WHAT IS *LIFE?*

A GUIDE TO BIOLOGY

WITH PHYSIOLOGY

WHAT IS *LIFE?*

A GUIDE TO BIOLOGY

WITH PHYSIOLOGY

FOURTH EDITION

Jay Phelan

UNIVERSITY OF CALIFORNIA, LOS ANGELES

New York

Vice President, STEM: Ben Roberts
Editorial Program Director: Brooke Suchomel
Executive Marketing Manager: Will Moore
Marketing Assistants: Cate McCaffery and Savannah DiMarco
Senior Development Editors: Beth Marsh, Debbie Hardin
Illustrations: Tommy Moorman
Assistant Editor: Kevin Davidson
Media Editor: Mike Jones
Photo Editor: Jennifer Atkins
Photo Researcher: Julia Phelan
Director of Design, Content Management: Diana Blume
Cover and Text Design: Dirk Kaufman
Director, Content Management Enhancement: Tracey Kuehn
Managing Editor: Lisa Kinne
Senior Content Project Manager: Liz Geller
Senior Workflow Project Manager: Paul Rohloff
Senior Media Project Manager: Chris Efstratiou
Composition and Layout: Sheridan Sellers, Lumina Datamatics, Inc.
Printing and Binding: LSC Communications
Cover Image: Thomas Jackson

Library of Congress Control Number: 2017959019
ISBN-13: 978-1-319-06544-7
ISBN-10: 1-319-06544-9

Printed in the United States of America

Second Printing 2018

W. H. Freeman and Company
One New York Plaza
Suite 4500
New York, NY 10004-1562
www.macmillanlearning.com

BRIEF CONTENTS

Part 5 Plant Life

Part 6 Health and Physiology

CONTENTS

Part 1 The Facts of Life

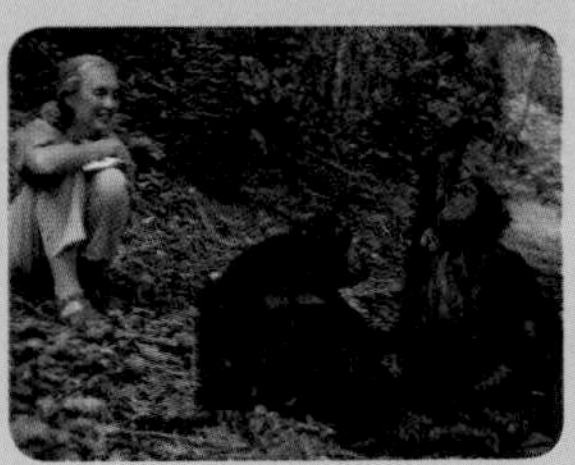

1 Scientific Thinking
Your best pathway to understanding the world

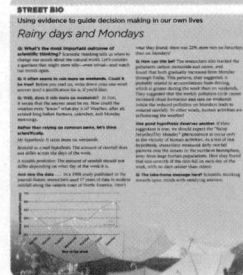

2 The Chemistry of Biology
Atoms, molecules, and their roles in supporting life

STREET BIO

Using evidence to guide decision making in our own lives

Bottled water or tap water

3 Molecules of Life

Macromolecules can store energy and information and serve as building blocks

STREET BIO

Using evidence to guide decision making in our own lives

Melt-in-your-mouth chocolate may not be such a sweet idea.

Food chemists have figured out how to make chocolate that melts in your mouth. Is that a good thing?

Q: Why are some fats "liquidy" like oil, and others solid?

Q: Can oils be made more solid?

Q: Does that improve their taste?

Q: Is there a downside?

Conclusion

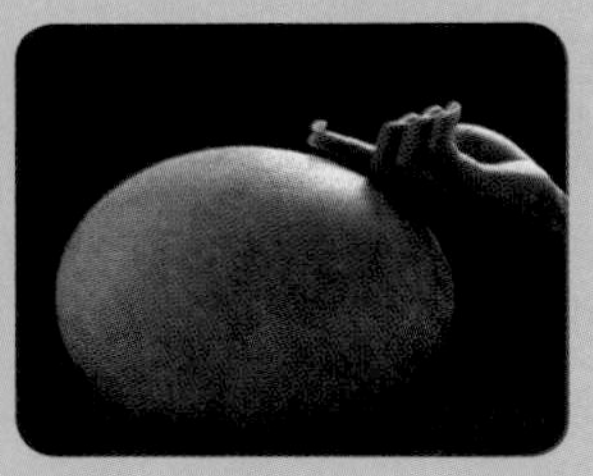

4 Cells
The smallest part of you

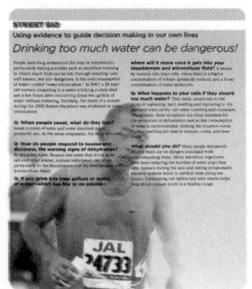

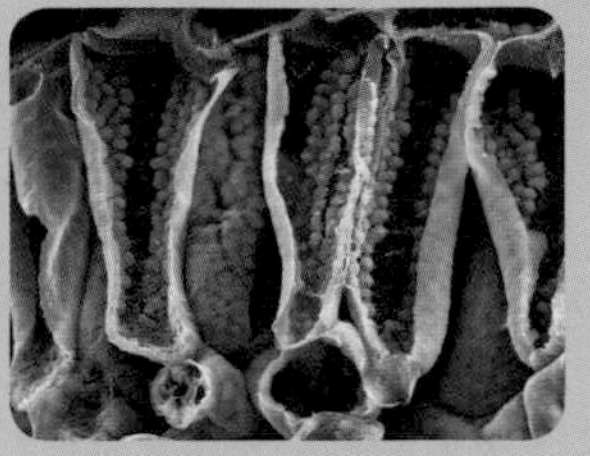

5 Energy
From the sun to you in just two steps

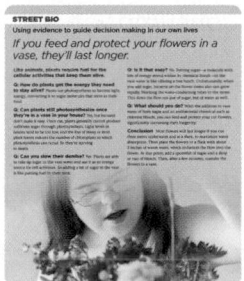

Part 2 Genetics, Evolution, and Behavior

6 DNA and Gene Expression
What is the genetic code and how does it work?

STREET BIO

Using evidence to guide decision making in our own lives

A faulty gene leads to unpleasant experiences from alcohol consumption. But maybe it's a beneficial gene?

7 Biotechnology
Harnessing the genetic code for medicine, agriculture, and more

8 Chromosomes and Cell Division
Continuity and variety

9 Genes and Inheritance
Family resemblance: how traits are inherited

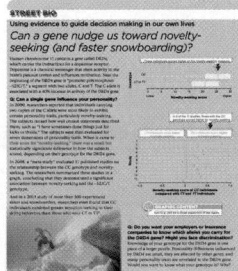

10 Evolution and Natural Selection
Darwin's dangerous idea

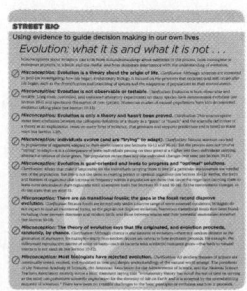

11 Evolution and Behavior
Communication, cooperation, and conflict in the animal world

Part 3 Evolution and the Diversity of Life

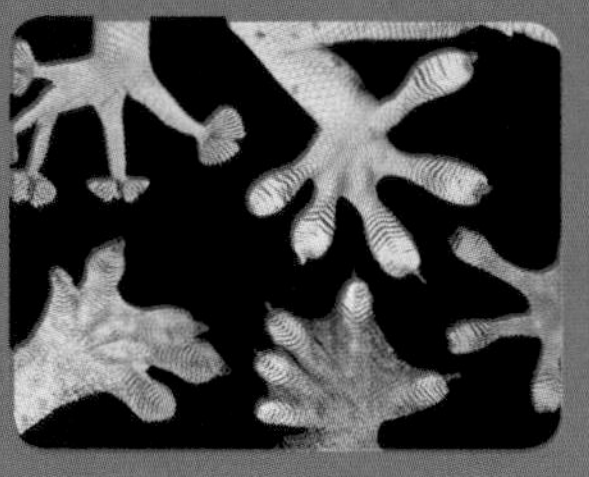

12 The Origin and Diversification of Life on Earth
Understanding biodiversity

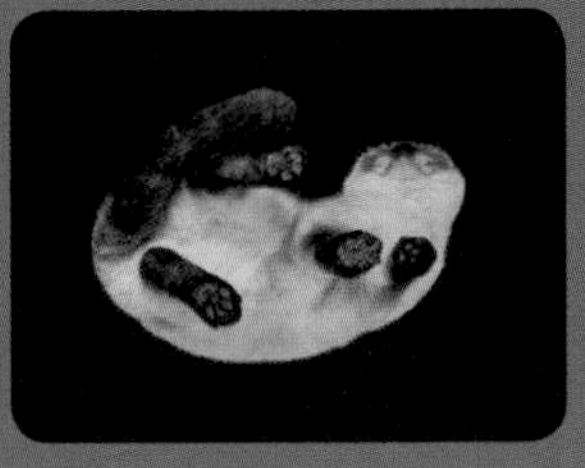

13 Animal Diversification

Visibility in motion

14 Plant and Fungi Diversification

Where did all the plants and fungi come from?

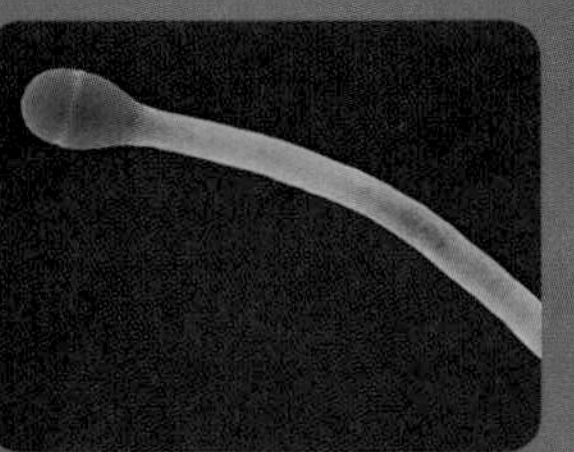

15 Evolution and Diversity Among the Microbes

Bacteria, archaea, protists, and viruses: the unseen world

STREET BIO

Using evidence to guide decision making in our own lives

The five-second rule: how clean is that food you just dropped?

Part 4 Ecology and the Environment

16 Population Ecology
Planet at capacity: patterns of population growth

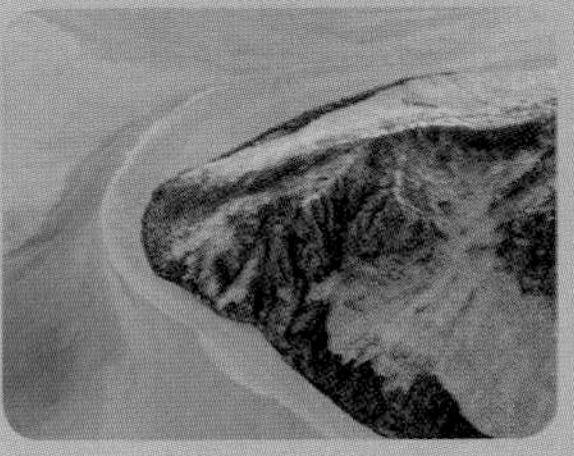

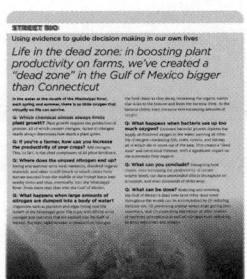

17 Ecosystems and Communities
Organisms and their environments

18 Conservation and Biodiversity
Human influences on the environment

Part 5 Plant Life

19 Plant Structure and Nutrient Transport
How plants function, and why we need them

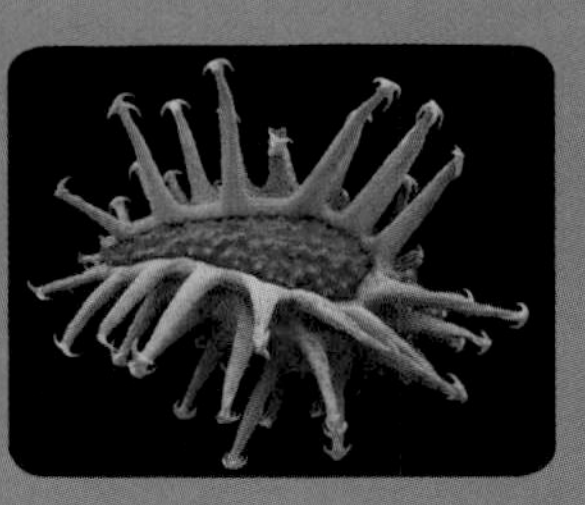

20 Growth, Reproduction, and Environmental Responses in Plants

Problem solving with flowers, wood, and hormones

21 Introduction to Animal Physiology
Principles of animal organization and function

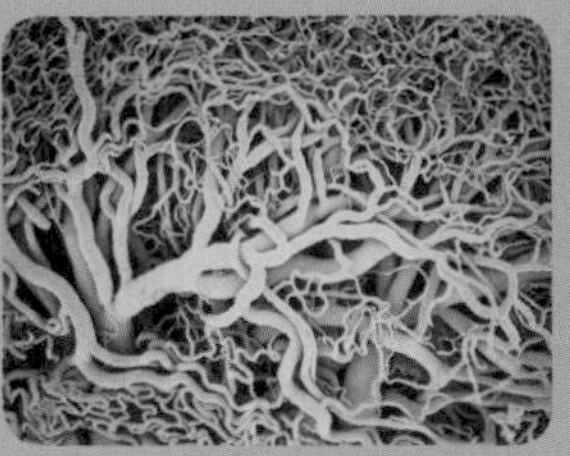

22 Circulation and Respiration

Transporting fuel, raw materials, and gases into, out of, and around the body

23 Nutrition and Digestion
At rest and at play: optimizing human physiological functioning

STREET BIO
Using evidence to guide decision making in our own lives
"An apple a day keeps the doctor away": Silly rhyme? Sound, data-backed advice? Or something in between?

24 Nervous and Motor Systems

Actions, reactions, sensations, and addictions: meet your nervous system

STREET BIO

Using evidence to guide decision making in our own lives

"We use only 10% of our brain!" Fact or fiction?

25 Hormones

Mood, emotions, growth, and more: hormones as master regulators

26 Reproduction and Development

From two parents to one embryo to one baby

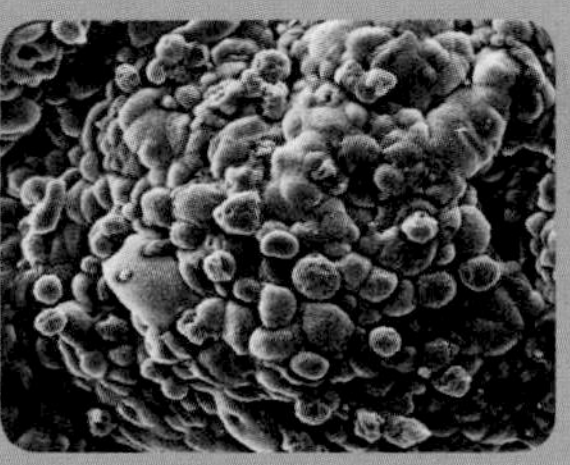

27 Immunity and Health
How the body defends and maintains itself

STREET BIO
Using evidence to guide decision making in our own lives
Facts or fallacies? How best to avoid getting sick

DEAR READER,

How many days do you wake up to breaking news about a scary-sounding virus, or a potential cause of cancer, or newly identified genes that make you better at math? In a world of easy access to information, it can be difficult to know how much confidence we should have about such reports. My mission is to help you evaluate the sometimes conflicting messages about science topics and science-related issues. In addition, I hope you will understand that ***biology is about you, and it touches every aspect of your life. It's creative. And it's fun.***

In these pages, you'll find an overview of the key themes in biology as well as detailed information about the natural world and its processes. I hope you will find answers to questions you're curious about and will be spurred to ask many more. You'll also find many red **Q** questions, such as:

- Do megadoses of vitamin C reduce cancer risk?
- An onion has five times as much DNA as a human. Why doesn't that make onions more complex than humans?
- Why doesn't natural selection lead to the production of perfect organisms?
- Why are big, fierce animal species so rare in the world?

The red **Q**s point toward passages that help uncover the answers. Often, the answer may not be apparent—but look again and think some more.

Within each chapter of *What Is Life?* you'll find a section called **This Is How We Do It**. In these sections we explore the diverse ways that scientists approach problems and how they go about finding answers. Example topics include *Why do we yawn?* and *Does sunscreen use reduce skin cancer risk?*

At the end of each chapter, you'll find a section called **Street Bio: Using evidence to guide decision making in our own lives**. These sections address issues that are particularly practical, such as *How clean is that food you just dropped?*

There's much more to biology than just words. Flip through *What Is Life?* and look at the photographs. Images can inspire and provide an alternative hook for remembering and understanding concepts. They can also challenge you to see ideas in new ways.

You'll also notice brief quotes from a variety of literary sources. It is my hope that as your scientific literacy increases, your experience and appreciation of literature also will be richer.

In organizing each chapter, I have broken down the topics into discrete, manageable sections. And at the end of each, I provide a **Take Home Message** that highlights and reinforces the section's most important ideas.

Included at the end of each chapter are summaries as well as multiple-choice and short-answer review questions, plus an exercise called **Graphic Content**. This critical thinking challenge will help you become adept at reading and analyzing visual displays of information, while identifying subtle assumptions, biases, and even manipulations.

This is just a sample of some of the features in *What Is Life?* I hope that you find this book stimulates new ways of thinking about and understanding the world.

Sincerely,

Jay Phelan

P.S. ABOUT THE COVER
I want to convey that biology isn't something that exists far away, separate from our personal lives. Rather, it intersects with our lives and is a central part of our world.

WHAT IS LIFE?
A GUIDE TO BIOLOGY WITH PHYSIOLOGY
FOURTH EDITION

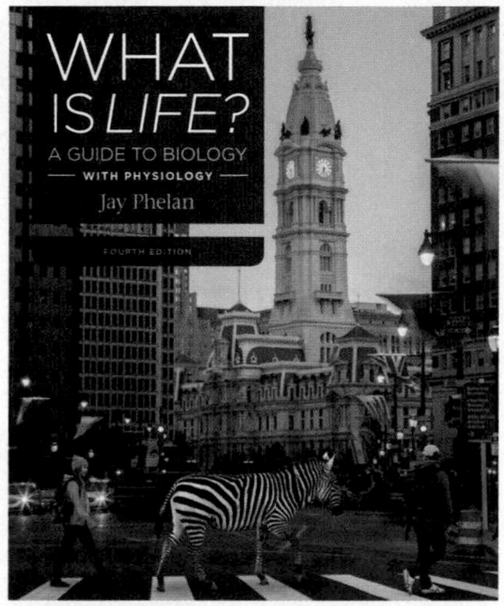

JAY PHELAN
University of California, Los Angeles

December 2017 (©2018)

Paperback with LaunchPad 978-1-319-15460-8
Paperback 978-1-319-06544-7
Loose-leaf with LaunchPad 978-1-319-15464-6
Loose-leaf 978-1-319-10632-4
LaunchPad only 978-1-319-10634-8

IN HIS POPULAR classes and best-selling textbooks, Jay Phelan is a master at captivating non-majors students with both the practical impact and awe-inspiring wonder of biological research. He also knows how to use the study of biology as a context for developing the critical thinking skills and scientific literacy students can draw on through college and beyond.

Phelan's dynamic approach to teaching biology is the driving force behind *What Is Life?*—the most successful new non-majors biology textbook of the millennium. The rigorously updated new edition brings forward the features that made the book a classroom favorite (chapters anchored to intriguing questions about life, spectacular original illustrations, and innovative learning tools) with a more focused and flexible presentation and enhanced art.

And more than ever, this edition seamlessly integrates multimedia resources with the text, with its dedicated version of LaunchPad, Macmillan's breakthrough online course space that brings together an interactive e-Book, all student and instructor media, and a wide range of assessment and course management features.

ALSO AVAILABLE

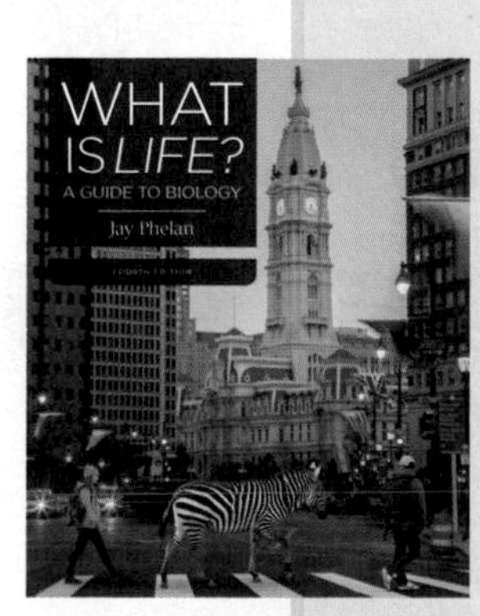

WHAT IS LIFE?
A Guide to Biology
Fourth Edition

Jay Phelan,
University of California, Los Angeles

December 2017 (©2018)

Paperback with LaunchPad 978-1-319-15465-3
Paperback 978-1-319-06545-4
Loose-leaf with LaunchPad 978-1-319-15463-9
Loose-leaf 978-1-319-10631-7
LaunchPad only 978-1-319-10636-2

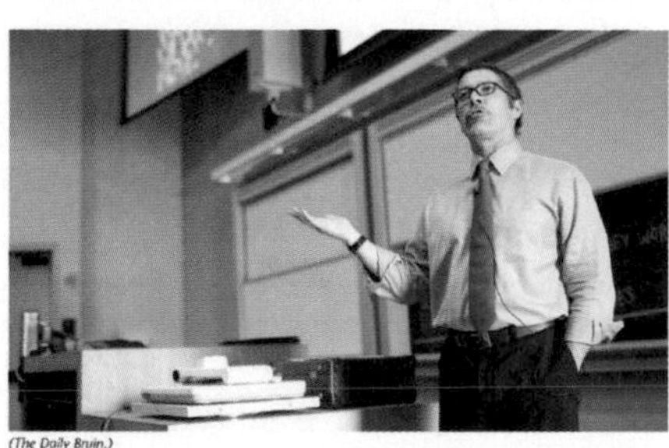
(The Daily Bruin.)

ABOUT THE AUTHOR
Jay Phelan is on the faculty of the UCLA Department of Life Sciences Core Education, where he has taught introductory biology in large lectures for majors and non-majors for 20 years. He received his PhD in evolutionary biology from Harvard in 1995, and his master's and bachelor's degrees from Yale and UCLA. His primary area of research is evolutionary genetics, and his original research has been published in *Evolution*, *Experimental Gerontology*, and the *Journal of Integrative and Comparative Biology*, among others. His research has been featured on *Nightline*, CNN, the BBC, and NPR; in *Science Times* and *Elle*; and in more than a hundred newspapers. He is the recipient of more than a dozen teaching awards, including UCLA's highest teaching honor, the Distinguished Teaching Award. With Terry Burnham, Jay is the coauthor of the international best-seller *Mean Genes: From Sex to Money to Food—Taming Our Primal Instincts*. Written for the general reader, *Mean Genes* explains in simple terms how knowledge of the genetic basis of human nature can empower individuals to lead more satisfying lives. Writing for a nonscientific audience has honed Phelan's writing style to one that is casual and inviting to students while also scientifically precise.

HOW DOES ***WHAT IS LIFE?*** GIVE NON-MAJORS

Brief Sections Make the Material Manageable

Each chapter is broken down into a series of short, accessible sections. This gives instructors flexibility in what they cover and makes the content easier for students to navigate and absorb.

Engaging Examples Showcase Biology in Everyday Life

What Is Life? A Guide to Biology threads fascinating, relevant, contemporary examples of the science throughout each chapter. Jay Phelan expertly uses analogies and interesting examples to inspire students to care about biology.

Intriguing, Often Surprising Q Questions Motivate Readers

Red Q questions spark students' interest and encourage critical thinking. These questions prompt students to consider how biology is a part of their lives and the world around them. Corresponding animations are available in LaunchPad.

5.5–5.11

Photosynthesis uses energy from sunlight to make food.

Large flat water lily leaves intercept sunlight.

5.5 Where does plant matter come from? Photosynthesis: the big picture.

Watching a plant grow over the course of a few years can seem like watching a magic trick. Of course it's not, but the process is nonetheless amazing. Consider that in five years a tree can increase its weight by 150 pounds (68 kg) (FIGURE 5-9). Where does that 150 pounds of new tree come from?

Q When humans grow, the new tissue comes from food we eat. When plants grow, where does the new tissue come from?

Our first guess might be the soil. It's easy enough to weigh the soil in a pot when first planting a tree. Five years later, however, if we weigh the soil again, we find that it has lost less than a pound—nowhere near enough to explain the massive increase in the amount of plant material. Perhaps the new growth comes from the water? Wrong again. Although the older and much larger tree holds more water in its many cells, the water provided to the plant does not account for the increase in the dry weight of the plant.

Most of the new material actually comes from an element present in an invisible gas in the air. During photosynthesis, plants capture this element, present as carbon dioxide gas (CO_2) in the atmosphere, and use the energy from sunlight, along with water and small amounts of chemicals usually found in soil, to produce solid, visible (and often tasty) sugars and other organic molecules. These molecules are used to make plant structures such as leaves, roots, stems, flowers, fruits, and seeds. In the process, the plants give off oxygen (O_2), a by-product that happens to be necessary for much of the life on earth—including all animal life!

Plants are not the only organisms capable of photosynthesis. Some bacteria and many other unicellular organisms,

FIGURE 5-9 **When plants grow, where does the new tissue come from?** From the soil? From thin air?

132 CHAPTER 5 • ENERGY

Vivid Photos Capture the Story of Biology

Striking images appear as unit openers and are combined with illustrations of biological processes, concepts, and experimental techniques to capture the imagination of the student.

BIOLOGY STUDENTS THE SUPPORT THEY NEED?

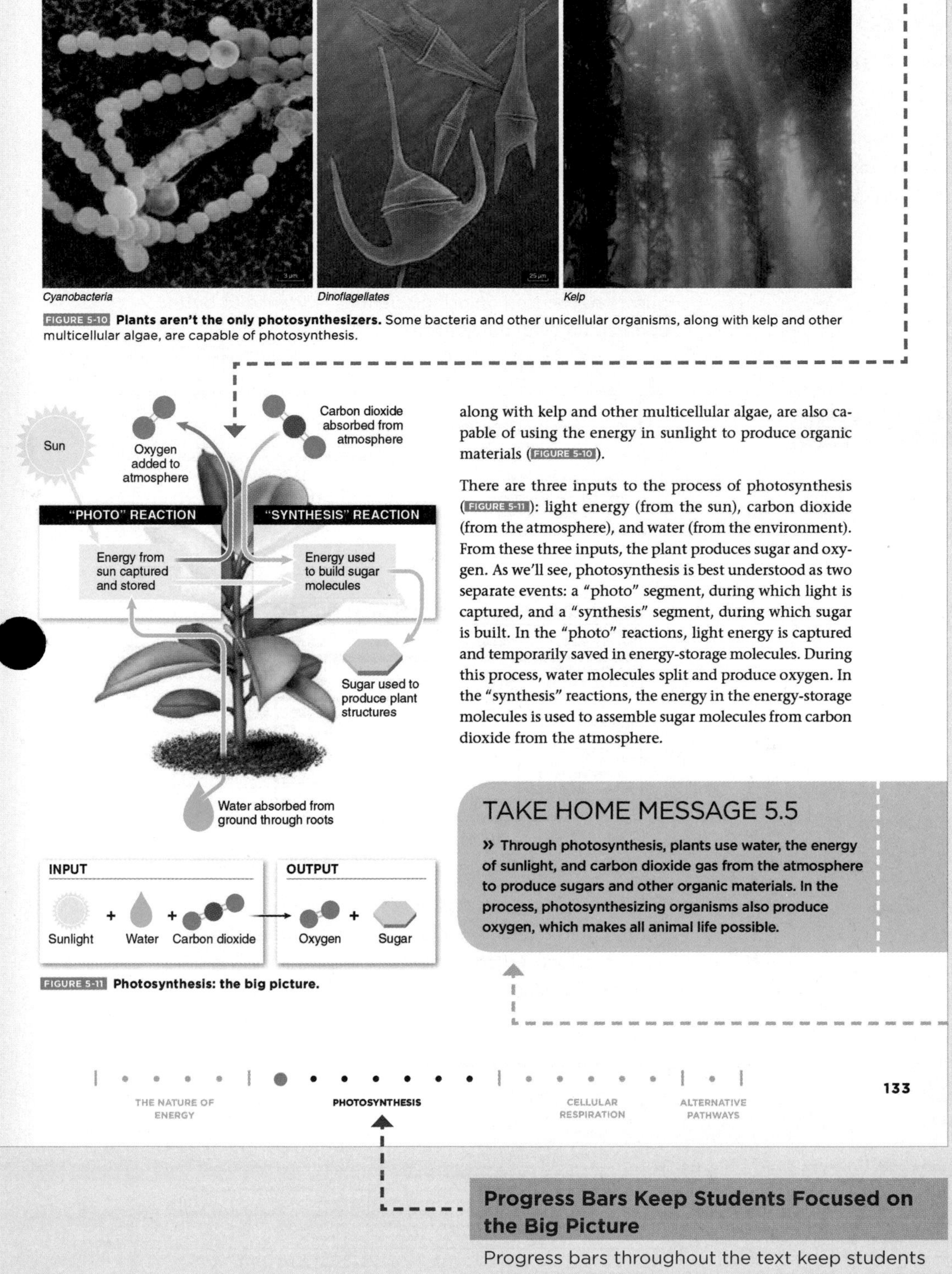

Cyanobacteria
Dinoflagellates
Kelp
FIGURE 5-10 Plants aren't the only photosynthesizers. Some bacteria and other unicellular organisms, along with kelp and other multicellular algae, are capable of photosynthesis.
Sun
Oxygen added to atmosphere
Carbon dioxide absorbed from atmosphere
"PHOTO" REACTION
"SYNTHESIS" REACTION
Energy from sun captured and stored
Energy used to build sugar molecules
Sugar used to produce plant structures
Water absorbed from ground through roots
INPUT
OUTPUT
Sunlight
Water
Carbon dioxide
Oxygen
Sugar
FIGURE 5-11 Photosynthesis: the big picture.
along with kelp and other multicellular algae, are also capable of using the energy in sunlight to produce organic materials (FIGURE 5-10).
There are three inputs to the process of photosynthesis (FIGURE 5-11): light energy (from the sun), carbon dioxide (from the atmosphere), and water (from the environment). From these three inputs, the plant produces sugar and oxygen. As we'll see, photosynthesis is best understood as two separate events: a "photo" segment, during which light is captured, and a "synthesis" segment, during which sugar is built. In the "photo" reactions, light energy is captured and temporarily saved in energy-storage molecules. During this process, water molecules split and produce oxygen. In the "synthesis" reactions, the energy in the energy-storage molecules is used to assemble sugar molecules from carbon dioxide from the atmosphere.
TAKE HOME MESSAGE 5.5
» Through photosynthesis, plants use water, the energy of sunlight, and carbon dioxide gas from the atmosphere to produce sugars and other organic materials. In the process, photosynthesizing organisms also produce oxygen, which makes all animal life possible.
THE NATURE OF ENERGY
PHOTOSYNTHESIS
CELLULAR RESPIRATION
ALTERNATIVE PATHWAYS
133

Clear and Compelling Illustrations

Fresh and easy-to-understand figures bring the concepts to life. Collaboratively developed by the author and scientific illustrators, the text and illustrations are seamlessly integrated, effective learning tools.

Recurring Chapter Features Develop Students' Scientific Skills

The following chapter features are built around key learning goals from Vision & Change:

- **Street Bio** demonstrates how students encounter biology every day, helping students apply the scientific knowledge they learn to improve their own lives.
- **This Is How We Do It** helps students develop the ability to understand and apply the process of science by showing how scientists have approached an intriguing question.
- **Graphic Content** helps students learn to interpret visual displays of quantitative information and think critically about evidence.

Take Home Messages Reinforce Key Concepts

Each section ends with a Take Home Message that highlights the most important content.

Progress Bars Keep Students Focused on the Big Picture

Progress bars throughout the text keep students aware of where they're headed and where they've been in the chapter, helping them to understand how individual topics relate to the bigger picture.

LAUNCHPAD

LAUNCHPADWORKS.COM

The full benefits of *What Is Life?* are realized through LaunchPad, Macmillan Learning's online companion platform.

WHAT IS LAUNCHPAD?

LaunchPad brings an interactive e-Book of *What Is Life?* together with book-specific media and assessments, including assignable versions of hallmark features of the text. Students can use self-study tools or complete instructor-created assignments that are automatically graded and tracked.

LearningCurve

Formative assessment in a game-like adaptive quizzing system, available for every chapter of the text.

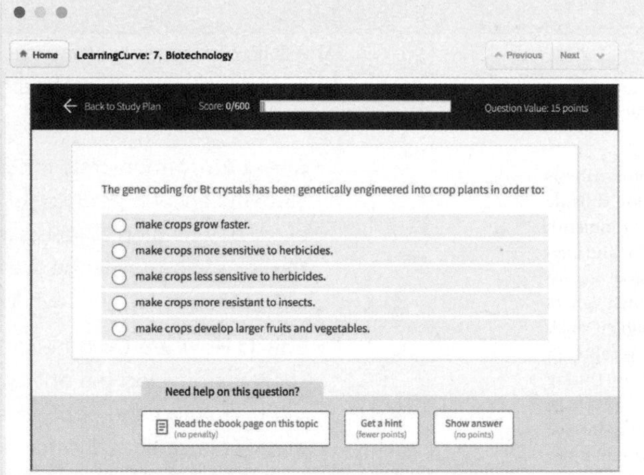

Graphic Content Assignments

Short, curated assessments featuring multiple-choice questions focused on interpreting visual displays of quantitative information.

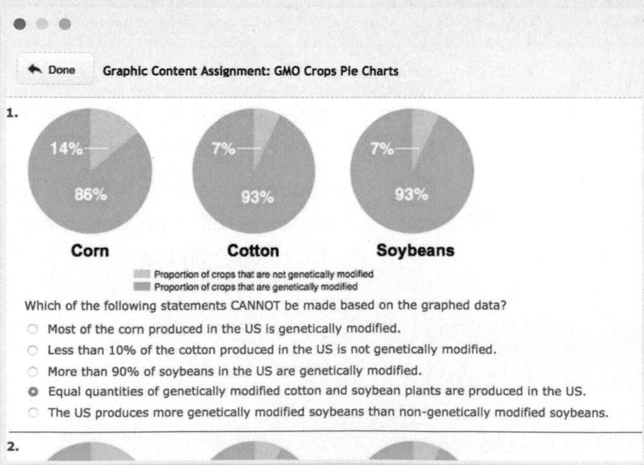

StreetBio Assignments

Short, curated assessments featuring multiple-choice questions focused on using scientific evidence to better understand everyday issues.

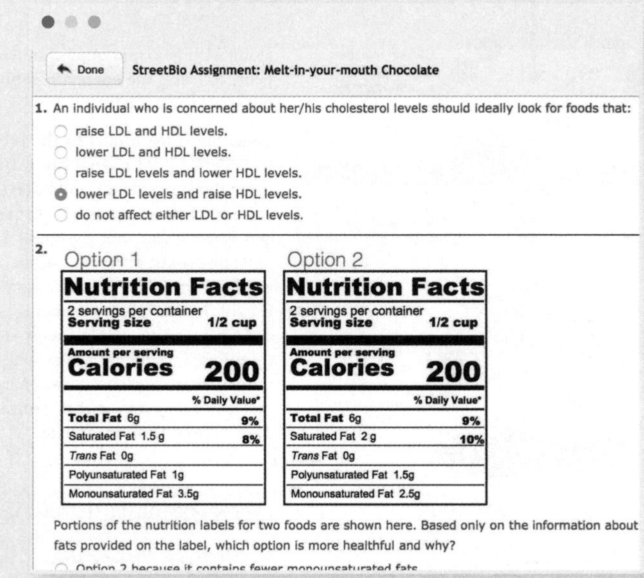

This Is How We Do It Assignments

Short, curated assessments featuring multiple-choice questions focused on understanding and applying the process of science.

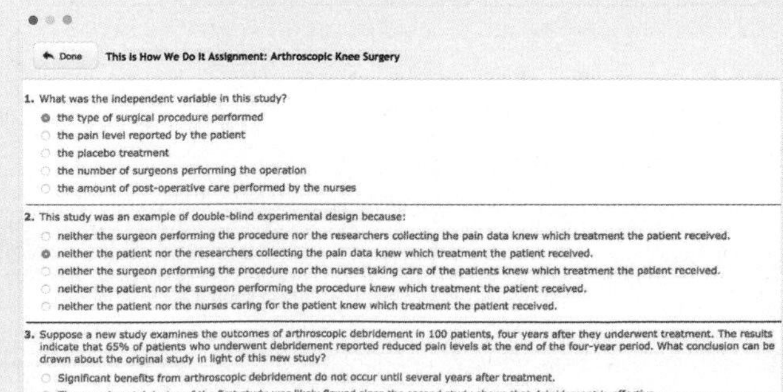

Tutorials

These brief activities, which feature drag/drop, click-to-match, and other interactive question types, are designed to help students review and learn core course concepts—particularly topics that are consistently difficult for students (ex: mitosis vs. meiosis, transcription vs. translation, and more).

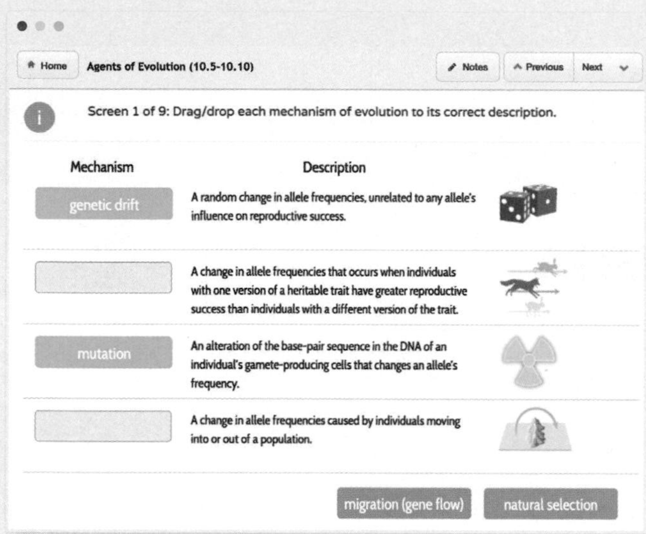

Animations

Available for select chapters, these animated videos bring to life key topics in non-majors biology (as identified by instructors) through a combination of chapter art and narration.

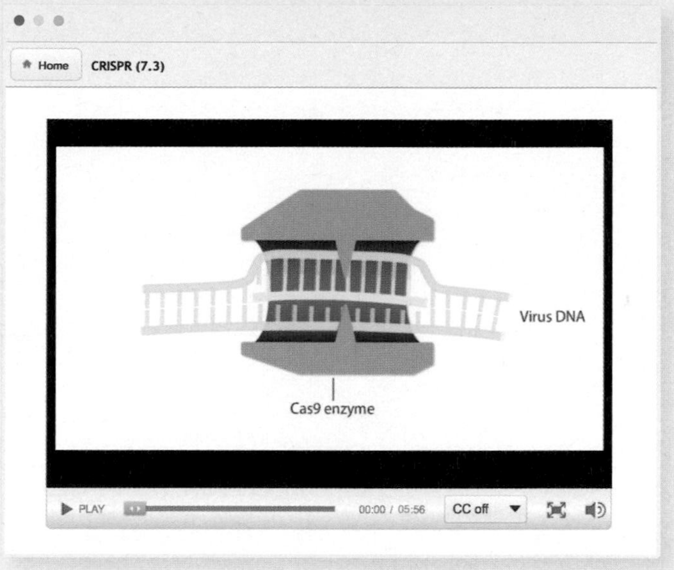

Active Learning: Visual Analysis Slides

Compatible with iClicker Cloud software, these clicker and discussion slides focus on checking understanding of important concepts through interpretation and analysis of select figures from the textbook.

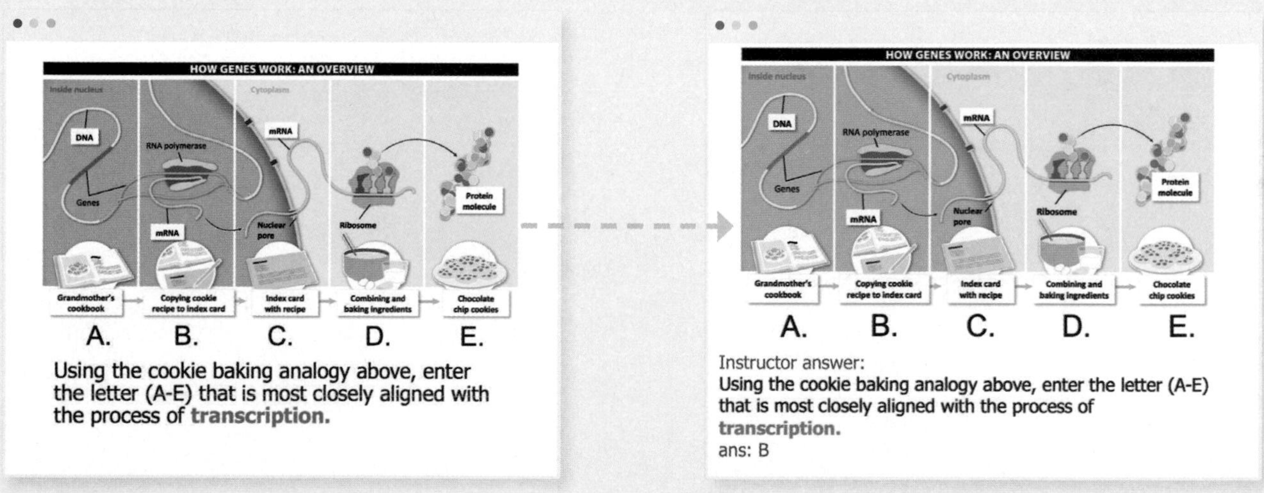

NEW E-BOOK ALTERNATIVE

Vitalsource™

VitalSource e-Book

For the first time, the VitalSource e-Book is also available through an app that allows students to read offline and have the book read aloud to them. These features are in addition to the highlighting, note-taking, and keyword search that VitalSource has long offered.

TO JULIA

ACKNOWLEDGMENTS

As a new graduate student at Harvard, I heard from experienced teaching fellows that if you were interested in learning how to be an effective teacher, it was essential to seek out extraordinary mentors. Based on word of mouth, I became involved with E. O. Wilson's course in Evolutionary Biology and Irv DeVore's course in Human Behavioral Biology. Both were known to be unusually provocative, challenging, and entertaining classes for non-science majors. Working under these legendary educators for twelve semesters, I was set on a course that inspired and prepared me to write this book.

The two courses were quite different from each other, but at their core both were built on two beliefs that are central to this book and to my thinking about education: (1) Biology is creative, interesting, and fun. (2) Biology is relevant to the daily life of every person. There was a palpable sense that, in teaching non-science majors especially, we had a responsibility to provide our students with the tools to thrive in a society increasingly permeated by scientific ideas and issues, and that one of our most effective strategies would be to convey the excitement we felt for biology and the enormous practical value it has to help us understand the world.

My development as a scientist and, particularly, my appreciation for rigorous and methodical critical thinking have been shaped by the kind support and wise guidance of Richard Lewontin. I have also been fortunate to have as a long-time mentor, collaborator, and friend Michael Rose, who has instilled in me a healthy skepticism about any observation in life that is not fivefold replicated. And for almost daily insightful input on matters relating to scientific content, teaching, writing, and more, I thank Terry Burnham.

There are many other friends and colleagues I wish to thank for helping me with *What Is Life?*

In researching and writing the book and in developing the numerous courses I teach, I have benefited from more than a decade of perceptive and valuable contributions from Glenn Adelson, Alon Ziv, and Michael Cooperson.

For a project covering so many topics and years, it is essential to have a close group of trusted and knowledgeable colleagues. I am grateful to Alicia Moretti, Harold Owens, Greg Graffin, Brian Swartz, Greg Laden, Jeff Egger, Joshua Malina, Melissa Merwin-Malina, Chris Bruno, and Michelle Richmond, who have offered advice, guidance, and support, far beyond the call of duty.

Numerous colleagues at UCLA provided assistance and support, including Cliff Brunk, Victoria Sork, Deb Pires, Gaston Pfluegl, Frank Laski, Erin Sanders, Liz Roth-Johnson, and Tracy Knox. As a result of their commitment to excellence in the UCLA Department of Life Science Core Education, I have been able to acquire a wealth of experiences that have helped me continue improving as a teacher.

I owe a tremendous debt to Sara Tenney and Peter Marshall, without whose encouragement and support this project could never have been begun or completed.

Macmillan Learning is an extremely author-centric publisher. Throughout the process of creating this book, Liz Widdicombe, John Sargent, Ken Michaels, Ben Roberts, and Susan Winslow have been tremendously supportive. I am grateful for their welcoming me into their publishing family. Editorial program director Brooke Suchomel has been an enthusiastic, wise, and skillful manager of the entire team. I am very fortunate to have such a committed and driven leader overseeing all aspects of this project. Development editor Beth Marsh oversaw every aspect of the writing of the book, while making insightful and helpful contributions throughout. I could not have completed this book without her commitment and guidance. I must also convey my gratitude to development editor Jane Tufts, whose meticulous attention to detail, commitment to accuracy, and almost obsessive drive to create a thorough and readable book are apparent on every page.

It is impossible to teach biology without illustrations. My deepest gratitude goes to Tommy Moorman for creating such innovative and effective figures for the book. Tommy's vision for an elegant and beautiful art program completely integrated with the text is apparent on every page. Working with him to develop each illustration in this book has been (and continues to be) one of my most enjoyable and satisfying professional collaborations. Thanks also go to Alison Kendall, Tamara Lau, and Erin Daniel for assisting with creation of the illustrations. And for excellent assistance with photo research thanks to Julia Phelan and Jennifer Atkins.

For creating the innovative media and print materials that accompany the book, I am thankful for the contributions of Mike Jones and Elaine Palucki. I thank all of the contributors and advisors who helped create the student and instructor resources; your efforts have been invaluable. Sheri Snavely provided significant input in developing pedagogical strategies throughout the book; I also appreciate her thoughtful and wise advice throughout the publishing process. Copyeditor Linda Strange helped to ensure consistency, accuracy, and readability throughout the text. I thank Robert Swanson (Arc Indexing) for compiling the thorough index. The rest of the life sciences editorial team at Macmillan Learning, too, have been knowledgeable and supportive, particularly Lisa Samols and Debbie Hardin. For their efficiency and commitment to producing a beautiful book, I am most grateful to the Macmillan Learning production team: Sheridan Sellers, Liz Geller, Paul Rohloff, Diana Blume, and Catherine Woods.

The people on the marketing team at Macmillan Learning have contributed enormously in helping with the challenging task of introducing this book to students and instructors across the country. Thanks especially to marketing assistants Cate McCaffery and Savannah DiMarco.

Finally, I thank my family—Kevin Phelan, Patrick Phelan, Erin Enderlin, and my parents—for their unwavering support and interest as I wrote this book. I thank Charlie, Jack, and Sam, too. Most of all, for her generous and passionate support of this project from day one, her substantive contributions to both the content and the presentation of ideas, and so much more, I thank Julia.

Contact the author with your feedback.

The content of this book has been greatly improved through the comments of reviewers and students. Your comments, suggestions, and criticism are also welcome; they are essential in guiding the book's ongoing evolution. Please contact the author at jay@jayphelan.com. I'm serious.

We thank the many reviewers who aided in the development of this text.

Fourth Edition Development

Christa Y. Adam, Prairie State College
Ana Maria Barral, National University
Tonya C. Bates, University of North Carolina at Charlotte
Mark W. Bland, University of Central Arkansas
Lisa L. Boggs, Southwestern Oklahoma State University
Mary E. Bonds, Northwest Mississippi Community College
Clay Britton, Methodist University
Dmitry Y. Brogun, CUNY–Brooklyn College
Sharon K. Bullock, University of North Carolina at Charlotte
Robert S. Byrne III, California State University, Fullerton
Carol L. Chaffee, California State University, Fullerton
Thomas T. Chen, Santa Monica College
Thomas Cholmondeley, Cincinnati State Technical and Community College
Carol Cleveland, Northwest Mississippi Community College
John G. Cogan, Ohio State University
Mary C. Colavito, Santa Monica College
Jennifer L. Cooper, University of Akron
Christina Couroux, Los Angeles Harbor College
William Crampton, University of Central Florida
Gregory A. Dahlem, Northern Kentucky University
Don C. Dailey, Austin Peay State University
Susan L. Dalterio, University of Texas at San Antonio
Ray Emmett, Daytona State College
Michele Engel, California State University, Bakersfield
Geneen Fitchett, Central Piedmont Community College
T. M. Gehring, Central Michigan University
Carrie L. Geisbauer, Moorpark College
Jenny Gernhart, Iowa Central Community College
Phoebe Gimple, College of the Canyons
Tammy Goulet, University of Mississippi
Tina T. Hopper, Missouri State University
John Hunt, University of Arkansas at Monticello
Meshagae Hunte-Brown, Drexel University
Alexander Imholtz, Prince George's Community College
Lisa A. E. Kaplan, Quinnipiac University
Jennifer M. Kilbourne, Community College of Baltimore County
Brian W. Kram, Prince George's Community College
Natasha LaMont, CUNY–Brooklyn College

Gary M. Lange, Saginaw Valley State University
Gerald Lasnik, Moorpark College
Debra A. Lewkiewicz, SUNY Sullivan
Suzanne Long, Monroe Community College
Cindy S. Malone, California State University, Northridge
Lisa Maranto, Prince George's Community College
Catarina Mata, Borough of Manhattan Community College
Amie Mazzoni, Fresno City College
Edward Mondor, Georgia Southern University
Britto P. Nathan, Eastern Illinois University
Lori Nicholas, New York University
Mary O'Sullivan, Elgin Community College
Murali Panen, Luzerne County Community College
Randi Papke, Southwestern Illinois College
James V. Price, Utah Valley University
Karen K. Resendes, Westminster College
Valerie Rosa, California State University, Fullerton
Lori Ann Rose, Sam Houston State University
Michael L. Rutledge, Middle Tennessee State University
Kim Cleary Sadler, Middle Tennessee State University
Michael Scott, Lincoln University
Pramila Sen, Houston Community College
Ryan Sessions, Florida State College at Jacksonville
Alka Sharma, University of Memphis
Mark Smith, Santiago Canyon College
Ashley Spring, Eastern Florida State College
Richard T. Stevens, Monroe Community College
Susan K. Sullivan, Louisiana State University of Alexandria
Kirsten Swinstrom, Santa Rosa Junior College
Kimberly Taugher, Diablo Valley College
Gail Anderson Tompkins, Wake Technical Community College
Robert Tompkins, Belmont Abbey College
Sue Trammell, John A. Logan College
Muatasem Ubeidat, Southwestern Oklahoma State University
Stephen C. Weeks, University of Akron
Carrie Wells, University of North Carolina at Charlotte
Wayne H. Whaley, Utah Valley University
Clayton J. White, Lone Star College–CyFair
Jennifer Wiatrowski, Pasco-Hernando State College

Active Learning Clicker Slides

J. Aaron Cassill, University of Texas at San Antonio
Michelle Cawthorn, Georgia Southern University
Stephen Chordas III, Ohio State University
Jennifer L. Cooper, University of Akron
Dani DuCharme, Waubonsee Community College
Lori A. Frear, Wake Technical Community College
Krista M. Henderson, California State University, Fullerton
Justin Hoshaw, Waubonsee Community College
Debra L. Linton, Central Michigan University
Mark Manteuffel, St. Louis Community College
Lisa Maranto, Prince George's Community College
Catarina Mata, Borough of Manhattan Community College

Maryanne Menvielle, California State University, Fullerton
Dana Newton, College of The Albemarle
Randi Papke, Southwestern Illinois College
Janet Ray, University of North Texas
Kimberly Regier, University of Colorado Denver
Valerie Rosa, California State University, Fullerton
Georgianna Saunders, Missouri State University
Jennifer Ripley Stueckle, West Virginia University
Heather Vance-Chalcraft, East Carolina University
Michael Wenzel, Folsom Lake College
Terry Williams, Community College of Denver
Jessica Wright, Southwestern Illinois College

Lecture Slides

Thomas T. Chen, Santa Monica College
Ana Esther Escandon, Los Angeles Harbor College
Tina T. Hopper, Missouri State University
Jessica Wright, Southwestern Illinois College

Media

Michael Black, California Polytechnic State University
J. Aaron Cassill, University of Texas at San Antonio
Michelle Cawthorn, Georgia Southern University
Julie Constable, California State University, Fresno
Dani DuCharme, Waubonsee Community College
Lori A. Frear, Wake Technical Community College
Tina T. Hopper, Missouri State University
Justin Hoshaw, Waubonsee Community College
Meshagae Hunte-Brown, Drexel University
Debra L. Linton, Central Michigan University
Lisa Maranto, Prince George's Community College
Dana Newton, College of The Albemarle
Randi Papke, Southwestern Illinois College
Ashley Ramer, University of Akron
Amy Reber, Georgia State University
Kimberly Reiger, University of Colorado Denver
Georgianna Saunders, Missouri State University
Jennifer Ripley Stueckle, West Virginia University
Michael Wenzel, California State University, Sacramento
Clayton J. White, Lone Star College–CyFair
Jessica Wright, Southwestern Illinois College

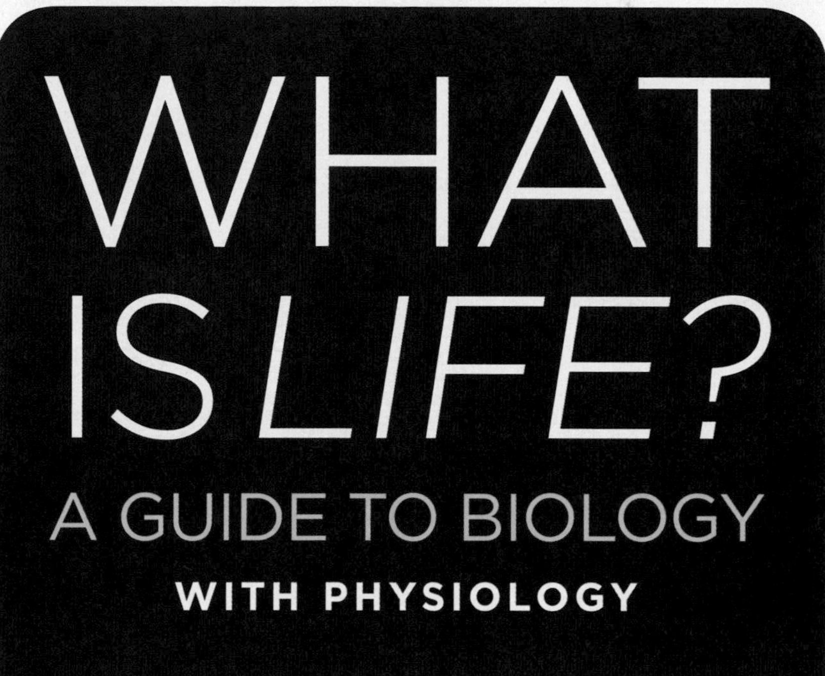
WHAT
IS LIFE?
A GUIDE TO BIOLOGY
WITH PHYSIOLOGY

Ch01

More than just a collection of facts, science is a process for understanding the world.

1.1 What is science? What is biology?
1.2 Biological literacy is essential in the modern world.

A beginner's guide to scientific thinking.

1.3 Thinking like a scientist: how do you use the scientific method?
1.4 Step 1: Make observations.
1.5 Step 2: Formulate a hypothesis.
1.6 Step 3: Devise a testable prediction.
1.7 Step 4: Conduct a critical experiment.
1.8 Step 5: Draw conclusions, make revisions.
1.9 When do hypotheses become theories, and what are theories?

Well-designed experiments are essential to testing hypotheses.

1.10 Controlling variables makes experiments more powerful.
1.11 *This is how we do it:* Is arthroscopic surgery for arthritis of the knee beneficial?
1.12 We've got to watch out for our biases.

Scientific thinking can help us make wise decisions.

1.13 Visual displays of data can help us understand and explain phenomena.
1.14 Statistics can help us in making decisions.
1.15 Pseudoscience and misleading anecdotal evidence can obscure the truth.
1.16 There are limits to what science can do.

On the road to biological literacy: what are the major themes in biology?

1.17 What is life? Important themes unify and connect diverse topics in biology.

Beginning in 1960 (and continuing for more than 50 years), Jane Goodall has studied the social and family interactions of chimpanzees in the wild, in Gombe Stream National Park, Tanzania.

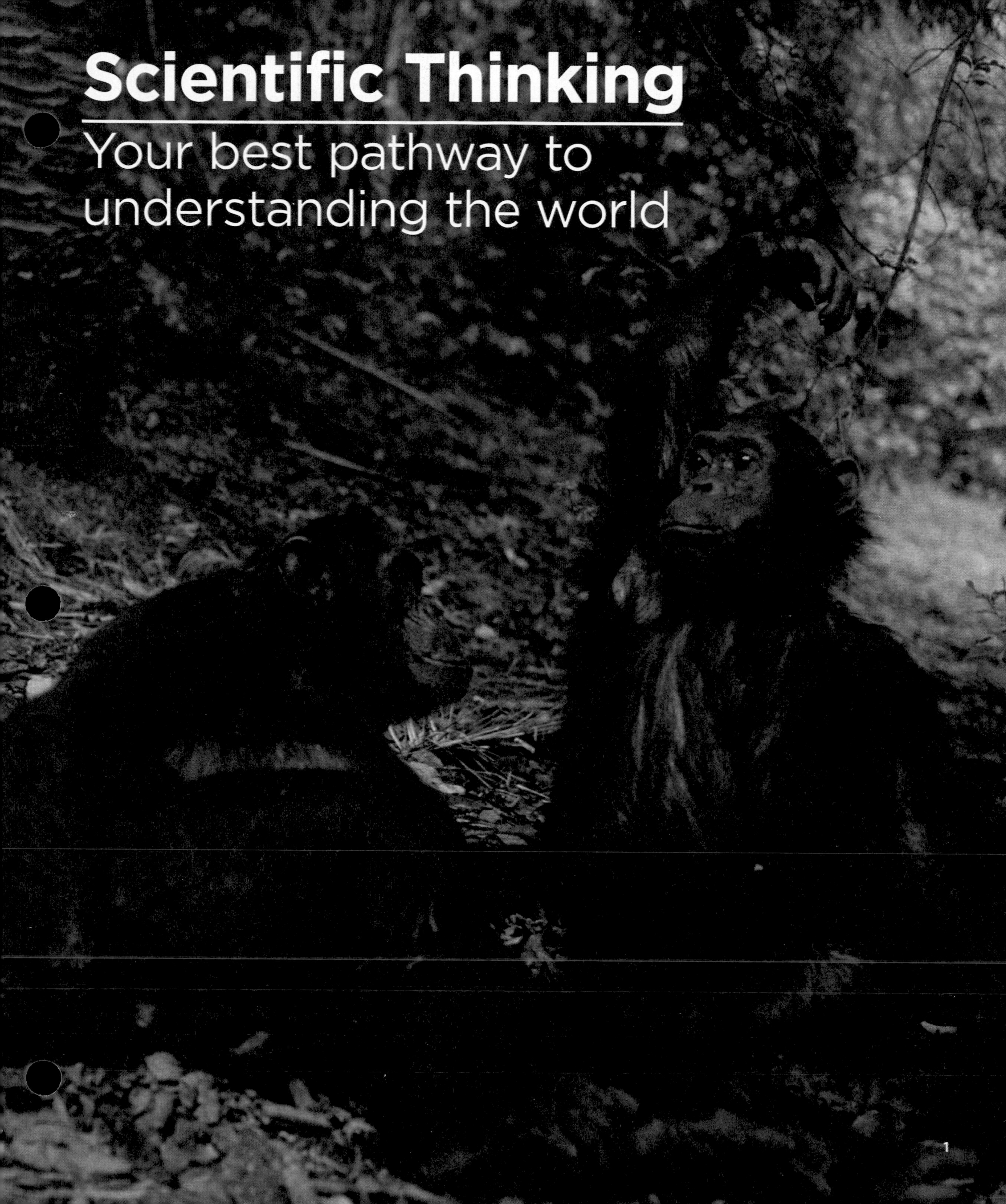

Scientific Thinking

Your best pathway to understanding the world

1.1–1.2

More than just a collection of facts, science is a process for understanding the world.

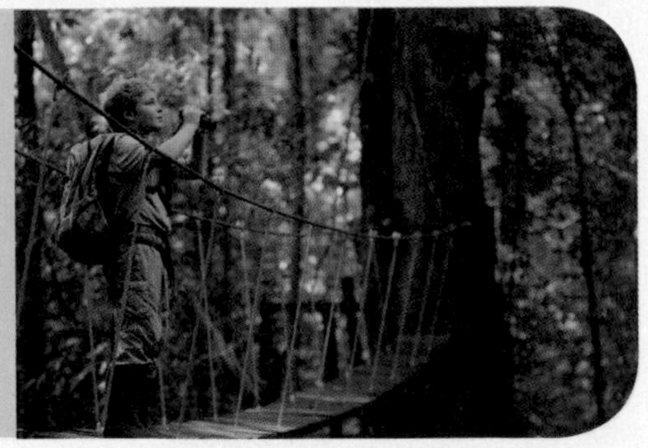

A boy (age 9) crossing the suspension bridge over the river at the research station in Gunung Palung National Park, Borneo, Indonesia.

1.1 What is science? What is biology?

You are already a scientist. You may not have realized this yet, but it's true. Because humans are curious, you have no doubt asked yourself or others questions about how the world works and wondered how you might find the answers.

- Does the radiation released by cell phones cause brain tumors?
- Are antibacterial hand soaps better than regular soap?

These are important and serious questions. But you've probably also pondered some less weighty issues, too.

- Why is morning breath so stinky? And can you do anything to prevent it?
- Why is it easier to remember gossip than physics equations?

And if you really put your mind to the task, you will start to find questions all around you whose answers you might like to know (and some whose answers you'll learn as you read this book).

- What is "blood doping," and does it really improve athletic performance?
- Why is it so much easier for an infant to learn a complex language than it is for a college student to learn biology?

Not convinced you're a scientist? Here's something important to know: science doesn't require advanced degrees or secret knowledge dispensed over years of technical training. It does, however, require an important feature of our species: a big brain, as well as curiosity and a desire to learn. But curiosity, casual observations, and desire can take you only so far.

Explaining how something works or why something happens requires methodical, objective, and rational observations and analysis that are not clouded by emotions or preconceptions. **Science** is not simply a body of knowledge or a list of facts to be remembered. It is an intellectual activity, encompassing methodical, objective observation, description, experimentation, and explanation of natural phenomena. Put another way, science is a pathway by which we can come to discover and better understand our world.

Later in this chapter, we explore specific ways in which we can most effectively use scientific thinking in our lives. But first let's look at a single powerful question that underlies scientific thinking:

How do you know that is true?

This is one question you can (and should) ask about the "scientific" claims that you see on products or read about in the newspaper or on the internet (FIGURE 1-1). For example, does Dannon yogurt help prevent colds and flu? Does Airborne? Contrary to the claims made by their manufacturers, there's no evidence to support this. The companies have been fined and the claims removed, all because people posed that important question—How do you know that is true?—and, with scientific thinking, the claims were put to the test.

Q Can we trust the packaging claims that companies make?

Our focus in this book is **biology**, the study of living things. Taking a scientific approach, we investigate the facts and ideas in biology that are already known and study the process by which we come to learn new things. You will learn exactly what it means to have scientific proof or evidence for something. As we move through the book, we'll explore the most important questions in biology.

FIGURE 1-1 **Some products claim to improve our health, but how do we know whether they work?**

- What is the chemical and physical basis for life and its maintenance?
- How do organisms use genetic information to build themselves and to reproduce?
- What are the diverse forms that life on earth takes, and how has that diversity arisen?
- How do organisms interact with each other and with their environment?

Because scientific thinking is important for the study of a wide variety of topics, we'll use a broad range of examples—including some from beyond biology—as we learn how to think scientifically. Although the examples vary greatly, they all convey a message that is key to scientific thinking: it's okay to be skeptical.

In this chapter, we explore the specific methods used in scientific thinking and how to make use of the knowledge we gain to make wise decisions. **Scientific literacy,** a general, fact-based understanding of the basics of biology and other sciences, is increasingly important in our lives, and literacy in matters of biology is especially essential.

TAKE HOME MESSAGE 1.1

» Through its emphasis on objective observation, description, and experimentation, science is a pathway by which we can discover and better understand the world around us.

1.2 Biological literacy is essential in the modern world.

A brief glance at any magazine or newspaper will reveal just how much scientific literacy has become a necessity (FIGURE 1-2). Many important health, social, medical, political, economic, and legal issues pivot on complex scientific data and theories. For example, why are unsaturated fats healthier for you than saturated fats? Why do allergies strike children from clean homes more than children from dirty homes? And why do new agricultural pests appear faster than new pesticides?

As you read and study this book, you will be developing **biological literacy,** the ability to (1) use the process of scientific inquiry to think creatively about real-world issues that have a biological component, (2) communicate these

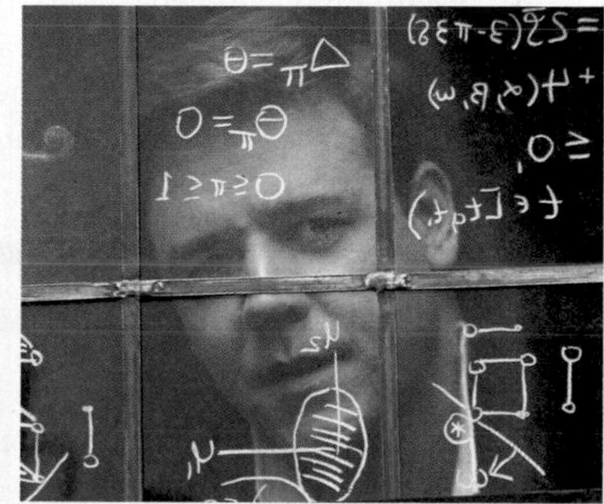

FIGURE 1-2 **In the news.** Every day, news sources report on social, political, medical, and legal issues related to science.

thoughts to others, and (3) integrate these ideas into your decision making.

Biological literacy doesn't involve just the big issues facing society or just abstract ideas. It also matters to you personally. Should you take aspirin when you have a fever? Are you using the wrong approach if you try to lose weight and, after some initial success, you find your rate of weight loss diminishing? Is it a good idea to consume moderate amounts of alcohol? Lack of biological literacy will put you at the mercy of "experts" who may try to confuse you or convince you of things in the interest of (their) personal gain. Scientific thinking will help you make wise decisions for yourself and for society.

Q Should you be swayed by the claim: "Experts say that ..."?

TAKE HOME MESSAGE 1.2

» Biological issues permeate all aspects of our lives. To make wise decisions, it is essential for individuals and societies to achieve biological literacy.

1.3–1.9

A beginner's guide to scientific thinking.

Eugenie Clark (at left), a pioneering investigator of shark behavior since the 1940s.

1.3 Thinking like a scientist: how do you use the scientific method?

Scientific method—this sounds like a rigid process to follow, much like following a recipe. By making the process sound inflexible, the phrase "scientific method" can also cause people to imagine—mistakenly—that it can be applied only to a narrow range of questions.

In practice, the scientific method is a highly flexible process that can be used to explore a wide variety of thoughts, events, or phenomena, not only in science but in other areas as well. For this reason, rather than framing it narrowly as the "scientific method," when it is appropriate, we will refer to the process by the more accurate term **scientific thinking.**

The basic elements of scientific thinking include:

- Make observations.
- Formulate a hypothesis.
- Devise a testable prediction.
- Conduct a critical experiment.
- Draw conclusions and make revisions.

Once begun, though, the process doesn't necessarily continue linearly through five non-overlapping steps until it is concluded (FIGURE 1-3). Sometimes, initial observations lead to more than one hypothesis and several testable predictions and experiments. Often, the conclusions drawn from experiments suggest new observations, refinements to hypotheses, and, ultimately, increasingly precise conclusions.

A key element of scientific thinking is that it is **empirical.** Empirical knowledge is based on experience and observations that are rational, testable, and repeatable. Another important element of scientific thinking is that it is self-correcting. As we continue to make new observations, a hypothesis about how the world works might change (FIGURE 1-4). If our observations do not support our current hypothesis, that specific hypothesis must be given up or, more typically, revised to one that is not contradicted

Q What should you do when something you believe in turns out to be wrong?

SCIENTIFIC THINKING

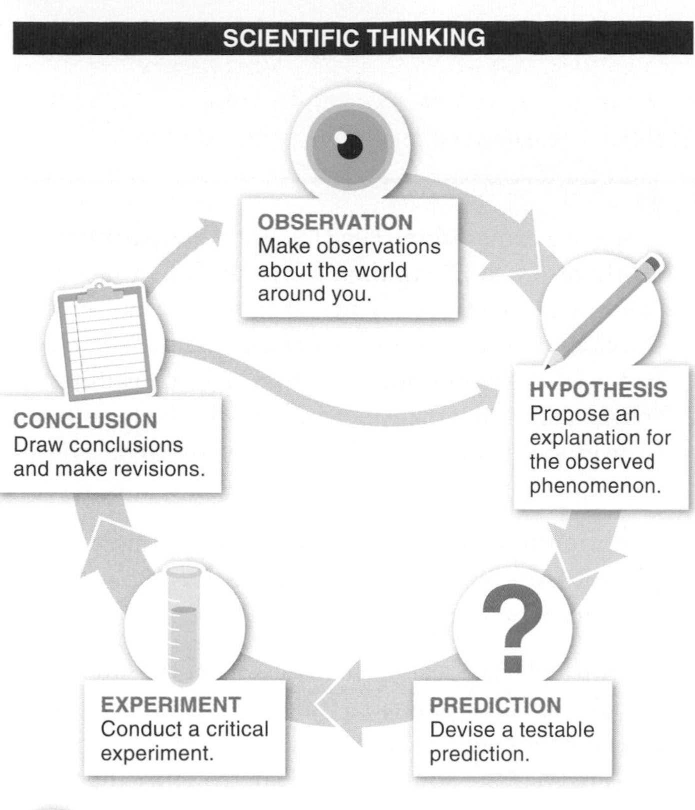
OBSERVATION
Make observations about the world around you.
HYPOTHESIS
Propose an explanation for the observed phenomenon.
PREDICTION
Devise a testable prediction.
EXPERIMENT
Conduct a critical experiment.
CONCLUSION
Draw conclusions and make revisions.
?

Scientific thinking rarely proceeds in a straight line. Conclusions, for example, often lead to new observations and refined hypotheses.

FIGURE 1-3 The scientific method: five basic steps and one flexible process.

by observations. This may be the most important feature of scientific thinking: *it tells us when we should change our minds.*

Because scientific thinking is a general strategy for learning, it needn't be used solely to learn about nature or scientific things. For example, we can analyze an important criminal justice question with scientific thinking:

- How reliable is eyewitness testimony in criminal courts?

> "If science proves some belief of Buddhism wrong, then Buddhism will have to change."
>
> — The 14th Dalai Lama, *New York Times,* **December 2005**

For more than 200 years, courts in the United States have viewed eyewitness testimony as unassailable. But is eyewitness identification always right? We will use scientific thinking to evaluate whether this perception is supported by evidence. We'll also look at how scientific thinking can be used to address a variety of other issues. And in addition to our criminal justice question, we'll take a detailed look at a question from biology:

- Does echinacea reduce the chances of catching the common cold or how long a cold lasts?

TAKE HOME MESSAGE 1.3

» Scientific thinking is an empirical process that incorporates making observations, articulating hypotheses, generating predictions, designing experimental tests, and drawing conclusions. It is a flexible, adaptable, and efficient pathway to understanding the world, in part because it tells us when we must change our beliefs.

CHANGING WHAT WE BELIEVE

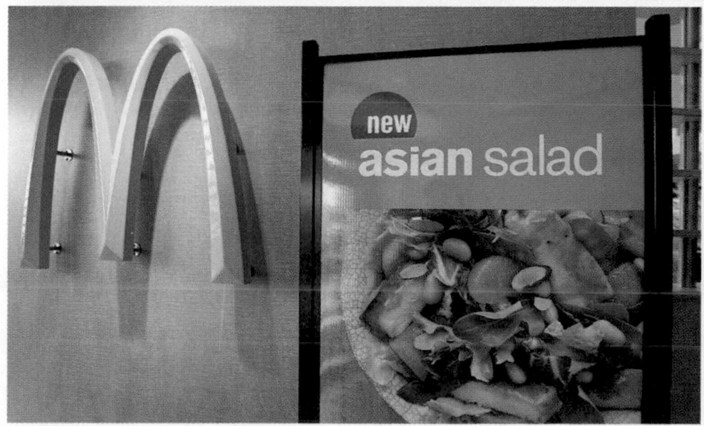
new
asian salad

Scientific thinking has helped us add more healthful options from the menu.

Can scientific thinking reveal problems with eyewitness testimony? (Spoiler alert: Yes!)

FIGURE 1-4 Scientific thinking is necessary in matters of health and criminal justice.

Scientific thinking can help us to understand the world and when we should change our minds.

1.4 Step 1: Make observations.

Scientific study always begins with observations: we simply look for interesting patterns or cause-and-effect relationships. This is where a great deal of the creativity of science comes from. The U.S. Justice Department recently reviewed 28 criminal convictions that had been overturned by DNA evidence. It found that in most of the cases, the strongest evidence against the defendant had been eyewitness identification. This constitutes an "observation" that many defendants who are later found to be innocent were initially convicted based on eyewitness testimony.

SCIENTIFIC THINKING: MAKE OBSERVATIONS

OBSERVATION
To many people, consuming echinacea extract seems to reduce the likelihood of catching the common cold or the duration of symptoms of a cold.

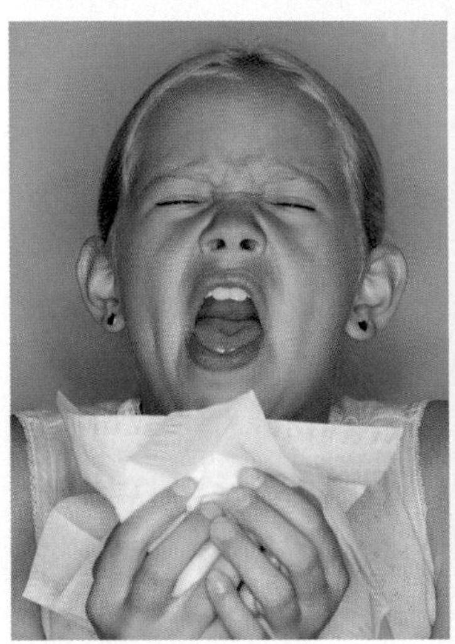

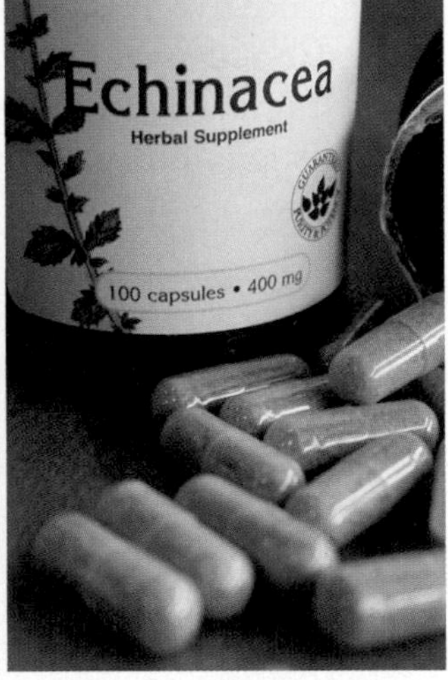

FIGURE 1-5 The first step of science: making observations about the world.

Opportunities for other interesting observations are unlimited. Using scientific thinking, we can (and will) also answer our other question.

Many people have claimed to observe that consuming extracts of the herb echinacea can reduce the likelihood of catching the common cold or the duration of cold symptoms (FIGURE 1-5). So let's investigate whether that is, in fact, true.

TAKE HOME MESSAGE 1.4

» Scientific thinking begins by making observations about the world, noting apparent patterns or cause-and-effect relationships.

1.5 Step 2: Formulate a hypothesis.

Based on observations, we can develop a **hypothesis** (*pl.* **hypotheses**), a proposed explanation for observed phenomena. What hypothesis could we make about the eyewitness-testimony observations described in the previous section? We could start with the hypothesis "Eyewitness testimony is always accurate." We may need to modify our hypothesis later, but this is a good start.

To be most useful, a hypothesis must accomplish two things.

1. It must establish an alternative explanation for a phenomenon. That is, it must be clear that if the proposed explanation is not supported by evidence or further observations, a different hypothesis is a more likely explanation.

2. It must generate testable predictions (FIGURE 1-6). This characteristic is important because we can evaluate the validity of a hypothesis only by putting it to the test. For example, we could disprove the "Eyewitness testimony is always accurate" hypothesis by demonstrating that, in certain circumstances, individuals who have witnessed a crime might misidentify the criminal.

Researchers often pose a hypothesis as a negative statement, proposing that there is no relationship between two factors, such as "Echinacea has no effect on the likelihood of catching a cold or the duration of cold symptoms." A hypothesis that states a *lack* of relationship between two factors is called a **null hypothesis.** Both types of hypothesis are equally valid, but a null hypothesis is typically easier to disprove. Any new observation that contradicts a null hypothesis can be sufficient for the researcher to reject it and conclude that an alternative hypothesis must be considered. Conversely, it is impossible to prove that a hypothesis is absolutely and permanently true: all evidence or further observations that support a hypothesis are valuable, but future evidence might show that the hypothesis is not true. Nonetheless, as more and more evidence supporting a hypothesis accumulates, our confidence increases that it is a true and accurate explanation for the phenomenon.

> "Scientific issues permeate the law. I believe [that] in this age of science we must build legal foundations that are sound in science as well as in law. The result, in my view, will further not only the interests of truth but also those of justice."
>
> — U.S. Supreme Court Justice Stephen Breyer, **at the annual meeting of the American Association for the Advancement of Science, February 1998**

For the question about echinacea and illness, we could state our hypothesis in either of these ways:

Hypothesis: Echinacea reduces the likelihood of catching the common cold and the duration of the symptoms of a cold.

Null hypothesis: Echinacea has no effect on the likelihood of catching the common cold or the duration of the symptoms of a cold.

SCIENTIFIC THINKING: FORMULATE A HYPOTHESIS

HYPOTHESIS
Echinacea reduces the likelihood of catching a cold and the duration of cold symptoms.

FIGURE 1-6 **Hypothesis: the proposed explanation for a phenomenon.**

Let's address one last thing about asking questions and proposing hypotheses. Although many of the questions scientists ask are *causal* and lead to hypotheses about what is responsible for observed phenomena, not all of the testable statements that scientists make are of this nature. Some questions, instead, are *descriptive*. They can be important questions and can lead to interesting discoveries, and the scientists may even be able to make predictions about them, but they're not answerable by a hypothesis. "What is the range of the bald eagle?" for example, is a descriptive question.

TAKE HOME MESSAGE 1.5

» A hypothesis is a proposed explanation for an observed phenomenon.

1.6 Step 3: Devise a testable prediction.

Not all hypotheses are created equal. A useful hypothesis helps us make predictions about novel situations; it guides us to knowledge about new situations.

When we devise a testable prediction from a hypothesis, our goal is to propose a situation that will give a particular outcome if our hypothesis is true, but will give a different outcome if our hypothesis is not true. Let's examine the two hypotheses we are considering.

Hypothesis: Eyewitness testimony is always accurate.

Prediction: Individuals who have witnessed a crime will correctly identify the criminal regardless of whether multiple suspects are presented one at a time or all at the same time in a lineup.

This is a good, testable prediction because, if our hypothesis is true, then our prediction will always be true. On the other hand, if one method of presenting suspects consistently causes incorrect identification of the criminal, our hypothesis cannot be true and must be revised or discarded.

Hypothesis: Echinacea reduces the likelihood of catching the common cold and the duration of the symptoms of a cold.

Prediction: If echinacea reduces the likelihood of catching the common cold and the duration of the symptoms of a cold, then individuals taking echinacea should get a cold less frequently than those not taking it, and when they do get sick, their illness should not last as long (FIGURE 1-7).

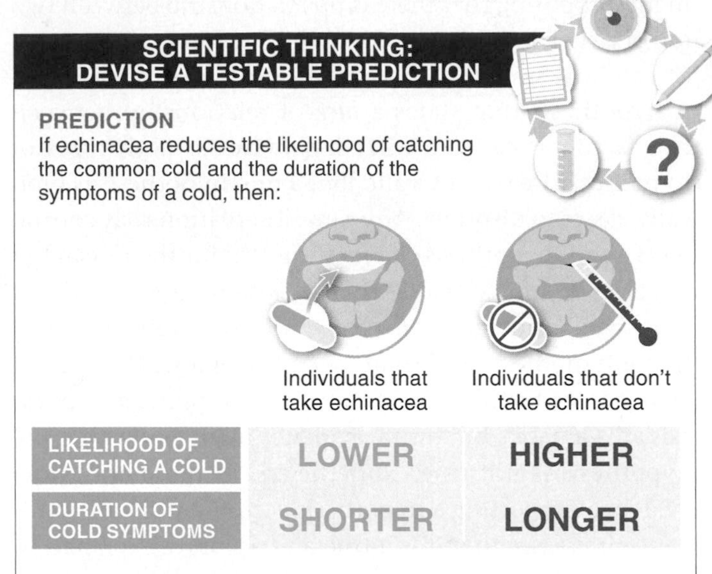
SCIENTIFIC THINKING:
DEVISE A TESTABLE PREDICTION
PREDICTION
If echinacea reduces the likelihood of catching the common cold and the duration of the symptoms of a cold, then:
Individuals that take echinacea
Individuals that don't take echinacea
LIKELIHOOD OF CATCHING A COLD
LOWER
HIGHER
DURATION OF COLD SYMPTOMS
SHORTER
LONGER
?

FIGURE 1-7 **Coming up with a testable prediction.** For a hypothesis to be useful, it must generate a testable prediction.

As you begin to think scientifically, you will find yourself making a lot of "if . . . then" types of statements: "*If* that is true," referring to some hypothesis or assertion someone makes, or perhaps a claim made by the manufacturer of a new health product, "*then* I would expect . . . ," proposing your own prediction about a related situation. Once you've made a testable prediction, the next step is to go ahead and test it.

TAKE HOME MESSAGE 1.6

» For a hypothesis to be useful, it must generate a testable prediction.

1.7 Step 4: Conduct a critical experiment.

Once we have formulated a hypothesis that generates a testable prediction, we conduct a **critical experiment,** an experiment that makes it possible to decisively determine whether a particular hypothesis is better than alternative hypotheses. There are many crucial elements in designing a critical experiment, and Section 1.10 covers the details of this process. For now, it is important just to understand that with a critical experiment we'll be able

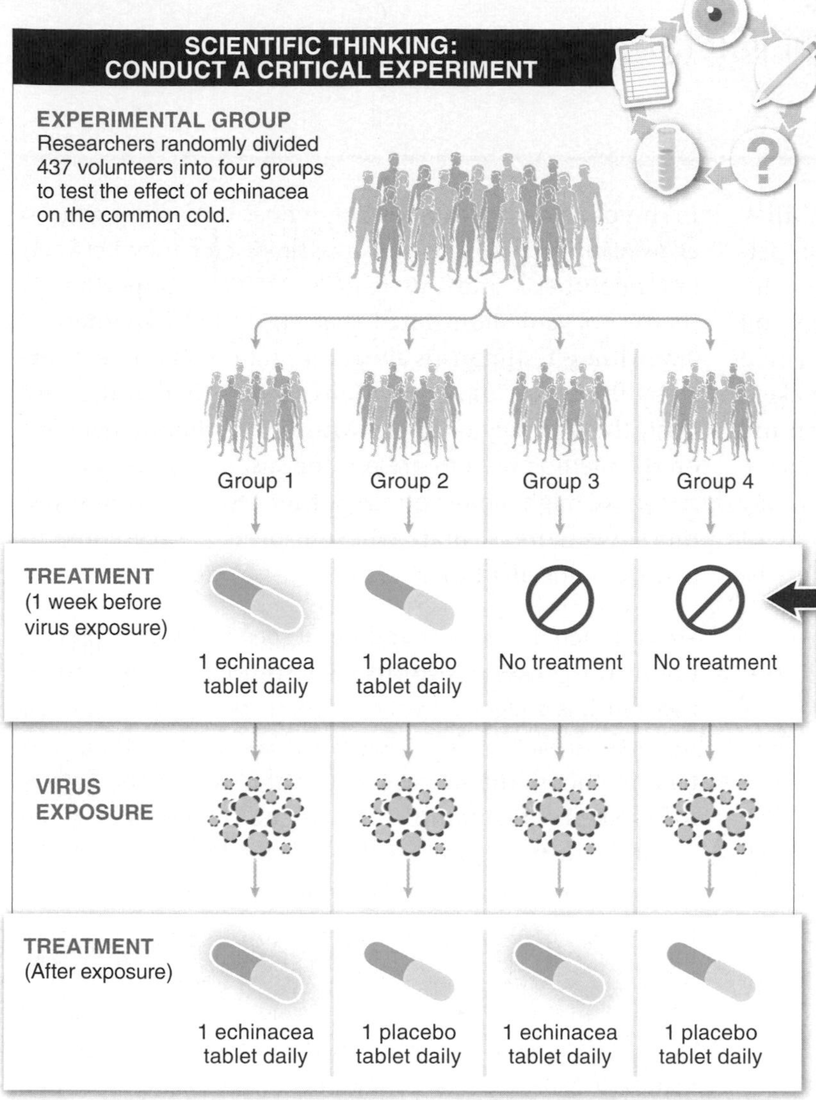
SCIENTIFIC THINKING:
CONDUCT A CRITICAL EXPERIMENT
EXPERIMENTAL GROUP
Researchers randomly divided 437 volunteers into four groups to test the effect of echinacea on the common cold.
Group 1
Group 2
Group 3
Group 4
TREATMENT
(1 week before virus exposure)
1 echinacea tablet daily
1 placebo tablet daily
No treatment
No treatment
To test whether treatment prior to exposure influences the development or duration of symptoms, groups 1 and 2 receive tablets for seven days prior to exposure, while groups 3 and 4 do not.
VIRUS EXPOSURE
TREATMENT
(After exposure)
1 echinacea tablet daily
1 placebo tablet daily
1 echinacea tablet daily
1 placebo tablet daily
HEALTH EVALUATION
For five days, doctors monitor all groups for cold symptoms.

FIGURE 1-8 **When you need to know . . .**

to make observations that compel us to reject alternative hypotheses, while not ruling out the particular hypothesis being tested, or vice versa.

In this step of the scientific method, a bit of cleverness can come in handy. Let's test our hypothesis "Eyewitness testimony is always accurate." First, we stage a mock purse snatching in front of a group of observers who do not know the crime is staged. Next, we ask observers to identify the criminal. To one group of observers we present six "suspects" all at once in a lineup. To another group of observers we present the same six suspects, but one at a time. The beauty of this experiment is that we actually know who the "criminal" is. We can evaluate with certainty whether an eyewitness's identification is correct or not.

This exact experiment has been done, but with an additional twist: the researcher did not include the actual "criminal" in any of the lineups or one-by-one presentations of the six "suspects." In the next section, we'll see what happened. Right now, let's consider a critical experiment for our other hypothesis.

To test the hypothesis "Echinacea reduces the likelihood of catching the common cold and the duration of cold symptoms," researchers performed an experiment with 437 people who volunteered to be exposed to viruses that cause the common cold. All of the volunteers had cold viruses (in a watery solution) dripped into their noses. The research participants were then secluded in hotel rooms for five days, and doctors examined them for the presence of the cold virus in their nasal cavity and for any cold symptoms (**FIGURE 1-8**). (We don't want to promulgate the movie stereotype of the "evil scientist." It is actually unusual to expose human participants to illness for research purposes, and they must always be informed of the risks and, by signing a form, give what is called "informed consent.")

Before the cold virus treatment, the volunteers were randomly divided into four groups. In two of the groups, each individual began taking a pill each day for a week prior to exposure to the virus. Those in one group received echinacea tablets, while those in the other took a **placebo,** a pill that looked identical to the echinacea tablet but contained no echinacea or other active ingredient. Neither the participant nor the doctor administering the tablets (and later checking for cold symptoms) knew what they contained. In the other two groups, the individuals began taking the tablets on the day of exposure to the cold virus. Again, one group got the echinacea and the other group got the placebo.

In the next section we'll evaluate how our hypotheses survive being confronted with the results of these critical experiments, and consider how we can move toward drawing conclusions.

TAKE HOME MESSAGE 1.7

» A critical experiment is one that makes it possible to decisively determine whether a particular hypothesis is correct.

1.8 Step 5: Draw conclusions, make revisions.

Once the results of a critical experiment are in, they are examined and analyzed. Researchers look for patterns and relationships in the evidence; they draw conclusions when possible and see whether their findings and conclusions support their hypothesis. If an experimental result differs from what is expected, that does not make it a "wrong answer." Science includes a great deal of trial and error. If the conclusions do not support the hypothesis, then it's necessary to revise the hypothesis, which may lead to more experiments. This step is a cornerstone of scientific thinking because it demands that you must be open-minded and ready to change what you think.

The results of the purse-snatching experiment were surprising. When the suspects (which, as you'll recall, did not include the actual "criminal") were viewed together in a lineup, the observers/witnesses erroneously identified someone as the purse snatcher about a third of the time. When the suspects were viewed one at a time, the observers made a mistaken identification less than 10% of the time.

Q Is eyewitness testimony in courts always right?

In this or any other experiment, it does not matter whether we can imagine a reason for the discrepancy between our hypothesis and our results. What is important is that we have demonstrated that our initial hypothesis—"Eyewitness testimony is always accurate"—is not supported by the data. Our observations suggest that, at the very least, the accuracy of an eyewitness's testimony depends on the method used to present the suspects. Based on this result, we might adjust our hypothesis to: "Eyewitness testimony is more accurate when suspects are presented to witnesses one at a time."

We can then devise new and more specific testable predictions. In the case of eyewitness testimony, further investigation has suggested that when suspects or pictures of suspects are placed side by side, witnesses compare them and tend to choose the suspect that *most resembles* the person they remember committing the crime. When viewed one at a time, suspects can't be compared in this way, and witnesses are less likely to make misidentifications.

In the echinacea study, the results were definitive (FIGURE 1-9). Those who took the echinacea were just as likely to catch a

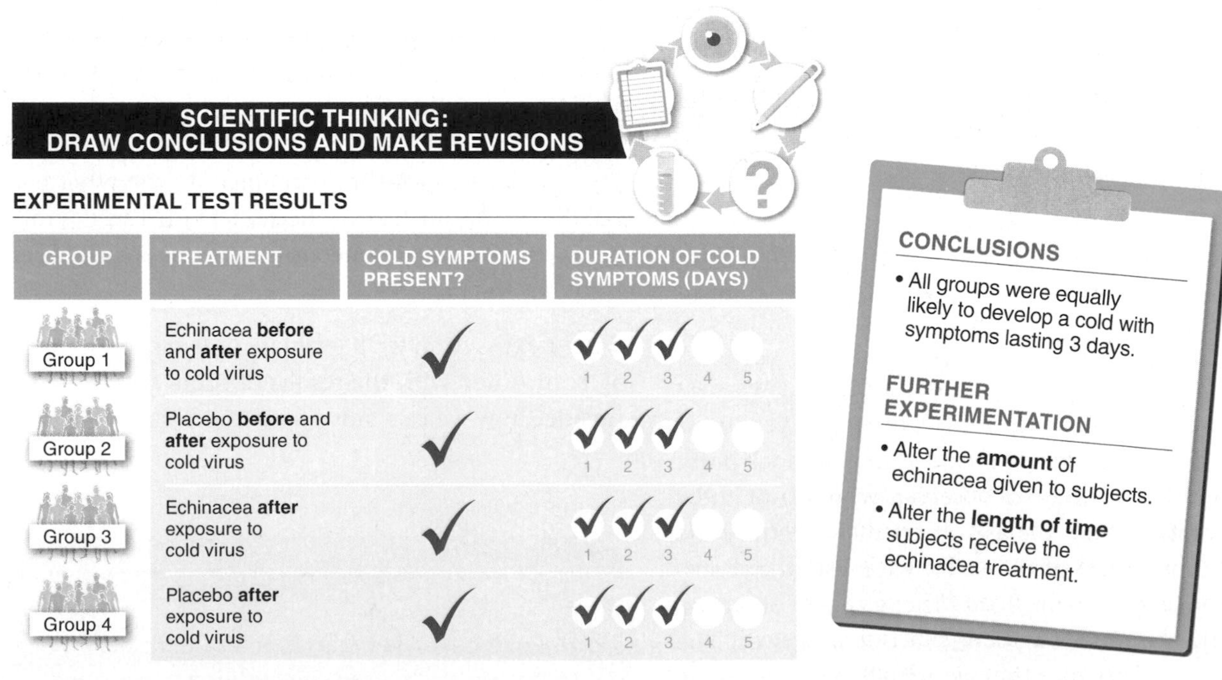

SCIENTIFIC THINKING:
DRAW CONCLUSIONS AND MAKE REVISIONS
EXPERIMENTAL TEST RESULTS
GROUP
TREATMENT
COLD SYMPTOMS PRESENT?
DURATION OF COLD SYMPTOMS (DAYS)
Group 1
Echinacea before and after exposure to cold virus
Group 2
Placebo before and after exposure to cold virus
Group 3
Echinacea after exposure to cold virus
Group 4
Placebo after exposure to cold virus
1 2 3 4 5
CONCLUSIONS
• All groups were equally likely to develop a cold with symptoms lasting 3 days.
FURTHER EXPERIMENTATION
• Alter the amount of echinacea given to subjects.
• Alter the length of time subjects receive the echinacea treatment.

FIGURE 1-9 **Using experimental conclusions to make revisions.**

Q Does echinacea help prevent the common cold or the duration of the symptoms of a cold?

cold as those who took placebo, and, once they caught the cold, the symptoms lasted for the same amount of time. In short, echinacea had no effect at all. Several similar studies have been conducted, all of which show that echinacea does not have any beneficial effect. As one of the researchers commented afterward, "We've got to stop attributing any efficacy to echinacea."

It seems clear that our initial hypothesis—"Echinacea protects people from catching colds and reduces the duration of cold symptoms"—is not correct. Further experimentation, however, might involve altering the amount of echinacea given to the research volunteers or the length of time they take echinacea before exposure to the cold-causing viruses. The outcomes in these studies illustrate that after the results of a critical experiment have been gathered and interpreted, it is important not just to evaluate the initial hypothesis but also to consider any necessary revisions or refinements to it. By revising a hypothesis, based on the results of experimental tests, we can explain the observable world with greater and greater accuracy.

TAKE HOME MESSAGE 1.8

» Based on the results of experimental tests, we can revise a hypothesis and explain the observable world with increasing accuracy. A great strength of scientific thinking, therefore, is that it helps us understand when we should change our minds.

1.9 When do hypotheses become theories, and what are theories?

It's an unfortunate source of confusion that, among the general public and in the popular media, the word "theory" is often used to refer to a hunch or a guess or speculation—that is, something we are not certain about. In fact, to scientists, the word means nearly the opposite: a hypothesis of which they are most certain. To reduce these common misunderstandings, we examine here how scientists describe our knowledge about natural phenomena.

Hypothesis As we have seen, hypotheses are at the very heart of scientific thinking. A hypothesis is a proposed explanation for a phenomenon. A useful hypothesis leads to testable predictions. Typically, when nonscientists use the word "theory"—as in, "I've got a theory about why there's less traffic on Friday mornings than on Thursday mornings"—they actually mean that they have a hypothesis.

Theory A **scientific theory** is an explanatory hypothesis for natural phenomena that is exceptionally well-supported by empirical data. A scientific theory can be thought of as a hypothesis that has withstood the test of time and is unlikely to be altered by any new evidence. Like a hypothesis, a scientific theory generates predictions and is testable; but because it has already been repeatedly tested and no experimental results have contradicted it, a scientific theory is viewed by the scientific community with nearly the same confidence as a fact. For this reason, it is inappropriate to describe something as "just a theory" as a way of asserting that it is not likely to be true.

Theories in science also tend to be broader in scope than hypotheses. In biology, two of the most important theories (which we explore in detail in Chapters 4 and 10) are the *cell theory,* that all organisms are composed of cells and all cells come from preexisting cells, and the *theory of evolution by natural selection,* that species can change over time and all species are related to one another through common ancestry.

TAKE HOME MESSAGE 1.9

» Scientific theories do not represent casual guesses about the natural world. Rather, they are hypotheses—proposed explanations for natural phenomena—that have been so strongly and persuasively supported by empirical observation that the scientific community views them as very unlikely to be altered by new evidence.

1.10–1.12

Well-designed experiments are essential to testing hypotheses.

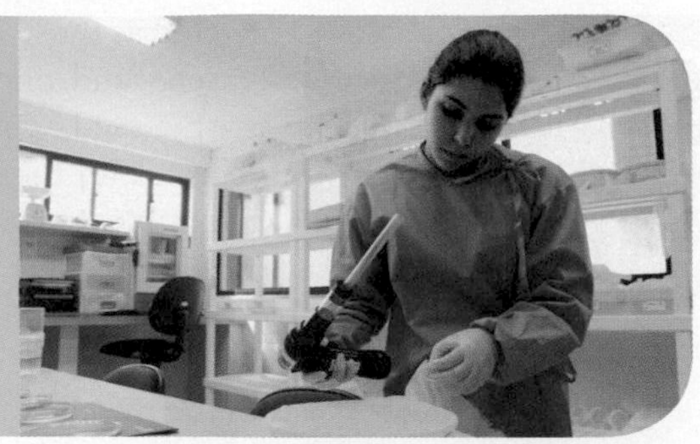

Controlled experiments increase the power of our observations. Here, biologist Danielle Varjal conducts research on mosquitoes that transmit the Zika virus.

1.10 Controlling variables makes experiments more powerful.

What are the elements that make up a well-designed experiment? In this section, we explore some of the ways to maximize an experiment's power to help us discern causes and effects.

First, let's consider some elements common to most experiments.

1. Treatment: any experimental condition applied to the research subjects. It might be the pattern used to show "suspects" (all at once or one at a time) to the witness of a staged crime, or a dosage of echinacea given to an individual.

2. Experimental group: a group of subjects who are exposed to a particular treatment—for example, the individuals given echinacea rather than placebo in the experiment described above. It is sometimes referred to as the "treatment group."

3. Control group: a group of subjects who are treated identically to the experimental group, with one exception—they are not exposed to the treatment. An example would be the individuals given placebo rather than echinacea.

4. Variables: the characteristics of an experimental system that are subject to change. In an experiment, the variables can be described as independent or dependent. An **independent variable** is some entity that can be observed and measured at the start of a process, and whose value can be changed as required. An independent variable might be, for example, the amount of echinacea a person is given. A **dependent variable** is one that can also be observed and measured, but whose response is created by the process being observed and depends on the independent variable. A dependent variable in the echinacea study might be how long the symptoms of a cold last.

When we speak of "controlling" variables we are describing the attempt to minimize any differences between a control group and an experimental group other than the treatment itself. That way, any differences between the groups in the outcomes we observe are most likely due to the treatment.

Let's look at a real-life example that illustrates the importance of considering all these elements when designing an experiment.

Stomach ulcers are erosions of the stomach lining that can be very painful. In the late 1950s, a doctor reported in the *Journal of the American Medical Association* that stomach ulcers could be effectively treated by having a patient swallow a balloon connected to tubes that circulated a refrigerated fluid. He argued that by super-cooling the stomach, acid production was reduced and the ulcer symptoms were relieved. He had convincing data to back up his claim: in all 24 of his patients who received this "gastric freezing" treatment, their condition improved (FIGURE 1-10). As a result, the treatment became widespread for many years.

Although there was a clear hypothesis ("Gastric cooling reduces the severity of ulcers") and some compelling observations (all 24 patients experienced relief), this experiment was poorly designed. In particular, there was no clear group with which to compare the patients who received the treatment.

A few years later, another researcher decided to do a more carefully controlled study. He recruited 160 patients with ulcers and gave 82 of them the gastric freezing treatment. The other 78 received a similar treatment in which they swallowed the balloon but had room-temperature water pumped in. The latter was an appropriate control group because the subjects were treated exactly like the experimental group, with the exception of a single difference between the groups—whether they experienced gastric freezing or not. The new experiment could test for an effect of the gastric freezing, while controlling for the effects of other, lurking variables that might affect the outcome.

The researcher did find that for 34% of those in the gastric freezing group their condition improved. Surprisingly,

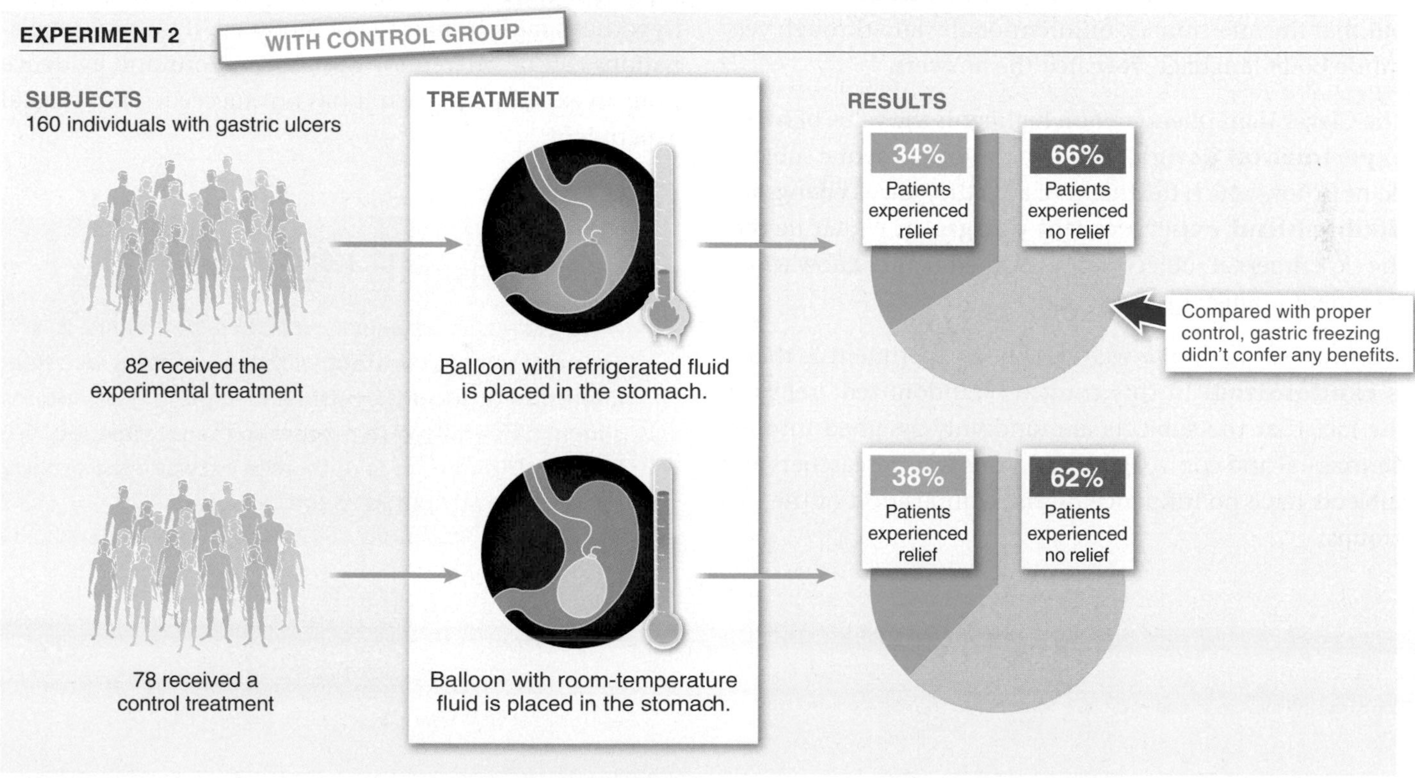

POOR EXPERIMENTAL DESIGN CAN LEAD TO FLAWED CONCLUSIONS
EXPERIMENT 1
WITHOUT CONTROL GROUP
SUBJECTS
24 individuals with gastric ulcers
24 received the experimental treatment (0 received a control treatment)
TREATMENT
Balloon with refrigerated fluid is placed in the stomach.
RESULTS
100% Patients experienced relief
Because all subjects experienced relief, it's tempting to conclude that the treatment worked.
EXPERIMENT 2
WITH CONTROL GROUP
SUBJECTS
160 individuals with gastric ulcers
82 received the experimental treatment
78 received a control treatment
TREATMENT
Balloon with refrigerated fluid is placed in the stomach.
Balloon with room-temperature fluid is placed in the stomach.
RESULTS
34% Patients experienced relief
66% Patients experienced no relief
Compared with proper control, gastric freezing didn't confer any benefits.
38% Patients experienced relief
62% Patients experienced no relief

FIGURE 1-10 **No controls.** Gastric freezing and stomach ulcers: a poorly designed experiment and a well-controlled "do-over."

however, he also found that 38% of those in the control group improved. In other words, gastric freezing didn't actually confer any benefit when compared with a treatment that did not involve gastric freezing. Not surprisingly, the practice was abandoned.

Another result from the gastric freezing study—and many other studies, as we'll see—was that it demonstrated the **placebo effect.** This is the frequently observed, but poorly understood, phenomenon in which people respond favorably to *any* treatment. The placebo effect amplifies the need for an appropriate control group. If the control group receiving the placebo or sham treatment has an outcome much like that of the experimental group, we can conclude that the treatment itself does not have an effect.

Another potential pitfall in designing an experiment is that the persons conducting the experiment may influence the experiment's outcome (even unintentionally). This is seen in the story of a horse named Clever Hans, whose owner claimed that Hans could perform remarkable intellectual feats, including multiplication and division. When given a numerical problem, the horse would tap out the answer number with his foot. Controlled experiments, however, demonstrated that Hans was able to solve problems only when he could see the person asking the question and when that person knew the answer (FIGURE 1-11). It turned out that the questioners, unintentionally and through very subtle body language, revealed the answers.

The Clever Hans phenomenon highlights the value of **blind experimental design,** in which the experimental subjects do not know which treatment (if any) they are receiving, and **double-blind experimental design,** in which neither the experimental subjects nor the experimenter know which treatment a subject is receiving.

Another hallmark of a well-designed experiment is that it is **randomized.** In this context, "randomized" refers to the fact that the subjects are randomly assigned into experimental and control groups. In this way, researchers and subjects have no influence on the composition of the two groups.

FIGURE 1-11 **Math whiz or ordinary horse?** The horse Clever Hans was said to be capable of mathematical calculations, until a controlled experiment demonstrated otherwise.

The use of "randomized, controlled, double-blind" experimental design can be thought of as an attempt to imagine all the possible ways that someone might criticize an experiment and to design the experiment so that the results cannot be explained by anything other than the effect of the treatment. In this way, the experimenter's results either support the hypothesis or invalidate it—in which case, the hypothesis must be rejected or modified. If multiple explanations can be offered for the observations and evidence from an experiment, then it has not succeeded as a critical experiment.

TAKE HOME MESSAGE 1.10

» To draw clear conclusions from experiments, it is essential to hold constant all those variables we are not interested in. Ideally, control and experimental groups should differ only with respect to the treatment of interest. Differences in outcomes between the groups can then be attributed to the treatment.

THIS IS HOW WE DO IT

Developing the ability to apply the process of science

1.11 Is arthroscopic surgery for arthritis of the knee beneficial?

Sometimes we may think that we "know" something when in fact there is no strong evidence supporting that belief. That's when the power of a randomized, well-controlled study can help us to gain a better understanding of our world—and to know when we should change our mind.

Consider, for example, the efficacy of a common knee surgery for arthritis. In arthroscopic surgery, surgeons make several small incisions and insert a tiny, flexible tube into the knee in order to see inside the joint. The surgeon may then remove or sand down debris and damaged cartilage. She may also flush out the debris and other materials from the knee joint.

Until very recently, this type of surgery was performed on more than 650,000 people in the United States each year. Some researchers asked a basic and reasonable question: Is this surgery actually beneficial for patients?

How could you determine whether a particular type of surgery is effective?

The general approach that the researchers took was straightforward. In a large group of volunteers who suffered from osteoarthritis of the knee, some received arthroscopic surgery, while others received a placebo surgery. The researchers then evaluated individuals' knee function and pain relief.

Wait a minute . . . what?

Yes. After recruiting 180 volunteer patients who were candidates for arthroscopic surgery for their arthritis, the researchers randomly assigned them to one of three groups. Two of the groups would undergo arthroscopic surgery, while the third group would receive the placebo surgery. The participants did not know which group they were in, but all of them understood that they might undergo only the placebo surgery. They all signed an "informed consent" form stating: "I realize that I may receive only placebo surgery. I further realize that this means that I will not have surgery on my knee joint. This placebo surgery will not benefit my knee arthritis."

How does general scientific literacy—particularly among nonscientists such as the volunteers in this study—help in advancing our knowledge and understanding about a particular phenomenon?

The patients were randomly assigned to one of the three groups only after they were in the operating room. The surgeon (one surgeon performed all the operations) was given an envelope indicating which treatment each patient was to receive:

1. *Arthroscopic surgery with debridement:* The surgeon made the incisions and irrigated the knee joint with 10 liters of fluid, and then shaved, trimmed, and smoothed (that is, debrided) the cartilage.

2. *Arthroscopic surgery with lavage:* The surgeon made the incisions and irrigated the knee joint with 10 liters of fluid, as in the first group, but did not debride. Any debris that could be flushed out was removed (the process of "lavage").

3. *Placebo surgery:* The surgeon made the same incisions as in the first two groups, and then asked for all instruments and manipulated the knee as if

(continued on following page)

performing arthroscopic surgery with debridement. The surgeon also splashed saline to simulate lavage.

In all three treatments, patients spent the night in the hospital, cared for by nurses who did not know which treatment group they were in.

How did the researchers decide whether the arthroscopic surgery was effective?

The researchers evaluated the effectiveness of the surgery at seven points over the next two years. The evaluations included patients' self-reports of knee pain and body pain and researchers' measurements of knee function.

Did patients improve in knee function or have less pain?

At no point following surgery was pain reduced or knee function improved in patients receiving either type of arthroscopic surgery relative to those undergoing the placebo surgery. Pain scores were on a scale of 0–100 (higher scores indicating greater pain), based on a 12-item evaluation called the Knee Specific Pain Scale. At two years, for example, the pain scores were:

Patient Group	Mean Pain Score (± standard deviation)
1. Debridement	51 ± 23
2. Lavage	54 ± 24
3. Placebo	52 ± 24

What conclusions can you draw from these results?

In the *New England Journal of Medicine,* the researchers summarized their conclusions succinctly: "This study provides strong evidence that arthroscopic lavage with or without débridement is not better than and appears to be equivalent to a placebo procedure in improving knee pain and self-reported function."

TAKE HOME MESSAGE 1.11

» In a well-controlled experiment, researchers demonstrated that arthroscopic knee surgery for osteoarthritis was no more beneficial for patients—in terms of knee pain and knee functioning—than a placebo surgery.

1.12 We've got to watch out for our biases.

In 2001, the journal *Behavioral Ecology* changed its policy for reviewing manuscripts that were submitted for publication. Its new policy instituted a double-blind process, whereby neither the reviewers' nor the authors' identities were revealed. Previously, the policy had been a single-blind process in which reviewers' identities were kept secret but authors' identities were known to the reviewers. An analysis of papers published between 1997 and 2005 revealed that, after 2001 (when the double-blind policy took effect), there was a significant increase in the number of published papers in which the first author was female (FIGURE 1-12). Analysis of papers published in a similar journal that maintained the single-blind process over that period revealed no such increase.

Q Can scientists be sexist? How would we know?

This study illustrates that people, including scientists, may have biases—sometimes subconscious—that influence their behavior. It also serves as a reminder of the importance of proper controls in experiments.

Another important element of scientific thinking is **replication**—the process of repeating a study. An experiment that can be done over and over again by a variety of researchers to give the same results is an effective defense against biases. Further, when experiments are repeated and the same results are obtained, our confidence in them—and the hypotheses the results support—is increased. Additionally, sometimes a tiny variation in the experimental design can lead to different results; this can help us isolate the variable that is primarily responsible for the outcome of the experiment.

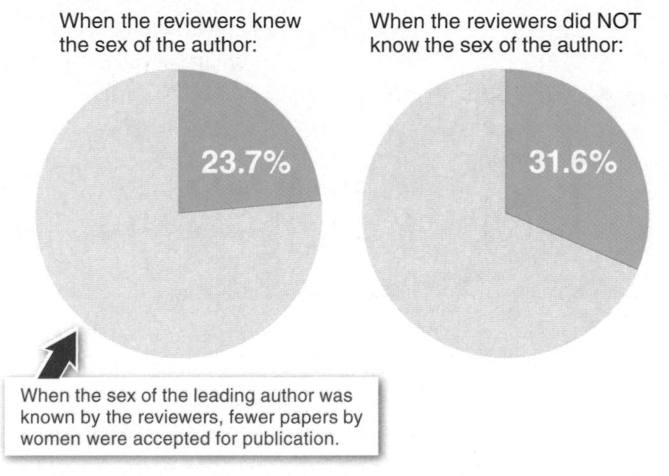

PERCENTAGE OF PAPERS PUBLISHED WITH FEMALE FIRST AUTHOR
When the reviewers knew the sex of the author:
23.7%
When the reviewers did NOT know the sex of the author:
31.6%
When the sex of the leading author was known by the reviewers, fewer papers by women were accepted for publication.

GRAPHIC CONTENT
Thinking critically about visual displays of data

Turn to p. 27 for a closer inspection of this figure.

FIGURE 1-12 **Bias against female scientists?** The journal *Behavioral Ecology* accepted more papers from female authors when the reviewers were not aware of the author's sex.

TAKE HOME MESSAGE 1.12

» **Biases can influence our behavior, including our collection and interpretation of data. With careful controls, it is possible to reduce the impact of biases.**

1.13–1.16

Scientific thinking can help us make wise decisions.

What can you believe? Reading labels is essential to evaluating products and the claims about them.

1.13 Visual displays of data can help us understand and explain phenomena.

"Let's look at the data." Whether making a point, illustrating an idea, or facilitating the testing of a hypothesis, visual displays of data condense large amounts of information into a more easily digested form. In doing so, they can help readers think about and compare data, ultimately helping them to synthesize the information and see useful patterns. Among the multitude of ways to display data, a few forms are used most frequently. These include bar graphs, line graphs, and pie charts (FIGURE 1-13).

Visual displays of data generally have a few common elements. Most have a title, which usually appears at the top and describes the content of the display. Bar graphs and line graphs include axes, usually a horizontal axis, also called the *x*-axis, and a vertical axis, called the *y*-axis. Each axis has a scale, generally labeled with some gradations, indicating one dimension by which the data can be described.

The *x*-axis of a line graph, for example, might be labeled "Time spent studying each day (hrs)," while the *y*-axis might be labeled "Performance on midterm exams (%)." Depending on the size of the data set, individual data points may be included on the graph, with additional information conveyed by the shape, color, or pattern of the data points (FIGURE 1-14). A line or curve may be used to connect data points or to illustrate a relationship between the two variables, and the axes must always be labeled and include the units of measure.

Rather than displaying individual data points, a bar graph has rectangular bars, each with a height proportional to the value being represented. In a pie chart, each "slice" is a proportion of the whole. A legend may be included for the graph, identifying which information is represented by which data point or bar or pie slice.

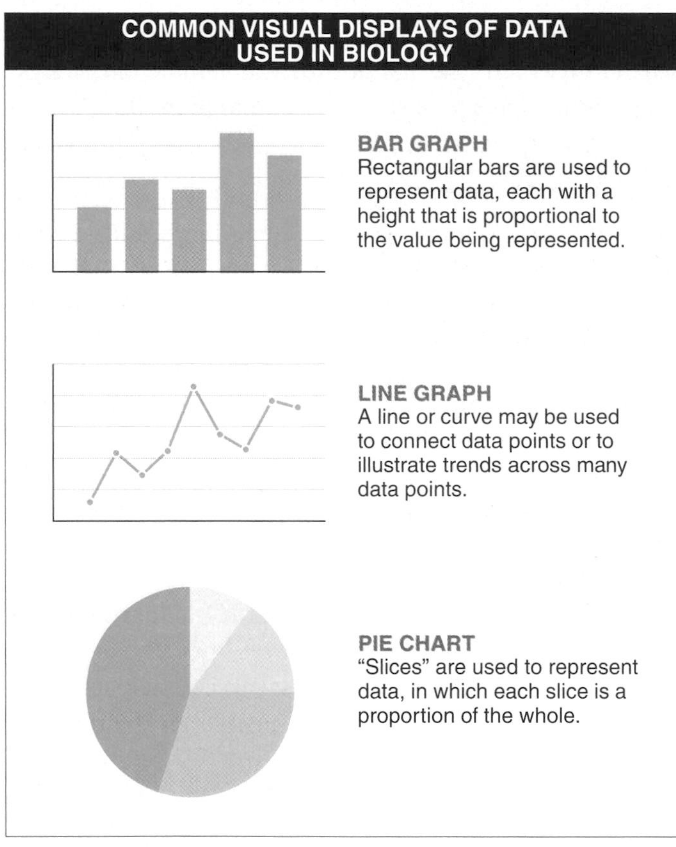

COMMON VISUAL DISPLAYS OF DATA USED IN BIOLOGY
BAR GRAPH
Rectangular bars are used to represent data, each with a height that is proportional to the value being represented.
LINE GRAPH
A line or curve may be used to connect data points or to illustrate trends across many data points.
PIE CHART
"Slices" are used to represent data, in which each slice is a proportion of the whole.

FIGURE 1-13 **Presenting what we have observed, precisely and concisely.**

One of the most common functions of visual displays of information is to present the relationship between two

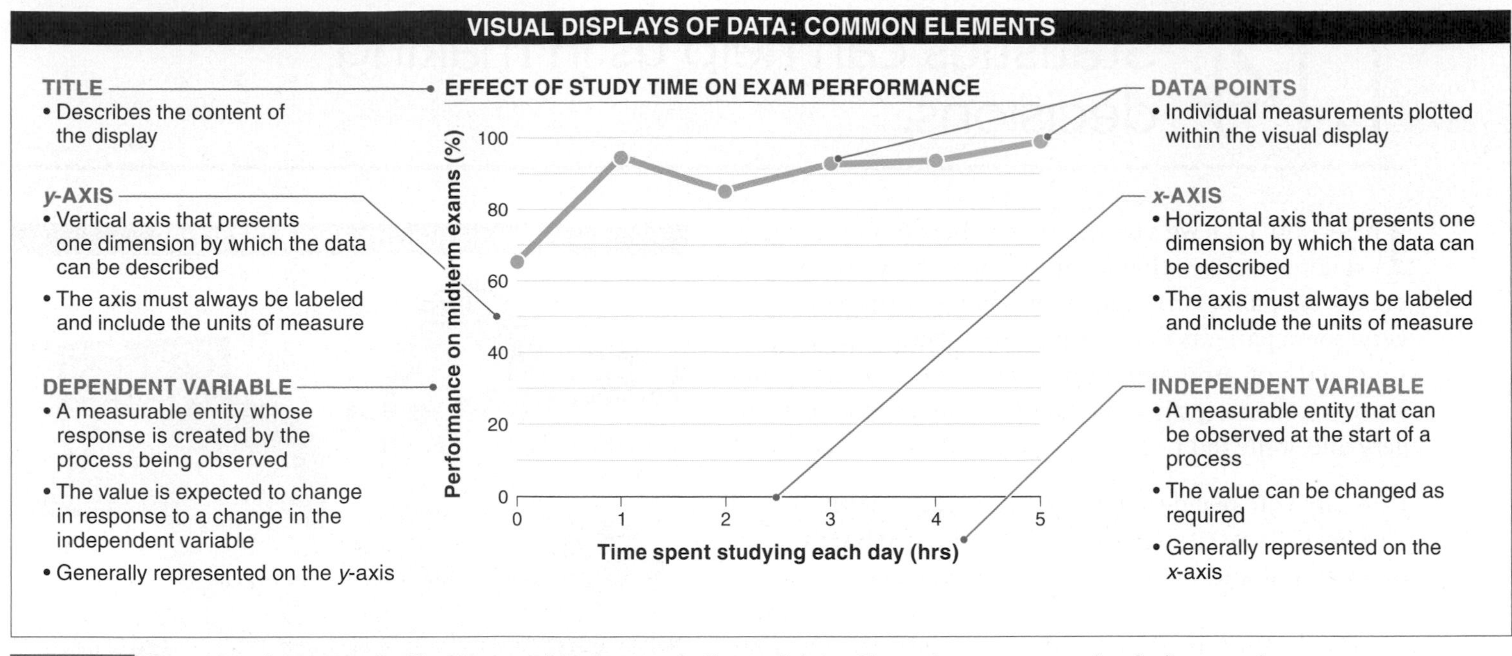

VISUAL DISPLAYS OF DATA: COMMON ELEMENTS
TITLE
• Describes the content of the display
EFFECT OF STUDY TIME ON EXAM PERFORMANCE
DATA POINTS
• Individual measurements plotted within the visual display
y-AXIS
• Vertical axis that presents one dimension by which the data can be described
• The axis must always be labeled and include the units of measure
x-AXIS
• Horizontal axis that presents one dimension by which the data can be described
• The axis must always be labeled and include the units of measure
DEPENDENT VARIABLE
• A measurable entity whose response is created by the process being observed
• The value is expected to change in response to a change in the independent variable
• Generally represented on the y-axis
INDEPENDENT VARIABLE
• A measurable entity that can be observed at the start of a process
• The value can be changed as required
• Generally represented on the x-axis
Performance on midterm exams (%)
100
80
60
40
20
0
0
1
2
3
4
5
Time spent studying each day (hrs)

FIGURE 1-14 **Common elements of effective graphical presentations of data.** Shown here: an example of a line graph.

variables, such as in a graph. As we saw in Section 1.10, an independent variable can be changed by the experimenter as required, while a dependent variable is one whose response is created by the process being observed and depends on the independent variable. The dependent variable is generally represented on the *y*-axis and is expected to change in response to a change in the independent variable, represented on the *x*-axis. The number of hours of sleep a student gets each night, for example, could be thought of as an independent variable, while some measure of academic performance—maybe grade point average—would be a dependent variable.

Visual displays of data may have features that reduce their effectiveness, or even cause them to be downright misleading. These difficulties can arise from ambiguity in the axis labels or scales, incomplete information on how each data point was collected (and how the points might have varied), biases or hidden assumptions in the presentation or grouping of the data, unknown or unreliable sources of data, or an insufficient or inappropriate context given for the data presentation.

Q Can a visual display "lie" to a reader, even though the data used to create it are true?

In each chapter of this book, you will see that one of the visual displays of data is labeled with the Graphic Content icon (you'll have noticed this on Figure 1-12). For each of these figures, in the Check Your Knowledge section at the end of the chapter you'll find several questions about the figure's content. These questions will guide you in evaluating aspects of the figure (such as: "What is its main point?") and in extracting information from it. They will also help you understand when it might be warranted to approach a data display with more than the usual amount of skepticism, and when greater trust is reasonable.

TAKE HOME MESSAGE 1.13

» Visual displays of data, which condense large amounts of information, can aid in the presentation and exploration of the data. The effectiveness of such displays is influenced by the precision and clarity of the presentation, and can be reduced by ambiguity, biases, hidden assumptions, and other issues that reduce a viewer's confidence in the underlying truth of the presented results.

1.14 Statistics can help us in making decisions.

In Section 1.11 we saw that researchers found, through experimentation, that arthroscopic surgery does not reduce knee pain. Two years after having arthroscopic surgery, some patients had less pain than before surgery and some did not. And among the patients in the group not receiving the surgery, some had less pain than before surgery and some did not.

How do you decide whether the surgery actually had an effect? This knowledge comes from **statistics,** a set of analytical and mathematical tools designed to help researchers gain understanding from the data they gather. To better understand statistics, let's start with a simple situation.

Suppose you measured the height of two people. One is a woman who is 5 feet 10 inches tall. The other is a man who is 5 feet 6 inches tall. If these were your only two observations of human height, you might conclude that female humans are taller than males. But suppose you measured the height of 100 women and 100 men, chosen randomly from a population. Then you find that for the 100 men, the average height is 5 feet 9.5 inches, and for the 100 women, the average height is 5 feet 4 inches.

Better still, with 100 measures for each sex, the data can illuminate not only the average but also some measure of how far the typical observation is from the average. One such measure is the standard deviation. You may find that not only is the average height for a man in this study 5 feet 9.5 inches, but that the standard deviation is 3 inches. This may be stated as "5 feet 9.5 inches ± 3 inches" ("plus or minus 3 inches"). The data might show for the women "5 feet 4 inches ± 2.5 inches" (FIGURE 1-15).

HUMAN HEIGHT AND STATISTICS

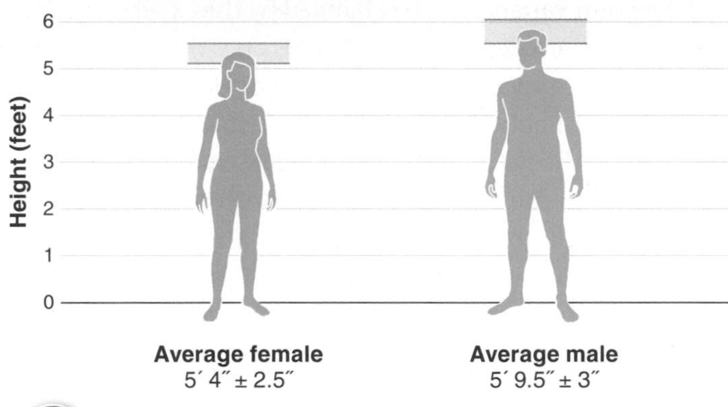

Don't get fooled! Making one or two observations isn't enough to draw general conclusions about natural phenomena, such as height.

FIGURE 1-15 **Drawing conclusions based on limited observations is risky.** Measuring a greater number of people will generally help us draw more accurate conclusions about human height.

One handy mathematical property of the standard deviation is that for a normally distributed set of observations, about two-thirds of the observations will be within one standard deviation above or below the average. In other words, about two-thirds of the men are between 5 feet 6.5 inches tall (3 inches less than the average) and 6 feet 0.5 inches tall (3 inches more than the average), and about two-thirds of the women are between 5 feet 1.5 inches and 5 feet 6.5 inches tall.

Using data to describe the characteristics of individuals participating in a study is useful, but often we want to know whether data support (or do not support) a hypothesis. For example, suppose we want to know whether having access to a textbook helps a student perform better in a biology class. Statistics can help us answer this question. After conducting a study, let's say we find that students who had access to a textbook scored an average of 81% ± 8% on their exams, while those who did not scored an average of 76% ± 7%.

In this example, it is difficult to distinguish between the two possible conclusions.

Possibility 1: Students having access to a textbook *do perform better* in biology class. Our sampling of this class reveals a true relationship between the two variables, textbook access and class performance. (After all, 81% is higher than 76%.)

Possibility 2: Students having access to a textbook *do not perform any better* in biology class. The variation in the scores for the two groups may be too large to allow us to conclude that there's any effect of having access to a textbook. Instead, the difference in average scores for the two groups might mean that more of the high-performing students just happened, through random chance, to be in the group given access to a textbook.

But what if the students with access to a textbook scored, on average, 95% ± 5%, while those without access scored only 60% ± 5%? In this case, we would be much more confident that there is a significant effect of having access to a textbook (FIGURE 1-16), because even with the large variation in the scores seen in each group, the two averages are still very different from each other.

Statistical methods help us to decide between these two possibilities and, equally important, to state how confident we are that one or the other is true. The greater the difference between two groups (95% vs. 60% is a greater difference than 81% vs. 76%), and the smaller the variation in each group (± 5% in the 95% and 60% groups vs. ± 8% and ± 7% in the 81% and 76% groups), the more confident we are of the conclusion that there is a significant effect of the treatment. In other words, in the case where the groups of students scored 95% or 60%, depending on whether they did or did not have access to a textbook, it still is possible that having a textbook did not actually improve performance and that this observed difference was just the result of chance. But this conclusion is very unlikely.

Statistics can also help us identify relationships (or lack of relationships) between variables. For example, we might note that when there are more firefighters at a fire, the fire is larger and causes more damage. This is a **positive correlation,** meaning that when one variable (the number of firefighters) increases, so does the other (the severity of the fire). Should we conclude that firefighters make fires worse? No. While correlations can reveal relationships between variables, they don't tell us *how* the variables are related or whether change in one variable *causes* change in another. (You may have heard or read the phrase "correlation is not causation," which refers to this sort of situation.) To estimate the effect of the number of firefighters on the amount of damage, we would need to compare the amount of damage from fires of similar size that are fought by different numbers of firefighters.

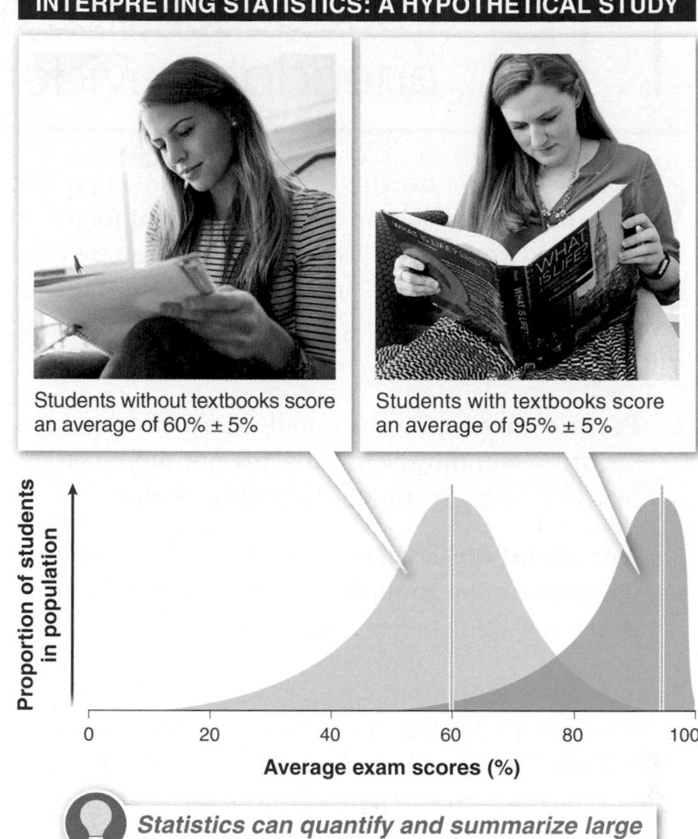
INTERPRETING STATISTICS: A HYPOTHETICAL STUDY
Students without textbooks score an average of 60% ± 5%
Students with textbooks score an average of 95% ± 5%
Proportion of students in population
0
20
40
60
80
100
Average exam scores (%)
Statistics can quantify and summarize large amounts of information, making it possible to draw more accurate conclusions.

FIGURE 1-16 **Drawing conclusions based on statistics.** Statistical analyses can help us evaluate the significance of our observations.

Ultimately, statistical analyses can help us organize and summarize the observations that we make and the evidence we gather in an experiment. These analyses can then help us decide whether any differences we measure between experimental and control groups are likely to be the result of the treatment, and how confident we can be in that conclusion.

TAKE HOME MESSAGE 1.14

» Because much variation exists in the world, statistics can help us evaluate whether any differences between a treatment group and a control group can be attributed to the treatment rather than chance.

1.15 Pseudoscience and misleading anecdotal evidence can obscure the truth.

One of the major benefits of scientific thinking is that it can prevent you from being taken in or fooled by false claims. Two types of "scientific evidence" are cited in the popular media frequently and are responsible for people erroneously believing that two things are linked, when in fact they are not.

1. **Pseudoscience,** in which individuals make scientific-sounding claims that are not supported by trustworthy, methodical scientific studies
2. **Anecdotal observations,** in which, based on just one or a few observations, people conclude that there is or is not a link between two things

Pseudoscience is all around us, particularly in the claims made on the packaging of consumer products and food (FIGURE 1-17). Pseudoscience capitalizes on a belief shared by most people: that scientific thinking is a powerful method for learning about the world. Beginning in the 1960s, for example, consumers encountered the assertion by the makers of a sugarless gum that "four out of five dentists surveyed recommend sugarless gum for their patients who chew gum." Maybe the statement is factually true, but the general relationship it implies may not be. How many dentists were surveyed? If the gum makers surveyed only five dentists, then the statement may not represent the proportion of *all* dentists who would make such a recommendation. And how were the dentists sampled? Were they at a shareholders' meeting for the sugarless gum company? What alternatives were given—perhaps gargling with a tooth-destroying acid? You just don't know. That's what makes it pseudoscience.

We are all familiar with anecdotal evidence. We may find compelling parallels between suggestions made in horoscopes and events in our lives, or we may think that we have a lucky shirt. That's how **superstitions** develop. Anecdotal observations can seem harmless and can be emotionally powerful. But because they do not include a sufficiently large and representative set of observations of the world, they can lead people to draw erroneous conclusions.

Q Why do people develop superstitions?

What do the scientific-sounding words "revive" and "essential" convey to you?
revive
fruit punch
essential
orange-orange
vitaminwater
vitaminwater

FIGURE 1-17 **Pseudoscientific claims are often found on food products.**

One such erroneous conclusion involves the developmental disorder autism and the vaccination for measles, mumps, and rubella (commonly called the MMR vaccine).

In 1998, the prestigious medical journal *The Lancet* published a report by a group of researchers that described a set of symptoms (diarrhea, abdominal pain, bloating) related to bowel inflammation in 12 children who exhibited the symptoms of autism. The parents or physicians of 8 of the children in the study said that the behavioral symptoms of autism appeared shortly after the children received MMR vaccination. For this reason, the authors of the report recommended further study of a possible link between the MMR vaccine, the bowel problems, and autism.

In a press conference, one of the paper's authors suggested a link between autism and the MMR vaccine, recommending single vaccines rather than the MMR triple vaccine until it could be proved that the MMR vaccine did not trigger autism. The press and many people took the claims made by the researcher at the press conference as evidence that the MMR vaccine causes autism. Over the course of the next few years, the number of children getting the MMR vaccine dropped significantly, as parents sought to reduce the risk of autism in their children (FIGURE 1-18).

Unfortunately, this is a notable case of poor implementation of scientific thinking. The study was small (only 12 children), the sample was not randomized (that is, the researchers selected the study participants based on the symptoms they showed), and no control group was included for comparison (for example, children who had been vaccinated but did not exhibit autism symptoms). As flawed as the study was, it did not actually find or even report a link between autism and the MMR vaccine. That purported link came only from the unsupported, suggestive statements made by one of the study's authors at the press conference.

Q Does the vaccine for measles, mumps, and rubella cause autism?

Those design flaws were not the only weaknesses in the research. It later was discovered that before the paper's publication, the study's lead author (the one who spoke at the press conference) had received large sums of money from lawyers seeking evidence to use in lawsuits against the MMR vaccine manufacturers. That author had also applied for a patent for a vaccine to compete with the most commonly used MMR vaccine. In the light of these initially undisclosed biases, 10 of the paper's 12 authors published a retraction of their original interpretation of their results. The *British Medical Journal* published an article in 2011 declaring that the "article linking MMR vaccine and autism was fraudulent."

In the years since the original paper was published, many well-controlled, large-scale critical experiments have been conducted. All have been definitive in their conclusion that there is no link between the MMR vaccine and autism. Here are a few of these studies.

1. A study of *all* children born in Denmark between 1991 and 1998 found no difference in the incidence of autism among the 440,655 children who were vaccinated with the MMR vaccine and the 96,648 children who were not vaccinated.

2. A 2005 study in Japan showed that after use of the MMR vaccine was stopped in 1993, the incidence of autism continued to increase.

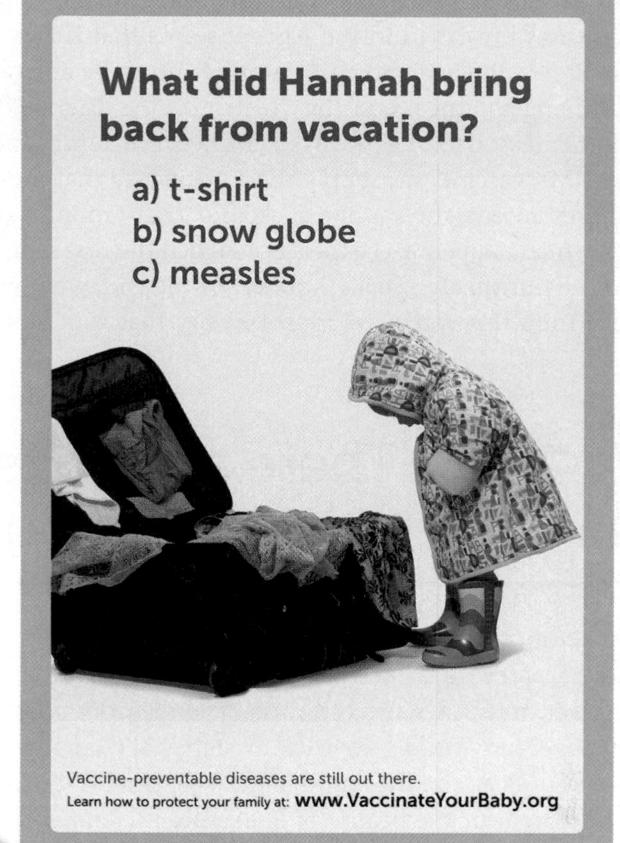

What did Hannah bring back from vacation?
a) t-shirt
b) snow globe
c) measles
Vaccine-preventable diseases are still out there.
Learn how to protect your family at: www.VaccinateYourBaby.org

A misguided fear of "catching" autism has caused some parents to decline immunizations for their children.

FIGURE 1-18 Bad science leads to dangerous behavior.

3. A study of 1.8 million children in Finland who were followed up for 14 years after getting the MMR vaccine found no link at all between the occurrence of autism and the vaccine.

At this point, the consensus of the international scientific community is that there is no scientific evidence for a link between the MMR vaccine and autism.

> "Science is a way to call the bluff of those who only pretend to knowledge. It is a bulwark against mysticism, against superstition, against religion misapplied to where it has no business being. If we're true to its values, it can tell us when we're being lied to."
>
> — Carl Sagan, in *The Demon-Haunted World: Science as a Candle in the Dark*, 1997

So what explains the observation that there are more autism cases now than in the past? It seems that this is a function of the better identification of autism by doctors and changes in the process by which autism is diagnosed. Another reason for the perceived link between autism and the MMR vaccine was simply the coincidence that most children receive the vaccine at around 18–19 months of age, which happens to be the age at which the first symptoms of autism are usually noticed. In the end, what we learn from this is that we must be wary that we do not generalize from anecdotal observations or let poorly designed studies obscure the truth.

TAKE HOME MESSAGE 1.15

» Pseudoscience and anecdotal observations often lead people to believe that links between two phenomena exist, when in fact there are no such links.

1.16 There are limits to what science can do.

Scientific thinking gives us a framework to make sense of what we see, hear, and read in our lives. There are limits, however, to what science can do. The scientific method will never prove or disprove the existence of God. It won't help us understand the mathematical elegance of Fermat's last theorem or the beauty of Shakespeare's sonnets. Nor can it address the question of whether we should allow the cloning of humans.

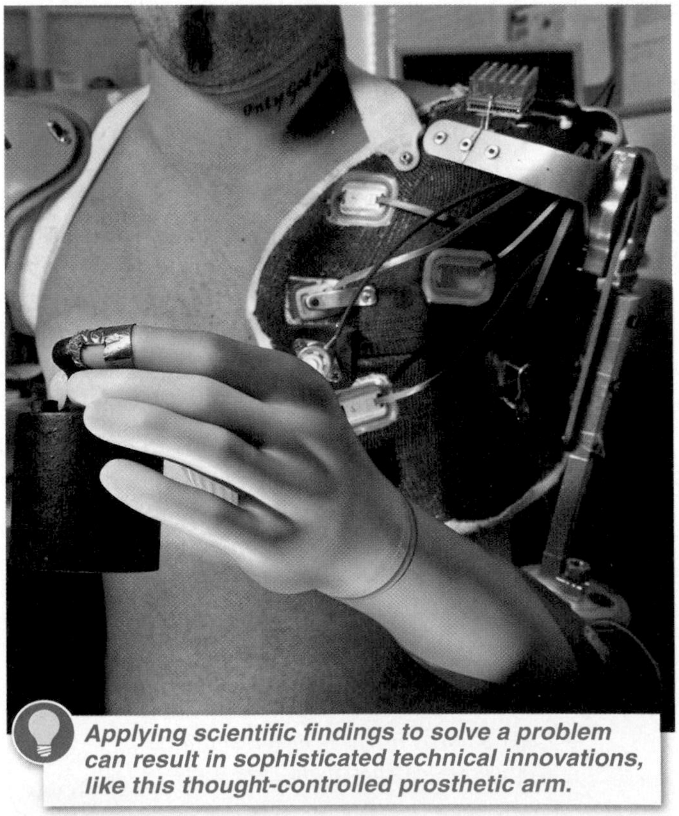

Scientific thinking differs from nonscientific approaches such as mathematics and logic and the study of artistic expression in that it relies on *measuring* phenomena in some way. The generation of value judgments and other types of non-quantifiable, subjective information—such as religious assertions of faith—falls outside the realm of science. Scientific thinking does not, for example, generate moral statements, and it cannot give us insight into ethical problems. What "is" (i.e., what we observe in the natural world) is not necessarily what "ought" to be (i.e., what is morally right). It may or may not be.

Also, much of what is commonly considered to be science, such as the construction of new engineering marvels or the heroic surgical separation of conjoined twins, is not scientific at all. Rather, these are technical innovations and developments. While **technology** relies on sophisticated scientific research, it represents the *application* of research findings to varied fields such as manufacturing and medicine to solve problems (FIGURE 1-19).

Applying scientific findings to solve a problem can result in sophisticated technical innovations, like this thought-controlled prosthetic arm.

FIGURE 1-19 The application of science. A prosthetic arm is tested in the lab.

TAKE HOME MESSAGE 1.16

» Although scientific thinking may be the most effective path toward understanding the observable world, it cannot give us insights into the generation of value judgments and other types of non-quantifiable, subjective information.

1.17

On the road to biological literacy: what are the major themes in biology?

Feather identification expert Roxie Laybourne, with a portion of the bird collection at the National Museum of Natural History.

1.17 What is life? Important themes unify and connect diverse topics in biology.

Biology is, literally, the study of life. But what is life? And how is it distinguished from non-life? Spend a few moments trying to define exactly what life is and you'll realize that it is not easily described with a simple definition. A useful approach is to consider the characteristics shared by all living organisms and living systems:

- **A complex, ordered organization consisting of one or more cells.** Cells carry out the functions necessary for life.
- **The use and transformation of energy to perform work.** Organisms can perform many reactions and activities, by acquiring, using, and transforming energy.
- **Sensitivity and responsiveness to the external environment.** Living organisms are able to respond to stimuli—such as light, moisture, or another organism.
- **Regulation and homeostasis.** Organisms are able to maintain relatively constant internal conditions that may differ from the external environment.
- **Growth, development, and reproduction.** Organisms can grow and develop, and they carry information relating to these and other processes that they can pass on to offspring.
- **Evolutionary adaptation leading to descent with modification over time.** Populations have the capacity to change over time. As a consequence of organisms' ability to reproduce and of evolutionary change, populations may become better adapted to their environments.

In this guide to biology, as we explore the many facets of biology and its relevance to life in the modern world, you will find two central and unifying themes recurring throughout.

Hierarchical organization. Life is organized on many levels within individual organisms, including atoms, cells, tissues, and organs. And in the larger world, organisms themselves are organized into many levels: populations, communities, and ecosystems within the biosphere.

The power of evolution. Evolution, the change in genetic characteristics of a population over time, accounts for the diversity of organisms and the unity among them.

These central unifying themes connect the diverse topics in biology, which include the chemical, cellular, and energetic foundations of life; the genetics, evolution, and behavior of individuals; the staggering diversity of life and the unity underlying it; and ecology, the environment, and the links between organisms and the world they inhabit. Let's continue our exploration of life!

TAKE HOME MESSAGE 1.17

» "Life" is not easily described with a simple definition. The characteristics shared by all living organisms include complex and ordered organization; the use and transformation of energy; responsiveness to the external environment; regulation and homeostasis; growth, development, and reproduction; and evolutionary adaptation leading to descent with modification.

STREET BIO

Using evidence to guide decision making in our own lives

Rainy days and Mondays

Q: What's the most important outcome of scientific thinking? Scientific thinking tells us when to change our minds about the natural world. Let's consider a question that might seem silly—even trivial—and watch our minds open.

Q: It often seems to rain more on weekends. Could it be true? Before you read on, write down your one-word answer (and a justification for it, if you'd like).

Q: Well, does it rain more on weekends? At first, it seems that the answer must be no. How could the weather even "know" what day it is? Weather, after all, existed long before humans, calendars, and Monday mornings.

Rather than relying on common sense, let's think scientifically.

My hypothesis: It rains more on weekends.

Restated as a null hypothesis: The amount of rainfall does not differ across the days of the week.

A testable prediction: The amount of rainfall should not differ depending on what day of the week it is.

And now the data . . . In a 1998 study published in the journal *Nature,* researchers used 17 years of data to analyze rainfall along the eastern coast of North America. Here's what they found: there was 22% more rain on Saturdays than on Mondays!

Q: How can this be? The researchers also tracked the pollutants carbon monoxide and ozone, and found that both gradually increased from Monday through Friday. This pattern, they suggested, is probably related to accumulations from driving, which is greater during the week than on weekends. They suggested that the weekly pollution cycle causes increased cloud formation and rain on weekends (while the reduced pollution on Mondays leads to reduced rainfall). In other words, human activities are influencing the weather!

One good hypothesis deserves another. If their suggestion is true, we should expect the "Rainy Saturday/Dry Monday" phenomenon to occur only in the vicinity of human activities. As a test of this hypothesis, researchers measured daily rainfall patterns over the oceans in the northern hemisphere, away from large human populations. Here they found that one-seventh of the rain fell on each day of the week, with no days rainier than others.

Q: The take-home message here? Scientific thinking rewards open minds with satisfying answers.

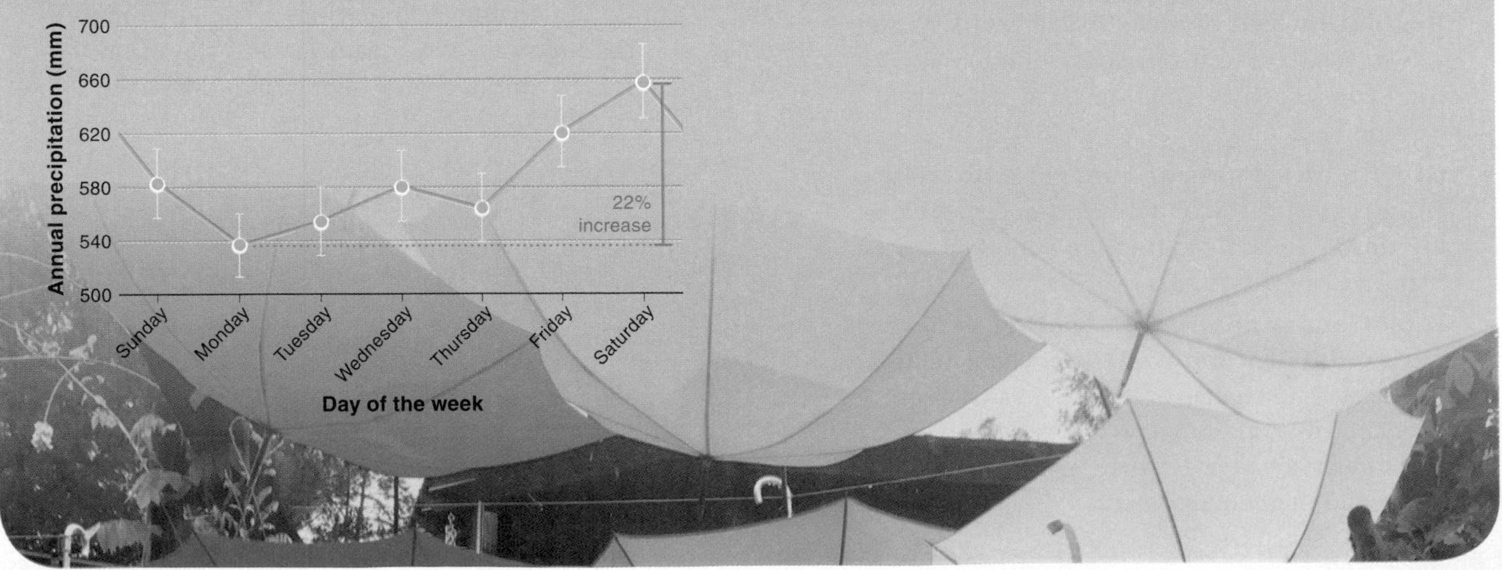

GRAPHIC CONTENT

Thinking critically about visual displays of data

1 What does the green portion of the pie chart represent? What does the gray portion represent? What does the "whole pie" represent?

2 Why are there two pie charts?

3 What can you conclude from this figure?

4 What might be an alternative explanation (other than bias against female scientists) for the phenomenon shown here?

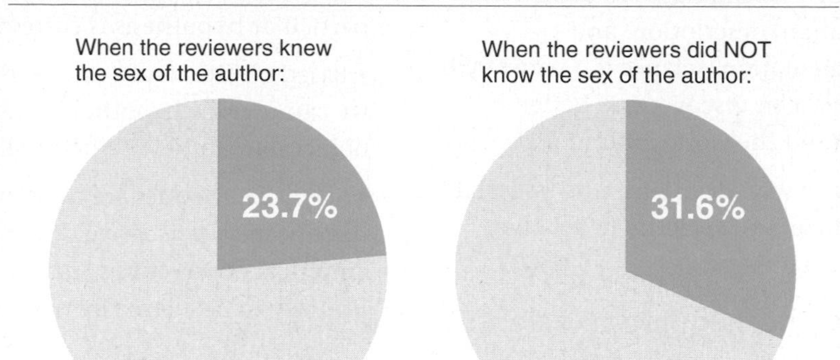

5 How big was the increase in number of papers published with a female first author following the institution of a double-blind review policy?

6 Why wasn't the percentage of papers published with a female first author 50%, even after the new review policy? Is that cause for concern? What data would help you to answer that question?

7 Do these graphs prove a general bias against female scientists?

See answers at the back of the book.

KEY TERMS IN SCIENTIFIC THINKING

anecdotal observation, p. 22
biological literacy, p. 3
biology, p. 2
blind experimental design, p. 14
control group, p. 12
critical experiment, p. 8
dependent variable, p. 12
double-blind experimental design, p. 14
empirical, p. 4
experimental group, p. 12
hypothesis (pl. hypotheses), p. 6
independent variable, p. 12
null hypothesis, p. 7
placebo, p. 9
placebo effect, p. 14
positive correlation, p. 21
pseudoscience, p. 22
randomized, p. 14
replication, p. 17
science, p. 2
scientific literacy, p. 3
scientific method, p. 4
scientific thinking, p. 4
scientific theory, p. 11
statistics, p. 20
superstition, p. 22
technology, p. 24
treatment, p. 12
variable, p. 12

BRIEF SUMMARY

More than just a collection of facts, science is a process for understanding the world.

• Through its emphasis on objective observation, description, and experimentation, science is a pathway by which we can discover and better understand the world around us.

• To make wise decisions, it is essential for individuals and societies to achieve biological literacy.

A beginner's guide to scientific thinking.

• Scientific thinking is an empirical process that incorporates making observations, articulating hypotheses, generating predictions, designing experimental tests, and drawing conclusions. It is a flexible, adaptable, and efficient process, and it can tell us when we must change our beliefs.

• Scientific thinking begins by making observations about the world, noting apparent patterns or cause-and-effect relationships.

• A hypothesis is a proposed explanation for an observed phenomenon.

• For a hypothesis to be useful, it must generate a testable prediction.

• A critical experiment is one that enables us to decisively determine whether a particular hypothesis is correct.

• Based on the results of experimental tests, we can revise a hypothesis and explain the observable world with increasing accuracy.

• Scientific theories are hypotheses that have been so persuasively supported by empirical observations that they are unlikely to be altered by new evidence.

Well-designed experiments are essential to testing hypotheses.

• To draw clear conclusions from experiments, it is essential to hold constant the variables we are not interested in. Control and experimental groups should differ only with respect to the treatment.

• Biases can influence our behavior, including our collection and interpretation of data. With careful controls, we can reduce the impact of biases.

Scientific thinking can help us make wise decisions.

• Visual displays of data can aid in the presentation and exploration of the data. Their effectiveness is influenced by the precision and clarity of the presentation.

• Statistics can help us evaluate whether differences between a treatment and a control group can be attributed to the treatment rather than chance.

• Pseudoscience and anecdotal observations often lead people to believe that links between two phenomena exist, when in fact there are no such links.

• Scientific thinking can't give us insights into value judgments and other non-quantifiable, subjective information.

On the road to biological literacy: what are the major themes in biology?

• The characteristics shared by all living organisms include complex and ordered organization; the use and transformation of energy; responsiveness to the external environment; regulation and homeostasis; growth, development, and reproduction; and evolutionary adaptation leading to descent with modification.

CHECK YOUR KNOWLEDGE

Short Answer

1. In nationwide advertisements, the Dannon Company claimed that its Activia yogurt relieved irregularity and helped with "slow intestinal transit time." Dannon also claimed that its DanActive dairy drink helped prevent colds and flu. These claims were based on no evidence. How would you design an experiment to try to test these statements?

2. Describe two examples of biological literacy.

3. Scientific thinking can be distinguished from alternative ways of acquiring knowledge in that it is empirical. What does empirical mean, and how does it relate to the study of biology?

4. An especially important feature of scientific thinking is that its steps are self-correcting. What does self-correcting mean in this context?

5. Describe something that you have observed in the world around you that you could study using scientific thinking. Describe something that you could not study scientifically.

6. A researcher hypothesizes that the more a person exercises, the less acne he or she will have. What would the null hypothesis be in this situation?

7. What would be a reasonable prediction for the hypothesis "Eating fresh fruit reduces the likelihood that you will get sick"?

8. Describe the key features of a critical experiment.

9. Many claims have been made concerning the health benefits of green tea. Suppose you read a claim that alleges drinking green tea causes weight loss. You are provided with the following information about the studies that led to this claim:

• People were weighed at the beginning of the study.

• People were asked to drink two cups of green tea every day for 6 weeks.

• People were weighed at the end of the study.

• People who drank green tea for 6 weeks lost some weight by the end of the study.

• It was concluded that green tea is helpful for weight loss.

This study obviously had some holes in its design. Assuming no information other than that provided above, indicate at least four things that could be done to improve the experimental design.

10. Following an experimental conclusion, what is a likely next step? Why?

11. Compare and contrast "theory" and "hypothesis."

12. A pharmaceutical company plans to test a potential anti-cancer drug on human subjects. The drug will be administered in pill form. How should this study be designed so that appropriate controls are in place?

13. Describe what is meant by a randomized, controlled, double-blind study.

14. Biases can influence our behavior, including our collection and interpretation of data. Which type of experimental design can help us eliminate biases?

15. When analyzing results of a study, what role do statistical analyses play?

16. You notice that all of the male students who signed up for French tutoring have blue eyes. What conclusions can you draw from this observation?

17. You have heard that people from Scandinavia usually have blond hair and fair skin. Your roommate is from Sweden and he has darker hair and more olive skin. Does this disprove the idea that Scandinavians are fair and blond? Why or why not?

18. What are the two major unifying themes in this guide to biology?

Multiple Choice

1. Science is:

a) a field of study that requires certain "laws of nature" to be taken on faith.

b) both a body of knowledge and an intellectual activity encompassing observation, description, experimentation, and explanation of natural phenomena.

c) a process that can be applied only within the scientific disciplines, such as biology, chemistry, and physics.

d) the only way to understand the natural world.

e) None of the above are correct.

0 EASY 8 HARD 100

2. Superstitions are:

a) evidence that not all phenomena can be understood purely through scientific thinking.

b) just one of many possible forms of scientific thinking.

c) true beliefs that have yet to be fully understood.

d) irrational beliefs that actions not logically related to a course of events influence its outcome.

e) proof that the scientific method is not perfect.

0 EASY 11 HARD 100

3. Empirical results:

a) rely on intuition.

b) are generated by theories.

c) are based on observation.

d) cannot be replicated.

e) must support a tested hypothesis.

4. Which of the following statements is correct?

a) A hypothesis that does not generate a testable prediction is not useful.

b) Common sense is usually a good substitute for scientific thinking when trying to understand the world.

c) Scientific thinking can be used only to understand scientific phenomena.

d) It is not necessary to make observations as part of scientific thinking.

e) All of the above are correct.

5. The placebo effect:

a) is the frequently observed phenomenon that people tend to respond favorably to any treatment.

b) reveals that sugar pills are more effective than actual medications.

c) reveals that experimental treatments cannot be proven effective.

d) demonstrates that most scientific studies cannot be replicated.

e) is an urban legend.

0 EASY 32 HARD 100

6. In controlled experiments:

a) one variable is manipulated while others are held constant.

b) all variables are dependent on each another.

c) all variables are held constant.

d) all variables are independent of each another.

e) all critical variables are manipulated.

0 EASY 24 HARD 100

7. If a researcher uses the same experimental setup as in another study, but with different research subjects, the process is considered:

a) an uncontrolled experiment.

b) intuitive reasoning.

c) extrapolation.

d) replication.

e) exploration.

8. An independent variable:

a) can cause a change in a dependent variable.

b) is generally less variable than a dependent variable.

c) is plotted on the *y*-axis in a line graph.

d) can be controlled less well than a dependent variable.

e) is typically more important than a dependent variable.

9. Statistical methods make it possible to:

a) prove that any hypothesis is true.

b) determine how likely it is that certain results have occurred by chance.

c) unambiguously learn the truth.

d) reject any hypothesis.

e) test non-falsifiable hypotheses.

10. Anecdotal evidence:

a) is the basis of scientific thinking.

b) tends to be more reliable than data based on observations of large numbers of diverse individuals.

c) is a necessary part of the scientific method.

d) is often the only way to prove important causal links between two phenomena.

e) can seem to reveal links between two phenomena, but the links do not actually exist.

11. Which of the following issues would be least helped by application of scientific thinking?

a) developing more effective high school curricula

b) evaluating the relationship between violence in video games and criminal behavior in teens

c) determining the most effective safety products for automobiles

d) formulating public policy on euthanasia

e) comparing the effectiveness of two potential antibiotics

Ch02

Atoms form molecules through bonding.

2.1 Everything is made of atoms.

2.2 An atom's electrons determine whether (and how) the atom will bond with other atoms.

2.3 Atoms can bond together to form molecules and compounds.

Water has features that enable it to support all life.

2.4 Hydrogen bonds make water cohesive.

2.5 Hydrogen bonds between molecules give water properties critical to life.

2.6 Living systems are highly sensitive to acidic and basic conditions.

2.7 *This is how we do it:* Do antacids impair digestion and increase the risk of food allergies?

Drops of dew bead on a dragonfly, magnifying parts of its compound eyes. These insects remain mostly immobile during the night, and water droplets collect all over their bodies.

The Chemistry of Biology

Atoms, molecules, and their roles in supporting life

2.1–2.3

Atoms form molecules through bonding.

Classical atomic models: a nucleus as a packed cluster of protons and neutrons, orbited by electrons.

2.1 Everything is made of atoms.

When most people think of a "chemical" they envision something toxic made in a laboratory, or maybe compounds such as paint or pesticides, produced in big factories. In fact, chemicals are everywhere. And many not only are nontoxic but are essential to life. They make up the air we breathe, the food we eat, and the soil that supports our crops. All organisms on earth are made up of interacting chemicals. By studying chemistry, we can better understand the properties and behavior of chemicals and the molecules of which they're made.

The chemistry that is most important in biology revolves around a few important elements, which are introduced at the end of this section. An **element** is a substance that cannot be broken down chemically into any other substances. Gold, carbon, and copper are elements you might be familiar with. Whatever the element, if you keep cutting it into ever smaller pieces, each of the pieces behaves exactly the same as any other piece. The smallest piece of pure gold will still have the softness, reflectivity, and malleability characteristic of that element (FIGURE 2-1).

If you could continue cutting, you would eventually separate the gold into tiny pieces that could no longer be divided without losing their gold-like properties. These individual component pieces of an element are called **atoms.** An atom is a bit of **matter** that cannot be subdivided any further without losing its essential properties. The word "atom" is from the Greek for "indivisible." (Of course, when breaking a substance into these tiny pieces, you would end up with something much smaller than you can see with your naked eye.)

Everything around us, living or not, can be reduced to atoms. All atoms—whether of the element gold or some other element such as oxygen or aluminum or calcium—have the same basic structure. At the center of an atom is a **nucleus,** which is usually made up of two types of particles, called

Gold

Carbon

Copper

FIGURE 2-1 **Familiar elements.**

Elements are substances that cannot be broken down chemically into any other substances.

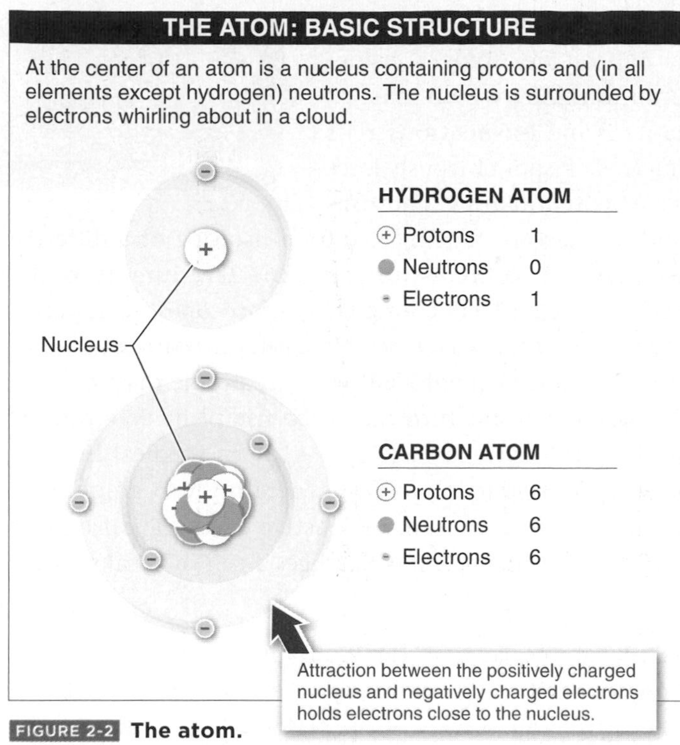

THE ATOM: BASIC STRUCTURE
At the center of an atom is a nucleus containing protons and (in all elements except hydrogen) neutrons. The nucleus is surrounded by electrons whirling about in a cloud.
HYDROGEN ATOM
Protons 1
Neutrons 0
Electrons 1
Nucleus
CARBON ATOM
Protons 6
Neutrons 6
Electrons 6
Attraction between the positively charged nucleus and negatively charged electrons holds electrons close to the nucleus.

FIGURE 2-2 **The atom.**

protons and neutrons. **Protons** are particles that have a positive electrical charge and **neutrons** are particles that have no electrical charge. The amount of matter in a particle is its mass; protons and neutrons have approximately the same mass (FIGURE 2-2).

Speeding around the nucleus of every atom are negatively charged particles called **electrons.** An electron weighs almost nothing—less than one-twentieth of one percent of the weight of a proton. (The mass of an atom—its **atomic mass**—is made up of the combined mass of all of its protons and neutrons; for our purposes here, electrons are so light that their mass can be ignored.)

Particles that have the same charge repel each other; those with opposite charges are attracted to each other. Because all electrons have the same charge, the electrons in an atom repel each other. But because they are negatively charged, they are attracted to the positively charged protons in the nucleus. This attraction holds electrons close enough to the nucleus to keep them from flying away, while the energy of their fast movement keeps them from collapsing into the nucleus. When the number of protons and electrons is equal, the charges in the atom are balanced.

Atoms are tiny. Enlarge an atom by a billion times and it would only be the size of a grapefruit. Paradoxically, most of the space taken up by an atom is empty. That is, because the nucleus is very small and compact, the electrons zip about relatively far from the nucleus. If the nucleus were the size of a golf ball, the electrons would be anywhere from half a mile to six miles away.

Elements Differ in Their Number of Protons What distinguishes one element, such as chlorine, from another, such as neon or oxygen? The number of protons in an atom's nucleus determines what element it is. As a rule, atoms of different elements have a different number of protons in the nucleus. A chlorine atom has 17 protons, a neon atom has 10 protons, and an oxygen atom has 8 protons. Each element is given a name (and an abbreviation, such as O for oxygen and C for carbon) and an **atomic number** that corresponds to how many protons it has (FIGURE 2-3).

The mass of an atom is often about double the element's atomic number. This is the case when the number of neutrons in the nucleus is equal to the number of protons, because protons and neutrons have approximately the same mass (which is designated as a mass of 1). The element oxygen, for example, has the atomic number 8 reflecting that it has 8 protons, and because it has 8 neutrons it has an atomic mass of 16, simply the mass of the 8 protons

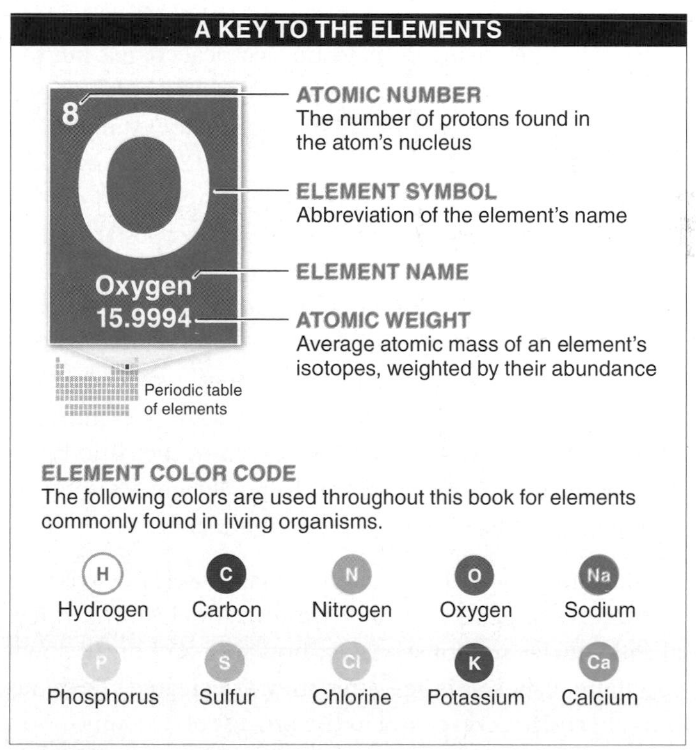

ELEMENT COLOR CODE
The following colors are used throughout this book for elements commonly found in living organisms.

FIGURE 2-3 **The vital statistics of atoms.** This key explains how to read a periodic table of elements. A full periodic table of the elements is at the back of the book.

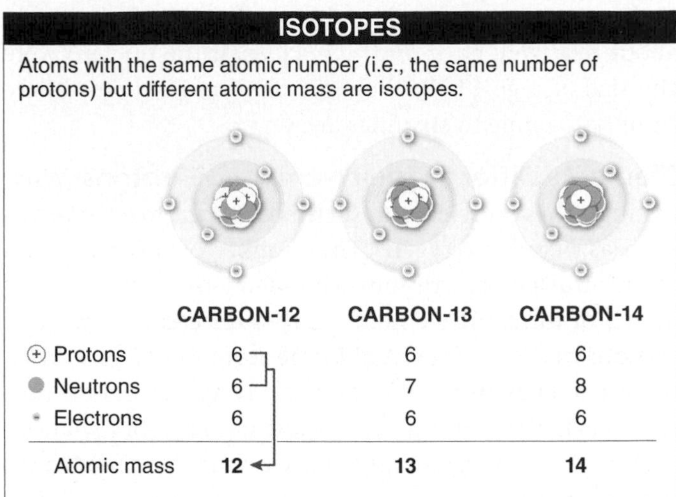

ISOTOPES
Atoms with the same atomic number (i.e., the same number of protons) but different atomic mass are isotopes.
CARBON-12
CARBON-13
CARBON-14
Protons 6 6 6
Neutrons 6 7 8
Electrons 6 6 6
Atomic mass 12 13 14

FIGURE 2-4 Isotopes are atoms that have the same number of protons but a different number of neutrons.

and the mass of the 8 neutrons added together. (As noted above, we can ignore the mass of the electrons.)

Atoms of the same element don't always have the same number of neutrons in their nucleus and electrons circling around it. For example, an atom may have more neutrons or fewer neutrons than the number of protons. Atoms with the same number of protons but different numbers of neutrons are called **isotopes.** An atom's charge doesn't change in an isotope, because neutrons have no electrical charge, but the atom's mass changes with the loss or addition of another particle in the nucleus (FIGURE 2-4). Carbon, for example, has 6 protons and so usually has a mass of 12. Occasionally, though, a rare carbon atom has an extra neutron or two and an atomic mass of 13 or 14. These isotopes are called carbon-13 (^{13}C) and carbon-14 (^{14}C) and are referred to as "heavy" carbon. In nature, we frequently see mixtures of several isotopes for a given element. So, although a sample of pure carbon consists mainly of ^{12}C atoms (with 6 protons and 6 neutrons), some ^{13}C and ^{14}C atoms are present in the sample, too. The **atomic weight** given in the periodic table is an average of the atomic mass of an element's isotopes, taking into account their different abundances.

Most elements and their isotopes have a stable nucleus that never loses or gains neutrons, protons, or electrons. A few atomic nuclei are not so stable, however, and break down spontaneously sometime after they are created. These atoms are **radioactive,** and in the process of decomposition they release, at a constant rate, a tiny, high-speed particle carrying a lot of energy. The particle may be a proton, neutron, or electron; sometimes, just energy is released and no particle.

Q Why is radioactivity hazardous to humans?

As an example of the decay of radioactive atoms, consider uranium-238 (which has 92 protons and 146 neutrons in its nucleus). It spontaneously loses a particle containing 2 protons and 2 neutrons, turning it into an isotope of a different element altogether—thorium-236. Thorium is radioactive as well, one in a long chain of isotopes, each decaying into another radioactive element, until finally producing the stable element lead (with an atomic mass of 206). Radioactive atoms turn out to be useful in determining the age of fossils (see Section 10.18), in medical imaging and cancer treatment, and in generating vast amounts of energy. The energy from radioactive decay can also cause adverse health effects if it damages a cell or breaks a molecule of DNA.

All the known elements can be arranged in a scheme, in the order of their atomic number, called the **periodic table** (see Figure 2-3). A copy of the periodic table is provided at the back of the book. So far, about 90 elements have been discovered that are present in nature, and about 28 others can be made in the laboratory. Everything you see around you is made up of some combination of the 90 or so naturally occurring elements.

GRAPHIC CONTENT
Thinking critically about visual displays of data
Turn to p. 50 for a closer inspection of this figure.

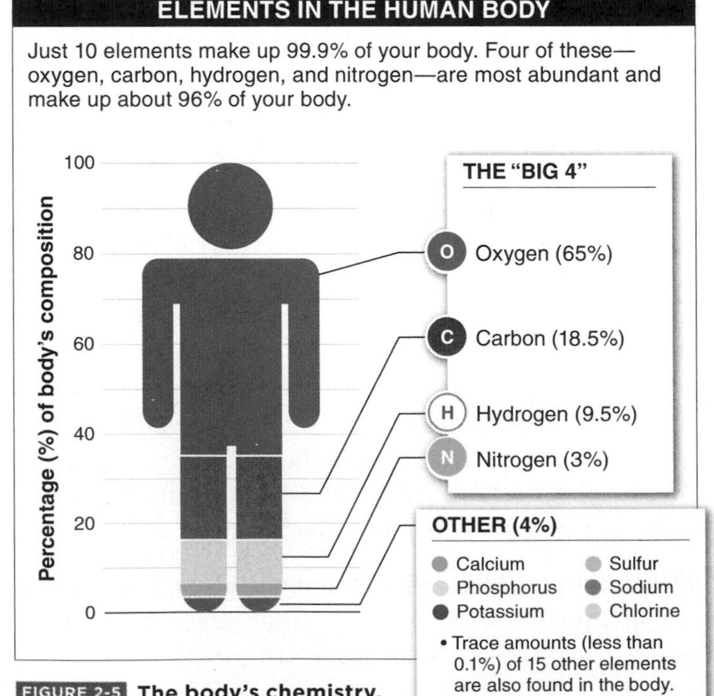

FIGURE 2-5 The body's chemistry.

Of all the elements found on earth, only 25 are found in your body. The "Top 10" most common of these make up 99.9% of your body mass, but they are not present in equal measures (FIGURE 2-5). The "Big 4" elements—oxygen, carbon, hydrogen, and nitrogen—predominate and make up more than 96% of your body mass. With knowledge about the Big 4, you can understand a huge amount about nutrition and physiology (how your body works), so in this chapter (and in Chapter 3) we'll focus on the properties of these four elements and the molecules they form.

TAKE HOME MESSAGE 2.1

» Everything around us, living or not, is made up of atoms, the smallest units into which an element can be divided without losing its essential properties. All atoms have the same general structure: protons and neutrons in the nucleus, and electrons circling, fast and far away, around the nucleus. Just four elements—oxygen, carbon, hydrogen, and nitrogen—make up the vast majority of your body mass.

2.2 An atom's electrons determine whether (and how) the atom will bond with other atoms.

Most substances are not pure elements but are compounds made of atoms from different elements that are joined by bonds. It is an atom's electrons that determine whether (and how) it bonds with other atoms. Electrons move so quickly that it is impossible to determine, at any given moment, exactly where an electron is. Electrons are not just moving about haphazardly, though. Speeding around the nucleus, they tend to stay within a prescribed area called an "electron shell." An atom may have several shells, each shell occupied by its own set of electrons. Within a shell, the electrons stay far apart because their negative charges repel one another. (Electron shells aren't actual physical tracks. Rather, they are simplified placeholders for the different possible energy levels of the electrons.)

The first electron shell is closest to the nucleus and can hold two electrons (FIGURE 2-6). If an atom has more than

ELECTRON SHELLS AND ATOM STABILITY

ELECTRON SHELLS
Electrons move around the nucleus in designated areas called electron shells.

First electron shell
(capacity: 2 electrons)
Second electron shell
(capacity: 8 electrons)
Vacancy
Oxygen atom

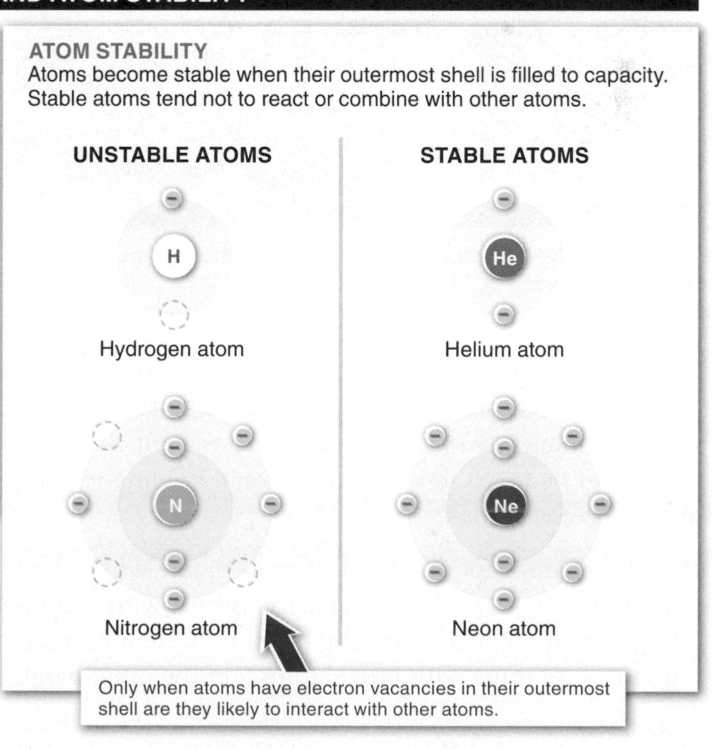

ATOM STABILITY
Atoms become stable when their outermost shell is filled to capacity. Stable atoms tend not to react or combine with other atoms.
UNSTABLE ATOMS
STABLE ATOMS
H
Hydrogen atom
He
Helium atom
N
Nitrogen atom
Ne
Neon atom
Only when atoms have electron vacancies in their outermost shell are they likely to interact with other atoms.

FIGURE 2-6 **Electron interactions.** The chemical characteristics of an atom depend upon the number of electrons in its outermost shell.

THE VERSATILITY OF CARBON

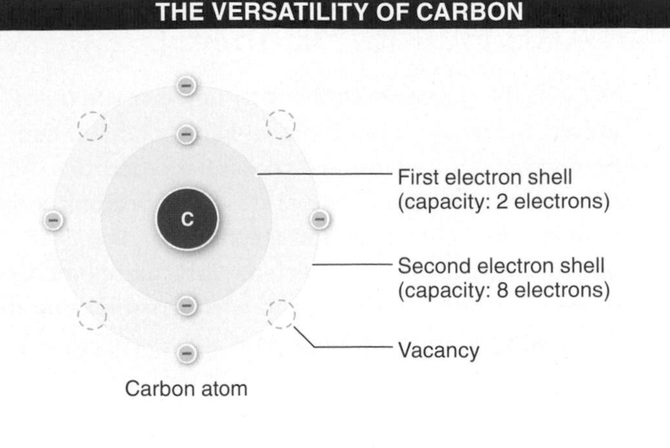

EXAMPLES OF THE WIDE VARIETY OF CARBON COMPOUNDS

Because carbon forms bonds so easily, it plays a role in many of the molecules central to our lives, including carbon dioxide, sugars, and proteins. How many more can you identify?

FIGURE 2-7 Carbon atoms can bond with other atoms in many different ways.

two electrons, as most atoms do, the other electrons are arranged in other shells. The second shell is a bit farther away from the nucleus and can hold as many as eight electrons. There can be up to seven shells in total, holding varying numbers of electrons.

Atoms become less reactive and more stable when their outermost shell is filled to capacity. Atoms with a completely filled outer shell behave like loners, neither reacting nor combining with other atoms. On the other hand, when atoms have outer shells with vacancies, they are likely to interact with other atoms, giving, taking, or sharing electrons to achieve that desirable state: a full outer shell of electrons. In fact, based on the number of electron vacancies in the outermost shell of an atom, it's possible to predict how likely that atom will be to bond, and even which other atoms its likely bonding partners will be.

Let's take a brief look at one element—carbon. Carbon's electron configuration gives it considerable versatility when it comes to bonding with other atoms and making important compounds. Carbon has 6 electrons overall: 2 in the first electron shell and 4 in the second electron shell. Because the second electron shell has a capacity of 8 electrons, carbon can share the 4 electrons in its outermost shell with other atoms to achieve that desirable 8. Having 4 electrons to share means a carbon atom can bond with other atoms in many ways—including in four different directions—and makes a huge variety of complex molecules possible (FIGURE 2-7). For example, methane, the chief component of natural gas, is formed when one atom of carbon covalently bonds with four atoms of hydrogen. It has the chemical formula CH_4.

Carbon atoms most commonly bond with oxygen, hydrogen, and nitrogen, and with other carbon atoms. Chains

IONS ARE CHARGED ATOMS

An atom that loses one or more electrons becomes positively charged; when an atom acquires electrons it becomes negatively charged. This electron transfer stabilizes atoms with incomplete outer electron shells.

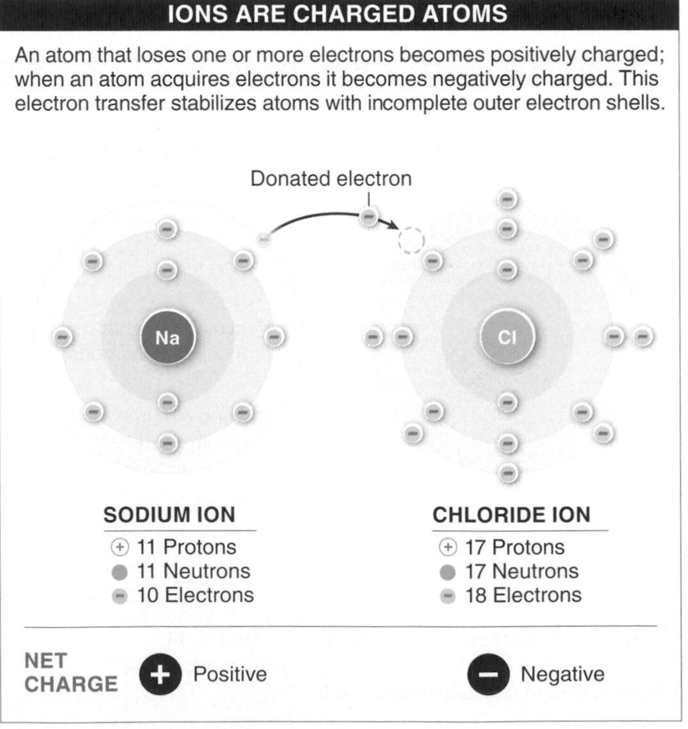

FIGURE 2-8 Ions are electrically charged atoms.

formed by carbon atoms bonded to other carbon atoms are very common—and each carbon bound to two other carbon atoms can still bind with one or two additional atoms. These carbon chains are common in the natural world.

Typically, as we've seen, an atom has the same number of electrons as protons. Sometimes an atom may have one or more extra electrons or may lack one or more electrons relative to the number of protons. An atom with extra electrons becomes negatively charged, and an atom lacking one or more electrons is positively charged. Such a charged atom is called an **ion** (FIGURE 2-8). Due to their electrical charge, ions behave very differently from the atoms that give rise to them. As we see later in the chapter, ions are more likely to interact with other, oppositely charged ions.

TAKE HOME MESSAGE 2.2

» The chemical characteristics of an atom depend on the number of electrons in its outermost shell. Atoms are most stable and least likely to bond with other atoms when their outermost electron shell is filled to capacity.

2.3 Atoms can bond together to form molecules and compounds.

When you eat a meal, it is like filling your car's tank with gasoline; you take in a source of energy that can be used to fuel activities like running, thinking, building muscle, and maintaining the machinery of life. That energy initially comes from the sun and is captured and stored by plants. When we eat plants or eat other animals that eat plants, we ingest the energy stored in the plant material. But how exactly is energy stored in plants? And how are we able to make use of that energy?

Groups of atoms held together by bonds are called **molecules.** It takes a certain amount of energy to break a bond between two atoms. The amount of this energy, called the **bond energy,** depends on the atoms involved. **Chemical reactions**—including those necessary for organisms to store and use chemical energy—involve the forming and breaking of chemical bonds. When chemical reactions occur, molecules called *reactants* are transformed into different molecules, called *products.*

If the bond energy of the new bonds formed in the reaction is less than the bond energy of the starting materials, the excess energy is released—and can be used to fuel the body's activities. In a sense, molecules are created as a short-term store of energy that can be harnessed later. As we will see in our investigation of how cells capture and make use of energy (Chapter 5), chemical reactions are central to nearly all the processes of life. We'll be exploring a great many of them.

Before looking at specific types of bonds that hold molecules together, let's look at how molecules are illustrated. In this book you will most often see molecular structures represented by the ball-and-stick and space-filling models.

DIFFERENT WAYS OF REPRESENTING MOLECULAR STRUCTURE

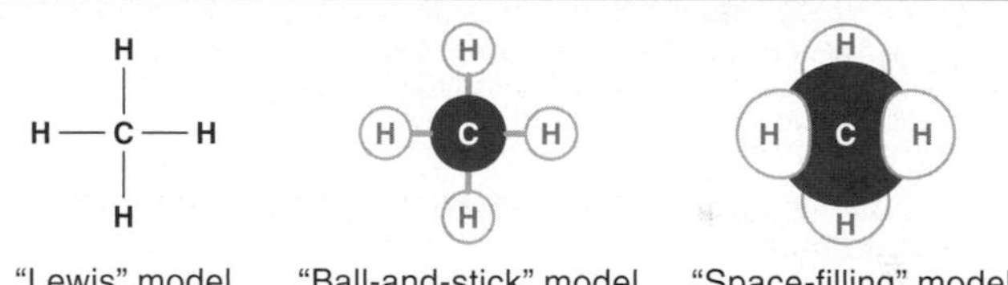

There are three principal types of bonds that hold atoms together: covalent, ionic, and hydrogen bonds. The type of bonding that any atom is likely to take part in depends almost entirely on the number of electrons in its outermost shell.

Covalent Bonds When two atoms share electrons, a strong **covalent bond** is formed. The simplest example of a covalent bond is the bonding of two hydrogen atoms to form a hydrogen molecule, H_2 (FIGURE 2-9), the simplest of all molecules. A hydrogen atom has an atomic number of 1: it has a single proton in its nucleus and a single electron circling around the nucleus in the first shell. Because the atom is most stable when the first shell has two electrons, two hydrogen atoms can each achieve a complete outermost shell by sharing electrons. The nuclei come close together

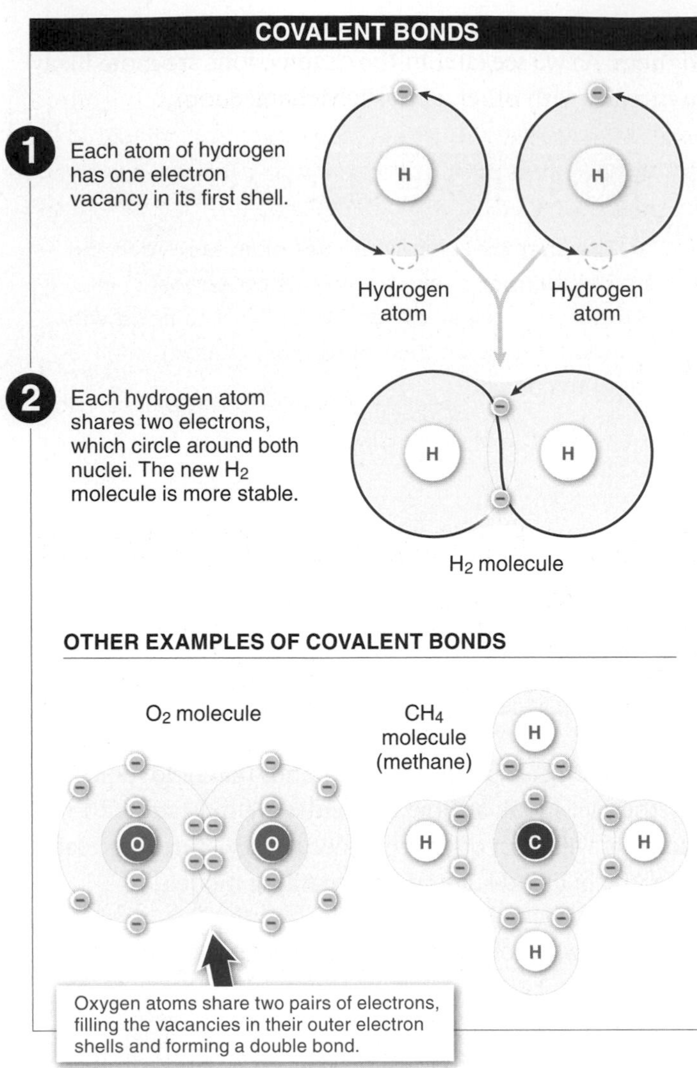

COVALENT BONDS
1
Each atom of hydrogen has one electron vacancy in its first shell.
H
H
Hydrogen atom
Hydrogen atom
2
Each hydrogen atom shares two electrons, which circle around both nuclei. The new H2 molecule is more stable.
H
H
H2 molecule
OTHER EXAMPLES OF COVALENT BONDS
O2 molecule
O
O
CH4 molecule (methane)
H
C
H
H
H
Oxygen atoms share two pairs of electrons, filling the vacancies in their outer electron shells and forming a double bond.

FIGURE 2-9 Covalent bonds: electron sharing that forms strong chemical bonds.

(but not too close, because they are positively charged and their mutual repulsion would destabilize the molecule), and the two electrons circle around both of the nuclei, almost in a figure 8. The H_2 molecule is very stable because, now that both atoms have two electrons in their outermost shell, they are no longer likely to bond with other atoms. The sharing of two electrons between two atoms forms a *single* covalent bond.

Oxygen, with an atomic number of 8, has two electrons in its innermost shell and six in its outermost shell. Consequently it needs to gain two electrons to fill its outermost shell. Sometimes two oxygen atoms join together to form O_2. Each oxygen atom shares a pair of electrons with the other, filling the outermost shell of both. The sharing of two pairs of electrons between two atoms is called a **double bond** (see Figure 2-9). O_2 is the most common form in which we find free oxygen in the world.

Carbon is a particularly "extroverted" atom. It has an atomic number of 6, meaning that two electrons fill its first shell and four remain to occupy the second shell. Because four electron vacancies are left, carbon can, and frequently does, form four covalent bonds, joining up with other atoms in a wide variety of molecules.

Ionic Bonds Atoms can also bond together without sharing electrons. When one atom transfers one or more of its electrons completely to another, each atom becomes an ion, since each has an unequal number of protons and electrons. The atom gaining electrons becomes negatively charged, while the atom losing electrons becomes positively charged. An **ionic bond** occurs when the two oppositely charged ions attract each other. Unlike in covalent bonds, in ionic bonds each electron circles around a single nucleus. Ions of two or more elements linked by ionic bonds form an **ionic compound.** Because the ions attracted to each other are of equal and opposite charge, the compound is neutral—that is, it has no charge (FIGURE 2-10). A common ionic compound you are familiar with is table salt (NaCl), composed of sodium and chloride ions.

"In the last third of his life, there came over Laszlo Jamf . . . a hostility, a strangely personal hatred, for the covalent bond . . . That something so mutable, so soft, as a sharing of electrons by atoms of carbon should be at the core of life, *his* life, struck Jamf as a cosmic humiliation. *Sharing?* How much stronger, how everlasting was the ionic bond—where electrons are not shared, but *captured. Seized!* and held! polarized plus and minus, these atoms, no ambiguities . . . how he came to love that clarity: how stable it was, in such mineral stubbornness!"

— THOMAS PYNCHON, in *Gravity's Rainbow*,

[Was Laszlo Jamf mistaken in his understanding of the relative strengths of ionic and covalent bonds?]

Hydrogen Bonds Ionic and covalent bonds are bonds *between atoms.* They link two or more atoms together within an ionic compound or a molecule. **Hydrogen bonds,** on the other hand, are important in *bonding molecules* together. A hydrogen bond is formed

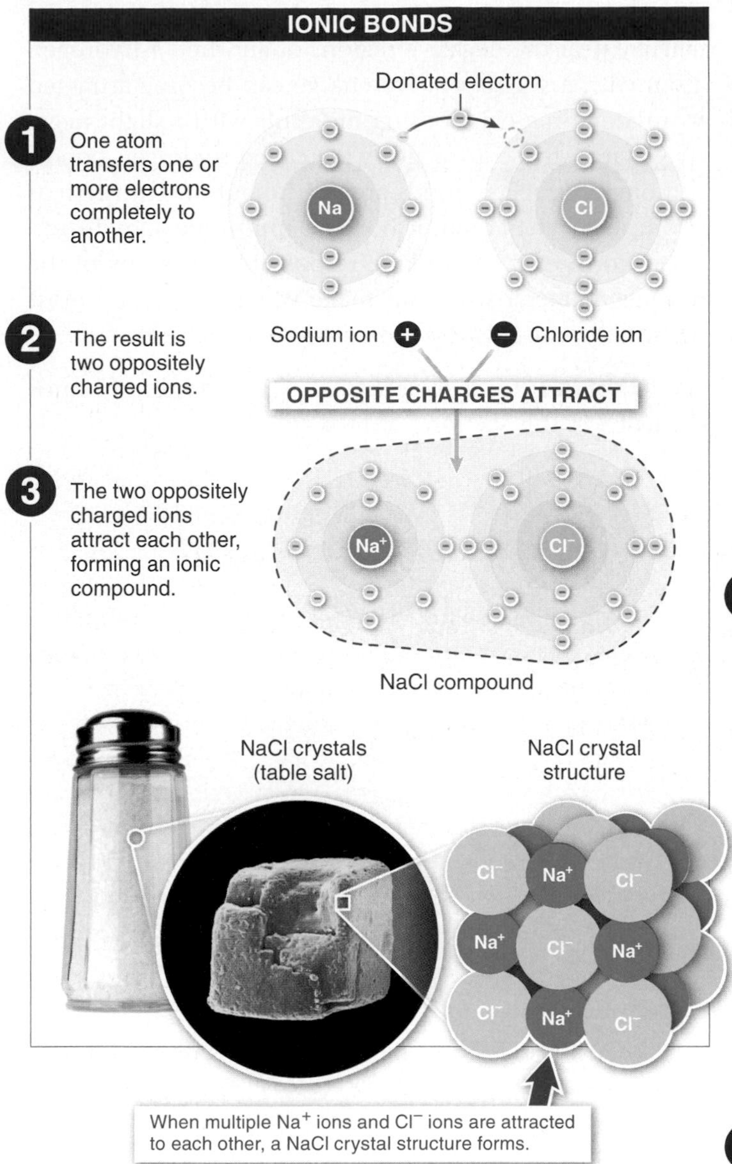

IONIC BONDS
Donated electron
1 One atom transfers one or more electrons completely to another.
Na
Cl
2 The result is two oppositely charged ions.
Sodium ion
Chloride ion
OPPOSITE CHARGES ATTRACT
3 The two oppositely charged ions attract each other, forming an ionic compound.
Na+
Cl−
NaCl compound
NaCl crystals (table salt)
NaCl crystal structure
When multiple Na+ ions and Cl− ions are attracted to each other, a NaCl crystal structure forms.

FIGURE 2-10 Ionic bonds: transfer of electrons from one atom to another.

between a hydrogen atom in one molecule and another atom, often an oxygen or nitrogen atom, in another molecule (or even, in larger molecules, in another part of the same molecule). This bond is based on the attraction between positive and negative charges.

The atoms taking part in hydrogen bonds are not necessarily ions, so where do the electrical charges come from? The hydrogen atom is already covalently bonded to another atom in the same molecule and shares its electron. That electron circles both the hydrogen nucleus and the nucleus of the other atom—but the electron is not shared equally. It's as if the nuclei of the atoms are playing tug-of-war with the electrons. Because the other atom always has more than the one proton found in the hydrogen nucleus, it is more positively charged. As a result, the electron contributed by hydrogen spends more of its time near the more positively charged nucleus than near the hydrogen nucleus. And so, the hydrogen atom is slightly positively charged, while the other is slightly negatively charged (FIGURE 2-11).

In a sense, molecules with slightly charged atoms become like a magnet, with distinct positive and negative sides (see Figure 2-11), and are said to be "polar." Polar molecules are attracted to other polar molecules, lining up in particular

HYDROGEN BONDS

1 In a water molecule, oxygen's eight positively charged protons attract electrons more readily than does the single proton in each hydrogen atom. As a result, the electrons are pulled toward the oxygen side of the molecule, making it slightly negative in charge, while the hydrogen side is slightly positive.

Polarity

O

H

Covalent bonds

Water molecule

2 Hydrogen bonds are formed between the slightly positively charged hydrogen atom of one molecule and the slightly negatively charged oxygen atom of another.

Hydrogen bond

FIGURE 2-11 Hydrogen bonds: attraction between a polar atom or molecule and a hydrogen atom. As we'll see later in the chapter, the hydrogen bonding between H_2O molecules gives water some of its characteristic properties.

SUMMARY: THREE TYPES OF BONDS

COVALENT BOND
A strong bond formed when atoms share electrons in order to become more stable, forming a molecule.

BOND STRENGTH: STRONG

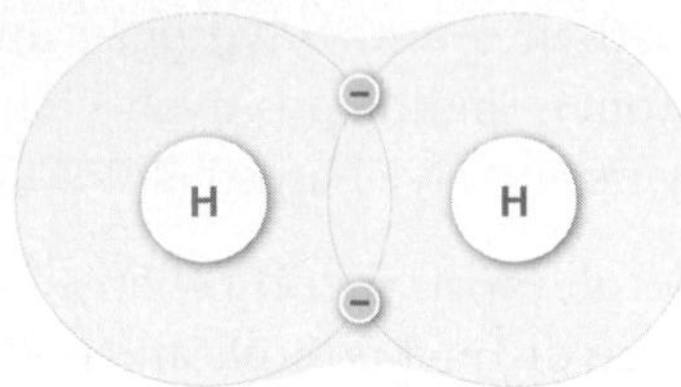

H_2 molecule

IONIC BOND
An attraction between two oppositely charged ions, forming an ionic compound.

BOND STRENGTH: STRONG

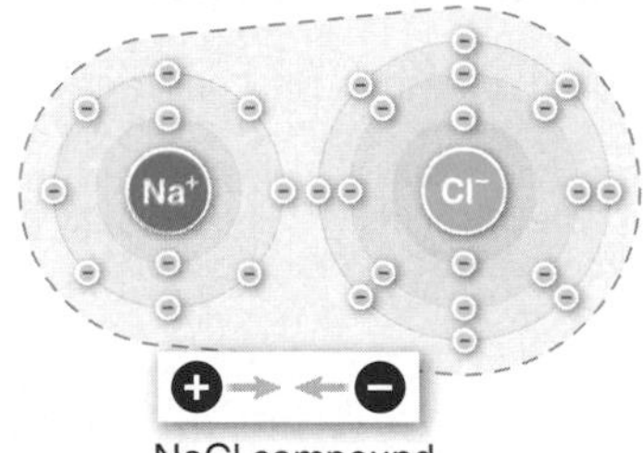

NaCl compound

HYDROGEN BOND
An attraction between the slightly positively charged hydrogen atom of one molecule and the slightly negatively charged atom of another.

BOND STRENGTH: WEAK

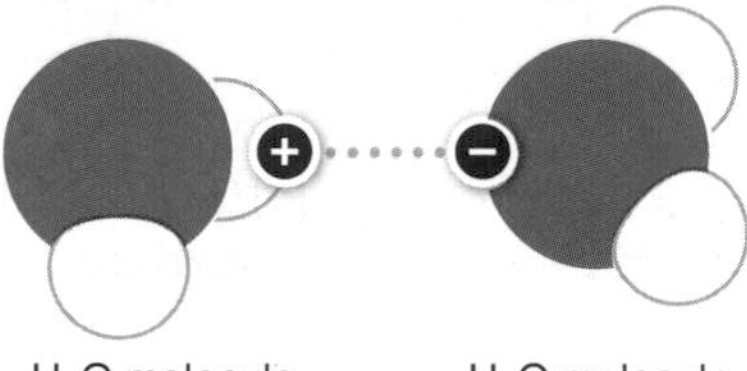

H_2O molecule H_2O molecule

FIGURE 2-12 The properties of covalent, ionic, and hydrogen bonds compared.

orientations such that the positive regions of one molecule are near the negative regions of another. A hydrogen atom with a slight positive charge can become attracted to an atom in a neighboring molecule with a slight negative charge; these attractions are called hydrogen bonds. Although hydrogen bonds are only about one-thirtieth as strong as covalent or ionic bonds, a multitude of hydrogen bonds form in water and are responsible for many of the unique characteristics that make water one of the most important molecules for life on earth.

See (FIGURE 2-12) for a review of the three types of bonds discussed in this section.

TAKE HOME MESSAGE 2.3

» Atoms can be bound together in three different ways. Covalent bonds are formed when atoms share electrons. In ionic bonds, one atom transfers its electrons to another atom and the two oppositely charged ions are attracted to each other, forming an ionic compound. Hydrogen bonds, which are weaker than covalent and ionic bonds, are formed from the attraction between a hydrogen atom in one molecule and another atom in another molecule that has a slight negative charge.

2.4–2.7 Water has features that enable it to support all life.

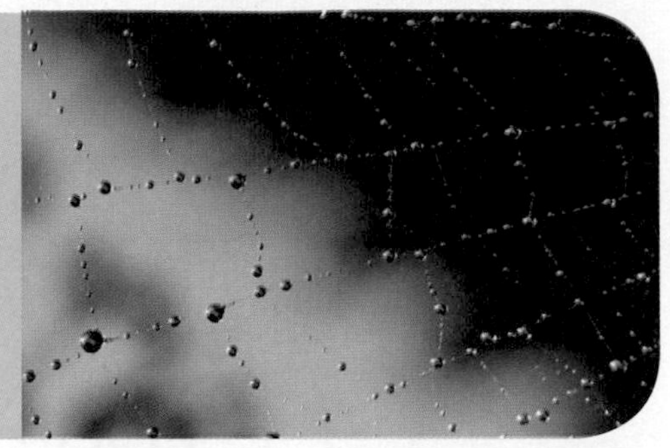

Drops of water adhere to the web of a spider.

2.4 Hydrogen bonds make water cohesive.

Every so often, a tadpole or small fish gets an unexpected—and life-ending—surprise from above as a giant spider moves quickly across the surface of the water, reaches down with its two front legs, and plucks the animal from the water. The spider, known as the fishing spider (*Dolomedes triton*), injects its prey with venom, carries it back to shore—moving across the water's surface as if across solid ground—and eats it. How can these spiders walk on water?

The fishing spider, like many other insects such as the water strider, makes use of the fact that water molecules have tremendous cohesion. That is, they stick together with unusual strength. This molecular cohesiveness is due to hydrogen bonds between the water molecules.

Each water molecule is V-shaped (FIGURE 2-13; see also Figure 2-11). The hydrogen atoms are at the ends of the two arms and the oxygen is at the bottom end of the V, between the two hydrogen atoms. Oxygen's strongly positively charged nucleus pulls the circling electrons toward itself and holds on to them for more than its fair share of the time. Consequently, the oxygen at the bottom of the V has a slight negative charge, and the top of the V, with the two hydrogen atoms, has a slight positive charge. Hydrogen bonds form between the relatively positively charged hydrogen atoms and the relatively negatively charged oxygen atoms of *adjacent* water molecules.

Hydrogen bonds, as we've noted, are much weaker than covalent and ionic bonds, and they don't last very

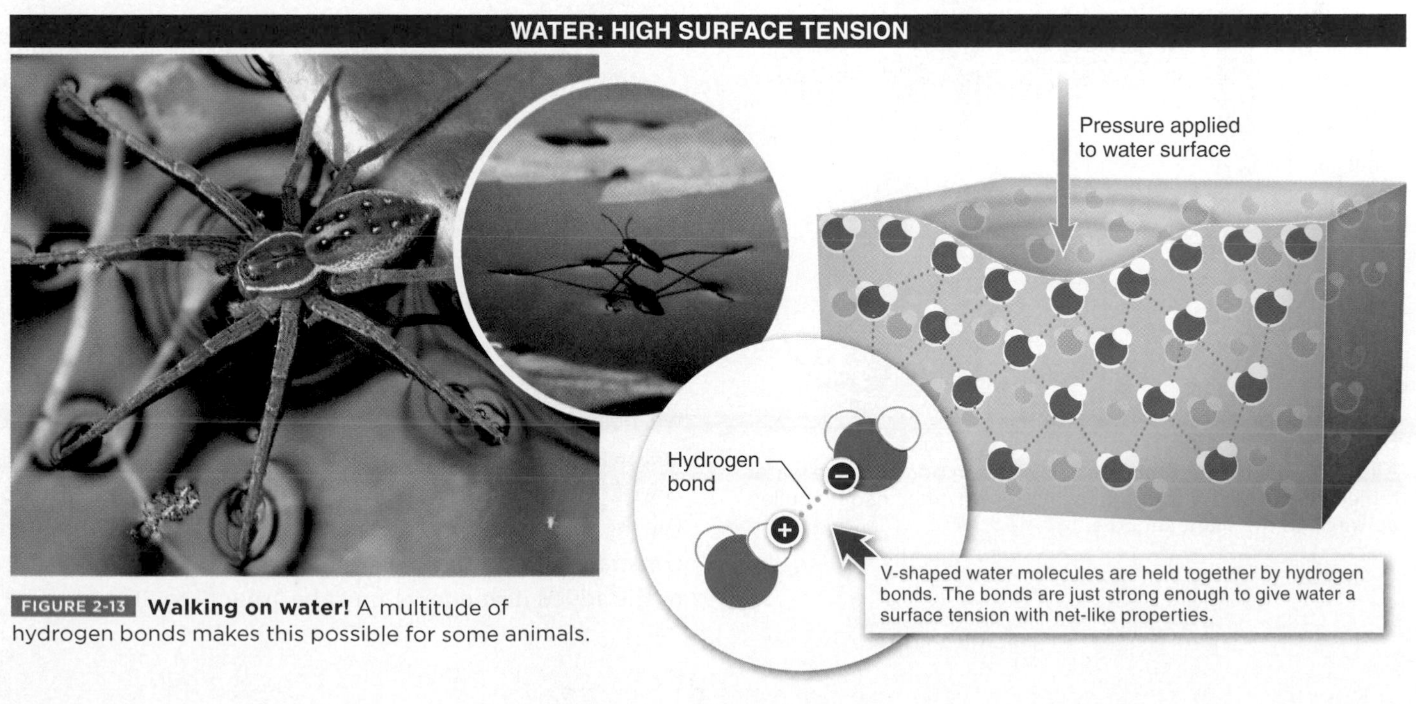

WATER: HIGH SURFACE TENSION
Pressure applied to water surface
Hydrogen bond
V-shaped water molecules are held together by hydrogen bonds. The bonds are just strong enough to give water a surface tension with net-like properties.

FIGURE 2-13 **Walking on water!** A multitude of hydrogen bonds makes this possible for some animals.

Q A spider can walk on water. Why can't a human?

long—constantly breaking and re-forming. Nonetheless, all the hydrogen bonds link together the water molecules in a chain or network just enough to give the water a surface tension with some net-like properties that can support a fishing spider.

TAKE HOME MESSAGE 2.4

» Water molecules easily form hydrogen bonds, giving water great cohesiveness.

2.5 Hydrogen bonds between molecules give water properties critical to life.

All life on earth depends on water; organisms are made up mostly of water and require it more than any other molecule. Hydrogen bonding among water molecules gives water several important properties that contribute to its crucial role in the biology of all organisms.

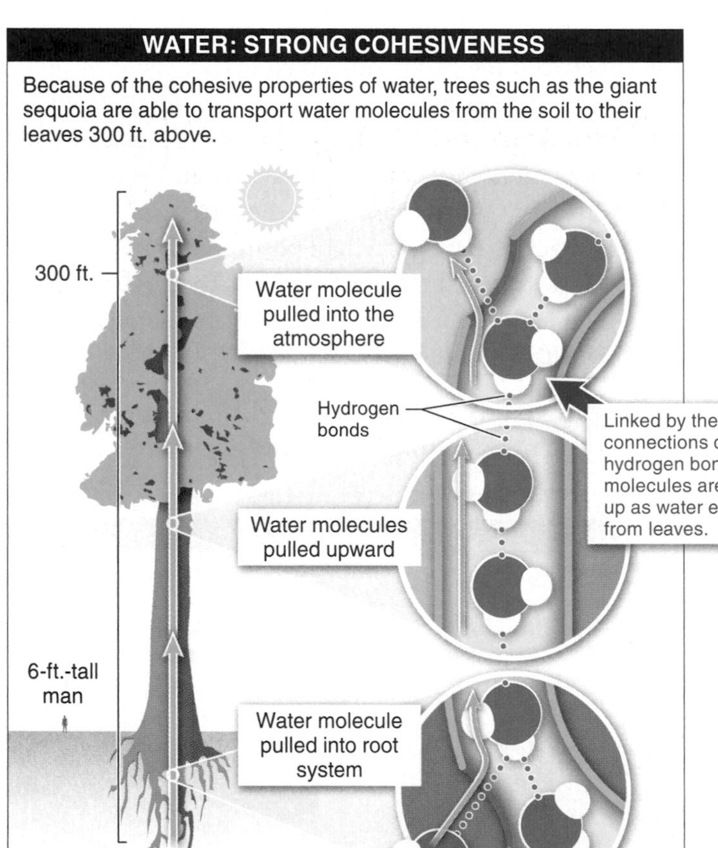
WATER: STRONG COHESIVENESS
Because of the cohesive properties of water, trees such as the giant sequoia are able to transport water molecules from the soil to their leaves 300 ft. above.
300 ft.
Water molecule pulled into the atmosphere
Hydrogen bonds
Linked by the "sticky" connections of their hydrogen bonds, water molecules are pulled up as water evaporates from leaves.
Water molecules pulled upward
6-ft.-tall man
Water molecule pulled into root system

FIGURE 2-14 **Like a giant straw.** Hydrogen bonds cause water molecules to "stick" together, which is why they can be pulled up through the giant sequoia.

1. Cohesion. We saw in the previous section how the interconnection of water molecules through hydrogen bonds makes water cohesive, resulting in, for example, high surface tension. The cohesiveness of water molecules also makes it possible for tall trees to exist (FIGURE 2-14). Leaves need water. Molecules of water in the leaves are continually lost to the atmosphere through evaporation or are used up in the process of photosynthesis. To get more water, plants must pull it up from the soil. The problem for many plants, such as the giant sequoia trees, is that the soil may be 300 feet below the leaf. However, water molecules can pull up adjacent water molecules to which they are hydrogen-bonded. The chain of linked molecules extends all the way from the leaf down to the soil, where another water molecule is pulled in via the roots each time a water molecule evaporates from the leaf far above.

2. Large heat capacity. Walking across a sandy beach on a hot day, you can feel how easily sand heats up. By comparison, the water is cool as you step from the beach into the ocean, because water resists warming. It takes a lot of energy to change the temperature of water even a small amount. Why? Again, we must look to hydrogen bonding for our answer.

The temperature of a substance is a measure of how quickly all of the molecules are moving. The molecules move more quickly when energy is added in the form of heat. When we heat water, the added energy can't immediately increase the movement of individual water molecules (which is what we would measure as temperature). Rather, because the molecules are cross-linked by hydrogen bonds, they are constrained from increased molecular motion. A portion of the added heat energy must be "spent" to disrupt some

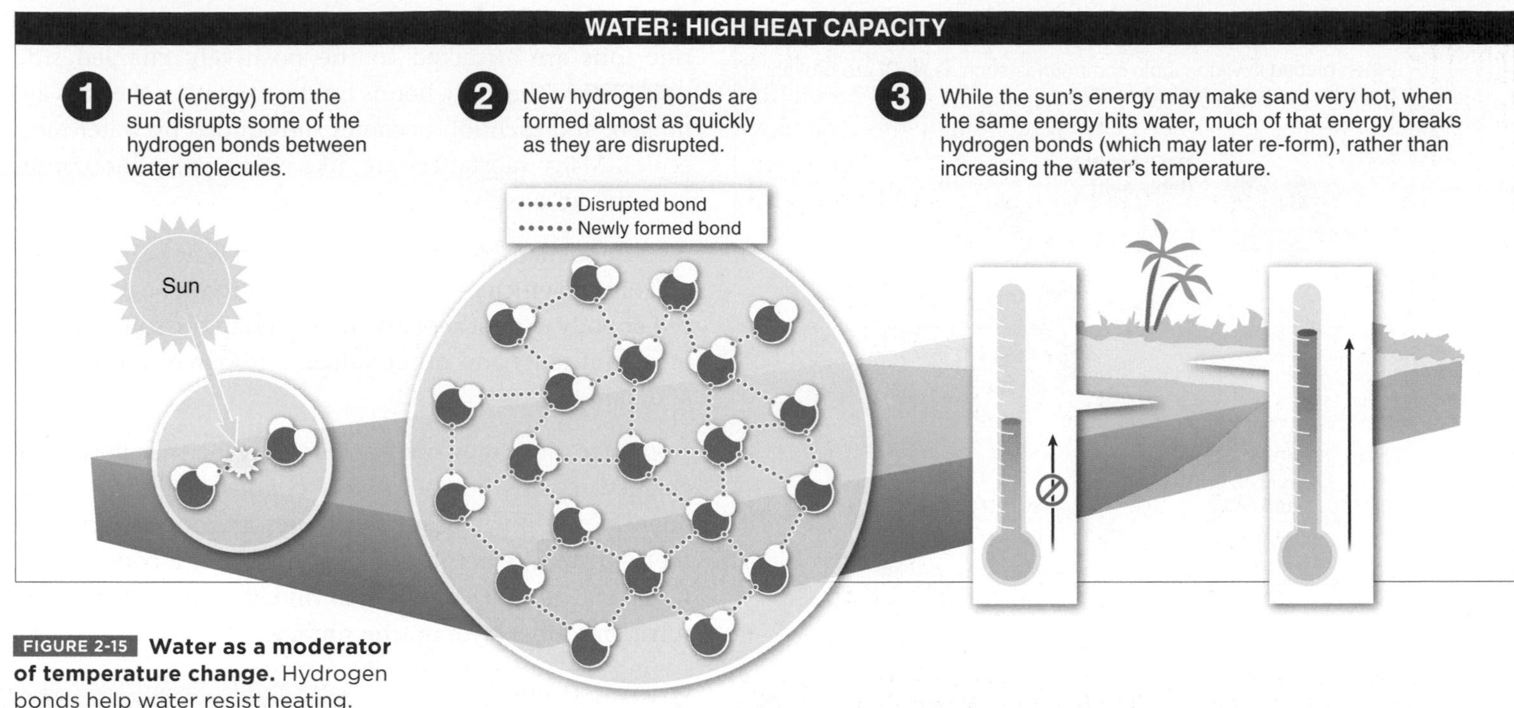

WATER: HIGH HEAT CAPACITY
1 Heat (energy) from the sun disrupts some of the hydrogen bonds between water molecules.
2 New hydrogen bonds are formed almost as quickly as they are disrupted.
3 While the sun's energy may make sand very hot, when the same energy hits water, much of that energy breaks hydrogen bonds (which may later re-form), rather than increasing the water's temperature.
Disrupted bond
Newly formed bond
Sun

FIGURE 2-15 **Water as a moderator of temperature change.** Hydrogen bonds help water resist heating.

of the many weak hydrogen bonds between the molecules (FIGURE 2-15). Even if you release a lot of energy into water, the temperature does not change very much. For this reason, because so much of your body is water, you are able to maintain a relatively constant body temperature.

Q Why do coastal areas have milder, less variable climates than inland areas?

Large bodies of water, especially oceans, can absorb huge amounts of heat from the sun during warm times of the year, reducing temperature increases on the coastland. During cold times of the year, the ocean cools slowly, giving off heat that reduces the temperature drop on shore.

3. Low density as a solid. Ice floats. This is unusual because most substances *increase* in density when changing from liquid to solid form as they cool (or freeze). As the cooling molecules slow down, they pack together more and more efficiently—and densely. Consequently, the solid sinks. Water, however, becomes less dense when solid, due to hydrogen bonding. As the temperature drops and water molecules slow down, each V-shaped molecule bonds with four partners, via hydrogen bonds, forming a crystalline lattice. The lattice holds the molecules slightly farther apart than in liquid form, causing ice to be less dense than water (FIGURE 2-16). This feature of water causes lakes and other bodies of water

FIGURE 2-16 **The lattice structure of ice allows it to float.**

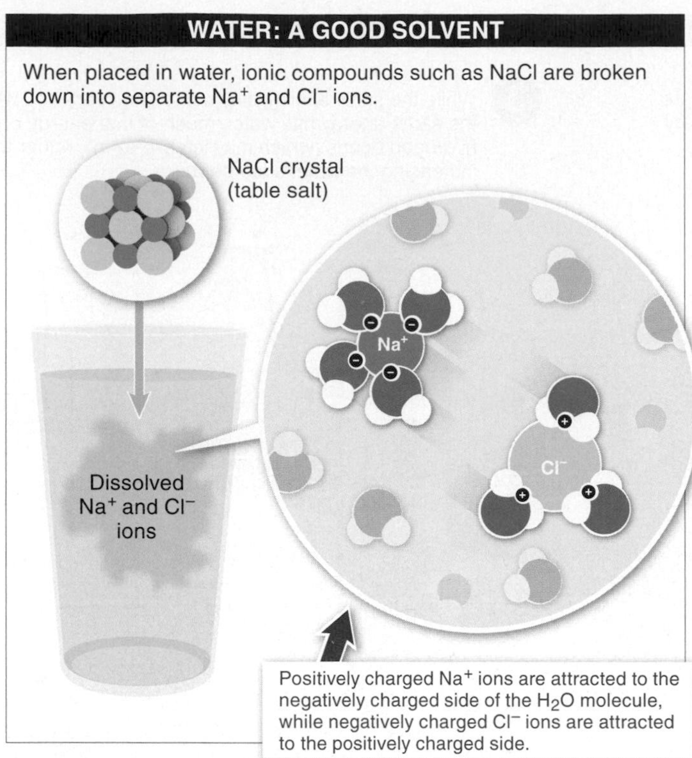

WATER: A GOOD SOLVENT
When placed in water, ionic compounds such as NaCl are broken down into separate Na^+ and Cl^- ions.
NaCl crystal (table salt)
Na^+
Cl^-
Dissolved Na^+ and Cl^- ions
Positively charged Na^+ ions are attracted to the negatively charged side of the H_2O molecule, while negatively charged Cl^- ions are attracted to the positively charged side.

FIGURE 2-17 **Solutions.** Water pries apart ionic bonds, dissolving ionic compounds.

to freeze from the top down. As a consequence, they are less likely to freeze completely, which is very good news for fish, aquatic plants, and other forms of life.

4. Good solvent. If you put a pinch of table salt (NaCl) into a glass of water, it will quickly dissolve. This means that all the charged sodium (Na^+) and chloride (Cl^-) ions in the compound become separated from one another. The sodium and chloride ions were initially attracted to each other because they carry small, opposing charges. Water is able to pry them apart because, as a polar molecule, it, too, carries charges. The positively charged sodium ions are attracted to the negatively charged side of the water molecule, and the negatively charged chloride ions are attracted to the positively charged side (FIGURE 2-17). The ionic bonds holding together the ions are broken, and each ion becomes surrounded by water molecules. Many substances are, like water, polar and consequently can dissolve easily in water.

Because so many substances dissolve in water, they can be quickly and efficiently moved about—from one part of the cell or body to another part, and through the soil—distributing nutrients and other valuable resources for animals and plants.

Nonpolar molecules such as oils have neither positively charged regions nor negatively charged regions. Consequently, the polar water molecules are not attracted to them. Instead, when an oil is poured into a container of water, the oil molecules remain in clumps that never dissolve, or form a layer on the surface.

Q Why don't oceans freeze as easily as freshwater lakes?

Because so much salt is dissolved in the oceans, many of the water molecules have their positively charged sides facing Cl^- ions and, simultaneously, many have their negatively charged sides facing Na^+ ions. Consequently, the orderly lattices of hydrogen bonds found in ice cannot form in salt water, and it does not freeze well.

TAKE HOME MESSAGE 2.5

» The hydrogen bonds between water molecules give water several of its most important characteristics, including cohesiveness, reduced density as a solid, the ability to resist temperature changes, and broad effectiveness as a solvent for ionic and polar substances.

2.6 Living systems are highly sensitive to acidic and basic conditions.

Have you ever wondered whether your drinking water was safe to consume? One thing you'd want to know is the acidic or basic composition of the water. Most water molecules are present as H_2O, but at any instant some of them are broken into two parts: H^+ and OH^-. In pure water, the amount of H^+ and OH^- must be exactly the same, since every time a molecule splits, one of each type of ion is produced. But in some fluids containing other dissolved materials, this balance is lost: the fluid can have more H^+ or more OH^-.

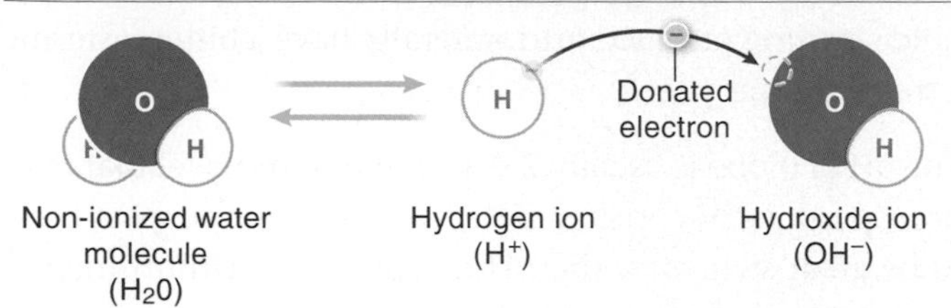

The amount of H^+ and OH^- in a fluid gives it some important properties. In particular, the amount of H^+ in a solution is a measure of its acidity and is called **pH.** The greater the number of free hydrogen ions floating around, the more acidic the solution is.

Pure water is in the middle of the pH scale, with a pH of 7.0. Any fluid with a pH below 7.0 has more H^+ ions (and fewer OH^- ions) and is considered an **acid.** Any fluid with a pH above 7.0 has fewer H^+ ions (and more OH^- ions) and is considered a **base.** The pH scale, like the Richter scale for earthquakes, is logarithmic, although in the case of pH, the *lower* the number the *greater* the acidity: a decrease of 1 on the pH scale represents a 10-fold increase in the hydrogen ion concentration (FIGURE 2-18). A decrease of 2 represents a 100-fold increase. This means that a cola, with a pH of about 3.0, is 10,000 times(!) more acidic than a glass of water, with a pH of 7.0.

Hydrogen ions are essentially free-floating protons. This is because hydrogen atoms have one proton and one electron (and no neutron). And when the atom loses the electron, just the proton remains. Acids can donate their H^+ ions to other chemicals. In fact, H^+ is a very reactive little ion. Its presence gives acids some unique properties. For instance, the hydrogen ions in acids can bind with atoms in metals, causing them to corrode. That's why you can dissolve nails by dumping them in a bucket of acid (or cola).

Q Is acid rain likely to have any impact on populations of microbes living in lakes, streams, and soil? Why?

Your stomach produces large amounts of hydrochloric acid (HCl dissolved in water) and has a pH between 1 and 3. (HCl dissolved in water is acidic because most of the hydrogen ions split off from the chlorine, raising the H^+ concentration of the fluid.) The acid in your stomach helps to kill most bacteria that you ingest. It also greatly enhances the breakdown of the chemicals in the food you eat, and increases the efficiency of digestion and absorption. You may have learned firsthand of the high acidity of your stomach fluids if you have experienced heartburn or the sour taste of vomit.

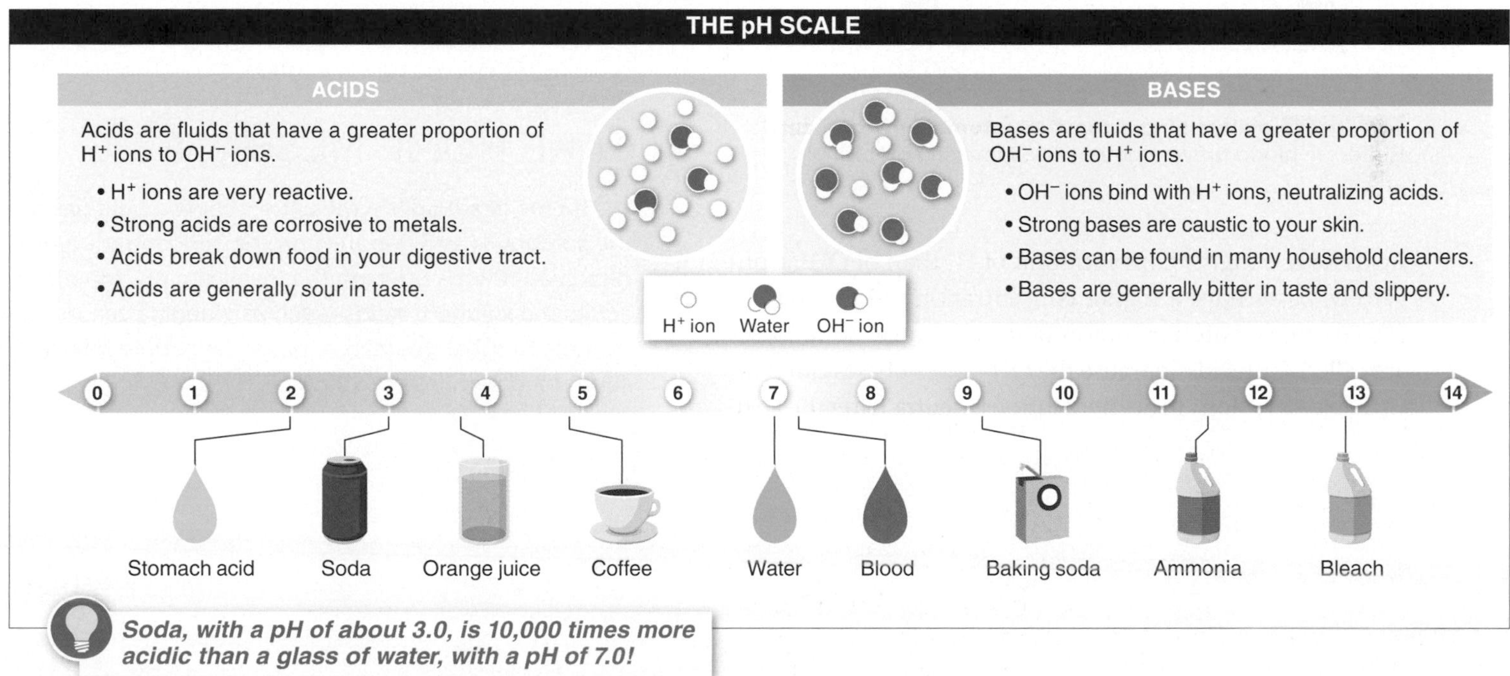

THE pH SCALE
ACIDS
Acids are fluids that have a greater proportion of H^+ ions to OH^- ions.
• H^+ ions are very reactive.
• Strong acids are corrosive to metals.
• Acids break down food in your digestive tract.
• Acids are generally sour in taste.
BASES
Bases are fluids that have a greater proportion of OH^- ions to H^+ ions.
• OH^- ions bind with H^+ ions, neutralizing acids.
• Strong bases are caustic to your skin.
• Bases can be found in many household cleaners.
• Bases are generally bitter in taste and slippery.
H^+ ion
Water
OH^- ion
0 1 2 3 4 5 6 7 8 9 10 11 12 13 14
Stomach acid
Soda
Orange juice
Coffee
Water
Blood
Baking soda
Ammonia
Bleach
Soda, with a pH of about 3.0, is 10,000 times more acidic than a glass of water, with a pH of 7.0!

FIGURE 2-18 The pH scale is a way of referring to the acidic, basic, or chemically neutral quality of a fluid.

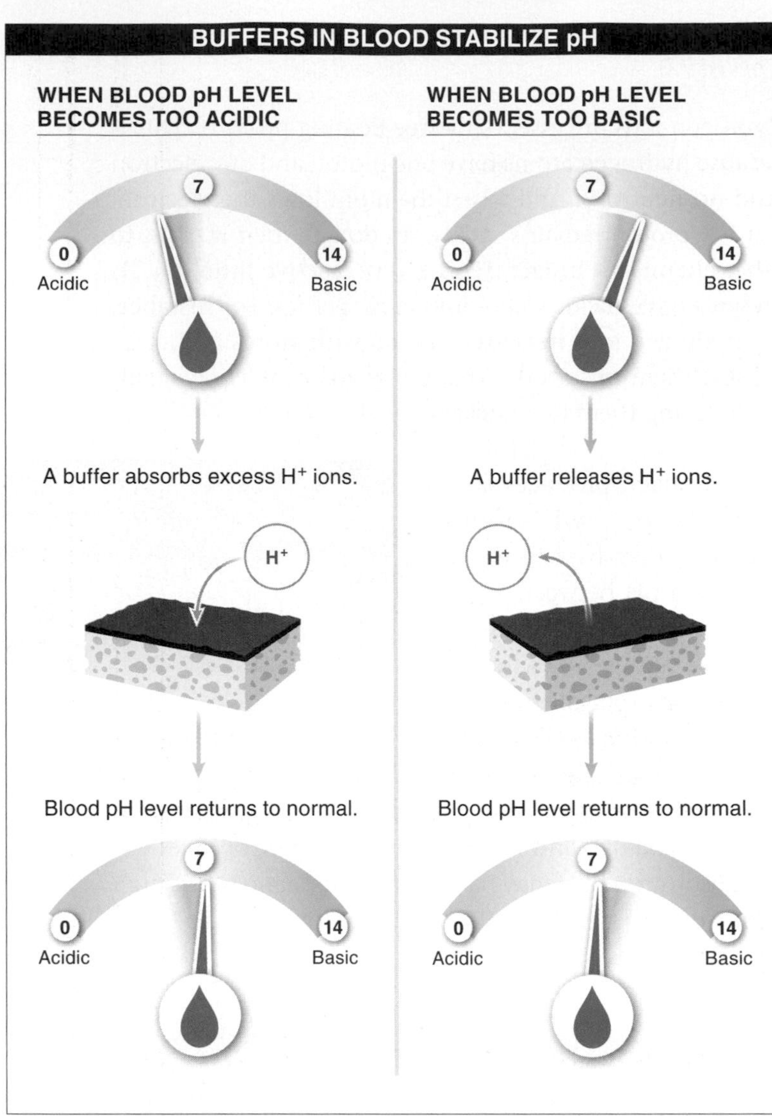

BUFFERS IN BLOOD STABILIZE pH
WHEN BLOOD pH LEVEL BECOMES TOO ACIDIC
WHEN BLOOD pH LEVEL BECOMES TOO BASIC
7
0
14
Acidic
Basic
A buffer absorbs excess H+ ions.
A buffer releases H+ ions.
H+
Blood pH level returns to normal.

FIGURE 2-19 **Maintaining a constant internal environment.** Buffers in blood prevent potentially damaging pH swings.

Acids have a higher concentration of H^+ than of OH^-. Conversely, bases have a higher concentration of OH^- than of H^+. Baking soda is a common basic substance. Some bases are called "antacids" because the OH^- ions in bases can bind with excess H^+ ions in acidic solutions, neutralizing the acid. Base-containing products such as Alka-Seltzer and Milk of Magnesia reduce the unpleasant feeling of heartburn and acid indigestion that sometimes arise from overproduction of acids by the stomach. Bases are commonly used in household cleaning products and generally have a bitter taste and a soapy, slippery feel.

The pH of blood is usually 7.4. Given that most cellular reactions produce or consume H^+ ions, you might expect there to be great swings in the pH of our blood. Unfortunately, our bodies can't tolerate such swings. Most of the chemical reactions in our blood and cells stop proceeding properly if the pH swings up or down, even by less than half a point. Fortunately, our bodies contain some chemicals that act like bank accounts for H^+ ions (FIGURE 2-19). Called **buffers,** these chemicals can quickly absorb excess H^+ ions to keep a solution from becoming too acidic, and they can quickly release H^+ ions to counteract any increases in OH– concentration. Buffers are chemicals that act to resist changes in pH.

Since the middle of the 18th century, the earth's oceans have become slightly more acidic, with the surface pH dropping from 8.25 to 8.14. This phenomenon, called ocean acidification, results from the absorption by the oceans of carbon dioxide in the atmosphere, and it is exacerbated by the high levels of carbon dioxide produced by human activities, including the burning of fossil fuels. The increasing ocean acidity may have harmful consequences for marine ecosystems, including corals, shellfish, and plankton, and the larger marine animals that feed on them.

TAKE HOME MESSAGE 2.6

» The pH of a fluid is a measure of how acidic or basic the solution is and depends on the concentration of dissolved H^+ ions present; the lower the pH, the more acidic the solution. Acids (such as vinegar) can donate protons to other chemicals; bases (including baking soda) bind with free protons.

THIS IS HOW WE DO IT

Developing the ability to apply the process of science

2.7 Do antacids impair digestion and increase the risk of food allergies?

Your stomach is a pretty extreme environment compared with the rest of your body—at least when it comes to acidity. In the previous section, we saw that hydrochloric acid produced in your stomach creates a highly acidic environment, with a pH between 1 and 3.

How does the stomach's acidic environment aid in digestion?

Your stomach's high acidity aids in breaking down chemicals in the food you eat—particularly proteins. (It also helps your body fight bacterial infections.) This is important because proteins that pass through the stomach undigested can trigger an allergic response: your body will treat them as a sort of foreign invader and health risk, mounting an immune response to them.

But for millions of people, the acidity in their stomach can lead to discomfort. For a wide variety of complaints—including ulcers, heartburn, and acid reflux—many people take medications (called antacids) that neutralize stomach acids or medications that reduce the stomach's production of acids or the major protein-digesting enzymes. In fact, one of these medications, Prilosec, is one of the best-selling drugs in the United States, with sales of more than $700 million in 2013.

Does reducing stomach acidity have health consequences? How could we evaluate whether acidity-reducing medications increase the potential for food allergy?

Researchers have suspected that antacid medications might reduce protein digestion, and that undigested proteins stimulate allergies. To test this idea, they explored two experimental strategies.

Strategy 1. First the researchers evaluated the effectiveness of protein-digesting enzymes at several different levels of acidity. For this, they simply combined these enzymes and common dietary proteins in test tubes, at body temperature.

The result? At the typical stomach pH of 2, they found that the proteins in milk are completely digested within 1 minute. Conversely, at less acidic pH values of 3 and 5, the milk proteins were largely undigested, even after 1 hour.

Based on the results of this first experiment, what might the researchers conclude about the role of stomach acidity in digesting proteins?

Strategy 2. In their second type of experiment, the researchers observed 152 people who were seeking treatment for an ulcer, before and after a three-month antacid treatment. They evaluated each participant for signs of allergic reactions to a variety of proteins.

The result? Before the three-month antacid treatment, 10% of the participants exhibited signs of having had an immune response to at least 1 of the 19 dietary proteins the researchers tested. After three months of antacid treatment, 26% of the participants exhibited signs of having had an immune response to at least 1 of the 19 tested dietary proteins.

(continued on following page)

Based on the results of the second experiment, what might the research team conclude about the role of antacids in causing normal dietary proteins to stimulate an allergic response?

The researchers also made observations on 50 control subjects who were not seeking anti-ulcer treatment. As with the other participants prior to antacid treatment, just 10% exhibited signs of having had an immune response to at least 1 of the 19 tested dietary proteins. (What do you think was the purpose of these controls? Suggest why the researchers might have included fewer than the 152 experimental subjects. Would the study have been better if they had used 152 control subjects?)

From their results, the researchers concluded that antacid drugs "hamper the biological gate-keeping function of the stomach." Do you think that such a strong conclusion is warranted from these two experiments? Did the researchers demonstrate a cause-and-effect relationship or just a correlation between antacids and allergic responses?

What follow-up experiments might be done—and what outcomes would you expect if the researchers' hypothesis were valid? Suppose, for example, you measured the incidence of protein allergies among individuals with a genetic disorder that causes their stomach pH to be higher than normal (due to impaired proton pump functioning in the stomach). What results would you expect?

TAKE HOME MESSAGE 2.7

» Dietary proteins are not digested as well by the stomach when the stomach's pH is increased as a result of medications taken to treat ulcers, heartburn, and other digestive problems. This can put people who take antacid medications at risk for developing allergic responses to common foods.

STREET BIO

Using evidence to guide decision making in our own lives

Bottled water or tap water

Drinking water is essential to good health. But where is the best place to get it?

Q: Which water is safer, tap or bottled? For safety, both bottled waters and tap water fare pretty well. The edge, however, goes to tap water.

The U.S. Environmental Protection Agency (EPA) oversees tap water quality. Each year, the EPA must send a water quality report to all customers, which includes the water source and the levels of any potential contaminants. Bottled water manufacturers, on the other hand, have no such requirement.

The U.S. Food and Drug Administration (FDA) oversees bottled water that is sold across state lines (but not the 60% of bottled water packaged and sold within the same state). Bottled water producers must follow numerous safety practices, but inspections are much less frequent and rigorous than those conducted on tap water by the EPA.

The FDA requires bottling plants to test for bacteria just once each week, while the EPA requires at least 100 tap water tests for bacteria each month, plus additional protection from bacteria due to the addition of small amounts of chlorine. Bottled water does not—and sitting on shelves can allow bacteria to grow.

The standard for lead contamination in bottled water, however, is stricter than for tap water—which takes into account that some homes have old pipes, causing higher lead levels. But the delivery system for bottled water has its own risks. And there may be some cause for concern about chemicals used in manufacturing the plastic bottles themselves, particularly polyethylene terephthalate (PET).

Testing 103 brands of bottled water, scientists at the Natural Resources Defense Council found that most are safe. But they also found that about 25% of bottled water was actually just bottled tap water!

Q: What about taste? On this question, it's a draw. The taste of tap water can suffer as a consequence of the small amounts of chlorine added for disinfecting. But bottled waters vary greatly in taste depending on their source and any treatment methods—such as distillation, filtration, and sterilization—which can make the water taste worse.

Q: Which is the best value? Tap water is the clear winner here. Typically, bottled water costs 2,000 to 3,000 times more than tap. And manufacturers usually draw the water from public sources—so they are selling something the public already owns.

Q: Is one water source more convenient? On convenience it's another draw. At home, tap water is as close as the nearest faucet. Away from home, it can be harder to get. Bottled water is easily transported (just throw a bottle in your backpack) and is sold everywhere.

Q: Which water has less environmental impact? Tap water is the clear winner when it comes to environmental impact. About 80% of recyclable bottles end up in landfills each year. Their breakdown is slow and can release contaminants into the environment.

The verdict? Both waters are largely safe. And so those willing to pay several thousands of times more for the convenience of bottled water, or those who prefer its taste, can continue to buy it. But they'll have to overlook the large and growing environmental toll, as well as the fact that most scientists, environmental agencies, and health agencies continue to recommend tap water.

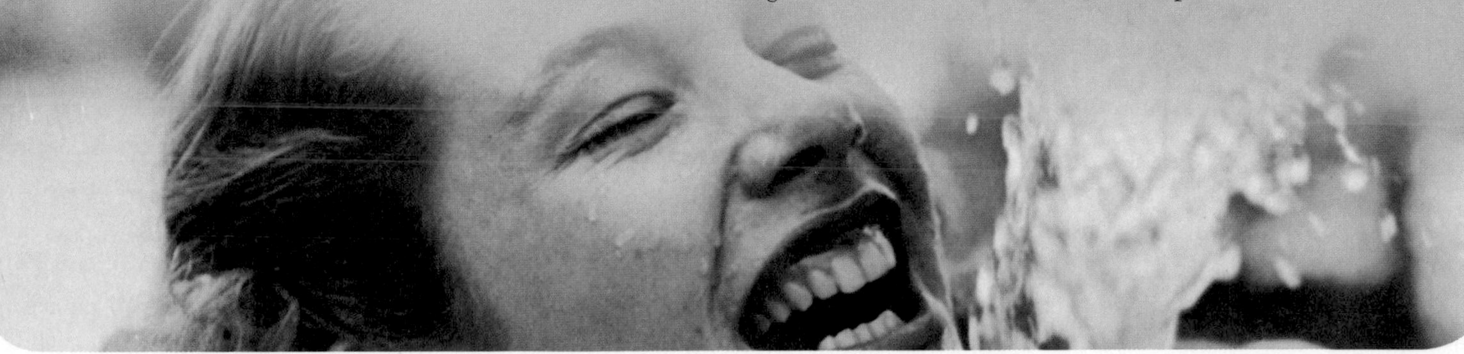

GRAPHIC CONTENT

Thinking critically about visual displays of data

1 Using a table rather than a graph/illustration, it would be possible to present more data in less space. What is gained by the display used here?

2 The percentage listed for carbon is 18.5%, but according to the *y*-axis, the top of the gray region lines up with about 36%. Why is there a discrepancy?

3 What data are presented in this figure? Why?

4 Present the same data as a bar graph and as a pie chart. What are the strengths and weaknesses of each, relative to the presentation here?

5 Could the positions of the colored regions for nitrogen, hydrogen, carbon, and oxygen be reversed? Why or why not?

6 Why is there no label for an *x*-axis? Is that appropriate in a graph?

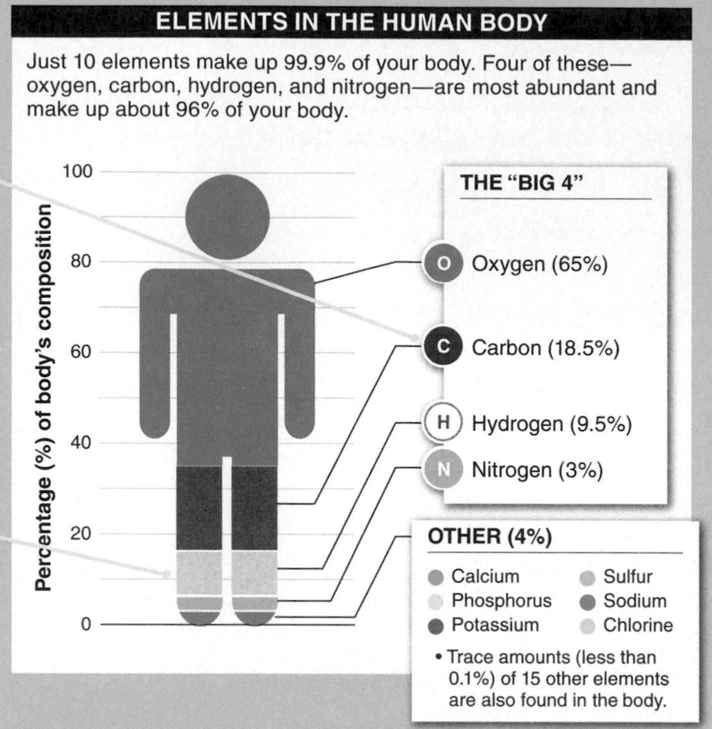

ELEMENTS IN THE HUMAN BODY
Just 10 elements make up 99.9% of your body. Four of these—oxygen, carbon, hydrogen, and nitrogen—are most abundant and make up about 96% of your body.
Percentage (%) of body's composition
100
80
60
40
20
0
THE "BIG 4"
O Oxygen (65%)
C Carbon (18.5%)
H Hydrogen (9.5%)
N Nitrogen (3%)
OTHER (4%)
Calcium
Sulfur
Phosphorus
Sodium
Potassium
Chlorine
• Trace amounts (less than 0.1%) of 15 other elements are also found in the body.

See answers at the back of the book.

KEY TERMS IN THE CHEMISTRY OF BIOLOGY

BRIEF SUMMARY

Atoms form molecules through bonding.

• Everything around us, living or not, is made up of atoms, the smallest unit into which material can be divided without losing its essential properties. All atoms have the same general structure: protons and neutrons in the nucleus, and electrons circling, fast and far away, around the nucleus. Just four elements—oxygen, carbon, hydrogen, and nitrogen—make up the vast majority of your body mass.

• The chemical characteristics of an atom depend on the number of electrons in its outermost shell. Atoms are most stable and least likely to bond with other atoms when their outermost electron shell is filled to capacity.

• Atoms can be bound together in three different ways. Covalent bonds are formed when atoms share electrons. In ionic bonds, one atom transfers its electrons to another and the two oppositely charged ions are attracted to each other, forming an ionic compound. Hydrogen bonds, which are weaker than covalent and ionic bonds, are formed from the attraction between a hydrogen atom of one molecule and another atom of (usually) another molecule with a slight negative charge.

Water has features that enable it to support all life.

• Water molecules easily form hydrogen bonds, giving water great cohesiveness.

• The hydrogen bonds between water molecules give water several of its most important characteristics, including cohesiveness (creating, for example, surface tension), reduced density as a solid, the ability to resist temperature changes, and broad effectiveness as a solvent for ionic and polar substances.

• The pH of a fluid is a measure of how acidic or basic the solution is and depends on the concentration of dissolved H+ ions present; the lower the pH, the more acidic the solution. Acids can donate protons to other chemicals; bases bind with free proton.

CHECK YOUR KNOWLEDGE

Short Answer

1. When we consider the mass of an atom, why do we ignore the weight of the electrons?

2. Why are atoms with complete outer shells not likely to bond with another atom?

3. Why is a hydrogen atom bonded to another hydrogen atom (to make an H_2 molecule) more stable than a hydrogen atom on its own?

4. Why are hydrogen bonds so much weaker than covalent or ionic bonds?

5. Why is water such a good solvent?

6. Why does vomit taste so sour?

Multiple Choice

1. The atomic number of carbon is 6. Its nucleus must contain:

a) 6 neutrons and 6 protons.

b) 3 protons and 3 neutrons.

c) 6 neutrons and no electrons.

d) 6 protons and no electrons.

e) 6 protons and 6 electrons.

0 EASY — 42 — HARD 100

2. The second orbital shell of an atom can hold ______ electrons.

a) 2 b) 3 c) 4 d) 6 e) 8

0 EASY — 20 — HARD 100

3. A covalent bond is formed when:

a) two nonpolar molecules associate with each other in a polar environment.

b) a positively charged particle is attracted to a negatively charged particle.

c) one atom gives up electrons to another atom.

d) two atoms share electrons.

e) two polar molecules associate with each other in a nonpolar environment.

0 EASY — 11 — HARD 100

4. Which of the following phenomena is most likely due to the high cohesiveness of water?

a) Lakes and rivers freeze from the top down, not the bottom up.

b) The fishing spider can walk across the surface of liquid water.

c) Adding salt to snow makes it melt.

d) The temperature of Santa Monica Bay, off the coast of Los Angeles, fluctuates less than the air temperature throughout the year.

e) All of the above are due to the cohesiveness of water.

0 EASY — 53 — HARD 100

5. Water can absorb and store a large amount of heat while increasing only a few degrees in temperature. Why?

a) The heat must first be used to break the hydrogen bonds rather than raise the temperature.

b) The heat must first be used to break the ionic bonds rather than raise the temperature.

c) The heat must first be used to break the covalent bonds rather than raise the temperature.

d) An increase in temperature causes an increase in adhesion of the water.

e) An increase in temperature causes an increase in cohesion of the water.

0 EASY — 24 — HARD 100

6. A chemical compound that releases H^+ into a solution is called:

a) a proton. b) a base. c) an acid.

d) a buffer. e) a hydrogen ion.

0 EASY — 32 — HARD 100

7. Ocean acidification results from:

a) acids released during the decay of dead fish and other marine organisms.

b) free electrons binding to water molecules.

c) free protons binding to water molecules.

d) absorption of carbon dioxide by chemicals in the ocean.

e) coral bleaching and the corresponding reduction of oceanic calcium.

0 EASY — 18 — HARD 100

Ch03

Macromolecules are the raw materials and fuel for life.

3.1 Carbohydrates, lipids, proteins, and nucleic acids are essential to organisms.

Carbohydrates are fuel for living machines.

3.2 Carbohydrates include macromolecules that function as fuel.

3.3 Glucose provides energy for the body's cells.

3.4 Many complex carbohydrates are time-release packets of energy.

3.5 Not all carbohydrates are digestible.

Lipids store energy for a rainy day.

3.6 Lipids are macromolecules with several functions, including energy storage.

3.7 Dietary fats differ in degrees of saturation.

3.8 *This is how we do it:* What is the impact of trans fatty acids on heart health?

3.9 Cholesterol and phospholipids are used to build sex hormones and membranes.

Proteins are versatile macromolecules that serve as building blocks.

3.10 Proteins are bodybuilding macromolecules.

3.11 Proteins are an essential dietary component.

3.12 A protein's function is influenced by its three-dimensional shape.

3.13 Enzymes are proteins that speed up chemical reactions.

3.14 Enzymes regulate reactions in several ways (but malformed enzymes can cause problems).

Nucleic acids store information on how to build and run a body.

3.15 Nucleic acids are macromolecules that store information.

3.16 DNA holds the genetic information to build an organism.

3.17 RNA is a universal translator, reading DNA and directing protein production.

An egret uses displays of its feathers (made from proteins) as part of a courtship display.

Molecules of Life

Macromolecules can store energy and information and serve as building blocks

3.1

Macromolecules are the raw materials and fuel for life.

Our food is made up of mixtures of carbohydrates, lipids, and amino acids.

3.1 Carbohydrates, lipids, proteins, and nucleic acids are essential to organisms.

A little bit of chemistry goes a long way in the study of biology and can help in understanding a great deal about your everyday life. For example, will the butter you spread on your toast sabotage your efforts to lose weight? Understanding something about the carbon-hydrogen connections in the fat molecules can help you decide.

Foods in our diet would seem to have nothing in common with a daisy or a Chihuahua, or bacteria. Yet they all are made from **organic molecules,** chemical compounds that contain both carbon and, usually, hydrogen.

In Chapter 2 we looked at tiny hydrogen (H) with its single proton and electron, as well as simple molecules such as water (H_2O). In this chapter, we'll look at much larger molecules. Because many of the organic molecules important to life are made from smaller building blocks and may have thousands of atoms in a single molecule, they are called **macromolecules.** Four types of macromolecule essential to the building and functioning of living organisms are carbohydrates, lipids, proteins, and nucleic acids. All organisms on earth consist largely of these same types of molecules.

TAKE HOME MESSAGE 3.1

» Macromolecules are large organic molecules made up from smaller building blocks, or subunits. Four types of macromolecule are essential to living organisms: carbohydrates, lipids, proteins, and nucleic acids.

3.2–3.5

Carbohydrates are fuel for living machines.

Worldwide, more corn is produced (by weight) than any other grain.

3.2 Carbohydrates include macromolecules that function as fuel.

Nearly every day, a story in a magazine or newspaper or on TV asks whether carbohydrates are good or bad for us. Meanwhile, nutritionists, doctors, and many diet programs tell people to eat more fiber—a type of carbohydrate. What exactly are they talking about?

Carbohydrates are the primary fuel for cells and form much of the structure of cells in all organisms. Carbohydrates always contain carbon, hydrogen, and oxygen, and sometimes contain atoms of other elements (FIGURE 3-1). A carbohydrate has approximately the same number of carbon atoms as it does H_2O units in its chemical formula. For instance, glucose has the composition $C_6H_{12}O_6$, and the carbohydrate maltose has the composition $C_{12}H_{22}O_{11}$.

As we will see in Section 3.4, carbohydrates are classified into several categories, based on their size and their composition. The simplest carbohydrates are the **monosaccharides,** or **simple sugars,** which contain anywhere from three to six carbon atoms and, when they are broken down, produce molecules that usually are not carbohydrates. Two common monosaccharides are glucose, found in the sap and fruit of many plants, and fructose, found primarily in fruits and vegetables, as well as in honey. Fructose is the sweetest of all naturally occurring sugars. The suffix *-ose* tells us that a substance is a carbohydrate.

SOME COMMON MONOSACCHARIDES

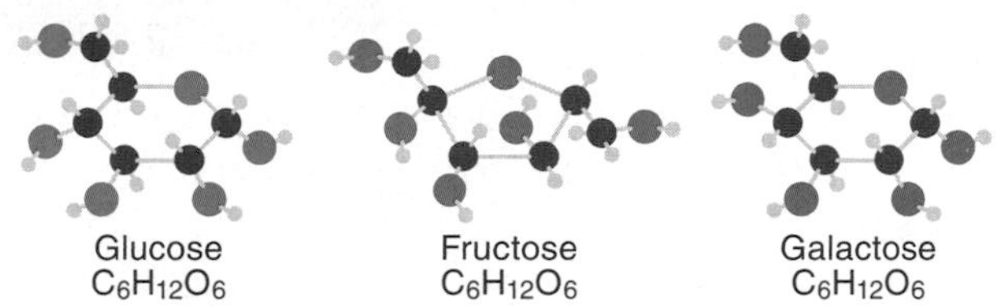
Glucose
$C_6H_{12}O_6$
Fructose
$C_6H_{12}O_6$
Galactose
$C_6H_{12}O_6$

Carbohydrates function well as fuels because of their many carbon-hydrogen bonds. As those bonds are broken down and other, more stable bonds (primarily between carbon and oxygen) are formed, a great deal of energy is released that organisms can use.

In the next section we investigate glucose, the chief carbohydrate that organisms use to fuel their activities.

CARBOHYDRATES

Carbohydrates, such as glucose (shown below), are composed primarily of three elements: carbon, hydrogen, and oxygen.

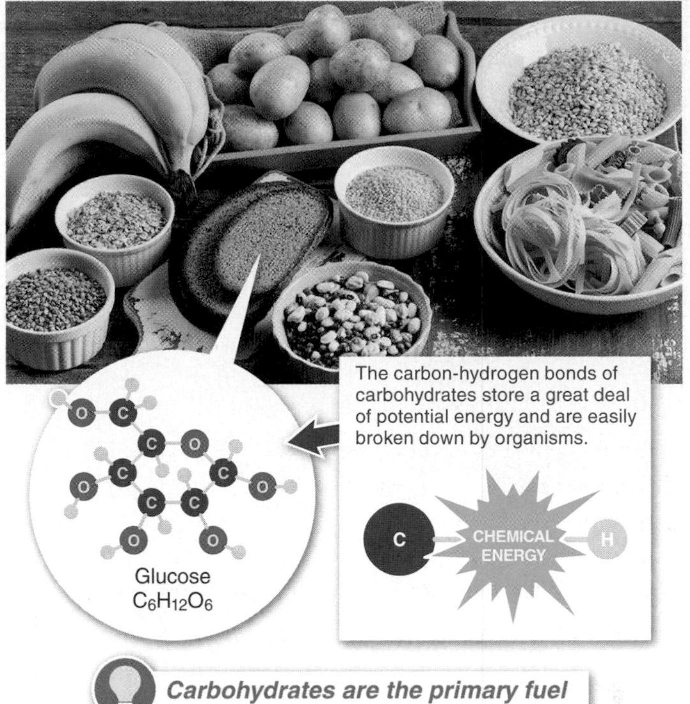
The carbon-hydrogen bonds of carbohydrates store a great deal of potential energy and are easily broken down by organisms.
Glucose
$C_6H_{12}O_6$
C
CHEMICAL ENERGY
H
Carbohydrates are the primary fuel source for cellular mechanisms.

FIGURE 3-1 All carbohydrates have a similar structure and function.

TAKE HOME MESSAGE 3.2

» Carbohydrates are the primary fuel for cells and also form much of the structure of cells in all organisms. The simplest carbohydrates are monosaccharides, or simple sugars. As the chemical bonds of carbohydrates are broken down and other, more stable bonds are formed, the process releases energy that the organism can use.

3.3 Glucose provides energy for the body's cells.

The carbohydrate of most importance to living organisms is glucose. This monosaccharide is found naturally in most fruits. Additionally, most of the carbohydrates that you eat, including table sugar (called sucrose) and the starchy carbohydrates found in bread and potatoes, are converted into glucose in your digestive system. The glucose then circulates in your blood at a concentration of about 0.1%. Circulating glucose, also called "blood sugar," has one of three fates (FIGURE 3-2).

1. Used as fuel for cellular activity. Once it arrives at and enters a cell, glucose can be used as an energy source.

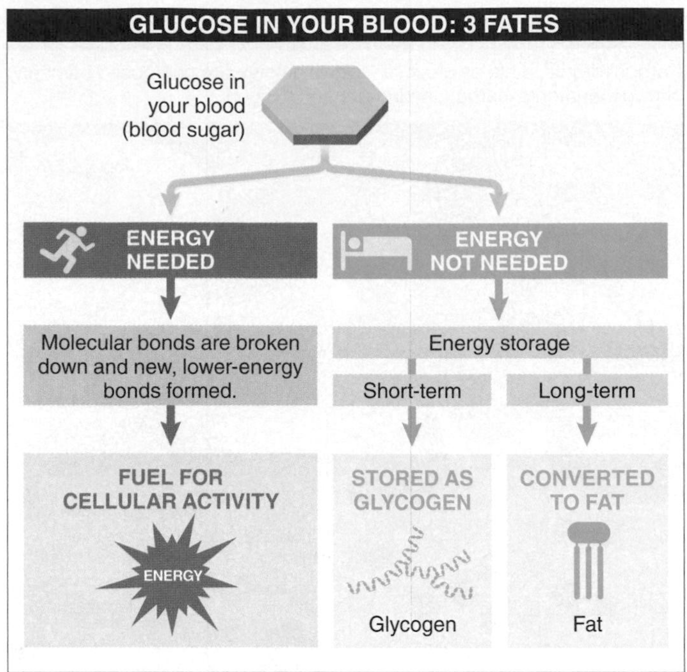

FIGURE 3-2 **What happens to sugar in your blood.** Depending upon whether energy is needed, glucose can be released for immediate use or stored for the short or long term.

The process that releases energy from the high-energy bonds of glucose is discussed in detail in Chapter 5.

2. Stored temporarily as glycogen. If glucose is circulating in your bloodstream in excess of your body's needs, it can be temporarily stored in various tissues, primarily your muscles and liver. The stored glucose molecules are linked together to form a large web of molecules called **glycogen.** When you need energy later, the glycogen can easily be broken down to release glucose molecules back into your bloodstream. Glycogen is the primary form of short-term energy storage in animals.

During the initial rapid weight loss that you experience if you're dieting, glycogen stores in your muscles and liver are the first, most accessible molecules that can be broken down for energy in the absence of sufficient sugar in your bloodstream. Large amounts of water are bound to glycogen. As that glycogen is removed from your tissue, so, too, is the water. This loss of water leads to the initial dramatic weight loss that occurs before your body resorts to using stored fat, and the rate of weight loss then slows considerably (FIGURE 3-3).

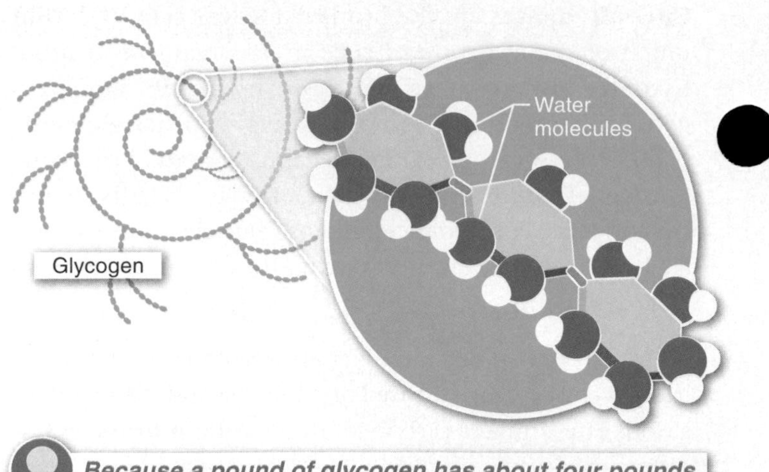

Water molecules
Glycogen
Because a pound of glycogen has about four pounds of water molecules bound to it, dieters lose a lot of "water weight" during the first days of a diet.

FIGURE 3-3 **Water weight.**

Q Dieters often experience an initial large loss of "water weight." Why?

3. Converted to fat. Finally, glucose circulating in your bloodstream can be converted into fat, a form of long-term energy storage.

TAKE HOME MESSAGE 3.3

» Glucose is the most important carbohydrate for living organisms. Glucose in the bloodstream can be used as an energy source, can be stored as glycogen in the muscles and liver for later use, or can be converted to fat.

3.4 Many complex carbohydrates are time-release packets of energy.

In contrast to the simple sugars, some carbohydrates contain more than one sugar unit or building block. A **disaccharide** forms when two simple sugars bond together. The joining of glucose and galactose, for example, makes the disaccharide lactose, which is the sugar found in milk. When large numbers of simple sugars are joined together—sometimes as many as 10,000—the resulting carbohydrate is called a **polysaccharide,** or a **complex**

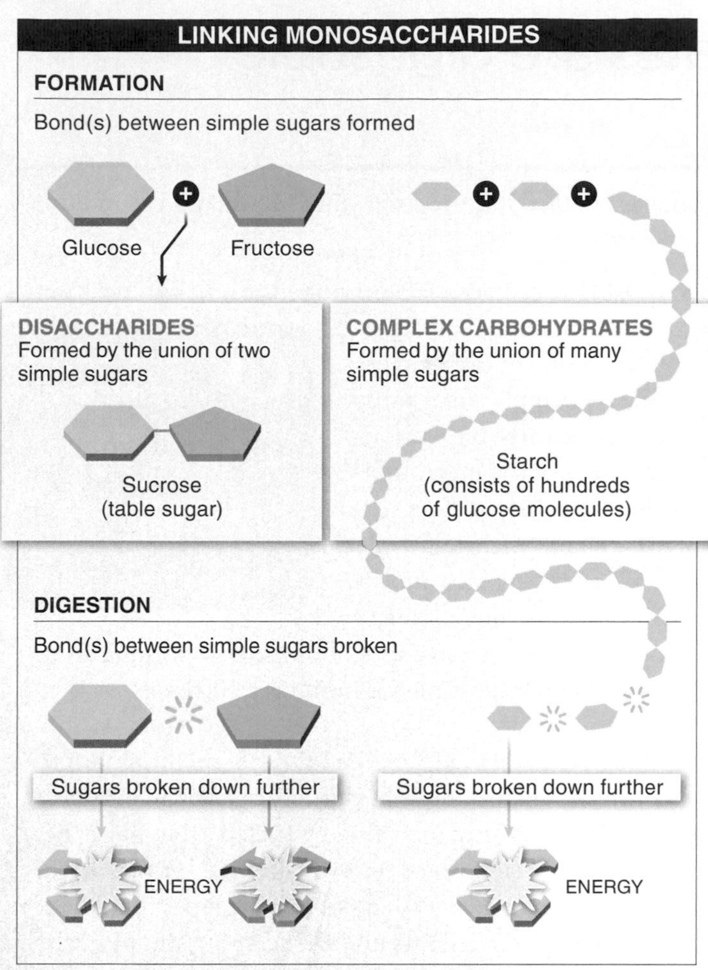

FIGURE 3-4 **Chains of sugars.** Polysaccharides are made from simple sugars bonded together.

FIGURE 3-5 **Short-term versus long-term energy.** Simple sugars and complex carbohydrates differ in the way they make energy available to you.

carbohydrate (FIGURE 3-4). Depending on how the simple sugars are bonded together, polysaccharides may function as "time-release" stores of energy or as structural materials for plants that may be completely indigestible by most animals.

Many disaccharides and polysaccharides are important sources of fuel for cells, but they must undergo some preliminary processing before the energy can be released from their bonds. Sucrose (common table sugar) is composed of the monosaccharides glucose and fructose linked together. To use sucrose, the body first breaks the bond linking the glucose and fructose. Then individual monosaccharides are broken down into their component atoms, and the energy that was stored in their chemical bonds is harvested and used.

In plants, the primary form of energy storage is a complex carbohydrate called **starch,** found in roots and other tissues (see Figure 3-4). Starch consists of linked glucose molecules. Grains such as barley, wheat, and rye are high in starch content, as are corn and rice, both of which are more than 70% starch. The glycogen that stores energy in your muscles and liver is also a complex carbohydrate, so it is sometimes referred to as "animal starch" (although it has a more branched structure than starch and carries more glucose units linked together).

Many complex carbohydrates—particularly starches that have not been refined or highly processed—are like "time-release" fuel pellets. The glucose molecules become available slowly, as the bonds between glucose units are broken. In contrast, many refined sugars and refined starches give a quick burst of energy because the sugars are more immediately available (FIGURE 3-5).

TAKE HOME MESSAGE 3.4

» Multiple simple carbohydrates are sometimes linked together into more complex carbohydrates. Complex carbohydrates include starch, which is the primary form of energy storage in plants, and glycogen, which is an important form of energy storage in animals.

3.5 Not all carbohydrates are digestible.

Despite their importance as a fuel source for humans, not all carbohydrates can be broken down by our digestive system. Two different complex carbohydrates—both indigestible by humans—serve as structural materials for invertebrate animals and plants, respectively: **chitin** (pronounced KITE-in) and **cellulose** (FIGURE 3-6). Chitin forms the rigid outer skeleton of most insects and crustaceans (such as lobsters and crabs). Cellulose forms a huge variety of plant structures. We find cellulose in trees and the wooden structures we build from them, in cotton and the clothes we make from it, in leaves and in grasses. In fact, it is the single most abundant compound on earth.

Surprisingly, cellulose is almost identical in composition to starch, but one small difference in the chemical bond between the simple sugar units changes the three-dimensional structure of the macromolecule. Even tiny differences in the shape of a molecule can affect its behavior. In this case, the difference in shape makes cellulose a sturdy structural material and makes it, unlike starch, impossible for humans to digest.

Q Cellulose is plentiful around the world. And, like starch, it's a polysaccharide. Why can't humans extract calories from it?

Although it is not digestible, cellulose is still important to our health. The cellulose in our diet is known as "fiber" (FIGURE 3-7). It is also appropriately called "roughage" because, as the cellulose of celery stalks and lettuce leaves passes through our digestive system, it scrapes the wall of the digestive tract. Its bulk and the scraping stimulate the rapid passage of food and of unwanted, possibly harmful products of digestion through our intestines. In this way, fiber reduces the risk of colon cancer and other diseases (but it is also why too much fiber can lead to diarrhea).

STRUCTURAL POLYSACCHARIDES

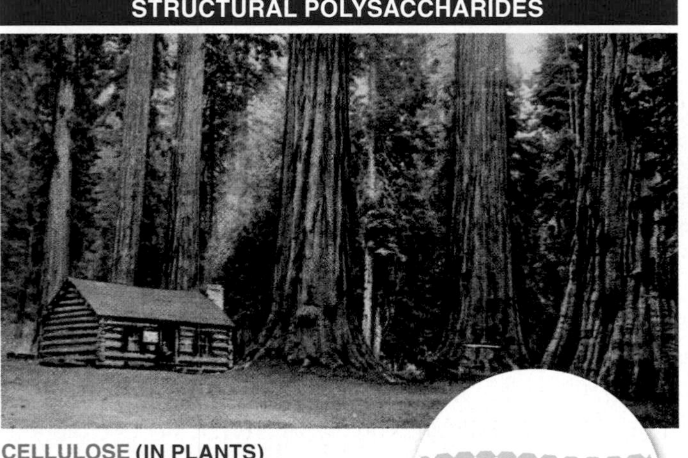

CELLULOSE (IN PLANTS)
Cellulose is made of chains of many linked glucose units. Hydrogen bonds between the parallel chains contribute to the rigidity of wood.

CHITIN (IN ANIMALS)
The linked sugar units of chitin contain nitrogen and have hydrogen bonds within and between chains, giving it strength.

FIGURE 3-6 **Carbohydrates can serve as structural materials.**

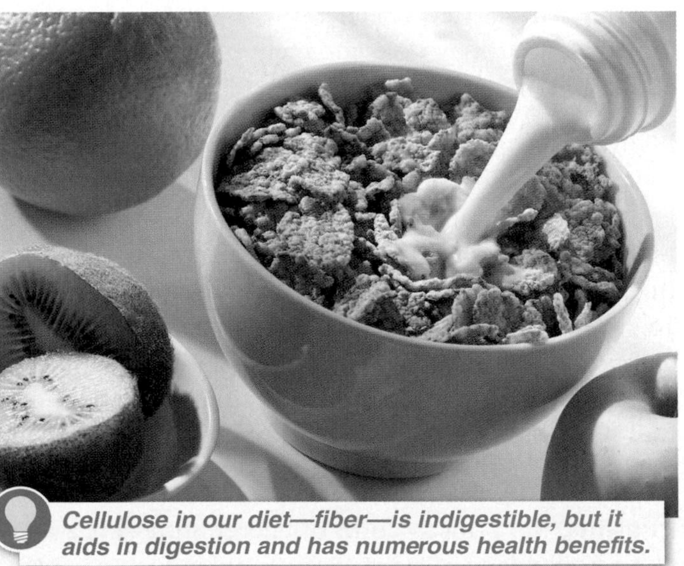

Cellulose in our diet—fiber—is indigestible, but it aids in digestion and has numerous health benefits.

FIGURE 3-7 **Fiber.** It's not digestible, but it's still important for our diet.

Unlike humans, termites and ruminant mammals (such as cows) have microorganisms living in their gut that are able to break down cellulose. That's how termites can chew on wood and ruminants can feed on plants and extract usable energy from the freed glucose molecules of cellulose.

TAKE HOME MESSAGE 3.5

» Some complex carbohydrates, including chitin and cellulose, cannot be digested by most animals. Such indigestible carbohydrates in the diet, called fiber, aid in digestion and have many health benefits.

3.6–3.9

Lipids store energy for a rainy day.

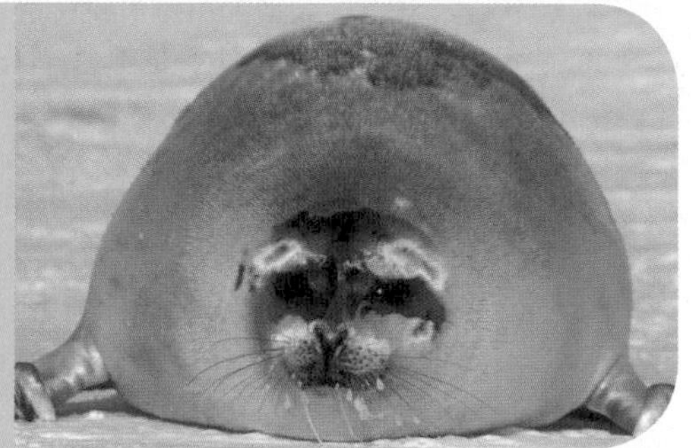

Protected from the cold: lipids help insulate seals.

3.6 Lipids are macromolecules with several functions, including energy storage.

Lipids are a second group of macromolecules important to all living organisms. Lipids, just like carbohydrates, are made primarily from atoms of carbon, hydrogen, and oxygen, but the atoms are in different proportions. The lipids in your diet, for example, tend to have significantly more energy-rich carbon-hydrogen bonds than do carbohydrates, and so contain significantly more stored energy.

Lipids are defined based on their physical characteristics (FIGURE 3-8). Most notably, lipids do not dissolve in water and are greasy to the touch—think of salad dressings.

Lipids are insoluble in water because they tend to have long chains consisting only of carbon and hydrogen atoms. In contrast to water, these chains of carbon and hydrogen atoms are nonpolar—there are no regions of positive or

INTRODUCTION TO LIPIDS

With so many energy-rich bonds, lipids are an important source of stored energy for organisms (including humans).

TYPICAL FEATURES OF LIPIDS
- Nonpolar molecules that do not dissolve in water
- Greasy to the touch
- Can be a significant source of energy storage

THREE TYPES OF LIPIDS

FATS — **Function:** Long-term energy storage and insulation

STEROLS — **Function:** Regulate growth and development

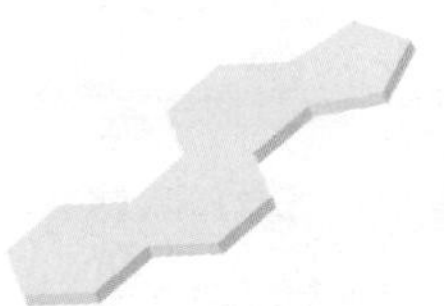

PHOSPHOLIPIDS — **Function:** Form cellular membranes

FIGURE 3-8 **Lipids serve many roles in the body.**

Q Why does a salad dressing made with vinegar and oil separate into two layers shortly after you shake it?

negative charge. Thus they do not form weak bonds with water and cannot dissolve in water. Nonpolar molecules (or parts of molecules) tend to minimize their contact with water and are considered **hydrophobic** ("water-fearing"). Lipids cluster together when mixed with water, never fully dissolving. Molecules that readily form hydrogen bonds with water, on the other hand, are considered **hydrophilic** ("water-loving").

One familiar type of lipid is *fat,* the type most important in long-term energy storage and insulation. (Penguins and walruses can maintain relatively high body temperatures, despite living in very cold habitats, due to their thick layer of insulating fat.) Lipids also include *sterols,* which include *cholesterol* and many of the sex hormones that play regulatory roles in animals, and *phospholipids,* which form the membranes that enclose cells.

TAKE HOME MESSAGE 3.6

» Lipids are insoluble in water and greasy to the touch. They are valuable to organisms for long-term energy storage and insulation, in membrane formation, and as hormones.

3.7 Dietary fats differ in degrees of saturation.

All fats have two distinct components: they have a "head" region and three long "tails" (FIGURE 3-9). The head region is a small molecule called **glycerol.** It is linked to "tail" molecules known as **fatty acids.** A fatty acid is simply a long hydrocarbon—that is, a chain of carbon atoms, often a dozen or more, linked together and with one or two hydrogen atoms attached to each carbon atom.

STRUCTURE OF FATS (TRIGLYCERIDES)

Fats are composed primarily of three elements: carbon, hydrogen, and oxygen.

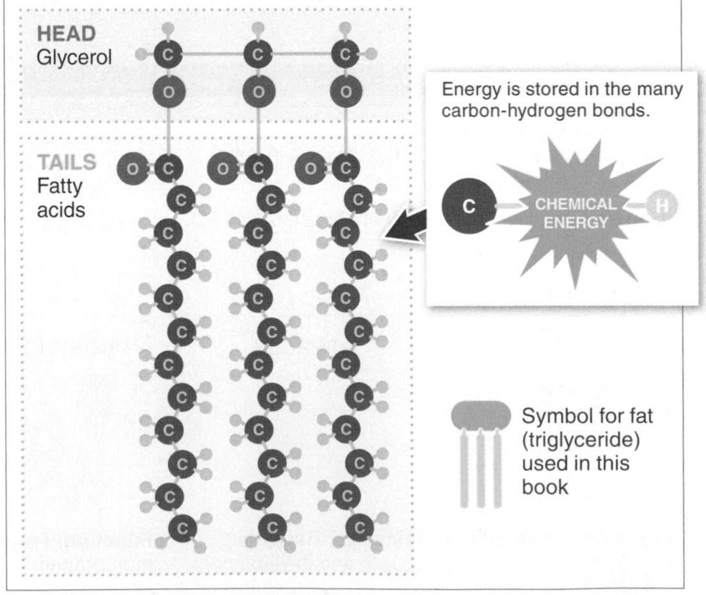

FIGURE 3-9 **Triglycerides have glycerol heads and fatty acid tails.**

The fats in most foods we eat are **triglycerides,** which are fats having three fatty acids linked to the glycerol molecule. For this reason, the terms "fats" and "triglycerides" are often used interchangeably. Triglycerides that are solid at room temperature are generally called "fats," while those that are liquid at room temperature are called "oils."

The chemical breakdown of fat molecules releases significantly more energy than the breakdown of carbohydrates. A single gram of carbohydrate stores about 4 calories of energy, while the same amount of fat stores about 9 calories—not unlike the difference between a $5 bill and a $10 bill. Because fats store such a large amount of energy, a strong taste preference for fats over other energy sources has evolved in animals (FIGURE 3-10). This taste preference helped the earliest humans to survive but today puts us in danger from the health risks of obesity when we overconsume fats.

The terms "saturated" and "unsaturated" in describing fats refer to the hydrocarbon chain of the fatty acids (FIGURE 3-11). If each carbon atom in the hydrocarbon chain of a fatty acid is bonded to two hydrogen atoms, the fat molecule carries the maximum number of hydrogen atoms and is said to be a **saturated fat.** Most animal fats, including those found in meat and eggs, are saturated. They are not essential to your health, and, because they accumulate in your bloodstream and can narrow the blood vessels, they can contribute to heart disease and strokes.

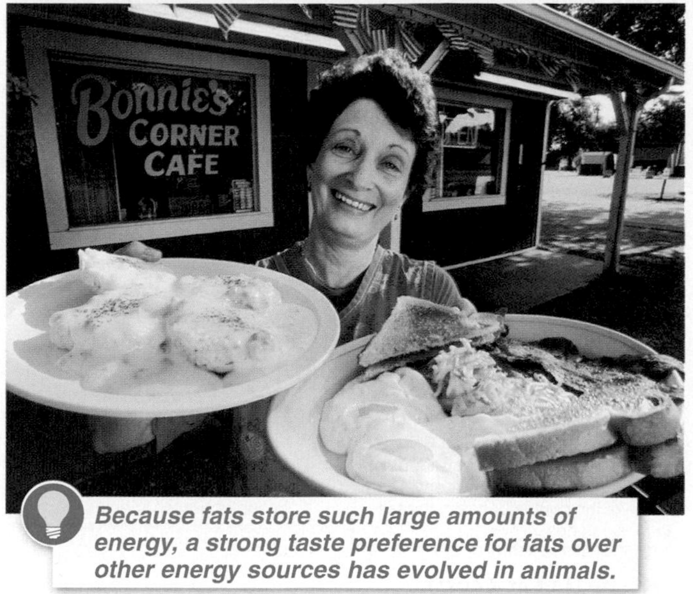

Because fats store such large amounts of energy, a strong taste preference for fats over other energy sources has evolved in animals.

FIGURE 3-10 **Animals (including humans) prefer the taste of fats.**

An **unsaturated fat** is one in which some of the carbon atoms are bound to only a single hydrogen atom (and pairs of neighboring carbons are connected to each other by a double bond). Most plant fats are unsaturated. Unsaturated fats may be *mono-unsaturated* or *polyunsaturated.* A mono-unsaturated fatty acid hydrocarbon chain has only one pair of carbon atoms in an unsaturated state—that is, has only one double bond. A *polyunsaturated* fatty acid hydrocarbon chain has more than one pair of carbons in an unsaturated state—there's more than one double bond. Unsaturated fats are still high in calories, but because they can lower cholesterol, they are generally preferable to saturated fats. Foods high in unsaturated fats include avocados, peanuts, and olive oil. Relative to other animals, fish tend to have less saturated fat.

The shapes of unsaturated fat molecules and saturated fat molecules are different. When saturated, the hydrocarbon tails of the fatty acids all line up very straight and the fat molecules can be packed together tightly. The tight packing causes the fats, such as butter, to be solid at room temperature. When

Q The "chewyness" of a cookie will differ depending on whether you make it with butter or vegetable oil as the lipid. Why?

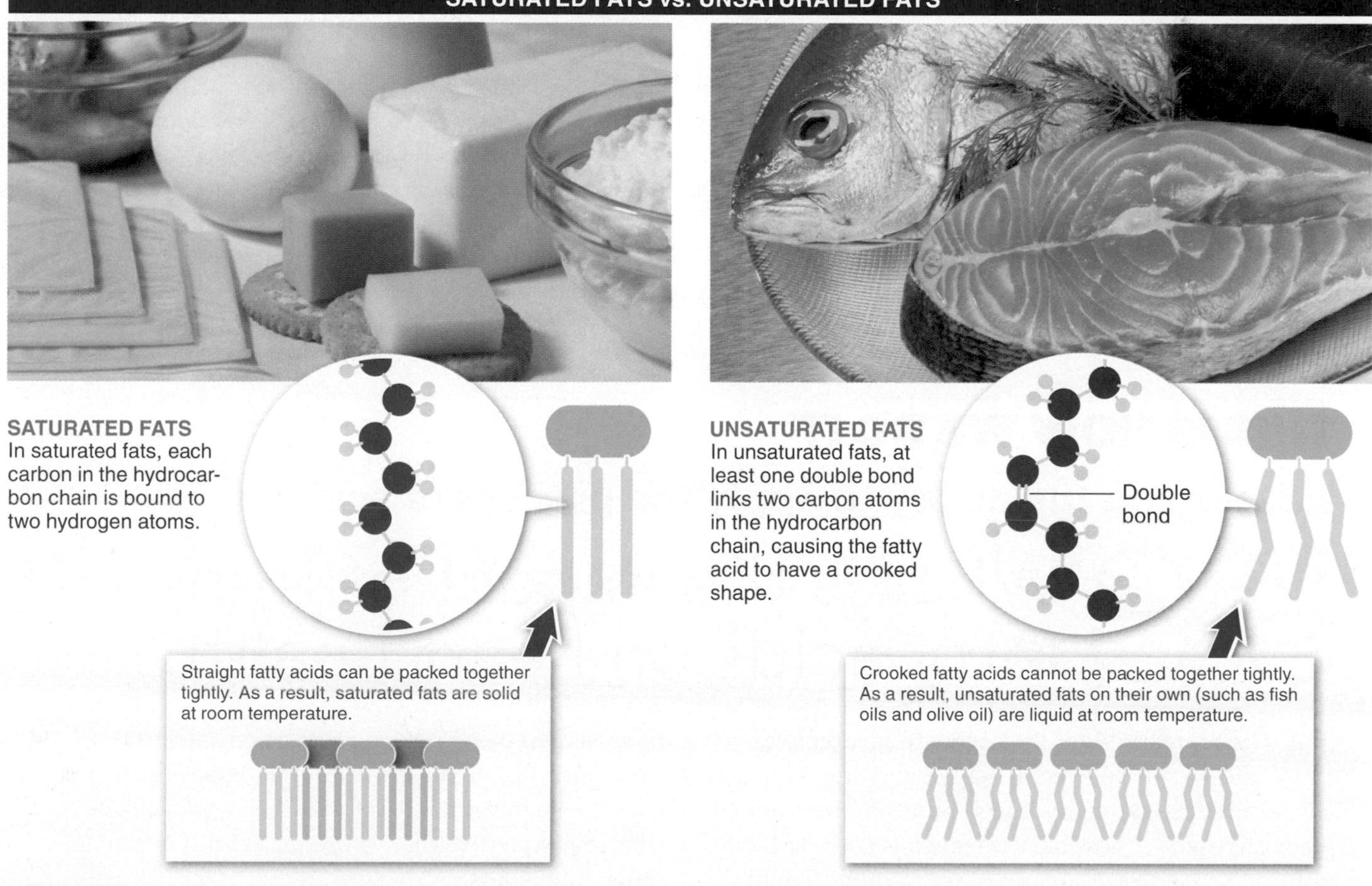

SATURATED FATS vs. UNSATURATED FATS
SATURATED FATS
In saturated fats, each carbon in the hydrocarbon chain is bound to two hydrogen atoms.
Straight fatty acids can be packed together tightly. As a result, saturated fats are solid at room temperature.
UNSATURATED FATS
In unsaturated fats, at least one double bond links two carbon atoms in the hydrocarbon chain, causing the fatty acid to have a crooked shape.
Double bond
Crooked fatty acids cannot be packed together tightly. As a result, unsaturated fats on their own (such as fish oils and olive oil) are liquid at room temperature.

FIGURE 3-11 **Degrees of saturation.** Fatty acids (and thus the fats that contain them) can be saturated or unsaturated.

HYDROGENATION

Hydrogenation can improve a food's taste, texture, and shelf-life, but it also makes the food less healthful.

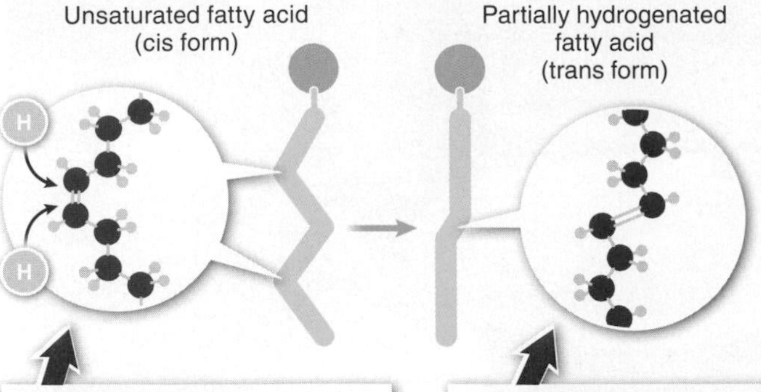
Hydrogenation is the artificial addition of hydrogen atoms to an unsaturated fat in order to make the fat more saturated.
Unsaturated fatty acid (cis form)
Partially hydrogenated fatty acid (trans form)
H
H
Hydrogenation leads to a straighter hydrocarbon tail shape by converting some of the double bonds in the tail to single bonds.
Some or all of the double bonds that remain are in the less healthful "trans" orientation.

FIGURE 3-12 **Hydrogenation improves a food's taste, texture, and shelf-life (but at a cost).**

unsaturated, the fatty acids have kinks in the hydrocarbon tails and the fat molecules cannot be packed together as tightly (see Figure 3-11). Consequently, unsaturated fats, such as canola oil and vegetable oil, do not solidify as easily and are liquid at room temperature.

Q Why should you minimize trans fats in your diet?

The ingredient list for many snack foods includes "partially hydrogenated" vegetable oils. The hydrogenation of an oil means that hydrogen atoms have been added to a liquid, unsaturated fat so that it becomes more saturated (FIGURE 3-12). Unfortunately, hydrogenation also makes the food less healthful because it also creates **trans fats,** the "trans" referring to the unusual orientation of some or all of the double bonds that remain following the addition of hydrogen atoms. The trans fats are less reactive—your body is less likely to break them down—and so are more likely to accumulate in your blood vessels. We explore the consequences of this in the next section.

TAKE HOME MESSAGE 3.7

» Fats, including the triglycerides common in the foods we eat, are one type of lipid. Characterized by long hydrocarbon tails, fats store energy effectively in the many carbon-hydrogen and carbon-carbon bonds. Their high calorie content per gram is responsible for humans' preferring fats to other macromolecules in the diet.

THIS IS HOW WE DO IT

Developing the ability to apply the process of science

3.8 What is the impact of trans fatty acids on heart health?

When a method for hydrogenating vegetable oils was discovered, it was an exciting advancement. Adding hydrogen bonds to unsaturated plant oils made it possible to create fats with desirable physical properties. Margarines, for example, could be spread on bread right out of the refrigerator—something not possible with butter. Foods made with hydrogenated vegetable oils had a longer shelf-life and were less expensive than those made with butter and other animal fats.

Over time, however, data accumulated indicating that trans fats increased the risk of heart disease. Researchers noted, for example, in a study of nurses, that those who consumed just 2% more trans fats experienced nearly double the risk of heart disease. Greater consumption of saturated fats was associated with a much smaller increase in heart disease.

The nurse study was large—more than 120,000 nurses—and spanned decades, but it was observational and did not include any controlled experimental manipulations. So the question "Do trans fats increase your risk of heart disease?" loomed ominous and unresolved.

How could you determine whether consumption of trans fats increases your risk of heart disease?

To examine heart disease risk and trans fat consumption, you could divide—randomly—a huge population into two groups, carefully control their diets, and monitor their heart health. One group would have a diet high in trans fats; the other, a diet with little or no trans fats, replacing those calories with calories from non-trans (called "cis") unsaturated fats. You would monitor each person in each group for their entire life, tracking heart attacks and other manifestations of heart disease.

Why is the ideal study not feasible?

As a thought experiment, such an approach is great. In reality, it's ridiculously impractical. Heart disease develops slowly—typically over decades. And not everyone gets heart disease, even if they have a high-fat diet. So you would need huge experimental groups. Moreover, it's unlikely that you could control what (and how much) the participants ate for a lifetime.

When the ideal study is impractical, what should you do?

When an ideal study is not possible, we seek a suitable compromise. We could observe fewer subjects, for a shorter period of time. Or we could measure outcomes that are related to the outcome we're actually interested in—development of heart disease. Let's look at one such approach.

Researchers recruited approximately 60 subjects. The average age was 25, and about 60% were women. The volunteers were restricted to identical diets *except* that 10% of the total calories came from (1) cis unsaturated fat, or (2) trans fat, or (3) saturated fat. Each diet was consumed by each person for three weeks, in random order, for a total of nine weeks.

How can you study the impact of trans fats on heart disease without measuring actual heart disease?

After each three-week period, the researchers analyzed blood samples taken from participants. They measured a lipoprotein (a mix of lipid and protein) called LDL that damages arteries and is associated with increased risk of heart disease. And they measured another lipoprotein, called HDL, which reduces damage to arteries and is associated with a reduced risk of heart disease.

The results? Blood levels of LDL—sometimes referred to as "bad cholesterol"—were significantly higher after three weeks on either the trans fat diet or the saturated fat diet, compared with the unsaturated fat diet. And perhaps more important, after three weeks on the trans fat diet, levels of HDL—"good cholesterol"—were significantly lower. This HDL decrease occurred only on the trans fat diet.

The ratio of LDL to HDL is one of the most useful measures of heart disease risk. By this measure, when subjects were consuming the trans fat diet, they had a significantly higher risk of heart disease than when consuming either the unsaturated or saturated fat diets.

What is the impact of trans fats on heart disease risk?

The researchers concluded that trans fats in the diet do cause harm. They increase the levels of the unhealthy LDL cholesterol, raising the risk of heart disease, and they reduce the body's production of beneficial HDL cholesterol, which protects against heart disease.

Based on their methods and results, are the researchers' conclusions appropriate? If not, how can you account for the results? In this light, it becomes easier to understand that it's often difficult to describe a hypothesis about the world as "true" or "false." It may make more sense to describe our level of confidence that a hypothesis accurately represents reality, and, importantly, to

(continued on following page)

discuss how we might conduct further investigations to increase that confidence. Can you think of any important changes to the experimental approach described here that might increase its power to change *your* mind about the influence of trans fats on heart disease?

TAKE HOME MESSAGE 3.8

» In short-term dietary manipulations, researchers found that trans fats increased the levels of unhealthy LDL cholesterol and decreased the levels of beneficial HDL cholesterol. These changes are known to increase the risk of heart disease.

3.9 Cholesterol and phospholipids are used to build sex hormones and membranes.

Not all lipids are fats, nor do lipids necessarily function in energy storage. A second group of lipids, called the **sterols,** play an important role in regulating growth and development (FIGURE 3-13). This group includes some very familiar lipids: cholesterol and the steroid hormones such as testosterone and estrogen. All of these molecules are variations on one basic structure formed from four interlinked rings of carbon atoms.

Cholesterol is an important component of most cell membranes. For this reason, it is an essential molecule for living organisms. Cells in our liver produce almost 90% of the circulating cholesterol by transforming the saturated fats in our diet. With excess cholesterol and fat, a substance called *plaque,* a mixture of cholesterol, fats, calcium, and clotting material, can cause hardening of the arteries and contributes to heart attack or stroke.

The steroid hormones estrogen and testosterone are built through slight chemical modifications to cholesterol. These hormones are among the primary molecules that direct and regulate sexual development, maturation, and sperm and egg production. In both males and females, estrogen influences memory and mood, among other traits. Testosterone has numerous effects, one of which is to stimulate muscle growth. Some athletes (particularly bodybuilders) take synthetic variants of testosterone to increase their muscularity (FIGURE 3-14). But the use of these supplements may cause dangerous side

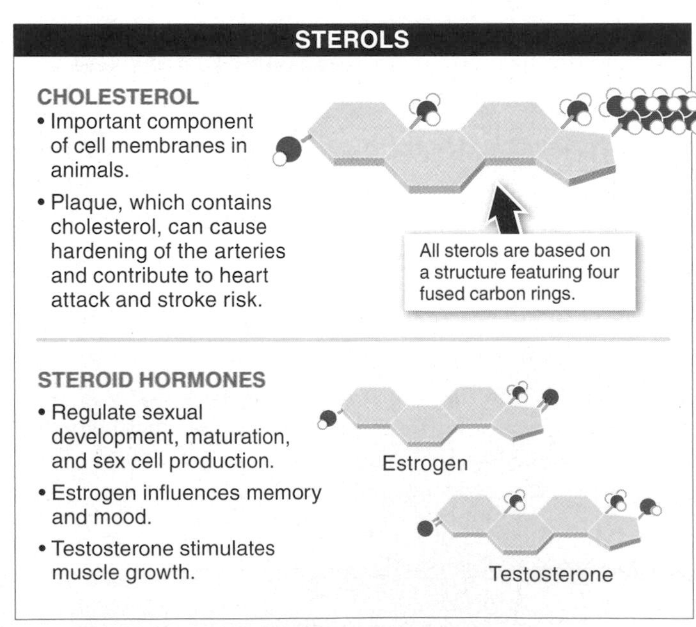

FIGURE 3-13 **Not all lipids are for energy storage.** Cholesterol, estrogen, and testosterone are all lipids.

FIGURE 3-14 **Dangerous bulk.** Steroids can increase muscularity, but sometimes have serious health consequences.

PHOSPHOLIPIDS

HYDROPHILIC HEAD (attracted to water)

Phosphate group

HYDROPHOBIC TAILS (not attracted to water)

Fatty acids

PHOSPHOLIPIDS IN WATER

Water

Hydrophilic heads

Hydrophobic tails

Phospholipids align with their hydrophilic heads extended toward water and their hydrophobic tails away from water.

Symbol for phospholipid used in this book

FIGURE 3-15 **Dual nature.** Phospholipids have a head region that is attracted to water and a tail that is not.

effects, including extreme aggressiveness ("'roid rage"), high cholesterol, and, following long-term use, cancer. As a consequence, nearly all athletic organizations have banned their use.

Phospholipids and waxes are also lipids. **Phospholipids** are the major component of the membrane that surrounds the contents of a cell and controls the flow of chemicals into and out of the cell (FIGURE 3-15). They have a structure similar to fats, but with two differences: they contain a phosphorus atom (hence *phospho*lipids), and they have two fatty acid chains rather than three. We explore the significant role of phospholipids in cell membranes in the next chapter.

Waxes resemble fats but have only one long-chain fatty acid linked to the glycerol head of the molecule. Because the fatty acid chain is highly nonpolar, waxes are strongly hydrophobic; that is, these molecules repel water. Their water resistance accounts for their presence as a natural coating on the surface of many plants and in the outer coverings of many insects. In both cases, the waxes prevent loss of the water essential to the organisms' life processes. Many birds, too, have a waxy coating on their wings, keeping them from becoming waterlogged when they get wet.

TAKE HOME MESSAGE 3.9

» Cholesterol and phospholipids are lipids that are not fats. Both are important components in cell membranes. Cholesterol also serves as a precursor to steroid hormones, important regulators of growth and development.

3.10–3.14

Proteins are versatile macromolecules that serve as building blocks.

In the cotinga, bright blue coloration comes from pockets of air in its feathers' protein coating.

3.10 Proteins are bodybuilding macromolecules.

You can't look at an animal and not see proteins (FIGURE 3-16). Inside and out, **proteins** are the chief building blocks of all life. They make up skin and feathers and horns. They make up muscles and are a significant component of bone. In your bloodstream, proteins fight invading microorganisms and stop you from bleeding to death from a shaving cut. Proteins control the levels of sugar and other chemicals in your bloodstream and carry oxygen from one place in your body to another. And in just about every cell in every living organism, proteins called **enzymes** initiate and assist every chemical reaction that occurs.

Although proteins perform several very different types of functions, all are built in the same way and from the same

PROTEIN DIVERSITY

Proteins perform a variety of different functions. They all, however, are built the same way and from the same raw materials in organisms.

Wing feathers on a Scarlet Macaw

STRUCTURE
Hair, fingernails, feathers, horns, cartilage, tendons

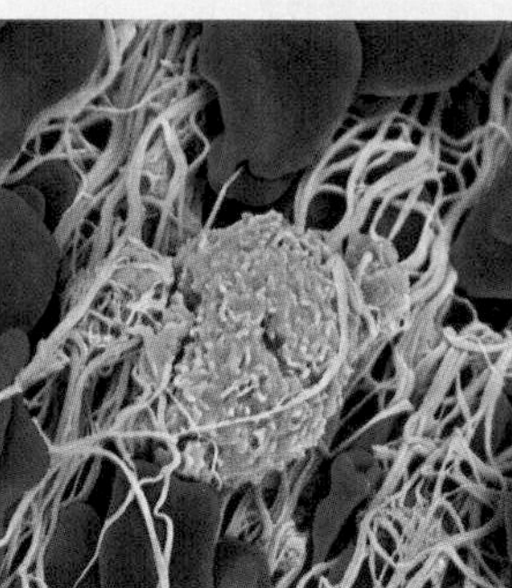

Blood clot

PROTECTION
Help fight invading micro-organisms, coagulate blood

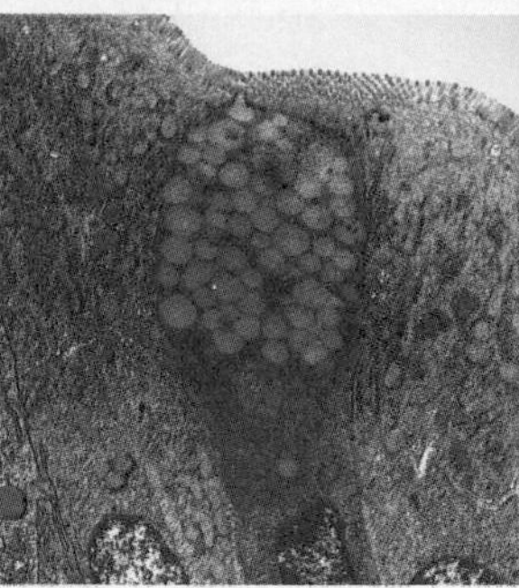

Goblet cell (pink and blue) in the mucosal lining of the small intestine

REGULATION
Control cell activity, constitute some hormones

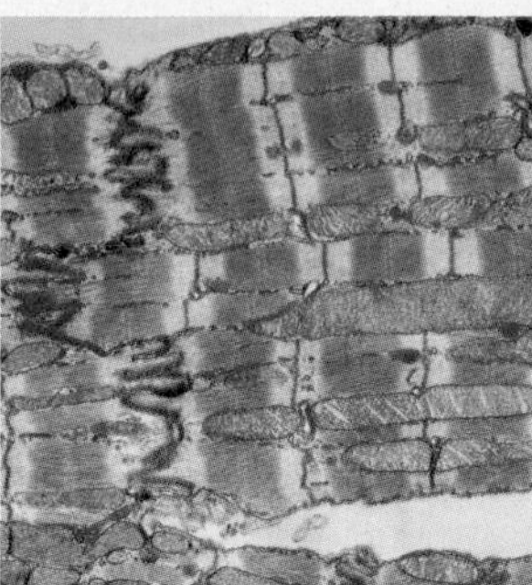

Heart muscle cells

CONTRACTION
Allow muscles to contract, heart to pump, sperm to swim

A model of hemoglobin molecules carrying oxygen

TRANSPORTATION
Carry molecules such as oxygen around your body

FIGURE 3-16 **Proteins everywhere!** Proteins are the chief building blocks of all organisms.

raw materials in all organisms. In the English language, every word is formed from one or more of the 26 letters of the alphabet. Proteins, too, are constructed from a sort of alphabet, known as **amino acids.** Unique combinations of 20 amino acids are strung together, like beads on a string, and the resulting protein has a unique structure and chemical behavior.

Amino acids contain the same familiar atoms as carbohydrates and lipids—carbon, hydrogen, and oxygen—but differ in an important way: they also contain nitrogen. At the center of every amino acid is a carbon atom, with its four covalent bonds (FIGURE 3-17). One bond attaches the carbon to a group of atoms called a **carboxyl group,** which is a carbon atom bonded to two oxygen atoms. The second bond attaches the central carbon to a single hydrogen atom. The third bond attaches the central carbon to an **amino group,** which is a nitrogen atom bonded to hydrogen atoms (usually two or three).

These components—the central carbon with its attached hydrogen atom, carboxyl group, and amino group—are the foundation that identifies a molecule as an amino acid. As multiple amino acids are joined together, the unit formed by these three components forms the “backbone” of the protein.

The fourth bond of the central carbon atom attaches to a side chain. This side chain is the unique part of each of the 20 amino acids. The side chain determines an amino acid's chemical properties, such as whether the amino acid molecule is polar or nonpolar. In the simplest amino acid, glycine, for example, the side chain is simply a hydrogen atom. In other amino acids, the side chain is a single CH_3 group or three or four such groups. Most of the side chains include both hydrogen and carbon atoms, and a few include nitrogen or sulfur atoms.

AMINO ACIDS

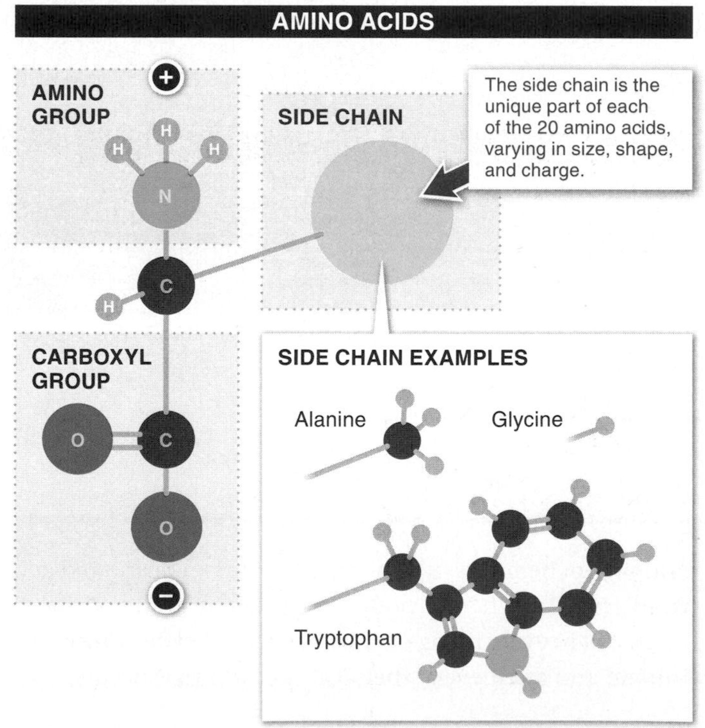

FIGURE 3-17 **Amino acid structure.** Amino acids are made up of a central carbon atom attached to a hydrogen atom, an amino group, a carboxyl group, and a side chain.

TAKE HOME MESSAGE 3.10

» Unique combinations of 20 amino acids give rise to proteins, the chief building blocks of the physical structures that make up all organisms. Proteins perform myriad functions, from assisting chemical reactions to causing blood clotting, building bones, and fighting microorganisms.

3.11 Proteins are an essential dietary component.

The atoms present in the proteins we eat—especially the nitrogen atoms—are essential to the growth, repair, and replacement that take place in our bodies. When we digest protein, breaking it down into its amino acids, our bodies use these amino acids for various building projects. Proteins also store energy in their bonds and, like carbohydrates and lipids, they can be used to fuel living processes.

The amount of protein we need depends on the extent of the building projects under way. Most individuals need 40–80 grams of protein per day. Bodybuilders, however, may need 150 grams or more a day to achieve the extensive muscle growth stimulated by their training; similarly, the protein needs of pregnant or nursing women are very high.

Contrary to the impression you might get from food labels, all proteins are not created equal. Every different protein has a different composition of amino acids. And while our bodies can manufacture certain amino acids as they are needed, many other amino acids must come from our diet. Those that we must get from our diet—about half of the 20 amino acids—are called "essential amino acids." For this reason, we shouldn't just speak of needing "*x* grams of protein per day." We need to consume all of the essential amino acids every day.

Foods that contain "complete proteins," have all of the essential amino acids. Animal products such as milk, eggs, fish, chicken, and beef tend to provide complete proteins. Vegetables, fruits, and grains often contain "incomplete proteins," which do not have all the essential amino acids. Two incomplete proteins that are "complementary proteins," when eaten together, can provide all the essential amino acids. Traditional dishes in many cultures often include such pairings. Examples are corn and beans in Mexico and rice and lentils in India (FIGURE 3-18).

Q Food labels list the amount of protein in a food item. This is insufficient for effectively guiding your protein intake. Why?

TAKE HOME MESSAGE 3.11

» Twenty amino acids make up all the proteins necessary for growth, repair, and replacement of tissue in living organisms. Of these amino acids, about half are essential for humans: they cannot be synthesized by the body so must be consumed in the diet. Complete proteins contain all essential amino acids, while incomplete proteins do not.

COMPLETE vs. INCOMPLETE PROTEINS

ESSENTIAL AMINO ACID CONTENT OF COMMON FOODS

	Apples	Almonds	White rice	Lentils	Milk - 2% fat	Tofu - firm (coagulated soy milk)	Chicken	Beef - T-bone steak
Histidine	●	●	●	●	●	●	●	●
Isoleucine	○	●	●	●	●	●	●	●
Leucine	●	●	●	●	●	●	●	●
Lysine	●	○	○	●	●	●	●	●
Methionine + Cysteine	○	○	●	○	●	●	●	●
Phenylalanine + Tyrosine	○	●	●	●	●	●	●	●
Threonine	○	●	●	●	●	●	●	●
Tryptophan	○	●	●	●	●	●	●	●
Valine	●	●	●	●	●	●	●	●

● Present at optimal levels
○ Not present at optimal levels

Optimal levels determined by the Institute of Medicine's Food and Nutrition Board

GRAPHIC CONTENT
Thinking critically about visual displays of data
Turn to p. 77 for a closer inspection of this figure.

Traditional dishes in many cultures combine proteins, bringing together all essential amino acids.

FIGURE 3-18 **All proteins are not created equal.** Some foods have "complete proteins" with all the essential amino acids. Other foods have "incomplete proteins," and we must consume proteins from multiple sources to get all the essential amino acids.

3.12 A protein's function is influenced by its three-dimensional shape.

Proteins are formed by linking individual amino acids together with a **peptide bond,** in which the amino group of one amino acid is bonded to the carboxyl group of another. Two amino acids joined together form a *dipeptide,* and several amino acids joined together form a *polypeptide.* The sequence of amino acids in the polypeptide chain is called the **primary structure** of the protein and can be compared to the sequence of letters that spells a specific word (FIGURE 3-19).

Amino acids in a polypeptide chain don't remain in a simple straight line like beads on a string. The chain folds as side chains come together and hydrogen bonds form between various atoms in the chain. The two most common patterns of hydrogen bonding between amino acids cause a segment of the chain to either twist in a corkscrew-like shape or form a zigzag folding pattern. The distribution of corkscrews and zigzags within a protein gives a protein its **secondary structure.**

The secondary structure itself continues to fold and bend, bringing together amino acids that then form bonds such as hydrogen bonds or covalent sulfur-sulfur bonds (see Figure 3-19). Eventually, the protein folds into a unique and complex three-dimensional shape called its **tertiary structure.**

Some protein molecules are made from more than one polypeptide chain. The polypeptide chains are held together by hydrogen bonds and bonds between amino acids from different chains, giving the protein a **quaternary structure.** An example of a protein with quaternary structure is hemoglobin, the protein molecule that carries oxygen from the lungs to the cells where it is needed (see Figure 3-16).

For proteins to function properly, they must retain their three-dimensional shape. When their shapes are deformed, they usually lose their ability to function. We can see proteins deforming when we fry an egg. The heat breaks the hydrogen bonds that give the proteins their natural shape. The proteins in the clear egg white unfold, losing their secondary and tertiary structure. This disruption of protein folding is called **denaturation** (FIGURE 3-20).

> **Q** Egg whites contain a lot of protein. Why does beating them change their texture, making them stiff?

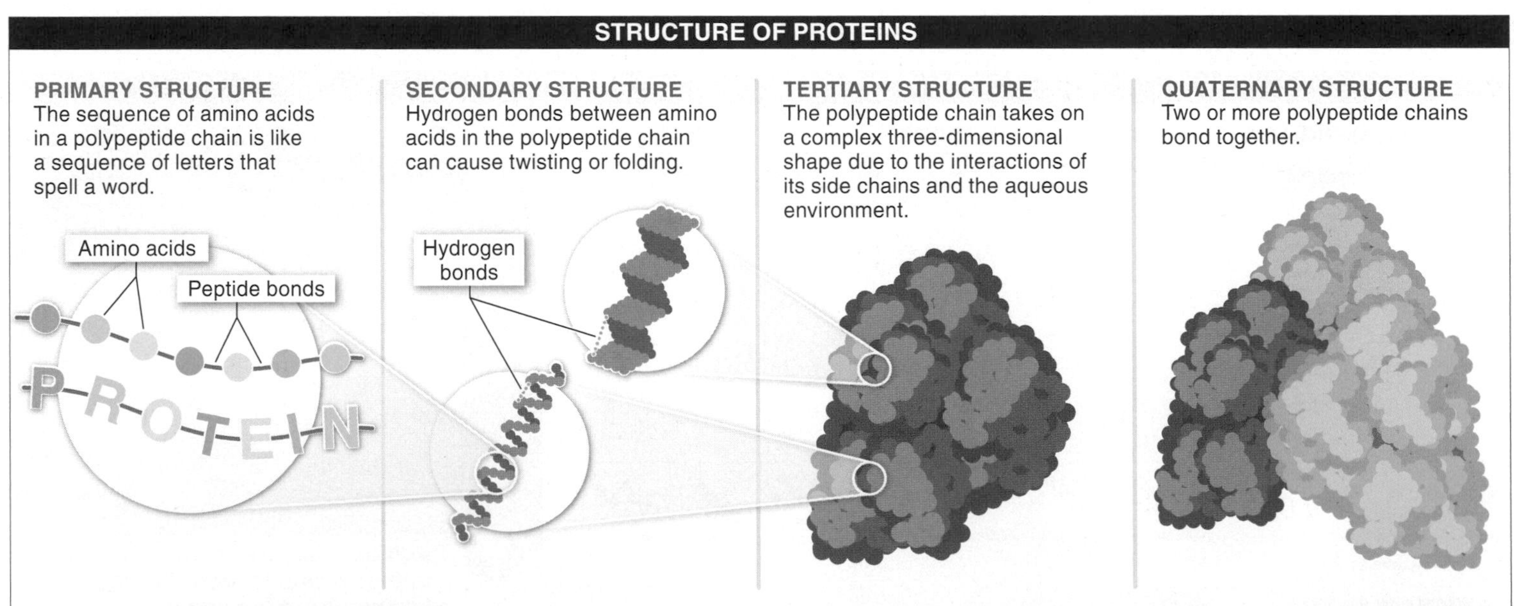

STRUCTURE OF PROTEINS
PRIMARY STRUCTURE
The sequence of amino acids in a polypeptide chain is like a sequence of letters that spell a word.
Amino acids
Peptide bonds
P-R-O-T-E-I-N
SECONDARY STRUCTURE
Hydrogen bonds between amino acids in the polypeptide chain can cause twisting or folding.
Hydrogen bonds
TERTIARY STRUCTURE
The polypeptide chain takes on a complex three-dimensional shape due to the interactions of its side chains and the aqueous environment.
QUATERNARY STRUCTURE
Two or more polypeptide chains bond together.

FIGURE 3-19 **The several levels of protein structure.** The functions of proteins are influenced by their three-dimensional shape.

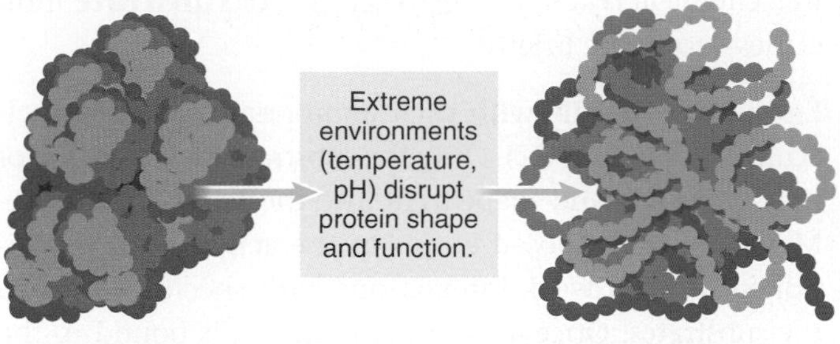

Extreme environments (temperature, pH) disrupt protein shape and function.
Normal protein
Denatured protein

When proteins are unfolded, they lose their function.

FIGURE 3-20 **Denaturation.**

Almost any extreme environment will denature a protein. Take a raw egg, for instance, and crack it into a dish containing baking soda or rubbing alcohol. Both chemicals are sufficiently extreme to turn the clear protein opaque white, as in fried egg whites.

Hair is a protein whose shape most of us have modified at one time or another. Styling hair—whether curling or straightening it—involves altering some of the hydrogen bonds between the amino acids that make up the hair protein, changing its tertiary structure. When your hair gets wet, the water can disrupt some of the hydrogen bonds, causing some amino acids in the protein to form hydrogen bonds with the water molecules instead. Thus, if you style your hair while it's wet, you can change your hair's shape—making it straighter or, if you manipulate it around curlers, making it curlier. The hair can then hold this shape when it dries because, as the water evaporates, the hydrogen bonds to water are replaced by hydrogen bonds between amino acids of the hair protein. Once your hair gets wet again, however, unless it is combed, brushed, or wrapped in a different style, it will return to its natural shape.

Whether your hair is straight or curly or somewhere in between depends on your hair protein's amino acid sequence and the three-dimensional shape it confers (FIGURE 3-21). This amino acid sequence is something you're born with (that is, it's genetically determined). The chains are more or less coiled, depending on the extent of covalent and hydrogen bonding between different parts of the coil. (Many hair salons make use of the ability to alter covalent bonds to change hair texture semi-permanently.)

FIGURE 3-21 **Curly or straight?** Proteins determine it!

Q Why do some people have curly hair and others have straight hair?

TAKE HOME MESSAGE 3.12

» A particular amino acid sequence determines how a protein folds into a particular three-dimensional shape. This shape determines many of the protein's properties, including which molecules it will interact with. When a protein's shape is deformed, the protein usually loses its ability to function.

3.13 Enzymes are proteins that speed up chemical reactions.

Protein shape is particularly critical for **enzymes,** molecules that help initiate and accelerate the chemical reactions in our bodies. Enzymes emerge unchanged—in their original form—when the reaction is complete and thus can be used again and again. Here's how they work.

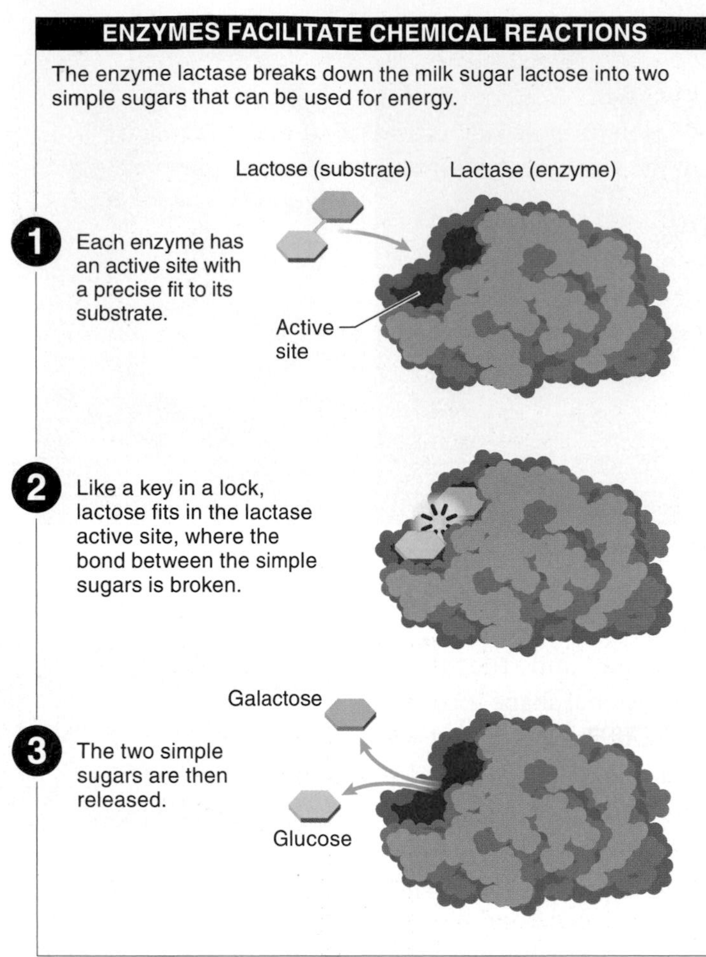

ENZYMES FACILITATE CHEMICAL REACTIONS
The enzyme lactase breaks down the milk sugar lactose into two simple sugars that can be used for energy.
Lactose (substrate)
Lactase (enzyme)
1 Each enzyme has an active site with a precise fit to its substrate.
Active site
2 Like a key in a lock, lactose fits in the lactase active site, where the bond between the simple sugars is broken.
Galactose
3 The two simple sugars are then released.
Glucose

FIGURE 3-22 **Lock and key.** Enzymes are very specific about which molecules and reactions they will catalyze.

Think of an enzyme as a big piece of popcorn. Its tertiary or quaternary structure gives it a complex shape with lots of nooks and crannies. Within one of those nooks is a small area called the **active site** (FIGURE 3-22). Based on the chemical properties of the atoms lining this pocket, the active site provides a place for the participants in a chemical reaction—the reactants, or **substrate** molecules—to nestle briefly.

Enzymes bind only with their appropriate substrate molecules (see Figure 3-22). Only the substrate molecules are of the correct size and shape to fit into the active-site groove. Moreover, the exposed atoms in the active site are positioned to form weak interactions with specific atoms in the substrates. Once a substrate molecule is bound to the active site, a reaction can take place—and usually does so very quickly.

The chemical reactions that occur in organisms can either release energy or consume energy. But in either case, a certain minimum energy—a little "push"—is needed to get the reaction going. This is called the **activation energy.** And although enzymes don't alter the amount of energy released by a reaction, they act as catalysts by lowering the activation energy, which causes the reaction to occur more quickly. For example, enzymes may lower the activation energy by holding substrate molecules in an orientation that stresses bonds that need to break, or by bringing together atoms that need to bond.

By virtue of their catalytic capacities, enzymes are at the heart of the chemistry of living organisms. Taken together, all of the chemical reactions in a living organism are its *metabolism*.

TAKE HOME MESSAGE 3.13

» Enzymes are proteins that help initiate and speed up chemical reactions. They aren't permanently altered in the process and can be used again and again.

3.14 Enzymes regulate reactions in several ways (but malformed enzymes can cause problems).

If not for enzymes, it's possible that almost nothing would get done in living organisms. Enzymes don't alter the outcome of reactions, but without the chemical "nudge" they supply—often increasing reaction rates to millions of times their uncatalyzed rate—the processes necessary to sustain life could not occur.

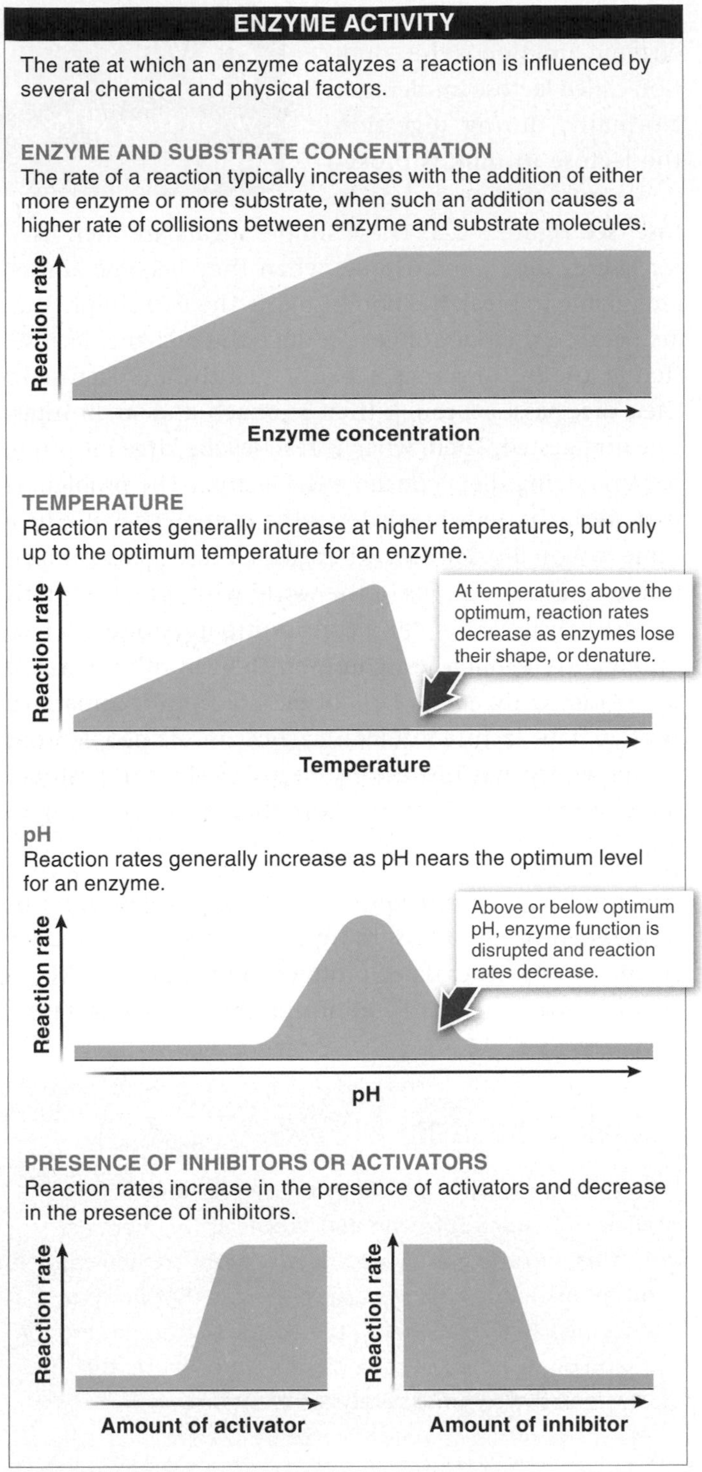

ENZYME ACTIVITY

The rate at which an enzyme catalyzes a reaction is influenced by several chemical and physical factors.

ENZYME AND SUBSTRATE CONCENTRATION

The rate of a reaction typically increases with the addition of either more enzyme or more substrate, when such an addition causes a higher rate of collisions between enzyme and substrate molecules.

Reaction rate
Enzyme concentration

TEMPERATURE

Reaction rates generally increase at higher temperatures, but only up to the optimum temperature for an enzyme.

Reaction rate
At temperatures above the optimum, reaction rates decrease as enzymes lose their shape, or denature.
Temperature

pH

Reaction rates generally increase as pH nears the optimum level for an enzyme.

Reaction rate
Above or below optimum pH, enzyme function is disrupted and reaction rates decrease.
pH

PRESENCE OF INHIBITORS OR ACTIVATORS

Reaction rates increase in the presence of activators and decrease in the presence of inhibitors.

Reaction rate
Amount of activator
Reaction rate
Amount of inhibitor

FIGURE 3-23 **Getting the job done.** Enzyme activity is influenced by physical factors such as temperature and pH and by chemical factors such as enzyme and substrate concentrations.

The rate at which an enzyme catalyzes a reaction is influenced by several chemical and physical factors (FIGURE 3-23). These include:

1. Enzyme and substrate concentration. The rate of a reaction typically increases with the addition of either more enzyme or more substrate, when such an addition causes a higher rate of collisions between enzyme and substrate molecules.

2. Temperature. Because molecules move faster as temperature increases, reaction rates generally increase similarly. Reaction rates continue to increase only up to the optimum temperature for an enzyme. At temperatures above the optimum, reaction rates decrease as enzymes lose their shape. Enzymes from different species can have widely differing optimum temperatures.

3. pH. As with temperature, enzymes have an optimum pH. Above or below this pH, excess hydrogen or hydroxide ions interact with amino acid side chains in the active site or elsewhere. These interactions disrupt enzyme function (and sometimes structure) and decrease reaction rates.

4. Presence of inhibitors or activators. One of the most common ways that cells can speed up or slow down their metabolic pathways is through the binding of other chemicals to enzymes. This binding can alter enzyme shape in a way that increases or decreases the enzyme's activity. **Inhibitors** reduce enzyme activity, and these come in two types. **Competitive inhibitors** bind to the active site, blocking substrate molecules from the site and thus from taking part in the reaction. **Noncompetitive inhibitors** do not compete for the active site but, rather, bind to another part of the enzyme, altering its shape in a way that changes the structure of the active site, thus reducing or blocking its ability to bind with substrate. Often, it is the very product of a metabolic pathway that acts as an inhibitor of enzymes early in the pathway, effectively shutting off the pathway when enough of its end product has been produced.

Just as a molecule can bind to an enzyme and inhibit the enzyme's activity, so can some cellular chemicals act as **activators,** altering the enzyme's shape or structure so that it can now catalyze a reaction.

Sometimes a cell produces a protein "word" that is misspelled—that is, the sequence of amino acids is incorrect. If an enzyme is altered even slightly, the active site or some other region of the protein molecule may change so that the enzyme will no longer function

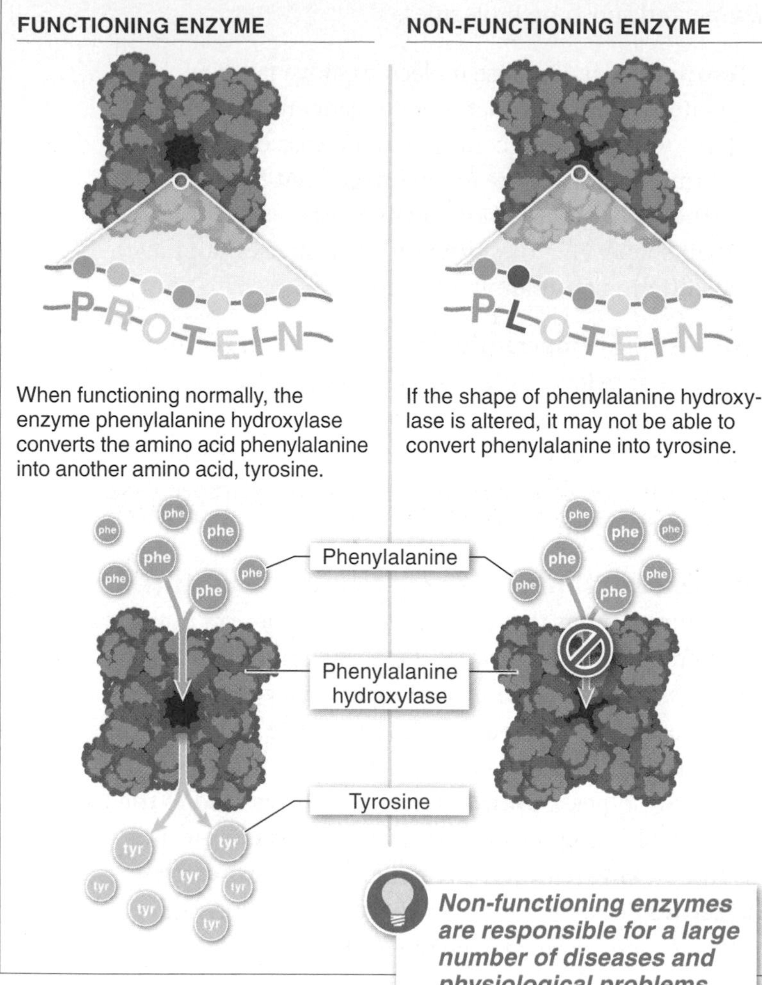
NON-FUNCTIONING ENZYMES
Sometimes a protein "word" is misspelled—that is, the sequence of amino acids is incorrect. If an enzyme is altered even slightly, particularly at the active site, it can cause the enzyme to no longer function.
FUNCTIONING ENZYME
NON-FUNCTIONING ENZYME
P-R-O-T-E-I-N
P-L-O-T-E-I-N
When functioning normally, the enzyme phenylalanine hydroxylase converts the amino acid phenylalanine into another amino acid, tyrosine.
If the shape of phenylalanine hydroxylase is altered, it may not be able to convert phenylalanine into tyrosine.
phe
Phenylalanine
Phenylalanine hydroxylase
Tyrosine
tyr
Non-functioning enzymes are responsible for a large number of diseases and physiological problems.

FIGURE 3-24 **When a protein is "misspelled."**

(FIGURE 3-24). Slightly modified, non-functioning enzymes are responsible for a large number of diseases and physiological problems (see Section 6.11). An example is the body's inability to break down the amino acid phenylalanine (in a condition known as phenylketonuria).

Q Why do some adults get sick when they drink milk?

One health issue influenced by enzyme function is the condition called lactose intolerance. Normally, during digestion, the lactose in milk is broken down into glucose and galactose (see Figure 3-22). These simple sugars are then used for energy. But some people, when they become adults, are unable to break the bond linking the two simple sugars because they no longer produce the enzyme, lactase, that assists in this process. Consequently, any lactose in their diet passes through their stomach and small intestine undigested. Then, when it reaches the large intestine, bacteria living there consume the lactose. The problem is that, as the bacteria break down the lactose, they produce some carbon dioxide and other gases, causing discomfort. Interestingly, in regions of the world with long traditions of pastoralism (raising and consuming livestock), lactose intolerance is much less common than in other parts of the world. Only about 10% of people from Denmark or Sweden have lactose intolerance, but among people from China, which has historically been largely non-pastoral, the vast majority of adults (more than 80% according to numerous published studies) are lactose intolerant.

The unpleasant symptoms of lactose intolerance can be avoided by not consuming milk, cheese, yogurt, ice cream, or any other dairy products, but they can also be avoided by taking a pill containing the enzyme lactase.

TAKE HOME MESSAGE 3.14

» **Enzyme activity is influenced by physical factors such as temperature and pH, as well as by chemical factors, including enzyme and substrate concentrations. Inhibitors and activators are chemicals that bind to enzymes and, by blocking the active site or altering the shape or structure of the enzyme, can change the rate at which the enzyme catalyzes reactions.**

Nucleic acids store information on how to build and run a body.

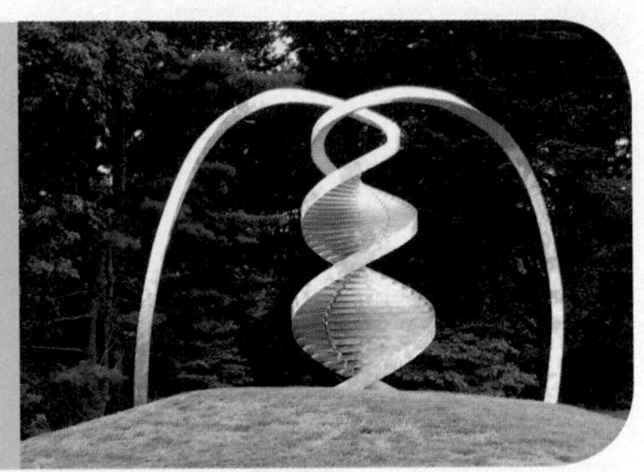

Spirals Time—Time Spirals: a sculpture inspired by DNA's double helix structure.

3.15 Nucleic acids are macromolecules that store information.

We have examined three of life's macromolecules: carbohydrates, lipids, and proteins. We turn our attention now to the fourth: **nucleic acids,** macromolecules that store information. There are two types of nucleic acids: **deoxyribonucleic acid (DNA)** and **ribonucleic acid (RNA).** Both play central roles in directing the production of proteins in living organisms and, by doing so, play a central role in determining all of the inherited characteristics of an individual.

Nucleic acids are made up of individual units called **nucleotides.** All nucleotides have three components: a molecule of sugar, a phosphate group (containing a phosphorus atom bound to four oxygen atoms), and a nitrogen-containing molecule (FIGURE 3-25). In both types of nucleic acids, nucleotides are linked in a series; a sugar molecule is attached to a phosphate group, which is attached to another sugar, which is attached to another phosphate, and so on. The linked series forms the backbone of the nucleic acid molecule. Attached to each sugar, and protruding from the backbone, is one of the nitrogen-containing molecules, the nucleotide **bases** (so named because of their chemical structure). A 10-unit nucleic acid strand therefore would have 10 bases, one attached to each sugar within the sugar-phosphate-sugar-phosphate backbone. But the base attached to each sugar is not always the same. It can be one of several different bases. For this reason, a nucleic acid is often described by its sequence of bases.

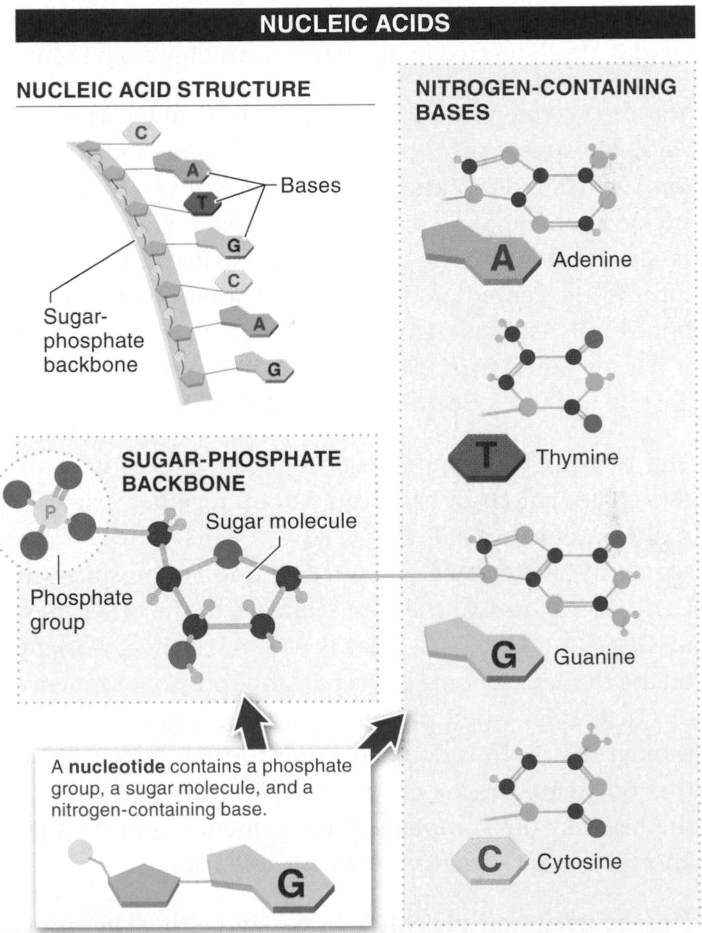

FIGURE 3-25 **The molecules that carry genetic information.** The nucleic acid shown here is DNA.

Nucleic acids store information in the order of bases attached at each position in the molecule's backbone. At each position in a molecule of DNA, for example, the base can be any one of four possible bases: adenine (A), thymine (T), guanine (G), or cytosine (C). Just as the meaning of a sentence is determined by which letters are strung together, the information in a segment of DNA is determined by its sequence of bases. One DNA segment may have the sequence adenine, adenine, adenine, guanine, cytosine, thymine, guanine, cytosine, thymine—abbreviated as AAAGCTGCT. Another DNA segment may have the sequence CGATTACCCGAT. Because the information differs in each case, so, too, does the polypeptide for which the sequence codes, as we'll see.

TAKE HOME MESSAGE 3.15

» The nucleic acids DNA and RNA are macromolecules that store information in their unique sequences of bases contained in nucleotides, their building-block molecules. Both nucleic acids play central roles in directing protein production in organisms.

3.16 DNA holds the genetic information to build an organism.

A molecule of DNA has two strands, each a sugar-phosphate-sugar-phosphate backbone with a base sticking out from each sugar molecule. The two strands wrap around each other, each turning in a spiral, and the two strands are connected by the bases protruding from them.

You can picture a molecule of DNA as a ladder. The two sugar-phosphate-sugar-phosphate backbones are like the long vertical sides of the ladder that give it height. A base sticking out represents a rung on the ladder (or, more accurately, half a rung). The bases protruding from each strand meet in the center and bind to each other (via hydrogen bonds). DNA differs slightly from a ladder, though, in that it has a gradual twist. The two spiraling strands together are said to form a **double helix** (FIGURE 3-26).

The two intertwining spirals fit together because only two combinations of bases can pair up together. The base A always pairs with T, and C always pairs with G. Consequently, if the base sequence of one of the spirals is CCCCTTAGGAACC, the base sequence of the other must be GGGGAATCCTTGG. That is why researchers working on the Human Genome Project describe only one sequence of nucleotides when presenting a DNA sequence—even though that DNA is double-stranded in our bodies. With that one sequence, we can infer the identity and order of the bases in the complementary sequence, and thus we know the exact structure of the nucleic acid.

The sequences of nucleotide bases containing the information about how to produce a particular protein have anywhere from a hundred to several thousand bases. In a human, all of the DNA in a cell, containing all of the instructions for every protein that a human must produce, contains about three billion base pairs. Almost all of this DNA is in the cell's nucleus.

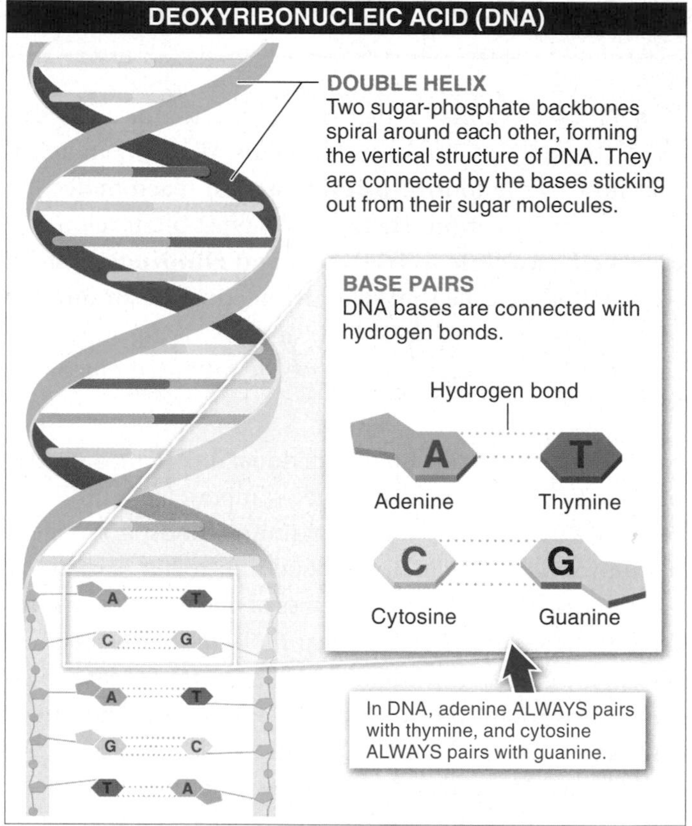

DEOXYRIBONUCLEIC ACID (DNA)
DOUBLE HELIX
Two sugar-phosphate backbones spiral around each other, forming the vertical structure of DNA. They are connected by the bases sticking out from their sugar molecules.
BASE PAIRS
DNA bases are connected with hydrogen bonds.
Hydrogen bond
A
T
Adenine
Thymine
C
G
Cytosine
Guanine
In DNA, adenine ALWAYS pairs with thymine, and cytosine ALWAYS pairs with guanine.

FIGURE 3-26 **A gradually twisting ladder.** The rungs of the DNA ladder are nucleotide base pairs, and each side of the ladder is made up of a sugar-phosphate backbone.

TAKE HOME MESSAGE 3.16

» DNA is shaped like a ladder in which the long, vertical sides are made from a sequence of sugar-phosphate-sugar-phosphate molecules and the rungs are pairs of nucleotide bases. The sequence of nucleotide bases contains the information about how to produce a particular protein.

3.17 RNA is a universal translator, reading DNA and directing protein production.

The process of building a protein from a DNA sequence is not a direct one. Rather, it incorporates a middleman, RNA (FIGURE 3-27). Although both RNA and DNA are built from nucleotides, the RNA nucleotide differs from the DNA nucleotide in three important ways. First, the sugar portion of the nucleotide contains an extra atom of oxygen. Second, while RNA has the bases A, G, and C, it does not have thymine (T); instead, it has a similar base called uracil (U). And third, RNA is single-stranded. Its bases do not bind with bases in another RNA strand to form the ladder-like structure of DNA.

When the cell needs to synthesize a protein, a short strip of RNA is produced using a segment of a DNA strand as a model. The RNA nucleotides are therefore complementary to the DNA nucleotides, so the RNA molecule contains all the information present in the order of nucleotides in that DNA segment. The RNA moves to another part of the cell and then directs the linking together of amino acids to form a polypeptide chain that folds into a three-dimensional protein. We explore this in greater detail in Chapter 6.

Whether we're looking at the nucleotides that make up RNA and DNA or the lipids used to build sex hormones and cell membranes, we see a recurring theme in the construction of biological macromolecules: from relatively simple sets of building blocks linked together, infinitely complex molecules can be formed.

TAKE HOME MESSAGE 3.17

» RNA acts as a middleman molecule—taking the instructions for protein production from DNA to another part of the cell, where, in accordance with the RNA instructions, amino acids are linked together into proteins.

RIBONUCLEIC ACID (RNA)

RNA STRUCTURE

There are three important structural differences between RNA and DNA.

- The sugar molecule in the RNA backbone contains an extra oxygen.
- Instead of thymine, RNA has a similar base called uracil.
- RNA has only one sugar-phosphate backbone, while DNA has two.

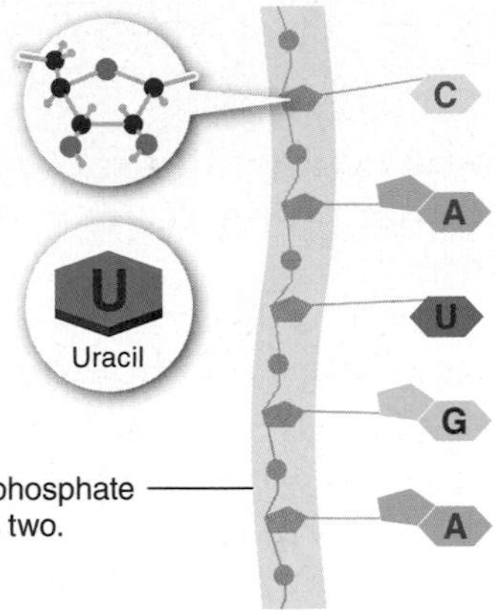

Uracil

C A U G A

RNA FUNCTION

RNA acts as a middleman molecule. It takes instructions for production of a protein from DNA, moves them to another part of the cell, and directs the building of a protein.

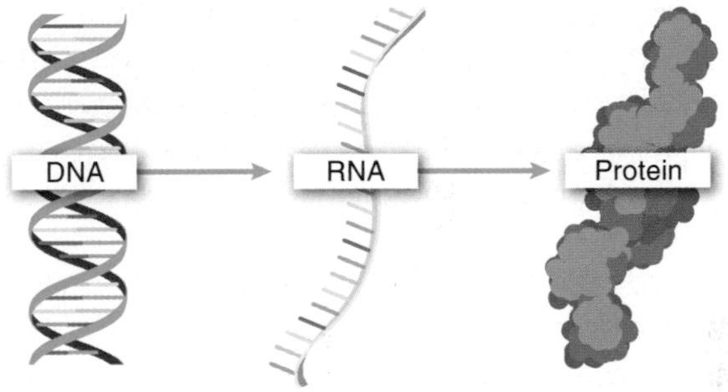

DNA → RNA → Protein

FIGURE 3-27 The middleman between DNA and protein. The structure of RNA.

STREET BIO

Using evidence to guide decision making in our own lives

Melt-in-your-mouth chocolate may not be such a sweet idea.

Food chemists have figured out how to make chocolate that melts in your mouth. Is that a good thing?

Q: Why are some fats "liquidy" like oil, and others solid? The less saturated a fat is, the more "liquidy" it is at room temperature. Most animal fats are saturated and are solid at room temperature. Best example: butter. Most plant fats are polyunsaturated and are liquid at room temperature. Best example: vegetable oils.

Q: Can oils be made more solid? As we saw in Section 3.8, it is possible to increase the saturation of plant fats. Just heat them up and pass hydrogen bubbles through the liquid. In creating partially hydrogenated plant oils, this process reduces the number of carbon-carbon double bonds and makes the oil more solid. (It's easy, it's cheaper than using butter, and it increases foods' shelf-life.)

Q: Does that improve their taste? By precisely controlling the level of saturation in plant fats, it is possible to create foods with a more desirable texture. They are solid, but have such a low melting point that they quickly melt on contact with the warmth of your mouth. This seems great, but . . .

Q: Is there a downside? The saturation of vegetable fats creates trans fats, which increase levels of LDL ("bad") cholesterol and decrease HDL ("good") cholesterol, and can narrow blood vessels and increase the risk of heart disease and strokes.

Conclusion Partially hydrogenated vegetable oils can give food a perfect texture and a pleasing feel in your mouth. But the creaminess comes with a high cost when it comes to your health. For this reason, the U.S. Food and Drug Administration is requiring that by 2018, food prepared in the United States must not include trans fats.

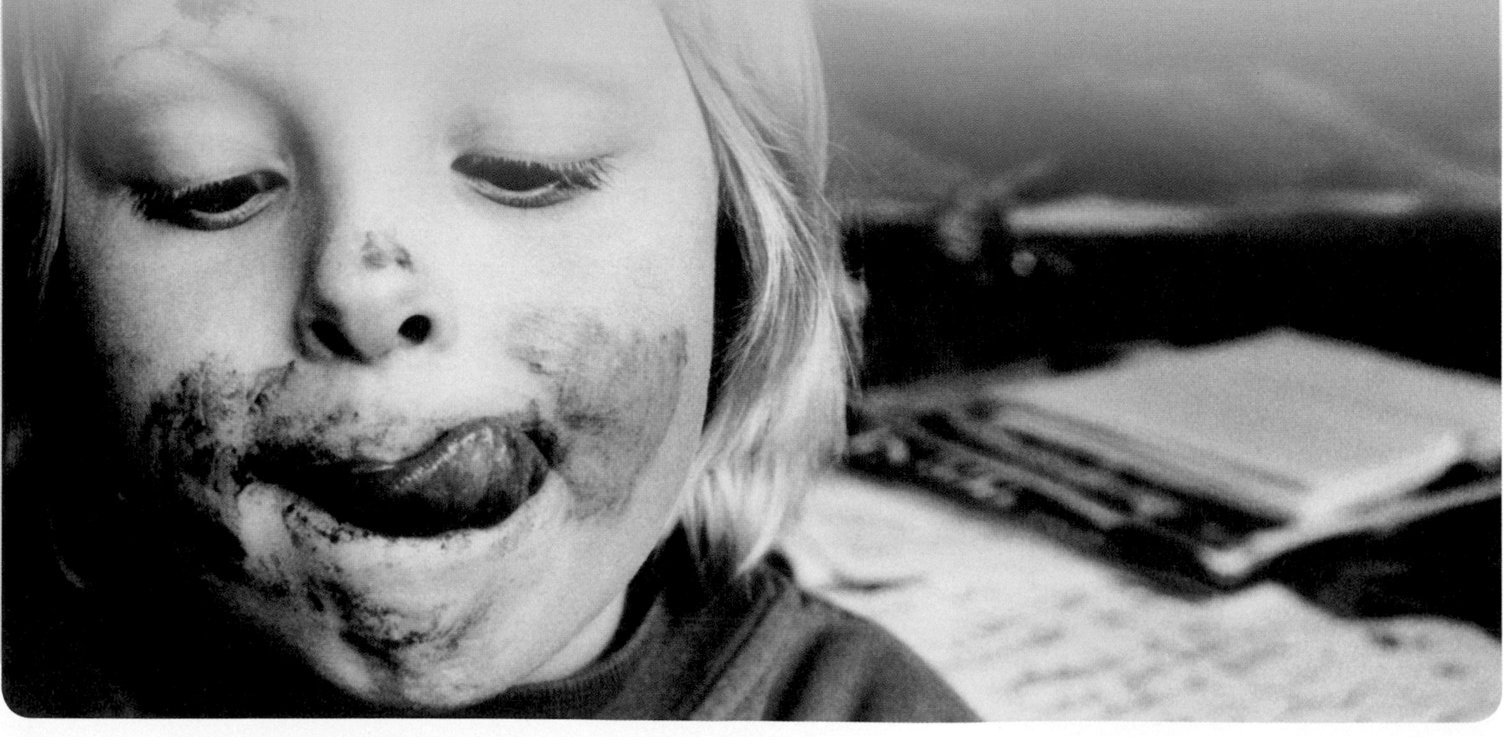

GRAPHIC CONTENT

Thinking critically about visual displays of data

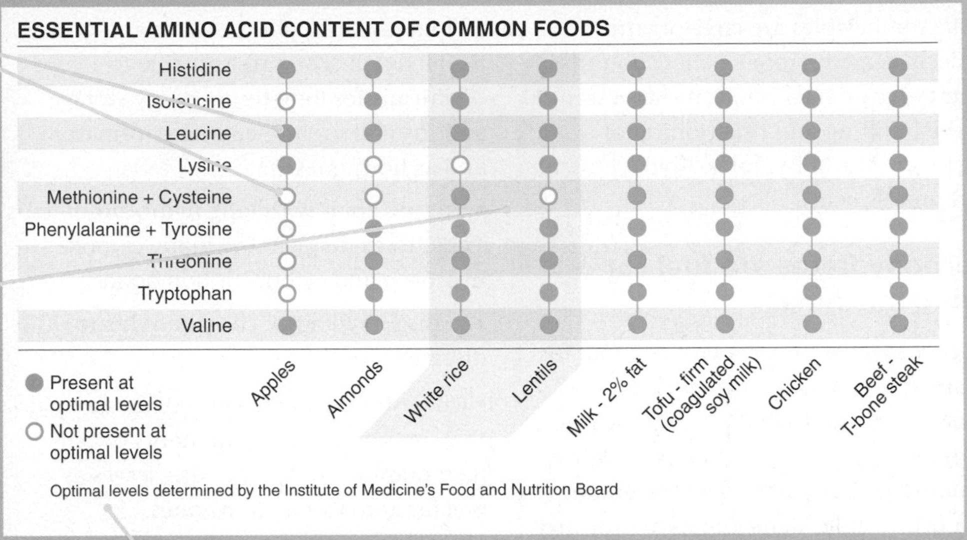

ESSENTIAL AMINO ACID CONTENT OF COMMON FOODS

	Apples	Almonds	White rice	Lentils	Milk - 2% fat	Tofu - firm (coagulated soy milk)	Chicken	Beef - T-bone steak
Histidine	●	●	●	●	●	●	●	●
Isoleucine	○	●	●	●	●	●	●	●
Leucine	●	●	●	●	●	●	●	●
Lysine	●	○	○	●	●	●	●	●
Methionine + Cysteine	○	○	●	○	●	●	●	●
Phenylalanine + Tyrosine	○	●	●	●	●	●	●	●
Threonine	○	●	●	●	●	●	●	●
Tryptophan	○	●	●	●	●	●	●	●
Valine	●	●	●	●	●	●	●	●

● Present at optimal levels
○ Not present at optimal levels

Optimal levels determined by the Institute of Medicine's Food and Nutrition Board

1 In this figure, what does a green dot indicate? What does a white dot indicate?

2 What can you conclude from this figure?

3 Why is there shading behind the data for white rice and lentils?

4 Should a food be avoided if it doesn't contain all of the essential amino acids at optimal levels? Why or why not?

5 Would a meal containing apples and white rice contain all of the essential amino acids? What about a meal of lentils and almonds?

6 Can you ascertain what "optimal levels" means in this figure?

7 What additional information would make this figure more helpful? Why?

See answers at the back of the book.

KEY TERMS IN MOLECULES OF LIFE

activation energy, p. 70
activator, p. 71
active site, p. 70
amino acid, p. 66
amino group, p. 66
base (nucleotide), p. 73
carbohydrate, p. 55
carboxyl group, p. 66
cellulose, p. 58
chitin, p. 58
cholesterol, p. 64
competitive inhibitor, p. 71
complex carbohydrate, p. 56
denaturation, p. 68
deoxyribonucleic acid (DNA), p. 73
disaccharide, p. 56
double helix, p. 74
enzyme, pp. 65, 69
fatty acid, p. 60
glycerol, p. 60
glycogen, p. 56
hydrophilic, p. 60
hydrophobic, p. 60
inhibitor, p. 71
lipid, p. 59
macromolecule, p. 54
monosaccharide, p. 55
noncompetitive inhibitor, p. 71
nucleic acid, p. 73
nucleotide, p. 73
organic molecule, 54
peptide bond, p. 68
phospholipid, p. 65
polysaccharide, p. 56
primary structure, p. 68
protein, p. 65
quaternary structure, p. 68
ribonucleic acid (RNA), p. 73
saturated fat, p. 60
secondary structure, p. 68
simple sugar, p. 55
starch, p. 57
sterol, p. 64
substrate, p. 70
tertiary structure, p. 68
trans fat, p. 62
triglyceride, p. 60
unsaturated fat, p. 61
wax, p. 65

BRIEF SUMMARY

Macromolecules are the raw materials and fuel for life.

- Macromolecules are large organic molecules made from smaller subunits. Four types of macromolecule are essential to the building and functioning of organisms: carbohydrates, lipids, proteins, and nucleic acids.

Carbohydrates are fuel for living machines.

- Carbohydrates are the primary fuel for organisms and also form much of cell structure. Carbohydrates contain carbon, hydrogen, and oxygen. As the chemical bonds of carbohydrates are broken down and other, more stable bonds are formed, energy is released that can be used by the organism.
- Glucose—the most important carbohydrate to organisms—can be used as an energy source, stored as glycogen for later use, or converted to fat.
- Simple carbohydrates can be linked into complex carbohydrates, including starch and glycogen.
- Some complex carbohydrates, including cellulose, cannot be digested by most animals. Such indigestible carbohydrates in the diet, called fiber, have health benefits.

Lipids store energy for a rainy day.

- Lipids are insoluble in water and greasy to the touch. They are valuable to organisms for long-term energy storage and insulation, in membrane formation, and as hormones.
- Fats are lipids with long hydrocarbon tails. They store energy in the many carbon-hydrogen and carbon-carbon bonds.
- Trans fats increase the risk of heart disease.
- Cholesterol and phospholipids are lipids that are important components of cell membranes. Cholesterol also serves as a precursor to steroid hormones.

Proteins are versatile macromolecules that serve as building blocks.

- Unique combinations of 20 amino acids give rise to proteins, the chief building blocks of organisms' physical structures. Proteins perform myriad functions, including assisting chemical reactions.
- Twenty amino acids make up all the proteins necessary for growth, repair, and replacement of tissue in living organisms. Of these, essential amino acids cannot be synthesized by humans and other animals so must be consumed in the diet.
- A protein's particular amino acid sequence determines how it folds into a particular three-dimensional shape. This shape determines many of the protein's properties.
- Enzymes are proteins that help initiate and speed up chemical reactions. They aren't permanently altered in the process and can be reused.
- Enzyme activity is influenced by physical factors such as temperature and pH, as well as by chemical factors, including enzyme and substrate concentrations.

Nucleic acids store information on how to build and run a body.

- The nucleic acids DNA and RNA are macromolecules that store information in their unique sequences of bases contained in nucleotides, their building-block molecules.
- DNA is shaped like a ladder; the sides are made from a sequence of sugar-phosphate-sugar-phosphate molecules, and the rungs are pairs of nucleotide bases. The sequence of bases contains information on how to produce a protein.
- RNA takes instructions for protein production from DNA to another part of the cell, where amino acids are linked together into proteins.

CHECK YOUR KNOWLEDGE

Short Answer

1. Why do carbohydrates function so well as a fuel for living organisms?

2. Describe the three possible "fates" of glucose in the body.

3. Why is it that if you take a piece of potato and leave it on your tongue for a while, it starts to taste sweet?

4. Why does cellulose pass through the digestive system unused?

5. Why do lipids contain so much more stored energy than carbohydrates?

6. Why do humans generally have a preference for fats in their diets?

7. Which types of lipids are important components of most cell membranes?

8. Describe two functions served by proteins in the body.

9. What does it mean to say that an amino acid is an "essential amino acid"?

10. How does the amino acid sequence of a protein affect its function?

11. Why is protein shape so important for enzymes?

12. How is information stored in the sequence of bases of a nucleic acid?

13. What determines the information in a segment of DNA? Why do researchers working on the Human Genome Project describe the sequence of nucleotides in only one strand when presenting a DNA sequence?

14. Describe the three ways in which RNA differs from DNA.

Multiple Choice

1. Which of the following foods is not a significant source of complex carbohydrates?

a) fresh fruit

b) rice

c) pasta

d) oatmeal

e) All of the above are significant sources of complex carbohydrates.

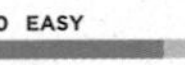

0 EASY HARD 100 40

2. Sucrose (table sugar) and lactose (the sugar found in milk) are examples of:

a) naturally occurring enzymes.

b) simple sugars.

c) monosaccharides.

d) disaccharides.

e) complex carbohydrates.

0 EASY HARD 100 45

3. Which of the following statements about starch is incorrect?

a) Starch is the primary form of energy storage in plants.

b) Starch consists of a hundred or more glucose molecules joined together in a line.

c) Starch tastes sweet because it is made from glucose.

d) Starch is a polysaccharide.

e) All of the above statements about starch are correct.

0 EASY HARD 100 66

4. Which of the following statements about fiber is incorrect?

a) Dietary fiber reduces the risk of colon cancer.

b) Fiber in the diet slows the passage of food through the intestines.

c) Humans cannot extract energy from fiber.

d) The cellulose of celery stalks and lettuce leaves is fiber.

e) Fiber scrapes the wall of the digestive tract, stimulating mucus secretion and aiding in the digestion of other molecules.

0 EASY HARD 100 47

5. A dietary fatty acid is liquid at room temperature (i.e., it has a low melting point) and contains carbon-carbon double bonds. It is most likely from:

a) a plant.

b) a cow.

c) a pig.

d) a chicken.

e) a lamb.

0 EASY HARD 100 36

6. In an unsaturated fatty acid:

a) carbon-carbon double bonds are present in the hydrocarbon chain.

b) the hydrocarbon chain has an odd number of carbons.

c) the hydrocarbon chain has an even number of carbons.

d) no carbon-carbon double bonds are present in the hydrocarbon chain.

e) not all carbons in the hydrocarbon chain are bonded to hydrogen.

0 EASY HARD 100 51

7. Which statement about phospholipids is incorrect?

a) They are used as organisms' chief form of short-term energy.

b) They are hydrophobic at one end.

c) They are hydrophilic at one end.

d) They are a major constituent of cell membranes.

e) They contain glycerol linked to fatty acids.

0 EASY HARD 100 45

8. Proteins are an essential component of a healthy diet for humans (and other animals). Their most common purpose is to serve as:

a) raw material for growth.

b) fuel for running the body.

c) organic precursors for enzyme construction.

d) long-term energy storage.

e) inorganic precursors for enzyme construction.

0 EASY HARD 100 58

9. Which of the following statements about enzymes is incorrect?

a) Enzymes can initiate chemical reactions.

b) Enzymes speed up chemical reactions.

c) Enzymes are proteins.

d) Enzymes contain an active site for binding of particular substrates.

e) Enzymes undergo a permanent change during the reactions they promote.

0 EASY HARD 100 47

10. Which of the following nucleotide bases are present in equal amounts in DNA?

a) adenine and cytosine

b) thymine and guanine

c) adenine and guanine

d) thymine and cytosine

e) adenine and thymine

0 EASY HARD 100 23

Ch04

What is a cell?

4.1 All organisms are made of cells.

4.2 Prokaryotic cells are structurally simple but extremely diverse.

4.3 Eukaryotic cells have compartments with specialized functions.

Cell membranes are gatekeepers.

4.4 Every cell is bordered by a plasma membrane.

4.5 Molecules embedded in the plasma membrane help it perform its functions.

4.6 Faulty membranes can cause diseases.

4.7 Membrane surfaces have a "fingerprint" that identifies the cell.

Molecules move across membranes in several ways.

4.8 Passive transport is the spontaneous diffusion of molecules across a membrane.

4.9 Osmosis is the passive diffusion of water across a membrane.

4.10 In active transport, cells use energy to transport molecules across the cell membrane.

4.11 Endocytosis and exocytosis are used for bulk transport of particles.

Cells are connected and communicate with each other.

4.12 Connections between cells hold them in place and enable them to communicate with each other.

Nine important landmarks distinguish eukaryotic cells.

4.13 The nucleus is the cell's genetic control center.

4.14 Cytoplasm and the cytoskeleton form the cell's internal environment, provide support, and can generate motion.

4.15 Mitochondria are the cell's energy converters.

4.16 *This is how we do it:* Can cells change their composition to adapt to their environment?

4.17 Lysosomes are the cell's garbage disposals.

4.18 In the endoplasmic reticulum, cells build proteins and lipids and disarm toxins.

4.19 The Golgi apparatus processes products for delivery throughout the body.

4.20 The cell wall provides additional protection and support for plant cells.

4.21 Vacuoles are multipurpose storage sacs for cells.

4.22 Chloroplasts are the plant cell's solar power plant.

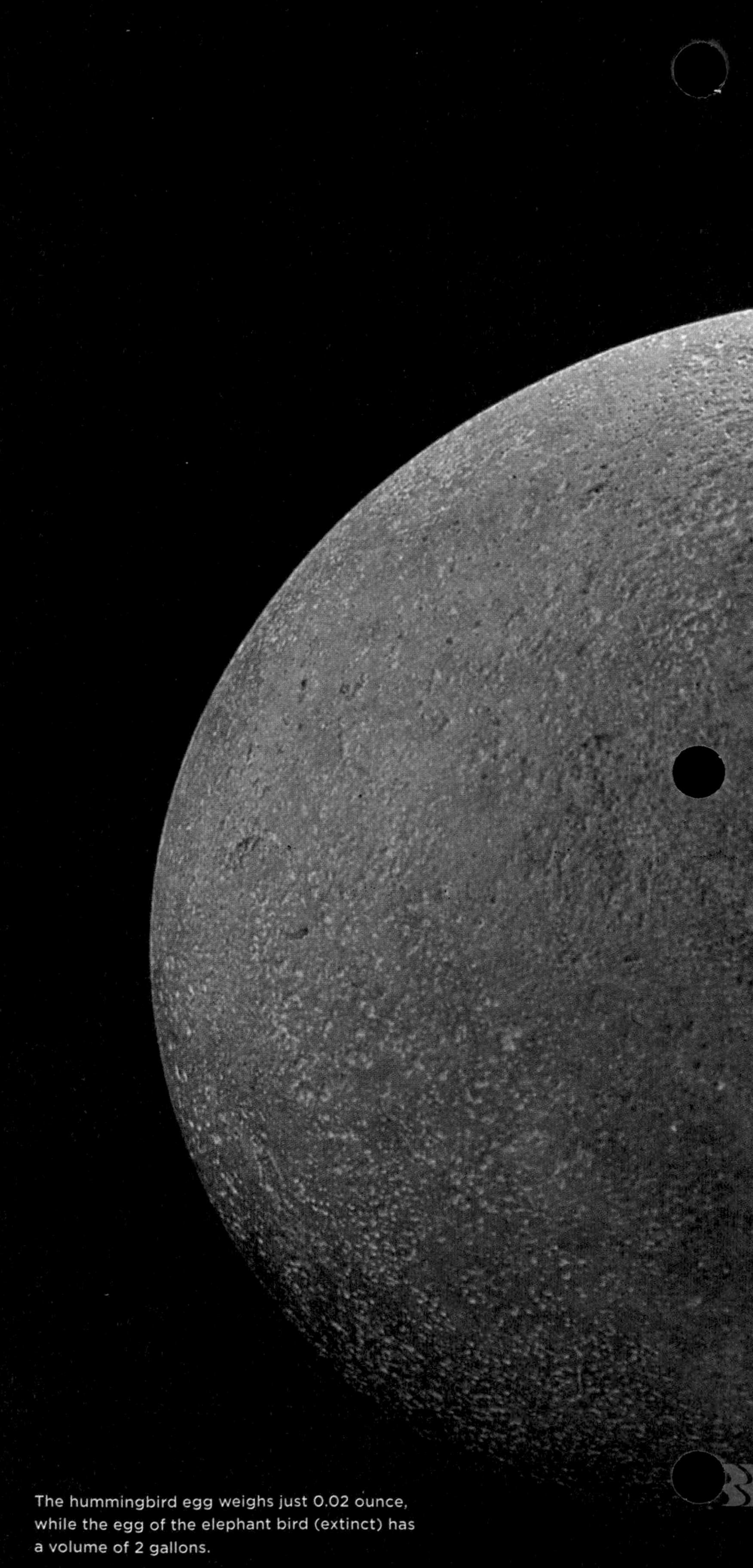

The hummingbird egg weighs just 0.02 ounce, while the egg of the elephant bird (extinct) has a volume of 2 gallons.

Cells

The smallest part of you

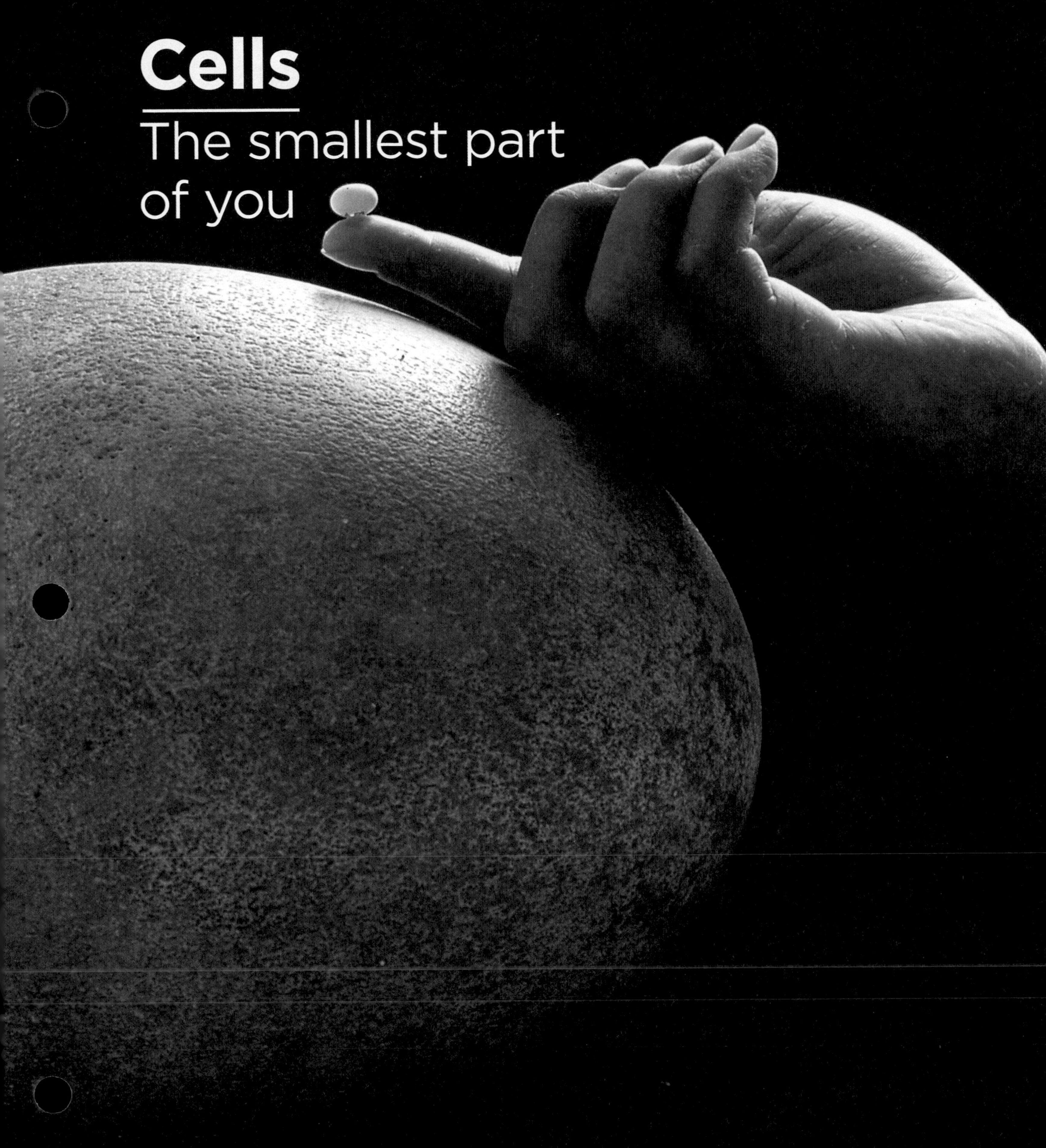

4.1–4.3

What is a cell?

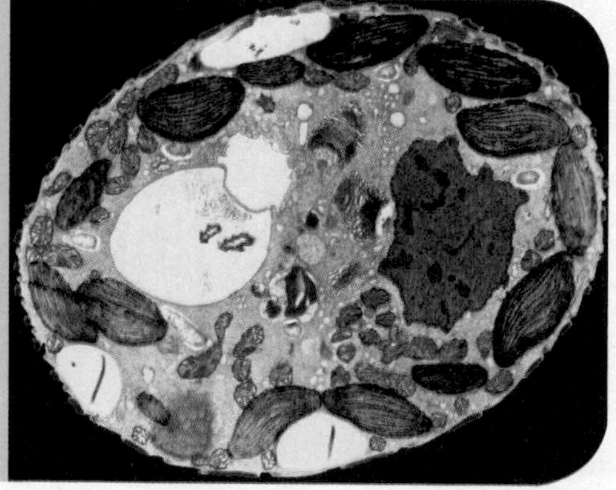

Plant cell packed with organelles.

4.1 All organisms are made of cells.

Whether we are studying a creature as small as a flea or as large as an elephant or giant sequoia, all organisms are made of smaller units that are more easily studied and understood (FIGURE 4-1). The most basic unit of any organism is the **cell,** the smallest unit of life that can function independently and perform all the necessary functions of life. Understanding cell structure and function is the basis for our understanding of how complex organisms are organized.

The term "cell" was first used in the mid-1600s by Robert Hooke, an English scientist who, as Curator of Experiments for the Royal Society of London, had access to many of the first microscopes. Hooke thought the magnified views of a very thin piece of cork resembled a mass of small, empty rooms. He named these compartments *cellulae,* Latin for "small rooms."

By the 1830s, improvements in microscopes helped scientists realize that all plants and animals were made entirely from cells. Subsequent studies revealed that every cell seemed to arise from the division of another cell. You, for example, are made up of at least 60 trillion cells, all of which came from just one cell: the single fertilized egg produced when an egg cell from your mother was fertilized by a sperm cell from your father.

The facts that (1) all living organisms are made up of one or more cells and (2) all cells arise from other, preexisting living cells are the foundations of **cell theory,** one of

Fly agaric mushroom

Gerenuk (also known as the giraffe gazelle)

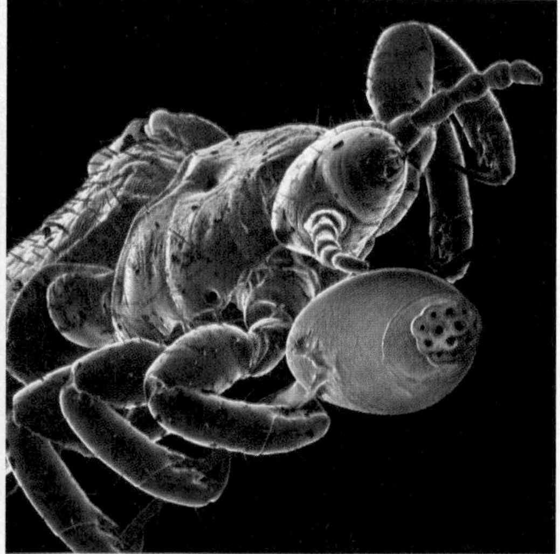

Head louse

FIGURE 4-1 **What do these diverse organisms have in common?** Cells.

Q Aristotle wrote that "living things form quickly whenever . . . air and . . . heat are enclosed in anything." Why is he wrong?

the unifying theories in biology, and one that is universally accepted by all biologists (FIGURE 4-2). As we see in Chapter 12, the origin of life on earth was a one-time deviation from cell theory: the first cells probably originated from free-floating molecules in earth's ancient oceans (about 3.5 billion years ago). Since that time, however, all cells and thus all life have been produced as a continuous line of cells, originating from these initial cells.

Today, we know that the cell is a three-dimensional structure, like a fluid-filled balloon, in which many of the essential chemical reactions of life take place. Generally, these reactions involve transporting raw materials and fuel into the cell and exporting finished materials and waste products out of the cell. In addition, most, but not all, cell types contain DNA (deoxyribonucleic acid), a molecule that contains the information that directs the formation of various cellular products within the cell, the chemical reactions in the cell, and the cell's ability to reproduce itself. Nearly all cells—even in multicellular organisms—contain the genetic information for the entire organism. We explore all of these features of cell functioning in this and the next four chapters.

Most cells are too small to see with the naked eye. At this moment, there are probably more than seven billion bacterial cells in your mouth—even if you just brushed your teeth! There are a few large-size exceptions, however, including hens' eggs from the supermarket. Each egg tends to contain just one cell (as long as the egg is unfertilized). The ostrich egg, weighing about three pounds, contains the largest of all animal cells. (We should note, however, that by the time an ostrich lays a fertilized egg, the embryo inside has already gone through multiple divisions.)

Eggs also tend to be the most valuable cells. Almas caviar, eggs from the beluga sturgeon, sells for nearly $700 per ounce. Human eggs currently fetch as much as $25,000 for a dozen or so eggs on the open market (FIGURE 4-3). (Human sperm cells command only about a penny per 20,000 cells!)

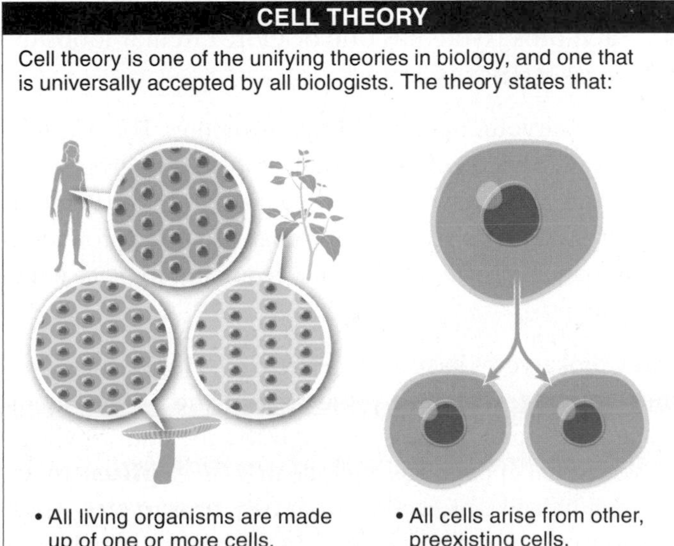
CELL THEORY
Cell theory is one of the unifying theories in biology, and one that is universally accepted by all biologists. The theory states that:
• All living organisms are made up of one or more cells.
• All cells arise from other, preexisting cells.

FIGURE 4-2 **The cell is the basic unit of life.**

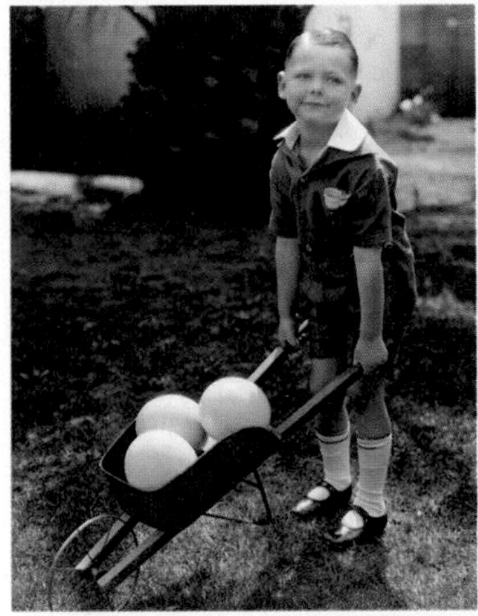

Ostrich eggs, weighing more than 3 pounds each

Beluga sturgeon eggs: $700 per ounce

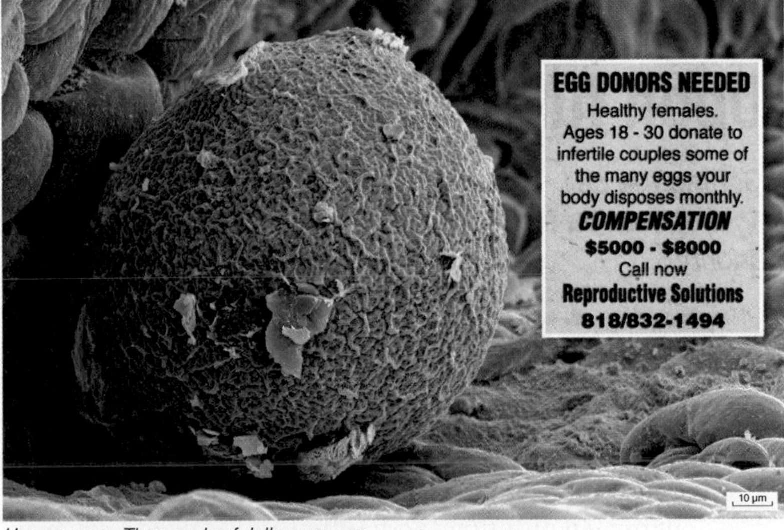
EGG DONORS NEEDED
Healthy females.
Ages 18 - 30 donate to infertile couples some of the many eggs your body disposes monthly.
COMPENSATION
$5000 - $8000
Call now
Reproductive Solutions
818/832-1494
10 μm

Human eggs: Thousands of dollars per egg

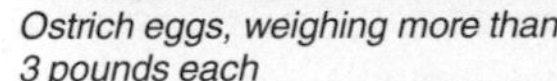

Eggs are the largest and most expensive cells in the world.

FIGURE 4-3 **Not all cells are tiny.** And some cells are extremely valuable!

In this chapter, we investigate the two different kinds of cells that make up all of the organisms on earth, the processes by which cells control how materials move into and out of the cell, and how cells communicate with each other. We also explore some of the important structures found in many cells and their functions. Along the way, we learn about some of the health consequences when cells malfunction.

TAKE HOME MESSAGE 4.1

» The most basic unit of any organism is the cell, the smallest unit of life that can function independently and perform all of the necessary tasks of life, including reproducing itself. All living organisms are made up of one or more cells, and all cells arise from other, preexisting cells.

4.2 Prokaryotic cells are structurally simple but extremely diverse.

Although millions of diverse species live on earth, and many of those organisms contain trillions of different cells, every cell falls into one of two basic categories.

A **eukaryotic cell** has a central control structure called a nucleus, which contains the cell's DNA. Organisms composed of eukaryotic cells are called **eukaryotes.** (The term "eukaryote" comes from the Greek words for "good" and "kernel," referring to the nucleus.)

A **prokaryotic cell** does not have a nucleus; its DNA simply resides in the cytoplasm. An organism consisting of a prokaryotic cell is called a **prokaryote.** All prokaryotes are one-celled organisms and are invisible to the naked eye.

The first cells on earth were prokaryotes, making their appearance about 3.5 billion years ago (the term "prokaryote" comes from the Greek for "before" and "kernel," referring to the evolutionary origin of these cells before the eukaryotes). For a long time (1.5 billion years), the prokaryotes had the planet to themselves.

Just two groups of prokaryotes exist: *bacteria* and *archaea.* All other organisms are eukaryotes. The largest group of prokaryotes is the bacteria. Bacteria are probably familiar to you for their ability to cause diseases such as pneumonia, to spoil food, and to decompose plant material in compost. They also play important and beneficial roles in human nutrition and health. Less familiar to you may be the archaea; these microorganisms inhabit some of the harshest environments on earth, thriving in extremes of temperature, salinity, and pH.

Prokaryotes have four basic structural features (FIGURE 4-4).

1. A **plasma membrane** encompasses the cell (and sometimes is simply called the "cell membrane"). Anything inside the plasma membrane is referred to as "intracellular," and everything outside the plasma membrane is "extracellular."
2. The **cytoplasm** refers to the cell's contents contained within the plasma membrane. This includes the jelly-like fluid, called the **cytosol,** and the cell's genetic material.
3. **Ribosomes** are little granular bodies where proteins are made; thousands of them are scattered throughout the cytoplasm.
4. Each prokaryote has one or more circular loops or linear strands of DNA.

Some prokaryotes have additional structures. For example, many have a rigid **cell wall** that protects and gives shape to the cell. Some have a sticky, sugary capsule as their outermost layer. This sticky outer coat provides protection and enhances prokaryotes' ability to anchor in place when necessary.

Many prokaryotes have a **flagellum** (*pl.* **flagella**), a long, thin, rotating, whip-like projection of the plasma membrane that moves the cell through the medium in which it lives. Other appendages include **pili** (*sing.* **pilus**), much thinner, hair-like projections that help prokaryotes attach to surfaces and can serve as "tubes" through which they exchange DNA.

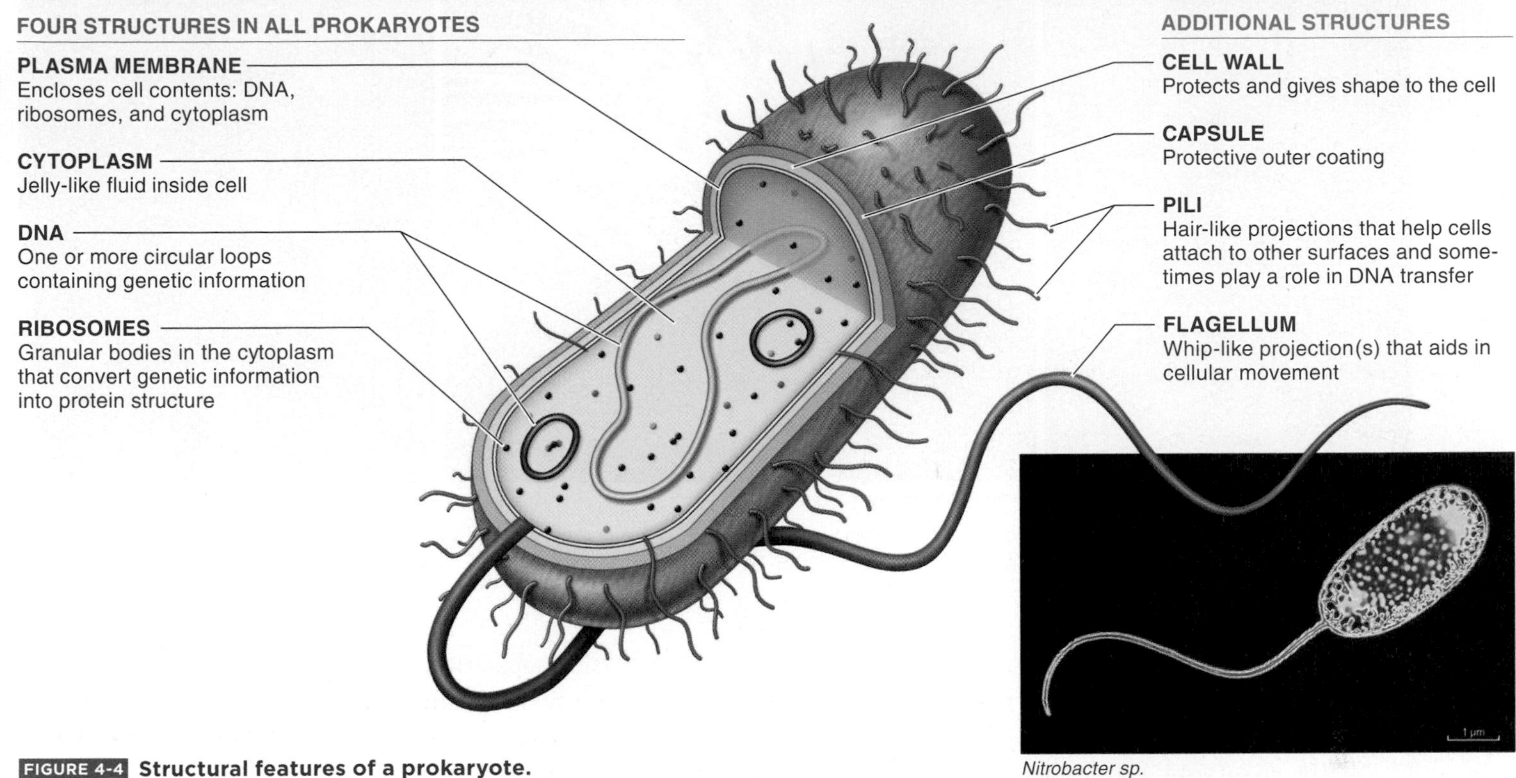
THE PROKARYOTE: BASIC STRUCTURE
FOUR STRUCTURES IN ALL PROKARYOTES
PLASMA MEMBRANE
Encloses cell contents: DNA, ribosomes, and cytoplasm
CYTOPLASM
Jelly-like fluid inside cell
DNA
One or more circular loops containing genetic information
RIBOSOMES
Granular bodies in the cytoplasm that convert genetic information into protein structure
ADDITIONAL STRUCTURES
CELL WALL
Protects and gives shape to the cell
CAPSULE
Protective outer coating
PILI
Hair-like projections that help cells attach to other surfaces and sometimes play a role in DNA transfer
FLAGELLUM
Whip-like projection(s) that aids in cellular movement
Nitrobacter sp.

FIGURE 4-4 **Structural features of a prokaryote.**

Although prokaryotes are smaller, evolutionarily older, and structurally simpler than eukaryotes, they are fantastically diverse metabolically (i.e., in the way they break down and build up molecules). For example, some can fuel their activities in the presence or absence of oxygen, and some prokaryotes can use the sulfur in deep-sea hydrothermal vents as an energy source, or hydrogen gas, or light from the sun.

TAKE HOME MESSAGE 4.2

» Every cell on earth is either a eukaryotic or a prokaryotic cell. Prokaryotes, which have no nucleus, were the first cells on earth. They are single-celled organisms. Prokaryotes include the bacteria and archaea and, as a group, are characterized by tremendous metabolic diversity.

4.3 Eukaryotic cells have compartments with specialized functions.

In the two billion years that eukaryotes have been on earth, they have evolved into dramatically different and interesting creatures, such as platypuses, dolphins, giant sequoias, and the Venus flytrap. Not all eukaryotes are multicellular, however. Many fungi are unicellular, and the Protista (or protists) are a huge group, of which most are single-celled organisms invisible to the naked eye. Nonetheless, because all prokaryotes are single-celled and thus invisible, all the plants and animals that we see around us are eukaryotic organisms (FIGURE 4-5).

Eukaryotic cells are about 10,000 times larger than prokaryotic cells in volume. It is easy to distinguish eukaryotes from prokaryotes under a microscope

A plant in the genus Lobelia

Malayan porcupine

Yellow-fuzz cone slime mold

FIGURE 4-5 **Diversity of the eukaryotes.** Every organism that we can see without magnification is a eukaryotic organism.

(FIGURE 4-6). Chief among the distinguishing features of eukaryotic cells is the presence of a **nucleus,** a membrane-enclosed structure that contains linear strands of DNA. Additionally, eukaryotic cells usually contain several other specialized structures in their cytoplasm. Called **organelles,** many of these structures are enclosed separately by their own lipid membranes.

The physical separation of compartments within a eukaryotic cell provides distinct areas in which different chemical reactions can occur at the same time. In the mostly non-compartmentalized interior of a prokaryotic cell, random molecular movements quickly blend the chemicals throughout the cell, reducing the ease with which different reactions can occur simultaneously.

EUKARYOTIC vs. PROKARYOTIC CELLS

Nucleus
Other organelles
1 μm

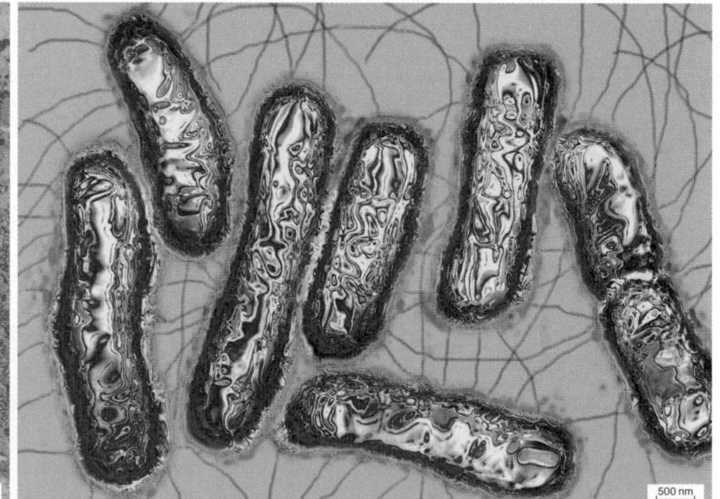

TYPICAL EUKARYOTIC CELL FEATURES
- DNA contained in nucleus.
- Cytoplasm contains specialized structures called organelles.
- Larger than prokaryotes—usually at least 10 times bigger.

TYPICAL PROKARYOTIC CELL FEATURES
- No nucleus—DNA is in the cytoplasm.
- Internal structures mostly not organized into compartments.
- Much smaller than eukaryotes.

FIGURE 4-6 **Comparison of eukaryotic and prokaryotic cells.**

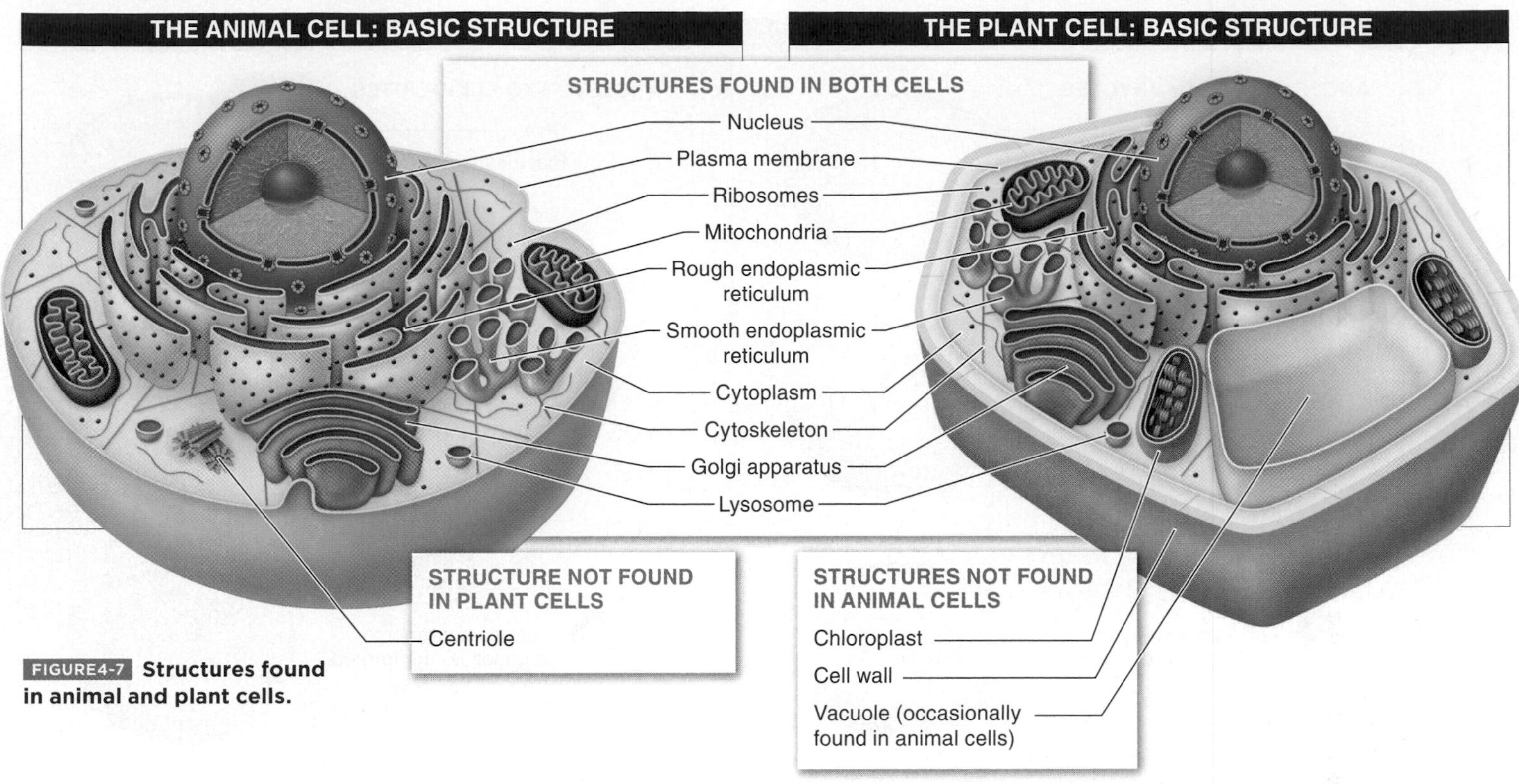

FIGURE4-7 **Structures found in animal and plant cells.**

FIGURE 4-7 illustrates a generalized animal cell and a generalized plant cell. Because they share a common, eukaryotic ancestor, they have much in common. We explore each of the illustrated animal and plant organelles in detail later in this chapter.

When you compare a complex eukaryotic cell with the structurally simpler prokaryotic cell, it's hard not to wonder about the origin of eukaryotic cells. The **endosymbiosis theory** provides the best explanation for the presence of two types of organelles in eukaryotes: chloroplasts in plants and algae, and mitochondria in plants and animals.

Q Humans—at a microscopic level—may be part bacteria. How can that be?

According to the theory of endosymbiosis, two different types of prokaryotes worked in close partnerships. For example, some small prokaryotes capable of performing photosynthesis (the process by which plant cells capture light energy from the sun and transform it into the chemical energy stored in food molecules) may have come to live inside a larger "host" prokaryote. The photosynthetic "boarder" may have made some of the energy that it captured in photosynthesis available for use by the host. After a long while, the two cells may have become more and more dependent on each other, until neither cell could live without the other (they became "symbiotic"). Eventually, the photosynthetic prokaryote evolved into a **chloroplast,** the organelle in plant and eukaryotic algae cells in which photosynthesis occurs. Similarly, a large host prokaryote may have engulfed a smaller prokaryote that was unusually efficient at converting food and oxygen into easily usable energy. This smaller prokaryote evolved into a **mitochondrion,** the organelle in plant and animal cells (as well as in protists and fungi) that converts the energy stored in food into a form usable by the cell (FIGURE 4-8).

The idea of the role of endosymbiosis in the evolution of eukaryotes is supported by several observations.

1. Chloroplasts and mitochondria are similar in size to prokaryotic cells and divide by splitting (fission), just like prokaryotes.
2. Chloroplasts and mitochondria have ribosomes, similar to those found in bacteria, that allow them to synthesize some of their own proteins; this ability is not found in other organelles, which rely on proteins made by the cell's cytoplasmic ribosomes.

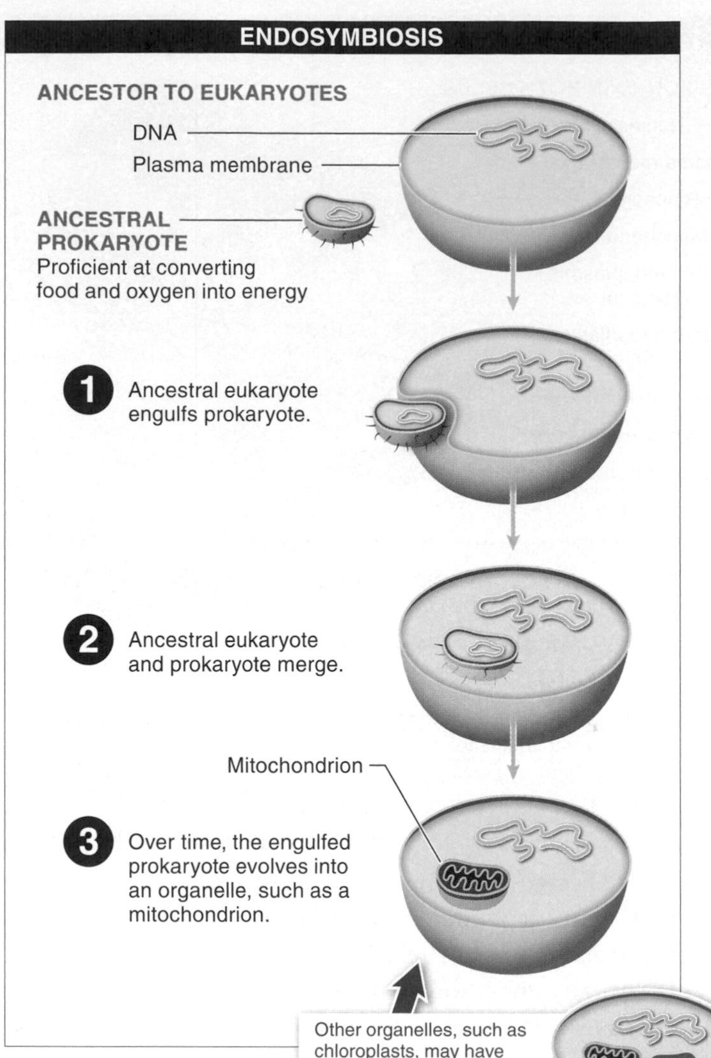
ENDOSYMBIOSIS
ANCESTOR TO EUKARYOTES
DNA
Plasma membrane
ANCESTRAL PROKARYOTE
Proficient at converting food and oxygen into energy
1 Ancestral eukaryote engulfs prokaryote.
2 Ancestral eukaryote and prokaryote merge.
Mitochondrion
3 Over time, the engulfed prokaryote evolves into an organelle, such as a mitochondrion.
Other organelles, such as chloroplasts, may have developed through a similar endosymbiotic partnership.

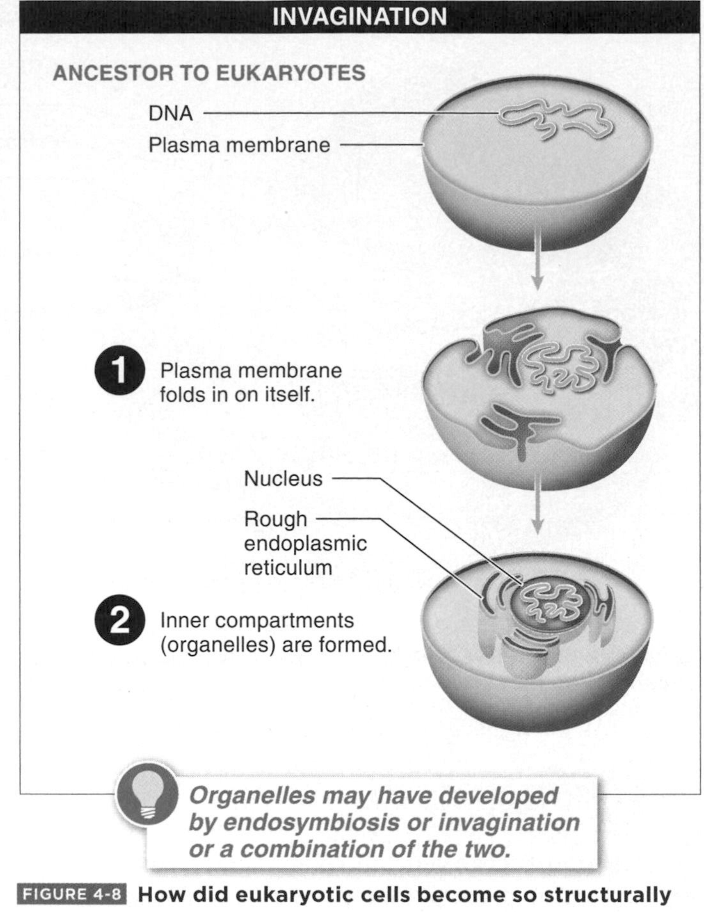
INVAGINATION
ANCESTOR TO EUKARYOTES
DNA
Plasma membrane
1 Plasma membrane folds in on itself.
Nucleus
Rough endoplasmic reticulum
2 Inner compartments (organelles) are formed.
Organelles may have developed by endosymbiosis or invagination or a combination of the two.

FIGURE 4-8 **How did eukaryotic cells become so structurally complex?** Two theories.

3. Chloroplasts and mitochondria have small amounts of circular DNA, similar to the circular DNA in prokaryotes and in contrast to the linear DNA strands found in a eukaryote's nucleus.

4. Analysis of chloroplast and mitochondrial DNA has revealed that it is highly related to bacterial DNA, much more closely than it is related to eukaryotic DNA.

The best current theory about the origin of the other organelles in eukaryotes is a process called **invagination.** The idea is that the plasma membrane around the cell may have folded in on itself to form inner compartments, which subsequently became modified and specialized (see Figure 4-8).

TAKE HOME MESSAGE 4.3

» Eukaryotes are single-celled or multicellular organisms consisting of cells with a nucleus that contains linear strands of genetic material. Most eukaryotic cells also have organelles throughout their cytoplasm; these organelles may have originated evolutionarily through endosymbiosis or invagination, or both.

4.4–4.7

Cell membranes are gatekeepers.

Like gatekeepers, cell membranes control the movement of material into and out of the cell.

4.4 Every cell is bordered by a plasma membrane.

Just as our bodies have a covering of skin, every cell of every living thing on earth is enclosed by a **plasma membrane,** a thin, flexible, two-layered membrane that holds the contents of a cell in place and regulates what enters and leaves the cell. Far from simple, the plasma membrane is filled with pores, outcroppings, channels, and complex molecules floating around within the two layers of the membrane (FIGURE 4-9).

Because cells are perpetually interacting with their external environment, plasma membranes must perform several critical functions.

- The membrane enables the cell to take in food and nutrients and dispose of waste products.
- It allows the cell to build and export molecules needed elsewhere in the body.
- It mediates communications with the external environment and with other cells, and adhesion to other cells or surfaces.
- Like a border control checkpoint, it controls the flow of molecules (including water) into and out of the cell.

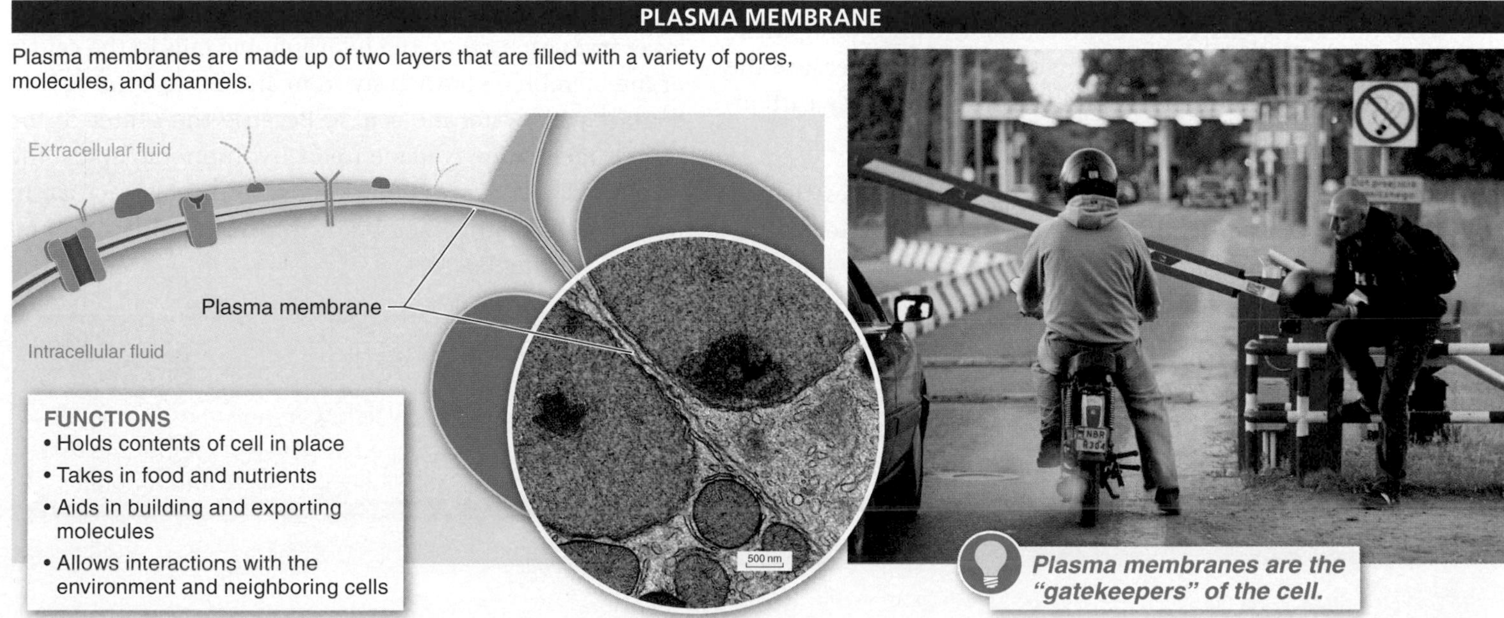

PLASMA MEMBRANE
Plasma membranes are made up of two layers that are filled with a variety of pores, molecules, and channels.
Extracellular fluid
Plasma membrane
Intracellular fluid
FUNCTIONS
• Holds contents of cell in place
• Takes in food and nutrients
• Aids in building and exporting molecules
• Allows interactions with the environment and neighboring cells
500 nm
Plasma membranes are the "gatekeepers" of the cell.

FIGURE 4-9 **More than just an outer layer.** The plasma membrane performs several critical functions beyond simply enclosing a cell's interior contents.

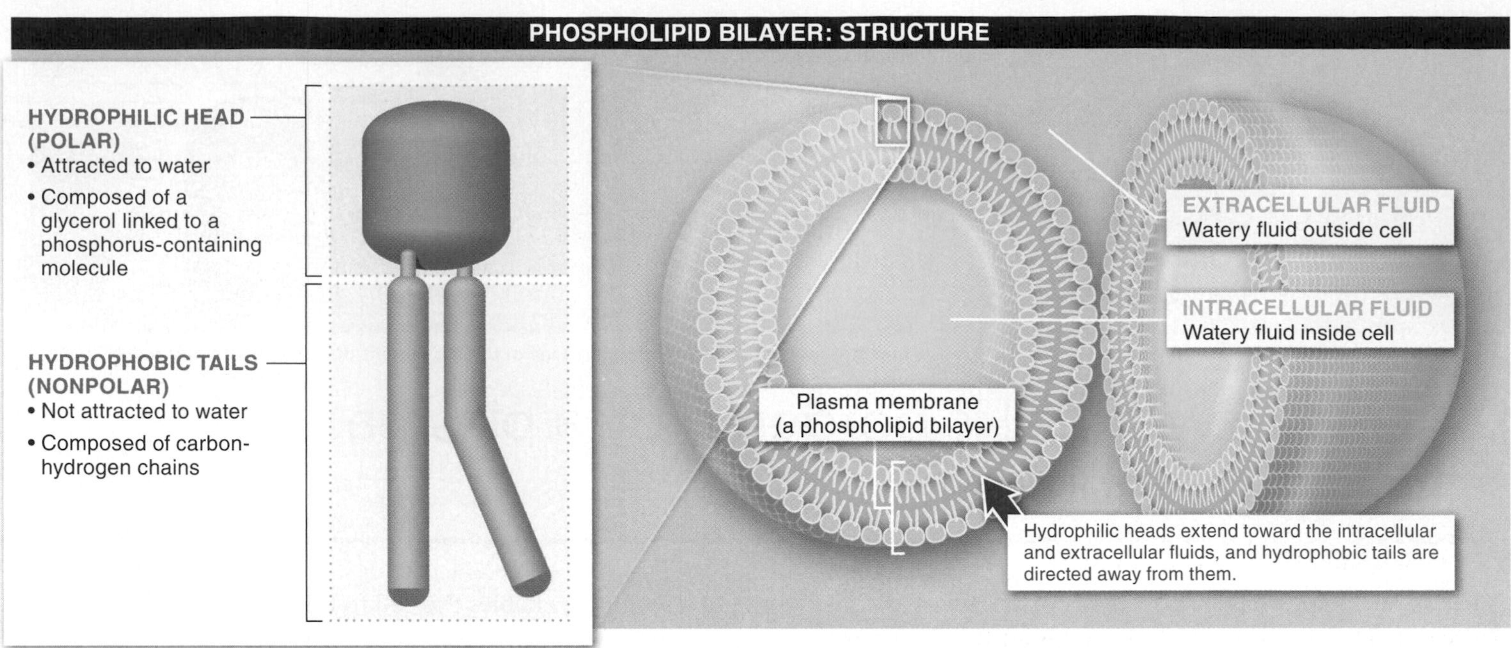
PHOSPHOLIPID BILAYER: STRUCTURE
HYDROPHILIC HEAD (POLAR)
• Attracted to water
• Composed of a glycerol linked to a phosphorus-containing molecule
HYDROPHOBIC TAILS (NONPOLAR)
• Not attracted to water
• Composed of carbon-hydrogen chains
EXTRACELLULAR FLUID
Watery fluid outside cell
INTRACELLULAR FLUID
Watery fluid inside cell
Plasma membrane (a phospholipid bilayer)
Hydrophilic heads extend toward the intracellular and extracellular fluids, and hydrophobic tails are directed away from them.

FIGURE 4-10 **Good membrane material.** The phospholipid bilayer of the plasma membrane prevents leakage of fluid from the cell.

The foundation of all plasma membranes is a layer of tightly packed lipid molecules, called **phospholipids.** You'll recall from Section 3-9 that phospholipids have what appear to be a head and two long tails. The head consists of a molecule of **glycerol** linked to a molecule containing phosphorus (FIGURE 4-10). This head region is said to be **polar,** because the electrons are not shared equally among the atoms, leading to regions of partial positive and partial negative charge. The two tails of the phospholipid are long chains of carbon and hydrogen atoms. Because the electrons in these bonds are shared equally, the carbon-hydrogen chains are **nonpolar.**

The chemical structure of phospholipids gives them a sort of split personality: their polar, hydrophilic ("water-loving") head region mixes easily with water, while their nonpolar, hydrophobic ("water-fearing") tail region does not mix with water.

The split personality of phospholipids makes them good membrane material. In the cell's plasma membrane, two sheets of phospholipids are arranged so that all the hydrophobic tails are in contact with one another in the center of the membrane and the hydrophilic heads are in contact with the watery solution outside and inside the cell (see Figure 4-10). This arrangement gives us another way to describe the structure of the plasma membrane: as a **phospholipid bilayer.**

The phospholipids are not locked in place; they just float around their side of the bilayer. They cannot pop out of the membrane or flop from one side of the membrane to the other. Resembling similarly charged magnets that push away from each other, the hydrophobic tails in the center of the membrane push away from and avoid coming into contact with water molecules. Because the center of the bilayer membrane is made up of hydrophobic lipids, the intracellular or extracellular fluids cannot leak across. In this way, the plasma membrane forms a boundary around the cell's contents.

TAKE HOME MESSAGE 4.4

» Every cell of every living organism is enclosed by a plasma membrane, a two-layered membrane that holds the contents of a cell in place and regulates what enters and leaves the cell.

4.5 Molecules embedded in the plasma membrane help it perform its functions.

The plasma membrane functions with the help of different types of protein, carbohydrate, and lipid molecules embedded within or attached to the phospholipid bilayer. Because many of these molecules float around, held in a proper orientation by hydrophobic and hydrophilic forces, the plasma membrane is often described as a **fluid mosaic** (FIGURE 4-11).

For every 50–100 phospholipids in the membrane, there is one protein molecule. Some of these proteins, called **transmembrane proteins,** penetrate right through the lipid bilayer, from one side to the other. Others, called **surface proteins,** or peripheral proteins, reside primarily on the inner or outer surface of the membrane. Hydrophobic and hydrophilic forces keep these membrane proteins properly oriented, while allowing them to float around without ever popping out.

There are several primary types of membrane proteins, each of which performs a different function (FIGURE 4-12).

MOLECULES WITHIN THE PLASMA MEMBRANE

Protein, carbohydrate, and lipid molecules are embedded in the plasma membrane, but many can float around because they are not always anchored in place.

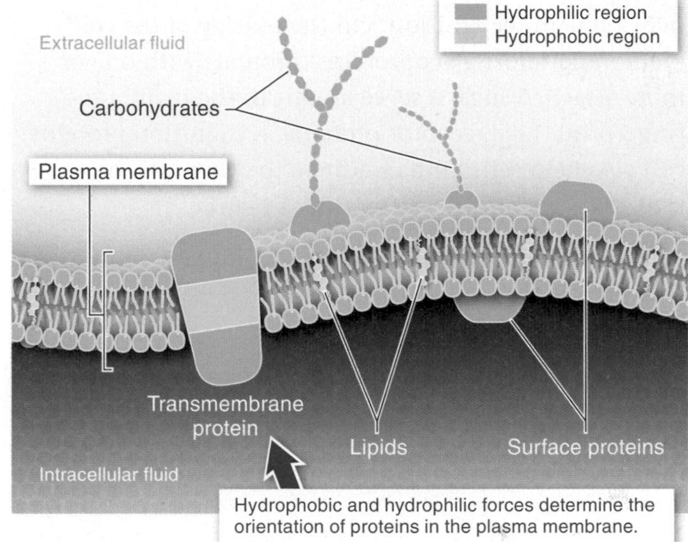

Hydrophilic region
Hydrophobic region
Extracellular fluid
Carbohydrates
Plasma membrane
Transmembrane protein
Lipids
Surface proteins
Intracellular fluid
Hydrophobic and hydrophilic forces determine the orientation of proteins in the plasma membrane.

FIGURE 4-11 **A fluid mosaic.**

1. Receptor proteins are surface or transmembrane proteins that bind to chemicals in the cell's external environment. In doing so, receptor proteins can attach a cell to the extracellular matrix or convey information from the outside to the inside of the cell.

Cells in the heart, for example, have receptor proteins within their membrane that bind to adrenaline, a chemical released into the bloodstream in times of extreme stress or fright. When adrenaline binds to

FUNCTION OF PLASMA MEMBRANE MOLECULES

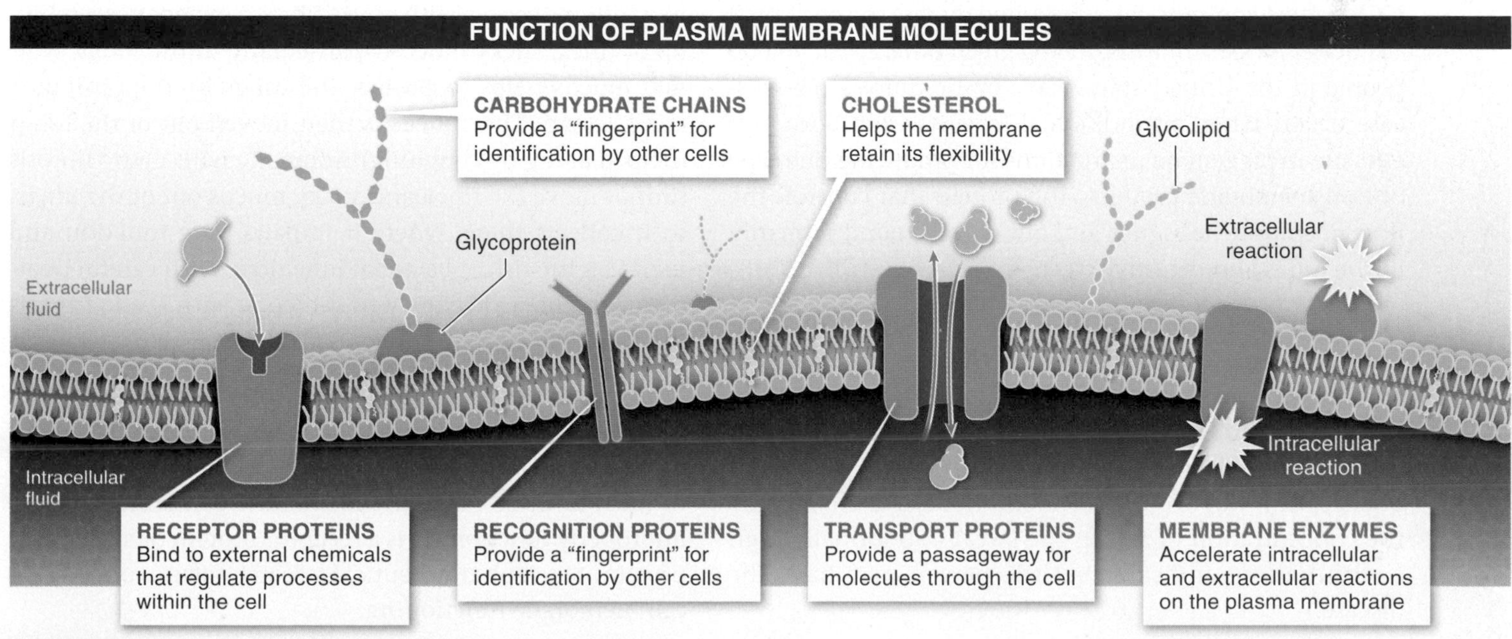

CARBOHYDRATE CHAINS
Provide a "fingerprint" for identification by other cells
CHOLESTEROL
Helps the membrane retain its flexibility
Glycolipid
Glycoprotein
Extracellular reaction
Extracellular fluid
Intracellular reaction
Intracellular fluid
RECEPTOR PROTEINS
Bind to external chemicals that regulate processes within the cell
RECOGNITION PROTEINS
Provide a "fingerprint" for identification by other cells
TRANSPORT PROTEINS
Provide a passageway for molecules through the cell
MEMBRANE ENZYMES
Accelerate intracellular and extracellular reactions on the plasma membrane

FIGURE 4-12 **Plasma membrane molecules serve diverse roles.**

these heart cells, the cells increase the heart's rate of contraction to pump blood through the body more quickly. You have experienced this reaction if you've ever been startled and felt your heart start to pound.

2. Recognition proteins are surface or transmembrane proteins that give each cell a "fingerprint" that makes it possible for the body's immune system (which fights off infections) to distinguish body cells from invaders. Carbohydrates also play a role in recognition. On the outside of the cell membrane, short glycoproteins (proteins with one or more attached sugars) serve as part of the membrane's fingerprint. Like receptor proteins, recognition proteins can also help cells bind or adhere to other cells or molecules within the extracellular matrix.

3. Transport proteins are transmembrane proteins that help polar or charged substances pass through the plasma membrane.

4. Membrane enzymes are surface or transmembrane proteins that accelerate chemical reactions on the plasma membrane's surface.

In addition, the lipid cholesterol can also be incorporated in a cell's plasma membrane. **Cholesterol** helps the membrane maintain its flexibility, preventing the membrane from becoming too fluid or floppy at moderate temperatures and acting as a sort of antifreeze, preventing the membrane from becoming too rigid at freezing temperatures. The membranes of some cells are about 25% cholesterol; other plasma membranes, such as those of most bacteria and plants, have no cholesterol.

TAKE HOME MESSAGE 4.5

» The plasma membrane is a fluid mosaic of proteins, lipids, and carbohydrates. Proteins in the membrane enable it to carry out most of its gatekeeping functions. In conjunction with carbohydrates, some plasma membrane proteins identify the cell to other cells. And, in addition to the phospholipids that make up most of the plasma membrane, the membrane lipid cholesterol influences fluidity.

4.6 Faulty membranes can cause diseases.

The most common fatal inherited disease in the United States is cystic fibrosis. At any given time, about 40,000 people in the United States have cystic fibrosis. The disease occurs when an individual inherits from both parents incorrect genetic instructions for producing one type of transmembrane protein—the protein that controls the flow of chloride ions into and out of cells found primarily in the lungs and digestive tract.

These genetic instructions, which can be defective in more than a thousand different ways, cause malfunction of chloride passageways and gradual accumulation of chloride ions within cells. In nearly all cases of cystic fibrosis, an improper salt balance in the cells occurs. (As a consequence, one way to test for cystic fibrosis is to measure the concentration of salt in the sweat—abnormally high concentrations indicate that the person may have the disease.)

In addition, people with cystic fibrosis experience a buildup of thick, sticky mucus—particularly in the lungs. Normal mucus helps to protect the lungs by trapping dust and bacteria. This mucus is then moved out of the lungs (helped along by coughing). Someone with cystic fibrosis cannot move the thick and sticky mucus out of the lungs, so it collects there, where it impairs lung function and increases the risk of bacterial infection. With careful treatment, the life expectancy of someone with cystic fibrosis can be 35–40 years or longer (FIGURE 4-13).

Faulty membranes also play a role in many other diseases, including heart disease (familial hypercholesterolemia), diabetes, and hormonal disorders (such as Graves disease). In most of these cases, the membrane component that is improperly functioning is a receptor protein or a transport protein. Much pharmaceutical research focuses on altering cell membrane functioning.

TREATING CYSTIC FIBROSIS

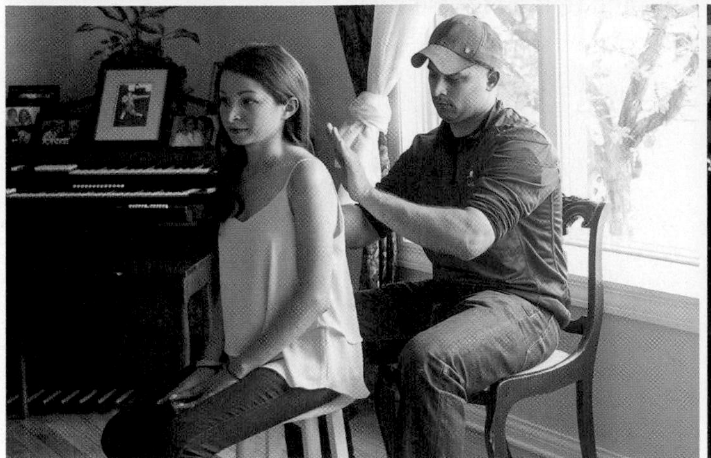

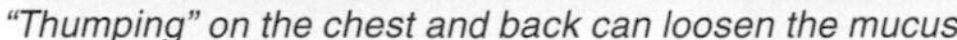

"Thumping" on the chest and back can loosen the mucus.

The vest, by inflating and deflating rapidly, can have a similar effect in the course of a 20-minute session.

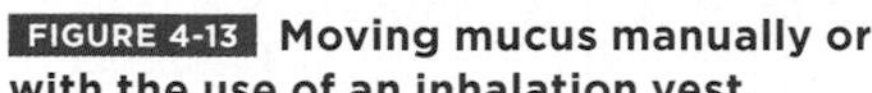

FIGURE 4-13 Moving mucus manually or with the use of an inhalation vest.

The thick and sticky mucus produced by someone with cystic fibrosis collects in the lungs, impairing lung function and increasing the risk of bacterial infection.

One group of drugs that alter membrane function—called beta-blockers—is extremely effective at reducing the symptoms of anxiety.

Here's how beta-blockers work. In stressful situations, your adrenal glands pump out adrenaline (FIGURE 4-14). Many cells in your body, particularly the cells of the heart, have receptor proteins called beta-adrenergic receptors (or beta-receptors) on their plasma membranes that can bind adrenaline. When adrenaline binds

Q Why do beta-blockers reduce anxiety?

BETA-BLOCKERS

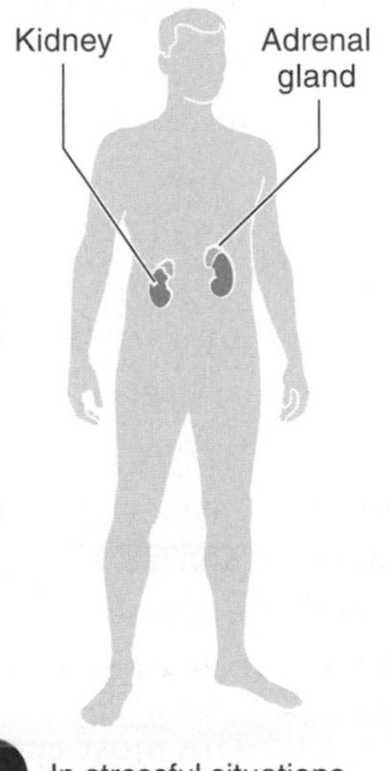

1 In stressful situations, the adrenal glands pump out adrenaline.

2 Adrenaline, the fight-or-flight chemical, cannot enter the cell when beta-blocking medications bind to the cell's beta-receptors.

FIGURE 4-14 Drugs can alter cell membrane function.

By binding to adrenaline receptors, beta-blockers reduce anxiety symptoms.

to beta-receptors, it promptly causes your heart to beat faster and more forcefully, increasing your blood pressure. This reaction is useful in a short-term fight-or-flight situation, but the increased pressure can damage blood vessels over the long run. Depending on its severity, this reaction can also be problematic if you are giving a presentation or taking a test, or if you're in any other anxiety-producing situation.

When you take a beta-blocker pill, the pill dissolves and the chemicals travel throughout your body until they encounter the beta-receptors. They bind to the receptors, hold on, and block the adrenaline from doing its job. This outcome slows your heart rate, causes a reduction in blood pressure, and can bring great relief to those suffering from the sweating and trembling associated with anxiety.

TAKE HOME MESSAGE 4.6

» Normal cell functioning can be disrupted when cell membranes—particularly the proteins embedded in them—do not function properly. Such malfunctions can cause health problems, such as cystic fibrosis. But intentional disruption of normal cell membrane function can have beneficial, therapeutic effects, such as in the treatment of high blood pressure and anxiety.

4.7 Membrane surfaces have a "fingerprint" that identifies the cell.

Every cell in your body has a "fingerprint" made from a variety of molecules located on the outer surface of the cell membrane. Some fingerprint molecules differ from cell to cell, depending on the specific function of the cell, and others are common to all of your cells. They tell your immune system, "I belong here." Cells with an improper fingerprint are recognized as foreign and are attacked by your body's defenses.

Throughout our evolutionary history, this system has helped our bodies fight infection. In some cases, however, this vigilance is a problem, like a car alarm that goes off even when you don't want it to. Suppose you receive a liver (or any other organ) transplant. Even if the donor is a close relative, the molecular fingerprint on the cells of the donated liver is not identical to your own. Your body sees the new organ as a foreign object and fights against it. To prevent organ rejection, doctors must administer drugs that suppress your immune system. Immune suppression helps you tolerate the new liver, but it leaves you without some of the defenses to fight off other foreign invaders, such as bacteria that may cause infection.

The AIDS-causing virus, HIV, uses molecular fingerprints to infect an individual's cells. These molecular markers belong to a group of identifying markers called "clusters of differentiation." Abbreviated as "CD markers" and having names such as CD1, CD2, and CD3, these marker molecules are proteins embedded in the plasma membrane. They enable the cell to bind to outside molecules and, sometimes, transport them into the cell.

Q Why is it extremely unlikely that a person will catch HIV from casual contact—such as shaking hands—with an infected individual?

One CD marker, called the CD4 marker, is found only on cells deep within the body and in the bloodstream, such as immune system cells and some nerve cells. HIV targets the CD4 marker, in conjunction with another receptor. If the virus locates a cell with a CD4 marker, it can infect that cell, and because CD4 markers never occur on the surface of skin cells, casual contact such as touching is very unlikely to transmit the virus (FIGURE 4-15). Even if millions of HIV particles are present on a person's hands, they just can't gain access into any of the other person's surface cells.

The most common ways for HIV to be transferred from one individual to another are through blood, semen, vaginal fluid, or breast milk. Particles of the virus, as well as cells infected by the virus, are present in these fluids. Consequently, the chief routes of transmission are from

HIV TRANSMISSION

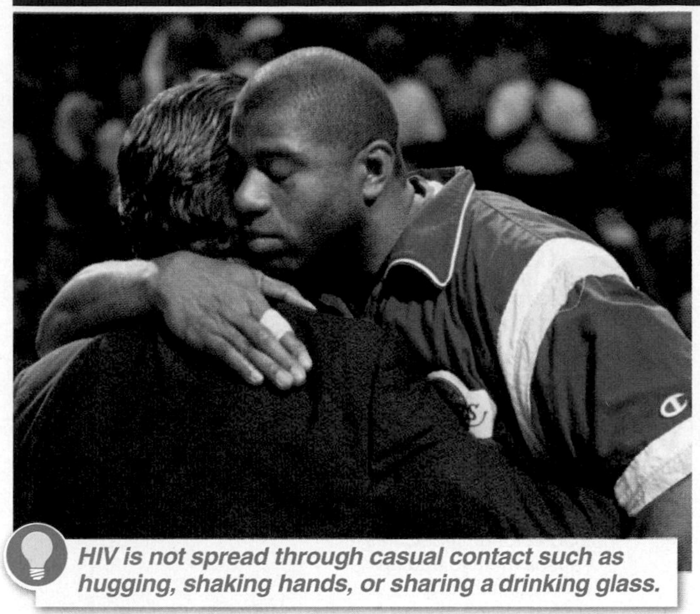

HIV is not spread through casual contact such as hugging, shaking hands, or sharing a drinking glass.

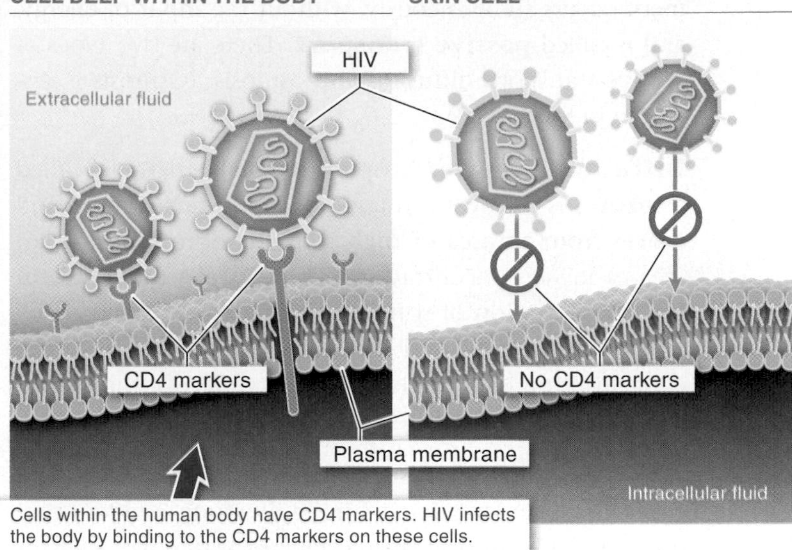

FIGURE 4-15 **HIV requires CD4 markers—not found on skin cells—to infect the body.**

an infected mother to her child in breast milk, from an infected mother to her baby at birth, the use of contaminated needles, and unprotected sexual intercourse. Because an open sore or cut might expose some of the cells in your bloodstream to the outside world, however, it is not impossible for casual transmission of HIV to occur.

TAKE HOME MESSAGE 4.7

» Every cell in your body has a "fingerprint" made from a variety of molecules on the outside-facing surface of the cell membrane. This molecular fingerprint is key to the function of your immune system.

4.8–4.11 Molecules move across membranes in several ways.

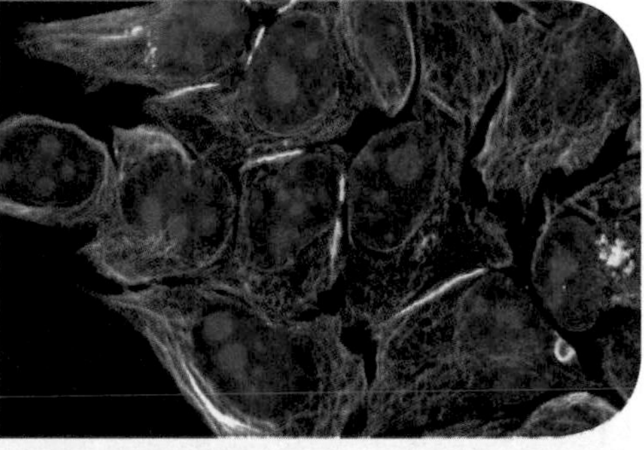

Passage of proteins between cells can be tracked (junctions between these human cancer cells are stained red and green).

4.8 Passive transport is the spontaneous diffusion of molecules across a membrane.

To function properly, cells must take in food and/or other necessary materials and must move out both metabolic waste and molecules produced for use elsewhere in the body. In some cases, this movement of molecules requires energy and is called active transport. (We cover active transport later in this chapter.) In other cases, the molecular movement occurs spontaneously, without the input of energy, and is called **passive transport.** There are two types of passive transport: diffusion and osmosis. (Osmosis is discussed in Section 4.9.)

Diffusion is passive transport in which a particle, called a **solute,** is dissolved in a gas or liquid (a **solvent**) and moves from an area of high solute concentration to an area of lower concentration (FIGURE 4-16). A difference in the concentration of solutes in two areas is called a *concentration gradient*—and the larger the difference in the concentration of the solutes in the two areas, the greater the concentration gradient.

We say that molecules tend to move "down" their concentration gradient. For a simple illustration, add a tiny drop of food coloring into a bowl of water and wait for a few minutes. The molecules of dye are initially clustered together in a very high concentration. Because the dye molecules are highly concentrated, they keep bumping into each other. Gradually, they disperse down their concentration gradient until the color is equally spread throughout the bowl.

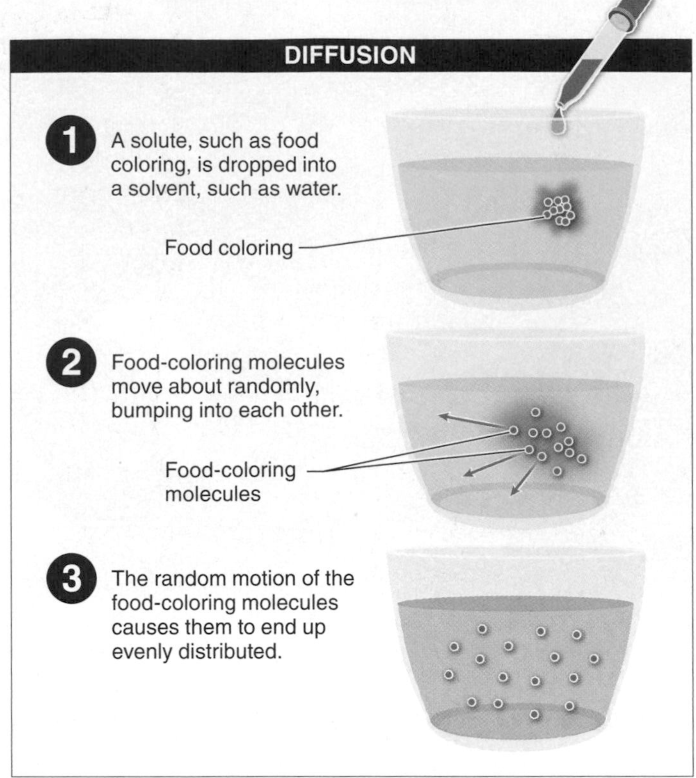

DIFFUSION
1 A solute, such as food coloring, is dropped into a solvent, such as water.
Food coloring
2 Food-coloring molecules move about randomly, bumping into each other.
Food-coloring molecules
3 The random motion of the food-coloring molecules causes them to end up evenly distributed.

FIGURE 4-16 **Diffusion: a form of passive transport that results in an even distribution of molecules.**

PASSIVE TRANSPORT

SIMPLE DIFFUSION
Molecules pass directly through the plasma membrane.

FACILITATED DIFFUSION
Molecules move across the plasma membrane with the help of a channel or carrier molecule.

Extracellular fluid

Channel or carrier molecule

Intracellular fluid

Oxygen (left) diffuses across membranes of lung cells, into the cells, while carbon dioxide (right) diffuses in the opposite direction.

FIGURE 4-17 **Simple and facilitated diffusion: no energy required.**

In cells, molecules such as oxygen (O_2) and carbon dioxide (CO_2) that are small and carry no charge can pass directly through the phospholipid bilayer of the membrane without the assistance of any other molecules, in a process called **simple diffusion.** Each time you take a breath, for example, there is a higher concentration of O_2 molecules in the air you pull into your lungs than in the blood in your lungs. And so O_2 diffuses across the plasma membranes of the lung cells and into your bloodstream, where red blood cells pick it up and deliver it to where it is needed. Similarly, because CO_2 in your bloodstream is at a higher concentration than in the air in your lungs, it diffuses from your blood into the cells of your lungs and is released to the atmosphere when you exhale (FIGURE 4-17).

Most molecules, however, can't get through plasma membranes by simple diffusion. Polar molecules are repelled by the hydrophobic middle region of the phospholipid bilayer. Other molecules may be too big to squeeze through the membrane. These molecules may still be able to diffuse across the membrane, down their concentration gradient, but only with the help of a transport protein. In a process called **facilitated diffusion** (see Figure 4-17), a transport protein spans the membrane and functions like a revolving door, allowing movement of molecules in either direction, depending on their concentration gradient.

Diffusion across membranes doesn't occur just in animals—we see it in all organisms. As we'll see in Chapter 5, one of the most important biological processes on earth is the diffusion of CO_2 from the atmosphere (an area of relatively high concentration) into the leaf cells of plants (areas of relatively low concentration), where it can be attached to other molecules, forming sugars through photosynthesis. At the same time, O_2 diffuses out of the leaf cells and into the atmosphere.

TAKE HOME MESSAGE 4.8

» For proper functioning, cells must acquire food molecules and/or other necessary materials from outside the cell. Similarly, metabolic waste molecules and molecules produced for use elsewhere in the body must move out of the cell. In passive transport—which includes simple and facilitated diffusion and osmosis—the molecular movement occurs spontaneously, without the input of energy. This generally takes place as molecules move down their concentration gradient.

4.9 Osmosis is the passive diffusion of water across a membrane.

Just as solute molecules will passively diffuse down their concentration gradients, water molecules will move from areas of high concentration to areas of low concentration to equalize the concentration of water inside and outside the cell. The diffusion of water across a membrane is a special type of passive transport called **osmosis** (FIGURE 4-18) (page 99). As solute molecules move across the membrane, molecules of water also move across the membrane, equalizing the water concentration inside and outside the cell. Although some water can pass through the lipid bilayer, the hydrophobic region severely limits this flow. Most of the rapid movement of water into and out of cells occurs through "water channels," called *aquaporins,* which are transmembrane proteins with hydrophilic channels through which water molecules pass in single file.

Osmosis can have some dramatic effects on cells. As we saw in Section 4.8, many molecules just can't move across a cell membrane. Water, however, can move into the cell down *its* concentration gradient. As a result, when water diffuses into or out of the cell, the cell will swell or shrink.

This is how osmosis works. In a fluid environment, dissolved substances are located inside as well as outside the cell. The relationship between the concentrations of solutes inside the cell and solutes outside the cell is referred to as **tonicity** (see Figure 4-18).

1. If the concentration of solutes outside the cell is *equal* to the concentration inside the cell, the outside solution is **isotonic.** Water still moves between the solution and the cell, but because it moves at the same rate in both directions, there is no net change in the amount of water inside versus outside the cell.

2. If the concentration of solutes outside the cell is *lower* than the concentration inside the cell, the outside solution is **hypotonic.** In this situation, if the cell membrane is not permeable to the solutes, more water molecules will move into the cell than out of it. The cell will swell.

3. If the concentration of solutes outside the cell is *higher* than the concentration inside the cell, the outside solution is **hypertonic.** In this situation, if the cell membrane is not permeable to the solutes, the concentrated solutes outside the cell will crowd and bump the water molecules and slow their entry into the cell membrane. More water will move out of the cell than into it. The cell will shrivel.

OSMOSIS

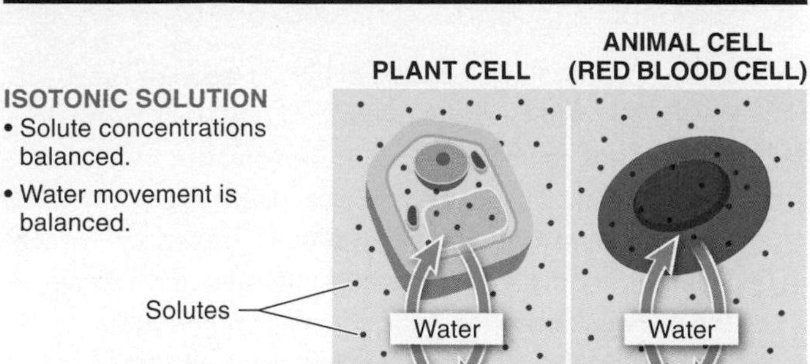
PLANT CELL
ANIMAL CELL (RED BLOOD CELL)
ISOTONIC SOLUTION
• Solute concentrations balanced.
• Water movement is balanced.
Solutes
Water
Water

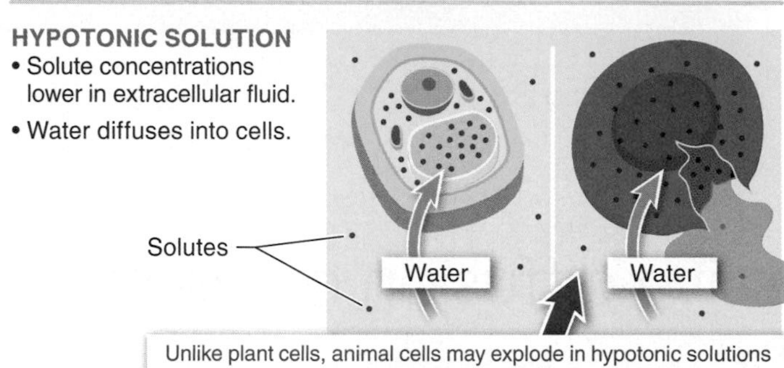
HYPOTONIC SOLUTION
• Solute concentrations lower in extracellular fluid.
• Water diffuses into cells.
Solutes
Water
Water
Unlike plant cells, animal cells may explode in hypotonic solutions because they don't have a cell wall to limit expansion.

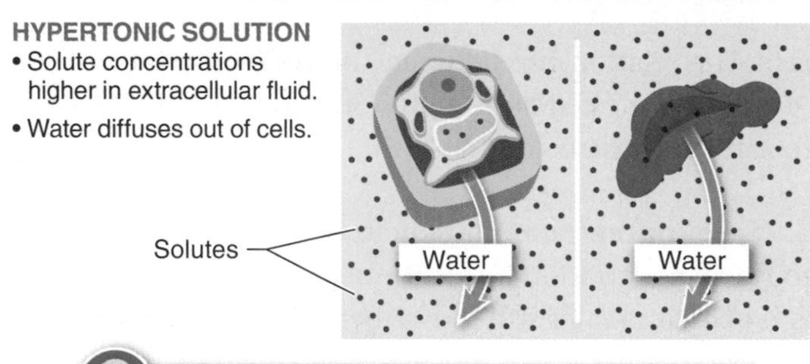
HYPERTONIC SOLUTION
• Solute concentrations higher in extracellular fluid.
• Water diffuses out of cells.
Solutes
Water
Water

Water will always move toward a region having a greater concentration of solutes.

FIGURE 4-18 **Osmosis overview.**

OSMOSIS IN ACTION

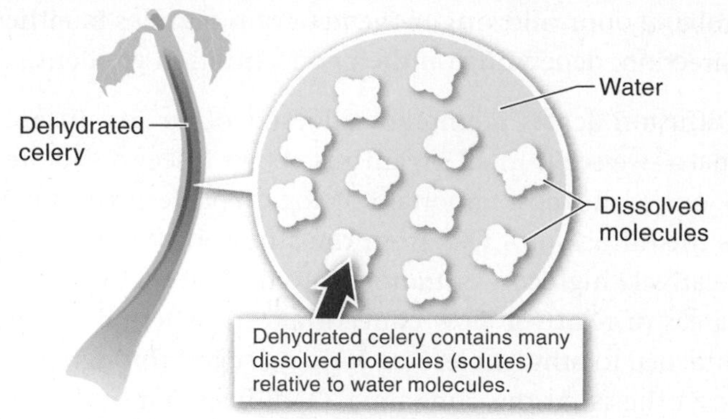
Dehydrated celery
Water
Dissolved molecules
Dehydrated celery contains many dissolved molecules (solutes) relative to water molecules.

WHEN PLACED IN DISTILLED WATER
Distilled water contains fewer dissolved molecules than the celery cells. Water molecules diffuse into the celery, equalizing the water concentration inside and outside the cells. The celery becomes crisp.

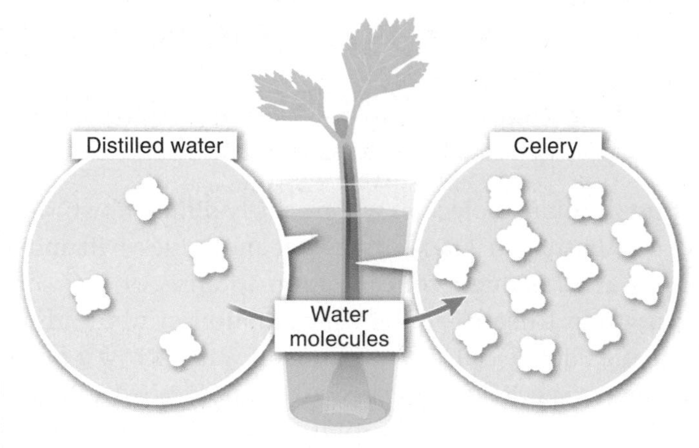
Distilled water
Celery
Water molecules

WHEN PLACED IN SALT WATER
Salt water contains more dissolved molecules than the celery cells. Water molecules diffuse out of the celery. The celery becomes even more shriveled.

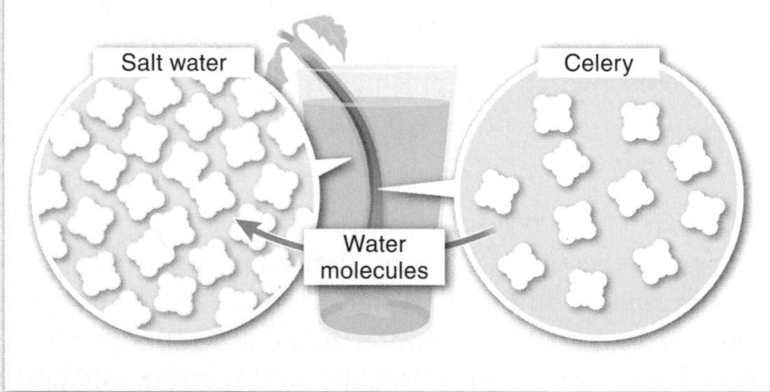
Salt water
Celery
Water molecules

FIGURE 4-19 **Osmosis in your kitchen.**

Q Drinking seawater can be deadly. Why?

You can see osmosis in action in your own kitchen (FIGURE 4-19). Take a stalk of celery and leave it on the counter for a couple of hours. As water evaporates from the cells o f the celery stalk, it will shrink and become limp. You can make it crisp again by placing it in a solution of distilled water. Why distilled water? Because it contains very few dissolved molecules—it is a hypotonic solution. The celery stalk cells, on the other hand, contain many dissolved molecules, such as salt. Because those dissolved molecules can't easily pass across the plasma membrane of the celery stalk cells, the water molecules move down their concentration gradient, from the distilled water into celery cells. Would the celery regain its crispness if you placed it in concentrated salt water? No, because there would be more dissolved solute molecules (and fewer water molecules) in the hypertonic salt solution than inside the celery cells. What little water remained in the celery cells would actually move, by osmosis, out of the celery, causing it to shrivel even more.

Q How do laxatives relieve constipation?

The laxative Milk of Magnesia makes use of osmosis to relieve constipation. This product contains magnesium salts, which are poorly absorbed from the digestive tract. After a dose of the laxative, water moves by osmosis from the surrounding cells into the intestines. The water softens the feces in the intestines and increases the fecal volume, thereby relieving constipation.

It's important to note that the direction of osmosis is determined only by a difference in the *total concentration* of all the molecules dissolved in the water: it does not matter what the solutes are, only how many molecules of solutes there are. To determine which way the water molecules will move, you need to determine the total amount of "dissolved stuff" on either side of the membrane. The water will move toward the side with the greater concentration of solute.

Because of osmosis, even small increases in the salinity of lakes can have disastrous consequences for the aquatic organisms living there, from fish to bacteria. Conversely, putting animal cells—such as red blood cells—in distilled water causes them to explode, because water will diffuse into the cell (which contains more solutes) and the cell will swell and burst. This does not generally happen to plant cells because (as we see later in this chapter) the plant cell plasma membrane is surrounded by a rigid cell wall that limits the amount of cell expansion possible when water moves in by osmosis.

TAKE HOME MESSAGE 4.9

» The diffusion of water across a membrane is a special type of passive transport called osmosis. Water moves from an area with a lower concentration of solutes to an area with a higher concentration of solutes. Water molecules move across the plasma membrane until the concentration of water inside and outside the cell is equalized.

4.10 In active transport, cells use energy to move small molecules into and out of the cell.

Sometimes the transport of molecules into and out of cells needs energy, in which case the process is called **active transport.** Such energy expenditures may be necessary if the molecules or ions to be moved are being moved against their concentration gradient. In all active transport, proteins embedded in the membrane act like motorized revolving doors, pushing molecules across cell membranes regardless of the concentration of those molecules on either side of the membrane. The two distinct types of active transport, primary and secondary, differ only in the source of the fuel that keeps the revolving doors spinning.

The process of digestion provides an example of **primary active transport,** the type of active transport that occurs when energy from ATP is used to fuel the transport of molecules.

To help break down food into more digestible bits, the cells lining your stomach create an acidic environment by pumping large numbers of H^+ ions (also called protons) into the stomach contents, against their concentration gradient (FIGURE 4-20). All of this H^+ pumping increases your ability to digest the food but requires the use of the high-energy molecule ATP—because protons would not normally flow into a region against their concentration gradient. (We explore ATP in more detail in Chapter 5.)

Many transport proteins use an indirect method of fueling their activities rather than using energy released directly from ATP. In the process of **secondary active transport,** the transport protein simultaneously moves one molecule against its concentration gradient while letting a second type of molecule flow down its concentration gradient. Although no ATP is used directly in this process, at some

ACTIVE TRANSPORT

Active transport occurs when the movement of molecules into and out of a cell requires the input of energy.

In secondary active transport, the transport protein moves one molecule against its concentration gradient, powered by the simultaneous flow of another molecule down its concentration gradient.

Active transport in the stomach increases your ability to digest food.

FIGURE 4-20 **With an input of energy, molecules can be moved against concentration gradients.**

other time and in some other location, energy from ATP was used to pump the second type of molecule against its concentration gradient. The process is like using energy to pump water to the top of a high water tower. Later, the water can be allowed to run out of the tower over a water wheel, which can, in turn, power a process such as grinding wheat into flour. Our bodies frequently use the energy from one reaction that occurs spontaneously to fuel another reaction that requires energy.

TAKE HOME MESSAGE 4.10

» In active transport, movement of molecules across a membrane requires energy. Active transport is necessary if the molecules to be moved are very large or if they are being moved against their concentration gradient. Proteins embedded in the plasma membrane act like motorized revolving doors to actively transport (pump) the molecules.

4.11 Endocytosis and exocytosis are used for bulk transport of particles.

Many substances are just too big to get into or out of a cell by passive or active transport. To absorb large particles, such as bacterial invaders, cells engulf them with their plasma membrane in a process called **endocytosis.** To export large particles, such as digestive enzymes manufactured for use elsewhere in the body, cells often use the process of **exocytosis.**

There are three types of endocytosis: phagocytosis, pinocytosis, and receptor-mediated endocytosis. In all three types of endocytosis, the plasma membrane oozes around an object outside the cell—surrounding the object—and forms a little pocket-like sac called a *vesicle*. The plasma membrane then pinches off the vesicle so that the vesicle is inside the cell but separated from the rest of the cell contents.

Phagocytosis and Pinocytosis Relatively large particles are engulfed in a process called **phagocytosis** (FIGURE 4-21). Amoebas and other unicellular protists, as well as white blood cells, use phagocytosis to consume entire organisms for defense or food. Whereas "phagocytosis" comes from the Greek for "eat" and "container," the term **pinocytosis** comes from the Greek for "drink" and "container," and describes the process of cells taking in dissolved particles and liquids. The two processes are largely the same, except that the vesicles formed during pinocytosis are generally much smaller than those formed during phagocytosis.

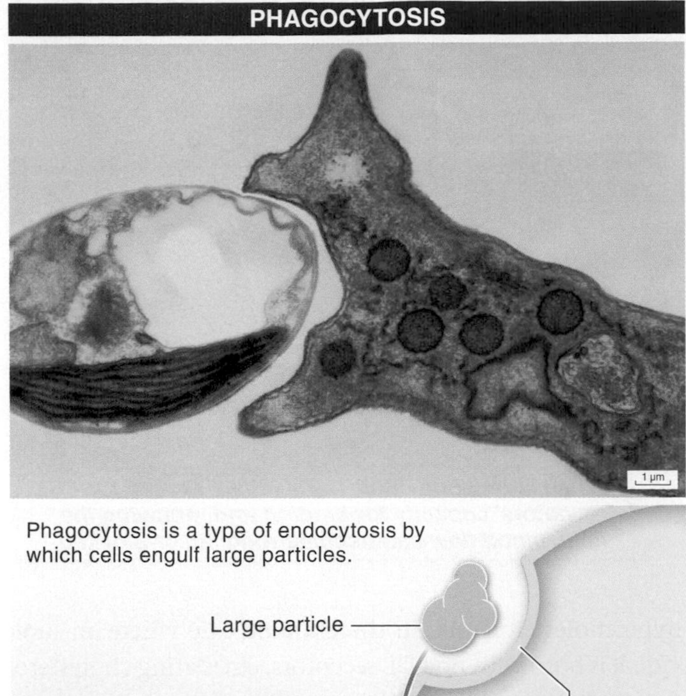

PHAGOCYTOSIS
1 µm
Phagocytosis is a type of endocytosis by which cells engulf large particles.
Large particle
Plasma membrane
1 The plasma membrane forms a pocket-like enclosure around a large particle.
Vesicle
2 The particle is transported into the cell in a vesicle.
Extracellular fluid
Intracellular fluid

FIGURE 4-21 **Through phagocytosis, amoebas and other unicellular protists, as well as white blood cells, consume other organisms for food or for defense.**

Receptor-Mediated Endocytosis

The third type of endocytosis, **receptor-mediated endocytosis,** is much more specific than either phagocytosis or pinocytosis. Receptor molecules on the surface of a cell recognize and bind one specific type of molecule. For one receptor it might be insulin, for another it might be cholesterol. When the appropriate molecule binds to each of the receptor proteins, the membrane begins to fold inward, first forming a little pit and then completely engulfing the molecules, which are still attached to their receptors.

One of the most important examples of receptor-mediated endocytosis involves cholesterol (FIGURE 4-22). Most cholesterol that circulates in the bloodstream is in the form of particles called low-density lipoproteins, or LDL. Each molecule of LDL is a cholesterol globule coated by phospholipids. Receptor proteins built into the plasma membranes of liver cells recognize and bind the proteins embedded within the LDL's phospholipid coat. Once the LDL is bound to the receptors, the cell consumes the LDL molecule by endocytosis, and the cholesterol is broken down and used to make a variety of other useful molecules, such as the hormones estrogen and testosterone.

Q Faulty cell membranes are a primary cause of cardiovascular disease. What modification to these membranes might be an effective treatment?

Circulating cholesterol collects on the walls of arteries, reducing blood flow and causing the artery to harden. Too much circulating cholesterol in LDL molecules can lead to cardiovascular disease and death (FIGURE 4-23). Individuals lucky enough to have large numbers of LDL receptors on the plasma membranes of their liver cells have a significantly lower risk of cardiovascular disease.

Conversely, some individuals have the misfortune of inheriting genes that code for faulty liver cell membranes that have few LDL receptors—a disorder called familial

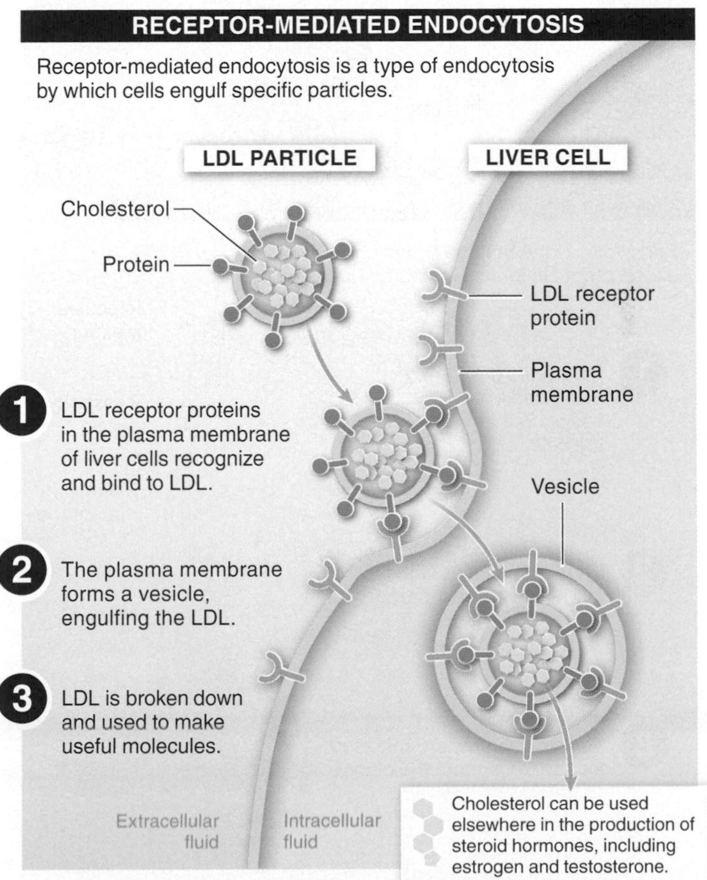

FIGURE 4-22 **Receptor proteins on liver cells bind cholesterol, helping remove it from the bloodstream.**

HEART DISEASE

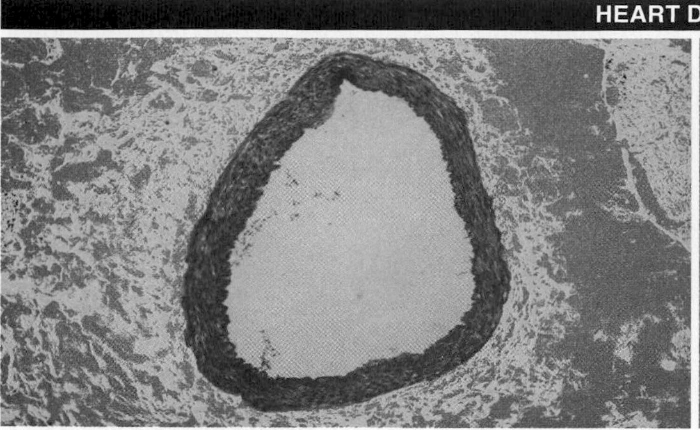

Healthy artery

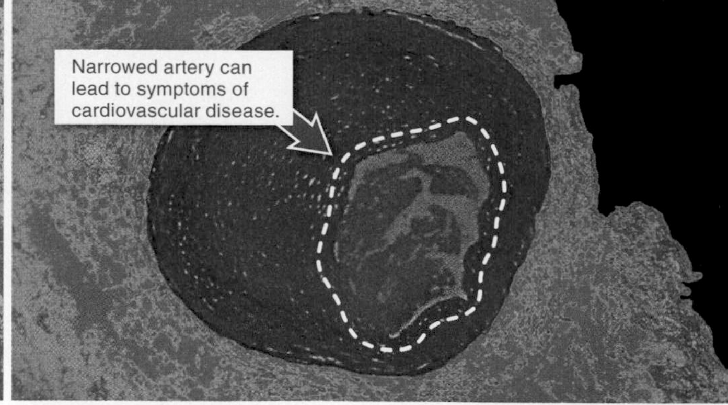

Narrowed artery can lead to symptoms of cardiovascular disease.

Blocked artery

FIGURE 4-23 Comparison of a healthy artery and an artery choked by the buildup of cholesterol.

When circulating cholesterol levels exceed the LDL receptors' capacity for binding and removing the lipid, blood flow can become reduced or blocked.

hypercholesterolemia. In the extreme case where an individual is born with no LDL receptors, circulating cholesterol accumulates in the arteries so rapidly that cardiovascular disease begins to develop even before puberty, and death from a heart attack can occur before the age of 30.

Exocytosis To shuttle molecules out of the cell, particles within the cell are first enclosed in a vesicle. The vesicle then moves through the cytoplasm to the plasma membrane, where the membrane of the vesicle merges with the plasma membrane, and the material in the vesicle is expelled from the cell. The hormone insulin, for example, is transported out of pancreatic cells in this way (FIGURE 4-24).

Exocytosis is not restricted to large molecules. In the brain and other parts of the nervous system, for example, communication between cells occurs as one cell releases large numbers of very small molecules, called neurotransmitters, by exocytosis.

EXOCYTOSIS

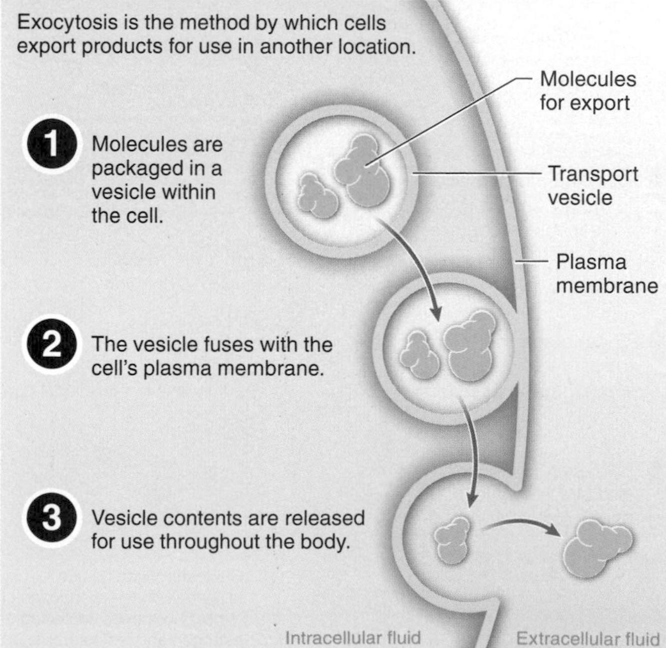

The white and red vesicles in this pancreas cell contain hormones—including insulin—for secretion into the bloodstream.
1 μm
Exocytosis is the method by which cells export products for use in another location.
Molecules for export
Transport vesicle
Plasma membrane
1 Molecules are packaged in a vesicle within the cell.
2 The vesicle fuses with the cell's plasma membrane.
3 Vesicle contents are released for use throughout the body.
Intracellular fluid
Extracellular fluid

FIGURE 4-24 Exocytosis moves molecules out of the cell. This is how insulin is exported from the cells where it is synthesized.

TAKE HOME MESSAGE 4.11

» When materials cannot get into a cell by diffusion or through a pump (for example, when the molecules are too big), cells can engulf the molecules or particles with their plasma membrane in a process called endocytosis. Similarly, molecules can be moved out of a cell by exocytosis. In endocytosis, the plasma membrane moves to surround the molecules or particles and forms a little vesicle that is pinched off inside the cell; in exocytosis, a vesicle inside the cell fuses with the plasma membrane and dumps its contents outside the cell.

4.12 Cells are connected and communicate with each other.

Cells reach out and connect; no telephone lines required.

4.12 Connections between cells hold them in place and enable them to communicate with each other.

So far we have examined the cell as if it were a free-living and independent entity. The majority of cells in any multicellular organism, however, are connected to other cells. We examine three primary types of connections between animal cells: (1) tight junctions, (2) desmosomes, and (3) gap junctions (FIGURE 4-25).

Tight junctions form continuous, water-tight seals around cells and anchor cells in place. Much like the caulking around a tub or sink that keeps water from leaking into the surrounding walls, tight junctions prevent fluid flow between cells. Tight junctions are particularly important in the small intestine.

THREE PRIMARY CONNECTIONS BETWEEN ANIMAL CELLS

TIGHT JUNCTIONS
Form a water-tight seal between cells, like caulking around a tub

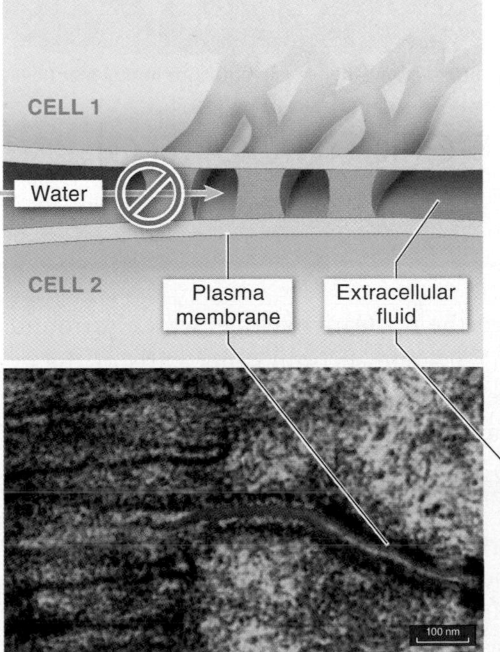

Tight junction between small intestine epithelial cells

DESMOSOMES
Act like Velcro and fasten cells together

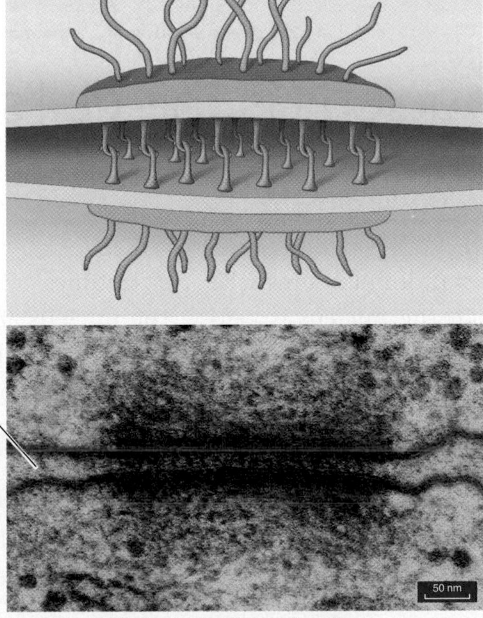

Desmosome between heart muscle cells

GAP JUNCTIONS
Act like secret passageways and allow materials to pass between cells

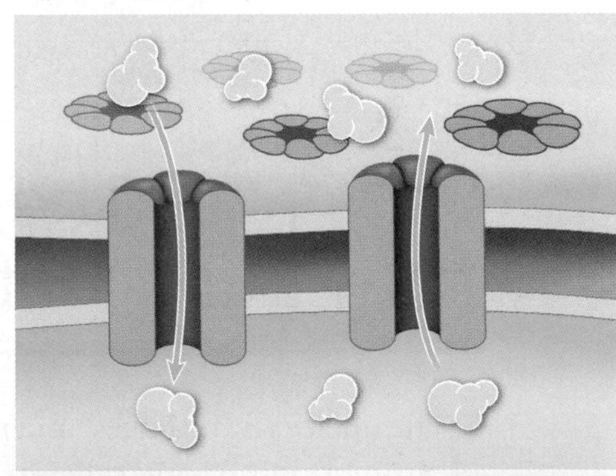

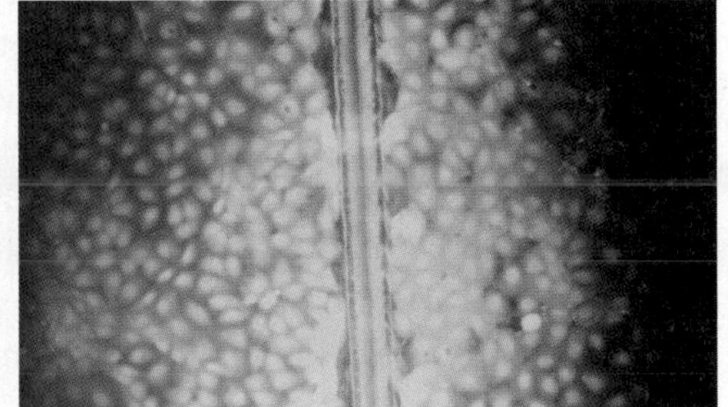

Fluorescent dye moving between cells via gap junctions

FIGURE 4-25 **Cell connections: tight junctions, desmosomes, and gap junctions.**

Desmosomes are like spot welds or rivets that fasten cells together into strong sheets. They occur at irregular intervals and function like fastened Velcro: they hold cells together but are not water-tight, allowing fluid to pass around them. Desmosomes and similar junctions are found in much of the tissue that lines the cavities of animals' bodies. They also are found in muscle tissue, holding fibers together.

Finally, **gap junctions** are pores surrounded by special proteins that form open channels between two cells (see Figure 4-25). Functioning like secret passageways, these junctions are an important mechanism for cell-to-cell communication. In the heart, for example, the electrical signal telling muscle cells to contract is passed from cell to cell through gap junctions.

Compared with normal cells, cancer cells have fewer gap junctions. And research suggests that the resulting reduction in intercellular communication among cancer cells may be important in the formation of masses of cells, called tumors, and the spread of cancer cells throughout the body. (Interestingly, treatments that cause an increase in the number of gap junctions tend to reduce tumor growth.)

Q Is a breakdown of cell-to-cell communication related to cancer?

TAKE HOME MESSAGE 4.12

» In multicellular organisms, most cells are connected to other cells. The connections can form a water-tight seal between the cells (tight junctions), can hold sheets of cells together while allowing fluid to pass between neighboring cells (desmosomes), or can function like passageways, allowing the movement of cytoplasm, molecules, and other signals between cells (gap junctions).

4.13–4.22

Nine important landmarks distinguish eukaryotic cells.

Cell of the voodoo lily. How many different organelles can you spot in this plant cell?

4.13 The nucleus is the cell's genetic control center.

The nucleus is the largest and most prominent organelle in most eukaryotic cells. In fact, the nucleus is generally larger than any prokaryotic cell. If a cell were the size of a large lecture hall or movie theater, the nucleus would be the size of an 18-wheel semi-trailer truck parked in the front rows. The **nucleus** is the genetic control center, directing most cellular activities by controlling which molecules are produced and in what quantity, and is the storehouse for hereditary information (FIGURE 4-26).

Three important structural components stand out in the nucleus (FIGURE 4-27). First is the **nuclear membrane,** sometimes called the nuclear envelope, which surrounds the nucleus and separates it from other parts of the cytoplasm. Unlike most plasma membranes, however, the nuclear membrane consists of two bilayers, one on top of the other, much like the double-bagging of groceries at the market. The nuclear membrane is a very leaky bag, though. It is perforated, covered with tiny pores that enable large molecules to pass between the nucleus and the cytosol. Made from multiple proteins embedded in the phospholipid membranes and spanning both bilayers, these pores also permit free diffusion of small molecules, such as water, sugars, and ions.

FIGURE 4-26 **The nucleus holds the genetic information that makes it possible to build organisms.** That's why identical twins, carrying the same genetic information, look so similar.

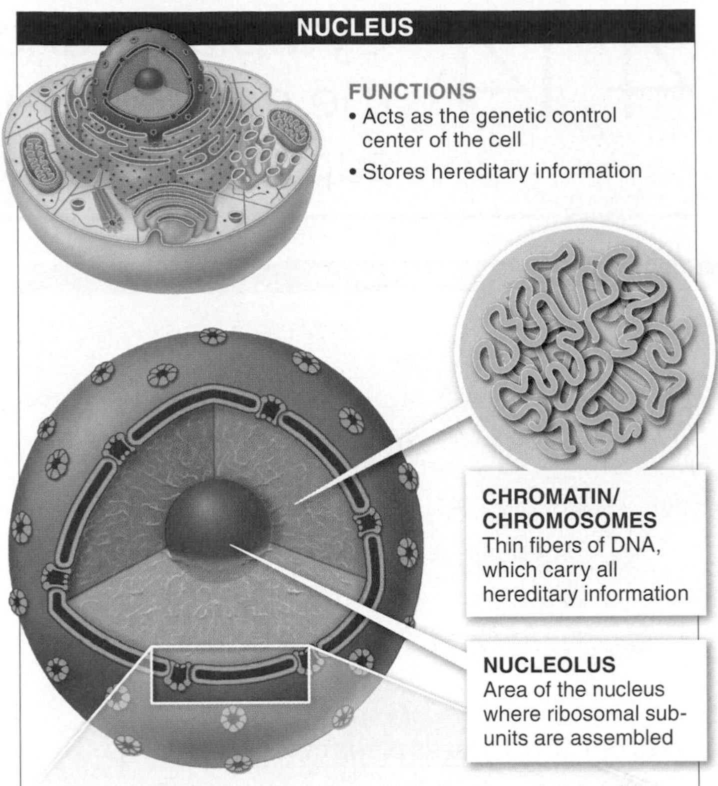

NUCLEUS
FUNCTIONS
• Acts as the genetic control center of the cell
• Stores hereditary information
CHROMATIN/ CHROMOSOMES
Thin fibers of DNA, which carry all hereditary information
NUCLEOLUS
Area of the nucleus where ribosomal subunits are assembled

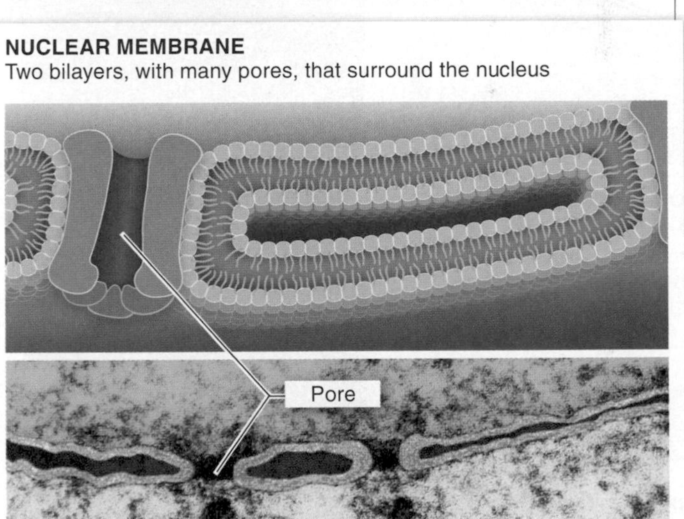

FIGURE 4-27 **The nucleus: the cell's genetic control center.**

The nuclear pore protein complexes regulate the transport of many molecules across the nuclear envelope in both directions. Two types of transport are particularly important. The nuclear membrane allows proteins that interact with DNA and catalyze nuclear activities to move from the cytosol into the nucleus. They also facilitate the movement of RNA and RNA-protein complexes from the nucleus to the surrounding cytosol.

The second prominent structure in the nucleus is the **chromatin,** a mass of long, thin fibers consisting of DNA with some proteins attached that keep the DNA from getting impossibly tangled. Most of the time, as the DNA directs cellular activities, the chromatin resembles a plate of spaghetti. When it's time for cell division (a process described in detail in Chapter 8), the chromatin coils and the threads become shorter and thicker until they become visible as chromosomes, the compacted, linear DNA molecules that carry hereditary information.

A third structure in the nucleus is the **nucleolus,** an area near the center of the nucleus where subunits of the ribosomes, a critical part of the cellular machinery, are assembled. Ribosomes are like little factories in which the information stored in the DNA is used to construct proteins. The ribosomes are built in the nucleolus but pass through the nuclear pores and into the cytosol before starting their protein-production work.

TAKE HOME MESSAGE 4.13

» The nucleus is usually the largest and most prominent organelle in the eukaryotic cell. It directs most cellular activities by controlling which molecules are produced and in what quantity. The nucleus is the storehouse for hereditary information.

4.14 Cytoplasm and the cytoskeleton form the cell's internal environment, provide support, and can generate motion.

If you imagine yourself inside a cell the size of a big lecture hall, it might come as a surprise that you can barely see that big rig of a nucleus parked in the front rows. Visibility is almost zero, not only because the room is filled with jelly-like cytoplasm, but also because there is a dense web of thick and thin, straight and branched ropes, strings, and scaffolding, running every which way throughout the room.

This inner scaffolding of the cell, which is made from proteins, is the **cytoskeleton** (FIGURE 4-28). It has three chief purposes. First, it gives animal cells shape and support—making red blood cells look like little round doughnuts (without the hole in the middle) and giving neurons their very long, thread-like appearance. Plant cells are shaped primarily by their cell wall (a structure we discuss later in the chapter), but they also have a cytoskeleton. Second, the cytoskeleton controls the intracellular traffic, serving as a series of tracks on which a variety of organelles and molecules are guided across and around the inside of the cell. And third, because the elaborate scaffolding of the cytoskeleton is dynamic and can generate force, it gives all cells some ability to control their movement.

Three types of protein fibers make up the cytoskeleton. **Microtubules,** the thickest, are linear fibers made from repeating units of protein, and look like rigid, hollow tubes. Like intracellular conveyor belts, they are the tracks to which molecules and organelles within the cell can become attached and moved along. Microtubules also help to pull chromosomes apart during cell division. Continuously built, disassembled, and rebuilt, microtubules rarely last more than about 10 minutes in a cell. **Intermediate filaments,** a second type of cytoskeleton fiber, are durable, rope-like systems of numerous different overlapping proteins. They give cells great strength. **Microfilaments** are the thinnest elements in the cytoskeleton. Long, solid, rod-like fibers, microfilaments help generate forces, including those important in cell contraction and cell division.

CYTOSKELETON

FUNCTIONS
- Acts as the inner scaffolding of the cell
- Provides shape and support
- Controls intracellular traffic flow
- Enables movement

THREE TYPES OF PROTEIN FIBERS IN THE CYTOSKELETON

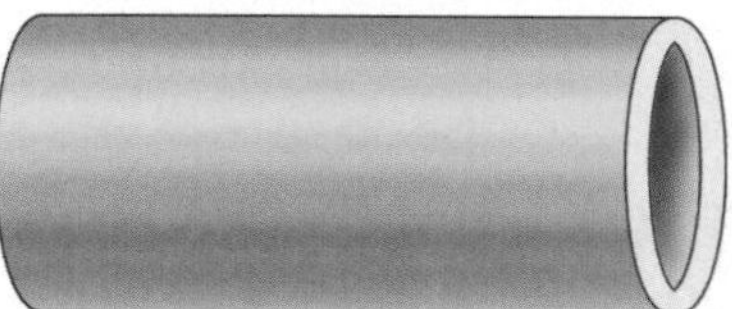

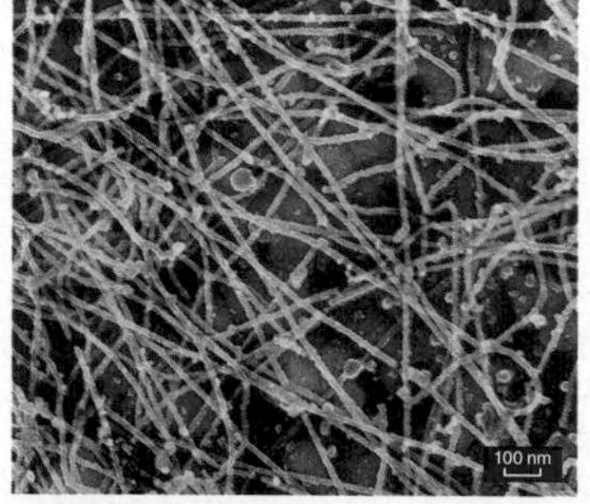

MICROTUBULES
- Thick, hollow tubes
- The tracks to which molecules and organelles within the cell may attach and be moved along

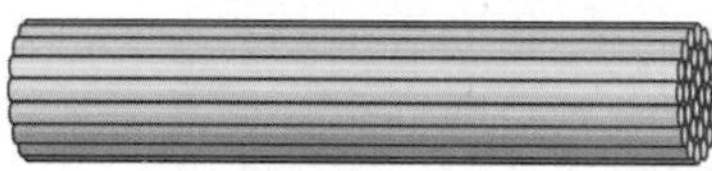

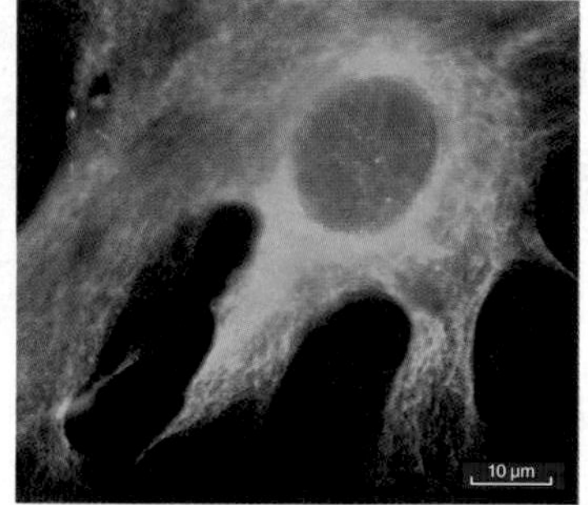

INTERMEDIATE FILAMENTS
- Durable, rope-like systems of numerous overlapping proteins
- Give cells great strength

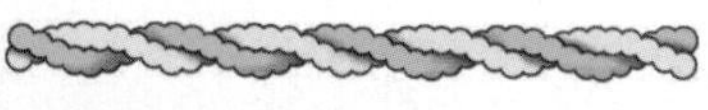

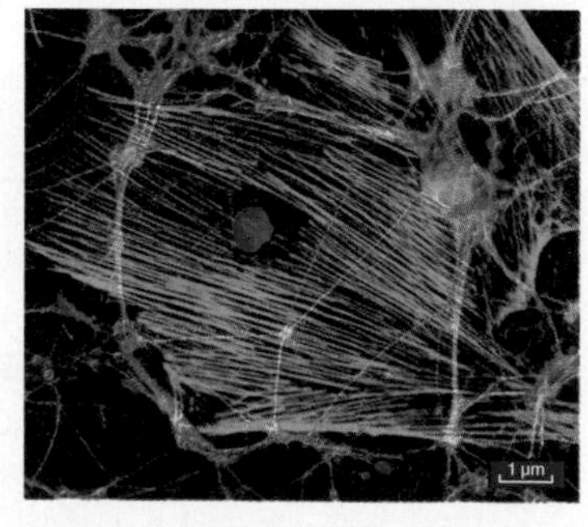

MICROFILAMENTS
- Long, solid, rod-like fibers
- Help with cell contraction and cell division

FIGURE 4-28 **The cytoskeleton: the cell's inner scaffolding.**

CELLULAR MOVEMENT

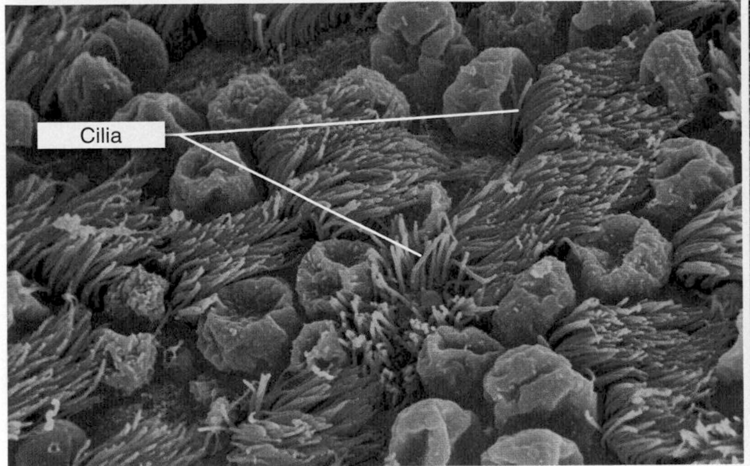

Within the lining of a human oviduct, cilia (orange) propel the egg to the uterus, while secretory cells (purple) nourish it.

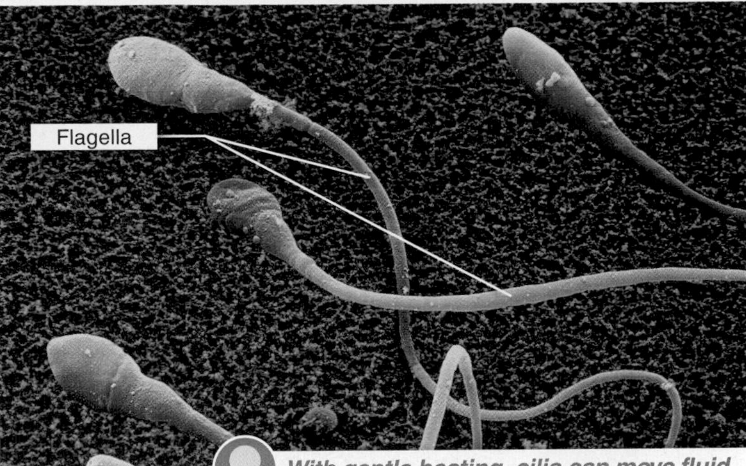

Human sperm cells

With gentle beating, cilia can move fluid past cells, whereas flagella, with whip-like motions, can move the cells themselves.

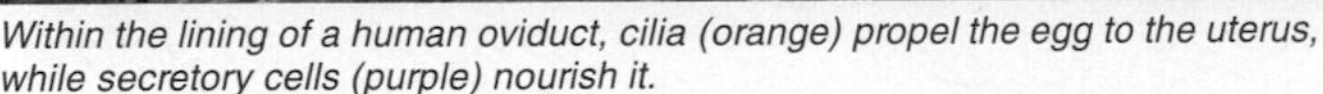

FIGURE 4-29 Cilia and flagella assist the cell with movement.

A couple of microtubule-based structures are sometimes present in cells and can help to move the cell through its environment (or, in stationary cells, can help move the environment past the cell). **Cilia** (*sing.* **cilium**) are short projections often found in large numbers on a single cell (FIGURE 4-29). Cilia beat swiftly, often in unison and in ways that resemble blades of grass in a field, blowing in the wind. Cilia can move fluid along and past a cell. This movement can accomplish many important tasks, including sweeping the airways to our lungs to clear them of debris (such as dust). Centrioles, found in animal cells, are made from bundles of microtubules and play a role in cell division.

Flagella are much longer than cilia. They occur in many prokaryotes and single-celled eukaryotes—although the flagella of prokaryotes differ structurally from those of eukaryotes. While many algae and plants have cells with one or more flagella, in animals, cells with one or more flagella are very rare. One of these cell types, however, has a critical role in every animal species: sperm cells. With a flagellum for a tail, sperm are among the most mobile of all animal cells. Some spermicidal birth control methods prevent conception by disabling the flagellum and immobilizing the sperm cells.

TAKE HOME MESSAGE 4.14

» The inner scaffolding of the cell, which is made from proteins, is the cytoskeleton. Consisting of three types of protein fibers—microtubules, intermediate filaments, and microfilaments—the cytoskeleton gives animal cells their shape and support, gives cells some ability to control their movement, and serves as a series of tracks on which organelles and molecules are guided across and around the inside of the cell.

4.15 Mitochondria are the cell's energy converters.

Cars generally run on a single type of fuel, and it is exactly the same fuel every time. With human bodies it's a different story. During some meals we put in meat and potatoes. Other times, fruit or bread or vegetables, popcorn or gummy bears. Yet our bodies can utilize the energy in these various foods to power all the reactions that make it

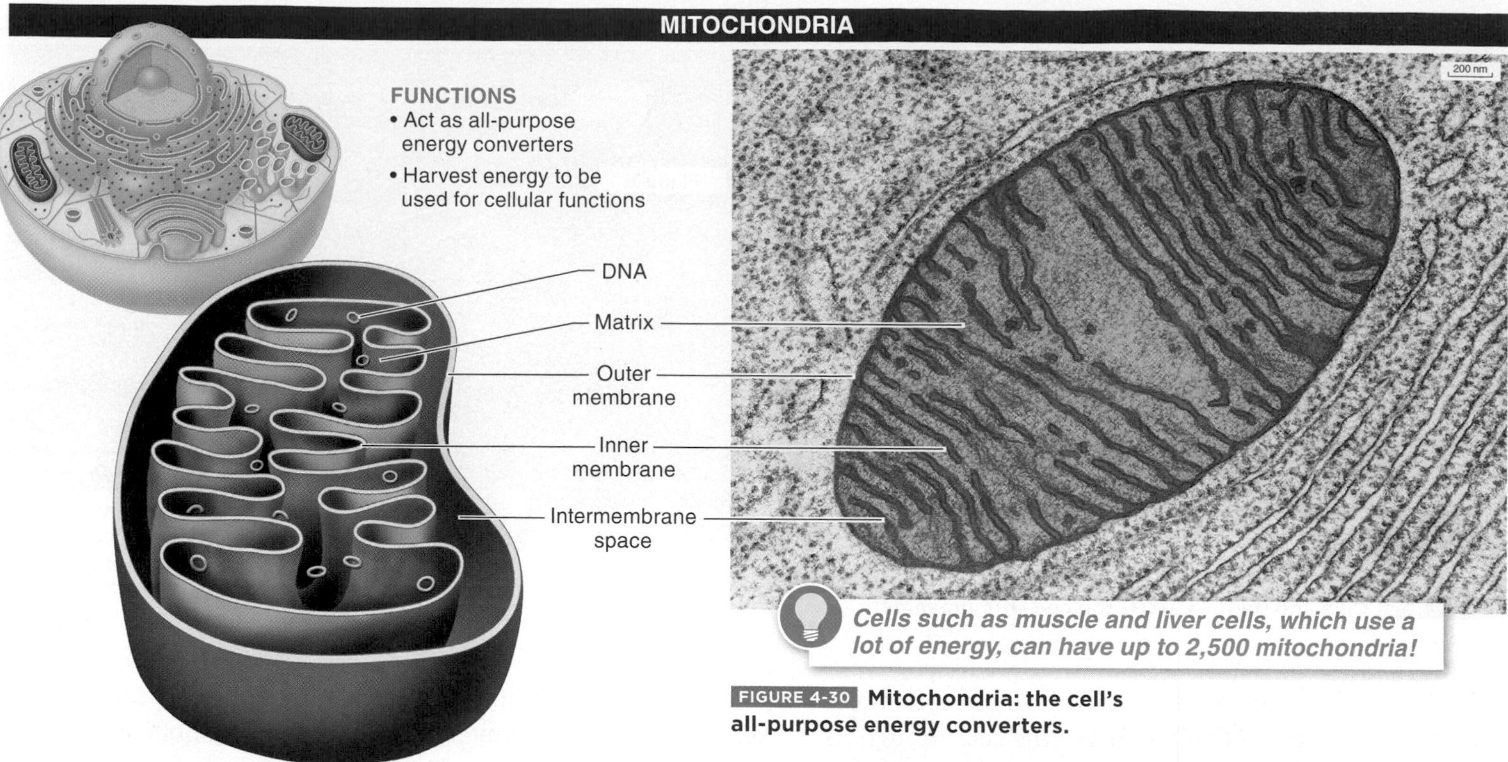

MITOCHONDRIA
FUNCTIONS
• Act as all-purpose energy converters
• Harvest energy to be used for cellular functions
DNA
Matrix
Outer membrane
Inner membrane
Intermembrane space
200 nm
Cells such as muscle and liver cells, which use a lot of energy, can have up to 2,500 mitochondria!

FIGURE 4-30 **Mitochondria: the cell's all-purpose energy converters.**

possible for us to breathe, move, and think. The mitochondria are the organelles that make this possible.

Mitochondria (*sing.* **mitochondrion**) are all-purpose energy converters that are present in nearly all plant cells, animal cells, and every other eukaryotic cell (FIGURE 4-30). Our mitochondria convert the energy contained in the chemical bonds of carbohydrates, fats, and proteins into carbon dioxide, water, and ATP. Cells use high-energy ATP molecules to fuel all their functions and activities. (ATP and how it works are described in detail in Chapter 5.) Because this energy conversion requires a significant amount of oxygen, mitochondria consume most of the oxygen used by each cell. In humans, for example, our mitochondria consume as much as 80% of the oxygen we breathe. Mitochondria give a significant return on this investment by producing about 90% of the energy our cells need to function.

Cells that are not very metabolically active, such as some fat storage cells in humans, have very few mitochondria. Cells that have large energy requirements, such as muscle, liver, and sperm cells in animals and fast-growing root cells in plants, are packed densely with mitochondria (FIGURE 4-31). Some of these cells have as many as 2,500 mitochondria! (See Section 4.16 for a description of an experimental approach to documenting how a cell's composition can change when the cell's function must change.)

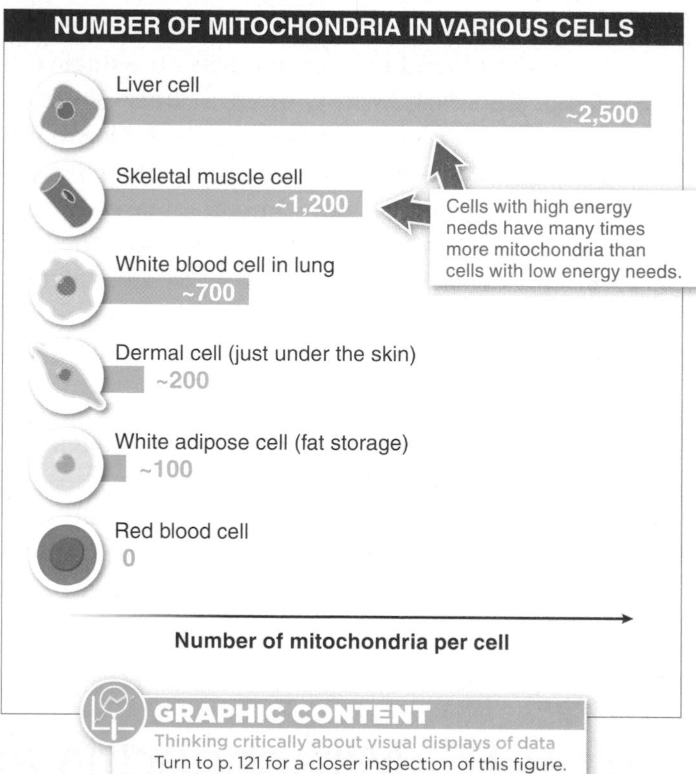

Dermal cell (just under the skin)

White adipose cell (fat storage)

Cells with high energy needs have many times more mitochondria than cells with low energy needs.

FIGURE 4-31 **How does the number of mitochondria vary among different types of cells?**

To visualize the structure of a mitochondrion, imagine a plastic sandwich bag. Now take another, bigger plastic bag and stuff it inside the sandwich bag. There is a smooth outer membrane and a scrunched-up inner membrane. This construction forms two separate compartments within the mitochondrion: a region outside the inner bag (called the **intermembrane space**) and another region, called the **mitochondrial matrix,** inside the inner bag. This bag-within-a-bag structure has important implications for energy conversion. And having a large, heavily folded inner membrane provides a huge amount of surface area on which to conduct chemical reactions. (We discuss the details of mitochondrial energy conversions and the role of mitochondrial structure in Chapter 5.)

As we learned in our earlier discussion of endosymbiosis (see Figure 4-8), mitochondria may very well have existed, billions of years ago, as separate, single-celled, bacteria-like organisms. Perhaps the strongest evidence for this is that mitochondria have their own DNA (see Figure 4-30). Anywhere from 2 to 10 copies of its own little ring-shaped DNA are mixed in among the approximately 3,000 proteins in each mitochondrion. This DNA carries the instructions for making 13 mitochondrial proteins essential for metabolism and energy production.

We're always taught that our mothers and fathers contribute equally to our genetic composition, but this isn't quite true. The mitochondria in every one of your cells (and the DNA that comes with them) come from the mitochondria that were initially present in your mother's egg, which, when fertilized by your father's sperm, developed into you. In other words, all of your mitochondria are descended from your mother's mitochondria. The tiny sperm contributes DNA, but no cytoplasm and, hence, no mitochondria. Consequently, mitochondrial DNA is something that we inherit exclusively from our mothers. This is true not only in humans but in most multicellular eukaryotes.

Q Who is the bigger contributor. Mom or dad? Why?

Given the central role of mitochondria in converting the energy in food molecules into a form that is usable by cells, you won't be surprised to learn that mitochondrial malfunctions can have serious consequences. Recent research has focused on possible links between defective mitochondria and diseases characterized by fatigue and muscle pain. It seems, for instance, that many cases of "exercise intolerance"—extreme fatigue or cramps after only slight exertion—may be related to defective mitochondrial DNA.

TAKE HOME MESSAGE 4.15

» In mitochondria, which are found in nearly all eukaryotic cells, the energy contained in the chemical bonds of carbohydrate, fat, and protein molecules is converted into carbon dioxide, water, and ATP, the energy source for all cellular functions and activities. Mitochondria may have evolutionary origins as symbiotic bacteria living inside other cells.

THIS IS HOW WE DO IT

Developing the ability to apply the process of science

4.16 Can cells change their composition to adapt to their environment?

Figuring out how to approach a question in biology can be one of the most important steps in trying to understand how things work. In most cases, the best strategy is to take the simplest approach. Researchers wondered, for example, how an animal's cells might respond to exposure to a much colder environment. Here's what they did.

(continued on following page)

They used 15 cats from five litters. From each litter, one cat served as a control, while the others were exposed to extreme cold—though within the range experienced by feral cats—for two periods of 1 hour per day. The rest of the time, all of the cats were kept at 20° C.

Why did the researchers use multiple cats from each litter?

After one week of this treatment, the team collected samples of fat tissue from all of the cats, which they examined microscopically. In the fat tissue, they measured the number and size of mitochondria, the number of tiny blood vessels (capillaries) per cell, and the size of the cells.

What was the purpose of counting the capillaries?

The results were clear and dramatic, as shown here (note that a micrometer, μm, also called a micron, is one-millionth of a meter).

	Control	Cold-stressed
Number of mitochondria (per μm^3)	1.48 ± 0.11	1.74 ± 0.10
Size of mitochondria (μm^3)	0.13 ± 0.04	0.48 ± 0.13
Number of capillaries (per cell)	0.34 ± 0.12	0.71 ± 0.2
Cell diameter (μm)	75 ± 2	18 ± 2

The experimental approach was extremely simple. The researchers used a single manipulation—exposure to cold—and then observed the consequences. This made it straightforward for drawing conclusions.
Cold stress caused a change in fat cells. In the animals exposed to cold, the mitochondria became more numerous, with each mitochondrion tripling in size. The number of blood vessels doubled. And the fat cell size was reduced to less than a quarter of the size found in the cells of the control group animals.

Why do you think the fat cells in the cats exposed to cold shrank so much?

Essentially, it appears that cells in the fat tissue of cold-exposed animals upgraded their heat-generating system. And by making use of stored lipids to fuel the intensive heat production, they depleted a significant portion of their energy reserves.

What change in the study would increase your confidence in the conclusions?

TAKE HOME MESSAGE 4.16

» Form follows function in an organism's cells and reflects their environment. When cells must perform intensive heat production, for example, they significantly increase the number and size of their mitochondria. They also increase the blood supply to the tissue and make use of existing stores of energy.

4.17 Lysosomes are the cell's garbage disposals.

Garbage. What does a cell do with all the garbage it generates? Mitochondria, for example, wear out after about 10 days of intensive activity. And white blood cells constantly track down and consume bacterial invaders, which they then have to dispose of. Similarly, the ongoing reactions of cellular metabolism produce many

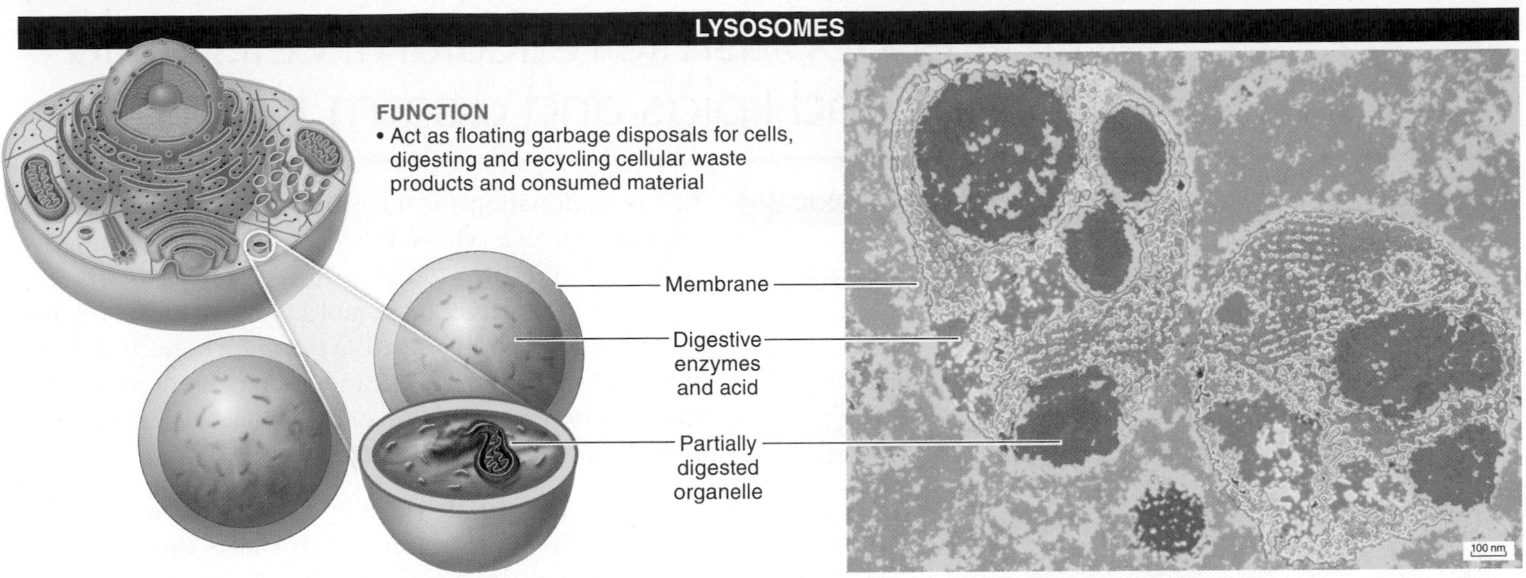

FUNCTION
• Act as floating garbage disposals for cells, digesting and recycling cellular waste products and consumed material
Membrane
Digestive enzymes and acid
Partially digested organelle
100 nm

FIGURE 4-32 **Lysosomes: digestion and recycling of the cell's waste products.**

waste macromolecules that cells must digest and recycle. To deal with this garbage, many eukaryotic cells maintain hundreds of versatile floating "garbage disposals" called lysosomes.

Lysosomes are round, membrane-enclosed, acid-filled vesicles that dispose of garbage (FIGURE 4-32). They are filled with about 50 different digestive enzymes and a super-acidic fluid, a corrosive broth so powerful that a burst lysosome would rapidly kill the cell by digesting many of its component parts. The broad spectrum of enzymes in the lysosome dismantle macromolecules that are no longer needed by the cell or are generated as by-products of cellular metabolism.

When a cell consumes a particle of food or even an invading bacterium via phagocytosis, the cell directs the material to lysosomes for dismantling. After this dismantling, the cell releases most of the component parts of molecules, such as the amino acids from proteins, into its cytoplasm, where they can be used by the cell as raw materials.

Some immune system cells tend to have particularly large numbers of lysosomes, most likely because of their great need for disposing of the by-products of disease-causing bacteria.

With 50 different enzymes necessary for lysosomes to function, sometimes problems occur. In a common genetic disorder called Tay-Sachs disease, a child inherits an inability to produce a critical lipid-digesting enzyme. Even though the lysosomes cannot digest certain lipids, the cells continue to send lipids to the lysosomes, where they accumulate, undigested. The lysosome swells until it bursts and digests the whole cell or chokes the cell to death. This process occurs in large numbers of cells within the first few years of life, and eventually leads to death.

Although the matter has long been debated among biologists, it does seem that plant cells also contain lysosomes, compartments with similar digestive broths that carry out the same digestive processes as in animals.

TAKE HOME MESSAGE 4.17

» Lysosomes are round, membrane-enclosed, acid-filled organelles that function as a cell's garbage disposals. They are filled with about 50 different digestive enzymes and enable a cell to dismantle macromolecules, including disease-causing bacteria.

4.18 In the endoplasmic reticulum, cells build proteins and lipids and disarm toxins.

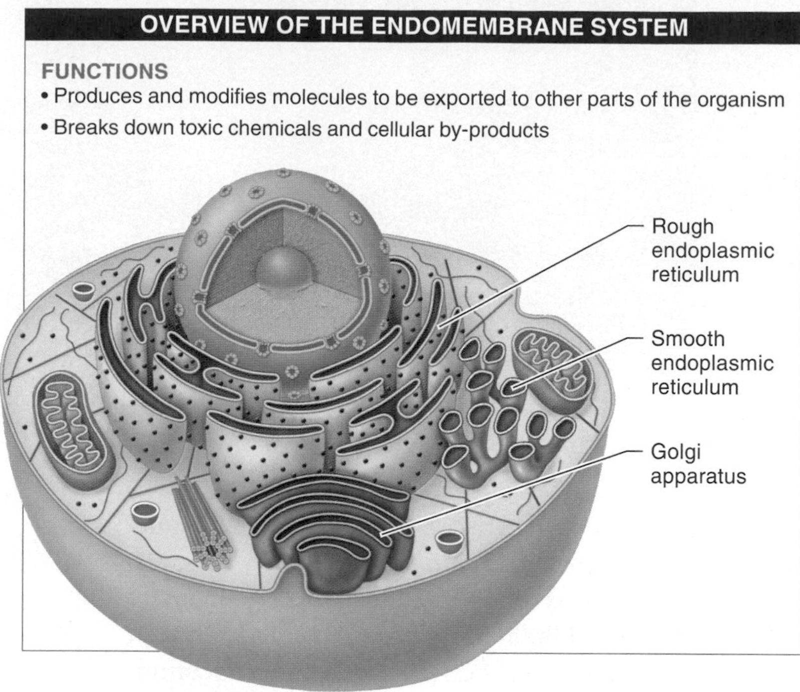

OVERVIEW OF THE ENDOMEMBRANE SYSTEM
FUNCTIONS
• Produces and modifies molecules to be exported to other parts of the organism
• Breaks down toxic chemicals and cellular by-products
Rough endoplasmic reticulum
Smooth endoplasmic reticulum
Golgi apparatus

FIGURE 4-33 The endomembrane system: the rough endoplasmic reticulum, smooth endoplasmic reticulum, and Golgi apparatus.

The information for how to construct the molecules essential to a cell's survival and smooth functioning is stored in the DNA found in the cell's nucleus. The energy used to construct these molecules and to run cellular functions comes primarily from the mitochondria. The actual production and modification of biological molecules, however, occurs in a system of organelles called the **endomembrane system** (FIGURE 4-33).

This mass of interrelated membranes spreads out from and surrounds the nucleus, forming chambers within the cell that contain their own mixtures of chemicals. The endomembrane system occupies as much as one-fifth of the cell's volume and is responsible for many of the fundamental functions of the cell.

Rough Endoplasmic Reticulum Perhaps the organelle with the most cumbersome name, the **rough endoplasmic reticulum,** or **rough ER,** is a large series of interconnected, flattened sacs that look like a stack of pancakes. These sacs are connected directly to the nuclear envelope. In most eukaryotic cells, the rough ER almost completely surrounds the nucleus (FIGURE 4-34). Its "rough"

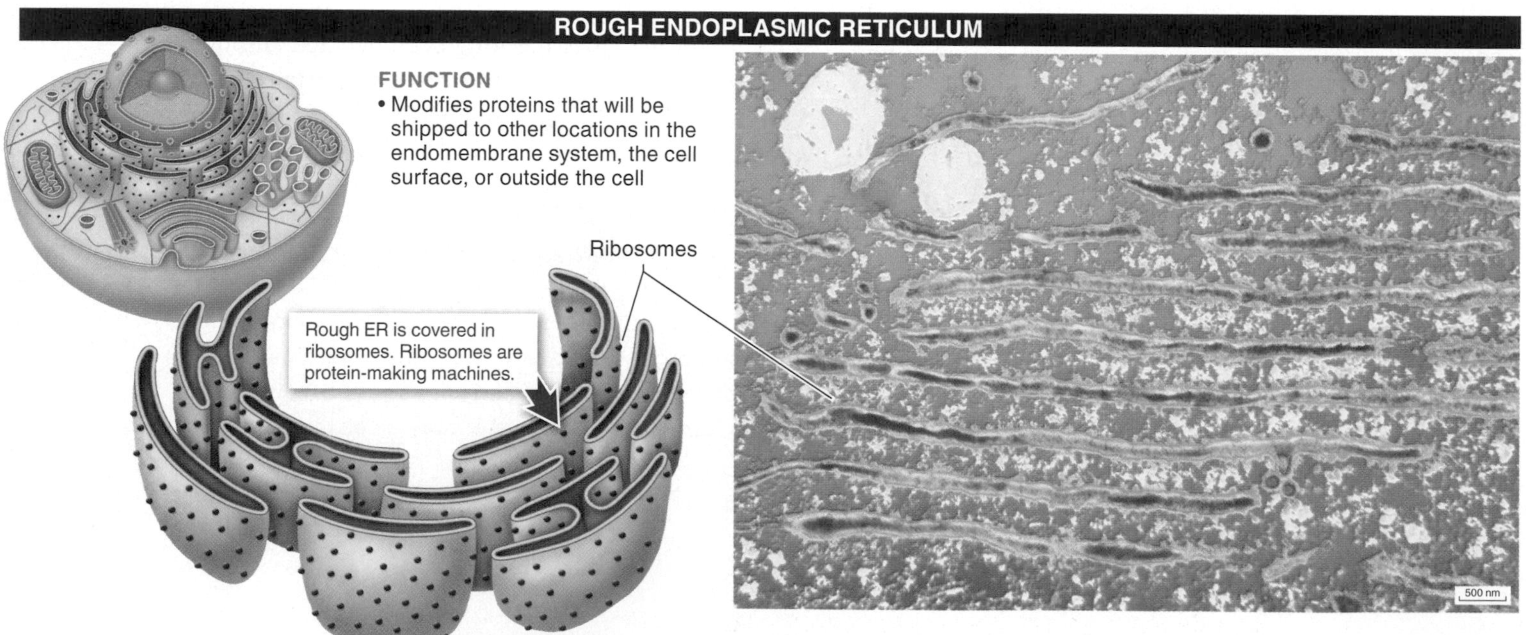

ROUGH ENDOPLASMIC RETICULUM
FUNCTION
• Modifies proteins that will be shipped to other locations in the endomembrane system, the cell surface, or outside the cell
Ribosomes
Rough ER is covered in ribosomes. Ribosomes are protein-making machines.
500 nm

FIGURE 4-34 The rough endoplasmic reticulum is studded with ribosomes.

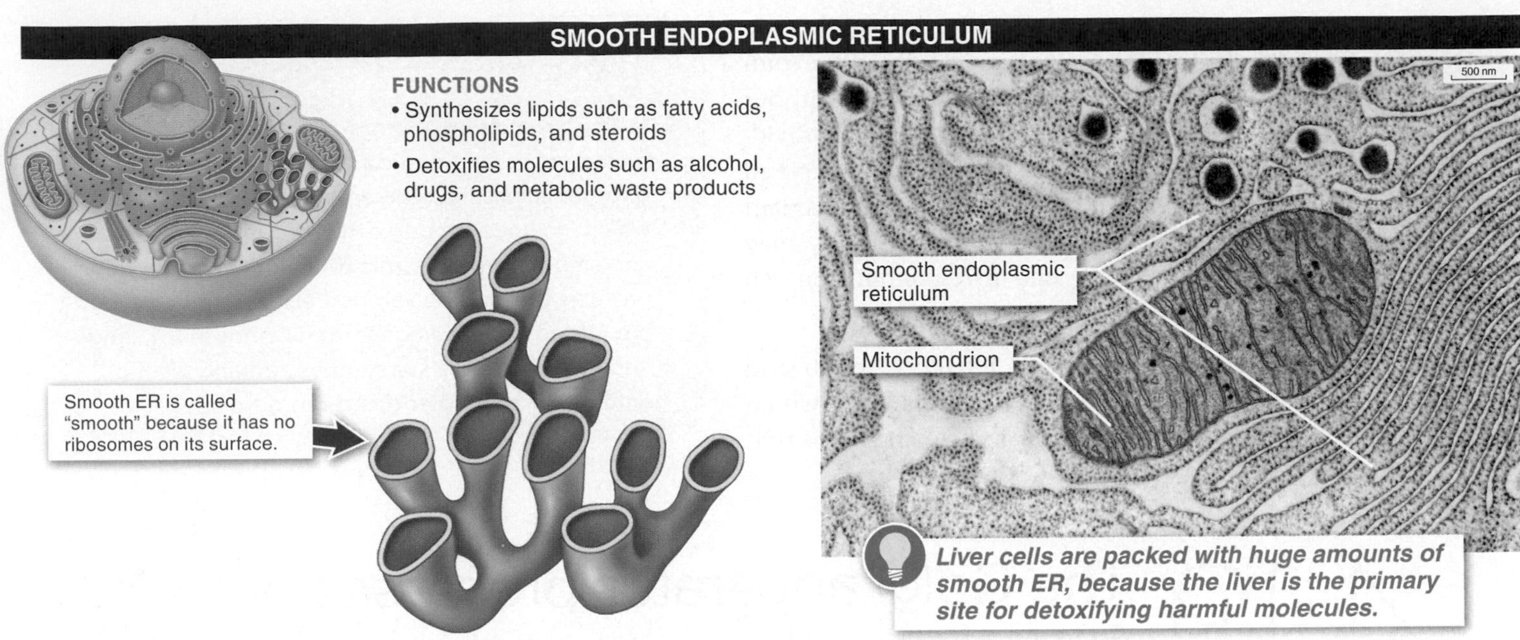

SMOOTH ENDOPLASMIC RETICULUM
FUNCTIONS
• Synthesizes lipids such as fatty acids, phospholipids, and steroids
• Detoxifies molecules such as alcohol, drugs, and metabolic waste products
Smooth ER is called "smooth" because it has no ribosomes on its surface.
500 nm
Smooth endoplasmic reticulum
Mitochondrion
Liver cells are packed with huge amounts of smooth ER, because the liver is the primary site for detoxifying harmful molecules.

FIGURE 4-35 In the smooth endoplasmic reticulum, lipids are synthesized and alcohol, antibiotics, and other drugs are detoxified.

surface is studded with ribosomes, the cell's protein-making machines. Cells with high rates of protein production generally have large numbers of ribosomes. We cover the details of ribosome structure and protein production in Chapter 6.

The primary function of the rough ER is to fold and package proteins for shipment to other locations in the endomembrane system, to the cell surface, or to the outside of the cell. (Proteins destined for use within the cell are generally produced on free-floating ribosomes in the cytoplasm.) Poisonous frogs, for example, package their poison in the rough ER of poison-producing cells before transporting it to the poison glands on their skin.

Smooth Endoplasmic Reticulum As its name advertises, the **smooth endoplasmic reticulum,** or **smooth ER,** is part of the endomembrane network that is smooth, because it has no ribosomes bound to it (FIGURE 4-35). Although connected to the rough ER, it is located farther from the nucleus. Whereas the rough ER looks like stacks of pancakes, the smooth ER sometimes looks like a collection of branched tubes.

The smooth ER is not extensively involved in folding or packaging proteins. Instead, it takes part in the synthesis of lipids such as fatty acids, phospholipids, and steroids, as well as carbohydrates. The exact lipids produced vary throughout the organism and across plant and animal species. The smooth ER of mammalian ovaries and testes, for example, produces the hormones estrogen and testosterone. The smooth ER of liver and fat cells produces other lipids. Following the same packaging process that occurs for proteins in the rough ER, lipids produced by the smooth ER are packaged in transport vesicles and then sent to other parts of the cell or to the plasma membrane for export.

A critical responsibility of the smooth ER—particularly in human liver cells—is to help protect us from the many dangerous molecules that get into our bodies. Detoxifying enzymes in the smooth ER reduce the harmfulness of alcohol, antibiotics, barbiturates, amphetamines, or other stimulants that we may consume, along with many toxic metabolic waste products formed in our bodies. And just as a bodybuilder's muscles get bigger in response to weightlifting, our smooth ER proliferates in cells that are exposed to large amounts of particular drugs. In liver cells, for example, smooth ER is abundant because the liver is the primary site of molecular detoxification.

Chronic exposure to many drugs (from antibiotics to heroin) can induce a proliferation of smooth ER and detoxification enzymes, particularly in the liver. This proliferation increases tolerance to the drugs, necessitating higher doses to achieve the same effect. This increased detoxification capacity of the

Q How can long-term use of one drug increase your resistance to another, different drug that you have never encountered?

cells often enables them to better detoxify other compounds, but not without consequences. An individual who has been exposed to large amounts of certain drugs, for example, may end up responding less well to antibiotics.

Other cells that are packed with ER (both rough and smooth) include blood cells called plasma cells, which produce immune system proteins, and pancreas cells that secrete large amounts of digestive enzymes.

TAKE HOME MESSAGE 4.18

» The production and modification of biological molecules in eukaryotic cells occurs in a system of organelles called the endomembrane system, which includes, among other organelles, the rough and smooth endoplasmic reticulum. In rough ER, proteins that will be shipped elsewhere in the body are folded and packaged. In the smooth ER, lipids and carbohydrates are synthesized and alcohol, antibiotics, and other drugs are detoxified.

4.19 The Golgi apparatus processes products for delivery throughout the body.

Moving farther outward from the nucleus, we encounter another organelle within the endomembrane system. The **Golgi** (GOHL-jee) **apparatus** (FIGURE 4-36) is a flattened stack of membranes (each of which is called a Golgi body), which are not connected to the endoplasmic reticulum. The job of the Golgi apparatus is to process molecules synthesized in the cell—primarily proteins and lipids—and to package those that are destined for use elsewhere in the body. The Golgi apparatus is also a site of carbohydrate synthesis, including the complex polysaccharides found in many plasma membranes.

To reach the Golgi apparatus, transport vesicles bud from the endoplasmic reticulum and move through the cytoplasm. The vesicles fuse with the Golgi apparatus membrane and dump their contents into a Golgi body. The molecules pass through about four successive Golgi body chambers. In each Golgi body, enzymes make slight modifications to the

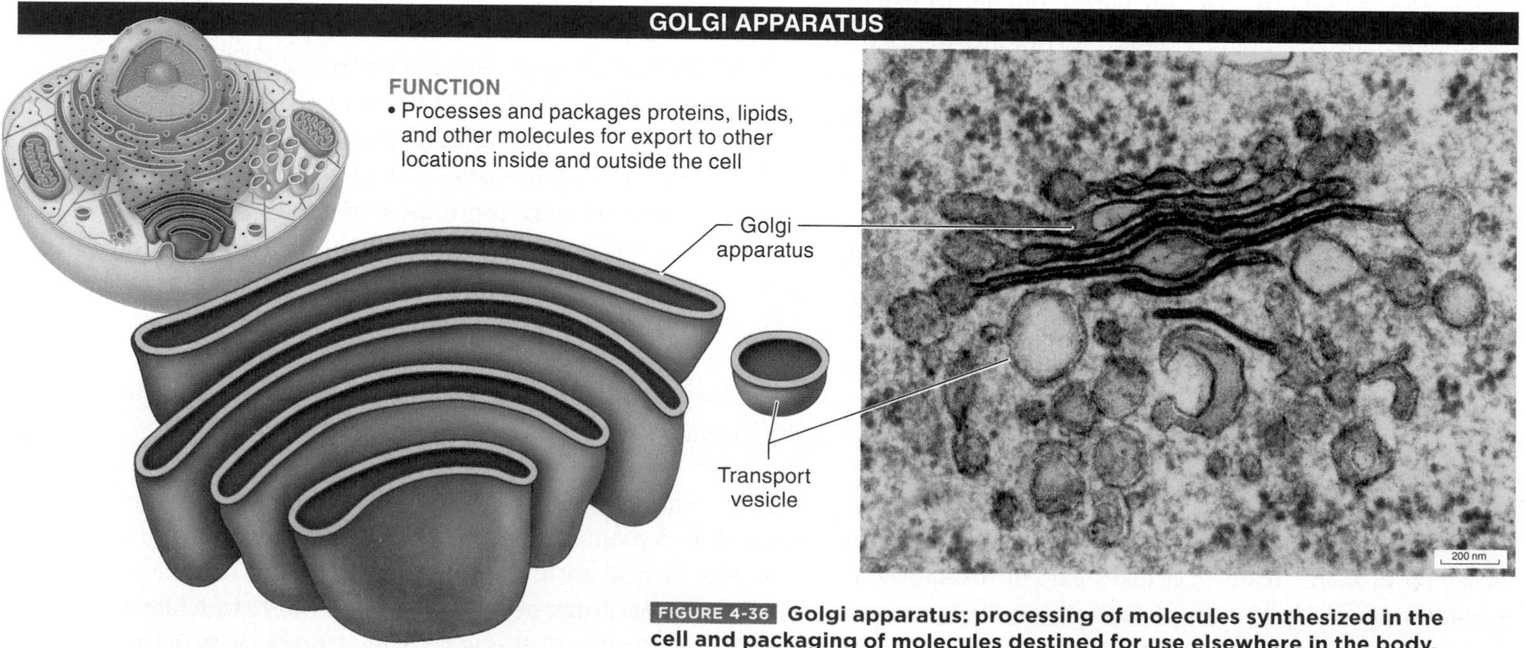

GOLGI APPARATUS
FUNCTION
• Processes and packages proteins, lipids, and other molecules for export to other locations inside and outside the cell
Golgi apparatus
Transport vesicle
200 nm

FIGURE 4-36 **Golgi apparatus: processing of molecules synthesized in the cell and packaging of molecules destined for use elsewhere in the body.**

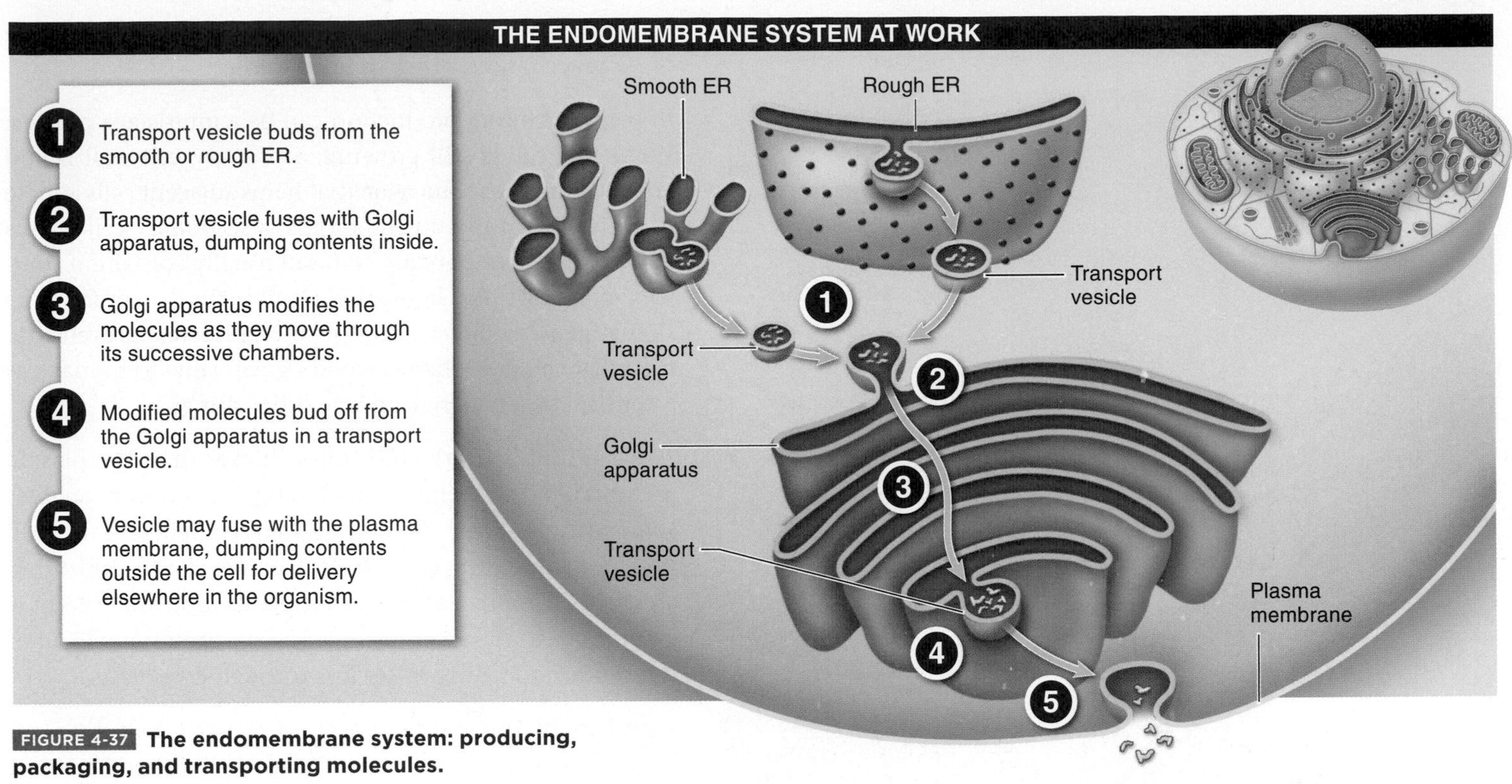
THE ENDOMEMBRANE SYSTEM AT WORK
1 Transport vesicle buds from the smooth or rough ER.
2 Transport vesicle fuses with Golgi apparatus, dumping contents inside.
3 Golgi apparatus modifies the molecules as they move through its successive chambers.
4 Modified molecules bud off from the Golgi apparatus in a transport vesicle.
5 Vesicle may fuse with the plasma membrane, dumping contents outside the cell for delivery elsewhere in the organism.
Smooth ER
Rough ER
Transport vesicle
Transport vesicle
Golgi apparatus
Transport vesicle
Plasma membrane

FIGURE 4-37 **The endomembrane system: producing, packaging, and transporting molecules.**

molecules (such as the addition or removal of phosphate groups or sugars). The processing that occurs in the Golgi apparatus often involves tagging molecules (much like adding a postal address or tracking number) to direct them to some other part of the organism.

After they are processed, the molecules bud off from the Golgi apparatus in a vesicle, which then moves into the cytosol. If the molecules are destined for delivery and use elsewhere in the body, the transport vesicle eventually fuses with the cell's plasma membrane and dumps the molecules into the bloodstream via exocytosis.

FIGURE 4-37 shows how the various parts of the endomembrane system work to produce, modify, and package molecules within a cell.

TAKE HOME MESSAGE 4.19

» The Golgi apparatus—another organelle within the endomembrane system—processes molecules synthesized in the cell and packages those that are destined for use elsewhere in the body.

4.20 The cell wall provides additional protection and support for plant cells.

Until this point, we've discussed organelles and structures that are common to both plants and animals. Now we're going to look at several structures that are not found in all eukaryotic cells. Because plants are rooted in the ground, unable to move, they have several special needs beyond those of animals. In particular, they can't outrun predators or outmaneuver their competitors to get more food. If they need more light, they have few options beyond simply growing larger or taller. And to resist plant-eating animals, they must simply reduce their edibility.

PLANT CELL WALL
FUNCTIONS
• Provides the cell with structural strength
• Gives the cell increased water resistance
• Provides some protection from insects and other animals that might eat plant parts
Plasmodesmata allow water and other molecules to pass between adjacent cells.
Cell 1
Cell 2
Cell 3
Primary cell wall
Secondary cell wall
Plasma membrane
Plasmodesma

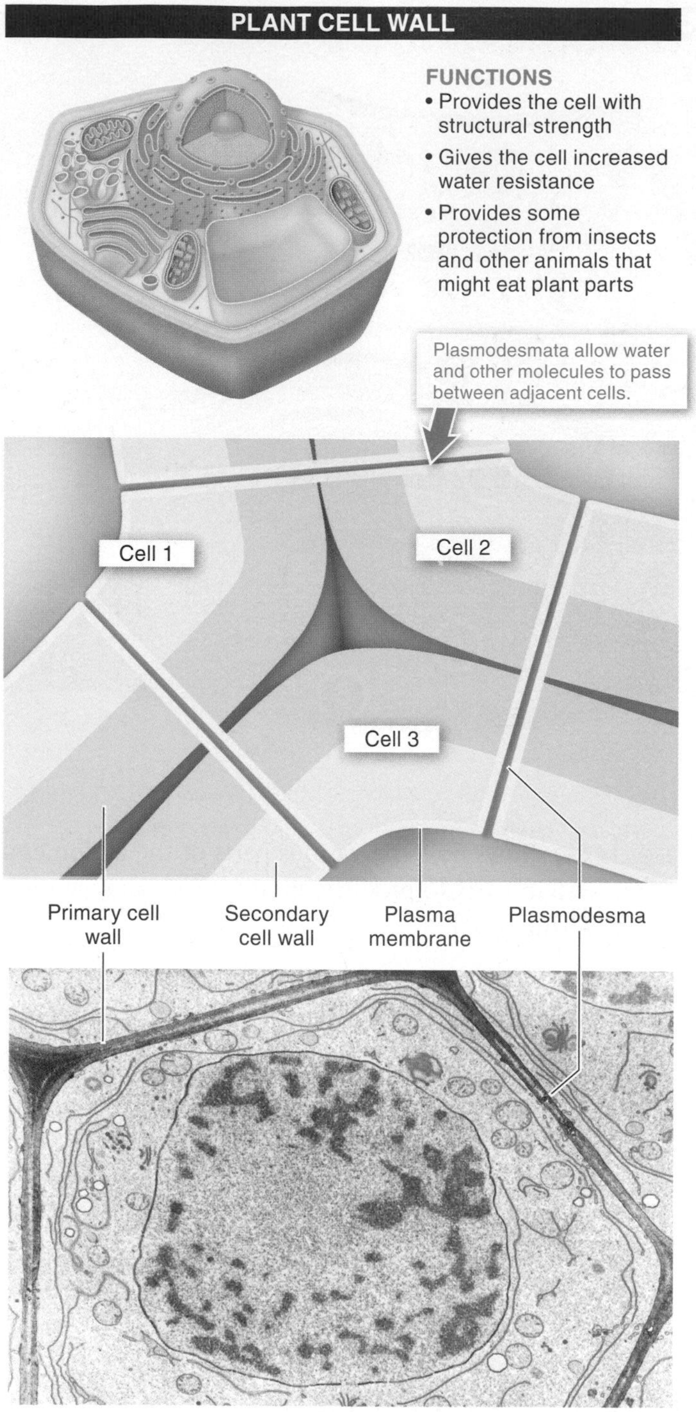

FIGURE 4-38 **The plant cell wall: providing strength.**

One structure that helps plants achieve these goals is the **cell wall,** which surrounds the plasma membrane (FIGURE 4-38). The plant cell wall is made largely from long fibers of a polysaccharide called cellulose. Animal cells do not have cell walls, but some non-plants such as archaea, bacteria, protists, and fungi do. Their cell walls are made of polysaccharides other than cellulose.

In plants, cell wall production can be a multistage process. When the cell is still growing, a primary cell wall is laid down, along with some glue that helps adjacent cells adhere to each other. Later, in some cells, a secondary cell wall is laid down. This secondary cell wall usually contains a complex molecule called lignin, which gives the wall much of its strength and rigidity. Oddly, this strength remains even after the plant cell dies—hence wood's great value to humans as a construction material, among other things.

The cell wall is nearly 100 times thicker than the plasma membrane. The tremendous strength it confers enables some plants to grow several hundred feet tall, and the wall provides some protection from feeding insects and other animals. Cell walls also help make plants more water-tight—an important feature in organisms that cannot move out of the hot sun to reduce water loss through evaporation.

Surprisingly, the great strength of the cell wall does not completely seal off plant cells from one another. Rather, it is porous, allowing water and solutes to reach the plasma membrane. Additionally, the cells of most plants have anywhere from 1,000 to 100,000 microscopic tube-like channels, called **plasmodesmata** (*sing.* **plasmodesma**) (see Figure 4-38), connecting the cells to each other and enabling communication and transport between them. In fact, because so many of the cells are connected to one another, sharing cytoplasm and other molecules, some biologists have wondered whether we should consider a plant as just one big cell. How would you respond to that idea?

TAKE HOME MESSAGE 4.20

» The cell wall is an organelle found in plants (and in some other non-animal organisms). It is made primarily from the carbohydrate cellulose, and it surrounds the cell's plasma membrane. The cell wall confers tremendous structural strength on plant cells, gives plants increased resistance to water loss, and provides some protection from insects and other animals that might eat the plant. Plasmodesmata connect the plant cells and enable communication and transport between them.

4.21 Vacuoles are multipurpose storage sacs for cells.

If you look at a mature plant cell through a microscope, one organelle, the **central vacuole,** usually stands out because it is so huge and appears empty (FIGURE 4-39). But it is not empty. Surrounded by a membrane, filled with fluid, and occupying from 50% to 90% of a plant cell's interior space, the central vacuole can play an important role in five different areas of plant life. (Vacuoles are also found in some other eukaryotes, including some protists, fungi, and animals, but they tend to be particularly prominent in plant cells.)

1. **Nutrient storage.** The vacuole stores hundreds of dissolved substances, including amino acids, sugars, and ions.

VACUOLE

FUNCTIONS
- Stores nutrients
- Retains and degrades waste products
- Accumulates poisonous materials
- Contains pigments, enabling plants to attract bird or insect pollinators
- Provides physical support

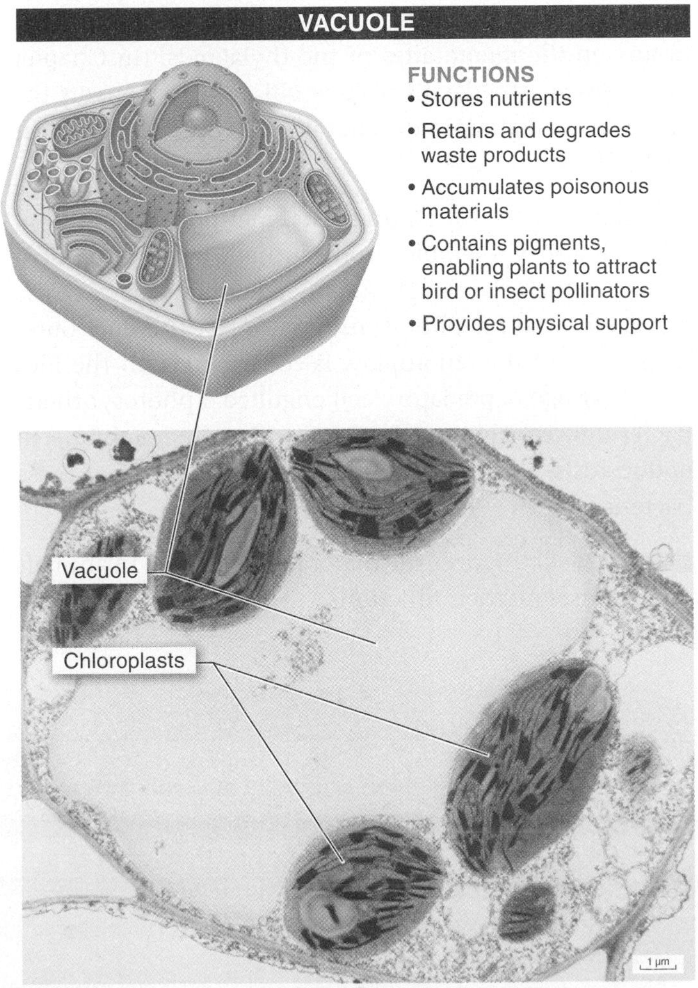

FIGURE 4-39 The vacuole: multipurpose storage.

2. **Waste management.** The vacuole retains waste products and degrades them with digestive enzymes, much like the lysosome in animal cells.
3. **Predator deterrence.** The poisonous, nasty-tasting materials that accumulate inside the vacuoles of some plants make a powerful deterrent to animals that might try to eat parts of the plant.
4. **Sexual reproduction.** The vacuole may contain pigments that give some flowers the colors that attract pollinators—birds and insects that help the plant reproduce by transferring pollen.
5. **Physical support.** High concentrations of dissolved substances in the vacuole can cause water to rush into the cells via osmosis. The increased fluid pressure inside the vacuole can cause the cell to enlarge and push against the cell wall. This process is responsible for the pressure (called **turgor pressure**) that allows stems, flowers, and other plant parts to stand upright. The ability of non-woody plants to stand upright is due primarily to turgor pressure. Wilting is the result of a loss of turgor pressure.

TAKE HOME MESSAGE 4.21

» In plants, vacuoles can occupy most of the interior space of the cell. Vacuoles, usually less prominent, are also present in some other eukaryotic species. In plants, they function as storage spaces and play roles in nutrition, waste management, predator deterrence, reproduction, and physical support.

4.22 Chloroplasts are the plant cell's solar power plant.

It would be hard to choose the "most important organelle" in a cell, but if we had to, the chloroplast would be a top contender. The **chloroplast,** an organelle found in all plants and eukaryotic algae, is the site of photosynthesis—the conversion of light energy into the chemical energy of food molecules, with oxygen as a by-product (FIGURE 4-40). Because all photosynthesis in plants and algae takes place in chloroplasts, these organelles are directly or indirectly responsible for everything we eat and for the oxygen we breathe.

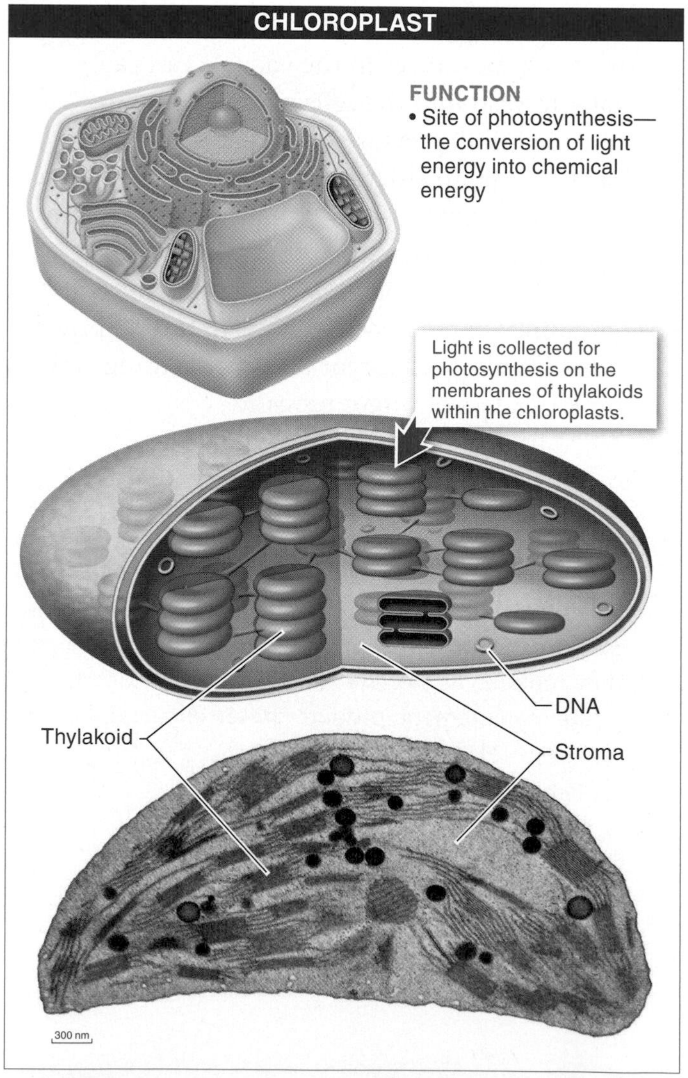

CHLOROPLAST

FUNCTION

- Site of photosynthesis—the conversion of light energy into chemical energy

Light is collected for photosynthesis on the membranes of thylakoids within the chloroplasts.

Thylakoid

DNA

Stroma

300 nm

FIGURE 4-40 **The chloroplast: location of photosynthesis.**

In a green leaf, each cell has about 40–50 chloroplasts. This means there are about 500,000 chloroplasts per square millimeter of leaf surface (that's more than 200 million chloroplasts in an area the size of a small postage stamp).

Chloroplasts are oval, somewhat flattened organelles. The outside of the chloroplast is encircled by two distinct layers of membranes (similar to the structure of mitochondria). Inside the compartment is a fluid called the **stroma,** which contains some DNA and much protein-making machinery.

With a simple light microscope, little spots of green are visible within the chloroplasts. Up close, these spots look like stacks of pancakes. Each stack consists of numerous interconnected little flattened sacs called **thylakoids.** The light-collecting process for photosynthesis occurs on the membranes of the thylakoids. (In Chapter 5, we discuss the details of how chloroplasts convert the energy in sunlight into the chemical energy stored in sugar molecules.)

A peculiar feature of chloroplasts, mentioned in Section 4.3, is that they resemble photosynthetic bacteria, particularly with their circular DNA (which contains many of the genes essential for photosynthesis). Also, the dual outer membrane of the chloroplast is consistent with the idea that, long ago, a predatory cell engulfed a photosynthetic bacterium, enveloping it with its plasma membrane in endocytosis, and that chloroplasts are derived from these bacteria.

FIGURE 4-41 summarizes the major parts of a cell, in animals and plants, and their functions.

TAKE HOME MESSAGE 4.22

» The chloroplast is the organelle of plants and algae that is the site of photosynthesis—the conversion of light energy into chemical energy, with oxygen as a by-product. Chloroplasts may originally have been bacteria that were engulfed by a predatory cell by endosymbiosis.

REVIEW OF CELL STRUCTURES

STRUCTURE and FUNCTION	ANIMALS	PLANTS
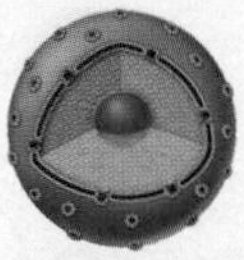 **NUCLEUS** Directs cellular activity and stores hereditary information	✓	✓
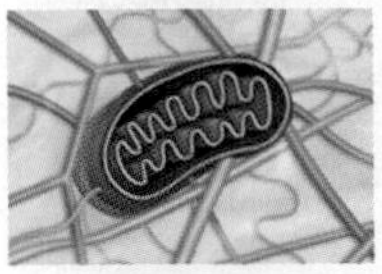 **CYTOSKELETON** Provides structural shape and support and enables cellular movement	✓	✓
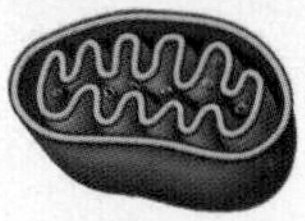 **MITOCHONDRION** Harvests energy for cellular functions	✓	✓
 **LYSOSOME** Digests and recycles cellular waste products and consumed material	✓	✓
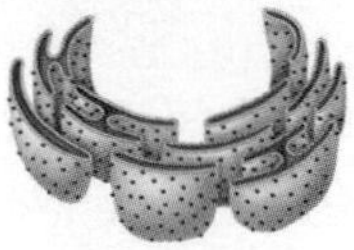 **ROUGH ER** Modifies proteins that will be shipped elsewhere in the organism	✓	✓
 **SMOOTH ER** Synthesizes lipids and detoxifies molecules	✓	✓
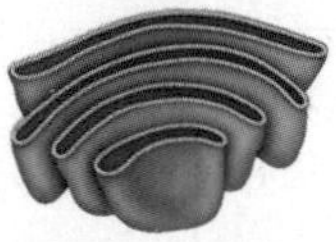 **GOLGI APPARATUS** Processes and packages proteins, lipids, and other molecules	✓	✓
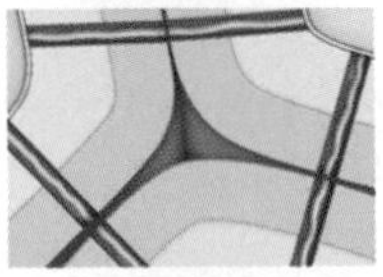 **CELL WALL** Provides structural strength, protection, and increased resistance to water loss	○	✓
 **VACUOLE** Stores nutrients, degrades waste, contains pigments, and provides structural support	✓ (Sometimes)	✓
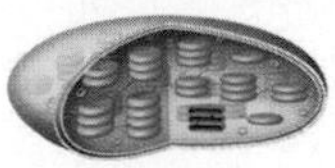 **CHLOROPLAST** Performs photosynthesis	○	✓

FIGURE 4-41 **Review of major cellular structures and their functions.**

STREET BIO

Using evidence to guide decision making in our own lives

Drinking too much water can be dangerous!

People have long understood the risks of dehydration, particularly during activities such as marathon running in which much fluid can be lost through sweating. Less well known, but also dangerous, is the overconsumption of water—called "water intoxication." In 2007, a 28-year-old woman competing in a water-drinking contest died just a few hours after consuming about two gallons of water without urinating. Similarly, the death of a runner during the 2002 Boston Marathon was attributed to water intoxication.

Q: When people sweat, what do they lose? Sweat consists of water and some dissolved solids, primarily salt. As the sweat evaporates, the body is cooled.

Q: How do people respond to nausea and dizziness, the warning signs of dehydration? By drinking water. Because the water they drink lacks salt and other solutes, sodium imbalances can occur, particularly in the bloodstream and the fluid around cells (extracellular fluid).

Q: If you drink lots (two gallons or more) of water—which has few or no solutes—where will it move once it gets into your bloodstream and extracellular fluid? It moves by osmosis into your cells, where there is a higher concentration of solutes (primarily sodium) and a lower concentration of water molecules.

Q: What happens to your cells if they absorb too much water? They swell, sometimes to the point of rupturing. Such swelling and rupturing in the enclosed brain cavity can cause vomiting and confusion. Disastrously, these symptoms are often mistaken for the symptoms of dehydration and so the consumption of water is recommended, making the situation worse. Further swelling can lead to seizures, coma, and even death.

What should you do? Many people mistakenly assume there are no dangers associated with overconsuming water. Many marathon organizers have been reducing the number of water stops they offer runners during the race and testing symptomatic runners' sodium levels in medical tents along the course. Consuming salt tablets and salty snacks helps keep blood sodium levels in a healthy range.

GRAPHIC CONTENT

Thinking critically about visual displays of data

1 What are the variables in this graph?

2 What additional information would make this figure more helpful? Why?

3 What can you conclude from this figure?

4 Is "number of mitochondria per cell" the best measure of a cell's "energy-generating capacity"? Can you think of a reason why this might not be a perfect measure? (Hint: muscle cells can be much, much larger than liver cells.)

5 Based on these data, can you make any of your own predictions about the energy requirements of particular cell types?

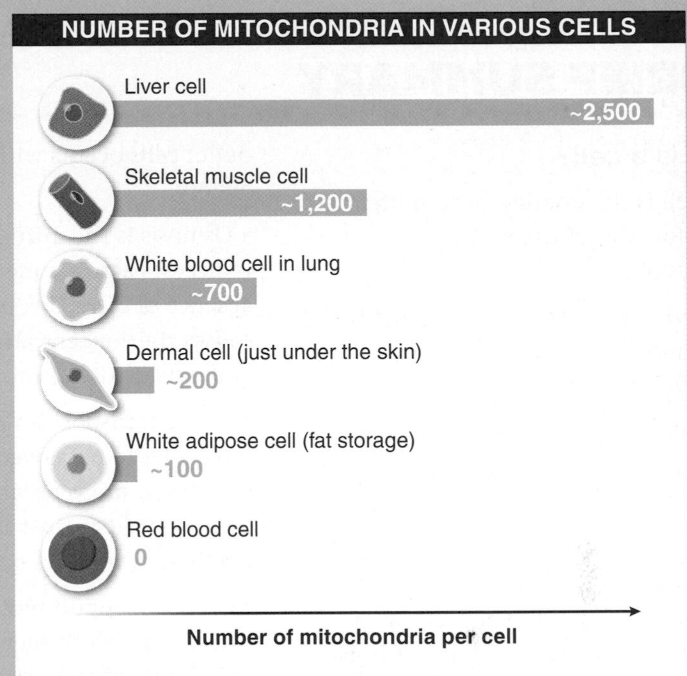

See answers at the back of the book.

KEY TERMS IN CELLS

active transport, p. 99
cell, p. 82
cell theory, p. 82
cell wall, pp. 84, 116
central vacuole, p. 117
chloroplast, pp. 87, 118
cholesterol, p. 92
chromatin, p. 105
cilium (*pl.* cilia), p. 107
cytoplasm, p. 84
cytoskeleton, p. 106
cytosol, p. 84
desmosome, p. 104
diffusion, p. 96
endocytosis, p. 100
endomembrane system, p. 112
endosymbiosis theory, p. 87
eukaryote, p. 84
eukaryotic cell, p. 84
exocytosis, p. 100
facilitated diffusion, p. 97
flagellum (*pl.* flagella), pp. 84, 107
fluid mosaic, p. 91
gap junction, p. 104
glycerol, p. 90
Golgi apparatus, p. 114
hypertonic, p. 97
hypotonic, p. 97
intermediate filaments, p. 106
intermembrane space, p. 109
invagination, p. 88
isotonic, p. 97
lysosome, p. 111
membrane enzymes, p. 92
microfilaments, p. 106
microtubules, p. 106
mitochondrial matrix, p. 109
mitochondrion (*pl.* mitochondria), pp. 87, 108
nonpolar, p. 90
nuclear membrane, p. 104
nucleolus, p. 105
nucleus, pp. 86, 104
organelle, p. 86
osmosis, p. 97
passive transport, p. 96
phagocytosis, p. 100
phospholipid, p. 90
phospholipid bilayer, p. 90
pilus (*pl.* pili), p. 84
pinocytosis, p. 100
plasma membrane, pp. 84, 89
plasmodesma (*pl.* plasmodesmata), p. 116
polar, p. 90
primary active transport, p. 99
prokaryote, p. 84
prokaryotic cell, p. 84
receptor-mediated endocytosis, p. 101
receptor protein, p. 91
recognition protein, p. 92
ribosome, p. 84

rough endoplasmic reticulum (rough ER), p. 112
secondary active transport, p. 99
simple diffusion, p. 96
smooth endoplasmic reticulum (smooth ER), p. 113
solute, p. 96
solvent, p. 96
stroma, p. 118
surface protein, p. 91
thylakoid, p. 119
tight junction, 103
tonicity, p. 97
transmembrane protein, p. 91
transport protein, p. 92
turgor pressure, p. 117
vacuole (central), p. 117

BRIEF SUMMARY

What is a cell?

• The cell is the smallest unit of life that can perform all of the necessary activities of life.

• All living organisms are made up of one or more cells, and all cells arise from other, preexisting cells.

• Prokaryotes are a metabolically diverse group of single-celled organisms, with no nucleus. They include the bacteria and archaea.

• Eukaryotes are single-celled or multicellular organisms consisting of cells with a nucleus that contains linear strands of genetic material, and most eukaryotic cells have organelles.

Cell membranes are gatekeepers.

• Every cell is enclosed by a plasma membrane, a two-layered membrane that holds the contents of a cell in place and regulates what enters and leaves the cell.

• The plasma membrane is a fluid mosaic of proteins, lipids, and carbohydrates. Membrane proteins act as receptors, help molecules enter and leave the cell, and catalyze reactions on the inner and outer cell surfaces.

• In conjunction with carbohydrates, some membrane proteins identify the cell to other cells, and help the immune system function.

Molecules move across membranes in several ways.

• In passive transport—which includes simple and facilitated diffusion and osmosis—molecular movement into and out of cells occurs without the input of energy.

• Osmosis is the diffusion of water molecules across a membrane. In the absence of other forces or barriers, it causes the concentration of water inside and outside the cell to be equalized.

• Active transport is necessary if molecules to be moved across a membrane are very large or if they are being moved against their concentration gradient. It requires energy.

• Cells can engulf molecules or particles with their plasma membrane in a process called endocytosis. Similarly, molecules can be moved out of a cell by exocytosis.

Cells are connected and communicate with each other.

• In multicellular organisms, most cells are connected to other cells by specialized structures—tight junctions, desmosomes, and gap junctions—that hold them in place and enable the cells to communicate with each other.

Nine important landmarks distinguish eukaryotic cells.

• The nucleus is the genetic control center of eukaryotic cells. It directs protein production and is the storehouse for hereditary information.

• The cytoskeleton is the inner scaffolding of a cell. Made from three types of protein fibers—microtubules, intermediate filaments, and microfilaments—it gives the cell shape and support and serves as a series of tracks on which organelles and molecules are guided.

• Mitochondria are found in virtually all eukaryotic cells and act as all-purpose energy converters, harvesting energy to be used for cellular functions.

• Lysosomes are acid- and enzyme-filled organelles that function as cellular garbage disposals.

• The production and modification of biological molecules in eukaryotic cells occurs in a system of organelles called the endomembrane system, which includes the rough and smooth endoplasmic reticulum.

• The Golgi apparatus—an organelle within the endomembrane system—processes molecules synthesized in the cell and packages those that are destined for use elsewhere in the body.

• The cell wall of plant cells is made primarily from cellulose and provides structural strength, increases resistance to water loss, and provides some protection from animals that might eat plant parts.

• Vacuoles are storage spaces found in several types of eukaryotes, and especially prominent in plants, that play a role in nutrition, waste management, predator deterrence, reproduction, and physical support.

• The chloroplast is the organelle in plants and algae that is the site of photosynthesis—the conversion of light energy into chemical energy, with oxygen as a by-product.

CHECK YOUR KNOWLEDGE

Short Answer

1. What are the key features of a cell?

2. Describe four structural features of prokaryotes.

3. Describe two evolutionary mechanisms by which organelles may have originated.

4. Why are phospholipids good membrane material?

5. Why is the plasma membrane referred to as a fluid mosaic?

6. Why do chloride ions accumulate within some cells of individuals with cystic fibrosis?

7. Describe two ways in which HIV is commonly transmitted.

8. Why are osmosis and simple and facilitated diffusion classified as passive transport?

9. Why is energy required for active transport to take place?

10. Where in the body would you expect to find cells with tight junctions? Why?

11. Why do mitochondria consume most of the oxygen used by each cell?

12. Describe how lysosomes function as a cell's "garbage disposal."

13. Why do liver cells contain more smooth ER than most other cells?

Multiple Choice

1. Which of the following statements about the cell theory is correct?

a) All living organisms are made up of one or more cells.

b) All cells arise from other, preexisting cells.

c) All eukaryotic cells contain symbiotic prokaryotes.

d) All prokaryotic cells contain symbiotic eukaryotes.

e) Both a) and b) are correct.

0 EASY 11 HARD 100

2. Which of the following statements about prokaryotes is incorrect?

a) Prokaryotes appeared on earth before eukaryotes.

b) Prokaryotes have circular pieces of DNA within their nuclei.

c) Prokaryotes contain cytoplasm.

d) Prokaryotes contain ribosomes.

e) Some prokaryotes can conduct photosynthesis.

0 EASY 51 HARD 100

3. Which of the following facts supports the claim that mitochondria developed from bacteria that, long ago, were incorporated into eukaryotic cells by the process of phagocytosis?

a) Mitochondria have flagella for motion.

b) Mitochondria have proteins for the synthesis of ATP.

c) Mitochondria are the "powerhouses" of the cell.

d) Mitochondria are small and easily transported across cell membranes.

e) Mitochondria have their own DNA.

4. Cellular "fingerprints":

a) are exposed on the cytoplasmic side of the membrane.

b) are made from cholesterol.

c) are "erased" by HIV.

d) can help the immune system distinguish "self" from "non-self."

e) All of the above are correct.

0 EASY 42 HARD 100

5. The movement of molecules across a membrane from an area of high concentration to one of low concentration is best described as:

a) active transport.

b) inactivated transport.

c) passive transport.

d) channel-mediated diffusion.

e) electron transport.

0 EASY 31 HARD 100

6. The cytoskeleton:

a) is a viscous fluid found in all cells.

b) fills a cell's nucleus but not the other organelles.

c) gives an animal cell shape and support, but cannot control movement.

d) helps to coordinate intracellular movement of organelles and molecules.

e) All of the above are correct.

0 EASY 55 HARD 100

7. Which of the following statements about mitochondria is correct?

a) Mitochondria are found in both eukaryotes and prokaryotes.

b) There tend to be more mitochondria in fat cells than in liver cells.

c) Most plant cells contain mitochondria.

d) Mitochondria may have originated evolutionarily as photosynthetic bacteria.

e) All of the above are correct.

0 EASY 70 HARD 100

8. Given that a cell's structure reflects its function, what function would you predict for a cell with a large Golgi apparatus?

a) movement

b) secretion of digestive enzymes

c) transport of chemical signals

d) rapid replication of genetic material and coordination of cell division

e) attachment to bone tissue

0 EASY 59 HARD 100

9. Cell walls:

a) occur only in plant cells.

b) are not completely solid, having many small pores.

c) confer less structural support than the plasma membrane.

d) dissolve when a plant dies.

e) are made primarily from phospholipids.

0 EASY 66 HARD 100

10. In plant cells, chloroplasts:

a) serve the same purpose that mitochondria serve in animal cells.

b) are the site of conversion of light energy into chemical energy.

c) play an important role in the breakdown of plant toxins.

d) have their own linear strands of DNA.

e) Both b) and d) are correct.

0 EASY 40 HARD 100

Ch05

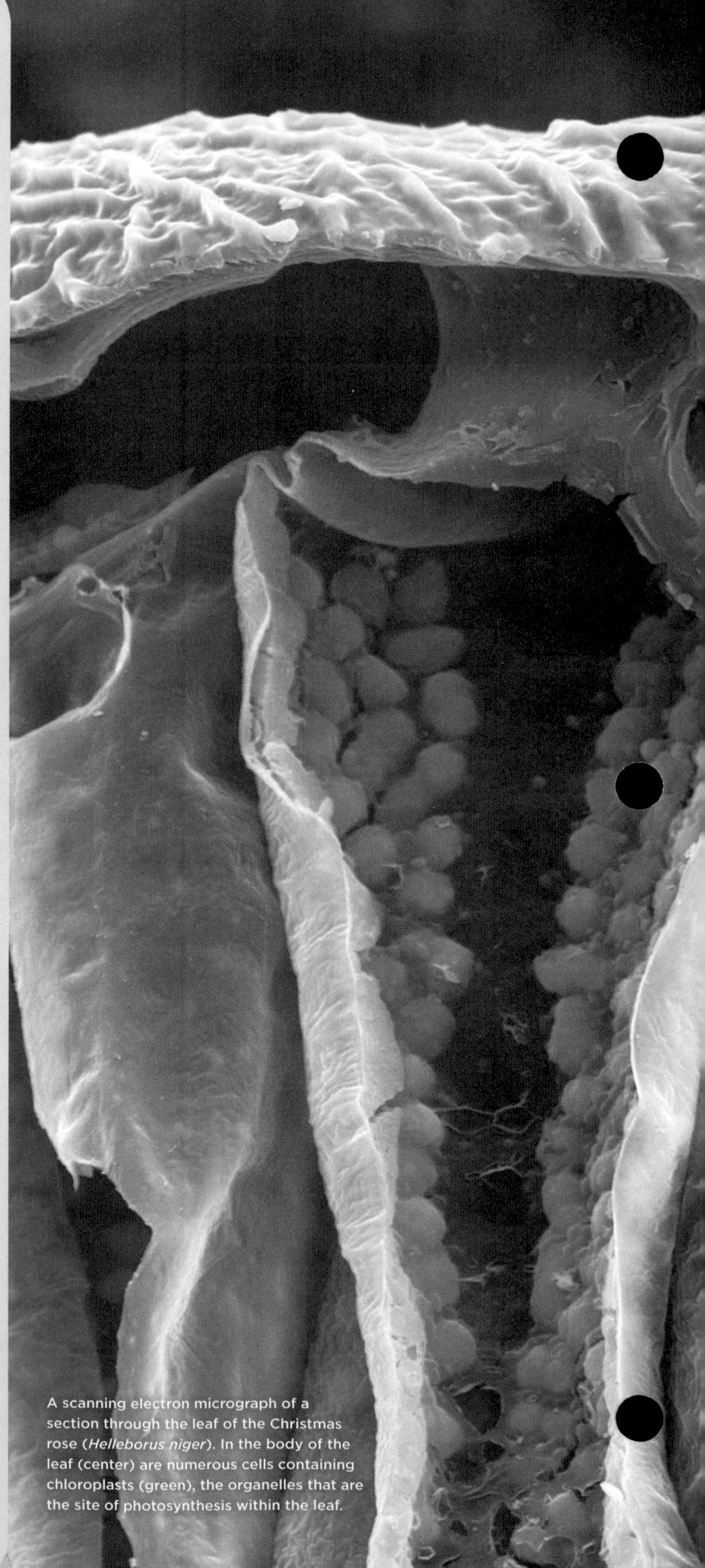

A scanning electron micrograph of a section through the leaf of the Christmas rose (*Helleborus niger*). In the body of the leaf (center) are numerous cells containing chloroplasts (green), the organelles that are the site of photosynthesis within the leaf.

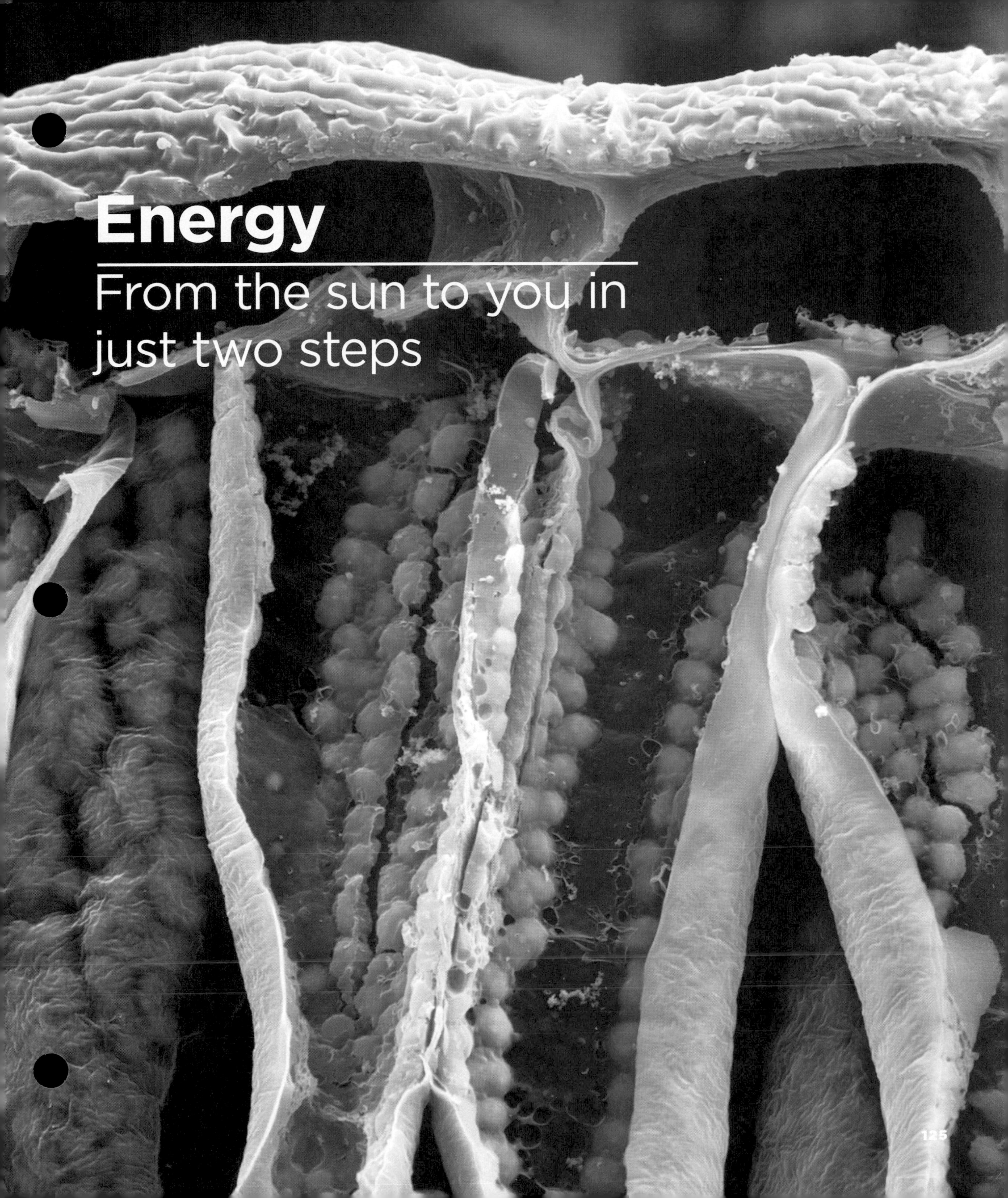

Energy

From the sun to you in just two steps

5.1–5.4

Energy flows from the sun and through all life on earth.

Sunlight energy fuels life.

5.1 Cars that run on french fry oil? Organisms and machines need energy to work.

Imagine that you are on a long road trip. Your car's fuel gauge is nearing empty, but instead of driving into a gas station, you head to the back of a fast-food restaurant and fill the fuel tank with used cooking grease. You head back to the highway, ready to drive several hundred miles before needing another pit stop.

Q Humans can get energy from food. Can machines?

Fast food for your car? Yes! The idea isn't as far-fetched as it sounds. In fact, many vehicles run on **biofuels,** fuels produced from plant and animal products (FIGURE 5-1). The production of biofuels requires only the plant or animal source, sunlight, air, water, and a relatively short amount of time—a few months or years, depending on the source.

In contrast, **fossil fuels,** such as gasoline, oil, natural gas, and coal, are produced from the decayed remains of plants and animals modified over millions of years by heat, pressure, and bacteria. It turns out that biofuels, fossil fuels, and the food fuels that supply energy to most living organisms are chemically similar. Energy from the sun is the source of the energy stored in the chemical bonds between the atoms in all these fuels. They contain chains of carbon and hydrogen atoms bound together, and breaking these bonds and forming new, lower-energy bonds releases large amounts of energy (and water and CO_2). If this released energy can be captured, it can be used not only to support life but to push pistons and turn car wheels.

Are we at the point where all our cars can run on biofuels? Not quite. Growing crops for biofuel reduces the land available for food crops and can cause the destruction of forests, wetlands, and other ecologically important habitats. Additionally, growing biofuel crops uses large amounts of fossil fuel and water. This is why the search for better fuels continues.

In this chapter we explore how plant, animal, and other living "machines" run on energy stored in chemical bonds. Nearly all life depends on capturing energy from the sun and

FIGURE 5-1 **Biofuel technology.** Biofuels are chemically similar to fossil fuels.

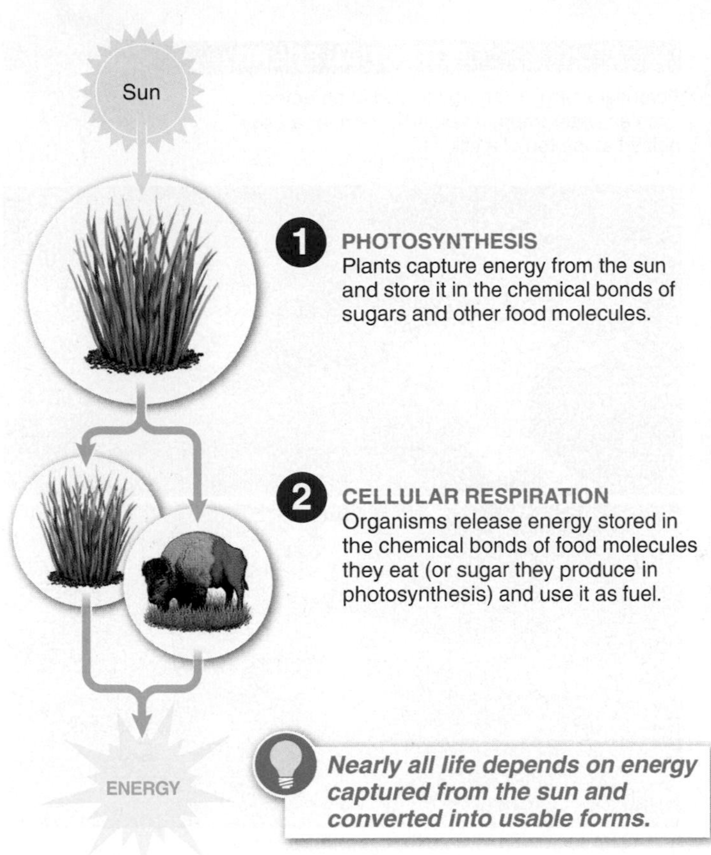

Sun
1 PHOTOSYNTHESIS
Plants capture energy from the sun and store it in the chemical bonds of sugars and other food molecules.
2 CELLULAR RESPIRATION
Organisms release energy stored in the chemical bonds of food molecules they eat (or sugar they produce in photosynthesis) and use it as fuel.
ENERGY
Nearly all life depends on energy captured from the sun and converted into usable forms.

FIGURE 5-2 **Photosynthesis and cellular respiration.** The sun to you in just two steps!

converting it into forms that living organisms can use. This energy capture and conversion occurs in two important processes that mirror each other: (1) **photosynthesis,** the process by which plants capture energy from the sun and store it in the chemical bonds of sugars, and (2) **cellular respiration,** the process by which all living organisms release the energy stored in the chemical bonds of food molecules and use it to fuel their lives (FIGURE 5-2). The sun to you in just two steps!

TAKE HOME MESSAGE 5.1

» The sun is the source of the energy that powers most living organisms and other "machines." The energy from sunlight is stored in the chemical bonds of molecules. When these bonds are broken, energy is released, regardless of whether the bond is in a molecule of food, of a fossil fuel, or of a biofuel such as the oil in which french fries are cooked.

5.2 Energy has two forms: kinetic and potential.

"Batteries not included." For a child, those are pretty depressing words. We know that many toys and electronic gadgets need energy—usually in the form of batteries. Generating ringtones, lights, and movement requires energy. The same is true for humans, plants, and all other living organisms: they need energy for their activities, from moving to reproducing to thinking.

Energy is the capacity to do work. And work is anything that involves moving matter against an opposing force. In the study of living things, we encounter two types of energy: kinetic and potential. **Kinetic energy** is the energy of motion. Legs pushing bike pedals and birds flapping wings are examples of kinetic energy (FIGURE 5-3). Heat, which results from lots of molecules moving rapidly, is another form of kinetic energy. Because it comes from the movement of high-energy particles, light is also a form of kinetic energy—probably the most important form of kinetic energy on earth.

An object does not have to be moving to have the capacity to do work; it may have **potential energy,** which is stored energy that results from an object's location, position, or composition. Water behind a dam, for example, has potential energy. If a hole is opened in the dam, the water can flow through, and perhaps spin a waterwheel or turbine. A concentration gradient, which we discuss in Chapter 4, also has potential energy: if molecules in an area of high concentration move toward an area of lower concentration, the potential energy of the gradient is converted to the kinetic energy of molecular movement, and this kinetic energy can do work. **Chemical energy,** the storage of energy in chemical bonds, is also a type of potential energy.

KINETIC ENERGY

Kinetic energy is the energy of motion, such as legs pushing pedals, birds flapping wings, and the rapidly moving molecules in a fire.

KINETIC ENERGY

POTENTIAL ENERGY

Potential energy is energy stored in an object, such as water trapped behind a dam, or a skier poised at the top of a hill.

POTENTIAL ENERGY

FIGURE 5-3 **Two forms of energy.**

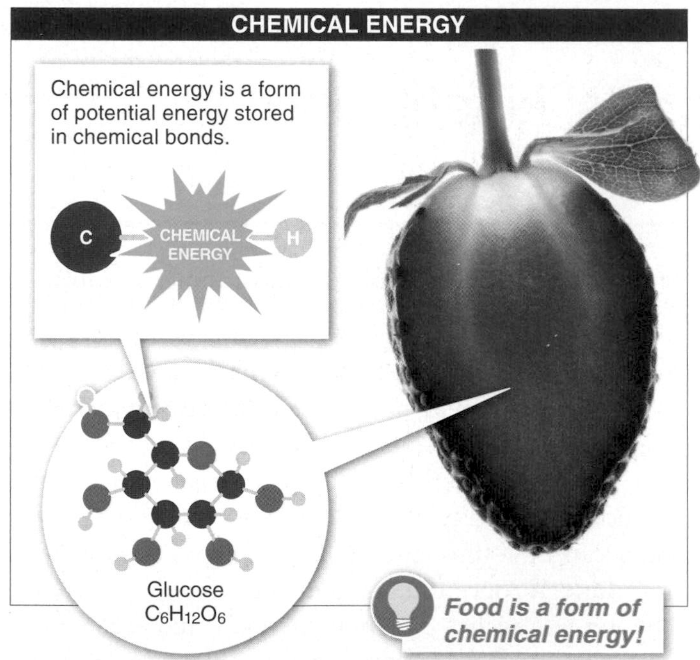

CHEMICAL ENERGY
Chemical energy is a form of potential energy stored in chemical bonds.
C
CHEMICAL ENERGY
H
Glucose
$C_6H_{12}O_6$
Food is a form of chemical energy!

FIGURE 5-4 **The energy of chemical bonds.** The chemical bonds in food molecules are a form of potential energy.

Because potential energy doesn't involve movement, it is a less obvious form of energy than kinetic energy. A strawberry has potential energy, as does any other type of food (FIGURE 5-4). Why? Because, during cellular respiration, the chemical energy stored in the chemical bonds making up the food can be released, enabling you to run, play, and work. We explore cellular respiration, the energy-releasing breakdown of molecules, later in this chapter. But first we need to know more about the nature of energy.

TAKE HOME MESSAGE 5.2

» Energy, the capacity to do work, comes in two forms. Kinetic energy is the energy of moving objects, while potential energy, such as chemical energy, is stored energy that results from the location, position, or composition of an object.

5.3 As energy is captured and converted, the amount of energy available to do work decreases.

Every minute of every day—even cloudy days—the sun shines brightly, releasing tremendous amounts of energy. Organisms on earth cannot capture all of the sun's energy that reaches the surface of the earth; indeed, most plants capture only a tiny fraction of the available energy. What happens to the rest? This unused energy does not simply disappear. All energy from the sun that reaches the earth can be accounted for. Some (probably less than 1%) is captured by organisms and transformed into usable chemical energy through photosynthesis (FIGURE 5-5). The rest is reflected back into space (probably about 30%) or is absorbed by land, the oceans, and the atmosphere (about 70%) and mostly transformed into heat.

The same accounting can be used on a smaller scale. If you eat a bowl of rice, some portion of the chemical energy stored in the bonds of the molecules that make up the rice grains is transformed into usable energy that can fuel your cells' activities. All the rest is transformed into heat and is ultimately lost into the atmosphere.

The fact that energy can change form but never disappear is an important feature of energy in the universe. Just as energy can never disappear or be destroyed, energy can never be created. In all our eating and growing, driving and sleeping, we are simply transforming energy. The study of the transformation of energy from one type to another, such as from potential energy to kinetic energy, is called **thermodynamics,** and the **first law of thermodynamics** states that energy can never be created or destroyed. It can only change from one form to another.

Because plants capture less than 1% of the sun's energy, it might seem that they are particularly inefficient. But

ENERGY TRANSFORMATIONS

KINETIC ENERGY TO POTENTIAL ENERGY

POTENTIAL ENERGY TO KINETIC ENERGY

FIGURE 5-5 **As energy is converted to do work, some energy is released as heat.**

we humans are also rather inefficient at extracting the chemical energy of plants that we eat. These inefficiencies occur because every time energy is converted from one form to another, some of the energy is converted to heat. For example, when a human converts the chemical energy in a plate of spaghetti into the kinetic energy of running a marathon, or when a car transforms the chemical energy of gasoline into the kinetic energy of forward motion, some energy is converted to heat, the least usable form of kinetic energy. In automobiles, for example, about three-quarters of the energy in gasoline is lost as heat (FIGURE 5-6).

Q If you ate a quarter-pound of food, the maximum amount of weight you could gain would be *less* than a quarter of a pound. Why?

FIGURE 5-6 **Inefficient conversion.** Much of the energy used to fuel a combustion engine is converted to heat rather than forward motion.

The **second law of thermodynamics** states that every conversion of energy is not perfectly efficient and invariably includes the transformation of some energy into heat. Although heat is certainly a form of energy, it is almost completely useless to living organisms for fueling their cellular activity because it is not easily harnessed to do work. (It can have other value for organisms, however, such as keeping them warm.) Put another way, the second law of thermodynamics tells us that although the quantity of energy in the universe is not changing, its quality is. Little by little, the amount of energy that is available to do work decreases.

TAKE HOME MESSAGE 5.3

» **Energy is neither created nor destroyed but can change form. Each conversion of energy is inefficient, and some of the usable energy is converted to heat, a less useful form of energy.**

5.4 ATP molecules are like free-floating rechargeable batteries in all living cells.

Much of the work that cells do requires energy. But even though light from the sun carries energy, as do molecules of sugar, fat, and protein, none of this energy can be used directly to fuel chemical reactions in organisms' cells. First it must be captured in the bonds of a molecule called **adenosine triphosphate (ATP),** a free-floating molecule found in cells that acts like a rechargeable battery, temporarily storing energy that can then be used for cellular work in plants, animals, and all other organisms on earth. The use of ATP solves an important timing and coordination problem for living cells: a supply of ATP guarantees that the energy required for energy-consuming reactions will be available when it's needed.

ATP is a simple molecule with three components (FIGURE 5-7). At the center of the ATP molecule is a small sugar molecule attached to a molecule called adenine and a chain of three negatively charged phosphate groups (hence the "tri" in "triphosphate"). Because the bonds between these three phosphate groups must hold the groups together in the face of the three negative electrical charges that all repel one another, each of these bonds contains a large amount of energy and is stressed and unstable. The instability of these high-energy bonds makes the three phosphate groups like a tightly coiled spring. With the slightest push, one of the phosphate groups will pop off. And in the process, a little burst of energy will be released that the cell can use.

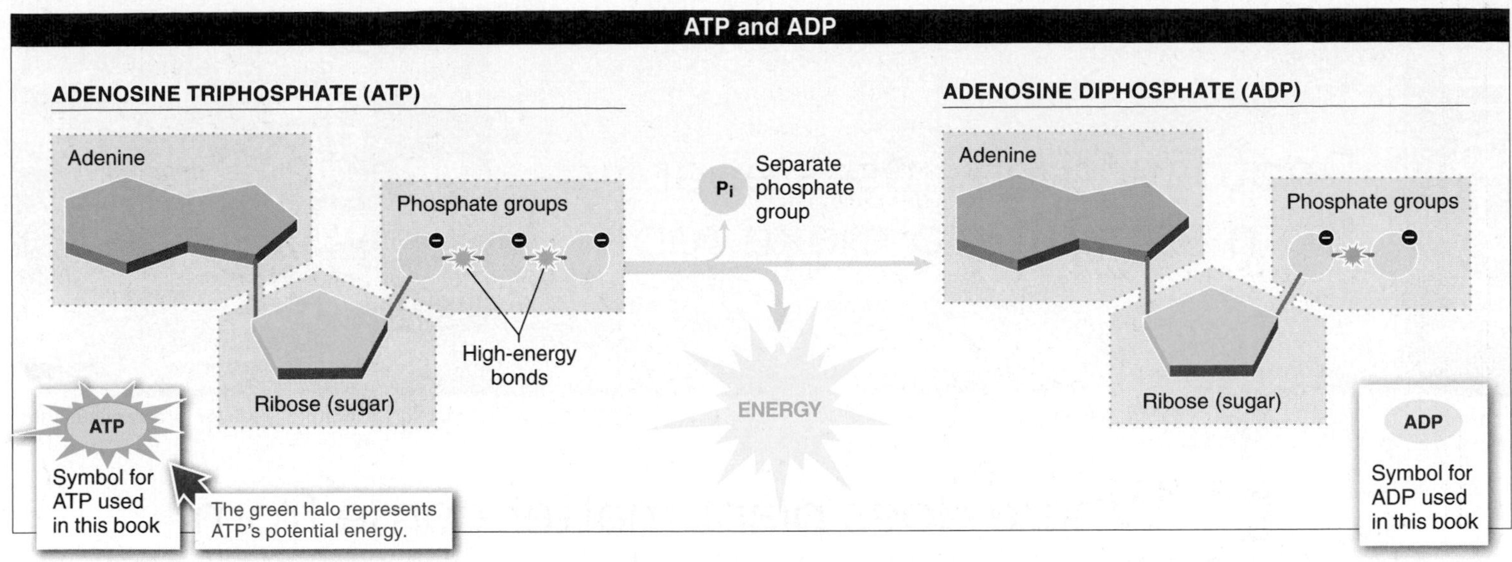

FIGURE 5-7 **The structure of ATP and ADP.** When ATP ejects one of its phosphate groups, energy is released as the ATP becomes ADP.

Each time a cell expends one of its ATP molecules to pay for an energetically expensive reaction, a phosphate is broken off and energy is released. What is left is a molecule with two phosphates, called ADP (adenosine *di*phosphate), and a separate phosphate group (labeled P_i).

An organism can then use ADP, a free-floating phosphate, and an input of kinetic energy to rebuild its ATP stocks (FIGURE 5-8). In this manner, ATP functions like a rechargeable battery. Where does the input of energy for recharging ATP come from? When we discuss photosynthesis, we'll see that plants, algae, and some bacteria can directly use light energy from the sun to make ATP from ADP and free-floating phosphate groups. And these photosynthetic organisms can also use the energy generated from breaking down sugar and other molecules. That's the same energy source that animals use, generating ATP from the energy contained in the bonds of their food molecules.

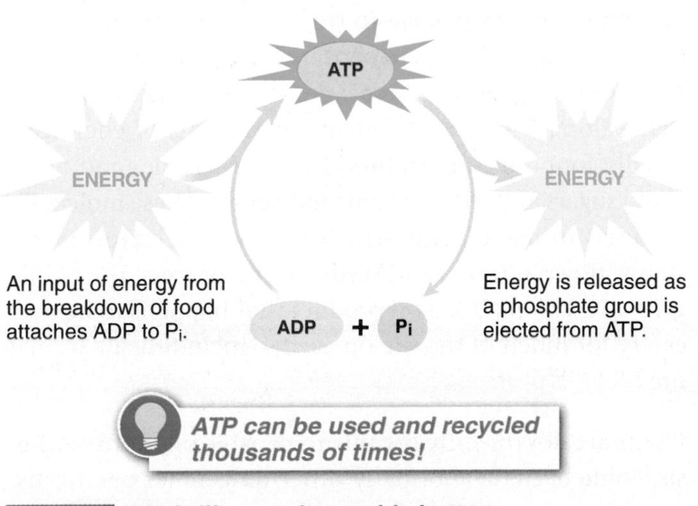

ATP
ENERGY
ENERGY
An input of energy from the breakdown of food attaches ADP to Pi.
ADP + Pi
Energy is released as a phosphate group is ejected from ATP.
ATP can be used and recycled thousands of times!

FIGURE 5-8 **ATP is like a rechargeable battery.**

Whether it comes from the sun or from the breakdown of molecules such as sugar, the energy is used to re-create the unstable bond in the triphosphate chain of ATP. When energy is needed, the organism can again release it by breaking the bond holding the phosphate group to the rest of the molecule.

Here's the ATP story in a nutshell. Breaking down a molecule of sugar—absorbed after drinking a glass of orange juice, for example—leads to a miniature burst of energy in one of your cells. The energy from many such mini-explosions is put to work building the unstable high-energy bonds that attach phosphate groups to ADP molecules, creating new molecules of ATP. Later—perhaps only a fraction of a second later—when an energy-consuming reaction is needed, your cells can release the energy stored in the new ATP molecules.

TAKE HOME MESSAGE 5.4

» Cells temporarily store energy in the bonds of ATP molecules. This potential energy can be converted to kinetic energy and used to fuel life-sustaining chemical reactions. In other reactions, inputs of kinetic energy are converted to the potential energy of the energy-rich but unstable bonds of ATP molecules.

5.5–5.11

Photosynthesis uses energy from sunlight to make food.

Large flat water lily leaves intercept sunlight.

5.5 Where does plant matter come from? Photosynthesis: the big picture.

Watching a plant grow over the course of a few years can seem like watching a magic trick. Of course it's not, but the process is nonetheless amazing. Consider that in five years a tree can increase its weight by 150 pounds (68 kg) (FIGURE 5-9). Where does that 150 pounds of new tree come from?

FIGURE 5-9 When plants grow, where does the new tissue come from? From the soil? From thin air?

Q When humans grow, the new tissue comes from food we eat. When plants grow, where does the new tissue come from?

Our first guess might be the soil. It's easy enough to weigh the soil in a pot when first planting a tree. Five years later, however, if we weigh the soil again, we find that it has lost less than a pound—nowhere near enough to explain the massive increase in the amount of plant material. Perhaps the new growth comes from the water? Wrong again. Although the older and much larger tree holds more water in its many cells, the water provided to the plant does not account for the increase in the dry weight of the plant.

Most of the new material actually comes from an element present in an invisible gas in the air. During photosynthesis, plants capture this element, present as carbon dioxide gas (CO_2) in the atmosphere, and use the energy from sunlight, along with water and small amounts of chemicals usually found in soil, to produce solid, visible (and often tasty) sugars and other organic molecules. These molecules are used to make plant structures such as leaves, roots, stems, flowers, fruits, and seeds. In the process, the plants give off oxygen (O_2), a by-product that happens to be necessary for much of the life on earth—including all animal life!

Plants are not the only organisms capable of photosynthesis. Some bacteria and many other unicellular organisms,

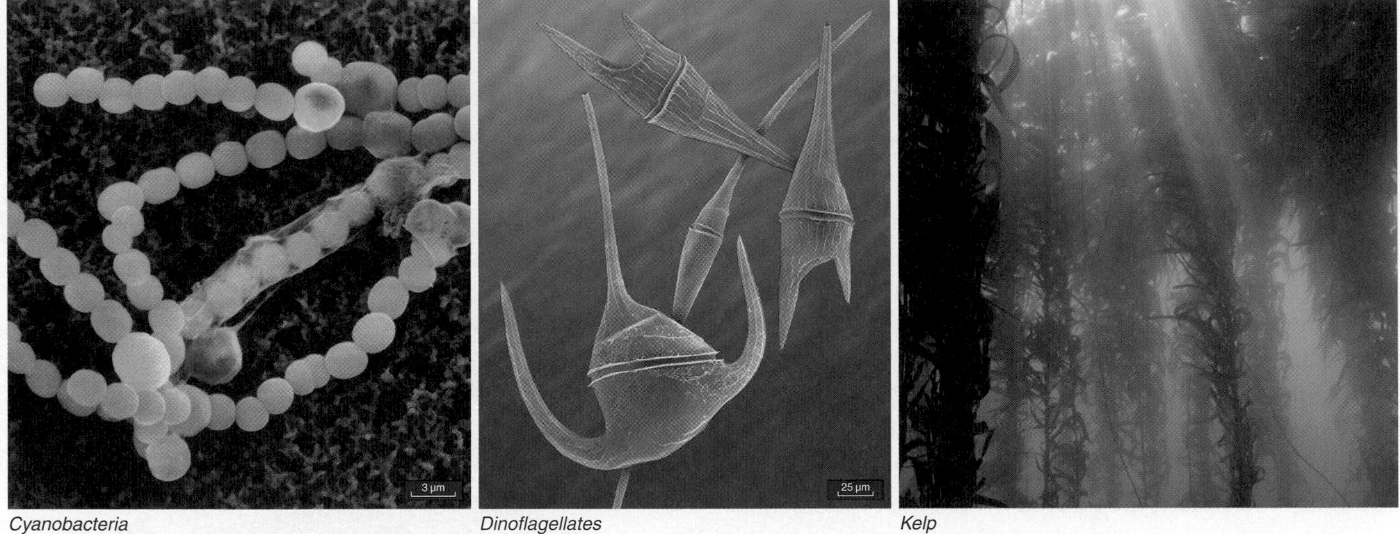

FIGURE 5-10 **Plants aren't the only photosynthesizers.** Some bacteria and other unicellular organisms, along with kelp and other multicellular algae, are capable of photosynthesis.

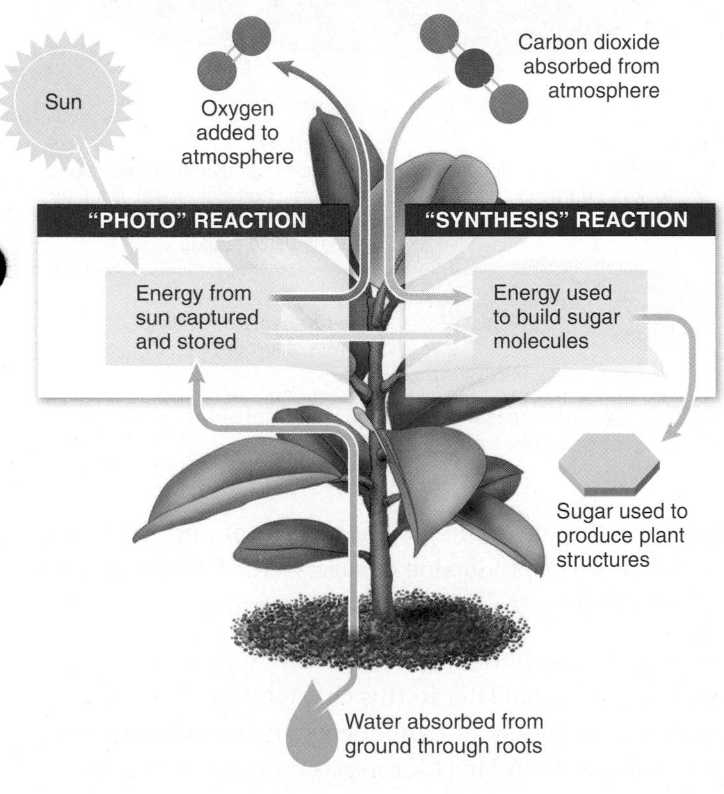

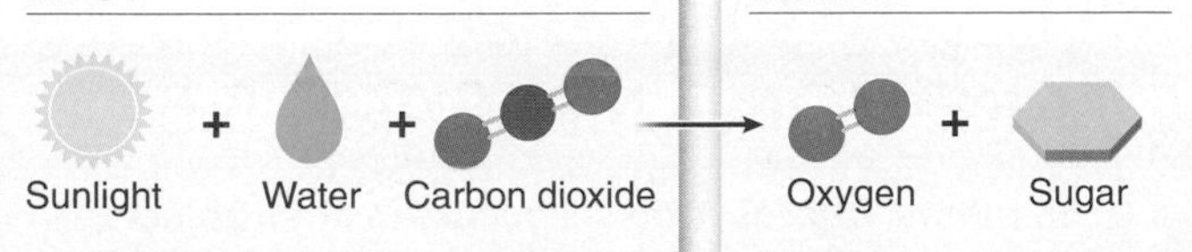

FIGURE 5-11 **Photosynthesis: the big picture.**

along with kelp and other multicellular algae, are also capable of using the energy in sunlight to produce organic materials (FIGURE 5-10).

There are three inputs to the process of photosynthesis (FIGURE 5-11): light energy (from the sun), carbon dioxide (from the atmosphere), and water (from the environment). From these three inputs, the plant produces sugar and oxygen. As we'll see, photosynthesis is best understood as two separate events: a "photo" segment, during which light is captured, and a "synthesis" segment, during which sugar is built. In the "photo" reactions, light energy is captured and temporarily saved in energy-storage molecules. During this process, water molecules split and produce oxygen. In the "synthesis" reactions, the energy in the energy-storage molecules is used to assemble sugar molecules from carbon dioxide from the atmosphere.

TAKE HOME MESSAGE 5.5

» Through photosynthesis, plants use water, the energy of sunlight, and carbon dioxide gas from the atmosphere to produce sugars and other organic materials. In the process, photosynthesizing organisms also produce oxygen, which makes all animal life possible.

5.6 Photosynthesis takes place in the chloroplasts.

If a plant part is green, then you know it is photosynthetic. Leaves are green because the cells near the surface are packed full of **chloroplasts,** light-harvesting organelles, which make it possible for the plant to use the energy from sunlight to make sugars (their food) and plant tissues (much of which animals use for food) (FIGURE 5-12). Other plant parts, such as stems, may also contain chloroplasts (in which case they, too, are capable of photosynthesis), but most chloroplasts are located within the cells of leaves.

Let's take a closer look at chloroplasts (FIGURE 5-13). The sac-shaped organelle is filled with a fluid called the **stroma.** Floating in the stroma is an elaborate system of interconnected membranous structures called **thylakoids,** which often look like stacks of pancakes. And embedded within the thylakoid membranes are many light-catching pigments organized into structures called **photosystems.**

Inside the chloroplast, there are two distinct places: in the stroma or inside the thylakoids. The conversion of light energy to chemical energy—the "photo" part of photosynthesis—occurs on the surface of and inside the thylakoids. The production of sugars—the "synthesis" part of photosynthesis—occurs within the stroma.

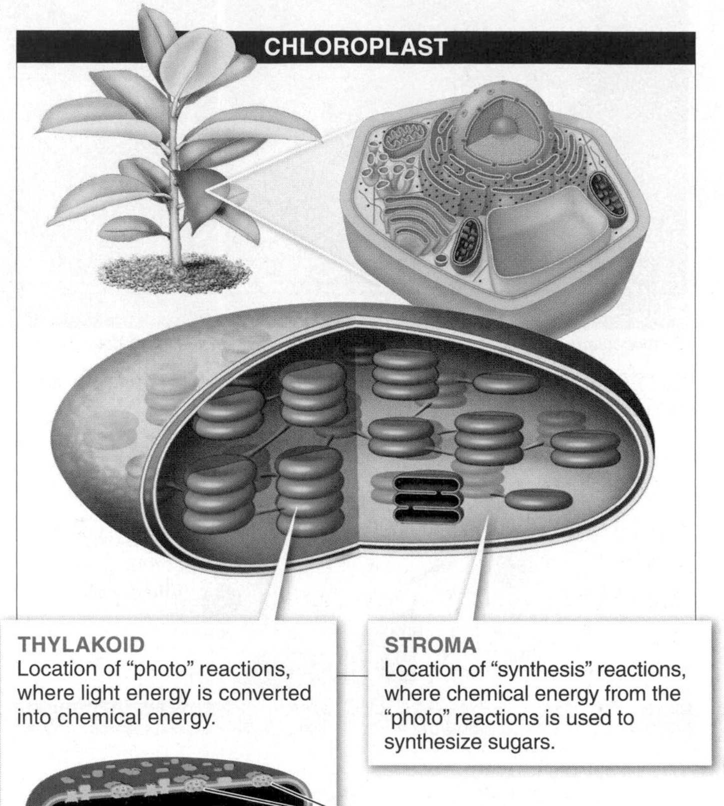

CHLOROPLAST
THYLAKOID
Location of "photo" reactions, where light energy is converted into chemical energy.
STROMA
Location of "synthesis" reactions, where chemical energy from the "photo" reactions is used to synthesize sugars.
Photosystems packed with chlorophyll

FIGURE 5-13 **Chloroplast structure.** The chloroplast is where photosynthesis takes place in a plant.

We examine both the "photo" and the "synthesis" processes in greater detail later in this chapter. First, however, we examine the nature of light energy and of **chlorophyll,** the molecule found in chloroplasts that makes the capture of light energy possible.

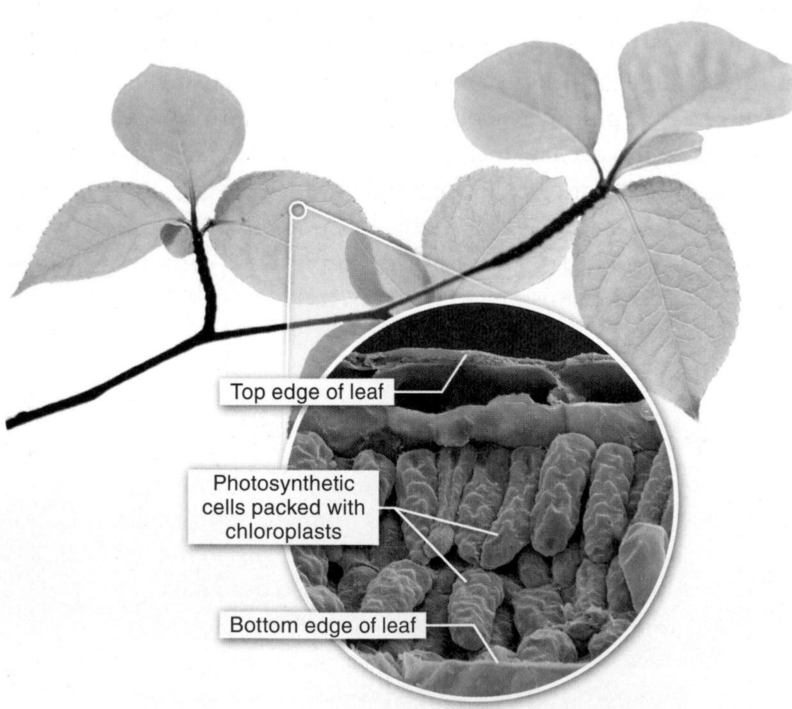

FIGURE 5-12 **Photosynthesis factories.** Cells near the leaf's surface are packed with photosynthetic chloroplasts.

TAKE HOME MESSAGE 5.6

» In plants, photosynthesis occurs in chloroplasts, green organelles packed in cells near the plants' surfaces, especially in the leaves.

5.7 Light energy travels in waves: plant pigments absorb specific wavelengths.

You can't eat sunlight. That's because sunlight is *light* energy rather than the chemical energy found in the bonds of food molecules. Photosynthesis is powered by **light energy,** a type of kinetic energy made up of little energy packets called **photons,** which are organized into waves. Photons can do work as they bombard surfaces such as your face (heating it) or a leaf (enabling it to build sugar from carbon dioxide and water).

Photons have various amounts of energy, and the length of the wave in which they travel corresponds to the amount of energy carried by the photon. The shorter the wavelength, the more energy the photon carries. Within a ray of light, there are super-high-energy photons (those with short wavelengths), relatively low-energy photons (those with longer wavelengths), and everything in between. This range, which is called the **electromagnetic spectrum,** extends from extremely short, high-energy gamma rays and X rays, with wavelengths as short as 1 nanometer (nm; a human hair is about 50,000 nm in diameter), to very long, low-energy radio waves, with wavelengths as long as a mile (FIGURE 5-14).

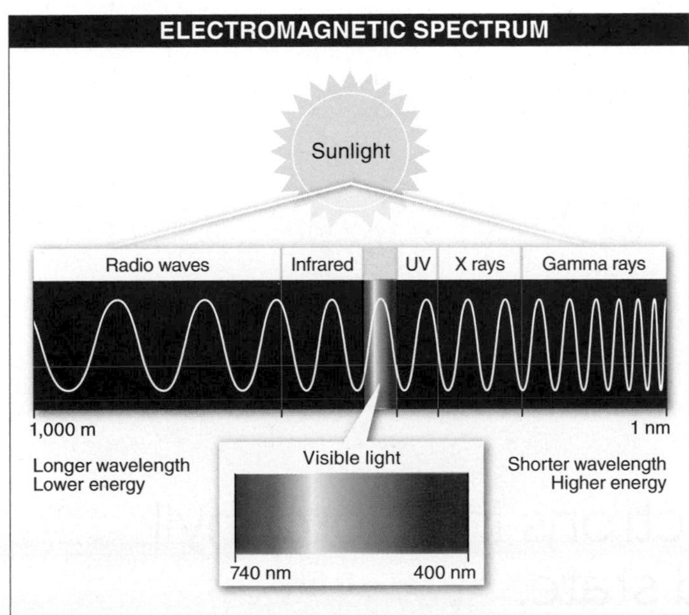

FIGURE 5-14 **A spectrum of energy.** A ray of light emits high-energy photons, low-energy photons, and everything in between. Plants use only a fraction of the light's available energy.

Just as we can't hear some high-pitched frequencies of sound that many dogs can hear, there are some wavelengths of light that we cannot see. The light that we can see, visible light, spans all the colors of the rainbow. We (and some other animals) can see colors because our eyes contain light-absorbing molecules called **pigments.** The energy in light waves excites electrons in the pigments, stimulating nerves in our eyes, which then transmit electrical signals to our brains. We perceive different wavelengths within the visible spectrum as different colors. The pigments in the human eye absorb many different wavelengths pretty well: that's why we can see so many colors. When plants use sunlight's energy to make sugar during photosynthesis, they also use the visible portion of the electromagnetic spectrum. Unlike the pigments in our eyes, however, plant pigments (the energy-capturing parts of a plant) absorb and use only a portion of the visible light wavelengths.

Chlorophyll is the main pigment molecule in plants that absorbs light energy from the sun. Chlorophyll molecules are embedded in the thylakoid membranes of chloroplasts. Just as light energy excites electrons in the pigments responsible for color vision in humans, electrons in a plant's chlorophyll can become excited by certain wavelengths of light and can capture a bit of this light energy.

Plants produce several different light-absorbing pigments (FIGURE 5-15). The primary photosynthetic pigment, called **chlorophyll *a*,** absorbs red and blue-violet wavelengths of light. Every other wavelength generally travels through or bounces off this pigment. Chlorophyll *a* cannot efficiently absorb green light and instead reflects those wavelengths. We perceive the reflected light waves as green, and so the pigment and the leaves that contain it appear green. Another pigment, **chlorophyll *b*,** is similar in structure but absorbs blue and red-orange wavelengths and reflects yellow-green wavelengths. Related pigments called **carotenoids** absorb blue-violet and blue-green wavelengths and reflect yellow, orange, and red wavelengths.

In the late summer, cooler temperatures cause some trees to prepare for the reduced daylight of winter by shutting

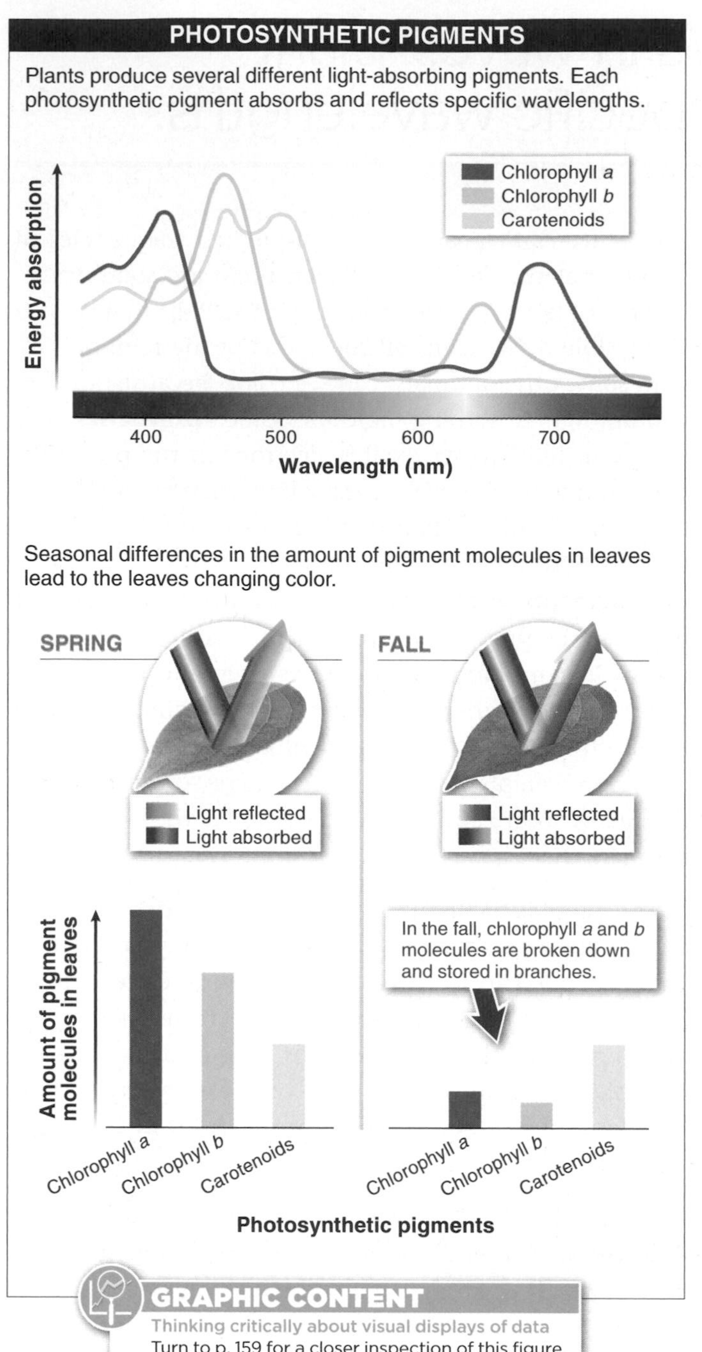
PHOTOSYNTHETIC PIGMENTS
Plants produce several different light-absorbing pigments. Each photosynthetic pigment absorbs and reflects specific wavelengths.
Energy absorption
Chlorophyll a
Chlorophyll b
Carotenoids
400
500
600
700
Wavelength (nm)
Seasonal differences in the amount of pigment molecules in leaves lead to the leaves changing color.
SPRING
FALL
Light reflected
Light absorbed
Light reflected
Light absorbed
Amount of pigment molecules in leaves
In the fall, chlorophyll a and b molecules are broken down and stored in branches.
Chlorophyll a
Chlorophyll b
Carotenoids
Chlorophyll a
Chlorophyll b
Carotenoids
Photosynthetic pigments
GRAPHIC CONTENT
Thinking critically about visual displays of data
Turn to p. 159 for a closer inspection of this figure.

FIGURE 5-15 **Plant pigments.** Each photosynthetic pigment absorbs and reflects specific wavelengths.

down chlorophyll production—which is energetically expensive—and reducing photosynthesis rates, much like an animal's hibernation. Gradually, the chlorophyll *a* and *b* molecules in the leaves are broken down and their chemical components are stored in the branches. As the amounts of chlorophyll *a* and *b* in the leaves decrease relative to the remaining carotenoids, the striking colors of the fall foliage are revealed (see Figure 5-15). During the rest of the year, chlorophylls *a* nd *b* are so abundant in leaves that green masks the colors of the other pigments, so that plants look green because they are reflecting green.

Q Why do the leaves of some trees turn beautiful colors each fall?

TAKE HOME MESSAGE 5.7

» Photosynthesis is powered by light energy, a type of kinetic energy made of energy packets called photons. Photons hit chlorophyll and other light-absorbing molecules in the chloroplasts of cells near the green surfaces of plants. These molecules capture some of the light energy and harness it to build sugar from carbon dioxide and water.

5.8 Photons cause electrons in chlorophyll to enter an excited state.

An organism can use energy from the sun only if it can convert the light energy of the sun into the chemical energy of the bonds between atoms. The most important molecule in this conversion is the pigment chlorophyll

ENERGY MOVEMENT THROUGH CHLOROPHYLL

1 Light energy bumps an electron in the chlorophyll molecule to a higher, excited energy level.

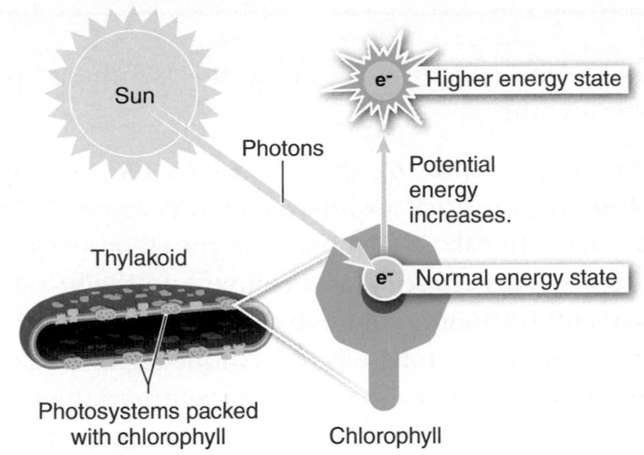

2 The excited electron generally has one of two different fates:

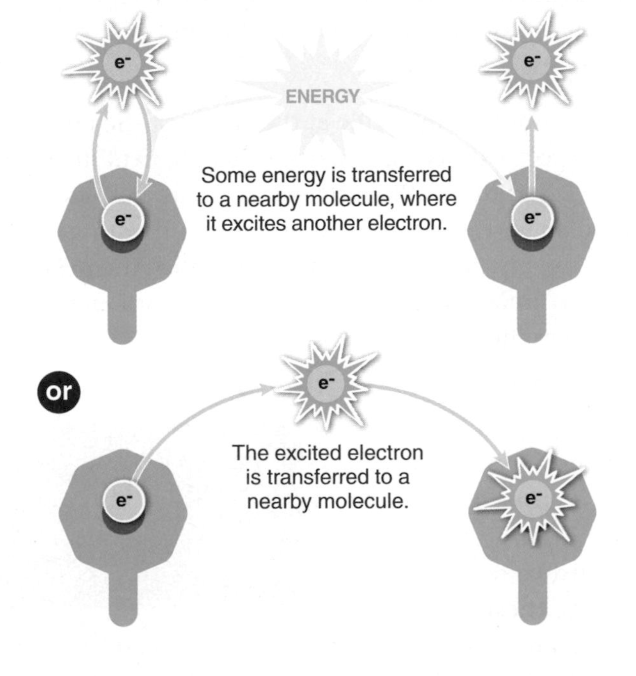

FIGURE 5-16 Capturing light energy as excited electrons. Chlorophyll electrons are excited to a higher energy state by light energy.

(FIGURE 5-16). When photons of certain wavelengths hit chlorophyll, the light energy bumps an electron (e^-) in the chlorophyll molecule to a higher energy level, an *excited* state. Upon absorbing the photon, the electron briefly gains energy, and the potential energy in the chlorophyll molecule increases.

The excited electron generally has one of two fates. (1) The electron returns to its resting, unexcited state, releasing energy in the process, some of which may bump electrons in a nearby molecule to a higher energy state (while the rest of the energy is dissipated as heat). Or (2) the excited electron itself is passed to another molecule.

The passing of electrons from molecule to molecule is an important way that energy moves through cells. Many molecules carry or accept electrons during cellular activities. All that is required is that the acceptor must have a greater attraction for electrons than does the donor molecule. A molecule that gains electrons always carries greater energy than it did before receiving the electron(s). For this reason, the passing of electrons from one molecule to another can be viewed as a passing of potential energy.

This transfer of electrons is one of the first steps of photosynthesis. As we see later in this chapter, the dismantling of food molecules such as glucose to generate energy is also a story of breaking and rearranging chemical bonds as electrons pass from one atom or molecule to another.

Q Suppose a large meteor hit the earth. How would smoke and soot in the atmosphere wipe out life far beyond the area of direct impact?

Because particles in the atmosphere can block light from the sun and reduce the excitation of electrons in chlorophyll molecules, photosynthesis can be hindered by light-blocking particles in the atmosphere. Any reduction in the available sunlight can have serious effects on plants. Scientists believe that if a large meteor hit the earth—as one did when the dinosaurs were wiped out 65 million years ago—smoke, soot, and dust in the atmosphere could block sunlight to such an extent that many plants in the region of the impact and beyond could not conduct photosynthesis at high enough levels to survive. And when plants die off, the animals and other species that rely on plants for energy die as well. Nearly all life on earth is completely dependent on the continued excitation of electrons by sunlight.

TAKE HOME MESSAGE 5.8

» When chlorophyll is hit by photons, the light energy excites an electron in the chlorophyll molecule, increasing the chlorophyll's potential energy. The excited electrons can be passed to other molecules, moving the potential energy through the cell.

5.9 Photosynthesis in detail: the energy of sunlight is captured as chemical energy.

Photosynthesis is a complex process, but our understanding can be greatly aided by remembering one phrase: FOLLOW THE ELECTRONS. Where are they coming from? What are they passing through? Where are they going? What will happen to them when they get there?

In the first part of photosynthesis, the "photo" part, sunlight hits the chloroplasts of a plant's leaves, and some of the energy in this sunlight is captured and stored in ATP and in another molecule, called NADPH, which stores energy by accepting high-energy electrons (FIGURE 5-17). After these transformations, the captured energy is ready to be used to make sugar molecules in the "synthesis" part of photosynthesis.

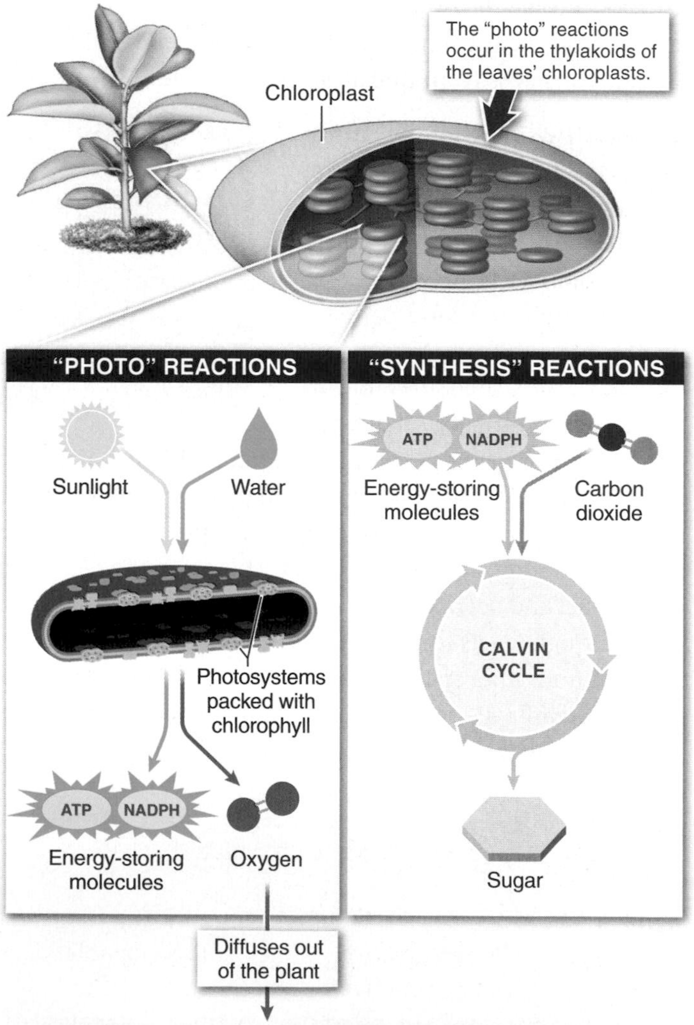

FIGURE 5-17 **Overview of the "photo" reactions.** Light energy is captured in the "photo" portion of photosynthesis. That energy is later used to power the building of sugar molecules.

The energy-capturing process occurs in two types of **photosystems** that are linked and work together. Embedded in the thylakoid membranes in the chloroplasts, these photosystems are structures composed of light-catching pigments (including chlorophyll) and proteins. As the pigments absorb photons, electrons in the pigments gain energy and become excited, and then return to their resting state, releasing energy. The released energy (but not the electrons) is transferred to neighboring pigment molecules. This process continues until the energy transferred among many pigment molecules makes its way to a chlorophyll *a* molecule at the center of the photosystem, and excites an electron there (FIGURE 5-18).

The chlorophyll *a* molecule at the center of the photosystems is special, differing from the other pigment molecules in one key feature. When its electrons are boosted to an excited state, they do not return to their resting, unexcited state. Instead, a nearby molecule, called the **primary electron acceptor,** acts like an electron vacuum, grabbing the excited electrons and leaving electron vacancies. This is where the electron journey begins.

Q Why must plants get water for photosynthesis to occur?

The electrons taken away from the special chlorophyll *a* molecule at the center of any photosystem must be replaced for photosynthesis to continue. Water supplies the replacement electrons in the first of the two linked photosystems. Molecules of water are continuously split near the special chlorophyll *a* molecule in this first photosystem. Four photons of light split two molecules of water into four electrons, four protons (H^+), and a molecule of O_2. A convenient and life-sustaining by-product of the splitting of water in photosynthesis is the oxygen that is released from the cell, a by-product essential for much of the life on earth, including all animal life.

Once the primary electron acceptor in the first photosystem grabs a high-energy electron from chlorophyll *a,* it passes it to another molecule, which passes it to another,

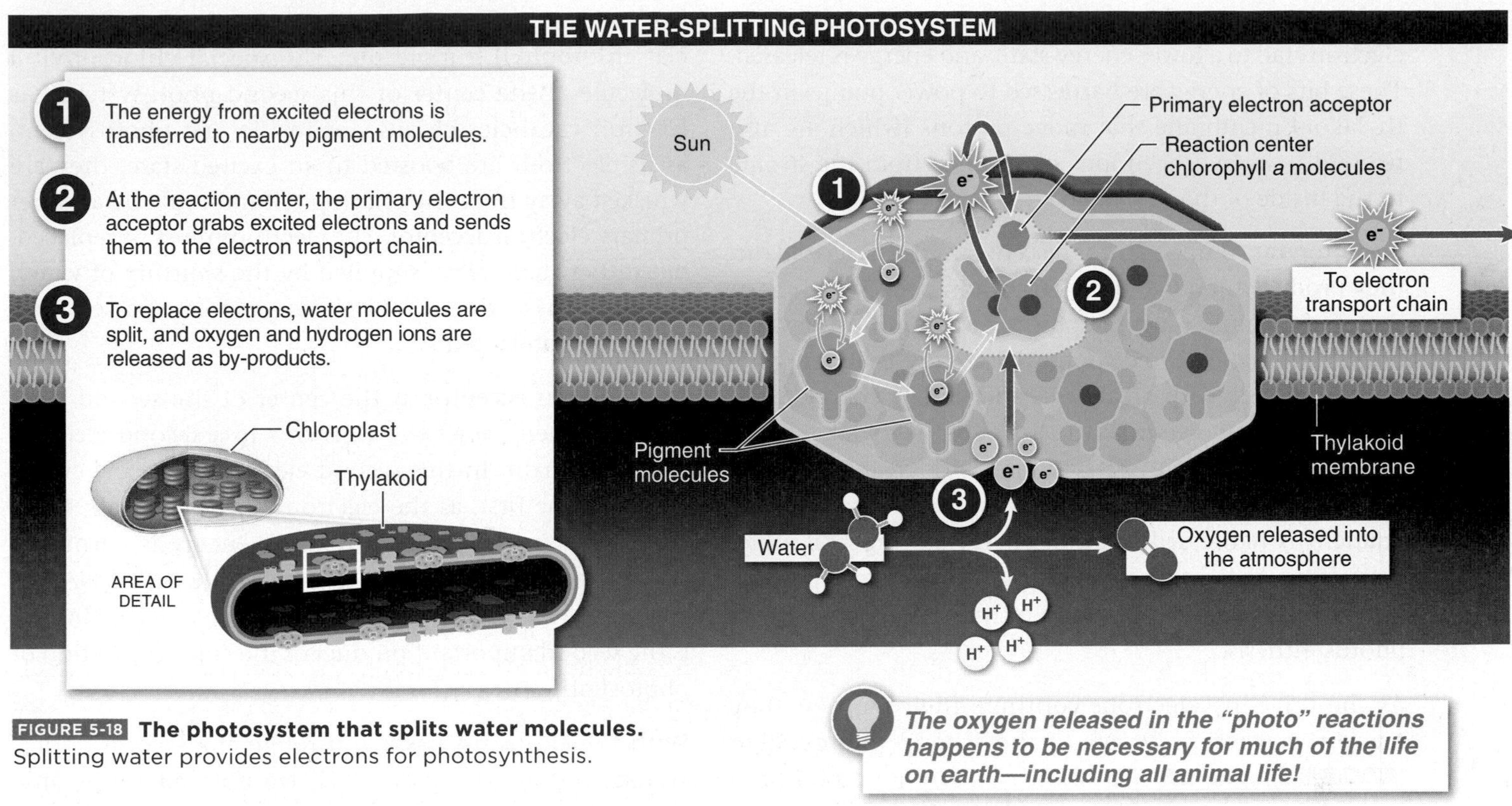

FIGURE 5-18 **The photosystem that splits water molecules.** Splitting water provides electrons for photosynthesis.

The oxygen released in the "photo" reactions happens to be necessary for much of the life on earth—including all animal life!

which passes it to yet another, in what is called an **electron transport chain** that links the first and second photosystems (FIGURE 5-19). This electron transport chain is embedded within the thylakoid membrane and consists of several protein complexes that hand off electrons from one to the next. At each step

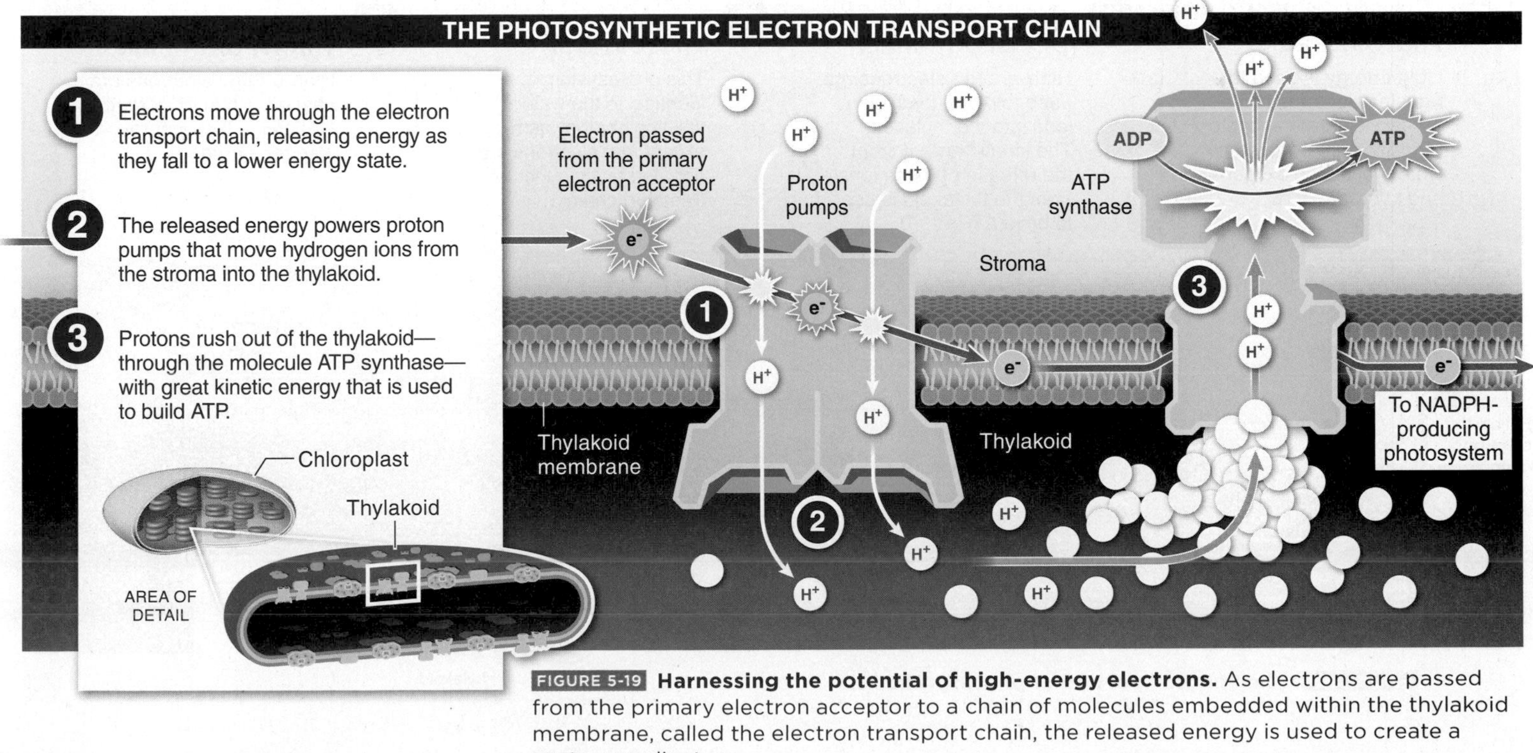

THE PHOTOSYNTHETIC ELECTRON TRANSPORT CHAIN
1 Electrons move through the electron transport chain, releasing energy as they fall to a lower energy state.
2 The released energy powers proton pumps that move hydrogen ions from the stroma into the thylakoid.
3 Protons rush out of the thylakoid—through the molecule ATP synthase—with great kinetic energy that is used to build ATP.
Chloroplast
Thylakoid
AREA OF DETAIL
Electron passed from the primary electron acceptor
Proton pumps
ATP synthase
ADP
ATP
Stroma
Thylakoid membrane
Thylakoid
To NADPH-producing photosystem

FIGURE 5-19 **Harnessing the potential of high-energy electrons.** As electrons are passed from the primary electron acceptor to a chain of molecules embedded within the thylakoid membrane, called the electron transport chain, the released energy is used to create a proton gradient.

in the electron transport chain's sequence of handoffs, the electrons fall to a lower energy state, and energy is released. These bits of energy are harnessed to power pumps in the thylakoid membrane that move protons (which are also referred to as hydrogen ions or H^+ ions) from the stroma to the inside of the thylakoid.

As protons are pumped into the center of the thylakoid sac, they produce higher and higher proton concentrations in that space. Think of a pump pushing water into an elevated tank, creating a store of potential energy that can gush out of the tank with great force and kinetic energy. Similarly, the protons eventually rush out of the thylakoid sacs, through the molecule ATP synthase, with great force—and the force of the protons moving down their concentration gradient is harnessed to build energy-storing ATP molecules from ADP and free-floating phosphate groups. This ATP is one of the two products of the "photo" portion of photosynthesis.

As the traveling electrons continue their journey, they reach the reaction center of the second photosystem (FIGURE 5-20). Like the first photosystem, the second photosystem has numerous pigments that harness photons from the sun and pass the light energy to another special chlorophyll *a* molecule. The special chlorophyll *a* molecule at the center of this second photosystem has electron vacancies because, as in the first photosystem, as its electrons are boosted to an excited state, they are whisked away from the chlorophyll molecule by another primary electron acceptor. The electrons must be replaced, but rather than being supplied by the splitting of water, the electrons in the second photosystem are passed from the first photosystem.

The electron acceptor at the center of the second photosystem then passes the electrons to a second electron transport chain. In this second electron transport chain, unlike in the first, as the electrons are handed off, there is no pumping of protons. Instead, energy is harnessed as the electrons are passed to a molecule called $NADP^+$, creating **NADPH,** a high-energy electron carrier. NADPH is the second important product of the "photo" portion of photosynthesis.

With the electrons' passage through the second photosystem and arrival in NADPH, we now have the final products of the "photo" part of photosynthesis (which is

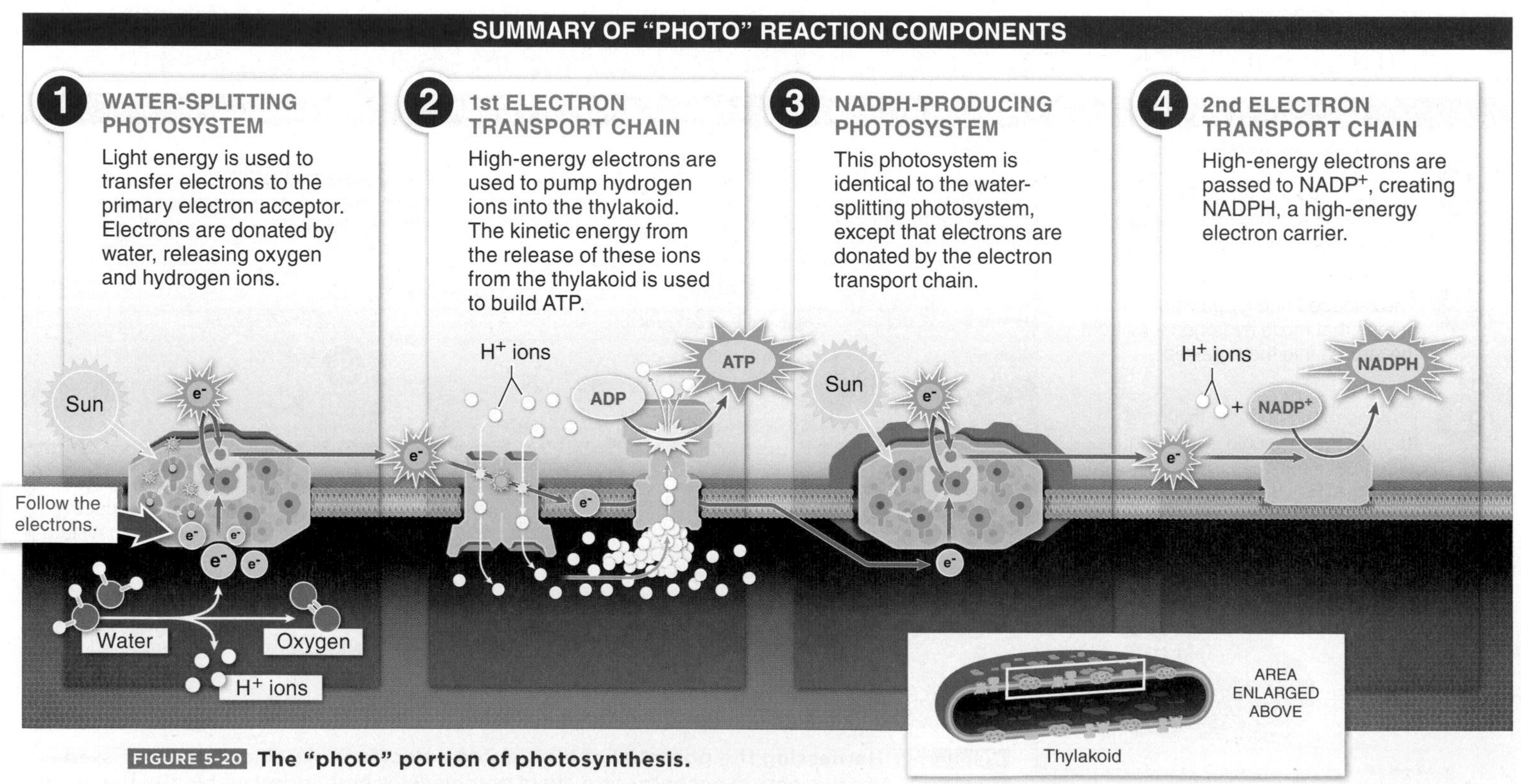

SUMMARY OF "PHOTO" REACTION COMPONENTS
1 WATER-SPLITTING PHOTOSYSTEM
Light energy is used to transfer electrons to the primary electron acceptor. Electrons are donated by water, releasing oxygen and hydrogen ions.
2 1st ELECTRON TRANSPORT CHAIN
High-energy electrons are used to pump hydrogen ions into the thylakoid. The kinetic energy from the release of these ions from the thylakoid is used to build ATP.
3 NADPH-PRODUCING PHOTOSYSTEM
This photosystem is identical to the water-splitting photosystem, except that electrons are donated by the electron transport chain.
4 2nd ELECTRON TRANSPORT CHAIN
High-energy electrons are passed to NADP+, creating NADPH, a high-energy electron carrier.
Sun
Follow the electrons.
Water
Oxygen
H+ ions
ADP
ATP
Sun
H+ ions
NADP+
NADPH
AREA ENLARGED ABOVE
Thylakoid

FIGURE 5-20 **The "photo" portion of photosynthesis.**

also called the "light-dependent reactions"): we've captured light energy from the sun and converted it to the chemical energy of ATP and the high-energy electron carrier NADPH (see Figure 5-20). But we haven't made any food yet. In the next section, we cover the "synthesis" part of photosynthesis and see how plants use the energy in ATP and NADPH to produce sugar from carbon dioxide.

TAKE HOME MESSAGE 5.9

» There are two parts to photosynthesis. The first is the "photo" part, in which light energy is transformed into chemical energy, while splitting water molecules and producing oxygen. The energy of sunlight is first captured when an electron in a chlorophyll molecule is excited. As this electron is passed from one molecule to another, energy is released at each transfer, some of which is used to build the energy-storage molecules ATP and NADPH.

5.10 Photosynthesis in detail: the captured energy of sunlight is used to make sugar.

The "synthesis" part of photosynthesis takes place in a series of chemical reactions called the **Calvin cycle.** If there is any part of photosynthesis that appears magical, it is the Calvin cycle. Just as a magician seems to make a rabbit appear from thin air, the Calvin cycle takes *invisible* gaseous molecules of CO_2 from the atmosphere and uses them to assemble molecules of sugar—sufficiently numerous that we can see them as tangible, solid, even edible, material.

All the Calvin cycle reactions occur in the stroma of the leaves' chloroplasts, outside the thylakoids. Plants carry out these reactions using the energy stored in the ATP and NADPH molecules that are built in the "photo" portion of photosynthesis. This dependency links the light-gathering ("photo") reactions with the sugar-building ("synthesis") reactions (FIGURE 5-21).

The processes in the Calvin cycle occur in three steps (FIGURE 5-22).

1. Fixation. First, using an enzyme called **rubisco,** plants pluck carbon from the air, where it occurs in the form of carbon dioxide (which has one carbon), and attach, or "fix," it to an organic molecule (which has five carbon atoms) within the chloroplast. Not surprisingly, given its role as the critical chemical that enables plants to build food molecules, rubisco is the most abundant protein on earth.

2. Sugar creation. The newly built molecule immediately splits in two, and each half is chemically modified. First a phosphate from ATP is added, and then each of the two molecules receives some high-energy electrons from NADPH. The result is two molecules of a three-carbon sugar called glyceraldehyde 3-phosphate (G3P). For the *net* synthesis of one molecule of G3P from carbon in the atmosphere, three "turns" of the Calvin cycle are required to fix three molecules of carbon dioxide.

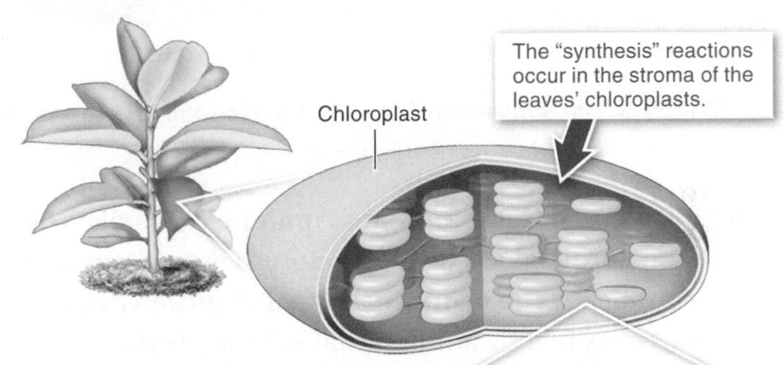

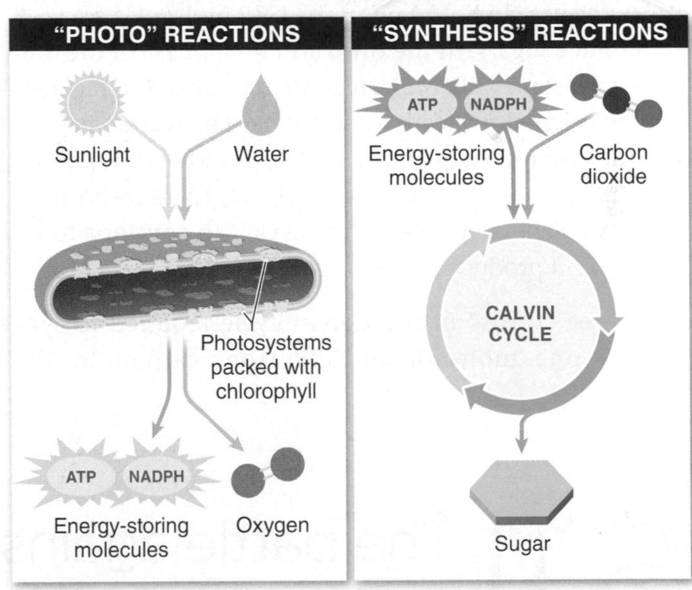

FIGURE 5-21 **Overview of the "synthesis" reactions of photosynthesis.**

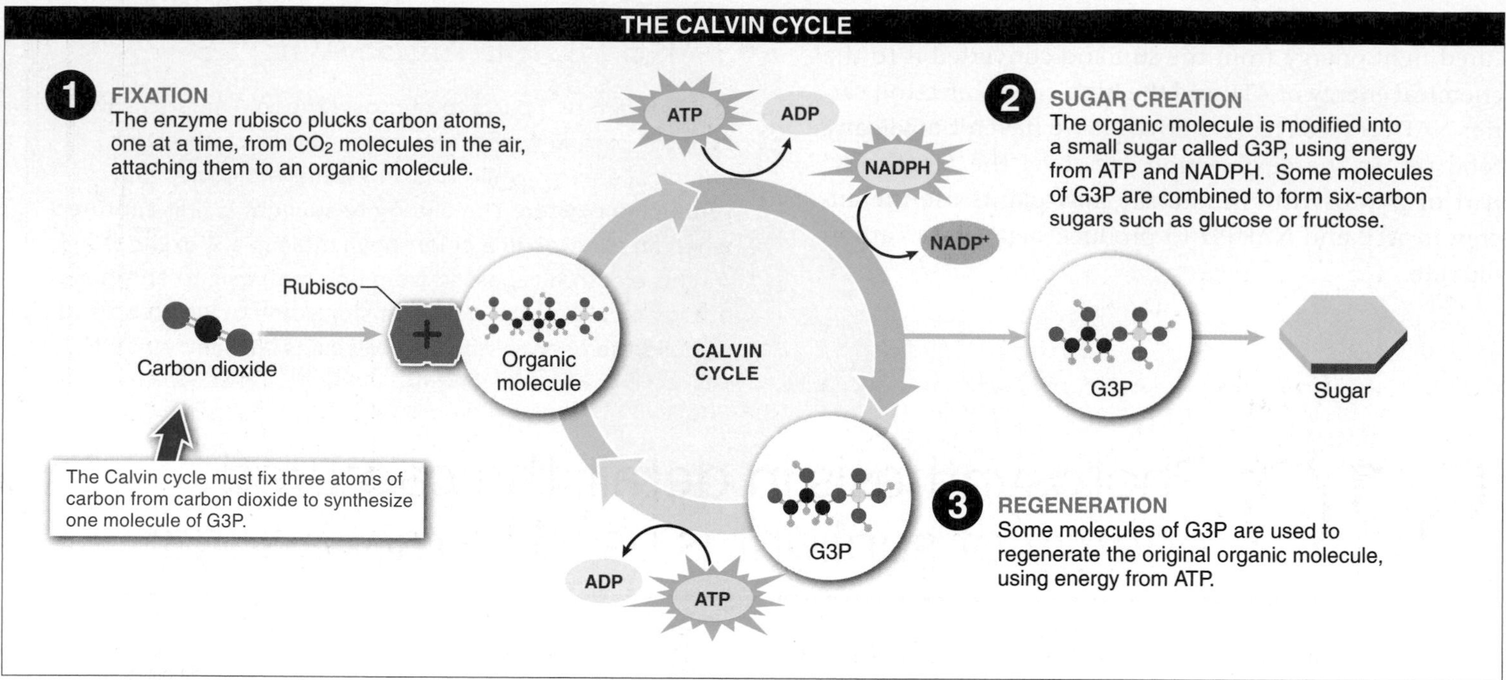

THE CALVIN CYCLE
1
FIXATION
The enzyme rubisco plucks carbon atoms, one at a time, from CO2 molecules in the air, attaching them to an organic molecule.
ATP
ADP
NADPH
NADP+
2
SUGAR CREATION
The organic molecule is modified into a small sugar called G3P, using energy from ATP and NADPH. Some molecules of G3P are combined to form six-carbon sugars such as glucose or fructose.
Rubisco
Carbon dioxide
Organic molecule
CALVIN CYCLE
G3P
Sugar
The Calvin cycle must fix three atoms of carbon from carbon dioxide to synthesize one molecule of G3P.
3
REGENERATION
Some molecules of G3P are used to regenerate the original organic molecule, using energy from ATP.
ADP
ATP
G3P

FIGURE 5-22 **Sugar synthesis.** The Calvin cycle—the "synthesis" portion of photosynthesis—produces sugars.

Some of the G3P molecules are combined to make the six-carbon sugars glucose and fructose. These sugars can be used as fuel by the plant, enabling it to grow. They can also be used as fuel by animals that eat the plant.

3. Regeneration. Not all of the G3P molecules are used to produce sugars. In the third and final phase of the Calvin cycle, some G3P molecules are rearranged to regenerate the original five-carbon molecule. The rearrangement of G3P requires energy from ATP produced in the "photo" reactions of photosynthesis. Once the five-carbon molecule is regenerated, the Calvin cycle can continue to fix carbon and produce molecules of G3P.

The three "turns" of the Calvin cycle required to synthesize one molecule of G3P from carbon in the atmosphere consume nine molecules of ATP and six molecules of NADPH generated in the "photo" reactions of photosynthesis.

TAKE HOME MESSAGE 5.10

» The "synthesis" part of photosynthesis is the Calvin cycle. During this phase, carbon from CO_2 in the atmosphere is attached to molecules in the stroma of chloroplasts, sugars are built, and molecules are regenerated to be used again in the Calvin cycle. The fixation, building, and regeneration processes consume energy from ATP and NADPH (the products of the "photo" part of photosynthesis).

5.11 The battle against world hunger can use plants adapted to water scarcity.

Sudan. Ethiopia. India. Somalia. Many of the world's regions with the highest rates of starvation are also places with the hottest, driest climates. This is not a coincidence. These climate conditions present difficult challenges for sustaining agriculture (FIGURE 5-23), and in the absence of stable crop yields, food production is unpredictable and the risk of starvation high. But evolutionary adaptations in some plants enable them to thrive in hot, dry conditions.

FIGURE 5-23 **Nowhere to hide.** Plants that lose too much water can't always survive in extremely hot, dry weather.

Recent technological advances in agriculture use these innovative evolutionary solutions to battle the problem of world hunger.

When it gets too hot and dry, animals can seek coolness in the shade. Plants, however, are anchored in place. Consequently, plants in hot, dry climates can lose significant amounts of water through evaporation. Evaporation is a problem for plants because water is essential to photosynthesis, growth, and the transport of nutrients. Without water, plants die.

To combat water loss through evaporation plants can close their **stomata** (*sing.* stoma), small pores usually on the underside of leaves (FIGURE 5-24). These openings are the primary sites for gas exchange in plants: carbon dioxide for photosynthesis enters, and oxygen generated as a by-product in photosynthesis exits through them. When open, the stomata also allow water to evaporate from the plant. Closing their stomata solves one problem for plants (too much water evaporation) but creates another: with the stomata shut, oxygen from the "photo" reactions of photosynthesis cannot be released from the chloroplasts, and carbon dioxide cannot enter. And without carbon dioxide molecules for sugar production, the Calvin cycle cannot proceed. Plant growth comes to a standstill and crops fail.

In some plants, including corn and sugarcane, a process has evolved that minimizes water loss but still enables the plants to make sugar when the weather is hot and dry. In the process called **C4 photosynthesis,** these plants add an extra set of steps to the usual process of photosynthesis, which is usually called C3 photosynthesis (FIGURE 5-25). In these steps, the plants use an enzyme that functions like the ultimate "CO_2-sticky tape." This enzyme has a tremendously strong attraction for carbon dioxide; it can find and bind carbon even when CO_2 concentration is very low. As a consequence, the plant's stomata can be opened just slightly, reducing evaporation and admitting a little CO_2. (In contrast, the enzyme rubisco, shown in Figure 5-22, functions poorly when CO_2 is scarce, necessitating greater stomata opening.)

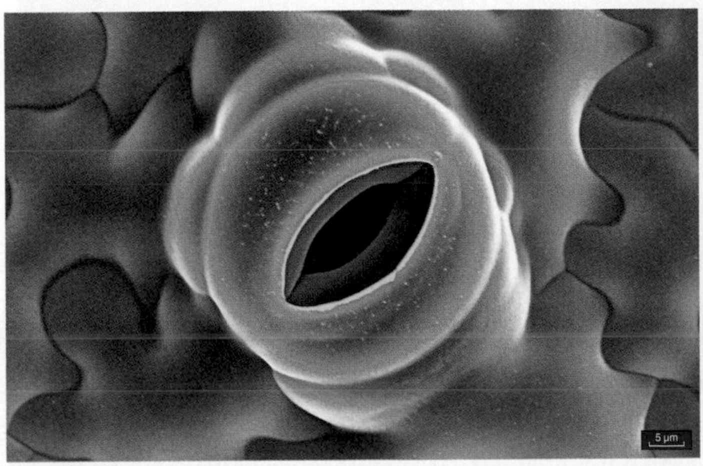

Stoma open

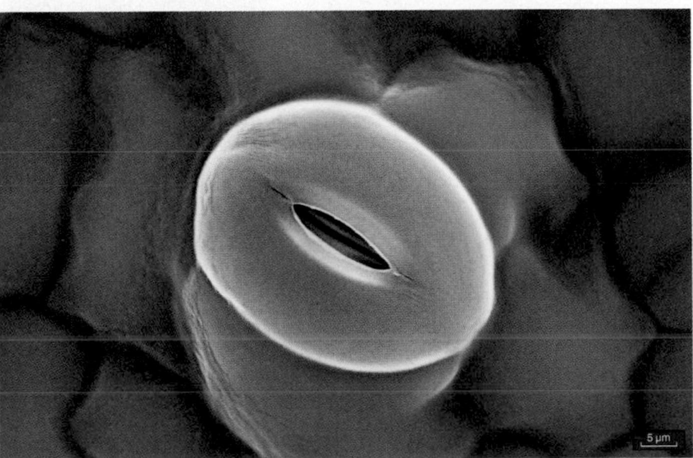

Stoma closed

FIGURE 5-24 **Plant stomata.** Carbon dioxide enters a plant through stomata, but water can be lost through the same openings.

COMPARISON OF PHOTOSYNTHESIS METHODS

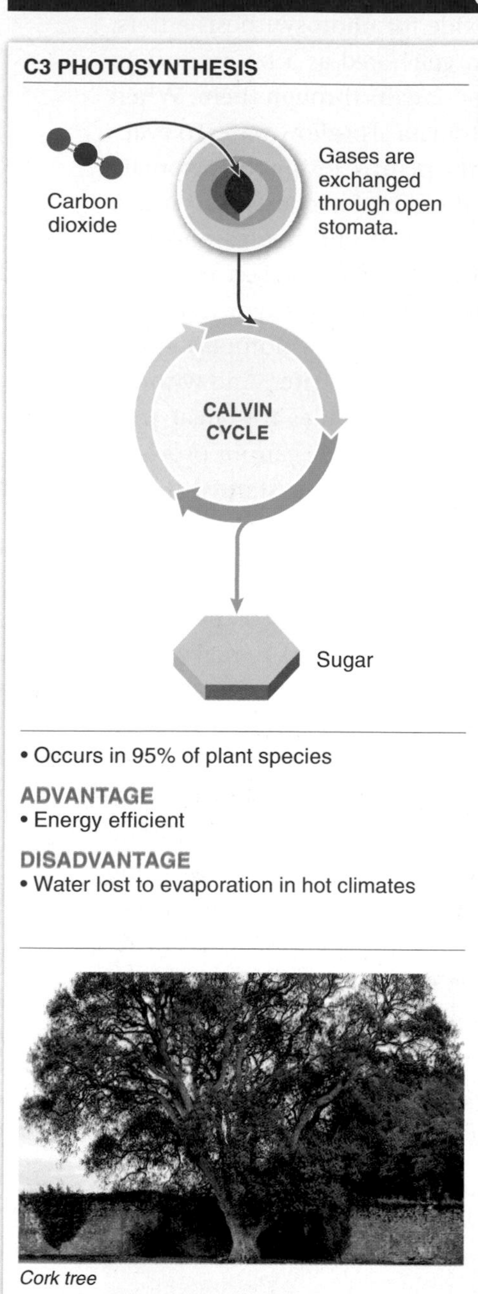

C3 PHOTOSYNTHESIS
Carbon dioxide
Gases are exchanged through open stomata.
CALVIN CYCLE
Sugar
• Occurs in 95% of plant species
ADVANTAGE
• Energy efficient
DISADVANTAGE
• Water lost to evaporation in hot climates

Cork tree

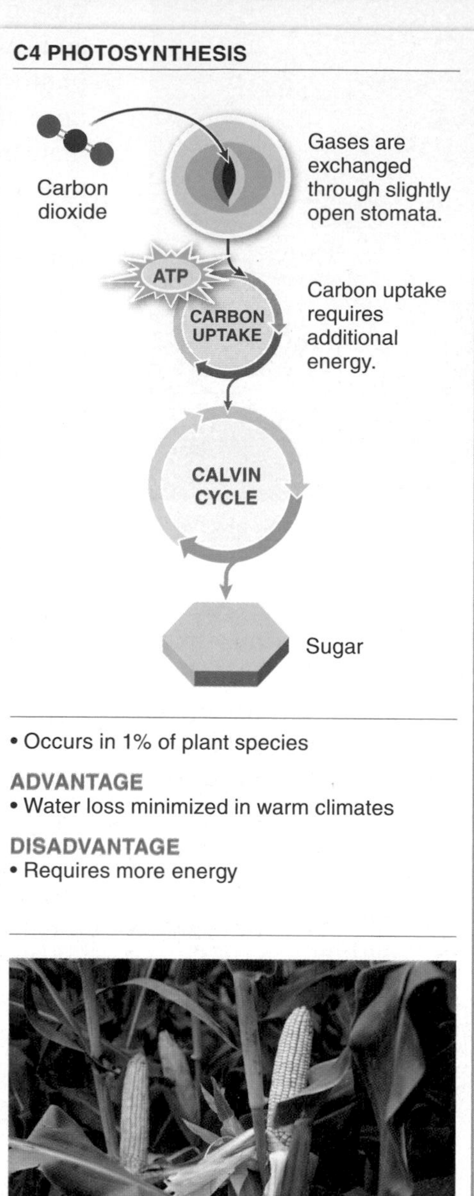

Corn

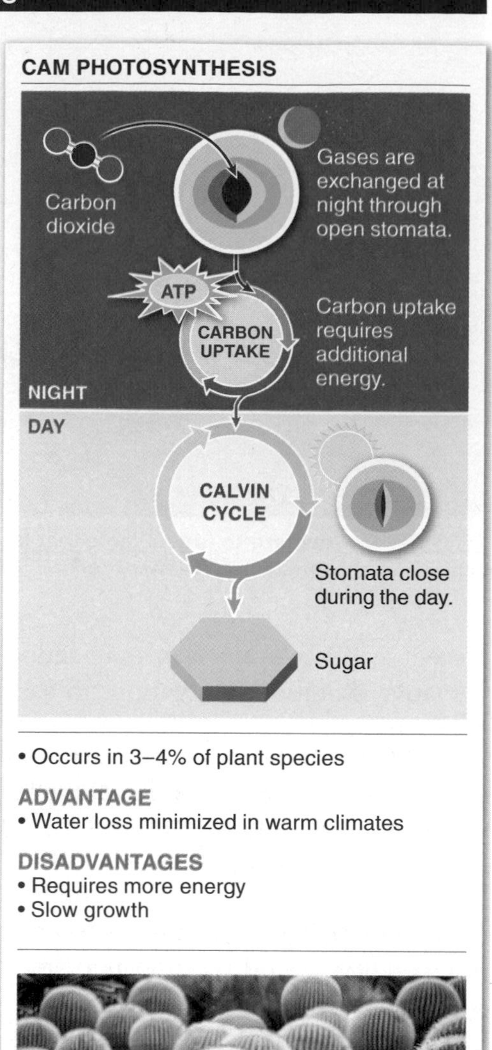

CAM PHOTOSYNTHESIS
Carbon dioxide
Gases are exchanged at night through open stomata.
ATP
CARBON UPTAKE
Carbon uptake requires additional energy.
NIGHT
DAY
CALVIN CYCLE
Stomata close during the day.
Sugar
• Occurs in 3–4% of plant species
ADVANTAGE
• Water loss minimized in warm climates
DISADVANTAGES
• Requires more energy
• Slow growth

Golden barrel cactus

FIGURE 5-25 **C3, C4, and CAM photosynthesis.**

> **Q** How might global warming be bad for agriculture?

This adaptation has a catch, though. The extra steps in C4 photosynthesis require the plant to expend additional energy. Specifically, every time the plant takes up a molecule of CO_2 with the "CO_2-sticky tape" enzyme, the plant must use an extra molecule of ATP (required for the production of that enzyme). It is acceptable to pay this energy cost only when the climate is so hot and dry that the plant would otherwise have to close its stomata and completely shut down all sugar production. In a mild climate, however, plants conducting C4 photosynthesis would be out-competed by the more efficient plants conducting standard C3 photosynthesis. Not surprisingly, we see few C4 plants in the temperate regions of the world and little C4 photosynthesis among photosynthetic organisms

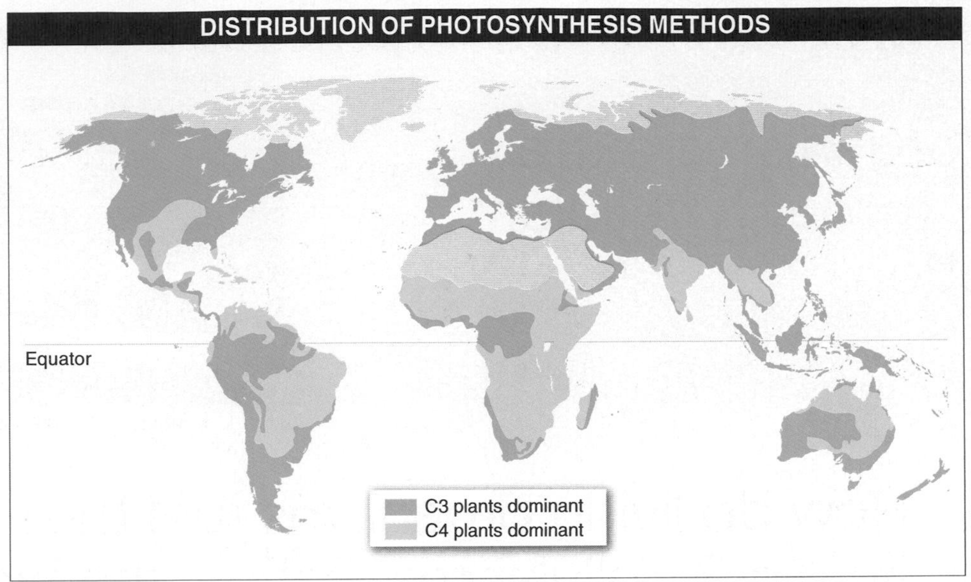

FIGURE 5-26 **Global distribution of C3 and C4 plants.**

living in the oceans. But in hot, dry regions, C4 plants are dominant and displace the C3 plants wherever both occur (FIGURE 5-26). With global warming, many scientists expect to see a gradual expansion of the ranges over which C4 plants grow, and non-C4 plants pushed farther and farther away from the equator.

A third and similar method of carbon fixation, called **CAM** (for "crassulacean acid metabolism"), is also found in hot, dry areas. In this method, used by many cacti, pineapples, and other fleshy, juicy plants, the plants close their stomata during hot, dry days. At night, they open the stomata and let CO_2 into the leaves, where it binds temporarily to a holding molecule. During the day, the holding molecule gradually releases CO_2, enabling photosynthesis to proceed while keeping the stomata closed to reduce water loss (see Figure 5-25). A disadvantage of CAM photosynthesis is that by completely closing their stomata during the day, CAM plants significantly reduce the total amount of CO_2 they can take in. As a consequence, they have much slower growth rates and cannot compete well with non-CAM plants under any conditions other than extreme dryness.

C4 and CAM photosynthesis originally evolved because they made it possible for plants to grow better in the world's hot and dry regions. Researchers are now experimenting with these adaptations as a way to fight world hunger. They have introduced into rice plants several genes from corn that code for the C4 photosynthesis enzymes. Once in the rice, these genes increase the rice plant's ability to photosynthesize, leading to higher growth rates and food yields. Whether the addition of C4 photosynthesis enzymes will make it possible to grow new crops on a large scale in previously inhospitable environments is not certain. Early results suggest, however, that this is a promising approach.

TAKE HOME MESSAGE 5.11

» C4 and CAM photosynthesis are evolutionary adaptations at the biochemical level that, although more energetically expensive than regular (C3) photosynthesis, allow plants in hot, dry climates to partly or fully close their stomata and conserve water without shutting down photosynthesis.

5.12–5.16 Living organisms extract energy through cellular respiration.

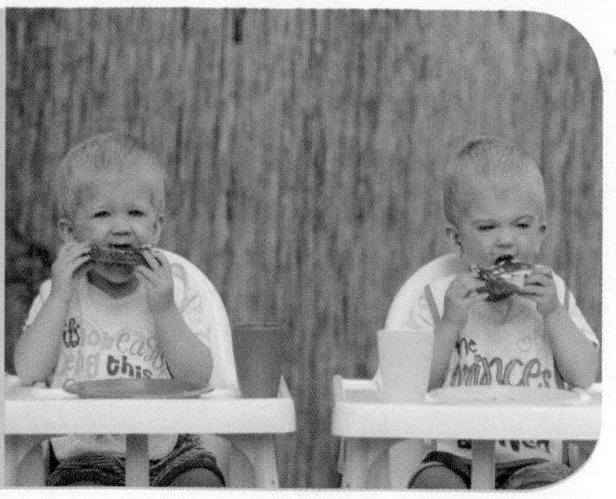
A pit stop at the toddler refueling station.

5.12 How do living organisms fuel their actions? Cellular respiration: the big picture.

Food is fuel. And all the activities of life—growing, moving, reproducing—require fuel. Plants, most algae, and some bacteria obtain their fuel directly from the energy of sunlight, which they harness through photosynthesis. Less self-sufficient organisms, such as humans, alligators, and insects, must extract the energy they need from the food they eat. This energy comes from photosynthetic organisms either directly (from eating plants) or indirectly (from eating animals that eat plants) (FIGURE 5-27).

All living organisms—including plants (a fact that is often overlooked)—extract energy from the chemical bonds of molecules (which can be considered "food") through a process called cellular respiration (FIGURE 5-28). This process is a bit like photosynthesis in reverse. In photosynthesis, the energy of the sun is captured and used to build molecules of sugars, such as glucose. In cellular respiration, plants and animals break down the chemical bonds of sugars and other energy-rich food molecules (such as fats and proteins) to release the energy that went into creating them. (Don't confuse cellular respiration with the act of breathing, which is also called *respiration*.) As energy is released, cells capture and store it in the bonds of ATP molecules. This plentiful, readily available stored energy can then be tapped as needed to fuel the work of the life-sustaining activities and processes of all living organisms.

Forest plants collecting light energy from the sun through photosynthesis

Caterpillar (herbivore) feeding on a leaf

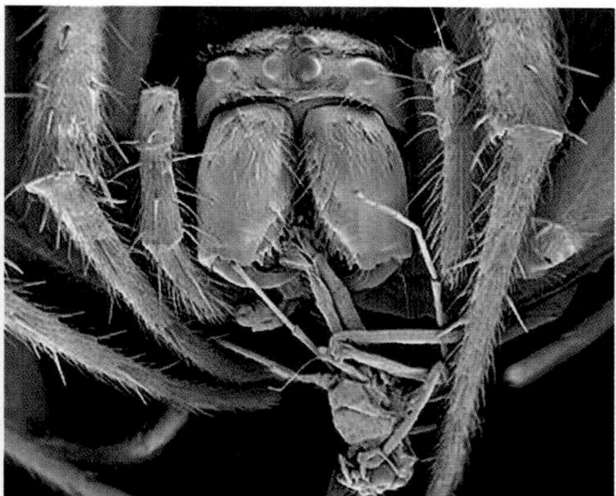
Garden spider (carnivore) feeding on an insect

FIGURE 5-27 **Living organisms require fuel (in one form or another).**

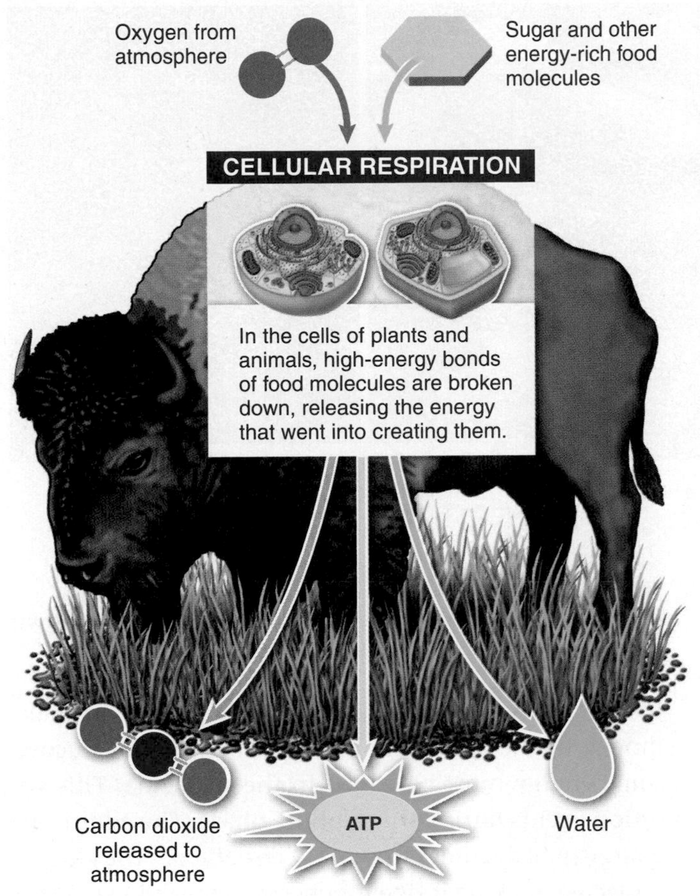

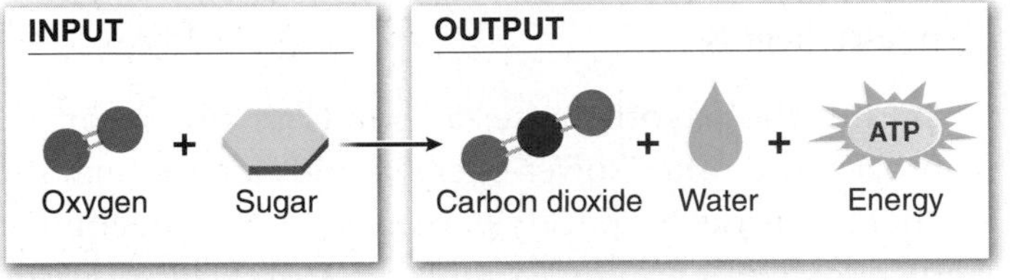

FIGURE 5-28 **Cellular respiration: the big picture.**

In humans, as in other animals, cellular respiration starts after we eat food, digest it, absorb the nutrient molecules into the bloodstream, and deliver them to the cells of our bodies. At this point, our cells begin to extract some of the energy stored in the bonds of the food molecules. We focus here on the breakdown of glucose, but later in this chapter we'll see that the process is similar for the breakdown of fats or lipids. Ultimately, when a food molecule has been completely processed, the cell has used the food molecule's stored energy (along with oxygen) to create a large number of high-energy-storing ATP molecules (which supply energy to power the cell's activities), water, and carbon dioxide (which is exhaled into the atmosphere).

TAKE HOME MESSAGE 5.12

» Living organisms extract energy through a process called cellular respiration, in which the high-energy bonds of sugars and other energy-rich molecules are broken, releasing the energy that went into creating them. The cell captures the food molecules' stored energy in the bonds of ATP molecules. This process requires fuel molecules and oxygen, and it yields ATP molecules, water, and carbon dioxide.

5.13 The first step of cellular respiration: glycolysis is the universal energy-releasing pathway.

To generate energy, fuels such as glucose and other carbohydrates, as well as proteins and lipids (after some preliminary modifications), are broken down in three steps: (1) glycolysis, (2) the citric acid cycle, and (3) the electron transport chain. **Glycolysis** means the splitting (*lysis*) of sugar (*glyco-*), and it is the first step that all organisms on the planet take to capture energy from the breakdown of food molecules; for many single-celled organisms, this one step is sufficient to provide all of the energy they need (FIGURE 5-29).

Every living organism, large or small, extracts energy through glycolysis!

FIGURE 5-29 Plants, animals, bacteria, and all other organisms use glycolysis to break down fuels.

Glycolysis is a sequence of chemical reactions (there are 10 in all), taking place in the cytoplasm of the cell, through which glucose is broken down to form two molecules of a substance called **pyruvate.** Glycolysis has two distinct phases: an "uphill" preparatory phase and a "downhill" payoff phase. A brief overview of the process is shown in FIGURE 5-30.

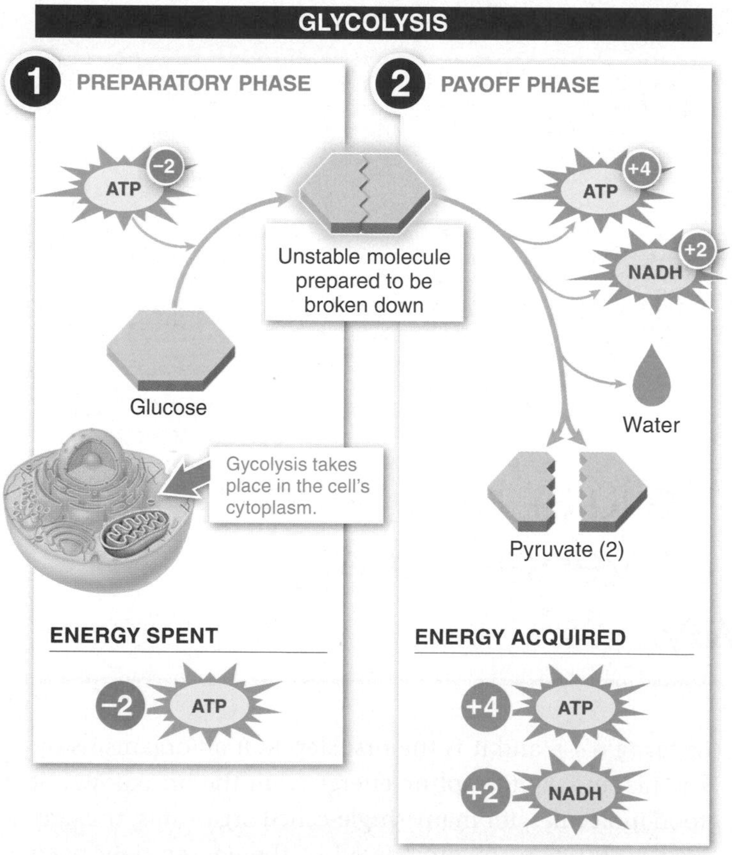

GLYCOLYSIS

1 PREPARATORY PHASE

ATP −2

Glucose

Unstable molecule prepared to be broken down

Gycolysis takes place in the cell's cytoplasm.

ENERGY SPENT

−2 ATP

2 PAYOFF PHASE

ATP +4

NADH +2

Water

Pyruvate (2)

ENERGY ACQUIRED

+4 ATP

+2 NADH

FIGURE 5-30 Glycolysis up close.

Just as you sometimes have to spend money to make money, before any energy can be extracted from glucose, some energy must be added to the molecule. This addition occurs during the "uphill" phase. The additional energy (which comes from ATP) destabilizes the glucose molecule, making it ripe for chemical breakdown. Once the glucose can be broken down chemically, the "payoff phase" begins.

During the payoff phase, as bonds in the sugar are broken and other, lower-energy bonds are formed, the energy released is quickly harnessed by the attachment of phosphate groups to molecules of ADP to create energy-rich ATP molecules. In another energy-yielding step of glycolysis, electrons originally from the glucose are transferred to NAD^+ to become the high-energy electron carrier NADH. Later (in an electron transport chain in the mitochondria), this energy will be converted to ATP.

The net result of glycolysis is that each glucose molecule is broken down into two molecules of pyruvate. During this breakdown, some of the released energy is captured in the production of energy-rich ATP molecules and molecules of the high-energy electron carrier NADH. Two molecules of water are also produced during glycolysis.

Glycolysis can proceed regardless of whether any oxygen is present. In the absence of oxygen, and in many yeasts and bacteria, glycolysis is the only means for fueling activity. Because single-celled organisms have much lower energy needs, they can function solely on the yields of glycolysis. When oxygen is present, however, glycolysis isn't the end of the story. For many organisms (including humans), glycolysis is a springboard to further energy extraction. The additional energy payoffs come from the citric acid cycle and the electron transport chain.

TAKE HOME MESSAGE 5.13

» Glycolysis is the initial phase in the process by which all living organisms harness energy from food molecules. Glycolysis occurs in a cell's cytoplasm and uses the energy released from breaking chemical bonds in food molecules to produce high-energy molecules, ATP and NADH.

5.14 The second step of cellular respiration: the citric acid cycle extracts energy from sugar.

Cells could stop extracting energy when glycolysis ends, but they rarely do so, because that would be like leaving most of your meal on your plate. In glycolysis, only a small fraction of the energy stored in sugar molecules is recovered and converted to ATP and NADH. Cells get much more of an "energy bang" for their "food buck" in the steps following glycolysis, which occur in the mitochondria, in the presence of oxygen. This is why mitochondria are considered ATP "factories."

In the mitochondria, the molecules produced from the breakdown of glucose during glycolysis are further broken down. The **citric acid cycle,** also known as the **Krebs cycle,** produces some additional molecules of ATP and, more importantly, captures a huge amount of chemical energy by producing high-energy electron carriers.

Q Aerobic training can cause our bodies to produce more mitochondria in cells. Why is this beneficial?

Before the citric acid cycle can begin, however, the end products of glycolysis—two molecules of pyruvate for every molecule of glucose used—must be modified. First, the pyruvate molecules move from the cytoplasm into the mitochondria, where they undergo three quick modifications that prepare them to be broken down in the citric acid cycle (FIGURE 5-31).

Modification 1. Each pyruvate molecule passes a pair of its high-energy electrons (and a proton) to the

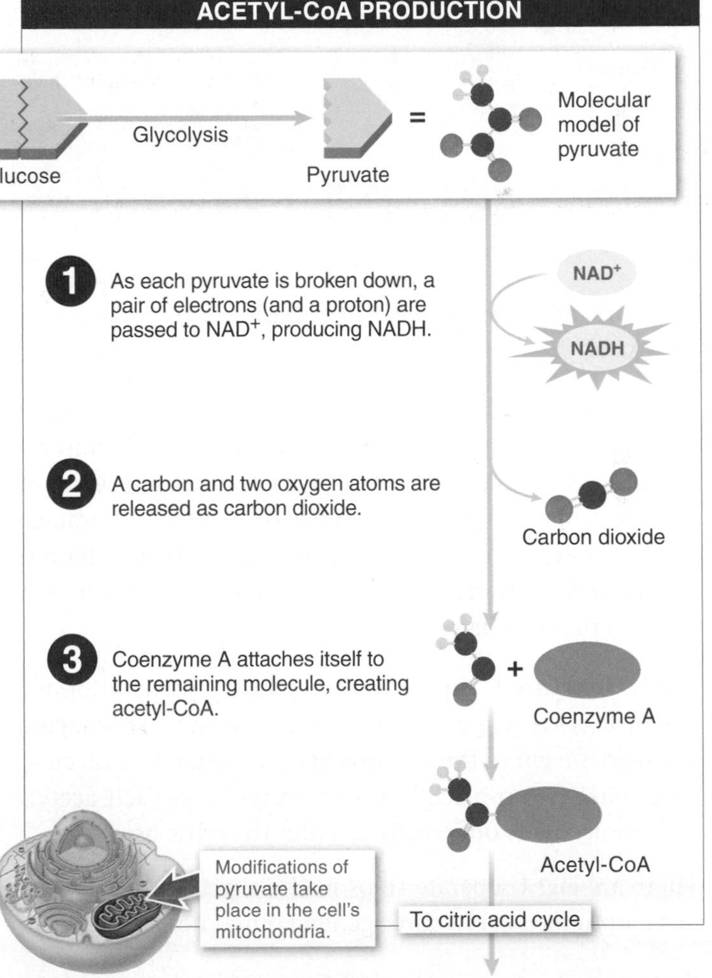

ACETYL-CoA PRODUCTION
Glucose
Glycolysis
Pyruvate
=
Molecular model of pyruvate
1 As each pyruvate is broken down, a pair of electrons (and a proton) are passed to NAD+, producing NADH.
NAD+
NADH
2 A carbon and two oxygen atoms are released as carbon dioxide.
Carbon dioxide
3 Coenzyme A attaches itself to the remaining molecule, creating acetyl-CoA.
+
Coenzyme A
Acetyl-CoA
Modifications of pyruvate take place in the cell's mitochondria.
To citric acid cycle

FIGURE 5-31 **Preparation of pyruvate.** In the mitochondria, pyruvate must be modified before it can be broken down in the citric acid cycle.

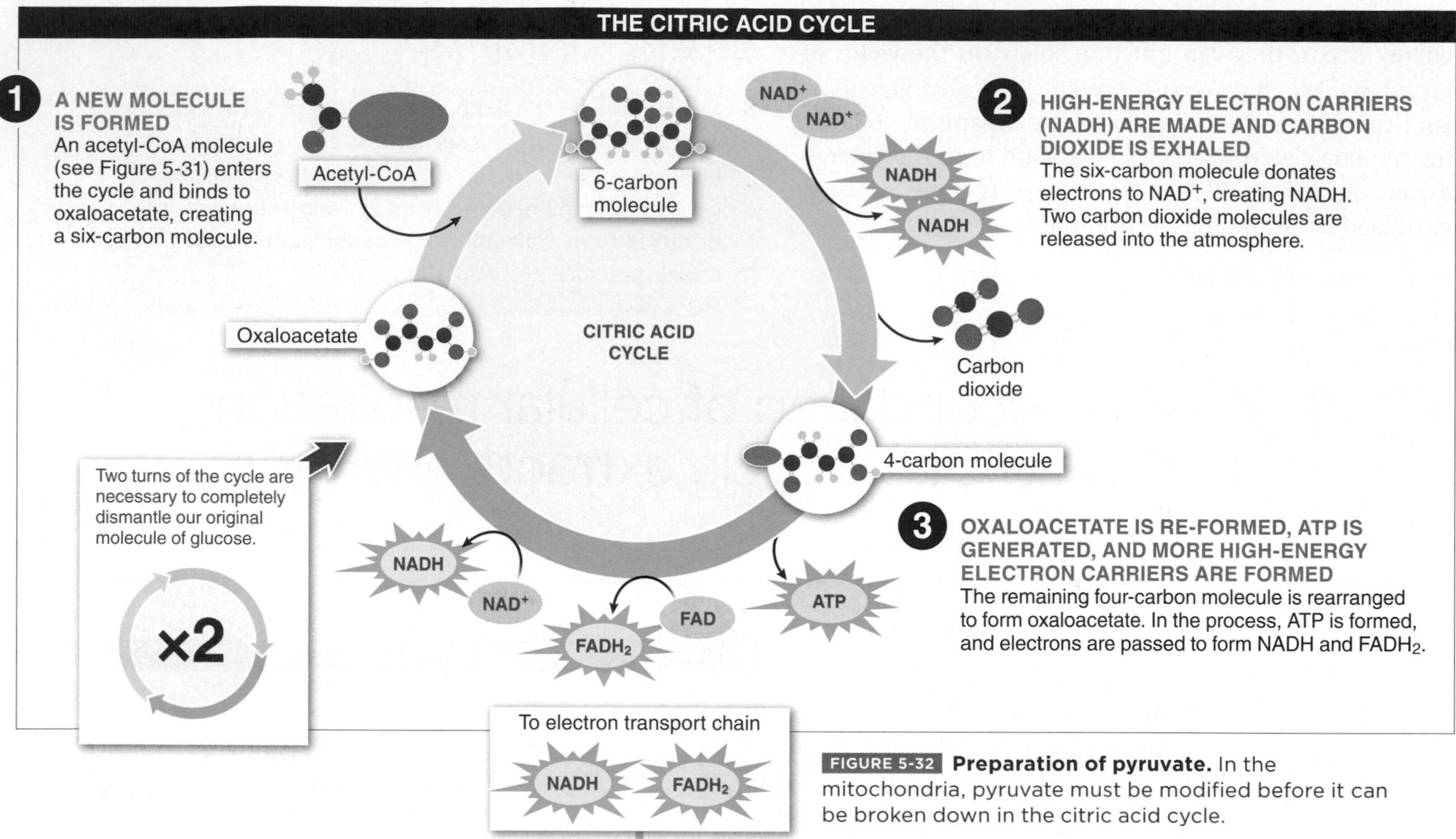

THE CITRIC ACID CYCLE
1 A NEW MOLECULE IS FORMED
An acetyl-CoA molecule (see Figure 5-31) enters the cycle and binds to oxaloacetate, creating a six-carbon molecule.
Acetyl-CoA
6-carbon molecule
NAD+
NAD+
NADH
NADH
2 HIGH-ENERGY ELECTRON CARRIERS (NADH) ARE MADE AND CARBON DIOXIDE IS EXHALED
The six-carbon molecule donates electrons to NAD+, creating NADH. Two carbon dioxide molecules are released into the atmosphere.
Oxaloacetate
CITRIC ACID CYCLE
Carbon dioxide
4-carbon molecule
Two turns of the cycle are necessary to completely dismantle our original molecule of glucose.
×2
NADH
NAD+
FAD
FADH2
ATP
3 OXALOACETATE IS RE-FORMED, ATP IS GENERATED, AND MORE HIGH-ENERGY ELECTRON CARRIERS ARE FORMED
The remaining four-carbon molecule is rearranged to form oxaloacetate. In the process, ATP is formed, and electrons are passed to form NADH and FADH2.
To electron transport chain
NADH
FADH2

FIGURE 5-32 **Preparation of pyruvate.** In the mitochondria, pyruvate must be modified before it can be broken down in the citric acid cycle.

electron-carrier molecule NAD+, building two molecules of NADH.

Modification 2. Next, a carbon atom and two oxygen atoms are removed from each pyruvate molecule and released as carbon dioxide. The CO_2 molecules diffuse out of the cell and, eventually, out of the organism. In humans, for example, these CO_2 molecules pass into the bloodstream and are transported to the lungs, from which they are eventually exhaled.

Modification 3. In the final step in the preparation for the citric acid cycle, a giant compound known as coenzyme A attaches itself to the remains of each pyruvate molecule, producing two molecules called acetyl-CoA. Each acetyl-CoA molecule is now ready to enter the citric acid cycle.

There are eight separate steps in the citric acid cycle, but our emphasis is on its three general outcomes (FIGURE 5-32).

Outcome 1. *A new molecule is formed.* Acetyl-CoA adds its two-carbon acetyl group to the starting material of the citric acid cycle, a four-carbon chemical called oxaloacetate, creating a six-carbon molecule (called *citrate*).

Outcome 2. *High-energy electron carriers (NADH) are made and carbon dioxide is exhaled.* The six-carbon molecule then gives electrons to NAD+ to make the high-energy electron carrier NADH. (Don't forget that the main purpose of the citric acid cycle is the capture of energy.) The six-carbon molecule also releases two carbon atoms along with four oxygen atoms to form two carbon dioxide molecules. In mammals (including humans), this CO_2 is carried by the bloodstream to the lungs.

Outcome 3. *The starting material of the citric acid cycle is re-formed, ATP is generated, and more high-energy electron carriers are formed.* After the CO_2 is released, the remaining four-carbon molecule is modified and rearranged to once again form oxaloacetate, the starting material of the citric acid cycle. In the process of this reorganization, one ATP molecule is generated, and more electrons are passed to NAD+ and a molecule called FAD to form NADH and $FADH_2$, both of which are high-energy electron carriers. The formation of these high-energy electron carriers increases the energy yield of the citric acid cycle. One oxaloacetate is re-formed, and the cycle is ready to break down the second molecule of acetyl-CoA. Two turns of the cycle are necessary to completely dismantle our original molecule of glucose.

Now that we have seen the citric acid cycle in its entirety, let's trace the path of the six carbons in the original glucose molecule. In a sense, the carbon atoms that were first plucked from the atmosphere to make sugar during photosynthesis have been exhaled back into the atmosphere as six molecules of carbon dioxide.

1. Glycolysis: the six-carbon starting point. Glucose is broken down into two molecules of pyruvate. No carbons are removed.
2. Preparation for the citric acid cycle: two carbons are released. Two pyruvate molecules are modified to enter the citric acid cycle, and both lose a carbon atom, forming two molecules of carbon dioxide.
3. Citric acid cycle: the last four carbons are released. A total of four carbon atoms enter the citric acid cycle in the form of two molecules of acetyl-CoA, entering one at a time. For each turn of the citric acid cycle, two molecules of carbon dioxide are released. So the two final carbons are released into the atmosphere during the second turn of the wheel. Poof! The six carbon atoms originally present in our single molecule of glucose are no longer present in the organism.

So we've come full circle. In photosynthesis, carbon atoms from the atmosphere were used to build sugar molecules, which had energy stored in the bonds between their carbon, hydrogen, and oxygen atoms. In cellular respiration, the energy previously stored in the bonds of the sugar is converted to molecules of ATP, NADH, and $FADH_2$. Carbon atoms from the sugar are exhaled back into the atmosphere as CO_2, and water is produced.

What happens to the high-energy electron carriers, NADH and $FADH_2$? They eventually give up their high-energy electrons to the final stage of cellular respiration, the electron transport chain. The energy released as those electrons pass through the transport chain is captured in the bonds of more ATP molecules. We explore that process in the next section.

Mitochondrial malfunctions have serious consequences for health. More than a hundred genetic mitochondrial disorders have been identified, all of which can lead to energy shortage, including muscle weakness, fatigue, and muscle pain. And, as noted in Section 4.15, many cases of extreme fatigue or cramps that occur after only slight exertion may be related to inherited mutations in the mitochondrial DNA.

Q Why might the malfunctioning of mitochondria play a role in lethargy or fatigue?

TAKE HOME MESSAGE 5.14

» Much additional energy can be harvested by cells after glycolysis. First, pyruvate is chemically modified. Then, in the citric acid cycle, the modified pyruvate is broken down. This breakdown releases carbon into the atmosphere (as CO_2) as bonds are broken and captures some of the released energy in two ATP molecules and many high-energy electron carriers for every glucose molecule.

5.15 The third step of cellular respiration: ATP is built in the electron transport chain.

How do we finally get a big payoff of usable energy from our glucose molecule? It is the energy held in the high-energy electron carriers NADH and $FADH_2$, formed in glycolysis and the citric acid cycle, that ultimately generates the largest amount of usable energy as ATP. In fact, almost 90% of the energy payoff from a molecule of glucose is harvested in the final step of cellular respiration, when the electrons from NADH and $FADH_2$ move along an electron transport chain. This process, like the citric acid cycle, takes place in the mitochondria.

In a manner similar to that seen in the chloroplast during photosynthesis, mitochondria convert kinetic energy (from electrons) into potential energy (a concentration gradient of protons). Two structural features of mitochondria are essential to their impressive ability to harness energy from molecules.

Feature 1. Mitochondria have a "bag-within-a-bag" structure. Consequently, material inside the mitochondrion can lie in one of two spaces: (1) in the intermembrane space, which is outside the inner bag, or (2) in the **mitochondrial matrix,** which is inside the inner bag. With two distinct regions separated by a membrane, the mitochondrion can create higher concentrations of molecules in one area or the other, creating a concentration gradient (FIGURE 5-33).

Feature 2. The inner bag of the mitochondrion is studded with molecules, mostly electron carriers, that are sequentially arranged as a "chain." This arrangement enables the molecules to hand off electrons in an orderly sequence.

Now let's explore how these features of mitochondria make it possible to harness energy from high-energy electron carriers (FIGURE 5-34).

Q Over-the-counter NADH pills provide energy to sufferers of chronic fatigue syndrome. Why?

Step 1 of the electron transport chain begins with NADH and $FADH_2$ in the mitochondrial matrix (inside the inner bag) moving to the membrane. There, the high-energy electrons they carry are transferred to molecules embedded within the membrane. After they donate their electrons, the molecules that remain, NAD^+ and FAD, are recycled back to the citric acid cycle.

The membrane-embedded molecules pass the electrons to the next carrier, which passes the electrons to the next, and

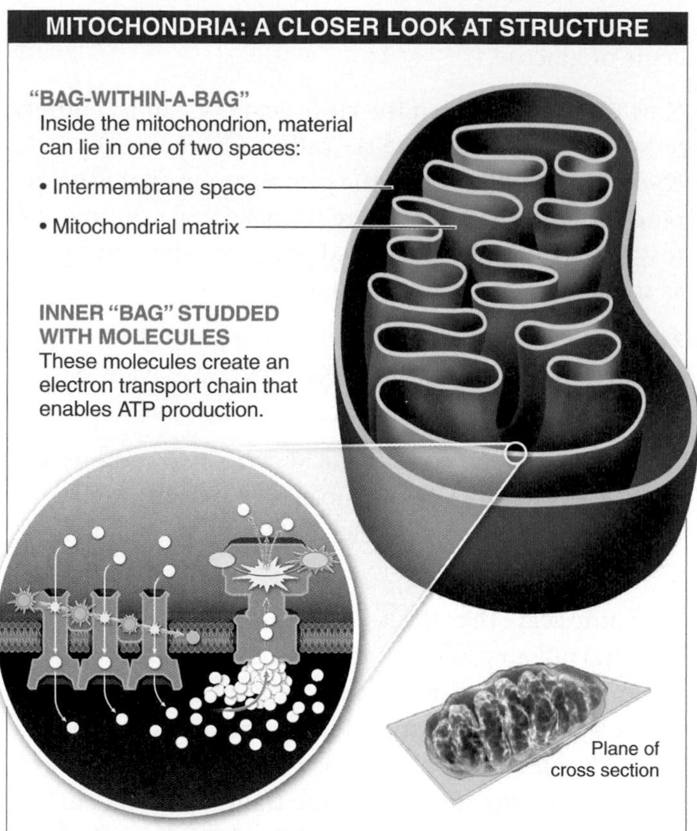

MITOCHONDRIA: A CLOSER LOOK AT STRUCTURE
"BAG-WITHIN-A-BAG"
Inside the mitochondrion, material can lie in one of two spaces:
• Intermembrane space
• Mitochondrial matrix
INNER "BAG" STUDDED WITH MOLECULES
These molecules create an electron transport chain that enables ATP production.
Plane of cross section

FIGURE 5-33 **"A bag-within-a-bag."** The structure of mitochondria makes possible their impressive ability to harness energy from food molecules.

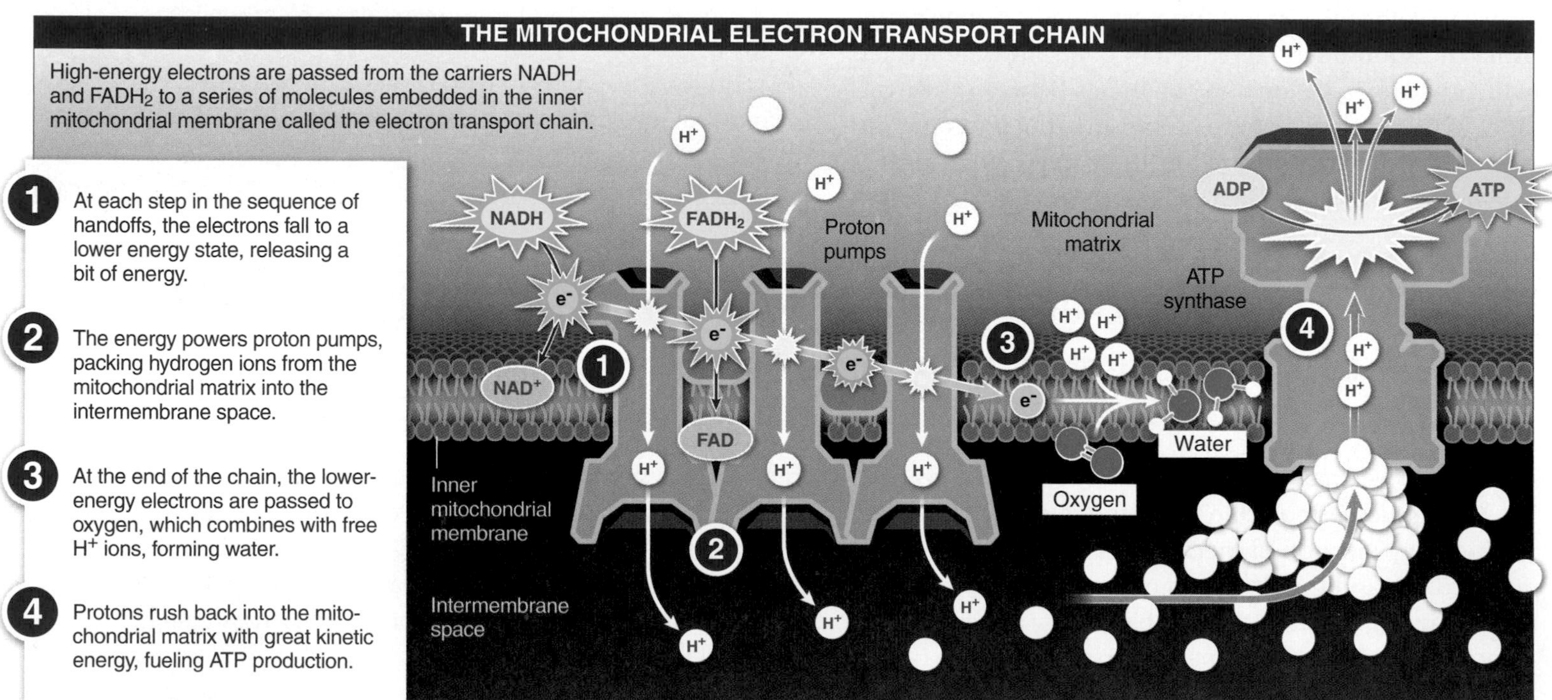

THE MITOCHONDRIAL ELECTRON TRANSPORT CHAIN
High-energy electrons are passed from the carriers NADH and $FADH_2$ to a series of molecules embedded in the inner mitochondrial membrane called the electron transport chain.
1 At each step in the sequence of handoffs, the electrons fall to a lower energy state, releasing a bit of energy.
2 The energy powers proton pumps, packing hydrogen ions from the mitochondrial matrix into the intermembrane space.
3 At the end of the chain, the lower-energy electrons are passed to oxygen, which combines with free H^+ ions, forming water.
4 Protons rush back into the mitochondrial matrix with great kinetic energy, fueling ATP production.
NADH
$FADH_2$
e^-
NAD^+
FAD
H^+
Proton pumps
Mitochondrial matrix
ATP synthase
ADP
ATP
Water
Oxygen
Inner mitochondrial membrane
Intermembrane space

FIGURE 5-34 **The big energy payoff.** Most of the energy harvested during cellular respiration is generated by the electron transport chain in the mitochondria.

so on. At each handoff, energy is released. Thus, as electrons move from one carrier to another, they lose energy.

At the end of the chain (step 3 in Figure 5-34), the lower-energy electrons are handed off to oxygen, which then combines with free H^+ ions in the mitochondrial fluid to form water.

As shown in step 2 of Figure 5-34, most of the energy released at each handoff from one electron carrier to another in the electron transport chain is used to pump protons (H^+ ions) from the mitochondrial matrix across the membrane and into the intermembrane space. As more and more protons are pumped across the membrane and packed into the intermembrane space, a concentration gradient is created. This gradient represents a significant source of potential energy.

This description may seem familiar. In chloroplasts, during photosynthesis, many protons are pumped from the stroma outside the thylakoid sacs to the inside of the thylakoids. We likened this potential energy to the potential energy of water in an elevated tower, which can be released with great force. Similarly, in step 4 of the mitochondrial electron transport chain, the protons pumped into the intermembrane space rush back into the mitochondrial matrix through channels in the inner mitochondrial membrane. And as the protons pass through, the force of their flow fuels the attachment of free-floating phosphate groups to ADP to produce ATP.

In the end, the number of ATP molecules generated from the complete dismantling of one molecule of glucose is about 36, most of which are produced with the energy harnessed from high-energy electron carriers as they pass their electrons down the electron transport chain (FIGURE 5-35).

Q Cyanide blocks the passage of electrons to oxygen in the electron transport chain. Why does this make it a toxic poison?

Given the central role of the electron transport chain in the generation of usable energy, any interference in its functioning has dire consequences. And in fact, murder by cyanide poisoning, an old tradition in detective stories, is just such an interference. When cyanide gets into the mitochondria, it binds to an enzyme in the electron transport chain, preventing it from accepting electrons. This halts the transfer of electrons and the pumping of protons across the mitochondrial membrane. ATP production ceases, removing the cell's energy source. For this reason, cyanide poisoning can cause rapid death.

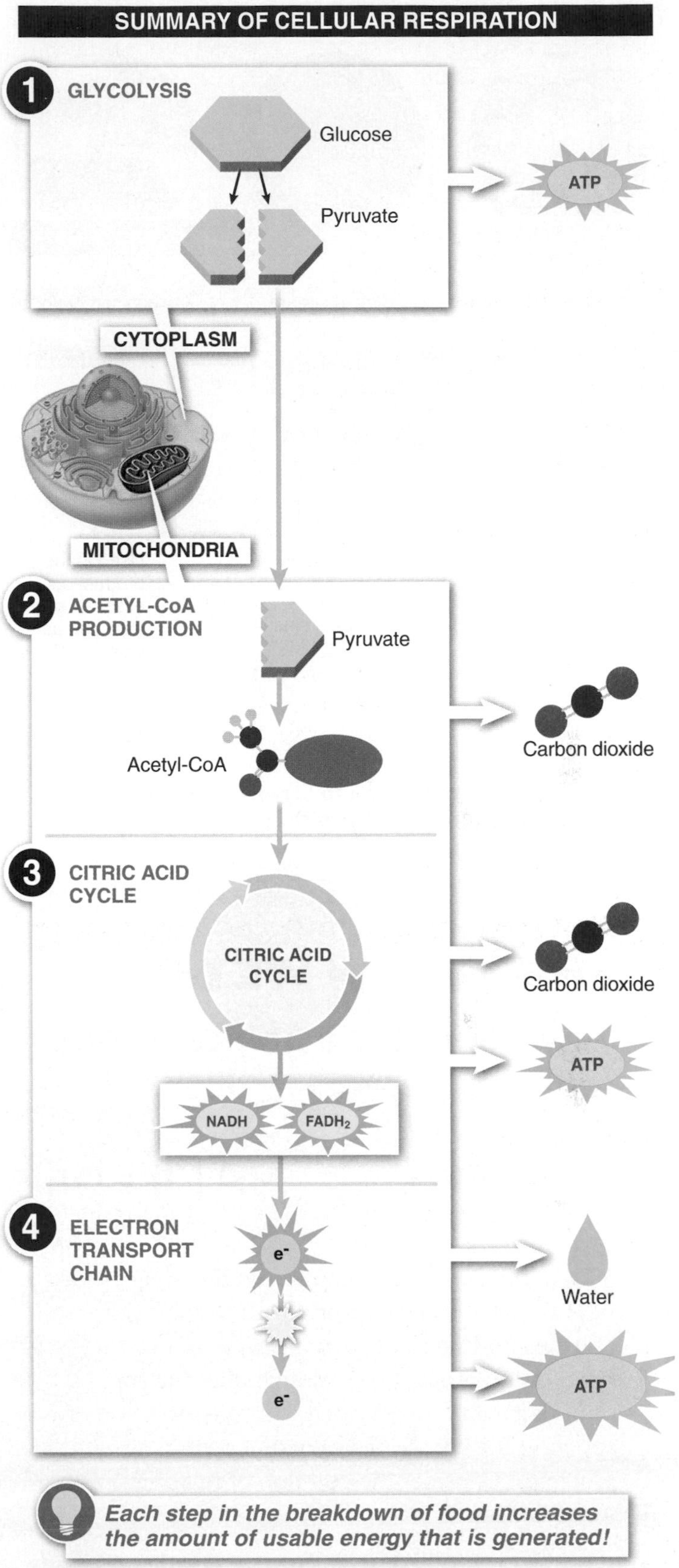

SUMMARY OF CELLULAR RESPIRATION

1 GLYCOLYSIS
Glucose
Pyruvate
ATP
CYTOPLASM
MITOCHONDRIA
2 ACETYL-CoA PRODUCTION
Pyruvate
Acetyl-CoA
Carbon dioxide
3 CITRIC ACID CYCLE
CITRIC ACID CYCLE
Carbon dioxide
ATP
NADH
$FADH_2$
4 ELECTRON TRANSPORT CHAIN
e^-
e^-
Water
ATP

Each step in the breakdown of food increases the amount of usable energy that is generated!

FIGURE 5-35 The steps of cellular respiration: from glucose to usable energy.

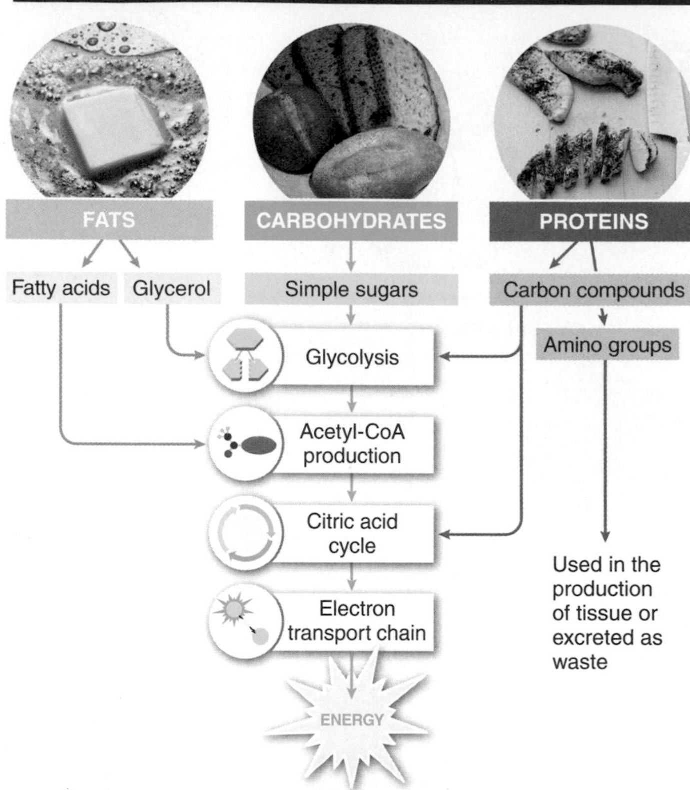

FIGURE 5-36 **Animals are able to harvest energy from proteins, carbohydrates, and lipids.**

Q **If a person has a low-carb diet, what provides the fuel for his body?**

Although we have examined just the breakdown of glucose, humans (and all other animals) are able to harvest energy from a variety of food sources. Whether a meal contains carbohydrates, lipids, proteins, or some combination thereof, the nutrients are chemically modified in some preliminary steps and then fed into one of the intermediate steps in glycolysis or the citric acid cycle to furnish usable energy for the organism (FIGURE 5-36).

TAKE HOME MESSAGE 5.15

» The largest energy payoff of cellular respiration comes as electrons from the NADH and $FADH_2$ produced during glycolysis and the citric acid cycle move along the electron transport chain. As the electrons are passed from one carrier to another, energy is released, pumping protons into the mitochondrial intermembrane space. As the protons rush back into the mitochondrial matrix, the force of their flow fuels the production of large amounts of ATP.

THIS IS HOW WE DO IT

Developing the ability to apply the process of science

5.16 Can we combat jet lag with NADH pills?

Often, scientific thinking is applied to questions about the natural world with the sole purpose of better understanding organisms and how they function. In other cases, research questions are formulated and investigated with the intention of applying the knowledge gained to address specific problems. Consider the question of treating the serious effects of jet lag.

What is jet lag and why is it of scientific interest?

Jet lag occurs when a person travels across several time zones and there is a mismatch between her body clock and the time of day or night at her destination. It is accompanied by a constellation of symptoms, including fatigue, gastrointestinal distress, memory loss, and reductions in cognitive performance. Jet lag affects a large number of travelers, including pilots and other air crew. It can have serious consequences because it hinders decision-making abilities, effective communication, and memory.

How is jet lag related to cellular respiration?

Researchers suspected that interventions targeting cellular respiration—with an eye toward increasing the rate at which cells generate ATP—might decrease the fatigue experienced in jet lag. In particular, they have focused on the high-energy electron carrier NADH. As we saw in the previous section, during cellular respiration energy is captured in the high-energy electron carriers NADH and $FADH_2$.

Why should NADH alleviate symptoms of jet lag?

If levels of NADH could be increased simply by taking the molecule in pill form, this might lead to increased production of usable energy through the electron transport chain. This hypothesis gave rise to a testable prediction: "Supplementing NADH should counteract some of the effects of jet lag, including reduced cognitive functioning and fatigue."

The experimental setup

Researchers used a randomized, controlled, double-blind experimental design. The participants were 36 volunteers, 35–55 years old, with at least 14 years of formal education and normal sleep schedules. During the study, the participants did not consume any caffeine, alcohol, or any medications known to affect nervous system functioning.

The volunteers were randomly assigned to one of two groups, placebo or NADH, and took a battery of tests to establish their baseline performance. They then took an overnight "red-eye" flight across four time zones, from California to Maryland. Arriving at 6 a.m., they were given breakfast and were then administered a pill containing either NADH or a placebo. At 9:30 a.m. and again at 12:30 p.m., they were given the same battery of tests.

Did NADH reduce the symptoms of jet lag?

Overall, the participants receiving NADH reported less sleepiness and performed significantly better on four tests of cognitive functioning.

What conclusions can we draw from these results?

The results reported in this well-designed study were clear and definitive. The placebo-receiving jet-lagged volunteers were more likely to make errors related to not paying attention, had greater difficulty with memory and concentration, and were less effective at multitasking. The researchers' conclusion, supported by the evidence, was that "NADH appears to be a suitable short-term countermeasure for the effects of jet lag on cognition and sleepiness."

What degree of confidence should we have that the question of whether NADH reduces jet lag is answered?

While the results do support the researchers' hypothesis, it still may be premature to consider the issue resolved. As they pointed out, the optimal doses of NADH still need to be investigated. Additionally, they examined the subjects' response to NADH only directly following the red-eye flight. It is not clear what the duration of the NADH effect might be on cognition and sleepiness.

It is wise, also, to be aware of any biases—even unconscious biases—that might influence researchers. In this study, for example, the researchers reported that "Menuco Corporation funded the study." A quick search reveals that Menuco Corporation was founded by one of the study's authors and is a for-profit company that markets and sells NADH supplements. The company and author hold the patent for the manufacturing process of the NADH supplement used in the study.

These facts do not invalidate the results, however. The study was carefully controlled and well designed. Most importantly, the researchers described their methods in such detail that the research can be replicated. Given the small number of subjects in their study, as well as the increasing relevance of jet lag in today's world, replication of these findings would be an important factor in increasing our confidence in the generalizability of the conclusions.

TAKE HOME MESSAGE 5.16

» The symptoms of jet lag—including fatigue, memory loss, and reductions in cognitive performance—can impair the performance of people in many professions. The results of a randomized, controlled, double-blind study support the hypothesis that an NADH supplement may be a suitable short-term countermeasure for these effects.

5.17

There are alternative pathways for acquiring energy.

Yeasts produce alcohol when they obtain energy in the absence of oxygen.

5.17 Beer, wine, and spirits are by-products of cellular metabolism in the absence of oxygen.

Every beer brewery, the entire wine industry, and all distilleries of vodka, tequila, and other alcoholic beverages owe their existence to microscopic yeast cells scrambling to break down their food for energy under stressful conditions. To better understand how yeast metabolism produces alcohol, it helps to begin by investigating what happens when humans and other animals try to metabolize energy from sugar molecules under some stressful conditions (FIGURE 5-37).

With rapid, strenuous exertion, our bodies soon fall behind in delivering oxygen from the lungs to the bloodstream to the cells and, finally, to the mitochondria. Oxygen deficiency then limits the rate at which the mitochondria can break down fuel and produce ATP (FIGURE 5-38). This slowdown occurs because the electron transport chain requires oxygen as the final acceptor of all the electrons generated during glycolysis and the citric acid cycle. If oxygen is in short supply, the electrons from NADH (and $FADH_2$) have nowhere to go. The whole process of cellular respiration can grind to a stop. Organisms don't let this interruption last long, though; most have a backup method for breaking down sugar.

During strenuous exertion, our muscles can require more oxygen than is available to them.

FIGURE 5-37 **Energy production without oxygen.** Organisms have a backup method for breaking down sugar when oxygen is not present.

In animals, there is an acceptor for the NADH electrons in the absence of oxygen: pyruvate, the end product of glycolysis. When pyruvate accepts the electrons, it forms lactic acid (see Figure 5-38). Once the NADH gives up its electrons, NAD^+ is regenerated, and glycolysis can continue. But as lactic acid builds up, it causes a burning feeling in our muscles. The next-day muscle soreness, though, is not due to this acid buildup, which goes away in a matter of minutes or hours, but rather to damage to the muscle fibers: they break down a bit before growing stronger. It's not ideal, but to escape a predator or to exercise strenuously, the two ATP molecules generated from each glucose molecule during glycolysis—which can serve as an immediate energy source—are better than nothing.

Like humans, yeasts normally use oxygen during their breakdown of food. And like humans, they have a backup method when oxygen is not available. But yeast cells make use of a different electron acceptor, and the resulting reaction leads to the production of all drinking alcohol.

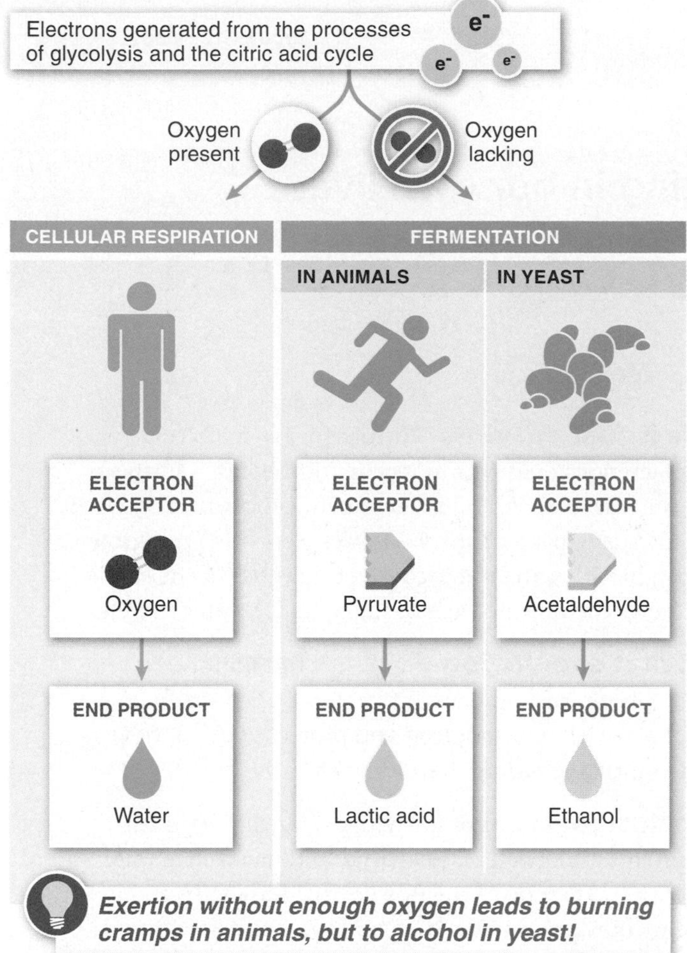

Oxygen lacking

Exertion without enough oxygen leads to burning cramps in animals, but to alcohol in yeast!

FIGURE 5-38 **Energy production with and without oxygen.**

After glycolysis in these single-celled organisms, pyruvate is usually converted to a molecule called acetaldehyde, releasing bubbles of CO_2 in the process (which, when yeast is used in baking, allows bread to rise). In the absence of oxygen, acetaldehyde accepts the electrons released from NADH, allowing glycolysis to resume. Acetaldehyde's acceptance of NADH's electrons results in the production of **ethanol,** the molecule that gives beer, wine, and spirits their kick.

Fermentation is the process by which cells obtain energy in the absence of oxygen. It occurs when, following glycolysis, alternative molecules are used as electron acceptors. Interestingly, although ethanol is always the alcohol produced by fermentation, the flavor of the output of fermentation depends on the source of the sugar metabolized by the yeast. Fruits, vegetables, and grains all give different results. If the sugar comes from grapes, wine is produced. If the sugar comes from a germinating barley plant, beer is produced. Potatoes, on the other hand, are the sugar source usually used to produce vodka.

Because yeasts prefer the more efficient process of aerobic respiration, they produce alcohol only in the absence of oxygen. Fermentation tanks used in producing wine, beer, and other spirits are built to keep oxygen out so that yeast cells are forced to use their backup pathway of fermentation (FIGURE 5-39).

TAKE HOME MESSAGE 5.17

» Oxygen deficiency limits the breakdown of fuel because the electron transport chain requires oxygen as the final acceptor of electrons during the chemical reactions of glycolysis and the citric acid cycle. When oxygen is unavailable, yeast cells resort to fermentation, in which they use a different electron acceptor, acetaldehyde, and in the process generate ethanol, the alcohol in beer, wine, and spirits.

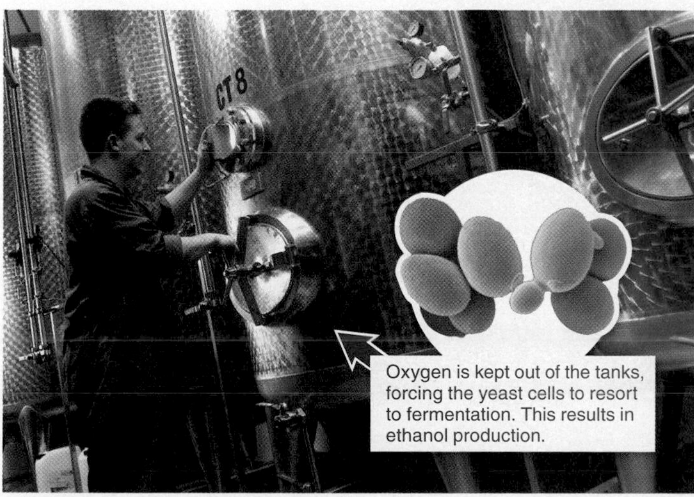

CT8
Oxygen is kept out of the tanks, forcing the yeast cells to resort to fermentation. This results in ethanol production.

FIGURE 5-39 **Just a by-product of metabolism under stressful conditions.** Beer, wine, and spirits are by-products of cellular metabolism in the absence of oxygen.

STREET BIO

Using evidence to guide decision making in our own lives

If you feed and protect your flowers in a vase, they'll last longer.

Like animals, plants require fuel for the cellular activities that keep them alive.

Q: How do plants get the energy they need to stay alive? Plants use photosynthesis to harness light energy, converting it to sugar molecules that serve as their food.

Q: Can plants still photosynthesize once they're in a vase in your house? Yes, but humans don't make it easy. Once cut, plants generally cannot produce sufficient sugar through photosynthesis. Light levels in houses tend to be too low, and the loss of many or most plant leaves reduces the number of chloroplasts in which photosynthesis can occur. So they're starving to death.

Q: Can you slow their demise? Yes. Plants are able to take up sugar in the vase water and use it as an energy source for cell activities. So adding a bit of sugar to the vase is like putting fuel in their tank.

Q: Is it that easy? No. Putting sugar—a molecule with lots of energy stored within its chemical bonds—in the vase water is like offering a free lunch. Unfortunately, when you add sugar, bacteria on the flower stems also can grow rapidly, blocking the water-conducting tubes in the stems. This slows the flow not just of sugar, but of water as well.

Q: What should you do? With the addition to vase water of both sugar and an antibacterial chemical such as chlorine bleach, you can feed and protect your cut flowers, significantly increasing their longevity.

Conclusion Most flowers will last longer if you cut their stems underwater and at a slant, to maximize water absorption. Then place the flowers in a flask with about 2 inches of warm water, which enhances the flow into the flower. At this point, add a spoonful of sugar and a drop or two of bleach. Then, after a few minutes, transfer the flowers to a vase.

GRAPHIC CONTENT

Thinking critically about visual displays of data

1 What are the axes of these two graphs?

2 What variable(s) is presented? How was it measured? What do the colors represent?

3 Do you know the source of the information in the graphs? Does that matter? Why or why not?

4 Why are there two graphs? What is the difference between them?

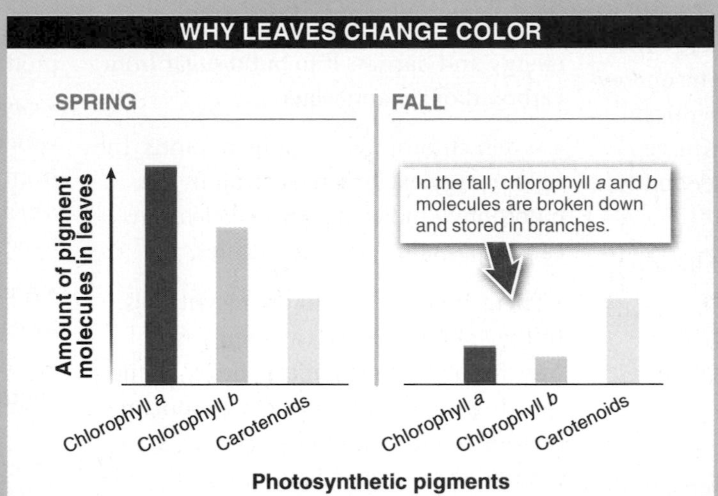

5 What can you conclude from this figure?

6 What additional information would make this figure more helpful? Why?

7 Do the data report experimental results? Is there a control group? An experimental group?

See answers at the back of the book.

KEY TERMS IN ENERGY

BRIEF SUMMARY

Energy flows from the sun and through all life on earth.

• The sun is the source of the energy that powers most living organisms. The energy from sunlight is stored in the chemical bonds of molecules. When these bonds are broken and new, lower-energy bonds are formed, energy is released.

• Kinetic energy is the energy of moving objects. Potential energy, such as chemical energy, is stored energy that results from the position, location, or composition of an object.

• Energy is neither created nor destroyed but can change form. Each conversion of energy is inefficient.

• Cells temporarily store energy in the bonds of ATP molecules. This potential energy can be converted to kinetic energy and used to fuel life-sustaining chemical reactions.

Photosynthesis uses energy from sunlight to make food.

• Through photosynthesis, plants use water, sunlight energy, and carbon dioxide gas to produce sugars and other organic materials, as well as oxygen.

• In plants, photosynthesis occurs in chloroplasts, green organelles packed in cells near a plant's surfaces, especially in the leaves.

• Photosynthesis is powered by energy packets called photons. Photons hit chlorophyll and other light-absorbing molecules in the chloroplasts of cells. These molecules capture some of the light energy and harness it to build sugar from carbon dioxide and water.

• When chlorophyll is hit by photons, the light energy excites an electron in the chlorophyll molecule. Excited electrons can be passed to other molecules.

• There are two parts to photosynthesis. In the "photo" part, light energy is transformed into chemical energy, while splitting water molecules and producing oxygen. Some of sunlight's energy is used to build the energy-storage molecules ATP and NADPH.

• In the second part of photosynthesis—the Calvin cycle—carbon from CO_2 is attached to molecules in chloroplasts, synthesizing sugars. These processes consume energy from ATP and NADPH generated in the "photo" part of photosynthesis.

• C4 and CAM photosynthesis are adaptations that are more energetically expensive than regular (C3) photosynthesis but allow plants in hot, dry climates to conserve water.

Living organisms extract energy through cellular respiration.

• In cellular respiration, organisms extract energy from the breakdown of the high-energy bonds of molecules, releasing their energy. The cell captures the released energy and stores it in the bonds of ATP molecules, and produces carbon dioxide as a by-product.

• Glycolysis is the initial process by which all living organisms harness energy from molecules. Glycolysis occurs in a cell's cytoplasm, and the released energy is captured as ATP and NADH.

• Additional energy can be harvested by cells after glycolysis. First, the end product of glycolysis, pyruvate, is chemically modified. Then, in the citric acid cycle, the modified pyruvate is broken down. This breakdown releases carbon (as CO_2) and captures some of the released energy in ATP and the high-energy electron carriers NADH and $FADH_2$.

• In the electron transport chain, electrons from the NADH and $FADH_2$ produced during glycolysis and the citric acid cycle are passed from one carrier to another, and energy is released, fueling the production of large amounts of ATP.

There are alternative pathways for acquiring energy.

• The electron transport chain requires oxygen as the final acceptor of electrons during glycolysis and the citric acid cycle. When oxygen is unavailable, yeast cells resort to fermentation, using a different electron acceptor and generating ethanol.

CHECK YOUR KNOWLEDGE

Short Answer

1. Why are fossil fuels considered a nonrenewable resource?

2. In terms of energy, describe what happens when a ball at the top of a steep ramp is released.

3. Why is the conversion of energy from one form to another considered inefficient.

4. Which chemical feature of ATP makes it effective in carrying and storing energy?

5. In which parts of a plant are most of its chloroplasts typically found? Why?

6. Why do leaves usually appear green?

7. Rubisco is the most abundant protein on earth. What does it do?

8. To reduce water loss via evaporation, plants may close their stomata. Although this action solves one problem for a plant, it creates another. Describe the new problem.

9. In which organelle does glycolysis occur?

10. Why are mitochondria considered to be ATP factories?

11. Describe how a structural feature of mitochondria helps them in the production of ATP.

12. A lack of oxygen limits the breakdown of fuel for most animals. Why?

Multiple Choice

1. A cyclist rides her bike up a very steep hill. Which of the following statements properly describes this activity in energetic terms?

a) Potential energy in food is converted to kinetic energy as the cyclist's muscles push her up the hill.

b) Kinetic energy is highest when the cyclist is at the crest of the hill.

c) The cyclist produces the most potential energy as she cruises down the hill's steep slope.

d) Potential energy is greatest when the cyclist is at the top of the hill.

e) Both a) and d) are correct.

2. The leaves of plants can be thought of as "eating" sunlight because:

a) light energy, like chemical energy released when the bonds of food molecules are broken, is a type of kinetic energy.

b) both light energy and food energy can be interconverted without heat loss.

c) the carbon-oxygen bonds within a photon of light release energy when broken by the enzymes in chloroplasts.

d) the carbon-hydrogen bonds within a photon of light release energy when broken by the enzymes in chloroplasts.

e) photons contain hydrocarbons.

3. A molecule of chlorophyll increases in potential energy:

a) when it binds to a photon.

b) when a photon strikes it, boosting electrons to a higher-energy excited state.

c) when it loses an electron.

d) only in the presence of oxygen.

e) None of the above. The potential energy of a molecule cannot change.

4. Plants rely on water:

a) to provide the protons necessary to produce chlorophyll.

b) to concentrate beams of sunlight on the reaction center.

c) to replenish oxygen molecules that are lost during photosynthesis.

d) to replace electrons that are excited by light energy and passed down an electron transport chain.

e) to serve as an energy source.

0 EASY 35 HARD 100

5. During C4 photosynthesis:

a) plants use less ATP in making sugar.

b) plants can produce sugars even when they close their stomata to reduce water loss on hot days.

c) plants are able to generate water molecules to cool their leaves.

d) plants produce more rubisco.

e) plants can produce sugars without any input of carbon dioxide.

6. During cellular respiration:

a) oxygen is used to transport chemical energy throughout the body.

b) metabolic oxygen is produced.

c) light is converted to kinetic energy.

d) ATP is converted to water and sugar.

e) energy from the chemical bonds of food molecules is captured.

7. Which of the following energy-generating processes is the only one that occurs in all living organisms?

a) the citric acid cycle

b) glycolysis

c) combustion

d) photosynthesis

e) None of the above. There are no energy-generating processes that occur in all living organisms.

8. All alcoholic beverages are produced as the result of cellular respiration by:

a) bacteria in the absence of oxygen.

b) bacteria in the absence of free electrons.

c) yeasts in the absence of free electrons.

d) all cells in the absence of sugar.

e) yeasts in the absence of oxygen.

0 EASY 18 HARD 100

Ch06

DNA: what is it, and what does it do?

6.1 Knowledge about DNA is increasing justice in the world.

6.2 The DNA molecule contains instructions for the development and functioning of all living organisms.

6.3 Genes are sections of DNA that contain instructions for making proteins.

6.4 Not all DNA contains instructions for making proteins.

6.5 How do genes work? An overview.

Information in DNA directs the production of the molecules that make up an organism.

6.6 In transcription, the information coded in DNA is copied into mRNA.

6.7 In translation, the mRNA copy of the information from DNA is used to build functional molecules.

6.8 Genes are regulated in several ways.

Damage to the genetic code has a variety of causes and effects.

6.9 What causes a mutation, and what are its effects?

6.10 *This is how we do it:* Does sunscreen use reduce skin cancer risk?

6.11 Faulty genes, coding for faulty enzymes, can lead to sickness.

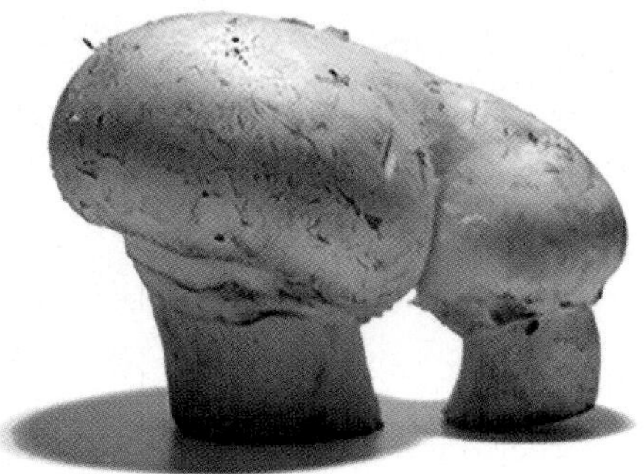

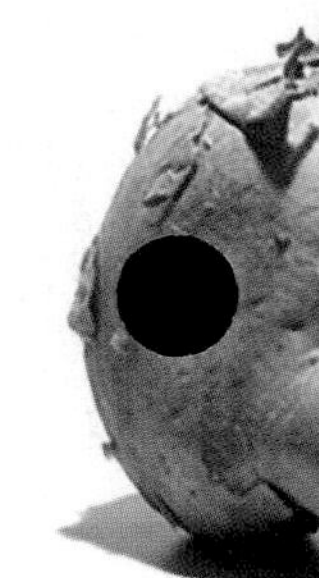

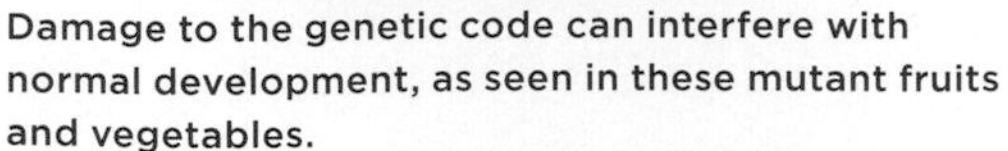

Damage to the genetic code can interfere with normal development, as seen in these mutant fruits and vegetables.

DNA and Gene Expression

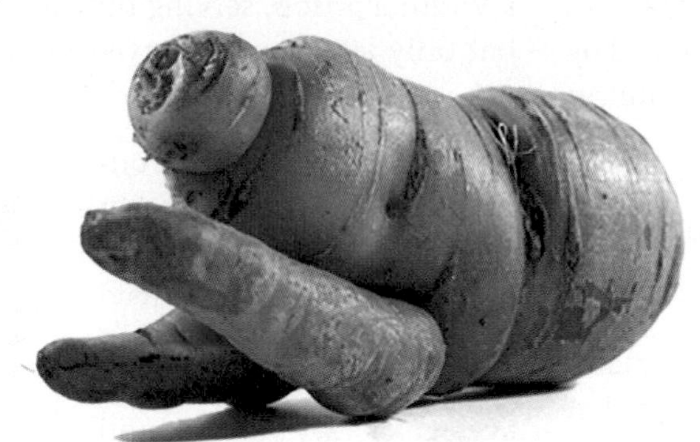
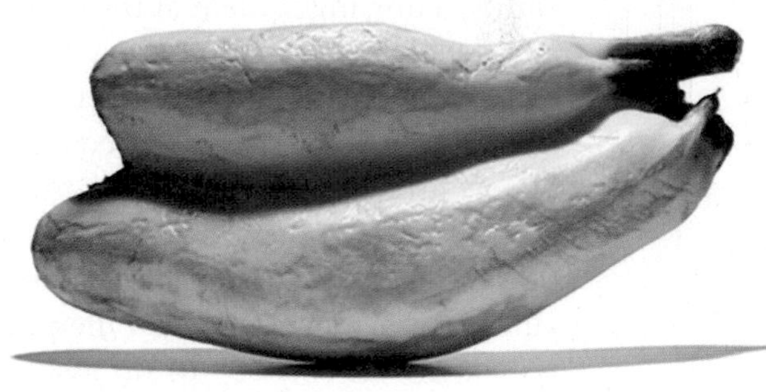

What is the genetic code and how does it work?

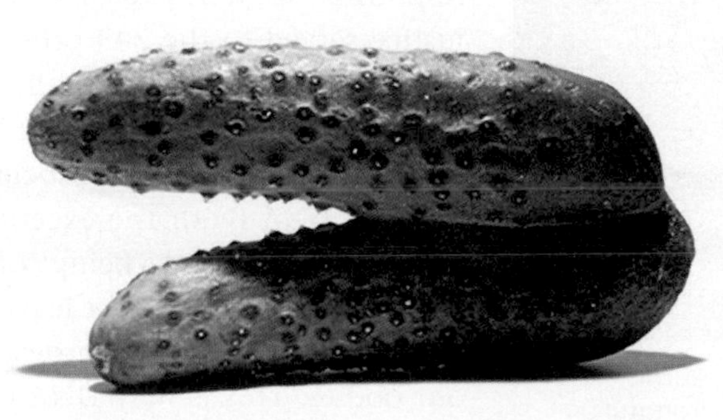

6.1–6.5

DNA: what is it, and what does it do?

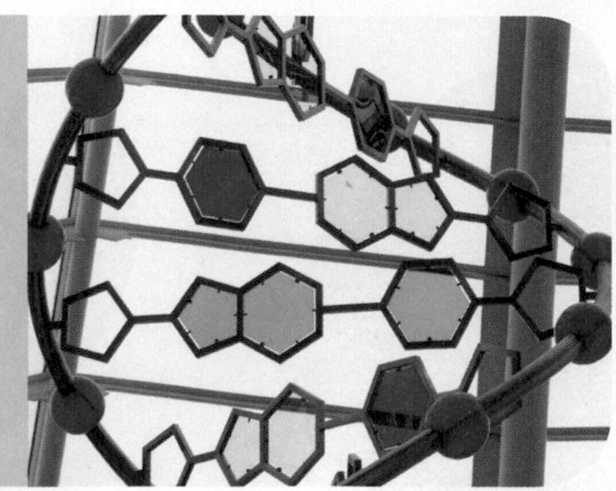

A model of the DNA helix on display at the City of Arts and Sciences, Valencia, Spain.

6.1 Knowledge about DNA is increasing justice in the world.

In 1981, Julius Ruffin, a 27-year-old maintenance worker at Eastern Virginia Medical School, got onto an elevator and lost more than 20 years of his life. Several weeks earlier, a nursing student at the school had been raped by an attacker who broke into her apartment. When Ruffin got on the elevator, the student thought she recognized him as her attacker and called the police, who immediately arrested him. Ruffin's girlfriend testified that he was with her at the time of the attack, but on the basis of the victim's eyewitness testimony, a jury found Ruffin guilty and sentenced him to life in prison. Ruffin maintained his innocence—to no avail, until 2003. At that time, the state's Division of Forensic Science performed a DNA analysis on a swab of evidence that remained from the investigation. The analysis revealed that the DNA perfectly matched that of a man who, in 2003, was already in a Virginia prison, serving time for rape. Ruffin was freed, but only after having served more than two decades in prison (FIGURE 6-1).

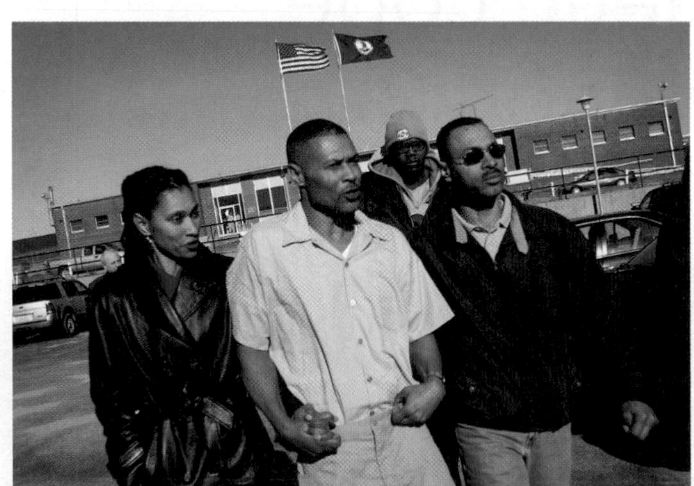

FIGURE 6-1 Vindicated by DNA evidence. Julius Ruffin was released from prison after two decades of wrongful incarceration.

Q What is the most common reason that DNA analyses overturn incorrect criminal convictions?

Julius Ruffin is one of 344 unjustly imprisoned people in the United States (as of late 2016) who have been freed from prison as a result of DNA analyses. They spent an average of 13.5 years behind bars. Eighty percent had been convicted of sexual assault; 28% had been convicted of murder. In three-quarters of the cases, inaccurate eyewitness testimony played an important role in the guilty verdict. (Recall, from Chapter 1, the experiments that revealed the unreliability of eyewitness identification.)

In this chapter, we take a close look at DNA, the molecule responsible for Julius Ruffin's exoneration and the deferred justice served to the 343 other people. All living organisms—people, plants, animals, bacteria, and otherwise—carry DNA in almost every cell in their body (with just a few exceptions). Like a social security number, every person's DNA (with the exception of identical twins) is unique. In addition to being contained in our living cells, our DNA exists in what we leave behind. It is in our saliva, hair, blood, and even the dead skin cells that fall from our bodies. This is why DNA can serve as an individual identifier.

The Economist
Genetics and music
Practice may not make perfect
Musical ability is in the DNA
TIME
The Man Behind Obama's Plan to Rescue the Economy
Why Banks Are Broke— And What to Do
Diabetes. Heart Disease. Parkinson's.
How the Coming Revolution in
STEM CELLS
The Telegraph
Home News World Sport Finance Comment Culture Travel Life
Health News
Born lazy: how genes dictate our love of exercise
Laziness runs in the family, a new study suggests.

FIGURE 6-2 **Genetic issues are in the news.**

The information carried within DNA is among the most important of all biological knowledge. It contains instructions for the function of every enzyme and cell in our bodies and carries a record of the evolutionary history of lineages of cells and organisms. And, as witnessed by the following excerpts, it is often in the news.

"Selfish dictators may owe their behaviour partly to their genes, according to a study that claims to have found a genetic link to ruthlessness."

— *Nature,* April 2008

"Too Many One-Night Stands? Blame Your Genes . . . according to a new study, it may be fair to say that while you jolly well could help cheating, your particular genes did make things more difficult."

— *Time* magazine, December 2010

In fact, it's nearly impossible to open a newspaper or watch a news report these days without being informed that yet another complex human characteristic or trait has been linked to a newly discovered gene (**FIGURE 6-2**). We'll explore these issues and more, beginning with a look at the structure of DNA and how it contains the information for producing organisms.

TAKE HOME MESSAGE 6.1

» DNA is a molecule carried by all living organisms in almost every cell of their body. It contains instructions for the functions of every cell. Because every person's DNA is unique, and because we leave a trail of DNA behind us as we go about our lives, DNA can serve as an individual identifier.

6.2 The DNA molecule contains instructions for the development and functioning of all living organisms.

Beginning in the 1900s and continuing through the early 1950s, a series of experiments revealed two important features of DNA. First, molecules of DNA are passed down from parent to offspring. Second, the instructions on how to create a body and control its growth, development, and behavior are encoded in the DNA molecule.

Given that DNA must hold the instructions for how to produce every possible type of structure in every living organism, scientists raced to determine the chemical structure of DNA and to understand how the molecule is assembled and shaped so that it can hold and transmit so much information.

Many formidable scientists rose to the challenge of determining the structure of DNA. When American Linus Pauling began investigating the structure of DNA, he had already won a Nobel prize in chemistry for his work on elucidating the structure of molecules. Simultaneously, in England, Maurice Wilkins and Rosalind Franklin were

James Watson (left) and Francis Crick figured out the exact structure of DNA

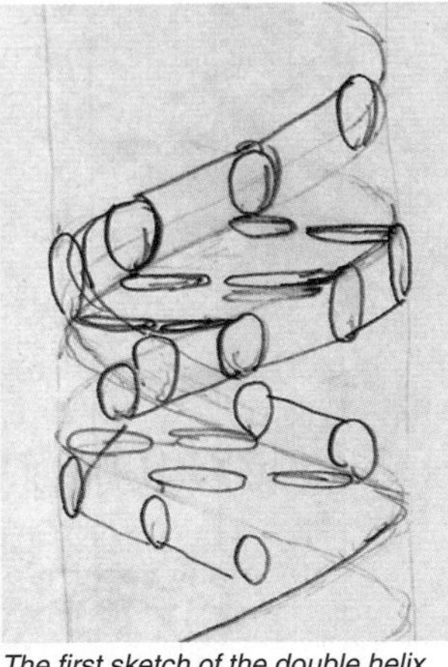

The first sketch of the double helix was rendered by Crick's wife, Odile

FIGURE 6-3 Watson and Crick.

devoting their research to this task and produced X-ray pictures of DNA that were critical to decoding its shape. But it was Englishman Francis Crick and American James Watson, working in Cambridge, England, who happened to put all the pieces together and deduce the exact structure of DNA (FIGURE 6-3).

As we'll see later in this chapter, their discovery was more than just a description of a molecule. As soon as they figured out DNA's structure, the answers to several other thorny problems in biology—such as how DNA might be able to duplicate itself—became apparent. We explore that process later, but first we need to examine the structure of DNA in more detail.

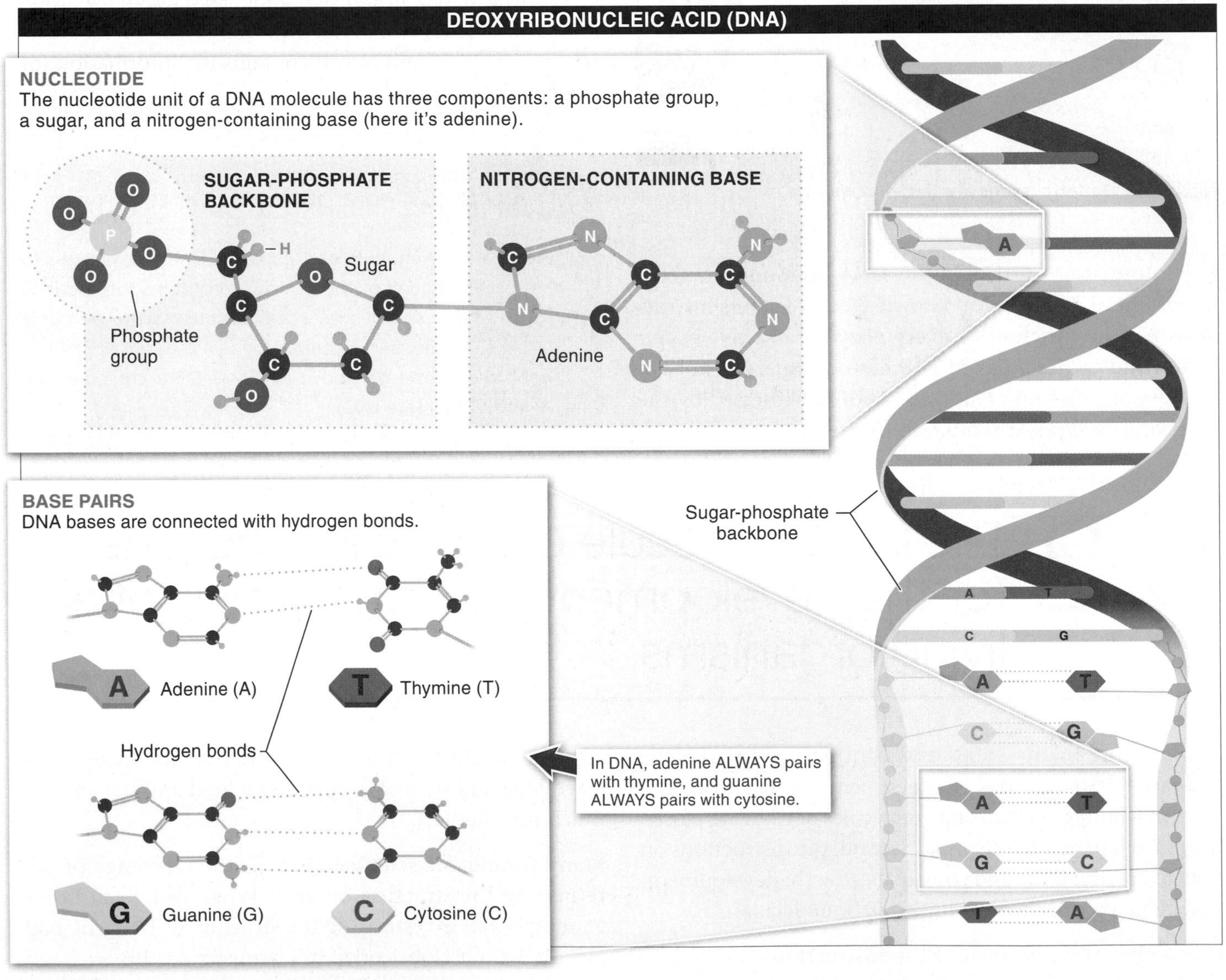

DEOXYRIBONUCLEIC ACID (DNA)

NUCLEOTIDE
The nucleotide unit of a DNA molecule has three components: a phosphate group, a sugar, and a nitrogen-containing base (here it's adenine).

SUGAR-PHOSPHATE BACKBONE

NITROGEN-CONTAINING BASE

Phosphate group

Sugar

Adenine

BASE PAIRS
DNA bases are connected with hydrogen bonds.

Adenine (A)

Thymine (T)

Hydrogen bonds

Guanine (G)

Cytosine (C)

Sugar-phosphate backbone

In DNA, adenine ALWAYS pairs with thymine, and guanine ALWAYS pairs with cytosine.

FIGURE 6-4 Overview of the structure of DNA.

DNA (deoxyribonucleic acid) is a **nucleic acid,** a macromolecule that stores information. It consists of individual units called **nucleotides,** which have three components: a molecule of sugar, a phosphate group (containing four oxygen atoms bound to a phosphorus atom), and a nitrogen-containing molecule called a **base.** The physical structure of DNA is frequently described as a "double helix." You can visualize the double helix as a long ladder twisted around like a spiral staircase (FIGURE 6-4).

The molecule has two distinct strands, like the vertical sides of a ladder. These are the "backbones" of the DNA molecule, and each is made up of two alternating molecules: a sugar, then a phosphate group, then a sugar, then a phosphate group, and so on. The sugar is always deoxyribose and the phosphate group is always the same, too. It is the shapes of the molecules in the backbone that cause the DNA "ladder" to twist.

The alternating sugars and phosphates hold everything in place, but the rungs of the ladder are where things get interesting. Attached to each sugar, and protruding inward like half of a rung on the ladder, is one of four nitrogen-containing bases: adenine, thymine, cytosine, and guanine. When discussing DNA, these bases are usually referred to by their first letter: A, T, C, and G.

Both backbones of the ladder have a base protruding from each sugar. The base on one side of the ladder—one strand of the DNA—binds, via hydrogen bonds, to a base on the other side, and together these **base pairs** form the rungs of the ladder. They pair up predictably. Every time a C protrudes from one side, it forms hydrogen bonds with a G on the other side (and vice versa: a G always bonds to a C). Similarly, each T forms hydrogen bonds with an A on the other side (and vice versa). For this reason, each DNA molecule always has the same number of Gs as Cs, and the same number of As as Ts. Because of these base-pairing rules, it also is true that if we know the base sequence for one of the strands in a DNA molecule, we know the complementary sequence in the other. This is why a DNA sequence is described by writing the sequence of bases in only one of the strands.

If a human DNA molecule were really a twisted ladder, it would be a very, very long one. One molecule of DNA can have as many as 200 million base pairs, or rungs. How does such a molecule fit into a cell? The rungs are small, and the twisting of the molecule—twists upon twists—shortens it considerably: think about how, if you repeatedly twist your shoelace around and around, it becomes shorter and shorter. That's what happens with DNA. Let's now investigate how DNA's structure—the rungs of the ladder, in particular—enables it to carry information.

TAKE HOME MESSAGE 6.2

» DNA is a nucleic acid, a macromolecule that stores information. It consists of individual units called nucleotides, which consist of a sugar, a phosphate group, and a nitrogen-containing base. DNA's structure resembles a twisted ladder, with the repeated sugar and phosphate groups serving as the two sides (backbones) and base pairs serving as the rungs. The sequence of bases on one side of the ladder-like molecule (one strand of the DNA) complements that of the bases on the other side (the second strand).

6.3 Genes are sections of DNA that contain instructions for making proteins.

Q Why is DNA considered the universal code for all life on earth?

One of DNA's most amazing features is that it embodies the instructions for building the cells and structures for almost every single living organism on earth (FIGURE 6-5). Thus, DNA is like a universal language, the letters of which are the bases A, T, C, and G. (Note that the sugar-phosphate backbone serves only to hold the bases in sequence, like the binding of a book. It does not convey genetic information.)

We've seen how the structure of DNA is like a spiral staircase. Another analogy may help you understand the information-containing aspect of DNA. You can think of an organism's DNA as a cookbook. Just as a cookbook contains detailed instructions on how to make a variety of

Human (Homo sapiens)

Onion (Allium cepa)

Fruit fly (Drosophila melanogaster)

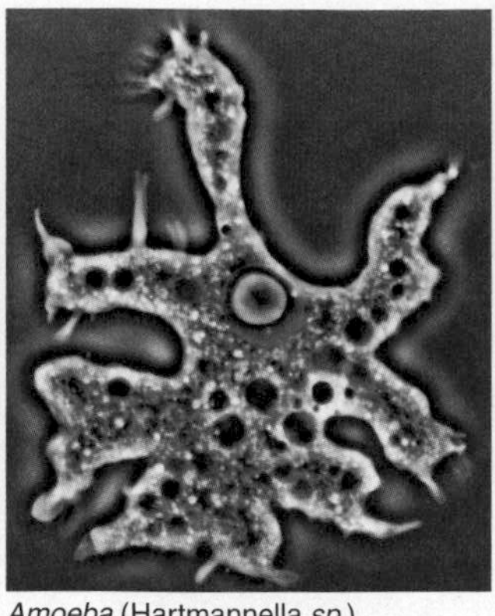
Amoeba (Hartmannella *sp.*)

Salamander (Ambystoma maculatum)

FIGURE 6-5 DNA is the universal code for all life on earth.

DNA provides the instructions for building virtually every organism on earth!

foods (such as French toast, macaroni and cheese, or chocolate chip cookies), an organism's DNA carries the detailed instructions to build the organism and keep it running. And just as a book can be viewed as a sequence of letters, with the book's meaning determined by which letters are strung together and in what order, a molecule of DNA can be viewed as a sequence of bases.

The sequence of bases in an organism's DNA is a **code** that specifies the detailed instructions for building polypeptides. Upon processing and folding, these polypeptides become the numerous and varied protein molecules that make up an organism—whether it is a one-celled amoeba, a giant oak tree, or a biology student.

The full set of DNA present in an individual organism is called its **genome** (FIGURE 6-6). In prokaryotes, including all bacteria, the information is contained within circular pieces of DNA. In eukaryotes, including humans, this information is laid out in long linear strands of DNA in the nucleus. Rather than being one super-long DNA strand, eukaryotic DNA exists as many smaller, more manageable pieces, called **chromosomes.** Humans, for example, have three billion base pairs, divided into 23 unique pieces of DNA. Because we have two copies of each (one from our mother, one from our father), we have 46 chromosomes in each cell.

Within the long sequences of bases in a cell's DNA molecules are relatively short sequences, on average about 3,000 bases long, called genes. You may hear the word "gene" misused in the media, as if a gene were a mysterious force controlling our bodies and behavior. Here we'll be precise: a **gene** is a sequence of bases (or, more precisely, base pairs) in a DNA molecule that carries the information necessary for producing a functional end-product, usually a polypeptide or an RNA molecule. Nothing more, nothing less. The

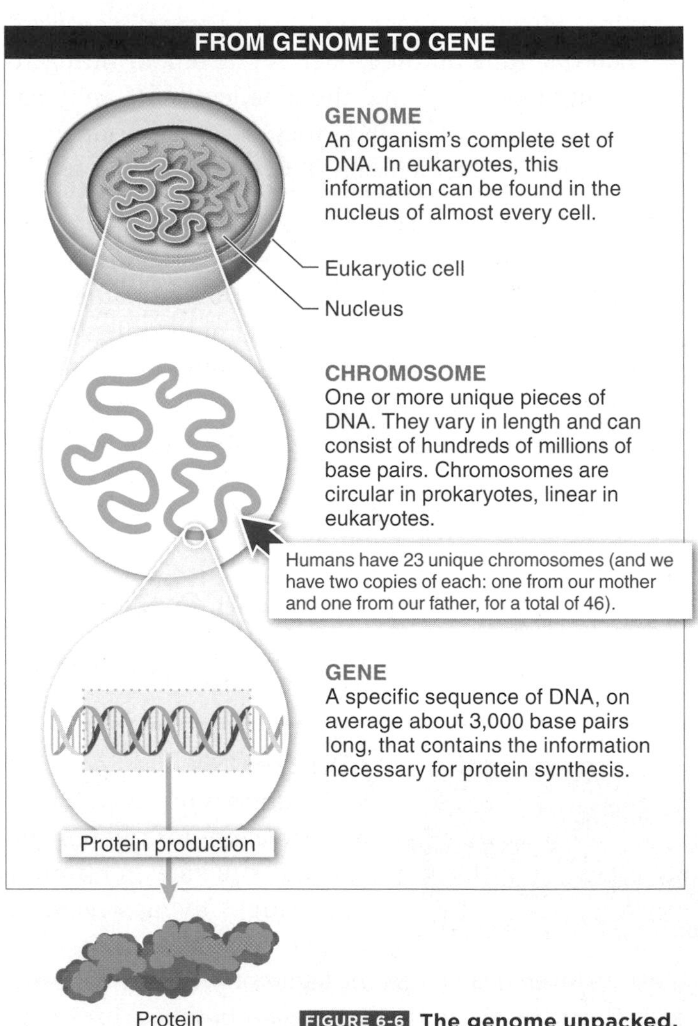

FROM GENOME TO GENE
GENOME
An organism's complete set of DNA. In eukaryotes, this information can be found in the nucleus of almost every cell.
Eukaryotic cell
Nucleus
CHROMOSOME
One or more unique pieces of DNA. They vary in length and can consist of hundreds of millions of base pairs. Chromosomes are circular in prokaryotes, linear in eukaryotes.
Humans have 23 unique chromosomes (and we have two copies of each: one from our mother and one from our father, for a total of 46).
GENE
A specific sequence of DNA, on average about 3,000 base pairs long, that contains the information necessary for protein synthesis.
Protein production
Protein

FIGURE 6-6 The genome unpacked.

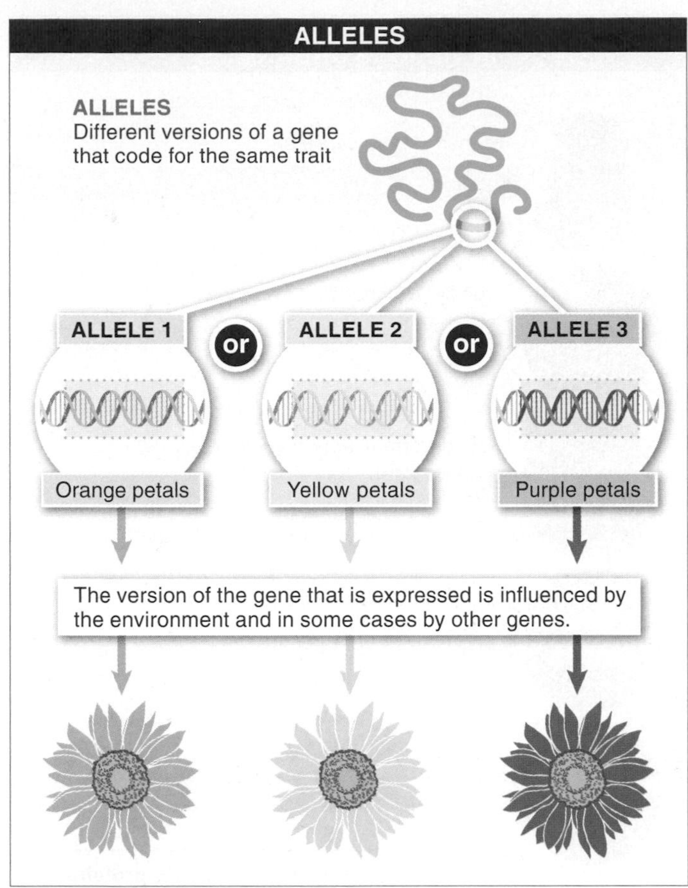
ALLELES
ALLELES
Different versions of a gene that code for the same trait
ALLELE 1
or
ALLELE 2
or
ALLELE 3
Orange petals
Yellow petals
Purple petals
The version of the gene that is expressed is influenced by the environment and in some cases by other genes.

FIGURE 6-7 **"Different versions of the same thing."** Alleles are alternative versions of a gene.

location or position of a gene on a chromosome is called a **locus** (pl. **loci**).

Each gene is the instruction set for producing one particular molecule, usually a protein. For example, there is a gene in silk moths that codes for fibroin, the chief component of silk. And, there is a gene in humans that codes for triglyceride lipase, an enzyme that breaks down dietary fat.

Within a species, individuals sometimes have slightly different instruction sets for a given protein, and these instructions can result in a different version of the same characteristic. These alternative versions of a gene that code for the same feature are called **alleles** (FIGURE 6-7)—and function like alternative recipes for chocolate chip cookies. Any single characteristic or feature of an organism is referred to as a **trait.**

A simple hypothetical example will clarify the meaning of these terms. The color of a daisy's petals is a trait. The instructions for producing this trait are found in a gene that controls petal color. This gene may have many different alleles within a population; one allele may specify the trait of orange petals, another may specify yellow petals, and yet another may specify purple petals (see Chapter 9). Similarly, one allele for eye color in fruit flies may carry the instructions for producing red eyes, while another, slightly different allele may have instructions for brown eyes. (Ultimately, though, the trait may be influenced not just by the genes an individual carries but by the way those genes interact with the environment, too.)

TAKE HOME MESSAGE 6.3

» DNA is a universal language that provides the instructions for building all the structures in all living organisms. The full set of DNA that an organism carries is called its genome. In prokaryotes, the DNA occurs in circular pieces. In eukaryotes, the genome is divided among smaller, linear strands of DNA. An organism's DNA pieces are generally called chromosomes. A gene is a sequence of bases in a DNA molecule that carries the information necessary for producing a functional product, usually a polypeptide or RNA molecule.

6.4 Not all DNA contains instructions for making proteins.

It is debatable whether humans are the most complex species on the planet, but surely we must be more complex than an onion. Comparing the amount of DNA present in various species, in terms of both numbers of chromosomes and numbers of base pairs, however, reveals a paradox: there does not seem to be any relationship between the size of an organism's genome and the organism's complexity (FIGURE 6-8).

Q An onion has five times as much DNA as a human. Why doesn't that make onions more complex than humans?

The description earlier in this chapter about what DNA is

and how genes code for proteins is logical and tidy, but it doesn't completely explain what we observe in cells. In humans, for example, genes make up only about 2% of the DNA (FIGURE 6-9). In many species, the proportion of the DNA that consists of genes is even smaller. In almost all eukaryotic species, the amount of DNA present far exceeds the amount necessary to code for all of the proteins in the organism. The fact is, a huge proportion of the base sequences in DNA do not code for proteins, and most of these have no known purpose. When it was first observed, some biologists even referred to this noncoding DNA as "junk DNA."

Bacteria and viruses tend to have very little noncoding DNA; genes make up 90% or more of their DNA. It is in the eukaryotes (with the exception of yeasts) that we see an explosion in the amount of noncoding DNA (FIGURE 6-10). Noncoding regions of DNA often take the form of sequences that are repeated, sometimes thousands (or even hundreds of thousands) of times; often these repeats exist because some DNA sequences can make copies of themselves and the copies can move about throughout the genome. Occasionally, the noncoding DNA consists of gene fragments, duplicate versions of genes, and "pseudogenes" (sequences that evolved from actual genes but accumulated mutations that made them lose their protein-coding ability).

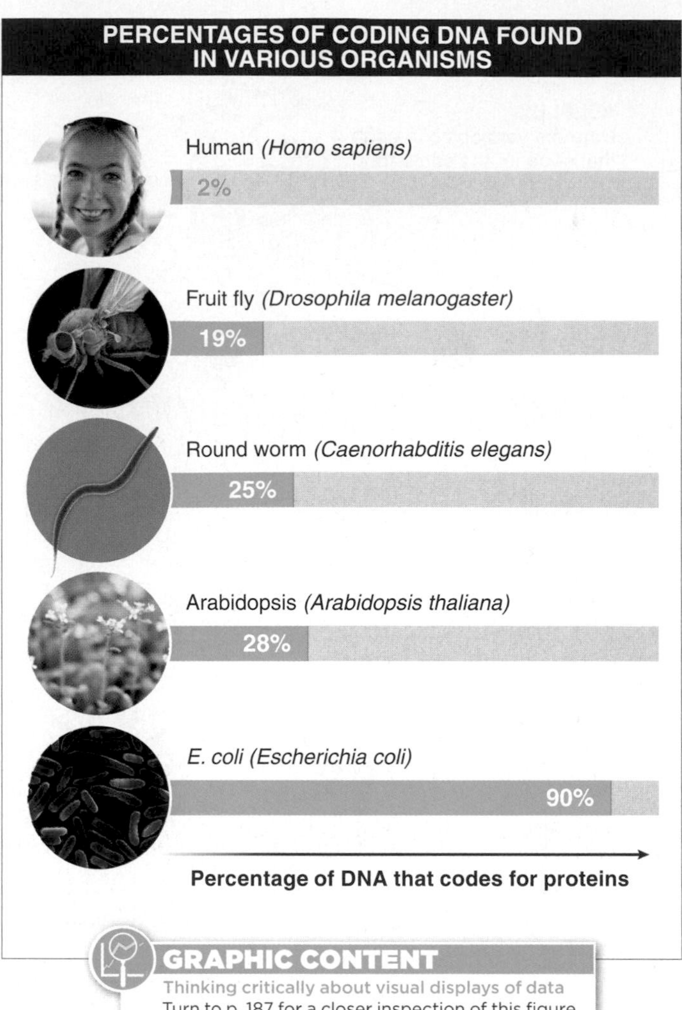

GRAPHIC CONTENT

Thinking critically about visual displays of data

Turn to p. 187 for a closer inspection of this figure.

FIGURE 6-9 **"Junk DNA."** The proportion of DNA that codes for proteins or RNA varies greatly among species. Of the species listed, which has the least "junk DNA"?

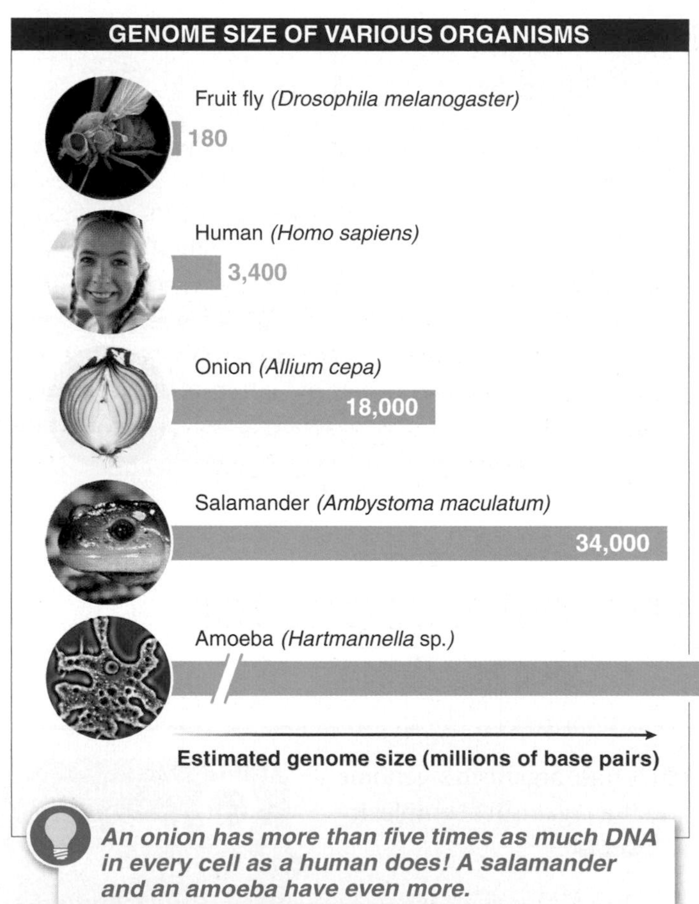

FIGURE 6-8 **Is the size of an organism's genome related to its complexity?**

About 25% of the noncoding regions occur within genes—in which case they are called **introns.** About 75% of the noncoding regions occur between genes (see Figure 6-10).

In the end, the presence of this noncoding DNA is still not completely understood. Recent evidence reveals, however, that some of the "noncoding" DNA does encode extremely

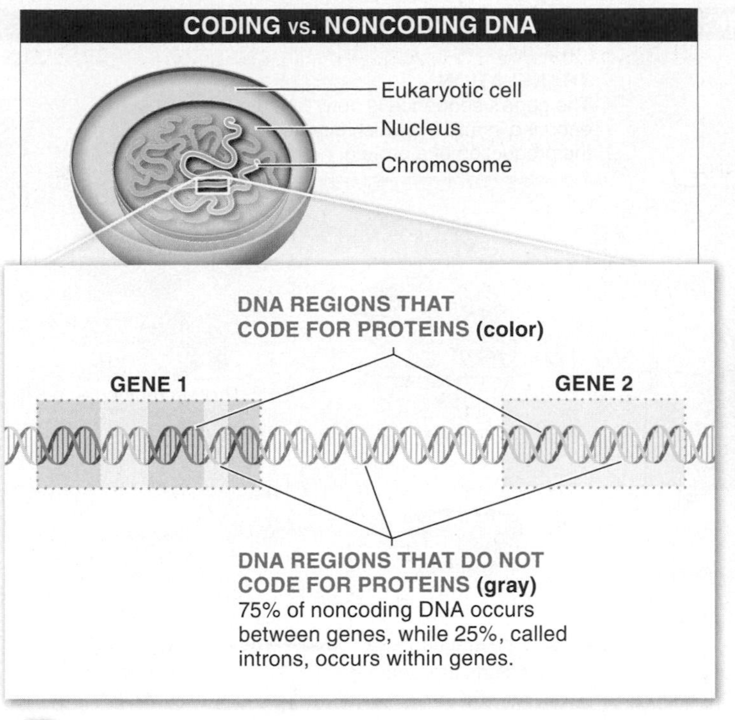
CODING vs. NONCODING DNA
Eukaryotic cell
Nucleus
Chromosome
DNA REGIONS THAT CODE FOR PROTEINS (color)
GENE 1
GENE 2
DNA REGIONS THAT DO NOT CODE FOR PROTEINS (gray)
75% of noncoding DNA occurs between genes, while 25%, called introns, occurs within genes.
Most DNA in eukaryotes does not code for any proteins!

FIGURE 6-10 **Noncoding regions of DNA.** These regions are found both between and within genes.

short RNA molecules (~20 nucleotides long) that function in gene regulation; they act as a "switch" that regulates when genes are turned on or off, or as a "volume knob" that influences the amount of the gene products produced. Noncoding DNA may also serve as a reservoir of potentially useful sequences. In any case, the label "junk DNA" is a bad description.

TAKE HOME MESSAGE 6.4

» Only a small fraction of the DNA in eukaryotic species is in genes that code for proteins; the function of much of the rest is still poorly understood, but at least some of it plays important roles in the cell, such as gene regulation.

6.5 How do genes work? An overview.

Having a recipe for chocolate chip cookies is not the same thing as having the actual cookies. Similarly, having a wealth of hereditary information about how to build muscle cells or leaf cells is not sufficient to produce an organism. Think about it: an organism's every cell contains all of the information needed to manufacture every protein in its body. This means that the skin cells on your arm contain the genes for producing proteins found only in liver cells or red blood cells or muscle tissue—but they don't produce those proteins. Having the instructions is not the same as having the products.

The genes in strands of DNA are a storehouse of information, an instruction book, but they are only one part of the process by which an organism is built. If the genes that an organism carries for a particular trait—its **genotype**—are like a recipe in a cookbook, the physical manifestation of the instructions—the organism's **phenotype**—is the cookie, or any of the other foods described by the recipes. And, just as you have to assemble the ingredients, mix them, then bake the dough to get a cookie, there are multiple steps in the production of the molecules, tissues, and even behaviors that make up a phenotype.

How does a gene (a sequence of bases in a section of DNA) affect a flower's color or the shape of a nose or the texture of a dog's fur (the phenotype)? The process occurs in two main steps: **transcription,** in which a copy of a gene's base sequence is made, and **translation,** in which that copy is used to direct the production of a polypeptide, which then, in response to a variety of factors, including the cellular environment, folds into a functional protein.

FIGURE 6-11 presents an overview of the processes of transcription and translation. In transcription, which in eukaryotes occurs in the nucleus, the gene's base sequence, or code, is copied into a middleman molecule called

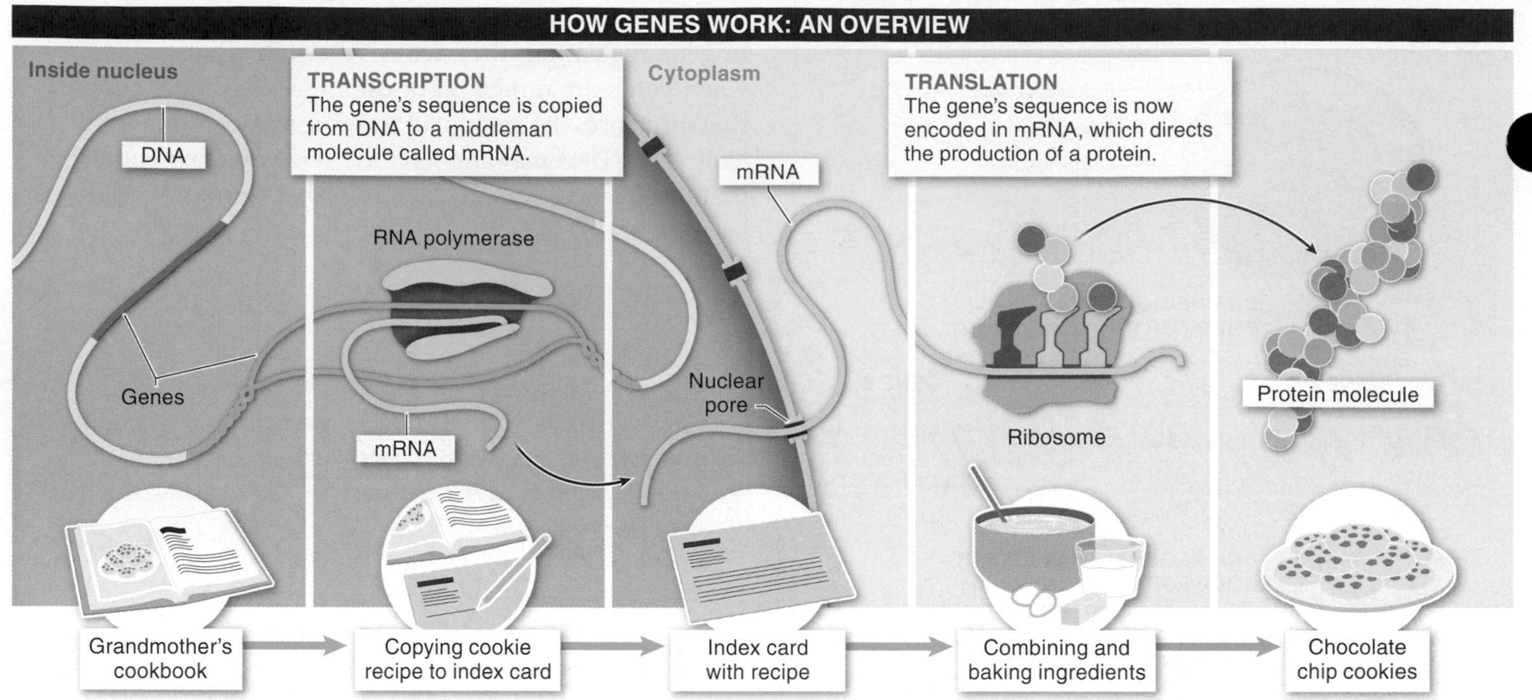

FIGURE 6-11 **Overview of the steps from gene to protein.**

messenger RNA (mRNA). (Because prokaryotes don't have a nucleus, transcription occurs in the cytoplasm.) This is like copying the information for the chocolate chip cookie recipe out of the cookbook and onto an index card. The mRNA then moves out of the nucleus into the cytoplasm, where translation allows the messages encoded in mRNAs to be used to build proteins.

TAKE HOME MESSAGE 6.5

» The genes in strands of DNA are a storehouse of information, an instruction book. The process by which this information is used to build an organism occurs in two main steps: transcription, in which a copy of a gene's base sequence is made, and translation, in which that copy is used to direct the production of a polypeptide.

6.6–6.8

Information in DNA directs the production of the molecules that make up an organism.

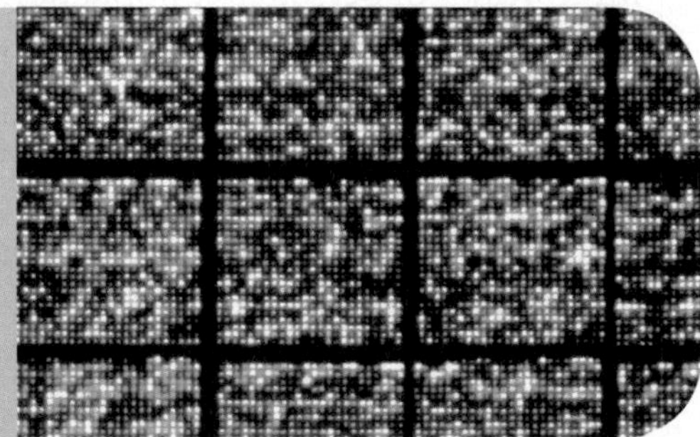

DNA microarray shows the expression levels of many genes simultaneously.

6.6 In transcription, the information coded in DNA is copied into mRNA.

If DNA is like a cookbook filled with recipes, transcription and translation are like cooking. In cooking, you use information about how to make chocolate chip cookies to produce the actual cookies. In an organism, the

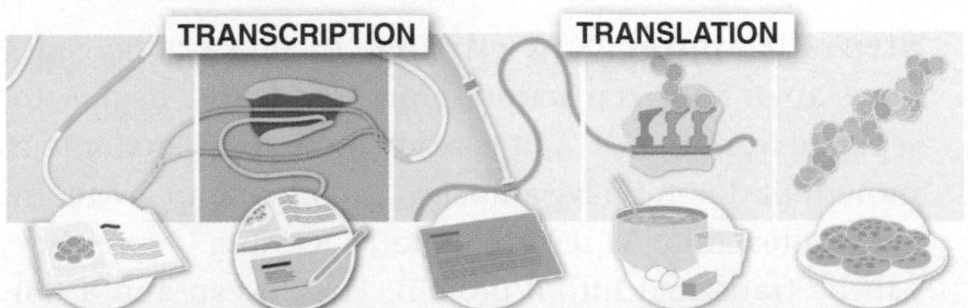

FIGURE 6-12 Transcription: copying the base sequence of a gene. This is the first step in a two-step process by which DNA regulates a cell's activity and synthesis of proteins. (Shown here is eukaryotic transcription.)

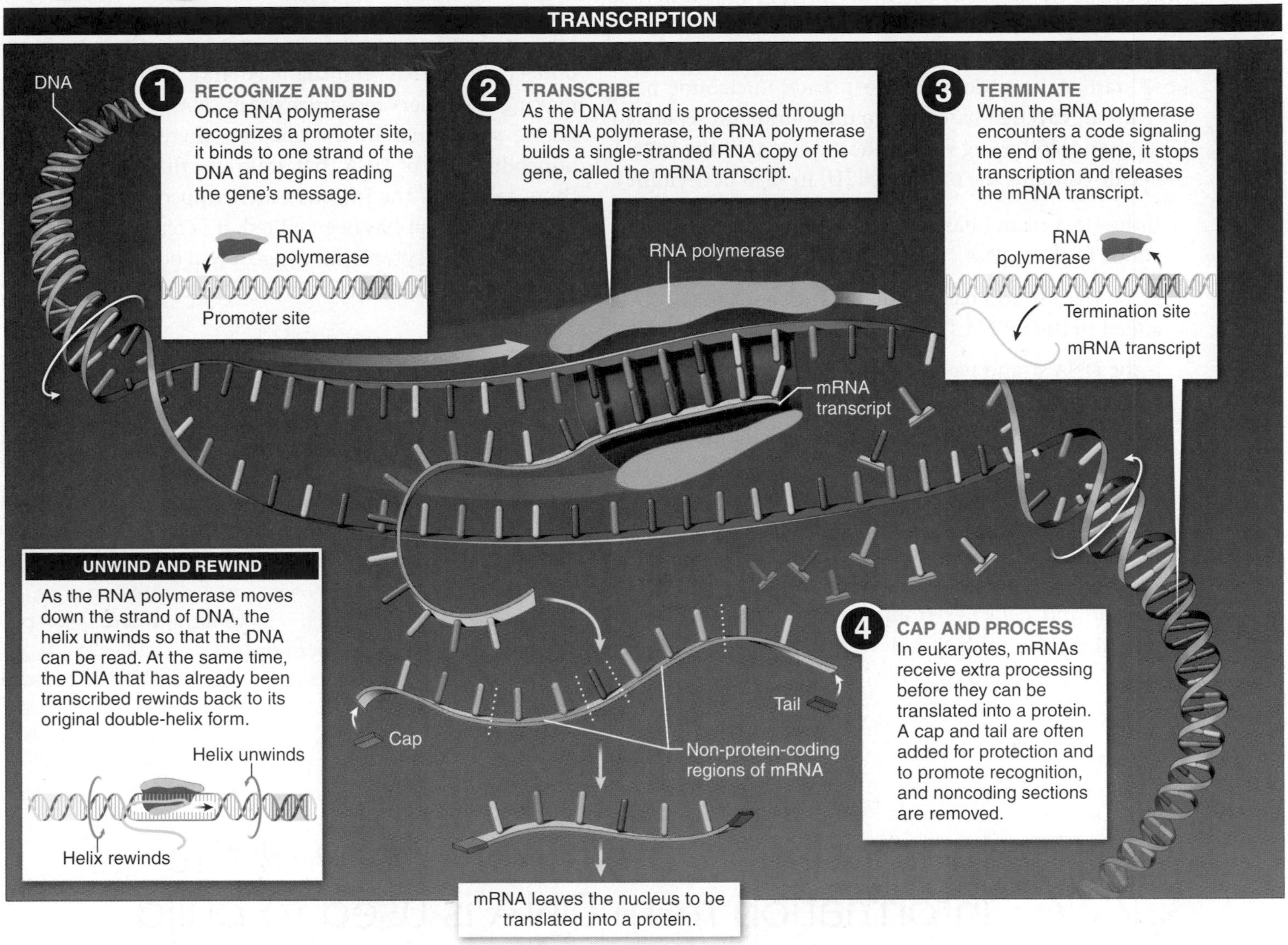

TRANSCRIPTION

DNA

1 RECOGNIZE AND BIND
Once RNA polymerase recognizes a promoter site, it binds to one strand of the DNA and begins reading the gene's message.

RNA polymerase

Promoter site

2 TRANSCRIBE
As the DNA strand is processed through the RNA polymerase, the RNA polymerase builds a single-stranded RNA copy of the gene, called the mRNA transcript.

RNA polymerase

mRNA transcript

3 TERMINATE
When the RNA polymerase encounters a code signaling the end of the gene, it stops transcription and releases the mRNA transcript.

RNA polymerase

Termination site

mRNA transcript

UNWIND AND REWIND

As the RNA polymerase moves down the strand of DNA, the helix unwinds so that the DNA can be read. At the same time, the DNA that has already been transcribed rewinds back to its original double-helix form.

Helix unwinds

Helix rewinds

4 CAP AND PROCESS
In eukaryotes, mRNAs receive extra processing before they can be translated into a protein. A cap and tail are often added for protection and to promote recognition, and noncoding sections are removed.

Cap

Tail

Non-protein-coding regions of mRNA

mRNA leaves the nucleus to be translated into a protein.

information about putting together proteins is used to build the proteins that the organism needs to function.

In this section, we examine transcription, the first step in the two-step process by which DNA regulates a cell's activity and its synthesis of proteins (see Figure 6-11). In transcription, a copy is made of one specific gene within the DNA. Continuing our cookbook analogy, transcription is like copying a single recipe from the cookbook onto an index card. Transcription happens in four steps (FIGURE 6-12).

Step 1. Recognize and bind. To start the transcription process, the enzyme RNA polymerase recognizes a **promoter site,** a base sequence in a gene that indicates the start of the gene and, in effect, tells the RNA polymerase, "Start here." RNA polymerase binds to the DNA molecule at the promoter site and unwinds it, so that only one strand of the DNA can be read. Throughout transcription, DNA is unwound ahead of the RNA polymerase so that a single strand of the DNA can be read, and it is rewound after the polymerase passes.

Step 2. Transcribe. As the DNA strand is processed through the RNA polymerase, the RNA polymerase builds a copy—called a "transcript"—of the gene from the DNA molecule, just as (to use another analogy) a court reporter transcribes and creates a record of everything that is said in a courtroom. This copy is called messenger RNA (mRNA) because, once this copy of the gene is created, it can move elsewhere in the cell and its message can be translated into a protein.

The mRNA transcript is constructed from four different nucleotides, which have a structure similar to that of DNA nucleotides but with a different kind of sugar ("ribose" rather than "deoxyribose"). Each nucleotide pairs up with an exposed base on the unwound and separated DNA strand, following these rules (notice that the mRNA molecule contains the base uracil (U) instead of thymine):

If the DNA strand has a thymine (T), an adenine (A) is added to the mRNA.

If the DNA strand has an adenine (A), a uracil (U) is added to the mRNA.

If the DNA strand has a guanine (G), a cytosine (C) is added to the mRNA.

If the DNA strand has a cytosine (C), a guanine (G) is added to the mRNA.

Step 3. Terminate. When the RNA polymerase encounters a sequence of bases on the DNA at the end of the gene (called a termination sequence), it stops creating the transcript and detaches from the DNA molecule. After termination, the mRNA molecule is released as a free-floating, single-strand copy of the gene.

Step 4. Capping and editing. In prokaryotic cells, once an mRNA transcript is produced and begins to separate from the DNA, it is ready to be translated into a protein (it doesn't have a nuclear membrane to cross). In eukaryotes, mRNAs receive extra processing before they can be translated into a protein. First, a cap and a tail may be added at the beginning and end of the transcript. Like the front and back covers of a book, these serve to protect the mRNA from damage and help the protein-making machinery recognize the mRNA. Second, because (as we saw earlier in the chapter) there may be some noncoding bits of DNA that are transcribed into mRNA, those sections—the introns—are snipped out. Once the mRNA transcript has been edited, it is ready to leave the nucleus for the cytoplasm, where it will be translated into a polypeptide that will fold into a protein.

TAKE HOME MESSAGE 6.6

» Transcription is the first step in the two-step process of producing proteins based on instructions contained in DNA. In transcription (which occurs in the nucleus in eukaryotic cells), a single copy of one specific gene in the DNA is made in the form of a molecule of mRNA. When the mRNA copy of a gene is completed and processed, it moves to the cytoplasm, where it can be translated into a polypeptide.

6.7 In translation, the mRNA copy of the information from DNA is used to build functional molecules.

Once the mRNA molecule has moved out of the cell's nucleus and into the cytoplasm, the translation process begins. In translation, the information carried by the mRNA is read, and its message can be translated into a protein.

To read an encrypted message, you need to know the code with which to translate each of the encrypted characters into a readable character. For mRNA, another, special type of RNA molecules hold the code. These molecules, called **transfer RNA (tRNA),** interpret the mRNA code, translating the language of DNA—coded in the linear sequence of bases—into the language of proteins, coded in the linear sequences of amino acids.

Picture a molecule with two distinct ends (FIGURE 6-13). On one end of the tRNA molecule is an attachment site consisting of a three-base sequence that matches up with a

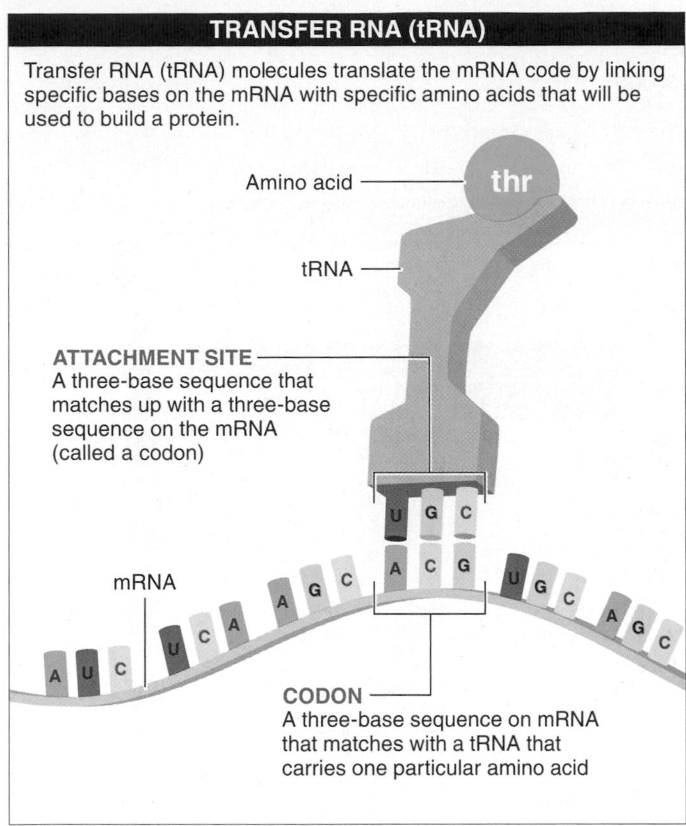

FIGURE 6-13 **Each transfer RNA attaches a particular amino acid to the mRNA.**

three-base sequence on the mRNA transcript. This match-up enables the tRNA molecule to attach to the mRNA. Attached to the other end of the tRNA molecule is an amino acid.

Each three-base sequence in mRNA—called a **codon**—matches with a tRNA molecule that carries a particular amino acid. The codon ACG, for example, is recognized by the tRNA molecule that carries the amino acid threonine. And the codon CAG is recognized by the tRNA molecule that carries glutamine. For every possible codon, there is one type of tRNA molecule that will recognize and bind to the mRNA at that point, and it will always carry the same amino acid.

The codon table (FIGURE 6-14), also called the "genetic code," describes which tRNA, and therefore which amino acid, is specified by each codon in the mRNA. With four possible bases in each of the three positions of a codon, 64 different codons are possible. Sixty-one of these codons specify amino acids, and 3 are "stop" sequences, indicating the end of translation. Because there are only 20 amino acids, many amino acids are specified by multiple codons. UGU and UGC, for example, both specify cysteine. With just a few small exceptions, the genetic code is the same in every organism on earth.

Several ingredients must be present in the cytoplasm for translation to occur. First, there must be a large supply of free amino acids. Recall from Chapter 3 that amino acids are the raw materials for building proteins and an essential component of our diet. Second, there must be **ribosomal subunits,** which are components of ribosomes, the protein-production factories where amino acids are linked together in the proper order to produce the protein.

The translation of an mRNA molecule into a sequence of amino acids (that will then fold into the complex three-dimensional shape of a protein) occurs in three steps (FIGURE 6-15).

Step 1. Recognize and initiate protein building. Translation begins in the cell's cytoplasm when the subunits of a ribosome, essentially a two-piece protein-building factory, recognize and assemble around a codon on the mRNA transcript called the start sequence. This start sequence is always the codon AUG. As the ribosomal subunits assemble themselves into a ribosome, the attachment site of a particular tRNA molecule also recognizes the start sequence on the mRNA and binds to it. This initiator tRNA carries the amino acid methionine (met). Thus, methionine is always the first amino acid in any protein that is produced. (Occasionally, in eukaryotes, this initial methionine is "edited out" later in the protein-building process.)

Step 2. Elongate. After the mRNA start sequence (AUG), the next three bases of the mRNA specify which tRNA molecule should bind to the mRNA. If the next three bases on the mRNA transcript are GUU, for example, a tRNA molecule that recognizes this sequence, and carries its particular amino acid (valine, or val), attaches to the mRNA at that point. The ribosome then facilitates the connection of this second amino acid to the first. After the first amino acid, carried by its tRNA molecule, is attached to the second amino acid, the first amino acid's tRNA molecule detaches from the mRNA and floats away (see Figure 6-15).

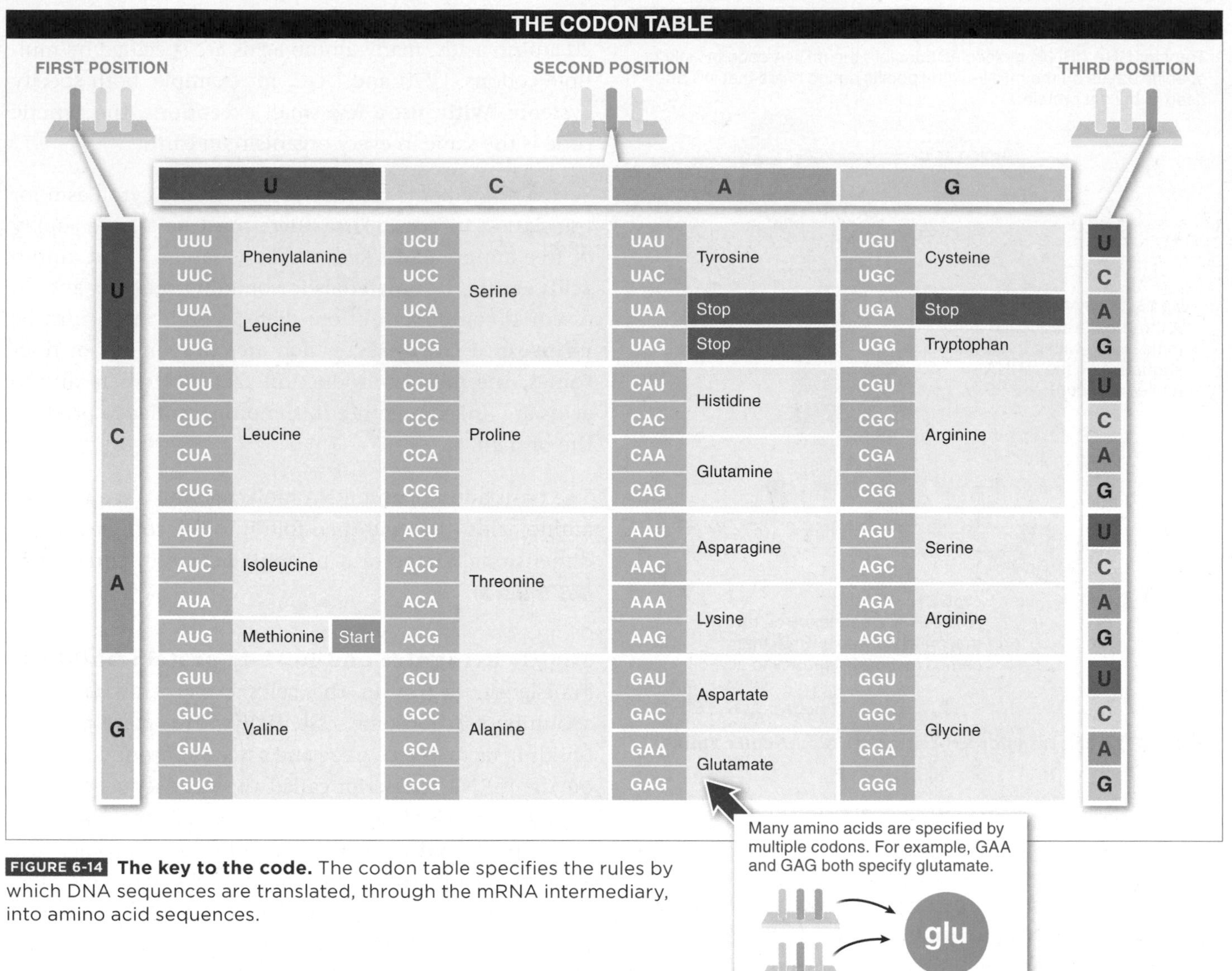

THE CODON TABLE

FIRST POSITION	SECOND POSITION: U		C		A		G		THIRD POSITION
U	UUU	Phenylalanine	UCU	Serine	UAU	Tyrosine	UGU	Cysteine	U
	UUC		UCC		UAC		UGC		C
	UUA	Leucine	UCA		UAA	Stop	UGA	Stop	A
	UUG		UCG		UAG	Stop	UGG	Tryptophan	G
C	CUU	Leucine	CCU	Proline	CAU	Histidine	CGU	Arginine	U
	CUC		CCC		CAC		CGC		C
	CUA		CCA		CAA	Glutamine	CGA		A
	CUG		CCG		CAG		CGG		G
A	AUU	Isoleucine	ACU	Threonine	AAU	Asparagine	AGU	Serine	U
	AUC		ACC		AAC		AGC		C
	AUA		ACA		AAA	Lysine	AGA	Arginine	A
	AUG	Methionine Start	ACG		AAG		AGG		G
G	GUU	Valine	GCU	Alanine	GAU	Aspartate	GGU	Glycine	U
	GUC		GCC		GAC		GGC		C
	GUA		GCA		GAA	Glutamate	GGA		A
	GUG		GCG		GAG		GGG		G

Many amino acids are specified by multiple codons. For example, GAA and GAG both specify glutamate.
glu

FIGURE 6-14 **The key to the code.** The codon table specifies the rules by which DNA sequences are translated, through the mRNA intermediary, into amino acid sequences.

As this process continues, the amino acid chain grows. The next three bases on the mRNA specify the next amino acid to be added to the first two. And the three bases after that specify the fourth amino acid, and so on. This process of progressively linking together the amino acids specified by an mRNA strand is called **protein synthesis,** because all proteins are chains of amino acids, like beads on a string. Precision in this process is essential. If the tRNAs were to misread or skip or double-read any bases, they could alter the amino acid sequence (and possibly the normal functioning) of the protein specified by the mRNA.

Step 3. Terminate. Eventually, the ribosome arrives at the codon on the mRNA that signals the end of translation. Once the ribosome encounters this stop sequence, the assembly of the amino acid chain is complete. Translation ends, and the amino acid chain and mRNA molecule are released from the ribosome. As it is being produced, the amino acid chain folds and bends, based on the chemical features in the amino acid side chains (as we saw in Chapter 3). Through the folding and bending, the protein acquires its three-dimensional structure. The completed protein—such as a membrane protein or insulin or a digestive enzyme—may be used within the cell or packaged for delivery via the bloodstream to somewhere else in the body where it is needed.

Following the completion of translation, the mRNA strand may remain in the cytoplasm to serve as the template for producing another molecule of the same protein. In

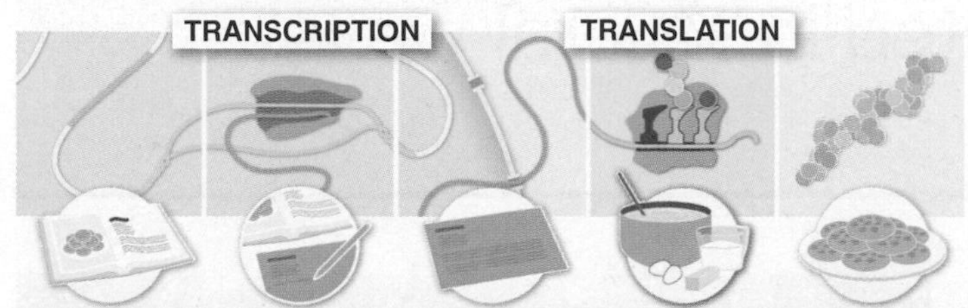

FIGURE 6-15 Translation: reading a sequence of nucleotides and producing protein. This is the second step in a two-step process by which DNA regulates a cell's activity and synthesis of proteins.

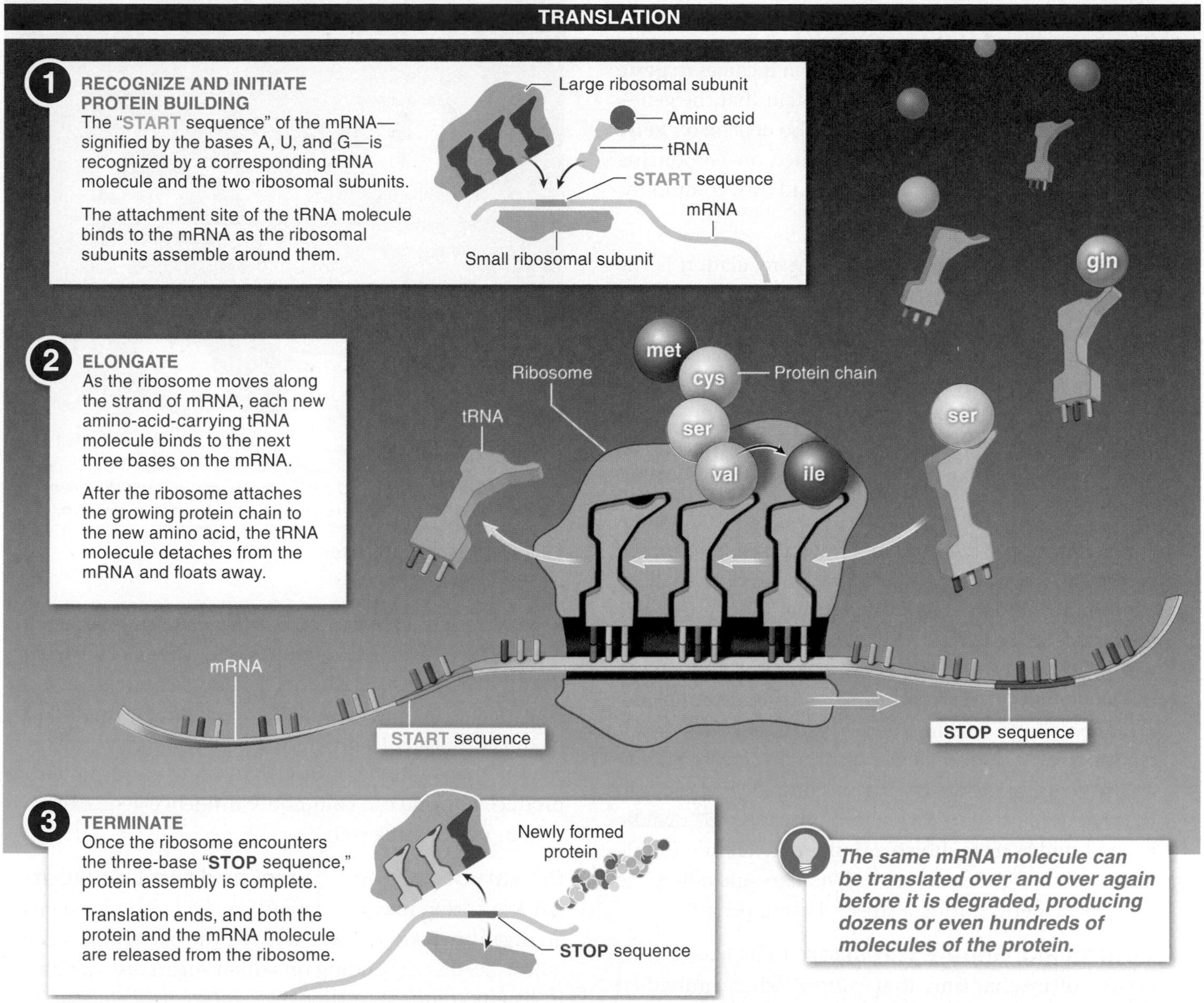

TRANSLATION

1 RECOGNIZE AND INITIATE PROTEIN BUILDING

The "START sequence" of the mRNA—signified by the bases A, U, and G—is recognized by a corresponding tRNA molecule and the two ribosomal subunits.

The attachment site of the tRNA molecule binds to the mRNA as the ribosomal subunits assemble around them.

Large ribosomal subunit
Amino acid
tRNA
START sequence
mRNA
Small ribosomal subunit

2 ELONGATE

As the ribosome moves along the strand of mRNA, each new amino-acid-carrying tRNA molecule binds to the next three bases on the mRNA.

After the ribosome attaches the growing protein chain to the new amino acid, the tRNA molecule detaches from the mRNA and floats away.

Ribosome
tRNA
met
cys
ser
val
ile
Protein chain
ser
gln
mRNA
START sequence
STOP sequence

3 TERMINATE

Once the ribosome encounters the three-base "**STOP** sequence," protein assembly is complete.

Translation ends, and both the protein and the mRNA molecule are released from the ribosome.

Newly formed protein
STOP sequence

The same mRNA molecule can be translated over and over again before it is degraded, producing dozens or even hundreds of molecules of the protein.

bacteria, an mRNA may last from a few seconds to more than an hour; in mammals, an mRNA may last several days. Depending on how long it lasts, the same mRNA strand may be translated hundreds of times. Eventually, it is broken down by enzymes in the cytoplasm.

TAKE HOME MESSAGE 6.7

» Translation is the second step in the two-step process by which information carried in DNA directs the synthesis of proteins. In translation, the information from a gene that has been encoded in the nucleotide sequence of an mRNA is read, and ingredients present in the cell's cytoplasm are used to produce a protein.

6.8 Genes are regulated in several ways.

Why do some people get sick more than others? Why is one person taller or shorter than another? It matters which genes a person has, but carrying a gene is not the only thing that matters when it comes to **gene expression**—production of the protein that the gene's sequence codes for. A person's traits also depend on **gene regulation,** whether a gene is turned on—producing its protein product—or turned off. And this, commonly, hinges on a person's environment.

A powerful tool in the study of gene regulation is the microarray, a small chip that looks like a microscope slide that can be used to monitor the expression levels of thousands of genes simultaneously. Taking cells from one part of the body, researchers examine gene expression in that tissue. Microarrays are particularly useful in exploring how gene expression differs in response to an illness or the treatment of an illness, or in response to aging.

For example, in a five-year medical study exploring the health consequences of loneliness, microarray analyses of immune system cells revealed that 209 genes were expressed differently in two groups of participants, those who were lonely, and those who were not. Individuals in the "lonely" group had impaired transcription (decreased activity) of some genes associated with an effective stress response. They also had increased activity in genes associated with inflammatory processes that are implicated in a variety of diseases (FIGURE 6-16). Related studies have shown a similarly reduced immune function among stressed-out individuals—including one study of students during a stressful exam period!

Lonely, isolated individuals have very different patterns of immune system gene expression.

FIGURE 6-16 **Social isolation.**

Controlling Gene Expression Cells have a wide variety of mechanisms that control when individual genes are expressed. One of the main ways of controlling expression is through proteins called transcription factors, which bind to specific sequences in the DNA called regulatory sites, often located in front (before the promoter site) of genes. The regulation may be "positive control," in which binding of the regulatory protein initiates or speeds up gene expression, or "negative control," in which the protein slows or blocks gene expression.

Regulatory proteins can be produced by the same cell whose DNA they are regulating, or they may come from nearby cells. During the initial development of an organism, for example, gene regulation is extremely important in directing cells to differentiate into one type of tissue—perhaps muscle, rather than bone—and proteins produced by one cell commonly influence gene expression in neighboring cells.

Prokaryotic Gene Control and the Lac Operon Regulation may also be influenced by the nutrients an organism takes in. For example, the bacterium *E. coli* adjusts its diet depending on which sugars are available in its environment. Although *E. coli* prefers glucose, in the absence of glucose, it will utilize lactose as an energy source. (The many *E. coli* cells that live in your intestine are likely to encounter lactose, which is the chief sugar in milk.)

The absorption and digestion of lactose, however, requires the expression of three genes, and bacteria do not devote energy to this unless necessary. The regulation of gene expression uses a set of regulatory sequences called an **operon,** which includes a group of several genes and the elements that control their expression as a unit.

The first regulatory site of the operon is the promoter. For a gene to be transcribed, RNA polymerase recognizes and binds to the promoter region—the specific nucleotide sequence that signals the beginning of the gene. The second regulatory site is the **operator.** A molecule called a repressor protein can bind at this site and block RNA polymerase from transcribing the genes necessary for lactose metabolism. The third regulatory element is a **regulatory gene.** This gene codes for the repressor protein that binds to the operator and blocks RNA polymerase.

Here's how this operon (called the "lac operon") functions. If lactose is not present: (1) The regulatory gene produces the repressor protein. (2) The repressor protein binds to the operator region. And (3) RNA polymerase is blocked from transcribing the lactose metabolism genes (FIGURE 6-17).

If, on the other hand, lactose is present: (1) The regulatory gene produces the repressor protein. (2) Lactose binds to the repressor protein, altering its shape so that it can no longer bind to the operator. And (3) with no repressor bound to the DNA, RNA polymerase binds to the promoter and transcribes the genes necessary for lactose metabolism.

Note that the lac operon is "switched on" only in the absence of glucose. In the absence of glucose, a regulatory protein called an activator helps the DNA unwind in the vicinity of the lac operon, enabling RNA polymerase to bind and transcribe genes. When glucose is present, however, the activator protein cannot bind to the DNA, and this prevents RNA polymerase from binding to and transcribing the lactose metabolism genes.

Eukaryotic Gene Control There are many other ways that genes can be regulated, besides operons. Many eukaryotic genes, for example, have associated nucleotide sequences called enhancer sequences. When a regulatory protein binds to these enhancer sequences, transcription rates for the associated gene are increased (a positive control).

Additional mechanisms of gene control include the binding of chemicals to DNA that can block genes from being transcribed (negative control) and the regulation of mRNA. Messenger RNA can be regulated by altering its processing and transport, or by altering the rate at which it is

THE OPERON

An operon is a group of several genes, along with the elements that control their expression as a unit, all within one section of DNA.*

OPERON ORGANIZATION

PROMOTER
The region of DNA that RNA polymerase recognizes and binds to in order to produce an mRNA transcript of the genes

OPERATOR
The region of DNA that a repressor protein can bind to and, by doing so, block RNA polymerase from transcribing the genes

GENES
The region of DNA that is transcribed by RNA polymerase

* Sizes of these elements, here and below, are not shown to scale. The promoter and operator regions are typically much shorter than the genes.

THE LAC OPERON IN ACTION

IF LACTOSE IS NOT PRESENT: OPERON OFF
A repressor protein binds to the operator, preventing RNA polymerase from binding to the promoter and transcribing the genes necessary for lactose metabolism.

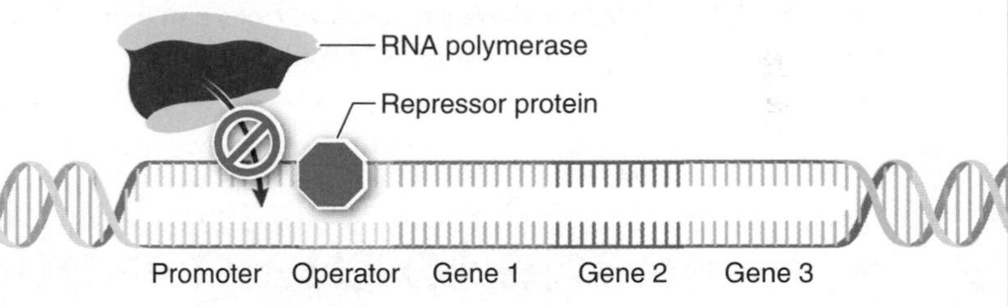

IF LACTOSE IS PRESENT: OPERON ON
Lactose (slightly modified) binds to the repressor protein, preventing it from binding to the operator. RNA polymerase can then bind to the promoter and transcribe the genes necessary for lactose metabolism.

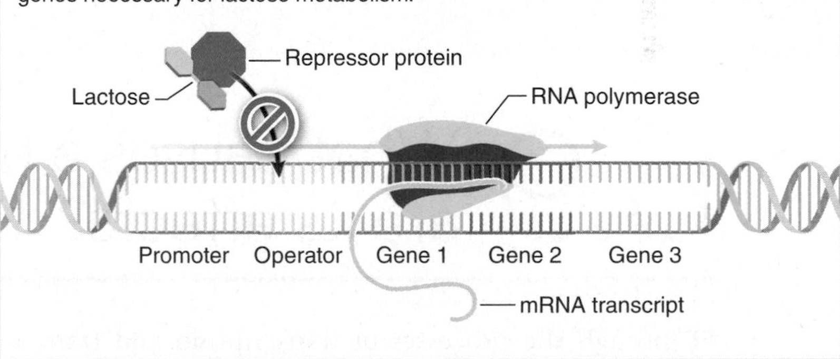

FIGURE 6-17 **Operon structure and action.** In *E. coli,* environmental conditions control the expression of the genes necessary for the metabolism of lactose.

broken down in the cytoplasm. Translation of mRNAs can be regulated, too, as can the processing of proteins following translation (FIGURE 6-18).

MECHANISMS OF REGULATING GENE EXPRESSION

TRANSCRIPTION REGULATION

POSITIVE CONTROL
- Activators can initiate or speed up gene expression.
- Enhancer sequences can speed RNA polymerase binding and gene transcription.

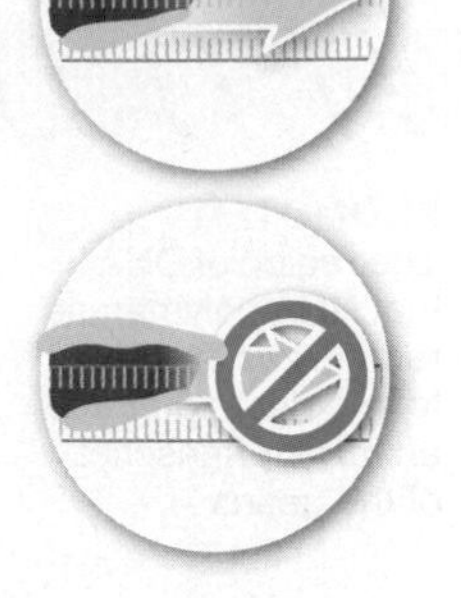

NEGATIVE CONTROL
- Repressors can block or slow down gene expression.
- Chemicals can bind to DNA and block gene transcription.

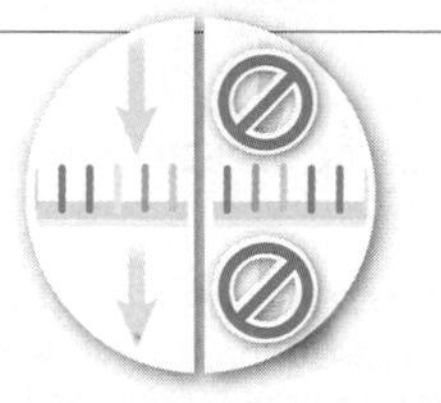

POST-TRANSCRIPTION REGULATION
- mRNA processing, transport, and enzymatic breakdown can be blocked or accelerated.
- mRNA translation and protein processing rates can be altered.

FIGURE 6-18 **Regulating gene expression.** Cells can increase or decrease gene expression in multiple ways.

TAKE HOME MESSAGE 6.8

» Environmental signals influence the expression—the turning on and turning off—of genes. By binding to DNA, regulatory proteins can block or facilitate the binding of RNA polymerase and the subsequent transcription of genes. Regulation of gene expression can also occur in a variety of other ways that enhance or impede transcription, or alter mRNA's longevity and rate of degradation, or influence translation or protein processing.

6.9–6.11

Damage to the genetic code has a variety of causes and effects.

This albino reticulated python *(Malayopython reticulatus)* lacks skin pigment.

6.9 What causes a mutation, and what are its effects?

Through the processes of transcription and translation, an organism converts the information held in its genes into the proteins necessary for life. But the process is only as good as the organism's underlying genetic information. Sometimes, something alters the sequence of bases in an organism's DNA. Such an alteration is called a **mutation,** and it can lead to changes in the structure and function of the proteins produced and have a range of effects. Sometimes mutations result in serious, even deadly, problems (FIGURE 6-19). Sometimes mutations have little or no detrimental effect. And occasionally—but very rarely—they may even turn out to be beneficial to the organism.

As an example of how mutations can affect organisms, consider breast cancer in humans. When two human genes called BRCA1 and BRCA2 are functioning properly, they reduce breast cancer risk by helping to repair DNA damage, which prevents accumulaton of changes that lead to cancer. If the DNA sequence of either of these genes is altered through mutation and the gene's normal function is lost, the person carrying the gene has a significantly increased risk of developing breast cancer. (Because a variety of factors, including environmental variables, are involved, it's impossible to know whether these individuals will develop breast cancer.)

Currently, more than 200 different mutations in the DNA sequences of these genes have been detected, each of which results in an increased risk of developing breast cancer.

Given the havoc mutations can cause for an organism, it might be surprising that most mutations are neutral, having neither a positive nor a negative effect on an organism. This happens when a mutation occurs in a noncoding region of DNA, or when a change in DNA within a gene doesn't alter the function of the protein produced. Researchers estimate that the rate of mutations in cells involved in reproduction is approximately 10^{-8} per base pair per generation.

Mutations are essential to evolution. Those mutations that don't kill an organism, or reduce its ability to survive and reproduce, can be beneficial. Every genetic feature in every organism was, initially, the result of a mutation. Most mutations you inherit from your parents will have no effect. And all of us are carrying mutations that we will never know about!

The type of cell in which a mutation occurs is important. Mutations in non-sex cells (such as skin or lung cells) can have bad health consequences for the person carrying them. Many forms of cancer, such as lung cancer and skin cancer, result from such mutations. But, non-sex-cell mutations are not passed on to your children.

Mutations in the sex cells (gametes), conversely, do not have any adverse health effects on those carrying them, but these mutations can be passed on to offspring. Individuals inheriting mutations from a parent can be at increased risk for diseases such as breast cancer or cardiovascular disease. Inherited mutations can also have an effect before birth, sometimes causing miscarriages or birth defects.

The changes to DNA caused by mutations are generally of two types: point mutations and chromosomal aberrations (FIGURE 6-20).

Point mutations occur when one base pair is substituted for another, or when a base pair is inserted or deleted. Insertions and deletions can be much more harmful than substitutions, because the amino acid sequence of a protein is certain to be affected. If a single base is added or removed, the three-base groupings in an mRNA get thrown off, and the entire sequence of amino acids "downstream" from that point will be wrong—the reading frame is shifted. It's like putting your hands on a computer keyboard, but offset by one key to the left or right, and then typing what should be a normal sentence. It comes out as gibberish.

Mutations can change the protein produced by the altered gene. Shown here: in fruit flies, a mutation in a gene for eye shape results in almost complete loss of the eye.

Normal fruit fly *Mutant fruit fly*

FIGURE 6-19 Wreaking havoc.

Chromosomal aberrations are changes to the overall organization of the genes on a chromosome. Chromosomal aberrations are like the manipulation of large chunks of text when editing a term paper. The aberrations can involve the deletion of an entire section of DNA or the relocation of a gene from one part of a chromosome to elsewhere on the same chromosome or to a different chromosome in the genome. It can involve the duplication of a gene, with the new copy inserted elsewhere in the genome. Whatever the type of aberration, a gene's expression—the protein that its sequence codes for—can be altered, as well as the expression of the genes around it.

There are three chief causes of mutation and, although one of them is beyond our control, the risk of occurrence of the other two can be significantly reduced (FIGURE 6-21).

1. Spontaneous mutations. Some mutations arise by accident as DNA is duplicating itself—at the rate of more than a thousand bases a minute in humans—when cells are dividing. DNA repair enzymes correct most errors, but not all.

2. Radiation-induced mutations. Ionizing radiation, such as X rays, has enough energy to disrupt atomic structure—even break apart chromosomes—by removing tightly bound electrons. Because ionizing rays

Q Why do dentists put a heavy apron over you when they X-ray your teeth?

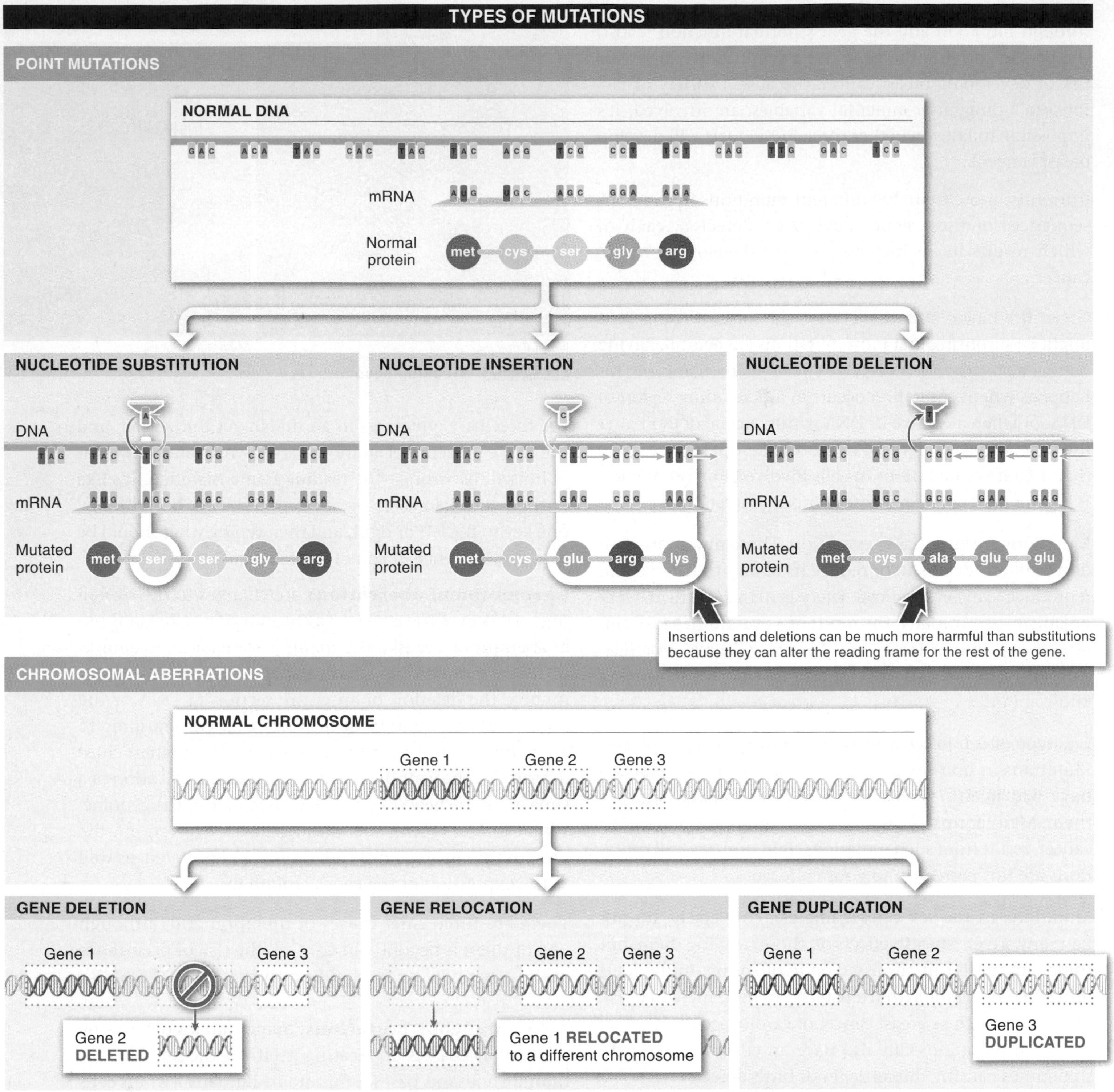

TYPES OF MUTATIONS

POINT MUTATIONS

NORMAL DNA

mRNA

Normal protein: met – cys – ser – gly – arg

NUCLEOTIDE SUBSTITUTION

DNA

mRNA

Mutated protein: met – ser – ser – gly – arg

NUCLEOTIDE INSERTION

DNA

mRNA

Mutated protein: met – cys – glu – arg – lys

NUCLEOTIDE DELETION

DNA

mRNA

Mutated protein: met – cys – ala – glu – glu

Insertions and deletions can be much more harmful than substitutions because they can alter the reading frame for the rest of the gene.

CHROMOSOMAL ABERRATIONS

NORMAL CHROMOSOME

Gene 1 Gene 2 Gene 3

GENE DELETION

Gene 1 Gene 3

Gene 2 **DELETED**

GENE RELOCATION

Gene 2 Gene 3

Gene 1 **RELOCATED** to a different chromosome

GENE DUPLICATION

Gene 1 Gene 2

Gene 3 **DUPLICATED**

FIGURE 6-20 Point mutations and chromosomal aberrations.

In point mutations, one nucleotide is changed, whereas in chromosomal aberrations, entire sections of a chromosome are altered.

cannot pass through lead, the lead apron you wear during a dental X ray protects your body from them. However, even non-ionizing, lower-energy radiation (which does not remove electrons) can damage DNA. Ultraviolet (UV) rays from the sun, for example, can be absorbed by bases in DNA, causing them to rearrange bonds. This can prevent them from pairing correctly with the complementary DNA strand and can transform a cell into a cancer cell. This is

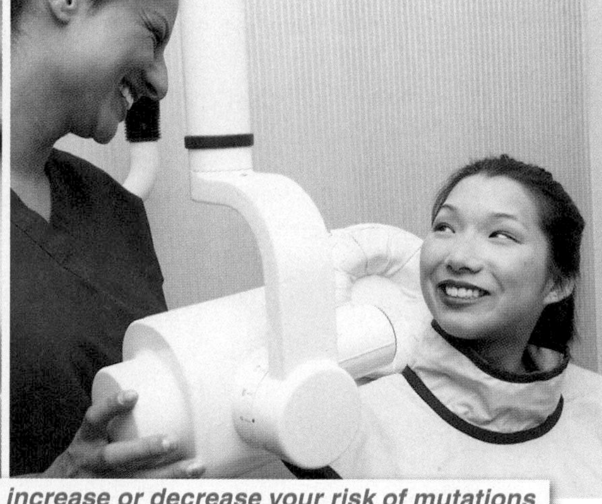

FIGURE 6-21 **Gambling with mutation-inducing activities.**

You can increase or decrease your risk of mutations with your behavior: smoking, tanning, and radiation (but notice the protective lead apron).

why long-term sun exposure can contribute to the development of skin cancer.

Another source of dangerous ionizing radiation is found in nuclear power plants, where radioactive atoms are used in energy-generating reactions. The high energy of the radioactivity can pass through the body and disrupt its DNA, causing point mutations and chromosomal aberrations. With safety precautions, however, nuclear power plant workers can minimize their exposure to harmful radiation.

3. Chemical-induced mutations. Chemicals, such as those found in cigarette smoke and engine exhaust, can also react with the atoms in DNA molecules and induce mutations.

In Section 6.11, we examine how even tiny changes in the sequence of bases in DNA can lead to errors in protein production and profound health problems.

TAKE HOME MESSAGE 6.9

» Mutations are alterations in a single base pair or changes in large segments of DNA that include several genes or more. They are rare, but may disrupt normal functioning of the body (although many mutations are neutral). Extremely rarely, mutations may have a beneficial effect. Mutations play an important role in evolution.

THIS IS HOW WE DO IT

Developing the ability to apply the process of science

6.10 Does sunscreen use reduce skin cancer risk?

Wear sunscreen!

Skin cells exposed to a type of ultraviolet radiation called UVB radiation can be damaged. Sunscreens contain chemicals that absorb or reflect some of the UV radiation. In doing so, they protect against many of the harmful effects of the sun, including sunburn. Nearly everyone assumed

(continued on the following page)

that sunscreen would also provide protection from skin cancer.

Why would anyone question whether sunscreen has a positive effect?

Sunscreens became widely available in the 1960s and 1970s. Since then, melanoma incidence has increased. Strangely, the greatest increase is in areas where sunscreen is most often used—North America, Australia, and Scandinavia. Deaths from melanoma in the United States more than doubled from 1950 to 1990. And eight studies published between 1979 and 1995 found a positive association between sunscreen use and melanoma.

Sunscreen users were at greater risk for cancer! How can that be?

Sunscreen users may have a false sense of security, and may not limit sun exposure or wear protective clothing. Further complicating matters, as recently as 2010, only one-third of sunscreens offered protection from all the wavelengths of UV radiation in sunlight. In other words, the sunscreen protected against "burns," but did not block harmful radiation.

Before throwing away your sunscreen, let's do some scientific thinking to be sure that it is of no benefit.

How can you figure out whether sunscreen users are actually protected from cancer?

The studies suggesting a positive relationship between sunscreen use and melanoma risk were "case-controlled" studies. In them, individuals developing melanoma were identified and then compared with individuals not having melanoma. The groups are analyzed to see whether they differ in some significant way—such as sunscreen use.

Such studies are undermined when the groups are heterogeneous—for example, if those in the melanoma group had more sun exposure than those in the healthy group. In fact, investigations suggested that the case-controlled studies were hindered by "confounding variables."

"Randomized controlled trials." Why are they better than case-controlled studies?

An alternative—and usually more powerful—study design is a randomized controlled study. In 2011, a study in Australia randomly assigned 1,621 adults to one of two groups: regular or discretionary sunscreen use. (Would it have been a better study if subjects in the second group had not been allowed to use sunscreen at all? Why didn't the researchers do that?)

For 5 years, those in the sunscreen group received unlimited sunscreen and were asked to apply it every morning (and after sweating, bathing, or long sun exposure). The discretionary sunscreen users were allowed to use sunscreen at their usual frequency (which included no use at all for some people). At the end of the 5-year treatment period, and for 10 additional years, the researchers noted the incidence of melanoma.

The results were dramatic. Among the 812 individuals in the sunscreen-use group, 11 new melanomas were identified. Among the 809 individuals in the discretionary-use group, twice as many new melanomas were identified. Invasive melanomas, the most severe type, occurred almost four times more frequently in the discretionary group. Based on questionnaires, there were no differences between the groups for the amount of time spent in the sun.

Cancer experts describe this first randomized, controlled study as a potential "game changer." They also view the results as a clear indication that melanoma-prevention strategies should incorporate efforts at increasing regular use of sunscreen.

TAKE HOME MESSAGE 6.10

>> Numerous case-controlled studies suggested that sunscreen use increases the incidence of melanoma, the most deadly type of skin cancer. But a more powerful, randomized controlled approach demonstrated that regular sunscreen use significantly reduces the risk of melanoma.

6.11 Faulty genes, coding for faulty enzymes, can lead to sickness.

Errors in the genetic code within a gene are responsible for many diseases. In Tay-Sachs disease (see Chapter 4), for example, an individual inherits genes with a mutation that causes an inability to produce a critical lipid-digesting enzyme in the lysosomes, the cellular garbage disposals. Because these organelles cannot digest certain lipids, the lipids accumulate, undigested. The lysosomes swell until they eventually choke the cell to death. This occurs in numerous cells in the first few years of life, and ultimately leads to the child's death.

Although the details differ from case to case, the overall picture is the same for many, if not most, inherited diseases. The pathway from mutation to illness includes just four short steps (FIGURE 6-22).

1. A mutated gene codes for a nonfunctioning protein, commonly an enzyme.
2. The nonfunctioning enzyme can't catalyze a reaction as it normally would, bringing the reaction to a halt.
3. The molecule on which the enzyme would have acted accumulates, just as half-made products would pile up on a blocked assembly line.
4. The accumulating chemical causes sickness and/or death.

The fact that many genetic diseases involve illnesses brought about by faulty enzymes suggests strategies for treatment. These include administering medications containing the normal-functioning version of the enzyme. For instance, lactose-intolerant individuals can consume the enzyme lactase, which gives them the ability to digest lactose. Alternatively, lactose-intolerant individuals can reduce their consumption of lactose-containing foods to keep the chemical from accumulating, reducing the problems that come from lactose overabundance.

FROM MUTATION TO ILLNESS
1 A mutated gene codes for a non-functioning protein, commonly an enzyme.
DNA
mRNA
Functioning enzyme should look like this:
Non-functioning enzyme
2 The non-functioning enzyme can't catalyze the reaction as it normally would.
3 The molecule it would have reacted with accumulates.
4 The accumulating chemical causes sickness or death (Tay-Sachs disease, for example).

FIGURE 6-22 Faulty enzymes can interfere with metabolism.

TAKE HOME MESSAGE 6.11

» Many genetic diseases result from mutations that cause a gene to produce a nonfunctioning enzyme, which blocks the functioning of a metabolic pathway.

STREET BIO

Using evidence to guide decision making in our own lives

A faulty gene leads to unpleasant experiences from alcohol consumption. But maybe it's a beneficial gene?

When we consume alcohol, our bodies start a two-step process to convert the alcohol molecules from their intoxicating form into harmless molecules. The steps are made possible by two different enzymes, whose assembly instructions are, for most people, coded in their DNA. Some problems can occur if these two enzymes don't function properly.

Q: What is "fast-flushing"? When some people consume alcohol, almost immediately they have some unpleasant symptoms: their face turns red, their heart races, and their head starts to pound. Worse still, they soon feel the need to vomit. This response is called "fast-flushing."

Q: What causes fast-flushing? Fast-flushers complete the first step of breaking down alcohol, but not the second. They carry defective genetic instructions for making aldehyde dehydrogenase, the enzyme that makes possible the second step of the alcohol-breakdown process. As a result, a poisonous substance accumulates in their body when they consume alcohol. This causes the symptoms of fast-flushing.

Q: Why is fast-flushing particularly common in Asia? Approximately half of the people living in Asia carry a non-functional form of the gene for aldehyde dehydrogenase. Individuals who cannot build this enzyme will experience fast-flushing when they consume alcohol.

Q: Why might there be a bright side to carrying this "faulty" gene? The inability to metabolize alcohol may actually confer more benefit than harm. In a study of 1,300 alcoholics in Japan, not a single one was a fast-flusher, even though half of all Japanese people are fast-flushers. The minor change in the genetic code that makes alcohol consumption such an unpleasant experience may be responsible for the lower incidence of alcoholism among Japanese and other Asian people.

GRAPHIC CONTENT

Thinking critically about visual displays of data

1 For the species listed, which has the largest percentage of "junk DNA"? Which has the smallest?

2 What does the gray section of each bar represent? What does the blue-green shaded part represent?

3 From the data presented, can you rank the species in order of how many coding DNA molecules are found in the species? If so, give the rankings. If not, explain why not.

4 Using the data presented, how would you refute the argument than having a greater proportion of coding DNA leads to a more complex organism?

5 What is the *x*-axis? What is the *y*-axis? Why might this graph be more persuasive than a table presenting the same data? How would your interpretation of this figure change if the axes were transposed?

6 The data presented do not give us information about the size of each genome. Why might this presentation—showing percentages of coding DNA—be useful?

7 What conclusions can you draw from this graph?

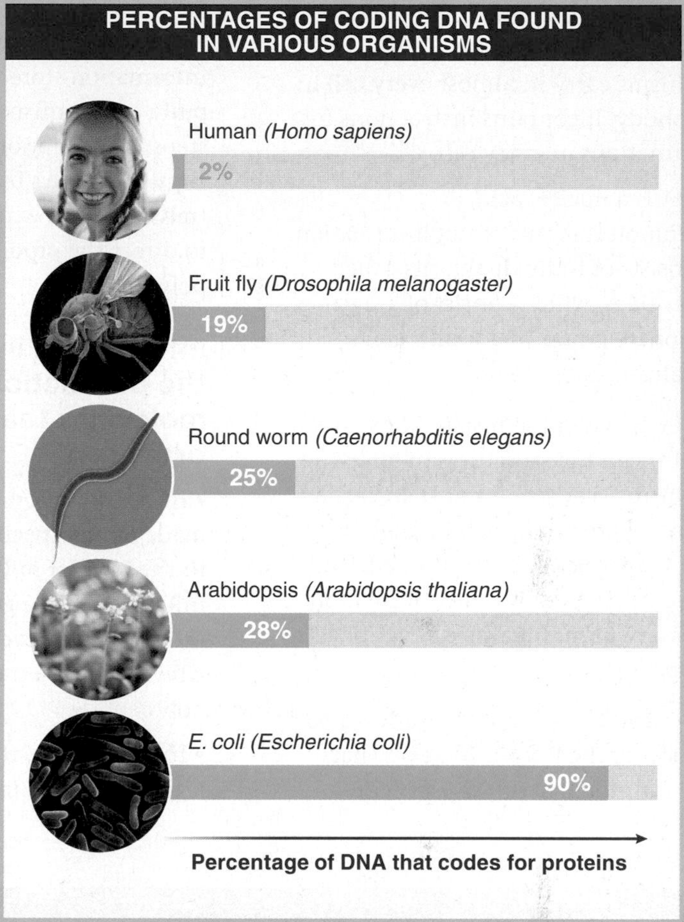

See answers at the back of the book.

KEY TERMS IN DNA AND GENE EXPRESSION

allele, p. 169
base, p. 167
base pair, p. 167
chromosomal aberration, p. 181
chromosome, p. 168
code, p. 168
codon, p. 175
deoxyribonucleic acid (DNA), p. 167
gene, p. 168
gene expression, p. 178
gene regulation, p. 178
genome, p. 168
genotype, p. 171
intron, p. 170
locus (*pl.* loci), p. 169
messenger RNA (mRNA), p. 172
mutation, p. 180
nucleic acid, p. 167
nucleotide, p. 167
operator, p. 179
operon, p. 179
phenotype, p. 171
point mutation, p. 181
promoter site, p. 173
protein synthesis, p. 176
regulatory gene, p. 178
ribosomal subunit, p. 175
trait, p. 169
transcription, p. 171
transfer RNA (tRNA), p. 174
translation, p. 171

BRIEF SUMMARY

DNA: what is it, and what does it do?

• DNA is a molecule that all living organisms carry in almost every cell in their body. It contains instructions for the functions of every cell.

• DNA is a nucleic acid, a macromolecule that stores information. It consists of individual units called nucleotides, which consist of a sugar, a phosphate group, and a nitrogen-containing base.

• DNA provides the instructions for building all the structures in all living organisms. The full set of DNA carried by an organism is called its genome. A gene is a sequence of bases in a DNA molecule that carries the information necessary for producing a functional product.

• In eukaryotic organisms, only a small fraction of the DNA is in genes that code for proteins; the function of much of the rest is still poorly understood.

• The process by which the information stored in DNA is used to build an organism occurs in two main steps: transcription, which produces a copy of a gene's base sequence, and translation, in which that copy is used to direct the production of a polypeptide.

Information in DNA directs the production of the molecules that make up an organism.

• In transcription, a single copy is made of one specific gene in the DNA, in the form of mRNA. When the mRNA copy of a gene is completed and processed, it moves to the cytoplasm, where it can be translated into a polypeptide.

• In translation, the information from a gene, encoded in the nucleotide sequence of an mRNA, is read, and ingredients present in the cell's cytoplasm are used to produce a protein.

• Environmental signals influence the turning on and turning off of genes.

Damage to the genetic code has a variety of causes and effects.

• Mutations are alterations in a single base or in large segments of DNA. They are rare, but when they do occur, they may disrupt normal functioning of the body (although many mutations are neutral). Extremely rarely, mutations may have a beneficial effect. Mutations play an important role in evolution.

• Many genetic diseases result from mutations that cause a gene to produce a non-functioning enzyme, which in turn blocks the functioning of a metabolic pathway.

CHECK YOUR KNOWLEDGE

Short Answer

1. How and why is DNA helpful in ensuring greater justice in our society?

2. After sequencing a molecule of DNA, you discover that 20% of the bases are cytosine. What percentage of the bases would you expect to be thymine? Why?

3. What function does the sugar-phosphate backbone of a DNA molecule serve?

4. Only a small fraction of the DNA in eukaryotic organisms codes for proteins. The remainder is sometimes referred to as "junk DNA." Why is this a poor description of the noncoding DNA?

5. What is the role of transcription? Why is this step necessary? Why isn't translation the first step in using information in DNA to build a new organism?

6. When the mRNA copy of a gene is completed in a eukaryotic cell, why does the mRNA move to the cytoplasm?

7. What is the function of tRNA in translation?

8. *E. coli* prefers glucose as an energy source, but if lactose, and no glucose, is present, the *E. coli* cells can survive. How is *E. coli* able to respond to this environmental change?

9. Mutations are often considered to be negative events with bad consequences. Why is this view of mutations somewhat of a paradox?

10. Which step in a metabolic pathway is disrupted in Tay-Sachs disease?

Multiple Choice

1. A person's DNA is carried in his or her:

a) muscle cells. b) hair.

c) saliva. d) skin cells.

e) All of the above contain a person's DNA.

0 EASY 23 HARD 100

2. Which of the following is/are always the same in every DNA molecule?

a) the sugar

b) the base

c) the phosphate group

d) Only a) and b) are always the same.

e) Only a) and c) are always the same.

3. The full set of an individual organism's DNA is called its:

a) complement.

b) genome.

c) nucleosome.

d) nucleotide.

e) chromosome.

0 EASY 15 HARD 100

4. In humans, genes make up _____ of the DNA.

a) about 75%

b) 100%

c) less than 5%

d) about 10%

e) about 50%

0 EASY 38 HARD 100

5. Genotype is to phenotype as:

a) cookie is to oven.

b) cookie is to recipe.

c) cookbook is to cookie.

d) recipe is to cookie.

e) oven is to cookie.

0 EASY 21 HARD 100

6. To start the transcription process, a large molecule, _____, recognizes a _____.

a) RNA polymerase; messenger RNA

b) DNA polymerase; termination site

c) DNA polymerase; promoter site

d) RNA polymerase; promoter site

e) DNA polymerase; messenger RNA

0 EASY 31 HARD 100

7. There are different _____ for each of the 20 different amino acids that are used in building proteins.

a) ribosomal subunits

b) tRNA molecules

c) mRNA molecules

d) DNA molecules

e) elongation factors

8. Deletions and substitutions are two types of mutations. Which type is more likely to cause mistranslations of proteins?

a) Substitutions, because they shift the reading frame and cause downstream amino acids to be changed.

b) Substitutions, because one protein is substituted for another protein.

c) Deletions, because they shift the reading frame and cause downstream amino acids to be changed.

d) Deletions, because one protein is deleted.

e) None of the above; mutations are not the cause of mistranslation.

0 EASY 37 HARD 100

9. Which of the following statements about mutations is incorrect?

a) Mutations are very rare.

b) A mutation in DNA always leads to changes in the structure and function of the protein produced.

c) When one nucleotide base pair is replaced by another, this constitutes a point mutation.

d) One general type of mutation is a change in the overall organization of chromosomal genes.

e) X rays can cause mutations.

10. Which of the following statements about the metabolism of ethanol (which is present in alcoholic beverages) is incorrect?

a) Individuals who produce non-functioning aldehyde dehydrogenase exhibit "fast-flushing."

b) The process requires two enzymes: alcohol dehydrogenase and isopropyl dehydrogenase.

c) Individuals who are "fast-flushers" are less likely to become alcoholics.

d) Approximately half of the people living in Asia carry a nonfunctional form of a gene necessary for ethanol metabolism.

e) None. All of the above statements are correct.

0 EASY 56 HARD 100

Ch07

Living organisms can be manipulated for practical benefits.

Biotechnology is producing improvements in agriculture.

Biotechnology has the potential for improving human health.

Biotechnology has value for criminal justice systems.

The double-helix sidewalks of McMillan Park in Huntsville, Alabama, will link businesses and a nonprofit educational and research center.

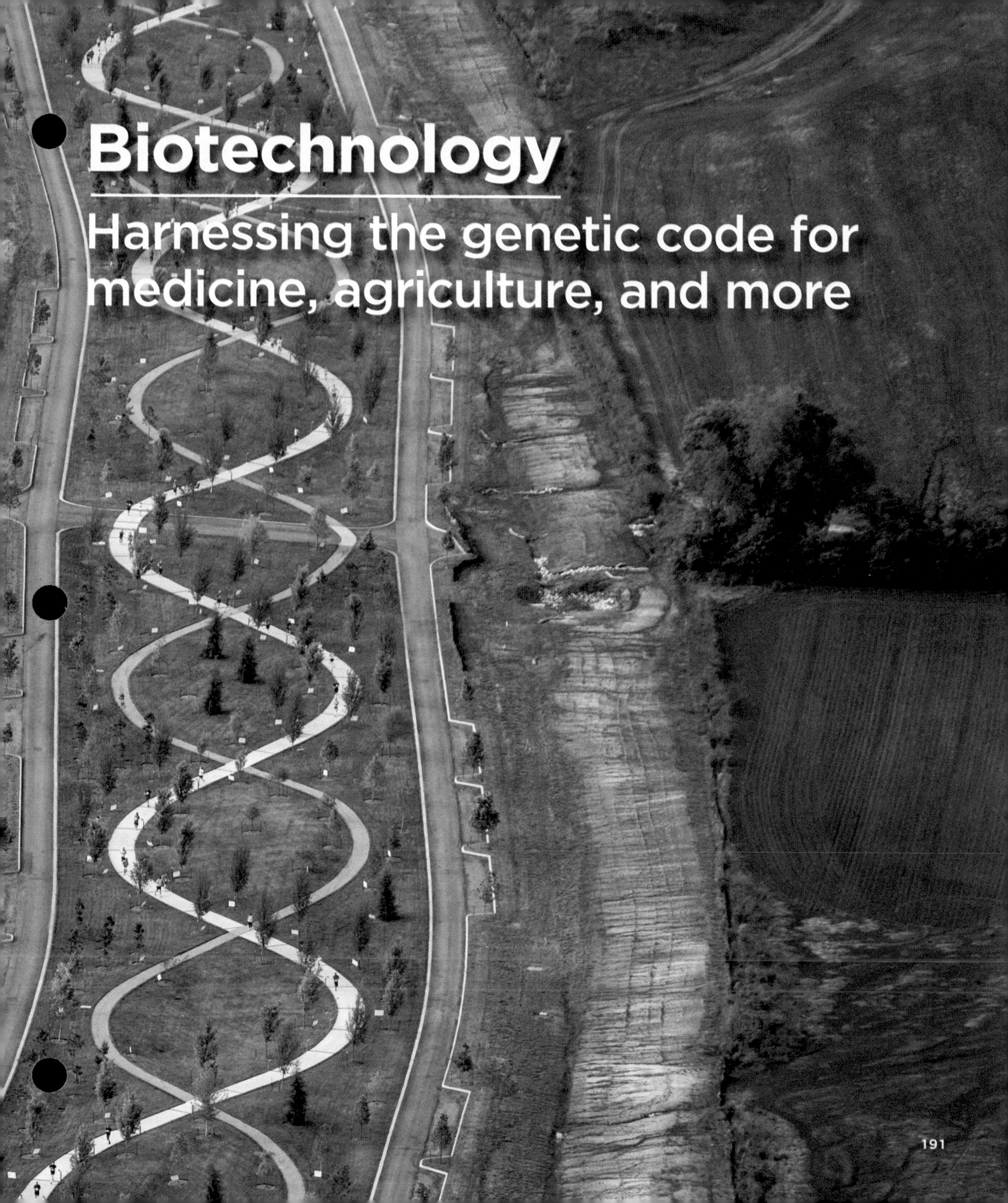

Biotechnology

Harnessing the genetic code for medicine, agriculture, and more

7.1–7.3

Living organisms can be manipulated for practical benefits.

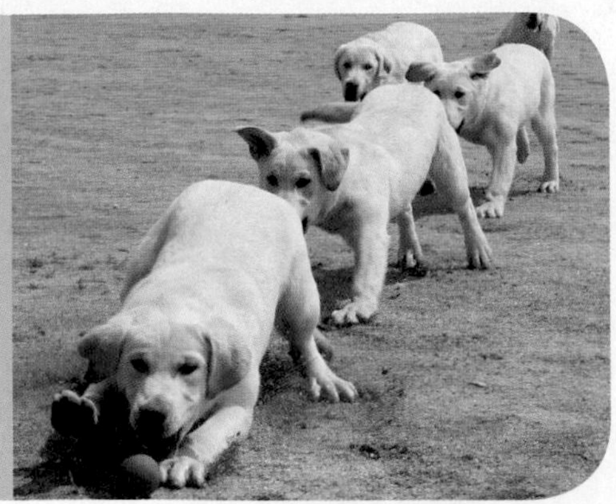

Cloned drug-sniffing Labrador puppies begin their training.

7.1 What is biotechnology and what does it promise?

Much interest surrounds the field of **biotechnology**, in which organisms, cells, and their molecules are modified to achieve practical benefits. Intriguing news stories about "biotech" announce attention-grabbing initial results:

- A kit that can test for cancer with just a drop of saliva
- A device that can stimulate regeneration of neurons following spinal injuries
- Targeted delivery of genetically engineered cells that can cure inherited diseases
- "Gene chips" that can identify your risk of developing any one of hundreds of diseases and highlight treatments personalized for your genome

Meanwhile, most scientists who conduct basic research are conservative when interpreting their results and cautious when drawing conclusions or when generalizing from limited observations to broader assertions about their significance.

Q Evaluating claims about biotechnology in the media requires more vigilance and skepticism than appraising claims about basic science. Why?

These differing points of view can be difficult to reconcile, but it helps to recognize that when scientific methods are used in the pursuit of practical applications, economic considerations can become important. And as entrepreneurs seek to establish the value of products they are developing, they may court the media with reports highlighting the *potential* of their products, which may have been demonstrated only in cell cultures, animal models, or extremely limited numbers of cases.

Let's look at an example. In 2008, *Good Morning, America* reported that "researchers have shown that gene therapy can be used to improve vision for blind children and young adults." The announcement was based on two research studies. In each, three patients received gene therapy treatments for an inherited disease that causes severe vision loss. Slight vision improvements occurred for four of the six patients. In the initial published reports, however, the researchers cautioned that "a placebo effect may have contributed to the improved measures and cannot be ruled out." They also noted that "we cannot be sure that the improvement reflects expression of the protein encoded [by the gene delivered in the therapy]."

The research studies were well done, but definitely preliminary. And in the researchers' subsequent report after following up the patients for three and a half years, it turns out that "no consistent improvement in visual acuity was evident."

There is nothing unusual about these studies; they simply reflect the typical process by which scientists come to understand phenomena. Hearing the report in the media, however, people might reasonably (but erroneously) have assumed that this was a victory for biotechnology rather than an early step in a lengthy process of discovery.

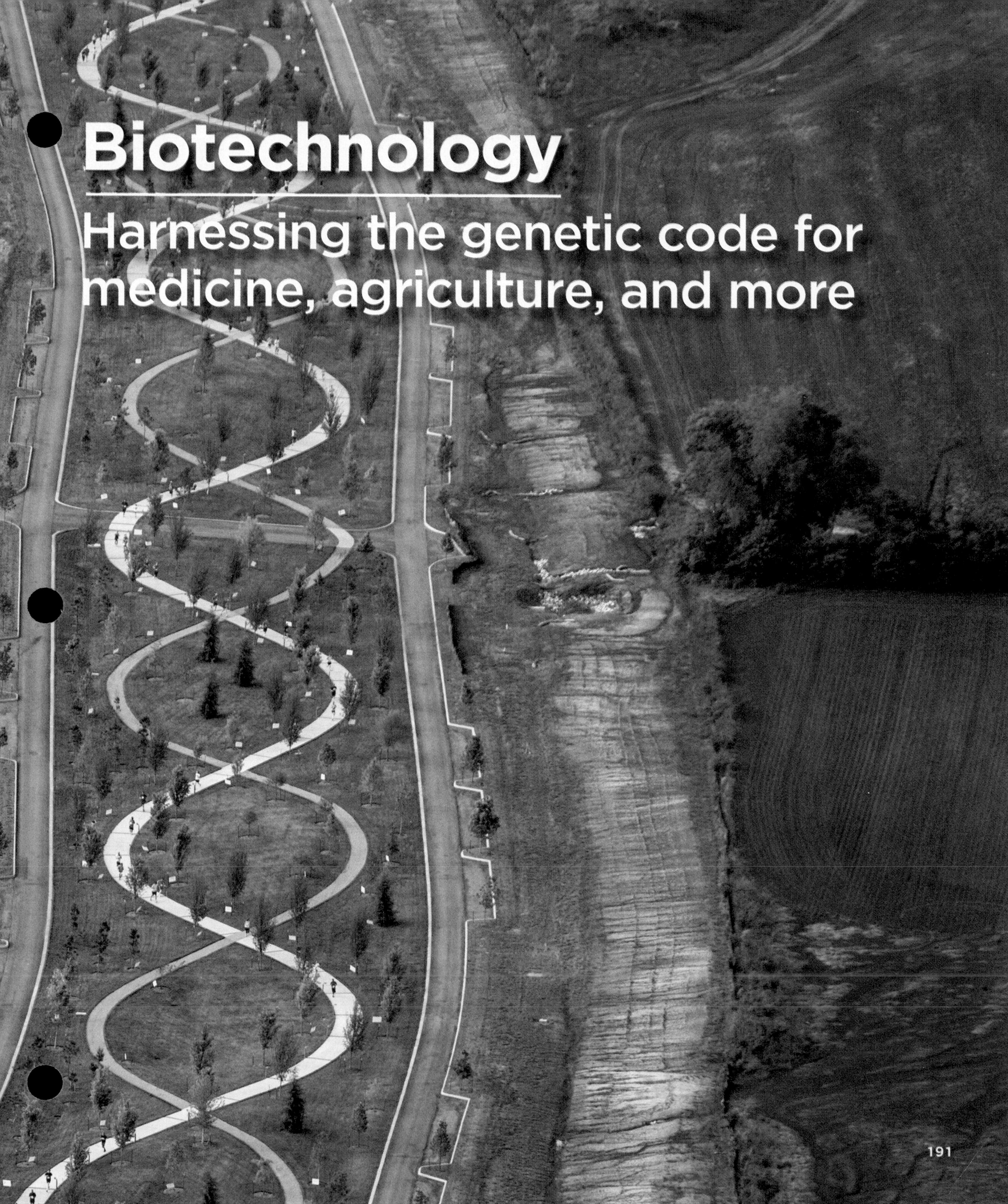

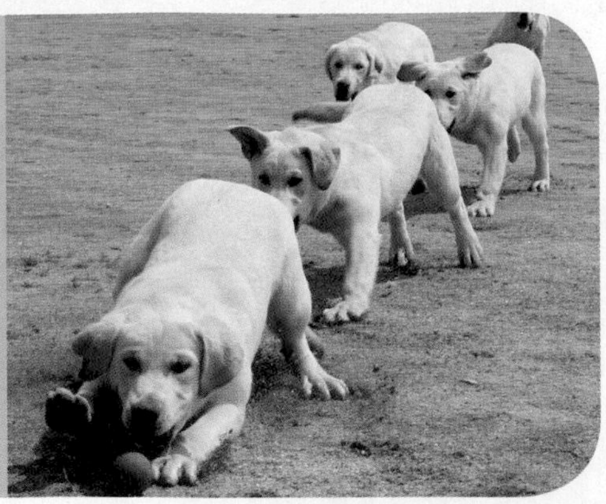

PRIMARY APPLICATIONS OF BIOTECHNOLOGY

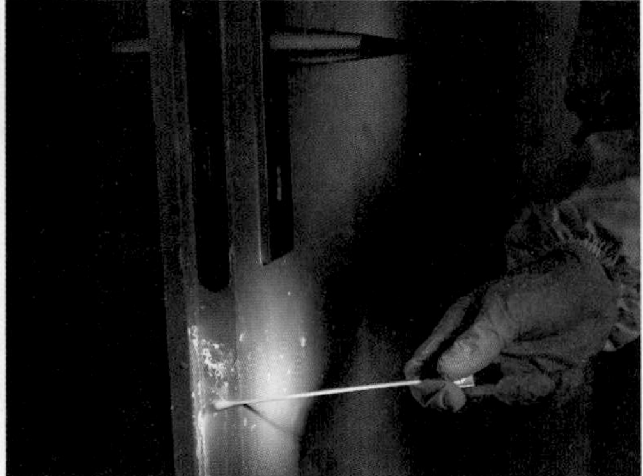

AGRICULTURE
Already changing our world significantly.
- Pest- and disease-resistant crops
- Dramatically higher crop yields
- Foods with enhanced nutrition

HUMAN HEALTH
Tremendous potential, but limited success so far.
- Improved treatment of disease through more effective medicine
- Improved diagnosis and screening for genetic diseases

FORENSIC SCIENCE
Advances already improving the justice system.
- Improved capabilities of law enforcement
- Important reforms to the criminal justice system

FIGURE 7-1 **Brave new world?** Biotechnology can change—and already has changed—our world.

Still, the steadfast work of biotechnology researchers has yielded successes that are robust and are changing our world in significant ways. This is particularly notable in agriculture, human health, and forensic science (FIGURE 7-1).

1. Agriculture. The development of crops resistant to pests and disease has increased yields dramatically while reducing costs and, sometimes, reducing the amount of pesticides that must be used. Moreover, in some cases, plants engineered to be resistant to pests can survive diseases that might otherwise make their cultivation impossible. The U.S. Department of Agriculture, for example, credits the genetic engineering of papayas resistant to the ringspot virus with saving the U.S. papaya industry. Additionally, biotechnology can be used to produce foods with enhanced nutrition.

2. Human health. Biotechnology has led to some notable successes in treating diseases—particularly juvenile (type 1) diabetes and several conditions that lead to anemia or insufficient production of human growth hormone. These advances have usually resulted from the ability to produce medicines more efficiently and effectively than they could be produced with traditional methods. As we'll see later in the chapter, biotechnology has also led to improvements in diagnosing and screening for genetic diseases.

3. Forensic science. Forensic biotechnologies have led to spectacular advances in the capabilities of law enforcement and improvements in the criminal justice system. The 2016 exoneration of Anthony Wright was the 344th case in which DNA evidence was used to free someone wrongly convicted. And perhaps equally important, the lessons we're learning from these sobering cases are helping us to reform the criminal justice system.

Any discussion of biotechnology must acknowledge that alongside the potential benefits come difficult ethical issues, potentially harmful environmental impacts, and potential health risks. These must be explored, debated, and addressed at every stage.

We begin our investigation of biotechnology by exploring some of the most important tools and techniques. These include several that are essential for **genetic engineering,** the manipulation of organisms' genetic material by adding, deleting, or transplanting genes from one organism to another.

TAKE HOME MESSAGE 7.1

» Biotechnology is the use of technology to modify organisms, cells, and their molecules to achieve practical benefits. The primary areas in which biotechnology is applied include agriculture, human health, and forensic science.

7.2 A few important processes underlie many biotechnology applications.

How would you create a plant resistant to being eaten by insects? Or a colony of bacteria that can produce human insulin? Surprisingly, although there are many different uses of biotechnology, a relatively small number of processes and tools are employed (FIGURE 7-2). These enable researchers to do the following:

1. **Chop** up the DNA from a donor organism that exhibits the trait of interest.
2. **Amplify** the small amount of DNA into larger quantities.
3. **Insert** pieces of the DNA into bacterial cells or viruses.
4. **Grow** separate colonies of the bacteria or viruses, each of which contains a different inserted piece of donor DNA.

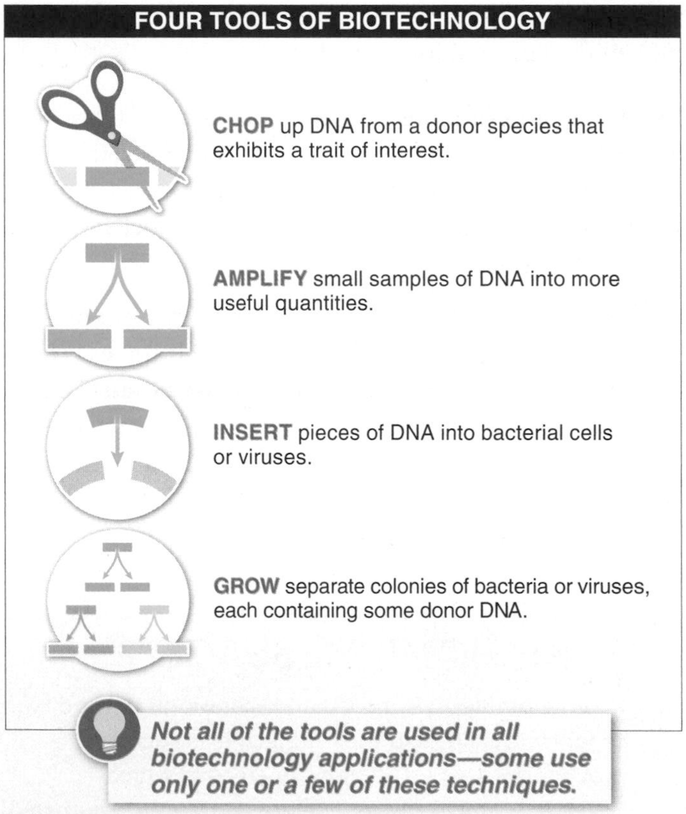

FOUR TOOLS OF BIOTECHNOLOGY

CHOP up DNA from a donor species that exhibits a trait of interest.

AMPLIFY small samples of DNA into more useful quantities.

INSERT pieces of DNA into bacterial cells or viruses.

GROW separate colonies of bacteria or viruses, each containing some donor DNA.

Not all of the tools are used in all biotechnology applications—some use only one or a few of these techniques.

FIGURE 7-2 Four important tools and techniques used in most biotechnology procedures.

These tools capture the essence of much modern biotechnology. Not all of the techniques are used in all biotech applications. Nonetheless, they all are mainstays of today's biotechnology world.

Tool 1. Chopping up DNA from a donor organism. To begin, researchers select an organism that has a desirable trait. For example, they might want to produce human growth hormone in large amounts. Their first step would be to obtain a sample of human DNA and cut it into smaller pieces. Cutting DNA into small pieces requires the use of **restriction enzymes** (FIGURE 7-3).

Restriction enzymes evolved in bacteria to defend against attack by viruses. Their job is to cut DNA into small pieces. A restriction enzyme recognizes a sequence of viral DNA and then binds to it and severs it, making it impossible for the virus to reproduce within the bacterial cell.

Dozens of different restriction enzymes exist, each of which recognizes and cuts a different, specific sequence of bases in DNA. Restriction enzymes cut DNA from any source, not just from viruses, as long as the specific four- to eight-base sequence is present.

Tool 2. Amplifying DNA pieces into larger quantities. Often in biotechnology, only a small sample of the DNA of interest is available. The **polymerase chain reaction (PCR)** is a laboratory technique that allows a tiny piece of DNA to be duplicated repeatedly (FIGURE 7-4).

The process involves briefly heating a solution containing the DNA of interest, which causes the two DNA strands to separate. The DNA solution is then cooled in the presence of certain enzymes, called DNA polymerases, that are specialized to be stable at high temperatures, along with special DNA fragments called primers, and plenty of nucleotides floating free in the solution. As the solution cools, an enzyme covalently links free nucleotides with their complementary bases on each of the single strands of DNA. Primers serve as starting points for DNA synthesis. And adding primers with sequences complementary to the region being amplified makes it possible to selectively amplify the regions of interest.

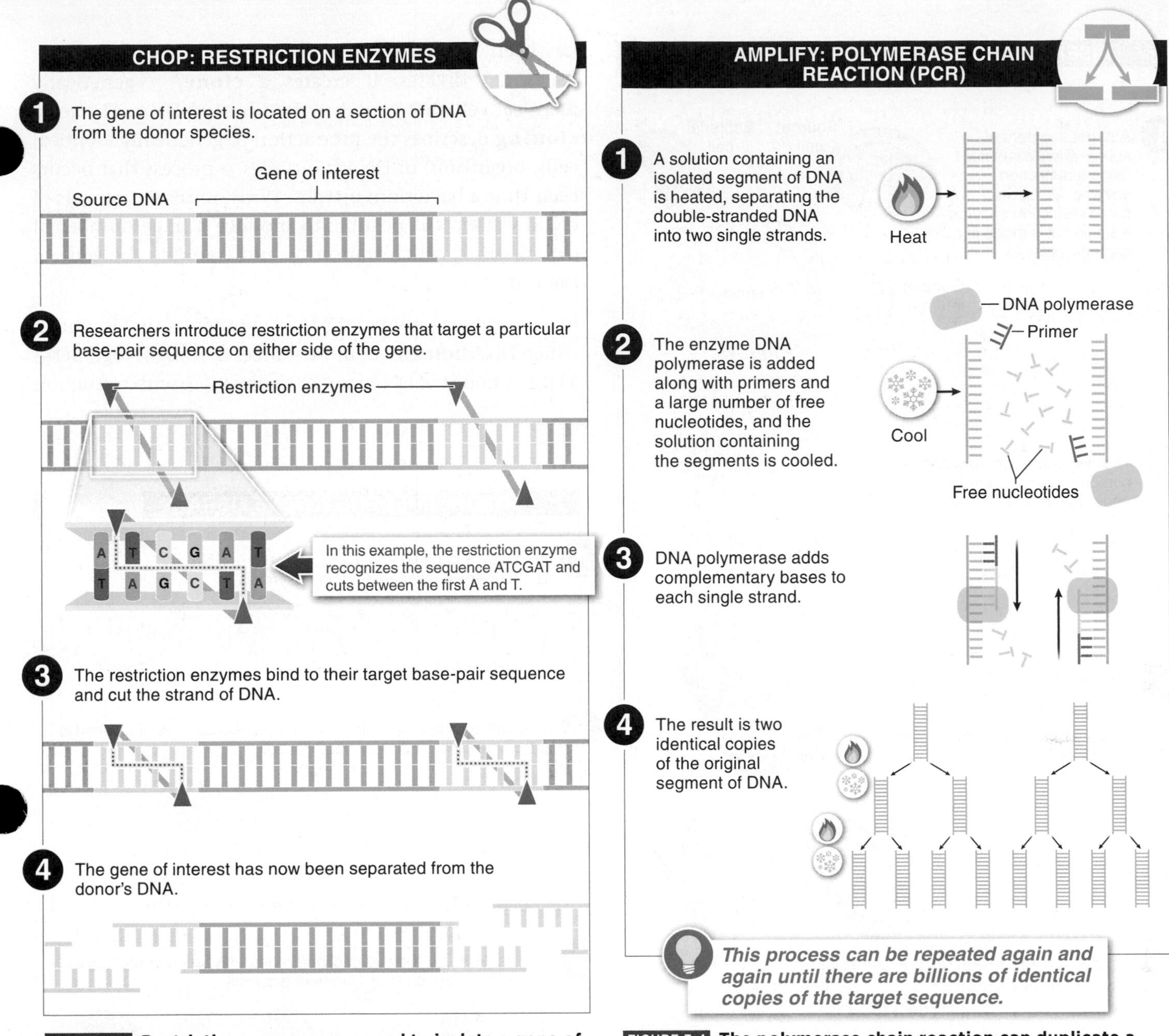

CHOP: RESTRICTION ENZYMES
1 The gene of interest is located on a section of DNA from the donor species.
Gene of interest
Source DNA
2 Researchers introduce restriction enzymes that target a particular base-pair sequence on either side of the gene.
Restriction enzymes
A T C G A T
T A G C T A
In this example, the restriction enzyme recognizes the sequence ATCGAT and cuts between the first A and T.
3 The restriction enzymes bind to their target base-pair sequence and cut the strand of DNA.
4 The gene of interest has now been separated from the donor's DNA.
AMPLIFY: POLYMERASE CHAIN REACTION (PCR)
1 A solution containing an isolated segment of DNA is heated, separating the double-stranded DNA into two single strands.
Heat
2 The enzyme DNA polymerase is added along with primers and a large number of free nucleotides, and the solution containing the segments is cooled.
DNA polymerase
Primer
Cool
Free nucleotides
3 DNA polymerase adds complementary bases to each single strand.
4 The result is two identical copies of the original segment of DNA.
This process can be repeated again and again until there are billions of identical copies of the target sequence.

FIGURE 7-3 **Restriction enzymes are used to isolate a gene of interest.**

FIGURE 7-4 **The polymerase chain reaction can duplicate a small strand of DNA repeatedly to form billions of copies.**

The result is two complete double-stranded copies of the DNA. This process of heating and cooling can be repeated again and again until there are billions of identical copies of the original sequence.

Tool 3. Inserting foreign DNA into the target organism. Gene editing (also referred to as genome editing) describes the alteration of an organism's genome using biotechnology. In the human growth hormone example mentioned above, the researchers might want to transfer the human growth hormone gene into the bacterium *E. coli,* creating **transgenic organisms** (that is, organisms with DNA inserted from a different species). The bacteria can then produce large amounts of the desired product.

To create a transgenic organism, researchers must physically deliver the DNA from a donor species into the recipient organism. This delivery requires a "vector"—something to carry the donor DNA—and is often accomplished using **plasmids,** circular pieces of DNA that can be incorporated into a bacterium's genome (FIGURE 7-5). A restriction

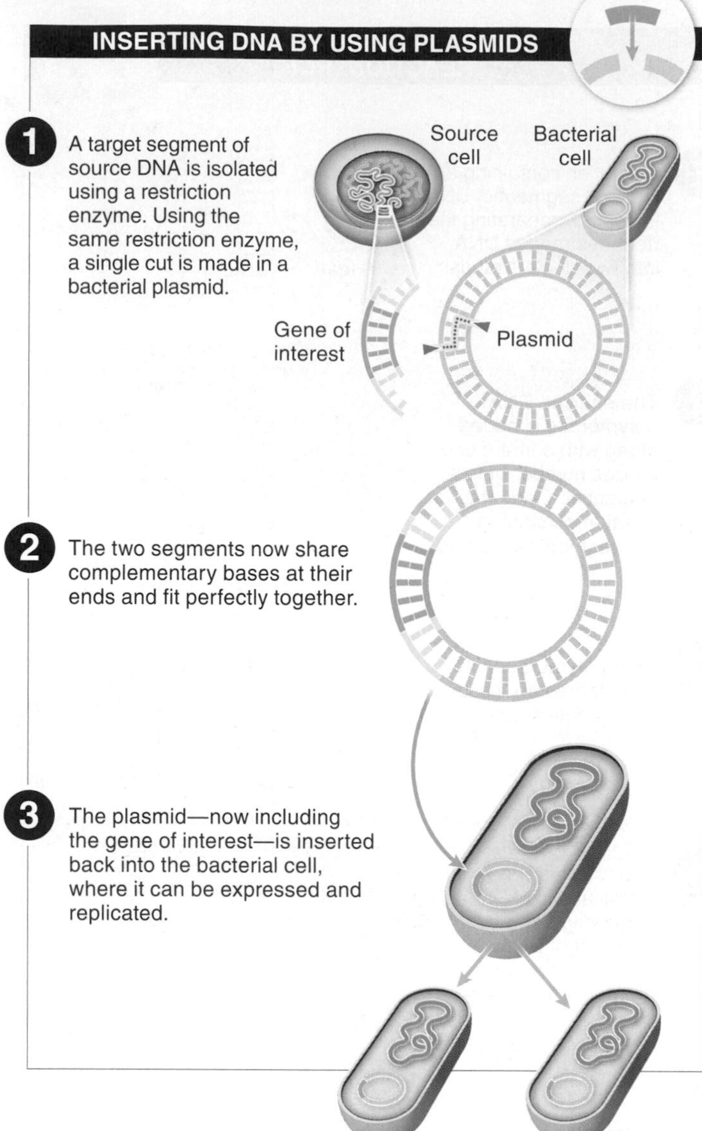

INSERTING DNA BY USING PLASMIDS
1
A target segment of source DNA is isolated using a restriction enzyme. Using the same restriction enzyme, a single cut is made in a bacterial plasmid.
Source cell
Bacterial cell
Gene of interest
Plasmid
2
The two segments now share complementary bases at their ends and fit perfectly together.
3
The plasmid—now including the gene of interest—is inserted back into the bacterial cell, where it can be expressed and replicated.

FIGURE 7-5 **A gene chauffeur.** Plasmids can transfer DNA from one species to another.

enzyme recognizes and binds to a particular sequence of four to eight bases of the plasmid and cuts there, thus allowing insertion of the matching sequence of donor DNA (matching because it was cut by the same restriction enzyme). After insertion of the plasmid into the bacterial cell, genes on the plasmid can be expressed in the bacterium and are replicated whenever the cell divides, so both of the new cells contain the plasmid with the donor gene. In other cases, genes are incorporated into viruses instead of plasmids. The viruses can then be used to infect organisms and transfer the genes of interest into those organisms.

Tool 4. Growing bacterial colonies that carry the DNA of interest: cloning. Once a piece of foreign DNA has been transferred to a bacterial cell, every time the bacterium divides, it creates a **clone,** a genetically identical cell that contains the inserted DNA. The term **cloning** describes the production of genetically identical cells, organisms, or DNA molecules, a process that occurs each time a bacterium divides. With numerous rounds of cell division, it is possible to produce a huge number of clones, all of which transcribe and translate the gene of interest.

In a typical recombinant DNA experiment—that is, one in which DNA from two or more sources is brought together—a large amount of DNA may be chopped up with restriction

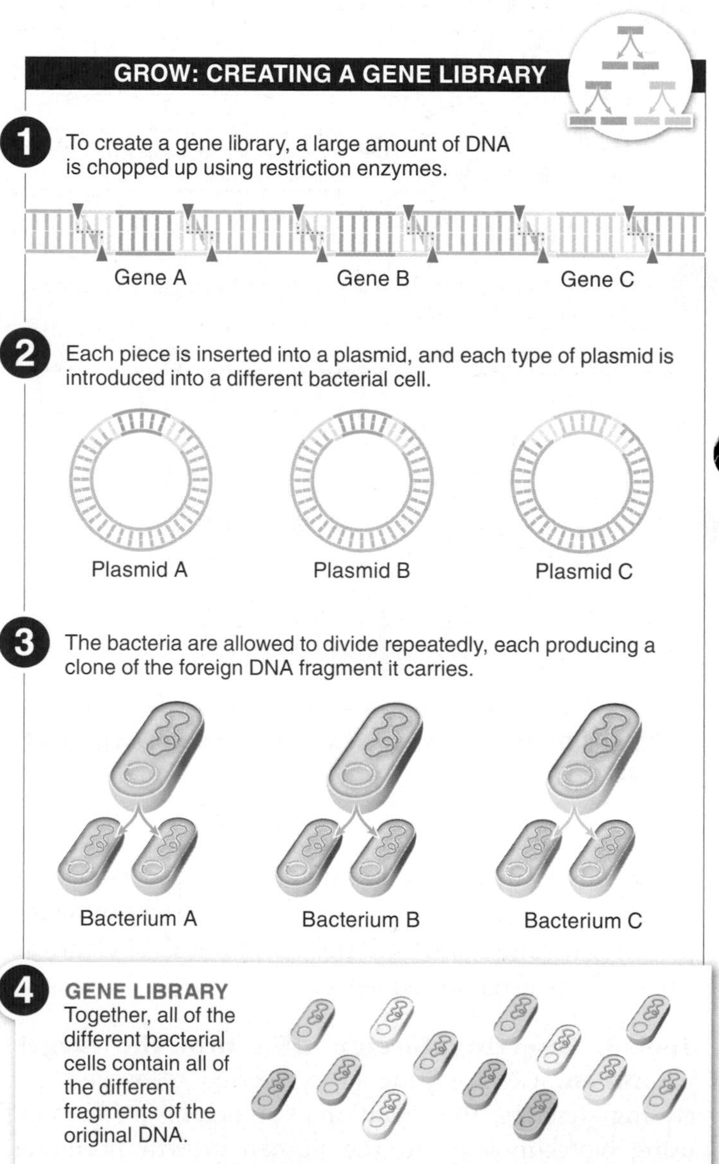

GROW: CREATING A GENE LIBRARY
1
To create a gene library, a large amount of DNA is chopped up using restriction enzymes.
Gene A
Gene B
Gene C
2
Each piece is inserted into a plasmid, and each type of plasmid is introduced into a different bacterial cell.
Plasmid A
Plasmid B
Plasmid C
3
The bacteria are allowed to divide repeatedly, each producing a clone of the foreign DNA fragment it carries.
Bacterium A
Bacterium B
Bacterium C
4
GENE LIBRARY
Together, all of the different bacterial cells contain all of the different fragments of the original DNA.

FIGURE 7-6 **A DNA archive.** A gene library (or clone library) is a collection of cloned DNA fragments.

enzymes, incorporated into plasmids, and introduced into bacterial cells. The bacteria are then allowed to divide repeatedly, with each bacterial cell producing a clone of the foreign DNA fragment it carries. Together, all of the different cells containing all of the different fragments of the original DNA are called a **clone library** or a **gene library** (FIGURE 7-6). Researchers can later identify those bacteria with the gene of interest and grow them in large numbers.

We turn next to how these tools and techniques are used to develop products. Keep in mind, however, that the field is still in its infancy.

TAKE HOME MESSAGE 7.2

» Modern molecular methods make it possible to cut and copy DNA from one organism and deliver it into another. The methods include the use of naturally occurring restriction enzymes for cutting DNA, the polymerase chain reaction for amplifying small amounts of DNA, insertion of the DNA into bacterial or viral vectors, and the cloning and identification of cells with the transferred DNA of interest.

7.3 At the cutting edge of biotech, CRISPR is a tool with the potential to revolutionize medicine.

Since 2012, one technology has generated almost unprecedented enthusiasm while racking up some impressive successes in delivering on its early promise. It goes by the acronym **CRISPR** (pronounced "crisper") and is a system for editing DNA. As we saw in the previous section, scientists already use several DNA editing techniques. But what makes CRISPR noteworthy is that it brings much greater precision and efficiency to gene editing, acting as a cut-and-paste tool that enables researchers to modify almost any gene in any organism.

CRISPR stands for "clustered regularly interspaced short palindromic repeats." The name describes the organization of DNA that originally came from viruses but now exists within the genomes of bacteria. It turns out that bacteria commonly incorporate virus DNA into their own DNA to serve as a sort of immune system that can help thwart infection by potentially lethal viruses.

Humans often develop immunity to illness-causing microbes after an initial encounter. Similarly, when a virus first infects a bacterium, the bacterium keeps a record of the encounter. It copies some of the virus's DNA and integrates it into its own DNA. When the same type of virus invades again, the bacterium uses the stored virus DNA to quickly produce an RNA molecule that acts as a sort of homing device, finding and binding to the infecting virus's DNA. After the RNA latches onto the virus DNA, the bacterium employs a DNA-cutting enzyme called Cas9 to cut the foreign DNA into harmless pieces (FIGURE 7-7). This effective way for a bacterium to remember and resist viral infections turns out to be present in almost half of all bacteria.

Adapting the CRISPR system for use in biotechnology is straightforward:

1. After identifying a particular DNA sequence of interest (perhaps to fix or inactivate a defective gene), researchers synthesize an RNA molecule—the "guide" molecule—with a sequence that matches the target gene to be sliced (following the complementary base-pairing rules described in Section 6-2).
2. The DNA sequences for the CRISPR RNA and the Cas9 enzyme are introduced to the target cells, using a plasmid as vector.
3. Within the target (host) cell, the RNA leads Cas9 to exactly the desired location on the cell's DNA, and the enzyme cuts the DNA there.
4. At that location, a sequence can then be inserted that repairs, or alters in some other way, the host cell's DNA.

CRISPR

CRISPR is a gene editing system that brings greater precision and efficiency to modifying genomes.

1. After identifying a particular DNA sequence of interest, researchers synthesize an RNA "guide" molecule with a sequence that matches the target gene to be sliced.

DNA sequence of interest

RNA guide

2. The sequences for the CRISPR RNA and Cas9 enzyme are introduced to target cells using a plasmid.

Plasmid

RNA sequence

Cas9 sequence

3. Within the cell, the RNA leads the Cas9 enzyme to exactly the desired location on the DNA, and Cas9 cuts the DNA there.

Cas9 enzyme

4. At the location where the DNA is cut, a sequence can be inserted that repairs or alters the host cell's DNA.

New DNA sequence

FIGURE 7-7 CRISPR! What it is and how it's used.

Mosquito DNA can be altered to block malaria from spreading.

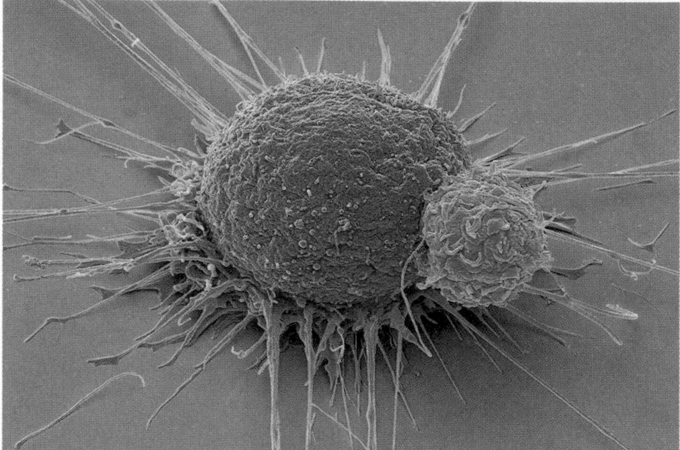

Human trials are under way in which CRISPR-altered white blood cells (in blue) are being used to fight cancer cells (red).

With CRISPR technology, beagles have been engineered to be born with twice as much muscle as a typical dog.

FIGURE 7-8 CRISPR! Initial successes and potential applications.

Q What makes CRISPR such a potentially powerful biotech tool?

CRISPR is considered a breakthrough in biotechnology because the ability to target and snip DNA precisely at a specific sequence opens the door to changing an organism's genes in almost any way imaginable. Just a few of the initial successful uses of CRISPR reveal the wide variety of potential applications (FIGURE 7-8). Researchers have achieved the following:

- Increasing significantly the muscle mass in the legs of a dog
- Inactivating immune system markers on tissue from a pig—by editing 62 genes simultaneously—in a way that could facilitate use of transplants from pigs into humans
- Introducing mutations into human stem cells to produce tissues with disease properties that can serve

as model systems for studying common human diseases

There is even more enthusiasm surrounding CRISPR's potential for fighting some of the most harmful diseases that affect humans. Significant work has already been done with CRISPR in altering the biochemistry of mosquitoes so that they cannot host or transmit the parasite that causes malaria. Eradicating malaria could save half a million lives every year.

With CRISPR, it might also be possible to inactivate the genes in disease-causing bacteria that confer resistance to antibiotics. Or to target invasive species that damage environments and threaten native species.

Without question, CRISPR has enormous potential. But some concerns (and other issues) remain. For starters, numerous legal proceedings are under way as several universities and companies fight over who invented the valuable techniques and has the rights to develop and profit from them. And perhaps a more significant impediment to unrestrained exploration of the possible applications of CRISPR relates to ethical issues. These include concern about the potential for editing human embryos or germline cells (sperm or eggs), because the gene changes could be passed to subsequent generations.

Critics have also noted that the consequences of introducing new or altered genes into the genomes of natural populations of organisms are difficult to predict. Even some desirable outcomes—eradicating the mosquito populations that transmit malaria, for example—might have secondary, harmful effects, such as for bird or bat populations that rely on mosquitoes as a food source.

TAKE HOME MESSAGE 7.3

» CRISPR is a gene (DNA) editing system that brings greater precision and efficiency to gene editing. It is considered a significant breakthrough in biotechnology because the ability to target and snip the DNA of almost any species, at a specific sequence, opens the door to changing an organism's genes in almost any way imaginable.

7.4–7.6 Biotechnology is producing improvements in agriculture.

Biologically engineered plants growing in a pink LED greenhouse.

7.4 Biotechnology can improve food nutrition and make farming more efficient and eco-friendly.

Your breakfast cereal is probably fortified with vitamins and minerals. And for snacking you may eat protein bars that have as much protein as a full chicken breast. It shouldn't come as a surprise, then, to learn that farmers use biotechnology to improve on the natural levels of vitamins, minerals, and other nutrients in the fruits, vegetables, and livestock they produce.

For thousands of years, humans have been practicing a relatively crude and slow form of **genetic engineering**—the

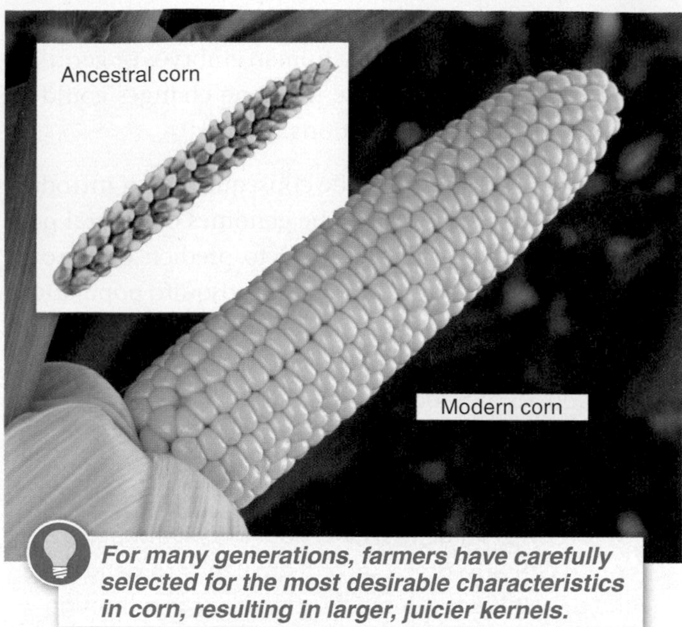

Ancestral corn
Modern corn
For many generations, farmers have carefully selected for the most desirable characteristics in corn, resulting in larger, juicier kernels.

FIGURE 7-9 **Ancestral corn and modern corn.**

manipulation of a species' genome in ways that do not normally occur in nature. In its simplest form, genetic engineering is the careful selection of the plants or animals to be used as the breeders for a crop or animal population. Through this process, farmers and ranchers have produced meatier turkeys, seedless watermelons, and big, juicy corn kernels (FIGURE 7-9). But what previously took many generations of breeding can now be accomplished in a fraction of the time, using **recombinant DNA technology,** the combination of DNA from two or more sources into a product.

In crop plants, for example, the process begins with the identification of a desirable characteristic, such as larger size or faster ripening time. Traditionally, breeders would then search for an organism of the same species that had the desirable trait, breed it with their crop organisms, and hope that the offspring would express the trait in the desired way. With recombinant DNA technology, the desired trait can come from *any* species, so the pool of organisms from which the trait can be taken becomes much larger. Organisms produced with recombinant DNA technology are referred to as **genetically modified organisms,** or **GMOs.**

Although difficult in practice, the results so far hint at a fruitful marriage of agriculture and technology. Here are some success stories.

Q How might a genetically modified plant help 500 million malnourished people?

Nutrient-Rich "Golden Rice" Almost 10% of the world's population suffers from vitamin A deficiency, which causes blindness in a quarter-million children each year and a host of other illnesses. These nutritional problems are especially severe in southern Asia and sub-Saharan Africa, where rice is a staple of most diets. Researchers have developed what may be the model for solving problems with biotechnology. It involves the creation of a new crop called "golden rice."

Humans and other mammals generally make vitamin A from beta-carotene, a substance found in abundance in most plants (it's what makes carrots orange), but not in the edible part of rice grains. To introduce beta-carotene into

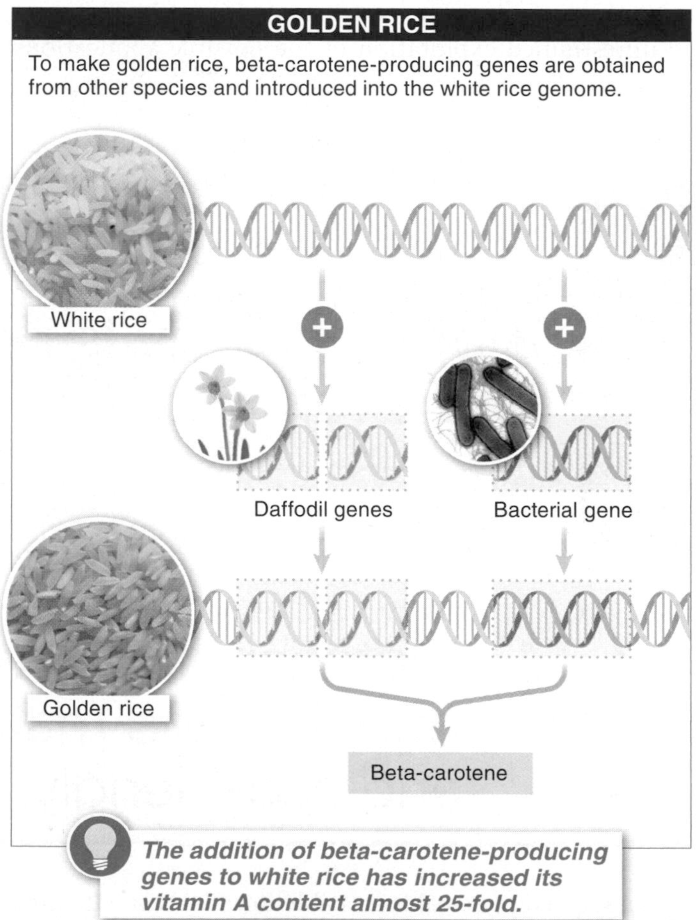

GOLDEN RICE
To make golden rice, beta-carotene-producing genes are obtained from other species and introduced into the white rice genome.
White rice
Daffodil genes
Bacterial gene
Golden rice
Beta-carotene
The addition of beta-carotene-producing genes to white rice has increased its vitamin A content almost 25-fold.

FIGURE 7-10 **The potential to prevent blindness in 250,000 people each year.** Rice can be engineered to prevent blindness by increasing its vitamin A content.

rice crops, researchers used recombinant DNA technology to insert three genes into the rice genome that code for the enzymes used in the production of beta-carotene. Two of the transplanted genes were from the daffodil plant and one was a bacterial gene (FIGURE 7-10). As the transplanted genes are expressed, the rice grain takes on a golden color from the accumulated beta-carotene. Since golden rice was first developed in 1999, new lines have been created that produce almost 25 times the vitamin A found in the original strains.

Field tests of golden rice are still in progress, and it is viewed as one of the most promising applications yet of biotechnology. Precautionary biosafety protocols and regulations, however, have restricted its cultivation by large-scale commercial farms. In June 2016, 110 Nobel laureates signed a letter urging an end to the opposition to GMOs, highlighting the potential value of golden rice, in particular.

Q What genetically modified foods do most people in the United States consume (usually without knowing it)?

Insect and Herbicide Resistance In the United States, biotechnology has already had a profound impact on agricultural practices. It is not a stretch to say that we are in the midst of a green revolution and that few people are aware of it.

The numbers are surprising: 86% of all corn grown in the United States is genetically modified; 93% of all cotton grown is genetically modified; and 93% of all soybeans grown are genetically modified (FIGURE 7-11). Two factors explain much of the extensive adoption of genetically modified plants in U.S. agriculture. (1) Many plants have insecticides engineered into them, which can reduce the amounts of insecticide that must be added during agriculture. (2) Many plants also have herbicide-resistance genes engineered into them. The cultivation of herbicide-resistant plants can reduce the amount of plowing needed to remove weeds. As a consequence, the use of genetically modified plants can reduce both the costs of producing food and the loss of topsoil to erosion.

Every year, about 40 million tons of corn are unmarketable as a consequence of insect damage. Increasingly, however, farmers have been enjoying greater success in their battles against insect pests. Farmers owe much of this success to soil-dwelling bacteria of the species *Bacillus thuringiensis* ("Bt"), which produce spores containing crystals that are fatally poisonous to insects but harmless to the crop plants and to people.

Beginning in 1961, the toxic Bt crystals were included in the pesticides sprayed on crop plants. Then in 1995, the gene coding for production of the Bt crystals was inserted

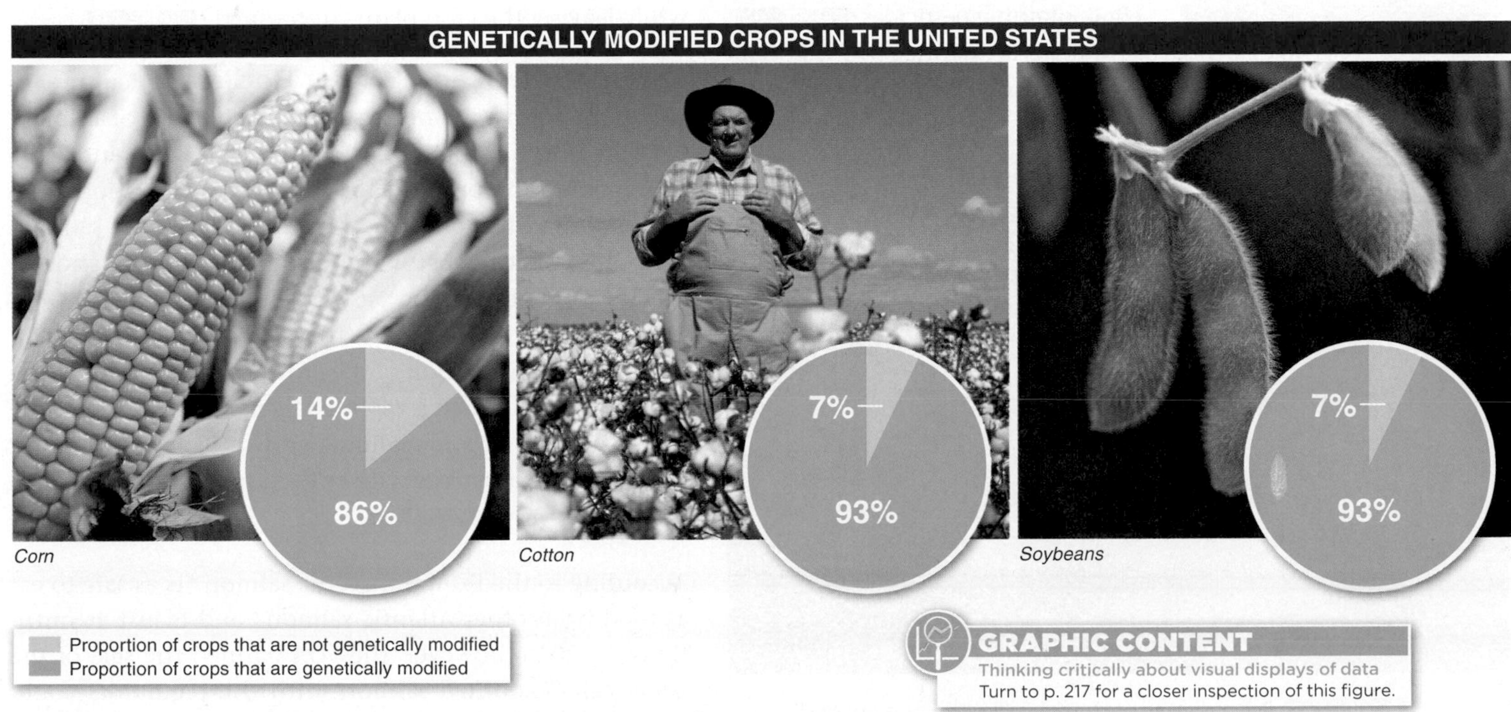

GRAPHIC CONTENT
Thinking critically about visual displays of data
Turn to p. 217 for a closer inspection of this figure.

FIGURE 7-11 **A significant percentage of crops grown in the United States are genetically modified.**

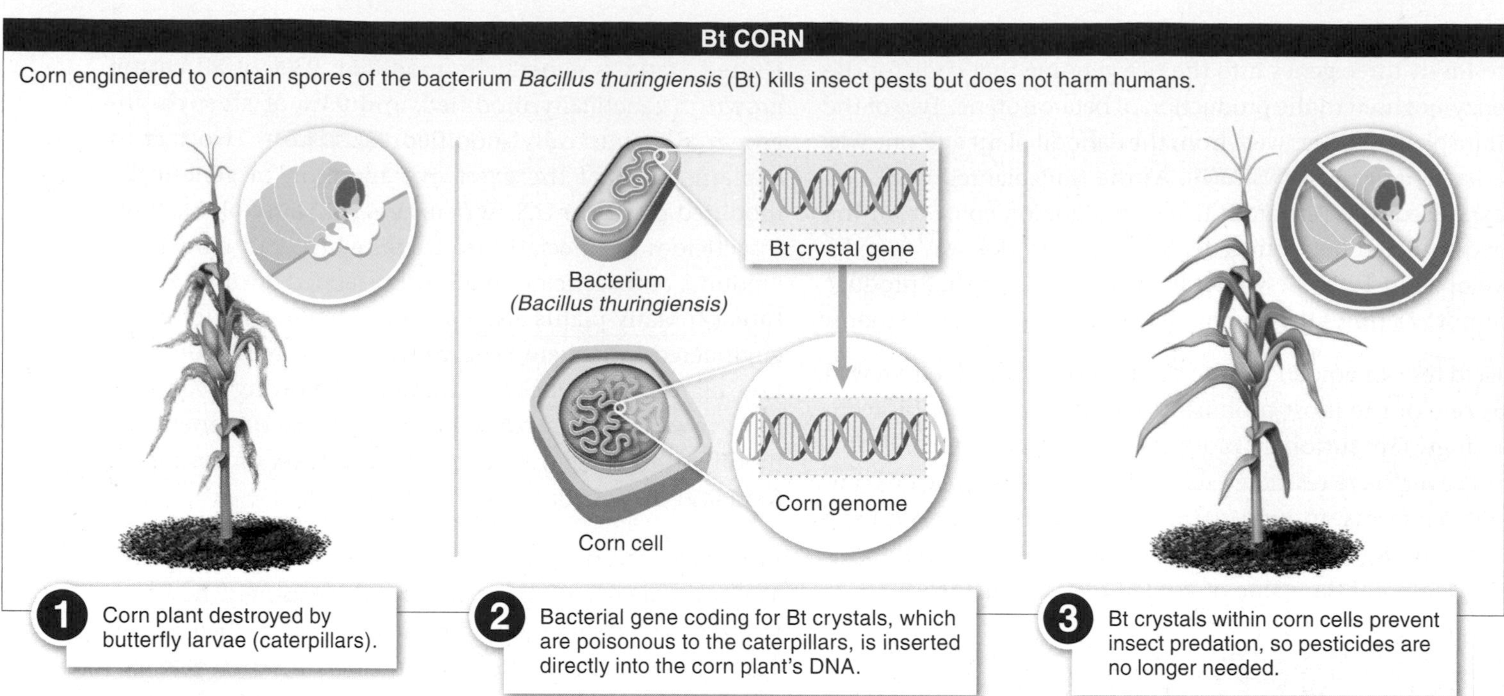

Bt CORN
Corn engineered to contain spores of the bacterium Bacillus thuringiensis (Bt) kills insect pests but does not harm humans.
Bt crystal gene
Bacterium
(Bacillus thuringiensis)
Corn genome
Corn cell
1 Corn plant destroyed by butterfly larvae (caterpillars).
2 Bacterial gene coding for Bt crystals, which are poisonous to the caterpillars, is inserted directly into the corn plant's DNA.
3 Bt crystals within corn cells prevent insect predation, so pesticides are no longer needed.

FIGURE 7-12 **Help from bacteria in growing disease-resistant corn.**

Q How can genetically modified plants lead to reduced pesticide use by farmers?

directly into the DNA of many crop plants, including corn, cotton, and potatoes. Because the plants themselves produce insect-killing Bt crystals, farmers no longer need to apply Bt-containing pesticides (FIGURE 7-12). There is no evidence that humans are harmed by the Bt crystals, even when they are exposed to very high levels.

FIGURE 7-13 **Crop duster.** Herbicides like the one applied by this crop-dusting drone must kill weeds while leaving the crop unharmed.

In the fight against weeds, bacteria have proven very useful. In the 1990s, researchers discovered a bacterial gene that confers resistance to herbicides. Integration of this gene into the plants' DNA gives the crops resistance to herbicides, allowing farmers to kill weeds with herbicides while leaving the crop plants unharmed (FIGURE 7-13).

Faster Growth and Bigger Bodies Agriculture includes the cultivation not just of plants but also of animals. And for the first time, in 2015, the U.S. Food and Drug Administration approved for human consumption a genetically modified animal—a transgenic Atlantic salmon. The salmon carries a growth hormone gene from another species (Chinook salmon), as well as a region of DNA from a third species (ocean pout) that acts as an "on" switch, facilitating transcription of the growth hormone gene. The transgenic fish, which is reported by its creators to taste the same as regular Atlantic salmon, grows much more quickly and reaches market size within 18 months rather than the usual three years (FIGURE 7-14).

According to the FDA, transgenic salmon "is as safe to eat as food from other Atlantic salmon" and is just as nutritious. In 2016, however, the FDA banned the sale of the genetically modified salmon until the regulators could publish labeling guidelines. Meanwhile, some fisheries experts, food safety experts, environmental groups, and

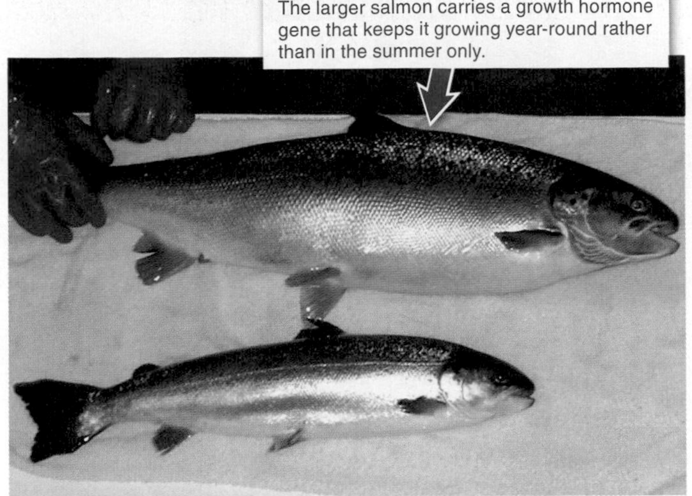

The larger salmon carries a growth hormone gene that keeps it growing year-round rather than in the summer only.

FIGURE 7-14 **Bigger salmon.**

consumer groups have expressed apprehensions that eating the salmon might cause increased rates of allergic reactions or have other adverse effects.

There is also the fear that the transgenic fish will escape from their enclosed breeding facilities into their natural habitat. If this occurred, the transgenic fish might outcompete wild populations for resources and grow too large to be consumed by natural predators. The wild salmon populations, many of which are listed as endangered, could suffer as a result.

TAKE HOME MESSAGE 7.4

» Biotechnology has led to important improvements in agriculture through the use of transgenic plants and animals to produce more nutritious food. Even more significant is the extent to which biotechnology has reduced the environmental and financial costs of producing food through the creation of herbicide-resistant and insect-resistant crops. The potential ecological and health risks of such widespread use of transgenic species are not yet fully understood.

7.5 Rewards, with risks: what are the possible risks of genetically modified foods?

Chickens without feathers look ridiculous (FIGURE 7-15). But such a genetically modified breed was developed to be easier and less expensive to prepare for market, benefiting farmers by lowering their costs and benefiting consumers by lowering prices. These "naked" chickens, however, have turned out to be unusually vulnerable to mosquito attacks, parasites, and disease, and are ultrasensitive to sunlight. They also have difficulty mating, because the males are unable to flap their wings.

Naked chickens teach us an important lesson about genetically modified plants and animals. Although the breed of featherless chickens was produced by relatively low-tech genetic engineering—the traditional animal husbandry method of crossbreeding—the desired trait of no feathers was accompanied by unintended and undesirable traits. Now, as more genetically modified foods are produced with modern methods of recombinant DNA technology, we must evaluate the risks of creating unintended and potentially harmful traits. For these and other reasons—some legitimate and rational, others irrational—many people

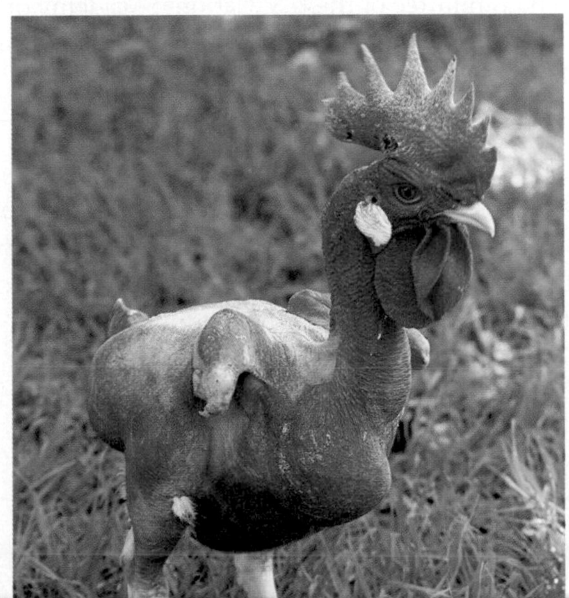

Featherless birds are cheaper for farmers and consumers. But there are unintended consequences, including vulnerability to mosquitoes and other parasites.

FIGURE 7-15 **"Naked" birds.**

have concerns about the production and consumption of genetically modified foods (FIGURE 7-16).

Q What are people's biggest fears about GMOs?

"Organisms that we want to kill may become invincible." Herbicide-resistant canola plants were cultivated in Canada, making it possible for farmers to apply herbicides freely to kill the weeds but not the canola crop. But the herbicide-resistant canola plants accidentally spread to neighboring farms and grew out of control, because traditional herbicides could not kill them. Similarly, there is the possibility that insect pests will develop resistance to the Bt produced by genetically modified crops, which will also make these pests resistant to Bt pesticides applied to crops that are not genetically modified.

"Genetically modified crops are not tested or regulated adequately." It is impossible to really know whether a new technology has been tested adequately. Scientists and lawmakers have been working toward a responsible set of policies designed to ensure that sufficient safety testing is done. For example, laboratory procedures for working with recombinant DNA have been established, and researchers have developed techniques that make it impossible for most genetically engineered organisms to survive outside the specific conditions for which they are developed.

Still, in a recent report on genetically modified animals, an expert committee of the U.S. National Academy of Sciences warned that GMOs pose risks that the government is unable to evaluate.

> "And he gave it for his opinion, 'that whoever could make two ears of corn, or two blades of grass, to grow upon a spot of ground where only one grew before, would deserve better of mankind, and do more essential service to his country, than the whole race of politicians put together.' "
>
> — JONATHAN SWIFT, *Gulliver's Travels,* 1726

"Eating genetically modified foods is dangerous." In the 1990s, a gene from Brazil nuts was used to improve the nutritional content of soybeans. The genetically modified soybeans were nutritious, but they acquired allergy-causing chemicals previously present in Brazil nuts but not in soybeans. This outcome illustrates the possibility that unwanted

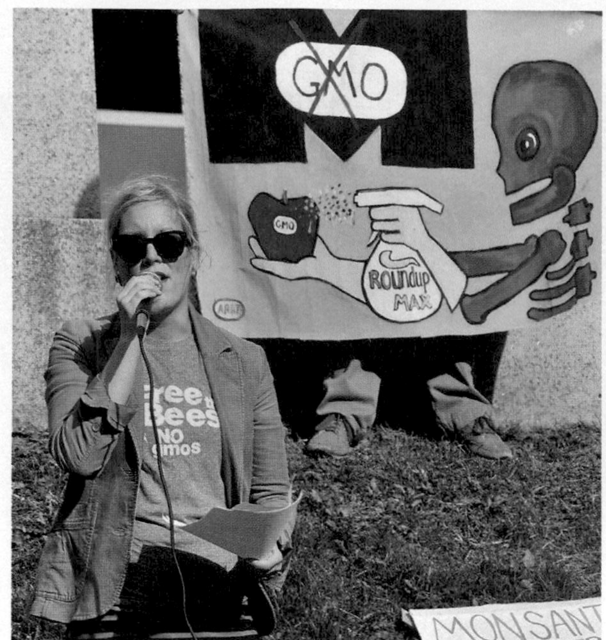

FIGURE 7-16 Consumer fears. Protesters voice opposition to the use of genetically modified organisms (GMOs).

features might be passed from species to species in the creation of transgenic organisms. In this case, all the genetically modified soybeans were destroyed and the research program was suspended. It is important to note that, to date, no evidence has appeared to suggest that consumption of any genetically modified foods is dangerous.

"Loss of genetic diversity among crop plants is risky." As increasing numbers of farmers favor one or a few genetically modified strains of crops, the genetic diversity of the crops declines. This can increase their vulnerability to environmental changes or pests. For example, in the mid-1800s, much of the population of Ireland depended on a diet of potatoes. Because most of the potato crops had been propagated from cuttings from the same plant, they were all genetically the same. When the crops were infected by a mold, most were wiped out, causing the Irish Potato Famine responsible for the deaths of more than a million people.

"Hidden costs may reduce the financial advantages of genetically modified crops." When seed companies create genetically modified seeds with crop traits desirable to farmers, the companies also engineer sterility into the seeds. As a consequence, the farmers are dependent on seed companies and must purchase new seeds for each generation of their crops.

Some argue that genetically modified foods are not "natural" and, for that reason, must be harmful. This is a flawed argument and should not be a cause for concern. Smallpox, HIV, poison ivy, and cyanide, after all, are natural. The

> **Q** Are "natural" things necessarily desirable? Does being "unnatural" make a product or technology harmful? (What about smallpox and its cure?)

smallpox vaccine, on the other hand, is unnatural. Innumerable other valuable technological developments are equally unnatural. There simply is no value in knowing whether something is natural or unnatural when evaluating whether or not it is good and desirable.

In the end, we must compare the risks of producing genetically modified foods with the benefits. These analyses will have to include the potential to reduce food costs, the ability to reduce environmental degradation by agriculture, and the safety of agricultural workers. For example, with genetically modified, pest-resistant crops, farmworkers will greatly benefit from spending less time applying pesticides. This is particularly important for workers in less-developed countries, where safety regulations for pesticide use are often ignored or unenforced. Unfortunately, establishing the environmental risks of genetically modified foods is difficult. This important issue must be evaluated closely and regulated carefully.

TAKE HOME MESSAGE 7.5

» **More and more genetically modified foods are being created through modern methods of recombinant DNA technology. For the public, however, numerous legitimate fears remain about the potentially catastrophic risks of these foods, given that their development relies on such new technology, and about the long-term financial advantages they offer.**

THIS IS HOW WE DO IT

Developing the ability to apply the process of science

7.6 How do we determine whether GMOs are safe?

On average, each American eats 193 pounds of genetically modified foods each year. (The World Health Organization defines GMOs as "those organisms in which the genetic material has been altered in a way that does not occur naturally.") It seems reasonable to evaluate their safety. In the United States, there is no mandatory review, regulation, or special safety testing of genetically modified foods. Rather, the producers take part in voluntary consultations with the FDA to determine whether genetically modified foods are "substantially equivalent" to non–genetically modified foods. As long as they are, no approval is required before bringing such foods to market.

While the political and ethical issues are contentious and complex, let's explore here potential approaches to answering the important biological question.

How can we evaluate whether the use of GMOs in foods is safe?

At first glance, this task seems straightforward. In practice, however, it turns out to be extremely difficult. To prove whether genetically modified foods are safe, we must first eliminate the possibility of danger.

How do you prove that something is *not* dangerous?

This is not possible.

The hypothesis "genetically modified foods are dangerous" cannot be falsified. For instance, one study may show that a GMO doesn't cause cancer, but that doesn't rule out the possibility that it may cause infertility. Another study may show that a GMO does

(continued on the following page)

not cause cancer or infertility, but that doesn't rule out the possibility that it may cause autism. The possibilities are, literally, endless.

From a scientific perspective, the question of whether genetically modified foods are safe is not the right question to ask. Instead, we must ask a specific question about a potential danger and figure out a way to approach it. Furthermore, no single research study that fails to find a danger of genetically modified food will conclusively resolve the issue. Rather, each such study adds to a body of evidence that, cumulatively, can give us confidence in GMO safety.

Humans live a long time. Is there a practical way to evaluate the risks of long-term consumption of genetically modified food?

It's just not feasible to conduct a whole lifespan study—70 years or more—involving hundreds or thousands of humans. So researchers used a shorter-lived mammal: rats. They randomly divided 180 rats into three groups. The first group consumed a diet high in a genetically modified rice (engineered to be insect-resistant); the second consumed a diet high in non–genetically modified rice, but otherwise equal in nutrients; and the third consumed a non–rice-based diet.

What health risks should be evaluated?

The researchers measured the mortality rates of the rat groups for 18 months and found no differences among the three groups. Similarly, they found no differences in the number of tumors or other types of organ damage. The researchers noted that the amount of genetically modified rice consumed by a rat in the first group was approximately 10 times greater (in proportion to body weight) than would be consumed by a human.

What if results from another study were different?

What happens when conflicting results are reported for other, seemingly similar studies? Let's look at an example.

For two years, another team of researchers fed rats a diet containing one of three different amounts of genetically modified corn, and evaluated health effects and mortality rates. The researchers also included groups of rats fed genetically modified corn that had been treated with a pesticide, and other groups of rats fed non–genetically modified corn that had been exposed to pesticide. The control group was fed non–genetically modified corn that had not been exposed to pesticide.

The reported results were dramatic and received widespread media coverage. The rats consuming the genetically modified corn were reported to have significantly higher mortality rates and higher incidence of tumors. But closer inspection of the study design generated a huge and critical response from other researchers.

How can small differences in experimental design compromise a study's results?

Critics focused on a couple of important methodological problems.

First, the researchers had used a strain of rat known to have an unusually high incidence of tumors. This made it difficult to know whether the differences between groups were caused by the genetically modified corn or were just normal variation.

Second, the large number of treatment (experimental) groups were made up of very small sample sizes: 10 different groups of males and 10 groups of females, with just 10 animals in each experimental group. As a result, the differences between groups reported by the researchers were not statistically significant. Furthermore, the researchers did not include the relevant statistical analyses.

When other researchers extracted data from figures in the published report and conducted their own careful statistical analyses, they concluded that the results provided no evidence at all that consumption of genetically modified corn had adverse effects on health. They even noted that by "cherry picking" the results to report—as the original researchers had done—they could state that males in the control groups had a mortality rate *three times higher* than the two groups consuming the most genetically modified corn. In response to reevaluations of the data, the editor-in-chief of the journal (*Food and Chemical Toxicology*) retracted the published article.

The significant media attention received by this study prior to its being discredited highlights the challenge in trying to prove that something is not dangerous. Just a single paper making an untrue assertion about the adverse effects of genetically modified food on health can arouse deep public fear.

The issue of the safety of foods containing GMOs continues to be debated. Research, and the accumulation of evidence that genetically modified foods do not pose health risks, continues. This process is central to the power of scientific thinking.

The consensus that genetically modified foods carry no more risk than non–genetically modified foods is endorsed by many organizations (that have no obvious financial stake in the outcome), including the U.S. National Academy of Sciences, the World Health Organization, the American Medical Association, and the British Royal Society.

TAKE HOME MESSAGE 7.6

» Determining whether genetically modified foods are safe is a complex and difficult challenge for scientists. We can, however, ask specific questions about potential dangers and use experimental approaches to collect evidence bearing on these questions. Although the issue of the safety of genetically modified food continues to be debated, there is growing consensus that GMOs carry no more risk than non-GMOs in the human diet.

7.7–7.9 Biotechnology has the potential for improving human health.

In 1996, Dr. Ian Wilmut created a sheep named Dolly, the first mammal to be cloned from an adult somatic cell.

7.7 The treatment of diseases and the production of medicines are improved with biotechnology.

In the best of all possible worlds, biotechnology would *prevent* debilitating human diseases. Next best would be to *cure* diseases forever. But these are difficult, long-term goals, so biotech often is directed at the more practical goal of *treating* diseases, usually by producing medicines more efficiently and effectively than with traditional methods. Biotechnology has had some notable successes in achieving this goal. The treatment of diabetes is one such success story.

Type 1 diabetes, often called juvenile diabetes, is a chronic disease in which the body cannot produce the hormone insulin. Without insulin, the body's cells are unable to take up and break down sugar from the blood. Type 2

FIGURE 7-17 **Lifesaving insulin.** Human insulin is engineered through recombinant DNA technology.

diabetes, which accounts for 90% of all cases of diabetes, is a metabolic disorder characterized by elevated blood sugar levels resulting from insulin resistance and insufficient insulin production. Complications from both types of diabetes can be deadly. Approximately one-third of all people with diabetes treat their condition with one or more daily injections of insulin. As recently as 1980, insulin necessary to treat diabetes was extracted from the pancreas of cattle or pigs that had been killed for meat. This process of collecting insulin was difficult and costly.

Q Why do some bacteria produce human insulin?

Everything changed in 1982, when a 29-year-old entrepreneur, Bob Swanson, joined scientists Herbert Boyer and Stanley Cohen to transform the potential of recombinant DNA technology. The team used restriction enzymes to snip out the human DNA sequence that codes for the production of insulin. They then inserted this sequence into the bacterium *E. coli,* creating a transgenic organism. After cloning the new, transgenic bacteria, the team was able to grow vats of the bacterial cells, all of which churned out human insulin (FIGURE 7-17). The drug could be produced efficiently in huge quantities and made available for patients with diabetes. This was the first genetically engineered drug approved by the FDA, and it continues to help millions of people every day.

This application of biotechnology revealed a generalized process for genetic engineering and, in doing so, started the biotech revolution. It instantly opened the door to a more effective method of producing many different medicines to treat diseases. Today, more than 1,500 companies work in the recombinant DNA technology industry, and their products generate more than $40 billion in revenues each year.

Several important achievements followed the development of insulin-producing bacteria. Here are just two examples.

1. Human growth hormone (HGH). Produced by the pituitary gland, human growth hormone stimulates protein synthesis, increases the utilization of body fat for energy to fuel metabolism, and stimulates the growth of virtually every part of the body (FIGURE 7-18). Insufficient growth hormone production, usually due to pituitary malfunctioning, leads to dwarfism.

When treated with supplemental HGH, individuals with dwarfism experience additional growth. Until 1994, however, HGH treatment was prohibitively expensive. HGH could be produced only by extracting and purifying it from the pituitary glands of human cadavers. Today, transgenic bacteria produce HGH in virtually unlimited supplies.

2. Erythropoietin. Produced primarily by the kidneys, erythropoietin (also known as EPO) is a hormone that regulates the production of red blood cells. Many clinical conditions (nutritional deficiencies and lung disease, among others) and medical treatments (such as chemotherapy) can cause a lower than normal number of red blood cells, a condition called anemia. Anemia reduces the transport of oxygen to tissues and cells, and this lack of sufficient oxygen causes a variety of symptoms, including weakness, fatigue, and shortness of breath.

Recombinant human erythropoietin (rhu-EPO), first cloned in 1985, is now produced in large amounts in cells originally derived from hamster ovaries. It is used to treat many forms of anemia. Worldwide sales of EPO are in the billions of dollars.

Q What is "blood doping"? How does it improve some athletes' performance

EPO has been at the center of several "blood doping" scandals in professional cycling.

Sylvester Stallone—age 61 in this photo—used human growth hormone (along with strength training) to develop larger muscles for a film role.

FIGURE 7-18 **Bulking up with a little help.**

Because this hormone increases the oxygen-carrying capacity of the blood, some otherwise healthy athletes have used EPO to improve their athletic performance. It can be very dangerous, though. By increasing the number of red blood cells, the blood can become much thicker, increasing the risk of heart attack.

The list of other medicines currently produced by transgenic organisms is long—including some, such as the hemophilia treatment, Factor VIII, that generate more than a billion dollars per year. Beyond these, plans are under way to create a variety of other useful products for treating diseases—including potatoes that produce antibodies that enable a more effective response to treatment of illness. In the next section we examine the strategies for preventing genetic diseases, as well as the less successful attempts to cure diseases through biotechnology.

TAKE HOME MESSAGE 7.7

» Biotechnology has led to some notable successes in treating diseases, usually by producing medicines more efficiently and effectively than they can be produced with traditional methods.

7.8 Gene therapy: biotechnology can help diagnose and prevent genetic diseases, but has had limited success in curing them.

Would you want to know? If you carried a gene that meant you were likely to develop a particular disease later in life, or if there was a large chance that your children would be born with a genetic disease—would you want to know? As biotechnology develops the tools to identify some of the genetic time-bombs that many of us carry, these are questions that we all must address.

Screening for genetic disorders generally focuses on three different scenarios.

1. Is a given set of parents likely to produce a baby with a genetic disease? Many genetic diseases occur only if an individual inherits two copies of the disease-causing gene, one from each parent. This is true for Tay-Sachs disease, cystic fibrosis, and sickle-cell anemia, among others. Individuals with only a single copy of the disease-causing gene never fully show the disease, but they may pass on the disease gene to their children. In these cases, it can be beneficial for potential parents to be screened to determine whether they carry a disease-causing copy of the gene. Such screening, combined with genetic counseling and testing of embryos following fertilization, can dramatically reduce the incidence of a genetic disease. Since screening began in 1969, the incidence of Tay-Sachs disease, for example, has fallen by more than 75% (FIGURE 7-19).

2. Will a baby be born with a genetic disease? Once fertilization has occurred, it is possible to test an embryo or developing fetus for numerous genetic problems. Prenatal genetic screening can detect cystic fibrosis, sickle-cell anemia, Down syndrome, and a rapidly growing list of other disorders.

To screen the fetus, doctors must examine some of the fetal cells and/or the amniotic fluid (the fluid that surrounds the fetus in the uterus and contains many chemicals produced

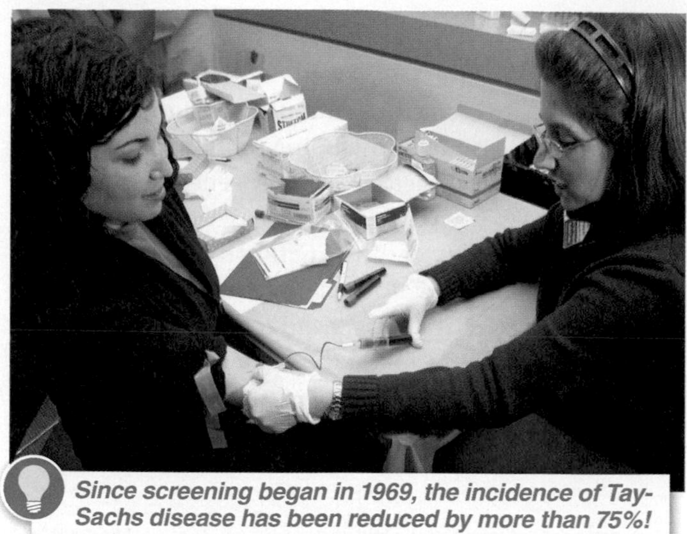

Since screening began in 1969, the incidence of Tay-Sachs disease has been reduced by more than 75%!

FIGURE 7-19 Genetic screening can determine the presence of the Tay-Sachs gene.

by the developing embryo). Cells and fluid are usually collected by amniocentesis or chorionic villus sampling (CVS), techniques that we explore in detail in Chapter 8. (Increasingly, these cell and fluid collection methods are being replaced by non-invasive methods requiring only a blood sample from the mother.)

3. Is an individual likely to develop a genetic disease later in life? DNA technology can be used to detect disease-causing genes in individuals who are currently healthy but are at increased risk of developing an illness later. Early detection of many diseases, such as breast cancer, prostate cancer, and skin cancer, greatly enhances the ability to treat the disease and reduce the risk of more severe illness or death.

These potential benefits of genetic technology come with potential costs. People who have a gene that puts them at increased risk of developing a particular disease, for example, might be discriminated against, even though they are not currently sick and may never suffer from the particular disease. Although the Genetic Information Nondiscrimination Act was signed into law in the United States in 2008, the law does not cover life insurance, disability insurance, and long-term care insurance. Insurance companies have already denied such coverage on discovering that an individual carries a gene that puts him or her at increased risk of disease. Another problem is that parents who discover that their developing fetus will develop a painful, debilitating, or fatal disease soon after birth are confronted with the difficult question of how to proceed.

When it comes to curing a disease by using biotechnology, there is good news and bad news. The good news is that, since the 1990s, a small number of humans with a usually fatal genetic disease called severe combined immunodeficiency disease (SCID) were completely cured through the application of biotechnology. The bad news is that it has not been possible to apply these promising techniques to other diseases.

It's not for lack of trying. There have been more than 500 other clinical trials for **gene therapies** designed to treat or cure a variety of diseases by inserting a functional gene into an individual's cells to replace a defective version of the gene. But no clear successes.

Let's look at what has been accomplished in treating SCID—a condition in which a baby is born with an immune system unable to properly produce a type of white blood cell. The infant is vulnerable to most infections and usually dies before the age of one (FIGURE 7-20). In gene therapy for SCID, researchers remove **stem cells** from an affected baby's bone marrow. These stem cells have the ability to develop into any type of cell in the body. In bone marrow, they normally produce white blood cells, but in individuals with SCID, a malfunctioning gene disrupts normal white blood cell production.

Next, in a test tube, the infant's bone marrow stem cells are infected with a transgenic virus carrying the functioning gene. The virus inserts the good gene into the DNA of the stem cells, which are then injected back into the baby's bone marrow. There, the cells can produce normal white blood cells, permanently curing the disease.

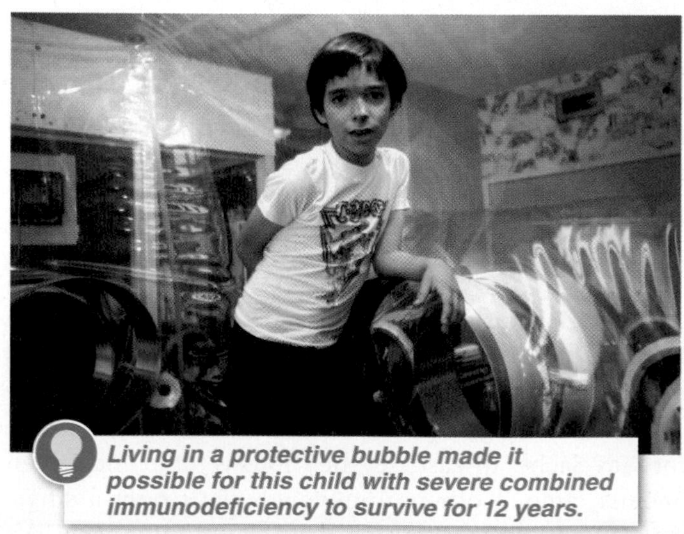

Living in a protective bubble made it possible for this child with severe combined immunodeficiency to survive for 12 years.

FIGURE 7-20 Protected by a bubble.

Several dozen cases of SCID have been cured, but two patients died from illness related to their treatment. As a result of recent advances, however, a variant of this treatment is pending approval in Europe. While no gene therapy treatments have been approved in the United States as of 2016, researchers are hopeful that the situation will change in the coming years. The most likely treatments are for hemophilia B, sickle-cell anemia, and a brain disease called cerebral adrenoleukodystrophy.

Q Why has gene therapy had such a poor record of success in curing diseases?

Difficulties with gene therapy usually relate to transfer of the normal-functioning gene into the cells of a person with a genetic disease, including:

1. Difficulty getting the working gene into the specific cells where it is needed.
2. Difficulty getting the working gene into enough cells and at the right rate to produce a physiological effect.
3. Difficulty arising from the transferred gene getting into unintended cells.
4. Difficulty regulating gene expression.

Beyond these technical problems—some of which may benefit from the CRISPR system described in Section 7-3—the malfunctioning gene has not been identified for most diseases, or the disease is caused by more than one malfunctioning gene. Additionally, it is important to keep in mind that gene therapy targets cells in the body other than sperm and eggs. Consequently, while a disease might, in theory, be cured in the individual receiving the therapy, he or she can still pass on the disease-causing gene(s) to offspring. It's not clear what the future holds for gene therapy, but a great deal of research is in progress.

TAKE HOME MESSAGE 7.8

» Biotech tools have been developed to reduce suffering and reduce the incidence of diseases, but their use comes with potential costs. Gene therapy has had limited success in curing human diseases, primarily because of technical difficulties in transferring normal-functioning genes into the cells of a person with a genetic disease.

7.9 Cloning—ranging from genes to organs to individuals—offers both opportunities and perils.

Cloning. Perhaps no scientific word is as emotionally loaded. Let's clarify what the word means. "Cloning" refers to a variety of different techniques. It can refer to the creation of new individuals that have exactly the same genome as the donor individual—a process called "whole organism cloning." The clone is like an identical twin, except that it may differ in age by years or even decades. It is also possible to clone tissues (such as skin) and entire organs from an individual's cells. And, as we saw in Section 7-2, it is possible to clone genes.

Cloning took center stage in the public imagination in 1997, when Ian Wilmut, a British scientist, and his colleagues announced that they had cloned a sheep—which they named Dolly. Their research was based on ideas dating back to 1938, when Hans Spemann first proposed the experiment of removing the nucleus from an unfertilized egg and replacing it with the nucleus from the cell of a different individual.

The process used by Wilmut and his research group was difficult and inefficient, but also simple in concept (FIGURE 7-21). They removed the nucleus from a mammary gland cell of a grown sheep and inserted it into an egg of another sheep (from which the nucleus had been removed). The egg was induced to divide as if it were a naturally fertilized egg, and they transplanted it into the uterus of a surrogate mother sheep. Out of 272 tries, they achieved just one success. But that was enough to show that cloning from an adult animal was possible.

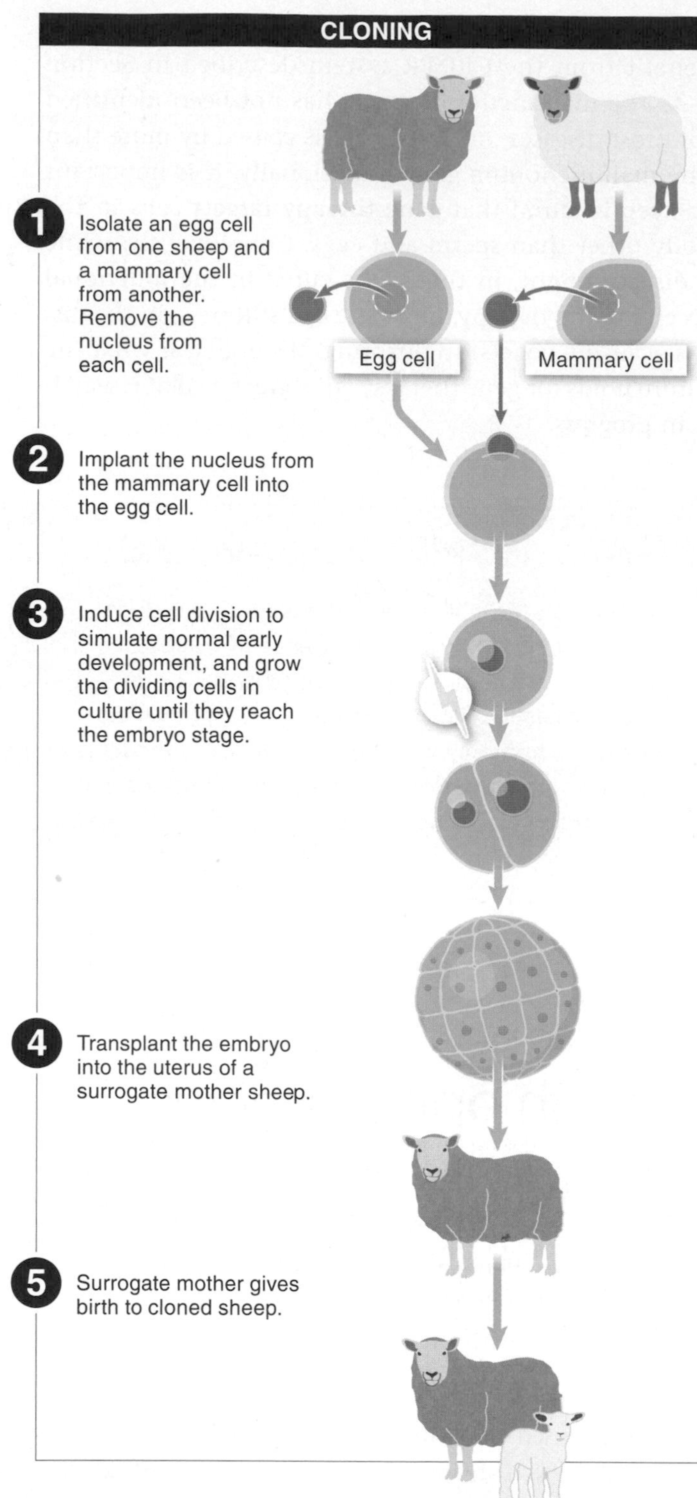

CLONING
1 Isolate an egg cell from one sheep and a mammary cell from another. Remove the nucleus from each cell.
Egg cell
Mammary cell
2 Implant the nucleus from the mammary cell into the egg cell.
3 Induce cell division to simulate normal early development, and grow the dividing cells in culture until they reach the embryo stage.
4 Transplant the embryo into the uterus of a surrogate mother sheep.
5 Surrogate mother gives birth to cloned sheep.

FIGURE 7-21 No longer science fiction. Steps used in the cloning of Dolly the sheep.

Shortly after news of Dolly's birth, teams set about cloning a variety of other species, including mice, cows, pigs, and cats (FIGURE 7-22). Not all of this work was driven by simple curiosity. For farmers, cloning could be an efficient method of producing animals with desirable traits, such as increased milk production in cows.

Q Are there any medical justifications for cloning?

Medical researchers, too, see much to gain from cloning. In particular, transgenic animals containing human genes—such as the hamster ovary cells producing rhu-EPO, discussed earlier—can be very valuable. But can a human be cloned? At this point, it is almost certain that the cloning of a human will be possible. There is near unanimity among scientists that human cloning to produce children should not be attempted. Some of the reasons cited relate to problems of safety for the mother and the

FIGURE 7-22 Genetically identical cloned animals. The cloning of animals can maintain desirable traits from generation to generation.

child, legal and philosophical issues relating to the inability of cloned individuals to give consent, problems of the exploitation of women, and concerns regarding identity and individuality. Governments are struggling to develop wise regulations for this new world.

TAKE HOME MESSAGE 7.9

» Cloning of individuals has potential benefits in agriculture and medicine, but ethical questions linger.

7.10 Biotechnology has value for criminal justice systems.

A DNA fingerprint is being evaluated in the forensic department of a police station.

7.10 DNA is an individual identifier: the uses and abuses of DNA fingerprinting.

In another time, Colin Pitchfork, a murderer and rapist, would have walked free. But in 1987, he was captured and convicted, betrayed by his DNA, and is now serving two life sentences in prison. Pitchfork raped and murdered two 15-year-old high school girls in a small village in England in the 1980s. The police thought they had the perpetrator when a man confessed, but only to the second murder. He denied any involvement in the first murder, which perplexed the police because the details of the two crimes strongly suggested a single culprit.

At the time, British biologist Alec Jeffreys made the important discovery that there are small stretches of DNA in human chromosomes that are tremendously variable in their base sequences. In much the same way that each person has a driver's license number or social security number that differs from everyone else's, it is extremely unlikely that two people would have identical sequences at these locations. Thus, investigators could compare a suspect's DNA sample against the DNA-containing evidence left at a crime scene.

Jeffreys analyzed DNA left by the murderer-rapist on the two victims and found that it did indeed come from a single person, and that person was *not* the man who had confessed to one of the crimes. The suspect was released and has the distinction of being the first person cleared of a crime through DNA fingerprinting.

To track down the criminal, police collected and analyzed more than 5,000 blood samples from all men in the area of the crimes who were between 18 and 35 years old—a practice that many viewed as an invasion of privacy. This procedure led them to Colin Pitchfork, whose DNA matched perfectly the DNA left on both victims, and ultimately was the evidence responsible for his conviction (FIGURE 7-23). (He almost slipped through, having persuaded a friend to give a blood sample in his name. But when the friend was overheard telling the story in a pub, police tracked down Pitchfork to get a blood sample.)

DNA fingerprinting is now used extensively in forensic investigations, in much the same way that regular fingerprints have been used for a century. DNA samples are frequently left behind, usually in the form of semen, blood, hair, skin, or other tissue. DNA fingerprinting has been directly responsible for bringing thousands of criminals to justice and exonerating the innocent. Let's examine how DNA fingerprinting is done, why it is such a powerful forensic tool, and why it is not foolproof.

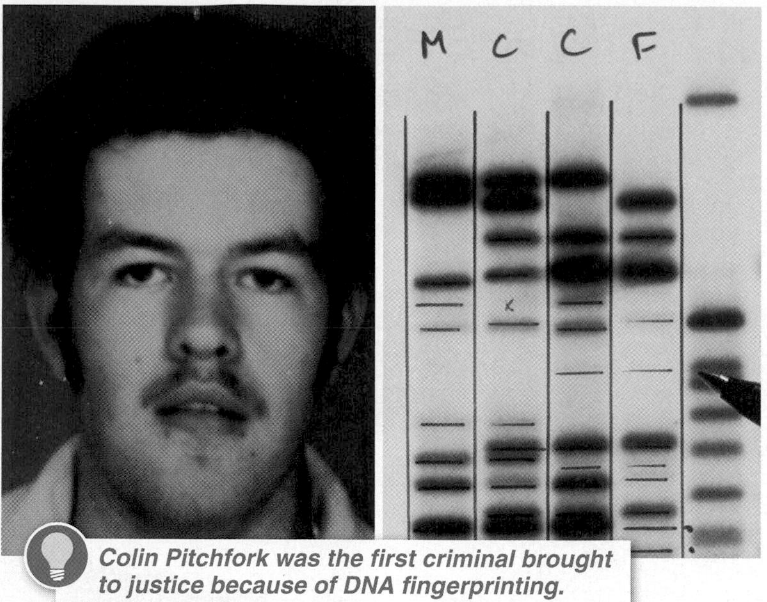

M C C F
Colin Pitchfork was the first criminal brought to justice because of DNA fingerprinting.

FIGURE 7-23 **Betrayed by his DNA.**

The DNA from all humans is almost completely identical. More than 99.9% of the DNA sequences of two individuals are the same, because we're all of the same species and thus share a common evolutionary history. Even so, in comparing two individuals' genomes of three billion base pairs each, a one-tenth of a percent difference still translates to about three million base-pair differences. These differences give individuals their own unique genome. (The lone exception? Identical twins, whose DNA is exactly the same.) Thus, the analysis of DNA from a crime scene focuses on the parts of our DNA that differ. There are thousands of these highly variable regions in the human genome.

Among these thousands of variable regions, one type is used for determination of a person's genetic fingerprint. These regions, called STRs (for short tandem repeats), are characterized by a short sequence (commonly four or five nucleotides long) that repeats over and over within a noncoding region of DNA. The number of repeats is what varies among individuals.

Here's an example. In Individual A, the number of times the sequence repeats at one STR region (say, on chromosome 2) is 3 times on the maternal copy of chromosome 2, and 14 times on the paternal copy. Individual A is said to have two different alleles for this STR region: 3 and 14. In contrast, in Individual B, in the same STR region on chromosome 2, the sequence repeats 5 times and 11 times. Individual B has alleles 5 and 11 (FIGURE 7-24).

For an STR region within the human genome, there are typically about 10 different alleles within a population, and each allele is shared by about 10% of all individuals in the population. So the likelihood that two individuals carry the same two alleles is about 1 in 100. This is unlikely, but given enough people, many are likely to carry the same alleles. If two different STR regions are analyzed, the likelihood that two individuals have the same four alleles is 1/10 × 1/10 × 1/10 × 1/10, or 1 in 10,000. Although rarer, multiple individuals within the same large city could carry the same four alleles. The real power of DNA fingerprinting comes from simultaneously determining the alleles an individual carries (that is, their genotype) not simply at one or two STR locations but at *13 different* STR locations. This is the number used by the FBI in constructing DNA fingerprints in the United States, and it makes the probability

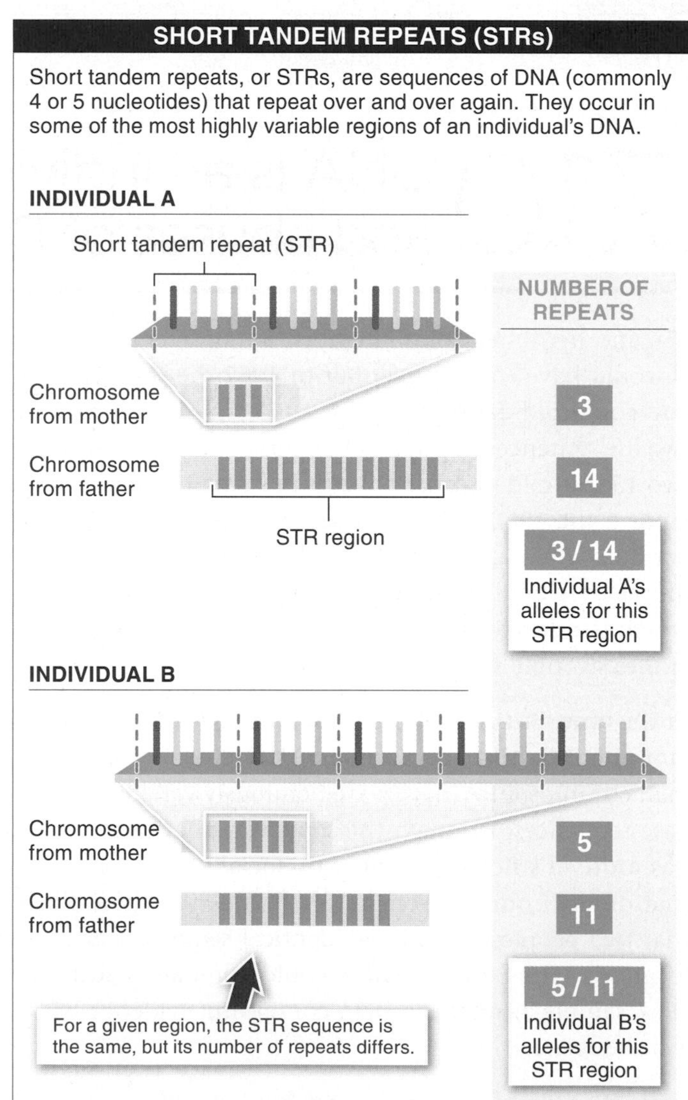

Chromosome from mother

Chromosome from father

Chromosome from mother

Chromosome from father

For a given region, the STR sequence is the same, but its number of repeats differs.

FIGURE 7-24 **Biotechnology in forensics.** Forensic scientists can use highly variable regions of DNA to genetically link a person to DNA-containing evidence left at a crime scene.

CREATING A DNA FINGERPRINT

A DNA fingerprint is created by determining which alleles an individual carries for 13 different STR regions.

1 AMPLIFY THE STR REGION
For each of the 13 STR regions used, the DNA fragment is amplified using PCR, resulting in huge numbers of those fragments.

The fragments differ in size depending on how many times the repeating unit of that STR is repeated.

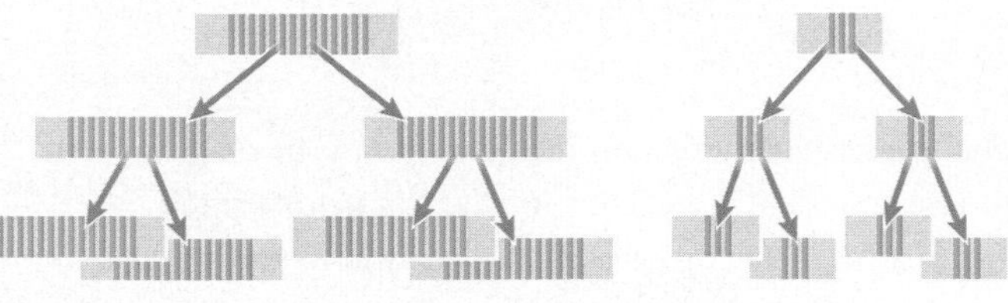

2 SORT THE FRAGMENTS BY SIZE
Amplified DNA fragments are poured into an electrophoresis gel and an electrical charge is applied.

Because DNA is a negatively charged molecule, the fragments move toward the positively charged electrode. Smaller pieces (having fewer repeats) move across the gel more quickly than larger pieces.

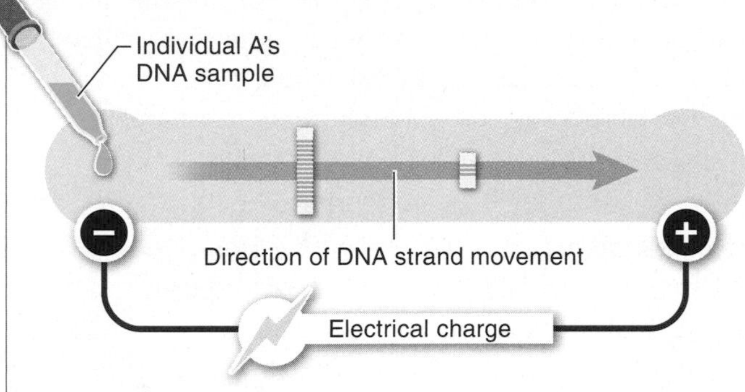

3 IDENTIFY THE GENOTYPE
The number of repeats within an STR region (indicating an individual's genotype) is determined by comparing the fragments with DNA fragments of known lengths.

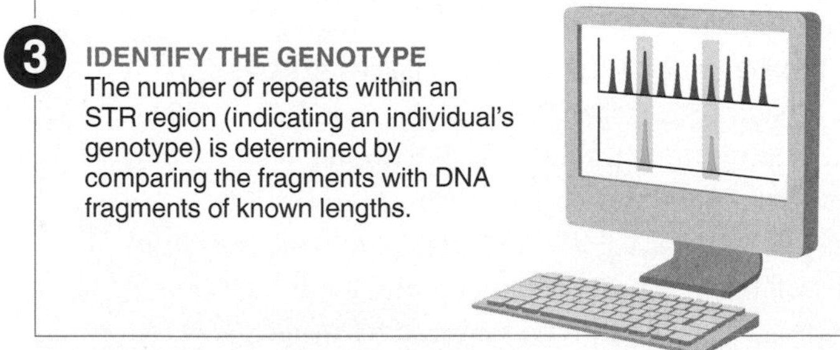

FIGURE 7-25 Your unique identifier: a DNA fingerprint. DNA fingerprints are being used to match a suspect's DNA to DNA found at the scene of a crime.

that two individuals would have exactly the same genotype extremely low (see Figure 7-24).

To produce a fingerprint, an individual's genotype is determined by using PCR to amplify the STR region. Using a technique called electrophoresis, the length of the STR region is measured. The length of the region is then used to determine the number of times the STR is repeated. For a single STR region, an individual's genotype is expressed by two numbers, as described above, reflecting the number of STR repeats in the copies inherited from the mother and from the father. And a person's full DNA fingerprint is a string of 26 numbers, consisting of the two numbers for each of 13 STRs.

In court, a suspect's genotype might be compared with the DNA fingerprint obtained from evidence found at the crime scene. DNA samples from different people produce different 26-number fingerprints, whereas different samples of DNA from the same person will have exactly the same genotype (FIGURE 7-25).

Despite universally accepted methods, DNA fingerprinting is not foolproof. Numerous incidences of human error—accidental as well as intentional—have been documented, ranging from mislabeled test tubes to tissue from a suspect being added to evidence from a crime scene. So we should not blindly draw conclusions solely from this one type of evidence. Nonetheless, DNA fingerprinting is an increasingly valuable tool for law enforcement, particularly because it is generally more reliable than eyewitness accounts. The FBI, for example, has reported that nearly one-third of their suspects are cleared immediately by DNA testing, and because of DNA fingerprinting, many more criminals now plead guilty to the crimes they have committed.

TAKE HOME MESSAGE 7.10

» Comparisons of highly variable DNA regions can be used to identify tissue specimens and determine the individual from whom they came.

STREET BIO

Using evidence to guide decision making in our own lives

Should you be considering genetic screening?

When it comes to genetic screening, the future is now. In fact, this may be one of the most valuable results of biotechnology.

Q: What can you be tested for? Most tests offered today are for diseases you might develop later in life. They include tests for the presence of the non-functional versions of the genes BRCA1 and BRCA2, which are associated with breast and ovarian cancer. Another test is for familial adenomatous polyposis, a condition that makes it likely that the person will develop colorectal cancer; another is for Huntington's disease.

Genetic tests also can identify an increased likelihood of having a child with a particular disease (that the person getting the test will not develop). These include Tay-Sachs disease, cystic fibrosis, thalassemia, Duchenne muscular dystrophy, hemophilia, and fragile X syndrome.

Q: How is genetic screening done? DNA for genetic screening can be isolated from a saliva or blood sample. Depending on how many tests are done, the price can range from about $100 to more than $2,000.

Q: What can you gain from these tests? Information from genetic screening can help with preparation and prevention. Suppose a person tests positive as a carrier of a disease; that is, he or she carries one copy of the gene associated with the disease, but the disease manifests only if an individual carries two copies of the gene. The person's partner may then elect to get tested. If both are carriers, each of their children will have a 1 in 4 chance of inheriting the disease. Couples at increased risk of having a child with a genetic disorder may explore the reproductive options available.

When genetic screening reveals that a person is at risk for developing a genetic disease later in life, the information can jumpstart lifestyle changes and/or early screening and preventive treatments.

Q: Are these tests risky? Several significant risks are associated with genetic screening, and because most are not related to the process of the screening test, they can be subtle.

- Not all causes for a specific disease can be detected. Consequently, a negative result can produce a false sense of security. For example, women carrying the non-functional versions of the BRCA1 and BRCA2 genes account for 5% to 10% of all breast cancer cases and 15% of all ovarian cancer cases. This means that the vast majority of all breast and ovarian cancers occur in women who would get a negative result on the BRCA1 and BRCA2 screenings.
- False positives can occur, sometimes with serious emotional and/or financial consequences, and possibly inappropriate treatments.
- There are risks (and fears of risks) of discrimination based on the test results, such as by life insurance providers.

Q: Are there any clear recommendations? In 2016, the American College of Obstetricians and Gynecologists advised screening "for a limited number limited testing is that with screening now possible for so many genetic disorders—many of which are extremely rare—the possibility that someone will test positive (as a carrier) for at least one condition is relatively high. This can lead to unnecessary anxiety, given the low probability that the person's partner would also be a carrier for the same rare gene.

Volunteers help collect DNA saliva samples for use in breast cancer research. The information can help researchers understand why some people get cancer and others do not.

GRAPHIC CONTENT

Thinking critically about visual displays of data

1 What proportion of corn grown in the United States is genetically modified?

2 From the figure, can you determine whether there is more genetically modified corn or genetically modified cotton produced in the United States? How do you do this? And if it is not possible, why isn't it?

3 What is the "take home message" from this figure? Does it influence your thoughts on genetically modified crops? Why or why not?

4 Why are data given for proportions of genetically modified crops rather than absolute amounts? Does this alter your interpretation of the graph?

5 Worldwide, the proportion of corn grown that is genetically modified is just 26%, while the proportion for soybeans is 77% and for cotton is 49%. How would this information add value to the figure?

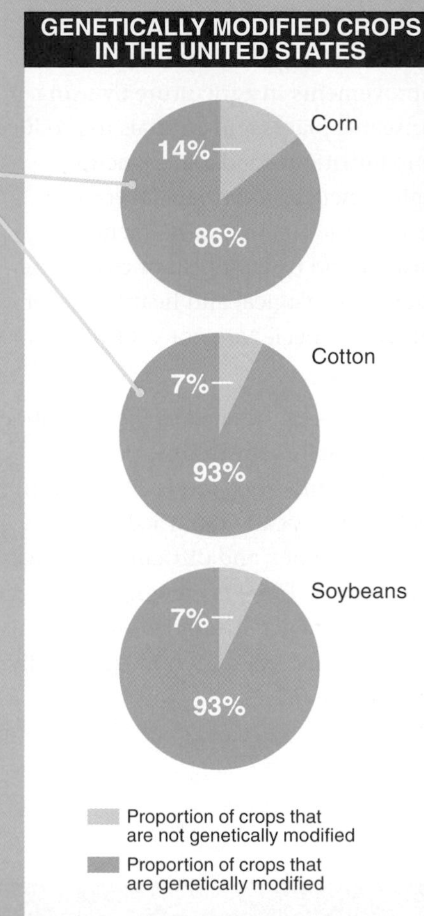

6 The proportion of corn grown in the United States that is genetically modified changed between 2001 and 2010 as follows:

Year	Proportion
2001	26%
2002	34%
2003	40%
2004	47%
2005	52%
2006	61%
2007	73%
2008	80%
2009	85%
2010	86% (shown in the graph)

Show two different ways that you could display these data. What is the clearest conclusion someone would draw from these figures?

See answers at the back of the book.

KEY TERMS IN BIOTECHNOLOGY

biotechnology, p. 192
clone, p. 196
clone library, p. 197
cloning, p. 196
CRISPR (clustered regularly interspaced short palindromic repeats), p. 197
gene library, p. 197
gene therapy, p. 210
genetic engineering, pp. 193, 199
genetically modified organism (GMO), p. 200
plasmid, p. 195
polymerase chain reaction (PCR), p. 194
recombinant DNA technology, p. 200
restriction enzyme, p. 194
stem cell, p. 210
transgenic organism, p. 195

BRIEF SUMMARY

Living organisms can be manipulated for practical benefits.

• Biotechnology is the use of technology to modify organisms, cells, and their molecules to achieve practical benefits. The primary areas in which biotechnology is applied include agriculture, human health, and forensic science.

• Modern molecular methods for transferring DNA from one organism into another include the use of naturally occurring restriction enzymes for cutting DNA, the polymerase chain reaction for amplifying DNA, insertion of the DNA into bacterial or viral vectors, and the cloning and identification of cells with transferred DNA.

• CRISPR is a gene editing system that brings greater precision and efficiency to gene (DNA) editing, allowing the targeting and cutting of DNA at a specific sequence in almost any species.

Biotechnology is producing improvements in agriculture.

• Biotechnology has led to important improvements in agriculture by using transgenic plants and animals to produce more nutritious food and reducing environmental and financial costs through the creation of herbicide-resistant and insect-resistant crops. The potential ecological and health risks of transgenic species are not yet fully understood.

• Numerous legitimate fears remain about the potentially catastrophic risks of genetically modified foods. Determining whether genetically modified foods are safe is a complex and difficult challenge for scientists. There is growing consensus, however, that genetically modified foods carry no more risk than non–genetically modified foods.

Biotechnology has the potential for improving human health.

• Biotechnology has led to some successes in treating diseases, usually by producing medicines more efficiently and effectively than they can be produced with traditional methods.

• Gene therapy has had a poor record of success in curing human diseases, primarily because of technical difficulties in transferring normal-functioning genes into the cells of a person with a genetic disease.

• Cloning of individuals has potential benefits in agriculture and medicine, but ethical questions linger.

Biotechnology has value for criminal justice systems.

• Comparisons of highly variable DNA regions can be used to identify tissue specimens and determine the individual from whom they came.

CHECK YOUR KNOWLEDGE

Short Answer

1. What is "CRISPR" and how is it used as a tool for editing an organism's genome?

2. Why is it impossible to falsify the hypothesis that "genetically modified foods are dangerous"? Is this a reasonable argument for permanently prohibiting their use? Why or why not?

3. Describe two ways in which advances in biotechnology have helped make farming more efficient.

4. The creation of genetically modified foods raises many concerns, including that it will lead to the loss of genetic diversity. Why is this a concern?

5. Describe three major applications of biotechnology and their impact on human health.

6. How has the advent of genetic screening for prospective parents led to decreases in the incidence of fatal genetic diseases?

7. What is a potential benefit to farmers of cloning technology?

8. What is an STR region? Describe the similarities and differences between an STR region and a gene for a structural trait (such as eye color), with respect to mechanisms of inheritance, genetic variation, and impact on an organism's phenotype.

Multiple Choice

1. "CRISPR" refers to repeated sequences located in:

a) the recipient cell of an organism that is to be genetically modified.

b) a virus's DNA.

c) a virus's RNA.

d) a eukaryote vector's genome.

e) bacterial DNA.

0 EASY — 37 — HARD 100

2. The polymerase chain reaction (PCR):

a) enables researchers to create huge numbers of copies of tiny pieces of DNA.

b) enables researchers to determine the sequence of a complementary strand of DNA when they have only single-stranded DNA.

c) utilizes RNA polymerase to build strands of DNA.

d) can create messenger RNA molecules from small pieces of DNA.

e) All of the above are correct.

3. Which of the following is not a difficulty that medicine has encountered in its attempts to cure human diseases through gene therapy?

a) The transfer organism—usually a virus—may get into unintended cells and cause disease.

b) It is difficult to get the working gene into the specific cells where it is needed.

c) It is difficult to get the working gene into enough cells at the right rate to have a physiological effect.

d) For many diseases, a malfunctioning gene has not been identified.

e) All of the above are difficulties encountered in attempts to cure human diseases through gene therapy.

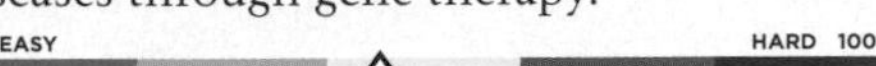

4. Golden rice:

a) grows without a husk, thereby reducing the processing required before the rice can be consumed.

b) can make vitamin A without beta-carotene.

c) could help prevent blindness due to vitamin A deficiency in 250,000 children each year.

d) supplies more vitamin A in one serving than an individual needs in a full week.

e) is one of the most recent developments in organic farming.

5. Which of the following statements about Bt crystals is correct?

a) They are produced by soil-dwelling bacteria of the species *Bacillus thuringiensis*.

b) The gene coding for the production of Bt crystals has been genetically engineered into the genome of dairy cows, increasing their milk production sixfold.

c) They are produced by the polymerase chain reaction (PCR).

d) They are produced by most weedy species of plants.

e) All of the above are correct.

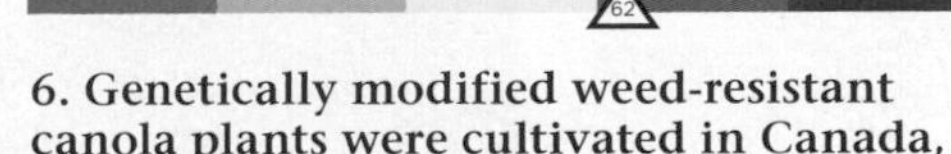

6. Genetically modified weed-resistant canola plants were cultivated in Canada, making it possible for farmers to apply herbicides freely to kill the weeds but not the canola plants. What went wrong with this scenario?

a) The weed-resistant canola plants spread by seed to neighboring farms that weren't growing canola. The weed-resistant canola grew out of control because traditional herbicides could not kill them.

b) The canola farmers applied the herbicide at such a great rate that it spread to other farms that were not growing weed-resistant canola and killed the crops on those farms.

c) Canola plants can't grow in Canada—it's too cold.

d) The genetic modification that made the canola plants weed-resistant caused them to become more vulnerable to some insect pests.

e) Farmers were so successful in growing canola plants that the market for canola crashed and the Canadian farmers went bankrupt.

7. Which of the following is not a potential problem caused by GMOs?

a) Some organisms, including weeds, might grow out of control and become impossible to kill with pesticides.

b) Genetically modified crops may turn out to be less expensive to cultivate than unmodified ones.

c) Some ecologically valuable organisms may unintentionally be killed.

d) The loss of genetic diversity in crop plants may make them more susceptible to pests.

e) We cannot completely evaluate the potential risks posed by GMOs.

0 EASY 51 HARD 100

8. Although genetic screening has contributed to a reduced incidence of Tay-Sachs disease:

a) genetic screening puts people at higher risk of developing a disease such as breast cancer.

b) genetic screening cannot reveal whether an individual is at increased risk of disease later in life.

c) genetic screening results can lead to cancellation of a life insurance policy.

d) genetic screening cannot be used in determining whether a fetus has a genetic disease.

e) genetic screening of prospective parents produces no useful information for couples wanting to have children.

0 EASY 61 HARD 100

9. Recombinant human ______ is/are at the center of the "blood doping" scandals in professional cycling.

a) erythropoietin (EPO)

b) growth hormone (HGH)

c) insulin

d) red blood cells

e) stem cells

0 EASY 47 HARD 100

10. Short tandem repeat sequences of DNA:

a) are characteristic of genes that code for biochemical traits rather than structural traits.

b) are used in biotechnology when creating a clone.

c) are produced when a mutation occurs in a non-sex cell.

d) can be used to produce a DNA fingerprint.

e) are produced when a mutation occurs in a sperm-producing or egg-producing cell.

0 EASY 64 HARD 100

11. DNA analyses can be used to overturn incorrect criminal convictions. What is the most common reason for this?

a) DNA evidence is a much more reliable identifier of an individual than are eyewitness accounts.

b) Science has not traditionally been used in courtroom prosecutions.

c) Researchers can now generate an accurate DNA fingerprint from surveillance photos of criminals.

d) The legal system now permits conviction of any suspects who refuse to have their DNA sampled.

e) Scientists can identify within a person's DNA whether he or she carries any of the most common "criminality" alleles.

Ch08

Sibling similarities—such as those seen among the Jackson brothers (shown here performing in 1973)—illustrate the phenotypic consequences of genetic similarity; but the subtle differences also reveal variability among offspring in sexually reproducing species.

Chromosomes and Cell Division

Continuity and variety

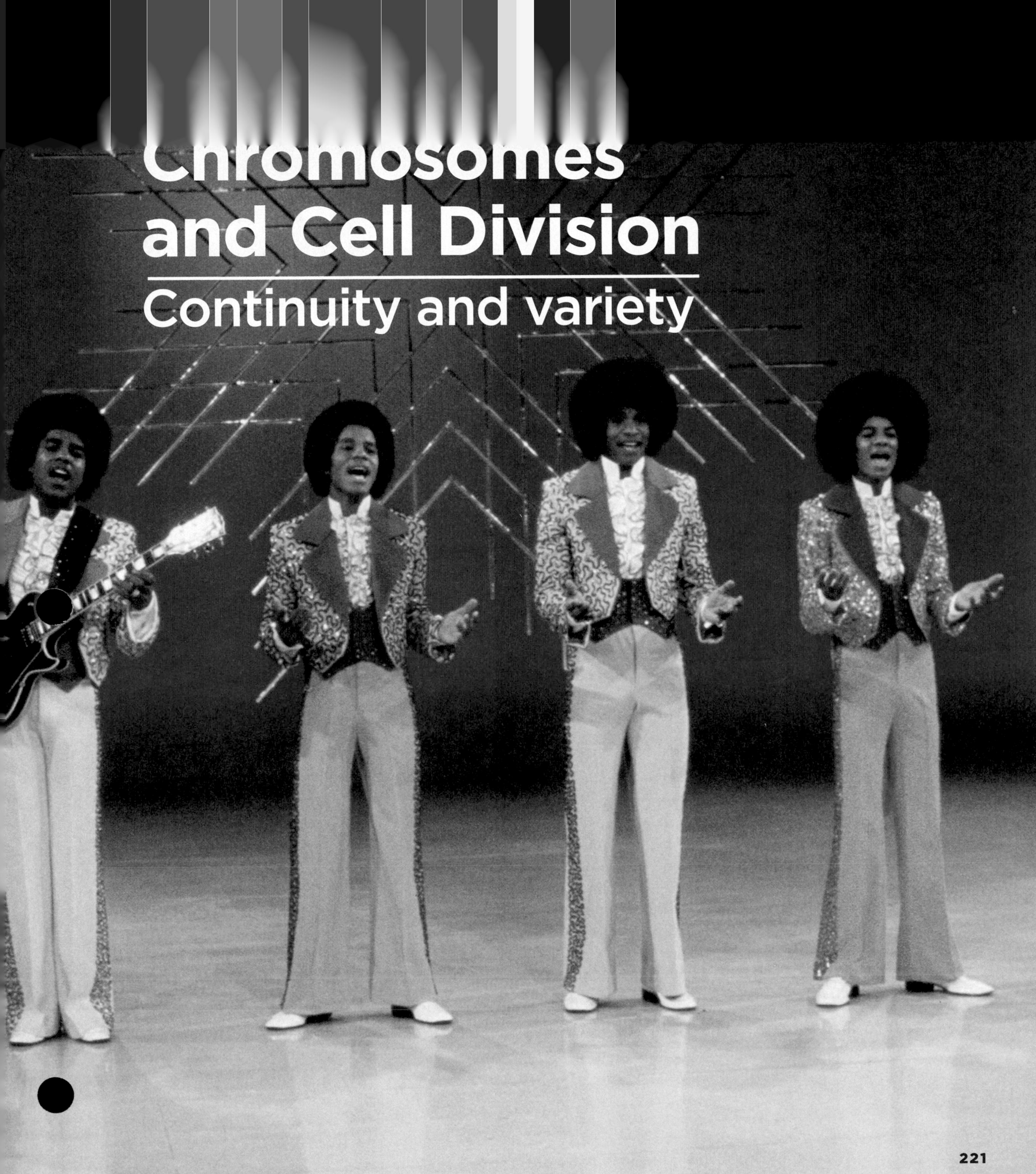

8.1–8.4 There are different types of cell division.

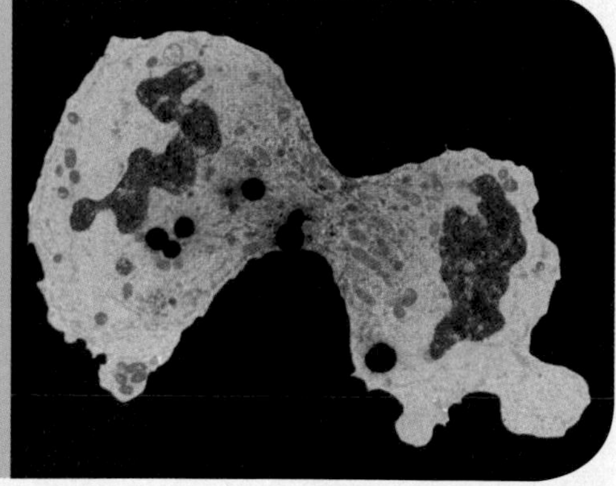

Cell division in a Ewing sarcoma bone tumor.

8.1 Immortal cells can spell trouble: cell division in sickness and in health.

Throughout most of your body your cells are continually dying off, and the ones that remain divide and replace the cells you've lost, in an ongoing process. But can this cell replacement go on forever? And does a cell even know how old it is?

Actually, a cell does have a feature that provides an approximate measure of how old it is. Just as a car comes with an odometer, which keeps track of how far the car has been driven, animal cells have a sort of counter that reflects how many times the cell has divided. This "counter" is a section of noncoding, repetitive DNA, called a **telomere,** that serves as a protective cap and is located at each tip of every chromosome, right next to the genes that direct the processes that maintain the organism (FIGURE 8-1).

Every time a cell divides, making an exact copy of itself, its DNA divides as well. However, each time the DNA divides, the process by which chromosomes are duplicated causes the telomere at each end of every chromosome to get a bit shorter. At birth, the telomeres in most human cells are long enough to support 80–90 cell divisions. In a 70-year-old person, the telomeres are only long enough for about 20–30 divisions. Eventually, the telomeres can become so short that additional cell divisions cause the loss of functional, essential DNA, and that means almost certain death for the cell.

Occasionally, individuals are born with telomeres that impair the functioning of a protein that helps maintain the nucleus of cells. This genetic condition, called Hutchinson-Gilford progeria syndrome, causes the cell's telomeres to be much shorter than normal. In addition, the normal functioning of many genes is disrupted, and consequently, cells and tissues begin to appear aged very soon after birth (FIGURE 8-2). Children with this disorder rarely live beyond the age of 13.

Given this information, you might wonder if, by rebuilding the telomere after each round of division, the cell and its descendants could function for a longer time than normal. Such a line of cells would never die. By extension, it's tempting to imagine that constantly rebuilding telomeres might act like a fountain of youth. Unfortunately, it does not.

Some cells do rebuild their telomeres after each cell division, restoring the chromosomes' protective caps. For single-celled eukaryotic organisms and for the cells that divide to produce sperm and eggs in multicellular organisms (cells that must go through many thousands of rounds of cell division), this telomere rebuilding is essential.

Unfortunately, for most of the other types of cells that rebuild their telomeres with each cell division, the telomere rebuilding presents a big problem: the cells are unable to stop dividing. Such cells commonly go by another name: cancer.

Because telomere rebuilding occurs in (and possibly is necessary to) many human cancers, researchers have hopes

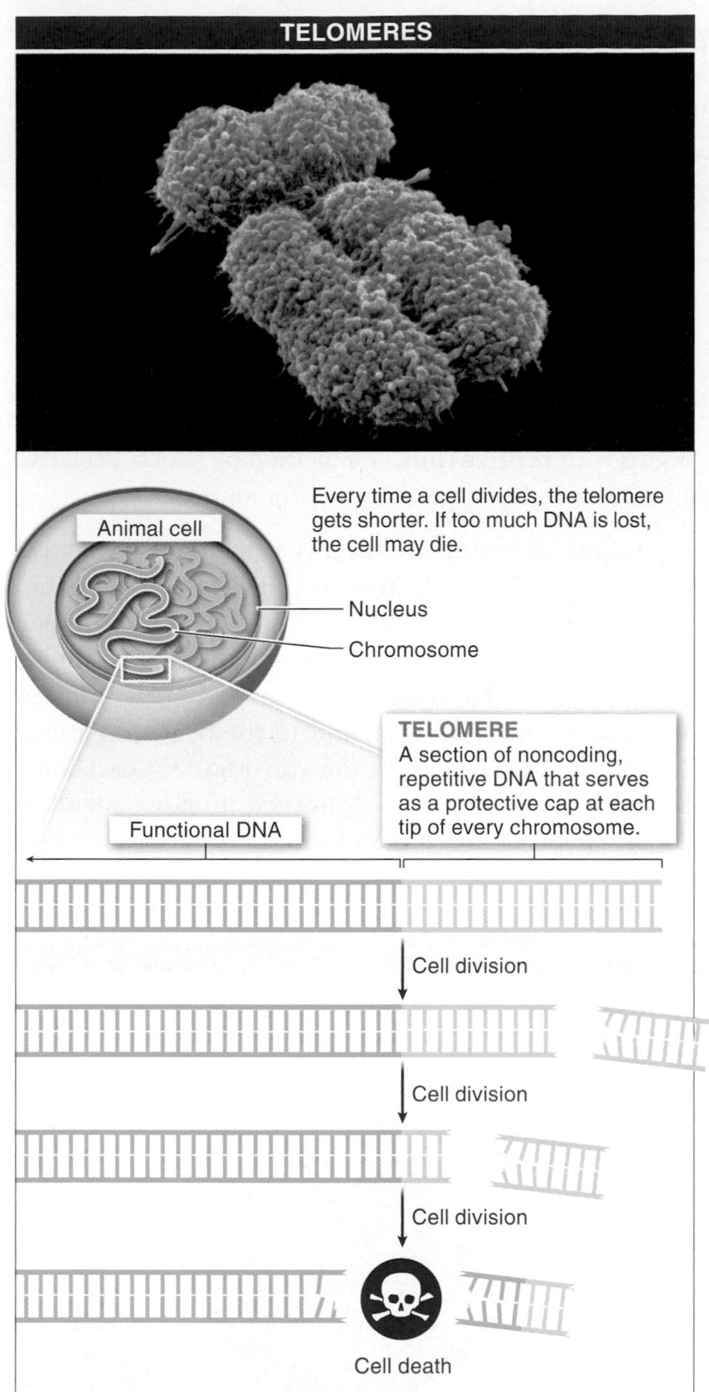

TELOMERES
Every time a cell divides, the telomere gets shorter. If too much DNA is lost, the cell may die.
Animal cell
Nucleus
Chromosome
TELOMERE
A section of noncoding, repetitive DNA that serves as a protective cap at each tip of every chromosome.
Functional DNA
Cell division
Cell division
Cell division
Cell death

FIGURE 8-1 **A cellular "odometer."** Telomeres limit cells to a fixed number of divisions.

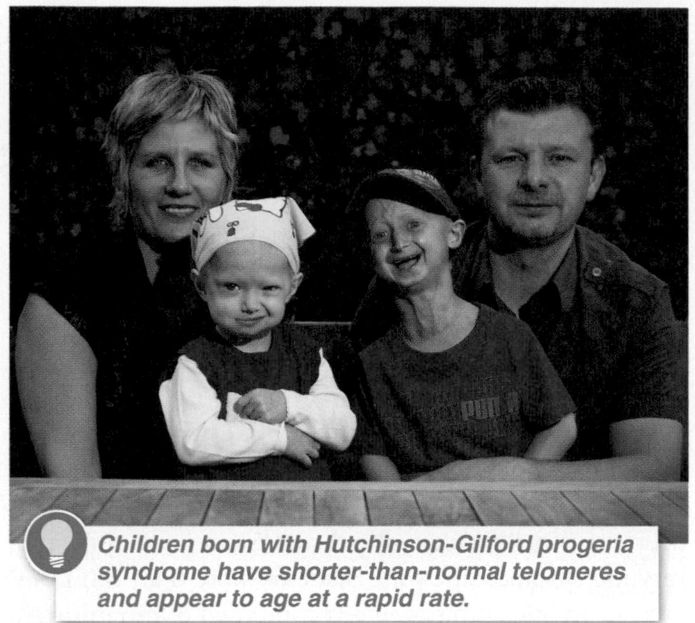

Children born with Hutchinson-Gilford progeria syndrome have shorter-than-normal telomeres and appear to age at a rapid rate.

FIGURE 8-2 **A chromosomal problem with serious health consequences.**

that inhibiting it might help fight cancer. In any case, because one of the defining features of cancer is runaway cell division, discovering a cure for cancer will necessarily involve a deep understanding of how cells divide.

In this chapter, we investigate the processes that enable cells to divide and create new cells. Prokaryotes have one method of cell division (called binary fission), and it serves all of their cell division needs. Eukaryotes have two methods of cell division, mitosis and meiosis, each of which has a specific purpose in a eukaryote's life cycle. We'll explore normal cell division and discuss what happens when cell division does not proceed in the normal way.

TAKE HOME MESSAGE 8.1

» Cell division is ongoing in most organisms and their tissues; disruptions to normal cell division can have serious consequences. In eukaryotic cells, a protective section of DNA called the telomere plays a role in keeping track of cell divisions, getting shorter every time the cell divides. If telomeres become too short, additional cell divisions cause loss of essential DNA and cell death. Some types of cells that rebuild their telomeres with each division can become cancerous.

8.2 Some chromosomes are circular, others are linear.

All life on earth uses DNA to store genetic information. This fact is remarkable considering the tremendous diversity of life on our planet—from single-celled bacteria to multicellular plants and animals. One way in which the DNA of different species varies is in how it is organized into chromosomes.

The most important part of a eukaryotic **chromosome** may be the DNA molecule, which carries information about how to accomplish the processes needed to support the life of the organism. But eukaryotic chromosomes (and some prokaryotic chromosomes) are made of more than just DNA. The eukaryotic chromosome is composed of **chromatin,** a linear DNA strand bound to and wrapped tightly around proteins called **histones,** which keep the DNA from getting tangled and enable it to be tightly and efficiently packed inside the nucleus. Plants and animals usually have between 10 and 50 chromosomes (although there are species with as few as 2 and others with more than a hundred).

Most prokaryotes—the bacteria and archaea—have less DNA than eukaryotes. They carry their genetic information in a single, circular chromosome, a closed loop of double-stranded DNA that is attached at one site to the cell membrane (FIGURE 8-3). And when it is time for them to reproduce, they use a method called **binary fission,** which means "division in two" (FIGURE 8-4). This process begins with **replication,** the method by which a cell creates an exact duplicate of each chromosome.

Replication in prokaryotes begins as the double-stranded DNA molecule unwinds from its coiled-up configuration. Once the strands are uncoiled, they split apart like a zipper, with bases exposed on each of the two separated, single-stranded, circular molecules of DNA. As the double-stranded molecule unzips, enzymes bind to the DNA and attach free-floating nucleotides to the growing DNA backbone, matching A to T and G to C, thus creating two identical double-stranded DNA molecules.

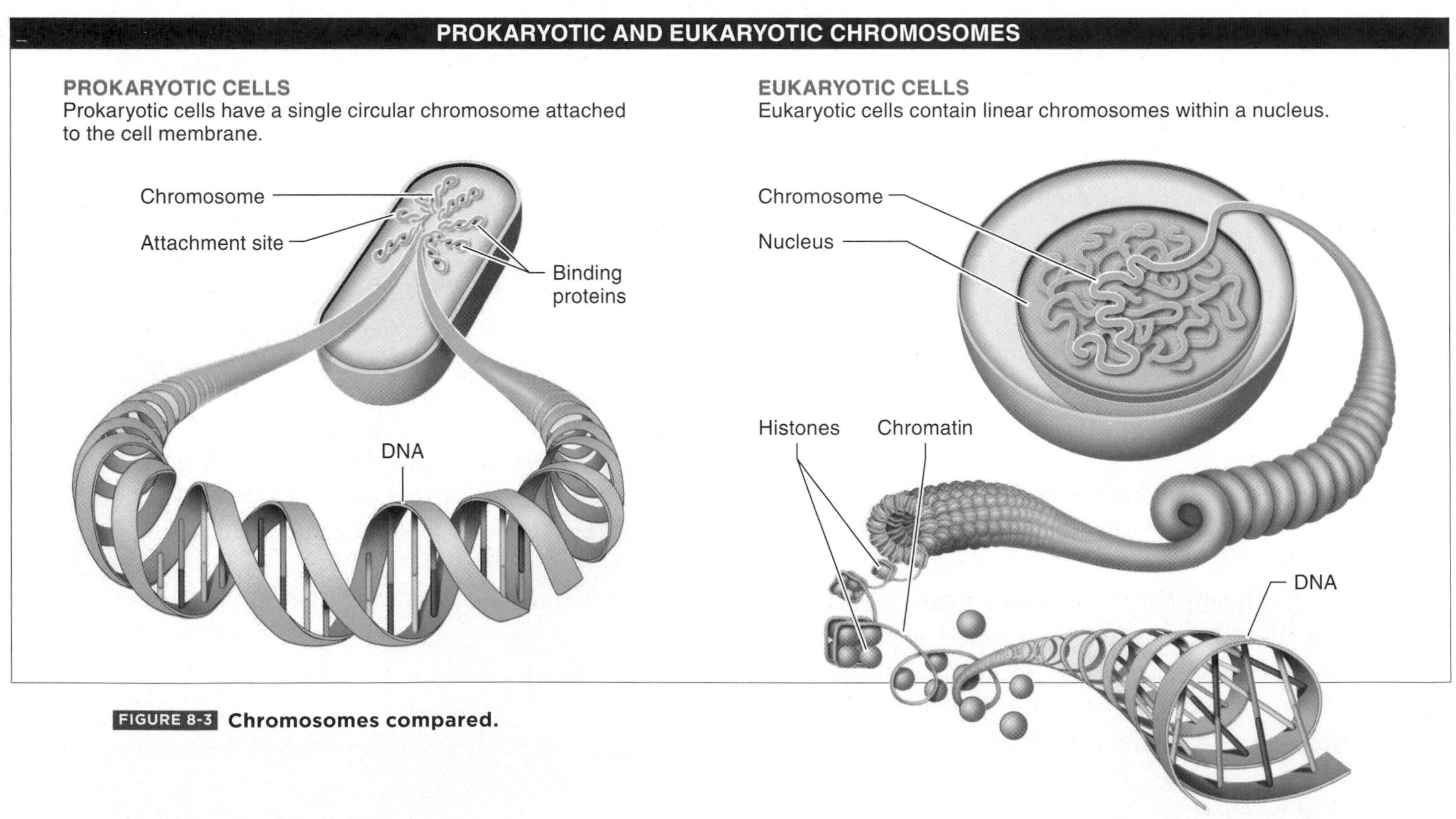

FIGURE 8-3 **Chromosomes compared.**

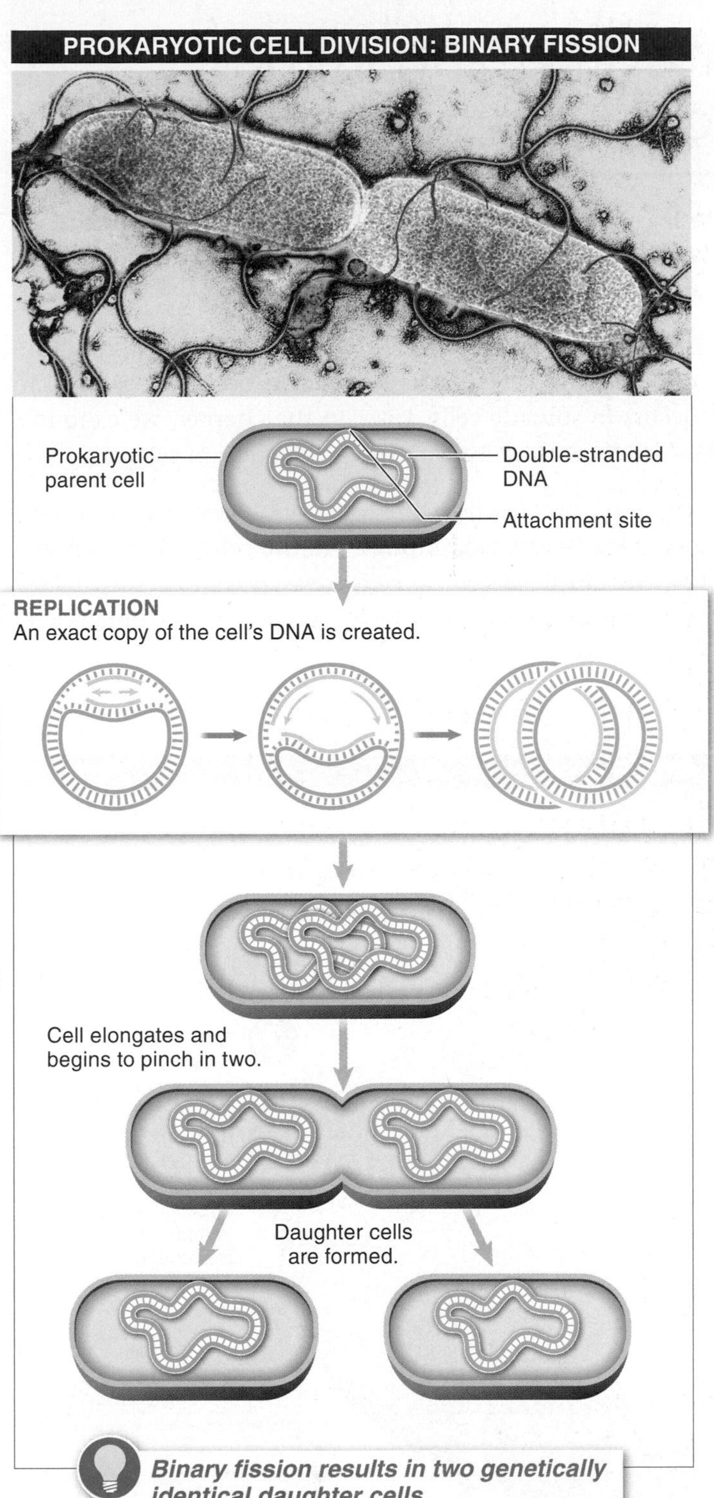
PROKARYOTIC CELL DIVISION: BINARY FISSION
Prokaryotic parent cell
Double-stranded DNA
Attachment site
REPLICATION
An exact copy of the cell's DNA is created.
Cell elongates and begins to pinch in two.
Daughter cells are formed.
Binary fission results in two genetically identical daughter cells.

FIGURE 8-4 **Dividing in two.** Asexually reproducing prokaryotes reproduce rapidly through binary fission.

The two newly created circular chromosomes attach to the inside of the plasma membrane, each at a different spot. The original cell, called the **parent cell,** then pinches in somewhere between these two spots until it divides into two new cells, called **daughter cells.** Each of the daughter cells has an identical two-stranded copy of the original two-stranded circular chromosome.

In some prokaryotes, such as the bacteria called *E. coli* that live in our digestive system, the complete process of binary fission can occur very quickly—often in as little as 20 minutes. Binary fission is considered **asexual reproduction,** because the daughter cells inherit their DNA from a single parent cell and thus are genetically identical to the parent.

TAKE HOME MESSAGE 8.2

» In most bacteria and archaea, the genetic information is carried in a single, circular chromosome, a closed loop of DNA that is attached at one site to the cell membrane. Eukaryotes have much more DNA than do bacteria and organize it into linear chromosomes within the nucleus. Bacteria divide by a type of asexual reproduction called binary fission: first, the circular chromosome duplicates itself, then the parent cell splits into two new, genetically identical daughter cells.

8.3 There is a time for everything in the eukaryotic cell cycle.

Eukaryotic cells don't do every job at once. They go through phases. They may, for example, spend long periods occupied with activities relating solely to their growth, and then may suspend those activities as they segue into a period devoted exclusively to reproducing themselves. This alternation of activities between processes related to growth and processes related to cell division is called the **cell cycle** (FIGURE 8-5).

Before we go any farther in discussing the cell cycle, we need to note an important distinction between the types of cells in the body. All the cells of a multicellular eukaryotic organism can be divided into two types: **somatic cells** are the cells forming the body of the organism; **reproductive cells** are the sex cells—that is, the **gametes** (sperm and eggs)—and the cells that give rise to them.

Somatic cells and reproductive cells use different methods of producing new cells. First we focus on cell division as it occurs in somatic cells. Later in the chapter, we examine cell division that leads to the production of sex cells.

The cell cycle describes the series of phases in somatic cell division. The two main phases in the cell cycle are **interphase,** during which the cell grows and prepares to divide, and the **mitotic phase** (or **M phase**), during which division occurs.

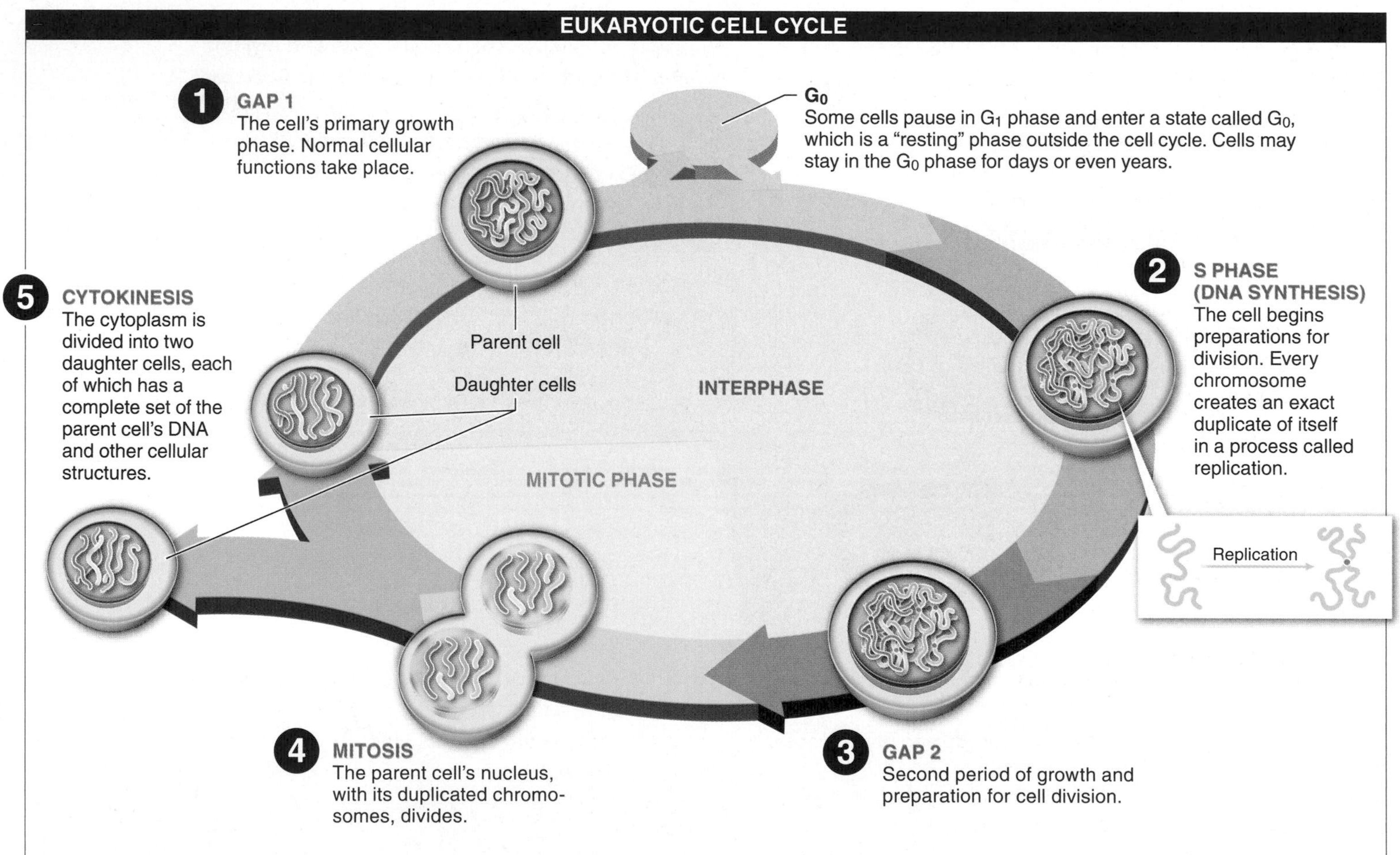

EUKARYOTIC CELL CYCLE
1 GAP 1
The cell's primary growth phase. Normal cellular functions take place.
G_0
Some cells pause in G_1 phase and enter a state called G_0, which is a "resting" phase outside the cell cycle. Cells may stay in the G_0 phase for days or even years.
2 S PHASE (DNA SYNTHESIS)
The cell begins preparations for division. Every chromosome creates an exact duplicate of itself in a process called replication.
Replication
5 CYTOKINESIS
The cytoplasm is divided into two daughter cells, each of which has a complete set of the parent cell's DNA and other cellular structures.
Parent cell
Daughter cells
INTERPHASE
MITOTIC PHASE
4 MITOSIS
The parent cell's nucleus, with its duplicated chromosomes, divides.
3 GAP 2
Second period of growth and preparation for cell division.

FIGURE 8-5 **The cycle of cellular activity.** The eukaryotic cell cycle is a series of events that results in the division of somatic cells (cells that make up the body of an organism).

Interphase Interphase is further divided into three distinct sub-phases, described below.

Gap 1 (G_1). During this period, a cell may grow and develop as well as perform its various cellular functions (making proteins, getting rid of waste, and so on). Most cells inhabit the Gap 1 phase the majority of the time. However, some cells enter a state called **G_0**, which is a "resting" phase outside the cell cycle in which no cell division occurs. The G_0 phase may last for days or even years—or permanently, as in the case of most neurons and heart muscle cells—before resuming cell division.

DNA synthesis (S phase). During this phase, the cell begins to prepare for cell division. First it creates an exact duplicate of each chromosome by replication. Before replication, each chromosome's DNA consists of a single long, linear molecule. After replication, each chromosome's DNA has become a pair of identical long, linear molecules, held together near the center; the region where the two pieces are in contact is called the **centromere.**

Gap 2 (G_2). Significant growth occurs in this phase, along with high rates of protein synthesis in preparation for division. Gap 2 differs from Gap 1 in its shorter duration and its genetic material now existing in duplicate.

Mitotic Phase This period begins with **mitosis,** a process in which the parent cell's nucleus, including its chromosomes, divides. Mitosis is generally followed by **cytokinesis** (which may begin before the end of mitosis), during which the cytoplasm is divided into two daughter cells. Each daughter cell has a complete set of the parent cell's DNA and other cellular structures. The mitotic phase is usually the shortest period in the eukaryotic cell cycle.

There is great variation in how cell types move through the phases of the cell cycle. Animal embryos, at one end of a continuum, may move through the cell cycle so quickly that they spend almost no time at all in the G_1 and G_2 phases. Heart muscle cells and brain neurons, at the other end of the continuum, may never pass through the cell division phase, remaining in a nondividing state for decades.

Rates of cell division are regulated by a **cell-cycle control system,** a group of molecules, mostly proteins, within a cell that coordinates the events of the cell cycle. This control system functions through a system of **checkpoints,** critical points in the cell cycle at which progress is blocked—and cells are prevented from dividing—until specific signals trigger continuation of the process. The signals that trigger cell-cycle transitions commonly consist of **growth factors,** which provide feedback about the cell's environment and can signal that division is appropriate.

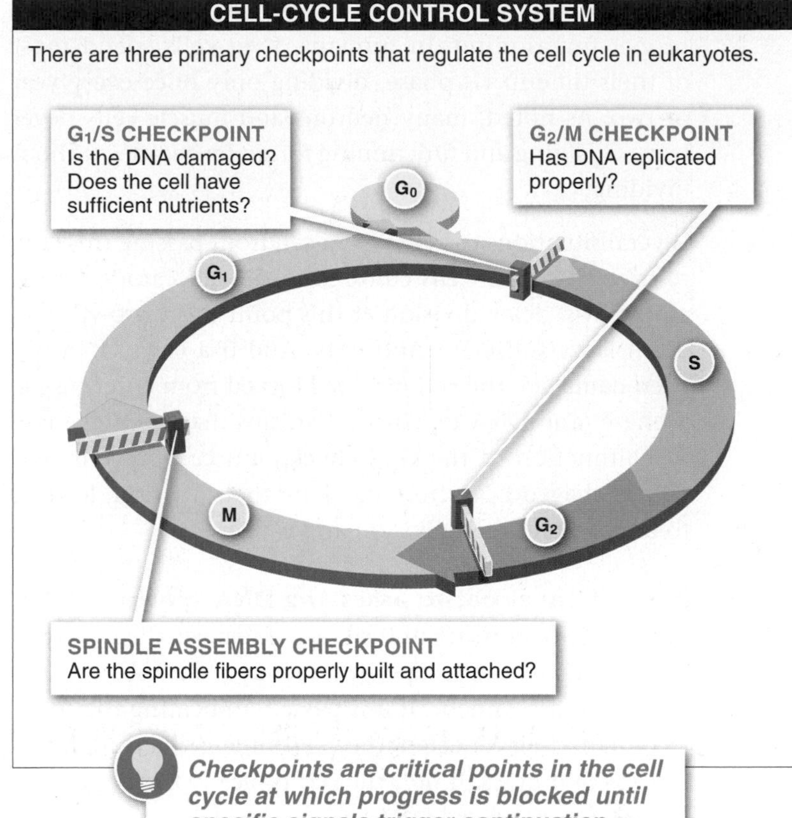

CELL-CYCLE CONTROL SYSTEM
There are three primary checkpoints that regulate the cell cycle in eukaryotes.
G1/S CHECKPOINT
Is the DNA damaged?
Does the cell have sufficient nutrients?
G2/M CHECKPOINT
Has DNA replicated properly?
G0
G1
S
M
G2
SPINDLE ASSEMBLY CHECKPOINT
Are the spindle fibers properly built and attached?
Checkpoints are critical points in the cell cycle at which progress is blocked until specific signals trigger continuation.

FIGURE 8-6 **Control of the cell cycle.** Cells use a system of checkpoints to ensure proper conditions have been achieved before the cell divides.

Checkpoints in the cell cycle make it possible for cells to (1) reduce the likelihood of completing cell division when errors have occurred in the process, and (2) respond to feedback conveying information about the cell's internal and external environments.

There are three primary checkpoints that regulate the cell cycle in eukaryotes (FIGURE 8-6).

1. G_1/S checkpoint: assessing DNA damage and cell growth. Occurring near the end of the G_1 phase, this is the point when a cell "decides" whether it will proceed to the S phase and complete cell division, or delay cell division, or enter into an extended "resting" phase, G_0.

Although often described as an extended "resting" phase, G_0 is, in fact, a nondividing, nongrowing phase that may include great metabolic activity critical to an organism's proper functioning. In humans, liver cells spend most of their time in G_0 phase, dividing only once every year or two. As noted, many neurons and muscle cells never leave G_0, living and functioning for many decades without dividing.

Several situations can prevent a cell from passing this G_1/S restriction point. Cells cultured in the laboratory, for example, will delay division at this point if the growth medium lacks sufficient nutrients. And if a cell's DNA has been damaged, the cell may be blocked from entering the S phase (and DNA repair mechanisms may be triggered). A malfunction in the G_1/S checkpoint can prevent cells with damaged DNA from blocking their division, leading to uncontrolled cell division and cancer.

2. G_2/M checkpoint: assessing DNA synthesis. Just before beginning mitosis, a cell reaches the G_2/M checkpoint. This checkpoint serves as a "mitosis-readiness" assessment. If it is passed, indicating that no DNA damage is detected, the cell initiates the complex process of mitosis. If not, the cell typically undergoes repair of its damaged DNA.

3. Spindle assembly checkpoint: assessing anaphase readiness during mitosis. This important cell-cycle checkpoint occurs during mitosis. At this point, cell-cycle control mechanisms assess whether the chromosomes have aligned properly at the metaphase plate and whether there is appropriate tension (pull) on them. If this checkpoint is passed, the cell completes cell division.

TAKE HOME MESSAGE 8.3

» Eukaryotic somatic cells alternate in a cycle between cell division and other cell activities. The cell division portion of the cycle is called the mitotic phase. The remainder of the cell cycle, called interphase, consists of two gap phases (during which cell growth and other metabolic activities occur) separated by a DNA synthesis phase, during which the genetic material is replicated. A cell-cycle control system functions through a series of checkpoints, critical points in the cell cycle at which progress is blocked—and cells are prevented from dividing—until specific signals trigger continuation of the process.

8.4 Cell division is preceded by chromosome replication.

In one of the great understatements in the scientific literature, James Watson and Francis Crick, in their paper describing the structure of DNA, wrote: "It has not escaped our notice that the specific pairing we have postulated immediately suggests a possible copying mechanism for the genetic material." The researchers' ability to explain how DNA copied itself was a critical feature of their description of DNA's structure. After all, every single time any cell in any organism's body divides, that cell's DNA must first duplicate itself so that each of the two new daughter cells has all of the genetic material of the original parent cell. The process of DNA duplication, as we've seen, is called replication.

What is it about DNA's structure that makes duplication so straightforward? Watson and Crick were referring to the feature of DNA called **complementarity,** meaning that in the double-stranded DNA molecule, the base on one strand always has the same pairing partner (called the **complementary base**) on the other strand: A pairs with T (and vice versa), and G pairs with C (and vice versa).

With this consistent pattern of pairing, one strand carries all the information needed to construct its complementary strand. Just before cells divide, the DNA molecule unwinds and "unzips," and each half of the unzipped molecule serves as a template on which the missing half is reconstructed. At the end of the process of reconstructing the missing halves, there are two DNA molecules—each identical to the original DNA molecule—one for each of the two new cells (FIGURE 8-7).

Before we describe the process by which DNA is duplicated, it is useful to revisit the structure of DNA because it dictates the directions along the linear molecule in which replication occurs. Recall from Chapter 6 that the individual units

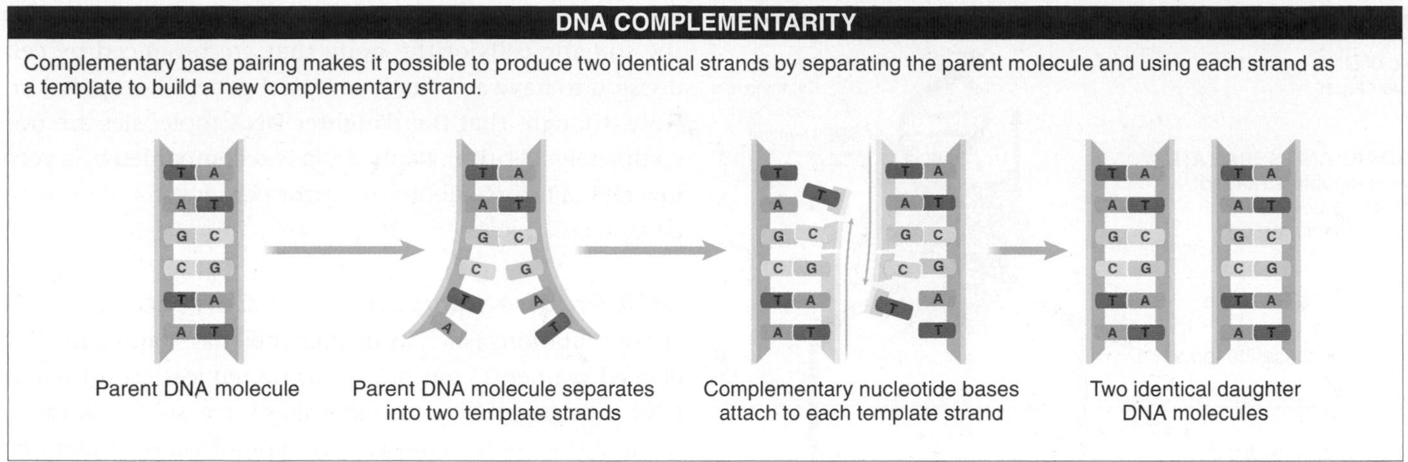

FIGURE 8-7 Meant for each other. The nucleotide base on one strand always has the same pairing partner on the other strand.

that make up DNA are nucleotides, which have three components: a nitrogen-containing molecule called a base, a phosphate group, and a molecule of a five-carbon sugar.

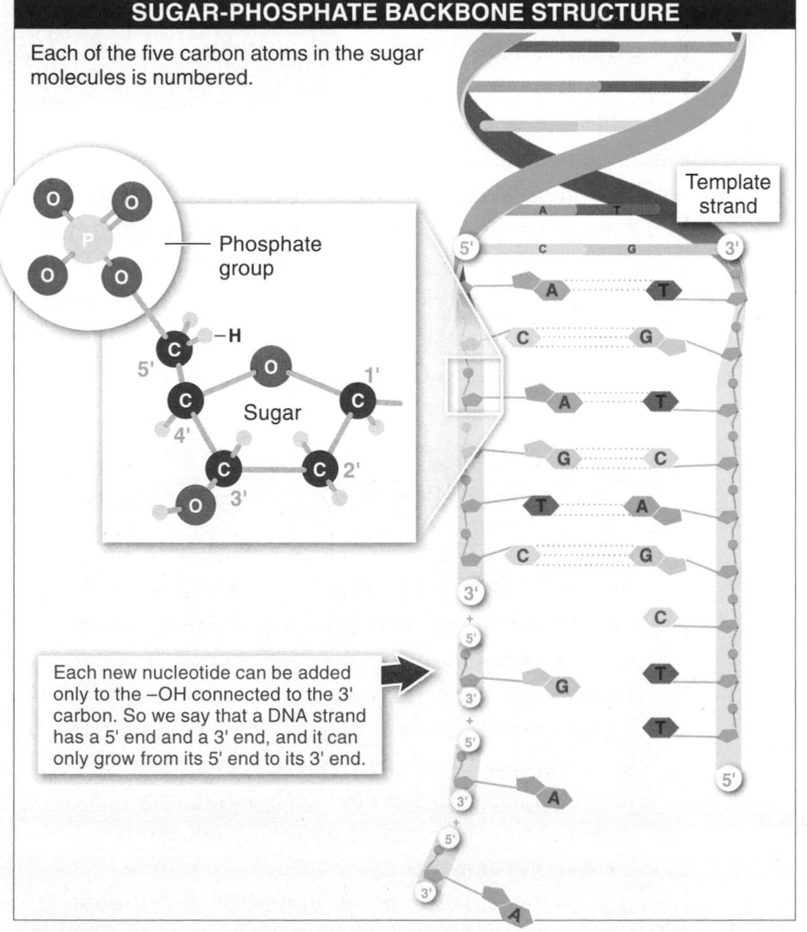

SUGAR-PHOSPHATE BACKBONE STRUCTURE
Each of the five carbon atoms in the sugar molecules is numbered.
Phosphate group
Sugar
Template strand
Each new nucleotide can be added only to the –OH connected to the 3' carbon. So we say that a DNA strand has a 5' end and a 3' end, and it can only grow from its 5' end to its 3' end.

FIGURE 8-8 A look at the structure of DNA. The DNA molecule has two strands of nucleotides held together by hydrogen bonds between bases. The two strands wind into a double helix.

Each of the five carbon atoms in the sugar molecules is given a number. The nitrogenous base is attached to the 1' (pronounced "one prime") carbon. An –OH group is attached to the 3' carbon. And the phosphate group is attached to the sugar's 5' carbon atom (FIGURE 8-8).

The process of DNA replication occurs in two steps: (1) unwinding and separation of the two strands, and (2) reconstruction and elongation of new, complementary strands (FIGURE 8-9). A feature of DNA that has important consequences for the process of replication is that its two strands run in opposite directions. The backbone of each strand of a DNA molecule is a long alternation of sugar, phosphate, sugar, phosphate, sugar, phosphate, and so on, going from one end to the other in a 5' to 3' pattern in one strand and a 3' to 5' pattern in the opposite strand.

1. Unwinding and separation. Replication begins at a specific site, called the **origin of replication,** where the coiled, double-stranded DNA molecule unwinds and separates into two strands, like a zipper unzipping. In prokaryotes, there is a single origin of replication, while eukaryotes have multiple origin sites on each chromosome. At the origin of replication, a complex of proteins binds to the DNA. One of the proteins, an enzyme called **DNA helicase,** unwinds the coiled DNA and separates the two complementary strands. The unwinding and separating of the two DNA strands creates what is called a **replication fork.**

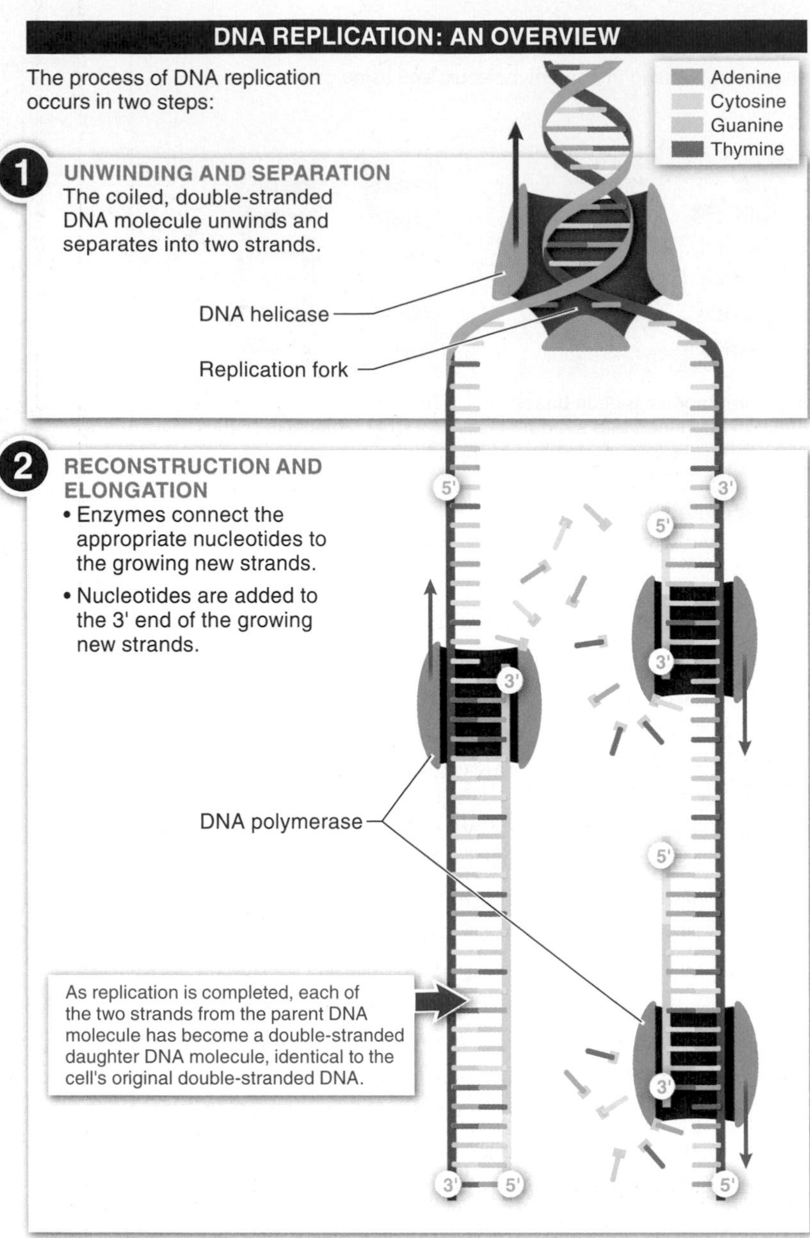

DNA REPLICATION: AN OVERVIEW
The process of DNA replication occurs in two steps:
Adenine
Cytosine
Guanine
Thymine
1
UNWINDING AND SEPARATION
The coiled, double-stranded DNA molecule unwinds and separates into two strands.
DNA helicase
Replication fork
2
RECONSTRUCTION AND ELONGATION
• Enzymes connect the appropriate nucleotides to the growing new strands.
• Nucleotides are added to the 3' end of the growing new strands.
DNA polymerase
As replication is completed, each of the two strands from the parent DNA molecule has become a double-stranded daughter DNA molecule, identical to the cell's original double-stranded DNA.
5'
3'
5'
3'
3'
5'
5'
3'
3'
5'
5'

FIGURE 8-9 **How DNA replicates: unwinding and rebuilding.**

2. Reconstruction and elongation. At the replication fork, a group of several proteins, called a replication complex, binds to each of the exposed strands. This complex includes the enzyme **DNA polymerase,** which adds DNA nucleotides with bases that are complementary to the bases on each exposed strand. Replication proceeds in both directions at once from each origin of replication.

DNA replication occurs in all types of cells, somatic and reproductive, before cell division. The result of replication is two double-stranded DNA molecules that carry virtually the same genetic information. This makes it possible for the somatic cells of the body that are produced by cell division to have a virtually identical genetic composition. Note, though, that the daughter DNA molecules are not completely identical. Replication is accompanied by a very low rate of errors—about one error per several billion base pairs in eukaryotes.

DNA Proofreading and Error Correction A variety of mutations (such as mismatched bases and added or deleted segments) can occur during replication (or when DNA is damaged by some external source, such as X rays). Because DNA polymerases perform proofreading functions as well as excision and repair functions, many of these errors are caught and repaired during or after replication or whenever they occur.

Q Errors sometimes occur when DNA duplicates itself. Why might that be a good thing?

Not all errors are corrected, however. And if an error remains, the sequences in a replicated DNA molecule (including the genes) can be different from those in the parent molecule. A changed sequence may ultimately lead to the production of a different protein. When an error is introduced into the DNA of a gamete-producing cell (i.e., one that divides to make sperm or eggs), a new gene can sometimes enter a population in the next generation and be acted on by evolution. We explore the relationship between mutation and evolution in more detail in Chapter 10.

TAKE HOME MESSAGE 8.4

» Every time a cell divides, that cell's DNA must first duplicate itself so that each of the two new daughter cells has all the genetic material of the original parent cell. The process of DNA duplication, called replication, is catalyzed by several different enzymes and occurs in two steps: unwinding and separation of the two strands, then reconstruction and elongation of the new, complementary strands. The end result is two double-stranded DNA molecules that carry virtually the same genetic information as the parent DNA. Although enzymes proofread and repair DNA during and after replication, some errors may remain.

8.5–8.8

Mitosis replaces worn-out old cells with fresh new duplicates.

For most of the cell cycle, chromosomes resemble a plate of spaghetti more than the familiar condensed, replicated X shape seen in photos.

8.5 Most cells are not immortal: mitosis generates replacements.

Q What is dust? Why is it your fault?

Look around your room. Dust is everywhere. What is it? It is primarily dead skin cells. In fact, you (and your friends and family) slough off millions of dead skin cells each day—yet your skin doesn't disappear. Why not? Because your body replaces the sloughed-off cells through the process of mitosis.

Mitosis has just one purpose: to enable existing cells to generate new, genetically identical cells. There are two different reasons for this need (FIGURE 8-10).

1. Growth. During development, organisms get bigger. Growth happens in part through the creation of new cells. A plant root, for example, may grow half an inch per day. The root is one of the fastest-growing parts of a plant.

2. Replacement. Cells must be replaced when they die. The wear and tear that come from living can physically damage cells. The daily act of shaving, for example, damages thousands of cells on a man's face. Fortunately, the layers under assault during shaving are made up primarily of dead cells. These dead cells help protect us from infection and reduce the rate at which the

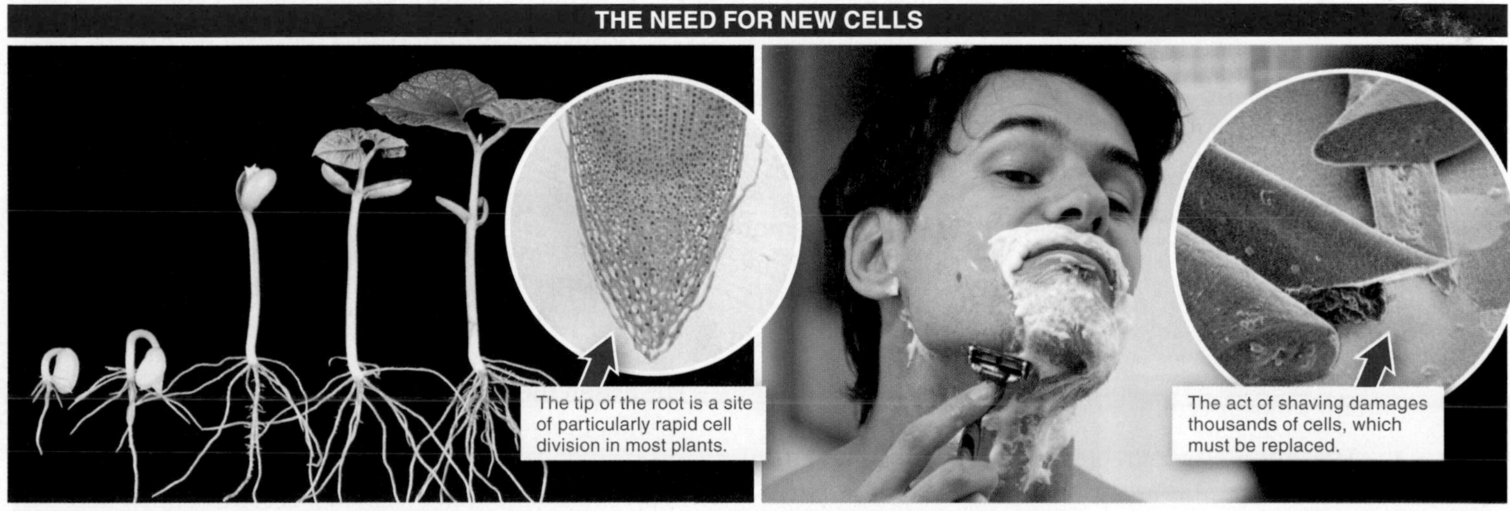

THE NEED FOR NEW CELLS
The tip of the root is a site of particularly rapid cell division in most plants.
The act of shaving damages thousands of cells, which must be replaced.

FIGURE 8-10 **Reasons for mitosis.** Mitosis is important in an organism's growth and the replacement of cells.

underlying living cells dry out. The cells living beneath the layers of dead cells are produced at a high rate by mitosis.

Every day, a huge number of cells in an organism must be replaced by mitosis. In humans, this number is in the billions. Nearly all the somatic cells of the body—that is, all cells except sperm- and egg-producing cells—undergo mitosis. There are a few notable exceptions, as we've already seen. Heart muscle cells and most neurons, in particular, do not seem to divide, or, if they do divide, they do it very slowly. (We don't know why this is so.)

The rate at which mitosis occurs in animals varies dramatically for different types of cells. The most rapid cell division takes place in the bone marrow as red blood cells are produced, as well as in the cells lining various tissues and organs. The average red blood cell, for example, circulates for two to four months and then must be replaced. The cells lining the intestines are replaced about every three weeks. Hair follicles, too, contain some of the most rapidly dividing cells.

Some cells die in a planned process of cell suicide called **apoptosis** (A-pop-TOE-siss). This seemingly counterproductive strategy takes place in cells that are likely to accumulate significant genetic damage over time and are therefore at high risk of becoming cancer cells (a process described later in this chapter). Cells targeted for apoptosis include many of the cells lining the digestive tract and cells in the liver, two locations where cells are almost constantly in contact with harmful substances.

TAKE HOME MESSAGE 8.5

» Mitosis enables existing cells to generate new, genetically identical cells. This makes it possible for organisms to grow and to replace cells that die.

8.6 Overview: mitosis leads to duplicate cells.

For mitosis to begin, the parent cell replicates its DNA, creating a duplicate copy of each chromosome. Once this task is completed, mitosis can proceed. During mitosis, newly duplicated chromosomes are separated into identical sets in two separate nuclei. After mitosis, the cytoplasm and the rest of the cell are divided into two cells that pinch apart (FIGURE 8-11). Where once there was one parent cell, now there are two identical daughter cells (FIGURE 8-12).

Throughout the cell cycle except for mitosis, chromosomes are uncoiled and spread out in a diffuse way, like a mass of spaghetti. And because they are so stretched out, they are not dense enough to be visible under a light microscope. However, just before mitosis begins, two important events occur.

1. The chromosomes replicate, becoming two identical linear DNA molecules. The two DNA molecules are held together at a region (on each molecule) called the **centromere.** Throughout mitosis, until the centromeres separate, each of the identical DNA molecules is called a **chromatid;** together, the two are called **sister chromatids** (FIGURE 8-13).
2. The sister chromatids begin the process of **condensation,** in which they coil tightly and become compact—in contrast to the uncondensed and tangled state of the chromosomes prior to replication, during most of interphase.

Q Animal chromosomes are linear. Why do they look like the letter X in pictures?

When the sister chromatids condense, they look like the letter X. Nevertheless, chromosomes are not X-shaped. They are *linear.* The reason the genetic material appears as X-shaped chromosomes in most photos is that once it has condensed in preparation for cell division, it is coiled tightly and only then is thick enough to be seen and photographed under the light microscope. And so the X-shaped image is not simply a chromosome, but a *replicated* chromosome.

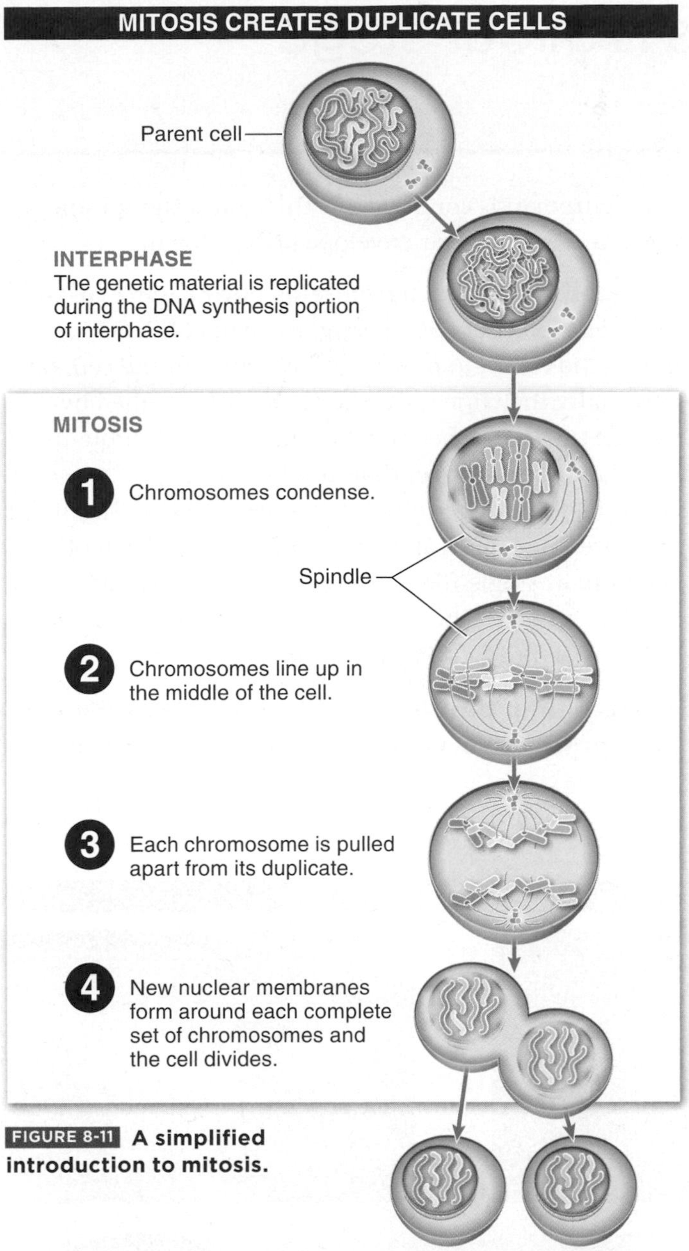

MITOSIS CREATES DUPLICATE CELLS

Parent cell

INTERPHASE
The genetic material is replicated during the DNA synthesis portion of interphase.

MITOSIS

1. Chromosomes condense.

Spindle

2. Chromosomes line up in the middle of the cell.

3. Each chromosome is pulled apart from its duplicate.

4. New nuclear membranes form around each complete set of chromosomes and the cell divides.

FIGURE 8-11 **A simplified introduction to mitosis.**

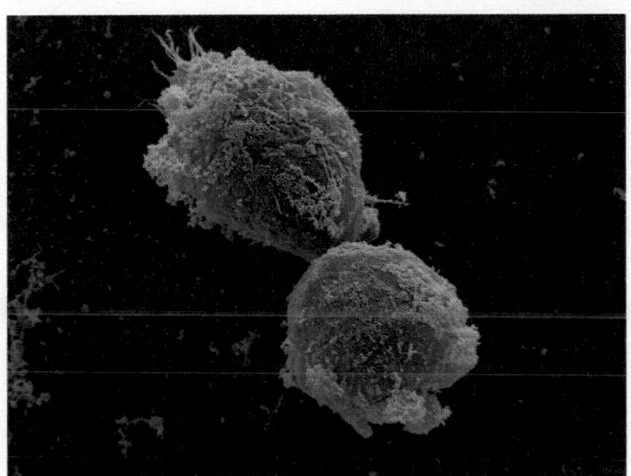

FIGURE 8-12 **These two human embryonic stem cells have just completed mitosis.**

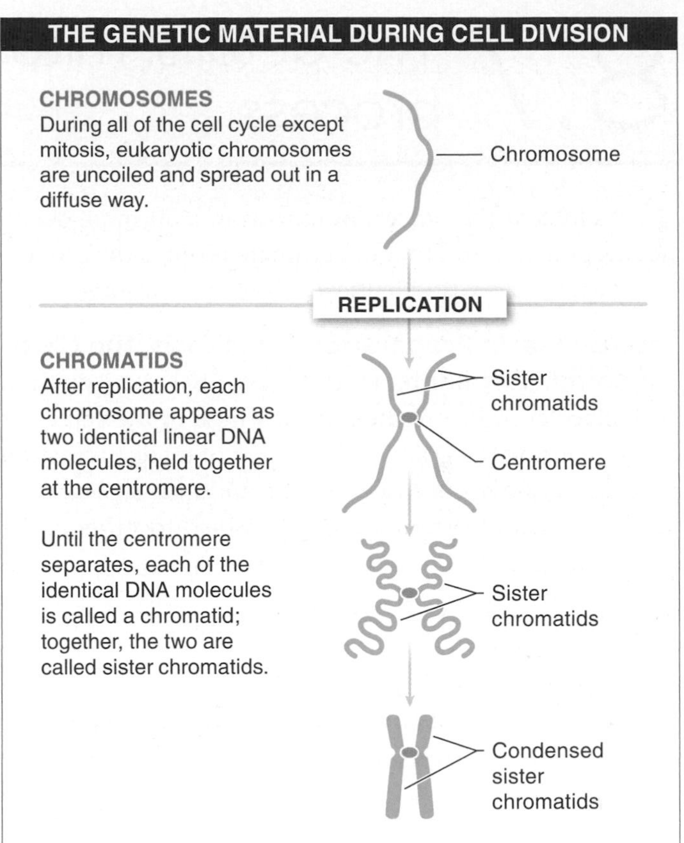
THE GENETIC MATERIAL DURING CELL DIVISION
CHROMOSOMES
During all of the cell cycle except mitosis, eukaryotic chromosomes are uncoiled and spread out in a diffuse way.
Chromosome
REPLICATION
CHROMATIDS
After replication, each chromosome appears as two identical linear DNA molecules, held together at the centromere.
Sister chromatids
Centromere
Until the centromere separates, each of the identical DNA molecules is called a chromatid; together, the two are called sister chromatids.
Sister chromatids
Condensed sister chromatids

FIGURE 8-13 **From one chromosome: two sister chromatids.**

Because the sister chromatids need space to separate, the membrane around the nucleus is dismantled and disappears early in mitosis. At the same time, a structure called the **spindle** is assembled. The spindle is made mostly of hollow tubes of protein called **microtubules,** which are part of the cell's cytoskeleton. They resemble a group of threads stretching across the cell between its two ends, or poles. In animal cells, the threads originate and spread out at each pole from structures called **centrosomes,** which contain a pair of **centrioles** and a mass of proteins that anchor the microtubules. These threads (known as **spindle fibers**) attach to the centromeres and pull the sister chromatids to the middle of the cell. During mitosis, they'll eventually pull the chromatids apart as cell division proceeds.

TAKE HOME MESSAGE 8.6

» Mitosis is the process by which cells duplicate themselves. Mitosis follows chromosome replication and leads to the production of two daughter cells from one parent cell.

8.7 The details: mitosis is a four-stage process.

Let's look at the process of mitosis in a bit more detail, keeping in mind that the ultimate result is the production of two cells with identical sets of chromosomes.

Interphase: In Preparation for Mitosis, the Chromosomes Replicate Processes essential to cell division take place even before the mitotic phase of the cell cycle begins. During the DNA synthesis part of interphase, sister chromatids are formed as every chromosome replicates itself. Each pair of sister chromatids is held together at the centromere.

Mitosis The actual process of cell division occurs in four stages (FIGURE 8-14).

1. Prophase: following replication, the sister chromatids condense. Prophase begins when the sister chromatids condense. At this point, the spindle forms and the nuclear envelope breaks down.

2. Metaphase: the chromatids congregate at the cell center. After condensing, the pairs of sister chromatids seem to move aimlessly around the cell, but eventually they line up at the cell's center, pulled by spindle fibers attached to a disk-like group of proteins, called a **kinetochore,** that develops on the centromeres. By the end of metaphase, all the chromatid pairs are lined up in an orderly fashion, straddling the center in a "single-file" congregation that is called the metaphase plate. The chromatids are at their most condensed during this part of mitosis.

3. Anaphase: the chromatids separate and move in opposite directions. In anaphase, the spindle

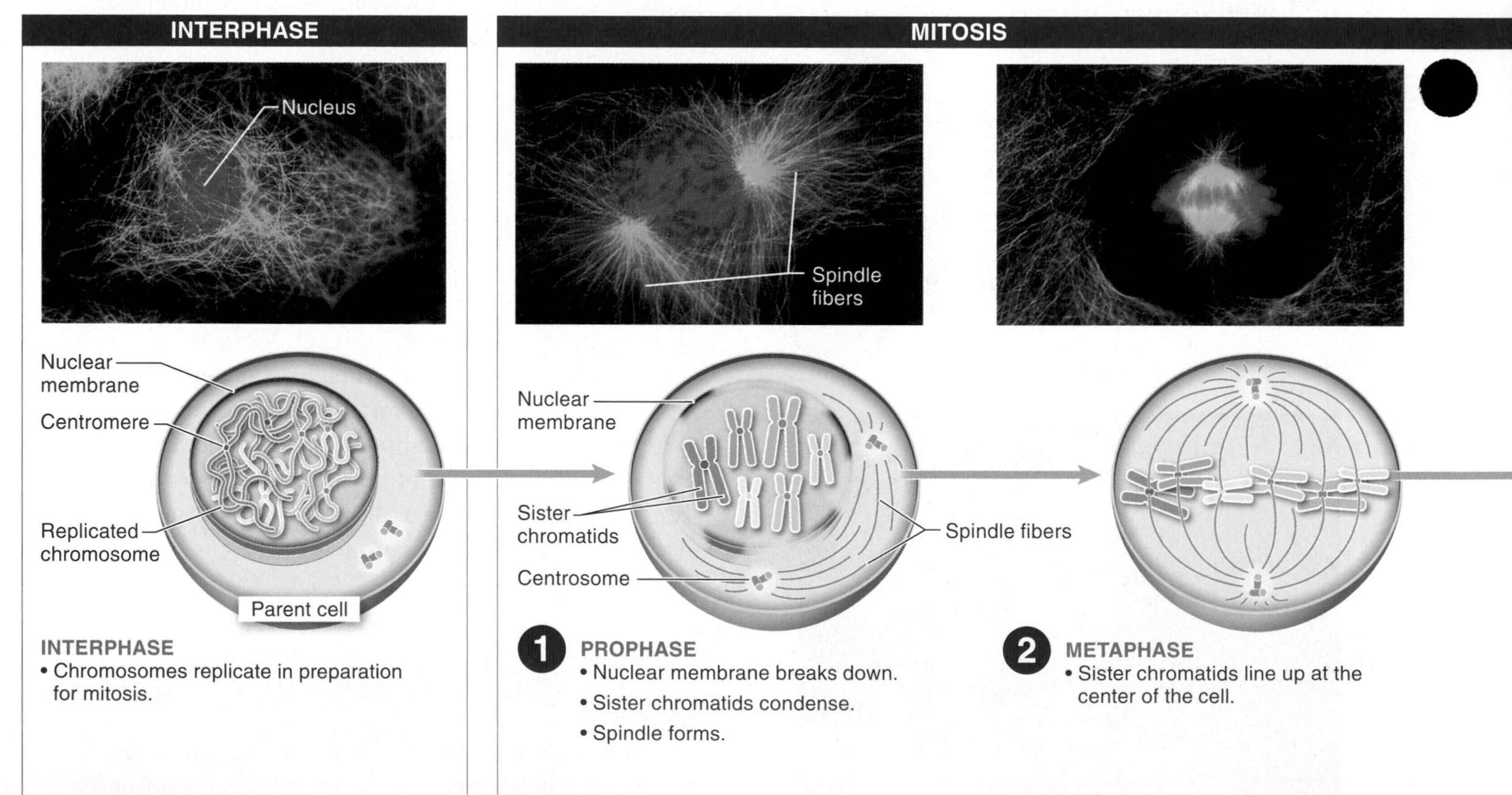

FIGURE 8-14 Mitosis: cell duplication, step by step.

microtubules attached to the centromeres begin pulling each of the chromatids in a sister chromatid pair toward opposite poles of the cell. In each pair of sister chromatids, the centromeres separate as one DNA molecule is pulled in one direction and the other, identical DNA molecule is pulled in the opposite direction. At the end of anaphase, one full set of chromosomes is at one end of the cell and another, identical full set is at the other end. These chromosome sets will eventually occupy the nucleus of each of the two new daughter cells that result from the cell division.

4. Telophase: new nuclear membranes form around the two complete chromosome sets. With two full, identical sets of chromosomes collected at either end of the cell, the parent cell is prepared to divide into two genetically identical cells. In this last stage, called telophase, the chromosomes begin to uncoil, fading from view, and nuclear membranes reassemble.

The process of mitosis is generally accompanied by cytokinesis, which typically begins during telophase. During cytokinesis, the cell's cytoplasm is divided into approximately equal parts and the cell divides, with some of the organelles going to each of the two new cells. When cytokinesis is complete, the two new daughter cells, each with an identical nucleus containing identical genetic material, enter interphase and begin the business of being cells.

Usually, mitosis occurs without errors and only when needed. In rare cases, however, something goes wrong that can cause cell division to proceed unchecked. In such cases, cancer can arise.

TAKE HOME MESSAGE 8.7

» **The ultimate result of mitosis and cytokinesis is the production of two genetically identical cells.**

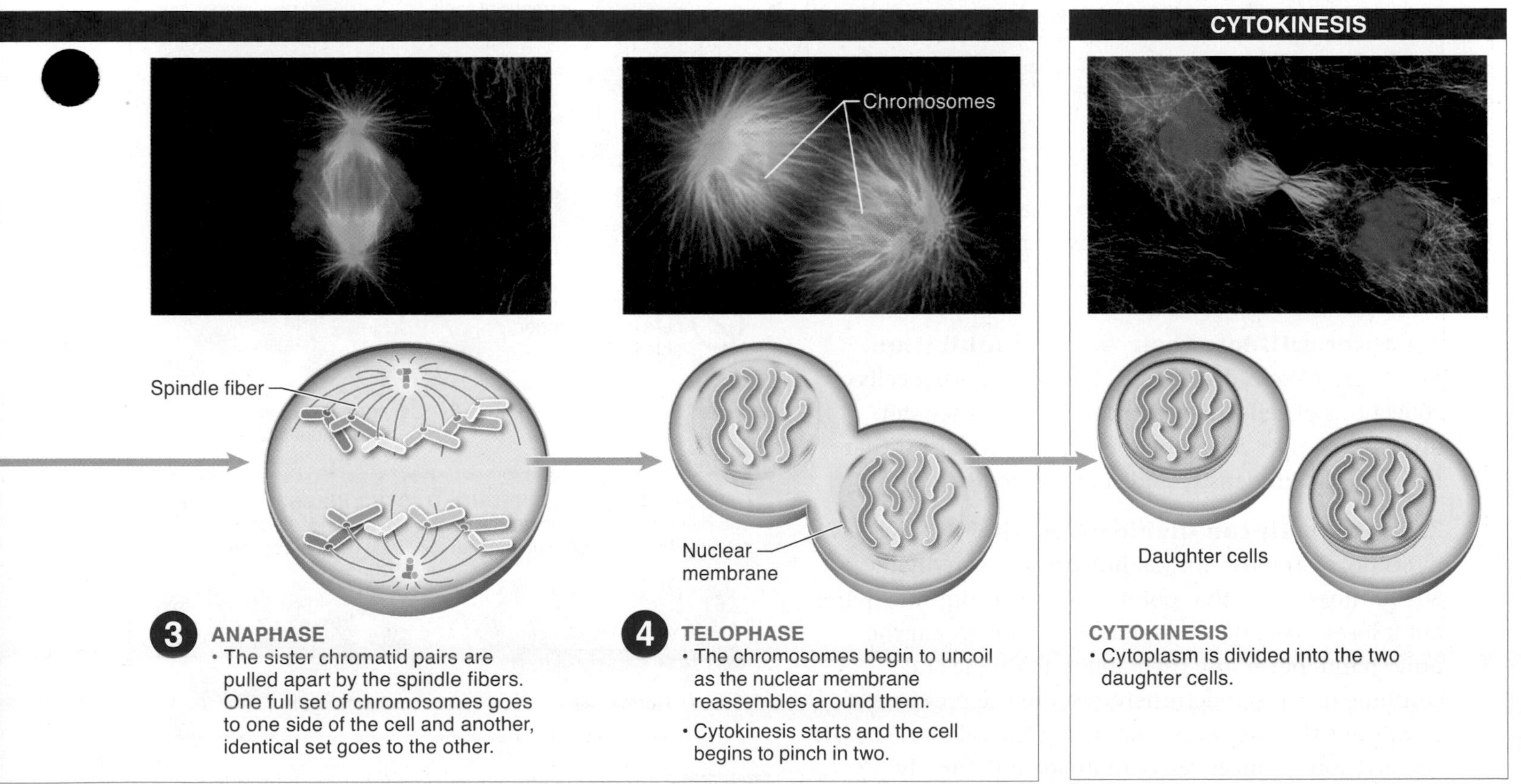

CYTOKINESIS
Chromosomes
Spindle fiber
Nuclear membrane
Daughter cells
3 ANAPHASE
• The sister chromatid pairs are pulled apart by the spindle fibers. One full set of chromosomes goes to one side of the cell and another, identical set goes to the other.
4 TELOPHASE
• The chromosomes begin to uncoil as the nuclear membrane reassembles around them.
• Cytokinesis starts and the cell begins to pinch in two.
CYTOKINESIS
• Cytoplasm is divided into the two daughter cells.

8.8 Cell division out of control may result in cancer.

Too much of a good thing can be bad. **Cancer** is characterized by unrestrained cell growth and division that can damage adjacent tissues. Some cancers can metastasize, or spread to other locations in the body. Cancer can cause serious health problems and is the second leading cause of death in the United States, responsible for more than 20% of all deaths. Only heart disease causes more deaths.

Cancer occurs when some disruption of the DNA in a normal cell interferes with the cell's ability to regulate cell division. DNA disruption can be caused by chemicals that mutate DNA or by sources of high energy such as X rays, the sun, or nuclear radiation. Cancer can even be caused by some viruses. Whatever the cause, once a cell loses control over its cell cycle, cell division can proceed unrestrained (FIGURE 8-15).

> "Cancer cells are those which have forgotten how to die."
>
> — HAROLD PINTER, playwright, from the poem "Cancer Cells," 2002

Cancer cells have several features that distinguish them from normal cells; the three most significant differences are the following:

1. Cancer cells lose their "contact inhibition." Most normal cells divide until they touch other cells or collections of cells (tissues). At that point, they stop dividing. Cancer cells, however, ignore the signal that they are at high density and continue to divide.

2. Cancer cells can divide indefinitely. As we saw in Section 8.1, most normal human cells can divide 80–90 times. After that point, a cell may continue living but it loses the ability to divide. Cancer cells, on the other hand, never lose their ability to divide and continue to do so indefinitely, even in the presence of conditions that would normally halt the cell cycle before cell division. (Cancer cells can divide indefinitely because they are able to rebuild their telomeres following each cell division, as we saw in Section 8.1.)

CANCER CELLS

Cancer cells have several features that distinguish them from normal cells.

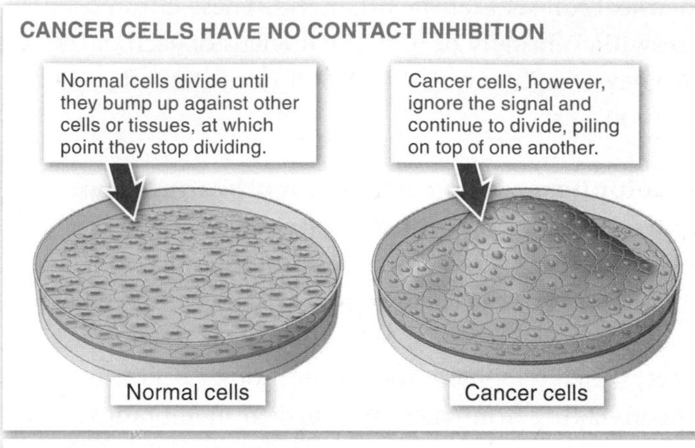

CANCER CELLS HAVE NO CONTACT INHIBITION
Normal cells divide until they bump up against other cells or tissues, at which point they stop dividing.
Cancer cells, however, ignore the signal and continue to divide, piling on top of one another.
Normal cells
Cancer cells

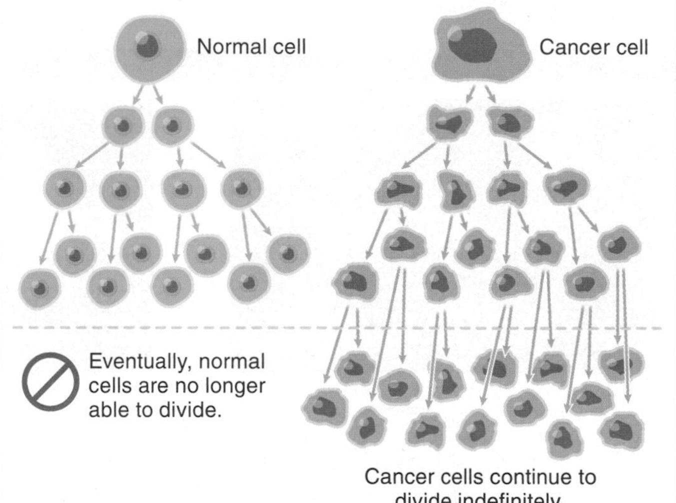

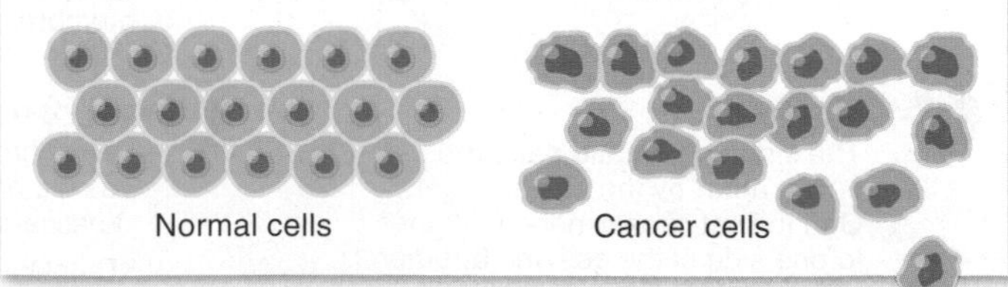

FIGURE 8-15 **Problem cells.** Cancer occurs when there is a disruption in cells' ability to regulate cell division and a reduced adhesion among the cells.

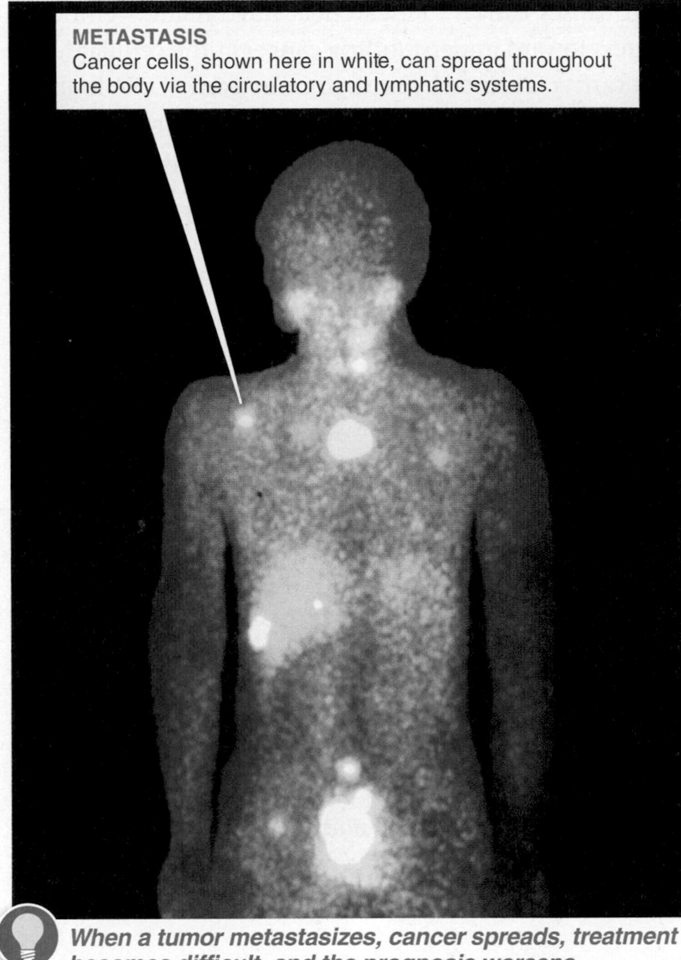

METASTASIS
Cancer cells, shown here in white, can spread throughout the body via the circulatory and lymphatic systems.
When a tumor metastasizes, cancer spreads, treatment becomes difficult, and the prognosis worsens.

FIGURE 8-16 **Spreading cancer.**

3. Cancer cells have reduced "stickiness." Cells are normally held together by adhesion molecules, proteins within cell membranes. And cancer cells, too, usually group together, forming a tumor. But the membranes of cancer cells tend to have reduced adhesiveness, causing them to stick to each other less than do non-cancerous cells.

Tumors caused by excessive cell growth and division are of two very different types: benign and malignant. Benign tumors, such as many moles, are just masses of normal cells that do not spread. They can usually be removed safely without any lasting consequences. Malignant tumors, on the other hand, are the result of unrestrained growth of cancerous cells (FIGURE 8-16). Malignant tumors shed and spread cancer cells, a process called metastasis (meh-TASS-tuh-siss). In this process, cancer cells separate from a tumor and invade the circulatory system and lymphatic pathways, then spread to different parts of the body where they can cause the growth of additional tumors.

Q What is cancer? How does it usually cause death?

How does cancer actually kill the organism? Somewhat surprisingly, it's not because of some toxic action of the cancer cells themselves. As a tumor gets larger, it uses up nutrients and energy, takes up more and more space, and presses against neighboring cells and tissues. Eventually, the tumor may block other cells and tissues from carrying out their normal functions and even kill them. This cell dysfunction or cell death can have disastrous consequences when the job of the affected normal tissue is to control critical processes such as breathing, heart function, or the detoxification processes in the liver.

To treat cancer, the rapidly dividing cells must be removed surgically or killed or, at least, their division slowed down. Currently, the killing and slowing down are done in two ways: by chemotherapy and by radiation.

In chemotherapy, drugs that interfere with cell division are administered, slowing down the growth of tumors. Because these drugs interfere with rapidly dividing cells throughout the body (not just the rapidly dividing cancer cells), they can have very unpleasant side effects. In particular, chemotherapy drugs disrupt normal systems that rely on the rapid and constant production of new cells. For instance, chemotherapy often causes extreme fatigue and shortness of breath because it reduces the rate at which red blood cells are produced, thus limiting the amount of oxygen that can be transported throughout the body. By interfering with the division of bone marrow stem cells, chemotherapy also reduces the production of platelets and white blood cells and thus increases bruising and bleeding, as well as a susceptibility to infection. The rapidly dividing cells within hair follicles are also affected by chemotherapy (and radiation). As a consequence, many people lose their hair when undergoing treatments. It usually grows back, however, when the treatments stop.

Like chemotherapy, radiation works by disrupting cell division. However, radiation therapy is more targeted than the drugs used in chemotherapy. It directs high-energy radiation only at the part of the body where a tumor is located. As with chemotherapy, the radiation process is not perfect, and nearby tissue is often harmed. The effects of chemotherapy and radiation treatment can initially

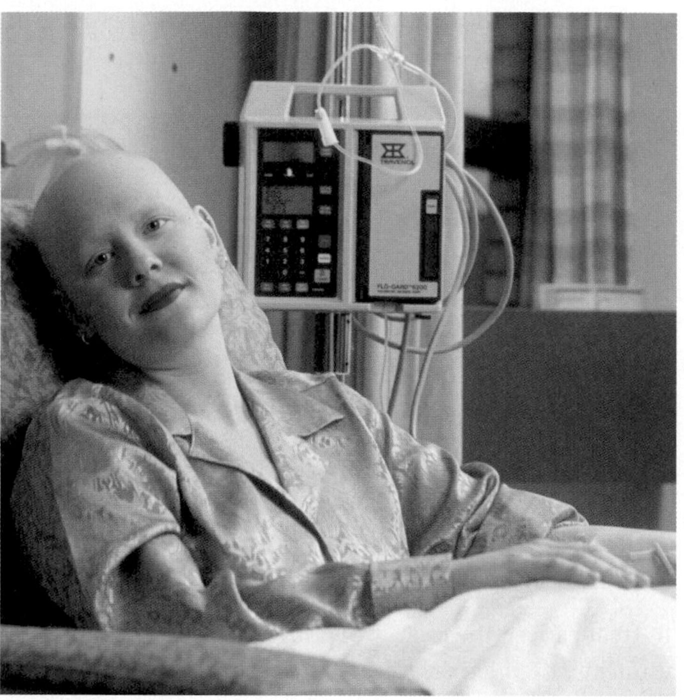

FIGURE 8-17 **Cancer treatments.** The treatments can be as painful and difficult as the disease.

feel worse than the disease itself—especially because patients may be asymptomatic at the time of treatment (FIGURE 8-17).

What causes cancer? Researchers have made significant progress toward understanding cancer at the cellular level. Most cancer is caused by mutations in a cell's DNA that disrupt the normal processes that control and regulate the cell cycle. More specifically, the cancer-causing mutations seem to affect two different types of genes: those that stimulate cell growth and those that restrain it.

Although there is no complete cure for cancer, many potentially successful therapies for treatment and prevention are on the horizon. Extensive research is being conducted on the mechanisms by which genes controlling the cell cycle are damaged in cancer cells, for example, and how such damage might be prevented or reversed.

TAKE HOME MESSAGE 8.8

» Cancer is unrestrained cell growth and cell division, which lead to large masses of cells that may cause serious health problems. It often results from mutations in genes important in controlling the cell cycle, thus reducing the effectiveness of the checkpoints. Treatment focuses on killing or slowing down the fast-growing and dividing cells, usually using chemotherapy and/or radiation.

8.9–8.13

Meiosis generates sperm and eggs and a great deal of variation.

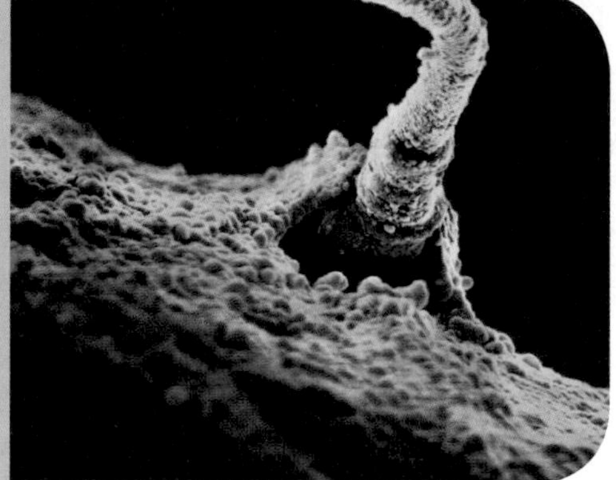

A sperm enters the egg: the moment of fertilization.

8.9 Overview: sexual reproduction requires special cells made by meiosis.

There are two ways in which an organism can reproduce. Many organisms, including bacteria, fungi, and even some plants and animals, undergo **asexual reproduction,** in which a single parent produces identical offspring. Other organisms, including most animals and plants, undergo **sexual reproduction,** in which offspring are produced by the fusion of two reproductive cells in the process of **fertilization.** And some species, particularly among plants, can use both methods. In sexual reproduction, because a combination of DNA from two separate individuals is passed on to offspring, the resulting offspring are genetically different from their parents and, for reasons explained later, from one another.

What would happen if sexually reproducing organisms, humans included, produced reproductive cells through mitosis? Both parents would contribute a full set of genes—that is, 23 *pairs* of chromosomes in humans—to create a new individual, and the new offspring would inherit 46 *pairs* of chromosomes in all. And when that individual reproduced, if he or she contributed 46 pairs of chromosomes and his or her mate also contributed 46 pairs, their offspring would have 92 pairs of chromosomes. Where would it end? The genome would double in size every generation.

In sexually reproducing organisms, the solution to chromosome overload is **meiosis.** This process enables organisms to make gametes, special reproductive cells that have half as many chromosomes as the rest of the cells in the organism's body (the somatic cells). In humans, for example, each gamete cell has only one set of 23 chromosomes, rather than two sets.

In genetics, the term **diploid** refers to cells that have two copies of each chromosome (in humans, two sets of 23 chromosomes, for 46 chromosomes in total), and the term **haploid** refers to cells that have one copy of each chromosome. Thus, somatic cells are diploid, and gametes, the cells produced in meiosis, are haploid.

At fertilization, two haploid cells, each with one set of 23 chromosomes, merge and create a new individual with the proper diploid human genome of 46 chromosomes. And this new individual, through meiosis, will also produce haploid gametes that have only a single set of 23 chromosomes. With sexual reproduction, then, diploid organisms produce haploid gametes that fuse at fertilization to restore the diploid state (FIGURE 8-18). In this way, meiosis maintains a stable genome size in a species.

Although there are some variations on this pattern of alternation between the haploid and diploid states, in most cases, multicellular animals produce haploid cells for reproduction. And after two haploid gametes come together to form a diploid fertilized egg, multiple cell divisions by mitosis produce a diploid, multicellular animal.

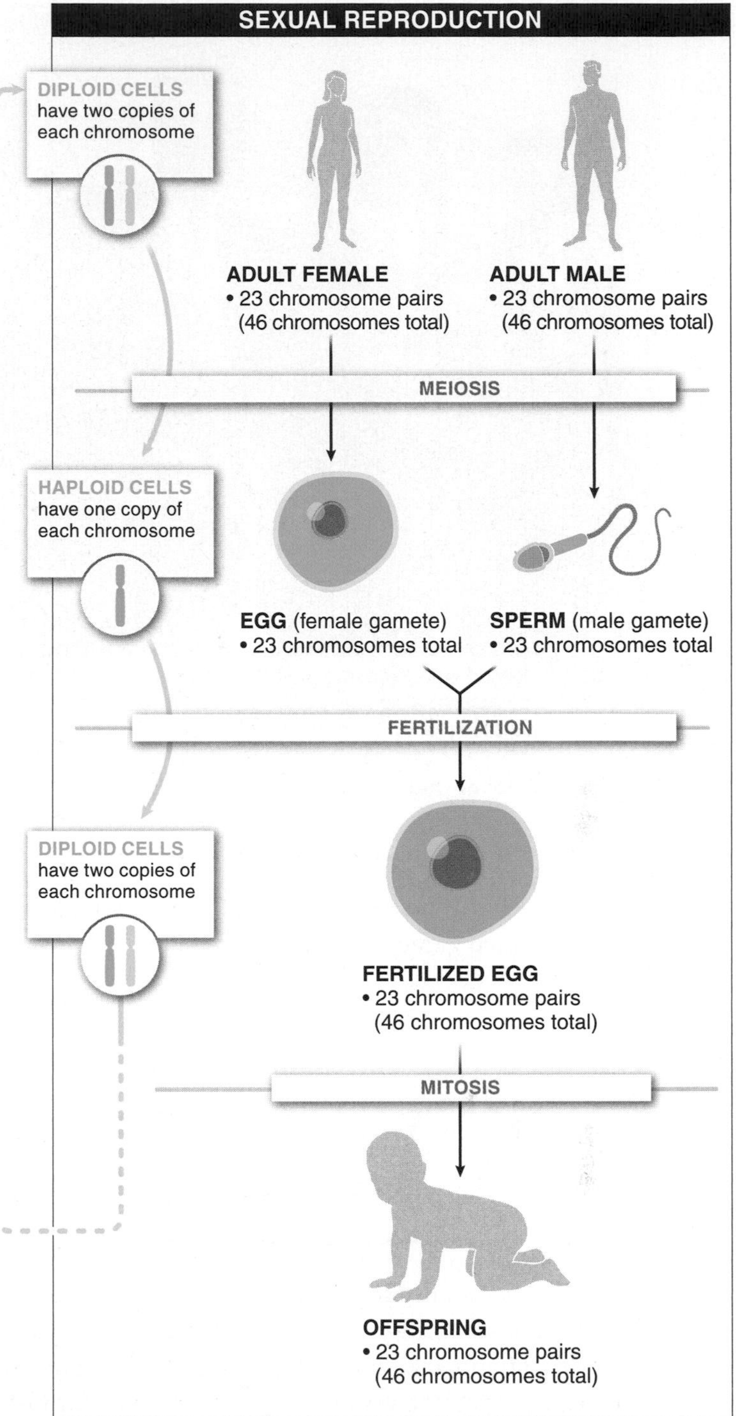

FIGURE 8-18 **Sexual reproduction.** Diploid organisms produce haploid gametes that fuse at fertilization and return to the diploid state.

Meiosis achieves more than just a reduction in the amount of genetic material in gametes. As a diploid individual, you have two copies (i.e., two alleles) of every gene in every somatic cell: one from your mother and one from your father.

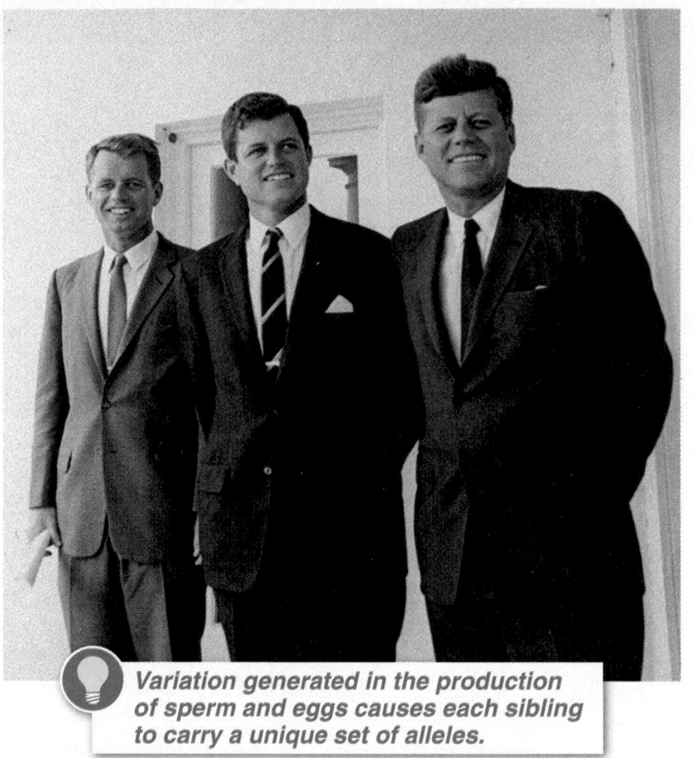

Variation generated in the production of sperm and eggs causes each sibling to carry a unique set of alleles.

FIGURE 8-19 **One mother, one father: varied offspring.**

As a consequence of meiosis, gametes have one allele for each trait rather than two. Which of these two alleles is included in each gamete cannot be predicted.

Each egg or sperm carries a unique set of alleles produced with a varied combination of maternal and paternal alleles. All offspring resemble their parents, yet none resemble them in exactly the same way. This variation among offspring can be seen among the five Jackson brothers in the photo that opens this chapter (and among the Kennedy brothers in FIGURE 8-19).

In all, then, meiosis has two important outcomes:

1. It reduces the amount of genetic material in gametes.
2. It produces gametes that differ from one another with respect to the combinations of alleles they carry.

In the next sections, we examine the exact steps that lead to these outcomes and investigate the consequences of all the variation that meiosis and sexual reproduction can generate.

TAKE HOME MESSAGE 8.9

» Meiosis is the process by which reproductive cells are produced in sexually reproducing organisms. It results in gametes that have only half as much genetic material as the parent cell and that differ from one another in the combinations of alleles they carry.

8.10 Sperm and egg are produced by meiosis: the details, step by step.

Mitosis is an all-purpose process for cell division. It occurs all through the body, all of the time. Meiosis, on the other hand, takes place in just a single place: the **gonads** (the ovaries and testes). And it takes place for just a single reason: the production of gametes (sperm and eggs).

Meiosis starts with a specialized diploid cell found in the gonads that is capable of undergoing meiosis. Thus, for humans, meiosis starts with a cell that has 46 chromosomes. These include a maternal copy and a paternal copy of each of 22 chromosomes—each of these pairs is called a **homologous pair**, or **homologues**—along with two additional chromosomes (one from each parent), called *sex chromosomes* (FIGURE 8-20).

Before meiosis can occur (and just as we saw with mitosis), each of these 46 chromosomes is duplicated during the cell's interphase. This means that when meiosis begins, after replication, each cell has 92 (2 × 46) strands of DNA.

Unlike mitosis, which has only one cell division, cells undergoing meiosis divide twice. In the first division, the homologues separate. In other words, for each of the 23 chromosome pairs, the maternal sister chromatid pairs

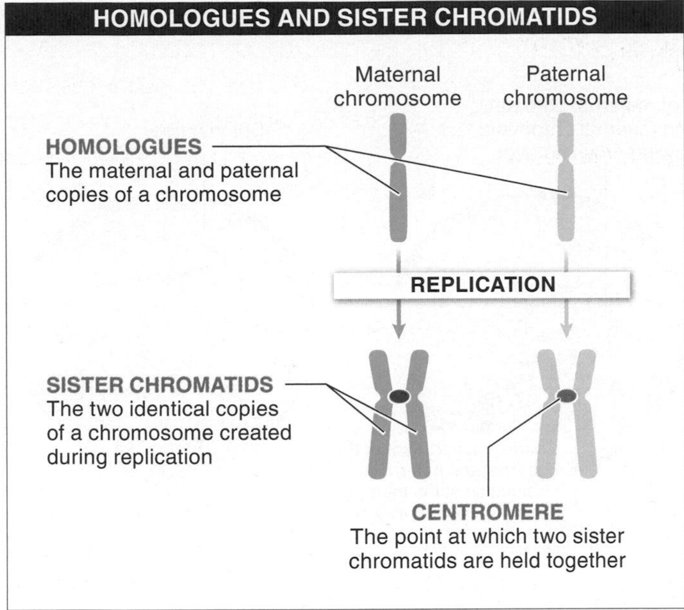

FIGURE 8-20 Chromosome vocabulary: homologues and sister chromatids.

and the paternal sister chromatid pairs separate (randomly, as we'll see) into two new cells. In the second division, each of the two new cells divides again, so that each of the four daughter cells contains a single chromosome from the homologous pair. At the end of meiosis, there are four new cells, each of which has 23 strands of DNA—that is, 23 chromosomes (FIGURE 8-21). Note that, in animals, none of these four cells will undergo any further cell division—they do not become parent cells for a new cycle of cell division.

Interphase: In Preparation for Meiosis, the Chromosomes Replicate Before meiosis begins, every chromosome creates an exact duplicate of itself by replication. The chromosomes that were each a single, long, linear piece of genetic material become a pair of identical long, linear pieces, held together at the centromeres.

Meiosis begins following replication and proceeds through eight steps (FIGURE 8-22).

Meiosis Division I: The Homologues Separate

The first meiotic division takes place in four stages.

1. Prophase I: chromosomes condense and crossing over occurs. This is by far the most complex of all the phases of meiosis. As in mitosis, it begins with all of the replicated genetic material condensing.

As the sister chromatids become shorter and thicker, the homologous chromosomes come together. This is where the process diverges from mitosis. The homologous chromosomes (each of which has become a pair of identical chromatids) line up, touching. Under a microscope, the two homologous pairs of sister chromatids appear as pairs of X's lying one on top of the other.

At this point, the sister chromatids that are next to each other do something that makes every sperm or egg cell genetically unique: they swap little segments of DNA. Some of the genes that you inherited from your mother may get swapped onto the strand of DNA you inherited from your father, and vice versa. This possible outcome of the **crossing over** of portions of the chromatids is called **genetic recombination** (or, more often, just **recombination**). It can take place at several spots (up to dozens) on each chromatid. As a result of recombination, every sister chromatid possesses a unique mixture of your genetic material. Note that crossing over takes place only during the

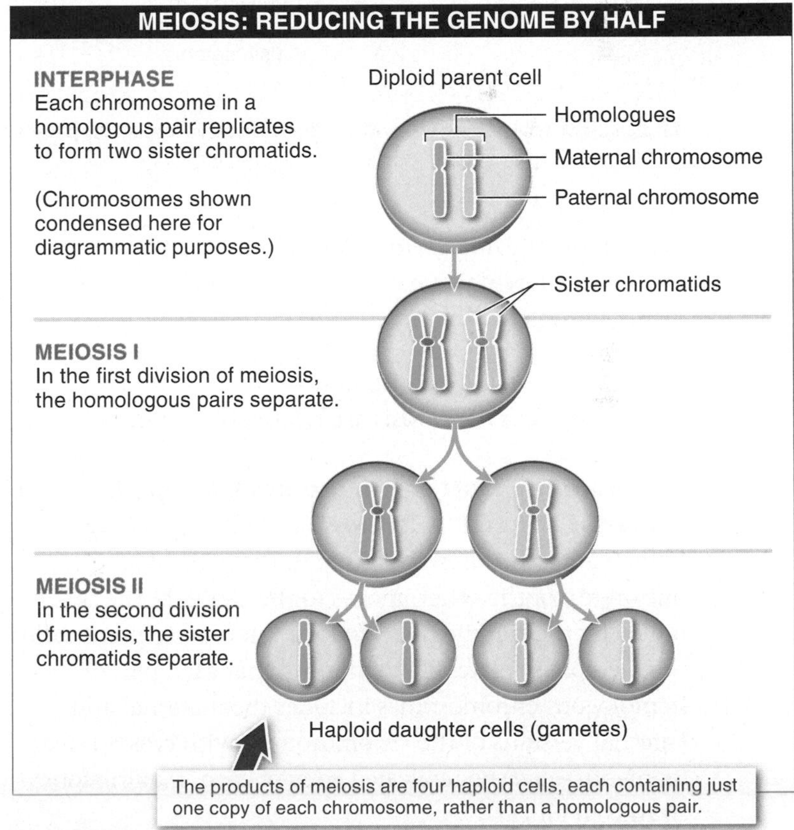

MEIOSIS: REDUCING THE GENOME BY HALF
INTERPHASE
Each chromosome in a homologous pair replicates to form two sister chromatids.
(Chromosomes shown condensed here for diagrammatic purposes.)
Diploid parent cell
Homologues
Maternal chromosome
Paternal chromosome
Sister chromatids
MEIOSIS I
In the first division of meiosis, the homologous pairs separate.
MEIOSIS II
In the second division of meiosis, the sister chromatids separate.
Haploid daughter cells (gametes)
The products of meiosis are four haploid cells, each containing just one copy of each chromosome, rather than a homologous pair.

FIGURE 8-21 Meiosis reduces the genome by half in anticipation of combining it with another genome.

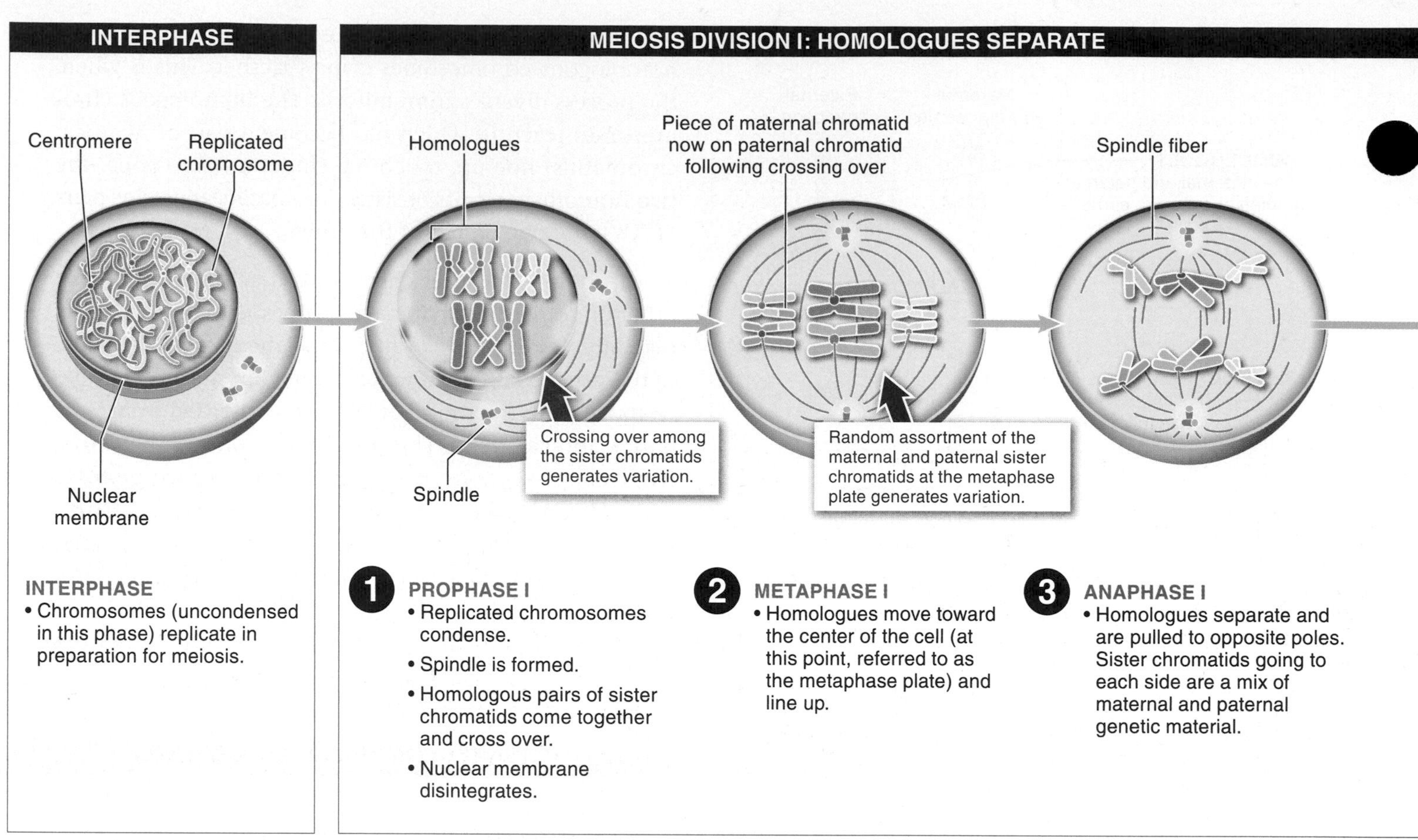

INTERPHASE
MEIOSIS DIVISION I: HOMOLOGUES SEPARATE
Centromere
Replicated chromosome
Nuclear membrane
Homologues
Spindle
Crossing over among the sister chromatids generates variation.
Piece of maternal chromatid now on paternal chromatid following crossing over
Random assortment of the maternal and paternal sister chromatids at the metaphase plate generates variation.
Spindle fiber
INTERPHASE
• Chromosomes (uncondensed in this phase) replicate in preparation for meiosis.
1 PROPHASE I
• Replicated chromosomes condense.
• Spindle is formed.
• Homologous pairs of sister chromatids come together and cross over.
• Nuclear membrane disintegrates.
2 METAPHASE I
• Homologues move toward the center of the cell (at this point, referred to as the metaphase plate) and line up.
3 ANAPHASE I
• Homologues separate and are pulled to opposite poles. Sister chromatids going to each side are a mix of maternal and paternal genetic material.

FIGURE 8-22 Meiosis: generating reproductive cells, step by step.

production of gametes, in meiosis. It does not occur during mitosis. Following crossing over, the nuclear membrane disintegrates.

We explore crossing over in more detail in Section 8.12. The remaining steps of meiosis are relatively straightforward.

2. Metaphase I: all chromosomes line up along the center of the cell. After crossing over, each pair of homologous chromosomes (that is, the pairs of X's lying one on top of the other) moves to the center of the cell, pulled by the spindle fibers to form the arrangement called the metaphase plate. (Keep in mind that each pair of homologous chromosomes includes the maternal and paternal versions of the chromosome—with crossed-over segments—*and* the replicated copy of each, making four strands in all.)

The maternal and paternal sister chromatid pairs line up at the metaphase plate in a random fashion, called **random assortment,** so that the pairs of sister chromatids pulled toward each pole are a mix of maternal and paternal sister chromatids. As a result of random assortment, all the products of meiosis are genetically unique.

Imagine all the different combinations that can occur in a species with 23 pairs of chromosomes. For chromosome pair 1, perhaps the maternal homologue will go to the "top" pole and the paternal to the "bottom" pole. And for pair 2, perhaps the maternal homologue also will go to the top and the paternal to the bottom. For each of the 23 pairs, you cannot predict which will go to the top and which will go to the bottom. There is a *huge* number of possible combinations. (Can you figure out how many?)

3. Anaphase I: homologues are pulled to either side of the cell. This phase is the beginning of the first cell division that occurs during meiosis. In anaphase, the spindle fibers pull the homologues apart toward opposite poles of the cell. One of the homologues (consisting of two sister chromatids) goes to one pole, the other to the opposite pole.

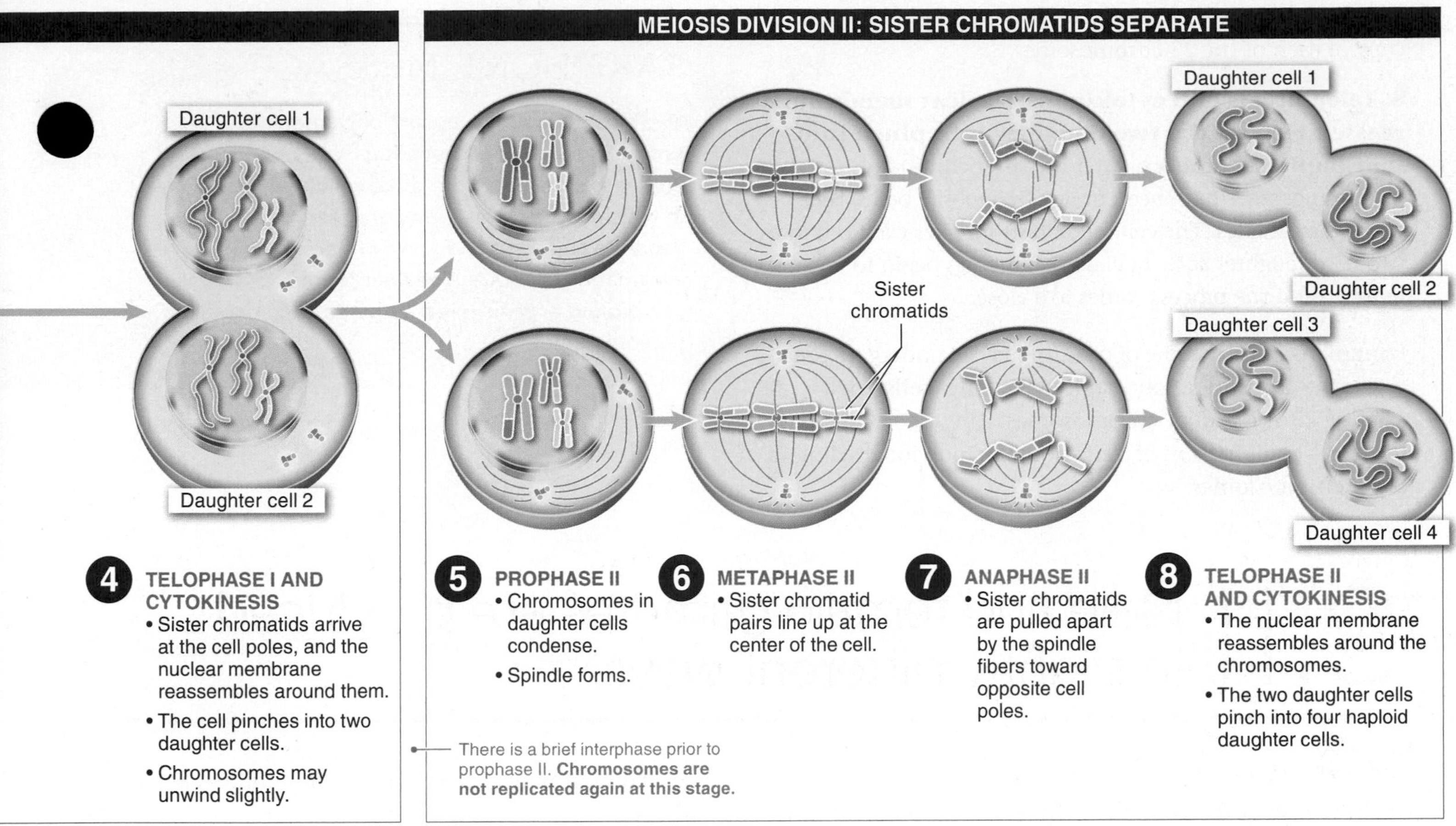

MEIOSIS DIVISION II: SISTER CHROMATIDS SEPARATE
Daughter cell 1
Daughter cell 2
Daughter cell 1
Daughter cell 2
Sister chromatids
Daughter cell 3
Daughter cell 4
4 TELOPHASE I AND CYTOKINESIS
• Sister chromatids arrive at the cell poles, and the nuclear membrane reassembles around them.
• The cell pinches into two daughter cells.
• Chromosomes may unwind slightly.
5 PROPHASE II
• Chromosomes in daughter cells condense.
• Spindle forms.
6 METAPHASE II
• Sister chromatid pairs line up at the center of the cell.
7 ANAPHASE II
• Sister chromatids are pulled apart by the spindle fibers toward opposite cell poles.
8 TELOPHASE II AND CYTOKINESIS
• The nuclear membrane reassembles around the chromosomes.
• The two daughter cells pinch into four haploid daughter cells.
There is a brief interphase prior to prophase II. Chromosomes are not replicated again at this stage.

4. Telophase I and cytokinesis: nuclear membranes reassemble around sets of sister chromatid pairs, and two daughter cells form. After the pairs of chromatids arrive at the two poles of the cell, nuclear membranes re-form, then cytokinesis occurs: the cytoplasm divides, and the cell membrane pinches the cell into two daughter cells. Each daughter cell has a nucleus that contains the genetic material—two sister chromatids for each of the 23 chromosomes in humans.

Meiosis Division II: Separating the Sister Chromatids There is a brief interphase after the first division of meiosis. In some organisms, the DNA molecules (now in the form of chromatid pairs) briefly uncoil and fade from view. In others, the second part of meiosis begins immediately. *It is important to note that in the brief interphase before prophase II, there is no replication of any of the chromosomes.* The second part of meiosis, like the first division of meiosis, is a four-phase process.

5. Prophase II: chromosomes re-condense. The second division of meiosis begins with prophase II. The genetic material in each of the two daughter cells once again coils tightly, making the pairs of chromatids visible under the microscope. (Unlike prophase I, no crossing over occurs during prophase II.)

6. Metaphase II: sister chromatid pairs line up at the center of the cell. In each of the two daughter cells, the sister chromatid pairs (each pair appearing as an X) are pulled to the center of the cell by spindle fibers attached to the centromeres. The congregation of all the genetic material in the center of each daughter cell is visible as a flat metaphase plate.

7. Anaphase II: sister chromatids are pulled to opposite sides of the cell. During this anaphase, the fibers attached to the centromeres begin pulling each chromatid in each of the 23 sister chromatid pairs toward opposite poles of the daughter cell. When anaphase II is finished, each of what will become the four daughter cells (you could think of them as granddaughter cells!) has one copy of each of the 23 chromosomes.

8. Telophase II and cytokinesis: nuclear membranes reassemble and the two daughter cells pinch into four haploid gametes. After the sister chromatids for all 23 chromosomes have been pulled to opposite poles, the cytoplasm divides. The cell membrane pinches each cell into two new daughter cells, nuclear membranes begin to re-form, and the process comes to a close.

In humans, the outcome of one diploid cell undergoing meiosis is the creation of four haploid daughter cells, each with a set of 23 individual chromosomes. These chromosomes contain a combination of traits from the individual's diploid set of chromosomes.

TAKE HOME MESSAGE 8.10

» Meiosis in animals occurs only in gamete-producing cells. It is preceded by DNA replication and consists of two rounds of cellular division, one in which homologous pairs of sister chromatids separate and a second in which the sister chromatids separate. The final product of meiosis in a diploid organism is four genetically dissimilar haploid gametes.

8.11 Male and female gametes are produced in slightly different ways.

For all sexually reproducing organisms there is just one way to distinguish males from females. Regardless of the species, the defining feature is always the same (and if you're thinking of anything visible to the naked eye, you're wrong). When there are two sexes—as in nearly every sexually reproducing animal and plant species—the females are the sex that produces the larger gamete, and the males produce the smaller, more motile, gamete (FIGURE 8-23).

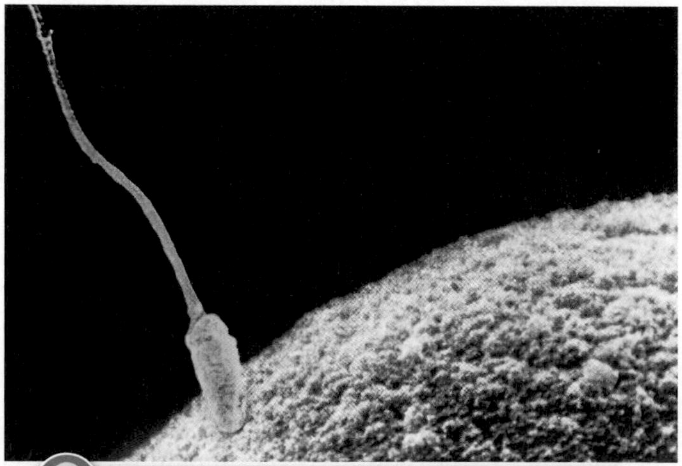

Females produce the larger gamete. Males produce the smaller, more motile gamete.

FIGURE 8-23 **Egg and sperm differences.**

Meiosis works differently in males and females. The female gamete is larger than the male gamete because it has more cytoplasm. During the production of sperm, meiosis takes place just as described in Section 8.10, resulting in four evenly sized cells that become sperm. During egg production, the cell divides in telophase I, and the genetic material is evenly divided, but nearly all of the cytoplasm goes to one of the cells and almost none goes to the other. The smaller cell is called a **polar body** and degrades almost immediately in most animals (although it sometimes goes through a second division, after which, in most animals, both of those polar bodies degrade). Then, in the second meiotic division of the larger cell, there is again an unequal division of cytoplasm.

As in the first division, one of the new cells gets nearly all of the cytoplasm and the other gets almost none, forming another polar body. The net result of meiosis in the production of eggs is one large egg with lots of cytoplasm and two or three small polar bodies with very little cytoplasm that degrade and never function as gametes (FIGURE 8-24).

Ultimately, whether egg or sperm, each gamete ends up with just one copy of each chromosome. That way, the fertilized egg that results from the fusion of sperm and egg carries two complete sets of chromosomes, so the developing individual will be diploid. The extra cytoplasm carried by the egg contains a large supply of nutrients and other chemical resources to help with initial development of the organism following fertilization.

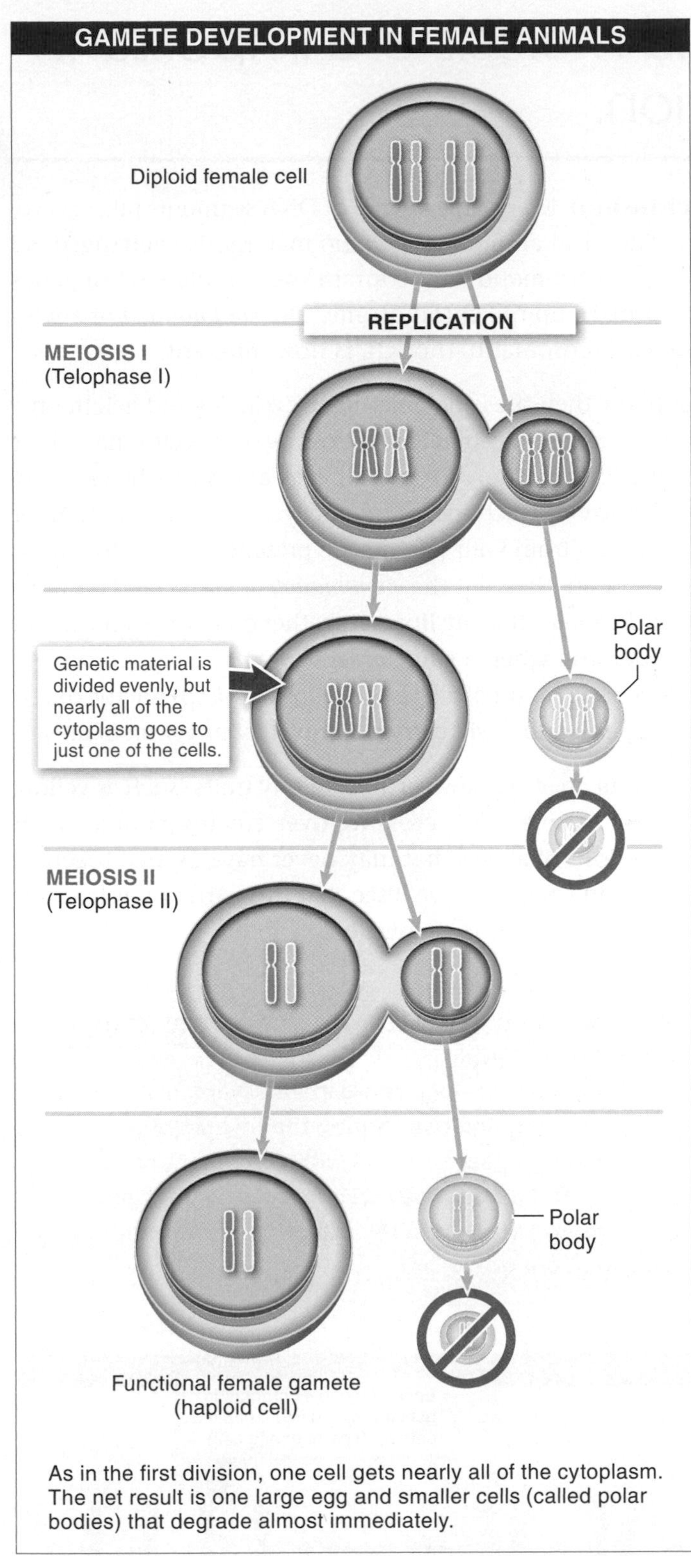

GAMETE DEVELOPMENT IN FEMALE ANIMALS
Diploid female cell
REPLICATION
MEIOSIS I
(Telophase I)
Polar body
Genetic material is divided evenly, but nearly all of the cytoplasm goes to just one of the cells.
MEIOSIS II
(Telophase II)
Polar body
Functional female gamete (haploid cell)
As in the first division, one cell gets nearly all of the cytoplasm. The net result is one large egg and smaller cells (called polar bodies) that degrade almost immediately.

FIGURE 8-24 Unequal distribution of cytoplasm results in one large egg.

TAKE HOME MESSAGE 8.11

» In species with two sexes—including nearly every sexually reproducing plant and animal species—females are the sex that produces the larger gamete, and males produce the smaller gamete. Whether it is male or female gametes that are being produced, each gamete ends up with just one copy of each chromosome.

8.12 Crossing over and meiosis are important sources of variation.

Genetically speaking, there are two ways to create unique individuals. The obvious way is for an organism to carry an allele that is not present in any other individuals. Alternatively—and equally successful in creating uniqueness—an individual can carry a *collection* of alleles that has never before existed in another individual. Both types of novelty introduce important variation into a population of organisms. The process of crossing over, or genetic recombination (FIGURE 8-25), which occurs during prophase I in meiosis, creates a significant amount of the second type of variation.

Let's look at crossing over more carefully. Take, for example, the homologous pair of human chromosome 15. It includes two copies of chromosome 15: one copy from your mother (which you inherited from the egg that was fertilized to create you) and one copy from your father (which you inherited from the sperm that fertilized the egg). Each chromosome in the pair carries the same genes, but because they came from different people, they don't necessarily have the same alleles.

Once the sister chromatids of the homologous chromosome pairs line up in prophase I (so that there are now four chromatids in two pairs lying very close together), regions that are close together can swap segments. A piece of one of the maternal chromatids—perhaps including the first 100 genes on the strand of DNA—may swap places with the same segment in a paternal chromatid. Elsewhere, a stretch of 20 genes in the middle may be swapped from the other maternal chromatid with one of the paternal chromatids. The points at which chromatids exchange genetic material during recombination are called **chiasmata** (*sing.* **chiasma**). Every time a swap of DNA segments takes place, an identical amount of genetic material is exchanged, so all four chromatids still contain the complete set of genes that make up the chromosome. The *combination* of alleles on each chromatid, though, is now different.

Suppose there are genes relating to eye color and height on a particular chromosome. After crossing over, a chromatid that carried instructions for brown eyes and short height may now carry instructions for brown eyes and tall height. All of the alleles from your parents are present on one DNA molecule or another. But the *combination of alleles (and the traits they determine)* that are linked together on a single chromatid is new. And when a gamete, let's say it's an egg, carrying a new combination of alleles is fertilized by a sperm, the developing individual will carry a completely novel set of alleles.

Without creating new versions of any traits (such as yellow eyes or purple hair), crossing over creates gametes with collections of alleles that may never have existed together before. In Chapter 9, we'll see that this variation is tremendously important for evolution.

TAKE HOME MESSAGE 8.12

» Although it doesn't create new versions of any alleles, crossing over during the first prophase of meiosis creates gametes with combinations of alleles (and of the traits they determine) that may never have existed before; this variation is important for evolution.

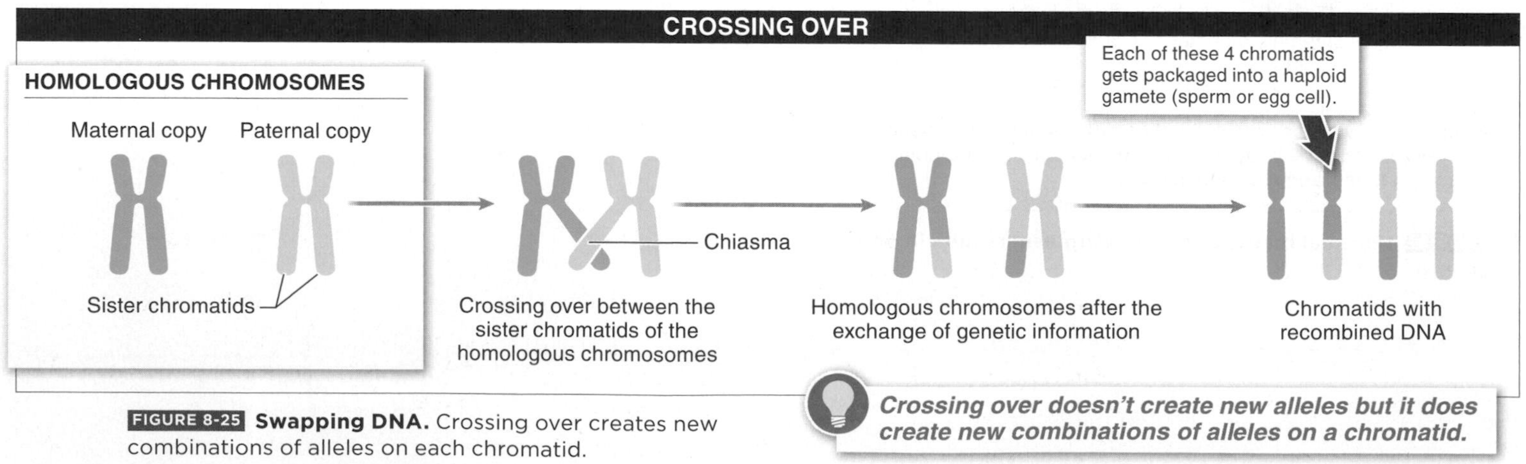

CROSSING OVER
HOMOLOGOUS CHROMOSOMES
Maternal copy
Paternal copy
Sister chromatids
Chiasma
Crossing over between the sister chromatids of the homologous chromosomes
Homologous chromosomes after the exchange of genetic information
Each of these 4 chromatids gets packaged into a haploid gamete (sperm or egg cell).
Chromatids with recombined DNA
Crossing over doesn't create new alleles but it does create new combinations of alleles on a chromatid.

FIGURE 8-25 **Swapping DNA.** Crossing over creates new combinations of alleles on each chromatid.

8.13 What are the costs and benefits of sexual reproduction?

As we've seen, cells and organisms can reproduce through asexual or sexual reproduction. Is one method better than the other? It depends. In fact, the more appropriate question is, what are the advantages and disadvantages of each method and under what conditions do the benefits outweigh the costs?

What Are the Advantages of Sexual Reproduction? Sexual reproduction leads to offspring that are genetically different from one another and from either parent, through three different processes (FIGURE 8-26).

1. **Combining alleles from two parents at fertilization.** First and foremost, with sexual reproduction, a new individual comes from the fusion of gametes from two different individuals. Each of these parents comes with his or her own unique set of genetic material.
2. **Crossing over during the production of gametes.** Crossing over during prophase I of meiosis causes every chromosome in a gamete to carry a mixture of an individual's maternal and paternal genetic material.
3. **Shuffling and reassortment of homologues during meiosis.** When homologues for each chromosome are pulled to opposite poles of the cell during the first division of meiosis (anaphase I), the maternal and paternal homologues are randomly separated. Many different combinations of maternal and paternal homologues could end up in each gamete.

Q Bacteria reproduce asexually, whereas most plants and animals reproduce sexually. Is either method better than the other?

The variability among the offspring produced by sexual reproduction enables populations of organisms to cope better with changes in their environment. After all, if the environment is gradually changing from one generation to the next, the production of many genetically different offspring increases the likelihood that some of the offspring will carry sets of genes particularly suited to the new environment. Over time, populations of sexually reproducing organisms can quickly adapt to changing environments. It's like buying different lottery tickets—the more tickets you buy, the more likely it is that one of them will be a winner.

SOURCES OF GENETIC VARIATION

There are multiple reasons why offspring are genetically different from their parents and from one another.

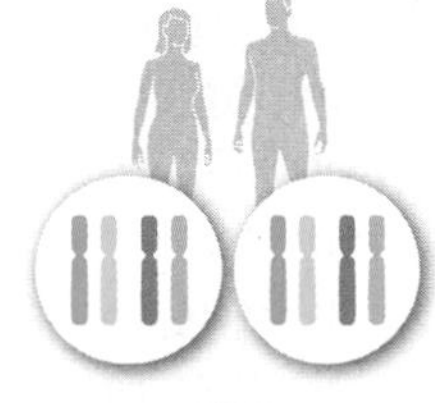

ALLELES COME FROM TWO PARENTS
Each parent donates his or her own set of genetic material.

CROSSING OVER
Crossing over during meiosis produces a mixture of maternal and paternal genetic material on each chromatid.

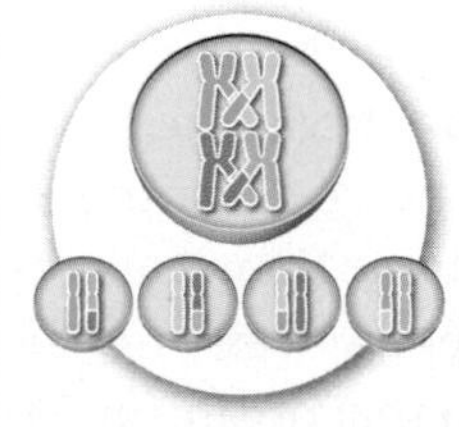

REASSORTMENT OF HOMOLOGUES
The homologues and sister chromatids distributed to each daughter cell during meiosis are a random mix of maternal and paternal genetic material.

FIGURE 8-26 Creating many different combinations of alleles.

FIGURE 8-27 **Sexual reproduction has its risks.** Female black widow spiders attempt to eat the male during or after mating.

What Are the Advantages of Asexual Reproduction? Because asexual reproduction involves only a single individual, it can be fast and easy. Some bacteria can divide, forming a new generation, every 20 minutes (FIGURE 8-28). And for organisms in isolated habitats or those establishing new populations, asexual reproduction can be advantageous as well. Asexual reproduction is efficient, too. Offspring carry all of the genes that their parent carried—they are genetically identical. If the environment is stable, it is beneficial for organisms to produce offspring as similar to themselves as possible.

What Are the Disadvantages of Sexual Reproduction? In the yellow dung fly (*Scathophaga stercoraria*), males sometimes wrestle each other for mating access to a female. The female awaits the outcome of the battle in a pile of dung. Occasionally, females drown in the dung pile as they wait. Males, too, can be at risk during reproduction. This is seen most dramatically in species such as praying mantises and black widow spiders, in which the female will attempt to eat the male (and often succeed in doing so) during or after mating, gaining not just gametes but a nutritious meal as well (FIGURE 8-27)!

Dangers associated with mating are one of the downsides to sexual reproduction. As we see with the dung flies, sex can be a risky proposition because organisms make themselves vulnerable to predation, disease, and other calamities during mating and reproduction. There are other risks, too. First, when an individual reproduces, only half of its offspring's alleles will come from that organism. The other half will come from the other parent. The complex cellular division required for sexual reproduction offers opportunities for mistakes, sometimes leading to chromosomal disorders. With asexual reproduction, an individual produces nearly identical offspring, so there is a very efficient transfer of genetic information from one generation to the next. Second, with sexual reproduction it takes time and energy to find a partner. This is energy that asexual organisms can devote to additional reproduction.

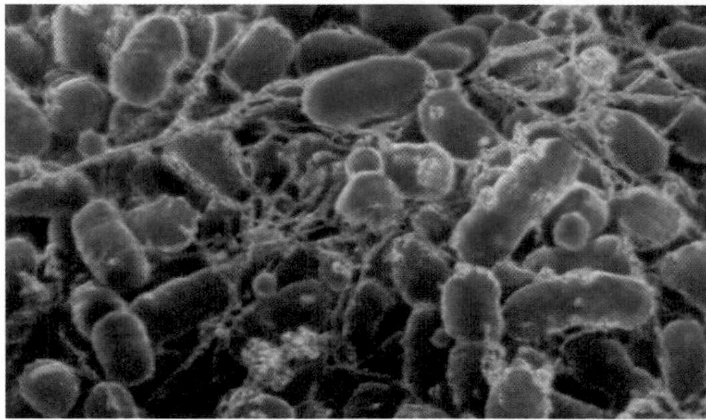

This bacterial colony can reproduce quickly and efficiently. But because the cell division is asexual, it doesn't produce genetic variety to enable adaptation should the environment change.

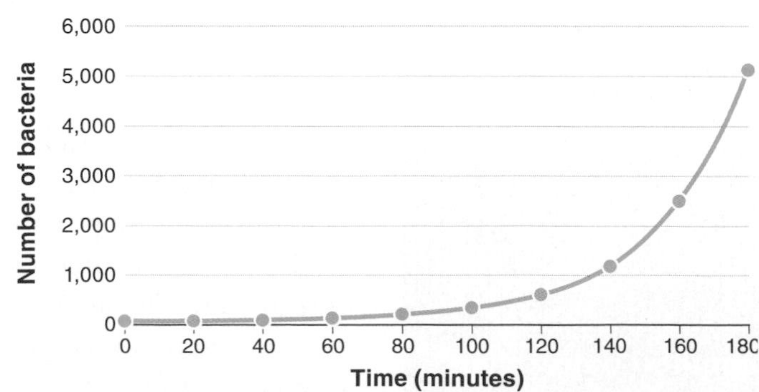

FIGURE 8-28 **Pluses and minuses.**

What Are the Disadvantages of Asexual Reproduction? The downside to asexual reproduction is that the more closely an offspring's genome resembles its parent's, the less likely it is that the offspring will be suited to the environment when it changes.

In the end, we still see large numbers of species that reproduce asexually and large numbers that reproduce sexually. It seems that conditions favoring one or the other occur in the great diversity of habitats in our world. That we see both sexual and asexual reproduction also highlights the recurring theme in biology that there often is more than one way to solve a problem.

TAKE HOME MESSAGE 8.13

» There are two fundamentally different ways that cells and organisms can reproduce: (1) mitosis and asexual reproduction, and (2) meiosis and sexual reproduction. Asexual reproduction can be fast and efficient, but it leads to genetically identical offspring that carry all of the genes that their parent carried, which could be disadvantageous in a changing environment. Sexual reproduction leads to offspring that are genetically different from one another and from either parent, but it takes more time and energy and can be risky.

8.14–8.15

There are sex differences in the chromosomes.

XX and XY: only one of the 23 chromosome pairs determines sex in humans.

8.14 How is sex determined in humans (and other species)?

Q Which parent determines a baby's sex? How?

In humans, the sex of a baby is determined by its father. The sequence of events involved in sex determination is instigated by one special pair of chromosomes, the sex chromosomes, which carry information that directs a growing embryo to develop as a male or as a female.

We've noted that there are 23 pairs of chromosomes in every somatic cell. These can be divided into two different types: 1 pair of sex chromosomes and 22 pairs of non-sex chromosomes. The human sex chromosomes are called the **X and Y chromosomes** (FIGURE 8-29).

How do the X and Y chromosomes differ from the other chromosomes? All of the genetic information is stored on the chromosomes in all the cells of an organism's body. But most of this information is not sex specific—that is, if you are building an eye or a neuron or a skin cell or a digestive

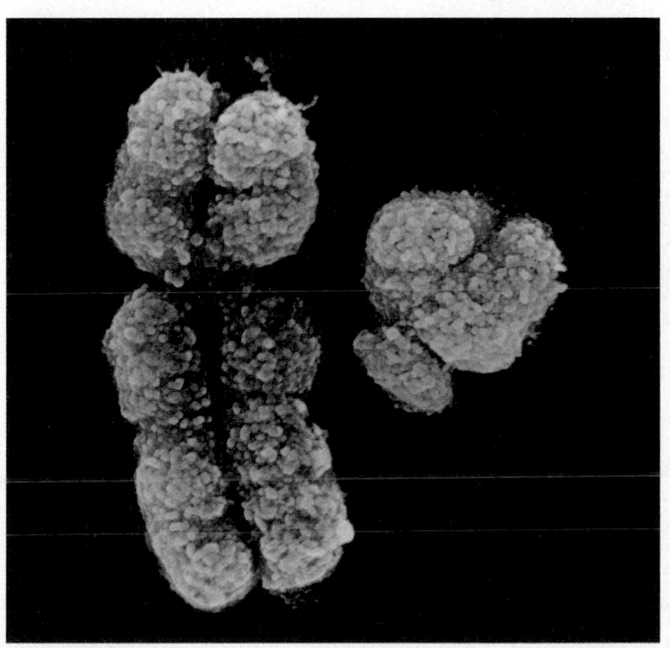

FIGURE 8-29 **X and Y: the human sex chromosomes.**

enzyme, it doesn't matter whether it is for a male or for a female; the instructions are the same for both sexes. Some genetic information, however, instructs the body to develop into one sex or the other. That information is found on the sex chromosomes.

An individual has two copies of all the non-sex chromosomes (called autosomes). One copy is inherited from the mother, one from the father. Individuals also have two copies of the sex chromosomes, but not always two copies of the same kind. Males have one copy of the X chromosome and one copy of the Y chromosome. Females, on the other hand, don't have a Y chromosome but instead have two copies of the X chromosome.

So how does the father determine the sex of the baby? During meiosis in females, the gametes that are produced carry only one copy of each chromosome, including the sex chromosomes. Half of the gametes receive a copy of one of the X chromosomes, and half receive a copy of the other X chromosome. Thus, every egg has an X for its one sex chromosome. During meiosis in males, the sperm that are produced also carry one copy of each chromosome, including the sex chromosomes, but in this case, half of the sperm inherit the X chromosome and the other half inherit the Y chromosome. At fertilization, an egg bearing a single X chromosome is fertilized by a sperm bearing *either* an X chromosome or a Y chromosome. When the sperm carries an X, the baby will have two X chromosomes and will develop as a female. When the sperm carries a Y, the baby will have an X and a Y and so will develop as a male (FIGURE 8-30).

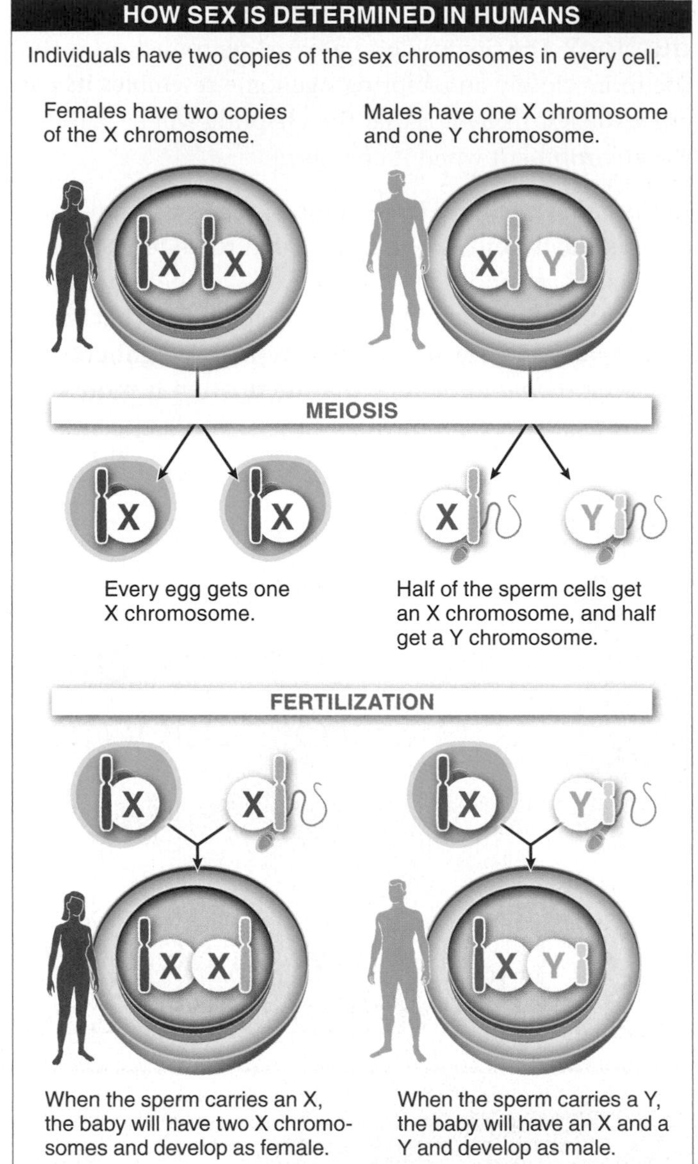

HOW SEX IS DETERMINED IN HUMANS
Individuals have two copies of the sex chromosomes in every cell.
Females have two copies of the X chromosome.
Males have one X chromosome and one Y chromosome.
X X
X Y
MEIOSIS
X X X Y
Every egg gets one X chromosome.
Half of the sperm cells get an X chromosome, and half get a Y chromosome.
FERTILIZATION
X X X Y
X X
X Y
When the sperm carries an X, the baby will have two X chromosomes and develop as female.
When the sperm carries a Y, the baby will have an X and a Y and develop as male.

FIGURE 8-30 Sex determination in humans. The sex of offspring is determined by the father.

Q We know that no genetic information on the Y chromosome is essential for producing a normally functioning human. Why?

Females don't have a Y chromosome in any of their cells, yet they are able to develop and live normal, healthy lives. For this reason, we know that nothing on the Y chromosome is absolutely necessary for the development of a normally functioning human.

Physically, the X and Y chromosomes look very different from each other. The X chromosome is relatively large and carries a great deal of genetic information relating to a large number of non–sex-related traits. The Y chromosome is tiny and carries genetic information about only a very small number of traits. The genetic instructions on the Y chromosome set off a cascade of gene activity, instructing the fetal gonads to develop as testes rather than ovaries. (In the absence of these instructions from a Y chromosome, the fetus develops as a female.) Once this is done, very little additional genetic input from the Y chromosome is necessary. Instead, the hormones produced by the testes or, in the absence of the Y chromosome, by the ovaries generally direct the rest of the body to develop as a male or female. Later, we investigate some of the ramifications of males having two different sex chromosomes.

A variety of other methods of sex determination are used in other species. In most plants, there aren't even distinct male and female individuals. Every individual produces male and female gametes in separate reproductive parts on the same plant. The same goes for all earthworms and garden snails. Such organisms are called **hermaphrodites,** because male and female gametes are produced by a single individual.

Among birds, it is the females that have one copy of two different sex chromosomes, while males have only one type, so the sex of bird offspring is determined by the female. In ants, bees, and wasps, sex is determined by the number of chromosome sets an individual possesses. Males are haploid and females are diploid. A female can fertilize an egg with sperm she has stored after mating, producing a female. Or she can lay the unfertilized egg, which develops into a male. And in some species, sex determination is controlled by the environment. In most turtles, for example, offspring's sex is determined by the temperature at which the eggs are kept. We explore this in the next section.

TAKE HOME MESSAGE 8.14

» A variety of methods are used for sex determination across the world of plant and animal species. In humans, the sex chromosomes carry information that directs a growing fetus to develop as either a male (if a Y chromosome is present) or a female (if no Y chromosome is present). As a consequence, sex determination depends on the sex chromosome inherited from the father.

THIS IS HOW WE DO IT

Developing the ability to apply the process of science

8.15 Can the environment determine the sex of a turtle's offspring?

Unexpected observations may be a sign that our ideas about how the world works are not quite right. As such, they provide a great opportunity for scientists to solve a problem. Figuring out how to approach the problem can be like solving a puzzle, and the solution can have far-reaching implications and unexpected importance.

In 1966, Madeleine Charnier reported a surprising observation in an obscure publication from West Africa. For a lizard species she observed, it seemed that the sex ratio of the offspring produced was influenced by the environment. In warmer temperatures, most of the eggs that hatched contained females. And in cooler temperatures, most of the hatched eggs contained males.

As members of a species in which our sex is determined by the *chromosomes* we inherit from our parents, it isn't surprising that we'd assume that sex determination would be the same for all animals. Charnier's observation, however, which spurred biologists to notice similar patterns of sex determination in other reptiles and some amphibian species, suggested that our understanding of sex determination wasn't quite right.

Can you propose how to test the "incubation temperature determines sex" hypothesis?

To test the effects of incubation temperature on offspring sex, researchers set up incubators at two different temperatures: one cool (25° C) and one warm (30.5° C). They then collected turtle eggs. From each clutch of eggs, they put half of the eggs in the cool incubator and half in the warm incubator. As the eggs hatched, the results were clear: at 25° C, all 210 offspring produced were males; at 30.5° C, all 211 offspring produced were females. The researchers concluded that—at least in the lab—temperature could influence the sex of the offspring.

When it comes to temperature, how does a lab differ from turtles' natural environment?

The turtles' natural habitat is cool at night and warmer in the day. In the initial lab experiment, however, the temperature was kept constant. To better approximate natural conditions, the researchers conducted two additional experiments. First, in the laboratory incubators they created fluctuating temperatures. In one, the temperature fluctuated from 20° to 30° C and back again each day. In the other, the temperature fluctuated

(continued on the following page)

similarly, but between 23° and 33° C. The results were identical to the first experiment: in the cooler incubator, 100% of the offspring hatching were male, and in the warmer incubator, 100% were female.

In the final study, the eggs were incubated in the turtles' natural habitat. Half were buried at a shaded nesting site that received little sun exposure (and rarely exceeded 30° C). The other half were buried at a nesting site that was exposed to the sun (and often exceeded 30° C). Of the 100 eggs hatching at the shaded site, all were males. Of the 127 eggs hatching at the exposed site, 123 were females.

Based on these results, would you be confident that incubation temperature does indeed influence the sex of the offspring? Can you think of any alternative explanations?

How might climate change have an impact on turtle populations?

Climate change is likely to bring an increase of several degrees centigrade in the next 100 years. This increase is likely to have adverse effects on species with temperature-dependent sex determination and may signal more and far greater impacts of climate change on biological systems.

TAKE HOME MESSAGE 8.15

» Observations of some lizard and turtle species reveal that sex determination is influenced by temperature during incubation of the eggs. Both in the lab and in natural habitats, at cooler temperatures more males develop, and at warmer temperatures more females develop. Climate change is likely to have adverse effects on species with temperature-dependent sex determination and may signal far greater impacts of climate change on biological systems.

8.16–8.17

Deviations from the normal chromosome number lead to problems.

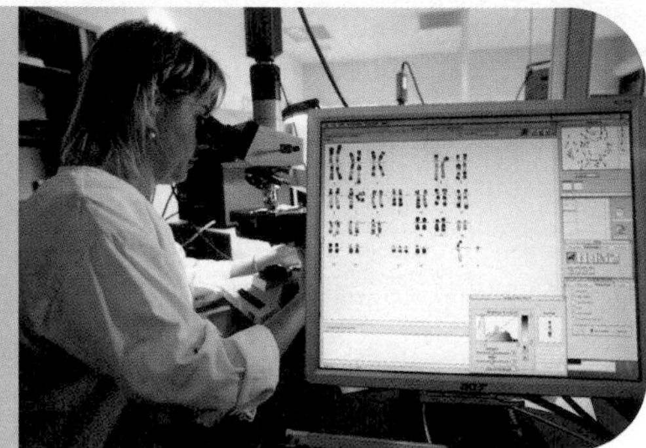
A karyotype displaying Down syndrome.

8.16 Down syndrome can be detected before birth: karyotypes reveal an individual's entire chromosome set.

Reproduction becomes riskier for women as they become older. Increasingly, their gametes contain incorrect numbers of chromosomes or chromosomes that have been damaged (FIGURE 8-31). These problems can lead to adverse effects on the offspring that range from minor to fatal.

Parents can request a quick test for some common genetic problems even before their baby is born. This information is available from an analysis of an individual's **karyotype,** a visual display of the complete set of chromosomes. A karyotype can be made for adults or children, but it is most

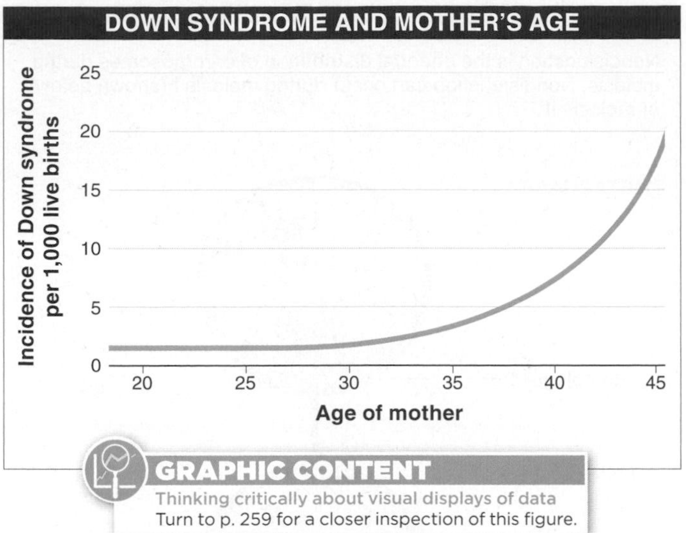

Turn to p. 259 for a closer inspection of this figure.

FIGURE 8-31 Reproduction risks and maternal age. Down syndrome incidence increases as the age of the mother increases.

commonly done for a fetus. A karyotype is a useful diagnostic tool because it can be prepared very early in the fetus's development to assess whether it has an abnormality in the number of chromosomes or in their structure. And because the test shows all of the chromosomes, even the sex hromosomes, it also reveals the sex of the fetus.

Preparing a karyotype takes five steps. (1) Cells are obtained from the individual. (2) The cells are cultured in a test tube with nutrients so that they divide. (3) After a period of cell division, the cells are treated with a chemical that stops them exactly midway through cell division—a time when the chromosomes are coiled thickly and are more visible than usual. (4) The cells are then placed on a microscope slide, and a stain is added that binds to the chromosomes, making them even more visible. (5) Finally, the chromosomes are arranged by size and shape and displayed on a monitor (FIGURE 8-32).

Collecting cells from a fetus is more challenging than taking a small blood sample, as is typical for testing adults and children. Two different methods can be used for collecting fetal cells.

1. Amniocentesis. This procedure can be done approximately three to four months into a pregnancy (FIGURE 8-33). In one quick motion (and without the use of anesthetic), a 3- to 4-inch needle is pushed through the abdomen, through the amniotic sac, and into the amniotic fluid that surrounds and protects the fetus. Using ultrasound for guidance, the doctor aims for a small pocket of fluid as far as possible from the fetus and withdraws about 2 tablespoons of fluid. This fluid contains many cells from the fetus, which can then be used for karyotype analysis. Any chromosomal abnormalities in the fetus will be present in these cells—and many of them can be seen in the karyotype.

2. Chorionic villus sampling (CVS). In this procedure, rather than sampling cells from the amniotic fluid, tissue is removed from the **placenta,** the temporary organ that allows the transfer of gases,

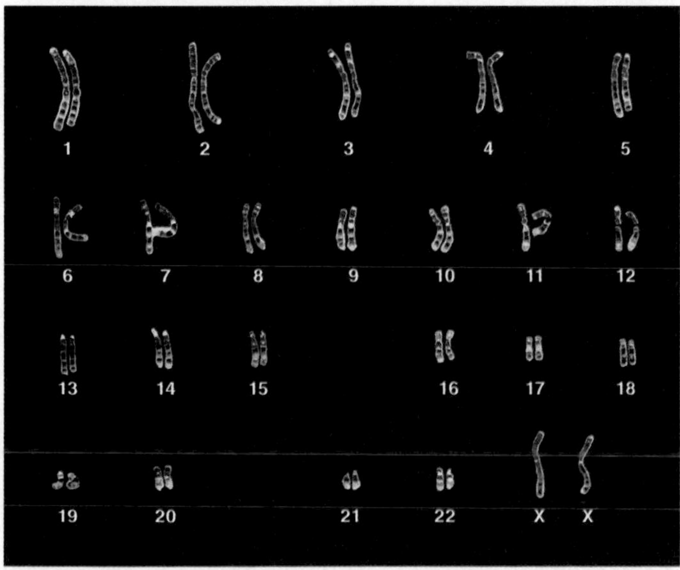

FIGURE 8-32 A human karyotype. A visual display of a complete set of chromosomes.

FIGURE 8-33 Amniocentesis. Fluid surrounding a fetus (and containing some of its cells) is extracted and analyzed to determine the genetic composition of the developing fetus.

nutrients, and waste products between mother and fetus. A needle is inserted either through the abdomen or through the vagina and cervix, again using ultrasound for guidance. Then a small piece of the finger-like projections from the placenta is removed by the syringe. Because much of the placenta develops from the fertilized egg, much of the placenta consists of cells with the same genetic composition as the fetus. The chief advantage of CVS over amniocentesis is that it can be done several weeks earlier in the pregnancy, usually between the 10th and 12th weeks.

The resulting karyotype, whether from amniocentesis or CVS, will reveal whether the fetus carries an extra copy of any of the chromosomes or lacks a copy of one or more chromosomes. Of all the chromosomal disorders detected by karyotyping, Down syndrome is the most commonly observed. Named after John Langdon Down, the doctor who first described it, in 1866, the syndrome is revealed by the presence of an extra copy of chromosome 21 (FIGURE 8-34). (For this reason, the condition that causes Down syndrome is also called "trisomy 21.") Striking about 1 in every 1,000 children born, Down syndrome is characterized by a suite of physical and mental characteristics that includes learning

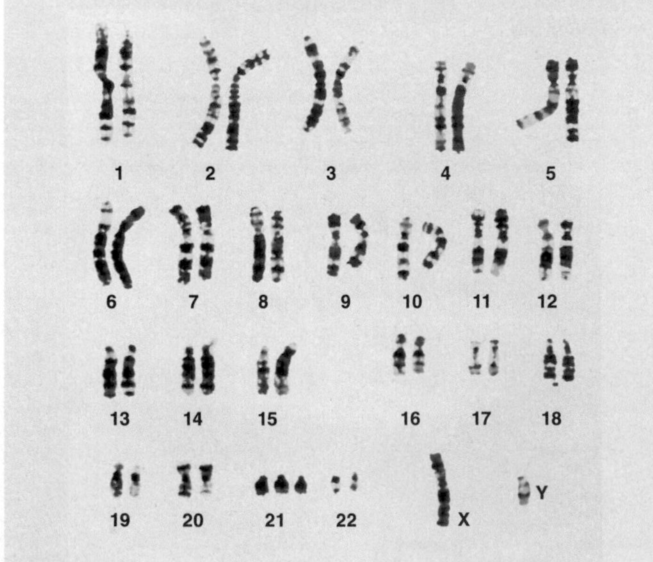

FIGURE 8-34 **Trisomy 21.** An extra copy of chromosome 21 causes Down syndrome.

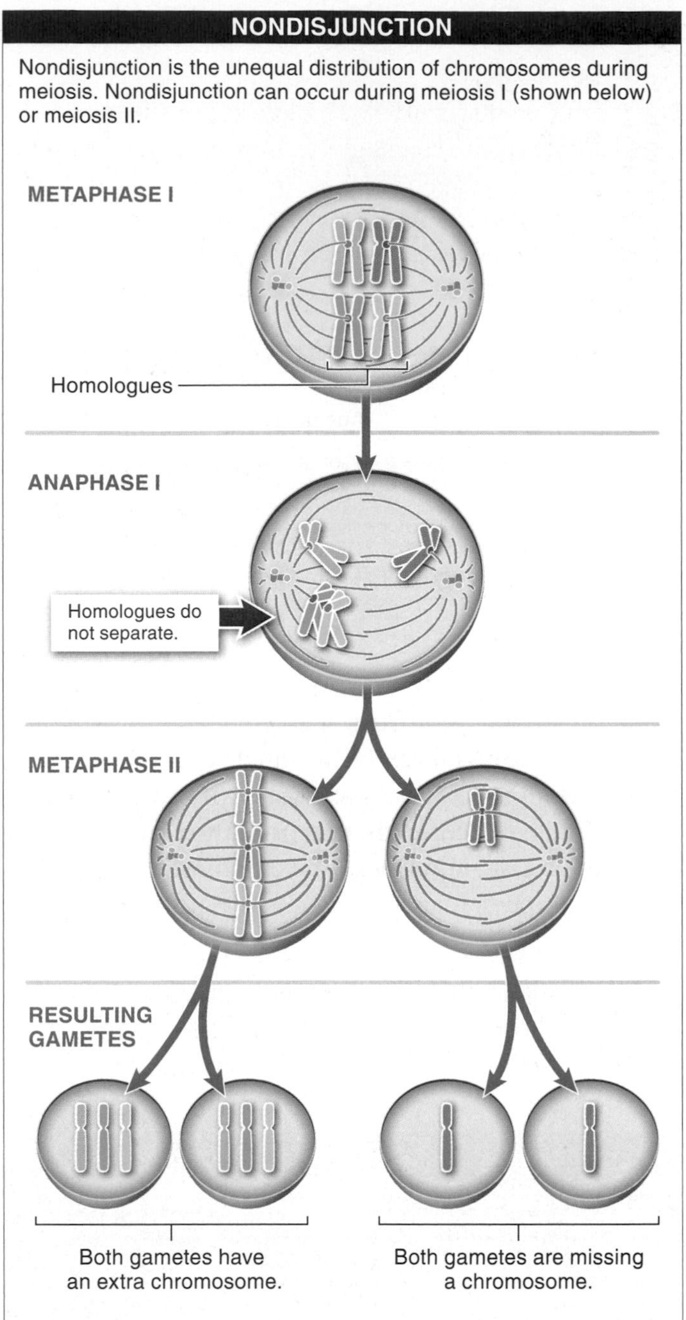

NONDISJUNCTION
Nondisjunction is the unequal distribution of chromosomes during meiosis. Nondisjunction can occur during meiosis I (shown below) or meiosis II.
METAPHASE I
Homologues
ANAPHASE I
Homologues do not separate.
METAPHASE II
RESULTING GAMETES
Both gametes have an extra chromosome.
Both gametes are missing a chromosome.

FIGURE 8-35 **An error during meiosis produces gametes with too many or too few chromosomes.**

disabilities, a flat facial profile, heart defects, and increased susceptibility to respiratory difficulties.

Down syndrome and other disorders caused by a missing chromosome or an extra copy of a chromosome are a consequence of **nondisjunction,** the unequal distribution of chromosomes during cell division. Nondisjunction can occur at two different points in meiosis: when homologues fail to separate during meiosis I, or when sister chromatids fail to separate during meiosis II (FIGURE 8-35). In both cases, nondisjunction results in an egg or sperm with zero or two copies of a chromosome rather than a single copy. Any of the chromosomes can fail to separate during cell division, but the ramifications of trisomy (having an extra copy of one chromosome in every cell) are greater for chromosomes with larger numbers of genes.

When trisomy occurs for chromosomes with greater numbers of genes, the likelihood that the developing embryo will survive to birth is reduced. Consequently, we tend to see cases of trisomy that involve only the chromosomes with the fewest genes, such as chromosomes 13, 15, 18, 21, and 22. In fact, observations show that trisomy 1 is never seen (all such fertilized eggs die before implantation in the uterus), trisomy 13 occurs in 1 in 20,000 newborns (and most die soon after birth), while trisomy 21, as we've seen, occurs in 1 in 1,000 newborns (many of whom live long lives).

As we mentioned at the beginning of this section, as women become older there are increased problems associated with reproduction. As women age, their gametes tend to have more errors. The reason is that the eggs began meiosis near the time the woman was born, and they may not complete it until she is 40 or more years old. During those decades, the cells may develop problems—including a reduction in the size and stability of the spindle fibers and reduced cohesion between sister chromatids—that interfere with normal cell division. In men, the cells that undergo meiosis are relatively young because new sperm-producing cells are produced every couple of weeks after puberty.

Q Why are babies born to older women more likely to have Down syndrome?

Whereas lacking a non-sex chromosome or having an extra chromosome usually has serious consequences for health, we'll see in the next section that lacking or having an extra X or Y chromosome has consequences that are much less severe.

TAKE HOME MESSAGE 8.16

» A karyotype is a visual display of a complete set of chromosomes. It is a useful diagnostic tool because it can be prepared early in fetal development to assess whether there is an abnormality in the number of chromosomes or in their structure, such as in Down syndrome. Down syndrome is caused by having an extra copy of chromosome 21.

8.17 Life is possible with too many or too few sex chromosomes.

It is usually fatal to have one too many or one too few of the non-sex chromosomes. However, many individuals are born lacking one of the sex chromosomes or having an additional X or Y chromosome, and they usually survive. We can get a glimpse into the role of the sex chromosomes by looking at the physical and mental consequences of having an abnormal number of these chromosomes. In each of the following cases, the condition is caused by nondisjunction of the sex chromosomes during the production of sperm or eggs (FIGURE 8-36).

Turner Syndrome: X_ Approximately 1 in 5,000 females carry only one X chromosome (and no Y chromosome), exhibiting a condition called Turner syndrome, denoted as X_ (or sometimes XO). This is the only condition in humans in which a person can survive without one of a pair of chromosomes. Instead of having 46 chromosomes in every cell, these individuals have only 45. As common as Turner syndrome is, in 98% of the fertilized eggs in which this condition occurs, the embryo is spontaneously aborted long before a fetus can come to term.

There are both physical and mental consequences of the absence of a second sex chromosome.

- Women with Turner syndrome are usually relatively short, averaging 4 feet 8 inches in height.

TOO MANY OR TOO FEW SEX CHROMOSOMES

GAMETE GENOTYPES	OFFSPRING GENOTYPE	SYNDROME CHARACTERISTICS

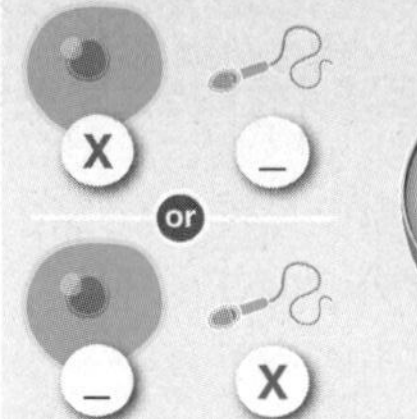

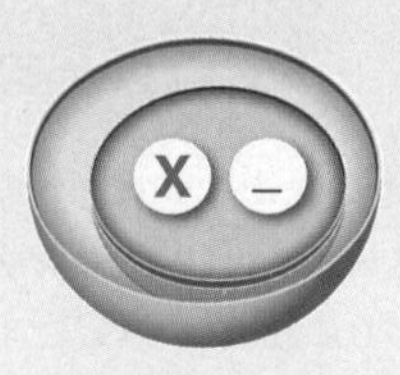

TURNER SYNDROME (FEMALE)
- Short height
- Web of skin between neck and shoulders
- Underdeveloped ovaries; often sterile
- Some learning difficulties

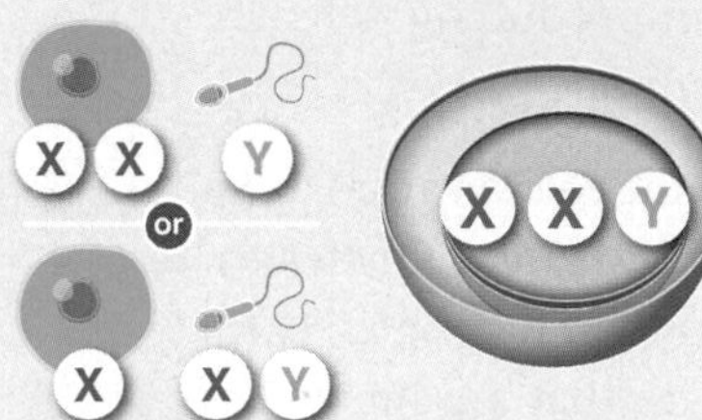

KLINEFELTER SYNDROME (MALE)
- Underdeveloped testes
- Lower testosterone levels; usually infertile
- Development of some female features
- Long limbs and slightly taller than average

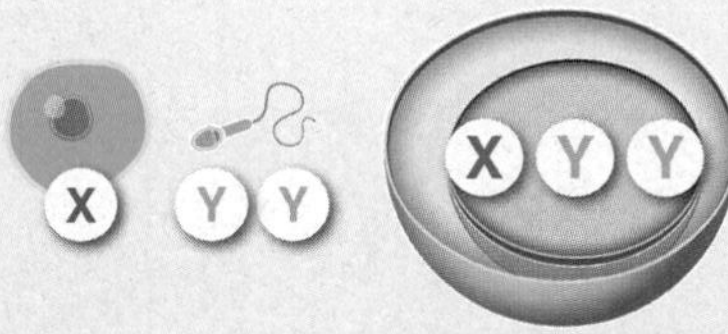

XYY MALE
- Taller than average
- Moderate to severe acne
- Intelligence may be slightly lower than average
- Sometimes called "super males"

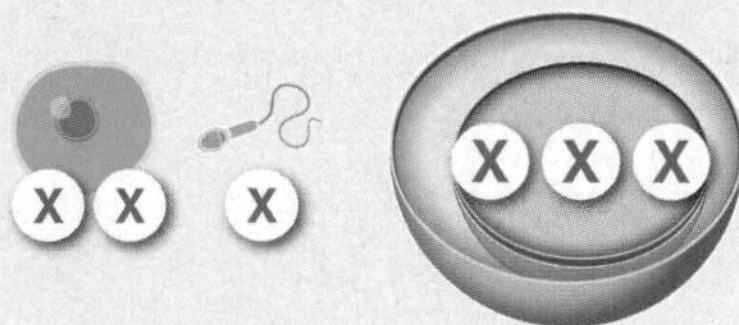

XXX FEMALE
- May be sterile
- No obvious physical or mental problems
- Sometimes called "metafemales"

FIGURE 8-36 **Characteristics of individuals with too many or too few sex chromosomes.**

- They develop a web of skin between the neck and shoulders.
- The ovaries never fully mature, so the women are almost always sterile.
- The breasts and other secondary sex characteristics develop incompletely.
- Intelligence is usually normal, but some learning difficulties are common.

Q If a person has two X chromosomes but also has a Y chromosome, is the individual male or female?

Klinefelter Syndrome: XXY An individual who has two X chromosomes is female; an individual with an X and a Y is male. But what happens when both of those conditions exist? An individual who carries two X chromosomes and a Y chromosome, a condition known as Klinefelter syndrome, develops as a male. This is because, as we saw earlier, the Y chromosome carries genetic instructions that cause fetal gonads to develop as testes; if these instructions are absent, the fetal gonads develop as ovaries. The extra X chromosome, however, does cause males with Klinefelter syndrome to be somewhat feminized, although this effect can be reduced through treatments such as testosterone supplementation. Approximately 1 in 1,000 males have the genotype XXY, making this one of the most common genetic abnormalities in humans.

Klinefelter syndrome has some physical and mental consequences.

- Men with this syndrome have testes that are smaller than average. Because the testes are small, levels of testosterone are low, and the men are almost always infertile.
- They develop some female features, including reduced facial and chest hair and some breast development.
- They have long limbs and are slightly taller than average (about 6 feet on average).
- They learn to speak at a later age than average and tend to have language impairments.
- Some individuals have further additional X chromosomes, with the karyotypes XXXY or even XXXXY. These males also exhibit the symptoms of Klinefelter syndrome but more frequently have mental retardation.

Q Do human hermaphrodites exist?

A hermaphrodite is an individual with functioning male and female reproductive organs capable of producing both male and female gametes. Often it is mistakenly assumed that a person with Klinefelter syndrome must be a hermaphrodite because he has both two X chromosomes (which would usually make an individual a female) and an X and a Y chromosome (usually making the individual a male). Hermaphroditism is common among invertebrates and occurs in some fish and other vertebrates, but contrary to urban legends, human hermaphrodites do not exist. Some men with Klinefelter syndrome may have some features of the opposite sex; however, they do not produce female gametes and so are not hermaphrodites. Individuals with sex characteristics that preclude definitive identification as male or female are called "intersex."

XYY Males There is no official name for the condition in which an individual has one X chromosome and two Y chromosomes, although such individuals are sometimes referred to as "super males." This chromosomal abnormality occurs in approximately 1 in 1,000 males (and in about 1 in 325 males who are 6 feet or taller). There are no distinguishing features at birth to indicate that an individual carries an extra Y chromosome, and the vast majority live their lives normally, without knowing they have an extra chromosome in every cell.

Several consequences of the XYY condition have been well documented.

- XYY males are relatively tall (about 6 feet 2 inches on average).
- They tend to have moderate to severe acne.
- Their intelligence usually falls within the same range as that of XY males, although the average may be slightly lower.

Because, in the United States, about three to five times as many XYY males are found in prisons (as a percentage of the prison population) as in the population as a whole, there has been tremendous controversy over whether the extra Y chromosome predisposes individuals to criminal behavior. These observations were not part of randomized, controlled, double-blind studies, however, so they are open to numerous alternative interpretations.

In one study, for example, researchers identified all male Danish citizens born in Copenhagen between 1944 and 1947 and obtained sex chromosome determinations for the 4,139 of these men who were taller than 6 feet 0.4 inches (184 cm). From this sample, they found that the level of intellectual functioning (as indicated by scores on an army selection test taken by the men) was significantly lower among the XYY males than among XY males. Based on this finding, the researchers suggested that the increased representation of XYY males in prison was more likely to be due to the fact that males of lower intelligence are more likely to be apprehended, regardless of which chromosomes they carry. The controversy remains unresolved, and the vast majority of XYY males are not in prison.

Although an error during meiosis in either the father or the mother can cause Klinefelter or Turner syndrome, it is only an error in meiosis in a male that can give rise to an XYY child. Why? Because only males carry Y chromosomes. Consequently, an XYY child must have received both of his Y chromosomes from his father. This means that the father's sperm cell must have contained two copies of the Y, a result of nondisjunction during meiosis in the father.

XXX Females Sometimes called "metafemales," individuals with three X chromosomes occur at a frequency of about 1 in 1,000 women. Very few studies of this condition have been completed, although initial observations suggest that XXX females are slightly taller than average but have no obvious physical or mental problems. In contrast to individuals who have Turner syndrome, most XXX females are fertile.

TAKE HOME MESSAGE 8.17

» Although it is usually fatal to have one too many or one too few of the non-sex chromosomes, individuals born with only a single sex chromosome that is an X, or with an additional X or Y chromosome, usually survive—though often with physical and/or mental problems.

STREET BIO

Using evidence to guide decision making in our own lives

Can you select the sex of your baby? (Would you want to?)

Some fertility clinics now promise that, for as little as $200, they can help a couple choose the sex of their baby.

Q: Given our knowledge of the behavior of the X and the Y chromosomes, is it more likely that these sex-selection techniques involve manipulations of sperm or of eggs? Why? Because all eggs have X chromosomes, the X chromosome doesn't play a role in determining the sex of a baby. Instead, fertility clinics must somehow sort and separate sperm cells based on whether they carry an X or a Y chromosome (men produce approximately equal numbers of each).

Q: What feature of sperm must be used in separating them? This process must be based on a way of distinguishing between sperm carrying an X chromosome and sperm carrying a Y chromosome. After the two types of sperm are distinguished, they must then be separated. Sperm with the desired sex chromosome are then used to fertilize the woman's egg (which may be done within her body, using artificial insemination, or in a Petri dish, after which a fertilized embryo is transferred into the woman's reproductive tract).

Q: What techniques might make it possible to separate sperm with an X chromosome from those with a Y chromosome? One method involves determining, by weighing sperm, which cells have more DNA. The heavier sperm must be carrying the X rather than the Y chromosome because the X is so much larger. Another method is to add a fluorescent dye to sperm that temporarily attaches to their DNA. Because sperm with an X chromosome have more DNA, more of the dye attaches to them and they are more fluorescent. A machine then sorts the sperm one by one, and at ovulation, insemination is performed using the sperm with the desired sex chromosome.

Q: Does it work? The technique is still fairly new, but initial results suggest that the procedures have a success rate of 70% to 90%.

Concerns for you to ponder. What are the ethical concerns raised by selecting the sex of one's baby? Are there some circumstances in which such a selection would be more acceptable than in others? Would it be acceptable to select for a baby with one eye color versus another? Why would that process be more difficult than selecting for sex?

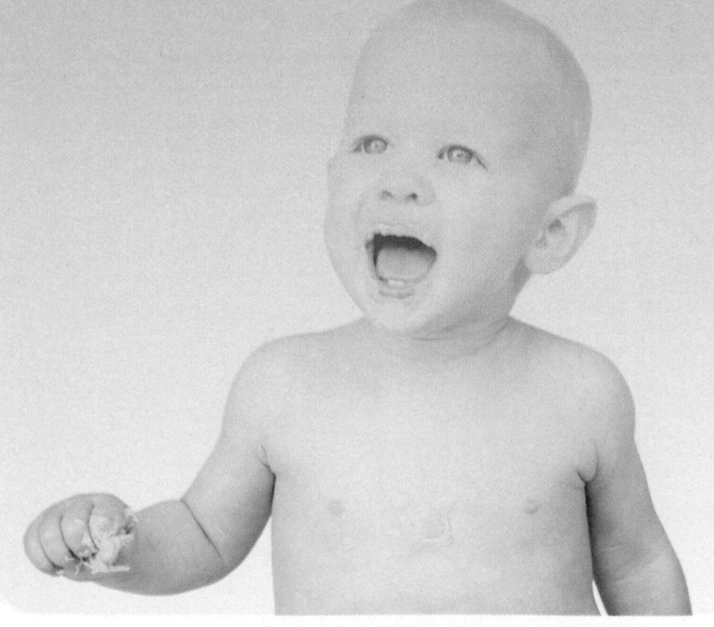

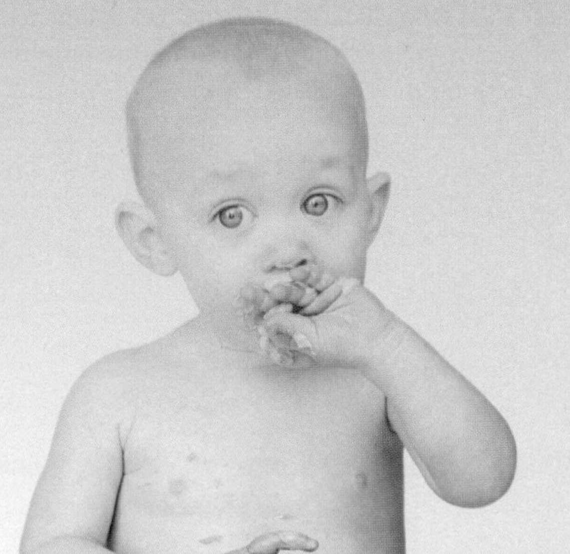

GRAPHIC CONTENT

Thinking critically about visual displays of data

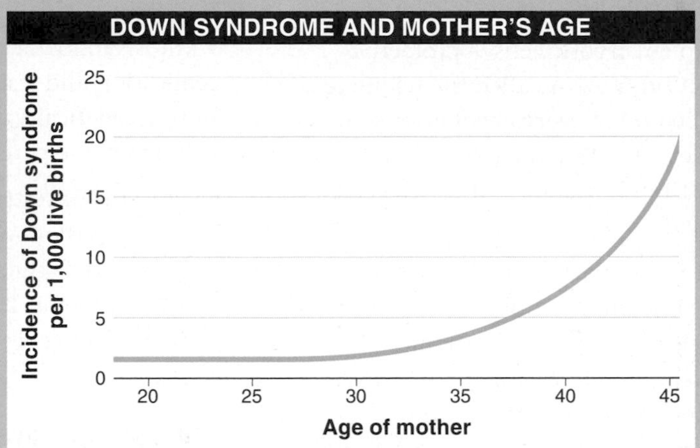

1 According to this graph, what is the probability (i.e., 1 in 800, or 1 in 256, etc.) that a baby born to a 35-year-old woman will have Down syndrome?

2 Is a woman of age 45 who gives birth more likely than not to have a baby with Down syndrome? What is the exact probability that her baby will not have Down syndrome?

3 From this graph, can you conclude that most babies with Down syndrome are born to women older than 35?

4 Can you determine from the graph what the chance of having a baby born with Down syndrome is after age 45? What factors might have been responsible for the graph not showing data for women older than 45?

5 Among women age 35 and older only, the incidence of fetuses produced that have an extra copy of chromosome 21 may be higher than the graph indicates. Why?

See answers at the back of the book.

KEY TERMS IN CHROMOSOMES AND CELL DIVISION

BRIEF SUMMARY

There are different types of cell division.

• In eukaryotic cells, a protective section of DNA called the telomere, at each end of every chromosome, plays a role in keeping track of cell division, getting shorter every time the cell divides.

• In most bacteria and archaea, the genetic information is carried in a single, circular chromosome. Eukaryotes have much more DNA, which is organized into linear chromosomes within the nucleus.

• Eukaryotic somatic cells alternate in a cycle between a mitotic phase, in which cell division occurs, and interphase, in which other cell activities take place. A cell-cycle control system functions through a series of checkpoints in the cell cycle at which specific signals trigger continuation of the process.

• Every time a cell divides, that cell's DNA must first duplicate itself, a process called replication. Although enzymes proofread and repair DNA during and after replication, some errors may remain.

Mitosis replaces worn-out old cells with fresh new duplicates.

• Mitosis enables existing cells to generate new, genetically identical cells. This makes it possible for organisms to grow and to replace cells that die.

• Mitosis follows chromosome replication and leads to the production of two genetically identical daughter cells from one parent cell.

• Cancer is unrestrained cell growth and cell division, which lead to large masses of cells that may cause serious health problems. It often results from mutations in genes controlling the cell cycle.

Meiosis generates sperm and eggs and a great deal of variation.

• Meiosis is the process by which reproductive cells are produced in sexually reproducing organisms. It results in gametes that have only half as much genetic material as the parent cell and that differ from one another and from the parents in the combinations of alleles they carry.

• Meiosis is preceded by DNA replication. The final product of meiosis in a diploid organism is four haploid gametes.

• In species with two sexes, females produce the larger gamete, and males produce the smaller, more motile gamete. Each gamete contains just one copy of each chromosome.

• Crossing over during the first prophase of meiosis creates gametes with combinations of alleles that may never have existed before; this variation is important for evolution.

• Asexual reproduction can be fast and efficient but leads to genetically identical offspring, which could be disadvantageous in a changing environment. Sexual reproduction leads to offspring that are genetically different from one another and from either parent, but it takes more time and energy and can be risky.

There are sex differences in the chromosomes.

• A variety of methods have evolved for sex determination in plant and animal species. In humans, the sex chromosomes carry information that directs a growing fetus to develop either as a male or as a female.

Deviations from the normal chromosome number lead to problems.

• A karyotype is a visual display of a complete set of chromosomes. It is used to assess whether there is an abnormality in the number of chromosomes or in their structure.

• It is usually fatal to have too many or too few of one of the non-sex chromosomes. Individuals with only a single X chromosome, or with an additional X or Y chromosome, usually survive, though often with physical and/or mental problems.

CHECK YOUR KNOWLEDGE

Short Answer

1. Each time a cell divides, telomeres get a little shorter. If a cell's telomeres get too short, this can interfere with genes and lead to cell death. Why would a therapy that helped cells rebuild their telomeres not be the answer to the quest for "eternal youth"?

2. Compare and contrast the genetic information found in prokaryotes and eukaryotes.

3. Describe the two main phases of the eukaryotic cell cycle and what occurs during each.

4. Describe the phase of the cell cycle during which the events make it possible to end up with sufficient genetic material for two identical daughter cells.

5. Describe three ways in which cancer cells differ from "typical" cells. Why is it that when someone has cancer, the properties of the cancer cells are not a direct cause of death?

6. Why do the gametes produced by meiosis have only half the genetic material of the parent cell?

7. One important type of genetic variation does not require the creation of a new allele. How can that be?

8. Why is crossing over considered such an important event with respect to evolution?

9. What are the differences between sex chromosomes and non-sex chromosomes?

10. Healthy individuals may have just one sex chromosome, as long as it is an X chromosome. Why can't a person survive with a Y chromosome and no X chromosome?

Multiple Choice

1. Which of the following statements about telomeres is incorrect?

a) They function like a counter, keeping track of how many times a cell has divided.

b) At birth, they are long enough to permit approximately 80–90 cell divisions in most cells.

c) They are slightly shorter in prokaryotic cells than in eukaryotic cells.

d) They function like a protective cap on chromosomes.

e) They contain no critical genes.

0 EASY — 62 — HARD 100

2. Prokaryotic cells can divide by:

a) mitosis.

b) binary fission.

c) meiosis.

d) both mitosis and binary fission.

e) none of the above.

0 EASY — 31 — HARD 100

3. Mitosis results in:

a) mitosis.

b) binary fission.

c) meiosis.

d) both mitosis and binary fission.

e) none of the above.

0 EASY — 45 — HARD 100

4. Using a light microscope, it is easiest to see chromosomes:

a) during mitosis and meiosis, because the condensed chromosomes are thicker and therefore more prominent.

b) during interphase, when they are concentrated in the nucleus.

c) in the mitochondria, because the chromosomes are circular.

d) during asexual reproduction.

e) during interphase, because they are uncoiled and more linear.

0 EASY — 40 — HARD 100

5. Which of the following statements about tumors is incorrect?

a) Benign tumors pose less of a health risk than malignant tumors.

b) Malignant tumors shed cancer cells that can spread in the body.

c) Tumors are caused by excessive cell growth and division.

d) Malignant tumor cells can travel to other parts of the body in a process called metastasis.

e) Tumors contain cells with abnormally high contact inhibition.

0 EASY — 48 — HARD 100

6. During meiosis but not during mitosis:

a) haploid gametes are produced that are identical in their allelic composition.

b) the cytoplasm divides.

c) chromosomes line up in the center of the cell during metaphase.

d) genetic variation among the daughter cells is increased.

e) two identical daughter cells are produced.

0 EASY — 52 — HARD 100

7. Sister chromatids are:

a) the result of crossing over.

b) identical molecules of DNA resulting from replication.

c) homologous chromosomes.

d) produced in meiosis but not in mitosis.

e) single-stranded.

0 EASY — 39 — HARD 100

8. A potential disadvantage of asexual reproduction is that:

a) it increases the time required to find a mate.

b) it allows perpetuation of a population even when the members are isolated.

c) it allows population size to increase rapidly.

d) it produces genetically uniform populations.

e) it requires additional rounds of meiosis, which is energetically much more costly than mitosis.

0 EASY — 76 — HARD 100

9. A karyotype of one of your skin cells would reveal a total of 46 chromosomes. How many of these are maternally inherited non-sex chromosomes?

a) 23

b) 20

c) 46

d) 22

e) 24

0 EASY — 76 — HARD 100

10. No karyotype has ever shown a person to have one Y chromosome and no X chromosome. Why?

a) No meiotic event could produce an egg without an X chromosome.

b) Individuals are probably born with this karyotype, but have such slight phenotypic abnormalities that they live out their lives normally and are unaware of this genetic anomaly.

c) Genes on the Y chromosome are expressed only when an X chromosome is present.

d) The X chromosome contains genes that are essential for life.

e) A karyotype would not reveal such an abnormality.

0 EASY — 65 — HARD 100

Ch09

Identical multiples (such as twins or triplets) come from a single fertilized egg and are genetically identical.

Genes and Inheritance

Family resemblance: how traits are inherited

9.1–9.5

Why do offspring resemble their parents?

These Japanese macaque parents and offspring have brown-gray fur and red faces.

9.1 Family resemblance: your mother and father each contribute to your genetic makeup.

Q How can a single bad gene make you smell like a rotten fish?

Can a gene be cruel? Consider this candidate: in humans, there is a gene for an enzyme called FMO3 (flavin-containing mono-oxygenase-3), which breaks down a chemical in our bodies that smells like rotting fish. Some unfortunate individuals inherit a defective FMO3 gene and can't break down the noxious chemical. Instead, their urine, sweat, and breath excrete it, causing them to smell like rotting fish. Worst of all, because the odor comes from within their bodies, they cannot wash it away. Called "fish odor syndrome," this disorder often causes those afflicted by it to suffer ridicule, social isolation, and depression.

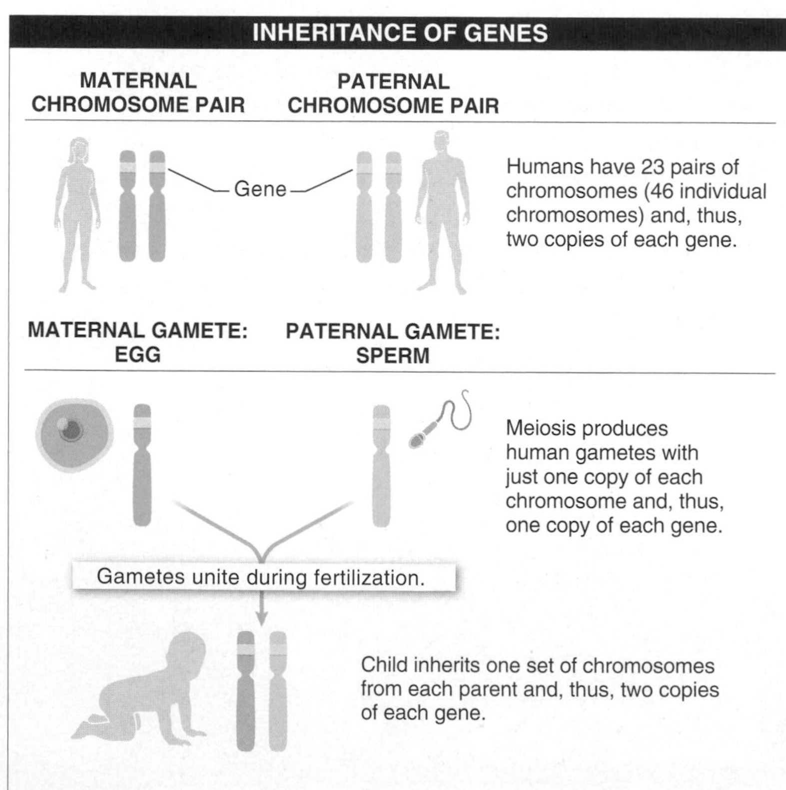

INHERITANCE OF GENES
MATERNAL CHROMOSOME PAIR
PATERNAL CHROMOSOME PAIR
Gene
Humans have 23 pairs of chromosomes (46 individual chromosomes) and, thus, two copies of each gene.
MATERNAL GAMETE: EGG
PATERNAL GAMETE: SPERM
Meiosis produces human gametes with just one copy of each chromosome and, thus, one copy of each gene.
Gametes unite during fertilization.
Child inherits one set of chromosomes from each parent and, thus, two copies of each gene.

FIGURE 9-1 **How we inherit our genes?** Each human offspring inherits one maternal set of 23 chromosomes and one paternal set.

Currently, there is no cure for the rare disorder, although there are a few ways of reducing the odor such as reducing consumption of eggs, fish, and some meats that contain certain chemicals, including sulfur.

For individuals born with this malady, beyond their own suffering there looms a scary question: "Will I pass this condition on to my children?" They might also wonder how they came to have the disorder, particularly if neither of their parents suffered from it (FIGURE 9-1).

To address this concern, let's recall from Chapter 8 that sexually reproducing organisms carry two copies of every chromosome in every cell (except their sex cells). Humans, for example, have 23 pairs of chromosomes (46 individual chromosomes). At each location—referred to as a locus (*pl.* loci)—on the two chromosomes of a pair is the same gene: one copy from the mother and one from the father. As we noted in Chapter 6, each of the two copies of the gene is called an **allele.**

The gene for FMO3 is on chromosome 1. There is a normal version—or allele—of the gene for FMO3, which most people carry, and there is a rare, defective version that is

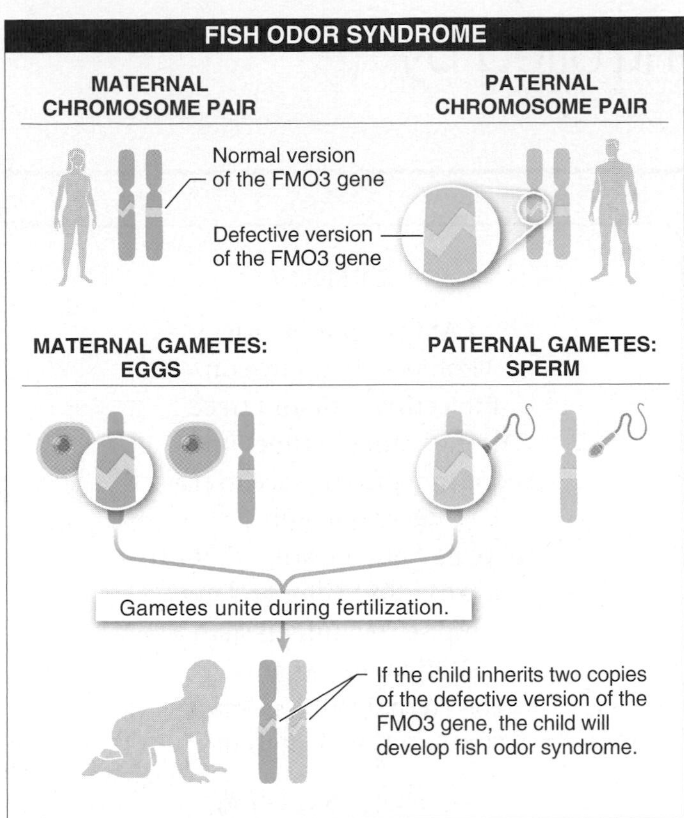

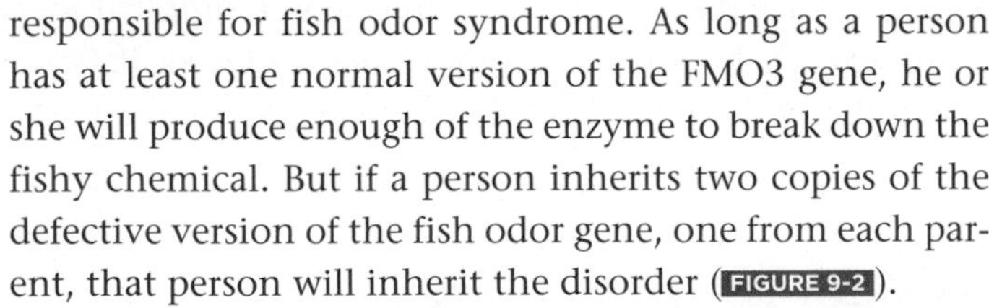

FIGURE 9-2 **Unlucky catch.** A baby inheriting two copies of the defective FMO3 gene will develop fish odor syndrome.

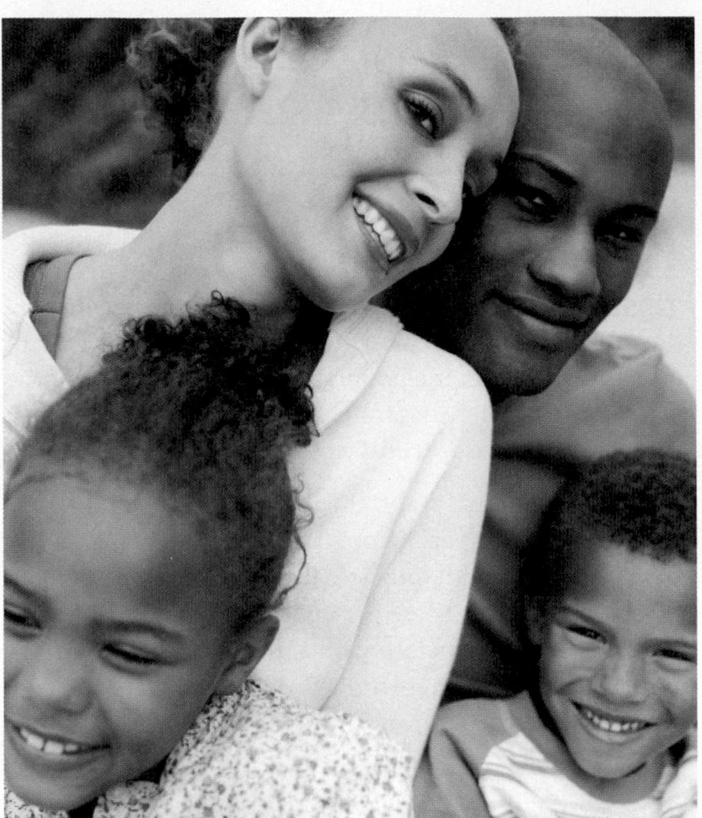

FIGURE 9-3 **Where do we begin and our parents end?** Family resemblances reveal that many traits are inherited.

responsible for fish odor syndrome. As long as a person has at least one normal version of the FMO3 gene, he or she will produce enough of the enzyme to break down the fishy chemical. But if a person inherits two copies of the defective version of the fish odor gene, one from each parent, that person will inherit the disorder (FIGURE 9-2).

Although individuals with fish odor syndrome will pass on one set of the bad instructions in every sperm or egg cell that they make, their children won't necessarily inherit the disease. As long as the other parent supplies a normal version of the FMO3 gene, the child will not have fish odor syndrome. The unaffected children, though, will carry a silent copy of the fish odor allele. If they have children, the fish odor trait will be expressed only if it comes together with a defective FMO3 allele from another person. In this way, some alleles can exist in a population without always revealing themselves.

In this chapter, we explore how heredity works. Inheritance follows some simple rules that allow us to make sense of patterns of family resemblance, such as facial features or hair texture, and even to predict the likelihood that an offspring will inherit a particular trait (FIGURE 9-3). We'll also examine why the behavior of some traits is easy to predict while many other traits have less straightforward patterns of inheritance, but still can be studied experimentally and yield predictable patterns.

TAKE HOME MESSAGE 9.1

» Offspring resemble their parents because they inherit genes—instructions for biochemical, physical, and behavioral traits, some of which are responsible for diseases—from their parents.

9.2 Some traits are controlled by a single gene.

Like father, like son. No one will ever seek a paternity test to prove that Jaden Smith is the son of Will Smith (FIGURE 9-4). We see all around us that offspring generally resemble their parents more than they resemble other random individuals in the population—a consequence of the passing of characteristics from parents to offspring through their genes. This is **heredity.**

Observing heredity is easy. Elucidating how it works is not. For thousands of years before the mechanisms of heredity were discovered and understood, plant and animal breeders recognized that there is a connection from parents to offspring across generations. In ancient Greece, for example, the poet Homer extolled the tremendous benefits to society that came from the skillful breeding of horses (FIGURE 9-5).

Once breeders recognized the existence of heredity, they began selecting individual plants or animals with the desired traits to breed with each other, in the hope that the offspring would also have those desirable traits. Since then, it has been possible to create a rich world of sweeter corn, loyal dogs, docile livestock, beautiful flowers, and more.

"HEREDITY

I AM the family face;
Flesh perishes, I live on,
Projecting trait and trace
Through time to times anon,
And leaping from place to place
Over oblivion.
The years-heired feature that can
In curve and voice and eye
Despise the human span
Of durance—that is I;
The eternal thing in man,
That heeds no call to die."

— THOMAS HARDY,
in *Moments of Vision and Miscellaneous Verses,* 1917

Practical successes were many, but people didn't understand exactly how the outcomes were achieved. Patterns of similarity among related individuals are impossible to

Offspring resemble their parents more than they resemble unrelated individuals in the population. This is heredity.

FIGURE 9-4 **Father and son: Will and Jaden Smith.**

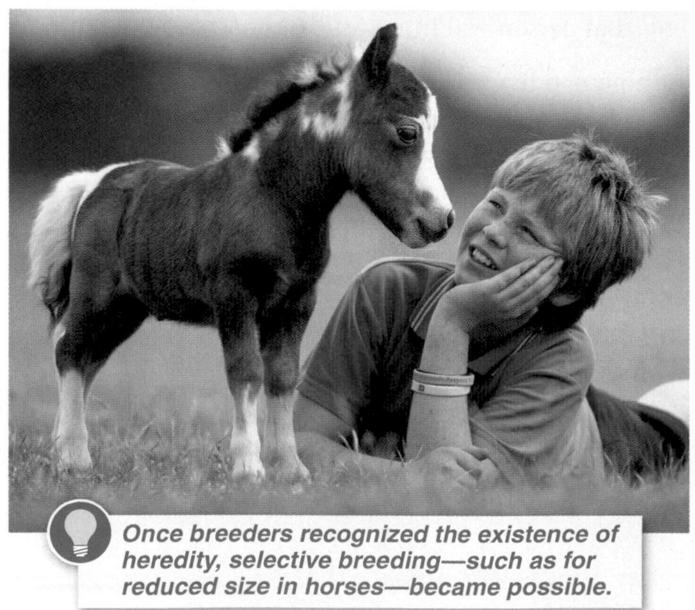

Once breeders recognized the existence of heredity, selective breeding—such as for reduced size in horses—became possible.

FIGURE 9-5 **Honey, I shrunk the horse!** Selective breeding was used to produce this tiny horse.

EXAMPLES OF SINGLE-GENE TRAITS

FUR LENGTH IN CATS

Long-haired cat

Short-haired cat

COAT COLOR IN CATS

White-haired cat

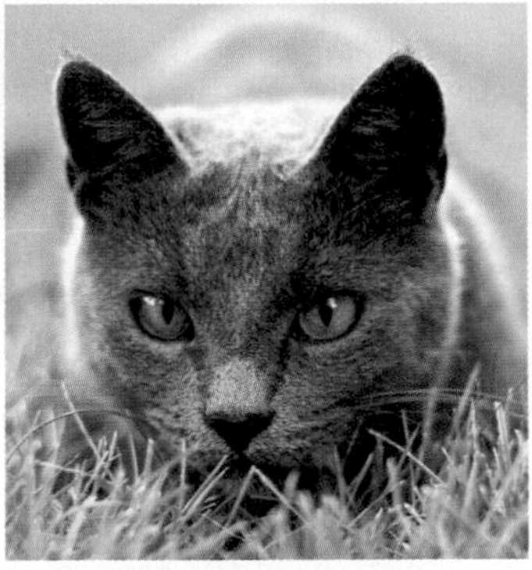
Colored-haired cat

FIGURE 9-6 **Fur length and coat color in cats: single-gene traits.** Some traits are determined by the instructions an organism carries on one gene.

make sense of without an understanding of how heredity works. This understanding is the province of the field of genetics.

Mechanisms of heredity are easier to grasp when we focus on visible traits with well-established inheritance patterns, such as coat color and fur length in cats. Virtually every short-haired cat, for instance, has at least one short-haired parent. And virtually every completely white-haired cat has at least one white-haired parent. In humans, many genetic diseases have similarly simple patterns of inheritance. Individuals who develop Huntington's disease, for example, have a parent who developed the disease. Traits that are determined by the instructions a person carries on one gene are called **single-gene traits** (FIGURE 9-6).

It is important to note here that most human characteristics are influenced by multiple genes, as well as by the environment. However, because the mechanism by which single-gene traits pass from parent to offspring is the easiest pattern of inheritance to decipher, we first explore this process.

TAKE HOME MESSAGE 9.2

» Some traits are determined by instructions that an individual carries on a single gene, and these traits exhibit straightforward patterns of inheritance.

9.3 Mendel learned about heredity by conducting experiments.

What do parents "give" their offspring that confers similarity? In the mid-1800s, Gregor Mendel, a monk living in what is now the Czech Republic, was the first to make headway in answering this question. Mendel conducted studies that not only shed light on the question but practically answered it completely.

When Mendel turned to questions about heredity, there were no obvious answers. A popular idea from the late 1600s suggested that an entire pre-made human—albeit a tiny one—was contained in every sperm cell (FIGURE 9-7). Though imaginative, this idea failed to explain why children resemble both their mothers and their fathers, not just their fathers. A "blended blood" idea suggested that traits are blended in the blood of parents and imparted to offspring. But how, then, could brown-eyed parents give birth to blue-eyed children?

Sometimes scientific breakthroughs are made because a

Q Individuals who share common ancestry are called "blood relatives," yet they don't actually share any blood. How might that phrase reflect early conceptions of inheritance?

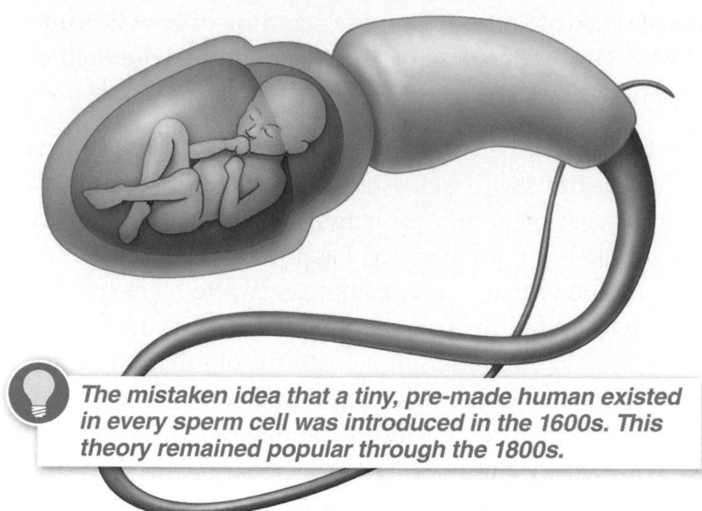

The mistaken idea that a tiny, pre-made human existed in every sperm cell was introduced in the 1600s. This theory remained popular through the 1800s.

FIGURE 9-7 **The "pre-made human."** A tiny human in every sperm cell? Why would children have any resemblance to their mothers?

new technique is invented or a lucky observation is made by chance. Neither played a role in Mendel's success. He simply applied the tried-and-true process of methodical experimentation and scientific thinking (as described in Chapter 1). In particular, three features of Mendel's research were critical to its success (FIGURE 9-8).

1. He wisely chose to study the garden pea. His goals were to understand inheritance in all organisms, not just plants. But cats and dogs or even mice would have been too hard to take care of in the large numbers he required—thousands and thousands of individuals. Humans, too, would have made a terrible study organism. We take too long to breed (and won't produce offspring on command). Pea plants, on the other hand, are relatively easy to fertilize manually by "pollen dusting." A single **cross**—the process in which male pollen (carrying sperm) is used to fertilize eggs—produces numerous offspring. In addition, pea plants reproduce quickly, so Mendel could conduct experiments that included multiple generations.

2. Mendel focused on easily categorized traits like shape and color. All peas of the variety that Mendel studied are round or wrinkled in shape, with nothing in between. In addition, all peas are yellow or green in color, never any intermediate shade. In all, Mendel looked at seven different traits of the pea plants, but for each trait only two—easily identified—variants ever appeared.

3. Mendel began his studies by repeatedly breeding together similar plants until he had many distinct populations, each of which was unvarying for a particular trait. He described these plants as **true-breeding** for that trait because they always produced offspring with the same variant of the trait as the parents. For example, when true-breeding round-pea plants were crossed

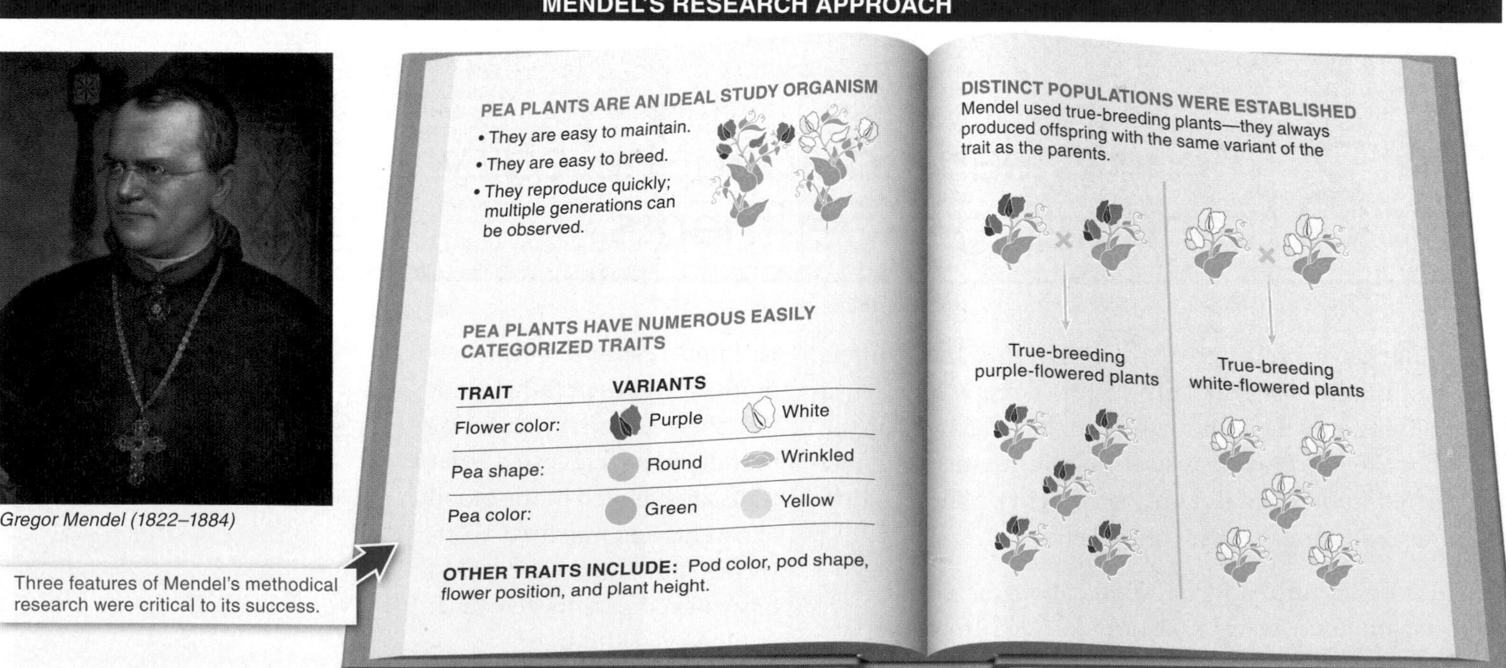

MENDEL'S RESEARCH APPROACH
Gregor Mendel (1822–1884)
Three features of Mendel's methodical research were critical to its success.
PEA PLANTS ARE AN IDEAL STUDY ORGANISM
• They are easy to maintain.
• They are easy to breed.
• They reproduce quickly; multiple generations can be observed.
PEA PLANTS HAVE NUMEROUS EASILY CATEGORIZED TRAITS
TRAIT
VARIANTS
Flower color: Purple White
Pea shape: Round Wrinkled
Pea color: Green Yellow
OTHER TRAITS INCLUDE: Pod color, pod shape, flower position, and plant height.
DISTINCT POPULATIONS WERE ESTABLISHED
Mendel used true-breeding plants—they always produced offspring with the same variant of the trait as the parents.
True-breeding purple-flowered plants
True-breeding white-flowered plants

FIGURE 9-8 **Mendel's pea plant experiments.** Careful planning and a well-selected study organism contributed to the success of Mendel's experiments.

together, they always produced plants with round peas. True-breeding purple-flowered pea plants always produced purple-flowered offspring. It took a lot of prep work, but once the true-breeding populations were established, Mendel could perform all of the crosses that enabled him to piece together the genetics puzzle.

Mendel began a straightforward process of experimentation with his groups of true-breeding plants. He crossed plants with different traits and observed large numbers of their offspring. For example, he crossed plants with green peas and plants with yellow peas and counted the number of offspring plants that produced green peas and the number that produced yellow peas. (This cross, written as "green × yellow cross," always resulted in plants that produced only yellow peas.) Mendel devised a hypothesis that would explain his observations and generate predictions about the outcomes of further crosses. Then he conducted those crosses to see whether the predictions generated by his hypothesis were borne out.

There was elegance to Mendel's work: from his predictions he articulated simple questions that could be answered by his experiments. For example, "If I cross a plant that produces wrinkled peas with a plant that produces smooth peas, will the offspring plants produce wrinkled or smooth peas?" His crosses would come out either one way or the other. This process of performing well-thought-out, rigorous experiments was a radical innovation for biology in Mendel's time. We explore his results in the next section.

TAKE HOME MESSAGE 9.3

» In the mid-1800s, Gregor Mendel conducted studies that help us understand heredity. He focused on easily observed and categorized traits in garden peas and applied methodical experimentation and rigorous hypothesis testing to determine how traits are inherited.

9.4 Segregation: you've got two copies of each gene but put only one copy in each sperm or egg.

Mendel was motivated by one odd and recurring result. Sometimes traits that weren't present in either parent pea plant would show up in their offspring—just as brown-eyed parents can have blue-eyed children. When plants with purple flowers (those *not* in his true-breeding group) were fertilized by pollen from other plants with purple flowers, they produced mostly purple-flowered offspring, but sometimes they produced plants with white flowers.

How was it possible to produce white flowers from a purple × purple cross? Where did the whiteness come from? Here's where Mendel's meticulous and methodical experiments paid off. First, he started with some true-breeding white-flowered plants. Then he got some true-breeding purple-flowered plants. He wondered: which color wins out when a white-flowered plant is crossed with a purple-flowered plant? The answer was definitive: purple wins (FIGURE 9-9).

All of the offspring from these crosses were purple, every time. For this reason, Mendel called the purple-flower trait **dominant,** and he considered the white-flower trait to be the **recessive** trait. In general, a dominant trait masks the effect of a recessive trait when an individual carries both the dominant and the recessive versions of the instructions for the trait.

Things got a bit more interesting when Mendel took the purple-flowered plants that came from the cross between true-breeding purple- and white-flowered plants and bred these purple offspring with each other. He found that these mixed-parentage plants were no longer true-breeding. Occasionally, they would produce white-flowered offspring. (To be exact, of the 929 offspring plants Mendel examined, 705 had purple flowers, and 224 had white flowers.) Apparently, the directions for building white flowers—last seen in one of their grandparents—were still lurking inside their purple-flowered parent plants. The existence of traits that could disappear for a generation and then show up again was perplexing.

Mendel devised a simple hypothesis to explain this pattern of inheritance. It incorporated three ideas that helped

DOMINANT AND RECESSIVE TRAITS

1 Mendel crossed true-breeding purple-flowered plants with true-breeding white-flowered plants.

All offspring had purple flowers.

The purple-colored flower is the dominant trait, while the white-colored flower is a recessive trait.

2 Then, Mendel crossed two of the purple-flowered offspring.

Most offspring had purple flowers, but some had white flowers.

The recessive trait for the white-colored flower must have been lurking in the previous generation, even though it was not visible.

FIGURE 9-9 **White or purple?** By careful and repeated crosses among pea plants, Mendel determined that there were "dominant" and "recessive" traits.

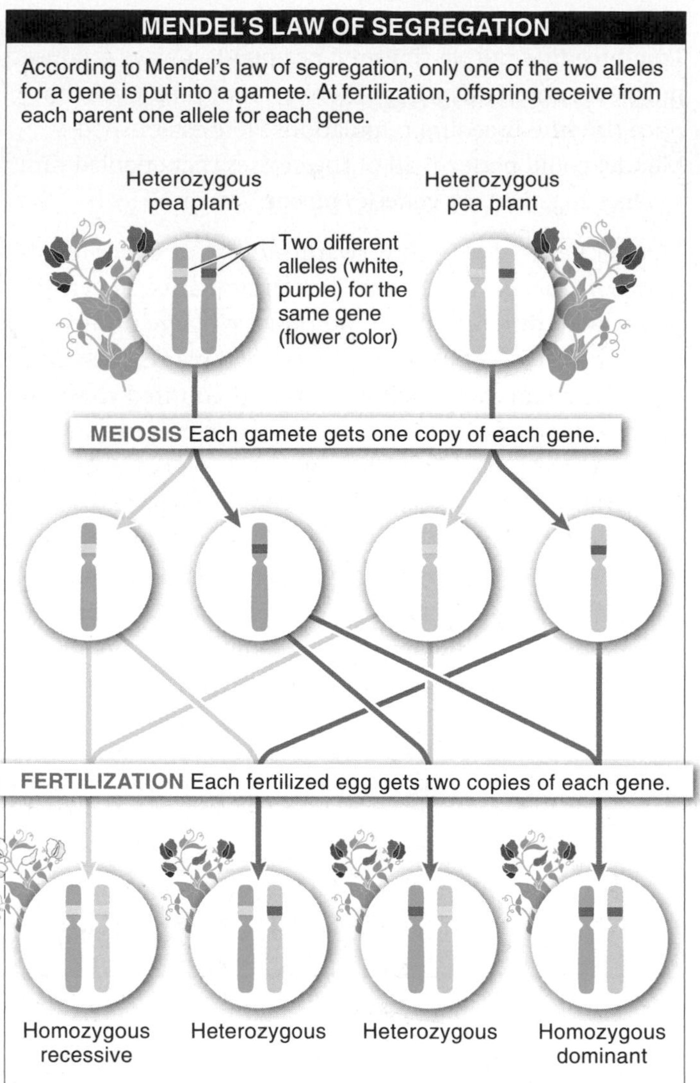

MENDEL'S LAW OF SEGREGATION
According to Mendel's law of segregation, only one of the two alleles for a gene is put into a gamete. At fertilization, offspring receive from each parent one allele for each gene.
Heterozygous pea plant
Heterozygous pea plant
Two different alleles (white, purple) for the same gene (flower color)
MEIOSIS Each gamete gets one copy of each gene.
FERTILIZATION Each fertilized egg gets two copies of each gene.
Homozygous recessive
Heterozygous
Heterozygous
Homozygous dominant

FIGURE 9-10 **Segregation of alleles in meiosis.** Organisms have two copies of each gene but place only one copy into each gamete during the process of meiosis (as described in Chapter 8).

him (and now help us) make predictions about crosses (**FIGURE 9-10**).

1. Rather than passing on the trait itself, **each parent puts into every sperm or egg it makes a single set of instructions for building the trait.** Today, we call that instruction set a gene.

2. Offspring receive two copies of the instructions for any trait. Often, both sets of instructions are identical, and the offspring produce the trait according to those instructions. Other times, though, each parent contributes a slightly different set of instructions—that is, a different allele—for that trait. So, for example, pea plants have two alleles for flower color: a purple-flower allele and a white-flower allele. Parents pass on the same gene (flower-color gene) but may contribute different alleles (purple or white).

3. The trait observed in an individual depends on the *two* copies of the gene it inherits from its parents. When an individual inherits the same two alleles for this gene, the individual's genotype for that gene is said to be **homozygous** and the individual shows the trait specified by the instructions embodied in those alleles.

When an individual inherits a different allele from each parent, the individual's genotype for that gene is said to be **heterozygous.** Dominant and recessive alleles are defined by their action when they are in the heterozygous state. A dominant allele and is said to "mask" the effect of the other allele, which is called the recessive allele. Note that this is the sole defining feature of "dominant" and "recessive." Describing an allele as dominant does not mean that it is more advantageous or more common than a recessive allele.

"There were no questions."

—Entry in meeting notes following Mendel's first public presentation of his ideas on how heredity works. (Unfortunately, no one in the audience of scientists had any idea what he was talking about. Not until about 40 years later did the world understand his discoveries.)

When an individual reproduces, it contributes just one of its two copies of a gene to its offspring. The other parent contributes the other allele. So, when sperm and eggs are made, each sex cell gets only one copy of a gene—as opposed to the two copies present in every other cell in the body. For a male who is heterozygous for a particular gene, for example, it means that half of the sperm he produces will have one of the alleles and half will have the other. The idea that, of the two copies of each gene everyone carries, only one of the two alleles gets put into each gamete is so important that it is called **Mendel's law of segregation.**

TAKE HOME MESSAGE 9.4

» Each parent puts a single set of instructions for building a particular trait into every sperm or egg that he or she makes. This instruction set is called a gene. The trait observed in an individual depends on the two copies (alleles) of the gene it inherits from its parents.

9.5 Observing an individual's phenotype is not sufficient for determining its genotype.

Things are not always as they appear. Take skin coloration, for example. Humans and many other animals have a gene that contains the information for producing melanin, one of the chemicals responsible for giving our skin its coloring (FIGURE 9-11). This gene is one of many that influence skin color. Unfortunately, there is also a defective, nonfunctioning version of the melanin gene that is passed along through some families. An individual who inherits two copies of the defective version of the gene cannot produce melanin and has a condition known as albinism, a disorder characterized by little or no pigment in the eyes, hair, and skin. A normally pigmented individual may carry one of these defective alleles; however, we would need to do a genetic analysis to discover this information.

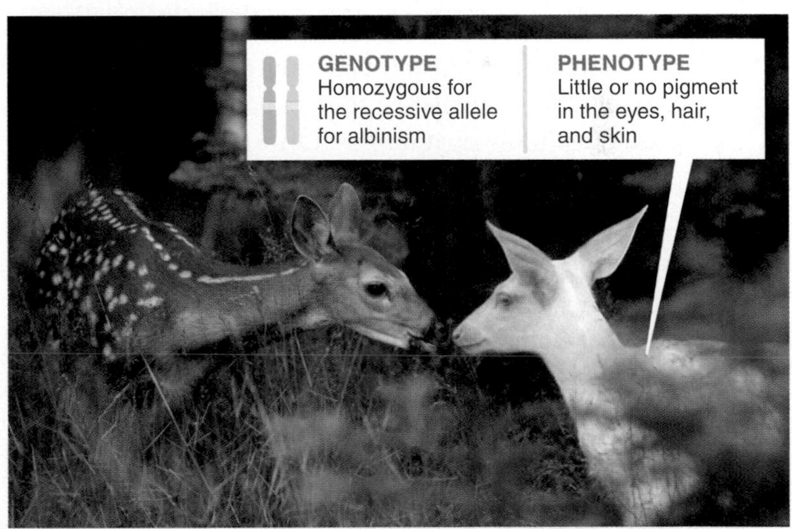

FIGURE 9-11 **One gene, much pigment.** Having two nonfunctioning versions of the gene carrying instructions for production of the pigment melanin causes albinism.

The outward appearance of an individual, such as skin pigment, is called its **phenotype.** A phenotype includes features visible to the naked eye, such as flashy coloration, height, or the presence of antlers. A phenotype also includes less easily visible characteristics such as the chemicals an individual produces to clot blood or digest lactose. An individual's phenotype even includes the behaviors it exhibits.

Underlying the phenotype is the **genotype.** This is an organism's genetic composition. We usually speak of an individual's genotype in reference to a particular trait. For example, an individual's genotype might be described as "homozygous for the recessive allele for albinism," meaning that the person has two recessive alleles of the gene. Another individual's genotype for the melanin gene might be described as "heterozygous" (a recessive allele and a dominant allele). Occasionally, the word "genotype" is also used as a way of referring to *all* of the genes an individual carries.

When an organism exhibits a recessive trait, such as albinism, we know with certainty what its genotype is for the melanin gene. When it shows the dominant trait, on the other hand, it's not possible to discern the genotype from the individual's appearance.

We can trace the possible outcomes of a cross between two individuals using a handy tool called the **Punnett square.** To see how this works, let's investigate the cross of an albino giraffe with a normally pigmented giraffe.

First, we assign symbols to represent the different variants of a gene. Generally, we use an uppercase letter for the dominant allele and lowercase letter for the recessive allele. In the case of pigmentation/albinism, we use the letter "m" for production of the pigment melanin: *M* for the dominant allele and *m* for the recessive allele. We represent the genotype of the albino giraffe as *mm,* because that individual must carry two copies of the recessive allele.

The pigmented giraffe must have the genotype *MM* or *Mm.* If we don't know which of the two possible genotypes the pigmented individual has, we can write *M_*, where _ is a placeholder for the unknown second allele.

In FIGURE 9-12 we illustrate the cross between a true-breeding pigmented individual, *MM,* and an albino, *mm.* Along the top of the Punnett square we list, individually, the two alleles that one of the parents produces, and along the left side of the square we list the two alleles that the other parent produces. We split up an individual's two alleles in this way because only one of the alleles is contained in each sperm or egg cell that it produces. The two gametes that come together at fertilization produce the genotype of the offspring.

In the four cells of the Punnett square, we enter the genotypes of all the possible offspring resulting from our cross. Each cell contains one allele given at the head of the column and one allele at the left of the row. In Cross 1

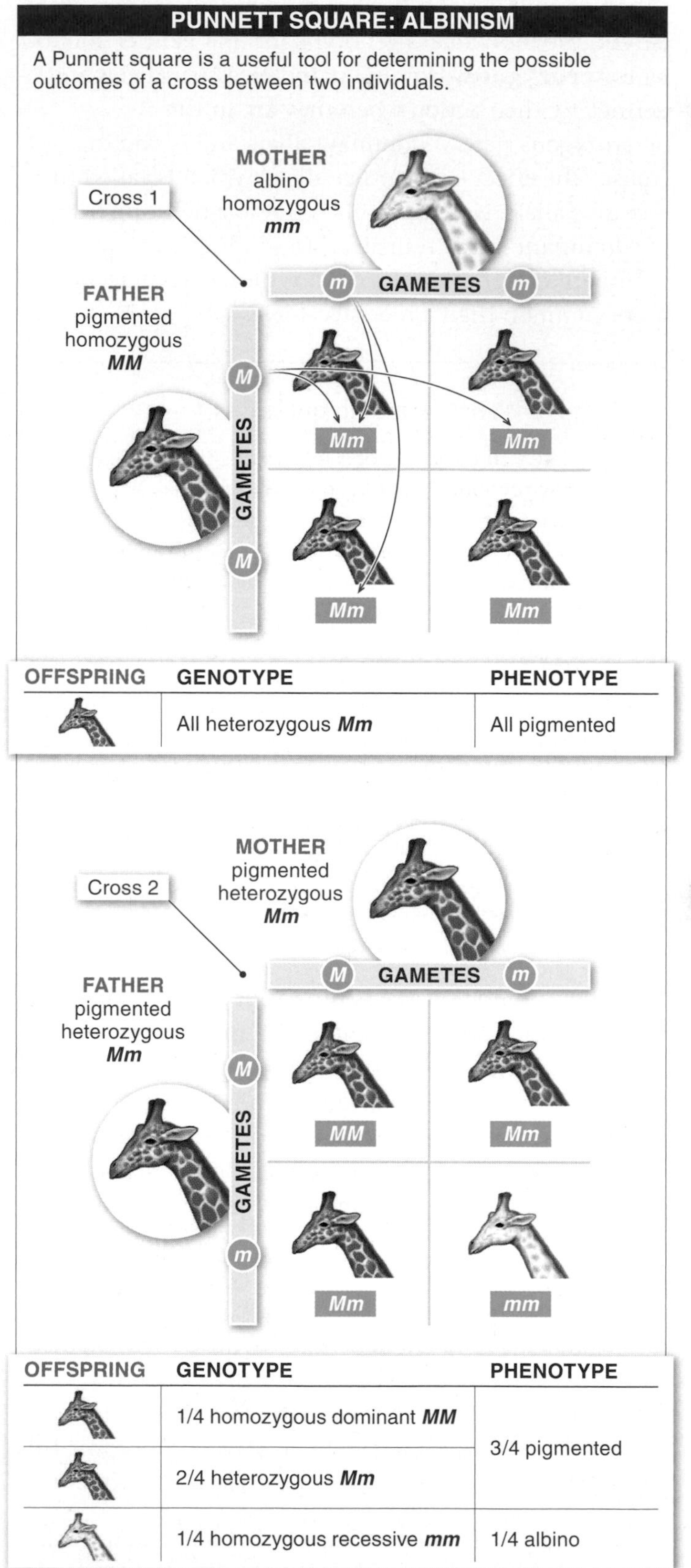

FIGURE 9-12 **Predicting the outcome of crosses.** A Punnett square shows us the likelihood of the albino genotype occurring in offspring.

illustrated in Figure 9-12, every possible offspring would be heterozygous and would be normally pigmented, because all offspring receive a dominant allele from the pigmented parent and a recessive allele from the albino parent.

In the bottom half of Figure 9-12 (Cross 2), we trace the cross between two heterozygous individuals. Note that each parent produces two kinds of gametes, one with the dominant allele and one with the recessive allele. This cross has three possible outcomes: one-quarter of the time the offspring will be homozygous dominant (*MM*), one-quarter of the time the offspring will be homozygous recessive (*mm*), and the remaining half of the time the offspring will be heterozygous (*Mm*). Phenotypically, three-quarters of the offspring will be normally pigmented (*MM* or *Mm*) and one-quarter will be albino (*mm*).

TAKE HOME MESSAGE 9.5

» It is not always possible to determine an individual's genetic makeup, known as its genotype, by observing the organism's outward appearance, known as its phenotype. For a particular trait, an individual may carry a recessive allele whose phenotypic effect is masked by the presence of a dominant allele. Much genetic analysis makes use of clever experiments and careful record-keeping, often using Punnett squares, to determine organisms' genotypes.

9.6–9.8 Probability and chance play central roles in genetics.

Because of the role of chance in genetics, we cannot always predict the exact outcome of a genetic cross.

9.6 Chance plays an important role in genetics.

Sometimes genetics is a bit like gambling. Even with perfect information, it can still be impossible to know a genetic outcome with certainty. It's like flipping a coin: you can know every last detail about the coin, but you still can't know whether the coin will land on heads or tails. The best you can do is define the probability of each possible outcome.

The rules of probability have a central role in genetics, for two reasons. First, Mendel described the process of segregation, in which each gamete receives only one copy of each gene present in somatic cells. As a result, for each gene the individual carries, the likelihood that the sperm or egg will include one allele is the same as the likelihood that it will include the other allele. It is impossible to know which allele it will be. Second, as we learned in Chapter 8, all of the sperm or eggs produced by an individual are genetically different from one another, and any one of those gametes may be the gamete involved in fertilization. Like segregation, fertilization is a chance event. Knowing everything about the alleles that a parent carries is not always enough to be able to determine with certainty which alleles his or her offspring will carry.

In games of chance, as well as in matings, we can make predictions based on probabilities. If an individual is homozygous for a trait, 100% of his or her gametes will carry that allele. An individual heterozygous for a trait has a

50% probability of carrying the dominant allele and a 50% probability of carrying the recessive allele. Let's use these probabilities in an example.

What is the probability that an offspring will be homozygous for albinism (*mm*) if the male parent is heterozygous for albinism (*Mm*) and the female parent is albino (*mm*)? To get the *mm* outcome, two events must occur. First, the father's gamete must carry the recessive allele (*m*), and second, the mother's gamete must carry the recessive allele (*m*). In this case, the probability of a homozygous recessive offspring is 0.5 (the probability that the father's gamete carries *m*) times 1.0 (the probability that the mother's gamete carries *m*), for a probability of 0.5, or 50%, or 1/2 (1 in 2). This is a general rule when determining the likelihood of a complex event occurring: if you know the probability of each component that must occur, you multiply all the probabilities together to get the overall probability of that complex event occurring (FIGURE 9-13).

Consider an example that involves Tay-Sachs disease. As described in Chapter 4, Tay-Sachs is caused by malfunctioning lysosomes that do not digest cellular waste properly. It leads to death in early childhood. Tay-Sachs occurs if a child inherits two recessive alleles (*t*) for the Tay-Sachs gene. If each parent is heterozygous for the Tay-Sachs gene (*Tt*), what is the probability that their child will have Tay-Sachs disease?

Half the father's sperm will have the dominant *T* allele and half will have the *t* allele. So the probability of the child inheriting the *t* allele is 0.5. Half of the mother's eggs will have the *T* allele, while the other half have the *t* allele, so the probability of the child inheriting the *t* allele from its mother is also 0.5. And the overall probability of the child having Tay-Sachs disease is $0.5 \times 0.5 = 0.25$, or 1 in 4 (FIGURE 9-14).

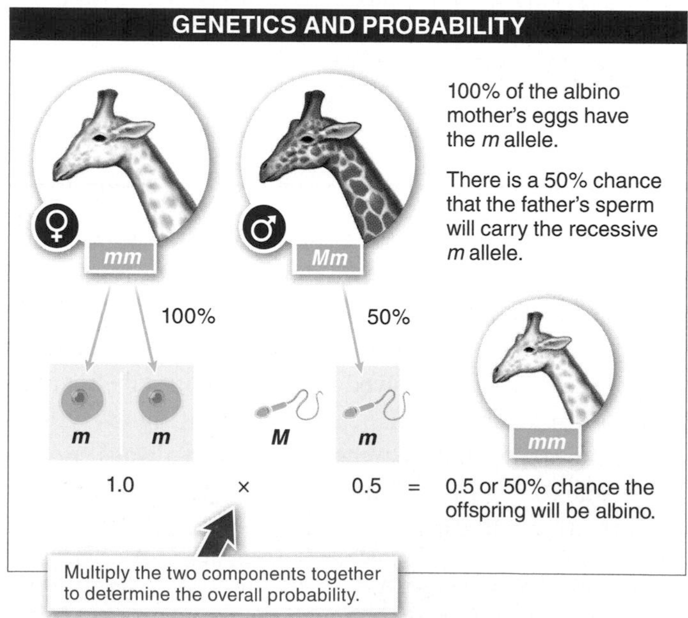

GENETICS AND PROBABILITY
100% of the albino mother's eggs have the m allele.
There is a 50% chance that the father's sperm will carry the recessive m allele.
mm
Mm
100%
50%
m
m
M
m
mm
1.0
×
0.5
=
0.5 or 50% chance the offspring will be albino.
Multiply the two components together to determine the overall probability.

FIGURE 9-13 **Using probability to determine the chance of inheriting a recessive trait such as albinism.**

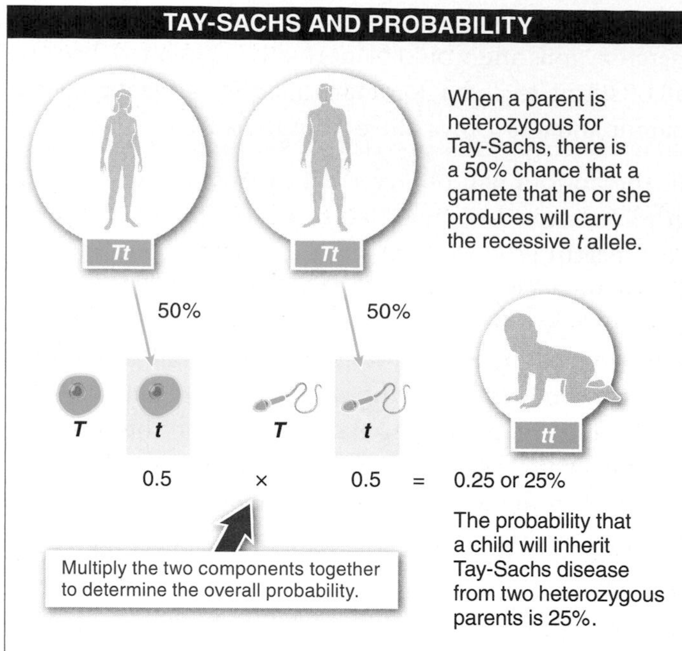

FIGURE 9-14 **The chance of inheriting Tay-Sachs disease.**

Of course, if the couple has only one child, we can't predict *with certainty* whether the child will have Tay-Sachs. The mathematical probability tells us, however, that if the couple had an infinite number of children, we could expect that one-fourth of them would have Tay-Sachs disease. This fact may not be much help for heterozygous couples trying to decide whether to have a child or not.

TAKE HOME MESSAGE 9.6

» Probability plays a central role in genetics. In segregation, each gamete that an individual produces receives only one of the two copies of each gene the individual carries in its other (somatic) cells, but it is impossible to know which allele goes into the gamete. Chance plays a role in fertilization, too: all of the sperm or eggs produced by an individual are different from one another, and any one of those gametes may be the gamete involved in fertilization.

9.7 A test-cross enables us to figure out which alleles an individual carries.

Genes are too small to be seen, and so determining an individual's genotype requires indirect methods. Suppose you are in charge of the alligators at a zoo. Some of your alligators come from a population in which white, albino alligators have occasionally occurred, although none of your alligators are white. Because white alligators are popular with zoo visitors, you would like to produce some at your zoo through a mating program. How would you go about producing white alligators?

The problem is that you cannot be certain of the genotype of your alligators. They might be homozygous dominant, *MM,* or they might be heterozygous, *Mm.* In either case, their phenotype is normal coloration. Determining the genotype of a particular alligator is a challenge to animal breeders, but not an insurmountable one. Genes may be invisible, but their identity can be revealed by a tool called the **test-cross.**

In the test-cross, you cross (i.e., mate) an individual exhibiting a dominant trait but whose genotype is unknown with an individual that is homozygous recessive. Then you examine the phenotypes of their offspring. To breed an albino alligator, you could borrow an albino alligator from another zoo and breed your unknown-genotype alligator (genotype: *M_*) with that albino alligator (genotype: *mm*). There are two possible outcomes, and they will reveal the genotype of your *M_* alligator (FIGURE 9-15). If your alligator is homozygous dominant (*MM*), it will contribute a dominant allele, *M,* to every offspring. Even though the albino alligator will contribute the recessive allele, *m,* to all of its offspring, all the offspring of this cross will be heterozygous, *Mm,* and none of them will be albino.

If, on the other hand, your *M_* alligator is heterozygous, *Mm,* half of the time it will contribute a recessive allele, *m,* to the offspring. In every one of those cases, the offspring will be homozygous recessive and thus albino.

So the cleverness of the test-cross is that when you cross your unknown-genotype organism with an individual

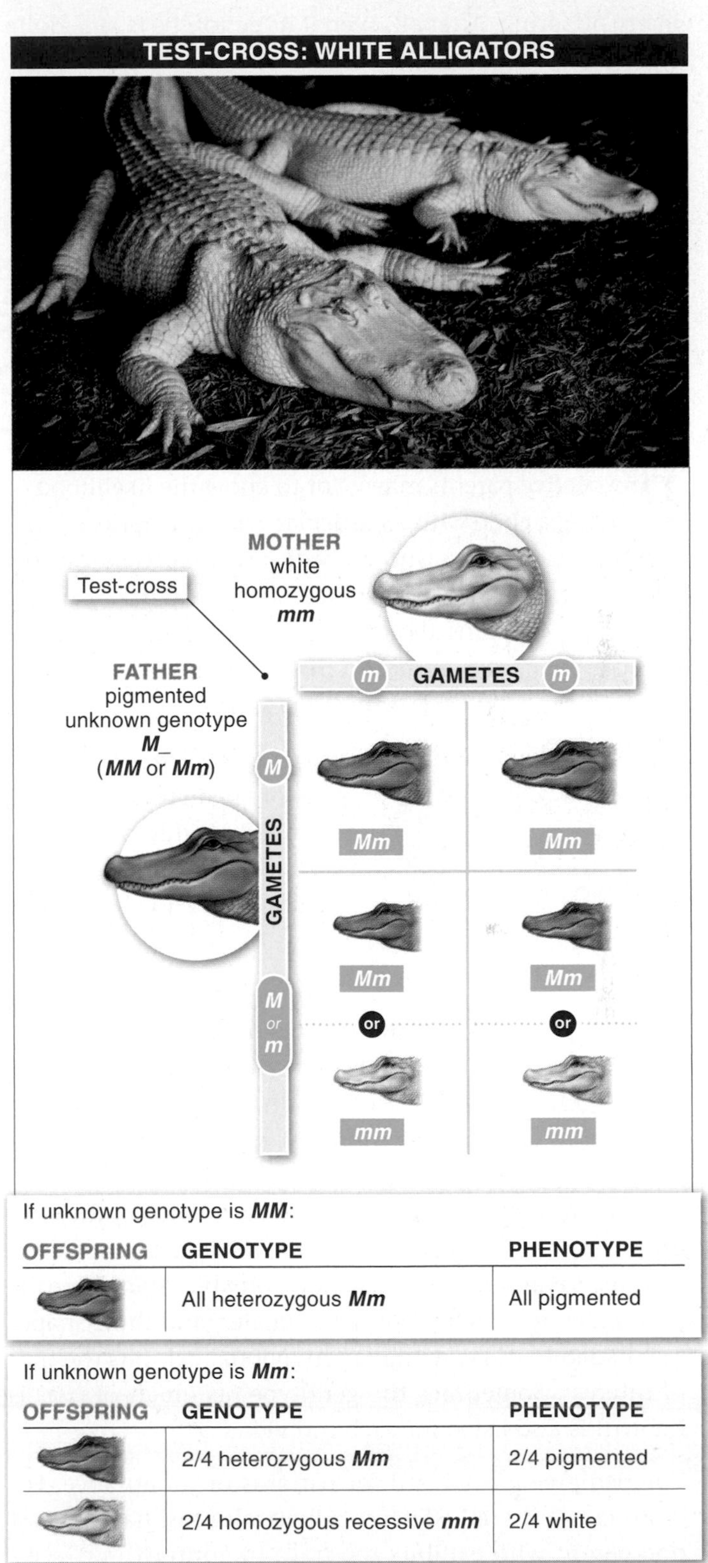

If unknown genotype is ***MM***:

OFFSPRING	GENOTYPE	PHENOTYPE
	All heterozygous ***Mm***	All pigmented

If unknown genotype is ***Mm***:

OFFSPRING	GENOTYPE	PHENOTYPE
	2/4 heterozygous ***Mm***	2/4 pigmented
	2/4 homozygous recessive ***mm***	2/4 white

FIGURE 9-15 A test-cross can reveal an unknown genotype. In this test-cross, a homozygous white female alligator is bred with a normally colored male of unknown genotype. The color of their offspring will help identify whether the male is homozygous dominant or heterozygous.

showing the recessive trait (and so having the known genotype of *mm*), the offspring will reveal the previously unknown makeup of the parent. To be confident in concluding that the unknown-genotype alligator has the *MM* genotype, though, you'd have to observe many, many offspring. After all, even if its genotype is *Mm,* quite a few offspring in a row might be normally pigmented, with the genotype *Mm.* Eventually, however, a heterozygous individual is likely to produce an offspring with the homozygous recessive genotype, *mm,* and the albino phenotype.

TAKE HOME MESSAGE 9.7

» In a test-cross, an individual that exhibits a dominant trait but has an unknown genotype is mated with an individual that is homozygous recessive. The phenotypes of the offspring reveal whether the unknown-genotype individual is homozygous dominant (all of the offspring exhibit the dominant trait) or heterozygous (half of the offspring show the dominant trait and half show the recessive trait).

9.8 We use pedigrees to decipher and predict the inheritance patterns of genes.

Prospective parents may want to know the likelihood of having a child with a particular genetic disease, say hemophilia. Geneticists who study diseases may want to know how a particular disease or trait is inherited. Is it recessive or dominant? Is it carried on the sex chromosomes or on one of the other chromosomes? A **pedigree** is a type of family tree that can help answer these questions.

Q Why do breeders value "pedigreed" horses and dogs so much?

Dog and horse breeders often speak of "pedigreed" animals, a feature that adds tremendous value to the animals. This is because, with knowledge of an animal's family tree, it is much less likely that any genetic surprises will occur as the animal develops and reproduces.

In a pedigree, information is gathered from as many related individuals as possible across multiple generations (FIGURE 9-16). Starting from the bottom, each row in the chart represents a generation, listing all of the children in their order of birth, their sex, and whether or not they have a particular trait. Working up the pedigree, the children's parents are indicated and, above them, their parents' parents, for as far back as data are available. Squares represent males and circles represent females, and these shapes are shaded to indicate that an individual exhibits the trait of interest. Sometimes the genotype (as much of it as is known) is also listed for each individual.

The pedigree is analyzed for patterns of inheritance. For dominant traits, all affected individuals must have at least one parent who exhibits the trait. In contrast, for recessive traits, an individual may exhibit a recessive trait even if neither parent exhibits the trait. In this case, the individual's parents must be heterozygous for that trait.

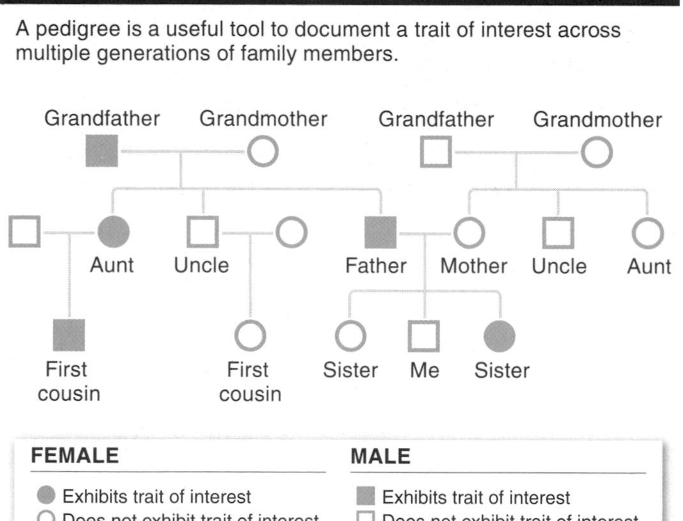

FIGURE 9-16 **Family tree.** A pedigree maps the occurrence of a trait in several generations of a family.

The pedigree can also help to determine whether a trait is carried on the sex chromosomes (X or Y, which we discuss in Section 9.13) or on one of the non-sex chromosomes (also called autosomes). Traits that are controlled by genes on the sex chromosomes are called **sex-linked traits.** Recessive sex-linked traits, for example, appear more frequently in males than in females, whereas dominant sex-linked traits appear more frequently in females. These patterns may become obvious only on inspection of a large pedigree.

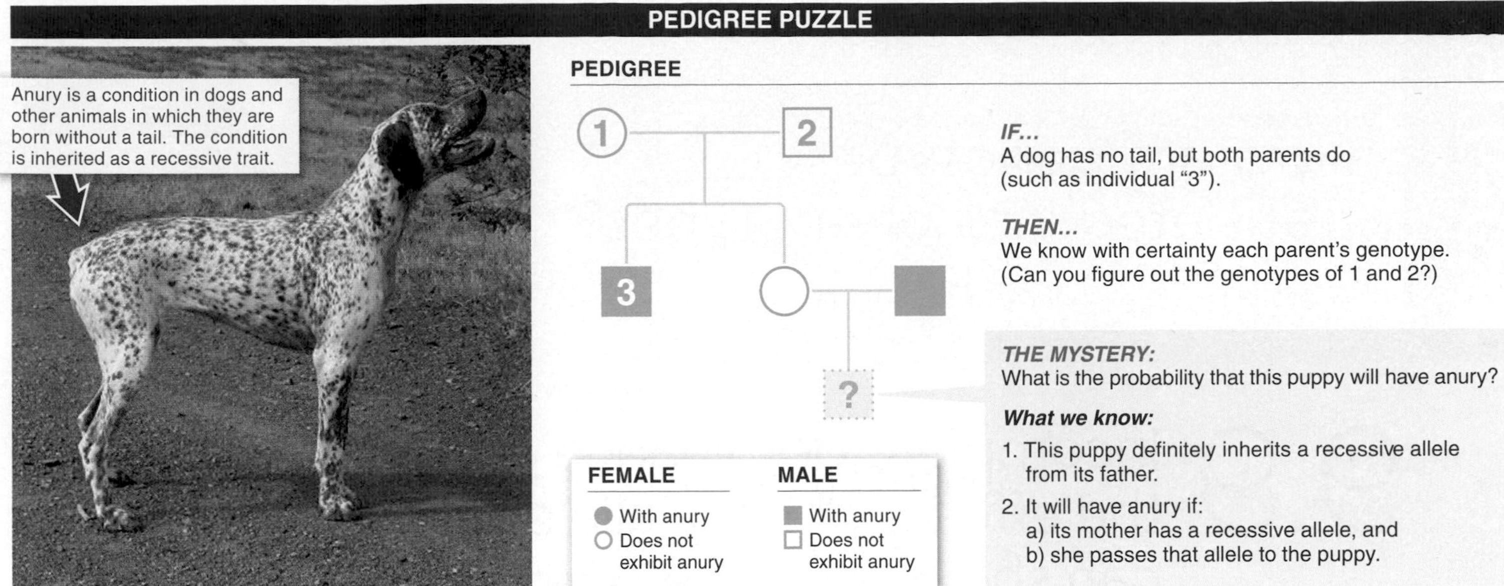

PEDIGREE PUZZLE
Anury is a condition in dogs and other animals in which they are born without a tail. The condition is inherited as a recessive trait.
PEDIGREE
1
2
3
?
IF...
A dog has no tail, but both parents do (such as individual "3").
THEN...
We know with certainty each parent's genotype. (Can you figure out the genotypes of 1 and 2?)
THE MYSTERY:
What is the probability that this puppy will have anury?
What we know:
1. This puppy definitely inherits a recessive allele from its father.
2. It will have anury if:
a) its mother has a recessive allele, and
b) she passes that allele to the puppy.
FEMALE
With anury
Does not exhibit anury
MALE
With anury
Does not exhibit anury

FIGURE 9-17 **Tracing the occurrence of anury in a family of dogs.**

An example of how pedigrees can help determine how traits are inherited is given in (FIGURE 9-17). Anury is a condition seen in dogs and some other animals in which the animal has no tail. The pedigree reveals that anury is inherited as a recessive trait, because unaffected parents can have offspring with the disorder.

Can you determine the genotype of the individual labeled "1" in Figure 9-17? Why must that female be heterozygous (*Aa*) for anury? (By the way, an individual that carries one allele for a recessive trait, and so does not exhibit the trait but can have offspring that do, is referred to as a **carrier** of the trait.) Must her mate have the same genotype?

On the same pedigree, note that two individuals in the second row of the pedigree have a puppy (indicated by "?"). What is the probability that this puppy has anury? Examine the pedigree carefully and see whether you can come up with the answer; the next paragraph will guide you through if you get stuck.

In the cross producing the mystery puppy, the father has anury and so must be homozygous recessive; he will definitely pass on one *a* to the son. The mother's genotype can be *AA*, *Aa*, or *aA*, given that both of her parents are heterozygous. Consequently, she has a 2/3 (2 in 3) probability of being a carrier of the *a* allele. If she is, then there is a 1/2 (50%) chance that she will pass the allele on to her offspring, and he will have anury. The probability, therefore, is $2/3 \times 1/2 = 1/3$, or a 1 in 3 chance that the puppy will have anury.

As we'll discuss in later sections of this chapter, some traits may not show complete dominance and many traits are also influenced by the environment, so it is not always obvious what a trait's mode of inheritance is. In such cases, the more individuals we can include in the pedigree, the more accurate the analysis. Some human pedigrees contain thousands of individuals and stretch back six or more generations. With some other species it is possible to analyze tens of thousands of individuals per generation for a dozen or more generations. This is why plants and small insects (among other organisms) are excellent for studying inheritance patterns.

Q Researchers often use plants and small insects to study inheritance patterns—even when they're more interested in how a trait works in humans. Why?

Once we have an idea about how a trait is inherited, we can identify individuals who are carriers of a recessive trait and make informed predictions about a couple's risk of having a child with a particular disorder. This knowledge can be useful in conjunction with prenatal testing and treatment.

TAKE HOME MESSAGE 9.8

» **Pedigrees help scientists, doctors, animal and plant breeders, and prospective parents determine the genes that individuals carry and the likelihood that the offspring of two individuals will exhibit a given trait.**

9.9–9.15

How are genotypes translated into phenotypes?

Phenotypic diversity—which is all around us—has multiple sources.

9.9 Incomplete dominance and codominance: the effects of both alleles in a genotype can show up in the phenotype.

As Mendel saw it, the world of genetics was straightforward. Each of the traits he studied was coded for by a single gene with two alleles—one completely dominant and one recessive—and with no environmental effects. We should be so lucky. In this and the following sections we build up a more complex model of how genes influence the building of bodies.

We begin with the observation that the phenotype of heterozygous individuals sometimes differs from that of either of the homozygotes for the trait, and instead reflects the influence of both alleles rather than a clearly dominant allele.

One situation in which complete dominance is not observed is called **incomplete dominance,** in which the phenotype of a heterozygote is intermediate between the phenotypes of the two homozygotes. An example of incomplete dominance that we can easily observe is the flower color of snapdragons (FIGURE 9-18).

We can obtain true-breeding (homozygous) lines of snapdragons with red flowers and true-breeding (homozygous) lines that produce only white flowers. When plants from these two populations are crossed, we would expect—if one allele were dominant over the other—a generation with either all red or all white flowers. Instead, such crosses always produce plants with pink flowers. Then, when we cross two plants with pink flowers, we get 1/4 red-flowered plants, 1/2 pink-flowered plants, and 1/4 white-flowered plants.

How can we interpret this cross? Here we use a slightly different way of denoting genotype. The plants with white flowers have the genotype C^WC^W and produce no pigment. At the other extreme, the plants with red flowers have the genotype C^RC^R and produce a great deal of pigment. The letter *C* indicates that the gene codes for color, and the superscript *W* or *R* refers to an allele producing no pigment (white) or red pigment. We use these designations for the genotypes because it isn't clear that either white or red is dominant over the other, and so neither should be represented by uppercase or lowercase. The pink flowers receive one of the pigment-producing C^R alleles and one of the no-pigment-producing C^W alleles, and so produce an intermediate amount of pigment. Ultimately, the intensity of pigmentation just depends on the amount of pigment chemical that is made by the flower-color genes.

An example of incomplete dominance in humans can be seen in the processing of cholesterol in the bloodstream. There is a membrane receptor that allows cells (chiefly those in the liver) to remove cholesterol from the bloodstream (see Section 4.11). Individuals who carry two copies of a mutant allele, called *FH,* for this gene produce almost no LDL receptors. Consequently, in these individuals, circulating cholesterol levels are high and cardiovascular disease develops at a very young age. Few individuals survive past 20 years of age. This is in sharp contrast to individuals carrying two copies of the allele for normal-functioning LDL receptors, who experience significantly lower levels of circulating cholesterol.

Heterozygotes, carrying one copy of the mutant *FH* allele and one copy of the allele for normal-functioning LDL receptors, represent an intermediate situation. They produce about half as many LDL receptors as do people who are homozygous for the normal-functioning receptor, leading

INCOMPLETE DOMINANCE: SNAPDRAGONS

Incomplete dominance occurs when a heterozygote exhibits an intermediate phenotype between the two homozygotes.

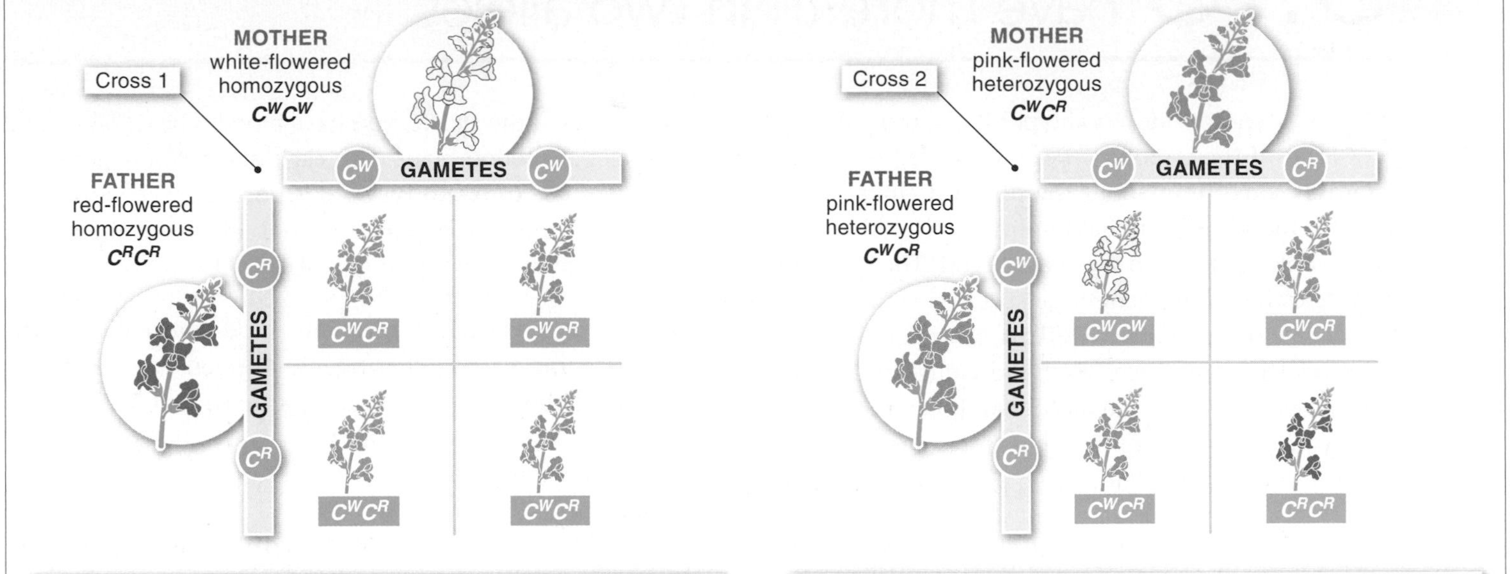

Cross 1

MOTHER white-flowered homozygous C^WC^W

FATHER red-flowered homozygous C^RC^R

GAMETES: C^W, C^W (mother); C^R, C^R (father)

	C^W	C^W
C^R	C^WC^R	C^WC^R
C^R	C^WC^R	C^WC^R

OFFSPRING	GENOTYPE	PHENOTYPE
	All heterozygous C^WC^R	All pink flowers

Because neither allele is dominant, we can't use uppercase and lowercase for the alleles. Instead, the superscript ***W*** represents the allele that produces the white flower, and ***R*** represents the red-flower allele.

Cross 2

MOTHER pink-flowered heterozygous C^WC^R

FATHER pink-flowered heterozygous C^WC^R

GAMETES: C^W, C^R (mother); C^W, C^R (father)

	C^W	C^R
C^W	C^WC^W	C^WC^R
C^R	C^WC^R	C^RC^R

OFFSPRING	GENOTYPE	PHENOTYPE
	1/4 homozygous C^RC^R	1/4 red flowers
	2/4 heterozygous C^WC^R	2/4 pink flowers
	1/4 homozygous C^WC^W	1/4 white flowers

FIGURE 9-18 **Pink snapdragons demonstrate incomplete dominance.** When true-breeding white and red snapdragons are crossed, offspring have pink flowers.

to higher-than-average levels of circulating cholesterol and 10 times the risk of death from cardiovascular disease.

A second situation in which complete dominance is not observed is called **codominance,** in which the heterozygote displays characteristics of both homozygotes—as playfully represented in **FIGURE 9-19** (although "shirt phenotype," of course, has no genetic basis). In codominance, neither allele masks the effect of the other. An example of codominance occurs with feather color in chickens. When white chickens are crossed with black chickens, all the offspring have both white and black feathers.

TAKE HOME MESSAGE 9.9

» Sometimes the effects of both alleles in a heterozygous genotype are evident in the phenotype for that trait. With incomplete dominance, the phenotype of a heterozygote appears to be an intermediate blend of the phenotypes of the two homozygotes. With codominance, a heterozygote has a phenotype that exhibits characteristics of both homozygotes.

CODOMINANCE

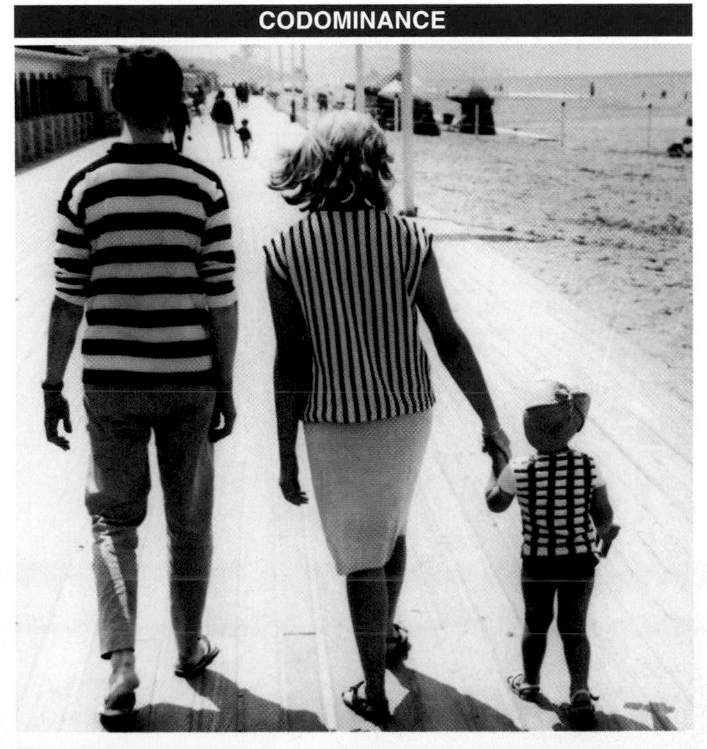

FIGURE 9-19 **With codominance, a heterozygous individual shows features of both alleles.**

9.10 What's your blood type? Some genes have more than two alleles.

Do you know your blood type? It can be O, A, B, or AB. Each of these blood types (also called blood groups) indicates something about the physical characteristics of your red blood cells and has implications for blood transfusions—both giving and receiving blood. The blood groups are interesting from a genetic perspective because they illustrate a case of **multiple allelism,** in which a single gene has more than two alleles. Each individual still carries only two alleles—one from the mother and one from the father. But if you surveyed all of the alleles for this gene *in the population,* you would find more than just two different alleles.

Inheritance of the ABO blood groups provides the simplest example of multiple allelism, because there are only three alleles. We can call these alleles I^A, I^B, and *i* (using both types of allele notation introduced earlier). The I^A and I^B alleles are both completely dominant to *i*, so individuals are considered to have blood type A whether they have the genotype I^AI^A or I^Ai (FIGURE 9-20). Similarly, an individual with the genotype I^BI^B or I^Bi is considered to have blood type B. If you carry two copies of the *i* allele, you have blood type O. The I^A and I^B alleles are codominant with each other, so the genotype I^AI^B gives rise to blood type AB. Consequently, with these three alleles in the population, individuals can be one of four different blood types: A, B, AB, or O.

These blood-type alleles direct the production of a specific set of chemicals, called antigens. Antigens are molecules (chiefly carbohydrates bound to protein) that jut from the surface of a cell and can "turn on" a body's defenses against foreign invaders.

The I^A allele directs the production of A antigens that cover the surface of red blood cells. Similarly, the I^B allele directs the production of B antigens on all red blood cells. Individuals with blood type AB have red blood cells with both A and B antigens. The *i* allele does not code for the A antigen *or* the B antigen. This means that individuals with blood type O have red blood cells that have neither A nor B antigens on their surface (FIGURE 9-21).

MULTIPLE ALLELISM: BLOOD TYPE

Multiple allelism occurs when there are three or more alleles for a gene within a population. An individual still inherits only two alleles—one from each parent.

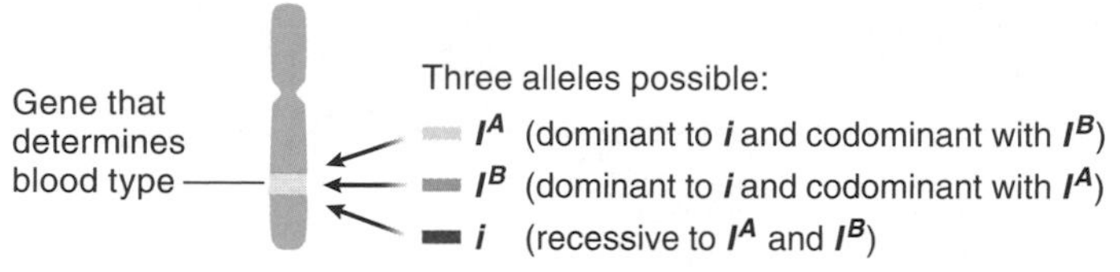

Red blood cells can have 6 different genotypes (I^AI^A, I^Ai, I^BI^B, I^Bi, I^AI^B, and *ii*). These genotypes result in 4 different phenotypes (type A, type B, type AB, and type O).

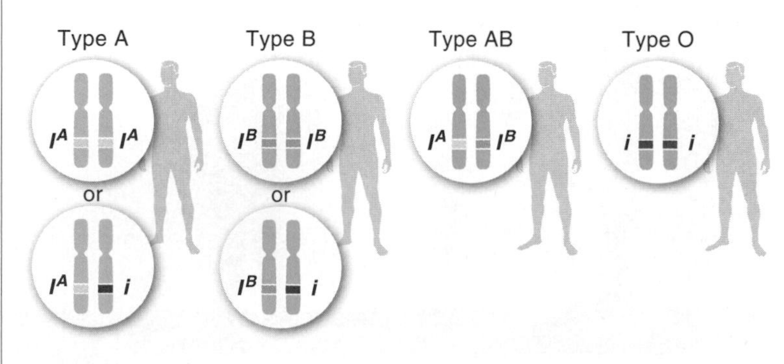

FIGURE 9-20 **Multiple allelism.** Three different alleles—I^A, I^B, and *i*—determine blood type in humans.

Antigens are like signposts in the body's disease-fighting immune system, telling the immune system whether a cell belongs in the body or not. If a red blood cell with the wrong antigens enters your bloodstream, your immune system recognizes it as a foreign invader and destroys it. Such an attack is initiated by molecules in the bloodstream called *antibodies*. Individuals with only A antigens on their red blood cells produce antibodies that attack B antigens. Under normal circumstances, antibodies do not encounter a red blood cell with foreign antigens. But such an event could occur if red blood cells with foreign antigens were accidentally injected into the person's bloodstream in a transfusion. An improper transfusion can cause destruction of red blood cells, low blood pressure, and even death.

Individuals with only B antigens on their red blood cells produce antibodies that attack A antigens. Individuals with blood type O, who have neither A nor B antigens on their red blood cells, produce antibodies that attack both A and B

Q Why are people with type O blood considered "universal donors"? Why are those with type AB considered "universal recipients"?

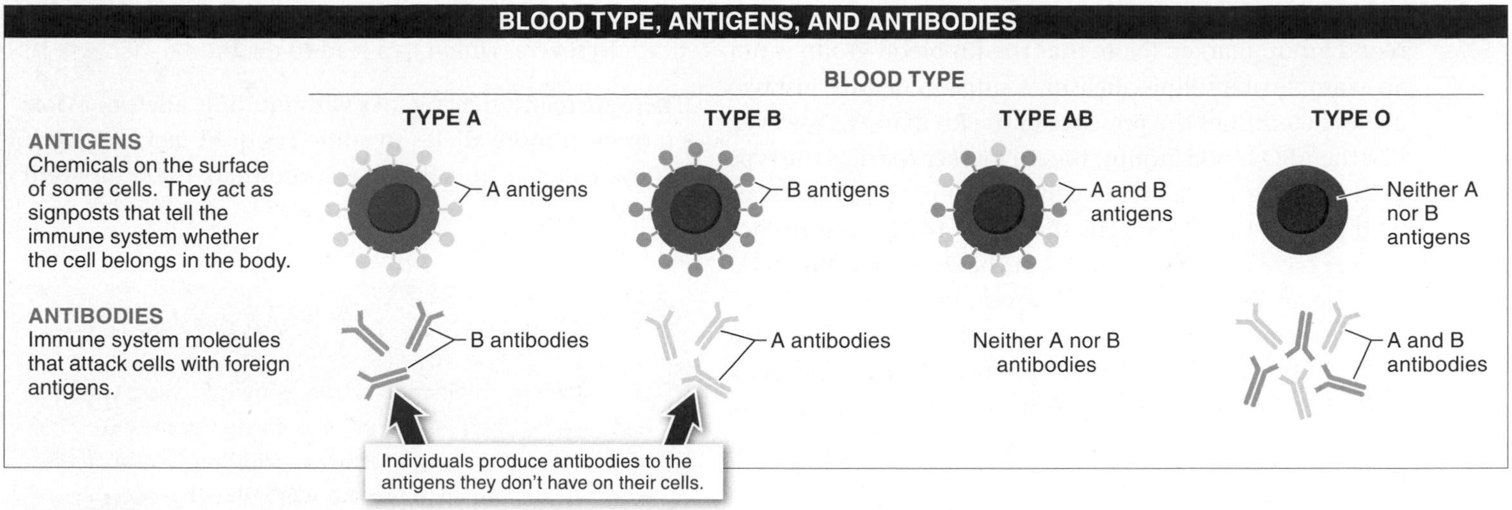

BLOOD TYPE, ANTIGENS, AND ANTIBODIES

BLOOD TYPE

TYPE A | TYPE B | TYPE AB | TYPE O

ANTIGENS
Chemicals on the surface of some cells. They act as signposts that tell the immune system whether the cell belongs in the body.

A antigens | B antigens | A and B antigens | Neither A nor B antigens

ANTIBODIES
Immune system molecules that attack cells with foreign antigens.

B antibodies | A antibodies | Neither A nor B antibodies | A and B antibodies

Individuals produce antibodies to the antigens they don't have on their cells.

FIGURE 9-21 **Friend or foe?** Antigens are signposts that tell the immune system whether or not a cell belongs in the body.

antigens. And individuals with blood type AB don't produce either type of antibody (or else they would have antibodies that attacked their own blood cells).

From this information, we can deduce which blood types can be used in transfusions. Individuals with blood type O are universal donors, because their red blood cells have no A or B antigens and so do not trigger a reaction from either type of antibody. And individuals with blood type AB are universal recipients, because they do not produce antibodies to either the A or B antigen. This is shown in FIGURE 9-22.

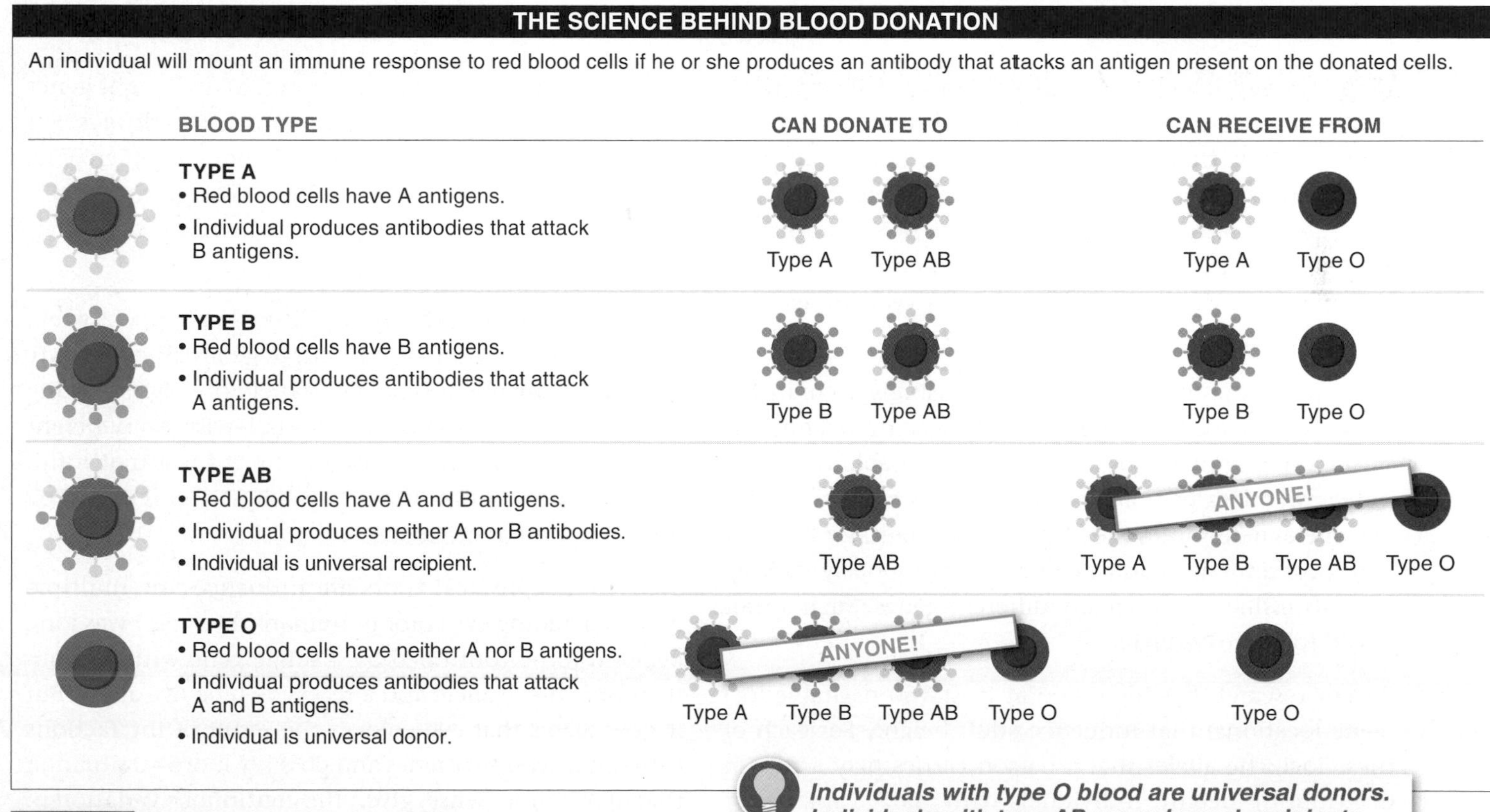
THE SCIENCE BEHIND BLOOD DONATION
An individual will mount an immune response to red blood cells if he or she produces an antibody that attacks an antigen present on the donated cells.
BLOOD TYPE
CAN DONATE TO
CAN RECEIVE FROM
TYPE A
• Red blood cells have A antigens.
• Individual produces antibodies that attack B antigens.
Type A
Type AB
Type A
Type O
TYPE B
• Red blood cells have B antigens.
• Individual produces antibodies that attack A antigens.
Type B
Type AB
Type B
Type O
TYPE AB
• Red blood cells have A and B antigens.
• Individual produces neither A nor B antibodies.
• Individual is universal recipient.
Type AB
ANYONE!
Type A
Type B
Type AB
Type O
TYPE O
• Red blood cells have neither A nor B antigens.
• Individual produces antibodies that attack A and B antigens.
• Individual is universal donor.
ANYONE!
Type A
Type B
Type AB
Type O
Type O

FIGURE 9-22 **Mapping blood compatibility.**

Individuals with type O blood are universal donors. Individuals with type AB are universal recipients.

Another marker on the surface of red blood cells is the Rh blood group marker. (Note that the Rh blood group is not an example of multiple allelism. A single gene with just two alleles determines the presence of the Rh marker. However, like the ABO blood groups, the Rh marker restricts the type of blood a person can receive in a transfusion.) Individuals who possess red blood cells that carry the Rh cell surface marker have one or two copies of the dominant Rh marker allele and are said to be "Rh-positive." This "positive" (or "+") is noted along with their ABO blood type, as in "O-positive." Individuals who have two copies of the recessive allele for this gene do not have any Rh markers, and they are described as "negative," as in "O-negative." If, during a blood transfusion, individuals who are Rh-negative are exposed to Rh-positive blood, their immune system attacks the Rh antigens as foreign invaders—an immune response that can vary from mild, which passes unnoticed, to severe, which can lead to death.

There are many, many genes with multiple alleles—a dozen or even more alleles in some cases. In fact, one gene for eye color in fruit flies has more than 1,000 different alleles!

TAKE HOME MESSAGE 9.10

» In multiple allelism, a single gene has more than two alleles. Each individual still carries only two alleles, but in the population, more than two alleles exist. This is the case for the ABO blood groups in humans.

9.11 Multigene traits: how are continuously varying traits such as height influenced by genes?

Old wives' tales suggest a couple of ways that parents can predict the ultimate height of their child. If the baby is a boy, they say to add 5 inches to the mother's height and average that with the father's height. If it is a girl, subtract 5 inches from the father's height and average that with the mother's height. Alternatively, the lore says, just take the child's height at two years of age and double it.

These prediction methods can be surprisingly accurate because genes play a strong role in influencing height; offspring do resemble their parents in measurable ways. But unlike Mendel's pea plants, in which a single gene with two alleles determines the height of the plant, for humans and most animals, adult height—a continuously varying trait—is influenced by many different genes. Such a trait is said to be **polygenic.**

Recent research has identified at least 180 heritable loci (gene locations) that influence adult height. For each of these loci, the alleles that a person carries play a role in determining height. Individuals with "tall" variants for more of the genes tend to be taller than those with the "tall" variants for fewer of the genes. Many height genes play roles in skeletal growth and hormone pathways. For example, a gene on chromosome 15 codes for an enzyme that converts testosterone to estrogen. This enzyme influences height because estrogen helps bones fuse at their ends and thus stop growing.

The term **additive effects** describes what happens when the effects of the alleles of multiple genes all contribute to the ultimate phenotype. The variety of heights among humans—from very short to very tall, with every height in between—reflects the fact that height is a trait influenced by contributions from multiple genes, as well as by the environment (FIGURE 9-23).

Many other physical traits are influenced by multiple genes, including eye color in humans. Eye color was long believed to be controlled by a single gene with a dominant brown-eye allele and a recessive blue-eye allele. But it now seems that eye color is the result of interactions between at least two genes and possibly more—a situation that makes more sense, given the continuous variation in eye color among adults.

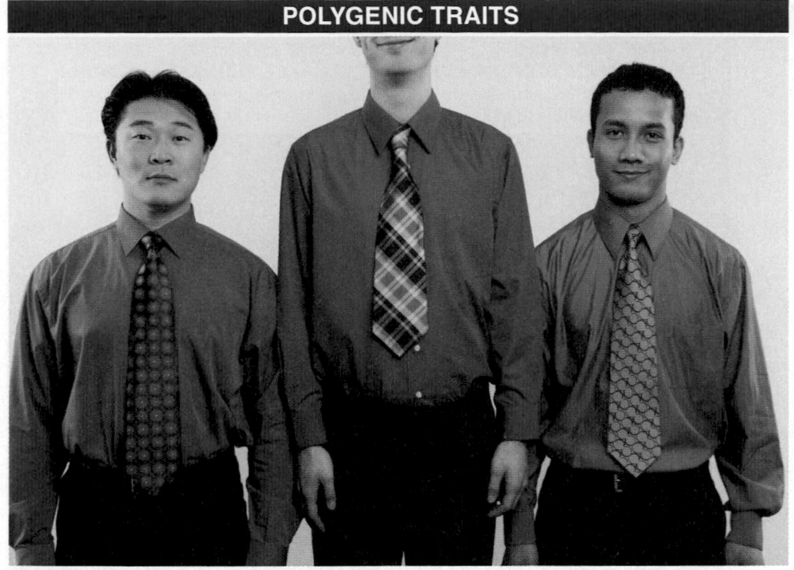

FIGURE 9-23 **From many genes, one trait.** Height and skin color are multigene traits.

Many behavioral traits are influenced by multiple genes. The developmental disorder known as autism, for example, seems to be the result of alterations in numerous—perhaps as many as 10 or even 20—genes. Individuals with autism have difficulty interacting with others, particularly in making emotional connections. They also tend to have narrowly focused and repetitive interests. Autism is notoriously difficult to study, because its symptoms are so varied, as is their intensity. These variations may result from different combinations of the many genes involved in autism.

The genes responsible for autism may also influence some desirable characteristics. This idea—called the "geek theory of autism"—arose when researchers documented an over-representation of autism among children of parents working in the fields of engineering, physics, computer science, and math. (The researchers point out, however, that children with autism are born to parents across all professions and socioeconomic backgrounds.) Perhaps alleles for genes that contribute to making individuals good at computer programming and solving complex technology problems can produce autism when someone carries too many of them.

TAKE HOME MESSAGE 9.11

» Many traits, including continuously varying traits such as height and eye color, are influenced by multiple genes.

9.12 Sometimes one gene influences multiple traits.

Q What is the benefit of "almost" having sickle-cell disease?

Just as multiple genes can influence one trait, some individual genes can influence multiple, unrelated traits, a phenomenon called **pleiotropy.** In fact, this may be true of nearly all genes. Consider sickle-cell disease (also called sickle-cell anemia), a potentially fatal condition in which individuals produce defective red blood cells that become sickle-shaped when they lose the oxygen they carry. The defective blood cells can't effectively transport oxygen to tissues and they accumulate in blood vessels, causing extreme pain. Individuals with sickle-cell disease suffer shortness of breath and numerous other problems that lead to a significantly reduced life span.

The gene responsible for sickle-cell disease codes for part of the hemoglobin molecule, the oxygen-carrying molecule in red blood cells: Hb^A is the allele for normal hemoglobin, and Hb^S is the abnormal, "sickle-cell" allele. Sickle-cell anemia occurs in individuals homozygous for the sickle-cell allele, Hb^SHb^S, but not in individuals carrying at least one copy of the normal allele, Hb^A. (Heterozygous individuals produce both normal and sickling red blood cells, but not enough sickle-shaped cells to cause sickle-cell anemia).

PLEIOTROPY: ONE GENE, MANY EFFECTS

The allele for sickle-cell disease is pleiotropic, influencing multiple traits. It causes red blood cells to form an unusual, sickled shape, and it also provides resistance to malaria.

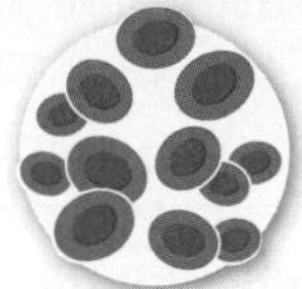

Hb^AHb^A **HOMOZYGOTE**
- Produces no sickling cells and so does not have sickle-cell disease
- Is susceptible to malaria

Hb^SHb^A **HETEROZYGOTE**
- Produces some sickling cells, but not enough to have sickle-cell disease
- Is immune to malaria

Hb^SHb^S **HOMOZYGOTE**
- Produces sickling red blood cells and so has sickle-cell disease
- Is immune to malaria

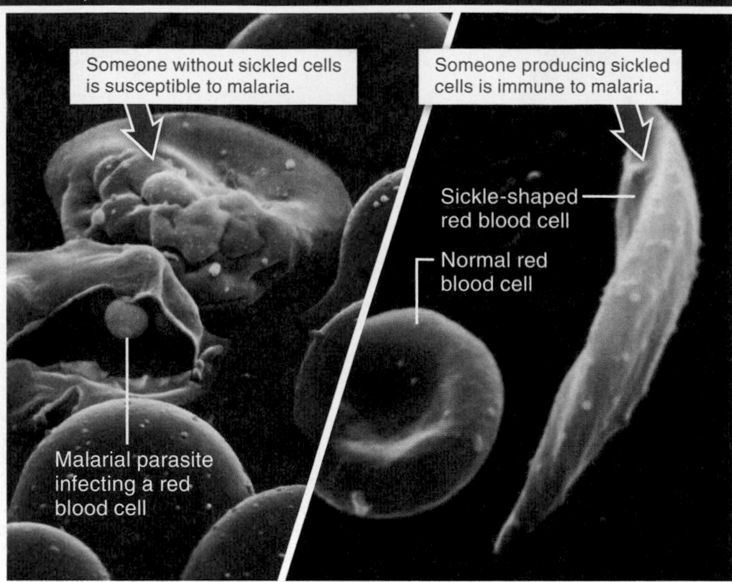
Someone without sickled cells is susceptible to malaria.
Someone producing sickled cells is immune to malaria.
Sickle-shaped red blood cell
Normal red blood cell
Malarial parasite infecting a red blood cell

FIGURE 9-24 From one gene, multiple traits. The allele for sickle-cell disease is pleiotropic: it causes red blood cells to form an unusual, sickled shape, and it also provides resistance to malaria.

This hemoglobin gene is pleiotropic because, although it is just one gene, it causes multiple phenotypic effects. Individuals homozygous or heterozygous for the sickle-cell allele, Hb^SHb^S or Hb^AHb^S, have abnormal hemoglobin and red blood cells, and circulatory problems. Moreover, individuals with these genotypes also are resistant to the parasite that causes malaria. The malarial parasite—which lives in red blood cells—cannot survive well in cells that carry the defective version of the hemoglobin gene (FIGURE 9-24).

Another example of pleiotropy is the gene called *CFTR,* which codes for a membrane protein that serves as a channel for chloride ions. Mutations that impair the functioning of these ion channels result in the disease cystic fibrosis, with its characteristic build-up of mucus in the lungs. They also give rise to diabetes resulting from reduced pancreatic functioning and, in males, cause a deterioration of the vas deferens, leading to infertility.

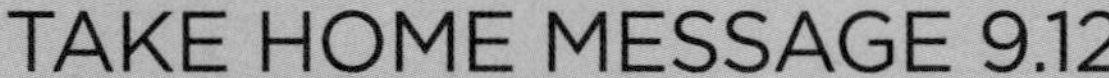

TAKE HOME MESSAGE 9.12

» In pleiotropy, one gene influences multiple, unrelated traits. Most, if not all, genes may be pleiotropic.

9.13 Why are more men than women color-blind? Sex-linked traits differ in their patterns of expression in males and females.

The patterns of inheritance of most traits do not differ between males and females. When a gene is on an autosome (one of the non-sex chromosomes), both males and females inherit two copies of the gene, one from their mother and one from their father. The likelihood that an individual inherits one genotype rather than another does not differ between males and females.

On the other hand, traits coded for by the sex chromosomes have different patterns of expression in males and females. One of the most easily observed examples of this phenomenon is red-green color-blindness. The X chromosome in humans has a gene that carries instructions for producing the light-sensitive proteins in the eye that make it possible to distinguish between the colors red and green. If an individual has at least one functioning copy of this gene, he or she produces sufficient amounts of the protein to have normal color vision.

There is a rare allele for this gene, however, that produces a nonfunctioning version of the light-sensitive protein. Having some of this nonfunctioning protein is not a problem if the person also carries another, normal version of the gene and produces some of the functioning protein.

Q If a man is color-blind, did he inherit this condition from his mother, his father, or both parents?

Here's the problem: men get only one chance to inherit the normal version of the gene that codes for red-green color vision. Because the gene is on the X chromosome, men inherit this chromosome only from their mother. Women get two chances; although a woman may inherit the defective allele from one parent, she still can inherit the normal allele from the other parent (and have normal color vision). If she inherits the defective allele from both parents, she will be red-green color-blind. As we would predict, then, the frequency of red-green color-blindness is significantly greater in males than in females (FIGURE 9-25). Approximately 7% to 10% of men exhibit red-green color-blindness, while fewer than 1% of women are red-green color-blind.

Although men exhibit sex-linked recessive traits more frequently than do women, the situation is reversed for sex-linked dominant traits. In these cases, because females have two chances to inherit the allele that causes the trait, they are more likely to have the allele and thus exhibit the trait than are males, who have only one chance to inherit the allele.

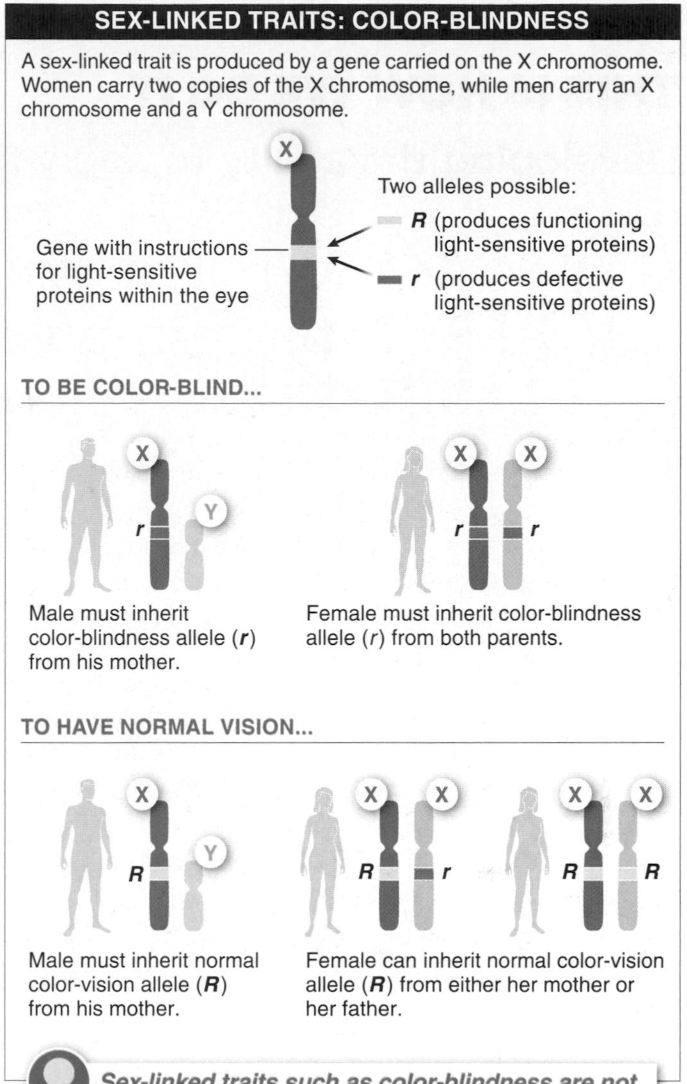

SEX-LINKED TRAITS: COLOR-BLINDNESS
A sex-linked trait is produced by a gene carried on the X chromosome. Women carry two copies of the X chromosome, while men carry an X chromosome and a Y chromosome.
X
Gene with instructions for light-sensitive proteins within the eye
Two alleles possible:
R (produces functioning light-sensitive proteins)
r (produces defective light-sensitive proteins)
TO BE COLOR-BLIND...
X
Y
r
Male must inherit color-blindness allele (r) from his mother.
X
X
r
r
Female must inherit color-blindness allele (r) from both parents.
TO HAVE NORMAL VISION...
X
Y
R
Male must inherit normal color-vision allele (R) from his mother.
X
X
R
r
X
X
R
R
Female can inherit normal color-vision allele (R) from either her mother or her father.
Sex-linked traits such as color-blindness are not expressed equally among males and females.

FIGURE 9-25 **Inheriting color-blindness.**

TAKE HOME MESSAGE 9.13

» The patterns of inheritance of most traits do not differ between males and females. However, when a trait is coded for by a gene on a sex chromosome, such as color vision on the X chromosome, the pattern of expression differs for males and females.

THIS IS HOW WE DO IT

Developing the ability to apply the process of science

9.14 What is the cause of male-pattern baldness?

It is a rare scientist who is interested only in finding a particular answer to a problem. Most are drawn to scientific thinking because it illuminates productive ways of *approaching* a problem. And the greatest advances in understanding often come about by exploring new approaches to old problems. Here we explore one of the oldest.

Why do men lose their hair? Conventional wisdom has long suggested that baldness in men is a trait they inherit from their mother. This hypothesis goes back to 1916, when Dorothy Osborn published the first scientific study putting forth heredity as a cause of baldness. In her paper, she challenged a widely held hypothesis of the time: that the wearing of hats caused baldness due to pressure on blood vessels that nourish the scalp.

Can you propose how to test the "hats cause baldness" hypothesis?

It was easy for Osborn to demonstrate that, as she put it, "the hat is not to blame." Numerous equally wrong hypotheses persist today: "The hair follicles are clogged from shampoo and too much washing." "Not enough blood is circulating around the scalp." "Hair gels and other products are toxic."

Are these hypotheses falsifiable? Propose a way of testing each of them?

What Osborn found, however, was that it's difficult to study the inheritance patterns of baldness carefully. For starters, it's very common—about 50% of men experience some balding by the age of 50, and 70% by age 70.

Why do these observations make conventional pedigree analysis difficult?

Also, because balding increases with age, it's hard to know whether younger, non-bald men will go bald later or never at all. And it's unclear whether all patterns of balding have the same underlying causes.

A modern approach: In 2005, researchers studied the DNA of 391 men—including 201 balding men—from 95 families in which at least two brothers exhibited early-onset male-pattern baldness. For comparison, they also examined the DNA of additional, unrelated men who were either under the age of 40 with male-pattern baldness or were over the age of 60 and unaffected by baldness.

The researchers found that the men with male-pattern baldness were significantly more likely to share a particular stretch of DNA on their X chromosome. The unaffected men were significantly more likely to share a different sequence of DNA.

If balding men commonly share a DNA sequence on their X chromosome, what does this tell us about the inheritance of this balding-associated DNA?

The finding that the DNA region implicated in baldness is on the X chromosome confirms the long-standing observation that male-pattern baldness is a trait passed down to men from their mothers. After all, men receive their sole X chromosome from their mother.

Future directions? This research resolved one long-standing debate. More important, perhaps, it illuminated an avenue for research on the *treatment* of male-pattern baldness. The DNA sequence associated with male-pattern baldness, it turned out, is located within the region of a single gene—a gene already suspected to play a role in triggering baldness. The sequence carries the instructions to produce a higher-than-typical amount of receptors for male sex hormones, such as testosterone. This is consistent with the finding that castrated males—who produce almost no testosterone at all—don't go bald. So, what strategies for treating male-pattern baldness do these observations suggest?

TAKE HOME MESSAGE 9.14

» Researchers' observations of male-pattern baldness within families and comparisons with unrelated individuals suggest that the baldness is caused by a sex-linked gene that codes for a male-hormone receptor. Males inheriting an allele on the X chromosome—always from their mother—for higher activity of this receptor gene are more likely to have male-pattern baldness than males inheriting an alternative allele.

9.15 Environmental effects: identical twins are not identical.

It is a very serious warning, in bold, capital letters:

PHENYLKETONURICS:
CONTAINS PHENYLALANINE

But it's not next to a skull and crossbones on a glass bottle in a chemistry lab. It's on cans of diet soda (FIGURE 9-26). Most of us don't notice the warning or we ignore it.

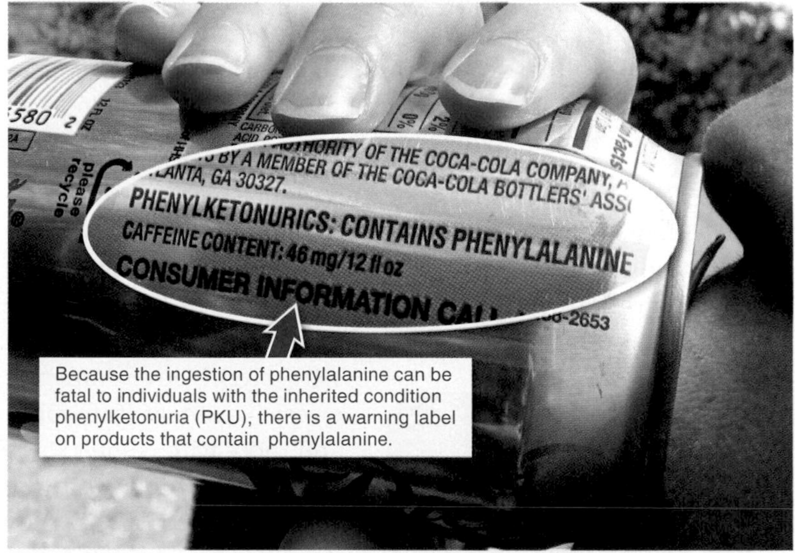

PHENYLKETONURICS: CONTAINS PHENYLALANINE
CAFFEINE CONTENT: 46 mg/12 fl oz
CONSUMER INFORMATION CAL
please recycle
Because the ingestion of phenylalanine can be fatal to individuals with the inherited condition phenylketonuria (PKU), there is a warning label on products that contain phenylalanine.

FIGURE 9-26 Change the environment, "cure" the disease.

The warning is for people who have a genotype that, in the presence of the amino acid phenylalanine, can be deadly. Sometimes our bodies use phenylalanine directly to build proteins, adding it to a growing amino acid chain. At other times, phenylalanine is chemically converted into another amino acid, called tyrosine. The body may then use tyrosine as it constructs proteins and in a variety of other functions.

The problem is this: at birth, some people carry two copies of a mutant version of the gene that is supposed code for the enzyme that converts phenylalanine into tyrosine. The mutant gene produces a malfunctioning enzyme, and none of the body's phenylalanine is converted into tyrosine. Little by little, as these individuals consume phenylalanine, it builds up in their bodies. Babies born with the two mutant alleles usually begin to show symptoms within 3–6 months. Within a few years, so much accumulates that it reaches toxic levels and poisons the brain, leading to mental retardation and other serious health problems. The disease is called phenylketonuria, or PKU.

Here's where the warning label comes in. By limiting the amount of phenylalanine consumed, individuals with the two mutant alleles for processing phenylalanine can avoid the toxic buildup of the amino acid in their brain. In essence, they modify their environment (their diet) so that it contains only a tiny, carefully monitored amount of phenylalanine. Because all newborn babies in the United States are screened for PKU at birth, they can be treated with the appropriate diet immediately. And in an environment free of excess phenylalanine, the PKU mutant alleles are harmless.

This example highlights the fact that genes interact with the environment to produce physical characteristics. Unless you have information about both the genes and the environment, it is not usually possible to know what the phenotype will be.

Q Could you create a temporarily spotted Siamese cat with an ice pack? How?

Siamese cats (as well as Himalayan rabbits) carry genes that produce dark pigmentation. These genes interact strongly with the environment and are heat-sensitive. Cold areas of the animal's body produce dark pigment, while warm areas remain light in color. The coldest parts of the body—the ears, paws, tail, and tip of the face—become the darkest, while the rest remains cream-colored or white (FIGURE 9-27). For Siamese cats living in cold climates and spending a lot of time outside, it's interesting to notice that they become significantly darker in color during the winter months. Those that lounge indoors all winter remain lighter in color.

Some pigment genes produce dark pigment only under cold conditions—such as on the tail, nose, ears, and feet of these animals.

FIGURE 9-27 **Heat-sensitive fur color.**

There are thousands of other cases in which genes' interactions with the environment influence their effects in the body. The scope of environmental influences ranges from traits with obvious environmental effects, such as body weight and its relationship with caloric intake; to those with slight environmental effects, such as fur color; and those with complex and subtle interactions with the environment, such as intelligence and personality.

Q James Watson, the co-discoverer of DNA, once wrote that when we completed the Human Genome Project, we would have "the complete genetic blueprint of man." Was that a good metaphor?

Given the role of environmental factors in influencing phenotypes, DNA is not like a blueprint for a house. There is nearly always significant interaction between the genotype and the environment that influences the exact phenotype produced. If this were not true, there would be no reason to invest in better schools, physical fitness regimens, nutritional monitoring, self-help efforts, or any other process by which individuals or societies try to improve people's lives (i.e., alter phenotypes) by enriching their environment.

TAKE HOME MESSAGE 9.15

» Genotypes are not like blueprints that specify phenotypes. Phenotypes are generally a product of the genotype and its interactions with the environment.

9.16–9.17

Some genes are linked together.

Different genes influence red hair and freckles, so why are they often inherited together?

9.16 Most traits are passed on as independent features: Mendel's law of independent assortment.

Sometimes you can be right about something for the wrong reason. This happened to Gregor Mendel. He didn't know that genes were carried on chromosomes. He believed that the units of heredity were just free-floating entities within cells. From this perspective, it made sense that the inheritance pattern of one trait wouldn't influence the inheritance of any other trait.

> ". . . the relation of each pair of different characters in hybrid union is independent of the other differences in the two original parental stocks."
>
> — Gregor Mendel, **clearly articulating the idea of independent assortment in his publication "Experiments in Plant Hybridization" (1865)**

Let's consider an example. Earlier in this chapter we saw that all cats with completely white fur have at least one parent that also has completely white fur. This is because white fur is coded for by a single dominant gene.

Imagine that you had a true-breeding population of cats with white fur (i.e., all offspring manifest the trait and, in this case, have the genotype *WW*). Now suppose that an individual in this population mated with a cat from a true-breeding population of cats that all had some colored fur (all individuals have the genotype *ww*). All of their offspring would have white fur, but they would be heterozygous (*Ww*). If two heterozygotes had offspring together, though, they would produce three-quarters white-furred and one-quarter non-white-furred offspring, with genotypes in the ratio of 1/4 *WW*, 1/2 *Ww*, and 1/4 *ww*. That is just what Mendel observed for traits in pea plants.

But what if we concurrently observed another characteristic of these cats? Suppose the original true-breeding population of white-furred cats (*WW*) was also true-breeding for long hair (all *ll*), a condition caused by carrying two recessive alleles for a single gene. And suppose that individuals in the colored-fur population (*ww*) were also true-breeding for short hair (all *LL*). The question is, do the alleles an individual inherits for the white-fur trait influence which alleles that individual inherits for fur length? And the answer is that they do not (FIGURE 9-28). Rather, the first cross of a long-haired, all-white cat with a cat having short, colored fur would result in offspring heterozygous for both traits—referred to as **dihybrid**—and expressing each of the dominant traits.

Phenotypically, all of the offspring from this cross would have short, completely white fur. In a mating between two of these doubly heterozygous individuals—referred to as a **dihybrid cross**—three-quarters of the offspring would have the dominant trait and one-quarter would have the recessive trait, regardless of which trait you are tallying. In other words, neither trait influences the inheritance pattern for the other trait; all traits are inherited independently of each other. This is known as **Mendel's law of independent assortment.**

In the next section, we'll see that, despite Mendel's correct understanding that separate traits are inherited

MENDEL'S LAW OF INDEPENDENT ASSORTMENT

Mendel's law of independent assortment states that one trait does not influence the inheritance of another trait.

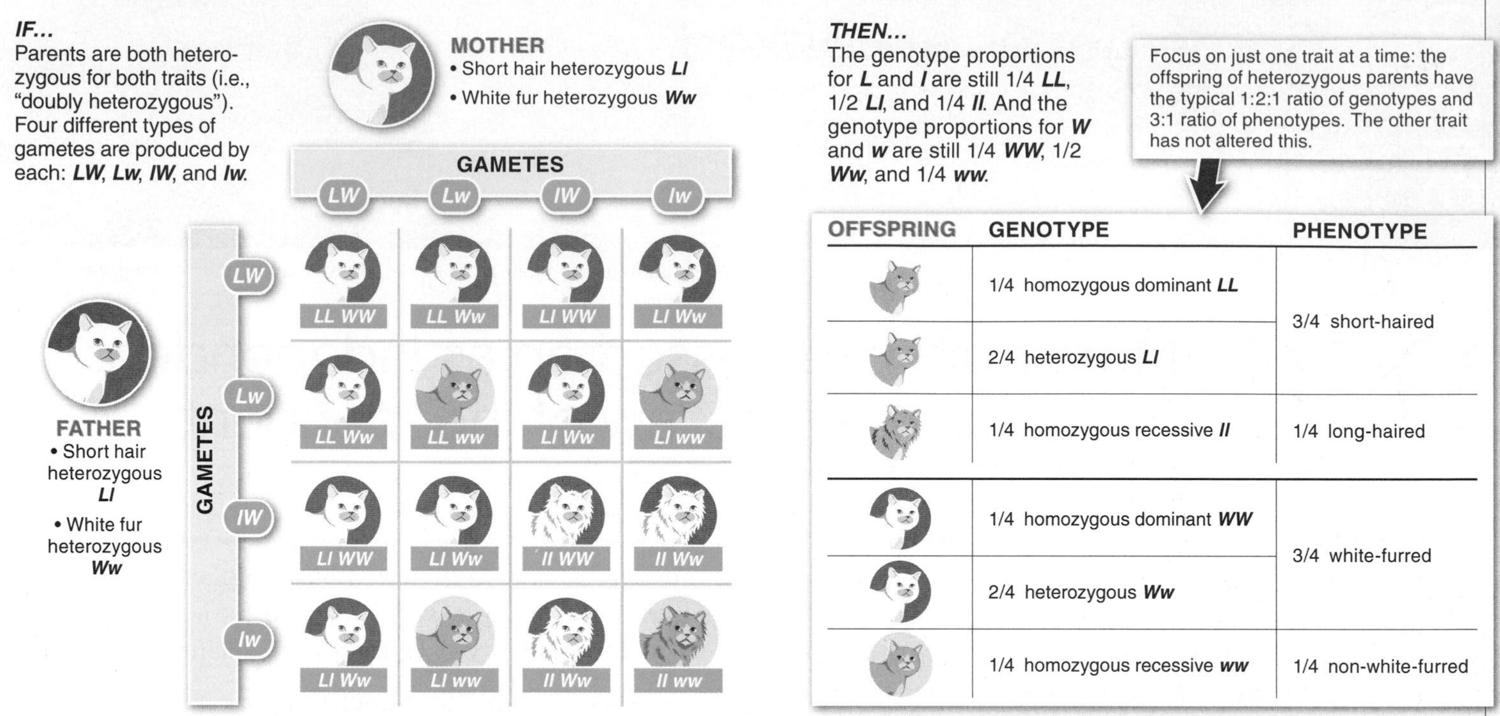

IF...
Parents are both heterozygous for both traits (i.e., "doubly heterozygous"). Four different types of gametes are produced by each: ***LW***, ***Lw***, ***lW***, and ***lw***.

MOTHER
- Short hair heterozygous ***Ll***
- White fur heterozygous ***Ww***

FATHER
- Short hair heterozygous ***Ll***
- White fur heterozygous ***Ww***

GAMETES	LW	Lw	lW	lw
LW	*LL WW*	*LL Ww*	*Ll WW*	*Ll Ww*
Lw	*LL Ww*	*LL ww*	*Ll Ww*	*Ll ww*
lW	*Ll WW*	*Ll Ww*	*ll WW*	*ll Ww*
lw	*Ll Ww*	*Ll ww*	*ll Ww*	*ll ww*

THEN...
The genotype proportions for ***L*** and ***l*** are still 1/4 ***LL***, 1/2 ***Ll***, and 1/4 ***ll***. And the genotype proportions for ***W*** and ***w*** are still 1/4 ***WW***, 1/2 ***Ww***, and 1/4 ***ww***.

Focus on just one trait at a time: the offspring of heterozygous parents have the typical 1:2:1 ratio of genotypes and 3:1 ratio of phenotypes. The other trait has not altered this.

OFFSPRING	GENOTYPE	PHENOTYPE
	1/4 homozygous dominant ***LL***	3/4 short-haired
	2/4 heterozygous ***Ll***	
	1/4 homozygous recessive ***ll***	1/4 long-haired
	1/4 homozygous dominant ***WW***	3/4 white-furred
	2/4 heterozygous ***Ww***	
	1/4 homozygous recessive ***ww***	1/4 non-white-furred

FIGURE 9-28 **Independent assortment of genes.**

independently, his belief that this happened because all genes just float freely around in the cell was not correct. The genes, as we now know, are carried on chromosomes. And this sometimes leads to situations in which independent assortment does *not* occur.

TAKE HOME MESSAGE 9.16

» Genes tend to behave independently, such that, for most traits, the inheritance pattern of one trait doesn't influence the inheritance of any other trait.

9.17 Red hair and freckles: genes on the same chromosome are sometimes inherited together.

Most redheads have pale skin and freckles. This simple observation is problematic for the law of independent assortment as Mendel imagined it. After all, he asserted that the inheritance of one trait does not influence the inheritance of another. But clearly, having red hair seems to influence the presence of another trait, pale skin (FIGURE 9-29). Strictly speaking, Mendel's second law is not true for *every* pair of traits. Sometimes the alleles for two genes are inherited together and expressed almost as a package.

We have about 21,000 genes in our genome. Yet we have only 23 unique chromosomes (two copies of each). Thus,

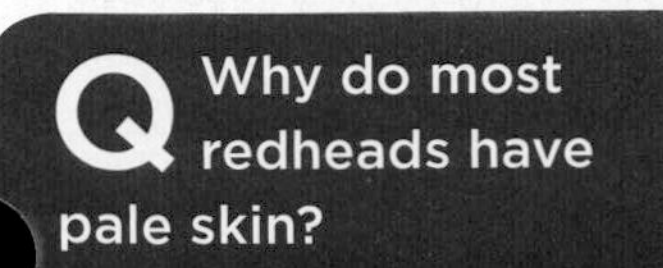

genes influencing different traits must sometimes be on the same chromosome, maybe even right next to each other. When they are close together, we say that they are **linked genes.** A 2008 study demonstrated a link between human genes that influence hair color and skin pigmentation (including freckles).

Why are linked genes inherited together? To answer this, we must revisit the behavior of chromosomes during the production of gametes, discussed in Chapter 8. When producing a sperm or egg by meiosis, only one of the two copies of each of your chromosomes ends up in the gamete. It may have been from your mother or it may have been from your father. All of the alleles on the chromosome from one parent get passed on as a group to the child at fertilization. This process continues generation after generation. The linked alleles remain together unless, during meiosis, recombination occurs, exchanging one or more alleles with the other chromosome in the pair so that they now become linked with the alleles on that chromosome (FIGURE 9-30).

When alleles are linked closely on the same chromosome, Mendel's second law doesn't hold true. It is surprising—and was fortunate for Mendel—that of the seven pea plant traits he chose for his studies, none were close together on the same chromosome. For this reason, they all behaved as if they weren't linked.

FIGURE 9-29 Violating the law of independent assortment. Red hair and freckles are often inherited as a package deal.

TAKE HOME MESSAGE 9.17

» Sometimes, having one trait influences the presence of another trait. This is because the alleles for two genes are inherited and expressed almost as a package deal when the genes are located close together on the same chromosome.

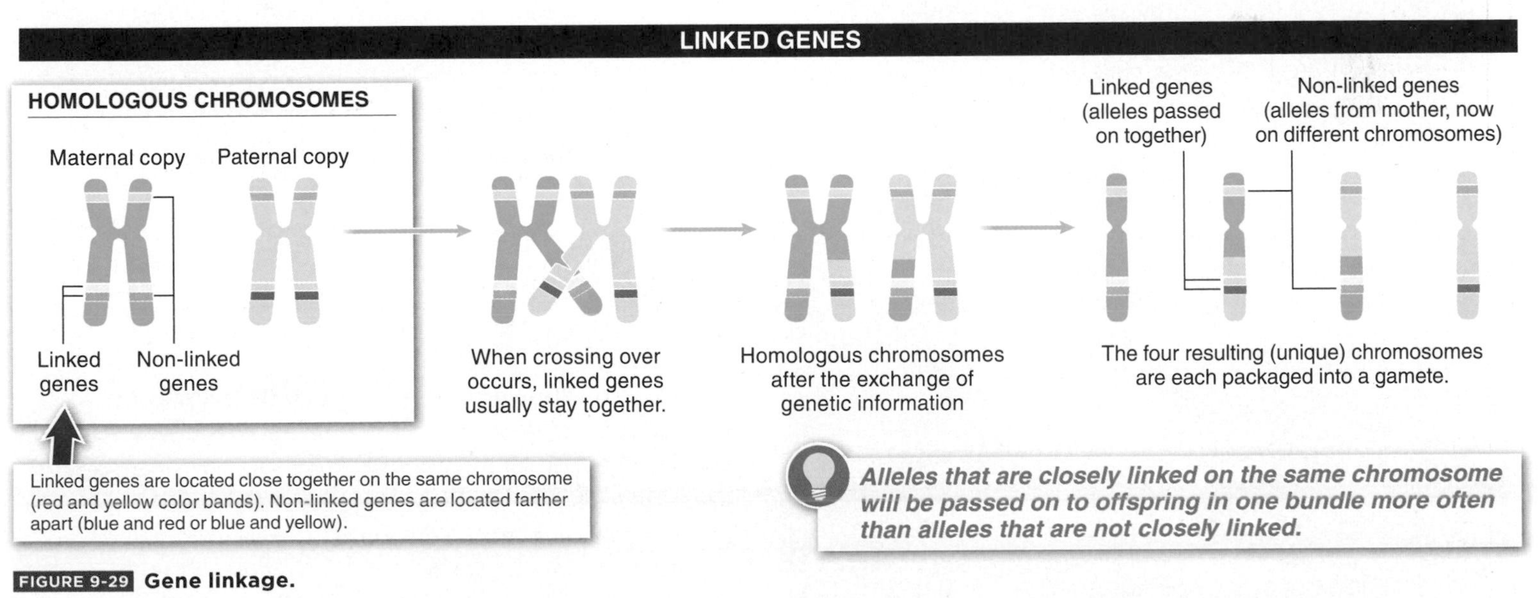

LINKED GENES
HOMOLOGOUS CHROMOSOMES
Maternal copy
Paternal copy
Linked genes
Non-linked genes
Linked genes are located close together on the same chromosome (red and yellow color bands). Non-linked genes are located farther apart (blue and red or blue and yellow).
When crossing over occurs, linked genes usually stay together.
Homologous chromosomes after the exchange of genetic information
Linked genes (alleles passed on together)
Non-linked genes (alleles from mother, now on different chromosomes)
The four resulting (unique) chromosomes are each packaged into a gamete.
Alleles that are closely linked on the same chromosome will be passed on to offspring in one bundle more often than alleles that are not closely linked.

FIGURE 9-29 Gene linkage.

STREET BIO

Using evidence to guide decision making in our own lives

Can a gene nudge us toward novelty-seeking (and faster snowboarding)?

Human chromosome 11 contains a gene called DRD4, which carries the instructions for a dopamine receptor. Dopamine is a chemical messenger that alters activity in the brain's pleasure centers and influences motivation. Near the beginning of the DRD4 gene is "promoter polymorphism –521C/T," a segment with two alleles, C and T. The C allele is associated with a 40% increase in activity of the DRD4 gene.

Q: Can a single gene influence your personality?
In 2000, researchers reported that individuals carrying two copies of the C allele were more likely to exhibit certain personality traits, particularly novelty-seeking. The subjects ranked how well certain statements described them, such as "I have sometimes done things just for kicks or thrills." The subjects were then evaluated for seven dimensions of personality traits. When it came to their score for "novelty-seeking," there was a small but statistically significant difference in how the subjects scored, depending on their genotype for the DRD4 gene.

In 2008, a "meta-study" evaluated 11 published studies on the relationship between the CC genotype and novelty-seeking. The researchers summarized these studies in a graph, concluding that they demonstrated a significant association between novelty-seeking and the –521C/T genotype.

And in a 2013 study of more than 500 experienced skiers and snowboarders, researchers even found that CC individuals exhibited greater sensation-seeking in their skiing behaviors than those who were CT or TT!

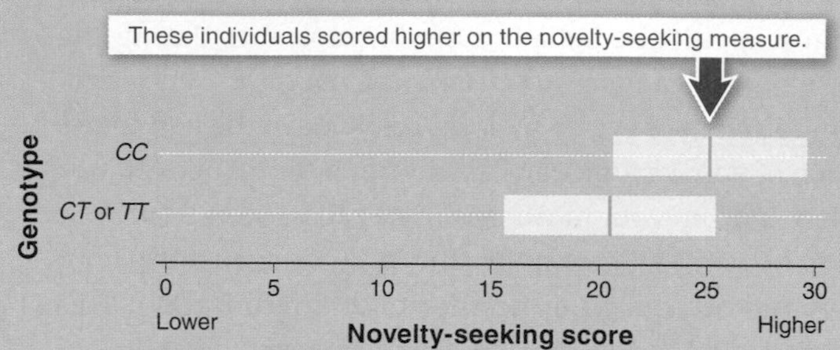

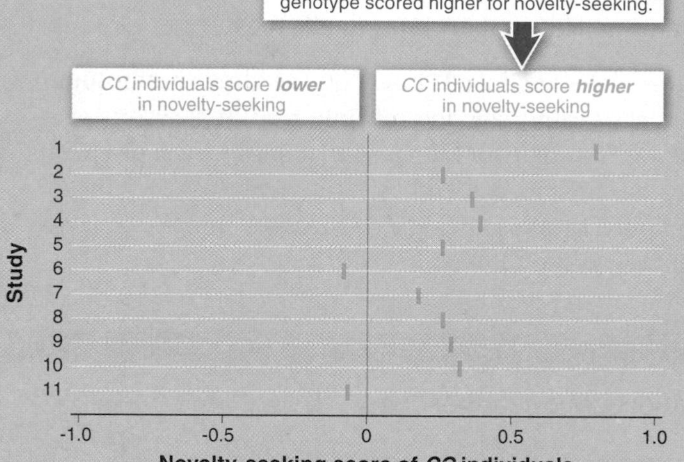

GRAPHIC CONTENT
Thinking critically about visual displays of data
Turn to p. 293 for a closer inspection of this figure.

Q: Do you want your employers or insurance companies to know which alleles you carry for the DRD4 gene? Might you face discrimination?
Knowledge of your genotype for the DRD4 gene is one piece of a larger puzzle. Personality differences influenced by DRD4 are small, they are affected by other genes, and many personality traits are unrelated to the DRD4 gene. Would you want to know what your genotype is? Why?

GRAPHIC CONTENT

Thinking critically about visual displays of data

1 The top graph displays the novelty-seeking scores of subjects who responded to a questionnaire on character and temperament. Based on this information, what was the average novelty-seeking score for individuals with the *CC* genotype?

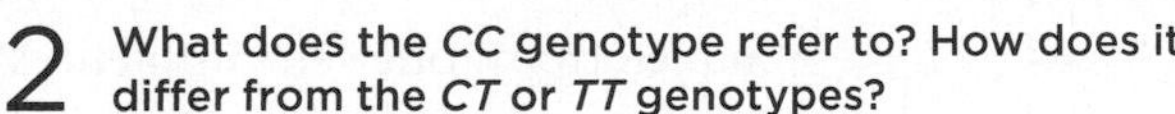

2 What does the *CC* genotype refer to? How does it differ from the *CT* or *TT* genotypes?

3 What information is conveyed by the light green bars? Why is this information helpful?

4 What would it mean if a green line were at 0?

5 The 11 studies shown all evaluated the same thing, yet their results are not identical. Why might that be?

6 What conclusion can you draw from this figure? How certain are you about that conclusion?

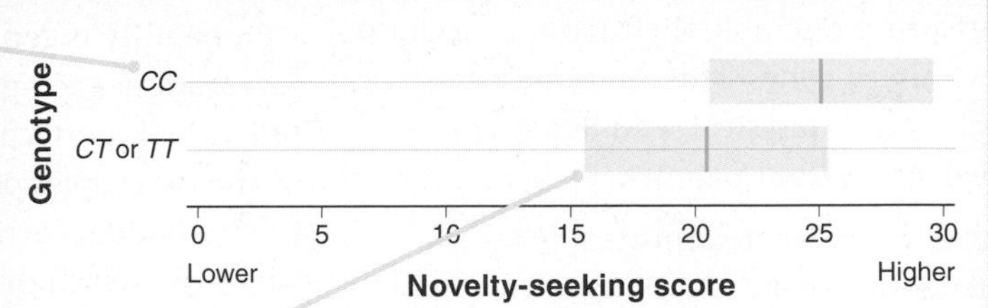

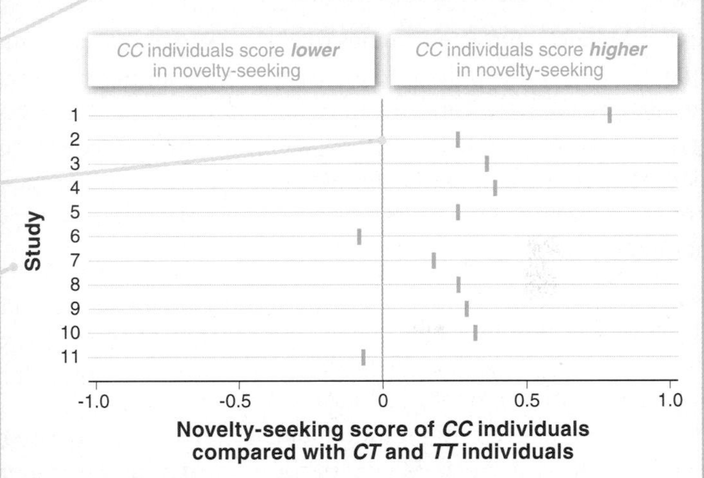

See answers at the back of the book.

KEY TERMS IN GENES AND INHERITANCE

BRIEF SUMMARY

Why do offspring resemble their parents?

• Offspring resemble their parents because they inherit genes—instructions for biochemical, physical, and behavioral traits—from their parents.

• Some traits are determined by instructions that an individual carries on a single gene, and these traits exhibit straightforward patterns of inheritance.

• Gregor Mendel conducted studies that help us understand heredity. He focused on easily observed and categorized traits in garden peas and applied methodical experimentation and rigorous hypothesis testing to determine how traits are inherited.

• Each parent puts a single set of instructions for building a particular trait into every sperm or egg that he or she makes. This instruction set is called a gene. The trait observed in an individual depends on the two copies (alleles) of the gene it inherits from its parents.

• It is not always possible to determine an individual's genetic makeup, known as its genotype, by observation of the organism's outward appearance, known as its phenotype. An individual may carry a recessive allele whose phenotypic effect is masked by the presence of a dominant allele.

Probability and chance play central roles in genetics.

• Probability is central to genetics. In segregation, each gamete that an individual produces receives only one of the two copies of each gene that the individual carries in its other cells. All of the sperm or eggs produced by an individual are different from one another.

• In a test-cross, an individual that exhibits a dominant trait but has an unknown genotype is mated with an individual that is homozygous recessive for that trait. The phenotypes of the offspring reveal the unknown genotype.

• Pedigrees help determine the genes that individuals carry and the likelihood that the offspring of two individuals will exhibit a particular trait.

How are genotypes translated into phenotypes?

• Sometimes the effects of both alleles in a heterozygous genotype are evident in the phenotype. With incomplete dominance, the phenotype of a heterozygote appears to be an intermediate blend of the phenotypes of the two homozygotes. With codominance, a heterozygote has a phenotype that exhibits characteristics of both homozygotes.

• In multiple allelism, a single gene has more than two alleles. Each individual still carries only two alleles, but in the population there are more than two alleles.

• Many traits, including continuously varying traits such as height and eye color, are influenced by multiple genes.

• In pleiotropy, one gene influences multiple, unrelated traits. Most, if not all, genes may be pleiotropic.

• When a trait is coded for by a gene on a sex chromosome, such as color vision on the X chromosome, the pattern of expression differs for males and females.

• Genotypes are not like blueprints that specify phenotypes. Phenotypes are generally a product of the genotype in combination with the environment.

Some genes are linked together.

• Genes tend to behave independently, such that the inheritance pattern of one trait doesn't usually influence the inheritance of any other trait.

• Sometimes, having one trait does influence the presence of another trait. This is because the alleles for two genes are inherited and expressed almost as a package deal when the genes are located close together (linked) on the same chromosome.

CHECK YOUR KNOWLEDGE

Short Answer

1. Why do offspring resemble their parents?

2. Describe the main focus of Mendel's research on pea plants.

3. Why does a sperm or egg have only one copy of a gene?

4. Compare and contrast the terms "phenotype" and "genotype."

5. In what situation is a test-cross helpful in determining an individual's genotype?

6. What does it mean to be the "carrier" of a trait?

7. Describe the phenotype of an individual heterozygous for an allele with incomplete dominance.

8. Describe why being heterozygous for the sickle-cell allele can confer resistance to malaria.

9. Why are sex-linked recessive traits more commonly observed in males than females?

10. How does the case of PKU support the notion that our phenotype is a product of our genotype in combination with the effect of our environment?

11. What is Mendel's second law? Why doesn't it apply for all traits?

Multiple Choice

1. Most genes come in alternative forms called:

a) alleles.

b) heterozygotes.

c) gametes.

d) chromosomes.

e) homozygotes.

2. Pea plants were well suited for Mendel's breeding experiments for all of the following reasons except:

a) Pea plants exhibit variations in a number of observable characteristics, such as flower color and seed shape.

b) Mendel could control the pollination between different pea plants.

c) It is easy to obtain large numbers of offspring from any given cross.

d) Many of the characteristics that vary in pea plants are not linked closely on the same chromosome.

e) Pea plants have a particularly long generation time.

3. The law of segregation states that:

a) the transmission of genetic diseases within families is always recessive.

b) an allele on one chromosome will always segregate from an allele on a different chromosome.

c) gametes cannot be separate and equal.

d) the number of chromosomes in a cell is always divisible by 2.

e) the two alleles for a given trait segregate into different gametes.

4. In pea plants, purple flower color is dominant to white flower color. If two pea plants that are true-breeding for purple flowers are crossed, in the offspring:

a) all of the flowers will be purple.

b) three-quarters of the flowers will be purple and one-quarter will be white.

c) half of the flowers will be purple and one-quarter will be white.

d) one-quarter of the flowers will be purple and three-quarters will be white.

e) all of the flowers will be white.

5. A test-cross:

a) makes it possible to determine the genotype of an individual of unknown genotype that exhibits the dominant version of a trait.

b) is a cross between an individual whose genotype for a trait is not known and an individual homozygous recessive for the trait.

c) sometimes requires the production of multiple offspring to reveal the genotype of an individual whose genotype is unknown (but who exhibits the dominant phenotype).

d) Only a) and b) are correct.

e) Choices a), b), and c) are correct.

6. All of the offspring of a black hen and a white rooster are gray. The simplest explanation for this pattern of inheritance is:

a) multiple alleles.

b) codominance.

c) incomplete dominance.

d) incomplete heterozygosity.

e) sex linkage.

7. A woman with type B blood and a man with type A blood could have children with which of the following phenotypes?

a) AB only

b) AB or O only

c) A, B, or O only

d) A or B only

e) A, B, AB, or O

8. A rare X-linked dominant condition in humans, congenital generalized hypertrichosis (CGH), is marked by excessive hair growth all over a person's body. Which of the following statements about this condition is incorrect?

a) The son of a woman with CGH has just slightly more than a 50% chance of having this disease.

b) All daughters of a man with CGH will have this disease.

c) The daughter of a woman with CGH has just slightly more than a 50% chance of having this disease.

d) Every son of a woman with CGH will have this disorder.

e) The son of a man with CGH is no more likely to have this condition than the son of a man who doesn't have CGH.

9. Individuals carrying two nonfunctioning alleles for the gene producing the enzyme that converts phenylalanine into tyrosine:

a) will develop the symptoms of phenylketonuria (PKU).

b) may or may not develop PKU symptoms, because diet also plays a role in determining their phenotype.

c) will develop both PKU and Tay-Sachs disease, because the genes causing the traits are linked.

d) will develop both PKU and Tay-Sachs disease, because one pleiotropic gene influences both traits.

e) do not need to worry about their phenylalanine consumption, because PKU depends on the environment.

10. Thousands (or even tens of thousands) of different traits make up an individual. For this reason:

a) in a species with 23 different chromosomes, some traits must be coded for by genes on the same chromosome.

b) the environment must influence more than half of our traits.

c) all genes must be pleiotropic.

d) knowing a person's phenotype is insufficient for determining his or her genotype.

e) All of the above are correct.

Ch10

Evolution is an ongoing process.

10.1 We can see evolution occurring right before our eyes.

Darwin journeyed to a new idea.

10.2 Before Darwin, many people believed that all species had been created separately and were unchanging.

10.3 A job on a 'round-the-world ship allowed Darwin to make observations that enabled him to develop a theory of evolution.

10.4 Observing geographic similarities and differences among fossils and living plants and animals, Darwin developed a theory of evolution.

Four mechanisms can give rise to evolution.

10.5 Evolution occurs when the allele frequencies in a population change.

10.6 Mutation—a direct change in the DNA of an individual—is the ultimate source of all genetic variation.

10.7 Genetic drift is a random change in allele frequencies in a population.

10.8 Migration into or out of a population may change allele frequencies.

10.9 When three simple conditions are satisfied, evolution by natural selection is occurring.

10.10 A trait does not decrease in frequency simply because it is recessive.

Through natural selection, populations of organisms can become adapted to their environments.

10.11 Traits causing some individuals to have more offspring than others become more prevalent in the population.

10.12 Organisms in a population can become better matched to their environment through natural selection.

10.13 Natural selection does not lead to perfect organisms.

10.14 Artificial selection is a special case of natural selection.

10.15 Natural selection can change the traits in a population in several ways.

10.16 *This is how we do it:* By picking taller plants, do humans unconsciously drive the evolution of smaller plants?

10.17 Natural selection can cause the evolution of complex traits and behaviors.

The evidence for evolution is overwhelming.

10.18 The fossil record documents the process of natural selection.

10.19 Geographic patterns of species distributions reflect species' evolutionary histories.

10.20 Comparative anatomy and embryology reveal common evolutionary origins.

10.21 Molecular biology reveals that common genetic sequences link all life forms.

10.22 Laboratory and field experiments, and antibiotic-resistant bacteria, reveal evolution in progress.

Eastern screech owl hidden in the hollow of a tree.

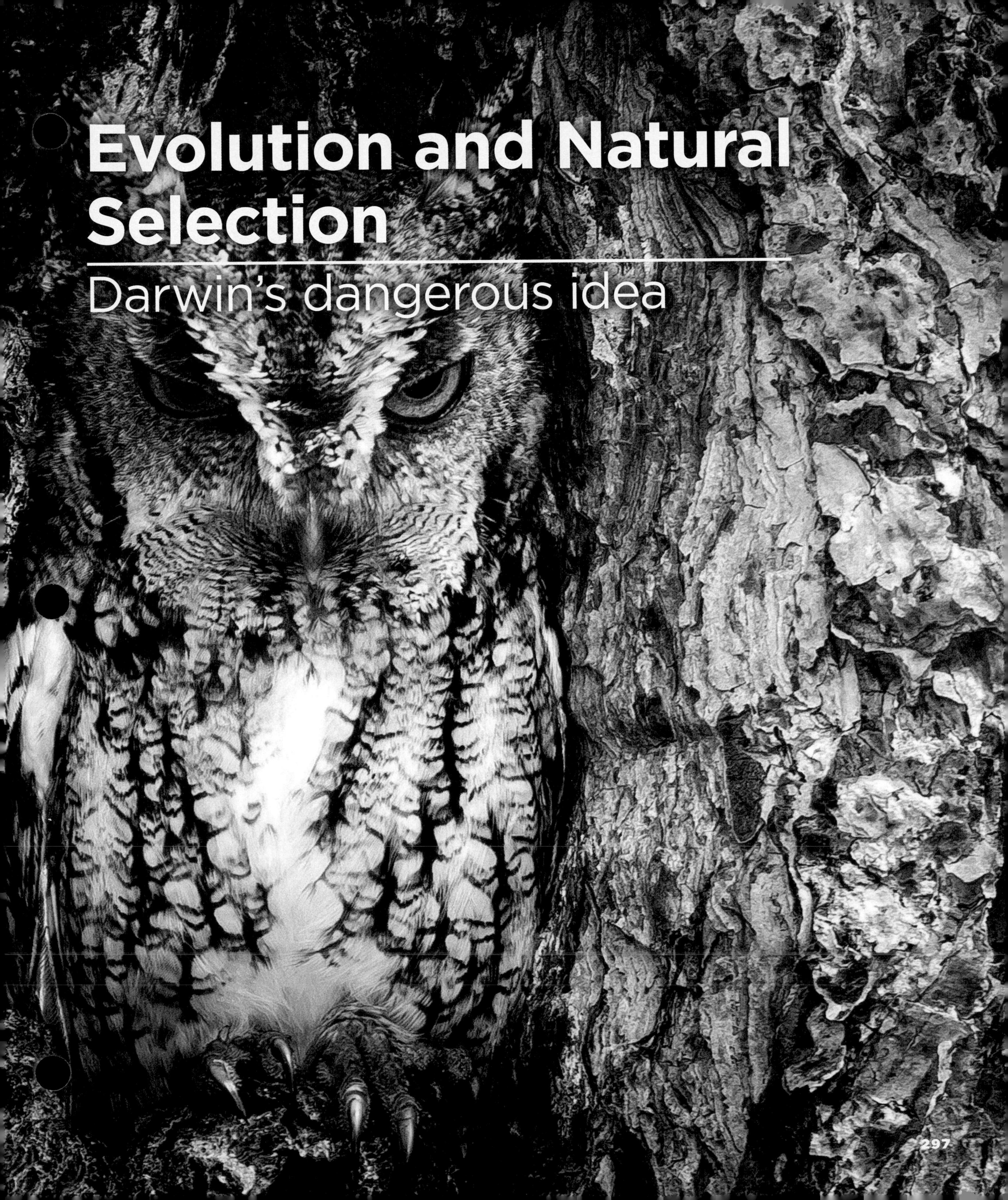

Evolution and Natural Selection

Darwin's dangerous idea

10.1

Evolution is an ongoing process.

Green coloration among plant-eating insects (such as the weevil shown here) provides effective crypsis.

10.1 We can see evolution occurring right before our eyes.

What's the longest you've gone without food? Twenty-four hours? Thirty-six hours? If you've ever gone hungry that long, you probably felt terrible. But humans can survive days, even weeks, without food. In 1981, for example, 27-year-old Bobby Sands went on a hunger strike. Forsaking all food and consuming only water, he gradually deteriorated and ultimately died—after 66 days without food.

A fruit fly, however, can last just under a day—20 hours or so (FIGURE 10-1). The tiny fruit fly body doesn't hold very large caloric reserves. But could you breed fruit flies that could live longer than 20 hours, on average? Yes.

First, start with a population of 5,000 ordinary fruit flies. In biology, the term **population** means a group of organisms of the same species living in the same geographic region. The region can be a small area such as a test tube or a large area such as a lake. In your experiment, your population occupies a cage in your laboratory.

From these 5,000 fruit flies, choose only the 20% of flies that can survive the longest without food.

1. Remove the food from the cage with the population of 5,000 flies.
2. Wait until 80% of the flies have starved to death, then put a container of food into the cage.
3. After the surviving flies eat, they'll have the energy to reproduce. When they do, collect and transfer the eggs to a new cage.

When these eggs hatch, the flies will show increased resistance to starvation. The average fly in the new generation can live for about 23 hours without food. And again, some of the new flies can survive for more than 23 hours, some for less (FIGURE 10-2). If you kept repeating these three steps, the average starvation-resistance time would increase, and by the fifth generation, the population's average survival time would be about 32 hours.

After 60 generations of allowing only the flies that are best at surviving without food to reproduce, how has the

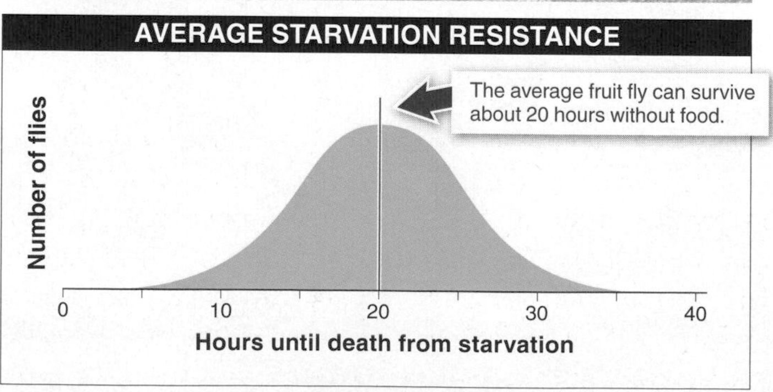

FIGURE 10-1 How long can a fruit fly survive without food?

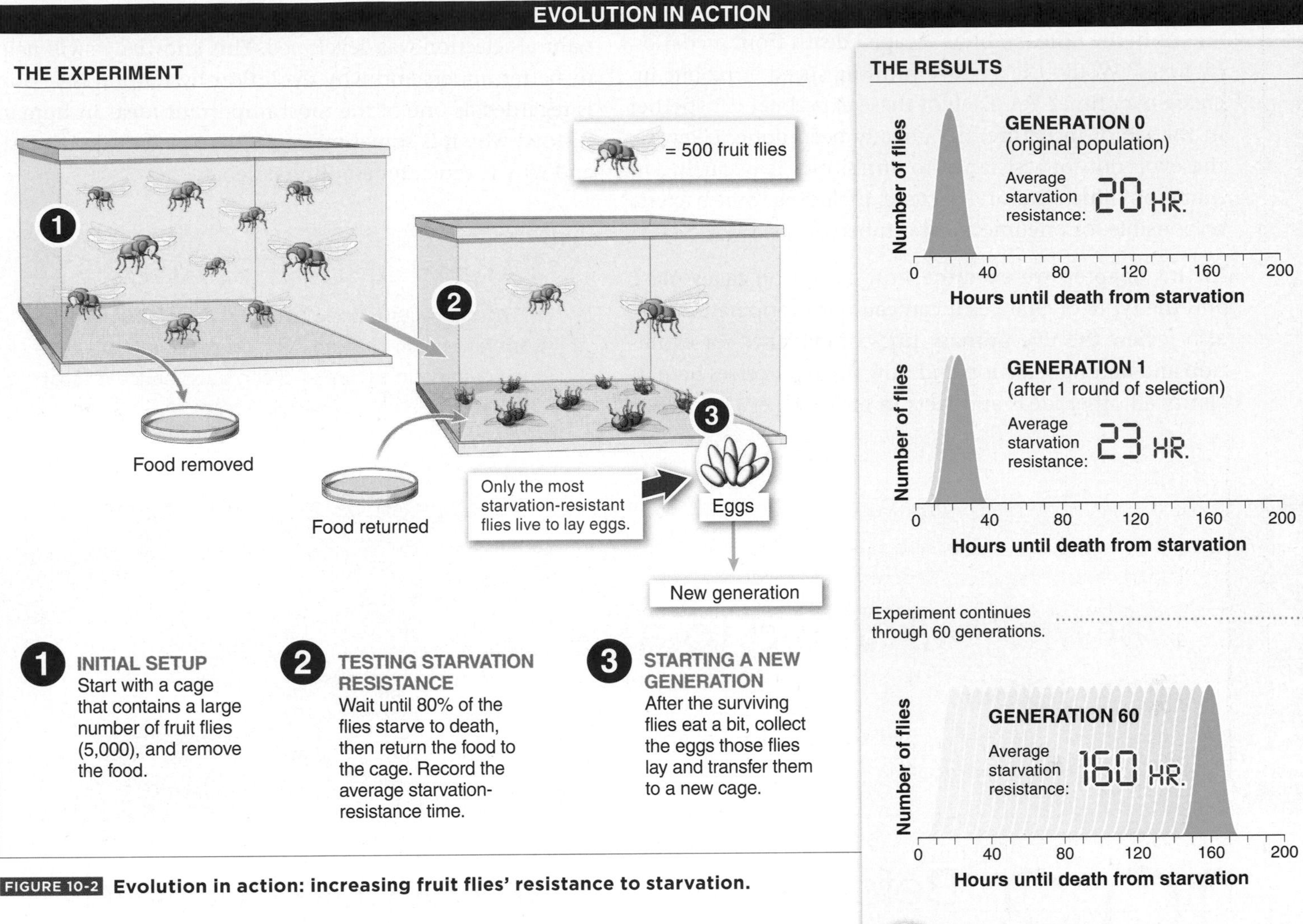

EVOLUTION IN ACTION
THE EXPERIMENT
= 500 fruit flies
1
2
3
Food removed
Food returned
Only the most starvation-resistant flies live to lay eggs.
Eggs
New generation
1 INITIAL SETUP
Start with a cage that contains a large number of fruit flies (5,000), and remove the food.
2 TESTING STARVATION RESISTANCE
Wait until 80% of the flies starve to death, then return the food to the cage. Record the average starvation-resistance time.
3 STARTING A NEW GENERATION
After the surviving flies eat a bit, collect the eggs those flies lay and transfer them to a new cage.
THE RESULTS
GENERATION 0
(original population)
Average starvation resistance: 20 HR.
Number of flies
0 40 80 120 160 200
Hours until death from starvation
GENERATION 1
(after 1 round of selection)
Average starvation resistance: 23 HR.
Number of flies
0 40 80 120 160 200
Hours until death from starvation
Experiment continues through 60 generations.
GENERATION 60
Average starvation resistance: 160 HR.
Number of flies
0 40 80 120 160 200
Hours until death from starvation

FIGURE 10-2 **Evolution in action: increasing fruit flies' resistance to starvation.**

Over many generations of natural selection, the population changes. Originally, the flies couldn't survive a day without food. Now they survive almost an entire week!

population changed? Amazingly, the *average* fly in the resulting population can survive for more than 160 hours without food (see Figure 10-2). Even the fly with the lowest starvation resistance is more than seven times better at resisting starvation than the best fly in the original population! At the point where the food is removed, the flies in generation 60 are noticeably fatter than the flies in generation 0.

What happened? In a word: **evolution.** That is, there was a genetic change in the population of fruit flies living in the cage.

Later in this chapter we'll fully discuss the mechanisms by which evolution can occur. In short, it is the consequence of certain individual organisms in a population being born with characteristics that make them more likely to survive and to reproduce than other individuals in the population. In the fruit fly experiment, the 20% of flies that were the most resistant to starvation could survive and reproduce, and the less starvation-resistant flies could not.

Recall from our discussion of scientific thinking (see Chapter 1) that to give us confidence in our hypotheses about anything, the hypotheses must be testable and reproducible. Researchers carried out the starvation-resistance experiment in fruit flies five separate times. The results were the same every time.

The fruit fly experiment answers what is sometimes perceived as a complex or controversial question: does evolution occur? The answer is an unambiguous *yes*. We can watch it happen in the lab whenever we want. What if we used dogs in our experiment, and we allowed only the smallest dogs to reproduce? Would the average body size of individual dogs in subsequent generations decrease over time? Yes. What if the "experiment" were done with rabbits in a natural habitat, in the absence of human intervention, and only the fastest rabbits escaped death from predation by foxes? Would the average running speed in rabbits increase over time? Yes. Each of these experiments, whether in the lab or in nature, has already been done. Likewise, the evolution of resistance to antibiotics is occurring in numerous illness-causing bacteria, including some bacteria responsible for pneumonia and tuberculosis.

In this chapter, we examine how evolution takes place and the types of changes it can cause in a population. We also review the five primary lines of evidence for evolution and natural selection and how these processes help us clarify all other ideas and facts in biology. Let's begin our investigation with a look at how the idea of evolution by natural selection was developed. This knowledge will help us better understand why evolution by natural selection is regarded as one of the most important ideas in human history, why it is sometimes considered a dangerous idea, and why it generates emotional debate.

TAKE HOME MESSAGE 10.1

» The characteristics of the individuals present in a population can change over time. We can observe such change in nature and can even cause it to occur.

10.2–10.4

Darwin journeyed to a new idea.

A Galápagos land iguana, *Conolophus subcristatus.*

10.2 Before Darwin, many people believed that all species had been created separately and were unchanging.

Charles Darwin grew up in an orderly world. In the Victorian British society he inhabited, many beliefs about humans and their place in the world had changed little over the previous two centuries. The Bible sufficiently explained most natural phenomena. Most people thought the earth was about 6,000 years old and was largely unchanging. People recognized that organisms existed in groups called species or kinds. However, most thought that all species, including humans, had been created simultaneously and never changed and never died out. (In Chapter 12, we discuss in more detail what a species is; for now, let's just say that individual organisms of a given species can interbreed with each other but not with members of another species.)

Darwin threw into question some dearly held, long-time beliefs about the natural world, and he forever changed our perspective on the origins of humans and their relationship to all other species. He didn't smash his society's worldview to pieces all at once, though, and he didn't do it by himself (FIGURE 10-3).

In the 1700s and 1800s, scientific thought was advancing rapidly. Darwin was influenced by the work of other scientists of this era. For example, in 1778, the respected French naturalist Georges-Louis Leclerc, Comte de Buffon, calculated that the earth must be about 75,000 years old—the minimum time required for the planet to cool from a molten state.

DARWIN'S INFLUENCES

GEORGES-LOUIS LECLERC, COMTE DE BUFFON
(1707–1788)
Suggested that the earth was much older than previously believed.

GEORGES CUVIER
(1769–1832)
By documenting fossil discoveries, showed that extinction had occurred.

JEAN-BAPTISTE LAMARCK
(1744–1829)
Suggested that living species might change over time.

CHARLES LYELL
(1797–1875)
Argued that geological forces had gradually shaped the earth and continue to do so.

Charles Darwin

FIGURE 10-3 **Scientists who shaped Darwin's thinking.**

In the 1790s, Georges Cuvier discovered fossil remains at the bottoms of coal and slate mines that were unlike any living species. Many people, believing that biblical accounts did not allow for species to be wiped out, found Cuvier's discoveries unthinkable. His publications documented giant fossils (including the Irish elk, the mastodon, and the giant ground sloth) that bore no resemblance to any currently living animals (FIGURE 10-4). Cuvier was not a proponent of the idea of evolution, but the fossils he discovered allowed only one explanation: extinction was a fact.

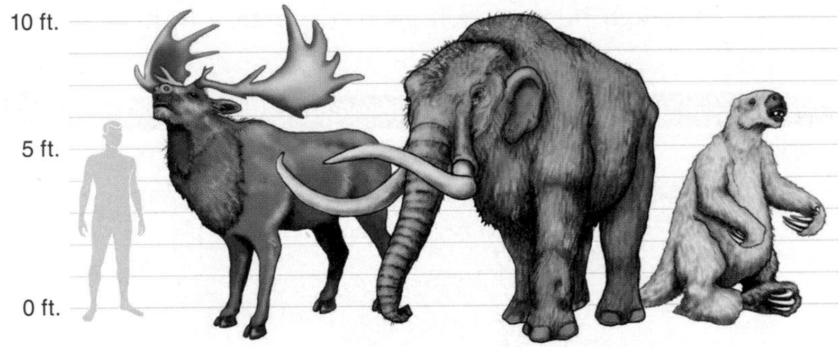

Fossils of organisms no longer found on earth mean that extinction occurs.

FIGURE 10-4 **Extinction occurs.** Deep in coal mines, Cuvier discovered the fossilized remains of very large animals no longer found on earth.

Not only was it starting to seem that species could disappear from the face of the earth, but several scientists, including Darwin's grandfather Erasmus Darwin, began to suggest that living species might change over time. And in the early 1800s, the biologist Jean-Baptiste Lamarck championed a popular idea that change came about chiefly through organisms' use or disuse of particular features. The idea was wrong, but the increased willingness among scientists to question previously sacred "truths" contributed to an atmosphere of unfettered scientific thought in which it was possible to challenge convention.

Perhaps the revolutionary ideas that most inspired Darwin were those of the geologist Charles Lyell. In his 1830 book *Principles of Geology,* Lyell argued that geological forces had shaped the earth and were continuing to do so, producing mountains and valleys, cliffs and canyons, through gradual but relentless change. This idea that the physical features of the earth were constantly changing would most closely parallel Darwin's idea that the living species of the earth, too, were gradually—but constantly—changing.

TAKE HOME MESSAGE 10.2

» In the 18th and 19th centuries, scientists began to overturn many commonly held beliefs in the Western world, including that the earth was only about 6,000 years old and that all species had been created separately and were unchanging. These gradual changes in scientists' beliefs helped shape Charles Darwin's thinking.

10.3 A job on a 'round-the-world ship allowed Darwin to make observations that enabled him to develop a theory of evolution.

Charles Darwin was born into a wealthy family in England in 1809. Never at the top of his class, Charles professed to hate schoolwork. Nonetheless, at 16, he went to the University of Edinburgh to study medicine, following in his father's footsteps. He was bored, though, and left at the end of his second year, when the prospect of watching gruesome surgeries (in the days before anesthesia) was more than he could bear.

At his father's urging, Darwin then pursued the ministry, studying theology at the University of Cambridge. Although he never felt great inspiration in his theology studies, Darwin was in heaven at Cambridge, where he could pursue his real love: the study of nature.

Shortly after graduation in 1831, Darwin landed his dream job, a position as "gentleman companion" to the captain of HMS *Beagle,* on a five-year, 'round-the-world surveying expedition.

Once on the *Beagle,* Darwin found that he didn't actually like sea travel. He was seasick for much of his time on board, writing to his cousin: "I hate every wave of the ocean with a fervor which you . . . can never understand." To avoid nausea, Charles spent as much time as possible on shore. It was there that he found fieldwork to be his true calling. At each stop, he investigated new worlds. In Brazil, he was enthralled by tropical forests. In Patagonia, he explored beaches and cliffs, finding spectacular fossils from huge extinct mammals. Elsewhere, he explored coral reefs and barnacles, always packing up specimens for museums and recording his observations for later use. He was like a schoolboy on permanent summer vacation (FIGURE 10-5).

Lyell's *Principles of Geology* was the only book Darwin took with him on the *Beagle.* He read it again and again, intrigued by the book's premise that the earth is constantly changing. Darwin's mind was ripe for fresh ideas—particularly as he began observing seemingly inexplicable things. How could he explain marine fossils high in the Andes, hundreds of miles from the nearest ocean? The idea that the earth was an ever-changing planet would serve Darwin well when, a few years into its journey, the *Beagle* stopped at the Galápagos Islands, off the northwest coast of South America.

This group of volcanic islands was home to many unusual species, from giant tortoises to extremely docile lizards, which made so little effort to run away that Darwin had to avoid stepping on them. Darwin was particularly intrigued by the wide variety of birds, especially the finches, which seemed dramatically more variable than those he had seen in other locations.

Darwin collected and donated finch specimens to the Zoological Society of London. Initially, Darwin thought that all the finches were of the same species but with different

DARWIN'S 'ROUND-THE-WORLD VOYAGE

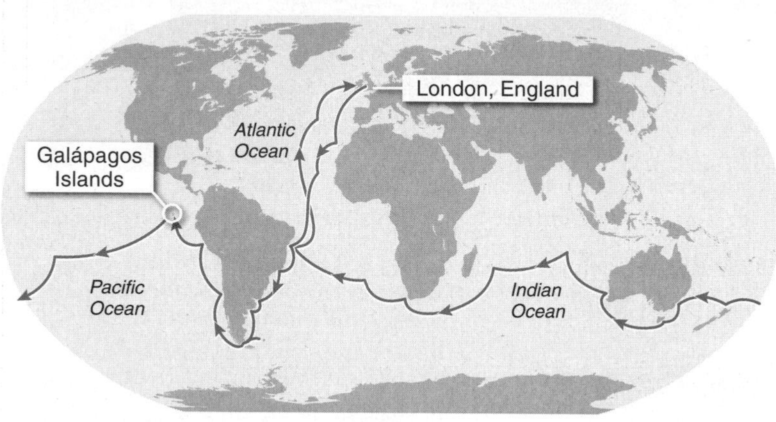

Galápagos Islands

FIGURE 10-5 **Like a schoolboy on permanent summer vacation.** Darwin set out on HMS *Beagle* on a five-year surveying expedition that took him around the globe.

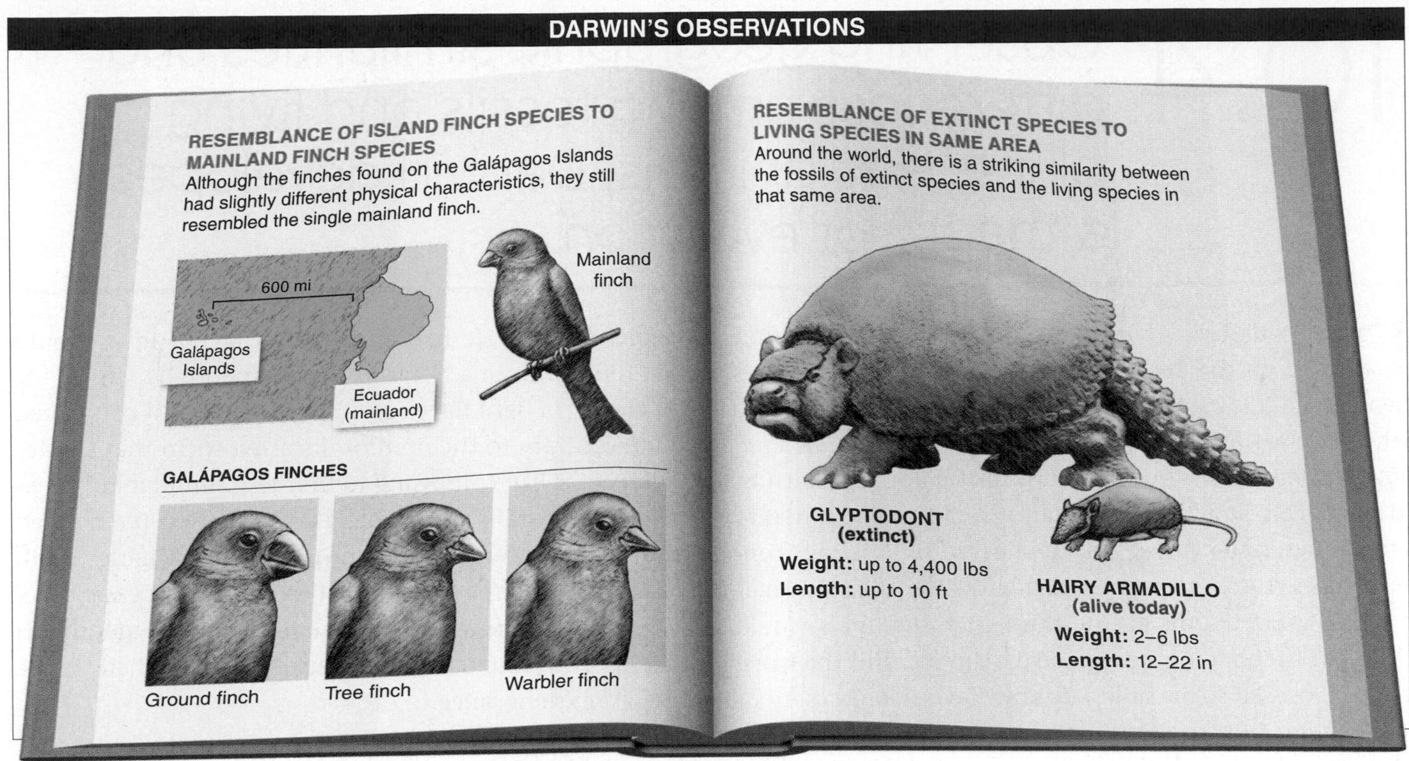
DARWIN'S OBSERVATIONS
RESEMBLANCE OF ISLAND FINCH SPECIES TO MAINLAND FINCH SPECIES
Although the finches found on the Galápagos Islands had slightly different physical characteristics, they still resembled the single mainland finch.
600 mi
Galápagos Islands
Ecuador (mainland)
Mainland finch
GALÁPAGOS FINCHES
Ground finch
Tree finch
Warbler finch
RESEMBLANCE OF EXTINCT SPECIES TO LIVING SPECIES IN SAME AREA
Around the world, there is a striking similarity between the fossils of extinct species and the living species in that same area.
GLYPTODONT (extinct)
Weight: up to 4,400 lbs
Length: up to 10 ft
HAIRY ARMADILLO (alive today)
Weight: 2–6 lbs
Length: 12–22 in

FIGURE 10-6 **Darwin observed unexpected patterns.**

physical characteristics or **traits,** such as body size, beak shape, or feather color. Back in London, however, ornithologists examining Darwin's specimens could see from the birds' physical differences that the birds were not reproductively compatible; there were unique lineages for every one of the Galápagos Islands that Darwin had visited. Moreover, although the birds were of 14 different species, they all resembled very closely the single species of finch living on the closest mainland, in Ecuador (FIGURE 10-6).

The finches were part of two important and unexpected patterns Darwin noticed on his voyage that would be central to his discovery of a mechanism for evolution.

This first pattern—that the different island finches all resembled the mainland species—seemed a suspicious coincidence to Darwin. Perhaps they used to be part of the same mainland population? Over time they may have separated and diverged from the original population and gradually formed new—but similar—species. Darwin's logic was reasonable, but his idea flew in the face of all the scientific thinking of the day.

The second pattern Darwin noted was that, throughout his voyage, there was a striking similarity between the fossils of extinct species and the living species in that same area. In Argentina, for instance, he found giant fossils from a group of organisms called "glyptodonts." The extinct glyptodonts, though huge, resembled armadillos, a species that still flourished in the same area (see Figure 10-6).

If glyptodonts had lived in South America in the past, and armadillos were currently living there, why was only one species living? And why were the glyptodont fossils found only in the same places where modern armadillos lived? Darwin deduced that glyptodonts resembled armadillos because they were their ancient relatives. Again, it was a logical deduction, but it contradicted the scientific dogma of the day that species were unchanging and extinction did not occur.

TAKE HOME MESSAGE 10.3

» Charles Darwin was able to study the natural world when he got a job on a ship conducting a five-year, 'round-the-world survey. Darwin noted unexpected patterns among fossils and living organisms. Fossils resembled but were not identical to living organisms in the same area. And finch species on each of the Galápagos Islands differed from one another in small but significant ways. These observations helped Darwin develop his theory of how species might change over time.

10.4 Observing geographic similarities and differences among fossils and living plants and animals, Darwin developed a theory of evolution.

Darwin had no "Eureka!" moment about evolution while on the *Beagle*. The wheels were turning in his head after his voyage, however, and inspiration struck when he was reading "for amusement" *Essay on the Principle of Population,* by the economist Thomas Malthus. Malthus calculated that populations had the potential to grow much faster than food supplies, with disastrous consequences. Darwin speculated that, rather than all sharing the same doom, maybe the best individuals would "win" in the ensuing struggle for existence, and the worst would "lose." If so, he suddenly saw, "favourable variations would tend to be preserved and unfavourable ones to be destroyed."

In 1842, Darwin prepared a draft of his ideas in a 35-page paper, written in pencil, and he fleshed it out over the next couple of years. He had an inkling that his ideas would rock the world. In a letter to a close friend, he wrote: "At last gleams of light have come, and I am almost convinced (quite contrary to the opinion I started with) that species are not (it is like confessing to murder) immutable." Inexplicably, he put his sketch into a drawer, where it remained for 14 years—even as his friends, including Charles Lyell, warned him that he should publish his ideas in case someone else came up with the same ideas independently. In 1858, Darwin found that his friends' warnings had "come true with a vengeance."

FIGURE 10-7 **Alfred Russel Wallace.** Darwin and Wallace independently identified the process of evolution by natural selection.

In a letter to Darwin, Alfred Russel Wallace (FIGURE 10-7), a young British biologist in the throes of malaria-induced mania in Malaysia, laid out a clear description of the process of evolution by natural selection, after having read Malthus's book. He asked Darwin to "publish it if you think it is worthy." Although ill, his thoughts were careful and precise, the result of many years of thought and exposure to the writings of many of the same scientists who influenced Darwin. Crushed at having been scooped, Darwin wrote that "all my originality will be smashed." Darwin's prominent scientist-friends, however, arranged for a joint presentation of Wallace's and Darwin's work to the Linnaean Society of London. As a result, both Darwin and Wallace are credited for the first description of evolution by natural selection.

Darwin then rapidly completed, 16 months later, a full book. In 1859, he published *The Origin of Species* (its full title is *On the Origin of Species by Means of Natural Selection, or the Preservation of Favoured Races in the Struggle for Life*).

The book was an instant hit, selling out on its first day, provoking public discussion and debate and ultimately causing a wholesale change in the scientific understanding of natural selection and many other important evolutionary ideas. Where once the worldview was of a young earth, populated by unchanging species all created at one time, with no additions or extinctions, now there was a new dynamic view of life on earth: descent with

EVOLVING BELIEFS IN DARWIN'S WORLD

BEFORE DARWIN	AFTER DARWIN
• All organisms were put on earth by a creator at the same time.	• Organisms change over time.
• Organisms are fixed: no additions, no subtractions.	• Some organisms have gone extinct.
• Earth is about 6,000 years old.	• Earth is more than 6,000 years old.
• Earth is mostly unchanging.	• The geology of earth is not constant, but always changing.

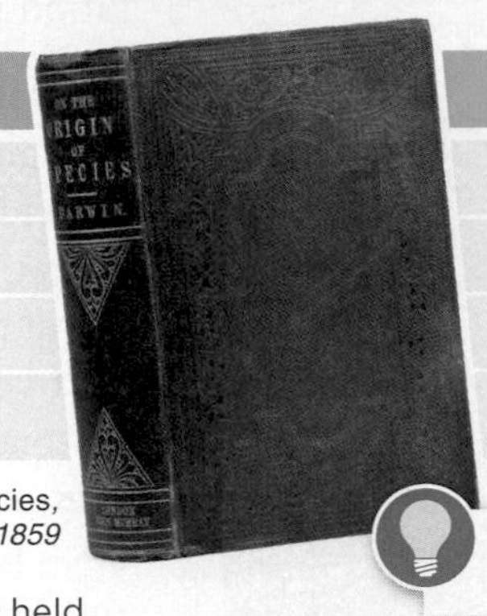

Darwin's The Origin of Species, *first published in 1859*

The ideas Darwin described in* The Origin of Species *reflected a new, dynamic view of life on earth.

FIGURE 10-8 **Reconsidering the world.** Some commonly held ideas before and after Darwin.

modification—species could and did change over time, and as some species split into new species, others became extinct (FIGURE 10-8).

Darwin's theory has proved to be among the most important and enduring contributions in all of science. It has stimulated an unprecedented diversity of theoretical and applied research, and it has withstood repeated experimental and observational testing. With the background of Darwin's elegant idea in hand, we can now examine its details.

TAKE HOME MESSAGE 10.4

» After putting off publishing his thoughts on natural selection for more than 15 years, Darwin did so only after Alfred Russel Wallace independently came up with the same idea. The two men published a joint presentation on their ideas in 1858, and Darwin published a much more detailed treatment in *The Origin of Species* in 1859, sparking wide debate and discussion about the processes and patterns of evolution.

10.5–10.10 Four mechanisms can give rise to evolution.

Harlequin ladybird beetles, *Harmonia axyridis.* Red in their wings signals to predators the presence of toxic chemicals.

10.5 Evolution occurs when the allele frequencies in a population change.

Suppose you are put in charge of a large population of tigers in a zoo. Almost all of them are orange and brown with black stripes. Occasionally, though, unusual all-white tigers are born. This white phenotype is the result of the presence of a rare pair of alleles that suppresses the tiger's production of most fur pigment (FIGURE 10-9). (Recall from

THE EVOLUTION OF POPULATIONS

Evolution is a change in the allele frequencies of a population over time.

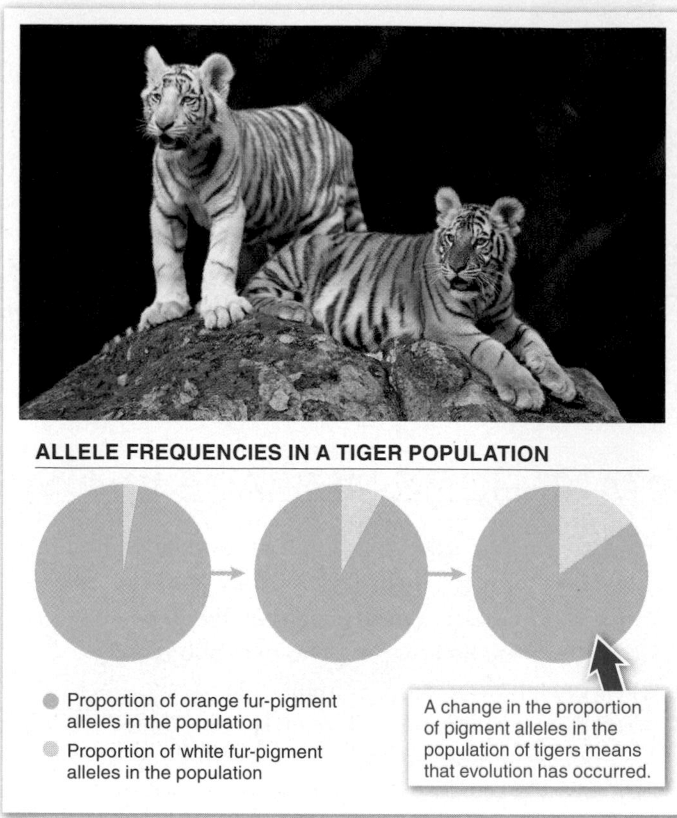
ALLELE FREQUENCIES IN A TIGER POPULATION
Proportion of orange fur-pigment alleles in the population
Proportion of white fur-pigment alleles in the population
A change in the proportion of pigment alleles in the population of tigers means that evolution has occurred.

FIGURE 10-9 **Evolution defined.**

Section 9-1 that alleles are variants of a gene and that most mammals inherit one allele from their mother and one from their father for each gene.) Because visitors flock to zoos to see white tigers, you want to increase the proportion of your tiger population that is white. How would you do so?

To produce more animals with the white phenotype, you could breed the white tigers with each other. And as the generations go by (this will take a while, though, since the generation time for tigers is about eight years), your population will include a higher proportion of white tigers. When this happens, you will have witnessed evolution: a change in the proportion of alleles for the pigment-suppression gene in the population.

Alternatively, you could increase the proportion of white tigers in your population by acquiring some white tigers from another zoo. Through this simple addition of white tigers to your population, the population will include a higher proportion of white tigers. When you add white tigers (or remove orange tigers), you will again witness evolution in your population.

As this example illustrates, evolution doesn't involve changing the genetics or physical features of *individuals*. Individuals do not evolve. Rather, you change the proportions of the alleles in the *population*. An allele's frequency is the proportion in which it is present in a population relative to the other alleles of the same gene, much like its "market share." In the tiger example, the white-fur alleles initially had little market share, perhaps as small as 1%. Over time, though, they came to make up a larger and larger proportion of the total fur-pigment alleles in the population. And as this happened, evolution occurred.

Any time an allele's market share changes, evolution is taking place. Perhaps the most dramatic way we see this is in acts of predation. When one organism kills another, the dead animal's alleles will no longer be passed on in the population. And if certain alleles alter coloration, or speed, or some other defense mechanism that makes an animal more likely to be killed by another, those alleles are likely to lose market share. The converse is equally true: evolution occurs when alleles increase in frequency.

THE MECHANISMS OF EVOLUTION

MUTATION
An alteration of the base-pair sequence in the DNA of an individual's gamete-producing cells that changes an allele's frequency.

GENETIC DRIFT
A random change in allele frequencies, unrelated to any allele's influence on reproductive success.

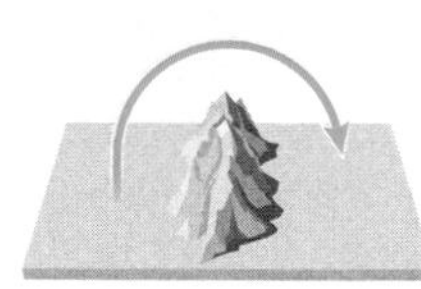

MIGRATION
A change in allele frequencies caused by individuals moving into or out of a population.

NATURAL SELECTION
A change in allele frequencies that occurs when individuals with one version of a heritable trait have greater reproductive success than individuals with a different version of the trait.

FIGURE 10-10 **Four ways in which evolution can occur.**

Keeping in mind that evolution is genetic change in a population, we'll now explore each of the four processes that lead to such genetic changes (FIGURE 10-10). The four evolutionary mechanisms are:

1. Mutation
2. Genetic drift
3. Migration
4. Natural selection

TAKE HOME MESSAGE 10.5

» Evolution is a change in allele frequencies within a population. It often occurs by four different mechanisms: mutation, genetic drift, migration, and natural selection.

10.6 Mutation—a direct change in the DNA of an individual—is the ultimate source of all genetic variation.

As we discuss this first of four mechanisms for evolution, it is helpful to keep in mind our precise definition of evolution: a change in the allele frequencies found in a population. **Mutation** is an alteration of the base-pair sequence of an individual's DNA, and when this alteration occurs in DNA that is part of a particular gene, the change in the DNA sequence may cause a change in an allele that has important consequences. (See Section 6-9 for a detailed discussion of the mutation process.)

Only mutations that affect reproductive cells can be passed from parent to offspring and have the potential to alter the allele frequencies within a population. Say, for example, that a mutation changes one of a person's two blue-eye alleles into a brown-eye allele. If this mutation occurs in the sperm- or egg-producing cells, it can be passed on to the next generation; the offspring may carry the brown-eye allele. When this happens, the proportion of blue-eye alleles in the population is slightly reduced, and the proportion of brown-eye alleles is slightly increased. Evolution has occurred.

Mutations can occur spontaneously during cell division or, as we saw in Chapter 6, can be induced by a variety of environmental factors—including radiation, such as X rays, and some chemicals. But although these factors can induce mutations and influence the rate at which mutations occur, they do not generally influence exactly *which* mutations occur. Thus, we tend to say that "mutations are random." What this means, more precisely, is that (1) we cannot predict which individuals will have which mutations, and (2) we cannot predict whether the consequences of a mutation will be neutral (no effect), harmful, or useful. Treatment of an agricultural pest with certain chemicals, for example, might increase the number of mutations in that pest population, but it does not increase the likelihood that a particular mutation is beneficial or detrimental. This doesn't mean, however, that all mutations have an equal probability of occurring. They do not. Some mutations are far more likely than others.

Mutation plays a critically important role relevant to all mechanisms of evolution, including natural selection: *mutation is the ultimate source of genetic variation in a population*. In the previous blue-eye and brown-eye example, we saw that a mutation may lead to the conversion of one allele to another that is already found within the population. More importantly, though, a mutation may create a completely novel allele that codes for the production of a new protein (FIGURE 10-11). That is, a change in the base-pair sequence of a person's DNA may cause the formation of a gene product that has never existed before in the human population: instead of blue or brown eyes, the mutated gene might code for yellow or red eyes. If such a new allele occurs in the reproductive cells, and if it does not significantly reduce an individual's "reproductive fitness" (which we consider below), the new allele can be passed on to offspring and remain in the population. At some future time, the mutation might even confer higher fitness, in which case natural selection could lead to an increase in frequency of this allele in the population. For this reason, mutation is critical to natural selection: all variation—the raw material for natural selection—must initially come from mutation.

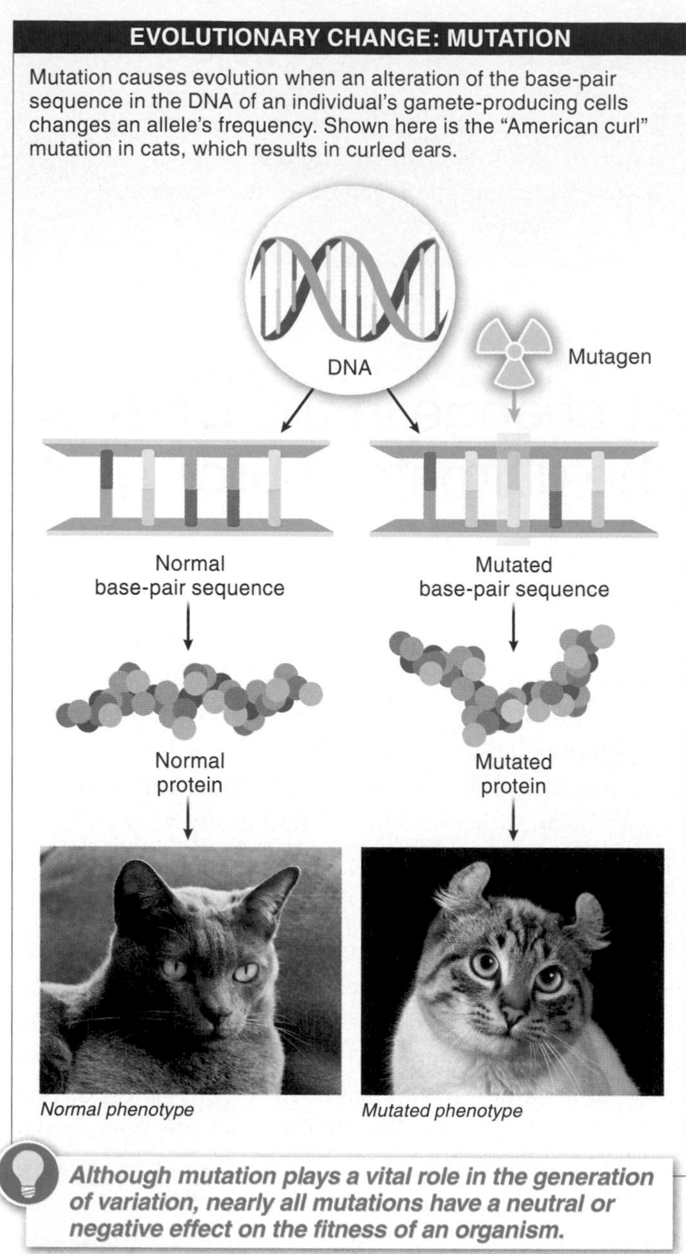

EVOLUTIONARY CHANGE: MUTATION
Mutation causes evolution when an alteration of the base-pair sequence in the DNA of an individual's gamete-producing cells changes an allele's frequency. Shown here is the "American curl" mutation in cats, which results in curled ears.
DNA
Mutagen
Normal base-pair sequence
Mutated base-pair sequence
Normal protein
Mutated protein
Normal phenotype
Mutated phenotype
Although mutation plays a vital role in the generation of variation, nearly all mutations have a neutral or negative effect on the fitness of an organism.

FIGURE 10-11 **Mechanisms of evolutionary change: mutation.**

Despite this vital role in generating variation, however, nearly all mutations either have no impact on an organism's fitness or have an adverse impact by causing early death or reducing the organism's reproductive success. Mutations in a normally functioning allele that codes for a normally functioning protein typically result in either an allele that codes for the same protein or a new allele that codes for a nonfunctioning protein. The latter case almost inevitably reduces an organism's fitness. For this reason, our bodies protect our sperm- or egg-producing DNA with a variety of built-in error-correction mechanisms. As a result, mutations are rare (in humans, on the order of 1 mutation in every 30 million base pairs in each generation).

TAKE HOME MESSAGE 10.6

›› Mutation is an alteration of the base-pair sequence in an individual's DNA. If such an alteration changes an allele in an individual's gamete-producing cells, the frequency of alleles in the population has changed, and this constitutes evolution within the population. Mutations can be caused by high-energy radiation or chemicals in the environment and also can appear spontaneously. Mutation is the only way that new alleles can be created within a population.

10.7 Genetic drift is a random change in allele frequencies in a population.

Another evolutionary mechanism is **genetic drift,** a random change in allele frequencies in a population. A change in allele frequencies due to genetic drift is not related to the alleles' influence on reproductive success. For example, consider two alleles for a particular trait such as white spotting in a cat's fur—a single-gene trait with two alleles, for which the white-spotting allele is dominant over the no-white-fur allele (FIGURE 10-12).

Now suppose that two heterozygous (*Ss*) parent cats have one offspring. Which combination of alleles will that kitten receive? This is impossible to predict because it depends completely on which sperm fertilizes which egg—the luck of the draw. If the parents' sole offspring inherits a recessive allele from each parent, will the *population's* allele frequencies change? Yes. The new individual's two recessive alleles (*ss*) increase the proportion of recessive alleles in the population.

GENETIC DRIFT

Genetic drift causes evolution when an allele's frequency changes for any of several reasons that are unrelated to the allele's influence on reproductive success.

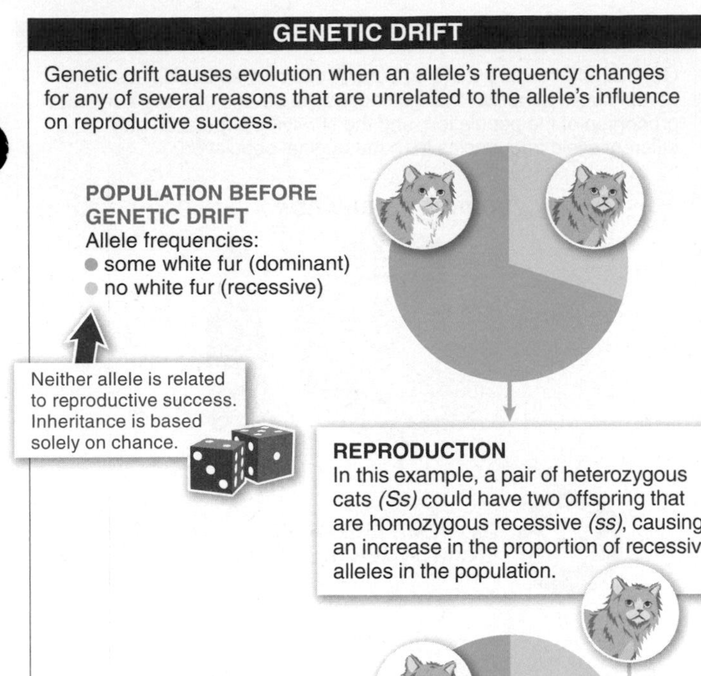

POPULATION BEFORE GENETIC DRIFT
Allele frequencies:
some white fur (dominant)
no white fur (recessive)
Neither allele is related to reproductive success. Inheritance is based solely on chance.
REPRODUCTION
In this example, a pair of heterozygous cats (Ss) could have two offspring that are homozygous recessive (ss), causing an increase in the proportion of recessive alleles in the population.

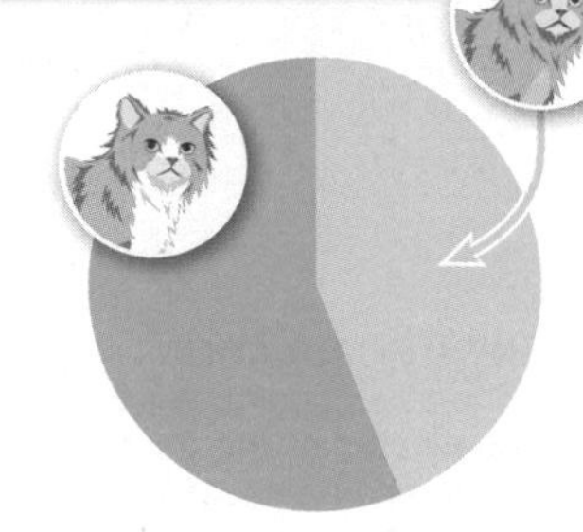

POPULATION AFTER GENETIC DRIFT
There are now more recessive alleles in the population than before.

FIXATION

Genetic drift leads to fixation when an allele's frequency becomes 100% in a population. If this occurs, there is no longer genetic variation for the gene.

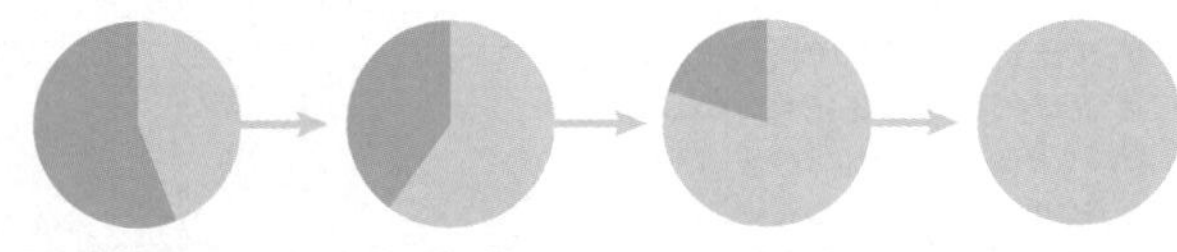

FIGURE 10-12 **Mechanisms of evolutionary change: genetic drift.** Genetic drift has the greatest impact in small populations.

It is equally likely that the parents' only offspring will receive two dominant alleles (*SS*). In either case, because a change in allele frequencies has occurred, evolution has happened.

This impact of genetic drift is much greater in small populations than in large populations. In a large population, any one couple having an offspring with two recessive alleles is likely to be offset by another couple having an offspring with two dominant alleles. When that happens, the overall allele frequencies in the population do not change, and no evolution has occurred.

One of the most important consequences of genetic drift is that it can lead to **fixation** for one allele of a gene in a population (see Figure 10-12). Fixation results when an allele's frequency in a population reaches 100% (and the frequency of all other alleles of that gene becomes 0%). If this happens, there is no more variability in the population for this gene; all individuals always produce offspring carrying only that allele (until new alleles arise through mutation). For this reason, genetic drift reduces the genetic variation in a population.

Multiple evolutionary mechanisms can occur simultaneously—an allele favored by natural selection, for example, may simultaneously increase or decrease in frequency due to genetic drift.

Two special cases of genetic drift, the founder effect and population bottleneck effect, are important in the evolution of many populations.

Founder Effect A small number of individuals may leave a population and become the founding members of a new, isolated population. The founder population may have different allele frequencies than the original, "source" population, particularly if the founders are a small group. If this new population does have different allele frequencies, evolution has occurred. Because the founding members of the new population will give rise to all subsequent individuals, the new population will be dominated by the genetic features that happened to be present in that group of founding fathers and mothers. This type of genetic drift is called the **founder effect.**

Q Why are Amish people more likely than other people to have extra fingers and toes?

The Amish population in the United States is thought to have been established by a small number of founders, some of whom happened to carry the allele for *polydactyly*—the condition of having extra fingers and toes. As a consequence, this trait, while rare, now occurs much more frequently among the Amish than it does worldwide (FIGURE 10-13).

Population Bottleneck Effect Occasionally, a famine, disease, or rapid environmental change causes the deaths of a large proportion of individuals in a population. Because the population is quickly reduced to a small fraction of its original size, this reduction is called the **bottleneck effect.** If the catastrophe is equally likely to strike any member of the population, the remaining members are essentially a random, small sample of the original population. For this reason, the remaining population may not possess the same allele frequencies as the original population. Thus, the consequence of such a population

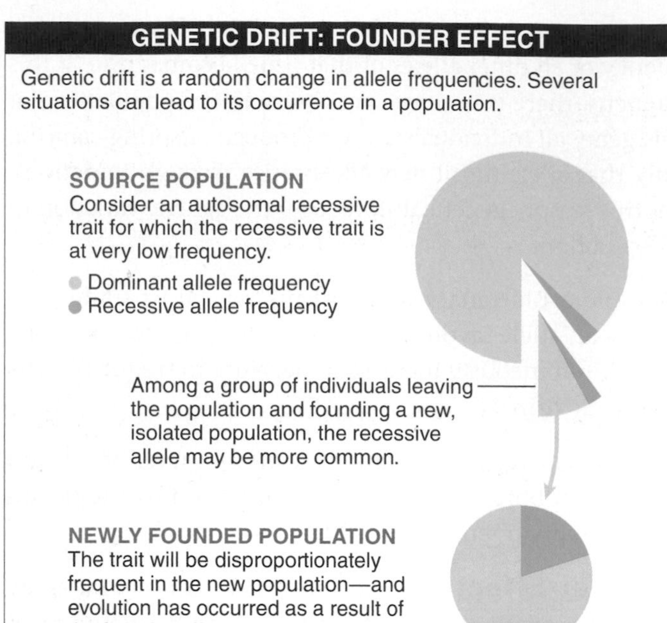

GENETIC DRIFT: FOUNDER EFFECT
Genetic drift is a random change in allele frequencies. Several situations can lead to its occurrence in a population.
SOURCE POPULATION
Consider an autosomal recessive trait for which the recessive trait is at very low frequency.
Dominant allele frequency
Recessive allele frequency
Among a group of individuals leaving the population and founding a new, isolated population, the recessive allele may be more common.
NEWLY FOUNDED POPULATION
The trait will be disproportionately frequent in the new population—and evolution has occurred as a result of the founder effect.

The condition of having extra toes or fingers (shown here) is called polydactyly. Typically extremely rare, it occurs at high frequency among the Amish of Pennsylvania (not shown), because the recessive allele causing it was carried by both a husband and wife in the founding population.

FIGURE 10-13 One way that genetic drift occurs: the founder effect.

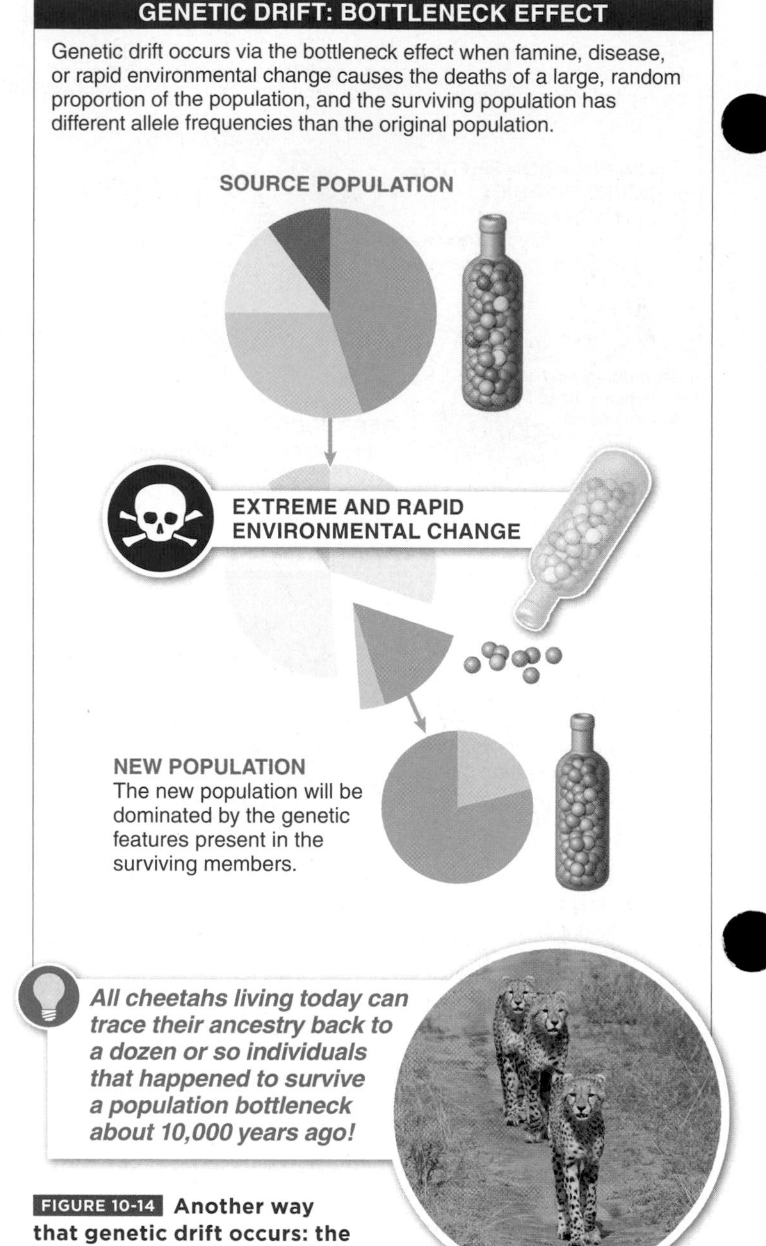

FIGURE 10-14 Another way that genetic drift occurs: the bottleneck effect.

bottleneck would be evolution through genetic drift (FIGURE 10-14).

Just such a population bottleneck occurred with cheetahs near the end of the last ice age, about 10,000 years ago. Although the cause is unknown—possibly environmental cataclysm or human hunting pressures—it appears that nearly all cheetahs died. And although the population rebounded, all cheetahs living today can trace their ancestry to a dozen or so lucky individuals that survived the bottleneck. As a result of this past instance of evolution by genetic drift, there is almost no genetic variation in the current population of cheetahs. (And, in fact, a cheetah is able to accept a skin graft from any other cheetah, much as identical twins can from each other.)

TAKE HOME MESSAGE 10.7

» Genetic drift is a random change in allele frequencies within a population, unrelated to the alleles' influence on reproductive success. Genetic drift is a significant mechanism of evolutionary change, primarily in small populations.

10.8 Migration into or out of a population may change allele frequencies.

The third mechanism of evolutionary change is **migration.** Migration, also called **gene flow,** is the movement of some individuals of a species from one population to another (FIGURE 10-15). This movement from population to population within a species distinguishes migration from the founder effect, in which individuals migrate to a new habitat previously unpopulated by that species. If migrating individuals survive and reproduce in the new population, and if they carry a different proportion of alleles than the individuals in their new home, then the recipient population experiences a change in allele frequencies and, consequently, evolution. And because alleles are simultaneously lost from the population that the migrants left behind, that population, too, experiences a change in its allele frequencies and thus evolves.

Gene flow between two populations is influenced by the mobility of the organisms and by environmental barriers, such as mountains or rivers. And as we'll see in Chapter 19, human activities can dramatically influence the migratory potential of organisms as we transport species from place to place. This can happen intentionally, as when people purchase exotic pets or non-native species for their gardens, or unintentionally, such as when marine species get trapped in the ballast tanks of cargo ships, later to be released in ports on the other side of the world when the tanks are flushed.

TAKE HOME MESSAGE 10.8

» Migration, or gene flow, leads to a change in allele frequencies in a population as individuals move into or out of the population.

MIGRATION (GENE FLOW)

Gene flow causes evolution if individuals move from one population to another, causing a change in allele frequencies in either population.

1 BEFORE MIGRATION
Two populations of the same species exist in separate locations. In this example, they are separated by a mountain range.

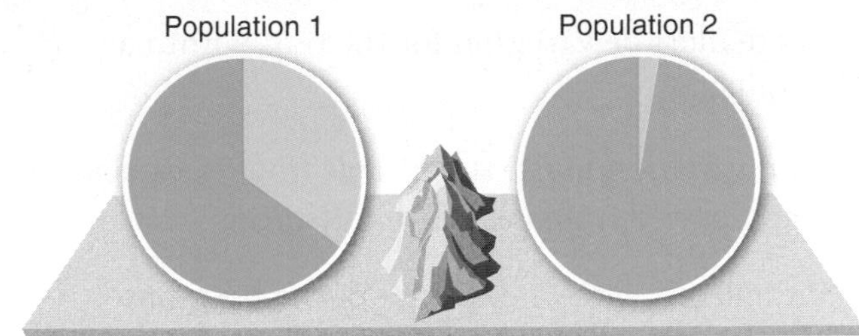

2 MIGRATION
A group of individuals from Population 1 migrates over the mountain range.

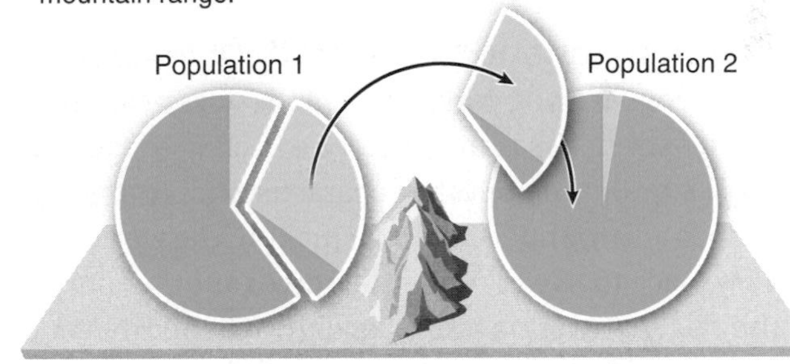

3 AFTER MIGRATION
The migrating individuals are able to survive and reproduce in the new population.

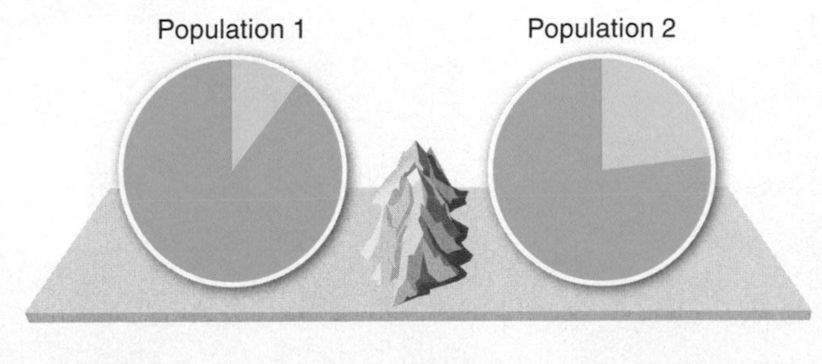

FIGURE 10-15 **Mechanisms of evolutionary change: migration (gene flow).**

10.9 When three simple conditions are satisfied, evolution by natural selection is occurring.

The fourth mechanism of evolutionary change is **natural selection.** This is the mechanism that Darwin identified in *The Origin of Species,* in which he noted that three conditions are necessary for natural selection to occur:

1. There must be variation for the trait within a population.
2. That variation must be heritable (that is, capable of being passed from parents to offspring).
3. Individuals with one version of the trait must produce more offspring than those with a different version of the trait.

Let's examine these conditions more closely.

Condition 1: Variation for a Trait Observing people with their dogs at a dog park, you will see dogs of all shapes, colors, and sizes. The humans, too, will vary in facial features and a multitude of other characteristics (FIGURE 10-16). Variation is all around us. Beyond making the world an interesting place to live in, variation serves another purpose: it is the raw material on which evolution works. Variation is not limited to physical features such as fur color or face shape. Individuals vary in physiological and biochemical ways, too. Some people can quickly and efficiently metabolize alcohol, for example. Others find themselves violently ill soon after sipping a glass of wine. Similarly, we vary in our susceptibility to poison ivy or to diseases such as malaria. Behavioral variation—from temperament to learning abilities to interpersonal skills—is dramatic and widespread, too. So impressed was Darwin with the variation he observed throughout the world that he devoted the first two chapters of *The Origin of Species* to a discussion of variation in nature and among domesticated animals.

Condition 2: Heritability For natural selection to act on a trait, offspring must inherit the trait from their parents. Although inheritance was poorly understood in Darwin's time, it was not hard to see that, for many traits, offspring look more like their parents than like some other, random individual in the population (FIGURE 10-17). Animal breeders had long known that the fastest horses generally give birth to the fastest horses. Farmers understood that the plants with the highest productivity

FIGURE 10-16 **The first condition for natural selection is variation for a trait.**

FIGURE 10-17 **The second condition for natural selection is heritability.** Goldie Hawn and daughter Kate Hudson resemble each other.

The tiniest pig in a litter has reduced differential reproductive success. Its more robust siblings prevent access to the food it needs to grow and thrive.

FIGURE 10-18 The third condition for natural selection is differential reproductive success.

generally produce seeds from which highly productive plants grow. We call the transmission of traits from parents to offspring through genetic information **inheritance** or **heritability.**

Condition 3: Differential Reproductive Success

Darwin derived the third condition for natural selection from three fairly simple observations. First, more organisms are born than can survive. Second, organisms are continually struggling for existence. Lastly, some organisms are more likely than others to survive and reproduce. In a world of limited resources, finding food or shelter is a zero-sum game: if one organism is feasting, another is likely to be starving.

This three-part observation led Darwin to his third condition for natural selection, which is called **differential reproductive success:** from all the variation existing in a population, individuals with traits more suited to survival and reproduction in their environment generally leave more offspring than do individuals with other traits (FIGURE 10-18). For example, in the food experiments with fruit flies, we saw that flies inheriting the ability to pack on fat when food was available had greater reproductive success than those inheriting a poor ability to pad their little fruit fly frames with fat deposits.

That's it. Natural selection—one of the most influential and far-reaching ideas in the history of science—takes place when three basic conditions are met (FIGURE 10-19):

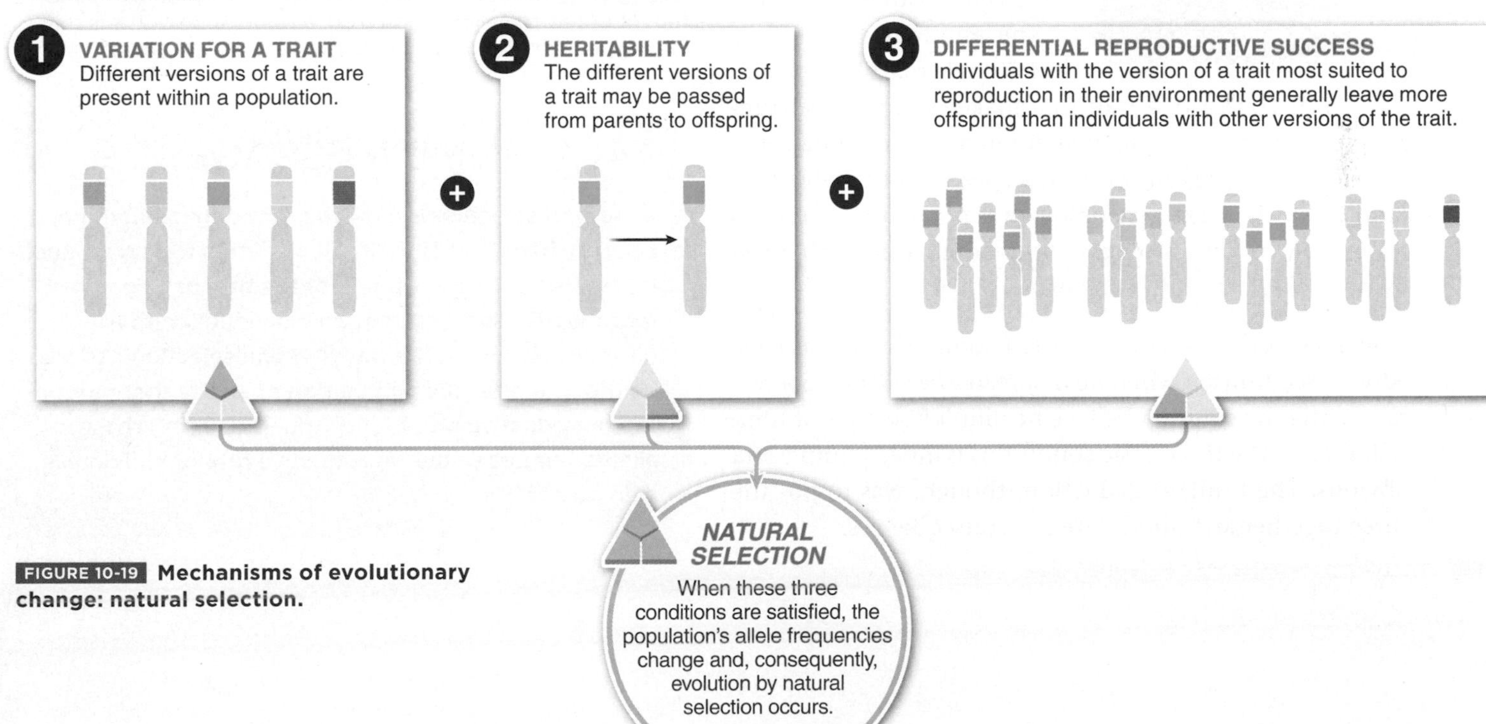

FIGURE 10-19 Mechanisms of evolutionary change: natural selection.

1. Variation for a trait
2. Heritability of that trait
3. Differential reproductive success based on that trait

When these three conditions are satisfied, evolution by natural selection is occurring. It's nothing more and nothing less. Over time, the traits that lead some organisms to have greater reproductive success than others will increase in frequency in a population, while traits that reduce reproductive success will become less and less common.

For some traits, the reason they specifically confer greater reproductive success is that they make the individual more attractive to the opposite sex. Such traits—including the brightly colored feathers of male peacocks, the large antlers of male red deer, and a variety of other "ornaments" that increase an individual's status or appeal—increase in frequency because they satisfy the three conditions for natural selection. This natural selection for mating success is called **sexual selection.**

Q In most agricultural pests treated with pesticides, resistance to the pesticides eventually evolves. How does this happen?

Another way of looking at natural selection is to focus not on the winners (the individuals who are producing more offspring) but on the losers. Natural selection can be viewed as the elimination of some heritable traits from a population. If you carry a trait that makes you a slower-running rabbit, for example, you are more likely to be eaten by a fox (FIGURE 10-20). If running speed is a heritable trait (and it is), the next generation in a population contains fewer slow rabbits. Over time, the population is changed by natural selection. It evolves.

One of Darwin's contemporaries, Thomas Huxley, supposedly cursed himself when he first read *The Origin of Species,* saying that he couldn't believe he didn't figure it out himself. Each of the three basic conditions is indeed simple and obvious. The brilliant deduction, though, was to put the three together and appreciate the consequences.

NATURAL SELECTION IN NATURE

1 VARIATION FOR A TRAIT
Running speed in rabbits can vary from one individual to the next.

2 HERITABILITY
The trait of running speed is passed on from parents to their offspring.

3 DIFFERENTIAL REPRODUCTIVE SUCCESS
In a population, rabbits with slower running speeds are eaten by the fox, and their traits are less likely to be passed on to the next generation.

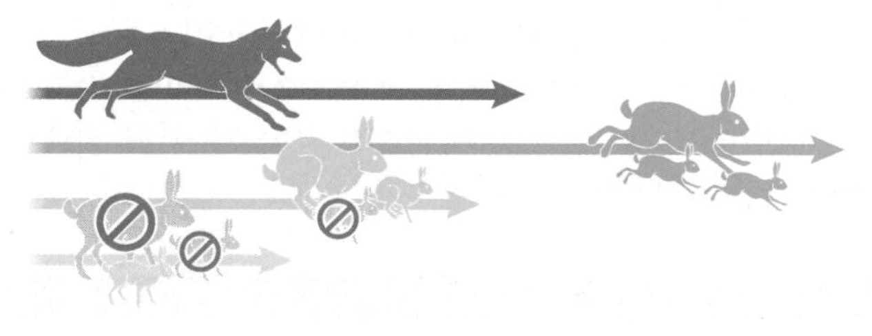

FIGURE 10-20 **Removing the losers.** Natural selection can be thought of as the elimination from a population of traits that confer poor reproductive success.

TAKE HOME MESSAGE 10.9

» Natural selection is a mechanism of evolution that occurs when there is heritable variation for a trait, and individuals with one version of the trait have greater reproductive success than do individuals with a different version of the trait. Natural selection can also be thought of as the elimination of alleles that reduce the reproductive rate of individuals carrying those alleles, relative to the reproductive rate of individuals who do not.

10.10 A trait does not decrease in frequency simply because it is recessive.

Q If a trait is recessive, does it eventually disappear from a population?

Do recessive traits gradually become less common in a population? In the early 1900s, there was much discussion among biologists about this question. They wondered: "If the allele for brown eyes is dominant over the allele for blue eyes, why doesn't a population eventually become all brown-eyed?"

R. C. Punnett (originator of the Punnett square, described in Chapter 9) posed this question to G. H. Hardy, a mathematician with whom Punnett played cricket. Hardy replied the next day, and he published his answer in *Science* in 1908. It turns out that Wilhelm Weinberg had answered the question independently six months earlier. But because he published his findings in an obscure journal, scientists didn't appreciate his result for another 35 years.

Their result underlies all evolutionary genetics and is referred to as the Hardy-Weinberg Law. It reveals several things, including that the answer to the question of whether recessive traits become rarer in a population is *no*. A trait does not decrease in frequency simply because it is recessive. Here's how Hardy and Weinberg demonstrated it.

First, we refer to the frequency of the dominant allele in the population, *A*, as "*p*" and the frequency of the recessive allele, *a*, as "*q*." (We're assuming here that the gene has only two alternative alleles in the population. The equations can be expanded for genes with more alleles, without changing the overall outcomes.) Since every allele in the population has to be either *A* or *a*, we can say that $p + q = 1$. The percentage of the two alleles must total 100%. If we know the frequency of either allele in the population, we can subtract it from 1 to calculate the frequency of the other allele.

Next, we can predict how common each genotype in the population will be. The frequency of *AA* is just the probability that an individual gets two copies of allele *A*, which is $p \times p$, or p^2. Applying the same math, the frequency of *aa* individuals in the population is simply $q \times q$, or q^2. Predicting the frequency of heterozygous individuals, *Aa*, is slightly more complicated. Because the dominant allele may come from either the mother or the father, the frequency of the *Aa* genotype is $2 \times p \times q$, or $2pq$. In other words, it's the sum of the frequency of the *Aa* genotype (i.e., getting the dominant allele from the mother) and the *aA* genotype (i.e., getting the dominant allele from the father), which is $p \times q$ plus $q \times p$, which simplifies to $2pq$.

Consider an example (FIGURE 10-21). Suppose a population of 1,000 kangaroo rats has the following phenotype and genotype frequencies: 16 are dark brown (*BB*), 222 are spotted (*Bb*), and 762 are light brown (*bb*). The trait shows incomplete dominance, with the allele for dark brown color, *B*, dominant over the allele for light brown color, *b*, and the heterozygote having a spotted phenotype. Because every individual in the population has two alleles for the coat-color gene (one from each parent), there are twice as many alleles as members of the population. So a population of 1,000 kangaroo rats has 2,000 alleles of the gene for coat color.

Now let's look at the allele frequencies. Each *BB* individual has two copies of *B*, and each *Bb* individual has one copy:

Frequency of $B = [(2 \times 16) + 222]/2{,}000 = 0.127 = p$

Frequency of $b = [(2 \times 762) + 222]/2{,}000 = 0.873 = q$

Since we know the allele frequencies, *p* (0.127) and *q* (0.873), we can determine the genotype frequencies of the offspring produced in this population. According to the equations above, they should be:

Frequency of $BB = p^2 = (0.127)^2 = 0.016$

Frequency of $Bb = 2pq = 2(0.127)(0.873) = 0.222$

Frequency of $bb = q^2 = (0.873)2 = 0.762$

If 1,000 kangaroo rats are produced, we expect to see genotype frequencies that are the same as in the parent generation: 16 *BB*, 222 *Bb*, and 762 *bb*. And from these genotype frequencies, we expect the following allele frequencies among the offspring they produce:

Frequency of $B = [(2 \times 16) + 222]/2{,}000 = 0.127 = p$

Frequency of $b = [(2 \times 762) + 222]/2{,}000 = 0.873 = q$

HARDY-WEINBERG EQUILIBRIUM

If a population is in Hardy-Weinberg equilibrium, and the allele frequencies of the population are known, predictions can be made about the genotypes of the offspring that the population produces.

PARENT GENERATION (Population of 1,000 kangaroo rats)

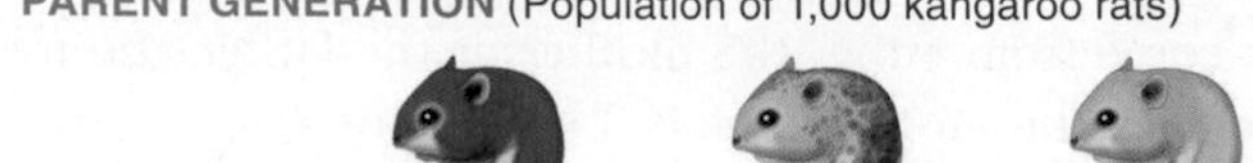

	Dark brown kangaroo rat	Spotted kangaroo rat	Light brown kangaroo rat
Genotype	*BB*	*Bb*	*bb*
# of individuals in population	16	222	762

ALLELE FREQUENCIES

B $p = [(2 \times 16) + 222] / 2{,}000 = \mathbf{0.127}$

- Each *Bb* individual has one copy of *B*.
- Each *BB* individual has two copies of *B*.

b $q = [(2 \times 762) + 222] / 2{,}000 = \mathbf{0.873}$

- Each *Bb* individual has one copy of *b*.
- Each *bb* individual has two copies of *b*.

REPRODUCTION (We assume random mating)

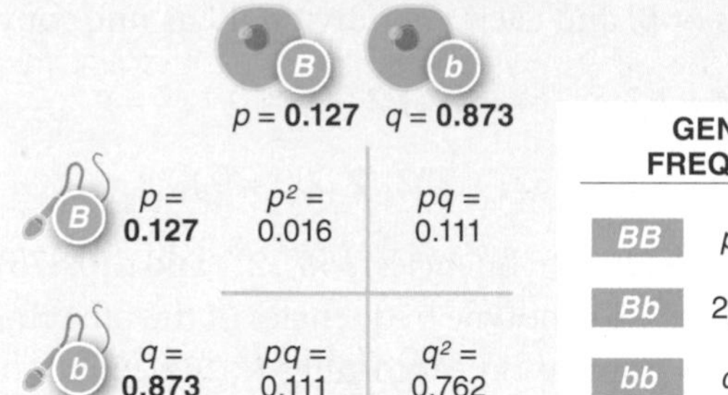

	B p = **0.127**	*b* q = **0.873**
B p = **0.127**	p^2 = 0.016	pq = 0.111
b q = **0.873**	pq = 0.111	q^2 = 0.762

GENOTYPE FREQUENCIES

BB	p^2 = 0.016
Bb	$2pq$ = 0.222
bb	q^2 = 0.762

NEXT GENERATION (If 1,000 kangaroo rat offspring are produced)

	Dark brown kangaroo rat	Spotted kangaroo rat	Light brown kangaroo rat
Genotype	*BB*	*Bb*	*bb*
# of individuals in population	16	222	762

As long as there is random mating and no evolution, the frequencies of recessive alleles and dominant alleles do not change over time.

FIGURE 10-21 Recessive alleles don't necessarily disappear from populations.

Notice that the frequencies are unchanged. The allele frequencies 0.127 and 0.873 will always produce the same genotype frequencies—0.16, 0.222, and 0.762—which, in turn, will always have the same allele frequencies. Put another way: the recessive allele *doesn't* decrease in the population over time!

Our example shows that if we know the frequency of each allele in a population, we can predict the genotypes and phenotypes we should see in that population. However, the Hardy-Weinberg equations and predictions are based upon two key assumptions: that random mating occurs, with alleles randomly coming together in all possible genotypes, and that there is no evolution. If individuals are dying off specifically because they carry the recessive allele (or because they carry the dominant), the allele frequencies would be changing due to natural selection. Likewise, other mechanisms of evolution—mutations, migration, or genetic drift—would alter the allele frequencies. In each of these exceptions, the allele frequencies would be changing through evolution.

What if we examine a population and observe genotype frequencies that are *not* those predicted by the Hardy-Weinberg equations? If this occurs, we say that the population is not in *Hardy-Weinberg equilibrium,* and we know that either evolution or non-random mating is occurring, or both. Our calculations can help us better understand the forces influencing the population and suggest further lines of investigation.

TAKE HOME MESSAGE 10.10

» Knowing the frequency of each allele in a population, we can predict the genotypes and phenotypes we should see. If the phenotypic frequencies in a population are not those predicted from the allele frequencies, the population is not in Hardy-Weinberg equilibrium, meaning that non-random mating or evolution, or both, is occurring. But as long as the Hardy-Weinberg assumptions are not violated, recessive alleles and dominant alleles do not change their frequencies over time.

10.11–10.17

Through natural selection, populations of organisms can become adapted to their environments.

Musk oxen, *Ovibos moschatus*, in a snowy landscape.

10.11 Traits causing some individuals to have more offspring than others become more prevalent in the population.

"Survival of the fittest." This is perhaps one of the most famous but most misunderstood and misused phrases. For starters, "survival of the fittest" was coined not by Darwin but by Herbert Spencer, an influential sociologist and philosopher. In fact, the phrase did not even appear in *The Origin of Species* when Darwin first published it. It wasn't until the fifth edition, 10 years later, that he used the phrase in describing natural selection.

As we'll see, in evolution, the word *fitness* is not defined by an organism's ability to survive or its physical strength or its health. Rather, **fitness** is a measure of the amount of reproduction (the reproductive output) of an individual with a particular phenotype relative to the reproductive output of individuals of the same species with alternative phenotypes.

Suppose there are two fruit flies. One fly carries the genes for a version of a trait that allows it to survive a long time without food. The other has the genes for a different version of the trait that allows it to survive only a short while without food. Which fly has the greater fitness? If the environment is one in which there are long periods without food, the fly that can live a long time without food is likely to produce more offspring than the other fly, and so over the course of its life it has greater fitness. The alleles carried by an individual with high fitness will increase their proportion in a population over time, and the population will evolve.

There are three important elements to an organism's fitness:

1. An individual's fitness is measured relative to other genotypes or phenotypes in the population. Those traits that confer the highest fitness will generally increase in frequency in a population, and their increase will always come at the expense of alternative traits that confer lower fitness.

2. Fitness depends on the specific environment in which the organism lives. The fitness value of having one trait versus another depends on the environment. A sand-colored mouse living in a beach habitat will be more fit than a chocolate-colored mouse. But that same sand-colored mouse will be vulnerable to predators if it lives in the darker brush away from the beach. An organism's fitness, although genetically based, can change over time and across habitats (FIGURE 10-22).

3. Fitness depends on an organism's reproductive success compared with other organisms in the population. If you carry an allele that gives you the trait of surviving for 200 years but also causes you to be sterile, your fitness is zero; that allele will never be passed down to future generations. On the other hand, if you inherit an allele that gives you a trait that causes you to die at half the age of everyone else but also causes you to have twice as many offspring as the average individual, your fitness is

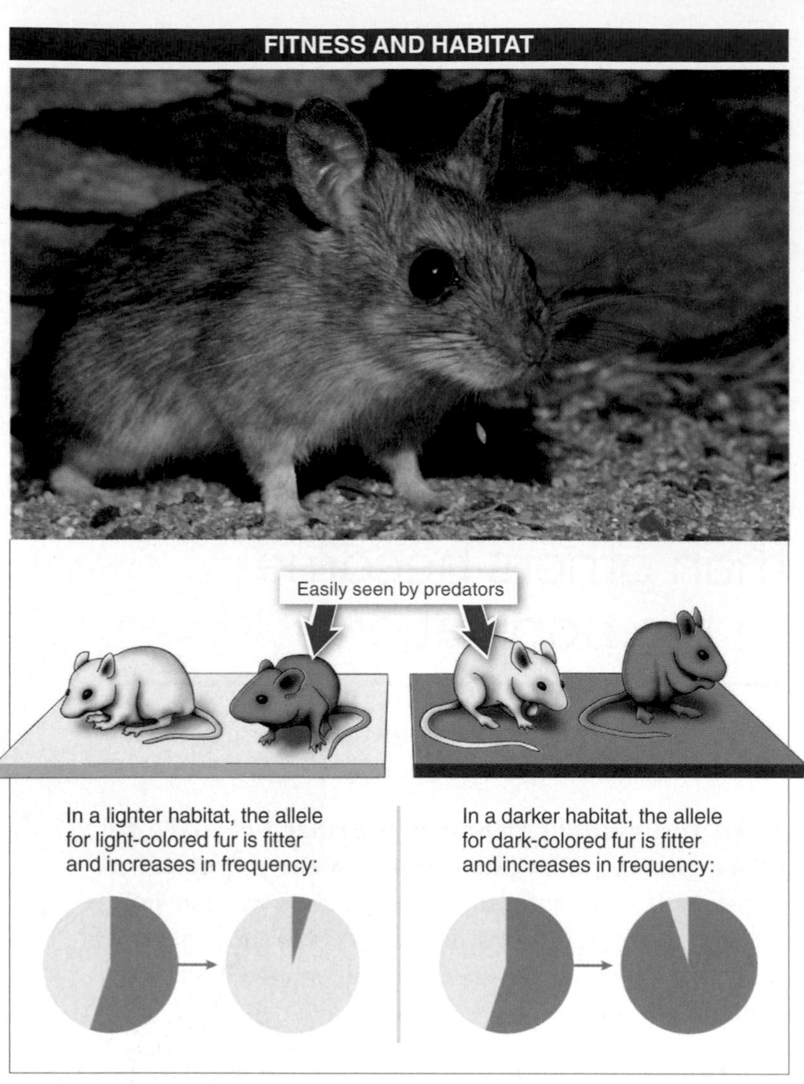

FITNESS AND HABITAT
Easily seen by predators
In a lighter habitat, the allele for light-colored fur is fitter and increases in frequency:
In a darker habitat, the allele for dark-colored fur is fitter and increases in frequency:

FIGURE 10-22 **An organism's fitness depends on the environment in which it lives.**

increased. It is reproductive success that is all-important in determining whether particular traits increase in frequency in a population.

The phrase "survival of the fittest" is misleading because it is the individuals with the greatest *reproductive output* that are the most fit in any population. It becomes a more meaningful phrase if we apply it to alleles rather than individual organisms: those alleles that increase an individual's fitness will "survive" in a population more than those that decrease an individual's fitness.

Q "Survival of the fittest" is a misleading expression. Why?

TAKE HOME MESSAGE 10.11

» **Fitness is a measure of the reproductive output of an individual with a particular phenotype, compared with the reproductive output of individuals with alternative phenotypes. An individual's fitness can vary, depending on the environment in which it lives.**

10.12 Organisms in a population can become better matched to their environment through natural selection.

If you put a group of humans on the moon, would they flourish? If you took a shark from the ocean and put it in your swimming pool, would it survive? In both cases, the answer is *no*. Organisms are rarely successful when put into novel environments. And the stranger the new environment, the less likely it is that the transplanted organism will survive. Why is that?

As Darwin noted over and over during his travels, the organisms that possess traits that allow them to better exploit the environment in which they live will tend to produce more offspring than the organisms with alternative traits. With passing generations, a population will be made up of more and more of these fitter organisms. And, as a consequence, populations of organisms will tend to be increasingly well matched or adapted to their environment.

Adaptation refers both to the process by which organisms become better matched to their environment *and* to the specific features that make an organism more fit.

FIGURE 10-23 **Adaptations increase fitness.** Quills are an adaptation that reduces (but doesn't eliminate) the risk of predation for porcupines.

Examples of adaptations abound. Bats have an extremely accurate type of hearing (called echolocation) for navigating and finding food, even in complete darkness. Porcupine quills make porcupines almost impervious to predation (FIGURE 10-23). Mosquitoes produce strong chemicals that prevent blood from clotting, so that they can extract blood from other animals.

TAKE HOME MESSAGE 10.12

» Adaptation, which refers both to the process by which organisms become better matched to their environment and to the specific traits that make an organism more fit, occurs as a result of natural selection.

10.13 Natural selection does not lead to perfect organisms.

In Lewis Carroll's *Through the Looking-Glass,* the Red Queen tells Alice that "in this place it takes all the running you can do, to keep in the same place." She might have been speaking about the process of evolution by natural selection. After all, if the least fit individuals are continuously weeded out of a population, we might logically conclude that, eventually, all organisms in all populations will be perfectly adapted to their environment. But this never happens. That is where the Red Queen's wisdom comes in.

Consider one of the many clearly documented cases of evolution in nature: the beak size of Galápagos finches. Over the course of a multi-decade study, biologists Rosemary Grant and Peter Grant closely monitored the average size of the finches' beaks. They found that the average beak size within a population fluctuated according to the food supply. During dry years—when the finches had to eat large, hard seeds—birds with bigger, stronger beaks were more successful and multiplied. During wet years, smaller-beaked birds were more successful, because there was a surplus of small, soft seeds.

The ever-changing "average" finch beak illustrates that adaptation does not simply march toward some optimal endpoint (FIGURE 10-24). Evolution in general, and natural selection specifically, does not guide organisms toward "betterness" or perfection. Natural selection is simply a process by which, in each generation, the alleles that cause organisms to have the traits that make them most fit in that environment at that time tend to increase in frequency. If the environment changes, the alleles that are most favored may change, too.

"I was taught that the human brain was the crowning glory of evolution so far, but I think it's a very poor scheme for survival."

— KURT VONNEGUT, American writer

Q Why doesn't natural selection lead to perfect organisms?

In the next to last paragraph of *The Origin of Species,* Darwin wrote: "as natural selection works solely by and for the good of each being, all corporeal [physical] and mental endowments will tend to progress towards perfection." In this passage, he overlooks several factors that prevent populations from progressing inevitably toward perfection:

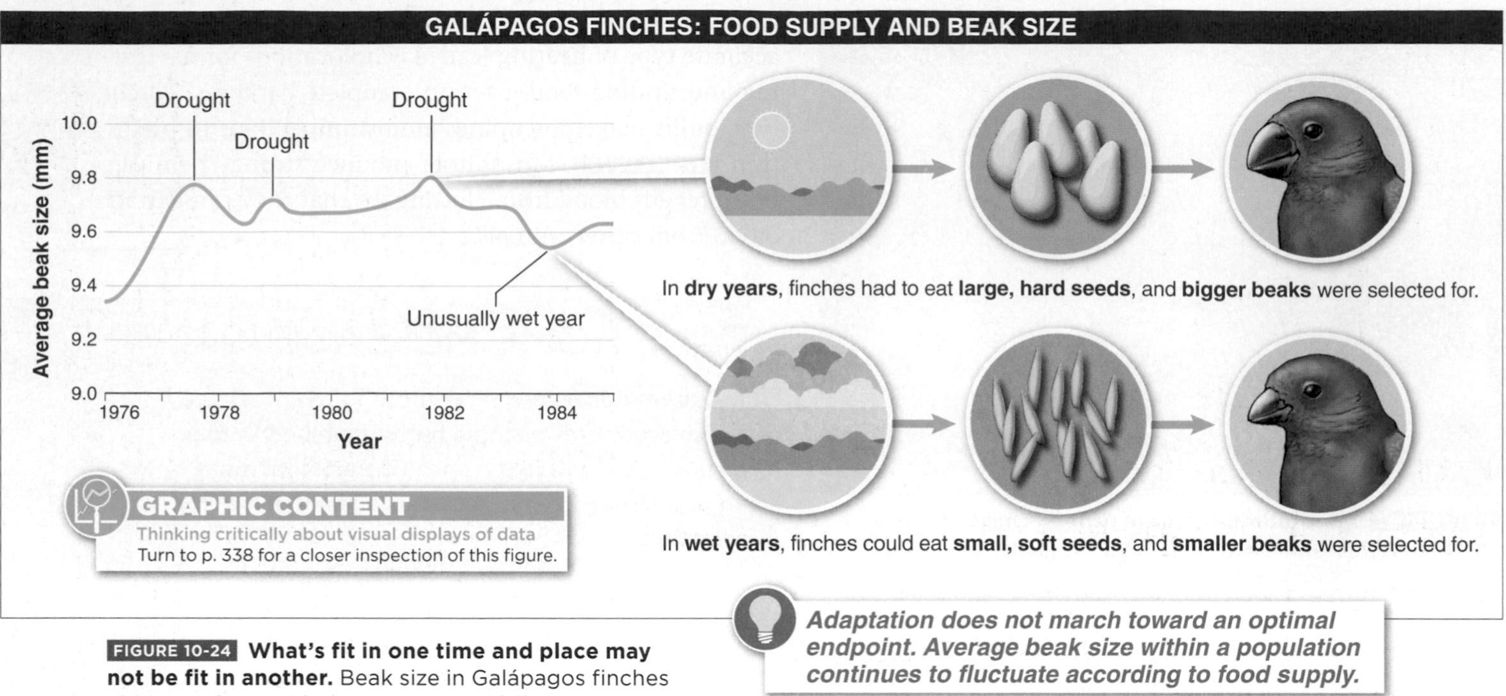
GALÁPAGOS FINCHES: FOOD SUPPLY AND BEAK SIZE
Drought
Drought
Drought
Unusually wet year
Average beak size (mm)
10.0
9.8
9.6
9.4
9.2
9.0
1976
1978
1980
1982
1984
Year
In dry years, finches had to eat large, hard seeds, and bigger beaks were selected for.
In wet years, finches could eat small, soft seeds, and smaller beaks were selected for.
GRAPHIC CONTENT
Thinking critically about visual displays of data
Turn to p. 338 for a closer inspection of this figure.
Adaptation does not march toward an optimal endpoint. Average beak size within a population continues to fluctuate according to food supply.

FIGURE 10-24 What's fit in one time and place may not be fit in another. Beak size in Galápagos finches changes along with the average rainfall.

1. Environments change quickly. Natural selection may be too slow to adapt the organisms in a population to such a constantly moving target.
2. Variation is needed as the raw material of selection. If a mutation creating a new, "perfect" version of a gene never arises, the individuals within a population will never be perfectly adapted.
3. There may be different alleles for a trait, each causing individuals to have the same fitness. In this case, each allele represents an equally fit "solution" to the environmental challenges.

TAKE HOME MESSAGE 10.13

» Natural selection does not lead to organisms perfectly adapted to their environment, because (1) environments can change more quickly than natural selection can adapt organisms; (2) mutation does not produce all possible alleles; and (3) there is not always a single, optimum adaptation for a given environment.

10.14 Artificial selection is a special case of natural selection.

In practice, plant and animal breeders understood natural selection long before Darwin did; they just didn't know that they understood it. Farmers bred crops for maximum yield, and dog, horse, and pigeon fanciers selectively bred animals with their favorite traits to produce more and more offspring with more and more exaggerated versions of those traits.

The process used by animal breeders and farmers is called **artificial selection,** but the underlying genetic process does not differ from natural selection, because the three conditions are satisfied. Artificial selection can be thought of as a special case of natural selection, however, because the differential reproductive success is being determined by humans rather than by nature (and artificial, unlike natural, selection is typically goal-oriented). Apple growers, for example, use artificial selection to produce the wide variety of apples now available. What is important is that it is still differential reproductive success, and the results are no different. It was a stroke of genius for Darwin to recognize that the same process farmers were using to develop new and better crop varieties is occurring naturally in every population on earth, and always has been.

TAKE HOME MESSAGE 10.14

» Animal breeders and farmers are making use of natural selection when they modify their animals and crops through selective breeding, because the three conditions for natural selection are satisfied. Because the differential reproductive success is determined by humans rather than by nature, this type of natural selection is also called artificial selection.

10.15 Natural selection can change the traits in a population in several ways.

Certain traits are easily categorized—white versus orange tiger fur, for example. Other traits, such as height, are influenced by many genes and environmental factors, so a continuous range of phenotypes occurs (FIGURE 10-25). Whether a given trait is influenced by one gene or by a complex interaction of many genes and the environment, it can be subject to natural selection and be changed in any of several ways.

TRAIT CATEGORIZATION

SPECIFIC CATEGORY
Some traits, such as white versus orange tiger fur, can be easily placed into one category or the other.

White tiger fur

Orange tiger fur

or

CONTINUOUS RANGE
Other traits, such as height, may fall within a wide range of values.

FIGURE 10-25 Some traits fall into clear categories; others range continuously.

Directional Selection In **directional selection,** individuals with one extreme of the range of variation in the population have higher fitness. Milk production in cows, for example, varies from one cow to another. Farmers select for breeding those cows with the highest milk production and have done so for many decades. As a result of such selection, between 1920 and 1945, average milk production increased by about 50% in the United States (FIGURE 10-26).

In contrast, cows that produce the smallest amounts of milk have reduced fitness, since farmers do not allow them to reproduce, thus limiting the number of "less milk"

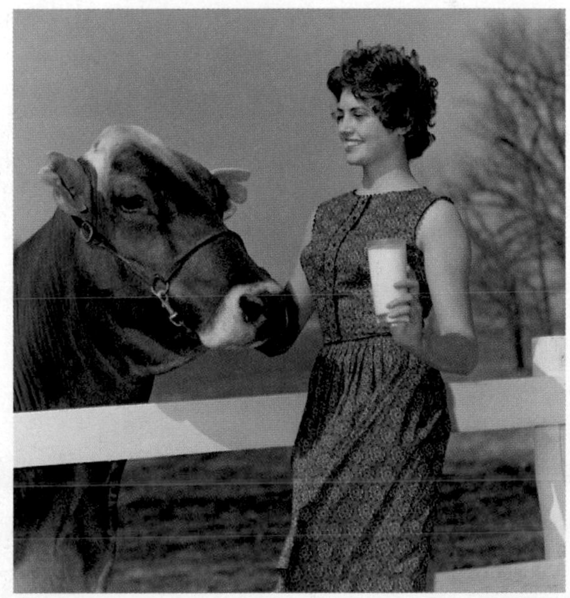

FIGURE 10-26 More milk? Cows have been selected for their ability to produce more and more milk.

DIRECTIONAL SELECTION

SELECTIVE PRESSURE
Babies with low birth weights are more vulnerable to complications.
Babies with high birth weights are more vulnerable to complications.
Farmers select against turkeys with smaller breast muscles because they are less valuable.

Population before selection

Population after selection

Change in average breast size

Proportion of individuals in population

Smaller

Larger

Relative size of breast muscles in farmed turkeys

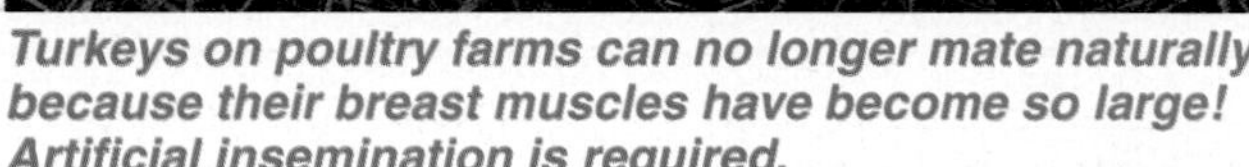

Turkeys on poultry farms can no longer mate naturally because their breast muscles have become so large! Artificial insemination is required.

FIGURE 10-27 Patterns of natural selection: directional selection.

alleles in the next generation. In subsequent generations, the average value for the milk-production trait increases. Note, however, that variation for the milk-producing trait in the cow population decreases a bit, too, because the alleles contributing to one of the phenotypic extremes are eventually removed from the population.

Hundreds of experiments in nature, in laboratories, and on farms demonstrate the power of directional selection. Turkeys, for instance, have been selected for increased size of their breast muscles, which makes them more profitable for farmers (FIGURE 10-27). Directional selection has been so successful that the birds' breast muscles are now so large

Q Turkeys on poultry farms have such large breast muscles that they can't get close enough to each other to mate. How can such a trait evolve?

STABILIZING SELECTION

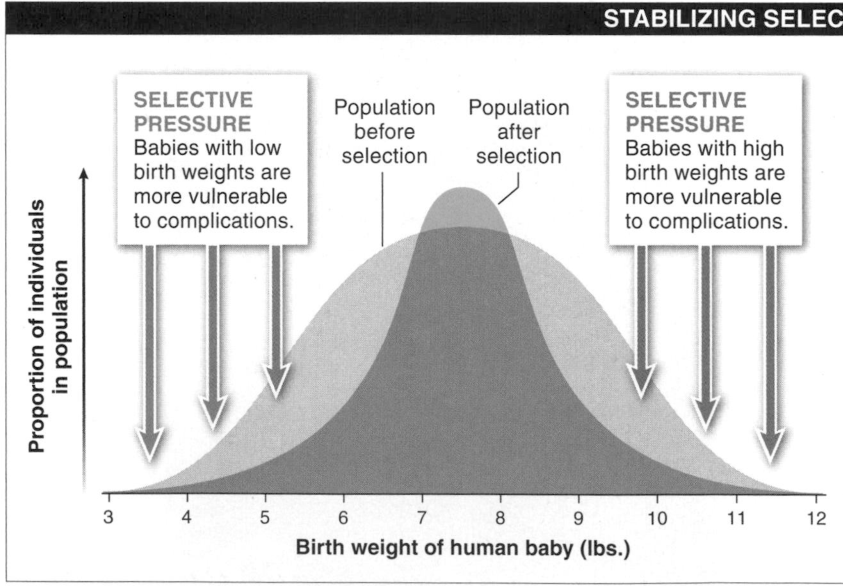

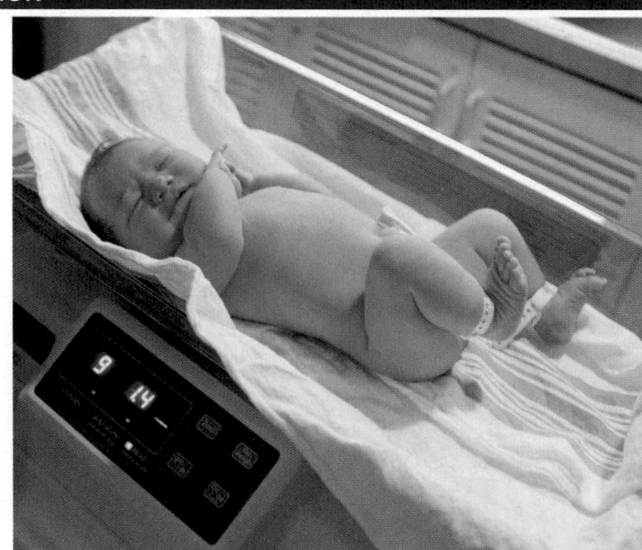

Birth weight of human baby (lbs.)

FIGURE 10-28 Patterns of natural selection: stabilizing selection.

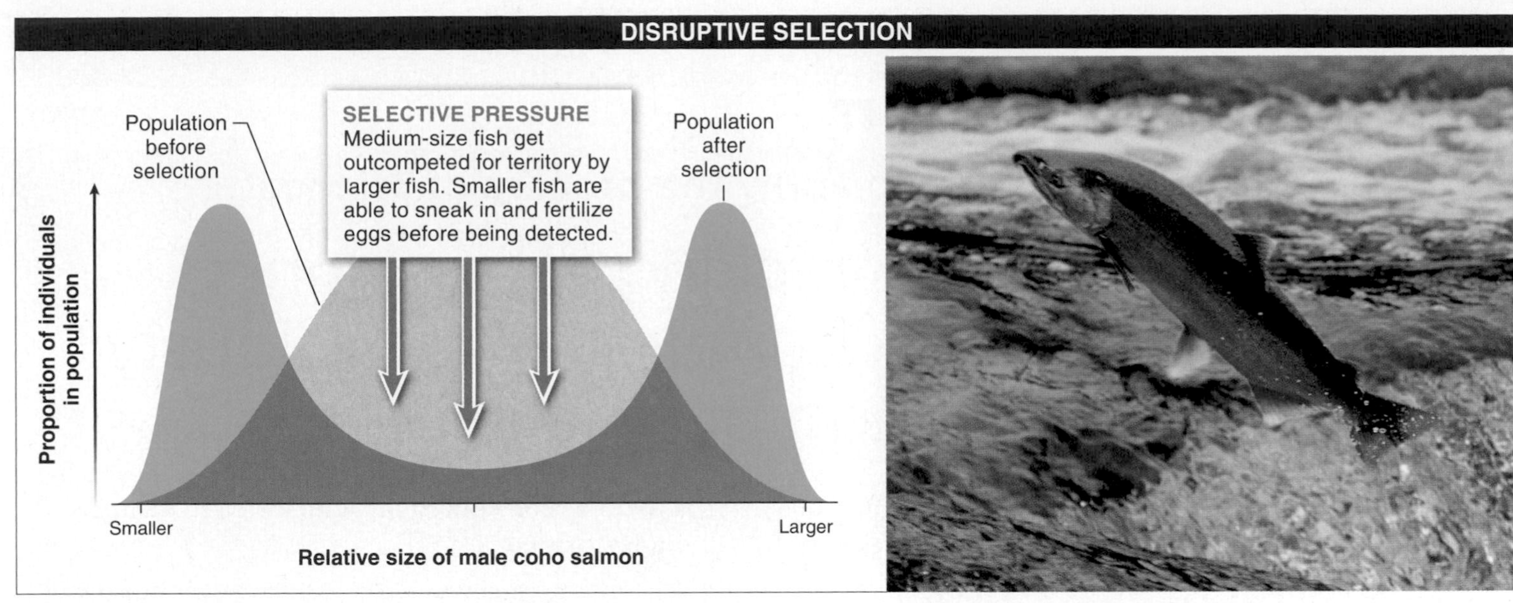

DISRUPTIVE SELECTION
Population before selection
SELECTIVE PRESSURE
Medium-size fish get outcompeted for territory by larger fish. Smaller fish are able to sneak in and fertilize eggs before being detected.
Population after selection
Proportion of individuals in population
Smaller
Larger
Relative size of male coho salmon

FIGURE 10-29 **Patterns of natural selection: disruptive selection.**

that it has become impossible for them to mate. All turkeys on large-scale, industrial poultry farms must now reproduce through artificial insemination.

Stabilizing Selection How much did you weigh at birth? Was it more than 10 pounds? Was it close to the 22½ pounds that Carmelina Fedele's child weighed when born in Italy in 1955? Or was it 5 pounds or less? Unlike directional selection, for which there is increased fitness at one extreme and reduced fitness at the other, **stabilizing selection** takes place when individuals with intermediate phenotypes are the most fit. The death rate among newborns, for example, is lowest between 7 and 8 pounds, and is higher for both lighter and heavier babies (FIGURE 10-28). The variation in birth weight has decreased as stabilizing selection has reduced the frequencies of genes associated with high and low birth weights.

Modern technologies, including Caesarean deliveries and neonatal intensive care units, allow many babies to live who would not have survived without such technology. This has the unintended consequence of reducing the selection against any alleles causing these "extreme" traits, and so reducing the rate of their removal from the population.

Disruptive Selection There is a third kind of natural selection, called **disruptive selection,** in which individuals with extreme phenotypes experience the highest fitness, and those with intermediate phenotypes have the lowest. Among some species of fish—the coho salmon, for instance—only the largest males acquire good territories. They generally enjoy relatively high reproductive success. While the intermediate-size fish regularly get run out of the good territories, some tiny males are able to sneak in and fertilize eggs before the territory owner detects their presence. Consequently, we see an increase in the frequency of small and large fish, with a reduction in the frequency of medium-size fish (FIGURE 10-29).

TAKE HOME MESSAGE 10.15

» Acting on traits for which populations show a large range of phenotypes, natural selection can change populations in several ways. These include directional selection, in which the average value for the trait increases or decreases; stabilizing selection, in which the average value of a trait remains the same while extreme versions are selected against; and disruptive selection, in which individuals with extreme phenotypes have the highest fitness.

THIS IS HOW WE DO IT

Developing the ability to apply the process of science

10.16 By picking taller plants, do humans unconsciously drive the evolution of smaller plants?

A clear and concise statement of a hypothesis often sheds light on how we might make the observations or collect the necessary data to test it. This step can help take a big idea or issue and break it down to a more manageable, answerable question. For example, let's consider a problem that conservation biologists wonder (and worry) about: the impact of human activities on plant populations.

Researchers investigated the Tibetan snow lotus, *Saussurea laniceps,* a rare plant that lives in rocky habitats high in the eastern Himalayas. In traditional Chinese and Tibetan medicine, snow lotus flowers are used to treat headaches and high blood pressure. The flowers are also valued as souvenirs. But they're not picked at random. Larger flowers (found on taller plants), which are believed to be more potent, are strongly preferred by the human harvesters.

One possible research question relating to these plants could be: *Is human harvesting of the snow lotus causing evolutionary change to the plants?*

How might we recast this question as a testable hypothesis?

We could restate this as: *Because humans like to pick the tallest of the wild-growing snow lotus plants, they are causing evolution.* Or we could get even more precise: *Human harvesting of the tallest snow lotus plants is causing evolution through selection of smaller plants.*

How could researchers determine whether the plants are becoming shorter?

For their first step, the researchers simply measured snow lotus plants from recent collections sold in China.

Is there a "control group" with which to compare today's plants?

The clever researchers used two different, but equally valuable, control groups. First, they measured snow lotus plants in numerous field collections and in eight herbaria, repositories of specimens acquired by explorers and cataloguers of biodiversity. They noted the date that each plant had been collected. What did their comparison reveal?

Before 1920, the average snow lotus plant height was *22 cm;* after 2000, the average height was just *14 cm.* The more recently the plant was collected, the smaller it was. (Statistically evaluating the relationship between the year of collection and the plant height, they found a significant negative correlation: $r^2 = 0.44$; $p < 0.0001$; $N = 218$.) In the same plant collections, the researchers also measured snow lotus plants of a closely related species that is not generally harvested. Over the course of 120 years, there was no decline in the average plant height for this relative. Snow lotus plants are getting smaller only for the human-harvested species.

Could another comparison be made?

As a second way of testing their hypothesis, the researchers compared snow lotus size at a heavily harvested site with a second control group: plants growing in protected, sacred areas of Tibet where flower harvesting is not permitted. In the protected area, the average plant height was 22.7 (± 2.0) cm. The plants in the heavily harvested area were about 40% shorter than this—average height 13.4 (± 0.4) cm.

Did the researchers prove their point?

The evidence strongly supports the researchers' hypothesis. Humans have inadvertently imposed directional selection on the snow lotus. Can you think of other naturally occurring species that humans value and harvest in ways that could lead to inadvertent evolution?

TAKE HOME MESSAGE 10.16

» Humans value the Tibetan snow lotus flower for medicinal and other purposes. Because people prefer to harvest the largest snow lotus plants, there has been selection for smaller plants. Data reveal a significant trend of decreasing height over the past hundred or so years.

10.17 Natural selection can cause the evolution of complex traits and behaviors.

We have seen that natural selection can change allele frequencies and modify the frequency with which simple traits, such as fur color or turkey-breast size or plant height, appear in a population. But what about complex traits, such as behaviors, that involve numerous physiological and neurological systems? For instance, can natural selection improve maze-running ability in rats?

The short answer is *yes*. Remember, evolution by natural selection, changing the allele frequencies for traits, is occurring whenever (1) there is variation for the trait, (2) that variation is heritable, and (3) there is differential reproductive success based on that trait. These conditions can easily be satisfied for complex traits, including behaviors.

> "All things must change
> To something new, to something strange."
>
> — HENRY WADSWORTH LONGFELLOW,
> American poet

In 1954, to address this question, William Thompson trained a group of rats to run through a maze for a food reward (FIGURE 10-30). He found a huge amount of variation in the rats' abilities: some rats learned much more quickly than others how to run the maze. Thompson then selectively bred the fast learners with each other and the slow learners with each other. Over several generations, he developed two separate populations: rats descended from a line of fast maze-learners and rats descended from a line of slow maze-learners. After only six generations, the slow learners made twice as many errors as the fast learners before mastering the maze, while the fast learners were adept at solving complex mazes that would give many humans some difficulty. More than 60 years later, it's still unclear which particular genes are responsible for maze-running behavior, yet the selection experiment still demonstrates a strong genetic component to the behavior.

Q How can a wing evolve if 1% of a wing is no use in helping an organism fly or glide?

Natural selection can also produce complex traits in unexpected, roundabout ways. One vexing case involves the question of how natural selection could produce an organ as complex as a fly wing, when 1% or 2% of a wing—that is, an incomplete structure—doesn't help an insect to fly (FIGURE 10-31).

The key to answering this question is that 1% of a wing doesn't need to function as a wing to increase an individual's fitness. Often, structures are enhanced or elaborated

NATURAL SELECTION FOR COMPLEX TRAITS

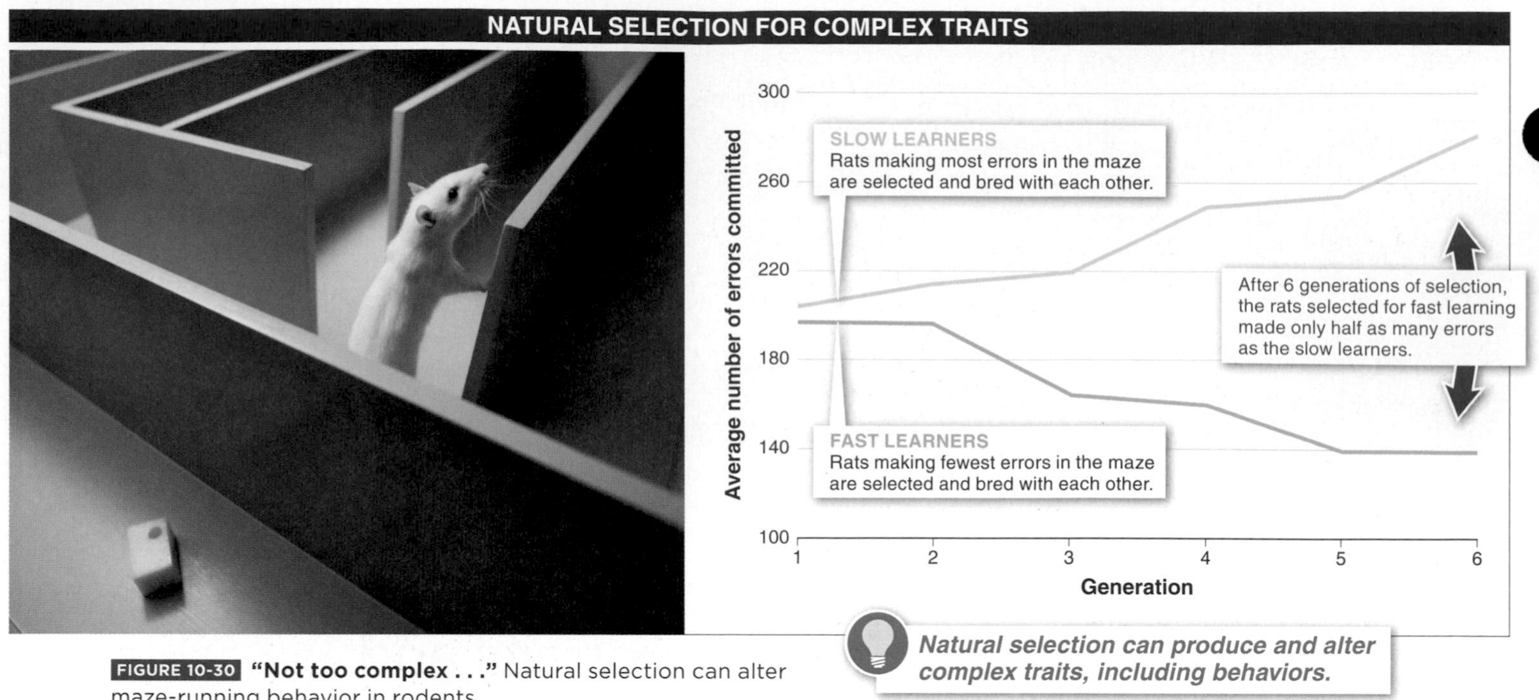

Average number of errors committed
300
260
220
180
140
100
SLOW LEARNERS
Rats making most errors in the maze are selected and bred with each other.
After 6 generations of selection, the rats selected for fast learning made only half as many errors as the slow learners.
FAST LEARNERS
Rats making fewest errors in the maze are selected and bred with each other.
1
2
3
4
5
6
Generation
Natural selection can produce and alter complex traits, including behaviors.

FIGURE 10-30 **"Not too complex . . ."** Natural selection can alter maze-running behavior in rodents.

on by natural selection because they enhance fitness by serving some other purpose. Experiments using models of insects demonstrated that a small percentage of a wing, in the form of a nub or "almost wing," confers no benefit at all when it comes to flying. (The nubs don't even help flies keep their orientation during a "controlled fall.") The incipient wings *do,* however, help the insect address a completely different problem. They allow much more efficient

NATURAL SELECTION CAN CO-OPT TRAITS FOR NEW FUNCTIONS

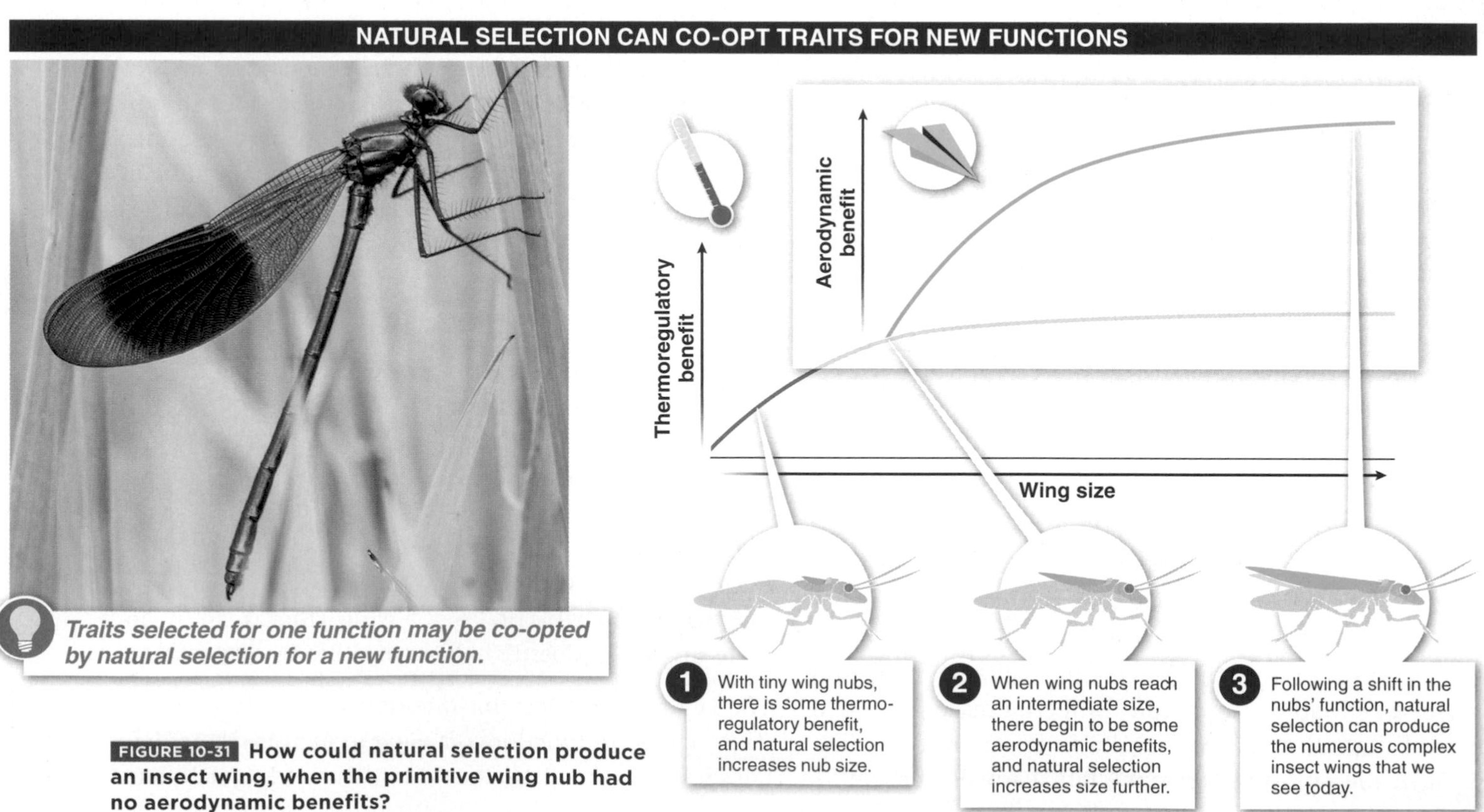

Traits selected for one function may be co-opted by natural selection for a new function.
Thermoregulatory benefit
Aerodynamic benefit
Wing size
1 With tiny wing nubs, there is some thermoregulatory benefit, and natural selection increases nub size.
2 When wing nubs reach an intermediate size, there begin to be some aerodynamic benefits, and natural selection increases size further.
3 Following a shift in the nubs' function, natural selection can produce the numerous complex insect wings that we see today.

FIGURE 10-31 **How could natural selection produce an insect wing, when the primitive wing nub had no aerodynamic benefits?**

temperature control, so that an insect can gain heat from the environment when the insect is cold and dissipate heat when the insect is hot. Experiments on heat-control efficiency, in fact, show that as small nubs become more and more pronounced, they are more effective—but only up to a point.

Eventually, the thermoregulatory benefit stops increasing, even if the nub length continues to increase. But right around this point, the proto-wing starts to confer some aerodynamic benefits (see Figure 10-31). Consequently, natural selection may continue to increase the length of this "almost wing," but now the fitness increase is due to a wholly different effect. Such functional shifts explain the evolution of numerous complex structures that we see today. These shifts may be common in the evolutionary process.

TAKE HOME MESSAGE 10.17

» Natural selection can change allele frequencies for genes involved in complex physiological processes and behaviors. Sometimes a trait that has been selected for one function is later modified to serve a completely different function.

10.18–10.22 The evidence for evolution is overwhelming.

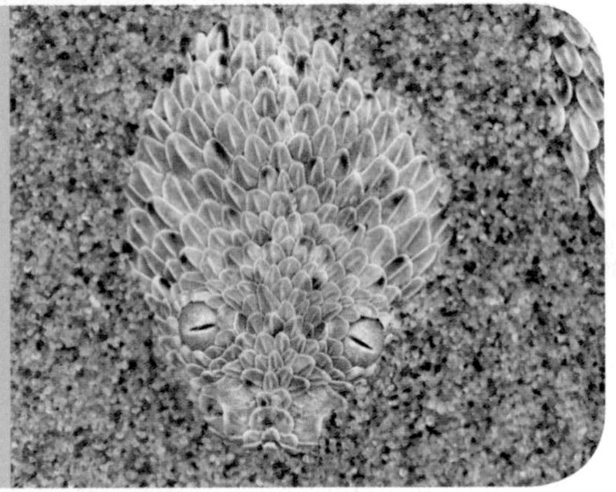

Namib desert sidewinding adder, *Bitis peringueyi*, Namibia.

10.18 The fossil record documents the process of natural selection.

> "It is indeed remarkable that this theory [evolution] has been progressively accepted by researchers, following a series of discoveries in various fields of knowledge. The convergence, neither sought nor fabricated, of the results of work that was conducted independently is in itself a significant argument in favor of this theory."
>
> — POPE JOHN PAUL II, 1996

In the 150 or so years since Darwin first published *The Origin of Species,* thousands of studies have been conducted on his theory of evolution by natural selection, both in the laboratory and in natural habitats. A wide range of modern methodologies—such as genome sequencing and algorithms for comparing genome similarities and constructing the most likely phylogenies—contribute to a much deeper understanding of the process of evolution. This ongoing accumulation of evidence overwhelmingly supports the basic premise that Darwin put forward, while filling in many of the gaps that frustrated him.

In the remainder of the chapter we review the five primary lines of evidence demonstrating the occurrence of evolution:

1. **The fossil record**—physical evidence of organisms that lived in the past

2. **Biogeography**—patterns in the geographic distribution of living organisms
3. **Comparative anatomy and embryology**—growth, development, and body structures of major groups of organisms
4. **Molecular biology**—examination of life at the level of individual molecules
5. **Laboratory and field studies**—implementation of the scientific method to observe and study evolutionary mechanisms

The first of the five lines of evidence is the fossil record. Although it has been central to much documentation of the occurrence of evolution, it is a very incomplete record. After all, the soft parts of an organism tend to decay rapidly and completely after death. And there are only a few environments (such as tree resin, tar pits, the bottom of deep lakes, and continental shelves) in which the processes of erosion and decomposition are so reduced that an organism's hard parts, including bones, teeth, and shells, can be preserved for thousands or even millions of years. These remains, called **fossils,** can be used to reconstruct what organisms looked like long ago. Such reconstructions often provide a clear record of evolutionary change, including insights into the patterns and processes of history (FIGURE 10-32).

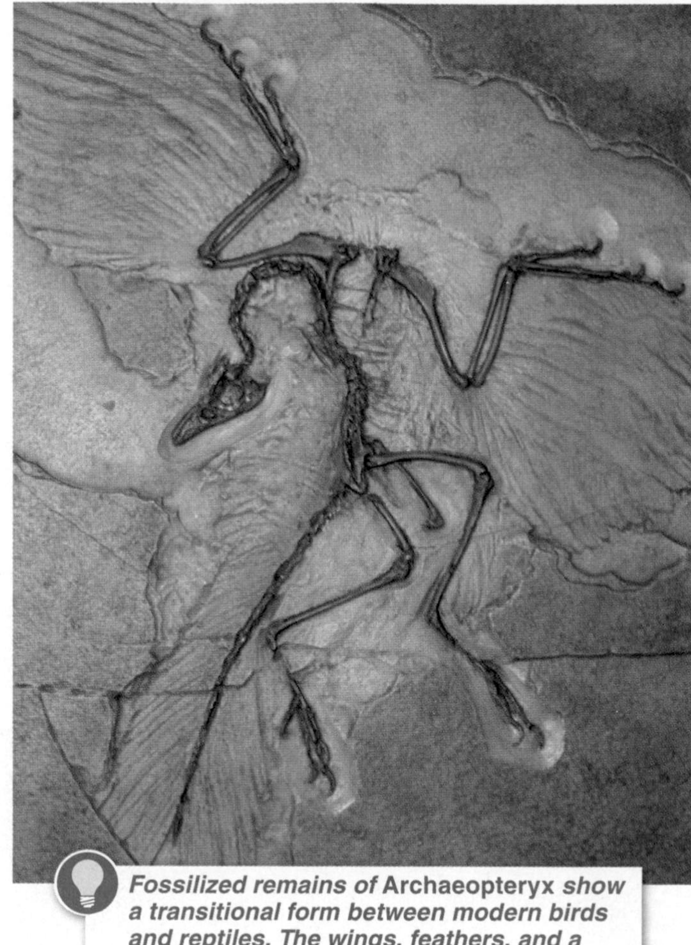

Fossilized remains of Archaeopteryx *show a transitional form between modern birds and reptiles. The wings, feathers, and a "wish bone" are features of modern birds.*

FIGURE 10-32 **Evidence for evolution: the fossil record.** Fossils can be used to reconstruct the appearance of organisms that lived long ago.

The use of **radiometric dating** paints a clearer picture of organisms' evolutionary history, revealing the age of the rock in which a fossil is found (FIGURE 10-33). In Darwin's time, it was assumed that the deeper in the earth a fossil was found, the older it was. Radiometric dating makes it possible to determine not just the *relative* age of fossils but also their *absolute, numerical* age. Dating is done by evaluating the amounts of certain radioactive isotopes present in the fossil-containing rocks or the layers above and below them. Radioactive isotopes in a rock begin breaking down into more stable compounds as soon as the rock is formed, and they do so at a constant rate. Nothing alters this. We can calculate the age of the rock, and thus of the fossil, by measuring the relative amounts of the radioactive isotope and its leftover decay product in the rock where a fossil is found.

Radiometric dating confirms that the earth is very old. Rocks more than 3.8 billion years old have been found on every continent, with the oldest so far found in northwestern Canada. By measuring the radioactive isotope uranium-238, with a half-life of 4.5 billion years, researchers have determined that the earth is about 4.6 billion years old and that the earliest organisms appeared at least 3.5 billion years ago. Radiometric dating also makes it possible to put the fossil record in chronological order. By dating all the fossils discovered in one locale, paleontologists can better evaluate whether the organisms are related to one another and how groups of organisms changed over time.

Paleontologists recognize that fossilization is an exceedingly unlikely event, and when it does occur, it represents only those organisms that (1) happened to live in that location, (2) could be preserved under certain chemical conditions, and (3) had physical structures that can leave fossils. For this reason, the fossil record is unavoidably incomplete. Entire groups of organisms have left no fossil record at all, and for others the record has numerous gaps. Still, for what there is, the fossil record can be incredibly

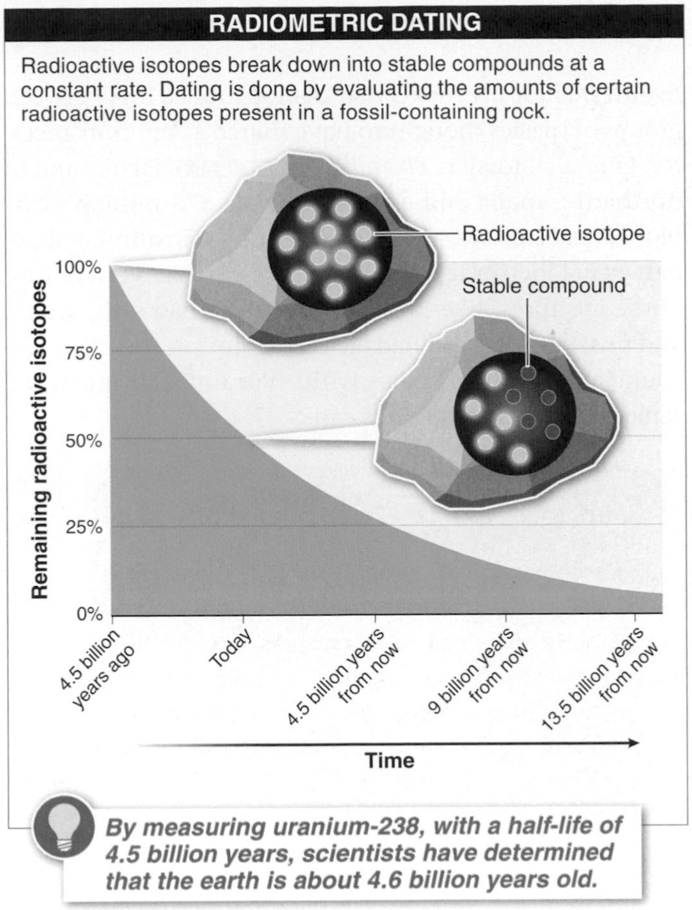

RADIOMETRIC DATING

Radioactive isotopes break down into stable compounds at a constant rate. Dating is done by evaluating the amounts of certain radioactive isotopes present in a fossil-containing rock.

Radioactive isotope
Stable compound
Remaining radioactive isotopes
100%
75%
50%
25%
0%
4.5 billion years ago
Today
4.5 billion years from now
9 billion years from now
13.5 billion years from now
Time

By measuring uranium-238, with a half-life of 4.5 billion years, scientists have determined that the earth is about 4.6 billion years old.

FIGURE 10-33 **How old is that fossil?** Radiometric dating helps to determine the age of rocks and the fossils in them.

detailed and interesting, and includes fossils that link every major group of vertebrates.

The evolutionary history of horses is among the most well-preserved in the fossil record. Horses first appeared in North America, about 55 million years ago, and radiated around the world, with more recent fossils appearing in Eurasia and Africa. These fossils exhibit distinct adaptations to those differing environments. Later, about 1.5 million years ago, much of the horse diversity—including all North American horse species—disappeared. Only a single genus, or group of species, remains on earth, called *Equus*. It is tempting to imagine a straight path from modern horses extending back through their 55-million-year evolutionary history. But that's just not how evolution works. Numerous branches of horses have split off over evolutionary time, flourished for millions of years, and only recently gone extinct. What we see living today is just a single branch of a greatly branched evolutionary tree (FIGURE 10-34).

The fossil record provides another valuable piece of evidence for evolution, in the form of fossils with transitional

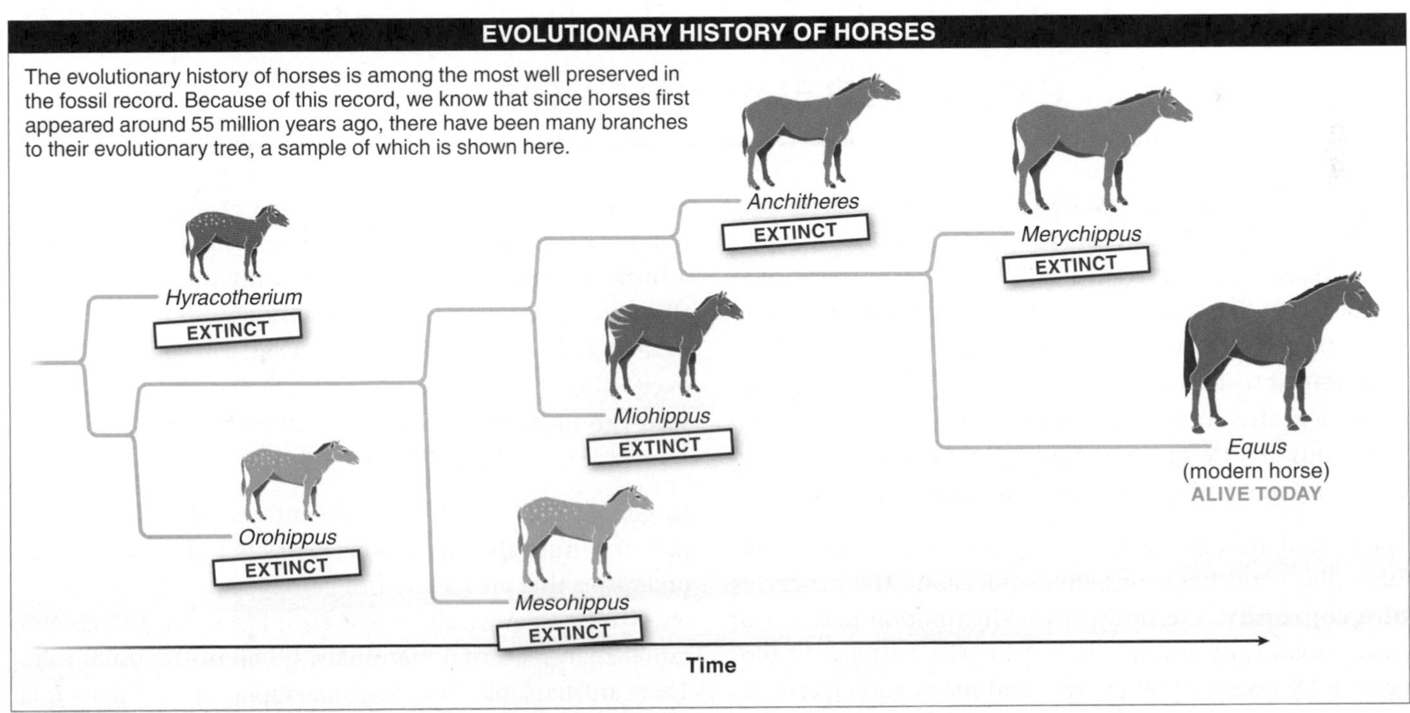
EVOLUTIONARY HISTORY OF HORSES
The evolutionary history of horses is among the most well preserved in the fossil record. Because of this record, we know that since horses first appeared around 55 million years ago, there have been many branches to their evolutionary tree, a sample of which is shown here.
Hyracotherium
EXTINCT
Orohippus
EXTINCT
Mesohippus
EXTINCT
Miohippus
EXTINCT
Anchitheres
EXTINCT
Merychippus
EXTINCT
Equus
(modern horse)
ALIVE TODAY
Time

FIGURE 10-34 **An evolutionary family tree.** The branching evolutionary tree of the horse.

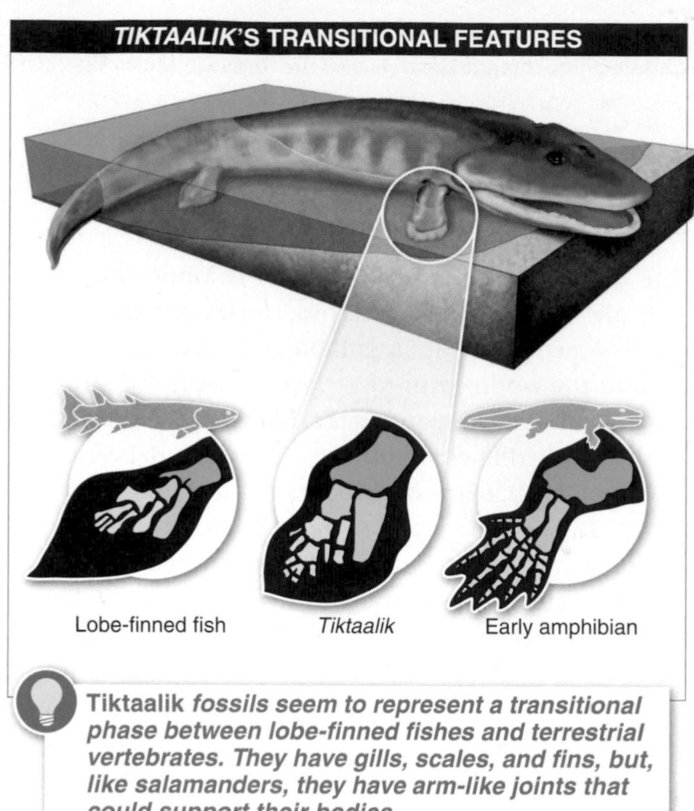
TIKTAALIK'S TRANSITIONAL FEATURES
Lobe-finned fish
Tiktaalik
Early amphibian
Tiktaalik fossils seem to represent a transitional phase between lobe-finned fishes and terrestrial vertebrates. They have gills, scales, and fins, but, like salamanders, they have arm-like joints that could support their bodies.

FIGURE 10-35 **A missing link?**

features. These are fossils that demonstrate a link between groups of species thought to have shared a common ancestor. One such fossil is *Tiktaalik* (FIGURE 10-35). First found in northern Canada and estimated to be 375 million years old, *Tiktaalik* fossils seem to represent a transitional phase between lobe-finned fishes and terrestrial vertebrates. These creatures, like lobe-finned fishes, had gills, scales, and fins, but they also had arm-like joints in their fins and could support their bodies with their limbs in much the same way that salamanders can.

TAKE HOME MESSAGE 10.18

» Radiometric dating confirms that the earth is very old and allows scientists to determine the age of fossils. By analyzing fossil remains, paleontologists can reconstruct what organisms looked like long ago, learn how organisms were related to each other, and understand how groups of organisms evolved over time.

10.19 Geographic patterns of species distributions reflect species' evolutionary histories.

When it comes to species distributions, history matters. The species that arrive in a geographic region first—usually coming from nearby—may take up numerous different lifestyles in numerous different habitats, and the populations adapt to and evolve in each environment. In Hawaii, it seems that a finch-like descendant of the honeycreepers arrived 4–5 million years ago and evolved into numerous diverse species. The same process occurred and continues to occur in all locales, not just on islands.

The second line of evidence that helps us see that evolution takes place and helps us better understand the process is **biogeography,** the study of the distribution patterns of living organisms around the world. The patterns of biogeography observed by Darwin and many subsequent researchers provide strong evidence that evolutionary forces are responsible for these patterns. For example, species often more closely resemble other species that live less than a hundred miles away but in radically different habitats than they resemble species living thousands of miles away in nearly identical habitats. In Hawaii, for example, nearly every bird is some sort of modified honeycreeper, from seed-eating honeycreepers to curved-bill nectar-feeding honeycreepers (FIGURE 10-36).

Large, isolated habitats also have interesting biogeographic patterns. Australia and Madagascar are filled with unique organisms that are clearly not closely related to organisms elsewhere. In Australia, for example, marsupial species, rather than placental mammals, fill all of the usual roles. There are marsupial "wolves," marsupial "mice," marsupial "squirrels," and marsupial "anteaters" (FIGURE 10-37).

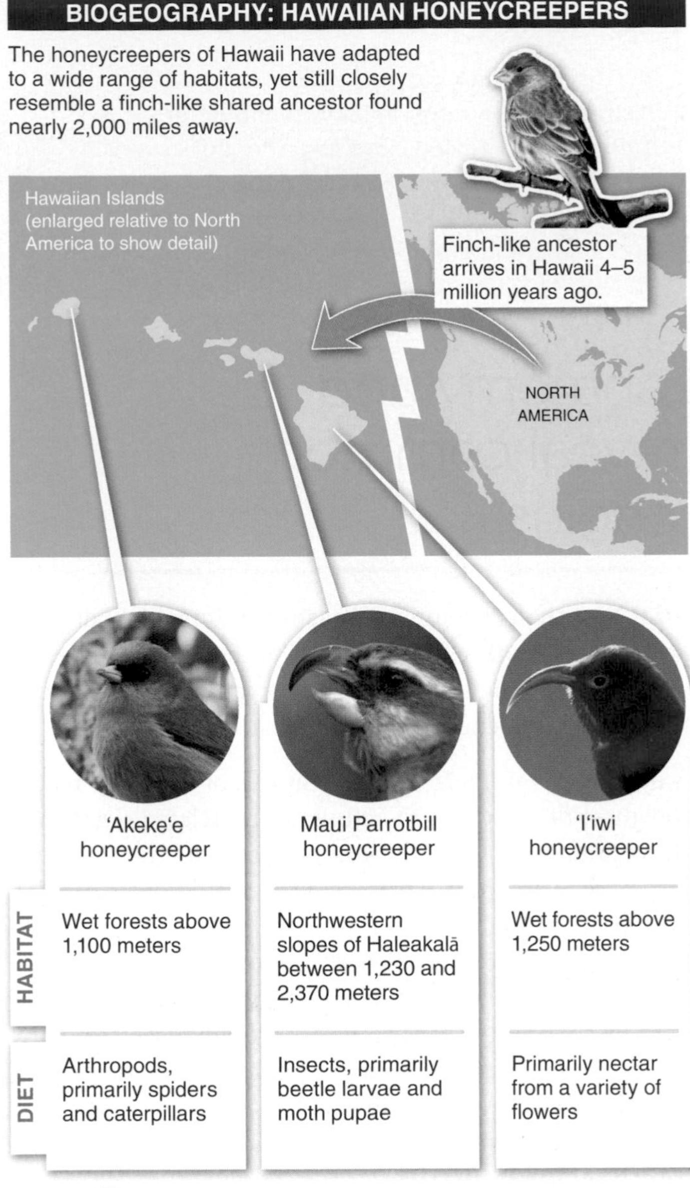

BIOGEOGRAPHY: HAWAIIAN HONEYCREEPERS
The honeycreepers of Hawaii have adapted to a wide range of habitats, yet still closely resemble a finch-like shared ancestor found nearly 2,000 miles away.
Hawaiian Islands (enlarged relative to North America to show detail)
Finch-like ancestor arrives in Hawaii 4–5 million years ago.
NORTH AMERICA
'Akeke'e honeycreeper
Maui Parrotbill honeycreeper
'I'iwi honeycreeper
HABITAT
Wet forests above 1,100 meters
Northwestern slopes of Haleakalā between 1,230 and 2,370 meters
Wet forests above 1,250 meters
DIET
Arthropods, primarily spiders and caterpillars
Insects, primarily beetle larvae and moth pupae
Primarily nectar from a variety of flowers

The Hawaiian honeycreepers resemble a common ancestor from mainland North America, but all have unique features.

FIGURE 10-36 **Evidence for evolution: biogeography.**

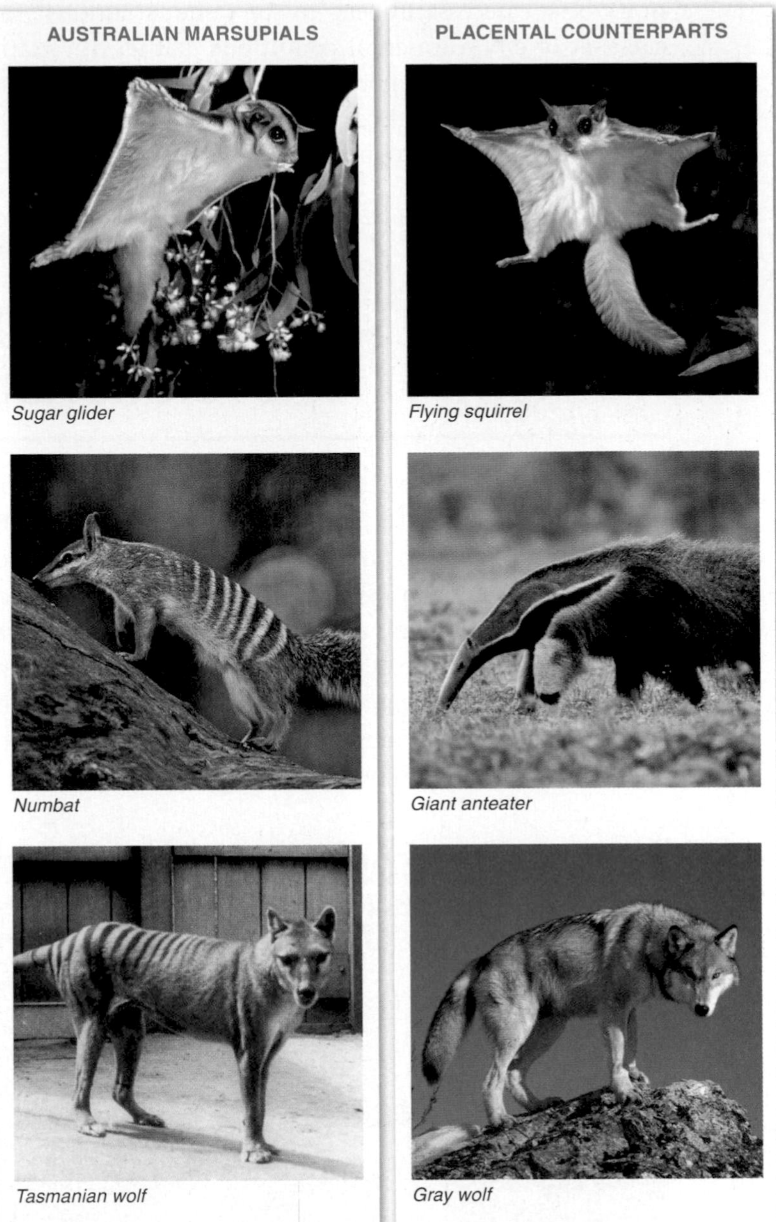

FIGURE 10-37 **Evidence for evolution: biogeography.**

The marsupials of Australia physically resemble their placental counterparts, but molecular analysis shows that they are more closely related to one another, sharing a common marsupial ancestor. Their relatedness is also revealed by similarities in their reproduction: females give birth to offspring at a relatively early stage of development, and the offspring finish their development in a pouch. The presence of marsupials in Australia does not simply mean that marsupials are better adapted than placentals to Australian habitats. When placental organisms are transplanted to Australia they do just fine, often thriving to the point of endangering the native species. Instead, it seems that the terrestrial placental mammals in Australia disappeared about 55 million years ago—for reasons that are not clear—giving the marsupials the chance to flourish and diversify.

Biogeographic patterns such as those seen in honeycreepers and in the marsupials of Australia illustrate that evolution doesn't necessarily lead to the same "solutions" for each set of environmental conditions. Rather, the traits in populations that happen to be in a particular location gradually change, and those species become better adapted to the habitats they occupy.

TAKE HOME MESSAGE 10.19

» Observing geographic patterns of species distributions—noting similarities and differences among species living close together but in very different habitats, and among species living in similar habitats but located far from one another—helps us understand the evolutionary histories of populations.

10.20 Comparative anatomy and embryology reveal common evolutionary origins.

If you observe any vertebrate embryo while it's developing, you will see that it passes through a stage in which it has little gill pouches on the sides of its neck. It will also pass through a stage in which it has a long bony tail. This is true whether it is a human embryo or that of a turtle or a chicken or a shark. The gill pouches disappear before birth in all but the fishes. Similarly, we don't find humans with tails. Why do these features exist during an embryo's development? Such shared embryological stages indicate that the organisms share a common ancestor, from which all have been modified (FIGURE 10-38). Study of these developmental stages and the adult body forms of organisms provides our third line of evidence for the occurrence of evolution.

HOMOLOGOUS STRUCTURES: EMBRYO STAGE

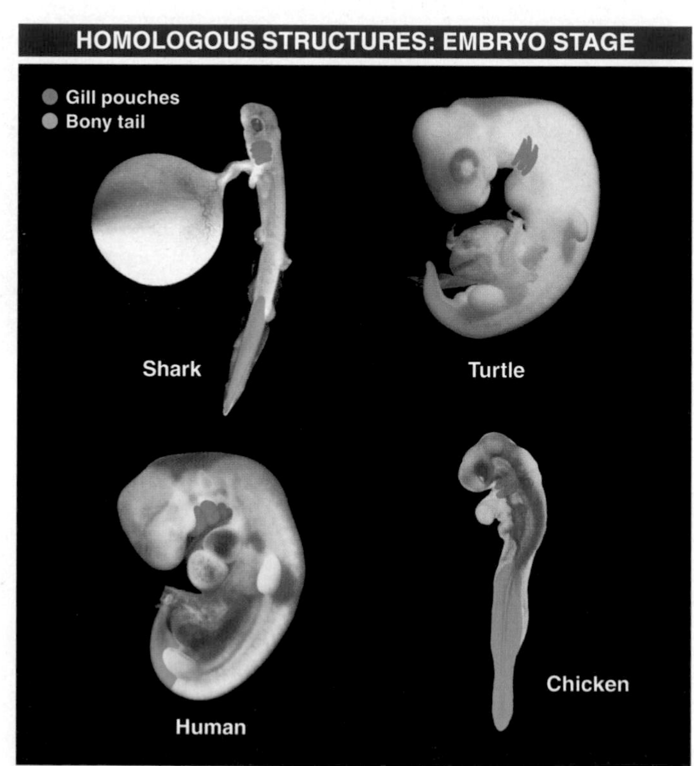

FIGURE 10-38 **Evidence for evolution: embryology.** Structures derived from common ancestry can be seen in embryos.

Among adult animals, several features of anatomy reveal the ghost of evolution in action. Many related organisms show similarities that can be explained only through evolutionary relatedness. The forelimbs of mammals such as bats, porpoises, horses, and humans are used for a variety of very different functions (FIGURE 10-39). If each had been designed for the uses necessary to that species—flying, swimming, running, grasping—we would expect dramatically different designs. And yet, all of these species possess the same bones—modified extensively—revealing that they share a common ancestor. These features are called **homologous structures.**

At the extreme, homologous structures sometimes come to have little or no function at all. Such evolutionary leftovers, called **vestigial structures,** exist because they had value in an ancestor species. Some vestigial structures in mammals include the molars that continue to grow in vampire bats, even though these bats consume a completely liquid diet (FIGURE 10-40); eye sockets with no eyes in some populations of cave-dwelling fishes; and in whales, pelvic bones that are attached to nothing (but in nearly all other mammals serve as important attachment points for leg bones). Even humans may have a vestigial organ—the appendix.

Q Why do vampire bats grow teeth specialized for grinding solid food, even though the bats have a completely liquid diet?

HOMOLOGOUS BONE STRUCTURES

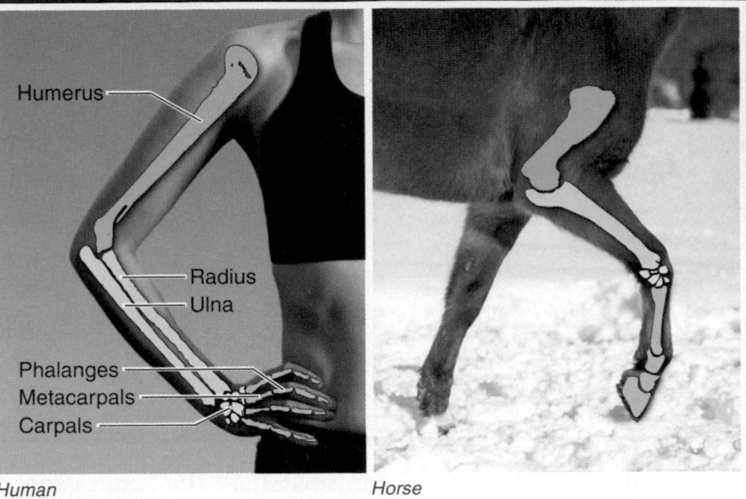

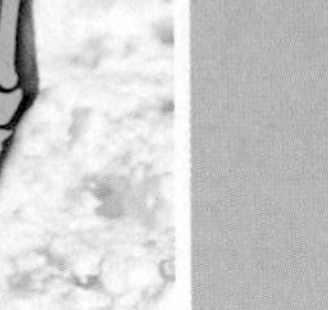

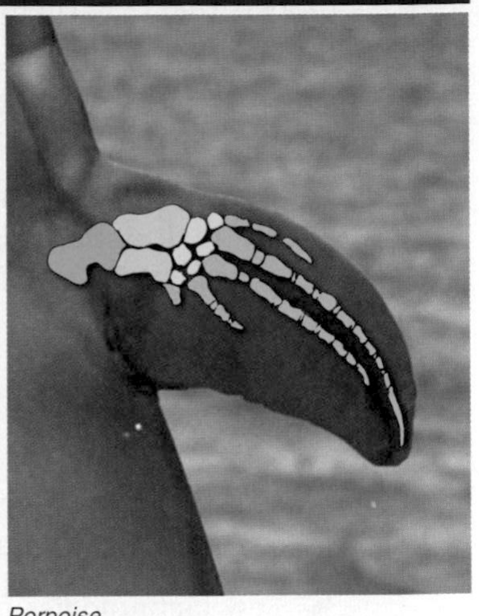

FIGURE 10-39 Evidence for evolution: comparative anatomy.

The similarities in the bone structure of the forelimbs of mammals demonstrate common ancestry.

It is greatly enlarged in our relatives the great apes, acting as host to cellulose-digesting bacteria that aid in breaking down plants in the apes' diet, but in humans it seems to serve no purpose (although recent studies suggest it provides some immune system support).

Not all organisms with similar-looking adaptations share recent ancestors. Consider, for example, flying mammals (bats) and flying insects (locusts) (FIGURE 10-41). Likewise, dolphins and penguins live in similar habitats and have

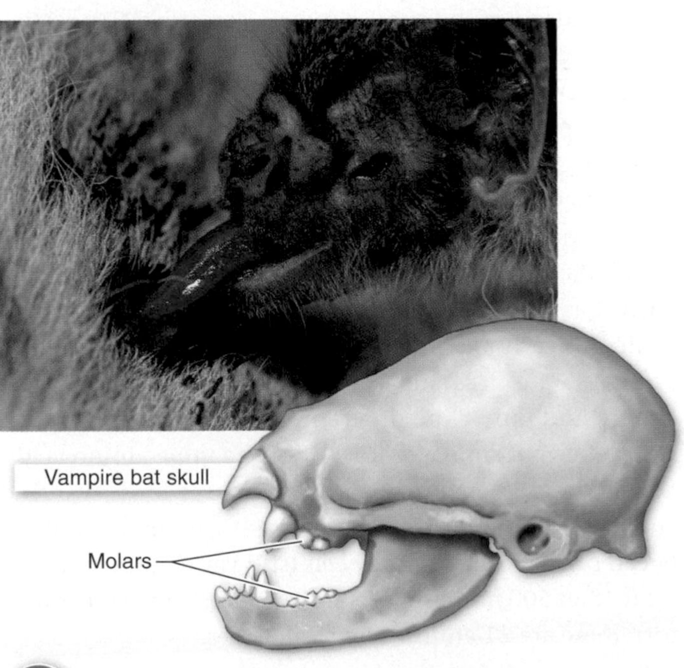

Vampire bats consume only a liquid diet.
Q: So why do they have molars for grinding food?
A: Because their ancestors did.

FIGURE 10-40 Evidence for evolution: vestigial structures.

CONVERGENT EVOLUTION

Different starting materials come to perform the same function through convergent evolution.

FIGURE 10-41 Evidence for evolution: convergent evolution and analogous structures.

flippers that help them swim. In both examples, however, the analogous structures developed from different original structures. Natural selection, in a process called **convergent evolution,** may act on different starting materials (such as a flipper or a forelimb) and modify them until they serve similar purposes—much as we saw in the marsupial and placental mammals in Figure 10-37.

TAKE HOME MESSAGE 10.20

» Similarities in the anatomy and development of different groups of organisms and in their physical appearance can reveal common evolutionary origins.

10.21 Molecular biology reveals that common genetic sequences link all life forms.

New technologies for deciphering and comparing the genetic code provide our fourth line of evidence for evolution. All living organisms share the same genetic code. From microscopic bacteria to flowering plants to insects and primates, the molecular instructions for building organisms and transmitting hereditary information are four simple bases, arranged in an almost unlimited variety of sequences.

GENETIC SIMILARITIES: HEMOGLOBIN

By comparing the similarity of their DNA sequences for individual genes—such as the gene that codes for a polypeptide used to build hemoglobin—we can estimate how closely related the species are and the amount of time that has passed since they had a common ancestor.

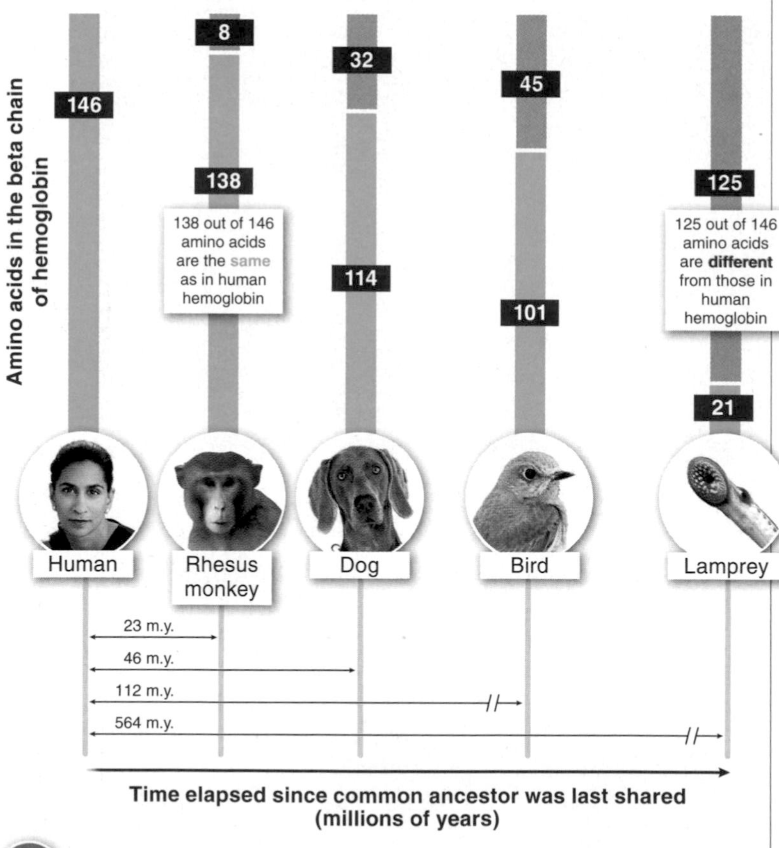

Amino acids in the beta chain of hemoglobin

146
8
138
138 out of 146 amino acids are the same as in human hemoglobin
32
114
45
101
125
125 out of 146 amino acids are **different** from those in human hemoglobin
21

Human
Rhesus monkey
Dog
Bird
Lamprey

23 m.y.
46 m.y.
112 m.y.
564 m.y.

Time elapsed since common ancestor was last shared (millions of years)

The longer two species have been evolving separately, the greater the number of genetic differences that accumulate.

FIGURE 10-42 **An evolutionary clock?** Genetic similarities (and differences) demonstrate species relatedness.

Related individuals within a species share a greater proportion of their DNA than do unrelated individuals. This is not unexpected; you and your siblings got all of your DNA from the same two parents, while you and your cousins got half of your DNA from the same two grandparents. The more distantly you and another individual are related, the more your DNA differs.

We can measure the similarity between the DNAs of two species by comparing their DNA sequences for individual genes. For example, let's look at the gene that codes for a polypeptide used to build the protein hemoglobin. In vertebrate animals, hemoglobin is found inside red blood cells, where its function is to carry oxygen throughout the body. It is made up of two types of chains of amino acids, the alpha chain and the beta chain. In humans, the beta chain has 146 amino acids. In rhesus monkeys, this beta chain is nearly identical in sequence: of the 146 amino acids, 138 are the same as those found in human hemoglobin, and only 8 are different. In dogs, the sequence is still somewhat similar, with 114 of the same amino acids. When we look at non-mammals such as birds, we find about 101 amino acid similarities with humans. And between humans and lamprey eels (still vertebrates, but with more than 500 million years since our last common ancestor), there are 21 amino acid similarities.

The differences in the amino acid sequence of the beta hemoglobin chain, and thus in the alleles that code for this structure, seem to indicate that humans have more recently shared a common ancestor with rhesus monkeys than with dogs. And that we have more recently shared an ancestor with dogs than with birds or lampreys. These findings are just as we would expect, based on estimates of evolutionary relatedness made from comparative anatomy and embryology, as well as from the fossil record. It is as if a molecular clock is ticking. The longer two species have been evolving separately, the greater the number of changes in amino acid sequences—or "ticks of the clock" (FIGURE 10-42).

TAKE HOME MESSAGE 10.21

» All living organisms share the same genetic code. The degree of similarity in the DNA of different species can reveal how closely related they are and the amount of time that has passed since their last common ancestor.

10.22 Laboratory and field experiments, and antibiotic-resistant bacteria, reveal evolution in progress.

A fifth line of evidence for the occurrence of evolution comes from multigeneration experiments and observations. We can observe and measure evolution as it is happening by choosing the right species—preferably organisms with very short life spans—and designing careful experiments.

In one clever study, researchers looked at populations of grass plants on golf courses—a habitat where lawn-mower blades represent a significant source of mortality. All the grass on the golf courses was of the same species, but on the putting greens it was cut very frequently, on the fairways it was cut only occasionally, and in the rough it was almost never cut at all (FIGURE 10-43).

Over just a few years, significant changes took place in these different grass populations. The grass plants on the putting greens rapidly developed to reproductive age and produced many seeds. For the plants in these populations, life was short and reproduction came quickly. Plants on the fairways had slightly slower development and a reduced seed output, while those in the rough had the slowest development and the lowest seed output of all.

When plants from each of the three habitats were collected and grown together in greenhouses under identical conditions (without mowing), the dramatic differences in growth, development, and life span remained. The frequencies of the alleles that control the traits of life span

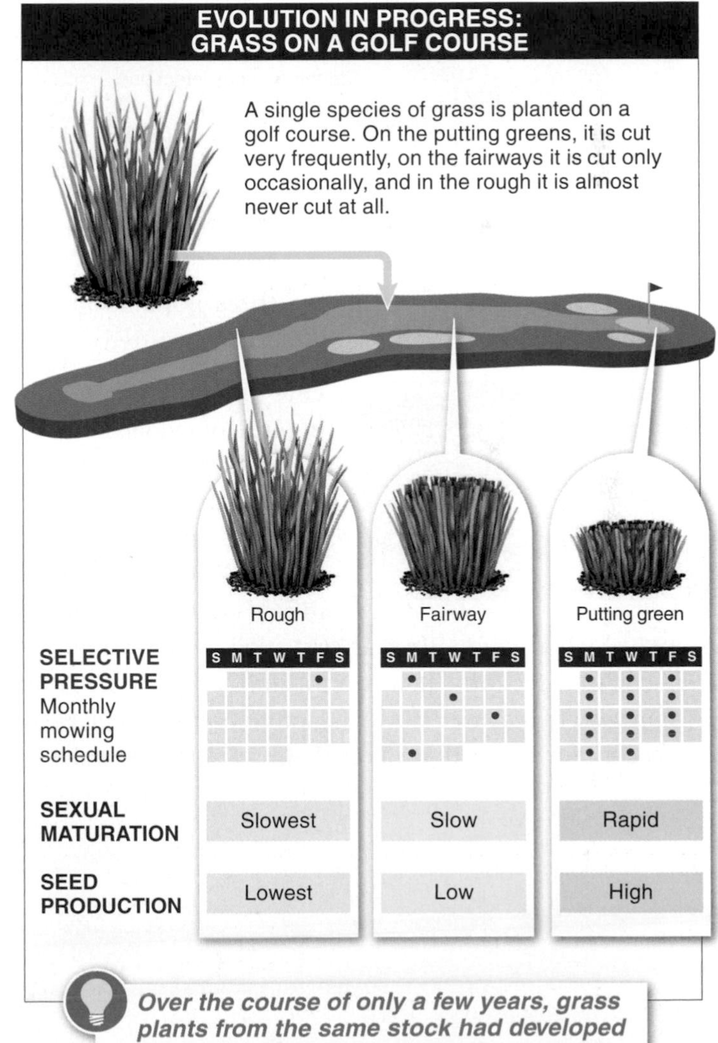

EVOLUTION IN PROGRESS: GRASS ON A GOLF COURSE
A single species of grass is planted on a golf course. On the putting greens, it is cut very frequently, on the fairways it is cut only occasionally, and in the rough it is almost never cut at all.
Rough
Fairway
Putting green
SELECTIVE PRESSURE Monthly mowing schedule
S M T W T F S
SEXUAL MATURATION
Slowest
Slow
Rapid
SEED PRODUCTION
Lowest
Low
High
Over the course of only a few years, grass plants from the same stock had developed into three distinct populations as a result of the frequency at which they were cut.

FIGURE 10-43 **Evolution in progress: grasses.**

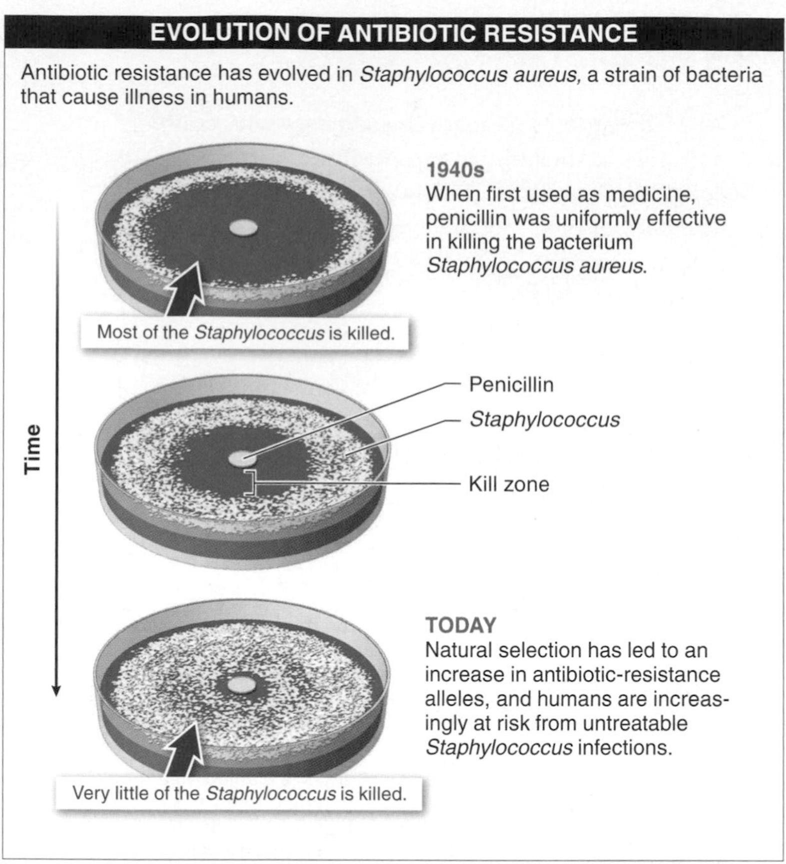

EVOLUTION OF ANTIBIOTIC RESISTANCE
Antibiotic resistance has evolved in Staphylococcus aureus, a strain of bacteria that cause illness in humans.
1940s
When first used as medicine, penicillin was uniformly effective in killing the bacterium Staphylococcus aureus.
Most of the Staphylococcus is killed.
Penicillin
Staphylococcus
Kill zone
Time
TODAY
Natural selection has led to an increase in antibiotic-resistance alleles, and humans are increasingly at risk from untreatable Staphylococcus infections.
Very little of the Staphylococcus is killed.

FIGURE 10-44 **Evolution in progress: disease-causing bacteria.**

and reproductive output in the three populations had changed. In other words, evolution had occurred.

Q Increasingly, bacterial infections are resistant to antibiotics that were once effective. Why?

One particularly disturbing line of evidence for the occurrence of evolution in nature comes from the evolution of antibiotic-resistant strains of bacteria that cause illness in humans. In the 1940s, when penicillin was first used as a treatment for bacterial infections, it was uniformly effective in killing *Staphylococcus aureus*. However, some strains of *Staphylococcus* contain alleles that make them resistant to penicillin. Because penicillin has become such a pervasive toxin in the environment of *Staphylococcus,* natural selection has led to an increase in the frequency of alleles that make these strains resistant to the drug.

Today, humans are increasingly at risk for becoming infected with *Staphylococcus* and getting diseases such as pneumonia and meningitis. Currently, more than 90% of isolated *S. aureus* strains are resistant to penicillin (FIGURE 10-44). In the 1960s, the antibiotics methicillin and oxacillin had nearly complete effectiveness against *Staphylococcus.* Now nearly a third of staph infections are resistant to these antibiotics as well. The meaning of such unintentional natural selection "experiments" is clear and consistent: evolution is occurring all around us.

Finally, let's return to the starvation-resistant fruit flies introduced at the beginning of this chapter. They are an unambiguous demonstration that natural selection can produce dramatic changes in a population, and that it can bring about these changes very quickly. And, perhaps most important, we saw that replicating the same evolutionary process over and over again produced the same predictable results. In the laboratory, on the farm, in the doctor's office, and in deserts, streams, and forests, evolution is occurring.

Reflecting on both the process and the products of evolution and natural selection, in the final paragraph of *The Origin of Species* Darwin eloquently wrote: "There is grandeur in this view of life . . . and that . . . from so simple a beginning endless forms most beautiful and most wonderful have been, and are being, evolved."

TAKE HOME MESSAGE 10.22

» Replicated, controlled laboratory selection experiments and long-term field studies of natural populations allow us to watch and measure evolution as it occurs.

STREET BIO

Using evidence to guide decision making in our own lives

Evolution: what it is and what it is not . . .

Misconceptions about evolution can arise from misunderstandings about subtleties in the process, from incomplete or erroneous accounts in schools and the media, and from deliberate interference with the understanding of evolution.

⊘ ***Misconception:* Evolution is a theory about the origin of life.** *Clarification:* Although scientists are interested in (and are investigating) how life began, evolutionary biology is focused on the processes that occurred (and still occur) *after* life began, such as the diversification and branching of species and the adaptation of populations to their environments.

⊘ ***Misconception:* Evolution is not observable or testable.** *Clarification:* Evolution is both observable and testable. Long-term, controlled, and replicated laboratory experiments on many species have demonstrated evolution (see Section 10-1) and speciation (formation of new species). Numerous studies of natural populations have also documented evolution taking place (see Section 10-13).

⊘ ***Misconception:* Evolution is only a theory and hasn't been proved.** *Clarification:* This misconception stems from confusion between the colloquial definition of a theory as a "guess" or "hunch" and the scientific definition of a theory as an explanation, based on many lines of evidence, that generates and supports predictions and is tested in many ways (see Section 1-10).

⊘ ***Misconception:* Individuals evolve (and are "trying" to adapt).** *Clarification:* Natural selection can lead to populations of organisms adapted to their environment (see Sections 10-12 and 10-16). But the process does not involve "trying" to adapt—it is a consequence of some individuals passing on their genes at a higher rate than individuals carrying alternative versions of those genes. The population rather than any one individual changes over time (see Section 10-11).

⊘ ***Misconception:* Evolution is goal-oriented and leads to progress and "optimal" solutions.** *Clarification:* Alleles that make it impossible for the individuals carrying them to live in a particular environment are weeded out of the population. But this is not the same as creating perfect or optimal organisms (see Section 10-13). Rather, the traits and features of organisms that increase in frequency within a population are those that cause the organisms carrying them to leave more descendants than organisms with alternative traits (see Sections 10-9 and 10-16). As the environment changes, so do the traits that are most fit.

⊘ ***Misconception:* There are no transitional fossils; the gaps in the fossil record disprove evolution.** *Clarification:* Because fossils are formed only under a narrow range of environmental conditions, biologists do not expect to find all transitional forms, so the gaps do not disprove evolution. Numerous transitional fossils *have* been found, including those between dinosaurs and modern birds, and those between whales and their terrestrial mammalian ancestors (see Section 10-18).

⊘ ***Misconception:* The theory of evolution says that life originated, and evolution proceeds, randomly, by chance.** *Clarification:* Although chance is one element of evolution—there is a random element to the generation of mutations, for example—many non-random factors are central to how evolution proceeds. For example, the differential reproductive success of some variants—such as bacteria with antibiotic-resistance genes—that leads to natural selection is not random (see Section 10-12).

⊘ ***Misconception:* Most biologists have rejected evolution.** *Clarification:* All modern theories of science are continually tested, verified, and modified as new and deeper understandings of the natural world emerge. The presidents of the National Academy of Sciences, the American Association for the Advancement of Science, and the National Science Teachers Association recently wrote a joint statement saying that "evolutionary theory has stood the test of time in serving as the most comprehensive scientific explanation for the diversity of life on Earth and it is accepted by the overwhelming majority of scientists." There have been no credible challenges to the basic principles of evolution and how it proceeds.

GRAPHIC CONTENT

Thinking critically about visual displays of data

1 What does the curvy line represent?

2 Finch beaks don't grow or shrink much once the bird reaches maturity. What is responsible for the average beak size changing from year to year?

3 Three droughts and one very wet year are highlighted. Why is that information included here?

4 If you wanted to make predictions about finches' beak size in the future, what information would be helpful?

5 Is there an "optimum" beak size for a finch? How does this graph help you answer that question?

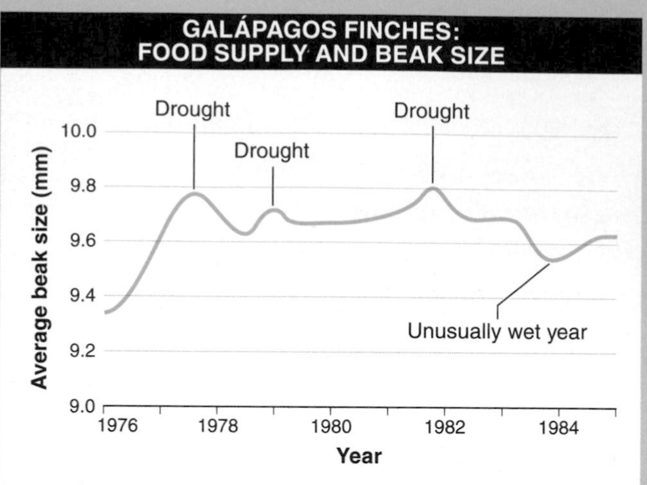

6 What do you think is the "Take Home Message" of this graph?

See answers at the back of the book.

KEY TERMS IN IN EVOLUTION AND NATURAL SELECTION

adaptation, p. 318
artificial selection, p. 320
biogeography, p. 330
bottleneck effect, p. 309
convergent evolution, p. 334
differential reproductive success, p. 313
directional selection, p. 321
disruptive selection, p. 323
evolution, p. 299
fitness, p. 317
fixation, p. 309
fossil, p. 328
founder effect, p. 309
gene flow, p. 311
genetic drift, p. 308
heritability, p. 313
homologous structure, p. 332
inheritance, p. 313
migration, p. 311
mutation, p. 307
natural selection, p. 312
population, p. 298
radiometric dating, p. 328
sexual selection, p. 314
stabilizing selection, p. 323
trait, p. 303
vestigial structure, p. 332

BRIEF SUMMARY

Evolution is an ongoing process.

• The characteristics of the individuals present in a population can change over time. We can observe such change in nature and can even cause it to occur.

Darwin journeyed to a new idea.

• In the 18th and 19th centuries, scientists began to overturn commonly held beliefs, including that the earth was only 6,000 years old and that all species had been created separately and were unchanging. These new ideas helped shape Charles Darwin's thinking.

• Darwin noted unexpected patterns among fossils and living organisms. Fossils resembled but were not identical to living organisms in the same area. And finch species on each of the Galápagos Islands differed from each other in small but significant ways.

• Darwin and Alfred Russel Wallace published a joint presentation on their ideas about natural selection in 1858, and Darwin published a much more detailed treatment in *The Origin of Species* in 1859, sparking debate and discussion about the processes and patterns of evolution.

Four mechanisms can give rise to evolution.

• Evolution is a change in allele frequencies within a population. It occurs by four different mechanisms: mutation, genetic drift, migration, and natural selection.

• Mutation is an alteration of the base-pair sequence in an individual's DNA. Mutations can be caused by high-energy radiation or chemicals in the environment, or can appear spontaneously. Mutation is the only way that new alleles can be created within a population.

• Genetic drift is a random change in allele frequencies within a population, unrelated to the alleles' influence on reproductive success.

• Migration, or gene flow, leads to a change in allele frequencies in a population as individuals move into or out of the population.

• Natural selection is a mechanism of evolution that occurs when there is heritable variation for a trait, and individuals with one version of the trait have greater reproductive success than do individuals with a different version of the trait.

• Knowing the frequency of each allele in a population, we can predict the genotypes and phenotypes we should see. If there is random mating and no evolution is occurring in a population, recessive and dominant alleles do not change their frequencies.

Through natural selection, populations of organisms can become adapted to their environments.

• Fitness is a measure of the relative amount of reproduction by an individual with a particular phenotype, compared with the reproductive output of individuals with alternative phenotypes. An individual's fitness can vary, depending on the environment in which it lives.

• Adaptation, which refers both to the process by which organisms become better matched to their environment and to the specific traits that make an organism more fit, occurs as a result of natural selection.

• Natural selection does not lead to organisms perfectly adapted to their environment.

• Natural selection can change allele frequencies for genes involved in complex physiological processes and behaviors. Sometimes a trait that has been selected for one function is later modified to serve a completely different function.

The evidence for evolution is overwhelming.

• Radiometric dating confirms that the earth is very old and allows scientists to determine the age of fossils. By analyzing fossilized remains in the fossil record, paleontologists can reconstruct what organisms looked like long ago, learn how organisms were related to each other, and understand how groups of organisms evolved.

• Observing geographic patterns of species distributions—noting similarities and differences among species living close together but in very different habitats, and among species living in similar habitats but located far from one another—helps us understand the evolutionary histories of populations.

• Similarities in the anatomy and development of different groups of organisms and in their physical appearance can reveal common evolutionary origins.

• All living organisms share the same genetic code. The degree of similarity in the DNA of different species can reveal how closely related they are and the amount of time that has passed since their last common ancestor.

CHECK YOUR KNOWLEDGE

Short Answer

1. Describe one benefit of conducting research on evolution in a population of short-lived species such as fruit flies.

2. Describe three commonly held Western beliefs about the natural world that were overturned in the 18th and 19th centuries.

3. How did Darwin's time on the *Beagle* help him develop his ideas on evolution?

4. What is evolution?

5. What does it mean when fixation for an allele occurs in a population?

6. Do recessive alleles tend to decrease in frequency in a population? Why or why not?

7. Describe three important components to an organism's evolutionary fitness.

8. How does the increasing frequency of antibiotic-resistant strains of bacteria represent an example of evolution in progress?

9. In *The Origin of Species,* Charles Darwin wrote: "We may look with some confidence to a secure future of great length. And as natural selection works solely by and for the good of each being, all corporeal and mental endowments will tend to progress towards perfection." Give three reasons why he was wrong.

10. How does modern medicine alter the selective pressures on birth weight in humans?

11. How is biogeography useful when studying the evolutionary history of a population?

12. Giving an example of each, compare and contrast homologous and analogous structures.

13. When evolutionary biologists speak of a "molecular clock," what do they mean?

Multiple Choice

1. Selecting for increased starvation resistance in fruit flies:

a) has no effect, because starvation resistance is not a trait that influences fruit flies' fitness.

b) has little effect, because ongoing mutation counteracts any benefits from selection.

c) cannot increase their survival time, because there is no genetic variation for this trait.

d) has no effect, because starvation resistance is dependent on the effects of too many genes.

e) can produce populations in which the average time to death from starvation is 160 hours.

2. Which of the following statements about mutations is incorrect?

a) Mutations are almost always random with respect to the needs of the organism.

b) A mutation is any change in an organism's DNA.

c) Most mutations are harmful or neutral for the organism in which they occur.

d) The origin of genetic variation is mutation.

e) All of these statements are correct.

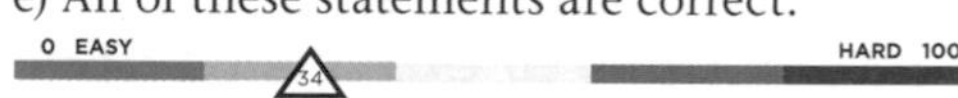

3. In a fish population in a shallow stream, the genotypic frequencies of yellowish-brown fish and greenish-brown fish changed significantly after a flash flood randomly swept away fish from the stream. This change in genotypic frequency was most likely attributable to:

a) gene flow.

b) disruptive selection.

c) directional selection.

d) stabilizing selection.

e) genetic drift.

4. When a small group of individuals colonizes a new habitat, an evolutionary event is likely to occur, because:

a) members of a small population have reduced rates of mating.

b) gene flow increases.

c) mutations are more common in novel environments.

d) new environments tend to be inhospitable, reducing survival there.

e) small founding populations are rarely genetically representative of the initial population.

0 EASY HARD 100 52

5. "Survival of the fittest" may be a misleading phrase to describe the process of evolution by natural selection, because:

a) it is impossible to determine the fittest individuals in nature.

b) survival matters less to natural selection than does reproductive success.

c) natural variation in a population is generally too great to be influenced by differential survival.

d) during population bottlenecks, it is the least fit individuals that have the greatest survival.

e) reproductive success, on its own, does not necessarily guarantee evolution.

6. Adaptations shaped by natural selection:

a) are magnified and enhanced through genetic drift.

b) are unlikely to be present in humans living in industrial societies.

c) may be out of date, having been shaped in the past under conditions that differed from those in the present.

d) represent perfect solutions to the problems posed by nature.

e) are continuously modified so that they are always matched to the environment in which a population lives.

0 EASY 77 HARD 100

7. In a population in which a trait is exposed to stabilizing selection:

a) neither the average value nor the variation for the trait changes.

b) both the average value and the variation for the trait increase.

c) the average value increases or decreases, and the variation for the trait decreases.

d) the average value for the trait stays approximately the same, and the variation for the trait decreases.

e) the average value for the trait stays approximately the same, and the variation for the trait increases.

0 EASY 46 HARD 100

8. Maze-running behavior in rats:

a) is too complex a trait to be influenced by natural selection.

b) is a heritable trait.

c) is not influenced by natural selection, because it does not occur in rats' natural environment.

d) shows no variation.

e) is influenced primarily by mutation.

0 EASY 52 HARD 100

9. What can be concluded from comparing differences in molecular biology among different species?

a) Extremely different species are fundamentally unrelated in any way.

b) Only DNA sequences can be used to compare species' relatedness.

c) Birds are more closely related to humans than dogs are.

d) Genetic similarities and differences demonstrate species relatedness.

e) The longer two species have been evolving on their separate paths, the fewer the genetic differences between them.

0 EASY 47 HARD 100

10. Evolution:

a) occurs too slowly to be observed.

b) can occur in the wild but not in the laboratory.

c) is responsible for the increased occurrence of antibiotic-resistant bacteria.

d) does not occur in human-occupied habitats.

e) None of these statements are correct.

0 EASY 48 HARD 100

Ch11

Behaviors are traits that can evolve.

11.1 Behavior has adaptive value, just like other traits.

11.2 Some behaviors are innate.

11.3 Some behaviors must be learned (and some are learned more easily than others).

11.4 Complex-appearing behaviors don't require complex thought in order to evolve.

Cooperation, selfishness, and altruism can be better understood with an evolutionary approach.

11.5 "Kindness" can be explained.

11.6 Apparent altruism toward relatives can evolve through kin selection.

11.7 Apparent altruism toward unrelated individuals can evolve through reciprocal altruism.

11.8 In an "alien" environment, adaptations produced by natural selection may no longer be adaptive.

11.9 Selfish genes win out over group selection.

Sexual conflict can result from disparities in reproductive investment by males and females.

11.10 There are big differences in how much males and females must invest in reproduction.

11.11 Males and females are vulnerable at different stages of the reproductive exchange.

11.12 Tactics for getting a mate: competition and courtship can help males and females secure reproductive success.

11.13 Tactics for keeping a mate: mate guarding can protect a male's reproductive investment.

11.14 *This is how we do it:* When paternity uncertainty seems greater, is paternal care reduced?

11.15 Monogamy versus polygamy: mating behaviors can vary across human and animal cultures.

11.16 Sexual dimorphism is an indicator of a population's mating behavior.

Communication and the design of signals evolve.

11.17 Animal communication and language abilities evolve.

11.18 Honest signals reduce deception.

Wolves communicate and bond through howling.

Evolution and Behavior

Communication, cooperation, and conflict in the animal world

11.1–11.4

Behaviors are traits that can evolve.

In Borneo, a macaque hangs upside down on a branch to take a sip of water.

11.1 Behavior has adaptive value, just like other traits.

What are your favorite things to eat? Perhaps donuts or hot fudge sundaes top your list. Or maybe bananas, or cheeseburgers and fries. One thing is almost certain about the food preferences of any human: the list is filled with calorie-rich substances. Put another way, when it comes to eating, humans show an almost complete aversion to dirt or pebbles or other substances from which no energy can be extracted (FIGURE 11-1).

In controlled studies, humans demonstrate a clear and consistent preference for sweet and fatty foods. Specifically, when people compare variations of a food dish that differ in the amount of sugar and/or fat, their preference for the dish increases as its sugar concentration increases—up to a point, after which adding more sugar makes the food less preferable. But their preference becomes stronger and stronger as the food's fat content increases. In short, we prefer sweet (but not too sweet) foods that are packed with as much fat as possible (FIGURE 11-2).

FIGURE 11-1 **Mud pie for dinner?** Humans show an aversion to eating dirt (and other substances from which we cannot extract nutrition).

Watch birds in a field and you'll see that they, too, have preferences. A starling, for example, walks through the grass, probing the ground and only occasionally picking up something to eat, such as a beetle larva. Shore crabs, too, have preferences. They favor the mussels that provide the greatest amount of energy relative to the energy required to open the mussel. The largest mussels take too long to open; the smallest mussels are easily opened but don't provide enough caloric content to make the effort worthwhile (FIGURE 11-3). Thus, the research shows, shore crabs prefer medium-size mussels.

Animals have taste preferences for a reason. Whereas plants can create their own food, all animals must consume materials from which they can extract the most energy and acquire essential nutrients. If they don't, they die. Taste preferences are a direct result of an interaction between the materials we eat and molecules within our taste buds. And those molecules are produced by a large number of genes that have been shaped by natural selection.

Taste preferences directly influence the feeding choices of animals, and these choices affect the number of offspring they can produce and their evolutionary fitness. In this way, natural selection shapes feeding behaviors just as it can bring about changes in physical structures.

Before we continue, let's define what we mean by "behavior." **Behavior** encompasses any and all of the actions performed by an organism, often in response to its environment or the actions of another organism. Feeding

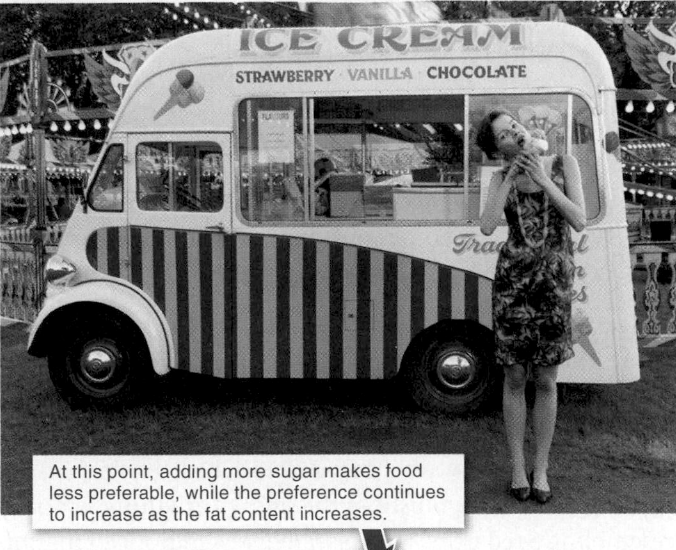

ICE CREAM
STRAWBERRY VANILLA CHOCOLATE
At this point, adding more sugar makes food less preferable, while the preference continues to increase as the fat content increases.

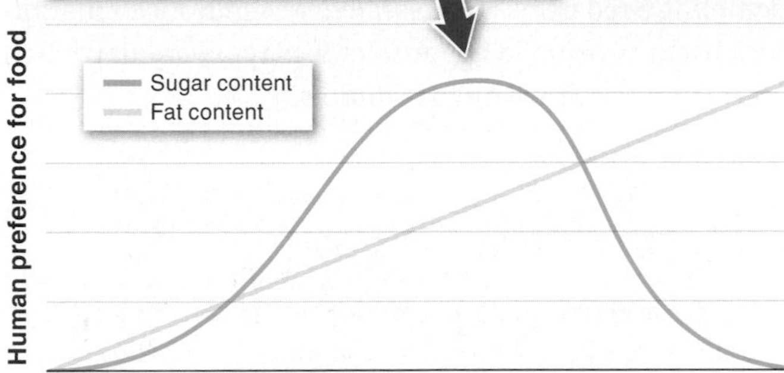

Across cultures, humans prefer sweet, but not too sweet, foods that are packed with fat.

FIGURE 11-2 Fatty and sweet.

ENERGY-EFFICIENT FEEDING BEHAVIOR HAS EVOLVED

A shore crab feeds on a mussel

Shore crabs preferentially choose mussels that provide the most energy relative to the effort it takes to open the shell.

FIGURE 11-3 Efficient eater.

behavior is only one of many behaviors influenced by natural selection.

The scope of animal behavior is vast and varied. Even a brief listing of well-studied topics in animal behavior is impressively long:

- Conflict, aggression, and territoriality
- Cooperation, alliance-building, and sociality
- Competing for food and avoiding predators
- Migration and navigation
- Behavioral control of body temperature
- Courtship and mate choice and fidelity
- Breeding and parental behavior
- Communication
- Learning and tool use

In Chapter 10, we learned about the evolutionary origins of many *physical traits* and the general process by which natural selection can lead to the evolution of organisms adapted to their environments. Similarly, in this chapter, we explore the evolution of *behavioral traits.*

Recall from the discussion of natural selection that when a heritable trait increases an individual's reproductive success relative to that of other individuals in the population, that trait tends to increase in frequency in that population. This is Darwin's mechanism for evolution by natural selection and its effects are evident all around us, from the fancy ornamentation of male peacocks' feathers produced by sexual selection to the cryptic coloration that camouflages many organisms.

In this chapter, we see that *behavior* is just as much a part of an organism's phenotype as is its anatomical structures, and that behavior is produced and shaped by natural selection.

We'll concentrate on three broad areas of animal behavior: cooperation and conflict, mating and parenting, and communication. Our focus is on why those behaviors evolved and how those behaviors contribute to an organism's fitness, survival, and reproduction.

TAKE HOME MESSAGE 11.1

›› Behavior encompasses any and all of the actions performed by an organism. Behavior is a part of an organism's phenotype, and as such it can be produced and shaped by natural selection.

11.2 Some behaviors are innate.

In his study of pea plants, Gregor Mendel described a single gene for plant height that caused a plant to be either tall or short. The production of a trait such as plant height, however, is not completely genetically determined. Certain environmental conditions, such as the type of soil and the availability of water, nutrients, and sunlight, also play roles. Nearly all physical traits of all organisms are the products not only of genes but also of environmental conditions. Likewise, when it comes to the production of behaviors, the environment also plays an important role.

The degree to which a behavior depends on the environment for expression, however, varies a great deal. At one extreme are behaviors—called **instincts** or **innate behaviors**—that don't require any environmental input to develop. Innate behaviors are present in all individuals in a population and do not vary much from one individual to another or over an individual's life span. An example of innate behavior is a **fixed action pattern,** a sequence of behaviors triggered in response to a specific signal called a **sign stimulus.** The behavior pattern requires no learning, does not vary among individuals, and, once started, runs to completion. Here are just two examples of fixed action patterns (FIGURE 11-4).

1. Egg retrieval in geese. When a goose spots an egg outside its nest (a sign stimulus), a fixed action pattern is triggered. The goose rolls the egg back to the nest, using a side-to-side motion, keeping the egg tucked underneath its bill. Once started, a goose continues the behavior to completion, even if the egg is taken away on the way to the nest.

2. Aggressive displays and attacks by stickleback fish. The bellies of male stickleback fish turn bright red when the breeding season arrives. During this time, the sight of a red belly (a sign stimulus) triggers a male stickleback to react aggressively to any other male stickleback. In fact, the males become antagonistic at the sight of anything remotely resembling a red belly. One researcher even noticed that his sticklebacks performed aggressive displays every day when a red mail truck drove by his window.

TAKE HOME MESSAGE 11.2

» Like any physical trait, behavior can depend on the environment for expression, though the degree of that dependence varies. Instincts, or innate behaviors, develop without any environmental input, are present in all individuals in a population, and vary little from one individual to another or over an individual's life span. A fixed action pattern, a type of innate behavior, is a sequence of actions that requires no learning, does not vary, and runs to completion once started.

SOME BEHAVIORS ARE INNATE

In geese, the sight of an egg outside the nest triggers a fixed action pattern: the goose uses a side-to-side egg-retrieval movement all the way back to the nest, even if the egg is taken away during the process.

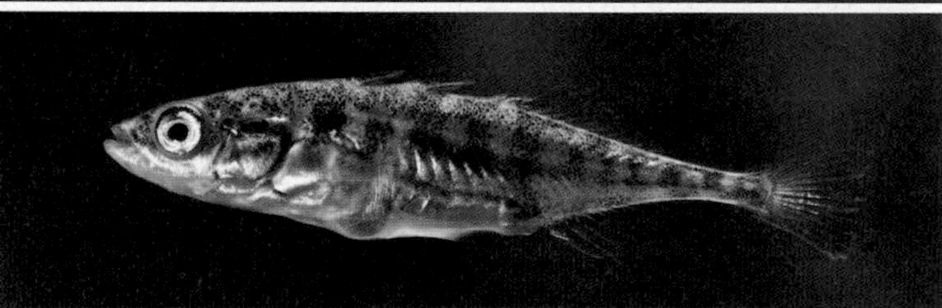

Male stickleback during non-breeding season

Male stickleback during breeding season

During the breeding season, a red belly on any other male stickleback fish triggers an aggressive response.

FIGURE 11-4 **No learning required.**

11.3 Some behaviors must be learned (and some are learned more easily than others).

In contrast to innate behaviors, some behaviors are acquired, altered, and modified over time in response to the individual's experiences. That is, they require **learning.** There can be tremendous variation among behaviors that require learning: some come relatively easily and are learned by most individuals in a population, while other behaviors are less easily learned.

Consider fear of snakes, a trait common to most primates, including humans. In the wild, rhesus monkeys are afraid of snakes. The sight of a snake causes the monkeys to engage in fear-related responses, including making alarm calls and rapidly moving away from the snake.

Observations of monkeys in captivity, however, reveal that they aren't born with a fear of snakes. Captive rhesus monkeys will reach over a plastic model of a snake to get a peanut. But, as experiments show, monkeys learn to fear snakes if they see another monkey terrified at the sight of a snake—even if they see the reaction on television (FIGURE 11-5). Ignoring the peanut, the frightened monkeys scream and move as far away from the artificial snake as possible. Studies on humans show an equally easily learned fear of snakes. We respond with sweaty palms and an increased heart rate.

Q Why is it so much easier for an infant to learn a complex language than for a college student to learn biology?

When behaviors are learned easily and by all (or nearly all) individuals, this is called **prepared learning.** In addition to the snake-fearing behavior of monkeys, examples of prepared learning abound. The acquisition of language in humans is a dramatic example. Most children begin to talk around one year of age; by age three, they understand most rules of sentence construction; and by age six, a native English-speaker has a vocabulary of about 13,000 words.

Let's return to our rhesus monkey experiments. Researchers altered the snake videotape to show a monkey having the same fear reaction in response to a flower or a toy rabbit. After watching this tape, the captive monkeys—which had never seen flowers or rabbits or snakes—did not learn to fear flowers or rabbits.

These observations point to an evolutionary basis for the acquisition of certain behaviors. Organisms are well-prepared to learn behaviors (such as snake avoidance) that were important to their ancestors' survival and reproductive success over the course of their evolutionary history. Organisms are less prepared to learn behaviors irrelevant to their evolutionary success (such as flower or rabbit avoidance).

Given that more than 30,000 people are killed by guns in the United States each year, while fewer than three dozen people are killed by snakes, it would seem that we ought to be very afraid of guns and relatively unconcerned about snakes. But we are built in just the opposite way

Q Human babies quickly and easily develop a fear of snakes. Yet they don't easily develop a fear of guns. Why?

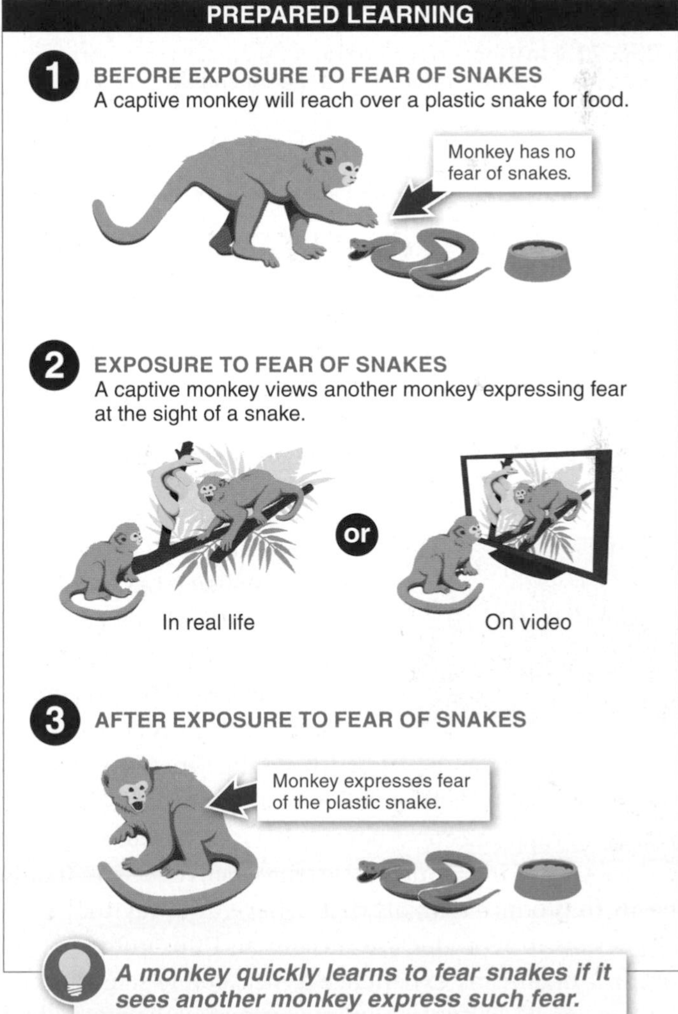

PREPARED LEARNING
1 BEFORE EXPOSURE TO FEAR OF SNAKES
A captive monkey will reach over a plastic snake for food.
Monkey has no fear of snakes.
2 EXPOSURE TO FEAR OF SNAKES
A captive monkey views another monkey expressing fear at the sight of a snake.
or
In real life
On video
3 AFTER EXPOSURE TO FEAR OF SNAKES
Monkey expresses fear of the plastic snake.
A monkey quickly learns to fear snakes if it sees another monkey express such fear.

FIGURE 11-5 **Not innate, but easily learned.** The monkey can learn to fear snakes by observing fear in other monkeys.

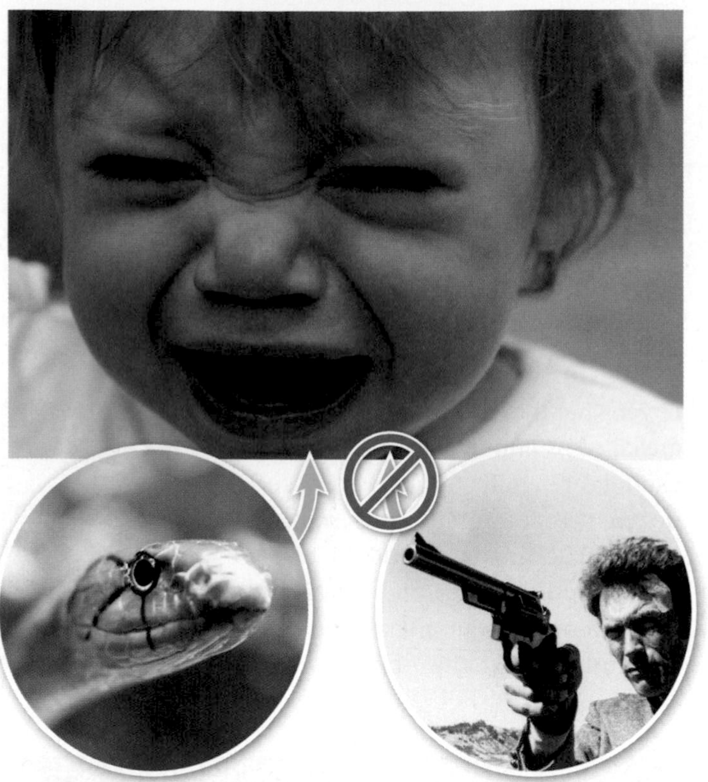

FIGURE 11-6 Instincts out of date. A human infant is born with a predisposition for developing a fear of snakes.

(FIGURE 11-6). The likely explanation is that evolution can be slow in producing populations that are adapted to their environments—that is, evolutionary change cannot always keep up with a rapidly (in evolutionary terms) changing environment.

Like all genes, any genes involved in behaviors have been handed down to us from our ancestors. Snakes caused many human deaths over the course of our evolutionary history. In contrast, guns didn't kill a single person until very recently on the evolutionary time scale. Accordingly, we still fear our ancient enemy, the snake, and have no instinctual response to novel threats, regardless of how deadly.

TAKE HOME MESSAGE 11.3

» In contrast to innate behaviors are behaviors that are influenced more by the individual's environment, requiring some learning, and are often modified over time in response to past experiences. Organisms are well prepared to learn behaviors that were important to the reproductive success of their ancestors, and less prepared to learn behaviors irrelevant to their species' evolutionary success.

11.4 Complex-appearing behaviors don't require complex thought in order to evolve.

Why do humans and other animals have sex? Is it because they are thinking, robot-like: "Must maximize reproductive success. Must maximize reproductive success"? Of course not. Yet they go about their lives behaving as if they were.

Q Do animals consciously act to improve their reproductive success?

An animal that experiences pleasure when it has sex has the incentive to seek out additional opportunities to experience that pleasure. As long as reproductive success is enhanced in the process, it is not necessary (from the fitness standpoint) for the animal to deliberately seek that outcome. In other words, natural selection doesn't have to produce animals consciously trying to maximize reproductive success. It only needs to produce animals that behave in a way that actually results in reproductive success, and this outcome will occur if organisms experience a pleasurable sensation by having sex. It's like an evolutionary shortcut. Behaviors that lead to a specific outcome that increases the animal's relative reproductive success will be favored by natural selection.

With a little trickery, we can investigate experimentally whether natural selection sometimes results in such evolutionary shortcuts. Consider the example of egg retrieval in geese. When an egg falls out of the nest, a goose retrieves it in exactly the same way every time. Is a goose able to do this because it vigilantly keeps tabs on all of its eggs? This seemed to be the case, until researchers decided to trick some geese. The researchers learned that putting *any* object that remotely resembles an egg near a goose's nest triggers the retrieval process. Geese will retrieve beer cans, door knobs, and a variety of other objects that will not increase the animal's reproductive success. Moreover, if multiple egg-like items are just outside the nest, the goose will retrieve the largest item first—even when it's an artificial egg the size of a basketball (FIGURE 11-7).

Thus, natural selection can produce organisms that exhibit silly and clearly maladaptive (i.e., fitness-decreasing) behaviors in experiments—the goose trying to retrieve a basketball-sized egg—if, in nature, individuals' performance of these simple actions nearly always leads to fitness-increasing behaviors—keeping eggs in the nest.

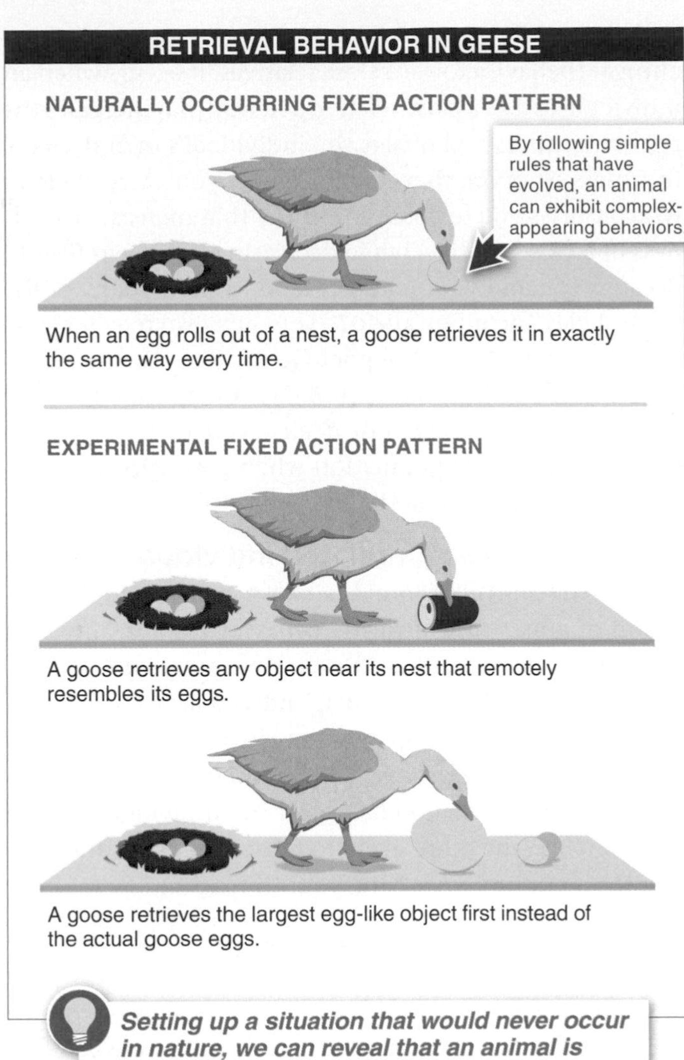

RETRIEVAL BEHAVIOR IN GEESE
NATURALLY OCCURRING FIXED ACTION PATTERN
By following simple rules that have evolved, an animal can exhibit complex-appearing behaviors.
When an egg rolls out of a nest, a goose retrieves it in exactly the same way every time.
EXPERIMENTAL FIXED ACTION PATTERN
A goose retrieves any object near its nest that remotely resembles its eggs.
A goose retrieves the largest egg-like object first instead of the actual goose eggs.
Setting up a situation that would never occur in nature, we can reveal that an animal is following simple behavioral rules.

FIGURE 11-7 Programmed to retrieve. The goose will retrieve any egg-like object outside its nest.

Taking an evolutionary approach to the study of the behavior of animals, including humans, is not new. Charles Darwin wrote in *On the Origin of Species:* "In the distant future I see open fields for far more important researches. Psychology will be based on a new foundation." And in the past several decades, Darwin's prediction has increasingly proved true.

In a synthesis of evolutionary biology and psychology, called "evolutionary psychology," researchers view the human brain and human behaviors, including emotions, as traits produced by natural selection. The evolutionary approach to studying behavior has also influenced the field of economics. The 2002 Nobel prize for economics was awarded to researchers who used insights from evolutionary biology and psychology in their analyses of human decision-making.

In the rest of this chapter, working from the understanding that organisms' traits, including their behaviors, have evolved by natural selection, we examine the question of why organisms behave as they do. We begin with an exploration of selfishness and cooperation in animals.

TAKE HOME MESSAGE 11.4

» If certain behavior in natural situations usually increases an animal's relative reproductive success, the behavior will be favored by natural selection. The natural selection of such behaviors does not require the organism to consciously try to maximize its reproductive success.

11.5–11.9 Cooperation, selfishness, and altruism can be better understood with an evolutionary approach.

An arctic ground squirrel in Alaska.

11.5 "Kindness" can be explained.

If we look closely, we see many behaviors in the animal world that *appear* to be **altruistic behaviors**—that is, they seem to come at a cost to the individual performing them, while benefiting a recipient. When discussing altruism, we define costs and benefits in terms of their contribution to an individual's fitness.

This mother spider is actually being selfish—not altruistic—when she lets her offspring eat her alive.

FIGURE 11-8 **Parental care to the extreme.** The female *Diaea ergandros* spider feeds her offspring with her own body.

Take the case of the Australian social spider (*Diaea ergandros*). After giving birth to about 50 hungry spiderlings, the mother's body slowly liquefies into a nutritious fluid that the newborn spiders consume. Over the course of about 40 days, her offspring literally eat her alive; they ultimately kill their mother, but start their lives well-nourished. The cost to the mother is huge—she is unable to produce any more offspring—but there is a clear benefit to her many offspring (FIGURE 11-8).

Such altruistic-appearing behavior puzzled Darwin. Natural selection, he believed, generally works to produce selfish behavior. Alleles that cause the individual carrying them to help increase other individual's reproductive success at the expense of their own success should decrease their relative frequency in a population. Put another way, natural selection should never produce altruistic behavior. Darwin worried that if the apparent instances of altruism were indeed truly altruistic, they would prove fatal to his theory.

As it turns out, Darwin's theory is safe. Virtually all of the apparent acts of altruism in the animal kingdom prove, on closer inspection, to be not truly altruistic; instead, they have evolved as a consequence of either **kin selection** or **reciprocal altruism.**

Kindness Toward Close Relatives: Kin Selection Kin selection can lead to the evolution of apparently-altruistic behavior toward close relatives. It occurs when an individual assists a close relative in a way that increases the relative's fitness and offsets the individual's own decrease in fitness. Suppose that, for one gene, you carry allele *K*, which causes you to behave in a way that increases the fitness of a close relative, while decreasing your own fitness. Because you and your relatives tend to share many of the same alleles, including allele *K*, your altruism may increase the frequency of *K* in the population. Thus the increased fitness of a relative you help might compensate for your own reduced fitness, because allele *K* will, overall, increase its frequency in the population when you help your relatives increase their reproductive output.

Kindness Toward Unrelated Individuals: Reciprocal Altruism Reciprocal altruism can lead to the evolution of apparently-altruistic behavior toward unrelated individuals. Suppose that, for another gene, you carry allele *R*, which causes you to help individuals to whom you are not related. This helps to increase their fitness and decreases yours in the process. Allele *R* might still increase its market share in the population, if the individuals whom you help become more likely to return the favor and help you in the future, increasing your fitness.

Seen in this light, both kin selection and reciprocal altruism can lead to the evolution of behaviors that are *apparently* altruistic but, in actuality, are beneficial—from an evolutionary perspective—to the individuals engaging in the behaviors.

In the next two sections, we explore kin selection and reciprocal altruism in more detail. We investigate some of the many testable predictions about when acts of apparent kindness should occur. It is important to note here that even as genes play a central role in cooperation and conflict, people's ability to override impulses toward selfishness is responsible for some of the rich diversity in human behavior that we see around us.

TAKE HOME MESSAGE 11.5

» Many behaviors in the animal world appear to be altruistic. In almost all cases, the apparent acts of altruism are not truly altruistic; they have evolved as a consequence of either kin selection or reciprocal altruism and, from an evolutionary perspective, are beneficial to the individual engaging in the behavior.

11.6 Apparent altruism toward relatives can evolve through kin selection.

The grasslands of the western United States are home to large colonies of Belding's ground squirrels, living in underground burrows. With hundreds or thousands of individuals, these colonies attract many birds of prey. But the squirrels have a system for reducing predation risk that resembles a neighborhood watch program. When an aerial predator approaches, one squirrel may stand on top of a burrow and produce a loud whistle-like alarm call, warning the colony of the impending danger. On hearing an alarm call, squirrels quickly take cover in their burrows, reducing their risk of death. Making an alarm call is a very dangerous activity: about half the time that an alarm call is made, the caller is killed by the predator (FIGURE 11-9).

Some squirrels see predators and make alarm calls, while other squirrels see predators and keep quiet. It turns out that about 80% of the squirrels making alarm calls are female. And older females are five times more likely than young females to make alarm calls.

Why are these particular squirrels making the alarm calls? Why would any squirrel make an alarm call? The age and sex of the squirrels reveal that alarm calling is about protecting relatives. The more kin an individual is likely to have, the more likely that individual is to sound the alarm. Males travel long distances to live in new colonies shortly after reaching maturity, so most adult males don't live near their parents, siblings, or other relatives except their own offspring. Females, on the other hand, remain near the area where they were born and are likely to have many close relatives nearby. Older females, then, are likely to have the largest number of relatives.

The biologist W. D. Hamilton expressed the idea of "kin selection," that one individual assisting another could compensate for its own decrease in fitness if it helped a close relative in a way that increased the relative's fitness. Benefactor and recipient share alleles that are preserved through the act of altruistic-appearing behavior. It seems as if individuals are acting selflessly/altruistically, but they are really acting in their genes' best interests.

Belding's ground squirrels are probably unaware of how closely they are related to the squirrels around them, but it's likely that natural selection has led to an evolutionary shortcut that allows them to behave as if they did know. In

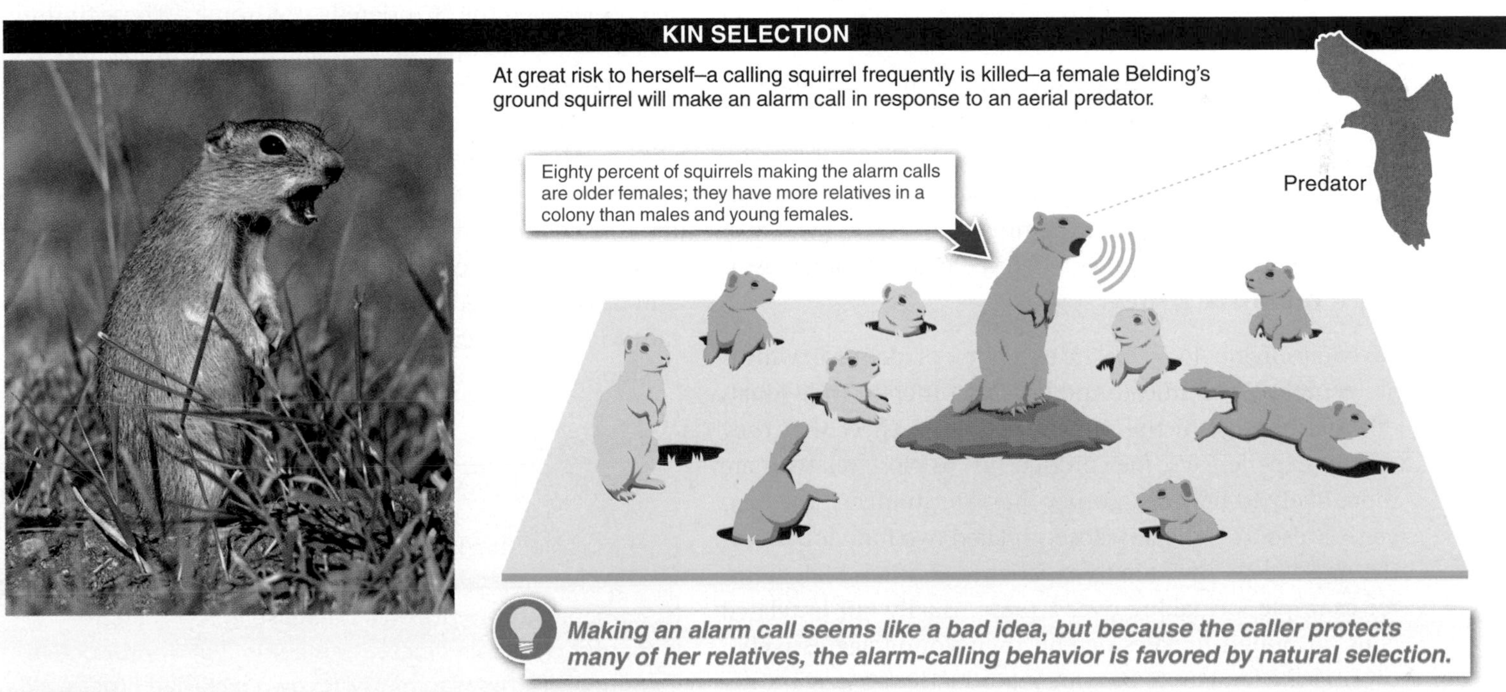

KIN SELECTION
At great risk to herself–a calling squirrel frequently is killed–a female Belding's ground squirrel will make an alarm call in response to an aerial predator.
Eighty percent of squirrels making the alarm calls are older females; they have more relatives in a colony than males and young females.
Predator
Making an alarm call seems like a bad idea, but because the caller protects many of her relatives, the alarm-calling behavior is favored by natural selection.

FIGURE 11-9 **Protecting relatives by making an alarm call.**

INAPPROPRIATE ALARM CALLING IN UNNATURAL SITUATIONS

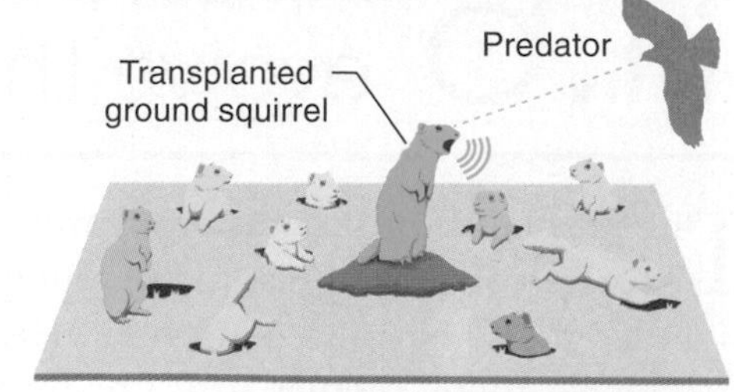

1 Researchers capture an older female Belding's ground squirrel (older females are the ones most likely to make alarm calls).

2 The captured ground squirrel is transported far away and released into a new colony, where she has no relatives.

3 The older female ground squirrel continues to make alarm calls, revealing that animals may sometimes follow simple behavioral rules.

FIGURE 11-10 **Following simple behavioral rules.**

A transplanted female Belding's ground squirrel that is not related to the squirrel colony behaves as though she is.

a clever test of this idea, researchers trapped adult female squirrels and relocated them to distant ground squirrel colonies in which they had no close relatives. If the transplanted females were able to determine exactly how closely they were related to the squirrels around them, they would not make alarm calls. To do so would put the caller at risk without benefiting her genes. What the researchers found, though, was that transplanted females were just as likely to sound the alarm in the colony of strangers as they were in their home territories, surrounded by close relatives (FIGURE 11-10). It seems that female ground squirrels have evolved to follow a simple rule that says, "If I am an older female, I behave as if I have many close relatives around me."

Based on the idea of kin selection, we need to redefine what is meant by an individual's fitness. An individual's fitness is not just his or her total reproductive output. Fitness also includes the reproductive output that individuals bring about through their seemingly altruistic behaviors toward their close kin. This redefined measure of fitness is called **inclusive fitness.**

No two individuals—with the exception of identical twins—are genetically identical. And because different individuals do not share all of the same alleles, we expect that they should experience some conflict. Just as close relatives are more likely to help each other than are strangers, the converse is also true: the less closely related two individuals are, the more likely they are to experience conflict. Male lions, for example, on taking over a pride usually kill unrelated cubs (although females defend their young aggressively), causing the females to become reproductively ready sooner than they would if they continued nursing their cubs.

The idea of kin selection gives rise to many predictions about conflict among humans. One of these is that child abuse, when it occurs, is more likely to be abuse by his or her stepparent than abuse by his or her biological parent. Is this the case?

Q Is a child living with one or more stepparents at greater risk of abuse than a child of the same age living with his or her biological parents?

Numerous studies across many different cultures, including an evaluation of 20,000 reports from the American Humane Association in the United States, support what has been called the "Cinderella syndrome." These findings included, for example, an estimate that the probability that a preschooler will be abused is about 1 in 3,000 for a child living with two biological parents (with whom he or she shares considerable genetic relatedness) and 40 in 3,000 for a child living with a stepparent (with whom he or she has no genetic relatedness). This difference in risk remains even when socioeconomic factors are taken into account and has been noted in multiple other cultures. It is important to note, however, that in the overwhelming majority of stepfamilies no abuse occurs.

TAKE HOME MESSAGE 11.6

» **Kin selection is apparently altruistic behavior in which an individual that assists a genetic relative compensates for its own decrease in reproductive output by helping increase the relative's reproductive output and, consequently, its own inclusive fitness.**

11.7 Apparent altruism toward unrelated individuals can evolve through reciprocal altruism.

As we saw in the previous section, when behavior appears to be altruistic, this is frequently because individuals are helping kin. But does any apparently-altruistic behavior occur toward non-relatives? Yes, but not much. Here we'll explore the conditions that give rise to reciprocal altruism, how it may have arisen, and why it is so common among humans but so rare among most other animal species.

We start by examining one species with well-documented altruistic-appearing behavior: the vampire bat. Vampire bats live in social groups of 8–12 mostly unrelated adults, roosting primarily in caves and hollow trees. They feed by landing on large mammals such as cattle and horses, piercing the skin with their razor-sharp teeth, and drinking blood that flows from the wound.

Because of their small body size (about the size of your thumb) and their high metabolic rate, vampire bats must consume almost their entire body weight in blood each night! If they go for more than about 60 hours without finding a meal, they are likely to die from starvation. Here's where the apparent altruism comes in: a bat that has not found food and is close to death will beg food from a bat that has recently eaten. In many cases, the bat that has just eaten will regurgitate some of the blood it has consumed into the mouth of the hungry bat, saving it from starvation. This act obviously has very high benefit for the recipient of the blood, but it comes at a cost to the sharing bat, which loses some of the caloric content of a meal it has just obtained (FIGURE 11-11).

RECIPROCAL ALTRUISM IN VAMPIRE BATS

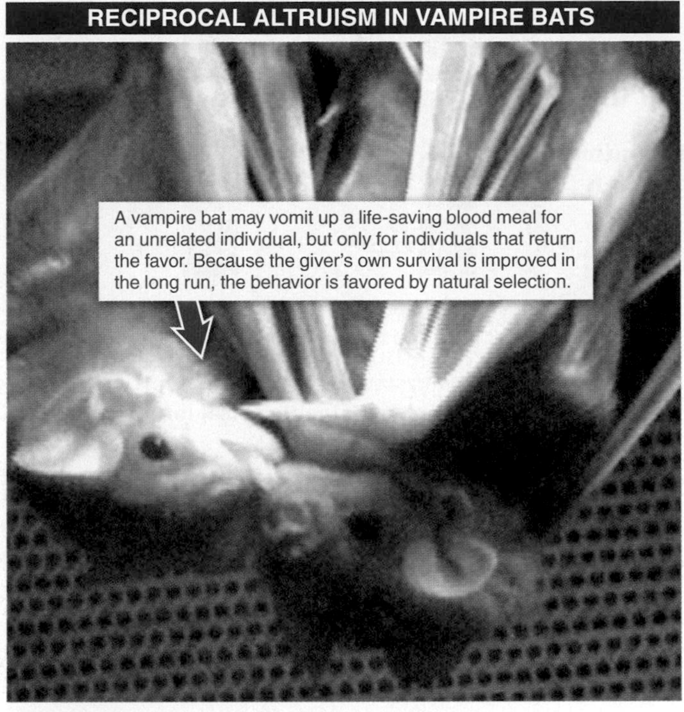

A vampire bat may vomit up a life-saving blood meal for an unrelated individual, but only for individuals that return the favor. Because the giver's own survival is improved in the long run, the behavior is favored by natural selection.

FIGURE 11-11 **Returning the favor.**

Kin selection is responsible for some of the blood sharing (females often regurgitate blood for their own offspring), but in many cases bats give blood to unrelated individuals. How might this behavior have arisen? To answer the question, it is important to note three other features of vampire bats. First, they can recognize more than a hundred distinct individual bats. Second, bats that receive blood donations from non-relatives reciprocate significantly more than average with the bats that have shared blood with them. And third, bats that do not have a history of helping each other generally do not regurgitate for each other.

The evolution of this apparent altruism may be due to the fact that the bats giving blood are repaid the favor when they are in need of blood. In other words, the act only *seems* to be selfless; in actuality it is selfish. With such reciprocal altruism, both individuals (at different encounters) give up something of *relatively* low value in exchange for getting something of great value at a later time when they need it most. In other words, they are storing goodwill in another individual and are protected from some of the world's uncertainties.

Taken together, studies of bats and other mammals show that reciprocal altruism can evolve if the following three conditions are met:

1. Repeated interactions among individuals, with opportunities to be both the donor and the recipient of altruistic-appearing acts
2. Benefits to the recipient that are significantly greater than the costs to the donor
3. The ability to recognize and punish cheaters, individuals that are recipients of altruistic-appearing acts but do not return the favor

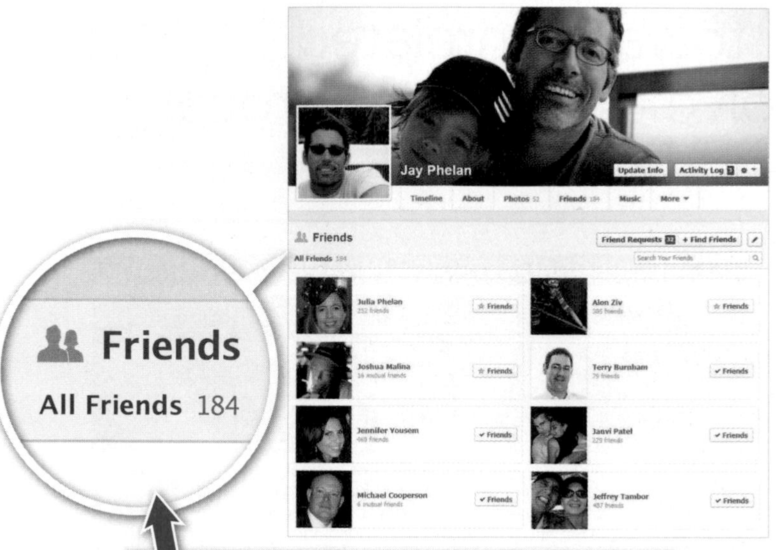
Jay Phelan
Friends
All Friends 184
The importance of keeping track of cheaters and of identifying good potential reciprocity partners may be why humans are so interested in social information and seemingly trivial gossip.

FIGURE 11-12 **Keeping track of valuable social information.**

In the absence of these conditions, selfishness is expected to be the norm among unrelated individuals. The three conditions required for the evolution of reciprocal altruism are not satisfied in many animal species, which may be why altruistic-appearing behavior among unrelated individuals is rare.

Q Why are humans among the few species to have friendships?

As rare as it is in other species, reciprocal altruism is very common among humans. Friendship is built on reciprocity and is almost universal. The opportunity for friendship may be enhanced by our long life span and our ability to recognize thousands of faces and keep track of cheaters. These features are essential for individuals engaging in reciprocal altruism, because an individual becomes very vulnerable when he or she acts in an altruistic manner toward an unrelated individual. The risk is that the altruism will not be repaid, in which case the cheater enjoys greater fitness than the altruist.

Q Why is it easier to remember gossip than physics equations?

While cooperators, those who repay altruistic-appearing behavior, have evolutionary advantages over loners, this advantage disappears if the interaction is unequal. The importance of keeping track of cheaters and of identifying good potential reciprocity partners (i.e., other cooperators) may be why humans are so interested in social information and seemingly trivial gossip. As with many of the complex behaviors we have explored, humans probably use some rules of thumb when making decisions about reciprocal altruism. This might include keeping track of *all* available social information to best identify any promising individuals with whom to engage in reciprocal altruism (FIGURE 11-12).

This includes information about health, generosity, social status, and reproduction. Researchers have even hypothesized that the evolution of maintaining an interest in social information, while valuable in most contexts, may now have some maladaptive manifestations. We may find ourselves interested in and distracted by the social lives of Angelina Jolie, Brad Pitt, or other individuals whom we'll most likely never meet.

TAKE HOME MESSAGE 11.7

» In reciprocal altruism, an individual engages in an altruistic-appearing act toward another individual. Although giving up something of value, the actor does so only when likely to get something of value at a later time. Reciprocal altruism occurs only if individuals have repeated interactions and can recognize and punish cheaters, conditions satisfied in humans but in few other species.

11.8 In an "alien" environment, adaptations produced by natural selection may no longer be adaptive.

The world in which alarm calling evolved did not include biologists with pickup trucks who trapped female squirrels, drove them long distances, and released them into colonies of unrelated individuals. So, when a female squirrel suddenly finds herself in this alien environment, her evolved behaviors can no longer be expected to be

MALADAPTIVE BEHAVIOR

How You Can Help › Syria Refugee Crisis

Syria Refugee Crisis

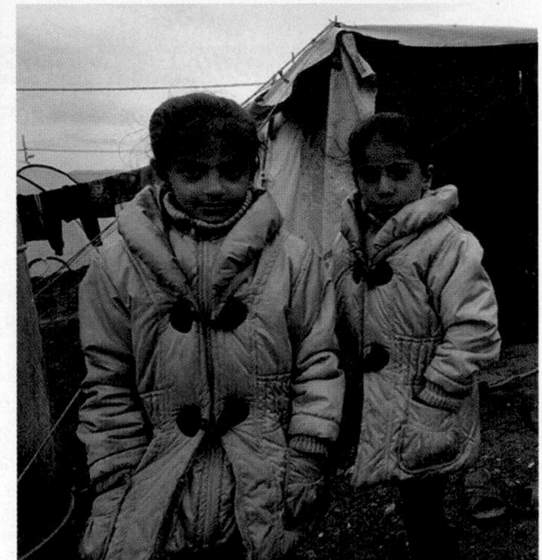

From Harm to Home
DONATE NOW.
Support our work ›
TAKE ACTION.
Advocate for change ›
SIGN UP
Get IRC News.
<your e-mail address>

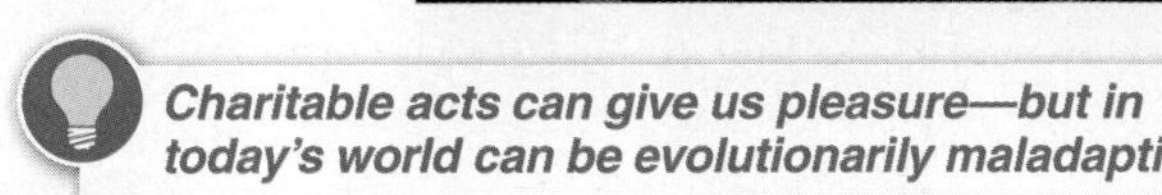

Charitable acts can give us pleasure—but in today's world can be evolutionarily maladaptive.

FIGURE 11-13 Helping those you may never meet.

adaptive—no more than a human could be expected to survive on the moon. And individuals' genes and all the traits they influence cannot change overnight. Adaptation to a new environment takes time, and the more quickly an environment changes, the more likely it is that the adaptations (including evolved behaviors) of a population will no longer be appropriate.

Let's consider another human example: donating money to refugees on another continent. To understand how natural selection could lead to this behavior, we need to understand a bit about human evolutionary history. We know from archaeological deposits that for more than two million years, our ancestors lived as hunter-gatherers in small groups of a few hundred people, at most. Their success depended on joint efforts against predators and in killing prey; being "nice" paid off when the prospect of hunting alone or sleeping outside the camp meant almost certain death. The loners died, so we are descended from those who could work well with others.

Only recently, the blink of an eye in evolutionary time, have humans invented agriculture, industrialization, and the means of food production and distribution. Now, most of us have easy access to unlimited amounts of food, and our typical group size has increased dramatically; on a given day, you may see 10 or even 100 *times* as many people as a hunter-gatherer ancestor might have seen. Additionally, if you were a hunter-gatherer, you would have many opportunities to help and, in turn, to be helped by everyone in your group. Reciprocity paid. And we evolved so that "altruistic" acts gave us pleasure, stimulating parts of our brain in ways that made us want to repeat such actions.

It is with this brain that you approach the issue of giving assistance to refugees halfway across the world or a homeless family somewhere in the United States. It is almost certain that you will not have repeated interactions with any of those people. From an evolutionary perspective, your action will almost certainly not increase your fitness. But in the world in which humans spent most of their evolutionary history, such altruistic-appearing behavior would most likely have been reciprocated at some future time—and thus your instincts guide you to, and reward you for, your kindness (FIGURE 11-13).

From the weight-control difficulties that come from easy access to food, to charitable contributions to people in faraway places, to alarm calling in transplanted squirrels, when organisms of any species find themselves in a situation where there is a **mismatch** between the environment they are in and the environment to which they are evolutionarily adapted, we expect (and see) behaviors and other traits that appear to be (and are) not evolutionarily adaptive. Still, understanding the process of natural selection can help us make sense of these behaviors (which might otherwise seem nonsensical).

TAKE HOME MESSAGE 11.8

» When there is a mismatch between the environment organisms are in and the environment to which they are adapted, the traits (including behaviors) they exhibit are not necessarily evolutionarily adaptive.

11.9 Selfish genes win out over group selection.

Casual observers of nature frequently see individuals acting in ways that appear altruistic, even though these individuals are truly acting—from the perspective of evolutionary fitness—in their own selfish interests. This raises the question of whether evolution ever leads to behaviors that are good for the species or population but detrimental to the individual exhibiting the behavior, a process called **group selection.**

It might seem that evolution would favor individuals that behave in a manner that benefits the group, even if it comes at a cost to the individual's own inclusive fitness. But this does not happen. Behaviors that reduce an individual's fitness (relative to that of other individuals in the population) are not likely to evolve.

Let's look at an example. Imagine a new allele in a population (perhaps arising through mutation) that causes the individual carrying the allele to double its reproductive output, even though this might spell doom for the species as the increasing numbers of individuals overuse their resources. An individual carrying this "selfish" allele will pass on more copies of the allele to its offspring than an individual carrying the alternative, "normal" allele. The selfish offspring, in turn, will pass on the selfish allele at a higher rate than the alternative allele is passed on. Excessive, "selfish," consumption of necessary resources will occur. This scenario may lead to extinction of the species, yet it still occurs (FIGURE 11-14).

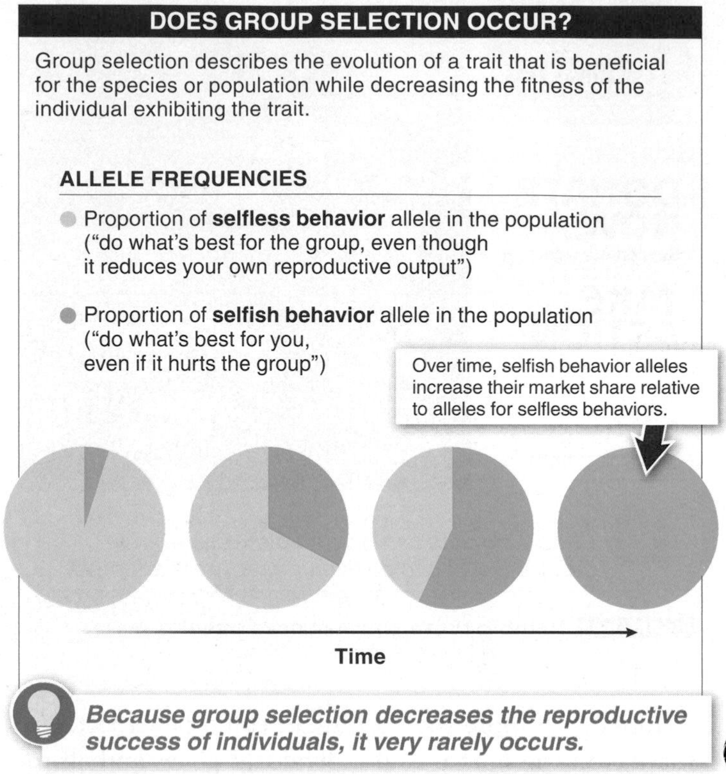

Now consider what happens when a new allele appears that causes an individual to reduce its reproductive output below what is best for that individual's fitness. Such a "selfless" allele does not increase its market share relative to the alternative "normal" or "selfish" allele even if it benefits the group by reducing the likelihood that the population goes extinct. Instead, natural selection generally causes increases in the frequencies of alleles that benefit the *individual* carrying them, even when this comes at the expense of the group. In some special situations, it is possible for natural selection to lead to group selection. But the stringent conditions necessary for this to occur are so rarely found in nature that we almost never see it.

DOES GROUP SELECTION OCCUR?

Group selection describes the evolution of a trait that is beneficial for the species or population while decreasing the fitness of the individual exhibiting the trait.

ALLELE FREQUENCIES

Proportion of **selfless behavior** allele in the population ("do what's best for the group, even though it reduces your own reproductive output")

Proportion of **selfish behavior** allele in the population ("do what's best for you, even if it hurts the group")

Over time, selfish behavior alleles increase their market share relative to alleles for selfless behaviors.

Time

Because group selection decreases the reproductive success of individuals, it very rarely occurs.

FIGURE 11-14 Can a "selfless" allele increase in frequency in a population?

TAKE HOME MESSAGE 11.9

» Behaviors that are good for the species or population but detrimental to the fitness of the individual exhibiting such behaviors are not generally produced in a population under natural conditions.

11.10–11.16

Sexual conflict can result from disparities in reproductive investment by males and females.

A red-crowned crane pair dances as they form and maintain a pair bond.

11.10 There are big differences in how much males and females must invest in reproduction.

As we've seen, many behaviors have evolved that influence how animals interact with each other. One aspect of life that necessarily involves interaction—for sexually reproducing species—is reproduction, from courtship and mating to parental investment and the forming and breaking of ties. In this and the next few sections, we'll explore these behaviors.

How many babies can a woman produce over her lifetime? The number is probably higher than you would guess: a Russian woman had 69 children (in 27 pregnancies). But even this high number is greatly exceeded by the 888 offspring produced by one man (Emperor of Morocco from 1672 to 1727). Among other species of mammals, the pattern is consistent: a male elephant seal can produce 100 offspring, while the maximum produced by a female over her lifetime is 8; a male red deer can produce 24 offspring, while the maximum produced by a female is 14. Here we examine the physical differences between males and females and how they lead to differences in sexual behavior.

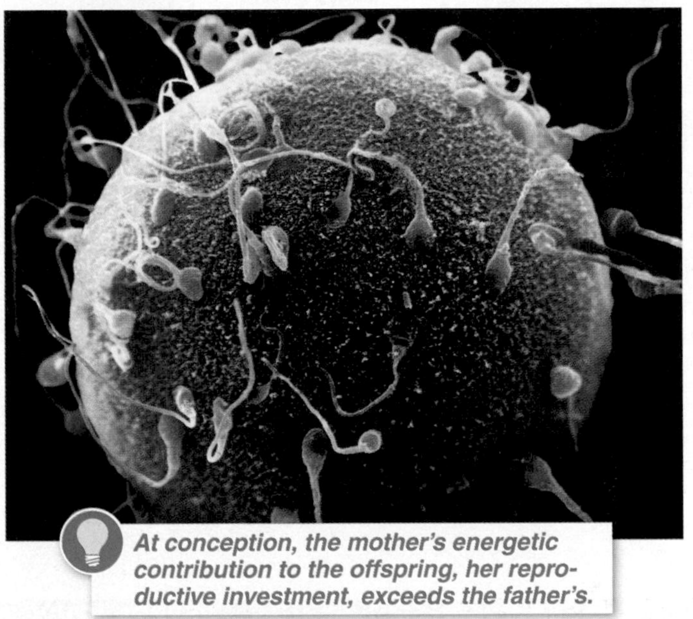

At conception, the mother's energetic contribution to the offspring, her reproductive investment, exceeds the father's.

FIGURE 11-15 **Small sperm, big egg.**

The very definition of "male" and "female" hinges on a physical difference between the sexes. Recall from Chapter 8 that in species with two distinct sexes, a **female** is defined as the sex that produces the larger gamete, while a **male** produces the smaller gamete (FIGURE 11-15). At conception, the mother's material and energetic contribution to the offspring—her **reproductive investment**—exceeds the father's. This is true for all animals, whether mammals, birds, insects, or sharks. (It is also true for plants.) Not only are female gametes larger, they tend to be relatively immobile and produced in smaller numbers, while male gametes are more plentiful and very motile.

The difference in the number of gametes that males and females can produce means that males have the potential to produce many, many more offspring than females. Put another way, a male's **total reproductive output,** the lifetime number of offspring he can produce, tends to increase as the number of females he is able to fertilize increases. A female, on the other hand, does not generally increase

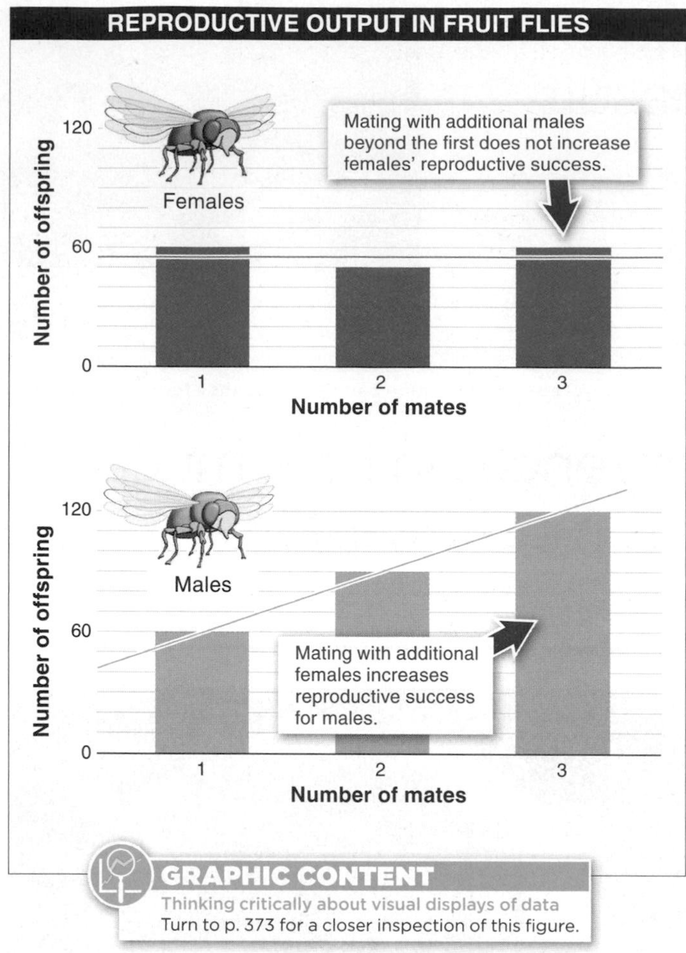

FIGURE 11-16 **Maximizing reproductive success in fruit flies.**

her reproductive success by mating with additional males beyond the first (FIGURE 11-16).

Because additional matings usually lead to greater increases in reproduction (and fitness) for males, selective pressure has resulted in the evolution of some differences in male and female reproductive behavior. For males, one of the most effective ways to maximize their reproductive success is to find and gain access to mating opportunities with additional females. For females, an effective way to maximize their reproductive success is to put more effort into parenting and less into mating.

Q Why do males usually compete for females rather than the other way around?

When it comes to putting more effort into parenting, two physical differences that can exist between males and females are particularly important. First, in species with internal fertilization, which includes most mammals, fertilization takes place in the female. The offspring also grow and develop within the female's body. The amount of energy females invest in reproduction is therefore much greater than males' investment. Females' reproductive investment also limits their reproductive output; they can be pregnant only once at a time.

A second important physical difference between females and males occurs in mammals: lactation takes place in females and not in males (with a very small number of exceptions). In these species, then, nurturing during both pregnancy and lactation can be accomplished only by the female. This difference in reproductive investment has led to the evolution of some very different reproductive behaviors in males and females.

Although across all animals the egg is bigger and more costly to make than the sperm, the physical differences between the sexes in the early nurturing (i.e., gestation and feeding) of offspring can vary considerably. In

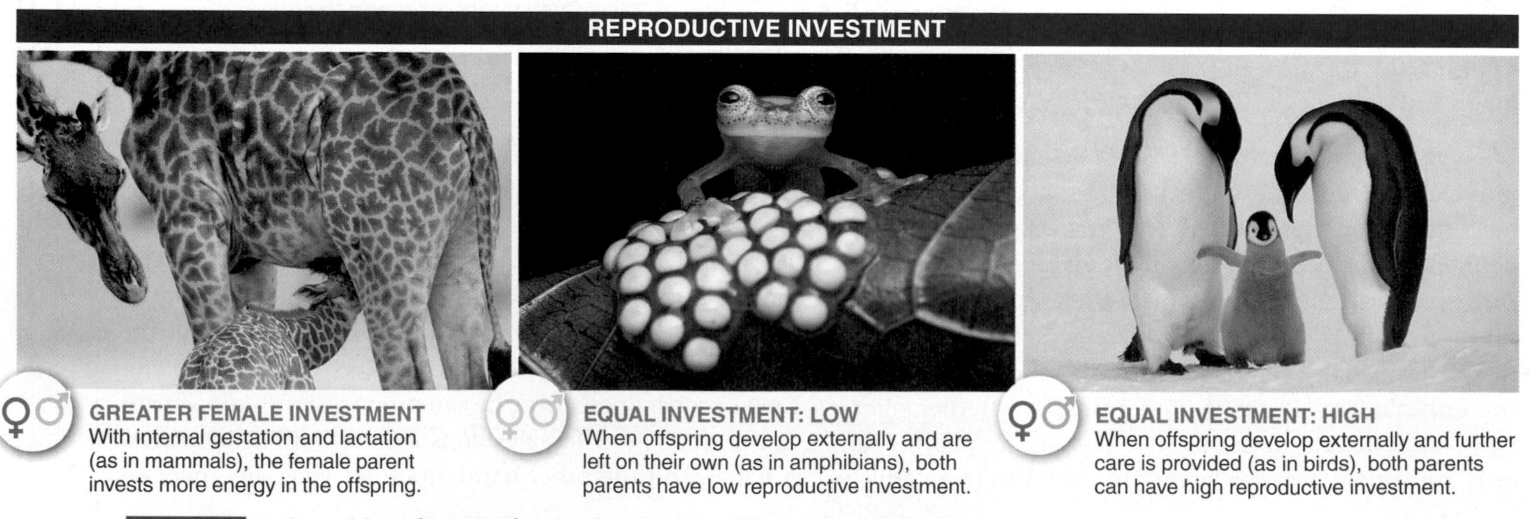

REPRODUCTIVE INVESTMENT
GREATER FEMALE INVESTMENT
With internal gestation and lactation (as in mammals), the female parent invests more energy in the offspring.
EQUAL INVESTMENT: LOW
When offspring develop externally and are left on their own (as in amphibians), both parents have low reproductive investment.
EQUAL INVESTMENT: HIGH
When offspring develop externally and further care is provided (as in birds), both parents can have high reproductive investment.

FIGURE 11-17 **Male and female reproductive investment differs across species.**

some cases, male and female investments become more nearly equal after fertilization. In birds, for example, before emergence of the chicks, much of the development of the fertilized egg is external: the female lays an egg, but the male or the female can protect and incubate the developing embryo. Further, birds do not lactate. Once hatched, the chicks must be fed—a task that can be done by both parents.

Accordingly, in many bird species, the maximum lifetime reproductive output of males and females is similar. In the kittiwake gull, for example, the largest number of offspring produced by a male is 26, while for a female the number is 28. External fertilization in fish and amphibian species further reduces the reproductive investment of the female—she does not spend any energy as the fertilized eggs develop into embryos (FIGURE 11-17).

Another profound consequence of internal fertilization is that a male cannot be 100% certain that any offspring a female produces are his progeny. Because it is possible for a female to mate with multiple males, any of whom could be the father, male mammals and birds will always have some degree of **paternity uncertainty.**

In the next few sections, we continue to explore how the physical differences between males and females in reproductive investment, along with paternity uncertainty, have led to the evolution of differences in male and female reproductive behavior.

TAKE HOME MESSAGE 11.10

» In mammals and many other animals, there are important physical differences between males and females relating to reproduction. Fertilization usually takes place in the female. Lactation (with few exceptions) takes place only in the female. And in species in which fertilization occurs inside the female, males cannot be certain that the offspring are their progeny. These physical differences have led to the evolution of differences in male and female reproductive behavior.

11.11 Males and females are vulnerable at different stages of the reproductive exchange.

Suppose you are on your college campus and a person of the opposite sex comes up to you and says, "Hi. I have been noticing you around campus. I find you very attractive. Would you go out with me tonight?" What percentage of men would answer *yes?* And women? In a study conducted at Florida State University in 1978 and 1982, the percentage answering *yes* was 50% for both males and females.

Now imagine the question was "Would you have sex with me tonight?" Among the men, 75% said *yes;* among the women, not a single one said *yes*. What could explain the difference between the responses of men and women? Studies of male-female differences in selectiveness about sexual partners, drawing on a wide range of human cultures—including many far removed from Western influence, such as the Trobriand Islanders in the South Pacific—reveal a consistent difference in men's and women's willingness to have sex.

Humans, like nearly all mammals, are characterized by a greater initial reproductive investment by females. For females, the cost of a poor mating choice can have significant consequences—pregnancy and lactation, with offspring from a low-quality male or a male who does not provide any parental investment or access to resources (FIGURE 11-18). For a male, the consequences of a poor choice are less dire—little beyond the time and energy involved in mating.

Two differences in the sexual behavior of males and females across the animal kingdom have evolved:

1. The sex with the greater energetic investment in reproduction is more discriminating about mating.
2. Members of the sex with the lower energetic investment in reproduction compete among themselves for access to the higher-investing sex.

When bush crickets mate, the male loses about a quarter of its body weight in contributing a massive ejaculate (the equivalent of nearly 50 pounds of semen in a human), which the female uses for energy (FIGURE 11-19). It can represent up to one-tenth of her lifetime caloric intake.

Not surprisingly, male crickets are very choosy when selecting a mate. They reject small females that would produce relatively few offspring. Females, on the other hand, spend much effort courting males.

FIGURE 11-18 **The choosier sex.**

Males and females differ in their attitude toward mating opportunities. For the female more than the male, the choice of the wrong mate could have expensive consequences.

The sex with the greater reproductive investment must be choosy: a poor choice of mate could be disastrous. The sex with the lower initial reproductive investment (usually males), on the other hand, is not made vulnerable through its mating choices; the matings are of little energetic consequence.

The point of greatest vulnerability for males comes when they provide parental care to offspring. Due to paternity uncertainty, there is some chance that the male may be investing in offspring that are not his own. This has significant evolutionary costs: rather than increasing his own fitness, he is increasing the fitness of another male. Females, conversely, are less vulnerable at the point of providing parental care; a female can be completely certain that the offspring she gives birth to are her own.

When bush crickets mate, the male loses about a quarter of his body weight in contributing a massive ejaculate that the female uses for energy. This makes males very choosy.

FIGURE 11-19 **A costly decision.**

As we explore some general patterns of reproductive behavior among males and females, it is important to keep two critical points in mind.

1. There is tremendous variability across species in male and female behaviors. We can see general behavioral patterns among animals, but these patterns are not necessarily universal features of their biology. A reasonable skepticism is important when identifying and interpreting broad trends across large groups of species.

2. Throughout history, there have been many cases of people using observations and scientific findings to justify a wide variety of discriminatory thoughts and behaviors—for example, that if male mammals "naturally" tend to be less faithful to their mates, this reduces humans' responsibility for their behavior. Such thinking ignores the tremendous variation in behavior among species and the power of cultural norms and socialization to influence human behavior.

TAKE HOME MESSAGE 11.11

» Differing patterns of investment in reproduction make males and females vulnerable at different stages of the reproductive process. This has contributed to the evolution of differences in their sexual behavior. The sex with greater energetic investment in reproduction is more discriminating about mates, and members of the sex with a lower energetic investment in reproduction compete among themselves for access to the higher-investing sex.

11.12 Tactics for getting a mate: competition and courtship can help males and females secure reproductive success.

Female choosiness (and the male-male competition it leads to) tends to increase the likelihood that a female will select only those males that have plentiful resources or relatively high-quality genes. Either is beneficial to the female, allowing her to produce more or better offspring—where "better" may mean increased disease resistance or physical traits that will be found attractive by future mates. Female choosiness is manifested by one or more of four general rules.

1. Mate only after subjecting a male to courtship rituals. In many bird species, the female requires the male to perform an elaborate and time-consuming courtship dance before she will mate with him. For the western grebe, for example, this dance involves fancy dives into water, graceful hovering, and flamboyant twists and turns. The courtship can go on for several days. But if the male passes the audition, he can generally be counted on to stick around to see a brood through hatching and early care (FIGURE 11-20).

2. Mate only with a male who controls valuable resources. Territorial defense is a common means by which males compete for access to females. Among arctic ground squirrels, for example, a female chooses a mate based, in part, on the territory he defends. With greater quality and quantity of resources in his territory, a male is better able to attract females. She will reside there after mating, and her reproductive success can be increased if the territory is rich in resources.

> "It is a truth universally acknowledged, that a single man in possession of a good fortune must be in want of a wife."
>
> —JANE AUSTEN, *Pride and Prejudice,* 1813

3. Mate only with a male who contributes a large parental investment up front. Better than a pledge to commit resources to future offspring is an actual exchange in which a male gives his parental investment up front, in the form of resources. In the hanging fly, for example, a female will not mate with a male unless he brings her a big piece of food, called a **nuptial gift**—usually a dead insect. The larger the food item, the longer she will mate; and the more she eats, the larger the number of eggs she will lay.

FACTORS IN MATE SELECTION

COURTSHIP RITUALS
A female grebe requires the male to perform a courtship dance before she will mate with him.

CONTROL OF VALUABLE RESOURCES
Female yellow-bellied marmots prefer rock outcroppings that provide protection from predators and for hibernation (and which are controlled by dominant males).

GIFTS UP FRONT
A female hanging fly will not mate with a male unless he brings her a large offering of food.

GOOD LOOKS
A female peacock is attracted to a male with the most beautiful tail feathers.

FIGURE 11-20 Four factors that influence a female's choice of mate.

After about 20 minutes of mating, though, when a male has transferred all of the sperm that he can, he is likely to break off the mating and take back whatever remains of the "gift," which he may use to try to attract another mate. Nuptial feeding is common among birds and insects.

4. Mate only with a male that has a valuable physical attribute. Male-male competition for the chance to mate with females can also take a more literal form: actual physical contests. Across the animal kingdom, from dung beetles to hippopotamuses, male-male contests determine the dominance rankings of males. Females then mate primarily with the highest-ranking males.

In a similar process, females sometimes base their choice on some physical attribute, such as antler size in red deer, the bright red chest feathers of frigate birds, or the elaborate tail feathers of the male peacock (see Figure 11-20). In each case, for the female, the physical feature serves as an indicator of the relative quality of the male, possibly because the feature is correlated with the male's health.

With so many examples of male-male fighting as part of courtship rituals, it is reasonable to ask: why is it so rare for females to fight? And why do females generally not have to advertise their health with flashy feathers or other ornamentation? The answer is that as long as females are making the greater investment in reproduction, nearly any male will mate with them. Consequently, there is nothing more to be gained by trying to outcompete other females or otherwise attract the attention of males.

Q Why do so few women get into barroom brawls?

Among humans, social and cultural values have powerful influences over mating behavior, complicating interpretation. We are not lumbering robots, destined to follow some genetic program. Researchers have noted some subtle manifestations of female health and fertility, including waist-to-hip ratios and patterns of facial and body symmetry. There is a rich and complex world of mating tactics, many of which we do not fully understand.

TAKE HOME MESSAGE 11.12

» As a consequence of male-female differences in initial reproductive investment, males tend to increase their reproductive success by mating with many females and have evolved to compete among themselves to get the opportunity to mate.

11.13 Tactics for keeping a mate: mate guarding can protect a male's reproductive investment.

If a male simply abandons a female after mating and searches for other mating opportunities, there is no risk of his investing further in offspring that are not his. This is one way to minimize the potential costs associated with paternity uncertainty. If you don't play, you can't lose.

This strategy, however, is not necessarily the most effective way for a male to maximize his reproductive success. If a male has multiple matings, but no offspring survive, that behavior is not evolutionarily successful. Consequently, in species for which offspring's survival can be enhanced with greater parental investment, there is incentive for males to invest in their offspring, even though such behavior makes him vulnerable to paternity uncertainty.

It is common for a male to reduce this vulnerability through some form of **mate guarding.** In contrast to a female, who can be certain that any offspring emerging from her body are hers, a male inhabits a "danger zone" that lasts as long as the female is fertile. If she mates with any other males during this time, the offspring she produces may not be his. He benefits by minimizing his risk during the danger zone. In this period, mate guarding is particularly common.

Mate-guarding methods range from the simple to the macabre. If a male wants to ensure that a female does not mate with another male, why bother stopping mating at all? Among house flies, even though the male completes the transfer of sperm to the female in 10 minutes of copulation, he does not separate from her for a full hour. Moths go even further and continue to mate for a full 24 hours. And in the extreme case of this strategy, certain frog species continue individual bouts of mating for several months (FIGURE 11-21). In humans, this would be equivalent to almost 10 years for a single round of intercourse.

Q Why do so few females guard their mates as aggressively as males do?

In a slightly subtler form of mate guarding that occurs in reptiles, insects, and many mammalian species, after copulation, males block the passage of additional sperm into the female by producing a copulatory plug. Formed in the female reproductive tract from coagulated sperm and mucus, copulatory plugs can be very effective. Male garter snakes that encounter a female snake with a copulatory plug, for example, do not court or mate with her, treating her instead as if she were not available.

A much more extreme form of mate guarding occurs in the black widow spider: the male breaks off his sexual organ inside the female, preventing her from ever mating again.

In some frog species, individual bouts of mating continue for several months. This is the equivalent in humans of ten-year-long episodes of intercourse!

FIGURE 11-21 Preventing paternity uncertainty.

EXTREME MATE GUARDING

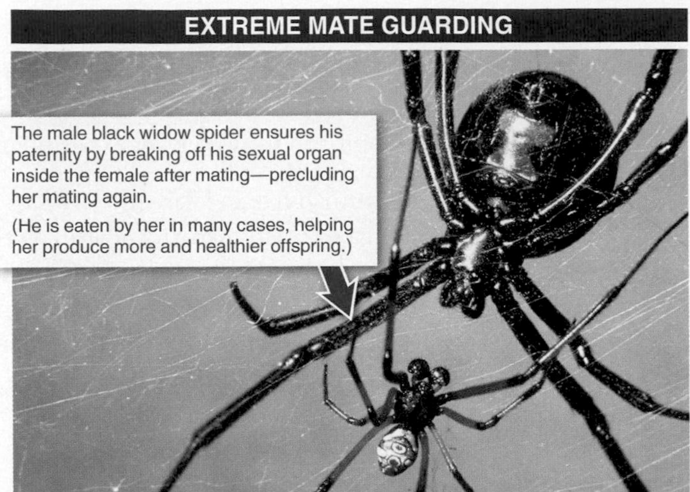

The male black widow spider ensures his paternity by breaking off his sexual organ inside the female after mating—precluding her mating again.
(He is eaten by her in many cases, helping her produce more and healthier offspring.)

FIGURE 11-22 A reasonable trade-off?

Interestingly, when the act is completed, the female usually kills and eats the male (FIGURE 11-22). In sealing his mate's reproductive tract, the male assures himself of fathering the offspring, and in consuming her mate's nutrient-filled body, the female gets resources that help her produce the offspring.

TAKE HOME MESSAGE 11.13

» Mate guarding can, in general, increase reproductive success by reducing additional mating opportunities for a partner, and it can improve a male's reproductive success by increasing his paternity certainty and thus reducing his vulnerability when he invests in offspring.

THIS IS HOW WE DO IT

Developing the ability to apply the process of science

11.14 When paternity uncertainty seems greater, is paternal care reduced?

In the previous section we saw that there should be a relationship between a male's certainty of paternity and his investment in the offspring. When paternity certainty is low, males should benefit by reducing their parental investment and, instead, seeking additional mating opportunities.

It can be difficult to test such predictions experimentally, however, and the results of a manipulation can be difficult to interpret. A powerful strategy to address this challenge is to use an experimental manipulation that makes contrasting

(continued on the following page)

predictions in two different situations. This serves as a sort of internal control in the experiment. Here's how a researcher used this approach in two clever experiments.

The system Bluegill sunfish live in lakes and rivers of North America. Most males reach maturity at age 7 years. During the breeding season, males create a depression in the sandy bottom and chase away almost everything that approaches. Females come and lay eggs—which the male fertilizes—and leave shortly afterward. The male remains to guard the eggs and the small offspring after they hatch. Typically, he doesn't even take a break to forage during this period of parental investment.

About 20% of the bluegill males in a population mature at age 2 years, at a much smaller size. These males, called "cuckold males," hide near the nests of other males and attempt to sneak into the nest, fertilize the eggs, and escape without being detected by the nest "owner."

Male bluegills are unable to distinguish between eggs they have fertilized and eggs fertilized by another male. They can, however, tell whether just-hatched bluegills are their biological offspring (from a chemical cue in the offspring's urine).

Experiment 1 The researcher randomly chose 34 nests. Around each nest he placed two glass containers, each containing two small cuckold males, and left them there for the duration of the egg-laying. The cuckold males couldn't fertilize the eggs, but they could be seen by the male at the nest. As a control, the researcher placed two empty glass containers around each of 20 other nests.

A day after the eggs were laid, the researcher placed a glass container with a predator fish (that eats eggs and just-hatched fish) near each nest, and then evaluated parental care by measuring how vigorously the nest owner defended his eggs over the course of two 30-second periods. Parental care was measured again after the offspring hatched.

In each case, the researcher allotted a "parental care score," reflecting the intensity of the male's guarding of the eggs and defense of the hatched offspring.

Would you expect the presence of cuckold males to influence a nest owner's perception of paternity certainty?

Prediction a: The presence of the cuckold males should reduce the nest owner's paternity certainty and therefore reduce his egg-guarding efforts.

Should the presence of cuckold males influence a nest owner's perception of paternity certainty after the offspring hatch? Why or why not?

Prediction b: After the offspring hatch, the nest owner can determine whether they are his genetic offspring, so he should not exhibit any reduction in parental care relative to males in the control group.

Results of Experiment 1 Manipulation: Cuckold males nearby, but all eggs fertilized by nest owner.

	Parental Care Score	
	Egg Guarding	Offspring Guarding
Prediction	Reduced	Unchanged
Actual results:		
No rivals (control)	80 ± 10	90 ± 10
Rivals present	52 ± 7	95 ± 10
Change in care	Reduced	Unchanged

Experiment 2 The researcher randomly chose 20 new nests and, the day after egg-laying and fertilization, he removed one-third of the eggs from each nest and replaced them with unrelated fertilized eggs from another male's nest. Then, as in the first experiment, the researcher placed a glass container with a predator fish near the nest and evaluated parental care.

Should a nest owner show reduced parental care of eggs that were swapped in from another nest? Why or why not?

Prediction a: Prior to hatching of the eggs, the nest owner should exhibit the same egg-guarding efforts regardless of whether or not the eggs were swapped.

Should a nest owner show reduced parental care of hatched offspring after eggs were swapped? Why or why not?

Prediction b: After the offspring hatch, because the nest owner can determine whether they are his genetic offspring, he should exhibit reduced parental care relative to the control males.

Results of Experiment 2 Manipulation: Eggs swapped with those fertilized by a different male.

	Parental Care Score	
	Egg Guarding	Offspring Guarding
Prediction	Unchanged	Reduced
Actual results:		
Eggs not swapped (control)	90 ± 10	73 ± 9
Eggs swapped	95 ± 10	50 ± 8
Change in care	Unchanged	Reduced

What conclusions can you draw from these results?

In each of the two experiments, males decreased their parental care relative to males in the control group in response to signs that the offspring were less likely to be their own genetic offspring. The experiments provide strong evidence that genetic relatedness to offspring plays an important role in parental care by a male bluegill sunfish.

TAKE HOME MESSAGE 11.14

» Experimental manipulations of the cues of paternity certainty in bluegill sunfish can increase or decrease a male's parental investment in accordance with the prediction that decision making about parental investment reflects perceptions of genetic relatedness.

11.15 Monogamy versus polygamy: mating behaviors can vary across human and animal cultures.

Continuing our tour of animal mating behavior, we turn again to the elephant seal. Each December, the males appear on islands off the coast of northern California, where they compete with each other for possession of the beach. Through bloody fights they establish dominance hierarchies, with the biggest males—which are 13 feet long and weigh more than 2 tons—generally winning.

In mid-January, the females arrive and are ready to mate. They congregate in large groups on just a few prime beaches. Because the females stick close together, the biggest males, who control the prime beaches, gain access to almost all of the females (FIGURE 11-23). In one study that observed 115 males, the 5 highest-ranking males fathered 85% of the offspring. While nearly every female will mate and produce offspring, the majority of males never get the chance to mate during the 10–20 years of their lives.

The elephant seals' mating pattern exemplifies **polygamy,** a system in which some individuals attract multiple mates while other individuals attract none. Polygamous mating systems can be subdivided into **polygyny,** in which individual males mate with multiple females, and **polyandry,** in which individual females mate with multiple males. Polygamy can be contrasted with **monogamy,** in which most individuals mate with and remain with just one other individual. Polygamy and monogamy are two types of **mating systems,** which describe the patterns of mating behavior in a species. In this section, we explore the features of environments and species that influence mating systems and survey the range of mating systems observed in nature.

As we have seen, throughout the animal world, as a result of their relative parental investment, females are choosy about which males they mate with, and males compete for access to mating opportunities. Not surprisingly, multiple females often end up selecting the same male—usually a male on a territory rich in resources or a male with unusually pronounced physical features, such as antler size.

POLYGAMY IN ELEPHANT SEALS

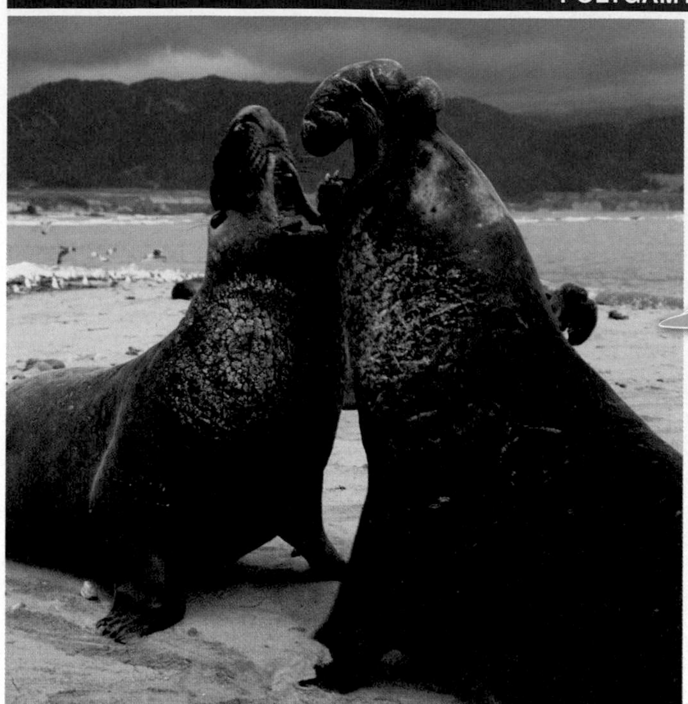

FIGURE 11-23 **King of the beach.** The biggest, best-fighting male elephant seals control the beach and mate with as many females as possible.

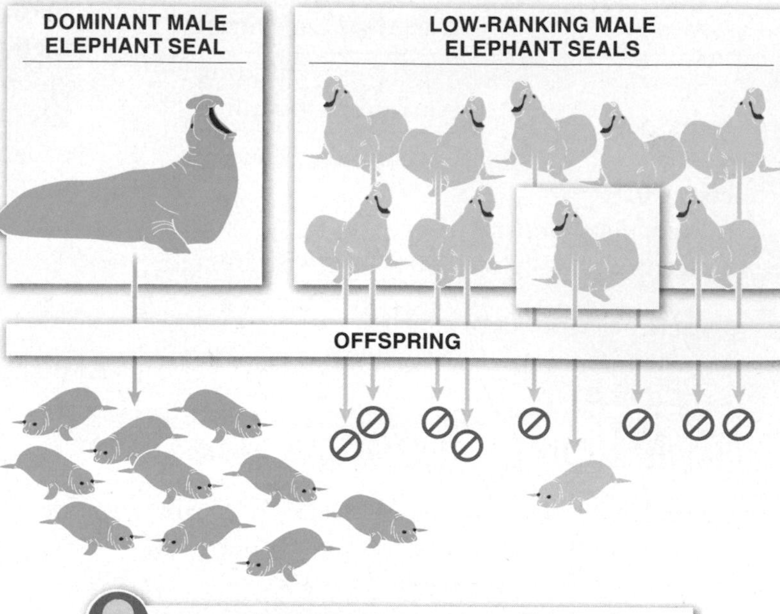

In a polygamous mating system, some individuals have multiple mates, while others have few or none.
DOMINANT MALE ELEPHANT SEAL
LOW-RANKING MALE ELEPHANT SEALS
OFFSPRING
In an extremely polygynous mating system, such as in elephant seals, the vast majority of males have no reproductive success at all.

Such female selection is the reason for the fancy ornamentation we see in male peacocks. Although the male peacock's tail is like a giant bull's-eye to predators, the number of eyespots on the tail is directly related to how well he can attract a mate: below 140 eyespots, he gets no mates; at 150, he gets two to three mates on average; with 160 eyespots, he gets six or more mates (see Figure 11-20).

Identifying a population's mating systems is not as easy as the elephant seal example might lead us to believe. Three issues, in particular, complicate the task. First, there are often differences between animals' mating behavior and their bonding behavior. It may seem that a male and female have formed a **pair bond**—in which they spend a high proportion of their time together, often over many years, sharing a "home" and contributing equally to parental care of offspring in what appears to be a monogamous relationship. Closer inspection (often including DNA analysis of the offspring), however, sometimes reveals that the male and/or the female may be mating with other individuals in the population, and perhaps the mating system is better described as a variation on polygamy. A second difficulty in defining a species' mating system arises because the mating system may vary within the species. Some individuals may be monogamous, while others are polygamous. And this may even change over the course of an individual's life. A third difficulty is that males and females often differ in their mating behavior. In the elephant seals described above, it could be said that the females are all mating monogamously, while the males are polygamous.

Examination of birds and mammals in general, however, reveals one sharp split. The vast majority of female mammals make a greater parental investment than male mammals. In birds, females and males have a more equal parental investment. Does the difference in parental investment patterns in birds and mammals lead to different mating systems? Yes. In mammals, polygyny is the most common mating system across all large groups, from rodents to primates. Polygyny is a consequence of the significant female investment and lesser male investment. In birds, more than 90% of the approximately 10,000 bird species we know about appear to be monogamous (although, in most cases, it is serial monogamy) (FIGURE 11-24).

Q Are humans monogamous or polygamous?

And what of humans? Across a variety of cultures, males consistently have greater variance in reproductive success than females—that is, some males have very high reproductive success and many others have

MONOGAMOUS BIRDS

Of the approximately 10,000 species of birds, more than 90% appear to be monogamous.

FIGURE 11-24 **Let's stay together.** Parental investment that is roughly equal often leads to monogamous mating behavior in birds.

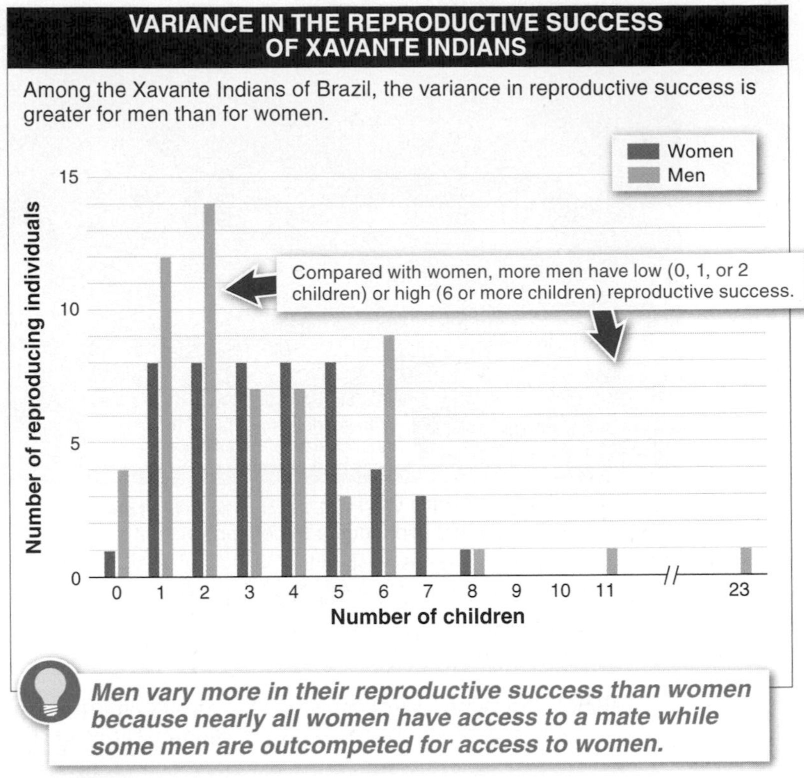

VARIANCE IN THE REPRODUCTIVE SUCCESS OF XAVANTE INDIANS
Among the Xavante Indians of Brazil, the variance in reproductive success is greater for men than for women.
Women
Men
Number of reproducing individuals
0
5
10
15
0 1 2 3 4 5 6 7 8 9 10 11 23
Number of children
Compared with women, more men have low (0, 1, or 2 children) or high (6 or more children) reproductive success.
Men vary more in their reproductive success than women because nearly all women have access to a mate while some men are outcompeted for access to women.

FIGURE 11-25 **Larger male variance in reproductive success.**

little or none. FIGURE 11-25, for example, presents data from one of the first investigations of this phenomenon. It is from a study of the Xavante Indians of Brazil, a pre-industrial population subsisting primarily as hunter-gatherers, with no access to reliable birth control and with almost no contact with Western cultures. The *average* number of children (3.6) does not differ between men and women, but the *range* does: some men have very large numbers of offspring (as many as 23!), and many have none. There is significantly less variability in reproductive success from one woman to another. A 2010 study using data from 7,710 women and men living in the contemporary United States similarly found significantly greater variance in reproductive success among men than among women.

Some differences in the variance in reproductive success among males and females, which can be significantly influenced by socialization and other cultural forces, are generally associated with polygynous mating systems. However, the difference between the sexes found in the Brazilian study is quite small when compared with many other mammalian species and is close to that seen in populations with a monogamous mating system. Humans, consequently, seem to have a mating system close to, but not completely, monogamous.

TAKE HOME MESSAGE 11.15

» Mating systems—monogamy, polygyny, and polyandry—describe the variation in number of mates and the reproductive success of males and females. They are influenced by the relative amounts of male and female parental investment.

11.16 Sexual dimorphism is an indicator of a population's mating behavior.

Male elephant seals are three to four times the size of female elephant seals. In contrast, the males and females of the majority of bird species (but certainly not all) are the same size. In some of these species, even expert bird-watchers have difficulty distinguishing between the sexes, except when the female bird is carrying eggs (FIGURE 11-26). When the sexes of a species do differ in size or appearance, this is called **sexual dimorphism.** Why do species have dramatic differences in the degree to which males and females resemble each other?

SEXUAL DIMORPHISM

With sexual dimorphism, the sexes differ in size or appearance.

SEXUAL MONOMORPHISM

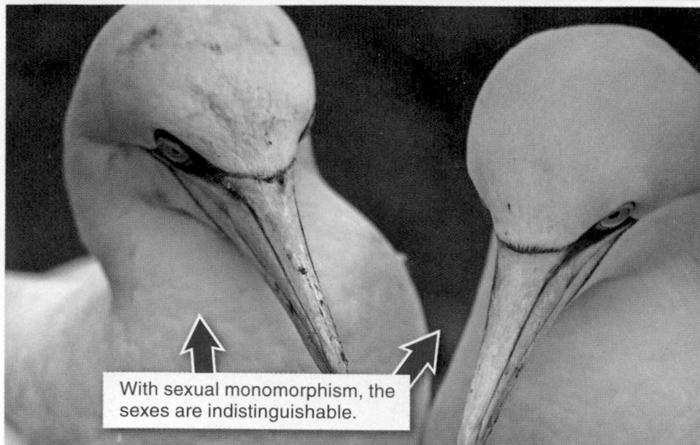
With sexual monomorphism, the sexes are indistinguishable.

BEHAVIORS ASSOCIATED WITH SEXUAL DIMORPHISM

- One parent invests more in caring for the offspring.
- Mating system tends toward polygamy.
- One sex (usually females) is choosier when selecting a mate.
- One sex (usually males) competes for access to mating opportunities with the other sex.

BEHAVIORS ASSOCIATED WITH SEXUAL MONOMORPHISM

- Both parents invest (approximately) equally in caring for the offspring.
- Mating system tends toward monogamy.
- Both sexes are equally choosy when selecting a mate.

FIGURE 11-26 **The same or not the same?** In monogamous species, the males and females are similar in appearance and behavior. In polygamous species, they tend to differ.

Body size is an important clue to mating behavior. We have seen that male elephant seals have a winner-take-all tournament for control of the beach, and as the winner, a male has access to mating opportunities. Because the largest individual will have the most offspring, there is selection for larger and larger body size. There is no selection for larger body size among female elephant seals, because females of any size can mate. Hence the dramatic size difference between the sexes.

Coloration, too, can be a clue to mating behavior. Because females of some polygynous species choose the males with the brightest or flashiest coloration rather than the largest males, male-male competition sometimes results in differences in physical appearance between the sexes.

Q It's almost impossible to distinguish males from females in most bird species. Why does this tell us they are monogamous?

What happens when there is little male-male competition for mates? Among bird species, males can provide significant investment in offspring. Eggs must be incubated, and because many chicks emerge in a poorly developed condition, it takes two parents to raise them. So a pair tends to stay together for a season or longer. If each female can pair up with only one male, there isn't much competition for mates. And without male-male competition, there is little selection for increased size. As a consequence, little sexual size dimorphism occurs in most bird species.

Q Men are bigger than women. What does that suggest about our evolutionary history of monogamy versus polygamy?

In turn, we can predict a bit about the parental practices of a species just by looking at a picture of a male and a female and examining the ratio of body sizes. If the two are dramatically different, as in elephant seals, we can hypothesize that the smaller sex is doing most of the care of the offspring and that the species is more likely to be polygamous than monogamous.

TAKE HOME MESSAGE 11.16

» Differences in the level of competition among individuals of each sex for access to mating opportunities can lead to the evolution of male-female differences in body size and other aspects of appearance. In polygynous species, this results in larger males than females. In monogamous species, there are few such sex differences.

11.17–11.18

Communication and the design of signals evolve.

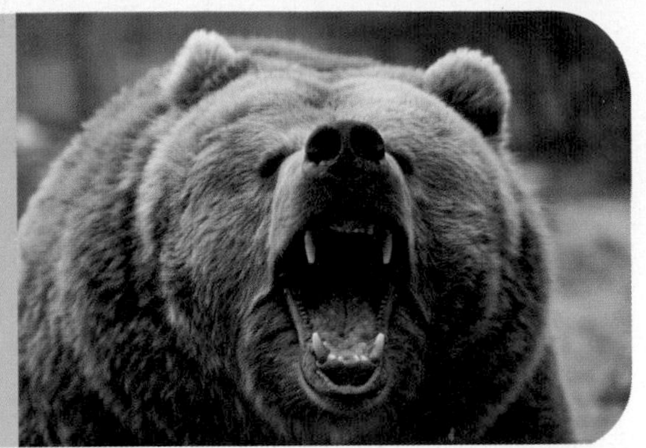

A grizzly bares its teeth in warning.

11.17 Animal communication and language abilities evolve.

The challenge of communication is central to the evolution of animal behaviors, whether they relate to cooperation, conflict, or selecting a mate. In this section and the next, we explore the evolution of a variety of methods of animal communication and how communication can influence fitness and the evolution of other behaviors.

Not all animals need to open their mouths when they have something important to communicate. When the female silkworm moth is ready to mate, for example, she releases a potent chemical called bombykol into the air. As molecules of the chemical come in contact with the bushy antennae of a male silkworm moth—who might be three miles away—he recognizes the mate attractant and responds by flying upwind, looking for the source of the chemical, until he zeroes in on the female and they mate.

This is just one example—a chemical one—of **communication,** an action or signal on the part of one individual that informs or alters the behavior of another individual (FIGURE 11-27). A huge range of animal behaviors require

FORMS OF ANIMAL COMMUNICATION

CHEMICAL COMMUNICATION
Pheromones released by one individual and detected by another, using its antennae, can trigger behavioral responses.

ACOUSTIC COMMUNICATION
Sounds, such as the roar of a lion, are a common method of triggering behavioral responses.

VISUAL COMMUNICATION
Organisms can convey information, such as threat or receptivity, with visual displays. Here, a balloonfish puffs up its body in response to a predator.

FIGURE 11-27 **Chemical, acoustic, and visual communication.** Many animal behaviors require communication: the ability to convey information to and receive information from other animals.

communication, including defending territory, alerting individuals to threats, announcing one's readiness for mating, establishing one's position in a dominance hierarchy, and caring for offspring.

Animals use many different types of signals to communicate. Three types are most common.

1. Chemical. Molecules released by an individual into the environment that trigger behavioral responses in other individuals are called **pheromones.** They include the airborne mate attractant of the silkworm moth, territory markers such as those in dog urine, and trail markers used by ants.

2. Acoustic. Sounds that trigger behavioral responses are abundant in the natural world. These include the alarm calls of Belding's ground squirrels described earlier, the complex songs of birds, whales, frogs, and crickets vying for mates, and the territorial howling of wolves.

3. Visual. Individuals often display visual signals of threat, dominance, or health and vigor. Examples include the male baboon's baring of his teeth and the vivid tail feathers of the male peacock.

With increasingly complex forms of communication, increasingly complex information can be conveyed. One extreme case is the honeybee **waggle dance.** When a honeybee scout returns to the hive after successfully locating a source of food, she performs a set of maneuvers on a honeycomb that resemble a figure 8, while wiggling her abdomen. Based on the angle at which she dances (relative to the sun's position in the sky) and the duration of her dance, the bees that observe the dance are able to leave the hive and locate the food source.

The complexity and power of the waggle dance raises an important question: at what point does communication become language? **Language** is a very specific type of communication in which arbitrary symbols represent concepts, and a system of rules, called grammar, dictates the way the symbols can be manipulated to communicate and express ideas (FIGURE 11-28).

It's not always easy to identify language. Consider one type of communication in vervet monkeys. Vervet monkeys live in groups, and when an individual sees a predator coming, it makes an alarm call to warn the others. Unlike in most other alarm-calling species, the vervet monkey's alarm call differs depending on whether the predator is a hawk, a snake, or a leopard. And the response of other vervet monkeys to the call is appropriate to the type of predator approaching. The snake alarm causes individuals to stand on their toes and look down. The leopard alarm causes individuals to quickly climb the nearest tree. And the hawk alarm causes vervet monkeys to look to the sky.

Does alarm calling in vervet monkeys or the honeybees' waggle dance constitute language? This is debatable. Is the form of American Sign Language taught to chimpanzees and gorillas a language (see Figure 11-28)? There are opinions on both sides. Interestingly, the great apes have shown a capacity for creating new signs, expressing abstract ideas, and referring to things that are distant in time or space—important features of human language. One thing that is certain, however, is that human language is near the most complex end of the communication continuum.

Language influences many of the behaviors discussed throughout this chapter. The evolution of reciprocal

COMPLEX FORMS OF ANIMAL COMMUNICATION

Honeybee waggle dance | *Orangutan and trainer giving hand signals* | *Children listening to a teacher*

FIGURE 11-28 **Dancing, signing, and speaking.** A variety of ways have evolved in animals to convey complex information to others.

altruism, for example, may be influenced by language, as this makes it easier for individuals to convey their needs and resources to each other. Similarly, in courtship and the maintenance of reproductive relationships, language gives individuals a tool for conveying complex information relating to the resources and the value they may bring to the interaction.

TAKE HOME MESSAGE 11.17

» Methods of communication—chemical, acoustic, and visual—have evolved among animal species, enabling them to convey information about their condition and situation. These abilities influence fitness and the evolution of almost all other behaviors.

11.18 Honest signals reduce deception.

Among Natterjack toads, found in northern Europe, males produce booming calls to attract females. Because females desire large males and size determines the volume of the calls, females seek the loudest croaker. They are drawn by calls that can be heard a mile away and are often louder than the legal noise limit for a car engine (FIGURE 11-29).

The Natterjack's call is an **honest signal,** a signal that cannot be faked and is given when both the individual making the signal and the individual responding to it have the same interests. An honest signal is one that carries the most accurate information about an individual or situation, and animals that respond to signals of any sort have evolved to value honest signals. A loud Natterjack's call, for example, cannot be faked by a small Natterjack toad.

When one animal can increase its fitness by deceiving another, deception can also be expected to evolve. An example might be baby birds begging for food from their parent. If an allele that causes a chick to exaggerate its need for food leads to faster growth and better health for that individual, the allele is likely to increase in frequency in the population.

Communication and signaling, therefore, are features of populations that are continually evolving. Over time, this evolutionary "arms race" between honest signals and deception can lead to both increasingly unambiguous signals and ever more sophisticated patterns of deception.

In this chapter, we've seen how our understanding of the behavior of animals, including humans, has been greatly expanded by applying scientific thinking, an experimental approach, and careful consideration of the process by which natural selection shapes populations. We return to an important point highlighted at the beginning of this chapter: an organism's phenotype is not limited to its physical traits but includes its behaviors as well. Consequently, behaviors respond to selective pressures and evolve.

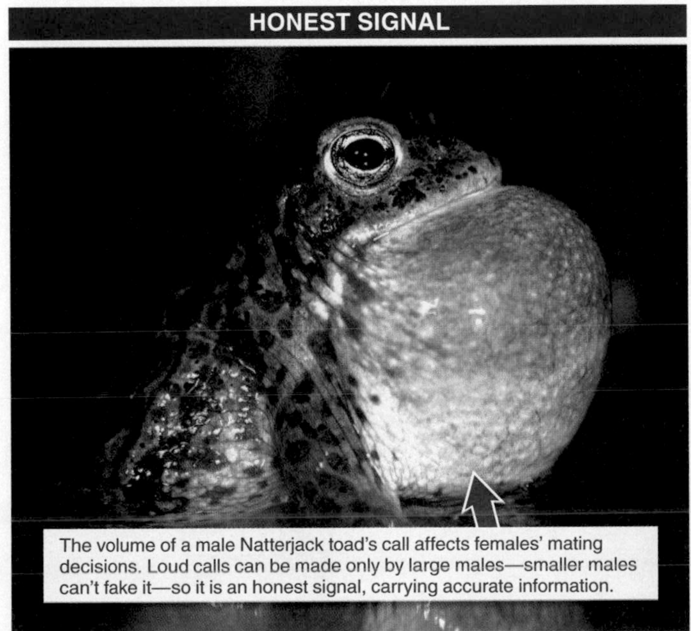

HONEST SIGNAL
The volume of a male Natterjack toad's call affects females' mating decisions. Loud calls can be made only by large males—smaller males can't fake it—so it is an honest signal, carrying accurate information.

FIGURE 11-29 **Loud and true.**

TAKE HOME MESSAGE 11.18

» Animals have evolved to rely primarily on signals that cannot easily be faked, in order to gain the maximum amount of information.

STREET BIO

Using evidence to guide decision making in our own lives

How to win friends and influence people

As we have seen, in the animal world, it is very rare to find individuals engaging in costly behaviors that benefit unrelated individuals; with few exceptions, we see little that resembles animal "friendship."

Humans are a rare exception—every day, in myriad ways, we see friendly behaviors. Just as we put money in the bank for a rainy day, we buffer ourselves from the world's uncertainties by storing goodwill in our neighbors. Cooperators have evolutionary advantages over loners.

Q: Why is "kindness" a risky behavior evolutionarily? With each costly act we perform for unrelated individuals, strangers and friends alike, there is the risk that our efforts will be in vain, our energies lost. For that reason, in our evolution as reciprocal altruists, we acquired a hesitancy to stick our necks out.

Q: How can we help others feel less vulnerable and more willing to cooperate? Our ability to override the impulses that often push us toward selfishness is one of the hallmarks of being human. By taking steps, often absurdly simple, to address the unconscious vulnerabilities that others feel, we can increase the likelihood that they choose cooperation, reciprocity, and friendship.

- Learn and use other people's names. It tells them you recognize them specifically and, consequently, that you understand exactly who it is that you "owe." Smile and make eye contact—even when attempting to change lanes in traffic. These gestures feel like the beginning of a relationship, thus stimulating favor-granting instincts.
- Embrace etiquette. Acknowledge your debts to others. Be effusive and public in your thanks for kindnesses done to you. Send thank-you notes.
- Take the first step in reciprocity. Human cooperation is so tied to reciprocal exchange that even tiny gestures of good faith can play an important role in building relationships. Whenever possible, give gifts. Even small ones. Even when they are not required. (Especially when they are not required.)
- Develop a good reputation. Become known for generosity, for loyalty, for remembering and acknowledging kindnesses done to you.

Can you think of other steps?

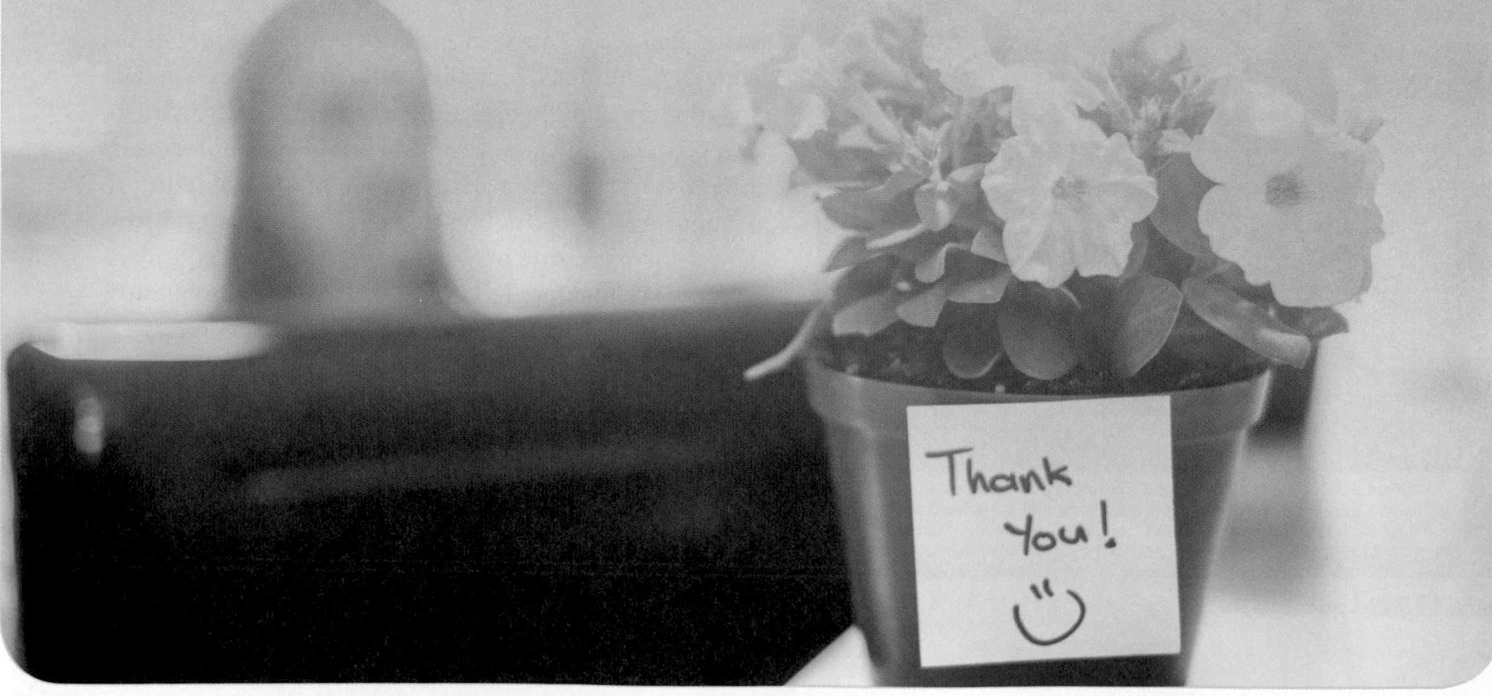

GRAPHIC CONTENT

Thinking critically about visual displays of data

1 According to the top bar graph, how many offspring does a female with one mate produce? How does this compare with the number of offspring produced by a male with one mate?

2 How many offspring are produced by females with two mates? Why is this lower than the number for females mating with just one male? What additional data would help you answer this question?

3 How many offspring do you think a female with four mates would have? What about a male with four mates?

4 According to this figure, what limits the reproductive output of a female fruit fly? What limits a male's reproductive output?

5 Why is it important to have the same axes on both graphs?

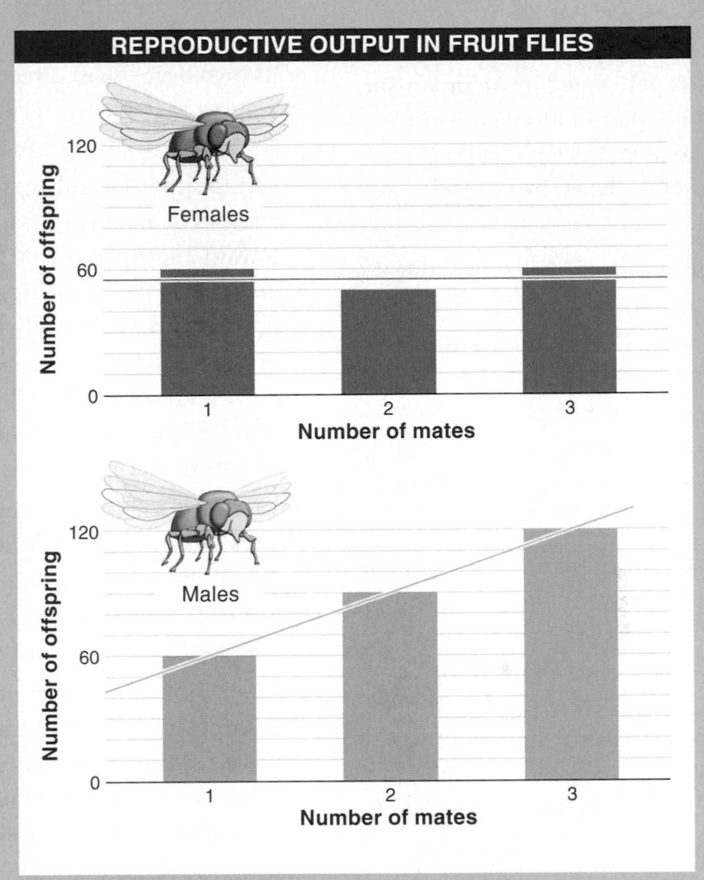

See answers at the back of the book.

KEY TERMS IN EVOLUTION AND BEHAVIOR

altruistic behavior, p. 349
behavior, p. 344
communication, p. 369
female, p. 357
fixed action pattern, p. 346
group selection, p. 356
honest signal, p. 371
inclusive fitness, p. 352
innate behavior, p. 346
instinct, p. 346
kin selection, p. 350
language, p. 370
learning, p. 347
male, p. 357
mate guarding, p. 362
mating system, p. 365
mismatch, p. 355
monogamy, p. 365
nuptial gift, p. 361
pair bond, p. 366
paternity uncertainty, p. 359
pheromone, p. 370
polyandry, p. 365
polygamy, p. 365
polygyny, p. 365
prepared learning, p. 347
reciprocal altruism, p. 350
reproductive investment, p. 357
sexual dimorphism, p. 367
sign stimulus, p. 346
total reproductive output, p. 357
waggle dance, p. 370

BRIEF SUMMARY

Behaviors are traits that can evolve.

• Behavior encompasses any and all of the actions performed by an organism. Behavior is a part of an organism's phenotype, and as such it can be produced and shaped by natural selection.

• Instincts, or innate behaviors, develop without any environmental input and are present in all individuals in a population, with little variation. A fixed action pattern, a type of innate behavior, is a sequence of acts that requires no learning, does not vary, and runs to completion once started.

• Many behaviors are influenced by an individual's environment, requiring learning, and are often modified in response to experiences. Organisms are well-prepared to learn behaviors that were important to the reproductive success of their ancestors, and less prepared to learn behaviors irrelevant to their species' evolutionary success.

• Natural selection of behaviors does not require the organism to consciously try to maximize its reproductive success.

Cooperation, selfishness, and altruism can be better understood with an evolutionary approach.

• Many behaviors in the animal world appear to be altruistic. In almost all cases, the behaviors have evolved as a consequence of either kin selection or reciprocal altruism and are beneficial to the individual's fitness.

• Kin selection is apparently-altruistic behavior in which an individual that assists a genetic relative compensates for its own decrease in direct fitness by helping increase the relative's fitness and, consequently, its own inclusive fitness.

• In reciprocal altruism, an individual engages in an altruistic-appearing act toward another individual. Although giving up something of value, the actor does so only when likely to get something of value at a later time. Reciprocal altruism occurs only if the individuals have repeated interactions and can recognize and punish cheaters.

• When there is a mismatch between the environment that organisms are in and the environment to which they are adapted, the traits they exhibit are not necessarily evolutionarily adaptive.

• Behaviors that are good for the species or population but detrimental to the fitness of the individual exhibiting such behaviors are not generally produced in a population under natural conditions.

Sexual conflict can result from disparities in reproductive investment by males and females.

• In mammals and many other animals, there are physical differences between males and females relating to fertilization, lactation, and paternity certainty. These differences have led to the evolution of differences in male and female reproductive behavior.

• The sex that makes a greater energetic investment in reproduction is more discriminating about mates, and members of the sex with a lower energetic investment in reproduction compete among themselves for access to the higher-investing sex.

• Mate guarding can increase reproductive success by reducing additional mating opportunities for a partner and can improve a male's reproductive success by increasing his paternity certainty.

• Mating systems—monogamy, polygyny, and polyandry—describe the variation in number of mates and the reproductive success of males and females. These systems are influenced by the relative amounts of male and female parental investment.

• Differences in the level of competition among individuals of each sex for access to mating opportunities can lead to the evolution of male-female differences in body size and appearance.

Communication and the design of signals evolve.

• Methods of communication—chemical, acoustic, and visual—have evolved among animal species, enabling them to convey information about their condition and situation.

• Animals have evolved to rely primarily on honest signals that cannot easily be faked, in order to gain the maximum amount of information.

CHECK YOUR KNOWLEDGE

Short Answer

1. Why is egg-retrieval behavior in geese referred to as a fixed action pattern?

2. Which type of behaviors are we most well prepared to learn?

3. Why is the risky behavior of alarm calling ever an evolutionarily advantageous behavior for a Belding's ground squirrel?

4. Describe two conditions that are essential if reciprocal altruism is to evolve.

5. An organism in an environment that differs from the environment to which it is evolutionarily adapted may exhibit maladaptive behaviors. Using an example, explain why.

6. Will natural selection favor a behavior that leads to a better outcome for the population but not for the individual? Explain why or why not.

7. Describe two ways in which, in mammals, males and females initially make unequal energetic investments in reproduction.

8. Compare and contrast the stages of the mammalian reproductive process at which males and females are most vulnerable.

9. In a polygynous mating system, many males produce no offspring. Why?

10. Describe the three types of mating systems.

11. Female moorhens are larger and more aggressive than males. They also compete among themselves for access to the smaller, fatter males. Which sex do you think provides more parental care? Explain your answer.

Multiple Choice

1. An animal will preferentially feed on:

a) the largest prey it can find.

b) the prey that provides the most energy relative to effort.

c) the smallest prey it can find.

d) the prey that provides the most calories relative to the prey size.

e) the prey that provides the most calories relative to the predator size.

2. From an evolutionary perspective, behavior can best be viewed as:

a) a trait that arises by learning, not by natural selection.

b) non-heritable.

c) a trait subject to drift and mutation, but not natural selection.

d) part of the phenotype.

e) All of the above are correct.

0 EASY 61 HARD 100

3. During the breeding season, the sight of a red belly on any other stickleback triggers aggressive behavior in a male stickleback. This is called:

a) a fixed action pattern. b) prepared learning.

c) aggressive conditioning. d) sexual selection.

e) male-male competition.

0 EASY 51 HARD 100

4. In Belding's ground squirrels, why are females much more likely than males to make alarm calls?

a) Belding's ground squirrels have a sex ratio that is biased toward females.

b) Females invest more in food storage, so they are more likely to lose their lives or their food if a predator attacks.

c) Belding's ground squirrels have a sex ratio that is biased toward males.

d) Females tend to remain in the area where they were born, so the females that call are warning their own kin.

e) Males forage alone, so their alarm calls would be useless.

0 EASY 18 HARD 100

5. Altruistic behavior in animals may result from kin selection, a process in which:

a) genes promote the survival of copies of themselves when behaviors by animals possessing those genes assist other animals that share those genes.

b) aggression within sexes increases the survival and reproduction of the fittest individuals.

c) companionship is advantageous to animals because, in the future, they can recognize and help individuals that have helped them.

d) aggression between the sexes increases the survival and reproduction of the fittest individuals.

e) companionship is advantageous to animals because, in the future, they can recognize individuals that have helped them and request help once again.

0 EASY 25 HARD 100

6. In a situation where males guard eggs and care for the young without help from the female, which of the following statements would most likely be correct?

a) The males are larger and more brightly colored in order to attract the very best females.

b) The males and females are equally brightly colored, but males court females aggressively.

c) The population is monogamous with no sexual dimorphism.

d) A single male controls a harem of females to which he has exclusive reproductive access.

e) The females are more brightly colored than males and court males aggressively.

0 EASY 44 HARD 100

7. Mate guarding is a reproductive tactic that functions to:

a) reduce paternity uncertainty.

b) increase the female's investment in offspring.

c) reduce the male's reproductive investment.

d) reduce the female's fitness.

e) increase the number of mates to which a male has access.

0 EASY 32 HARD 100

8. Relative to birds, more mammalian species are:

a) polygynous. b) monogamous.

c) polyandrous. d) hermaphroditic.

e) sexually monomorphic.

0 EASY 59 HARD 100

9. In a species such as pigeons, in which males are almost indistinguishable in appearance from females, the most likely mating system is:

a) monomorphism. b) monogamy.

c) polygyny. d) polyandry.

e) It is impossible to predict the mating system with only this information.

0 EASY 37 HARD 100

10. Polygynous species:

a) usually employ external fertilization.

b) are usually sexually dimorphic, with males larger and more highly ornamented.

c) are usually sexually dimorphic, with females larger and more highly ornamented.

d) usually have males and females that are physically indistinguishable.

e) are more commonly found among birds than among mammals.

Ch12

Life on earth most likely originated from non-living materials.

12.1 Complex organic molecules arise in non-living environments.

12.2 Cells and self-replicating systems evolved together to create the first life.

12.3 *This is how we do it:* Could life have originated in ice, rather than in a "warm little pond"?

Species are the basic units of biodiversity.

12.4 What is a species?

12.5 How do we name species?

12.6 Species are not always easily defined.

12.7 How do new species arise?

Evolutionary trees help us conceptualize and categorize biodiversity.

12.8 The history of life can be imagined as a tree.

12.9 Evolutionary trees show ancestor-descendant relationships.

12.10 Similar structures don't always reveal common ancestry.

Macroevolution gives rise to great diversity.

12.11 Macroevolution is evolution above the species level.

12.12 The pace of evolution is not constant.

12.13 Adaptive radiations are times of extreme diversification.

12.14 There have been several mass extinctions on earth.

An overview of the diversity of life on earth: organisms are divided into three domains.

12.15 All living organisms are classified into one of three groups.

12.16 The bacteria domain has tremendous biological diversity.

12.17 The archaea domain includes many species living in extreme environments.

12.18 The eukarya domain consists of four kingdoms: plants, animals, fungi, and protists.

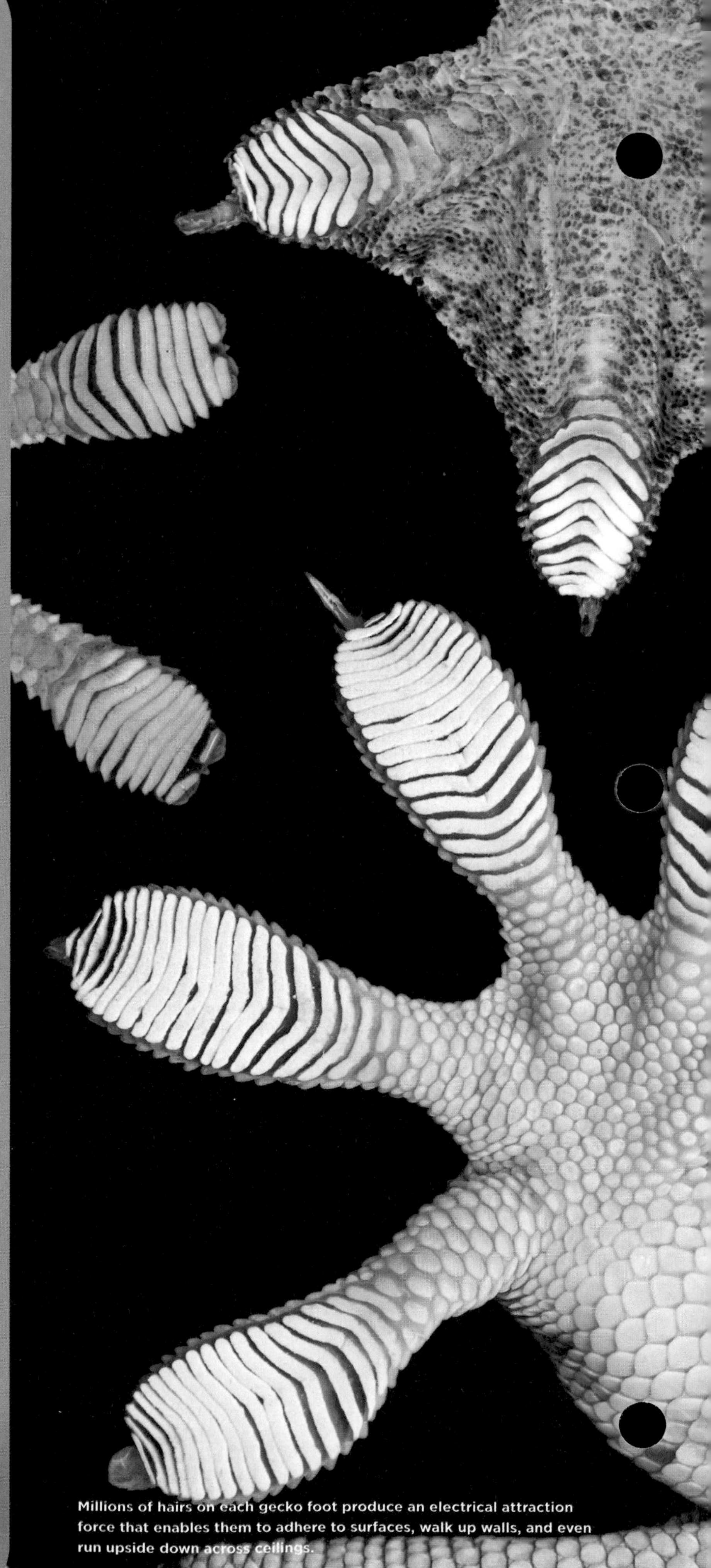

Millions of hairs on each gecko foot produce an electrical attraction force that enables them to adhere to surfaces, walk up walls, and even run upside down across ceilings.

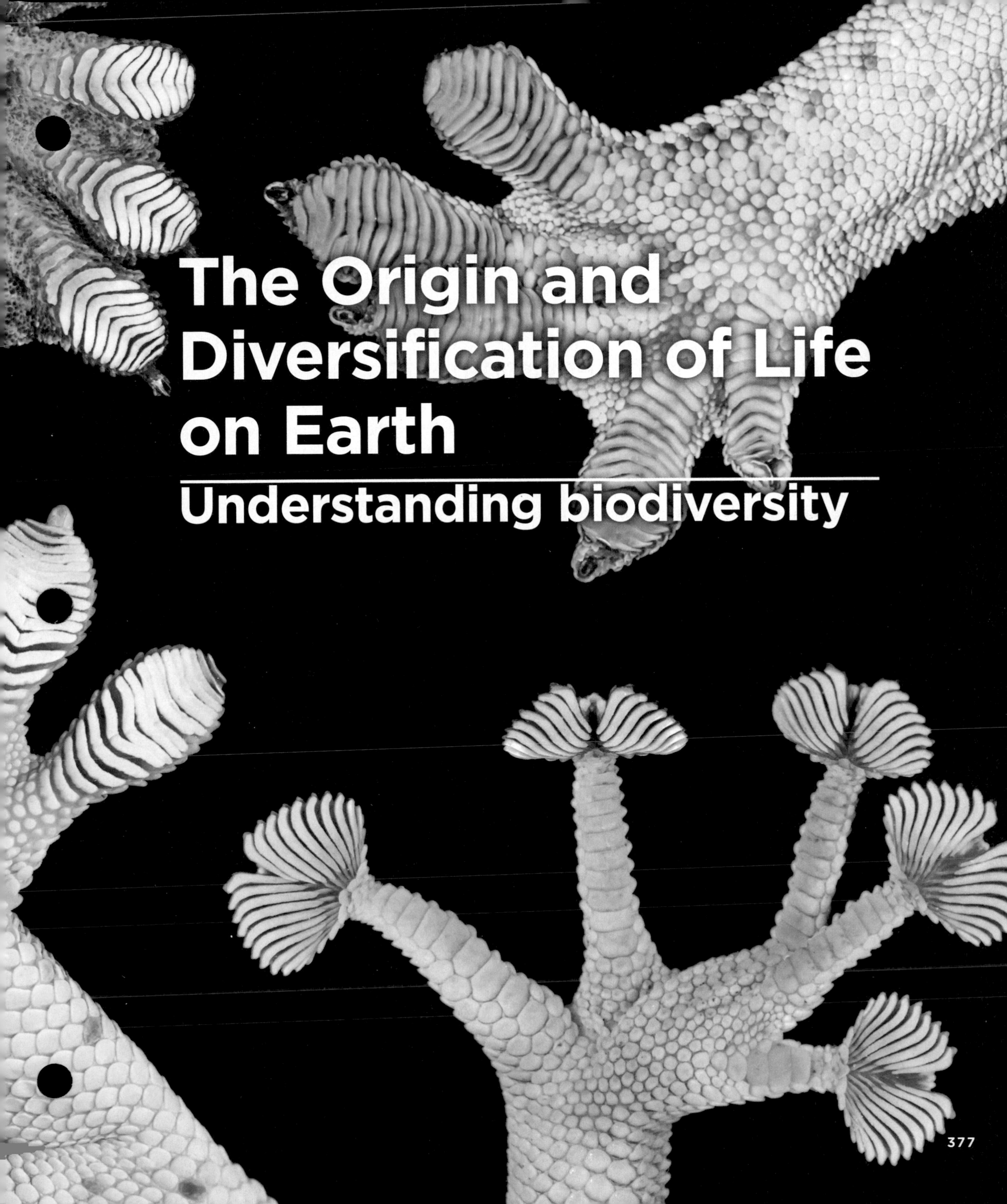

The Origin and Diversification of Life on Earth

Understanding biodiversity

12.1–12.3

Life on earth most likely originated from non-living materials.

Life on earth arose in a very different environment than we experience today. (Shown here: the formation of volcanic rock in Hawaii.)

12.1 Complex organic molecules arise in non-living environments.

In the beginning, there was nothing. Now there is something. That, in a nutshell, describes one of the most important, yet difficult to resolve, questions in science: how did life on earth begin? First, let's give a basic definition of what we mean by "life." **Life** is defined by the ability to replicate and by the presence of some sort of metabolic activity (the chemical processes by which molecules are acquired and used and energy is transformed in controlled reactions).

Earth formed about 4.5 billion years ago from clouds of dust and gases. As it very gradually cooled, a crust formed at the surface and condensing water formed the oceans. This probably took several hundred million years. The oldest rocks, found in Canada, are about 3.8 billion years old. And the earliest life forms appeared not long after these first rocks formed: fossilized bacteria-like cells have been found in rocks that are 3.4 billion years old.

From that initial point in earth's formation, tremendous **biodiversity** arose—variety and variability among all genes, species, and ecosystems. In this chapter, we explore how this biodiversity might have come to be, how we name groups of organisms, and how we determine the relatedness of these groups to one another. We begin by returning to the question of the origin of life on earth.

How did these first organisms arise? Some have suggested that life may have originated elsewhere in the universe and traveled to earth, possibly on a meteor. It is hotly debated, however, whether microbes could even survive the multimillion-year trip to earth in the cold vacuum of space with no protection from ultraviolet and other forms of radiation. Experimental data have been unable to answer this question definitively. The vast majority of scientists believe, instead, that life originated on earth, probably in several distinct phases, as described here and in the next section.

Phase 1: The formation of small molecules containing carbon and hydrogen. We know from chemical analyses of ancient rock that no oxygen gas was present on earth around the time of the origin of life. The atmosphere included large amounts of carbon dioxide, nitrogen, methane, ammonia, hydrogen, and hydrogen sulfide, mostly produced by volcanic eruptions. It was this environment that probably served as the cradle of life, or what Darwin called the "warm little pond" (FIGURE 12-1).

CONDITIONS ON EARTH AT THE TIME LIFE BEGAN

Earth's early atmosphere had almost no oxygen and was filled with many compounds released during volcanic eruptions.

Small organic molecules eventually formed, providing the building blocks of life.

FIGURE 12-1 **Darwin's "warm little pond."** The first life on earth tolerated an atmosphere without oxygen.

THE UREY-MILLER EXPERIMENT

Stanley Miller (shown) and Harold Urey developed a simple four-step experiment that demonstrated how complex organic molecules could have arisen in earth's early environment.

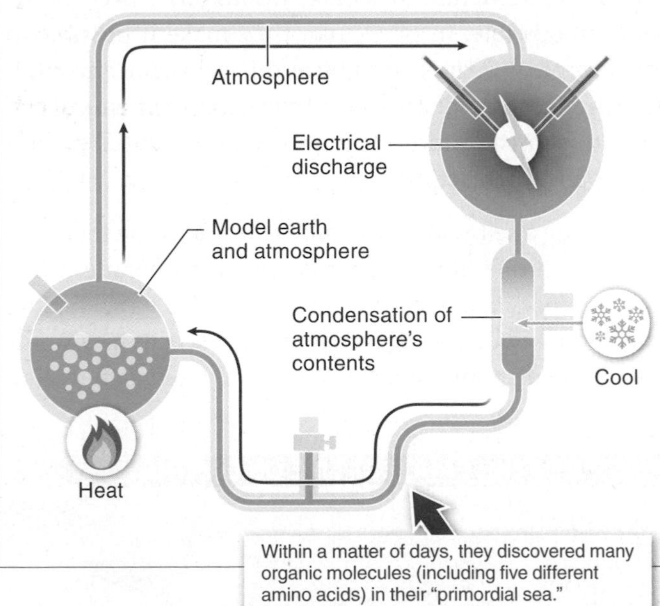

Atmosphere
Electrical discharge
Model earth and atmosphere
Condensation of atmosphere's contents
Cool
Heat
Within a matter of days, they discovered many organic molecules (including five different amino acids) in their "primordial sea."

FIGURE 12-2 **A promising first step.** The Urey-Miller experiment generated organic molecules from hydrogen, methane, and ammonia.

Critical to the origin of life was the formation of small molecules containing carbon and hydrogen. As we saw in Chapter 2, carbon readily bonds with hydrogen and with other carbon atoms, creating molecules in a variety of forms that can interact with each other in processes essential to life. There are several plausible scenarios for how these small organic molecules might have formed. The most likely comes from some simple but revealing four-step experiments done in 1953 by a 23-year-old student named Stanley Miller and his advisor, Harold Urey (FIGURE 12-2).

1. They created a model of the "warm little pond" and their best estimate of earth's early atmosphere: a flask of water with H_2, CH_4 (methane), and NH_3 (ammonia).
2. They subjected this mini-world to sparks, to simulate lightning.
3. They cooled the atmosphere so that any compounds formed in it would rain back down into the water.
4. They waited, and then they examined the contents of the water to see what happened.

They didn't have to wait long to get exciting results. Within days—not millions of years or even a few months—they discovered many organic molecules, including five different amino acids, in their primordial sea. (Using more sensitive equipment, recent re-analyses reveal that all 20 amino acids present in living organisms were produced during these experiments.) This was the first demonstration that complex organic molecules could have arisen in earth's early environment. Questions remain, however, such as whether it is reasonable to think that the environment Urey and Miller assumed to exist on the early earth is actually likely to have existed. Nonetheless, the experiments are a promising first step, suggesting that complex organic compounds—including amino acids, the primary constituents of proteins and an essential component of living systems—could have been produced from inorganic chemicals and lightning energy in the primitive environment of earth.

TAKE HOME MESSAGE 12.1

›› Under conditions similar to those thought to have existed on early earth, small organic molecules can form, and these molecules have some chemical properties of life.

12.2 Cells and self-replicating systems evolved together to create the first life.

After the generation of organic molecules such as amino acids, the second phase in the generation of life from non-life was probably the assembly of these building block molecules into self-replicating, information-containing molecules. This is where things get a bit more speculative. It's complicated to generate a complex organic molecule, but even more complicated to generate an organic molecule that can replicate (make copies of) itself. Researchers believe that to get to the replication phase, enzymes, or something with the catalytic activity of enzymes, were required.

Phase 2: The formation of self-replicating, information-containing molecules. Recently, researchers discovered that the nucleic acid RNA can do what proteins do—catalyze reactions necessary for replication—meaning that this single, relatively simple molecule could have been a self-replicating system and a precursor to cellular life.

In the early world, self-replicating nucleic acid molecules probably carried the information on how to replicate *and* served as the machinery to carry out the replication. Is that enough to be considered living?

Based on our definition of life—the ability to replicate and the ability to carry out some sort of metabolism—these early self-replicating RNA molecules were close to being considered living. They were able to replicate, but they could not carry out metabolism. What, then, were the first truly *living* organisms on earth?

Fossils of 3.4-billion-year-old cells have been found in rocks in South Africa and Australia. These cells appear to be prokaryotic cells, similar to living bacterial cells, with no nucleus, no organelles, and a circular strand of genetic information (FIGURE 12-3). Some even look as if they were in the process of dividing. Several lines of evidence support the idea that they are indeed remnants of cells: (1) the age of the rocks themselves has been reliably determined, (2) the size of the circular impressions in the rocks is similar to that of modern-day prokaryotes, and (3) the ratio of two carbon isotopes (^{12}C/(^{13}C) is more characteristic of fossilized organisms than of typical rocks that do not contain fossils. (Some sediments in Greenland that are 3.8 billion years old also have carbon isotope ratios that suggest biological activity.)

But many questions remain, including: Were these cells the first living organisms on earth? And were they descendants of earlier, self-replicating molecules of RNA? Because no earlier fossils have been found, we are not certain about the answers to these questions.

We now explore the critical third phase in the generation of life from non-life: the development of a membrane separating these self-replicating small molecules from their surroundings, thus forming cells and facilitating metabolic activity.

Phase 3: The development of a membrane, enabling metabolism and creating the first cells. As we saw in Chapter 4, cell membranes are semipermeable barriers that separate the inside of cells from their external environment. Membranes make numerous aspects of metabolism possible. In particular, they make it possible for chemicals inside the cell to be at different concentrations than they are outside the cell. Differences in chemical concentrations inside and outside a cell are essential to most life-supporting reactions.

Combining a self-replicating molecule and some metabolic chemicals into a unit, surrounded by a membrane, would make life possible. But how could this have happened initially? Some evidence suggests that the first cells may simply have formed spontaneously. Specifically, researchers have

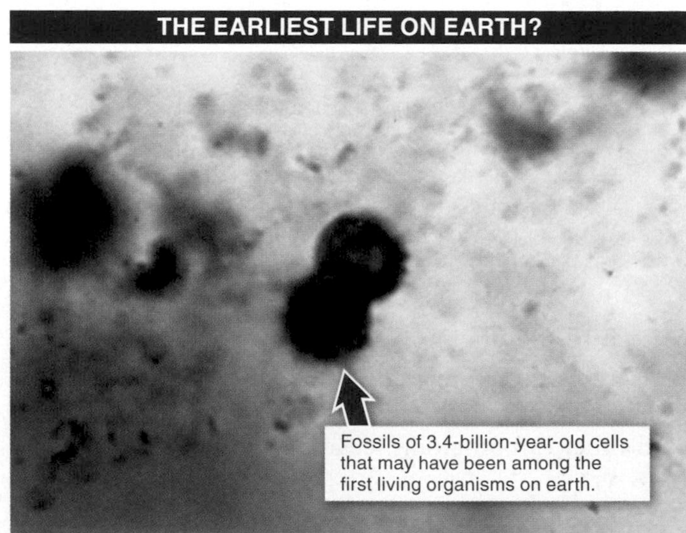
THE EARLIEST LIFE ON EARTH?
Fossils of 3.4-billion-year-old cells that may have been among the first living organisms on earth.

FIGURE 12-3 **Ancient prokaryotic cells.**

THE ORIGIN OF LIFE?

How did the first organisms on earth arise? Most scientists believe that life originated on earth, probably in several distinct phases.

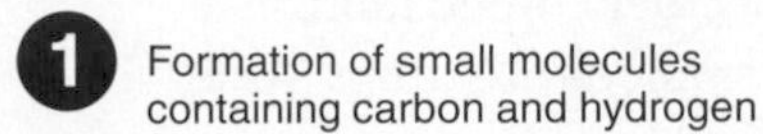
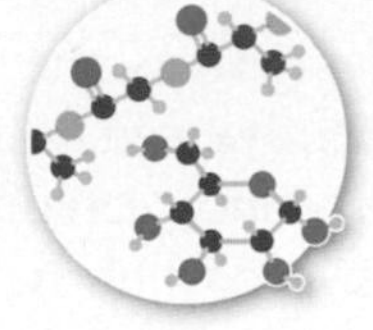

1 Formation of small molecules containing carbon and hydrogen

FIGURE 12-4 **How did life arise?**

found that mixtures of phospholipids placed in water or salt solutions tend to spontaneously form small spherical units that resemble living cells. These units may even "sprout" new buds at their surface, appearing to divide. Because these cell-like units don't have any genetic material, however, they cannot be considered to be alive. But if, at some point, units like these incorporated some self-replicating molecules inside, maybe by forming around them, such **microspheres** might have been important in the third phase in the generation of life from non-life: the compartmentalization of self-replicating, information-containing molecules into cells. If this did occur, the final step in the creation of something from nothing would be complete (FIGURE 12-4).

The exact process by which life on earth originated is still uncertain, and research continues. What is clear, however, is that life now abounds in great diversity. In this chapter we explore how this diversity is generated, as well as what a species is. We also consider how individual species may split and create additional species.

TAKE HOME MESSAGE 12.2

» The earliest life on earth, which resembled bacteria, appeared about 3.4 billion years ago (and maybe as early as 3.8 billion years ago), not long after the earth was formed. Evidence supports the idea that self-replicating molecules, possibly RNAs, may have formed in earth's early environment and later acquired or developed membranes, enabling them to replicate and making metabolism possible—the two conditions that define life.

THIS IS HOW WE DO IT

Developing the ability to apply the process of science

12.3 Could life have originated in ice, rather than in a "warm little pond"?

In science, when do we need to think outside the box?

Phenomena in the natural world don't always give up their secrets easily. When trying to better understand some process in nature, applying a new experimental technique occasionally allows a breakthrough. Other times, it may be the discovery of new or unexpected evidence that provides the breakthrough. But sometimes progress requires a more radical approach and thinking outside the box.

Researchers have been questioning some of the most basic assumptions about life's origin on earth, such

(continued on the following page)

as the widely held view that life emerged from an environment that was warm or hot and was wet—something along the lines of the "warm little pond" Darwin speculated about in *The Origin of Species*. Evidence from the experiments of Miller and Urey and others supports this view.

But what if icy baths, not warm ponds, were the "incubator" of life?

That's thinking outside the box. And researchers have provided some intriguing evidence and proposed some clever ideas to support it. Their ideas build upon the broad consensus about the most important physical and chemical requirements for the initial generation of self-replicating molecules.

Chemical requirement 1 Precursor molecules need to persist and need to come in contact with each other.

Conventional assumption: In living cells today, compartments make this duration and closeness of contact possible. But before life appeared on earth, under warm, wet conditions, some sort of chambers or microspheres may have spontaneously formed and served this purpose.

Novel approach: As water freezes, tiny compartments form within the ice. On early earth, low temperatures may have slowed the degradation of any precursor molecules—including RNA—that formed. And the tiny compartments may have held those precursor molecules close together, making it possible for them to react with each other.

Chemical requirement 2 Precursor molecules need to exhibit catalytic properties.

Conventional assumption: At warm or hot temperatures—but not at colder temperatures—molecules move quickly and collide frequently, enhancing reaction rates.

Novel approach: Although reactions usually slow down as the temperature drops, some actually speed up. As water freezes, the ice crystals form only from pure water. If there are any impurities present—such as salt or cyanide—they are excluded from the crystals and concentrated in small chambers of liquid water within the ice. At these higher concentrations, the molecules collide more frequently, even as the temperature drops. As a consequence, certain reactions can occur more rapidly—possibly including the creation and elongation of the first RNA chains.

Is it even feasible that ice was present on early earth and that precursor molecules could have formed in it?

Intriguing observations and evidence
Researchers have carried out experiments in which they prepare—and then freeze—tubes containing seawater and the building blocks of RNA. After thawing out the tubes, they find numerous RNA molecules, some long enough to be able to act as enzymes.

And recent evaluations of glacier-encased land north of the Arctic Circle suggest that 4 billion years ago, the sun may have been dimmer, and the earth may have cooled so much that ice covered the oceans. Some scientists have even gone so far as to describe earth as a "giant snowball" at that time.

Has exploration of the plausibility of ice as the initial medium of RNA replication answered the questions about how life on earth originated?

There is still plenty of skepticism about the idea that the primordial soup was a cold soup. Many researchers suspect that reported evidence of RNA chains forming under freezing temperatures may reflect accidental contamination, or that the chains actually formed during the thawing-out process.

As a case study of scientific thinking in action, however, this example illuminates the importance of evaluating the assumptions underlying our hypotheses.

Is there any value to false starts (and even dead ends) encountered in research investigations?

Keeping an open mind is more important than rigidly holding onto an idea. Observations and evidence must take the central role in guiding our understanding of natural processes.

TAKE HOME MESSAGE 12.3

» Some researchers question the long-held assumptions that self-replicating molecules with catalytic properties are most likely to have formed in a warm, wet environment. They've proposed that the laws of chemistry and the properties of water as it freezes may actually favor ice as the initial incubator of life. The answer is unclear, but the process of scientific thinking is guiding investigators.

12.4–12.7

Species are the basic units of biodiversity.

A world of beauty: there are more than 280,000 species of plants.

12.4 What is a species?

A cat and a mouse are different things. Oprah Winfrey and Bill Gates, on the other hand, are different *versions* of the same thing. That is, although they are different individuals, both are humans. How do we know when to classify two individuals as members of different groups and when to classify them as members of the same group (FIGURE 12-5)?

THE SAME
We easily recognize Cristiano Ronaldo of Real Madrid and Kerry Washington as members of the same species.

DIFFERENT
The Persian buttercup and the pink rose or the Bornean orangutan and the golden lion tamarin look similar but are not of the same species.

FIGURE 12-5 **Appearance is not always an accurate identifier of species.**

Biologists use the word **species** to label different kinds of organisms. According to the **biological species concept,** species are populations of organisms that interbreed, or could possibly interbreed, with each other under natural conditions, and that cannot interbreed with organisms from other such groups (FIGURE 12-6).

Notice that the biological species concept completely ignores physical appearance when defining a species. Instead it emphasizes **reproductive isolation,** the inability of individuals from two populations to produce fertile offspring with each other, thereby precluding gene exchange between the populations.

Let's clarify two important features of the biological species concept. First, it says that members of a species are actually interbreeding or *could possibly* interbreed. Just because two individuals are physically separated, they aren't necessarily of different species. Two people living in different countries, for example, may not be able to mate because of the distance, but if they were brought to the same location, they could. So, although the people are not in the same population, we do not consider them to be reproductively isolated. Second, our definition refers to "natural" conditions. This distinction is important because occasionally, in captivity, individuals may interbreed that would not interbreed in the wild, such as zebras and horses (FIGURE 12-7).

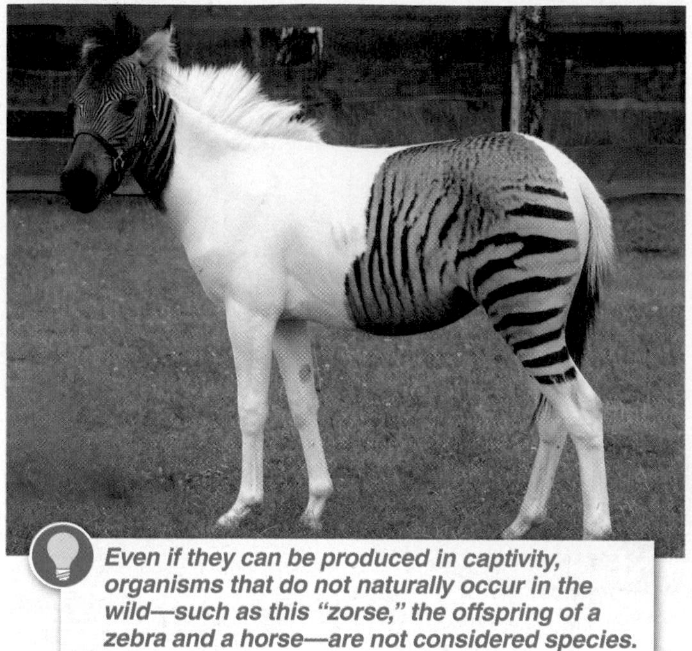

Even if they can be produced in captivity, organisms that do not naturally occur in the wild—such as this "zorse," the offspring of a zebra and a horse—are not considered species.

FIGURE 12-7 **A species? Interbreeding is not enough.**

WHAT MAKES A SPECIES?

Herd of guanacos

SPECIES ARE...

- Populations of organisms that interbreed with each other
- Or could possibly breed, under natural conditions
- And are reproductively isolated from other such groups

FIGURE 12-6 **An interbreeding population.**

Two types of barriers prevent individuals of different species from reproducing: prezygotic and postzygotic barriers (FIGURE 12-8). (Remember, an egg that has been fertilized by a sperm cell is a *zygote.*)

Prezygotic barriers make it impossible for individuals to mate with each other or, if they can mate, make it impossible for the male's reproductive cell to fertilize the female's reproductive cell. These barriers include situations in which the members of the two species have different courtship rituals or have physical differences that prevent mating or fertilization.

Postzygotic barriers occur after fertilization and generally prevent the production of fertile offspring from individuals of two different species. These barriers are responsible for the production of **hybrid** individuals that either do not survive long after fertilization or, if they do, are infertile or have reduced fertility. Mules, for example, are the hybrid offspring of horses and donkeys, and although they can survive, they cannot produce offspring.

At first glance, the biological species concept seems to make it possible to determine unambiguously whether individuals belong to the same or different species. For plants and animals this is usually true. For many other organisms,

REPRODUCTIVE ISOLATION: KEEPING SPECIES SEPARATE

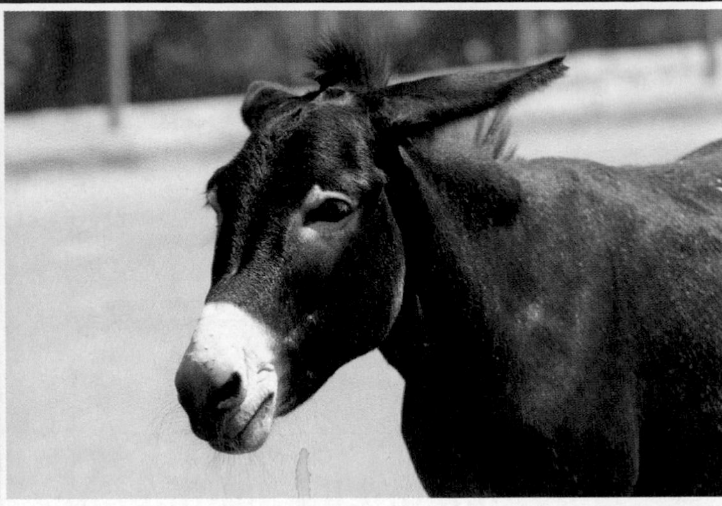

PREZYGOTIC BARRIERS

- Individuals are physically unable to mate with each other.

OR

- If individuals are able to mate, the male's reproductive cell is unable to fertilize the female's reproductive cell.

POSTZYGOTIC BARRIERS

- Matings produce hybrid individuals that do not survive long after fertilization.

OR

- If hybrid offspring survive (such as the mule above), they are infertile or have reduced fertility.

FIGURE 12-8 **Barriers to reproduction.**

however, it is impossible to apply the biological species concept. We examine those species in Section 12-6 and explore some alternative methods of identifying species. Even though it is not perfect, however, the biological species concept gives us an important tool for conceptualizing and categorizing the tremendous diversity of organisms found on earth.

TAKE HOME MESSAGE 12.4

» Species are generally defined as populations of individuals that interbreed with each other, or could possibly interbreed, and that cannot interbreed with organisms from other such groups. This concept can be applied to most plants and animals, but is not applicable for many other types of organisms.

12.5 How do we name species?

Keeping track of a large group of anything requires an organizational system. To keep track of earth's rich diversity, biologists use the system developed by the Swedish biologist Carolus Linnaeus in the mid-1700s. Here's how it works (FIGURE 12-9). Every species is given a scientific name that consists of two parts, a **genus** (*pl.* genera) and a **specific epithet.** Linnaeus named humans *Homo sapiens,* meaning "wise man." *Homo* is the genus and *sapiens* is the specific epithet, and *Homo sapiens* is the species name. The genus is capitalized, and both genus and specific epithet are italicized. The redwood tree has the name *Sequoia sempervirens.*

The strength of Linnaeus's system is that it is hierarchical—that is, each element of the system falls under a single element in the level just above it. The narrowest classification in the Linnaean system is the species. The name for every species within a genus is unique, but many different species may be in the same genus. Similarly, many genera are grouped within a **family.** And many families are grouped within an **order.** Orders are grouped within a **class.** Classes are grouped within a **phylum.** And, as Linnaeus originally set it up, all phyla were classified under one of three **kingdoms:** the animal kingdom, the plant kingdom, or the "mineral kingdom."

THE ORGANIZATION OF LIFE

Each species on earth, like *Equus quagga*, is given a scientific name and is categorized according to hierarchical groups.

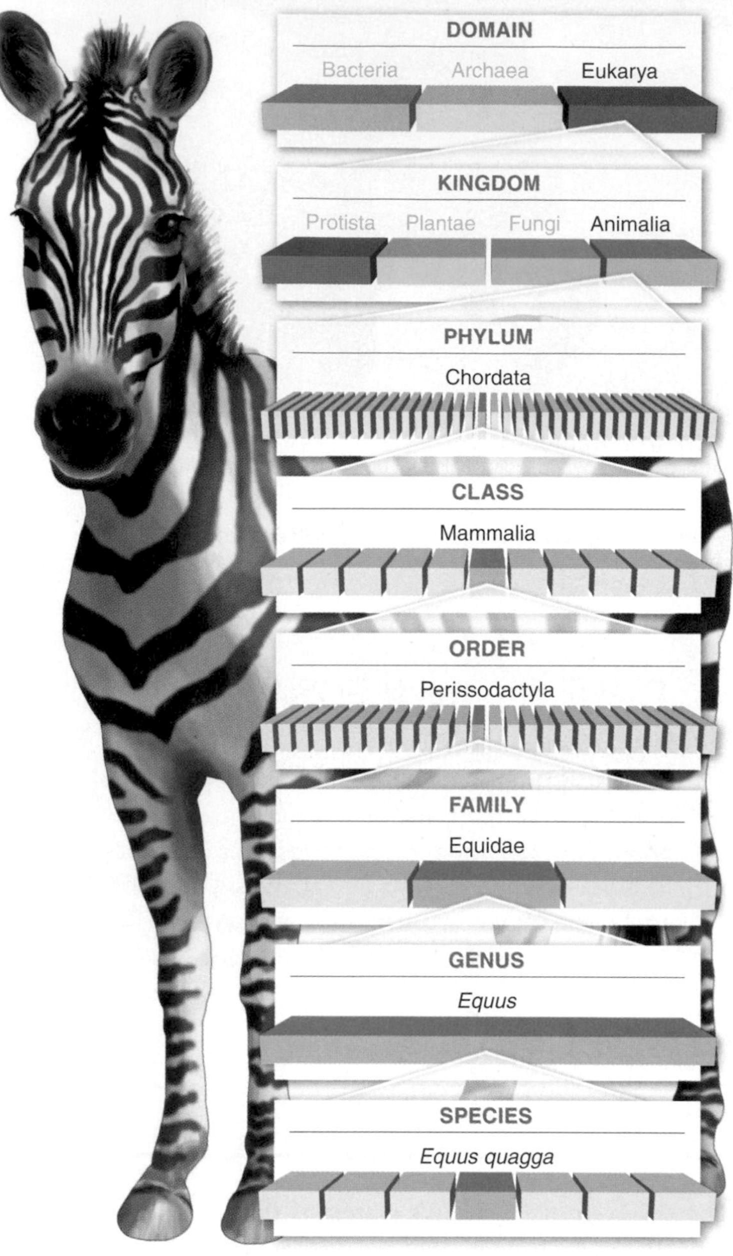

FIGURE 12-9 **Name that zebra.**

Today, many of the species classifications that Linnaeus described have been revised. Some of his designations, such as the "mineral kingdom," are now left out, while two other kingdoms (fungi and protists) have been added. Also, all of the kingdoms are now classified under an even higher order of classification, the **domain.** But Linnaeus's basic hierarchical structure remains, and all life on earth is still named using this system, with all organisms belonging to one of the three domains: the bacteria, the archaea, and the eukarya.

When new species are discovered, they are classified based on the Linnaean system. And when it comes to assigning a specific epithet, scientists frequently have a little fun. In recent years, for example, an amphibian fossil was named *Eucritta melanolimnetes,* which translates roughly into "creature from the black lagoon." And in honor of Elvis Presley, a wasp species was named *Preseucoila imallshookupis*. A variety of other celebrities also have had species named after them, including the Beatles, Mick Jagger, Kate Winslet, and Steven Spielberg.

Plants and animals often are referred to by common names, which are based on similarities in appearance. These common names, however, can cause confusion. "Fish" for example is used as part of the names of jellyfish, crayfish, and silverfish, none of which is closely related to the vertebrate group of fishes that includes salmon and tuna. Later in this chapter we'll see how an important goal of modern classification and the naming of organisms is to link an organism's classification more closely with its evolutionary history than with its physical characteristics.

TAKE HOME MESSAGE 12.5

» **Each species on earth is given a unique name, using a hierarchical system of classification. Every species is classified in one of three domains.**

12.6 Species are not always easily defined.

Biologists, like all humans, can be biased. When investigating the natural world, for example, they often focus on plants and animals, to the exclusion of the rest of the earth's rich biodiversity. This is problematic when it comes to the biological species concept. While the biological species concept is remarkably useful when describing most plants and animals, it falls short of representing a universal and definitive way of distinguishing many life forms (FIGURE 12-10).

Difficulties in Classifying Asexual Species The biological species concept defines species as populations of interbreeding individuals. But this is a useless distinction for the asexual species of the world. Recall from Chapter 8 that asexual reproduction, common among single-celled organisms (including all bacteria), many plants, and some animals, is a form of reproduction that doesn't involve fertilization or even two individuals. Rather, the cells (or cell) of an individual simply divide, creating new individuals. Because asexual reproduction does not involve partners or interbreeding, the concept of reproductive isolation is not meaningful, and it might seem that every individual should be considered a separate species. Clearly, that's not a helpful rule to follow.

Difficulties in Classifying Fossil Species When classifying fossil species, differences in the size and shape of fossil bones from different individuals can never definitively reveal whether there was reproductive isolation between those individuals. This makes it impossible to apply the biological species concept.

Difficulties in Determining When One Species Has Changed into Another Based on fossils, it seems that modern-day humans, *Homo sapiens,* probably evolved from a related species called *Homo heidelbergensis* about 250,000–400,000 years ago. This seems reasonable, until you consider that your parents—who are in the species *H. sapiens*—were born to your *H. sapiens* grandparents, who were born to your *H. sapiens* great-grandparents, and so on. If humans evolved from *H. heidelbergensis,* at what *exact* point did *H. heidelbergensis* turn into *H. sapiens?* It may not be possible to identify the exact point at which this change occurred.

Q Chihuahuas and Great Danes generally can't mate. Does that mean they are different species?

Difficulties in Classifying Ring Species Living in central Asia are some small, insect-eating songbirds called greenish warblers. Unable to live at higher elevations of the Tibetan mountain range, the warblers live in a ring around it. At the southern end of the mountain range, in northwest India, the warblers interbreed with each other. Along either side of the range, the warbler population is split, and warblers on one side do not interbreed with those on the other, because the mountain range separates them. Where the two "side" populations meet up again at the northernmost end of the mountain range, in the forests of Siberia, they can no longer interbreed.

What happened? Gradual changes in the warblers on each side of the mountain range accumulated so that the two populations that meet up in Siberia are sufficiently different, physically and behaviorally, that they have become reproductively incompatible. But because the two non-interbreeding populations in the north are connected by gene flow through other populations farther south, there is no exact point at which one species stops and the other begins. So where do you draw the line? The greenish warblers are just one example of the more than 20 such **ring species** that biologists have described.

THE BIOLOGICAL SPECIES CONCEPT DOESN'T ALWAYS WORK

The biological species concept is remarkably useful when describing most plants and animals, but it doesn't work for distinguishing all life forms.

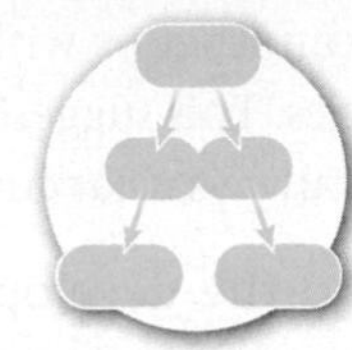

CLASSIFYING ASEXUAL SPECIES
Asexual reproduction does not involve interbreeding, so the concept of reproductive isolation is no longer meaningful.

CLASSIFYING FOSSIL SPECIES
Differences in size and shape of fossil bones cannot reveal whether there was reproductive isolation between the individuals from which the bones came.

DETERMINING WHEN ONE SPECIES HAS CHANGED INTO ANOTHER
There is rarely a definitive moment marking the transition from one species to another.

CLASSIFYING RING SPECIES
Two non-interbreeding populations may be connected to each other by gene flow through another population, so there is no exact point where one species stops and the other begins.

CLASSIFYING HYBRIDIZING SPECIES
Hybridization—the interbreeding of closely related species—sometimes occurs and produces fertile offspring, suggesting that the borders between the species are not clear cut.

FIGURE 12-10 A useful concept that can't always be easily applied.

Difficulties in Classifying Hybridizing Species

Increasingly, **hybridization**—the interbreeding of closely related species—has been observed among plant and animal species. And in some cases, such as among butterflies in the genus *Heliconius,* the hybrids have high survival rates *and* are fertile, whether interbreeding with one another or with individuals of either of the two parental species. This suggests that the borders between the species are not always clear cut.

All of these shortcomings have prompted the development of alternative approaches to defining what a species is. These alternatives tend to focus on aspects of organisms other than reproductive isolation as defining features. The most commonly used alternative is the **morphological species concept,** which characterizes species based on physical features such as body size and shape. Although the choice of which features to use is subjective, an important aspect of the morphological species concept is that it can be used effectively to classify asexual species. And because it doesn't require knowledge of whether individuals can actually interbreed, the morphological species concept is a bit easier to use when observing organisms in the wild.

Although the biological species concept is the most widely used and can be applied without difficulty to most plants and animals, there will probably never be a universally applicable definition of what a species is. From asexual species to ring species to hybridizing species, the diversity of the natural world is simply too great to fit into neat, completely defined and distinct little boxes. Nonetheless, scientists can generally use a species definition that is satisfactory for a particular situation.

TAKE HOME MESSAGE 12.6

» The biological species concept is useful when describing most plants and animals, but it falls short as a way of distinguishing many life forms. Difficulties arise when trying to classify asexual species, fossil species, species arising over long periods of time, ring species, and hybridizing species. In these cases, alternative approaches to defining species can be used.

12.7 How do new species arise?

We don't know how many species there are on earth. Estimates vary tremendously, from 5 million to 100 million. Biologists do, however, know the process by which these species arose.

The process of **speciation,** in which one species splits into two distinct species, occurs in two phases and requires more than just evolutionary change in a population. The first phase of speciation is *reproductive isolation,* in which two populations are separated from one another and so come to have independent evolutionary fates. The second phase is *genetic divergence,* in which two populations evolving separately accumulate physical and behavioral differences over time. These differences may arise as random, neutral mutations, or through each population adapting differently to features of their separate environments, including to different predators and different types of food. When physical and behavioral differences have accumulated that prevent members of the two populations from interbreeding, we say that speciation has occurred.

Often, the initial reproductive isolation necessary for speciation comes about when two populations are geographically separated. Although this is an effective and common way for speciation to occur, speciation can also occur without it.

Speciation with Geographic Isolation: Allopatric Speciation

Suppose the local climate grows wetter and a river forms that splits one population of squirrels into two separate populations. Because the squirrels cannot cross the river, the populations on either side are reproductively isolated from each other. Over time, the two populations have different evolutionary paths as they accumulate different mutations and adapt to particular features of their separate habitats, which may differ. Eventually, the two populations might genetically diverge so much that even if the populations came back into contact, squirrels from the two groups could no longer interbreed. In fact, two species of antelope ground squirrels have formed on the north and south rims of the Grand Canyon as a result of this type of speciation. Speciation that occurs as a result of geographic isolation is known as **allopatric speciation** (FIGURE 12-11).

Another example of allopatric speciation is seen in the various finch species of the Galápagos Islands, the same finch

Allopatric speciation occurs when a geographic barrier causes one group of individuals in a population to be reproductively isolated from another group.

1 INITIAL POPULATION

2 REPRODUCTIVE ISOLATION
Suppose a river forms through the squirrels' habitat, dividing the population. Because the squirrels cannot cross the river, the two groups are now reproductively isolated.

3 GENETIC DIVERGENCE
Over time, the populations diverge enough genetically that they wouldn't be able to interbreed if they came in contact.

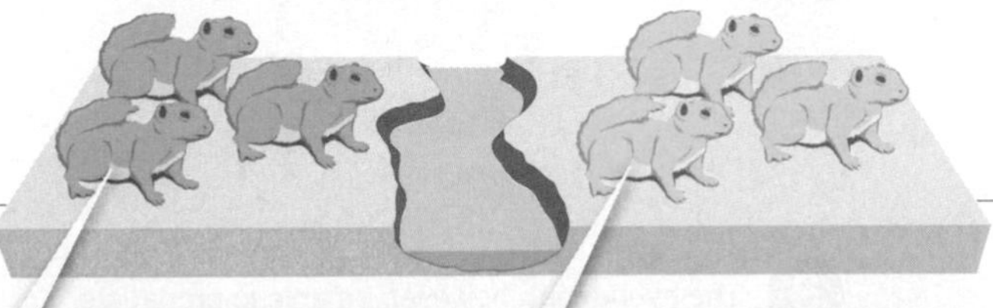

Harris's antelope ground squirrel

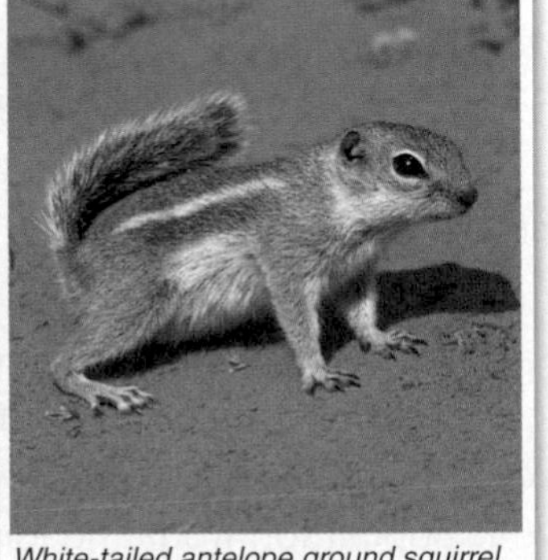
White-tailed antelope ground squirrel

FIGURE 12-11 Geographic isolation can result in genetic divergence. Two species of antelope ground squirrel that have evolved on the north and south rims of the Grand Canyon.

ALLOPATRIC SPECIATION: GALÁPAGOS ISLAND FINCHES

Due to allopatric speciation, 14 different species of finches have evolved in the Galápagos Islands. Only one species of finch is found on the nearest mainland.

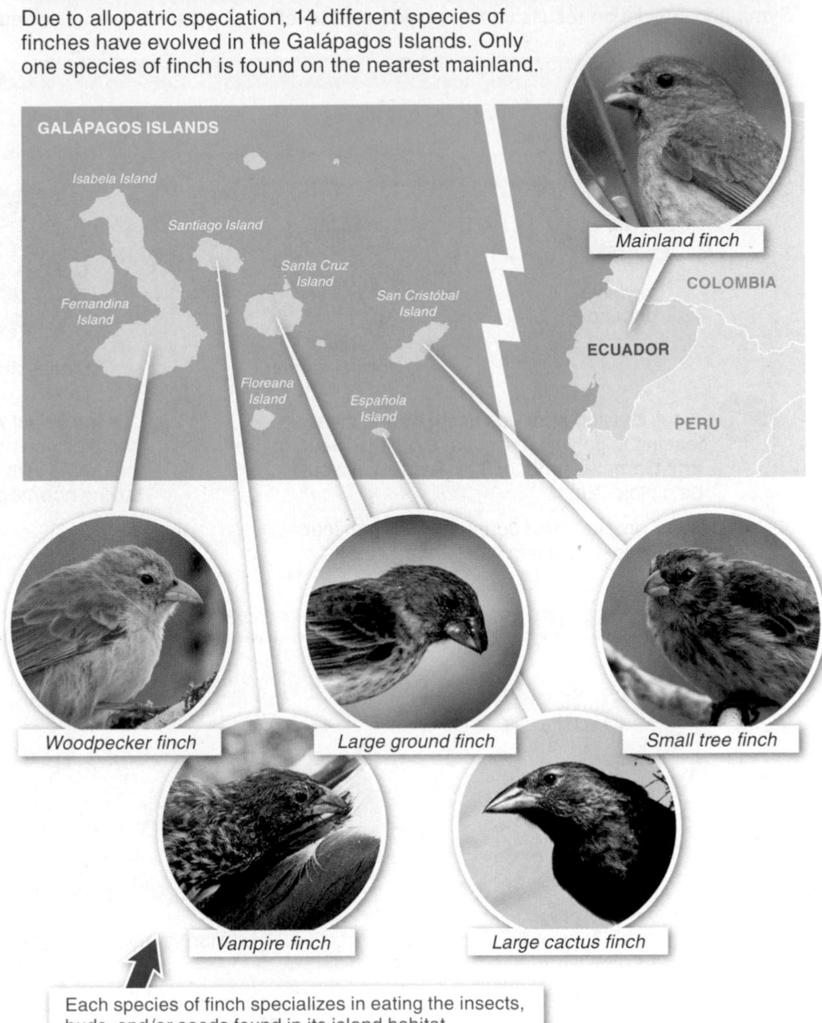

FIGURE 12-12 Five of the 14 Galápagos Island finch species.

species that Darwin observed and collected while on his 'round-the-world trip on HMS *Beagle*. Individual finches from the nearest mainland, now Ecuador, originally colonized one or more of the islands. (And later, additional islands may have been colonized by birds from the mainland or from previously colonized islands.) But because the islands are far apart, the finches tended not to travel between them, and the populations remained reproductively isolated from one another. Consequently, 14 different finch species have evolved on the Galápagos Islands, each species adapting to the predominant food source on its particular island and now having features that allow it to specialize in eating certain of the wide range of insects, buds, and seeds found on the islands (FIGURE 12-12). Only one species of finch is found in mainland Ecuador.

The barrier doesn't have to be an expanse of water. A forming glacier could split a population into two or more isolated populations. Or a drop in the water level of a lake might expose strips of land that divide the lake into detached,

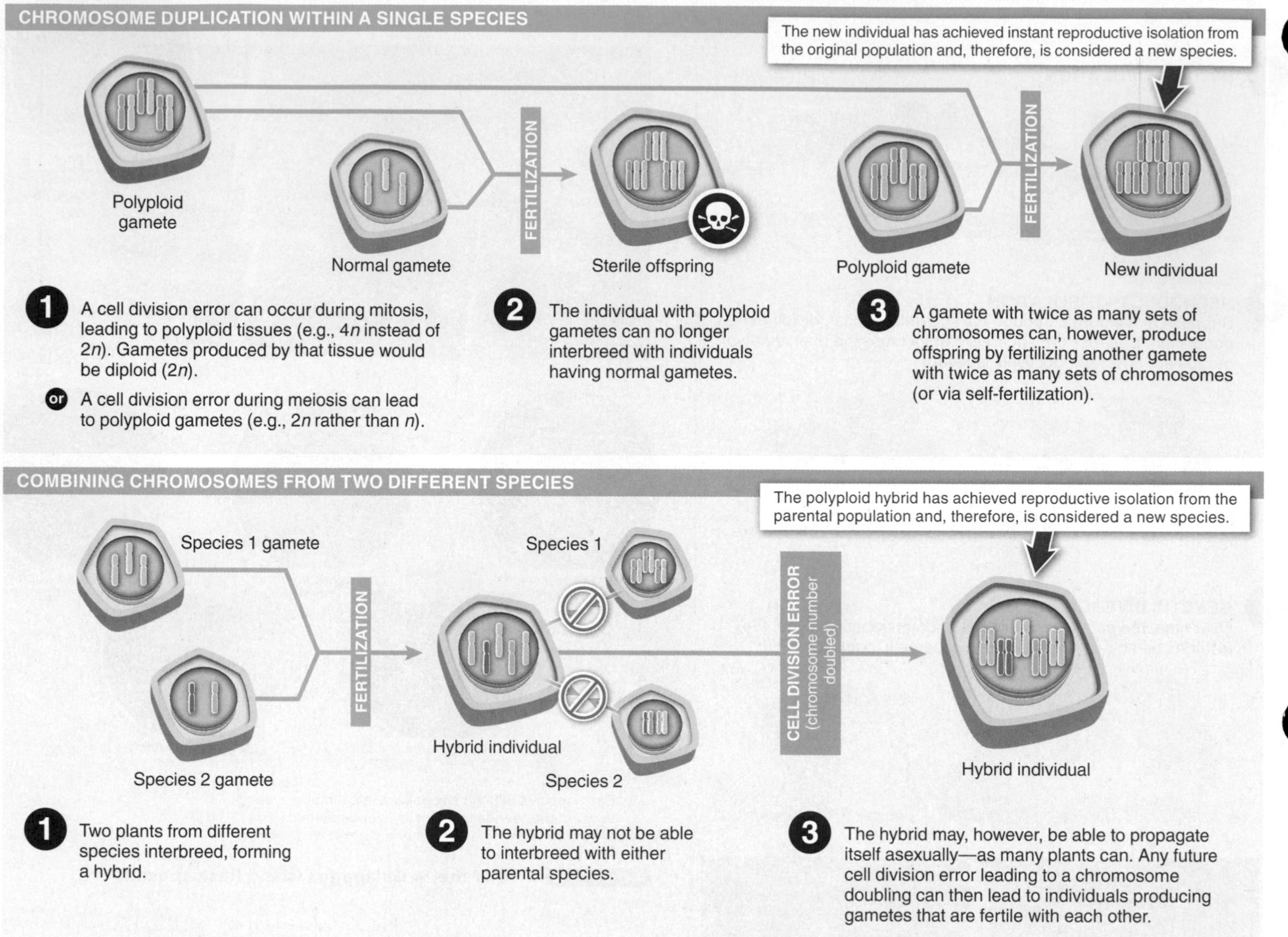

SYMPATRIC SPECIATION
Sympatric speciation results in the reproductive isolation of populations that coexist in the same area. Two scenarios lead to this method of speciation.
CHROMOSOME DUPLICATION WITHIN A SINGLE SPECIES
The new individual has achieved instant reproductive isolation from the original population and, therefore, is considered a new species.
Polyploid gamete
Normal gamete
FERTILIZATION
Sterile offspring
Polyploid gamete
FERTILIZATION
New individual
1 A cell division error can occur during mitosis, leading to polyploid tissues (e.g., 4n instead of 2n). Gametes produced by that tissue would be diploid (2n).
or A cell division error during meiosis can lead to polyploid gametes (e.g., 2n rather than n).
2 The individual with polyploid gametes can no longer interbreed with individuals having normal gametes.
3 A gamete with twice as many sets of chromosomes can, however, produce offspring by fertilizing another gamete with twice as many sets of chromosomes (or via self-fertilization).
COMBINING CHROMOSOMES FROM TWO DIFFERENT SPECIES
The polyploid hybrid has achieved reproductive isolation from the parental population and, therefore, is considered a new species.
Species 1 gamete
Species 1
FERTILIZATION
CELL DIVISION ERROR (chromosome number doubled)
Hybrid individual
Species 2 gamete
Species 2
Hybrid individual
1 Two plants from different species interbreed, forming a hybrid.
2 The hybrid may not be able to interbreed with either parental species.
3 The hybrid may, however, be able to propagate itself asexually—as many plants can. Any future cell division error leading to a chromosome doubling can then lead to individuals producing gametes that are fertile with each other.

FIGURE 12-13 **Speciation without geographic isolation.** Plants may genetically diverge through sympatric speciation.

smaller bodies of water, separating one large population of fish into two distinct populations. In each case, the result is the same: geographic isolation that enforces reproductive isolation.

Q Could you create a new species in the laboratory? How?

Researchers can easily create new species in the lab, using an analogous strategy. In one experiment, a single population of fruit flies (*Drosophila pseudoobscura*) was divided into two. The two populations were then maintained on different diets—one on the sugar maltose and the other on a starch-based food. After only eight generations of enforced reproductive isolation and adaptation to their differing nutritional environments, the populations had diverged sufficiently to form separate species. When the populations were mixed, the fruit flies from one population would no longer interbreed with flies from the other population.

Speciation without Geographic Isolation: Sympatric Speciation Speciation can also occur among populations that overlap geographically. This is called **sympatric speciation.** Among vertebrates, it is rare for populations of the same animal to become reproductively isolated when they coexist in the same area, so this method of speciation is relatively uncommon. But it *is* common among plants, and it occurs in one of two ways.

During cell division in plants (both in reproductive cells and in other cells of the plant body), an error sometimes occurs in which the chromosomes are duplicated but a cell does not divide. This creates a new cell that can then grow into an individual with twice as many sets of chromosomes as the parent from which it came. The new individual may have four sets of chromosomes, for example, while the original individual had two. This doubling of the number of sets of chromosomes is called **polyploidy** (FIGURE 12-13).

The individual with four sets of chromosomes can no longer interbreed with individuals having only two sets, because their offspring would have three sets (two sets from the parent that had four, and one set from the parent that had two), which could not divide evenly during cell division. The individual with four sets can, however, propagate through self-fertilization or by mating with other individuals that also have four sets. As a consequence, the individuals with four sets of chromosomes have achieved instant reproductive isolation from the original population and are therefore considered a new species.

Although rare in animals, speciation by polyploidy has occurred several times among some species of tree frogs.

A much more common method of sympatric speciation occurs when plants of different (but closely related) species interbreed, forming a hybrid. The hybrid may not be able to interbreed with either of the parental species, but it may be able to propagate asexually—as many plants can. And meiosis often is disrupted in hybrids, causing a chromosome doubling (polyploidy). When this occurs, the hybrids with a doubled number of chromosome sets can interbreed with each other (see Figure 12-13). This method of speciation has led to the production of many important crop plants, including wheat, bananas, potatoes, and coffee.

Whether populations are separated from each other allopatrically or sympatrically, speciation is not considered complete until sufficient differences have evolved in the two populations that they could no longer interbreed even if they did come in contact.

Separation of two populations of a species can occur relatively quickly—as long as it takes for a new river (or freeway) to divide one large population into two—but the genetic divergence that causes true reproductive isolation can take a very long time, sometimes thousands of years. For this reason, speciation can be difficult to study and observe.

TAKE HOME MESSAGE 12.7

» Speciation is the process by which one species splits into two distinct species that are reproductively isolated. It can occur by polyploidy or by a combination of geographic isolation and genetic divergence.

12.8–12.10 Evolutionary trees help us conceptualize and categorize biodiversity.

The diversity of life on earth can be thought of as branching like a tree.

12.8 The history of life can be imagined as a tree.

Although Carolus Linnaeus's method for naming species was an important step in categorizing and cataloguing earth's biodiversity, his classification of organisms into hierarchical groups was based on nothing more than his own evaluation of how physically similar various organisms appeared.

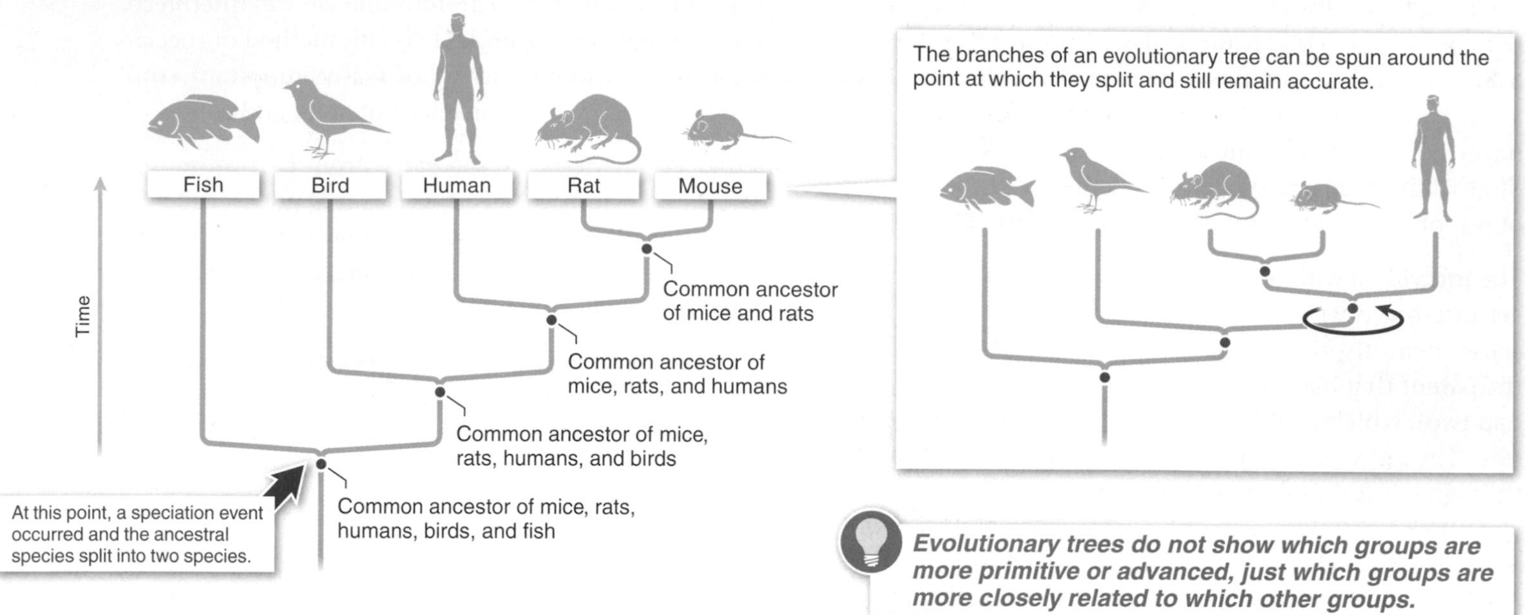
HOW TO READ AN EVOLUTIONARY TREE
Evolutionary trees reveal the evolutionary history of species and the sequence of speciation events that gave rise to them.
Fish
Bird
Human
Rat
Mouse
Time
Common ancestor of mice and rats
Common ancestor of mice, rats, and humans
Common ancestor of mice, rats, humans, and birds
Common ancestor of mice, rats, humans, birds, and fish
At this point, a speciation event occurred and the ancestral species split into two species.
The branches of an evolutionary tree can be spun around the point at which they split and still remain accurate.
Evolutionary trees do not show which groups are more primitive or advanced, just which groups are more closely related to which other groups.

FIGURE 12-14 **Mapping common ancestors.**

A hundred years later, however, Charles Darwin proposed and documented that species could change and give rise to new species. With Darwin, the classification of species acquired a new goal and a more important function. In *The Origin of Species* Darwin wrote: "Our classifications will come to be, as far as they can be so made, genealogies." With these words, Darwin was the first to link classification with evolution.

The modern incarnation of Darwin's vision of classification is called **systematics** and has the broader goal of reconstructing the **phylogeny,** or evolutionary history, of organisms. That is, through systematics, all species—even extinct species—are named and arranged in a manner that indicates the common ancestors they share and the points at which the species diverged. A complete phylogeny of all organisms is like a family tree for all species, past and present.

A phylogenetic tree not only shows the relationships among organisms but also presents a hypothesis about evolutionary history (FIGURE 12-14). At the beginning of life on earth there was the first living organism, one that could replicate itself. Then a **speciation event** occurred, after which the population of the first living organism split into independent evolutionary lineages. The phylogenetic tree had its first branch, and there was biodiversity. Over hundreds of millions of years, speciation events continued to occur, and today the tree has branches with millions of tips that represent all the species on earth. Moving up a branch of any evolutionary tree from its trunk toward its tips, we can see when groups split, with each branching point representing a speciation event.

It is important to remember that because phylogenies are hypotheses they are subject to revision and modification. In the next section, we explore how phylogenetic trees are constructed and look at the rich information they convey about the history of life on earth. We also investigate the modern methods of molecular systematics and see how, with new molecular and DNA evidence, biologists are revising many earlier phylogenies that were based on organisms' physical features.

TAKE HOME MESSAGE 12.8

» The history of life can be visualized as a tree; by tracing from the branches back toward the trunk, we can follow the pathway back from descendants to their ancestors. The tree reveals the evolutionary history of all species and the sequence of speciation events that gave rise to them.

12.9 Evolutionary trees show ancestor-descendant relationships.

Many human traditions have a sacred "Tree of Life." The bodhi tree that the Buddha sat beneath to achieve the ultimate truth is the Tree of Enlightenment; the apple tree in the Garden of Eden is the Tree of Knowledge. The history of the relationship between all organisms that constitute life on earth is another tree of life.

The trunk and branches of an evolutionary tree represent ancestor-descendant relationships that link living organisms with all life that has ever existed on earth. The evolutionary tree of life can be thought of as one giant tree, but as a practical matter, biologists often study only particular branches. These branches can be illustrated as big trees or little trees and can be expanded to include whatever organisms the biologist is studying: the tree might, for example, include all animals or just the rodents.

In evolutionary trees, it does not matter on which side of the tree you put a particular group of organisms. Any branches can be spun around the **nodes** at which they split (see Figure 12-14). This pinwheel effect means that you cannot assume that rats are more evolutionarily advanced than mice, or vice versa. They are equally advanced in the sense that both groups derived from the same speciation event.

Q Are humans more advanced, evolutionarily, than cockroaches? Can bacteria be considered "lower" organisms?

Evolutionary trees do not tell us which groups are most "primitive" and which are most "advanced." This property of phylogenetic trees can serve to undermine some of our most sacred beliefs, such as the one that humans are the pinnacle of evolution. Many trees can be drawn to support this idea, including Figure 12-14. But notice that if you rotate a part of this tree around any one of its nodes, you can get a number of different trees.

What trees do tell us is which groups are most closely related to which other groups. One of the most interesting revelations of "tree thinking" is that—despite appearances—fungi, such as mushrooms, yeasts, and molds, are more closely related to animals than to plants (FIGURE 12-15). We investigate this surprising fact in Chapter 13.

Biologists use the term **monophyletic** to describe any group in which all of the individuals are more closely related to each other than to any individuals outside that group. Monophyletic groups are determined by looking at the

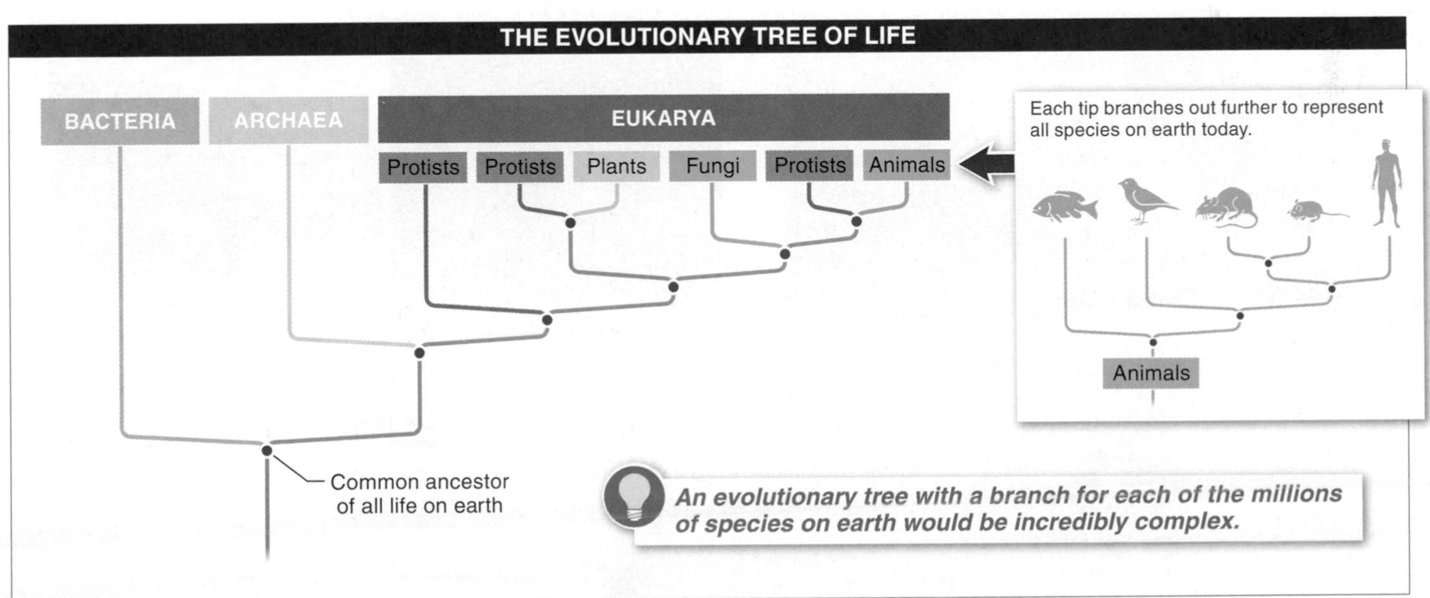
THE EVOLUTIONARY TREE OF LIFE
BACTERIA
ARCHAEA
EUKARYA
Protists
Protists
Plants
Fungi
Protists
Animals
Each tip branches out further to represent all species on earth today.
Animals
Common ancestor of all life on earth
An evolutionary tree with a branch for each of the millions of species on earth would be incredibly complex.

FIGURE 12-15 **A growing tree.** The evolutionary tree of life has branches with millions of tips representing all species on earth. As speciation events occur, new branches are added to the tree.

MONOPHYLETIC GROUPS

MONOPHYLETIC GROUP
Members of a monophyletic group share a common ancestor, and the group contains all of the descendants of that ancestor.

Lizards and crocodiles together do not make a monophyletic group: although they share a common ancestor, some descendants of that ancestor are in another group (the birds).

Birds
Crocodiles
Lizards
Mammals
A
B

Members of a monophyletic group are more closely related to each other than to any individuals outside that group.

FIGURE 12-16 **Similar appearance doesn't necessarily mean groups are monophyletic.**

nodes of the trees. For example, birds and crocodiles, taken together, compose a monophyletic group because they share a more recent common ancestor (designated by node A in FIGURE 12-16) than either group shares with lizards or mammals. Lizards and crocodiles, taken together, do not compose a monophyletic group, because their common ancestor (at node B) is also shared by birds. But birds, crocodiles, and lizards, all taken together, *do* compose a monophyletic group, by virtue of all three sharing the common ancestor at node B. Similarly, birds, crocodiles, lizards, and mammals also compose a monophyletic group. Notice also, in Figure 12-15, that the **protists** (or "Protista"), a kingdom of mostly single-celled eukaryotes, are not actually a monophyletic group, contrary to original expectations.

Constructing evolutionary trees by comparing similarities and differences among organisms. Reading an evolutionary tree reveals which groups are most closely related and approximately how long ago they shared a common ancestor. But how are evolutionary trees—which might hypothesize historical events such as the cat-dog split that happened 60 million years ago—constructed in the first place?

Until recently, these trees were assembled by looking carefully at numerous physical features, or traits, of species and generating tables that compared these features across the species. FIGURE 12-17 is a simple example of such a table, showing a clear split between the characteristics of the lion and the hyena, on the one hand, and those of the wolf and the bear, on the other. For most of the 20th century, biologists classifying organisms would often use 50 or more traits to generate a tree.

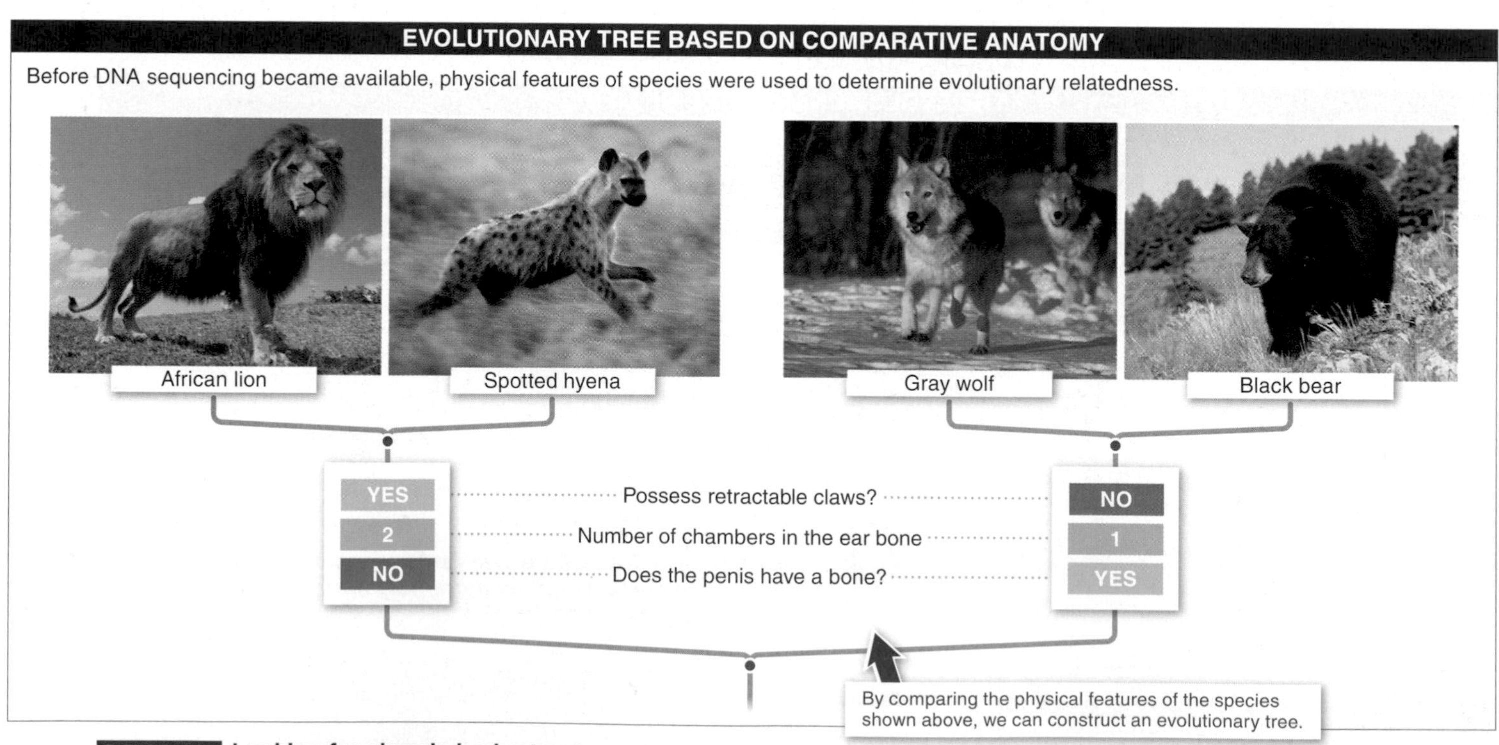

EVOLUTIONARY TREE BASED ON COMPARATIVE ANATOMY
Before DNA sequencing became available, physical features of species were used to determine evolutionary relatedness.
African lion
Spotted hyena
Gray wolf
Black bear
YES
2
NO
Possess retractable claws?
Number of chambers in the ear bone
Does the penis have a bone?
NO
1
YES
By comparing the physical features of the species shown above, we can construct an evolutionary tree.

FIGURE 12-17 **Looking for clues in body structures.**

> "We all should know that diversity makes for a rich tapestry, and we must understand that all the threads of the tapestry are equal in value no matter what their color."
>
> — MAYA ANGELOU, in *Wouldn't Take Nothing for My Journey Now*, 1993

Then, beginning in the 1980s, biologists began using molecular sequences rather than physical traits to generate evolutionary trees. The rationale for this approach is that organisms inherit DNA from their ancestors and so, as species diverge, their DNA sequences also diverge, becoming increasingly different. As more time passes following the splitting of one species into two, the differences in their DNA sequences become greater. By comparing how similar the DNA sequences are between two groups, it is possible to estimate how long ago they shared a common ancestor (FIGURE 12-18).

In theory, using DNA sequences to construct evolutionary trees is not really different from using physical traits. Each DNA base pair can be thought of as a "trait," and many, many more such "traits" can be compared.

Although most evolutionary trees are now produced by using molecular sequence comparisons, when it comes to actually naming species we still use Linnaeus's system, assigning a species name and fitting the new species within the other categories such as kingdom, class, and family. This can be difficult, because there is no objective way to assign groups above the level of species—for example, deciding whether multiple genera should be grouped in the same or different families, or whether multiple families should be grouped in the same or different orders.

DNA-BASED EVOLUTIONARY TREES

By comparing how similar the DNA sequences are between two groups, we can estimate how long it has been since they shared a common ancestor.

Species 1 and 2 are more similar to each other than they are to either species 3 or 4. This is because they probably share a more recent common ancestor.
Difference in DNA base pairs
Species 1
Species 2
Species 3
Species 4
A
T
C
G
The more similar their DNA sequences, the more closely two species are related.

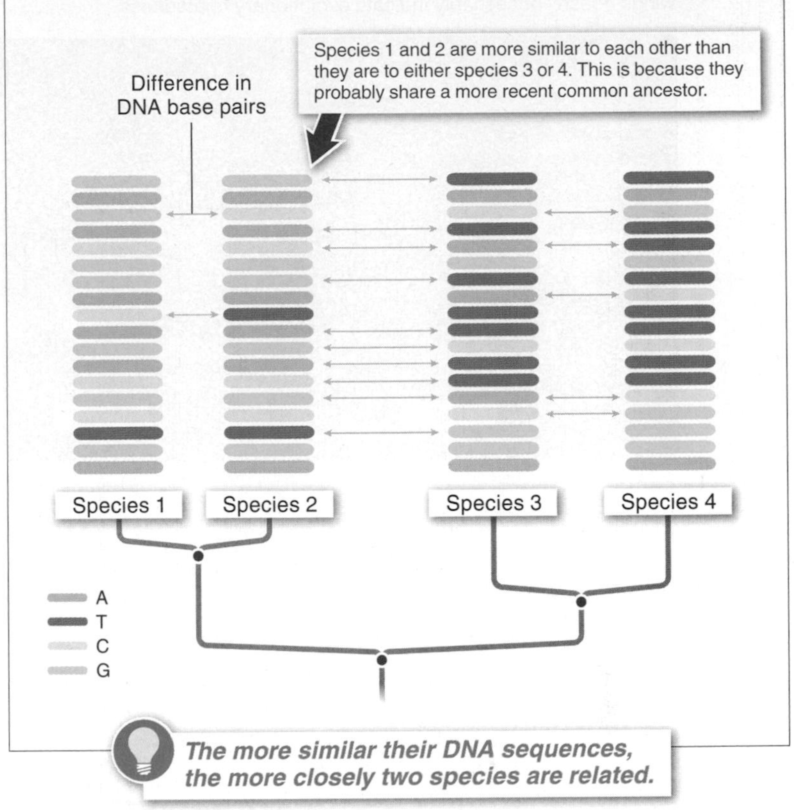

FIGURE 12-18 DNA sequences reveal evolutionary relatedness.

TAKE HOME MESSAGE 12.9

» Evolutionary trees constructed by biologists are hypotheses about the ancestor-descendant relationships among species. The trees represent an attempt to describe which groups are most closely related to which other groups.

12.10 Similar structures don't always reveal common ancestry.

Whales and sharks have fins; bats and many insects have wings. Do bats have wings because they are most closely related to insects and inherited their wings from a common, winged ancestor? Or did insects' and bats' wings evolve independently? As the evolutionary tree in FIGURE 12-19 makes clear, insects' wings and bats' wings evolved independently—an adaptation that arose separately on more than one occasion.

Mapping species' characteristics onto phylogenetic trees provides us with the story of evolution. As we've seen, before the 1980s, biologists tried to decode the process of

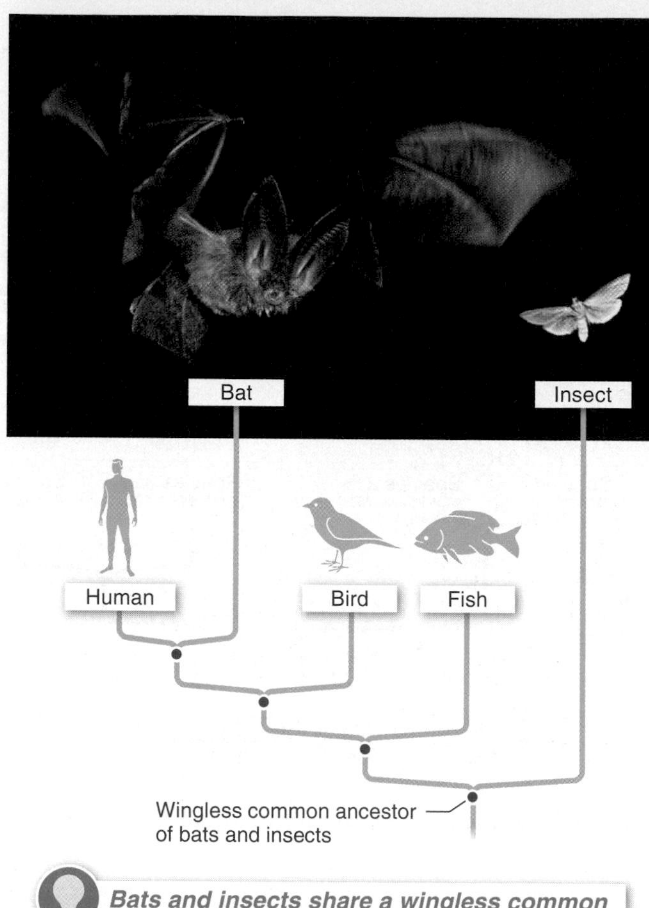

FIGURE 12-19 Separate paths to flight.

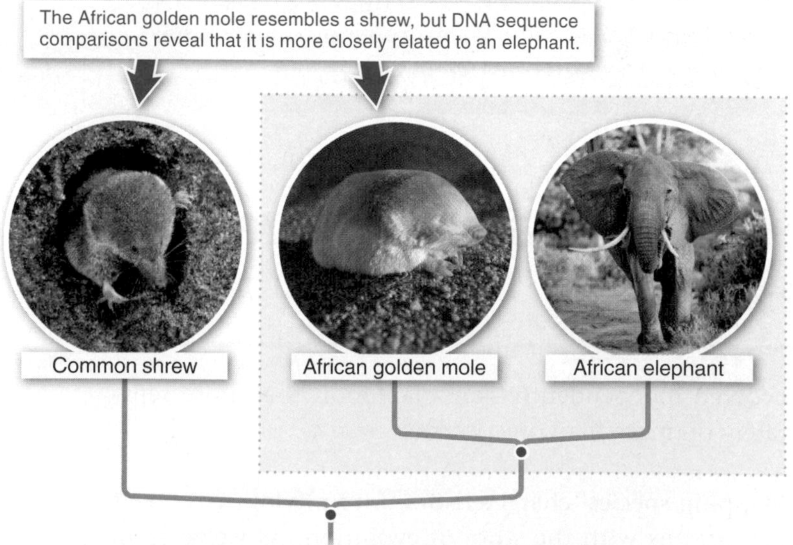

The African golden mole resembles a shrew, but DNA sequence comparisons reveal that it is more closely related to an elephant.
Common shrew
African golden mole
African elephant

FIGURE 12-20 Looks can be deceiving. Which two of these three animals are more closely related? Can you tell just by looking?

evolution by comparing physical features of organisms. Then, with the advent of methods for comparing DNA sequences, tracing the history of life on earth became a more rigorous science. Let's look at a case that illuminates why the original methods were weaker (FIGURE 12-20).

Initially, biologists thought that African golden moles belonged in the order known as insectivores, which includes shrews, hedgehogs, and other moles. Moles have many characteristics in common with these other animals: they are small, they have long, narrow snouts, their eyes are tiny, and they live in underground burrows. Biologists thought that this group of characteristics evolved just once, and that all species in the insectivore order possessed these characteristics because they inherited them from a common ancestor.

Q Humans have several fused vertebrae at the bottom of their spinal cord that look like a tiny, internal "tail." Are we evolving a tail or losing a tail?

Recent DNA evidence reveals that African golden moles are more closely related to elephants than to the insectivores, including all of the other mole species! Why do they look so similar to the insectivores? Because of a phenomenon called **convergent evolution,** which occurs when populations of different organisms live in similar environments and so experience similar selective forces. **Analogous traits** are characteristics (such as bat wings and insect wings) that are similar because they were produced by convergent evolution, not because they descended from a common structure in a shared ancestor. Features that are inherited from a common ancestor are called **homologous features.** All mammals have hair, for example, because they inherited this trait from a common ancestor.

Analogous features can cause problems when biologists are constructing evolutionary trees, because they are the result of natural selection rather than relatedness. But we can determine whether traits are homologous or analogous by using DNA analysis. DNA sequences do not become increasingly similar during convergent evolution, so molecular phylogenies cannot be fooled. This is an important reason why molecular-based evolutionary trees are preferable to trees based on physical features.

Although analogous features are not helpful in constructing evolutionary trees, convergence provides some of the best evidence for the power of natural selection. For example, from the 19th to the 20th centuries in industrialized parts of Europe, a large number of distinct butterfly and moth species became darker and darker. This change was not the result of their sharing a common ancestor, but was due to natural selection. The light-colored moths and butterflies were easy targets for predators when they landed on trees that were covered by dark, industry-produced soot. As a consequence, natural selection led to a rapid evolution of body color, with many species converging on a darker coloration at the same time.

TAKE HOME MESSAGE 12.10

» Evolutionary trees are best constructed by comparing organisms' DNA sequences rather than physical similarities, because convergent evolution can cause distantly related organisms to appear closely related, but it doesn't increase their DNA sequence similarity.

12.11–12.14

Macroevolution gives rise to great diversity.

More than 400 Anolis lizard species exist. (Shown here: A blue-throated anole, *Anolis chrysolepsis*, from Ecuador.)

12.11 Macroevolution is evolution above the species level.

When water runs over rocks, it wears them away. The process is simple and slow, yet powerful enough to have created the Grand Canyon. No additional physical processes are necessary. The process of evolution has a lot in common with a stream of water wearing away solid rock: in the short term, it produces small changes in a population, yet the accumulation of these changes over the long term can be "canyonesque" (FIGURE 12-21). Let's consider some examples.

The production of 200-ton dinosaurs from rabbit-size reptile ancestors. The diversification from a single species of flowering plant into more than 230,000 species. These examples of large-scale changes, the products of evolutionary change involving the origins of entirely new groups of organisms, are referred to as **macroevolution.** This can be contrasted with phenomena that involve changes in allele frequencies within a population—referred to as **microevolution**—such as the increase in milk production by cows during the 20th century, or the gradual

No additional processes necessary! Just as a stream of water running over rocks can create the Grand Canyon, accumulated microevolutionary changes can lead to dramatic macroevolutionary changes.

FIGURE 12-21 From a trickle of water to the Grand Canyon.

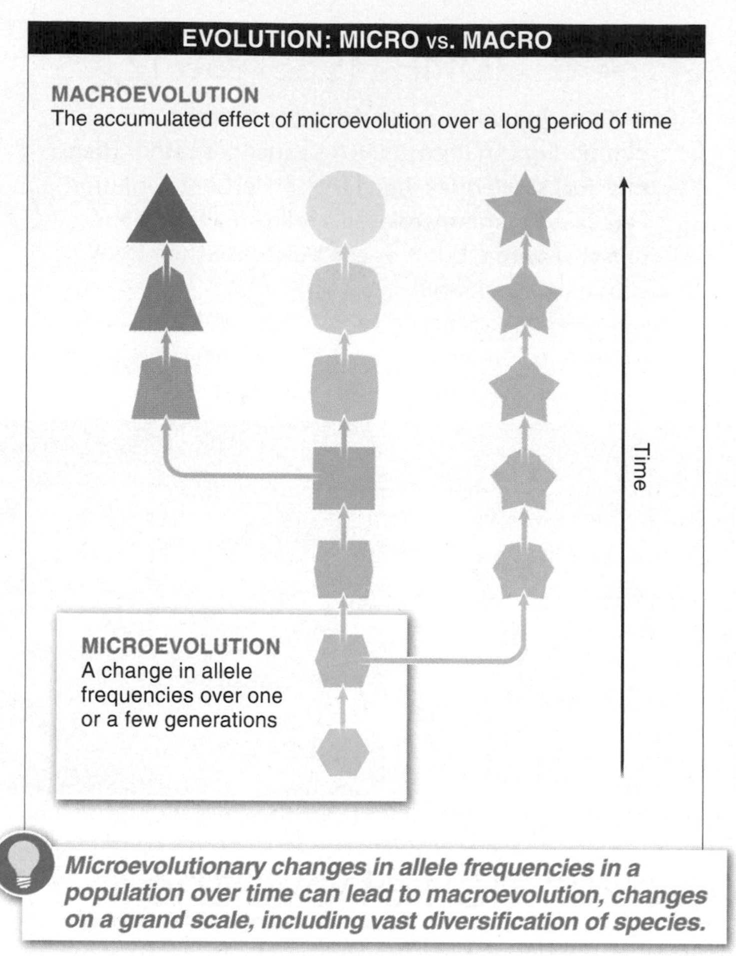

FIGURE 12-22 **The scale of evolution.**

change in the average beak size of birds with changing patterns of rainfall (FIGURE 12-22).

These micro and macro events might seem like two very different processes, but they are not. Evolution, whether at the micro or macro level, is one thing only: a change in allele frequencies in a population. In the short term, over one or a few generations, evolution can appear as a slight and gradual change within a species. Over the long term, these gradual, accumulating changes may combine with reproductive isolation, causing the dramatic phenomenon described as macroevolution. Just as a trickle of water, given enough time, produced the Grand Canyon, so evolution has created the endless forms of diversity on earth. In a sense, microevolution is the process and macroevolution is the result.

TAKE HOME MESSAGE 12.11

» The process of evolution—changes in allele frequencies within a population—in conjunction with reproductive isolation is sufficient to produce speciation and the rich diversity of life on earth.

12.12 The pace of evolution is not constant.

If you listened to some scientists debating their research, you might hear one side describing "evolution by jerks" and the other side saying "evolution by creeps." They're not gearing up for a fight out in the schoolyard. They're just debating the pace of evolution.

The traditional model of evolutionary change recognizes that populations slowly but surely accumulate sufficient genetic differences for speciation—hence the phrase "evolution by creeps." Spurred on by findings from the fossil record that do not always support this view, however, researchers have come to believe that evolution may often occur in a different way: brief periods of rapid evolutionary change immediately after speciation, followed by long periods with relatively little change—hence "evolution by jerks." This newer view of the pace of evolution, in which long periods of relatively little evolutionary change are punctuated by bursts of rapid change, is called **punctuated equilibrium** (FIGURE 12-23).

In nature, we find examples of both gradual change and the irregular pattern of punctuated equilibrium. For many groups of organisms, such as mammals, the fossil record reveals a long period in which particular species seem to change very little. This period is followed by the appearance of a large number of newer, but clearly related, species—but no record of fossils that transition from the old to the new. For other groups, such transitional fossils do exist. For example, very complete sets of fossils reveal the transitional sequence from fox-like ancient terrestrial mammals to modern whales.

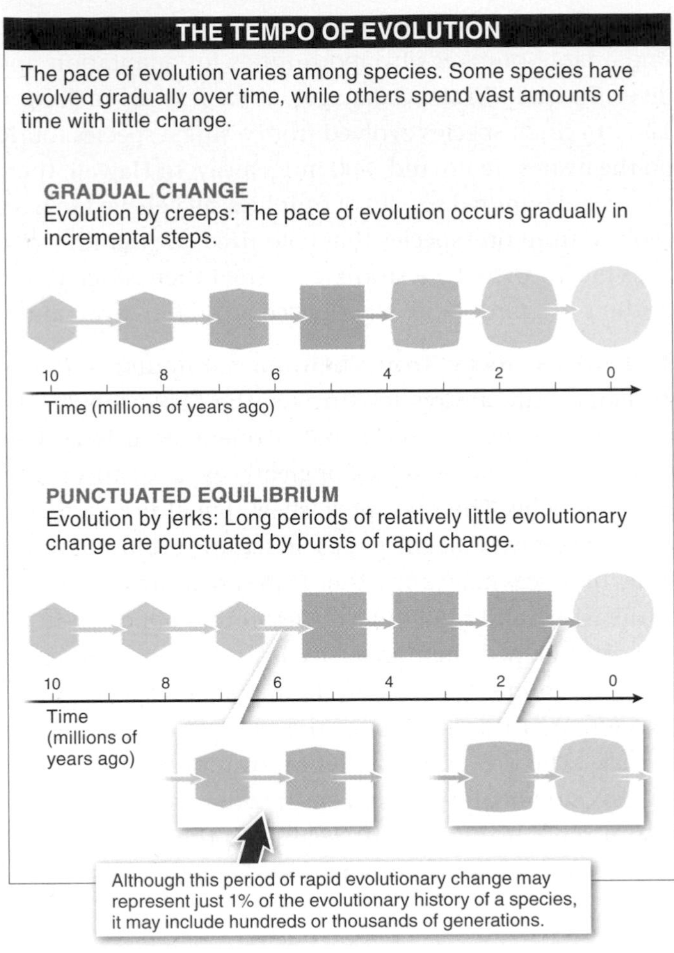

THE TEMPO OF EVOLUTION
The pace of evolution varies among species. Some species have evolved gradually over time, while others spend vast amounts of time with little change.
GRADUAL CHANGE
Evolution by creeps: The pace of evolution occurs gradually in incremental steps.
10
8
6
4
2
0
Time (millions of years ago)
PUNCTUATED EQUILIBRIUM
Evolution by jerks: Long periods of relatively little evolutionary change are punctuated by bursts of rapid change.
10
8
6
4
2
0
Time (millions of years ago)
Although this period of rapid evolutionary change may represent just 1% of the evolutionary history of a species, it may include hundreds or thousands of generations.

FIGURE 12-23 **The pace of evolution can be rapid or slow.**

In the end, there is no single rate at which evolution occurs across all species. Some species have spent vast periods of time with little change at all. The fish species called coelacanths, for example, seem to have undergone almost no physical changes over more than 300 million years. Other species have changed gradually, for their entire evolutionary history. For any organism, the rate of evolutionary change depends on the selective forces acting on the population. Strong and directional forces may act over long periods of time and lead to rapid change, whereas stabilizing selection may lead to very little change.

Darwin foreshadowed the discussion about the pace of evolution in *The Origin of Species,* writing that "the periods during which species have undergone modification, though long as measured in years, have probably been short in comparison with the periods during which they retain the same form."

TAKE HOME MESSAGE 12.12

» The pace at which evolution occurs can be rapid or very slow. In some cases, the fossil record reveals long periods with little evolutionary change punctuated by rapid periods of change. In other cases, species may change at a more gradual but consistent pace.

12.13 Adaptive radiations are times of extreme diversification.

As mammals, it might seem that we owe a little of our success to luck. Sixty-five million years ago, our mammalian ancestors were rodent-size, insect-eating, nocturnal creatures that didn't remotely possess the dominant position we hold today. It was the dinosaurs' time, and the giant reptiles dominated the earth.

All that changed in an instant when the earth was struck by an asteroid, about 6 miles (10 km) in diameter, near what is now the eastern part of Mexico. Environmental conditions on earth changed quickly and drastically. We'll explore the details of this catastrophic event in the next section, but one outcome was extreme: almost all of the dinosaur species were wiped out. Our mammalian ancestors, though, survived and found themselves living on a planet where most of their competitors had suddenly disappeared. We were in the right place at the right time.

What followed was an explosive expansion of mammalian species. In a brief period of time, a small number of species diversified into a much larger number of species, able to live in a wide diversity of habitats. Called an **adaptive radiation,** such large, rapid diversifications have occurred many times throughout history (FIGURE 12-24).

Three different types of phenomena tend to trigger adaptive radiations.

1. Mass extinction events. With the near-total disappearance of the dinosaurs, mammals suddenly had few competitors. Not surprisingly, the number of mammalian

ADAPTIVE RADIATION

Adaptive radiation occurs when a small number of species diversify into a larger number of species. Three phenomena tend to trigger adaptive radiation.

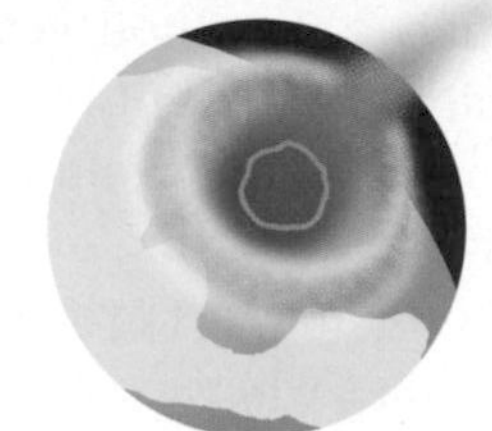

MASS EXTINCTION EVENTS
With their competition suddenly eliminated, remaining species can rapidly diversify.

COLONIZATION EVENTS
Moving to a new location with new resources (and possibly fewer competitors), colonizers can rapidly diversify.

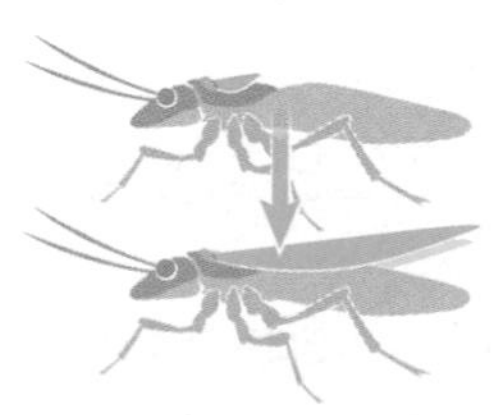

EVOLUTIONARY INNOVATIONS
With the evolution of an innovative feature that increases fitness, a species can rapidly diversify.

FIGURE 12-24 **A rapid diversification of species.**

species increased from perhaps just a few hundred to more than 4,000 species in about 130 genera. This happened over about 10 million years, barely the blink of an eye by geological standards. Following other large-scale extinctions, numerous other groups that suddenly lost most of their competitors experienced similar adaptive radiations.

2. Colonization events. In a rare event, one or a few birds or small insects will fly off from a mainland and end up on a distant island group, such as Hawaii or the Galápagos Islands. Once there, the new arrivals tend to find a large number of opportunities for adaptation and diversification. In the Galápagos, as we learned in Section 12-7, 14 finch species evolved from a single species found on the nearest mainland, 600 miles away. In Hawaii, there are several hundred species of fruit flies, all believed to have evolved from one species that colonized the islands—perhaps blown there by a storm, or carried there stuck in the feathers of a bird—and experienced an adaptive radiation.

3. Evolutionary innovations. Computer software developers are always looking for the "killer app"—the new application so useful that it opens up a large new niche in the software market or greatly expands an already-existing niche. The first spreadsheet, email program, and web browser were all killer apps. In nature, evolution sometimes produces killer apps, too. These are innovations such as the wings and rigid outer skeleton that appeared in insects and helped them diversify into the most successful group of animals, with more than 800,000 species today (more than a hundred times the number of mammalian species). The flower is another innovation that propelled an explosion of diversity and ensured the evolutionary success of the flowering plants relative to the non-flowering plants, such as ferns and pine trees. Today, about 9 out of 10 plant species are flowering plants.

TAKE HOME MESSAGE 12.13

›› Adaptive radiations—brief periods of time during which a small number of species diversify into a much larger number of species—tend to be triggered by mass extinctions of potentially competing species, colonizations of new habitats, or the appearance of evolutionary innovations.

12.14 There have been several mass extinctions on earth.

"Forests keep disappearing, rivers dry up, wild life's become extinct, the climate's ruined and the land grows poorer and uglier every day."

— ANTON CHEKHOV, *Uncle Vanya*, 1899

If the past is a guide to the future, we know this: no species lasts forever. Speciation is always producing new species, and **extinction,** the complete loss of all individuals in a species population, takes them away. Extinction is always occurring and is faced by all species.

For any given time in earth's history, scientists can estimate the rate of extinctions. This rate can be expressed in several different ways, such as the number of species that become extinct over a given period of time. And the evidence reveals that these rates are far from constant (FIGURE 12-25). Although the details differ in most cases, extinctions generally fall into one of two categories: "background" extinctions or mass extinctions.

Background extinctions are those that occur at lower rates, during periods other than times of mass extinctions. Background extinctions occur mostly as the result of natural selection. Competition with other species, for example, may reduce a species' population size or the range over which it can roam or grow. Or a species might be too slow to adapt to gradually changing environmental conditions and become extinct as its individuals die off.

Mass extinctions are periods during which a large number of species on earth become extinct over a relatively short period. There have been at least five mass extinctions on earth and, during each of these, 50% or more of the animal species living at the time became extinct. Mass extinctions are due to extraordinary and sudden changes to the environment (such as an asteroid impact). As a consequence, nothing more than bad luck is responsible for the extinction of species during mass extinctions; fit and unfit individuals alike perish.

Sixty-five million years ago, an asteroid 6 miles in diameter wiped out three-fourths of all species on earth, including almost all the dinosaurs.

FIGURE 12-26 **Global climate change!** Worldwide mass extinction? Catastrophic events set the stage for explosive speciation of the remaining groups, including the mammals.

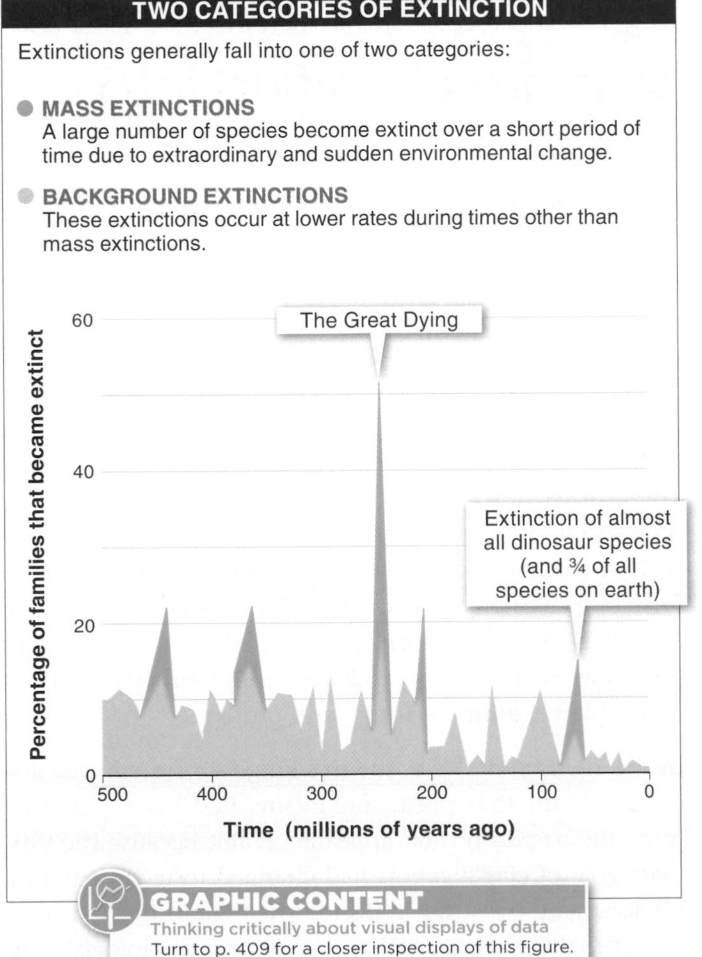

FIGURE 12-25 **Extinction never sleeps.**

Of the five mass extinctions during the past 500 million years, the most recent is also the best understood: the asteroid crash, described in the previous section, that cleared the way for our mammalian ancestors to diversify. This massive asteroid smashed into the Caribbean near the Yucatán Peninsula of Mexico 65 million years ago (FIGURE 12-26). The impact left a crater more than 100 miles wide and probably created an enormous fireball that caused fires worldwide, followed by a cloud of dust and debris that blocked all sunlight from the earth and disturbed the global climate for months. In the aftermath of this catastrophe, about 75% of all species on earth were wiped out, including almost all dinosaurs—a tremendously successful group that had been thriving for 150 million years. (Among the rare surviving dinosaurs were those that would become the birds.)

As bad as the dinosaur-devastating asteroid event was, from a biodiversity perspective it is not the worst catastrophe in earth's history. That distinction falls to a mass extinction that took place 250 million years ago, called the Great Dying (see Figure 12-25). Although the cause is not clear—hypotheses include an asteroid impact, continental drift, a supernova, or extreme volcanic eruptions—more than 95% of all marine life became extinct, along with almost 75% of all terrestrial vertebrates. The causes of the three other mass extinctions also are poorly understood, but the evidence in the fossil record of the tremendous upheavals is clear.

An important question now being debated is whether we are currently in the midst of a human-caused mass extinction event. We'll explore whether this is true—and why—in Chapter 18.

TAKE HOME MESSAGE 12.14

» As new species are being created, others are lost through extinction, which may be a consequence of natural selection or large, sudden changes in the environment. Mass extinctions are periods during which a large number of species on earth become extinct over a short period of time.

12.15–12.18 An overview of the diversity of life on earth: organisms are divided into three domains.

The Gouldian finch (*Erythrura gouldiae*) is a colorful bird native to Australia. There is evidence of population decline in the wild. Large numbers are bred in captivity, particularly in Australia.

12.15 All living organisms are classified into one of three groups.

Biological diversity can be humbling. At the very least, it makes it harder to believe that humans are particularly special. We are not at the center of earth's "family tree." Nor are we at its peak. We are simply one branch.

When Linnaeus first put together his system of classification, he saw a clear and obvious split: all living organisms were either plant or animal. Plants could not move and could make their own food. Animals could move but could not make their own food. So in Linnaeus's original classification, all organisms were put in either the animal kingdom or the plant kingdom (his third, "mineral" kingdom, now abandoned, included only nonliving matter).

With the refinement of microscopes and subsequent discovery of the rich world of **microbes**—microscopic organisms—the two-kingdom system was inadequate. Where did the microbes belong? Some could move, but many of those could also make their own food, seeming to put them somewhere between plants and animals. Furthermore, mushrooms and molds, among other organisms originally categorized as plants, didn't move but they didn't make their own food either—they digested the decaying plant and animal material around them.

The two-kingdom system gave way in the 1960s to a five-kingdom system. At its core, the new system was a division based on the distinction between *prokaryotic* cells (those without nuclei) and *eukaryotic* cells (those with nuclei). The prokaryotes were put in one kingdom, in which the only residents were the bacteria: single-celled organisms with no nucleus, no organelles, and genetic material in the form of a circular strand of DNA. The eukaryotes—having a nucleus, compartmentalized organelles, and individual, linear pieces of DNA—were divided into four separate kingdoms: plants, animals, fungi, and protists.

In the 1970s and 1980s, the five-kingdom system was discarded. Until that point, organisms had been classified primarily based on their appearance. But because the ultimate goal of classification had changed to reconstructing phylogenetic trees that reflected the evolutionary history of earth's diversity, Carl Woese, an American biologist, and his colleagues began classifying organisms by their nucleotide sequences.

Q How can one molecule support the claim that all living organisms probably evolved from a single common ancestor?

Woese assumed that the more similar the genetic sequences were between two species, the more closely related they were, and he built phylogenetic trees accordingly. The only way Woese could compare the evolutionary relatedness of all the organisms present on earth today was by examining one molecule that was found in *all* living organisms and looking at the degree to which it differed from species to species. That molecule is ribosomal RNA. Ribosomal RNA helps translate genes into proteins (see Chapter 6), and it carries out this function in all organisms on earth, almost certainly because it comes from a common ancestor. Over time, however, its genetic sequence (i.e., the DNA that codes for the RNA) has changed a bit. Tracking these changes makes it possible to reconstruct the process of diversification and change that has taken place.

The genetic sequence data gathered by Woese and colleagues generated some surprising evolutionary trees. First and foremost, the biggest division in the diversity of life on earth was not between plants and animals or between prokaryotes and eukaryotes. The new trees revealed instead that the diversity among microbes was much, much greater than ever imagined. In fact, a completely new group of prokaryotes was identified. Called **archaea** (*sing.* archaeon), they thrive in some of the most extreme environments on earth and differ greatly from bacteria. The tree of life was revised to show three primary branches, called domains: the bacteria, the archaea, and the eukarya (FIGURE 12-27). The domain names are often capitalized: Bacteria, Archaea, and Eukarya.

In Woese's new system, which is the most widely accepted classification scheme today, the domains are above the kingdom level in the Linnaean system. The bacteria and archaea domains each have one kingdom and the eukarya domain has four.

Because both bacteria and archaea are microscopic, it can be hard to believe that the two domains are as different from each other as either domain is from the eukarya. However, each of the three domains is monophyletic, meaning that each contains species that share a common ancestor and includes all descendants of that ancestor. Close inspection even reveals that the archaea are more closely related to the eukarya than to the bacteria.

The three-domain, six-kingdom approach is not perfect and is still subject to revision. For example, within the eukarya, the kingdom of single-celled protists has turned out to be much more diverse than initially thought, and protists are not a monophyletic group. Increasingly, scientists recognize that the protists should be split into multiple kingdoms. Another problem is that bacteria sometimes engage in **horizontal gene transfer,** which means that, rather than passing genes simply from "parent" to "offspring," they transfer genetic material directly into another species. This process complicates attempts to determine phylogenies based on sequence data, because it creates situations in which two organisms might have a similar genetic sequence, not because they share a common ancestor, but as a result of a direct transfer of the sequence from one species to another.

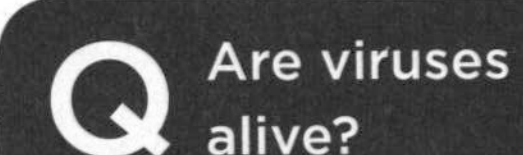

Additionally, a fourth group of incredibly diverse and important biological entities, the **viruses,** is not even included in the tree of life, because they are not considered to be living organisms. Viruses can replicate, but they can have metabolic activity only by taking over the metabolic processes of another organism. Their lack of metabolic activity puts viruses just outside the definition

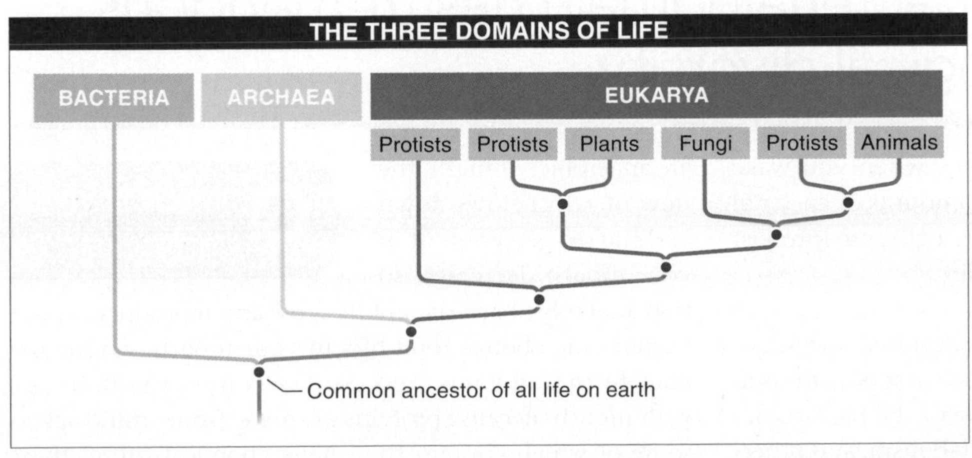

FIGURE 12-27 All living organisms are classified into one of three groups.

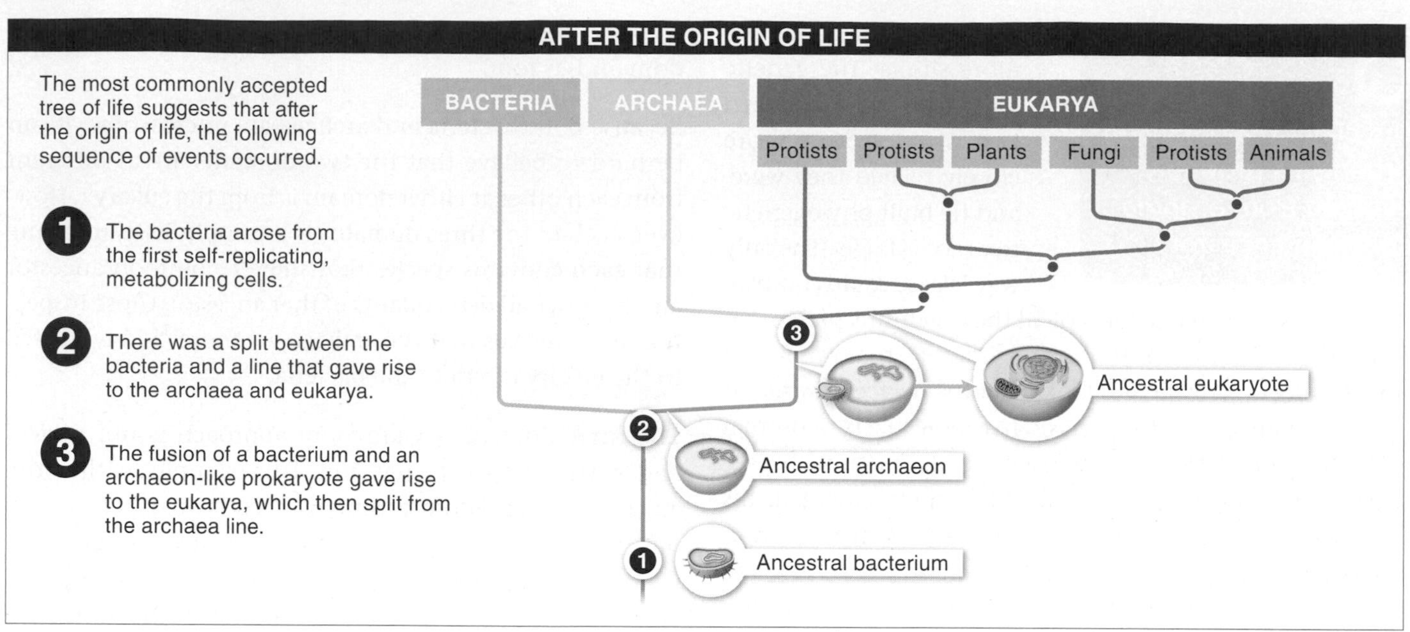
AFTER THE ORIGIN OF LIFE
The most commonly accepted tree of life suggests that after the origin of life, the following sequence of events occurred.
1 The bacteria arose from the first self-replicating, metabolizing cells.
2 There was a split between the bacteria and a line that gave rise to the archaea and eukarya.
3 The fusion of a bacterium and an archaeon-like prokaryote gave rise to the eukarya, which then split from the archaea line.
BACTERIA
ARCHAEA
EUKARYA
Protists
Protists
Plants
Fungi
Protists
Animals
Ancestral eukaryote
Ancestral archaeon
Ancestral bacterium

FIGURE 12-28 From self-replicating, metabolizing cells to complex organisms. The biggest branches on the tree of life separate the three domains.

of life that we use in this book, but some scientists do view viruses as living.

The most commonly accepted tree of life suggests that, after the origin of life, the following sequence of events occurred (FIGURE 12-28):

1. The bacteria arose from the first self-replicating, metabolizing cells.
2. There was a split between the bacteria and a line that gave rise to the archaea and eukarya.
3. The fusion of a bacterium and an archaeon-like prokaryote gave rise to the eukarya, which then split from the archaea line.

The next three sections introduce each of the three domains, surveying the broad diversity that has evolved within each domain and the common features that link all the members of each. Chapters 13, 14, and 15 cover the domains in greater detail.

TAKE HOME MESSAGE 12.15

» All life on earth can be divided into three domains—bacteria, archaea, and eukarya—which reflect species' evolutionary relatedness to one other. Plants and animals are just two of the four kingdoms in the eukarya domain, encompassing only a small fraction of the domain's diversity.

12.16 The bacteria domain has tremendous biological diversity.

Morning breath is stinky. You see, when you wake up, your mouth contains huge amounts of bacterial waste products. Bad breath can give us a glimpse into just how diverse and resourceful bacteria are.

At any given time, there are several hundred species of bacteria in your mouth—mostly on your tongue—all competing for resources (FIGURE 12-29). Some of the bacteria are aerobic, requiring oxygen for their metabolism, and others are anaerobic. At night, the flow of saliva slows down and the oxygen content of your mouth decreases, so that anaerobic bacteria rapidly grow and reproduce. These bacteria metabolize food bits in your mouth, plaque on your teeth and gums, and dead cells from the lining of your mouth. Because proteins are made from amino acids, some of which contain the smelly chemical sulfur, their

Q Why is morning breath so stinky?

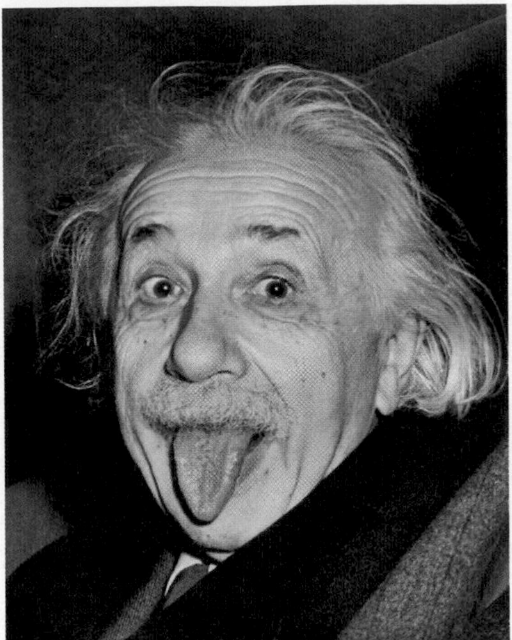

FIGURE 12-29 **Bacteria thrive on your tongue and in your mouth.**

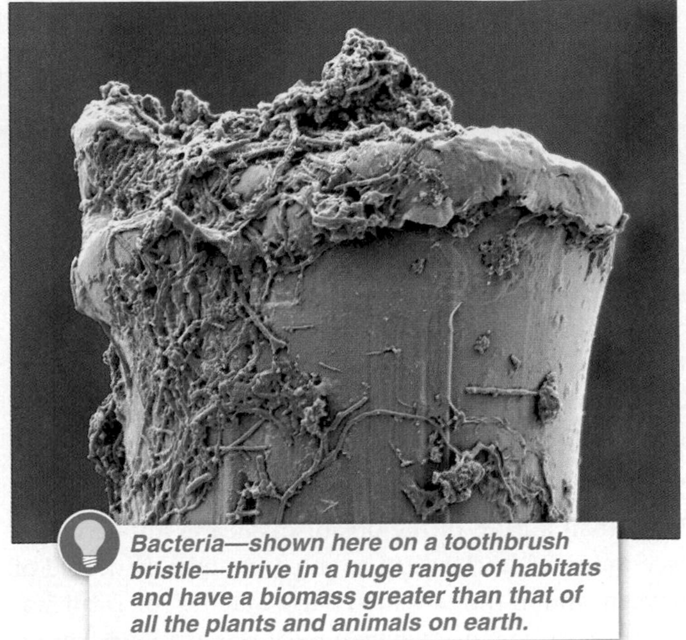

Bacteria—shown here on a toothbrush bristle—thrive in a huge range of habitats and have a biomass greater than that of all the plants and animals on earth.

FIGURE 12-30 **We're outnumbered.**

breakdown leads to the odor in the accumulating waste products.

Once you wake up, you breathe more and produce more saliva, both of which increase the oxygen level in your mouth, making the environment favorable for aerobic bacteria. Because aerobic bacteria prefer carbohydrates as their energy source, and because carbohydrates don't contain sulfur, the sulfur smell goes away as the aerobic bacteria start to outcompete the anaerobic bacteria.

On a small scale, your mouth reveals some of the tremendous biological versatility of the bacteria: hundreds of species can live in a tiny area—a teaspoon of soil, for example, is home to more than a billion bacteria—they can thrive in a variety of unexpected habitats, they can utilize a variety of food sources, and they can survive and thrive with or without oxygen. By any measure, this is their planet: the biomass of bacteria (if they were all collected, dried out, and weighed) exceeds that of all the plants and animals on earth (FIGURE 12-30).

While the various species differ in many ways, the bacteria are a monophyletic group, sharing a common ancestor. For this reason, they all have a few features in common. All bacteria are single-celled organisms with no nucleus or organelles, with one or more circular molecules of DNA as their genetic material, and using several methods of exchanging genetic information. Because they are asexual—they reproduce without a partner, just by dividing—the biological species concept cannot be applied to bacteria. As a consequence, bacteria are classified on the basis of physical appearance or, preferably, genetic sequences.

Although bacteria are responsible for many diseases, including strep throat, pneumonia, anthrax, leprosy, and tuberculosis, disease-causing bacteria are only a small fraction of the domain. Bacteria are essential to our lives. Bacteria (*E. coli*) living in your gut help your body digest the food you eat and, in the process, make certain vitamins your body needs. Other bacteria produce antibiotics such as streptomycin. Still others live symbiotically with plants as small fertilizer factories, converting nitrogen into a form that is usable by the plant. Bacteria also give taste to many foods, from sour cream to cheese, yogurt, and sourdough bread. Increasingly, bacteria are used in medical and industrial applications, as described in Chapter 7.

We explore the great diversity of the bacteria in more detail in Chapter 15.

TAKE HOME MESSAGE 12.16

» All bacteria share a common ancestor and have a few features in common. They are prokaryotic, asexual, single-celled organisms with no nucleus or organelles, with one or more circular molecules of DNA as their genetic material, and using several methods of exchanging genetic information. Bacteria have a much broader diversity of metabolic and reproductive abilities than do the eukarya.

12.17 The archaea domain includes many species living in extreme environments.

In a bubbling hot spring in Yellowstone National Park, the temperature ranges from boiling water, at 212° F (100° C), down to a relatively cool 165° F (74° C) at the surface. It would seem a most inhospitable place for life. Yet researchers have found 38 different species of archaea thriving there. Perhaps even more surprising, the genetic differences among these species are more than double the genetic differences between plants and animals.

In the freezing waters of Antarctica, too, archaea abound. More than a third of the organisms in the Antarctic surface waters are archaea. In swamps, completely devoid of oxygen, and in the extremely salty water of the Dead Sea, the story is the same: where once it was assumed that no life could survive, the archaea not only exist but thrive and diversify (FIGURE 12-31).

Champagne pool,
Waiotapu, New Zealand
Geogemma barossii, Strain 121,
from a 121° C hydrothermal vent
in the Pacific Ocean

FIGURE 12-31 Archaea can thrive in a diversity of environments, including the most inhospitable-seeming places.

Q Will life be found elsewhere in the universe? How does the discovery of archaea alter that likelihood?

Until relatively recently, our perception of life on earth was that there were bacteria and there were eukaryotes, such as the plants and animals. But as researchers have explored some of the most unlikely of habitats, they have found archaea thriving.

Analyses of genetic sequences indicate that the archaea and the bacteria diverged about 3 billion years ago and that the eukarya split off from the archaea approximately 2.5 billion years ago. The archaea are grouped in one kingdom within the domain archaea, but we have no idea how many species exist. Given that they are the dominant microbes in the deep seas, it may very well be that archaea are the most common organisms on earth.

We do know that, like bacteria, all archaea are single-celled prokaryotes. For that reason, under a microscope they look very similar to bacteria. Several physical features distinguish them from the bacteria, however. Specifically, the archaeal cell walls contain polysaccharides not found in either bacteria or eukaryotes. The archaea also have cell membranes, ribosomes, and some enzymes similar to those found in the eukarya.

The archaea are divided into five groups based on their physiological features.

1. Thermophiles ("heat lovers"), which live in very hot places
2. Halophiles ("salt lovers"), which live in very salty places
3. High- and low-pH-tolerant archaea
4. High-pressure-tolerant archaea, found as deep as 4,000 meters (about 2.5 miles) below the ocean surface, where the pressure is almost 6,000 pounds per square inch (compared with an air pressure of less than 15 pounds per square inch at sea level)
5. Methanogens, which are anaerobic and produce methane

There also seem to be large numbers of archaea living in relatively moderate environments that are also commonly home to bacteria. We explore the great diversity of the domain archaea in more detail in Chapter 15.

TAKE HOME MESSAGE 12.17

» Archaea, many of which are adapted to life in extreme environments, physically resemble bacteria but are more closely related to eukarya. Because they thrive in many habitats that humans have not yet studied well, including the deepest oceans, they may turn out to be more common than currently believed.

12.18 The eukarya domain consists of four kingdoms: plants, animals, fungi, and protists.

After exploring the archaea and bacteria domains, turning to the eukarya feels like coming home. We are most familiar with organisms in this domain: plants, mushrooms, insects, fish, birds, and, of course, mammals, including humans. Three kingdoms of eukarya can be seen with the naked eye: plants, animals, and fungi (FIGURE 12-32). All are made up of eukaryotic cells—they have a membrane-enclosed nucleus—and each kingdom is almost entirely multicellular.

A fourth kingdom contains the protists, which are often too small to be seen by the naked eye, and is a sort of grab bag that includes a wide range of mostly single-celled eukaryotic organisms, including amoebas, paramecia, and algae. Discovery of new species of protists, like that of the other microscopic organisms on earth (the bacteria and archaea), continues at a very high rate. As we noted earlier, biologists now know that the protists are not a monophyletic group, and they are increasingly splitting them into multiple kingdoms within the eukarya.

Because they are so much easier to see than bacteria and archaea, a disproportionate number of the named species on earth are in the domain eukarya. In fact, of the 1.5 million named species, the majority are eukarya, with about half being insects. This is more a result of the biases of biologists than a reflection of the relative numbers of actual species in the world. We explore the great diversity of the eukarya in Chapters 13 (on animals) and 14 (on plants and fungi).

TAKE HOME MESSAGE 12.18

» All living organisms that we can see with the naked eye (and many that are too small to be seen) are eukarya, including all plants, animals, and fungi. The eukarya are unique among the three domains in having cells with organelles.

THE EUKARYA

Slime mold

Freesia

Mushrooms

Chacma baboon

FIGURE 12-32 **All shapes and sizes.** There is a huge diversity of eukaryotic organisms.

STREET BIO

Using evidence to guide decision making in our own lives

Do racial differences exist on a genetic level?

In a fraction of a second, we are aware of someone's size, gender, approximate age—and race. But what is race? Does race have biological meaning?

Q: What is race, biologically speaking? The concept of race is not straightforward. Some biologists have proposed that races are groups of interbreeding populations that have different and distinctive traits but are reproductively compatible. But it is important to note that, in humans—beyond phylogenetic distinctions—religion, culture, ethnicity, and geography are common aspects of many legal, social, and political conceptions of race. So do the many social constructions of "race" reflect meaningful divisions?

Q: What do human racial groupings reflect? Are blacks and whites genetically different? The answer is obviously yes; black-skin alleles differ from white-skin alleles. Moreover, the prevalence of some genetic diseases varies by race and ethnicity. Sickle-cell anemia, for example, is relatively common among Africans and Southeast Asians (and people of African and Southeast Asian descent) because the genes that cause it also improve a person's resistance to malaria, a disease prevalent in Africa and Southeast Asia.

Yet Africans from the southernmost part of the continent have no greater risk of sickle-cell anemia than do the Japanese, because malaria is similarly rare in both of their homelands. Should we use this trait rather than skin color when determining a person's race? If we did, southern Africans would be grouped with the Japanese and northern Africans with southern Asians (who share their genetic resistance to malaria and the higher risk of sickle-cell anemia).

This observation reveals an important problem with racial groupings, one that DNA sequence comparisons also show: similarity in skin color between two people doesn't tell you anything about overall genetic similarity. There is, for example, a huge variation in blood type among Africans: some are type O, some AB, others A or B. But the same goes for any population in the world, Asians and Turks, Russians and Spaniards. Yet we do not group people according to blood type or any other physiological or biochemical variations.

Q: What is the future of race? Racial groupings such as "Hispanic" or "black" or "Asian" or "white" do not reflect consistent genetic differences. Still, race remains an important part of individual, social, cultural, and ethnic identity for many, and its use is mandated by federal agencies. For these reasons, it will remain a powerful social construct, even as scientific consensus emerges that racial groupings do not reflect meaningful genetic distinctions.

GRAPHIC CONTENT

Thinking critically about visual displays of data

1 What is the approximate rate of background extinctions over the time period shown here?

2 What distinguishes the mass extinctions from the background extinctions in this figure?

3 Why do you think the graph records the percentage of families becoming extinct rather than the percentage of species becoming extinct?

4 Why isn't the light green line indicating the rate of background extinctions a straight line?

5 Rates of extinction are generally estimated based on the fossil record. Why might this not be an accurate measure of the magnitude of a mass extinction?

6 Conservation biologists believe it is important that we keep a close watch on the current rate of extinctions. Why might that be?

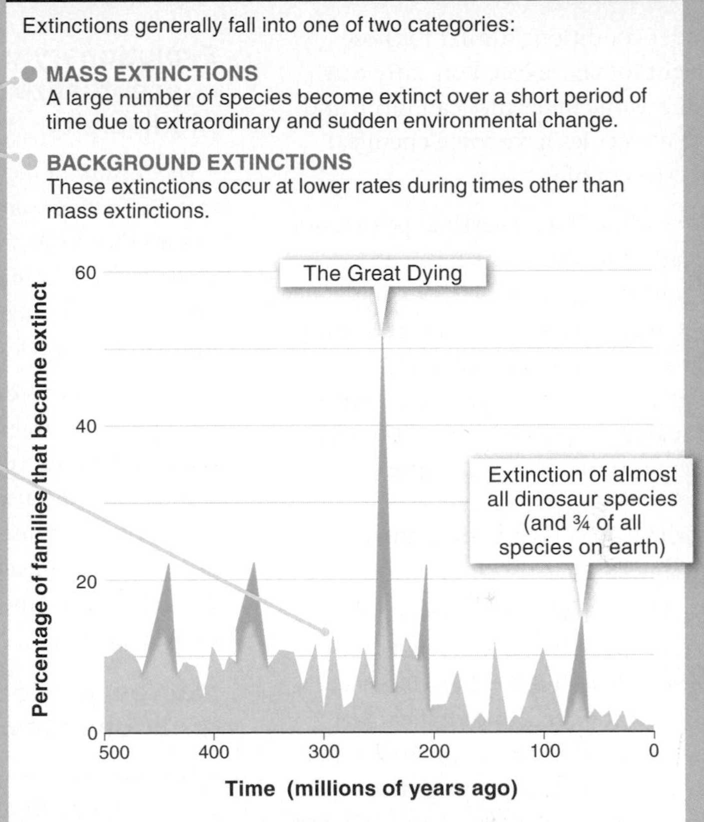

TWO CATEGORIES OF EXTINCTION
Extinctions generally fall into one of two categories:
MASS EXTINCTIONS
A large number of species become extinct over a short period of time due to extraordinary and sudden environmental change.
BACKGROUND EXTINCTIONS
These extinctions occur at lower rates during times other than mass extinctions.
The Great Dying
Extinction of almost all dinosaur species (and ¾ of all species on earth)
Percentage of families that became extinct
60
40
20
0
500
400
300
200
100
0
Time (millions of years ago)

See answers at the back of the book.

KEY TERMS IN THE ORIGIN AND DIVERSIFICATION OF LIFE ON EARTH

BRIEF SUMMARY

Life on earth most likely originated from non-living materials.

• Under conditions similar to those thought to have existed on early earth, small organic molecules can form, and these molecules have some chemical properties of life.

• The earliest life on earth appeared not long after the earth was formed. Self-replicating molecules, possibly RNAs, may have formed in earth's early environment and later acquired membranes, enabling them to replicate and making metabolism possible—the two conditions that define life.

Species are the basic units of biodiversity.

• Species are generally defined as populations of individuals that interbreed with each other, or could possibly interbreed, and that cannot interbreed with individuals from other such groups. This concept can be applied to most plants and animals, but is not applicable for many other types of organisms.

• Each species on earth is given a unique name, using a hierarchical system of classification.

• Difficulties arise when trying to classify asexual species, fossil species, species arising over long periods of time, ring species, and hybridizing species.

• Speciation is the process by which one species splits into two distinct species that are reproductively isolated. It can occur by polyploidy or a combination of geographic isolation and genetic divergence.

Evolutionary trees help us conceptualize and categorize biodiversity.

• The history of life can be visualized as a tree. By tracing from the branches back toward the trunk, we can follow the pathway back from descendants to their ancestors, revealing the evolutionary history of all species.

• Evolutionary trees are best constructed by comparing organisms' DNA sequences rather than physical similarities, because convergent evolution can cause distantly related organisms to appear closely related but doesn't indicate increased similarity in DNA sequences.

Macroevolution gives rise to great diversity.

• The process of evolution in conjunction with reproductive isolation is sufficient to produce speciation and the rich diversity of life on earth.

• The pace at which evolution occurs can be rapid or very slow. In some cases, the fossil record reveals long periods with little evolutionary change punctuated by rapid periods of change.

• Adaptive radiations—brief periods of time during which a small number of species diversify into a much larger number of species—tend to be triggered by mass extinctions of potentially competing species, colonizations of new habitats, or the appearance of evolutionary innovations.

• As new species are being created, others are lost through extinction, which may be a consequence of natural selection or large, sudden changes in the environment. Mass extinctions are periods during which a large number of species become extinct over a short period of time.

An overview of the diversity of life on earth: organisms are divided into three domains.

• All species are classified in one of three domains—bacteria, archaea, and eukarya—which reflect species' evolutionary relatedness to each other. Plants and animals are just two of the four kingdoms in the eukarya domain.

• All bacteria share a common ancestor and are prokaryotic, asexual, single-celled organisms with no nucleus or organelles. Bacteria have a much broader diversity of metabolic and reproductive abilities than do the eukarya.

• Archaea physically resemble bacteria but are more closely related to eukarya. They thrive in many habitats that humans have not yet studied well, including the deepest seas and oceans.

• All living organisms that we can see with the naked eye (and many that are too small to be seen) are eukarya, including all plants, animals, and fungi.

CHECK YOUR KNOWLEDGE

Short Answer

1. Why is the development of a membrane system an important phase in the origin of life?

2. Using the biological species concept, define a species.

3. Why must the first phase of speciation involve reproductive isolation?

4. Phylogenetic trees show the relationships among organisms. What else can we glean from phylogenetic trees?

5. Why do analogous features (such as bat wings and insect wings) pose problems for biologists as they construct phylogenetic trees?

6. Compare and contrast microevolution and macroevolution.

7. What sort of evidence do we use to help develop models of evolutionary change?

8. What is the fundamental difference between background extinctions and mass extinctions?

9. Why did the species classification system created in the 1960s have to be revised in the 1970s and 1980s?

10. Archaea are single-celled prokaryotes, and under a microscope they look very similar to bacteria. Which physical features distinguish them from the bacteria?

11. Which features make the domain eukarya unique?

Multiple Choice

1. In a set of classic experiments performed in the early 1950s, Urey and Miller subjected an experimental system composed of H_2, CH_4 (methane), NH_3 (ammonia), and water to electrical sparks. A few days later, they found in their system:

a) amino acids. b) DNA.
c) microspheres. d) cells.
e) RNA.

0 EASY — 13 — HARD 100

2. What kind of molecule is thought to be the most likely to have been the first genetic material?

a) protein b) DNA
c) carbohydrate d) RNA
e) microsphere

0 EASY — 25 — HARD 100

3. In animals, it is believed that the most common mode of speciation is:

a) autopolyploidy. b) chromosomal.
c) directional. d) allopatric.
e) sympatric.

0 EASY — 54 — HARD 100

4. Polyploidy:

a) arises only when an error in meiosis results in diploid gametes instead of haploid gametes.

b) is a common method of sympatric speciation for animals.

c) arises when allopatric speciation causes plants to have fewer sets of chromosomes than their parent plants.

d) is an increased number of sets of chromosomes.

e) always results in allopatric speciation.

0 EASY — 46 — HARD 100

5. Phylogenetic trees should be viewed as:

a) true genealogical relationships among species.

b) the result of vertical, but never horizontal, gene transfer.

c) intellectual exercises, not to be interpreted literally.

d) representations of allopatric speciation events.

e) hypotheses regarding evolutionary relationships among groups of organisms.

0 EASY — 34 — HARD 100

6. The idea of punctuated equilibrium challenges which component of Darwin's theory of evolution?

a) steady change

b) gradualism

c) species stasis

d) Both a) and b) are correct.

e) None of the above are correct.

0 EASY — 45 — HARD 100

7. Which of the following scenarios would best facilitate adaptive radiation?

a) A population of birds native to an island archipelago is forced to relocate to the mainland by a storm.

b) A population of cheetahs goes through an event in which all genetic diversity in the population is wiped out.

c) Darker-colored moths have a selective advantage over lighter-colored moths due to industrial soot on trees.

d) A population of birds becomes stranded on an island archipelago.

e) All of the above are equally likely to facilitate adaptive radiation.

0 EASY — 78 — HARD 100

8. The mass extinction that occurred on earth 65 million years ago was immediately followed by:

a) the rise of archaea.

b) the emergence of the first non-photosynthetic organisms.

c) the rise of the reptiles, including the dinosaurs.

d) an increase in atmospheric oxygen levels.

e) the rapid divergence and radiation of modern mammals.

0 EASY — 46 — HARD 100

9. Prokaryotes are classified into ______ domain(s).

a) 1 b) 2
c) 3 d) 4
e) 5

0 EASY — 42 — HARD 100

10. Which of the following pairs of domains share the most recent common ancestor?

a) archaea and eukarya

b) bacteria and eukarya

c) archaea and bacteria

d) None of the above; all three domains evolved from different ancestors.

e) None of the above; all three domains are equally related to each other.

0 EASY — 57 — HARD 100

11. Which of the following groups would be placed nearest the fungi in an evolutionary tree based on DNA sequences?

a) plants b) bacteria
c) animals d) archaea
e) protists

0 EASY — 41 — HARD 100

Ch13

Animals are just one branch of the eukarya domain.

13.1 What is an animal?
13.2 There are no "higher" or "lower" species.
13.3 Four key distinctions divide the animals.

Invertebrates—animals without a backbone—are the most diverse group of animals.

13.4 Sponges are animals that lack tissues and organs.
13.5 Jellyfishes and other cnidarians are among the most poisonous animals in the world.
13.6 Flatworms, roundworms, and segmented worms come in all shapes and sizes.
13.7 Most mollusks live in shells.
13.8 Are some animals smarter than others?
13.9 Arthropods are the most diverse group of all animals.
13.10 *This is how we do it:* How many species are there on earth?
13.11 Flight and metamorphosis produced the greatest adaptive radiation ever.
13.12 Echinoderms are vertebrates' closest invertebrate relatives.

The phylum Chordata includes vertebrates, animals with a backbone.

13.13 All vertebrates are members of the phylum Chordata.
13.14 The evolution of jaws and fins gave rise to the vast diversity of vertebrate species.
13.15 The movement onto land required lungs, a rigid backbone, four legs, and eggs that resist drying.

All terrestrial vertebrates are tetrapods.

13.16 Amphibians live a double life.
13.17 Birds are reptiles in which feathers evolved.
13.18 Mammals are animals that have hair and produce milk.
13.19 Humans tried out different lifestyles.
13.20 How did we get here? The past 200,000 years of human evolution.

Cats are digitigrades, mammals that walk on their tippy toes. Walking and running on toes makes cats fast and quiet, and excellent hunters.

Animal Diversification

Visibility in motion

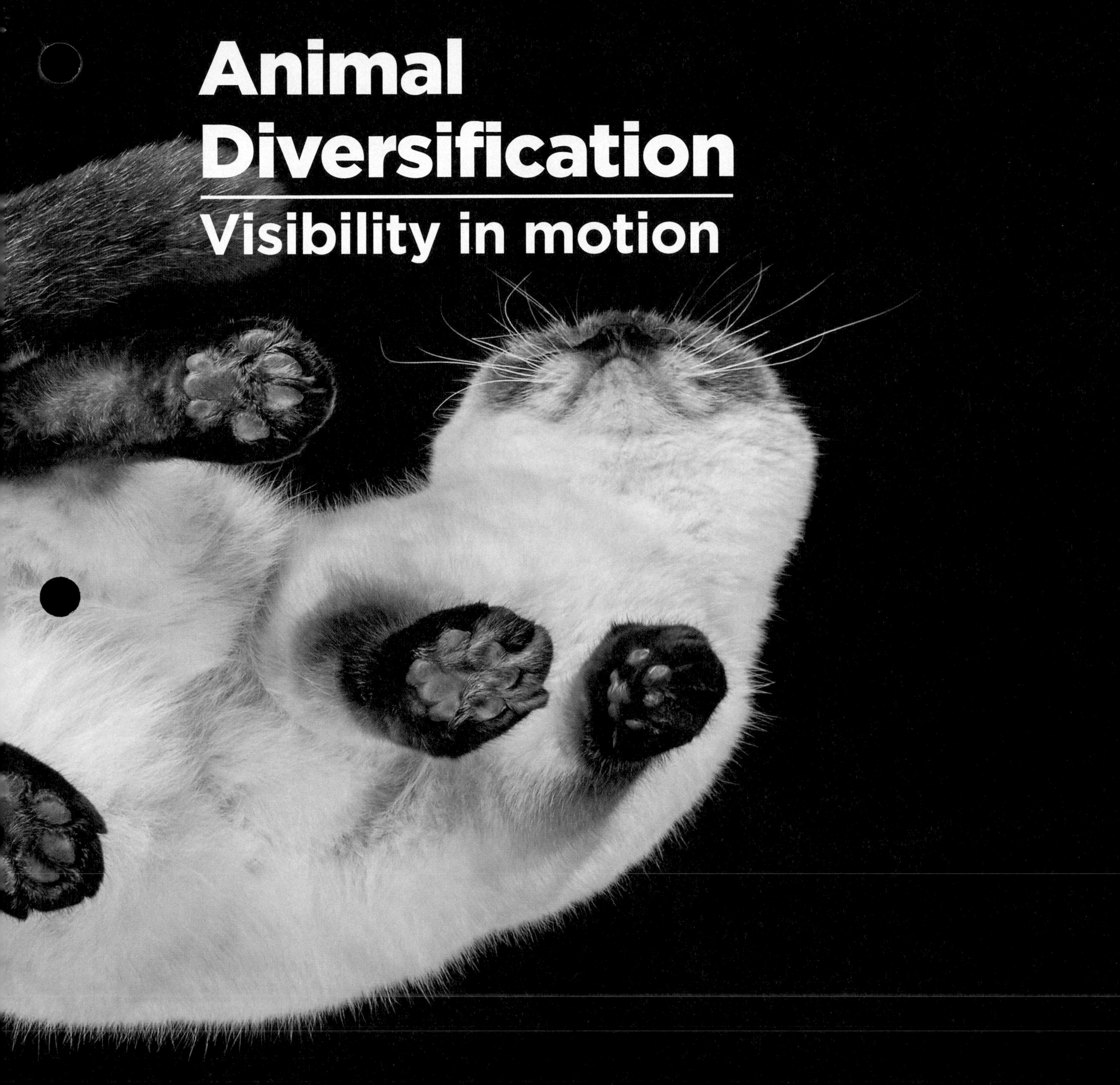

Animals are just one branch of the eukarya domain.

A group of the highly sociable southern carmine bee-eaters (*Merops nubicoides*) in the Luangwa Valley, Zambia.

13.1 What is an animal?

Looking at a kangaroo, it's obvious that it is an animal. The same goes for a mosquito, squid, or earthworm. It gets a bit more difficult, though, when you look at a sponge. Is it an animal? Or is it a plant? Or is it something else? This is when it is helpful to have some guidelines for identifying and organizing the biodiversity we can see on earth.

The sponge is classified as an animal because it shares certain features common to all animals. Surprisingly, considering the enormous differences among even the few animals mentioned above, just three characteristics are generally sufficient to define **animals** (FIGURE 13-1).

1. All animals eat other organisms. In contrast, plants and some bacteria manufacture their own food through photosynthesis. And because the best way to catch other organisms is to move toward them or chase after them, that brings us to the second characteristic that sets animals apart from plants.

2. All animals move—*at least, at some stage of their life cycle.* That qualifying phrase is important because some animals move only during a period called the "larval stage" that comes early in their lives, and when they become adults they fasten themselves to a surface and stay there. Organisms that are fastened in place, such

WHAT IS AN ANIMAL?

Rock agama (Agama agama) *male eating a grasshopper*

ANIMALS EAT OTHER ORGANISMS
All animals acquire energy by consuming other organisms.

Blackbucks (Antilope cervicapra)

ANIMALS MOVE
All animals have the ability to move—at least at some stage of their life cycle.

Spanish shawl nudibranch (Flabellina iodinea), *a type of sea slug*

ANIMALS ARE MULTICELLULAR
Animals consist of multiple cells and have body parts that are specialized for different activities.

FIGURE 13-1 **Characteristics of an animal.**

as adult mussels and barnacles living on rocks in the tidal zone, are said to be **sessile,** but even animals that are sessile as adults moved when they were larvae.

3. All animals are multicellular, and most animals have body parts that are specialized for different activities. For example, most animals possess sensory organs (eyes, ears, nose, tongue, antennae, whiskers) enabling them to acquire information about their environment and, among other things, detect potential prey and predators.

There are a few other general features of animals—most reproduce sexually, for example—but these three criteria are the easiest to apply when judging whether a particular organism is an animal.

TAKE HOME MESSAGE 13.1

» Animals are organisms that share three characteristics: all of them eat other organisms, all can move during at least one stage of their development, and all are multicellular.

13.2 There are no "higher" or "lower" species.

Humans often reveal their biases when speaking of other organisms. There are two possible states for a species. It can be *extant,* meaning that it currently exists. Or it can be *extinct.* From an evolutionary perspective, referring to any extant species as "lower" or "primitive" does not make sense. Rather, these labels usually reflect nothing more than how similar a species is to humans. Evolution can lead to greater adaptation between an organism and its environment, but it does not necessarily lead to greater complexity or features that can be identified as "higher" or more advanced.

Darwin wrote a note to himself in the margin of a book that proposed a theory of cosmic and biological evolution: "Never use the word higher or lower."

Throughout his writings, for the many species he described, Darwin emphasized the process of adaptation and the relationship between populations of organisms and the environments in which they lived. One species was never better or worse than another. Rather, each species was differentiated from the others, with specializations that adapted individuals of that species to the particular niche in which they lived. The fact that a species exists means it can do the things all organisms must do: find food, escape predators, and reproduce (FIGURE 13-2).

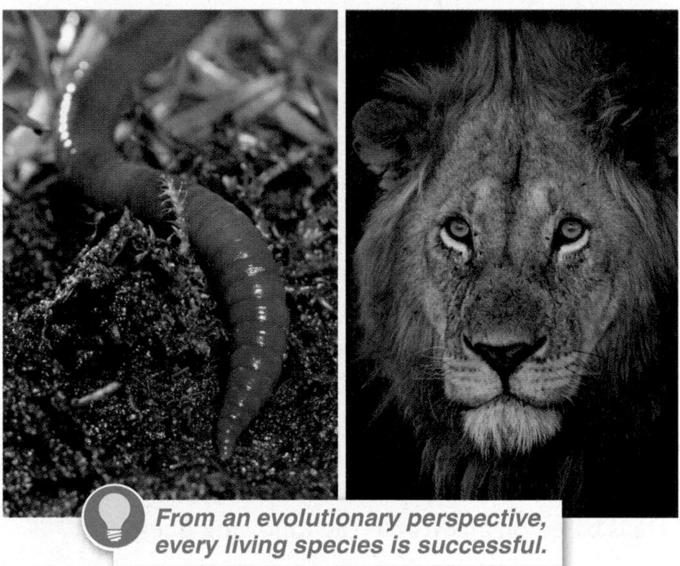

From an evolutionary perspective, every living species is successful.

FIGURE 13-2 Two equally successful organisms: the earthworm and the lion.

Our tour of the animal kingdom begins with animals having a simple body plan, such as sponges, jellyfishes, and worms. It continues through the mollusks and arthropods—complex animals that do not have a backbone. And it culminates with the vertebrates: mammals (including humans), amphibians, reptiles (including birds), and fishes. Taking this path allows us to explore the major evolutionary transitions in the chronological order in which they occurred. This order does not provide any information about the relative success of any species. Nor does it imply that humans are the "crown of creation."

> "I have been studying the traits and dispositions of the lower animals (so-called), and contrasting them with the traits and dispositions of man. I find the result humiliating to me."
>
> — MARK TWAIN, in *The Lowest Animal* (written ~1897, first published 1962)

Certain adaptations have led to the rapid and extensive diversification of particular *groups* of species. We investigate nine of the phyla (there are about three dozen animal phyla in all) with the most species (see Figure 13-3) and the evolutionary innovations that contributed to their success. These groups represent only a quarter of all animal phyla, but they account for about 99% of the animal species on earth today.

TAKE HOME MESSAGE 13.2

» From an evolutionary perspective, no species is "higher" or "lower." Certain adaptations have led to the rapid and extensive diversification of particular groups of species. Of the approximately three dozen animal phyla, nine phyla account for more than 99% of described animal species.

13.3 Four key distinctions divide the animals.

About two-thirds of the almost two million identified species on earth are animals. These animals share a common ancestor, a free-living unicellular organism resembling a sperm in size and shape.

Analyses of DNA and RNA sequences have helped biologists identify species that share a common ancestor. These analyses have made it possible to better classify the animals into monophyletic groupings. Each of these groupings hinges on a particular adaptation and on the question of whether or not the animals of a species have descended from an ancestor with that adaptation (FIGURE 13-3). Generally, the appropriate placement of species in a phylogenetic tree is based on answers to the following questions:

1. Does the animal have specialized cells that form defined tissues? Animals having no tissues—groups of cells with a common structure and function—belong to the phylum Porifera, the sponges. Sponges are aggregations of similar cells, with little coordination of activities among the cells. All other animals form a monophyletic group with clearly defined tissues. Humans, for example, have highly specialized cells such as skin, muscle, and sensory cells. For all non-sponge animals, we can ask a second question.

2. Does the animal develop with radial symmetry or bilateral symmetry? All animals with defined tissues develop either radial or bilateral symmetry. **Radial symmetry** describes animals with bodies structured like pies, such as jellyfishes, corals, and sea anemones. Slow-moving or free-floating, radially symmetrical animals have no front or back ends. It is possible to make multiple slices through the center that divide the organism into identical pieces.

Other animals develop with **bilateral symmetry.** These organisms (such as humans, cows, and scorpions) have left and right sides that are mirror images. The animals can move adeptly, searching for food and avoiding predators. In some animals, such as starfish, embryonic development is bilaterally symmetrical, but the animal becomes radially symmetrical as an adult. Any animal showing bilateral symmetry at any life stage is considered bilaterally symmetrical.

Radial symmetry

Bilateral symmetry

Among the monophyletic group of animals with defined tissues and bilateral symmetry, there is a further phylogenetic division based on the answer to a third question.

3. During gut development, does the mouth or anus form first? Early in the evolution of animals—about 630 million years ago—a major split occurred between the

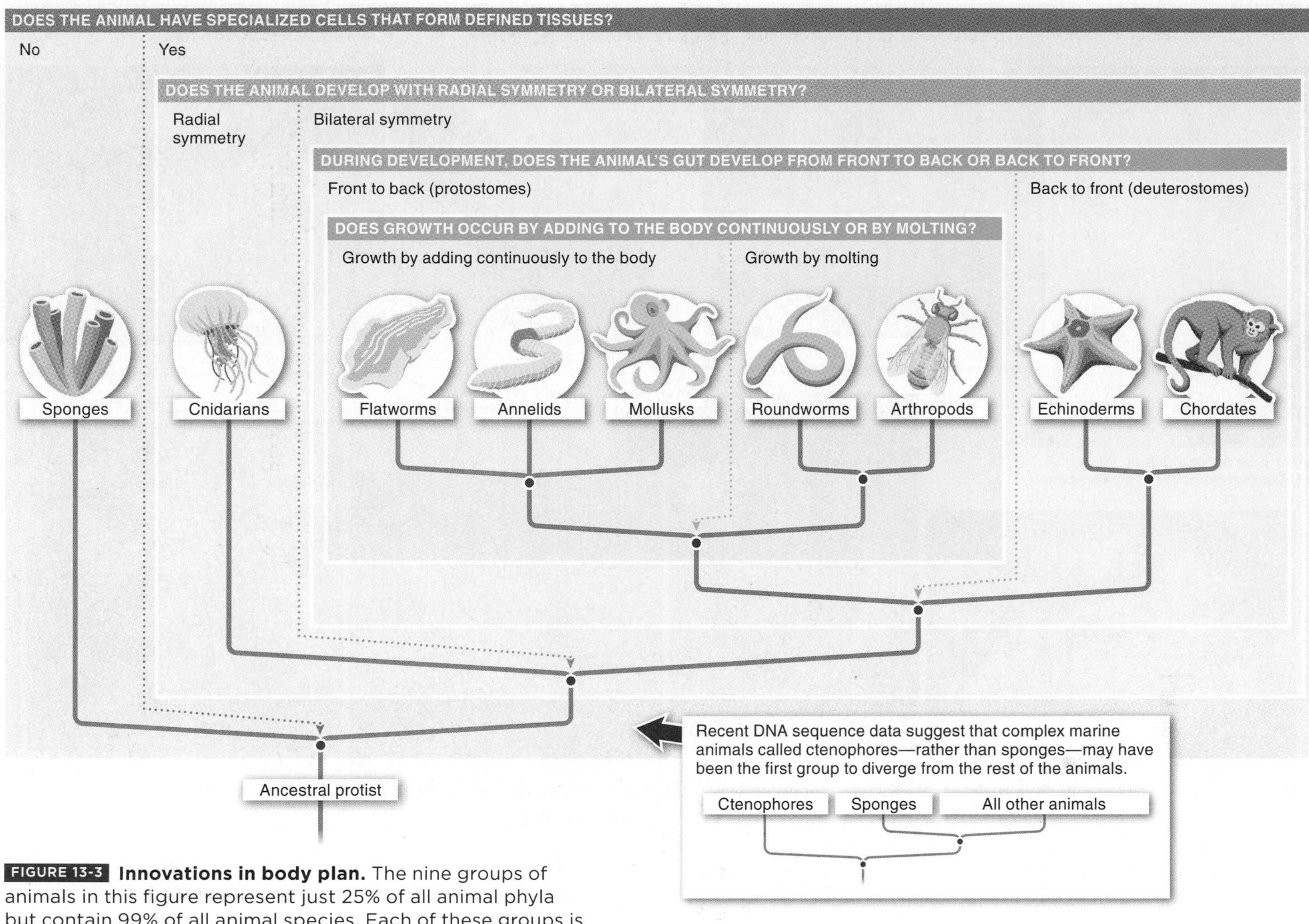

FIGURE 13-3 **Innovations in body plan.** The nine groups of animals in this figure represent just 25% of all animal phyla but contain 99% of all animal species. Each of these groups is explored in this chapter.

protostomes (which can be translated as "mouth first") and the **deuterostomes** ("mouth second").

These names refer to the way the gut develops—from front to back or from back to front. In protostomes, the gut develops from front to back, so the first opening that forms becomes the mouth of the adult animal, and the last opening becomes the anus. In deuterostomes, the gut develops from back to front. The anus is the first opening to form, and the mouth is the second. Although it's something you can see only when closely observing animals' embryonic development, the protostome-deuterostome split is one of the most basic divisions of animals, in terms of revealing their evolutionary relatedness to one another.

In building an overall phylogeny of the animals, we can further subdivide into monophyletic groups the animals that (1) have defined tissues, (2) develop with bilateral symmetry, and (3) are protostomes. To do this, we need to ask a fourth question.

4. Does growth occur by molting or by continuous addition to the animal's body? Animals that molt (such as lobsters and insects) shed their exoskeleton (hard outer layer), replacing it with a larger one, at intervals during development. Other animals (such as clams and octopuses) grow by adding to the size of their body in a continuous manner.

These four key distinctions help place the approximately three dozen phyla of living animals into monophyletic groups. Within any of the groups, we know that individuals are more closely related to one another than to any individuals outside that group.

Besides being organized within monophyletic groups, animals are grouped according to the presence or absence of a backbone. **Invertebrates**—including sponges—are animals that do not have a backbone; **vertebrates** are animals that do have a backbone.

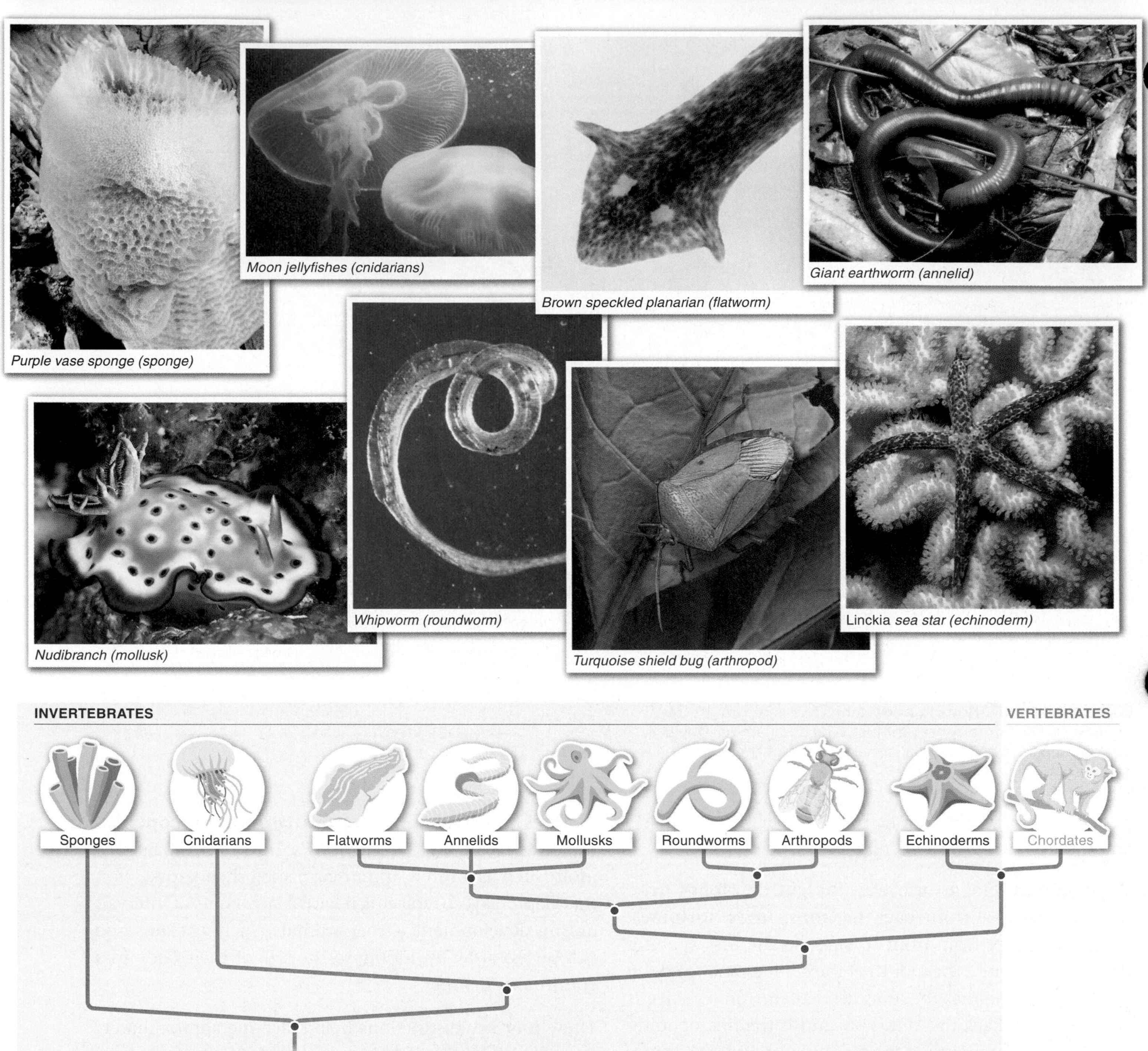

FIGURE 13-4 **Phylogeny of the major groups of invertebrates.**

Invertebrates comprise more than 95% of all living animal species (FIGURE 13-4). But the invertebrates are not a monophyletic group. Until the advent of molecular methods for building phylogenetic trees, we had been fooled by convergent evolution into believing that organisms that *appear* similar also share close evolutionary relatedness.

In fact, the invertebrates include eight separate evolutionary lineages with an enormous diversity of body forms, body sizes, habitats, and behaviors. All invertebrates, however, are protostomes—except for one large group, whose members are all deuterostomes. Called the echinoderms, and including sea stars, sea urchins, and sand dollars, this animal phylum shows the same from-back-to-front gut development as the vertebrates, and these animals are the vertebrates' closest relatives. We examine them in detail in the following sections.

It's important to note that any phylogeny—including the one shown in Figure 13-3—is a hypothesis and, as such, subject to revision as we accumulate more evidence. Recently, for example, researchers have made the surprising suggestion that perhaps the sponges are not the first lineage of organisms to branch off from the rest of the animals (as shown in Figure 13-3).

DNA sequence data from a group called the ctenophores—previously believed to be most closely related to the cnidarians—indicate that they may have been the first to diverge from the rest of the animals. This is surprising because, unlike the sponges, the ctenophores, marine organisms called comb jellies, have a nervous system (and several types of defined tissues). If the ctenophores were the first animal group to diverge, then the ancestor of all animals would have had a nervous system, and this was subsequently lost by the sponges. Or, the nervous system evolved independently in the ctenophores. In either case, this proposed new phylogeny may change our understanding of the earliest events in animal evolution.

TAKE HOME MESSAGE 13.3

» Animals probably originated from an ancestral protist. Four key distinctions divide the extant animals into monophyletic groups: (1) tissues or not, (2) radial or bilateral symmetry, (3) protostome or deuterostome development, and (4) growth through molting or through continuous addition to the body.

13.4–13.12 Invertebrates—animals without a backbone—are the most diverse group of animals.

The most diverse animals on earth? More than 25% of all described species are beetles. Shown here is a wood-boring beetle (*Madecassia rothschildi*) from Madagascar.

13.4 Sponges are animals that lack tissues and organs.

We begin our investigation of animal diversity with the group that may be the least "animal-like," and the simplest, of all the animals: the sponges (phylum Porifera) (FIGURE 13-5).

Sponges have no specialized tissues or organs, and most lack any symmetry. Despite their simplicity, sponges are remarkably efficient at obtaining food. Living as sessile suspension-feeders, sponges strain suspended matter and food particles from water. Sponges have several different cell types. Epithelial cells—flattened, skin-like cells, connected to each other—cover the outside of the sponge. Collar cells, each of which has a long, whip-like flagellum, cover the inside. The middle layer is a gel, incorporating a diversity of cell types involved in reproduction, digestion, and growth, and embedded fibers that stiffen the sponge's body.

The beating of each collar cell's flagellum creates a current that carries water upward and out of the body. An inward flow of water through the pores on the sides of the sponge carries bacteria and algae and other microscopic organisms into the sponge's body. In the "collars" of the collar cells, sticky mucus traps some of this material, and some particles are taken up into the cells by endocytosis (see Section 4.11). The volume of water moving through its body

THE SPONGES

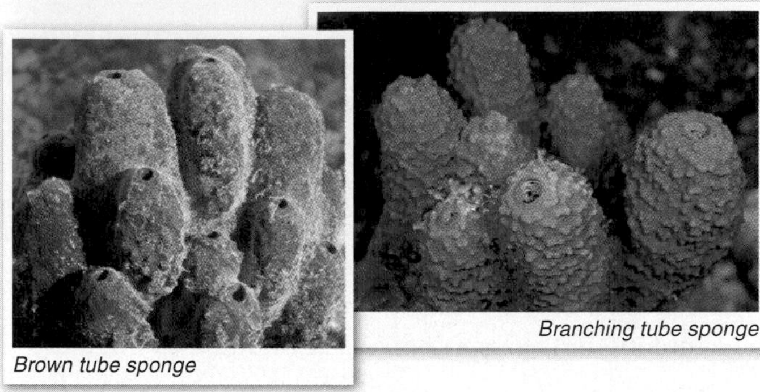

COMMON CHARACTERISTICS
- No tissues or organs
- Body consists of a hollow tube or network of canals with pores in their walls
- Feed by pumping in water, along with bacteria, algae, and small particles of organic material, through their pores
- Free-swimming larvae
- Sessile as adults

MEMBERS INCLUDE
- ~5,000 species

FIGURE 13-5 **An overview of the sponges.**

THE ANATOMY OF A SPONGE

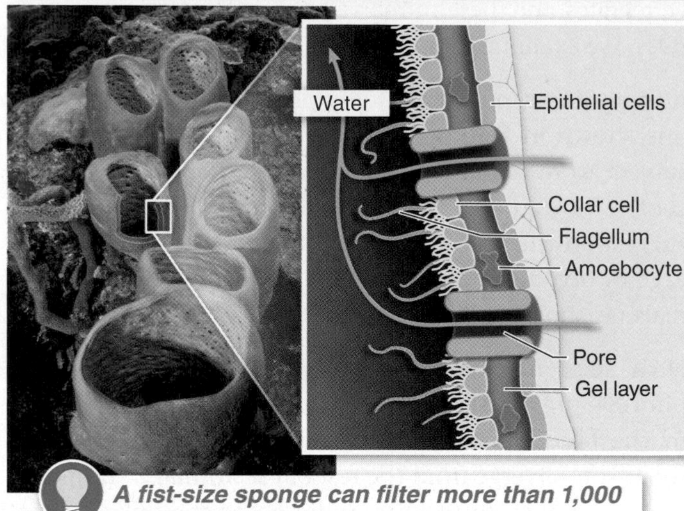

A fist-size sponge can filter more than 1,000 gallons (5,000 liters) of water every day!

FIGURE 13-6 **Sponge structure.**

daily can be as much as 20,000 times the volume of the sponge (FIGURE 13-6).

Sponges are hermaphrodites. That is, each individual contains both male and female reproductive organs, but it produces only one kind of gamete (eggs or sperm) at a time. Sponges that are acting as males release a cloud of sperm that swim to other sponges, which are acting as females, and fertilize their eggs. The eggs develop into larvae, which are released into the water and then drift for a few days before settling on a rock or coral outcrop and developing into a sessile adult sponge.

Sponges also reproduce asexually: small buds form on the exterior and eventually break off, settle to the bottom, and may grow into new sponges. Most remarkably, sponge cells are able to reassemble. Picture this experiment. Puree a living sponge in a food blender, then strain the liquid through a fine sieve to remove the chunks, leaving a suspension of individual cells, and dump that into an aquarium. Within a day you'll see small clumps of sponge forming as individual cells move around and attach to each other, and within a week a new sponge will form. You can repeat the experiment with two sponge species of different colors. Not only do the sponges reassemble, but the cells from each species assemble only with other cells of the same species, so you'll have two sponges, each with only the cells of its own species.

Because the dried skeletons of some sponges are soft and absorbent, humans have long made use of them—as padding in helmets, as tools for painting, and, in ancient Rome, as toilet paper. But these sponges are far from living. They are just a matrix of soft, elastic fibers of a protein called spongin, isolated from dried sponges. Because over-fishing has reduced sponge populations, they are no longer widely available. Rather, the "sponges" used in most kitchens and bathrooms today are nearly always produced from synthetic materials. Loofah sponges, contrary to their name, are made from plant material.

Q **Is your kitchen sponge an animal? Is it alive?**

TAKE HOME MESSAGE 13.4

» Sponges are among the simplest of the animal lineages. A sponge has no tissues or organs. Sponges reproduce sexually (by eggs and sperm) and asexually (by budding). Fertilized eggs grow into free-swimming larvae that settle and develop into sessile, filter-feeding adult sponges.

13.5 Jellyfishes and other cnidarians are among the most poisonous animals in the world.

If you feel a sting when swimming in the ocean, you have met a jellyfish. If you are lucky, it will be the mild sting of one of thousands of species that feed on tiny floating organisms called plankton. Other jellyfish species have powerful stings that can cause extreme pain or even death.

The jellyfishes, along with sea anemones and corals, belong to the phylum Cnidaria (pronounced nigh-DARE-ee-ah). As with all animals except the sponges, the 11,000 species in this phylum have defined tissues. The cnidarians have a simple, radially symmetrical body plan (FIGURE 13-7).

There are two types of cnidarian bodies: a sessile polyp and a free-floating medusa. In some species, individuals spend part of their life cycle as a polyp and part as a medusa. Other species exist only as medusas or, as in corals and sea anemones, only as polyps.

Polyp

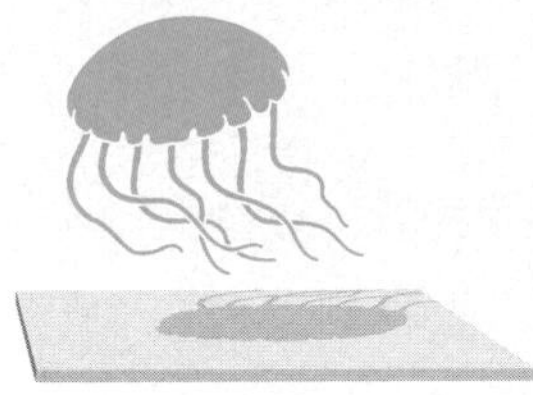

Medusa

Cnidarians are carnivores, using tentacles to capture and feed on marine organisms, from protists to fish and shellfish. Their method of capturing prey relies on a stinging cell called a cnidocyte that is unique to the cnidarians. All cnidarians have tentacles, located near their mouth, that are armed with rows of stinging cnidocytes. Each cnidocyte has a coiled thread with barbs inside and a "trigger" on the outside. When prey comes in contact with the trigger, the coiled thread is ejected and, like a harpoon, can penetrate the prey, often injecting a toxin (FIGURE 13-8). The Portuguese man-o'-war is one of the dangerously poisonous species of cnidarians. Being stung by a Portuguese man-o'-war is painful (extremely painful if the unlucky swimmer becomes entangled in the tentacles and receives hundreds of stings), but these encounters are rarely fatal.

We explore here some of the great diversity among three familiar groups of cnidarians.

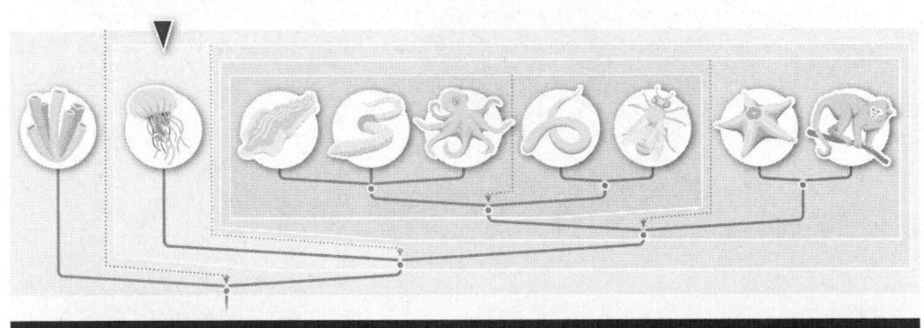

THE CNIDARIANS

Mosaic jellyfish

Sun coral

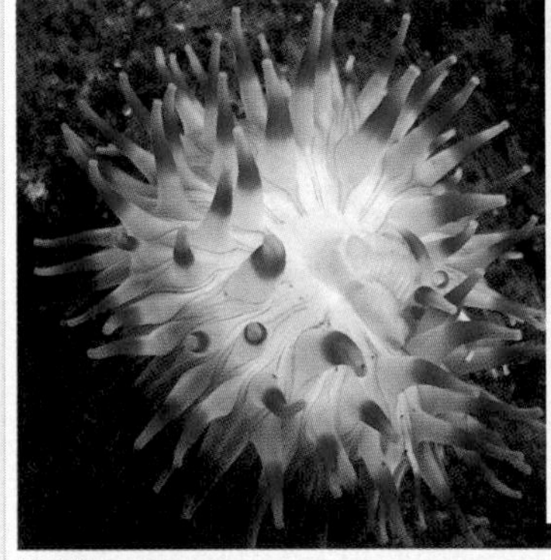

Dahlia anemone

COMMON CHARACTERISTICS
- Radially symmetrical
- Tentacles armed with rows of stinging cells, used to paralyze prey

MEMBERS INCLUDE
- Jellyfishes
- Sea anemones
- Corals

FIGURE 13-7 An overview of the cnidarians.

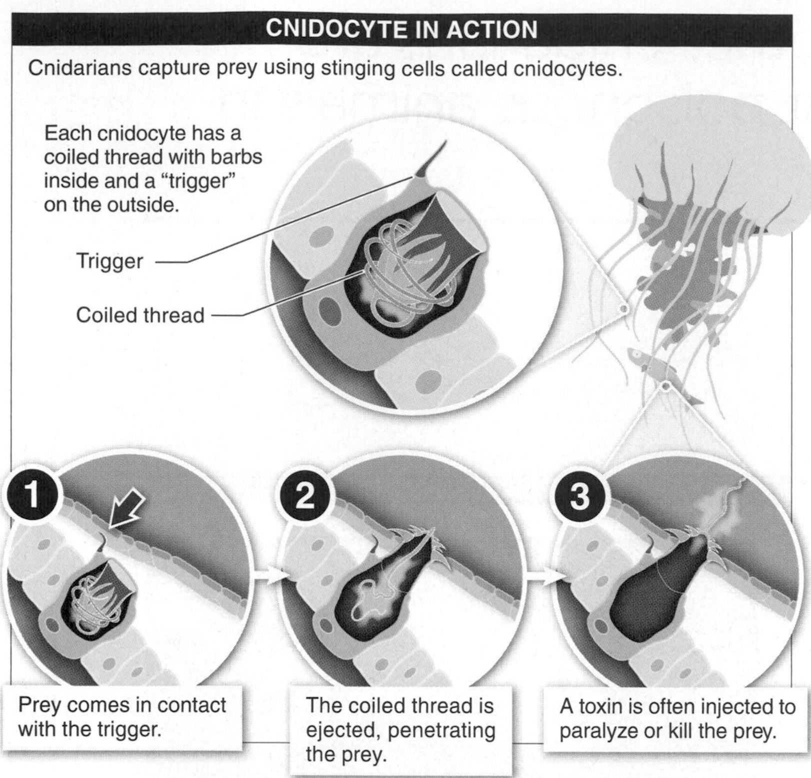

CNIDOCYTE IN ACTION
Cnidarians capture prey using stinging cells called cnidocytes.
Each cnidocyte has a coiled thread with barbs inside and a "trigger" on the outside.
Trigger
Coiled thread
1
2
3
Prey comes in contact with the trigger.
The coiled thread is ejected, penetrating the prey.
A toxin is often injected to paralyze or kill the prey.

FIGURE 13-8 **Stinging cells are used to capture prey.**

Corals The corals live as small (just a few millimeters long) polyps in large colonial groups, familiar as the beautiful structures found in coral reefs. Corals sting prey and use their tentacles to catch plankton (and even, on occasion, small fish), which are directed into the mouth by the tentacles and then digested in the stomach. Corals most commonly reproduce sexually, with both external fertilization (releasing sperm and eggs into the water, where they fuse and form larvae) and internal fertilization (in which only sperm are released and can fertilize eggs within female corals). Corals can also reproduce asexually.

Corals grow in a variety of shapes—from free-standing spherical or branched forms to crusts on rocks or on other corals. This is possible because polyps secrete calcium carbonate, creating hard shells on which other corals can live as a colony. Such assemblies of giant calcium carbonate skeletons and corals are called **coral reefs.** Coral reefs provide an environment that is home to a greater diversity of species than any other marine habitat. The Great Barrier Reef, located in the Coral Sea, off the coast of Australia, extends more than 1,500 miles (about 2,600 km) from north to south. An awesome display of biological productivity, this reef is the largest biological structure in the world and is visible from the International Space Station.

Q How is global warming affecting the coral reefs of the world?

Coral reefs are among the first casualties of global warming. Coral polyps obtain most of their nutrition from symbiotic algae, called zooxanthellae, living within the polyps. The algae use carbon dioxide produced by the polyps to photosynthesize, and the polyps gain oxygen and nutrients produced by the algae. Strangely, when coral polyps get too hot they expel their zooxanthellae. Because zooxanthellae are responsible for the colors of many corals, as well as much of their nutrition, when coral polyps expel their zooxanthellae, the coral appears white—a phenomenon called coral bleaching. The zooxanthellae may return to the polyps when the water cools, but repeated bleaching events often lead to the death of the coral polyps. The world's oceans are already warming, and coral bleaching events are becoming more frequent (FIGURE 13-9).

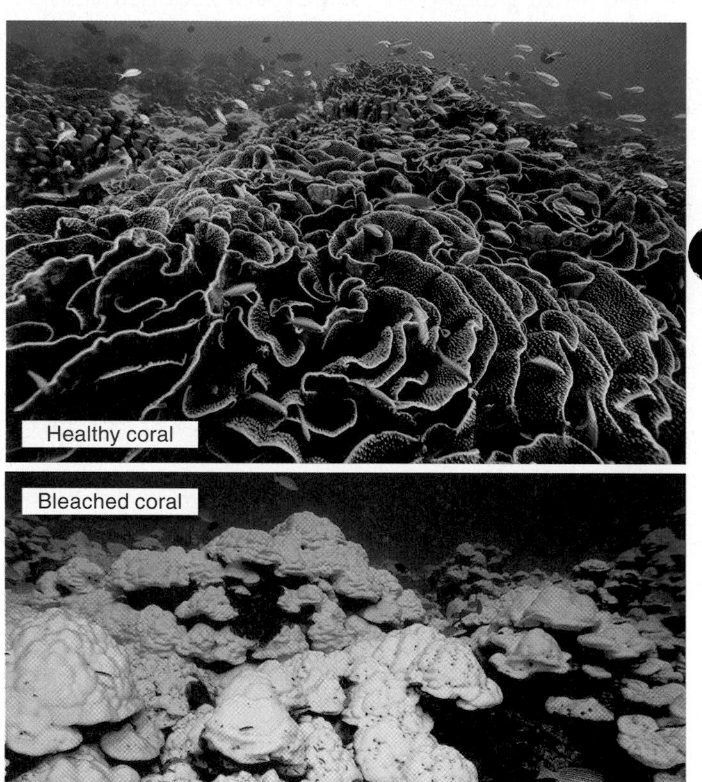

With warming water, corals eject their photosynthetic symbionts, causing coral "bleaching." Without the colored algae, the corals appear white and may die, endangering the coral reef and the diversity it supports.

FIGURE 13-9 **Coral bleaching.**

Sea Anemones Sea anemones have polyp bodies that resemble flowers. The anemone has tentacles with stinging cells at the top of its body, surrounding its mouth. Sea anemones have a larval stage that swims freely, then settles on a rock and metamorphoses into the adult form. Even as adults, many sea anemones can crawl a few inches a day, and wandering sea anemones are often found in seaweed.

Jellyfishes The jellyfishes (which are not fish!) range tremendously in size. The Asian giant jellyfish is more than 6 feet (about 2 m) across and weighs nearly 500 pounds (more than 200 kg). Swarms of these giants clog nets of fishing boats with sticky masses of toxic stingers. At the opposite end of the size scale is the Irukandji jellyfish, about the size of a hen's egg. This species is so deadly that, in 2007, the sighting of just five Irukandji off Australia halted the filming of a Hollywood movie (*Fool's Gold*), which had to be completed in the safety of a studio.

TAKE HOME MESSAGE 13.5

» Corals, sea anemones, and jellyfishes are radially symmetrical animals with defined tissues, in the phylum Cnidaria. All cnidarians are carnivores and use specialized stinging cells located in their tentacles to capture prey.

13.6 Flatworms, roundworms, and segmented worms come in all shapes and sizes.

It's a good thing that biologists don't run bait shops—if they did, a simple trip to buy worms for fishing would be too complicated. The name "worm" is commonly applied to a long, skinny, slimy animal without a backbone, but you can find animals fitting that description in eight different phyla—so worms do not make up a monophyletic group. Their general body plan has evolved several times independently. Before their evolutionary relationships were understood, these groups were all classified as they first were by Linnaeus—as worms, simply because of similarities in their appearance.

We will consider three phyla that illustrate most of the diversity of the animals we call worms. These include **flatworms** (phylum Platyhelminthes), **roundworms** (phylum Nematoda), and **segmented worms,** or **annelids** (phylum Annelida). Keep in mind, however, that from an evolutionary perspective, "worm" is an obsolete—and misleading—label. We know today that the segmented worms (annelids) are probably more closely related to mollusks (including snails, clams, and octopuses) than to roundworms. And roundworms probably are more closely related to arthropods (insects) than to flatworms or segmented worms (see Figure 13-3).

The worms have defined tissues and are protostomes—the gut develops from front to back. Unlike the cnidarians, however, the worms (along with the remaining animals discussed in this chapter) develop with a body plan characterized by bilateral symmetry, which adapts them for forward movement.

Flatworms Unlike the roundworms and segmented worms, flatworms have no body cavity. That is, the space between the body wall and the digestive tract is not a fluid-filled cavity. Possibly as a consequence of this, the flatworms have no specialized respiratory or circulatory organs and so have a reduced capacity for diffusion of gases and nutrients.

Flatworms grow by adding body mass rather than by molting (FIGURE 13-10). Flatworms have well-defined head and tail regions, with clusters of light-sensitive cells called eyespots at the head end, used for navigation. Most flatworms are hermaphrodites, each individual producing both male and female gametes, and they engage in both sexual and asexual reproduction. Flatworms include more than 20,000 species and can be parasites (flukes and tapeworms) or brilliantly colored, free-living aquatic creatures.

Many flatworms have a digestive system in which the gut has only one opening, requiring them to consume food and eliminate waste through the same opening. The digestive systems of roundworms and segmented worms, in contrast, have two openings. One group of flatworms lacking a digestive system is the tapeworms. These parasitic worms

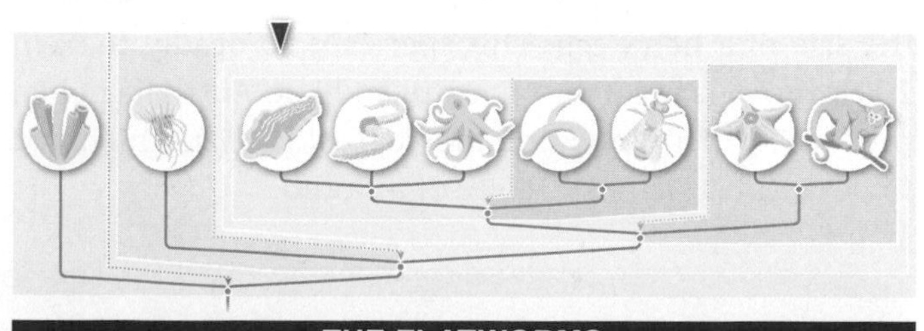

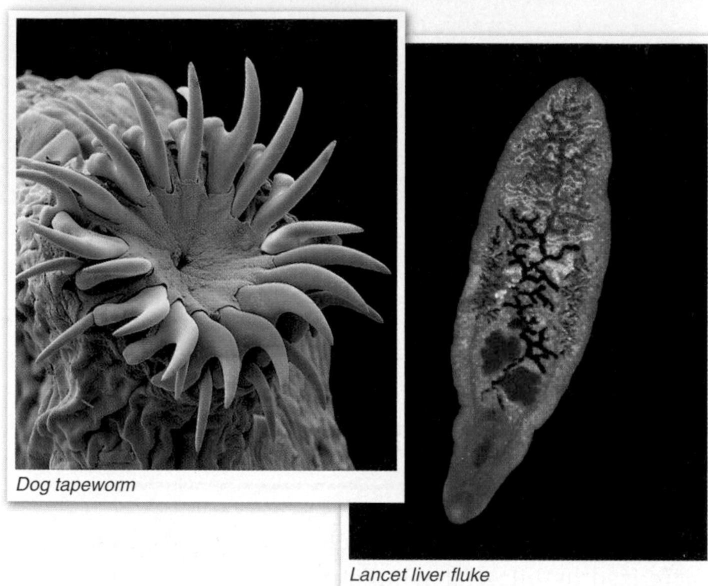
Dog tapeworm

Lancet liver fluke

COMMON CHARACTERISTICS
- Well-defined head and tail regions
- Grow by adding body mass
- Hermaphroditic and can engage in both sexual and asexual reproduction
- Some have a single opening for their digestive tract, which serves as a mouth and anus

MEMBERS INCLUDE
- Tapeworms
- Flukes

FIGURE 13-10 An overview of the flatworms.

live in their host's gut and absorb nutrients from the host directly through their body wall.

All 5,000 species of tapeworms are parasites, most having a two-stage life cycle split between two different host species. For example, the common tapeworm that infects dogs spends half its life cycle in fleas. The human pork tapeworm splits its life between pigs and humans. Tapeworms have long, flat bodies made up of repeated segments, each of which is a reproductive unit. Mature tapeworms spread by breaking off segments, which are then shed in the feces of infected hosts.

Roundworms The roundworms, also called nematodes, are probably the most abundant animals on earth—a spoonful of garden soil contains several thousand individuals, and some species produce more than 200,000 eggs every day. All roundworms grow by molting. More than 90,000 roundworm species have been named, and there may be five times as many species not yet identified (FIGURE 13-11).

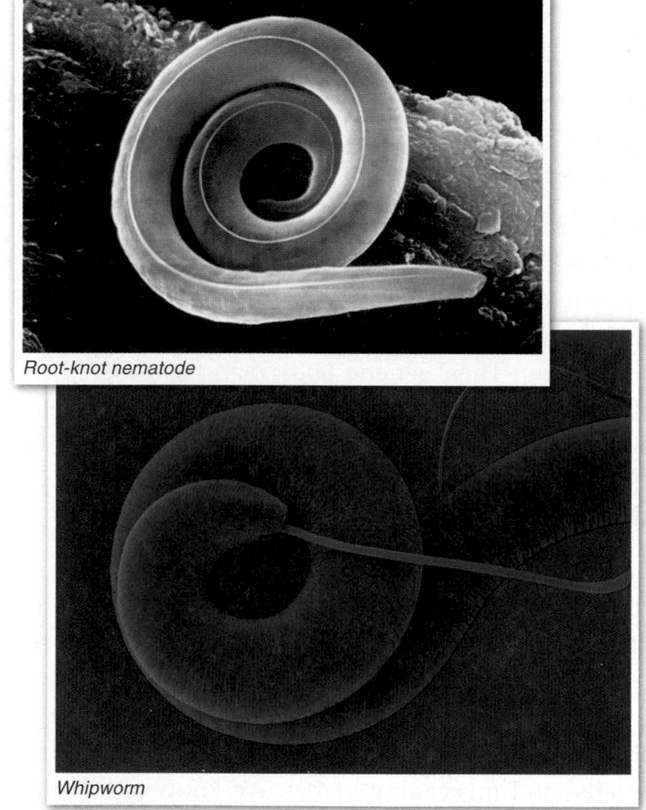
Root-knot nematode

Whipworm

COMMON CHARACTERISTICS
- Long, narrow, unsegmented body
- Surrounded by a strong, flexible cuticle
- Must molt in order to grow larger

MEMBERS INCLUDE
- More than 90,000 species (scientists estimate there may be five times as many species not yet identified)

FIGURE 13-11 An overview of roundworm diversity.

Many soil-dwelling roundworms live in the roots of plants, damaging or even killing the plant. About 15,000 species of roundworms are parasites of other animals, and roundworms are responsible for a large number of human diseases. Most roundworms are transmitted by fecal contamination of soil or food.

Tiny parasitic roundworms called filariae (*sing.* filaria) are responsible for several devastating tropical diseases. Elephantiasis, for example, is a disease in which filariae transmitted by mosquito bites block the lymph ducts so that fluid accumulates in the limbs or scrotum, causing grotesque swelling. These filariae are found in Africa, South Asia, Pacifica, and tropical regions of the Americas.

Segmented Worms (Annelids) The segmented worms, or annelids, number about 13,000 species and are easy to recognize by the grooves running around the body that mark the divisions between segments. The segmented worms are protostomes with defined tissues, and they do not molt. They are organized into three different groups: marine polychaetes (pronounced POL-lee-keets), terrestrial oligochaetes, and leeches (FIGURE 13-12).

Polychaetes are marine worms, living on the seafloor. *Polychaete* means "many bristles," and the combination of segments and bristles makes polychaetes easy to recognize. Some species burrow through the mud and extract organic material from it. Tube worms use sand grains or limestone to make a tube in which they live, with just their waving tentacles exposed. Small particles of food are trapped in mucus on the tentacles and transported to the worm's mouth.

Earthworms, which belong to the *oligochaetes* ("few bristles"), are the annelid worms you are most likely to have seen. The night crawler, a typical earthworm, gets its name from its habit of emerging from its burrow on rainy nights to crawl across the surface of the ground. More than 4,000 earthworm species have been named, ranging in size from less than half an inch to the giant Gippsland earthworm at 7–10 feet (2–3 m)!

Earthworms consume particles of soil and organic material. The organic material is digested as it passes through the worm's one-way gut, and fecal plus inorganic soil material is excreted as feces, called castings. Gardeners value earthworm castings as soil supplements, and earthworms can produce their own weight in castings every day. Earthworms' activities mix the soil components, create a more uniform mixture of nutrients, and expedite the breakdown of organic materials in the soil, making the

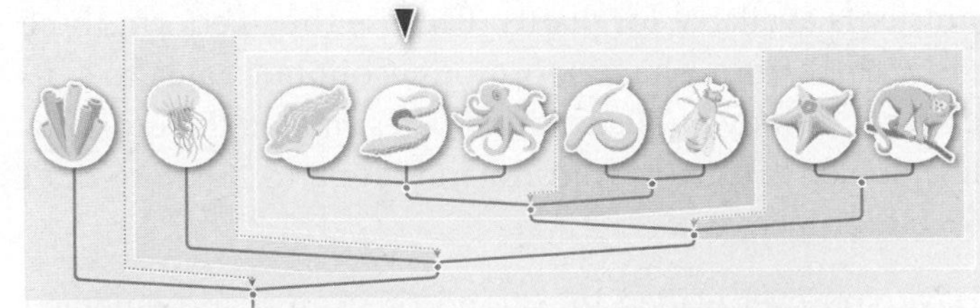

THE ANNELIDS

Fireworm

Earthworm

Leech

COMMON CHARACTERISTICS
- Segmented body
- Grow by adding body mass

MEMBERS INCLUDE
- Marine polychaetes (~9,000 species)
- Earthworms (more than 4,000 species)
- Leeches (~500 species)

FIGURE 13-12 **An overview of annelid diversity.**

nutrients available for plants. Because they perform these tasks, earthworms have enormous economic value to agriculture. A study published in 2008 estimated the economic value of earthworms in Ireland alone to be more than $1 billion per year.

Have you ever waded in a pond or swamp and emerged to find a long, dark brown worm clinging to your leg? If so, you have met the third major group of annelids, the *leeches*. Probably you pulled the leech off (not easy, because leeches are both slippery and stretchy) and threw it as far as you could. If you had looked carefully, though, you'd have noticed that, like earthworms, leeches have segmented bodies.

Not all leeches are blood-suckers—in fact, more than half the leech species are predators. The horse leech, for example, which reaches a length of 8 inches (about 20 cm), feeds on smaller annelid worms, snails, and aquatic insect larvae.

TAKE HOME MESSAGE 13.6

» Worms are found in several different phyla and are not a monophyletic group. All are bilaterally symmetrical protostomes with defined tissues. Flatworms and segmented worms (annelids) do not molt; the roundworms do. Flatworms include parasitic flukes and tapeworms, many of which infect humans. Many roundworms are parasites of plants or animals and are responsible for several widespread human diseases. Earthworms are annelids that play an important role in recycling dead plant material.

13.7 Most mollusks live in shells.

Just because they are related, that doesn't mean organisms will have much of a physical resemblance. Consider these animals. A colossal predatory squid, more than 40 feet long, with eyes the size of beach balls and a razor-sharp beak. A small snail—*escargot* in French—one of the more than 700 million consumed in France each year. An oyster, with its nearly 2,000-year-old reputation as an aphrodisiac (unsupported by any scientific proof, it turns out). All of these are mollusks, the second most diverse group of animals.

More than 100,000 mollusk species have been named (and many more have not yet been named). The members of this large phylum (Mollusca) live in the ocean, in fresh water, and on land, and they include clams, scallops, mussels, and octopuses. They are so diverse that it is difficult to describe any single defining characteristic. The position of mollusks within the animal phylogeny, however, reflects several important characteristics that they all share: defined tissues, bilateral symmetry, and protostome development. Further, all mollusks grow by adding tissue rather than by molting.

Several additional features are common to many mollusks. Some groups of mollusks have a shell that protects the soft body, a mantle (the tissue that secretes calcium carbonate to form the shell), and a sandpaper-like tongue structure, called the radula, that is used during feeding.

Here, we examine three of the major groups of mollusks. Animals in these groups share most of the features common to mollusks, but with very different body plans: **gastropods, bivalve mollusks,** and **cephalopods** (FIGURE 13-13).

Gastropods Snails and slugs are gastropod mollusks. *Gastropod* means "belly foot," and these mollusks get their name from the expanded foot on the bottom of their body, which allows them to climb a vertical surface and glide across a horizontal one. Snails have a one-piece, curled shell, and slugs have just a tiny remnant of a shell covered by the mantle. Found in both aquatic and terrestrial environments, snails and slugs account for three-quarters of all mollusks.

The snail's shell is its primary protection against predators, but for terrestrial slugs and sea slugs, which have very little shell material, other defense mechanisms are necessary. A terrestrial slug relies on slime for defense: when a slug is attacked, it secretes slime that sticks to the predator. Worse still, anything that touches the slime coating sticks to it, so a bird that attacks a slug quickly finds that its beak and face are covered by debris. Sea slugs have other methods of defense. Some species synthesize toxic chemicals, while

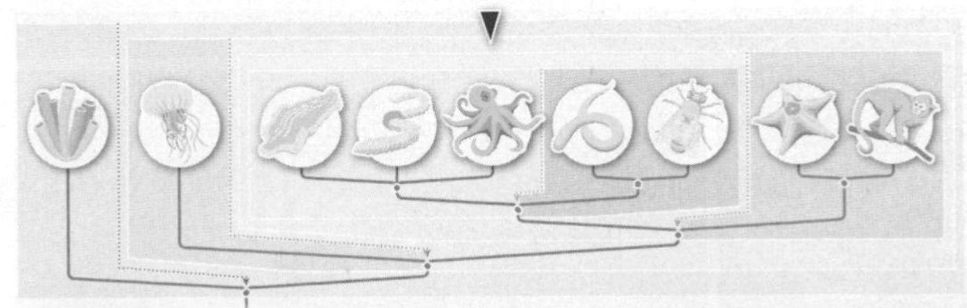

THE MOLLUSKS

COMMON CHARACTERISTICS

- Most have a shell that protects the soft body
- Mantle (tissue that forms the shell)
- Radula (sandpaper-like tongue structure used during feeding); found in all mollusks except bivalves
- Grow by adding body mass

MEMBERS INCLUDE

- Gastropods (~35,000 species)
- Bivalves (~8,000 species)
- Cephalopods (~700 species)

Southern blue-ringed octopus (cephalopod)

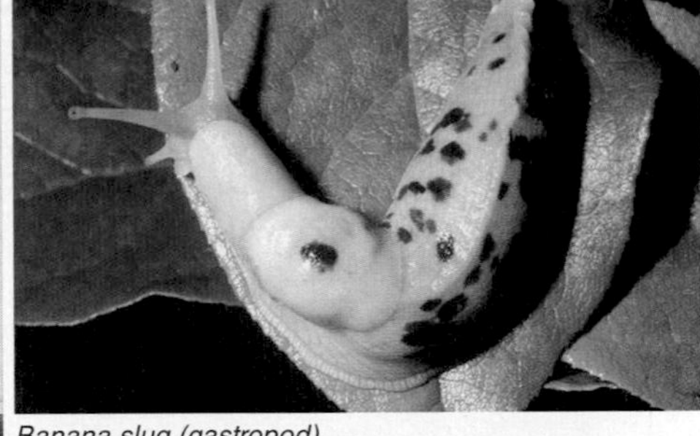

Banana slug (gastropod)

Clam (bivalve)

Swollen phyllidia nudibranch (gastropod)

FIGURE 13-13 An overview of mollusks.

those feeding on sponges can recycle sponge toxins into their own slime. Sea slugs that consume sea anemones can transfer the anemone's stinging cells to the sea slug's own skin, enabling it to sting other creatures.

Bivalve Mollusks The bivalve mollusks are soft-bodied animals protected by a pair of shells hinged together by a ligament. Clams, scallops, oysters, and mussels are examples of bivalves. There are about 8,000 bivalve species and most of them live in the ocean, although there are some freshwater species. Clams spend their lives buried in mud or sand, scallops live on the seafloor, and oysters and mussels fasten themselves to underwater objects such as rocks, the pilings that support piers, and the hulls of boats. All bivalves are filter-feeders that draw a current of water in through a tube called the "incurrent siphon," across their gills, where tiny food particles are captured, and out through the "excurrent siphon."

When a grain of sand is trapped in the mantle tissue of a bivalve (or between the mantle and the shell), the mantle may secrete layer after layer of a shell-like protein material that covers the sand grain and forms the iridescent gem called a pearl. Oysters are the best known source of pearls, but clams and other bivalve mollusks also form pearls.

Cephalopods The third major group of mollusks is the cephalopods (FIGURE 13-14). It includes 6 nautilus species and more than 600 species of squids and octopuses. The nautilus has an external shell, squids have very small shells that are covered by the mantle, and octopuses have lost the shell entirely. In these species, tentacles appear to grow directly from the head, which explains the name: *cephalopod* translates as "head-footed."

Cephalopods use their tentacles to walk, swim, and capture prey. Squids—many of which are ferocious predators—have

GROUPS OF CEPHALOPODS

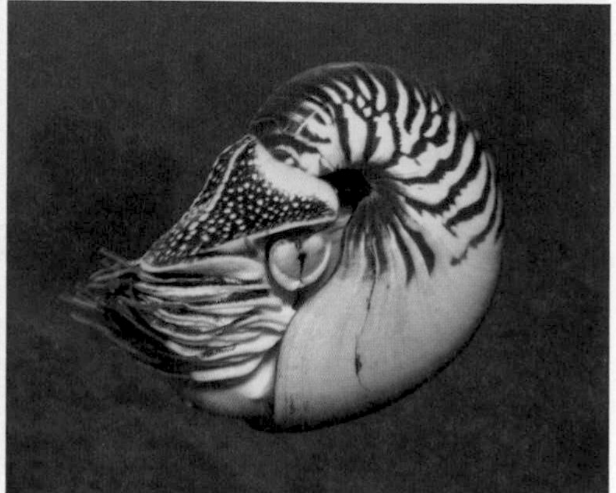

SQUIDS
- 8 short tentacles and 2 long sucker-bearing tentacles
- Free-swimming
- ~300 species

OCTOPUSES
- 8 short tentacles called arms
- Bottom-dwellers, living in coral reefs and on rocky coasts
- ~300 species

NAUTILUSES
- Chambered shell used for protection
- Free-swimming
- 6 species

FIGURE 13-14 An overview of cephalopod diversity.

eight short tentacles called arms and two long, sucker-bearing tentacles used to capture prey.

A squid can propel its tentacles forward with astonishing speed—accelerations of more than 800 feet (about 250 m) per second have been measured, the equivalent of your car accelerating from 0 to 6,000 mph in 10 seconds! The suckers on the tentacles adhere to the prey and draw it back toward the squid, where the arms take over, turning and manipulating the prey as it is bitten by the squid's sharp beak and pulled into the mouth by the tongue-like radula.

Many squids are more than 3 feet (about 1 m) long, and members of some species may exceed 40 feet (about 13 m). When a squid is in a hurry, it swims tail-end first, using jet propulsion. Water is expelled through the siphon at the tentacle end of the body, shooting the animal backward.

Octopuses, which can have tentacles that spread 12 feet (about 4 m), are bottom-dwellers that live in coral reefs and on rocky coasts. Having no shell at all, octopuses can squeeze through astonishingly small openings, and their ability to escape from sealed tanks is a perpetual challenge for the keepers of zoos and aquaria, who are accustomed to arriving in the morning and finding an octopus wandering around the room.

In the next section we consider whether their skills of manual dexterity make them the smartest of the invertebrates.

TAKE HOME MESSAGE 13.7

>> Mollusks are protostome invertebrates that do not molt. They are the second most diverse phylum of animals and include snails and slugs, clams and oysters, and squids and octopuses. Most mollusks have a shell for protection, a mantle of tissue that wraps around their body, and a specialized tongue called a radula.

13.8 Are some animals smarter than others?

If you consider yourself a good multitasker, consider that an octopus can capture prey with one of its tentacles, use a second tentacle to hold something else, and keep the remaining six tentacles in motion, searching for more prey.

FIGURE 13-15 **An octopus attempts to open a jar.**

Octopuses are very good at manipulating things with their tentacles. It doesn't take a captive octopus long to learn how to unscrew tops on glass jars to reach fish swimming inside—a minor variation on twisting a clam to open it (FIGURE 13-15).

Because of their manipulative skills and exploratory behavior, octopuses are sometimes featured in online videos, TV shows, and magazine articles, where they are often described as the smartest invertebrates and their intelligence is compared to that of mammals. But this is the sort of claim that we must view skeptically.

Q Are octopuses smart?

What is intelligence, after all? The concept of intelligence loses its meaning when we are comparing animals that live in completely different environments and respond to entirely different stimuli. We can say that octopuses are very good at doing the things that octopuses need to do—searching for prey, capturing it, and manipulating it, all within their marine environment, for example. But is following a complex sequence of actions the same thing as intelligence? Spiders, for example, construct elaborate webs that are strategically placed in the flight paths of insects; they detect the impact of an insect on the web and rush out to wrap the insect in silk before it can escape. Does that require more or less intelligence than opening a glass jar to eat a fish? And what about the red squirrels of northern Europe? Every fall a squirrel can hide more than 3,000 acorns, cones, and nuts, then recover more than 80% of its hidden nuggets of food over the course of the winter.

If intelligence is defined, in part, as the ability to do some problem-solving task valued by humans, the concept of "intelligence" may be relevant only to humans. Applying the question of intelligence to non-human species may be an assessment of how well they can accomplish tasks essential to their survival. For this reason, questions of comparative animal intelligence are not generally useful. Rather, animals' abilities should be considered as evolutionary responses to particular selective pressures imposed by their environments, and the fact that a species survives is an indication of some mastery of its particular niche. As stated earlier in this chapter, from an evolutionary perspective, any species alive today must be considered a success.

TAKE HOME MESSAGE 13.8

» The predatory behavior of octopuses involves exploration and manipulation, behaviors that humans often consider to be intelligent. But the concept of intelligence cannot be applied objectively to other species, which have evolved in response to the selective forces at work in their own particular niches.

13.9 Arthropods are the most diverse group of all animals.

When it comes to species diversity, the **arthropods** take the cake, outnumbering all other forms of life. The phylum contains almost one million recognized species, or approximately 75% of all animal species and 60% of *all species* on earth (FIGURE 13-16). And about 80% of these are insects. There are about six times as many species of beetles as species of all birds, mammals, fishes, amphibians, and reptiles combined.

Not only are the arthropods diverse, they're also extremely abundant. Taken together, there are more than a billion billion (1 followed by 18 zeroes) arthropods alive at any given

THE ARTHROPODS

COMMON CHARACTERISTICS
- Body with distinct segments
- Exoskeleton made of chitin
- Jointed appendages

MEMBERS INCLUDE
- Insects (more than 800,000 species)
- Arachnids (~60,000 species)
- Crustaceans (~52,000 species)
- Millipedes and centipedes (~10,000 species)

Inchworm moth (insect)

Polydesmid millipede

Spiny orb-weaver spider (arachnid)

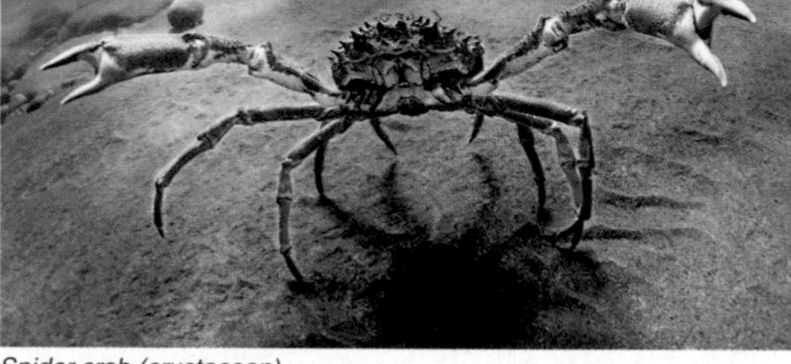

Spider crab (crustacean)

Arthropods make up about 75% of the animal species on earth.

FIGURE 13-16 **Arthropod diversity: insects, millipedes and centipedes, arachnids, and crustaceans.**

time! Put another way, right now there are more than 150 million insects for every single human.

The phylum of arthropods (Arthropoda) contains bilaterally symmetrical, protostome invertebrates that have distinct body segments, a rigid external covering (called an **exoskeleton**) made of a stiff carbohydrate known as chitin, and legs with joints (think of all the joints in a lobster's claw) (FIGURE 13-17). The four major lineages of arthropods are:

- Millipedes and centipedes
- Chelicerates (including horseshoe crabs, spiders, mites, ticks, and scorpions)
- Crustaceans (including lobsters, crabs, shrimp, and barnacles)
- Insects

The arthropod phylum is so large that, even if we excluded all the insects—the most diverse arthropod group—it would still contain more species than any other phylum.

Millipedes and Centipedes *Millipede* and *centipede* translate to "thousand feet" and "hundred feet" and refer to the enormous numbers of legs on these long, skinny animals. They don't really have a thousand—or even a hundred—feet, but millipedes do have more legs than centipedes, and that is the easiest way to tell them apart. Millipedes and centipedes have long, segmented bodies that seem almost worm-like, but their jointed legs and hard exoskeletons are characteristics of arthropods. Both live among fallen leaves, especially in forests where the leaf litter layer is cool and moist. Millipedes feed on decaying plant material, while centipedes are predators that use fangs equipped with venom to kill insects and even snakes

GROUPS OF ARTHROPODS

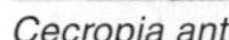

Cecropia ant

INSECTS
- Three pairs of walking legs
- Legs located on the thorax
- Life cycle with separate life stages

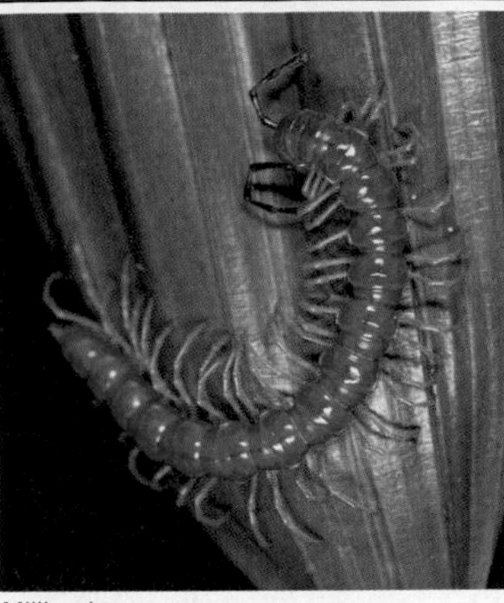

Millipede

MILLIPEDES AND CENTIPEDES
- Many pairs of legs (two pairs per segment in millipedes; one pair per segment in centipedes)
- Long, segmented body

Jumping spider

ARACHNIDS
- Usually four pairs of walking legs
- Legs located on a fused head-midsection
- Specialized mouthparts
- Predators

Sally lightfoot crab

CRUSTACEANS
- Many pairs of legs
- Usually five pairs of appendages that extend from the head
- Mostly aquatic

FIGURE 13-17 Exploring the arthropods.

and small mammals. Centipedes use their jaws to tear prey into pieces small enough to swallow.

Chelicerates There are about 60,000 species of chelicerates, a group that includes four horseshoe crab species and the arachnids, land-dwelling arthropods that include spiders, scorpions, mites, and ticks. Most have four pairs of walking legs and a specialized feeding apparatus.

Arachnids are predators. Many spiders construct webs by spinning silk fibers (consisting of interconnected protein molecules). Orb-weaving spiders construct intricate spiral-shaped webs that trap (and sometimes even lure) flying insects. When a spider captures prey, it uses its fangs to inject venom that contains two types of enzymes. The first set of enzymes disrupts the prey's nervous system and paralyzes it, then other enzymes dissolve its internal organs. The spider sucks out the liquefied contents.

The venom of most spiders is usually harmless to animals as large as humans, but two North American spiders are dangerous. The black widow spider is identified by a red hourglass marking on the bottom surface of the female. The brown recluse spider lacks distinctive markings, so it is harder to identify—and it is more dangerous. Both species often move indoors, constructing webs in closets and behind furniture. Brown recluse bites can create wounds that remain open for months and sometimes require repeated skin grafts to close.

Scorpions look dangerous to humans. The sting of North American scorpions, however, is no worse than a bee's. (The sting of African scorpions, on the other hand, can be lethal to humans.) Scorpions are deadly for insects, though. They seize insects with their claws and use their stinger only if the prey struggles. Some of the scorpion's mouth structures are covered with spines that grind the prey, while glands secrete enzymes that liquefy it, and the scorpion sucks up the soup.

Crustaceans In the United States, the top three fishery crops, by annual market value, are crustaceans: shrimp, crabs, and lobsters. There are about 52,000 species of crustaceans, and although crustacean species can look very different—from sessile barnacles to mobile, tiny krill—they have one feature in common: five pairs of appendages extending from the head. Three of these appendages are used for feeding, and two are antennae that sense the environment.

Crustaceans have many pairs of legs, which can be modified for many purposes. Shrimp and barnacles (which are closely related to shrimp but spend their adult lives within a shell, fastened to a rock) have legs with comb-like projections that they use to capture plankton. Free-living crustaceans (shrimp, crabs, lobsters, crayfish) have modified limbs in the abdominal region used to hold eggs or newly hatched young.

Although most crustaceans are aquatic, the inconspicuous wood lice (also known as pill bugs, sow bugs, or roly-polies) are terrestrial forms that you can find when you turn over a rock in your garden. Wood lice are herbivores and play an important role in recycling dead plant material.

Insects Mostly terrestrial, insects have three pairs of walking legs, and most also have one or two pairs of wings extending from the midsection. Insects abound in nearly all terrestrial habitats and are, by far, the most diverse group of arthropods. Every species of plant, for example, is eaten by at least one insect species. Wings and the ability to fly, as well as the developmental process of metamorphosis, have played important roles in the great success and diversification of insects. We explore these features in more detail in Section 13.11.

TAKE HOME MESSAGE 13.9

» The arthropods are protostome invertebrates; with nearly one million species (and at least as many more yet to be identified), they outnumber all other forms of life in species diversity. Centipedes are predators with fangs that inject venom, and millipedes are herbivores that feed on dead plant material. Spiders and scorpions are predatory arthropods that eat insects and, occasionally, small vertebrates. Lobsters, crabs, shrimp, and barnacles are predatory marine crustaceans. With adaptations that include the ability to fly and metamorphosis, insects are the most diverse group of arthropods.

THIS IS HOW WE DO IT

Developing the ability to apply the process of science

13.10 How many species are there on earth?

When contemplating a question in biology, the most useful approach often resembles a sort of "back-of-the-envelope" calculation. This requires gross simplification, but it can illuminate the information we will need to come to a more accurate answer, while revealing assumptions that may have a large influence on the answer we obtain.

As we consider the diversity of life on earth, one seemingly straightforward question comes to mind.

How many distinct species are there on our planet?

We just don't know. Some scientists put the number at around 3 million, while others believe it could be greater than 100 million.

In a famous paper published in 1982—a paper short enough to fit on the back of an envelope—the biologist Terry Erwin laid out a simple approach to estimating the answer. His strategy was two-pronged. It began by posing a question that he could answer with a high degree of accuracy.

How many species of beetles are there in one species of tree?

Erwin set out to get a definitive count of all the beetle species living in the canopy of a single species of tree (*Luehea seemannii*) in a tropical forest in Panama. His methods were not elegant. Over the course of three seasons, he sprayed a fog of pesticide into the canopy of 19 trees of this species, then collected and identified the beetles that fell to the ground.

He found 1,143 different species of beetles that fell from the trees. Because other researchers had reported collecting additional beetle species from the same species of tree, he rounded off the number to 1,200 species. Next, Erwin used these data, in which he had high confidence, to extrapolate the numbers of other insect species living in the diverse tree species in the tropics.

How is the number of beetle species in one tree species related to the number of all insect species in all types of trees?

Erwin made several educated guesses. His back-of-the-envelope calculations included estimates of the proportion of beetle species that are host-specific (13.5%), the number of unique trees per hectare of tropical forest (~70), the proportion of all arthropod species that are beetles (40%), the proportion of arthropods in the forest canopy as opposed to elsewhere (67%), and the number of unique species of tropical trees (50,000). When he multiplied everything together, his estimates led him to conclude that there were almost 30 million species of arthropods in tropical forests. This number was remarkable because it was about 10 times higher than most previous estimates of the total number of species on earth!

Understanding the self-correcting and collaborative nature of scientific thinking, Erwin said, "I would hope someone will challenge these figures with more data." Erwin had, after all, made numerous assumptions. (Which of these seem likely to have had a significant influence on his final answer?)

Taking Erwin's ideas as a starting point, many researchers have noted (and tried to improve on) his assumptions. In 2010, researchers came up with a better way to estimate. They used probability distributions for each parameter (such as for the proportion of beetle species specific to each species of tree).

This enabled the researchers to come up with a range of estimates, each having a likelihood associated with it, and their estimates were much lower than Erwin's 30 million. One of their models generated an estimate of the median number of tropical arthropod species as 3.7 million, with a 90% confidence interval that the true number was between 2.0 and 7.4 million.

These estimates and the process by which they've been generated give us greater confidence that we know the answer. Unfortunately, this isn't an answer to our original question. It's an answer to an easier question: how many *arthropod* species are there in the tropics?

The reason for limiting the scope of the question is that it is much easier to define and identify beetle or other arthropod species. But does our answer bring us closer to knowing how many species, in total, there are on earth? Consider another question.

What types of species are the most difficult to estimate (or count)?

Given the impossibility of applying the biological species concept to asexual species, for example, how should we define and identify species of prokaryotes and viruses? And, are the numbers of marine species proportional to the numbers of arthropod species?

The back-of-the-envelope approach described here can be a useful starting point for these new calculations as well.

TAKE HOME MESSAGE 13.10

» Determining how many distinct species there are on earth is a challenge. By counting the exact number of beetle species in one species of tropical tree and estimating several parameters that can be used to extrapolate this number to the number of species of all arthropods across the entire tropics, it is possible to make headway on this challenge. Initial estimates suggest there may be many more species on earth than previously believed.

13.11 Flight and metamorphosis produced the greatest adaptive radiation ever.

Two characteristics of insects that have been central to their success are the ability to fly and the way they cope with change in body size as they grow. Wings are adaptations that first appeared in insects. With flight, insects can avoid many predators and can efficiently search for food and for mates. Flight also enables organisms to more quickly leave one location and disperse to another that is potentially more hospitable.

METAMORPHOSIS

COMPLETE METAMORPHOSIS

Complete metamorphosis is the division of an organism's life history into three completely different stages (occurs in 83% of insect species).

LARVA
An egg hatches into a larva, which eats (and eats), growing as it molts several times until large enough to enter the pupal stage.

PUPA
The larva encloses itself in a case, body structures are broken down, and new structures are assembled into the adult form.

ADULT
The adult emerges from the pupa and no longer grows. Its primary function is to reproduce.

INCOMPLETE METAMORPHOSIS

Incomplete metamorphosis is the pattern of growth and development in which an organism does not pass through separate, dramatically different life stages (occurs in 17% of insect species).

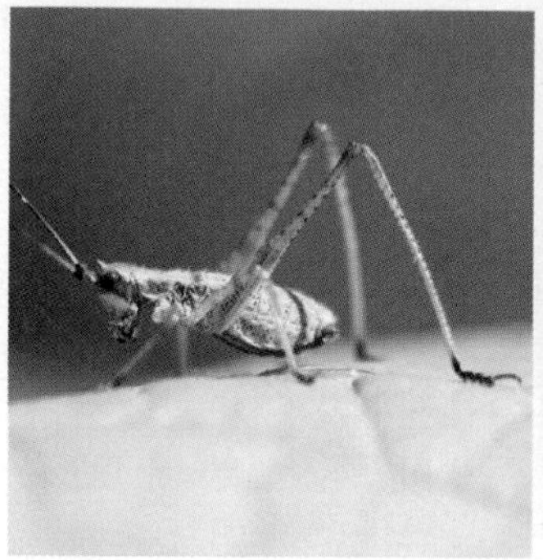

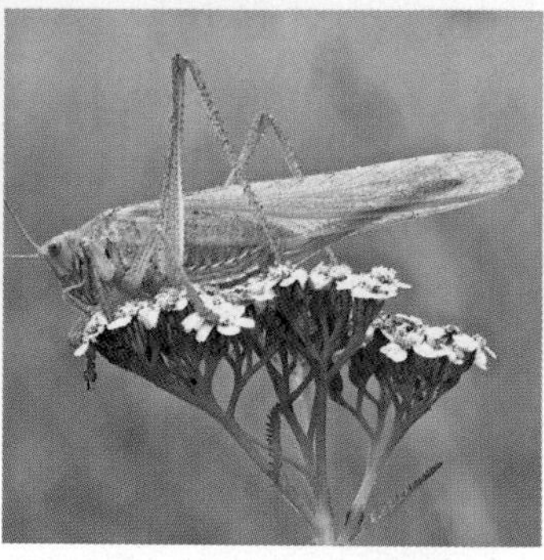

NYMPH
An egg hatches into a nymph, resembling a small version of the adult (but without wings or reproductive organs), growing as it molts several times.

ADULT
Upon reaching adult size, the nymph stops molting and the adult forms.

The separation of life stages has contributed to the enormous ecological diversity of insects.

FIGURE 13-18 Metamorphosis. Some insects have a life history divided into unique stages, each with a distinct purpose.

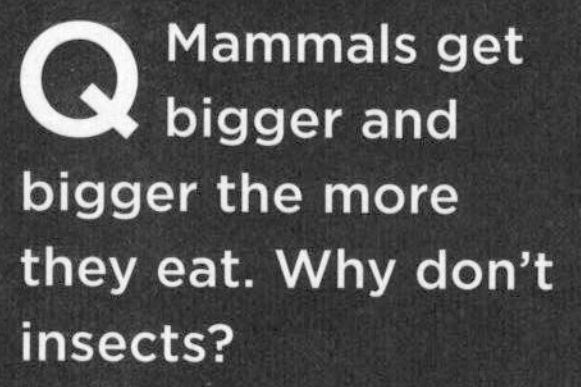

Q
Mammals get bigger and bigger the more they eat. Why don't insects?

As arthropods, insects have a rugged exoskeleton, to which their muscles are attached. Like a suit of armor, the exoskeleton provides excellent protection from injury and predators. It also helps individuals conserve water and resist drying out. The exoskeleton, however, prohibits growth. To overcome this constraint, all arthropods (as well as the roundworms, of phylum Nematoda) have a process of growth that includes molting, which is very different from the growth pattern in other animals. After hatching, their life is divided into three completely different stages (FIGURE 13-18).

1. Larva. An egg hatches into a **larva** (*pl.* larvae), which looks completely different from the adult form. A caterpillar is the larval stage of a butterfly or moth, and a grub is the larval stage of a beetle. A larva's job is to eat and grow large enough to enter the next life stage, the pupa. Larvae grow by passing through numerous stages (usually 4–8) in which they shed their exoskeleton, increase in size, and develop a new, slightly larger exoskeleton in the process of molting.

2. Pupa. Eventually, a larva completely covers itself with a casing (such as cocoon or chrysalis), at which point the insect is called a **pupa.** In the pupa, genes are activated that code for the adult body form, and the proteins that made up the larva are broken down to amino acids, which are recycled to synthesize adult proteins and reassembled into the adult form. This rebuilding process is called **metamorphosis,** from two Greek words meaning "change of form." The entire process requires several days, and by the end of the pupal stage, the adult insect is curled tightly inside the pupal case.

3. Adult. When the **adult** form hatches from the pupal case, it is at its full size. For this reason, the adults of some insects do not even eat. Rather than eating and growing, the job of an adult insect is reproduction.

The developmental process called **complete metamorphosis,** which includes these three stages, occurs only in insects and allows the larva and adult to act as if they were animals from different species, each optimized to perform very specialized tasks. Caterpillars eat leaves, for example, and spend their entire larval period on a single plant. In contrast, butterflies feed on nectar that they collect by flying from flower to flower, saving up enough resources to lay or fertilize eggs. As a larva, the animal feeds and grows. As an adult, it reproduces. The genes that control the larval body form are different from the genes that determine the adult form, so natural selection can act on the larval and adult stages independently.

About 83% of insect species go through complete metamorphosis, and this separation of the life stages has been an important factor in helping insects diversify into nearly

20 times as many named species as vertebrates. Among those insect species that do not undergo complete metamorphosis, typically the eggs hatch as nymphs (the juvenile form), which resemble a smaller version of the adult. These juveniles do not have wings or reproductive organs, but they live in the same habitats as adults and eat the same foods. They then undergo several molts as they grow, and stop molting when they reach adult size. This pattern of growth and development, called **incomplete metamorphosis,** occurs in grasshoppers and cockroaches, among many other insect species.

TAKE HOME MESSAGE 13.11

» The ability to fly and the developmental process of metamorphosis as a means of overcoming the constraints imposed by a rugged exoskeleton have contributed to the enormous ecological diversity of insects. The life cycle of most insects includes a larval stage devoted to feeding and growth, a pupal stage during which metamorphosis occurs, and an adult stage in which the insect reproduces.

13.12 Echinoderms are vertebrates' closest invertebrate relatives.

Appearances can be deceiving. Sea stars, sand dollars, and sea urchins are members of the phylum Echinodermata and live in the oceans. They are radially symmetrical (as adults) and are covered by spiny plates. And yet, because they are deuterostomes (their gut forms from back to front), it turns out that they are more closely related to humans than to any other invertebrate group. This unexpected relationship reveals the force of natural selection and the dramatic effects it can have on body form as populations of organisms become adapted to their environment.

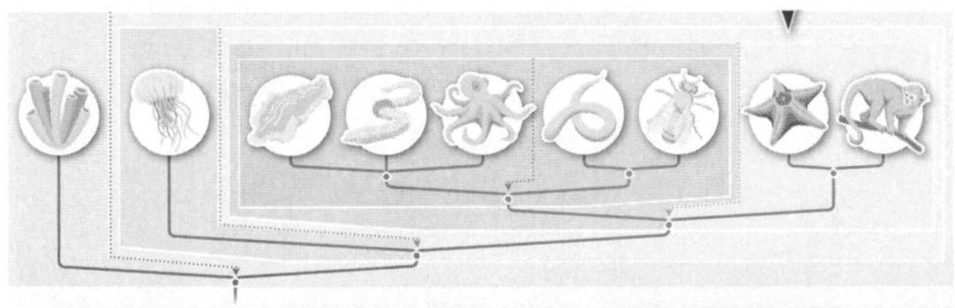

The echinoderms include about 6,000 species of marine animals, most of which are enclosed by a hard skeleton of spiny plates (FIGURE 13-19). Adult sea stars (starfishes and brittle stars), probably the most recognizable echinoderms, have five or more appendages evenly distributed around

FIGURE 13-19 **The echinoderms.**

THE ECHINODERMS

COMMON CHARACTERISTICS
- Enclosed by a hard skeleton of spiny plates
- Larvae are bilaterally symmetrical and share some anatomical features with chordates
- Adults are radially symmetrical
- Undersides covered with tube feet that aid in locomotion and grasping

MEMBERS INCLUDE
- Sea stars (~1,600 species)
- Sea urchins and sand dollars (~940 species)
- Sea cucumbers (~1,100 species)

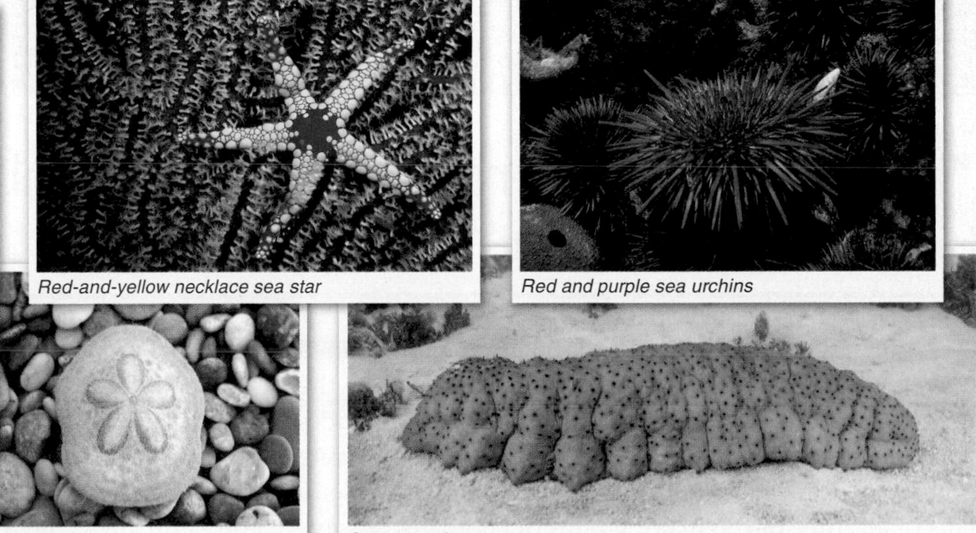

Red-and-yellow necklace sea star

Red and purple sea urchins

Sand dollar

Sea cucumber

their body circumference. They move with equal ease in any direction, and their sensory organs are distributed around their circumference. Adult sea urchins and sand dollars don't have projecting arms, but they also are radially symmetrical.

Although radial symmetry is characteristic of adult echinoderms, their larvae are bilaterally symmetrical. Bilaterally symmetrical larvae are evidence that echinoderms evolved from bilaterally symmetrical ancestors, and their radial symmetry as adults is an adaptation associated with their mode of locomotion and feeding specializations.

Echinoderms do not have a brain. They have a nervous system that consists of a central ring of nerves, with branches that extend into each of their appendages and help them gain information about and respond to their environment. Echinoderms creep on little tube feet that are extensions of an internal system of water-filled canals radiating throughout the body. Tube feet carpet the undersides of sea stars' arms and extend up and around the bodies of sea urchins and sand dollars, extending and contracting in waves to grasp and release the surface on which they move as the animal glides along.

Although each tube foot is tiny, there are thousands of them, and their combined force allows a sea star to pull apart the two shells of a clam or mussel. Once the shells are apart, the sea star pushes its stomach out through its mouth and inserts it into the opened shells. The stomach then secretes digestive enzymes that break down the clam or mussel tissue, and the sea star absorbs the soup that results. Sea urchins feed on algae, scraping them loose from rocks with sharp tooth-like surfaces made of calcium. Sand dollars capture floating particles of algae and other organic matter by trapping them in streams of mucus, which are moved toward their mouth by beating cilia.

TAKE HOME MESSAGE 13.12

» Because they are deuterostomes (as are vertebrates), echinoderms are the invertebrates that are the closest evolutionary relatives to the vertebrates. Their aquatic larvae are bilaterally symmetrical and share some anatomical features with chordates, but adult echinoderms are radially symmetrical.

13.13–13.15

The phylum Chordata includes vertebrates, animals with a backbone.

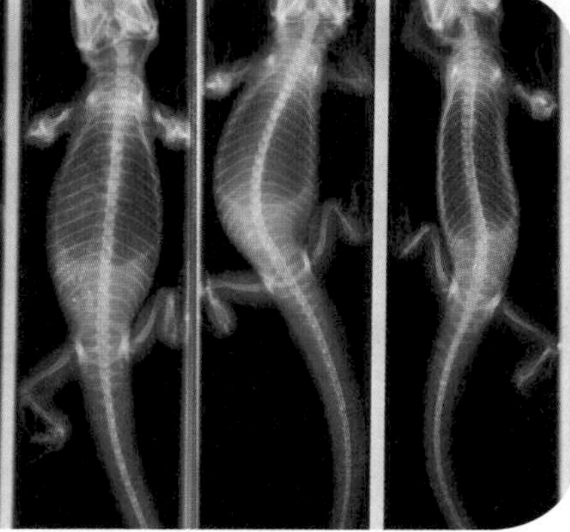

Animal in motion: these X-ray images show a tuatara (a reptile from New Zealand) walking.

13.13 All vertebrates are members of the phylum Chordata.

Vertebrates are deuterostome animals, having defined tissues and bilateral symmetry. They are part of the phylum called Chordata (the chordates), which includes three major groups: the vertebrates and two invertebrate groups, the tunicates and the lancelets (FIGURE 13-20). Members of all three of the major groups of chordates possess, for at least a portion of their life cycle, four distinctive chordate body structures that are found in no other animal groups (FIGURE 13-21).

1. The **notochord,** a rod of tissue extending from head to tail, is the structure that gives chordates their name (from the Latin for "cord"). The notochord stiffens the body when muscles contract during movement. The simplest

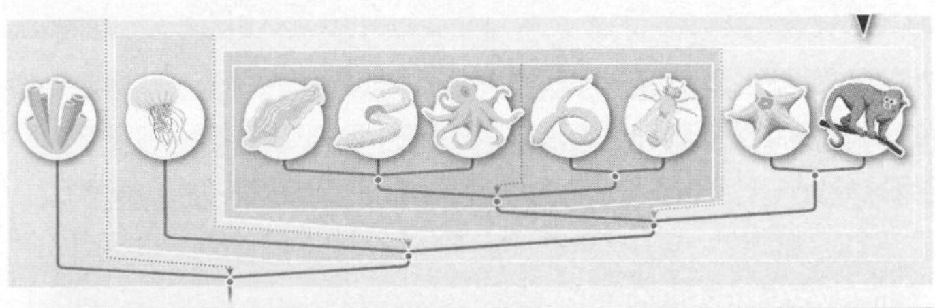

THE CHORDATES

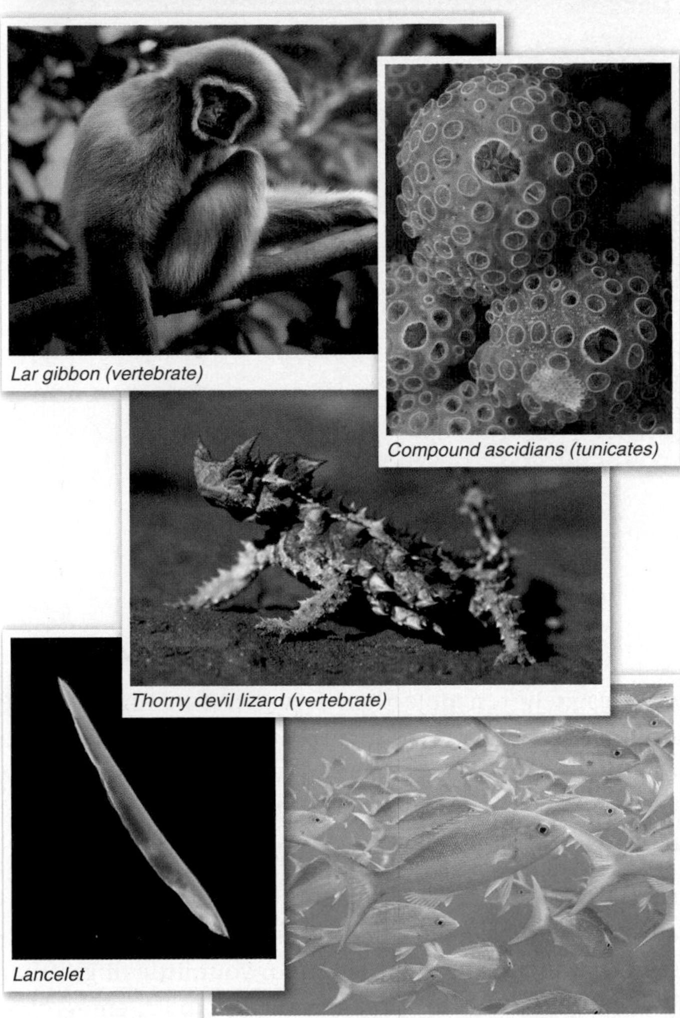

Lar gibbon (vertebrate)

Compound ascidians (tunicates)

Thorny devil lizard (vertebrate)

Lancelet

Schooling yellowtail snapper (vertebrates)

COMMON CHARACTERISTICS
All chordates possess four distinctive body structures (sometimes present only during specific life stages).

- Notochord
- Dorsal hollow nerve cord
- Pharyngeal slits
- Post-anal tail

MEMBERS INCLUDE

- Tunicates (~2,000 species)
- Lancelets (~20 species)
- Vertebrates (~56,000 species)

FIGURE 13-20 **An overview of chordates.**

chordates—those most distantly related to the vertebrates—retain the notochord throughout life. In the more complex chordates such as vertebrates, however, the notochord is present only in the early embryo and is replaced by the backbone (vertebral column) as the embryo develops.

2. A **dorsal hollow nerve cord** extends along the animal's back (its dorsal side) from its head to its tail. In vertebrates, this nerve cord eventually forms the central nervous system, which consists of the spinal cord and the brain. Other kinds of animals (worms, insects, and so on) also have a nerve cord, but it lies in the lower portion of the front (ventral) part of the body and is solid instead of hollow.

3. Pharyngeal slits, or gill slits, are present in the embryos of all chordates, but in many chordates (including humans) the slits disappear as the animal develops. The earliest chordates were aquatic, and to breathe and feed they passed water through slits in the pharyngeal region (the area between the back of the mouth and the top of the throat).

THE DISTINCTIVE BODY STRUCTURES OF CHORDATES

The members of Chordata are defined by four distinctive body structures present in all chordates but in no other animal groups.

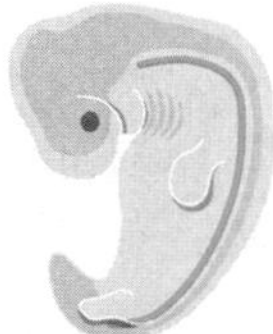

NOTOCHORD
A rod of tissue extending from the head to the tail

- Simpler chordates retain the notochord throughout life
- In more complex chordates, the notochord in early embryos is replaced by a backbone

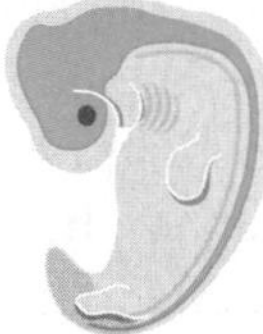

DORSAL HOLLOW NERVE CORD
Nerve cord that extends along the animal's back (its dorsal side)

- In vertebrates, the nerve cord eventually forms the spinal cord and brain

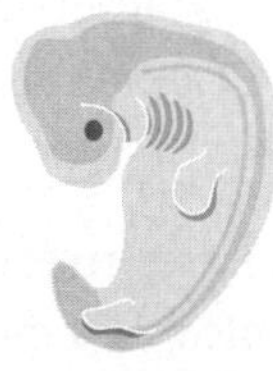

PHARYNGEAL SLITS
Slits through which water is passed in order to breathe and feed

- In many chordates (including humans), the slits disappear as the embryo develops

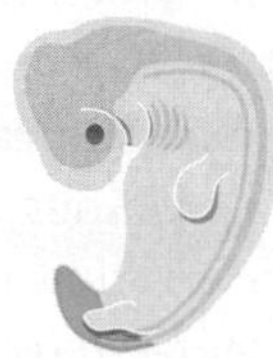

POST-ANAL TAIL
Tail that extends beyond the posterior (back) end of the digestive system

- Some vertebrates (including humans) have a tail only briefly, during embryonic development

FIGURE 13-21 **Characteristic chordate body structures.**

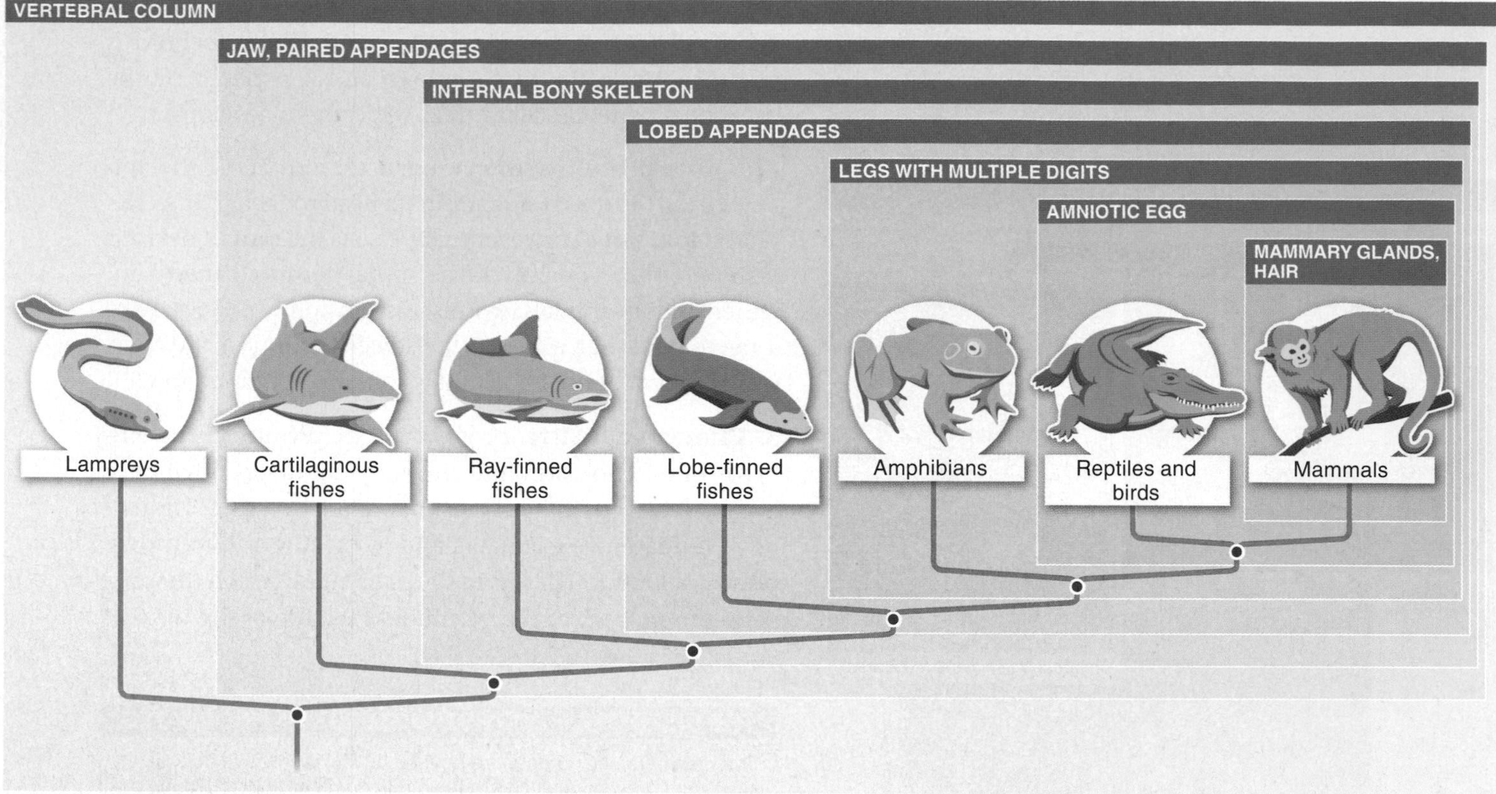

FIGURE 13-22 **Vertebrate phylogeny.**

In humans, pharyngeal slits are present only in the embryo; our gills were lost far back in evolutionary time.

4. A **post-anal tail** is another chordate characteristic. The posterior (back) end of the digestive system is the anus. The region of the body extending beyond the opening of the anus is referred to as "post-anal." All vertebrates have a tail in this location, but some, including humans, have a tail only for a brief period, during embryonic development.

Although all chordates share these four characteristic structures, the animals in this phylum exhibit tremendous physical and ecological diversity.

Tunicates (Urochordata; about 2,000 species) are invertebrate marine animals that have defined tissues, bilateral symmetry, and deuterostome development, in which the gut develops from back to front. The adults are thumb-sized and look like balls of brownish green jelly. You can find them attached to docks and the mooring lines of boats. It is the free-swimming larvae of tunicates that reveal their chordate characteristics. Tunicate larvae have a notochord, dorsal hollow nerve cord, and tail, all of which disappear during development. The sessile adult tunicates only vaguely resemble their larvae. Adult tunicates are filter-feeders: cilia draw a current of water through the mouth into the pharynx and out through the pharyngeal slits. Microscopic food items are trapped by a layer of mucus stretched across the slits.

Lancelets (Cephalochordata; about 20 species), also bilaterally symmetrical deuterostomes, are slender, eel-like invertebrate animals, about the length of your little finger, that live in coastal waters. Unlike tunicates, both embryonic and adult lancelets have all the chordate characteristics. Lancelets are also filter-feeders.

Vertebrates (Vertebrata; about 56,000 species) are the most diverse group of chordates (FIGURE 13-22). Because they are deuterostomes, they are bilaterally symmetrical, with defined tissue and from-back-to-front gut development. Vertebrates differ from the other chordates in two important ways.

1. They have a backbone, formed when a column made from hollow bones (or cartilage in some species), called vertebrae, forms around the notochord. This backbone surrounds and protects the dorsal hollow nerve cord.

2. They have a head, at the front (anterior) end of the organism, containing a skull, a brain, and sensory organs.

Blue whales, opossums, tree frogs, and finches are all vertebrate animals. Although they vary tremendously in size, shape, and locomotion, they nonetheless share the four chordate characteristics at some stage of their lives. In the remaining sections, we focus on the diversity of the vertebrates.

TAKE HOME MESSAGE 13.13

» All chordates have four characteristic structures: a notochord, a dorsal hollow nerve cord, pharyngeal slits, and a post-anal tail. The three major groups of chordates, though superficially very different, are united by possessing these four structures at some stage of their life cycle.

13.14 The evolution of jaws and fins gave rise to the vast diversity of vertebrate species.

The earliest vertebrates were fish-like animals that lived more than 500 million years ago and did not have jaws. Today, just two kinds of jawless vertebrates exist: the lampreys (with 41 species) and the hagfishes (with 43 species). Most lamprey species do not feed as adults; instead, they live off energy stores built up by filter-feeding in their larval stage. Some lamprey species, however, are parasitic and attach to other fishes by creating suction with a circular oral disk that is studded with sharp spines. The lamprey emits a protein that keeps the wound from closing, as it feeds on the seeping blood and body fluids. Hagfishes feed on dead animals, embedding two spiny dental plates in their prey.

Fishes swim in the same way lampreys and hagfishes swim—that is, by bending the body from side to side to create an S-shaped wave that moves from head to tail. Unlike the simple, tube-like, limbless, and fin-less body of hagfish and lamprey, a typical fish has seven fins serving different purposes: to drive the fish forward, to minimize rolling from side to side, and for steering and stopping. The combined effects of the seven fins allow the fish to swim rapidly in a straight line in open water or to weave its way through a dense stand of plants or corals. Fins propel the organism to its prey, and jaws capture and kill it (FIGURE 13-23). The evolution of fins and jaws is what set the stage for the explosion of diversity in the vertebrates.

Q Why are fins and jaws needed?

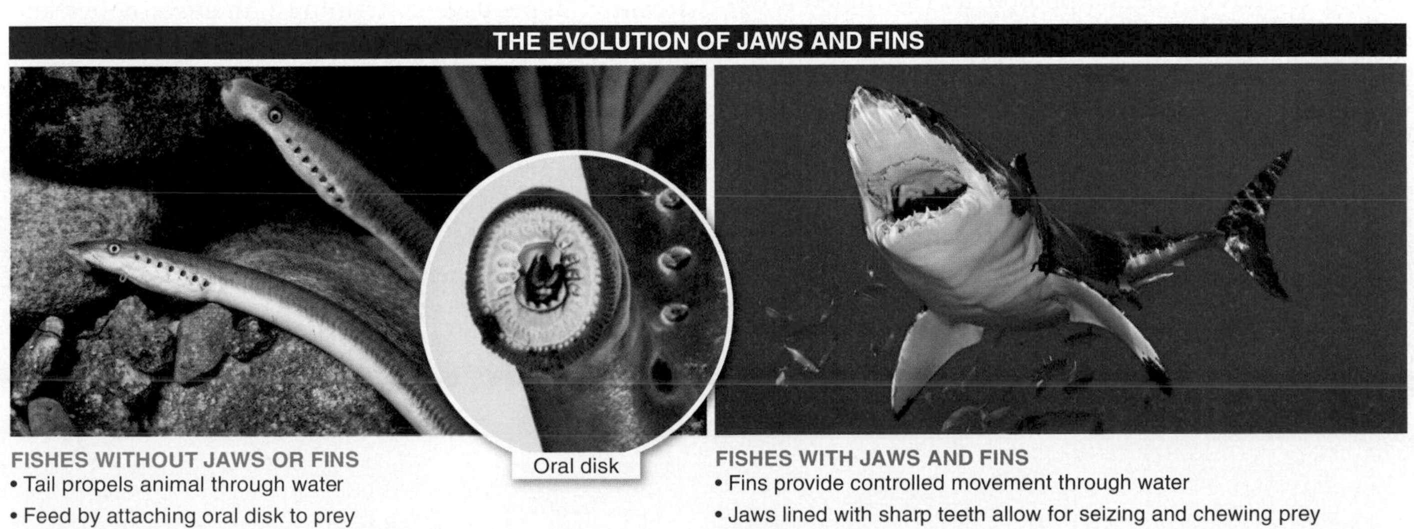

THE EVOLUTION OF JAWS AND FINS
Oral disk
FISHES WITHOUT JAWS OR FINS
• Tail propels animal through water
• Feed by attaching oral disk to prey
FISHES WITH JAWS AND FINS
• Fins provide controlled movement through water
• Jaws lined with sharp teeth allow for seizing and chewing prey
The development of fins and jaws set the stage for the evolutionary explosion of vertebrate diversity.

FIGURE 13-23 The lamprey and the shark.

JAWED FISHES

Stingray

Coral grouper

Australian lungfish

CARTILAGINOUS FISHES

- Characterized by a skeleton made completely from cartilage
- ~880 species, including sharks and rays

RAY-FINNED FISHES

- Characterized by rigid bones and fins lined with hardened rays
- Possess a swim bladder, which aids in flotation
- ~27,000 species, including almost everything we think of as "fish"

LOBE-FINNED FISHES

- Characterized by two pairs of sturdy fins on the underside of the body
- 6 species of lungfish and 2 species of coelacanths

FIGURE 13-24 Fishes with jaws: cartilaginous, ray-finned, and lobe-finned fishes.

There are three categories of jawed fish. **Cartilaginous fishes,** with about 880 species, include the sharks and rays (FIGURE 13-24). Cartilaginous fishes are characterized by a skeleton made entirely of cartilage, a solid but slightly flexible connective tissue—the same tissue that gives your nose and ears their shape.

Ray-finned fishes have a rigid skeleton made from bone, which, like cartilage, is a solid connective tissue consisting of specialized cells and an extracellular material (matrix) that the cells secrete. Bone, however, is much less flexible than cartilage because its extracellular matrix is mineralized by crystals of calcium phosphate. Ray-finned fishes have a mouth at the narrow tip, or apex, of the body, and their fins, made from webs of skin, are supported with hardened rays made of bone.

Ray-finned fishes are the largest group of jawed fishes and the most diverse group of vertebrates, with about 27,000 species. This group includes almost everything you think of as "fish," from salmon to goldfish.

In addition to a bony skeleton, an important evolutionary development in the ray-finned fishes was the swim bladder, a gas-filled organ that keeps the fish from sinking. This bladder is believed to be homologous to the lungs of land-dwelling (terrestrial) vertebrates, with both evolving from simple air sacs, involved in gas exchange, connected to the gut. Cartilaginous fishes, which have no swim bladder, must constantly move through the water or they will sink.

The third group of jawed fishes, the **lobe-finned fishes,** are represented by just eight species. These fishes have sturdy pelvic and pectoral fins on the underside of their body (see Figure 13-25), which, unlike other fins, have a central appendage containing numerous bones and muscles that connect the fins to the body. As we see in the next section, lobe fins were useful in initiating the move onto land.

TAKE HOME MESSAGE 13.14

» The development of two structures in fishes—fins and jaws—set the stage for the enormous diversity of modern vertebrates.

13.15 The movement onto land required lungs, a rigid backbone, four legs, and eggs that resist drying.

Movement from water onto land required four major evolutionary adaptations. All four characteristics are necessary for a land animal, and all four evolved in predatory fishes that lived in shallow water—the immediate ancestors of terrestrial vertebrates. The four adaptions were lungs, a backbone, four legs, and eggs that won't dry out.

Q Where did legs and lungs come from?

As we've noted, lungs probably evolved from the air sacs that were also evolutionary precursors to the swim bladder found in ray-finned fishes. All of the terrestrial vertebrates—amphibians, reptiles (including birds), and mammals—are descendants of the lobe-finned fishes that lived during the Devonian period, some 400 million years ago. At this time, the lobe-finned fishes had lungs and the rudiments of limbs, but they were still fully aquatic and living near the shore. The four sturdy fins on the underside of their body helped them move through shallow water, and lungs allowed them to breathe air when the oxygen concentration in the warm, stagnant water was low. Thus, these fishes already had some of the basic characteristics of terrestrial animals. The jointed bones in the fins of lobe-finned fishes don't look much like your arms and legs, but they are homologous structures.

To move onto land, a vertebrate needed more than just legs and lungs. It needed structural support to resist the pull of gravity. Each vertebra of a terrestrial vertebrate has projections that interlock with projections from the vertebra ahead of it and the vertebra behind it. These interlocking projections prevent the backbone from sagging under the pull of gravity, and body weight is transmitted through the limbs to the ground.

The last innovation necessary to move onto land was an egg that resists drying out (FIGURE 13-25). When eggs are deposited on land, they are exposed to air and lose water by evaporation. The eggs of terrestrial animals need a waterproof, protective covering—a membrane and a shell. The

FROM WATER TO LAND

The transition of vertebrates from life in water to life on land required overcoming three main obstacles. Four major evolutionary innovations allowed this transition.

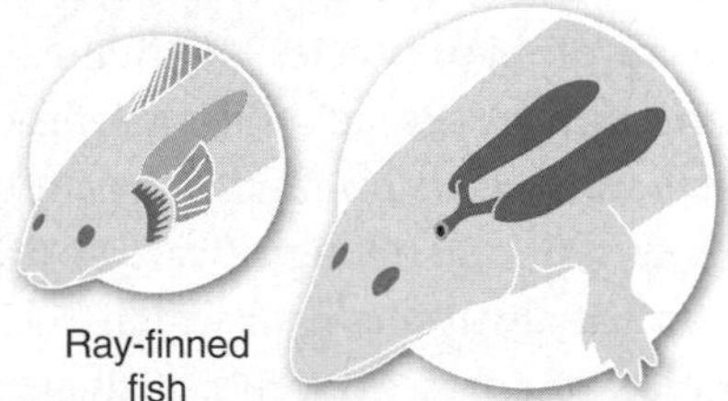

PROBLEM: RESPIRATION
Aquatic animals use gills to acquire dissolved oxygen from water. The transition onto land required the ability to breathe air.

SOLUTION: LUNGS
Gas exchange occurs in lungs, which evolved from the air sacs (the evolutionary precursors to the swim bladder found in ray-finned fishes).

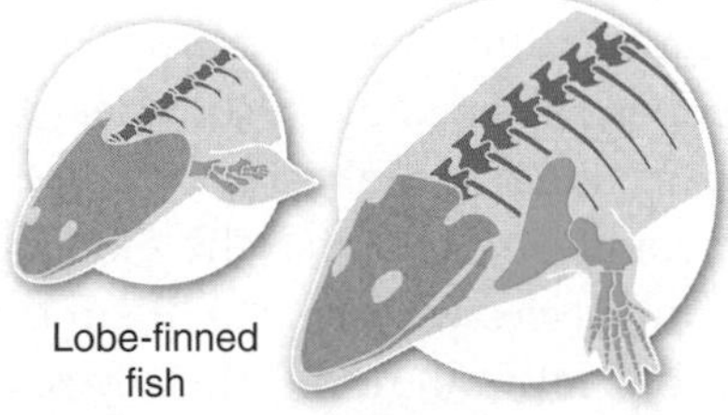

PROBLEM: GRAVITY
The transition onto land from a buoyant water environment required structural support to resist the pull of gravity.

SOLUTIONS: LIMBS and MODIFIED VERTEBRAE
Limbs evolved from the jointed fins on the underside of lobe-finned fishes.

Vertebrae were modified to transmit body weight through the limbs to the ground.

PROBLEM: EGG DESICCATION
On land, eggs must resist the drying effects of air.

SOLUTION: AMNIOTIC EGG
A waterproof eggshell prevents eggs from drying before they hatch.

FIGURE 13-25 **How did vertebrates make the transition from life in water to life on land?**

shell appeared about 380 million years ago and further facilitated the evolution of entirely terrestrial vertebrates, the groups that evolved into mammals and reptiles (including birds). But these initial egg coverings were soft and probably still required a watery environment. The eggs of terrestrial vertebrates acquired a water-tight membrane that keeps the embryo surrounded by a bath of fluid called amniotic fluid.

TAKE HOME MESSAGE 13.15

» Four adaptations were important in the transition of life from water to land. Fins were modified into limbs. Vertebrae were modified to transmit body weight through the limbs to the ground. The site of gas exchange was transferred from gills and swim bladders to lungs. And terrestrial vertebrate eggs have membranes and a shell, and resist drying out.

13.16–13.20

All terrestrial vertebrates are tetrapods.

Capybaras (*Hydrochoerus hydrochaeris*), rodents found in South America. They can reach more than 200 pounds (91 kg)!

13.16 Amphibians live a double life.

The ancestors of all vertebrates that live on land had lungs and four legs. Some modern vertebrates, such as whales and snakes, have evolved so that their limbs are reduced to a few shrunken and unused bones. Whales and some snakes have returned to living in water, and some salamanders have lost lungs and breathe through their skin. But these are recent changes, occurring in species whose ancestors had lungs and four legs. Taxonomically, all terrestrial vertebrates are **tetrapods** (*tetra* = four; *poda* = feet).

Terrestrial vertebrates are divided into two main groups: (1) animals called **non-amniotes,** such as amphibians, that reproduce in water and do not have desiccation-proof amniotic eggs; and (2) animals called **amniotes,** such as reptiles, birds, and mammals, that have amniotic eggs. We discuss amphibians here, then reptiles and mammals in the next two sections.

The first terrestrial vertebrates were **amphibians,** from the Greek word *amphibios,* meaning "living a double life." There are two stages in the life of most amphibians: a water-breathing juvenile form and an air-breathing adult form. The adults of many amphibian species lay their eggs in water (FIGURE 13-26). Amphibian eggs lack the waterproof layers that allow amniotes to lay eggs on land, so most of the 6,000 species in this group are still significantly tied to life in the water. They must always be near water to lay their eggs, which are simple structures, not unlike fish eggs. Frogs and toads make up the vast majority of amphibians (about 5,400 species). Other amphibians include salamanders (550 species) and caecilians, a group of legless, burrowing animals (170 species).

The juvenile stage of frogs, called a tadpole, lives in water, lacks legs, and eats algae. A few weeks (or up to many months in some species) after hatching, metamorphosis occurs, and the tadpole develops legs, lungs, and a digestive system fit for its adult life as a carnivore. Adult frogs have thin, moist skin through which gas (oxygen and carbon dioxide) exchange

THE AMPHIBIAN LIFE CYCLE

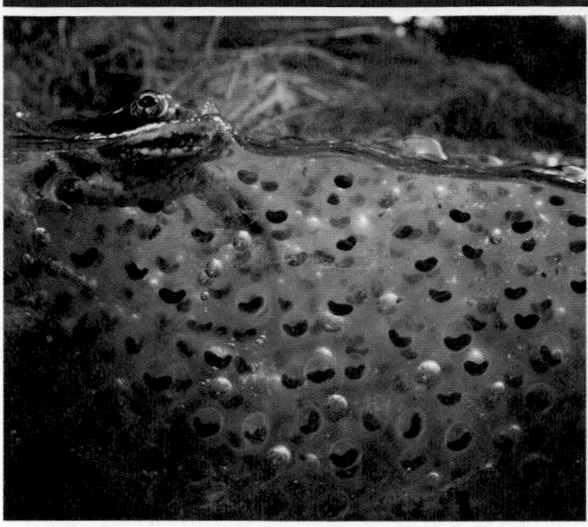

EGGS
Amphibians have non-amniotic eggs, which must be laid in water to prevent desiccation.

JUVENILES
Amphibians spend their juvenile stage underwater and undergo metamorphosis to develop legs and lungs.

ADULTS
Only the adults are true land animals; however, most amphibians stay close to water to lay their eggs.

FIGURE 13-26 From egg to larva to adult: the amphibian life cycle.

Most species of amphibians, such as frogs, toads, and salamanders, live on land as adults, but develop in water.

with the air can take place. When a frog is underwater, oxygen diffuses directly from the water into the blood.

The past two decades have seen a stunning decline in amphibian populations throughout the world, and almost a third of all amphibian species are endangered or threatened. This situation is attributable to a combination of causes: climate change, habitat degradation, fungal diseases, and increased pollution.

TAKE HOME MESSAGE 13.16

» Amphibians are terrestrial vertebrates, but the adults of most species still lay eggs in water. The eggs hatch into aquatic juveniles.

13.17 Birds are reptiles in which feathers evolved.

Soon after amniotic vertebrates appeared, two different evolutionary lineages began to diverge (FIGURE 13-27). One of these lineages is the mammals, the group to which humans belong. The other lineage is the one referred to in this chapter as "reptiles (including birds)." That is an awkward term, but the evolutionary tree shows why it is necessary. Surprisingly, birds (about 9,700 species) are one branch of the reptile lineage that also includes snakes and lizards (about 8,000 species), turtles (about 300 species), crocodiles and alligators (23 species), the New Zealand tuatara (2 species)—and the dinosaurs.

The characteristics that hold the bird-crocodile-dinosaur group together are mostly similarities in bones (especially the bones of the skull and legs) and in DNA sequences. Reptiles are amniotes and thus have amniotic eggs. You are familiar with the hard-shelled eggs of birds, and some lizard, snake, and turtle eggs also have hard shells. Other species in these groups have eggs with a paper-like shell.

Most people find it confusing that birds are grouped with turtles, lizards and snakes, and alligators and crocodiles, and that confusion is easy to understand (FIGURE 13-28). After all, birds are covered by feathers and can fly, whereas the other reptiles have bare skin and do not fly. And there is an important

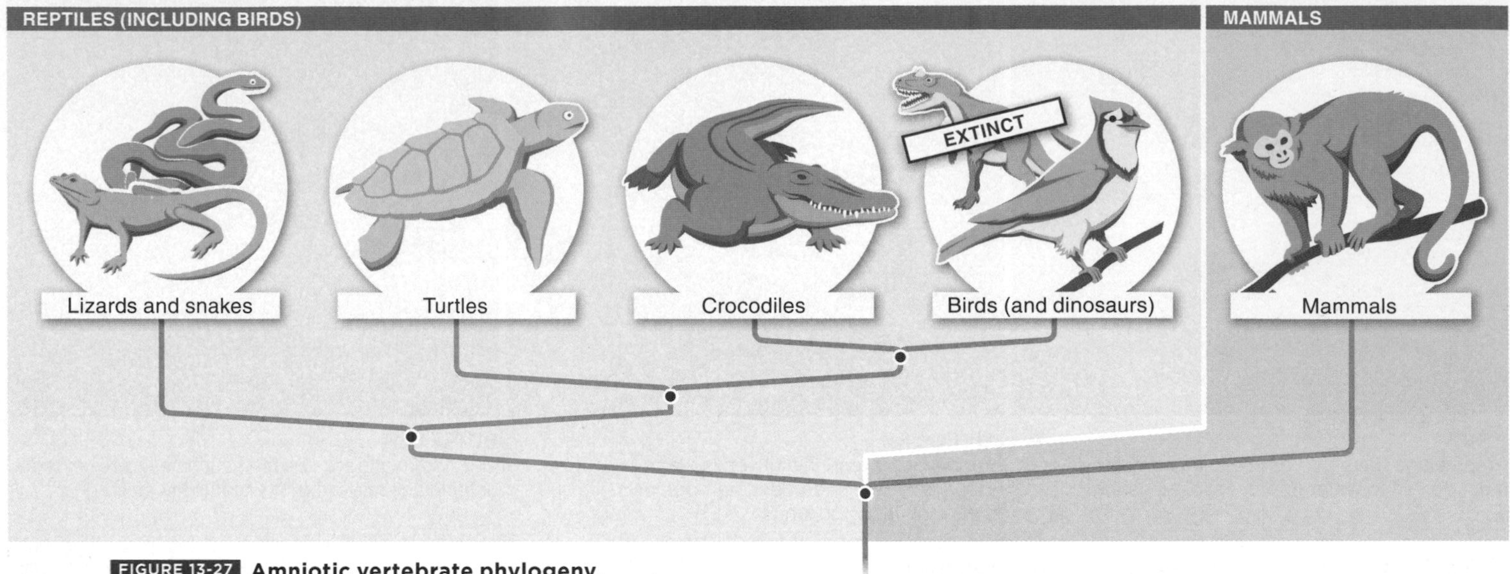

FIGURE 13-27 Amniotic vertebrate phylogeny.

physiological difference as well: birds are **endotherms,** meaning they use the heat produced by cellular respiration to raise their body temperature above air temperature, whereas other reptiles are **ectotherms**—they bask in the sun to raise their body temperature and seek the shade when the air is too warm.

Birds are, in fact, just one group of dinosaurs. During the Mesozoic era, from about 250 million to 65 million years ago, dinosaurs were the dominant terrestrial vertebrates. The fossil record shows a remarkably clear series of animal forms that bridge the transition between bare-skinned dinosaurs and feathered birds.

REPTILES vs. BIRDS

REPTILES
- Skin is covered in scales
- Body temperature is controlled by external conditions, such as air temperature (exothermic)
- Include snakes and lizards (~8,000 species), turtles (~300 species), crocodiles and alligators (23 species), and tuatara (2 species)

BIRDS
- Have feathers and wings, providing insulation and enabling flight
- Body temperature is maintained by heat generated from cellular respiration (endothermic)
- ~9,700 species

FIGURE 13-28 Comparison of reptiles and birds.

Q What were the first feathers used for?

The most apparent difference between reptiles (including most dinosaurs) and birds is the presence of feathers in birds. The fossil record reveals that feathers evolved before birds. Fossil evidence also suggests that the initial evolution of feathers probably had nothing to do with flight.

Among reptiles, many different species had feathers. A small, agile dinosaur called *Sinosauropteryx,* for example, living about 120 million years ago, had feathers that were simple spiky filaments on its neck, back, and tail, along with shorter filaments covering its body (FIGURE 13-29). The 2010 discovery of fossilized pigment-packed organelles—called melanosomes—within these feathers indicates that the feathers and filaments were probably brightly colored. Researchers believe that they may have been used by male *Sinosauropteryx* for courtship displays to females, as well as for aggressive displays to other males. *Sinosauropteryx* and many of the other feathered dinosaurs were latecomers, however. The pigeon-sized *Archaeopteryx,* which was among the first bird species, was already, by 147 million years ago, using its feathers to fly. By 140 million years ago, the skies were filled with birds.

Another big difference between birds and reptiles is that, as noted earlier, birds use internally generated heat to maintain a high body temperature, whereas reptiles rely on the sun to heat their bodies. This evolutionary change may have been related to feathers. As feathers evolved for display and flight, they may also have provided some insulation. This insulation may then have enabled the evolution of high rates of cellular respiration and the maintenance of a high and constant body temperature.

Colorful feathers originally may have been used for behavioral displays. In modern birds, feathers provide insulation and aid in flight.

FIGURE 13-29 ***Sinosauropteryx:* an early feathered, flightless dinosaur.**

TAKE HOME MESSAGE 13.17

» Birds are a branch of the reptile lineage, but unlike other reptiles, they possess feathers and can generate body heat. The complex anatomical and physiological systems in extant animals, such as feathers and endothermy in birds, are the products of hundreds of millions of years of step-by-step changes that began with simple structures. Feathers were originally colorful structures, possibly used for behavioral displays; additional functions such as insulation and flight evolved later.

13.18 Mammals are animals that have hair and produce milk.

As we saw in the previous section, two different evolutionary lineages began to diverge after amniotic vertebrates appeared. One gave rise to the reptiles, and the other gave rise to the mammals. Originally small, nocturnal insect-eaters, the early mammals remained small. And, judging from the scarce fossil record, they were never in great abundance while the dinosaurs were around. Early mammals had several features in common that are still present in all mammals today. Two important features are **hair**—dead cells filled with the protein keratin—which serves as an insulator, and **mammary glands** in female mammals, which enable them to produce nutritious, calorie-rich milk and nurse their young.

As the earliest mammals evolved, there was a gradual transition from bodies with short legs that projected out horizontally from the trunk, like the legs of an alligator, to bodies with long legs held vertically beneath the trunk, like those of a dog (FIGURE 13-30). These anatomical changes would have allowed the earliest mammals to run faster and farther to capture prey. But an increase in muscle activity for running had to be accompanied by an increase in the rate of cellular respiration to supply energy for active muscles. With this increase in metabolism, along with the evolution of hair, the early mammals could trap the heat produced and use it for temperature regulation, setting the stage for endothermy (but through a different sequence of events than in the birds).

One feature common to most mammals is **viviparity,** or giving birth to babies ("live birth") rather than laying eggs, but it is not a defining mammalian characteristic. **Monotremes** retain the ancestral condition of laying eggs. Monotremes do produce milk, but they do not have nipples—their mammary glands open directly onto the skin, so babies can lap up milk from the skin and hair.

Today, only five species of monotremes survive: the platypus and four species of spiny animals called echidnas. The platypus lives in streams and rivers in eastern Australia. It uses its broad, leathery, electrosensitive bill to find aquatic insect larvae and crayfish, which it locates by sensing the electrical activity of their muscles as they try to hide. The echidnas—three species in Australia and one in New Guinea—also have a leathery electrosensitive snout.

The remaining two lineages of mammals, the **marsupials** and the **placentals,** are viviparous, but the newborn young of marsupials and placentals are quite different (FIGURE 13-31). Marsupials are called the "pouched mammals," because the female of most marsupial species has

EARLY MAMMALS

Early mammals, such as the mouse-sized *Morganucodon*, underwent two major changes in body structure that led to endothermy.

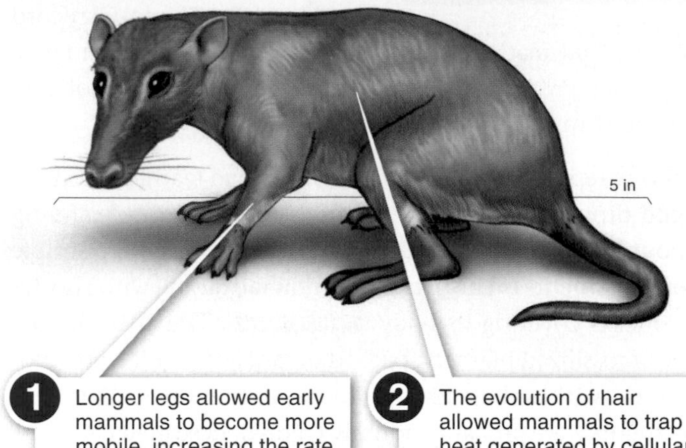

5 in
1 Longer legs allowed early mammals to become more mobile, increasing the rate of cellular respiration and thus generating more heat.
2 The evolution of hair allowed mammals to trap heat generated by cellular respiration and use it for temperature regulation.

Endothermy evolved independently in mammals and birds, and by different paths.

FIGURE 13-30 **How did endothermy evolve in early mammals?**

GROUPS OF MAMMALS

Mammals are endothermic vertebrates that have hair and produce milk for their young.

Platypus

MONOTREMES
- Females lay eggs
- Females produce milk, but have no nipples; babies suck milk from hairs on mother's chest
- Only 5 species survive, including the platypus

Kangaroo

MARSUPIALS
- Females give birth after short gestation period
- In most species, young complete development in the mother's pouch
- ~300 species, including kangaroos, koalas, and possums

African elephant

PLACENTAL MAMMALS
- Females have a placenta that provides oxygen and nutrients to embryos in the uterus
- ~4,500 species

FIGURE 13-31 **Three groups of mammals: monotremes, marsupials, and placentals.**

a pouch on her abdomen in which the young complete their development, following a short period of embryonic life in the uterus.

Placental mammals (including humans) take their name from the *placenta,* which is the structure responsible for the transfer of nutrients, respiratory gases, and metabolic waste products between the mother and the developing fetus. The placental fetus has a much longer period of development in the uterus than marsupials.

TAKE HOME MESSAGE 13.18

» Hair and mammary glands are defining characteristics of mammals. Monotremes are egg-laying mammals. Marsupial mammals give birth after a short period of development in the uterus, and the newborn completes its development in the mother's pouch. Placental mammals have a placenta that provides oxygen and nutrients to the fetus as it undergoes a longer development in the uterus.

13.19 Humans tried out different lifestyles.

The primates, the evolutionary lineage to which humans belong, originated about 55 million years ago. We can get an idea of what the ancestral primate looked like from the modern species of tree-living (arboreal) mammals commonly called prosimians—a group that includes the lemurs. This is because many of the anatomical characteristics of humans and the other primates can be traced to our arboreal origin. Our forward-directed eyes and binocular vision that allow us to judge distances accurately, our shoulder and elbow joints that allow our arms to rotate, and our fingers and opposable thumbs, and our toes, that allow us to grasp objects—all these are traits we inherited from our arboreal ancestors (FIGURE 13-32).

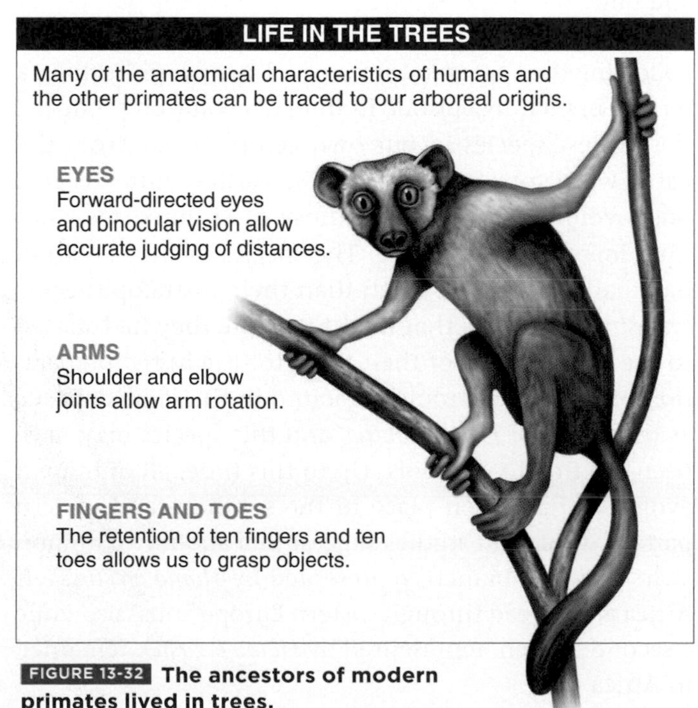

LIFE IN THE TREES
Many of the anatomical characteristics of humans and the other primates can be traced to our arboreal origins.
EYES
Forward-directed eyes and binocular vision allow accurate judging of distances.
ARMS
Shoulder and elbow joints allow arm rotation.
FINGERS AND TOES
The retention of ten fingers and ten toes allows us to grasp objects.

FIGURE 13-32 **The ancestors of modern primates lived in trees.**

Humans are part of the primate lineage that includes the groups commonly referred to as the New World and Old World monkeys (both of which have tails) and the apes (which lack tails). Within the apes, genetic and anatomical characteristics show that chimpanzees are our closest living relatives (FIGURE 13-33). Human and chimpanzee genes are very similar: their base sequences differ only by about 1%, and one-third of human and chimpanzee genes are identical. The amount of genetic difference between humans and chimpanzees indicates that the chimpanzee and modern human lineages separated only five or six million years ago.

Humans differ from chimpanzees in three major anatomical characteristics: humans are bipedal (we normally walk on two legs, whereas chimpanzees usually walk on four legs), humans are bigger than chimpanzees, and the human brain is about three times the size of the chimpanzee brain.

When we trace the appearance of these human characteristics through the fossil record, we find that humans did not become bipedal, big, and brainy all at once. Bipedality

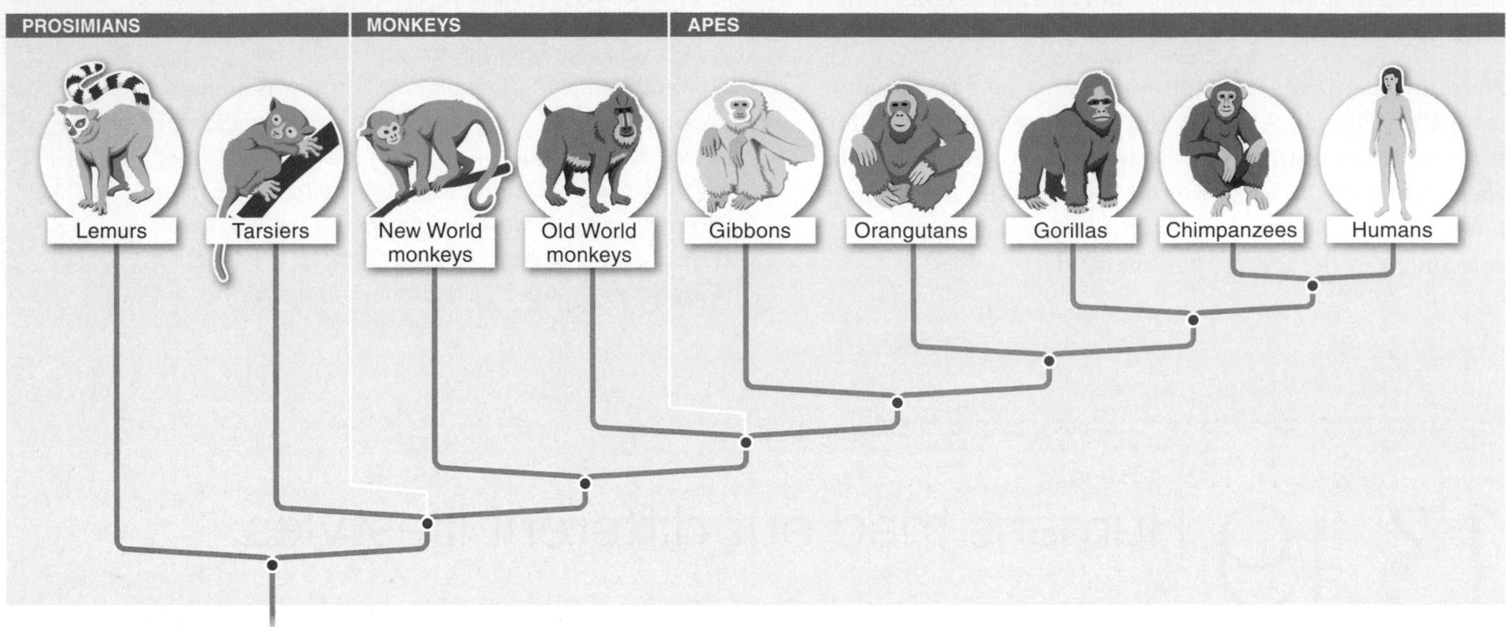

FIGURE 13-33 **The phylogeny of primates.**

evolved first, then brain volume increased, and finally body size increased, accompanied by a further increase in brain size.

Shifting from walking on four legs to walking on two legs required changes in several parts of the skeleton. The evolution of bipedalism was not a simple process, but it was one of the first important changes in the human lineage, and it seems to have set the stage for all the changes that followed. The primary advantage of bipedal locomotion for early humans was probably energy efficiency, though the issue is still debated as research continues. Bipedal locomotion at walking speed uses less energy than quadrupedal locomotion, and it also frees the hands for carrying and for tool use.

About 3.5 or 4 million years ago, several bipedal groups, the australopithecines, appeared (FIGURE 13-34). They were no larger than chimpanzees and had the same brain volume, just a bit larger than a pint jar (350–400 cc). The fossil known as Lucy was the first of many australopithecines to be discovered. An adult female, Lucy was only 3 feet (about 1 m) tall and probably weighed no more than 60 pounds (a bit less than 30 kg). The group of australopithecines to which Lucy belonged lived in grassy habitats with scattered trees; their hands and feet retained the curved finger and toe bones that are characteristic of arboreal animals. Lucy and other members of the closely related species of australopithecines probably foraged on the ground for food and climbed into trees to escape predators and perhaps, at night, to sleep. With jaws and teeth much like the jaws and teeth of chimpanzees, they had a diet probably much like that of chimpanzees—a mixture of leaves, soft fruits, and nuts.

Following the appearance of bipedal australopithecines, a second branching episode in human evolution produced the earliest species in our own genus, *Homo* (from the Latin word for "human"). These earliest humans had brain volumes about twice those of chimpanzees, but little increase in body size. The species in this radiation had markedly smaller teeth than their australopithecine ancestors, a change that might indicate they had started to use tools instead of their teeth for the initial preparation of food. Stone tools are found in the same deposits as the fossils of *Homo habilis,* and this species may have been the first to use tools. Up to this time, all of human evolution had taken place in the southern and eastern parts of Africa, but studies suggest that about two million years ago, one branch, represented by *Homo erectus,* left Africa and spread through eastern Europe and Asia, while a second branch, represented by *Homo ergaster,* remained in Africa.

THE EVOLUTIONARY HISTORY OF HUMANS

The evolutionary history of humans did not follow a straight line. Since australopithecines first appeared around 4 million years ago, there have been many branches to the human evolutionary tree, a sample of which is shown here.

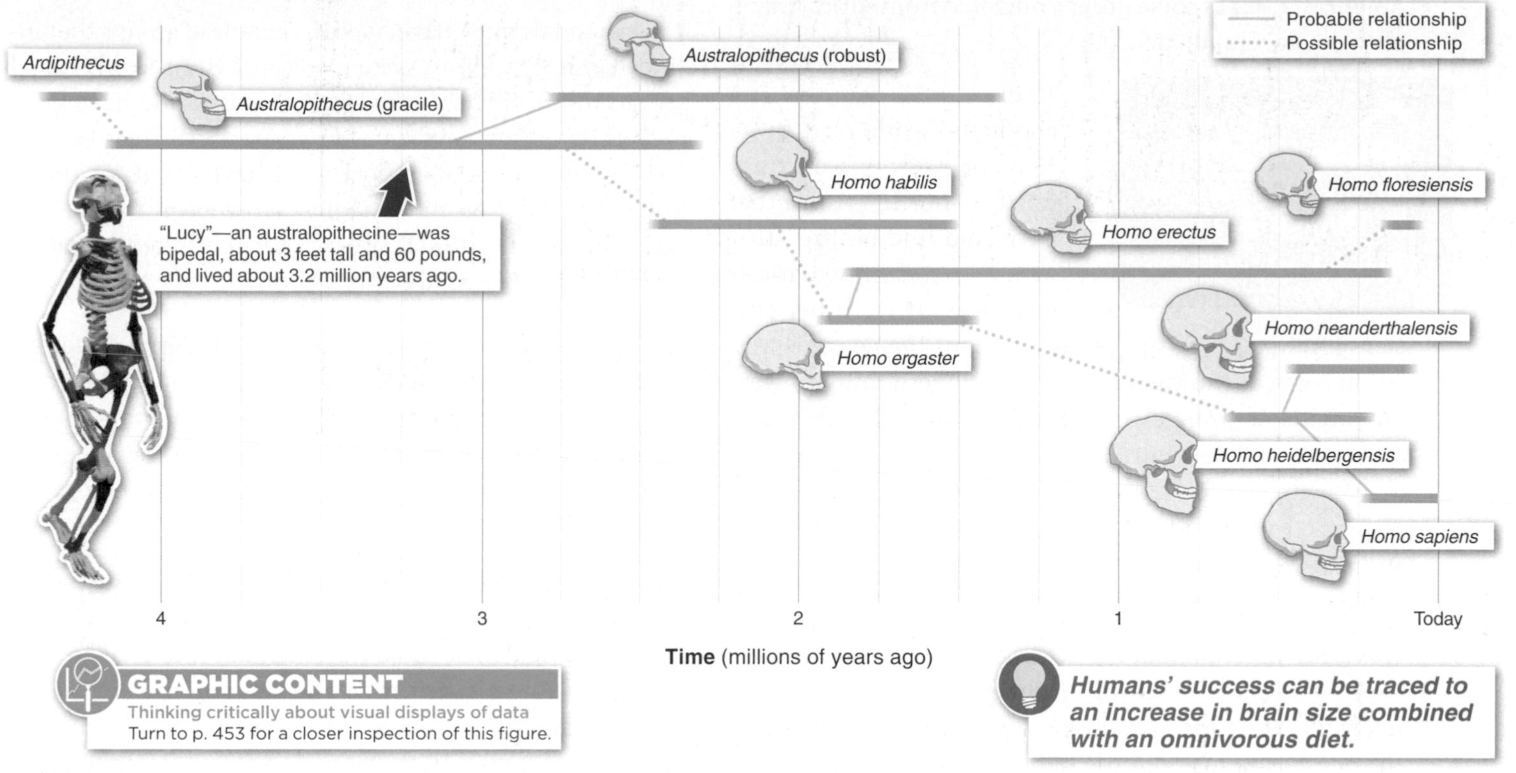

Probable relationship
Possible relationship
Ardipithecus
Australopithecus (gracile)
Australopithecus (robust)
Homo habilis
Homo floresiensis
Homo erectus
"Lucy"—an australopithecine—was bipedal, about 3 feet tall and 60 pounds, and lived about 3.2 million years ago.
Homo ergaster
Homo neanderthalensis
Homo heidelbergensis
Homo sapiens
4
3
2
1
Today
Time (millions of years ago)
GRAPHIC CONTENT
Thinking critically about visual displays of data
Turn to p. 453 for a closer inspection of this figure.
Humans' success can be traced to an increase in brain size combined with an omnivorous diet.

FIGURE 13-34 **Branching episodes in human evolution.**

Additional branching episodes in human evolution (see Figure 13-34) gave rise to several species of humans, including Neandertals (*Homo neanderthalensis*), "Flores man" (*Homo floresiensis*), and our own species (*Homo sapiens*). This radiation coincided with an increase in body size to approximately the height and weight and body form of modern humans, as well as an increase in brain volume to nearly twice that of earlier ancestors.

TAKE HOME MESSAGE 13.19

» Humans' forward-looking eyes, hands and feet with 10 fingers and 10 toes, and shoulder and elbow joints that allow the arms to rotate are characteristics retained from our arboreal ancestors. The early ancestors of humans left the trees and took up life on the ground, where they walked on two legs.

13.20 How did we get here? The past 200,000 years of human evolution.

About 200,000 years ago, a new human species branched off the *Homo ergaster* lineage in Africa. This was the first modern *Homo sapiens*. Then, about 60,000 years ago, a small group—probably no more than 100—of these modern humans left Africa and ultimately spread across the earth.

Mitochondrial DNA analyses show that the initial human migration out of Africa followed three major pathways (FIGURE 13-35). One path turned west and spread into Europe. A second path turned southeast and spread into southern Asia and through the Indo-Australian Archipelago to

Australia. The third migration went northeast, populating northern Asia. This group crossed the Bering Straits bridge about 15,000 years ago and spread southward through the Americas. Even as some groups migrated from Africa, many individuals remained.

Q Were two species of humans ever alive at the same time? If so, what happened?

There was something about the world that those *H. sapiens* groups encountered as they left Africa that we would find bizarre—*they were not alone.* Populations of at least three other species of humans (i.e., species in the genus *Homo*) were present in some of the areas into which modern humans were moving. How strange and different that world must have been with multiple human species! What would it be like to live with species that were as similar to us as dogs are to wolves, coyotes, and jackals? How would you interact with an individual who was clearly human, but not human in exactly the same way that you are?

This is the situation that the first modern *H. sapiens* encountered when they left Africa. Neandertals (*H. neanderthalensis*) had already spread across Europe and the Middle East. Neandertals were about the same size as modern humans, but more robust and muscular. Fossils of Neandertals often include bones that had been broken and healed. These injuries provide two types of information about Neandertals.

1. Neandertals must have lived in organized groups that included a social support system, because these fossils reveal serious injuries. The injured individuals would have been incapacitated for days or even weeks until they healed, and family or clan members must have cared for them. Ritual burials of Neandertals have also been found, and these provide additional support for the hypothesis that they lived in organized social groups.

2. The pattern of injuries found in Neandertal skeletons closely resemble the injuries of professional rodeo bull and bronco riders who are trampled by large, angry animals. Neandertals probably hunted large mammals (bison, mammoths, wooly rhinoceroses) with short spears that were used for close-up jabbing.

The *H. sapiens* who migrated through the Indo-Australian Archipelago encountered two species of humans: *H. erectus* and *H. floresiensis*. *Homo erectus* had migrated to Asia at least a million years earlier and was still present on the island of Java when modern humans arrived, about 50,000 years ago. Although *H. erectus* was the same size as modern

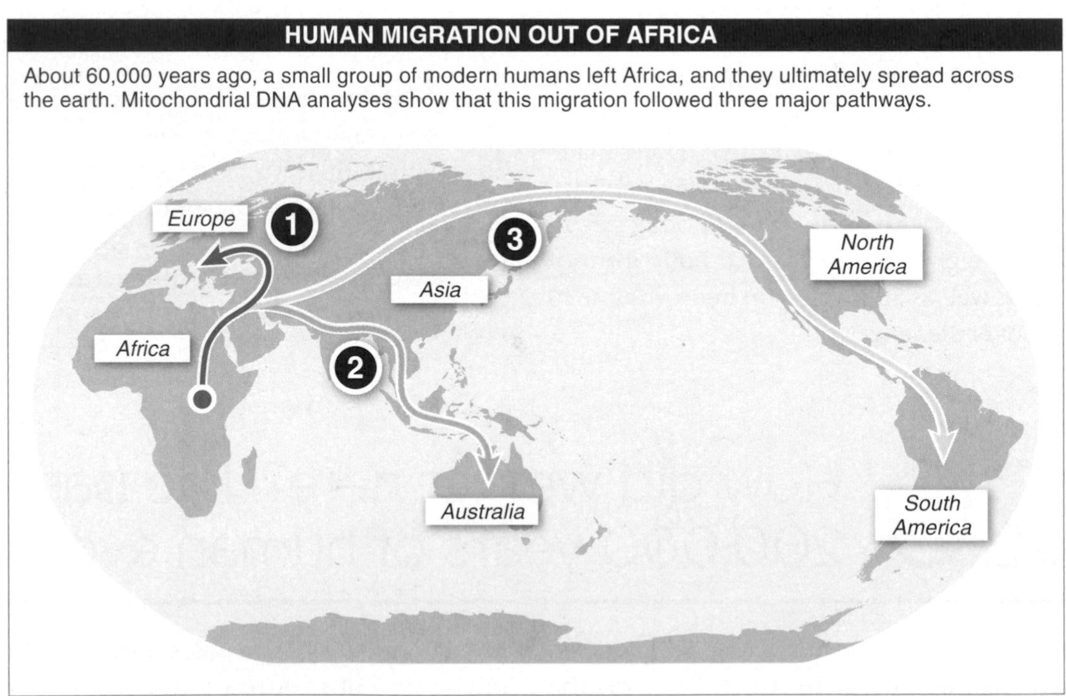
HUMAN MIGRATION OUT OF AFRICA
About 60,000 years ago, a small group of modern humans left Africa, and they ultimately spread across the earth. Mitochondrial DNA analyses show that this migration followed three major pathways.
Europe
1
3
Asia
North America
Africa
2
Australia
South America

FIGURE 13-35 ***Homo sapiens*** **on the move: three migratory paths out of Africa.**

humans, it had a smaller brain—an average brain volume of just over a quart (about 1,000 cc) for *H. erectus* compared with a quart and a half (about 1,400 cc) for *H. sapiens*. Nonetheless, *H. erectus* may have been the first humans to use fire (the evidence on this is not definitive), and they almost certainly built boats that allowed them to move along coasts and from island to island.

In 2003, researchers discovered stone tools and fossils of a species of human only 3 feet (about 1 m) tall and with a brain volume of just over a pint (about 350 cc) on Flores Island, which lies east of Java. The paleontologists who discovered this species gave it the scientific name *Homo floresiensis* ("Flores man"), but the world press promptly called it the Hobbit because of its tiny size. When the dwarf human *H. floresiensis* lived, Flores Island was also home to a dwarf elephant, which *H. floresiensis* may have eaten, and a giant monitor lizard, which may have eaten *H. floresiensis*.

After modern *H. sapiens* groups spread into the areas occupied by the three other human species, these three other species disappeared. Neandertals became extinct about 30,000 years ago, *H. erectus* about 27,000 years ago, and *H. floresiensis* about 12,000 years ago. We're not sure why they vanished. Previously, researchers thought that modern *H. sapiens* exterminated the other species, either by monopolizing access to food and living space or by killing them in battles for these resources. But recent results from DNA sequencing suggest that, just as they emerged from Africa, *H. sapiens* interbred with Neandertals. This conclusion is based on the finding that at least 1% to 4% of the genetic makeup of most modern humans includes Neandertal DNA.

Now we are the only extant human species, a situation that is unique in the evolutionary history of humans. Ever since the first branching event, about four million years ago, multiple, closely related species of humans had coexisted. We are the first human species to be alone.

TAKE HOME MESSAGE 13.20

» Modern humans (*Homo sapiens*) evolved in Africa about 200,000 years ago, and all living humans are descended from that evolutionary radiation. About 60,000 years ago, a small group of modern humans moved out of Africa, and the descendants of this group ultimately populated Europe, Asia, and the Americas. Three other species of humans were living at this time, all of which became extinct between 30,000 and 12,000 years ago, after modern humans had spread into the areas where they were living.

STREET BIO

Using evidence to guide decision making in our own lives

Where are you from? "Recreational genomics" and the search for clues to your ancestry in your DNA

How much do you know about your ancestral heritage? More than two dozen companies now offer a promise to help you discover your genetic ancestry—usually for $100 to $1,000. Can they deliver what they promise?

Q: Are clues to your ancestry hidden within your DNA? The short answer is yes. Researchers use comparisons of DNA sequences to explore personal ancestries *within* our species, even preparing pie charts that purport to identify a person's ancestral proportions (for example, 10% West African, 70% Middle Eastern/North African, and 20% European).

Q: How is DNA used to build a family tree? Most tests are based on one of three types of analysis. (1) Mitochondrial DNA tests evaluate highly variable sequences in maternally inherited mitochondrial DNA. (2) Y-chromosome tests look at variation in the paternally inherited Y chromosome. (3) Autosomal DNA tests examine highly variable sections of DNA scattered throughout the chromosomes. In each case, a sample of cells is collected (usually by swabbing the inside of the cheeks), then technicians determine the base sequences at a number of locations (a few dozen to several hundred) and compare them with sequences from thousands of individuals in populations around the world. The matching of your sequences with those of other samples is then interpreted as potential evidence of shared ancestry.

Q: Genetic ancestry testing may be good fun, but is it good science? Unfortunately, the limitations of commercially available tests of genetic ancestry are significant. These are among the most important limitations:

- High genetic variation among individuals within most populations makes it difficult to identify specific sequences that can reliably indicate membership in a population. As an analogy: if Native Americans are highly variable for a trait (e.g., height), we can't use that trait to prove someone's Native American ancestry.
- High rates of gene flow between populations reduce the reliability with which any sequence can demonstrate membership in one particular population. As an analogy: dark hair is extremely common in Asia, but having dark hair doesn't necessarily indicate Asian ancestry.
- Evaluating too few genetic loci, of which just a small number happen to be similar, can lead to the conclusion that individuals are much more genetically similar than they actually are.
- A DNA match between two individuals living today is not a match with an ancestor. Rather, it suggests that the two people may have inherited the DNA sequence from a common ancestor.

What can you conclude? Until we have a much more comprehensive catalog of genetic variation—including precise estimates of how much variation exists within populations and how much exists between populations—the results of genetic ancestry testing are speculative and insufficiently reliable. Conclusions may be incorrect and misleading. But, because the conceptual underpinnings are solid, the future of genetic ancestry testing is promising.

GRAPHIC CONTENT

Thinking critically about visual displays of data

THE EVOLUTIONARY HISTORY OF HUMANS

The evolutionary history of humans did not follow a straight line. Since australopithecines first appeared around 4 million years ago, there have been many branches to the human evolutionary tree, a sample of which is shown here.

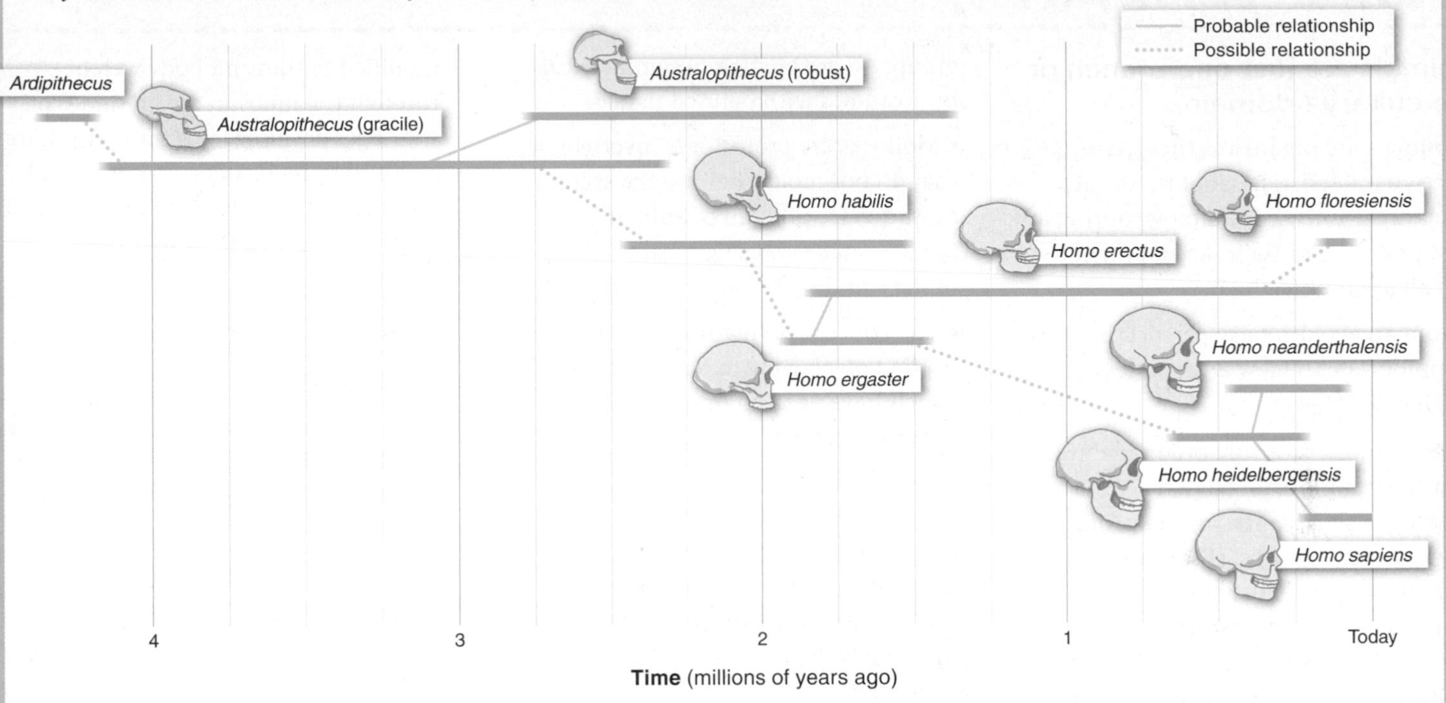

Probable relationship
Possible relationship

Ardipithecus
Australopithecus (gracile)
Australopithecus (robust)
Homo habilis
Homo erectus
Homo floresiensis
Homo ergaster
Homo neanderthalensis
Homo heidelbergensis
Homo sapiens

4 3 2 1 Today

Time (millions of years ago)

1 Moving back in time from today, which species are the most likely to be the direct ancestors of modern humans? Which are not? Why?

2 When did the genus *Homo* first appear? How did it differ from earlier species in this evolutionary tree?

3 What is the difference between the dotted and the solid yellow-green lines? Why might that information change in the future?

4 What is the largest number of different species on the human "family tree" that were alive at the same time? When was that?

5 How long ago did *Homo sapiens* become the only living representative of the genus *Homo*?

See answers at the back of the book.

KEY TERMS IN ANIMAL DIVERSIFICATION

lobe-finned fishes, p. 440
mammary glands, p. 445
marsupial, p. 446
metamorphosis, p. 434
monotreme, p. 446
non-amniote, p. 442
notochord, p. 436
pharyngeal slits, p. 437
placental, p. 446
post-anal tail, p. 438
protostome, p. 417
pupa, p. 434
radial symmetry, p. 416
ray-finned fishes, p. 440
roundworm, p. 423
segmented worm, p. 423
sessile, p. 415
tetrapod, p. 442
vertebrates, p. 417
viviparity, p. 446

BRIEF SUMMARY

Animals are just one branch of the eukarya domain.

• Animals are organisms that share three characteristics: all of them eat other organisms, all can move during at least one stage of their development, and all are multicellular.

• From an evolutionary perspective, it is inappropriate to view any species as "higher" or "lower."

• The animals probably originated from an ancestral protist. Four key distinctions divide the animals into monophyletic groups: (1) tissues or not, (2) radial or bilateral symmetry, (3) protostome or deuterostome development, and (4) growth through molting or through continuous addition to the body.

Invertebrates—animals without a backbone—are the most diverse group of animals.

• Sponges are among the simplest of the animal lineages. A sponge has no tissues or organs. Free-swimming sponge larvae settle and develop into sessile, filter-feeding adults.

• The cnidarians—corals, sea anemones, and jellyfishes—are radially symmetrical animals with defined tissues. All cnidarians are carnivores and use specialized stinging cells located in their tentacles to capture prey.

• Worms are found in several different phyla and are not a monophyletic group. All are bilaterally symmetrical protostomes with defined tissues.

• Mollusks are protostome invertebrates that do not molt. They are the second most diverse phylum of animals and include snails and slugs, clams and oysters, and squids and octopuses.

• The concept of intelligence cannot be applied objectively to other species, which have evolved in response to the selective forces at work in their own particular niches.

• The arthropods are protostome invertebrates, and with nearly one million species, they outnumber all other forms of life in species diversity. Insects are the most diverse group of arthropods.

• Because they are deuterostomes, echinoderms are the invertebrates that are the closest evolutionary relatives to the vertebrates.

The phylum Chordata includes vertebrates, animals with a backbone.

• All chordates have four characteristic structures: a notochord, a dorsal hollow nerve cord, pharyngeal slits, and a post-anal tail.

• The development of two structures in fishes—fins and jaws—set the stage for the enormous diversity of modern vertebrates.

• Four adaptations were important in the transition of life from water to land. Fins were modified into limbs. Vertebrae were modified to transmit body weight through the limbs to the ground. The site of gas exchange was transferred from gills and swim bladders to lungs. And terrestrial vertebrate eggs have membranes and a shell, and resist drying out.

All terrestrial vertebrates are tetrapods.

• Amphibians are terrestrial vertebrates, but the adults of most species still lay eggs in water. The eggs hatch into aquatic juveniles.

• Birds are a branch of the reptile lineage but, unlike other reptiles, possess feathers and can generate body heat.

• Hair and mammary glands are defining characteristics of mammals.

• Humans' forward-looking eyes, hands and feet with 10 fingers and 10 toes, and shoulder and elbow joints are characteristics retained from our arboreal ancestors. The early ancestors of humans left the trees and took up life on the ground, where they walked on two legs.

• Modern humans (*Homo sapiens*) evolved in Africa about 200,000 years ago, and all living humans are descended from that evolutionary radiation.

CHECK YOUR KNOWLEDGE

Short Answer

1. Why did Darwin write the note to himself, "Never use the word higher or lower"?

2. All animals share a common ancestor—what was it? How do we know this is true?

3. What important ecological role do roundworms play?

4. What is the specialized tongue found in mollusks called? How does it function?

5. Why are questions of comparative animal intelligence not generally useful?

6. Compare and contrast the benefits and costs of having an exoskeleton.

7. Which two features of insects contributed to their great adaptive radiation?

8. Describe the four distinctive chordate body structures.

9. Which two evolutionary innovations in vertebrates resulted in their eventual domination among the large animals?

10. Why does an ectotherm, such as a snake, require only an occasional meal to survive, while an endotherm, such as a mouse, requires far more frequent meals?

11. In mammals, what is the main function of the placenta?

12. Describe the three major anatomical differences between humans and chimpanzees.

Multiple Choice

1. The type of adaptation most likely to occur over evolutionary time is one that:

a) increases the physiological complexity and sophistication of the individuals in a population.

b) increases the physiological simplicity of the individuals in a population.

c) causes the individuals in a population to have greater intelligence.

d) causes the individuals in a population to more closely resemble humans.

e) causes the individuals in a population to more efficiently reproduce in the environment in which they live.

0 EASY HARD 100
45

2. In cnidarians, cnidocytes are primarily used for:

a) creation of water flow across the body wall.

b) formation of free-living medusas.

c) secretion of digestive enzymes.

d) prey capture and defense.

e) muscular contraction during movement.

3. The mollusk's mantle is used primarily for:

a) feeding.

b) gas exchange.

c) producing the shell.

d) excretion.

e) reproduction.

0 EASY HARD 100
36

4. The phylum Arthropoda includes all of the following kinds of animals except:

a) snails. b) crabs.

c) crayfish. d) butterflies.

e) scorpions.

0 EASY HARD 100
54

5. Which of the following characteristics distinguishes all chordates from all other animals?

a) a vertebral column

b) a dorsal hollow nerve cord

c) collar cells

d) bilateral symmetry during embryonic or larval development

e) an amniotic egg

0 EASY HARD 100
50

6. Which of these animals is a tetrapod that does not produce amniotic eggs?

a) salamander

b) human

c) monkey

d) elephant

e) python

0 EASY HARD 100
48

7. Why is the amniotic egg considered a key evolutionary innovation?

a) It prohibits external fertilization, thereby facilitating the evolutionary innovation of internal fertilization.

b) It has an unbreakable shell.

c) It greatly increases the likelihood of survival of the eggs in a terrestrial environment.

d) It enables eggs to float in an aquatic medium.

e) It extends the time of embryonic development.

0 EASY HARD 100
37

8. Which came first, the chicken or the egg?

a) The chicken, because the amniotic egg did not evolve until the first chicken appeared.

b) The egg, because the amniotic egg evolved well before the first birds.

c) The chicken, because, during speciation, the adult stage always precedes the juvenile stage.

d) The egg, because the chicken is not a real species.

e) It is impossible to determine because eggs leave no fossils.

0 EASY HARD 100
42

9. Marsupials and which of the following groups combine to make a monophyletic group?

a) birds b) carnivores

c) primates d) monotremes

e) placental mammals

0 EASY HARD 100
43

10. According to the fossil record, the first modern humans, *Homo sapiens*, appeared approximately ______ years ago

a) 6 million b) 6,000

c) 190 million d) 200,000

e) 1 million

0 EASY HARD 100
55

Plants are just one branch of the eukarya.

14.1 What makes a plant?

The first plants had neither roots nor seeds.

14.2 Colonizing land brings new opportunities and new challenges.

14.3 Mosses and other non-vascular plants lack vessels for transporting nutrients and water.

14.4 The evolution of vascular tissue made large plants possible.

The advent of the seed opened new worlds to plants.

14.5 What is a seed?

14.6 With the evolution of the seed, gymnosperms became the dominant plants on earth.

14.7 Conifers include the tallest and longest-living trees.

Flowering plants are the most diverse and successful plants.

14.8 Angiosperms are the dominant plants today.

14.9 A flower is nothing without a pollinator.

14.10 Angiosperms improve seeds with double fertilization.

Plants and animals have a love-hate relationship.

14.11 Fleshy fruits are bribes that flowering plants pay to animals to disperse their seeds.

14.12 Unable to escape, plants must resist predation in other ways.

Fungi and plants are partners but not close relatives.

14.13 Fungi are more closely related to animals than they are to plants.

14.14 Fungi have some structures in common but exploit an enormous diversity of habitats.

14.15 Most plants have fungal symbionts.

14.16 *This is how we do it:* Can beneficial fungi save our chocolate?

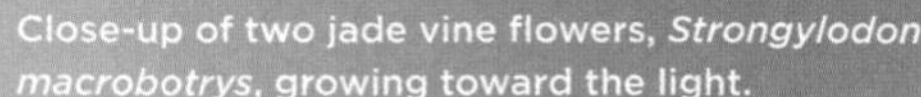

Close-up of two jade vine flowers, *Strongylodon macrobotrys*, growing toward the light.

Plant and Fungi Diversification

Where did all the plants and fungi come from?

14.1 Plants are just one branch of the eukarya.

Succulents often have thick and fleshy leaves that store water.

14.1 What makes a plant?

An animal can move from place to place to get food or water, to avoid being eaten, or to find a mate. It's harder for a plant. It must do all the things an animal does—obtain food and water, protect itself from predators, and reproduce—*but* a plant can't move a centimeter in the process. Wherever a seed puts down roots, that's where the plant stays anchored for its entire life.

Immobility is only one aspect of being a plant. A **plant** is a multicellular eukaryote that produces its own food by carrying out photosynthesis and has an embryo that develops within the protected environment of the female parent. Plants occur almost exclusively on land, and they vary in size from less than 0.04 inch (1 millimeter) to 380 feet (116 meters) tall (FIGURE 14-1). (There are a few exceptions to our definition of a plant, but it applies to the vast majority of the more than 280,000 species of plants on earth today.)

The closest relatives of land plants are multicellular and photosynthetic eukaryotes that include many species of algae, such as seaweed. They differ from plants, however, in that they live only in water or on very moist land surfaces. This is in sharp contrast to plants, which can live in a wide range of habitats from moist places to deserts.

Most plants can make their own food—carbohydrates—by photosynthesis. But a plant also needs nitrogen to build proteins, phosphorus to make ATP, and salts to create concentration gradients between the inside and outside of cells. Plants use **roots,** the part of a plant below ground, to obtain these needed substances from the soil. Above ground, plants have a **shoot** that consists of a stem and leaves. The stem is the structure that supports the main photosynthetic organ of a plant: its leaves.

WHAT IS A PLANT?

Almost all plants share the following three characteristics:

- **PLANTS CREATE THEIR OWN FOOD**
  Almost all plants carry out photosynthesis, using energy from sunlight to convert carbon dioxide and water into sugar.
- **PLANTS ARE SESSILE AND (MOSTLY) TERRESTRIAL**
  Plants are anchored in place at their bases and occur almost exclusively on land.
- **PLANTS ARE MULTICELLULAR**
  Plants consist of multiple cells and have structures that are specialized for different functions.

FIGURE 14-1 **The defining characteristics of a plant.**

STATIONARY LIVING CREATES CHALLENGES

Moving the light source every few days, growers can make "lucky bamboo" grow in spirals.

OBTAINING FOOD
Because plants can't move to reach sunlight, they bend in place and grow toward light.

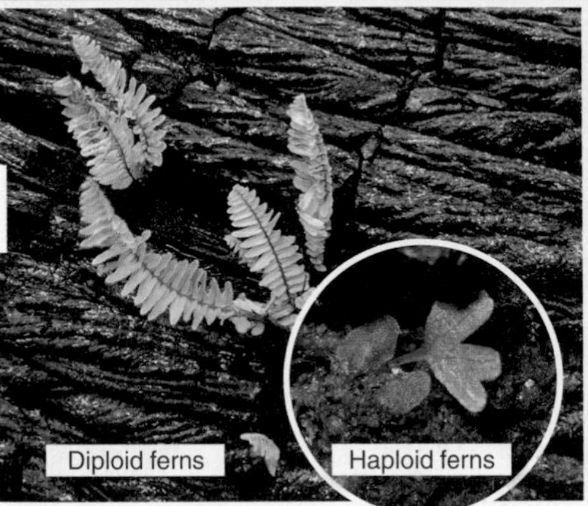

FINDING A MATE
Male and female plants can't meet to reproduce, but have alternative solutions, including alternating haploid and diploid life stages and using other organisms to transport the male gametes.

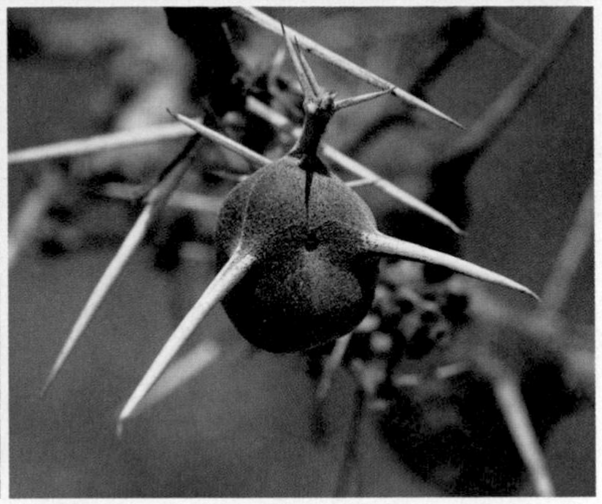

RESISTING PREDATION
Plants can't run from predators, so they have developed adaptations such as thorns to defend themselves.

FIGURE 14-2 Plants must overcome the constraints of being immobile.

When we think of plants, we often think of them as "chlorophyll-containing," but some plants have no chlorophyll—their ancestors had chlorophyll (which is why they are classified as plants), but some plants have lost almost all of it over evolutionary time—and they can't carry out photosynthesis. Instead, they live as parasites that steal nutrients from other plants. Dodder is an example of a parasitic plant that almost completely lacks chlorophyll and gets its sugar from the host plant it grows on. You can probably find dodder growing on plants in a nearby vacant lot or at the roadside.

Q If a plant growing in the shade can't move to a new, sunnier location, how can it reach the available sunlight?

Plants that carry out photosynthesis need sunlight. This creates a challenge for plants that grow under the tree canopy, with just a few sunny spots where light penetrates the tree branches overhead (FIGURE 14-2). If a seed sprouts (germinates) in the shade, the young plant needs to somehow get to the nearest sunny spot. Plants can do that by growing toward the light. You may have seen a "lucky bamboo" plant (it's not really bamboo), which can grow in spirals or loops. The growers move the light source every few days, forcing the plant to bend to follow the light. If you buy lucky bamboo, you must continue rotating the light source or it will start to grow straight, like any other plant.

Sex can also be a challenge for an organism that is anchored in place by its roots. Not surprisingly, plants have a sex life very different from that of animals. Consider humans as a typical animal. As soon as an egg is fertilized by a sperm, a human has the diploid number of chromosomes—that is, two sets of chromosomes, one set from each parent. Only the egg and sperm are haploid—an egg or a sperm has only one set of chromosomes. As we will see, plants have haploid *and* diploid life stages. The haploid stage allows the male and female to reproduce, even though the plants will never meet because they can't move. Some plants enlist the help of animals to carry the male gamete to the female gamete.

Resisting predators is another challenge for organisms that cannot move, but plants have ways to defend themselves. Thorns, for example, are an anatomical defense that plants use to avoid predation (see Figure 14-2). Plants also use chemicals to deter predators, as many people learn when they develop an itchy rash from poison ivy.

The earliest land plants were the first multicellular organisms to live on land. They were **non-vascular,** meaning that they had no tube-like vessels to transport water and nutrients (FIGURE 14-3). The subsequent evolution of land plants was a series of radiations of forms with characteristics that made them increasingly independent of water. The evolutionary tree of plants shows these stages clearly: first the development of vessels to conduct water from the

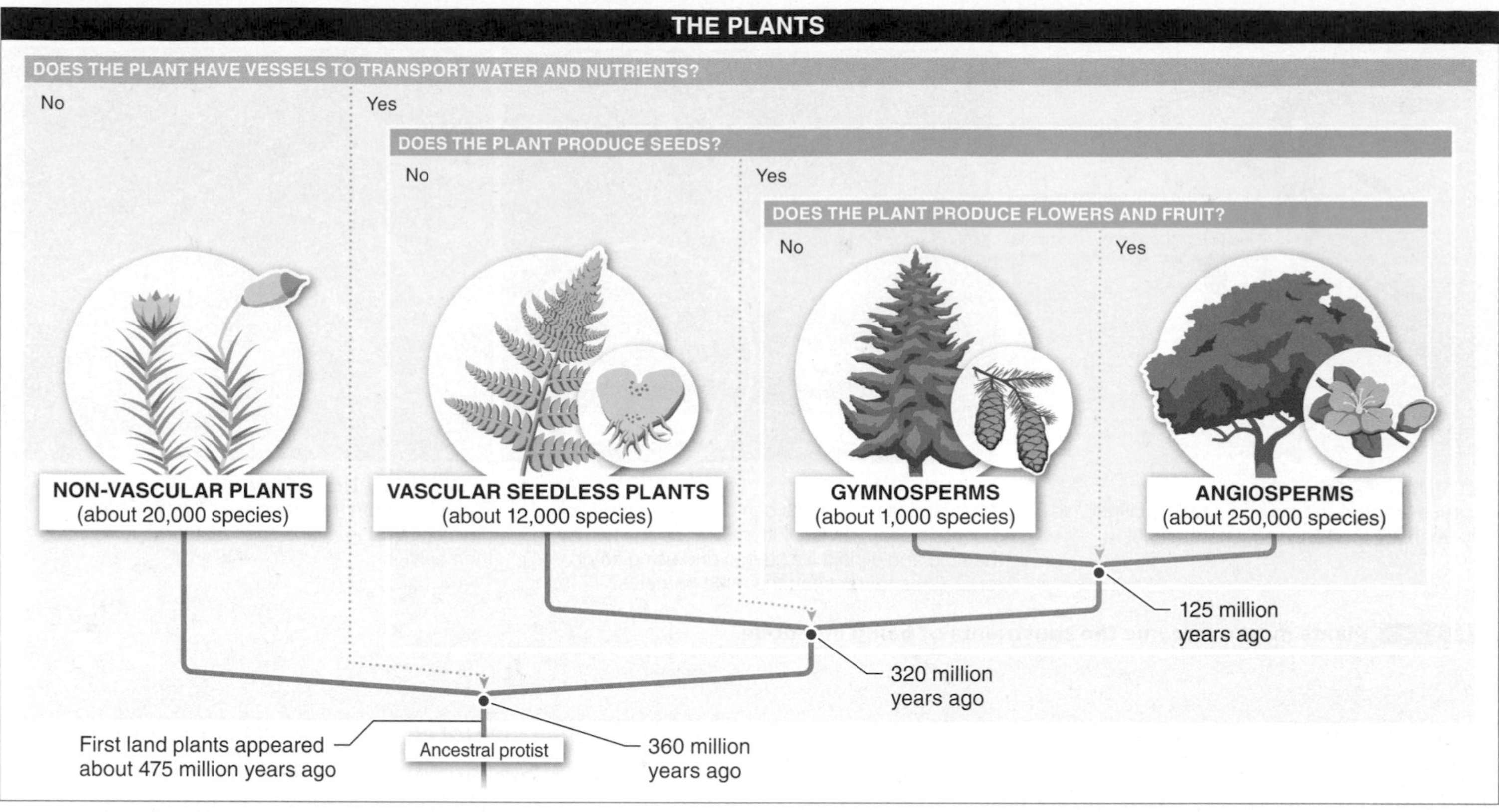

FIGURE 14-3 **Phylogeny of the plants.**

soil through the plant (the vascular plants), then seeds that provide nutrients to get the next generation off to a good start (the gymnosperms), and finally flowers that allow plants to entice or trick insects and birds into spreading the plant's male gametes (the angiosperms). We look at each of these plant adaptations in this chapter. We also explore the ecologically important kingdom of fungi. Although fungi are not plants, they are very closely associated with plants.

TAKE HOME MESSAGE 14.1

» Plants are multicellular organisms that spend most of their lives anchored in one place by their roots. Characteristics evolved that made it possible for plants to successfully obtain food, reproduce, and protect themselves from predation on land, despite their immobility.

The first plants had neither roots nor seeds.

Uluhe ferns (*Dicranopteris linearis*) growing in Hawaii Volcanoes National Park.

14.2 Colonizing land brings new opportunities and new challenges.

The aquatic ancestors of land plants were a group of protists called green algae. Like the land plants, green algae are multicellular, photosynthetic eukaryotes, but they live only in water or on very moist land surfaces. As water dwellers, green algae do not require specialized structures to obtain water and nutrients; water simply enters their cells by osmosis, and the nutrients they require are in solution in the water that surrounds them.

Taxonomists consider some green algae that look like slime on rocks—organisms called coleochaetes (pronounced KOH-lee-oh-keets)—to be the closest relatives of plants (FIGURE 14-4). An individual coleochaete is about the size of a pinhead and is only one cell-layer thick. Coleochaetes can withstand exposure to air, so they survive when the water level in a lake falls and leaves them high and dry. This resistance to drying out was the first evolutionary step taken by plants as they moved from water to land.

About 472 million years ago, the first land plants appeared as patches of low-growing green stems at the water's edge. They had no roots, leaves, or flowers. But from an evolutionary perspective, those early plants were enormously important. Until terrestrial plants evolved, there was nothing on land for other land organisms to eat. These first land plants not only set the stage for the tremendous diversity of plant life that we know today but also paved the way for the evolution and diversification of land animals.

As land plants emerged from water, they faced the same two challenges that were to confront the first terrestrial animals, some 25 million years later: supporting themselves against the pull of gravity and retaining moisture. The second of these problems was the more urgent. The earliest plants did not have to grow upward—they could creep along the ground—but they *did* have to avoid drying out (FIGURE 14-5). The material that protects all land plants from drying is a shiny, waxy layer on the stem and leaves called the **cuticle.**

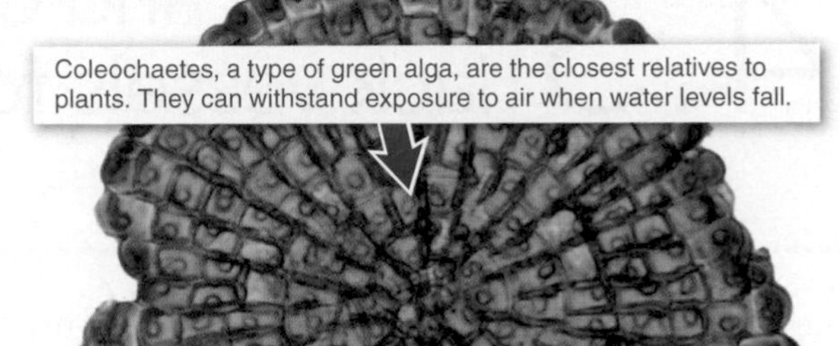

Coleochaetes, a type of green alga, are the closest relatives to plants. They can withstand exposure to air when water levels fall.

FIGURE 14-4 Plants' closest relative: a green alga called a coleochaete.

TAKE HOME MESSAGE 14.2

» The first land plants were small; had no leaves, roots; or flowers; and could grow only at the water's edge. Nonetheless, they set the stage for the enormous diversity of terrestrial plants and animals on earth today.

MOVING ONTO LAND PRESENTS CHALLENGES

When plants emerged onto land, they faced the same two challenges that terrestrial animals faced 25 million years later.

PROBLEM: GRAVITY

ADAPTATION: The earliest plants grew very close to the ground, as mosses do today, in order to resist the pull of gravity.

PROBLEM: DESICCATION

ADAPTATION: Plants developed an outer waxy layer called a cuticle that covers their entire surface.

FIGURE 14-5 Leaving the water.

14.3 Mosses and other non-vascular plants lack vessels for transporting nutrients and water.

The earliest land plants were low-growing for a reason: they had no structures that could transport water and nutrients from the soil upward into the plant. The only way that these substances could move within the plant was by slowly diffusing from one cell into an adjacent cell, and so on.

Despite the limitations of diffusion, three groups of plants—liverworts, hornworts, and mosses—all known as **bryophytes,** still use diffusion to move substances through their bodies, rather than having any sort of "circulatory system."

More than 12,000 species of mosses grow in habitats extending from arctic and alpine regions to the tropics. Liverworts and hornworts are small (less than an inch in height), simple plants that grow in moist and shady places and resemble flattened moss. All of these plants are referred to as "non-vascular" because they do not have vessels to transport water and food. Projections that penetrate just a few micrometers into the soil absorb water and nutrients into the outermost layer of cells. Because these projections are so short, non-vascular plants must either live in places where the soil is always moist or become dormant when the soil surface dries out (FIGURE 14-6).

To adapt to land, the non-vascular plants had to develop a method of reproduction that protected the plant embryo from drying out and provided it with a source of nutrients. A life cycle of alternating haploid and diploid generations made this possible. In animals, the haploid gametes, at fertilization, produce a new, diploid cell that becomes a multicellular organism, which in the adult stage starts the process all over. All plants exhibit an "alternation of generations."

The life cycle of a moss provides a useful example of how the alternation of generations works. When you look at a spongy mass of moss (or any other non-vascular plant), what you see is the *haploid* (or **gametophyte**) part of the life cycle, rather than the *diploid* (or **sporophyte**) part of the life cycle (FIGURE 14-7). For this reason, mosses are described as having a dominant gametophyte. The adult moss plants are haploid plants—that is, all the cells have only one set of chromosomes. There are male and female

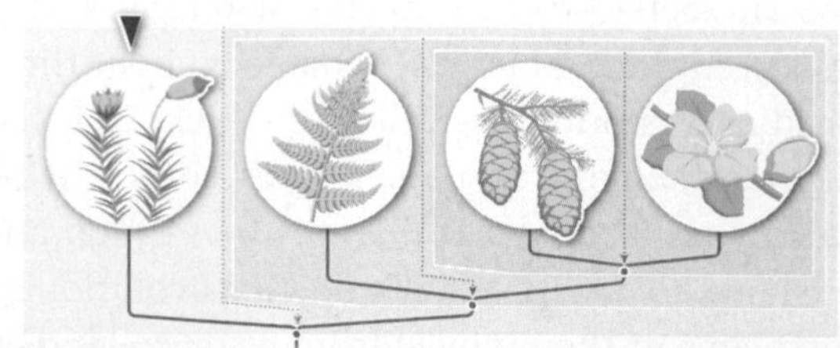

THE NON-VASCULAR PLANTS

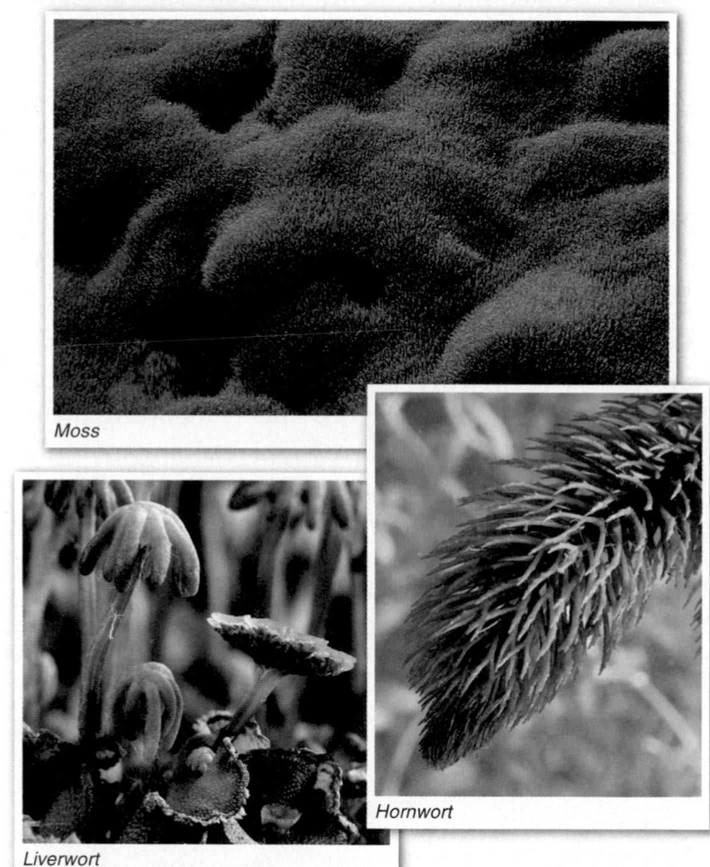

COMMON CHARACTERISTICS
- Distribute water and nutrients throughout the plant by diffusion
- Release haploid spores, which grow and produce gametes
- Life cycle with multicellular haploid and diploid phases

MEMBERS INCLUDE
- Mosses (~12,000 species)
- Liverworts (~8,000 species)
- Hornworts (~100 species)

FIGURE 14-6 **Overview of the non-vascular plants.**

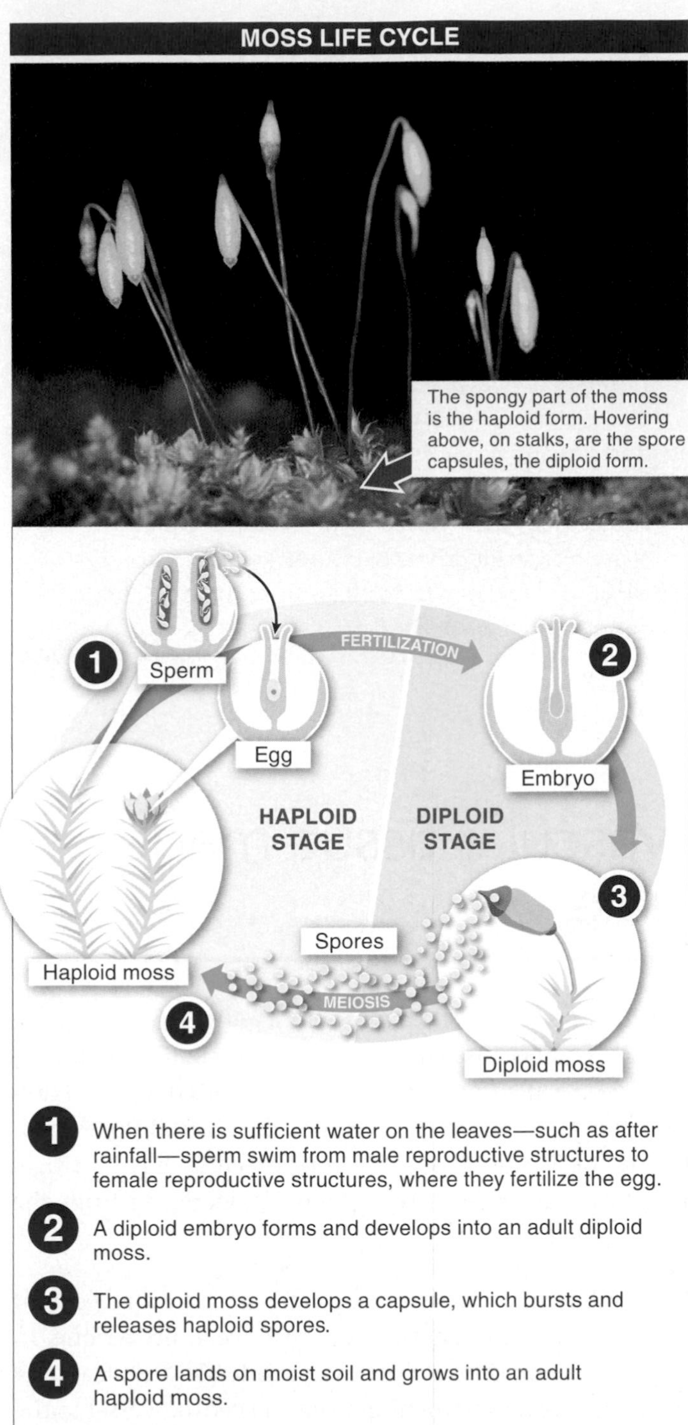

MOSS LIFE CYCLE
The spongy part of the moss is the haploid form. Hovering above, on stalks, are the spore capsules, the diploid form.
1
Sperm
FERTILIZATION
2
Egg
Embryo
HAPLOID STAGE
DIPLOID STAGE
3
Spores
Haploid moss
MEIOSIS
4
Diploid moss
1 When there is sufficient water on the leaves—such as after rainfall—sperm swim from male reproductive structures to female reproductive structures, where they fertilize the egg.
2 A diploid embryo forms and develops into an adult diploid moss.
3 The diploid moss develops a capsule, which bursts and releases haploid spores.
4 A spore lands on moist soil and grows into an adult haploid moss.

FIGURE 14-7 **Alternation of generations in mosses.**

moss plants, which have male or female reproductive structures located at the tips of the stems. Water collects here during rainstorms, allowing the sperm to "swim" (or be splashed) from the male structures to fertilize eggs in the female structures. Once an egg is fertilized, a diploid zygote is formed within the female structure and divides to become an embryo.

The embryo is sheltered within the female reproductive structure, which provides it with water and nutrients. Eventually, the female structure elongates to such a degree that it breaks in two and forms a capsule extending over the top of the plant like a raised fist. Inside a capsule are hundreds of haploid **spores**—single cells, containing DNA, RNA, and a few proteins. When a capsule ruptures, spores that land in moist, sheltered spots grow into new (haploid) male or female moss plants.

Some non-vascular plants are economically or ecologically important. "Peat moss" (*Sphagnum* moss), found in many parts of the world, has several uses. It is sold as a soil enhancer for gardening. In areas where trees are scarce, peat is dried and burned as fuel. In places with monsoon seasons, such as Malaysia and Indonesia, peat bogs—consisting of partially decayed moss—are important in flood control, because they can absorb enormous amounts of water and release it over a period of months. Mounds of burning peat are also used to dry the barley used in producing Scotch whisky, giving Scotch its distinctive smoky taste (FIGURE 14-8).

Unlikely as it seems, non-vascular plants can grow in deserts, and mosses (along with lichens and cyanobacteria) are important components of the biological crust that holds desert soils in place. The crust cements the soil particles together and allows the soil to resist wind erosion, but this crust is

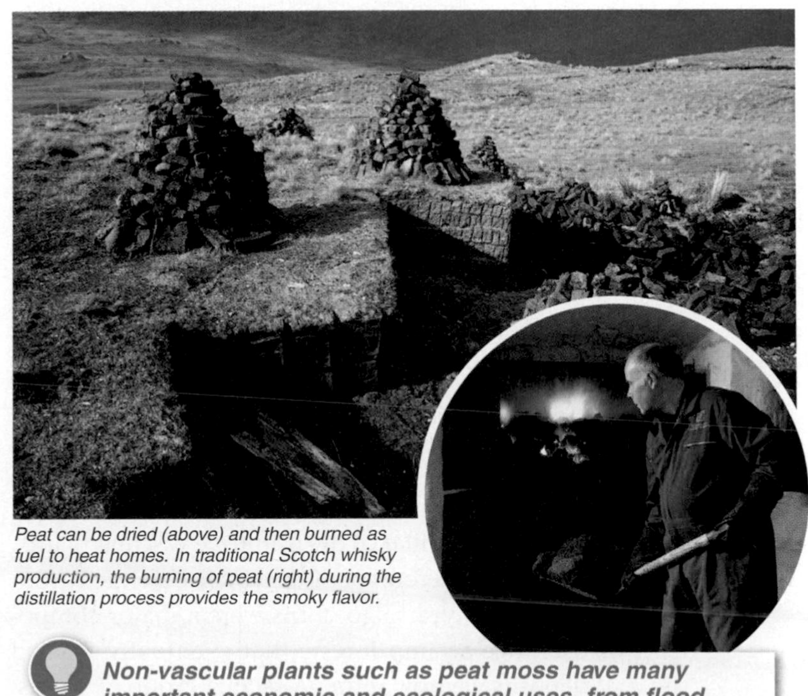

Peat can be dried (above) and then burned as fuel to heat homes. In traditional Scotch whisky production, the burning of peat (right) during the distillation process provides the smoky flavor.

Non-vascular plants such as peat moss have many important economic and ecological uses, from flood control to gardening to the production of Scotch whisky.

FIGURE 14-8 **Uses of peat.**

extremely fragile and very slow to regenerate. When people or cattle walk over the crust, they break it into small pieces that cannot resist wind, and erosion can then strip a meter or more of soil in a few decades, leaving tree trunks supported in mid-air by their roots (FIGURE 14-9).

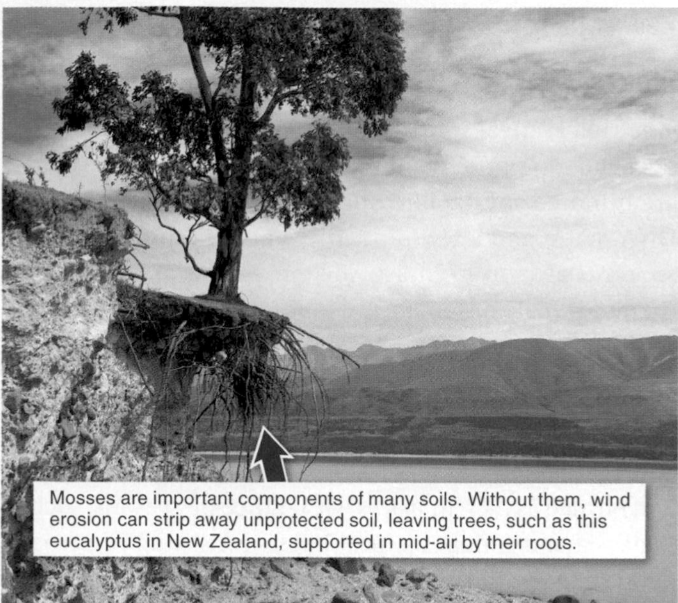

Mosses are important components of many soils. Without them, wind erosion can strip away unprotected soil, leaving trees, such as this eucalyptus in New Zealand, supported in mid-air by their roots.

FIGURE 14-9 Non-vascular plants reduce erosion.

TAKE HOME MESSAGE 14.3

» Non-vascular plants—mosses, liverworts, and hornworts—have scarcely evolved beyond the stage of the earliest land plants. They lack roots and vessels to move water and nutrients from the soil into and throughout the plant body, and they reproduce with spores formed when a sperm from a male reproductive structure "swims" through a drop of rainwater to fertilize the egg in a female reproductive structure.

14.4 The evolution of vascular tissue made large plants possible.

What a difference some tubes make. That's all vascular tissue is—an infrastructure of tubes that begins in a plant's roots and extends up its stem and out to the tips of its leaves. The evolution of vascular tissue allowed early land plants to transport water and nutrients faster and more effectively than does the cell-to-cell diffusion that non-vascular plants must rely on. The roots of vascular plants penetrate far enough into the soil to reach moisture even when the soil surface is dry. Roots that reach deep into the soil also provide the support that a plant needs to grow upward without falling over. This is why **vascular plants** can grow taller than non-vascular plants and are more successful in areas where the surface of the ground dries out between rains (FIGURE 14-10).

Ferns are the most familiar of the primitive vascular plants. During the Carboniferous period, which extended from 360 to 300 million years ago, ferns were a major component of the huge swamp forests that eventually formed the coal deposits we now mine. A related group of plants, the horsetails, like their Carboniferous ancestors, grow in wet habitats where the oxygen concentration around the roots is low because the spaces between soil particles are filled with water (rather than air). Horsetails developed hollow stems that allow oxygen from the air to diffuse down to the roots. The thin leaves of horsetails have a single nutrient- and water-carrying vessel extending from the base to the tip.

Simply having vessels, as horsetails do, is an important evolutionary innovation, but ferns have a further one. In addition to the central vessel in each leaflet, the leaves of ferns have vessels branching from the central vessel to the edges of the leaflet. This arrangement places a channel for the movement of water and nutrients close to each cell in the leaf.

Like non-vascular plants, horsetails and most ferns reproduce with spores. In horsetails, the central hollow stem is topped by a conical structure where the haploid spores are produced and released. Many ferns have **sporangia** (*sing.* **sporangium**) on the undersides of the leaves, where spores

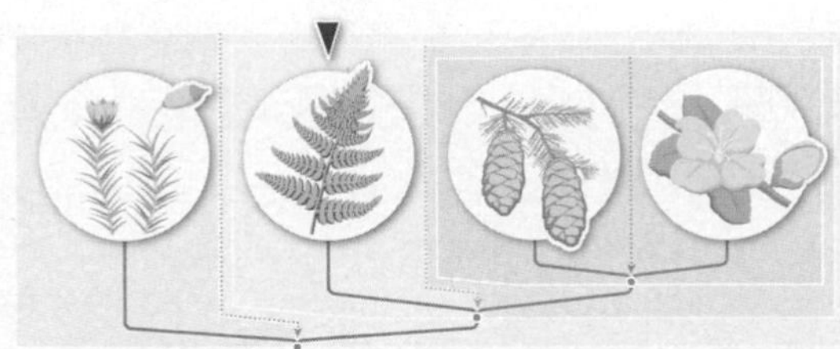

THE VASCULAR SEEDLESS PLANTS

Katote tree ferns

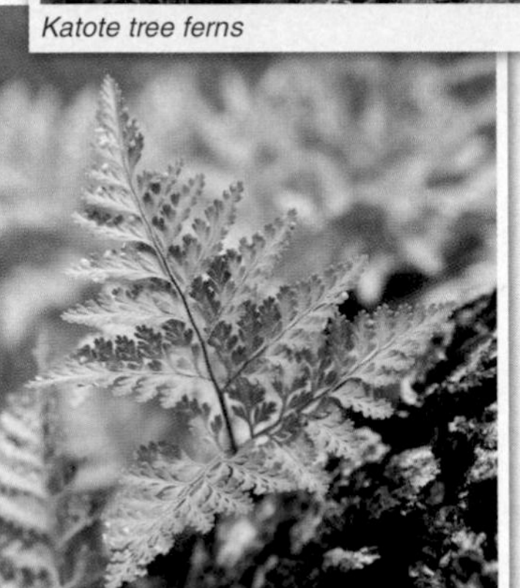

Canary Island hare's foot fern

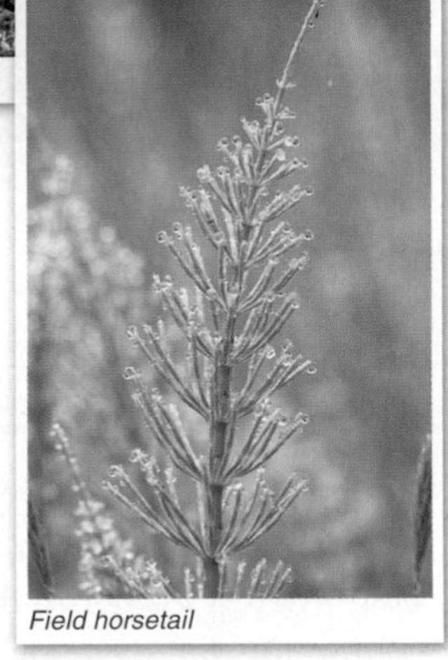

Field horsetail

COMMON CHARACTERISTICS

- Distribute water and nutrients throughout the plant with a "circulatory system" of vascular tissue
- Release haploid spores, dispersed by the wind, which grow and produce gametes
- Life cycle (unlike in animals) with multicellular haploid and diploid phases

MEMBERS INCLUDE

- Ferns (~12,000 species)
- Horsetails (~15 species)

FIGURE 14-10 Snapshot of the vascular seedless plants: ferns and horsetails.

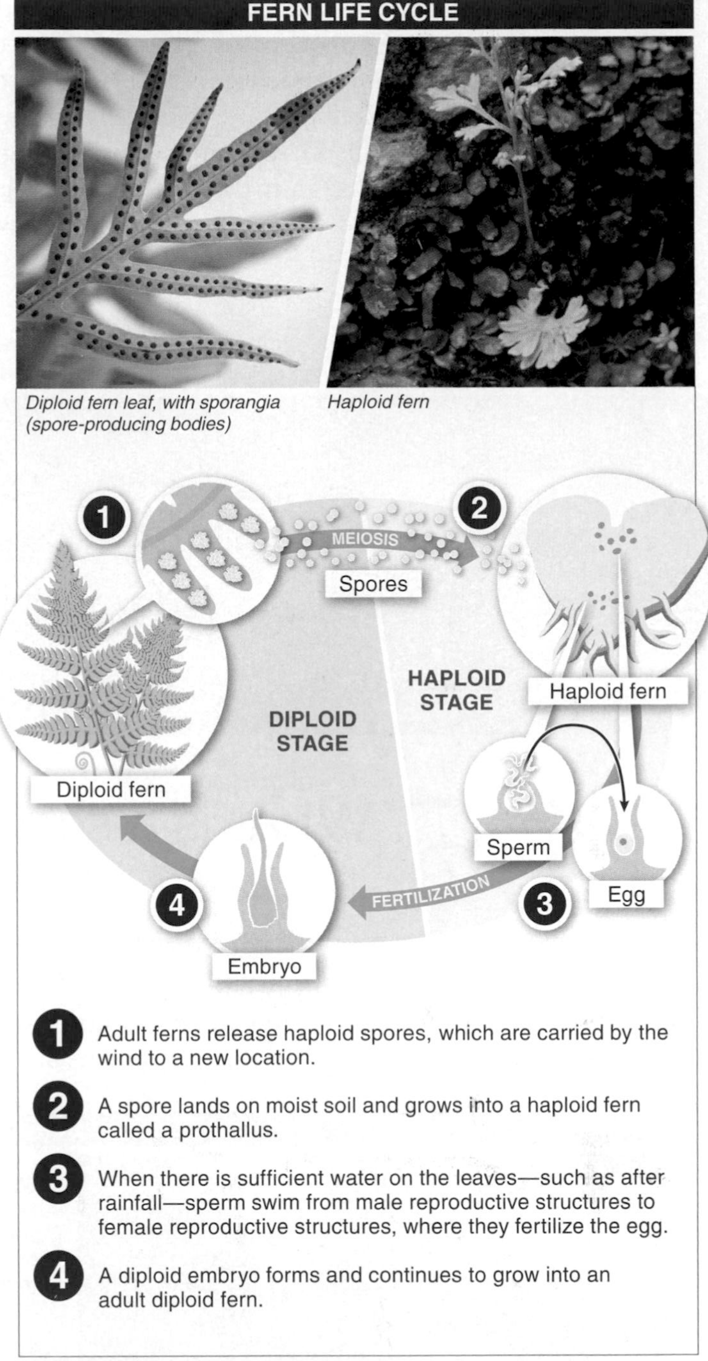

FERN LIFE CYCLE
Diploid fern leaf, with sporangia (spore-producing bodies)
Haploid fern
1
MEIOSIS
2
Spores
HAPLOID STAGE
Haploid fern
DIPLOID STAGE
Diploid fern
Sperm
Egg
FERTILIZATION
3
4
Embryo
1 Adult ferns release haploid spores, which are carried by the wind to a new location.
2 A spore lands on moist soil and grows into a haploid fern called a prothallus.
3 When there is sufficient water on the leaves—such as after rainfall—sperm swim from male reproductive structures to female reproductive structures, where they fertilize the egg.
4 A diploid embryo forms and continues to grow into an adult diploid fern.

FIGURE 14-11 Haploid and diploid life stages in ferns.

are produced (FIGURE 14-11). Because the haploid gametophyte is much smaller and simpler than the diploid sporophyte, ferns are described as having a dominant sporophyte.

Horsetails and ferns are taller than non-vascular plants, so their spores can be blown by the wind when they are released, and they may settle some distance from the parent plant. This increased dispersal ability was an important adaptation—non-vascular plants are so low-growing that wind can't play much of a role in moving their spores.

A spore that lands on moist soil grows into a tiny, heart-shaped structure called a **prothallus,** which is the free-living haploid stage of a fern (see Figure 14-11). The prothallus produces the haploid gametes: some cells produce eggs and others produce sperm. A sperm "swims" through drops of rainwater to fertilize an egg, and the fertilized egg (a diploid zygote with two sets of chromosomes) grows into an adult fern.

TAKE HOME MESSAGE 14.4

» Vessels are an effective "circulatory system" to carry water and nutrients up from the soil to a plant's leaves. The first vascular plants—including the earliest ferns and horsetails—were able to grow much taller than their non-vascular predecessors.

14.5–14.7

The advent of the seed opened new worlds to plants.

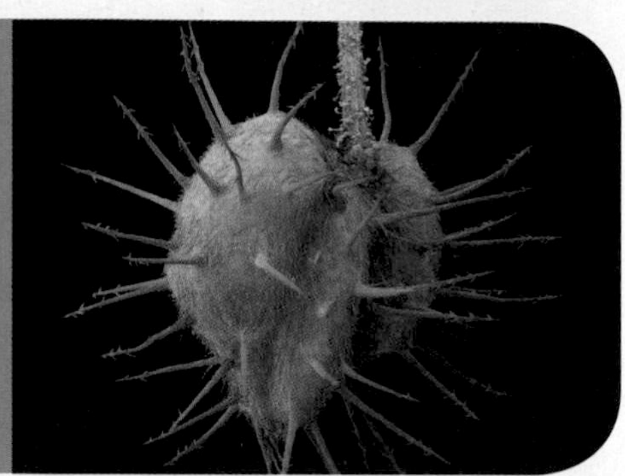

A seed is a plant embryo packed with its own food supply.

14.5 What is a seed?

As we've seen, the evolutionary development of vascular tissue allowed plants to grow larger. Vascular plants could colonize areas where the soil at the surface of the ground was not always wet, because their roots could penetrate the soil to reach water and nutrients. The next big innovation in plant evolution was the **seed,** an embryonic

SEEDS: STRUCTURE AND GROWTH

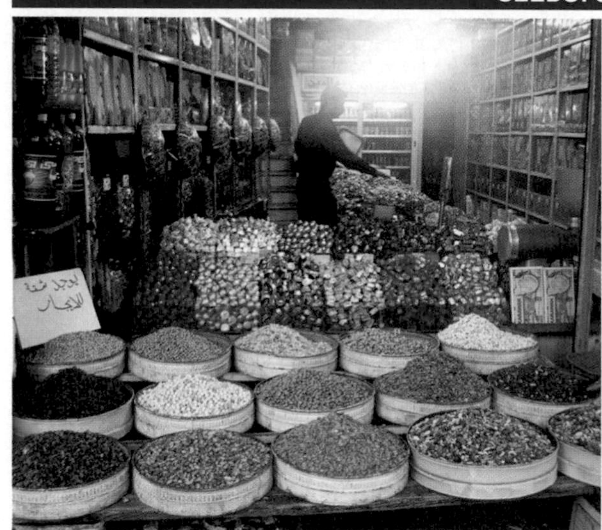

Due to their store of nutrients, seeds are a desirable food source. Shown here: seeds for sale at a market in Jordan.

STRUCTURE
Fertilization produces a diploid seed, which contains a multicellular embryo and a store of carbohydrate (endosperm) to fuel its initial growth.

GROWTH
A seedling draws energy from the endosperm while it extends its leaves upward to begin photosynthesis and its roots downward into the soil to reach water and nutrients.

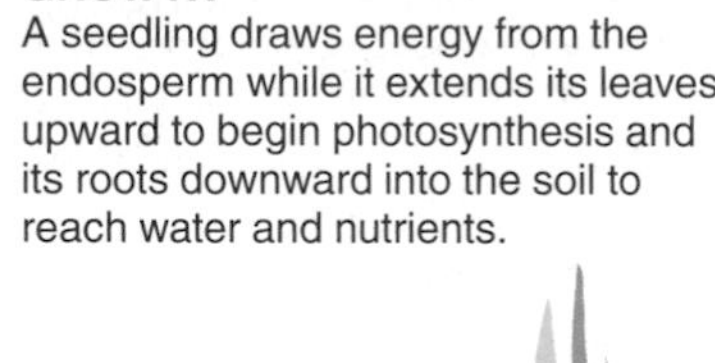

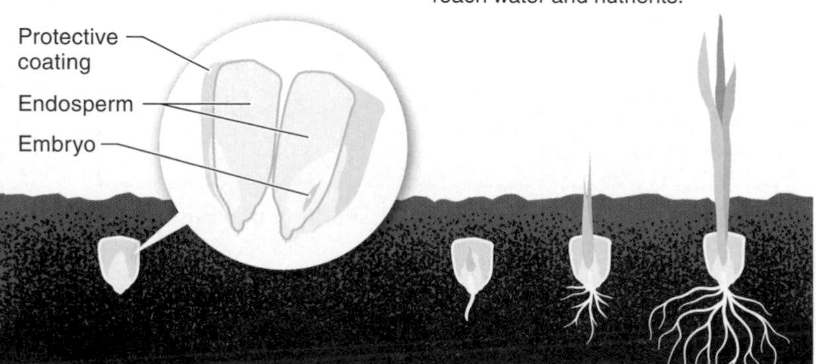

FIGURE 14-12 **What is a seed?**

plant with its own supply of water and nutrients encased within a protective coating (FIGURE 14-12).

Unlike spores, which are single cells that contain only DNA, RNA, and a few proteins, seeds contain both a multicellular embryo and a store of nutrients, mostly starch. This nutritive tissue—called **endosperm** in flowering plants and trees—can fuel the seed's initial growth. A seedling draws energy from the endosperm while it extends its leaves upward to begin photosynthesis and its roots downward to reach water and nutrients in the soil. There are two modern groups of seed-producing plants: **gymnosperms** (including pines, firs, and redwoods) and **angiosperms** (all flowering plants and trees).

Q Do plants have sex? How are seeds formed?

Like plants that do not produce seeds, seed plants have a life stage, called the gametophyte, that produces haploid gametes (sperm and egg). This is analogous to humans' production of haploid gametes. Because the gametophyte is much smaller than the sporophyte in seed plants, these plants are considered sporophyte dominant. **Pollen grains** and **ovules** are the male and female gametophytes, respectively, of seed plants. In brief (we expand on this later in the chapter), a haploid female gamete (egg) forms inside the ovule. When a pollen grain lands near the ovule, it produces a tube that grows into the ovule. Sperm from the pollen grain move through the pollen tube into the ovule and fertilize the egg. The external layer of the ovule then forms the seed coat.

Another of the challenges confronting plants is dispersal of their seeds. As we'll see later in the chapter, many methods for dispersing seeds have evolved.

TAKE HOME MESSAGE 14.5

» Seeds are a way that plants can give their offspring a good start in life. A seed contains a multicellular embryo plus a store of carbohydrate and other nutrients. Seeds are distributed by wind, animals, or water.

14.6 With the evolution of the seed, gymnosperms became the dominant plants on earth.

Gymnosperms include four major groups: conifers, cycads, gnetophytes, and ginkgo (just one species) (FIGURE 14-13). All of the 900 or more species of gymnosperms are seed-bearing plants that produce ovules on the edge of a cone-like structure. During the middle of the Mesozoic era, 160 million years ago, the forests that dinosaurs walked through consisted entirely of gymnosperms—pine trees, redwoods, cycads, and their relatives. Flowering plants (angiosperms) did not appear for another 35 million years. Indeed, when it comes to population size and habitat range, pine trees and their relatives are among the most evolutionarily successful groups of plants on earth, growing on every continent except Antarctica and extending from sea level to the tree line on mountains (see Figure 14-13).

Pines, spruces, firs, redwoods, and their relatives are familiar to the residents of temperate regions. Many of these conifers have needle-like leaves—but that is not true of all gymnosperms. The ginkgo has distinctive fan-shaped leaves, and cycads have palm-like fronds with many small leaflets (see Figure 14-13). Both are nearly identical to fossils from the Mesozoic era.

The reproductive structures of gymnosperms—the cones—are male or female. The pine cones you are probably familiar with are the female cones, which produce the ovules and, eventually, the seeds (FIGURE 14-14). The male cones are smaller and release pollen that is blown by the wind, and some of it reaches the ovules, which lie beneath the protruding scales of the female cones. The quantity of pollen released by conifers is beyond imagination—pollen-heavy air in a pine forest becomes hazy, and every surface is coated with yellow. Wind dispersal is clearly an inefficient method of **pollination** (getting

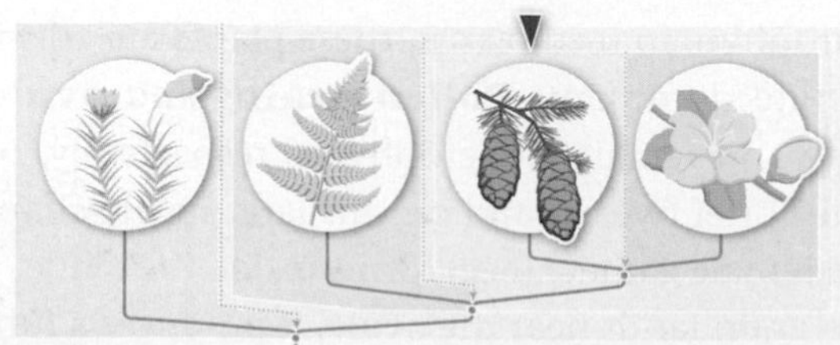

THE GYMNOSPERMS

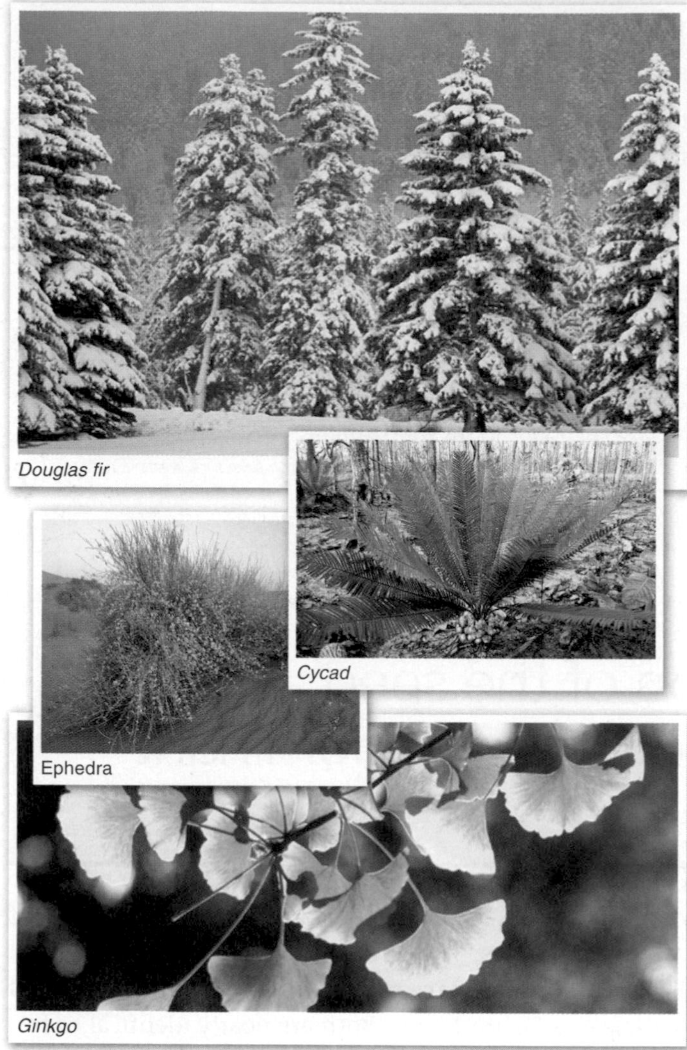

COMMON CHARACTERISTICS

- Distribute water and nutrients with a "circulatory system" of vascular tissue
- Reproductive structures called cones produce the gametes
- Fertilization produces seeds

MEMBERS INCLUDE

- Conifers (~600 species)
- Cycads (~300 species)
- Gnetophytes (~65 species)
- Ginkgo (1 species)

FIGURE 14-13 Overview of the gymnosperms: non-flowering plants with seeds.

CONES

MALE CONE
The male cone releases pollen grains that require wind to reach a female cone.

FEMALE CONE
The female cone has ovules on the protruding scales. They produce seeds when fertilized by pollen.

Although billions of pollen grains are wasted for each one that fertilizes an egg, gymnosperms have successfully used wind pollination for more than 200 million years.

FIGURE 14-14 Cones are the reproductive structures of gymnosperms.

the pollen to the vicinity of the ovule): billions of pollen grains are wasted for each grain that lands on a female cone and produces sperm to fertilize an egg in an ovule. Nonetheless, this "brute force" method of ensuring fertilization works: for more than 200 million years, pines have been successfully pollinated by wind.

Q Why does so much tree pollen coat the windshields of parked cars in the spring?

When the pollen arrives at the female cone, a pollen tube forms and transports the haploid sperm to the ovule, where fertilization occurs and a diploid embryo begins to grow. The embryo develops slowly within the female cone over the course of many months, until the scales of the cone open and release the seed, ready to sprout into a new plant (FIGURE 14-15). With the evolution of seeds, gymnosperms developed a life cycle with no free-living haploid stage such as we see in non-vascular plants.

This evolutionary change in life history probably reflects the different degrees of evolutionary fitness of haploid

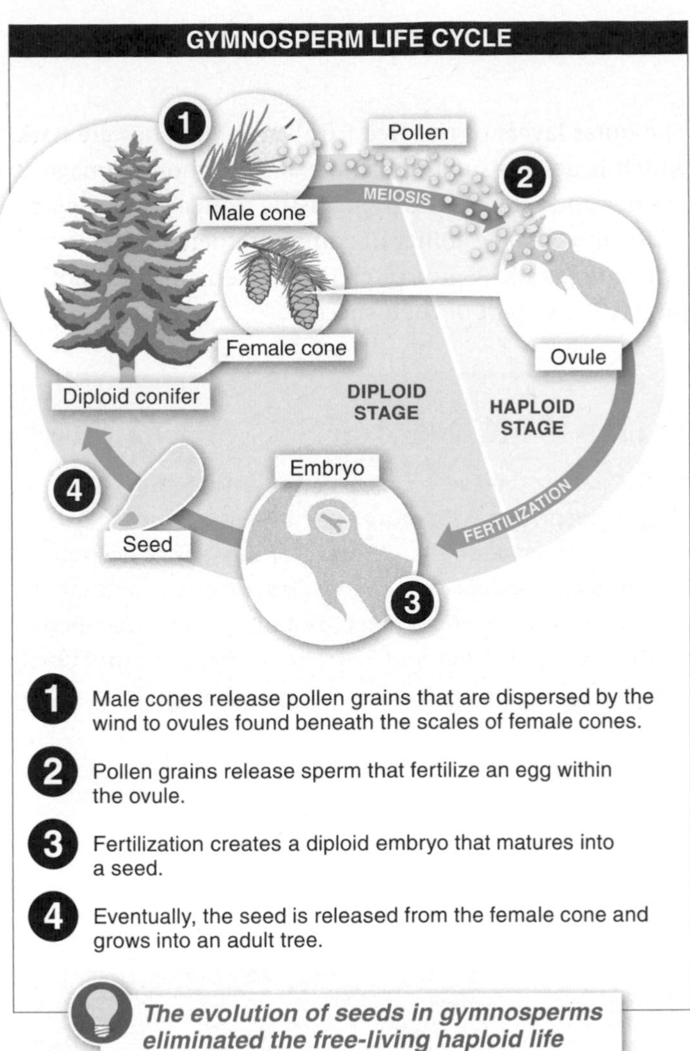

GYMNOSPERM LIFE CYCLE
1
Pollen
2
MEIOSIS
Male cone
Female cone
Ovule
Diploid conifer
DIPLOID STAGE
HAPLOID STAGE
Embryo
4
Seed
FERTILIZATION
3
1 Male cones release pollen grains that are dispersed by the wind to ovules found beneath the scales of female cones.
2 Pollen grains release sperm that fertilize an egg within the ovule.
3 Fertilization creates a diploid embryo that matures into a seed.
4 Eventually, the seed is released from the female cone and grows into an adult tree.
The evolution of seeds in gymnosperms eliminated the free-living haploid life stage seen in mosses and ferns.

FIGURE 14-15 Assisted by the wind: the life cycle of gymnosperms.

versus diploid organisms. A haploid organism has just one copy of each chromosome, which means it has just one copy of each allele. If the haploid organism carries a defective allele for a critically important gene, the organism is doomed, because it has only that one defective version of the gene. In contrast, diploid organisms have two sets of chromosomes and thus two copies of each allele. If one copy is defective, the second copy can function as a backup, so that mutations are much less likely to be lethal for diploid organisms than for haploid organisms.

TAKE HOME MESSAGE 14.6

» Gymnosperms (pine trees and their relatives) were the earliest plants to produce seeds. This mode of reproduction offered advantages over the spores of earlier plants and gave gymnosperms the boost they needed to become the dominant plants of the early and middle Mesozoic era. Gymnosperms depend on wind to carry their pollen. Conifers protect the developing seeds in the female cone.

14.7 Conifers include the tallest and longest-living trees.

In addition to having the widest geographic range and the greatest number of species among the gymnosperms, the cone-bearing trees—the conifers—include both the tallest and the oldest living organisms on earth (FIGURE 14-16). The four tallest trees in the world are conifers: a coast redwood that is 380 feet (116 m) tall, a Douglas fir and a Sitka spruce, each 318 feet (97 m), and a Sierra redwood at 311 feet (95 m). Miniature species of conifers exist, too. The shore pine, for example, can be just 20 centimeters tall. The oldest tree trunk belongs to a Great Basin bristlecone pine, the Methuselah tree, which has lived for more than 4,800 years (a slightly older tree was cut down in 1964). There's more to a tree than its trunk, though, and trunks can be replaced. That is what a Norway spruce in Sweden has done: the current trunk is about 600 years old, but the roots are 9,550 years old. This tree has persisted by sending up a new trunk each time the old one died.

Structural characteristics of trees allow them to grow large and live to great ages. Rigidity is provided primarily by the heartwood, which is a core of dead tissue that contains complexly cross-linked molecules.

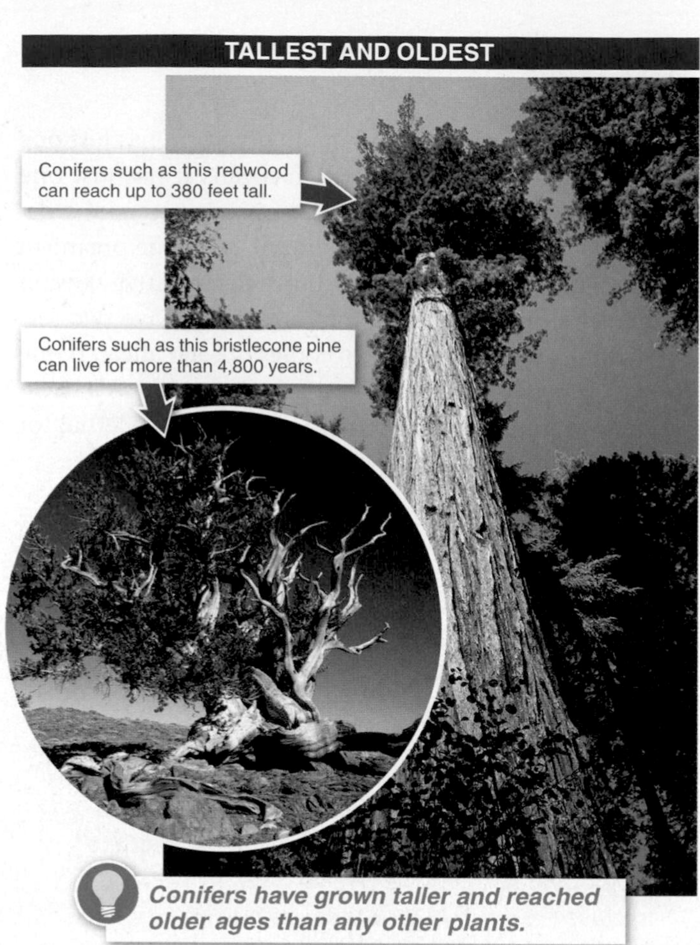

TALLEST AND OLDEST
Conifers such as this redwood can reach up to 380 feet tall.
Conifers such as this bristlecone pine can live for more than 4,800 years.
Conifers have grown taller and reached older ages than any other plants.

FIGURE 14-16 **Towering and ancient gymnosperms.**

The outer layers of the tree trunk and branches are bark, which is dead tissue that can be shed without damage to the tree and protects the living tissue from attack by plant-eating insects. The ability of conifers to defend themselves by exuding a sticky pine pitch that can engulf and smother insects is also part of their success.

TAKE HOME MESSAGE 14.7

» Conifers are the success stories among the gymnosperms, with more species and a larger geographic range than the rest of the gymnosperms combined. Rigidity, an exterior layer of bark, and the ability to exude sticky pitch protect conifers, allowing them to grow taller and live longer than any other plants.

14.8–14.10

Flowering plants are the most diverse and successful plants.

A South African pink daisy is in full bloom.

14.8 Angiosperms are the dominant plants today.

The appearance of flowering plants (angiosperms) about 135 million years ago in the Cretaceous period set the stage for the botanical world we know today, with flowering trees, flowering bushes, and all the grasses and herbaceous (non-woody) plants we see around us. The vast majority of plants on earth are flowering plants in the angiosperm group (FIGURE 14-17). Many of the early flowering plants would look familiar to us, and angiosperms dominate the plant world now, with some 250,000 species, compared with approximately 1,000 species of gymnosperms.

Flowers come in a bewildering variety of sizes, shapes, and colors, but all have similar structures: a supporting stem with modified leaves—the flashy petals and the sepals, which

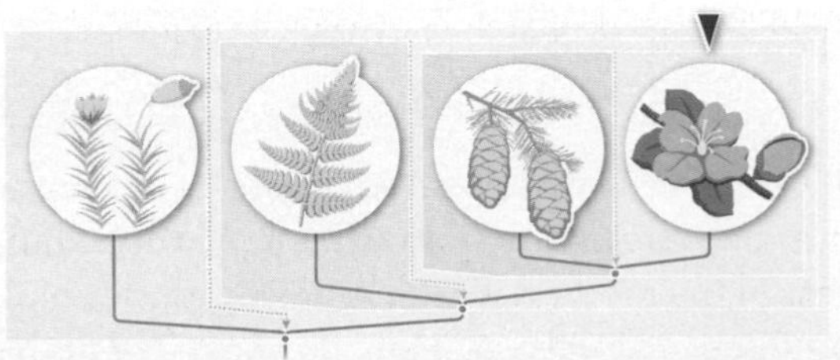

FIGURE 14-17 **Snapshot of the angiosperms.**

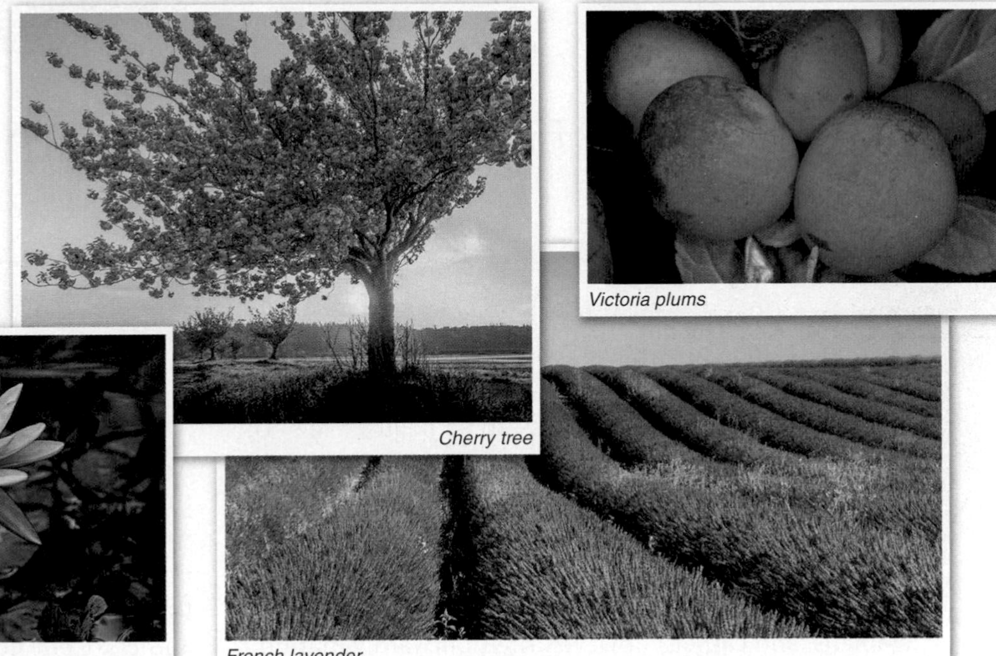

form a (usually) green wrapping that encloses the flower while it is in bud. Most angiosperms combine the male and female reproductive structures in the same flower. The male structure is the **stamen** and includes the **anther,** which produces the pollen, and its supporting stalk, the **filament.** The female reproductive structure is the **carpel.** It has an enclosed **ovary** at its base, which contains one or more ovules in which eggs develop; a stalk, called the **style,** extending from the ovary; and a sticky tip, the **stigma** (FIGURE 14-18).

THE ANGIOSPERMS
COMMON CHARACTERISTICS
• Distribute water and nutrients throughout the plant with a "circulatory system" of vascular tissue
• Produce flowers, which produce gametes
• Seeds are enclosed within an ovary
MEMBERS INCLUDE
• Flowering trees, bushes, herbs, and grasses (~250,000 species)
Cherry tree
Victoria plums
Water lily
French lavender

Pollination in angiosperms is the transfer of pollen from the male reproductive structures to the female reproductive structures of a flower—sometimes the same flower, and sometimes a different flower (on the same plant or on a different plant). Pollination is a multi-step process in which a pollen grain sticks to the stigma and forms a tube that grows until it reaches the ovule, thus providing a route for sperm to travel to fertilize the egg. In the next section, we explore the tremendously varied methods by which angiosperms get the male gametes to the female gametes.

Although most flowers contain both male and female structures, several thousand species of plants produce flowers

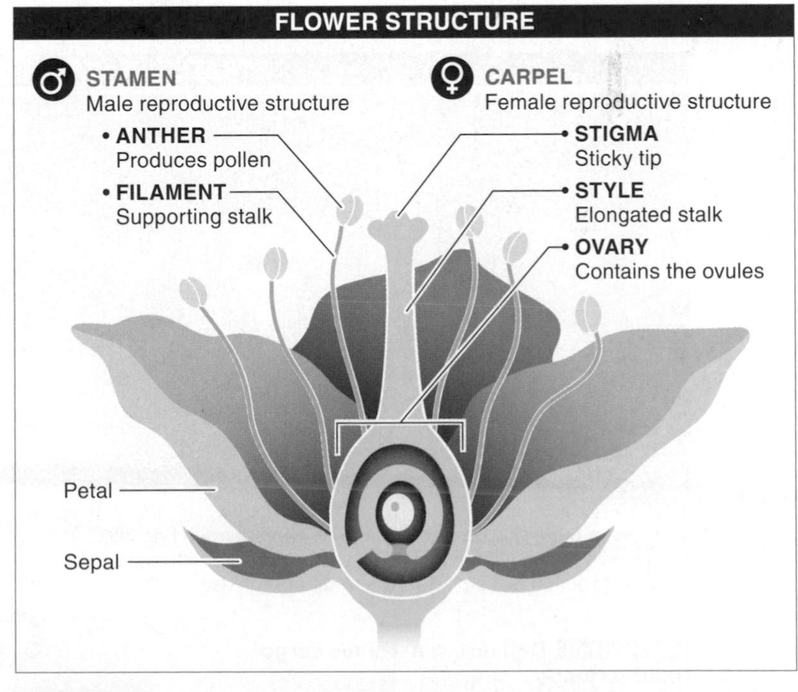

FIGURE 14-18 **A flower houses a plant's reproductive structures.**

that are only male or only female. Sometimes, male and female flowers are borne on different individual plants, but in quite a few angiosperm species, the same plant has some male flowers and some female flowers, often side by side. Maize (corn) is a familiar plant with male and female flowers in different places on the same plant, although you might not recognize the reproductive structures as flowers. The tassel at the top of the plant is the male flower, and the ear is formed by female flowers.

TAKE HOME MESSAGE 14.8

» Flowering plants appeared about 135 million years ago and diversified rapidly to become the dominant plants in the modern world. A flower houses a plant's reproductive structures. Most flowers have both male and female reproductive structures, but some species have flowers with only male or only female structures.

14.9 A flower is nothing without a pollinator.

Picture a flower. What does it look like? Flowers vary widely and we all have our favorites. There are tiny violets, giant sunflowers, sweet-smelling roses, and some rotten-smelling orchids and lilies, to name just a few. The label "flower" would seem to indicate little beyond the fact that flowers are usually brightly colored, frequently have a prominent odor, and generally sit at the end of a plant's branches. With very few exceptions, all flowers have the same four fundamental structures: sepals, petals, stamens, and carpels; however, their outward variation is important in sexual reproduction.

Fertilization occurs when the male gamete merges with the female gamete. But that's a lot easier said than done: the male gamete must travel to the female gamete. This voyage is pollination—the journey of a pollen grain to the stigma.

A small number of angiosperms achieve pollination by releasing tremendous amounts of pollen into the wind, as do the gymnosperms, or into water—on the slim chance that some pollen will land on the female reproductive organs of another plant of the same species. Most angiosperms, though, have a different way of moving pollen from the anthers of one flower to the stigma of another: they use animals to carry it (FIGURE 14-19).

Q Why are some flowers so flashy?

To ensure that animals will visit a flower, and thus pick up

STRATEGIES FOR ATTRACTING POLLINATORS

TRICKERY
This isn't a bee! This orchid flower deceives the male bee into carrying its pollen. When the bee attempts to mate with the flower it picks up pollen, which it delivers to other orchids.

BRIBERY
Some plants offer something of value to an animal, bribing the animal to carry pollen from one plant to another. Here, a bee, covered in pollen, flies from flower to flower in search of nectar.

FIGURE 14-19 Delivering precious cargo (inadvertently).

Angiosperms have developed an efficient way to transfer pollen from one flower to another: get an animal to carry it!

COEVOLUTION: FLOWERS AND THEIR POLLINATORS

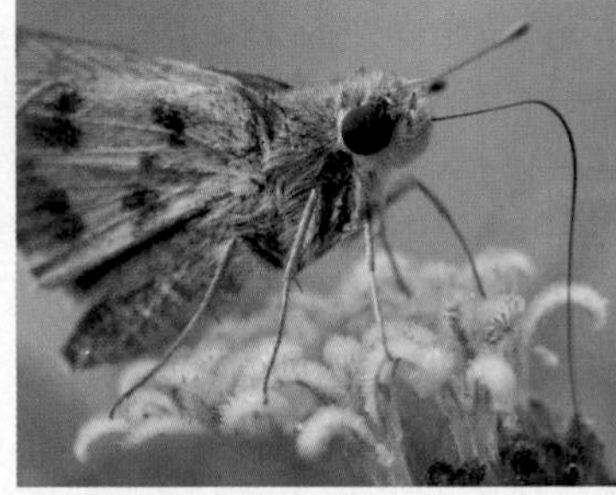

COLORS AND PATTERNS

- **WHITE:** Nocturnal pollinators, such as moths and bats
- **BRIGHT COLORS:** Visually oriented diurnal pollinators, such as birds, butterflies, and bees

FLOWER STRUCTURE

- **TUBE:** Pollinators with long tongues, such as moths
- **INTRICATE/CLOSED:** Pollinators such as bees

ODORS

- **SWEET:** Pollinators with a good sense of smell, such as moths, butterflies, and bees
- **STINKY:** Pollinators looking for rotten meat on which to lay eggs, such as flies
- **NO ODOR:** Pollinators with a poor sense of smell, such as birds

NECTAR

- **ABUNDANT:** Pollinators with high energy needs, such as bees, birds, and butterflies
- **NECTAR ABSENT:** Pollinators, such as flies, looking for a place to lay eggs, or beetles, looking for petals, pollen, and other parts to eat

FIGURE 14-20 **Evolving together: plants and pollinators.** You can often determine the type of animal that pollinates a flower just by examining the features of the flower.

and deliver their pollen cargo, two strategies for achieving pollination have evolved among the flowering plants.

1. Trickery. The plant deceives some animals into carrying its pollen from one plant to another. Some orchid species, for example, produce flowers that resemble female wasps. The mimicry is so good that male wasps mount the flower and attempt to mate with it. The male wasp twirls wildly on the flower, like a cowboy on a bucking bronco, repeatedly whacking his head against the strategically located anthers and getting pollen stuck all over his head and body. That is not enough for the plant to achieve pollination—but it's a start. If that male wasp gets fooled again by another orchid flower and mounts it to mate, he will inadvertently deposit some of the pollen from his body onto the also strategically placed stigma of that flower. In the end, the wasp does not gain from his actions, but the orchids have an effective system of pollination.

2. Bribery. The plant bribes some animals to carry its pollen from one plant to another. Rather than just using trickery, the plant offers something of value to the animal. For this mutually beneficial method (a "mutualism"; see Section 17-14) to work, the plant must produce (a) a sticky pollen; (b) a flower that catches the attention of the pollinator; and, most important, (c) something of value to the pollinator, such as nutritious nectar rich in sugars and amino acids, or perhaps a safe, hospitable location for an insect to lay its eggs.

The variety of flower structures is tremendous. They differ in shape, color, and smell; and in the time of day they are open; the ability to produce nectar; and whether their pollen is edible. Pollinators likewise vary, and include birds (mostly hummingbirds), bees, flies, beetles, butterflies, moths, and even some mammals (mostly bats) (FIGURE 14-20). In each case, there has been strong coevolution between the plants and their pollinators: the plants become increasingly effective at attracting the pollinators and deterring other species from visiting the flower, while the pollinators become increasingly effective at exploiting the resources offered by the plants. Because of this strong coevolution between plant and animal species, for most flowers we can determine, just by examining the features of the flower, the type of animal that will pollinate it.

Pollination is just one step toward fertilization. Fertilization itself doesn't happen until the male and female sex cells meet and fuse. We explore this process in the next section.

TAKE HOME MESSAGE 14.9

» Angiosperms rely on animals to carry pollen from the anthers of one flower to the stigma of another. Flowers are conspicuous structures that advertise their presence with colors, shapes, patterns, and odors. With their flowers, plants can trick or bribe animals into transporting male gametes to female gametes, where fertilization can occur.

14.10 Angiosperms improve seeds with double fertilization.

Fertilization—the fusion of two gametes to form a zygote—*begins* when a pollen grain lands on a stigma of a flower, but that is far from the end. In angiosperms, the process is called **double fertilization,** because there are two separate fusions of male nuclei (each carrying a complete set of the parent organism's genetic material) from the pollen grain with female nuclei in an ovule. Double fertilization is a more efficient system than the type of fertilization in gymnosperms, because whenever (and only when) an embryo is produced at fertilization, so, too, is a more substantial, ready-made food source.

When we left off in the previous section, a pollen grain—which will deliver the haploid male gamete—had just arrived on the female's stigma. The pollen grain forms a tube that extends downward through the stigma and style to the ovary, ultimately entering an ovule. As the pollen tube grows, two haploid sperm are formed within it by mitosis, and both move down the pollen tube (FIGURE 14-21).

While the pollen tube is growing, a cell within the ovule undergoes meiosis, forming four haploid cells, called spores. Three of these usually degenerate. The surviving haploid spore enlarges and undergoes mitosis, forming an embryo sac. A large, central cell in the embryo sac has two haploid nuclei. One of the other cells in the embryo sac is the haploid female gamete (the egg). The pollen tube will be guided into the ovule by two other haploid cells within the embryo sac.

When the pollen tube reaches the ovule, one of the two sperm fuses with the egg to form a zygote. The other sperm fuses with the two nuclei in the middle of the embryo sac to form the endosperm, which has *three sets* of chromosomes (i.e., is triploid). The process is called double fertilization because two sperm enter the ovule and combine with female cells in two separate events, forming (1) a zygote (with two sets of chromosomes) and (2) an endosperm (with three sets of chromosomes).

The final steps in producing a seed occur as the diploid zygote cell undergoes multiple mitotic divisions to form an embryo, while the triploid cells multiply mitotically to produce the endosperm that will provide nutritional support for the seedling during its initial growth stages. The outer layers of the ovule form the seed coat that will protect the seed until it sprouts. The enclosure of the seed within the ovary is a distinction between angiosperms and gymnosperms. The seeds of a gymnosperm are unenclosed and are sometimes referred to as "naked."

Q What are the advantages of double fertilization for angiosperms?

In gymnosperms, a single sperm fuses with an egg to form a zygote in an ovule that contains hundreds of other female cells that are used as a source of nutrients for the resulting embryo. In angiosperms, however, double fertilization offers two important advantages.

1. Double fertilization initiates formation of endosperm *only* when an egg is fertilized. Angiosperms do not waste energy forming endosperm in ovules that will not contain embryos. Gymnosperms invest that energy up front, and nutrients in ovules that are not fertilized are wasted because those "seeds" do not contain an embryo.

ANGIOSPERM LIFE CYCLE

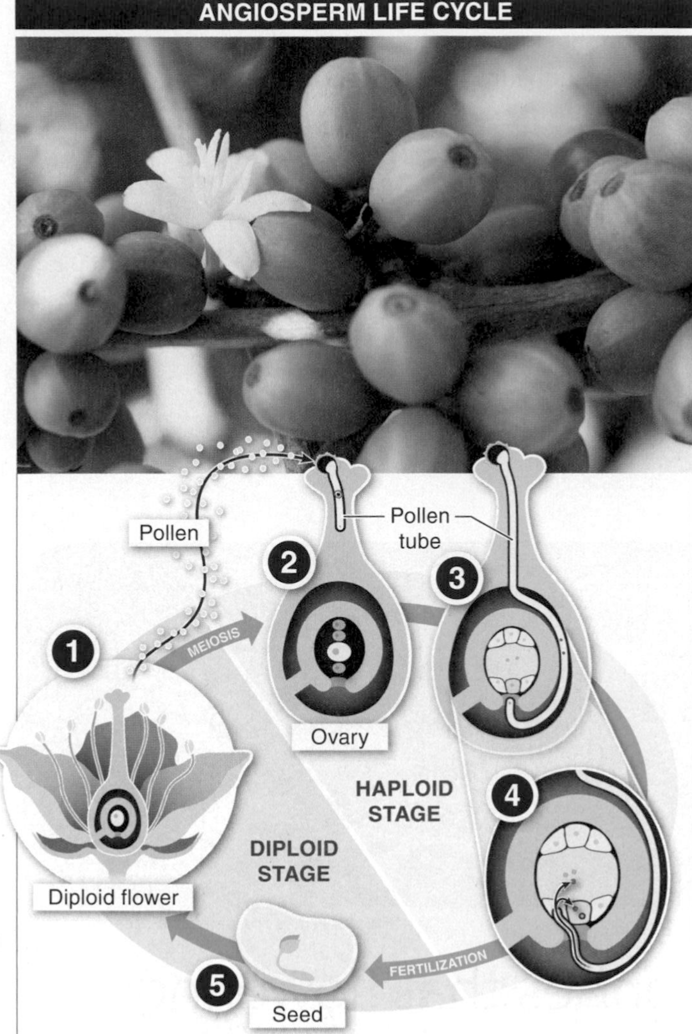

1. Pollen grains are released from the male anthers of a flower, delivering the male haploid gamete to the stigma—the sticky tip of the female reproductive structure.
2. The pollen grain produces a tube that extends through the stigma to the ovary. Meanwhile, a cell within the ovule undergoes meiosis, forming four haploid cells, called spores. Three of these usually degenerate.
3. The surviving haploid spore undergoes mitosis, forming an embryo sac that contains seven haploid cells. One cell becomes the egg, another cell—with two nuclei—will form the endosperm following fertilization by a sperm cell.
4. The pollen tube reaches the ovule, and two sperm are released. One fuses with the egg to form a zygote. The other fuses with two nuclei in the embryo sac to form the endosperm, which nourishes the embryo. The process is called double fertilization.
5. The zygote and the endosperm continue to develop within the ovule, forming a seed that will eventually be released and grow into a mature plant.

FIGURE 14-21 Double fertilization fortifies the seeds of angiosperms.

2. Angiosperms can produce smaller gametes, because the large energetic reserves will be produced only *after* fertilization occurs. The small size of the male and female gametes of angiosperms ensures that seeds are produced quickly. FIGURE 14-22 shows how, as different reproductive strategies have evolved in plants, their gametes have become progressively smaller. Rapid production of seeds allows angiosperms to grow as annual plants (plants that complete their life cycle from sprouting to seed production in one growing season), which is something gymnosperms cannot do.

Outbreeding, the combination of haploid cells from two different individuals, produces offspring with greater genetic diversity (that is, carrying a greater

Q Do flowers with both male and female structures fertilize themselves?

HAPLOID AND DIPLOID LIFE STAGES

HAPLOID STAGE	DIPLOID STAGE

NON-VASCULAR PLANTS
The majority of the life cycle is spent in the haploid stage.

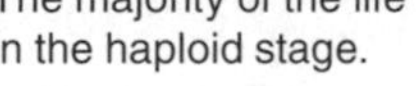

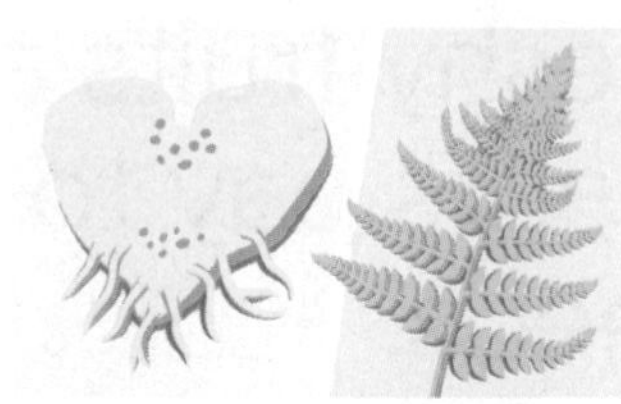

VASCULAR SEEDLESS PLANTS
The haploid and diploid stages are both multicellular and are physically independent of one another.

GYMNOSPERMS
The evolution of seeds almost completely eliminates the prominent haploid stage seen in mosses and ferns.

ANGIOSPERMS
Haploid gametes are further reduced in size, enabling more rapid seed production.

As plants have developed different reproductive strategies, they have progressed from having a prominent haploid stage of life to simply having haploid gametes.

FIGURE 14-22 Overview of the haploid and diploid stages of plant life cycles.

diversity of alleles) than offspring that result from inbreeding, the combination of a male and female gamete from the same individual. Angiosperms have a variety of ways to increase the chance of outbreeding. When male and female reproductive structures are in different flowers or even on different individuals, for example, this increases the chance that a pollen grain landing on a stigma comes from a different plant. But many flowers with both male and female parts also have mechanisms to prevent self-fertilization. For example, the anthers may mature before the stigma, so the stigma is not ready to receive pollen when the anthers are active. And when the stigma does become functional, the anthers are no longer producing pollen. Other plants use a molecular recognition system to prevent inbreeding: proteins on the surface of the stigma will not allow pollen from the same individual to form a pollen tube.

TAKE HOME MESSAGE 14.10

» Angiosperms undergo double fertilization, which ensures that a plant does not invest energy in forming endosperm for an ovule that has not been fertilized. Angiosperms have also developed methods to reduce the occurrence of self-fertilization and increase genetic variation among offspring.

14.11–14.12

Plants and animals have a love-hate relationship.

Fleshy fruits for sale at a street market.

14.11 Fleshy fruits are bribes that flowering plants pay to animals to disperse their seeds.

Leaving home is an inevitable part of growing up. But as difficult as it may be, imagine you had to move away but didn't have a car or moving van—or even legs. This is yet another issue facing plants. Ingenious methods have evolved for transporting the male gametes to the female sex organs for fertilization, but what would happen if all the fertilized eggs and all the developed offspring remained on the female plant—or nearby, if they fall off the parent plant? Parent and offspring would end up competing for the same light, space, soil nutrients, and other resources.

The distribution solution is the fruit, a structure that aids in dispersing seeds—the reproductive packets made up of the embryo, some food reserves, and a hard, protective coat. Pea pods, sunflower seeds, corn kernels, and hazelnuts are a few examples of fruits. These are "dry fruits," and the seeds they contain are transported by wind, water, or animals (FIGURE 14-23).

Many angiosperms, though, make **fleshy fruits** that consist of the ovary plus some additional parts of the flower. For example, the core of an apple is the ovary and the flesh of the apple is derived from adjacent parts of the flower. Blueberries, watermelons, oranges, tomatoes, and peaches are other examples of fleshy fruits. The fruit usually develops from the ovary (and sometimes from nearby tissue), right around the seeds, which develop from the ovules. Given that the ovary is part of a flower, after pollination and fertilization that flower will turn into a fruit. This should cause you to look at a field of flowers differently. Every fruit you eat was once a flower.

SEED DISPERSAL

HITCHING A RIDE
Some seed pods have spines or projections that attach them to passing animals.

FLYING AND FLOATING
The structure of some seeds allows them to be carried away from the parent plant by wind or water.

PROVIDING A FOOD SOURCE
Fleshy fruit is a form of bait that lures an animal to eat the seed and carry it far from the parent plant before eliminating it.

FIGURE 14-23 Methods of dispersing seeds.

The fleshy part of a fruit is often larger than the ovary, and an angiosperm invests a lot of energy in producing fleshy fruits. The payoff comes when an animal eats the fruit and then defecates the seeds at a location far from the parent plant. In other words, fleshy fruit is the bait that some flowering plants use to get animals to disperse their seeds.

When you look at fruits, you can see several characteristics that help this system work.

- Fruits are colorful—red is the most common color, followed by yellow and orange. Bright colors contrast with green foliage and make fruits conspicuous.
- Fruits typically taste good—plants pour sugars into fruits, and their sweetness appeals to many animals.
- Fruit is good for animals, serving as a significant source of sugar and enabling animals to produce ATP for their needs. Additionally, fruit has other nutritional value. Many birds incorporate the red and yellow pigments into their own color patterns, and colorful male birds are appealing to female birds. Thus, male birds that eat fruit are likely to produce a lot of offspring who also eat fruit—a good deal for the plant.

Q Can seeds still sprout after being eaten by an animal?

To germinate, seeds must pass through the animal's digestive tract and survive. That's not a problem—seeds are fully viable when they emerge. You may test this yourself. The next time you eat fresh tomatoes, check the toilet about 24 hours later. The seeds that have passed through your digestive system are still intact. If you want to test their viability, recover a few, rinse them off, and put them on a wet paper towel in a sealed plastic bag. If you want to do a controlled experiment, you can take some seeds directly from a tomato and put them in another bag. Put both bags in a warm, dark place, and check them in a couple of days. You'll find that both sets of seeds have sprouted. (Or, you can just take our word for it.)

Seeds not only survive being eaten by animals—some seeds will not germinate *unless* they are eaten. For some plants, the pulp that surrounds the seeds in the fruit inhibits germination, and it must be digested away before the seed will sprout. The seeds of other plants have chemicals in the seed coats that must be removed by the acidity of an animal's stomach before the seeds will germinate. And there's another plus to this system: the seeds are deposited with a bit of manure that provides nutrients for the seedlings.

TAKE HOME MESSAGE 14.11

» Following pollination and fertilization, plants often enlist animals to disperse their fruits, which contain the fertilized seeds; the seeds are deposited at a new location where the seedlings can grow. Fruits are made from the ovary and, occasionally, some surrounding tissue.

14.12 Unable to escape, plants must resist predation in other ways.

The interactions of plants and animals are not simple. On the one hand, plants depend on animals to pollinate their flowers and disperse their seeds. On the other hand, many kinds of animals eat plants, and plants are vulnerable because they can't run away. Plants, however, are not defenseless. A host of devices have evolved in plants that give them some protection against being eaten. These defenses fall into two categories: anatomical structures, such as thorns, and chemical compounds, including hallucinogens (FIGURE 14-24).

Spines, prickles, and thorns are structures that discourage herbivores, and some are impressively large. The acacias, a group of plants that grow as large bushes or small trees, are notoriously spiny. The deterrent effects of acacia spines are sometimes enhanced by ants that live in the swollen bases of the spines and fight off other animals—from insects to mammals, including humans—that would feed on their host. Spines on palm fronds carry bacteria that infect spine-inflicted wounds, providing a long-lasting reminder to avoid palm fronds. In most plants, the surfaces of leaves and stems have microscopic hairs that protect the plant against insect herbivores by exuding a sticky fluid trap. Stinging plants such as nettles use larger hairs to inject a potent fluid into any animal that brushes against a leaf.

Many chemical defenses make plants bitter or otherwise unpalatable, but some plants produce potent chemicals that affect animals' nervous systems. Locoweeds, which grow throughout western North America, contain alkaloids that can cause permanent damage to the nervous system. When cattle and horses eat locoweed, they tire quickly and stop eating. Other defensive chemicals are hallucinogens. Tetrahydrocannabinol, or THC, in the leaves and buds of marijuana plants disorients mammals and causes them to stop eating.

Many of the chemicals that plants produce to protect themselves have physiological effects in humans, and many have been used for medicinal purposes. The aboriginal peoples of North America and Eurasia, for example,

PLANT DEFENSES

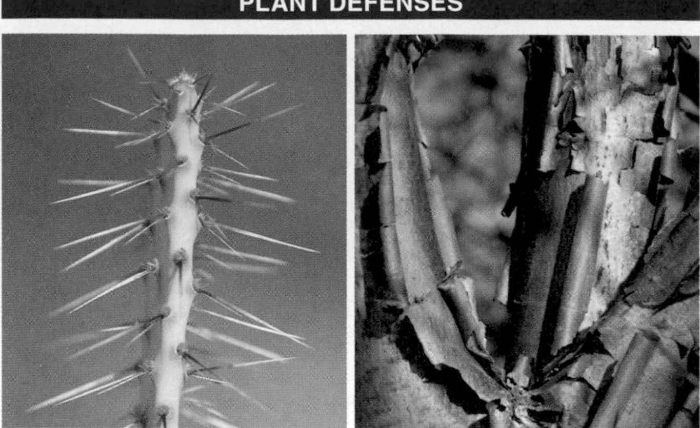

ANATOMICAL STRUCTURES
Some plants have spines, spikes, and thorns that deter predators. Some trees have thick layers of bark that are shed in order to get rid of attacking insects.

STICKY TRAPS
Conifers exude pitch, a sticky substance that can engulf and smother attacking insects.

Locoweed contains substances that, when eaten, can cause cattle and horses to become lethargic and stop feeding.
Stinging nettles have needle-like hairs that inject a chemical that causes a stinging sensation when contacted by an animal.

CHEMICAL COMPOUNDS
Some plants synthesize chemicals that induce physiological and behavioral changes in the animals eating them.

FIGURE 14-24 **Spines, sticky traps, and toxic compounds.** Plants have developed a range of defenses against predation.

knew that extracts of willow or aspen bark could relieve pain. Salicin is the compound in the bark that relieves pain, and we still use acetylsalicylic acid (aspirin) for pain relief. Opium poppies produce a more powerful pain reliever; foxglove flowers contain a compound that slows the heart rate; and a compound from the ipecacuanha plant can induce vomiting.

Plants with medicinal uses are much sought after, and representatives of pharmaceutical companies trudge through rain forests and deserts to find them (a process called "bioprospecting"). The rewards—medicinal as well as financial—from such a discovery can be great.

There's another role that chemicals can play in a plant's defense systems. Although plants can't move, they do send messages. They release airborne chemicals that can warn nearby plants of an impending insect threat or call insects to their aid. When insects nibble on plants, the plants release the chemical methyl jasmonate (MJ), which drifts on the breeze to nearby plants. The drifting MJ causes plants to ramp up production of their own defensive chemicals so that they are prepared if the insects invade them. Meanwhile, insects that consume other insects are attracted to plants emitting MJ, because it signals the presence of prey. When corn and cotton plants are attacked by caterpillars, release of MJ summons parasitic wasps that deposit their eggs inside the caterpillars. When the eggs hatch, the wasp larvae eat the caterpillars from the inside out.

About 900 species of plants turn the tables—and eat insects! They use the protein in their insect meals to supplement the nitrogen compounds they get from the soil. Insectivorous plants are most common in boggy areas, because boggy soil often has low nitrogen concentrations (FIGURE 14-25). Pitcher plants are an example—their name comes from conspicuous pitcher-shaped structures that capture insects. It's easy for an insect to enter the open top of a pitcher, but hard to get out, because the inner walls are lined with hair-like structures that point downward, forcing the insect into a pool of liquid. Some pitcher plants merely absorb the nitrogen that is released when the trapped insects decay, but other species secrete digestive enzymes into the pool. These enzymes, which are basically the same as the enzymes in your stomach, digest the insects and make the nitrogen available more rapidly.

Some plants living in nitrogen-deficient soil have turned the tables, becoming predators on insects.

A bug nymph trapped in the northern pitcher plant (Sarracenia purpurea) *in an acidic bog in Michigan.*

FIGURE 14-25 Carnivorous plants feed on insects.

TAKE HOME MESSAGE 14.12

» Plants have a wide range of defenses against herbivorous animals, from physical defenses such as thorns to chemicals that have complex effects on animals' physiology. Plants respond to insect attack by synthesizing chemicals that make the plant less palatable. Some plants living in soil that is deficient in nitrogen have switched roles, preying on insects.

14.13–14.16

Fungi and plants are partners but not close relatives.

A pixies parasol fungi (*Mycena interrupta*) growing in Tasmania, Australia.

14.13 Fungi are more closely related to animals than they are to plants.

Fungi make up their own monophyletic kingdom (see Chapter 12) within the eukarya domain. Most fungi are multicellular, sessile decomposers (FIGURE 14-26). Although they were originally thought to be plants lacking chlorophyll, it turns out that they have little in common with plants. In fact, DNA sequence comparisons reveal that fungi are more closely related to animals than they are to plants (FIGURE 14-27). As eukaryotes, fungi have all the basic cellular components you would expect to find: nuclei, mitochondria, an endomembrane system, and a cytoskeleton. They also have cell walls, but the cell walls, instead of including cellulose, as in plants, are made of the carbohydrate chitin, a chemical also important in producing the exoskeleton of insects.

WHAT IS A FUNGUS?

Fungi share the following three characteristics:

- **FUNGI ARE DECOMPOSERS OR SYMBIONTS**
  Fungi acquire energy by breaking down the tissues of dead organisms or by absorbing nutrients from living organisms.
- **FUNGI ARE SESSILE**
  Fungi are anchored to the organic material on which they feed.
- **FUNGI HAVE CELL WALLS MADE OF CHITIN**
  The chitin component of fungal cell walls is the same chemical that is important in producing the exoskeleton of insects.

FIGURE 14-26 **The defining characteristics of fungi.**

THE FUNGI

PLANTS

FUNGI

ANIMALS

Ancestral protist

Fungi are more closely related to animals than they are to plants.

FIGURE 14-27 **Where the fungi fit in.**

FAMILIAR FUNGI

Portobello mushrooms

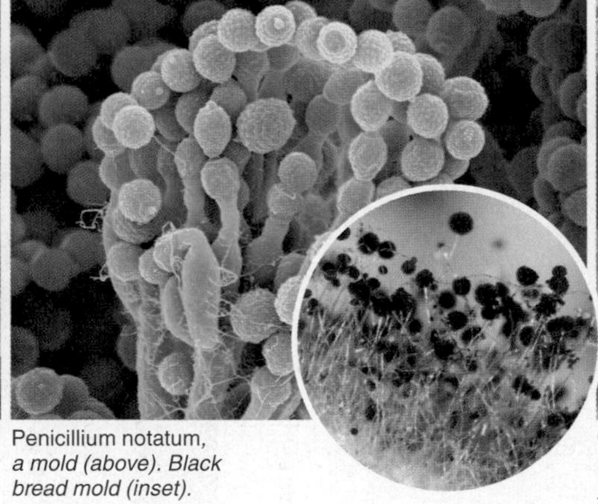

Penicillium notatum, a mold (above). Black bread mold (inset).

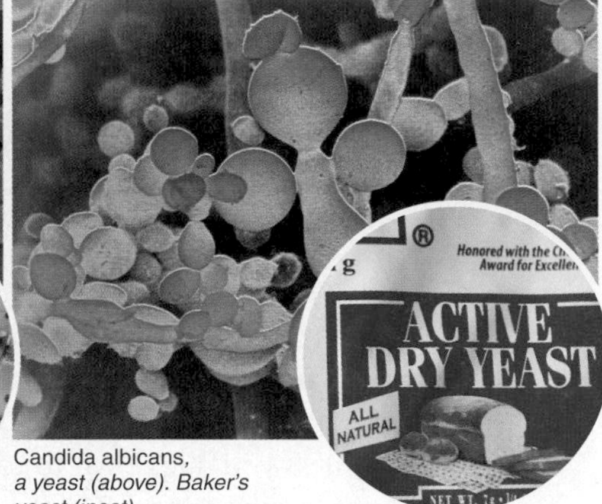

Candida albicans, a yeast (above). Baker's yeast (inset).

FIGURE 14-28 **The fungi that humans know best.** There are more than 1.5 million fungal species, usually divided among seven phyla. Those most recognizable to humans are mushrooms of the Basidiomycota (basidiomycetes), yeasts of the Ascomycota (ascomycetes), and molds (a term that does not refer to a monophyletic group), which occur in multiple phyla.

Fungi most likely arose from a unicellular, flagellated, aquatic protist more than 500 million years ago (and possibly as long as 1.3 billion years ago). Close to 100,000 species of fungi have been described, but the total number of species is estimated to be about 1.5 million. These species are divided into about seven phyla, but the overall phylogenetic classification is still very much in flux, with many of the relationships unresolved. Molecular evidence is, increasingly, guiding the process. About 98% of the described species belong to two monophyletic phyla: Ascomycota (about 64,000 species) and Basidiomycota (about 31,000 species). Besides these phyla, the two phyla with the largest numbers of described species are Microsporidia (about 1,300 species) and Chytridiomycota (about 700 species). These two groups, however, are not monophyletic.

The most commonly encountered fungi are of three types: (1) yeasts (the only single-celled fungi), truffles, and morels of the phylum Ascomycota; (2) mushrooms, of the phylum Basidiomycota; and (3) molds (not a phylogenetic grouping), such as *Penicillium* (source of the antibiotic penicillin) of the phylum Ascomycota, and black bread mold of the monophyletic phylum Zygomycota (FIGURE 14-28).

Bread rises because yeast cells are mixed into the dough, which is then kept in a warm place while the yeast consumes sugar and produces carbon dioxide through fermentation. The carbon dioxide released by the yeast makes the dough swell into a loaf.

The fungus that causes athlete's foot is multicellular and consists of thread-like structures made up of long strings of cells called **hyphae** (pronounced HIGH-fee). Because of their thinness and length, hyphae have an enormous surface area. This means they are very good at taking up nutrients from your skin, but it also makes them very vulnerable to drying out. The fungus responsible for athlete's foot grows best in moist places—like on the skin between your toes.

TAKE HOME MESSAGE 14.13

» Fungi are eukaryotes with the same internal cellular elements as other eukaryotes—and one distinctive feature: a cell wall formed from the carbohydrate chitin. Some fungi, the yeasts, live as individual cells; most other fungi are multicellular.

14.14 Fungi have some structures in common but exploit an enormous diversity of habitats.

As we've seen, most fungi are multicellular and are composed of long strings of barely visible, thread-like hyphal cells. The hyphae interconnect to form a mass of tissue called a **mycelium,** the form in which a fungus spends most of its time, usually underground or in a decaying tree or log. Unless you dug down to reach the mycelium, you would not know a fungus was there.

The structure that most people associate with fungi is the mushroom. But a mushroom is just a temporary reproductive structure (or "fruiting body"), part of a complex reproductive cycle that includes both sexual and asexual reproduction in some fungi (FIGURE 14-29). This cycle consists of several steps.

1. Underground, genetically distinct haploid hyphal cells join together. But their nuclei *don't* fuse, so instead of being diploid, they are "dikaryotic," meaning that each cell in the hypha has two nuclei.
2. The dikaryotic mycelium can grow and spread for years.
3. At some point, tightly packed dikaryotic hyphae may form a mushroom.
4. The haploid nuclei in the dikaryotic hyphae now fuse in some cells, putting the mushroom into a diploid state.
5. In meiosis, the diploid cells produce huge numbers of haploid spores (up to a billion in a single mushroom!), which are dispersed by wind or water or on the bodies of animals.
6. After landing in a hospitable place, the spores grow as haploid hyphae, and the cycle begins anew.

Fungi have an unusual and effective method of getting nutrition: they digest their food outside their "body." While growing underground, hyphae secrete strong enzymes that break down the organic molecules around them, and the hyphae then absorb the nutrient-rich fluid. They are extremely effective at absorbing nutrients and growing. In fact, in eastern Oregon, a yellow honey mushroom fungus covers an area of about 4 square miles (nearly 10 square kilometers)

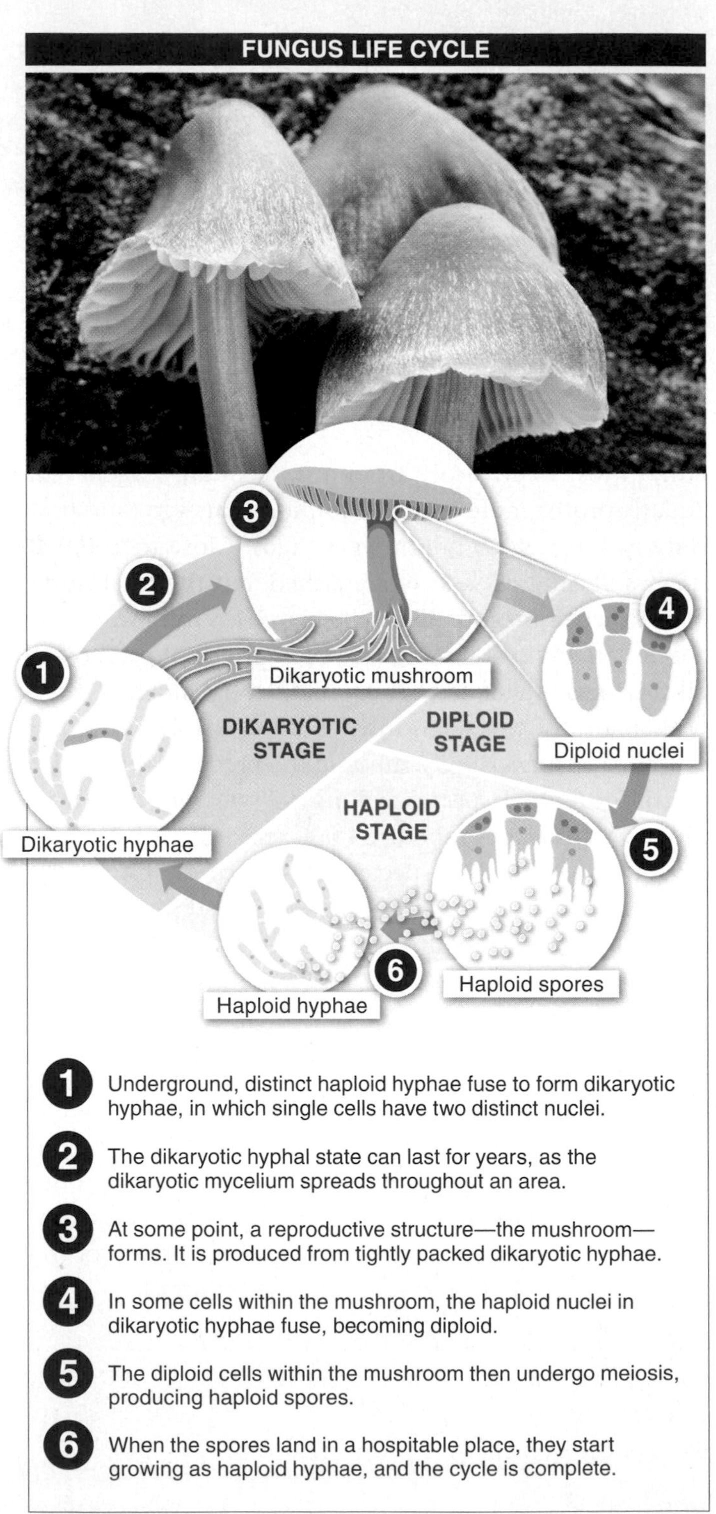

FUNGUS LIFE CYCLE
Dikaryotic mushroom
DIKARYOTIC STAGE
DIPLOID STAGE
Diploid nuclei
HAPLOID STAGE
Dikaryotic hyphae
Haploid spores
Haploid hyphae
1 Underground, distinct haploid hyphae fuse to form dikaryotic hyphae, in which single cells have two distinct nuclei.
2 The dikaryotic hyphal state can last for years, as the dikaryotic mycelium spreads throughout an area.
3 At some point, a reproductive structure—the mushroom—forms. It is produced from tightly packed dikaryotic hyphae.
4 In some cells within the mushroom, the haploid nuclei in dikaryotic hyphae fuse, becoming diploid.
5 The diploid cells within the mushroom then undergo meiosis, producing haploid spores.
6 When the spores land in a hospitable place, they start growing as haploid hyphae, and the cycle is complete.

FIGURE 14-29 **The three-stage reproductive cycle of a fungus.**

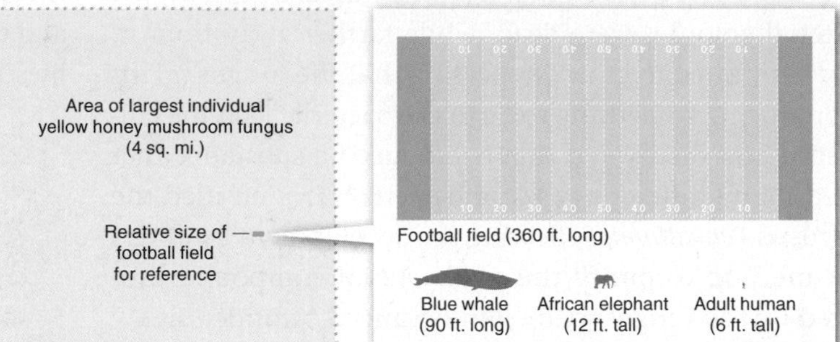

FIGURE 14-30 **The largest organism in the world: a yellow honey mushroom fungus.**

Q What is the largest living organism in the world?

(FIGURE 14-30). This fungus is estimated to be at least 2,400 years old, and it may be more than 8,000 years old, which could make it the oldest living organism. Moreover, by area, it is the largest. But, huge as it is, the fungus is hard to detect because the mycelium is underground. The fungus appears on the surface only as mushrooms (edible, but not particularly tasty) and as hyphae in dying and dead trees. The discovery of this enormous fungus occurred in 2000, when Oregon foresters sought an explanation for an epidemic of dead and dying trees in the Malheur National Forest. (They discovered that the fungus was killing trees by causing their roots to rot.)

As **decomposers,** fungi play an enormously important ecological role in speeding the decay of organic material in forests. All they need for their nutrient supply is some sort of organic material that they can break down. They don't need light, so they can grow underground or inside dead trees and logs.

Besides being able to live off dying trees, fungi can also grow in and on people. "Mycosis" is a general term for any disease that is caused by a fungus, such as athlete's foot, or the related fungi that cause jock itch, beard itch, scalp itch, ringworm, and toenail fungus. These fungi digest some of the organic molecules of your body! And because they are spread by spores, these fungal diseases are quite contagious. The fungi can linger on moist surfaces and on clothing and towels, so they tend to be a problem in places like dormitories and fitness centers.

Fungi can also thrive in poorly ventilated spaces in buildings (FIGURE 14-31). Molds are multicellular fungi that are responsible for many unpleasant effects. People exposed to a "sick building" can experience burning or watering eyes, a runny nose, and itchy skin—allergic reactions to the proteins in fungal spores. More severe effects, such as cancers and miscarriages, are probably produced by toxins released when the fruiting bodies of the fungi disintegrate. Curing a sick building can be so expensive that in some cases the entire building is destroyed and rebuilt.

It wouldn't be fair to mention so many undesirable effects of fungi without describing some of the many fungi that are beneficial to humans, particularly one that has benefited millions of humans: the fungus that produces penicillin. Alexander Fleming was a researcher studying bacteria. In 1928, just before taking a month-long vacation, he stacked some of his used Petri dishes containing bacterial colonies

Mold contaminations can cause a variety of irritating and potentially dangerous health problems.

FIGURE 14-31 **Cleaning out toxic mold.** A building overrun with mold spores may cause its occupants to become ill.

in a corner of his lab. When he returned from his vacation, he found that some of the dishes had been contaminated by a fungus. He noticed that near the fungal infection, the bacterial colonies were dead, while farther away from it, they continued to grow. Suspecting that the fungus might be producing something toxic to the bacteria, Fleming cultured it and discovered that it produced a substance that killed many different types of bacteria. He identified the fungus as *Penicillium*. Following the development by others of a method to purify the antibacterial compound produced by the fungus, penicillin became a "wonder drug," responsible for saving millions of lives by helping to treat bacterial infections.

Many mushrooms are gastronomic delicacies; commercially grown portobello and shiitake mushrooms, for example, command high prices. Truffles—the underground reproductive structures of a fungus—can sell for as much as $3,500 per pound. Because truffles grow underground, there is nothing on the surface to indicate their presence, and truffle hunters use trained pigs or dogs to locate them by scent.

TAKE HOME MESSAGE 14.14

» Fungi are decomposers, and all they need to thrive is organic material to consume and a moist environment so their hyphae don't dry out. Fungi have complex life cycles, with both sexual and asexual phases, and the parts of a fungus that are most often visible are its temporary spore-producing bodies. Fungi can cause a variety of health problems, but also are responsible for antibacterial medicines such as penicillin.

14.15 Most plants have fungal symbionts.

If you examine the roots of a plant with a microscope, you will find round structures and fibers closely associated with the fine rootlets and root hairs. These are root fungi, or **mycorrhizae** (pronounced my-koh-RYE-zee). Some mycorrhizae have hyphae that press closely against the walls of root hair cells. Others send hyphae through the root cell walls and into the space between the cell wall and the plasma membrane (FIGURE 14-32).

Beneficial associations between roots and fungi have been found in 400-million-year-old fossils of early land plants—and nearly all species of modern plants have

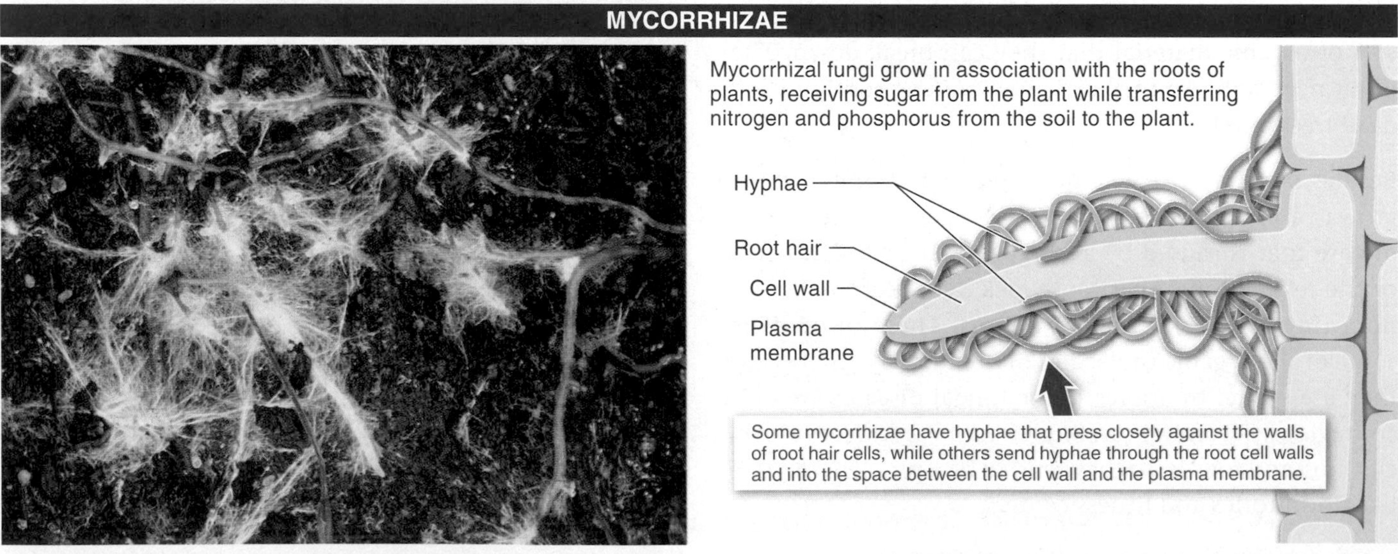

MYCORRHIZAE
Mycorrhizal fungi grow in association with the roots of plants, receiving sugar from the plant while transferring nitrogen and phosphorus from the soil to the plant.
Hyphae
Root hair
Cell wall
Plasma membrane
Some mycorrhizae have hyphae that press closely against the walls of root hair cells, while others send hyphae through the root cell walls and into the space between the cell wall and the plasma membrane.

FIGURE 14-32 **The association between mycorrhizal fungi and plants is beneficial to both.**

SYMBIOTIC FUNGI ARE BENEFICIAL TO THEIR HOSTS

Researchers investigated the benefits of root fungi in the invasive velvetleaf. They compared the survival of offspring from parents grown with and without symbiotic root fungi:

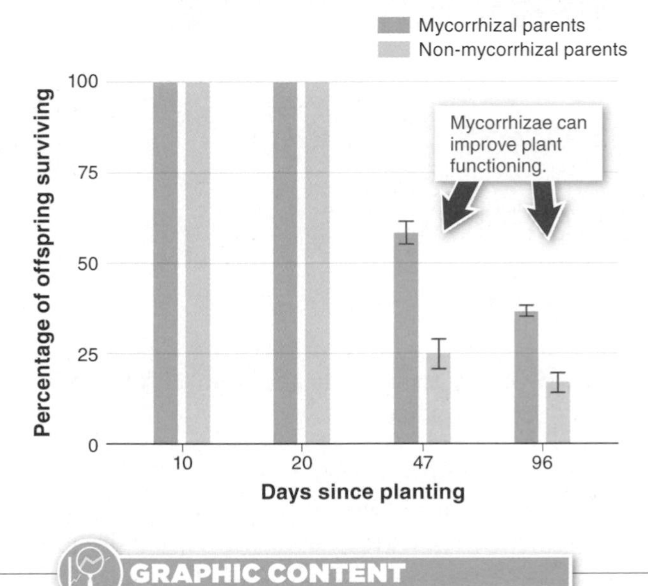

GRAPHIC CONTENT

Thinking critically about visual displays of data

Turn to p. 489 for a closer inspection of this figure.

FIGURE 14-33 **Plant partners.** Root fungi (mycorrhizae) can improve plant functioning.

them. Lumber companies have found that they have greater success when replanting clear-cut areas if the conifer seedlings are first grown in soil that has been inoculated with mycorrhizal fungi. Home gardeners can buy potting soil with mycorrhizal spores, or can buy spores compressed into tablets that can be placed in the soil when seedlings are transplanted. The association between mycorrhizal fungi and plants is beneficial to both partners. The mycorrhizal fungus benefits by drawing sugar from the plant, which it uses to support its own cellular respiration. The plant benefits as the fungus extracts phosphorus and nitrogen from the soil and releases them inside the roots, allowing it to grow faster and larger. Research on the impact of root fungi on their plant hosts even demonstrated that the *offspring* of parental plants with mycorrhizae had significantly better survival than the offspring of plants without mycorrhizae (FIGURE 14-33). (In this case, unfortunately, the fungi were enhancing the growth and survival of an invasive plant called velvetleaf, which causes hundreds of millions of dollars of damage each year to corn crops in the United States.)

Some plants, however, take nutrients from the fungus and give nothing in return. One example is the mycorrhizal parasite called ghost pipe (named for its ghostly white appearance), which grows in forests in the northern hemisphere (FIGURE 14-34). Ghost pipe is an angiosperm related to heaths and heathers. Its roots are associated with mycorrhizal fungi that are also connected to the roots of trees. The ghost pipe obtains all its nutrients from or through the mycorrhizae: nitrogen and phosphorus are provided by the mycorrhizal fungus, and sugar travels through the fungus from the tree roots to the ghost pipe.

Some plants manipulate mycorrhizae to gain a competitive advantage over other plants. Garlic mustard, for example, was brought to North America by European colonists to add a tangy flavor to salads. Initially it was planted in kitchen gardens, but it escaped into the wild, where it thrives as an invasive species. Garlic mustard excretes compounds from its roots that interfere with the partnership between mycorrhizal fungi and local native plants. Because

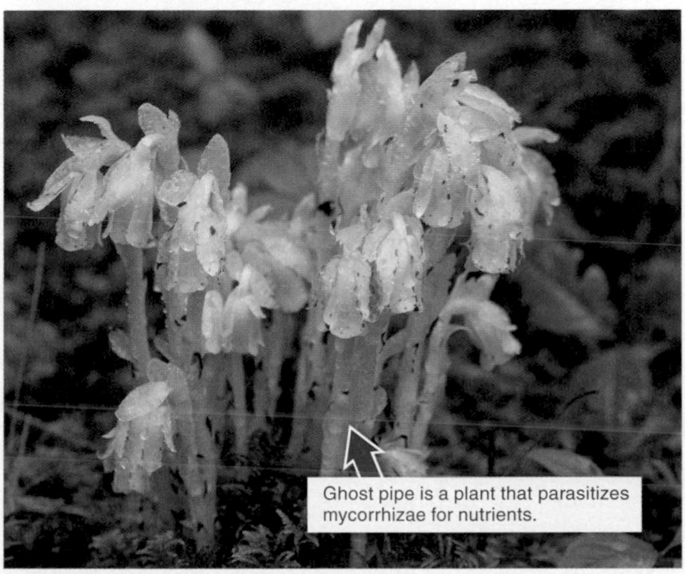

Ghost pipe is a plant that parasitizes mycorrhizae for nutrients.

FIGURE 14-34 **Not all plants play fair.**

the native plants require mycorrhizal fungi to prosper, the destruction of these fungi by garlic mustard weakens the plants. This chemical warfare is so successful that garlic mustard has wiped out entire populations of native woodland plants and is threatening many forests in the central United States.

Fungi also form mutually beneficial relationships with chlorophyll-containing bacteria and/or algae. These two-way or three-way partnerships are called **lichens,** and they grow on surfaces such as tree trunks and rocks. The fungus is fed by its photosynthetic partner and helps it absorb water and nutrients. In addition, the fungus provides nutrients for its partners by excreting enzymes that dissolve organic material and excreting acid that dissolves rock—which is important in soil formation.

TAKE HOME MESSAGE 14.15

» Plants and fungi have a close and mutually beneficial association in mycorrhizae. Mycorrhizal fungi grow in intimate association with the roots of most plants, receiving sugar from the plant and transferring nitrogen and phosphorus to the plant.

THIS IS HOW WE DO IT

Developing the ability to apply the process of science

14.16 Can beneficial fungi save our chocolate?

Chocolate is a plant product at the heart of a huge economic industry. More than 3.5 million tons of cocoa beans are produced each year, and the global chocolate market is valued at close to $100 billion per year.

Chocolate is made from the seeds of the cacao tree (*Theobroma cacao*), native to Central and South America. Twenty to 60 seeds are harvested from the large fruits, called cacao pods. The seeds are fermented, dried, roasted, cracked, and crushed, after which they may be combined with other ingredients, depending on the product.

Chocolate production is affected each year by fungal diseases. The fungi reduce the ability of trees to grow and to produce pods, and they rot the interior of the seed pods. In Peru, for example, the "frosty pod rot" fungus alone has reduced cacao production by one-third.

Pesticides that are used to fight the fungus are not very effective, and researchers are looking for other ways to reduce the impact of these fungal pathogens.

How could you approach the problem of finding ways to fight fungal pathogens?

Fungi often live within plants. In fact, *every single plant species examined* harbors fungi inside its tissues. These fungi are extremely diverse and, in many cases, do not seem to harm their plant host. For this reason, researchers are looking for species of fungi that might be able to outcompete or even harm the pathogenic fungi responsible for three of the most common fungal diseases. To get started, the cacao researchers identified three questions requiring investigation.

1. Are there fungal species with anti-pathogen activity?

2. Can those species be introduced into cacao plants (without harming the plants)?

3. Within cacao plants, do these fungi limit damage by the pathogenic fungi and improve a plant's productivity?

To investigate the first question, researchers isolated 110 different types (called *morphospecies*) of fungi that live in cacao trees without harming them. They then cultivated these fungi in Petri dishes, along with the pathogenic fungi—pairing one harmless and one pathogenic species in each dish. In each case, they evaluated which of the two interacting species survived.

Why are simple laboratory experiments a good place to start?

This approach made it possible to efficiently identify fungi with anti-pathogen activity that could serve as "biocontrol agents." In fact, the researchers discovered 53 different fungi that showed antagonism toward the major pathogenic fungi. Of these, they identified three candidates that had both anti-pathogen activity and good colonizing ability.

The next step was to attempt to deliver the helpful fungi into cacao seedlings. The researchers first germinated cacao seeds in sterile soil. They prevented exposure to airborne pathogenic fungi by growing the seedlings in plastic shade houses. Some of the seedlings were sprayed with a suspension containing spores of a potentially beneficial fungus (while other plants were not sprayed, serving as controls).

Why is it necessary to document an effective method of inoculating seedlings with the beneficial fungus?

Evaluating leaf tissue after 10, 25, and 33 days, the researchers identified fungal growth and estimated the percentage of fungal colonization. After 33 days, they found that colonization had occurred in 39%, 62%, and 92% of the cacao plants, depending on which of the three potentially beneficial fungi they had used for inoculation.

Might beneficial fungi harm pathogenic fungi in Petri dishes in the lab, but not in trees in the field? Why?

Researchers then examined whether the beneficial species would reduce the impact of the pathogenic species on cacao trees' productivity. Here, the results were a bit mixed.

In a greenhouse study, the researchers found 25% leaf mortality among trees infected with one of the pathogens. Among trees infected with both the pathogenic and the potentially beneficial fungus, leaf mortality was just under 10%, a statistically significant reduction.

Results were different in the field. Across four farms, researchers observed 320 experimental trees (inoculated with the potentially beneficial fungus) and 320 control trees (not inoculated with the potentially beneficial fungus). They found no reduction in the incidence of pod loss due to the pathogenic fungi. This may have been a consequence of an unusually low rate—less than 20%—of infection by the pathogenic fungi on these farms. The researchers noted, however, that in plants treated with the potentially beneficial fungus, there was no reduction in plant growth and there was a 10% reduction in *lesions* caused by the pathogenic fungus.

What can we conclude from these results?

Biocontrol isn't a magic bullet that wipes out the major fungal diseases of cacao trees. But researchers are optimistic, for several reasons. (1) They could identify potentially "good" fungi quickly in lab experiments; (2) they could inoculate cacao trees with "good" fungi; and (3) there seems to be at least some inhibition of pathogen activity in trees inoculated with "good" fungi.

The researchers are continuing their work, evaluating the impact of other factors on the anti-pathogen impact of the "good" fungi. These factors include (1) the timing of the spraying of seedlings to inoculate them, (2) the impact of using multiple potentially "good" fungal morphospecies simultaneously, (3) the amount of shade versus sun the trees are exposed to, and (4) the extent of pathogen infection.

TAKE HOME MESSAGE 14.16

» Pathogenic fungi significantly reduce the productivity of cacao plants—with a significantly adverse effect on the world chocolate market. Other fungi have been identified that are antagonistic to the pathogens when cultured in Petri dishes. Inoculation with these potentially "good" fungi reduces the pathogenic fungi's impact on plant growth in greenhouse experiments and shows promise when used in the field.

STREET BIO

Using evidence to guide decision making in our own lives

Yams: nature's fertility food?

Have you ever wondered how common twins are? In the United States and Europe, about 14 in every 1,000 births are twin births. In Asia, the number is slightly lower, at 8 sets of twins per 1,000 births. Elsewhere, the rates are similar—except for southwest Africa, particularly in Nigeria, which has been called "The Land of Twins." There, the rate of twin births among the Yoruba people is more than four times that in the United States, with about 50 pairs of twins per 1,000 births. (Triplet births are unusually common, too, occurring 16 times more frequently in Nigeria than in the United States.)

Q: Why are so many twins born in Nigeria? Is it something in the water? Nope. But that's not too far off. It's the diet of the Yoruba people. A staple in their diet is the white yam, a starchy vegetable that looks a bit like a potato; many people eat yams several times a day.

Q: What's in the yams? Some preliminary studies suggest that an estrogen-related compound in yams is responsible for stimulating the ovaries, increasing the likelihood, in any given month, that more than one egg will be released.

Q: How can we be sure it's the yams? At this point, the relevant data are still being collected. In one intriguing laboratory study, rats fed a diet rich in yams *doubled* their litter size. In another, the circulating levels of the follicle-stimulating hormone (which stimulates growth and maturity of follicles in the ovaries) and estradiol were both 32% higher in rats fed diets rich in yams than in rats fed a standard diet, while the levels of luteinizing hormone (which triggers ovulation) were 158% higher in the yam-fed rats.

Q: So, will it work for you? It's unclear. Anecdotes abound about women having twins after purposely increasing their yam consumption. But a randomized, controlled, double-blind study has yet to be conducted. How would you set up and analyze a study like that? (It is important to note that the yams consumed in Nigeria are not the same as American yams, which are sweet potatoes and do not contain the steroid levels of yams grown in Nigeria.)

GRAPHIC CONTENT

Thinking critically about visual displays of data

1 What can you conclude from this graph?

2 What percentage of the offspring from parents with mycorrhizal fungi were surviving after 47 days? What percentage were surviving from parents without mycorrhizal fungi?

3 There are error bars for the percentage of offspring surviving at 47 days and at 96 days after planting. Why are there no error bars at 10 days and 20 days after planting?

4 What does it mean to say that the differences in survival of offspring from mycorrhizal and non-mycorrhizal parents are "statistically significant"?

5 If you could have offspring survival data for one additional point in time, when would it be (i.e., how many days after planting)? Why?

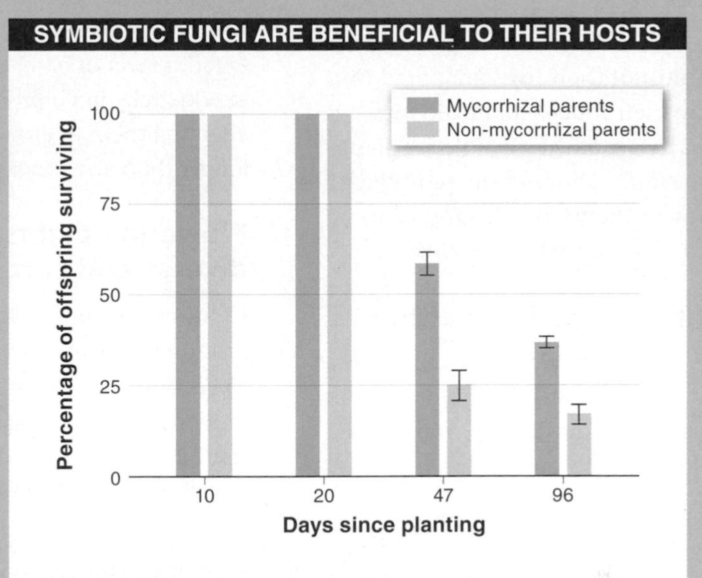

See answers at the back of the book.

KEY TERMS IN PLANT AND FUNGI DIVERSIFICATION

BRIEF SUMMARY

Plants are just one branch of the eukarya.

• Plants are multicellular organisms that spend most of their lives anchored in place by their roots. Characteristics evolved that made it possible for plants to successfully obtain food, reproduce, and protect themselves from predation on land, despite their immobility.

The first plants had neither roots nor seeds.

• The first land plants were small, had no leaves, roots, or flowers, and could grow only at the water's edge.

• Non-vascular plants, including mosses, lack roots and vessels to move water and nutrients from the soil into and throughout the plant, and they reproduce with spores.

• Vessels are an effective "circulatory system" to carry water and nutrients up from the soil to a plant's leaves. The first vascular plants, including ferns, were able to grow much taller than their non-vascular predecessors.

The advent of the seed opened new worlds to plants.

• Seeds contain a multicellular embryo plus a store of carbohydrates and other nutrients. They are distributed by wind, animals, or water.

• Gymnosperms were the earliest seed-producing plants and the dominant plants of the early and middle Mesozoic era. Gymnosperms depend on wind to carry their pollen.

• Conifers have more species and a larger geographic range than the rest of the gymnosperms combined. Rigidity, an exterior layer of bark, and the ability to exude sticky pitch protect conifers, allowing them to grow taller and live longer than any other plants.

Flowering plants are the most diverse and successful plants.

• Flowering plants appeared about 135 million years ago and diversified rapidly to become the dominant plants in the modern world. A flower houses a plant's reproductive structures. Most flowers have both male and female reproductive structures.

• Angiosperms rely on animals to carry pollen from one flower to another. Flowers advertise their presence with colors, shapes, patterns, and odors. With flowers, plants can trick or bribe animals into transporting male gametes to female gametes, where fertilization can occur.

• Angiosperms undergo double fertilization, which ensures that a plant does not invest energy in forming endosperm for an ovule that has not been fertilized.

Plants and animals have a love-hate relationship.

• Plants often enlist animals to disperse their fruits, which contain the fertilized seeds. The seeds are deposited at a new location where the seedlings can grow. Fruits are made from the ovary and, occasionally, some surrounding tissue.

• Plants have a wide range of defenses against herbivorous animals, from physical defenses such as thorns to chemicals that make the plant less palatable. Some plants living in soil that is deficient in nitrogen have switched roles, preying on insects.

Fungi and plants are partners but not close relatives.

• Fungi are eukaryotes with the same internal cellular elements as other eukaryotes—and one distinctive feature: a cell wall formed from the carbohydrate chitin.

• Fungi are decomposers, and all they need to thrive is organic material to consume and a moist environment so their hyphae don't dry out. Fungi have complex life cycles, with both sexual and asexual phases. The fungus parts most often visible are temporary spore-producing bodies.

• Fungi can cause health problems, but also are responsible for antibacterial medicines such as penicillin.

• Plants and fungi have a close and mutually beneficial association in mycorrhizae. Mycorrhizal fungi grow in intimate association with the roots of most plants, receiving sugar from the plant and transferring nitrogen and phosphorus to the plant.

CHECK YOUR KNOWLEDGE

Short Answer

1. Describe two challenges facing plants as a consequence of their lack of mobility.

2. The earliest land plants grew very close to the ground. Why?

3. How is fertilization in mosses dependent on water?

4. What distinguishes a vascular plant from a non-vascular plant? How does this difference affect their geographic distribution?

5. What source of energy do seedlings rely on for their initial growth?

6. Describe an advantage that seeds have over the spores of earlier land plants.

7. If all animals became extinct, what changes would occur in plant communities? Why?

8. Why is double fertilization considered more efficient than the type of fertilization that occurs in gymnosperms?

9. How does development of a fleshy fruit benefit plants?

10. Describe two features that distinguish fungi from other eukarya.

11. How do mycorrhizal fungi benefit from growing in association with plant roots?

Multiple Choice

1. Which of the following adaptations arose first in early plants?

a) vessels to transport water and food

b) seeds

c) roots

d) pollen

e) resistance to drying out

0 EASY 59 HARD 100

2. Which is the best brief description of the vascular system of the very first terrestrial plants?

a) The first plants developed specialized vessel cells that conducted water.

b) The first plants did not possess a vascular system.

c) The first plants had a very basic vascular system with a simple method of internal transport.

d) The first plants had only long, needle-like leaves, from which water could evaporate easily.

e) None of the above are correct.

0 EASY 56 HARD 100

3. Which of the following statements about ferns is incorrect?

a) Their sporophyte is dominant.

b) They require liquid water for fertilization.

c) Their seeds are dispersed by the wind.

d) They have vascular tissue for distributing water and nutrients throughout the plant.

e) Their spores are contained in sporangia.

0 EASY 64 HARD 100

4. In terms of their adaptation to living on land, how are reptiles similar to the seed plants?

a) Both reptiles and seed plants became completely independent of water.

b) Reptiles eat plants.

c) Seed plants and reptiles have developed structures such as cuticles and impermeable skin to minimize desiccation.

d) Reptiles and seed plants have developed structures that house their gametes and protect them from the surrounding environment.

e) Both c) and d) are correct.

0 EASY 35 HARD 100

5. Unlike higher plants such as angiosperms and gymnosperms, all bryophytes lack:

a) stomata.

b) cuticles.

c) alternation of generations.

d) water transport mechanisms.

e) roots.

0 EASY 69 HARD 100

6. Anthers and stigmas are found on:

a) bryophytes.

b) fungi.

c) angiosperms.

d) gymnosperms.

e) all of the above.

0 EASY 35 HARD 100

7. Over the evolutionary history of plants:

a) the sporophyte has become smaller, though more independent.

b) the gametophyte and sporophyte have grown increasingly independent of each other.

c) there has been a trend toward gametophyte dominance.

d) there has been a trend toward gametophyte dependence on the sporophyte.

e) the gametophyte has become larger, though more dependent.

0 EASY 73 HARD 100

8. Which of the following comparisons and contrasts between fungi and plants is incorrect?

a) Fungi cannot photosynthesize, but plants can.

b) Both fungi and plants use chitin as a structural stabilizer.

c) Fungi are heterotrophs (i.e., cannot fix carbon and so must use organic carbon for growth), but plants are not.

d) Both fungi and plants have cell walls.

e) Both fungi and plants have a sexual stage in their reproductive cycle.

0 EASY 62 HARD 100

9. Dispersal of fungal spores is typically done by:

a) movement of cilia.

b) insects.

c) hummingbirds.

d) wind.

e) movement of flagellas.

0 EASY 25 HARD 100

10. In most cases, the relationship between roots and fungi in mycorrhizae can best be described as:

a) mycelium.

b) competition.

c) trickery.

d) symbiosis.

e) parasitism.

0 EASY 39 HARD 100

Ch15

There are microbes in all three domains.

Bacteria may be the most diverse of all organisms.

In humans, bacteria can have harmful or beneficial health effects.

Archaea exploit some of the most extreme habitats.

Most protists are single-celled eukaryotes.

Viruses are at the border between living and non-living.

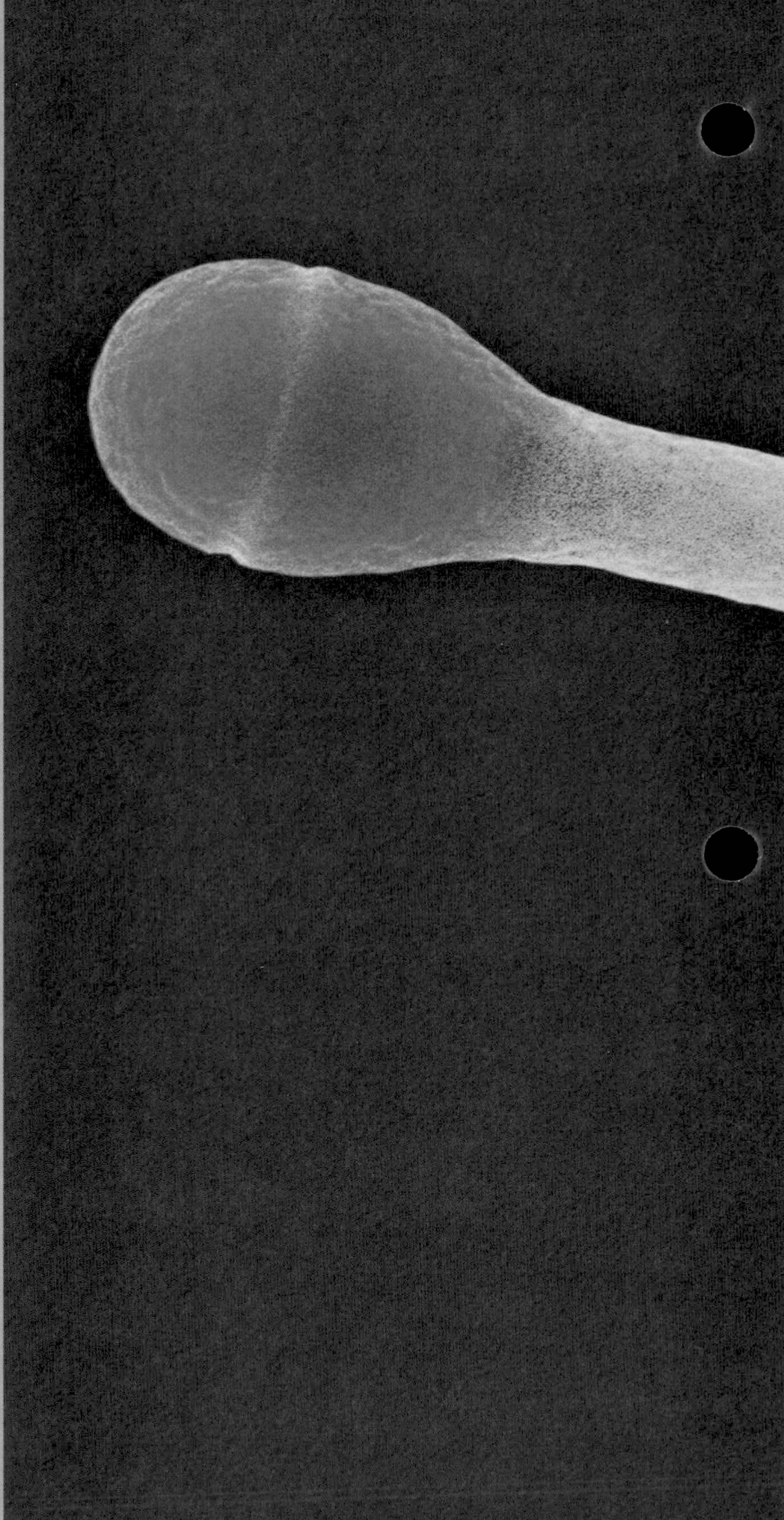

Clostridium phytofermentans is a bacterium found in forest soil. Capable of metabolizing plant cellulose and producing ethanol, it may have value in the production of biofuels.

Evolution and Diversity Among the Microbes

Bacteria, archaea, protists, and viruses: the unseen world

15.1–15.2

There are microbes in all three domains.

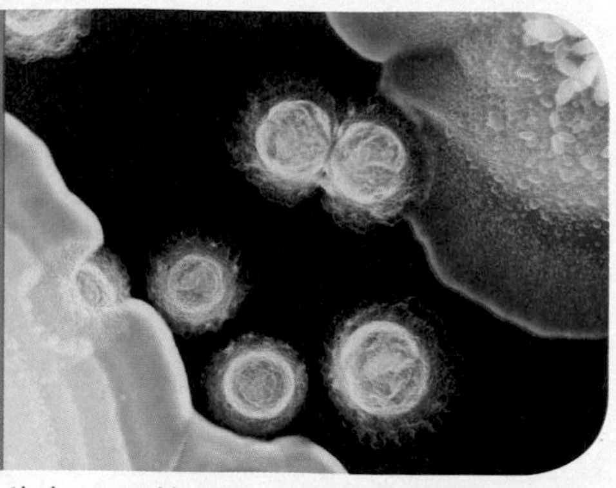

Dental plaque, a film that forms on teeth, is caused by the growth of bacterial colonies (shown here). It can generally be removed by brushing, but, if left, can lead to tooth decay.

15.1 Not all microbes are closely related evolutionarily.

Microbe is an appropriately descriptive name, but it's sloppy. The word has Greek origins and means "small life," but we could apply the label "microbe" to any one of many different types of organisms too small to see without magnification. In Chapters 4 and 12 we introduced two very different kinds of microbes: bacteria and viruses. Here we explore bacteria and viruses in more detail, as well as two other kinds of microbes: the protists and the archaea. The amoeba, a kind of protist, may be familiar to you, but the archaea are probably unfamiliar to you. That's not surprising, because archaea were only recently recognized as a distinct group of microbes. The group doesn't even have a common name (unless you like "the group formerly known as archaebacteria"). We'll just call them archaea.

It is hard to generalize about microbes. They are grouped together simply because they are small, not because they all share a recent common ancestor. In fact, microbes occur in all three domains of life—bacteria, archaea, and eukarya—and so the various types of microbes could not be

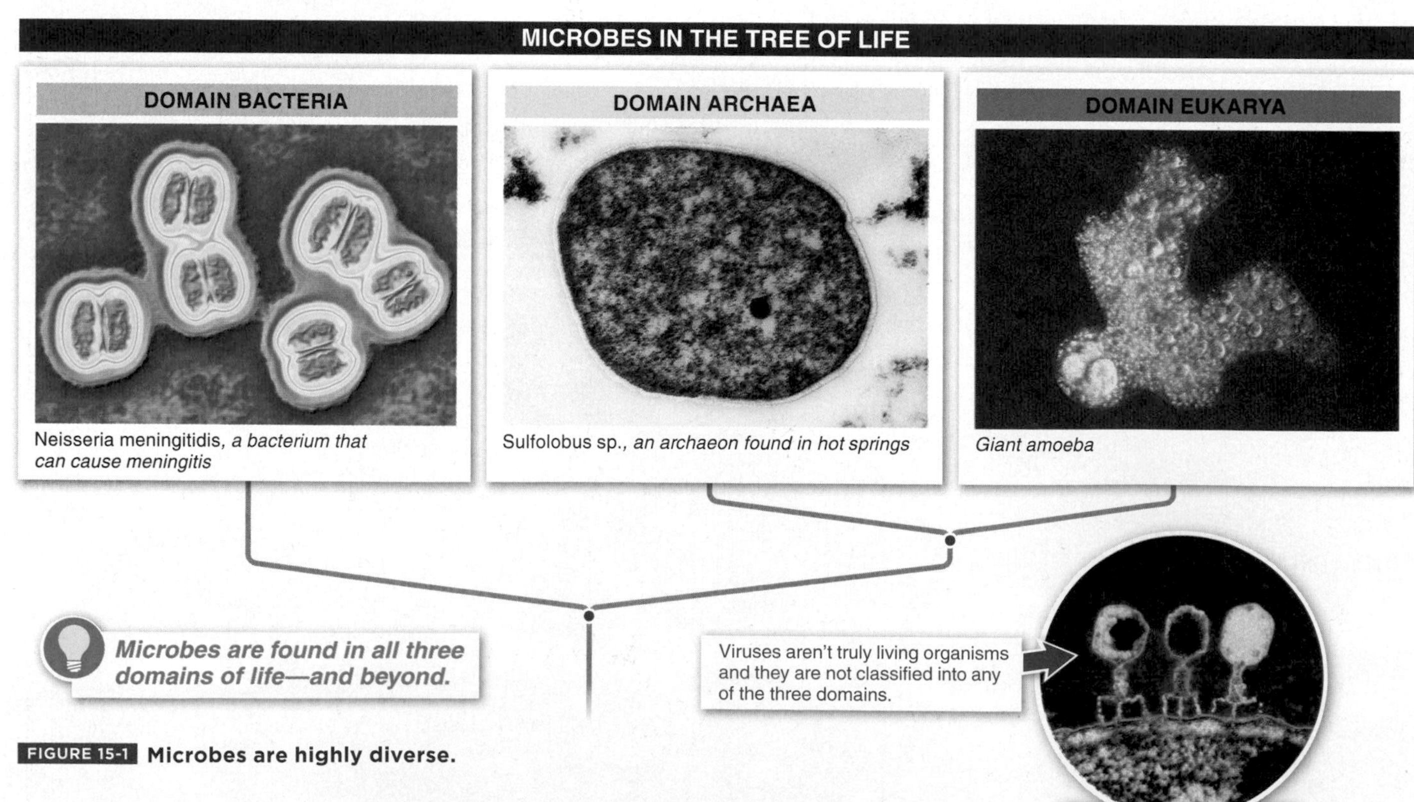
MICROBES IN THE TREE OF LIFE
DOMAIN BACTERIA
DOMAIN ARCHAEA
DOMAIN EUKARYA
Neisseria meningitidis, *a bacterium that can cause meningitis*
Sulfolobus sp., *an archaeon found in hot springs*
Giant amoeba
Microbes are found in all three domains of life—and beyond.
Viruses aren't truly living organisms and they are not classified into any of the three domains.
Bacteriophage virus

FIGURE 15-1 **Microbes are highly diverse.**

more widely separated (FIGURE 15-1). In this chapter we focus on the tiny organisms from each of these domains. The microbes in the domains bacteria and archaea are prokaryotic, although archaea have some characteristics like those of bacteria and some like those of eukaryotes. Protists are the mostly microbial members of the domain eukarya. And viruses, another type of microbe, are not classified into any domain at all because they are only at the borderline of life.

TAKE HOME MESSAGE 15.1

» Microbes are grouped together only because they are very small, not because of evolutionary relatedness. They occur in all three domains of life and also include the viruses, which are not classified in any of the domains.

15.2 Microbes are the simplest but most successful organisms on earth.

Humans are large organisms, and our size comes with some "baggage." We need a skeletal system to support our weight and a respiratory system to take in oxygen and get rid of carbon dioxide. We need a circulatory system to move oxygen, carbon dioxide, and other molecules around our bodies, and a digestive system to take in food and break it down. We even need a nervous system so that our brain knows what distant parts of our body are doing.

Q How can a microbe function when its body is just a single cell?

Microbes—the most abundant organisms on earth—are too small to need these types of systems. The volume of an amoeba, for example, is about a million billion (10^{15}) times smaller than that of a human. At that size, the force of gravity is trivial, so the amoeba needs no skeleton to support it. Plenty of oxygen diffuses inward across its cell membrane, and carbon dioxide diffuses outward, so an amoeba does not need a respiratory system. And because every part of its interior is close to the body surface, it doesn't need a circulatory system to transport gases. An amoeba doesn't need a digestive system, either: it eats by enclosing food items in a piece of its cell membrane and digests the food with the same enzymes it uses to recycle its own proteins, lipids, and carbohydrates. And no part of an amoeba is far enough from any other part to require a specialized nervous system for communication.

Most microbes are even smaller than an amoeba: a typical bacterium or archaeon is about one thousand million billion (10^{18}) times smaller than a human, and an influenza virus is about one thousand billion trillion (10^{24}) times smaller than you are (FIGURE 15-2). Although they may be invisible, microbes are highly successful in almost every imaginable way.

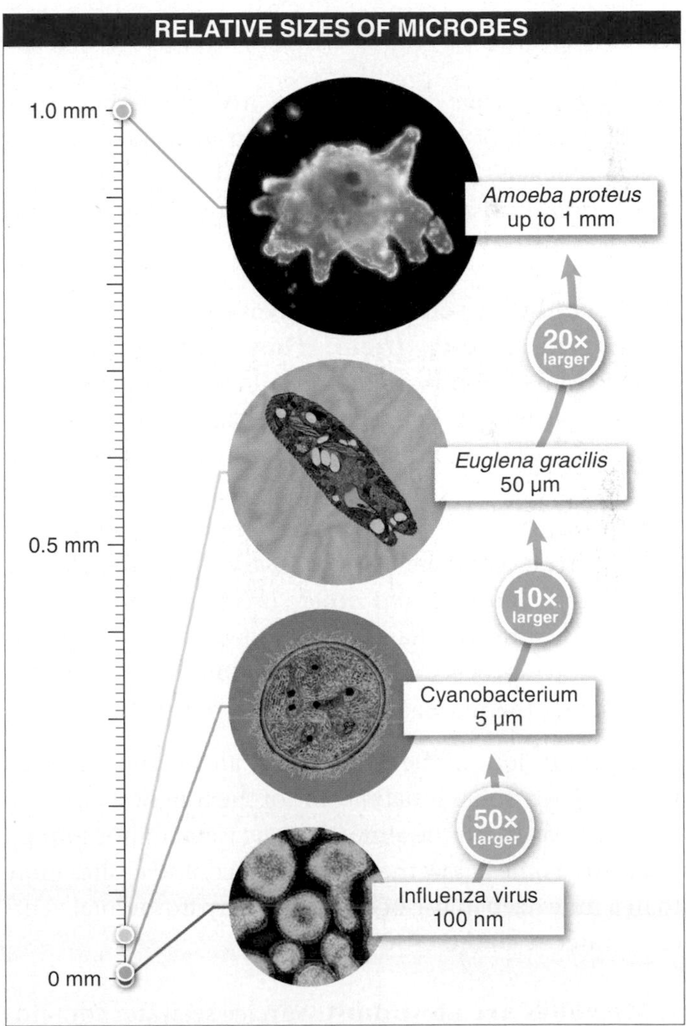

FIGURE 15-2 The most abundant organisms on earth are too small to see.

MICROBES CAN LIVE ALMOST ANYWHERE

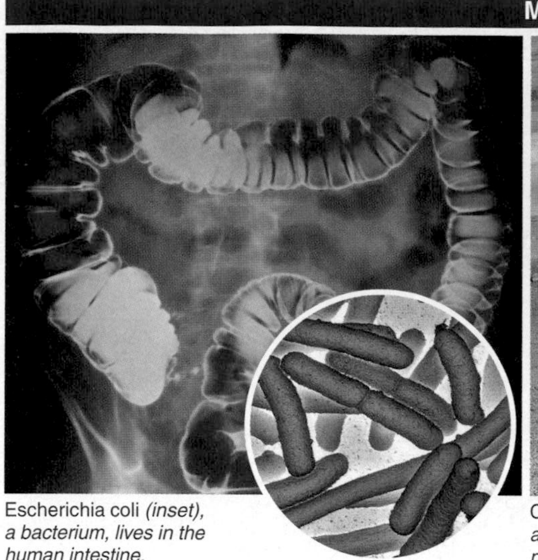

Escherichia coli *(inset), a bacterium, lives in the human intestine.*

Chlorobium tepidum *(inset), a bacterium, lives in geothermal pools that can reach temperatures greater than 160° F (71° C).*

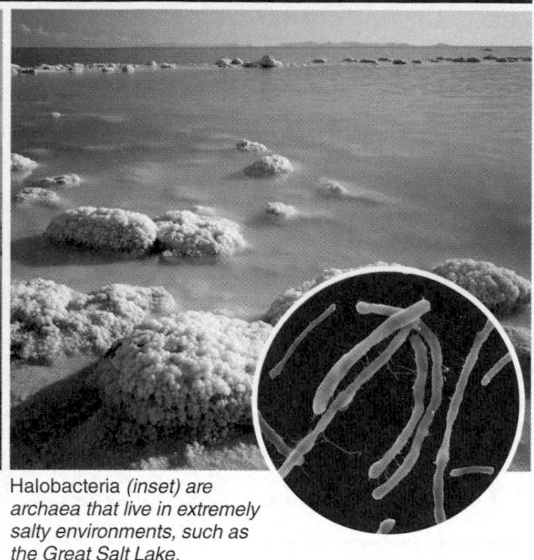

Halobacteria *(inset) are archaea that live in extremely salty environments, such as the Great Salt Lake.*

FIGURE 15-3 **Microbes are everywhere on earth.**

Microbes live in nearly every kind of environment, including water at temperatures of up to 750° F and as low as 5° F!

1. Microbes are genetically diverse. More than 500,000 kinds of microbes have been identified by their unique nucleotide sequences, and further studies will almost certainly distinguish millions of additional microbial species.

2. Microbial species live in almost every habitat on earth; among them, they can eat almost anything. Right now, more than 400 species of microbes are thriving in your intestinal tract (FIGURE 15-3), 500 more species thrive in your mouth, and nearly 200 species call your skin home. The microbes that live in and on you eat mostly what you eat—some of the bacteria in your mouth and intestine compete with you, trying to digest your food before you can, and others use the waste products you release after you have broken down the food. Others feed on the leftovers released by the breakdown of your cells during the normal process of cell renewal.

Living conditions in the human body are relatively moderate. Other microbes inhabit some of the toughest environments on earth—in the almost boiling water of hot springs, at depths a mile below the earth's surface, and at sites more than a mile deep in the oceans, where hydrothermal vents emit water at 400° C (750° F).

3. Microbes are abundant. Surface seawater contains more than 100,000 bacterial cells per milliliter, and diatoms (protists in the eukarya domain) are as abundant

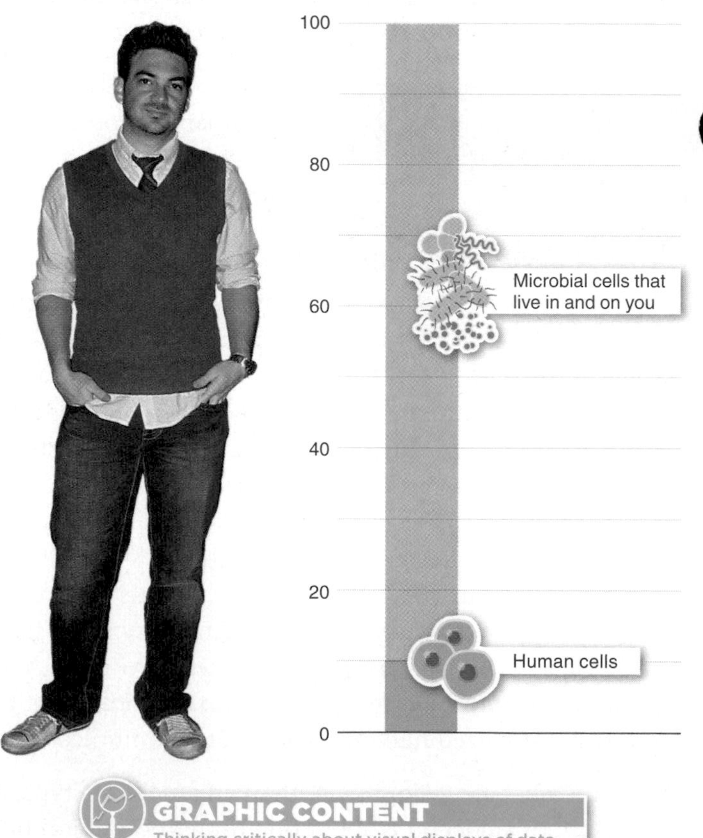

GRAPHIC CONTENT
Thinking critically about visual displays of data
Turn to p. 523 for a closer inspection of this figure.

FIGURE 15-4 **Microbe majority.** Your body has more microbial cells than human cells.

there as are the bacteria. These densities translate to about 8,000 million billion trillion (8×10^{30}) individuals of just these two kinds of microbes in the world's oceans. Your own body contains about 100 trillion cells, but only one-tenth of those cells are human cells—the remaining 90 trillion cells are the microbes that live in and on you (FIGURE 15-4). You're a minority in your own body.

TAKE HOME MESSAGE 15.2

» Microbes are very small, simple organisms, but they are capable of the same life processes as larger, multicellular organisms. They can live anywhere, from moderate to extreme environments. There are millions of different kinds of microbes on earth, in enormous numbers.

15.3–15.5

Bacteria may be the most diverse of all organisms.

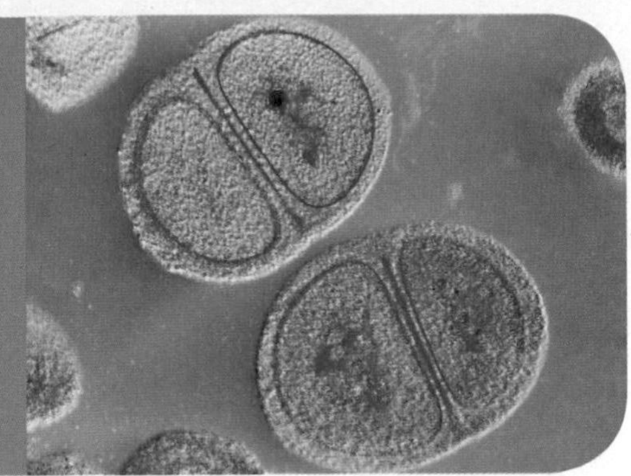

The bacterium *Staphylococcus epidermidis* is part of the normal flora on human skin and many mucous membranes.

15.3 What are bacteria?

A bacterium has a simple structure. It has a cell envelope, consisting of a plasma membrane and, usually, a cell wall, that maintains conditions inside the cell that are different from conditions outside. The cell envelope surrounds a cytoplasm—the substance that fills all kinds of cells (including your own). Because bacteria are prokaryotes, they have no organelles. Proteins in the cytoplasm carry out essential functions such as digesting molecules of food and transferring the energy gained to ATP. DNA in the cytoplasm carries the instructions for making those proteins, and messenger RNAs carry this information to ribosomes, where the proteins are synthesized. That's all a bacterium needs.

Bacteria may be classified by their shape: some are spherical cells (known as the cocci), some are rod-shaped (the bacilli), and others are spiral-shaped (the spirilla) (FIGURE 15-5). Bacteria usually reproduce by binary fission. As bacterial cells divide, the number of cells doubles every generation, producing a colony of cells—each of which is a clone of the original cell. Within a few hours, a single cell can form

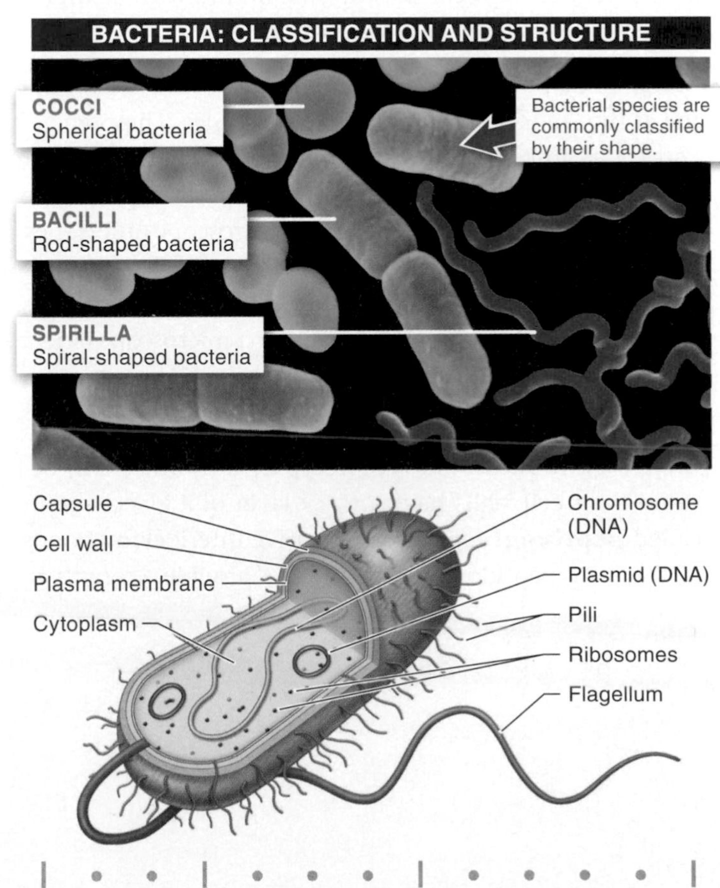

FIGURE 15-5 **Bacteria basics.** Bacteria are single-celled organisms that lack a nucleus.

METHODS OF IDENTIFYING BACTERIA

APPEARANCE
Some bacteria can be identified by looking at the colors and shapes of their colonies.

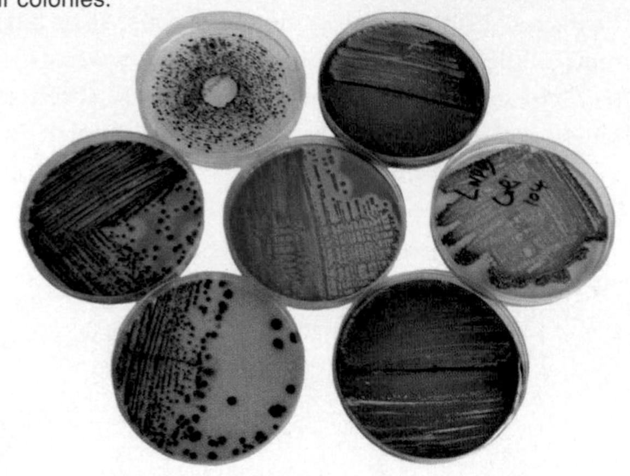

GRAM STAINING

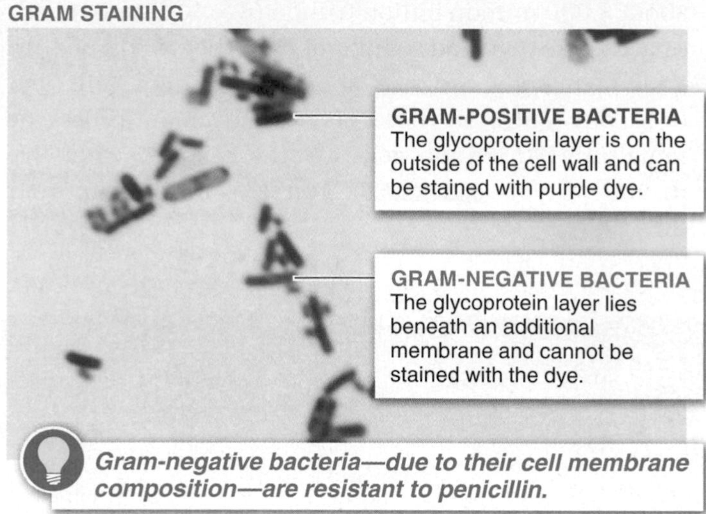

GRAM-POSITIVE BACTERIA
The glycoprotein layer is on the outside of the cell wall and can be stained with purple dye.
GRAM-NEGATIVE BACTERIA
The glycoprotein layer lies beneath an additional membrane and cannot be stained with the dye.

Gram-negative bacteria—due to their cell membrane composition—are resistant to penicillin.

FIGURE 15-6 **Bacterial IDs.** Bacteria are often identified by the appearance of an entire colony or by the cells' response to Gram staining.

a culture containing thousands of cells. Colonies of different species of bacteria look different. The familiar human intestinal bacterium *Escherichia coli,* for example, forms beige or gray colonies that have smooth margins and a shiny, mucus-like covering. Species of *Proteus,* which are often responsible for spoiling food because they can grow at refrigerator temperatures, form colonies with a surface that looks like a contour map.

Microbiologists can often identify bacteria simply by looking at the colors and shapes of their colonies. They can get additional information by examining a single cell under a microscope, but living bacterial cells are transparent, so you can't see them with an ordinary microscope unless they have been dyed. In 1884, Hans Christian Gram, a Danish microbiologist, described a method of staining the cell walls of bacteria to make them visible under a microscope, and a **Gram stain** is still the first test microbiologists use when identifying an unknown bacterium (FIGURE 15-6).

Gram-positive bacteria are colored purple by the stain, because their cell walls have a thick layer of a glycoprotein called **peptidoglycan.** The extensive interlocking bonds of the long peptidoglycan molecules provide strength to the cell wall. In *Gram-negative bacteria,* the layer of peptidoglycan is thinner and lies beneath an additional membrane, and these cells are not stained by the dye. Penicillin is effective in treating infections by Gram-positive bacteria because it interferes with the formation of peptidoglycan cross-links. Penicillin is less effective on Gram-negative bacteria because it does not pass through the outer membrane that covers the peptidoglycan layer.

Many bacteria also have a **capsule** that lies outside the cell wall. This capsule can restrict the movement of water out of the cell and thus allow bacteria to live in dry places, such as on the surface of your skin. For some bacteria, the capsule is important in allowing them to bind to solid surfaces such as rocks or to attach to human cells.

TAKE HOME MESSAGE 15.3

» Bacteria are single-celled organisms, with an envelope surrounding a cytoplasm that contains the DNA. (Bacteria have no nucleus and no organelles.) A single bacterial cell can grow into a colony of cells.

15.4 Bacterial reproduction and growth are fast and efficient.

A bacterium is a rapid reproducer; most bacteria have generation times of between 1 and 3 hours, and some are even shorter. *Escherichia coli,* for example, has a generation time of 20 minutes under optimal conditions, so a single *E. coli* cell could give rise to a population of 20 billion cells in less than 12 hours.

Bacteria carry genetic information in two structures: the chromosome and plasmids. The chromosome, a circular DNA molecule, carries the genes that provide instructions for the cell's basic life processes. Most bacteria have just one chromosome, but some have more than one. A bacterial chromosome is organized more efficiently than a eukaryotic chromosome, in two ways. First, in bacteria, the genes that code for proteins with related functions—such as enzymes that play a role in a pathway that breaks down food for energy—are often situated next to one another on the chromosome. This organization allows transcription of all the genes together. Second, almost all the DNA in a bacterial chromosome codes for proteins, so bacteria do not use time and energy transcribing mRNA that will not be translated. As we saw in Chapter 6, as much as 90% of the DNA in eukaryotic chromosomes does not code for proteins, and after transcription, this is edited out of the mRNA before it is translated.

Plasmids are a second type of information-carrying structure in bacteria. These circular DNA molecules carry genes for specific functions. For example, *metabolic plasmids* carry genes enabling bacteria to break down specific substances, such as toxic chemicals; *resistance plasmids* carry genes enabling bacteria to resist the effects of antibiotics; and *virulence plasmids* carry genes that control how sick an infectious bacterium makes its host. Many bacteria have one or more (sometimes more than a hundred) plasmids. The strain of *E. coli* that has sickened patrons of some fast-food restaurants carries a virulence plasmid that magnifies the effects of a gene for a sometimes-lethal toxin. (*E. coli* strains without this virulence plasmid are normal components of the bacterial community of the human intestine.)

When a bacterium divides, it creates two new daughter cells, each "offspring" carrying the genetic information that was present in the chromosome of the mother cell. Thus, binary fission transmits genetic information from one bacterial generation to the next (FIGURE 15-7). However, bacteria can also transfer genetic information laterally—to other individuals within the same generation—through any

Q What would be the benefit of being able to transfer genetic information directly from one adult human to another?

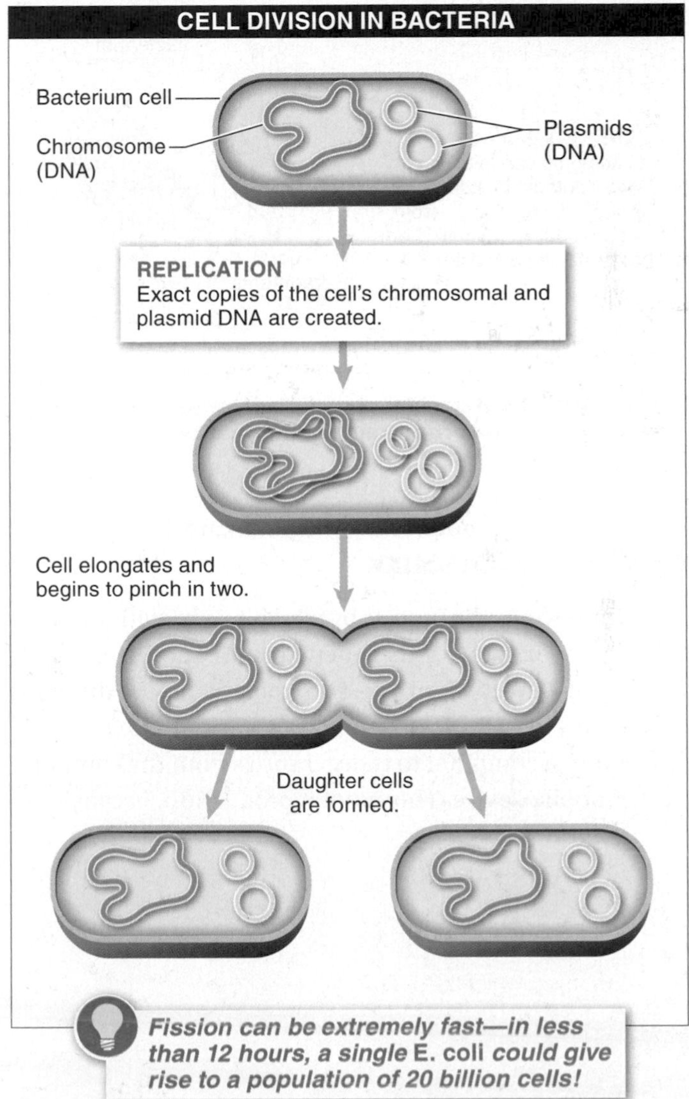

CELL DIVISION IN BACTERIA
Bacterium cell
Chromosome (DNA)
Plasmids (DNA)
REPLICATION
Exact copies of the cell's chromosomal and plasmid DNA are created.
Cell elongates and begins to pinch in two.
Daughter cells are formed.
Fission can be extremely fast—in less than 12 hours, a single E. coli could give rise to a population of 20 billion cells!

FIGURE 15-7 **Binary fission.** This asexual cell division method is used by prokaryotes.

METHODS OF GENETIC EXCHANGE IN BACTERIA WITHIN THE SAME GENERATION

CONJUGATION
A bacterium transfers a copy of some or all of its DNA to another bacterium—genetic information the recipient may not have had.

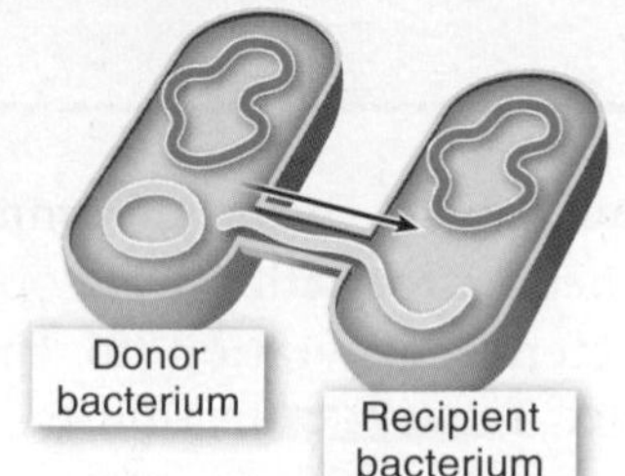

TRANSDUCTION
A virus containing pieces of bacterial DNA inadvertently picked up from its previous host infects a new bacterium—passing on genetic information the recipient may not have had.

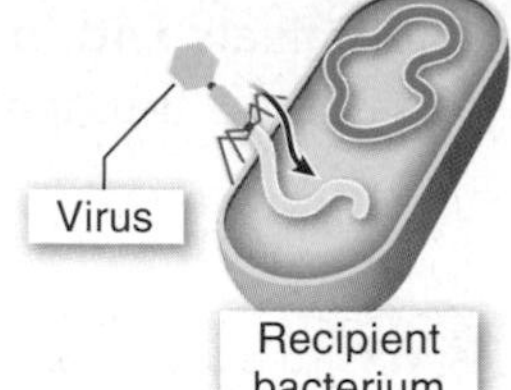

TRANSFORMATION
A bacterium can take up DNA—potentially including alleles it did not carry—from its surroundings (usually from bacteria that have died).

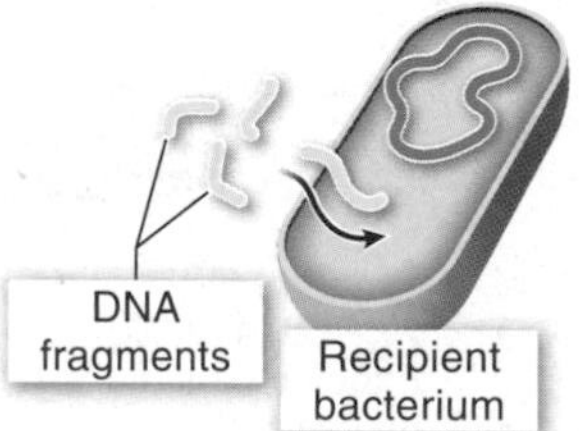

FIGURE 15-8 Lateral transfer of genetic information: conjugation, transduction, and transformation.

of three different processes: conjugation, transduction, and transformation (FIGURE 15-8).

Conjugation is the process by which one bacterium transfers a copy of some of its genetic information to another bacterium—even when the two bacteria are of different species. It's very much like plugging your phone or music player into a computer to transfer songs from the computer to the mobile device. This is not reproduction, because you start with one iPod and one computer, and that's what you have when you finish. But the music player now contains songs that it did not have before. Similarly, plasmid transfer gives the recipient bacterium genetic information that it did not have before. These genes could enable the bacterium to make an enzyme that allows it to metabolize a new chemical or to defend itself against a new antibiotic.

Transduction occurs when a kind of virus called a bacteriophage (one type is shown in Figure 15-1) infects a bacterial cell. The virus reproduces inside the bacterial cell, and sometimes, inadvertently, the new virus particles contain pieces of the bacterium's DNA in addition to or instead of the viral DNA. When these viruses are released and infect new bacterial cells, the bacterial DNA that they carry can be inserted into the host bacterium's chromosome, passing new bacterial genes to that bacterium.

Transformation is the process by which bacterial cells scavenge DNA from their environment. This DNA comes from other bacterial cells that have burst open, releasing their cellular contents. The released chromosomes break into short lengths of DNA, which can then be taken up by living bacterial cells and inserted into their own chromosomes, potentially adding genes they did not originally have.

TAKE HOME MESSAGE 15.4

» Bacteria undergo binary fission and grow rapidly in number to form colonies. Bacterial genes are efficiently organized in groups with related functions. Virtually all the DNA codes for proteins. Bacteria sometimes carry genes for specialized traits on plasmids (small DNA molecules), which can be transferred from one bacterial cell to another by conjugation. DNA can also be transferred laterally between bacterial cells by transduction or transformation.

15.5 Metabolic diversity among the bacteria is extreme.

One important attribute that makes bacterial diversity possible is that bacteria can metabolize almost anything. (Not all bacteria can metabolize everything. Rather, there is a huge variety of bacteria, each type with its own set of metabolic specializations.) Some can even use energy from light to make their own food, just as plants do. Microbiologists place bacteria into different "trophic" (feeding) categories that reflect their metabolic specialization.

Chemical organic feeders (**chemoorganotrophs**) are bacteria that consume organic molecules, such as carbohydrates. You probably see the products of organic feeders every time you take a shower—they are responsible for the pink deposits on the shower curtain and other discolorations on shower tiles (FIGURE 15-9). Most of the bacteria that live in and on your body are also organic feeders. Some compete with you to metabolize the food you eat. Others digest things you can't digest.

Chemical inorganic feeders (**chemolithotrophs,** meaning "rock feeders") can use inorganic molecules such as ammonia, hydrogen sulfide, hydrogen, and iron. The most common inorganic feeders are the iron bacteria responsible for the brown stains that form on plumbing fixtures where tap water contains high levels of iron. Sulfur bacteria are associated with iron bacteria and leave the slimy black deposits that you may find if you lift the stopper out of the drain in your bathroom sink.

On a larger scale, inorganic feeders are responsible for the acidic drainage that is a by-product of mining. Once ore is extracted from rock, the piled-up remainders are called "tailings." This material is often rich in minerals such as pyrites (iron sulfides). Inorganic feeders can gain energy by oxidizing these minerals, and in the process they release compounds that combine with rainwater to produce strong acids, such as sulfuric acid. When this acidic water drains into streams, it can kill fish and aquatic plants and insects.

Bacteria that use the energy from sunlight (**photoautotrophs,** or "light self-feeders") contain chlorophyll and use light energy to convert carbon dioxide to glucose by photosynthesis. The floating mats of gooey green material growing on ponds or in ditches are a type of photoautotroph called cyanobacteria.

The cyanobacteria living today closely resemble the first photosynthetic organisms that appeared on earth about

METABOLIC DIVERSITY AMONG BACTERIA

CHEMOORGANOTROPHS
Feed on organic molecules

CHEMOLITHOTROPHS
Feed on inorganic molecules (shown here: rusticles—created as bacteria break down iron—on the wreck of the *Titanic*)

PHOTOAUTOTROPHS
Use energy from sunlight to produce glucose via photosynthesis

FIGURE 15-9 **A broader palate than yours.** Bacteria can metabolize almost anything.

Bacteria are resourceful and can extract food from a huge range of sources: the sun, inorganic molecules in your drainage pipes, and even your shower curtain/door!

2.6 billion years ago. These organisms could use solar energy to build organic compounds from carbon dioxide, and in the process they broke down water molecules to release free oxygen. Before cyanobacteria, the earth's atmosphere consisted almost entirely of nitrogen and carbon dioxide. The accumulation of oxygen released by cyanobacteria is called the **Oxygen Revolution.** Oxygen—which, like all animals, humans depend on—now makes up about 21% of the volume of air, and cyanobacteria still release important quantities of oxygen into the atmosphere.

One of the common ways that bacteria are classified is as aerobic or anaerobic, depending on whether they require or do not require oxygen for growth—although some bacteria, called facultative anaerobes, utilize oxygen if it is present but can also switch to anaerobic respiration when oxygen is absent.

TAKE HOME MESSAGE 15.5

» Some bacteria eat organic molecules, some eat minerals, and still others carry out photosynthesis. About 2.6 billion years ago, photosynthesizing bacteria were responsible for the first appearance of free oxygen in the earth's atmosphere.

15.6–15.10

In humans, bacteria can have harmful or beneficial health effects.

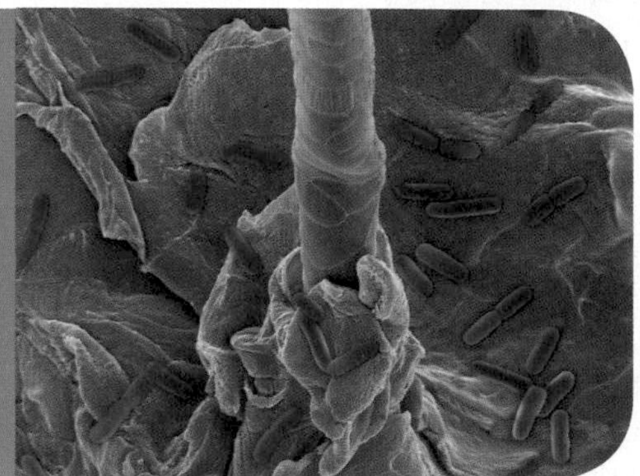

Escherichia coli (E. coli) bacteria (pink) on the surface of human skin and hair follicle.

15.6 Many bacteria are beneficial to humans.

Do you like yogurt? You can thank bacteria for it—*Lactobacillus acidophilus* and several other species of bacteria are added to milk to create yogurt. As the bacterial cells use the lactose (milk sugar) for energy, the by-product is lactic acid, which reacts with the milk proteins to produce the characteristic taste and texture of yogurt. If you eat yogurt that is labeled as containing "live cultures," you are consuming living bacterial cells that may take up residence in your digestive tract and improve your extraction of nutrients from food (FIGURE 15-10). Bacteria are also used to produce many other foods, such as cheeses, and they (along with yeasts) are used in the production of beer, wine, and vinegar. Industrial microbiology is a multi-billion-dollar industry.

Hundreds of species of bacteria grow in and on your body; these microbes are called your "normal flora." They take up every spot on your body that a disease-causing bacterium could adhere to, and they consume every potential source of nutrition, making it difficult for a disease-causing bacterium to gain a foothold. Thus, maintaining a robust population of these benign bacteria is your first line of defense against infection by harmful bacteria.

Probiotic therapy is a method of treating infections by introducing benign bacteria in numbers large enough to swamp the harmful forms. *Lactobacillus acidophilus,* which is a normal inhabitant of the human body, is used to treat gastrointestinal upsets, such as traveler's diarrhea, and

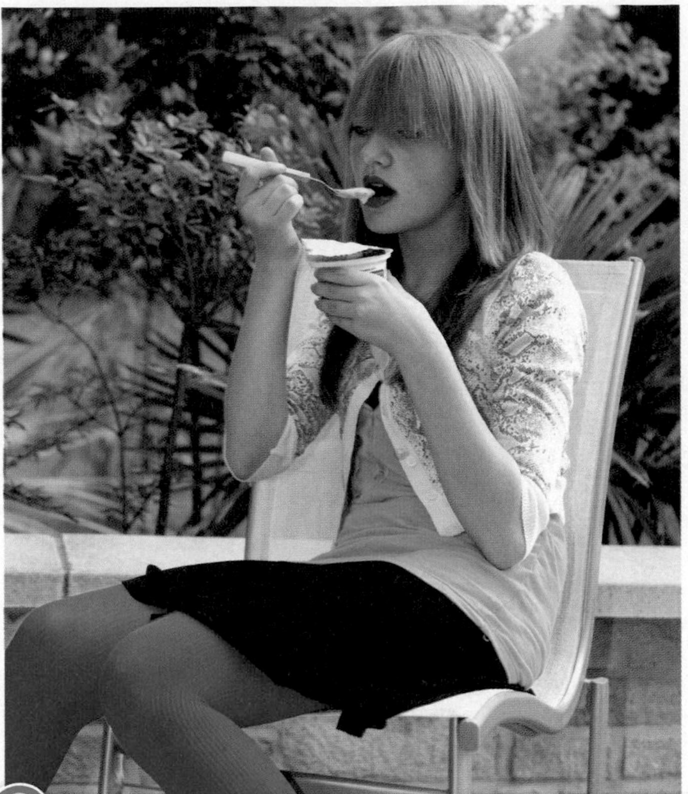

Many bacteria are beneficial. Those living in yogurt, for example, can take up residence in your digestive tract and improve your extraction of nutrients from food.

FIGURE 15-10 **Yogurt contains beneficial bacteria.**

urinary tract infections. In addition to replicating so vigorously that it crowds out harmful bacteria, *L. acidophilus* releases lactic acid, which interferes with the growth of other bacteria and prevents them from adhering to the walls of the urinary tract and bladder.

TAKE HOME MESSAGE 15.6

» A disease-causing bacterium must colonize your body before it can make you sick, and your body is already covered with harmless bacteria. If the population of harmless bacteria is dense enough, it will prevent invading bacteria from gaining a foothold.

THIS IS HOW WE DO IT

Developing the ability to apply the process of science

15.7 Are bacteria thriving in our offices, on our desks?

Bacteria can live in a huge variety of habitats. As humans create new, "artificial" environments, do bacteria thrive in these habitats as well? Some researchers wondered, in particular, about the extent to which bacteria might be thriving in office buildings, with shared bathrooms, meeting rooms, and other common areas.

To evaluate the abundance of office bacteria, where could you take samples?

Because they wanted to draw some general conclusions, the researchers decided to collect samples in three different cities. They chose

(continued on the following page)

Tucson, Arizona (where one of the researchers lived), New York, and San Francisco. And to avoid being overly influenced by any one office—which might be unusually hospitable (or inhospitable) to bacteria—they selected 30 office buildings in each city. They noted the gender of the occupant of each office where they collected a sample.

The samples were collected by wiping a cotton swab across a small (13 cm^2) area of the desk. Each swab was stored in a sterile tube on ice and sent to the laboratory in Arizona. In the lab, the researchers swirled the swab in a sterile saline (saltwater) solution, and then rubbed it across the surface of a culture medium on a small Petri plate. The culture medium was a gel containing nutrients in which bacteria could grow. They then kept the plate at 30° C (86° F) for five days, allowing any bacteria present to grow.

How could the researchers evaluate the abundance of bacteria in each office?

The researchers simply counted the number of colonies growing on each plate. To evaluate any large-scale patterns, they calculated the average number of colonies that grew on the 30 plates from each city. They also compared the average number of colonies from women's desks and men's desks.

What did they find?

The researchers plotted the average number of colonies on the plates from each city:

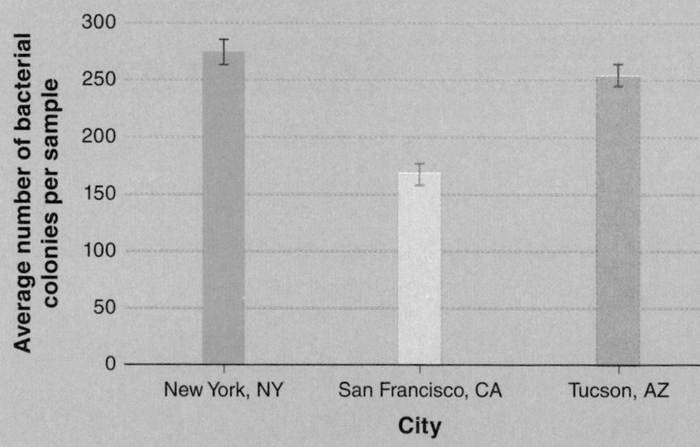

They found a significant difference among the cities: San Francisco had the lowest number of colonies per sample (about 170), and New York had the highest (more than 275).

The researchers found an even more striking difference when they compared men's and women's desks:

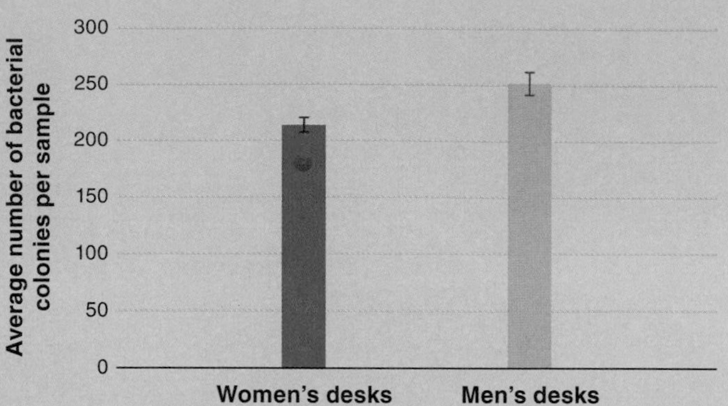

The samples from men's desks contained significantly more bacterial colonies (about 250, on average) than those from women's desks (about 215).

What can we conclude from these results?

When these research results were first published in 2012, many newspapers proclaimed that "men are grubbier than women." This conclusion seemed to confirm previously published research reporting that men wash their hands and brush their teeth less frequently than women. But the researchers were more conservative in the conclusions they drew.

Why might these results reflect something other than male hygiene issues?

The researchers pointed out that men are bigger than women, on average. Men have more skin surface and larger nasal and oral cavities: more space for bacteria to colonize. And that, in turn, may lead to more bacteria being shed onto their desk surfaces.

What interesting or useful observations could you make from this experimental approach?

This study also looked at the variety of bacteria in each office. They found a broader diversity of bacterial types in Tucson than in the other two cities, but no difference in the diversity of bacteria living on the desks of men and women

Are there any practical consequences of these results?

Most of the bacteria the researchers identified typically live on or in humans and are not harmful. The presence or absence of these bacteria probably has no health consequences.

Although the Clorox Corporation provided some of the funding for this study, the researchers indicated that it had no role in the study design or analysis. Perhaps these methods will be used in the future to evaluate the effectiveness of various cleaning products or regimens.

TAKE HOME MESSAGE 15.7

» The abundance of bacteria in human habitats can be evaluated and compared, using simple swabbing and culturing methods. In office buildings, bacteria are extremely common, but with significant differences in abundance depending on the location of the office and the gender of its occupant.

15.8 Bacteria cause many human diseases.

The number of **pathogenic** (disease-causing) bacteria is very small compared with the total number of bacterial species, but some pathogens kill millions of people annually, despite advances in medicine and sanitation. Some bacteria are always pathogenic, such as those that cause cholera, plague, and tuberculosis. Others, however, such as the ones responsible for acne, strep throat, scarlet fever, and "flesh-eating" necrotizing fasciitis, are normal parts of the communities of bacteria that live in or on humans. These bacteria in our "normal flora" become pathogenic only under special circumstances.

The cholera epidemic that devastated London in 1854 became a milestone in epidemiology (the study of the occurrence of disease outbreaks), when Dr. John Snow, by mapping the pattern of deaths, identified the Broad Street pump as the source of infection (FIGURE 15-11). When Dr. Snow persuaded the authorities to shut down the pump by removing the handle, new cases of cholera in the area dropped sharply.

Cholera epidemics are a continuing threat, especially where sanitation is compromised during war or following a natural disaster. The strains of cholera found in these areas are especially potent; as the bacteria multiply inside their hosts, they cause severe diarrhea with a massive loss of water. Victims of these severe strains of cholera are incapacitated and soon die, but before they die they release

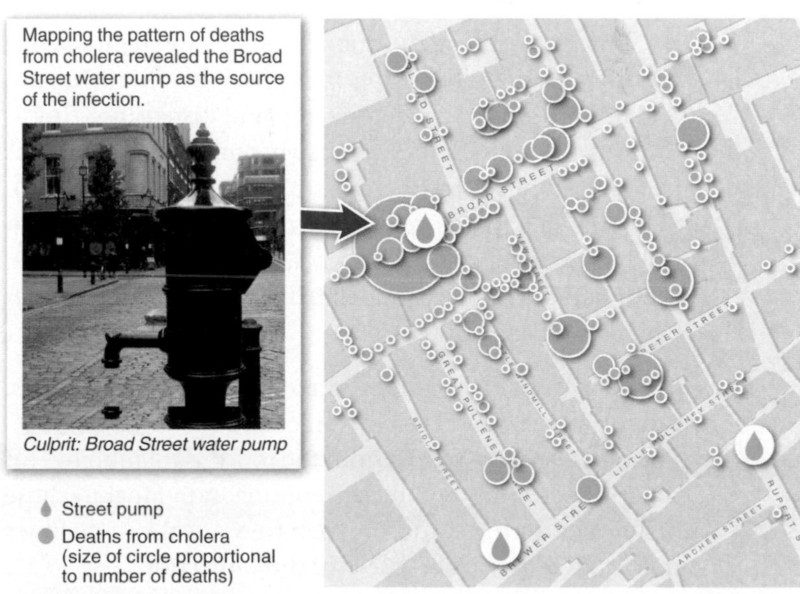

Mapping the pattern of deaths from cholera revealed the Broad Street water pump as the source of the infection.
Culprit: Broad Street water pump
Street pump
Deaths from cholera (size of circle proportional to number of deaths)
BROAD STREET

FIGURE 15-11 **Contaminated!**

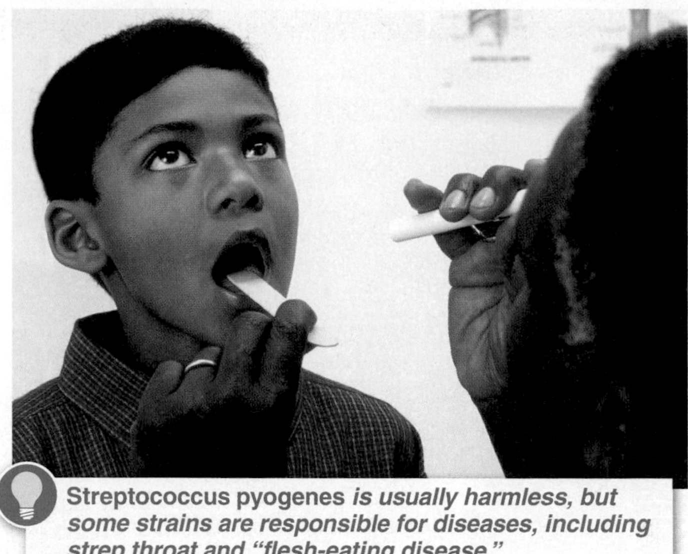

Streptococcus pyogenes is usually harmless, but some strains are responsible for diseases, including strep throat and "flesh-eating disease."

FIGURE 15-12 Strep throat is the most common disease caused by *S. pyogenes*.

billions of cholera bacteria in diarrhea—an effective strategy for the bacteria, which contaminate the water used for drinking or bathing and rapidly infect new hosts.

Streptococcus pyogenes is a part of the bacterial community of your nose and mouth. Normally it is harmless. But when the population of *S. pyogenes* is not held in check by competition with other members of the bacterial community, it can become a pathogen. Strep throat (more formally known as "streptococcal pharyngitis") is the most common disease caused by *S. pyogenes;* it produces a severe sore throat and a distinctive rash of red abscesses with white pus at the top of the throat (FIGURE 15-12). Some strains of *S. pyogenes* produce a toxin that is released into the bloodstream and produces a red rash that spreads across the skin—the disease called scarlet fever. Most threatening of all is an infection caused by strains of *S. pyogenes* that have a toxin that allows them to enter body tissues, where they produce necrotizing fasciitis. These strains of *S. pyogenes* are known by their well-earned name, "the flesh-eating bacteria."

TAKE HOME MESSAGE 15.8

» Some bacteria always cause disease, and others are harmful only under certain conditions. For example, *Streptococcus pyogenes* can be harmless, but under some conditions it releases toxins that are responsible for strep throat, scarlet fever, and necrotizing fasciitis.

15.9 Sexually transmitted diseases reveal battles between microbes and humans.

To many microbes, the human genitals and reproductive tract are desirable places to find shelter, nourishment, and opportunities for reproducing and dispersing. Unfortunately, these microbes can cause problems for humans in the form of **sexually transmitted diseases (STDs).** STDs produce symptoms of varying severity, ranging from mild to extreme discomfort, to sterility, or even death. It is estimated that more than 300 million new cases of STDs occur each year worldwide.

STDs are caused by all types of microbes—bacteria, viruses, fungi, protists—and even by some arthropods (FIGURE 15-13). The organisms are passed from the mucous membranes (of the genitals, anus, and mouth) of one individual to those of another during sexual contact; they are also sometimes transmitted by needles used for drug injections.

Although most STDs are curable with drugs, two characteristics of STDs make them nearly impossible to completely eradicate from a population: (1) their symptoms may be mild or absent, causing many people to unwittingly pass an infection to their partners, and (2) to prevent reinfection, both partners must be treated simultaneously. Furthermore, because most microbes have such high reproductive rates, populations of a microbe can evolve quickly and become resistant to existing drugs. The treatment of STDs is one of the most pressing public health issues in the world today.

TAKE HOME MESSAGE 15.9

» Sexually transmitted diseases (STDs) are caused by a variety of organisms, including bacteria, viruses, protists, fungi, and arthropods. Worldwide, more than 300 million people are newly infected each year.

SEXUALLY TRANSMITTED DISEASES			
CAUSE	EXAMPLES	SYMPTOMS	TREATMENT
BACTERIUM	• Gonorrhea	Often none; sometimes painful urination, genital discharge, or irregular menstruation	Several antibiotics can successfully cure gonorrhea; however, drug-resistant strains are increasing.
	• Syphilis	Often no symptoms for years; eventual sores, skin rash, and if untreated, organ damage	Penicillin, an antibiotic, can cure a person in the early stages of syphilis.
	• Chlamydia	Often none; sometimes painful urination, genital discharge	Chlamydia can be easily treated and cured with antibiotics.
VIRUS	• HIV/AIDS	Initial symptoms range from none to flu-like; late stages involve severe infections and death	Currently no cure. Antiretroviral treatment can slow progression. Drug-resistant strains occur.
	• Genital herpes	Often none; outbreaks include sores on genitals, flu-like symptoms	Currently no cure. Antiviral medications can shorten and prevent outbreaks.
	• Human papilloma virus (HPV)	Often none; some types can lead to genital warts, others can cause cervical cancer	A vaccine prevents HPV, and is recommended for boys and girls 11–12. Warts and cancerous lesions can be removed.
PROTIST	• Trichomoniasis	Painful urination and/or vaginal discharge in women; often no symptoms in men	Trichomoniasis can usually be cured with prescription drugs.
FUNGUS	• Yeast infections	Genital itching or burning and/or vaginal discharge in women; genital itching in men	Yeast infections can usually be cured with antifungal suppositories or creams.
ARTHROPOD	• Crab lice	Visible lice eggs or lice crawling or attached to pubic hair; itching in the pubic and groin area	Crab lice can be treated with over-the-counter lotions.

FIGURE 15-13 **The most common STDs.**

15.10 Bacteria's resistance to drugs can evolve quickly.

Penicillin was the first antibiotic to be manufactured and used widely against illness-inducing bacteria. It came into use during the Second World War and caused a revolution in the care of the wounded. But antibiotic-resistant infections soon appeared, and the number of resistant bacteria has increased rapidly ever since (FIGURE 15-14). Now, almost 80 years after the first use of antibiotics, many bacteria are resistant to many, or even to most, antibiotics. In the United States, more people now die of *Staphylococcus aureus* infections that are resistant to many different antibiotics (MRSA infections) than die of HIV/AIDS. A few years ago, antibiotic-resistant staph infections were acquired only in hospitals, but now these infections have spread to the community at large.

Microbes live everywhere they can, and compete for the best places to attach themselves and for the richest sources of food to eat. This competition takes many forms: rapid growth to grab space, superior ability to seize nutrients and starve competitors, and, importantly, the production of chemicals to kill other microbes, or at least stop them from growing. Microbes produce antibiotics to help them compete with other microbes, and most of the antibiotics we use today are derived from microbes.

Q Where do antibiotics come from, and why do they so quickly lose their effectiveness?

Bacteria and other microbes have developed a variety of ways to resist antibiotics. For example, some bacteria pump an antibiotic out of their cells as fast as it enters, so it never reaches a lethal concentration inside the bacterial cell. Other bacteria block its lethal effect with proteins that bind to the antibiotic molecule. Some bacteria have enzymes that break down antibiotic molecules, which are then used as fuel to help the bacteria grow faster! Antibiotic resistance within a population of bacteria can spread quickly, because many

ANTIBIOTIC RESISTANCE IN BACTERIA

PATIENT A
PATIENT B
1 INFECTION
Patient A and Patient B both have an infection caused by a harmful strain of bacteria.
Harmful bacteria
Harmless bacteria
2 ANTIBIOTICS
Both patients are prescribed an antibiotic to treat the infection, which reduces the initial number of harmful bacteria.

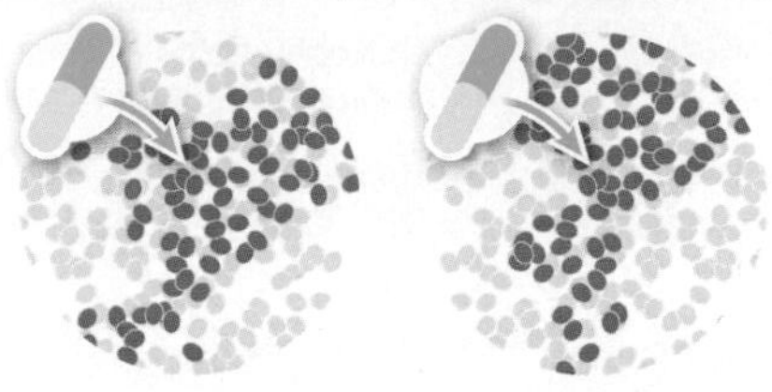

3 OUTCOMES
a Patient A continues to take the antibiotic as prescribed until all of the pills have been consumed.
b Patient B stops taking the antibiotic before finishing all of the prescribed amount.

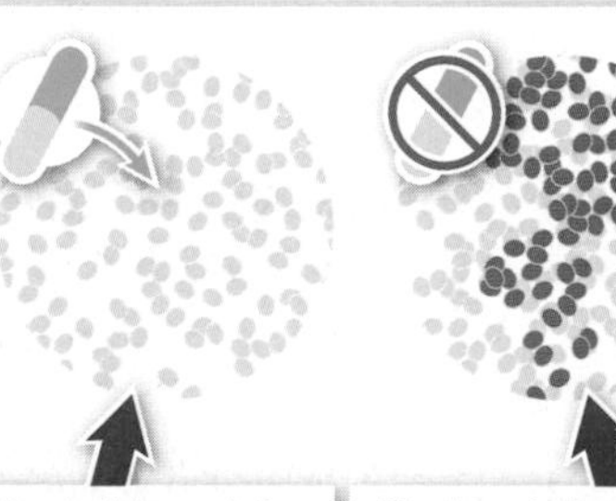

Harmful bacterial population is significantly reduced (and can be held in check by competition with other, harmless bacteria).
The harmful bacterial cells still alive are the most resistant to the antibiotic and can proliferate, leading to future health problems.

FIGURE 15-14 Evolution of resistance. Almost 80 years after the appearance of antibiotics, many bacteria are resistant to many, or even most, antibiotics.

FIGURE 15-15 Antibiotics are used in agriculture.

of the genes that code for resistance are on plasmids. This means that a bacterium carrying a resistance gene can transmit the gene to other bacteria by conjugation; there's no need to wait for natural selection over multiple generations.

Q Why is it essential to take every dose of an antibiotic prescribed by a doctor?

When an antibiotic is taken as prescribed—that is, at the times specified on the label and until all the pills have been consumed—the population of target bacteria is greatly reduced. Some target bacteria are resistant to the antibiotic, but they are few. The growth of these resistant bacteria will be held in check by competition with other types of bacteria. But, if you do not complete the antibiotic treatment as prescribed, many of the target bacterial cells will survive, including the ones that are most resistant to the antibiotic. These resistant cells will be the founders of a new population of bacteria in your body, so the next time you take that drug it will be ineffective. Even worse, taking antibiotics when they are not needed—to treat a viral infection, for example—selects for resistant bacteria without providing any benefit. Antibiotics have no effect on viruses.

The use of antibiotics in agriculture is another reason for the spread of antibiotic resistance. Low concentrations of antibiotics are routinely added to the feed for cattle, hogs, chickens, and turkeys. This can be beneficial in the short term, promoting growth and minimizing disease in the crowded conditions of commercial meat and milk production (FIGURE 15-15). But in the long run it can have disastrous consequences, as the practice can lead to selection for bacteria resistant to the antibiotics.

The antibiotics can also pass through the food chain to humans. Data gathered by the Union of Concerned Scientists indicate that agriculture in the United States uses about 25 million pounds of antibiotics each year—about eight times more than is used for all human medicine!

TAKE HOME MESSAGE 15.10

» Antibiotic resistance routinely evolves in microbes, and plasmid transfer allows an antibiotic-resistant bacterium to pass that resistance to other bacteria. Excessive use of antibiotics in medicine and agriculture has made several of the most important pathogenic bacteria resistant to every known antibiotic.

15.11–15.12

Archaea exploit some of the most extreme habitats.

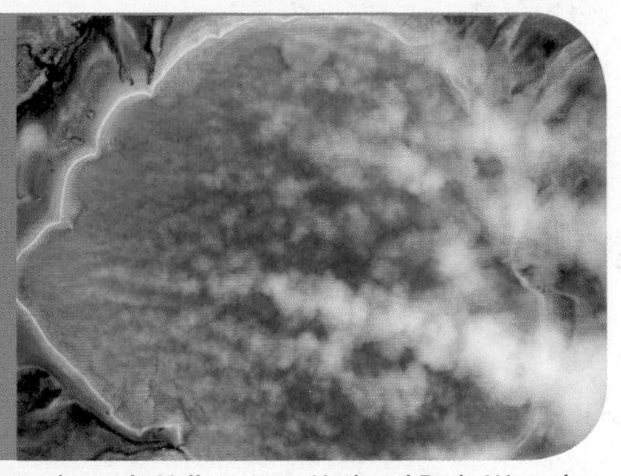

Grand Prismatic Spring hydrothermal vent in Yellowstone National Park, Wyoming. Hot springs are home to many microbes, including thermophilic archaea.

15.11 Archaea are profoundly different from bacteria.

When you look at the photograph of archaea in FIGURE 15-16, you may find it hard to believe that they are even a little bit different from bacteria, let alone *profoundly* different. Both live as single cells or colonies of cells, both are surrounded by a plasma membrane, and both have species with flagella that twirl like propellers. In fact, until the 1970s, biologists considered the archaea to be bacteria. Comparisons of DNA then revealed that archaea are as different from bacteria as humans are.

Archaea look very much like bacteria. But closer inspection—of their physiology, biochemistry, and DNA—reveals them to be profoundly different from all bacteria.

FIGURE 15-16 Appearances are deceiving.

Those studies of archaeal nucleotide sequences stimulated other comparisons, which identified additional differences among bacteria, archaea, and eukarya. For example, the chemical compositions of the plasma membranes, cell walls, and flagella of archaea are qualitatively different from those of bacteria. And beyond the large DNA sequence differences and the differences in plasma membranes, cell walls, and flagella, a third difference reflects the phylogenetic positioning of archaea between bacteria and eukarya on the tree of life (see Figure 15-1): eukaryotes have a distinct cell nucleus that is separated from the cytoplasm by a nuclear membrane, whereas bacteria and archaea have neither a nucleus nor a nuclear membrane. The presence of a nucleus protects the chromosomes of a eukaryotic cell and allows the cell to control what molecules interact with its DNA.

TAKE HOME MESSAGE 15.11

» Archaea possess characteristics that place them between bacteria and eukaryotes on the tree of life. Archaea and bacteria may look similar, but they have significant differences in their DNA sequences, as well as in their plasma membranes, cell walls, and flagella. Furthermore, neither archaea nor bacteria have a distinct cell nucleus and nuclear membrane.

15.12 Archaea thrive in habitats too extreme for most other organisms.

Archaea are famous for their ability to live in places where life would seem to be impossible, such as in water around hydrothermal vents that emerge from the seafloor. At 2,000 meters below the surface, the water pressure reaches 200 atmospheres, or almost 3,000 pounds per square inch, and the 400° C water coming from the vent cannot boil. Archaea thrive there. Most organisms die at temperatures between 40° and 50° C because their protein molecules are denatured, so the ability of archaea to survive at temperatures above 100° C is truly remarkable.

Equally impressive is the ability of archaea to live in water as acidic as pH 0 or as salty as a saturated NaCl solution, and to obtain energy from an extraordinary range of compounds, including sulfur, iron, and hydrogen gas (FIGURE 15-17). Some bacteria also can tolerate extreme physical and chemical conditions, and organisms that can live in these conditions, both bacteria and archaea, are called **extremophiles** ("lovers of extreme conditions").

The extremophile nature of so many archaea makes it difficult to study them. How do you grow an organism in your lab if it requires 200 atmospheres of pressure and a water temperature of 100° C or higher and eats a gas that is toxic to humans? It's not easy. Consequently, only about one-quarter of the identified archaeal species have been cultured in the lab. We know that the other species exist, because their transfer RNA has been identified in samples taken from the environment, and biologists are confident that thousands, perhaps millions, of species of archaea await discovery.

Extremophile archaea have important applications in bioengineering and environmental remediation. For example, *Thermus aquaticus,* an extremophile archaeon that lives in hot springs where the water is nearly boiling, produces an enzyme important in making PCR possible (see Chapter 7), and this enzyme is now sold for laboratory use under the name *Taq* polymerase. Far from being destroyed at high temperatures, *Taq* polymerase works best at 75° C, a temperature at which most human enzymes would simply fall apart.

> "He was a killer, a thing that preyed, living on the things that lived, unaided, alone, by virtue of his own strength and prowess, surviving triumphantly in a hostile environment where only the strong survive."
>
> — JACK LONDON,
> *The Call of the Wild,* 1903

Extremophile archaea and bacteria that thrive at high temperatures and pressures and metabolize toxic substances may have enormous potential for industry. Recent

FIGURE 15-17 **Living in the harshest environments.** Archaea thrive in many extreme environments, such as this salt pond.

FIGURE 15-18 **Practical applications for archaea.** The metabolic activity of archaea has potential for use in the cleanup of oil spills.

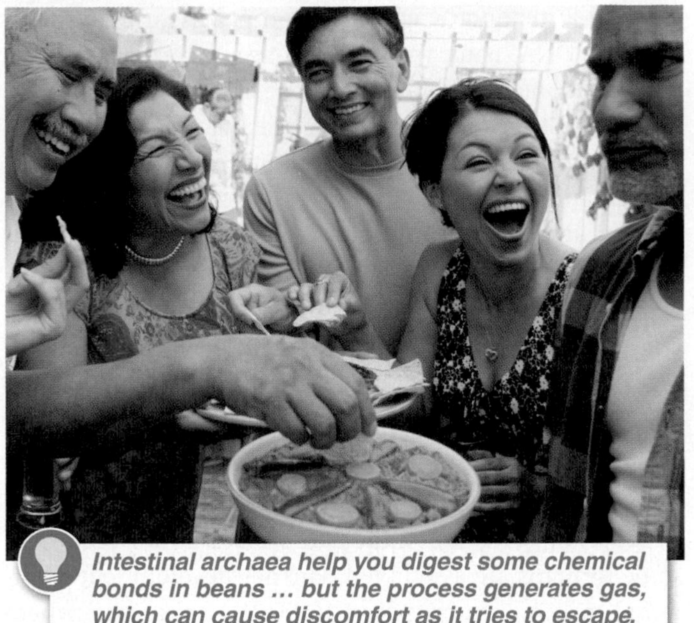

Intestinal archaea help you digest some chemical bonds in beans ... but the process generates gas, which can cause discomfort as it tries to escape.

FIGURE 15-19 **Helping you digest tough bonds.**

experiments have demonstrated the ability of some archaea to efficiently degrade hydrocarbons, making it possible for them to be used in the removal of sludge that accumulates in oil refinery tanks and, potentially, in the cleanup of contaminated environments such as oil slicks (FIGURE 15-18). Naturally occurring archaea are helping to break down some of the more than 200 million gallons of oil released into the Gulf of Mexico following an oil rig explosion and subsequent massive leak in 2010. Other archaea show promise in clearing mineral deposits from pipes in the cooling systems of power plants.

But not all archaea are extremophiles. Many of them live in places you would find comfortable. For example, beans are notorious for their tendency to produce gas in the intestine—but archaea are the culprits (FIGURE 15-19). Methane-producing archaea produce an enzyme that targets the carbohydrate bonds in beans that are not broken down very well by any human enzymes. As a result, the archaea in your intestine digest most of these carbohydrates. But in the process of breaking these bonds, they produce gases that, as they escape the digestive system, can cause considerable distress.

TAKE HOME MESSAGE 15.12

» Archaea can tolerate extreme physical and chemical conditions. Archaea are hard to study, however, because many require extreme heat or pressure to grow in the laboratory. Archaea are potentially valuable for industrial and environmental purposes. Not all archaea are extremophiles; some live in moderate conditions, including the human intestine.

15.13–15.15 Most protists are single-celled eukaryotes.

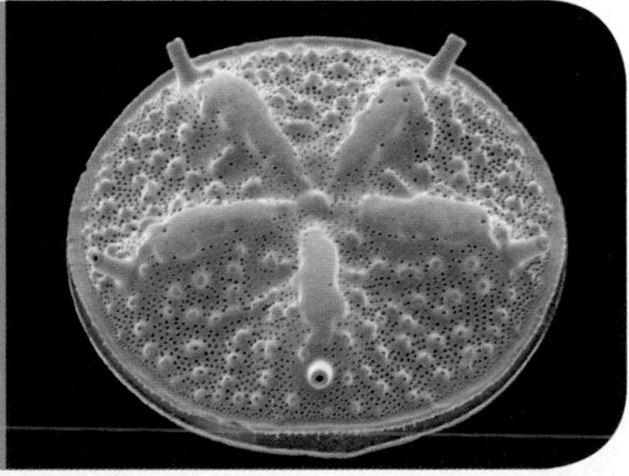

Diatoms, single-celled photosynthetic algae that form an important part of the plankton, have mineralized cell walls that provide protection and support.

15.13 The first eukaryotes were protists.

For the first 2 billion years of life on earth, organisms were extremely small—bacteria and archaea less than 10 micrometers across, one-eighth the diameter of a human hair. But in rocks about 1.9 billion years old, we find

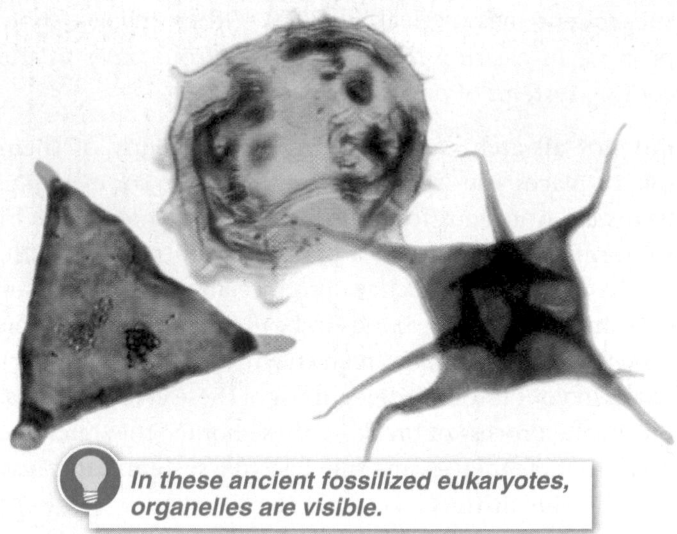

In these ancient fossilized eukaryotes, organelles are visible.

FIGURE 15-20 **Ancient protists.** The oldest acritarch fossils—found in 1.9-billion-year-old rocks—represent the first eukaryotes.

fossils of new kinds of organisms that are 10 times larger (FIGURE 15-20). These are a group of organisms called acritarchs (a name that can be translated as "confusing old things"). They were the first eukaryotes.

The larger size is the first thing you notice about these fossil cells, but the internal changes were the basis for the success of the entire eukaryote lineage, including humans. For the first time in the history of life, cells had internal structures that carried out specific functions. These structures, the cellular organelles, perform the specialized activities that make eukaryotic cells more complex than prokaryotes.

The nucleus is an evolutionary innovation that first appeared in protists. Infoldings of the plasma membrane that surrounded the cell probably fused around the DNA, creating a membrane-enclosed compartment containing the cell's DNA. Thus, the cell nucleus was formed, separated by a double membrane from the cytoplasm of the cell. Many modern prokaryotes have plasma membranes with complex infoldings that increase the total surface area of the cell and allow better exchange of material between the internal and external environments. Membrane infoldings of this type may also have played a role in endosymbiosis and formation of the first organelles (see Figure 4-8).

Subsequently, the first nuclear membranes developed two specializations: they incorporated proteins that controlled the movement of molecules into and out of the nucleus, and they extended outward from the nucleus to form a folded membrane called the endoplasmic reticulum. The endoplasmic reticulum is the part of a eukaryotic cell where some proteins are assembled. Further development of internal membranes produced the Golgi apparatus, where newly synthesized proteins are given some final processing steps, and sac-like structures called lysosomes, which contain enzymes that break down damaged molecules. Finally, a lineage of protists took in a guest—a bacterial cell that subsequently became the mitochondrion, the organelle that produces most of the ATP synthesized by a eukaryotic cell.

TAKE HOME MESSAGE 15.13

» The nucleus is an evolutionary innovation that appeared for the first time in protists. Early protists took in a bacterial cell that subsequently became the mitochondrion, an organelle in eukaryotic cells that produces ATP.

15.14 There are animal-like protists, fungus-like protists, and plant-like protists.

Six major lineages of protists have been named, but even that classification does not capture the huge diversity of the group, and some of the best-known protists do not fit into any of the six lineages. Protists include forms that are very much like animals, others that seem a lot like fungi, and still others that look like plants (FIGURE 15-21).

Animal-like Protists Some protists propel themselves quickly around their environment and appear to hunt for prey. These animal-like protists, which include *Paramecium,* are the ciliates. They get their name from the cilia (hair-like projections) that cover their body surfaces and propel the cells through water. *Paramecium* feeds by the process of **phagocytosis:** cilia in a funnel-shaped structure called the gullet create an inward flow of water that carries bacteria and other small particles of food with it.

Fungus-like Protists Some protists resemble fungi: living as heterotrophs, establishing sheet-like colonies of

DIVERSITY OF PROTISTS

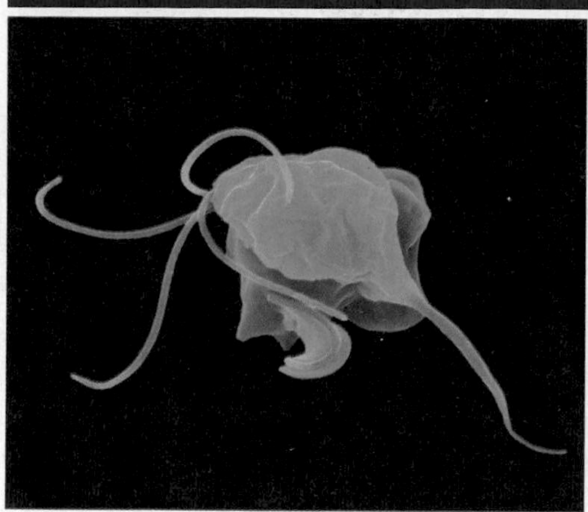

ANIMAL-LIKE PROTISTS
Some protists, such as *Trichomonas vaginalis* shown here, move around and hunt for prey like an animal.

FUNGUS-LIKE PROTISTS
Some protists, such as the carnival candy slime mold shown here, live as heterotrophs and form sheet-like colonies of cells like a fungus.

PLANT-LIKE PROTISTS
Some protists, such as the kelp forest shown here, are multicellular and carry out photosynthesis like a plant.

FIGURE 15-21 **Protists come in all shapes and sizes.**

cells on surfaces (such as the grout in shower stalls), using spores to reproduce, and sometimes producing fruiting bodies. These are the slime molds. They generally spread without any individual cells moving, but rather by adding new cells at the edges of the colony. Some slime molds, called plasmodial slime molds, are oozing masses that flow along a surface, engulfing bacteria, fungi, and small bits of organic material as they go. The streaming of a slime mold in its feeding phase is easy to observe with a microscope, and such a slime mold can flow around, over, or through almost anything—it can even flow through a window screen and reassemble itself on the other side! You may have seen an irregularly shaped blob of yellow material in a moist, shaded garden—that was probably a plasmodial slime mold.

Remarkably, a plasmodial slime mold is a very large, single cell (but has multiple nuclei). Slime molds divide by binary fission, just like other cells, but in this case a single cell may cover an area of several square centimeters. All of the nuclei in the cell undergo mitosis simultaneously.

Plant-like Protists Other members of the protists grow in water and resemble plants. These include the protists referred to colloquially as algae and seaweeds. The term "seaweed" generally refers to the macroscopic, multicellular marine algae, many of which are used as a source of food for humans. And although all are protists, seaweeds encompass several groups that do not share a common multicellular ancestor. These include some red algae and some green algae (from which the land plants most likely evolved), as well as some of the brown algae—such as the giant kelp that grows in water 30 meters deep. (The seaweeds, of course, represent species of protists that obviously are not "microbial.")

Brown algae cover large portions of the rocks in the intertidal zone. Giant kelp, growing in temperate regions of the North and South Pacific and off the Atlantic coast of South Africa, are among the fastest growing organisms on earth, with some growing as much as 60 meters in a single year. Kelp forests are an enormously diverse habitat; more than 1,000 species of fishes, crustaceans, snails, and mammals make their homes in these marine "forests." Sea otters wrap a kelp frond around their waist when they sleep, and gray whales hide in kelp forests to escape killer whales.

Although many protists are unicellular, there are many exceptions. As we've seen, many green and brown algae are multicellular, composed of many different cells and cell types that perform different functions. Some other protists, such as spirogyra, are colonial, living as collections of cells, each of which can carry out all of its life processes independent of the other cells.

Also among the plant-like protists are the diatoms (FIGURE 15-22). Diatoms are unicellular organisms that live in ponds, lakes, and rivers as well as in the oceans. They are so small that, for some species, 30 individuals could be lined up across the width of a human hair. A characteristic of the diatoms is that they are enclosed in a shell made of silica. Many species of diatoms float in the water, forming part of the phytoplankton—the collection of microscopic organisms that fix carbon dioxide and release oxygen. The phytoplankton can reach densities of hundreds of

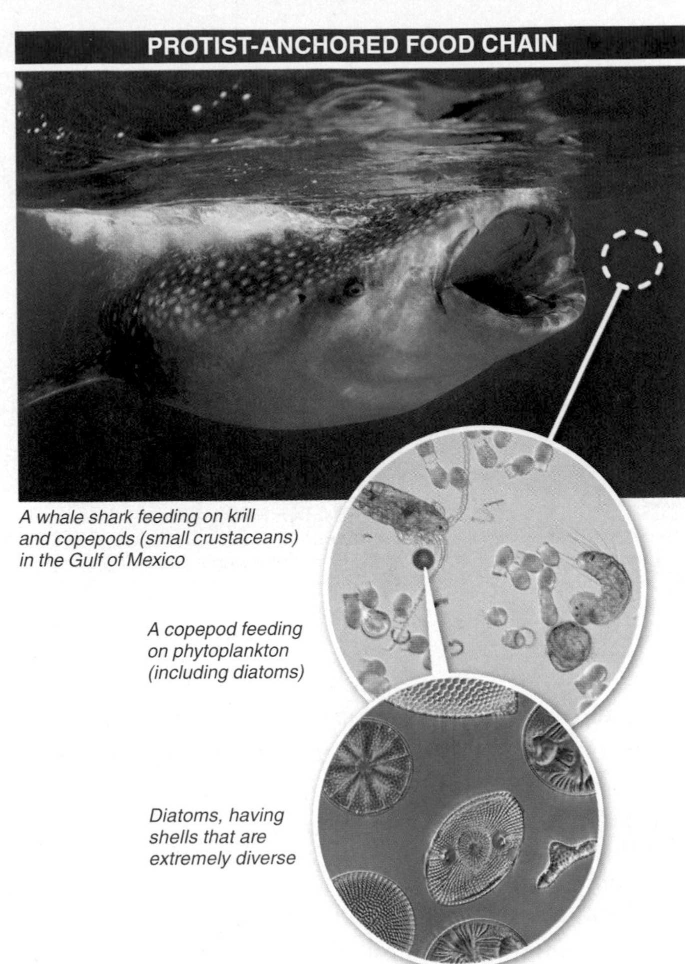

PROTIST-ANCHORED FOOD CHAIN
A whale shark feeding on krill and copepods (small crustaceans) in the Gulf of Mexico
A copepod feeding on phytoplankton (including diatoms)
Diatoms, having shells that are extremely diverse

FIGURE 15-22 Diatoms are aquatic photosynthetic protists. Diatoms are food for copepods, which are an important source of food for many larger predators.

thousands of cells per liter, and it accounts for about one-quarter of the photosynthetic production of oxygen on earth, occupying a critical position in marine food chains. Small fishes and shrimp-like organisms called copepods feed on phytoplankton, and these small predators are eaten by larger predators, which are eaten by still larger predators. Thus the diversity of life in the marine habitat relies on diatoms and the other microbes that make up the phytoplankton.

Q The health of protist populations can serve as a bioindicator of marine ecosystems. Why?

TAKE HOME MESSAGE 15.14

» Protists are a diverse group of mostly unicellular eukaryotic organisms. The ciliates, such as *Paramecium,* are animal-like protists. Plasmodial slime molds are fungus-like protists. Colonial protists and multicellular protists such as the giant kelp can be enormous and are plant-like in appearance.

15.15 Some protists can make you very sick.

A **parasite** is an organism that lives in or on another organism, called a **host,** and damages it. A parasitic protist called *Plasmodium* that is transmitted by a mosquito is responsible for malaria. Malaria occurs in tropical parts of Africa, Asia, and Latin America, and it is common in the eastern Mediterranean as well. Between 350 million and 500 million people have clinical cases of malaria, and about 1 million people die of it each year. Malaria is the leading cause of death for children younger than 5 years old in sub-Saharan Africa. Somewhere on the African continent, about every 30 seconds, a child dies of malaria.

Neither the incidence of malaria nor the rate of mortality has changed very much since *Plasmodium* was identified as the cause of malaria nearly a century ago. The reason for the lack of progress is the changing nature of *Plasmodium* (FIGURE 15-23). The human immune system has a difficult time fighting a malarial infection because, to survive from one generation to the next, *Plasmodium* parasites go through a series of distinct developmental stages. And as the *Plasmodium* cells change from one stage to another, the parasite produces different cell surface proteins. *Plasmodium* stays ahead of the human immune system by constantly changing the way it appears to immune system cells.

Although the immune system doesn't usually have much success fighting malarial infection, some individuals produce red blood cells that are inhospitable to *Plasmodium,* conferring a resistance to the malaria-causing protist. These

MALARIA: A PROTIST-CAUSED ILLNESS

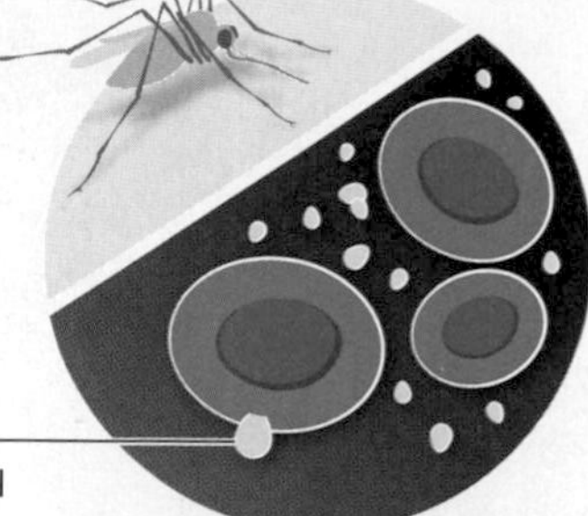

1 Following the bite of a *Plasmodium*-infected mosquito, malaria-causing protists take up residence in healthy red blood cells.

Plasmodium parasite infecting a red blood cell

2 Once inside a red blood cell, *Plasmodium* cells modify the cell's surface proteins, making it difficult for the immune system to fight the malarial infection.

RESISTANCE TO MALARIA
In individuals with sickle-cell trait, the sickle shape of some red blood cells makes the cells inhospitable to the protist. This makes these individuals resistant to malaria.

FIGURE 15-23 Malaria-infected blood cells and sickled blood cells.

individuals carry an allele, called Hb^S, which (as we learned in Chapter 9) codes for a variant of adult hemoglobin that causes hemoglobin molecules to stick together in long chains that distort the red blood cell (see Figure 15-23), giving the cell a sickled shape. The sickled red blood cells leak substances that are essential to *Plasmodium,* making these cells inhospitable to the parasite. Although carrying a copy of the Hb^S allele confers resistance to malaria, individuals who carry two copies of the Hb^S allele suffer from sickle-cell anemia—a severely debilitating genetic disease. Individuals with a single Hb^S allele are said to have "sickle-cell trait."

TAKE HOME MESSAGE 15.15

» Some protists cause debilitating diseases. *Plasmodium,* the protist responsible for malaria, is one of these. *Plasmodium* has characteristics that protect it from the human immune system, but some humans do have a defense against malaria: people with sickle-cell trait make red blood cells that are inhospitable to the parasite.

15.16–15.19

Viruses are at the border between living and non-living.

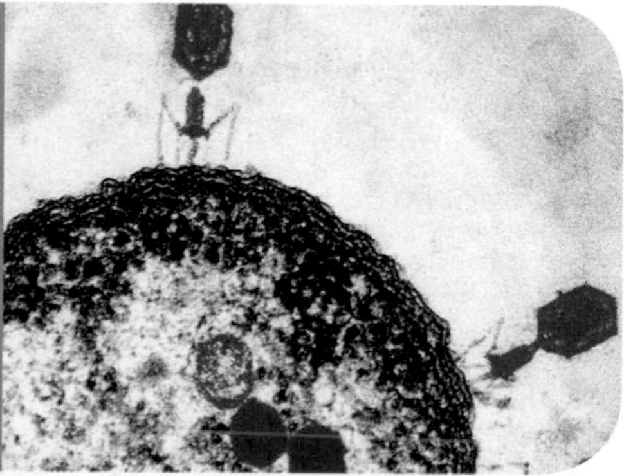

Colored transmission electron micrograph of T2 bacteriophage viruses (red) attacking an *E. coli* bacterium.

15.16 Viruses are not exactly living organisms.

A virus is not a cell, and that is why viruses do not fit into any of the three domains of life. A virus particle (called a virion) consists of genetic material inside a container made of protein. Some viruses also contain a few enzymes. That's all there is to a virus. The protein container is called the **capsid,** and the genetic material can be either

BASIC STRUCTURE OF A VIRUS

All viruses have a container (the capsid) that holds their genetic material, and sometimes the capsid is wrapped in a membrane (an envelope).

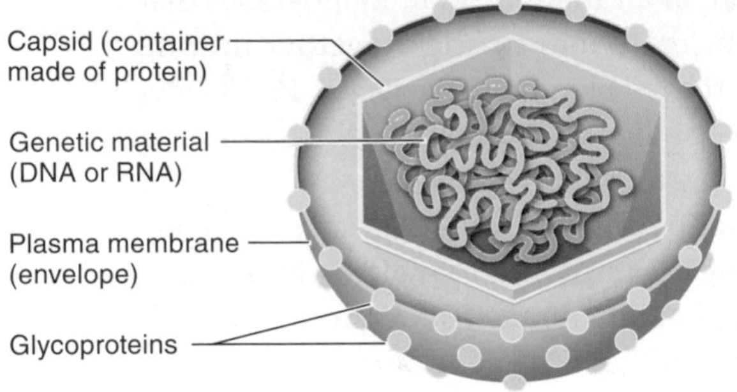

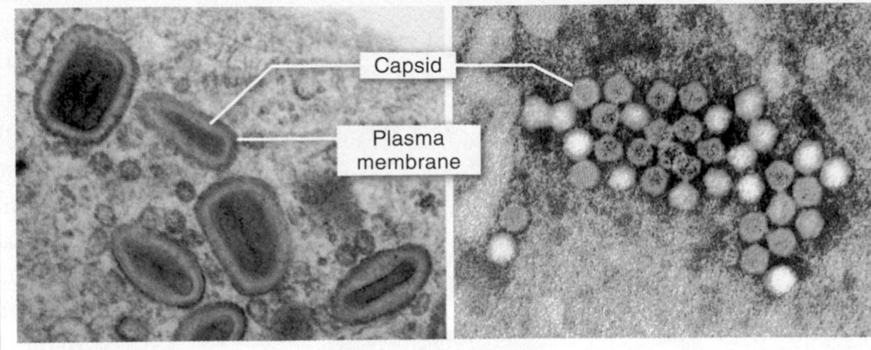

ENVELOPED VIRUS
Enveloped viruses wrap themselves in a bit of the plasma membrane of the host cell as they are released.

NON-ENVELOPED VIRUS
Non-enveloped viruses are enclosed only by a capsid.

FIGURE 15-24 **The simple structure of viruses.**

DNA or RNA. Some viruses wrap themselves in a bit of the plasma membrane of the host cell as they are released. A virus of this type is called an enveloped virus—the flu virus is an example. Non-enveloped viruses are enclosed only by the protein container; the virus that causes the common cold is an example of a non-enveloped virus (FIGURE 15-24).

There is almost nothing inside the capsid of a virus except DNA or RNA. A virus particle does not carry out any metabolic processes, and it does not control the inward or outward movement of molecules to make conditions inside the virus particle different from conditions outside. Viruses just wait for a chance to insert their genetic material into a living cell.

We speak of "catching" a cold, but in reality, a virus catches us. Viruses identify the cells that they can infect by the specific glycoprotein molecules on the surfaces of those cells. Every cell in your body has glycoprotein molecules that are embedded in the plasma membrane and extend

VIRAL REPLICATION

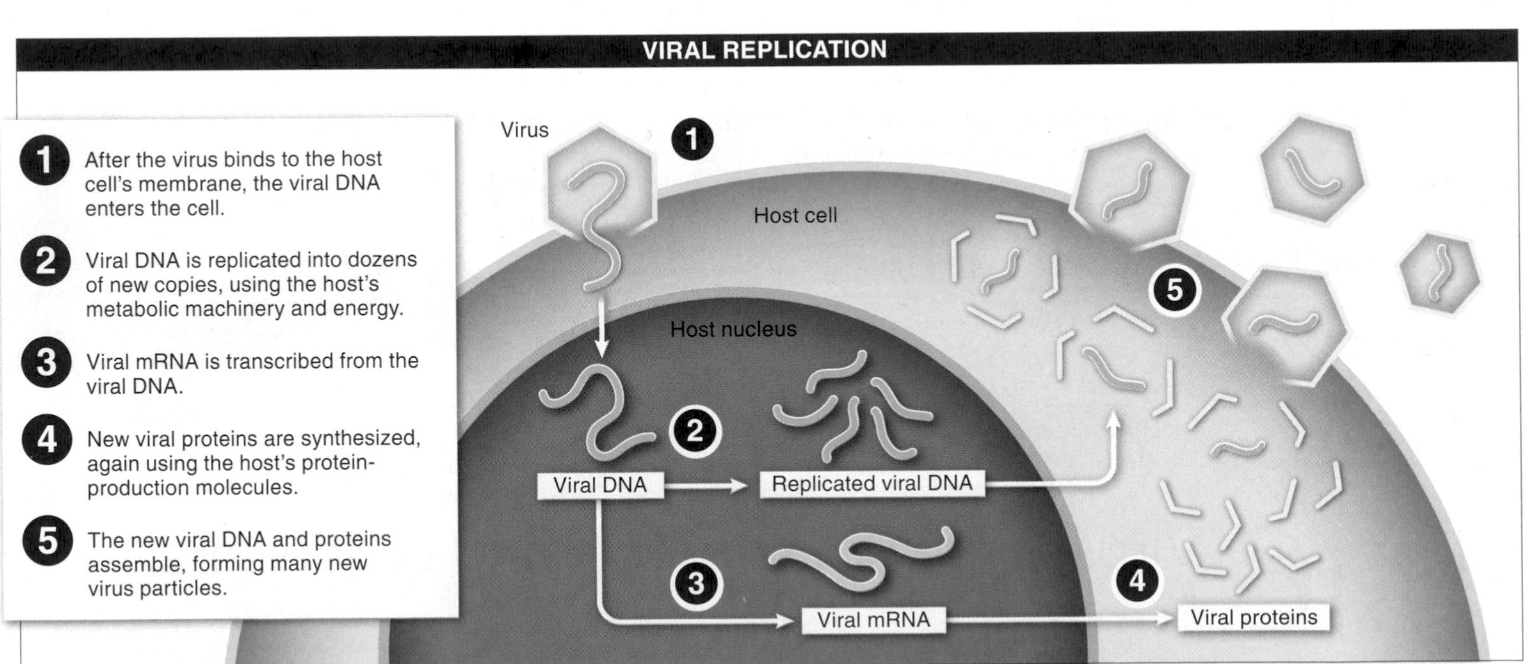

1 After the virus binds to the host cell's membrane, the viral DNA enters the cell.
2 Viral DNA is replicated into dozens of new copies, using the host's metabolic machinery and energy.
3 Viral mRNA is transcribed from the viral DNA.
4 New viral proteins are synthesized, again using the host's protein-production molecules.
5 The new viral DNA and proteins assemble, forming many new virus particles.
Virus
Host cell
Host nucleus
Viral DNA
Replicated viral DNA
Viral mRNA
Viral proteins

FIGURE 15-25 **Making more viruses.** A virus duplicates its own genetic material (DNA in the example shown here) by taking over the resources and metabolic machinery of a host cell.

Because they are dependent on their hosts' metabolic machinery for replication, viruses are not considered "living."

outward. Your immune system uses these proteins to identify the cells as part of you. When viruses find a cell with the appropriate glycoprotein on its surface, they bind to that cell's plasma membrane and insert their genetic material into the cell (FIGURE 15-25). The viral DNA or RNA takes over the cellular machinery and uses it to produce more viruses.

Viruses carry out most activities by hijacking materials and organelles in the host cell. Viral proteins are synthesized in the same way as host cell proteins—mRNA binds to the cell's ribosomes, and tRNA matches the correct amino acid to each mRNA codon. The mRNA is produced from the virus's DNA or RNA, but all the protein-building machinery comes from the host cell, as does the ATP required to synthesize the new viral protein.

Besides viruses, there are some other types of non-living infectious agents. Prions, for example, are misfolded proteins that form plaques and interfere with normal tissue. Usually acquired through ingestion of an infected animal (or its bodily fluids), prions can cause a variety of degenerative neurological diseases in humans. One of these, Creutzfeldt-Jakob disease, is sometimes referred to as a human form of mad cow disease: it is characterized by the development of holes in brain tissue, causing the tissue to appear sponge-like. It leads to dementia, memory loss, and disruption of balance and coordination. No cure or treatment is known, and few victims survive more than a year following the appearance of symptoms.

TAKE HOME MESSAGE 15.16

» A virus is not alive, but it can carry out some of the same functions as living organisms if it can get inside a cell. A virus takes over the protein-making machinery of the host cell to produce more viral genetic material (RNA or DNA) and protein. The viral proteins and genetic material are assembled into new virus particles and released from the cell.

15.17 Viruses are responsible for many health problems.

Many diseases are caused by viruses. Some viral diseases have been responsible for worldwide epidemics, called pandemics. The influenza pandemic of 1918–1919 killed at least 20 million people, and possibly as many as 50 million. In the current HIV/AIDS pandemic, more than 75 million people have been infected and about 35 million people have died of AIDS. Worldwide, about 1.1 million people died of AIDS-related illnesses in 2015.

Other viral diseases, such as the common cold, are not usually serious. Herpes is another common viral infection in humans, caused by two related viruses. Oral herpes is an infection near the mouth and can cause cold sores lasting two to three weeks. More than three-quarters of all adults in the United States are infected with the virus, with more than 50 million people experiencing outbreaks each year. Genital herpes, a sexually transmitted disease, affects about one in six people between the ages of 14 and 49 in the United States. The viruses causing herpes are present in and released from sores and spread by skin-to-skin contact, but an infected person can also pass on the virus even when she or he has no symptoms. Although a variety of treatments are available for both types of herpes, including antiviral drugs that can reduce the severity and duration of symptoms, there is currently no cure.

DNA viruses have base-pair sequences that are stable over time, because the enzymes that replicate the DNA check for errors and correct them during replication. From a health perspective, the fact that DNA viruses do not change rapidly makes it much easier to fight and treat diseases caused by these viruses, using vaccines. Vaccination against a DNA virus such as smallpox, for example, provides years of protection.

Q Why do flu viruses change so quickly?

RNA viruses, by contrast, change quickly, because the enzymes that carry out RNA replication do not have error-checking mechanisms. Because the RNA-replicating enzymes make errors as they assemble new RNA molecules, RNA viruses are continuously mutating into new forms. The common influenza virus that causes outbreaks of flu every year, for example, is an RNA virus. Flu viruses mutate so fast that the virus

VIRAL OUTBREAKS

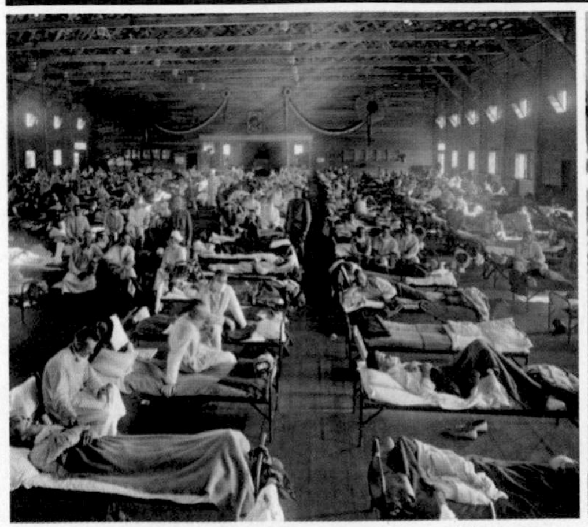

The influenza pandemic of 1918–1919 killed millions.

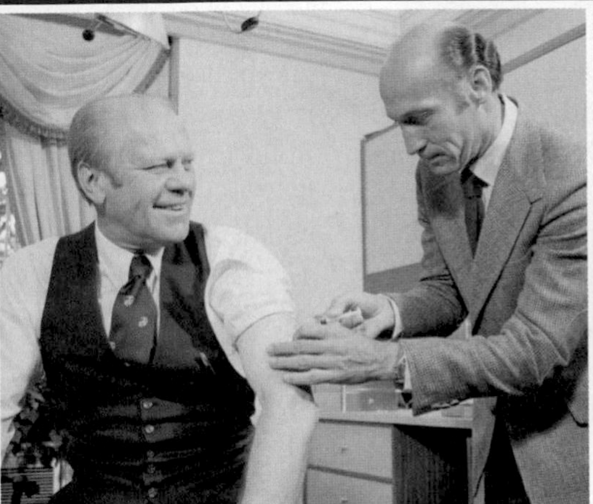

President Gerald Ford receives the swine flu vaccine in the White House in 1976.

In 2002–2003, the virus that causes severe acute respiratory syndrome (SARS) infected several thousand people around the world.

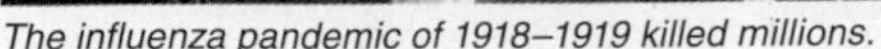

FIGURE 15-26 **Viral outbreaks can threaten large populations.**

changes from one flu season to the next, necessitating a different flu vaccine every year.

About every 50 years, a new variety of influenza causes a pandemic. The 20th century had three major influenza pandemics. In each case, a bird flu virus gained the ability to infect human cells by passing through pigs, and then the virus spread rapidly through the human population worldwide. The most famous of these pandemics was that of 1918–1919, mentioned above (FIGURE 15-26). It was called the Spanish influenza, because it seemed to enter Europe through Spain, but it originated in Asia. The Asian flu and Hong Kong flu pandemics also originated in Asia, as did smaller viral outbreaks, such as Korean flu (1947), swine fever (1976), and SARS (severe acute respiratory syndrome, 2002–2003), which did not become pandemics.

TAKE HOME MESSAGE 15.17

» Many diseases are caused by viruses. DNA viruses are relatively stable, because DNA-replicating enzymes check for errors and correct them during replication. RNA viruses change quickly, however, because RNA-replicating enzymes do not have error-checking mechanisms.

15.18 Viruses infect a wide range of organisms.

Viruses are nearly everywhere. Viruses infect animals and plants, and even infect bacteria. Most viruses infect just one species, or only a few closely related species, and enter only one kind of cell in that species. But some viruses can infect a wide range of hosts—the rabies virus, for example, can infect any mammal. Pet dogs and cats are routinely immunized against it, but wild mammals, such as bats, skunks, and raccoons, can get rabies. That is why, if you are bitten by one of those wild animals, your treatment includes rabies shots as a precaution.

The glycoproteins on the surface of a virus determine which host species the virus can infect and which tissues of the host it can enter. Influenza A viruses (the ones that cause flu outbreaks every year), for example, have two types of glycoprotein that have different functions. One glycoprotein matches that of a host cell and allows the virus to enter the cell. The other glycoprotein allows virus particles to get back out of the cell—new virus particles that can infect other cells (FIGURE 15-27).

Influenza A is an example of a virus that can move from one species to another, and as we noted, all influenza pandemics in the past century began when a bird flu virus infected humans. Here's how "species jumping" can occur.

Viruses that infect birds don't bind well to glycoproteins in human cells, making it difficult for bird flu viruses to infect humans. However, the cells of pigs have glycoproteins that allow both human and bird flu viruses to bind to them.

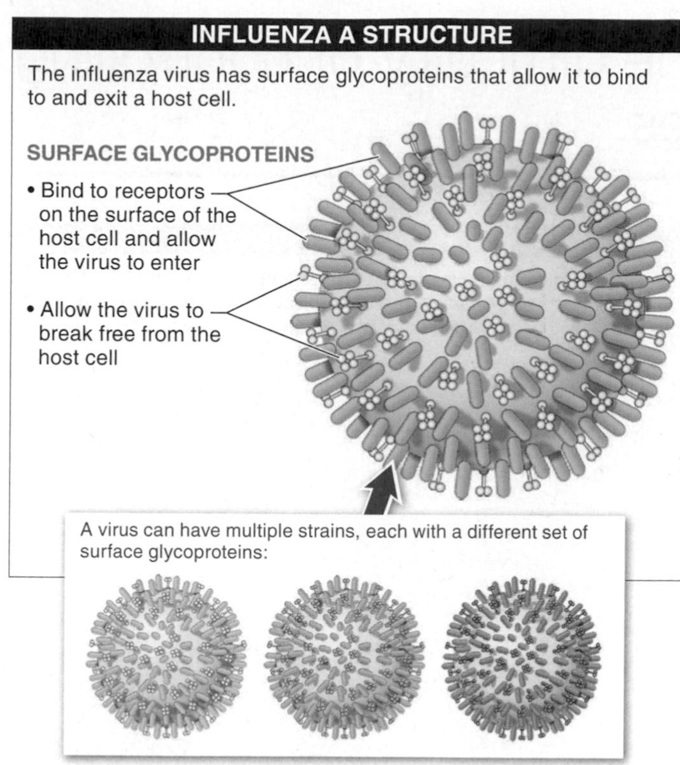

INFLUENZA A STRUCTURE
The influenza virus has surface glycoproteins that allow it to bind to and exit a host cell.
SURFACE GLYCOPROTEINS
• Bind to receptors on the surface of the host cell and allow the virus to enter
• Allow the virus to break free from the host cell
A virus can have multiple strains, each with a different set of surface glycoproteins:

FIGURE 15-27 **A virus's surface proteins.**

newly synthesized RNA into new particles. If RNA strands from a human flu virus and a bird flu virus happen to be in a pig cell at the same time, some virus particles released from the cell might include RNA from both the bird and the human virus. These influenza viruses are a new strain, and they may be able to infect humans (FIGURE 15-28).

Since 1997, one strain of bird flu has spread from Hong Kong through most of Asia and into Turkey, France, Germany, and England. This avian influenza virus can also infect humans, not needing pigs as intermediary. The virus readily infects birds, and tens of millions of chickens, turkeys, and ducks have been killed in attempts to eradicate the infection. Because the virus's host-entry glycoproteins do not bind well to human cells, the virus does not easily infect people and spread from human to human. So far, nearly all of the people infected by the avian influenza virus have gotten it through close contact with infected flocks of birds or by eating birds that died of the viral infection. Only a half-dozen cases are known in which the virus seems to have been transmitted from one person to another. However, avian flu can be deadly when it does infect humans—more than half of the 650 human cases reported as of 2014 have ended in death.

Q **What role does a pig play in the transmission of virus from a bird to a human?**

Thus, a pig cell can be infected by a human flu virus and a bird flu virus at the same time. But influenza viruses are not very careful when they incorporate

TAKE HOME MESSAGE 15.18

» Glycoproteins on the surfaces of viruses determine the types of cells a virus can invade. Most viruses infect just one species, or only a few closely related species, and enter only one kind of cell in that species.

VIRAL TRANSMISSION THROUGH "SPECIES JUMPING"

Viruses that infect birds don't bind well to glycoproteins in human cells, making it difficult for bird viruses to infect humans.

However, a virus that infects humans and a virus that infects birds can meet, if both have infected a pig cell.

Because viral RNA replication is error-prone, the two viral RNA molecules can get packaged together into a new virus particle. The new virus can have features from the bird virus and be capable of infecting humans.

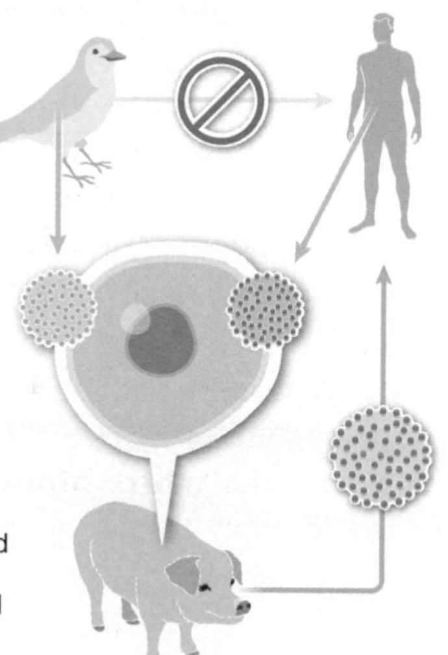

FIGURE 15-28 **Inside a pig, bird viruses can gain the ability to infect humans.**

15.19 HIV illustrates the difficulty of controlling infectious viruses.

New infectious diseases—some caused by bacteria and others by viruses—emerge quite frequently. Many diseases originate in an animal and subsequently acquire the ability to infect humans. There are enough of these new diseases to fill the monthly issues of the journal *Emerging Infectious Diseases,* which is published by the Centers for Disease Control and Prevention. Even so, **acquired immunodeficiency syndrome (AIDS)** stands out. AIDS is caused by the **human immunodeficiency virus (HIV),** which is derived from a strain of the simian immunodeficiency virus (SIV) that jumped from chimpanzees to humans in the early 1900s. HIV has the hallmark qualities that make viral diseases hard to control—plus an additional characteristic of its own.

1. HIV mutates easily. HIV is a **retrovirus,** an RNA-containing virus that also contains a viral protein called reverse transcriptase, an enzyme that uses a strand of viral RNA as a template to synthesize a single strand of DNA. That DNA strand, in turn, is used as a template to make a complementary strand of DNA, and the resulting double-stranded DNA is integrated into the host cell's DNA. From here, the DNA is used by the host cell's machinery to make more viral RNA (FIGURE 15-29).

Reverse transcriptase is so error-prone, however, that virtually every copy of HIV in an infected individual's body has a different mutation.

This enormous genetic variation in the HIV particles circulating in an infected person's body makes the infection hard to treat. Virus particles with different mutations can have different proteins on their surfaces, and these surface proteins change each time the virus replicates inside a host cell. Each new generation of HIV in the infected individual contains viruses with surface proteins that his or her immune system has never seen. Furthermore, some of the HIV mutations confer resistance to the drugs that are being used to treat the patient, and new drugs must be used.

2. HIV attacks white blood cells. All of those problems would apply to any disease caused by a retrovirus,

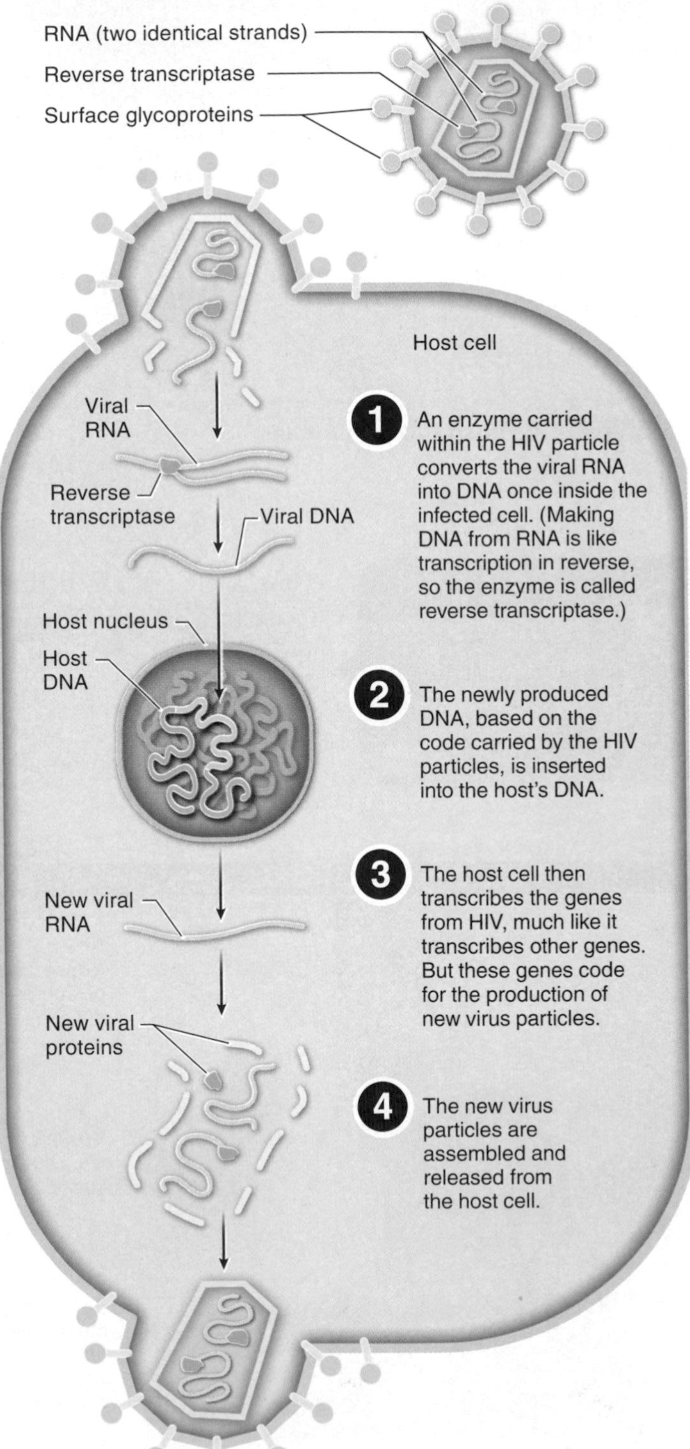

HIV REPLICATION
The HIV particle carries its genetic instructions in the form of RNA rather than DNA. The virus takes over a white blood cell's replicating machinery to produce a new generation of HIV particles.
RNA (two identical strands)
Reverse transcriptase
Surface glycoproteins
Host cell
Viral RNA
Reverse transcriptase
Viral DNA
1 An enzyme carried within the HIV particle converts the viral RNA into DNA once inside the infected cell. (Making DNA from RNA is like transcription in reverse, so the enzyme is called reverse transcriptase.)
Host nucleus
Host DNA
2 The newly produced DNA, based on the code carried by the HIV particles, is inserted into the host's DNA.
New viral RNA
3 The host cell then transcribes the genes from HIV, much like it transcribes other genes. But these genes code for the production of new virus particles.
New viral proteins
4 The new virus particles are assembled and released from the host cell.

FIGURE 15-29 **HIV infection.**

but HIV offers an additional challenge: it targets cells in the host's immune system, especially white blood cells, and particularly those that search for and attack invading bacteria or viruses. During the incubation period, which can last for many years, HIV infects white blood cells. New ones are produced to replace those killed by the virus so that the infected person has virtually no symptoms. Nonetheless, HIV is present in the individual's body fluids during the incubation period and can be transmitted to other individuals. HIV testing is used to detect infections during this stage.

Each time HIV infects another white blood cell, the reverse transcriptase makes errors in transcribing the RNA to DNA, and eventually one of the mutations allows the virus to bind to the glycoprotein on the surface of a specialized type of white blood cell—a bacteria- and virus-hunting white blood cell that is critically important in identifying disease-causing threats. A suitable mutation may occur in a couple of years, though more often it takes about 10 years or longer. But when it does happen, it signals a new stage in the HIV infection: the development of AIDS (FIGURE 15-30).

3. The immune system collapses. Normally, white blood cells all work together to identify and destroy cells that have been infected by a virus. When HIV begins to kill the cells that hunt for viruses and bacteria, the immune system begins to fail—it can no longer respond to HIV, or to any other infectious agent. Patients with AIDS develop multiple infections, bacterial and viral, as well as cancers, because they have lost the immune system cells that would normally have marked infected and cancerous cells for destruction.

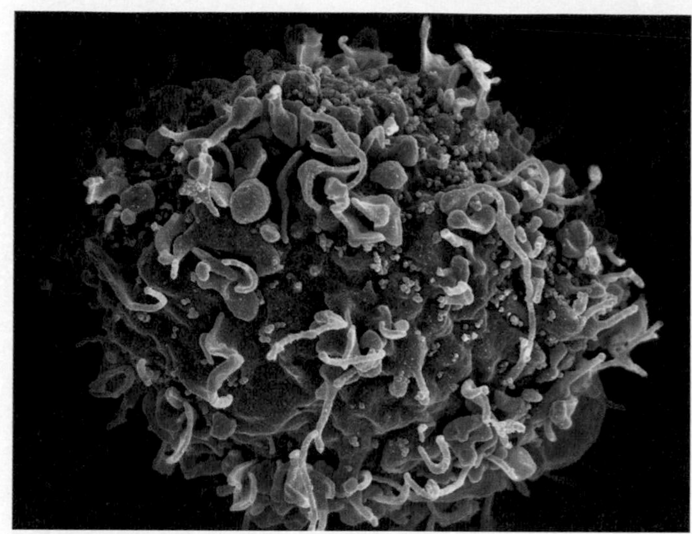

FIGURE 15-30 HIV attacks white blood cells essential for identifying foreign invaders.

TAKE HOME MESSAGE 15.19

» HIV is especially difficult to control. Mutations change the properties of the retrovirus so that it is hard for the immune system to recognize it, and they produce variants that are resistant to the drugs used to treat the HIV infection.

STREET BIO

Using evidence to guide decision making in our own lives

The five-second rule: how clean is that food you just dropped?

In 2007, two students at Connecticut College decided to test the "five-second rule": the idea that it's safe to eat food you've dropped on the ground if you pick it up within 5 seconds. They dropped apple slices and Skittles candies on the floor of the cafeteria and a snack bar, let the foods lie there for 5, 10, 30, or 60 seconds, and then tested them for bacteria.

Q: Is the five-second rule valid? The students found no bacteria on the food picked up within 30 seconds. After a minute, the apple slices had picked up some bacteria, but the Skittles had none (in a later experiment, they found bacteria on Skittles only after 5 minutes). The students concluded that the five-second rule should get an extension: they proposed that you have 30 seconds to pick up moist foods and more than a minute to pick up dry foods without risk of bacterial contamination.

Q: Should we worry less about bacterial contamination of food? This question is a serious one and warrants more study. Each year in the United States there are more than 76 million cases of illness caused by contaminated food, of which more than 5,000 are fatal.

Q: Can you generalize from a small number of observations to all possible situations? In the Connecticut College study, the students examined two food types, dropped in just two locations, and onto surfaces with unknown concentrations of bacteria. In another study, published in the *Journal of Applied Microbiology,* researchers focused on *Salmonella* bacteria. Because it has been documented that bacteria can survive on clothes, hands, sponges, cutting boards, and utensils for several days, the researchers decided to drop food on surfaces known to be covered in bacteria.

Q: Which factor is more important for dropped food: location or duration? Armed with slices of bologna and pieces of bread, the researchers found that when dropped on surfaces covered with *Salmonella* and left for a full minute, both types of food took up 1,500–80,000 bacteria. And although picking up the food quickly—within just 5 seconds—reduced by 90% the number of bacteria present, 150–8,000 bacteria still had time to hitch a ride on the food in that first 5 seconds. These experiments were replicated, with no significant differences, three times, on separate days.

A grubby conclusion. It's probably safe to say that if you're in the Connecticut College cafeteria, you can eat your dropped food as long as you pick it up quickly. But for all other situations, a bit more caution is advised—particularly, taking notice of the "drop zone." Unsanitary surfaces likely to have microbes will contaminate your food almost immediately.

GRAPHIC CONTENT

Thinking critically about visual displays of data

1 What can you conclude from this figure?

2 How many human cells are there in a human?

3 How many microbial cells are there in a human?

4 How can bacterial cells outnumber human cells in a human, yet a person still look human rather than bacterial?

5 How were the data obtained for the graph? Are they from an experiment? If so, what additional information would have been useful?

6 Create two alternative ways of conveying the information in this figure. What are the relative strengths and weaknesses of your representations compared with this figure?

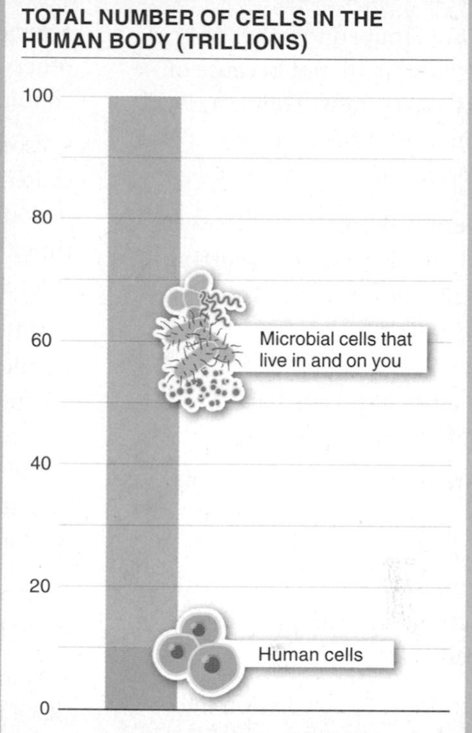

See answers at the back of the book.

KEY TERMS IN EVOLUTION AND DIVERSITY AMONG THE MICROBES

BRIEF SUMMARY

There are microbes in all three domains.

• Microbes are grouped together only because they are small, not because of evolutionary relatedness. They occur in all three domains of life and also include the viruses.

• Microbes are very small, and they are an extremely diverse group, occurring in all types of environments, from moderate to extreme.

Bacteria may be the most diverse of all organisms.

• Bacteria are single-celled organisms with an envelope surrounding the cytoplasm, which contains the DNA (they have no nuclei and no organelles). A single bacterial cell can grow into a colony of cells.

• Bacteria undergo binary fission and grow rapidly in number. Virtually all bacterial DNA codes for proteins, and three mechanisms exist for genetic transfer among bacteria.

• Some bacteria eat organic molecules, some eat minerals, and still others carry out photosynthesis. Photosynthesizing bacteria were responsible for the first appearance of free oxygen in the earth's atmosphere.

In humans, bacteria can have harmful or beneficial health effects.

• A disease-causing bacterium must colonize your body before it can make you sick, and your body is already covered with harmless bacteria.

• Some bacteria always cause disease, and others are harmful only under certain conditions.

• Sexually transmitted diseases (STDs) are caused by a variety of organisms, including bacteria, viruses, protists, fungi, and arthropods, and infect more than 300 million people each year.

• Antibiotic resistance routinely evolves in microbes, and plasmid transfer allows an antibiotic-resistant bacterium to pass that resistance to other bacteria. Excessive use of antibiotics in medicine and agriculture has led to pathogenic bacteria resistant to every known antibiotic.

Archaea exploit some of the most extreme habitats.

• Archaea possess characteristics that place them between bacteria and eukaryotes on the tree of life. Archaea and bacteria may look similar, but they have significant differences in their DNA sequences, as well as in their plasma membranes, cell walls, and flagella.

• Archaea can tolerate extreme physical and chemical conditions that are intolerable for most other living organisms, but this makes them hard to study.

Most protists are single-celled eukaryotes.

• The nucleus is an evolutionary innovation that appeared for the first time in protists. Early protists took in a bacterial cell that subsequently became the mitochondrion.

• Protists are a diverse group of mostly unicellular eukaryotic organisms, some of which can cause debilitating diseases.

Viruses are at the border between living and non-living.

• A virus is not alive, but it can carry out some of the same functions as living organisms, provided that it can get inside a cell. A virus takes over the protein-making machinery of the host cell to produce more viral genetic material and protein.

• Many diseases are caused by viruses. DNA viruses are relatively stable, while RNA viruses change quickly.

• Glycoproteins on the surfaces of viruses determine the types of cells a virus can invade. Most viruses infect just one or a few species.

• HIV is especially difficult to control. Mutations change the properties of the retrovirus so that it is hard for the immune system to recognize it, and they produce variants that are resistant to the drugs used to treat the HIV infection.

CHECK YOUR KNOWLEDGE

Short Answer

1. Why doesn't an amoeba need a respiratory system?

2. It is difficult to make generalizations about the group of organisms sometimes referred to as "microbes." Why?

3. Describe three types of lateral transfer of genetic information in bacteria.

4. What produces the characteristic taste and texture of yogurt?

5. In the mid-1800s, how was the spread of cholera slowed in London?

6. Why do more people die from *Staphylococcus aureus* infections than from HIV/AIDS?

7. Describe two characteristics of STDs that make it nearly impossible to completely eliminate these diseases.

8. Describe two differences between archaea and bacteria.

9. Why might scientists eager to learn more about archaea have a hard time doing so?

10. Which group of protists is referred to as the "animal-like protists"? Why?

11. Which feature of *Plasmodium*, the protist responsible for malaria, helps it stay ahead of the human immune system?

12. Why are viruses not considered "living"?

Multiple Choice

1. Prokaryotes are found in ______ domain(s).

a) 1 b) 2

c) 3 d) 4

e) 5

0 EASY HARD 100

30

2. "Establishing a phylogeny for bacteria is more difficult than establishing a phylogeny for plants or animals." This statement is:

a) correct, because of the quick generation time of bacteria.

b) incorrect, because lateral gene transfer makes creating a phylogeny simpler.

c) incorrect, because the metabolic diversity of bacteria makes classification impossible.

d) correct, because it is difficult to obtain enough morphological data from bacteria.

e) correct, because bacteria can engage in lateral gene transfer.

0 EASY HARD 100

62

3. Plasmids containing genes for antibiotic resistance can be exchanged between bacterial cells by:

a) conjugation.

b) artificial exchange.

c) cloning.

d) transduction.

e) conduction.

48

4. Which group of organisms utilizes the largest variety of energy sources?

a) prokaryotes

b) fungi

c) protists

d) invertebrate animals

e) vertebrate animals

53

5. Which of the following statements about antibiotics is incorrect?

a) Penicillin was the first antibiotic widely used to fight bacterial infections.

b) Antibiotics help microbes compete with other microbes.

c) Antibiotic-resistant microbes are selected for in humans who are taking antibiotics.

d) Antibiotics, though effective against viruses, are not effective against bacteria.

e) Antibiotics are used not just in human health care but also in agriculture.

44

6. Which of the following pairs of domains are the most closely related, in that they share a unique common ancestor?

a) archaea and bacteria

b) archaea and eukarya

c) bacteria and eukarya

d) None of the above; all three domains evolved from different ancestors.

e) None of the above; all three domains are equally related to one another.

7. Most archaeal species:

a) have yet to be discovered.

b) are pathogenic.

c) are bacteriophages.

d) are animal parasites.

e) are beneficial to humans.

8. Many biologists do not consider viruses to be alive. Which of the following characteristics of viruses leads to this conclusion?

a) Viruses lack a metabolic system.

b) Viruses do not respond to external stimuli.

c) Viruses are unable to reproduce on their own.

d) All of the above are correct.

e) Only a) and c) are correct.

44

9. The genetic information in all viruses is:

a) DNA.

b) RNA.

c) protein.

d) a polymerase enzyme.

e) DNA or RNA.

10. Which of the following statements about HIV is incorrect?

a) It is a virus.

b) It attacks red blood cells.

c) It mutates frequently.

d) It is derived from a simian immunodeficiency virus.

e) It contains RNA but not DNA.

Ch16

Population ecology is the study of how populations interact with their environments.

16.1 What is ecology?

16.2 A population perspective is necessary in ecology.

16.3 Populations can grow quickly for a while, but not forever.

16.4 A population's growth is limited by its environment.

16.5 Some populations cycle between large and small.

16.6 "Maximum sustainable yield" is useful but nearly impossible to implement.

A life history is like a species summary.

16.7 Life histories are shaped by natural selection.

16.8 There are trade-offs between growth, reproduction, and longevity.

16.9 *This is how we do it:* Life history trade-offs: rapid growth comes at a cost.

16.10 Populations can be described quantitatively in life tables and survivorship curves.

Ecology influences the evolution of aging in a population.

16.11 Things fall apart: what is aging and why does it occur?

16.12 What determines the average longevity in different species?

16.13 Can we slow down the process of aging?

The human population is growing rapidly.

16.14 Age pyramids reveal much about a population.

16.15 As less-developed countries become more developed, a demographic transition often occurs.

16.16 Human population growth: how high can it go?

Because of the limited availability of land for housing, Hong Kong is home to thousands of tall skyscrapers that house its densely packed population.

Population Ecology

Planet at capacity: patterns of population growth

16.1–16.6

Population ecology is the study of how populations interact with their environments.

African elephants (*Loxodonta africana*) at the Chobe River in Botswana.

16.1 What is ecology?

Lobster tastes delicious. Many people consider it one of the finest delicacies. It's not surprising, then, that catching and selling lobsters is big business. Generating almost $200 million a year for the state of Maine alone, it is among the most economically important businesses in New England. Now imagine for a minute that you were in charge of the lobster industry. The 6,000 lobstermen in Maine depend on your managing the industry so that not only can they catch and sell enough lobsters to survive, but sufficient numbers of lobsters are left to ensure there will be lobsters to catch in all the years to come. How would you do it? You would be faced with some tough questions.

- How many lobsters should you allow each lobsterman to take each day? Each year?
- Should you require that certain lobsters be thrown back? If so, should they be the biggest or the smallest? Males or females?
- Is it better to increase the size of the lobster population or maintain it at current levels?
- But wait a minute. Forget about these obviously complex questions and start with a seemingly simpler one: how many lobsters currently live in the waters off the coast of Maine? Is this even possible to determine (FIGURE 16-1)?

Ecologists have a challenging task managing natural populations when it's difficult to count the number of individuals present or determine how many the environment can support.

FIGURE 16-1 **Managing valuable resources.** Population growth models can help.

Welcome to ecology.

These difficult questions, and many others like them, are all part of **ecology,** a subdiscipline of biology defined as the study of the interactions between organisms and their environments. But don't be fooled by the simple definition. Ecology encompasses a very large range of interactions and units of observation and is studied at different levels (FIGURE 16-2). These include:

- *Individuals:* How do individual organisms respond biochemically, physiologically, and behaviorally to their environments?
- *Populations:* How do groups of interbreeding individuals change over time, in terms of growth rates, distributions, and genetic compositions?

WHAT IS ECOLOGY?

Ecology is the study of interactions between organisms and their environments. It can be studied at many levels, including:

INDIVIDUALS
Individual organisms

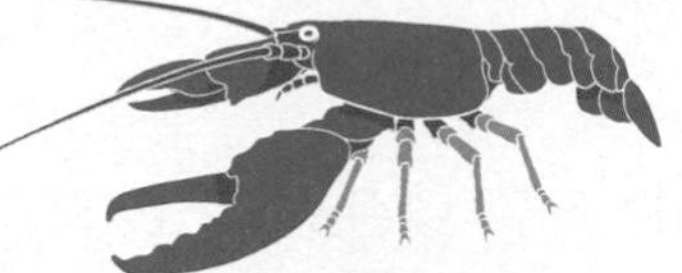

POPULATIONS
Groups of individual organisms that interbreed with each other

COMMUNITIES
Populations of different species that interact with each other within a locale

ECOSYSTEMS
All living organisms, as well as non-living elements, that interact in a particular area

FIGURE 16-2 **From individuals to ecosystems.** Ecologists study living organisms and their relationship to the environment.

- *Communities:* How do populations of different species in a locale interact?
- *Ecosystems:* How do the living and non-living elements interact in a particular area, such as a forest, desert, or wetland?

We explore ecology at all of these levels throughout this and the next two chapters. We focus in this chapter on **population ecology,** a subfield of ecology that focuses on populations of organisms of a species and how they interact with the environment. We also examine the special case of human population growth and its impact on the environment, as well as the ways in which ecological knowledge can contribute to effective conservation policies.

TAKE HOME MESSAGE 16.1

» **Population ecology is the study of the interactions between populations of organisms and their environments, particularly their patterns of growth and how they are influenced by other species and by environmental factors.**

16.2 A population perspective is necessary in ecology.

As we begin our study of population ecology, a subtle but critical shift in perspective is required. It is a shift from the individual to the population—a group of organisms of the same species living in a particular geographic region—as the primary focus. Most ecological processes cannot be observed or studied within an individual. Rather, they emerge when considering the entire group of individuals that regularly exchange genes in a particular locale (FIGURE 16-3).

Population ecology examines features that cannot be studied within an individual organism, such as population size.

FIGURE 16-3 **A change in perspective.** Ecology requires a focus on populations rather than on isolated individuals.

As an example of how a population perspective is needed, consider the case of adaptations produced by natural selection. Genetic changes don't occur within an individual. That is, a longer neck did not evolve in one individual giraffe. Instead, the genetic changes occurred within a population of giraffes over time. As a consequence of differential reproductive success among the individuals of a population with different neck lengths, there came to be more individuals with longer necks over time. Birth rates, death rates, and immigration and emigration rates are also features not of individuals but of populations.

In the next sections, we explore how populations grow, a question important both to the management of consumable natural resources—as in the lobster fisheries described above—and to the conservation of rare and endangered species.

TAKE HOME MESSAGE 16.2

» Most ecological processes cannot be observed or studied within an individual. Rather, we need to consider the entire group of individuals that regularly exchange genes in a particular locale.

16.3 Populations can grow quickly for a while, but not forever.

In a stable population, how many of the five million eggs laid by a female cod over the course of her life survive and grow to adulthood? How many of an elephant's babies survive to adulthood and reproduce? These may seem like difficult questions to answer, requiring complex mathematical models of population growth and special knowledge about cod or elephants. But actually, the most fundamental fact about population growth is ridiculously simple.

How many of the female cod's eggs and the female elephant's babies will, on average, make it to adulthood? Just two.

Q Who leaves more surviving offspring, a pair of elephants or a pair of rabbits?

If a population is not growing, each individual is ultimately replacing itself with a new individual. This is true for all species. Two of those cod eggs will make it. And two of those baby elephants will make it. For a stable population, on average two survive, rather than one, because both a male and a female contributed to each offspring. If each pair produced only one surviving offspring, the population would get smaller and smaller. But what happens to the rest of the cod eggs and elephant offspring? Along the way from gamete to adult, they may not survive. For example, an egg may not get fertilized. Or a fertilized egg may not survive to give rise to an offspring. Predation and other factors may prevent a youngster's surviving to adulthood. Or an adult may not find a mate. There are too many risks to name them all.

What would happen if more than two cod or two elephants survived? If there were more than one offspring per individual, the population would grow until the earth was covered with cod, elephants, and every other species. But that hasn't happened, and it can't. Let's investigate why.

> "There is no exception to the rule that every organic being naturally increases at so high a rate that, if not destroyed, the Earth would soon be covered by the progeny of a single pair."
>
> — CHARLES DARWIN
>
> ***The Origin of Species*, 1859**

We start by figuring out how a population grows if each individual does more than just "replace" itself. To calculate the **growth rate** of a population—the change in the number of individuals in the population in some unit of time, such as a year—we need two pieces of information: the number of individuals in the population now, N, and the per capita growth rate, r, which is simply the birth rate minus the death rate. The growth of the population in a year is the growth rate times the number of individuals present to start with: $r \times N$.

If the birth rate is greater than the death rate, the population gets bigger. If the death rate is bigger than the birth rate, the population gets smaller. For example, if there are 500 individuals in a population ($N = 500$) and, over the course of a year, 125 offspring are born, the birth rate is 125/500, or 0.25 births per individual. And if 25 of the 500 individuals die during the same year, the death rate is 25/500, or 0.05 deaths per individual. In this population, the growth rate, r, is $0.25 - 0.05$, or 0.20 individuals per individual per year. But how many individuals is that?

Since there were 500 individuals to start with, the population would increase by 0.20×500, or 100 individuals, during a year (the time period observed). After a second year, if it grew at the same rate, the population growth would be 0.20×600, or 120 new individuals. This would mean the population now has 720 individuals.

With the same calculations, we could predict the population size for the next 10, 50, or even 100 years (FIGURE 16-4). When a population's size increases at a rate proportional to its current size—in other words, the bigger the population, the faster it grows—the growth is called **exponential growth.** As the J-shaped graph reveals, the size of a population growing exponentially becomes astronomical very quickly. In 10 years, the population of 500 elk would grow to 2,580. In 50 years, it would reach almost four million, and after 80 years it would surpass a billion individuals. It's clear that unchecked exponential population growth ends badly.

But the world isn't overrun with cod or elephants or elk, so, clearly, exponential growth doesn't occur for long. In the next section, we explore why populations cannot keep growing unchecked forever.

EXPONENTIAL GROWTH

Exponential growth occurs when each individual produces more than the single offspring necessary to replace itself. According to realistic (and moderate) estimates of birth and death rates, a population of just 500 elk would grow to more than a billion individuals within 80 years and eventually would cover the earth.

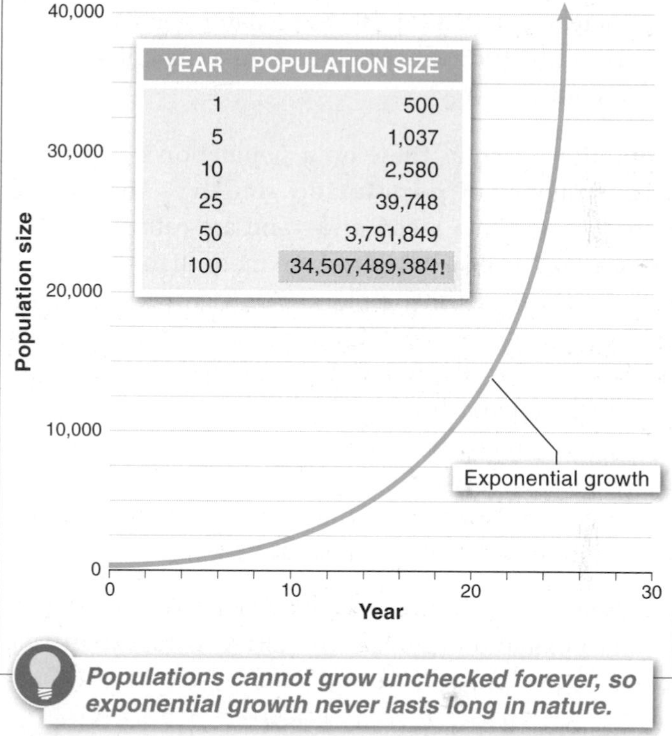

YEAR	POPULATION SIZE
1	500
5	1,037
10	2,580
25	39,748
50	3,791,849
100	34,507,489,384!

Populations cannot grow unchecked forever, so exponential growth never lasts long in nature.

FIGURE 16-4 **Unchecked growth.** Any animal or plant species that grew exponentially would eventually cover the planet.

TAKE HOME MESSAGE 16.3

» Populations tend to grow exponentially, but this growth is eventually limited.

16.4 A population's growth is limited by its environment.

Life gets harder for individuals when it's crowded. Whether you are an insect, a plant, a small mammal, or a human, difficulties arise in conjunction with increasing competition for limited resources. In particular, as population size increases, organisms experience:

- Reduced food supplies, due to competition
- Diminished access to places to live and breed, also due to competition
- Increased incidence of parasites and disease, which spread more easily when their hosts live at higher density
- Increased predation risk, as predator populations grow in response to the increased availability and visibility of prey

Limitations such as these on a population's growth are a consequence of **population density**—the number of individuals in a given area—and are called **density-dependent factors.** They cause more than discomfort: with increased density, a population's growth is reduced as limited resources slow it down. This ceiling on growth is the **carrying capacity, *K*,** of the environment. As a population approaches the carrying capacity, death rate increases, emigration rate increases (as individuals seek more hospitable places to live), and a reduction in birth rate usually occurs, as low food supplies result in poor nutrition, which, in turn, reduces fertility (FIGURE 16-5).

Here's how the carrying capacity of an environment influences a population's growth. We start with our exponential growth equation:

$$\text{Population growth} = r \times N$$

Then we multiply by a term that can slow down exponential growth:

$$\text{Population growth} = (r \times N) \times [(K - N)/K]$$

When *N*, the population size, is small relative to the carrying capacity, *K*, it means there are plenty of resources to go around. In this situation, the term $[(K-N)/K]$ is close to 1. To illustrate this, let's choose a large number for *K*, say, 100,000, and a small number for *N*, say, 1,000. This would mean $[(K-N)/K]$ equals [(100,000 – 1,000)/100,000], or 99,000/100,000, which equals 0.99.

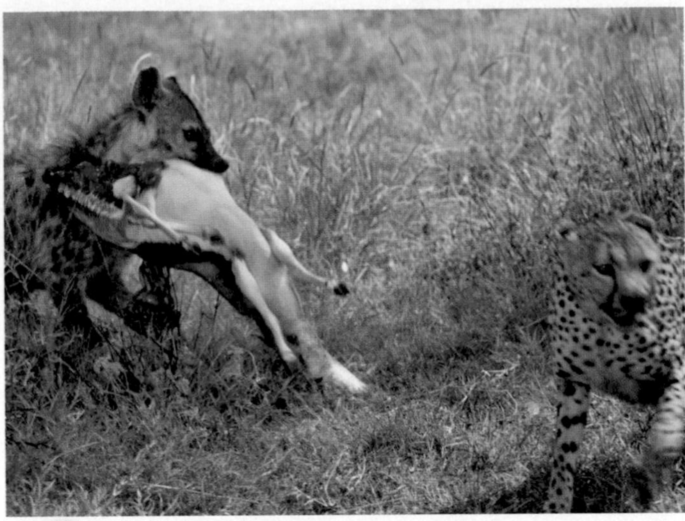

With increased density, a population's growth can be reduced by limited resources.

FIGURE 16-5 **Fighting over scarce resources.**

So in this situation, we are multiplying normal exponential growth by 0.99. This is almost the same as multiplying it by 1, which doesn't change it. As a consequence, the population is growing exponentially (or very close to exponentially).

On the other hand, if the population size, *N*, is close to the value of *K*, it means that the environment is nearly full to capacity, and the term $[(K - N)/K]$ is close to 0. Let's plug in some numbers again. We can use 100,000 again for *K*. But let's use 99,000 for *N*. In this case, $[(K - N)/K]$ equals [(100,000 − 99,000)/100,000], or 1,000/100,000, which equals 0.01. In this situation, we multiply normal exponential growth by 0.01, which reduces it to almost zero. In other words, population growth slows more and more as the population size nears the environment's carrying capacity.

When a population grows exponentially at first, but then slows as the population size approaches the carrying capacity, the growth is called **logistic growth** and makes an S-shaped curve (FIGURE 16-6).

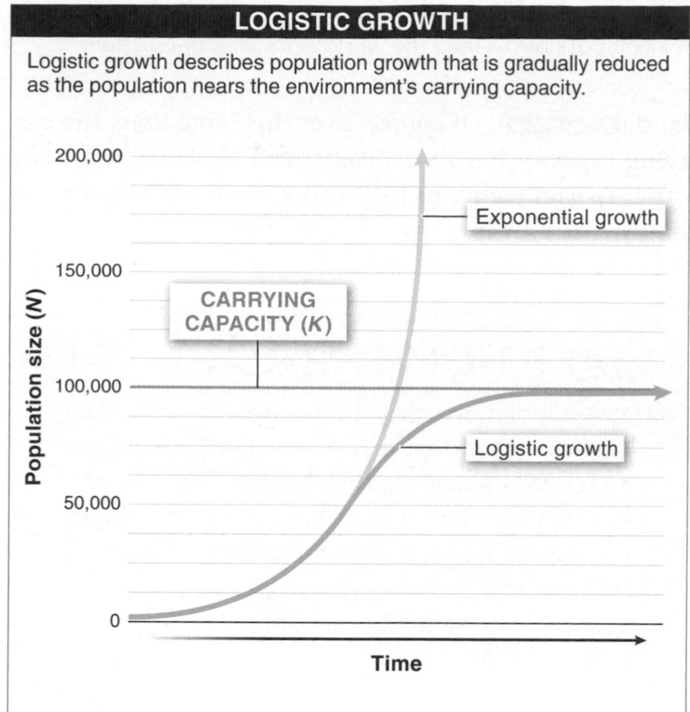

FIGURE 16-6 **Lack of resources limits growth.** Logistic growth is population growth that has stabilized because of limited resources.

DENSITY-INDEPENDENT FACTORS LIMIT POPULATION SIZE

Density-independent forces—including fires, floods, and earthquakes—can dramatically reduce population size.

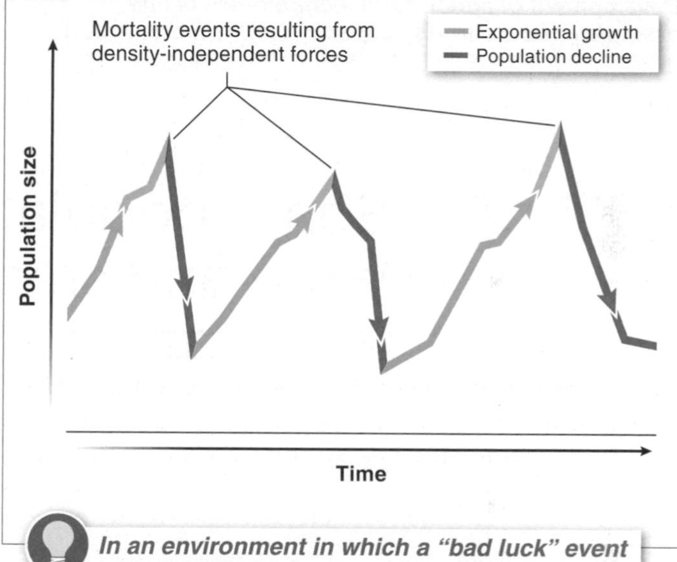

In an environment in which a "bad luck" event repeatedly occurs, a population can be in a perpetual state of exponential growth with periodic massive mortality events.

FIGURE 16-7 **Events can limit population growth.**

Density-independent factors can also knock a population down, acting like "bad luck" limits to population growth. These forces strike populations without regard for population size, by increasing the death rate or decreasing the number and rate of offspring produced. They are mostly weather- or geology-based, including calamities such as floods, earthquakes, and fires. They also include habitat destruction by other species, such as humans. The population hit by the disaster may be at the environment's carrying capacity, or it may be in the initial stages of exponential growth, before its growth is slowed by the carrying capacity. In either case, the density-independent force simply knocks down the size. The population then resumes logistic growth.

In an environment where these "bad luck" events repeatedly occur, a population might never have time to reach the carrying capacity. Instead, as the series of jagged curves in FIGURE 16-7 show, the population might be perpetually in the exponential growth portion of the logistic growth curve, with periodic massive mortality events.

The growth of populations doesn't always appear as a smooth S-shaped logistic growth curve. For some populations, particularly humans, the carrying capacity of an environment is not set in stone. Consider that, in 1883, an acre of farmland in the United States produced an average of 11.5 bushels of corn. By 1933, this was up to 19.5 bushels per acre. By 1992, it had increased to 95 bushels per acre. And by 2013, it had increased to 159 bushels per acre! How did this happen? The development of agricultural technologies—including the use of vigorous hybrid varieties of corn, rich fertilizers, crop rotation, and effective pest management—has allowed farmers to produce more and more food from the same amount of

Development of agricultural technologies is one example of how carrying capacity can be increased.

FIGURE 16-8 Efficient crop production. Advances in agriculture make it possible to feed the world's growing population.

land (FIGURE 16-8). Of course, over this same time, the carrying capacity has most likely been *decreased* for many other species trying to live in the same environment as humans and their crops.

TAKE HOME MESSAGE 16.4

» A population's growth can be constrained by density-dependent factors: as density increases, a population reaches the carrying capacity of its environment, and limited resources put a ceiling on growth. It can also be reduced by density-independent factors such as natural or human-caused environmental calamities.

16.5 Some populations cycle between large and small.

Nature is not always tidy. Exponential and logistic growth equations, for instance, help us understand the concept of population growth, but real populations don't always show such "textbook" growth patterns. Sometimes they vary greatly in their rates and patterns of growth.

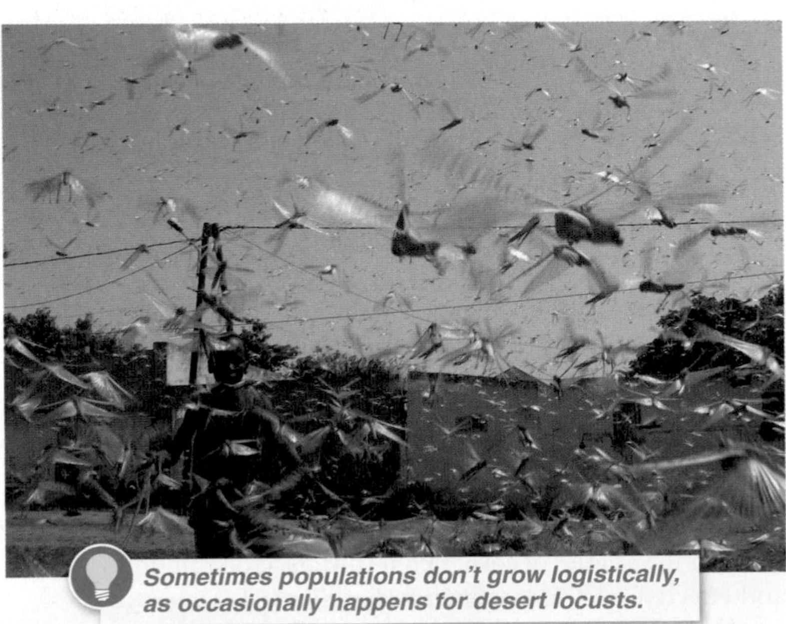

Sometimes populations don't grow logistically, as occasionally happens for desert locusts.

FIGURE 16-9 Population explosion!

Locust swarms of biblical proportions certainly attest to the unpredictability of population growth (FIGURE 16-9). In northwest Africa, the desert locusts (migrating grasshoppers) normally live as solitary individuals in relatively small, scattered populations. In 2004, however, the population size increased rapidly, probably due to unusually good rains and mild temperatures. As the rainy season ended and green areas gradually shrank, the locusts became progressively concentrated in smaller and smaller areas. At this point, for reasons related to overcrowding but not completely understood, the locusts behaved like a mob, rather than solitary individuals. Giant swarms—some including tens of millions of insects—began flying across huge expanses of land in search of food. They completely consumed the crops on giant swaths of farmland, causing more than $100 million in damage.

There is more than one way for population growth to deviate from the standard pattern. The explosive locust population growth, for example, occurs at unpredictable intervals. Another unusual pattern is the oscillations of the lynx and snowshoe hare populations of Canada. As seen in FIGURE 6-10, rather than undergoing smooth logistic growth, the populations of both the snowshoe hare and its predator, the lynx, cycle regularly between increases to very large numbers and crashes to much smaller numbers. Thanks to the Hudson's Bay Company, which kept detailed records for decades on the number of pelts it purchased from fur trappers, this population cycling is well documented. Although its cause is not fully certain, to some extent, the predator and the prey cause their own cycling:

1. The hare population size grows,
2. providing more food for the lynx,
3. which then reproduce at a higher rate,
4. causing them to eat great numbers of the hares,
5. thereby reducing the hare population (and the lynx's food source),
6. causing the lynx population to crash,
7. enabling the hare population to grow, and the cycle to begin anew.

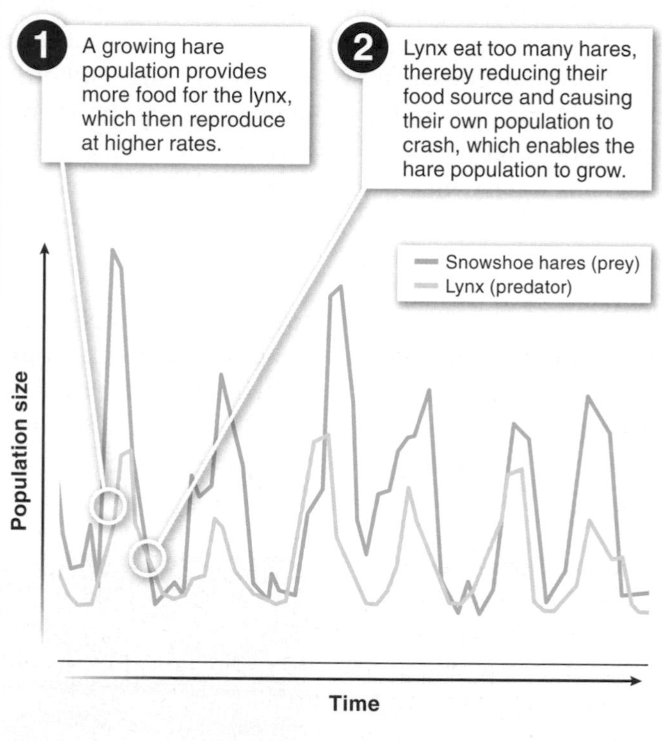

POPULATION OSCILLATIONS
The snowshoe hare and lynx populations rise and crash in a 10-year cycle.
1 A growing hare population provides more food for the lynx, which then reproduce at higher rates.
2 Lynx eat too many hares, thereby reducing their food source and causing their own population to crash, which enables the hare population to grow.
Snowshoe hares (prey)
Lynx (predator)
Population size
Time

FIGURE 16-10 **Predator and prey.** Population numbers—based on the number of pelts sold to the Hudson's Bay Company by trappers—reveal cycles.

In discussing these examples of unusual population growth, we must not lose sight of the fact that, for the most part, regularly cycling populations and populations with periodic outbreaks of huge numbers are more the exception than the rule. In general, the logistic growth pattern describes populations better than any other model.

Q Do lemmings commit suicide by jumping off cliffs when their populations get too big?

One population-growth myth that demands debunking is the story of lemmings and their supposed suicides. Do lemmings jump off cliffs when their population becomes too large? In a word: *no*. Lemming populations do occasionally experience large increases in size. Then, just as locust behavior changes when population density gets too great, lemming behavior changes, and many individuals migrate in search of less crowded habitats and more food. As they enter unfamiliar territory, some lemmings may suffer increased rates of death. But these deaths are not suicides, and they do not occur in large groups.

How did the myth of lemming mass suicides become so prevalent? In the filming of a 1958 Disney documentary, *White Wilderness,* many lemmings were placed on a giant snow-covered turntable so that it would appear they were migrating. In a later scene, lemmings were filmed on a cliff overlooking a river and were then chased into the water as if in a mad rush. None of these scenes were of actual lemming migrations, and in nature, lemmings never behave like this.

There are practical reasons for predicting population sizes and growth rates, as we'll see. But it can be difficult to translate simple growth models into feasible management practices.

TAKE HOME MESSAGE 16.5

» Although the logistic growth pattern is better than any existing model for describing the general growth pattern of populations, some populations cycle between periods of rapid growth and rapid shrinkage.

16.6 "Maximum sustainable yield" is useful but nearly impossible to implement.

Suppose you worked for a logging company and were responsible for selecting which trees in an eastern hardwood forest to cut down. Your responsibilities would be similar to those you'd have if you were in charge of the lobster fisheries discussed at the beginning of this chapter. In the case of the hardwood forest, how many trees would you cut down for maximum wood yield?

The maximum yield on any given day would be obtained by cutting *everything* down (known as clear-cutting). No other strategy could yield more immediately. But such a harvest can be done only once, so it isn't really a management strategy. A more sensible strategy would include harvesting some individuals and leaving others for the future. With that in mind, which trees (or lobsters) would you harvest to minimally affect the population's growth? Ideally, you would harvest those individuals that are no longer contributing to the population growth. But there may not be many post-reproductive individuals in the population, or it may be too hard to identify them.

What's the better solution? For long-term management, it is better to harvest some but leave others still growing and reproducing for harvest at a later time, so the population can persist indefinitely. The special case in which as many individuals as possible are removed from the population without impairing its growth is called the **maximum sustainable yield** (FIGURE 16-11). The value in such a harvest is that it can be carried out repeatedly, clearly yielding more in the long run than the shortsighted strategy of a complete, one-time harvest.

Your first step as manager is to determine the maximum sustainable yield for the resource, calculated as the point at which the population is growing at its fastest rate. In Figure 16-6, we can see that in logistic growth, the population is getting larger at the fastest rate when its size is equal to half the carrying capacity. At this midpoint, scarcity of mates is not a problem, as it can be at low population levels, nor is competition a problem, as it can be near the carrying capacity—one of the reasons that the population doesn't grow at all when it reaches the environment's carrying capacity.

Maximum sustainable yield is a useful concept, applicable not just to tree or lobster harvesting but also to managing nearly all natural resources. In fact, there are 31 official U.S. government agencies mandated to utilize the maximum sustainable yield concept in determining harvest levels. This is, however, nearly always an impossible task. Why is that so?

Q Almost all natural resource managers working for the U.S. government fail to do their job exactly as mandated. Why?

There are numerous reasons. For starters, if maximum sustainable yield occurs when a population is at half its carrying capacity, do we first wait until the population stabilizes at its carrying capacity to determine what half of that carrying capacity would be? But then isn't it inefficient to sit around waiting for the population to reach its carrying capacity, when we want to maintain it at half that size? Or can we just estimate carrying capacity and then calculate half of it? But, for lobsters, won't that be difficult, given that they live underwater? Put simply, we rarely know the value of K.

MANAGING NATURAL RESOURCES

FIGURE 16-11 **Tree harvest.** Wood is a renewable resource, but how many trees can we cut down (and which ones?) without irreversibly depleting the supply?

Effective and sustainable management of natural resources requires the determination of a population's maximum sustainable yield, the point at which the maximum number of individuals are being added to the population (and so can be harvested or utilized).

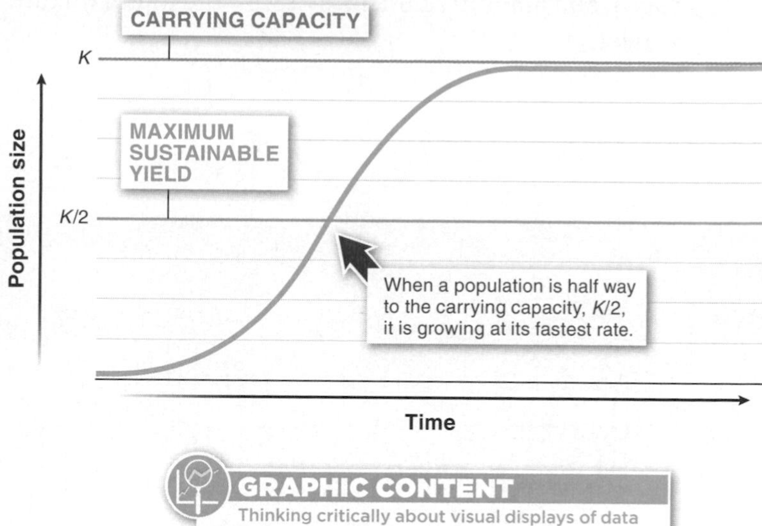

GRAPHIC CONTENT
Thinking critically about visual displays of data
Turn to p. 559 for a closer inspection of this figure.

And the problem gets worse. For one thing, with many species, not only do we not know the carrying capacity, but we don't even know the number of individuals currently living. It is difficult to accurately count humans, so imagine how hard it is to count individuals of species that live underwater, fly, or are microscopic.

And if we were to solve the mysteries of counting individuals and knowing the carrying capacity, we would still have to figure out whether carrying capacity is stable from year to year. For instance, if it depends on levels of resources, it may be cyclic. And even if we can figure out the carrying capacity and the population size, we would not be certain which individuals to harvest. Often, not all individuals in a population are contributing equally to population growth. The post-reproductive individuals, as mentioned above, do not contribute to this growth.

As it turns out, harvesting natural resources for maximum efficiency generates insights into how to fight biological pests, such as cockroaches or termites (FIGURE 16-12). The problem is just turned on its head: which individuals in the pest population would you concentrate on killing to most effectively slow population growth? Those at the age of maturity, with the highest reproductive value, contribute most to the growth of the population. Similarly, because a population can still grow very rapidly with only a few males—given that any one male can fertilize a large number of females—it is effective to focus pest control on females.

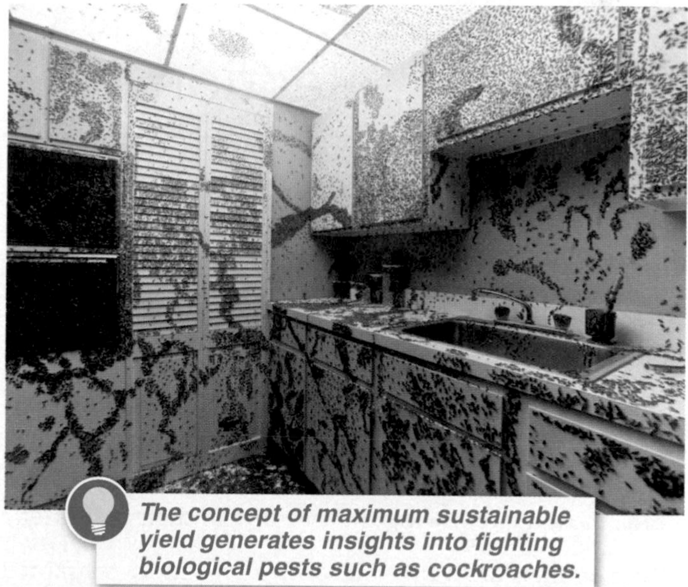

The concept of maximum sustainable yield generates insights into fighting biological pests such as cockroaches.

FIGURE 16-12 **Population explosion—pests!** Population-growth models don't just guide us to maximum production of valuable resources. They can help us efficiently reduce pest populations, too.

Patterns of population growth and the environmental features that influence them can lead to evolutionary changes in populations, shaping features such as life span, age of reproduction, number of offspring produced, and amount of offspring care—the topics we turn to next.

TAKE HOME MESSAGE 16.6

» Based on models of population growth, it might seem easy to utilize natural resources efficiently and sustainably. In practice, however, difficulties such as estimating population size and carrying capacity complicate the implementation of such strategies.

16.7–16.10

A life history is like a species summary.

An American Oystercatcher feeding its young.

16.7 Life histories are shaped by natural selection.

Some animals reproduce with a "big bang." *Antechinus* is an Australian mouse-sized marsupial, and the males are classic big-bang reproducers (FIGURE 16-13). At one year of age, they enter a two- to three-week period of intense mating activity, copulating for as long as 12 hours at a time. Shortly after this, the males undergo rapid physical deterioration—they lose weight, much of their fur falls out, their resistance to parasites falls—and then they die. (Although occasionally a female will live for two or three years, most die following the weaning of their first litter.)

Other animals are a bit—but only a bit—more restrained in their reproduction. The house mouse reaches maturity in about one month and produces litters of 6–10 offspring nearly every month, sometimes generating more than a hundred offspring in its first year of life.

And some animals could not be farther from big-bang reproducers. The little brown bat is also mouse-sized, yet does not reach maturity until one year of age and typically produces only a single offspring each year. It can, however, live more than 33 years in its natural habitat.

Why all the variation? And is one strategy better than the others, evolutionarily? One of the most important recurring themes in biology is that evolution nearly always finds more than one way to solve a problem. As the marsupial mouse, house mouse, and little brown bat illustrate, there are many possible responses to the challenge of when to reproduce, how often to reproduce, and how much to reproduce in any given episode. The answers to these questions make up an organism's **life history**—the species' vital statistics, including age at first reproduction, probabilities of survival and reproduction at each age, litter size and frequency, and longevity. Life histories tell us as much about a species as possible with a small amount of information.

Life histories vary from the big-bang reproductive strategy in *Antechinus,* in which the male's **reproductive investment**—all of the material and energetic contribution that an individual will make to its offspring—is made in a single episode, to strategies, such as that of humans, in which organisms have repeated episodes of reproduction. Plants also have life histories, and, like animals, they

VARIATION IN LIFE HISTORIES

Antechinus

BIG-BANG REPRODUCTION
- Reaches sexual maturity at one year
- Mates intensely over a three-week period
- Males die shortly after mating period
- Females usually die after weaning their first litter

House mouse

FAST, INTENSIVE REPRODUCTIVE INVESTMENT
- Reaches sexual maturity at one month
- Produces litters of six to ten offspring every month

Little brown bat

SLOW, GRADUAL REPRODUCTIVE INVESTMENT
- Reaches sexual maturity at one year
- Produces about one offspring per year

FIGURE 16-13 **Reproductive strategies.**

have a range of strategies. Annuals, such as corn, wheat, rice, peas, marigolds, and cauliflower, usually reproduce once and then die. Perennials, such as apples, raspberries, grapes, ginger, and garlic, on the other hand, reproduce repeatedly, often living for many years.

Returning to the question posed above—which life history strategy is best?—we need to consider two questions.

1. What is the cost of the reproductive investment during any reproductive episode? Producing offspring is risky in several ways. The act of mating can be risky (by increasing the likelihood of an individual's being eaten by a predator, for example). And the wear and tear of reproduction on an individual's body takes its toll. Thus, the number of offspring an organism produces can only go so high before it becomes detrimental. An organism might produce four offspring in an episode, for example, without significantly increasing its risk of dying from the effort. But if, by chance, it produces five or six, this may take such a toll that the individual is unlikely to live to have another litter. Thus, for many organisms, a smaller litter is favored by natural selection because it enables the individual to have additional litters in the future, maximizing its *lifetime* reproductive success.

2. What is an individual's likelihood of surviving to have future reproductive episodes? If predation rates or other sources of mortality are very high, an individual might not be alive in one month. If so, it makes less sense to defer reproduction: why save for a future unlikely to come?

Taken together, the answers to these questions help us understand the variety of life history strategies we see in nature.

Rodents are toward the fast extreme of reproductive strategy. They experience very high mortality rates in the wild, and natural selection has consequently favored early and heavy reproductive investment. Rodents provide relatively little parental care but produce many litters, each with many offspring. Organisms with this type of life history tend to have relatively poor competitive ability but a high rate of population growth; each individual's likelihood of surviving is low, but so many are produced that the odds of two offspring surviving are high.

Humans are at the slow extreme of life histories, waiting two and sometimes several more decades before reproducing. (And they generally have just one offspring at a time.)

FIGURE 16-14 With such a low probability of dying in any given year, humans can defer reproduction.

At the slow extreme, humans have such a low probability of dying in any given year that they can defer reproduction, or at least reproduce in small amounts—once every few years—without much risk, enabling them to have just one offspring at a time but to invest significant parental effort in maximizing its likelihood of surviving. Organisms with this type of life history generally have evolved in such a way that their competitive ability—the likelihood of an individual surviving—is high, although their rate of population growth is low. That is, not many offspring are produced, but those that are have a high likelihood of surviving (FIGURE 16-14).

Q Why do humans defer reproducing so much longer than do cats or mice?

TAKE HOME MESSAGE 16.7

» An organism's investment pattern in growth, reproduction, and survival is described by its life history. Very different strategies can achieve the same outcome in which a mating pair of individuals produces at least two surviving offspring.

16.8 There are trade-offs between growth, reproduction, and longevity.

If you were designing an organism, how would you structure its life history for maximum fitness? (Recall from Chapter 10 that fitness is an organism's reproductive output relative to that of other individuals in the population.) Ideally, you would create an organism that (1) produces many offspring (beginning just after birth and continuing every year), while (2) growing tremendously large, to reduce predation, and (3) living forever.

Unfortunately, evolution operates with some constraints. Some of these traits are just not possible, because selection that changes one feature tends to adversely affect others. When evolution increases one life history characteristic, another characteristic is likely to decrease. Life history trade-offs can be better understood in the context of the three areas to which an organism allocates its resources: growth, reproduction, and survival. When resources are limited, increased allocation to one of these areas tends to reduce allocation to one or both of the others. Some examples of the best-studied trade-offs follow.

1. Reproduction and survival. Among big-bang reproducers such as the marsupial mouse (*Antechinus*) and salmon, investment in reproduction is exceedingly

LIFE HISTORY TRADE-OFFS

REPRODUCTION AND SURVIVAL
Big-bang reproducers such as salmon make a single, exceptionally high investment in reproduction, then die shortly afterward.

REPRODUCTION AND GROWTH
Beech trees grow much more slowly in the years when they produce many seeds than they do in years when they produce few seeds.

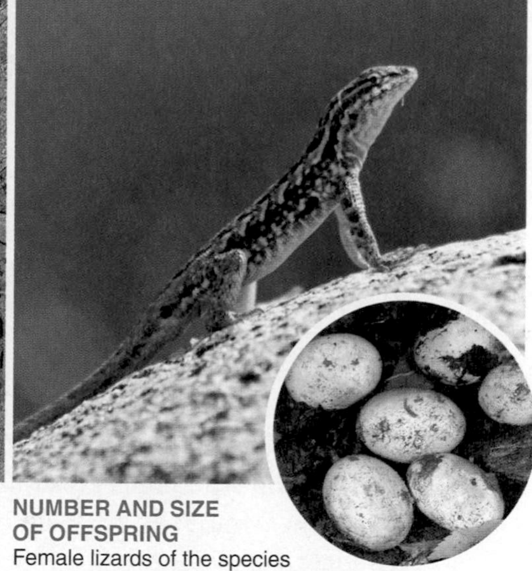

NUMBER AND SIZE OF OFFSPRING
Female lizards of the species *Uta stansburiana* produce medium-size eggs as a compromise between a large number of small eggs (with poor survival of offspring) and few large eggs (with better survival of offspring).

FIGURE 16-15 **Making trade-offs.**

Q With one simple surgical procedure, most men could add more than 10 years to their life span. Why doesn't anyone opt for it?

high, followed shortly by death (FIGURE 16-15). If individuals are physically prevented from reproducing, however, they can live many more years.

2. Reproduction and growth. Beech trees grow more slowly during years when they produce many seeds than in years when they produce no seeds. Similarly, female red deer kept from reproducing in their first year of life grow significantly larger during that year.

3. Number and size of offspring. The females of one lizard species, *Uta stansburiana,* can lay more eggs if the eggs are smaller, but a higher proportion of the offspring survive if the eggs are larger. Evolution has led to a medium-size egg that tends to maximize the number of surviving offspring.

There are many other trade-offs, such as the number of offspring produced and amount of parental care given. Additionally, among mammals, litter size increases with latitude (distance from the equator). This trend may be due to the trade-off between litter size and frequency. Closer to the equator, animals can breed for a longer portion of the year (year-round near the equator). Farther from the equator, the breeding season is shorter. To compensate for having fewer litters at higher latitudes, mammals produce more offspring per litter.

Can you think of other trade-offs?

TAKE HOME MESSAGE 16.8

» Because constraints limit evolution, life histories are characterized by trade-offs between investments in growth, reproduction, and survival.

THIS IS HOW WE DO IT

Developing the ability to apply the process of science

16.9 Life history trade-offs: rapid growth comes at a cost.

Sometimes it's tricky to collect experimental evidence supporting a theoretical prediction, even when it seems extremely likely that the prediction is accurate. This is the case for the idea that there must be a trade-off between growth and longevity.

Collecting the appropriate evidence is challenging due to several confounding factors.

1. If you look at the slowest-growing organisms in a population, they may turn out not to have the greatest longevity, simply because their slow growth is due to poor nutrition—which tends to reduce longevity.
2. If you look at the fastest-growing organisms, they may have the greatest longevity simply because their faster growth reflects access to better nutrition.
3. And, because growth rate is often positively correlated with adult body size—which is, in turn, positively correlated with longevity—the fastest-growing organisms may end up having the greatest longevity.

Why is it useful to randomize subjects to experimental treatments?

In each of these cases, the data aren't appropriate for evaluating whether there is a trade-off specifically between growth and longevity. Testing that prediction requires extremely well-controlled experiments.

In a 2013 paper in the *Proceedings of the Royal Society of London,* some researchers reported a clever experimental approach that enabled them to evaluate, in a rigorous way and for the first time, the relationship between growth rate and life span. Here's how they did it.

Working with small (~5 cm long) fish called three-spined sticklebacks (*Gasterosteus aculeatus*), the researchers were able to alter the growth trajectories of the fish by manipulating the temperature of their aquarium water. The sticklebacks are ectothermic, meaning that, unlike us, they don't maintain a constant body temperature. Rather, their body temperature decreases or increases with the temperature of the water in which they are living.

The researchers randomly assigned groups of fish to one of three temperature conditions: normal (6° C), warmer (10° C), or cooler (4° C). They maintained them at this temperature for 4 weeks, designating this as Period 1. All of the groups were then maintained at 6° C (normal temperature) for the remainder of the fishes' lives, designated Period 2. Because the median life span of these sticklebacks is about two years—with many living three years or longer—Period 1 represented less than 4% of the median life span.

What happened to fish growth when their water was warmer or cooler than usual?

The effect of the temperature manipulation was to reduce growth at the cooler temperature and increase growth at the warmer temperature during Period 1. As expected, at the end of this period there were significant differences in the average size of the sticklebacks in the three experimental groups.

The temperature perturbations, however, had another effect: they altered growth rates during Period 2, when all were kept at the normal temperature of their environment. Sticklebacks from the populations maintained at a cooler temperature in Period 1 (which had grown more slowly) experienced more rapid, "catch-up" growth, and those maintained at a warmer temperature in Period 1 exhibited slowed-down growth. As a result of the catch-up or slowed-down growth, after 3–4 months there were no differences among the three groups in the average size of the fish.

Is there a cost to growing more quickly? What must an organism give up in exchange?

As noted above, the researchers maintained all of the sticklebacks until they died, and they monitored their longevity. When they compared the average life spans across the three groups, the results were dramatic.

- Groups that spent Period 2 in catch-up growth had a decreased median life span—14.5% lower than that of the normal-temperature group.
- Groups that spent Period 2 in slowed-down growth had an increased life span—30.6% higher than that of the normal-temperature fish.

What can we conclude from these results?

The researchers concluded that the results demonstrated the existence of a growth–life span trade-off.

They also reported additional measurements that supported their conclusion and provided evidence that the experiment was very well-controlled. The life span differences were independent of (1) the eventual size attained by the fish, and (2) the reproductive investment made by the fish during Period 2. Also, life span was unrelated to growth rate during the temperature manipulation (Period 1).

In discussing their results, the researchers noted that the sticklebacks grew more quickly or slowly during Period 1 as a consequence of digesting and processing food more quickly or slowly. And that depended solely on the temperature. Why didn't this affect longevity? Because the differences in growth rate were not a consequence of diverting resources away from maintaining their bodies.

During Period 2, however, all the fish were living at the same temperature, so those growing more slowly were able to divert more resources to maintenance or reproduction. Those growing more quickly, on the other hand, could achieve that increased growth only by diverting valuable resources away from maintenance and reproduction.

The experimental design did not allow the researchers to determine the mechanisms for the longevity differences. They speculated, however, that more rapid growth during Period 2 may have led to more cellular damage, resulting from stresses associated with increased metabolic activity.

TAKE HOME MESSAGE 16.9

» Three-spined sticklebacks exposed to relatively cold or warm temperatures early in life have, respectively, reduced or increased growth rates. Returned to normal temperature, they exhibit catch-up or slowed-down growth. Catch-up growth reduces longevity, while slowed-down growth increases longevity, reflecting a trade-off between growth and life span.

16.10 Populations can be described quantitatively in life tables and survivorship curves.

Biologists have learned some important lessons from insurance agents. Because insurance companies need to estimate how long an individual is likely to live, they invented something called the **life table** (FIGURE 16-16). It ought to be called a death table, though, because the table tallies the number of people in a population within a certain

LIFE TABLE

AGE (beginning of interval)	NUMBER ALIVE (beginning of interval)	PROPORTION ALIVE (beginning of interval)	DEATHS DURING INTERVAL	PROBABILITY OF DYING DURING INTERVAL
0	210	1.00	140	0.67
3	70	0.33	28	0.40
6	42	0.20	28	0.67
9	14	0.07	10	0.71
12	4	0.02	4	1.00
15	0	0.00	n/a	n/a

In species with type II survivorship curves, individuals experience approximately the same risk of death at any age.

Cactus ground finch

FIGURE 16-16 **Summarizing life and death in a table.** Shown here: a life table for the cactus ground finch on one of the Galápagos Islands. (The finch has one of the three types of "survivorship curves" discussed in the text.)

SURVIVORSHIP CURVES

Survivorship curves show the proportion of individuals of a particular age that are alive in a population.

TYPE I

High survivorship until old age, then rapidly decreasing survivorship

Giant tortoise

TYPE II

Survivorship decreases at a steady, regular pace

Common kingfisher

TYPE III

High mortality early in life, but those that survive the early years live long lives

Mackerel

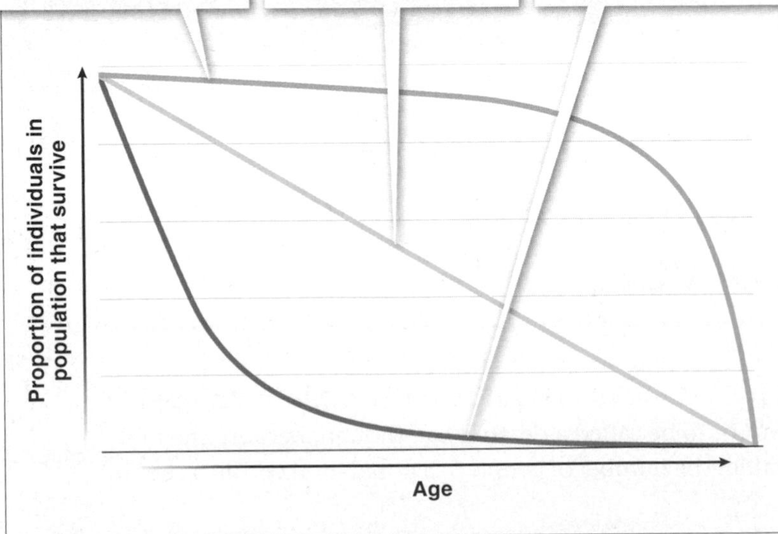

FIGURE 16-17 **Three types of survivorship.**

age group, say 10–20 years old, and the number of individuals within that age range who die. From these numbers, insurers can predict an individual's likelihood of dying or surviving within a particular age interval. For the insurance companies, making money hinges on making accurate life span predictions. For biologists, a life table is a quick window into the lives of the individuals of a population, showing how long they're likely to live, when they'll reproduce, and how many offspring they'll produce.

From life tables, biologists create **survivorship curves,** graphs showing the proportion of individuals of a particular age that are now alive in a population. Survivorship curves indicate an individual's likelihood of surviving through a particular age interval. And survivorship curves reveal a huge amount of information about a population, such as whether most offspring die shortly after birth—think back to the five million eggs produced by some fish—or whether most survive to adulthood and are likely to live long lives, such as humans in the United States.

Three distinct types of survivorship curves are shown in FIGURE 16-17. The curves plot the proportion of individuals surviving at each age, across the entire range of ages seen for that species.

Type I. At the top of the graph in Figure 16-17, in blue, is the survivorship curve seen in most human populations, and shared by the giant tortoise. We and the tortoise have a very high probability of surviving every age interval until old age, then the risk of dying increases dramatically.

Species with a type I survivorship curve, which includes most large mammals, usually have a few features in common. They have few natural predators, so they are likely to live long lives. They tend to produce only a few offspring—after all, most will survive—and they invest significant time and effort in each offspring.

Type II. In the middle of the graph, in green, is a survivorship curve seen in many bird species, such as the common kingfisher, and in small mammals such as squirrels. The straight line indicates that the proportion alive in each age interval drops at a steady, regular pace. In other words, with a type II survivorship curve, the likelihood of dying in any age interval is the same, whether the bird is between 1 and 2 years old, or between 10 and 11.

Type III. At the bottom of the graph, in purple, is a survivorship curve in which most of the deaths occur in the youngest age groups. Common in most plant and insect species, as well as in many marine species such as oysters and fish, the type III curve describes populations in which the few individuals lucky enough to survive the first few age intervals are likely to live a much longer time. Species with this type of survivorship curve tend to produce very large numbers of offspring, because most will not survive. They also tend not to provide much parental care, if any. A classic example is the mackerel. One female might produce a million eggs! Obviously, most of these (on average, 999,998) do not survive to adulthood, or the planet would be overrun with mackerel.

In reality, most species don't have survivorship curves that are definitively type I, II, or III: they may be anywhere in between. These three types, though, represent the extremes and help us make predictions about reproductive rates and parental investment without extensive observations of individual behavior. Survivorship curves can also change over time or location. Humans in developing countries, for example, tend to have higher mortality rates in all age intervals—particularly in the earliest intervals—relative to individuals in industrialized countries.

TAKE HOME MESSAGE 16.10

» **Life tables and survivorship curves summarize the survival and reproduction patterns of the individuals in a population. Species vary greatly in these patterns: the highest risk of mortality may occur among the oldest individuals or among juveniles, or mortality may strike evenly at all ages.**

16.11–16.13

Ecology influences the evolution of aging in a population.

Protected from predators by their quills, North American porcupines are extremely long-lived.

16.11 Things fall apart: what is aging and why does it occur?

Picture your grandparents or people in their seventies or eighties. Clearly they have aged. But what exactly does that mean? And does it differ from the way that people in their thirties or forties have aged? Sometimes aging involves sagging skin. Sometimes it involves weakened muscles and bones. Sometimes it involves senility.

AGING

Aging differs from one individual to another, but we recognize individuals of different ages without any difficulty. Shown here: group portraits of the same 16 women, taken 26 years apart. (The inscription in the upper left reads "For the future.")

1973

1999

FIGURE 16-18 **What does "aging" mean?**

The difficulty comes when we realize that no two people age in exactly the same way, yet without fail everyone ages (FIGURE 16-18). So we must retreat to a somewhat vague definition. For an individual, aging is the gradual breakdown of the body's machinery. With this definition, we see how each individual can experience aging differently. But aging can be seen much more clearly by examining a population as a whole.

From a population perspective, **aging** emerges as a definitive and measurable feature: it is simply an increased risk of dying with increasing age, after reaching the age of maturity. For example, among humans, people are more likely to die between the ages of 70 and 71 than between 60 and 61. Similarly, they are more likely to die between the ages of 44 and 45 than between 34 and 35. The causes of death at these ages often differ, but the shuffle toward death becomes more and more dire, with no respite. This is the most useful definition of aging for ecologists, and it doesn't apply only to humans. Aging can be measured and assessed for any species in the same way.

Jeanne Louise Calment lived for more than 122 years, the longest life span ever documented for a human. While this is spectacularly long relative to most organisms on earth (and is significantly longer than the average human life span in the United States, 78 years), it pales in comparison with some of the longest-lived organisms. Bristlecone pine trees are the longest-lived, surviving for thousands of years. The growth rings of one bristlecone pine showed that it was more than 7,000 years old!

Among animals, lobsters and quahogs (a type of clam) are the longest-lived invertebrates (up to 100 and 200 years, respectively). Long-lived vertebrates include striped bass, which can exceed 120 years, and tortoises, which can reach 150 years. Toward the other end of the spectrum, rodents may live just a couple of years, and fruit flies and many other insects generally live and die within the span of a few weeks. Interestingly, bird and bat species live 5–10 times longer than similarly sized rodents. These data prompt the questions: Why is there so much variation? And what controls how long the individuals of a species live? We explore these questions in the next section, but first we consider the more general question of why organisms age at all.

> "Things fall apart; the centre cannot hold;
> Mere anarchy is loosed upon the world,
> The blood-dimmed tide is loosed, and everywhere
> The ceremony of innocence is drowned;
> The best lack all conviction, while the worst
> Are full of passionate intensity."
>
> — WILLIAM BUTLER YEATS
> ***The Second Coming*, 1919**

Aging, in fact, is not a mystery. The question of why we age is one of the great solved problems in biology. Oddly, though, it may also be one of the best-kept secrets in science. Let's investigate.

It is easier to *state* the evolutionary explanation for why we age than to *understand* it. But that's where we start. The explanation is based on one fundamental fact: the force of natural selection lessens with advancing age. So what does this mean?

Q Many genetic diseases kill old people, but almost none kill children. Why not?

1. Imagine a mutation that causes a person carrying it to die at age 10. Will that person pass the mutation on to many offspring? Of course not. The carrier of that mutant allele will die before she gets a chance to pass it on. Alternative versions of the gene that don't cause death will be the only ones that persist.

2. Now imagine a mutation that causes a person carrying it to die at the age of 150. Will that person pass it on? Yes! The carrier of this fatal allele will already have had children, passing on the mutation to them long before she even knows she carries it. In fact, she will no doubt die long before this mutation has the opportunity to exert its disastrous effect. The same thing happens if the mutation has its negative effect at age 100 . . . or 70 . . . or even 50 (though it may be the cause of death at these "younger old" ages).

In the first example, we see that natural selection "weeds out" mutant alleles that cause sickness and death early in life. Those alleles never get passed on (FIGURE 16-19). No one inherits them, and they disappear from the population without fail. Individuals with the alternative versions of the gene pass on the "good" versions at a higher rate. Consequently, those "good" alleles end up as the only versions of the gene in the population. These are the genes that we inherited from our ancestors, and this is why very few people have genetic diseases that kill them in their teens. We sometimes think of natural selection as only acting positively, to improve or increase a trait—the length of a giraffe's neck, for example. But natural selection can also select *against* a trait—in this case, disease.

Natural selection isn't so effective with some other genes. The second example describes mutations that arise and only later in life make their carriers more likely to die. Examples include mutations that increase the risk of cancers, heart disease, or other ailments. Unfortunately, the carriers of these mutant alleles will already have reproduced and

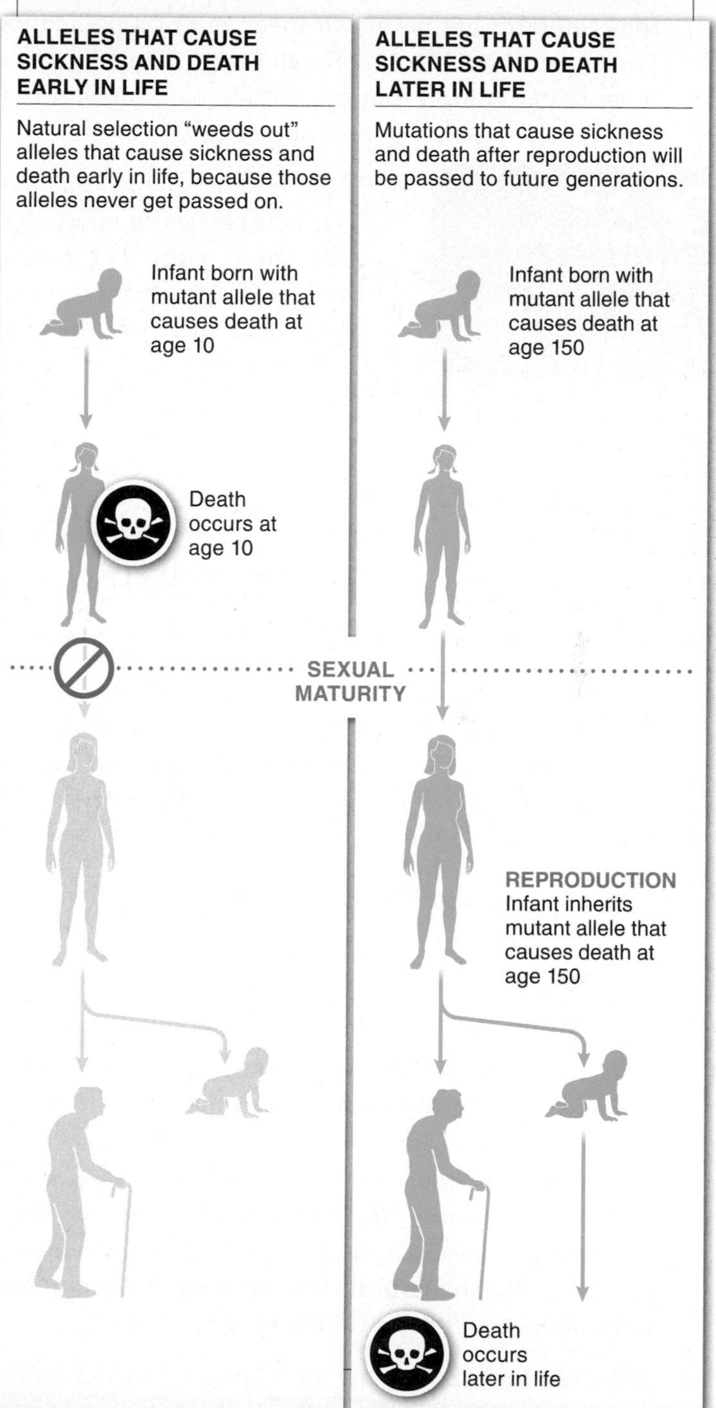
GENETIC DISEASE AND AGING
Natural selection can't weed out genetic defects that wait until we've already reproduced to harm our health. Those mutations that act early in life will be passed on to subsequent generations at a greatly reduced frequency.
ALLELES THAT CAUSE SICKNESS AND DEATH EARLY IN LIFE
Natural selection "weeds out" alleles that cause sickness and death early in life, because those alleles never get passed on.
Infant born with mutant allele that causes death at age 10
Death occurs at age 10
ALLELES THAT CAUSE SICKNESS AND DEATH LATER IN LIFE
Mutations that cause sickness and death after reproduction will be passed to future generations.
Infant born with mutant allele that causes death at age 150
SEXUAL MATURITY
REPRODUCTION
Infant inherits mutant allele that causes death at age 150
Death occurs later in life
The cumulative effects of the build-up of late-acting harmful alleles are responsible for what we see as aging.

FIGURE 16-19 **Carrying a time bomb.**

passed on the alleles before they have had their negative effect. In other words, it doesn't matter whether you carry one of these mutants or an alternative allele that does not harm you, the number of offspring you produce—your **reproductive output**—is the same. Consequently, these mutants are never "cleaned" out of a population. The practical result is that we all have inherited many of these mutant alleles that have arisen over the past hundreds and thousands of generations.

Q A cure for cancer may be discovered, but not for aging. Why the difference?

These mutant alleles that have their adverse health effects later in life are responsible for aging. The later in life they have their effect—and therefore the more likely it is that they are passed on to offspring—the more common they will be in the population. This is why there are *many* causes of death, and why it seems as if all of our bodily systems fall apart as we get older.

TAKE HOME MESSAGE 16.11

» Natural selection cannot weed out harmful alleles that do not diminish an individual's reproductive output. Consequently, these mutant alleles accumulate in the genomes of individuals of nearly all species. This leads to the physiological breakdowns that we experience as we age.

16.12 What determines the average longevity in different species?

Can individuals live forever? No. Just because they are not at risk from internal, genetic sources of mortality that doesn't mean there aren't things in the world around them that might kill them—for example, predators. lightning, drought, or disease. And for different species, these risks are more or less serious. Rodents, for example, are at *very* high risk of predation nearly every minute of every day, so they are likely to be dead within a few years (FIGURE 16-20).

Because death from some external source is likely to come so early, organisms living in high-risk worlds must reproduce early: natural selection favors this reproductive strategy because otherwise, organisms wouldn't leave any descendants. Now, when a harmful mutation arises that causes its damage at an early age—say, one or two years of age—the rodent has probably already reproduced and passed the mutation on. And so that harmful allele increases in frequency in the population, causing individuals to age earlier.

Tortoises, by contrast, live in relatively low-risk worlds. Because they have armor-like shells that protect them from danger, risk of death from external sources is low. Early, intensive investment in reproduction is not necessary and so has not been favored by natural selection. In such species, when a harmful mutation arises that has its effect at five or ten years of age, the individual carrying it is likely to die before it has reproduced. Consequently, the mutation isn't passed on; natural selection weeds it out of the population.

The age at time of reproduction, then, is a key factor determining longevity. Factors that favor early reproduction also favor early aging; factors that don't favor early reproduction—or actively favor later reproduction—favor later aging. But then, what determines when an organism reproduces?

We can think of each population as evolving in a world with a specific **hazard factor** (see Figure 16-20). This factor includes the risk of death from all types of external forces. When the hazard factor is low, as it is for tortoises and humans, individuals of a species tend to reproduce later, and so natural selection can weed out all harmful mutations except those that have their effects late in life. Individuals of these species will live a long time before they succumb to aging. A high hazard factor, on the other hand, will lead to earlier reproduction and therefore early aging and shorter life span.

Which species would you expect to age more slowly in captivity, a porcupine or a guinea pig? The difference is large,

ENVIRONMENTAL RISKS AND AGING

Tortoise

HIGH HAZARD FACTOR

- Relatively high risk of death at each age
- Individuals tend to reproduce earlier
- Earlier aging
- Shorter life spans

LOW HAZARD FACTOR

- Relatively low risk of death at each age
- Individuals tend to reproduce later
- Later aging
- Longer life spans

150 years

Longevity

In environments characterized by low mortality risk, populations of slowly aging individuals with long life spans evolve. In environments characterized by high mortality risk, the opposite occurs.

FIGURE 16-20 **Hazard factor.** The mouse and the tortoise face different risks of dying.

revealing unambiguously that it's good to be a porcupine. Few predators want to eat you. Or rather, few predators *can* eat you. Because of their sharp quills, porcupines have evolved in a world with little risk of predation and with a lower hazard factor.

Meanwhile, guinea pigs, who have no protective quills, live with a high risk of predation. Because they've been evolving in a world characterized by a high hazard factor, they reproduce much sooner, age more quickly, and die younger. A guinea pig in captivity lives less than 10 years, whereas a captive porcupine can live 28 years. In the wild the difference is even greater: 15 years for the porcupine and only 3 or 4 years for the guinea pig.

What difference in longevity would you expect between poisonous and non-poisonous snakes, or between bats (which can fly) and mice?

TAKE HOME MESSAGE 16.12

» The rate of aging and pattern of mortality are determined by the hazard factor of the organism's environment. In environments characterized by low mortality risk, populations of slowly aging individuals with long life spans evolve. In environments characterized by high mortality risk, populations of early-aging, short-lived individuals evolve.

16.13 Can we slow down the process of aging?

Life extension is possible. Real, honest-to-goodness doubling of the normal life span. Not only is it possible, researchers have demonstrated it repeatedly . . . in fruit flies.

The results of these life-extension studies are astonishing, but the methods by which they were achieved are simple. Researchers kept several large populations of fruit flies in cages in the laboratory. The flies would feed and lay eggs on a small plate of food. Every few days, a fresh plate would be put in the cage and the old one discarded. At 2 weeks, eggs were collected and used to start the next generation of flies. Propagated in this way, the fly populations were maintained for hundreds of generations. When the longevity of flies in these populations was measured, the researchers found that the flies started to experience the physiological breakdowns associated with aging after a few weeks, and after about a month the flies died.

Q Is it possible, with our current knowledge, to double longevity in humans?

One researcher thought that changing the force of natural selection on the flies might produce an evolution in their aging pattern. So, instead of collecting eggs at 2 weeks to start the new generation, he cleverly waited longer and longer (FIGURE 16-21). The first new generation had parents that had lived for 3 weeks. If any fly carried a mutation that caused it to die between 2 and 3

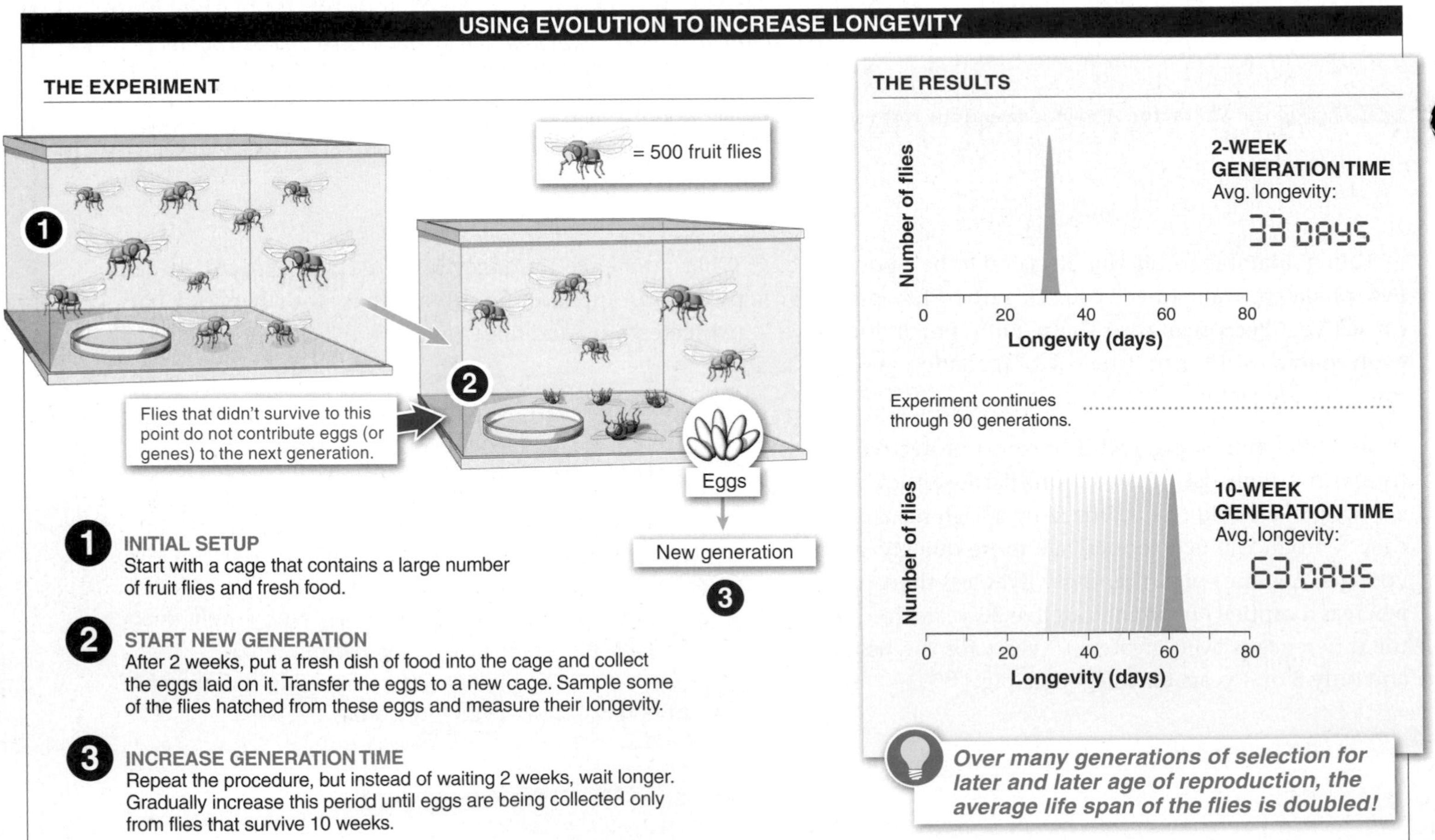

USING EVOLUTION TO INCREASE LONGEVITY
THE EXPERIMENT
= 500 fruit flies
Flies that didn't survive to this point do not contribute eggs (or genes) to the next generation.
Eggs
New generation
1 INITIAL SETUP
Start with a cage that contains a large number of fruit flies and fresh food.
2 START NEW GENERATION
After 2 weeks, put a fresh dish of food into the cage and collect the eggs laid on it. Transfer the eggs to a new cage. Sample some of the flies hatched from these eggs and measure their longevity.
3 INCREASE GENERATION TIME
Repeat the procedure, but instead of waiting 2 weeks, wait longer. Gradually increase this period until eggs are being collected only from flies that survive 10 weeks.
THE RESULTS
Number of flies
Longevity (days)
0
20
40
60
80
2-WEEK GENERATION TIME
Avg. longevity:
33 DAYS
Experiment continues through 90 generations.
10-WEEK GENERATION TIME
Avg. longevity:
63 DAYS
Over many generations of selection for later and later age of reproduction, the average life span of the flies is doubled!

FIGURE 16-21 **Creating longer-lived organisms through evolution in the laboratory.**

weeks, that fly did not contribute to the next generation and that mutation was not passed on.

After several generations, eggs were collected at 4 weeks instead of 3 weeks. At this point, all the flies making up the new generation had parents that had survived for 4 weeks. Any flies that died before 4 weeks of age didn't contribute to the new generation. Natural selection was now able to reduce the prevalence of mutant genes responsible for early aging.

The experiment continued, progressively increasing the generation time to 5 weeks, then 6, 7, and ultimately 10 weeks—equivalent to decreasing the hazard factor of the fly population. At this point, only those flies with a genome free of mutations that might cause death in the first 10 weeks of life could contribute to the new generation. The net result was the creation of a population of flies with much "cleaner" genomes. When put under ideal conditions outside the cages, these "super" flies did not experience aging until long after the original flies would have died. In fact, the average life span of the flies hatching from eggs harvested from parents of 10 weeks of age was more than 60 days—double that of the original flies!

TAKE HOME MESSAGE 16.13

» By increasing the strength of natural selection later in life, it is possible to increase the mean and maximum longevity of individuals in a population. This occurs in nature (as in porcupines and bats) and has also been achieved under controlled laboratory conditions.

16.14–16.16

The human population is growing rapidly.

High-density human habitations in Hong Kong.

16.14 Age pyramids reveal much about a population.

The "baby boom." It happened half a century ago, yet young people in the United States today may end up paying a price for it. What was it? And why does it still matter? Increasing birth rates caused the baby boom. Beginning just after World War II, and continuing through the early 1960s, families in the United States had about 30% more babies per month than they would have had if they had followed their parents' pattern. Then the birth rates began to return to their earlier levels, where they have remained since. As the babies from the boom years grew to school age, schools had to increase in size to accommodate all the children. Then the schools had to downsize once the baby boomers had graduated. Now, as these individuals enter or approach retirement age, the question of how their retirement and health care needs will be met is one of the biggest issues facing American society.

Q What is the baby boom? Why is it bad news for young people today?

Populations often vary across space—dense in the cities, more sparse in rural areas. The baby boom illustrates that populations have an "age distribution" as well. Just as there may be more individuals in some areas and fewer in others, there may be more individuals in certain age groups and fewer in others.

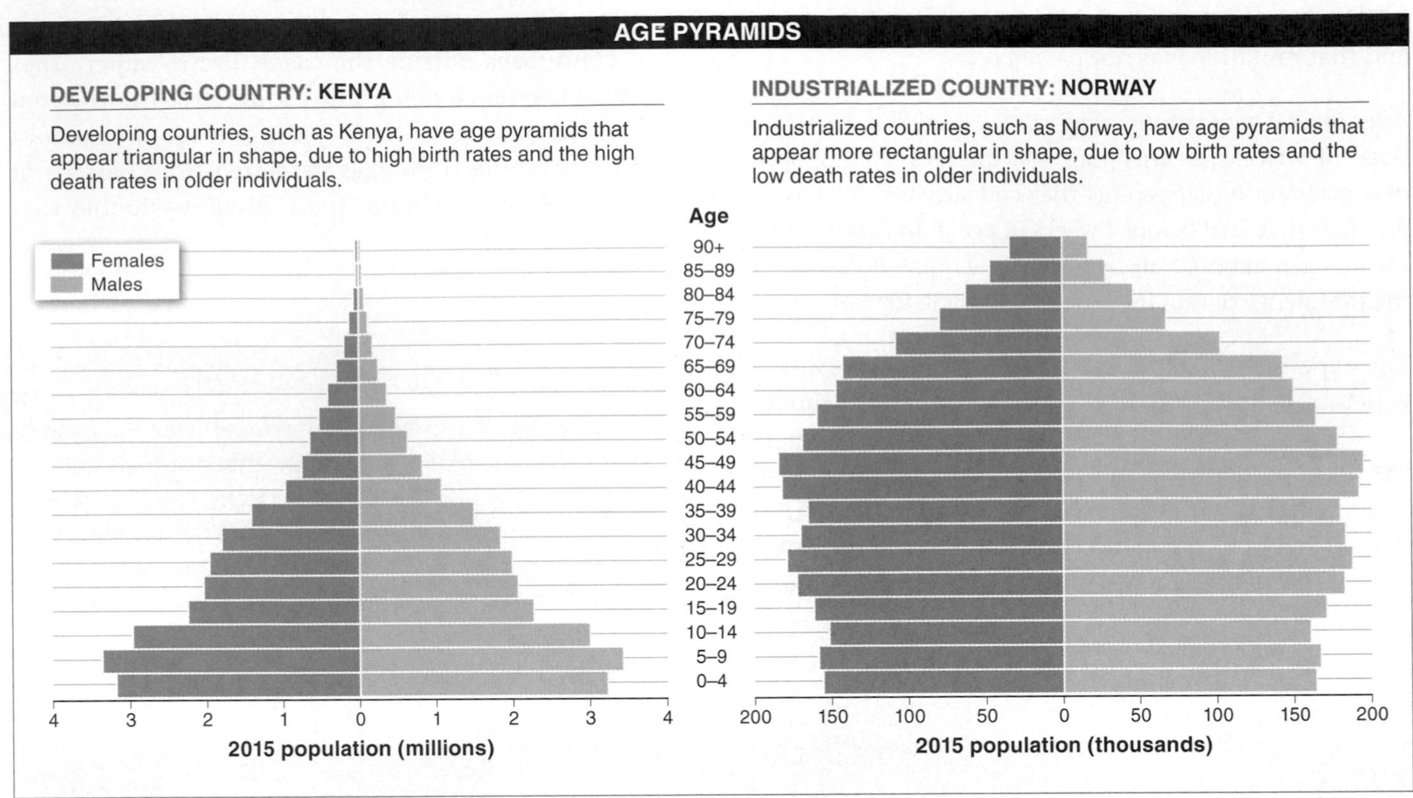

FIGURE 16-22 **A visual representation of population growth.** Age pyramids go from triangular to rectangular when population growth stabilizes.

Describing populations in terms of the proportion of individuals in each age group reveals interesting population features. A population can be divided into the percentages of individuals that are in specific age groupings, called *cohorts,* such as 0–4 years, 5–9 years, 10–14 years, and so on, in an "age pyramid." People have radically different likelihoods of dying or reproducing, for instance, depending on their age. A 10-year-old isn't going to produce any offspring, but a 30-year-old has a relatively high likelihood of reproducing. And a 10-year-old is less likely to die than an 80-year-old.

It can be useful, therefore, for a society to know the relative numbers of 10-, 30-, and 80-year-olds in its population. The age data can be used to determine whether a society would be better off investing in new schools or in fertility wards or in convalescent homes, among other social issues. Companies, too, rely on such demographic data to best plan their strategies for producing the goods that people will want and need.

Around the world, countries vary tremendously in the age pyramids describing their populations. If two populations are the same size but have different age distributions, they will have some very different features. Most industrialized countries are growing slowly or not at all; most of the population is middle-aged or old. In these countries, such as Norway, the age pyramid is more rectangular than pyramid-shaped (FIGURE 16-22). Because birth rates are low, the bottom of the pyramid is not very wide. And because death rates are low, too, the higher age classes in the pyramid don't get significantly narrower. Instead, the cohorts remain about the same size all the way into the late sixties and seventies, at which point high mortality rates finally cause them to narrow. This gives the pyramid a more or less rectangular shape.

A more triangular shape is seen in the age pyramids of developing countries. Kenya, for example, has very high birth rates, reflected as a large base in its age pyramid (see Figure 16-22). But high mortality rates, usually due to poor health care, cause a large and continuous reduction in the proportion of individuals in older age groups. In these countries, most of the population is in the younger age groups.

Notice the "bulge" in the U.S. age pyramid (FIGURE 16-23). The shape of the U.S. age pyramid has economists worried that the social security system (including Social Security and Medicare) will not be able to offer older citizens sufficient benefits in the next 10–30 years. Because of the unusually

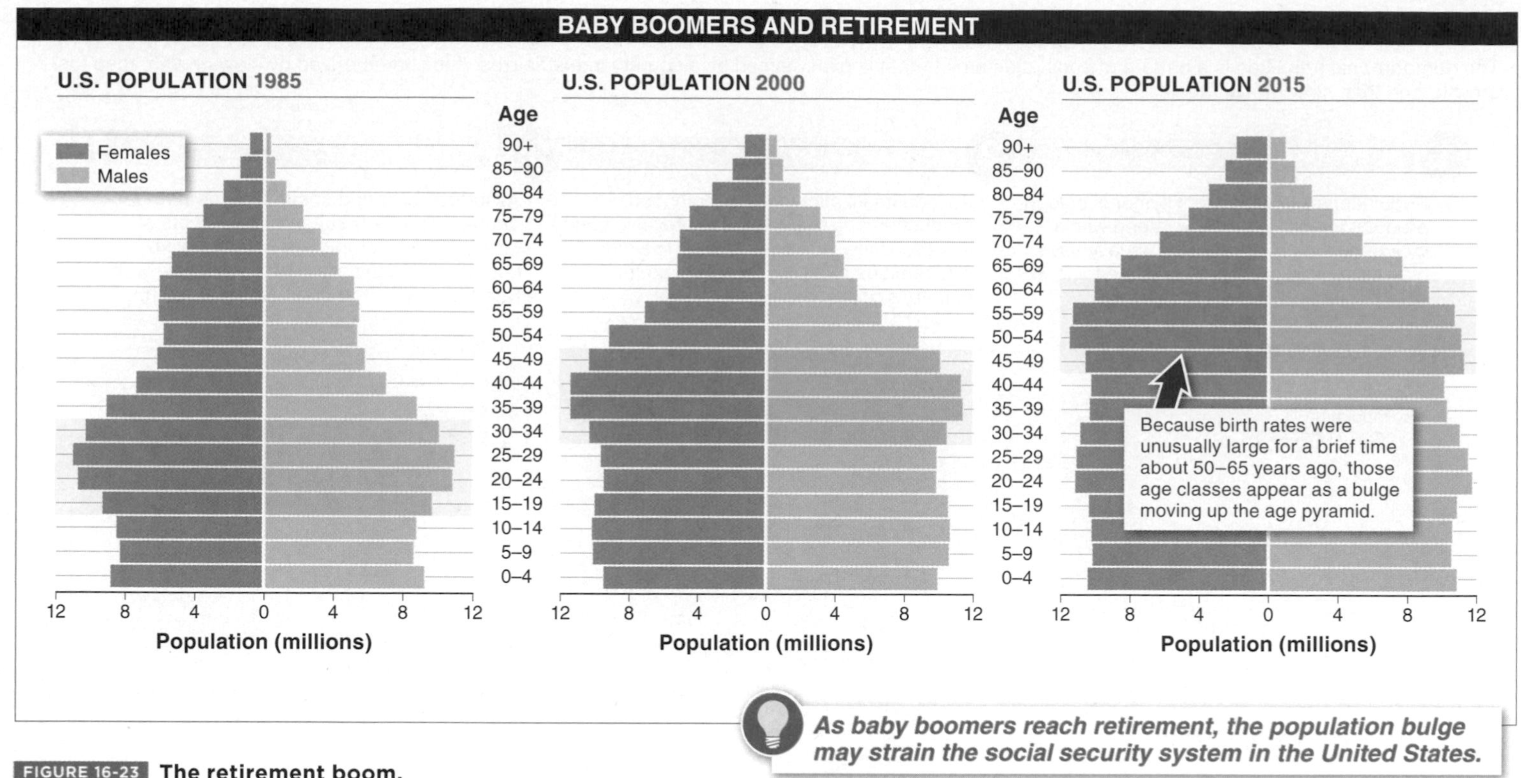

FIGURE 16-23 **The retirement boom.**

large number of babies born about 50–70 years ago, an unusually large number of people are now reaching retirement age. Since the baby boomers were born, there haven't been any years with such a large cohort. This means that the current numbers of working individuals who contribute to the social security system are not sufficient to cover the payouts promised to the large number of retirees, and the baby boomers will be expensive to support as they reach retirement.

TAKE HOME MESSAGE 16.14

» Age pyramids show the number of individuals in a population within any age group. They allow us to estimate birth and death rates over multiyear periods.

16.15 As less-developed countries become more developed, a demographic transition often occurs.

Populations can sometimes seem as strange and stubborn as individuals. Around the world, governments encourage (or sometimes even coerce) their citizens to reduce their reproductive rates in efforts to check population growth, but they are rarely successful. Yet it seems that when governments stop trying, their countries' population growth rates slow down all on their own.

Many countries have undergone industrialization, with increases in health, wealth, and education. And in the process, their patterns of population growth follow common paths, marked first by periods of faster growth and later by slower growth. The sequence of changes is remarkably consistent.

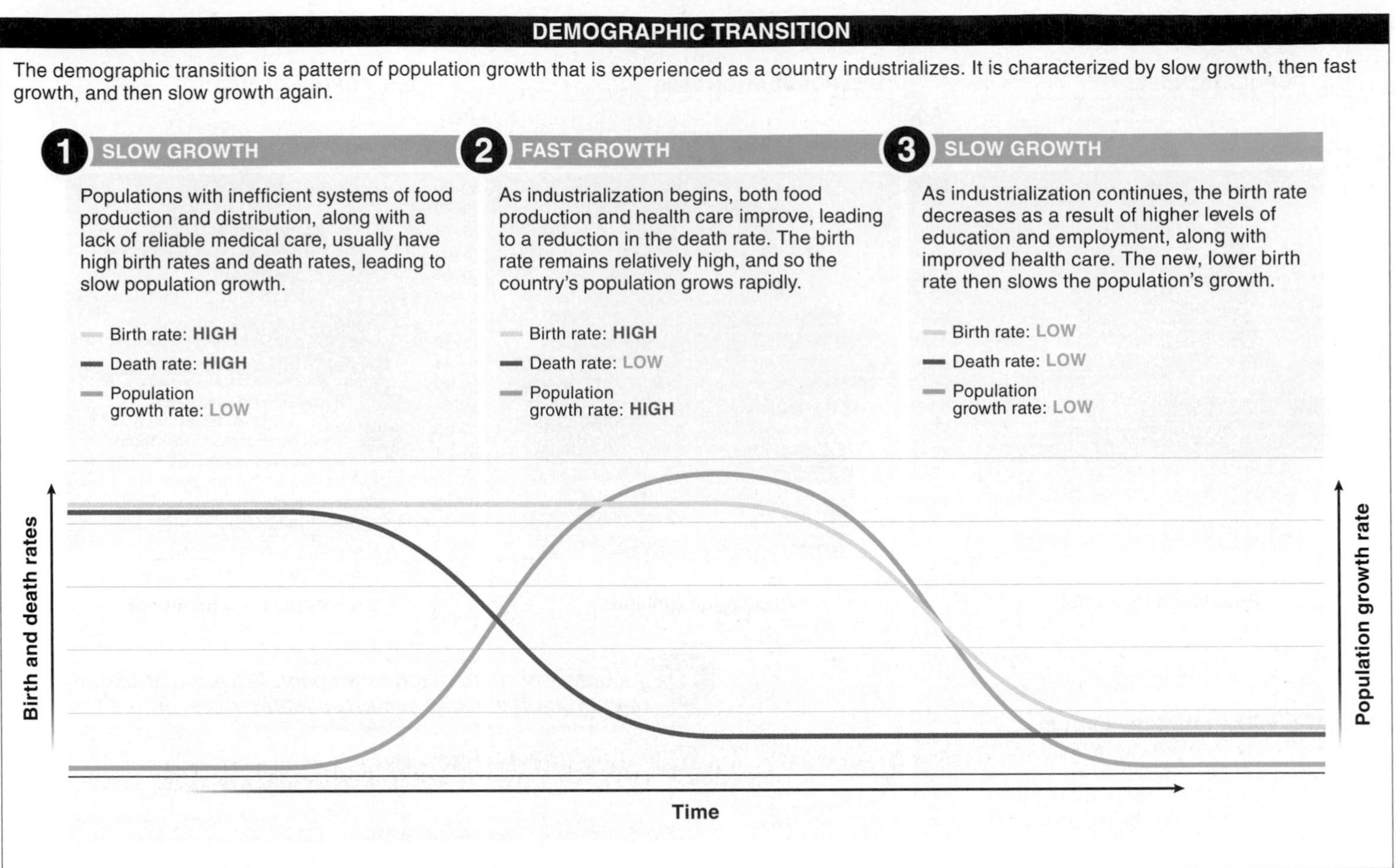

DEMOGRAPHIC TRANSITION
The demographic transition is a pattern of population growth that is experienced as a country industrializes. It is characterized by slow growth, then fast growth, and then slow growth again.
1 SLOW GROWTH
Populations with inefficient systems of food production and distribution, along with a lack of reliable medical care, usually have high birth rates and death rates, leading to slow population growth.
Birth rate: HIGH
Death rate: HIGH
Population growth rate: LOW
2 FAST GROWTH
As industrialization begins, both food production and health care improve, leading to a reduction in the death rate. The birth rate remains relatively high, and so the country's population grows rapidly.
Birth rate: HIGH
Death rate: LOW
Population growth rate: HIGH
3 SLOW GROWTH
As industrialization continues, the birth rate decreases as a result of higher levels of education and employment, along with improved health care. The new, lower birth rate then slows the population's growth.
Birth rate: LOW
Death rate: LOW
Population growth rate: LOW
Birth and death rates
Population growth rate
Time

FIGURE 16-24 **With industrialization, death rates drop and, later, birth rates drop, too.**

Start with a country prior to industrialization. Such countries usually have high birth rates and high death rates, resulting from poor and inefficient systems of food production and distribution, along with a lack of reliable medical care. Food production and health care typically improve as industrialization begins. These improvements inevitably lead to a reduction in the death rate. The birth rate, however, remains relatively high, so the country's population grows rapidly.

As industrialization continues, further changes occur. Most important, the standard of living increases. This results from higher levels of education and employment, and, in conjunction with improved health care, this finally causes a reduction in the birth rate. The new, lower birth rate then slows the population's growth. The progression from

1. high birth rates and high death rates (slow population growth) *to*
2. high birth rates and low death rates (fast population growth) *to*
3. low birth rates and low death rates (slow population growth)

is called the **demographic transition** (FIGURE 16-24).

The demographic transition can take decades to complete, so it is not always easy to identify it as it occurs. A survey of countries around the world reveals that many are at different points along the transition. Sweden, for example, has a low fertility rate (1.7 children born per woman) and a low death rate (10.4 deaths per 1,000 people). At the other extreme, Nigeria has a very high fertility rate (5.5 children born per woman) and death rate (17.2 deaths per 1,000). Mexico, in the midst of a clear demographic transition, has a moderately high fertility rate (2.45 children born per woman) yet a very low death rate (4.7 deaths per 1,000).

Q **Population growth is alarmingly slow in Sweden and alarmingly fast in Mexico. Why this difference?**

In the past few decades, the demographic transition has been completed in Japan, Australia, the United States, Canada, and most of Europe, leading to a slowing of population growth. In Mexico, Brazil, Southeast Asia, and most of Africa, on the other hand, the transition is not complete and population growth is still dangerously fast.

The demographic transition illustrates how health, wealth, and education can lead to a reduction in the birth rate without direct government interventions. But because more than three-quarters of the world's population lives in developing countries and less than a quarter in developed countries, the slowed population growth that generally accompanies the demographic transition is unlikely to be sufficient to keep the world population at a manageable level. Instead, world population growth will continue to rise quickly. What are the potential consequences of such explosive growth? We explore this next.

TAKE HOME MESSAGE 16.15

» The demographic transition tends to occur with the industrialization of countries. It is characterized by an initial reduction in the death rate, followed later by a reduction in the birth rate.

16.16 Human population growth: how high can it go?

Humans are a phenomenally successful species. There are more than 7 billion people alive today and we add 80 million people to the total each year, because birth rates greatly exceed death rates (FIGURE 16-25). But for all of our success, the laws of physics and chemistry still apply. We all need food for energy and space to live. We need other resources, too, and we need the capability of processing and storing all of the waste our societies generate. Because of these limits to perpetual population growth, we may become victims of our own success.

This we know for certain: human population growth cannot continue forever at the current rate. As is the case for every other species, our environment has a carrying capacity beyond which the population cannot be maintained indefinitely. The question is, how high can it go? What is the earth's carrying capacity for humans? This, unfortunately, is a difficult question to answer.

For more than 300 years, biologists have been making estimates, starting when Antonie van Leeuwenhoek (the inventor of the microscope) made an estimate of just over 13 billion people. The median of all the estimates is just over 10 billion, and the United Nations suggests that the carrying capacity is somewhere between 7 and 11 billion.

There is huge variation from one biologist's estimate to the next. Why is it so hard to figure this out? After all, it's important that we know so that we can work to avoid a global catastrophe. The problem, it seems, goes back to the reasons behind the tremendous human success in the first

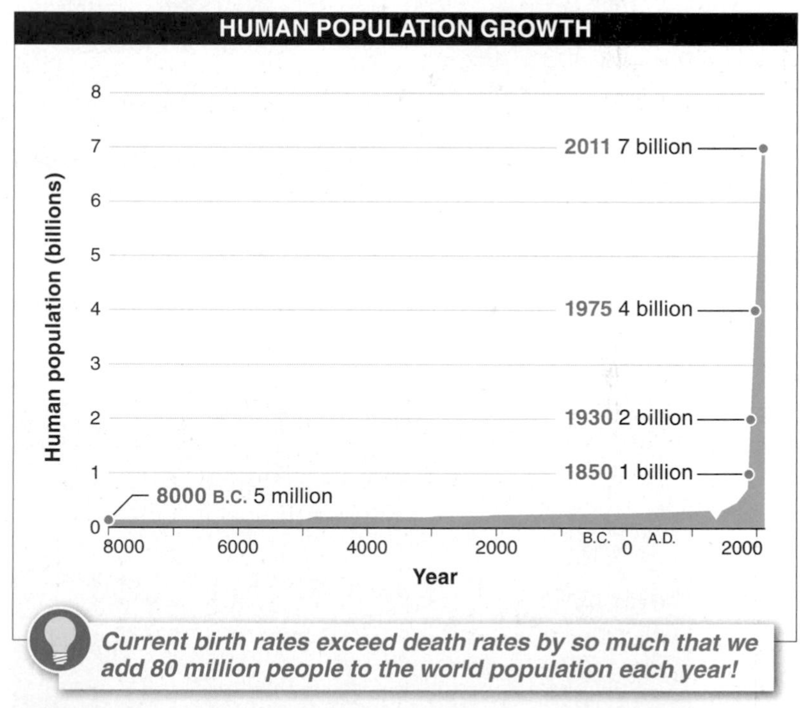

FIGURE 16-25 The world's human population: a slow start, but rapidly picking up steam.

LIMITLESS CARRYING CAPACITY?

One reason that the carrying capacity for the human population is difficult to estimate is that we can increase it in a variety of ways.

Dubai, United Arab Emirates

EXPANDING INTO NEW HABITATS
With fire, tools, shelter, and efficient food distribution, we can survive almost anywhere on earth.

Combine harvesting wheat

INCREASING THE AGRICULTURAL PRODUCTIVITY OF THE LAND
With fertilizers, mechanized agricultural methods, and selection for higher yields, fewer people can now produce much more food than was previously thought possible.

Skyscrapers in Hong Kong

FINDING WAYS TO LIVE AT HIGHER DENSITIES
Public health and civil engineering advances make it possible for higher and higher densities of people to live together with minimal problems from waste and infectious diseases.

FIGURE 16-26 Three ways of increasing the carrying capacity for the human population.

place. We are so clever that we seem to keep increasing the carrying capacity before we ever bump into it. We do this in a variety of ways, all made possible by various technologies that we invent. In particular, we make advances on three fronts (FIGURE 16-26).

1. **Expand into new habitats.** With fire, tools, shelter, and efficient food distribution, we can survive almost anywhere on earth.
2. **Increase the agricultural productivity of the land.** With fertilizers, mechanized agricultural methods, and selection for higher yields, fewer people can now produce much more food than was previously thought possible.
3. **Circumvent the problems that usually accompany life at higher densities.** Public health and civil engineering advances make it possible for higher and higher densities of people to live together with minimal problems from waste and infectious diseases.

But the question remains: how high can the population go? The most difficult problem in determining the earth's carrying capacity for humans may be assessing just how many resources each person needs. It is possible to estimate the **ecological footprint** of an individual or an entire country by evaluating how much land, food, water, and fuel, among other things, are used. This method reveals that although some countries (e.g., New Zealand, Canada, and Sweden) have more resources available than are required to support the needs of their population, resource use in the world as a whole is significantly greater than the resources available, implying that we are already at our planet's carrying capacity. The populations of many countries—including the United States, Japan, Germany, and England—currently consume an unsustainable level of resources (FIGURE 16-27).

Another difficulty in estimating the earth's carrying capacity is that populations can and do occasionally alter their fundamental growth properties. Relatively small changes in the worldwide birth rate can have dramatic effects on the ultimate population size of the planet.

The difficulties in estimating the earth's carrying capacity illuminate an important issue. We must be more specific when we ask how high it can go. We must add: and at what level of comfort, security, and stability, and with what impact on the other species on earth? There are trade-offs.

FIGURE 16-27 **How high can the population go?** As human population growth continues, so, too, does our consumption of resources. Shown here: a poultry plant in China that processes and packages 375,000 chickens every day.

> "There is in every American, I think, something of the old Daniel Boone—who, when he could see the smoke from another chimney, felt himself too crowded and moved further out into the wilderness."
>
> — HUBERT HUMPHREY
>
> **38th U.S. Vice President, 1966**

The ecological footprint of the 1.25 billion people currently living in India, for example, is much smaller than that of the populations of Japan, Norway, or Australia. It is possible, we know, to live with much less impact on the environment per person. But how much do we want to sacrifice to enable a larger number to live stably on the planet? Is our goal to maximize the number? Or to reduce the number and increase the resources available to each person? Or should our goal be something else entirely? The answers to these questions will influence whether the ultimate carrying capacity is even higher than 11 billion, or perhaps lower than the current 7.3 billion people now alive.

The problem gets even more difficult, though, because the world population currently has significant momentum. That is, even if, today, we immediately and permanently reduced our fertility rate to the replacement rate of just two children per couple, the world population would continue to grow for more than 40 years, putting us up to at least 8 billion. This is because there are so many young people alive that, each year, more and more of them will enter the reproductive population. As we get closer to (or exceed) the earth's carrying capacity for humans, we should be better able to recognize it, but by then it may be too late to take preemptive measures—to avoid the resource depletion that may doom us to the much more unpleasant experience of natural population controls.

TAKE HOME MESSAGE 16.16

» The world's human population is currently growing at a very high rate, but limited resources will eventually limit this growth, most likely at a population size between 7 and 11 billion.

STREET BIO

Using evidence to guide decision making in our own lives

Life history trade-offs and a mini-fountain of youth: what is the relationship between reproduction and longevity?

It's well documented that there is a trade-off between reproduction and longevity. In a wide variety of animals, decreasing reproductive effort has been shown to increase longevity.

Q: Does this have any practical applications? Yes: spay or neuter your pet to increase its life span!

Q: Does this work? Yes! Data from more than 1,000 cat autopsies showed that non-sterilized females lived a mean of 3.0 years, whereas sterilized females lived significantly longer, with a mean life span of 8.2 years. The results for males were similar. (Perhaps the most dramatic example of the reproduction–longevity trade-off is that of the marsupial mouse, *Antechinus stuartii,* in which castration led to a doubling or even tripling of the usual life span. Similarly, following castration, Pacific salmon lived up to 8 years, double their usual life span.)

Not that you asked . . . In a study in the early 1900s of men institutionalized for intellectual and developmental disabilities, those who were castrated lived 13 years longer, on average, than non-castrated men, matched for age and intelligence, at the same institution—69.3 years vs. 55.7 years.

Q: Does a vasectomy have the same effect as castration? No! The life-extending effect of sterilization occurs only when the ovaries or testes are removed; "tying the tubes" and vasectomy leave the gonads intact. And there is no longevity increase. Why do you think that is the case?

Q: What is responsible for the trade-off between reproduction and longevity? If it were a question of sterilized individuals living longer simply because they don't have to expend energy on reproduction, perhaps similar life span increases could be achieved by simply giving animals access to more energy. In practice, however, this doesn't work (and usually decreases life span). Rather, it seems that a significant part of the "cost" of reproduction, at least in mammals, is an increased incidence of cancer, caused by the higher levels of circulating hormones in "reproductively ready" (that is, fertile and receptive) animals.

Possibilities to think about. The link between maintaining reproductive readiness and reduced longevity has led some researchers to contemplate the design of birth control pills that might have the additional effect of increasing longevity. Stay tuned.

GRAPHIC CONTENT

Thinking critically about visual displays of data

1 What is the purpose of this graph?

2 What factors contribute to the blue line becoming flat?

3 When the population reaches the carrying capacity, it appears to persist indefinitely. How is that possible? Do organisms stop dying?

4 What data are represented in the graph?

5 What could be added to this graph to improve it?

6 Describe two assumptions about population growth that are implicit in this figure.

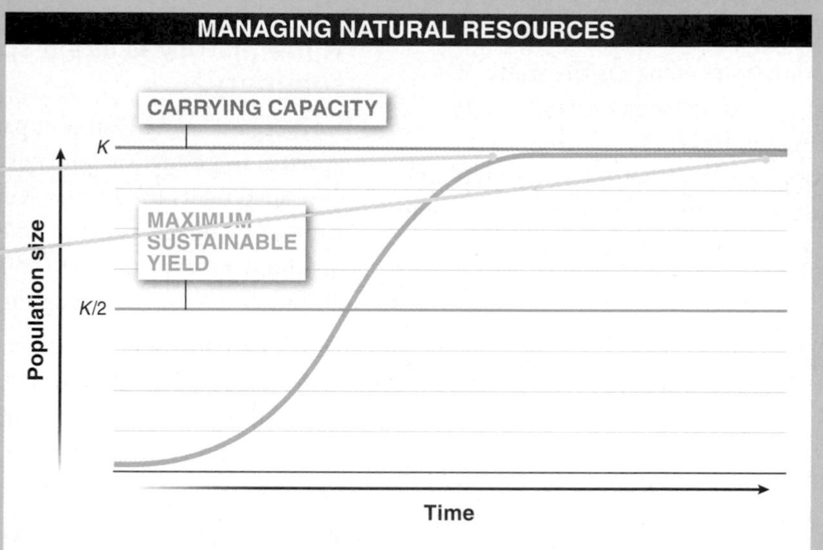

7 Create an alternative graph showing the relationship between population size (on the *x*-axis) and growth rate (on the *y*-axis). Label carrying capacity and the point of maximum sustainable yield on your graph.

See answers at the back of the book.

KEY TERMS IN POPULATION ECOLOGY

BRIEF SUMMARY

Population ecology is the study of how populations interact with their environments.

• Population ecology is the study of interactions between populations of organisms and their environments, particularly their patterns of growth and how they are influenced by other species and by environmental factors.

• Most ecological processes cannot be observed or studied within an individual. Rather, we need to consider the entire group of individuals in a population.

• Populations tend to grow exponentially, but this growth is eventually limited.

• A population's growth can be constrained by density-dependent factors: as density increases, limited resources constrain the population's growth. It can also be reduced by density-independent factors such as natural or human-caused environmental calamities.

• Some populations cycle between periods of rapid growth and rapid shrinkage.

• Based on models of population growth, it might seem easy to utilize natural resources efficiently and sustainably. In practice, however, difficulties such as estimating population size and carrying capacity complicate the implementation of such strategies.

A life history is like a species summary.

• An organism's investment pattern in growth, reproduction, and survival is described by its life history. Very different strategies can achieve the same outcome in which a mating pair of individuals produces at least two surviving offspring.

• Because constraints limit evolution, life histories are characterized by trade-offs between investments in growth, reproduction, and survival.

• Life tables and survivorship curves summarize the survival and reproduction patterns of the individuals in a population. Species vary greatly in these patterns.

Ecology influences the evolution of aging in a population.

• Natural selection cannot weed out harmful alleles that do not diminish an individual's reproductive output. Consequently, these mutant alleles accumulate in the genomes of individuals, leading to the physiological breakdowns we experience as we age.

• The rate of aging and pattern of mortality are determined by the hazard factor of the organism's environment. In environments characterized by low mortality risk, populations of slowly aging individuals with long life spans evolve. In environments characterized by high mortality risk, populations of early-aging, short-lived individuals evolve.

• By increasing the strength of natural selection later in life, it is possible to increase the mean and maximum longevity of individuals in a population. This occurs in nature and has also been achieved under laboratory conditions.

The human population is growing rapidly.

• Age pyramids show the number of individuals in a population within any age group. They allow us to estimate birth and death rates over multi-year periods.

• The demographic transition tends to occur with the industrialization of countries. It is characterized by an initial reduction in the death rate, followed later by a reduction in the birth rate.

• The world's human population is currently growing at a very high rate, but limited resources will eventually limit this growth, most likely at a population size between 7 and 11 billion.

CHECK YOUR KNOWLEDGE

Short Answer

1. Why can't we study ecological processes within an individual?

2. Explain the differences between density-dependent and density-independent factors that can limit a population's growth. Give an example of each.

3. Explain the difference between maximum yield and maximum sustainable yield in a population.

4. In a situation where an organism has a short breeding season, how would you expect the number of offspring per litter to compare with that of organisms with longer breeding seasons?

5. Explain how natural selection "weeds out" alleles that cause sickness and death early in life.

6. Why do some species have individuals with very long lives, while others do not? How could the life span of any species be extended?

7. Describe the process by which researchers were able to increase the life span of fruit flies.

8. Why is it useful for a society to know the relative numbers of 10-, 30-, and 80-year-olds in its population?

9. What is meant by a "demographic transition"?

10. Why is the ecological footprint of the 1.25 billion people currently living in India much smaller than that of the population of Japan?

Multiple Choice

1. In a population exhibiting logistic growth, the rate of population growth is greatest when *N* is:

a) 0.5 *K*.

b) 0.

c) above the carrying capacity.

d) *K*.

e) All of the above are correct; the rate of population growth is constant in logistic growth.

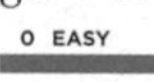

2. In a population, as *N* approaches *K*, the logistic growth equation predicts that:

a) the carrying capacity of the environment will increase.

b) the growth rate will approach zero.

c) the population will become monophyletic.

d) the population size will increase exponentially.

e) the growth rate will not change.

3. Which of the following statements about maximum sustainable yield is incorrect?

a) The maximum sustainable yield for a population is the population's growth rate at *K*/2.

b) The maximum sustainable yield for a population can be difficult to determine because it is not always possible to accurately measure *N*.

c) The maximum sustainable yield for a population can be difficult to determine because it is not always possible to accurately measure *K*.

d) The maximum sustainable yield for a population is a useful management guideline for harvesting plant products such as timber, but is not helpful for managing animal populations.

e) The concept of maximum sustainable yield can generate useful information for fighting the growth of pest species.

0 EASY HARD 100 58

4. Dr. David Reznick has studied the evolution of life history in guppies that live in streams in Trinidad. Guppies are found in two different types of habitat: sites where predation is high and sites where predation is low. Which of the following life history characteristics would you expect to evolve in a guppy population living in a high-predation site?

a) bright colors and courtship displays

b) increased egg number

c) a female-biased sex ratio

d) increased egg size

e) delayed sexual maturation

0 EASY HARD 100 33

5. Which of the following is a major trade-off in life histories?

a) size of offspring for amount of parental investment

b) size of offspring for number of reproductive events

c) growth for reproduction

d) size for life span

e) number of reproductive events for number of offspring per reproductive event

6. Natural selection:

a) does not influence aging, because aging is determined by an individual's environment.

b) cannot reduce the frequency of alleles that cause mortality among individuals who have not yet reached maturity.

c) cannot weed out from a population any alleles that do not reduce an individual's relative reproductive success, even if these alleles increase an individual's risk of dying.

d) can influence aging but not longevity.

e) leads to an increase in the frequency of any illness-inducing alleles that have their effect when an organism can reproduce.

7. Which of the following statements about the hazard factor of a population is incorrect?

a) It is a measure of organisms' risk of death resulting from external sources.

b) It is lower for a population of porcupines than for a population of guinea pigs.

c) It is a measure of the ratio of mortality risk due to external (environmental) causes relative to internal (genetic) causes.

d) It is responsible for the rate of aging among individuals in the population.

e) It is a measure of how quickly the individuals in a population age.

0 EASY HARD 100 55

8. Life extension:

a) is not possible, because natural selection cannot weed out disease-causing alleles that have an effect only at an age when reproduction is no longer possible.

b) can be achieved by selectively breeding those individuals that have the earliest age of maturity.

c) works in insects but could not work in humans.

d) has been achieved using laboratory selection for delayed reproduction.

e) None of the above are correct.

9. A population pyramid:

a) represents the number of individuals in various age groups in a population.

b) can be constructed from the data in a life table.

c) directly predicts future age distributions of the population.

d) shows the current birth and death rates of a population.

e) predicts survival and mortality rates for an individual at a given age.

10. A primary difference between the age pyramids of industrialized and developing countries is that:

a) mean longevity is significantly greater in developing countries.

b) in developing countries, much larger proportions of the population are in the youngest age groups.

c) developing countries show a characteristic "bulge" that indicates a baby boom.

d) in developing countries, females live significantly longer than males, whereas in industrialized countries the reverse is true.

e) developing countries have significantly more individuals than industrialized countries.

Ch17

Ecosystems have living and non-living components.

Interacting physical forces create weather.

Energy and chemicals flow within ecosystems.

Species interactions influence the structure of communities.

Communities can change or remain stable over time.

An aerial view of the Landmannalaugar region in southern Iceland, known for its unusual geological features.

Ecosystems and Communities

Organisms and their environments

Ecosystems have living and non-living components.

Flamingoes feeding at Lake Nakuru National Park, Kenya.

17.1 What are ecosystems?

Picture a lush nature scene: some greenery, a bit of rotting wood, and abundant wildlife. Grazing animals abound, while predators feed on other animals and their eggs. Parasites are poised, looking for hosts, and, just below the surface, scavengers find meals among the organic detritus. It would seem to be the quintessential **ecosystem,** a community of biological organisms plus the non-living components with which the organisms interact. But now imagine that the entire scene gets up and walks away! The "camera" in your mind pulls back to reveal that the scene is playing out on the back of a beetle no more than 2 inches (5 cm) long (FIGURE 17-1).

The host of this mini-ecosystem is a beetle from New Guinea called the large weevil. The weevil is camouflaged from its predators by lichens—which consist of fungi and photosynthetic algae living together—while the lichens are given a safe surface on which to live. And the garden of lichens supports a wide range of other organisms, from tiny mites to a variety of other microscopic invertebrates, some free-living and others parasitic.

Not all ecosystems are the obvious assemblages of plants and animals that we usually picture—ponds, deserts, or tropical forests. A similar scene can just as easily be found in your large intestine, where several hundred bacterial species flourish. These ecosystems are contained within ecosystems that are contained within ecosystems. The scale can vary tremendously. The closer you look, the more you find.

What is important is that the two essential elements of an ecosystem are present: the biotic environment and the physical (abiotic) environment (FIGURE 17-2).

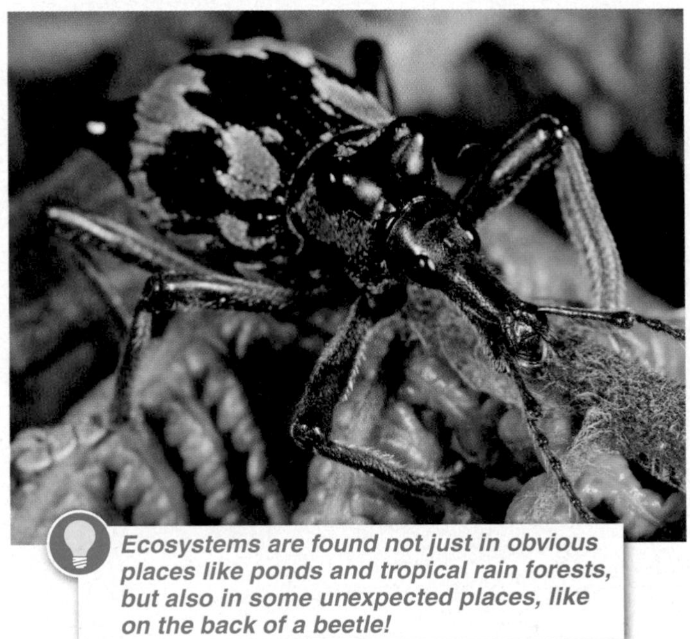

Ecosystems are found not just in obvious places like ponds and tropical rain forests, but also in some unexpected places, like on the back of a beetle!

FIGURE 17-1 **"Be it ever so humble . . ."** A small-scale ecosystem can exist on the large weevil of New Guinea.

1. The **biotic** environment consists of all the living organisms within an area and is often referred to as a **community.**

2. The physical, or **abiotic,** environment, often referred to as the organisms' **habitat,** consists of:

   • the *chemical resources* of the soil, water, and air, such as carbon, nitrogen, and phosphorus, and

   • the *physical conditions,* such as the temperature, salinity (salt level), moisture, humidity, and energy sources.

ECOSYSTEMS

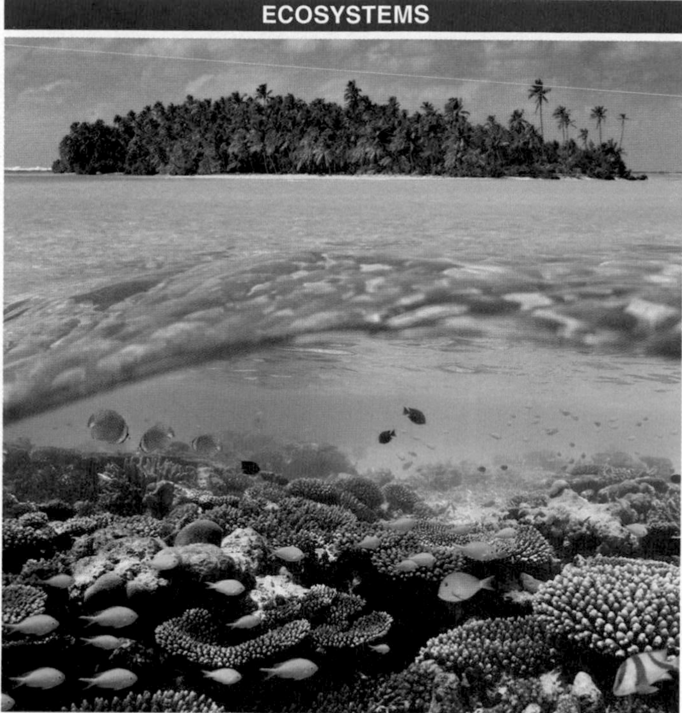

All ecosystems share two essential features:

BIOTIC ENVIRONMENT	PHYSICAL (ABIOTIC) ENVIRONMENT
• The living organisms within an area • Often referred to as a community	• The chemical resources and physical conditions within an area • Often referred to as the organisms' habitat

FIGURE 17-2 **What makes up an ecosystem?**

Biologists view communities of organisms and their habitats as "systems" in much the same way engineers might, hence the term eco*system*. Biologists monitor the inputs and outputs of the system, tracing the flow of energy and various molecules as they are captured, transformed, and utilized by organisms and later exit the system or are recycled. They also study how the activities of one species affect the other species in the community—whether the species have a conflicting relationship, such as predator and prey, or a complementary relationship, such as flowering plants and their pollinators. On a small scale, such as the back of a beetle, making some of these measurements can be easy. But there are also some well-studied giant ecosystems, such as the Hubbard Brook Experimental Forest in New Hampshire, which covers 7,600 acres. The same principles apply to the study of ecosystems regardless of size: observe and analyze organisms and their environments, while monitoring everything that goes into and out of the system.

Why should researchers bother with such a methodical—and tedious—analysis of ecosystems? Using the principles of scientific thinking and carrying out experimental tests of hypotheses have led to numerous valuable discoveries, from understanding how the clear-cutting of forests dramatically reduces soil quality to understanding the link between the use of fossil fuels and the creation of destructive acid rain. As we see in this chapter and the next, humans, perhaps more than any other species in history, are significantly affecting most of the ecosystems on earth. And, in addition to improving our understanding of environmental issues, ecosystem studies can also tell us about the microbial ecosystems living inside humans and thus lead to advances in public health.

TAKE HOME MESSAGE 17.1

» An ecosystem is a community of biological organisms plus the non-living components in the environment with which the organisms interact. Ecosystems are found not just in obvious places such as ponds, deserts, and tropical rain forests but also in some unexpected places, such as in the digestive tracts of organisms or on the shell of a beetle.

17.2 Biomes are large ecosystems that occur around the world, each determined by temperature and rainfall.

Dense vegetation surrounds you. Above you is a canopy of evergreen trees, 30–40 meters (100–130 feet) tall. Climbing vines hang from virtually all the trees. And dozens or even hundreds of species of insects are flourishing around you. Even if you've never been there, the description of a tropical rain forest is easy to recognize. But

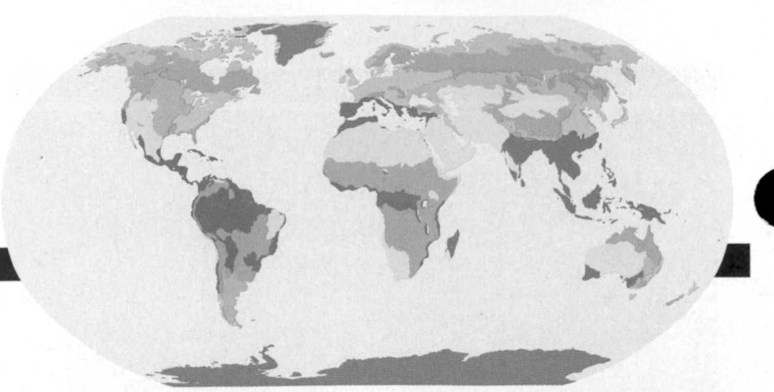

TERRESTRIAL BIOMES

Biomes are large ecosystems that cover huge geographic areas. The nine chief terrestrial biomes are determined by the temperature and amounts of precipitation in conjunction with the magnitude of seasonal variation in these factors.

TROPICAL FOREST

DESERT

SAVANNA

TEMPERATE GRASSLAND

TEMPERATE DECIDUOUS FOREST

CHAPARRAL

CONIFEROUS FOREST

TUNDRA

POLAR ICE

FIGURE 17-3 **Terrestrial ecosystem diversity.**

where exactly would you be if you were in this scenario? It could be South America or Africa or Southeast Asia. The species are different, but the general pattern of life forms is the same. The same holds for arctic tundra: whether you were in northern Asia or North America, the view would be similar. These are examples of the largest of the earth's ecosystems, the **biomes.**

Biomes cover huge geographic areas of water or land—the deserts that stretch almost all the way across the northern part of Africa, for example. Terrestrial (land) biomes are defined and usually described by the predominant types of plant life in the area. But looking at a map of the world's terrestrial biomes, it is clear that they are mostly determined by the weather. Specifically, when defining terrestrial biomes, we ask four questions about the weather:

1. What is the average temperature?

2. What is the average rainfall (or other precipitation)?

AQUATIC BIOMES

Aquatic biomes are determined by physical features, including salinity, water movement, and depth.

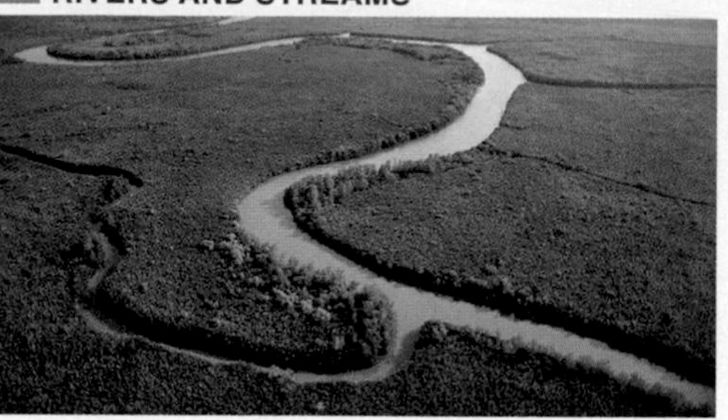

FIGURE 17-4 **Aquatic ecosystem diversity.**

3. Is the temperature relatively constant or does it vary seasonally?
4. Is the rainfall relatively constant or does it vary seasonally?

For example, where it is always moist and the temperature does not vary across the seasons, **tropical rain forests** develop. And where it is hot but with strong seasonality that brings a "wet" season and a "dry" season, **savannas** or **tropical seasonal forests** tend to develop. At the other end of the spectrum, in dry areas with a hot season and a cold season, **temperate grasslands** or **deserts** develop. FIGURE 17-3 shows examples of the nine chief terrestrial biomes; all are determined, in large part, by the precipitation and temperature levels.

Aquatic biomes are defined a bit differently, usually based on physical features such as salinity, water movement, and depth. Chief among these environments are (1) lakes and ponds, with non-flowing fresh water; (2) rivers and streams, with flowing fresh water; (3) **estuaries** and wetlands, where salt water and fresh water mix in a shallow region characterized by exceptionally high productivity; (4) open oceans, with deep salt water; and (5) coral reefs, highly diverse and productive regions in shallow oceans (FIGURE 17-4).

If the terrestrial biomes of the world are determined by the temperature and rainfall amounts and seasonality, what determines those features? In other words, what makes the weather? We investigate next how the geography and landscape of the planet—the shape of the earth and its orientation to the sun, and patterns of ocean circulation—cause the specific patterns of weather that create the different climate zones and the biomes characteristic of each. Then we'll see how energy and chemicals are made available for life to flourish in these biomes.

TAKE HOME MESSAGE 17.2

›› Biomes are the major ecological communities of earth, characterized mostly by the vegetation present. Different biomes result from differences in temperature and precipitation, and the extent to which both vary from season to season.

Interacting physical forces create weather.

Storm clouds gather over a Utah cornfield.

17.3 Global air circulation patterns create deserts and rain forests.

"There was a hot desert wind blowing that night. It was one of those hot dry Santa Anas that come down through the mountain passes and curl your hair and make your nerves jump and your skin itch. On a night like that every booze party ends in a fight. Meek little wives feel the edge of a carving knife and study their husbands' necks. Anything can happen."

— **RAYMOND CHANDLER**, *Red Wind*, **1938**

Temperature and rainfall, that's all. The type of terrestrial biome depends on little else besides these two aspects of the weather. But what determines the temperature and rainfall in a particular part of the world? Differences in both ultimately result from one simple fact: the earth is round. As we'll see, the earth's curvature influences temperature, and rainfall patterns are an inevitable consequence of the variation in temperature across the globe.

Wherever you are, begin walking toward the equator. As you get closer, it gets hotter. Nearly everyone is aware of this universal trend (and the reverse as well: it gets colder as you move away from the equator and approach the North or South Pole). What is responsible for the increased warmth at the equator?

The sun shines most directly on the equator (FIGURE 17-5). At the equator, solar energy hits the earth and spreads out over a relatively small area. Away from the equator, the earth curves toward the North and South Poles. Because of this curvature, the same amount of solar energy hitting the earth at the Poles is spread out over a much larger area and also travels a greater distance through the atmosphere, which absorbs or reflects much of the heat. With the energy dispersed over a large area, there is less warmth at any one point on the earth's surface. It's similar to the fact that at noon, the sun's rays hit the earth at a more direct angle than they do later in the day. That is why the sun provides less warmth late or early in the day. It is also why the risk of sunburn and skin cancer is greatest around noon and the nearer you are to the equator.

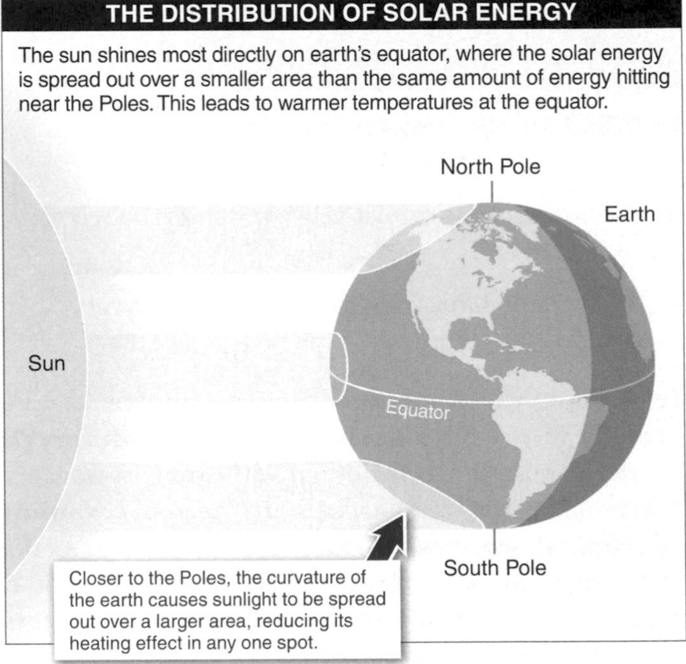

THE DISTRIBUTION OF SOLAR ENERGY
The sun shines most directly on earth's equator, where the solar energy is spread out over a smaller area than the same amount of energy hitting near the Poles. This leads to warmer temperatures at the equator.
North Pole
Earth
Sun
Equator
South Pole
Closer to the Poles, the curvature of the earth causes sunlight to be spread out over a larger area, reducing its heating effect in any one spot.

FIGURE 17-5 Why is it warmer at the equator than at the Poles?

FORMATION OF RAIN

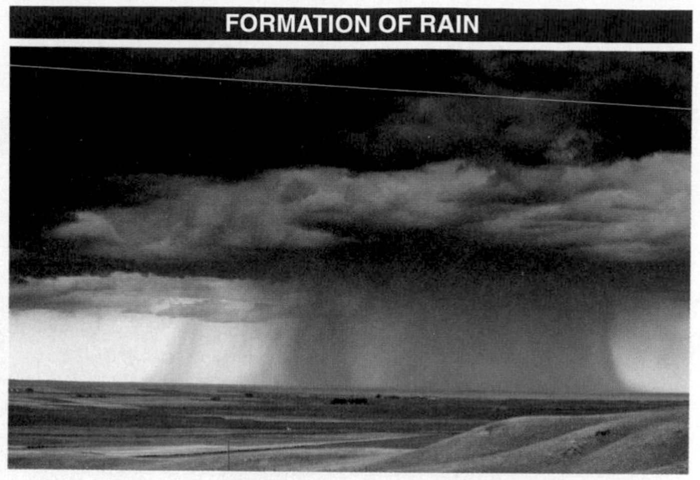

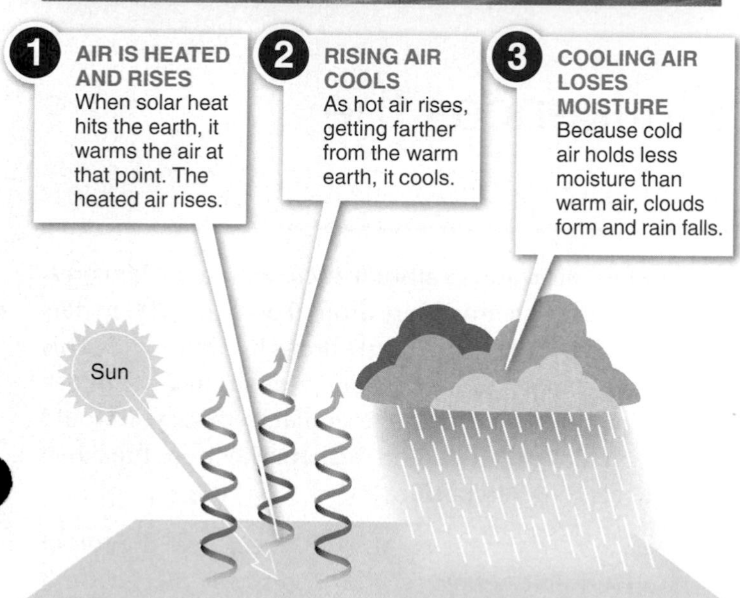

1 AIR IS HEATED AND RISES
When solar heat hits the earth, it warms the air at that point. The heated air rises.
2 RISING AIR COOLS
As hot air rises, getting farther from the warm earth, it cools.
3 COOLING AIR LOSES MOISTURE
Because cold air holds less moisture than warm air, clouds form and rain falls.
Sun

FIGURE 17-6 **Rainmaking.**

Global patterns of rainfall can be predicted just as easily, by taking into account that warm air holds more moisture than cold air. We'll start at the equator again, where the greatest warming power of the sun hits the earth, and some of that energy radiates back, warming the air. This starts a three-step process: (1) hot air rises; (2) as it rises high into the atmosphere, it cools; and (3) because cool air holds less moisture, as the air rises, clouds form, and the moisture that can no longer be held in the air falls as rain (FIGURE 17-6). The equator is hot, but it is also very wet.

The high, cold air in the atmosphere over the equator expands outward, to about 30° north and south. Here, the cold air, which is heavier than warm air, begins to fall downward toward the earth and is warmed by heat radiated off the earth's surface. As the air gets warmer, it can hold more and more moisture, rather than releasing it as rain. In fact, the falling and rapidly heating air can hold so much moisture that it sometimes sucks up moisture from the land itself. For this reason, at about 30° north and south, around the world, there is very little rainfall, the ground is very dry, and deserts form (FIGURE 17-7). The Atacama Desert of South America, which lies at approximately 30° south of the equator, is an extreme example. In some parts of this large desert, *no rainfall* has ever been recorded. Other great deserts of the world are also at 30° latitude, including the Sahara, Kalahari, Mojave, and Australian deserts (although some deserts do occur at other latitudes). These areas stand in

Q Nearly all of the world's deserts occur a third of the way from the equator toward the Poles, at 30° latitude. Is this just a coincidence?

THE FORMATION OF DESERTS

Sand dunes in Namibia

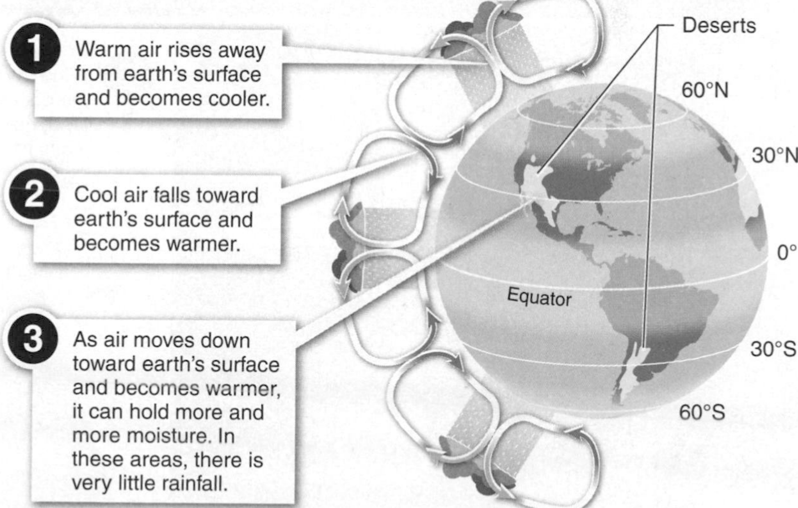

1 Warm air rises away from earth's surface and becomes cooler.
2 Cool air falls toward earth's surface and becomes warmer.

FIGURE 17-7 **Desert formation.** Circulating masses of air, caused by solar energy hitting different parts of the earth at different angles, determine rainfall patterns throughout the world.

stark contrast to the equator, where it is not uncommon for an area to receive more than 3 meters (almost 120 inches) of rain in a year.

As the air falls near 30° north and south and hits the earth, it spreads equally toward the equator and toward the Poles, gradually getting warmer, until, at around 60° latitude, it begins to rise because of its accumulated heat. Again, the rising air loses its moisture as rainfall. So at 60° latitude, two-thirds of the way toward the Poles from the equator, it's not particularly warm but there is a great deal of rain. Not surprisingly, at these latitudes lie huge temperate forests with extensive plant growth. And finally, around the Poles, air masses again descend. As they do, because they can hold more moisture, very little rain falls. The Poles are cold, but with little precipitation—they resemble frozen desert.

TAKE HOME MESSAGE 17.3

» Global patterns of weather are largely determined by the earth's round shape. Solar energy hits the equator at a more direct angle than at the Poles, leading to warmer temperatures at lower latitudes. This temperature gradient generates atmospheric circulation patterns that result in heavy rain at the equator and many deserts at 30° latitude.

17.4 Local topography influences the weather.

Why is it so windy on the sidewalk around tall buildings? Why does it rain all the time on one side of some mountain ranges, while deserts form right on the other side? And is it actually warmer in the city than in the country? The answers reflect how the physical features of land, its **topography**—including features created by humans—can have dramatic but predictable effects on temperature, precipitation, and wind.

High altitudes have lower temperatures. With increasing elevation, the air pressure drops, because the weight of the atmosphere decreases as altitude increases. And when pressure is lower, the temperature drops. For each 1,000 meters above sea level, the temperature drops by about 6° C. This is why the changes in weather and vegetation that you see while climbing a mountain are similar to those you would see as you moved farther and farther away from the equator—it gets colder in both cases.

Rain shadows create deserts. Air rises to get over the top of a mountain. But because the air cools as it rises, it can't hold much moisture. So clouds form and rain falls. As the

FORMATION OF RAIN SHADOWS

Mountain range

Rain shadow desert

The Andes Mountains, near the west coast of South America, cause a rain shadow effect.

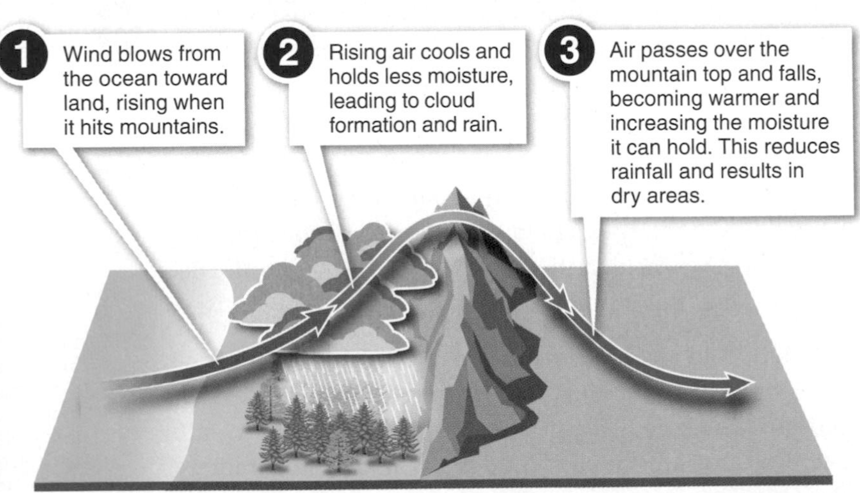

1 Wind blows from the ocean toward land, rising when it hits mountains.
2 Rising air cools and holds less moisture, leading to cloud formation and rain.
3 Air passes over the mountain top and falls, becoming warmer and increasing the moisture it can hold. This reduces rainfall and results in dry areas.

FIGURE 17-8 **The rain shadow effect.**

MODERN LANDSCAPES INFLUENCE WEATHER
122
Temperature (°F)
32
Asphalt, cement, and building tops absorb heat, raising the temperature.
Tall buildings force wind downward.

FIGURE 17-9 Human engineering. Unintended consequences can occur when humans alter the land, such as changes in temperature and in wind speed and direction.

air eventually passes over the top of the mountain, it falls and warms. And because warm air holds onto moisture, there is rarely any rain on the back side of the mountain. In fact, often the air will pull moisture from the ground, intensifying the already dry conditions, creating **rain shadow** deserts (FIGURE 17-8). Along the west coast of the United States, the Sierra Nevada and the Cascade mountain range are responsible for the Mojave Desert in California and the Great Sandy Desert in Oregon.

Q Is it warmer or cooler in urban areas relative to nearby rural areas?

Asphalt, cement, and tops of buildings absorb heat, raising the temperature. Modern landscapes also influence the weather, creating "urban heat islands." When energy from the sun hits concrete, pavement, or the dark roof of a building, some of it is reflected, heating the air around it, and most of the rest is absorbed by these man-made surfaces and held until night. In the darkness, the surfaces lose heat to the sky, which is colder by comparison, through radiation. In contrast, when sunlight hits trees or other plant life, the solar energy evaporates water in the leaves. The ground surface doesn't get much hotter, and neither does the air. It's not surprising, then, that cities tend to be 1° to 6° C warmer than surrounding rural areas. And not only is it hotter in cities, but the rising warm air also alters rainfall patterns both in and around cities.

A variety of steps are being taken to create "greener" cities that absorb and release less heat from the sun, and require less energy for air conditioning. These methods include the creation of buildings with lighter-colored rooftops, the planting of trees around buildings and along roads, and the development of rooftop gardens rich with vegetation.

Q Why is it so windy on streets with tall buildings?

Tall buildings channel wind downward. Tall buildings are responsible for the perpetual winds you feel when walking on a city sidewalk. Here's why. Winds blow more strongly when they are higher above the earth, freed from the earth's frictional drag (from plants, dirt, rocks, and water) that slows wind as it gets near the surface. When these strong, elevated winds suddenly encounter tall buildings, they are deflected. Some of the wind goes up and over the building, but much of it is pushed downward, reaching double or even triple its initial speed by the time it reaches street level (FIGURE 17-9).

TAKE HOME MESSAGE 17.4

» Local features of topography influence the weather. With higher altitude, the temperature drops. On the windward side of mountains, rainfall is high; on the back side, descending air reduces rainfall, causing rain shadow deserts. Urban development increases the absorption of solar energy, leading to higher temperatures, and creates wind near the bottom of tall buildings.

17.5 Ocean currents affect the weather.

Weather is affected not just by circulating air masses. It is also affected by the oceans, which are vast and deep, warmed almost exclusively by the sun. And water is continuously moving and mixing, due to a combination of forces, including wind, the earth's rotation, the gravitational pull of the moon, temperature, and salt concentration. These forces create several large, circular patterns of flow in the world's oceans, as illustrated in FIGURE 17-10.

Q Why are beach communities cooler in summer and warmer in winter than inland communities?

Much of water's effect on weather stems from its great heat capacity. For equal volumes of water and air, water can absorb and hold 10,000 times more heat than can air. This means that temperatures fluctuate much more in air than in water. At the beach, the air temperature can go from mild to very hot and back to mild over the course of a day, while the heat of the sun will have a tiny, almost negligible effect on the water temperature. Thus, during summers, much of the heat of the sun is absorbed and held by the ocean water in coastal towns, rather than causing hot air temperatures. Conversely, during cold winter months, heat from the water can reduce the coldness of the air.

One of the strongest ocean currents is the Gulf Stream. It travels north through the Caribbean, carrying warm water up the east coast of the United States and across the Atlantic Ocean toward Europe. Because the current begins in a warm part of the globe, close to the equator, it is still warm when it reaches the east coast of the United States and then Europe. The warm water also warms the climate in these areas. In fact, if it weren't for the Gulf Stream, much of Europe—given its high latitudes—would have a climate more like Canada's. Because all ocean currents in the northern hemisphere rotate in a clockwise direction, water reaching the beaches of California, unlike water reaching east coast beaches, has just come from the north, near Alaska, where it gets very cold.

Every two to seven years, a dramatic climate change driven by ocean currents, called **El Niño,** causes a sustained surface temperature change in the central Pacific Ocean. It is blamed for flooding, droughts, famine, and a variety of other extreme climate disruptions. We can more easily understand El Niño if we contrast it with the more common climate pattern, as illustrated in FIGURE 17-11.

These predictable changes during an El Niño event can start a global pattern of unusual weather. Although the mechanisms aren't clear, in El Niño years, more rain falls in the Midwestern United States and storms form off the coast of California, while South Africa, Japan, and Canada enjoy warmer weather than usual. The counterpart of the El Niño phenomenon is called **La Niña.** During La Niña periods, ocean surface temperatures are lower than usual, and the climate effects are approximately opposite to El Niño effects.

OCEAN CIRCULATION PATTERNS

There are several large, circular patterns of flowing water in the oceans due to a combination of forces, including wind, the earth's rotation, the gravitational pull of the moon, temperature, and salt concentration.

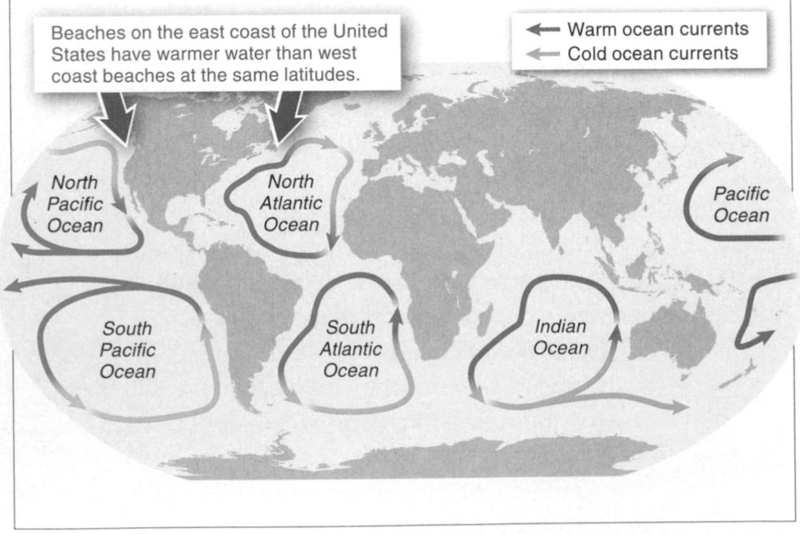

FIGURE 17-10 **Ocean currents influence water temperatures.**

TAKE HOME MESSAGE 17.5

» Oceans have global circulation patterns. Disruptions in these patterns occur every few years and can cause extreme climate disruptions around the world.

EL NIÑO

El Niño occurs every two to seven years and is blamed for flooding, droughts, famine, and a variety of other extreme climate disruptions.

USUAL CONDITIONS

1. Steady winds blow westward across the Pacific Ocean from South America toward Southeast Asia.
2. These winds push warm surface water away from the coast of South America, heating the air above it, which causes air to rise and produce tropical rainstorms.

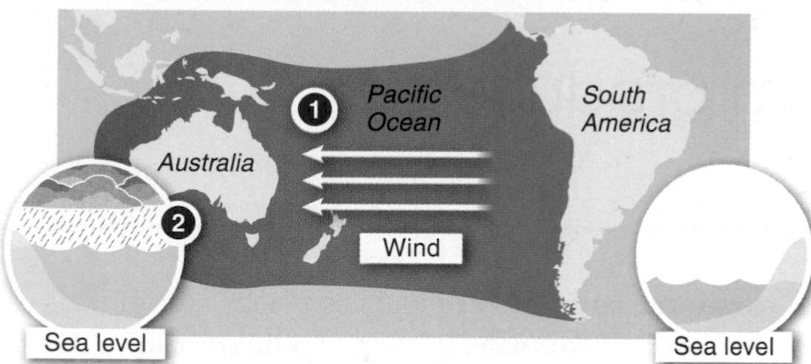

3. Off the coast of South America, colder water upwells from the ocean depths. The cold water cools the air above, causing extremely dry weather.
4. Water from the depths of the ocean brings up rich nutrients, enabling plankton to flourish and feed the huge populations of fish.

EL NIÑO CONDITIONS

1. The usual South America–to–Southeast Asia winds ease just a bit.
2. Without the push of the wind, warm water flows back toward South America. The cooled water around Australia and the Philippines cools the air and causes dry weather that can lead to droughts.

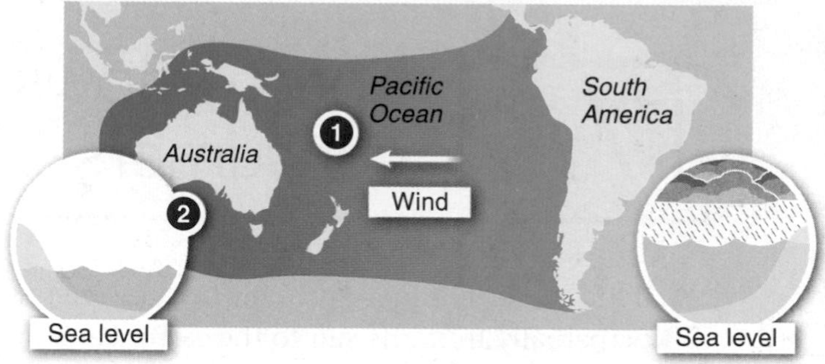

3. Without the warm surface water being blown west, colder water from the ocean depths cannot upwell. The warm surface water warms the air, resulting in rainfall.
4. With no upwelling of cold water and nutrients, plankton levels drop dramatically and fish stocks are wiped out.

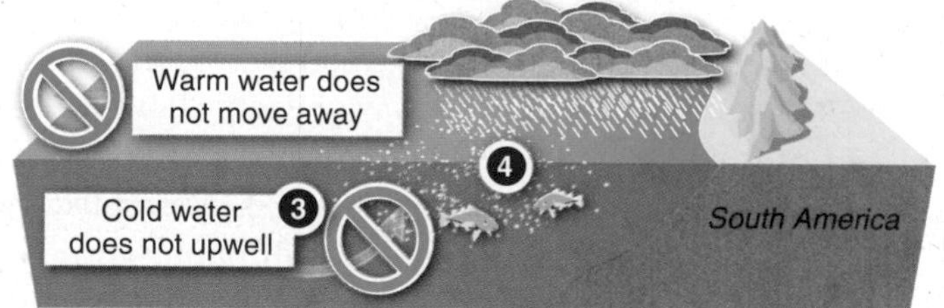

FIGURE 17-11 **Domino effect.** A reduction in the usual east-to-west ocean breeze can cause a cascade of disastrous weather.

Changes during an El Niño event start a chain reaction of unusual weather around the globe: increased rainfall in the U.S. Midwest, storms in coastal California, and warmer weather in South Africa and Japan.

17.6–17.8

Energy and chemicals flow within ecosystems.

A Parson's chameleon (*Calumma parsonii*) eating a grasshopper in Madagascar.

17.6 Energy flows from producers to consumers.

All life on earth is made possible because energy flows perpetually from the sun to the earth. Looking at an ecosystem such as a desert savanna, for example—trying to understand how all the species, from grasses and trees to birds and mammals and worms, interact with one another, and what role each plays within the system—can seem overwhelming. But if we focus on just one aspect of the ecosystem—the pathways energy takes as it flows through the system—a simple and logical order becomes clear.

The sun is where our pathway of energy flow begins. Most of the energy is absorbed or reflected by the earth's atmosphere or surface, but about 1% of it is intercepted and converted to chemical energy through photosynthesis. That intercepted energy is then transformed again and again by living organisms, making about four stops as it passes through an ecosystem. Let's examine what happens at each of the stops, known as **trophic levels** (FIGURE 17-12).

First stop: producers. When it comes to energy flow, all the species in an ecosystem can be placed in one of two groups: producers and consumers. Plants (along with some algae and bacteria) are the **producers.** They convert the sun's light energy into chemical energy through photosynthesis, as discussed in Chapter 5. We use another word to describe that chemical energy: food. The amount of organic material produced in a biome is called its **primary productivity** level.

Second stop: primary consumers—the herbivores. Cattle grazing in a field, gazelles browsing on herbs, insects devouring the leaves of a crop plant—these are the **primary consumers** in an ecosystem, the animals that eat plants. Plant material such as cellulose can be difficult to digest. Consequently, most **herbivores,** the animals that eat plants, need a little help in digesting their food. Primary consumers, from termites to cattle, often have bacteria living in their digestive system. These microorganisms benefit the organism in which they live by breaking down the cellulose, enabling the herbivore to harness the energy held in the chemical bonds of the plants' cell walls.

Third stop: secondary consumers—the carnivores. The energy that the herbivore harnesses fuels its growth, reproduction, and movement, but that energy doesn't remain in the herbivore forever. **Carnivores,** such as cats, spiders, and frogs, are animals that feed on herbivores. They are also known as **secondary consumers.** As they eat their prey, some of the energy stored in the chemical bonds of carbohydrate, protein, and lipid molecules is again captured and harnessed for their own movement, reproduction, and growth.

Fourth stop: tertiary consumers—the "top" carnivores. In some ecosystems, energy makes yet another stop: the **tertiary consumers,** or "top carnivores." These are the "animals that eat the animals that eat the animals that eat the plants." They are several steps removed from the initial capture of solar energy by a plant, but the general process is the same. A top carnivore, such as a tiger, eagle, or great white shark, consumes other carnivores, breaking down their tissues and releasing energy stored in the chemical bonds of the cells. As in each of the previous

ENERGY FLOW WITHIN AN ECOSYSTEM

Energy from the sun is intercepted and converted into chemical energy, which passes through an ecosystem in about four stops.

Sun

1 PRODUCERS
Plants convert light energy from the sun into food through photosynthesis.

2 PRIMARY CONSUMERS
Herbivores are animals that eat plants.

3 SECONDARY CONSUMERS
Carnivores are animals that eat herbivores.

4 TERTIARY CONSUMERS
Top carnivores are animals that eat other carnivores.

The food chain is a simplified pathway. In actuality, **food webs** are often a better representation, because many organisms can occupy more than one position in the chain.

FIGURE 17-12 **Follow the fuel.**

steps, the top carnivores harness this energy for their own physiological needs.

This path from producers to tertiary consumers is called a **food chain.** We see later in this chapter why a food chain almost never extends to a fifth stop.

EARTH'S RECYCLERS

When they die, organisms from every level in the food chain provide sustenance for decomposers and detritivores, and important chemical components are recycled through the food chain.

Mold decomposes an orange.

The dung beetle (a detritivore) feeds on decomposing matter.

Decomposers and detritivores break down organic wastes, releasing chemical components that can then be reused by plants and other producers.

FIGURE 17-13 **Nothing is wasted.**

The food chain pathway from photosynthetic producers through the various levels of animals is a slight oversimplification. In actuality, food chains are better thought of as **food webs,** because many organisms are **omnivores** and can occupy more than one position in the chain (see Figure 17-12). When you eat a simple meal of chicken and vegetables, after all, you're simultaneously a carnivore and an herbivore. On average, about 30% to 35% of the human diet comes from animal products and the remaining 70% to 65% from plant products. Many other animals, from bears to cockroaches, also have diets that involve harvesting energy from multiple stops in the food chain.

In every ecosystem, as energy is transformed through the steps of a food chain, organic material accumulates in the form of animal waste and dead plant and animal matter. **Decomposers,** usually bacteria or fungi, and **detritivores,** including scavengers such as vultures, worms, and a variety of arthropods, break down the organic material, harvesting energy still stored in the chemical bonds (FIGURE 17-13). Decomposers are distinguished from detritivores by the decomposers' ability to break down a much larger range of organic molecules. But both groups release many important chemical components from the organic material, which can eventually be recycled and used by plants and other producers.

Energy flows from one stop to the next in a food chain, but not in the way that a baton is passed by runners in a relay race. The difference is that at every step in the food chain, much of the usable energy is lost as heat. An animal that eats five pounds of plant material doesn't convert that into five new pounds of body weight. Not by a long shot. In the next section, we'll see how this inefficiency of energy transfers ensures that most food chains are very short.

TAKE HOME MESSAGE 17.6

» Energy from the sun passes through an ecosystem in several steps known as trophic levels: (1) producers convert light energy to chemical energy in photosynthesis; (2) herbivores then consume the producers; (3) the herbivores are consumed by carnivores; and (4) the carnivores may be consumed by top carnivores. Detritivores and decomposers extract energy from organic waste and the remains of dead organisms. At each step in a food chain, some usable energy is lost as heat.

17.7 Energy pyramids reveal the inefficiency of food chains.

Look out of the nearest window. What organisms can you see? Almost without fail, you will see green plant life. Maybe some trees, possibly bushes and grasses as well. You'll have to look longer and harder to see any animals, but you'll probably see a few, most likely small animals and various insects that eat plants. On the other hand, you might stare out of the window all day and not see any animals (other than some fellow humans) that eat other animals. Why? And why are big, fierce animals so rare? Also, why are there so many more plants than animals?

The answers to these questions are closely related to our observation that an animal consuming five pounds of plant material does not gain five pounds in body weight from its meal. The actual amount of growth such a meal can support is far, far less—about 10%—and this is fairly consistent across all levels of the food chain. So the herbivore consuming five pounds of plant material is likely to gain only about half a pound in new growth, while the remaining 90% of the meal is either expended in cellular respiration or lost as feces. Similarly, a carnivore eating the herbivore converts only about 10% of the mass it consumes into its own body mass. Again, 90% is lost to metabolism and feces. Additionally, non-predatory deaths reduce the transfer of energy from one trophic level to the next. And the same inefficiency holds for a top carnivore as well. Let's explore how this 10% rule limits the length of food chains and is responsible for the rarity of big, fierce animals outside your window and across the world.

Q Why are big, fierce animal species so rare in the world?

Biomass is the total weight of living or non-living organic material in a given volume, such as a single organism, or, on a larger scale, the weight of all plant and animal matter in an ecosystem. Given the 10% efficiency with which herbivores convert plant biomass into their own biomass, how much plant biomass is necessary to produce a single 1,200-pound (500 kg) cow? On average, that cow would need to eat about 12,000 pounds (5,000 kg) of grain in order to grow to weigh 1,200 pounds. But that 1,200-pound cow, when eaten by a carnivore, could only add about 120 pounds of biomass to the carnivore, and only 12 pounds to a top carnivore. That's a huge amount of plant biomass required to generate a tiny amount of our top carnivore, which explains why big, fierce animals are so rare (and why vegetarianism is more energetically efficient than meat-eating). Multiply that 5,000 kg of grain by several hundred—or more appropriately, thousands—and you can see that millions of kilograms of grain are required to support only a few top carnivores.

How much plant biomass would be required to support an even higher link on the food chain? Ten times as much—so much that there might not be enough land in the ecosystem to produce enough plant material. And even if there were, the area required would be so large that the "top, top carnivores" might be so spread out and so busy trying to eat enough that they'd be unlikely to encounter each other

ENERGY PYRAMID

"The 10% rule": only about 10% of the biomass from each trophic level is converted into biomass in the next trophic level. The rest of the available energy is lost to the environment, a consequence of several factors, including non-predatory deaths, incomplete digestion of prey/food, and respiration.

TERTIARY CONSUMERS

SECONDARY CONSUMERS

10% converted to biomass

PRIMARY CONSUMERS

10% converted to biomass

PRODUCERS

10% converted to biomass

Inefficiencies in the transfer of energy from one trophic level to the next explain why there are so many more plants than animals.

GRAPHIC CONTENT
Thinking critically about visual displays of data
Turn to p. 597 for a closer inspection of this figure.

FIGURE 17-14 **The 10% rule.**

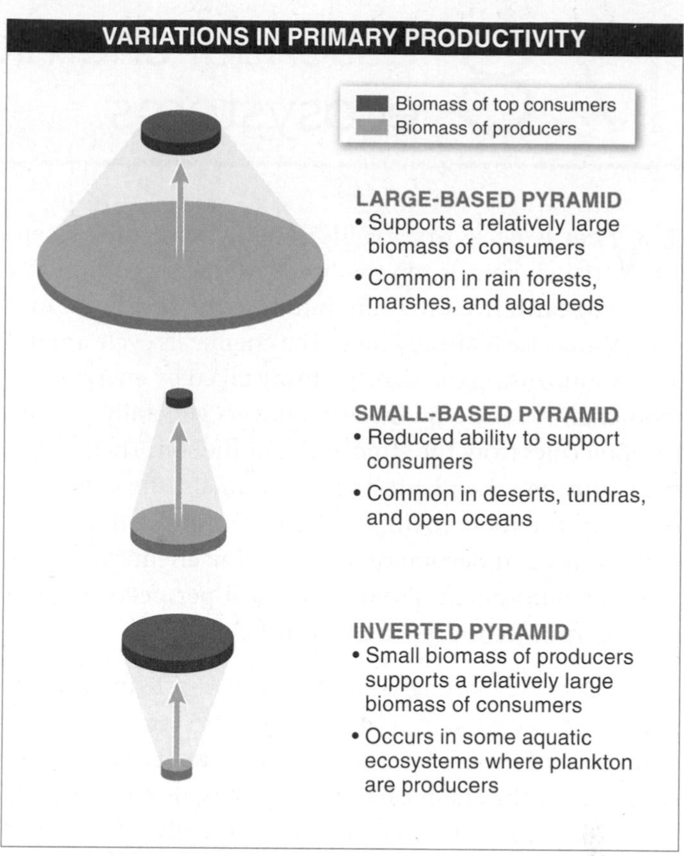

VARIATIONS IN PRIMARY PRODUCTIVITY
Biomass of top consumers
Biomass of producers
LARGE-BASED PYRAMID
• Supports a relatively large biomass of consumers
• Common in rain forests, marshes, and algal beds
SMALL-BASED PYRAMID
• Reduced ability to support consumers
• Common in deserts, tundras, and open oceans
INVERTED PYRAMID
• Small biomass of producers supports a relatively large biomass of consumers
• Occurs in some aquatic ecosystems where plankton are producers

FIGURE 17-15 **Relative biomass of producers and consumers.** Across ecosystems, there is huge variation in primary productivity.

in order to mate. Hence, the 10% rule limits the length of food chains.

We can illustrate the path of energy through the organisms of an ecosystem with an **energy pyramid,** in which each layer of the pyramid represents the biomass of a trophic level. In FIGURE 17-14, we can see that for terrestrial ecosystems, the biomass (in kilograms per square meter) found in photosynthetic organisms, at the base of the pyramid, is reduced significantly at each step, given the incomplete utilization by organisms higher up the food chain. FIGURE 17-15 illustrates the huge variation in primary productivity across a variety of ecosystems. It is highest in tropical rain forests, marshes, and algal beds, and lowest in deserts, tundra, and the open ocean. In each case, the shapes of the energy and biomass pyramids are similar. With a smaller base, though, the ability of an ecosystem to support higher levels in the food chain is reduced. One dramatic exception is seen in some aquatic ecosystems where the producers are plankton. Because plankton have such short life spans and rapid reproduction rates, a relatively small biomass can support a large biomass of consumers, giving rise to an inverted pyramid (see the bottom pyramid in Figure 17-15). However, if you quantified the amount of energy available to consumers (rather than measuring biomass), the pyramid would resemble those seen in terrestrial ecosystems.

TAKE HOME MESSAGE 17.7

» Energy pyramids reveal that the biomass of producers in an ecosystem tends to be far greater than the biomass of herbivores. Similarly, the biomass transferred at each successive step in the food chain tends to be only about 10% of the biomass of the organisms consumed. Due to this inefficiency, food chains rarely exceed four levels.

17.8 Essential chemicals cycle through ecosystems.

What is necessary for life? Energy and some essential chemicals top the list. New energy continually comes to earth from the sun, fulfilling the first need. And everything else is already here. The chemicals cycle around and around, using the same pathway taken by energy—the food chain. Plants and other producers generally take up the molecules from the atmosphere or the soil. Then as animals consume the plants or other animals, the chemicals move up the food chain. As the plants and animals die, detritivores and decomposers return the chemicals to the abiotic environment. From a chemical perspective, life is just a continuous recycling of molecules.

Chemicals cycle through the living and non-living components of an ecosystem. Each chemical is stored in a non-living part of the environment called a "reservoir." Organisms acquire the chemical from the reservoir, the chemical cycles through the food chain, and eventually it is returned to the reservoir.

We can get a deeper appreciation of the functioning of ecosystems and the ecological problems that can occur when they are disturbed—particularly by humans—by investigating a few of these cycles in more detail. Here we explore three of the most important chemical cycles: carbon, nitrogen, and phosphorus.

Carbon Carbon is found largely in four compartments on earth: the oceans, the atmosphere, terrestrial organisms, and fossil deposits. Plants and other photosynthetic organisms obtain most of their carbon from the atmosphere, where carbon is in the form of carbon dioxide (CO_2). As we saw in Chapter 5, photosynthetic organisms use carbon dioxide in photosynthesis, separating the carbon molecules from CO_2 and using them to build sugars and other macromolecules. Carbon then moves through the food chain as organisms eat plants and are themselves eaten (FIGURE 17-16). The oceans contain most of the earth's carbon. Here, many organisms use dissolved carbon to build shells (which later dissolve back into the water, after the organism dies).

Most carbon returns to its reservoir as a consequence of organisms' metabolic processes. Organisms extract energy from food by breaking carbon-carbon bonds, releasing the energy

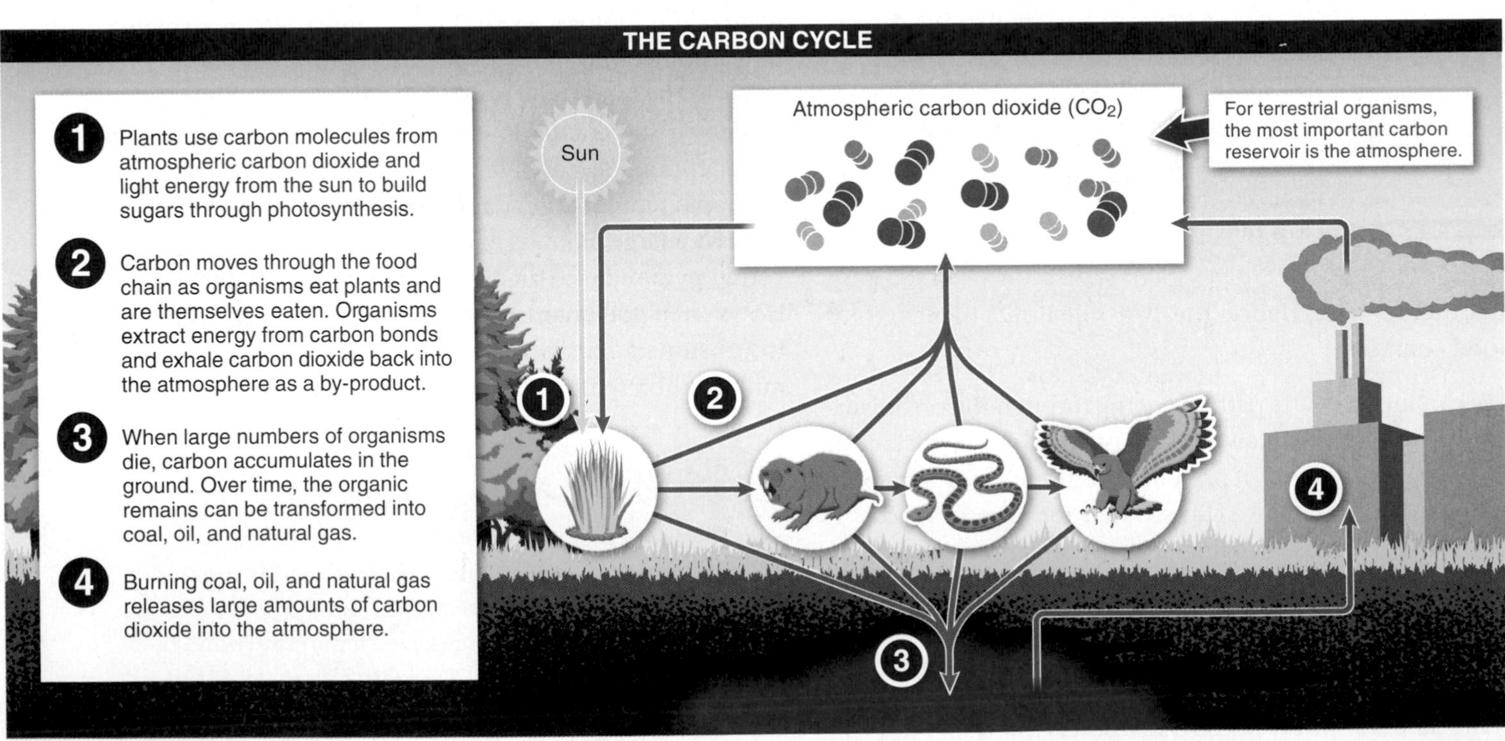

THE CARBON CYCLE
1 Plants use carbon molecules from atmospheric carbon dioxide and light energy from the sun to build sugars through photosynthesis.
2 Carbon moves through the food chain as organisms eat plants and are themselves eaten. Organisms extract energy from carbon bonds and exhale carbon dioxide back into the atmosphere as a by-product.
3 When large numbers of organisms die, carbon accumulates in the ground. Over time, the organic remains can be transformed into coal, oil, and natural gas.
4 Burning coal, oil, and natural gas releases large amounts of carbon dioxide into the atmosphere.
Sun
Atmospheric carbon dioxide (CO_2)
For terrestrial organisms, the most important carbon reservoir is the atmosphere.
1
2
3
4

FIGURE 17-16 **Element cycling: carbon.**

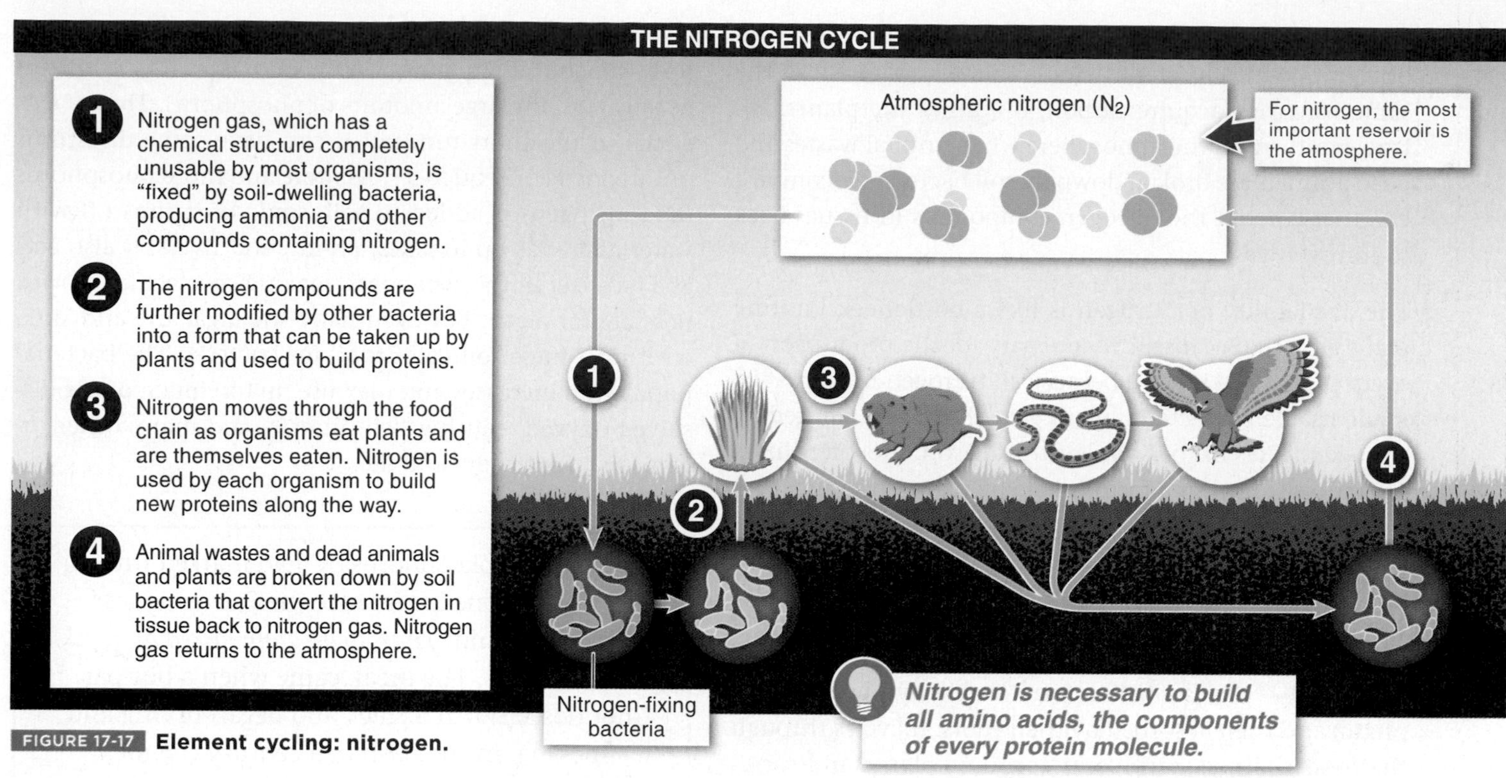

THE NITROGEN CYCLE
1 Nitrogen gas, which has a chemical structure completely unusable by most organisms, is "fixed" by soil-dwelling bacteria, producing ammonia and other compounds containing nitrogen.
2 The nitrogen compounds are further modified by other bacteria into a form that can be taken up by plants and used to build proteins.
3 Nitrogen moves through the food chain as organisms eat plants and are themselves eaten. Nitrogen is used by each organism to build new proteins along the way.
4 Animal wastes and dead animals and plants are broken down by soil bacteria that convert the nitrogen in tissue back to nitrogen gas. Nitrogen gas returns to the atmosphere.
Atmospheric nitrogen (N_2)
For nitrogen, the most important reservoir is the atmosphere.
Nitrogen-fixing bacteria
Nitrogen is necessary to build all amino acids, the components of every protein molecule.

FIGURE 17-17 **Element cycling: nitrogen.**

stored in the bonds, and combining the released carbon atoms with oxygen. They then exhale the end product as CO_2.

The burning of fossil fuels is adding significantly to the atmospheric carbon reservoir. Fossil fuels are created when large numbers of organisms die and are buried in sediment lacking oxygen. In the absence of oxygen, at high pressures, and after very long periods of time, the organic remains are transformed into coal, oil, and natural gas. Trapped underground or in rock, these sources of carbon played little role in the global carbon cycle until humans in industrialized countries began using fossil fuels to power their technologies. Burning coal, oil, and natural gas releases large amounts of carbon dioxide, thus increasing the average CO_2 concentration in the atmosphere—the current level of CO_2 in the atmosphere is the highest it has been in almost half a million years. This has potentially disastrous implications, as we'll see in Chapter 18.

Q Why are global CO_2 levels rising?

Q Global CO_2 levels are rising overall, but they also exhibit a sharp rise and fall over the course of each year. Why?

Although, on average, the level of CO_2 in the environment is increasing, there is a yearly cycle of ups and downs in the CO_2 levels in the northern hemisphere. This is due to fluctuations in the ability of plants to absorb CO_2. Many trees lose their leaves each fall and, during the winter months, relatively low rates of photosynthesis lead to low rates of CO_2 consumption, causing an annual peak in atmospheric CO_2 levels. During the summer, trees have their leaves, sunlight is strong, and photosynthesis (consuming CO_2) occurs at much higher levels, causing a drop in the atmospheric CO_2 level.

Nitrogen Nitrogen is necessary to build a variety of molecules critical to life, including all amino acids, the components of every protein molecule. The chief reservoir of nitrogen is the atmosphere, but even though more than 78% of the atmosphere is nitrogen gas (N_2), for most organisms this nitrogen is completely unusable. The problem is that atmospheric nitrogen consists of two nitrogen atoms bonded tightly together, and these bonds need to be broken to make the nitrogen usable for living organisms. Only through the metabolic magic (chemistry, actually) of some soil-dwelling bacteria, the nitrogen-fixers, can most nitrogen enter the food chain (FIGURE 17-17).

These bacteria chemically convert, or "fix," nitrogen by attaching it to other atoms, including hydrogen, to produce ammonia and related compounds. These compounds are then further modified by other bacteria into a form that can be taken up by plants and used to build proteins. And once nitrogen is in plant tissues, animals acquire it in the same way they acquire carbon: by eating the plants. Nitrogen returns to the atmosphere when animal wastes and dead animals are broken down by soil bacteria (decomposers) that convert the nitrogen compounds in tissues back to nitrogen gas.

The availability of nitrogen is like a bottleneck limiting plant growth. Because it is necessary for the production of every plant protein, and because all nitrogen must first be made usable by bacteria, plant growth is often limited by nitrogen levels in the soil. For this reason, most fertilizers contain nitrogen in a form usable by plants.

Phosphorus Every molecule of ATP and DNA requires phosphorus. But no (or barely any) phosphorus is available in the atmosphere. Instead, soil serves as the chief reservoir. Like nitrogen, phosphorus is chemically converted—"fixed"—into a form usable by plants (phosphate) and then absorbed by their roots. It cycles through the food chain as herbivores consume plants and carnivores consume herbivores. As organisms die, their remains are broken down by bacteria and other organisms, returning the phosphorus to the soil. The pool of phosphorus in the soil is also influenced by the much slower process of formation of rock on the seafloor, its uplifting into mountains, and its eventual weathering, releasing its phosphorus (FIGURE 17-18).

Like nitrogen, phosphorus is often a limiting resource in soils, constraining plant growth. Consequently, fertilizers usually contain large amounts of phosphorus. This is beneficial in the short run, but it can have some disastrous unintended consequences. As more and more phosphorus (and nitrogen) is added to soil, some of it runs off with water and ends up in lakes, ponds, and rivers. It also acts as a fertilizer in these habitats, making spectacular growth possible for algae. But eventually the algae die and sink, creating a huge source of food for bacteria. The bacterial population increases and may use up too much of the dissolved oxygen, causing fish, insects, and many other organisms to suffocate and die.

> "An atom of phosphorus 'X' had marked time in the limestone ledge since the Paleozoic seas covered the land. *Time, to an atom locked in a rock, does not pass.* The break came when a bur-oak root nosed down a crack and began prying and sucking. In the flash of a century the rock decayed, and X was pulled out and up into the world of living things. He helped build a flower, which became an acorn, which fattened a deer, which fed an Indian all in a single year."
>
> — **ALDO LEOPOLD,** *A Sand County Almanac,* **1949**

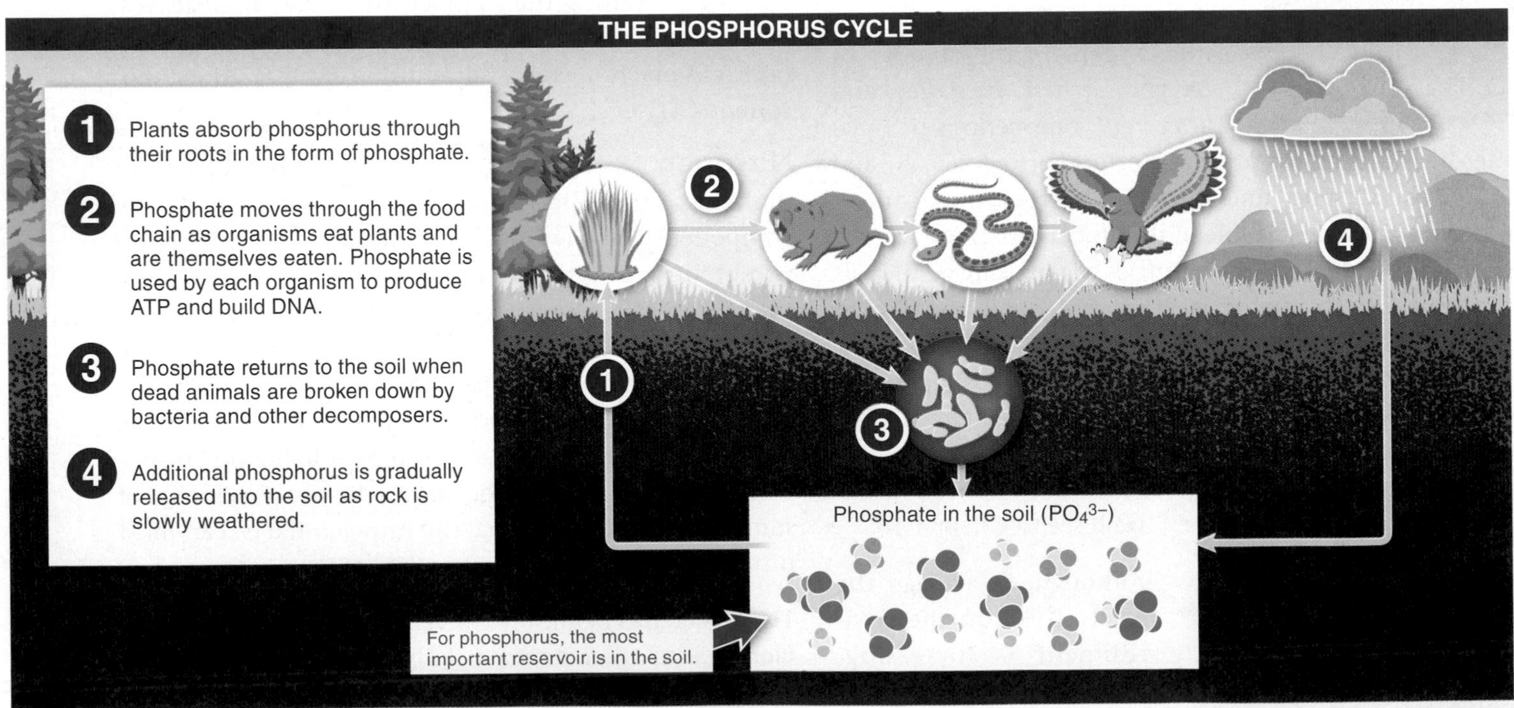

THE PHOSPHORUS CYCLE
1 Plants absorb phosphorus through their roots in the form of phosphate.
2 Phosphate moves through the food chain as organisms eat plants and are themselves eaten. Phosphate is used by each organism to produce ATP and build DNA.
3 Phosphate returns to the soil when dead animals are broken down by bacteria and other decomposers.
4 Additional phosphorus is gradually released into the soil as rock is slowly weathered.
Phosphate in the soil (PO_4^{3-})
For phosphorus, the most important reservoir is in the soil.

FIGURE 17-18 **Element cycling: phosphorus.**

EUTROPHICATION

Eutrophication—the increase in nutrients in an ecosystem, particularly nitrogen and phosphorus—often leads to the rapid growth of algae and bacteria in aquatic ecosystems. These organisms then consume much of the oxygen, leading to large die-offs of animals.

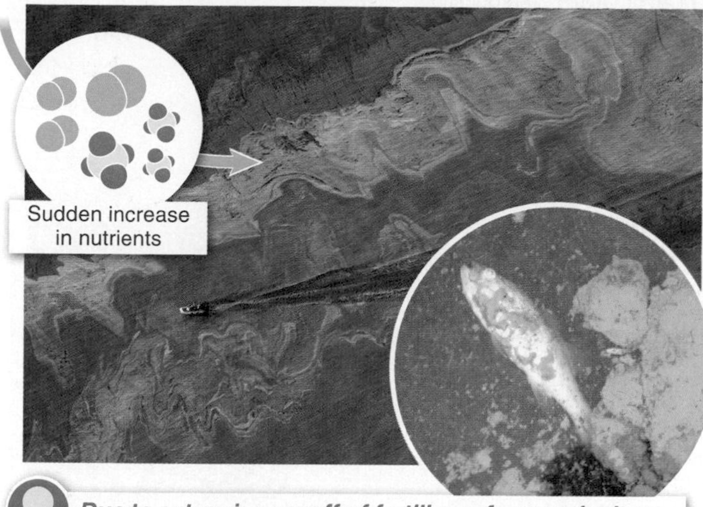

Due to extensive runoff of fertilizers from agriculture, eutrophication now affects more than half of the lakes in Asia, Europe, and North America.

FIGURE 17-19 **Too much of a good thing?**

The increase in nutrients, particularly nitrogen and phosphorus, in an ecosystem, is called **eutrophication.** The consequent rapid growth of algae and bacteria, followed by large-scale die-offs of other organisms, is increasingly a problem in both small and large bodies of water, including more than half of the lakes of Asia, Europe, and North America (FIGURE 17-19). Lake Erie, on the U.S.–Canadian border, for example, has experienced eutrophication as a result of all the phosphorus- and nitrogen-containing wastewater that drains into the lake from the extensive surrounding farmlands. A eutrophication-caused algae bloom there in August 2014 was responsible for the growth of toxic bacteria in municipal water supplies. The governor of Ohio was forced to declare a state of emergency, banning half a million people from using water for drinking, cooking, or even bathing. Given the lower use of fertilizers in South America and Africa, eutrophication is less common there.

TAKE HOME MESSAGE 17.8

» Chemicals essential to life—including carbon, nitrogen, and phosphorus—cycle through ecosystems. They are usually captured from the atmosphere, soil, or water by growing organisms, passed from one trophic level to the next as organisms eat other organisms, and returned to the environment through respiration, decomposition, and erosion. These cycles can be disrupted as human activities significantly increase the amounts of chemicals released to the environment.

17.9–17.15 Species interactions influence the structure of communities.

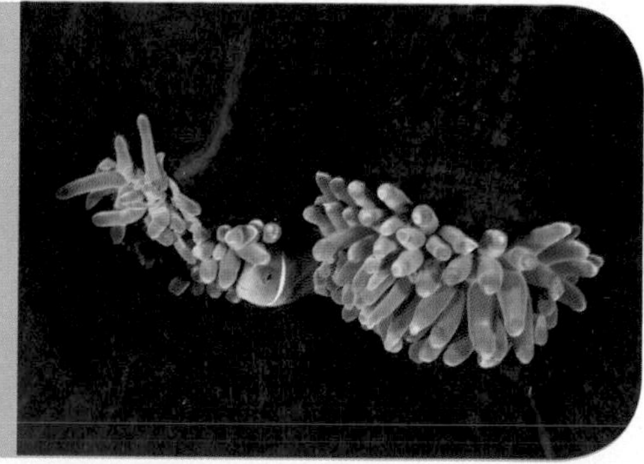

The sea anemone (green) and the pink anemonefish live together. The anemone's sting warns predators away, while the fish eats parasites and drives off fish that feed on anemones.

17.9 Each species' role in a community is defined as its niche.

Within a society, most humans seem to find their niche. Each person plays a particular role, defined by the nature of his or her work, activities, and interactions with others. Other species do the same thing. Within their

NICHE

FIGURE 17-20 **A way of living.** "Niche" describes all of the ways in which an organism utilizes the resources of the environment.

NICHE FEATURES

- The space an organism requires
- The type and amount of food an organism utilizes
- The timing of an organism's reproduction
- An organism's temperature and moisture requirements and other necessary living conditions
- The organisms for which it is a food source
- Its influence on competitors

REALIZED NICHE vs. FUNDAMENTAL NICHE

An organism's fundamental niche—the full range of conditions under which it may live—may be larger than its realized niche, due to competition or other factors.

For example, urban areas reduce suitable habitat within the geographic range of a bald eagle. Likewise, interference from other species can limit access to habitat within its geographic range.

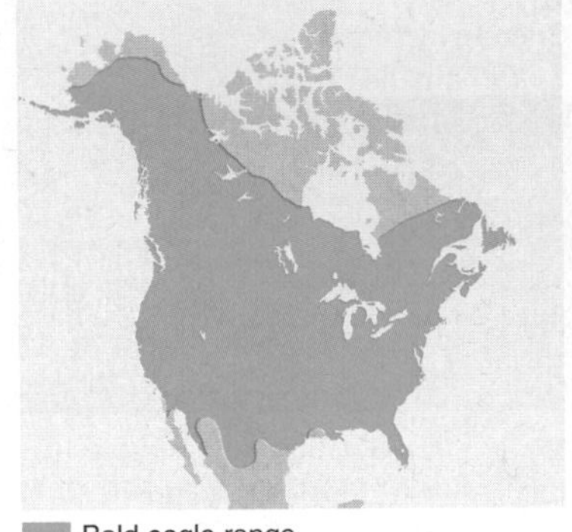

communities—geographic areas defined as loose assemblages of species, sometimes interdependent, with overlapping ranges—each species has its own niche.

We can think of an organism's **niche** as its place in the community. More than just a *place* for living, however, a niche is a complete *way* of living. In other words, an organism's niche encompasses (1) the space it requires, (2) the type and amount of food it consumes, (3) the organisms for which it is a food source, (4) its influence on competitors, (5) the timing of its reproduction (its life history), (6) its temperature and moisture requirements—and almost every other aspect that describes the way the organism uses its environment and influences the other organisms in that environment (FIGURE 17-20).

Although a niche describes the role a species *can* play within a community, the species doesn't always get to have that exact role. It is common for species to find themselves competing with other species for parts of a niche, and for one of the species to be restricted from its full niche (see Figure 17-20). Consider the rats of Boston. Until the 1990s, they lived in relative peace in the sewers beneath the city's streets. But when the city embarked on the largest underground highway construction project in U.S. history, engineers displaced and forcibly drove out thousands of rats from much of their habitat. In essence, there was suddenly an overlap between the rat niche and the human niche, and the rats were now restricted to just some portions of the sewers. As a result, the rats' **realized niche,** the narrower role that they may occupy in a community, is now just a subset of their **fundamental niche,** the full range of environmental conditions under which they can live.

TAKE HOME MESSAGE 17.9

» A population of organisms in a community fills a unique niche, defined by the manner in which the population utilizes the resources and influences the other organisms in its environment. Organisms do not always completely fill their niche; competition with other species within overlapping niches can reduce their range.

17.10 Interacting species evolve together.

There is a moth in Madagascar with a tongue that is 11 inches long! This might seem absurd—until the moth approaches a similarly odd-looking orchid. The orchid's flower has a very long tube, also about 11 inches long, with a bit of nectar at the very bottom. The moth's tongue, although usually rolled up, straightens out as fluid is pumped into it, and the moth can then insert it into the long nectar tube. As its tongue reaches the bottom, the moth slurps up the nectar. The moth also gets a bit of pollen stuck to it, which gets brushed onto the reproductive parts of the next orchid flower it visits (FIGURE 17-21).

Q Which came first, the long-tongued moths or the long-tubed flowers?

It's clear that having an 11-inch tongue is useful, even necessary, to extract nectar from an 11-inch nectar tube. And it's clear that putting nectar at the bottom of an 11-inch tube is a strategy to restrict access to only those pollinators that will reliably pass pollen from plant to plant of the same species. But how did such a system originate? Each trait seems to make sense only if the other already exists. The answer is that the two traits evolved—**coevolved**—together.

As a consequence of natural selection, populations of organisms commonly become better adapted to their environment. As long as there is variation for a trait and the trait is heritable, differential reproductive success will lead to a change in the population. It is easy to imagine populations becoming more and more efficient at making use of non-living resources. But natural selection does not distinguish between biotic and abiotic resources as selective forces. Either can cause individuals with certain traits to reproduce at a higher rate than others, so either can cause evolution: small changes in the moth lead to small changes in the orchid, which select for further small changes in the moth . . . and the process continues. In the end, species become adapted not just to their physical environment but to the other species around them.

FIGURE 17-21 **A perfect match?** Long-tubed orchids and long-tongued moths in Madagascar each influence the evolution of the other.

TAKE HOME MESSAGE 17.10

» In producing organisms better adapted to their environment, natural selection does not distinguish between biotic and abiotic resources as selective forces.

17.11 Competition can be hard to see, yet it influences community structure.

Interacting species' agendas are not always aligned, as they are for plants and their pollinators. Often, species interact because both are trying to exploit the same resources. In such cases—that is, when species have similar

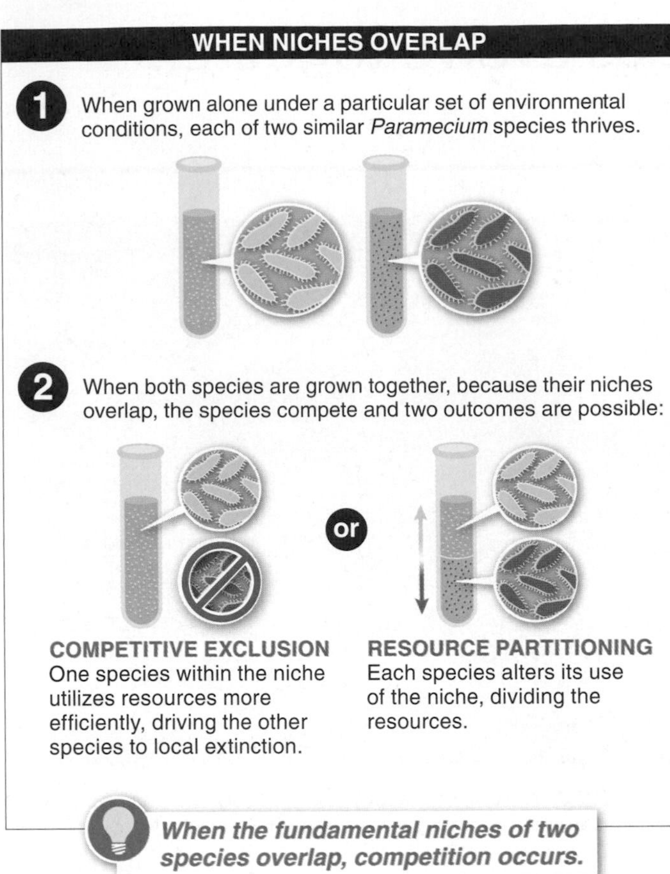

FIGURE 17-22 Overlapping niches: competitive exclusion and resource partitioning.

niches—conflict usually occurs. This competition doesn't last forever, though. Inevitably, one of two outcomes occurs: competitive exclusion or resource partitioning (FIGURE 17-22).

1. In **competitive exclusion,** two species battle for resources in the same niche until the more efficient one wins and the other species is driven to extinction in that location ("local extinction"). In the 1930s, this was demonstrated in simple laboratory experiments using *Paramecium,* a single-celled organism. Populations of two similar *Paramecium* species were grown either separately or together in test tubes containing water and their bacterial food source. When grown separately, each species thrived. When grown together, one species always drove the other to extinction.

2. Resource partitioning is an alternative outcome of niche overlap. Individual organisms and species can adapt to changing environmental conditions, and resource partitioning can result from an organism's behavioral change or a change in its structure. When this occurs, one or both species become restricted in some aspect of their niche, dividing the resource. In other experiments with *Paramecium,* for example, researchers used one of the two species from the initial experiment and a new, third species. Again, each species thrived when grown alone. But when the two species were grown together in the same test tube, they ended up dividing the test tube "habitat." One species fed exclusively at the bottom of the test tube, and the other fed only at the top. Simple behavioral change made coexistence possible.

In many situations, resource partitioning is accompanied by **character displacement,** an evolutionary divergence in one or both of the species that leads to a partitioning of the niche. A clear example occurs in two species of seed-eating finches on the Galápagos Islands. On islands where both species live, their beak sizes differ significantly. One species has a deeper beak, better for large seeds, while the other has a shallower beak, better for smaller seeds, and they do not compete. On islands where either species occurs alone, beak size is intermediate between the two sizes (FIGURE 17-23).

COMPETITION AND CHARACTER DISPLACEMENT

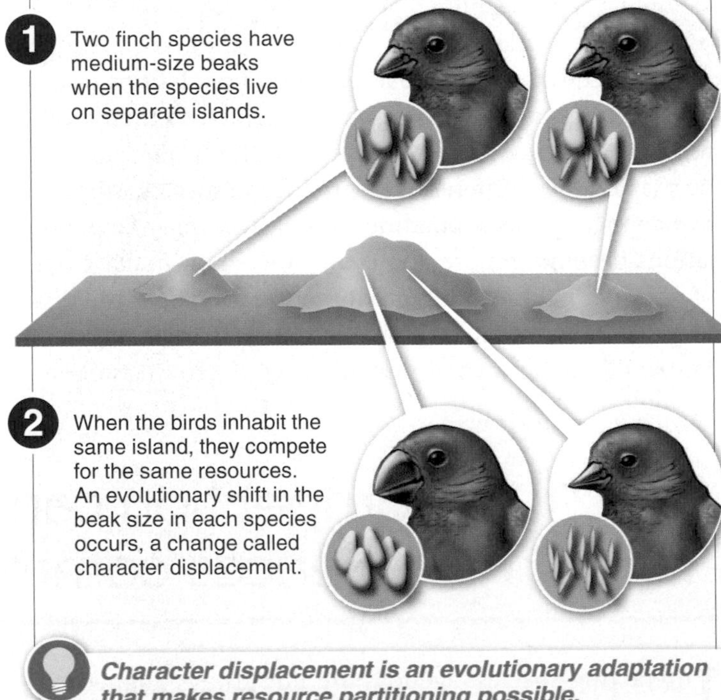

Character displacement occurs when natural selection reduces the competition between two species by producing an evolutionary divergence in one or both species.
1 Two finch species have medium-size beaks when the species live on separate islands.
2 When the birds inhabit the same island, they compete for the same resources. An evolutionary shift in the beak size in each species occurs, a change called character displacement.
Character displacement is an evolutionary adaptation that makes resource partitioning possible.

FIGURE 17-23 Allowing organisms to divide resources.

Competition between species has one very odd feature: it is hard to actually see it happening because it causes itself to disappear. After only a short period of competition, one of the species becomes locally extinct or leaves the area where the niches overlap, or character displacement occurs, largely reducing the competition. In either case, the net result is that the level of competition is significantly reduced or wiped out altogether. For this reason, biologists often have to look for character displacement—the "ghost of competition past"—to identify areas where competition once took place. Moreover, even while it is occurring, competition tends to be indirect, rather than head-to-head battles. Like a game of musical chairs, it's more a question of both species trying to use a particular resource, with one being a bit better at it.

Biologists do have the opportunity to see competition in occasional natural experiments in which a new species is introduced to a community. For example, the American mink, a weasel-like mammal, was introduced into Europe, where previously only the European mink had lived. Within 10 years, in the area where their ranges overlapped, the European mink increased in size while the American mink decreased in size, as biologists documented the evolution. In the next chapter, we investigate in greater depth why newly introduced species frequently win the competition with native species.

TAKE HOME MESSAGE 17.11

» Populations with completely overlapping niches cannot coexist forever. Competition for resources occurs until one or both species evolve in ways that reduce the competition, through character displacement, or until one species becomes extinct in that location.

17.12 Predation produces adaptation in both predators and their prey.

Some words of advice in case you ever think about quickly approaching a horned lizard: be afraid. Be very afraid. Here's why: as you get close to the lizard, it may zap you with streams of blood squirted from its eyes. In all likelihood, you will flinch. And as you flinch, the horned lizard will scurry away. The display is shocking, but the fact that evolution has produced extreme and effective anti-predator adaptations is not.

Predation—an interaction between two species in which one species eats the other—is one of the most important forces shaping the composition and abundance of species in a community. Predation, though, isn't restricted to the obvious interactions involving one animal chasing down and killing another. Herbivores' eating of leaves, fruits, or seeds is a form of predation, even though it doesn't necessarily kill the plant. And predators are not necessarily physically imposing. Each year more than a million humans die as a result of disease from mosquito bites, compared with fewer than a dozen from shark attacks.

Predators are a potent selective force: organisms eaten by predators tend to have reduced reproductive success. Consequently, in prey species, a variety of features have evolved (and continue to evolve)—including the blood-squirting-eyeball effect—that reduce the organisms' predation risk. But as prey evolve, so do predators. This coevolution is a sort of arms race with ever-changing and escalating predation-effectiveness adaptations causing more effective predator-avoidance adaptations, and vice versa. In this light, it may seem unexpected that exotic species often flourish when released into novel habitats, even though natural selection has not adapted them to their new environment. As it turns out, just as these species are not fully adapted to their new environment, they also have few predators there. And with low predation risk, they often can flourish. We'll examine some common adaptations of both predators and prey.

Q Why do exotic species often flourish when released into novel habitats, even though natural selection has not adapted them to this new environment?

Prey Adaptations for Reducing Predation There are two broad categories of defenses against predators: physical and behavioral.

PHYSICAL DEFENSES FOR REDUCING PREDATION

MECHANICAL DEFENSES
Physical structures—such as the quills of the Malayan porcupine—can help protect an organism from predation.

CHEMICAL DEFENSES
Toxins—such as those found within the strawberry poison dart frog—can make an organism poisonous or unpalatable to a predator.

WARNING COLORATION
Organisms that produce toxic chemicals—such as monarch butterflies—frequently have bright color patterns, warning potential predators to stay away.

CAMOUFLAGE
Cryptic coloration—such as that seen in the praying mantis—can enable an organism to blend into its surroundings and avoid predation.

FIGURE 17-24 **Prey defenses: some physical means for avoiding predation.**

Physical defenses include mechanical, chemical, warning coloration, and camouflage mechanisms (FIGURE 17-24).

1. Mechanical defenses. Predation plays a large role in producing adaptations such as the sharp quills of a porcupine, the prickly spines of a cactus, or the tough armor protecting an armadillo or sow bug. These, as well as claws, fangs, stingers, and other physical structures, can reduce predation risk.

2. Chemical defenses. Further prey defenses can include chemical toxins that make the prey poisonous or unpalatable. Plants can't run from their predators, so chemical defenses are especially important to them. Almost all plants produce some chemicals to deter organisms that might eat them. The toxins can be severe, such as strychnine from plants in the genus *Strychnos,* which kills most vertebrates, including humans, by stimulating nonstop convulsions and other extreme and painful symptoms leading to death. At the other end of the spectrum are chemicals toxic to some insects but relatively mild to humans, such as those found in cinnamon, peppermint, and jalapeño peppers. Ironically, many plants that evolved to produce such chemicals to deter predators are now cultivated and eaten specifically *for* these chemicals. One organism's toxic poison is another's spicy flavor.

Some animals can also synthesize toxic compounds (or perhaps, more commonly, can safely acquire them from the organisms they eat). The poison dart frog has poison glands all over its body, making it toxic to the touch. The fire salamander is also toxic, with the capacity to squirt a strong nerve toxin from poison glands on its back. Some animals, including milkweed bugs and monarch butterflies, can safely consume toxic chemicals from plants and sequester them in their tissues, becoming toxic to predators who try to eat them.

3. Warning coloration. Species protected from predation by toxic chemicals frequently have bright color patterns to warn potential predators, essentially carrying a sign that says "Warning: I'm poisonous—keep away." To make it as easy as possible for predators to learn, different poisonous species often have the same color patterns.

In a clever twist on this, some species that are perfectly edible to their predators also have the same bright colors, in a phenomenon known as **mimicry.** As long as they are relatively rare compared with the toxic individuals they mimic—reducing the chance that predators might catch on to their trickery—the evolutionary ruse is quite successful.

4. Camouflage. An alternative to warning coloration, and one of the most effective ways to avoid being eaten, is simply to avoid being seen. An adaptation in many organisms is patterns of coloration that enable them to blend into their surroundings. Examples include insects that look like leaves or twigs and hares that are brown for most of the year but turn white when there is snow on the ground.

Behavioral defenses include both seemingly passive and active behaviors: hiding or escaping, and alarm calling or fighting back (FIGURE 17-25).

1. Hiding or escaping. Anti-predator adaptations need not involve toxic chemicals, physical structures, or special coloration. Many species excel at hiding or running, or both. With vigilance, it is possible to get advance warning of impending predator attacks, and then quickly and effectively avoid the predator. A variation of this strategy comes from safety in numbers: many species, including schooling fish and emperor penguins, travel in large groups to reduce their predation risk.

Q On islands, animals frequently have no fear of humans. Why?

On many islands, animals show no fear of humans. Rather than encountering skittish lizards, for example, a human visitor to the Galápagos Islands must be careful not to step on them. The animals are not afraid because, given the small size of most islands, the number of predators is restricted. As a consequence, there has been no selection for skittishness and hypervigilance, traits that normally evolve in response to predation risk.

2. Alarm calling and fighting back. In many species, especially birds and mammals, individuals warn others with an alarm call. Although risky for the caller, such alarm calling can give other individuals—often close kin that are nearby—enough advance warning to escape (recall, from Chapter 11, the warning calls of Belding's ground squirrels). Some prey species also turn the tables, mobbing predators to keep them from successfully completing their task. This category might also include the blood-squirting lizard described above, or the fulmar, a seabird that defends its nest from attacks with projectile vomiting aimed at the intruder.

Predator Adaptations for Enhancing Predation

Just as prey use physical and behavioral features to reduce their risk of predation, predators evolve in parallel ways to increase their efficiency. As milkweed plants have evolved to produce toxic chemicals, sequestered within the structures of the plant, milkweed bugs have evolved to be able to eat the toxic plants without suffering harm. And as prey have become better at hiding and escaping, predators have developed better sensory perception to help them detect hiding prey and faster running ability to catch them. Predators, too, make use of mimicry. The angler fish, tasseled frogfish, and snapping turtle all have physical structures that mimic something—usually a food item—of interest to potential prey. As a prey animal comes closer to inspect, the predator snaps it up.

Although natural selection leads to predators with effective adaptations for capturing prey, the adaptations are rarely so efficient that the prey species are driven to extinction. The explanation for this is referred to as the "life-dinner hypothesis." Selection for "escape ability" in the prey is stronger

BEHAVIORAL DEFENSES THAT REDUCE PREDATION

HIDING OR ESCAPING
Some prey species excel at hiding or running, or both, effectively avoiding predation. A variation of this strategy comes from safety in numbers: many species, including schooling fish, travel in large groups to reduce their predation risk.

FIGHTING BACK
Some prey species fight back against their attacker, effectively avoiding predation. For example, the fulmar, a seabird, defends its nest from attacks with projectile vomiting aimed at the intruder.

FIGURE 17-25 **Prey defenses: behavioral means for avoiding predation.**

Q Why don't predators become so efficient at capturing prey that they drive the prey to extinction?

than selection for "capture ability" in the predator. When, for example, a rabbit can't escape from a fox, the cost to the rabbit is its life—and it will never reproduce again. On the other hand, when a fox can't keep up with a rabbit, all it loses is a meal; it can still capture prey and reproduce in the future. In other words, the cost of losing in the interaction is much higher for the rabbit.

TAKE HOME MESSAGE 17.12

» Predators and their prey are in an evolutionary arms race: as physical and behavioral features evolve in prey species to reduce their predation risk, predators develop more effective and efficient methods of predation. This coevolutionary process can result in brightly colored organisms, alarm calling, and many types of mimicry.

17.13 Parasitism is a form of predation.

For most people, the word "predation" conjures images of a large animal such as a cheetah, chasing and killing another large animal such as a gazelle. But there is another world of predation, largely unseen but equally deadly: **parasitism,** defined as a symbiotic relationship—that is, a close and commonly long-term relationship between individuals of different species—in which one organism (the **parasite**) benefits while the other (the **host**) is harmed. In fact, this hidden world of parasites is thriving with activity: parasites are probably the most numerous species on earth, with three or four times as many species as non-parasites!

Two general types of parasites make life difficult for most organisms. *Ectoparasites* (*ecto* = outside) include organisms such as lice, leeches, and ticks. One species of Mexican parrot is all too aware of ectoparasites: it has 30 different species of mites living on its feathers alone. And many of the parasites even have parasites themselves! *Endoparasites* are parasites that live inside their hosts (*endo* = inside). They are equally pervasive. Endoparasites infecting vertebrates include many different phyla of both animals and protists, the single-celled eukaryotes (FIGURE 17-26). In all of these parasite-host interactions, as with all predator-prey

PARASITES

Parasites are predators that benefit from a symbiotic relationship with their hosts.

ECTOPARASITES
Ectoparasites—such as the bedbug (*Cimex lectularius*), an ectoparasite of humans—live *on* their host.

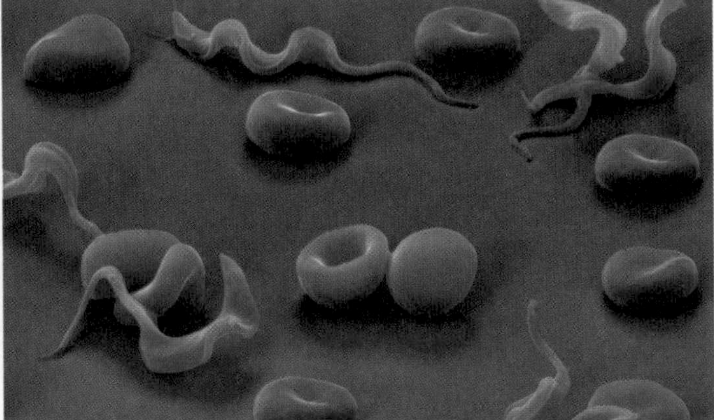

ENDOPARASITES
Endoparasites—such as *Trypanosoma brucei*, an endoparasite that invades a human host and is the cause of African sleeping sickness—live *inside* their host.

FIGURE 17-26 **Parasites: predators dwarfed by their prey.**

interactions, the predator or parasite benefits and the prey or host is harmed.

Even though they are considered predators, parasites have some unique features and face some unusual challenges in comparison with other predators. The most obvious of these features is that the parasite generally is much smaller than its host and stays in contact with it for extended periods of time, normally not killing the host but weakening it as the parasite uses some of the host's resources. Being located right on your food source all the time can be advantageous. But this also leads to what is perhaps the greatest challenge that parasites face: how to get from one individual host to another—after all, a parasite can't survive long once its host dies. The methods by which parasites accomplish such dispersal are surprising. Many of their complicated life cycles involve passing through two (or more) different host species (and could have come about only through coevolution with each of the host species). These life cycles are likely to give us a new appreciation for the ingenuity of these microorganisms—or rather, for the evolutionary process that produced them. Let's look at a few representative examples.

PARASITE INGENUITY

Rabies parasite is more effectively passed from one animal to another by causing its host to foam at the mouth and aggressively bite.

Parasites have surprising ways of solving a challenging problem: how to get from one individual host to another.

FIGURE 17-27 **Spreading disease.** Rabies symptoms aid the disease-causing virus in spreading to new hosts.

Case 1: Parasites can induce foolish, fearless behavior in their hosts. During their evolution, rats have developed a protective wariness of cats, as well as areas in which cats have been roaming. Toxoplasma is an organism that changes this. This single-celled parasite of rats must also spend part of its life cycle in cats. It does this by altering the brain of its rat host so that the rat no longer fears cats, but are attracted to them instead. This behavioral change, while quite dangerous for the rat, is exactly the change that increases the likelihood that the rat will be attacked by a cat, spreading the parasite in the process.

Case 2: Parasites can induce inappropriate aggression in their hosts. Rabid animals don't behave normally. They froth at the mouth and become unusually aggressive (FIGURE 17-27). Is this an accident? No. Rabies is caused by a virus that infects warm-blooded animals, mostly raccoons, skunks, foxes, and coyotes. It is passed from one host to another via saliva. Inducing these "rabid" behaviors is the change in behavior that will most help the virus to spread.

Case 3: Parasites can induce bizarre and risky behavior in their hosts. The lancet fluke is a parasitic flatworm. It has also been described as a "zombie-maker." This fluke spends most of its life in sheep and goats, but the fluke's eggs pass into snails that graze on vegetation contaminated by sheep and goat feces. Once inside the snail, the fluke eggs grow and develop, eventually forming cysts that the snail surrounds with mucus and then excretes. Continuing on their complex life cycle, the fluke cysts find their way into ants that eat the snail mucus. The flukes' journey back to a sheep or goat is now expedited by the so-called zombie-making. In an infected ant, the flukes grow into the ant's brain, altering its behavior. Whereas ants normally remain low to the ground, when infected by the lancet fluke, an ant climbs to the top of a grass blade or plant stem and clenches its mandibles on a leaf. This behavioral change puts the ant at greater risk of being eaten by a grazing mammal—a bad outcome for the ant, but just what the parasite needs to complete its life cycle.

These parasite-induced behavior modifications are among the most dramatic, but there are many others that are more subtle yet equally effective at allowing a parasite to thrive attached to or within its host's body. This subset of predation is an active and exciting area of ecological and physiological investigation.

TAKE HOME MESSAGE 17.13

» Parasitism is a symbiotic relationship in which one organism benefits while the other is harmed. Parasites face some unusual challenges relative to other predators, particularly in how to get from one individual host to another, and some complex parasite life cycles have evolved.

17.14 Not all species interactions are negative: mutualism and commensalism.

It's easy to get the idea that all species interactions are harsh and confrontational, marked with a clear winner and loser. However, not all species interactions are combative. Every flower should be a reminder that evolution produces beneficial species interactions as well. These types of interactions fall into two categories: mutualism and commensalism.

Mutualism: Everybody Wins Corals that gain energy from photosynthetic algae living inside their tissues. Termites capable of subsisting on wood, but only with the assistance of cellulose-digesting microbes living in their digestive system. Flowers pollinated by animals that are nourished by nectar. Each of these relationships is an example of **mutualism,** an interaction in which both species gain and neither is harmed (FIGURE 17-28). Mutualism is common in almost all communities. Plants, particularly, have numerous such interactions: with nutrient-absorbing fungi, nitrogen-fixing bacteria, and animals that pollinate them and disperse their fruits.

Mutualistic interactions are so widespread and ecologically important that they are described in detail throughout this book. These include the relationship between plants and mycorrhizal bacteria (Section 14-15), between plants and their animal pollinators and fruit dispersers (Sections 14-9 and 14-11), and lichens, the symbiotic association of fungi and photosynthetic algae or bacteria (Section 14-15).

Commensalism: An Interaction with a Winner but No Loser Some species interactions are one-sided. The cases in which one species benefits and the other neither benefits nor is harmed are called commensal relationships, or **commensalism.** Cattle egrets have just such a relationship with grazing mammals such as buffalo and elephants (see Figure 17-28). As the large mammals graze through grasses, they stir up insects. The birds, which feed near the mammals, are able to catch more insects with less effort this way. The grazers are neither helped nor harmed by the presence of the birds.

TAKE HOME MESSAGE 17.14

» Not all species interactions are combative: beneficial species interactions evolve as well. Mutualism, in which both species benefit from the interaction, is widespread and critically important to all ecosystems. In commensalism, one species benefits and the other is neither harmed nor helped.

NON-NEGATIVE SPECIES INTERACTIONS

MUTUALISM
In mutualism, both species benefit from an interaction. For example, the swamp aster flower is pollinated by a bumblebee that is nourished by the plant's nectar.

COMMENSALISM
In commensalism, one species benefits from an interaction and the other neither benefits nor is harmed. For example, the cattle egret feeds on insects stirred up by a grazing Cape buffalo. The buffalo is neither helped nor harmed in the interaction.

FIGURE 17-28 **Not always "red in tooth and claw."** Mutualisms and commensalisms abound in nature.

THIS IS HOW WE DO IT

Developing the ability to apply the process of science

17.15 Investigating ants, plants, and the unintended consequences of environmental intervention.

The complexity of the webs of interactions within ecosystems makes it difficult to predict the impact of individual changes to a system. This has important implications as we try to protect valuable environments and natural resources.

Consider a situation that many natural resource managers encounter. Suppose that a population of organisms with great value appears to be threatened by the predatory behavior of another species. To protect the valuable population, the natural resource manager might be tempted to intervene, blocking the predator from harming the valuable population. This seems like a reasonable plan. But how can we be sure?

In fact, our plan, with all its good intentions, may not work. Worse, it may have exactly the opposite effect of what we intend. Let's look at a situation that, surprisingly (but unambiguously), revealed a scary truth: seemingly helpful, straightforward manipulations may have significant, unintended, and negative consequences.

It started when a biologist visiting Africa observed an intervention that was in place to help a plant species—acacia trees. The intervention: large groups of these trees were fenced off to protect them from the destructive herbivory of elephants and giraffes.

What do you think was the intended consequence of putting enclosures around acacia trees?

The natural resource managers assumed that the enclosures, by protecting the trees from herbivores, should benefit the plants. The acacia trees ought to grow faster and live longer. But the visiting biologist noticed that the several hundred plants within the enclosures—distributed across six 10-acre plots of land that had been fenced off for more than 10 years—weren't doing very well. Compared with unprotected acacia plants in adjacent plots, most of the protected plants looked unhealthy, and many were dying. Researchers decided to find out why.

With an intervention already in place, how could the researchers figure out what was responsible for the poor health of the enclosed acacia trees?

The researchers began by observing acacia trees closely to figure out what normally occurs (that is, outside enclosures) in these plants. Here's what they found out.

1. Acacia plants and certain types of ants have a mutualistic relationship. The plant makes hollow thorns, in which the ants live, and produces sugary nectar, which the ants consume.

2. When something touches the tree—such as an elephant eating leaves—the ants swarm out and defend the tree. This reduces herbivory on the plant.

The biologists' next step was to make careful observations of the acacia trees within the enclosures to figure out what was happening as a result of the intervention to keep out elephants and giraffes. Here's what they found.

1. Reduced predation → reduced investment in defense against predators. When protected from elephants and giraffes, the acacias somehow detected the reduced herbivory and reduced their investment in supporting the symbiotic ants: they decreased by one-third the number of nectar-producing sites, and they produced fewer swollen thorns in which the ants could live.

2. Reduced support of ants → reduced defense behavior by ants. When the nectar reward dwindled, the ants played a lesser role in plant defense: there was a 30% reduction in the number of ants remaining on the plants, and among the remaining ants, 50% fewer individuals responded to disturbances of the plant—and those that did were significantly less aggressive in their response.

These two changes, on their own, should not have been bad for the enclosed plants. Why?

3. Reduced number of ants → opportunity for competitors to get in. When the number of mutualistic ants dwindled, this created an opportunity for competing (non-mutualistic) species of ants to take up residence in the acacias. The proportion of trees occupied by these competing, non-mutualistic ants doubled.

4. Increased number of competitor ants → increase in stem-boring beetles and plant damage. The competitor, non-mutualistic ants encouraged beetles to infect the plants: the beetles bored cavities in the acacia stems, where the competing ants could lay eggs. And when beetles bore cavities, plants suffer. The researchers documented that acacia growth was much slower, and mortality was doubled.

In just a few short steps, the researchers were able to piece together what was happening.

1. Big herbivores eat acacia plants.

2. Restricting herbivores causes acacia plants to invest less in mutualistic ants.

3. Reduced support of the ants helps competing ants and stem-boring beetles to invade.

4. Formation of bore holes by the beetles slows plant growth and increases plant mortality.

In other words, the researchers learned why restricting big herbivores by placing enclosures ultimately harmed the acacia trees.

What can we conclude from these results?

Beyond this specific message, the researchers noted that their results illustrate the extreme complexity of community interactions. Within an ecosystem, seemingly helpful, straightforward manipulations may have significant, negative consequences. Or, in the words of poet Robert Burns, "The best-laid schemes o' mice an' men / Gang aft agley [often go awry]."

What does this study suggest about the possible consequences of the loss of large herbivores from the world's ecosystems? Why?

TAKE HOME MESSAGE 17.15

» Study of acacia–ant mutualisms and attempts to manipulate them reveals that community interactions are very complex. Within an ecosystem, seemingly helpful, straightforward manipulations may have significant, negative consequences.

Communities can change or remain stable over time.

Buffalo grazing in a South Dakota grassland.

17.16 Many communities change over time.

Human "progress" and development can transform an environment. A patch of barren land, for example, may be turned into a productive agricultural field. Or an urban landscape may obliterate any signs of the nature that was once there. This is why it can be surprising (and heartening) to observe what happens when humans abandon an area. Little by little, nature reclaims it. The area doesn't necessarily recover completely, and change is slow. Still, this process is almost universal and virtually unstoppable. Nature responds similarly to other disturbances, too, from a single tree falling in a forest, to a massive flood or fire, to enormous volcanic eruptions.

The process of nature reclaiming an area and of communities gradually changing over time is called **succession.** It is defined as a change in the species composition over time, following a disturbance. There are two types of succession. *Primary succession* is when the process starts with no life and no soil. *Secondary succession* is when an already established habitat is disturbed, but some life and some soil remain.

Primary Succession Primary succession can take thousands or even tens of thousands of years, but it generally occurs in a consistent sequence. It always begins with a disturbance that leaves an area barren of soil and life. The huge volcanic eruption on Krakatau, Indonesia, in 1883, for example, completely destroyed several islands and wiped out all life and soil on others. Primary succession has also begun, in a less dramatic fashion, in regions where glaciers have retreated, such as Glacier Bay, Alaska. Although succession does not occur in a single, definitive order, several steps are relatively common (FIGURE 17-29).

- The first arrivals—called **colonizers** or pioneer species—in a lifeless, soil-less area are usually bacteria or fungi or other photosynthetic microorganisms, floating in on the air.
- Lichens—symbiotic associations between a fungus and a photosynthetic alga or bacterium—are also common first colonizers. They can grow on rocks and can generate energy through photosynthesis. As they grow, lichens change their environment, secreting acids that break down the rocks and release useful minerals, making the terrain more hospitable for other organisms.
- Mosses also tend to arrive in the early stages, using many of the nutrients freed up by lichens. They trap moisture and can provide a hospitable site for germination of the seeds of other plants.
- Following mosses, small herbs often arrive. Later, shrubs may begin to thrive, shading out the mosses and herbs.
- Shrubs, in turn, are eventually outcompeted and shaded out by small trees.
- And the first trees generally are outcompeted by taller, faster-growing trees.

An important feature of the colonizers seen in the earliest stages of primary succession—whether plants, animals, or other species—is that while they are good **dispersers,** able to move away from their original home (hence their early arrival in a new locale), they are not particularly good *competitors*. That's why they are gradually replaced.

Ultimately, it is the longer-living, larger species that tend to outcompete the initial colonizers and persist within a stable and self-sustaining community, called a

SUCCESSION

Succession is the change in species composition in a community over time.

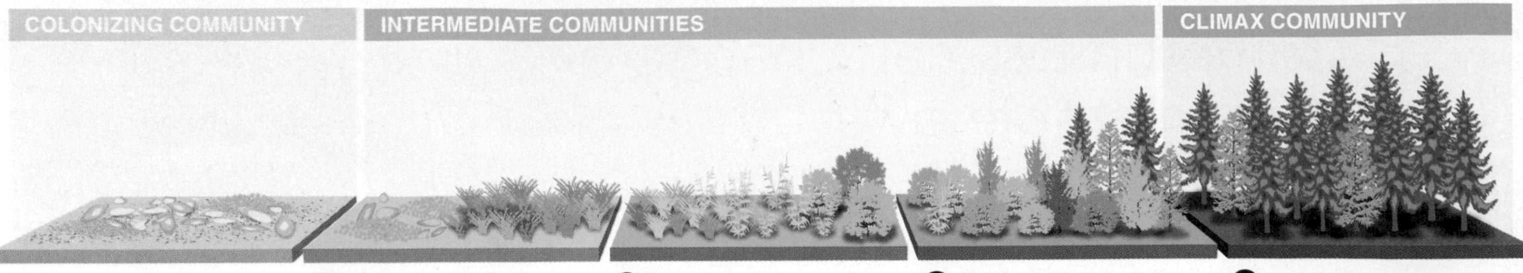

1. Fungi, bacteria, lichens, and seeds are often among the earliest colonizers.
2. Mosses begin to grow and trap moisture, allowing the seeds of other plants to germinate.
3. Small herbs arrive, and shrubs grow, shading out the mosses and herbs.
4. Small trees grow and outcompete the shrubs.
5. Longer-living, larger species outcompete the initial colonizers and persist as a stable and self-sustaining community.

PRIMARY SUCCESSION
Primary succession begins after a disturbance leaves an area barren of soil and with no life.

SECONDARY SUCCESSION
Secondary succession is like primary succession with a head start. It begins when a disturbance opens up part of a community to the development and growth of species previously outcompeted by other species in the area.

Disturbance is a fundamental part of most ecosystems and can repeatedly set a community back to an earlier stage of succession.

FIGURE 17-29 **The species composition of a community changes over time.**

climax community. The particular species present in the climax community depend on physical factors such as temperature and rainfall.

Secondary Succession Secondary succession is much faster than primary succession, because life and soil are already present. Rather than the thousands of years that primary succession may take, secondary succession can happen in a matter of centuries, decades, or even years. It frequently begins with organisms colonizing the decaying remains of dead organisms. This may involve fungi establishing themselves in the decaying trunk of a fallen tree, and these being replaced over time by different species of fungi.

Secondary succession may also begin with weeds springing up in formerly cultivated land. If they are allowed to grow, the weeds are eventually outcompeted by perennial species, then shrubs, and eventually larger trees. The process is similar to primary succession, but with a head start. Ultimately, if undisturbed, secondary succession also leads to establishment of a stable, self-sustaining climax community.

Given that primary and secondary succession both progress toward climax communities, it might seem surprising that so many communities are in states far from climax. This paradox reflects the fact that succession leads to climax communities in the *absence of disturbance*. But disturbance is a fundamental part of most ecosystems. Massive disturbances, such as a wildfire that blazed across more than 500,000 acres in Oregon in 2012 or the volcanic eruptions of Mount St. Helens, can send a community back to square one. Minor disturbances, on the other hand, may undo just a small degree of the succession that has already occurred. Interestingly, it is at intermediate levels of disturbance that communities tend to support the greatest number of species. With very high or very low disturbance levels, the community is restricted to the top colonizers or the top competitors, respectively. At the intermediate level, a larger variety of species with intermediate combinations of features can persist.

Q Why doesn't succession occur on front lawns?

TAKE HOME MESSAGE 17.16

» Succession is the change in the species composition of a community over time, following a disturbance. In primary succession, the process begins in an area with no life and no soil; in secondary succession, the process occurs in an area where life is already present. In both, the process usually takes place in a predictable sequence.

17.17 Some species are more important than others within a community.

In this chapter, we have examined the ways in which species interact with one another and with their environments, noting the dependence of species on each other. But not all species have equal dependence on or influence over others. Within a community, the presence of some species, called **keystone species,** greatly influences which other species are present and which are not (FIGURE 17-30). If a keystone species is removed, the species mix in the community changes dramatically. The removal of other species causes relatively little change.

Q When it comes to conservation, are some species more valuable than others? Why?

Bison are a keystone species. In a Kansas prairie, experiments were set up in which a herd of bison was allowed to graze in some areas but excluded from others. The areas without the bison were soon dominated by a single species of tall grass that was like a bully to other plant species. The areas where bison grazed, on the other hand, had significantly greater species diversity. Kept in check by the grazing bison, the bully grass was unable to outcompete the other plant species, which were able to thrive.

Sea stars, too, are a keystone species (see Figure 17-30). On the rocky seashores of the Pacific Northwest, they consume mussels, which tend to crowd out other species. In a five-year study, sea stars were removed and kept out of certain areas of the intertidal zone but allowed in other areas. As with the bison study, species diversity dropped dramatically where sea stars were excluded. Ultimately, just the mussels remained. In the areas with sea stars, mussels also remained but in reduced numbers. More important, 28 other species of animals and algae were also able to live there.

Keystone species make it possible to get more "bang for your buck" when conserving biodiversity. Consequently, identifying keystone species is an important part of conservation biology. Besides bison and sea stars, some other keystone species are dam-building beavers, elephants of the African savanna, and desert lichens.

TAKE HOME MESSAGE 17.17

» Keystone species have a relatively large influence on which other species are present in a community and which are not. Unlike other species, when a keystone species is removed from the community, the species mix changes dramatically. For this reason, protecting keystone species is an important strategy in preserving biodiversity.

KEYSTONE SPECIES

A keystone species, such as the sea star, has an unusually large influence on the presence or absence of numerous other species in the community.

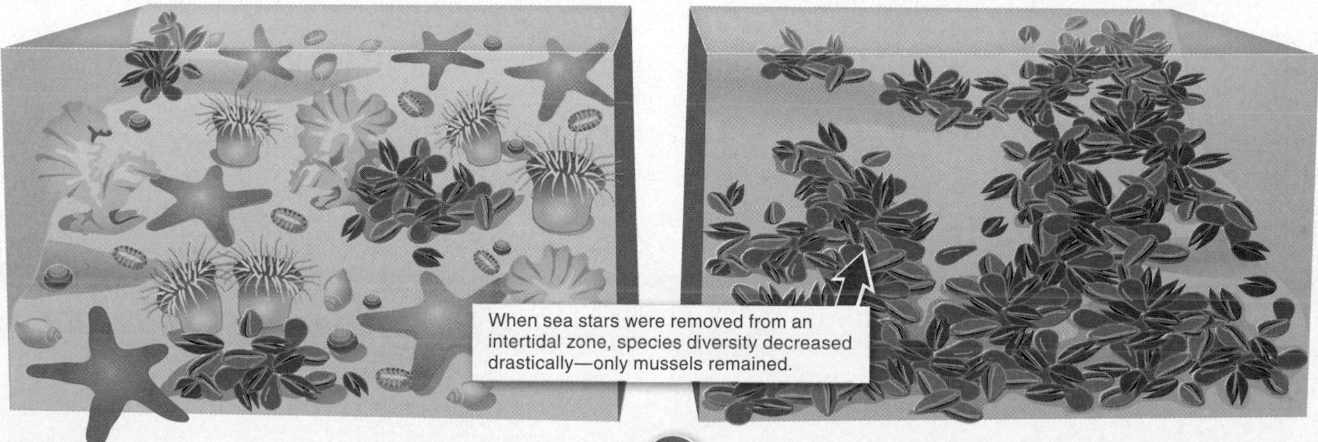

COMMUNITY WITH KEYSTONE SPECIES
COMMUNITY WITH KEYSTONE SPECIES REMOVED
When sea stars were removed from an intertidal zone, species diversity decreased drastically—only mussels remained.

FIGURE 17-30 **Preserving biodiversity.** Keystone species can keep aggressive species in check, allowing more species to coexist.

Preserving just one keystone species has the effect of preserving many additional species at the same time.

STREET BIO

Using evidence to guide decision making in our own lives

Life in the dead zone: in boosting plant productivity on farms, we've created a "dead zone" in the Gulf of Mexico bigger than Connecticut

In the water at the mouth of the Mississippi River, each spring and summer, there is so little oxygen that virtually no life can survive.

Q: Which chemical almost always limits plant growth? Plant growth requires the production of proteins, all of which contain nitrogen. Access to nitrogen nearly always determines how much a plant grows.

Q: If you're a farmer, how can you increase the productivity of your crops? Add nitrogen. This, in fact, is the chief component of all plant fertilizers.

Q: Where does the unused nitrogen end up? Spring and summer rains wash nutrients, dissolved organic materials, and other runoff (much of which comes from human sources) from the middle of the United States into nearby rivers and thus, eventually, into the Mississippi River. From there they flow into the Gulf of Mexico.

Q: What happens when large amounts of nitrogen are dumped into a body of water? Organisms such as plankton and algae living near the mouth of the Mississippi grow like crazy with all the extra nitrogen and nutrients that are washed into the Gulf of Mexico. But their rapid increase in productivity disrupts the food chain as they decay, increasing the organic matter that sinks to the bottom and feeds the bacteria there. As the bacteria thrive, they consume ever-increasing amounts of oxygen.

Q: What happens when bacteria use up too much oxygen? Excessive bacterial growth depletes the supply of dissolved oxygen in the water, starving all other life of oxygen—including fish, crabs, oysters, and shrimp, all of which die or move out of the area. This creates a "dead zone" and ruins local fisheries, with a significant impact on the economies they support.

Q: What can you conclude? Disrupting food chains, even increasing the productivity of certain trophic levels, can have unintended effects throughout an ecosystem, and even thousands of miles away.

Q: What can be done? Reducing and reversing the Gulf of Mexico's dead zone (and other dead zones throughout the world) can be accomplished by (1) reducing fertilizer use, (2) preventing animal wastes from getting into waterways, and (3) controlling the release of other sources of nutrients (phosphorus as well as nitrogen) from industrial facilities into rivers and streams.

GRAPHIC CONTENT

Thinking critically about visual displays of data

1 Why is the drawing of the hawk so much smaller than those of the other organisms?

2 What does the size of each "pie" represent? Why aren't they all the same size?

3 What does the slice of the pie represent in each drawing?

4 Why isn't there a slice taken from the top pie?

5 Would this figure be different if it included data from consumers that ate plants and other animals? How?

6 What can you conclude from this figure about the relative numbers of predators versus herbivores in any environment?

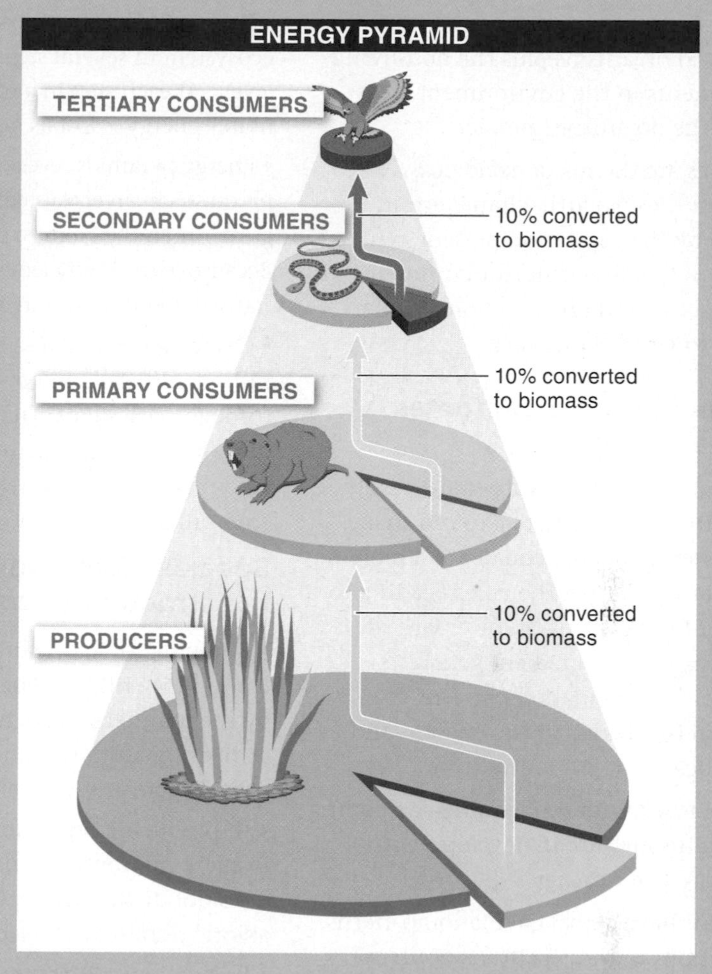

See answers at the back of the book.

KEY TERMS IN ECOSYSTEMS AND COMMUNITIES

BRIEF SUMMARY

Ecosystems have living and non-living components.

• An ecosystem is a community of biological organisms plus the non-living components in the environment with which the organisms interact.

• Biomes are the major ecological communities of earth, characterized mostly by the vegetation present. Different biomes result from differences in temperature and precipitation and the extent to which both vary.

Interacting physical forces create weather.

• Global patterns of weather are largely determined by the earth's round shape. Solar energy hits the equator at a more direct angle than at the Poles, leading to warmer temperatures at lower latitudes. This temperature gradient generates atmospheric circulation patterns that result in heavy rain at the equator and many deserts at 30° latitude.

• Local features of topography—including mountains and urban developments—influence the weather.

• Oceans have global circulation patterns. Disruptions in these patterns occur every few years and can cause extreme climate disruptions around the world.

Energy and chemicals flow within ecosystemss.

• Energy from the sun passes through an ecosystem in several steps known as trophic levels. At each step in a food chain, some usable energy is lost as heat.

• Energy pyramids reveal that the biomass of producers in an ecosystem tends to be far greater than the biomass of herbivores. Because of the inefficiency of biomass transfer, food chains rarely exceed four levels.

• Chemicals essential to life—including carbon, nitrogen, and phosphorus—cycle through ecosystems. These cycles can be disrupted as human activities increase the amounts of chemicals released to the environment.

Species interactions influence the structure of communities.

• A population of organisms in a community fills a unique niche, defined by the manner in which the population utilizes the resources and influences the other organisms in its environment.

• In producing organisms better adapted to their environment, natural selection does not distinguish between biotic and abiotic resources as selective forces.

• Populations with completely overlapping niches cannot coexist forever. Competition for resources occurs until one or both species evolve in ways that reduce the competition.

• Predators and their prey are in an evolutionary arms race. This coevolutionary process can result in brightly colored organisms, alarm calling, and many types of mimicry.

• Parasitism is a symbiotic relationship in which one organism benefits while the other is harmed.

• Not all species interactions are combative. Mutualism, in which both species benefit from the interaction, is widespread and critically important to all ecosystems. In commensalism, one species benefits and the other is neither harmed nor helped.

Communities can change or remain stable over time.

• Succession is the predictable pattern of change in the species composition of a community over time, following a disturbance. In primary succession, the process begins in an area with no life and no soil; in secondary succession, the process occurs in an area where life is already present.

• Keystone species have a relatively large influence on which other species are present or absent in a community. Protecting keystone species is an important strategy in preserving biodiversity.

CHECK YOUR KNOWLEDGE

Short Answer

1. What changes in climate occur on moving from the equator toward the Poles? Which two aspects of weather determine the type of biome found at any location?

2. Why are areas at lower latitudes typically warmer than high-latitude areas?

3. You are charged with creating a "greener" city that will be cooler and use less air conditioning. Describe three strategies you could use to help achieve this goal.

4. Why is the term "food web" more accurate than "food chain" when describing the pathway of energy flow?

5. Describe the energy flow through an ecosystem. How does this energy flow influence the number of top carnivores?

6. Why do levels of CO_2 in the northern hemisphere undergo ups and downs throughout the year?

7. What is a niche? What are the potential fates of two species that occupy the same niche?

8. When the niches of different species overlap, the result is competition. Describe the two possible outcomes from competition.

9. Why are plants more likely than animals to use chemical defense mechanisms to reduce predation?

10. Which feature of colonizers is crucial in the earliest stages of primary succession?

Multiple Choice

1. Earth's largest terrestrial ecosystems, the biomes, are defined primarily by:

a) the average rainfall.

b) the average temperature.

c) the seasonal variability in temperature.

d) the seasonal variability in rainfall.

e) All of the above are correct.

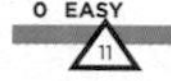

2. Why is it hotter at the equator than at the Poles?

a) Because of the angle at which sunlight hits the earth, solar energy is spread over a smaller area at the equator than at the Poles.

b) The increased rotational speed of the earth at the Poles creates strong winds that increase radiant cooling.

c) Greater cloud cover at the equator leads to a greater "greenhouse" effect than at the Poles.

d) The water in the oceans has a high heat capacity, and there is significantly more water at the equator than at the Poles.

e) All of the above are responsible for the higher average temperature at the equator than at the Poles.

3. When a moist ocean wind blows onshore toward a mountain range:

a) as the air rises, it pulls moisture from the ground, causing the higher elevations to be drier.

b) as the air goes over the top of the mountains and falls back down toward lower elevations, it holds less moisture, creating a "rain shadow" zone of unusually high precipitation.

c) as the air rises, it holds less moisture, causing dissipation of all the clouds.

d) as the air goes over the top of the mountains and falls back down to lower elevations, it holds even more moisture, creating a "rain shadow" desert.

e) None of the above are correct.

0 EASY HARD 100 42

4. "Top" carnivores:

a) are more common than secondary consumers.

b) rely directly on producers for energy.

c) consume primarily herbivores.

d) consume primarily carnivores.

e) rely on symbiotic bacteria to help them digest cellulose.

5. The 10% rule of energy-conversion efficiency:

a) explains why big, fierce animals are so rare.

b) explains why the biomass of herbivores must exceed that of carnivores.

c) limits the length of food chains.

d) suggests that 90% of what an organism eats is used in cellular respiration or is lost as feces.

e) All of the above are correct.

6. Nitrogen enters the food chain:

a) primarily through soil-dwelling bacteria that "fix" nitrogen by attaching it to other atoms.

b) from the atmosphere when "fixed" by the photosynthetic machinery of plants.

c) when rocks dissolved by rainwater become soil, which is then utilized for plant growth.

d) through soil erosion followed by runoff into streams and ponds.

e) through methane, produced by herbivores as a by-product of the breakdown of plant material.

7. The "ghost of competition past" refers to the fact that:

a) competition often leads to character displacement, which remains even after direct competition is reduced.

b) competition cannot be seen in nature.

c) competition inevitably leads to extinction of one of the competitors.

d) competition inevitably leads to extinction of both competitors.

e) the fossil record is a record of competitive interactions.

0 EASY HARD 100 60

8. Chemical defenses are more common among plants than animals because:

a) plants cannot move to escape predators and so must develop other deterrents.

b) the cell wall can contain the chemicals more effectively than can a simple plasma membrane.

c) mechanical defenses against predators can evolve only in animals.

d) parasite loads are significantly higher in plants than in animals.

e) All of the above are correct.

0 EASY HARD 100 30

9. Coevolution:

a) is responsible for all the beautiful flowers in the world.

b) is responsible for nectar production by plants.

c) reveals that both biotic and abiotic resources can serve as selective forces.

d) can produce an insect with a tongue as long as its body.

e) All of the above are correct.

10. Keystone species:

a) occur only in intertidal zones.

b) play an unusually important role in determining the species composition in a habitat.

c) can be removed from a habitat without any impact on the remaining species in the community.

d) are producers and therefore usually are plants.

e) are more expendable than commensal species, from a conservation perspective.

Ch18

Biodiversity—of genes, species, and ecosystems—is valuable in many ways.

18.1 Biodiversity can have many types of value.

18.2 *This is how we do it:* When 200,000 tons of methane disappears, how do you find it?

18.3 Biodiversity occurs at multiple levels.

18.4 Where does the greatest biodiversity occur?

Extinction reduces biodiversity.

18.5 There are multiple causes of extinction.

18.6 We are in the midst of a mass extinction.

Human activities can have disruptive environmental impacts.

18.7 Some ecosystem disturbances are reversible, and others are not.

18.8 Human activities can damage the environment: 1. Introduced non-native species may wipe out native organisms.

18.9 Human activities can damage the environment: 2. Acid rain harms forests and aquatic ecosystems.

18.10 Human activities can damage the environment: 3. The release of greenhouse gases influences the global climate.

18.11 Human activities can damage the environment: 4. Deforestation of rain forests causes loss of species and the release of carbon.

We can develop strategies for effective conservation.

18.12 Reversal of ozone layer depletion illustrates the power of good science, effective policymaking, and international cooperation.

18.13 With limited conservation resources, we must prioritize which species should be preserved.

18.14 There are multiple effective strategies for preserving biodiversity.

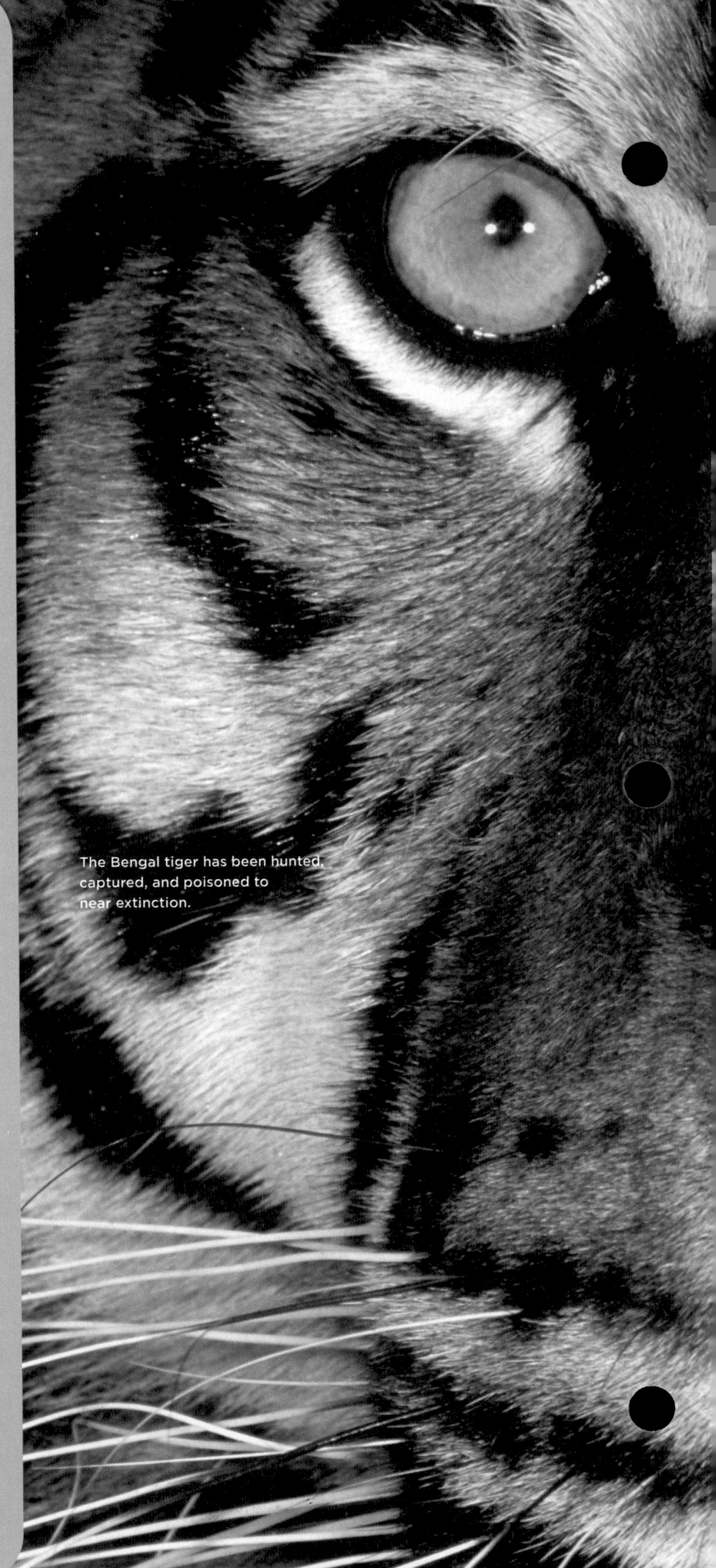

The Bengal tiger has been hunted, captured, and poisoned to near extinction.

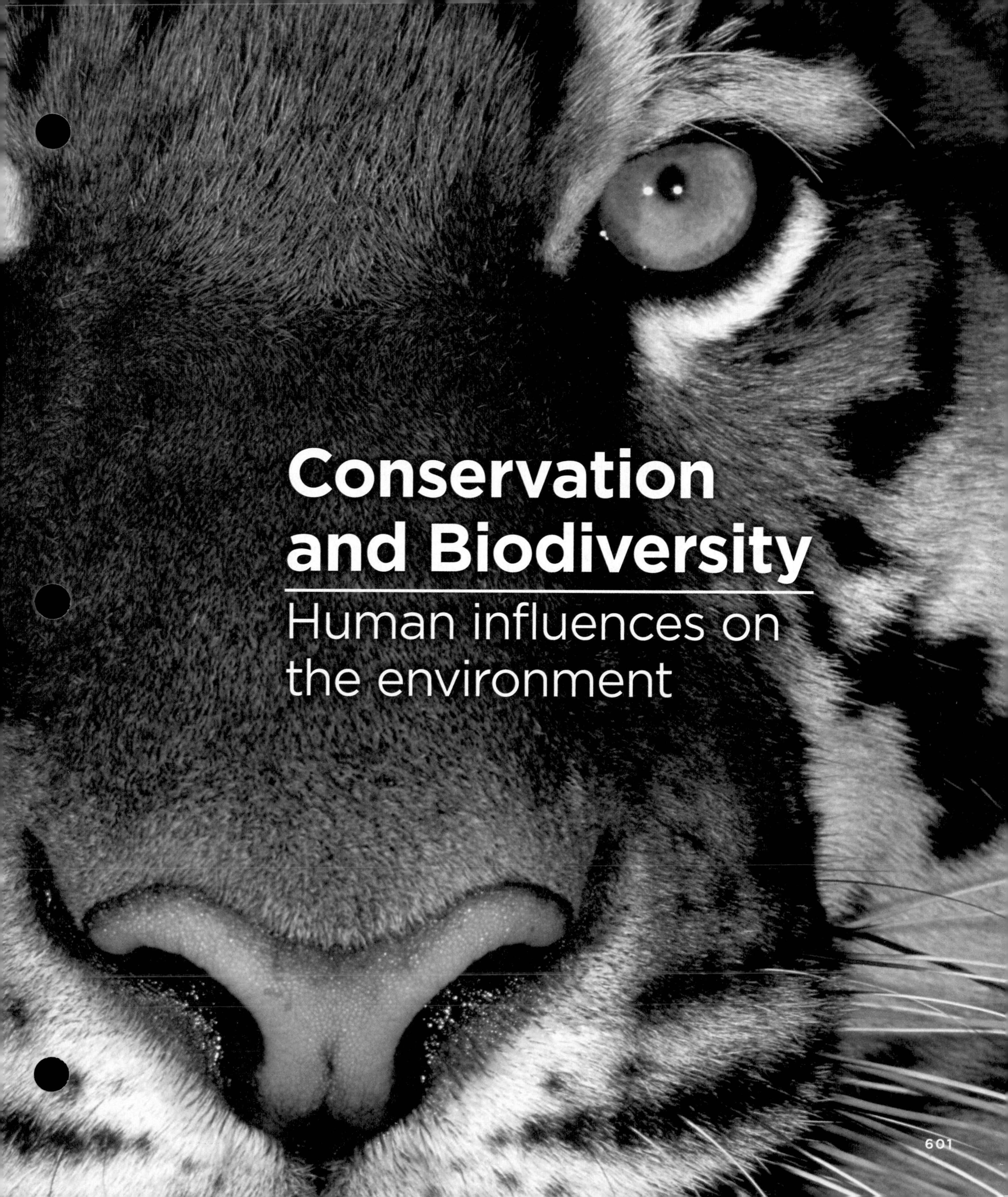

Conservation and Biodiversity

Human influences on the environment

18.1–18.4

Biodiversity—of genes, species, and ecosystems—is valuable in many ways.

Endangered Nilgiri tahr (*Nilgiritragus hylocrius*) visit a grassy alpine plateau in India.

18.1 Biodiversity can have many types of value.

In the Pacific Northwest of the United States, a conifer grows. This tree, the Pacific yew (*Taxus brevifolia*), isn't the biggest or most common tree in the forest. But within the bark of the Pacific yew there is a remarkable chemical called taxol, which acts as an anti-cancer agent, interfering with the division of cells that encounter it (FIGURE 18-1). Although we don't fully understand the role taxol plays for the Pacific yew—it may reduce the rate at which other organisms feed on the plant—in humans, taxol is effective in the treatment of ovarian cancer, breast cancer, and lung cancer (generating more than $1 billion a year in pharmaceutical sales).

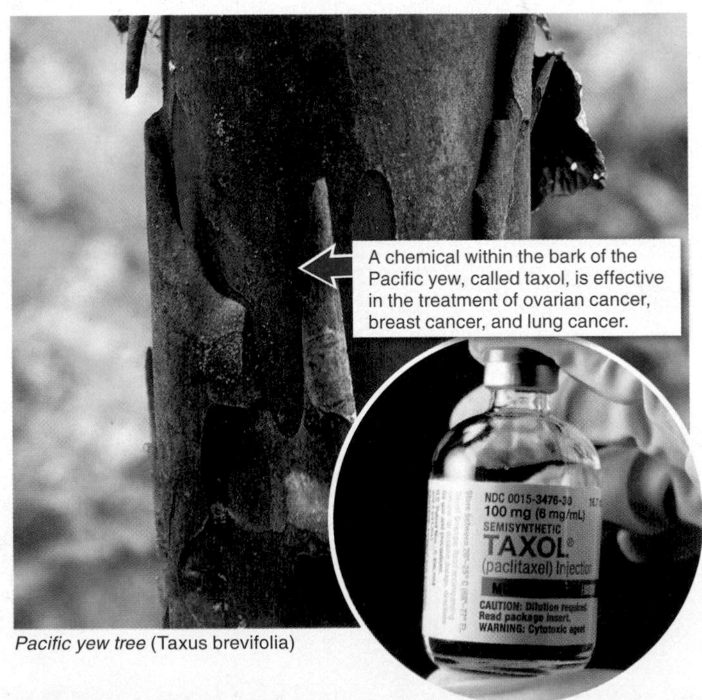

A chemical within the bark of the Pacific yew, called taxol, is effective in the treatment of ovarian cancer, breast cancer, and lung cancer.
NDC 0015-3476-30
100 mg (6 mg/mL)
SEMISYNTHETIC
TAXOL®
(paclitaxel) Injection
CAUTION: Dilution required.
Read package insert.
WARNING: Cytotoxic agent
Pacific yew tree (Taxus brevifolia)

FIGURE 18-1 **Taxol is cancer medicine derived from a tree.**

Consider the following as well:

- The chemicals vinblastine and vincristine, isolated from the Madagascar periwinkle (*Catharanthus roseus*), are so effective in treating leukemia and Hodgkin's lymphoma that both diseases, formerly incurable, are now highly curable.
- The chemical ancrod, found only in the Malayan pit viper snake (*Calloselasma rhodostoma*), dissolves blood clots and is effective in treating some heart attack and stroke patients.
- Epibatidine, a poisonous compound in the saliva of a small frog (*Epipedobates tricolor*) that lives in Ecuador, is 200 times more effective than morphine in relieving pain and is non-addictive.

Q Why would another species produce a chemical that fights cancer in humans?

These are just a few of the many examples of how medically important compounds often come from organisms living in various locations around the world. Toxin production commonly evolves in these organisms as protection against predators. Our ability to co-opt these chemicals for human use reveals that one important value of the world's diverse plants and animals is as a sort of universal medicine cabinet.

Plants and animals aren't the only species with great value to humans. Microbes, too, have great utility. In the next section, for example, we see that hydrocarbon-consuming bacteria can play a significant role in cleaning up oil spills.

Humans use the chemicals derived from plants and animals and the metabolic processes of organisms such as oil-removing microbes for many purposes. The plants, animals, and microbes are examples of **biodiversity,** which is defined as the variety of genes, species, and ecosystems on earth. And, as we've seen, earth's biodiversity has tremendous practical value, and it often comes from unexpected places. For these reasons, the loss of biodiversity—including genes, species, and ecosystems—can be hugely detrimental to humans.

But viewing the value of biodiversity only in the context of our abilities to extract medicines or to better process the messes we make is to take much too narrow a view. The United Nations has asserted that biodiversity has an intrinsic value—that is, an inherent value independent of its value to humans—stating that "every form of life is unique, warranting respect regardless of its worth to man."

And, in describing the extrinsic or utilitarian value of biodiversity, scientists and economists use the concept of **ecosystem services** to highlight the benefits to humans of biodiversity and natural ecosystems. In a 2005 United Nations study that involved more than 1,300 scientists, the researchers identified four distinct categories of ecosystem services (FIGURE 18-2).

- **Provisioning services.** These include the many useful products that humans obtain from nature, including foods, spices, minerals, and pharmaceutical and industrial products.
- **Regulating services.** Ecosystems provide numerous services relating to the regulation of our environment. These include climate regulation through carbon storage in plants; pollination; waste decomposition and detoxification; protection from natural disasters; and pest and disease control.
- **Habitat services.** Significant value from biodiversity and natural ecosystems also comes in the form of habitats for species and the maintenance of genetic diversity they provide.
- **Cultural services.** Much of what humans gain from biodiversity isn't easily assigned a dollar value—but that value is extremely important, nevertheless. We think of a field of flowers or a deer in the woods as beautiful to look at and pleasant to experience, for example. Or we make use of images of animals and plants to convey meaningful abstract ideas: the dove as a symbol of peace; the oak as a symbol of strength;

BIODIVERSITY VALUES AND ECOSYSTEM SERVICES

INTRINSIC VALUE
The inherent value of biodiversity, independent of its value to humans

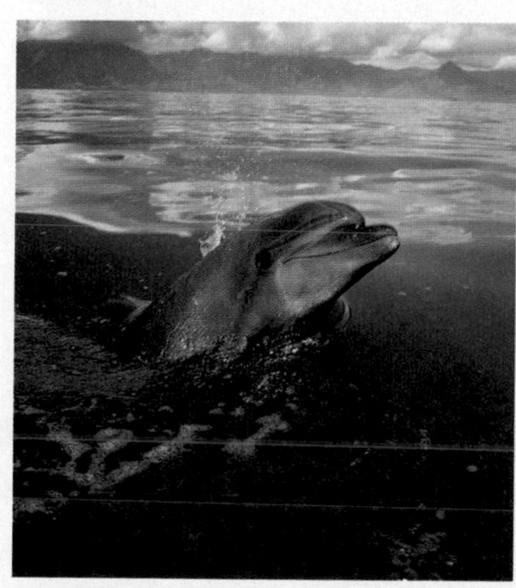

EXTRINSIC (UTILITARIAN) VALUE
The value of biodiversity to humans, often described as ecosystem services, can be grouped in four categories of services:

PROVISIONING SERVICES
- Food
- Raw materials
- Fresh water
- Medicines

HABITAT SERVICES
- Habitats for species
- Maintenance of genetic diversity

REGULATING SERVICES
- Air, water, and soil quality
- Carbon storage and climate regulation
- Pollination and biological control

CULTURAL SERVICES
- Aesthetics and inspiration
- Recreation and mental health
- Tourism
- Spiritual experience

FIGURE 18-2 **Ecosystem services categorize the many different ways that biodiversity has value to humans.**

the butterfly as a symbol of transformation; and the owl as a symbol of wisdom. Beyond the aesthetic, symbolic, and spiritual values of biodiversity, cultural services also include recreational experiences, as well as scientific and other educational value.

When we recognize that people can value the same organism or gene or ecosystem in different ways, we are on our way to making wiser conservation decisions. This knowledge, for example, can help us anticipate potential conflicts when it comes to allocating conservation resources. We explore some strategies for making these difficult decisions later in this chapter.

TAKE HOME MESSAGE 18.1

» Biodiversity—the variety of genes, species, and ecosystems on earth—has intrinsic value as well as extrinsic value, or value to humans. The value of biodiversity to humans is often described in terms of four categories of ecosystem services: provisioning, regulating, habitat, and cultural services. These categories help distinguish among various utilitarian values, such as the production of food and medicines, the regulation and support of our environments, and aesthetic and symbolic value.

THIS IS HOW WE DO IT

Developing the ability to apply the process of science

18.2 When 200,000 tons of methane disappears, how do you find it?

Even during an ecological catastrophe, there can be opportunities for scientific thinking and problem solving. Such situations, however, require fast thinking to reduce the impact of the catastrophe and to guide management strategies.

For example, in April 2010, an oil rig in the Gulf of Mexico was drilling a well about a mile under the surface of the water when the rig exploded. It took nearly three months to cap the well, during which time more than 200 million gallons of crude oil poured into the Gulf.

The oil spill heavily damaged the coastal and marine environments and killed thousands of animals, including birds, dolphins, sea turtles, mollusks, and crustaceans. The longer-term consequences of the oil spill for the region—including the eight U.S. national parks threatened—are not yet clear.

Crude oil wasn't the only hydrocarbon released into the Gulf. Methane gas, too—almost 200,000 tons of it—was also released by the spill. As scientists worried about the impact of the gas and oil on the ecosystem, they tracked the movement of the methane.

Can you propose a way to keep track of oil a mile below the water surface?

To build a three-dimensional map of the affected Gulf regions, researchers lowered sensors into the water at 207 locations and collected water samples at many depths, from the surface down to the Gulf floor. The samples enabled researchers to identify the location and concentration of methane in the water column.

Writing in the journal *Science,* the team reported something quite unexpected. Just a couple of months after the leak had been sealed, a huge proportion of the methane released in the spill was gone.

Most of the methane was gone from the area of the spill. What could have happened to it?

Two explanations for the methane disappearance seemed possible. Perhaps the researchers simply lost track of the flowing gas and oil underwater. Or, could populations of bacteria be consuming the methane?

How could the scientists figure out whether bacteria were eating the methane or the methane was lost?

The researchers made a testable prediction. If bacteria were consuming the methane, there would be a chemical trace of their metabolic activities. The bacteria, working on such a huge scale, would consume large amounts of oxygen as they burned the methane for fuel. Consequently, the amount of dissolved oxygen in the water should have decreased significantly. If, on the other hand, the methane had just drifted off in the Gulf, there would be no drop in the dissolved oxygen.

As the researchers sampled the water and evaluated oxygen concentrations, they noted an unprecedented, significant drop in oxygen saturation—down from 67% to 59%. This suggested that bacterial populations were indeed proliferating rapidly and "mopping up" the oil.

Researchers calculated that the amount of missing oxygen was almost exactly equal to the oxygen required by bacteria to consume the amount of methane that had leaked from the well. "The math worked out scary good," is how one researcher put it.

The diversity of bacteria in the world is huge. How could the scientists figure out which species were responsible?

Twenty years ago, researchers had to culture any bacteria they found to identify them. Today scientists can bypass this difficult step. Using DNA sequencing methods, the researchers could test the microbes (and pieces of microbes) in the Gulf. In these sequences, they found genetic fragments previously identified in methane-eating microbes.

Moreover, the bacterial DNA they found was related to several previously observed oil-eating bacterial species from the Gulf of Mexico. These bacteria live in the Gulf because natural oil—equivalent to as much as a million barrels of oil per year—seeps out of the ground. The researchers hypothesized that because the bacteria were already present, the populations of microbes could quickly respond to the spill.

Should we be confident that we can rely on microbes to clean up any oil that we spill?

By having methane-consuming microbes already in place, the Gulf of Mexico was particularly suited to a rapid, effective response to an oil spill. In other places, however, such as the numerous Arctic drilling locations, there is not the same sort of natural oil seepage. And with no similar populations of methane-eating bacteria already in place, the impact of an oil spill in these habitats might be quite different.

TAKE HOME MESSAGE 18.2

» Following a massive oil leak in the Gulf of Mexico, researchers noted a rapid disappearance of methane. Finding a simultaneous drop in the oxygen saturation of the water, the researchers were able to determine that populations of methane-eating bacteria already living in the area grew rapidly in response to the new food source.

18.3 Biodiversity occurs at multiple levels.

Which habitat has greater biodiversity, one with three or four species each of birds, reptiles, mammals, plants, and insects, or one with 38 unique species of fruit flies, or one with 100 radically different species of prokaryotes (FIGURE 18-3)?

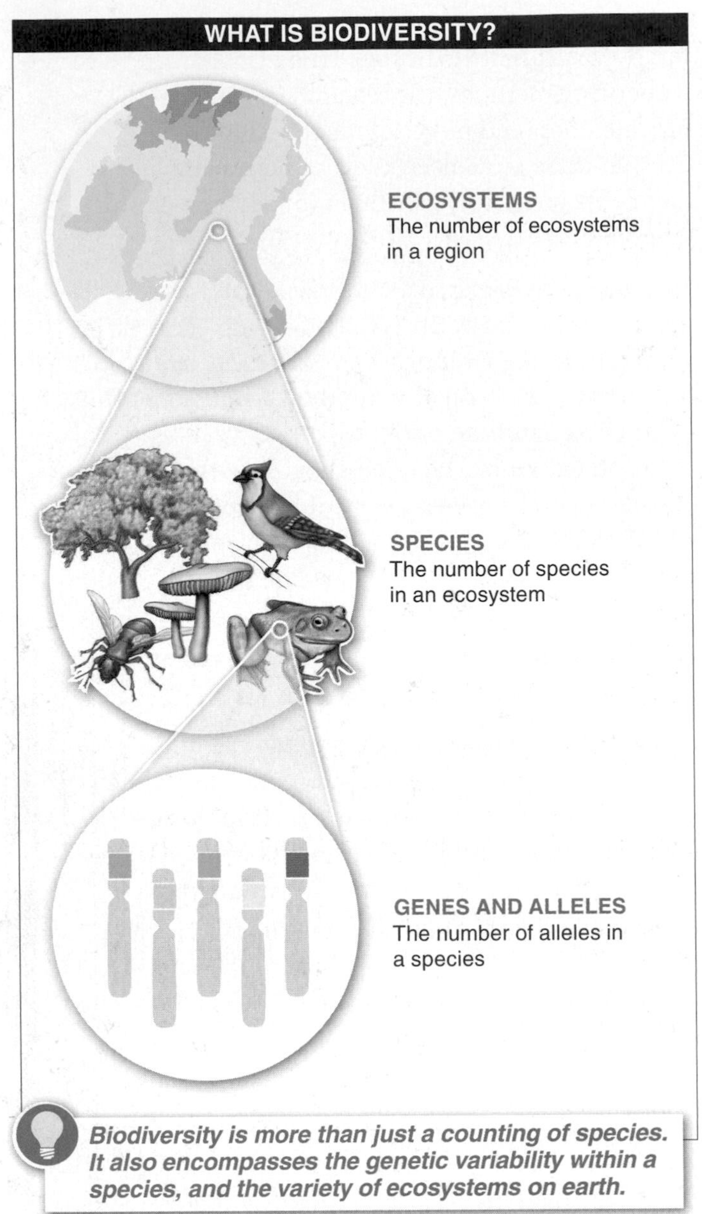

WHAT IS BIODIVERSITY?
ECOSYSTEMS
The number of ecosystems in a region
SPECIES
The number of species in an ecosystem
GENES AND ALLELES
The number of alleles in a species
Biodiversity is more than just a counting of species. It also encompasses the genetic variability within a species, and the variety of ecosystems on earth.

FIGURE 18-3 **Biodiversity is found on many different scales.**

If you were given the task of overseeing the conservation of biodiversity in a region where these three habitats occurred, which habitat would you protect first? In making your decision, you might prioritize the number of different classes or families of organisms in a habitat, or maybe the number of distinct species in the habitat—which is, in fact, the most commonly used measure of biodiversity. In doing so, however, you'd be acknowledging that biodiversity can be prioritized in many ways and that any assessment is a function of your values, biases, and interests. But this recognition helps us to discuss, manage, use, and conserve biodiversity in a practical and less ambiguous way.

No Vehicles
Off Road
NO
SNOWMOBILES
Conservation biology is a multidisciplinary science that addresses how to preserve natural resources and protect biodiversity.

FIGURE 18-4 **Preserving the earth's natural resources.**

Conservation biology is the interdisciplinary field that addresses how to understand and preserve the natural biological resources of earth, including its biodiversity at all levels (FIGURE 18-4). In this chapter, we explore how the actions of humans can have significant, even disastrous, consequences for biodiversity and natural resources, and we also explore the strategies that have been developed for effective conservation.

Underpinning many of the difficulties in conservation biology is that biodiversity, as we've noted, can have value for humans in many ways. Throughout the chapter, we'll see how this complexity makes it difficult to balance competing interests when they are in conflict.

TAKE HOME MESSAGE 18.3

›› Assessing biodiversity can be difficult because it must be considered at multiple levels, from entire ecosystems to species to genes and alleles. In practice, biodiversity is most often defined as the number of distinct species in a habitat, which has important implications for conservation biology—the field that addresses questions of how to preserve the natural resources of earth.

18.4 Where does the greatest biodiversity occur?

Dwelling at the equator in South America are about 400 different species of land mammals. If you started walking southward, away from the equator, and assessed the diversity of land mammals at 20° latitude, you would find about 350 different species. Continuing your long walk south, at around 30° latitude you would find only 200 different species. And the trend would continue for as far south as you could go, with 100 species at 40° latitude and fewer than 50 species at 50° latitude. On a similar walk northward from the equator, through North America, you would observe the same trend, with more than 300 species of land mammals at 20° latitude, reduced to only 100 species in northern Canada (FIGURE 18-5). Biodiversity is not evenly distributed throughout the world. As you move away from the equator in either direction, diversity is reduced with increasing latitude.

The strong biodiversity gradient from the tropics to the Poles occurs for nearly any group of plants or animals observed; recent surveys have documented the trend in woody plants, herbaceous plants, and mangrove trees. Interestingly, even the diversity of organisms in the oceans follows this trend. One survey of marine copepods—tiny crustacean animals, including plankton—found 80 species present in the tropical region of the Pacific Ocean and, moving northward, fewer than 10 species in the Arctic Ocean. Similar patterns have been documented in fishes, mollusks, marine and terrestrial arthropods, corals, and marine protists.

Q Why are there more species in an acre of tropical rain forest than in an acre farther from the equator, such as in a temperate forest or prairie?

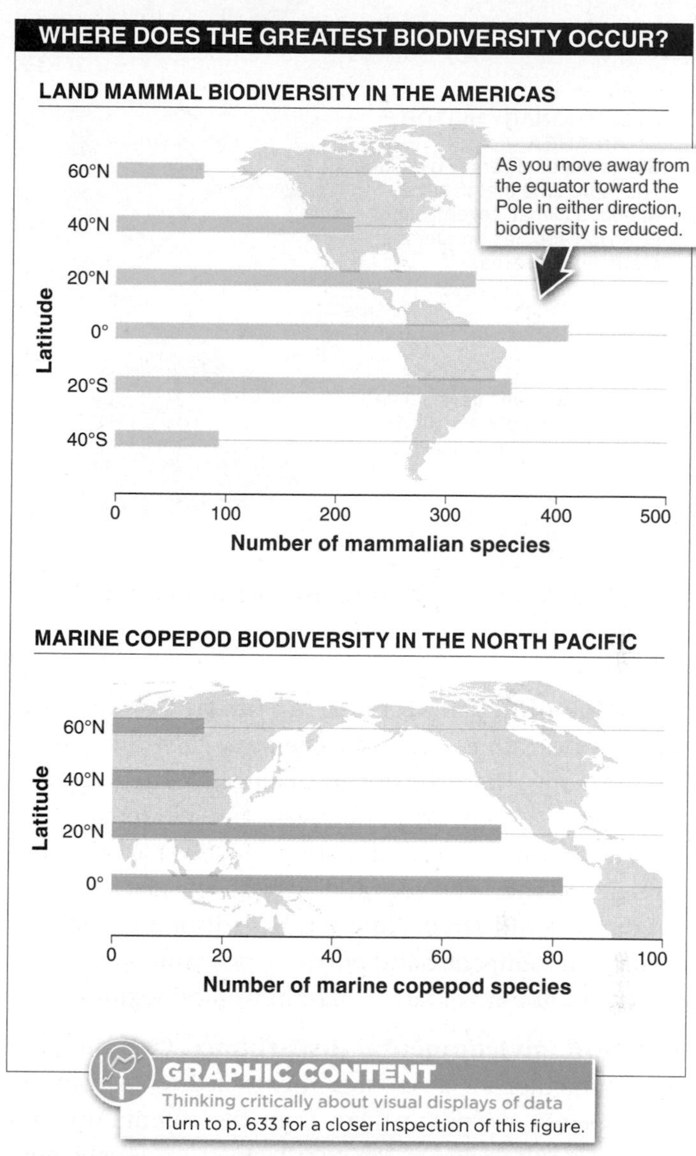

Turn to p. 633 for a closer inspection of this figure.

FIGURE 18-5 **Biodiversity is unevenly distributed.**

Numerous factors that influence **species richness**—that is, the number of species in an area—are responsible for producing the tropical-to-temperate-to-polar gradient in biodiversity. Three factors play strong roles (FIGURE 18-6).

1. Solar energy available. Perhaps the simplest predictor of species diversity in an area is the amount of solar energy, the ultimate fuel for nearly all life, that is available in that area. As we saw in Figure 17-5, the sun shines most directly on the equator, and, as we move away from the equator, the earth's curvature causes the solar energy hitting the earth to be spread out over greater areas. This leads to less solar energy at any one point as latitude increases. In a variety of species of plants and animals, researchers have documented strong relationships between energy availability and species richness.

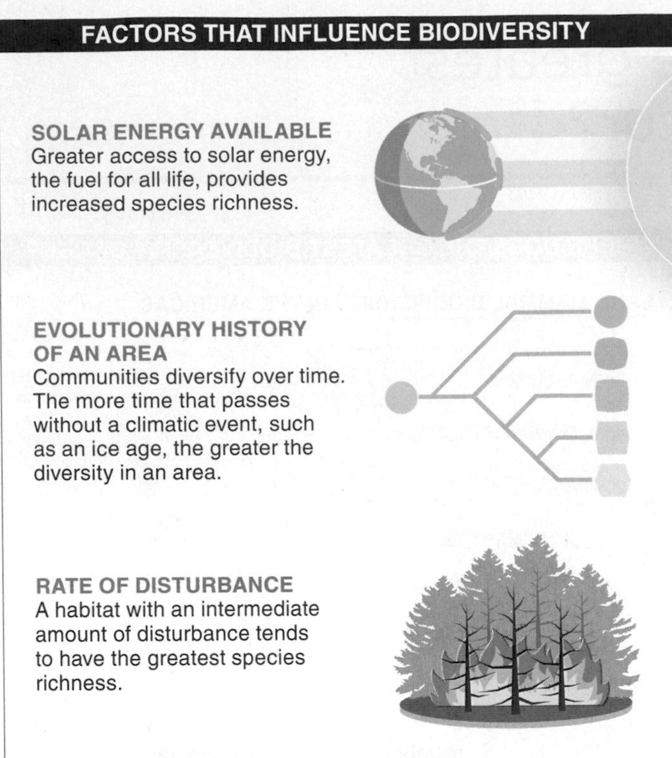

FACTORS THAT INFLUENCE BIODIVERSITY
SOLAR ENERGY AVAILABLE
Greater access to solar energy, the fuel for all life, provides increased species richness.
EVOLUTIONARY HISTORY OF AN AREA
Communities diversify over time. The more time that passes without a climatic event, such as an ice age, the greater the diversity in an area.
RATE OF DISTURBANCE
A habitat with an intermediate amount of disturbance tends to have the greatest species richness.

FIGURE 18-6 **What determines species richness in an area?**

2. Evolutionary history of an area. Communities diversify over time. Generally, the more time that passes, the greater is the diversity in an area. A high level of biodiversity in an area, however, can be reduced by climatic disasters such as glaciations. Biodiversity declines caused by the advance and retreat of glaciers over hundreds of millions of years may have disproportionately occurred in temperate and polar regions (and on the tops of some mountains), rather than in tropical regions.

3. Rate of environmental disturbance. Over time, the best competitor for a resource is expected to outcompete other species, excluding them from the community and reducing species richness. However, factors such as predation, or environmental disturbances such as floods, fires, and volcanic eruptions, may keep communities from reaching this species-depleted state. For example, a habitat with an *intermediate* amount of environmental disturbance—that is, not so high that many or all species are wiped out and not so low that organisms with a competitive advantage drive less-competitive species from the community—is expected to have the greatest species richness.

Conservation biologists are increasingly interested in **biodiversity hotspots,** regions of the world having significant reservoirs of biodiversity that are under threat of destruction. Twenty-five biodiversity hotspots have been identified around the globe (FIGURE 18-7). These cover less than 1% of the world's area but have 20% or more of the world's species. Habitats included among the biodiversity hotspots are tropical rain forests, coral reefs, and islands. While not considered hotspots, the deepest regions of the oceans have recently been found to hold great biodiversity.

Tropical Rain Forests These habitats are common near the equator in Asia, Australia, Africa, South America, and Central America, as well as on many Pacific islands. An important but still poorly studied region of tropical rain forests is the canopy, the uppermost region of a forest, where nearly all of the sunlight is intercepted. The canopy, which can be up to 30 feet (almost 10 m) thick, contains the most biodiversity in tropical rain forests; as many as half of all species on earth reside there. Due to its inaccessibility—65–135 feet (20–40 m) above the ground—the canopy is not easily studied, and the extent of its biodiversity is still not fully known.

Coral Reefs Built by corals, members of the animal kingdom, coral reefs occur in ocean areas with low levels of nutrients. Corals produce calcium carbonate structures that are habitat for a wide variety of species. The Great Barrier Reef (which is not listed among the 25 biodiversity hotspots because it currently is a focus of extensive conservation efforts) is the largest coral reef system in the world and covers such a large area (almost the size of California) off the northeast coast of Australia that it can be seen from space. It is home to numerous species of whales, dolphins, sea turtles, sharks, and stingrays, as well as 1,500 species of fishes and 5,000 species of mollusks. Five hundred species of algae are found there, too. Above the Great Barrier Reef, more than 200 species of birds are supported by the rich diversity of life below. Other coral reefs are not as well protected by laws and conservation efforts, and the world's second largest coral reef, in New Caledonia in the southwestern Pacific, is one of the 25 biodiversity hotspots.

Islands Because of their small landmass, which limits the population size of organisms living on them, islands can be regions where biodiversity is at risk. Madagascar, located in the Indian Ocean off the southeastern coast of Africa, is the fourth largest island in the world (just a bit smaller than Texas). It is home to 5% of the world's species, 80% of which are **endemic species** (exclusively native to a

BIODIVERSITY HOTSPOTS

There are 25 biodiversity hotspots around the world; they cover less than 1% of the world's area, but have 20% or more of the world's species.

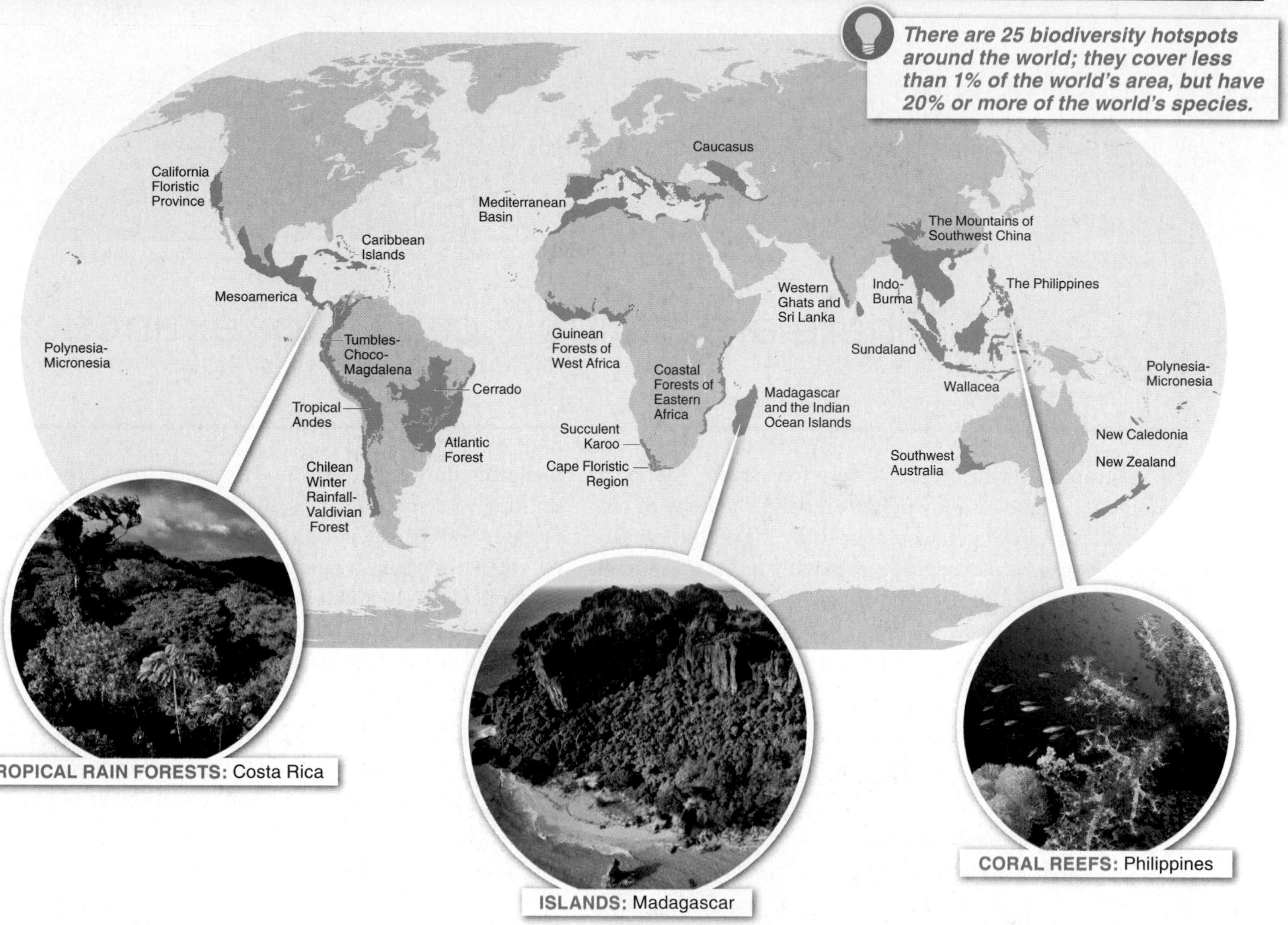

FIGURE 18-7 Biodiversity hotspots are significant reservoirs of biological diversity.

place)—a higher percentage of endemic plants and animals than in any comparably sized area on earth. One of the 25 biodiversity hotspots, Madagascar is an area unusually rich in species diversity.

Deep Oceans The oceans, covering two-thirds of the earth's surface, can be much more difficult to explore than terrestrial habitats. In particular, deep ocean regions are much less studied than terrestrial regions. Once thought to be lifeless, the deepest ocean regions are now known to be teeming with biodiversity. One survey of the deepest parts of the Antarctic Ocean, sampling at a depth of more than 18,000 feet (6,000 m), discovered 585 new species of crustaceans alone.

TAKE HOME MESSAGE 18.4

» Biodiversity is not evenly distributed over the earth. For nearly all groups of plants and animals, both marine and terrestrial, biodiversity is greatest near the equator and gradually decreases toward the North and South Poles. Factors that influence the species richness in an area include the amount of solar energy available, the area's evolutionary history, and the rate of environmental disturbance. Biodiversity hotspots are regions of significant biodiversity under threat of destruction.

18.5–18.6

Extinction reduces biodiversity.

A mother rhinoceros (*Ceratotherium simum*) with her offspring.

18.5 There are multiple causes of extinction.

Imagine that you are on a deserted island. Alone. It would be hard and it would be lonely. The hope of rescue someday, however, would probably make things bearable. "Martha," a passenger pigeon living in the Cincinnati zoo in the early 1900s, found herself in a situation of this sort—minus the hope. Passenger pigeons in the wild had been completely wiped out by a combination of factors, including hunting of the birds for meat, loss of their habitat due to deforestation in North America, and possibly disease. And so, except for the small flock that Martha was part of, the species was gone. One by one, most of Martha's flock died as well, and when the second-to-last passenger pigeon died in 1912, Martha was all alone. She lived for two more years, with no possibility for ever reproducing, and died on September 1, 1914 (FIGURE 18-8).

The aftermath of a pigeon-hunting expedition; in a single day, as many as 50,000 birds could be killed.

Stuffed passenger pigeon

Once numbering in the millions, passenger pigeons were hunted to extinction.

FIGURE 18-8 **Once plentiful, now extinct.**

On average, species persist for about 10 million years, but there is huge variation. Some may last for hundreds of millions of years, but for many others the number of years is much lower than the average. An extinction occurs when all individuals in a species have died. And whether you look at biodiversity from a utilitarian, aesthetic, or symbolic perspective, extinction is a tragic loss.

As we discussed in Chapter 12, extinctions can be divided into two general categories: mass extinctions and background extinctions (FIGURE 18-9), which are distinguished by the numbers of species affected. A mass extinction occurs when a large proportion of species on earth are lost in a short period of time (typically less than two million years, and in some cases much shorter)—such as when an asteroid struck earth 65 million years ago and about 75% of all species were wiped out. These extinctions are above and beyond the normal rate of extinctions that occur in any given period, referred to as the background extinction rate.

Q Does extinction happen only to weak, unfit species?

Both background and mass extinctions result in the same outcome for the species involved, but the causes tend to differ. In catastrophic mass extinction events, such as an asteroid impact, the

TWO CATEGORIES OF EXTINCTIONS

Extinctions generally fall into two categories:

- **MASS EXTINCTIONS**
A large number of species (even entire families) become extinct over a short period of time due to extraordinary and sudden environmental change.
- **BACKGROUND EXTINCTIONS**
These extinctions occur at lower rates during times other than mass extinctions.

Extinction rate (percentage of families that became extinct)

60
40
20
0

500 400 300 200 100 0

Time (millions of years ago)

FIGURE 18-9 Mass extinctions and background extinctions.

FACTORS THAT INFLUENCE THE RISK OF BACKGROUND EXTINCTION

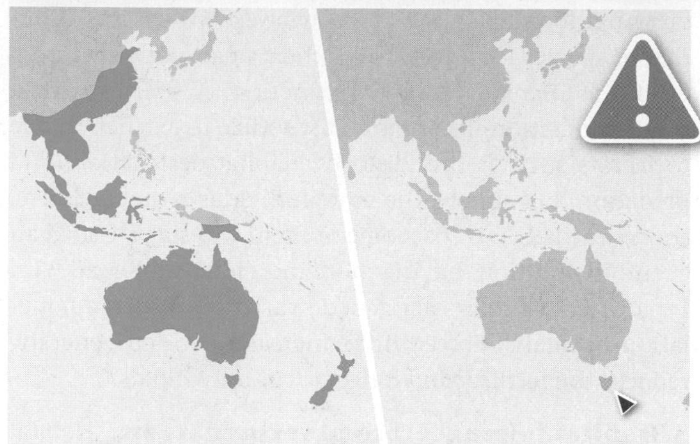

GEOGRAPHIC RANGE: EXTENSIVE vs. RESTRICTED
Species restricted in their range are more vulnerable than those with extensive ranges.

LOCAL POPULATION SIZE: LARGE vs. SMALL
Species with small population sizes are at increased risk of extinction.

HABITAT TOLERANCE: BROAD vs. NARROW
Species with narrow habitat tolerances are at greater risk of extinction than species with broader habitat tolerances.

FIGURE 18-10 Range, population size, and habitat tolerance influence risk of extinction.

features of a species' biology—its biochemistry, physiology, behavior—don't necessarily play a role. It's more like bad luck. Background extinctions, by contrast, tend to be a consequence of one or more aspects of a species' biology.

For any species, there is always a risk of extinction. Here we look at three important aspects of a species' biology that can influence its extinction risk (FIGURE 18-10).

1. Geographic range: extensive versus restricted. Species restricted in their range—including those limited to small bodies of water and those confined to islands—are more vulnerable than species with extensive ranges. The Tasmanian devil is a marsupial carnivore about the size of a dog. Although these animals once thrived in Australia, they now are confined to the island of Tasmania, which is smaller than the state of Maine. Unable to expand their range, Tasmanian devils are more vulnerable to extinction than if their range were not limited. The house mouse, on the other hand, is found nearly everywhere in the world and so has a much lower risk of extinction.

2. Local population size: large versus small. Tigers and California condors, along with *Welwitschia,* a slow-growing, long-lived plant in southwest Africa, are examples of species that, because of their small population sizes, are at increased risk of extinction. A species with a small population size is more susceptible to extinction due to factors such as fire, diseases, habitat destruction, and predation. Put simply, the more individuals that are alive, the more likely it is that some of them will survive such an event. In addition, having more individuals to breed with typically leads to greater genetic variation. When a population is small, inbreeding is increased, which generally reduces the fertility and longevity of individuals.

3. Habitat tolerance: broad versus narrow. "Habitat tolerance" describes the breadth of habitats in which a species can survive. Some plant species, for example, can tolerate large swings in water availability or soil pH or temperature. Others are limited to very narrow habitat ranges. The now-extinct passenger pigeons could build nests only in a specific type of forest and needed large numbers of individuals, breeding communally, to breed successfully. As forests were cut down and as the birds were hunted, the size of their flocks diminished. Their narrow habitat tolerance then made them extremely vulnerable to extinction. In general, because species with narrow habitat tolerance cannot adapt in the face of habitat degradation and loss, they are at greater risk than species with broader habitat tolerance.

Humans are enormously successful, thanks in part to our extensive, expanding geographic range, large local population sizes, and broad habitat tolerance. Our success, unfortunately, is having an increasingly negative effect on the species with which we share the planet, causing extinctions at a significantly higher rate than ever before. We next explore in more detail the conflict between humans' success as a species and the survival of other species.

TAKE HOME MESSAGE 18.5

» Mass extinctions, which can destroy many or all of the species in an area, may reflect bad luck more than the particulars of a species' biology, including its biochemistry, physiology, and behavior. Background extinctions, on the other hand, tend to be a consequence of one or more features of the species' biology. Small population size, limited habitat range, and narrow habitat tolerance contribute to background extinctions.

18.6 We are in the midst of a mass extinction.

Sometimes you can be so close to something that it's difficult to really see it. That may be the situation that humans experience when it comes to the global loss of biodiversity. But we are becoming increasingly aware that we are amid a mass extinction. In a recent survey by the American Museum of Natural History, 70% of biologists indicated that they believed this to be so and that steps must be taken by governments and individuals to stop this massive loss of species.

Data on current rates of extinction in every well-studied group of plants and animals support the hypothesis that a mass extinction is under way. Among the mammals, more than 10% of all species are currently *endangered,* under imminent threat of extinction, and approximately 21% are endangered or *threatened,* vulnerable to extinction. Among birds, 4% of all species are endangered and approximately 13% are endangered or threatened. Close to 50% are in decline. Almost one-third of amphibian species are endangered or threatened. Among fishes, mollusks, insects, fungi, and plants, too few species have been evaluated to make accurate estimates of the proportion of all species that are endangered or threatened. But for the species that have been evaluated, the situation is similarly dire.

Q If mass extinctions are a natural part of earth's history, should we be concerned about our effects on extinction rates?

Historically, evidence from the fossil record suggests that background extinction rates are about one extinction per million species per year. Currently, as the numbers above suggest, extinction rates may be 1,000 times (or more) greater than this. It is difficult to know the exact magnitude of the problem because, as we saw in Chapter 12, biologists don't have much of an idea about how many species there are on earth—the estimates range from 5 million to 100 million.

Unlike the last mass extinction event, in which the dinosaurs and most other species were wiped out following an asteroid's collision with the earth, this current mass extinction is the result of the activities of one species—humans. Ironically, it is the unprecedented success of humans that is responsible for this situation. The resource needs of so many people have inevitably interfered with the ability of other species to coexist with us.

The chief reason for the loss and impending loss of so many species is habitat loss and habitat degradation. Particularly harmful is habitat loss in earth's tropical rain forests, where biodiversity is greatest. In the past 25 years, half of the world's tropical rain forests have been destroyed, usually by burning to make way for agricultural use of the land or by logging (FIGURE 18-11). Urban development, too, is responsible for destruction of rain forests, as growing human populations continue to expand. Intensive agriculture, livestock grazing, and the development of urban centers have led to the destruction and fragmentation of habitats worldwide.

The loss of biodiversity-supporting habitat is not restricted to the tropics. In the Pacific Northwest, logging of forests has occurred at an unsustainable rate, leading to significant reductions in a diverse range of populations. Another factor that reduces biodiversity is overexploitation of species, such as the killing of animals for food, pelts, tusks or other body parts, and medicinal products. Bonobos (*Pan paniscus*), for example, a type of chimpanzee and our closest living relatives, are almost never seen in the wild. Because of habitat loss and their exploitation as a source of food in parts of Central Africa, bonobos have been driven to near extinction (FIGURE 18-12).

Bonobos, the closest living relative of humans, have been driven close to extinction, in part due to hunting.

FIGURE 18-12 Exploitation of species: the bonobo.

Additionally, the introduction of exotic species into habitats where they would not naturally be found adversely affects biodiversity. We explore this in Section 18.8. The consequences of the loss of biodiversity, beyond the values of biodiversity discussed earlier, are largely unknown, but most likely include a serious reduction in the capacity of the environment to provide clean air and water and to recover from environmental and human-induced disasters.

Later in this chapter, we discuss some strategies by which the current high rate of biodiversity loss can be reduced.

DEFORESTATION

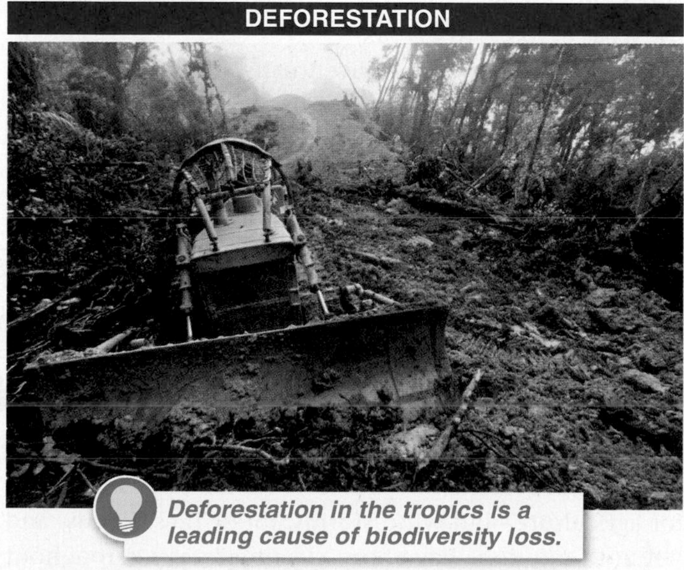

Deforestation in the tropics is a leading cause of biodiversity loss.

FIGURE 18-11 Removing habitat through deforestation.

TAKE HOME MESSAGE 18.6

›› Most biologists believe that we are currently in the midst of a mass extinction, that it is the result of human activities, and that it poses a serious threat to our future survival.

18.7–18.11

Human activities can have disruptive environmental impacts.

A dense green jungle surrounds Hong Kong.

18.7 Some ecosystem disturbances are reversible, and others are not.

When you fly over southern New England today, you look down on an undulating carpet of forest that covers hills and valleys. Roads traverse the forest, and here and there you see towns and cities or cultivated land and pastures, but mostly what you see is the tops of trees. The same route 200 years ago would have looked very different. By the early 1800s, most southern New England forests had been cleared. Homesteads consisting of farmhouses, barns, stables, and other outbuildings were scattered across the countryside. Stone walls extended across hills and valleys, separating cropland from pastures.

Four hundred years ago that same land was mostly forested, and a few Native American villages stood among cleared fields, but most of what you saw then would have looked similar to what you see today.

The change from forest 400 years ago to cleared land 200 years ago was a major ecosystem disturbance, but now the area has returned to forest. The species compositions are not identical—today's forests are not the same as those of 400 years ago. Nonetheless, stone fences found deep in the New England woods today are the only obvious trace of the agricultural history of the area. But they run through solid forest instead of between fields and pastures.

By the middle 1800s, New England settlers were abandoning farms on marginal land and moving west. And, once abandoned, the fields previously cultivated began a process of ecological succession that returned them to forests (see Section 17.16).

In the first year after a field is abandoned, its bare soil provides a harsh environment for plants. Sunlight heats the ground surface and bakes moisture from the soil. Only tough plants sprout and grow under such conditions. We call these pioneering species of plants "weeds," and we see them growing in disturbed habitats everywhere—along roadsides, in vacant lots, and even in cracks in pavement.

As weeds grow and die each year, their organic matter enriches the soil and helps it retain moisture. And every year, the plants grow more densely and shade the ground, which no longer gets quite so hot and dry. Additional species of plants can grow in these milder conditions. These plants are taller, so they provide still more shade, and the ground becomes still cooler and moister. At this stage, the seeds of bushes and small trees can sprout. As these woody plants grow, they make the ground cool and shady enough to allow the seeds of forest trees (oak, maple, or beech) to sprout. Eventually, forest trees grow so closely packed that little sunshine penetrates their leafy canopies to reach ground level. Undisturbed, the forest stops changing at this point, because the mature trees have created conditions in which only their own seeds can sprout and grow.

Q Once land is cultivated or developed, can it ever return to its previous state?

An ecosystem disturbance is reversible when it alters the biotic (living organisms) and abiotic (decaying organisms, soil, rocks) components of habitats but does not cause the complete extinction of any species. The species may recover and reestablish their populations. The disturbance of the New England forests was caused by humans—clearing land for agriculture—but wind storms, forest fires, floods, and volcanic eruptions have been clearing forests throughout the history of life (FIGURE 18-13).

SOME DISTURBANCES ARE REVERSIBLE: MOUNT ST. HELENS

In 1980, a destructive eruption of Mount St. Helens, in Washington State, wiped out the dense surrounding forest. Since then, the ecosystem has rebounded, and the recovery continues today.

MAY 1980
Seen here on the day before the eruption, the volcano is surrounded by coniferous forests.

SEPTEMBER 1980
The eruption flattened surrounding vegetation and left the area coated in a layer of ash.

MAY 2015
Shrubs and small trees have returned to the plain at the foot of the active volcano.

FIGURE 18-13 **The forest returning.**

SOME DISTURBANCES ARE IRREVERSIBLE

Tasmanian wolves

Chinese river dolphin

Blue whale

Once lost, a species will never exist again. The blue whale—the largest animal ever to have lived on earth—is at risk of being lost forever.

FIGURE 18-14 **The end of a species.**

> "Any fool kid can step on a beetle. But all the professors in the world can't make one."
>
> — ARTHUR SCHOPENHAUER
> ***German philosopher (1788–1860)***

If an ecosystem disturbance causes a species' extinction, however, the disturbance is irreversible. Each species is the result of a long and uninterrupted evolutionary history, involving an interplay of random and selective forces and producing a unique genome. Once lost, a species can never exist again (FIGURE 18-14). For this reason, ecosystem disturbances that involve the loss of species can have more devastating consequences than those that do not lead to extinctions, no matter how great the changes to the abiotic environment might be.

TAKE HOME MESSAGE 18.7

» An ecosystem disturbance is reversible as long as it does not include the complete extinction of any species, so members of the species can reestablish their populations. An ecosystem disturbance that involves the complete loss of a species to extinction is irreversible because a species, once lost, can never exist again.

18.8 Human activities can damage the environment: 1. Introduced non-native species may wipe out native organisms.

In the rain and sleet of a Chicago winter, you might see colorful birds flying about. These small green members of the parrot family are about a foot (30 cm) from head to tail tip. And if they look out of place in Chicago, it's because they are. Native to South America, monk parakeets were imported to the United States as pets in the 1960s and 1970s. Some of those pet birds escaped, survived, and bred. Free-living populations of monk parakeets now live in 11 states across the country, including Oregon, Florida, and New York, as well as Illinois (FIGURE 18-15).

Monk parakeets are just one example of an **exotic species** (also called an **introduced species**), a species introduced by human activities, intentionally or accidentally, to areas other than the species' native range.

The phrase "exotic species" suggests colorful birds and butterflies winging through a forest, but the ecological reality of such species can be much darker. Although some introduced species do not harm their new habitat, in many cases they cause economic or environmental harm, or harm to human health. Harmful introduced species are referred to as **invasive species.**

The arrival of free-living monk parakeets in the United States triggered alarm because they are regarded as agricultural pests in South America. Records from the Inca civilization that flourished before Pizarro's invasion of Peru in 1532 attest to crop damage by monk parakeets. And when Charles Darwin visited Uruguay in 1833, he was told that monk parakeets attack fruit orchards and grain fields. Fortunately, parakeets have not devastated crops in the United States, but in Chicago, the monk parakeets build massive nests on utility poles, transformers, and floodlights. The nests, when soaked with rain, are heavy enough to bring down wires, interrupting electrical service to entire neighborhoods.

Q Why should we worry about exotic species?

While monk parakeets were initially introduced to the United States intentionally, in many cases the introduction of an exotic species has been unintentional. The brown tree snake, which reached Guam sometime before 1952, stowed away in shipments of military equipment just after the Second World War. Whether introduced intentionally or unintentionally, two characteristics of introduced species are troubling (FIGURE 18-16).

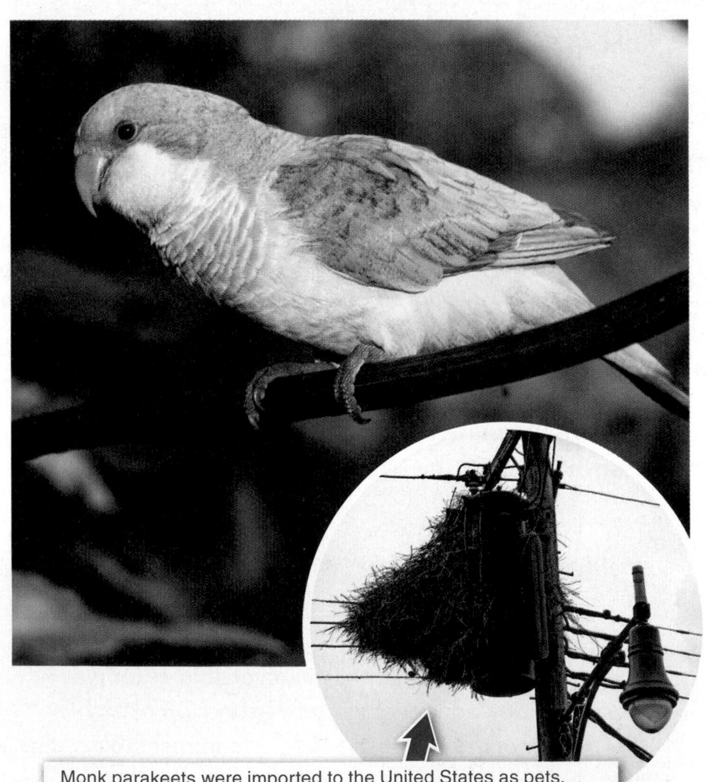
Monk parakeets were imported to the United States as pets. Some of those birds escaped, survived, and bred. They now are a nuisance in cities, where they build huge nests on utility poles.

FIGURE 18-15 **A parakeet in the Midwest.**

DISRUPTIVE IMPACTS — INVASIVE SPECIES

PROBLEM
Introduced species can harm habitats and their native populations.

CAUSE
When non-native species are introduced–accidentally or intentionally–and cause harm, they are called invasive species. Invasive species can multiply, unchecked by predation, overwhelming competitors and irreversibly altering ecosystems.

STRATEGIES FOR SOLUTION
Better regulation and restriction of intentional introductions; better vigilance against accidental introductions

FIGURE 18-16 **Understanding the impact of invasive species.**

INVASIVE SPECIES: ACCIDENTAL INTRODUCTIONS

BROWN TREE SNAKE
- Introduced to the island of Guam
- Eradicated most of the native bird species, which had never been preyed upon by snakes and had not developed defense mechanisms

PURPLE LOOSESTRIFE
- Introduced to the U.S. as a garden plant
- Invades wetlands and outgrows native grasses, sedges, and flowering plants in every state except Florida

ZEBRA MUSSELS
- Introduced to the Great Lakes
- Cause damage to industrial facilities by clogging water-intake pipes, and deplete resources available to small fish, eliminating the food supply for larger game fish

FIGURE 18-17 **Unwanted guests.** Invasive species sometimes get into natural habitats accidentally. The results can be devastating.

1. Exotic species may have no predators or pathogens in their new habitat, so their populations may grow unchecked.
2. Native plants and animals may have no mechanisms to compete with or defend themselves against invading exotic species.

The island of Guam lies in the middle of the South Pacific Ocean. It has no native snakes or predators that eat snakes. Guam was once home to magnificent native birds that had evolved in isolation for thousands of years. These bird species had never experienced predation by snakes, so they had no defense mechanisms against them. Invasive brown tree snakes have now eradicated most of the native species of birds. Extraordinary efforts are being made to prevent brown tree snakes from spreading to other Pacific islands, such as Hawaii. Like Guam, these islands are home to species of birds that occur nowhere else in the world, and lack native species of snakes and snake predators.

One current promising strategy, used by the U.S. Department of Agriculture, is to airdrop dead mice, their bodies packed with tablets of acetaminophen—the active ingredient in Tylenol. The brown tree snakes feed on the dead mice, and because the acetaminophen interferes with the oxygen-carrying capability of the snake's hemoglobin, the snakes die. While initial results are promising, it is too early to predict the long-term success of this program.

Invasive plants can also cause massive ecological shifts and economic costs. Purple loosestrife is an attractive flowering plant that is native to Eurasia. It was imported to the United States in the 1800s as a garden plant. Once here, it rapidly escaped from cultivation and invaded wetlands in every state except Florida. It produces thousands of seeds a year and also spreads by sending out underground stems. This aggressive growth overwhelms native grasses, sedges, and flowering plants, replacing diverse wetland communities with monocultures of loosestrife that provide poor-quality habitats for bog-dwelling insects, birds, reptiles, amphibians, and mammals (FIGURE 18-17).

The title of "Most Destructive Invaders in North America" may belong to zebra mussels and quagga mussels. These thumbnail-size freshwater mussels are native to eastern Europe and western Asia. In the early 1800s, the mussels spread to western Europe. They unexpectedly came to North America in the 1980s—in the ballast water of ships traveling through the Saint Lawrence Seaway on their way from the Atlantic Ocean to the Great Lakes. Female mussels produce more than a million eggs per year, and the larvae settle on any solid surface. Sometimes they settle so thickly that they clog the intake pipes of water-treatment plants and factories and the cooling systems of power plants.

The ecological threat that zebra mussels represent is far more serious than the damage they cause to industrial facilities. The Great Lakes fisheries are based on game species (lake trout, salmon, walleye) and commercial species

INVASIVE SPECIES: INTENTIONAL INTRODUCTION

Introduced to Australia to control agricultural pests, cane toad populations exploded and have become an ecological nightmare, killing indigenous snakes, lizards, birds, and even crocodiles.

FIGURE 18-18 **Cane toad infestation.**

(whitefish, perch, herring), and they produce revenues of more than $1.5 billion annually. Most of the commercially valuable fish feed on small fish species, such as smelt, which, in turn, feed on microscopic plants and animals (phytoplankton and zooplankton). Zebra and quagga mussels are exceptionally efficient at filtering phytoplankton and zooplankton from the water—so good that they are depleting the resources available to the small fish. Without enough food, the populations of small fish are crashing and eliminating the food supply for the large game and commercial fish. The Great Lakes contain nearly one-fifth of the fresh water in the world, but this enormous ecosystem is being irreversibly degraded by just two exotic species.

Another legendary pest is the cane toad. Ironically, it was imported into Australia to control a native sugarcane pest, the cane beetle. Unfortunately, the cane toads were not successful at controlling the pest populations. And worse, the cane toad populations exploded—preying on native amphibians, reptiles, mammals, and even birds. They spread throughout much of the continent and now number more than 200 million. Meanwhile, conservation workers struggle to control their numbers (FIGURE 18-18). The cane toad illustrates a common problem caused by the deliberate introduction of non-native species to control pests: the effects of a species on an ecosystem in which it does not naturally occur are unpredictable and often terrible.

Controlling or eradicating invasive species once they have become established and have spread can be difficult and costly. Consequently, the best approach when dealing with invasive species is early detection and a rapid response. Toward these goals, the U.S. National Invasive Species Council spends approximately $1.3 billion per year coordinating the efforts of 38 federal departments and agencies to prevent and control invasive species.

TAKE HOME MESSAGE 18.8

» Exotic (or introduced) species are species intentionally or accidentally introduced into a new habitat. They are considered invasive species if they cause harm in their new habitat. Invasive species can dominate and irreversibly alter communities and entire ecosystems.

18.9 Human activities can damage the environment: 2. Acid rain harms forests and aquatic ecosystems.

"What goes up must come down." That's a saying that applies to molecules just as much as it does to larger objects. When molecules come down, however, they may not be in the same form as when they went up. Some chemicals that rise into the atmosphere as gases return to earth as acidic precipitations, in the form of fog, sleet, snow, or rain.

We use the term "fossil fuels" to describe oil, natural gas, and coal. Composed largely of carbon and hydrogen, these substances form carbon dioxide (CO_2) and water (H_2O) when they burn. Although they are referred to as hydrocarbons, fossil fuels are not pure carbon and hydrogen. They also contain other elements, including sulfur and nitrogen (FIGURE 18-19).

When fossil fuels are burned, the gases sulfur dioxide (SO_2) and nitrogen dioxide (NO_2) are produced. These gases react with water vapor in the atmosphere to produce sulfuric acid (H_2SO_4) and nitric acid (HNO_3), and these acids fall to earth as acid precipitation (FIGURE 18-20).

In North America, precipitation has an average pH as low as 4.3 in some parts of the Northeast, more than 10 times as acidic as clean rain, which has a pH of about 5.6. (Clean rain is more acidic than pure water because carbon dioxide in the air dissolves in water to form carbonic acid, making the rain slightly acidic.) High concentrations of sulfuric and nitric acids in the atmosphere are to blame, and several factors play a role. The Northeast has more people per square mile than the rest of the United States, and that means there are more houses, factories, and automobiles burning the oil, coal, and gasoline that create acid precipitation (FIGURE 18-21).

Not all pollution that causes acid precipitation is local. In the United States, the Midwest and Southeast have many electric power plants that burn coal, and wind currents

THE CHEMISTRY OF ACID RAIN

1 Burning of fossil fuels releases the gases sulfur dioxide and nitrogen dioxide.

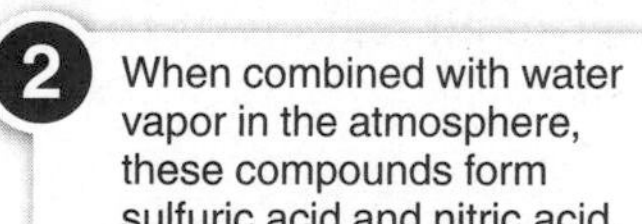

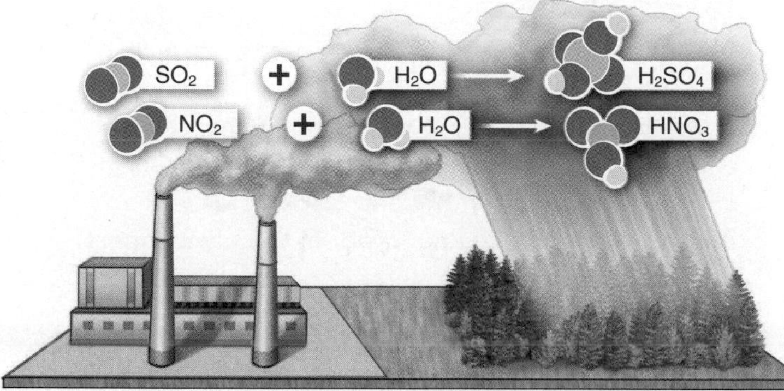

3 The acids fall to earth as acid precipitation.

FIGURE 18-19 **Formation of acid rain.**

DISRUPTIVE IMPACTS ACID RAIN

PROBLEM
Acid precipitation kills plants and aquatic animals directly, and also acts indirectly via changes in soil and water chemistry.

CAUSE
Burning of fossil fuels releases sulfur dioxide and nitrogen dioxide. The compounds form sulfuric and nitric acids when combined with water vapor.

STRATEGIES FOR SOLUTION
Tighter regulation and reduction of sulfur dioxide and nitrogen dioxide emissions

FIGURE 18-20 **Acid precipitation: the problem and its cause.**

ACID RAIN DISTRIBUTION

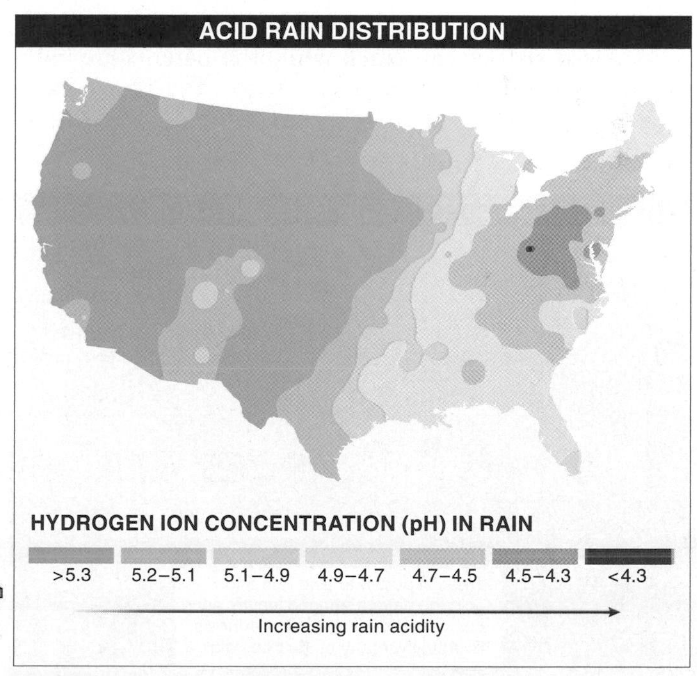

FIGURE 18-21 **Acid precipitation across the United States.** Pure water has a neutral pH of 7.

Q Why is acid rain an *international* issue?

carry the gases produced from this combustion across the Northeast. The precipitation in the western states is not as acidic as that in the Northeast (but it's still significantly more acidic than clean rain). Much of this region's acid precipitation comes from sulfur dioxide and nitrogen dioxide from Asia, which are converted to acids as they blow eastward across the Pacific Ocean. Precipitation in Europe and Asia is also acidified by local and distant sources of pollution.

Both terrestrial and aquatic organisms are harmed by acid precipitation. For example, when mountain forests are blanketed by fog year after year, their leaves are damaged, their rates of photosynthesis decrease, and eventually, many of the trees die (FIGURE 18-22). The acidic water directly affects the living tissues of the trees.

The indirect effects of acid precipitation on forests are not as conspicuous as the direct effects, but they are more far-reaching. As acid falls on the soil, it carries away calcium, potassium, magnesium, and sodium ions. These ions are essential nutrients for plants, and when they are leached from the soil, the rate of plant growth is reduced. Some forest animals are harmed by a change in soil chemistry—when calcium is depleted, for example, the shells of snails are thinner than usual. Because of the thinning of snail shells, birds that eat snails receive less calcium than normal, and these birds lay eggs with shells so fragile that they can crack while the parents are incubating them.

Acid precipitation affects lakes and streams directly and indirectly. Some lakes in the northeastern United States and in northern Europe have pH values as low as 4.0–4.3, which is at least 100 times more acidic than most lakes (which typically have a pH between 6 and 8). Very few species of insects, mollusks, crustaceans, fishes, or amphibians can live in such acidic water. Still worse, acid precipitation dissolves aluminum in soil and carries the dissolved metal (as aluminum ions), which is toxic to animals, into lakes and streams, where it can kill most forms of aquatic animal life.

Acid rain is a solvable problem. The deposition of acids from acid rain can be both prevented and cleaned up. By improving energy efficiency, reducing the use of coal, switching to the use of natural gas, and increasing the use of renewable wind and solar energy resources, we can prevent or reduce the emissions of sulfur dioxide and nitrogen dioxide. In the United States, many coal-burning plants have now installed special "scrubbers" on their smokestacks. The scrubbers extract hot gases from the power plant and function like sponges, removing sulfur dioxide from the smoke before it is released into the atmosphere and converting the gas to a less hazardous form that can be physically removed from the tower for disposal.

In the United States, implementation of regulations requiring that emissions of sulfur dioxide and nitrogen oxides be reduced by 2010 to approximately one-half the levels in the 1980s has improved air quality. Ambient sulfate concentrations during 2007–2009, for example, were 40% lower in all regions of the country relative to 1989–1991 levels.

EFFECTS OF ACID PRECIPITATION

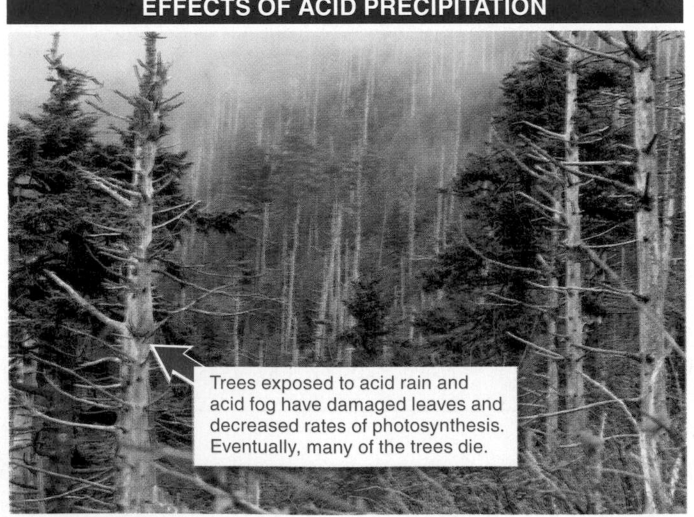

Trees exposed to acid rain and acid fog have damaged leaves and decreased rates of photosynthesis. Eventually, many of the trees die.

FIGURE 18-22 **Acid precipitation can kill trees.**

TAKE HOME MESSAGE 18.9

» The burning of fossil fuels releases the gases sulfur dioxide and nitrogen dioxide, and these compounds form sulfuric and nitric acids when they combine with water vapor in the atmosphere. Rain, fog, sleet, and snow that contain these acids can be more than 10 times more acidic than clean rain. Acid precipitation kills plants and aquatic animals directly, by contact with living tissues, and indirectly, through changes in soil and water chemistry.

18.10 Human activities can damage the environment: 3. The release of greenhouse gases influences the global climate.

Not long ago, debates about global climate change focused on two questions: "Is it real?" and "Is it caused by human activities?" Now, the vast majority of scientists believe that the answer to the first question is *yes* and, most likely, the answer to the second question is also *yes*. Consequently, the scientific debate has increasingly moved to questions such as "What effect will global climate change have on the world we know?" and "How can we stop it?" Nonetheless, in part because of the significant political implications of global climate change—including, but not limited to, global warming—and because of the complex nature of the data establishing its occurrence, global climate change and the role of human activities remains a controversial topic.

Some of the evidence that the earth's atmosphere is warming comes from weather records that extend back into the 1700s. The records show that the average temperature has increased rapidly over the past 50 years and that 18 of the 20 hottest years on record have occurred since 1990 (FIGURE 18-23). Three centuries, though, are just a blink of an eye in the history of the earth. Data from a report on Antarctic ice cores, published in the journal *Science* in 2007, extend the temperature records back to more than 800,000 years ago.

These ice cores reveal regular cycling between periods of cold climate ("glaciations") and of warmer climate and show that we are currently nearing the end of a colder period. But while they document that the current temperatures are not the highest on record, the Antarctic ice cores reveal that the current levels of carbon dioxide and methane in the atmosphere are far higher than *any* that have occurred during the 800,000-year span of the cores.

Q Why do we blame global warming on humans?

The term "greenhouse effect" describes the process by which energy from the sun warms the earth's atmosphere. The light we see lies in the visible portion of the light spectrum, and visible wavelengths of light pass easily through the atmosphere to warm the earth's surface. Some energy in the infrared part of the spectrum flows from earth back toward space, and several gases in the atmosphere absorb this energy, heating the air (FIGURE 18-24).

DISRUPTIVE IMPACTS **INCREASED GREENHOUSE GAS EMISSIONS**

PROBLEM
The average temperature has increased rapidly during the past 50 years, affecting both the physical environment and the biological world.

CAUSE
Burning of fossil fuels and clearing of land to cultivate crops have significantly increased levels of greenhouse gases in the atmosphere.

STRATEGIES FOR SOLUTION
Reduced emissions of greenhouse gases (particularly from the burning of fossil fuels)

FIGURE 18-23 Increasing greenhouse gases: the problem and its cause.

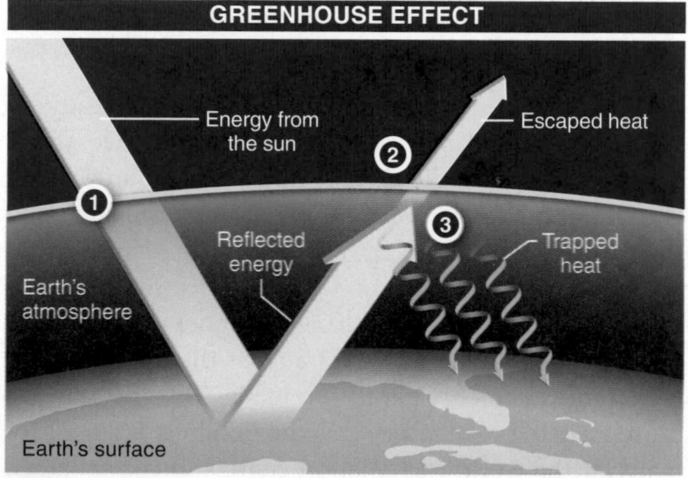

1 Energy from the sun passes easily through the atmosphere to warm earth's surface.

2 Some energy is reflected back toward space and escapes the atmosphere.

3 Some energy is absorbed by greenhouse gases and remains trapped in the atmosphere, heating the air.

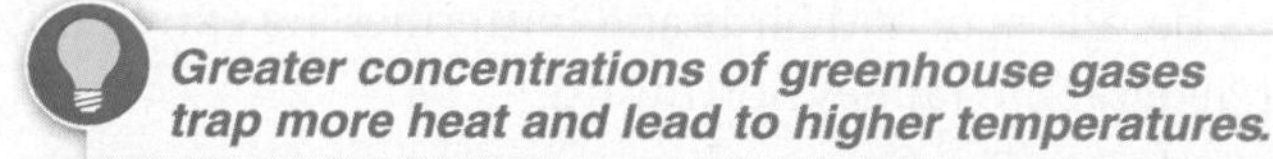

Greater concentrations of greenhouse gases trap more heat and lead to higher temperatures.

FIGURE 18-24 The earth is warmed by the greenhouse effect.

These gases act like the glass panels that make up the roof of a greenhouse: they allow light energy to pass through, but they keep heat from escaping—hence the name "greenhouse gases." Carbon dioxide and methane are among the most important of the greenhouse gases. Without them, heat would escape into space and the earth's average temperature would be lower.

Humans have increased the amount of greenhouse gases in the atmosphere. And we've done it primarily by the burning of fossil fuels—coal, oil, gasoline, and natural gas—for much of the energy we use. The burning of fossil fuels, which have been stored deep in the earth for millions of years, releases carbon dioxide. Clearing land and plowing soil to cultivate crops, too, releases both carbon dioxide and methane.

The Intergovernmental Panel on Climate Change, a scientific body set up in 1988 by the World Meteorological Organization and the United Nations, has predicted an increase in the average annual temperature of 2° to 12° F (1° to 6.4° C) during the 21st century. Sea levels will rise by 7–24 inches (18–59 cm) as warmer temperatures accelerate the melting of glaciers and ice sheets near the North and South Poles. Rainfall is predicted to decrease overall, and droughts will become more frequent and more severe. The impact of droughts will be most severe in areas already on the borderline of aridity, such as Australia, the American Midwest, and sub-Saharan Africa.

Both environmental and biological evidence already reveal the effects of global warming. The Arctic Ocean ice cap is melting, and the glaciers that cover Antarctica and Greenland have also been melting at unprecedented and increasing rates (FIGURE 18-25). In 1996, 21.6 cubic miles (90 cubic kilometers) of the Greenland ice sheet melted, and by 2010, the rate of melting was more than 57 cubic miles per year—faster than in any of the 53 years over which measurements have been made. This melting will cause a rise in ocean levels, creating flooding problems for many of the world's population centers in coastal areas, including New York City, Tokyo, and Amsterdam, and in the world's small island nations.

Biological systems are also showing the effects of climate change. Trees and flowers in northern latitudes are blooming earlier in the spring than they used to; migratory birds are arriving at their summer ranges earlier than they did even a decade ago; and birds and butterflies are extending their geographic ranges northward.

ICE CAP MELTING

1979

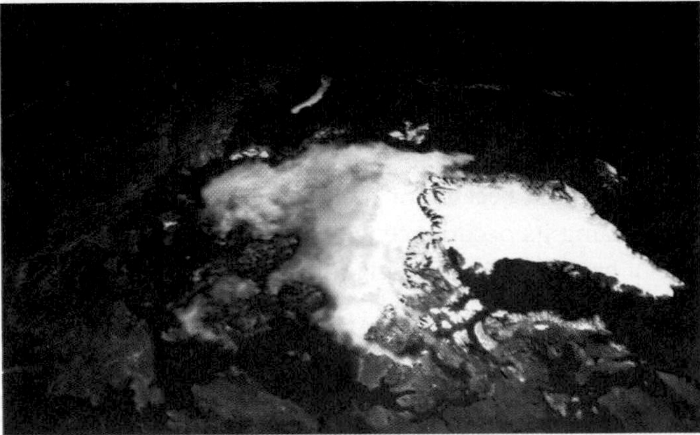

2016

FIGURE 18-25 **Disappearing ice pack.** The satellite image at the top shows the minimum concentration of Arctic sea ice in 1979. The lower image shows the concentration of sea ice recorded on September 10, 2016.

The changes to the physical environment caused by global warming, such as the thinning and shrinking of the Arctic ice or the flooding of coastal estuaries, can have profound effects on biodiversity at the species and ecosystem levels. For example, in 2004, a research ship in the Arctic Ocean found nine walrus calves swimming alone in deep water. Normally, adult female walruses leave their calves on ice floes in shallow water while they dive to the seafloor to feed on clams and crabs, and then return to the ice to nurse their calves. But the Arctic ice cap over shallow water has melted, driving the female walruses to the remaining ice, which is in deep water. The mother walruses cannot dive deep enough to reach the sea bottom, and they seem to be abandoning their calves.

Other consequences of global warming reveal the tremendous interconnectedness of so many ecosystem elements.

Many bird and butterfly species' migratory patterns have been altered as they search for lower temperatures. Many populations of small mammals living near high-altitude peaks are increasingly constrained as they attempt to move toward milder temperatures during the summer months—if they move to lower altitudes it gets hotter; moving to higher altitudes brings them to milder temperatures, but if they are already at the peak, they've got nowhere to go when it gets too hot. Massive croplands throughout the world may no longer be able to support the crops previously grown there.

Infectious disease experts have warned that global climate change is likely to alter the distribution of disease vectors—especially arthropods such as mosquitoes and ticks, which are particularly sensitive to temperature—changing and potentially increasing the incidence of diseases such as malaria and dengue fever.

Global warming is a difficult issue to tackle because it is so large in scope; it encompasses every biome, every ecosystem, every species, and every body of water, and its solutions require individual efforts in all nations as well as international cooperation among governments. Fundamentally, however, the science is straightforward: carbon emissions must be reduced significantly. This can be done by individuals who choose to cut down on their own fossil fuel use, improve the energy efficiency of their homes and cars, and support the development of carbon-free renewable energy sources. Internationally, it can be done through efforts to reduce deforestation while increasing the replanting of forests, and by developing and implementing transportation and industrial innovations that reduce dependence on the fuels that produce greenhouse gases. Innovative technologies, such as the underground sequestration of carbon, may also contribute to the reversal of global warming.

TAKE HOME MESSAGE 18.10

» Carbon dioxide and methane are called "greenhouse gases" because they trap heat in the earth's atmosphere. As humans burn fossil fuels and clear forests, the concentrations of greenhouse gases have been increasing and global temperatures have been rising. Ecological changes in plant and animal communities have already been observed and are likely to become more serious unless there is a global reduction in emissions of greenhouse gases.

18.11 Human activities can damage the environment: 4. Deforestation of rain forests causes loss of species and the release of carbon.

Towering trees, colorful birds and butterflies, maybe a glimpse of Tarzan swinging past on a hanging vine—that is the popular image of a tropical rain forest. And it is a reasonably accurate picture (minus the Tarzan part). But it is also a picture that is rapidly fading, as agriculture, logging, gold mines, and oil wells destroy tropical forests (FIGURE 18-26).

Tropical rain forests grow in a region extending just a bit north and south of the equator, in South America, Africa, Asia, and Australia. As recently as a few centuries ago, that belt of rain forest covered about 6 million square miles (15.5 million square kilometers). More than half of that

DISRUPTIVE IMPACTS DEFORESTATION

PROBLEM
Tropical rain forests are being cleared at unprecedentedly high rates, endangering countless species and increasing the concentration of greenhouse gases in the atmosphere.

CAUSE
The land is cleared for agriculture, logging, gold mines, and oil wells.

STRATEGIES FOR SOLUTION
Reduced destruction of high-biodiversity habitats, particularly tropical rain forests

FIGURE 18-26 **Deforestation: the problem and its cause.**

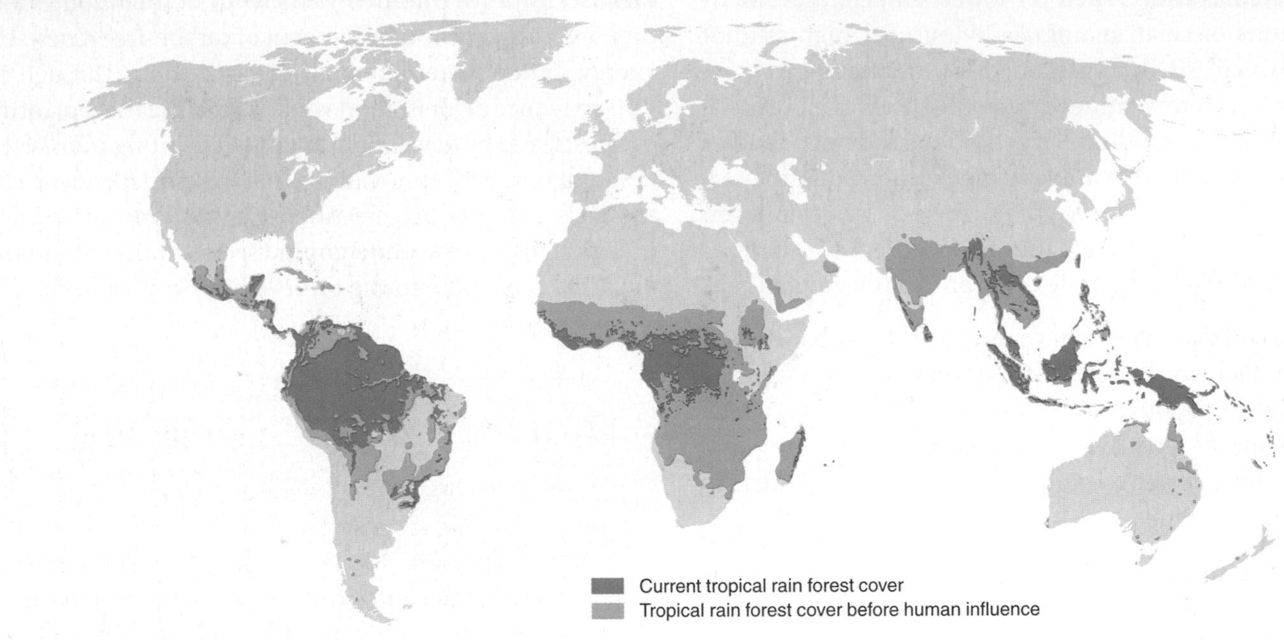

FIGURE 18-27 **Worldwide loss of tropical rain forests.**

forest has already been destroyed, most of it in just the past 200 years (FIGURE 18-27).

The destruction of a tropical rain forest means an enormous loss of species. More than 170,000 species of plants grow in tropical rain forests—more than 60% of the total number of living plant species in the world. A survey of a Brazilian rain forest found 487 different species of trees in just 2.5 acres (1 hectare), an area less than half the size of one city block in Manhattan. To put that number in perspective, consider that only 700 species of trees are found in the 5 *billion* acres (2 billion hectares) of the United States and Canada combined!

Agriculture—on both large and small scales—is responsible for the greatest loss of tropical forests. In "slash-and-burn agriculture," usually, just a few acres of forest are cut and burned, and crops are planted in the clearing. But the land is fertile for only a few years after it has been cleared. Then that plot is abandoned and more forest is cleared. Working on a large scale, multinational corporations clear hundreds of acres of forest to plant bananas, coffee, or oil palms, and clear thousands of acres of forest at a time to create pastures for cattle.

Pollution from oil wells and mining in tropical rain forests has an extensive impact. Leaking oil contaminates streams and groundwater, and acidic water drains from mines. Even worse, gold miners use mercury to extract gold, and some of the mercury enters streams, where it is converted to methyl mercury—a toxic compound that accumulates in plant and animal tissues as it moves up the food chain.

Agriculture and mining both require roads to bring equipment to the sites and take crops or minerals to market. Roads have a huge impact on forests because they make access easy. Traveling through a virgin tropical rain forest is difficult—bogs and natural tree falls often prevent travel in a straight line; slopes are often steep and the wet soil is slippery; and the ground-level vegetation can be both dense and thorny. No wheeled vehicle larger than a motorcycle can penetrate most rain forests, and traveling by motorcycle is typically slower than walking. These difficulties protect rain forests, but as soon as a road is bulldozed through a forest, people flock to it. They and their activities then spread into the forest on both sides of the road, creating new forest clearings.

Destruction of tropical rain forests has two serious environmental impacts: reducing the earth's biodiversity and increasing the concentration of greenhouse gases in the atmosphere (which, in turn, affects global climate change).

1. Reducing biodiversity. Tropical rain forests of Africa, Asia, the Pacific region, and Central and South America, as we've seen, contain an unusually large number of species of plants and animals, and probably other groups of organisms. Half of the world's biodiversity hotspots are in tropical rain forests.

2. Increasing greenhouse gases. Photosynthesis in tropical rain forests removes an estimated 610 billion tons (550 trillion kg) of carbon dioxide from the atmosphere each year. Accumulation of carbon dioxide in the atmosphere is the major cause of global warming. Thus, the continued photosynthetic activity of tropical rain forests is important in slowing the rate of warming. When forests are cleared and burned, however, that carbon is released into the atmosphere. And tropical forests are being cleared at a frightening rate—more than 50,000 square miles (an area larger than New York State) each year between 2000 and 2010.

Because deforestation alters the characteristics of land cover, changing the proportion of solar radiation reflected and absorbed, destruction of the rain forests can have additional significant effects on climate—including rainfall amounts and seasonality. These changes can further accelerate climate change, with potentially significant impacts on ecosystems.

The problem of tropical deforestation is one of the most difficult environmental problems to solve, and solving it will rely on international cooperation and require multiple strategies: (1) identifying and protecting the most diverse areas; (2) addressing the poverty that drives the need to destroy rain forests for human activities; (3) developing alternative sources of food and income; (4) reducing population growth; and (5) making education about the value of biodiversity a central part of these solutions.

In spite of these varied and complex difficulties, numerous programs around the world have had some success (FIGURE 18-28). In the state of Pará, in the Brazilian Amazon, for example, some encouraging progress has been made. As of 2012, this region, which is three times the size of California, had nearly eliminated illegal deforestation by adopting several new forest management practices.

- **Accountability and protection:** establishing a local system that requires landowners to register their holdings and an environmental police force to identify and punish illegal deforestation.
- **Education:** creating training programs for ranchers and farmers to help them make more efficient use of already-deforested land.
- **Better forest management:** developing environmentally sound infrastructure, such as reducing the building of new roads, long associated with increasing deforestation, in favor of improving existing roads and thus enhancing the ability to bring renewable forest products to market.

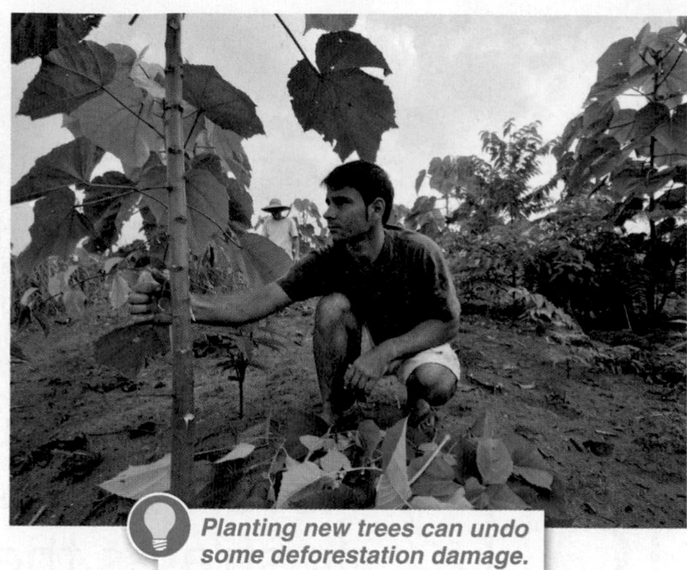

Planting new trees can undo some deforestation damage.

FIGURE 18-28 Cultivating seedlings for reforestation in the Amazon rain forest.

The initial transition is difficult, particularly because it can be accompanied by job losses. But efforts are under way to offset the loss of jobs associated with deforestation with increased economic development in non-deforestation-related areas. These areas include ecosystem services such as water management and a focus on non-wood forest products, including medicinal plants.

Still, many worry that this progress is not secure. In 2015 and 2016, for example, deforestation in Brazil increased. Conservationists attribute this reversal, after 10 years of reductions in deforestation, to growing worldwide demand for agricultural products, coupled with economic and political difficulties in Brazil.

TAKE HOME MESSAGE 18.11

» Tropical rain forests contain more species of plants and animals than all other terrestrial habitats combined, and half of the earth's biodiversity hotspots are in these forests. Tropical rain forests remove more carbon dioxide from the atmosphere than any other terrestrial habitats, and they are enormously important in limiting global warming. These forests, however, are being destroyed at an alarming rate.

18.12–18.14

We can develop strategies for effective conservation.

A worker carries a tree to plant at the edge of Maowusu Desert, China. Trees are planted to prevent the desert from expanding.

18.12 Reversal of ozone layer depletion illustrates the power of good science, effective policymaking, and international cooperation.

You breathe in oxygen molecules that consist of two oxygen atoms. The chemical formula is O_2. Ozone is a different molecular arrangement of oxygen that packs three oxygen atoms into a molecule; its chemical formula is O_3. Ozone irritates the respiratory pathways and lungs, causing coughing and wheezing, and exposure to ozone can induce asthma attacks.

During summer in the city, you are likely to hear radio and television warnings of high ozone levels. These warnings announce that ozone is expected to exceed safe levels that day. Children and adults with lung disease are advised to remain indoors during an ozone alert.

If ozone makes people sick, why would anyone worry about *depletion* of ozone in the atmosphere? The answer, like the adage about the value of real estate, depends on "location, location, location." At ground level, ozone is bad for you. Higher up, in the lower part of the stratosphere, about 30 miles (50 km) above the earth's surface, ozone protects you from ultraviolet radiation that causes sunburn and skin cancer. Ozone is a Jekyll-and-Hyde molecule. One set of environmental regulations tries to reduce the formation of ozone at ground level, while a different set of regulations tries to protect ozone in the stratosphere.

There are two types of "ozone depletion." One type is the general reduction in the amount of ozone in the stratosphere. The second type is the formation of areas, called "ozone holes," that have very low ozone concentration. These holes occur over the North and South Poles every winter. Synthetic chemicals known as chlorofluorocarbons (CFCs) cause both forms of ozone depletion. CFCs, which were developed in the 1930s, have a wide range of applications, including use as coolants in refrigeration systems. When CFC molecules leak into the atmosphere, they rise to the stratosphere, where sunlight knocks a chlorine atom off the CFC molecule. These free chlorine atoms then catalyze the breakdown of ozone to oxygen (FIGURE 18-29).

The cold temperatures and circular flows of air that develop over the Poles during the winter concentrate CFCs, forming ozone holes in those locations. The Antarctic ozone hole,

DISASTROUS IMPACTS — IN RECOVERY — OZONE LAYER DEPLETION

PROBLEM
Increased levels of ultraviolet light reach the earth's surface, leading to a greater incidence of health problems in animals and decreased rates of photosynthesis in plants.

CAUSE
Synthetic chemicals known as chlorofluorocarbons (CFCs) leak into the atmosphere, where they cause the breakdown of ozone.

STRATEGIES FOR SOLUTION
Reduced production and emission of CFCs

FIGURE 18-29 Depletion of the ozone layer: the problem and its cause.

ANTARCTIC OZONE DEPLETION

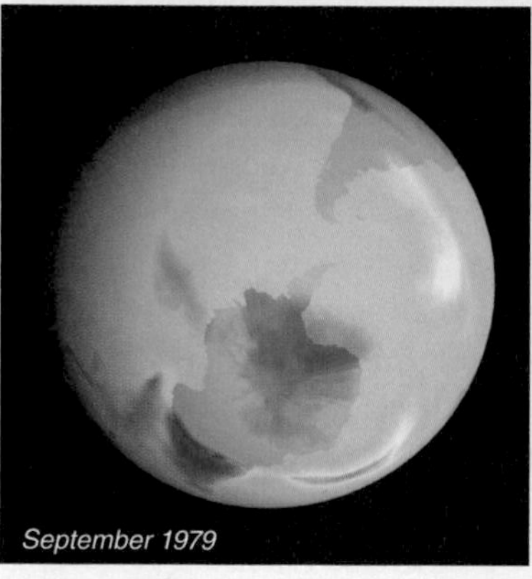

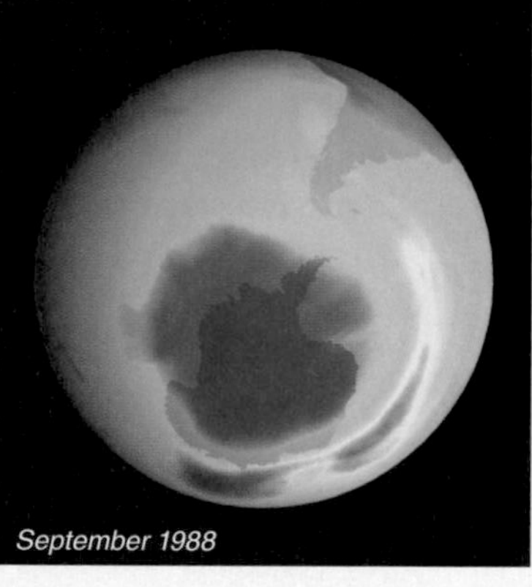

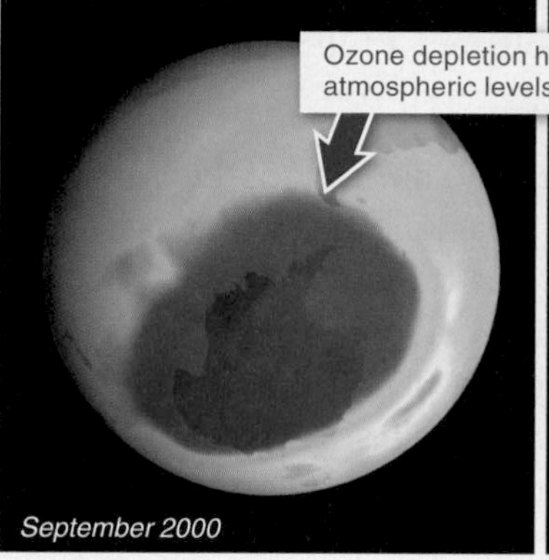

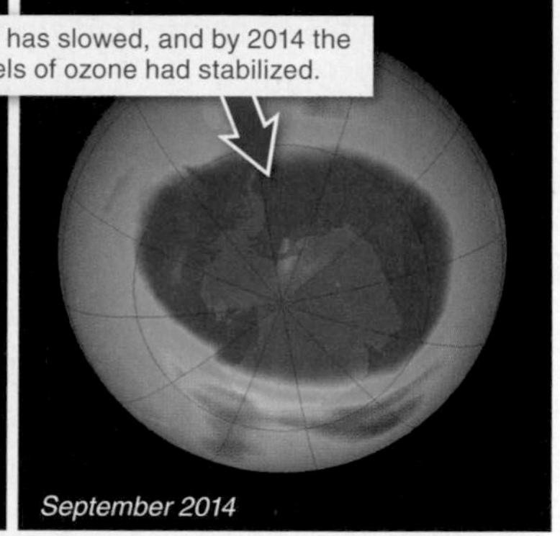

FIGURE 18-30 **Monitoring the ozone hole.** The satellite images show changes in the ozone hole over the Antarctic during the past 35 years. The darkest blue color indicates the lowest ozone concentration.

Depletion of the ozone layer and the formation of an ozone "hole" over earth's polar regions are the result of human activities.

which lasts for several months a year, covers the entire continent of Antarctica and extends northward to include the southern tips of South America and Australia (FIGURE 18-30). The Arctic ozone hole is smaller than the Antarctic hole and does not last as long, but it is large enough to extend southward into northern Europe, Asia, and North America.

Ozone depletion allows short-wavelength ultraviolet light (UVB light, with wavelengths of 270–315 nanometers) to reach the earth's surface. In the 1970s, scientists noted that the amount of ozone in the stratosphere was decreasing by about 4% per decade. The decline in global stratospheric ozone intensified UVB radiation everywhere on earth, particularly under the ozone holes. An increase in UVB intensity of 1% increases the incidence of skin cancer by 2% to 3%. Exposure to UVB radiation also promotes the formation of cataracts and reduces the effectiveness of our immune system. Domestic animals suffer the same kinds of damage that humans do, and a few studies indicate that wild animals are also affected.

Q **Should we be concerned about ozone holes in our atmosphere?**

Beyond the health risks associated with the higher UVB radiation, the damage to ecosystems is an even more serious consequence of ozone depletion. UVB radiation reduces the rate of photosynthesis by plants, and reduced photosynthesis in agricultural ecosystems means that crop yields decline.

Fortunately, the worldwide response to ozone depletion is an encouraging example of effective integration of science and policymaking. When CFCs were first invented, they were considered harmless, and so their use as coolants, as aerosol propellants, and in the production of materials such as Styrofoam became widespread. But the recognition of their effects in the atmosphere prompted relatively quick reaction and efforts to find solutions (FIGURE 18-31). In the 1980s, ozone depletion was slowed when most

REDUCING CFCs

Products made from ozone-depleting CFCs, such as Styrofoam food containers and hair sprays with damaging propellants, have been phased out, contributing to the ozone layer recovery.

FIGURE 18-31 **Taking simple steps to reduce ozone depletion.**

countries adopted an agreement to discontinue the use of CFCs. And in 2010, scientists found that the atmospheric levels of ozone had stabilized. A full recovery could occur by 2050.

It's important to clarify that depletion of the ozone layer is not the mechanism of global warming. Ultraviolet radiation represents less than 1% of the sun's energy. Greenhouse gases (Section 18.10) do not reduce the atmosphere's ozone layer. Global warming and ozone depletion are largely unrelated environmental issues, linked only by their common cause: human activities. The response to ozone depletion, however, represents an encouraging example of how nations can work together to address and correct human-caused environmental problems.

TAKE HOME MESSAGE 18.12

» Ozone in the stratosphere prevents short-wavelength ultraviolet light (UVB) from reaching the earth's surface, but for many decades the amount of ozone was decreasing, largely due to synthetic chemicals called chlorofluorocarbons (CFCs). An increase in UVB light reaching the earth's surface can seriously damage ecosystems and can harm the health of humans and other organisms. A decline in the production and use of CFCs is reducing atmospheric ozone depletion, and a full recovery of the ozone layer appears possible.

18.13 With limited conservation resources, we must prioritize which species should be preserved.

Is it preferable to save one beautiful, well-studied bird species or 1,000 species of unnamed, undescribed bacteria? And what about the 1,200 species of beetles in Panama that live in the evergreen tree *Luehea seemannii,* almost 200 of which live *only* in this one species of tree—how should we prioritize these species for conservation relative to, say, a single primate species? From locale to locale these questions may change, but the underlying issue remains: in a world where species are being driven to extinction faster than we can save them, which should be singled out for preservation and which should we leave to almost certain extinction (FIGURE 18-32)? This is a question that biologists, policymakers, and, ultimately, citizens must address.

In an ideal world, a conservation goal might be to protect "all biodiversity." Barring that, options include protecting "most biodiversity"—that is, the most *diverse* subset of biodiversity (in terms of genes, species, and ecosystems)—or protecting the most *valuable* biodiversity. There are numerous pros and cons for each goal, but once a goal is set, a plan is needed that outlines priorities for achieving

WHAT SHOULD WE PROTECT?

Egyptian tortoise hatchling

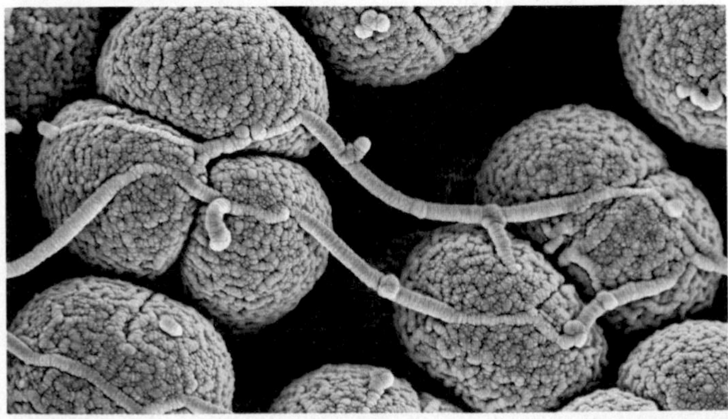

Soil bacteria

FIGURE 18-32 **Prioritizing conservation efforts.** How do we decide among the many options?

PRESERVING SPECIES vs. PRESERVING HABITATS

Conservation efforts can focus on preserving individual species (such as orangutans) or important habitats (such as rain forests).

FIGURE 18-33 **Which to preserve?** Limited resources and great demand make preservation choices difficult.

it. Frequently, such a plan involves an assessment of the various threats to biodiversity. We can rank species, for example, as relatively intact, relatively stable, vulnerable, endangered, or critical. These rankings can then be used, in conjunction with measures of the biological value of the biodiversity to humans, in formulating a plan.

As we saw earlier in the chapter, biodiversity can be valued in many ways, and its worth is not easily quantified or weighed. Sooner or later, many difficult and subjective decisions must be made in the goal-setting and prioritizing process. The undiscovered bacterial species may harbor the metabolic secrets to curing a devastating medical condition in humans, but the beautiful bird species may be much loved by bird-watchers or may serve as a powerful icon of strength or freedom, inspiring generations of people. The decisions are difficult, but not facing them is, in most cases, the same as deciding.

In the United States, much conservation policy involves response to the **Endangered Species Act (ESA),** a law that defines **endangered species** as those in danger of extinction throughout all or a significant portion of their range. The law is designed to protect these species from extinction. As we noted earlier in the chapter, species that are dwindling are listed as endangered or threatened, according to an assessment of their risk of extinction. Once a species is listed, legal tools are available to help rebuild its populations and protect the habitat critical to its survival.

Seemingly straightforward, the ESA has had the effect of focusing most conservation efforts on the preservation of individual species (and populations), sometimes at the expense of other elements of biodiversity and sometimes at the expense of efforts to reduce the loss of ecologically important habitats (FIGURE 18-33). Other difficulties are also associated with the ESA. Consider the task of determining the critical population size below which a population is endangered—is it 500 individuals or 5,000 or 50,000? With its emphasis on preservation of single species, the ESA does not effectively address ecosystem decline, which is an equally urgent and serious problem.

Despite these difficulties, the species-focused approach to conservation has had some success in the United States. Since it became law in 1973, approximately 40 species that were listed as endangered—including the bald eagle, the peregrine falcon, the gray whale, and the grizzly bear—have been taken off the list as their numbers have recovered. In the final section, we examine other strategies that have been used in efforts to preserve biodiversity.

TAKE HOME MESSAGE 18.13

» Effective conservation requires the setting of goals on the elements of biodiversity (genes, species, or ecosystems) that should be conserved and priorities among those elements. The U.S. Endangered Species Act has focused much conservation effort on the preservation of individual species.

18.14 There are multiple effective strategies for preserving biodiversity.

In a survey asking people which animal they would like to be, the top answer among men was eagle, followed by tiger, lion, and dolphin. Women chose cat, followed by butterfly, swan, and swallow. It's just a silly survey, but the differences in men's and women's selections illustrate how people have very different preferences when it comes to preserving biodiversity. People differ in the way they value species or habitats.

Most approaches to conservation biology, as we've seen, have focused on the preservation of individual species. Increasingly, however, conservation biology is shifting toward the preservation of important habitats, focusing on conserving communities and ecosystems. This habitat-conserving approach also leads to the preservation of individual species, but it has the added benefit of preserving many different species living within a habitat, including many that have not yet been identified, particularly microbes and fungi.

One effort aimed at the preservation of important habitat is the World Wildlife Fund's "Global 200," an identification of the most biologically distinct habitats on earth. The identification of these habitats, which include terrestrial, freshwater, and marine habitats, is part of an innovative strategy, sometimes called **landscape conservation,** that is geared toward conserving not just species but habitats and ecological processes (e.g., large-scale migrations and predator-prey interactions). In an age of scarce conservation resources and limited time, this approach prioritizes the conservation of biologically unique habitats on earth.

The habitat conservation approach is not new. The very first attempts at conservation in the United States were the creation of Yellowstone Park in 1872 and Yosemite in 1890, both large-scale efforts geared not simply toward the preservation of one or a few species but rather toward the preservation of the "wilderness." In the time since then, significant efforts have been made to establish national parks, wilderness areas, and recreation areas (FIGURE 18-34).

Today, conservation biologists integrate knowledge of population dynamics and biogeography into the design of natural preserves. Modern preserve design maximizes the variety of habitats and the efficiency of the preserves (FIGURE 18-35). The design includes:

FIRST ATTEMPTS AT PRESERVING WILDERNESS

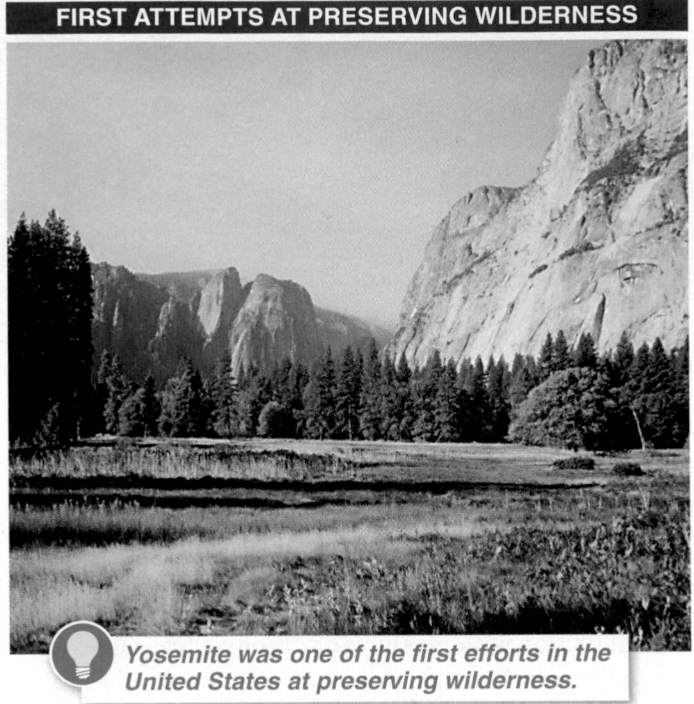

Yosemite was one of the first efforts in the United States at preserving wilderness.

FIGURE 18-34 **The natural beauty of Yosemite.**

- Larger, rather than smaller, preserves (including a preference for a single, undivided preserve over several smaller preserves)
- Circular, rather than linear, preserves to reduce the negative effects at the preserve edges
- Corridors—even just relatively narrow strips of land—that connect larger natural preserves, allow gene flow, and reduce inbreeding among distinct populations within the individual preserves
- Buffer zones (which permit limited amounts of human activity) around core areas (which contain the habitat to be conserved)

As conservation plans aimed at preserving habitats increase, efforts focusing on individual species continue to be effective. Several strategies targeting individual species for conservation have been particularly successful at preserving large amounts of biodiversity beyond that single species.

1. Flagship species. Some species, because they are particularly charismatic, distinctive, vulnerable, or otherwise

DESIGNING EFFECTIVE NATURE PRESERVES

Modern preserves use design features that maximize their efficiency, including larger, rather than smaller, preserves; circular, rather than linear, preserves; and:

CORRIDORS
Strips of land that allow gene flow and reduce inbreeding among distinct populations within the larger, individual natural preserves

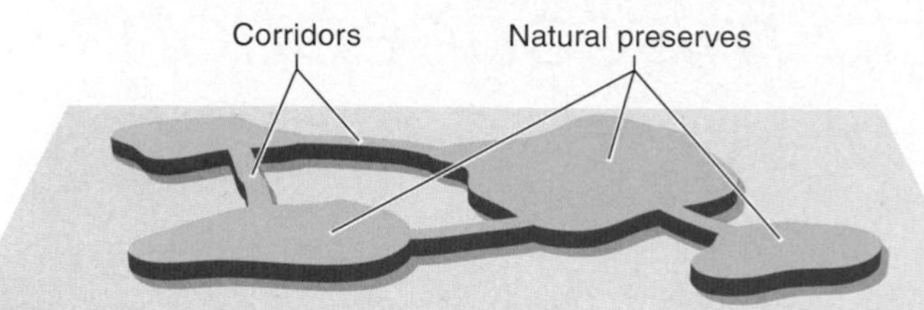

BUFFER ZONES
Areas where limited amounts of human use are permitted, surrounding a core natural preserve

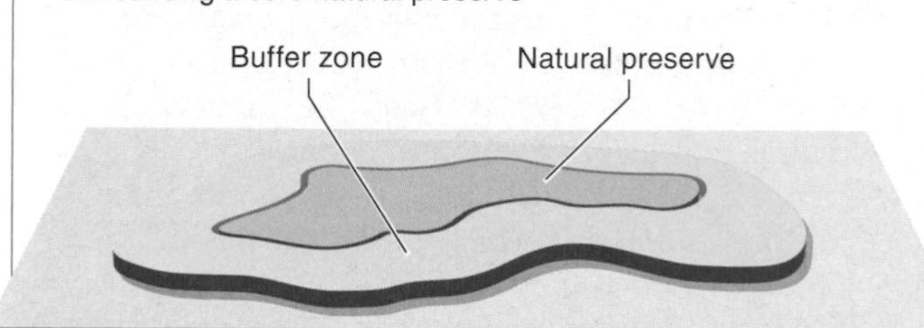

FIGURE 18-35 **Carefully planned nature preserves focus on features that maximize biodiversity.**

appealing, can engender significant public support. Examples include the giant panda of China, the golden lion tamarin of Brazil's coastal forest, the mountain gorilla of Central Africa, the orangutan of Southeast Asia, the leatherback sea turtle, the Indian tiger, and the African elephant. Preservation of these species, given their habitat needs, can serve to preserve many other species as well.

2. Keystone species. As we saw in Chapter 17, keystone species have a disproportionate effect on the biodiversity of a community. Their removal can lead to massive changes in the composition of species in an ecosystem, often causing huge loss of biodiversity. Examples include kelp, California mussels, grizzly bears, beavers, and sea stars.

3. Indicator species. These are species whose presence within an ecosystem indicates the presence or well-being of a large range of other species. For example, lichens are an indicator of air quality, because they are sensitive to sulfur dioxide, a component of industrial fumes. When lichens disappear from trees, it is usually an indicator of increased pollution, which endangers the entire ecosystem. Conversely, when lichens are present and healthy, they indicate that the ecosystem is healthy.

4. Umbrella species. These wide-ranging species require large habitats and other resources so that their preservation simultaneously protects numerous other species within that same habitat. Umbrella species tend to be large, wide-ranging vertebrates.

Captive breeding programs and habitat restoration have had some success in bringing back biodiversity from the brink of extinction. Zoos and botanical gardens have taken the lead in many of these efforts. By the 1980s, for example, the population of the California condor had dropped to 22 individuals, due to poaching, lead poisoning, and habitat loss. All of the birds were caught and taken to zoos, where a captive breeding program began. By 2016 the population had reached 446, including 276 birds in the wild and the remainder in zoos. Breeding programs are not a complete conservation solution on their own, however. It is essential that the species' habitats are not destroyed or altered if the species are to flourish again under natural conditions.

Restoration ecology, which uses the principles of ecology to restore degraded habitats to their natural state, has also been an important tool of conservation biologists. Wetlands that have been degraded by dredging and development have benefited from such programs. Reintroduction of native plant species and restoration of water and stream flow patterns can be instrumental in restoring these habitats.

Taken together, the strategies used by conservation biologists represent an important step toward reducing the adverse effects on many species of one hugely successful (from a population growth perspective) species—humans. Continued conservation efforts offer our best hope for a sustainable future.

TAKE HOME MESSAGE 18.14

» Conservation biology has focused, in the past, on preserving individual species. Increasingly, there is a shift toward the preservation of important habitats, focusing on conserving communities and ecosystems. Several methods focusing on single species remain useful, however, particularly when preserving the selected species requires the preservation of an amount and type of habitat that simultaneously protects many other species.

STREET BIO

Using evidence to guide decision making in our own lives

Climate change: clearing up misconceptions that cloud the issue

Misconceptions about climate change arise from its scale and complexity. And addressing it requires cooperation among many countries, corporations, and individuals, with varied interests and priorities. Nonetheless, the science is straightforward, as are the necessary courses of action. The misconceptions can (and must) be clarified and corrected.

Misconception: The climate changes naturally. It has changed before and will change again. Analyses of ice cores, tree rings, and lake sediments have helped us understand long-term historical patterns of climate change. These reveal changes that occurred over thousands or even millions of years, rather than decades, as we're observing now. Also, past climate changes were not accompanied by the large increase in atmospheric carbon dioxide seen over the past 100 years. Recent increases in CO_2 levels are due to human activities—80% due to fossil fuel burning and 20% to deforestation—with no changes in natural sources of CO_2 emission.

Misconception: We had more snow last winter than ever before, so there can't be any global warming. Anecdotal observations are appealing, but also misleading. Weather events are short-term phenomena, influenced by regional factors. Climate, on the other hand, is the long-term average of the weather. Cold spells or unusually heavy snowfalls or storms do occur, but the benchmark measure is *average* temperature. This has increased consistently since the 1950s.

Misconception: It's the ozone hole that causes global warming, and we've fixed that. Depletion of ozone in the atmosphere—much of which was human-caused—increased the UV radiation reaching earth. This radiation causes harmful mutations in organisms, but has little impact on temperatures. Several other human-caused environmental issues—including acid rain, nuclear waste, and heat from car exhausts—also don't have a significant impact on earth's climate.

Misconception: Plants will grow faster, winters will be milder. Global warming isn't such a bad thing. Yes, the impact of global warming will vary by location, and in some places may seem positive. But globally, the negative consequences will vastly outweigh any positive changes. Coral reef die-offs, for example, will damage the entire food webs of the oceans. Melting polar ice will flood coastal cities and small islands around the world, displacing many millions of people. Rising sea levels will also lead to longer droughts and altered wind patterns that may produce intense storms, tornadoes, and other extreme weather events.

Misconception: So, maybe climate change is occurring, but there's nothing we can do. We can reduce greenhouse gas emissions by burning fewer fossil fuels. We can develop new, renewable energy sources. However, because CO_2 emissions diffuse evenly throughout the earth's atmosphere, all nations must work together. The Kyoto Protocol is an international agreement created to address climate change. Lack of full commitment from the countries responsible for most greenhouse gas emissions (including the United States) highlights the difficulty in securing international cooperation.

Misconception: The sun is to blame. It has gotten hotter, or gotten closer to earth. Solar output—which has been closely monitored by NASA and other independent organizations for several decades—has not increased. A peer-reviewed summary of published research on solar energy concluded that fluctuations in solar output have had a negligible impact on climate and cannot explain the magnitude (1° F) or direction (consistently upward) of observed climate change over the past hundred years.

Misconception: Not everyone agrees; there's no consensus among the experts. There is debate among scientists about how best to model the climate and predict changes. And in the 1970s, there were even some predictions of "global cooling." As data accumulate, hypotheses are refined and improved. Today, there is near-universal agreement among climate scientists that climate change is occurring and is the result of human activities. In a recent analysis of the publications of 1,372 climate researchers, more than 97% of the researchers were in agreement about the existence and cause of global climate change.

Shown in background: a representation of what has been called the "hockey stick graph." The red line reflects earth's temperature in each year relative to the long-term historical average.

GRAPHIC CONTENT

Thinking critically about visual displays of data

1 What do the axes of these graphs represent? What are the variables being observed?

2 Which variable is independent? Which is dependent? Do these graphs conform to typical norms for the placement of the independent and dependent variables? Why or why not?

3 At what latitude do the largest numbers of species occur?

4 Why are data presented both for mammalian species and for marine copepod species (rather than just one or the other)? Why aren't data for all species presented?

5 At 20° north latitude, are there more mammalian or more marine copepod species? What aspect of the graphs may be responsible for someone answering this question incorrectly?

6 Bar graphs often have error bars, indicating some measure of variance in the data. Why do the data bars in these graphs have no such error bars?

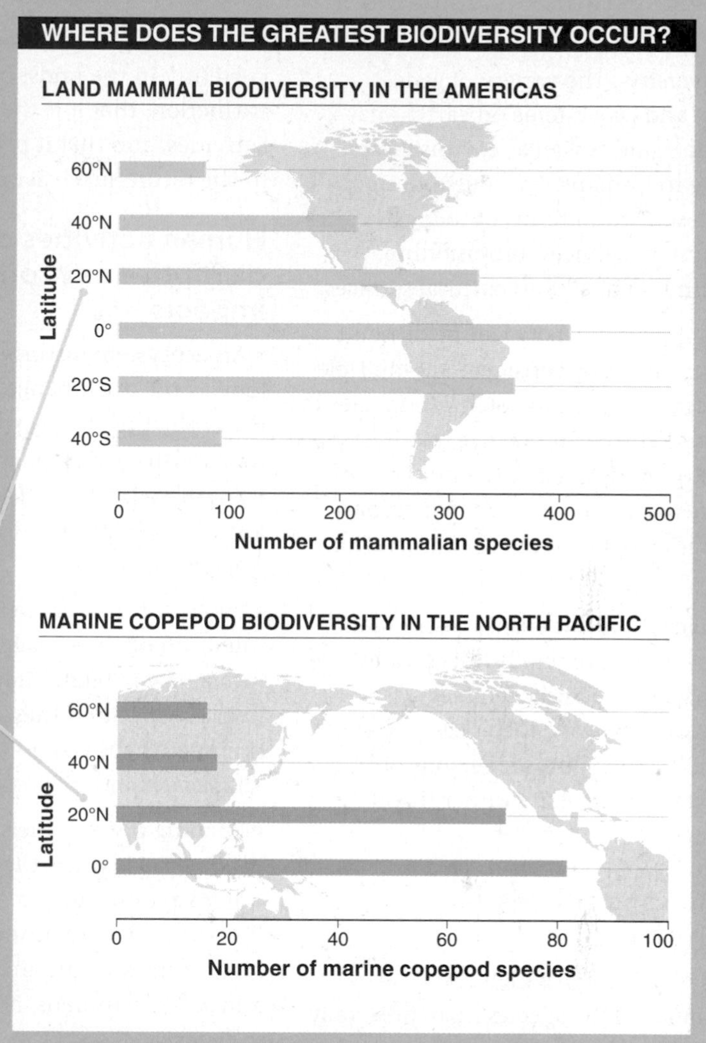

See answers at the back of the book.

KEY TERMS IN CONSERVATION AND BIODIVERSITY

biodiversity, p. 603
biodiversity hotspots, p. 608
conservation biology, p. 606
ecosystem services, p. 603
endangered species, p. 629
Endangered Species Act (ESA), p. 629
endemic species, p. 608
exotic species, p. 616
introduced species, p. 616
invasive species, p. 616
landscape conservation, p. 630
species richness, p. 607

BRIEF SUMMARY

Biodiversity—of genes, species, and ecosystems—is valuable in many ways.

• Biodiversity—the variety of genes, species, and ecosystems on earth—has intrinsic value as well as extrinsic value, or value to humans. Its value to humans is often described in terms of four categories of ecosystem services: provisioning, regulating, habitat, and cultural services.

• Assessing biodiversity can be difficult because it must be considered at multiple levels, from entire ecosystems to species to genes and alleles. In practice, biodiversity is most often defined as the number of distinct species in a habitat, which has important implications for conservation biology.

• Biodiversity is not evenly distributed over the earth. Generally, it is greatest near the equator and decreases progressively toward the Poles. Biodiversity hotspots are regions of significant biodiversity under threat of destruction.

Extinction reduces biodiversity.

• Mass extinctions, which can destroy many or all of the species in an area, may reflect bad luck more than the particulars of species' biology. Background extinctions, on the other hand, tend to be a consequence of one or more features of the species' biology, such as small population size, limited habitat range, and narrow habitat tolerance.

• Most biologists believe that we are currently in the midst of a mass extinction, that it is the result of human activities, and that it poses a serious threat to the future survival of humans.

Human activities can have disruptive environmental impacts.

• An ecosystem disturbance is reversible as long as the disturbance does not include the extinction of any species, so they can reestablish their populations. An ecosystem disturbance that involves the loss of a species to extinction is irreversible.

• Exotic (or introduced) species are species intentionally or accidentally introduced into a new habitat. They are considered invasive species if they cause harm in their new habitat, and they can dominate and irreversibly alter entire ecosystems.

• Burning of fossil fuels releases gases that become acids when they combine with water vapor in the atmosphere. Precipitation containing these acids can be extremely acidic and can kill plants and aquatic animals, both directly and indirectly.

• Carbon dioxide and methane are called "greenhouse gases" because they trap heat in the atmosphere. As humans burn fossil fuels and clear forests, greenhouse gas concentrations increase and global temperatures rise. Unless there is a global reduction in greenhouse gas emissions, the ecological consequences will be significant.

• Tropical rain forests contain more species of plants and animals than all other terrestrial habitats combined. These forests, which are enormously important in limiting global warming, are being destroyed at an alarming rate.

We can develop strategies for effective conservation.

• Ozone in the stratosphere prevents ultraviolet light (UVB) from reaching the earth's surface. For decades, ozone was decreasing—due to chlorofluorocarbons (CFCs)—putting ecosystems and human health at risk. Reduction in CFC use is reducing ozone depletion, and a full recovery of the ozone layer appears possible.

• Effective conservation requires the setting of goals and priorities on the elements of biodiversity (genes, species, or ecosystems) that should be conserved. The U.S. Endangered Species Act has focused much conservation effort on the preservation of species.

• Although conservation biology has long focused on individual species, there is an increasing shift toward the preservation of important habitats. However, several methods focusing on single species remain useful.

CHECK YOUR KNOWLEDGE

Short Answer

1. Following a massive oil leak in the Gulf of Mexico, scientists observed a drop in the oxygen saturation of the water and disappearance of the leaked methane. What were they able to conclude from this observation?

2. How is biodiversity defined?

3. What makes studying biodiversity particularly difficult?

4. Describe three factors that influence the species richness in an area.

5. Describe a feature of tigers that renders their species susceptible to extinction.

6. What is the chief reason for the current loss and impending loss of so many species?

7. Why are ecosystem disturbances that involve the loss of species considered so devastating?

8. Which gases are produced when fossil fuels are burned?

9. Why are carbon dioxide and methane called "greenhouse gases"?

10. Describe two important ecological effects of deforestation.

11. Describe the significance of keystone species such as kelp.

Multiple Choice

1. Biodiversity is considered important because of:

a) its direct economic benefits to humans, such as medicines and food.

b) the symbolic value that elements of the natural world can provide.

c) its aesthetic value.

d) its potential for helping humans understand the processes that gave rise to the diversity we see.

e) All of the above are correct.

0 EASY — 40 — HARD 100

2. Biodiversity hotspots are defined by which two criteria?

a) species richness and ecosystem integrity

b) size and distance from nearest adjacent hotspot

c) species endemism and degree of threat

d) species richness and population size

e) ecological diversity and species diversity

0 EASY — 40 — HARD 100

3. Madagascar is unusually important to conservation because it:

a) has more species per unit area than any other place on earth.

b) is the fourth largest island in the world.

c) is home to more endangered species than any other country.

d) has a higher percentage of endemic plants and animals than any comparably sized area on earth.

e) is the native habitat of the rosy periwinkle, a rain forest plant that helps cure childhood lymphocytic leukemia.

0 EASY — 37 — HARD 100

4. How many species do biologists estimate to be currently existing on earth?

a) fewer than 1 million

b) between 3 million and 5 million

c) more than 1 billion

d) between 5 million and 100 million

e) between 1 million and 2 million

0 EASY — 45 — HARD 100

5. Which of the following is currently the leading cause of extinction?

a) pollution

b) habitat loss

c) disease

d) exotic species

e) overexploitation

0 EASY — 29 — HARD 100

6. Currently, the major threat to most large land mammals is:

a) invasive species.

b) loss of genetic variation.

c) overexploitation.

d) habitat loss.

e) mutation.

0 EASY — 33 — HARD 100

7. Exotic species can disrupt ecosystems because:

a) they frequently have no predators in their new habitat and can grow unchecked.

b) they are favored by ecotourists.

c) they have no natural prey and so must rely on humans for their survival.

d) they have better dispersal capability than endemic species.

e) All of the above are correct.

0 EASY — 42 — HARD 100

8. Even though there is a carbon cycle, carbon dioxide levels around the world seem to be rising. Which of the following best explains why this is so?

a) Animals give off CO_2 during their normal metabolism.

b) As the atmosphere heats up, it can contain more CO_2.

c) The destruction of coral reefs leads to increased levels of CO_2.

d) More CO_2 is being given off by ocean waters as they heat up.

e) The burning of fossil fuels releases more CO_2 into the atmosphere.

0 EASY — 19 — HARD 100

9. A charismatic species that can engender significant public support for conservation of other species and of the ecosystem they all inhabit is called:

a) an indicator species.

b) a keystone species.

c) a flagship species.

d) a phylogenetic species.

e) an endangered species.

0 EASY — 52 — HARD 100

10. In conservation biology, what does the term "corridor" refer to?

a) a section of habitat that organisms use to travel between two or more otherwise isolated patches of habitat

b) the edge of a given patch of habitat

c) the minimum range required by a keystone species in any habitat

d) the path a conservation biologist takes through a habitat to minimize any negative effects on the habitat's biodiversity

e) the ancestral geographic range of a species

0 EASY — 31 — HARD 100

Three basic tissue types give rise to diverse plant characteristics.

19.1 Older, taller, bigger: plants are extremely diverse (but share a basic structural organization).

19.2 Flowering plants are divided into two major groups: the monocots and the eudicots.

19.3 The plant body is organized into tissues, each with specific functions.

Most plants have common structural features.

19.4 Roots anchor the plant and take up water and minerals.

19.5 Stems are the backbone of the plant.

19.6 Leaves feed the plant.

19.7 Several structures help plants resist water loss.

Plant nutrition: plants obtain sunlight and usable chemical elements from the environment.

19.8 Four factors are necessary for plant growth.

19.9 Nutrients cycle from soil to organisms and back again.

19.10 Plants acquire essential nitrogen with the help of bacteria.

19.11 *This is how we do it:* Carnivorous plants can consume prey *and* do photosynthesis. Why are they confined to bogs and other nutrient-poor habitats?

Plants transport water, sugar, and minerals through vascular tissue.

19.12 Plants take up water and minerals through their roots.

19.13 Water and minerals are distributed through the xylem.

19.14 Sugar and other nutrients are distributed through the phloem.

A red maple tree (*Acer rubrum*), the most abundant native tree in eastern North America.

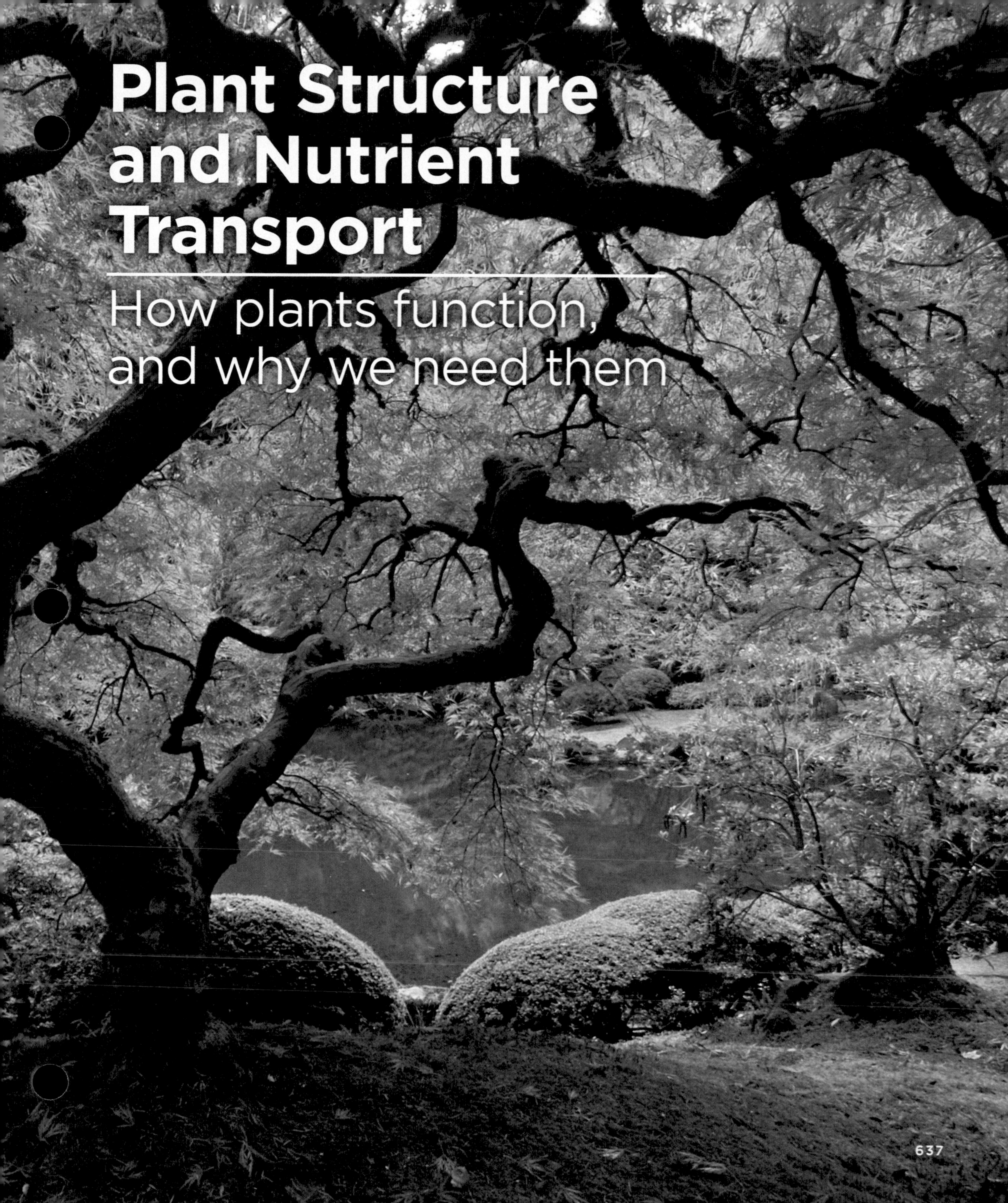

Plant Structure and Nutrient Transport

How plants function, and why we need them

19.1–19.3

Three basic tissue types give rise to diverse plant characteristics.

A yellow calla lily, prepared to reveal all of its structures.

19.1 Older, taller, bigger: plants are extremely diverse (but share a basic structural organization).

Plants are better than animals. Actually, that's an absurd thing to say. But only because biology isn't a contest, and from the perspective of evolution, any species that is not extinct can be considered a success. By many measures, however, plants have developed a stunning diversity of structures and features during their evolution.

Consider some vital statistics on plants and animals (FIGURE 19-1).

Q Plants versus animals: Which can live longer? Grow larger?

1. **Longevity.** The longest-living plants live 10 times longer than the longest-living animals. Bristlecone pines, found in the western United States, can live for more than 4,800 years. Among animals, the ocean quahog (a clam-like mollusk) can live about 200 years, and giant tortoises can live almost 200 years.
2. **Height.** The tallest plants are 20 times taller than the tallest animals. Coast redwoods, found along the west coast of North America, have grown to more than 379 feet (115 m). The tallest giraffe grew to just under 20 feet (6 m).
3. **Weight.** The heaviest plants weigh almost 20 times as much as the heaviest animals. The largest giant sequoias are estimated to weigh more than 3,300 tons (almost 3 million kg). The largest blue whales—probably the largest animals ever to inhabit the earth—weigh in at approximately 190 tons (154,000 kg).
4. **Energy acquisition.** Most important is that plants can make their own food by capturing energy from the sun and converting it into a usable chemical form, but animals cannot. Consequently, animals rely on producers such as plants for energy. This is why the total amount of plant matter on earth is more than 10 times the total biomass of all the animals.

Plants have a variety of strategies for growth, competition, defense, and reproduction, all vastly different from strategies that have evolved in animals. The difference in the strategies used by plants and animals illustrates a fundamental truth of biology: there are multiple pathways to evolutionary success.

In this chapter, we explore the structural and physiological features of plants and how they represent adaptive

EXTRAORDINARY ATTRIBUTES OF PLANTS

LONGEVITY
The longest-living plants live 10 times longer than the longest-living animals. Bristlecone pines have been documented to live for more than 4,800 years.

HEIGHT
The tallest plants are 20 times taller than the tallest animals. Coast redwoods have grown to be more than 379 feet.

WEIGHT
The heaviest plants weigh almost 20 times as much as the heaviest animals. The largest giant sequoias are estimated to weigh more than 3,300 tons.

ENERGY
Plants can make their own food by capturing energy from the sun and converting it into a usable chemical form, but animals cannot.

FIGURE 19-1 **In many important attributes, plants surpass animals.**

VASCULAR PLANT STRUCTURE

LEAVES
- The primary site of photosynthesis, the conversion of energy from the sun into food for the plant

STEMS
- Provide structural support for the plant
- Position leaves so that they can be exposed to sunlight for photosynthesis
- Conduct food, water, and nutrients throughout the plant

ROOTS
- Absorb water and minerals from the soil
- Anchor the plant in place

Shoot system

Root system

FIGURE 19-2 **Three distinct plant structures: roots, stems, and leaves.**

solutions to the constraints imposed by their environments. We begin by discussing the three main vegetative structures in vascular plants (FIGURE 19-2) and how their vascular (or "circulatory") system works.

TAKE HOME MESSAGE 19.1

» Plants are an extremely diverse and successful group of organisms, with distinct strategies for growth, competition, defense, and reproduction. Plants are generally composed of three distinct parts: roots, stems, and leaves.

19.2 Flowering plants are divided into two major groups: the monocots and the eudicots.

Based on several prominent structural features, the flowering plants are classified into two major groups. The groups are named according to the structure of the plant embryo within the seed, called a **cotyledon** (pronounced COT-uh-LEE-din), which usually becomes a plant's first embryonic leaf or leaves. Plants in which one cotyledon forms are called **monocots,** and those in which two cotyledons form are called dicots.

Unlike the monocots, the dicots are not a monophyletic group and include many different lineages. A large subset of the dicots, however, called the **eudicots** (you-DIE-cots), is a monophyletic group.

In addition to having seeds with only one cotyledon, the monocots (about 70,000 species) differ from the eudicots (about 200,000 species) with respect to four other common structural features (FIGURE 19-3).

1. *Leaves.* Monocots generally have parallel veins in their leaves, while eudicots have branching veins.

2. *Stems.* Within the stems of monocots, the vascular tissue—the tissue responsible for nutrient transport—is arranged as numerous, randomly scattered bundles. You can see the scattered vascular bundles if you cut through an asparagus spear (FIGURE 19-4). In eudicots, such as a sunflower, the vascular tissue is arranged in an orderly ring.

3. *Flowers.* The flower parts (such as petals) of monocots typically occur in multiples of three, while in eudicots they occur in fours or fives.

4. *Roots.* Monocots typically have many fibrous roots branching from the stem. Eudicots usually have a taproot, a single primary root with smaller roots branching from it.

Some common monocots are palm trees, orchids, lilies, and all of the grasses, including most grains used in food

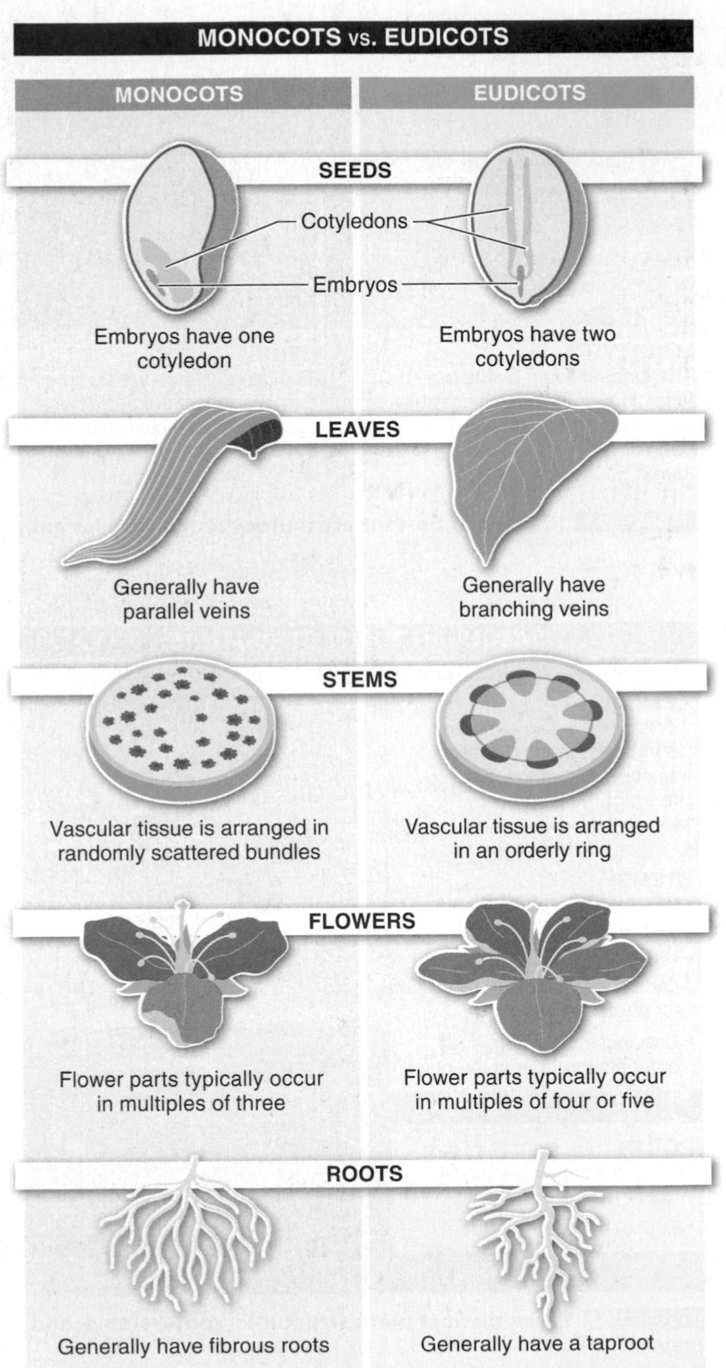

FIGURE 19-3 Monocots and eudicots: two major groups of flowering plants that differ in several structural features.

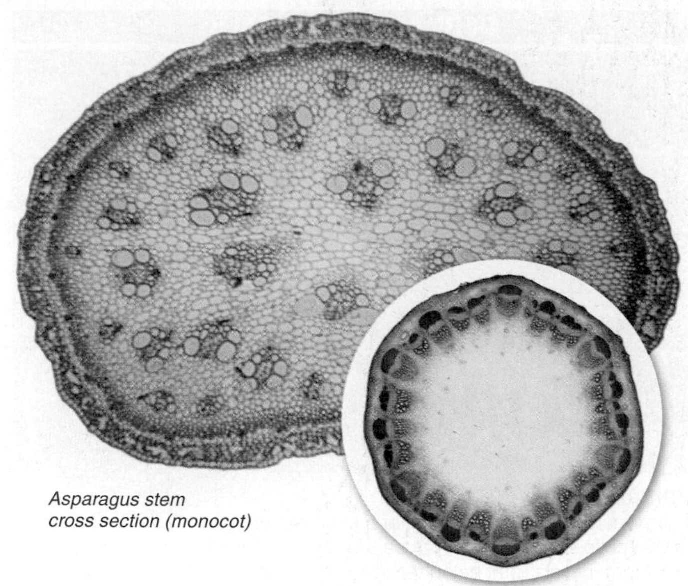

FIGURE 19-4 **Vascular tissues of monocot and eudicot stems.** By cutting through an asparagus spear you can see the vascular tissue (shown here in purple), which is arranged in bundles and scattered throughout the stem.

products. Common eudicots include roses, daisies, coffee, potatoes, apples, strawberries, and most large trees, including maples and oaks.

TAKE HOME MESSAGE 19.2

» The flowering plants are divided into two major groups, the monocots and the eudicots, based on structural features of their seeds, leaves, stems, flowers, and roots.

19.3 The plant body is organized into tissues, each with specific functions.

A rosebush, a cactus, and a redwood tree are recognizable plants. All are rooted in the ground and have green parts. But beyond these basic characteristics, they don't look much like one another. If we examined their internal structure, however, we would find great similarities. These similarities are common to all the vascular plants. Recall from Chapter 14 that the vast majority of all plant species are vascular, having a sort of circulatory system for moving water and nutrients into and throughout the organism.

In this chapter, we focus on vascular plants. All vascular plants are organized around the same basic body plan and built up from the same three types of tissues. Leaves, stems, roots, and flowers—all consist of these tissue types (FIGURE 19-5).

1. **Dermal tissue** covers and protects the surface of the plant body, much like skin covers the human body.
2. **Vascular tissue** transports water and nutrients throughout the plant body, acting much like the circulatory system in vertebrates. (It's important to note, however, that plants have no heart or pump of any kind.)
3. **Ground tissue,** which makes up the bulk of the plant, is where most of the plant's metabolic activities are carried out.

Each of the three tissue types is composed of one or more different types of cells. Recall from Chapter 4 that plant cells commonly differ from other eukaryotic cells: they may contain chloroplasts, they may contain a large vacuole for storage, and they have a cellulose-containing cell wall surrounding the cell membrane.

Dermal Tissue Like an animal's skin, a plant's dermal tissue covers and protects the entire plant body (FIGURE 19-6). Dermal tissue usually consists of a single layer of tightly packed, thin cells, called the **epidermis.** The epidermis produces a waxy covering, called the **cuticle,** that helps reduce water loss due to evaporation. The cuticle also offers some protection from pathogens and from microorganisms that might attempt to eat the plant. The epidermal cells of roots, however, do not secrete a cuticle, which would reduce the root cell's ability to absorb water.

Specialized cells in the epidermis, called **guard cells,** help regulate the gases coming into and out of leaves.

TYPES OF TISSUE IN VASCULAR PLANTS

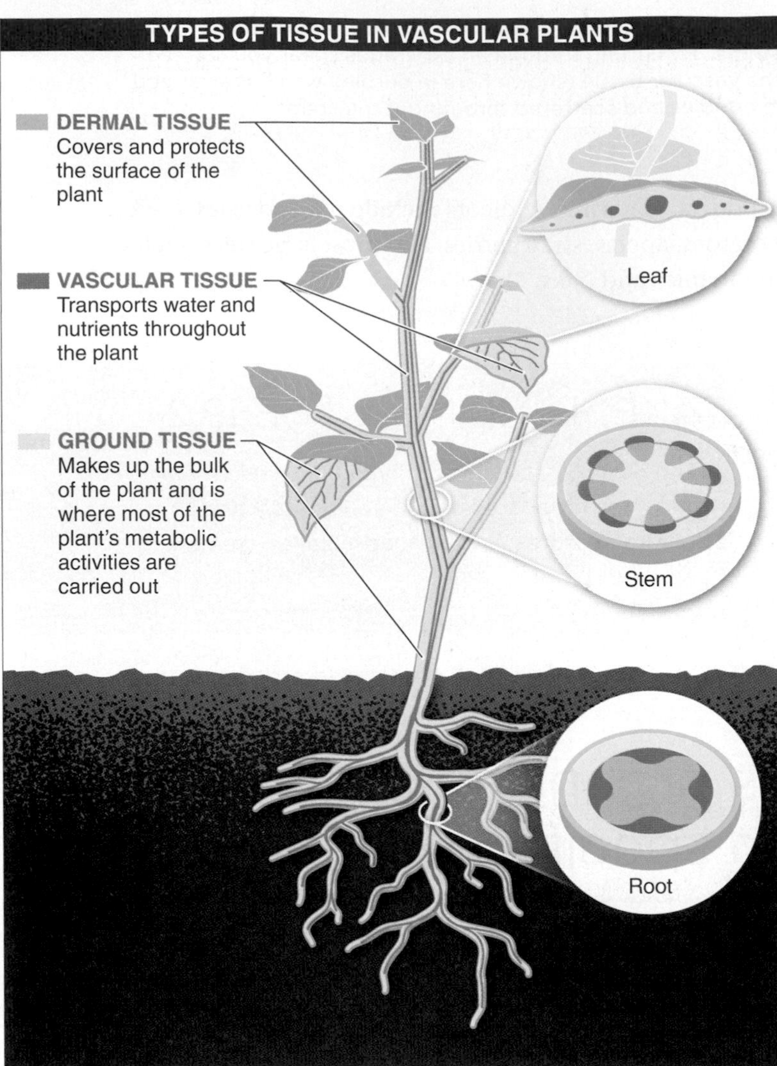
DERMAL TISSUE
Covers and protects the surface of the plant
VASCULAR TISSUE
Transports water and nutrients throughout the plant
GROUND TISSUE
Makes up the bulk of the plant and is where most of the plant's metabolic activities are carried out
Leaf
Stem
Root

FIGURE 19-5 **Basic tissue types found in all vascular plants.** Plants come in a huge variety of shapes and sizes, but inside, all are composed of the same three types of tissue.

DERMAL TISSUE CELLS

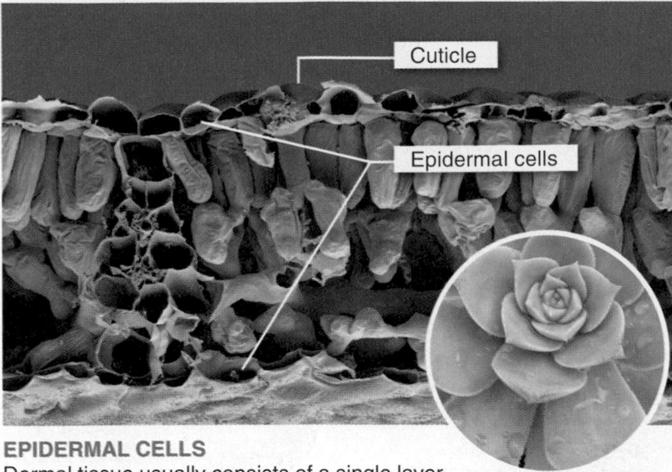
Cuticle
Epidermal cells

EPIDERMAL CELLS
Dermal tissue usually consists of a single layer of thin cells. These cells produce a waxy substance, called the cuticle, that helps reduce water loss.

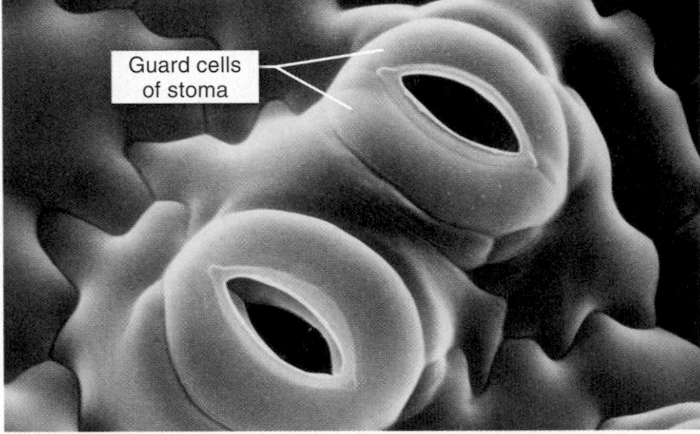
Guard cells of stoma

GUARD CELLS
Guard cells are specialized epidermal cells that create openings for gas exchange during photosynthesis. They can also close the openings, reducing water loss by evaporation.

CORK CELLS
As a plant grows, cork cells replace epidermal cells, providing a thicker and more protective covering. On trees, this protective covering forms the outer layer of bark.

FIGURE 19-6 **Dermal tissue protects the entire plant.**

Two guard cells surround a pore in the leaf, and by changing their shape the guard cells create an opening through which gases (including carbon dioxide and oxygen) can be exchanged during photosynthesis. The guard cells can also change shape to close the opening, sealing off the inside tissues and minimizing water loss by evaporation (see Section 5.11). The two guard cells and pore are called a **stoma,** from the Greek word for "mouth."

In many vascular plants, the epidermis covering the stems and roots is replaced by a thicker, more protective dermal covering as the plant grows. This covering consists mostly of dead **cork cells.** Cork cells contain a waxy, fatty substance that makes the tissue impermeable to water and resistant to fire and decay. Small pores interspersed among the cork cells allow the gas exchange essential for living cells. On trees, this covering forms the outer layer of bark and protects the inner, living tissue from risks such as fire and predation. The covering also allows the entry of oxygen for cellular respiration in the interior tissues and the release of carbon dioxide.

Vascular Tissue Because plants can grow to be very large, they must have a system for delivering nutrients to and removing waste products from all parts of the plant. These functions are carried out by vascular tissue, which plays a role analogous to that of the circulatory system in animals (FIGURE 19-7). Most sugar, for example, is produced in the leaves, while most water is absorbed through the roots. Because all plant cells require both water and sugar to survive, the physical separation of the leaves and roots is overcome by the vascular tissue that links them.

Vascular plants possess two parallel transport systems, made of two distinct vascular tissues—xylem and phloem—that transport a fluid called **sap.** Unlike animal circulatory systems, plant systems simply transport fluids from one place to another anywhere in the plant, rather than circulating them in one direction. **Xylem** (pronounced ZY-luhm) conducts xylem sap, containing water and dissolved minerals, from the roots to the rest of the plant body. **Phloem** (FLOW-uhm) conducts phloem sap, consisting mostly of water but also containing sugar—the plant's fuel source—and some minerals, to the tissues needing fuel for their activities.

Q Cells that have already died are among the most important parts of a plant. Why?

Plants build some structures in ways not found in animals. For example, the tubes of the xylem are made from cells that are *dead*. The cells grow close together and, after their death, become connected, end to end, to form the outer surface of a pipeline. Water flows through these many parallel pipelines and reaches all the living cells in the plant through tiny holes in the cell walls, in much the same way that a gardener's soaker hose, with hundreds of small holes, waters a garden. The rigid cell walls that make up the pipelines also give the plant structural support.

The phloem transport system is built from living cells that function as another sort of pipe. These cells are arranged end to end to form pipe-like **sieve tubes.** Unlike pipes, however, sieve tube cells have openings in their side walls through which sugar passes to cells outside the phloem.

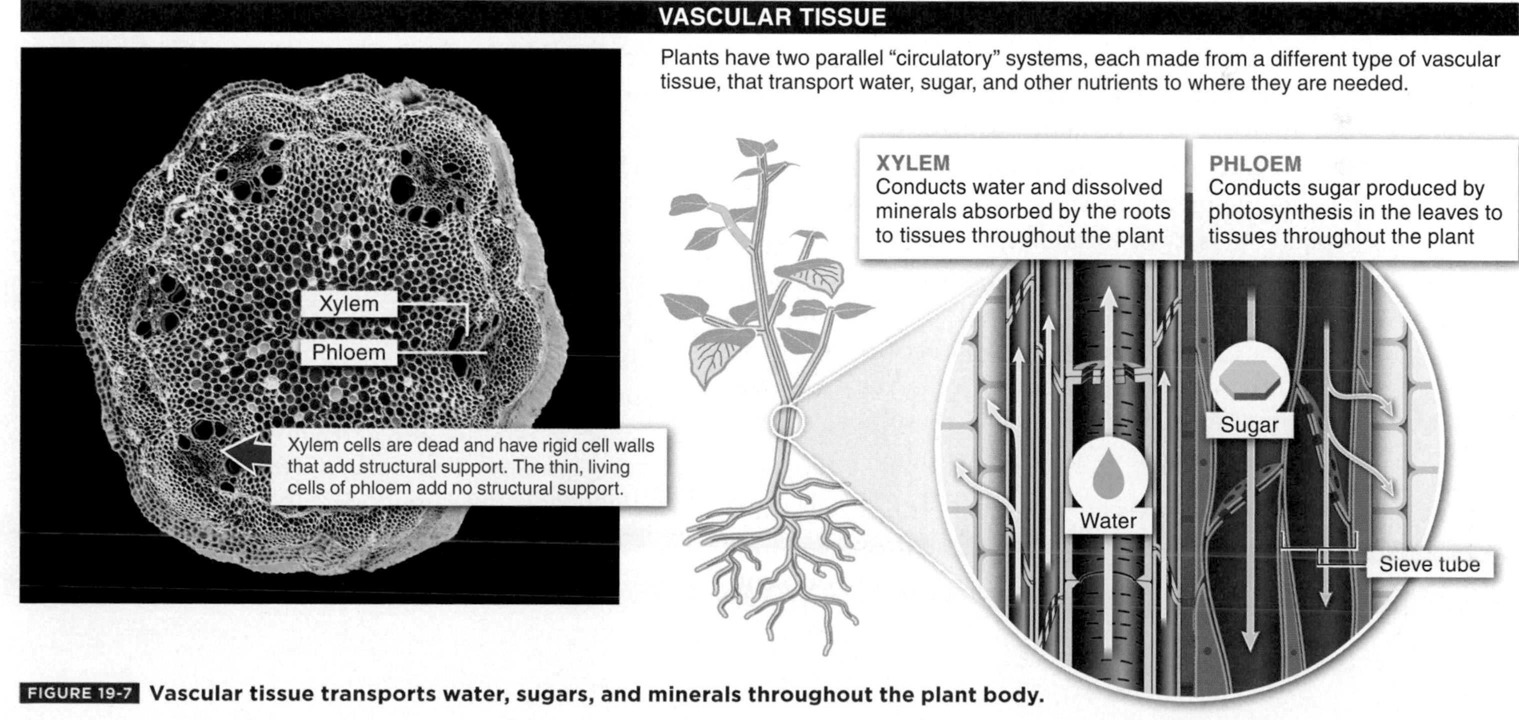

VASCULAR TISSUE
Plants have two parallel "circulatory" systems, each made from a different type of vascular tissue, that transport water, sugar, and other nutrients to where they are needed.
Xylem
Phloem
Xylem cells are dead and have rigid cell walls that add structural support. The thin, living cells of phloem add no structural support.
XYLEM
Conducts water and dissolved minerals absorbed by the roots to tissues throughout the plant
PHLOEM
Conducts sugar produced by photosynthesis in the leaves to tissues throughout the plant
Sugar
Water
Sieve tube

FIGURE 19-7 **Vascular tissue transports water, sugars, and minerals throughout the plant body.**

GROUND TISSUE CELLS

Parenchyma cells of an apple

PARENCHYMA CELLS

- Make up the majority of plant tissue, including most of the soft, flexible tissue found in leaves, flowers, stems, roots, and fruits
- Responsible for photosynthesis, food storage, and the production and release of hormones

Collenchyma cells of a sunflower

COLLENCHYMA CELLS

- Elongated, stringy cells with thickened cell walls
- Give the plant flexibility, enabling it to twist and bend

Sclerenchyma cells of a balsa tree

SCLERENCHYMA CELLS

- Have very thick cell walls containing lignin, an important component of wood
- Function like the steel girders of a building, enabling plants to resist the force of gravity and grow tall
- Not living when mature

FIGURE 19-8 **Ground tissue makes up most of the plant body.**

Because phloem cells have much thinner cell walls than xylem, they don't offer much structural support to the plant.

Ground Tissue The third type of tissue found in plants is the ground tissue. Because ground tissue includes everything that is neither the outer covering (dermal tissue) nor the inner vascular tissue (xylem and phloem), it makes up most of the plant body. There are three different types of ground tissue (FIGURE 19-8).

1. **Parenchyma cells** make up most plant tissue, including most of the soft, flexible tissue found in leaves, flowers, stems, roots, and fruits. Parenchyma cells are the metabolic workhorses of the plant body. Depending on their location, parenchyma cells have the capacity to photosynthesize, store food molecules, produce ATP through cellular respiration, or produce and release hormones. When you eat a carrot or a potato, the carbohydrate-packed cells you consume are primarily parenchyma cells that were storing energy for the plant to use at a later time.
2. **Collenchyma cells** are elongated, stringy cells with thickened cell walls. This might at first seem like sloppy cell construction, but it actually gives a plant great flexibility, enabling it to twist and bend. An example of collenchyma tissue is the strings on the outside of a celery stalk.
3. Unlike the other types of ground tissue, **sclerenchyma cells** are not living when they are mature. Sclerenchyma cells have thick cell walls containing **lignin,** a substance that is one of the chief chemical components of wood and that makes the cell walls in woody plants indigestible to nearly all organisms. Sclerenchyma cells are strong and can function like steel girders in a highrise, enabling plants to resist the force of gravity and grow very tall. Humans use fibrous sclerenchyma cells to make cloth, rope, and paper.

TAKE HOME MESSAGE 19.3

» All vascular plants have a similar body plan and are made from the same three types of tissue. Dermal tissue covers and protects the surface of the plant. Vascular tissue transports water, sugars, and minerals throughout the plant body. Ground tissue, which makes up the bulk of the plant body, is where most of the plant's metabolic activities take place.

19.4–19.7

Most plants have common structural features.

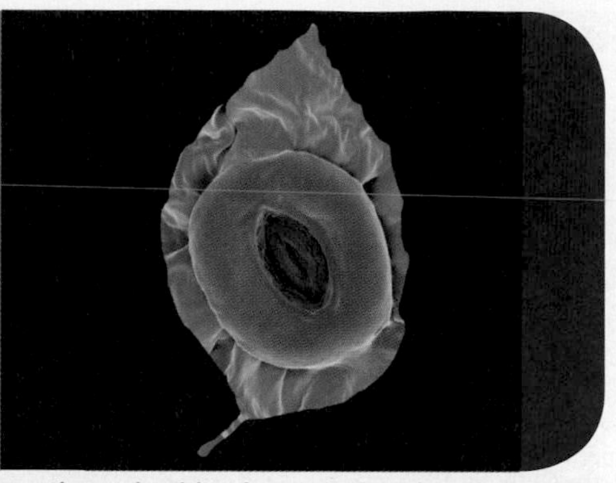

Take a deep breath: a stoma—where carbon dioxide is taken in and water is lost—from the underside of an American beech tree leaf.

19.4 Roots anchor the plant and take up water and minerals.

Roots don't garner the attention enjoyed by flowers, fruits, and tree trunks. But they are essential to plant growth and survival, and they perform three primary functions (FIGURE 19-9).

1. *Absorption.* Roots take up water and dissolved minerals, as well as oxygen, from the soil. In deserts, mesquite plants send roots straight down as deep as 175 feet to reach water, while some cacti send roots outward 50 feet in all directions to collect moisture.
2. *Anchorage.* Roots secure the plant in place. Also, without this anchorage, plants would not be able to stand upright.
3. *Storage.* Sugar produced in the leaves can be sent to fleshy roots, where it is converted to starch and packed away for later use, much like humans store energy as lipids in fat cells. Sweet potatoes, carrots, radishes, and turnips are examples of starch-filled roots. Water, too, is sometimes stored in roots.

THE PRIMARY FUNCTIONS OF ROOTS

ABSORPTION
Roots take up water, minerals, and oxygen from the soil for use by the plant.

ANCHORAGE
Roots secure the plant in place.

STORAGE
Sugar produced by photosynthesis and converted to starch, as well as water absorbed by roots, can be packed away in the roots for later use.

FIGURE 19-9 **Mostly unseen, but essential.** Roots serve functions that are necessary for growth and survival.

> "If the roots are not severed, all is well—
> and all will be well—in the garden."
>
> —JERZY KOSINSKI
> *Being There* (1971)

There are two types of root systems: taproots and fibrous roots (FIGURE 19-10). **Taproots** typically grow deep down into the soil. A taproot is a thick primary root with many smaller roots branching out from it. Found primarily in eudicots, taproot systems can be thought of as more expensive to produce (energetically) but providing greater storage and depth of penetration. In residential areas, taproots can clog sewer pipes or burrow into house foundations, causing damage.

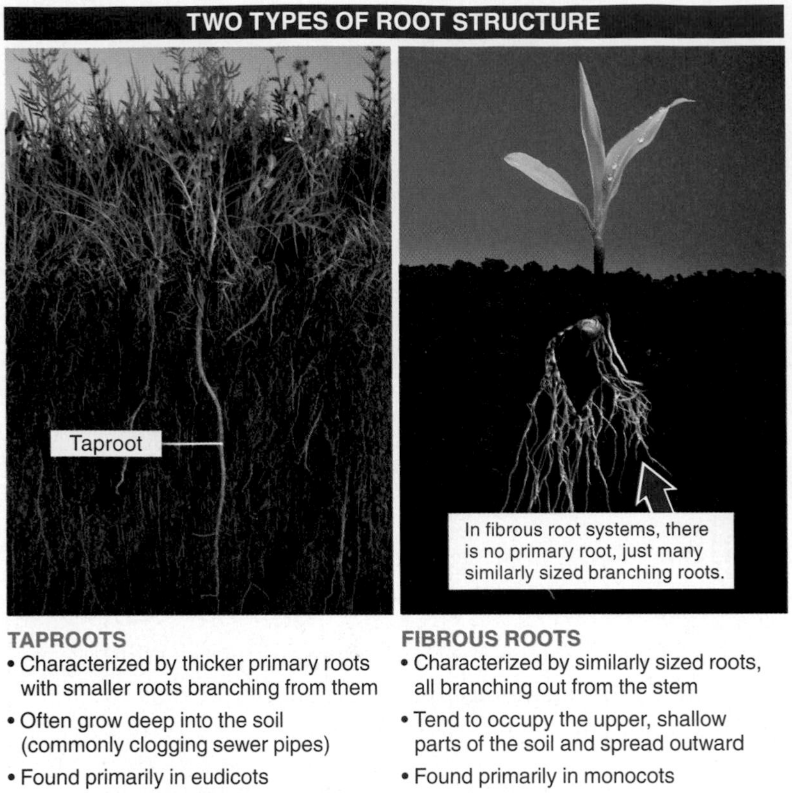

TWO TYPES OF ROOT STRUCTURE

Taproot

In fibrous root systems, there is no primary root, just many similarly sized branching roots.

TAPROOTS
- Characterized by thicker primary roots with smaller roots branching from them
- Often grow deep into the soil (commonly clogging sewer pipes)
- Found primarily in eudicots

FIBROUS ROOTS
- Characterized by similarly sized roots, all branching out from the stem
- Tend to occupy the upper, shallow parts of the soil and spread outward
- Found primarily in monocots

FIGURE 19-10 Taproots and fibrous roots.

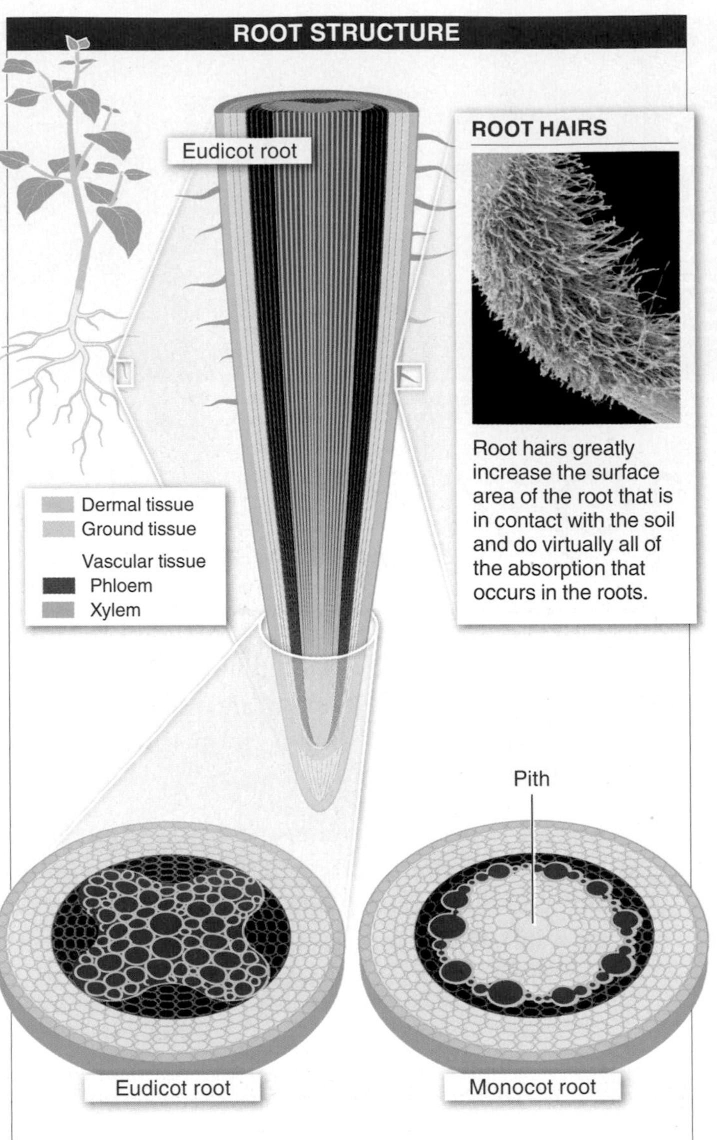
ROOT STRUCTURE
Eudicot root
ROOT HAIRS
Root hairs greatly increase the surface area of the root that is in contact with the soil and do virtually all of the absorption that occurs in the roots.
Dermal tissue
Ground tissue
Vascular tissue
Phloem
Xylem
Pith
Eudicot root
Monocot root

FIGURE 19-11 Organization of tissues within the root. In eudicots, the vascular tissue of the stem is arranged in a ring. In monocots, vascular tissue is dispersed throughout the ground tissue in the stem.

Fibrous roots are more commonly associated with monocots (but also found in some eudicots). They consist of numerous, similarly sized roots that rise and branch from the stem. They tend to occupy the upper, shallow parts of the soil and spread outward. Fibrous root systems are energetically less expensive to produce and allow greater spread and faster growth.

The organization of all root tissue is simple, and the three tissue types are revealed in a cross section (FIGURE 19-11). In most eudicots, the vascular tissue (xylem and phloem) is a central column, surrounded by ground tissue, and at the outer edge is dermal tissue. The xylem is in the center of the root, with several evenly spaced bulges swelling outward. Nestled between the bulges are phloem strands. Most monocot roots have a different organization, with a central core of soft, spongy parenchyma tissue, called **pith,** surrounded by a ring of vascular tissue.

Q How are sweet potatoes like the "love handles" on a person's waist?

When the ground tissue in a root is used to store energy in the form of starch, the root becomes enlarged and fleshy. Plants store excess energy for the times when rapid growth or flower production demands a lot of energy. Animals are often tempted to find such storehouses and eat them. Humans, for instance, like to dig up and eat large roots such as turnips, carrots, and sweet potatoes.

The outer surface of the root is epidermis, including epidermal cells that elongate and become tiny hairs. These hairs greatly increase the surface area of the root that is in contact with the soil and accomplish most absorption that occurs in the roots. The porous epidermal cells easily absorb water, which then moves through and between the cells of the root's ground tissue until it reaches the xylem, from where it is distributed throughout the plant.

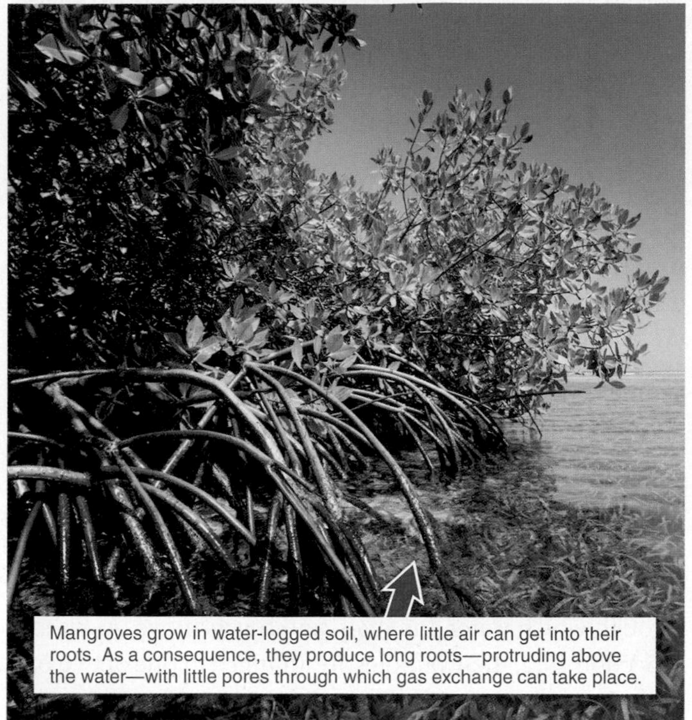

FIGURE 19-12 **"Snorkel roots."**

The spaces between root cells allow oxygen to diffuse in, enabling the root's cells to respire (make ATP and exchange gases with the environment). Mangroves grow in water-logged soil, so very little air can get into the submerged roots directly from outside. Consequently, mangroves produce odd-looking long roots—with numerous little pores in them—that protrude above the water (FIGURE 19-12). The pores open into loosely packed cells, creating a column of air that extends down to the underwater roots.

TAKE HOME MESSAGE 19.4

» Roots have three primary functions in plants: (1) absorption, the uptake of water and dissolved minerals from soil; (2) anchorage, securing the plant in place; and (3) storage of water and excess starch for future use.

19.5 Stems are the backbone of the plant.

When an organism must reach toward the sun for the energy to make its own food, gravity can be the enemy. Plants compete with each other, and taller plants can leave shorter plants in the shade. The plant body is one continuous structure, with the root system below ground and a portion called the shoot system above ground. The **shoot** is divided into two parts: stems and leaves. In some species, the shoot system also includes flowers and fruits, which we discuss in Chapter 20.

Serving as the backbone of the plant, the **stem** has three main functions. First, it provides structural support for the plant. Second, in conjunction with the roots, the stem enables the plant to grow and orient its leaves toward the sun. The stem's third function is housing the plant's transport infrastructure—the xylem and the phloem—through which water, sugars, and other nutrients are distributed throughout the plant (FIGURE 19-13).

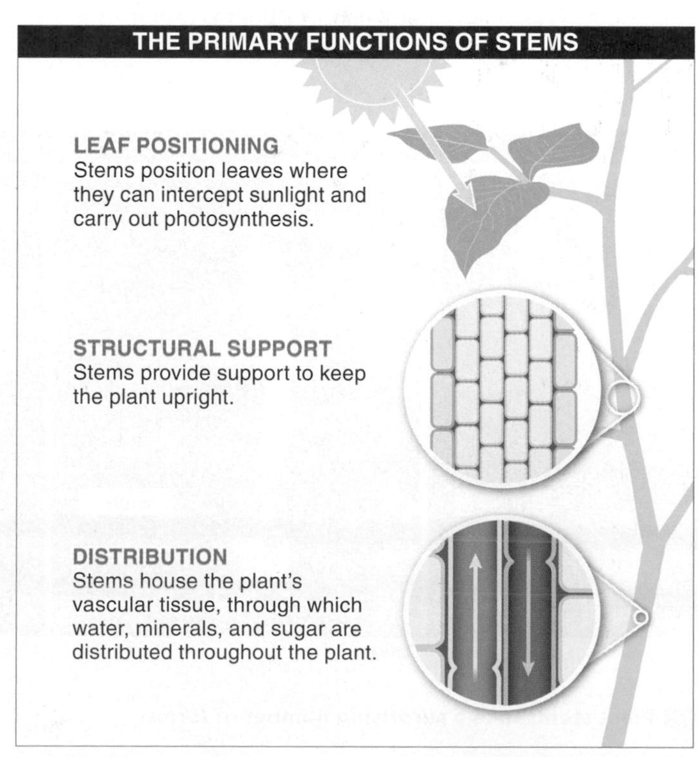

THE PRIMARY FUNCTIONS OF STEMS
LEAF POSITIONING
Stems position leaves where they can intercept sunlight and carry out photosynthesis.
STRUCTURAL SUPPORT
Stems provide support to keep the plant upright.
DISTRIBUTION
Stems house the plant's vascular tissue, through which water, minerals, and sugar are distributed throughout the plant.

FIGURE 19-13 **Stems provide support and the transport infrastructure needed for the distribution of water, sugars, and other nutrients.**

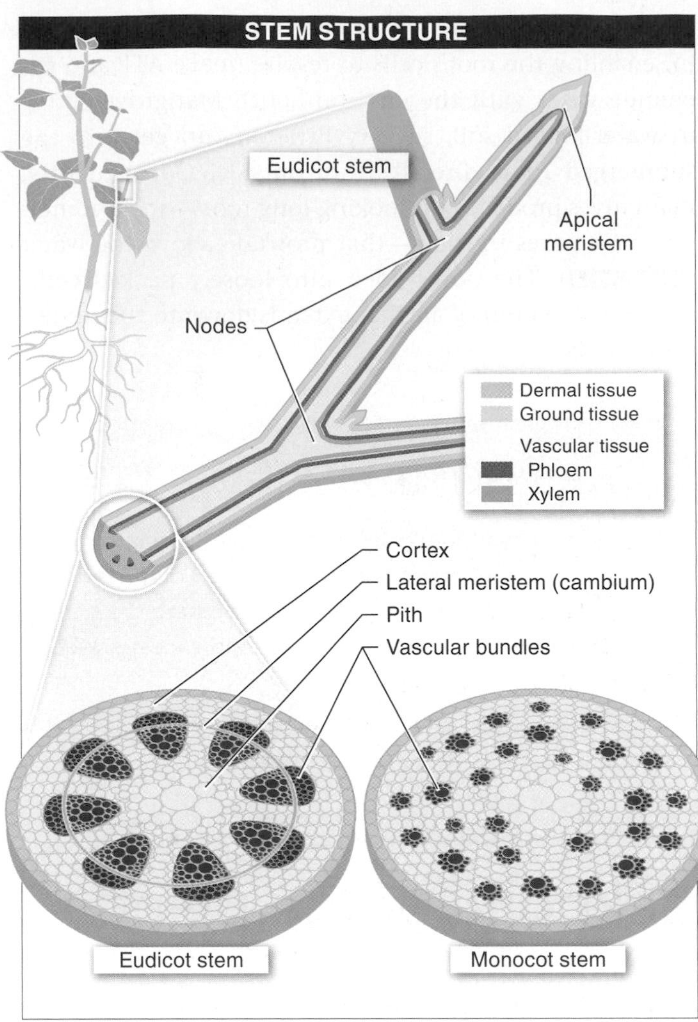

FIGURE 19-14 **Organization of tissues within the stem.**

In plants, growth occurs in regions called **meristems,** made up of undifferentiated cells. These cells divide, often at very high rates, generating additional meristem cells along with cells that differentiate into epidermis, vascular tissue, or ground tissue. At the tip of a root or stem, the meristem is called an **apical meristem;** division of cells there causes the root or stem to increase in length. **Lateral meristems** consist of a layer of cells, called cambium, that divide to make a stem or root thicker, as opposed to longer. We discuss plant growth in more detail in Chapter 20.

Stem growth and root growth are similar, except that stem growth is segmented (FIGURE 19-14). That is, the stem grows a bit and then produces **nodes,** little lumps of tissue from which leaves, flowers, cones, or additional stems (or a branch) may grow.

The stems of flowering plants contain vascular bundles—through which water, minerals, and sugars move. As we saw in Section 19.2, these bundles are arranged in one of two patterns. In eudicots, a ring of vascular bundles is arranged near the outside of the stem. A cross section of the stem reveals a ring of bundles, like a clock face. Where each of the numbers might be, there is a double pipeline. The outer pipe is the phloem and the inner pipe is the xylem. This ring of vascular bundles divides the stem's ground tissue into two areas: one inside and one outside the ring. The inner tissue is the pith and the outer tissue is the **cortex**. In monocots, the vascular bundles are dispersed throughout the stem. Each node of the stem may be the site of a leaf or a branch; if so, some vascular bundles turn outward and enter the leaf or branch.

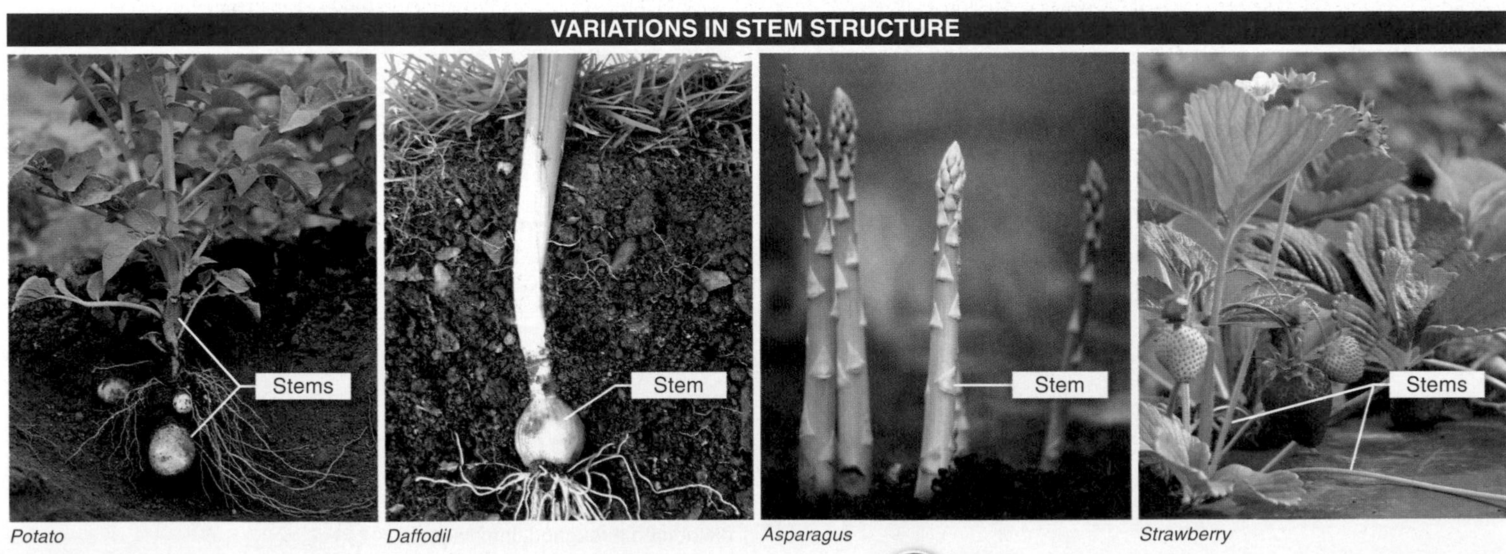

FIGURE 19-15 **Plant stems take a surprising number of forms.**

Stems can become modified—sometimes into structures almost unrecognizable as stems.

Although the stems of all flowering plants are variations on the two patterns described here, they vary tremendously (FIGURE 19-15). Most bulbs that gardeners plant each year, such as those for daffodils and lilies, are stems. Many modified leaves are attached to the bulb stems, forming the bulk of the bulb, and, after planting, roots emerge from the bottom.

Many foods come from modified stem tissue. Asparagus spears, for example, are fleshy stems. The scales along the spear are tiny leaves that haven't branched far from the stem. The horizontal runners of a strawberry plant are stems, from which roots emerge at intervals to anchor the plant to the ground. And, although they grow underground, potatoes, too, are highly modified stems—not roots. When potatoes sprout, the "eyes" contain lateral buds that branch out. Similar-appearing foods that are true roots, such as sweet potatoes, never form such "eyes."

As stems grow and branch, they increase in girth and length, becoming stronger and better able to position leaves for exposure to the sun for photosynthesis. In the next chapter, we investigate the process by which stems develop into tree trunks and, in doing so, produce wood—perhaps, for humans, the most important non-food plant product on earth.

TAKE HOME MESSAGE 19.5

» Stems serve three chief functions in plants: (1) they put leaves in positions where they can intercept sunlight and carry out photosynthesis; (2) they provide structural support for the plant; and (3) they contain the xylem and phloem, through which water, minerals, and sugars are distributed throughout the plant body.

19.6 Leaves feed the plant.

A recurring theme when comparing plants and animals is that plants must acquire energy and nutrients without being able to move. **Leaves** solve this problem handily. They have one primary purpose—producing food for the plant—and a dazzling array of forms has evolved that allow them to accomplish this purpose. Leaves intercept sunlight and, by putting chloroplast-containing photosynthetic cells in the path of that light, convert the potential energy of the sun into the usable chemical energy of sugars.

Cells inside the leaf are organized for capturing solar energy to make sugar and transporting that food out of the leaf to the rest of the plant. We describe leaf structure starting at the surface and moving inward (FIGURE 19-16).

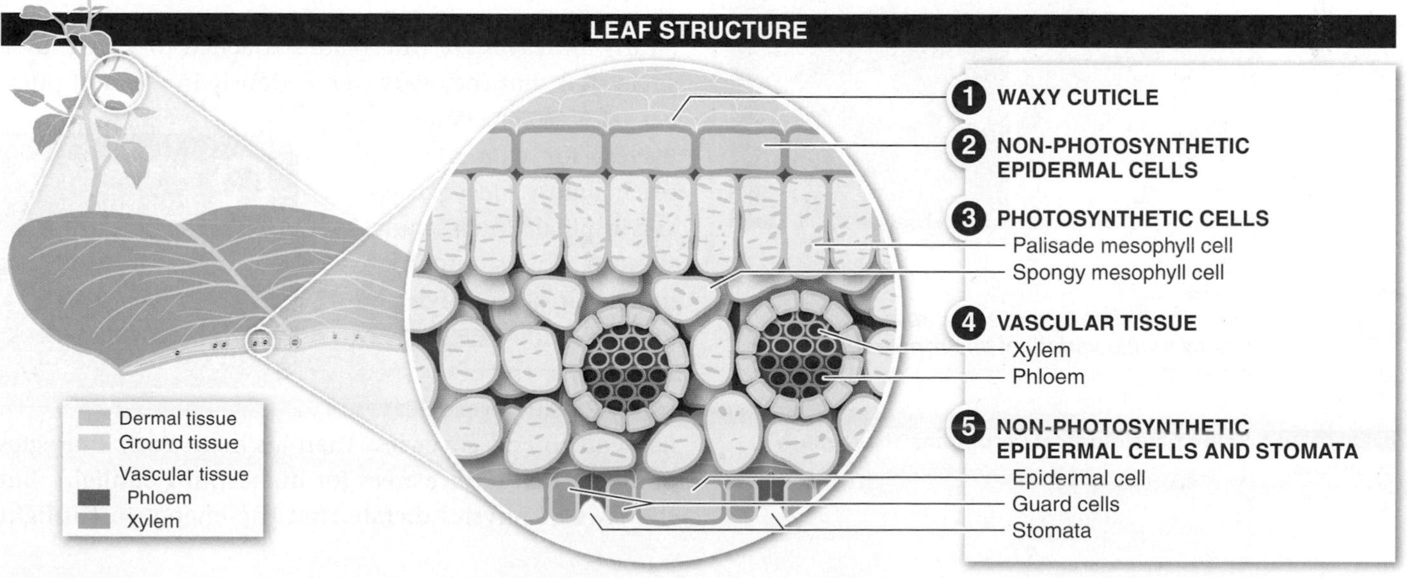

FIGURE 19-16 **A five-part "photosynthetic-cell sandwich."** In a leaf cross section we see five layers of structure, each with a distinct function.

VARIATIONS IN LEAF STRUCTURE

CONIFER LEAVES

Serbian spruce

Monkey puzzle tree

MONOCOT LEAVES

Hosta

Bamboo

EUDICOT LEAVES

Swiss chard

Pincushion cactus

Leaves come in all shapes and sizes, revealing adaptations to a wide variety of environments.

FIGURE 19-17 Some physiological "problems" have many solutions.

1. A waxy cuticle on the upper surface protects the leaf from damage and reduces the rate at which water evaporates in the hot sun.
2. The epidermis—cells forming the upper surface—is non-photosynthetic and protective. These cells secrete the cuticle.
3. Below the epidermis are a few layers of photosynthetic cells. The cells in the first layer, called the palisade mesophyll, are closely packed parenchyma cells, rich in chloroplasts. Rates of photosynthesis are highest in this part of the leaf. In the next layer or two, parenchyma cells are irregularly shaped and more loosely arranged, forming numerous air pockets that make the layers spongy—hence this portion of the leaf is called spongy mesophyll. Most cells in the spongy layer are surrounded by air pockets, enhancing the diffusion of gases, including the CO_2 essential to photosynthesis, from the air and into the plant's photosynthetic cells.
4. Veins—composed of xylem and phloem—run through the middle of the leaf. Much like blood vessels in humans, they reach out to all the cells to deliver water and essential minerals and, in plants, haul away the sugar products of photosynthesis for use elsewhere in the plant body.
5. The bottom layer is composed of non-photosynthetic epidermal cells. Interspersed among these cells are **stomata** (see Section 19.3), the pore-like openings through which the plant takes in CO_2 and releases O_2, one of the by-products of photosynthesis.

Nearly all leaves are thin blades attached to a stem by a small stalk, but they vary tremendously in size and number (FIGURE 19-17). Water lily leaves, for example, can reach almost 7 feet across—and inspired Monet's mural-sized paintings, which are more than 6 feet high and 42 feet wide (FIGURE 19-18). At the other end of the spectrum, 25 duckweed leaves laid end to end cover less than an inch. Both strategies produce large surface areas for intercepting sunlight. But the laws of physics dictate that the energy in sunlight

Q Leaves come in a huge number of shapes, but nearly all leaves are very, very thin. Why?

Monet was inspired by the large leaves of water lilies and painted them on a huge canvas.

FIGURE 19-18 **Water lilies produce some of the largest leaves of any plants.**

can't pass through very many layers of cells without being absorbed or reflected. For this reason, leaves are always thin.

Among the flowering plants, many monocots can be identified by their long, flat leaves that are oriented vertically, as in the grasses, in a pattern that keeps them from shading each other. Eudicot leaves come in two types: simple leaves, each with just a single undivided blade, and compound leaves, having several leaflets attached to a central support structure. Each of these two types can vary greatly (see Figure 19-17).

Perhaps the most extremely modified plant leaves are the hard, dry spines of most cacti. The spines don't contribute to photosynthesis—that occurs in the stems, which we see as the prominent green parts of the plant. Instead, the spines help adapt cacti for desert living. They slow the speed of warm winds passing over the plant, reducing evaporation. They help moisture from the cooler night air condense, enabling the plant to capture a bit of water. Spines also discourage birds and mammals from eating the plant. In Chapter 20, we explore some of the chemicals that plants sequester in their leaves to further protect them from being eaten.

TAKE HOME MESSAGE 19.6

» In leaves, plants convert the potential energy of sunlight into the usable chemical energy of sugar. Leaves are thin and have a layered structure that enables them to effectively capture energy and transport water and nutrients. The vascular tissue of leaves carries food to the rest of the plant body, and brings in water and minerals.

19.7 Several structures help plants resist water loss.

When we're lying on a beach or working in a field, the sun's heat can feel relentless. But we can seek shade when the heat becomes too intense and drink water whenever we're thirsty. Plants don't have these options. Consequently, they are perpetually at risk of losing too much water as it evaporates from their tissues. Because water is essential to nearly every chemical reaction in every living organism, water loss is a serious issue. Not surprisingly, plants have multiple adaptations that keep them from becoming dangerously dehydrated. We see three of these important adaptations on the surfaces of most leaves (**FIGURE 19-19**).

1. **The cuticle.** The top layer of most leaves, secreted by their epidermal cells, is the cuticle. This waxy, water-repelling substance keeps water inside the leaf from diffusing through the leaf surface and evaporating. It seals in water almost completely. Carnauba wax, from the cuticle of Brazilian palm trees, is an ingredient in many sunscreen products, as well as in car polishes.

LEAF STRUCTURES THAT HELP REDUCE WATER LOSS

CUTICLE
This waxy, water-repelling substance keeps water inside the leaves from diffusing through the leaf and evaporating into the atmosphere.

LEAF HAIRS
These tiny hairs can reflect sunlight and reduce the speed at which wind moves over the leaf's surface, decreasing water loss through evaporation.

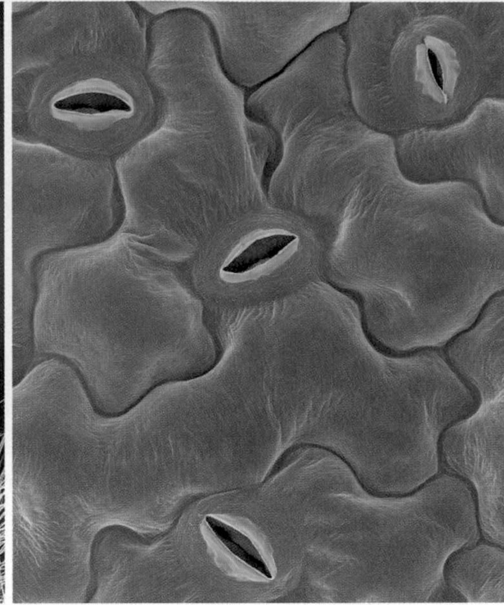

STOMATA
Guard cells swell or shrink to control the size of the stomatal opening. The stomata block or permit water passage and gas exchange.

FIGURE 19-19 Plants have multiple adaptations that keep them from becoming dangerously dehydrated.

2. **Leaf hairs.** Some of the surface cells on leaves become modified as tiny hairs. These hairs can reduce water loss by reflecting some of the sunlight (thus reducing the temperature inside the leaf) and by reducing the speed at which breezes move over the leaf's surface, taking water, through evaporation, with them.
3. **Guard cells.** The cuticle would solve the problem of water loss in plants almost completely if it covered the entire leaf. Unfortunately, the cuticle is also impervious to CO_2, an essential ingredient for photosynthesis. So, just as a castle needs a front door—which then becomes a point of vulnerability to attack—leaves must have openings. These openings are the stomata. There are as many as 10,000 stomata per square centimeter on the underside of leaves (and, to a lesser extent, on stems). Their structure reflects an important water-conservation adaptation. A pair of guard cells (seen in Figure 19-6) surrounds each pore, one cell on either side. The guard cells can increase in size to create a small opening through which CO_2 can enter and water can be lost, or they can decrease in size to seal off the opening. This opening and closing is controlled by osmotic pressure in the guard cells and occurs in response to changes in temperature, humidity, light intensity, or CO_2 concentration.

Stomata are the chief sites where plants "breathe." If you were to do a cruel experiment and coat the underside of the leaves of some plants with Vaseline, what do you think would happen? With no way for air to enter, the plants would become starved for CO_2 and the leaves would die.

Q How could smearing Vaseline on leaves cause a plant to "suffocate."

TAKE HOME MESSAGE 19.7

» Plants have multiple adaptations that enable them to resist becoming dangerously dehydrated. These adaptations include the cuticle, leaf hairs, and guard cells.

19.8–19.11

Plant nutrition: plants obtain sunlight and usable chemical elements from the environment.

Sprouting coconut on a beach in Bora Bora, Tahiti.

19.8 Four factors are necessary for plant growth.

Plants need sunlight, water, and air to thrive. But that's not enough. Just as humans need a nutritious diet, plants will get sick, wither, and even die if they don't have adequate nutrition (FIGURE 19-20).

Plants generally require four things for proper nutrition. The first three are (1) sunlight for the energy to build molecules of sugar; (2) water, for numerous reasons, including as a source of electrons for photosynthesis; and (3) air as a source of carbon dioxide and oxygen. The carbon dioxide is a source of carbon atoms used in the construction of sugar during photosynthesis, and the oxygen serves as an electron acceptor in cellular respiration as a plant breaks down sugar for energy (just as animals and other organisms do; see Chapter 5). And (4) plants usually need soil, from which they get essential minerals.

For a plant to grow, it must build new cells, which are then assembled into new tissues. The raw materials for growth, aside from the carbon and oxygen atoms obtained from CO_2 in the air and the hydrogen atoms from water, come from the soil. In all, there are 13 different minerals (each of which is an element) that plants require for proper survival and growth.

Six of these 13 essential elements are required in relatively large amounts and are particularly important for producing proteins and the enzymes that catalyze all the reactions taking place in plants:

Nitrogen, for making proteins and DNA

Phosphorus, for making ATP, DNA, and cell membranes

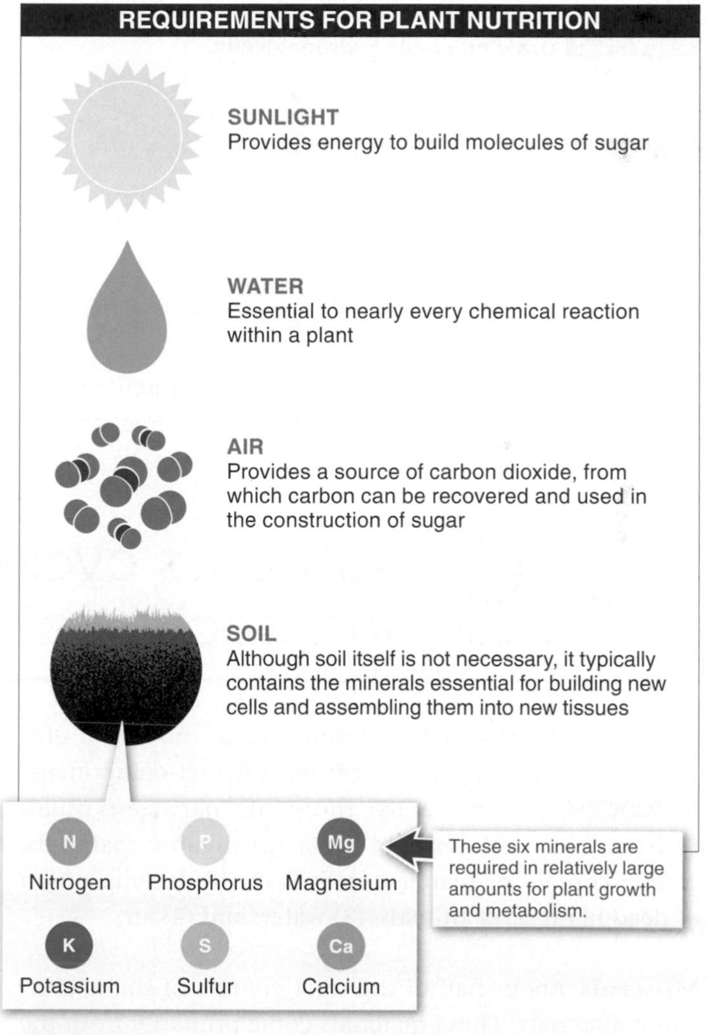

REQUIREMENTS FOR PLANT NUTRITION
SUNLIGHT
Provides energy to build molecules of sugar
WATER
Essential to nearly every chemical reaction within a plant
AIR
Provides a source of carbon dioxide, from which carbon can be recovered and used in the construction of sugar
SOIL
Although soil itself is not necessary, it typically contains the minerals essential for building new cells and assembling them into new tissues
N
Nitrogen
P
Phosphorus
Mg
Magnesium
K
Potassium
S
Sulfur
Ca
Calcium
These six minerals are required in relatively large amounts for plant growth and metabolism.

FIGURE 19-20 The components necessary for maintaining plant health.

Plants can grow just fine without soil if the essential nutrients are added to their water.

FIGURE 19-21 **Growing plants hydroponically.**

Magnesium, for making chlorophyll

Potassium, for making enzymes and controlling water balance

Sulfur, for making proteins and vitamins

Calcium, for a large variety of cell functions

Plants need seven additional elements in much smaller amounts: chlorine, iron, boron, manganese, zinc, copper, and molybdenum. How small are the amounts needed? A plant could get all the chlorine it needed for the day just from being touched by a human with sweaty fingertips. Similarly, a tiny amount of one of these minerals present in the seed that the plant grows from may be enough to supply the plant for its entire life. In other cases, the growing plant roots "reach out" to the necessary elements.

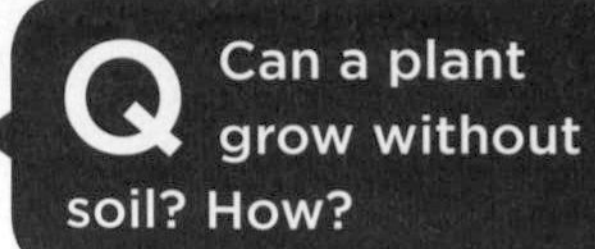

Q Can a plant grow without soil? How?

Much of what we know about plant nutrition comes from experiments with **hydroponically grown plants** (FIGURE 19-21). These plants are grown without soil. Instead, they are grown in water that has chemical elements added. By adding or withholding exact amounts of certain elements, the researcher can determine which are essential and in what amounts. These findings are important in helping farmers determine what types of fertilizers they need for crops, to coax more growth out of each acre of land than would occur naturally.

TAKE HOME MESSAGE 19.8

» Plant growth depends on four important factors: (1) sunlight for energy, (2) water, (3) air as a source of carbon dioxide and oxygen, and (4) minerals, usually obtained from soil.

19.9 Nutrients cycle from soil to organisms and back again.

Soil is more than just a homogeneous mound of dirt. It is a complex mixture of four distinct components (FIGURE 19-22): (1) minerals: inorganic particles, usually from the breakdown of rock; (2) organic materials: carbon-containing matter, usually from the decomposition of dead plants and animals; (3) water; and (4) air.

Minerals About half of the total volume of soil is inorganic materials. These materials come primarily from the weathering of rocks, which decay physically and chemically over time, releasing minerals. The inorganic particles are grouped into three sizes: sand, silt, and clay. As you may know from feeling sand at the beach, there's plenty of air between the particles, so sand doesn't hold water very well. That's why the grains fall through your fingers. Silt particles, such as those from a riverbed, are smaller. Clay particles, the smallest of all, pack together very densely. Because minerals passing through the soil cling to clay, it holds the essential minerals in place for plants to absorb as needed. However, too much densely packed clay in soil can limit the supply of air available for absorption by plant roots, leading to poor plant growth.

COMPOSITION OF SOIL

Soil, a mixture of minerals, organic matter, air, and water, is a source of nutrients that are critical to a plant's health and survival.

MINERALS: 50%
- Inorganic particles
- Formed from the breakdown of weathering rock
- Grouped into three sizes: sand, silt, and clay

WATER AND AIR: 45–50%
- Fill the space between particles of inorganic and organic matter

ORGANIC MATERIALS: 1–5%
- Carbon-containing matter
- Formed from the decomposition of dead plants and animals

GRAPHIC CONTENT
Thinking critically about visual displays of data
Turn to p. 669 for a closer inspection of this figure.

FIGURE 19-22 The components of soil.

The best soils have about equal amounts of each type of particle. The minerals present in soil determine the pH of the soil and can affect how well plants grow; they can even influence the color of flowers (FIGURE 19-23).

Organic Materials Leaves fall and animals die, but **decomposition** returns their chemicals to the soil, making reuptake by plant roots possible. Animal droppings, too, are important to plants. They serve as concentrated sources of nitrogen, perhaps the most important of the essential plant nutrients. Called **humus** (pronounced HY-oo-muhss), these organic decay products make up 1% to 5% of the soil (FIGURE 19-24). Humus also absorbs water and nutrients easily and can release them as needed. With too little humus, the soil may be deficient in nutrients. With too much, it may retain too much water.

> "I bequeath myself to the dirt
> to grow from the grass I love,
> If you want me again
> look for me under your boot-soles."
>
> — WALT WHITMAN
> *Leaves of Grass* (1855)

SOIL pH AND FLOWER COLOR

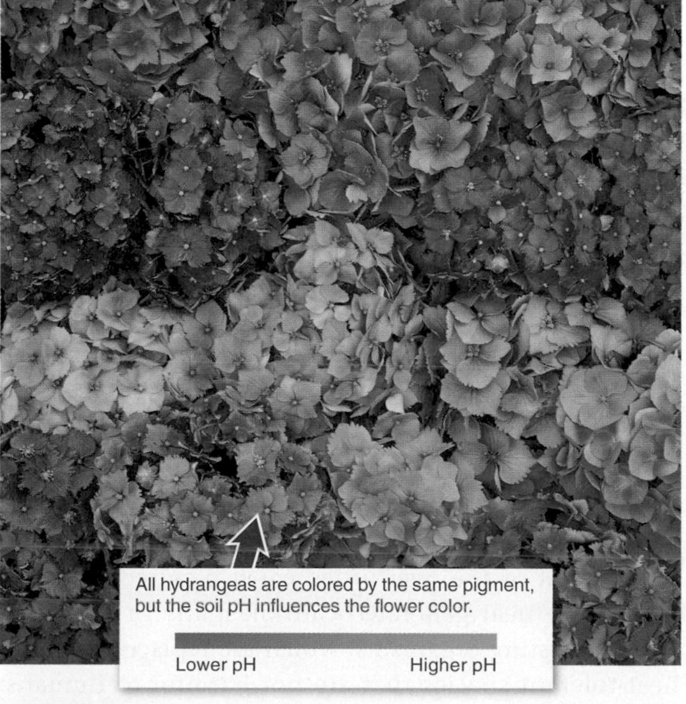

All hydrangeas are colored by the same pigment, but the soil pH influences the flower color.
Lower pH
Higher pH

FIGURE 19-23 Multicolored hydrangeas.

DECOMPOSITION OF ORGANIC MATTER

The decomposition of organic matter is facilitated by huge populations of bacteria, fungi, insects, and earthworms that live in the soil.

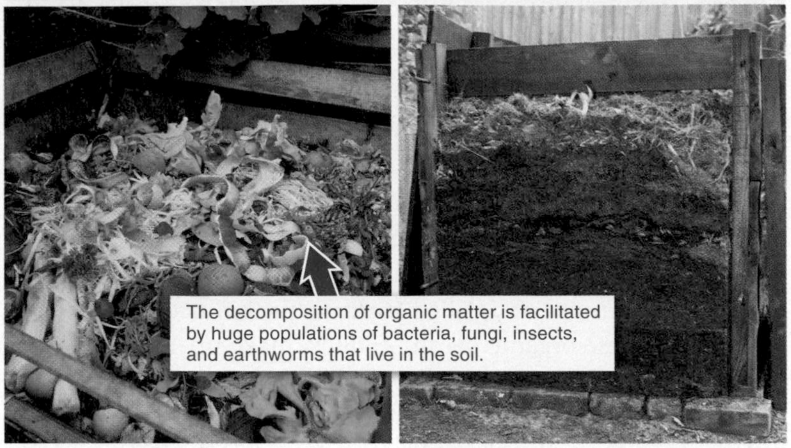

BEFORE: *Decaying organic matter* ***AFTER:*** *Humus*

FIGURE 19-24 Decomposition (such as in the compost bin shown here) returns the chemicals inside animal and plant remains to the soil.

Composting can greatly reduce the volume of waste that cities must process and dispose of, while generating a product that can return important nutrients to the soil.

FIGURE 19-25 The practice of composting converts organic waste into useful fertilizer.

Most of the decomposition of organic matter is carried out by huge populations of bacteria, fungi, insects, and earthworms that live in soil. Aristotle called earthworms "the intestines of the earth." Passing soil through their guts as they feed on organic matter, earthworms increase tremendously the amount of nitrogen available for plants.

A practical application of the decomposition of organic matter is **composting**, used to decompose organic wastes—such as kitchen garbage, manure, yard clippings, and even sewage. In composting, bacteria metabolize the material, converting it into fertilizer (FIGURE 19-25). Compost heaps can speed organic breakdown and greatly reduce the volume of waste that cities must process and dispose of, while generating a product that can be distributed in gardens and croplands to help return important plant nutrients to the soil.

Q What is composting? Why is it useful? Is it dangerous?

The growing populations of bacteria in compost generate large amounts of heat—a compost heap can reach temperatures above 60° C (140° F). This is why steam is coming out of the overturned earth pictured in Figure 19-25. The heat generated can kill many of the original composting organisms, which are replaced by more heat-tolerant species that are not harmful to humans. For many centuries, up until the middle of the 20th century, the use of human feces, called "night soil," as fertilizer was a common practice. In much of rural Japan and China, the purchase, collection, and distribution of night soil was a large industry. The night soil from wealthier villages even sold for higher prices; these villagers had better diets, so their waste products contained more nutrients (particularly nitrogen from protein). Because human waste may carry pathogens, however (and because of the increased availability of chemical fertilizers), the use of night soil has become significantly less common and its use has been strongly discouraged.

Q Does human waste have value as fertilizer?

Water and Air Water and air, the final components of soil, fill the spaces between the particles of inorganic and organic matter. They account for about half of the total volume of soil.

TAKE HOME MESSAGE 19.9

» Soil is a mixture of minerals, organic materials, air, and water that serves as an almost perpetual source of nutrients critical to a plant's health and survival.

19.10 Plants acquire essential nitrogen with the help of bacteria.

There is a sad irony to floating in a lifeboat on the ocean. Despite being surrounded by water, a person is likely to die from lack of water, because salt water just isn't drinkable for a human. Plants encounter a similar situation when it comes to a key component of their diet: nitrogen.

Q Burying dead fish in their fields can help farmers produce more food. Why?

Almost 80% of the air around us is nitrogen. But despite its abundance, a lack of nitrogen, after lack of water, is the factor most likely to limit plant growth. Early farmers recognized this fact and would bury dead fish along with their corn crops. As the fish decomposed, the rotting organic material increased the soil nitrogen and significantly improved the size and health of the corn plants. Today, people regularly treat their crops, flowers, and lawns with nitrogen-rich fertilizers.

Adding nitrogen to the soil increases a plant's growth because every protein, every bit of DNA, and every molecule of chlorophyll it synthesizes contains nitrogen. Plants need nitrogen to survive and grow. If nitrogen is so plentiful in the air, why can't plants make use of it directly?

The problem is that nitrogen in the atmosphere exists as a molecule of two nitrogen atoms (N_2) bound very tightly and stably together. The nitrogen becomes usable for plants and animals only when it is converted to a molecule with a single nitrogen atom, such as ammonium (NH_4^+) or nitrate (NO_3^-). Plant evolution did not produce the metabolic know-how to break apart nitrogen molecules. Bacteria, on the other hand, are metabolic wizards; several species produce and use an enzyme, called nitrogenase, that changes the nitrogen into a plant-usable form—a process called **nitrogen fixation** (FIGURE 19-26). Most nitrogen-fixing bacteria are free-living in the soil. Some, however, live within plants in a mutually beneficial relationship that enables plants to gain access to the nitrogen fixed by these bacteria.

The alliance between plants and nitrogen-fixing bacteria can occur in several ways. In legumes—alfalfa, peas, beans, and soy, for example—the symbiosis develops in several simple steps (FIGURE 19-27).

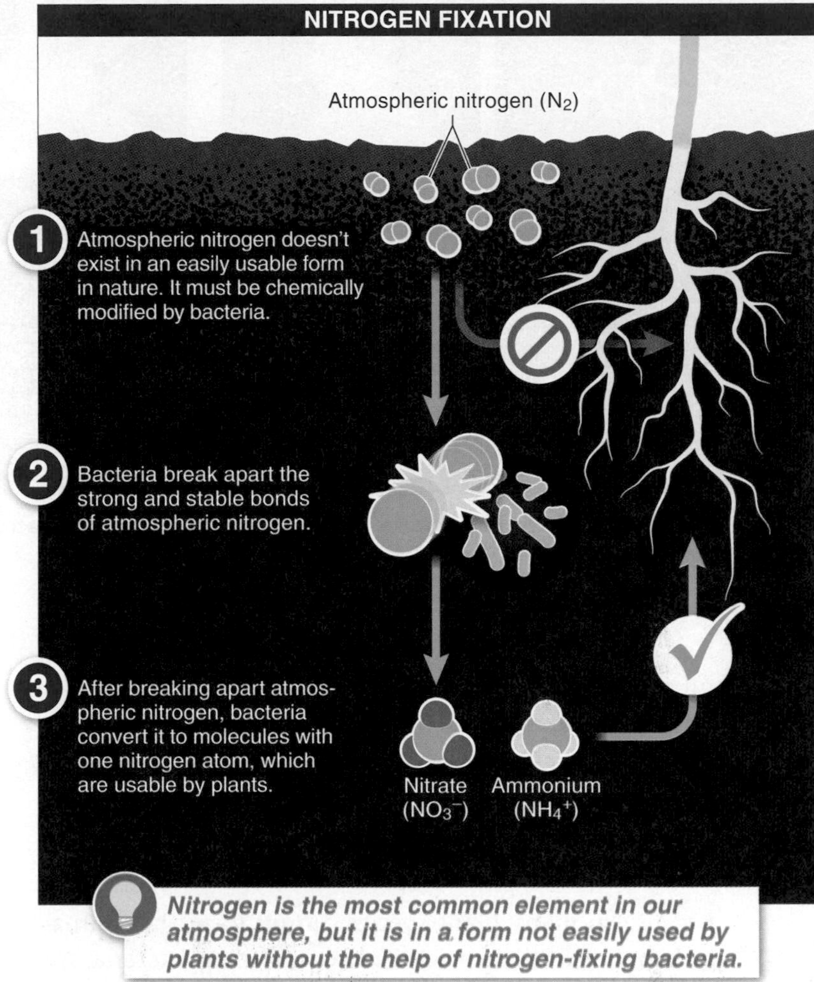

NITROGEN FIXATION
Atmospheric nitrogen (N_2)
1 Atmospheric nitrogen doesn't exist in an easily usable form in nature. It must be chemically modified by bacteria.
2 Bacteria break apart the strong and stable bonds of atmospheric nitrogen.
3 After breaking apart atmospheric nitrogen, bacteria convert it to molecules with one nitrogen atom, which are usable by plants.
Nitrate (NO_3^-)
Ammonium (NH_4^+)
Nitrogen is the most common element in our atmosphere, but it is in a form not easily used by plants without the help of nitrogen-fixing bacteria.

FIGURE 19-26 The conversion of atmospheric nitrogen into a form that plants can use.

1. A plant's roots secrete a bacteria-attracting compound.
2. Bacteria enter the roots' inner cells, multiplying to form, in some cases, a big, round lump called a **nodule.**
3. The bacteria in the nodule split apart nitrogen molecules absorbed from the soil—an energetically expensive process that is fueled by the plant's sugars.
4. The bacteria convert the nitrogen into a usable form and release this into the plant. Any excess usable nitrogen, not required by the plant, is released into the soil.
5. The nitrogen in this form is absorbed throughout the plant body and used for growth. (And, we should add here, animals eat plants, from which all of their nitrogen ultimately comes.)

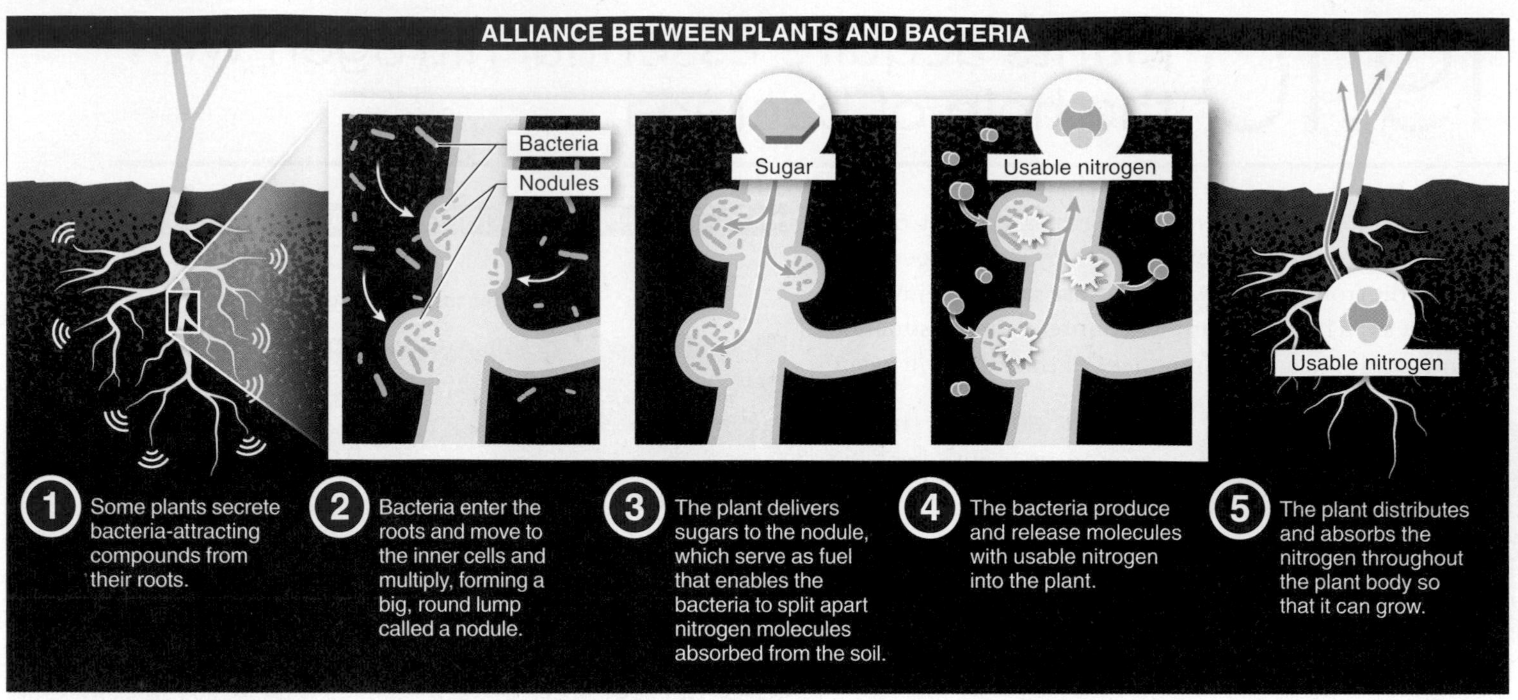

ALLIANCE BETWEEN PLANTS AND BACTERIA
Bacteria
Nodules
Sugar
Usable nitrogen
Usable nitrogen
1 Some plants secrete bacteria-attracting compounds from their roots.
2 Bacteria enter the roots and move to the inner cells and multiply, forming a big, round lump called a nodule.
3 The plant delivers sugars to the nodule, which serve as fuel that enables the bacteria to split apart nitrogen molecules absorbed from the soil.
4 The bacteria produce and release molecules with usable nitrogen into the plant.
5 The plant distributes and absorbs the nitrogen throughout the plant body so that it can grow.

FIGURE 19-27 Nitrogen fixation: a product of the symbiotic relationship between plants and bacteria.

Not all soil conditions are conducive to the growth of nitrogen-fixing bacteria, and some plants have alternative ways to acquire nitrogen. Insect-eating plants, for example, generally grow in very acidic soils in which nitrogen-fixing bacteria do not thrive. The Venus flytrap (FIGURE 19-28) captures insects rich in nitrogen-containing molecules, digests them, and recovers the nitrogen for its own uses.

Because so many plant species cannot form the mutually beneficial alliance with nitrogen-fixing bacteria and instead must scavenge for usable nitrogen in the soil, farmers commonly rotate their crops every few seasons.

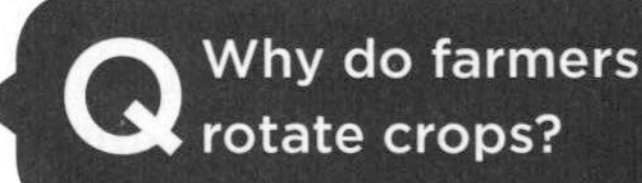

They alternate between growing plants that have nodules of nitrogen-fixing bacteria (such as peas, soybeans, and alfalfa) and those that don't. As excess nitrogen fixed by the bacteria is released, usable nitrogen in the soil is replenished like a natural fertilizer. Also, because these plants are often plowed under before the next crop is planted, they decay and add even more nitrogenous compounds to the soil. Called "green manure," these compounds can reduce the need for additional fertilizer (and the costs to farmers). The soil is then ready for a non-nitrogen-fixing crop.

Some plants are carnivorous and can harvest nitrogen from insects.

FIGURE 19-28 Another way to acquire nitrogen: the Venus flytrap.

TAKE HOME MESSAGE 19.10

» Nitrogen, required by nearly all plant cells, is the mineral that most often limits plant growth. Most nitrogen does not exist in an easily usable form in nature. Several species of bacteria can chemically modify nitrogen molecules into a plant-usable form—a process called nitrogen fixation. Some of these species are free-living in soil, while others live symbiotically within the root nodules of plants.

THIS IS HOW WE DO IT

Developing the ability to apply the process of science

19.11 Carnivorous plants can consume prey *and* do photosynthesis. Why are they confined to bogs and other nutrient-poor habitats?

"Murderous plants." This is how many botanists (and others) described Venus flytraps, pitcher plants, and other carnivorous plants in the 19th century. But Linnaeus called them "miracles of nature," and Charles Darwin spent 15 years writing an entire book about them.

What sets carnivorous plants apart from other plants is their ability to trap, digest, and absorb nutritionally beneficial substances from animal prey. They have a second energy-acquisition system *in addition to* photosynthesis.

Often, the best way to approach a question in biology—such as why carnivorous plants are confined to nutrient-poor habitats—is with simple experimental manipulation. This can be one of the most important steps in trying to understand how things work.

Why don't carnivorous plants thrive in all habitats?

Carnivorous plants occur on every continent except Antarctica, but they compete successfully *only* in nutrient-poor environments. Why don't they dominate in all habitats? After all, they can make their own food by capturing energy from the sun and converting it into a usable chemical form. *And* they can capture and kill many types of animals, primarily insects, thus gaining access to nutrients in these organisms.

How can we explain the distribution patterns we see for carnivorous plants?

Researchers have used some logic and a clever experimental manipulation to address this question.

First: the logic Noting that carnivorous plants are almost exclusively found in nutrient-poor habitats, researchers concluded that *non*-carnivorous plants struggle to survive and grow in nitrogen-poor soils. Carnivorous plants "win" in these environments because their carnivory provides them with the nitrogen necessary to compensate for the poor soil.

But why don't carnivorous plants also outcompete other plants everywhere else? The researchers proposed that the structures necessary for carnivory—flytraps or pitchers, for example—might be costly to produce and maintain, putting carnivorous plants at a competitive disadvantage anywhere the soil is sufficiently nourishing.

Is there experimental evidence to support this speculation? Yes.

Second: the manipulation Give some plants extra nitrogen.

The setup. For their experiments, researchers chose pitcher plants growing in bogs in the northeastern United States. Every two weeks during the growing season (June through September), for three consecutive years, they added 5 milliliters of a solution directly to each open pitcher on each plant, using one of three types of solution:

Solution 1: distilled water only

Solution 2: low concentration of nitrogen

Solution 3: high concentration of nitrogen

The researchers measured (1) the number, size, and shape of pitchers (which are necessary for catching insects) and (2) the photosynthetic rate (as micromoles of CO_2 used per square meter per second) of the largest leaf on each plant.

The results. So, what did they find?

- Plants that received either distilled water or a *low* concentration of nitrogen produced normal carnivorous pitchers with large tubes.
- Plants that received *high* concentrations of nitrogen produced leaves with small carnivorous tubes (less than half the size of those of typical pitchers). In some cases, no tubes were produced at all, and all of the leaves were entirely non-carnivorous.
- Photosynthetic rate increased as leaves became more flattened and less carnivorous.
- Photosynthetic rate decreased when plants invested more in making pitchers.

Conclusions The results led to two general conclusions.

1. When nitrogen is limited—as in bogs and other nutrient-poor habitats—carnivorous plants invest more in the structures necessary for carnivory. When there is an excess supply of nitrogen, these plants reduce their investment in carnivory.

Why are carnivorous plants outcompeted where nitrogen isn't limited?

2. There is a photosynthetic cost to producing carnivorous structures. Although pitchers are good for capturing prey, they are less efficient at photosynthesis than other leaves. Consequently, unless a habitat is nutrient-poor, carnivorous plants are less efficient than (and thus outcompeted by) non-carnivorous plants.

Broader applications: Could researchers use these methods to make fast, inexpensive, and accurate evaluations of the nitrogen status of a habitat? How?

TAKE HOME MESSAGE 19.11

» Carnivorous plants can trap, digest, and absorb nutritionally beneficial substances from animal prey. They can also do photosynthesis. With low availability of nitrogen in nutrient-poor habitats, carnivorous plants invest more in the structures necessary for carnivory and can thrive. But carnivorous structures are less efficient at photosynthesis, reducing carnivorous plants' competitive ability in habitats with sufficient nitrogen.

19.12–19.14

Plants transport water, sugar, and minerals through vascular tissue.

A mature cypress sits beside a country road in Tuscany.

19.12 Plants take up water and minerals through their roots.

Plant roots absorb water from the soil. The cell membranes of the root hair cells are permeable to water, and when a root reaches water in the soil that has a lower concentration of dissolved minerals than the fluid inside the cells, water moves into the root through osmosis (FIGURE 19-29).

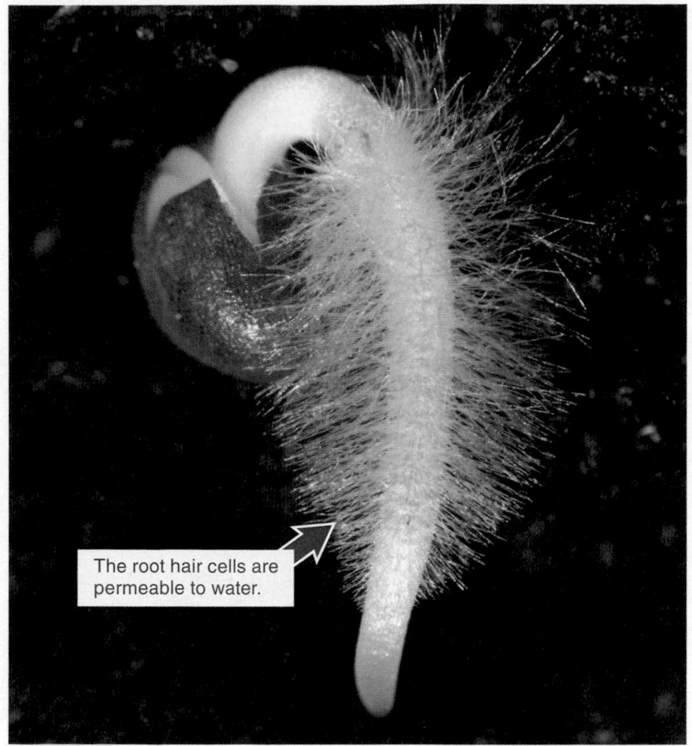
The root hair cells are permeable to water.

FIGURE 19-29 **Plants absorb water from the soil into their root hairs through osmosis.** The absorption of minerals requires an intervening step.

Plant roots also absorb minerals (FIGURE 19-30). Because most minerals are chemically charged, plants take up minerals from the soil only with the help of transport proteins embedded in their root cell membranes. If a mineral is at a higher concentration outside a root cell than inside, the mineral can move across the membrane by simple facilitated diffusion (a process that does not require the plant cell to expend any energy). If a mineral is at a higher concentration inside the root cell, it can still be taken up by the plant, but only through active transport (see Section 4.10), in which the plant expends energy to move the mineral into the root cell against its concentration gradient.

When it comes to acquiring water and minerals, plants benefit from collaboration with fungi (FIGURE 19-31). Most plants have fungi growing all around and even within their roots, forming mutualistic associations known as **mycorrhizae**. Tiny, thread-like fungi trap water like a sponge and hold it around the roots. The fungi's huge surface area dramatically increases the amount of water and minerals that can be absorbed. The fungi don't do all that work for free, of course. In exchange, they receive sugars, amino acids, and

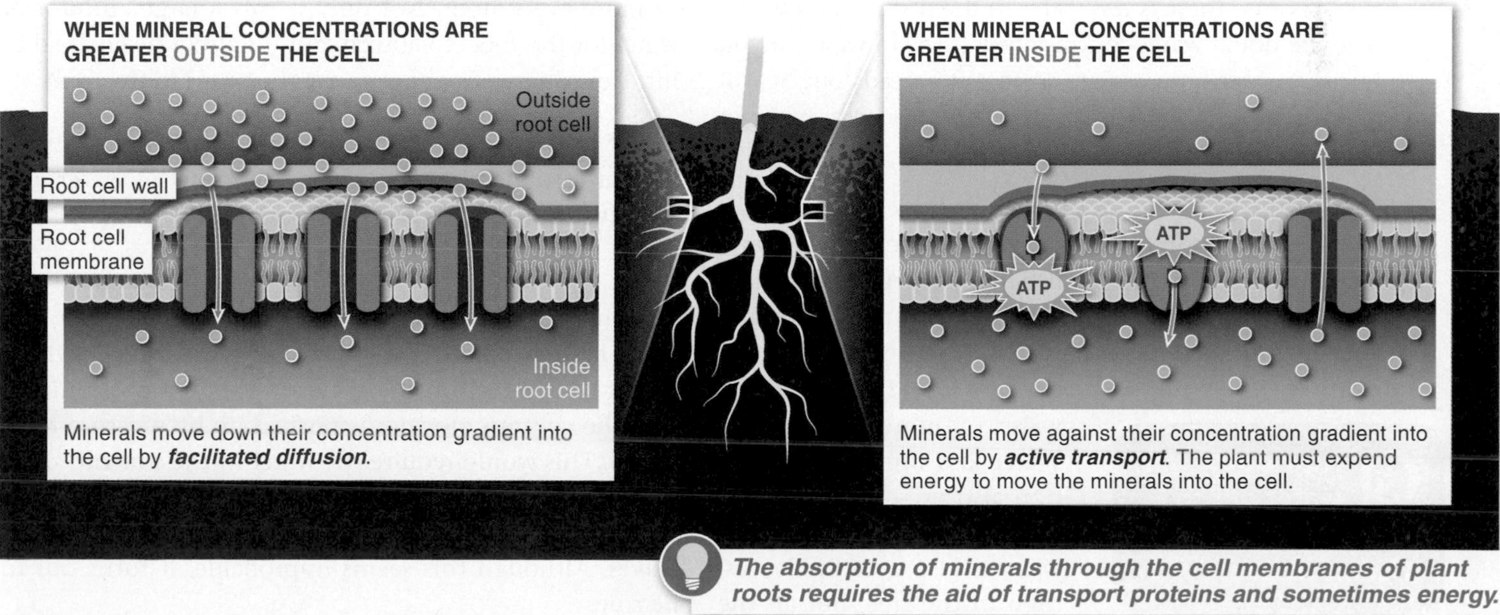
ABSORBING MINERALS
Because most minerals are chemically charged, they must pass through transport proteins embedded within the membranes of root cells.
WHEN MINERAL CONCENTRATIONS ARE GREATER OUTSIDE THE CELL
Outside root cell
Root cell wall
Root cell membrane
Inside root cell
Minerals move down their concentration gradient into the cell by *facilitated diffusion*.
WHEN MINERAL CONCENTRATIONS ARE GREATER INSIDE THE CELL
ATP
ATP
Minerals move against their concentration gradient into the cell by *active transport*. The plant must expend energy to move the minerals into the cell.
The absorption of minerals through the cell membranes of plant roots requires the aid of transport proteins and sometimes energy.

FIGURE 19-30 **Taking up minerals into the plant.**

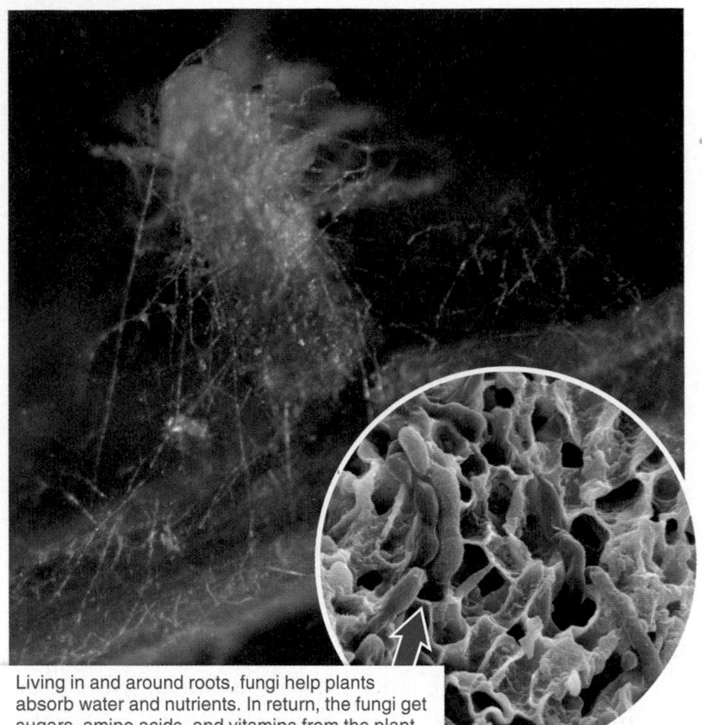

Living in and around roots, fungi help plants absorb water and nutrients. In return, the fungi get sugars, amino acids, and vitamins from the plant.

FIGURE 19-31 **Mutualistic relationship between plants and fungi.**

vitamins from the plant. The association between plants and fungi is so important to plants that many trees can't grow at all if they don't have fungi around their roots.

TAKE HOME MESSAGE 19.12

» Plants absorb water from the soil into their root hairs through osmosis. Minerals are absorbed into the roots with the help of transport proteins in the root cell membranes. In mutualistic associations called mycorrhizae, fungi growing into and around plant roots increase the plant's water and mineral absorption, while gaining access to energy and nutrients from the plant.

19.13 Water and minerals are distributed through the xylem.

Evaporation is relentless. Leave a glass of water on your desk for a day or two, and the water will evaporate away. Because there is less water in the dry air than in the glass, the liquid water continuously turns to vapor. If you pour the water onto a plate so that it spreads out, it will evaporate even faster, because more of the water is in direct contact with the air.

Evaporation steals the moisture from leaves, too. And although evaporation can cool leaves and help move water and ions throughout a plant body, perpetual water loss to the atmosphere is a problem for plants, especially trees. In one day, a big tree in a North American forest can lose 100 gallons of water or more (FIGURE 19-32). Water loss through evaporation also presents a physical challenge for the plant. Most water is lost from the leaves, high in the tree, but all the water comes from the ground. Somehow, water must be constantly delivered to the leaves—dozens of gallons or more every hour. Not only is this a large amount of water for the roots to absorb, but it's also a heavy load to lift. How do trees manage to get the water to the top?

Q How does water get to leaves at the top of a giant tree?

Water transport is the job of the xylem, the transport system responsible for directing the flow of water and dissolved minerals from the roots to all of the plant's cells.

How, in the absence of a pump like an animal's heart, can water be transported from the roots to the leaves—which might be 300 feet off the ground? The first possibility is that the column of water is pushed up by pressure from below. This would require the water in the soil and roots to be under very high pressure, which it isn't. The alternative explanation is that water is *pulled* up to the highest leaves. Although this seems improbable, it turns out to be true.

Evaporation is relentless. Large trees can lose a hundred gallons of water or more in a single day.

FIGURE 19-32 Trees can lose significant amounts of water to the atmosphere through evaporation.

> "To believe that columns of water should hang in [the xylem] . . . and should . . . transmit downwards the pull exerted on them . . . by the transpiring leaves, is to some of us equivalent to believing in ropes of sand."
>
> — FRANCIS DARWIN
> botanist, Charles Darwin's son (ca. 1900).
> (His disbelief in Henry Dixon's 1896 cohesion-tension theory of water transport in plants is understandable, but he was *wrong*.)

Three properties of water aid its movement throughout a plant (FIGURE 19-33).

1. *Evaporation*. Because of low water concentration in the air relative to that in the leaf, molecules of water, one by one, are vaporized and lost from the leaf.
2. *Tension*. Recall from Chapter 2 that hydrogen bonds cause water molecules to stick together. As one water

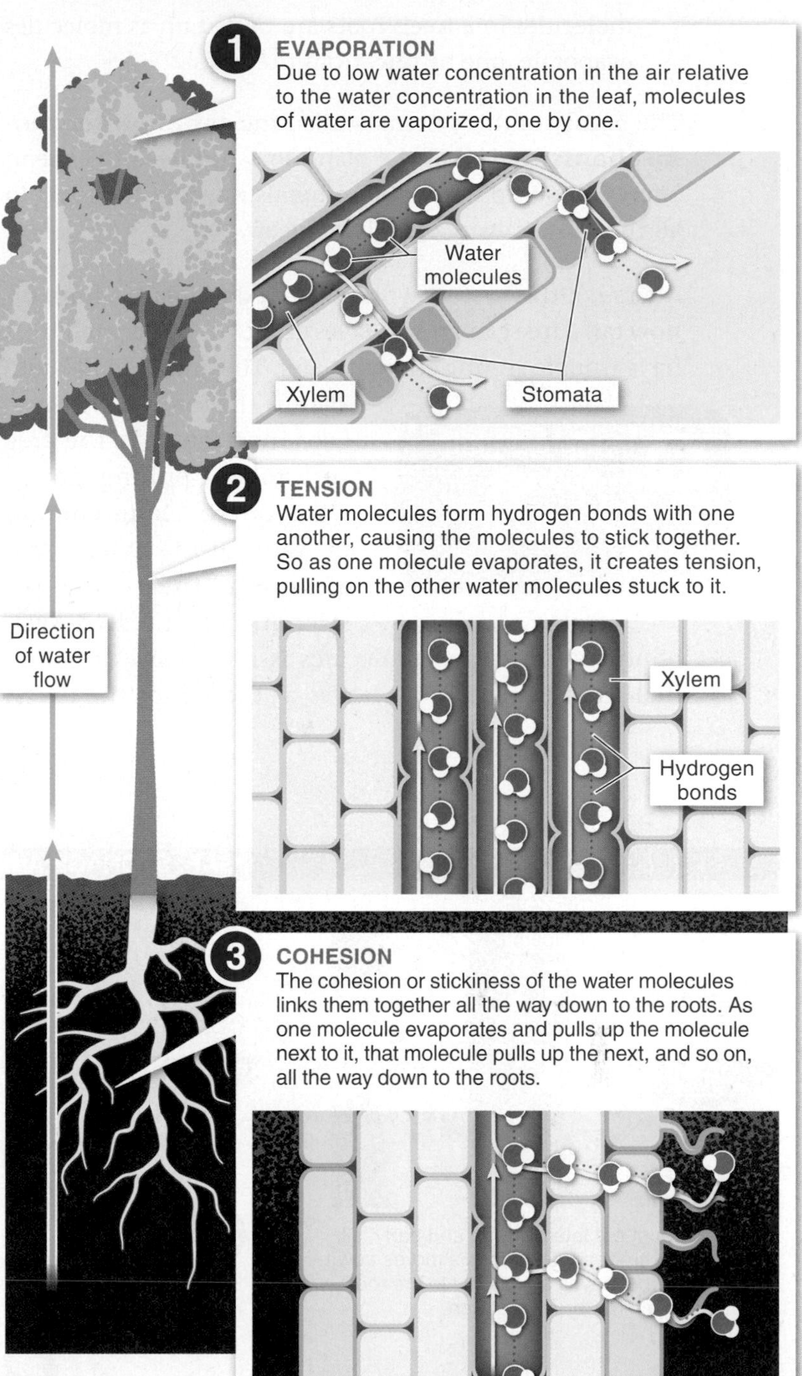

WATER TRANSPORT
Plants use a cohesion-tension mechanism to transport water and dissolved minerals from the roots and circulate them throughout the plant.
1 EVAPORATION
Due to low water concentration in the air relative to the water concentration in the leaf, molecules of water are vaporized, one by one.
Water molecules
Xylem
Stomata
2 TENSION
Water molecules form hydrogen bonds with one another, causing the molecules to stick together. So as one molecule evaporates, it creates tension, pulling on the other water molecules stuck to it.
Xylem
Hydrogen bonds
Direction of water flow
3 COHESION
The cohesion or stickiness of the water molecules links them together all the way down to the roots. As one molecule evaporates and pulls up the molecule next to it, that molecule pulls up the next, and so on, all the way down to the roots.

Moving heavy fluid—water and nutrients—around a plant could be energetically expensive. However, the cohesion of water molecules allows evaporation to do this work, and the fluids are "pulled" through the plant, from roots to leaves.

FIGURE 19-33 The cohesion-tension mechanism of fluid movement in the xylem.

molecule evaporates from the leaf, it creates a tension, pulling the chain of water molecules stuck to it.

3. *Cohesion*. Through this cohesion, or stickiness, water molecules in a tree's roots are pulled up as molecules evaporate, one by one, from the leaf.

The beauty of this system—called the **cohesion-tension mechanism**—is that the plant does not need to expend energy to pump water and minerals up from the roots to all the cells in every branch and leaf.

Although this system works well, it does place a limit on how tall a tree can grow. Leaves at the top of very tall trees have trouble getting enough water. They must "pull" it up against the force of gravity—ever-increasing as the tree gets taller—pulling downward on the long column of water in the xylem. And although the collective strength of the hydrogen bonds linking water molecules is impressive, it is limited. Consequently, the leaves at the top of the tallest trees resemble desert leaves: they are small, grow slowly, photosynthesize at a low rate, and otherwise behave like a parched leaf. Researchers estimate that the physical constraint of lifting water against the force of gravity restricts trees to a maximum height of about 400 feet.

Q Is there a limit to how tall a tree can grow? Why?

The sap in the xylem doesn't always contain just water and dissolved minerals. Sometimes the fluid contains sugars, too. (Recall that phloem is generally the sugar-transporting pipeline.) Throughout the late summer and early fall, just before maple trees (*Acer saccharum*) lose their leaves, they begin storing the sugar from photosynthesis as starch in their roots. In late winter, the temperature rises and the starch in the roots begins to break down into sugar and is released into the xylem (FIGURE 19-34). The sap moves upward toward the newly forming leaf buds, which need energy to develop. By drilling into the xylem and inserting a small tube, we can collect some of the sugary sap, which simply drips out. At about 1% to 5% sugar,

Q Maple syrup comes from trees. What is it?

SAP: STORAGE AND TRANSPORT

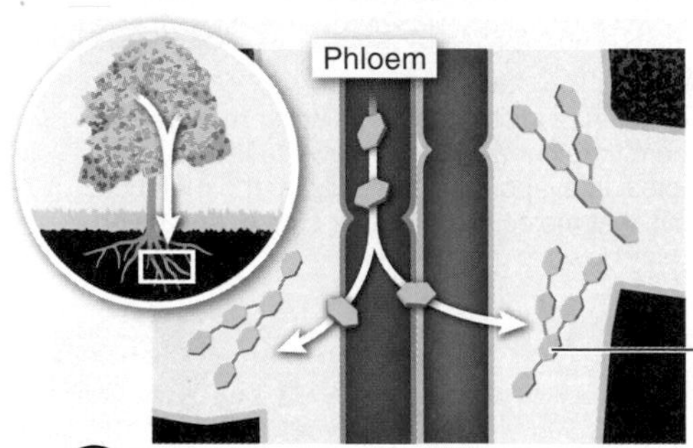

1 Throughout late summer and early fall, sugar from photosynthesis moves down from the leaves and stems to the roots, where it is stored as starch.

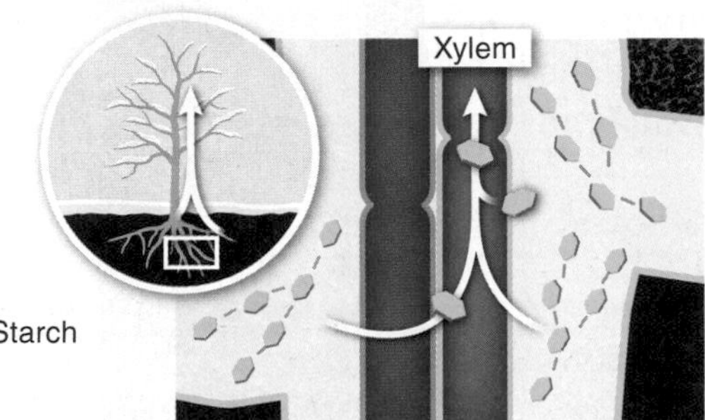

2 As the temperature rises in late winter, starch in the roots begins to break down, and the sugar is released into the xylem.

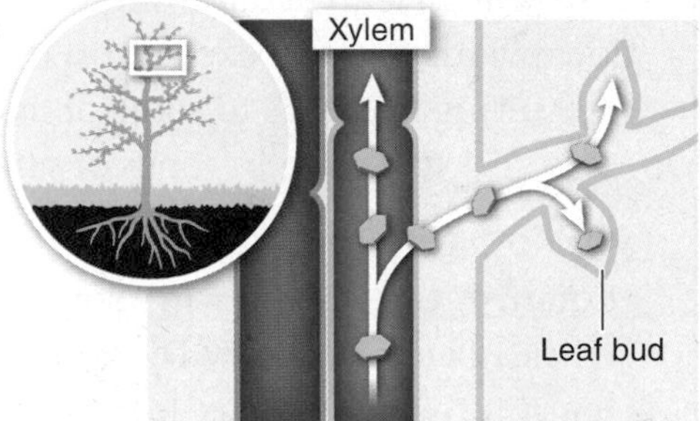

3 Increased pressure caused by the added sugar moves sap upward toward newly forming leaf buds.

Sugar from photosynthesis can be stored in the roots as starch. Later, the sugar can provide energy for new leaf buds.

FIGURE 19-34 **In late winter and early spring, sugar-carrying sap moves upward from the plant roots to nourish the forming leaf buds.**

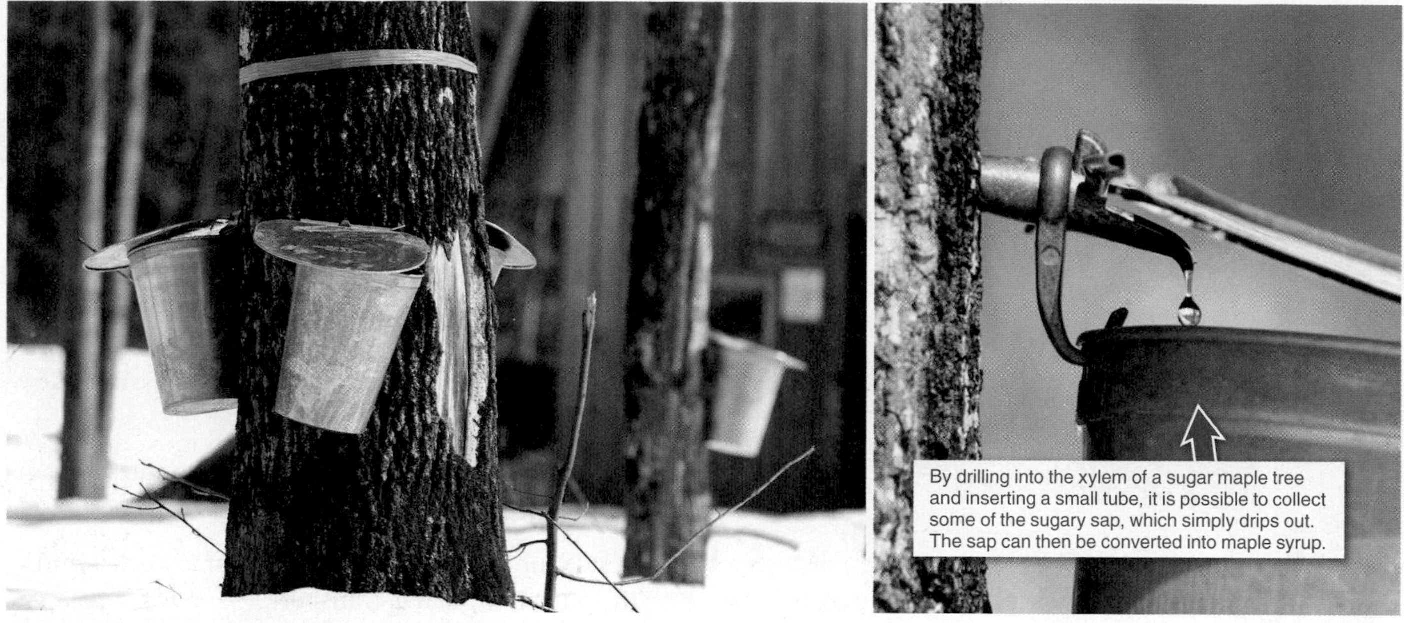

By drilling into the xylem of a sugar maple tree and inserting a small tube, it is possible to collect some of the sugary sap, which simply drips out. The sap can then be converted into maple syrup.

FIGURE 19-35 **Tapping a maple tree.**

the sap is not sweet enough to pour on your pancakes, so it is boiled to evaporate much of the water and concentrate the sugar. When the sap reaches about 85% sugar, it makes a tasty maple syrup. About 1 gallon of sap can be collected from a tree over the several weeks during which it flows, not enough to harm the tree. But it takes about 50 gallons of sap to make 1 gallon of syrup (FIGURE 19-35).

TAKE HOME MESSAGE 19.13

» Xylem directs the flow of water and dissolved minerals from the roots to all the tissues of the plant. The force driving the flow of this fluid (the xylem sap) comes from evaporation of water from the leaves, which pulls water up from the roots.

19.14 Sugar and other nutrients are distributed through the phloem.

Phloem is the chief food-delivery service of plants. Through a branching network of highways, sugars produced in the leaves (the "source") are transported throughout the plant to places (called "sinks") where they are needed. The sinks may be regions of rapid growth such as roots or leaf buds, or developing fruits, or roots and stems where sugars can be stored.

When a leaf is young, it cannot produce adequate food for itself; instead, it receives food supplied through the phloem. Once the leaf matures and makes more food than it uses, the transport direction is reversed. Sugar moves out of the leaf, into the phloem, and to other places in the plant. So, unlike mammalian circulatory systems, the direction of flow depends on the plant's needs at the time.

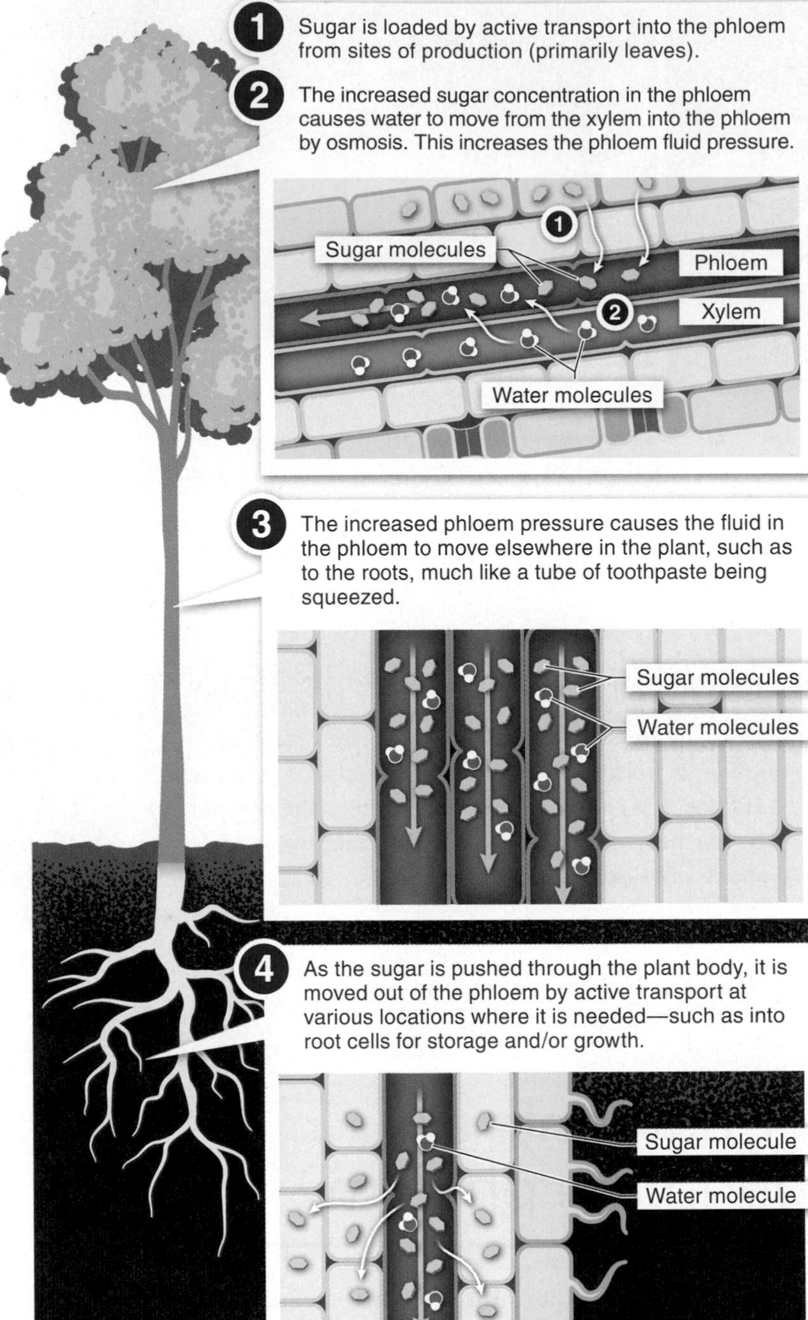

SUGAR TRANSPORT
The phloem controls sugar movement throughout the plant in four steps.
1 Sugar is loaded by active transport into the phloem from sites of production (primarily leaves).
2 The increased sugar concentration in the phloem causes water to move from the xylem into the phloem by osmosis. This increases the phloem fluid pressure.
Sugar molecules
Phloem
Xylem
Water molecules
3 The increased phloem pressure causes the fluid in the phloem to move elsewhere in the plant, such as to the roots, much like a tube of toothpaste being squeezed.
Sugar molecules
Water molecules
4 As the sugar is pushed through the plant body, it is moved out of the phloem by active transport at various locations where it is needed—such as into root cells for storage and/or growth.
Sugar molecule
Water molecule

FIGURE 19-36 **The pressure-flow mechanism of fluid movement in the phloem.**

Phloem vessels are visible to the naked eye. Along with the xylem vessels, they make up the veins in a leaf. As discussed in Section 19.3, the living cells of the phloem vessels are called sieve tubes and have many small openings in their walls (primarily in the ends of the cells). Sugar can quickly move through sieve tubes, from the leaves throughout the plant to roots, stems, buds, flowers, and fruits. The process is called the **pressure-flow mechanism.**

Phloem controls sugar movement throughout the plant in four steps (FIGURE 19-36).

1. At its source (primarily leaves), sugar is loaded into the phloem by active transport.
2. Water moves from the xylem into the phloem by osmosis, because of the increased sugar concentration in the phloem. This increases the pressure inside the phloem.
3. The increased pressure (particularly higher up in the plant, where photosynthesis is occurring) causes the fluid in the phloem to move to other parts of the plant.
4. Sugar is moved out of the phloem at various locations where it is needed (the sinks) by active transport.

Sometimes, important science questions can be answered with unexpected methods. Consider this question: what exactly is the fluid circulating in the phloem? To answer this, researchers used aphids—insects that use their mouthparts to pierce the plant stem and gain access to the sugar-rich sap in the phloem (FIGURE 19-37). The fluid simply flows into their gut.

The researchers selected plants on which aphids were feeding, then removed the aphids' bodies, leaving their mouthparts sticking into the phloem like tiny syringes. As the sap dripped out of the severed mouthparts, the

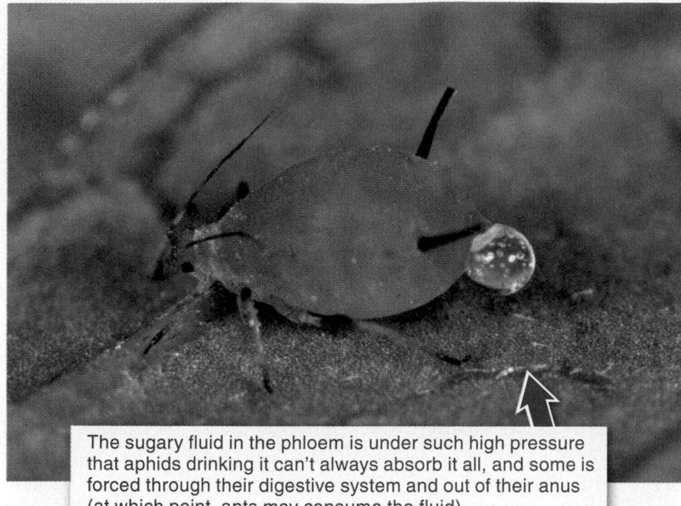
The sugary fluid in the phloem is under such high pressure that aphids drinking it can't always absorb it all, and some is forced through their digestive system and out of their anus (at which point, ants may consume the fluid).

FIGURE 19-37 Like drinking from a fire hose.

researchers collected and analyzed it. As it turns out, phloem sap is a sugar solution, of which about 10% to 25% is dissolved solids that include some lipids and amino acids. In most plants, about 90% of the phloem sugar is sucrose, the same molecule we use as table sugar.

Aphids can also be a nuisance. If you have ever found your car covered in a sticky substance when parked under some trees, you can blame aphids. When aphids are attacking a tree, puncturing holes in the plant tissues to feed on the phloem sap, the pressure in the phloem fluid can cause more of the fluid to come out than the aphids can use. The excess sugary (and sticky) fluid secreted by the aphids, called honeydew, drops onto the ground (or your car) below.

In the next chapter, we see that as a tree grows, phloem also takes on a secondary, protective function.

TAKE HOME MESSAGE 19.14

» The phloem consists of a branching network of vessels made from living cells lined up end to end to form sieve tubes, with small openings in their end walls and side walls. Sugar, usually sucrose, is moved in the phloem from sites of production (sources) to sites of use or storage (sinks). The direction of flow can change to meet the plant's needs.

STREET BIO

Using evidence to guide decision making in our own lives

Organic foods: are they worth the price?

Q: What are "organic" foods? To be labeled and sold as "organic," a food must be grown or raised without the use of chemical fertilizers or non-organic pesticides. They also cannot have been genetically modified or irradiated, or had hormones added.

Q: Don't be fooled by sneaky labeling. Products labeled "100% organic" must contain only organic ingredients with the exception of water and salt, according to the U.S. Department of Agriculture. But be careful: products can be labeled "organic" if they contain at least 95% organic ingredients. And a product can be labeled "made with organic ingredients" if it is made with at least 70% organic ingredients.

Q: Are there downsides to organic foods? Organic fruits and vegetables cost more than other produce, typically 13 to 36 cents more per pound. And hormone- and antibiotic-free milk sells for $6 per gallon, on average, versus $3.50 per gallon for non-organic-certified milk.

Q: Are organic foods nutritionally superior to conventionally produced foods? In an exhaustive 2009 review of 55 published articles on the nutritional content of organic foods, researchers funded by the U.K. Food Standards Agency examined 13 nutrient categories and found no evidence that organic foods were superior in nutritional content when compared with conventionally produced foods.

Q: Might organic foods have some other benefits? Although the results of the review seem to suggest that it might not be worth paying a premium for organic foods, there is reason to withhold final judgment. The study examined only nutritional content—but there may be other benefits:

- *Safety.* There may be a reduced risk of *E. coli* contamination of meat from grass-fed cattle, which are maintained with less crowding. When it comes to pesticide exposure, however, it is less clear whether there are benefits. In its *Guidelines on Nutrition for Cancer Prevention,* the American Cancer Society states: "At present there is no evidence that residues of pesticides and herbicides at the low doses found in foods increase the risk of cancer."
- *Heart health.* Beef from grass-fed cattle has been shown to have less fat and higher levels of omega-3 fatty acids, which may reduce the risks of certain cancers and heart disease.
- *Environmental degradation.* Organic farming reduces the addition of synthetic pesticides to the environment, potentially improving water supplies and long-term soil quality.
- *Risks to farmworkers.* Reduced exposure to synthetic pesticides and chemical fertilizers benefits farmworkers, decreasing their contact with potential carcinogens and other harmful chemicals.

Because extensive research continues, it isn't yet possible to say definitively whether there are benefits to organic foods that justify their increased cost.

GRAPHIC CONTENT

Thinking critically about visual displays of data

1 What is the purpose of this pie chart?

2 What is the source of the data in the chart? What is an assumption underlying this presentation of data?

3 Are the data presented in the chart valid for all types of soil? How might such comprehensive data be presented?

4 Bar graphs can present not just an average value for data but also an estimate of variation. Can that be done with a pie chart? How?

5 What is the benefit to presenting these data as a pie chart rather than a bar graph with three bars? If they were presented as a bar graph, what would the axis labels be?

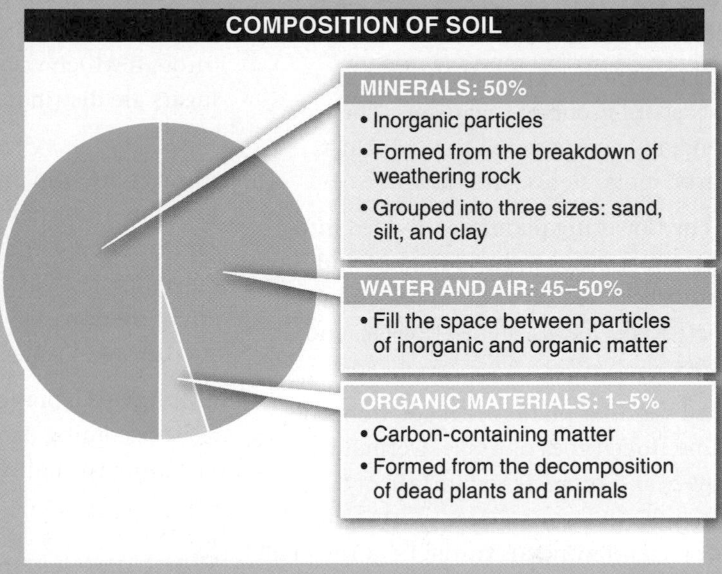

COMPOSITION OF SOIL
MINERALS: 50%
• Inorganic particles
• Formed from the breakdown of weathering rock
• Grouped into three sizes: sand, silt, and clay
WATER AND AIR: 45–50%
• Fill the space between particles of inorganic and organic matter
ORGANIC MATERIALS: 1–5%
• Carbon-containing matter
• Formed from the decomposition of dead plants and animals

6 Pie charts often are shown in three dimensions, as shown here. In this chart, the "water and air" portion is rotated toward the front, causing more blue coloring than turquoise coloring to be visible. Does that mean the blue "slice" represents a larger number than the turquoise "slice"? Could this method of data presentation be used to mislead viewers?

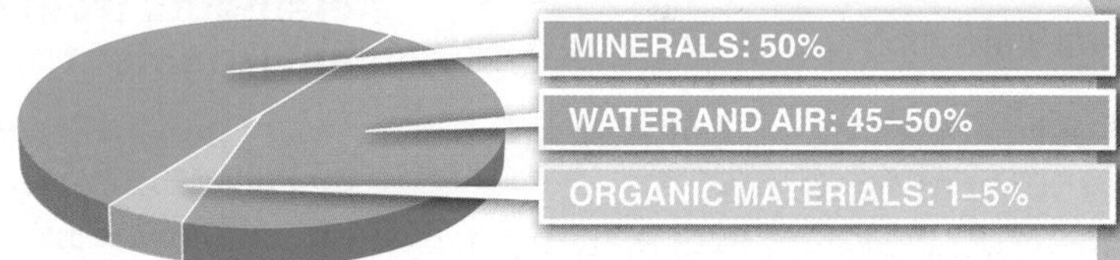

See answers at the back of the book.

KEY TERMS IN PLANT STRUCTURE AND NUTRIENT TRANSPORT

apical meristem, p. 648
cohesion-tension mechanism, p. 664
collenchyma cell, p. 644
composting, p. 656
cork cell, p. 643
cortex, p. 648
cotyledon, p. 640
cuticle, p. 641
decomposition p. 655
dermal tissue, p. 641
epidermis, p. 641
eudicot, p. 640
fibrous root, p. 646
ground tissue, p. 641
guard cell, p. 641
humus, p. 655
hydroponically grown plant, p. 654
lateral meristem, p. 648
leaf, p. 649
lignin, p. 644
meristem, p. 648
monocot, p. 640
mycorrhizae, p. 661
nitrogen fixation, p. 657
node, p. 648
nodule, p. 657
parenchyma cell, p. 644
phloem, p. 643
pith, p. 646
pressure-flow mechanism, p. 666
root, p. 645
sap, p. 643
sclerenchyma cell, p. 644
shoot, p. 647
sieve tube, p. 643
stem, p. 647
stoma (*pl.* stomata), pp. 642, 650
taproot, p. 645
vascular tissue, p. 641
xylem, p. 643

BRIEF SUMMARY

Three basic tissue types give rise to diverse plant characteristics.

• Plants are an extremely diverse and successful group of organisms. They are generally composed of three distinct parts: roots, stems, and leaves.

• The flowering plants are divided into two major groups, the monocots and the eudicots, based on structural features of their seeds, leaves, stems, flowers, and roots.

• All vascular plants are made from the same three types of tissue. Dermal tissue covers and protects the surface of the plant. Vascular tissue transports water, sugars, and minerals throughout the plant body. Ground tissue, which makes up the bulk of the plant, is where most of the plant's metabolic activities occur.

Most plants have common structural features.

• Roots have three primary functions: (1) absorption, the uptake of water and dissolved minerals from soil; (2) anchorage, securing the plant in place; and (3) storage of water and excess starch.

• Stems serve three chief functions: (1) they put leaves in positions where they can intercept sunlight and carry out photosynthesis; (2) they provide structural support for the plant; and (3) they contain the xylem and phloem, through which water, minerals, and sugars are distributed through the plant body.

• Leaves are thin and have a layered structure that enables them to effectively capture energy and transport water and nutrients. The vascular tissue carries food out of the leaves to the rest of the plant and brings in water and minerals.

• Multiple adaptations enable plants to resist becoming dangerously dehydrated, including the cuticle, leaf hairs, and guard cells.

Plant nutrition: plants obtain sunlight and usable chemical elements from the environment.

• Plant growth depends on (1) sunlight for energy; (2) water; (3) air as a source of carbon dioxide and oxygen; and (4) minerals, usually obtained from soil.

• Soil is a mixture of minerals, organic materials, air, and water that serves as a source of nutrients critical to a plant's health and survival.

• Nitrogen is the mineral that most often limits plant growth, but does not exist in an easily usable form in nature. Several species of bacteria—living in the soil or symbiotically within plant roots—can chemically modify nitrogen molecules into a plant-usable form, in a process called nitrogen fixation.

Plants transport water, sugar, and minerals through vascular tissue.

• Plants absorb water from the soil into their root hairs through osmosis. Minerals are also absorbed into the roots, with the help of transport proteins in root cell membranes. Fungi growing into and around plant roots increase the plant's water and mineral absorption, while gaining access to energy and nutrients from the plant.

• Xylem directs the flow of water and dissolved minerals to all the tissues of the plant. The force driving this flow comes from evaporation from the leaves, which pulls water up from the roots.

• The phloem consists of a branching network of vessels made from living cells lined up end to end to form sieve tubes, with small openings in their ends and side walls. Sugar is moved in the phloem from sites of production to sites of use or storage.

CHECK YOUR KNOWLEDGE

Short Answer

1. Describe the three distinct parts of vascular plants.

2. Why would a monocot be more likely to survive having a tight wire wrapped around its stem, causing a shallow cut, than a eudicot plant suffering the same injury?

3. Which plant tissue functions much like the circulatory system in vertebrates? Why is the function of this tissue not exactly like that of the vertebrate circulatory system?

4. What happens to the ground tissue in a root when it is used to store energy in the form of starch?

5. Describe one similarity and one difference in the ways that apical meristems and lateral meristems contribute to plant growth.

6. Describe three important adaptations that help reduce water loss in plants.

7. How does soil contribute to the nutrition of a plant?

8. Why does adding nitrogen to soil increase plant growth?

9. Why is an association with fungi so important to many plants?

10. How does the cohesion-tension mechanism allow vascular plants to transport water without some type of pump? How does the use of this mechanism affect the maximum height that can be attained by trees?

11. Which plant structure could be referred to as the "food-delivery service of plants"? Why?

Multiple Choice

1. The embryonic precursors to leaves in a seedling are called:

a) seedling leaves.

b) cotyledons.

c) angiosperms.

d) needles.

e) monocots.

2. Which of the following is most important in providing structural support in vascular plants?

a) xylem

b) apical meristem

c) phloem

d) epidermis

e) cortex

3. What role does soil play in the life of a terrestrial plant?

a) The plant is anchored in the soil.

b) The plant obtains water and nutrients from the soil.

c) The roots of the plant obtain oxygen from the soil.

d) All of the above are correct.

e) Only a) and b) are correct.

4. Which of the following tissues is found in both the stem and the roots of a plant?

a) vascular tissue

b) dermal tissue

c) ground tissue

d) meristematic tissue

e) All of the above tissues are found in both the stem and the roots.

5. Leaves are generally thin, with a three-layered structure that includes all of the following except:

a) veins of vascular tissue.

b) non-photosynthetic epidermal cells.

c) a few layers of photosynthetic cells.

d) apical meristem cells.

e) stomata.

6. The cuticle of a leaf is:

a) responsible for delivering water to the leaves.

b) responsible for much of the photosynthesis in plants.

c) the site where most of the photosynthetic products of plants are used.

d) a waxy, non-living layer that minimizes water loss.

e) All of the above are correct.

7. The enzyme known as ______ catalyzes the fixation of nitrogen in nitrogen-fixing bacteria.

a) nitrogen reductase

b) nitrogenase

c) rhizobiase

d) ammoniase

e) nitrogen fixase

0 EASY HARD 100 (29)

8. All of the following statements about carnivorous plants are correct except:

a) The carnivorous structures (such as pitchers) contain chloroplasts and are photosynthetic.

b) The carnivorous structures are less efficient at photosynthesis than typical leaves.

c) Increasing the amount of nitrogen available to a pitcher plant causes a decrease in the plant's investment in carnivorous structures.

d) Carnivorous plants usually cannot outcompete non-carnivorous plants in nutrient-rich soils.

e) Carnivorous plants thrive in nitrogen-rich tropical habitats.

0 EASY HARD 100 (18)

9. When the concentration of a mineral is higher inside root cells than in the soil:

a) ATP is used to import more of the mineral into the root cells.

b) the mineral moves by diffusion from the soil into the root cells.

c) energy is used to export the mineral into the soil.

d) the absorption of water is blocked.

e) the plant loses the mineral passively to the environment.

10. Evaporation from the leaves of a tree pulls water up from the roots as an unbroken column through the entire height of the tree. This feat is possible because of which characteristic(s) of water?

a) high heat capacity

b) cohesion and kinetic energy

c) cohesion

d) low heat capacity

e) absorption

Ch20

Plants can reproduce sexually and asexually.

20.1 Plant evolution has given rise to two methods of reproduction.

20.2 Many plants can reproduce asexually when necessary.

20.3 Plants can reproduce sexually, even though they cannot move.

Flowers have several roles in plant reproduction.

20.4 The flower is the chief structure for sexual reproduction.

20.5 The male reproductive structure produces pollen grains.

20.6 Female gametes develop in embryo sacs.

Pollination, fertilization, and seed dispersal often depend on help from other organisms.

20.7 Plants need help getting the male gamete to the female gamete for fertilization.

20.8 *This is how we do it*: Does it matter how much nectar a flower produces?

20.9 Fertilization occurs after pollination.

20.10 Most plants can avoid self-fertilization.

20.11 Following fertilization, the ovule develops into a seed.

20.12 Fruits are a way for plants to disperse their seeds.

Plants have two types of growth, usually enabling lifelong increases in length and thickness.

20.13 How do seeds germinate and grow?

20.14 Plants grow differently from animals.

20.15 Primary plant growth occurs at the apical meristems.

20.16 Secondary growth produces wood.

Hormones regulate growth and development.

20.17 Hormones help plants respond to their environments.

20.18 Seed germination and stem elongation are stimulated by gibberellins.

20.19 Seedlings grow and orient themselves under the direction of auxins.

20.20 Other plant hormones regulate flowering, fruit ripening, and responses to stress.

External cues trigger internal responses.

20.21 Tropisms influence plants' direction of growth.

20.22 Plants have internal biological clocks.

20.23 With photoperiodism and dormancy, plants detect and prepare for winter.

20.24 Plants actively resist being eaten.

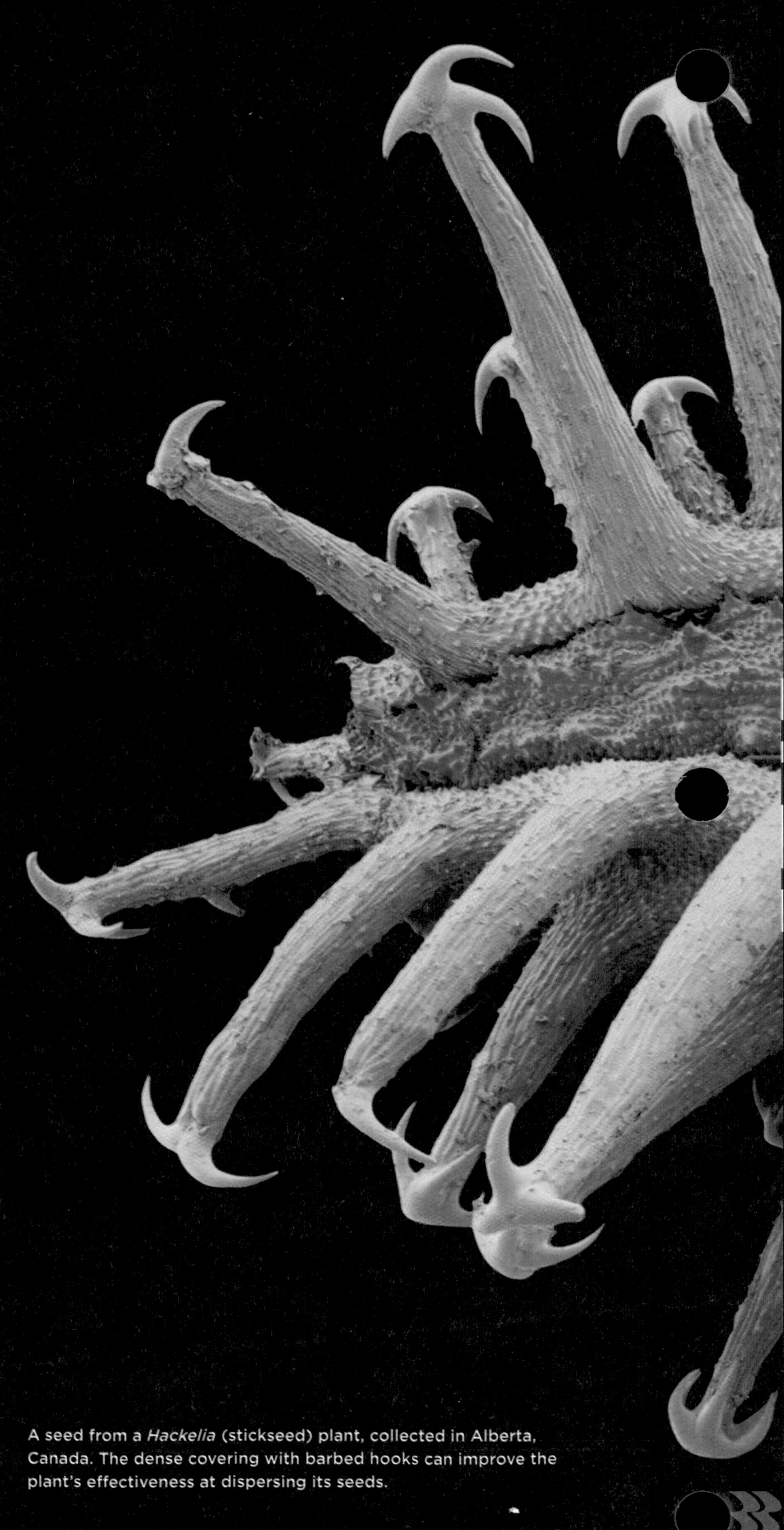

A seed from a *Hackelia* (stickseed) plant, collected in Alberta, Canada. The dense covering with barbed hooks can improve the plant's effectiveness at dispersing its seeds.

Growth, Reproduction, and Environmental Responses in Plants

Problem solving with flowers, wood, and hormones

20.1–20.3

Plants can reproduce sexually and asexually.

A rainbow of flowers.

20.1 Plant evolution has given rise to two methods of reproduction.

Why do flowers have such a hold on people? Brides clutch roses during wedding ceremonies. Suitors and partners give flowers during courtships and in relationships. We send flowers to those in mourning and to those celebrating. And many people spend countless hours and dollars tending to flowering plants in their gardens.

Floral displays, however, are not produced for us. They are intended for a wide range of other animals. They are the means by which plants enlist the assistance of animals in transporting male gametes to female gametes, thus using sexual reproduction to produce seeds and increase genetic variability among their offspring.

Take the carrion flower, *Rafflesia arnoldii*, for example (FIGURE 20-1). Some insects feed on feces, rotting flesh, and other decaying organic matter, and they lay their eggs in damp, stinky places. Carrion flowers lure these insects into their foul-smelling blossoms and, in doing so, ensure the transport of male gametes to female gametes.

Sexual reproduction isn't the only method of reproduction for flowering plants. Many are able to make use of an alternative method, producing offspring asexually under certain conditions. Aspens, for example (FIGURE 20-2), are a species

Carrion flowers lure insects that feed on decaying organic matter into their foul-smelling blossoms. In doing so, they ensure the transport of male gametes to female gametes.

FIGURE 20-1 **The carrion flower, *Rafflesia arnoldii*.**

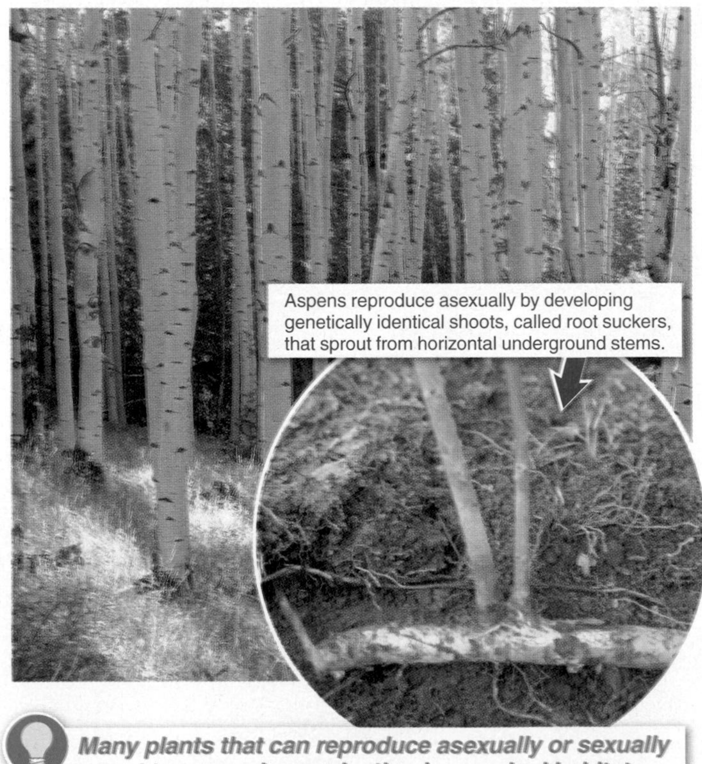

Aspens reproduce asexually by developing genetically identical shoots, called root suckers, that sprout from horizontal underground stems.
Many plants that can reproduce asexually or sexually resort to asexual reproduction in marginal habitats.

FIGURE 20-2 **Aspen trees.**

of tree that thrives in the cool regions of Europe, Asia, and North America. Occasionally, however, aspens take root in inhospitable places. Consider the aspen populations on the Orkney Islands, at the northernmost tip of Scotland. Unlike aspen populations farther south, aspens on the Orkneys produce seeds that seem to be unable to germinate and generate new trees in the harsh climate, possibly due to inadequate soil moisture and limited light availability. This inability to reproduce sexually would seem to be a fatal limitation to aspens' survival on these islands.

On the Orkneys, however, the aspens reproduce only asexually. They develop shoots, called suckers, that sprout from horizontal underground stems and can give rise to new, genetically identical trees. One tree can give rise to dozens or even hundreds of new trees, all genetically identical.

Such asexual reproduction is rare among animals, but it's common among plants growing in marginal habitats. Such habitats may limit the survival of seeds or, because of small population sizes, may reduce the likelihood that an individual can find a sexual partner.

In this chapter, we examine why evolution may have led to most plants having two very different options for reproduction. We also investigate how plants grow and how they interact with their environment. We begin by exploring the process of asexual reproduction and its advantages and disadvantages.

TAKE HOME MESSAGE 20.1

» Most plants have two very different options for reproducing: asexual and sexual reproduction.

20.2 Many plants can reproduce asexually when necessary.

Cloning isn't new. Gardeners have been doing it for centuries. Cloning is just a more modern term to describe the asexual reproductive process by which one organism gives rise to another, genetically identical individual. And most of the trees or shrubs we see in gardens are, in fact, the result of asexual reproduction. So, too, are nearly all houseplants purchased from nurseries.

The process of asexual reproduction in plants is simple. New individual plants grow directly from the tissue of established plants through mitosis. There is no meiosis to generate gametes and no fertilization of one gamete by another. Asexual reproduction is also called vegetative reproduction, because it starts with a vegetative structure of the plant (such as a stem or root or leaf), which becomes separated from the parent plant and continues to grow to form a new individual.

Raspberry plants, for example, can be propagated from sprouts cut from their roots (FIGURE 20-3). And potato plants can be generated from their underground storage stems (that is, from the potato itself). Thus, a new potato plant can be generated just by planting a potato and waiting for it to sprout into a genetically identical individual. We consider the new plant and the "parent" plant distinct individuals because they have become separated and can grow and function independent of one another. In a group of plant species called *Bryophyllum,* small "plantlets" grow right along the edges of the leaves, eventually dropping to the soil, taking root, and growing as independent plants—a reproductive pattern responsible for the plants' common name: "mother of millions."

The critical feature in every case of asexual reproduction is that the plant part that is used to start a new individual must have some undifferentiated cells. Such cells—like stem cells in humans—are **totipotent,** meaning that they have the potential to develop into any type of cell and tissue.

The ability to reproduce asexually is an evolutionary adaptation that helps plants overcome their immobility—they are rooted in the ground and cannot easily seek out sexual partners. In most situations, however, there is a critical disadvantage to asexual reproduction. The advantages and disadvantage can be summarized quite simply.

ASEXUAL REPRODUCTION

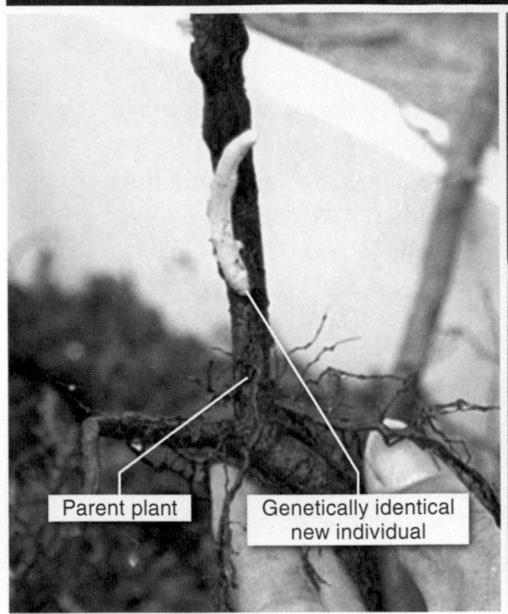

Raspberry root with sprout

Potato with sprouts

Bryophyllum *leaf with plantlets*

FIGURE 20-3 Reproduction without a partner.

The process of asexual reproduction in plants is simple. It always involves the growth of new, genetically identical plants directly from the established plant's tissue via mitosis.

Advantages of Asexual Reproduction There are three main advantages to reproducing asexually.

1. *It is energetically efficient.* No energy is wasted in producing flowers, pollen, and fruits—plant tissues used exclusively for reproduction. Instead, the plant's energy can be invested in producing roots, support and nutrient-transporting tissues, leaves, and other vegetative matter for growth and increased photosynthetic capacity.

2. *It is faster.* Participation of another individual is not required, nor is a means of bringing together the male and female gametes. As a result, new individuals can be produced at a much faster rate, which enhances an individual's reproductive success and fitness.

3. *It preserves winning allele combinations.* If a plant is well adapted to a particular environment, all the individuals it produces by asexual reproduction will carry the same, well-adapted set of alleles and are likely to flourish just as the parent plant did.

Disadvantage of Asexual Reproduction Asexual reproduction has one huge disadvantage relative to sexual reproduction: *It leads to no new genetic variation among the offspring.* Asexually reproduced individuals are identical to their parents (with the rare exception of any mutations that might occur). If the environment is changing—perhaps the climate is changing, or a new variety of pest species is on the increase—the new individuals are less likely to be adapted to the changed environment. Moreover, if individuals in an area are produced by asexual reproduction, and those individuals then reproduce together sexually, there will be less genetic variation among their offspring. Each parent plant will simply draw from the same set of alleles when producing gametes. This reduced variation in genetic material can leave plant populations with a reduced capacity to evolve.

TAKE HOME MESSAGE 20.2

» Many plants can reproduce asexually. This process involves the growth of new, individual plants directly from the tissue of an established plant through mitosis. Asexual reproduction can be energetically efficient and fast and can preserve successful genetic combinations, but it does not lead to genetic variability among an individual's offspring.

20.3 Plants can reproduce sexually, even though they cannot move.

For all its efficiency and advantages for plants, asexual reproduction just doesn't measure up to sexual reproduction. Plants' ability to reproduce sexually is at the heart of their evolutionary success. Because sexually reproducing populations generate large amounts of genetic diversity, they are better at keeping up with changing conditions by adapting to new environments.

Sexual reproduction (and the resulting genetic variation in a population) also turns out to be crucial in agriculture, because plant breeders can selectively breed plants for the most sought-after features—the size of the corncob, the speed at which the oranges ripen, or the scent or color of a rose. Tremendous gains in agricultural productivity have come from selective breeding programs, particularly with major crops such as wheat, rice, corn, and soybeans (FIGURE 20-4).

The process of sexual reproduction has some common steps in plants and animals. Meiosis is an essential part of the process by which individuals produce gametes. The gametes come together to produce a fertilized egg, which develops and grows by mitosis. Finally, the fertilized egg is transported to a place where the embryo can survive and grow. These steps make up what is called a **life cycle.**

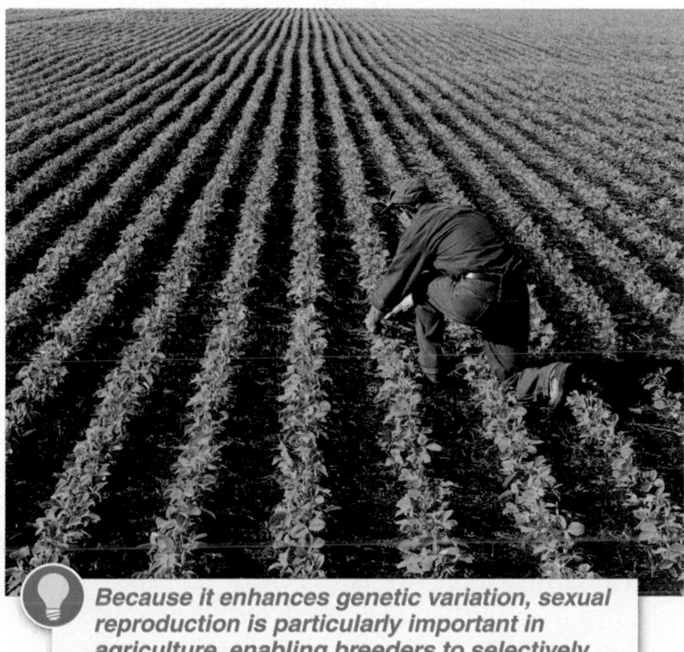

Because it enhances genetic variation, sexual reproduction is particularly important in agriculture, enabling breeders to selectively breed plants with the most desirable features.

FIGURE 20-4 **Breeding a better crop of soybeans.**

Because plants cannot move, the male gamete must reach the female gamete without the parent plant uprooting itself to find an individual to mate with. Thus, the plant life cycle includes steps that increase a plant's ability to safely deliver the male gamete to the female gamete. We trace the life cycles of the non-flowering, non-seed plants and the flowering plants—the angiosperms (which all have seeds)—in FIGURE 20-5. (For more details, see Sections 14.6 and 14.10.)

Life Cycle of Non-flowering, Non-seed Plants (Such as Ferns)

1. One form of the plant "body" that we see is diploid (but there are also multicellular haploid forms).
2. Production of gametes:
   a. Through meiosis, the diploid body, called a **sporophyte,** produces haploid cells called **spores.** Although spores are haploid, they *are not* gametes.
   b. The haploid spores are carried by wind or water to new soil, where they **germinate**—that is, after a period of dormancy, they begin to grow.
   c. Through mitosis, the haploid spores develop and grow into the second form of the plant "body" that we see: a new, haploid plant, called a **gametophyte.**
   d. Through mitosis, the new haploid body—the gametophyte—produces gametes.
3. Fertilization: Male and female gametes come together at fertilization.
4. The fertilized egg develops and grows, by mitosis, into a new diploid body, and the cycle begins again.

Note that in non-flowering, non-seed plants, spores are carried by wind or water to soil where they can germinate and grow into a gamete-producing form. Also, the male and female gametes are assisted, usually by rain or wind, in coming together for fertilization. The flowering plants have a slightly different strategy for dealing with immobility. The spores never leave the body of the diploid plant; rather, it is the fertilized egg inside a seed that is dispersed to a new location, as described in Section 20.12. The flowering plants also find a way for fertilization to occur, as described in Section 20.7.

Life Cycle of Flowering Plants

1. The plant "body" that we see is diploid.
2. Production of gametes:

LIFE CYCLES OF NON-FLOWERING, NON-SEED PLANTS AND FLOWERING PLANTS

NON-FLOWERING, NON-SEED PLANTS

1 DIPLOID PLANT "BODY"
The plant "body" that we are most familiar with exists in a diploid form.

2 GAMETE PRODUCTION

- Haploid spores are produced through meiosis.
- The spores are carried by wind or water to new soil.
- The haploid spores grow through mitosis into the haploid form of the plant body.
- Through mitosis, the haploid body produces gametes.

3 FERTILIZATION
Male and female gametes come together at fertilization.

4 DEVELOPMENT OF NEW INDIVIDUAL
The fertilized egg develops into a new diploid body and the cycle begins again.

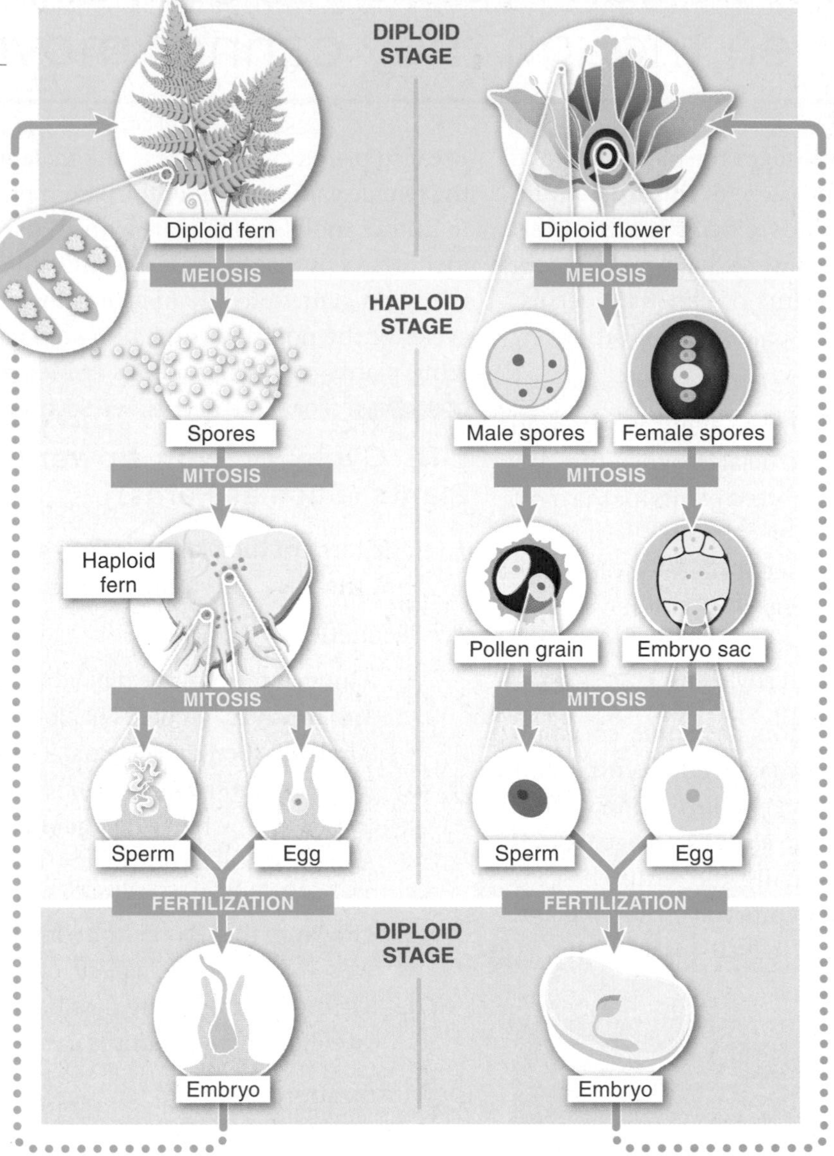

FLOWERING PLANTS

1 DIPLOID PLANT "BODY"
The plant "body" that we are most familiar with exists in a diploid form.

2 GAMETE PRODUCTION

- Haploid spores are produced through meiosis.
- The spores do not leave the diploid plant body (and can't survive away from it).
- The haploid spores divide by mitosis into either a female embryo sac or a male pollen grain.
- Through mitosis, the male gamete (sperm) and female gamete (egg) are produced.

3 FERTILIZATION
Male and female gametes come together at fertilization.

4 DEVELOPMENT OF NEW INDIVIDUAL
The fertilized egg develops into a new diploid body and the cycle begins again.

FIGURE 20-5 **Life cycles of plants.**

The life cycle in plants is called an alternation of generations because a period in which a plant is in a multicellular haploid form alternates with a period in which it is in a multicellular diploid form.

a. Through meiosis, the diploid body—the sporophyte—produces haploid spores. Although they are haploid, spores *are not* gametes.

b. The haploid spores do not leave the diploid plant body.

c. Within a flower on the plant body, the spores divide by mitosis to form a gametophyte that is either a haploid female embryo sac or a haploid male pollen grain.

d. Through mitosis, the haploid embryo sac produces a haploid female gamete (an egg), and the haploid pollen grain produces a haploid male gamete (a sperm).

3. Fertilization: Male and female gametes come together at fertilization. Following fertilization, a seed develops from the ovule (see Section 20.11.)

4. The fertilized egg (which typically has been moved away from the plant on which fertilization occurred; see Section 20.12 for a discussion of dispersal methods) develops and grows, by mitosis, into a new diploid body, and the cycle repeats.

The life cycle in plants is called an **alternation of generations,** because a period in which a plant is in a multicellular haploid form alternates with a period in which it is in a multicellular diploid form. If the same thing happened in humans, it would be the stuff of science fiction: the cells in our testes or ovaries, instead of producing sperm or eggs, would produce a special creature that would leave our bodies, live elsewhere on its own for a while, and then produce sperm or eggs. The eggs and sperm from one individual creature would then find some way to come together with sperm or eggs from another such creature, undergo fertilization, and produce babies.

In this chapter, we focus on the sexual reproduction of flowering plants (the angiosperms), because they are the predominant group of plants on earth. All angiosperms produce flowers, the structures in which the gametes develop and in which much of the reproductive activity takes place. First, we see how flowers ensure that male gametes travel to female gametes, even though individual plants are rooted in the ground.

TAKE HOME MESSAGE 20.3

» Many plants can benefit from producing genetically varied offspring through sexual reproduction, with the flower as the chief structure for sexual reproduction.

20.4–20.6 Flowers have several roles in plant reproduction.

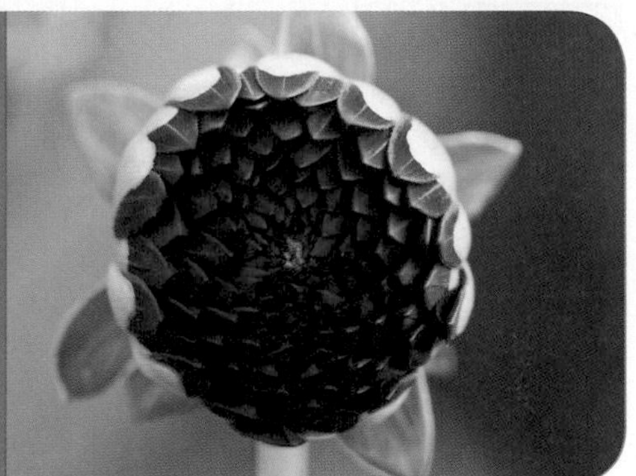

An emerging Withypitts dahlia, also known as a black wizard.

20.4 The flower is the chief structure for sexual reproduction.

Take a look at FIGURE 20-6, which showcases the diversity of flower shapes and colors. Despite appearances, all flowers, with only a few exceptions, have the same fundamental structures. Flowers typically have four distinct parts (FIGURE 20-7). (We'll see later, however, that not all flowers have all four parts. Some may have only male

FLOWER SHAPE DIVERSITY

Water lily

Daffodil

Kalanchoe

Angel's trumpet

FIGURE 20-6 The captivating and varied colors, textures, and shapes of flowers.

Flowers can vary greatly in shape, but, with only a few exceptions, they all have the same fundamental structures.

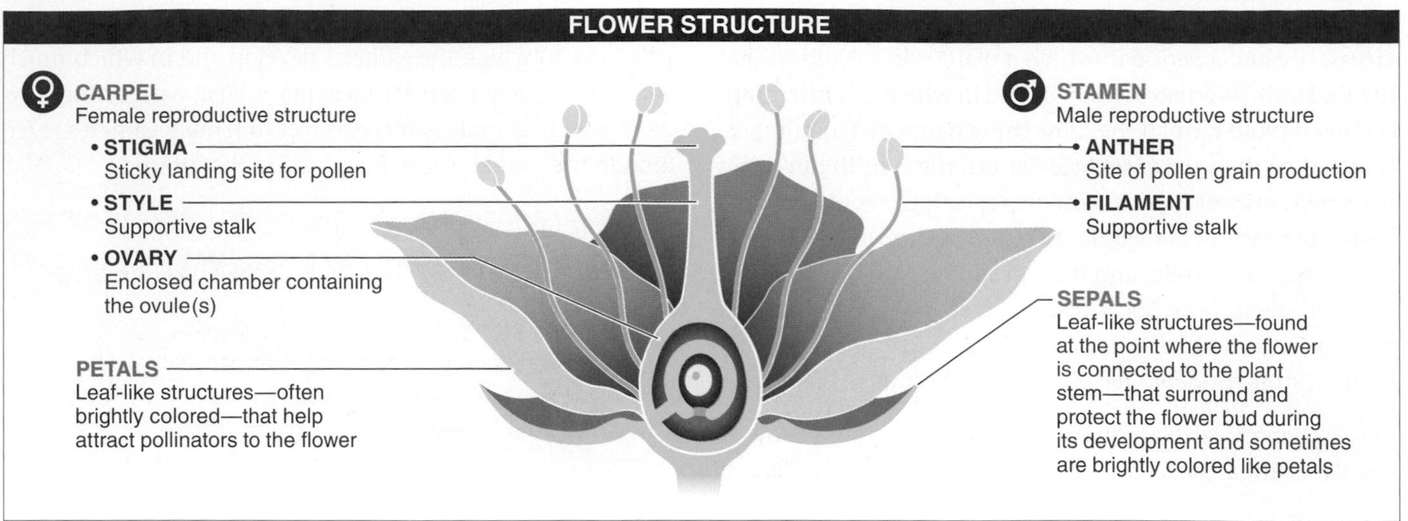
FLOWER STRUCTURE
CARPEL
Female reproductive structure
• STIGMA
Sticky landing site for pollen
• STYLE
Supportive stalk
• OVARY
Enclosed chamber containing the ovule(s)
PETALS
Leaf-like structures—often brightly colored—that help attract pollinators to the flower
STAMEN
Male reproductive structure
• ANTHER
Site of pollen grain production
• FILAMENT
Supportive stalk
SEPALS
Leaf-like structures—found at the point where the flower is connected to the plant stem—that surround and protect the flower bud during its development and sometimes are brightly colored like petals

FIGURE 20-7 **Most flowers are organized on the same general body plan.**

or only female reproductive parts, and some may lack sepals or petals.)

1. The **sepals** are leaf-like structures located at the point where the flower is connected to a main support structure of the plant, such as a branch, stem, or stalk. Sepals grow in a ring surrounding the outside of the flower, and they protect the flower bud during its development. Although sepals are usually green, in some flowers they are brightly colored and closely resemble the petals.

2. Just inside the ring of sepals is another ring of leaf-like structures, the **petals.** They are usually brightly colored—helping the flower to attract pollinators—and frequently have unusual shapes, from long and thin to short and broad.

3. Moving inward toward the center of the flower, the next structures are the **stamens,** the male reproductive parts. There are usually several stamens, each of which has a "head" on top of a long, thin stalk. The stalk is called the **filament,** and the head-like top structure is called the **anther.** The anthers are the sites where the **pollen grains**—the structures that contain the male gametes, or sperm—are produced. The typical anther on a cornflower produces about 3,000 pollen grains.

4. In the center of the flower is the **carpel,** the female reproductive structure. There is usually only one carpel, and it is usually shaped like a long-necked vase. At the very top of the carpel is a flat, sticky surface called the **stigma,** which functions as a landing pad for pollen. The stigma is held high, out of the center of the flower, by the **style,** a long thin structure that leads down to an enclosed chamber, the **ovary.** Within the ovary are one or more **ovules,** which produce the female gametes (eggs). After an egg is fertilized, the ovule develops into a seed.

The prize for the most valuable stigma probably goes to *Crocus sativus,* a flowering plant native to Southwest Asia that is more commonly known as the saffron crocus (FIGURE 20-8). Collected and dried, the stigma and style

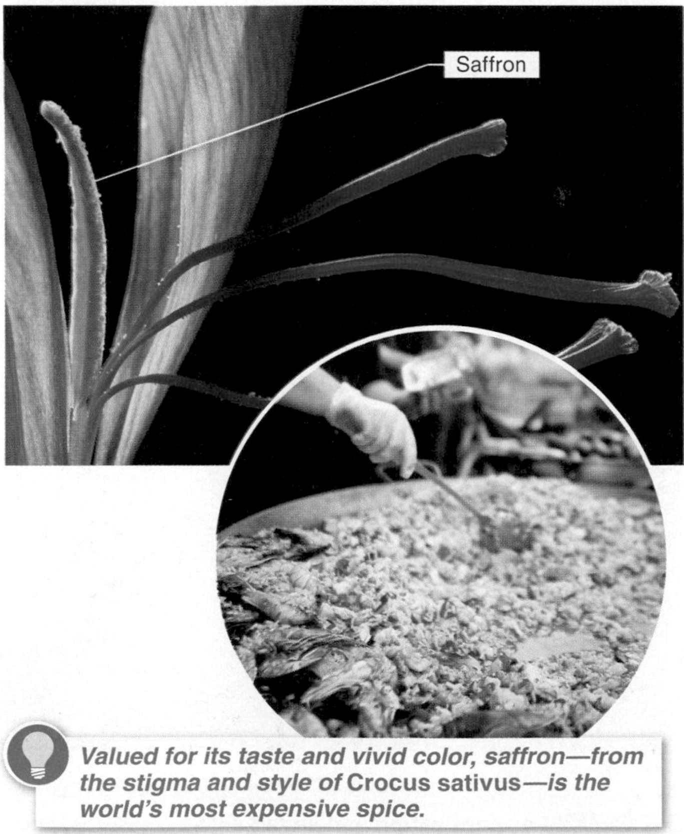
Saffron

Valued for its taste and vivid color, saffron—from the stigma and style of Crocus sativus*—is the world's most expensive spice.*

FIGURE 20-8 **Saffron: the world's most expensive spice.**

produce a deep yellow powder, valued by food lovers for its taste, sometimes described as slightly bitter and earthy. Common in Persian and Indian dishes, saffron is the world's most valuable spice: a pound (requiring the stigma and style from about 75,000 flowers) sells for more than $2,000.

TAKE HOME MESSAGE 20.4

» Flowers are plant structures specialized for sexual reproduction. Most flowers have the same fundamental structures: sepals, petals, stamens, and a carpel.

20.5 The male reproductive structure produces pollen grains.

Pollen—so annoying to allergy sufferers—is the male gametophyte and is essential to plants' sexual reproduction. To form pollen, cells in the anther of a plant undergo meiosis to form haploid cells called spores. As the spores are produced, the anther grows and forms four chambers, sometimes called "spore sacs." Each chamber is filled with diploid cells called microspore mother cells. (To see the four spore sacs, slice the anther of a flower in half.) The microspore mother cells divide by meiosis, each cell producing four haploid microspores. The microspores then quickly divide by mitosis, forming two-celled grains of pollen with very complex, water-tight, sticky surfaces. This two-celled structure, the pollen grain, is a reproductive packet containing two haploid cells (FIGURE 20-9).

Q Is pollen the equivalent of plant sperm?

One of the two cells in a pollen grain will eventually grow to form a **pollen tube** (FIGURE 20-10), aiding in fertilization. The other will divide once to produce two **sperm** cells. But those steps don't usually occur until the pollen grain has made its way from the anther into the environment and come to rest on a stigma.

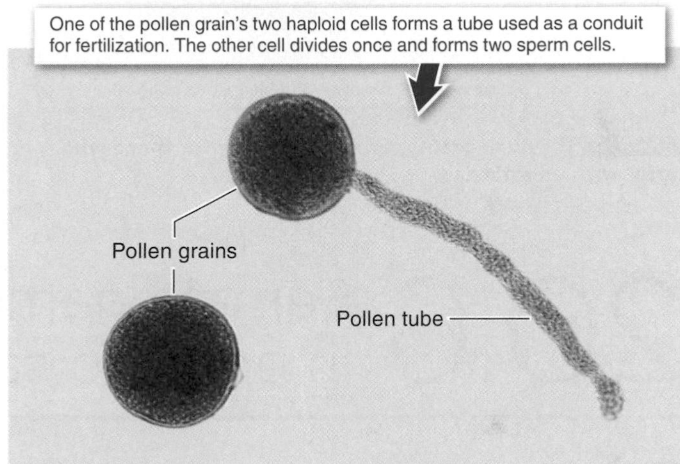

One of the pollen grain's two haploid cells forms a tube used as a conduit for fertilization. The other cell divides once and forms two sperm cells.
Pollen grains
Pollen tube

FIGURE 20-10 **A pollen grain contains two haploid cells.**

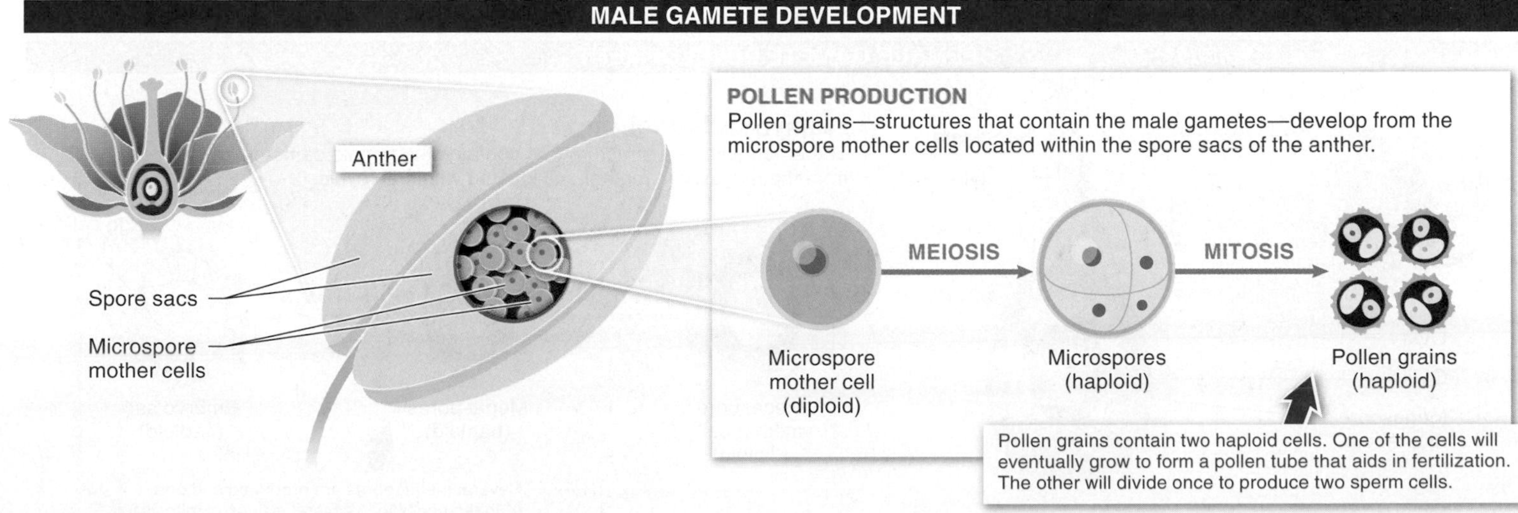

FIGURE 20-9 **The anther of a flower produces pollen grains containing the male gametes.**

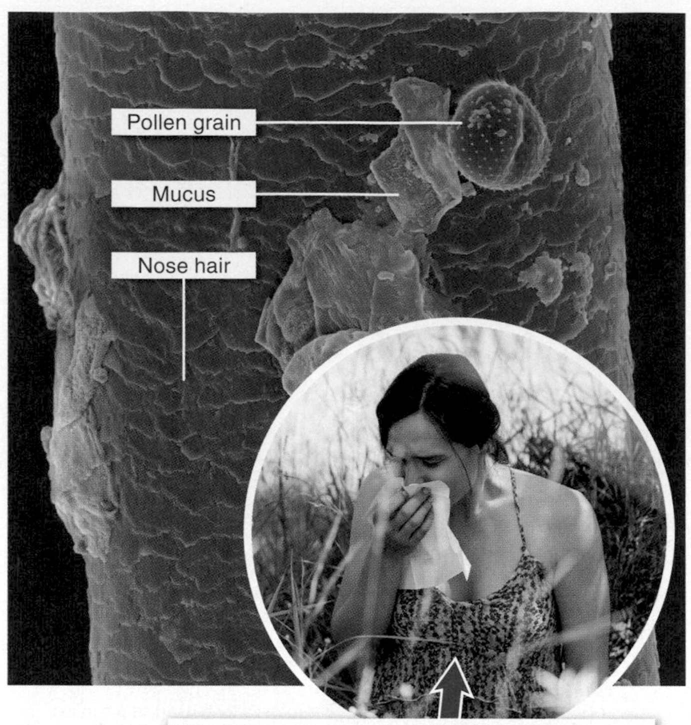

Pollen grain
Mucus
Nose hair
Identifying pollen grains as foreign invaders, our bodies may mount an immune response that includes a runny nose, sneezing, coughing, and congestion.

FIGURE 20-11 Pollen grains can cause misery for those who suffer with allergies.

Tens of millions of people suffer sneezing, coughing, and watery eyes from allergies to certain types of pollen. The trouble begins each spring and summer, as plants release billions and billions of tiny pollen grains into the air. Proteins project from the cell wall of every one of these pollen grains, increasing the pollen's stickiness. Inevitably, we breathe in some pollen grains, and our bodies immediately detect the pollen's surface proteins and identify them as foreign invaders (FIGURE 20-11). In response to this invasion, our immune system attacks and tries to flush the pollen from our bodies. Mucus on nose hairs helps move the pollen out of the nose. However, sometimes the response is extreme. The runny nose, sneezing, coughing, and congestion common to pollen allergies are all manifestations of the body's over-reaction.

TAKE HOME MESSAGE 20.5

» The male reproductive structure of flowering plants produces pollen grains. Each grain is a two-celled structure that is water-tight and sticky. One of the cells in the pollen grain will form a pollen tube, and the other will divide to produce two sperm cells.

20.6 Female gametes develop in embryo sacs.

Production of a plant's female gametes (eggs) doesn't cause the suffering in some humans that pollen grains do, primarily because the production is confined to the closed structure of the ovary. Within the ovary, one or more diploid cells differentiate into ovules. Each ovule is made up of outer protective cells that surround a diploid, egg-producing

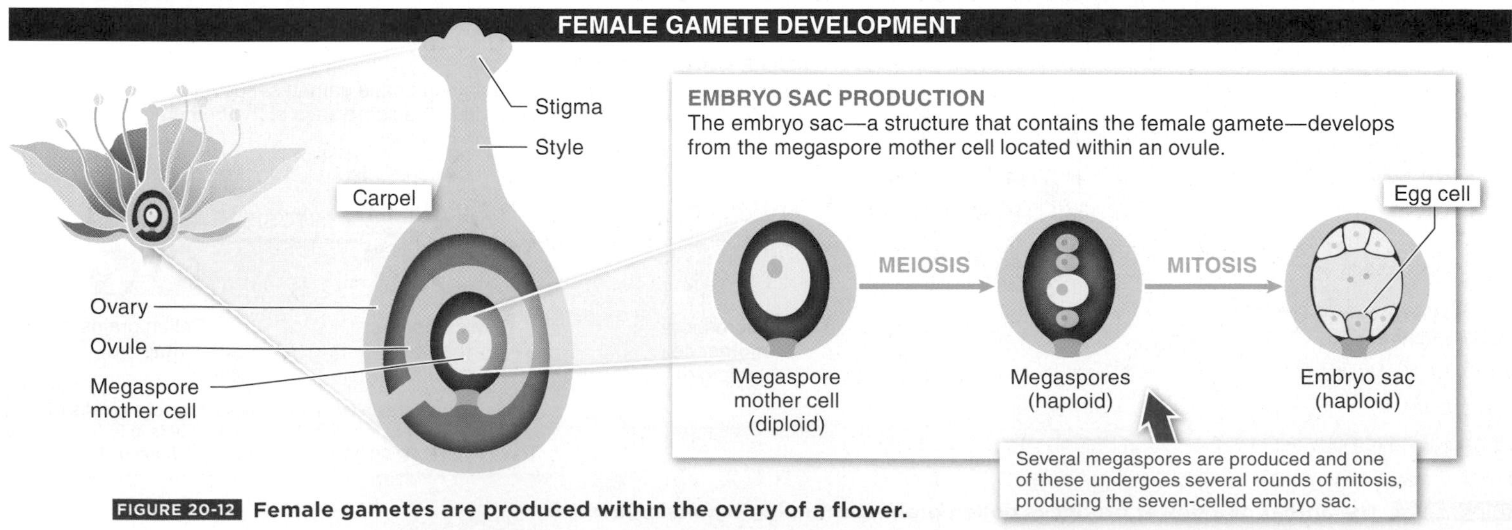

Several megaspores are produced and one of these undergoes several rounds of mitosis, producing the seven-celled embryo sac.

FIGURE 20-12 Female gametes are produced within the ovary of a flower.

cell called a megaspore mother cell. The megaspore mother cell undergoes meiosis to produce four haploid megaspores. Within the ovule, three of the haploid megaspores disintegrate, and the fourth undergoes mitosis several times to produce the **embryo sac,** the structure in which fertilization will occur (FIGURE 20-12). The embryo sac is an unusual collection of seven cells. Six of these cells—including the one that is the egg—have haploid nuclei, and the seventh cell, called the central cell, has two distinct haploid nuclei. The embryo sac waits for a male gamete to arrive.

TAKE HOME MESSAGE 20.6

» Within the ovary, diploid cells differentiate into ovules. Each ovule is a group of protective cells surrounding a diploid egg-producing cell, which produces four haploid megaspores. One of these megaspores produces the embryo sac, the structure that contains the egg and is the place where fertilization occurs.

20.7–20.12 Pollination, fertilization, and seed dispersal often depend on help from other organisms.

A honeybee coated with pollen can transport male gametes.

20.7 Plants need help getting the male gamete to the female gamete for fertilization.

For sexual reproduction to occur, a pollen grain from a plant must journey to the stigma of another plant of the same species. This process is called **pollination.**

About 10% of plant species achieve pollination by simply releasing their pollen into the wind (grasses and pine trees, for example) or into water (eelgrass, for example), on the slim chance that—through luck—some of the pollen will land on the female reproductive organs of another plant of the same species. To increase their odds of pollination, wind- and water-pollinated plants produce tens of millions of pollen grains per plant. Wind pollination is not particularly efficient, but it works.

In the other 90% of plants, animals act as "go-betweens," moving pollen from one plant to another. A flowering plant attracts the animal with its flowers' visual cues (color, shape), olfactory cues (smell), or even tactile cues (soft, bristly, hard, rough, smooth). Regardless of the method of attraction, the goal is always the same: to attach some male gametes to an animal's body so that they will rub off on the female reproductive parts of another plant, where fertilization can then occur.

Most animal-pollinated plants are pollinated by insects, although numerous other species are pollinated by birds or mammals (mostly bats), and a few by lizards (FIGURE 20-13). It is important to note that the data in Figure 20-13 are very imprecise estimates. There are several difficulties in establishing the exact proportion of plants pollinated by any group of animals. One difficulty is that the vast majority (~90%) of animal-pollinated flowering plant species are visited by more than one animal pollinator species. Another is that it is difficult to demonstrate that a particular animal is actually a pollinator of the plant (rather than a predator or just a harmless visitor).

How does a flowering plant get an animal to transport its gametes? Plants use two different, clever strategies for achieving pollination (FIGURE 20-14): (1) bribery—bribing

MEANS OF POLLINATION

Sexual reproduction requires two individuals. Many solutions have evolved that help plants to bridge the (sometimes very large) gap between plants.

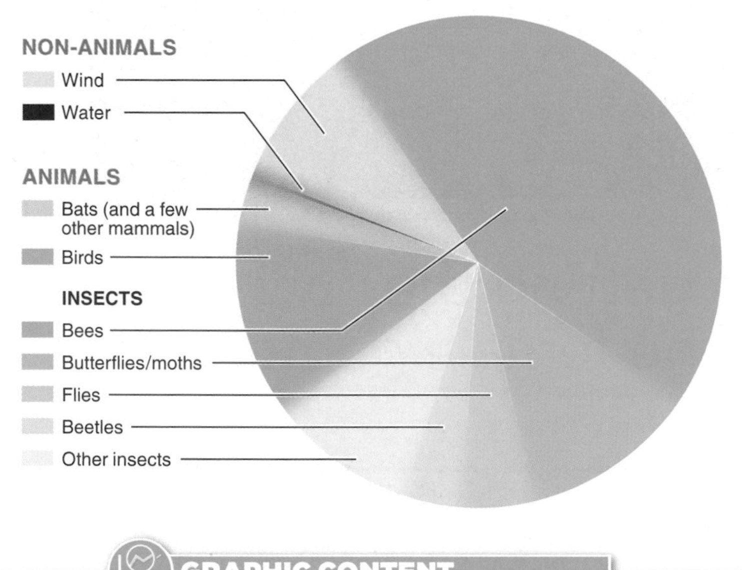

GRAPHIC CONTENT
Thinking critically about visual displays of data
Turn to p. 715 for a closer inspection of this figure.

FIGURE 20-13 Options for getting the male gamete to the female gamete.

STRATEGIES FOR ACHIEVING POLLINATION

BRIBERY
Plants offer something of value to an animal so that it carries pollen from one plant to another.

Cycad plants "trick" insects called thrips into pollinating them. With coordinated changes in smell and temperature, the plant induces thrips to move from male to female reproductive structures.

TRICKERY
Plants use methods of deception to trick animals into carrying pollen from one plant to another.

FIGURE 20-14 Gamete couriers. Plants have two general strategies for getting animals to transport pollen.

Q Why do some flowers smell nice while others don't smell at all?

animals to carry pollen from one individual plant to another, such as with **nectar,** a solution rich in sugars; and (2) trickery—deceiving animals into doing the job.

In each case, there has been strong coevolution between the plants and their pollinators: the plants have become more and more effective at attracting their pollinators and deterring other species from visiting the flower, while the pollinators have become more and more effective at exploiting the resources offered by the plants (see Figure 14-19).

TAKE HOME MESSAGE 20.7

» Most plants employ trickery or bribery to get the assistance of animals in carrying the male gametes to the female gametes. There has been strong coevolution between plants and their animal pollinators.

THIS IS HOW WE DO IT

Developing the ability to apply the process of science

20.8 Does it matter how much nectar a flower produces?

Most animal-pollinated plants use bribery: they produce nectar for consumption by visiting animals and get pollination in return. Would you expect this arrangement to lead to perpetual directional selection for plants that produce more and more nectar to attract more pollinators? At first that might seem reasonable. Within a plant population, an individual's fitness is closely linked to its relative success in reproducing.

But one of the most common themes in nature is the pervasiveness of trade-offs. Does producing more nectar always lead to higher fitness for the plant? Does nectar production have significant costs? These are important questions. How could we investigate them experimentally?

Researchers devised a powerful experiment to do this. By crossing petunia plants from two closely related species that differ in the average volume of nectar they produce, the researchers generated two distinct petunia populations. One population produced flowers with a low volume of nectar (low-nectar plants). Plants in the other population produced flowers with a high volume of nectar (high-nectar plants). The petunia populations resembled each other in all traits except nectar production. The low-nectar plants produced only 30% as much nectar as the high-nectar plants.

Armed with large numbers of plants from each of the two populations, the researchers evaluated whether nectar volume influenced the behavior of the plants' pollinator, a hawkmoth.

Are pollinators more attracted to plants that produce more nectar?

The researchers conducted 37 "choice" experiments in which they simultaneously presented one of each type of plant to an individual hawkmoth. They found that the hawkmoths showed no preference for either plant, approaching each with equal likelihood.

This finding didn't change when the researchers artificially increased the amount of nectar in the low-nectar flowers (by adding nectar with an eyedropper), creating plants with equal amounts of nectar. What additional observations would be helpful?

Do pollinators behave differently when probing flowers with different volumes of nectar?

What the researchers did notice was that hawkmoths spent significantly less time probing the low-nectar flowers (6.9 seconds, on average) compared with the high-nectar flowers (11.1 seconds). This 47% reduction in probing time on the low-nectar flowers disappeared when the researchers supplemented the flowers with nectar to the same level as the high-nectar flowers.

Does increased pollinator "probing time" benefit a plant?

From an evolutionary perspective, perhaps the most important observation in these experiments was this: The average number of seeds produced by each low-nectar plant was 328 (± 22), while the average number for the high-nectar plants was significantly higher, at 424 (± 25).

Is more nectar unambiguously a better strategy for petunias?

Based on these results, it appears that from an evolutionary perspective, it benefits a plant to produce flowers with a greater volume of nectar. But things are not quite so simple. Remember: benefits often come with costs.

In a clever twist, the researchers took samples of low- and high-nectar plants and hand-pollinated them. In this experiment, they discovered that the low-nectar plants were more effective at producing seeds—an average of 480 (± 20) seeds per plant, compared with the high-nectar plants' 383 (± 20) seeds.

It seems that producing more nectar reduces a plant's ability to produce seeds, given equal amounts of pollination. The optimum nectar volume, rather than being "as much as a plant can produce," more likely reflects a balance based on a trade-off between the ability to produce seeds and the need to attract pollinators that spend time probing the flowers.

TAKE HOME MESSAGE 20.8

» In plants producing nectar for their animal pollinators, the optimal nectar volume reflects a trade-off between the benefit of attracting pollinators that spend a long time at each flower and the cost of nectar production—a cost that may reduce a plant's ability to produce seeds following pollination.

20.9 Fertilization occurs after pollination.

Pollination is a bit like sexual intercourse in mammals: it brings the male and female gametes close to each other, but it isn't quite fertilization. Fertilization requires that the male and female gametes fuse so that their genetic material can be combined. We examine here the male gamete's last steps on its journey to the female gamete.

Recall that the pollen grain consists of two cells encased in a sticky, spiky wall. When the pollen grain lands on a stigma of a flower of its own species, it sticks to the stigma through interactions between the pollen grain's outer coating and the cells of the stigma. Pollen grains from other species have the wrong identifiers on the surface and slide off the stigma easily, keeping it open to the "right" pollen grains (FIGURE 20-15).

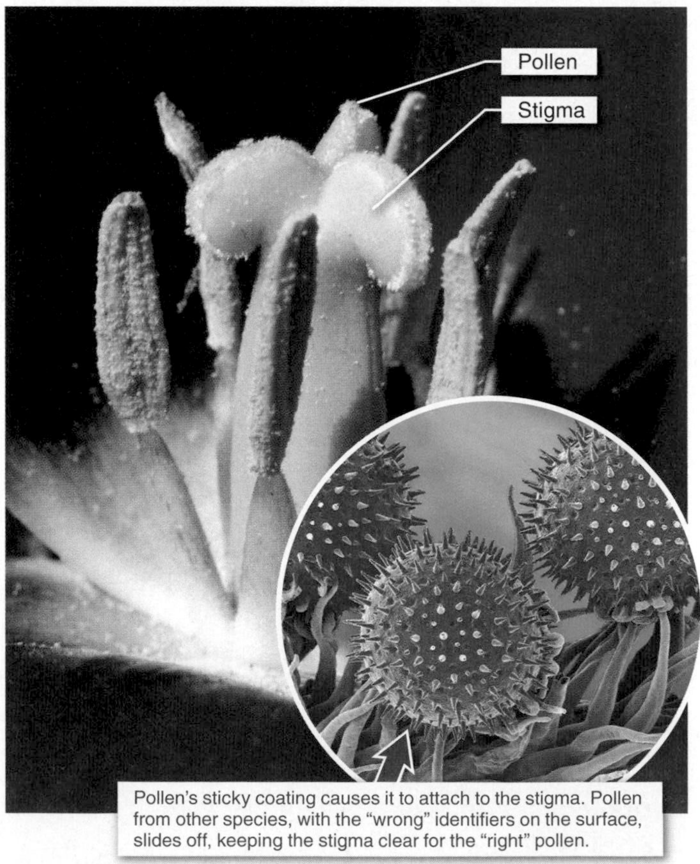

Pollen
Stigma
Pollen's sticky coating causes it to attach to the stigma. Pollen from other species, with the "wrong" identifiers on the surface, slides off, keeping the stigma clear for the "right" pollen.

FIGURE 20-15 Built to stick. Pollen grains adhere to the stigma of a flower.

Within 12–36 hours of the pollen landing on a stigma, one cell in the pollen grain starts to grow into a pollen tube, which elongates and pushes the sperm-producing cell downward, moving closer and closer to the ovary. This growth isn't an easy stroll down the style, however. It's more like running a gauntlet. The cells within the style test whether the tube cell is too closely related to the plant on which the pollen landed (as it would be if the plant self-pollinated). If the tube cell is too closely related, it is killed—often by a chemical reaction that breaks down the cytoskeleton of the pollen tube. Otherwise, the pollen tube is allowed to continue growing

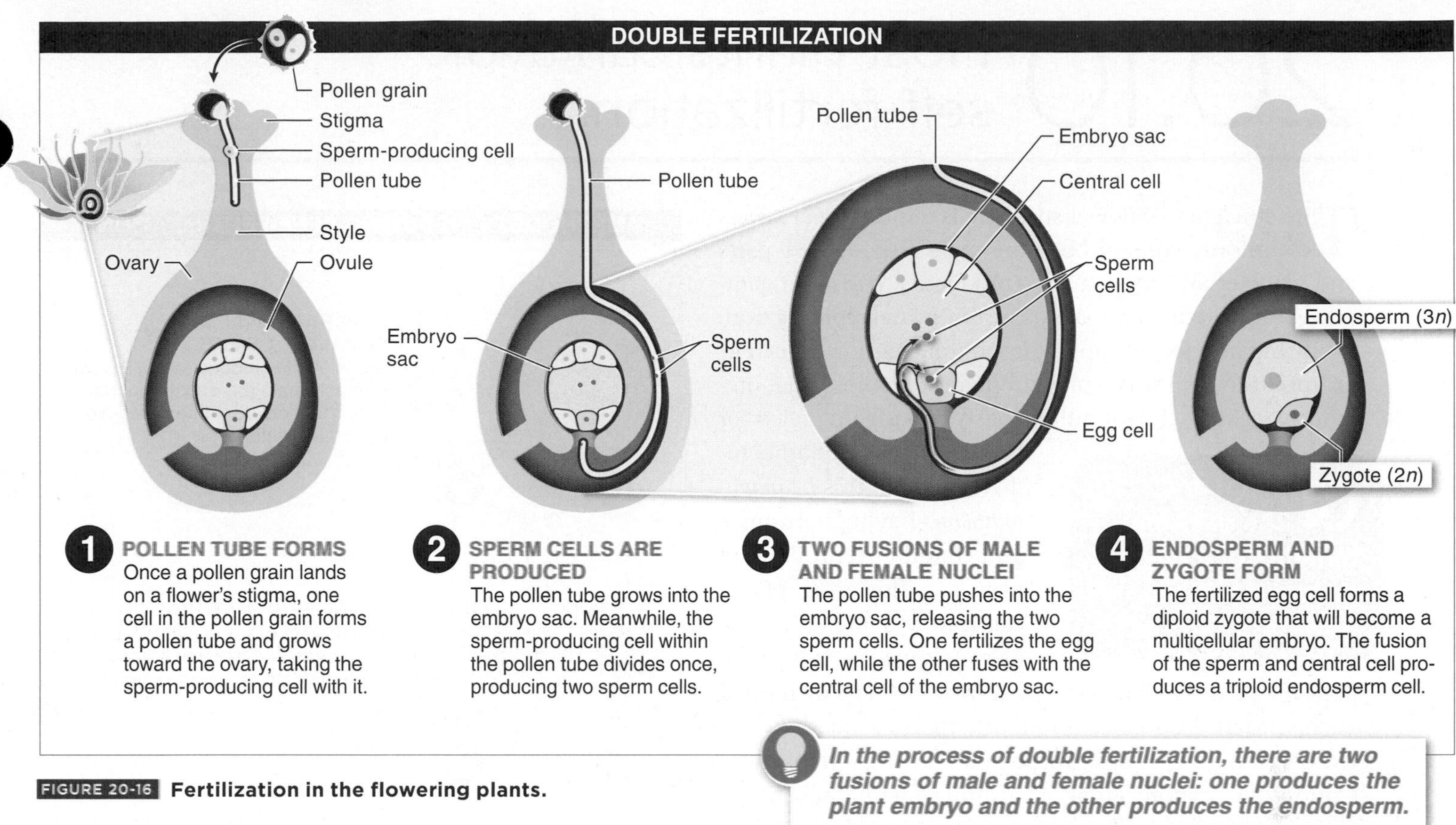

DOUBLE FERTILIZATION
Pollen grain
Stigma
Sperm-producing cell
Pollen tube
Style
Ovary
Ovule
Embryo sac
Pollen tube
Sperm cells
Pollen tube
Embryo sac
Central cell
Sperm cells
Egg cell
Endosperm (3n)
Zygote (2n)
1 POLLEN TUBE FORMS
Once a pollen grain lands on a flower's stigma, one cell in the pollen grain forms a pollen tube and grows toward the ovary, taking the sperm-producing cell with it.
2 SPERM CELLS ARE PRODUCED
The pollen tube grows into the embryo sac. Meanwhile, the sperm-producing cell within the pollen tube divides once, producing two sperm cells.
3 TWO FUSIONS OF MALE AND FEMALE NUCLEI
The pollen tube pushes into the embryo sac, releasing the two sperm cells. One fertilizes the egg cell, while the other fuses with the central cell of the embryo sac.
4 ENDOSPERM AND ZYGOTE FORM
The fertilized egg cell forms a diploid zygote that will become a multicellular embryo. The fusion of the sperm and central cell produces a triploid endosperm cell.
In the process of double fertilization, there are two fusions of male and female nuclei: one produces the plant embryo and the other produces the endosperm.

FIGURE 20-16 **Fertilization in the flowering plants.**

toward the ovary until it grows right into the embryo sac (FIGURE 20-16).

Meanwhile, the sperm-producing cell within the pollen tube divides once, producing two sperm cells. As the pollen tube pushes its way into the embryo sac, it releases the two sperm cells, and one of them fertilizes the egg cell inside the embryo sac. When the sperm and egg fuse in fertilization, a diploid cell called the **zygote** results. The zygote will develop by mitosis into the new diploid plant.

What happens with the other sperm cell? This cell fuses with the central cell of the embryo sac, which, as we learned earlier, has two haploid nuclei. This second fusion produces a triploid cell, with three sets of genes. The triploid cell is called the **endosperm.** As the embryo starts to grow and divide, so, too, does the endosperm. It continues to do so at a high rate, turning into nutritional tissue that nourishes the developing embryo.

Because, in this fertilization, there are two separate fusions of male nuclei with female nuclei—one producing the embryo and the other producing the endosperm—the whole process is called **double fertilization.** This process, which is unique to flowering plants, is an efficient system because when, and only when, an embryo is produced, so, too, is a ready-made food source. A large proportion of the calories we ingest when we eat plants comes from endosperm, including rice, the flour in pasta and bread, and the sweet material in corn on the cob.

Q What part of a plant supplies the majority of calories when eaten by humans?

TAKE HOME MESSAGE 20.9

» Pollination is necessary but not sufficient for achieving fertilization. Following pollination, a pollen tube must grow down the style and into the ovule, during which time the sperm-producing cell divides once to form two sperm cells. Released into the embryo sac, one sperm cell fertilizes the egg cell to form the zygote, and the other fuses with the central cell, which contains two haploid nuclei, to form the endosperm, the tissue that will nourish the developing embryo.

20.10 Most plants can avoid self-fertilization.

The efficiency of flower structure is remarkable. Flowers commonly contain both the male reproductive parts and gametes and the female reproductive parts and gametes, so a pollinator can easily assist in sexual reproduction. A visiting bee, for example, can pollinate a flower such as an apple blossom with pollen from another blossom it previously visited, while simultaneously picking up pollen for later fertilization of other individuals. But this *hermaphroditism*—having both male and female reproductive parts—in flowering plants can lead to some unintended consequences. What happens if a plant's male gamete fertilizes one of its own eggs? Such self-fertilization leads not only to less genetically varied offspring but also to offspring that are more likely to express one or more lethal or harmful genes because of the extreme inbreeding. How can plants avoid self-fertilization?

Q What are the risks and benefits of a plant's producing "bisexual" flowers?

Plants have many mechanisms for avoiding self-fertilization, including (as we saw in Section 20.9) chemical reactions within the cells of the style that destroy a growing pollen tube if it is too closely related to the plant being pollinated. Three of the most effective general mechanisms for reducing self-fertilization are described here (FIGURE 20-17).

1. **Separate male and female flowers.** In some plants, an individual produces some flowers with only male reproductive parts and others with only female reproductive parts. Overall, the plant is still a hermaphrodite, but these unisexual flowers minimize the likelihood of self-pollination and self-fertilization. Examples of plants with separate male and female flowers include corn and most fig species.

2. **Staggered maturation of male and female reproductive parts.** To prevent self-fertilization in hermaphrodite flowers, some plants have flowers with male parts that develop before the female parts, or vice versa. This pattern occurs, for example, in geraniums.

3. **Separate male and female plants.** The most extreme method of avoiding self-fertilization is for some plants within a population to produce only flowers with male reproductive parts, while others produce only flowers with female reproductive parts. This makes self-fertilization impossible, but it also reduces the likelihood that any given visit by a pollinator will result in pollination and fertilization. Examples of plants with separate male and female plants include poplars, American holly, and willows.

METHODS OF REDUCING SELF-FERTILIZATION

SEPARATE MALE AND FEMALE FLOWERS
Some plants produce some flowers with only male reproductive parts and other flowers with only female reproductive parts.

STAGGERED MATURATION
The flowers of some plants have male parts that develop before the female parts, or vice versa.

SEPARATE MALE AND FEMALE PLANTS
Some plants within a population produce only flowers with male reproductive parts, and other plants produce only flowers with female reproductive parts.

Plants have several effective mechanisms for reducing the incidence of potentially detrimental self-fertilization.

FIGURE 20-17 **Avoiding inbreeding.**

TAKE HOME MESSAGE 20.10

» **Plants can reduce the likelihood of self-fertilization in several ways: producing separate male and female flowers, or staggering the time of maturation of male and female reproductive parts, or having separate male and female plants.**

20.11 Following fertilization, the ovule develops into a seed.

After fertilization within the ovule, numerous cell divisions occur in rapid succession. The developing embryo forms a **root meristem** and a **shoot meristem,** each of which is a cluster of active, dividing embryonic cells that will generate new tissue. (We discuss the unique qualities of meristems in more detail in Section 20.14.) Also, one or two **cotyledons** form, structures in the embryo that usually become the embryonic leaves (see Section 19.2). As the embryo matures, the outer cells of the ovule develop into a hard casing. The result is a **seed**—a sort of "plant in waiting" that consists of the embryo (meristems and cotyledons) and any remaining endosperm, surrounded by the casing (FIGURE 20-18).

At the same time that the seed is forming, the ovary wall—surrounding one or more seeds—develops into a **fruit,** which in many species is fleshy, juicy, and edible, but in other species may be dry or inedible. The fruit both protects the seed or seeds inside and, as we'll see, aids in dispersal. Metabolism and oxygen consumption grind to a halt, and the seed dries out, reducing its water content to about 15% of its total weight. Then the waiting begins.

Humans have learned a trick that turns one type of seed into an unusual snack. Each kernel of popcorn comes from a single kernel of corn on a cob and is the equivalent of a fruit: it includes a fertilized egg surrounded by a nutritious, edible outer layer, formed from the ovary wall. (If you plant some popcorn kernels, they will grow!) There is also moisture within the kernel—some oil and some water—and the kernel is surrounded by a moisture-proof coat. Heating the kernel turns the water inside to steam and chemically alters the starch and protein in the seed, making it soft and a bit gelatinous. With continued heating, the pressure from the steam rises and ultimately blasts through the seed coat with a pop. The fluffy white material is the starch and protein that surrounded the seed.

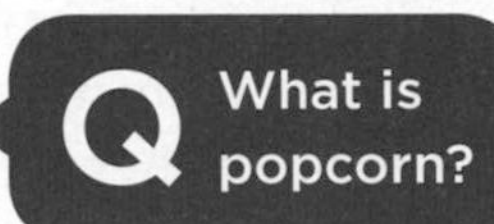

Before a seed can grow into a new plant, it must leave the parent plant. In the next section we look at how ever-immobile plants ensure that their seeds are dispersed and prepared for growth.

TAKE HOME MESSAGE 20.11

» Following fertilization, the ovule develops into a seed, containing an embryo having a root meristem, a shoot meristem, and one or two cotyledons, and also perhaps containing some endosperm, all surrounded by a hard casing. The seed is protected within a fruit, which aids in its dispersal.

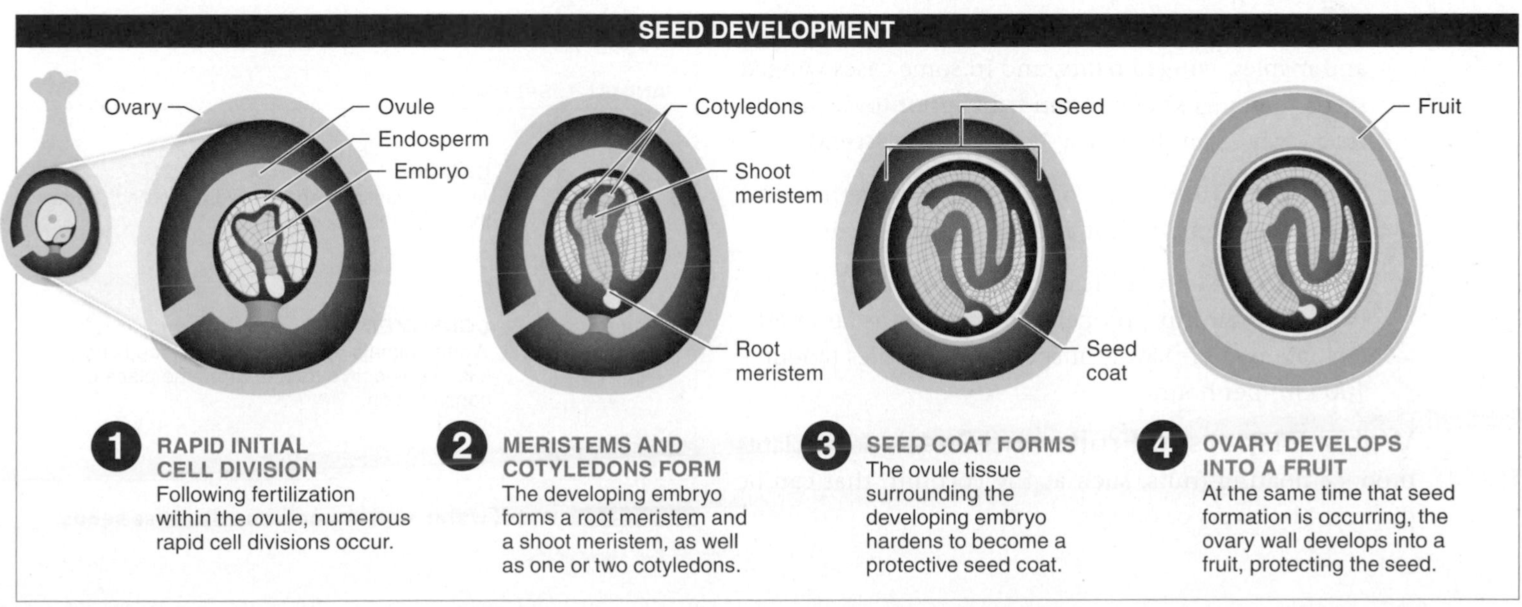

SEED DEVELOPMENT
Ovary
Ovule
Endosperm
Embryo
Cotyledons
Shoot meristem
Root meristem
Seed
Seed coat
Fruit
1 RAPID INITIAL CELL DIVISION
Following fertilization within the ovule, numerous rapid cell divisions occur.
2 MERISTEMS AND COTYLEDONS FORM
The developing embryo forms a root meristem and a shoot meristem, as well as one or two cotyledons.
3 SEED COAT FORMS
The ovule tissue surrounding the developing embryo hardens to become a protective seed coat.
4 OVARY DEVELOPS INTO A FRUIT
At the same time that seed formation is occurring, the ovary wall develops into a fruit, protecting the seed.

FIGURE 20-18 **Protected for later dispersal.** Seeds are protected by a casing and nestled within a fruit.

20.12 Fruits are a way for plants to disperse their seeds.

Every fruit you eat used to be a flower. Fruits are produced by plants to aid in dispersing their seeds, which will grow into new offspring. By encasing seeds in a juicy, nutritious structure—the fruit—plants create a package that entices hungry animals. Animals eat the fruit, gaining valuable nutrients and energy. Then, at a later point—and in a location far from the parent plant—the animals eliminate the seeds, which have passed unharmed through their digestive system. At the new location, often surrounded by a bit of feces that serves as fertilizer, the seed can grow.

There are many fruits beyond those adapted for consumption by animals. Just as with dispersal of pollen, for example, plants can also use wind or water to transport their seeds. FIGURE 20-19 presents some of the rich variety of fruits and seeds produced by plants. The term "fruit" applies to a much broader range of structures than we typically associate with this term. And the makeup of each fruit clearly reflects the method by which it is dispersed—wind, water, or animals.

Wind-Dispersed Fruits and Seeds Several characteristics of fruits and seeds make them well adapted for wind dispersal.

1. **Hairy.** Bushy hairs allow the fruits of dandelions and milkweed to float in the air.
2. **Winged.** Usually released from tall trees such as elms and maples, winged fruits, and in some cases winged seeds that are expelled from fruits prior to dispersal, float away from the tree as they slowly descend.
3. **Tiny, dust-like.** Many orchids have seeds that are as tiny and light as dust and so float in the air.
4. **Explosive.** Many fruits, such as those of mistletoe, explode when ripe, propelling the seeds as far as 50 feet (about 15 m) away at speeds of 60 miles (about 100 km) per hour.

Water-Dispersed Fruits and Seeds Some plants produce floating fruits, such as the coconut, that can be dispersed by rivers or oceans.

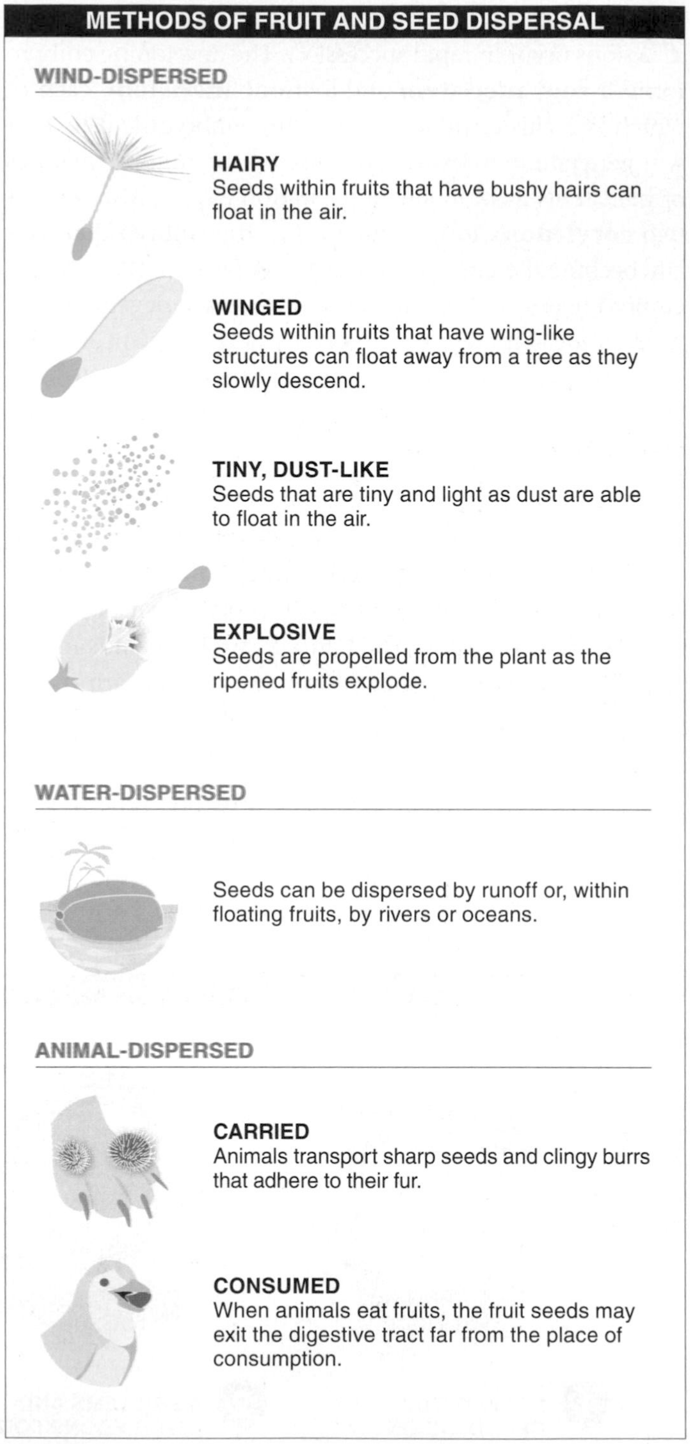

METHODS OF FRUIT AND SEED DISPERSAL

WIND-DISPERSED

HAIRY
Seeds within fruits that have bushy hairs can float in the air.

WINGED
Seeds within fruits that have wing-like structures can float away from a tree as they slowly descend.

TINY, DUST-LIKE
Seeds that are tiny and light as dust are able to float in the air.

EXPLOSIVE
Seeds are propelled from the plant as the ripened fruits explode.

WATER-DISPERSED

Seeds can be dispersed by runoff or, within floating fruits, by rivers or oceans.

ANIMAL-DISPERSED

CARRIED
Animals transport sharp seeds and clingy burrs that adhere to their fur.

CONSUMED
When animals eat fruits, the fruit seeds may exit the digestive tract far from the place of consumption.

FIGURE 20-19 **Wind, water, and animals can disperse seeds.**

Animal-Dispersed Fruits and Seeds Animals act as dispersers simply by carrying or eating fruits.

1. Carried. Some plants produce fruits that have sharp or clingy burrs that catch on the legs or fur of animals (including humans). The fruit is carried away until the animal grooms itself and removes it, leaving the seeds to germinate and grow in the new location. Burrs were the inspiration for Velcro.

2. Consumed. Many fruits are sugar- and nutrient-laden fleshy structures eaten by animals, usually vertebrates. In some cases (such as the peach), the fruit develops from a single ovary; in others (such as blackberries or pineapple), it forms from multiple ovaries from numerous flowers. Because it does not benefit a plant to have its seeds dispersed before they are ready, fruits generally do not ripen and become sweet and edible until the seeds are fully developed. In fact, most fruits remain green until the seeds are ready for dispersal, at which point the fruit sweetens and turns a bright color to attract the attention of dispersers.

Q Why do fruits taste so bad before they are ripe?

The seed's coating is tough enough to pass through an animal's digestive tract without being destroyed, and the seed can germinate and begin growing wherever the animal defecates (or, in the case of birds, regurgitates).

The various animal-dispersed fruits include tomatoes, grapes, cherries, olives, apples, pears, squash, zucchini, all beans and peas, corn kernels, and avocados. Because there appears to be no species of animal alive today that disperses avocado seeds, researchers have speculated that the avocado plant evolved in concert with some large animal species that has since become extinct. Today, avocado plants rely on humans to disperse their seeds. Humans are also the sole dispersers of the seed of the Osage-orange tree, whose fruit was once eaten (and seeds dispersed) by the now-extinct mastodon (FIGURE 20-20).

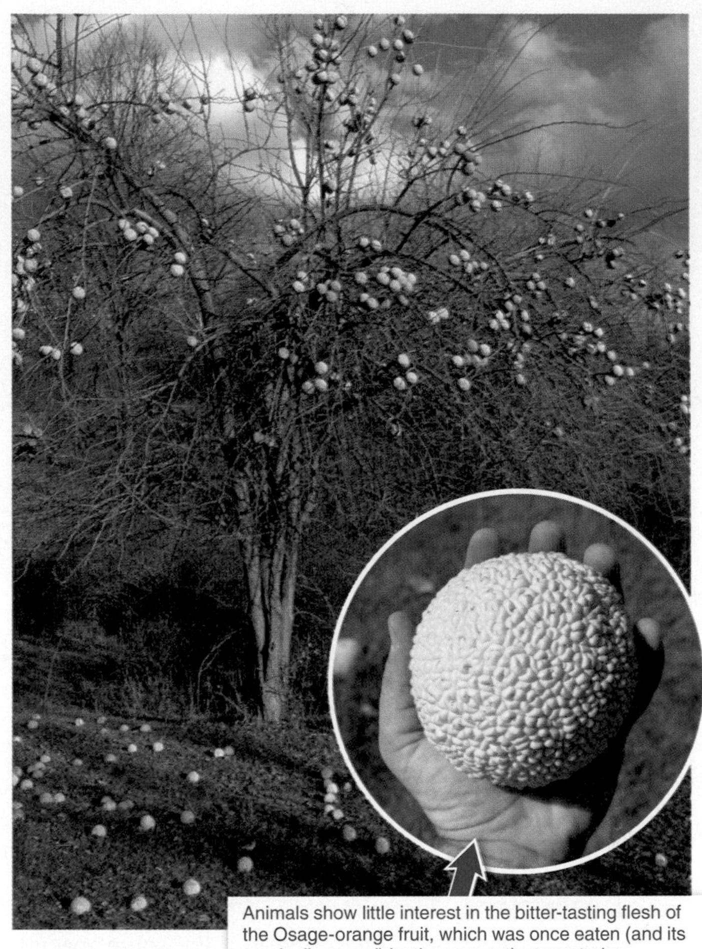

Animals show little interest in the bitter-tasting flesh of the Osage-orange fruit, which was once eaten (and its seeds dispersed) by the now-extinct mastodon.

FIGURE 20-20 Fruits of the Osage orange, *Maclura pomifera*, lie rotting beneath the tree.

TAKE HOME MESSAGE 20.12

» Seeds and fruits form following pollination and fertilization. Plants use the assistance of animals, water, or wind to disperse their fruits and seeds, which are deposited at new locations where the seeds can germinate and new plants can grow.

20.13–20.16

Plants have two types of growth, usually enabling lifelong increases in length and thickness.

Trees on a hill grow in a variety of shapes and sizes.

20.13 How do seeds germinate and grow?

Within the seed, the embryo patiently waits until the conditions for its new life are just right and then bursts forth with growth.

The seed does not start to grow until the water, temperature, and oxygen conditions are good for germination and growth (FIGURE 20-21). Sometimes this means waiting a matter of days or weeks before germination. For some species, though, the seeds remain dormant even in the face of ideal environmental conditions. They may require passage through the gut of a bird or mammal to roughen up and weaken the seed coat. Or they may require exposure to fire. Such seeds can remain dormant for dozens or even hundreds of years before finally germinating. Conversely, plant breeders are likely to have selected seeds that germinate quickly and have the fastest initial growth.

> **Q** The seeds of plants used in agriculture have thinner coats and more stored energy than the seeds of wild, naturally occurring plants. Why is this so?

There can be benefits to producing seeds with coats that require extra "processing" before they can germinate. (1) The need for extra processing can prevent the germination of seeds that have not been dispersed widely enough. (2) The need to pass through an animal gut can ensure that seeds are always left in a dollop of natural fertilizer. Or (3) exposure to fire may ensure that the seed germinates in clearings where competing trees have been burned down and light is available.

Seed germination occurs and growth begins when the water, temperature, and oxygen conditions for life are just right.

FIGURE 20-21 The emerging plant.

> "Though I do not believe that a plant will spring up where no seed has been, I have great faith in a seed. Convince me that you have a seed there, and I am prepared to expect wonders."
>
> — HENRY DAVID THOREAU, ***Faith in a Seed***

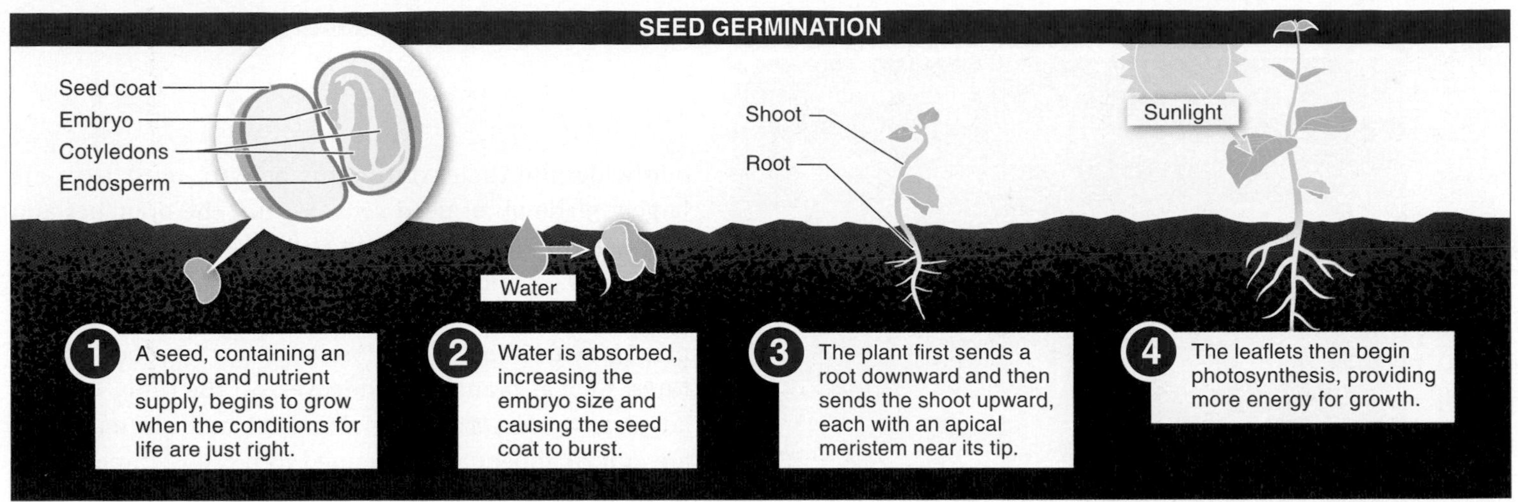

SEED GERMINATION
Seed coat
Embryo
Cotyledons
Endosperm
Water
Shoot
Root
Sunlight
1 A seed, containing an embryo and nutrient supply, begins to grow when the conditions for life are just right.
2 Water is absorbed, increasing the embryo size and causing the seed coat to burst.
3 The plant first sends a root downward and then sends the shoot upward, each with an apical meristem near its tip.
4 The leaflets then begin photosynthesis, providing more energy for growth.

FIGURE 20-22 From seed to seedling: the initial growth of a plant.

Such requirements, though, can also precariously link the evolutionary success of the plant to the behavior of another species or to specific ecological conditions. For example, some of the plants producing seeds that require extreme heat or smoke from a fire to induce germination are in trouble. As we have reduced the incidence of wildfires, so, too, have we reduced the ability of many plants to germinate.

Q Animals (including humans) love to eat seeds. Why?

Consider another risk that plants face in using seeds for offspring dispersal. The whole purpose of a seed is to package an embryo with a ready-made supply of nutrients and energy, so that it has the fuel to start growing before it can photosynthesize for itself. This high-energy package is exactly what animals are looking for in a food source, too. Consequently, seeds can be calorically rich snacks for animals, who may digest and absorb seeds rather than simply eliminating them intact, following passage through the gut.

But let's return to the seed as it is ready to begin its growth. With the proper processing and sufficient warmth, oxygen, and water, germination can finally begin. As germination begins, water is absorbed, causing the embryo to enlarge and the seed coat to burst. Utilizing the fat and starch reserves of the endosperm and embryo, metabolic activity increases dramatically and the plant begins growing (FIGURE 20-22). It first sends down a root and then sends up the shoot. In many plants, the cotyledons can photosynthesize after their energy reserves are used, providing even more energy for the plant.

TAKE HOME MESSAGE 20.13

» A seed, containing an embryo and a supply of nutrients, begins to grow only when the water, temperature, and oxygen conditions for life are just right. Seeds sometimes must pass through an animal's digestive system or be exposed to fire before they can germinate. Initial growth utilizes fat and starch reserves stored in the endosperm and embryo.

20.14 Plants grow differently from animals.

Plants and animals grow very differently. Three particular features stand out. First, in most animals, growth is determinate. In other words, after a period of maturation and "growing up," the growth more or less comes to an end. Most plants, though, keep growing taller and thicker their whole lives. Second, perhaps more

FEATURES OF PLANT GROWTH

Several features of plant growth are very different from growth in most animals.

- **INDETERMINATE GROWTH**
  Most plants continue to grow—both taller and thicker—for their entire life.
- **CONSISTING OF BOTH LIVING AND DEAD CELLS**
  Many plant cells, including water-conducting cells and those that make up wood, are dead.
- **LOSS OF STRUCTURES**
  Plants are able to lose relatively large structures, such as limbs and leaves, without adverse effects.

FIGURE 20-23 Defining characteristics of plant growth.

oddly, despite their continuous growth, most trees are largely made up of dead cells. Within the branches and roots, all the water-conducting cells are actually dead (see Section 19.3), and the majority of a large tree trunk—the wood—is dead, too. And finally, a distinctive aspect of plant growth is that plants lose relatively large body structures, such as branches and leaves, all the time, without adverse effects (FIGURE 20-23). Whereas limbs in mammals are crucial appendages intended to last a lifetime, many plant structures function more like our skin cells, which are regularly sloughed off and replaced by new ones.

Plants have two different methods of growing, each for a different direction of growth, which we explore in more detail in the next two sections. **Primary growth** causes plants to get taller, and **secondary growth** makes them thicker (and stronger). For both types, the cellular processes are similar and begin with regions called **meristems,** clusters of active, dividing cells. Acting like human stem cells, the meristems contain perpetually youthful cells that can (1) repeatedly divide and (2) develop into any type of plant tissue. Nowhere else in an adult plant body are there cells with these unlimited developmental options.

The two different types of plant growth depend on the activities of two different types of meristem (FIGURE 20-24).

TYPES OF PLANT GROWTH

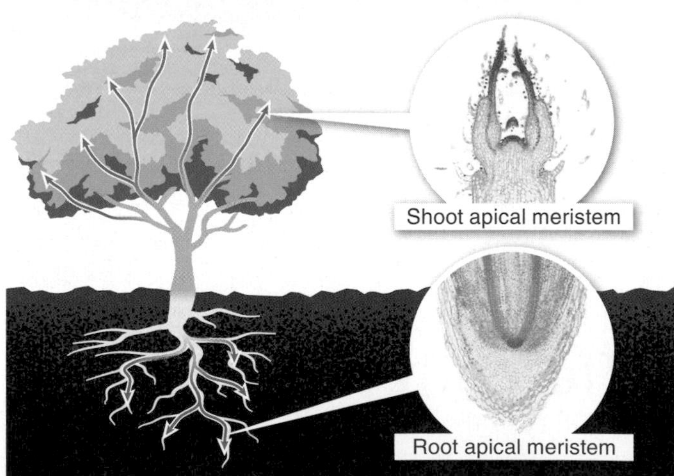

PRIMARY GROWTH
- Makes a plant taller and roots and branches longer
- Due to cell division within apical meristems present at the tips of shoots and roots

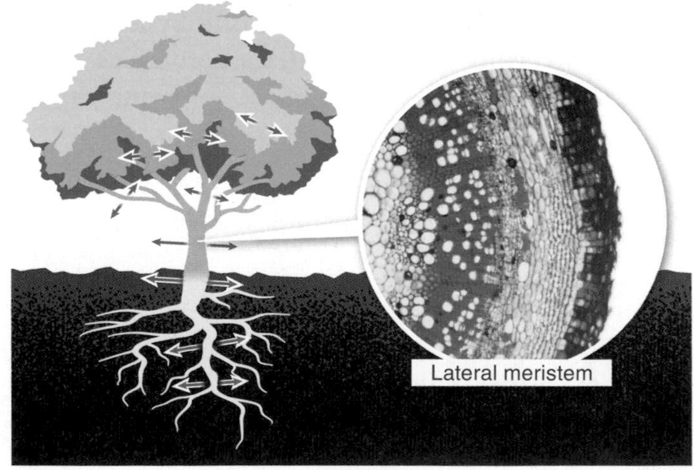

SECONDARY GROWTH
- Makes a plant stronger and roots and branches thicker
- Due to cell division within lateral meristems

FIGURE 20-24 Apical and lateral meristems are the source of plant growth.

FIGURE 20-25 **Taken to extremes.**

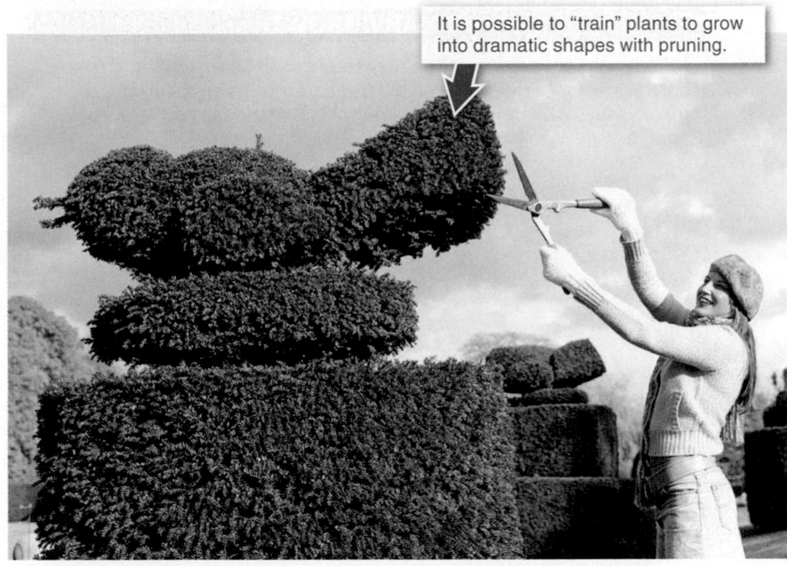

It is possible to "train" plants to grow into dramatic shapes with pruning.

1. *Apical meristems* are clusters of meristem cells at the ends of roots and shoots or branches that repeatedly divide to produce primary growth, making the plant taller and plant parts (roots and branches) longer. This process is in dramatic contrast to human growth. It would be as if we grew by continuously adding more material on top of our heads and at the ends of our hands and feet.

   Axillary buds that are lower on the plant shoot are inhibited from growth as long as the apical meristem is present. If the top meristem of the shoot is cut off, axillary buds lower on the shoot increase their growth. This causes the plant to grow "out" more than "up." Taking this approach to the extreme, gardeners have found that they can cause plants to grow into a variety of shapes through constant pruning (FIGURE 20-25).

2. *Lateral meristems*—not present in all plant species—give rise to secondary plant growth, the thickening of trunks and branches and the formation of wood and bark.

TAKE HOME MESSAGE 20.14

» Plants generally grow for their entire lives, using two types of growth. Primary growth makes plants taller and plant parts longer. Secondary growth makes plants thicker and sturdier.

20.15 Primary plant growth occurs at the apical meristems.

So, as we've seen, primary growth makes shoots taller and roots and branches longer. It also forms new tissues such as buds and leaves.

All primary growth occurs as a result of cell division in apical meristems (FIGURE 20-26). Meristem cells are totipotent—they have the potential to develop into any type of cell the plant is capable of producing. Growth occurs as a meristem cell divides and creates two new cells. One differentiates into a specific type of tissue cell, such as xylem, phloem, or perhaps a storage cell. The other cell remains a meristem cell. Thus, growth and differentiation can continue while the meristem remains a perpetual source of new cells. Apical meristem cells are present at all root tips and shoot tips, just behind a small cap of cells, at the very end of the tip of the root or shoot, that protects the meristem.

Apical meristem cells are also responsible for the production of branches as a plant grows. In the shoot, as the meristem cells divide, pushing the top of the plant higher and higher, some meristem cells are left behind at regular intervals. These meristem cells left behind may divide at any time—usually after stimulation by hormones. They then push outward to form a branch. As the branch grows, the meristem again leaves behind some cells near each bud, allowing further branching. Some of the new cells become xylem and phloem, and the newly formed vessels join up with the central circulatory vessels within the stem.

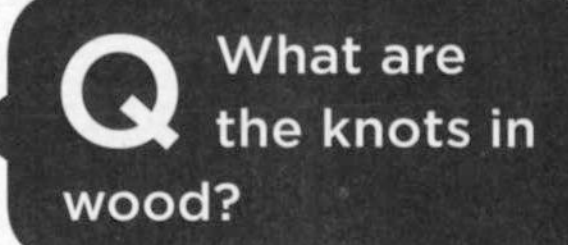

Q What are the knots in wood?

Most timber used in construction comes from pieces of wood cut from the tree along the length of its trunk. Knots

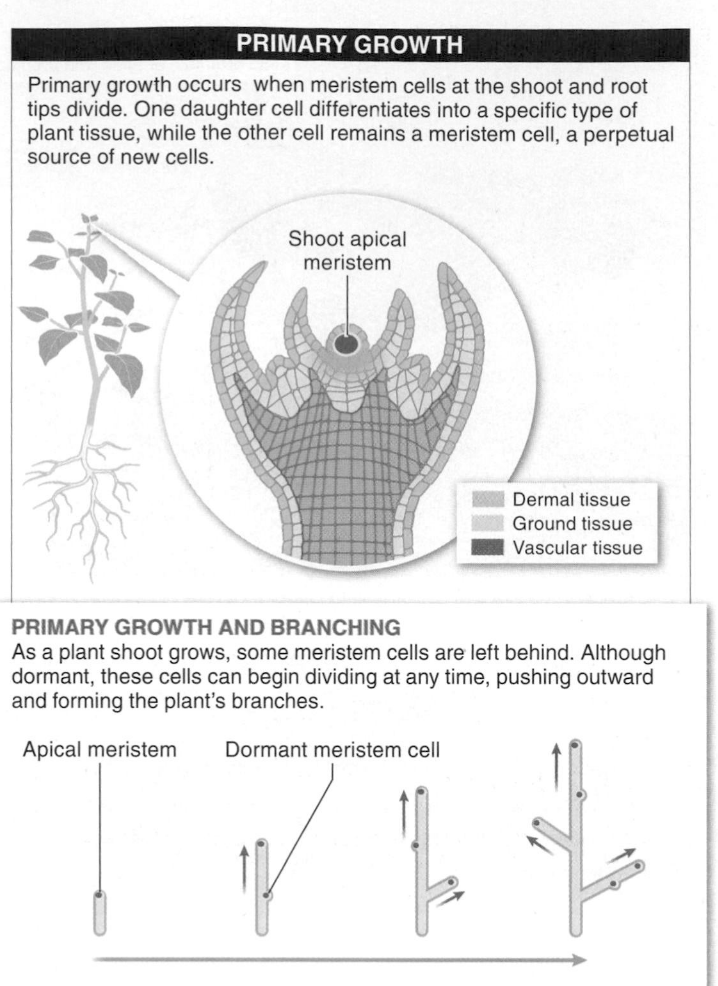

PRIMARY GROWTH
Primary growth occurs when meristem cells at the shoot and root tips divide. One daughter cell differentiates into a specific type of plant tissue, while the other cell remains a meristem cell, a perpetual source of new cells.
Shoot apical meristem
Dermal tissue
Ground tissue
Vascular tissue
PRIMARY GROWTH AND BRANCHING
As a plant shoot grows, some meristem cells are left behind. Although dormant, these cells can begin dividing at any time, pushing outward and forming the plant's branches.
Apical meristem
Dormant meristem cell

FIGURE 20-26 **Growing taller: primary growth.**

are places where branches were connected to the trunk and where their vascular tissue merged with that of the trunk. Knots become more and more deeply embedded within the trunk as secondary growth makes the trunk thicker.

TAKE HOME MESSAGE 20.15

» Plant growth occurs as a result of cell division in meristems, small collections of totipotent cells. Primary growth—the lengthening of stems, branches, and roots and the formation of new tissues such as buds and leaves—results from the division of apical meristem cells.

20.16 Secondary growth produces wood.

Wood is amazing. And not just from an aesthetic or utilitarian perspective—although it is quite remarkable in that way, too (FIGURE 20-27). The beauty of wood is that it both confers strength, keeping the tree from toppling over, and serves as the plumbing conduit for water and minerals. And it does this while being tremendously lightweight.

Once the apical meristems have moved past a location in a stem or trunk during primary growth, that part of the plant can no longer elongate or become taller. It can, however, grow outward and become thicker through secondary growth.

How does secondary growth occur? Imagine a cross section of a young woody plant. It is packed with concentric cylinders, one inside another (FIGURE 20-28). At the center, extending about halfway to the outer circumference, is the **pith,** consisting of soft spongy parenchyma cells. Just beyond the pith is a thin cylinder of vascular tissue, the xylem, which is near another cylinder of vascular tissue, the phloem. And beyond that is the epidermis and outer covering of the trunk. Sandwiched between the xylem and phloem is the **vascular cambium,** a layer of lateral meristem cells. Recall that meristems contain perpetually youthful cells that can repeatedly divide and develop into any type of plant tissue.

As the cells of the vascular cambium divide, they produce two rings of cells—one toward the outer circumference of the trunk and one toward the center. The

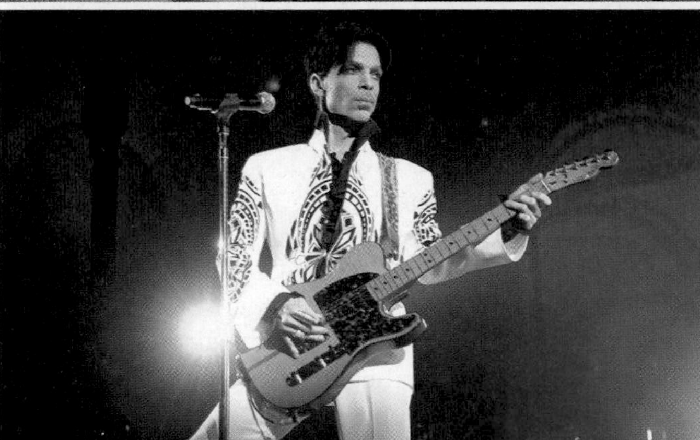

Humans utilize the strength, durability, and beauty of wood in a variety of applications.

FIGURE 20-27 **Wood is of great value to humans.**

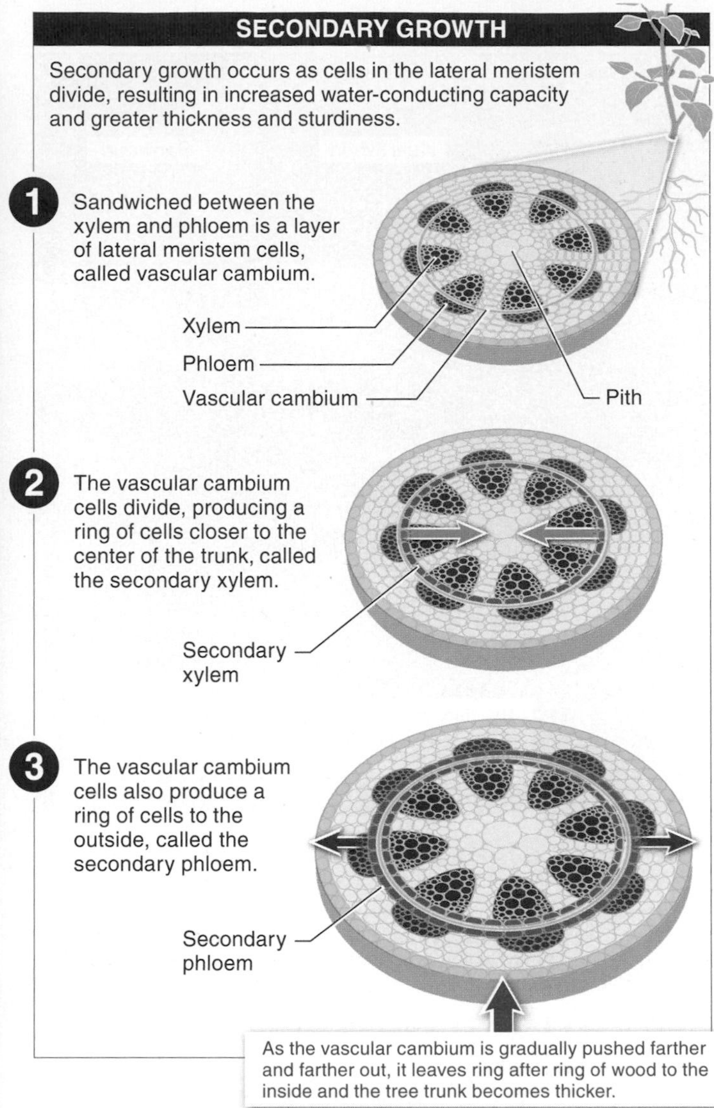

SECONDARY GROWTH
Secondary growth occurs as cells in the lateral meristem divide, resulting in increased water-conducting capacity and greater thickness and sturdiness.
1
Sandwiched between the xylem and phloem is a layer of lateral meristem cells, called vascular cambium.
Xylem
Phloem
Vascular cambium
Pith
2
The vascular cambium cells divide, producing a ring of cells closer to the center of the trunk, called the secondary xylem.
Secondary xylem
3
The vascular cambium cells also produce a ring of cells to the outside, called the secondary phloem.
Secondary phloem
As the vascular cambium is gradually pushed farther and farther out, it leaves ring after ring of wood to the inside and the tree trunk becomes thicker.

FIGURE 20-28 **Becoming thicker and sturdier: secondary growth.**

cells produced toward the outside are called **secondary phloem** and conduct sugars throughout the plant body. The cells produced closer to the center are **secondary xylem** and conduct water and minerals, while also providing structural support to the plant. We call these secondary xylem cells, collectively, "wood." Because the vascular cambium is producing cells toward the inside of the stem or trunk, it is gradually pushed farther and farther out, away from the center. As it is pushed farther out, it leaves ring after ring of wood to the inside, thickening the tree trunk.

Although most xylem initially acts like a long water pipe, older xylem is more like a clogged artery through which less and less blood can flow. Because the clogging of xylem is a progressive condition, the older the xylem, the more likely it is to be clogged. And the closer it is toward the center of the trunk, the older the xylem is. The clogged xylem is called "heartwood" and is darker, due to accumulated resins and other non-soluble metabolic waste molecules. Although heartwood cannot conduct water, it is valuable to the plant for the structural support it provides. Farther toward the outside, closer to the vascular cambium, is the xylem that was more recently created and, consequently, is

WOOD STRUCTURE

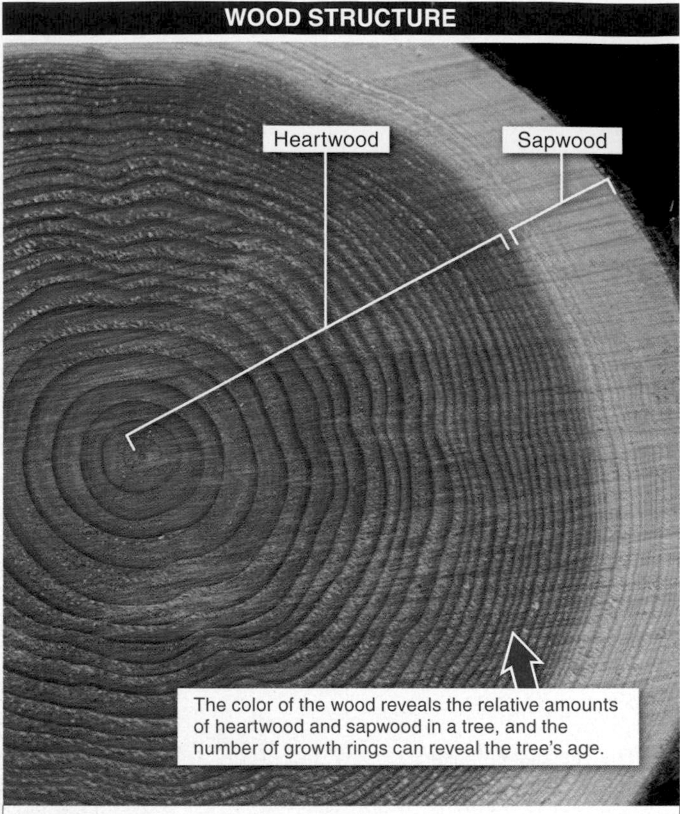

Heartwood
Sapwood
The color of the wood reveals the relative amounts of heartwood and sapwood in a tree, and the number of growth rings can reveal the tree's age.

STRUCTURAL COMPOSITION OF A TREE TRUNK

- Most of the cells are secondary xylem, also known as wood, and are dead.
- Secondary growth occurs in a ring called the vascular cambium, a lateral meristem, which produces new xylem on the inside and new phloem to the outside.

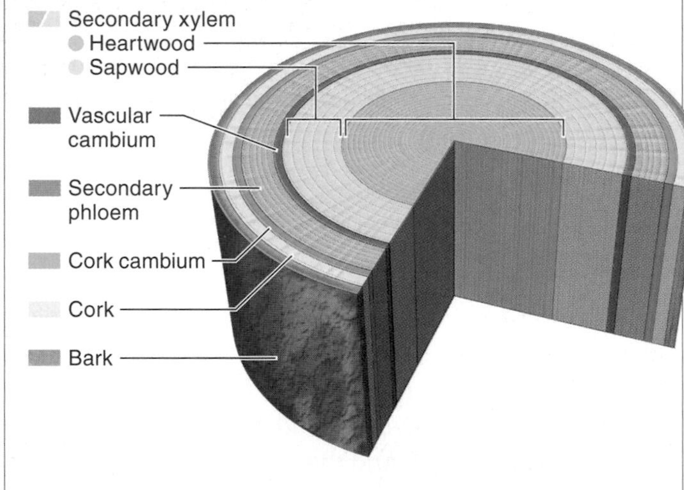

FIGURE 20-29 **Cross section of a tree trunk.**

better at conducting water. This xylem is called "sapwood" (FIGURE 20-29).

Looking at a cross section of a tree, we find that the color of the wood reveals the relative amounts of heartwood and sapwood. It can also reveal the age of the tree, in the growth rings. In temperate climates, more growth—that is, greater rates of cell division in the vascular cambium—occurs during the spring and summer, causing a wider, lighter band. During winter and fall, growth is slower and produces a darker band. Wider bands also indicate wetter years, and narrower bands drier years. By counting the number of rings, we can accurately determine the age of the tree. In the tropics, there is less variation in growth from one season to another, so trees don't always lay down rings, and it isn't always possible to accurately establish their ages.

Q Why can you tell a tree's age by counting its rings?

As the vascular cambium increases the girth of a tree and is pushed farther and farther out, the epidermal cells that make up the bark continuously split and fall from the tree. This tissue is replaced by the **cork cambium**—a cylinder of lateral meristem cells close to the outer edge of the trunk. The cork cambium plays an important role in the perpetual task of maintaining the protective covering of bark. As these cambium cells divide, they produce a cylinder of waxy cork cells that protects the outer surface of the trunk from water loss, fire, and infection by microbes. Secondary phloem, too, is pushed toward the outer surface of the trunk and eventually is sloughed off. For this reason, at any given time, although there may be a huge amount of functioning xylem, there is only a very thin layer of functioning phloem, and it is very close to the outer surface of the trunk. Often, only the phloem produced within the past year is still actively conducting sugars throughout the plant.

Secondary growth increases the girth of a tree but doesn't elongate it. So if you were to carve your initials in the trunk of a tree, the section below your initials would never increase in height. Moreover, because increases in girth can cause the outer layers of bark and epidermis to fall off, to

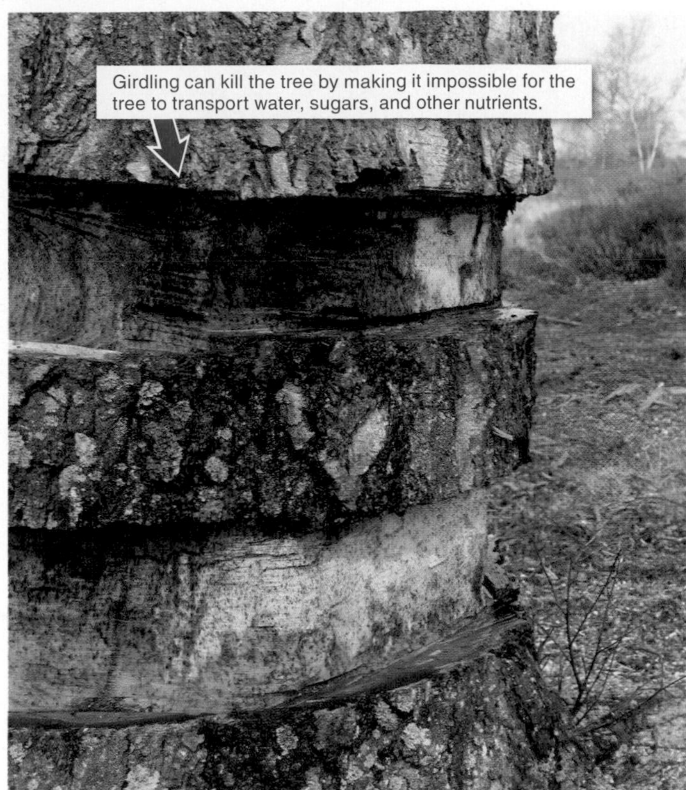
Girdling can kill the tree by making it impossible for the tree to transport water, sugars, and other nutrients.

FIGURE 20-30 **Girdling a tree can kill it.**

Q If you thoughtlessly carved your initials in the trunk of a tree 4 feet above the ground and came back in 10 years, would your initials be at the same height or higher? Why? What if you came back in 20 years?

be replaced by new cells pushing outward, your initials would eventually completely disappear from the tree trunk.

An action that damages the bark of a tree all the way around its circumference is called "girdling." It is possible to girdle a tree by accident with a "weed whacker" or a lawn mower. Girdling can kill a tree, even if the tree is very thick and the damage is not particularly deep. Because the phloem and vascular cambium layers are so close to the outer surface of the trunk, relatively shallow injuries to the trunk can destroy the secondary phloem and, if the damage goes just a bit deeper, the cells that give rise to new phloem.

Q Bumping into a tree repeatedly with a lawn mower or "weed whacker" can kill the tree. Why?

When such damage occurs, the tree is still able to transport water from the roots to all of the tissues, because the xylem is much closer to the center of the trunk and isn't damaged. But damage to the phloem makes it impossible for the tree to transport sugars and other nutrients from the sites of photosynthesis down to the roots where they are needed and stored. Shortly after the damage occurs, the tree may die (FIGURE 20-30).

Roots can be girdled, too. This kind of girdling can happen when trees are planted with wire or mesh netting around their roots. As the roots increase in girth, the metal or nylon can gradually destroy the vascular cambium as the roots grow outward, killing the tree about 5–10 years after it was planted.

It is an odd fact about plants, especially trees, that most of their cells are dead. Remember (from Section 19.3) that all xylem cells, including the secondary xylem cells that make up wood, are dead at maturity. In a huge oak tree, for example, most of the tree's mass is in the wood, and more than 98% of the cells may be dead. Nonetheless, the tree has buds that keep forming and growing, and it has vascular cambium cells that continue to divide to produce new xylem and phloem. As long as there are living cells in a tree, it can grow and is considered alive.

TAKE HOME MESSAGE 20.16

» Secondary growth results from cell divisions in a thin cylinder of tissue between the primary xylem and the primary phloem—the vascular cambium, a lateral meristem. As this tissue divides, it produces a ring of non-living xylem cells, close to the center of the trunk, that conduct water and minerals, while also providing structural support to the plant. We call these cells, collectively, wood.

20.17–20.20

Hormones regulate growth and development.

A weightlifter hoists the winning marrow (120 pounds) at a "giant vegetable" competition.

20.17 Hormones help plants respond to their environments.

Nature versus nurture. The complex interactions between nature and nurture that influence phenotypes are just as important to plants as they are to animals. How does the plant change its growth patterns to respond to different environments? Through the use of hormones.

Plant **hormones** are chemical signals that enable plants to respond quickly and appropriately to changing environmental variables (such as amount of moisture, amount and direction of sunlight, and temperature). These chemicals convey information about the physiological state of the plant's tissues or the environment in which the plant finds itself and then regulate some aspect of the metabolism of the plant's cells. Sometimes, a hormone may stimulate a certain response—such as growth—in the target cell. Other times, the hormone may suppress an action in the target cell.

The hormones in humans and other animals are generally produced in specific glands or tissues and then transported (commonly in the bloodstream) to different locations to exert their influence. In contrast, many plant hormones are produced in numerous places throughout the plant body and may have their effects in those same places or may be transported to another part of the plant before taking effect. As we will see, the effects of plant hormones are usually predictable and easily observed.

PLANT HORMONES

HORMONE	FUNCTION	LOCATION
GIBBERELLINS	Increase the speed of seed germination; promote stem elongation; induce early blooming of flowers; increase fruit size	Shoot and root apical meristems; seeds
AUXINS	Stimulate stem elongation; control seedling orientation; stimulate root branching; promote fruit development	Apical meristems; immature plant tissue
ETHYLENE	Increases the speed at which fruit ripens; stimulates leaf dropping and the death of flowers	All parts of the plant, including the fruits
ABSCISIC ACID	Inhibits growth and reproduction; inhibits seed germination; stimulates closure of stomata	Leaves; fruits; root tips; seeds
CYTOKININS	Cause rapid cell division, in conjunction with auxin; induce seed germination; initiate new branches from lateral buds	Roots and fruits, primarily

FIGURE 20-31 The location and function of five types of plant hormones.

There are many plant hormones, many of which fall into one of five major groups: gibberellins, auxins, ethylene, abscisic acid, and cytokinins. FIGURE 20-31 provides a brief summary of the functions of these hormones. We explore them in greater detail in the next three sections.

TAKE HOME MESSAGE 20.17

» **Plant hormones are chemical signals produced by plant cells that enable the plant to respond to environmental variables (such as amount of moisture, amount and direction of sunlight, and temperature) and that influence its growth and development.**

20.18 Seed germination and stem elongation are stimulated by gibberellins.

Gibberellins are a group of about 125 related hormones that regulate a plant's growth, primarily by stimulating cell division and cell elongation. Gibberellins are produced in growing areas of a plant and are found throughout all plant structures. In some plants, only two or three different gibberellins have been reported, while in others, more than four dozen have been found. Gibberellins occur in the largest amounts in seeds, but they are also found in high concentrations in the shoot and root apical meristems. The areas with the greatest concentrations of gibberellins experience the most dramatic growth.

Gibberellins have four main types of effects (FIGURE 20-32).

1. Initiating seed germination. Gibberellins within the seed stimulate the production of enzymes that break down and metabolize nutrients stored in the endosperm. When gardeners apply additional gibberellins to seeds, the seeds can more quickly and efficiently use their energy reserves to germinate.

2. Elongating stems. Some gibberellins increase stem elongation by spacing the nodes farther apart, thus increasing the distance between branch points.

3. Inducing flowering. Some plants do not produce flowers until the nights are sufficiently short or long (as we discuss in Section 20.23), or until the plant is exposed to a certain degree of coldness. The application of gibberellins to such plants can cause flower production in the absence of the triggering event. In other plant species, including rose, gibberellins have been shown to play roles in both the initiation and regulation of flowering under normal conditions.

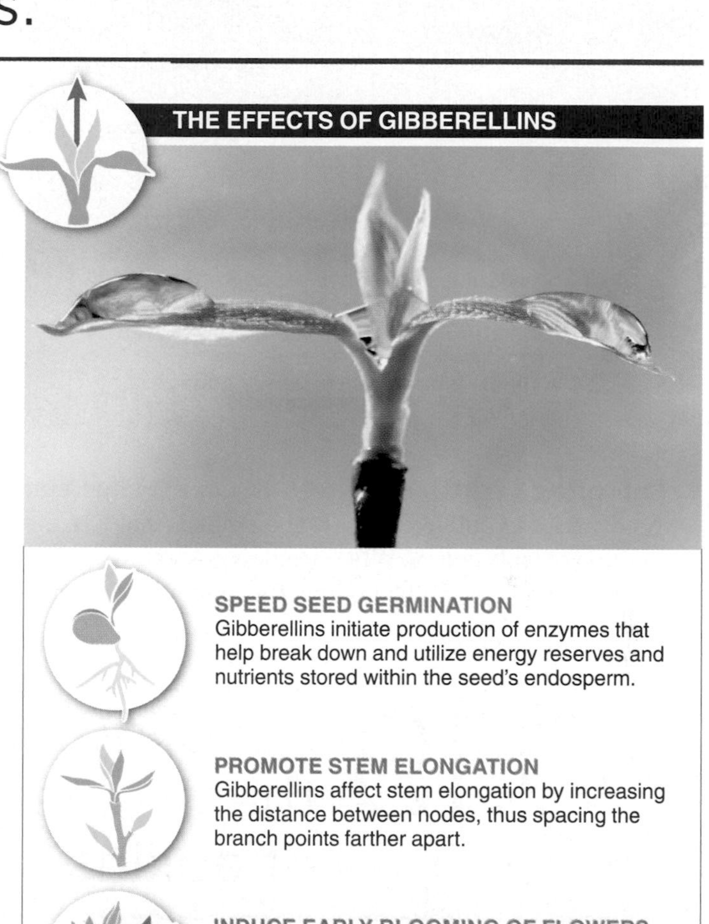

THE EFFECTS OF GIBBERELLINS
SPEED SEED GERMINATION
Gibberellins initiate production of enzymes that help break down and utilize energy reserves and nutrients stored within the seed's endosperm.
PROMOTE STEM ELONGATION
Gibberellins affect stem elongation by increasing the distance between nodes, thus spacing the branch points farther apart.
INDUCE EARLY BLOOMING OF FLOWERS
Gibberellins can cause flower production in the absence of a triggering event from the external environment.
PROMOTE FRUIT ENLARGEMENT
Seedless grapes sprayed with gibberellins grow larger and, due to stem elongation, have more space between the grapes on the bunch.

FIGURE 20-32 **Physiological effects of gibberellins.**

Gibberellins are powerful growth stimulators and, when applied in unnaturally large concentrations, can produce giant plants!

FIGURE 20-33 **A giant, gibberellin-treated cabbage.**

4. Enlarging fruits. One of the most important economic uses of gibberellins in the United States is the production of table grapes. When seedless grapes are sprayed with large amounts of gibberellins, the grapes grow larger, and due to the effect of gibberellins on stem elongation, there is more space between grapes on the bunch. Both of these effects increase the attractiveness of a bunch of grapes to shoppers. (Normally, plants don't ripen their ovary walls, producing fruits, unless they have set seeds, which need to mature. After all, that would be a waste of energy.)

Gibberellins are about as close to "miracle grow" chemicals as exist. For example, if a type of gibberellin is sprayed on cabbage plants, the plants produce huge cabbages (FIGURE 20-33) and the plants can grow to more than 10 feet tall!

Q How can farmers grow massive cabbages?

If plants grow more when they have greater amounts of gibberellins, why don't all plants simply produce more gibberellins so they can outgrow other plants competing for sunlight? The answer is that, as in most aspects of biology, there are trade-offs. The fast-growing cabbages, for instance, can grow to great heights only if humans secure them to a stake for support. Under natural conditions, their stems would not have the strength to stand at a height of 10 feet: the cabbage plants would fall over, wilt, and die.

TAKE HOME MESSAGE 20.18

» **Gibberellins are a group of about 125 related hormones, produced primarily in meristems and seeds, that regulate a plant's growth, mainly by stimulating cell division and cell elongation.**

20.19 Seedlings grow and orient themselves under the direction of auxins.

Auxins are a small group of naturally occurring hormones (and a larger group of synthetic variants that chemists can produce) that play several important roles in stimulating and regulating a plant's growth and development. They are found primarily in shoot tips and immature plant tissue, such as young leaves. The primary role of auxins in plant growth is to stimulate the expression of genes that promote cell division, stem elongation, the formation of roots, and the formation of vascular cambium. Auxins also influence a plant's orientation—its growth in response to directional light and gravity.

The chief effects of auxins are of four types (FIGURE 20-34).

1. Stimulating shoot elongation. Auxins enhance the effect of gibberellins in shoot elongation. As we saw in Section 20.14, primary growth in plants causes apical

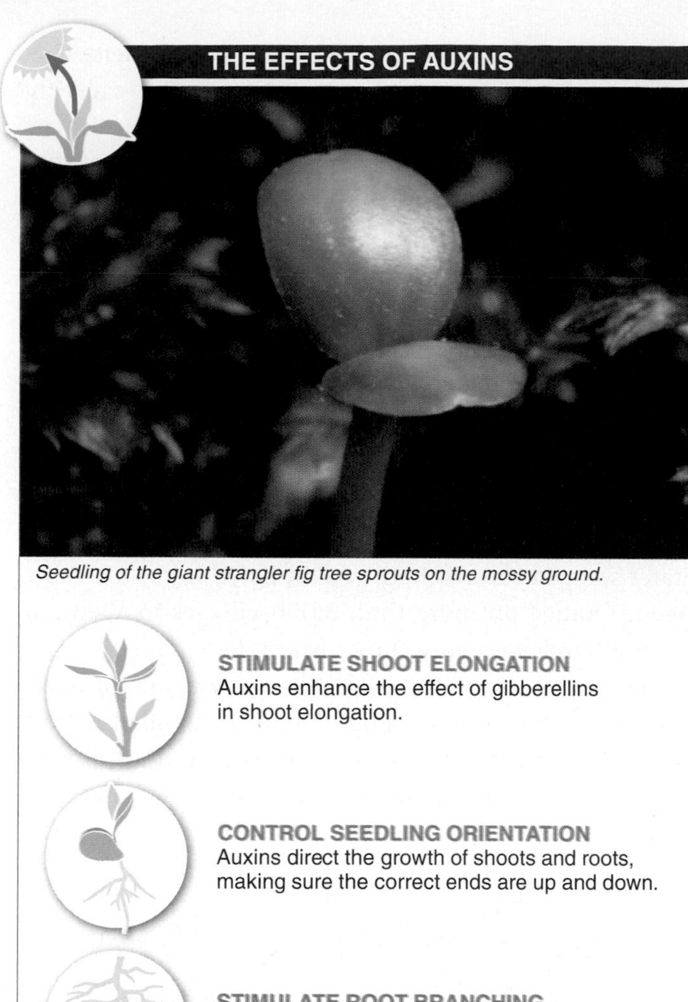

Seedling of the giant strangler fig tree sprouts on the mossy ground.

STIMULATE SHOOT ELONGATION
Auxins enhance the effect of gibberellins in shoot elongation.
CONTROL SEEDLING ORIENTATION
Auxins direct the growth of shoots and roots, making sure the correct ends are up and down.
STIMULATE ROOT BRANCHING
Auxins induce the formation of roots.
PROMOTE FRUIT DEVELOPMENT
Auxins produced within an embryo promote the maturation of the ovary wall and development of the fruit.

FIGURE 20-34 Physiological effects of auxins.

meristems to repeatedly divide. This causes plants to grow "up." Lateral buds, lower on the stem are also capable of rapid cell division and growth, but typically these lateral buds remain dormant.

When the top meristem of a shoot is cut off, however, lateral buds increase their growth. The inhibition of the lateral buds by the apical meristem is called *apical dominance* and is caused by the large amount of auxin produced by the apical bud. Cutting off the top meristem removes the inhibiting source of auxin, which is why the lateral buds then grow. This is the basis for "pruning" and makes it possible to control the branching pattern and shape of plants (see Figure 20-25).

2. Controlling seedling orientation. Immediately after emerging from the seed, the shoot grows as if it senses which way is up. And, as Charles Darwin and his son Francis were the first to document, seedlings exposed to light bend in whichever direction the light is coming from. Whether a shoot grows down into the earth or up toward the light actually depends on (1) where the auxins are located and (2) how the auxins influence the cells in those specific locations. Like gibberellins, auxins are produced near the growing tips of shoots and branches, but unlike gibberellins, they don't remain there. Auxin molecules move in two directions within a cell: they are pulled downward by gravity, and they move laterally away from light. These two movements distribute the auxin molecules unevenly within the plant. In regions of higher auxin concentration, cells in the stem elongate more rapidly than those in regions of lower auxin concentration (FIGURE 20-35). This difference in growth occurs because auxins increase the flexibility of the usually rigid cell wall (so that it can grow) and increase its permeability (so that it can take on more water and expand). The overall effect on growth is best seen when a plant is tipped onto its side. First, the auxins flow downward to the bottom side of the now-horizontal shoot. The bottom side then elongates more rapidly than the top side, causing the shoot to bend upward, against gravity.

3. Stimulating root branching. Auxins induce the formation of roots. In fact, you can buy auxin powders that can be dusted onto the bottom of plant stem cuttings. This causes the stems to send out numerous roots (transforming some stem cells into root cells), so that the cuttings can be planted and will form new plants.

4. Promoting fruit development. Auxins produced within an embryo have several roles in development of the fruit. First, they promote maturation of the ovary wall, and then they promote several steps in full development of the fruit. It is even possible to trick plants into producing fruits by applying auxins to unfertilized flowers. Because there has been no fertilization, such fruits are always seedless (for example, seedless tomatoes).

Q How can we trick plants into producing seedless fruits for our convenience?

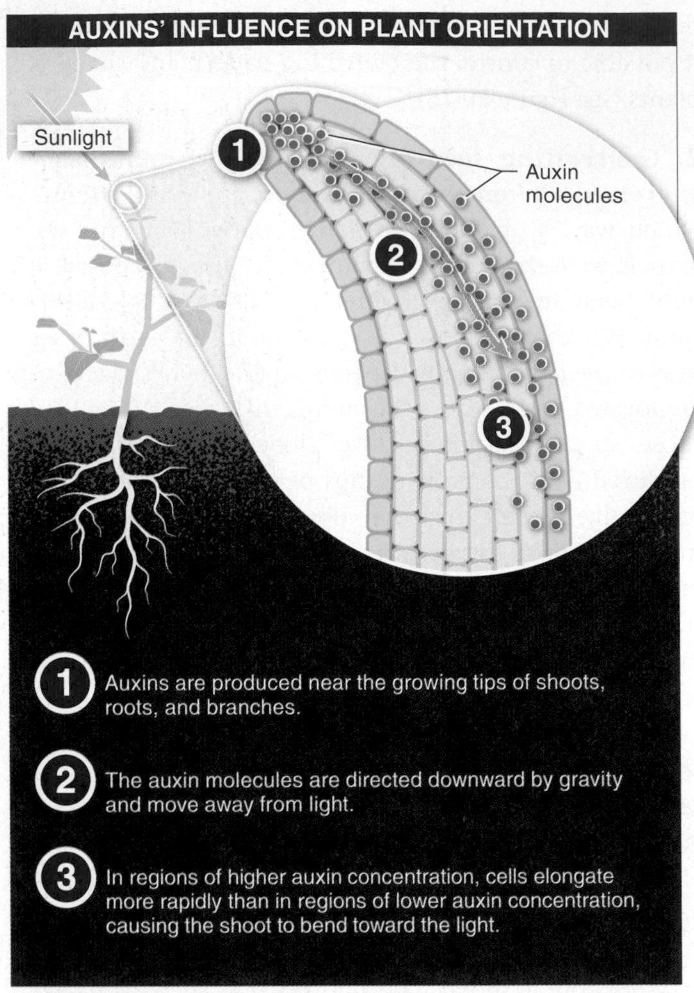

AUXINS' INFLUENCE ON PLANT ORIENTATION
Sunlight
Auxin molecules
1
2
3
Auxins are produced near the growing tips of shoots, roots, and branches.
The auxin molecules are directed downward by gravity and move away from light.
In regions of higher auxin concentration, cells elongate more rapidly than in regions of lower auxin concentration, causing the shoot to bend toward the light.

FIGURE 20-35 **Auxins cause a plant to grow toward light.**

Auxins applied to fruiting plants can also prevent the fruits from prematurely dropping from the tree or vine. These fruits are easier and less expensive to harvest, because they can all be picked at one time.

Synthetic auxins are sometimes used as weed killers. They are sprayed on plants in high concentrations and cause the plants to begin growing uncontrollably. Like a mismanaged start-up company, the plants devote so much of their energy budget to growth that they quickly find themselves without sufficient energy for essential metabolic maintenance functions. And then they die. One type of synthetic auxin, called 2,4-D, is a particularly useful weed killer because it kills only eudicots—such as dandelions—while not harming monocots, including grasses and cereals such as the valuable crop plants corn, rice, wheat, and barley. (The reason for this difference is not fully understood.)

The plant-destroying properties of synthetic auxins have even found military uses. During the 1960s, the United States sprayed the synthetic, auxin-based herbicide called Agent Orange on more than 3,000 villages in Vietnam, in an attempt to reduce the brush under which opposing military forces could hide. But the manufacture of Agent Orange creates a highly toxic by-product called dioxin. Agent Orange contaminated with dioxin was unexpectedly responsible for causing birth defects, leukemia, liver diseases, and other disorders. Its use has since been stopped.

TAKE HOME MESSAGE 20.19

» Auxins are a small group of naturally occurring hormones found primarily in shoot tips and immature plant tissue (and a larger group of synthetic auxin variants) that play several important roles in stimulating and regulating a plant's growth and development, often by increasing the usually rigid cell wall's flexibility and permeability.

20.20 Other plant hormones regulate flowering, fruit ripening, and responses to stress

Auxins and gibberellins are the primary plant regulators when it comes to growth and orientation. Three other types of hormones—ethylene, abscisic acid, and cytokinins—also play critical roles in directing germination, sprouting, fruiting, and other activities when the environmental conditions are favorable.

Q How can farmers ensure that every banana they ship to market ripens at exactly the right time?

Ethylene Suppose you wanted to pick 10,000 bananas in Central America and deliver them to market in the United States so that they all turned yellow at just the right time. What would you do? The hormone **ethylene** is a gas produced in every part of a plant, and it has several important effects, including speeding up the rate at which many fruits ripen. Thus, ethylene can help you solve the banana problem: pick the fruits in Central America before they are ripe (when they are green), and just prior to their arrival in the United States, inject ethylene into the cargo hold of the ship and initiate the ripening of all the bananas simultaneously (FIGURE 20-36).

Not all fruits ripen in response to ethylene—strawberries, for example, do not, and so they must be picked from the vine only when ripe. Still, the use of ethylene for ripening fruits is probably the single most important agricultural use of any plant hormone. You can see this effect on a smaller scale by putting one ripe or rotting fruit (which produces very large amounts of ethylene) in a bag or container with fruits that haven't yet ripened. The ripe or rotting fruit will cause all of the other fruits to ripen very quickly. Alternatively, you can delay the ripening of fruits by separating them from those that are already ripe.

Ethylene also hastens the aging and dropping of leaves from trees and the death of flowers at specific times. A plant has to maintain a flower's ovary until the seeds and fruit are mature, but it does not maintain the petals or stamens after fertilization. Most plants use ethylene to abort some flowers to reduce the number of fruits the plant must mature. Some flower merchants fight the effects of ethylene by briefly soaking cut-flower stems in a chemical solution of silver salts, which inhibits the effect of ethylene on the petals (FIGURE 20-37).

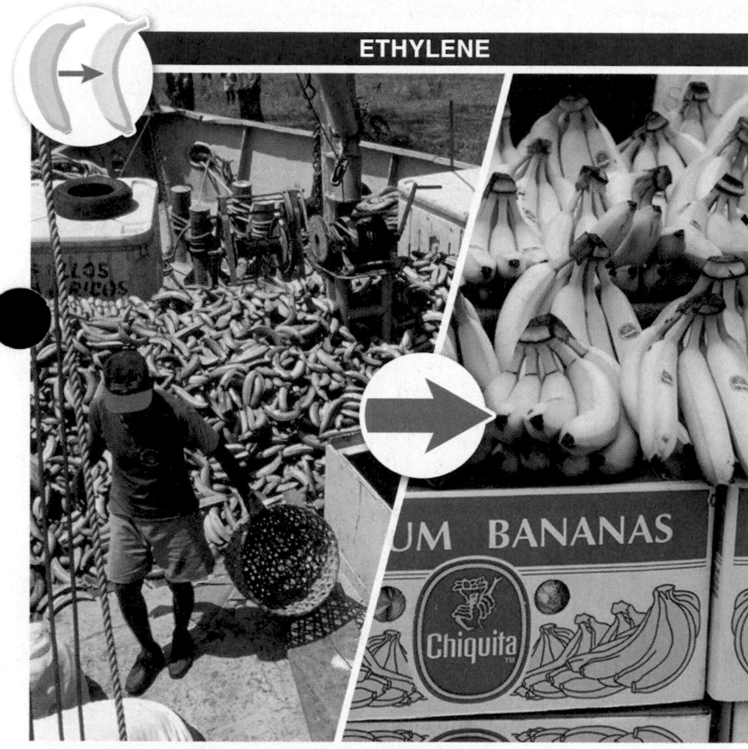
ETHYLENE
UM BANANAS
Chiquita

THE PRIMARY EFFECTS OF ETHYLENE
- Speeds up the rate at which many fruits ripen
- Hastens the dropping of leaves from trees and the death of flowers at specific times

Bananas are picked prior to ripening and exposed to ethylene gas just before delivery to market, so that all ripen simultaneously.

FIGURE 20-36 **Ethylene speeds fruit ripening.**

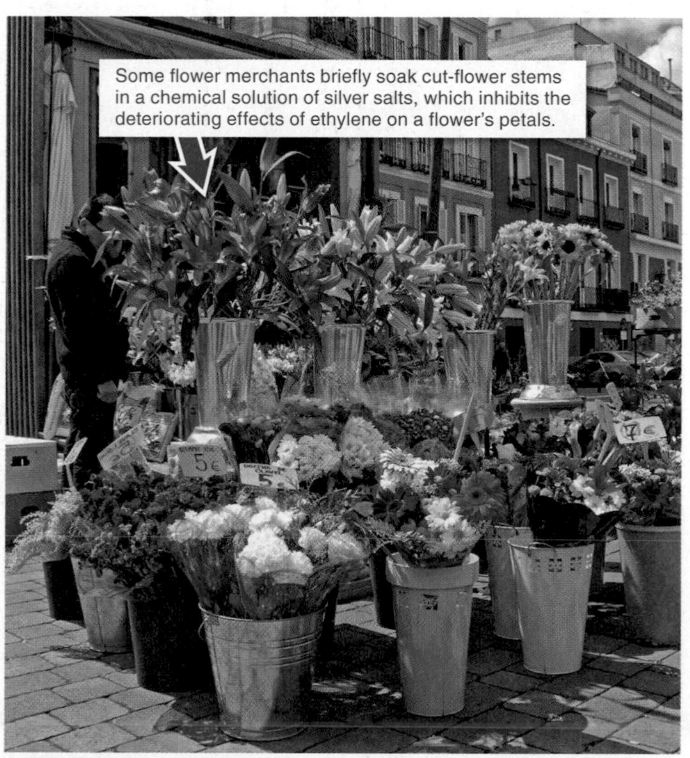
Some flower merchants briefly soak cut-flower stems in a chemical solution of silver salts, which inhibits the deteriorating effects of ethylene on a flower's petals.

FIGURE 20-37 **Flower merchants fight the withering effects of ethylene on flowers.**

Abscisic Acid Under stressful conditions, such as water shortage, the plant hormone **abscisic acid** is produced in relatively large amounts. Synthesized primarily in leaves, fruits, and root tips (but also in other plant parts), this hormone has the general effect of inhibiting growth and reproductive activities under adverse environmental conditions (FIGURE 20-38). In seeds, for example, abscisic acid can inhibit germination.

In effect, abscisic acid causes a plant to hunker down and ride out difficult conditions before embarking on new growth. When roots encounter unusually dry or cold or salty conditions, the production of abscisic acid in the root increases, and the hormone travels through the plant body, inhibiting growth. It also signals the stomata on the plant's leaves to close, thereby conserving water.

Cytokinins The **cytokinins** are primarily stimulators of cell division throughout the plant body and throughout the lifetime of the plant. Although they are produced primarily in the roots and fruits, cytokinins exert their influence in all parts of the plant. They usually work in conjunction with auxins to produce four main types of effects (FIGURE 20-39).

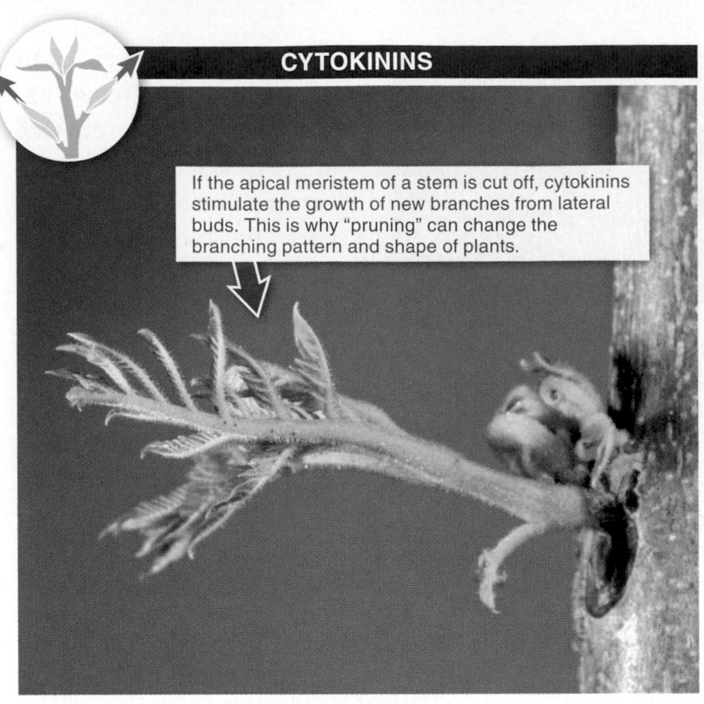

CYTOKININS
If the apical meristem of a stem is cut off, cytokinins stimulate the growth of new branches from lateral buds. This is why "pruning" can change the branching pattern and shape of plants.

THE PRIMARY EFFECTS OF CYTOKININS
- Cause rapid cell division, in conjunction with auxins
- Initiate new branches from lateral buds
- Induce seed germination
- Retard leaf aging and death

FIGURE 20-39 **Cytokinins stimulate cell division.** New leaf shoots are growing from a pruned buddleia ("black night") stem.

1. **Causing rapid cell division and promoting primary growth.**
2. **Initiating new branches from lateral buds.**
3. **Inducing seed germination.** Application of cytokinins will even cause seeds that normally germinate only in the light to germinate in the dark.
4. **Retarding leaf aging and death.** Even when applied to a leaf that has already fallen off the tree, cytokinins will cause the leaf to remain green longer than it otherwise would.

THE PRIMARY EFFECTS OF ABSCISIC ACID
- Inhibits growth and reproductive activities when environmental conditions are stressful
- Signals the stomata on a plant's leaves to close, increasing water conservation

FIGURE 20-38 **Not a good time to invest in growth.** Abscisic acid inhibits plant growth in times of stress, such as during a spring ice storm.

TAKE HOME MESSAGE 20.20

» In addition to auxins and gibberellins, three other types of hormones play critical roles in plant regulation. Among their multiple functions, ethylene induces and speeds fruit ripening, abscisic acid inhibits growth and reproduction under stressful environmental conditions, and cytokinins stimulate cell division throughout the plant body.

20.21–20.24

External cues trigger internal responses.

Tendrils of American vetch, *Vicia americana*, climbing a fence.

20.21 Tropisms influence plants' direction of growth.

A plant's immobility affects virtually every aspect of its structural design and function. Whereas animals can move from a problematic, changing environment to one with more suitable conditions, plants instead grow toward or away from various environmental stimuli such as light, gravity, and physical obstacles. In responding to these stimuli, plants use a variety of growth patterns, known as **tropisms**—such as bending, curving, and twisting. Three of the most common tropisms are phototropism, gravitropism, and thigmotropism.

Phototropism If you have indoor plants, you've probably noticed that they always seem to grow toward a window. This tendency to grow toward a directional light source is **phototropism.** It occurs when the cells in a plant's stem grow unevenly, at different rates, in such a way that the stem bends toward the light (FIGURE 20-40). The reason that such a pattern of growth has evolved is not surprising: if a plant can orient itself so that its photosynthesizing cells (in leaves and stems) can intercept more light, the plant can photosynthesize more efficiently and generate more energy for growth and reproduction.

It was the study of phototropism that led to the discovery of the plant hormone that was named "auxin" (which turned out to be a group of similar hormones). When light hits a plant from a particular direction, auxins produced in the plant cells move away from the light source and end up on the opposite, shaded side of the stem. There, the auxins stimulate a slightly greater rate of growth than on

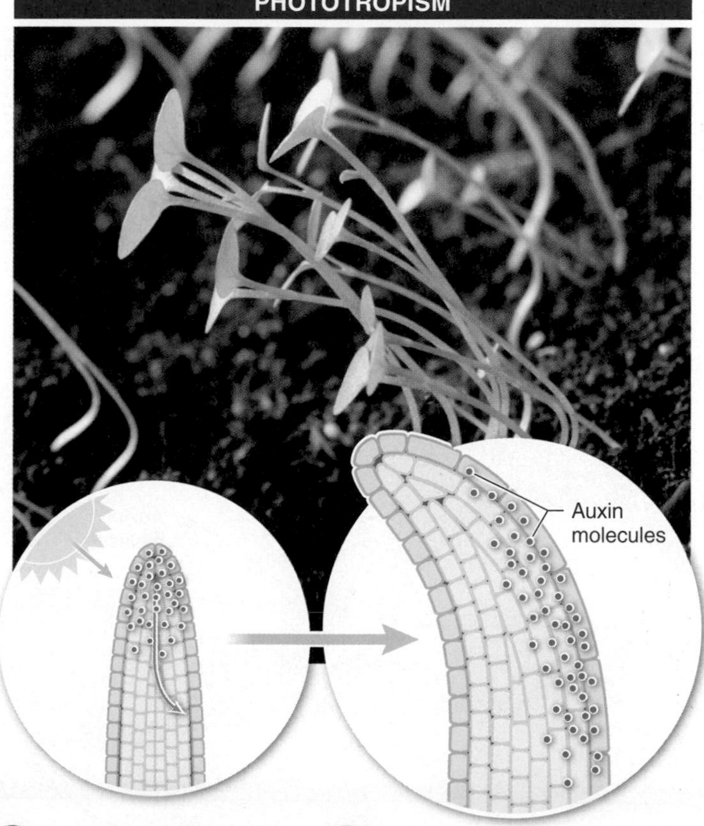

PHOTOTROPISM
Auxin molecules
1 Auxins produced in the plant move away from the light source to the shaded side of a stem.
2 Auxins on the shaded side of the stem stimulate a greater rate of growth than on the side with less auxin, causing the plant to bend toward the light.

FIGURE 20-40 Phototropism is plant growth that is influenced by the presence of light.

the lighter side, where there is less auxin. The elongation of cells on the shaded side of the stem causes the stem to bend toward the light.

Phototropism appears to be a response simply to "light," but that's not exactly right. The pigments in plants that are responsible for phototropism respond only to part of the full spectrum of light—the wavelengths that correspond to blue light.

One special type of phototropism, first described by Leonardo da Vinci, is called **heliotropism**—growth or movement in response to the position of the sun. "Heliotropic" leaves and flowers, such as the alpine buttercup, change orientation as they track the sun's movement across the sky each day. Cells in regions of the plant away from the light elongate as potassium ions are pumped into them and the subsequent movement of water into the cells increases turgor pressure (see Section 4.21), which changes the orientation of the flowers or leaves.

GRAVITROPISM

In response to gravity, a root grows downward, where it is most likely to find water. A stem grows upward, where it is most likely to find light.

Root

Stem

Auxin molecules

1 Starches within the cells of the stem sink downward in response to gravity, triggering the movement of auxin toward them.

2 Auxin then stimulates faster growth where it occurs in higher concentration, causing the stem to bend upward.

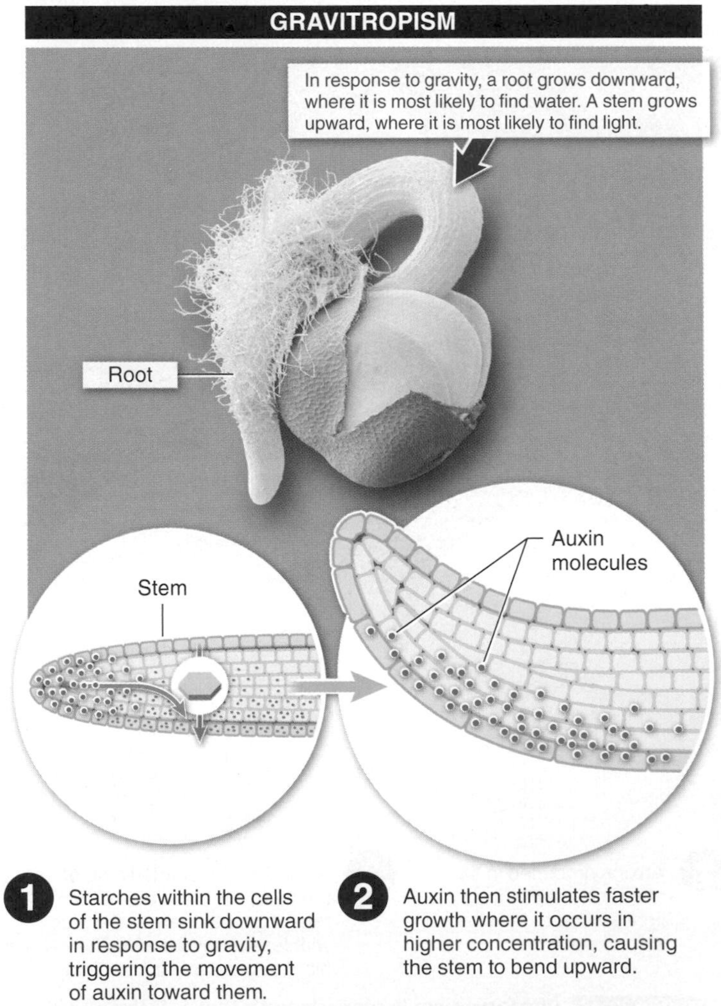

FIGURE 20-41 Plant stems grow away from the force of gravity.

Gravitropism Plants' growth is also guided by gravity. Plant response to gravity, known as **gravitropism** (also first described by Leonardo), is the reason that stems grow upward and roots grow downward. You can observe gravitropism if you take a potted houseplant and tip it on its side. Within a matter of days, the plant will grow upward. It doesn't matter what direction the pot faces (it can even be suspended upside down): roots will grow downward, in response to the force of gravity—and in the direction in which they are most likely to find water—and stems will grow upward, in the opposite direction, which will reliably put them in position to intercept maximal amounts of light (FIGURE 20-41).

Gravitropism occurs as a result of the uneven distribution of auxins, much as in the case of phototropism. How do the auxins detect the force of gravity? Small bodies that contain starches are present within plant cells. Like marbles in a bottle of fluid, these starch-containing bodies, pulled by gravity, sink toward the bottom of the cell, regardless of the plant's orientation. Once there, the starch bodies trigger the migration of auxin molecules toward them, and the auxin again causes uneven cell growth so that the stems and roots bend in the right direction.

Thigmotropism In **thigmotropism,** plant growth occurs in response to touch or physical contact with an

THIGMOTROPISM

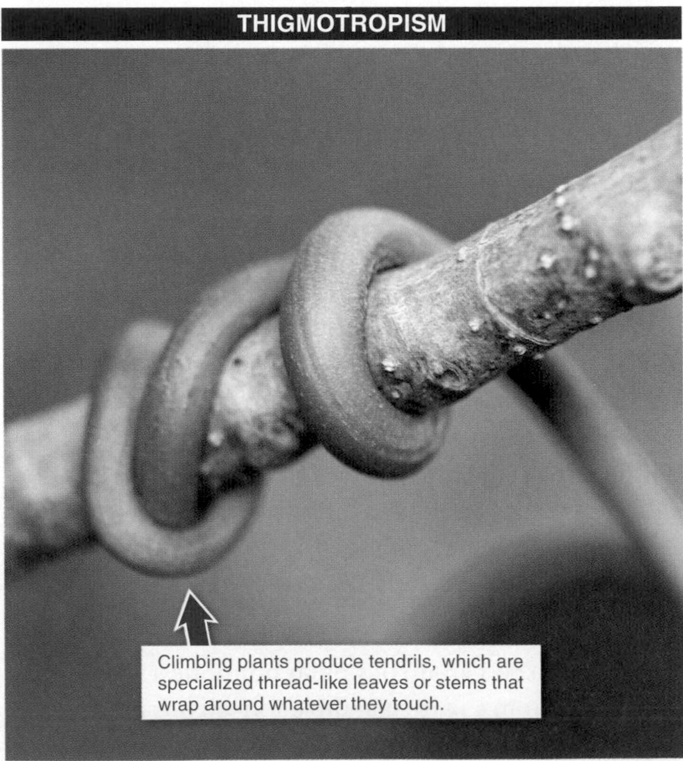
Climbing plants produce tendrils, which are specialized thread-like leaves or stems that wrap around whatever they touch.

FIGURE 20-42 Plant growth can be affected by touch or contact with physical objects.

object (*thigmo-* derives from the Greek word for "touch"). For example, many climbing plants, such as cucumbers and morning glories, will wrap around structures—other plants, wires, posts, trellises, or anything that might support them as they grow upward. Climbing plants produce **tendrils,** which are specialized thread-like leaves or stems or branches that wrap around whatever they touch (FIGURE 20-42). In some species, a tendril can grow so fast that it wraps completely around something in less than an hour.

As in the case of gravitropism and phototropism, thigmotropism occurs through uneven growth: cells on one side of a shoot (the side in physical contact with the object) elongate less quickly than cells on the opposite side. This, again, involves auxins. The cells in contact with the object produce auxins, and these auxins are transported to cells not touching the object, where they induce the cells to elongate, causing the tendril to coil around the object.

TAKE HOME MESSAGE 20.21

» Plants have a variety of growth patterns, known as tropisms, by which they grow toward or away from various environmental stimuli. Phototropism is growth in response to directional light, gravitropism is growth in response to gravity, and thigmotropism is growth in response to touch or physical contact.

20.22 Plants have internal biological clocks.

If you were to photograph or film a plant over a 24-hour period, you would observe many slow but sure movements. The common bean plant, for example, opens its leaves and orients them so that they face the sun during the day. Then each night, the plant pulls its leaves close to the stem (FIGURE 20-43). Similarly, in many moth-pollinated tobacco

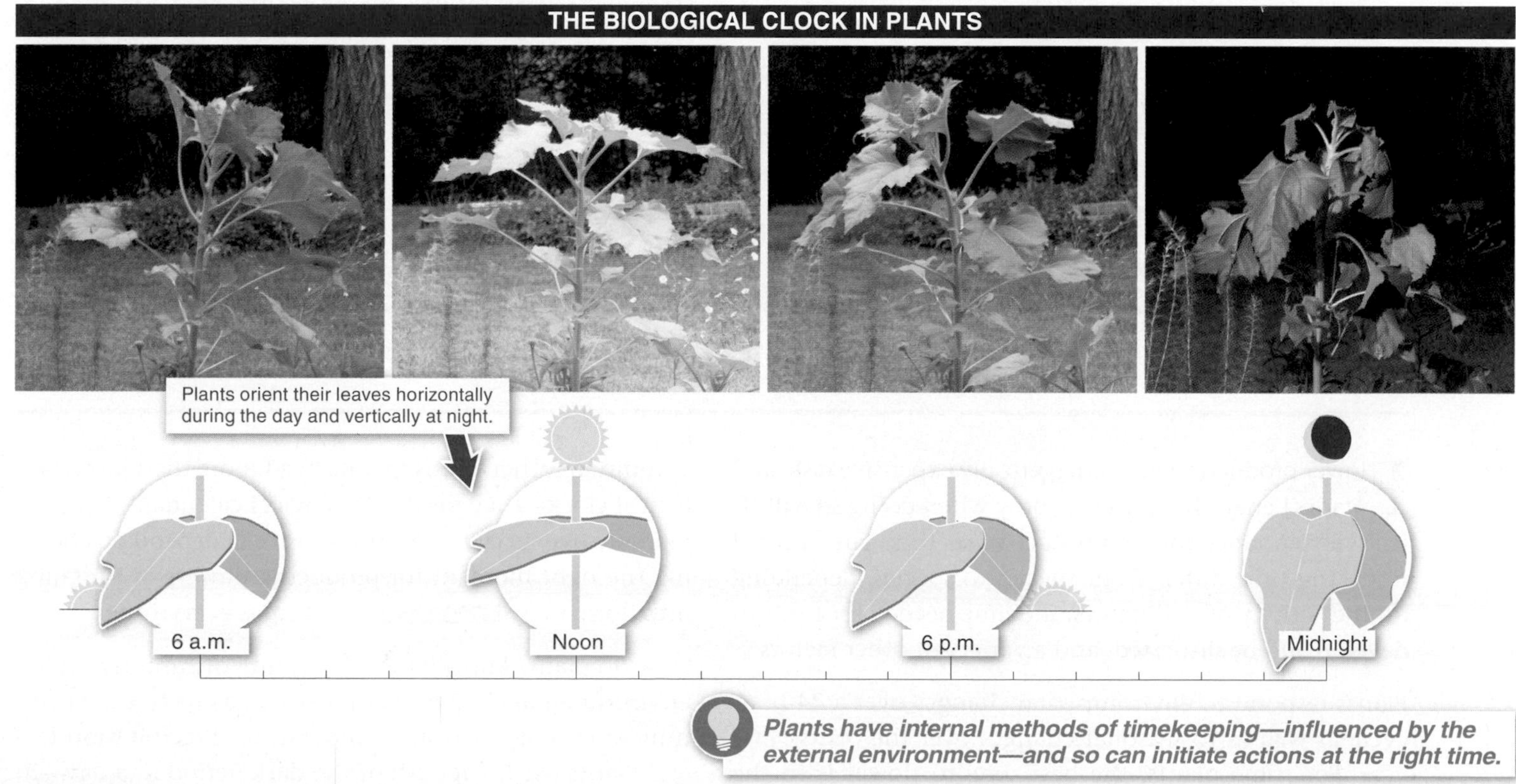

THE BIOLOGICAL CLOCK IN PLANTS
Plants orient their leaves horizontally during the day and vertically at night.
6 a.m.
Noon
6 p.m.
Midnight
Plants have internal methods of timekeeping—influenced by the external environment—and so can initiate actions at the right time.

FIGURE 20-43 **Great timing.** Plants adjust their biological clocks in response to environmental cues.

plants, the white trumpet-shaped flowers open each evening for their night-time pollinators and close again in the morning. Do the plants respond to environmental cues to schedule their daily activities at the proper times? Or do they have some sort of built-in alarm clock?

The answers are *yes* and *yes*. Plants do have a "biological clock," an internal method of keeping time that tells the plant when to turn on or off the expression of certain genes. This internal clock makes possible the plant's perpetually precise behavior—such as properly orienting leaves for efficient photosynthesis or making flowers accessible when pollinators are available. We know this because, even if the bean plants or tobacco plants described above are taken indoors and maintained in a room with constant, dim light, they still open their leaves or flowers on a daily schedule.

But this biological clock is not like a regular watch. The internal clock of plants is continuously adjusted by environmental cues. When maintained under constant light conditions, in the absence of external cues, plants have anywhere from a 21-hour to a 27-hour cycle. Consequently, after a while, they tend to get out of synch with the natural pattern of daylight and darkness. However, left outside, plants constantly adjust their clocks to these external cues and maintain a cycle that always corresponds to the daily light-dark cycle. It's as if plants sense that the environment is always the "correct" time and, if their clock is out of synch, they change it accordingly.

The most important environmental cues are light-dark cycles (but with a bit of input from temperature cycles). The precise mechanism by which plants keep track of time is not known for all species. In many plants, however, the mechanism seems to be regulated by pigments in the leaves that are sensitive to particular wavelengths of light. Because the specific wavelengths present in sunlight vary throughout the day, plants reliably have more of the pigments that are activated during the day, and different pigments are activated as the sun sets and night begins.

In the mid-1700s, Swedish botanist Carolus Linnaeus used his understanding of plants' biological clocks to design what he called a "floral clock," made of flowers that opened and closed at specific times of day. By planting a garden with 41 species that differed in their flowers' opening and closing times, Linnaeus was able to tell the time of day just by noting which flowers were open and which were closed.

Q Could we tell time accurately just by looking at some flowers?

TAKE HOME MESSAGE 20.22

» Plants have internal methods of keeping time that enable them to initiate various biochemical and physiological actions at the appropriate times. They constantly adjust their biological clocks to environmental cues, such as light-dark cycles and temperature cycles.

20.23 With photoperiodism and dormancy, plants detect and prepare for winter.

Flower production is an energetically expensive task, and plants benefit by flowering only when doing so will effectively enhance their reproductive success. The optimal flowering time differs from species to species, depending on the activity of pollinators, the time needed for fruits to develop and be dispersed, and a variety of other factors.

Plants respond to environmental changes over a 24-hour cycle as well as to seasonal changes over the course of a year. For some plants, the best time to flower is in the spring or summer. For others, it is best to wait until early autumn. In other words, plants need more than just a biological clock—they need a "biological calendar." A plant's reproductive success or even survival can depend on choosing the right moment for producing flowers or for going into dormancy (FIGURE 20-44). Timing is everything.

How do plants know when it is spring or summer? What environmental change do they use as a cue to the beginning or end of a season? Temperature? Precipitation levels? Plants use the length of the dark period in a day—in a process known as **photoperiodism**—to regulate their flowering time and numerous other responses to seasonal changes.

> "Art is the unceasing effort to compete with the beauty of flowers—and never succeeding."
>
> — MARC CHAGALL

All flowering plants fall into one of three categories when it comes to regulating flower production. There are (1) long-day plants, (2) short-day plants, and (3) day-neutral plants (FIGURE 20-45). The groups are so named because the amount of daylight seems to determine when they produce flowers. Long-day plants begin producing flowers only when the length of daylight exceeds a critical amount. Short-day plants are the opposite: they flower only when the length of daylight becomes less than some threshold amount. The critical amounts of daylight generally occur in spring for long-day plants and in late summer or fall for short-day plants, when conditions are best suited for reproduction for that particular species. Day length has no effect on day-neutral plants, which flower when they reach a sufficient state of growth and maturity.

Although short-day and long-day plants are named for the amount of daylight that seems to determine when they flower, experiments have shown that they are actually sensitive to the length of the night rather than the day. That is, flowering in short-day plants is triggered when the nights become longer than some threshold period, and flowering in long-day plants is triggered by the onset of short nights. This seemingly unimportant distinction had disappointing implications for the California Department of Transportation. The department wanted to line a highway with bright-red flowering poinsettias. Unfortunately, these short-day plants never flowered as expected, because the car headlights kept interrupting the plants' measurement of the night length. As a consequence, the poinsettia plants—which require at least 14 uninterrupted hours of darkness to flower—never detected an appropriately long night (and thus short day), and they remained green.

How do plants detect night length? As noted earlier, phototropism occurs as photoreceptors in plants respond to

When it comes to producing flowers—an energetically expensive task—a plant's life can depend on choosing the right moment.

FIGURE 20-44 Plants flower and become dormant in accordance with the seasons.

PHOTOPERIODISM

All flowering plants fall into one of three categories when it comes to regulating their flower production.

LONG-DAY PLANTS
- Flower production is triggered by decreasing periods of darkness (generally in spring)
- Species include: carnations and clovers

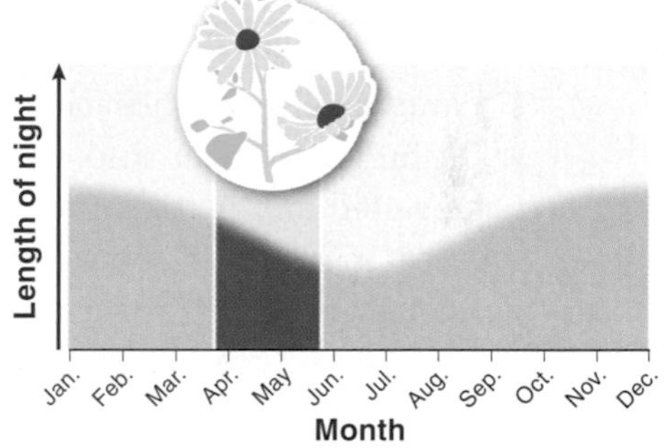

SHORT-DAY PLANTS
- Flower production is triggered by increasing periods of darkness (generally in late summer or fall)
- Species include: poinsettias and strawberries

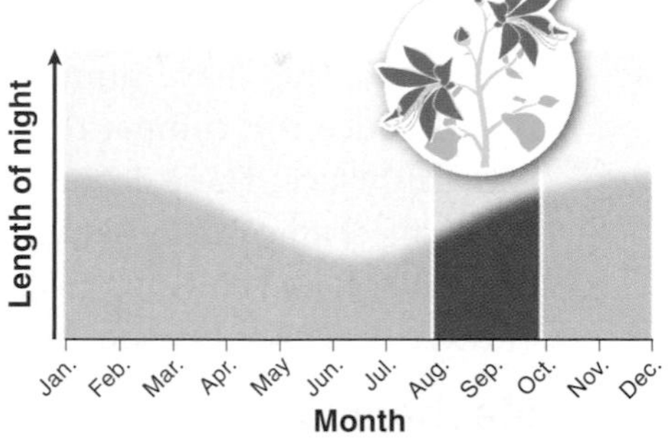

DAY-NEUTRAL PLANTS
- Flower production is triggered by a sufficient state of maturity and not by periods of darkness
- Species include: roses and tomatoes

Some plants are triggered to produce flowers when the nights are long (and the amount of daylight is relatively small). Others are triggered when nights are shorter and the daylight lasts longer.

FIGURE 20-45 Day length and flowering.

blue light. Photoperiodism, on the other hand, is regulated in large part by plant pigments in the leaves that are sensitive to red and far-red wavelengths of light. The ratio of activated red-sensitive pigments to activated far-red-sensitive pigments varies at different points during the day and night. These pigments also play important roles in regulating seed germination and plant elongation, as well as in several aspects of leaf growth and shape.

Flowering may be disrupted, or its timing altered, in plants cultivated indoors. It can be particularly difficult to trigger flowering in short-day plants, as people who keep poinsettias as houseplants discover. Plants growing under the artificial lights in houses don't usually experience the long, uninterrupted period of darkness required for blooming. Long-day flowers, on the other hand, are easier to cultivate indoors, because flashes of light or continuous light during the night can trick the plants into behaving as if the night is short, leading to flowering. Day-neutral plants are not influenced by artificial lights.

TAKE HOME MESSAGE 20.23

» Plants exhibit photoperiodism, responding to seasonal changes over the course of a year. They may time their production of flowers or initiation of winter dormancy, for example, according to environmental factors such as the length of darkness each night.

20.24 Plants actively resist being eaten.

Plants are under almost constant attack. From microbes to fungi, animals, and even other plants, organisms make varied and persistent efforts to access the valuable resources in plants. In some cases, the attackers can kill the plant outright. In most cases, though, the attackers simply siphon off resources—such as stored chemical energy.

Rooted in the ground, plants are unable to employ many of the strategies used by animals and other organisms for resisting predation. Nevertheless, plants have evolved to fight back. They have numerous defenses that can help them reduce the competitive and predatory "tax" they constantly face. These defenses fall into four general categories: mechanical defenses, chemical defenses, mimicry or camouflage, and enlisting help from elsewhere. And some of these have uses beyond defending against herbivores.

Mechanical Defenses Several types of plant defenses take the form of physical structures or movements (FIGURE 20-46).

Thorns, spines, and hairs. If you've ever tried to pick blackberries or raspberries, you probably encountered sharp prickles on the stems. And while these may have been only a nuisance to you, they can be enough of a deterrent to many herbivores—in some cases, harming or even killing them—to significantly reduce herbivory.

Waxes and saps. Producing leaves covered with waxy compounds can also reduce herbivory. Researchers have documented, for example, that beetles spend more time slipping and falling off waxy leaves—such as holly leaves—than they do off non-waxy leaves.

Defensive (and offensive) movements. The Venus flytrap is the most dramatic example of a plant making use of rapid movement, closing together quickly as a method of acquiring nitrogen (released as insect prey are digested by the plant). Many plants use a similar movement mechanism to rapidly fold their leaves in response to touch, decreasing the surface area available to potential pests.

Chemical Defenses One of the most common methods that plants use to fight herbivory is the production of chemical defenses (called "secondary compounds") that make the plant toxic, reduce palatability, or reduce digestibility.

One common strategy, used by more than 3,000 species of plants, is to produce cyanide-containing molecules called cyanogenic glycosides. When consumed by herbivores, these molecules break down into cyanide, which blocks electron transport in cellular respiration—killing or seriously harming the animal.

As the consequence of a sort of evolutionary arms race between plants and herbivores, the individuals of some herbivore species have the ability to eat toxic secondary compounds without suffering ill effects. Monarch butterfly caterpillars, for example, can safely consume the toxic cardiac glycosides produced by milkweed plants. Moreover, the monarchs store the toxic chemicals within their cell vacuoles and then become poisonous to their predators.

MECHANICAL DEFENSES IN PLANTS

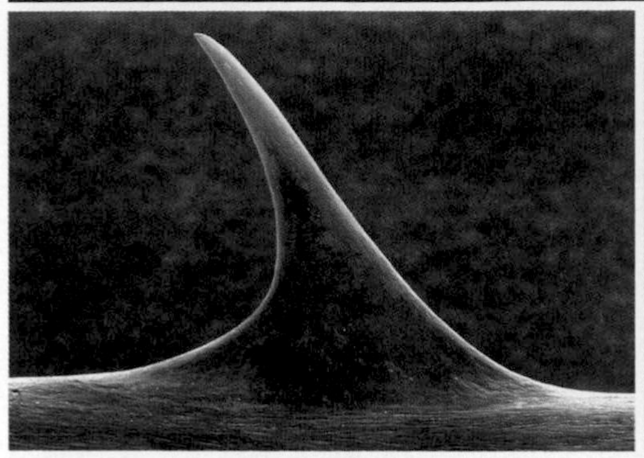

THORNS, SPINES, AND HAIRS
Structures such as sharp spines or fine hairs can significantly reduce herbivory.

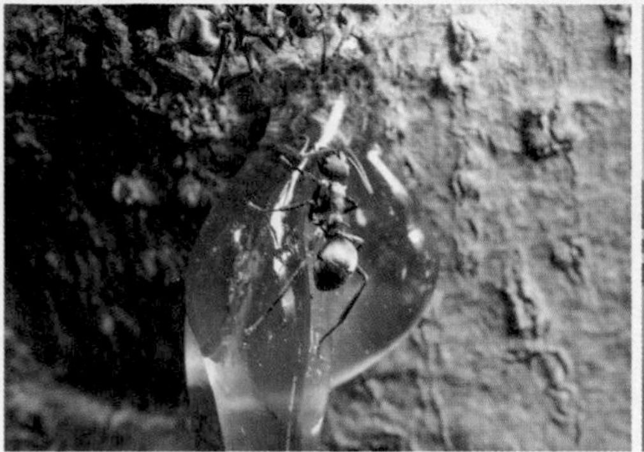

WAXES AND SAPS
Leaf secretions such as slippery waxy compounds or sticky saps significantly reduce herbivory.

DEFENSIVE MOVEMENTS
Rapid movements, such as folding leaves in response to touch, can decrease the available surface area and significantly reduce herbivory.

FIGURE 20-46 **Plants have physical defenses to ward off predation by insects and other animals.**

Many plants secrete toxic chemicals from their roots that block the germination of seeds and reduce the growth of nearby plants, conferring a competitive advantage.

Q Can humans benefit from plants' attempts to avoid being eaten?

Humans have long made use of some of the "defensive" compounds produced in or by plants.

Spices. Spicy foods, such as mustard, usually get their flavor from secondary plant compounds that may make the plant toxic or unpalatable to an insect but are not toxic to humans.

Medicines. Many of our medicines come from compounds isolated from plants. After all, a chemical that disrupts a microbe's metabolism can have its effect whether the microbe is on a plant or inside a human. These medicines include taxol, a compound used to treat a wide variety of cancers (see Section 18.1), and quinine, an anti-herbivore compound isolated from the bark of *Cinchona* trees in South America, which is an effective treatment for malaria. There has been a large increase in the number of researchers—sometimes called "bio-prospectors"—looking for plant compounds with potential medicinal uses (FIGURE 20-47).

Mimicry and Camouflage Some plants resist predation through mimicry and camouflage. One example is a type of passion flower plant. Many species in the *Heliconius* group of butterflies lay their eggs on the leaves of passion flower plants. On hatching, the insects begin eating the plant, often causing significant damage. By producing leaves that are dotted with spots and appear to be already covered with butterfly eggs, one species of passion flower reduces the likelihood that the butterflies will lay eggs on its leaves.

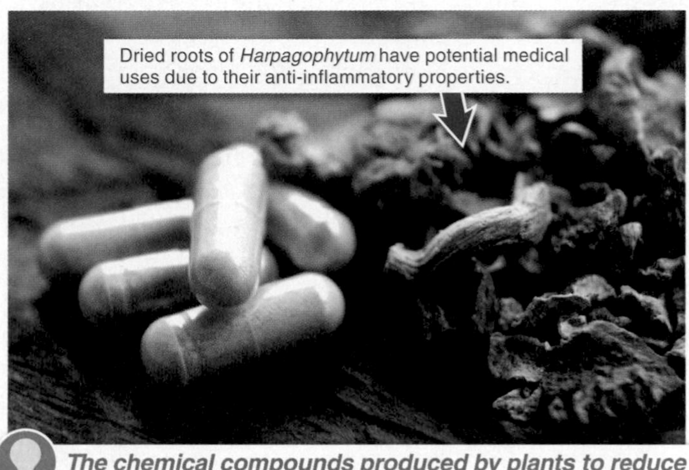

Dried roots of *Harpagophytum* have potential medical uses due to their anti-inflammatory properties.

The chemical compounds produced by plants to reduce herbivory can also have medicinal effects in humans.

FIGURE 20-47 **"Bio-prospectors" search for plant chemicals that have medicinal properties.**

Enlisting Other Organisms for "Security"

Researchers have even discovered situations in which plants outsource some of their defenses to another species. Acacia plants produce a sweet nectar that attracts ants. The ants hollow out thorns in the plants and live inside them. In exchange for their "room and board," the ants quickly and aggressively attack leaf-eating insects.

TAKE HOME MESSAGE 20.24

» Rooted in the ground, plants are targets for predators and pathogens. They actively resist these challenges by several methods, including mechanical and chemical defenses that make them distasteful, poisonous, or difficult to consume.

STREET BIO

Using evidence to guide decision making in our own lives

Practical botany and Aloe vera*: can* Aloe vera *lessen the pain or duration of burns?*

Q: What is *Aloe vera*? *Aloe vera* is a species of succulent plant (also known as *Aloe barbadensis*) widely distributed in hot, dry areas, particularly in Africa.

Q: Why the interest in *Aloe vera*? When *Aloe vera* leaves are cut, sap oozes from the inner tissues of the leaves. This clear fluid has long been touted as having a variety of medicinal properties—including providing relief from the pain of burns (including sunburn).

Q: Does it work? Despite a long history of the use of *Aloe vera* to treat burns, the scientific evidence has often been conflicting. Numerous published research studies on *Aloe vera* have reported (1) antibacterial properties, (2) anti-inflammatory activity, and (3) direct stimulation of the immune system. In a 2008 report in the *New England Journal of Medicine,* for example, doctors suggested that *Aloe vera* may be used in lieu of topical anti-inflammatory drugs to reduce the pain of first-degree burns. And a review of clinical trials of the use of *Aloe vera* for burn-wound healing in burn patients found an average reduction in healing time of almost nine days when compared with patients not treated with *Aloe vera*.

Still, there is no consensus at this point, with several published research reports finding no increase in burn healing or skin regeneration, and no reductions in bacterial counts. One study even found that *Aloe vera* "hindered the healing process" when compared with an antibacterial cream.

Q: Why would a plant's compound soothe human sunburn? *Aloe vera* contains a large number of compounds that are toxic to many herbivores, making the plant resistant to most insect pests. Researchers believe that these molecules, particularly some organic compounds called anthraquinones, salicylic acid (an aspirin-like compound), and some sterols with anti-inflammatory effects, are responsible for the soothing, healing effects of *Aloe vera* when applied to the skin. So far, though, researchers have been unable to figure out exactly how the process works.

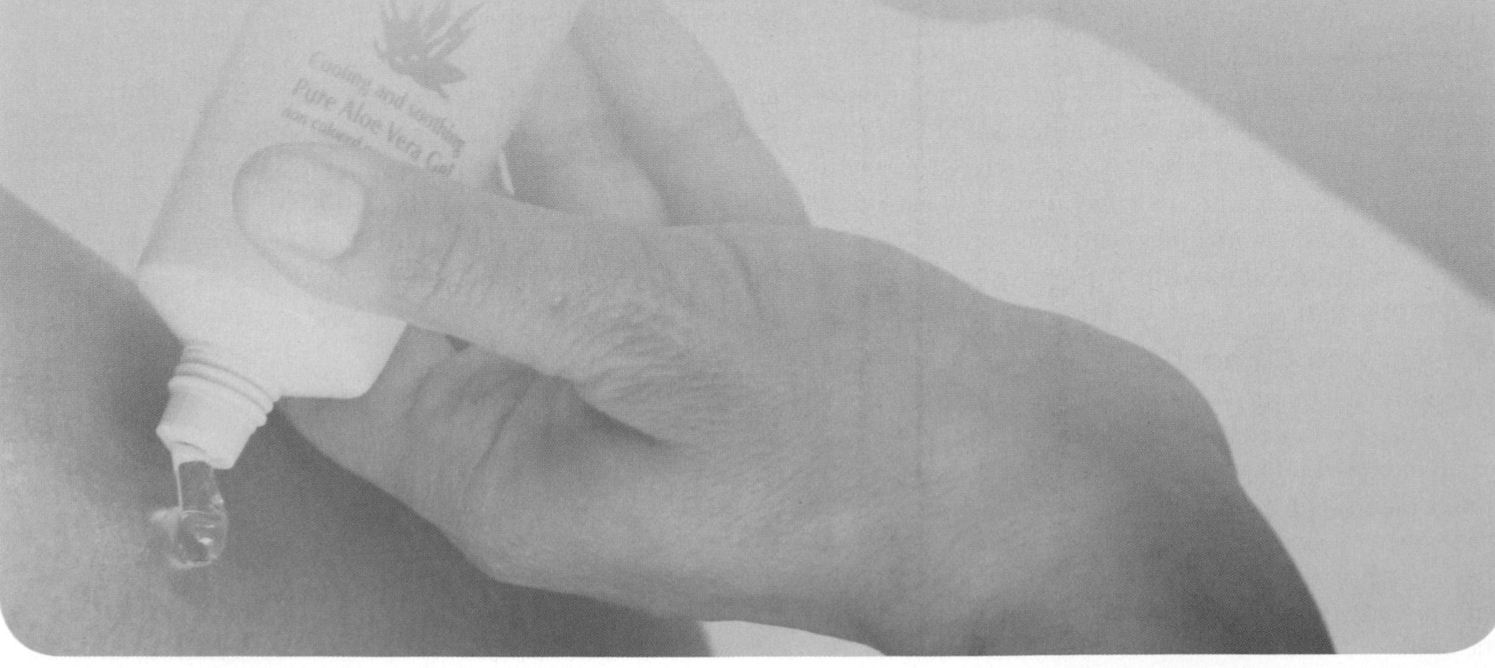

GRAPHIC CONTENT

Thinking critically about visual displays of data

1 What idea is conveyed by the "fuzzy" borders to each slice of the pie? How else could these data (and the uncertainty associated with them) be conveyed?

2 What is the reason for the color scheme used in this graph (two different shades of blue, multiple shades of green, and yellow and red)?

3 Would it be helpful to have actual percentages included with each slice of the pie? Why might percentages have been excluded?

4 Which group of pollinators is used by the largest number of flowering plant species? Approximately what proportion of plant species are pollinated by this group of pollinators? Which pollinators come in second?

5 Why might it be useful to know the specific habitats surveyed in collecting the data reported here?

6 What are two important conclusions you can draw from this pie chart?

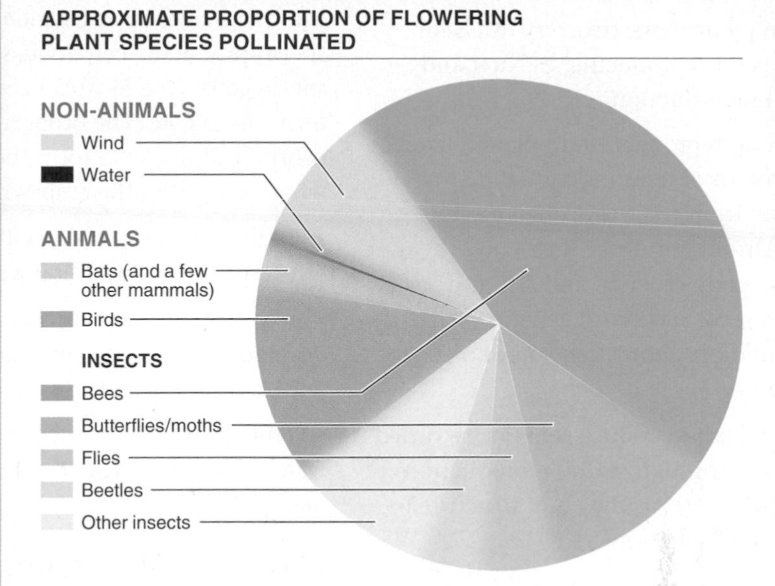

7 The total pie represents "all flowering plant species." Why would the pie chart differ if it instead reported the proportion of all plants (individuals, rather than species) that are pollinated by each type of pollinator?

8 What are the risks of reporting and interpreting data described as "loose approximations"? Are there benefits that outweigh the risks?

See answers at the back of the book.

KEY TERMS IN GROWTH, REPRODUCTION, AND ENVIRONMENTAL RESPONSES IN PLANTS

abscisic acid, p. 706
alternation of generations, p. 679
anther, p. 680
auxins, p. 702
carpel, p. 680
cork cambium, p. 698
cotyledon, p. 689
cytokinins, p. 706
double fertilization, p. 687
embryo sac, p. 683
endosperm, p. 687
ethylene, p. 705
filament, p. 680
fruit, p. 689
gametophyte, p. 677
germinate, p. 677
gibberellins, p. 701
gravitropism, p. 708
heliotropism, p. 708
hormone, p. 700
life cycle, p. 677
meristem, p. 694
nectar, p. 684
ovary, p. 680
ovule, p. 680
petal, p. 680
photoperiodism, p. 710
phototropism, p. 707
pith, p. 696
pollen grain, p. 680
pollen tube, p. 681
pollination, p. 683
primary growth, p. 694
root meristem, p. 689
secondary growth, p. 694
secondary phloem, p. 694
secondary xylem, p. 697
seed, p. 689
sepal, p. 680
shoot meristem, p. 689
sperm, p. 681
spore, p. 677
sporophyte, p. 677
stamen, p. 680
stigma, p. 680
style, p. 680
tendril, p. 709
thigmotropism, p. 708
totipotent, p. 675
tropism, p. 707
vascular cambium, p. 696
zygote, p. 687

BRIEF SUMMARY

Plants can reproduce sexually and asexually.

• Most plants have two very different options for reproducing: asexual and sexual reproduction.

• Asexual reproduction in plants involves the growth of new, individual plants directly from the tissue of an established plant through mitosis. It can be energetically efficient and fast and can preserve successful genetic combinations, but it generates no genetic variability.

• Many plants produce genetically varied offspring through sexual reproduction, with the flower as the chief structure for sexual reproduction.

Flowers have several roles in plant reproduction.

• Flowers are plant structures specialized for sexual reproduction. Most flowers have the same fundamental structures: sepals, petals, stamens, and a carpel.

• The male reproductive structure of flowering plants produces two-celled pollen grains. One of the cells will form a pollen tube, and the other will divide to produce two sperm cells.

• Within the ovary, diploid cells differentiate into ovules; these consist of outer protective cells surrounding a diploid egg-producing cell, which produces four haploid megaspores. One of these megaspores produces the embryo sac, containing the egg, where fertilization occurs.

Pollination, fertilization, and seed dispersal often depend on help from other organisms.

• Plants usually employ trickery or bribery to get the assistance of animals in carrying the male gametes to the female gametes.

• Following pollination, a pollen tube must grow down through the style and into the ovule, during which time the sperm-producing cell divides to form two sperm cells. One fertilizes the egg cell to form the zygote. The other fuses with two central cell nuclei to form the endosperm, which nourishes the embryo.

• Plants can reduce the likelihood of self-fertilization in several ways.

• Following fertilization, the ovule develops into a seed. The seed is protected within a fruit, which aids in its dispersal.

• Plants use the assistance of animals, water, or wind to disperse their fruits and seeds.

Plants have two types of growth, usually enabling lifelong increases in length and thickness.

• A seed, containing an embryo and a supply of nutrients, begins to grow only when the water, temperature, and oxygen conditions for life are just right.

• Plants generally grow for their entire lives, using two types of growth. Primary growth makes plants taller and plant parts longer. Secondary growth makes plants thicker and sturdier.

• Plant growth occurs as a result of cell division in meristems, small collections of totipotent cells. Primary growth results from the division of apical meristem cells. Secondary growth results from cell divisions in a lateral meristem.

Hormones regulate growth and development.

• Plant hormones are chemical signals produced by cells that enable the plant to respond to environmental variables and that influence its growth and development.

• Gibberellins, produced primarily in meristems and seeds, regulate a plant's growth, mainly by stimulating cell division and cell elongation.

• Auxins, found primarily in shoot tips and immature plant tissue, play important roles in stimulating and regulating a plant's growth and development, often by increasing the cell wall's flexibility and permeability.

• Ethylene induces and speeds fruit ripening. Abscisic acid inhibits growth and reproduction under stressful environmental conditions. And cytokinins stimulate cell division throughout the plant body.

External cues trigger internal responses.

• Plants have a variety of growth patterns, known as tropisms, by which they grow toward or away from various environmental stimuli.

• Plants have internal methods of keeping time that enable them to initiate various biochemical and physiological actions at the appropriate times. They adjust their biological clocks to environmental cues.

• Plants exhibit photoperiodism, responding to seasonal changes over the course of a year.

• Rooted in the ground, plants are targets for predators and pathogens. They actively resist these challenges through several types of defense mechanisms.

CHECK YOUR KNOWLEDGE

Short Answer

1. Potato plants can reproduce sexually, producing flowers, fruits, and seeds. A potato farmer can collect the seeds and plant them. Alternatively, she can plant some of the potatoes instead. Why might it be preferable to plant potatoes instead of seeds? And which potatoes should be chosen for planting?

2. Why is sexual reproduction in plants especially important in agriculture?

3. Which feature of pollen contributes to its "success" as an allergen?

4. Describe the coevolution between plants and their animal pollinators.

5. Why is the process of double fertilization so efficient?

6. Describe some of the evolutionary pros and cons of self-fertilization.

7. Why is it that fruits do not taste good until the seeds inside are fully developed?

8. If you wanted to induce a plant to grow taller, which part(s) of the plant would you cut off?

9. The term "miracle grow" best suits which plant hormone? Why?

10. Which plant hormone or group of hormones plays a role in the ripening of fruit picked while it is still green? Why is this information valuable in the agricultural industry?

11. Why don't plants rely on temperature or rainfall levels to help determine whether it is spring or summer?

12. Why might it be problematic to keep poinsettias inside a well-lit house?

Multiple Choice

1. Asexual reproduction in plants:

a) tends to be particularly common in a population growing in a marginal habitat.

b) like sexual reproduction requires meiosis.

c) requires reproductive cells that are produced in flowers.

d) is energetically more expensive than sexual reproduction.

e) Both b) and c) are correct.

0 EASY 56 HARD 100

2. "Double fertilization" in angiosperms refers to the:

a) release of two sperm from a single pollen grain.

b) production of two or more eggs in the ovary.

c) fusion of the ovary with the anther.

d) fusion of a sperm cell with two nuclei of the endosperm-forming cell.

e) fusion of one sperm cell with the egg and another with two nuclei of the endosperm-forming cell.

3. Which of the following is a likely way in which plants increase dispersal of their fruit?

a) Fruits are conspicuously colored.

b) Fruits taste good.

c) Fruit stems are highly elastic.

d) Both a) and b) are correct.

e) Both a) and c) are correct.

0 EASY 53 HARD 100

4. For angiosperms, which of the following statements about growth is the most accurate?

a) Vascular cambium increases girth; cork cambium increases length.

b) Apical meristems increase length; vascular cambium increases girth.

c) Cork cambium increases length; apical meristems increase girth.

d) Apical meristems increase length; apical meristems increase girth.

e) Apical meristems increase girth; vascular cambium increases length.

0 EASY 32 HARD 100

5. The seed of a certain plant remains dormant until it is exposed to fire. What would be the greatest advantage of this for the survival of this plant species?

a) avoiding germination during cold winters

b) germination under ideal soil conditions

c) avoiding competition with other plants

d) germination timed to coincide with autumn

e) utilization of heat energy to break down starch to glucose to feed the embryo

0 EASY 62 HARD 100

6. When you are seven years old, you scratch your name, 4 feet above the ground, into a 12-foot-tall bamboo plant. You spend the next 10 years in a special home for vandals. When you get out, the bamboo is 24 feet tall. At that time, how far above the ground will your name be?

a) 16 feet

b) 4 feet

c) 8 feet

d) 12 feet

e) 2 feet

0 EASY 67 HARD 100

7. Which plant hormone or group of hormones do grape-growers spray on their crop in order to get seedless grape plants to grow fruits that are as large as grapes with seeds?

a) gibberellins

b) ethylene

c) cytokinins

d) auxins

e) abscisic acid

0 EASY 38 HARD 100

8. High concentrations of ________ stimulate root production.

a) gibberellins

b) cytokinins

c) auxins

d) ethylene

e) abscisic acid

0 EASY 58 HARD 100

9. Which hormone or group of hormones stimulates the ripening of fruit?

a) auxins

b) ethylene

c) cytokinins

d) gibberellins

e) abscisic acid

0 EASY 23 HARD 100

10. Phototropism in plants is:

a) a process that produces glucose.

b) caused by ethylene.

c) mediated by cytokinins.

d) the bending of roots away from the light.

e) the bending of a shoot tip toward a light source.

Ch21

Animal body structures reflect their functions.

21.1 Most animal bodies are organized in a hierarchy from cells to tissues, organs, and organ systems.

21.2 Connective tissue provides support.

21.3 Epithelial tissue covers most interior and exterior surfaces of the body.

21.4 Muscle tissue enables movement.

21.5 Nervous tissue transmits information.

21.6 Each organ system performs special tasks.

Animals have an internal environment.

21.7 Our bodies function best within a narrow range of internal conditions.

21.8 Animals regulate their internal environment through homeostasis.

How does homeostasis work?

21.9 Negative and positive feedback systems influence homeostasis.

21.10 Temperature control is a component of homeostasis.

21.11 *This is how we do it:* Why do we yawn?

21.12 Animals must balance their water content within a narrow range.

21.13 In humans, the kidneys regulate water balance.

A bison crossing Mammoth Hot Springs in Yellowstone National Park. When exposed to extreme low temperatures (as low as −30° C) and windy conditions, bison increase their heart rate slightly, enabling them to maintain a relatively constant body temperature.

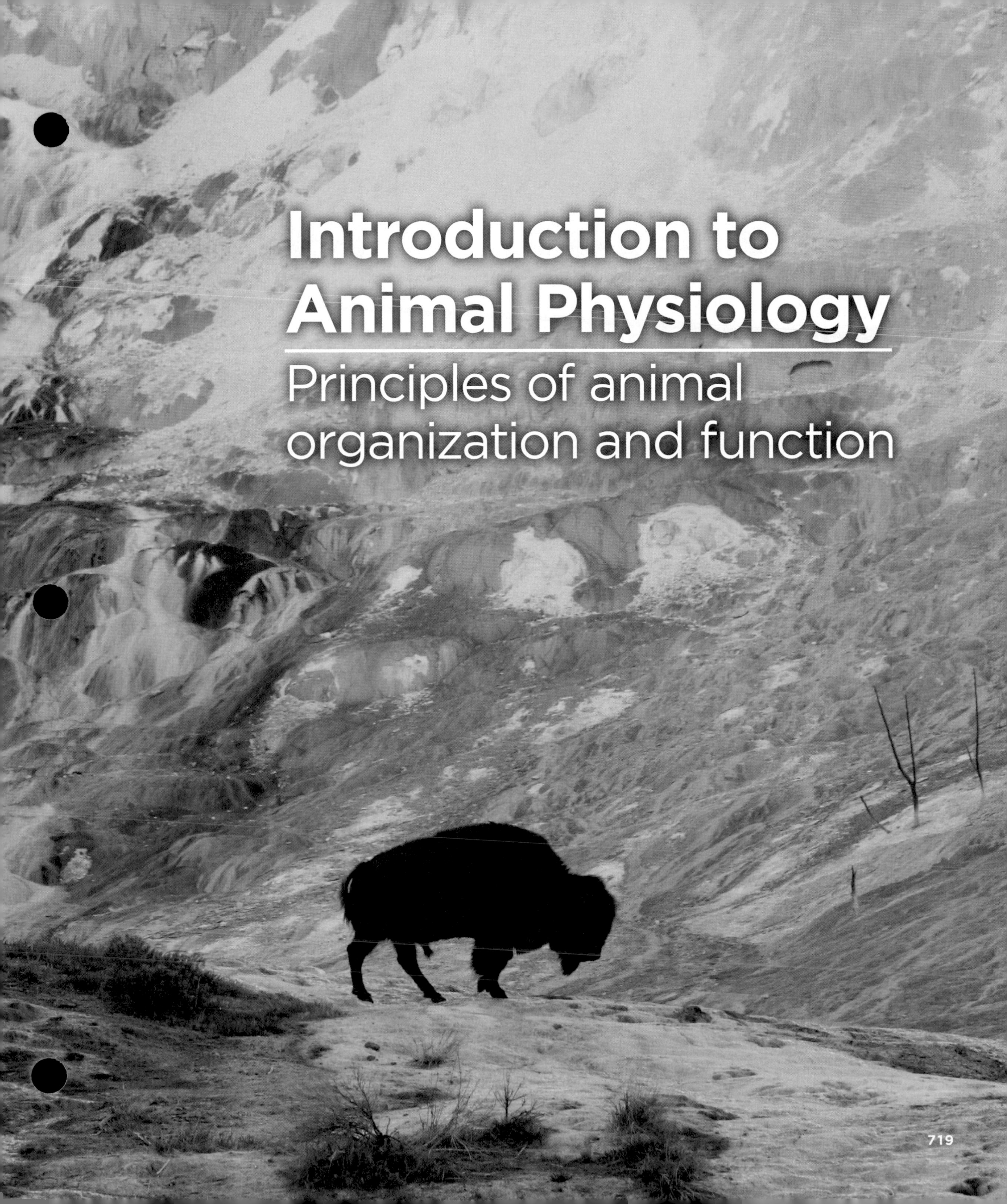

Introduction to Animal Physiology

Principles of animal organization and function

21.1–21.6

Animal body structures reflect their functions.

The teeth of a great white shark make it capable of consuming large prey.

21.1 Most animal bodies are organized in a hierarchy from cells to tissues, organs, and organ systems.

Form follows function. This simple statement captures one of the most universal relationships in the living world. For example, fast-swimming organisms, from penguins to tuna to sharks, share a common streamlined body shape (FIGURE 21-1). If a structure is adaptive—the product of natural selection—then its physical features closely reflect its function. And just as form follows function for large animal structures, the relationship also holds true for molecules, organelles, and cells. Form fits function in each.

FORM FOLLOWS FUNCTION

If a structure is the product of natural selection, the structure usually closely reflects its function. For example, fast-swimming organisms have streamlined body shapes.

FIGURE 21-1 Adapted for the life aquatic: the streamlined body of a penguin.

Animals are multicellular organisms. And multicellularity makes it possible for animals to attain much larger sizes and much greater physiological complexity than single-celled organisms. Increased size and complexity bring many benefits, including fewer potential predators, more potential prey, and, more generally, some protection from the influence of the external environment. For these reasons, among others, the transition from a single-celled to a multicellular body is one of the most important evolutionary transitions.

Perhaps the chief benefit of multicellularity is that it makes possible a division of labor and specialization at the cellular level. It is unnecessary for each cell to carry out every single process (such as generating movement, detoxifying harmful chemicals, digesting biological molecules, and sensing and responding to environmental changes). Instead, cells can be organized into groups, and groups organized into larger groups, to carry out specific life-sustaining functions such as exchanging gases between the organism and its environment, thinking and feeling, and fighting pathogens (FIGURE 21-2).

One of the fundamental features of the animal body is its hierarchical organization. Cells, as you'll recall from Chapter 4, are the smallest units of organization in all organisms. Even sponges, the structurally simplest of all animals, have cells specialized to perform a few distinct tasks. Some cells form the covering of the sponge's body, and others obtain food to provide nutrition and energy to sustain the sponge's activities. At the other end of the spectrum are humans—with more than 200 different cell types.

ORGANIZATION OF ANIMAL BODIES

One of the fundamental features of the organization of the animal body is that it is hierarchical, containing four levels of organization.

CELL
The smallest unit of organization in all organisms

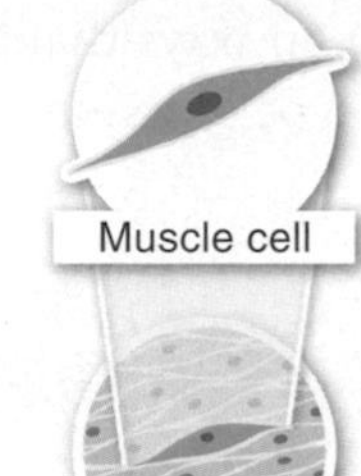

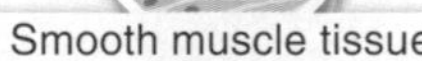

TISSUE
Group of cells that share similar structure and function

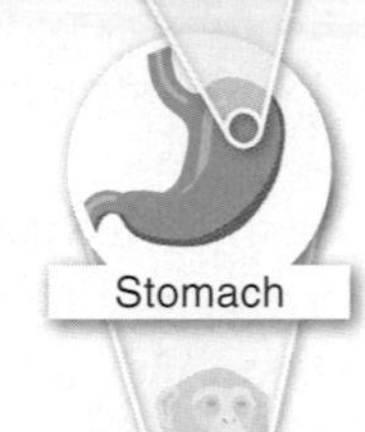

ORGAN
Group of tissues that perform specialized functions (most organs, including the stomach, contain all four tissue types)

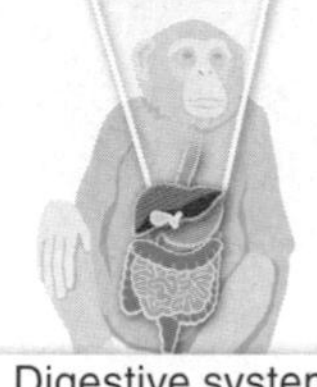

ORGAN SYSTEM
Group of organs that work together to accomplish one or a few, usually related, physiological functions

Multicellularity allows larger body size and greater physiological complexity through the specialization of cell and tissue functions.

FIGURE 21-2 **From cells to organ systems.**

In most animals, groups of cells with similar structure, along with some products of those cells, form **tissues,** in which the cells act together to perform specific functions in the body. Adult animals generally have four main types of tissue (FIGURE 21-3). We'll discuss them in more detail in Sections 21.2 through 21.5, but we introduce them here.

Connective tissue consists of cells embedded in a large amount of extracellular material, called **matrix,** which together contribute to body structure and support. Found throughout the human body, connective tissue can also serve to anchor cells, regulate communication between cells, and influence growth and wound healing. Some of the most important structures formed from connective tissue include tendons, cartilage, blood, adipose tissue (fat), and bone.

Epithelial tissue covers and lines most exterior and interior surfaces of the body. Skin is an epithelial tissue, as are the tissues that line the nose, throat, lungs, blood vessels, and digestive tract.

Muscle tissue is made up of cells that can contract. This characteristic gives muscle tissue the ability to generate movement or pump fluids through the body.

TYPES OF ANIMAL TISSUE

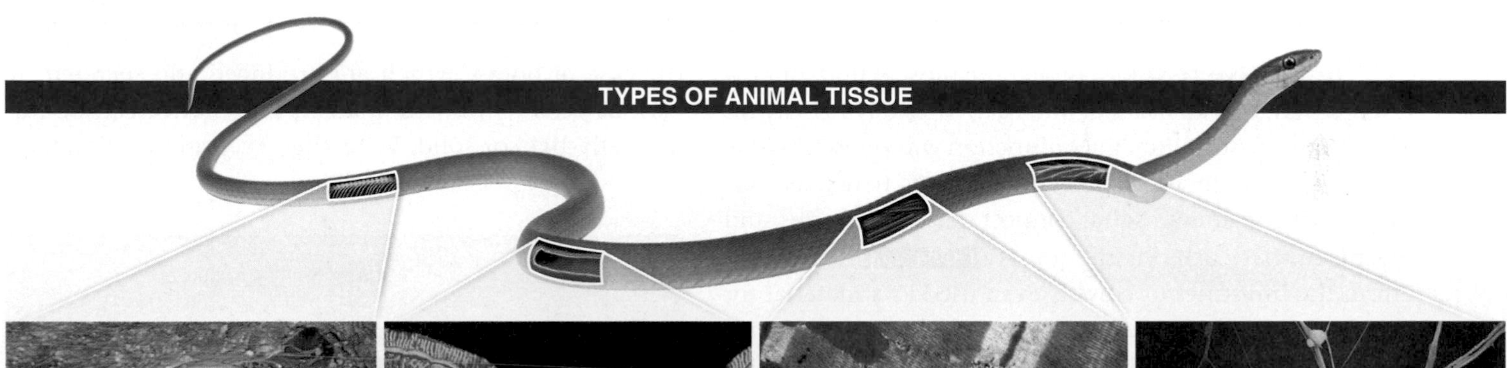

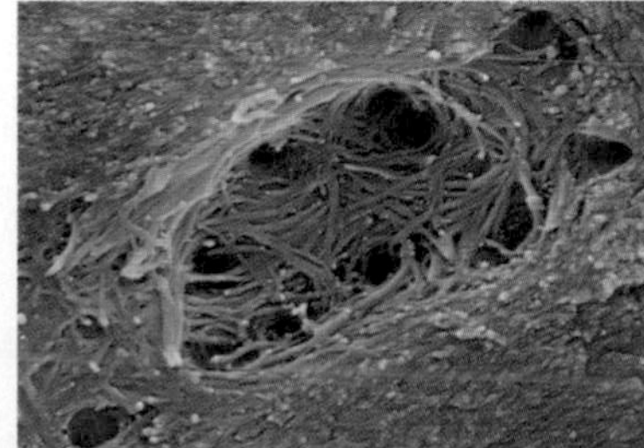

CONNECTIVE TISSUE
- Composed of cells interspersed throughout a matrix
- Provides structure and support, anchors cells, and regulates communication between cells

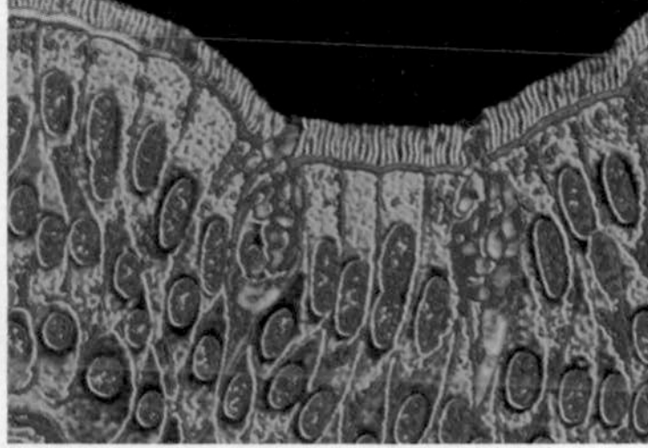

EPITHELIAL TISSUE
- Composed of cells that cover and line most surfaces of animal bodies
- Forms the skin and the lining of the lungs, digestive tract, and blood vessels

MUSCLE TISSUE
- Composed of cells that can contract
- Generates movement, pumps fluid, and moves substances

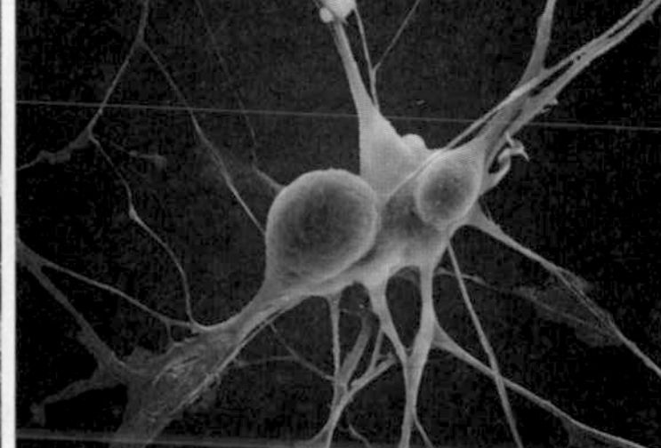

NERVOUS TISSUE
- Composed of specialized cells that send and receive electrical signals
- Stores and transmits information

FIGURE 21-3 **Tissues perform specific functions in the body.**

Nervous tissue, found throughout the human body, is specialized to send and receive electrical and chemical signals and, in doing so, can store and transmit information. The brain and spinal cord are made up of large amounts of nervous tissue.

Just as cells with similar functions are grouped into tissues, so tissues are often grouped into organs or organ systems. **Organs** are structures that serve specialized functions, and they can contain several (or even all four) types of tissue. The heart, liver, kidneys, and brain are examples of organs in the human body. **Organ systems** are groups of organs that work together to accomplish one or more, usually related, functions. The circulatory system, for example, includes the heart, blood vessels, and blood.

In the remaining chapters of the book, we explore animal **physiology**—the complex functions carried out by each of the organ systems in animals. We describe the structures that are part of each system, how they function, and how they have evolved. As we do this, we see that an organism is greater than the sum of its parts. The working together of cells to form tissues, and of tissues to form organs and organ systems, enables multicellular organisms to reproduce, defend themselves, and communicate (among many other abilities), in ways that are not possible for a single cell or a single type of tissue.

TAKE HOME MESSAGE 21.1

» Animal bodies are highly organized, and at all levels of organization, the physical features are closely related to function. In most animals, cells with similar structure and function are organized into tissues of four types: connective tissue, epithelial tissue, muscle tissue, and nervous tissue. Tissues are often organized into organs, which serve specialized functions and can contain several types of tissue. In turn, organs can be organized into organ systems that accomplish highly complex tasks.

21.2 Connective tissue provides support.

Connective tissue is usually the most abundant type of tissue in an animal. Most activities rely heavily on connective tissue, often in conjunction with muscle tissue. Playing tennis, for example, relies on blood to deliver oxygen to muscle, tendons that connect muscles to bones, and cartilage that cushions the joints (FIGURE 21-4). (Collagen from the intestines of cows is even used as a material for producing the strings in some tennis rackets.) Connective tissue is sometimes called "cellular glue," because it holds cells together and, as bone and cartilage, for example, gives shape, structure, and support to other tissues, structures, and organs throughout the body.

Regardless of the function it performs or the form it takes, all connective tissue consists of cells that are embedded in matrix—a mass of non-living extracellular material (FIGURE 21-5). Matrix, in vertebrates, consists chiefly of polysaccharides and protein (and some minerals, in the case of bone), which are produced and secreted by various cells within the matrix. The matrix can be liquid, jelly-like, or solid. With the exception of blood, all

FIGURE 21-4 Most activities, such as tennis, rely heavily on connective tissue, often in conjunction with muscle tissue.

CONNECTIVE TISSUE STRUCTURE

Connective tissue is a collection of cells arranged within an extracellular matrix that gives shape, structure, and support to other body tissues.

FIBROBLASTS
Cells that produce and secrete the proteins collagen and elastin (present in every type of connective tissue except blood)

MATRIX
A non-living, extracellular mass of protein fibers and surrounding liquid, jelly-like, or solid material

- Collagen protein
- Elastin protein
- Surrounding material

FIGURE 21-5 **Connective tissues consist of cells embedded in an extracellular matrix.**

connective tissue contains cells called *fibroblasts,* which produce and secrete the matrix proteins **collagen** and (in most cases) **elastin.** So common is connective tissue that collagen, which often forms a sort of net surrounding organs, is the most abundant protein in humans and other mammals.

A type of tissue called **connective tissue proper** functions much like the body's packing material. It can be loosely or densely packed. In loose connective tissue, the cells are in a semi-fluid, flexible matrix that generally has many fibers, usually collagen, embedded in it (FIGURE 21-6). Loose connective tissue includes the soft padding under your skin, the tissue surrounding most organs, and adipose (fat) tissue, which aids in cushioning, lubricating, and insulating other tissues.

Dense connective tissue is stronger than loose connective tissue. It also has collagen as its chief matrix component, but it has many more tightly packed collagen fibers than loose connective tissue. Examples of dense connective tissue include **tendons,** which connect muscle to bone,

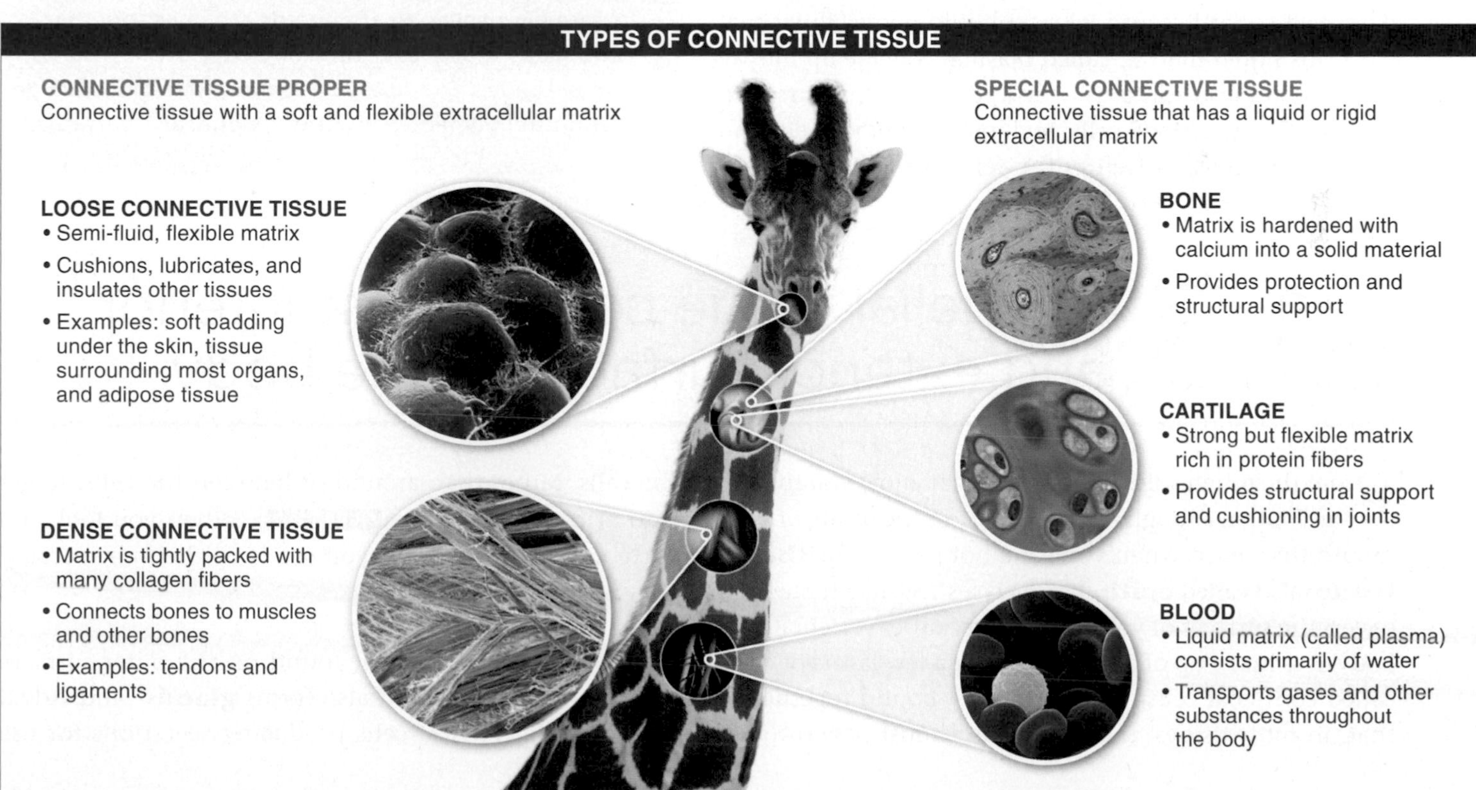

TYPES OF CONNECTIVE TISSUE
CONNECTIVE TISSUE PROPER
Connective tissue with a soft and flexible extracellular matrix
LOOSE CONNECTIVE TISSUE
• Semi-fluid, flexible matrix
• Cushions, lubricates, and insulates other tissues
• Examples: soft padding under the skin, tissue surrounding most organs, and adipose tissue
DENSE CONNECTIVE TISSUE
• Matrix is tightly packed with many collagen fibers
• Connects bones to muscles and other bones
• Examples: tendons and ligaments
SPECIAL CONNECTIVE TISSUE
Connective tissue that has a liquid or rigid extracellular matrix
BONE
• Matrix is hardened with calcium into a solid material
• Provides protection and structural support
CARTILAGE
• Strong but flexible matrix rich in protein fibers
• Provides structural support and cushioning in joints
BLOOD
• Liquid matrix (called plasma) consists primarily of water
• Transports gases and other substances throughout the body

FIGURE 21-6 **Connective tissues: the most abundant tissues in most animals.**

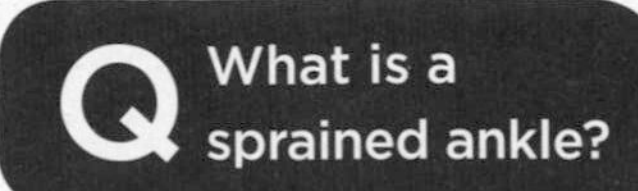

and **ligaments,** which bind bone to bone. A sprained ankle is a common injury in which a twisting of the ankle overstretches and sometimes tears part of a ligament in the foot.

The second type of connective tissue, called **special connective tissue,** differs from connective tissue proper in that it is rigid or liquid. Bone, cartilage, and blood (see Figure 21-6) are examples.

In **bone,** the mineral calcium is incorporated into the extracellular matrix, which then hardens into a solid material. Bones can give significant protection to the body (the skull protects the brain, the ribs protect the lungs and heart) or can provide structural support, as the backbone does.

Cartilage, a dense connective tissue with an extracellular matrix rich in collagen, elastin, and proteins bound to long carbohydrate chains, has a hardness between that of bone and that of tendons. Strong, but also flexible, cartilage is found in the ears and the tip of the nose in humans, and it cushions the bones in joints throughout the body. In some vertebrates, including sharks, the entire skeleton is made of cartilage.

Blood differs from the other types of connective tissue described so far because it has a liquid extracellular matrix. This liquid matrix, called plasma, is made up mostly of water, but also contains dissolved proteins, sugars, and other molecules. Blood cells, including red blood cells and white blood cells, as well as the cellular fragments called platelets (see Chapter 22), are suspended within plasma as they transport gases and other substances throughout the body.

All connective tissues in the body, as we've seen, have essentially the same structure: cells embedded within an extracellular matrix that may be solid, soft and flexible, or fluid. An inflammation or malfunction in connective tissue is the cause of many diseases with widespread effects, including scleroderma and Marfan syndrome. The most common type of arthritis, osteoarthritis, results from the breakdown of connective tissue, usually the cartilage, around joints. The reduced cushioning often leads to inflammation of tissue around the joint, with stiffness and pain, and is commonly treated with anti-inflammatory drugs, including aspirin and ibuprofen.

Q The most common form of arthritis is called "wear and tear" arthritis. Why?

TAKE HOME MESSAGE 21.2

» The most abundant type of tissue in most animals is connective tissue. Connective tissue is a collection of cells embedded within an extracellular matrix, usually containing collagen, that holds the cells together and gives shape, structure, and support to other body tissues. Examples of connective tissue include tendons, ligaments, fat, blood, bone, and cartilage.

21.3 Epithelial tissue covers most interior and exterior surfaces of the body.

From the tough soles of your feet that allow you to walk barefoot on a rough, hot sidewalk, to the lining of your mouth that burns when you eat a hot pepper, **epithelial tissue** (also called **epithelium**) is a sheet-like tissue that covers the surfaces of your—and every other vertebrate's—body. It is made up of a single layer or a few layers of cells, called epithelial cells, that are tightly bound together so that, in most tissues, fluids and gases must pass through the cells, rather than around or between the cells, to get into or out of the body (FIGURE 21-7). When you look at a vertebrate, most of what you see is epithelium, because skin is an epithelial tissue.

Epithelial tissue is not just found on the outer surfaces of organisms, however. It also forms **glands**—individual cells or collections of cells producing secretions for use

FUNCTIONS OF EPITHELIAL TISSUE

Epithelium, a sheet-like tissue that separates different parts of the body, plays important roles in protection, transport, and secretion.

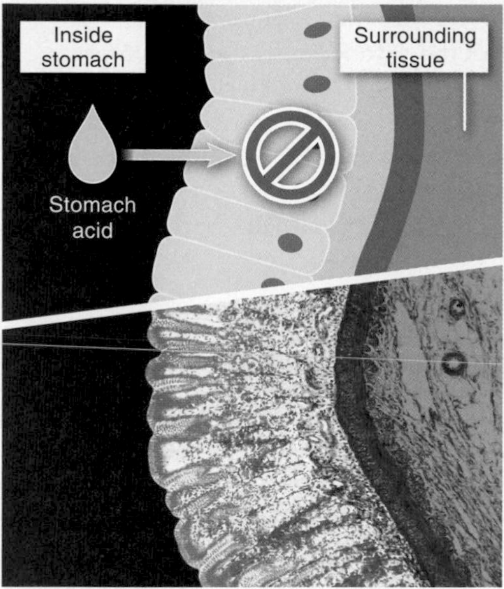

PROTECTION

- Acts as a barrier between the inside and outside of an organism
- Keeps fluids from leaking into or out of tissue

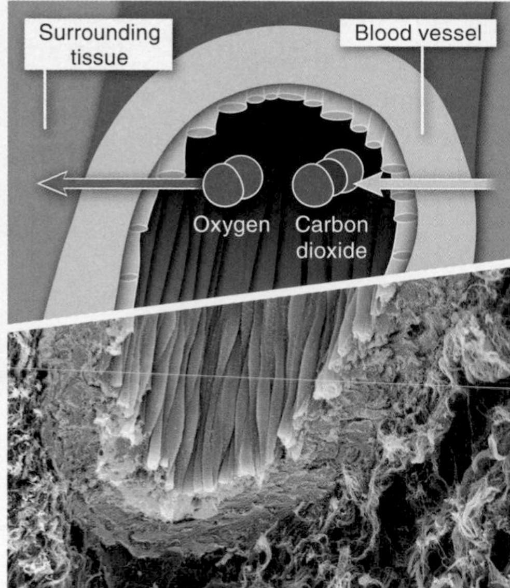

TRANSPORT

- Regulates the movement of nutrients and other molecules into and out of body tissues

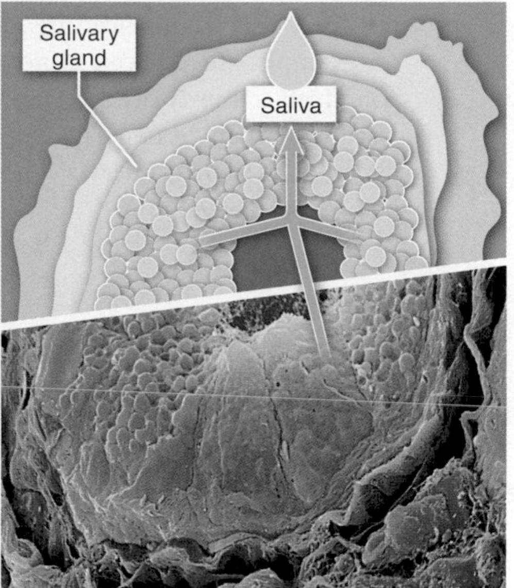

SECRETION

- Can form exocrine glands, which secrete products such as saliva, sweat, and mucus
- Can form endocrine glands, which secrete hormones

FIGURE 21-7 The multiple roles of epithelial tissue.

elsewhere in the body—and lines the internal tubes and cavities of the body, such as the stomach and intestines, lungs, and blood vessels. In each location, epithelium always has two distinctive sides. The "outside" can be in contact with the outside of the body or with an internal cavity such as the stomach, where it plays a protective role. The "inside" (or underside) faces away from the surface and is generally secured to underlying tissues. Epithelium plays multiple roles in organisms. Three of its most important functions are protection, transport, and secretion (see Figure 21-7).

1. Protection. Acting as a barrier, epithelial cells are linked together closely by tight junctions and desmosomes (see Section 4.12), which keep fluids from leaking into or out of tissue. If the strong acids in the stomach, for example, were to leak into surrounding tissue, they would seriously damage the tissue. This is what occurs in individuals with stomach ulcers, open sores in the lining of the stomach. The tight fit between epithelial cells is also why the skin of terrestrial vertebrates is usually waterproof.

Q Damage to the lining of the stomach can have painful consequences. Why?

2. Transport. Epithelium forms small finger-like projections in the lining of the small intestine, where nutrients are absorbed from digested food into the bloodstream. In blood vessels, epithelium controls which molecules can enter other tissues of the body. And in the kidneys, epithelium helps to regulate which molecules from the bloodstream are eliminated in the urine.

3. Secretion. Epithelium can also form glands. **Exocrine glands** generally secrete products—including earwax, sweat, saliva, milk, mucus, and, in the case of some frogs, toxic poisons—into ducts that lead to the external environment. **Endocrine glands,** on the other hand, are ductless glands that produce hormones, which are released into the fluid surrounding the glands and, from there, usually enter the bloodstream for distribution to other parts of the body, where they act as chemical messengers (see Chapter 25).

Because epithelial cells are often in contact with a large variety of materials, they tend to experience more damage than other tissues. For this reason, they tend not to last long and are replaced frequently. Human skin cells, for example, are replaced approximately every two weeks. (Dandruff is mostly dead skin cells.) Cells lining the digestive tract last about five days, and because liver epithelium experiences slightly less wear and tear, it lasts a year or two before replacement.

TAKE HOME MESSAGE 21.3

» Epithelium is a very thin, sheet-like tissue that covers most of the exterior and interior surfaces of an animal's body. Made up of a single layer or a few layers of cells that are tightly bound together, epithelium acts as a barrier between the inside and outside of the organism.

21.4 Muscle tissue enables movement.

Most animals move. And **muscle tissue,** made up of elongated cells capable of generating force by contracting, is responsible for much of that movement. Most muscle tissue cells are packed with protein filaments that slide past one another as they break down ATP, causing the entire cell to shorten and thus the muscle to contract. The action of muscle cells enables organisms to generate force and motion. There are three types of muscle tissue: skeletal, cardiac, and smooth muscle (FIGURE 21-8).

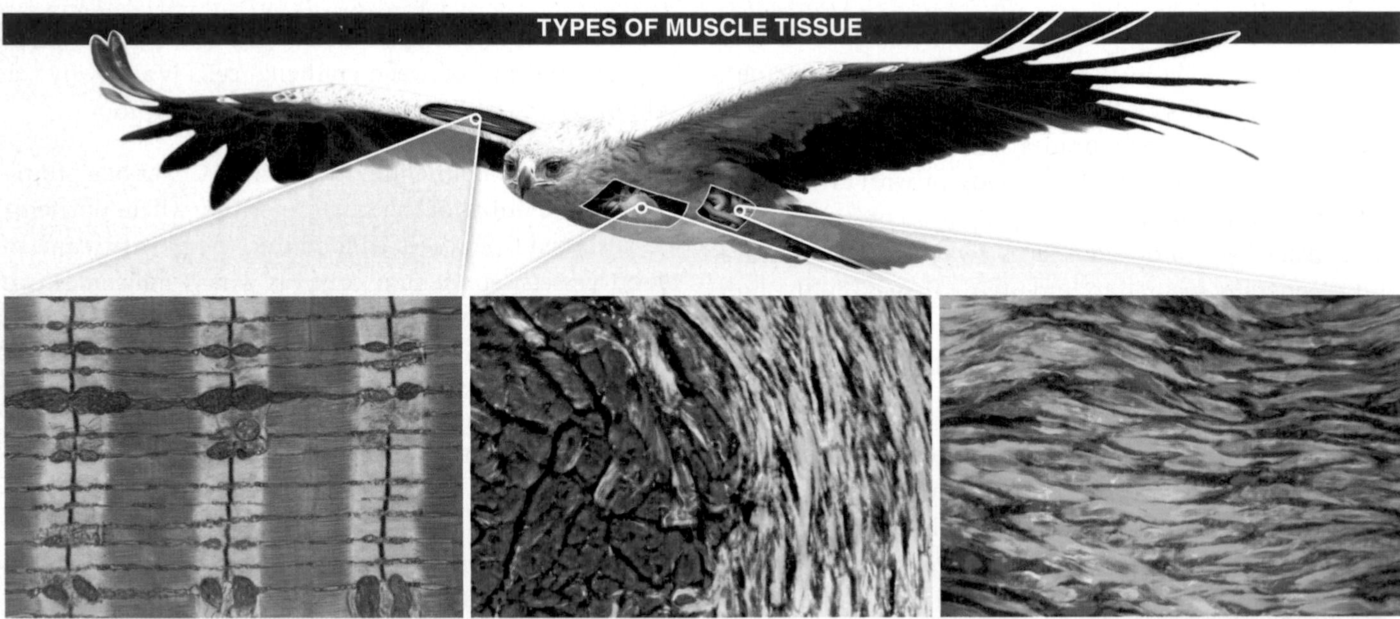

TYPES OF MUSCLE TISSUE
SKELETAL MUSCLE
• Generates most of the movement seen in animals
• Under both conscious and unconscious control
CARDIAC MUSCLE
• Muscle in the heart that pumps blood through the body
• Under unconscious control
SMOOTH MUSCLE
• Generates slow, rhythmic contractions that can gradually move food or other substances through the body or alter blood pressure
• Under unconscious control and can generate contractions without nervous stimulation

FIGURE 21-8 **Muscle tissue is made up of elongated cells capable of generating force when they contract.**

Skeletal muscle (sometimes called voluntary muscle) is usually attached to bones and is responsible for generating most movement we see in animals, including facial expressions and breathing. Skeletal muscles can be under conscious control, such as when you choose to flex your biceps, or unconscious control, such as is often the case for the muscles that control breathing or moving your eyes. Muscles account for about 40% of human body weight. The individual skeletal muscle cells are called **muscle fibers.** The muscle fibers are very long and contain multiple nuclei, and the repeating units of protein filaments in the cells give the fibers a striped, or striated, appearance. Skeletal muscles are controlled by the nervous system, and the individual nerve cells (neurons) connected to each muscle fiber stimulate its contraction.

Cardiac muscle, as the name indicates, is located only in the heart. It is not under conscious control and causes the heart to pump blood through the body. Because of the tremendous amount of energy used by cardiac muscles, constantly contracting throughout our life, the cells contain many more mitochondria than do other types of muscle cells. As with skeletal muscle, cardiac muscle has striations. The cells of heart muscle are fused together and connected by gap junctions (see Section 4.12) that allow the electrical signals that initiate each contraction to pass through.

Smooth muscle is found in the walls surrounding blood vessels, the stomach and intestines, the bladder, and many other organs and internal tubes within the body. It gets its name from the fact that it lacks obvious striations. Not under conscious control (and so sometimes called involuntary muscle), smooth muscle generates slow, rhythmic contractions that can gradually move food or other substances through the organ or tube or alter blood pressure by increasing or decreasing the diameter of blood vessels. Smooth muscle can generate contractions without nervous stimulation—for example, through the opening and closing of ion channels, such as those controlling the movement of calcium ions.

TAKE HOME MESSAGE 21.4

» Muscle tissue consists of elongated cells capable of generating force when they contract. Skeletal muscle is responsible for generating most of the movement we see in animals. Cardiac muscle causes the heart to pump blood through the body. Smooth muscle, surrounding blood vessels and many internal organs, generates slower contractions that can gradually move food or other substances and alter blood pressure.

21.5 Nervous tissue transmits information.

The fourth type of animal tissue is **nervous tissue,** specialized to store and transmit information. It is responsible for much of the communication that occurs within an animal's body. Nervous tissue enables animals to sense and respond to stimuli such as the smell of food; the sight, smell, or sounds of a predator or a potential mate; or the heat of a fire.

Neurons and glial cells are two types of nervous tissue cells (FIGURE 21-9). **Neurons** are the "excitable" cells that receive and transmit a signal. Most neurons have three distinctive elements: dendrites, a cell body, and an axon. **Dendrites** are a bit like an antenna system, specialized for receiving signals from the external environment or from other neurons. The **cell body** contains the nucleus and other cellular machinery found in eukaryotic cells. And the **axon** is a single projection that transmits impulses away from the cell body and can extend over very long distances—sometimes 3 feet (1 m) or more!

Glial cells, also called **neuroglia,** are like the support staff to neurons. They do not carry signals but assist and nourish the neurons. There are numerous types of glial cells, and they vastly outnumber neurons. Together, the

STRUCTURE OF NERVOUS TISSUE

A neuron from the part of the brain called the hippocampus

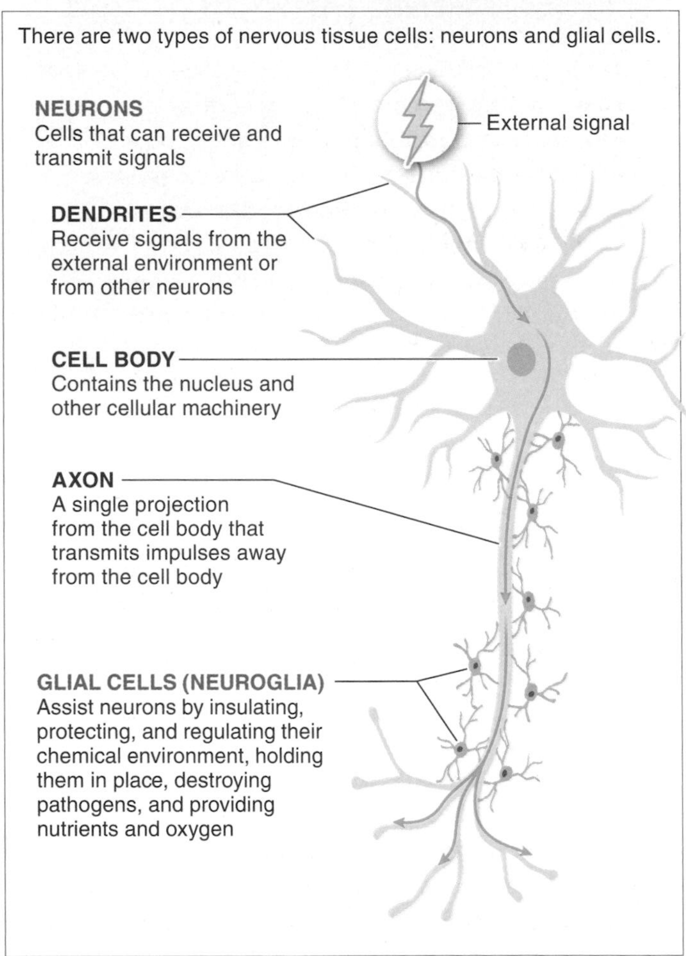
There are two types of nervous tissue cells: neurons and glial cells.
NEURONS
Cells that can receive and transmit signals
External signal
DENDRITES
Receive signals from the external environment or from other neurons
CELL BODY
Contains the nucleus and other cellular machinery
AXON
A single projection from the cell body that transmits impulses away from the cell body
GLIAL CELLS (NEUROGLIA)
Assist neurons by insulating, protecting, and regulating their chemical environment, holding them in place, destroying pathogens, and providing nutrients and oxygen

FIGURE 21-9 Neurons and glial cells.

ORGANIZATION OF THE VERTEBRATE NERVOUS SYSTEM

In vertebrates, the nervous system is divided into the central nervous system and the peripheral nervous system.

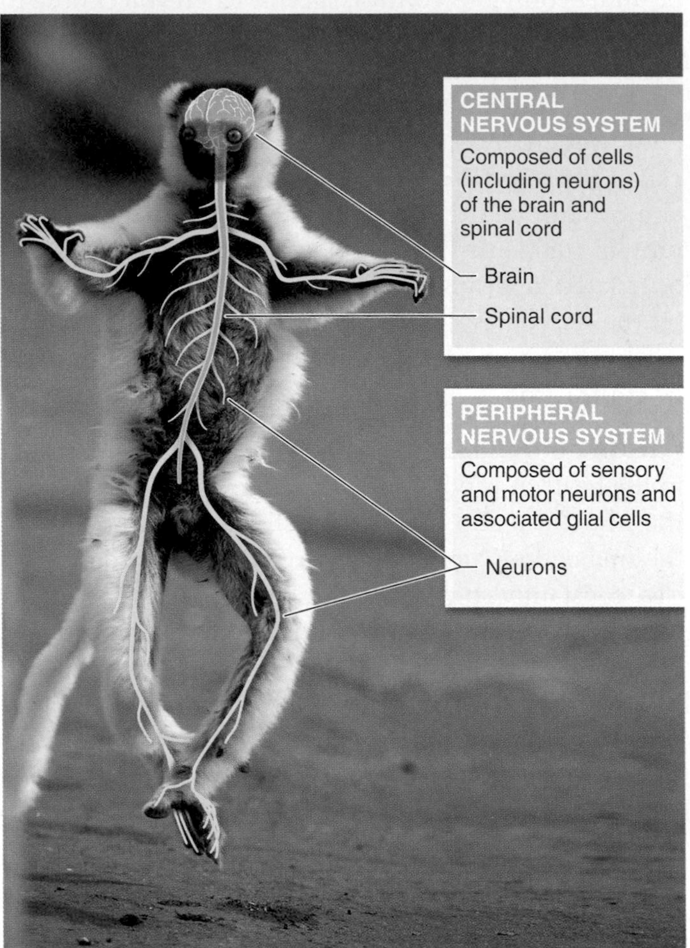
CENTRAL NERVOUS SYSTEM
Composed of cells (including neurons) of the brain and spinal cord
Brain
Spinal cord
PERIPHERAL NERVOUS SYSTEM
Composed of sensory and motor neurons and associated glial cells
Neurons

FIGURE 21-10 The vertebrate nervous system: brain, spinal cord, and neurons.

various types of glial cells produce insulation for neurons, protect and regulate the chemical environment around neurons, hold the neurons in place, destroy pathogens, and provide nutrients and oxygen.

All of the nervous tissue of an organism makes up its nervous system (FIGURE 21-10). In vertebrates, the nervous system is divided into the **central nervous system,** which includes the brain and spinal cord, and the **peripheral**

At the point where the ends of a motor neuron axon are in contact with muscle cells, signals from the neuron can initiate contractions in the muscle cells.

FIGURE 21-11 **Motor neurons of the peripheral nervous system.**

nervous system, which includes the neurons that detect a stimulus (sensory neurons) and the neurons that transmit signals to the muscles (motor neurons) and glands in response to that stimulus (FIGURE 21-11). (The nervous system is discussed in detail in Chapter 24.)

TAKE HOME MESSAGE 21.5

» Nervous tissue is specialized to store and transmit information. There are two types of nervous tissue cells: neurons, which can receive and transmit a signal, and glial cells, which assist and provide nutrients for neurons.

21.6 Each organ system performs special tasks.

Most animals have some tissues that are organized into organs (the exceptions are sponges and some cnidarians). Organs are structures, such as the heart, brain, lungs, and liver, that serve specialized functions and consist of multiple tissue types. And just as cells make up tissues and tissues make up organs, organs, too, are part of larger functional units, the organ systems, which carry out the various physiological processes necessary for the growth, development, maintenance, and reproduction of organisms.

We briefly describe 11 animal organ systems in FIGURE 21-12, and we'll discuss them in greater detail in Chapters 22–28. It's important to keep in mind that these systems do not operate in isolation: in carrying out its tasks, each system depends on the proper functioning of one or more of the other systems and, in turn, often influences the functioning of other systems. The act of running, for example, might be initiated when the *nervous system* detects a threat and sends impulses to the *musculoskeletal system* (which causes the body to move), to the *respiratory system* (which increases the rate of oxygen consumption), and to the *circulatory system* (which increases the rate at which the heart pumps blood and thus increases the oxygen available to muscle tissue).

As we explore organ systems in more detail in the following chapters, we'll pay particular attention to the ways in which the form and function of the systems' components are intimately related and what this relationship reveals about how these systems (and the organisms of which they are a part) evolved.

THE ORGAN SYSTEMS

DIGESTIVE SYSTEM
Disassembles and absorbs food so the body can acquire the nutrients it needs to function

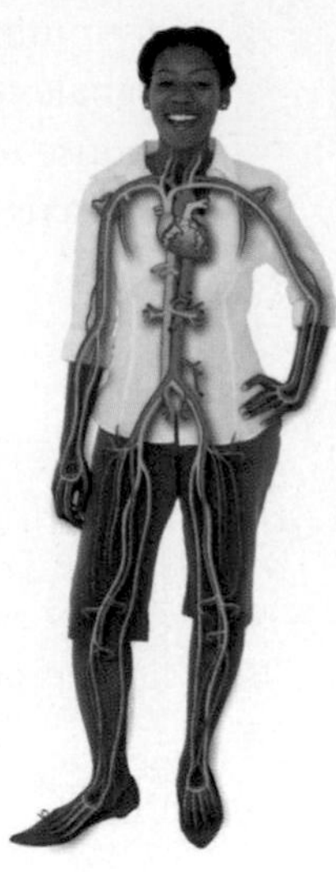

CIRCULATORY SYSTEM
Transports nutrients and respiratory gases to the tissues and eliminates wastes from the tissues

RESPIRATORY SYSTEM
Provides a site for gas exchange between the external environment and an organism's circulatory system

REPRODUCTIVE SYSTEM (MALE)
Produces sperm and delivers them to the female reproductive system, where fertilization may occur

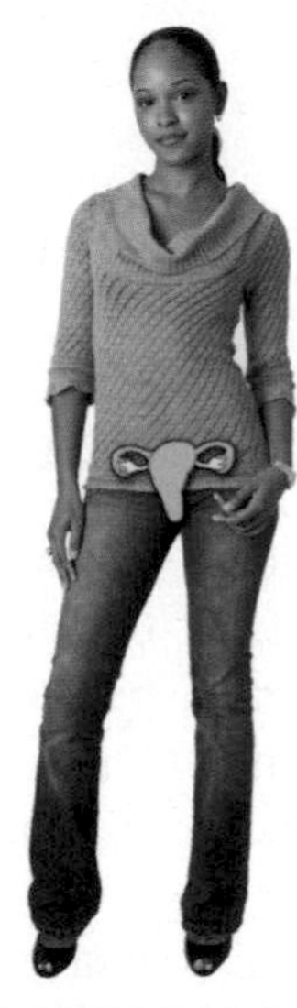

REPRODUCTIVE SYSTEM (FEMALE)
Produces eggs and provides an environment that can nurture a developing embryo and fetus, if fertilization occurs

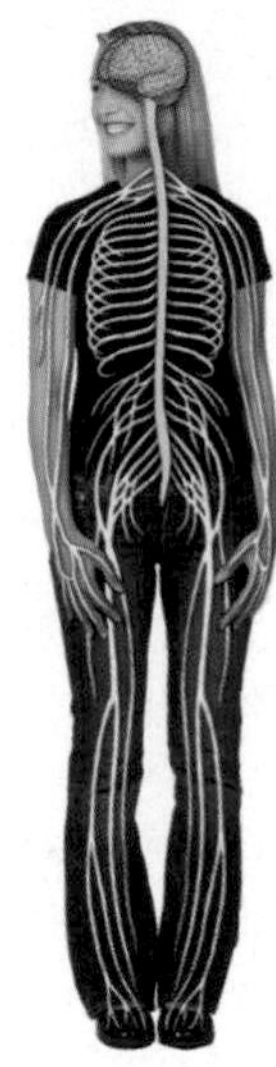

NERVOUS SYSTEM
Acts as the control center of the body and interprets, stores, and transmits information, using electrical impulses and chemical signals

FIGURE 21-12 **The major organ systems of vertebrates.**

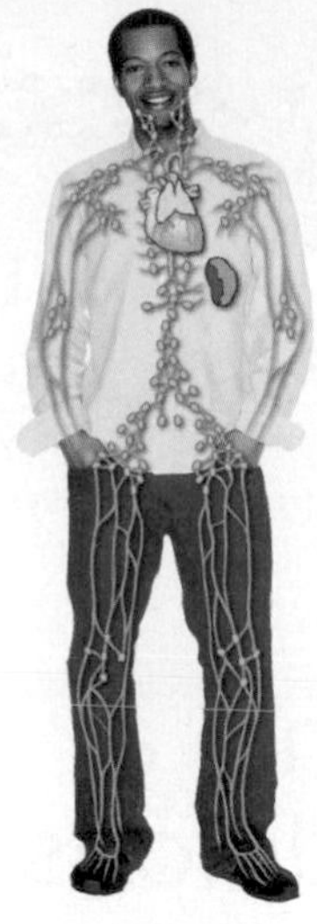

IMMUNE AND LYMPHATIC SYSTEM
Attacks pathogens that threaten the body and plays a supporting role in circulation by recycling fluid that leaks from the circulatory system

URINARY/EXCRETORY SYSTEM
Purifies the blood by filtering out wastes and transports wastes out of the body

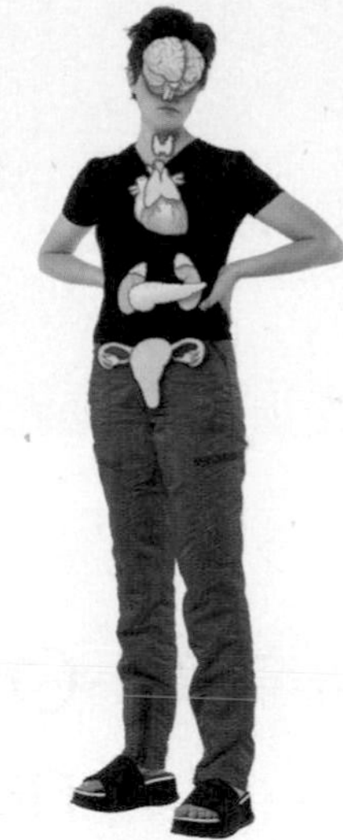

ENDOCRINE SYSTEM
Regulates body activities by releasing hormones that travel through vessels in the circulatory system to reach target cells

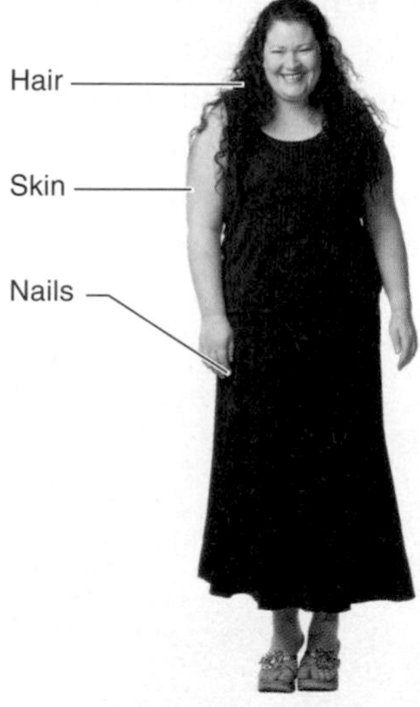

INTEGUMENTARY SYSTEM
Provides protection by forming a barrier between the inside and outside of an organism and can aid in the secretion and transport of molecules

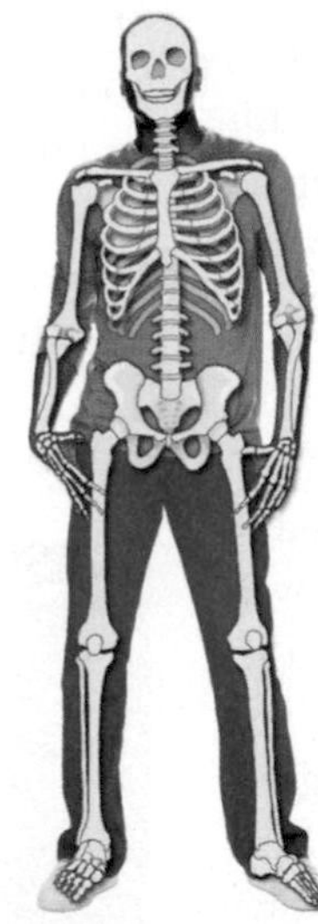

SKELETAL SYSTEM
Supports and protects the body and internal organs, manufactures blood cells, and provides a surface for muscle attachment, creating a foundation for movement

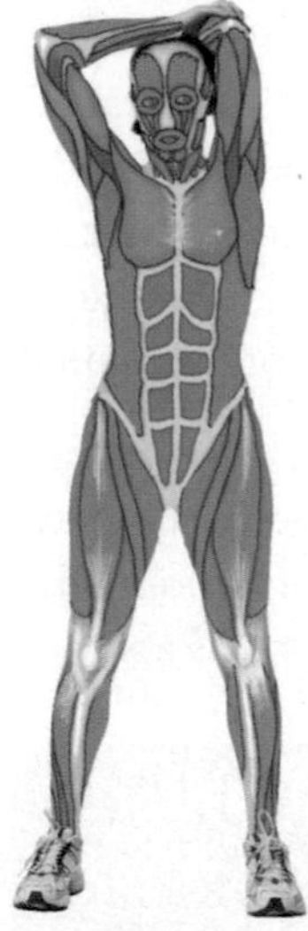

MUSCULAR SYSTEM
Generates force through contraction, which enables movement of the body and of blood, food, and other substances throughout the body

TAKE HOME MESSAGE 21.6

» In nearly all animals, some tissues are organized into organs that serve specialized functions and consist of multiple tissue types. These organs are part of organ systems that carry out the various physiological processes necessary for the organism's growth, development, maintenance, and reproduction.

21.7–21.8

Animals have an internal environment.

Heat radiates from warm bodies in the cold.

21.7 Our bodies function best within a narrow range of internal conditions

Following an extremely rigorous practice in the summer of 2001, 27-year-old Minnesota Vikings football player Korey Stringer walked into an air-conditioned tent to recover. Almost immediately, however, the 6 foot 4 inch, 335-pound all-pro player began to feel weak and dizzy; his breathing became rapid and his blood pressure dropped. He was rushed to the hospital, but as a consequence of his body temperature allegedly reaching 108.8° F (42.7° C), many of his organs failed, and he died that night.

Normal body temperature is approximately 98.6° F (37° C). **Hyperthermia** occurs when too much heat is produced or when high environmental temperature and humidity overwhelm the body's ability to dissipate heat. (It is unlike a fever, in which the body sets its core temperature slightly higher in response to an infection by pathogens.) Korey Stringer's death is an example of the disastrous consequences that can result from extreme hyperthermia, called **heat stroke,** and illustrates the danger when organisms fail to maintain an internal environment within a safe range.

Q What is heat stroke? Why is it dangerous?

Although the environmental temperature range you experience every day may involve a 20°, 30°, or even 40° F difference between high and low, humans cannot tolerate such large internal temperature swings. Even deviations of less than 10° F can lead to a cascade of biochemical problems with serious health implications. For mammals, the risks of getting too hot include reduced enzyme activity and, at extreme temperatures, the complete breakdown of enzymes. Excessive water loss, too, can occur, leading to numerous problems—including organ failure. For these reasons, heat stroke can be life-threatening (FIGURE 21-13).

At the other extreme, **hypothermia,** or extremely low body temperature, can cause difficulties in coordination and movement, along with disorientation and confusion, irregular heartbeat, and, ultimately, death. Similarly disastrous consequences can result from the failure to maintain a consistent internal environment in other ways, including

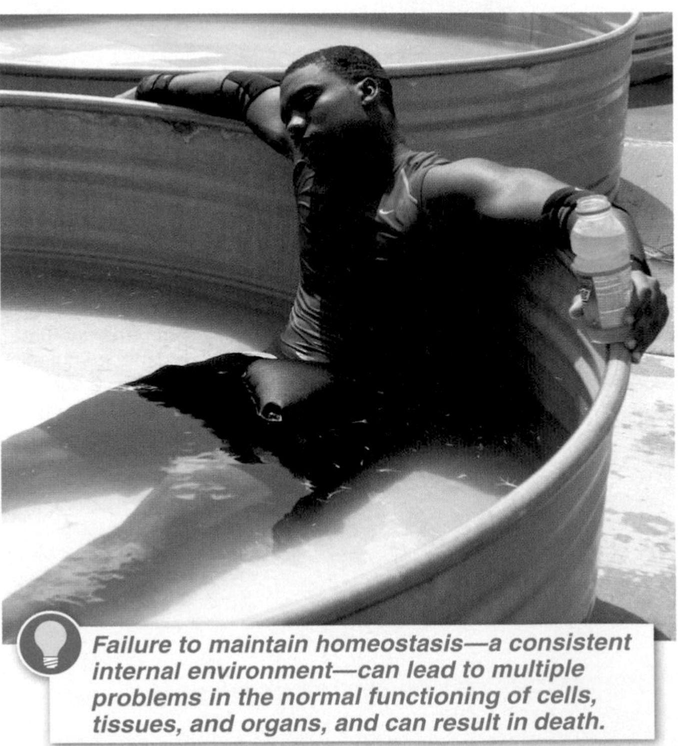

Failure to maintain homeostasis—a consistent internal environment—can lead to multiple problems in the normal functioning of cells, tissues, and organs, and can result in death.

FIGURE 21-13 A cooling soak after a tough workout can prevent hyperthermia.

blood sugar levels, blood pH, and tissue concentrations of oxygen and carbon dioxide.

In the remainder of this chapter, we explore a fundamental characteristic of organism function: **homeostasis,** the body's use of physical and chemical processes to maintain a consistent internal environment, even in the face of changing external and internal environmental forces.

TAKE HOME MESSAGE 21.7

» Failure to maintain a consistent internal physical and chemical environment can lead to multiple problems in the normal functioning of cells, tissues, and organs, and can result in death.

21.8 Animals regulate their internal environment through homeostasis.

All of the functions and activities of a single-celled organism are straightforward, thanks to the advantages of living close to the external environment. Food and other materials necessary for life are nearby, just across a plasma membrane. Getting rid of waste products from the cell is similarly simple. But direct contact with the external environment has its costs, too. A single-celled organism is, to a large extent, at the mercy of its external environment. Changes in the environment—such as in temperature or pH—can have a great impact on the cell itself.

Like single-celled organisms, multicellular animals also must acquire food and other materials, as well as get rid of waste. These tasks become increasingly complex when cells are not in direct contact with the external environment. On the other hand, a cell can be protected from harsh or changing environmental conditions if it's not in direct contact with the environment.

With multicellularity and increasing size, then, an animal's internal environment takes on greater importance than the external environment in influencing cell functioning. In vertebrates, this internal environment consists of the **extracellular fluid** (also called the **interstitial fluid**) that fills the space between cells, bathing the cells. This fluid is primarily water, but it also contains nutrients and raw materials for growth and development, as well as waste products that have diffused out of or been removed from cells. The volume of interstitial fluid is not insignificant. In fact, one-third of the water in the human body is in interstitial fluid, with more than two and a half gallons (10 liters—picture five 2-liter soda bottles!) surrounding and bathing the cells.

One of the hallmarks of animal physiology, however, is that organisms generally maintain homeostasis, or the ability to remain within a narrow range of physical and chemical conditions, even in the face of changing external environmental forces. In this relatively constant, steady, internal environment, variables such as temperature, water-solute balances, pH, blood sugar levels, and O_2 and CO_2 concentrations in blood and other tissues are maintained within narrow ranges by the activities of cells within the organism's tissues, organs, and organ systems (FIGURE 21-14).

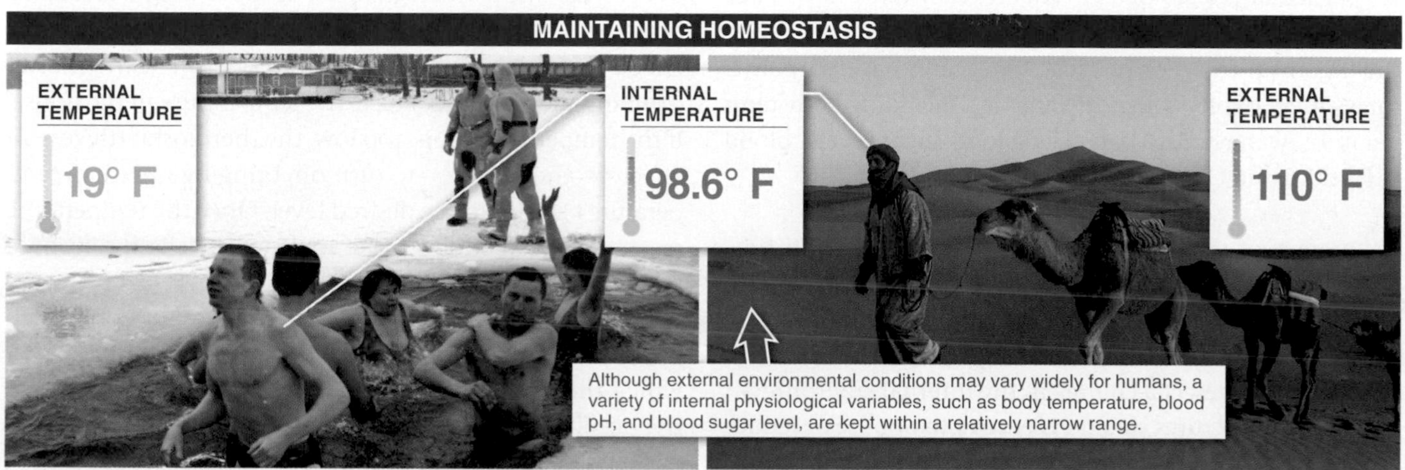

MAINTAINING HOMEOSTASIS
EXTERNAL TEMPERATURE
19° F
INTERNAL TEMPERATURE
98.6° F
EXTERNAL TEMPERATURE
110° F
Although external environmental conditions may vary widely for humans, a variety of internal physiological variables, such as body temperature, blood pH, and blood sugar level, are kept within a relatively narrow range.

FIGURE 21-14 **Homeostasis: maintaining a stable internal environment.**

Through homeostasis, organisms can maintain optimum metabolic functioning. For example, the work of cells, from DNA replication to protein production to intercellular communication, depends in large part on the activities of enzymes. In turn, an enzyme's activity depends critically on the temperature and other characteristics of its immediate environment. Even slight changes in these variables can disrupt, reduce, or even stop enzyme functioning. And this can have catastrophic consequences, such as in the case of heat stroke described in Section 21.7.

Enzyme activity is just one of many facets of cell function influenced by temperature and other physical features of the cell's internal environment; membrane permeability and the rates at which materials diffuse across membranes also respond to changes in the environment. In the following sections, we explore how homeostasis is maintained.

TAKE HOME MESSAGE 21.8

» Although the internal environment of multicellular animals is continuously influenced by their external environment, many animals maintain homeostasis: they keep a variety of internal physiological variables—including temperature, water-solute balances, pH, blood sugar levels, and blood gas concentrations—within a relatively constant range.

21.9–21.13

How does homeostasis work?

Why do we yawn? (Hint: It's about homeostasis.)

21.9 Negative and positive feedback systems influence homeostasis.

For animals to maintain homeostasis with regard to a particular physiological variable, they must have a **set point** for that variable, a target value or range to which the organism can return. Most vertebrates have set points for a variety of physiological variables, including body temperature, water-solute balances, blood sugar levels, blood pH, and tissue O_2 and CO_2 concentrations.

In the face of changes in its external environment, how does an organism maintain its homeostasis, its constant internal environment? The most common method involves the regulatory mechanism of **negative feedback,** in which sensors detect a change in the internal environment and trigger structures called **effectors** to oppose or reduce the change. This cycle of detection, response, and change is called a negative feedback loop (FIGURE 21-15).

An example of a negative feedback loop that we encounter in everyday life is the regulation of room temperature in a house, which involves a sensor (the thermostat) and two effectors (a furnace and an air conditioner). The thermostat contains a sensor that detects the temperature in the house. If the temperature drops too low, the thermostat triggers an effector—the furnace—to turn on, bringing the room temperature back up to the desired level. Once the temperature reaches a specified level, the thermostat tells the furnace to shut off. Similarly, if the temperature gets too high, the thermostat can trigger the air conditioner to turn on, which brings the temperature back down to the desired level, after which the air conditioner shuts off. In both cases, a negative feedback system senses a change in the temperature, triggers events that reverse the change, and restores the house's internal environment to its desired level.

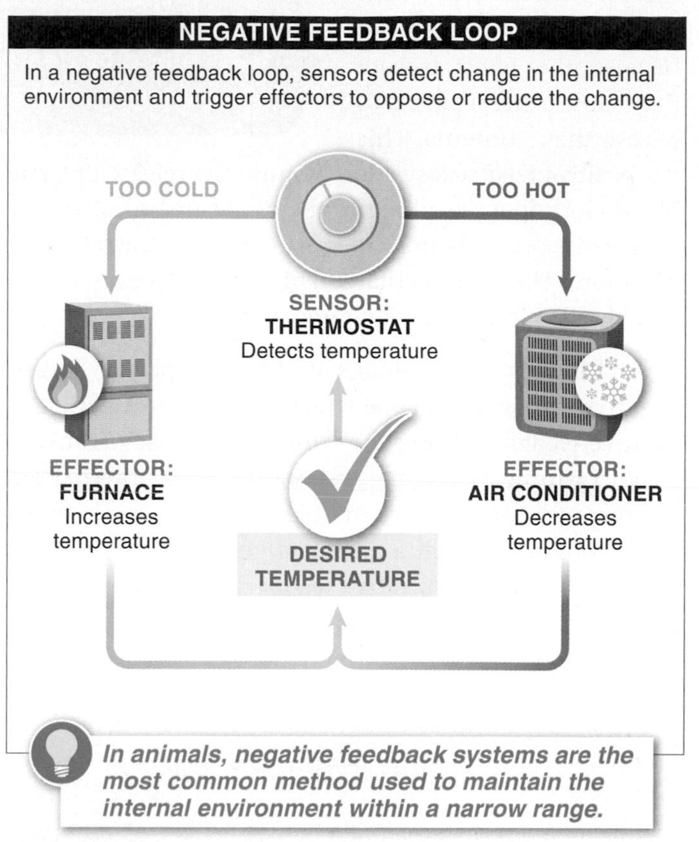

NEGATIVE FEEDBACK LOOP
In a negative feedback loop, sensors detect change in the internal environment and trigger effectors to oppose or reduce the change.
TOO COLD
TOO HOT
SENSOR:
THERMOSTAT
Detects temperature
EFFECTOR:
FURNACE
Increases temperature
EFFECTOR:
AIR CONDITIONER
Decreases temperature
DESIRED TEMPERATURE
In animals, negative feedback systems are the most common method used to maintain the internal environment within a narrow range.

FIGURE 21-15 A negative feedback loop restores the internal environment to a set point.

In the human body, negative feedback systems are the most common method used to maintain the internal environment within a narrow range. These systems include those that maintain body temperature, regulate levels of sugar in the bloodstream, and control the levels of salt and other solutes in body fluids. We explore these systems in detail later in this chapter.

It is important to note that while some groups of organisms, called **regulators,** maintain homeostasis for a certain variable (as we've just described for temperature), other organisms, called **conformers,** may have no set point for that variable, and the variable may fluctuate with external changes. Animals differ in the traits for which they are regulators and those for which they are conformers. Most fishes, for example, are conformers with regard to the temperature of the surrounding water but are regulators when it comes to the concentration of salt in their blood and tissues (FIGURE 21-16).

In addition to negative feedback systems, organisms also have a small number of **positive feedback** systems, in which deviations from conditions normally found in the internal environment cause an increase in or acceleration of the change, in the same direction. Such systems push the body away from homeostasis and often initiate cascading processes that increase the rate at which the system deviates

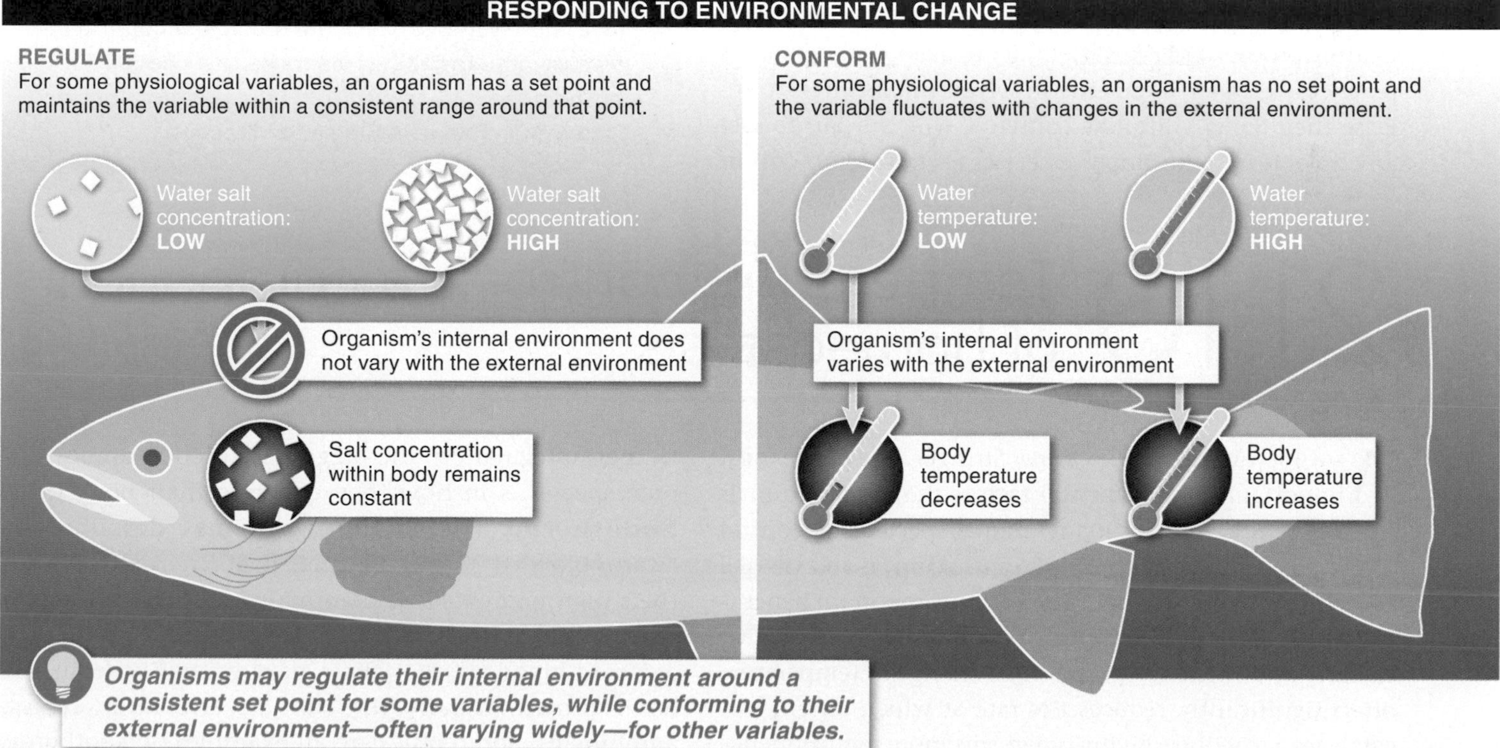

RESPONDING TO ENVIRONMENTAL CHANGE
REGULATE
For some physiological variables, an organism has a set point and maintains the variable within a consistent range around that point.
Water salt concentration: LOW
Water salt concentration: HIGH
Organism's internal environment does not vary with the external environment
Salt concentration within body remains constant
CONFORM
For some physiological variables, an organism has no set point and the variable fluctuates with changes in the external environment.
Water temperature: LOW
Water temperature: HIGH
Organism's internal environment varies with the external environment
Body temperature decreases
Body temperature increases
Organisms may regulate their internal environment around a consistent set point for some variables, while conforming to their external environment—often varying widely—for other variables.

FIGURE 21-16 Regulate or conform? Two strategies to cope with a changing environment.

POSITIVE FEEDBACK SYSTEM

In positive feedback systems, such as in the formation of a blood clot after an injury, a change away from conditions normally found in the body causes an increase or acceleration of the change.

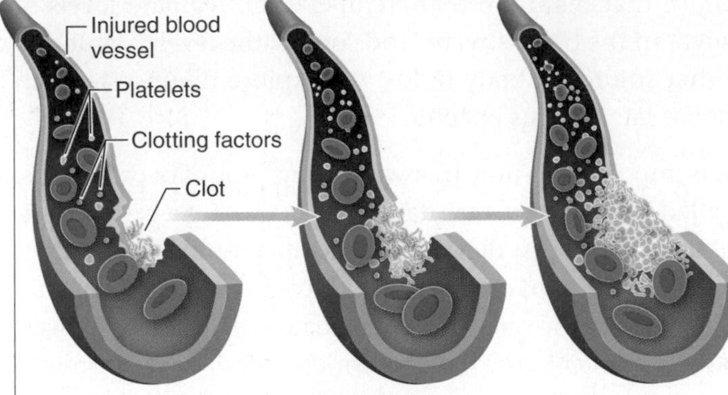

An injured blood vessel triggers platelets to form a clot.

The release of blood-clotting molecules into the bloodstream causes the release of more such molecules, which causes the release of still more, to aid in forming the clot.

Positive feedback is generally part of a larger, negative feedback system; after a period of positive feedback, the variable is brought back within its typical range.

FIGURE 21-17 **Patching a tear.** The blood clot that forms after an injury is a result of a positive feedback system.

from the normal range. For this reason, positive feedback systems, if unchecked, can become highly unstable.

An example of a positive feedback system is the blood-clotting process (FIGURE 21-17). Injury to a blood vessel causes platelets—cellular fragments involved in clotting—to release their Super Glue–like contents, which begin to seal any rip or tear in the blood vessel (see Section 22.8). Within the bloodstream, the contents released from a platelet have another effect as well: they cause other platelets to release their contents. This is a positive feedback system, because the release of some blood-clotting molecules into the bloodstream causes the release of more such molecules, which causes the release of still more. The result is that a dangerous situation, injury to a blood vessel, is quickly remedied.

Q Why don't we bleed to death when we get a cut?

Generally, positive feedback systems are part of larger, negative feedback systems, which, after a short period of positive feedback, bring the variable (and the organism) back to within its original, normal range of internal environmental conditions. After blood clotting has stopped the loss of blood from a damaged vessel, other chemical signals interrupt the positive feedback loop, stop the further release of blood-clotting molecules, and even begin to break down blood clots that no longer serve a function.

TAKE HOME MESSAGE 21.9

» For animals to maintain homeostasis with regard to a particular physiological variable, they must have a set point for that variable to which the organism can return. Through negative feedback, sensors detect a change in the internal environment and trigger effectors to oppose or reduce the change. Positive feedback systems, which are much less common, oppose homeostasis in response to a change, pushing the body away from normal conditions and increasing change in the same direction.

21.10 Temperature control is a component of homeostasis.

As we saw in the case of Korey Stringer, one of the most important environmental factors that affect animals is temperature. And for many animal species, the control of temperature, called **thermoregulation,** is an essential component of homeostasis. Some of the important benefits to controlling body temperature include enhancing enzyme stability and efficiency (just a small change in temperature often significantly reduces the rate at which an enzyme catalyzes a reaction), giving organisms more independence in the timing of their daily activities, and resisting freezing.

In thermoregulation, the biggest distinction among animal species is in how they generate their body heat. **Endotherms** (which are sometimes described as "warm-blooded") generate their heat internally, within their own bodies. Most mammals and birds are endotherms. **Ectotherms** (which are sometimes described as "cold-blooded") get their heat primarily from the environment, usually the sun. Invertebrates, fishes, amphibians, and reptiles are all examples of ectotherms (FIGURE 21-18).

GENERATING BODY HEAT

ENDOTHERMS

- Animals that generate body heat internally
- Sometimes described as "warm-blooded"
- Include most mammals and birds

ECTOTHERMS

- Animals that get their heat primarily from the environment
- Sometimes described as "cold-blooded"
- Include invertebrates, fishes, amphibians, and reptiles

FIGURE 21-18 Strategies to heat the body.

ENDOTHERMS vs. ECTOTHERMS

Endotherms generate their heat within their own bodies. Ectotherms get their heat primarily from the environment, usually the sun.

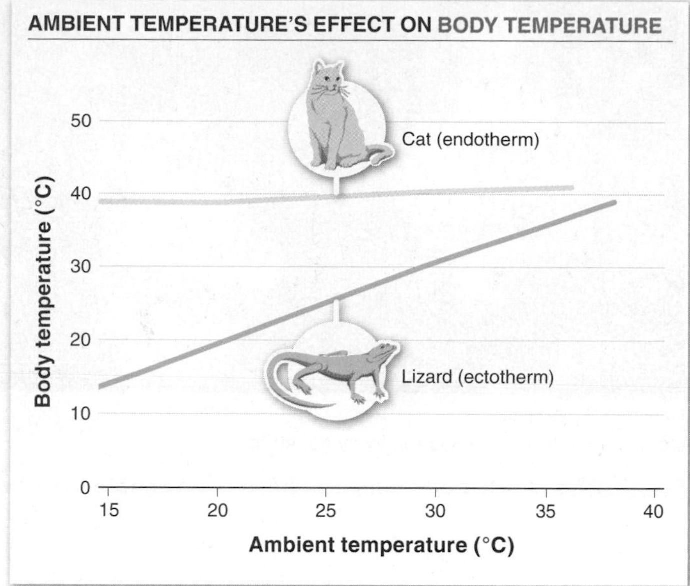

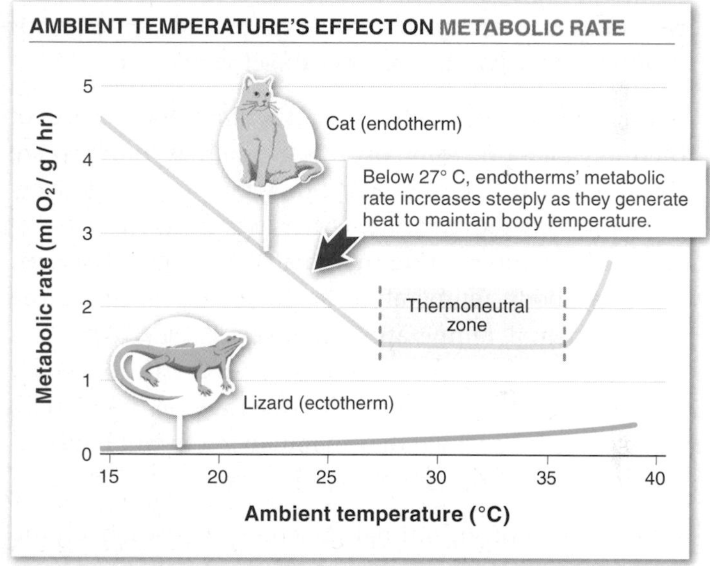

Turn to p. 749 for a closer inspection of this figure.

FIGURE 21-19 Energy expenditure and thermoregulation.

Each means of generating body heat has its costs and benefits. Between the temperatures of 27° and 36° C, for example, endotherms are able to maintain a constant internal body temperature without any energy expenditure, using passive methods of regulation such as heat loss through the skin. Below this temperature zone (called the **thermoneutral zone**), however, their metabolic rate increases significantly as they use active heating to maintain their body temperature. By contrast, the metabolic rate of ectotherms increases continuously across this range of temperatures (27° to 36° C), and in temperatures below 27° C, their metabolic rate decreases as they let their body temperature drop along with the ambient temperature (FIGURE 21-19).

As we just noted, endotherms and ectotherms are often distinguished as "warm-blooded" and "cold-blooded," but on closer investigation, such terms do not always accurately describe the two groups. Some animals, including many mammalian species, **hibernate**, going into a state of reduced metabolic activity for days or weeks at a time, during which their body temperature can drop considerably. The body temperature of some hibernating ground squirrels, for example, can drop below

Q Why do some cold-blooded animals have "hot" blood and some warm-blooded animals have "cold" blood?

MAINTAINING BODY TEMPERATURE

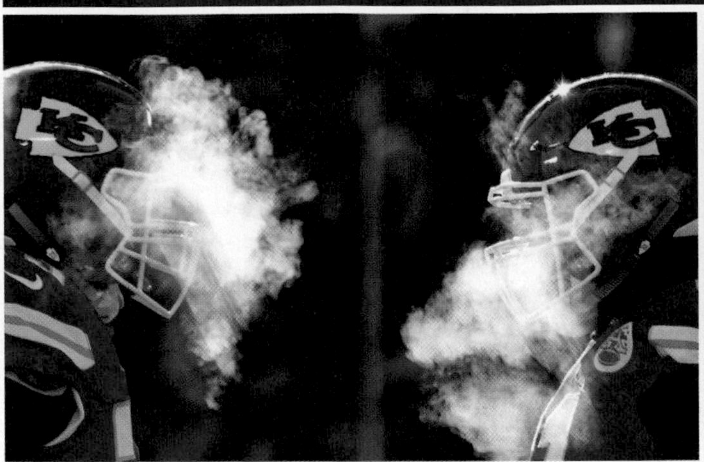

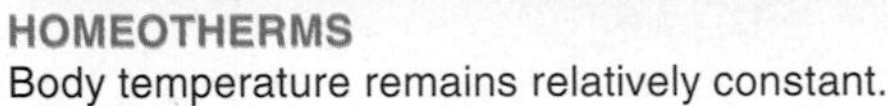

HOMEOTHERMS
Body temperature remains relatively constant.

Reptiles can often be found on the edges of roads in the afternoon and early evening, using the warmth of the road to elevate their body temperature.

HETEROTHERMS
Body temperature fluctuates as environmental temperatures change.

FIGURE 21-20 Stable body temperature versus variable body temperature.

freezing for more than three weeks. And, while some hibernating mammals can become quite cold, some lizards heating themselves in the sun can get very warm.

These observations highlight another important aspect of thermoregulation. Some organisms (such as humans) are **homeotherms,** meaning that they maintain a relatively constant body temperature. Other organisms (such as lizards and the hibernating mammals) are **heterotherms,** and their body temperature, at times, fluctuates as the environmental temperature changes (FIGURE 21-20). As a consequence, animals are most accurately categorized by describing both their source of heat (internal or external) *and* the degree to which they maintain a constant temperature or have a temperature that fluctuates.

Whether or not an animal generates its own heat and maintains a constant body temperature, all animals exchange heat with the environment. Ultimately, an organism's temperature is a function of the heat produced by its body in conjunction with the heat transferred from the environment to its body or from its body to the environment. This heat exchange can occur in four ways.

1. *Conduction* is the transfer of heat that occurs when two objects at different temperatures come into contact. Holding an ice cube or sitting on a hot car seat dramatically reveals conduction in action.
2. *Convection* is the transfer of heat from an object to a medium such as water or air as it passes next to the object. This is particularly noticeable to us when a cold breeze blows.
3. *Radiation* is the transfer of heat, without direct contact, from a warmer object to a colder object. You are experiencing radiant heat when you feel the warmth of the sun on your face or the heat from a fireplace on your feet.
4. *Evaporation* is the loss of heat that occurs as a liquid substance, such as liquid water, turns to a gas. When you sweat and the sweat evaporates, it cools you off.

We'll find further examples of each of these mechanisms as we look at four of the methods—physical, behavioral, physiological, and cellular—by which animals regulate body temperature (FIGURE 21-21).

Numerous physical features have evolved in organisms that influence their body temperature. These include body size, surface area, and levels of insulation. The thick, lipid-rich coat of blubber in walruses, whales, and many other marine mammals, for example, can account for up to 50% of their body weight and provides effective insulation, helping to maintain a constant body temperature.

An organism can also use behavioral strategies to regulate its body temperature. Most lizards, for example, heat themselves in the morning sun, increasing their body temperature. Later, when the day is at its hottest, they may retreat to a burrow to reduce their body temperature. Many large mammals that cannot get to shade orient their bodies to minimize the angle at which the sunlight strikes them. The African ground squirrel even shades itself with its tail as it forages.

METHODS FOR REGULATING BODY TEMPERATURE

PHYSICAL METHODS
The walrus has a thick coat of blubber that provides insulation from the external environment.

BEHAVIORAL METHODS
The African ground squirrel shades itself with its tail while foraging, to minimize heat from the sun.

PHYSIOLOGICAL METHODS
By panting, dogs make use of the efficient loss of heat due to evaporation.

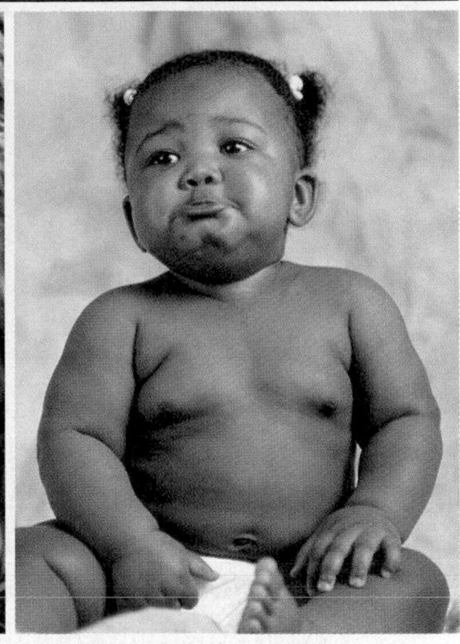

CELLULAR METHODS
Human infants have a special type of fat that produces heat, rather than ATP, when broken down.

FIGURE 21-21 **Adaptations that aid in temperature regulation.**

Animals also employ many physiological methods of thermoregulation. One specific example of this occurs as cells maintain concentration gradients of various ions, such as potassium (K^+) and sodium (Na^+). Many cell membranes are somewhat permeable to K^+ and Na^+ ions. The ions "leak" across the membrane, requiring active transport to restore and maintain the concentration gradients necessary for some cellular processes. Because the reactions involved in active transport are not perfectly efficient in their use of energy, some energy is released as heat with each reaction. This heat can help maintain an organism's body temperature. Cells may use up to 40% of their available energy in maintaining Na^+ and K^+ ion concentration gradients through the process of active transport. Interestingly, cell membranes in endotherms, such as humans, are leakier than cell membranes in ectotherms, such as fishes, suggesting that such "leakiness" is an adaptation that helps endotherms generate heat internally. Heat generation is also the likely reason endotherms have, on average, three to four times as many mitochondria per cell as ectotherms, because mitochondria are the sites of greatest heat generation.

Another physiological method by which animals—both ectotherms and endotherms—can regulate their body heat is by controlling the flow of blood to the skin. To lose heat, such as during periods of extreme exertion or at extremely high external temperatures, animals increase blood flow to the skin, allowing greater heat loss by convection. Through panting and sweating, too, many animals make use of the efficient loss of heat due to evaporation (FIGURE 21-22). Alternatively, in cold environments, the loss of heat can be reduced by reducing the flow of blood to the skin and

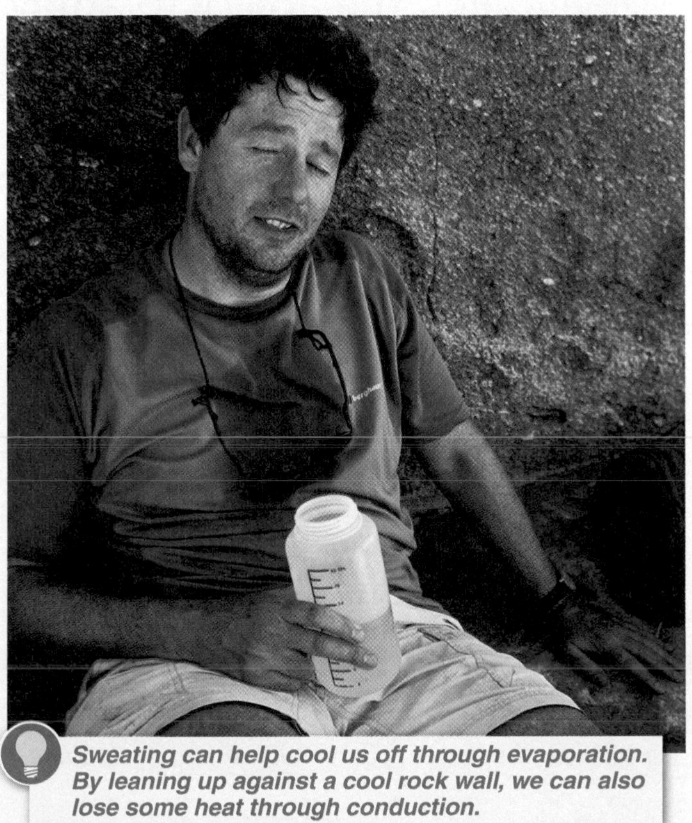

Sweating can help cool us off through evaporation. By leaning up against a cool rock wall, we can also lose some heat through conduction.

FIGURE 21-22 **Heat exchange.**

by shivering, through which animals can increase their heat production.

> **Q** Why does "baby fat," unlike regular fat, act like a built-in heating pad?

One cellular method of temperature regulation is particularly important to human babies. It takes place in a special type of connective tissue called "brown fat." Unlike the cells of "white fat," or adipose tissue (most of the fat in adults' bodies), which have few mitochondria, the cells of brown fat have a high density of mitochondria along with the stored fat. When brown fat cells oxidize their fat, however, they don't generate ATP. Instead, a special protein causes protons to leak directly across the mitochondrial membrane, rather than passing through the enzyme that synthesizes ATP (see Section 5.15), causing the production of heat rather than ATP from the fat breakdown. Human infants have significantly more brown fat than adults. This difference probably evolved because infants' small body size gives them a large surface-area-to-volume ratio, which results in a relatively large surface area over which they can lose heat and a relatively small body mass in which they can generate heat.

TAKE HOME MESSAGE 21.10

» The control of body temperature, called thermoregulation, is an important component of homeostasis. Body temperature is a function of internal heat production and heat transfer between an organism and its environment. Heat transfer to and from the environment is regulated physically, behaviorally, physiologically, and at the cellular level.

THIS IS HOW WE DO IT

Developing the ability to apply the process of science

21.11 Why do we yawn?

The purpose of yawning has long vexed scientists. It's an almost universal behavior among vertebrates, yet its purpose has never been understood. Thanks to some recent creative hypothesis testing, that may be changing.

Have you ever heard any explanations for why we yawn?

Speculation about the function of yawning has been wide-ranging. Two thousand years ago, Hippocrates proposed that it served to rid the body of illness-causing "bad air" that accumulates inside us. This idea morphed into a widely held belief:

- *Yawning is triggered by insufficient oxygen in the brain.*

Widely held, but wrong.

Propose some testable predictions of the hypothesis that yawning is triggered by too little oxygen in the brain.

The evidence against the "insufficient oxygen" hypothesis is clear.

1. People don't yawn more when they exercise, although they require more oxygen.

2. When they are exposed to a gas mixture high in carbon dioxide and low in oxygen, people aren't more likely to yawn.

3. Breathing air with a higher-than-typical concentration of oxygen doesn't reduce the rate of yawning.

On top of these findings, there is no evidence that yawning actually increases brain oxygenation.

Moreover, an organism can increase the supply of oxygen to its brain by simply breathing more rapidly or deeply.

Some other persistent but unsupported hypotheses include:

• *Yawning reduces ear pressure.* This does, in fact, occur. But there is no evidence that yawning rates increase in response to rapidly changing air pressure. And many other behaviors, including swallowing and chewing, have the same pressure-reducing effect.

• *Yawning counteracts drowsiness, increasing arousal.* Although we do yawn more frequently when we are tired, further observations contradict this hypothesis. For example, measurements of brain activity after yawning show the opposite response: increased drowsiness and reduced vigilance.

Recently, a new idea has generated some interesting experiments and promising results—simultaneously illustrating how researchers approach a question. The idea is this:

• *Yawning is a mechanism for cooling the brain.*

Several observations prompted this line of thinking. (1) Yawning—at least in humans and rats—is commonly preceded by a brief increase in brain temperature; (2) following a yawn, brain temperature usually decreases; and (3) yawning is associated with drowsiness, a physiological state correlated with increases in body temperature.

How might yawning cool the brain?

Researchers proposed that yawning increases cerebral blood flow and—as the sinus walls expand and contract like a bellows—pumps air onto the brain, lowering its temperature. They predicted that, if yawning serves as a mechanism for cooling the brain, the effectiveness of yawning will depend on ambient air temperature. In other words, when the air temperature is high, there will be little benefit to yawning—the air pumped onto the brain won't provide cooling. And when the ambient air temperature drops below some critical point, cooling should be unnecessary. The researchers used three extremely simple but powerful experiments to test their predictions.

If yawning is for brain cooling, would you expect more yawning in winter or summer?

Experiment 1

What they measured: Yawning frequency in mild versus hot ambient temperatures. During winter (February) and summer (June), the researchers approached pedestrians in outdoor public areas in Arizona and asked them to participate in the study. Each participant (80 individuals participated during each of the two data-collection periods) was shown 20 images of people yawning and then asked to fill out a survey, reporting (among other things) whether they had yawned while looking at the images. (They also reported how long they had slept the previous night and how long they had been outdoors before participating in the study.) While the participant was filling out the survey, the researchers measured the temperature. During the winter data collection, the average temperature was 22° C (71° F). During the summer, it was 37° C (98° F).

Prediction: The researchers predicted that a higher percentage of individuals would yawn during the winter portion of the study than the summer portion.

Results: The findings provided strong support for the hypothesis. During the winter data collection, 45% of the participants reported yawning, while during the summer—when ambient temperature equaled or exceeded body temperature—only 24% reported yawning.

Experiment 2

What they measured: Yawning frequency in mild versus cold ambient temperatures. Using the same methods as in Experiment 1, during winter and summer the researchers approached pedestrians in outdoor public areas and asked them to participate in the study. This time, however, the study was conducted in Vienna, Austria. During the winter data collection, the average temperature was 1° C (33° F). During the summer, it was 19° C (66° F).

As in the first experiment, participants were shown 20 images of people yawning and then asked to fill out a survey, which included reporting whether they had yawned while looking at the images.

Prediction: The researchers predicted that a higher percentage of individuals would yawn during the summer portion of the study than the winter portion.

Results: Again, the researchers found strong support for the hypothesis. During the winter data collection, just 18% of the participants reported yawning, while during the summer, 42% reported yawning.

Experiment 3

What they measured: Forehead cooling and yawning frequency. Participants held a warm pack (46° C/114° F) or a cold pack (4° C/39° F) to their forehead while watching videotapes of people yawning, and the researchers observed whether they yawned.

Prediction: The researchers predicted that holding a cold pack to the forehead would reduce brain temperature, thereby reducing the need for the thermoregulatory benefit of yawning.

Results: The results strongly supported the hypothesis. Among the participants that held a warm pack to their forehead, 41% yawned while watching the videotape. Among those holding a cold pack to their forehead, just 9% yawned.

Taken together, the results of these experiments provide compelling support for the idea that yawning's function is to cool the brain. Further, the researchers found that temperature was the only significant predictor of yawning; other factors, including sex, age, hours of sleep on the previous night, and air humidity did not have a significant effect on the incidence of yawning.

Can you think of another experiment for testing the "yawning for brain cooling" hypothesis?

At this point, the issue is not completely resolved. A variety of observations suggest that other factors may also influence yawning. The contagious nature of yawning, in particular, suggests a possible link to empathy. Researchers have found, for example, that the closer two people are genetically or emotionally, the more likely they are to "catch" each other's yawns. They've also observed that yawning rates are reduced among people with schizophrenia and autism. As additional evidence accumulates, a fuller picture will emerge. And, unlike over the past two thousand years, progress is likely to be much more rapid given the results presented here.

TAKE HOME MESSAGE 21.11

» Accumulating experimental evidence supports the hypothesis that yawning in humans and other vertebrates functions as a mechanism for brain cooling.

21.12 Animals must balance their water content within a narrow range.

Organisms, whether they live in the water or on land, require water. But they can't take in too much or they will burst. And they can't lose too much or they will shrivel up and die. **Osmoregulation,** an important component of homeostasis in animals, is the regulation of water content and of the concentrations of dissolved solutes that influence osmosis. The most abundant solutes in animal fluids are sodium, chloride, potassium, magnesium, and calcium ions.

Although a variety of mechanisms of water regulation have evolved in animals, there are just four ways that animals get water: by drinking it, by eating foods that contain it, by absorbing it through osmosis, and as a by-product of cellular respiration. Similarly, animals lose water in four ways: urination, defecation, evaporation (including panting, breathing, and sweating), and osmosis.

Osmoregulation revolves around controlling and regulating these mechanisms of water loss and gain. Through osmosis (see Section 4.9), water flows from areas of low solute concentration to areas of high solute concentration. Thus organisms often control their water content indirectly by regulating their solute content. They generally balance the total amount of solutes—which influences the direction and magnitude of osmosis—and the concentration of each solute individually. Imbalances in the concentrations of certain solutes (minerals and electrolytes) can cause serious health problems, from muscle spasms to confusion to paralysis, and even death.

Because animals live in such a wide range of habitats, they encounter a similarly wide range of osmoregulatory challenges. The challenges are very different for aquatic organisms than for terrestrial organisms, for example. And among the aquatic organisms, the challenges faced by organisms living in saltwater habitats are very different from—almost the opposite of—those faced by organisms living in fresh water.

Most invertebrates that live in salt water are **osmoconformers,** meaning that they let the solute concentration of their body fluids reflect that of their environment. Most vertebrates, on the other hand, are **osmoregulators,** maintaining their fluids and solute concentrations within narrow ranges that differ from those of their environment (FIGURE 21-23; see also Figure 21-16).

A variety of structures have evolved that enable animals to regulate their water balance. In insects, for example, very small tubes branch off the digestive tract, near its end at the rectum. These extensions from the gut, called Malpighian tubules, function as insects' chief excretory organs. Potassium ions (K^+) and cellular waste products move by active transport from the body cavity of the insect—where its blood is—into the Malpighian tubules. The increase in solutes in the tubules attracts water by osmosis. Together, K^+ ions, water, and waste products move down the digestive tract until, near its end, the K^+ ions and water are mostly reabsorbed into the circulatory system. This is a very effective system for conserving water, leaving mostly waste products to be excreted as feces.

OSMOREGULATORY STRATEGIES

OSMOCONFORMERS
Osmoconformers are organisms that let the composition of their body fluids reflect that of their environment.

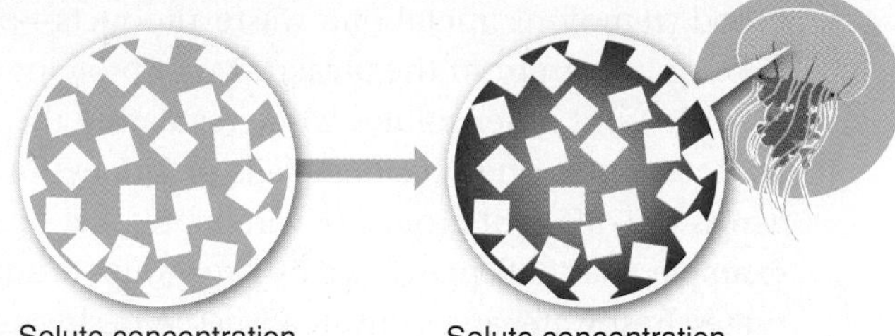

OSMOREGULATORS
Osmoregulators are organisms that maintain their fluids and solute concentrations within narrow ranges that differ from those of their environment.

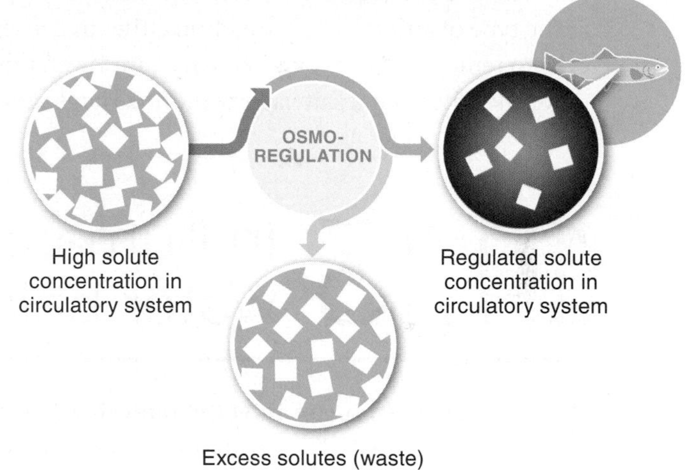

OSMOREGULATORY STRUCTURES

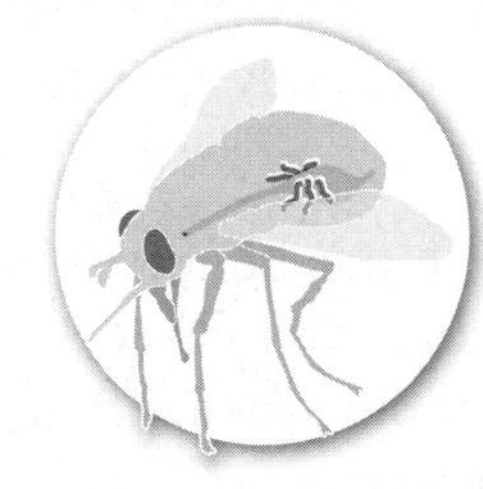

MALPIGHIAN TUBULES (IN INSECTS)
These small tubes regulate osmotic balance in insects by removing excess solutes from the circulatory system.

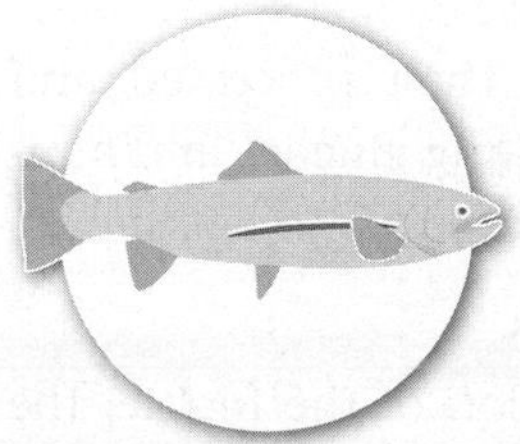

KIDNEYS (IN VERTEBRATES)
This complex organ regulates osmotic balance in vertebrates by removing either excess solutes or excess water from the circulatory system, depending on the organism's external environment.

FIGURE 21-23 Two strategies for regulating the amount of salt and water in the body—and some structures that make osmoregulation possible.

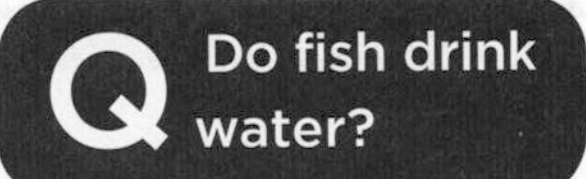

Another osmoregulatory structure is the vertebrate kidney. The kidney, as we shall see, is a complex organ that filters blood, removing metabolic waste products—particularly excess nitrogen from the breakdown of proteins and nucleic acids—and other solutes while regulating the organism's water balance. The specific details of kidney function vary across the different groups of vertebrates. Let's consider, for example, the challenges faced by two groups of aquatic vertebrates: freshwater and saltwater fishes. A freshwater fish faces the problems of high concentrations of water entering its body and of solutes leaving its body. A saltwater fish, on the other hand, tends to lose water to its environment while taking in high concentrations of solutes from the environmental water. As a consequence, vertebrate kidneys—depending on the type of animal—may function either to conserve water or to remove it. (This explains why a freshwater fish does not drink water, but a saltwater fish drinks large amounts.)

In the next section, we examine the details of kidney functioning, focusing specifically on the human kidney and how it can produce urine that has a solute concentration more than four times that of blood, and how it can eliminate the many potentially harmful molecules contained in the food and drink that we consume.

TAKE HOME MESSAGE 21.12

» Many organisms maintain their water content within a narrow range. To maintain osmotic balance, organisms must be able to take up water and get rid of water, and they must be able to regulate concentrations of ions in their body fluids. Various mechanisms and strategies have evolved for coping with these challenges.

21.13 In humans, the kidneys regulate water balance.

The kidney is an organ in vertebrates that helps maintain homeostasis by regulating water balance and solute concentration in body fluids. It accomplishes this by filtering blood—as much as 2,000 liters (or almost 400 times the total volume of blood in your body!) passes through the human kidneys each day—and reabsorbing water and other substances needed by the body. The kidneys accomplish three primary functions.

1. Filtration. As blood passes through the kidneys, it is filtered. During this process, water, ions, and other dissolved substances are removed from the blood as a filtrate, leaving behind just cells and large proteins.

2. Reabsorption. The filtrate is then processed and modified, as water, nutrients (including glucose and amino acids), and some ions are reabsorbed into the blood, concentrating the filtrate.

3. Secretion. Following concentration of the filtrate, the substances not needed by the body are excreted in the urine.

The amount of urine excreted each day depends on an individual's fluid intake and water loss (such as due to perspiration) and typically ranges from 0.7 to 2.0 liters. In controlling the amount of water excreted in urine or reabsorbed, the kidneys play an important role in preventing dehydration, even as fluid consumption and fluid loss may vary tremendously from day to day.

The filtering of blood by the kidneys begins as blood flows into blood vessels within each kidney via a renal artery. The filtered blood leaves the kidney via a renal vein, while waste products isolated during filtration and any excess water removed from the bloodstream pass to the bladder, and from there are excreted as urine (FIGURE 21-24).

A human has two kidneys, each about the size of a fist, located on either side of the spine, just above the waist. Each kidney is made up of approximately a million nephrons, the structural units that accomplish the work of the kidneys. A **nephron** consists of two basic components: a

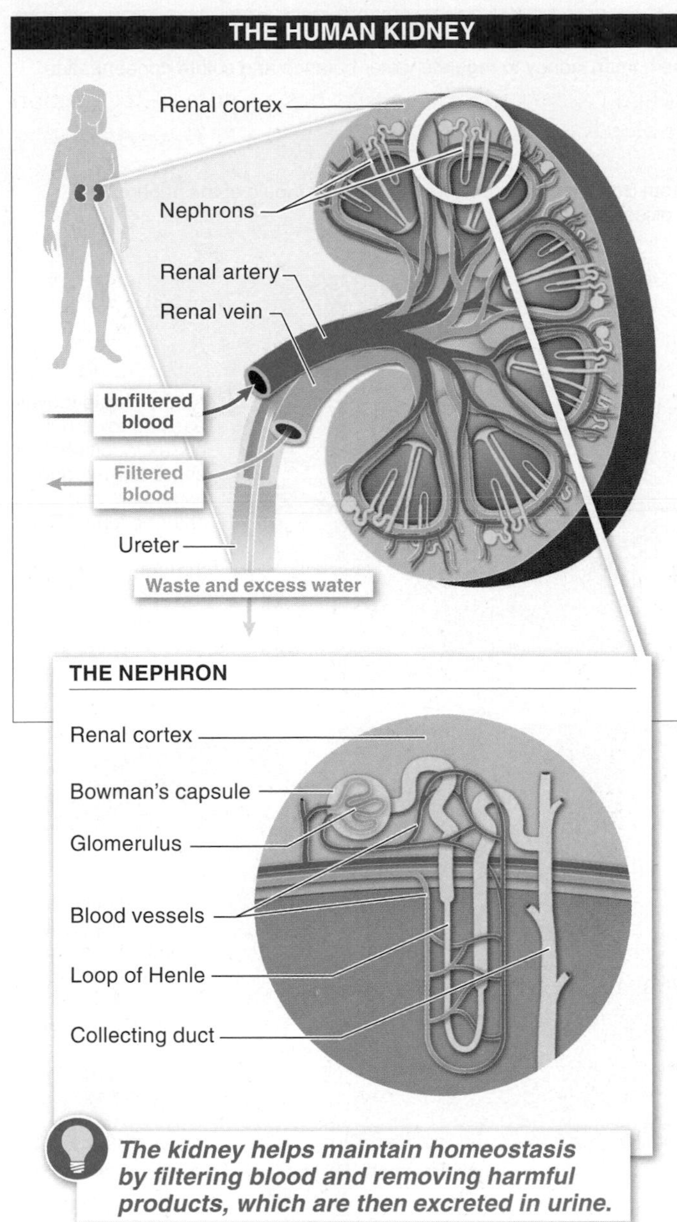

FIGURE 21-24 **Structure of the kidney.** Human kidneys filter blood, reabsorb water and solutes, and excrete waste.

nephron tubule and a mass of blood vessels that work together to accomplish the tasks of filtration, reabsorption, and secretion (**FIGURE 21-25**). The blood-filtering unit of the nephron is a mass of capillaries called a **glomerulus,** and the ball-like structure that surrounds it is called **Bowman's capsule.** Each Bowman's capsule is connected to a single, long, urine-producing tube that excretes its filtered fluid into a collecting duct.

The capillaries in the glomerulus are porous, and blood pressure forces out water and small molecules and ions (but not blood cells or most proteins) through the capillary walls. Fluid that accumulates in Bowman's capsule, called **filtrate,** contains salts, sugars, amino acids, vitamins, and many other molecules, all at the same concentration as in the blood.

As the filtrate moves from Bowman's capsule through the urine-collecting tubule of the nephron, the vast majority of its water and dissolved solutes must be reabsorbed. In fact, of the 2,000 liters of blood that pass through the kidneys each day, only about 1.5 liters of urine are produced and excreted.

Reabsorption in the long tubule of the nephron is a complex process. It begins with the active transport of sodium ions out of the tubule, which causes water and chloride ions to follow, moving out by passive transport.

The tubule (and the filtrate it contains) then loops from the outer part of the kidney, called the cortex (see Figure 21-24), down into the innermost part of the kidney and back again, in a path called the loop of Henle. As the tubule passes to the innermost part of the kidney, it encounters a higher solute concentration in the interstitial fluid, outside the tube. This creates an osmotic gradient, which causes more and more water to be drawn out of the tubule, moving from the filtrate into the interstitial fluid.

Then, as the filtrate moves up toward the kidney cortex again, more salt is lost, through diffusion and active transport. In the cortex, the tubule empties into the collecting duct, which passes through the innermost part of the kidney again, returning additional water to the interstitial fluid and further concentrating the urine for excretion. The collecting duct eventually merges with collecting ducts from other nephrons, ultimately forming the ureter, which empties into the urinary bladder. From the bladder, urine passes through the urethra and is excreted from the body.

Filtering the blood and producing urine not only follows a complicated path but is very energetically expensive. Considerable amounts of energy are expended in the active transport of solutes that leads to the recovery of water. Nonetheless, some animals, such as kangaroo

Q Why do some desert mammals *never* need to drink water?

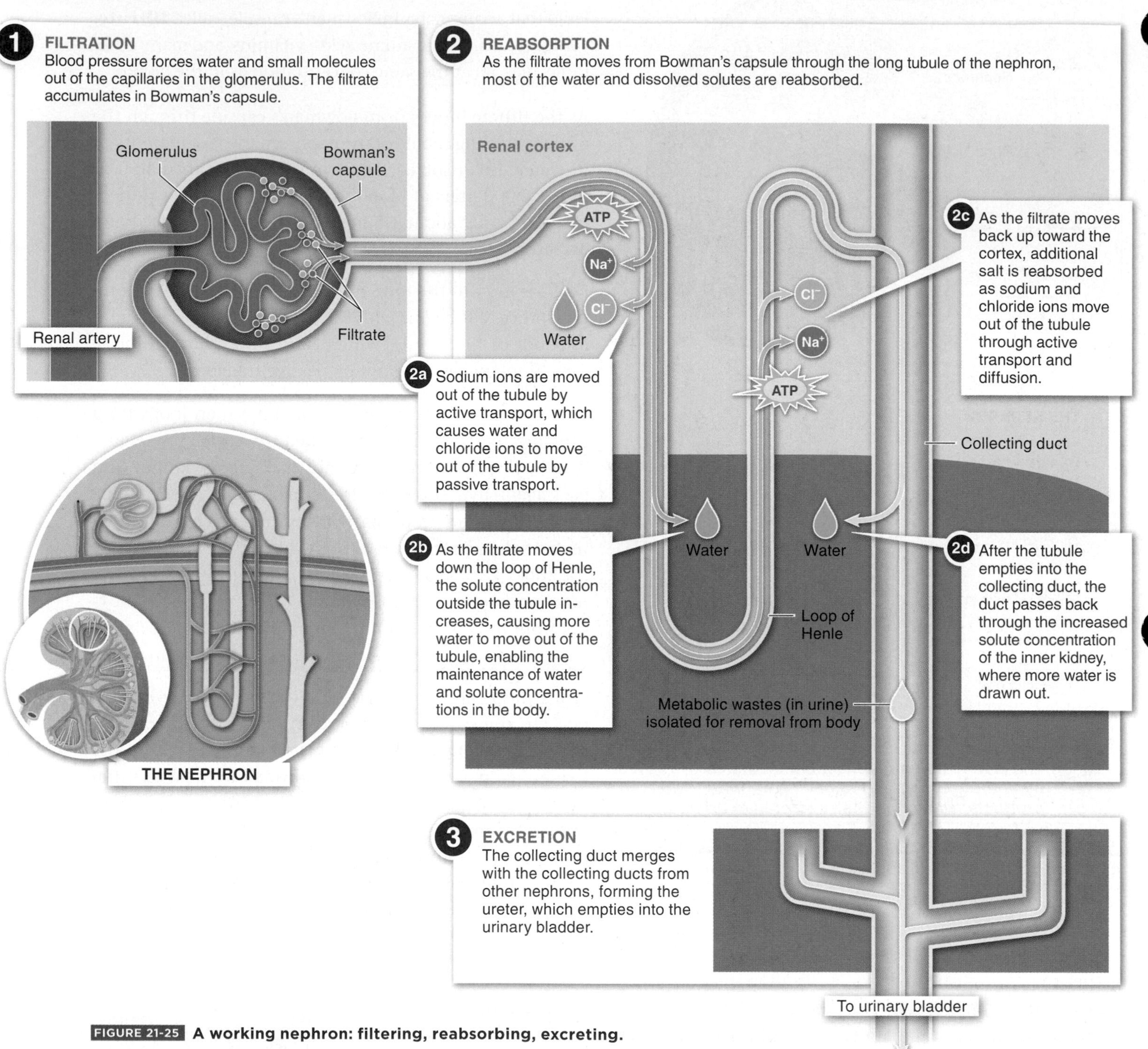

HOW THE HUMAN KIDNEY WORKS
The nephron is the fundamental unit of the human kidney. One million nephrons enable the human kidney to regulate water balance and solute concentration in body fluids, while producing urine.
1
FILTRATION
Blood pressure forces water and small molecules out of the capillaries in the glomerulus. The filtrate accumulates in Bowman's capsule.
Glomerulus
Bowman's capsule
Renal artery
Filtrate
2
REABSORPTION
As the filtrate moves from Bowman's capsule through the long tubule of the nephron, most of the water and dissolved solutes are reabsorbed.
Renal cortex
ATP
Na+
Cl−
Water
2a Sodium ions are moved out of the tubule by active transport, which causes water and chloride ions to move out of the tubule by passive transport.
2b As the filtrate moves down the loop of Henle, the solute concentration outside the tubule increases, causing more water to move out of the tubule, enabling the maintenance of water and solute concentrations in the body.
2c As the filtrate moves back up toward the cortex, additional salt is reabsorbed as sodium and chloride ions move out of the tubule through active transport and diffusion.
2d After the tubule empties into the collecting duct, the duct passes back through the increased solute concentration of the inner kidney, where more water is drawn out.
Collecting duct
Loop of Henle
Metabolic wastes (in urine) isolated for removal from body
THE NEPHRON
3
EXCRETION
The collecting duct merges with the collecting ducts from other nephrons, forming the ureter, which empties into the urinary bladder.
To urinary bladder

FIGURE 21-25 **A working nephron: filtering, reabsorbing, excreting.**

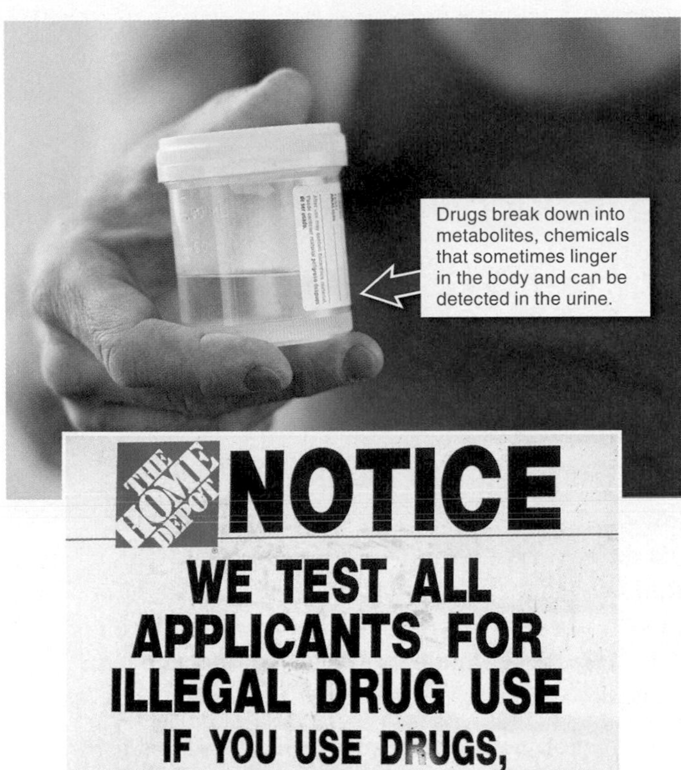

Drugs break down into metabolites, chemicals that sometimes linger in the body and can be detected in the urine.
THE HOME DEPOT
NOTICE
WE TEST ALL APPLICANTS FOR ILLEGAL DRUG USE
IF YOU USE DRUGS, DON'T BOTHER TO APPLY!

FIGURE 21-26 **Detectable remainders.**

rats, are so efficient at reabsorbing water that they never have to drink water at all. They can recover nearly all of the water filtered by their kidneys so that the water contained in their food and generated as a by-product during cellular respiration is sufficient.

Nitrogen-containing compounds are among the most important of the metabolic waste products that are removed from the blood by the kidneys. Produced from the breakdown of proteins and nucleic acids, this nitrogen tends to be in the form of ammonia, which is generally very toxic to organisms. Some organisms—mostly aquatic organisms—are able to rid their bodies of excess nitrogen by simply excreting the ammonia. Terrestrial animals (and many marine animals), however, cannot consume sufficient water to safely dilute toxic ammonia. Instead, these organisms use an energetically expensive process that combines ammonia with carbon dioxide to produce urea, which can be stored for longer periods of time and at higher concentrations (thereby requiring much less water). Insects, terrestrial snails, birds, and many reptiles use the least water but the most energy in removing nitrogenous wastes, which are excreted as a paste called uric acid.

The human kidneys are so effective at filtering blood and concentrating the many waste products of metabolism in urine that it is possible to detect the use of many drugs through urinalysis. Most drug-screening urinalysis does not involve testing for the presence of the drugs themselves, which may remain in the body for only a short time. Instead, the tests look for the presence of chemicals, called metabolites, that result from breakdown of the drugs (FIGURE 21-26). Metabolites have been identified for many drugs, including Ecstasy, cocaine, methamphetamines, and marijuana. In the case of marijuana, the metabolites are fat-soluble. This means that they are stored in fat cells indefinitely and released only when fat from those cells is metabolized for energy. As a consequence, marijuana can be detected a month or longer after the last use.

Q How can some drugs be detected by urinalysis even months after the last intake?

TAKE HOME MESSAGE 21.13

» The kidney is the organ in vertebrates that helps maintain homeostasis by regulating water balance and solute concentrations in body fluids, filtering blood, and removing potentially harmful ions and metabolic waste products, excreting them in the urine.

STREET BIO

Using evidence to guide decision making in our own lives

Your body sometimes deliberately upsets homeostasis. (And you might not want to fight it.)

Q: What do you usually do when you have a fever? At the first sign of fever, many people take aspirin, Tylenol, Advil, or Motrin.

Q: What is the result? Aspirin, acetaminophen (Tylenol), and ibuprofen (Advil and Motrin) quickly reduce a fever.

Q: Why did you have a fever? Generally, fever is not itself an illness. Rather, it is your body's response to cues that there is a bacterial or viral infection. In response to infection, your temperature set point is raised, because pathogens are more easily brought under control by the body's defenses at higher temperatures. (Ectotherms use a similar strategy, moving to warmer areas when they have an infection!)

Q: How does reducing a fever interfere with your body's defenses? Blocking a fever by taking aspirin or other medication may reduce your body's ability to fight infection. A recent well-controlled study demonstrated, for example, that chicken pox lasts longer, on average, when aspirin is used to treat it, compared with placebo.

Q: Is the lesson here about more than fever? Yes. There are other signs of infection that, as defenses rather than part of the illness itself, maybe shouldn't be fought. For example: (1) *Coughing:* the use of codeine to block coughing after surgery increases the risk of pneumonia; the coughing is *helpful.* (2) *Diarrhea:* anti-diarrhea medications delay recovery and slow the eradication of bacteria from the digestive tract; the diarrhea is *helpful.* (3) *Vomiting* and *inflammation,* too, appear to be important parts of our evolved defenses and, as such, blocking them can have serious health consequences.

What can you conclude? *Is this the dawn of Darwinian medicine?* This perspective on when to treat and when not to treat symptoms is called "Darwinian medicine." It represents a newfound appreciation for the fact that many protective responses have evolved that may be useful to the organism. Of course, suffering is not always the solution. There can be real costs to vomiting, diarrhea, coughing, fever, and other body defenses. Opting not to treat them isn't necessarily the best solution if, for example, antibiotics can bring an infection under control easily. Either way, bringing an evolutionary perspective to medical decision making can be valuable.

GRAPHIC CONTENT

Thinking critically about visual displays of data

1 Is the body temperature of cats influenced by the ambient temperature? Provide data to support your answer.

2 What would you expect the lizard body temperature to be when the ambient temperature is 10° C? Why?

3 What is the unit of measure for metabolic rate? Is the rate higher in cats or lizards?

4 Body weight in cats is typically much greater than in lizards. How is this difference accounted for in the lower graph?

5 In lizards, what is the percentage increase in metabolic rate between 27° and 36° C? What is the percentage increase for cats across that range?

6 Suggest a way to re-plot these data that would better highlight the greater rate of increase in the metabolic rate of lizards relative to cats between 27° and 36° C. What is a drawback of your alternative version of the graph?

7 Based on this graph, what is meant by the phrase "thermoneutral zone"? Is there a thermoneutral zone for lizards? Why or why not?

8 From the perspective of relative metabolic costs, under what conditions is it more efficient to be an endotherm? Under what conditions is it more efficient to be an ectotherm?

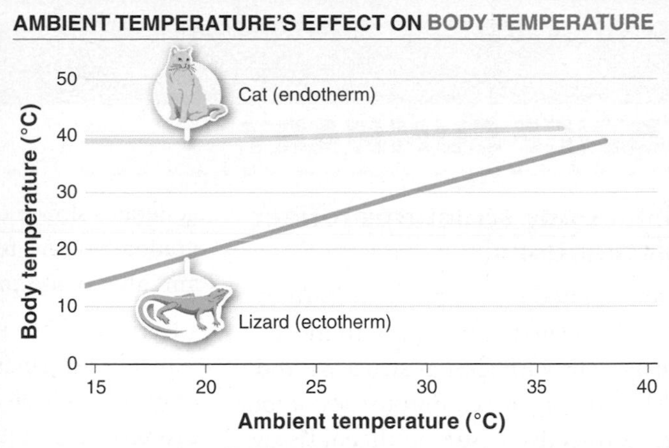

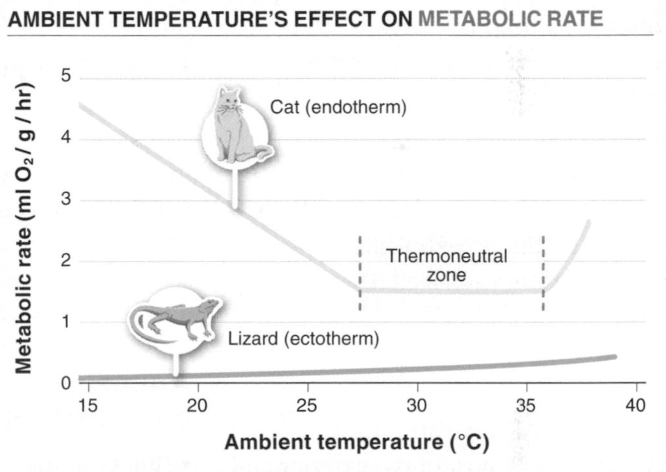

See answers at the back of the book.

KEY TERMS IN INTRODUCTION TO ANIMAL PHYSIOLOGY

hypothermia, p. 732
interstitial fluid, p. 733
ligament, p. 724
matrix, p. 721
muscle fiber, p. 727
muscle tissue, pp. 721, 726
negative feedback, p. 734
nephron, p. 744
nervous tissue, pp. 722, 727
neuroglia, p. 727
neuron, p. 727
organ, p. 722
organ system, p. 722
osmoconformer, p. 745
osmoregulation, p. 742
osmoregulator, p. 743
peripheral nervous system, p. 728
physiology, p. 722
positive feedback, p. 735
regulator, p. 735
set point, p. 743
skeletal muscle, p. 727
smooth muscle, p. 727
special connective tissue, p. 724
tendon, p. 723
thermoneutral zone, p. 737
thermoregulation, p. 736
tissue, p. 721

BRIEF SUMMARY

Animal body structures reflect their functions.

• In animal bodies, the physical features are closely related to function. In most animals, cells with similar structure and function are organized into tissues of four types: connective tissue, epithelial tissue, muscle tissue, and nervous tissue. Tissues are often organized into organs, and organs can be organized into organ systems.

• The most abundant type of tissue in most animals is connective tissue. Connective tissue is a collection of cells embedded within an extracellular matrix, usually containing collagen, that holds the cells together and gives shape, structure, and support to other body tissues.

• Epithelium is a very thin, sheet-like tissue that covers most of the exterior and interior surfaces of an animal's body. Epithelium acts as a barrier between the inside and outside of the organism and can aid in the secretion and transport of molecules.

• Muscle tissue consists of elongated cells capable of generating force when they contract. Skeletal muscle generates most of the movement we see in animals. Cardiac muscle causes the heart to pump. Smooth muscle, surrounding blood vessels and many internal organs, generates slower contractions that can gradually move food or other substances and alter blood pressure.

• Nervous tissue is specialized to store and transmit information. There are two types of nervous tissue cells: neurons, which can receive and transmit a signal, and glial cells, which assist and provide nutrients for neurons.

• In nearly all animals, some tissues are organized into organs that serve specialized functions and consist of multiple tissue types. These organs are part of organ systems that carry out the various physiological processes necessary for growth, development, maintenance, and reproduction.

Animals have an internal environment.

• Failure to maintain a consistent internal physical and chemical environment can lead to multiple problems in the normal functioning of cells, tissues, and organs, and can result in death.

• Many animals maintain homeostasis: they keep a variety of internal physiological variables—including temperature, water-solute balances, pH, blood sugar levels, and blood gas concentrations—within a relatively constant range.

How does homeostasis work?

• To maintain homeostasis for a physiological variable, animals must have a set point for that variable to which the organism can return. Through negative feedback, sensors detect a change in the internal environment and trigger effectors to oppose or reduce the change. Positive feedback systems, which are much less common, oppose homeostasis in response to a change, pushing the body away from normal conditions.

• The control of body temperature, called thermoregulation, is an important component of homeostasis, reflecting internal heat production and heat transfer between an organism and its environment. Heat transfer to and from the environment is regulated physically, behaviorally, physiologically, and at the cellular level.

• Many organisms maintain their water content within a narrow range. To maintain osmotic balance, organisms must be able to take up water and get rid of water, and to regulate concentrations of ions in their body fluids.

• In vertebrates, the kidney regulates water balance and solute concentrations in body fluids, filtering blood and removing potentially harmful ions and metabolic waste products, excreting them in the urine.

CHECK YOUR KNOWLEDGE

Short Answer

1. **What is the major benefit to organisms of multicellularity?**

2. **Why is connective tissue sometimes referred to as "cellular glue"?**

3. **Describe one way that epithelium functions in the secretion of molecules.**

4. **Describe the three types of muscle tissue in vertebrates and their similarities and differences.**

5. **What is a possible consequence of damage to your glial cells?**

6. **Which organ system regulates the rate at which the heart pumps blood?**

7. **What is hyperthermia? How does it reflect an inability of the body to maintain homeostasis?**

8. **Multicellularity and large body size typically lead to the increased importance of the internal (relative to the external) environment in influencing cell functioning. Why?**

9. **In organisms, why are negative feedback systems more common than positive feedback systems?**

10. **An organism's temperature is a function of the heat produced by its body in conjunction with the heat transferred from the environment to its body or from its body to the environment. Describe two ways in which heat exchange occurs.**

11. **Does an invertebrate osmoconformer living in salt water have a set point with respect to the solute concentrations in its body fluids? Why or why not?**

12. **Describe the factors that influence the amount of urine produced by a human in one day.**

Multiple Choice

1. Which of the following statements is not an extension of the adage "form follows function"?

a) Penguins are streamlined like fish because both types of animals swim through water.

b) A neuron is a long, wire-like cell because it connects structures that are far apart in the body.

c) Red-colored muscle cells are found in the heart because hearts must be red overall.

d) Skin cells are stacked in layers because the outer coating of the body needs to be waterproof.

e) Bones are hard and stiff so that they can resist bending.

0 EASY 30 HARD 100

2. Which of the following is not a type of connective tissue?

a) bone

b) cartilage

c) blood

d) collagen

e) ligament

0 EASY 54 HARD 100

3. Neurons generally have all of the following components except:

a) dendrites.

b) a cell body.

c) an axon.

d) glial processes.

e) a nucleus.

0 EASY 31 HARD 100

4. Which of the following explains why complex multicellular organisms face a major challenge in maintaining homeostasis?

a) Most of the body cells are not in direct contact with the external environment.

b) Most of the body cells are protected from harsh or changing environmental conditions.

c) There is less extracellular fluid from which to gather nutrients.

d) There are more chemical reactions per cell that require energy and nutrients.

e) Nutrients and waste materials enter or leave the cells too quickly.

0 EASY 32 HARD 100

5. Interstitial fluid:

a) is found exclusively within the spinal cord, surrounding the nerve bundles.

b) is found exclusively within the skull, surrounding the brain.

c) is mostly water.

d) is found exclusively within the skull and spinal cord, surrounding nervous tissue.

e) occurs within the organelles of all eukaryotic cells.

0 EASY 57 HARD 100

6. Negative feedback loops:

a) generally lead to highly unstable internal physiological conditions.

b) cause internal conditions to deviate from the normal range.

c) are part of larger, positive feedback systems.

d) rely on sensors to trigger effectors to alter an organism's internal environment.

e) None of the above are properties of negative feedback.

0 EASY 37 HARD 100

7. A set point:

a) is the target value or range for a physiological variable, to which it generally returns following a perturbation.

b) is a physiological state that occurs in animals called "conformers," but not in "regulators."

c) can occur in fishes but not in terrestrial animals.

d) is the target value or range for a physiological variable regulated through positive feedback, but not negative feedback.

e) Both a) and c) are correct.

0 EASY 44 HARD 100

8. It is not necessarily accurate to refer to endotherms as "warm-blooded," because:

a) some endotherms allow their body temperature to drop significantly during hibernation.

b) on average, their body temperature is colder than that of "cold-blooded" animals.

c) they generate their heat primarily within their own bodies.

d) most do not actually have blood in their bodies.

e) some are homeotherms rather than heterotherms.

0 EASY 44 HARD 100

9. The glomerulus is ______ that is/are a key component of the filtration process in the nephron.

a) a system of ducts

b) a renal tubule

c) a layer of connective tissue

d) the terminal end of the nephron

e) a bed of capillaries

0 EASY 62 HARD 100

10. In homeostasis terms, a fever is classified as:

a) a rise in internal temperature in response to foreign invaders.

b) a sudden drop in internal temperature.

c) a crazed reaction toward some object.

d) any general rise in internal temperature.

e) All of the above are correct.

Ch22

The circulatory system is the chief route of distribution in animals.

22.1 What is a circulatory system, and why is one needed?

22.2 Circulatory systems can be open or closed.

22.3 Vertebrates have several different types of closed circulatory systems.

The human circulatory system consists of a heart, blood vessels, and blood.

22.4 Blood flows through the four chambers of the human heart.

22.5 Electrical activity in the heart generates the heartbeat.

22.6 Blood flows out of and back to the heart in blood vessels.

22.7 *This is how we do it:* Does thinking make your head heavier?

22.8 Blood is a mixture of cells and fluid.

22.9 Blood pressure is a key measure of heart health.

22.10 Cardiovascular disease is a leading cause of death in the United States.

22.11 The lymphatic system plays a supporting role in circulation.

The respiratory system enables gas exchange in animals.

22.12 Oxygen and carbon dioxide must get into and out of the circulatory system.

22.13 Oxygen is transported while bound to hemoglobin.

22.14 Gas exchange takes place in the gills of aquatic vertebrates.

22.15 Respiratory systems of terrestrial vertebrates move oxygen-rich air into and carbon-dioxide-rich air out of the lungs.

22.16 Birds have unusually efficient respiratory systems.

22.17 Muscles control the flow of air into and out of the lungs.

Evolutionary adaptations maximize oxygen delivery.

22.18 Animals living at high elevations have special adaptations to the low-oxygen conditions.

22.19 Humans become acclimated to low-oxygen conditions.

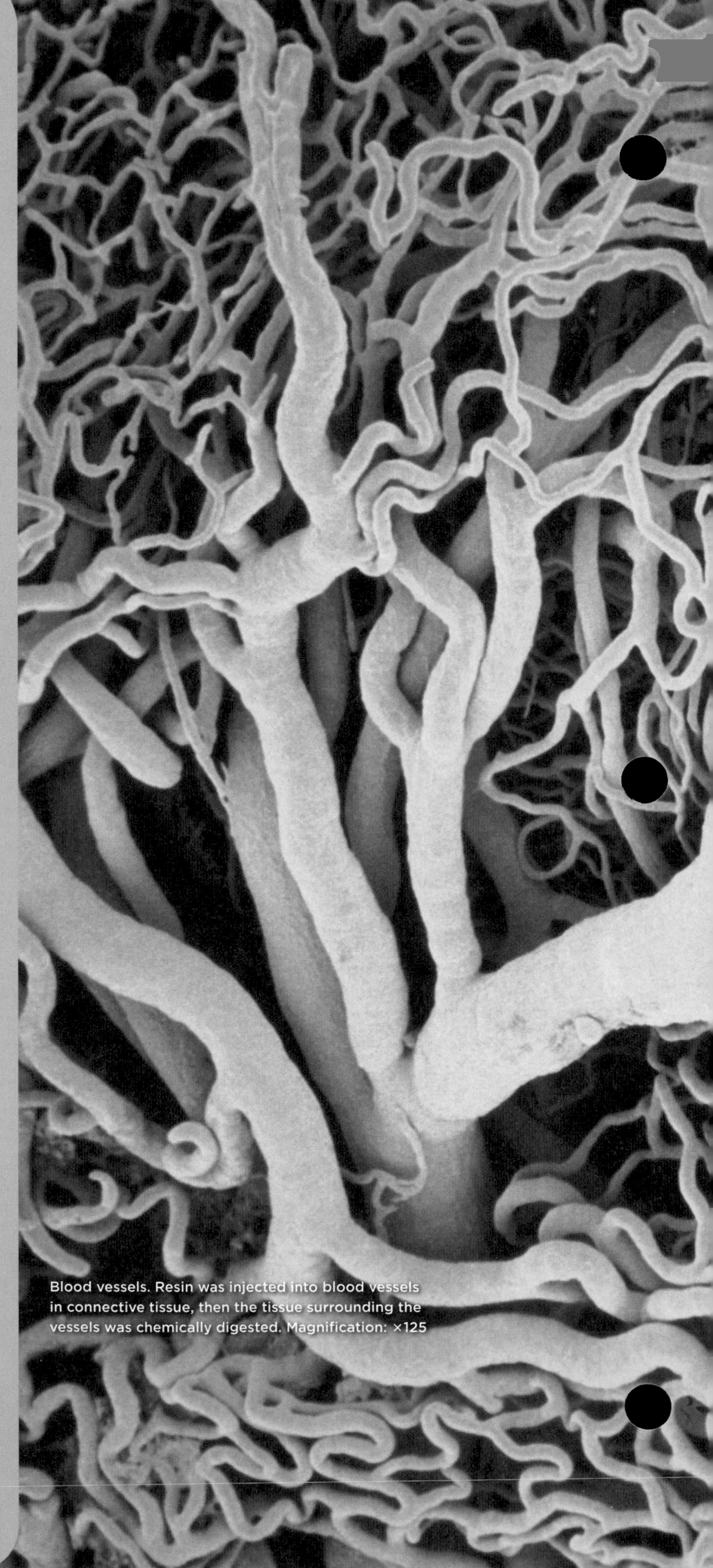

Blood vessels. Resin was injected into blood vessels in connective tissue, then the tissue surrounding the vessels was chemically digested. Magnification: ×125

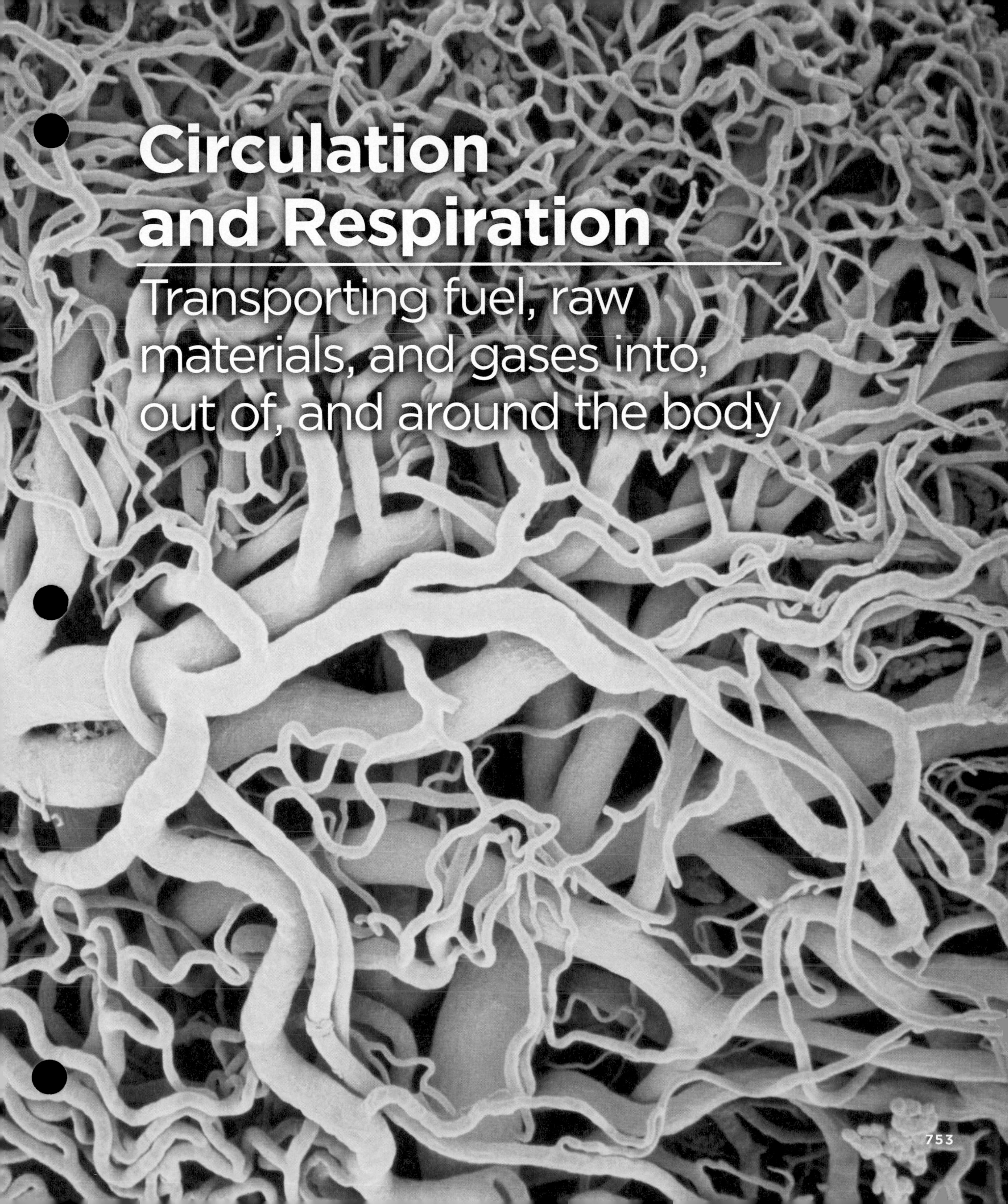

Circulation and Respiration

Transporting fuel, raw materials, and gases into, out of, and around the body

22.1–22.3

The circulatory system is the chief route of distribution in animals.

Wildebeests in Kenya bounding through open plains.

22.1 What is a circulatory system, and why is one needed?

Size matters. In single-celled organisms and small multicellular organisms, all cells are in contact with (or just a few cells away from) the external environment and can acquire raw materials and dispose of metabolic waste by simple diffusion.

The evolution of large body sizes created a host of new challenges. With increasing size, most of an animal's cells were no longer in direct contact with the outside world, the environment from which the animal obtained the oxygen, nutrients, water, and other substances it needed to survive. As body size increased, dedicated delivery and removal systems became a necessity.

In animals, the primary distribution system is the circulatory system, which consists of tens of thousands of miles of hollow, tube-like vessels that reach all tissues of the body. In vertebrates, circulatory systems have three principal functions: transport, body temperature regulation, and protection (FIGURE 22-1).

1. Transport. Like a system of highways for delivering important goods and removing garbage, the circulatory system transports oxygen, nutrients, waste products, hormones, and immune system cells in the blood throughout the body.

- In vertebrates, blood vessels take oxygen from the lungs or gills and deliver it to the tissues for energy-releasing cellular respiration. Blood arriving from the lungs or gills is loaded with oxygen and is said to be oxygenated or oxygen-rich.

FUNCTIONS OF THE CIRCULATORY SYSTEM

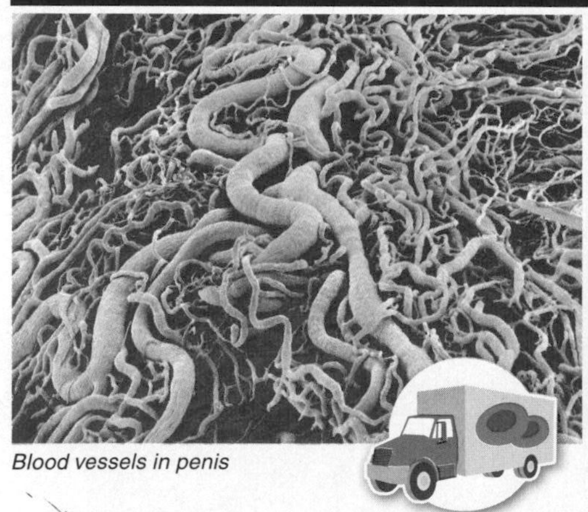
Blood vessels in penis

TRANSPORT
The circulatory system transports oxygen, nutrients, waste products, immune system cells, and hormones in the blood throughout the body.

Black-tailed jackrabbit

TEMPERATURE REGULATION
The circulatory system helps to maintain body temperature within the optimum range for metabolic functioning.

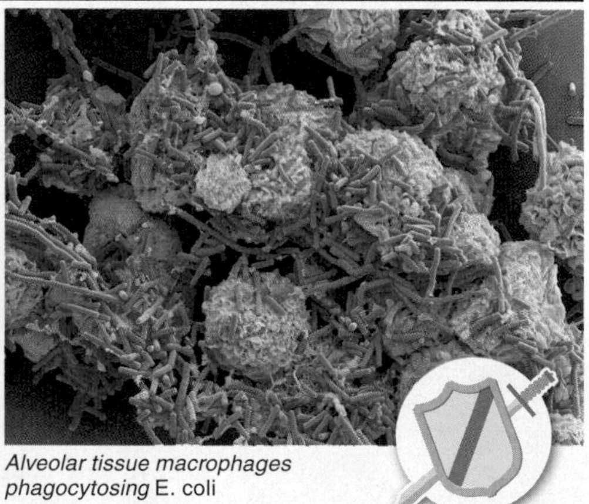
Alveolar tissue macrophages phagocytosing E. coli

PROTECTION
The circulatory system contains a variety of cells and chemicals that contribute to the individual's defenses against infection by pathogens.

FIGURE 22-1 **Like a set of highways for the body.** Circulatory system functions: transport, temperature regulation, and protection.

- Simultaneously, blood vessels whisk away carbon dioxide and other metabolic wastes that are produced by cellular respiration and other cell processes and must be removed from the body. Blood that carries a lot of carbon dioxide is said to be deoxygenated or oxygen-poor.
- Once food particles are digested, the nutrients must be absorbed and delivered to all the tissues of the body, for activity, growth, and reproduction.
- The circulatory system delivers hormones (chemicals produced throughout the body by glands and other tissues; see Chapter 25) to target tissues to regulate growth, development, and reproduction.

2. Body temperature regulation. By expanding or contracting the blood vessels closest to the exterior of the body, animals can release or absorb (or at least minimize the loss of) heat to maintain body temperature within the optimum range for metabolism.

3. Protection. A variety of cells and chemicals can defend against infection by pathogens. White blood cells, or leukocytes, can engulf and destroy many disease-causing microorganisms. Platelets and certain chemicals in the blood provide protection by limiting blood loss and infection when the skin or other tissues are damaged.

TAKE HOME MESSAGE 22.1

» In animals, the circulatory system is the chief distribution system. The circulatory system transports gases, nutrients, waste products, and hormones to and from tissues throughout the body. The circulatory system also helps animals regulate their body temperature and protect against infection.

22.2 Circulatory systems can be open or closed.

There's more than one way to build a circulatory system. And, in fact, not all multicellular organisms even need a circulatory system. For example, although flatworms and cnidarians such as jellyfish are large and multicellular, every cell lies close to the external surface of the body or to its internal cavity, called the *gastrovascular cavity* (FIGURE 22-2). The gastrovascular cavity, while not a true circulatory system, serves many digestive ("gastro") and circulatory ("vascular") functions. It directs water into the central mouth and through an elaborate system of channels. The cells that line

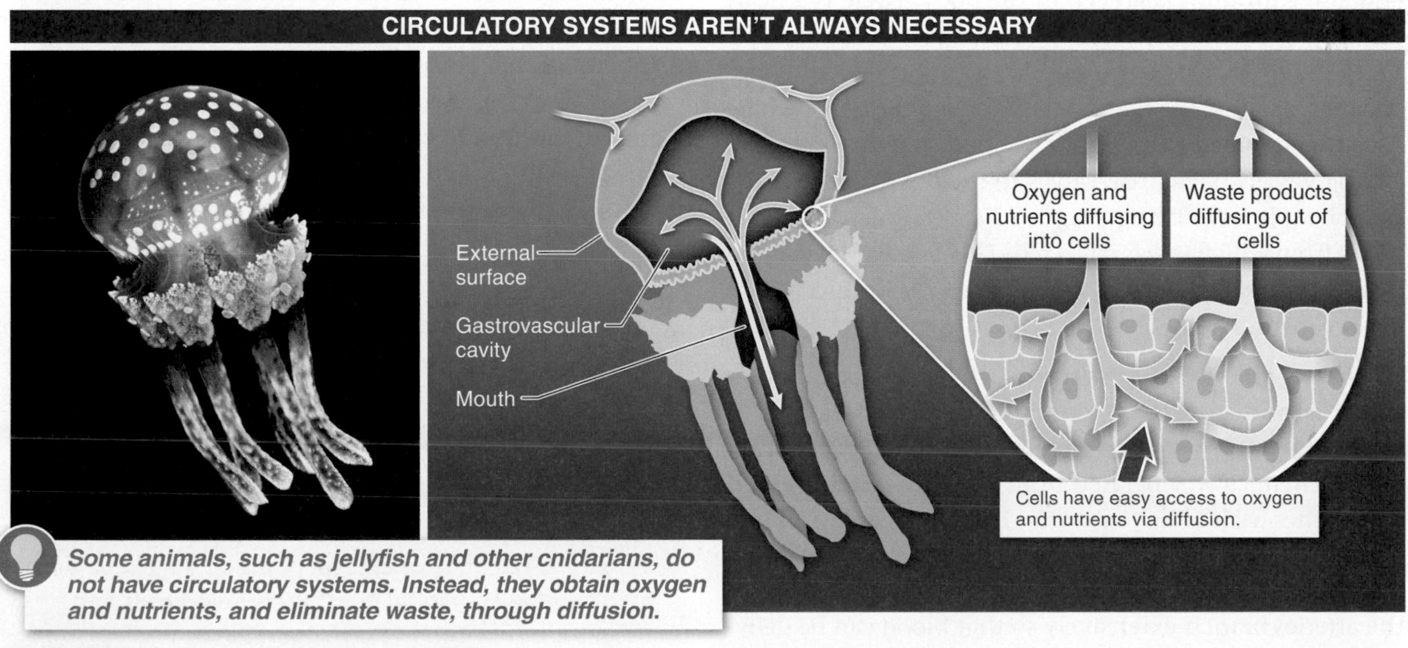
CIRCULATORY SYSTEMS AREN'T ALWAYS NECESSARY
External surface
Gastrovascular cavity
Mouth
Oxygen and nutrients diffusing into cells
Waste products diffusing out of cells
Cells have easy access to oxygen and nutrients via diffusion.
Some animals, such as jellyfish and other cnidarians, do not have circulatory systems. Instead, they obtain oxygen and nutrients, and eliminate waste, through diffusion.

FIGURE 22-2 No circulatory system required.

OPEN CIRCULATORY SYSTEMS

In open circulatory systems, there is no clear distinction between the circulating fluid and interstitial fluid. The heart (or hearts) pumps the fluid mixture—called hemolymph—throughout the extracellular spaces inside the body.

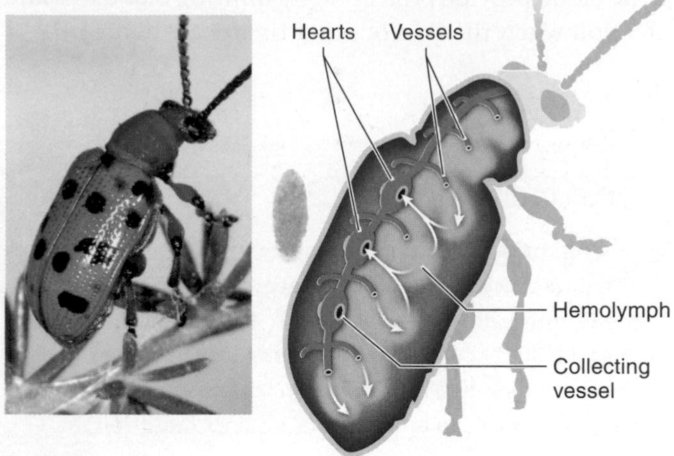

CLOSED CIRCULATORY SYSTEMS

In closed circulatory systems, blood is contained within vessels that separate it from interstitial fluid. A muscular heart propels blood through vessels to tissues throughout the body.

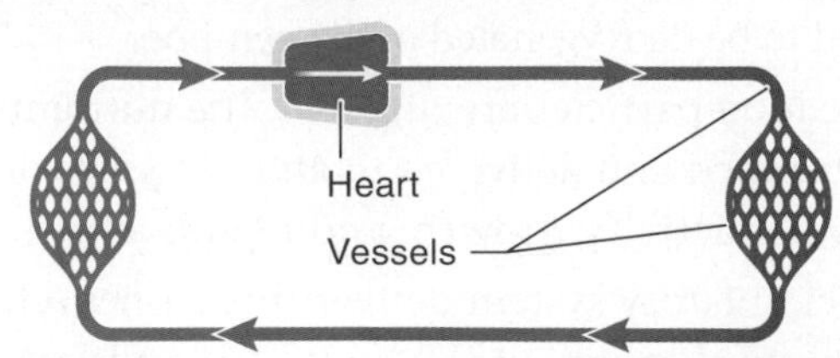

FIGURE 22-3 There are open and closed circulatory systems.

BLOOD VESSELS IN A CLOSED CIRCULATORY SYSTEM

ARTERIES
Vessels that carry blood away from the heart and to the capillaries

CAPILLARIES
Tiny porous vessels that bring blood close to tissue, enabling the diffusion of gases, nutrients, and other molecules into and out of the tissue

VEINS
Vessels that carry blood away from the capillaries and back toward the heart

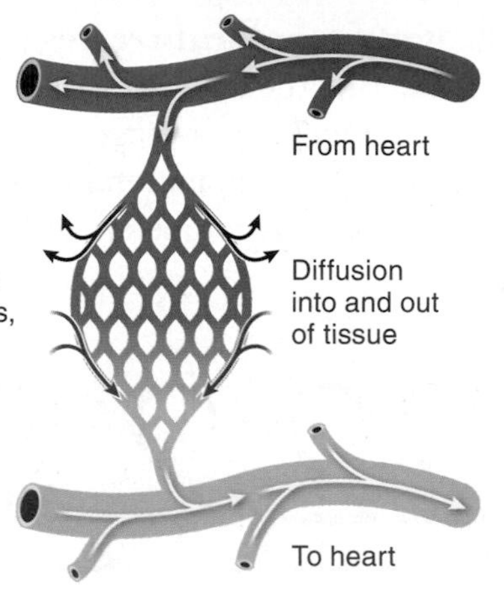

FIGURE 22-4 In closed circulatory systems, blood flows from arteries to capillaries to veins.

the mouth and the channels can absorb nutrients and exchange gases by diffusion. The absorbed nutrients diffuse to nearby cells. Once nutrients have been extracted (and waste products picked up), fluid in the gastrovascular cavity is flushed back out of the mouth, and the process is repeated.

Open and closed circulatory systems are found among other multicellular animals (FIGURE 22-3). Insects and most mollusks have open systems, and all vertebrates have closed circulatory systems. The defining feature of an **open circulatory system** is that it has one fluid, called **hemolymph,** that circulates to transport nutrients, gases, and waste products and also surrounds each cell in the body. There is a heart (or sometimes many hearts) that pumps the hemolymph throughout the extracellular spaces. Large collecting vessels then channel the hemolymph back to the heart, where it can be pumped throughout the body again. These collecting vessels have little valves that close to prevent the hemolymph from being pumped back through the same vessel from which it was collected. This one-way collection system directs the circulation of hemolymph throughout the organism's body.

In **closed circulatory systems,** the circulating fluid—called **blood**—is always contained in a vessel as it is pumped throughout the animal's body, and it is physically and chemically separated from the **interstitial fluid** that bathes each cell. A muscular **heart** serves as a pump, and with each contraction it propels blood at high pressure through vessels called **arteries.** Like a highway system, the arteries branch extensively so that blood can be delivered to all the tissues of the body.

At the organs and tissues, arteries branch into thousands of ever-smaller vessels, eventually becoming the narrowest and thinnest-walled: the **capillaries.** The thin walls of the capillaries allow gases, nutrients, and other molecules to diffuse easily into and out of the tissue, down their concentration gradients (FIGURE 22-4). After passing through the capillaries, blood returns to the heart in vessels called **veins,** which increase in diameter as they near the heart. We discuss the structure and function of blood vessels in more detail in Section 22.6.

TAKE HOME MESSAGE 22.2

» Animals (such as flatworms and cnidarians) that can acquire all the nutrients and oxygen they need by diffusion do not have circulatory systems. Among animals that do have circulatory systems, the system can be open, with no clear distinction between the circulating fluid and the interstitial fluid, or closed, in which the circulating fluid and the interstitial fluid surrounding cells and tissues are kept completely separate.

22.3 Vertebrates have several different types of closed circulatory systems.

Among the vertebrates, there are several variations of the closed circulatory system. These variations evolved in concert with numerous adaptations that made possible the transition from life in the seas to life on land. As some vertebrates developed lungs and the ability to extract oxygen from the air rather than water, their circulatory systems increased in complexity. Here we give a broad overview of the structures and functions of the various types of closed circulatory systems.

In fishes, blood flow follows a circular path (FIGURE 22-5). We can trace its flow, beginning at the two-chambered heart. Blood first passes into the **atrium,** the collecting chamber, and from there is pumped into the **ventricle,** the chamber from which it is pumped to the gills. Blood flows through the gills, where it picks up oxygen. Flowing through the millions of blood vessels in the gills, blood encounters friction and loses the pressure it had as it was pumped from the heart. From the gills, the blood flows to the tissues of the body, delivering the oxygen. After passing through the body tissues, the oxygen-poor blood flows back to the heart, and the cycle begins again.

In contrast to the two-chambered heart of fishes, birds and mammals have a four-chambered heart with two atria and two ventricles (FIGURE 22-6). Twice as many chambers are required because birds and mammals have two circuits of flow. The two circuits can be visualized as a figure 8. Oxygen-poor blood flows through the body into the right atrium and through to the right ventricle (see Figure 22-6). From there it is pumped out to the lungs, where it picks up oxygen, and then returns to the heart. This first circuit of blood flow, between the heart and the lungs, is called the **pulmonary circuit.**

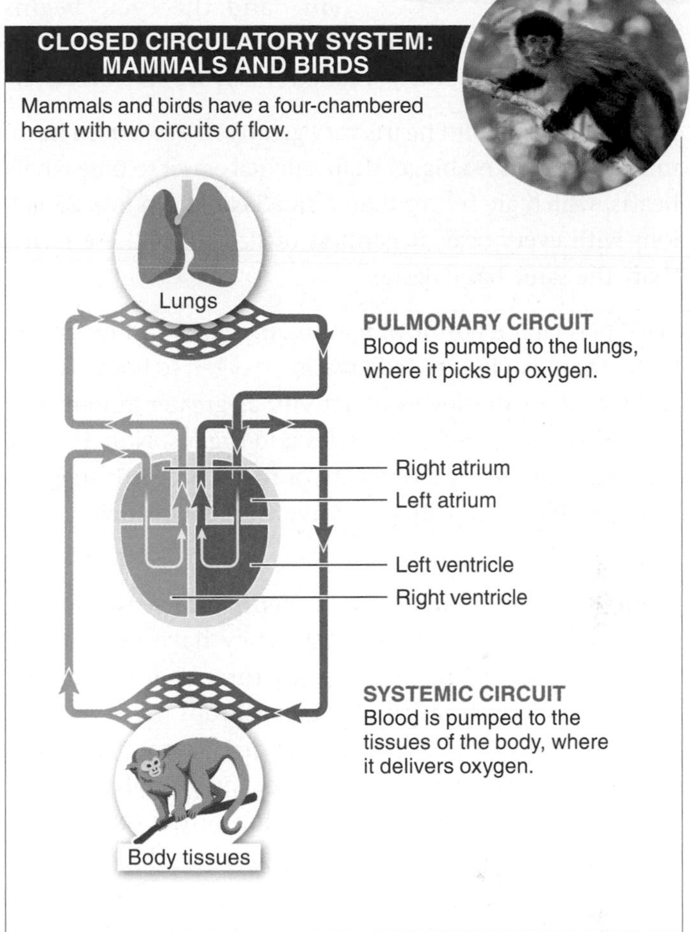
CLOSED CIRCULATORY SYSTEM: MAMMALS AND BIRDS
Mammals and birds have a four-chambered heart with two circuits of flow.
Lungs
PULMONARY CIRCUIT
Blood is pumped to the lungs, where it picks up oxygen.
Right atrium
Left atrium
Left ventricle
Right ventricle
SYSTEMIC CIRCUIT
Blood is pumped to the tissues of the body, where it delivers oxygen.
Body tissues

FIGURE 22-6 **Blood flow in mammals and birds.** Note that the right side of the animal's heart is on the left side in this diagram, and the left side is on the right. This is because the heart is drawn as if the animal were facing you.

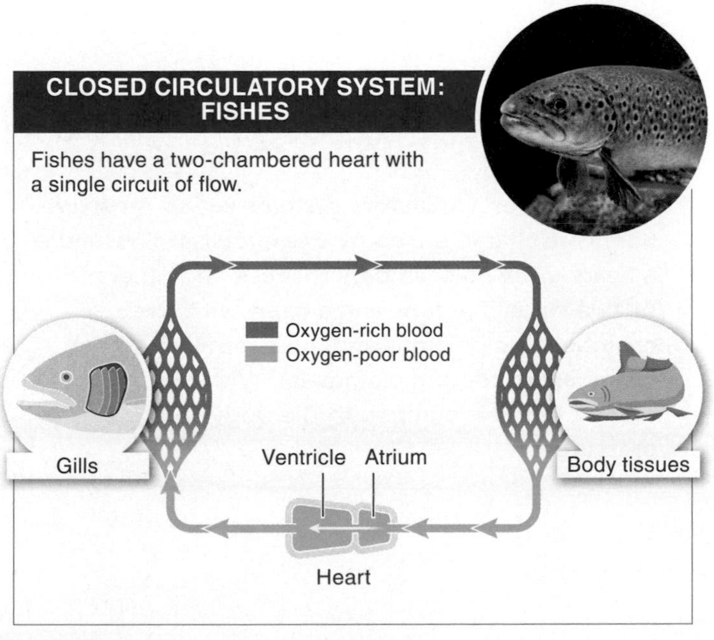
CLOSED CIRCULATORY SYSTEM: FISHES
Fishes have a two-chambered heart with a single circuit of flow.
Oxygen-rich blood
Oxygen-poor blood
Gills
Ventricle
Atrium
Body tissues
Heart

FIGURE 22-5 **Blood flow in fishes.**

The oxygen-rich blood that returns from the lungs to the heart collects in the left atrium. The blood immediately passes into the muscular left ventricle and from there is

> **Q** Why is the left side of mammalian and bird hearts always bigger than the right?

pumped at great pressure to the rest of the body. After delivering its nutrients and oxygen to body tissues, the now oxygen-poor blood travels back to the right atrium, and the cycle begins again. This second circuit of blood flow, between the heart and the rest of the body, is called the **systemic circuit.**

Mammalian and bird hearts vary greatly in size—from hummingbird hearts no bigger than a pencil eraser to blue whale hearts, which are bigger than a small car and pump 25 gallons with every beat. Regardless of size, all of these hearts share the same basic design.

With two circuits of flow, mammalian and bird hearts are better than fish hearts at delivering oxygen to body tissues, sustaining greater levels of activity as greater amounts of oxygen are delivered to muscles and organs. Nonetheless, the large number and diversity of fishes on earth suggests that they have been able to thrive with their single-circuit system.

Amphibians have circulatory systems similar to those of mammals and birds (FIGURE 22-7). They have two circuits of flow, but have a heart with only three chambers rather than four. Blood from the lungs and from the rest of the body collects in the left and right atria, respectively, but it then flows into a single ventricle. Surprisingly, though, in the ventricle there is little mixing of the oxygenated blood from the lungs and the deoxygenated blood from the rest of the body. Rather, the two types of blood flow side by side and are pumped into arteries that direct the oxygenated blood to the body and the deoxygenated blood to the lungs.

With the exception of birds, most reptiles also have three-chambered hearts, although the ventricle is partially divided in two. In one group of reptiles, however—the crocodilians—the division is complete. These reptiles also have an extra little artery through which blood can be pumped from the right ventricle to the rest of the body rather than to the lungs. This adaptation allows the crocodile to bypass sending blood to the lungs when the animal is underwater. Instead, the blood is sent back to the body, where the remaining oxygen in the blood can be utilized.

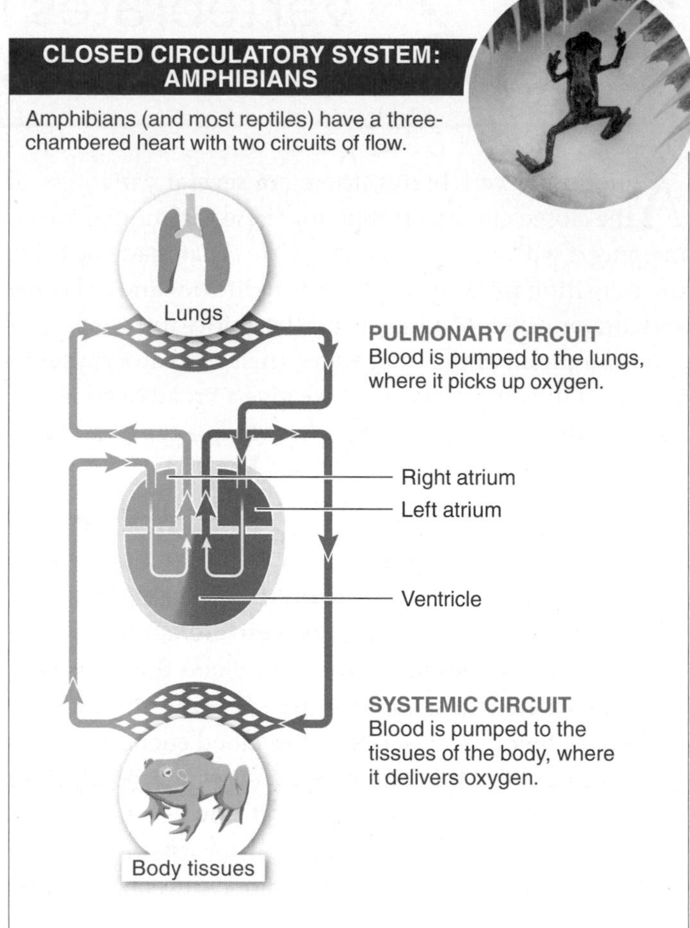
CLOSED CIRCULATORY SYSTEM: AMPHIBIANS
Amphibians (and most reptiles) have a three-chambered heart with two circuits of flow.
Lungs
PULMONARY CIRCUIT
Blood is pumped to the lungs, where it picks up oxygen.
Right atrium
Left atrium
Ventricle
SYSTEMIC CIRCUIT
Blood is pumped to the tissues of the body, where it delivers oxygen.
Body tissues

FIGURE 22-7 **Blood flow in amphibians.** (Note that some deoxygenated blood is sent to the skin, rather than the lungs, where exchange of respiratory gases can also occur.)

TAKE HOME MESSAGE 22.3

» Vertebrates' circulatory systems vary in structure. Some are characterized by one circuit of flow and a heart with two chambers (fishes), and others by two circuits of flow and a heart with three chambers (amphibians and most reptiles) or four chambers (birds and mammals). With two circuits of flow, blood is pumped to the body at higher pressure.

The human circulatory system consists of a heart, blood vessels, and blood.

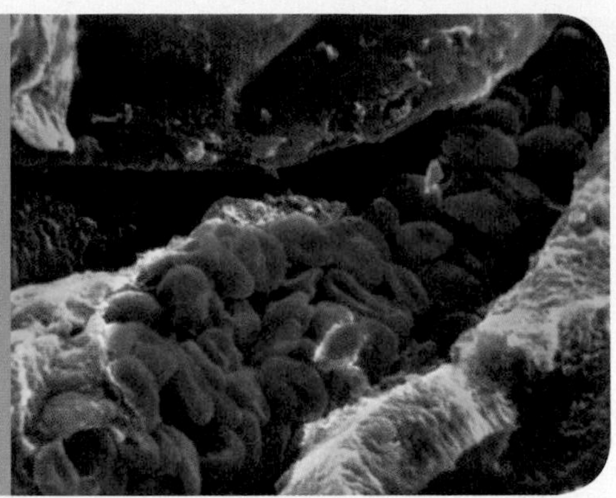

Red blood cells are packed in a capillary.

22.4 Blood flows through the four chambers of the human heart.

Clench your fist. That is the size of your heart. Now clench your fist and relax it a hundred thousand times. That is what your heart does every day for 70 or more years. The human heart is at the center of our circulatory system, and it's one of the most durable and reliable pumps ever produced (FIGURE 22-8).

The four-chambered heart sends blood on a figure 8, two-circuit path through the body, first to the lungs for loading up with oxygen and then, on its second circuit, to the tissues and organs. Let's trace the blood flow as it cycles through the heart, lungs, and tissues of the body (FIGURE 22-9). We start with the arrival of oxygen-depleted blood from the organs and tissues.

1. Deoxygenated blood from the organs and tissues enters the right atrium. The blood arriving from the lower half of the body enters through a large vein called the inferior vena cava. Blood arriving from the head and arms enters through the superior vena cava.
2. Most of the blood passes directly through the right atrium into the right ventricle. A contraction pushes the remainder of the blood in the right atrium into the right ventricle.
3. The contraction continues, pumping the blood out of the right ventricle and into the pulmonary artery. This large artery immediately forks, sending half of the blood to the left lung and half to the right lung.

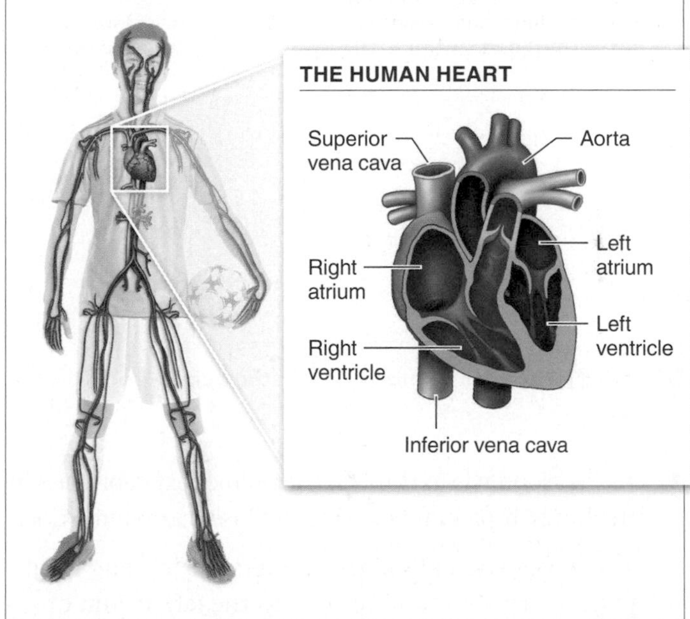

THE HUMAN CIRCULATORY SYSTEM
The human circulatory system is composed of a fist-sized heart and an intricate system of blood vessels that transport respiratory gases, nutrients, and waste products throughout the body.
THE HUMAN HEART
Superior vena cava
Aorta
Left atrium
Right atrium
Left ventricle
Right ventricle
Inferior vena cava

FIGURE 22-8 **Heart and blood vessels.** An overview of the human circulatory system.

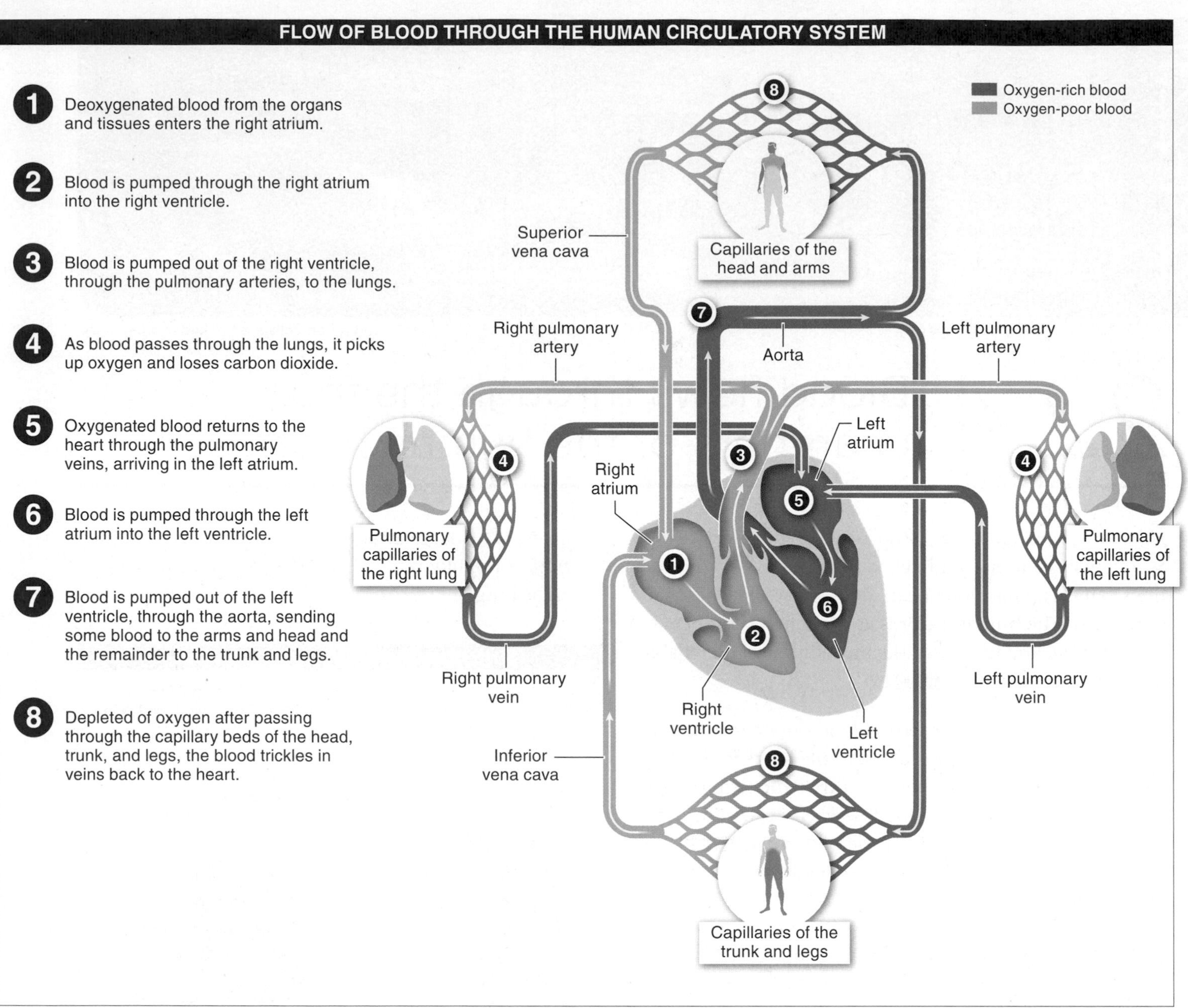

FLOW OF BLOOD THROUGH THE HUMAN CIRCULATORY SYSTEM
1 Deoxygenated blood from the organs and tissues enters the right atrium.
2 Blood is pumped through the right atrium into the right ventricle.
3 Blood is pumped out of the right ventricle, through the pulmonary arteries, to the lungs.
4 As blood passes through the lungs, it picks up oxygen and loses carbon dioxide.
5 Oxygenated blood returns to the heart through the pulmonary veins, arriving in the left atrium.
6 Blood is pumped through the left atrium into the left ventricle.
7 Blood is pumped out of the left ventricle, through the aorta, sending some blood to the arms and head and the remainder to the trunk and legs.
8 Depleted of oxygen after passing through the capillary beds of the head, trunk, and legs, the blood trickles in veins back to the heart.
Oxygen-rich blood
Oxygen-poor blood
Superior vena cava
Capillaries of the head and arms
Right pulmonary artery
Aorta
Left pulmonary artery
Left atrium
Right atrium
Pulmonary capillaries of the right lung
Pulmonary capillaries of the left lung
Right pulmonary vein
Left pulmonary vein
Right ventricle
Left ventricle
Inferior vena cava
Capillaries of the trunk and legs

FIGURE 22-9 **The path of blood flow in the human body.**

4. As the blood passes through the pulmonary capillaries in the lungs, it picks up oxygen and loses carbon dioxide.
5. The oxygenated blood then enters the left and right pulmonary veins and returns to the left atrium of the heart.
6. Most of the blood passes directly through the left atrium and into the left ventricle. A contraction pushes the remaining blood from the left atrium into the left ventricle.
7. As the contraction continues, the oxygenated blood is pumped up and out of the ventricle at tremendous pressure into the aorta, the largest artery in the body. After making a sharp turn (called the aortic arch), the aorta forks and sends some blood to the arms and head and the remainder to the trunk and legs.
8. After passing through the capillary beds in the tissues of the head, arms, trunk, and legs, the blood

FIGURE 22-10 **Audible heartbeats.**

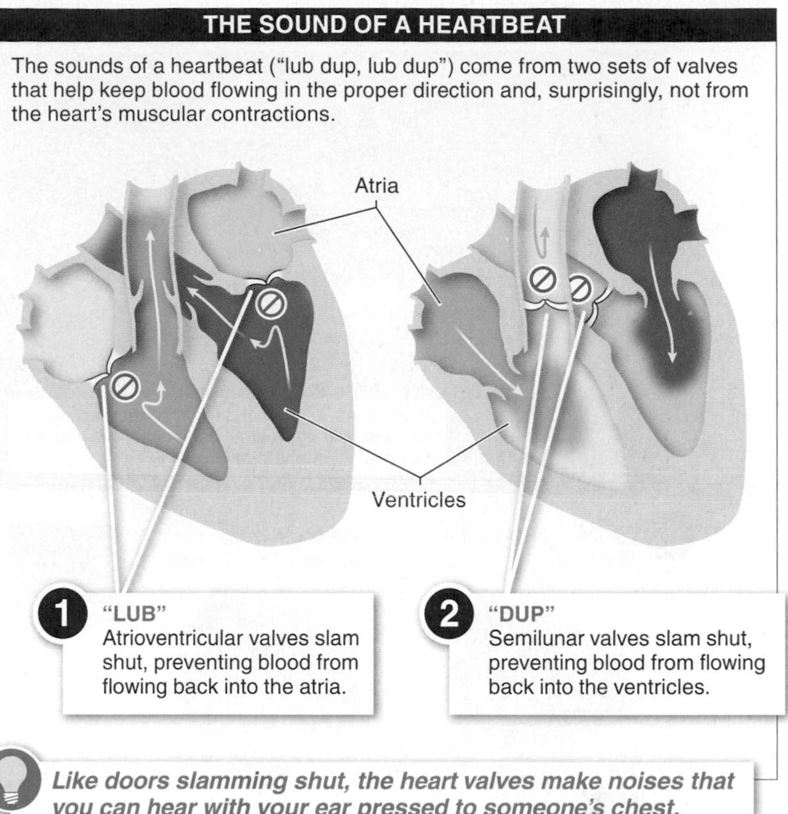

THE SOUND OF A HEARTBEAT
The sounds of a heartbeat ("lub dup, lub dup") come from two sets of valves that help keep blood flowing in the proper direction and, surprisingly, not from the heart's muscular contractions.
Atria
Ventricles
1 "LUB"
Atrioventricular valves slam shut, preventing blood from flowing back into the atria.
2 "DUP"
Semilunar valves slam shut, preventing blood from flowing back into the ventricles.
Like doors slamming shut, the heart valves make noises that you can hear with your ear pressed to someone's chest.

moves at low pressure through the veins. This deoxygenated blood returns to the heart by way of the superior and inferior venae cavae and collects in the right atrium, and the cycle repeats.

Surprisingly, the muscular contractions don't make much noise. Rather, the characteristic "lub dup" sounds come from two sets of valves that help keep blood flowing in the proper direction (FIGURE 22-10).

The two atrioventricular (AV) valves, located between the atrium and ventricle on each side of the heart, allow blood to flow from the atrium to the ventricle. When the ventricles contract, these flaps of tissue slam shut, preventing backflow into the atria. This makes the "lub" sound. With no other escape, blood flows out through the pulmonary arteries on the right side and through the aorta on the left. At these two primary exits from the heart are two more valves, the semilunar valves (so called because of their half-moon shape), which close and prevent backflow of blood into the ventricles. The closing of these valves makes the "dup" sound.

Q What is the source of heart murmurs?

When the atrioventricular or semilunar valves do not completely close, some blood can flow back in the wrong direction. The blood moving backward through the valve can be heard with a stethoscope as a buzzing or swishing noise. These noises are called **heart murmurs.** Most are not life-threatening, and the individual suffers no ill effects.

TAKE HOME MESSAGE 22.4

›› The human heart, at the center of our circulatory system, is an extremely durable pump. It sends blood on a figure 8, two-circuit path through the body, first to the lungs for loading up with oxygen, and then, on its second circuit, to the tissues and organs of the body. Valves in the heart keep blood flowing in one direction.

22.5 Electrical activity in the heart generates the heartbeat.

Vertebrate hearts have a small piece of modified muscle tissue, the sinoatrial (SA) node, that initiates the regular, rhythmic contractions of the heart. Unlike most muscular tissue, which must be stimulated by a nerve before it contracts, the sinoatrial node spontaneously fires. This spontaneous firing begins early in fetal development and initiates every heartbeat for your entire life. As the pacemaker of the heart, the SA node, through its repeated firing, ensures the efficient pumping of blood to the lungs and to the rest of the body.

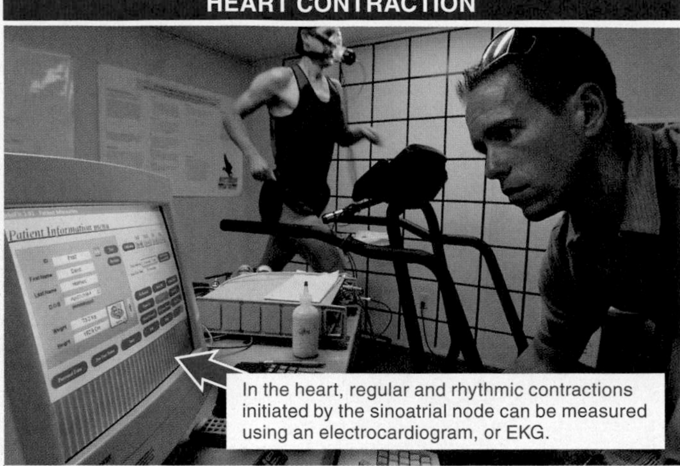

HEART CONTRACTION
In the heart, regular and rhythmic contractions initiated by the sinoatrial node can be measured using an electrocardiogram, or EKG.

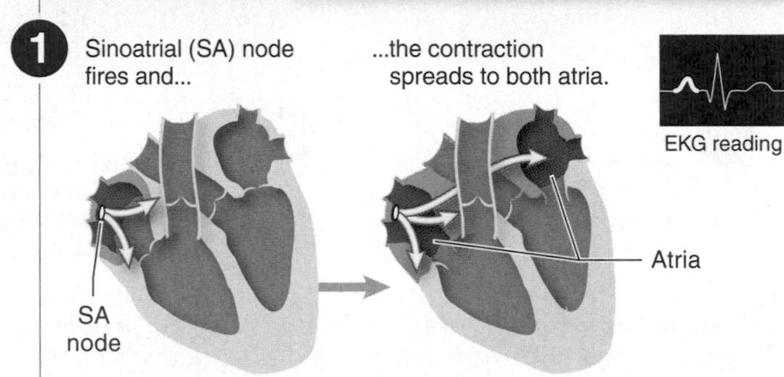

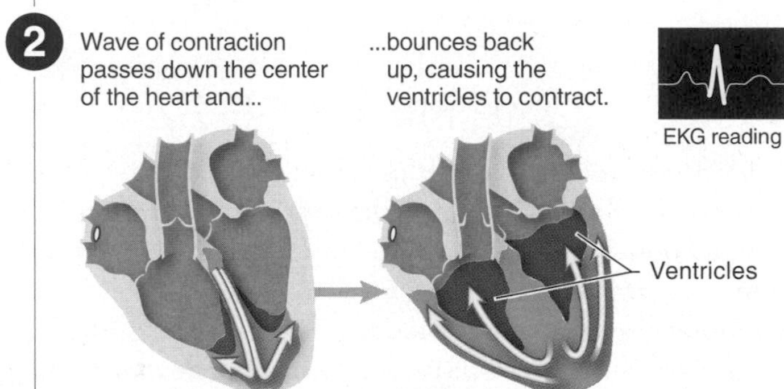

2
Wave of contraction passes down the center of the heart and...
...bounces back up, causing the ventricles to contract.
EKG reading
Ventricles

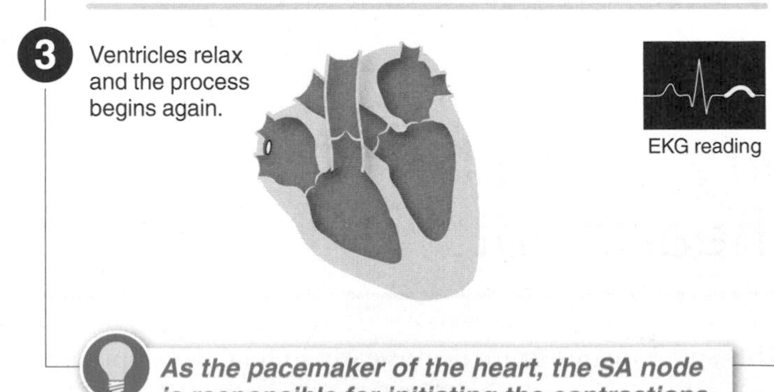

3
Ventricles relax and the process begins again.
EKG reading

As the pacemaker of the heart, the SA node is responsible for initiating the contractions that pump blood through the body.

FIGURE 22-11 Rhythmic contractions of the heart.

Beginning just above the right atrium, the electrical impulse in the SA node quickly spreads to the left atrium. As the atria contract, blood is pushed into the ventricles. The atria then enter a "relaxation" phase, and the wave of contraction continues, traveling down the center of the heart and pausing briefly as it passes between the two ventricles. On reaching the bottom of the heart, the contraction appears to almost "bounce" back upward, causing a deep contraction that pushes the blood up from the bottom of both ventricles and into the pulmonary arteries (from the right ventricle) and the aorta (from the left ventricle), much like squeezing a tube of toothpaste from the bottom up. The ventricles then enter a relaxation phase, and the SA node starts the contraction anew (FIGURE 22-11).

The contraction of muscle tissue is a powerful electrical event that can be recorded by electrodes placed on the skin. The charges are traced into an electrocardiogram, or EKG (from the German *Elektrokardiogramm*). EKG readings allow quick and easy display and analysis of the cardiac cycle.

When the SA node does not function properly, the heart may beat too slowly or erratically. An artificial pacemaker is a battery-operated electronic device that generates stimulation to the heart, causing a more regular heartbeat.

TAKE HOME MESSAGE 22.5

» The sinoatrial node, modified muscle tissue in the vertebrate heart, initiates regular, rhythmic contractions. A heart contraction begins with an electrical impulse in the SA node in the right atrium. The contraction quickly spreads to the left atrium, passes down the center to the bottom of the heart, then moves upward, pushing blood from both ventricles out through the pulmonary arteries and aorta.

22.6 Blood flows out of and back to the heart in blood vessels.

Blood flows from the heart to the tissues of the body and back to the heart again through a system of blood vessels. There are three distinct types of blood vessels: arteries, capillaries, and veins (FIGURE 22-12).

1. Arteries. Leaving the heart, blood flows through arteries. The pulmonary arteries, arising from the right ventricle, carry deoxygenated blood to the lungs. The arteries extending from the left ventricle carry oxygenated blood, first to the aorta, then to the brain and the other tissues.

Arteries are lined with endothelium (a type of epithelial tissue) and surrounded by smooth muscle and connective tissue. These layers of tissue, along with collagen and other, elastic connective tissue fibers, enable arteries to stretch with the pressure generated by each beat of the heart, and then to recoil, pushing the blood forward. The thickness of the artery walls minimizes the exchange of materials between the blood and the tissues. Arteries branch repeatedly, becoming arterioles, which become smaller and smaller in diameter as they reach into all the tissues and organs. Blood then flows into capillaries.

2. Capillaries. The tiniest blood vessels are the capillaries, which bring blood close to all cells in the body and enable molecules to diffuse between the blood and cells. Each capillary has an inner diameter so narrow that blood cells must pass through in single file. In contrast to arteries, capillaries have walls that are thin and somewhat porous, permeable to water, ions, and some small molecules.

Capillaries are the last branch of the circulatory system to carry nutrients to the interstitial fluid and thus into the cells. Capillaries are also the first branch to carry blood back from cells toward the heart—transporting the carbon dioxide and waste products that diffuse from cells into the interstitial fluid and then through capillary walls.

Capillaries are arranged in capillary beds, networks of vessels that supply blood to tissues and organs. Although the diameter of each capillary is tiny, cumulatively their volume is much greater than that of the arterioles and arteries. The network of capillaries in an adult human covers more than 50,000 miles! And this number fluctuates: the gain

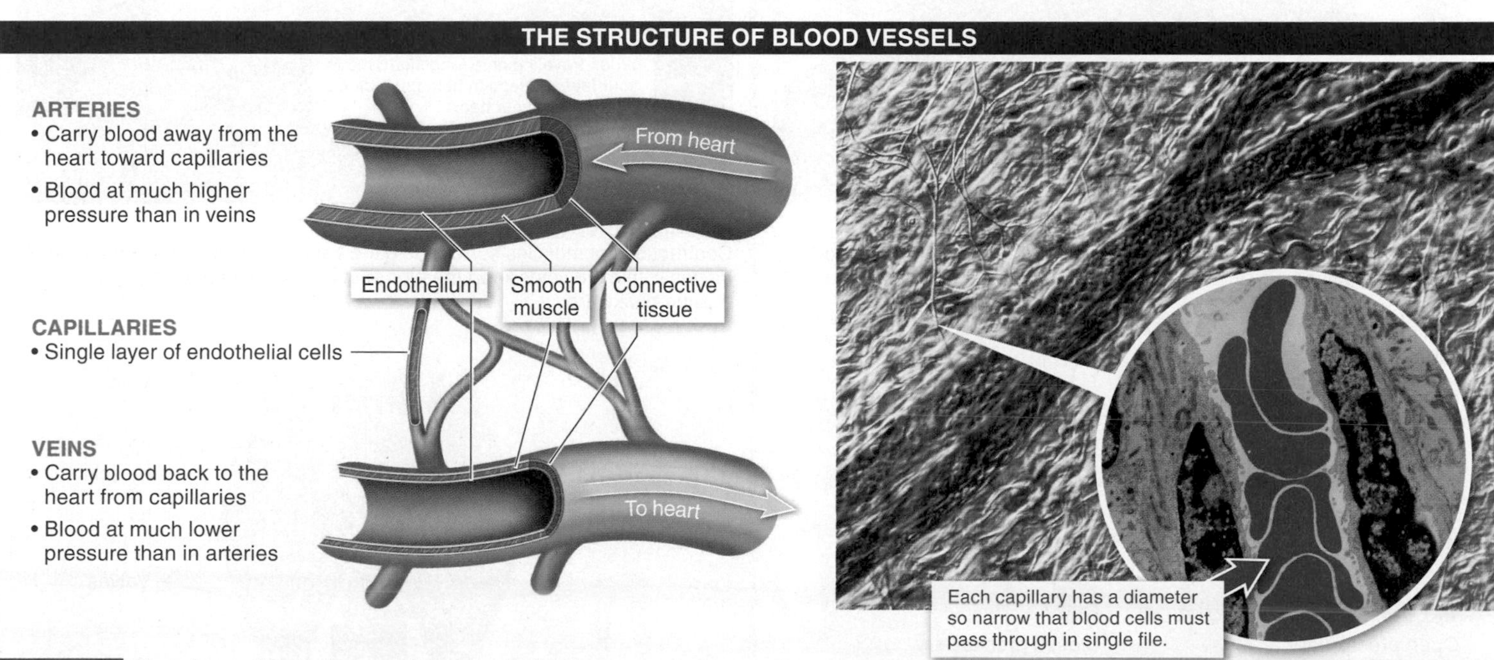
THE STRUCTURE OF BLOOD VESSELS
ARTERIES
• Carry blood away from the heart toward capillaries
• Blood at much higher pressure than in veins
From heart
Endothelium
Smooth muscle
Connective tissue
CAPILLARIES
• Single layer of endothelial cells
VEINS
• Carry blood back to the heart from capillaries
• Blood at much lower pressure than in arteries
To heart
Each capillary has a diameter so narrow that blood cells must pass through in single file.

FIGURE 22-12 **Structure of arteries, capillaries, and veins.**

of one pound of fat, for example, is accompanied by the addition of more than a mile of new capillaries.

Surprisingly, most capillaries have little or no blood flowing through them at any given time, and the body is constantly directing blood flow to the tissues where it is most needed. In frigid weather, your face or hands may feel cold and turn bluish in color. This happens because smooth muscle around arterioles, called precapillary sphincters, contracts and reduces blood flow (and the heat it brings) to the capillaries of the face or hands, where blood flow could lead to wasteful heat loss (FIGURE 22-13). Simultaneously, these muscles shunt blood to critical parts of the body, including the brain, heart, liver, and kidneys. Blushing is the opposite situation: precapillary sphincters relax and blood flow to the face and neck increases.

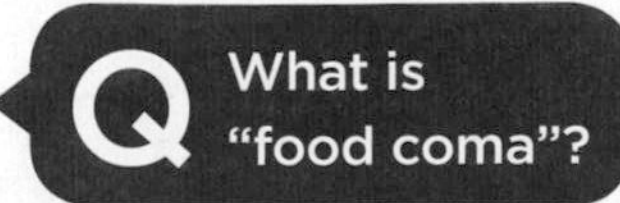

Q
What is "food coma"?

"Food coma," that feeling of lethargy following a large meal, results when more blood flows through the capillaries surrounding the digestive tract, reducing flow to other parts of the body. In contrast, during strenuous exercise, more blood is directed toward the skeletal muscles. This accounts for the feeling of being "pumped up" during and shortly after a weightlifting session.

3. Veins. Once blood has passed through the capillaries, it flows back to the heart through veins. Like arteries, veins are lined with endothelium and surrounded by smooth muscle and connective tissue. The layer of smooth muscle surrounding veins, however, is much thinner than that surrounding arteries, making the walls of veins much more

DIRECTING THE FLOW OF BLOOD

Precapillary sphincters can reduce blood flow to hands or feet in cold weather, saving the blood's warmth for vital organs.

PRECAPILLARY SPHINCTERS
In the arterioles, precapillary sphincters can contract and cut off blood flow to the capillaries in order to shunt it elsewhere in the body.

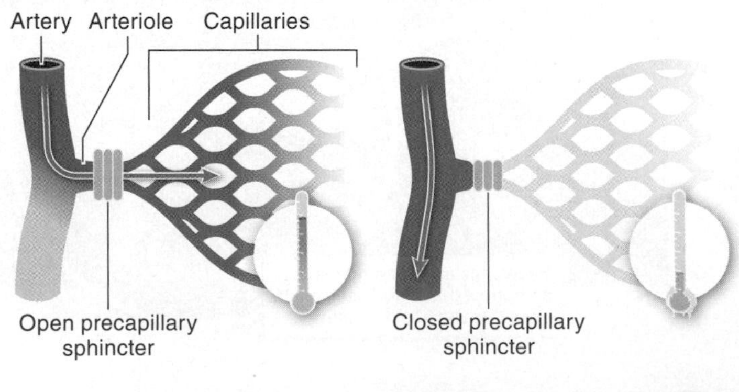

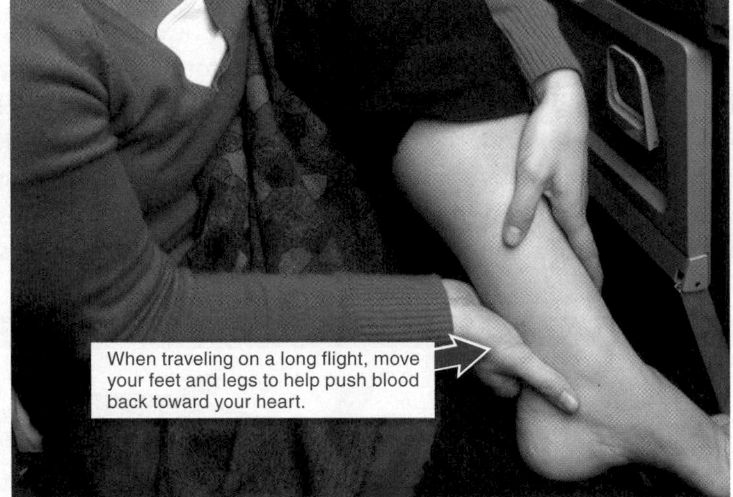

When traveling on a long flight, move your feet and legs to help push blood back toward your heart.

MUSCLE CONTRACTIONS AND VALVES
Contractions of muscles surrounding the veins push blood toward the heart. Valves within the veins keep the blood on course by preventing it from moving backward.

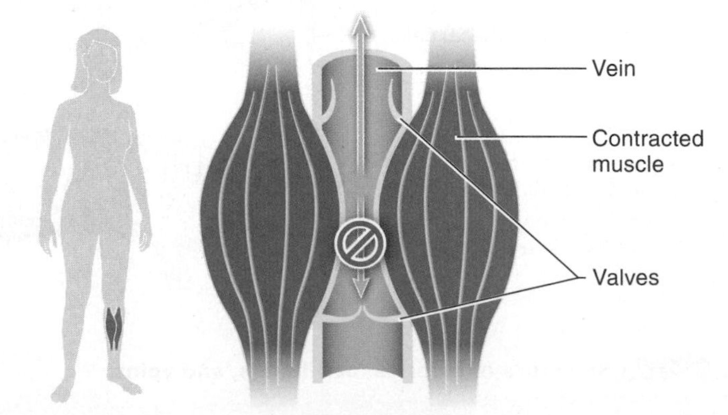

Muscle contractions help direct blood flow—blocking blood from capillaries or pushing blood through veins back toward the heart.

FIGURE 22-13 **Controlling blood flow.**

expandable. In a person at rest, as much as 60% of the total blood volume may be in his or her veins.

The pressure of blood in the veins is only about one-tenth of the pressure in the arteries. This is because as blood flows through capillaries, an increasing proportion is in direct contact with the inside surface of the vessels and the increased friction slows the blood flow. Moving from the capillaries through the veins, the pressure of the blood is reduced further. For this reason, although blood will spurt and gush when an artery is cut, it will only trickle from a cut vein. It is not surprising that most arteries are not as close to the skin as are veins.

Like the capillaries, veins also aid in the control of blood flow. After sitting for a very long time, such as on a long plane ride, you may notice that your feet become swollen. What causes this swelling? The answer has to do with the way blood flows back to the heart. It returns to the heart only with the assistance of two important features. First, during normal movement, muscles that surround the veins squeeze the veins, pushing the blood through. And second, within veins, at regular intervals, are one-way valves that allow blood to flow in only one direction—toward the heart (see Figure 22-13).

Regulating blood flow by means of valves in the veins and the contractions of muscles surrounding the veins usually works fine. If you sit for several hours, however, you may encounter a problem. In the absence of contractions of your calf and thigh muscles, blood in your feet can't climb up your legs, and it pools in your feet. This causes them to swell (and increases the risk of blood clots). It is good to get up and move around occasionally or to exercise your leg muscles as you sit.

Varicose veins are a painful condition that occurs when the valves preventing backflow of blood malfunction and blood pools in the veins, stretching them. Standing for long periods of time can cause or exacerbate the condition. Laser surgery or injections are effective for treating varicose veins. The veins slowly fade and disappear, with deeper veins taking over the circulation in that region.

Q **Some individuals develop a painful condition called varicose veins—enlarged, twisted, and visible veins. Why might this occur?**

TAKE HOME MESSAGE 22.6

›› Blood flows from the heart to the tissues of the body and back to the heart through endothelium-lined blood vessels. Arteries are elastic and carry blood at high pressure from the heart to capillaries. Capillaries are tiny, thin-walled vessels, the site of diffusion of nutrients, gases, waste products, and other materials between tissues and blood. Blood returns to the heart in veins, at low pressure; this movement is aided by contractions of skeletal muscles.

THIS IS HOW WE DO IT

Developing the ability to apply the process of science

22.7 Does thinking make your head heavier?

We sometimes speak of thoughts that "weigh heavy" on our mind, or brain. Brain activity—particularly in response to cognitively complex tasks or emotionally engaging stimuli—can seem taxing. But is there a literal, measurable cost to brain activity?

In 1884, an Italian researcher named Angelo Mosso pondered these questions and proposed a testable hypothesis:

To supply the necessary fuel and oxygen for increased brain activity, more blood is pumped to the brain, causing the head to become heavier.

After all, we know that blood can be shunted to where it is needed most. Does the same thing occur in response to increased brain activity?

Mosso's hypothesis is a reasonable one. But, of course, for a hypothesis to be useful, we must be able to test it.

Toward this end, Mosso invented an apparatus to measure the weight of a human head, called a "human circulation balance," and reported some preliminary observations. Due to some confounding variables and the limitations of his apparatus, however, it was difficult to draw definitive conclusions.

In 2014, a student and her professor decided to recreate Mosso's balance and to rigorously test his hypothesis by carefully evaluating the weight of brain activity.

How can you detect when the head gets heavier?

The human circulation balance is essentially a long wooden board, balanced like a see-saw. (Mosso's daughter said that he called it the "metal cradle" and "the machine to weigh the soul.") A person lies on the board, and a scale is placed under the end where his or her head is, to register any downward force. Small weights are added or removed from the other end of the board, where the person's feet are, until the scale registers just the slightest downward force—approximately 2 newtons—which indicates that the balance is slightly tipped toward the head end. (A newton is a unit of force, equivalent to about 0.22 pounds.) The scale can then register any additional downward force due to increased blood flow to the head.

How can you stimulate brain activity?

In the experimental design, each participant (14 in all) lay on the balance and was exposed to a stimulus for 2 seconds, followed by 23 seconds of rest. During the stimulus period, the participant was exposed to (1) music alone, or (2) music and a visual display of colorful geometric shapes synchronized to the music, or (3) no stimulation, as a control. The researchers recorded the force on the scale during the stimulation (or control) period. Any increase in head weight was attributed to increased mental activity due to increased blood flow. For each participant, the experiment was repeated 22 times.

How can you reduce the effects of other stimuli during this experiment?

Some fine-tuning of the apparatus was necessary to eliminate the impact of participants' breathing and heartbeats on the force measured. And the experimental protocol was designed to minimize any extraneous stimuli: each participant rested on the balance for 1 hour before measurements were made, and all wore headphones and kept their eyes shut during the experiment—except in the visual stimulation trials, in which they looked at a display on a computer screen above them.

Did heads get heavier with brain activity?

Yes! Listening to music only, the force increased by 0.61% relative to when no stimulus was used. And music and a visual display together produced a 1.21% increase. Although these amounts are very small—representing only about 0.005 newtons—the differences among the three stimuli were consistent across the research participants and were statistically significant.

Can you suggest alternative stimuli that might produce even greater brain activity and blood flow to the head?

Mosso also experimented with having subjects read complex math texts and a letter from an upset creditor (for which, he wrote, "the balance fell all at once"). Would you expect the addition of exposure to odors to have any effect?

What can we conclude from these results?

With a clearly stated, testable hypothesis and a simple experimental setup—straightforward enough to have been devised 130 years ago—it is possible to generate interesting observations and draw conclusive results. And brain activity does, indeed, cause increased blood flow to your head (though perhaps not enough for you to notice).

What further information would increase our confidence in these conclusions?

Sophisticated brain imaging technologies used today, including MRI and PET scans—which can detect movement of blood and metabolic activity within the brain—have also documented greater blood flow in response to neural activity.

TAKE HOME MESSAGE 22.7

» In response to increased brain activity, the circulatory system pumps additional blood to the brain, thus supplying the necessary fuel and oxygen. Using a sensitive balance, it's possible to detect an increase in the weight of the head due to this influx of additional blood.

22.8 Blood is a mixture of cells and fluid.

If circulatory systems are like highways throughout our bodies, then blood is the traffic. Blood is a salty, protein-rich mixture of cells and fluid, with a consistency like that of motor oil. The average human body has 4–5 quarts (3.7–4.7 liters) of blood, which makes up just under 10% of our total body weight. Blood's many functions include the transport of (1) respiratory gases such as oxygen and carbon dioxide, (2) vitamins and minerals, (3) nutrients, (4) hormones, (5) immune system cells, and (6) metabolic wastes. Blood also helps to maintain body temperature and homeostasis (see Chapter 21).

Blood has several components, which we can identify by putting a sample of blood in a test tube and spinning it rapidly in a centrifuge (FIGURE 22-14). The lightest-weight part of the blood, the upper, creamy yellow layer in the test tube, is the **plasma,** the liquid part of the blood. Plasma is 90% salty water. Dissolved within this water is a huge variety of molecules: metabolites and wastes, salts and ions, and hundreds of plasma proteins that transport lipids, vitamins, and a host of other chemicals to the tissues where they are required. Most of the carbon dioxide produced in tissues as a by-product of cellular respiration is carried to the lungs dissolved in the plasma.

A thin middle layer in the centrifuged tube, called a *buffy coat,* contains white blood cells mixed with platelets. The heaviest components of blood are at the bottom of the tube. This layer consists of tightly packed blood cells. The proportion of the blood that consists of cells is called the **hematocrit.** In humans, a hematocrit of about 45% is normal. Individuals living at high elevations for a few weeks or longer have hematocrits of around 48% or 49%. The body adapts to the reduced oxygen concentrations in air at high elevations by producing more blood cells to carry oxygen.

Blood cells are made in the bone marrow (the material that fills the interior of our bones) by specialized cells, called **stem cells,** that are able to develop into a diverse range of cell types. Stem cells throughout the bones in the human body produce blood cells at a rate of about two million cells per second. There are two types of blood cells suspended in the plasma, red blood cells and white blood cells, as well as platelets, which are cellular fragments (FIGURE 22-15).

CONSTITUENTS OF BLOOD

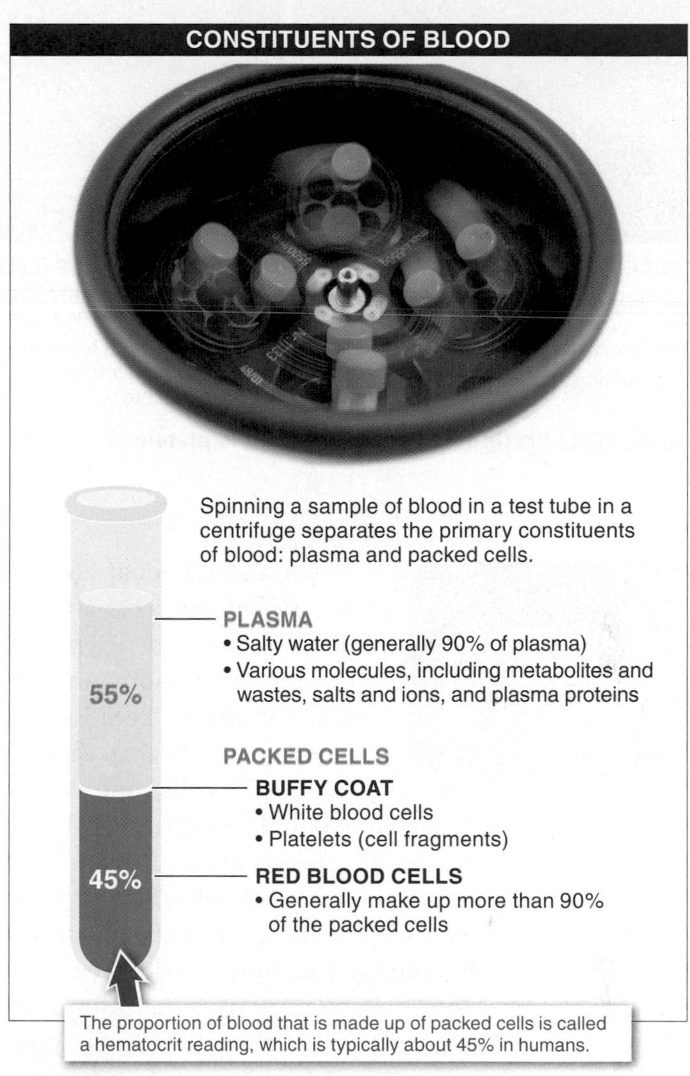

Spinning a sample of blood in a test tube in a centrifuge separates the primary constituents of blood: plasma and packed cells.
55%
45%
PLASMA
• Salty water (generally 90% of plasma)
• Various molecules, including metabolites and wastes, salts and ions, and plasma proteins
PACKED CELLS
BUFFY COAT
• White blood cells
• Platelets (cell fragments)
RED BLOOD CELLS
• Generally make up more than 90% of the packed cells
The proportion of blood that is made up of packed cells is called a hematocrit reading, which is typically about 45% in humans.

FIGURE 22-14 **What makes up blood?**

1. Red blood cells, also called **erythrocytes,** are the workhorses of the circulatory system and the most common blood cells. In a human, about 95% of the circulating blood cells are red blood cells. They are oxygen-transporting specialists. Shaped like flexible disks, they can squeeze through capillaries in single file. They have no nucleus, mitochondria, or protein-making machinery. Instead, each red blood cell contains about 250 million molecules of **hemoglobin,** an oxygen-carrying protein molecule (which we discuss in detail in Section 22.13). Lacking the cellular machinery for repair and upkeep, a red blood lasts only about four months.

CELLS AND PLATELETS IN BLOOD

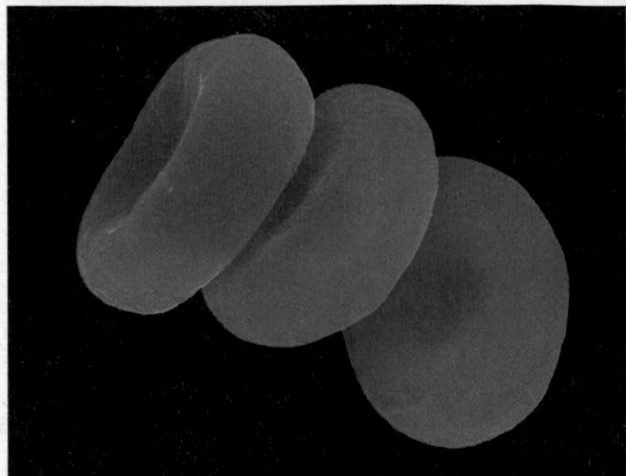

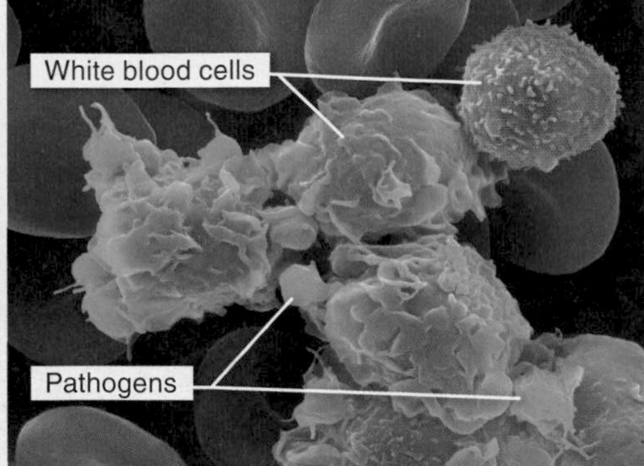

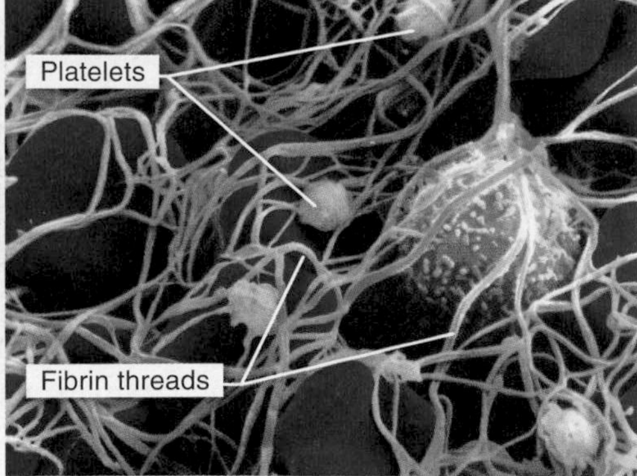

RED BLOOD CELLS (ERYTHROCYTES)
- Transport oxygen from the lungs to the rest of the body
- Flexible disks containing few organelles
- Packed full of hemoglobin

WHITE BLOOD CELLS (LEUKOCYTES)
- Destroy pathogens and foreign organisms in the bloodstream and interstitial fluid
- There are several types of white blood cells that differ in their methods of fighting disease and responding to foreign materials

PLATELETS
- Slow blood loss by initiating the constriction of blood vessels and the formation of a clot
- Composed of small pieces of cytoplasm
- Contain no organelles

FIGURE 22-15 **Erythrocytes, leukocytes, and platelets.**

Q What is anemia? Why are women more susceptible than men?

When an individual has too few red blood cells, anemia develops. Iron deficiency is the most common cause of this condition. Because red blood cells deliver oxygen to the body's cells, one consequence of anemia is a reduction in the oxygen available to cells. This causes people with anemia to feel tired and rundown. A reduced number of red blood cells is also associated with an increased susceptibility to infection, apparently by weakening the immune system. Iron is a critical element that enables red blood cells to carry oxygen. If iron is in short supply, red blood cells can't deliver sufficient oxygen to the tissues where it's needed. Both men and women can become anemic, but anemia affects women much more commonly, because of the blood lost during menstruation. An old folk remedy for anemia—which actually worked—was to sip liquid daily from a jug containing rusty iron nails and water. Even food cooked in an iron skillet has a significantly increased iron content (FIGURE 22-16).

FIGURE 22-16 **Cooking in a cast-iron skillet can significantly increase the iron content of foods!**

2. White blood cells, also called **leukocytes,** are the defenders of the body and the primary components of the body's immune response system. Five different types of white blood cells can be found circulating in the blood (see Chapter 27 for more on these white blood cells and the immune system). Like red blood cells, white blood cells arise from stem cells in bone marrow. In the bloodstream, they patrol for **pathogens,** disease-causing foreign organisms circulating in the blood. White blood cells also diffuse out of capillaries and into the interstitial fluid, where they may encounter and destroy cancerous body cells or cells infected by pathogens. The number of leukocytes circulating in the blood can vary greatly, depending on a person's health. Under normal conditions, there is approximately one leukocyte for every thousand red blood cells, but the number of leukocytes can increase up to threefold during an infection.

3. With more than 50,000 miles of blood vessels in our bodies, it is inevitable that there will be occasional cuts or

FIGURE 22-17 **Fighting blood loss.** Platelets patch damaged blood vessels. Shown here: Elena Reid during a world championship boxing bout.

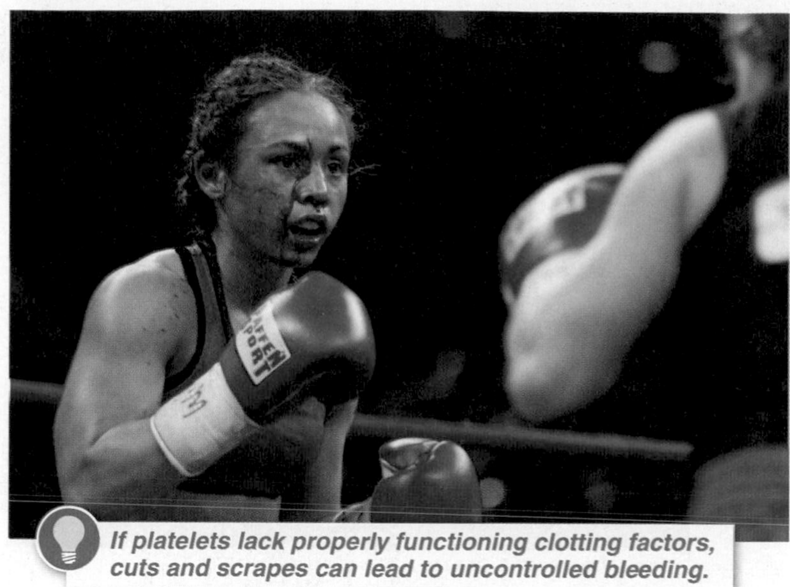

If platelets lack properly functioning clotting factors, cuts and scrapes can lead to uncontrolled bleeding.

punctures (FIGURE 22-17). Fortunately, the blood contains platelets, which can patch punctures. **Platelets** are cellular fragments rather than full-fledged cells. In the bone marrow, large cells called megakaryocytes repeatedly pinch off fragments of cytoplasm that have no nuclei or other organelles. These cell fragments—the platelets—are filled with enzymes and chemicals for patching damaged blood vessels. Hundreds of thousands of platelets circulate at any given time, with each platelet generally lasting about a week. You can envision platelets as tiny jars full of Super Glue. When they bump into the edge of a cut in a blood vessel, they release their contents, initiating constriction of the blood vessel and production of fibrin threads that form a blood clot to reduce blood loss.

Individuals who have platelets that lack properly functioning clotting factors can experience uncontrolled bleeding from even minor cuts or scrapes. These problems can be due to an inherited malfunctioning gene (as in hemophilia), or they can be environmental. If the liver, where many of the clotting enzymes are produced, is damaged by disease (such as from alcoholism), uncontrolled bleeding can occur. Conversely, problems caused by blood clotting *too* readily can also lead to health problems. Thrombosis, for example, is the formation of clots of coagulated blood within a blood vessel. When such clots block circulation in vessels that supply blood to the muscle tissue of the heart, a heart attack occurs.

TAKE HOME MESSAGE 22.8

» Blood is a salty, protein-rich mixture of cells and fluid that transports (1) respiratory gases, (2) vitamins and minerals, (3) nutrients, (4) hormones, (5) components of the immune system, and (6) metabolic wastes. Blood also helps maintain a constant internal environment, including body temperature. Plasma contains two types of cells, with different functions—red blood cells (oxygen transport) and white blood cells (defense from infections)—as well as cellular fragments called platelets (repair).

22.9 Blood pressure is a key measure of heart health.

Feel your pulse. A person's pulse provides a quick and easy determination of the rate and rhythm of the heartbeats. For some arteries, such as those on the underside of your wrist or the side of your neck, you can feel with your fingers the pressure increase as a blood surge stretches the arteries with each contraction of the heart.

Additional information about heart health can be gained by measuring **blood pressure,** the force with which blood flows through a person's arteries. There are two different parts to a blood pressure reading (FIGURE 22-18). The first, called **systolic pressure,** is the pressure when the heart contracts, momentarily stretching arteries as they accommodate the large pulse of blood. The second blood pressure reading is called **diastolic pressure.** This is a measure of the force that blood exerts on the artery walls while the heart is between beats. Because blood isn't being actively pumped at that moment, the diastolic pressure is always lower than the systolic pressure.

Blood pressure can be measured in four easy steps, using a blood pressure cuff.

BLOOD PRESSURE

Boxer Muhammad Ali at a pre-fight medical exam, 1970

Blood pressure readings consist of two measurements:

SYSTOLIC PRESSURE

- The force that blood exerts on the artery walls when the heart contracts and pumps blood into the arteries
- Typical range is between 90 and 140 mm Hg

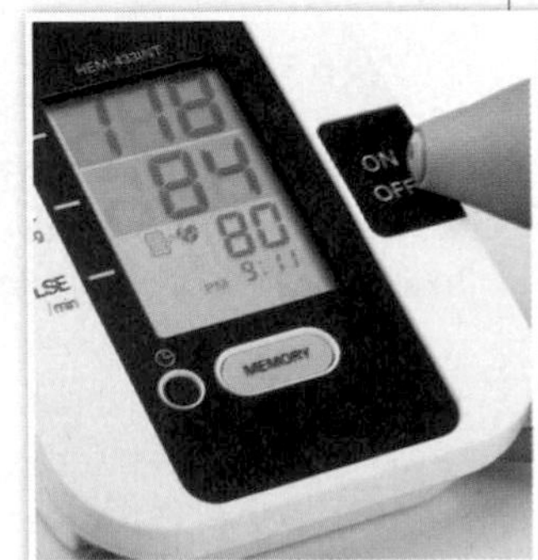

DIASTOLIC PRESSURE

- The force that blood exerts on the artery walls while the heart is between beats
- Typical range is between 60 and 90 mm Hg

BLOOD PRESSURE GUIDELINES
According to the United States National Heart, Lung, and Blood Institute

	Systolic pressure (mm Hg)	Diastolic pressure (mm Hg)
Normal	<120	<80
Prehypertension	120–139	80–89
Stage 1 Hypertension	140–159	90–99
Stage 2 Hypertension	≥160	≥100

With a blood pressure above 120/80, your heart must work harder at all times and your arteries can lose some of their elasticity. This increases your health risks.

FIGURE 22-18 Blood pressure readings can reveal heart health.

1. The cuff is fastened around the upper arm and pumped up, clamping off the arteries in the arm so that no blood gets through.
2. Gradually, pressure on the cuff is released.
3. When the pulsing of blood getting pushed through the arteries under the cuff can first be heard with a stethoscope—as a little squirt—the pressure reading is noted. This is the systolic pressure. Each contraction of the heart is just strong enough to push blood through the barrier of that much pressure.
4. Additional pressure in the cuff is released until the squirting sound disappears. The pressure at that point is the diastolic pressure. Blood is flowing through the arteries with this amount of pressure between heart contractions.

If your blood pressure is "120 over 80," it means that the systolic pressure is 120 and the diastolic pressure is 80. This is written as 120/80, and the units of measure are millimeters of mercury (mm Hg), representing how high a column of mercury could be lifted by such pressure.

Blood pressure above 120/80 is a potential health hazard. In its guidelines concerning high blood pressure (also called **hypertension**), the U.S. National Heart, Lung, and Blood Institute included the following definitions:

Normal blood pressure: systolic = less than 120, *and* diastolic = less than 80 mm Hg

Prehypertension: systolic = 120–139, *or* diastolic = 80–89 mm Hg

Stage 1 hypertension: systolic = 140–159, *or* diastolic = 90–99 mm Hg

Stage 2 hypertension: systolic = 160 or greater, *or* diastolic = 100 or greater mm Hg

Why is high blood pressure cause for concern? Imagine a pair of shorts with an elastic waistband. If you stretch the waistband and hold it in that position for a long time, the waistband loses its elasticity. This is similar to what happens to your arteries if they are stretched by high pressures for long periods of time. With high blood pressure, not only must your heart work harder at all times, but your arteries have a reduced ability to expand and accommodate the increasing pulses of blood during times of exertion. Furthermore, cholesterol more easily sticks to artery walls when they are rigid than when they are elastic. If cholesterol buildup narrows the blood vessel, it further taxes the heart. These problems increase the risk of heart attacks and strokes, which we discuss in Section 22.10. Although both systolic pressure and diastolic pressure are important, the systolic blood pressure is more important in identifying and controlling hypertension.

At the other extreme, low blood pressure, or **hypotension**, is defined as a pressure of 90/60 or lower. It can cause

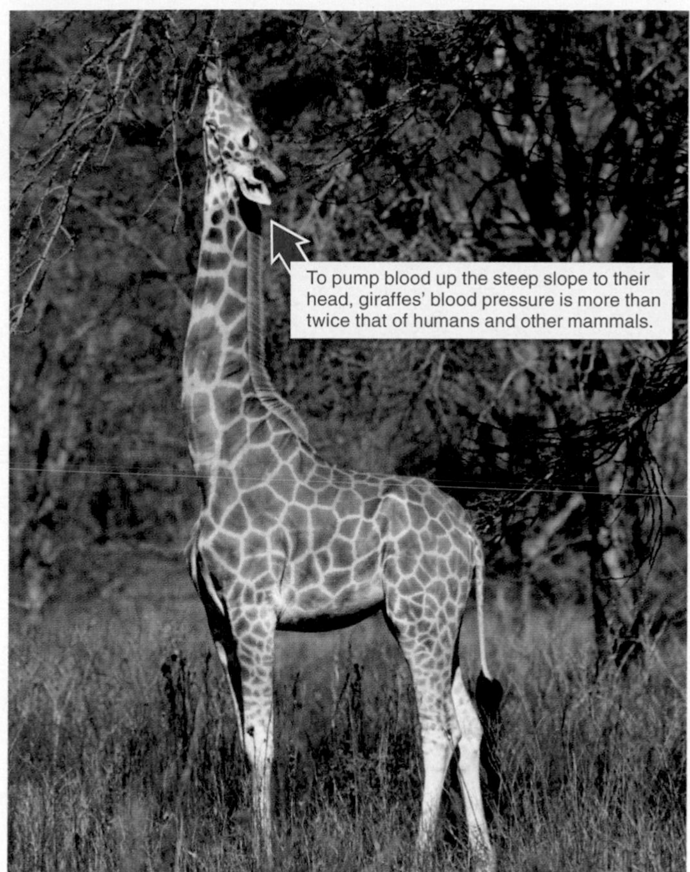

To pump blood up the steep slope to their head, giraffes' blood pressure is more than twice that of humans and other mammals.

FIGURE 22-19 **Contents under pressure.** The large heart of a giraffe pumps blood through its long neck all the way to its head.

symptoms such as dizziness, particularly just after a person stands up, due to inadequate blood flow to the brain. Sometimes caused by medications, low blood pressure can also be associated with weakness or depression. Most often it is not a problem, and it rarely carries long-term risks.

Giraffes' blood pressure is unusual among mammals. With their necks stretching 8 feet (2.5 m) or more above the heart (and with their extremely long legs), giraffes must pump blood up a steep slope and retrieve it from far below (FIGURE 22-19). Although the size of their heart relative to their body size is similar to that of most mammals, giraffes have an average blood pressure double that of humans and other mammals (and at times their blood pressure is more than five times that of humans). Scientists study giraffe circulation in the hope of learning physiological secrets to help humans cope with high blood pressure. Why, for example, with such high blood pressure, doesn't a giraffe's head explode when it is lowered to drink from a pond? (At least part of the answer may be that giraffes' blood vessels are extremely thick.)

TAKE HOME MESSAGE 22.9

» Blood pressure measurement gives clues about an individual's cardiovascular health. A blood pressure reading consists of systolic pressure, the force that blood exerts on the artery wall when the heart contracts and pumps blood into the arteries, and diastolic pressure, the force that blood exerts on the artery wall between heartbeats. With high blood pressure, the heart must work harder, arteries can lose some of their elasticity, and health risks are increased.

22.10 Cardiovascular disease is a leading cause of death in the United States.

Many people take their heart for granted. But diseases of the heart, including heart attacks, result in more deaths in the United States every year than any other single cause. They also are among the most avoidable of all causes of death. In 2014, just over 23% of the deaths in the United States were caused by diseases of the heart. Heart attacks are brought on by an interruption in the flow of blood through one of the **coronary arteries**—the blood vessels that deliver oxygen and nutrients to the heart muscle itself. When heart muscle cells are deprived of oxygen, the heart may beat irregularly or cease to beat, and those cells may die. The long-term implications are serious, because very few heart cells divide and renew each year. Although the heart attack itself is a sudden event, it usually occurs after decades of progressive deterioration of the circulatory system, collectively called **cardiovascular disease** (FIGURE 22-20).

CARDIOVASCULAR DISEASE

Cardiovascular disease includes all of the diseases of the heart and blood vessels, including heart attacks and strokes. It generally begins with the development of fatty deposits on the inner walls of arteries that can disrupt blood flow and interfere with proper circulation.

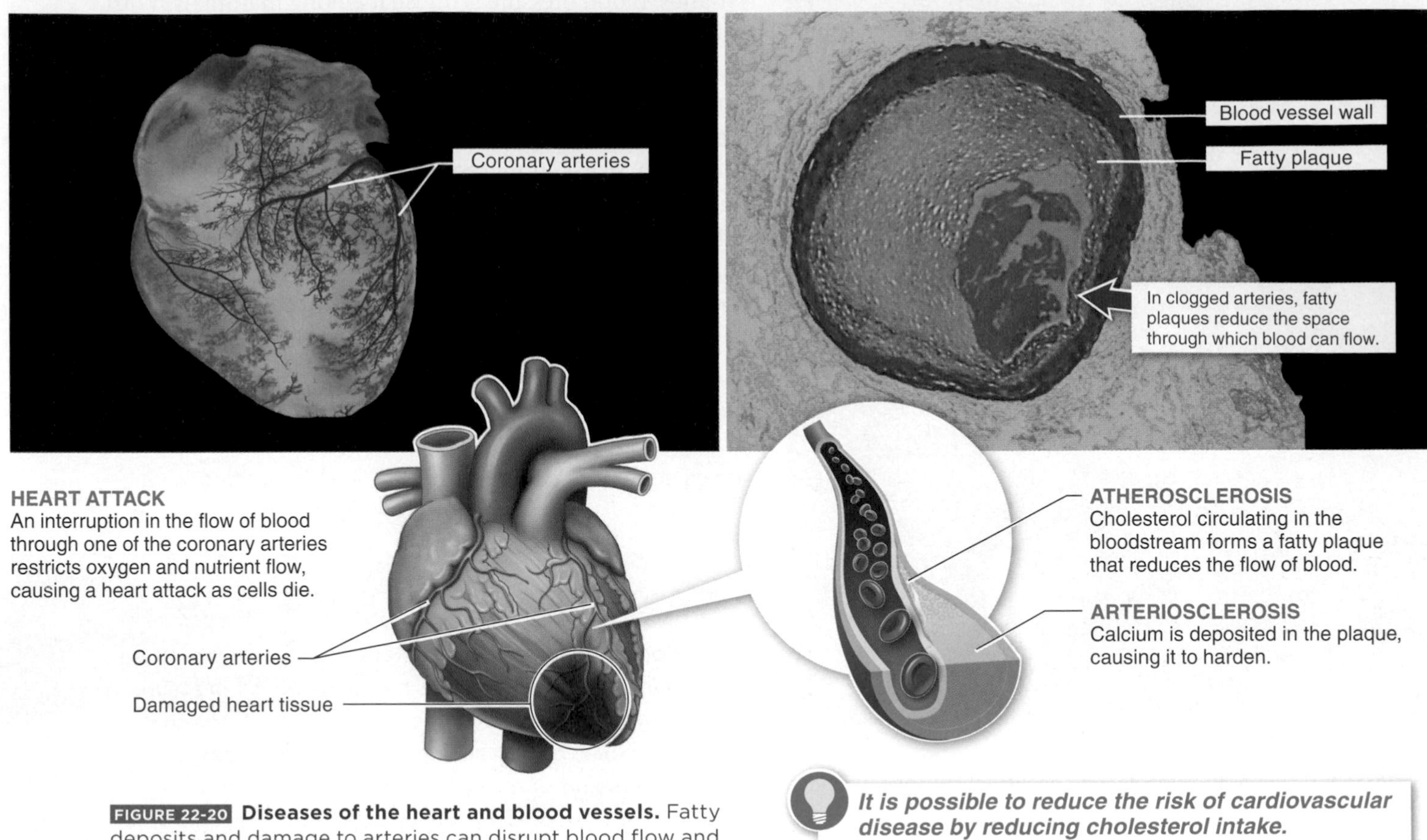

Coronary arteries
Blood vessel wall
Fatty plaque
In clogged arteries, fatty plaques reduce the space through which blood can flow.
HEART ATTACK
An interruption in the flow of blood through one of the coronary arteries restricts oxygen and nutrient flow, causing a heart attack as cells die.
Coronary arteries
Damaged heart tissue
ATHEROSCLEROSIS
Cholesterol circulating in the bloodstream forms a fatty plaque that reduces the flow of blood.
ARTERIOSCLEROSIS
Calcium is deposited in the plaque, causing it to harden.

FIGURE 22-20 **Diseases of the heart and blood vessels.** Fatty deposits and damage to arteries can disrupt blood flow and interfere with proper circulation.

It is possible to reduce the risk of cardiovascular disease by reducing cholesterol intake.

> "Broken heart. A pump after all, pumping thousands of gallons of blood every day. One fine day it gets bunged up and there you are."
>
> — JAMES JOYCE, *Ulysses*, 1922

In addition to heart attacks, strokes are a common outcome of advanced cardiovascular disease. Caused by blocked arteries or blood clots in the brain, strokes starve tissues of oxygen, killing cells in the brain. Although still a major health threat, the death rate from heart disease has been declining steadily for the past 50 years. These reductions are mostly due to improvements in diet and exercise as well as advances in diagnosis and preventive medicine.

Cardiovascular disease generally begins when fatty deposits called plaques build up on the inner walls of arteries. Plaques increase the risk of blood clot formation and, by narrowing the artery, reduce the flow of blood. This narrowing of arteries, called **atherosclerosis,** is often followed by deposition of calcium at the plaques, causing them to harden in a process known as **arteriosclerosis.** The initial formation of plaques is usually a consequence of circulating cholesterol in the bloodstream.

All cholesterol is the same chemically, but you may hear references to "good" cholesterol and "bad" cholesterol. Why is this? As we discussed in Chapter 4, most cholesterol circulating in the bloodstream is packaged as LDL, or low-density lipoproteins. These molecules consist of thousands of molecules of cholesterol surrounded by a phospholipid coat. Because the LDL particles are sticky, they adhere to artery walls and can initiate the buildup of dangerous plaques. Other circulating particles, high-density lipoproteins (HDL), are considered "good" cholesterol. These particles seem to remove cholesterol from arteries and deliver it to liver cells, where it can be broken down. This process can actually reduce the progression of cardiovascular disease. By including in your diet fish and other foods that contain a specific type of fatty acids called omega-3 fatty acids, you can increase your HDL levels.

Both nature and nurture play a role in cardiovascular disease. As we saw in Chapter 4, the tendency to develop cardiovascular disease is inherited. Individuals differ in the number of LDL receptors they produce on their liver cells, and the more receptors an individual has, the better that individual is able to remove "bad" cholesterol from circulation. You cannot alter the genes you inherit for LDL receptor production. But you *can* alter the amount of cholesterol or type of cholesterol (LDL vs. HDL) that is circulating in the first place. Several different behavioral changes, such as increasing aerobic exercise, not smoking, and eating a low-fat diet, can reduce the risk of cardiovascular disease (FIGURE 22-21).

With aerobic training, such as running, the muscle fibers of the heart get bigger and cardiovascular health is improved. When people are sedentary, though, their hearts may have to work harder for reasons such as hypertension or poor diet. The increased load on the heart causes the heart to get bigger, but in a pathological manner that increases the risk of heart failure, rather than in a manner that increases strength and efficiency.

In summarizing its research-based recommendations for reducing risk factors for heart disease, heart attack, and stroke, the American Heart Association suggests a focus on "A, B, and C":

Avoid tobacco

- Stop using tobacco products and minimize your exposure to tobacco smoke. Because cigarette smokers are two to three times more likely to die from heart disease, this change can lead to significant improvements in health and longevity.

Be more active

- Participate in at least 30 minutes of moderate-intensity physical activity (such as brisk walking) on five or more days each week (or vigorous-intensity activity, such as jogging, on three or more days). Additionally, every adult should perform activities that increase or maintain muscle strength, including progressive weight training and/or stair climbing, on a minimum of two days each week.

Choose good nutrition

- Eat a diet that balances energy intake with exercise to prevent weight gain. If you are overweight, take steps to increase physical activity and decrease energy intake to establish a healthy body weight.

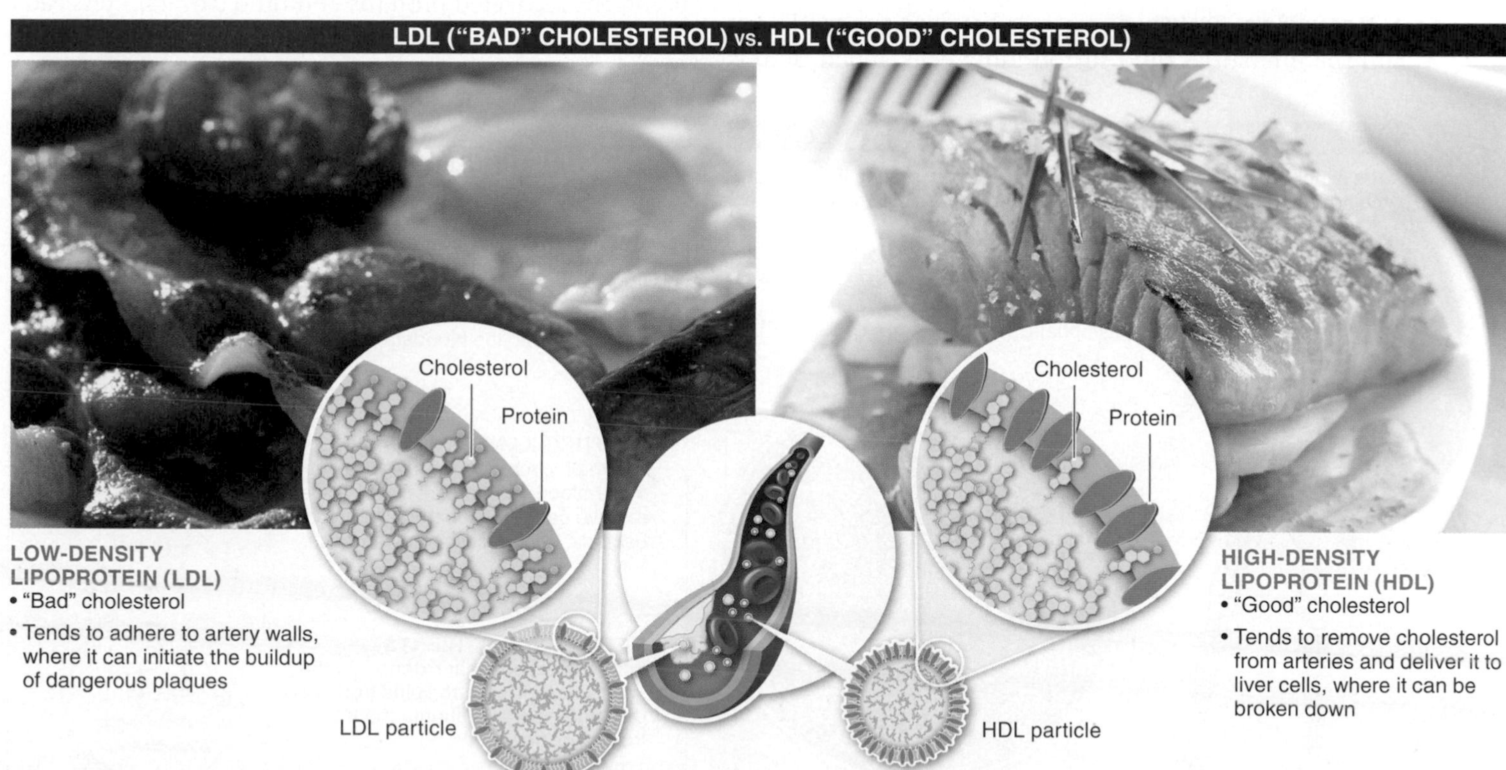

LDL ("BAD" CHOLESTEROL) vs. HDL ("GOOD" CHOLESTEROL)
Cholesterol
Protein
Cholesterol
Protein
LOW-DENSITY LIPOPROTEIN (LDL)
• "Bad" cholesterol
• Tends to adhere to artery walls, where it can initiate the buildup of dangerous plaques
LDL particle
HDL particle
HIGH-DENSITY LIPOPROTEIN (HDL)
• "Good" cholesterol
• Tends to remove cholesterol from arteries and deliver it to liver cells, where it can be broken down

FIGURE 22-21 **Cholesterol can be helpful or harmful.** Foods high in saturated fats can increase your levels of "bad" cholesterol, while foods high in omega-3 fatty acids can increase your levels of "good" cholesterol.

- Choose a diet rich in vegetables, fruits, and whole grains, particularly those that are high in fiber.
- Limit your intake of saturated fat and trans fat by choosing lean meats, vegetables, and low-fat dairy products.
- Limit alcohol to one drink daily for women and two drinks daily for men.

These behavioral changes are much easier said than done, but because they include the most effective strategies for reducing the risk of cardiovascular disease, the payoffs are significant.

TAKE HOME MESSAGE 22.10

» Cardiovascular disease includes all diseases of the heart and blood vessels, including heart attacks and strokes, and is the leading cause of death in the United States. It generally begins with the development of fatty deposits on the inner walls of arteries that increase the risk of blood clots and reduce the flow of blood in coronary arteries. A healthy lifestyle can minimize the risk of cardiovascular disease.

22.11 The lymphatic system plays a supporting role in circulation.

Our bodies have what appears to be another circulatory system (FIGURE 22-22). In addition to the cardiovascular system, and running close to it throughout the body, is the lymphatic system. The lymphatic system is a network of vessels—though not a closed system—that has a supporting role in the process of circulation. It has three important functions (FIGURE 22-23).

1. Recycling. As diffusion occurs between the capillaries and the interstitial fluid surrounding cells, much fluid is lost from the blood. The lymphatic system recovers this fluid. Intertwined around blood vessels, the lymphatic capillaries take in, by diffusion, fluid, proteins, and other substances that have leaked into the interstitial fluid from the blood. Once recovered, this fluid, now called **lymph,** travels through progressively larger lymphatic vessels that eventually join up with veins in the shoulders. At this point, the recovered proteins and fluid (several liters each day) are returned to the blood as it makes its way back to the heart.

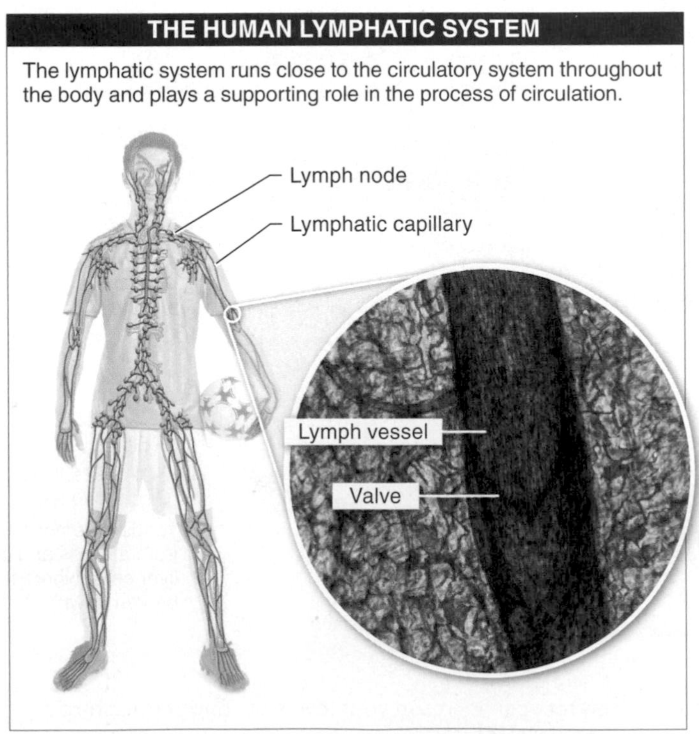

FIGURE 22-22 **The "other" circulatory system.**

FUNCTIONS OF THE LYMPHATIC SYSTEM

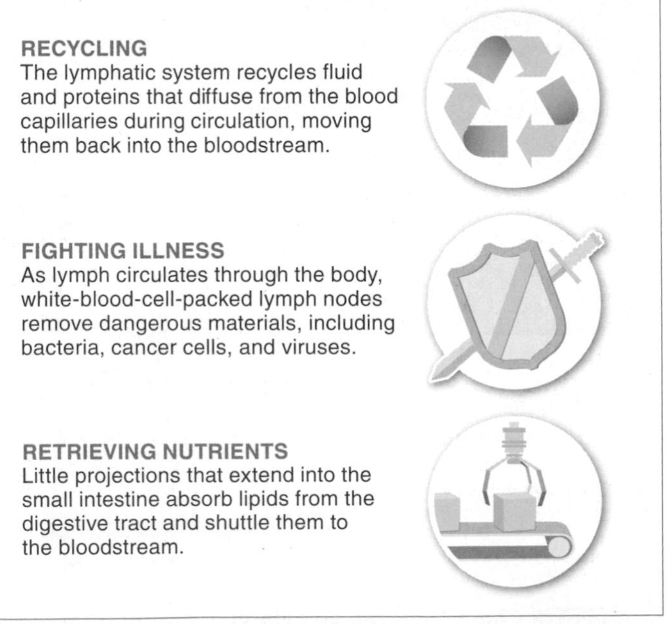

RECYCLING
The lymphatic system recycles fluid and proteins that diffuse from the blood capillaries during circulation, moving them back into the bloodstream.

FIGHTING ILLNESS
As lymph circulates through the body, white-blood-cell-packed lymph nodes remove dangerous materials, including bacteria, cancer cells, and viruses.

RETRIEVING NUTRIENTS
Little projections that extend into the small intestine absorb lipids from the digestive tract and shuttle them to the bloodstream.

FIGURE 22-23 **The lymphatic system supports the circulatory system while fighting illness.**

Q Why is it not necessarily a good idea to have your tonsils taken out, even if they are painfully swollen?

2. Fighting illness. As it moves through the lymphatic system, lymph passes through patches of connective tissue called **lymph nodes.** These compartmentalized sacs are filled with pathogen-fighting white blood cells that remove dangerous materials (including bacteria, cancer cells, and viruses) from the body. This is why your lymph nodes—including your tonsils, the largest lymph nodes of all—become swollen when your body is fighting an infection.

3. Retrieving nutrients. The lymphatic system has numerous little projections that extend into the small intestine and absorb lipids from the food you have eaten and shuttle them from the digestive tract to the bloodstream.

In humans, the lymphatic system accomplishes these tasks without a pump. Lymph is pushed through the system when muscles adjacent to lymph vessels contract and squeeze the fluid onward. Lymph vessels (like veins) have valves that keep the lymph flowing in one direction. When you sit for an extended period of time, lymph can accumulate in vessels. You can help move it along by contracting the muscles in your extremities and progressively contracting muscles closer and closer to your shoulders.

Infection by some parasitic worms can cause the scarring of lymph vessels. When lymph vessels are scarred, fluid recovered by the lymphatic system cannot be returned to the circulatory system, and the extremities swell, often tremendously. This disease is called elephantiasis. Because the parasitic worms causing elephantiasis are transmitted by mosquitoes, the condition is most common in tropical regions. It can be treated with antibiotics that kill the symbiotic bacteria necessary for the parasitic worm to live.

TAKE HOME MESSAGE 22.11

» The lymphatic system runs close to the circulatory system throughout the body and plays a supporting role in the process of circulation, by performing three main functions: recycling fluid that leaks out of the capillaries of the circulatory system, marshaling white blood cells to help fight dangerous cells and pathogens, and absorbing nutrients from the digestive system.

22.12–22.17

The respiratory system enables gas exchange in animals.

An African elephant swimming across a river lifts its trunk to breathe.

22.12 Oxygen and carbon dioxide must get into and out of the circulatory system.

Our circulatory systems are like trucking systems and the highways on which they move. Of the substances they transport, among the most important are the respiratory gases. Body cells must take up oxygen and release carbon dioxide. In the rest of this chapter, we investigate the structures where gas exchange occurs and the transport molecules that make it possible.

In single-celled and very small multicellular organisms, gas exchange can occur by direct diffusion. In larger

GAS EXCHANGE IN ANIMALS

In large, multicellular organisms, gas exchange is a two-stage process.

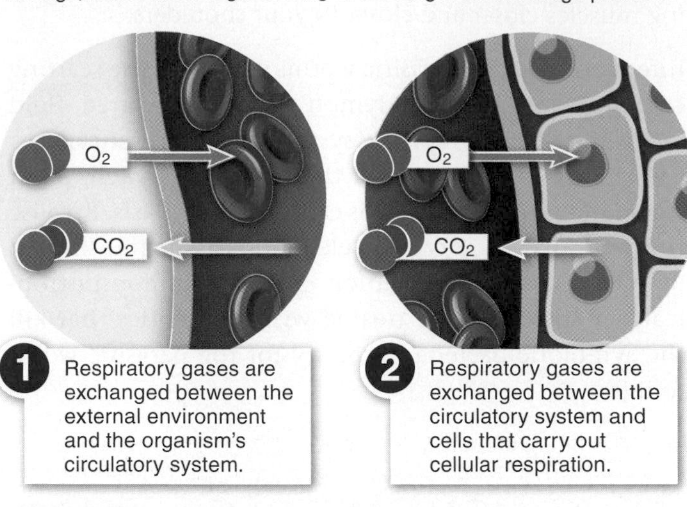

O2
CO2
O2
CO2
1 Respiratory gases are exchanged between the external environment and the organism's circulatory system.
2 Respiratory gases are exchanged between the circulatory system and cells that carry out cellular respiration.

FIGURE 22-24 **Overview of respiratory gas exchange in animals.**

multicellular organisms, however, gas exchange becomes a two-stage process (FIGURE 22-24). First is the exchange between the external environment and the organism's circulatory system. Later comes the exchange between the circulatory system and the cells involved in cellular respiration.

The first of these two stages can occur in several different types of organs specialized for respiration, such as lungs or gills. In all cases, however, it requires a respiratory medium—air or water—that serves as a reservoir for the gases and a moist respiratory surface across which the gas exchange can occur. The many systems that have evolved for gas exchange fall into five categories (FIGURE 22-25).

1. Direct diffusion. Organisms with low metabolic demands, such as marine flatworms, can accomplish respiration by direct diffusion between the cells and the environment.

2. Protruding respiratory sacs. Sea stars and other echinoderms have little balloon-like sacs that protrude from their skin—greatly increasing the surface area—and exchange gases between the body cavity and the environment.

3. Gills. Fishes and many aquatic invertebrates, such as lobsters and clams, have **gills.** These extensions of the body are tremendously elaborate structures in which the large surface area allows extensive exchange of gases between the water and the blood vessels of the circulatory system.

4. Tracheae. Although insect bodies look rather solid, they have a huge number of tiny openings—the spiracles—leading to tubes called tracheae that branch extensively throughout the body. These inner tubes make it

DIVERSITY IN GAS-EXCHANGE SYSTEMS

DIRECT DIFFUSION

- Gas exchange occurs directly between cells and the environment
- Occurs in single-celled organisms and small multicellular organisms with low metabolic demands

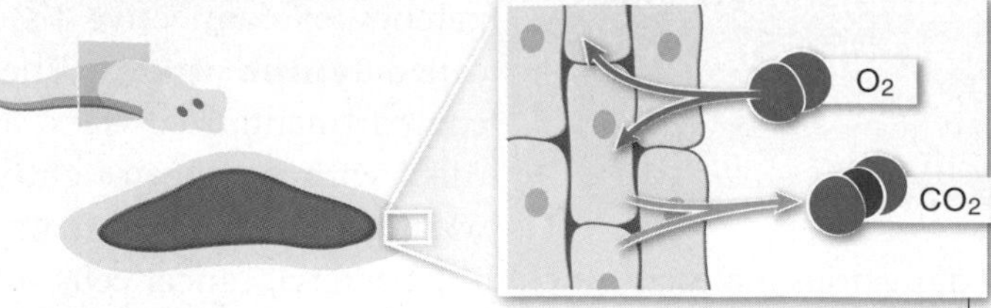

PROTRUDING RESPIRATORY SACS

- Balloon-like sacs that increase surface area for gas exchange
- Occur in sea stars and other echinoderms with low metabolic demands

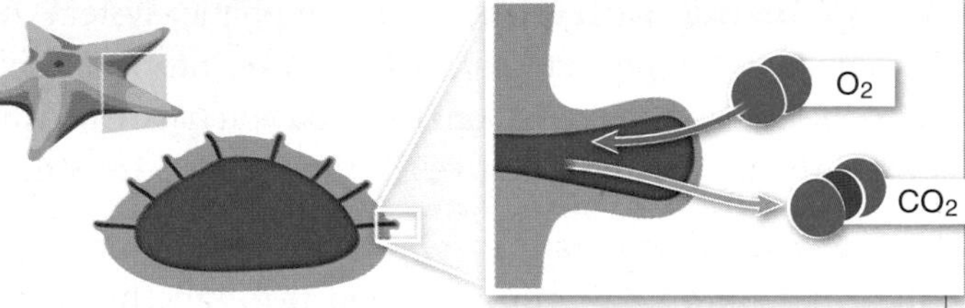

GILLS

- Elaborate extensions of the body that exchange significant amounts of gases dissolved in water
- Occur in fishes and many marine invertebrates such as lobsters and clams

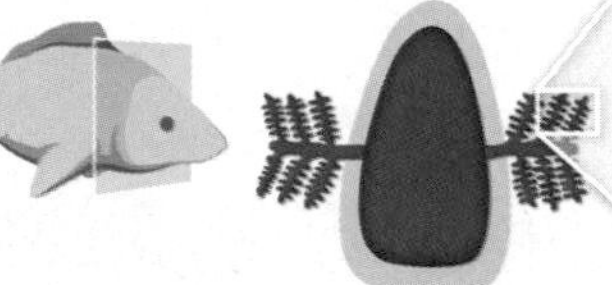

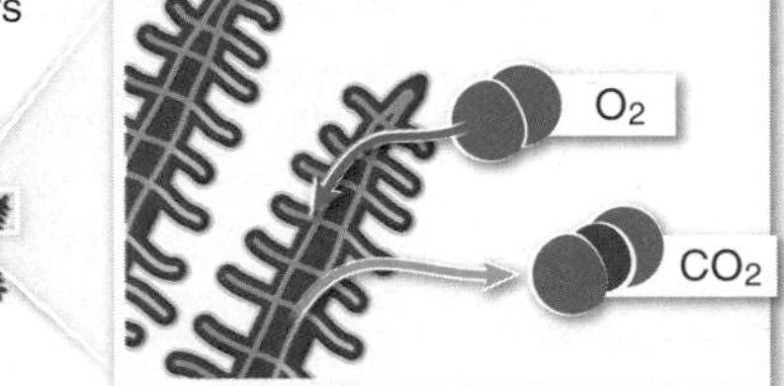

TRACHEAE

- Network of branching tubes connected to tiny openings on the body called spiracles
- Occur in most terrestrial insects

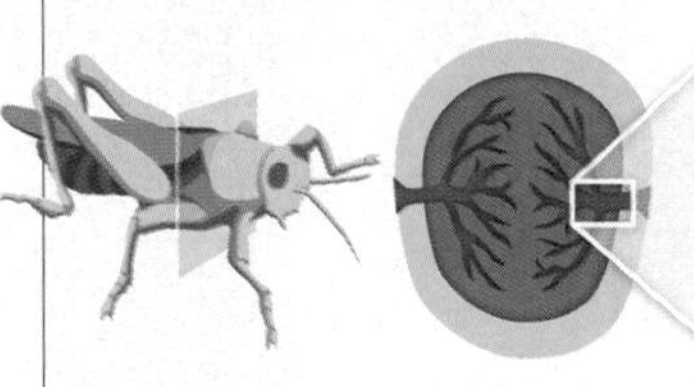

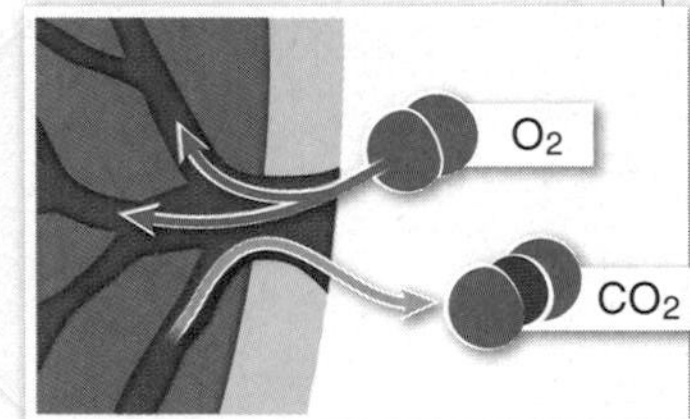

LUNGS

- Internal organs with highly branched, moist surfaces
- Occur in most land vertebrates

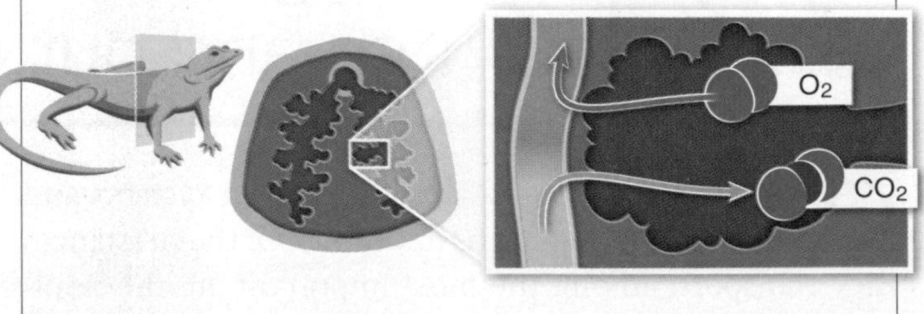

FIGURE 22-25 **Gas-exchange systems and the body structures that support them.**

possible for gases in the air to come in direct contact with most of the organism's cells.

5. Lungs. Most land vertebrates have **lungs.** These are internal organs, characterized by highly branched, moist respiratory surfaces, across which gases in the air breathed in are exchanged with gases dissolved in the blood circulating through the lung tissue. Birds, reptiles, and mammals do virtually all of their respiration through their lungs. Amphibians also exchange gases through their skin. Their skin stays moist because of their largely aquatic lifestyle.

TAKE HOME MESSAGE 22.12

» In single-celled and very small multicellular organisms, gas exchange can occur by direct diffusion. In larger multicellular organisms, gas exchange is a two-stage process: (1) exchange between the external environment and the organism's circulatory system, which usually takes place in lungs, tracheae, or gills, and (2) exchange between the circulatory system and the cells involved in cellular respiration.

22.13 Oxygen is transported while bound to hemoglobin.

In vertebrates, oxygen is transported to body tissues within red blood cells, bound to a molecule called hemoglobin. Hemoglobin is like a "shuttle bus," transporting oxygen around the body. In the lungs, it picks up O_2. Later, when it reaches tissues that are in need of oxygen, the hemoglobin release its O_2 "passengers." The hemoglobin then returns to the lungs, where it can load up on oxygen again.

Hemoglobin is a tiny molecule; each red blood cell holds about 250 million copies of it. Hemoglobin is built right inside the blood cell as it is being formed and remains there for the cell's entire life. Each hemoglobin molecule is a tangled mass of four polypeptide chains. Nestled within the molecule are four cozy compartments, each of which can carry one molecule of oxygen gas on a seat of iron. This iron attaches to the O_2 that diffuses into the red blood cell, temporarily making it part of the hemoglobin molecule (FIGURE 22-26). As we saw earlier, a shortage of iron in your diet can lead to anemia. When iron is limited, less oxygen can be bound by hemoglobin and transported by each red blood cell. Muscles and organs become starved of oxygen, causing fatigue and weakness.

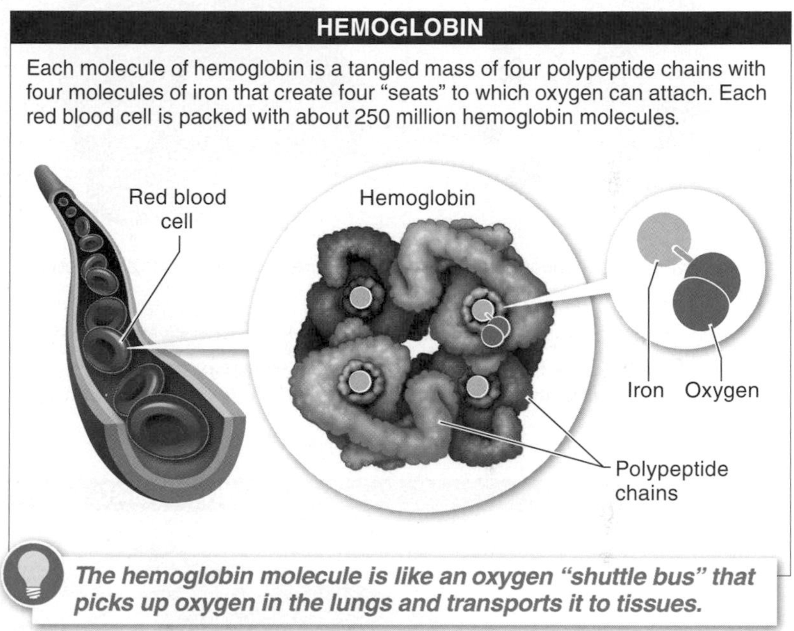

HEMOGLOBIN
Each molecule of hemoglobin is a tangled mass of four polypeptide chains with four molecules of iron that create four "seats" to which oxygen can attach. Each red blood cell is packed with about 250 million hemoglobin molecules.
Red blood cell
Hemoglobin
Iron
Oxygen
Polypeptide chains
The hemoglobin molecule is like an oxygen "shuttle bus" that picks up oxygen in the lungs and transports it to tissues.

FIGURE 22-26 **Hemoglobin: the oxygen transporter.**

Although "blood doping" is banned by most sports organizations, some athletes use it to improve their performance. One method involves withdrawing red blood cells during the weeks and months leading up to a big competition, storing them, and re-injecting them in the few days before the competition. Because red blood cells are filled with hemoglobin, blood doping can increase the athlete's capacity for delivering O_2 to body tissues. In addition to being against the rules in most competitions, blood doping also carries health risks because it increases the viscosity of the blood. In the 1990s, dozens of apparently healthy, elite cyclists died inexplicably from heart failure. It was suspected that their blood had become so thick with red blood cells from blood doping that the burden on the heart to pump their sludge-like blood became too great.

Oxygen binds to hemoglobin, but not too tightly. Like Post-it notes—useful because they are sticky enough to attach to surfaces but not so sticky as to become permanently affixed—hemoglobin functions as if it "knows" when to bind to O_2 and when to release it. This function hinges on something called the partial pressure of oxygen (denoted as Po_2), the force of oxygen particles in the air pressing against

HEMOGLOBIN BINDS AND RELEASES OXYGEN

HIGH PARTIAL PRESSURE OF OXYGEN: Hemoglobin *binds* oxygen

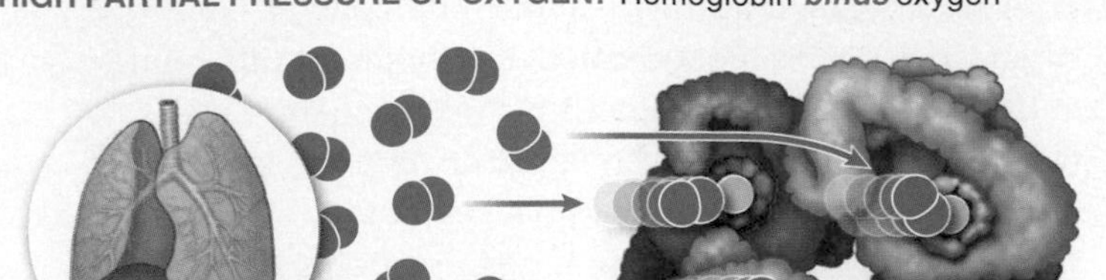

When hemoglobin encounters a high partial pressure of oxygen, such as in inhaled air in the lungs, the hemoglobin gets packed with oxygen.

LOW PARTIAL PRESSURE OF OXYGEN: Hemoglobin *releases* oxygen

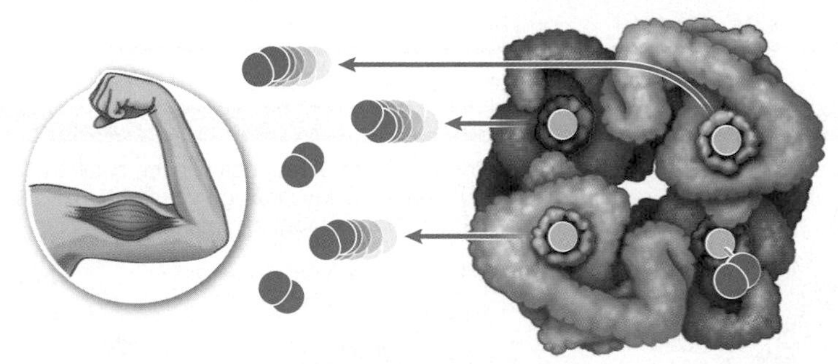

When hemoglobin encounters a low partial pressure of oxygen, such as in active muscle tissue in the body, the hemoglobin releases oxygen.

FIGURE 22-27 **Hemoglobin binds and releases oxygen, depending on the partial pressure of oxygen in the vicinity.**

the body. The partial pressure of oxygen can be thought of as a measure of the amount of oxygen present.

When you breathe in air that has a large amount of oxygen, all four O_2 compartments in hemoglobin bind oxygen molecules. Deep in the tissues of your body, though, oxygen is not in great supply—especially if you are exerting yourself and your muscles have been consuming oxygen. Any hemoglobin in the vicinity encounters a reduced amount (a low partial pressure) of oxygen (low Po_2). Because the oxygen it carries is not held too tightly, some of it is released (FIGURE 22-27). This oxygen is quickly soaked up by the tissue, which can then continue to generate ATP.

Sitting at your desk, your tissues have plenty of oxygen. In these circumstances, hemoglobin typically gives up only one of its four molecules of bound oxygen gas before returning to the lungs to load up again. It cycles back and forth between getting packed with four oxygen molecules in your lungs and being reduced to three oxygens in your tissues.

If your hemoglobin usually oscillates between picking up a single molecule of oxygen in the lungs and dropping off that O_2 in the organs or muscles, why carry the other three oxygen molecules? These oxygen reserves are needed to fuel energetic activity. If you are exercising vigorously, for instance, the amount of oxygen in your tissues can drop so low that hemoglobin gives up another of its oxygen molecules. In extreme cases of exertion, it might give up three or even all four of the oxygens it carries. FIGURE 22-28 shows

OXYGEN BINDING CURVE FOR HEMOGLOBIN

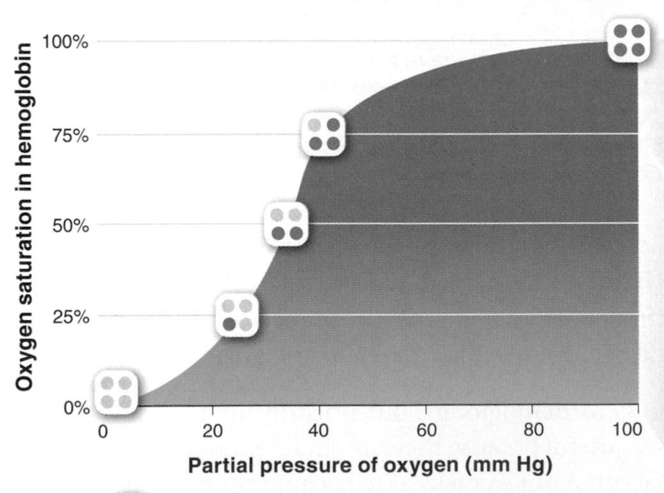

OXYGEN DELIVERED DURING NORMAL METABOLISM

- Oxygen saturation is high.
- Hemoglobin gives up only one oxygen molecule in the body tissues before returning to the lungs.

OXYGEN DELIVERED DURING HIGH PHYSICAL EXERTION

- Oxygen saturation is lower.
- Hemoglobin dips into its reserves, releasing two, three, or even all four of its oxygen molecules before returning to the lungs.

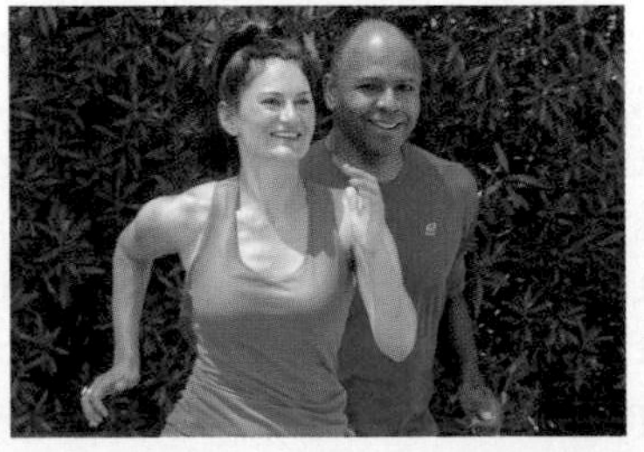

GRAPHIC CONTENT
Thinking critically about visual displays of data
Turn to p. 789 for a closer inspection of this figure.

FIGURE 22-28 **"Sticky, but not too sticky."** A graph showing hemoglobin's affinity for oxygen.

an oxygen binding curve for hemoglobin, illustrating the relationship between the amount of oxygen (measured as Po_2) and the proportion of oxygen molecules that hemoglobin holds on to.

When a woman is pregnant, the fetus does not breathe air, but it needs a great deal of oxygen. How does it get the O_2 it needs? It scavenges oxygen molecules released by the mother's hemoglobin. The fetus and mother have their own, separate circulatory systems, and there is no intermingling of blood. At the placenta, however, the fetal and maternal blood vessels are highly branched and come into extremely close contact. Here, the fetus is able to get access to oxygen. It does this by producing a special type of hemoglobin that is a bit stickier than normal adult hemoglobin. At a Po_2 that is low enough for the mother's hemoglobin to release oxygen, the fetal hemoglobin—with its greater stickiness (or oxygen affinity)—binds the oxygen. It can then deliver that oxygen to its own, fetal tissues (FIGURE 22-29).

A single change in the genetic instructions for building hemoglobin causes a malfunction in hemoglobin molecules, which causes sickle-cell disease (see Chapter 9). When the abnormal hemoglobin molecules lose their bound oxygen molecules—such as when an individual is exercising—the hemoglobin molecules become misshapen and stick together. The entire red blood cell collapses into a sharply pointed sickle. Sickled red blood cells cause numerous problems. Many break open, which can cause anemia if the cells aren't replaced promptly. Others clump together, often blocking capillaries where, normally, they can pass through one at a time. This leads to intense pain and can cause a stroke if it occurs in the brain. About 70,000 people in the United States live with sickle-cell disease. They can minimize the effects by avoiding strenuous activity or other situations in which the amount of oxygen in their muscles and other tissues drops too low.

Q What is carbon monoxide poisoning?

In an unfortunate coincidence, carbon monoxide (CO) also binds to hemoglobin, but with a higher affinity than oxygen. In areas with high carbon monoxide concentrations—such as around a faulty furnace or a kerosene heater without adequate ventilation—the carbon monoxide will outcompete oxygen for hemoglobin's binding sites. Even worse, hemoglobin binds the CO very tightly. When the hemoglobin travels to the body tissues, it has no oxygen to release. Consequently, the tissue is suffocated, even as the person takes deeper and deeper breaths.

FETAL HEMOGLOBIN

Mother and fetus do not share a blood supply. Their separate circulatory systems come close together in the placenta, however, and the fetus produces hemoglobin that is able to bind oxygen released by the mother's hemoglobin.

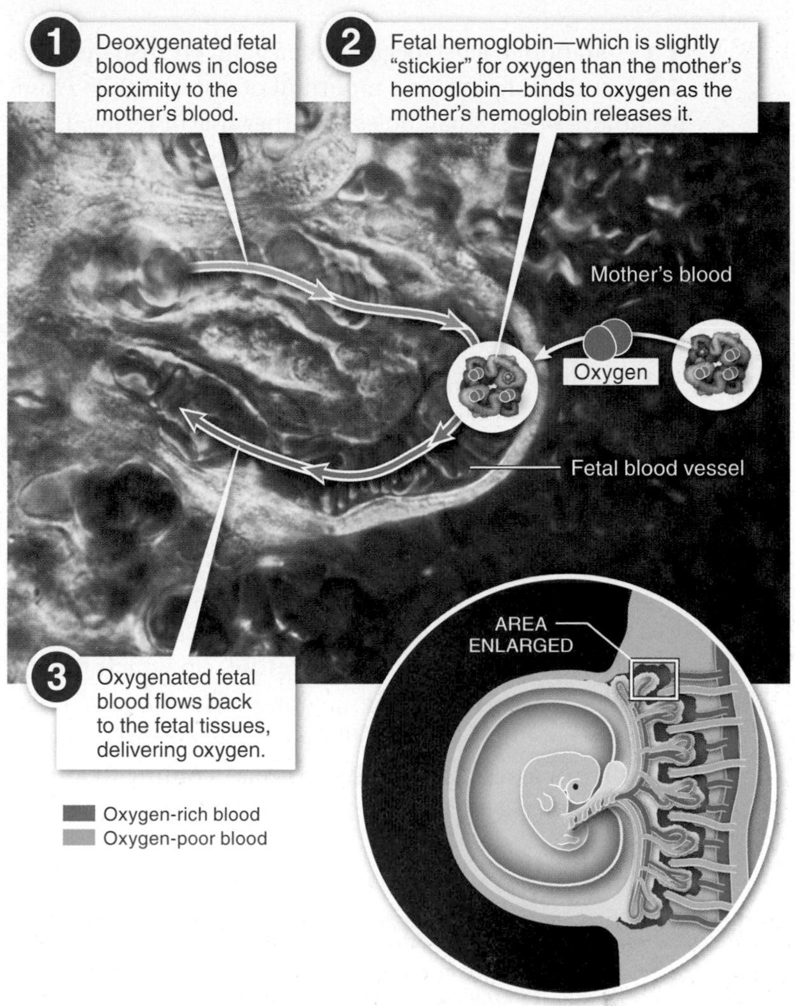

1 Deoxygenated fetal blood flows in close proximity to the mother's blood.
2 Fetal hemoglobin—which is slightly "stickier" for oxygen than the mother's hemoglobin—binds to oxygen as the mother's hemoglobin releases it.
Mother's blood
Oxygen
Fetal blood vessel
AREA ENLARGED
3 Oxygenated fetal blood flows back to the fetal tissues, delivering oxygen.
Oxygen-rich blood
Oxygen-poor blood

FIGURE 22-29 How a fetus gets oxygen.

TAKE HOME MESSAGE 22.13

» Red blood cells are filled with hemoglobin, a molecule that picks up oxygen in the lungs and transports it around the body. Hemoglobin releases its oxygen in organs and tissues, such as muscles, where it is needed for cellular respiration.

22.14 Gas exchange takes place in the gills of aquatic vertebrates.

Respiration in a fish begins with a gulp. The fish opens its mouth and sucks in a mouthful of water. It then closes its mouth, opens small holes on either side of its head, and releases, or "exhales," the water. Unlike the air that humans and other mammals breathe in and out, water follows a one-way path into and out of the fish, never changing direction.

Gas exchange takes place as the water passes through the gills, complex structures adapted to extract as much O_2 as possible from water (FIGURE 22-30). This extraction is a difficult task, because water has only 5% of the oxygen concentration found in air. The gills generally consist of bony or cartilaginous gill arches on either side of the head. Similar in appearance to the teeth of a comb, these arches support the gill. Long filaments of tissue extend like an accordion from each gill arch, spreading out and creating a large surface area. The filaments are stacks of hundreds of disk-like structures, called **lamellae,** on which the gas exchange takes place. Each membranous lamella is a disk of elaborately branched capillaries. As water rushes across the gills, it passes between the lamellae, coming in almost direct contact with the capillaries. Dissolved O_2 can pass from the inhaled water to the blood by direct diffusion, and dissolved CO_2 can pass by direct diffusion from the blood to the exhaled water.

Blood circulation in gills is set up in a simple pattern that is highly efficient at extracting oxygen from the water. In each filament, the blood vessels are arranged so that the blood is moving in the opposite direction from the water flowing past the gills. Called a "countercurrent exchange system," this layout is much more efficient than if the blood flowed in the same direction as the water. If the blood and water flowed in the same direction, the gills would extract a maximum of about 50% of the oxygen in the water; with the countercurrent system, the gills extract as much as 85% of the oxygen in the water.

Fishes are not the only aquatic animals with gills. Many other groups of vertebrates and invertebrates have gills, including mollusks, such as clams, and arthropods, such as lobsters. Gills range from simple to complex, and there is a great deal of variation in the associated structures. Most sharks, for example, move water past their gills, not by gulping in water as do most fishes but just by swimming

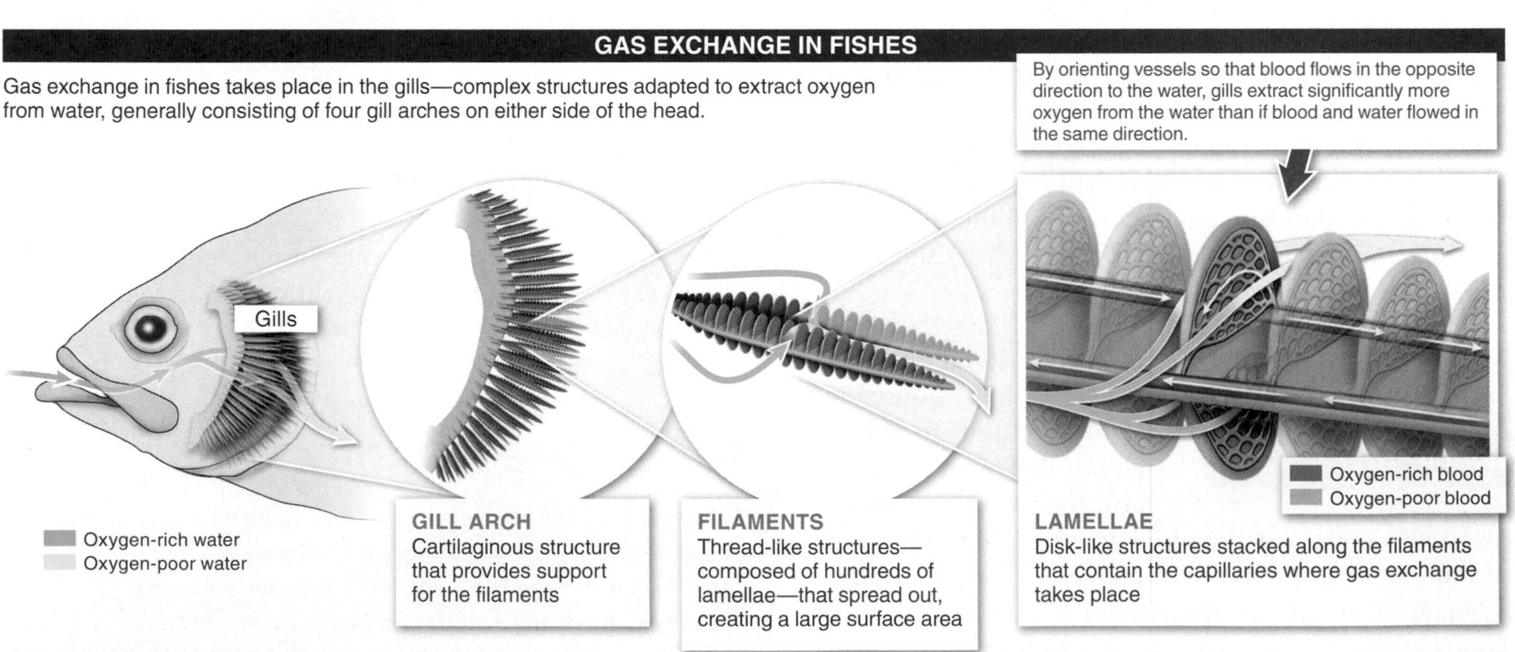

GAS EXCHANGE IN FISHES
Gas exchange in fishes takes place in the gills—complex structures adapted to extract oxygen from water, generally consisting of four gill arches on either side of the head.
By orienting vessels so that blood flows in the opposite direction to the water, gills extract significantly more oxygen from the water than if blood and water flowed in the same direction.
Gills
Oxygen-rich water
Oxygen-poor water
GILL ARCH
Cartilaginous structure that provides support for the filaments
FILAMENTS
Thread-like structures—composed of hundreds of lamellae—that spread out, creating a large surface area
Oxygen-rich blood
Oxygen-poor blood
LAMELLAE
Disk-like structures stacked along the filaments that contain the capillaries where gas exchange takes place

FIGURE 22-30 **Gills.** The remarkable structure of gills makes gas exchange possible in fishes.

forward all the time. If the shark stops moving forward, the oxygen in the water surrounding its gills is quickly depleted and the shark will suffocate. Consequently, most sharks spend their whole lives moving forward.

"Relationships are like sharks; they have to keep moving or they die."

— WOODY ALLEN, *Annie Hall*

TAKE HOME MESSAGE 22.14

» In aquatic vertebrates, respiration begins when an organism opens its mouth, takes in water, and moves the water out through its gills. Gas exchange takes place in the gills, which extract as much as 85% of the oxygen from the water.

22.15 Respiratory systems of terrestrial vertebrates move oxygen-rich air into and carbon-dioxide-rich air out of the lungs.

Life on land is very different from life underwater, but the fundamental energetic needs are the same. Cells need oxygen to produce ATP to fuel the reactions necessary for life. Consequently, an organism must enable oxygen to get into the cells and carbon dioxide to get out. These are universal challenges facing all terrestrial animals.

Terrestrial vertebrates have a general solution to these challenges. First, they suck in air through their mouth or nose. The air moves down a trachea into lungs. In the lungs, O_2 diffuses from air to blood, while CO_2 diffuses from blood to air. Finally, the oxygen-depleted air is exhaled and the process begins again.

Mammalian respiration begins with a deep breath (FIGURE 22-31). Air enters through the nose, filling the nasal cavity, where it becomes warm and moist. Additional air can be taken in through the mouth. These two entry points for air join at the throat (also called the pharynx), at the back of the mouth. The air then moves into the trachea (also called the windpipe), a long tube that takes the air into the chest cavity. Once there, the trachea branches

THE HUMAN RESPIRATORY SYSTEM

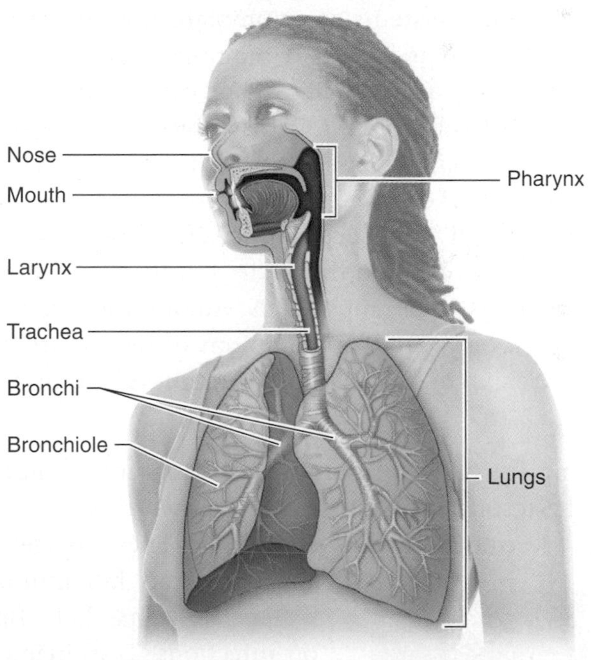

FUNCTIONS OF THE RESPIRATORY SYSTEM
- Acquires the oxygen necessary for cellular respiration
- Removes carbon dioxide, a waste product of cell functions

In terrestrial animals, the respiratory system provides the route for inhaled air to meet the blood vessels of the body.

FIGURE 22-31 **A terrestrial mammal.** Overview of the human respiratory system.

into two smaller tubes called bronchi. One bronchus goes to the left lung and the other goes to the right lung.

When the bronchi enter the lungs, which are like stretchy, elastic bags, they branch again. And again. And again. With each successive branching, the bronchi get smaller. Under a certain size they are called bronchioles. Eventually, the bronchioles reach dead ends—tiny elastic sacs, the alveoli, where the air meets the blood vessels (FIGURE 22-32).

There are about 300 million alveoli in each human lung, with a total surface area roughly the size of a movie screen. Alveoli are made up of the most delicate cells in our bodies and have ultra-thin walls. Completely surrounding the alveoli, the way your fingers might surround a small ball, are tiny capillaries; like the alveoli, the capillaries have extremely thin walls. Oxygen in the air in the alveoli dissolves in moisture on the cells lining them. Oxygen then passes through the two sets of thin membranes—alveolar and capillary—and is picked up by the bloodstream. Simultaneously, carbon dioxide can diffuse from the blood into the alveoli. Instantly, the breath you inhaled is changed. When exhaled, it is depleted of O_2 and laden with CO_2.

The respiratory systems of amphibians and reptiles are almost identical to those of mammals. The lungs of amphibians are a bit smaller in relation to body size, but amphibians make up for this reduced lung capacity by conducting gas exchange through their skin. This is why they must keep their skin moist at all times. Reptiles are generally too thick-skinned and scaly to achieve any respiration through their skin, but they have slightly larger lungs than amphibians. Among the terrestrial vertebrates, birds are the champions of respiratory efficiency. We explore some of their unique adaptations in Section 22.16.

Smoking introduces thousands of different chemicals into the respiratory system, many of which have powerfully destructive effects on its cells. These dangerous chemicals can kill immune system cells that help fight off infections. The chemicals in smoke also trigger mucous secretions that can block airways and lead to other respiratory difficulties. After

Q How does smoking damage the lungs? Can the damage be reversed?

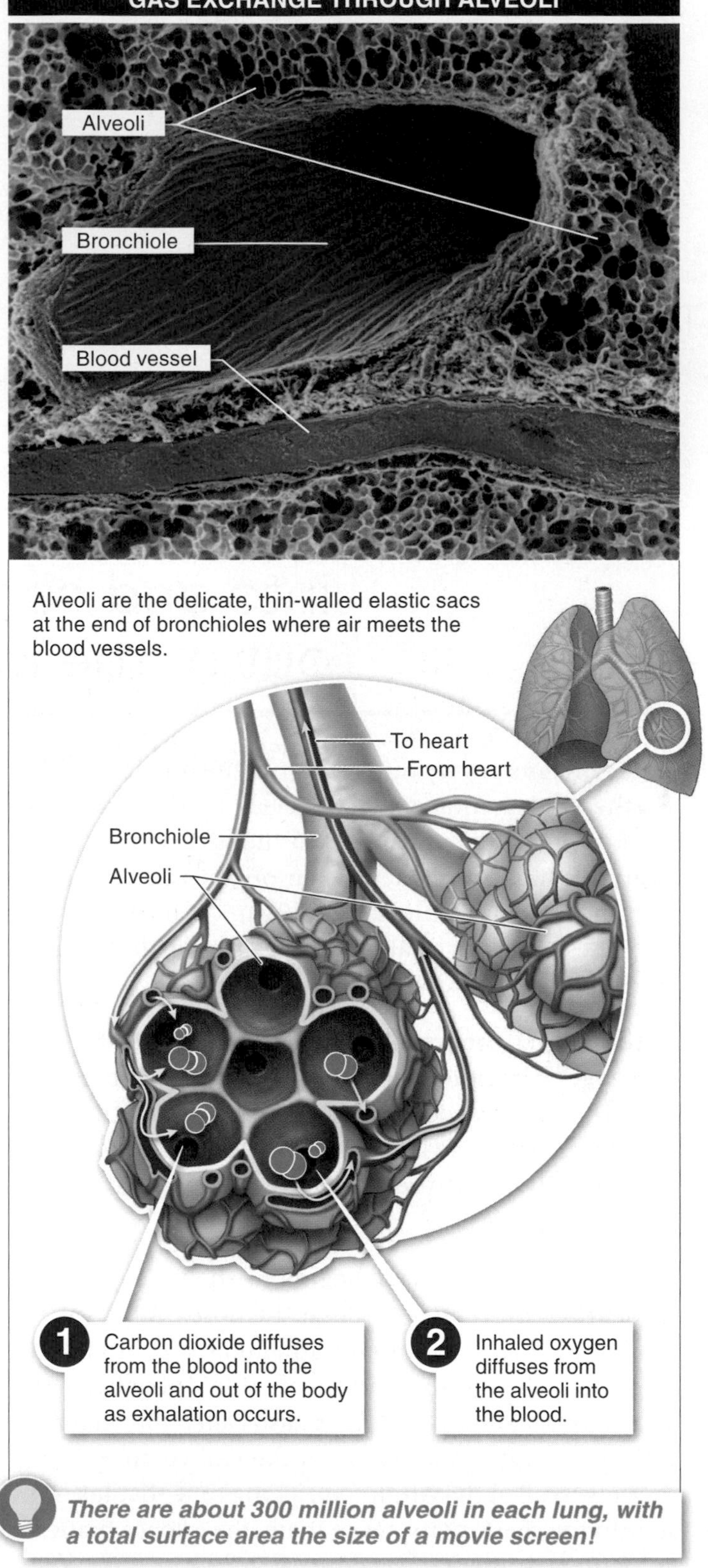
GAS EXCHANGE THROUGH ALVEOLI
Alveoli
Bronchiole
Blood vessel
Alveoli are the delicate, thin-walled elastic sacs at the end of bronchioles where air meets the blood vessels.
To heart
From heart
Bronchiole
Alveoli
1 Carbon dioxide diffuses from the blood into the alveoli and out of the body as exhalation occurs.
2 Inhaled oxygen diffuses from the alveoli into the blood.
There are about 300 million alveoli in each lung, with a total surface area the size of a movie screen!

FIGURE 22-32 **Alveoli in the lungs: where air meets blood.**

chronic exposure to smoke, the walls of the alveoli become brittle, reducing respiratory capacity. Toxic particles in tobacco smoke can also damage the tiny, hair-like cilia on the cells lining the trachea, reducing their ability to filter out dirt and microorganisms from the air we breathe. Finally, and perhaps most significantly, carcinogenic chemicals in tobacco smoke can trigger unrestrained cell multiplication in lung tissues, causing cancer.

Although smoking is destructive in numerous ways—it causes almost half a million deaths in the United States every year—stopping smoking at any point can begin the process of reversing some of the damage. By the end of the first year of non-smoking, the risk of death from lung cancer and heart disease begins to decrease, and after 15 years of non-smoking, the risk of death from these causes returns to the same levels as for individuals who have never smoked.

TAKE HOME MESSAGE 22.15

» In terrestrial vertebrates, respiration begins as air is sucked in through the mouth or nose and down the trachea into the lungs. There, O_2 diffuses from the air into capillaries and the bloodstream, while CO_2 diffuses from blood to air. Finally, the oxygen-depleted air is exhaled and the process begins again.

22.16 Birds have unusually efficient respiratory systems.

Birds take breathing to new heights. Because they often spend time in high-altitude, low-oxygen habitats and fly for long periods of time, they need *a lot* of oxygen for ATP production (FIGURE 22-33). Several key evolutionary adaptations make it possible for birds to exchange gases much more efficiently than other terrestrial vertebrates.

For starters, birds' respiratory adaptations enable them to keep oxygen-rich air flowing through the lungs twice as long as in mammals. In mammals, air is inhaled and reaches a dead end at the alveoli. On the exhale, it changes direction and is breathed out. During the exhale, no new oxygen reaches the lungs, a problematic situation during times of exertion. In birds, air never runs into a dead end and the lungs never experience "stale" air. Here's how they do it (FIGURE 22-34).

When a bird inhales, some of the air passes through the lungs, where oxygen can diffuse into the blood. The rest of the air fills temporary holding structures called the posterior air sacs. Then, when the bird exhales, the posterior

Whooper swans, which have been recorded migrating at altitudes up to 8,200 m, where the ambient temperature is – 40° C.

Respiratory adaptations have evolved that enable birds to exchange gases with great efficiency—even at high altitudes where oxygen availability is low.

FIGURE 22-33 High-altitude respiration.

CIRCULAR RESPIRATORY SYSTEMS IN BIRDS

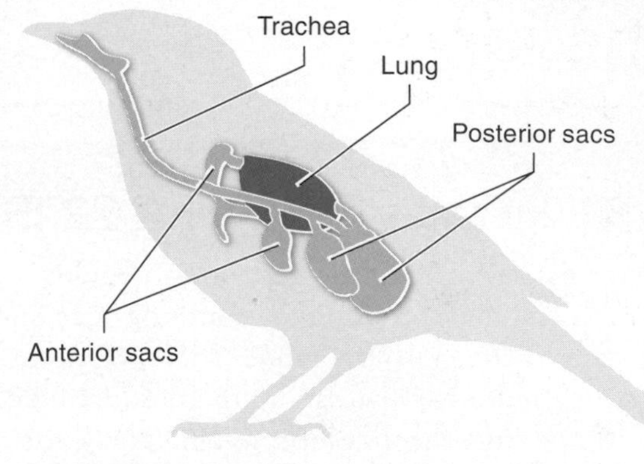

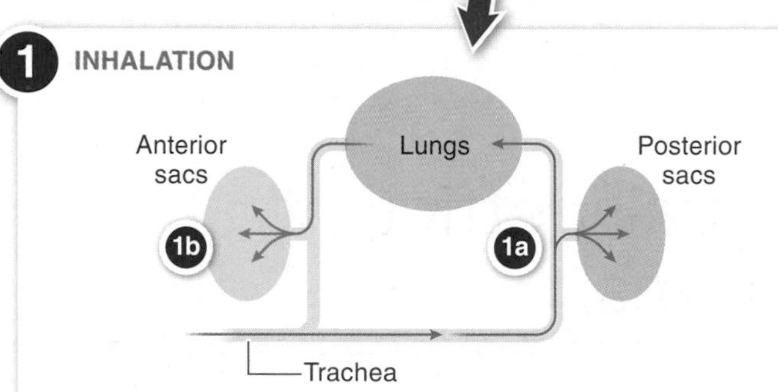

a Fresh, oxygen-rich air moves down the trachea, inflating the posterior "waiting room" sacs, as well as the lungs.

b At the same time, oxygen-poor air is expelled from the lungs, inflating the anterior sacs.

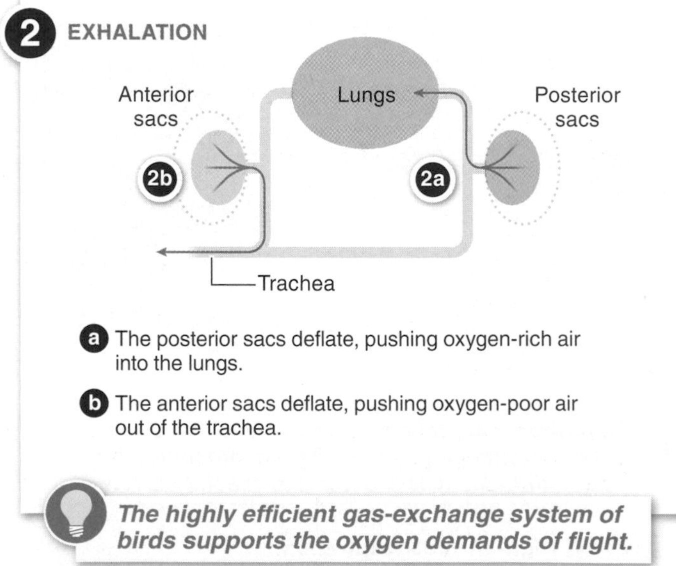

a The posterior sacs deflate, pushing oxygen-rich air into the lungs.

b The anterior sacs deflate, pushing oxygen-poor air out of the trachea.

The highly efficient gas-exchange system of birds supports the oxygen demands of flight.

FIGURE 22-34 **Meeting extreme needs.**

air sacs contract and push more oxygen-rich air through the lungs. With this system, even during exhales, when no "new" air is being breathed in, oxygen-rich air is passing through the lungs.

When a bird is inhaling, air moving through the lungs passes into other temporary holding structures, called the anterior air sacs. Then, as the bird exhales, oxygen-poor air passing from the lungs *and* oxygen-poor air from the anterior air sacs together move into the trachea, and this mix is expelled from the bird's body.

Q How is bird breathing less like mammalian breathing and more like fish "breathing"?

With their circular respiratory system, made possible by the air sacs and by the lungs' unique pass-through channels, birds maintain a continuous and unidirectional flow of oxygen-rich air through the lungs. These adaptations make it possible for geese to fly over Mount Everest, nearly six miles above sea level. A mammal deposited at that height would pass out almost instantly from insufficient oxygen.

TAKE HOME MESSAGE 22.16

›› Birds often spend time in high-altitude, low-oxygen habitats and may fly for long periods of time, both of which require a great deal of oxygen. These needs are met by a circular system of air flow, which makes it possible for birds to exchange gases more efficiently than other terrestrial vertebrates.

22.17 Muscles control the flow of air into and out of the lungs.

Breathe in. Breathe out. You don't even have to think about it. But over the course of your life, this simple process will happen 500 million times without fail. How does it work?

THE MECHANICS OF BREATHING

Breathing is made possible by the following structures in the chest cavity:

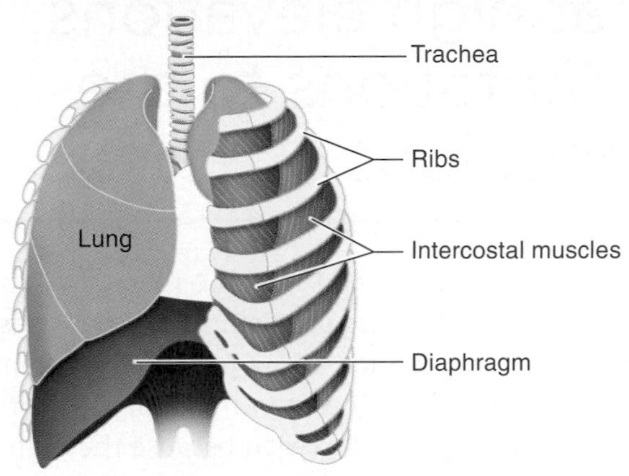

In reptiles, birds, and mammals, breathing occurs in two steps. (Note the difference in the size of the lungs, outlined by the dashed lines, during inhalation and exhalation.)

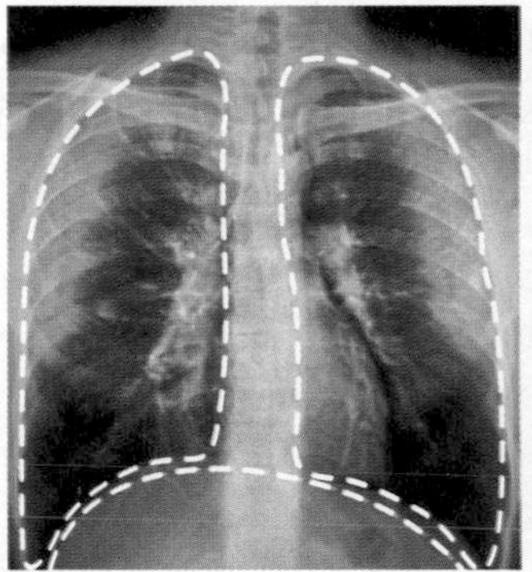

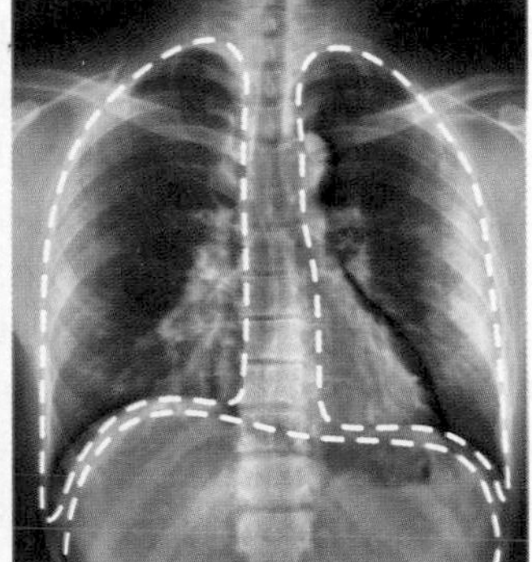

1 INHALATION
- Diaphragm and intercostal muscles contract
- Diaphragm is pulled lower and rib cage expands
- Air is sucked into the lungs

2 EXHALATION
- Diaphragm and intercostal muscles relax
- Chest cavity returns to its original size
- Air is forced back out to the trachea

FIGURE 22-35 Body structures that make breathing possible.

In reptiles, birds, and mammals, breathing occurs in two steps: inhalation and exhalation (FIGURE 22-35). At the bottom of the chest cavity is a large sheet of muscle called the **diaphragm,** which separates the chest cavity from the abdominal cavity. The rest of the chest cavity is surrounded by the rib cage and the intercostal muscles between the ribs. During inhalation, the diaphragm and external intercostal muscles contract; the diaphragm is pulled down and the intercostal muscles lift the ribs and expand the rib cage. This causes a rapid increase in the volume of the chest cavity and lungs, which causes air to be sucked into the lungs. When the diaphragm and intercostal muscles relax, the chest cavity returns to its original size. This reduction in volume compresses the lungs and forces air back out.

"I'm going to hold my breath until I die." As a child, you may have threatened your parents this way. Although we can consciously control our breathing to some extent, chemical sensors in the body detect when carbon dioxide levels rise dangerously high, and our brain responds by sending signals to the muscles that control our breathing, spurring them to override our efforts to stop breathing.

TAKE HOME MESSAGE 22.17

» In reptiles, birds, and mammals, breathing occurs in two steps: inhalation and exhalation. During inhalation, muscles contract, pulling the diaphragm down, expanding the rib cage, and increasing the volume of the chest cavity and lungs, which causes air to be sucked into the lungs. When the muscles relax, the chest cavity returns to its original size and air is forced out of the lungs.

22.18–22.19

Evolutionary adaptations maximize oxygen delivery.

A yak at the base camp of Mt. Everest.

22.18 Animals living at high elevations have special adaptations to the low-oxygen conditions.

Breathing is more difficult in some places than others. One of the most important constraints on the rate at which oxygen diffuses into the blood is air pressure. At high elevations much less oxygen is available. At 15,000 feet above sea level, for example, the air pressure is half of what it is at sea level. This is because there is less atmosphere above the air, pushing down on it. With lower pressure from above, less oxygen is squeezed into a given volume. The lower air pressure makes it hard to push air into lungs and to drive the oxygen across the alveolar and capillary membranes and into the blood. We struggle at high elevations because our hemoglobin isn't designed to pick up oxygen at such a low pressure; the hemoglobin isn't sticky enough. Thus, mountain climbers often carry canisters of pressurized oxygen because the high-pressure oxygen can more easily diffuse into the bloodstream.

Animals living at high elevations are adapted to survive with less available oxygen (FIGURE 22-36). For example, living at elevations of about 3.1 miles (5,000 m) above sea level, llamas produce a slightly different form of hemoglobin. Under the low-oxygen conditions of high elevations, llama hemoglobin has a higher affinity for oxygen than human hemoglobin. It's "stickier." This stickiness enables the hemoglobin to become saturated with four molecules of oxygen even when the llama breathes in relatively "oxygen-poor" air. Because oxygen is being consumed by cellular respiration, the oxygen bound to the hemoglobin is still released to muscles to allow normal activity.

BREATHING AT HIGH ALTITUDES

Llamas can thrive in high-elevation, low-oxygen environments, because they produce a "stickier" form of hemoglobin that has a higher affinity for oxygen.

FIGURE 22-36 Gas exchange is more difficult in some environments than others.

TAKE HOME MESSAGE 22.18

» At high elevations, the amount of oxygen present is lower: breathing and activity are difficult. Animals living at high elevations solve this problem by producing a form of hemoglobin that has a higher affinity for oxygen.

22.19 Humans become acclimated to low-oxygen conditions.

Although llamas are adapted for high elevations from the day they are born, most humans are not—but we acclimate well, and this has a significant impact on our strength, speed, and stamina.

Training for three to five weeks at high elevations, usually 6,000–7,000 feet above sea level, triggers several physiological changes—a process called *acclimation*. And, as with llamas, some of the human acclimation to low-oxygen conditions comes from modifications (though they are acquired rather than inherited) to hemoglobin. Other changes to the human body include stimulating the production of additional red blood cells, increasing blood and capillary volume, and increasing the number of mitochondria.

High-elevation training causes an increase in a chemical called diphosphoglyceric acid (DPG) in red blood cells. This acid combines with hemoglobin in the red blood cells and alters the protein's shape ever so slightly. In doing so, it reduces hemoglobin's stickiness, giving it a lower affinity for oxygen. With this reduced oxygen affinity, the DPG-modified hemoglobin releases—at any elevation—more of the O_2 that it carries (FIGURE 22-37).

Q How can high-elevation training help athletes?

Athletes who train at high elevations can improve their performance by about 3%. This difference was manifested dramatically during the 1968 Olympics in Mexico City (elevation 7,500 feet) when the top five finishers in the 10,000-meter race were all year-round high-elevation residents. The benefits of high-elevation training remain after an athlete returns to sea level—increasing performance significantly. But just as humans become acclimated to high elevations, we also become acclimated to low elevations, and DPG (and the performance enhancement it can bring) returns to low-elevation levels about three to five weeks after returning to sea level.

HIGH-ELEVATION TRAINING

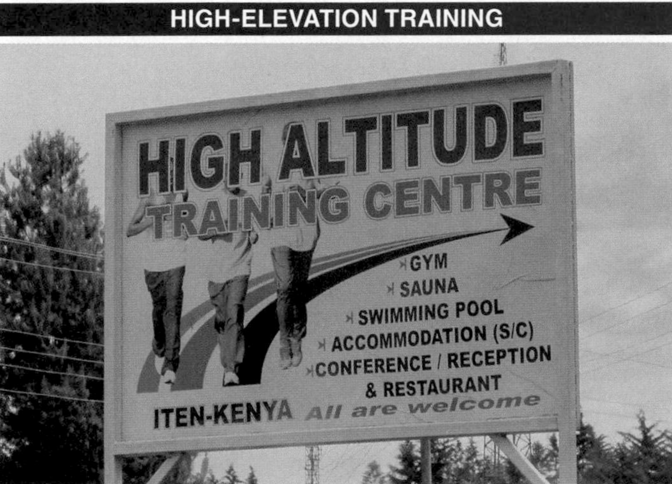

BENEFITS OF HIGH-ELEVATION TRAINING

- Additional red blood cell production
- Increased blood and capillary volume
- Increased number of mitochondria
- Increased diphosphoglyceric acid (DPG) in red blood cells

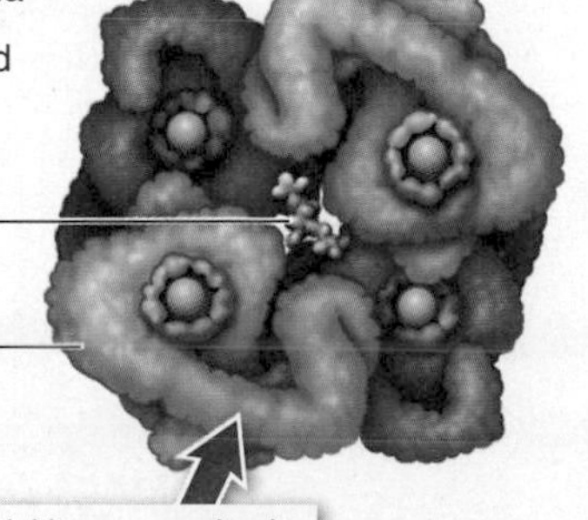

FIGURE 22-37 **High-elevation training.**

TAKE HOME MESSAGE 22.19

» **Humans living at high elevations become acclimated to low-oxygen conditions over the course of three to five weeks. This acclimation includes increased production of diphosphoglyceric acid (DPG) in red blood cells, thereby reducing hemoglobin's affinity for oxygen and leading to release of higher levels of oxygen to muscles during exertion.**

STREET BIO

Using evidence to guide decision making in our own lives

"Contents under pressure": recognizing and avoiding the bends

Q: What do scuba divers and champagne bottles have in common? Under high pressure, large amounts of a gas can be dissolved in fluid. This is why huge amounts of carbon dioxide can be present in a bottle of champagne. It is also what happens to divers who, with tremendous water pressure on their bodies, must breathe highly pressurized air in order for oxygen to dissolve in their bloodstream.

Q: Why are bubbles bad news for divers? When a champagne bottle is uncorked, the pressure is suddenly released. Under low pressure, not nearly as much carbon dioxide can be held in the fluid, so it quickly bubbles out. Similarly, when a scuba diver returns too quickly to the surface, the pressure on the body is rapidly reduced, as is the capacity of the blood to hold gases. And, like champagne after the cork is popped, many gases in the diver's blood—particularly nitrogen, which is the biggest component of air—quickly bubble out and can almost cause the diver to explode from within, in a condition known as "the bends." Something similar to the bends can occur if someone flies in an unpressurized plane that suddenly climbs very high to where the air pressure is lower. This was sometimes reported by pilots in World War II.

Q: What can you conclude about avoiding the bends? Just as the solution to opening a bottle of champagne without it bubbling over is to release the pressure very slowly by removing the cork slowly, a diver must reduce the pressure on his or her blood very slowly by rising gradually from the depths of the water. Alternatively, a diver can be placed in a chamber (called a hyperbaric chamber) to be gradually depressurized.

GRAPHIC CONTENT

Thinking critically about visual displays of data

1 What is the purpose of this graph?

2 Why is the lower portion of the graph shaded blue and the top portion shaded red?

3 Are data presented in this graph? If so, what are they? If not, what is presented?

4 What is the significance of the curve's S shape? How would an organism's physiology differ if it were a straight line (at a 45° angle) instead?

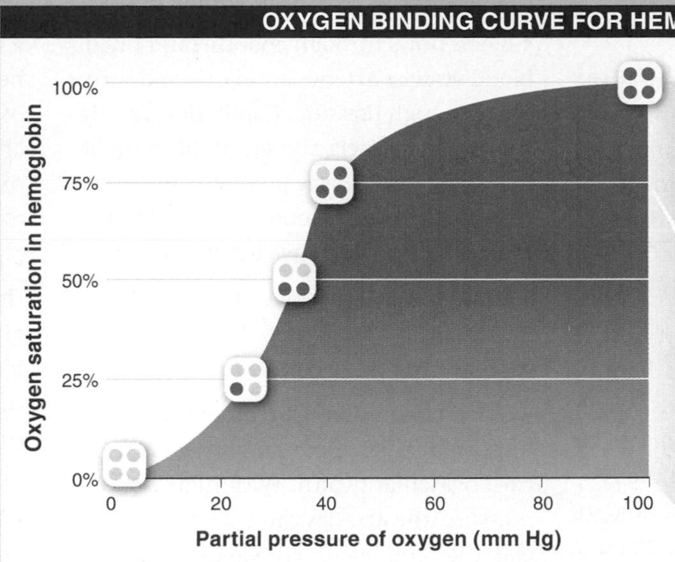

OXYGEN BINDING CURVE FOR HEMOGLOBIN
Oxygen saturation in hemoglobin
100%
75%
50%
25%
0%
0
20
40
60
80
100
Partial pressure of oxygen (mm Hg)
OXYGEN DELIVERED DURING NORMAL METABOLISM
• Oxygen saturation is high.
• Hemoglobin gives up only one oxygen molecule in the body tissues before returning to the lungs.
OXYGEN DELIVERED DURING HIGH PHYSICAL EXERTION
• Oxygen saturation is lower.
• Hemoglobin dips into its reserves, releasing two, three, or even all four of its oxygen molecules before returning to the lungs.

5 What is the oxygen saturation in hemoglobin at an oxygen partial pressure of 60 mm Hg? Does your answer make sense, given that a molecule of hemoglobin can carry either one or two or three or four molecules of oxygen (but nothing in between)?

6 For what 20-mm-Hg change in the partial pressure of oxygen is there the greatest release of oxygen from hemoglobin?

7 How does this graph help you to understand why people become short of breath when climbing high-elevation mountains?

See answers at the back of the book.

KEY TERMS IN CIRCULATION AND RESPIRATION

arteriosclerosis, p. 772
artery, p. 756
atherosclerosis, p. 772
atrium, p. 757
blood, p. 756
blood pressure, p. 769
capillary, p. 756
cardiovascular disease, p. 771
closed circulatory system, p. 756
coronary artery, p. 771
diaphragm, p. 785
diastolic pressure, p. 769
erythrocytes, p. 767
gills, p. 776
heart, p. 756
heart murmur, p. 761
hematocrit, p. 767
hemoglobin, p. 767
hemolymph, p. 756
hypertension, p. 770
hypotension, p. 770
interstitial fluid, p. 756
lamella, p. 780
leukocytes, p. 768
lungs, p. 777
lymph, p. 774
lymph node, p. 775
open circulatory system, p. 756
pathogen, p. 768
plasma, p. 767
platelet, p. 769
pulmonary circuit, p. 757
red blood cells, p. 767
stem cells, p. 767
systemic circuit, p. 758
systolic pressure, p. 769
vein, p. 756
ventricle, p. 757
white blood cells, p. 768

BRIEF SUMMARY

The circulatory system is the chief route of distribution in animals.

• In animals, the circulatory system is the chief distribution system. It transports gases, nutrients, waste products, and hormones to and from tissues throughout the body. The circulatory system also helps animals regulate their body temperature and protect against infection.

• Animals that can acquire all the nutrients and oxygen they need by diffusion do not have circulatory systems. In animals that do have circulatory systems, the system can be open, with no distinction between the circulating fluid and the interstitial fluid, or closed, in which the circulating fluid and interstitial fluid are separate.

• Vertebrates' circulatory systems vary in structure, characterized by one circuit of flow and a heart with two chambers (fishes) or by two circuits of flow and a heart with three chambers (amphibians and most reptiles) or four chambers (birds and mammals). With two circuits of flow, blood is pumped to the body at higher pressure.

The human circulatory system consists of a heart, blood vessels, and blood.

• The human heart is a pump that sends blood on a figure 8 circuit, first to the lungs for loading up with oxygen and then to the tissues and organs of the body. Valves keep blood flowing in one direction.

• The sinoatrial node in the vertebrate heart initiates regular, rhythmic contractions.

• Blood flows through endothelium-lined blood vessels. Arteries are elastic and carry blood at high pressure. Capillaries are tiny, thin-walled vessels, the site of diffusion of nutrients, gases, waste products, and other materials between blood and cells. Blood returns to the heart in veins at low pressure.

• Blood is a mixture of cells and fluid important for transport, defense, and maintaining homeostasis.

• Blood pressure measurement gives important clues about an individual's cardiovascular health. With high blood pressure, the arteries can lose some of their elasticity and health risks are increased.

• Cardiovascular disease includes all diseases of the heart and blood vessels, including heart attacks and strokes. A healthy lifestyle can minimize the risk of cardiovascular disease.

• The lymphatic system runs close to the circulatory system throughout the body and helps in recycling circulatory fluids, marshaling white blood cells, and absorbing nutrients.

The respiratory system enables gas exchange in animals.

• In single-celled and very small multicellular organisms, gas exchange can occur by direct diffusion. In larger multicellular organisms, gas exchange is a two-stage process: (1) between the external environment and the circulatory system, and (2) between the circulatory system and the body's cells.

• Red blood cells are filled with hemoglobin, a molecule that picks up oxygen in the lungs and transports it around the body. Hemoglobin releases its oxygen in organs and tissues, where the oxygen is needed for cellular respiration.

• In aquatic vertebrates, water moves in through the mouth and out through the gills, where gas exchange occurs. In terrestrial vertebrates, air moves down the trachea into lungs, where O_2 diffuses into the bloodstream and CO_2 diffuses from blood to air.

• Birds often spend time at high altitudes and experience low-oxygen conditions. Their oxygen needs are met by a more efficient, circular system of air flow.

• In reptiles, birds, and mammals, breathing occurs in two steps: inhalation and exhalation. It is coordinated by muscle contractions in the diaphragm.

Evolutionary adaptations maximize oxygen delivery.

• At high elevations, the amount of oxygen in the air is lower. Animals living at high elevations produce a form of hemoglobin that has a higher affinity for oxygen.

• Humans living at high elevations become acclimated to low-oxygen conditions, with hemoglobin developing a reduced affinity for oxygen, leading to greater release of oxygen during exertion.

CHECK YOUR KNOWLEDGE

Short Answer

1. Why has the evolution of multicellular organisms led to the development of circulatory systems?

2. What is the difference between single- and double-circuit cardiovascular systems? Which would be better suited for a higher energy demand? Why?

3. Why is it that if you cut into an artery you are more likely to have blood spurt all over you than if you cut into a vein?

4. Describe some of the health risks associated with high blood pressure.

5. Describe three behavioral changes everyone can make to reduce their risk of heart disease and heart attacks.

6. How does lymph travel around the body?

7. How is gas exchange accomplished in animals with gills and in those with lungs?

8. How does hemoglobin "know" when to release the oxygen it is carrying?

9. Explain how the countercurrent exchange system increases the efficiency of gas exchange in fishes.

10. Why don't birds pass out from lack of oxygen when they fly at high altitudes?

11. How does low air pressure at higher altitudes affect the availability of oxygen for breathing?

12. Would you expect hemoglobin to be more "sticky" or less "sticky" in individuals living at higher altitudes? Explain your answer.

Multiple Choice

1. How can flatworms, simple animals with a flat body shape, survive without a circulatory system?

a) Flatworms do not have internal body fluids that need to be sectioned off and transported by vessels.

b) Because all cells in flat-bodied animals are in direct contact with the external environment, direct diffusion is sufficient for all their respiratory needs.

c) In flatworms, all cells are located near central body branches or near the body surface, so diffusion alone can transport important nutrients.

d) Flatworms use the water currents in their aquatic environment to circulate the necessary nutrients throughout their body.

e) Flatworms have multiple hearts and so can pump blood directly into tissues without needing a circulatory system.

0 EASY 25 HARD 100

2. Blood pressure tends to fall with increasing distance from the heart. Based on this information, which of the following lists the types of blood vessel, in order, from highest blood pressure to lowest?

a) veins, arteries, capillaries

b) arteries, capillaries, veins

c) veins, capillaries, arteries

d) capillaries, veins, arteries

e) arteries, veins, capillaries

0 EASY 60 HARD 100

3. The heartbeat in a vertebrate:

a) is initiated by modified muscle tissue, the sinoatrial (SA) node, that contracts without nerve stimulation.

b) is triggered by rhythmic stimulation from the cardiac nerve.

c) begins at the bottom of the ventricles and moves upward through the heart.

d) cannot be recorded by an EKG, but the activity of neurons that control it can be.

e) is initiated by the atrioventricular (AV) node.

0 EASY 16 HARD 100

4. In a person at rest, most of his or her blood is likely to be in the:

a) heart.

b) capillaries.

c) veins.

d) arteries.

e) interstitial fluid.

0 EASY 25 HARD 100

5. Which of the following components of human blood is a cell fragment?

a) spiracle

b) red blood cell

c) white blood cell

d) plasma

e) platelet

0 EASY 16 HARD 100

6. Breathing in birds is more like "breathing" in fishes than in mammals, because:

a) water and high-altitude air have similar viscosity.

b) oxygen never runs into a dead end.

c) both birds and fishes have gill arches.

d) the blood vessels in bird lungs, like those in fish gills, are insulated with a fatty coating, reducing heat loss.

e) in birds and fishes, but not mammals, the heart pumps in a two-circuit system of blood flow.

0 EASY 70 HARD 100

7. Under resting metabolic conditions, what percentage of oxygen is released from the blood to the tissues?

a) 1%

b) 75%

c) 25%

d) 100%

e) 50%

0 EASY 32 HARD 100

8. Why is carbon monoxide (CO) so dangerous?

a) CO prevents hemoglobin from binding to and transporting O_2 to the body tissues, resulting in oxygen starvation and death.

b) CO prevents CO_2 from dissolving in the blood, and CO_2 builds up to toxic levels.

c) CO modifies the structure of hemoglobin, making it more difficult for O_2 to be transported and released in the body.

d) CO coats the inside of the alveoli, preventing diffusion of O_2 from the lungs into the blood.

e) CO binds to hemolymph, so tissues cannot remove O_2 from the red blood cells and the tissue dies.

0 EASY 26 HARD 100

9. Which of the following has the highest binding affinity for oxygen?

a) hemolymph

b) human hemoglobin

c) llama hemoglobin

d) plasma

e) All have equal binding affinity for oxygen.

0 EASY 41 HARD 100

10. Diphosphoglyceric acid (DPG):

a) causes the release of more O_2 to body tissues.

b) makes it possible for blood to carry more CO_2.

c) is the main reason that breathing is difficult at high elevations.

d) increases the binding affinity of adult hemoglobin for O_2.

e) encourages the production of more hemoglobin.

0 EASY 41 HARD 100

Ch23

Food provides the raw materials for growth and the fuel to make it happen.

Nutrients are grouped into six categories.

We extract energy and nutrients from food.

What we eat profoundly affects our health.

The individual foods used to make a hamburger are broken down into their component macromolecules during digestion, yielding energy and nutrients.

Nutrition and Digestion

At rest and at play: optimizing human physiological functioning

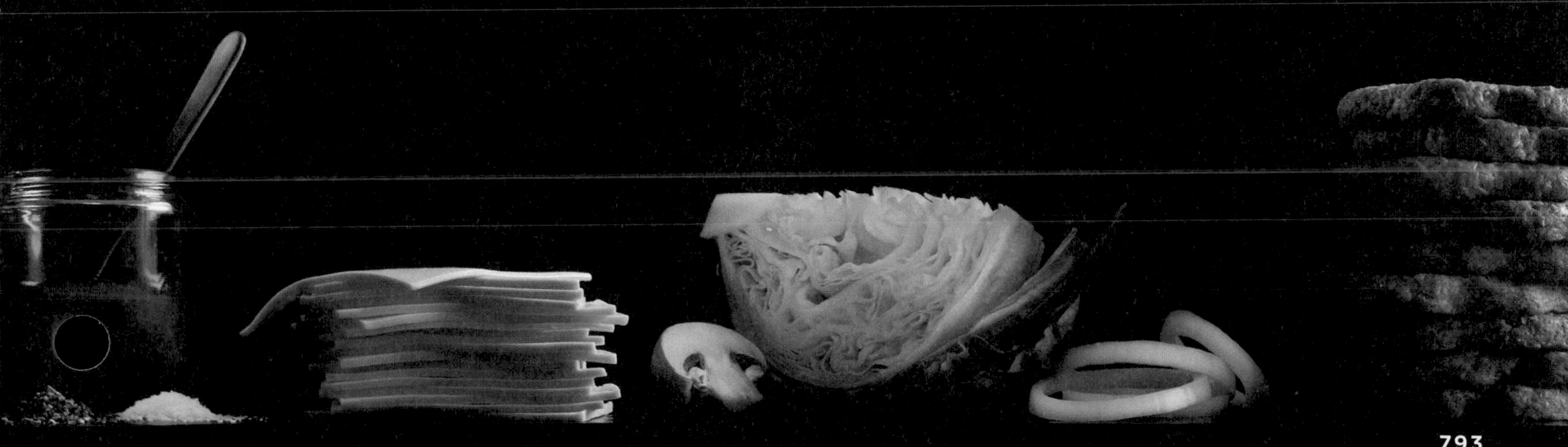

23.1–23.3

Food provides the raw materials for growth and the fuel to make it happen.

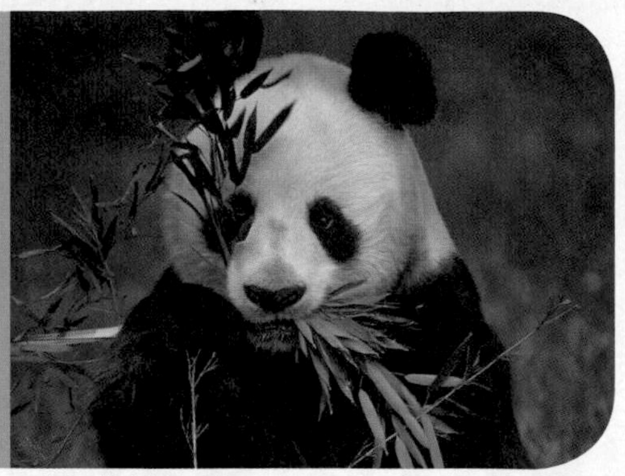

A panda's diet is made up almost entirely of bamboo.

23.1 Why do organisms need food?

Lamb roulade with leek basil stuffing and a red wine sauce. Chili fries. Lime-marinated tofu kabobs. A big vanilla milkshake. Although we dress food up in an almost infinite variety of combinations, it really boils down to two simple biological needs: raw materials and fuel. Just as a carpenter needs wood and a car needs gas, living organisms need raw materials and fuel to function.

Our need for fuel is one of the two reasons why we must eat: building a body, moving around, reproducing, and just staying alive all require energy. Food provides that energy. The other reason is that to grow and to build the myriad complex molecules required for life, we need raw materials: molecules containing carbon and nitrogen and phosphorus, to name just a few. Food provides these raw materials.

The food we eat is physically and chemically broken down into its fundamental macromolecular components in the process of **digestion** (FIGURE 23-1). Our body quickly breaks down food and separates it into the usable and unusable. The usable materials are carbohydrates, lipids, proteins, vitamins, minerals, and water and are referred to as **nutrients.** These six crucial substances are used for energy, raw materials, and maintenance of the body's systems. The unusable materials pass through the digestive system and are eliminated.

Ultimately, an organism's body weight reflects a balance between the energy carried within the molecular bonds of the food it consumes and the energy burned in the processes of living. Any surplus calories are stored, usually as fat or as glycogen (see Section 3.3), a form of carbohydrate stored primarily in muscle and liver tissues.

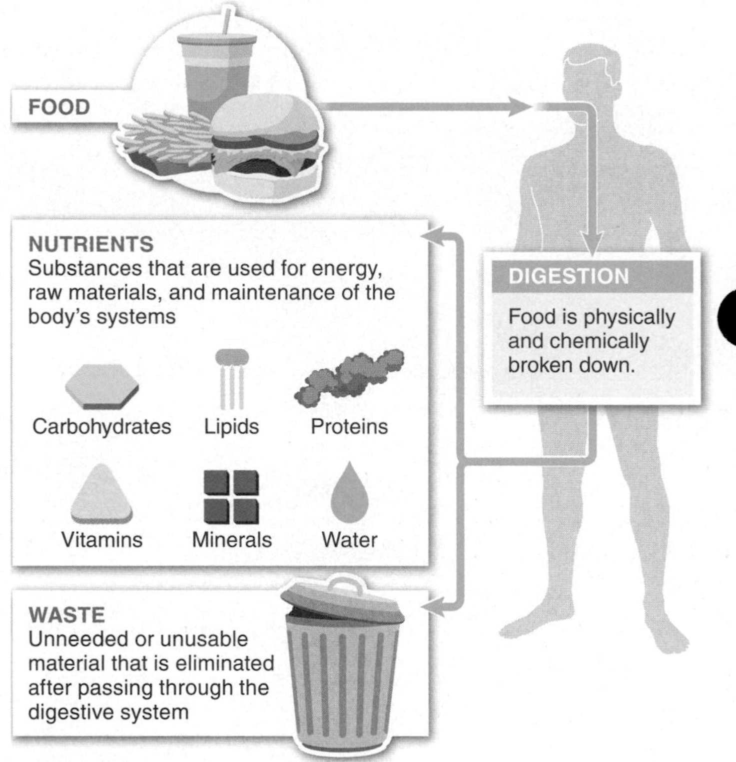

FOOD
NUTRIENTS
Substances that are used for energy, raw materials, and maintenance of the body's systems
Carbohydrates
Lipids
Proteins
Vitamins
Minerals
Water
DIGESTION
Food is physically and chemically broken down.
WASTE
Unneeded or unusable material that is eliminated after passing through the digestive system

FIGURE 23-1 **Made possible by food.** Living organisms need raw materials and fuel to function.

TAKE HOME MESSAGE 23.1

» Animals must eat for two reasons: to acquire the energy needed for all growth and activity, and to acquire the raw materials required for life.

23.2 What's on the menu? Animals have a variety of diets.

All organisms need food—but there are some very different ways to get it. Plants and other photosynthetic organisms (including many bacteria and protists) make their own food, through photosynthesis. But animals do not have the cellular machinery to do this, and so they must eat other organisms. Most animals eat plants, while some animals eat other animals. But in every case, though some organisms are able to capture solar energy to make food, all animals must acquire that energy indirectly by consuming other organisms.

In the animal world, species fall into three groups based on their diets (FIGURE 23-2).

Carnivores Predatory animals that consume only other animals are **carnivores.** They include spiders and snakes; mammalian species such as wolves, seals, bats, and cats; and hawks, owls, and other birds of prey. Some carnivores don't actually kill their prey but instead just suck nutrient-rich fluids from them. Mosquitoes and ticks are carnivorous fluid-feeders.

Herbivores Food that is easy to get, but hard to digest—that sums up life as an **herbivore,** an animal that consumes only plants. Because plants are plentiful in most habitats and can't run away, they are easy targets for predation. To protect themselves, however, plants tend to carry many toxic compounds that are difficult or impossible for animals to break down chemically. Nonetheless, many herbivores have developed digestive adaptations to overcome these difficulties. We'll explore a couple of them in detail in Section 23.13. Examples of herbivores include beavers, tortoises, and caterpillars, as well as many species of birds, squirrels, and large grazing mammals. And just as there are fluid-feeding carnivores, there are fluid-feeding herbivores. Aphids, for example, pierce the surface of plants and suck their sugary sap.

Omnivores Most humans, as **omnivores,** eat plants and animals and can digest both efficiently. We share this diet with numerous other species, including cockroaches. Omnivores come in all sizes, from bears to raccoons, from chickens to flies and wasps.

TYPES OF ANIMAL DIETS

CARNIVORES
Animals that consume only other animals. (Shown here: A leopard seal feeding on a penguin)

HERBIVORES
Animals that consume only plants. (Shown here: A desert cottontail feeding on a flower)

OMNIVORES
Animals that consume both plants and animals. (Shown here: A western scrub-jay feeding on an acorn (left) and an insect larva (right))

FIGURE 23-2 **Animal diets: carnivores, herbivores, and omnivores.**

TAKE HOME MESSAGE 23.2

» Animals require food. Plants and other photosynthetic organisms produce food through photosynthesis, and animals have one of three types of diet. Carnivores consume only other animals. Herbivores consume only plants. And omnivores consume both plants and animals.

23.3 Calories count: organisms need sufficient energy.

Have you ever gone a whole day without eating? Or have you ever spent a few weeks trying to reduce your caloric intake? Living in a state of hunger can be torturous. In the early 1990s, eight human "guinea pigs" learned this the hard way. Living in the Biosphere 2 dome—a self-contained, 3.2-acre world filled with plants and animals that was designed to explore sustainable living with minimal environmental impact—the group took part in an experiment on the effects of a low-calorie diet, among other things. The findings were not surprising: all the participants lost weight, but they also became very unhappy. They argued constantly and frequently squabbled over meal portions in what they dubbed "the hunger dome" (FIGURE 23-3). Put simply, humans (and all other living organisms) need a steady and sufficient flow of energy to function well.

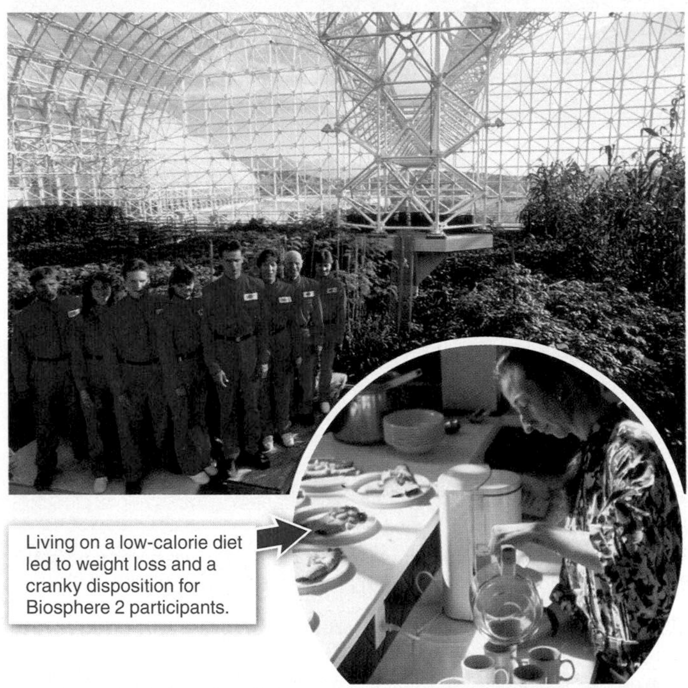

Living on a low-calorie diet led to weight loss and a cranky disposition for Biosphere 2 participants.

"If we ever all start talking to each other, that would be a major accomplishment."

— ONE OF THE EIGHT PARTICIPANTS OF BIOSPHERE 2

FIGURE 23-3 **Hungry Biosphere 2 crew.**

So, how much energy do we actually need to survive? The energetic value of food is measured in very small units called **calories,** with a single calorie defined as the energy required to raise the temperature of 1 gram of water by 1° C. However, use of this word can be a bit confusing: in discussing human consumption, the word "calorie" is often used to refer to a **kilocalorie (kcal),** which is 1,000 calories.

To estimate how many kilocalories an individual expends each day, we first establish an energy baseline. This is called the **basal metabolic rate,** or **BMR,** and refers to the amount of energy the individual expends with no food in the digestive tract, in a neutral-temperature environment, doing nothing more than the equivalent of sitting on a couch all day. For humans, the BMR is approximately 1 calorie per hour per gram of body weight, or about 1,400 kcal/day for a woman (weighing 120 pounds) and about 1,700 kcal/day for a man (weighing 160 pounds). The difference between the sexes is mostly a function of the difference in body size.

The BMR allows easy comparisons across species because it is so clearly defined, but to accurately evaluate organisms' energy needs, we must account for how active they are (FIGURE 23-4). Individuals need about 50% to 100% more kilocalories per day than their BMR. A 120-pound woman requires 1,800–2,400 kcal/day, and a 160-pound man requires about 2,400–3,200 kcal/day.

From one animal species to another, basal metabolic rates differ tremendously. The BMR of a tiny shrew, for example, is about 35 times higher than that of a human. The shrew's heart beats more than 500 times per minute *at rest!* Let's estimate what a shrew needs to eat each day. If it weighs 5 grams and has a BMR of 35 calories per hour per gram, we can calculate as follows:

Body weight	×	*Energy needed each hour*	×	*Hours per day*	=	*Animal's caloric needs*
5 grams	×	35 cal/ gram/hr	×	24 hr/ day	=	4,200 cal/day (4.2 kcal/day)

Thus, the shrew would need 4.2 kcal/day if it were at rest. But if it needed another 4 kcal/day for its normal activities, it would have to eat about 8 kcal/day. A pure source of

COMPONENTS OF ENERGY EXPENDITURE

BASAL METABOLIC RATE (BMR)
The minimal energy expenditure of an organism at rest
ENERGY REQUIRED FOR ALL ACTIVITY
Individuals generally need about 50% to 100% more kilocalories per day than their BMR.

Female
Male
0
1,000
2,000
3,000
ACTUAL DAILY ENERGY EXPENDITURE
Average kcal/day to cover the energetic needs of an organism

FIGURE 23-4 **How many calories do you need?** Energy expenditure varies depending on body size and activity level.

carbohydrate or protein carries about 4 kcal/gram. Thus, every day, the shrew must find and consume at least 2 grams of food. This is a challenge when you weigh only 5 grams. The task is equivalent to a 200-pound man finding and eating 80 pounds of food every day.

Animals that fast or hibernate for long periods of time must also prepare by consuming many kilocalories. During the breeding season, male elephant seals eat little or nothing; they spend much of their time battling with other males for dominance and (if they are among the most dominant) mating with females. Over these 100 days, a male that begins the season at 6,600 pounds (3,000 kg) may lose one-third of his body weight. Consequently, the male seal's caloric requirements are much higher in the months leading up to the breeding season.

TAKE HOME MESSAGE 23.3

» To function well, living organisms need sufficient energy, measured in kilocalories. The minimal energy needed by an individual not engaged in any activity is called its basal metabolic rate, or BMR.

23.4–23.7

Nutrients are grouped into six categories.

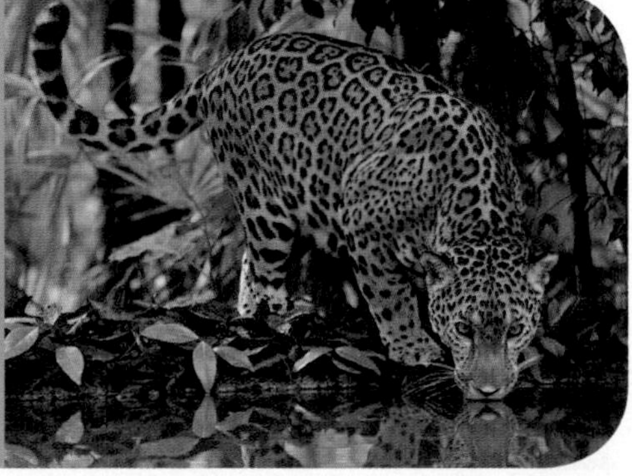

A jaguar hydrating.

23.4 Water is an essential nutrient.

Water constitutes about 65% of the body weight of most mammals and is considered an essential nutrient. This water plays important roles in both the intracellular fluid and the extracellular fluids, including blood (FIGURE 23-5). Water in body fluids serves a variety of critical purposes, all of which can be impaired when the animal becomes dehydrated.

Water is essential for intracellular and extracellular fluids (including blood), and for nearly every system in the body.

FIGURE 23-5 **Water is an essential component of the diet.**

Water has the following functions:

- Transports nutrients and waste in the body
- Takes part in chemical reactions
- Serves as a solvent for many vitamins and minerals, amino acids, and sugars
- Lubricates many joints, the spinal cord, and the eyes
- Helps regulate body temperature

A person expending about 2,000 kcal/day needs about 2–3 liters of water each day. Water intake can come from drinking water, milk, and other beverages and from water in foods, and water is released as a by-product of many chemical reactions. Water intake must offset the water lost in urine, feces, respiration, and sweating. It is important to remember, too, as we saw in the Chapter 4 StreetBio, that drinking too much water too quickly can lead to water intoxication—which sometimes happens to marathon runners during a race. When coupled with the loss of salt through sweating, over-consumption of water can lead to severe sodium imbalance. This imbalance causes dizziness, nausea, confusion, and swelling of the extremities. It can even lead to swelling of the brain, causing death.

Water usage varies among animal species. Tremendous water efficiency has evolved in some desert mammals—the kangaroo rat, for example, gets all of the water it needs not from drinking but from food and metabolic processes. Among the marine birds and reptiles, most are able to drink salt water. They can do this with the aid of salt glands that remove and excrete the excess salt they consume.

Q If water is so important, why do some desert animals never need to drink?

TAKE HOME MESSAGE 23.4

›› Water is probably the single most important component of an animal's diet. It constitutes about 65% of the body weight of most mammals. It transports nutrients and waste materials throughout the body, takes part in metabolic reactions, serves as a solvent, lubricates many body parts, and helps regulate body temperature.

23.5 Proteins in food are broken down to build proteins in the body.

For optimum health, organisms must consume many types of nutrients. In this section and the next, we investigate the nutritional features of the three types of calorie sources: proteins, carbohydrates, and fats.

Proteins: Raw Material for Growth *What are proteins, and what is their role in the diet?* Protein in our diet is principally a building material. As described in detail in Sections 3.10–3.12, protein molecules are made of long chains of amino acids linked together, like beads on a string. Once eaten, protein molecules are broken down into the individual amino acids, much like removing the beads from the string one at a time. It's only in the form of individual amino acids (and, occasionally, chains of two or three amino acids) that proteins can be absorbed by the digestive system.

Our bodies can then reorder the beads, making proteins that are different from those we ate.

Besides serving as building materials, the proteins we build from amino acids function as enzymes, catalyzing chemical reactions (see Sections 3.13 and 3.14). Proteins, whether from our diet or from our own body tissues, can also be broken down to release energy or to be converted to and stored as fat. Protein breakdown happens, for example, when we are consuming too few (or too many) calories to sustain necessary growth and activities. It also happens during a long exercise session. Our bodies can use protein as fuel because we have a variety of methods of converting one type of chemical to another to release or store energy (see Figure 5-36). The breakdown of proteins for energy generates 4 kilocalories per gram. However, when a protein is broken down for energy, the nitrogen it carries cannot be used to build new proteins; instead, it is excreted in our urine. In essence, burning protein for fuel wastes valuable nitrogen (which is not found in carbohydrates and lipids) that could be used for building new protein.

There are many sources of dietary protein, from both plants and animals (FIGURE 23-6). Animal sources include egg whites, shrimp, tuna, chicken, turkey, and all meat products. Plant sources of protein include grains, vegetables, nuts, seeds, and beans.

Are all proteins the same, or do they vary in important ways? Animals, including humans, require 20 different amino acids to make proteins. Most animals, however, can produce only about half of these amino acids themselves—for example, under most normal conditions humans can make only 11 of the 20 amino acids. The 9 amino acids we cannot produce we must get from the food we eat; we say that these 9 amino acids are **essential amino acids,** and the remaining 11 are **non-essential amino acids.**

All proteins are not created equal. Animal products vary in their amino acid composition. All nine essential amino acids are found in milk, eggs, meat, poultry, cheese, and fish; these animal products have "complete" proteins (FIGURE 23-7). Most proteins that we get from eating plants are not complete. Corn proteins, for example, have only six of the essential amino acids, as do beans (but not the same six). In fact, with only a few exceptions (including soybeans and quinoa), plant proteins do not have all nine. A healthy diet is one containing not just a sufficient *amount of protein* but a sufficient amount of each of the nine essential amino acids that we need.

Q Why is it beneficial for vegetarians to eat beans and corn over the course of the same day?

How and where do we store proteins? Proteins and the amino acid pool they generate on digestion cannot be stored very long in our bodies, usually less than a day. By that point, they are broken down and the nitrogen we don't need is excreted. Because our bodies are making proteins constantly, we must take in all of the essential amino acids over the course of each day.

PROTEIN: RAW MATERIAL FOR GROWTH

FUNCTION
Once proteins are broken down, the amino acids are used as the raw materials to build new complex proteins, such as hemoglobin and muscle.

SOURCE
- Animals: egg whites, shrimp, tuna, poultry, and meat
- Plants: grains, vegetables, nuts, seeds, and legumes

STORAGE
- Amino acids usually stored for less than half a day before being reassembled into proteins throughout the body
- Can be converted to fat and stored in fat cells

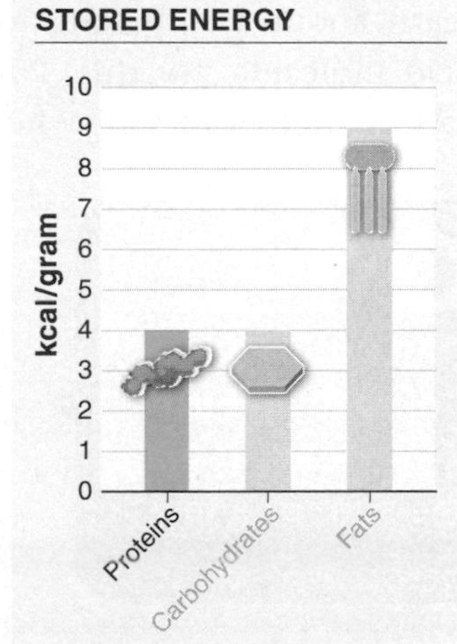

FIGURE 23-6 **Versatile proteins.** Besides serving as building materials, proteins can be broken down to release energy or can be stored as fat.

COMPLETE vs. INCOMPLETE PROTEINS

Sushi with salmon, tuna, and shrimp

COMPLETE PROTEINS

- Contain all nine essential amino acids
- Almost all animal proteins are complete

Rice and beans (part of a meal from the Federal Detention Center in Portland, Oregon)

INCOMPLETE PROTEINS

- Do not contain all nine essential amino acids
- Almost all plant proteins are incomplete

FIGURE 23-7 **Animal vs. plant proteins.**

A healthy diet contains sufficient amounts not just of protein but of each of the nine essential amino acids.

The recommended daily intake of protein is approximately 0.8 grams per kilogram of body weight. This translates to about 45 grams of protein for a 120-pound woman (and about 25 additional grams per day for pregnant or lactating women) and 60 grams for a 160-pound man. Regular muscle-building exercise can double or even triple this requirement, but it is important to remember that protein on its own does nothing to stimulate muscle growth; it simply serves as the source for the raw materials.

In our diet, linked amino acids don't necessarily function as tissue-building molecules or energy sources. To learn about a surprising role played by a two-amino-acid molecule, try this. Put one can of Coke and one can of Diet Coke somewhere hot and leave them for a few days. Then taste them. The Diet Coke will lose its flavor and taste bitter, even if you cool it down, while the Coke will taste fine. What's going on? Diet Coke, but not regular Coke, is sweetened with aspartame (NutraSweet is the brand name). Aspartame is made by linking together two amino acids, aspartic acid and phenylalanine—each of which tastes bitter by itself. When the two are joined as aspartame, however, the taste is very sweet. But the reaction by which aspartic acid and phenylalanine are linked together is reversed by heat. So in a hot area, the aspartame in Diet Coke decomposes to the bitter amino acids. They aren't harmful, but they aren't very tasty either.

Q Why does Diet Coke (but not regular Coke) taste bitter if you leave it in the sun for a few days?

Every organism—even plants—must build enzymes and so needs protein. This is dramatically apparent in the Venus flytrap, a plant that captures and consumes insects in order to supply itself with protein. The plant releases digestive juices that break down the insect into its usable nutrients. Venus flytraps grow in places where the soil is of poor quality, particularly with respect to nitrogen; they can extract nitrogen from the animal prey they capture. The green leaves and stems of Venus flytraps, however, should serve as a reminder that these plants are also photosynthetic. The nitrogen they need (and resort to carnivory to obtain) helps them build the enzymes necessary to conduct photosynthesis.

TAKE HOME MESSAGE 23.5

» Animals consume three different types of macromolecules for calories: proteins, carbohydrates, and fats. Proteins provide raw materials for growth and for the production of enzymes. Food sources of protein vary in amino acid composition. Humans require 20 amino acids for building proteins, and 9 of these, called essential amino acids, must be consumed in the diet.

23.6 Carbohydrates and lipids provide bodies with energy and more.

Reading these words requires carbohydrates in your body—obtained from your diet—to fuel your brain (and some muscles as well). Food fat, too, provides energy for your body's functioning (see Sections 3.6–3.9).

Carbohydrates: Fuel for Living Machines

What are carbohydrates, and what is their role in the diet? Carbohydrates are the primary fuel on which animal bodies run. Nearly all of the energy used by our brain comes from the carbohydrate glucose. As we saw in Chapter 3, all carbohydrates are made primarily from carbon, hydrogen, and oxygen. When the bonds between these atoms are broken and other bonds are formed, energy is released that can be captured by the body and used to fuel movement, growth, and all the cellular activities that require energy.

We get the majority of our dietary carbohydrates from fruits, vegetables, and grains (FIGURE 23-8). And, as with proteins, the breakdown of carbohydrates for energy generates 4 kilocalories per gram.

Are all carbohydrates the same, or do they vary in important ways? All carbohydrates are formed from carbon, hydrogen, and oxygen in the approximate proportions of CH_2O. But they vary considerably in their complexity and break down in the body at different rates (FIGURE 23-9). Animals consume carbohydrates in the form of simple sugars (monosaccharides), disaccharides and digestible complex sugars (starch and glycogen), and indigestible complex sugars (fiber).

Simple sugars. These are linear or ring structures with three to seven carbon atoms, and include glucose and fructose. Animals can break them down directly through the steps of glycolysis (see Section 5.13), rapidly releasing the stored energy from the bonds of the sugar.

Disaccharides and digestible complex sugars. Multiple simple sugars can bond together to form complex but digestible molecules. Disaccharides, such as sucrose (table sugar), are just two simple sugars joined together. Complex sugars, such as starch and glycogen, are large molecules that may consist of hundreds or thousands of glucose molecules connected in dense, branching patterns. For an animal to have access to the energy stored in the bonds of the individual simple sugars, it must first break the bonds that link the sugars together. As these bonds are broken, simple sugars become available for the energy-releasing reactions of glycolysis.

Fiber. This is a complex carbohydrate, such as cellulose, that forms the structural parts of plants. Fiber differs from

CARBOHYDRATES: FUEL FOR RUNNING YOUR BODY

FUNCTION
Carbohydrates provide energy to fuel movement, growth, and all cellular activities in the body.

SOURCE
Fruits, vegetables, and grains

STORAGE

- Carbohydrates stored in the liver and muscle cells as glycogen for about a day before being broken down to provide energy
- Can be converted to fat and stored in fat cells

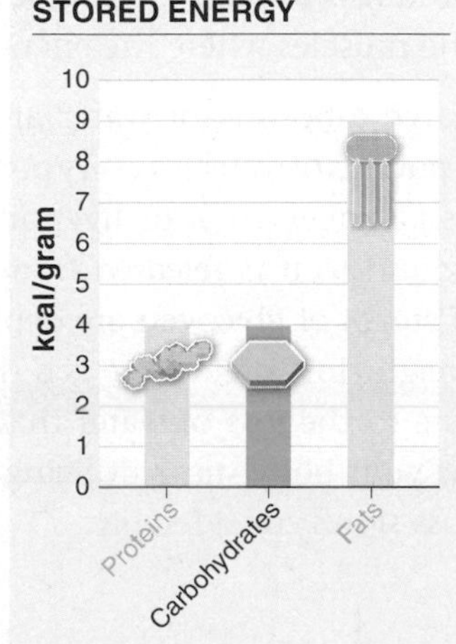

FIGURE 23-8 **Carbohydrate is the primary fuel on which animal bodies run.**

TYPES OF CARBOHYDRATES

SIMPLE SUGARS

- Glucose, fructose, and others
- Breakdown by glycolysis releases energy rapidly

COMPLEX SUGARS

DIGESTIBLE COMPLEX SUGARS (STARCH)

- Simple sugars bonded together, such as sucrose (table sugar) or starch
- Bonds between simple sugars must be broken before the energy-releasing reactions of glycolysis occur

INDIGESTIBLE COMPLEX SUGARS (FIBER)

- Complex carbohydrate that forms structural parts of plants
- Indigestible in humans, but has a significant role in digestion

FIGURE 23-9 **Carbohydrate complexity.** Some carbohydrates are readily broken down for fuel, but fiber passes through the human body undigested.

starch and other digestible complex sugars in the type of bond connecting the simple sugars together. Because this bond cannot be broken by any digestive enzyme in humans, fiber is indigestible. As we'll see later, it still plays an important role in digestion and is necessary in the diet to maintain health.

How and where do humans store carbohydrates? Carbohydrates in our body are stored mostly as glycogen (see Section 3.3) in liver and muscle cells. At any given time, we can store only about one day's worth of energy. When our bodies need energy for an activity, a signal is sent that causes the release of enzymes that break the bonds holding together the highly branched glycogen. This glycogen breakdown produces a flood of glucose into the bloodstream and in the muscles where the energy is needed.

Large amounts of water are bound to stored glycogen: 4 pounds of water for every pound of glycogen. Consequently, as glycogen in your liver and muscles is used, the water bound to it is released from the tissue and lost as urine. If stores of glycogen are depleted in the initial stages of a weight-loss diet, there is a dramatic initial loss of weight due to the loss of water that was bound to the glycogen. As your body starts utilizing stored fat, the rate of weight loss slows considerably.

Fats: Long-Term Energy-Storage Experts

What are fats, and what is their role in the diet? Fats in the animal diet function primarily as a dense source of energy that can be efficiently stored in the body (FIGURE 23-10). The average person has about four or five weeks' worth of stored energy in the form of fat. Compared with carbohydrates or proteins, a given amount of fat contains more than twice as much stored energy: 1 gram of fat produces 9 kilocalories. Another feature that makes fats particularly efficient as energy-storage molecules is that, because they are hydrophobic, they are stored without binding to water. (Because fats are poor conductors of heat, fats stored in a layer just beneath the skin can also help to keep the body warm.)

Are all fats the same nutritionally, or do they vary in important ways? In our diet, fats usually come in the form of fatty acids, long chains of carbon atoms with hydrogens attached. With proteins, we saw that there are essential and non-essential amino acids. Similarly, with dietary fats, there are essential and non-essential fatty acids. The essential fatty acids—including linoleic acid, in the omega-6 family of fatty acids—are those that humans cannot produce and must be consumed. Linoleic acid is essential as a building block for signaling molecules, such as some hormones; a deficiency can lead to infertility and difficulty lactating. Another

FATS: LONG-TERM ENERGY STORAGE

FUNCTION
Fats provide a dense source of energy that can be efficiently stored in the body, and they aid in keeping the body warm.

SOURCE
Butter, cheese, oils, eggs, and meat

STORAGE
Fats stored in fat cells throughout the body

STORED ENERGY

kcal/gram
10
9
8
7
6
5
4
3
2
1
0
Proteins
Carbohydrates
Fats

At 9 kilocalories per gram, fats are rich in stored energy.

FIGURE 23-10 **Energy-rich fats.**

essential fatty acid is linolenic acid, essential for normal growth and development, especially in the eyes and brain.

An important distinction among dietary fats is between saturated and unsaturated fats (FIGURE 23-11). If each carbon within the fatty acid chain is bonded to two hydrogen atoms, the molecule carries the maximum number of hydrogen atoms and is said to be saturated. (For a refresher, see Figure 3-11.) Conversely, if some of the carbons are bound to only a single hydrogen, the fatty acid is unsaturated.

When saturated, fatty acids are very straight and the fat molecules can be packed together tightly. This causes saturated fats to be solid at room temperature, like butter. When unsaturated, the fatty acids have kinks in the hydrocarbon tail and cannot be packed together as tightly. Consequently, unsaturated fats do not solidify so easily and tend to be liquid at room temperature (think: vegetable oil). Because unsaturated fatty acids can accept one or more hydrogen atoms, they are a bit less stable and more reactive than saturated fatty acids—that isn't a bad thing, it just means they will take part in a greater variety of chemical reactions, making them less likely to be stored as body fat.

Dietary trans fats have been in the news because of their tendency to raise levels of low-density lipoprotein cholesterol (LDL), increasing the risk of coronary heart disease. Because trans fats are no longer recognized as safe, the U.S. Food and Drug Administration is taking steps to have them removed from processed foods.

SATURATED vs. UNSATURATED FATS

SATURATED FATS
- Fatty acids have straight tails and can be packed together tightly
- Tend to be solid at room temperature
- More likely to be stored as fat in the body

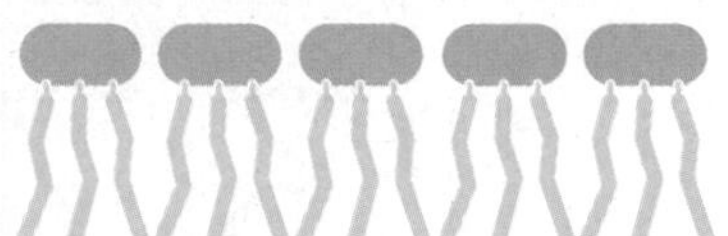

UNSATURATED FATS
- Fatty acids have kinked tails and cannot be packed together tightly
- Tend to be liquid at room temperature
- Less likely to be stored as fat in the body

FIGURE 23-11 **Saturated and unsaturated fats compared.**

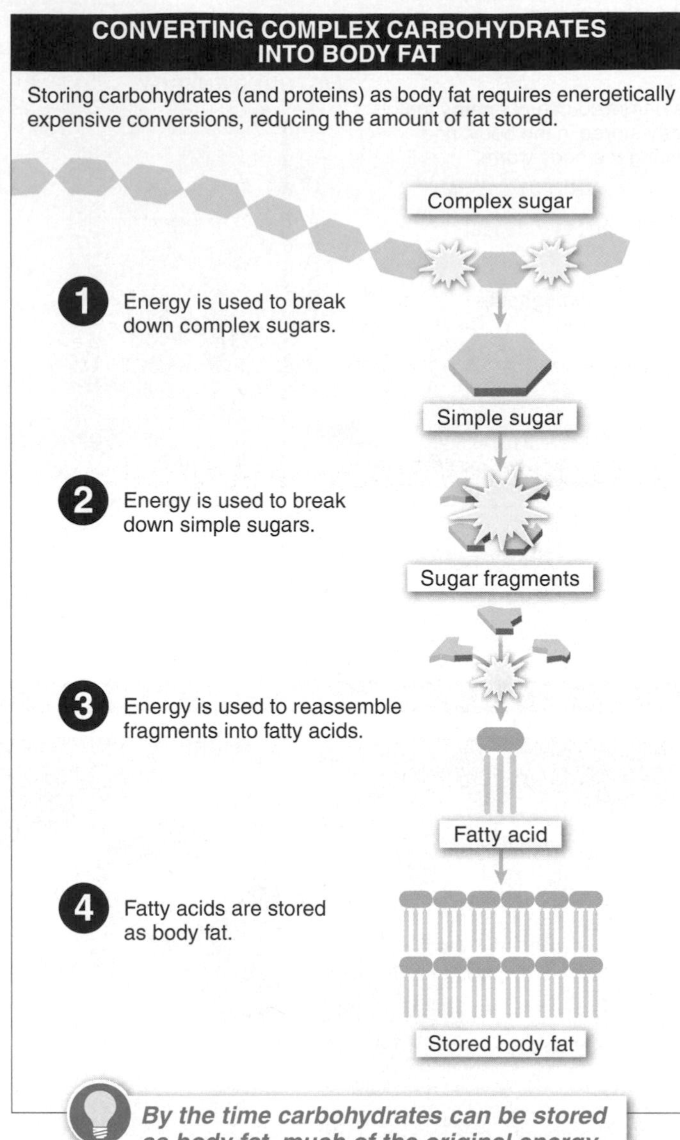
CONVERTING COMPLEX CARBOHYDRATES INTO BODY FAT
Storing carbohydrates (and proteins) as body fat requires energetically expensive conversions, reducing the amount of fat stored.
Complex sugar
1 Energy is used to break down complex sugars.
Simple sugar
2 Energy is used to break down simple sugars.
Sugar fragments
3 Energy is used to reassemble fragments into fatty acids.
Fatty acid
4 Fatty acids are stored as body fat.
Stored body fat
By the time carbohydrates can be stored as body fat, much of the original energy has been lost in fueling the conversion.

FIGURE 23-12 **Inefficient conversion to body fat.**

How and where do we store fats? If a human consumes more calories than he or she burns, most of the excess (regardless of whether these calories were consumed as carbohydrate, fat, or protein) gets converted to fat and stored in fat cells distributed throughout the body. A pound of body fat holds 3,600 kilocalories worth of energy. It is like a savings account for an uncertain future.

Why is it an effective weight-management strategy to minimize dietary fat intake? The answer rests in the number of chemical conversions carbohydrates and proteins must undergo to become body fat. In the case of carbohydrates, complex sugars must first be broken down into simple sugars—a process that requires energy. Then the simple sugars must be broken down and the fragments reassembled as fatty acids—another process that requires energy. By the time those fatty acids can be stored as body fat, much of the original energy stored in the carbohydrates has been lost in fueling all the chemical conversions (FIGURE 23-12). Storing protein as body fat also requires several energetically expensive conversions that reduce the ultimate quantity of fat stored.

TAKE HOME MESSAGE 23.6

» **Carbohydrates are the primary fuel on which animal bodies run. Dietary fats function primarily as a dense source of energy that can be efficiently stored in the body.**

23.7 Vitamins and minerals are necessary for good health.

> "Vitamins can double your health!"
>
> — Dr. ATKINS,
>
> *on* The Larry King Show *(making perhaps the most nonsensical nutritional claim ever)*

It would be nice if we could take a pill that would make us more healthy or fit. Vitamin supplements are not such miracle pills. They're a bit like security guards at a museum: above a certain number, they don't make the museum any better, but in their absence, the museum is likely to become a whole lot worse. And while vitamin supplements are not a health panacea, vitamins and minerals do have important roles in nutrition.

Vitamins are organic compounds required by the body in small amounts for normal growth and health. **Minerals** are chemical elements, other than those commonly found in organic molecules (carbon, hydrogen, oxygen, and nitrogen), some of which are required in the diet in small amounts. There are three features common to all vitamins and minerals.

1. They don't yield any usable energy. Rather, they serve as collaborators with enzymes to enable the processing of the proteins, carbohydrates, and fats we eat, and they catalyze a wide range of other chemical reactions around the body.
2. They need to be consumed in much smaller amounts than proteins, carbohydrates, and fats. This is because they tend to serve as reaction catalysts and so can be recycled and reused again and again.
3. If we have a healthy diet, we tend to consume sufficient quantities of vitamins and minerals in our food.

Although their biological roles are frequently similar, vitamins and minerals have a fundamental chemical difference. Vitamins are organic molecules—they contain carbon—and are more fragile, easily destroyed by heat and other chemical or physical extremes. Minerals, on the other hand, are elements, so they can't be broken down further or they lose their chemical identity. They stay in your body until they are excreted. Seventeen minerals have been identified that are essential to the human diet. They are described in FIGURE 23-13.

ESSENTIAL VITAMINS AND MINERALS IN THE HUMAN DIET

VITAMIN	MAJOR FUNCTION	SOURCES
WATER-SOLUBLE		
B_1 (thiamin)	Coenzyme in cellular respiration	Liver, legumes, whole grains
B_2 (riboflavin)	Coenzyme in FAD	Dairy foods, meat, eggs, green leafy vegetables
B_6 (pyridoxine)	Coenzyme in amino acid metabolism	Liver, whole grains, dairy foods
B_{12} (cobalamin)	Formation of nucleic acids, proteins, and red blood cells	Liver, meat, dairy foods, eggs
Biotin	Found in coenzymes	Liver, yeast, bacteria in gut
C (ascorbic acid)	Formation of connective tissues; antioxidant	Citrus fruits, tomatoes, potatoes
Folic acid	Coenzyme in formation of heme and nucleotides	Vegetables, eggs, liver, whole grains
Niacin	Coenzyme in NAD and NADP	Meat, fowl, liver, yeast
Pantothenic acid	Found in acetyl-CoA	Liver, eggs, yeast
FAT-SOLUBLE		
A (retinol)	Found in visual pigments	Fruits, vegetables, liver, dairy foods
D (cholecalciferol)	Absorption of calcium and phosphate	Fortified milk, fish oils, sunshine
E (tocopherol)	Muscle maintenance; antioxidant	Meat, dairy foods, whole grains
K (menadione)	Blood clotting	Intestinal bacteria, liver

MINERAL	MAJOR FUNCTION	SOURCES
MAJOR MINERALS (essential minerals found in the human body in amounts larger than 5 g)		
Calcium (Ca)	Found in bones and teeth; blood clotting; muscle action	Dairy foods, eggs, green leafy vegetables, whole grains
Chlorine (Cl)	Water balance; principal negative ion in extracellular fluid	Table salt (NaCl), meat, eggs, vegetables, dairy foods
Magnesium (Mg)	Required by many enzymes; found in bones and teeth	Green vegetables, meat, whole grains, nuts, milk, legumes
Phosphorus (P)	Found in nucleic acids, ATP, and phospholipids	Dairy foods, eggs, meat, whole grains, legumes, nuts
Potassium (K)	Nerve and muscle action; principal positive ion in cells	Meat, whole grains, fruits, vegetables
Sodium (Na)	Nerve and muscle action; water balance	Table salt, dairy foods, meat, eggs
Sulfur (S)	Found in proteins and coenzymes; detoxification	Meat, eggs, dairy foods, nuts, legumes
TRACE MINERALS (essential minerals found in the human body in amounts smaller than 5 g)		
Chromium (Cr)	Glucose metabolism	Meat, dairy foods, whole grains, legumes, yeast
Cobalt (Co)	Found in vitamin B_{12}; formation of red blood cells	Meat, tap water
Copper (Cu)	Found in enzymes and electron carriers; hemoglobin production	Liver, meat, fish, shellfish, legumes, whole grains, nuts
Fluorine (F)	Found in teeth; helps prevent tooth decay	Most water supplies
Iodine (I)	Found in thyroid hormones	Fish, shellfish, iodized salt
Iron (Fe)	Found in many enzymes, hemoglobin, and myoglobin	Liver, meat, green vegetables, eggs, whole grains, legumes
Manganese (Mn)	Activates many enzymes	Organ meats, whole grains, legumes, nuts, tea, coffee
Molybdenum (Mo)	Found in some enzymes	Organ meats, dairy foods, whole grains, green vegetables
Selenium (Se)	Fat metabolism	Meat, seafood, whole grains, eggs, milk, garlic
Zinc (Zn)	Found in some enzymes and some transcription factors	Liver, fish, shellfish, and many other foods

FIGURE 23-13 Overview of essential vitamins and minerals.

More than half of the U.S. population takes regular vitamin supplements; however, nearly all adults get all of the nutrients they need from the food they eat.

FIGURE 23-14 Is supplementation necessary?

Thirteen vitamins essential to humans, also described in Figure 23-13, have been discovered. They fall into two groups, based on whether they are soluble in fat or water. The water-soluble vitamins include eight vitamins that are part of the vitamin B complex, plus vitamin C. There is some variation among species in the ability to manufacture (rather than needing to ingest) vitamins. Cats and dogs, for instance, can make vitamin C and so don't need to consume it.

There are four types of fat-soluble vitamins: A, D, E, and K. Because they are stored in fatty tissue and the liver until they are needed, they don't need to be consumed as regularly as water-soluble vitamins. Because excess water-soluble vitamins can be removed by the kidney and excreted in urine, they are less likely to reach toxic levels than fat-soluble vitamins.

Currently, more than half of the U.S. population takes regular vitamin supplements (FIGURE 23-14). The consumption of vitamins above their required amount, however, has no additional benefits. Usually, taking vitamin supplements is a costly but harmless behavior, but in some cases, it can be harmful. Excess consumption of fat-soluble vitamins, for example, can lead to toxicity, with potential health problems. Moreover, research has definitively discredited claims that vitamin supplementation slows or prevents aging or improves physical performance. And no supplement has been demonstrated to "relieve stress"—despite the extravagant claims of marketers.

While most people in the United States receive all of the nutrients they need from the foods they eat, under special circumstances, some individuals may need a supplement.

- Women who lose unusually large amounts of blood during menstruation may need iron.
- Post-menopausal women and those allergic to milk may not get enough calcium in their diet to prevent bone degeneration.
- Pregnant and breast-feeding women may need additional vitamins and minerals to support a developing fetus or rapidly growing infant.
- Pregnant women, or women who may become pregnant, may need additional folate, which helps prevent neural tube defects in the fetus.
- People on extremely low-calorie diets may need vitamin and mineral supplements.
- People with limited consumption of milk and extremely low sun exposure may need additional vitamin D.
- People with intestinal absorption problems (such as when taking antibiotics) may need vitamin and mineral supplementation.

Individuals who do not fall into one of these groups, however, generally get all of the essential vitamins and minerals simply by eating a varied diet, rich in nutritious foods.

TAKE HOME MESSAGE 23.7

» Vitamins and minerals are organic and inorganic molecules, respectively, needed in the diet. They are used in the production and action of enzymes and other molecules involved in the processing of food and other biochemical reactions. While vitamins and minerals are essential in small amounts, most people in the United States do not benefit from taking them as supplements.

23.8–23.13

We extract energy and nutrients from food.

A woodland kingfisher eating a frog.

23.8 We convert food into nutrients in four steps.

The human digestive system is like an assembly line running backward. Imagine starting with an assembled car and dismantling it into its many parts: the tires, doors, windshield, steering wheel. The food entering the digestive system is like the intact car. It enters the assembly line and then passes through four distinct phases, during which the food is progressively chewed up and broken down, the nutrients absorbed by the body, and the non-usable portion of the raw materials discarded as waste products (FIGURE 23-15).

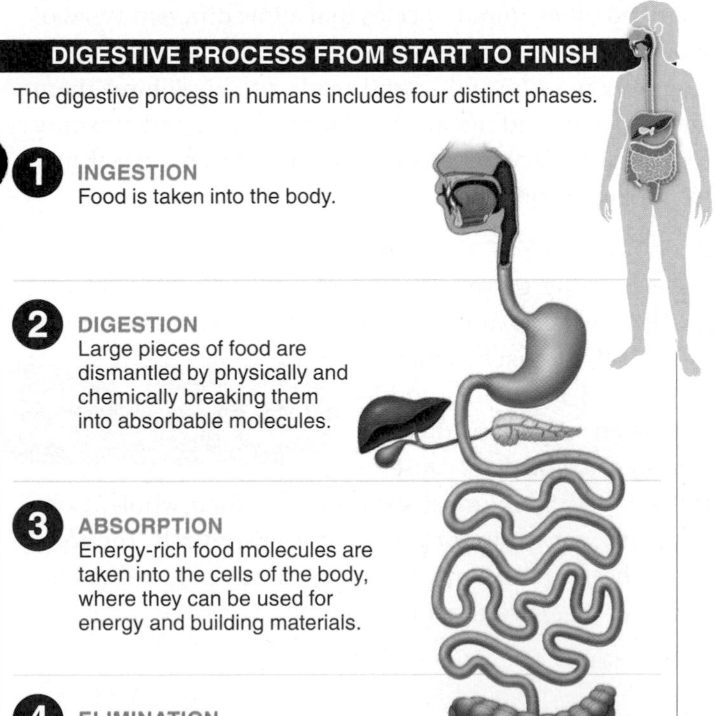

DIGESTIVE PROCESS FROM START TO FINISH
The digestive process in humans includes four distinct phases.
1 INGESTION
Food is taken into the body.
2 DIGESTION
Large pieces of food are dismantled by physically and chemically breaking them into absorbable molecules.
3 ABSORPTION
Energy-rich food molecules are taken into the cells of the body, where they can be used for energy and building materials.
4 ELIMINATION
The remaining parts of the consumed foods—mostly indigestible materials—are discarded as waste products, and much water reabsorption occurs.

FIGURE 23-15 There are four steps in the body's processing of food.

In the next several sections, we examine in detail the four stages in the processing of food: ingestion, digestion, absorption, and elimination. Throughout these sections, it will be useful to picture the digestive system as a long tube with an opening at each end, the mouth and the anus, and some glands along the way that produce the necessary chemicals to help the process along.

TAKE HOME MESSAGE 23.8

» The digestive process in humans includes four distinct phases during which food is progressively chewed up, broken down, and absorbed by the body, after which the non-usable portion of the raw materials is discarded as waste.

23.9 Ingestion is the first step in the breakdown of food.

The intake of an item of food into your body, called **ingestion**, is quick (FIGURE 23-16). Typically lasting less than a minute, ingestion involves your mouth, teeth, tongue, and esophagus, as well as your salivary glands. (In fact, just thinking about food can stimulate saliva secretion!)

Salivary glands secrete mucus through tiny ducts throughout your mouth. Mucus lubricates the food to help it pass into your stomach. The glands also secrete an enzyme called alpha-amylase that initiates the process of digestion. Alpha-amylase breaks the bonds holding together the starch molecule and releases a bit of glucose that can be used for energy. About 20% of the ingested starch is broken down in the mouth. Little or no breakdown of protein and fat molecules occurs in the mouth.

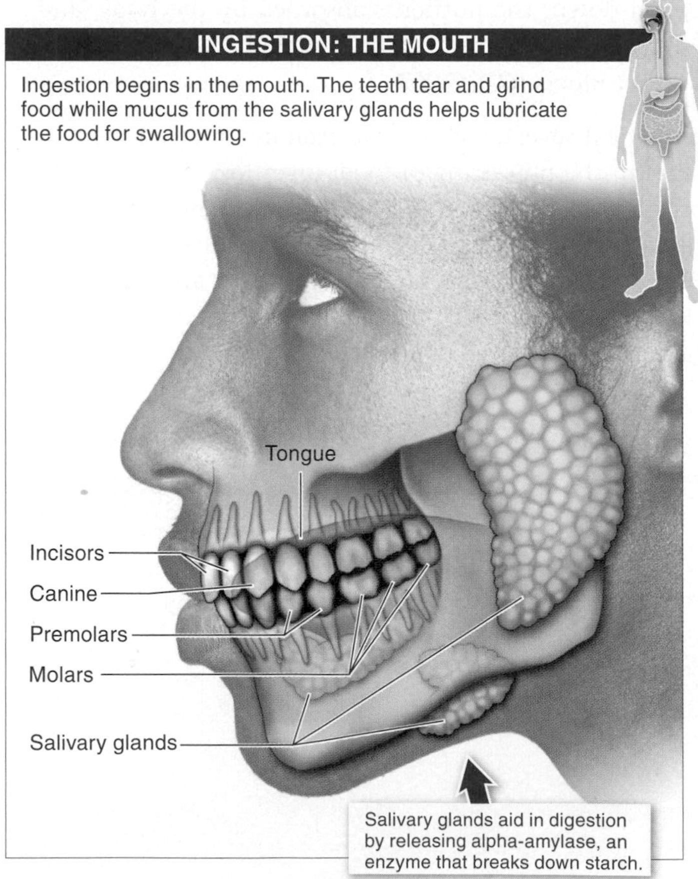

INGESTION: THE MOUTH
Ingestion begins in the mouth. The teeth tear and grind food while mucus from the salivary glands helps lubricate the food for swallowing.
Tongue
Incisors
Canine
Premolars
Molars
Salivary glands
Salivary glands aid in digestion by releasing alpha-amylase, an enzyme that breaks down starch.

Interestingly, across cultures, humans have developed common ways of preparing their food, including using heat and marinating food in acidic solutions, such as vinegar or lemon juice. Because harsh conditions such as heat and acid disrupt the tissue of food items, they increase the efficiency with which digestive enzymes can make contact with the food and break it down.

Q Humans use heat for cooking and often marinate foods in vinegar or lemon juice. How do these processes help with digestion?

You start chewing food in your mouth to tear and grind it into little bits. This is a first important step toward harvesting the energy stored in the chemical bonds of the food. Several different types of teeth have evolved in mammals and other animal species that allow different types of food items to be processed. Incisor and canine teeth, in the front, are used for biting and tearing food. Behind them are premolars and molars, used for grinding and crushing food. Just by looking at the type of teeth an animal has, we can learn a lot about its diet.

Birds have no teeth. This explains why they can sometimes be seen eating gravel. The gravel has no nutritional value, but it collects in the gizzard, one of the two chambers of a bird's stomach, where it helps to grind up food. This is an important digestive step for birds because, lacking teeth, they must swallow their food whole. Consequently, when it first arrives in the stomach, the food hasn't been ground up at all, which can reduce the efficiency of digestive enzymes.

Q Why do many birds eat gravel?

FIGURE 23-16 Ingestion, the intake of food into the body, begins in the mouth.

FIGURE 23-17 Ingestion is complete when food passes from the mouth through the esophagus to the stomach.

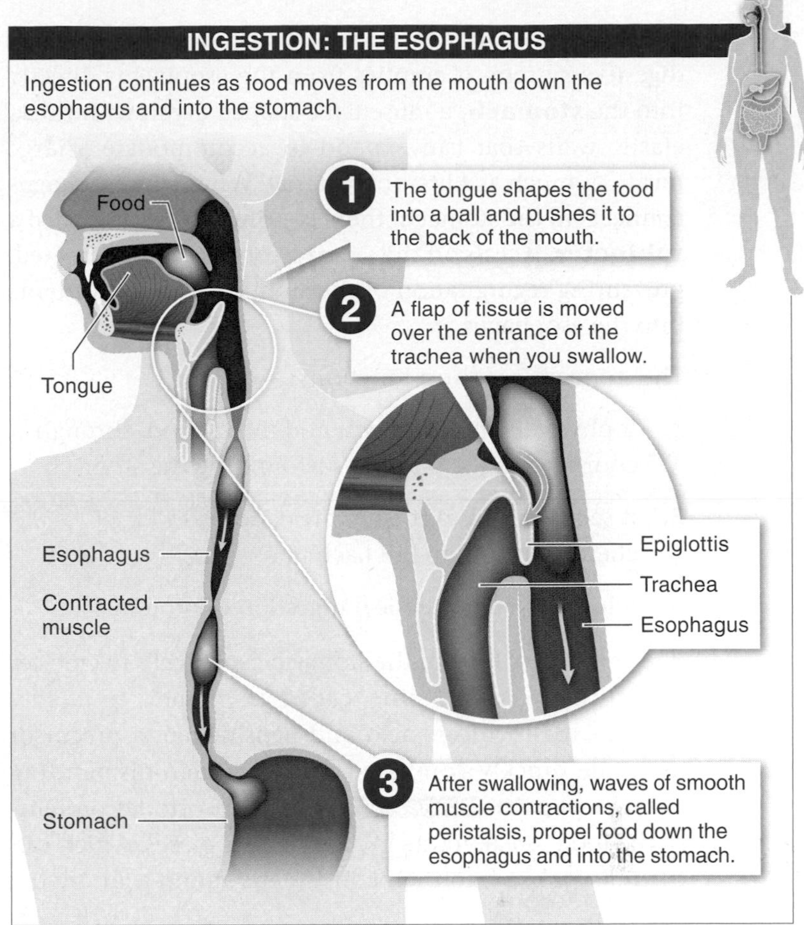

INGESTION: THE ESOPHAGUS
Ingestion continues as food moves from the mouth down the esophagus and into the stomach.
Food
Tongue
1 The tongue shapes the food into a ball and pushes it to the back of the mouth.
2 A flap of tissue is moved over the entrance of the trachea when you swallow.
Epiglottis
Trachea
Esophagus
Esophagus
Contracted muscle
Stomach
3 After swallowing, waves of smooth muscle contractions, called peristalsis, propel food down the esophagus and into the stomach.

While you are chewing your food, your tongue forms the food into a ball shape that can be swallowed. The saliva-coated food is pushed to the back of your mouth by your tongue. There, your throat opens to two passageways, the **trachea** and the **esophagus** (FIGURE 23-17). The trachea, also called the windpipe, connects to your lungs. The esophagus connects to your stomach. Food destined for your stomach is kept from entering your trachea by a fast but critical maneuver in which your voice box moves up (due to muscle contractions), causing a flap of tissue (called the epiglottis) to be pushed over the entrance to the trachea just as you begin to swallow. If the ball of food you swallow is too big, it can get stuck at the beginning of the esophagus, wedging against the flap of tissue in front of the trachea—and this can cause you to choke by blocking air from getting into your lungs.

Once the ball of chewed food makes its way into the esophagus, waves of smooth muscle contractions, called **peristalsis,** propel the food down the esophagus and into the stomach, where digestion continues. Because of peristalsis, it's not necessary for you to be sitting upright or standing when you eat. Even if you are standing on your head, the food is pushed down the esophagus to the stomach.

TAKE HOME MESSAGE 23.9

» Ingestion is the first phase of the digestive process. Usually lasting less than a minute, ingestion involves tearing and grinding a food item in preparation for passing it to the stomach. Digestion also begins during this phase, with some starch being broken down by enzymes in saliva.

23.10 Digestion dismantles food into usable parts.

Food is fuel. But first, it must be completely broken down into its fundamental chemical constituents. Only then can it be absorbed into the bloodstream and used constructively. The process of dismantling large pieces of food, physically and chemically breaking them down into absorbable molecules, is digestion. It occurs primarily in the stomach and small intestine.

The Stomach As a ball of food moves through the digestive system, it empties from the esophagus directly into the **stomach,** a muscular J-shaped organ with thick, elastic walls that can expand to accommodate a large meal—as much as 4 liters of material! Where the esophagus connects to the stomach, there is a ring of muscle, called a **sphincter.** It seals off the stomach once food has entered, preventing regurgitation of the stomach's acidic contents into the esophagus.

The stomach has three functions:

1. It physically breaks down and mixes food, through churning of the muscles surrounding the stomach.
2. It secretes acid to further break down food chemically and to kill bacteria.
3. It begins some chemical digestion of proteins.

The presence of food in the stomach causes cells in crevices within the stomach lining, called gastric pits, to rapidly produce hydrochloric acid and pepsinogen, a precursor molecule quickly converted to the protein-dismantling enzyme **pepsin** (FIGURE 23-18). The acid corrodes proteins in the food, breaking them into smaller pieces, which can then be broken into their constituent amino acids by the pepsin.

The burning sensation of indigestion occurs when the sphincter between the esophagus and the stomach leaks, and acidic contents of the stomach move back up the esophagus. It usually occurs when a person eats too much or too quickly. Antacids can neutralize acid from the stomach. Remember, though, that the high acidity in the stomach is important for digestion, so while reducing it may alleviate heartburn, antacids can reduce the digestive efficiency of the stomach.

Q What is indigestion? How do antacids cure it?

Besides helping to chemically break down food in your stomach, the strong acid also kills most of the bacteria that you might consume. Generally, bacteria can't survive when the pH gets so acidic (although some microbes occasionally sneak through in higher-pH pockets of food passing through the stomach).

The end result of all the churning and digesting is that the food you have eaten is no longer recognizable. It becomes a creamy, acidic liquid called **chyme.** While some carbohydrates and proteins have been partly or completely digested in the stomach, fiber is undigested, as are the lipids, which are suspended as greasy droplets within the chyme. At the other end of the stomach is another ring of muscle. About every 20 seconds, it opens just a tiny bit and squirts a few tablespoons of chyme into the small intestine.

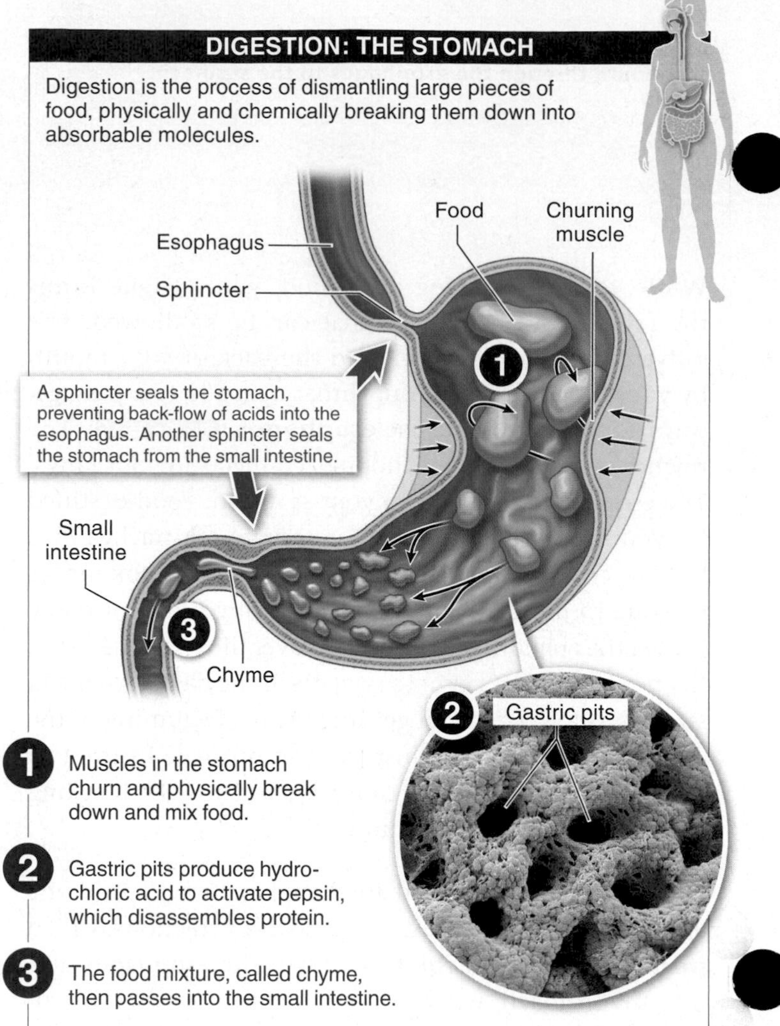

DIGESTION: THE STOMACH
Digestion is the process of dismantling large pieces of food, physically and chemically breaking them down into absorbable molecules.
Esophagus
Sphincter
Food
Churning muscle
A sphincter seals the stomach, preventing back-flow of acids into the esophagus. Another sphincter seals the stomach from the small intestine.
Small intestine
Chyme
Gastric pits
1 Muscles in the stomach churn and physically break down and mix food.
2 Gastric pits produce hydrochloric acid to activate pepsin, which disassembles protein.
3 The food mixture, called chyme, then passes into the small intestine.

FIGURE 23-18 **The role of the stomach in digestion.**

The Small Intestine Digestion only *begins* in the mouth and stomach. Most chemical digestion actually occurs in the **small intestine,** a long thin tube connected to the stomach. It is about 20 feet long and winds all

DIGESTION: THE SMALL INTESTINE

Although digestion begins in the mouth and stomach, most of it occurs in the small intestine.

1. As a sphincter at the end of the stomach relaxes, small amounts of chyme are squirted into the small intestine.
2. The liver produces bile, which travels from the gallbladder, where it is stored, to the small intestine, where it separates globules of fat into tiny, more easily digested and absorbed droplets.
3. The pancreas secretes pancreatic juice, which neutralizes the chyme and aids in the digestion of carbohydrates, proteins, and fats.
4. Cells within the walls of the small intestine produce enzymes that further digest fats, carbohydrates, and proteins.

Digestion is completed in the small intestine, at which point carbohydrates, proteins, and fats have been broken down into simple sugars, amino acids, and fatty acids.

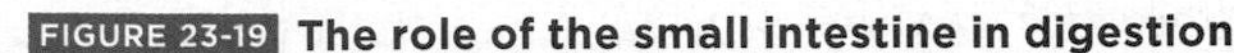

FIGURE 23-19 The role of the small intestine in digestion.

around your abdominal cavity (how else could 20 feet of tubing be packed into such a small area?). The small intestine begins its job of chemical digestion once the sphincter at the end of the stomach relaxes, creating an opening through which small amounts of chyme are squirted. And as the creamy substance is pushed through the small intestine by the rhythmic contractions of peristalsis, the macromolecules are gradually digested, with help from the pancreas and the liver (FIGURE 23-19).

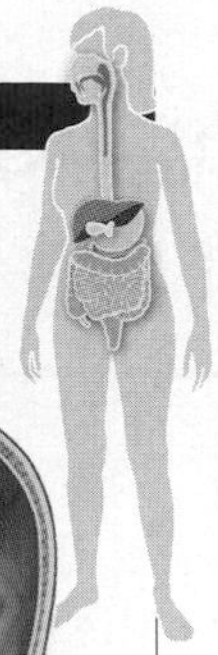

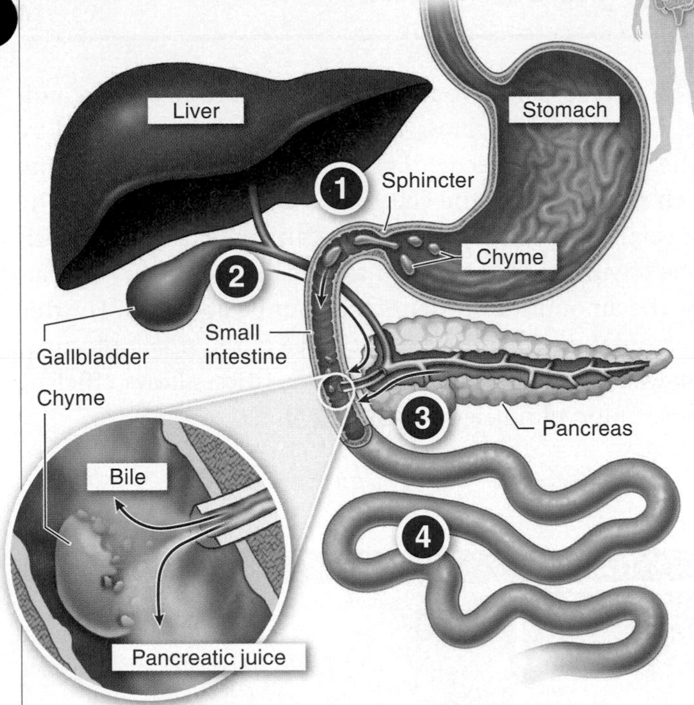

The **pancreas,** nestled at the point where the stomach connects to the small intestine, plays a central role in digestion by secreting pancreatic juice through a duct into the beginning portion of the intestine. This juice is a mixture of chemicals that neutralize the acidic chyme and enzymes that digest carbohydrates, proteins, and fats. Another organ that assists the small intestine with digestion is the **liver,** which produces **bile,** a juice that aids in the breakdown of fats. The liver sends the bile to the gallbladder, and from there it passes through a small duct into the small intestine, where the bile initiates the first step in fat digestion.

Lipids are not water-soluble, so bile initiates the first step in fat digestion by separating globules of ingested fat into tiny, more easily digested and absorbed droplets that pancreatic enzymes can successfully break down. Enzymes generated by cells within the walls of the small intestine further digest fats, carbohydrates, and proteins.

Digestion is completed in the small intestine when the consumed carbohydrates, proteins, and fats have been broken down into their component parts: simple sugars, amino acids, and fatty acids. These simple molecules can then be absorbed into the bloodstream, a process that we explore in the next section.

TAKE HOME MESSAGE 23.10

» Digestion—the process of dismantling large pieces of food, physically and chemically breaking them down into absorbable molecules—is the second phase of the breakdown of food in animals. It occurs primarily in the stomach and small intestine.

23.11 Absorption moves nutrients from your gut to your cells.

Chewing, tearing, churning, acidifying, dismantling. You put a lot of energy into reducing food to its simplest component molecules. Eventually, though, there is a payoff. That payoff is **absorption,** the process by which the energy-rich food particles are taken from the digestive tract into the bloodstream and then into cells throughout the body, where they can be used for energy and building materials. Absorption occurs mainly in the small intestine; however, a few molecules are absorbed in the stomach, including aspirin and alcohol, which explains why drinking alcohol on an empty stomach can lead to unexpectedly rapid inebriation.

The secret to effective absorption is surface area. For a molecule such as a sugar or an amino acid to be absorbed by the body, the molecule must be in direct contact with the cell membrane of the cell that is going to absorb it. The greater the number of cells that can come in contact with the chyme passing through the small intestine, the greater the amount of nutrients that can be absorbed. For this reason, the tremendous surface area of the small intestine, as well as other structural characteristics, allows efficient absorption of nutrients (FIGURE 23-20).

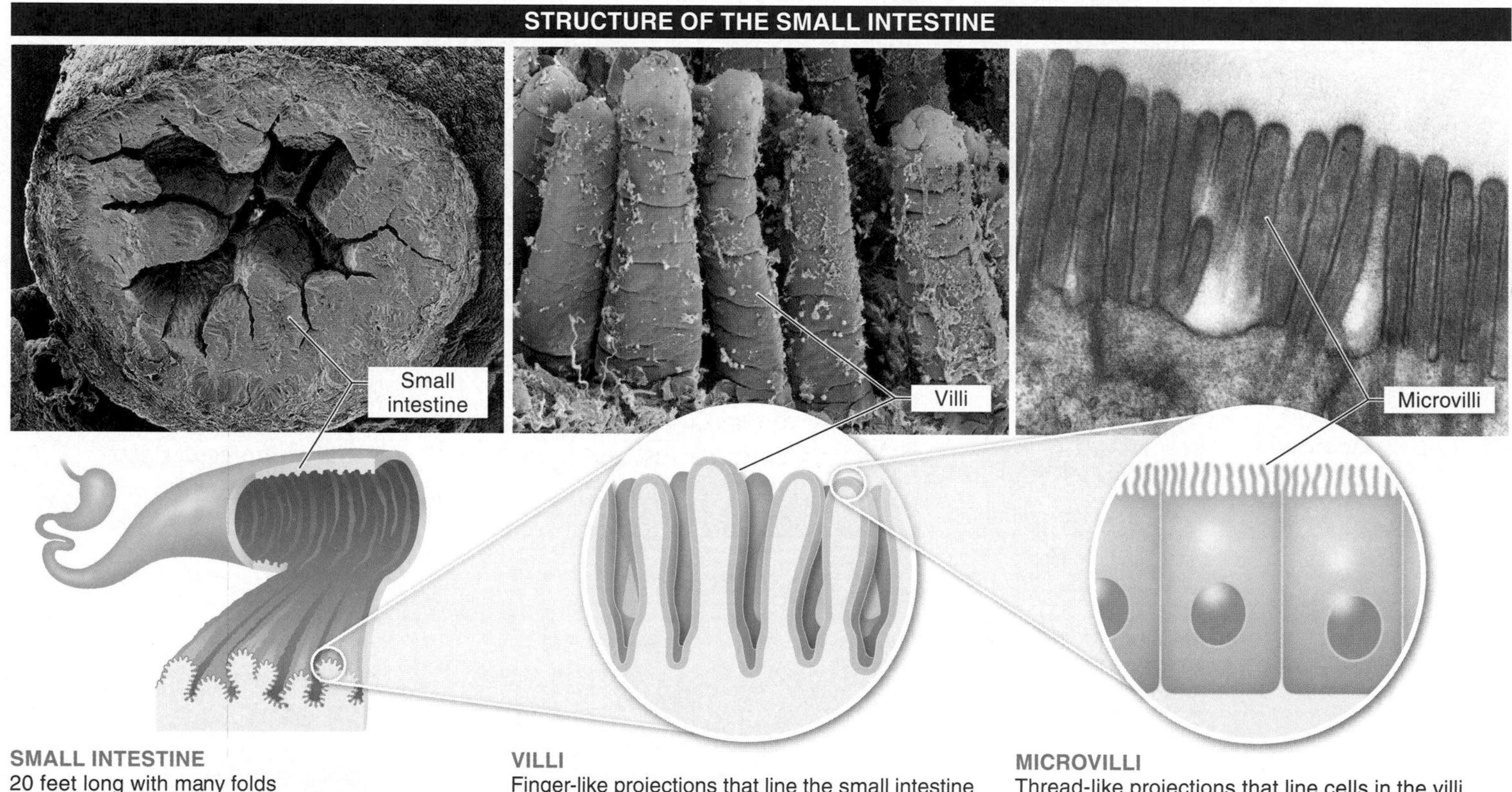

FIGURE 23-20 **The small intestine has a very large area for absorbing nutrients.**

The tremendous surface area of the small intestine allows very efficient absorption of nutrients.

1. As we noted in Section 23.10, the small intestine is long—about 20 feet.

2. Rather than being a straight, smooth tube, it has many folds.

3. The interior lining of the small intestine is made up of thousands of small finger-like projections, called villi. These create lots of nooks and crannies where chyme can come in contact with absorptive cells.

4. Each of the cells along the villi has hundreds of its own tiny thread-like projections, called microvilli. These create micro-nooks and micro-crannies that allow more of the molecules in chyme to directly touch the membranes of absorptive cells.

The molecules that can be absorbed in the small intestine include simple sugars, individual amino acids (and, occasionally, short chains of two or three amino acids), fatty acids, vitamins, and minerals. When a nutrient molecule is small enough to be absorbed, it gravitates toward the villi lining the interior of the small intestine. The tiny particles become trapped in the microvilli and are drawn across the cell membranes into the cells (FIGURE 23-21). The nutrients then diffuse out of the cells and into the interstitial fluid bathing the cells. From here, the nutrients are picked up by capillaries and move into the bloodstream, where they can be delivered to the organs and tissues that need them.

Q Is it true that certain food combinations should not be eaten together?

There is a food myth that if certain foods are combined, digestion and absorption are impaired. This is not true. The myth ignores a couple of facts: the body is able to produce its digestive enzymes for fats, proteins, and carbohydrates simultaneously, and it can absorb nutrients regardless of which other nutrients are also in the digestive tract. In fact, the contrary is often true. Many foods enhance the absorption of nutrient molecules in other foods. Vitamin C in citrus fruits, for example, increases the efficiency of iron absorption from a meal of beef or beans.

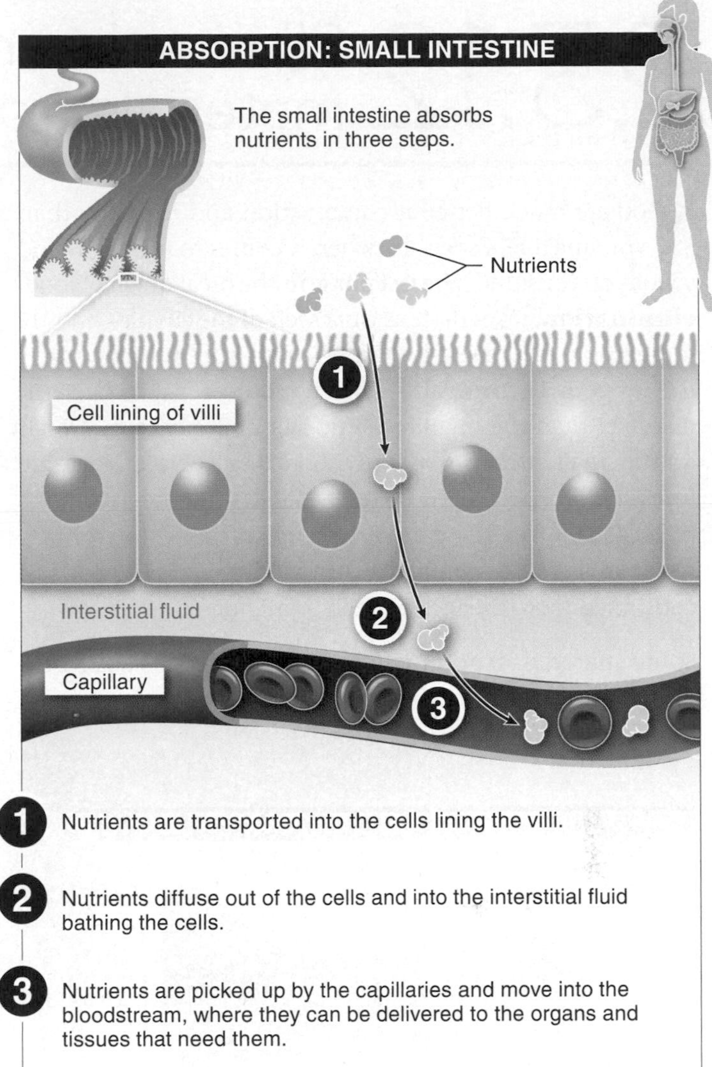
ABSORPTION: SMALL INTESTINE
The small intestine absorbs nutrients in three steps.
Nutrients
Cell lining of villi
Interstitial fluid
Capillary
1 Nutrients are transported into the cells lining the villi.
2 Nutrients diffuse out of the cells and into the interstitial fluid bathing the cells.
3 Nutrients are picked up by the capillaries and move into the bloodstream, where they can be delivered to the organs and tissues that need them.

FIGURE 23-21 Absorption moves nutrients into the bloodstream.

TAKE HOME MESSAGE 23.11

» Absorption is the process by which energy-rich food particles are taken up from the digestive tract into the cells of the body, where they can be used for energy and building materials. Absorption takes place primarily in the small intestine.

23.12 Elimination removes unusable materials from your body.

You are much better at conservation and recycling than you imagine. Especially when it comes to the fluids and solids you consume. The last phase in the breakdown of food, **elimination,** takes place as what's left of the chyme—mostly indigestible materials—leaves the small intestine and enters the **large intestine,** also called the **colon.** Much larger in diameter than the small intestine (about 3 inches vs. 1 inch for the small intestine) but only about 3–6 feet long, the large intestine serves to absorb water, salts, and some vitamins (FIGURE 23-22). The last part of the large intestine, the rectum, serves as a storage compartment for the remaining parts of consumed food, the feces, which is later defecated.

Some material in food cannot be digested or absorbed. This material is called fiber and includes plant gums and cellulose. For this reason, when fiber is consumed, it increases fecal mass. With additional mass, more water is attracted, speeding the movement of chyme through the colon, softening the feces, and making it easier to eliminate (that's why too much fiber can lead to diarrhea). Fiber also binds to bile, causing some of it to be eliminated, reducing the body's ability to absorb cholesterol from food. It is important to achieve just the right balance of water absorption in the large intestine. If too much is absorbed, the remaining indigestible material becomes too solid and can't move easily through the last part of the digestive system, causing constipation. Alternatively, if too little water absorption occurs, the body loses more water through diarrhea.

Q Fiber is an indigestible carbohydrate. If we can't digest it, how can it be essential in our diet?

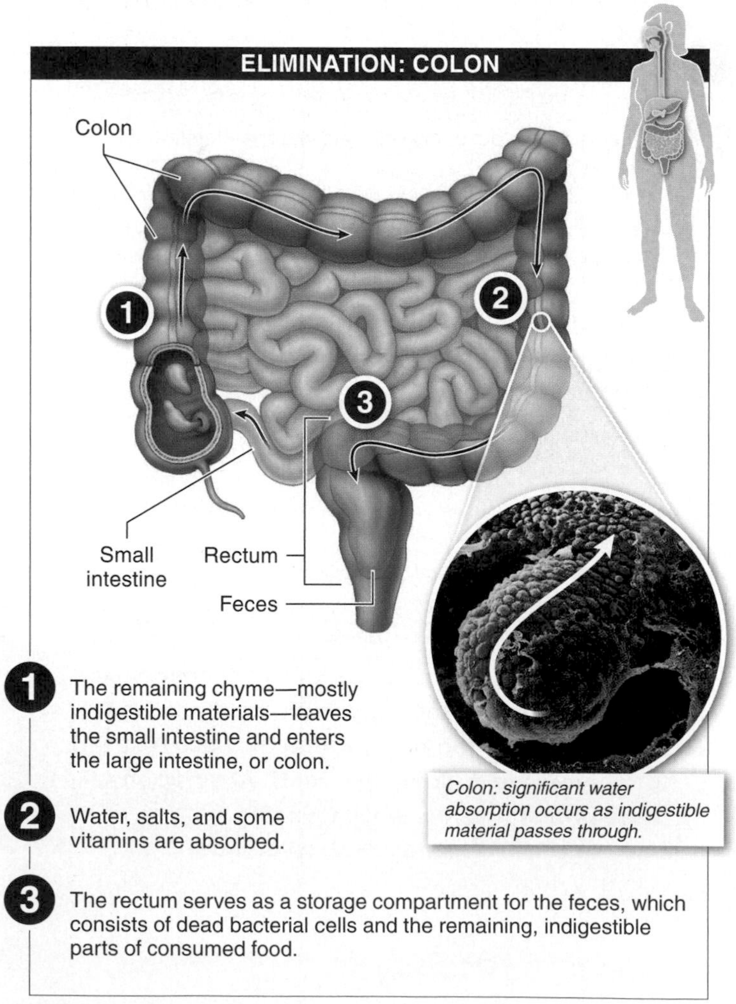

ELIMINATION: COLON

Colon
Small intestine
Rectum
Feces
1 The remaining chyme—mostly indigestible materials—leaves the small intestine and enters the large intestine, or colon.
2 Water, salts, and some vitamins are absorbed.
3 The rectum serves as a storage compartment for the feces, which consists of dead bacterial cells and the remaining, indigestible parts of consumed food.
Colon: significant water absorption occurs as indigestible material passes through.

FIGURE 23-22 **Elimination is the final step in the digestion process.**

Huge colonies of bacteria live in the colon. Before we are born, our guts are free of bacteria. During birth, we acquire some of our mother's vaginal and fecal bacteria. By the age of two, all of the numerous bacterial species present in adults have been acquired from food and the environment. At any given time, we generally contain more bacterial cells than our own body cells (see Figure 15-4). The bacteria are usually harmless and live off the undigested materials that end up in the colon. Bacteria also release important metabolic by-products, such as vitamin K and one of the B vitamins, biotin. About half of the feces that we excrete each day is made up of dead bacterial cells; the rest is mostly indigestible materials such as cellulose and other types of fiber.

Antibiotics frequently (and unintentionally) kill a large proportion of the colon bacteria, in addition to whatever illness-causing microbe they were prescribed to kill. This can have negative side effects. First, the transit of undigested materials through the colon may not be slowed down as much as usual, in which case less water is removed and diarrhea results. Also, a reduction in the production of vitamin K and biotin can lead to deficiencies.

Q Why can taking antibiotics lead to vitamin deficiencies?

In the colon, water is removed from feces through osmosis. First, salts are pumped out of the colon into surrounding cells, and then water moves out by simple diffusion (toward the area with more ions). This process can be manipulated as

Q Laxatives contain salts. How can this help them reduce constipation?

a treatment for constipation. Laxatives contain magnesium salts, which are so slowly absorbed that they are still largely intact when they reach the colon. Laxatives work by increasing the salt concentration in the colon, so that more water remains. This additional water makes it easier to excrete feces (but can sometimes cause diarrhea).

TAKE HOME MESSAGE 23.12

» The last phase of food breakdown takes place as the mostly indigestible materials leave the small intestine and enter the large intestine. There, water and ions are absorbed before the remaining materials are defecated.

23.13 Some animals have alternative means for processing their food.

Cellulose could feed the world. It is everywhere—it's the major carbohydrate making up the cell wall around plant cells—and a tremendous amount of energy is stored in its chemical bonds. But most animals don't produce (or acquire in any other way) the enzymes that break down cellulose. Those with the appropriate enzymes, however, gain access to one of the most plentiful sources of chemical energy on the planet. Let's investigate how some animals do it.

Ruminant animals, including cows, bison, deer, goats, and sheep, have complex four-part stomachs in which they can digest plant matter that humans cannot (FIGURE 23-23). First, the grazing animals chew on the plant material for a while, grinding the tough cell walls. Then they swallow it, and it passes into the first part of their stomach, which contains a huge pool of enzymes and cellulose-digesting bacteria. There, the plant material gets broken down, and much of the cellulose is digested. To increase their energy-extraction efficiency, the animals regurgitate the food back into the mouth, chew it some more, and swallow it again.

Called "chewing the cud," the additional chewing further breaks up the plant cell walls so that the bacteria have easier access to the cellulose, digesting more of it. From the first part of the stomach, the food then passes through the remaining chambers, where some additional digestion takes place, before moving into the small intestine and continuing the usual path of digestion. Although cellulose is the primary component of the ruminant diet, ruminants also get a significant amount of protein every day by digesting many of the bacteria working in their gut.

FOUR-PART STOMACH

In ruminant animals, a complex four-part stomach has evolved that can digest the plant matter that humans cannot digest.

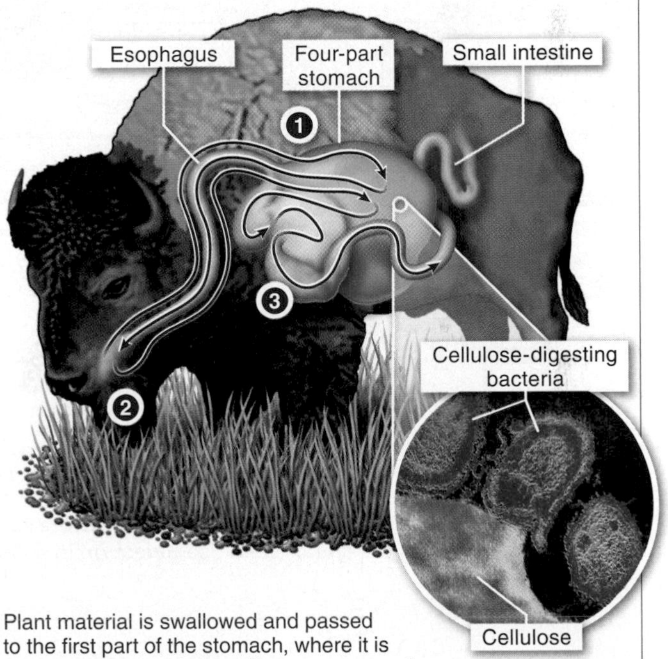

1. Plant material is swallowed and passed to the first part of the stomach, where it is broken down and digested by enzymes and cellulose-digesting bacteria.
2. Food is regurgitated, chewed more, then swallowed and passed to the first chamber again.
3. Food passes through the remaining chambers, where additional digestion takes place before it moves into the small intestine.

FIGURE 23-23 **Ruminants can digest plant material that humans cannot.**

THE CECUM

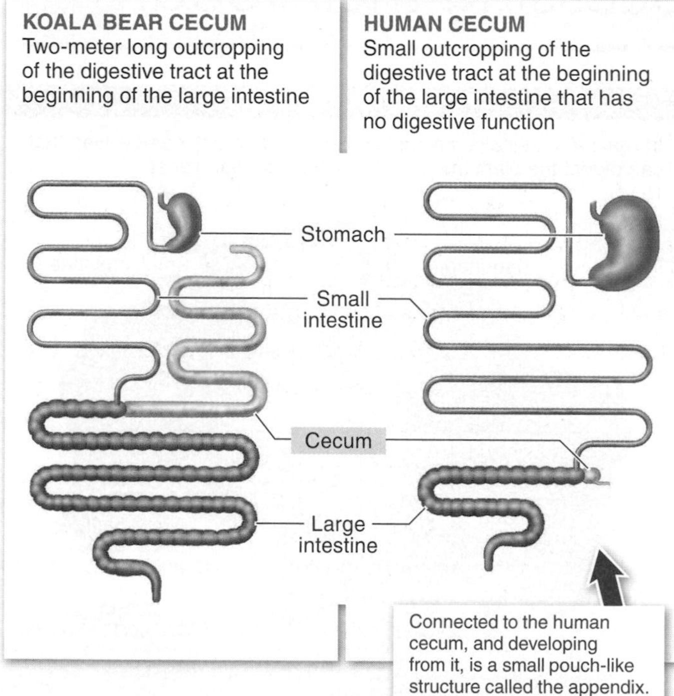

KOALA BEAR CECUM
Two-meter long outcropping of the digestive tract at the beginning of the large intestine
HUMAN CECUM
Small outcropping of the digestive tract at the beginning of the large intestine that has no digestive function
Stomach
Small intestine
Cecum
Large intestine
Connected to the human cecum, and developing from it, is a small pouch-like structure called the appendix.

FIGURE 23-24 **Koalas can eat foods that are indigestible for humans.** The cecum is the site of cellulose digestion in some animals.

Some insects, including silverfish, produce cellulose-digesting enzymes. These enzymes make it possible for them to eat books and paper, which are made of plant products containing cell walls made from cellulose.

Q Silverfish are insect pests in libraries rather than in kitchens. Why might this be? (Hint: they are one of the few species of animals to produce enzymes that digest cellulose.)

Other animals, lacking the large bacteria-filled stomach of ruminants, also employ cellulose-digesting bacteria to enable them to extract energy from cellulose. Horses, rodents, rabbits, and koalas have an outcropping of the digestive tract, called the cecum, in which bacteria live. Usually considered the beginning of the large intestine, the cecum is a separate chamber where the food goes for a while and the cellulose gets digested, before the food continues down the intestine. In animals that don't break down cellulose, the cecum is much smaller. The cecum is almost nonexistent in the meat-eating coyote, for example, while the similarly sized koala has a cecum that is 2 meters long. Within this lengthy chamber, bacteria convert shredded eucalyptus leaves into usable food (FIGURE 23-24). In humans, no cellulose-digesting bacteria live in the cecum. (Connected to the human cecum, and developing from it, is a small pouch-like structure called the appendix.)

Rabbits and rodents have mastered another way to increase their ability to extract energy from their food: they pass it through their entire digestive system twice. Eating some of their feces—a process called **coprophagy**—allows them to significantly increase their nutritional intake from their cellulose-laden diet. Numerous other species—including pigs, elephants, cats, gorillas, and chimps—also resort to coprophagy from time to time, although the reasons are not always clear and may be related to vitamin and mineral absorption or the acquisition of bacteria to aid in digestion.

TAKE HOME MESSAGE 23.13

» Most animals do not produce enzymes that break down cellulose. In ruminant animals, complex four-part stomachs have evolved with which the animals can digest plant matter that humans cannot, in part due to the presence of symbiotic cellulose-digesting bacteria. Other animals have alternative methods of utilizing cellulose-digesting bacteria.

23.14–23.18

What we eat profoundly affects our health.

What makes some foods better than others?

23.14 What constitutes a healthy diet?

At the most basic level, just two requirements—quality and quantity—must be satisfied in the design of a healthy diet (FIGURE 23-25). First, a diet must contain sufficient amounts of each of the six categories of nutrients: water, proteins, carbohydrates, lipids, vitamins, and minerals. And second, a healthy diet must contain sufficient energy to support an individual's metabolic needs without containing a surplus of calories. But it is not easy to find the best strategy to satisfy these requirements.

Balance in a diet is important, because no one food is completely adequate. Milk, for example, is a good source of protein and calcium but does not contain sufficient iron for most adults. Meats, on the other hand, are rich in iron (as well as protein) but generally are poor sources of calcium. For this reason, nutritionists recommend consuming a variety of foods from each of the basic food groups—(1) grains, (2) vegetables, (3) fruits, (4) milk and other dairy products, and (5) meats, poultry, and beans—and small quantities of unsaturated fats, such as olive oil and canola oil, and exercising daily to manage body weight.

The U.S. Department of Agriculture (USDA) has established dietary guidelines to help people plan a healthy diet. The USDA updates the guidelines every five years in order to incorporate advances in nutritional science. In addition to urging people to take steps to ensure that their food is safe to eat, the USDA also recommends the following:

1. Keep your weight within the recommended range.
2. Be physically active.
3. Choose a variety of fruits, vegetables, grains—particularly whole grains—and nonfat or low-fat milk and milk products.
4. Keep your diet low in saturated fat, cholesterol, and total fat.
5. Keep your diet low in sugars relative to complex carbohydrates and fiber.

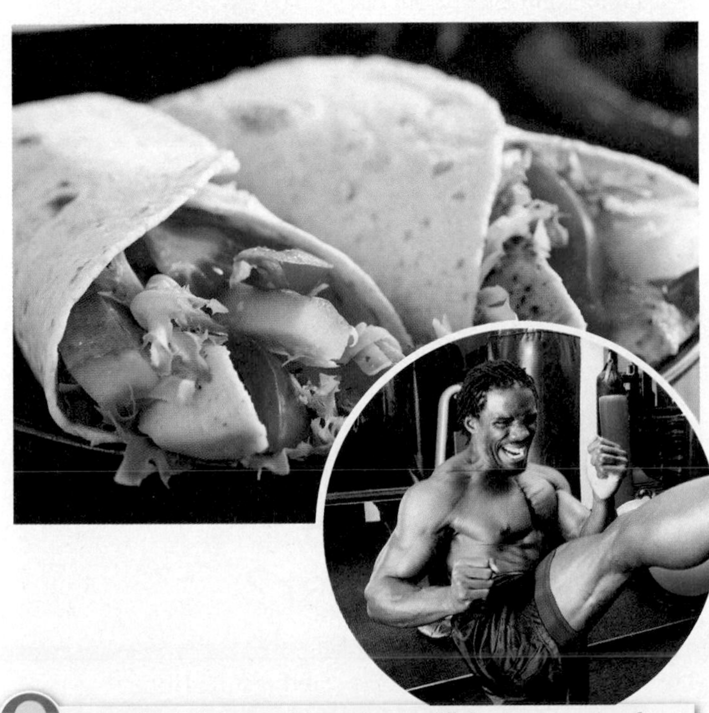

A healthy diet should contain sufficient amounts of each of the six categories of nutrients and just enough energy to support an individual's metabolic needs.

FIGURE 23-25 Food quality and quantity are the main factors to consider in choosing a healthy, balanced diet.

6. Keep your diet low in sodium.
7. If you consume alcoholic beverages, do so in moderation.

Food labels are a useful tool that can help in adhering to these guidelines and selecting a balanced diet (FIGURE 23-26). Ingredient lists, too, are very valuable, listing every ingredient in order of amount (by weight). Ultimately, following these guidelines can help reduce the incidence and severity of chronic diseases.

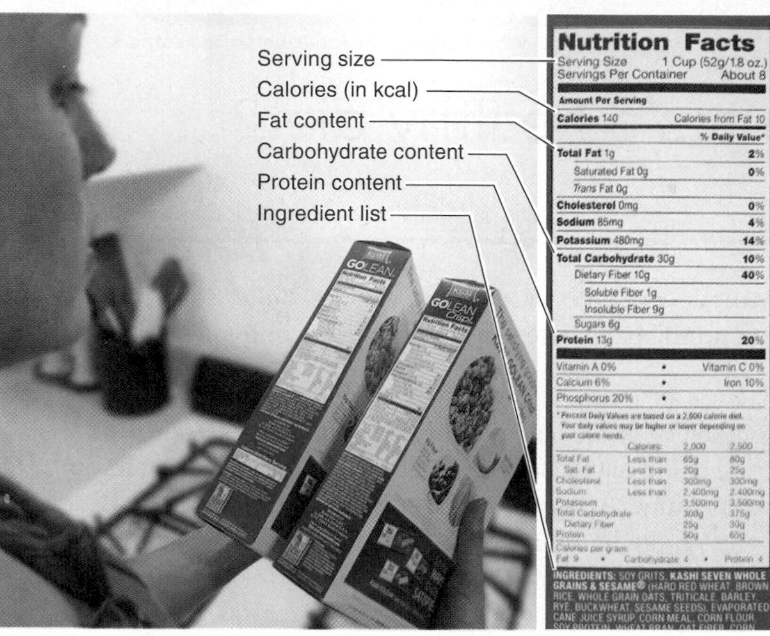

FIGURE 23-26 Food labels provide a wealth of nutritional information.

Q Why is it increasingly difficult to maintain a healthy weight as we get older?

Complicating the USDA's dietary guidelines is the fact that we need to reduce our caloric intake as we get older. Metabolic rate falls slowly but surely, beginning around age 30. Consequently, without an increase in activity, eating the same amount of food—even if it is a healthy diet—leads to weight gain. Most adults gain about half a pound every year throughout their thirties, forties, and fifties.

It is important to note that people who do not consume meat, poultry, fish, and/or milk products can still have a balanced diet. Legumes, seeds, and nuts can provide many of the same nutrients found in meat, including protein. Dark leafy vegetables can provide iron, another nutrient plentiful in meat. And in fact, because such diets tend to be lower in fat content, they can be valuable in helping to maintain a healthy body weight.

TAKE HOME MESSAGE 23.14

» A balanced diet contains adequate amounts of essential nutrients and energy, but not surplus amounts, and is low in substances—including saturated fats, cholesterol, sugar, salt, and alcohol—that can have adverse health effects when consumed in greater quantities.

THIS IS HOW WE DO IT

Developing the ability to apply the process of science

23.15 Does human judgment depend on blood sugar?

A balanced diet is essential for health—from simply surviving to maximizing physiological functioning. But what about for decision making? To what extent is brain functioning influenced by our diet and nutrition?

Three researchers recently reported results from a clever study designed to evaluate some of the factors that may influence cognitive reasoning. Their approach is an elegant reminder that rigorous scientific thinking and investigation doesn't necessarily involve a clear "control group" versus "experimental group" comparison.

The researchers examined real-world data on judicial rulings, to address the question: "Does the outcome of legal cases depend solely on laws and facts?"

Can you rephrase that question as a null hypothesis?

Specifically, they set out to determine whether "extraneous factors can sway highly consequential decisions of expert decision makers." They evaluated all the decisions made by judges presiding over two parole boards in Israel, over the course of 50 days during a 10-month period. This included 1,112 judicial rulings. Serving four major prisons, these two parole boards process approximately 40% of all parole requests in Israel.

When a prisoner appeared before a parole board, the judge was given information on (1) the number of previous incarcerations, (2) the seriousness of the crime, and (3) the number of months already served in prison.

Each day, each parole board heard 14–35 cases (mean = 24) in succession. The cases were presented in a random order, and the judge deliberated for an average of 6 minutes per case. The judge then ruled to either "accept" or "reject" each parole request, and the time of the decision was recorded. The default decision was to reject: 64.2% of requests were denied.

This is a chapter on "Nutrition and Digestion." Where's the connection here?

The routine of the parole board hearings is very regimented (and is recorded by a clerk). In the study period, the judges had a late-morning snack (starting between 9:49 and 10:27 A.M. and lasting an average of 38 minutes) and a lunch break (starting between 12:46 and 2:10 P.M. and lasting an average of 57 minutes), so each day's hearings were divided into three decision sessions.

Why would anyone care about the "feeding schedule" of the judges?

The researchers were interested in the food breaks because many published research reports have presented evidence that mental fatigue can interfere with brain functioning. For example, one study reported that car buyers, after being asked to make numerous consecutive decisions about which options they preferred, became more likely to simply accept the default option in subsequent choices. And we know that consumption of food increases blood glucose levels, which can replenish mental resources.

Back to the hypothesis. How did the researchers evaluate whether extraneous factors influenced the judges' decisions?

Presenting their results in the *Proceedings of the National Academy of Sciences*, the researchers plotted the proportion of favorable decisions as a function of the time of day the case was heard, as shown below.

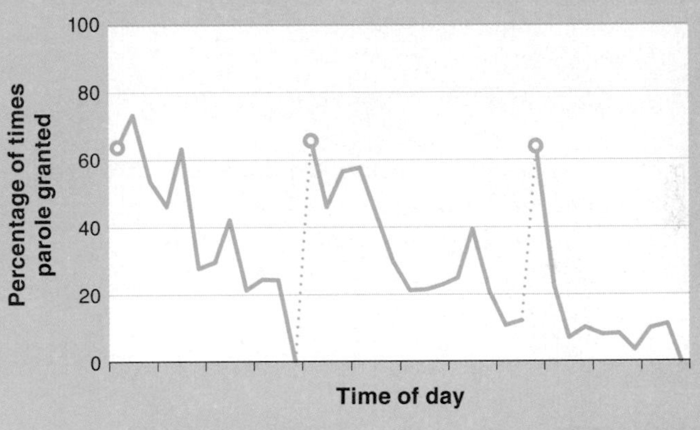

Turn to p. 827 for a closer inspection of this figure.

They found a clear and dramatic pattern: a parole request was much more likely to be granted at the beginning of the day or just after a food break—indicated by a dotted line in the graph. As each session wore on, the probability of a favorable ruling decreased, until it was close to 0%. Following a food break, the probability of a favorable ruling returned to about 65%. The first decision of each session is circled in the graph.

If the null hypothesis were true, what would the graph look like?

Because the order in which cases were presented was random, if the cases were being decided based solely on law and the facts of the case, the probability of a favorable decision should have been the same regardless of what time of day the case was heard.

What can you conclude from these results?

The researchers concluded that mentally fatigued, hungry judges fall back on the easier default position of denying a request for parole. Engaging in cognitive reasoning depletes blood sugar, making additional mental effort difficult and unpleasant; ingesting food restores one's decision-making abilities. (To further support their conclusions that favorable decisions required greater mental effort, the researchers noted that those decisions took 41% longer and the written decisions were almost twice as long.)

What other possible explanations should be explored?

In their analyses, the researchers tested several additional hypotheses. And, while they found that prisoners without previous incarceration were more likely to be granted parole, they also observed that the likelihood of getting parole was *not* influenced by the severity of the crime, the amount of time already served, or the prisoner's sex or ethnicity.

Do these results matter except for prisoners up for parole?

Viewed more broadly—as an investigation of the impact of nutritional state on decision making—these results may have numerous important implications. Can you see their relevance for doctors? Or stockbrokers? Or university admissions committees?

Can you propose some next steps that should be taken in studying this topic?

TAKE HOME MESSAGE 23.15

» Proper diet and nutrition are necessary not just for health and fitness but also for effective brain functioning. Human cognitive reasoning, as demonstrated by judicial decision making, is significantly influenced by hunger and food intake.

23.16 Obesity can result from too much of a good thing.

Jerry and Louis Kahn are two brothers who really love to eat. The pair tip the scales at about 500 pounds each—and their hunger never subsides. Doctors have pinpointed the exact physiological source of their problem. They both possess an altered gene that renders their bodies unable to recognize the internal measure of food intake and register "enough," much like a broken thermostat that never shuts off a heater. Jerry and Louis are eternally hungry because their brains never get the "we're full" message from their stomachs.

Like the Kahns, we all have inherited genes from our parents that significantly influence our body weight. All humans inherit dozens (possibly hundreds) of genes that influence body weight, and even in their "normal" condition, these genes are likely to lead to weight problems (FIGURE 23-27). But this raises a couple of questions. Why would it be in our genetic interests to have huge appetites? Why would natural selection favor such genes? To answer these questions, we must consider the harsher, less predictable environment in which *Homo sapiens* evolved.

Our ancestors lived off the land by hunting animals and gathering plants. Under such conditions, it was unclear where the next meal was coming from. Powerful, instinctual hunger kept our ancestors going in that tough, energetically demanding world. Those individuals who were enthusiastic rather than restrained eaters were most likely to survive periods of food shortage and live to reproduce. Recent studies using PET scans (positron emission tomography, a type of imaging) have supported this, demonstrating that just the sight of food sets off activity in the brain's pleasure centers—a mechanism that generally leads to consumption of food when it is available.

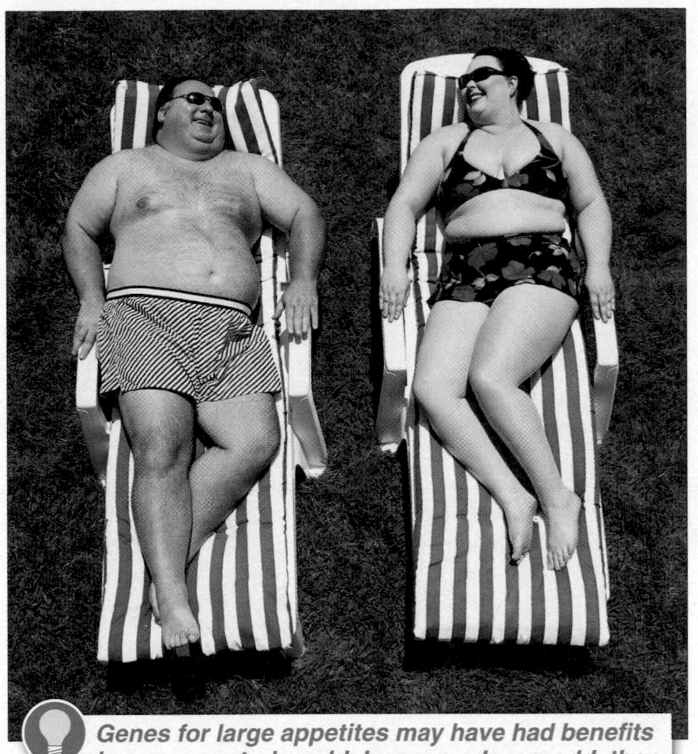

Genes for large appetites may have had benefits in our ancestral world. In our modern world, they may lead to weight-management problems.

FIGURE 23-27 Are your genes making it difficult for you to maintain a healthy weight?

The consequence of our perpetual hunger is not news: one of every four Americans is obese. In terms of size, plumpness gets labeled "obesity" when our body mass index hits 30. **Body mass index,** or **BMI,** equals body weight in kilograms divided by height in meters squared (kg/m^2). A BMI of 30 translates to about 209 pounds if you are 5 feet 10 inches tall, and 180 pounds if you are 5 feet 5 inches. With a BMI of 25 or higher but under 30, you are considered "overweight." BMI isn't the only way to evaluate obesity, and it has some important problems as a measure of body fat. Because the measure does not consider body composition, professional athletes and bodybuilders with unusually large muscle mass, for example, may have BMIs that fall within the obese range. Still, it is a useful and easily obtained measure (FIGURE 23-28). Repeatedly, around the world, as societies get richer and as individuals age, they tend to become fatter. Most of us would reduce our risk of heart disease, stroke, and diabetes if we lost even as little as 10 pounds. But because of the "famine-fearing" genes we carry, this is easier said than done.

BODY MASS INDEX (BMI)

1 Find your height on the *y*-axis.

2 Move right across the row until you get to the column with your weight.

3 Move up to the top of the row to find out what your BMI is.

The BMI indicates the ratio between height and body weight; the chart shows healthy body-weight ranges.

	UNDER-WEIGHT (<18.5)	HEALTHY WEIGHT (18.5–24.9)						OVERWEIGHT (25–29.9)					OBESE (>30)					
Height / BMI	**18**	**19**	**20**	**21**	**22**	**23**	**24**	**25**	**26**	**27**	**28**	**29**	**30**	**31**	**32**	**33**	**34**	**35**
6′3″	144	152	160	168	176	184	192	200	208	216	224	232	240	248	256	264	272	279
6′2″	141	148	155	163	171	179	186	194	202	210	218	225	233	241	249	256	264	272
6′1″	136	144	151	159	166	174	182	189	197	204	212	219	227	235	242	250	257	265
6′0″	132	140	147	154	162	169	177	184	191	199	206	213	221	228	235	242	250	258
5′11″	129	136	143	150	157	165	172	179	186	193	200	208	215	222	229	236	243	250
5′10″	126	132	139	146	153	160	167	174	181	188	195	202	209	216	222	229	236	243
5′9″	122	128	135	142	149	155	162	169	176	182	189	196	203	209	216	223	230	236
5′8″	118	125	131	138	144	151	158	164	171	177	184	190	197	203	210	216	223	230
5′7″	115	121	127	134	140	146	153	159	166	172	178	185	191	198	204	211	217	223
5′6″	112	118	124	130	136	142	148	155	161	167	173	179	186	192	198	204	210	216
5′5″	108	114	120	126	132	138	144	150	156	162	168	174	180	186	192	198	204	210
5′4″	105	110	116	122	128	134	140	145	151	157	163	169	174	180	186	192	197	204
5′3″	102	107	113	118	124	130	135	141	146	152	158	163	169	175	180	186	191	197
5′2″	98	104	109	115	120	126	131	136	142	147	153	158	164	169	175	180	186	191

Body Weight (Pounds)

FIGURE 23-28 One indicator of healthy body weight: body mass index (BMI).

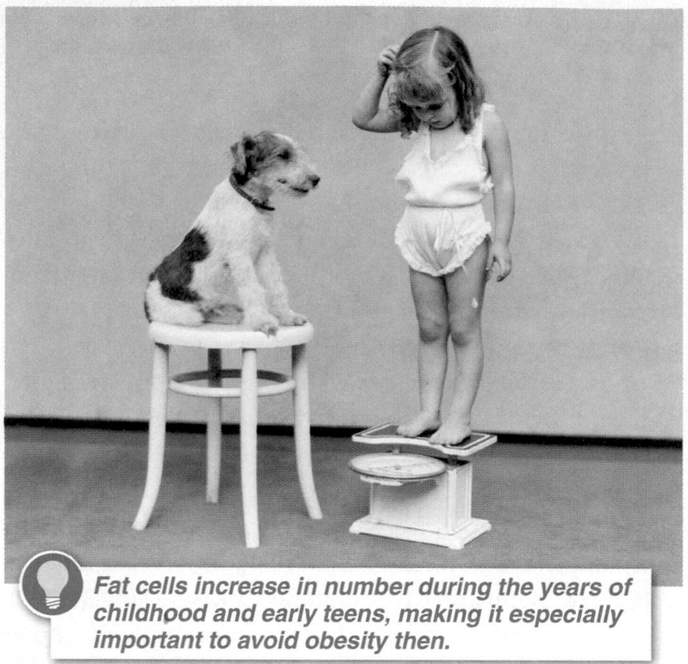
Fat cells increase in number during the years of childhood and early teens, making it especially important to avoid obesity then.

FIGURE 23-29 It's important to maintain a healthy body weight during childhood.

From the cellular perspective, the specific causes of obesity are an increase in the size and number of fat cells an individual carries. Excess calories can be converted to fat regardless of whether they come from fats, carbohydrates, or proteins. Whatever the source, the fat molecules are added to the fat cells in the body's adipose tissue. Fat cells increase in number primarily during the late years of childhood and in the early teens (FIGURE 23-29). After that, the cells tend to grow in size rather than number with excess caloric intake, although when a fat cell becomes too large it will divide. The reverse, unfortunately, is not true. When people lose weight, their fat cells become smaller but are never lost completely. Consequently, it is especially important to avoid obesity early in life.

The problems of obesity are not restricted to humans. Most primates in zoos are also overweight, as are many pets and most laboratory research animals. In each case, the reason is the same as for humans. In their natural world, food is limited and unpredictable. Animals overeat when they do not need to expend much energy to acquire food and food is plentiful. In the next section, we explore some weight-loss diets and why they are almost universally unsuccessful.

TAKE HOME MESSAGE 23.16

» Food supplies were unpredictable for our early ancestors, leading to the evolution of strong appetites. In the modern industrial world, such instincts have problematic consequences, leading most people to overeat and to weigh more than they would prefer and more than is healthy. Zoo animals, laboratory animals, and domesticated animals have a similar problem.

23.17 Weight-loss diets are a losing proposition.

"I can reason down or deny everything, except this perpetual Belly: feed he must and will, and I cannot make him respectable."

— RALPH WALDO EMERSON, ***Representative Men, 1850***

Weight loss is both a simple and a complicated problem. It is simple because there is one complete plan that guarantees success, requiring only five words of description: "Eat less. Move around more." Any animal of any species will lose weight when it expends more calories than it consumes.

But weight loss is also much more complex than this, and in the long run almost all "diets" fail. Current interventions designed to facilitate weight loss range from mild to extreme. They fall into three categories: drugs, surgery, and behavior modification. Each has both promising and problematic elements (FIGURE 23-30).

Drugs and Other Chemical Interventions

Xenical. One of the more promising of recently developed weight-control drugs is Xenical, a product that interferes

INTERVENTIONS TO FACILITATE WEIGHT LOSS

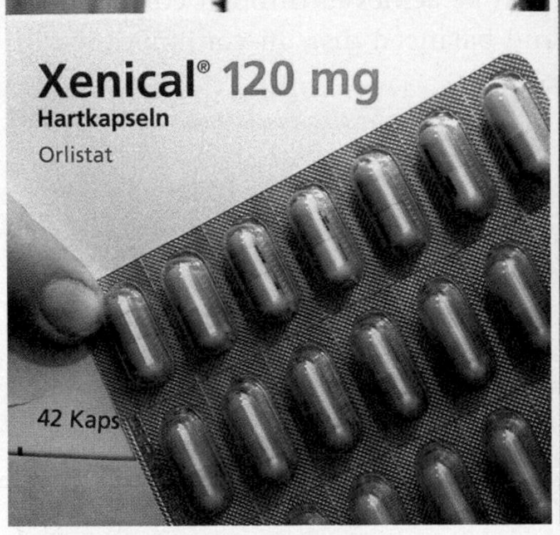

DRUGS AND OTHER CHEMICAL INTERVENTIONS

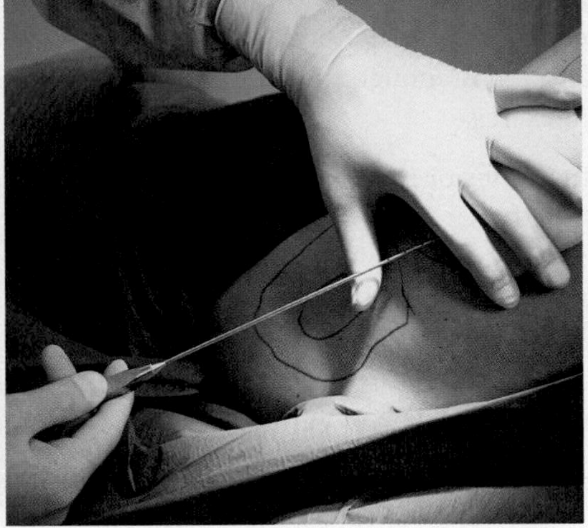
SURGERY

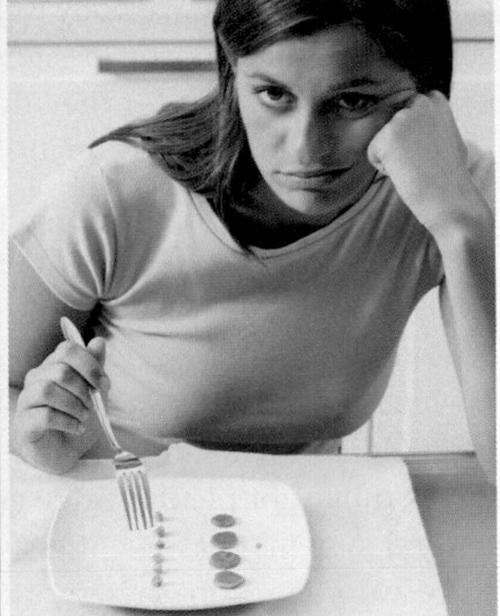
BEHAVIOR MODIFICATIONS: WEIGHT-LOSS DIETS

FIGURE 23-30 A wide variety of strategies are employed in the quest for weight loss.

with fat digestion by binding to lipases (fat-digesting enzymes) and blocking them from doing their job. As a consequence, some of the fat in the digestive system passes through the body without absorption.

Olestra and NutraSweet. Some artificially created molecules, such as Olestra and NutraSweet, can cause the same taste perception as if a molecule of fat or sugar, respectively, were present. Still, in several published research studies, subjects who unknowingly ingested fewer calories as a consequence of consuming products containing either NutraSweet or Olestra responded by increasing their caloric consumption in the following days. We're not tricked quite so easily.

Caffeine and other stimulants. A variety of products claim to increase the expenditure of energy without an offsetting increase in appetite. Clinical trials demonstrate that stimulants such as caffeine can produce short-term weight loss in the range of 5–10 pounds, but long-term studies reveal that most of the short-term losses do not last.

Surgery

Liposuction. In this procedure, doctors directly remove fat cells from various parts of the body, using a hollow tube and a suction device. Over time, however, individuals tend to regain all the lost weight (or more).

Bariatric surgery and stomach banding. In bariatric surgery, surgeons bypass a significant portion of the small intestine and seal off most of a person's stomach. This reduces the amount of food people can eat before becoming full as well as the ability of their small intestine to digest and absorb nutrients. The surgery carries significant risks, though, and leads to nutritional deficiencies in almost a third of all cases. Still, something can be said for bariatric surgery that cannot be said unequivocally for any drug or diet: it works. One study of more than 600 individuals who had undergone this surgery found that after 14 years, the average weight loss was 100 pounds!

Behavior Modification: Weight-Loss Diets

Portion control and general caloric restriction. Can we change our eating habits permanently? To answer this question, researchers put a group of monkeys on a very-low-calorie diet. The monkeys shed pounds initially, then stabilized at much lower weights for two years. Then they were given unlimited access to food. Did they maintain their new weights? No. These monkeys quickly returned to their original, pre-diet weights. Similarly, researchers observed a group of successful human dieters. The newly skinny people had lost an average of 70 pounds through a comprehensive program. Three years after completing the program, however, the participants had, on average, regained all of their lost weight.

General Problems with Weight-Loss Diets

The main problems with most popular weight-loss diets are that (1) they focus on reducing weight (even weight due to water) rather than reducing body fat; (2) they reduce muscle mass, the body tissue best able to burn fat; (3) because they reduce body weight too rapidly, they trigger defense mechanisms designed to preserve the body's energy reserves; and (4) they don't focus enough on the other side of the energy equation: exercise.

An evaluation by the National Institutes of Health of *all* the major diet programs concluded that there was no good evidence that any of the programs reliably led to long-term weight loss. A large study published in the *New England Journal of Medicine* came to a similar conclusion in 2009. Unfortunately, living in our modern, zoo-like environment of plenty, we're going to struggle with our natural systems that relentlessly seek out and efficiently store calories. Nonetheless, the National Institutes of Health notes that even modest weight loss (5% to 10% of total body weight) is likely to produce significant health benefits, and it can be achieved through consumption of a moderate and balanced diet, in conjunction with moderate exercise.

TAKE HOME MESSAGE 23.17

» Weight loss is both a simple and a complicated problem. There is only one plan that guarantees success: reduced caloric intake and increased caloric expenditure. Interventions designed to facilitate weight loss involve drugs, surgery, or behavior modification, none of which are reliably successful and safe.

23.18 Diabetes is caused by the body's inability to regulate blood sugar effectively.

When it functions properly, the human body can resemble a finely tuned, responsive machine. After you digest and absorb food, for example, there's an increase in the amount of glucose circulating in your bloodstream. This triggers the release of insulin by your pancreas. And insulin causes your body's cells, especially muscle cells and fat cells, to pull the glucose in. They can then either use it for energy or convert it to glycogen or fat for storage until the energy is needed some other time. This leads to a reduction in glucose in the bloodstream and brings your blood sugar level back down.

Foods differ in the extent to which they cause a surge in blood sugar and subsequent release of insulin. Foods that cause a rapid and large surge—such as orange juice, honey, and white potatoes—are classified as having a high **glycemic index** and are less desirable in the diet. More desirable are foods having a low glycemic index, those that cause only a slow, moderate increase in release of insulin. These include whole grains and beans. With reduced insulin response comes a more efficient utilization of the sugar and lipids in the bloodstream and a reduction in fat storage. Insulin surges can also be reduced by eating smaller meals or, if you consume foods that have a high glycemic index, eating them with other foods as part of a meal.

Even finely tuned machines can malfunction. More than 10 million Americans (and 100 million people worldwide) have problems with their insulin-response systems, a condition referred to as **diabetes.** These problems are chiefly of two types (FIGURE 23-31): either (1) the pancreas doesn't secrete enough insulin in response to an increase in blood sugar, or (2) the pancreas secretes plenty of insulin but the cells of the body don't respond to it, usually due to a deficiency in glucose receptors on their cell membranes. In both cases, blood sugar remains high, causing a host of problems.

Q Why must people who have diabetes inject insulin rather than taking it in pill form?

The first type of diabetes (called type 1) is believed to be an autoimmune disease and is most often diagnosed in children or young adults. Generally, people with type 1 diabetes have a pancreas that doesn't secrete enough insulin, and thus they can be treated by insulin injections (FIGURE 23-32). (Unfortunately, because insulin is a protein, it is digested by stomach acids and the enzymes of the small intestine, so insulin must be injected in order to get

DIABETES: TYPE 1 vs. TYPE 2

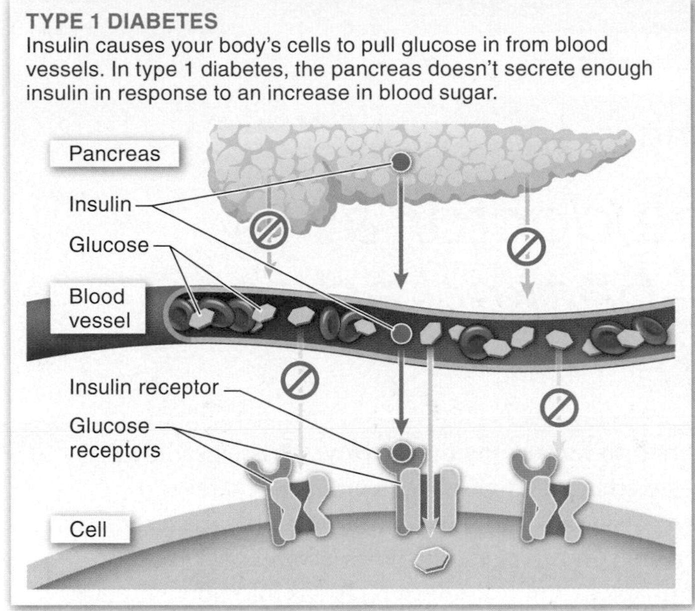
TYPE 1 DIABETES
Insulin causes your body's cells to pull glucose in from blood vessels. In type 1 diabetes, the pancreas doesn't secrete enough insulin in response to an increase in blood sugar.
Pancreas
Insulin
Glucose
Blood vessel
Insulin receptor
Glucose receptors
Cell

TYPE 2 DIABETES
The pancreas secretes plenty of insulin, but the cells of the body don't respond to it, usually due to a deficiency in glucose receptors on the cell membranes.

Pancreas
Insulin
Glucose
Blood vessel
Insulin receptor
Deficient glucose receptors
Cell

FIGURE 23-31 Diabetes is a disruption in the body's regulation of blood sugar.

it into the bloodstream intact.) The second type of diabetes (type 2) is about 10 times more common. It generally develops after the age of 40, and in about 90% of cases it is a consequence of obesity. It appears that chronic and excessive amounts of sugar in the diet (and, later, in the bloodstream) reduce the sensitivity and/or number of cellular insulin receptors that help control blood sugar. This causes even greater releases of insulin, which can, ultimately, wear out the insulin-producing cells of the pancreas. Type 2 diabetes is treated by minimizing the glucose fluctuations in the bloodstream. Weight loss is also encouraged, as it reduces insulin resistance.

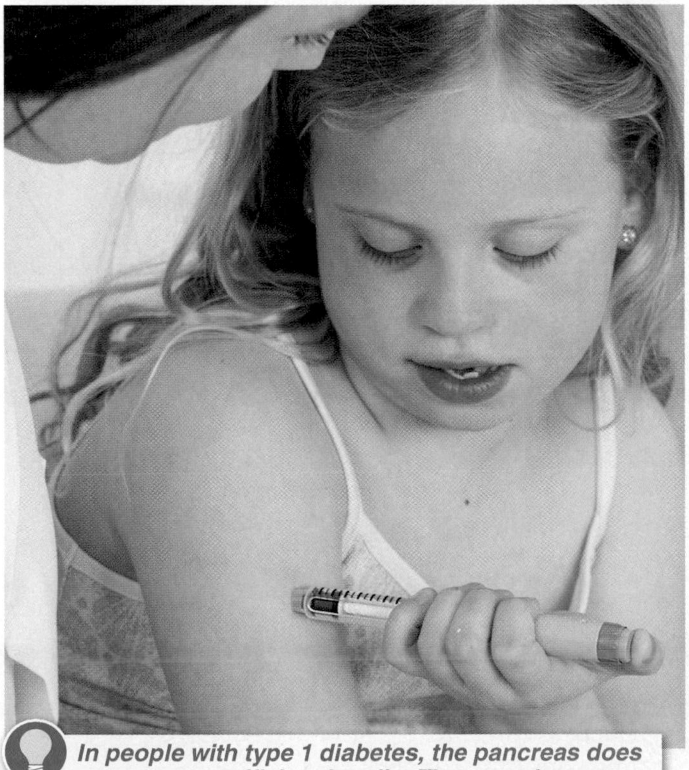

Because chronically high levels of blood sugar affect nearly all the cells of the body, the health effects can be far-reaching and severe. Cells in the eye lens can become deformed, causing blurry vision; blood vessels can be damaged, causing circulatory system and kidney malfunction; and nerves can also be damaged. Taken together, the varied problems resulting from diabetes make it the sixth leading cause of death in the United States.

In people with type 1 diabetes, the pancreas does not secrete sufficient insulin. They can, however, modulate their blood sugar by injecting insulin.

FIGURE 23-32 Living with diabetes.

TAKE HOME MESSAGE 23.18

» Digesting and absorbing food leads to an increase in the amount of glucose circulating in the bloodstream, which triggers the release of insulin by the pancreas; insulin causes the body's cells to pull the glucose in for energy or storage. Problems with regulation of blood sugar, called diabetes, affect millions of people and are caused by heredity or by poor diet.

STREET BIO

Using evidence to guide decision making in our own lives

"An apple a day keeps the doctor away": Silly rhyme? Sound, data-backed advice? Or something in between?

Q: What is the claim? A 2011 research study—described by media around the world, including *Time,* MSNBC, and CBS—touted the first experimental support for the adage "an apple a day keeps the doctor away."

Q: What is the purported apple advantage? The media stories reported that participants in the study experienced the following benefits from eating apples:

1. A 23% decrease in LDL (low-density lipoprotein, associated with a higher risk of cardiovascular disease).
2. A 4% increase in HDL (high-density lipoprotein, associated with a reduced risk of cardiovascular disease).
3. A 32% decrease in C-reactive protein (a protein associated with inflammation and a marker for cardiovascular disease).
4. A 33% decrease in lipid hydroperoxide (a chemical associated with plaque formation in arteries).
5. A loss of 3.3 pounds of body weight, on average.

Q: How did the researchers evaluate the benefits of apples? The study lasted for one year and included 160 post-menopausal women, aged 45–65, who were randomly assigned to one of two groups. In the "treatment" group, participants were assigned to eat 75 grams (2.7 ounces) of dried apples each day. In the "control" group, participants were assigned to eat 100 grams of prunes each day.

Q: What do the results mean for you? The headlines suggested that "an apple a day" confers a general and dramatic benefit—and they are correct in that regard. Unfortunately, this story illustrates a common difficulty in translating technical research reports into popular news stories. The research paper, published in *Journal of the Academy of Nutrition and Dietetics,* reports a rigorous study, notable for randomizing participants to the experimental groups and for including individuals—post-menopausal women—frequently overlooked in research. But how might this reduce the generalizability of the results? Why?

Also, of the five important results listed above, the researchers found statistically significant differences for only two findings: the LDL decrease and the lipid hydroperoxide decrease. Do you think the media should make claims about the 3.3-pound weight loss and the 4% increase in HDL levels when those results were not sufficiently different between treatment and control groups? Is there another way the media could present those results for people without statistical training?

Q: What can you conclude? Be cautious when reading sensational claims. When a study is presented in the form of a media-friendly sound bite, much can be lost in translation. Because there are limits to what researchers can definitively conclude from any study, read with caution. In this case, there are indeed two promising results. In post-menopausal women, daily apple consumption improved two measures of heart health: (1) LDL level, an important measure of the risk of cardiovascular disease, was reduced, and (2) lipid hydroperoxide, an indicator of plaque formation in arteries, was reduced. Though encouraging, these results fall short of the claims made by news reports that apples are a "miracle fruit" and that "everyone can benefit from consuming apples."

GRAPHIC CONTENT

Thinking critically about visual displays of data

1 Identify two aspects of this graph that support the hypothesis that judges' blood sugar level influences their decision making. Explain your answer.

2 If judges' blood sugar level did not affect their decision making, what would you expect this graph to look like?

3 What do the two dotted green lines represent?

4 Why are there no numbers on the "Time of day" axis?

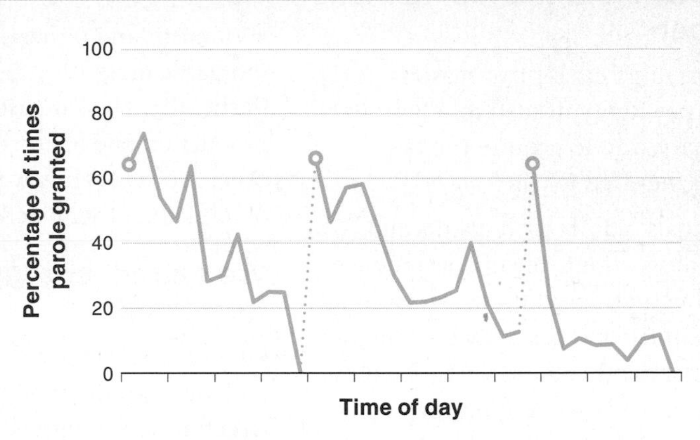

5 What do the green circles represent?

6 Why are there no error bars above and below the green circles (or other data points)?

7 After the third green circle, there is a steeper drop off than after either of the first two green circles. Why might this be?

See answers at the back of the book.

KEY TERMS IN NUTRITION AND DIGESTION

BRIEF SUMMARY

Food provides the raw materials for growth and the fuel to make it happen.

• Animals must eat for two reasons: to acquire the energy needed for all growth and activity, and to acquire the raw materials required for life.

• All animals require food. Plants and other photosynthetic organisms produce food through photosynthesis, and animals have one of three types of diet. Carnivores consume only other animals. Herbivores consume only plants. And omnivores consume both plants and animals.

• To function well, living organisms need sufficient energy, measured in kilocalories. The minimal energy needed by an individual not engaged in any activity is called its basal metabolic rate, or BMR.

Nutrients are grouped into six categories.

• Water is probably the single most important component of an animal's diet. It transports nutrients and waste materials throughout the body, takes part in metabolic reactions, serves as a solvent, lubricates body parts, and helps regulate body temperature.

• Animals consume three different types of macromolecules. Proteins provide raw materials for growth and the production of enzymes.

• Carbohydrates are the primary fuel on which animal bodies run. Dietary fats function primarily as a dense source of energy that can be efficiently stored in the body.

• Vitamins and minerals are organic and inorganic molecules, respectively, needed in the diet. They are used in the production and action of enzymes and other molecules involved in the processing of food and other biochemical reactions.

We extract energy and nutrients from food.

• The digestive process in humans includes four distinct phases during which food is progressively chewed up, broken down, and absorbed by the body, after which the non-usable portion of the raw materials is discarded as waste.

• Ingestion, the first phase of the digestive process, involves tearing and grinding food in preparation for passing it to the stomach.

• Digestion, the process of physically and chemically breaking down food into absorbable molecules, is the second phase of the breakdown of food. It occurs primarily in the stomach and small intestine.

• Absorption is the process by which energy-rich food particles are taken up from the digestive tract into the cells of the body, where they can be used for energy and building materials. Absorption takes place primarily in the small intestine.

• The last phase of food breakdown takes place as the mostly indigestible materials leave the small intestine and enter the large intestine. There, water and ions are absorbed before the remaining materials are defecated.

• Most animals do not produce enzymes that break down cellulose. In ruminant animals, complex four-part stomachs have evolved with which the animals can digest plant matter, in part due to the presence of symbiotic cellulose-digesting bacteria.

What we eat profoundly affects our health.

• A balanced diet contains adequate amounts of essential nutrients and energy, but not surplus amounts, and is low in substances that can have adverse health effects.

• Food supplies were unpredictable for our early ancestors, leading to the evolution of strong appetites. In the modern industrial world, such instincts have problematic consequences.

• Weight loss is both a simple and a complicated problem. Interventions designed to facilitate weight loss involve drugs, surgery, or behavior modification, none of which are reliably successful and safe.

• Digesting and absorbing food leads to an increase in the amount of glucose in the bloodstream, which triggers the release of insulin by the pancreas; insulin causes the body's cells to pull the glucose in for energy or storage. Problems with regulation of blood sugar, called diabetes, affect millions of people and are caused by heredity or by poor diet.

CHECK YOUR KNOWLEDGE

Short Answer

1. Why must herbivores spend a large portion of each day eating?

2. What is basal metabolic rate?

3. Why can it be dangerous to consume too much water?

4. What is meant by a "complete" protein?

5. Animals, including humans, require 20 different amino acids to make proteins. How many of these can be made in our body? Where do the others come from?

6. Why does over-consumption of fat-soluble vitamins pose a greater health risk

than over-consumption of water-soluble vitamins?

7. What are the four distinct phases used by humans to extract nutrients from food? During each phase, what significant activity contributes to nutrient harvesting and absorption?

8. Besides helping to chemically break down food in your stomach, what other function do stomach acids serve?

9. Which structural feature of the small intestine is crucial in helping the body get nutrients from food into the bloodstream?

10. Describe how some mammals, which do not produce enzymes to break down cellulose, are able to extract nutrients from cellulose-containing food.

11. Describe the role of insulin in human digestion and absorption.

Multiple Choice

1. Basal metabolic rate:

a) depends to a large degree on how active an individual is.

b) does not vary across mammalian species.

c) is a measure of the minimal energy needs of an individual not engaged in any activity.

d) is the same for males and females of any given species.

e) All of the above are correct.

2. Water has many functions in animals' bodies. These include all of the following except:

a) lubricating many joints, the spinal cord, and the eyes.

b) serving as a solvent for many vitamins and minerals.

c) transporting nutrients and waste materials throughout the body.

d) regulating growth and development.

e) All of the above are functions of water in animals' bodies.

0 EASY HARD 100 44

3. "Fiber" refers to a type of:

a) carbohydrate.

b) nucleic acid.

c) lipid.

d) amino acid.

e) protein.

0 EASY HARD 100 32

4. Proteins are an essential component of a healthy diet for humans (and other animals). Their most common purpose is to serve as:

a) fuel for running the body.

b) raw material for growth.

c) inorganic precursors for enzyme construction.

d) organic precursors for membrane construction.

e) long-term energy storage.

0 EASY HARD 100 59

5. Why do birds eat gravel?

a) Gravel contains most of the essential minerals for a bird's diet.

b) Birds have poor vision and have difficulty distinguishing gravel from small seeds.

c) By chewing on gravel, birds are able to sharpen their teeth, increasing their ability to crack open hard nuts or catch their prey.

d) The gravel collects in the gizzard (one of the two chambers of the bird stomach), where it helps to grind up the food they eat.

e) Because they have such a small brain relative to body size, birds tend to be the least intelligent of the vertebrates.

6. Cross-culturally, humans have developed ways of preparing food, including using heat and marinating food in acidic solutions, such as lemon juice. How might these be adaptations that help with digestion?

a) Heat and acid help disrupt the tissue of food items. This causes digestive enzymes to have increased contact with the food molecules and more easily break them down.

b) By reducing the need for chyme, these methods of food preparation increase the caloric value of food.

c) Because even the weakest acids are toxic to all bacteria, these methods of food preparation reduce the incidence of dietary-induced bacterial infection.

d) These methods increase the body's ability to extract energy from normally indigestible cellulose.

e) These food preparation methods reduce the need for water consumption.

0 EASY HARD 100 53

7. Which of the following statements about the small intestine is incorrect?

a) It is the primary site of digestion and of absorption of nutrients into the bloodstream.

b) It is the chief site of absorption of water by the digestive system.

c) It is the longest part of the digestive tract.

d) It receives secretions from the gallbladder.

e) It is the part of the digestive tract where all food macromolecules can be broken down into absorbable monomers.

0 EASY HARD 100 38

8. To leave the digestive tract and enter the cells of the body, a substance must cross a cell membrane. During which stage of digestion does this take place?

a) peristalsis

b) absorption

c) elimination

d) chemotaxis

e) ingestion

0 EASY HARD 100 33

9. Digestion and absorption:

a) involve the breakdown of food into small nutrient molecules (absorption) and passage of those molecules into the bloodstream (digestion).

b) both occur primarily in the large intestine, or colon.

c) are terms that describe the same process.

d) both occur primarily in the stomach.

e) involve the breakdown of food into small nutrient molecules (digestion) and passage of those molecules into the bloodstream (absorption).

10. Cows have large populations of bacteria in their digestive systems. Which of the following best explains why?

a) The mutualistic microbes combat the harmful microbes that may enter a cow's body with its food.

b) Cows are able to use cellulose-producing bacteria to help them digest their food.

c) Most cows do not actually have large populations of bacteria in their digestive systems. Only infected cows have these microbes.

d) Scientists put the microbes there so that they can study how the microbes affect cows' digestion.

e) The microbes metabolize the cellulose in the plants that cows eat.

0 EASY HARD 100 41

Ch24

What is the nervous system?

24.1 Why do we need a nervous system?

24.2 Neurons are the building blocks of all nervous systems.

24.3 The vertebrate nervous system consists of the peripheral and central nervous systems.

How do neurons work?

24.4 Dendrites receive external stimuli.

24.5 The action potential propagates a signal down the axon.

24.6 At the synapse, a neuron interacts with another cell.

24.7 There are many types of neurotransmitters.

Our senses detect and transmit stimuli.

24.8 Sensory receptors are our windows to the world around us.

24.9 Taste: an action potential serves up a taste sensation to the brain.

24.10 Smell: receptors in the nose detect airborne chemicals.

24.11 Vision: seeing is the perception of light by the brain.

24.12 Hearing: sound waves are collected by the ears and stimulate auditory neurons.

24.13 Touch: the brain perceives pressure, temperature, and pain.

The muscular and skeletal systems enable movement.

24.14 Muscles generate force through contraction.

24.15 Skeletal systems enable movement, among several other important functions.

The brain is organized into distinct structures dedicated to specific functions.

24.16 The brain has several distinct regions.

24.17 Specific brain areas are involved in the processes of learning, language, and memory.

24.18 *This is how we do it:* Can intense cognitive training induce brain growth?

Drugs can hijack pleasure pathways.

24.19 Our nervous system can be tricked by chemicals.

24.20 A brain slows down when it needs sleep. Caffeine wakes it up.

24.21 Alcohol interferes with many different neurotransmitters.

This blackfaced spider monkey of the Amazon rain forest is a fruit eater and spends most of its time in the upper canopy of the forest.

Nervous and Motor Systems

Actions, reactions, sensations, and addictions: meet your nervous system

24.1–24.3

What is the nervous system?

Huskies that work as sled dogs—which are descended from lineages that were crossed with sighthounds—have excellent vision.

24.1 Why do we need a nervous system?

Imagine a world without pain. Think of all that you could achieve. You could work harder, run farther, and just plain feel better. Maybe that's why so many products are marketed as "painkillers."

But would it really be a better world? Think again.

The case of Gabby Gingras tells us that we wouldn't really be happier in a pain-free world. Gabby feels no pain. She was born that way. At first it seemed like a blessing. But rather than being a gift that makes her life easier, Gabby's inability to feel pain is a crippling curse. She inadvertently damaged her eyes—permanently scratching one of her corneas—by poking her fingers into them. Chewing on plastic toys, she cracked most of her teeth. This condition, called "heritable sensory autonomic neuropathy," reveals that, no matter how unpleasant pain is to experience, it does have a very large benefit. The pains we feel alert us to the need to extricate ourselves from a dangerous situation and can prevent much greater suffering in the long run (FIGURE 24-1).

The survival of organisms depends on their awareness of the world and their ability to avoid physically harmful situations. Our **nervous system** accomplishes these tasks by letting us see, hear, feel, taste, smell, remember (and forget!), think about, act on, and react to various events and stimuli around us. Present in all multicellular animals other than sponges (and not found in plants), a nervous system is a network of cells that collects information about an organism's internal and external environments, processes that information, and sends signals to effectors, the muscles and glands that are capable of responding to the information.

Nervous systems have three critical features.

1. They receive input (stimuli) from the surrounding world.
2. They process that information.
3. They initiate responses to the internal and external environments, when necessary.

The ability to feel pain, no matter how unpleasant, makes us aware of the world around us and can prevent much greater damage in the long run.

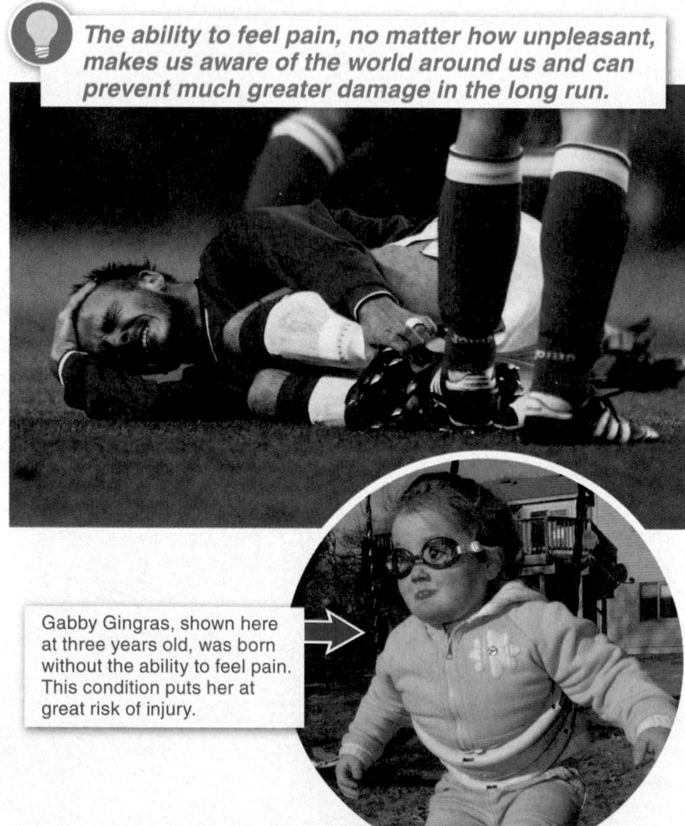

Gabby Gingras, shown here at three years old, was born without the ability to feel pain. This condition puts her at great risk of injury.

FIGURE 24-1 **Avoiding danger.** The sensation of pain gives us important information about the world.

In this chapter, we examine the organization of and structures found in nervous systems and the diversity of nervous system specializations that have evolved in the animal world.

TAKE HOME MESSAGE 24.1

» Present in all multicellular animals other than sponges, a nervous system is a network of cells that collects information about the organism's internal and external environments, processes that information, and responds by sending signals to muscles and glands.

24.2 Neurons are the building blocks of all nervous systems.

The **neuron** is a type of cell that is specialized for generating and conducting electrical impulses and is found in all animals, with the exception of sponges. It is the building block of all nervous systems (FIGURE 24-2). Each neuron is very small and has but a single option at any point in time: "fire" or "don't fire" (meaning that it either generates an electrical signal that can convey information to nearby cells, or it does not).

Collectively, groups of neurons bundled together with connective tissue—into structures called **nerves**—connect us to our world by enabling us to sense light, sound, touch, tastes, and smells and to respond to all of that sensory information. The accumulated information carried by all of our neurons—whether they are firing or not firing—is also responsible for information storage and retrieval and all thought in our brain.

Every day, about 9,000 of our neurons die—significantly more if you consume any alcohol or happen to inhale any gasoline fumes—and are not replaced. The 100 billion to one trillion you were born with, however, are more than enough for you to keep your wits about you.

Even though they perform such varied and significant functions, neurons are relatively standard eukaryotic cells. Each has a **cell body** that contains a nucleus, mitochondria, endoplasmic reticulum, and so on. But neurons also have two specialized structures that make them unusually adept at interacting with the external environment and with other cells.

The first of these two important specializations is the **dendrite.** A dendrite is the part of a neuron that is like an antenna: it senses and responds to stimulation from outside the cell and sends that information toward the cell body of the neuron. Numerous dendrites may branch out from the neuron cell body.

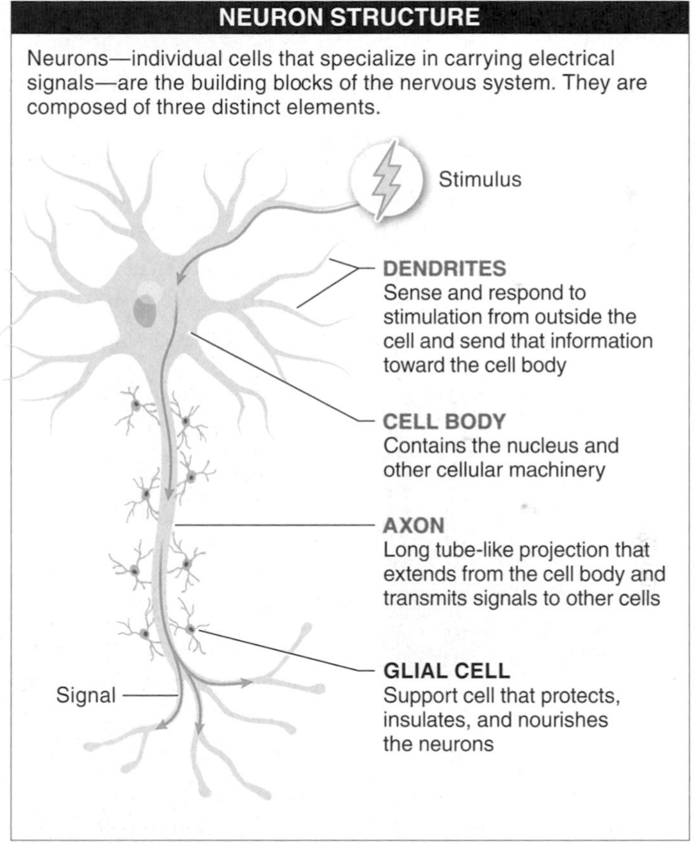

NEURON STRUCTURE
Neurons—individual cells that specialize in carrying electrical signals—are the building blocks of the nervous system. They are composed of three distinct elements.
Stimulus
DENDRITES
Sense and respond to stimulation from outside the cell and send that information toward the cell body
CELL BODY
Contains the nucleus and other cellular machinery
AXON
Long tube-like projection that extends from the cell body and transmits signals to other cells
GLIAL CELL
Support cell that protects, insulates, and nourishes the neurons
Signal

FIGURE 24-2 **The neuron is the building block of the nervous system.**

The neuron's second important specialization is the **axon.** An axon can transmit a signal, much as an electrical wire does, over great distances. Like dendrites, the axon is an extension of the plasma membrane from the main cell body. Sometimes *very* long, the axon transmits the signals picked

TYPES OF NEURONS

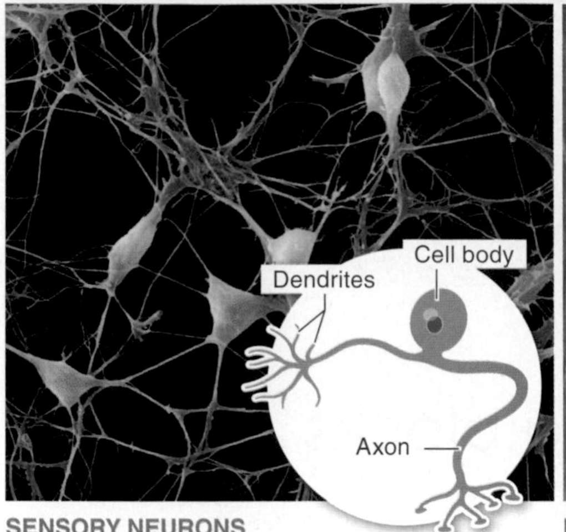

SENSORY NEURONS
- Collect information from an animal's environment
- Dendrites modified to respond to external stimulation, such as temperature, touch, taste, smell, light, or sound

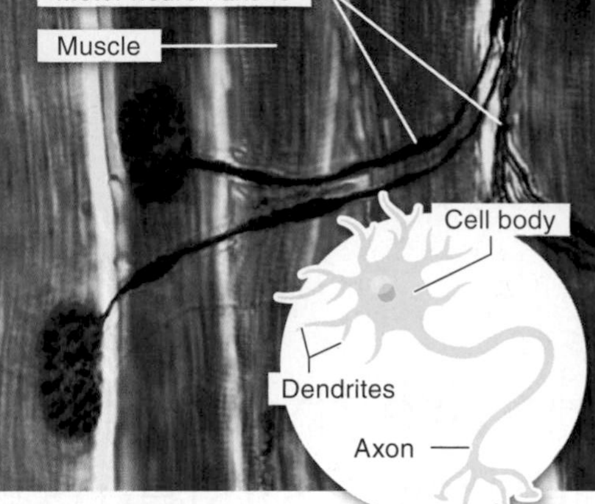

MOTOR NEURONS
- Initiate an animal's response to stimuli
- Stimulate action by conveying signals to muscles or glands

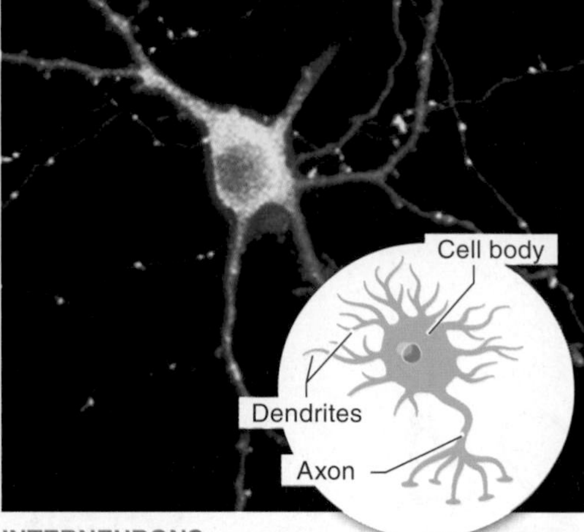

INTERNEURONS
- Interpret signals coming in from sensory neurons and relay them to motor neurons
- Located only in the brain and the spinal cord

FIGURE 24-3 **Three types of neurons work together in the vertebrate nervous system.**

up by the dendrites to the rest of the organism's body. The end of the axon is specially modified in a way that allows it to transmit the signals to another cell. Some neurons are several feet long. The sciatic (sigh-At-ick) nerve, for example, runs from your spinal cord all the way to the tips of your toes. Most of this distance is covered by a single axon.

> **Q** If neurons rarely (if ever) divide, how can people get brain tumors, which are the result of unstoppable cell division?

Although neurons do all the actual work of the nervous system, they do it with a lot of help from other cells. Large numbers of non-neuronal cells, called **glial cells,** function like a support staff to protect, insulate, and nourish the neurons. In the human brain, there are approximately nine times as many glial cells as there are neurons. And unlike most neurons, glial cells regularly divide. Not surprisingly, then, virtually all brain tumors in adults are formed from glial cells.

Just as corporations keep their big computers locked up in super-clean, secure rooms to keep them functioning at an optimal level, the brain is protected from potentially harmful molecules in the blood. Glial cells line the blood vessels that supply the brain, forming a semi-permeable barrier called the "blood-brain barrier." This barrier allows essential nutrients and gases to pass through, but it bars harmful molecules such as metabolic wastes produced throughout the body. The blood-brain barrier is not perfect, though. Many small molecules, including anesthetics and alcohol, can make it through to influence the brain. The barrier also can be broken down by hypertension, radiation, and a variety of infectious organisms.

To carry out the activities of the nervous system, there are three functional types of neurons (FIGURE 24-3).

1. **Sensory neurons** collect information from an animal's environment and have dendrites modified to respond directly to internal and external stimulation. This stimulation can include temperature, touch, taste, smell, light, or sound.
2. **Motor neurons** stimulate action by conveying signals to muscles or glands and initiating a body's response to stimuli.
3. **Interneurons** integrate the signals coming in from the sensory neurons and relay them to the motor neurons. These "middlemen" are located only in the brain and the spinal cord.

TAKE HOME MESSAGE 24.2

›› The neuron is a type of cell specialized for carrying electrical signals and is the building block of all nervous systems. Each neuron is very small, but groups of neurons bundled together enable us to sense light, sound, touch, tastes, and smells, and to respond to these sensations.

24.3 The vertebrate nervous system consists of the peripheral and central nervous systems.

In most animals, including all vertebrates, the nervous system is divided into the peripheral nervous system and the central nervous system (FIGURE 24-4). The **peripheral nervous system (PNS)** is the network of (1) sensory cells modified to receive information from the environment and (2) motor pathways that transmit signals to effectors, the muscles and glands that are capable of responding to that stimulus. But the information received by the body's sensory cells first passes through the **central nervous system (CNS),** which is made up of the spinal cord and brain. Not directly connected to sensory organs or to muscles, the central nervous system processes information that it receives from sensory cells about the organism's surroundings and sends out instructions to other nervous tissue to act in response to that sensory information.

The motor pathways of the peripheral nervous system have two major divisions: the somatic and the autonomic nervous systems. The **somatic nervous system** relays signals to your skeletal muscles and enables you to contract those muscles and move your limbs consciously, such as when you pick up a fork. The **autonomic nervous system** relays signals to your glands and your smooth muscle tissue and cardiac muscle. These signals, which are not under conscious control, cause the muscles surrounding your stomach and intestine, for example, to contract and to move food through the digestive tract. The central nervous system controls both the somatic and the autonomic nervous systems.

Because it enables you to initiate movements consciously, the somatic nervous system is commonly referred to as the voluntary nervous system. When a sensory neuron in the peripheral nervous system senses an irritation (such as a mosquito landing on your hand), it sends a signal to convey this information to the spinal cord, where the sensory neuron connects with an interneuron that extends to the brain (FIGURE 24-5). The signal is interpreted and a response is determined (you want to make the mosquito go away), and this information is sent through another interneuron back down your spinal cord to a motor neuron that is attached to a muscle. The signal from the brain to the muscle causes you to swat at the mosquito—removing the source of irritation. This whole process takes only a tiny fraction of a second.

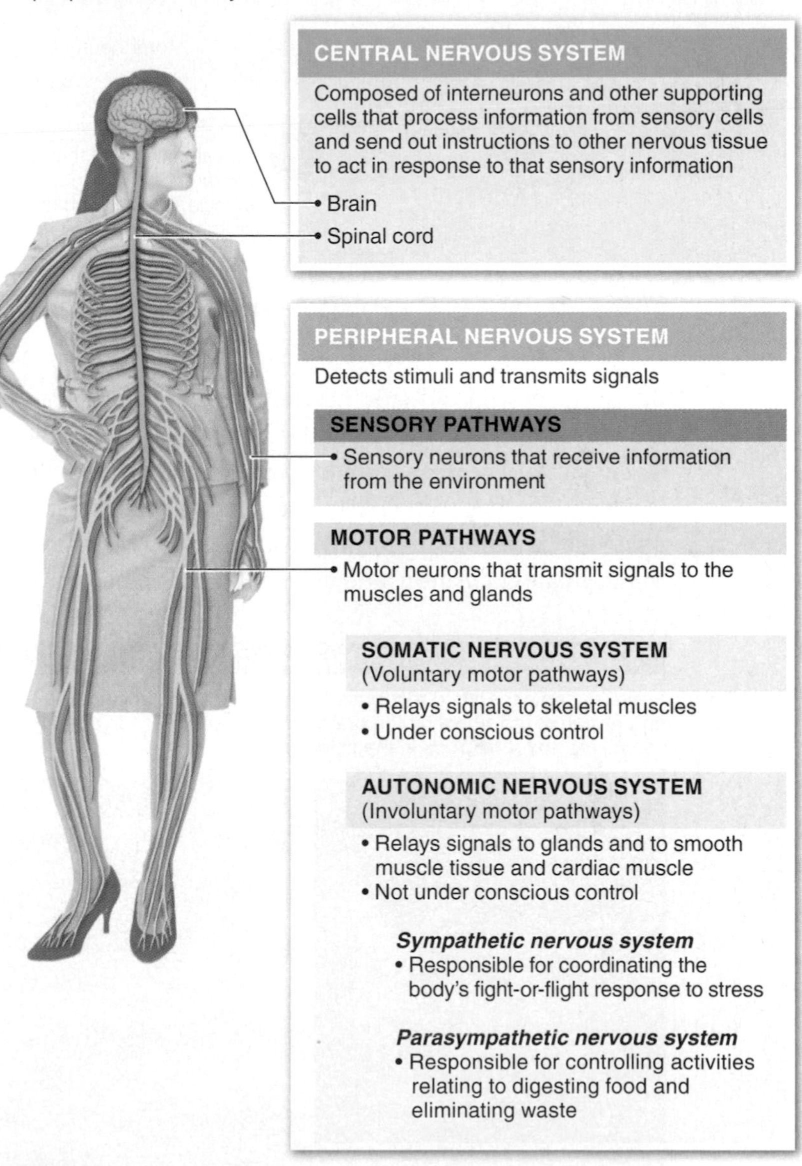

THE VERTEBRATE NERVOUS SYSTEM
In vertebrates, the nervous system is divided into the central nervous system and the peripheral nervous system.
CENTRAL NERVOUS SYSTEM
Composed of interneurons and other supporting cells that process information from sensory cells and send out instructions to other nervous tissue to act in response to that sensory information
• Brain
• Spinal cord
PERIPHERAL NERVOUS SYSTEM
Detects stimuli and transmits signals
SENSORY PATHWAYS
• Sensory neurons that receive information from the environment
MOTOR PATHWAYS
• Motor neurons that transmit signals to the muscles and glands
SOMATIC NERVOUS SYSTEM
(Voluntary motor pathways)
• Relays signals to skeletal muscles
• Under conscious control
AUTONOMIC NERVOUS SYSTEM
(Involuntary motor pathways)
• Relays signals to glands and to smooth muscle tissue and cardiac muscle
• Not under conscious control
Sympathetic nervous system
• Responsible for coordinating the body's fight-or-flight response to stress
Parasympathetic nervous system
• Responsible for controlling activities relating to digesting food and eliminating waste

FIGURE 24-4 **Organization of the vertebrate nervous system.**

Within the somatic nervous system, some signals can bypass the brain, with sensory information stimulating the motor neuron through the spinal cord. This kind of direct sensory-motor response is called a **reflex.** Reflexes enable an organism to respond faster to an imminent danger (FIGURE 24-6). Your brain learns about it just a bit later.

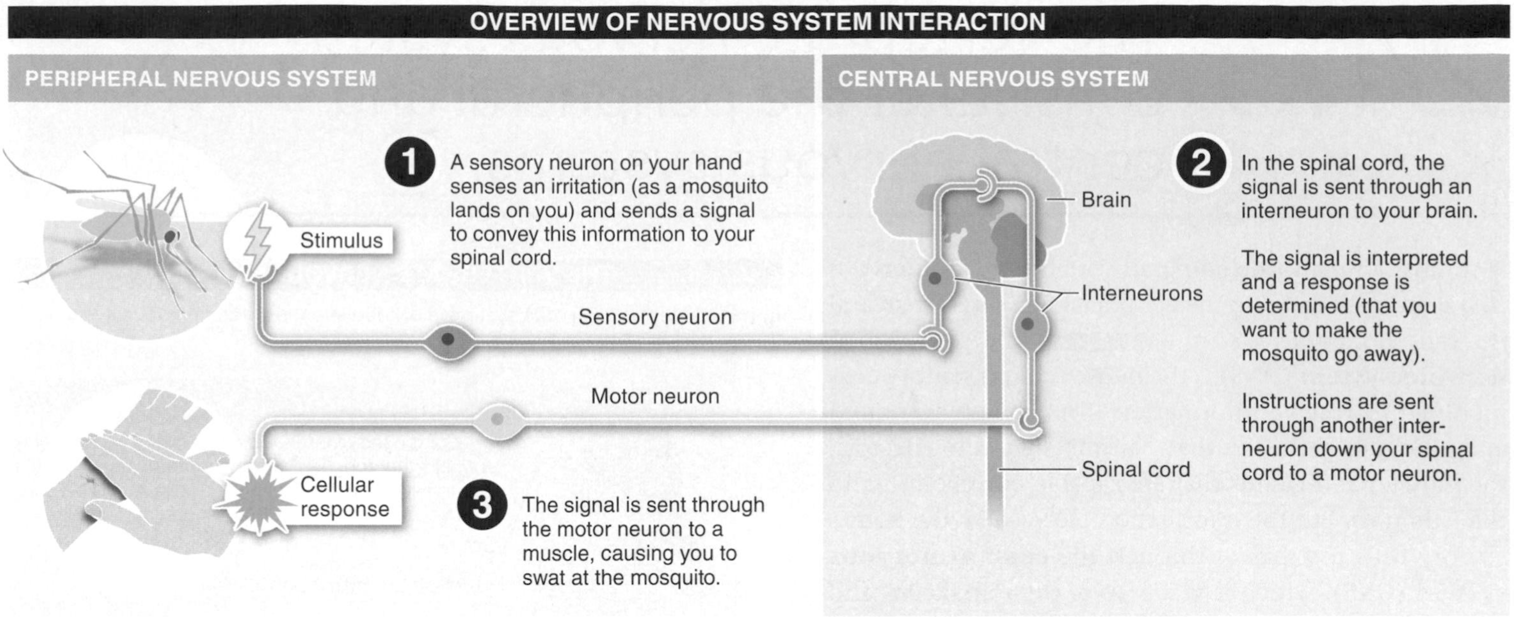

OVERVIEW OF NERVOUS SYSTEM INTERACTION

PERIPHERAL NERVOUS SYSTEM

CENTRAL NERVOUS SYSTEM

Stimulus

1 A sensory neuron on your hand senses an irritation (as a mosquito lands on you) and sends a signal to convey this information to your spinal cord.

Sensory neuron

Brain

Interneurons

Spinal cord

2 In the spinal cord, the signal is sent through an interneuron to your brain.

The signal is interpreted and a response is determined (that you want to make the mosquito go away).

Instructions are sent through another interneuron down your spinal cord to a motor neuron.

Motor neuron

Cellular response

3 The signal is sent through the motor neuron to a muscle, causing you to swat at the mosquito.

FIGURE 24-5 **The peripheral nervous system interacts with the central nervous system.**

In contrast to the somatic nervous system, the autonomic nervous system—which can be thought of as the "automatic" or involuntary nervous system—carries the signals by which the central nervous system regulates the heartbeat, breathing, glandular secretions, the muscles surrounding blood vessels that regulate circulation, and the movement of food through the digestive tract. It is through the autonomic nervous system, in large part, that the central nervous system regulates homeostasis.

Two components make up the autonomic nervous system: the sympathetic nervous system and the parasympathetic nervous system. The **sympathetic nervous system** is responsible for coordinating the body's fight-or-flight response to stress, including increasing the heart and breathing rates and providing muscles with additional blood flow. The **parasympathetic nervous system,** in contrast, is responsible for controlling activities related to digesting food and eliminating waste, tending to slow the heart and breathing rates as these processes occur. Although the two systems tend to coordinate opposing effects, most muscles, organs, and glands receive input from both the sympathetic and the parasympathetic nervous systems.

REFLEXES

Reflexes can generate a response to a stimulus without need for processing of the signal by the brain—sensory information directly stimulates motor neurons in the spinal cord.

Reflexes enable an organism to respond faster to an imminent danger.

FIGURE 24-6 **Quick reaction.**

TAKE HOME MESSAGE 24.3

» In all vertebrates, the nervous system is divided into the peripheral nervous system and the central nervous system. The central nervous system is made up of the spinal cord and brain. The peripheral nervous system carries signals to and from the sensory and motor pathways. The motor pathways carry signals from both the somatic and autonomic nervous systems, which relay signals that can be controlled consciously and other signals that cannot be controlled consciously.

How do neurons work?

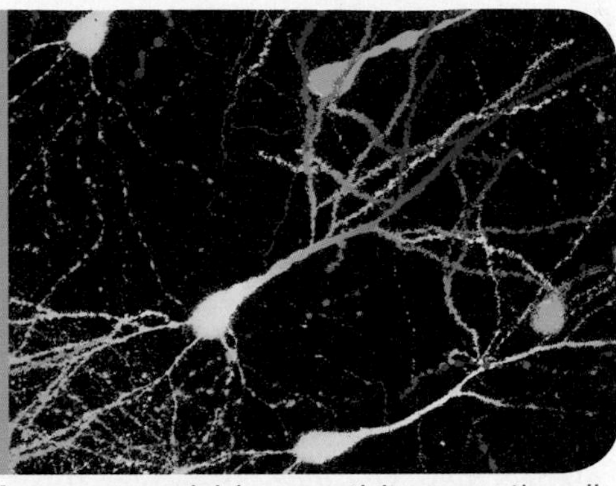

Neurons in the brain of a mouse reveal rich connectivity among the cells.

24.4 Dendrites receive external stimuli.

Fire!" "Don't fire!" "Fire!" "Don't fire!" Dendrites are on the receiving end of an almost constant barrage of (sometimes conflicting) signals. A neuron may have hundreds or even thousands of dendrites, which give it a huge amount of surface area over which to make connections with other neurons and receive signals.

Dendrites receive stimuli in one of two ways. Those on motor neurons and interneurons generally connect with and receive signals from other neurons. Sensory neuron dendrites, on the other hand, are modified to respond to a specific external stimulus such as a touch or sound, light, or a chemical (FIGURE 24-7).

When a neuron is not transmitting a signal, the inside of the cell has a negative charge relative to the outside of the cell. This difference in charge, referred to as a membrane potential, is the neuron's **resting potential.** The resting potential is produced as proteins within the neuron's plasma membrane pump sodium ions (Na^+) out of the cell and potassium ions (K^+) into the cell. Three sodium ions are pumped out of the cell for every two potassium ions moved in, contributing to the establishment of a greater positive charge outside the cell than inside the cell. The cell is described as "polarized" and can be thought of as "ready to fire."

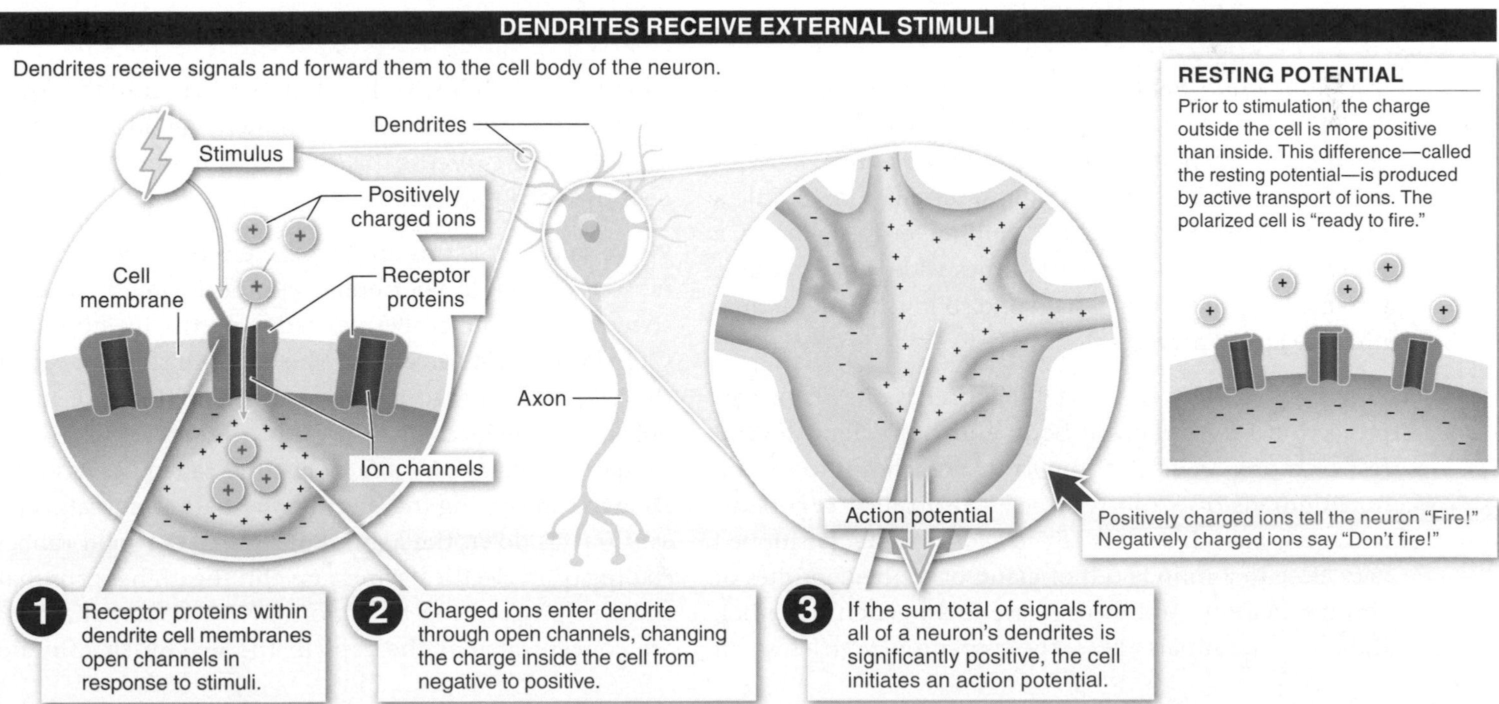

DENDRITES RECEIVE EXTERNAL STIMULI
Dendrites receive signals and forward them to the cell body of the neuron.
Stimulus
Dendrites
Positively charged ions
Cell membrane
Receptor proteins
Ion channels
Axon
Action potential
RESTING POTENTIAL
Prior to stimulation, the charge outside the cell is more positive than inside. This difference—called the resting potential—is produced by active transport of ions. The polarized cell is "ready to fire."
Positively charged ions tell the neuron "Fire!" Negatively charged ions say "Don't fire!"
1 Receptor proteins within dendrite cell membranes open channels in response to stimuli.
2 Charged ions enter dendrite through open channels, changing the charge inside the cell from negative to positive.
3 If the sum total of signals from all of a neuron's dendrites is significantly positive, the cell initiates an action potential.

FIGURE 24-7 **"Fire!" (or "Don't fire!).**

On stimulation, receptors within the cell membrane of the dendrite respond by briefly opening up ion channels (usually sodium channels). These channels, made from proteins, allow the passage of charged ions (usually Na^+ ions) down their concentration gradient. This ion flow momentarily alters the negative electrical charge inside the cell—increasing it or decreasing it, depending on the type of ion flow. When opening the channels makes the neuron more negatively charged (that is, the electrical charge inside the cell relative to that in the fluid surrounding the cell becomes even more negative), this says, "Don't fire!" When it makes the neuron more positively charged, such as when sodium channels open and sodium ions rush into the cell, this says, "Fire!"

"The trees grew heavy with blackbirds, branches like dendrites of the nervous system fattening, deep in twittering nerve dusk, waiting for some important message."

— THOMAS PYNCHON,
Gravity's Rainbow, 1973

As stimuli cause channels to open in the dendrites of a neuron—some making the neuron more negatively charged, others making the neuron more positively charged—the changes in the cell's electrical charge, occurring in all these dendrites, converge at the cell body. The cell body then integrates the signals, much like tallying the votes in an election. If the sum total of signals coming in is sufficiently positive—exceeding a threshold favoring "Fire!"—then the neuron initiates an **action potential,** an electrical signal that travels down its axon. If the sum total is negative, no action potential is generated.

TAKE HOME MESSAGE 24.4

» Dendrites receive external stimuli in one of two ways. Dendrites on motor neurons and interneurons generally connect with and receive signals from other neurons, whereas sensory neuron dendrites are modified to respond to a specific external stimulus such as a touch or sound, light, or a chemical.

24.5 The action potential propagates a signal down the axon.

Giant squids happen to have giant neurons. And because neurons function almost exactly the same in all animals, much of what we know about human nervous systems has come from studying squid neurons. While a human neuron might be 0.02 millimeter in diameter, a squid's may be more than 100 times wider, up to double the thickness of a human hair, and so much easier to manipulate and observe.

Regardless of the species, each neuron has one axon. This projection leaves the cell body and can extend several feet or more. At its end, the axon branches into several (or hundreds of) axon terminals, also called **terminal buttons,** knob-like ends of the axon that are positioned very close to a muscle cell or gland or to the dendrites of another neuron. And in response to an action potential, these axon terminals release the contents of vesicles, small sacs of chemicals inside the axon terminal, into the space between the cells, potentially influencing adjacent cells (FIGURE 24-8). We discuss this process in more detail in the next two sections.

We've said that axons are like the electrical wires of the nervous system. In your home, electrical wires are usually covered with rubber. Besides reducing the likelihood of someone receiving an electric shock, this insulation causes the electrical charge to move through the wire more quickly and reduces dissipation of the signal. Axons are similarly insulated, by a fatty coating called the **myelin sheath,** preventing the action potential from weakening as it travels down the axon. But, unlike the solid rubber coating of an electrical wire, a myelin sheath has gaps that assist in propagating an action potential. Ion channels are concentrated at the gaps in the myelin sheath and

THE AXON PROPAGATES A SIGNAL

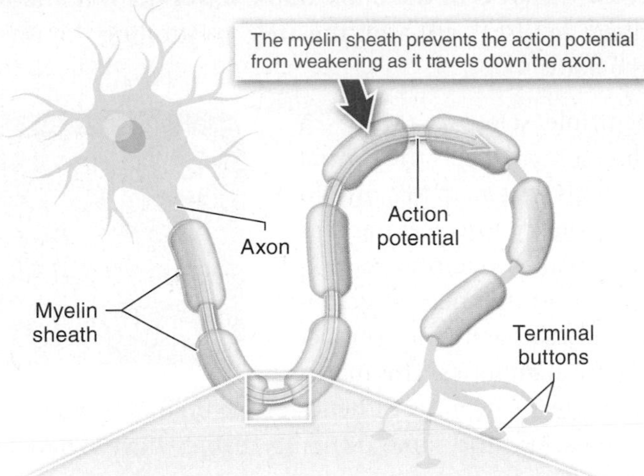

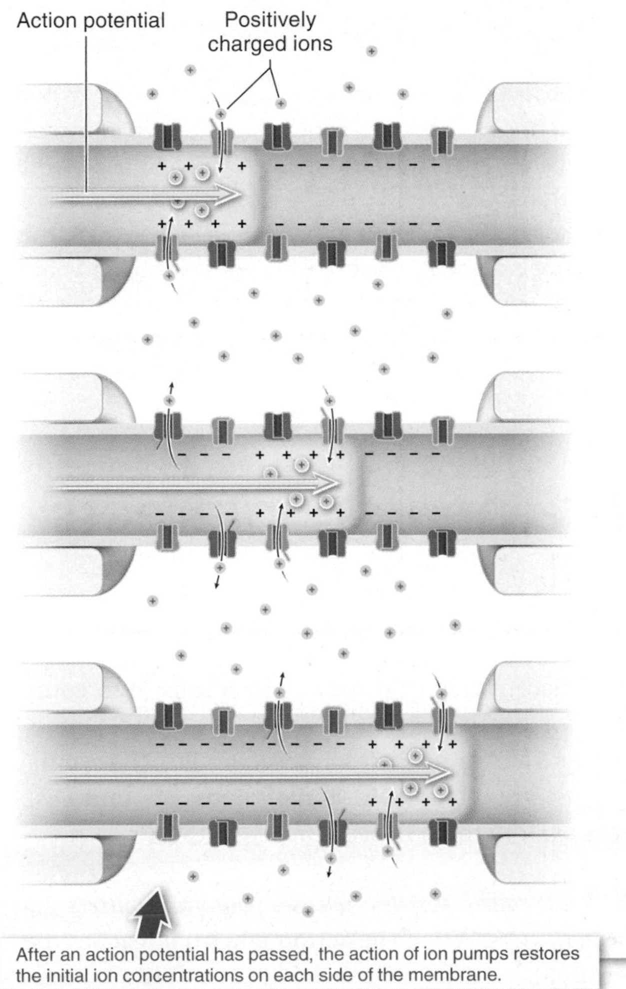

As the action potential moves down an axon, ion channels allow positively charged ions (typically sodium ions) to rush in, changing the charge to positive in that region and propagating the action potential.
Other ion channels then allow positively charged ions (typically potassium ions) to rush out, restoring the resting potential.
Action potential
Positively charged ions
After an action potential has passed, the action of ion pumps restores the initial ion concentrations on each side of the membrane.

FIGURE 24-8 **An action potential moves down an axon.**

allow sodium ions to rush into the axon. The rushing ions make the cell's charge sufficiently positive in that region to cause the opening of ion channels at the next gap in the myelin sheath, thereby propagating the action potential along the axon.

The fatty myelin is white, so, in cross sections of the brain, those areas where axons are densely packed together appear white (FIGURE 24-9). Regions of the brain with many cell bodies and dendrites appear gray. These different regions are often referred to as white matter and gray matter.

What would happen to signals traveling down an axon in your leg if the myelin sheath insulator were not there? You can observe this effect by watching babies when they first start trying to walk, at around 10 or 11 months of age. At

NEURONS IN THE BRAIN

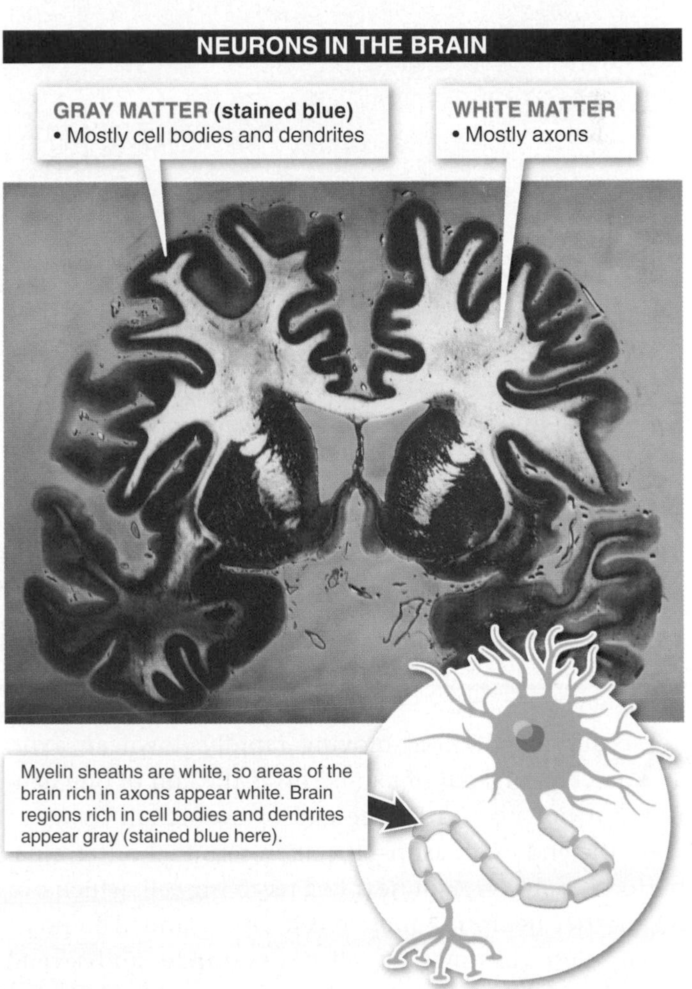

GRAY MATTER (stained blue)
• Mostly cell bodies and dendrites
WHITE MATTER
• Mostly axons
Myelin sheaths are white, so areas of the brain rich in axons appear white. Brain regions rich in cell bodies and dendrites appear gray (stained blue here).

FIGURE 24-9 **White and gray matter.**

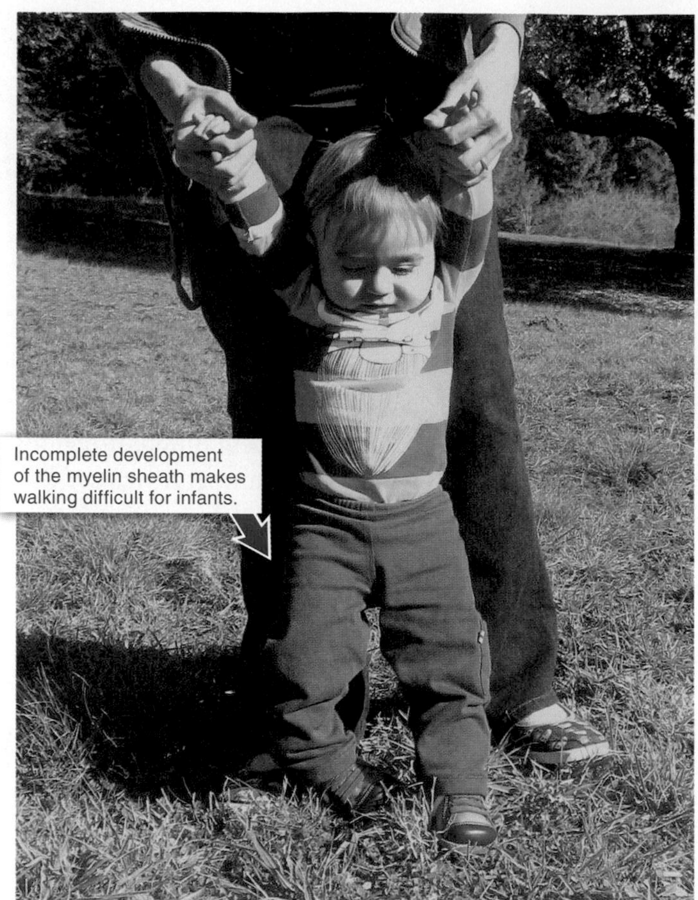

Incomplete development of the myelin sheath makes walking difficult for infants.

FIGURE 24-10 **Unsteady on his feet.**

that time, myelin hasn't completely formed around all of their axons, and consequently, their gross motor control isn't very good (FIGURE 24-10). So, although the baby's brain may be saying "Walk!" the signal never makes it to the leg muscles. Walking is especially difficult to master because the areas of the brain that control the feet and legs don't develop until after the areas controlling the rest of the body.

Multiple sclerosis (MS) is a disease that affects the central nervous system. In MS, myelin is gradually lost, leaving scar tissue called sclerosis. As myelin is lost, the neuron gradually loses its ability to conduct electrical impulses. This makes it progressively more difficult for the brain to send signals to muscles—the same difficulty babies have when first learning to walk—and can lead to a gradual loss in motor control, balance and coordination, and bladder and bowel control.

Q Some symptoms of multiple sclerosis resemble difficulties that babies have when learning to walk. Why?

TAKE HOME MESSAGE 24.5

» Each neuron has one axon, a projection that extends from the cell body, sometimes several feet or more. At its end, the axon branches into numerous terminal buttons, positioned close to a muscle cell or a gland or to the dendrites of another neuron. In response to an action potential, which travels down the axon from the cell body, the axon terminals release chemicals into the extracellular space, potentially influencing adjacent cells.

24.6 At the synapse, a neuron interacts with another cell.

An action potential moving rapidly down an axon quickly runs out of axon. This is the end of one neuron, but not necessarily the end of the signal. As we've seen, the end of an axon—the axon terminal or terminal button—is always right next to a receiving cell, which can be another neuron, a muscle cell, or a gland. The point where these cells meet is called a **synapse,** and several possible things can happen there. The signal arriving at the end of the axon may stimulate an action potential in the adjacent cell, or it may cause a muscle to contract or relax, or it may initiate a secretion by a gland. Or the signal may end altogether. In each case, though, the events at the synapse are remarkably similar across all animal species. Let's explore them in sequence (FIGURE 24-11).

1. *Sacs called vesicles release neurotransmitters into the synaptic cleft.* When the action potential reaches the axon terminal, it causes little sacs called **vesicles** to merge with

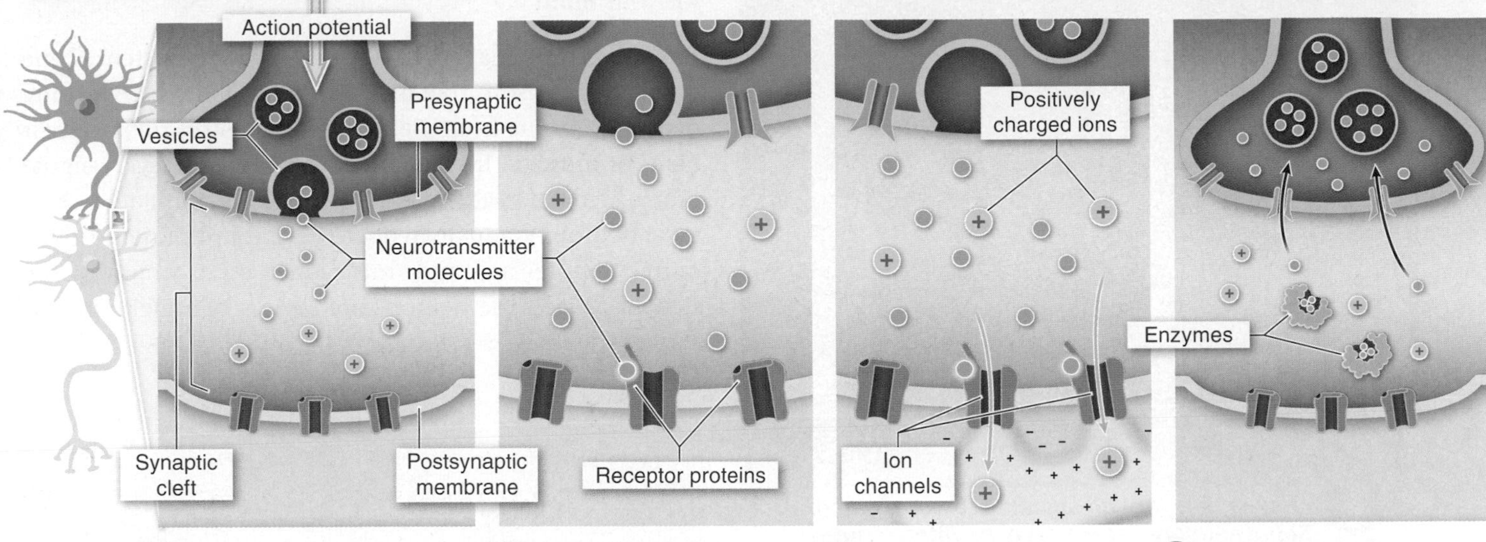
THE SYNAPSE
Action potential
Vesicles
Presynaptic membrane
Neurotransmitter molecules
Synaptic cleft
Postsynaptic membrane
Receptor proteins
Positively charged ions
Ion channels
Enzymes
1 When the action potential reaches the terminal button, vesicles fuse with the presynaptic membrane, releasing neurotransmitter molecules into the synaptic cleft.
2 The neurotransmitter molecules diffuse away and bind to receptor sites on the postsynaptic membrane of an adjacent cell.
3 Binding to a receptor, a neurotransmitter causes a channel to open, allowing ions to flow into the adjacent cell, stimulating or inhibiting it.
4 The neurotransmitters are then released from postsynaptic receptors. They are cleared from the synapse by enzymatic breakdown or reuptake by the axon terminal.

FIGURE 24-11 **The sequence of events when an action potential reaches the synapse.**

the axon's cell membrane, called the **presynaptic membrane.** The sacs open up and release their contents, chemical messengers called **neurotransmitters,** into the **synaptic cleft,** the space between the axon and the cell receiving the signal (muscle cell, gland, or neuron).

2. *Neurotransmitter diffuses away and binds to nearby receptor sites.* As the neurotransmitter molecules float around in the fluid in the synaptic cleft, they diffuse away from the axon until they bump into the adjacent neuron, muscle cell, or gland. Some of the neurotransmitter molecules attach to receptor sites on the **postsynaptic membrane** of the adjacent cell (called the *postsynaptic cell*).

3. *Gates open in the postsynaptic cell membrane, and the signal enters the postsynaptic cell.* When neurotransmitter binds to a receptor in the postsynaptic cell membrane, this causes a gate to open, which allows ions (often Na^+ ions) to flow into the cell. The resulting chemical change in the postsynaptic cell can cause an electrical change and, consequently, initiate an action potential (if the postsynaptic cell is a neuron), a contraction (if it's a muscle cell), or a secretion (if it's a gland).

4. *Neurotransmitter is released from the postsynaptic cell receptors and recycled or broken down.* The receptors then release the bound neurotransmitter molecules back into the fluid in the synaptic cleft. Eventually, the neurotransmitter molecules return to the presynaptic membrane and are recycled, or they are broken down within the synaptic cleft by enzymes, clearing out the area so that the synapse can be used again.

A contraction may occur if the postsynaptic cell is a muscle cell, and a secretion may occur if the postsynaptic cell is a gland, but how do neurotransmitters affect a neuron? In some cases, the released neurotransmitters are excitatory and excite the next neuron, increasing the likelihood that it will fire its own action potential. In other cases, the neurotransmitters are inhibitory and reduce the likelihood that the next cell will produce an action potential. And for some neurotransmitters, whether it is excitatory or inhibitory depends on the receptor. With the giant web of connections between neurons—each neuron synapses with hundreds or thousands of other neurons—the ultimate outcome of whether or not an action potential is initiated depends on a neuron democratically weighing all of its inputs and assessing whether most of its synapses are urging it to fire (pass on the signal) or not to fire (stop the signal from getting through).

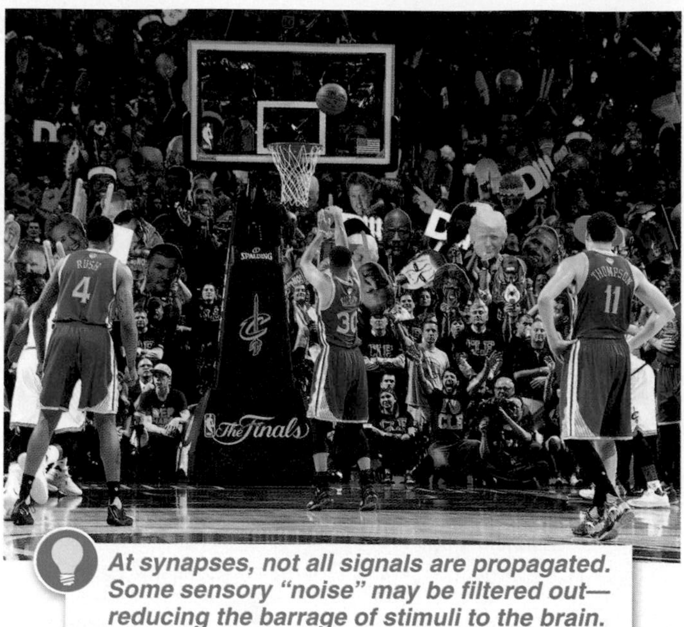

At synapses, not all signals are propagated. Some sensory "noise" may be filtered out—reducing the barrage of stimuli to the brain.

FIGURE 24-12 **Like call-screening for your brain.**

It might seem odd that the total of all the signals might direct the cell not to fire. However, the capacity to stop the signal enables the nervous system to modulate and filter some of the overwhelming amount of sensory information coming into the brain. It's like call-screening for your brain.

This "filtering" characteristic of the nervous system is one of the main reasons that organisms don't have single, long neurons running from their sensory receptors right to their brain or muscles. It's not always best to have every signal propagated (FIGURE 24-12). With continuous stimulation, too, most neurons gradually reduce the amount of neurotransmitter they release, and thus reduce the strength of the signal. It's as if the neurons are saying, "Enough already. We get the message."

TAKE HOME MESSAGE 24.6

» At the synapse, a neuron interacts with other cells. In response to an action potential, neurotransmitters are released into the synaptic cleft, diffuse, and may bind to receptors on an adjacent neuron, muscle cell, or gland, potentially stimulating an action potential, muscle contraction, or secretion. Neurotransmitters may then be recycled or broken down.

24.7 There are many types of neurotransmitters.

The neurotransmitters released by neurons can be thought of as the chemicals that initiate or modify a wide variety of actions, moods, feelings, or other sensations. In all, several dozen neurotransmitters have been identified, each of which has a sort of "personality" based on the effects associated with its release. A discussion of all the neurotransmitters is beyond the scope of this book. Here we discuss four that are particularly important.

Acetylcholine Acetylcholine is the neurotransmitter released by motor neurons at the point where they synapse with muscle cells. When enough acetylcholine binds to a muscle cell, the muscle contracts. Botulinum toxin—a protein produced by several species of bacteria—is the most toxic substance known: less than a microgram is lethal if injected into a person or inhaled. One type of botulinum toxin, known commercially as Botox, is an increasingly popular drug used in cosmetic procedures (and as part of several medical treatments), with more than a billion dollars spent on it worldwide each year (FIGURE 24-13). What does this have to do with neurotransmitters? When injected into muscles, Botox blocks the fusion of acetylcholine-filled vesicles with the presynaptic neuron's membrane, preventing release of acetylcholine into the synapse and thus paralyzing the muscle. This smooths lines in the forehead and can reduce other wrinkles. A tiny amount of Botox can prevent a muscle from contracting for three or four months!

Q Why would someone pay to get injections of the most toxic substance known? (Hint: it's also known as Botox.)

Glutamate Its mechanism of action as a neurotransmitter isn't well understood, but glutamate appears to be involved with learning and memory. Mice that were genetically engineered to have more sensitive glutamate receptors, for instance, learned tasks better and became much better at running mazes than normal mice.

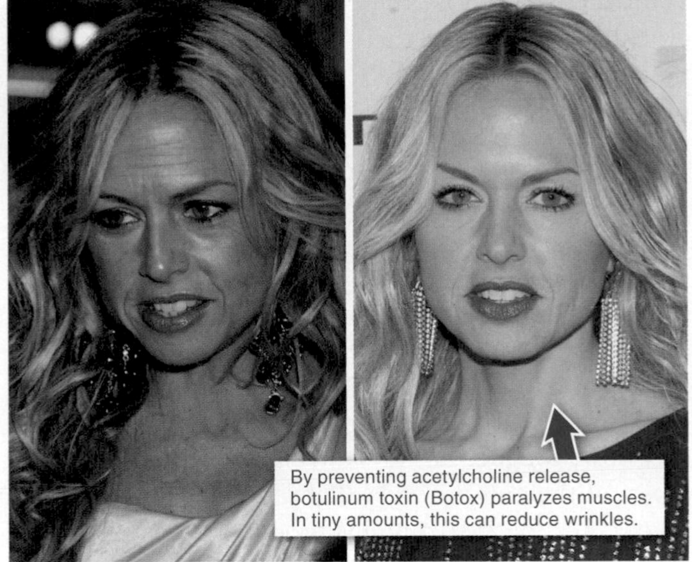

By preventing acetylcholine release, botulinum toxin (Botox) paralyzes muscles. In tiny amounts, this can reduce wrinkles.

Fashion stylist Rachel Zoe, in 2006 (left) and 2009.

FIGURE 24-13 Controlled paralysis.

Dopamine Dopamine is important in initiating and coordinating movement. Loss of dopamine neurons may be responsible for Parkinson's disease, for which the symptoms include tremors of a hand or foot, stiffness or inflexibility of muscles, and impaired balance and coordination. Dopamine is also one of the body's chief "happiness" neurotransmitters. Its release in certain parts of the brain is associated with feelings of intense pleasure.

Serotonin Serotonin generally functions as an inhibitory neurotransmitter. It affects appetite, sleep, anxiety, and mood, and produces feelings of contentment and satiation when released. Women make serotonin only about two-thirds as quickly as men, an observation that may be related to why depression is twice as common among women as men. As we see later in the chapter, antidepressant medications such as Prozac increase the amount of serotonin in the synapses and are effective at elevating the mood of someone with depression.

TAKE HOME MESSAGE 24.7

» Many neurotransmitters have been identified, each of which has several actions, depending on the synapse where it acts. Acetylcholine is released by motor neurons where they synapse with muscle cells. Glutamate is involved with learning and memory. Dopamine is important in initiating and coordinating movement and in producing feelings of intense pleasure. Serotonin affects appetite, sleep, anxiety, and mood, and produces feelings of contentment and satiation.

Our senses detect and transmit stimuli.

A mongoose lemur stands alert.

24.8 Sensory receptors are our windows to the world around us.

As animals move through their environment, their "front" part encounters new things in the environment first; with its sensory equipment up front, the organism is able to decide what to do—eat or run—as quickly as possible, which may have conferred evolutionary benefits. And as the decisions to be made get

THE FIVE SENSES

SIGHT

TOUCH

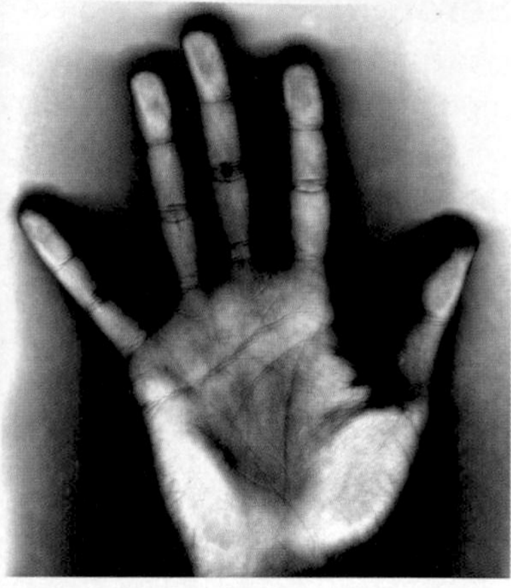

TASTE

SMELL

HEARING

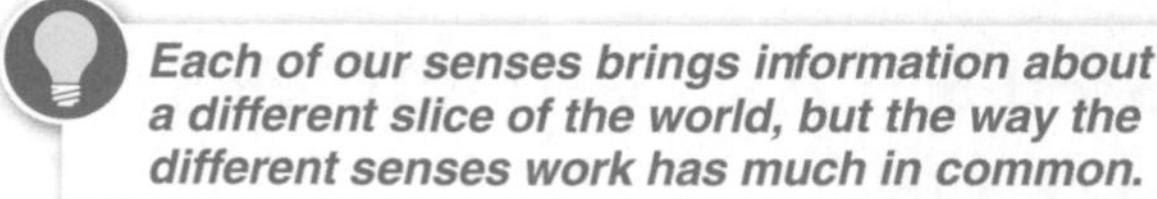

Each of our senses brings information about a different slice of the world, but the way the different senses work has much in common.

FIGURE 24-14 Making us aware of our environment: the senses. (Note that most of the body parts shown here—such as hands and tongue—may have more than one function.)

increasingly complex, the circuitry required for the animal to "make up its mind" becomes more complex. In fact, only then does it really become necessary to have a mind. And so the brain and head (which always go hand in hand) are prominent structures.

Q Why do animals have only one head? Why is it in the front?

Our senses—sight, hearing, smell, taste, and touch—make us physically aware of the environment around us and have much in common (FIGURE 24-14). A receptor—commonly a modified dendrite on a sensory neuron (in the eye, nose, tongue, ear, or skin)—is stimulated by some aspect of the outside world (light, odor, taste, sound, or touch). This outside stimulus produces a change in the neuron, causing the sensory neuron to either (1) fire an action potential itself, which shoots down the axon and ultimately reaches a part of the brain where the signal is perceived as a particular smell or sound, for example, or (2) alter its rate of neurotransmitter secretion so that it increases or reduces the rate of firing of action potentials in a neighboring neuron.

How is it possible to experience variations in the intensity of sensation, when an action potential is, as we've seen, an all-or-nothing event? It either occurs or doesn't occur as the result of dendrite stimulation. After all, our experience of sensations is not "all or nothing." We are able to feel gradations of intensity. A 10-pound bowling ball resting in your left hand will feel very heavy (and perhaps cooler and smoother) compared with a lightweight rubber ball in your right hand (FIGURE 24-15). How can the "fire" versus "don't fire" event produce such a range of sensations? The intensity of the sensation is modulated by the number of action potentials per unit of time and the number of neurons stimulated. The rubber ball causes a small number to fire, and the bowling ball causes many more to fire.

VARIATIONS IN SENSATION INTENSITY

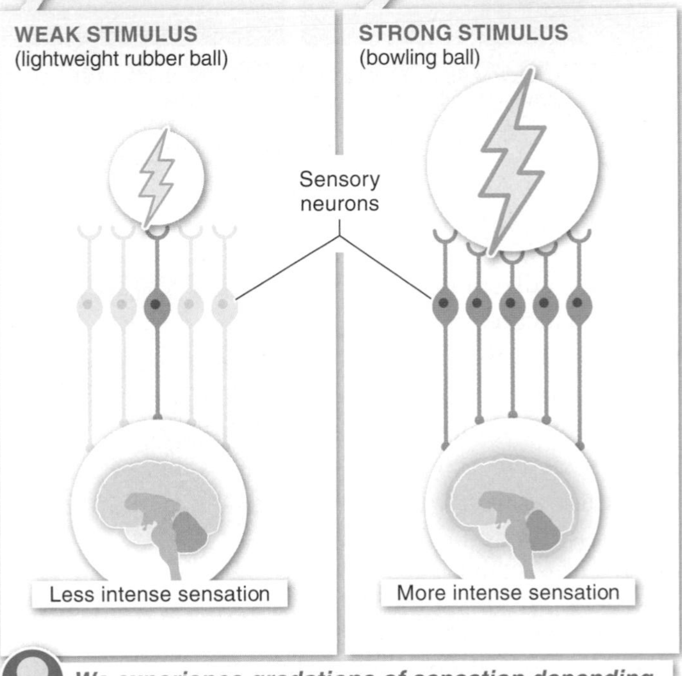

We experience gradations of sensation depending on the number of neurons that fire.

FIGURE 24-15 **Heavy or light?**

In the following sections, we present information about each of the five primary senses, exploring how the specific senses are variations on a theme, fine-tuned by natural selection to give an animal specific information about its environment that helps it respond appropriately. (Among animals, there are several senses in addition to taste, smell, vision, hearing, and touch, including those that perceive and respond to balance and motion, electricity, magnetism, and heat production in the animal's environment.)

TAKE HOME MESSAGE 24.8

» The process by which all our senses work is basically the same. A modified dendrite on a sensory neuron is stimulated, causing the sensory neuron either to fire an action potential (which shoots down the axon and ultimately reaches a part of the brain where the signal is perceived as a particular smell or sound, for example) or to alter its rate of neurotransmitter secretion (which increases or reduces the rate of action potential firing in a neighboring neuron).

24.9 Taste: an action potential serves up a taste sensation to the brain.

Your tongue is covered with bumps. Embedded within these bumps are taste buds—more than 10,000 in all. Within each taste bud are 60–80 sensory receptor cells, called **chemoreceptors** (which are modified epithelial cells that synapse with sensory neurons), that are stimulated when particular chemicals in food dissolve in saliva

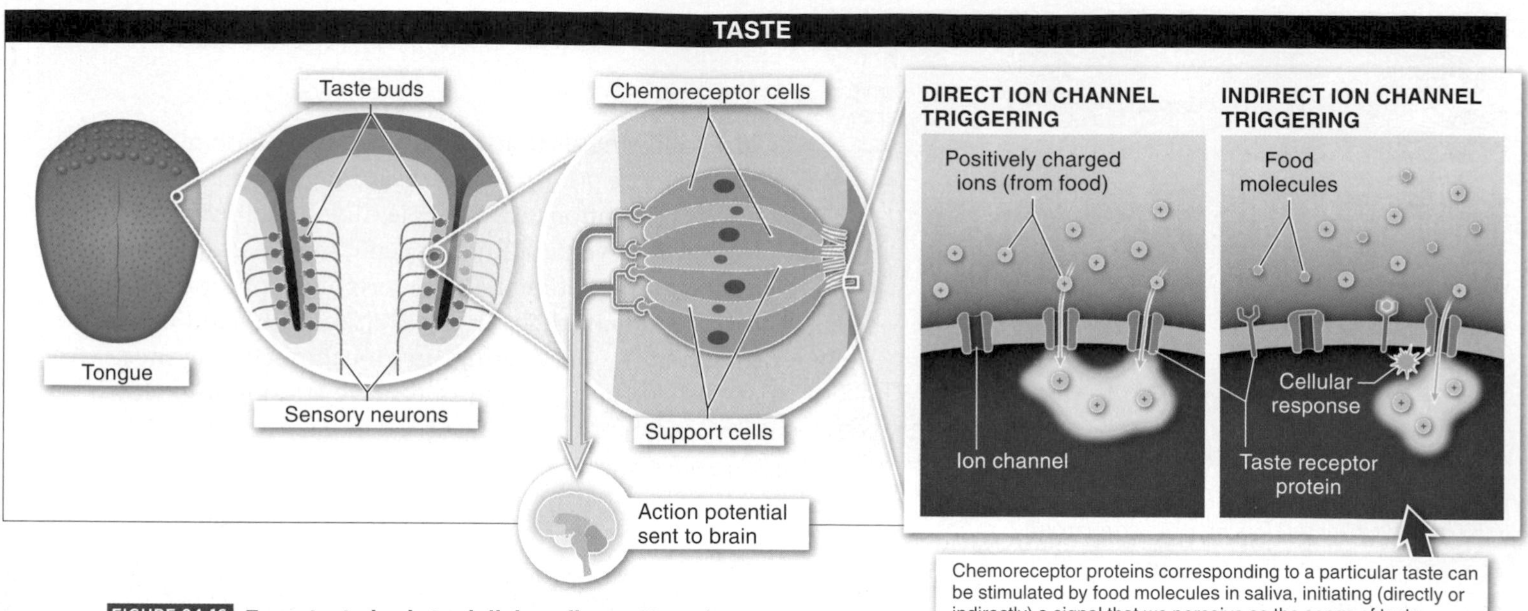

FIGURE 24-16 **From taste buds to delicious flavor.** How the sensation of taste is generated in the brain.

and bind to proteins on the receptor cell surface. Particular taste receptors allow only specific food molecules, with exactly the right shape, to bind (FIGURE 24-16).

Taste chemoreceptors fall into five groups, depending on which type of molecule chemically stimulates them: sweet, salty, sour, bitter, or a recently discovered, but difficult to describe, savory taste called umami. These different types of chemoreceptors occur all across the tongue (although the different groups tend to be concentrated in different regions of the tongue). A rich variety of tastes is possible because most foods stimulate unique combinations of the different taste receptors. Foods also release molecules into the air in the mouth, which stimulate smell receptors within the nasal cavity.

Not all animals use their mouth to taste things. Some insects have chemoreceptors on their legs, and they "taste" things just by touching them. Other animals have taste receptors on their antennae or tentacles.

Your brain never actually "knows" the true identity of the food on your tongue. Rather, it senses a particular taste based solely on the combination of receptor cells that are stimulated. If a molecule that is not sugar has a chemical structure closely resembling sugar, it can stimulate the same taste-bud receptors and be perceived by the brain as sugar. Many non-nutritive sugar-substitute molecules, such as saccharin and sucralose, do exactly this. They have three-dimensional arrangements of atoms that are similar to sugar molecules such as sucrose, glucose, or fructose. Consequently, when we consume them, we sense the sweet taste of sugar but don't actually derive any energy from the molecules. Instead, they pass through our digestive tract unaltered.

Q How can artificial sweeteners taste like sugar while not actually being sugar?

Aspartame also resembles sugar closely enough to stimulate sugar receptors in taste buds. Molecularly, however, it is quite different: it is made from two amino acids. Unlike the other artificial sweeteners, it can be broken down and releases energy (and hence contains calories), but it is so efficient at stimulating sweet-taste receptors that it can generate a strong sugary taste even when used in tiny amounts.

TAKE HOME MESSAGE 24.9

» Your tongue has about 10,000 taste buds, each containing 60–80 chemoreceptors, which are stimulated when particular chemicals in food bind to receptor proteins on the cell surface, triggering a signal to the brain.

24.10 Smell: receptors in the nose detect airborne chemicals.

Our sense of smell works in almost exactly the same way as our sense of taste. Neurons that can detect smells have dendrites modified with tiny, hair-like projections. These dendrites—densely packed within the nasal cavity—are covered with chemoreceptors. Airborne chemicals move through mucus in the nasal cavity and bind to the smell receptors, triggering action potentials that shoot down the axon, all the way into the smell center of the brain, where the signal is perceived as a particular odor. More than a thousand different types of receptors—modified neurons—are present in a human nose, each capable of detecting a different scent (FIGURE 24-17).

Q Why are your senses of smell and taste dulled when you have a cold?

In humans, the senses of smell and taste are closely connected, because the air in the mouth, throat, and nasal passages circulates around all these areas. You may have noticed that when you have a cold, your senses of taste and smell are dulled. The dulling occurs because mucus in your nasal passages reduces the rate at which airborne chemicals can reach the smell receptors and taste receptors on dendrites in your nose and on your tongue.

Dogs are among the most smell-sensitive vertebrates, having as many as 40 times more smell receptors than a human—hence their tremendous proficiency at detecting drugs or explosives (FIGURE 24-18). Some moths, such as gypsy moths and silkworm

Dogs have as many as 40 times more smell receptors than a human—hence their keen sense of smell, which gives them tremendous proficiency at detecting drugs or explosives.

FIGURE 24-18 **Smelling champ: the bomb-sniffing dog.**

Q People are much worse than dogs at sniffing out drugs. Why?

SMELL

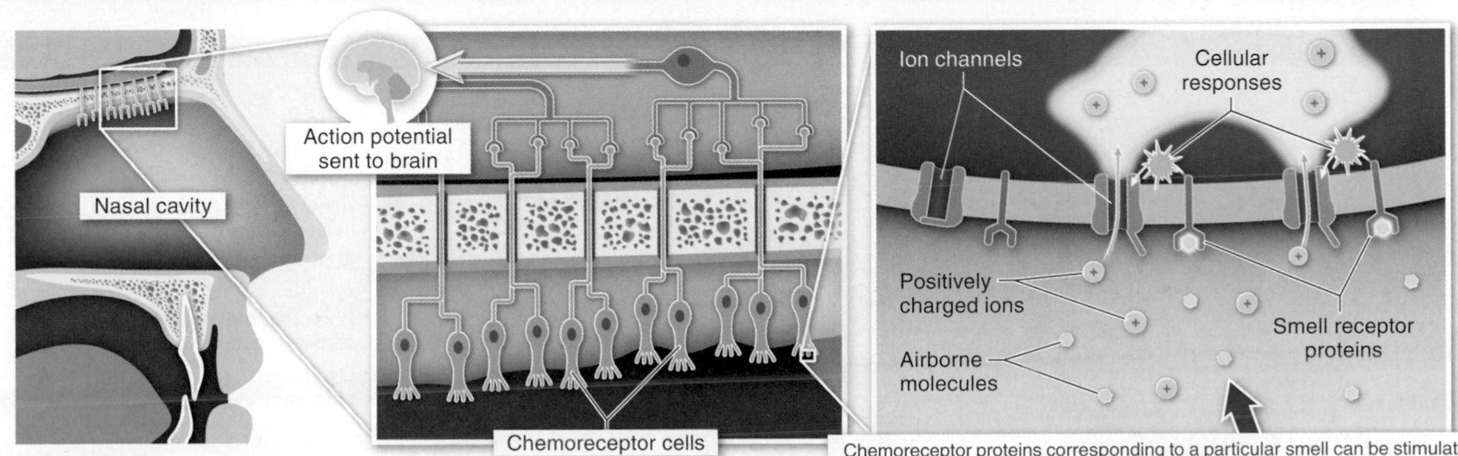

Action potential sent to brain
Nasal cavity
Chemoreceptor cells
Ion channels
Cellular responses
Positively charged ions
Airborne molecules
Smell receptor proteins
Chemoreceptor proteins corresponding to a particular smell can be stimulated by airborne molecules in the nasal cavity. This triggers ion channels (indirectly), initiating a signal that we perceive as the sense of smell.

FIGURE 24-17 **How the sensation of smell is generated.**

moths, may be the champions of chemoreception. The tiny hairs on the males' antennae can detect just a few molecules of the sex attractant released by females. By moving in the direction of increasing concentration of the airborne molecules, a male can track down a female two miles away. Snakes, as they stick their tongue out, are actually catching the odor molecules that enable them to "smell" their environment.

TAKE HOME MESSAGE 24.10

» Neurons that can detect smells have dendrites modified with tiny, hair-like projections covered with chemoreceptors, densely packed within the nasal cavity.

24.11 Vision: seeing is the perception of light by the brain.

What we lack in smell proficiency, we more than make up for in visual acuity, a consequence of our image-forming, **single-lens eyes.** (This type of eye has evolved independently twice: in vertebrates and in a group of mollusks that includes the squid.) Light hits the eye and enters the eye's interior through an **iris.** The pupil, an opening in the iris, opens and closes to control the amount of light that gets into the eye. Once the light is through the pupil, a lens focuses it onto the **retina,** nervous tissue containing light-sensitive cells called **photoreceptor cells** (FIGURE 24-19).

The retina lines the inner surface of the eye and transmits impulses to the vision center of the brain via the **optic nerve.** The two types of photosensitive cells in the retina are **rods,** which are highly sensitive to even tiny amounts of light and make it possible to see at night and in low-light situations, and **cones,** which are less sensitive to light and so are more effective during daylight. Humans have three kinds of cone cells: red-, green-, and blue-sensitive. Our ability to detect color is based on the combination of cones being stimulated. Some individuals (usually men) produce non-functioning red or green cones—the result of mutant genes. In either of these cases, the individual cannot distinguish red from green, so the person is said to be **color-blind.**

Vision is one of the senses in which our capabilities exceed those of most other animals. Still, as with nearly every trait, there are plenty of species—including many birds and some spiders—that have abilities that exceed ours. Despite the great diversity of light-sensing capabilities, the basic functioning of light-absorbing cells is quite consistent among all animals. Here's how these cells work (FIGURE 24-20).

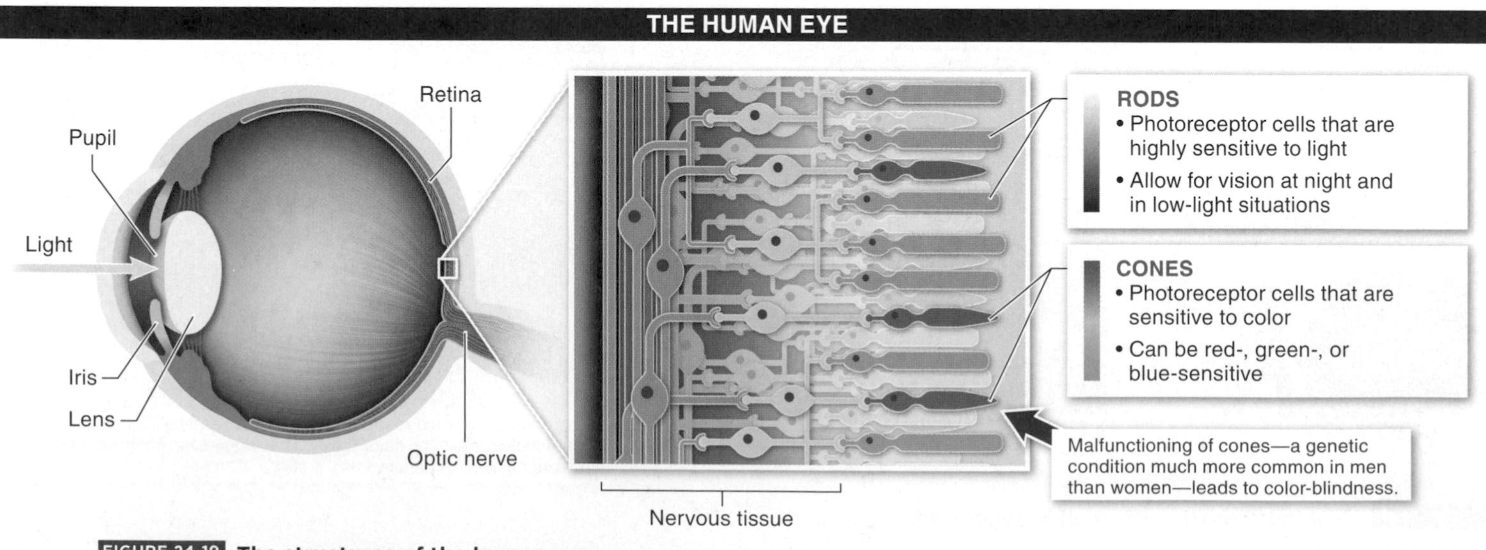

THE HUMAN EYE
Retina
Pupil
Light
Iris
Lens
Optic nerve
Nervous tissue
RODS
• Photoreceptor cells that are highly sensitive to light
• Allow for vision at night and in low-light situations
CONES
• Photoreceptor cells that are sensitive to color
• Can be red-, green-, or blue-sensitive
Malfunctioning of cones—a genetic condition much more common in men than women—leads to color-blindness.

FIGURE 24-19 **The structures of the human eye.**

HOW PHOTORECEPTOR CELLS WORK

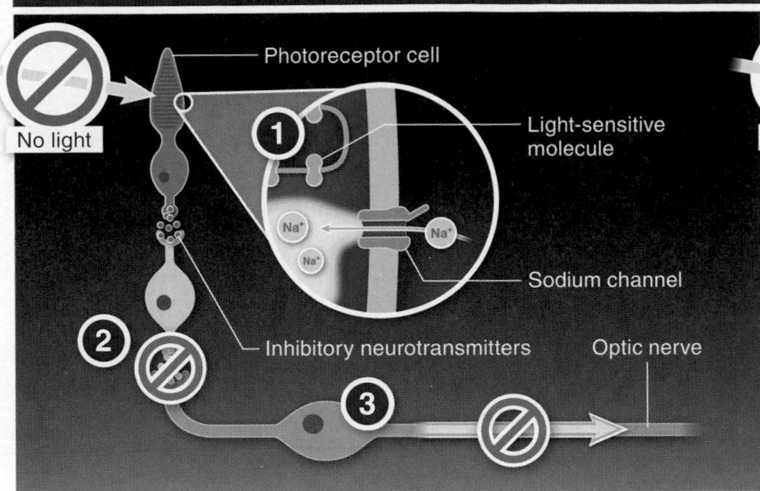

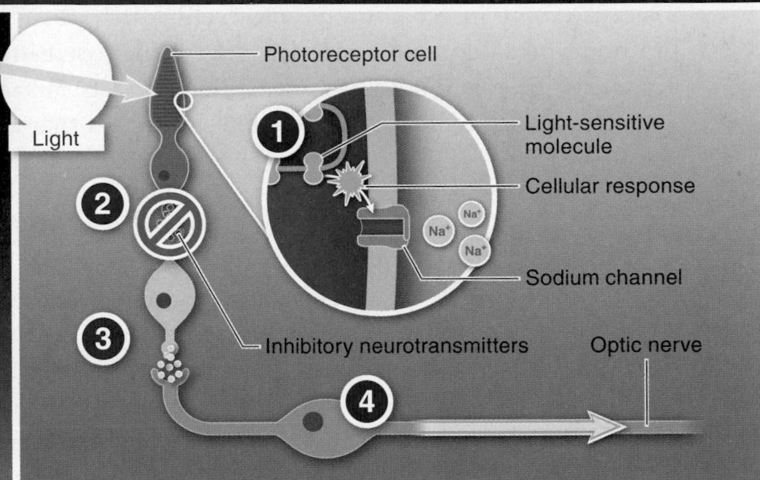

IN THE DARK

1. Sodium channels, located within photoreceptor cell membranes, are open, allowing the photoreceptor cells to continuously release inhibitory neurotransmitters.
2. The inhibitory transmitters reduce the ability of adjacent neurons to excite the neurons that signal the brain.
3. An action potential is not sent through the optic nerve and light is not perceived by the brain.

IN LIGHT

1. Light energy causes light-sensitive molecules to change shape, closing sodium channels in the photoreceptor cell membranes.
2. The closing of sodium channels reduces the amount of inhibitory neurotransmitter released by the photoreceptor cells.
3. No longer inhibited, adjacent neurons excite the neurons that signal the brain.
4. An action potential is sent through the optic nerve and light is perceived by the brain.

FIGURE 24-20 **Photoreceptors responding to light enable vision.**

Within the eye or other light-detecting structure, there are photoreceptor cells, with light-sensitive molecules embedded in their cell membranes. In the dark, a molecule binds to these photoreceptor cells, causing them to depolarize. They also continuously release a neurotransmitter that, like the depressing of a brake pedal, inhibits the ability of many adjacent neurons to send information to the brain. In a sense, darkness blocks the ability of optic nerves to send visual signals to the brain.

When light hits one of the light-sensitive molecules in the membrane of a photoreceptor cell, the light energy causes some of the chemical bonds in the molecule to become stretched. (A similar capture of light energy occurs in plants during photosynthesis.) This closes sodium channels in the photoreceptor cell membranes and reduces the amount of inhibitory neurotransmitter that is released. Without the inhibitory neurotransmitter at the synapse, the adjacent neurons are able to excite the neurons that send visual signals to the brain. The particular wavelength perceived by the brain depends on the particular light-sensitive molecules stimulated.

Insects have much more refined visual capabilities (FIGURE 24-21). They possess **compound eyes** made of dozens to thousands of separate light-sensing units, each with its own image-forming lens that directs light onto about a dozen photoreceptors. The photoreceptors then send signals along neurons. When the signals reach the

INSECT VISION

HOW YOU SEE IT
Silverweed flowers in daylight

HOW AN INSECT SEES IT
Silverweed flowers in ultraviolet light

Some flowers build ultraviolet-light-reflecting petals that guide insects—which see this part of the light spectrum—to the nectar, ensuring pollination of the flowers.

FIGURE 24-21 **Insects see things differently than we do.**

brain, the brain interprets these signals as an image. Because some insect photoreceptor cells—such as those in honeybees—contain a broader range of light-sensitive pigments than is found in humans, these insects can see wavelengths of light in the ultraviolet spectrum that are invisible to us.

Q Insects don't just fly into spiders' webs accidentally. Sometimes they do it on purpose. Why?

Plants sometimes capitalize on this ability of insects to detect the ultraviolet spectrum by constructing flowers with UV-reflecting petals that look like the landing lights at an airport, guiding insects to their nectar reward (while ensuring pollination of the flowers). Spiders capitalize on this ability, too. Some build webs with UV-reflecting threads that trick insects into thinking the web is a flower.

TAKE HOME MESSAGE 24.11

» Vision results from the stimulation of light-sensitive sensory neurons, called photoreceptor cells. The photoreceptor cells have a variety of molecules, embedded within their membranes, that are chemically altered by light. Signals are conveyed to the brain and interpreted as an image. The particular wavelength perceived depends on which version of the light-sensitive molecules in the photoreceptor cell membranes is stimulated.

24.12 Hearing: sound waves are collected by the ears and stimulate auditory neurons.

While tastes and smells are detected by chemoreceptors and sight is made possible by photoreceptors, hearing is a result of the stimulation of **mechanoreceptors,** specialized neurons with receptors that respond to mechanical pressure. The stimuli from the outside world are sound waves, tiny fluctuations in air pressure, collected and amplified by the ears, which then pass information to the brain.

Although the details of how hearing works vary across different species, a general, six-step hearing model applies in most cases (FIGURE 24-22).

1. The outer part of the ear collects sound waves and, because of its shape, funnels them down the **ear canal,** a channel that conducts sound waves.
2. At the end of the ear canal, the sound waves bang into the **eardrum**—a thin membrane that divides the outer ear from the middle portion of the ear—causing it to vibrate.
3. Small bones on the other (inner) side of the eardrum pass the vibrations on to the inner ear membrane.
4. The vibrations are conducted by fluid inside the inner ear, which consists of fluid-filled canals in the bones of the skull. The semicircular canals function in balance, and the coiled cochlea is involved in hearing. This fluid movement bends hair-shaped receptor cells, which release neurotransmitters. (The more vigorous their bending—in response to stronger sound waves—the greater the amount of neurotransmitter released.)
5. Auditory neurons respond to the neurotransmitters released from the hair cells by firing an action potential.
6. The signal continues along a path of neurons to the brain, where it is interpreted as sound.

Q How can you get motion sickness without moving at all?

The fluid in the semicircular canals of the inner ear also acts like an inner motion detector, telling your body about its orientation, speed, and direction. It's not a foolproof system, though. Sitting in an IMAX theater watching a movie filmed from the seat of a roller coaster, your eyes tell your brain that you are moving, while your inner ear senses no motion at all. The conflicting signals can confuse your

HEARING

1. Sound waves are funneled down the ear canal.
2. Sound waves cause the eardrum to vibrate.
3. Vibrations pass to the inner ear membrane.
4. Fluid in the inner ear conducts the vibrations, bending hair-shaped receptor cells and releasing neurotransmitters.
5. The neurotransmitters cause auditory neurons to fire an action potential.
6. The signal is passed to the brain, where it is interpreted as sound.

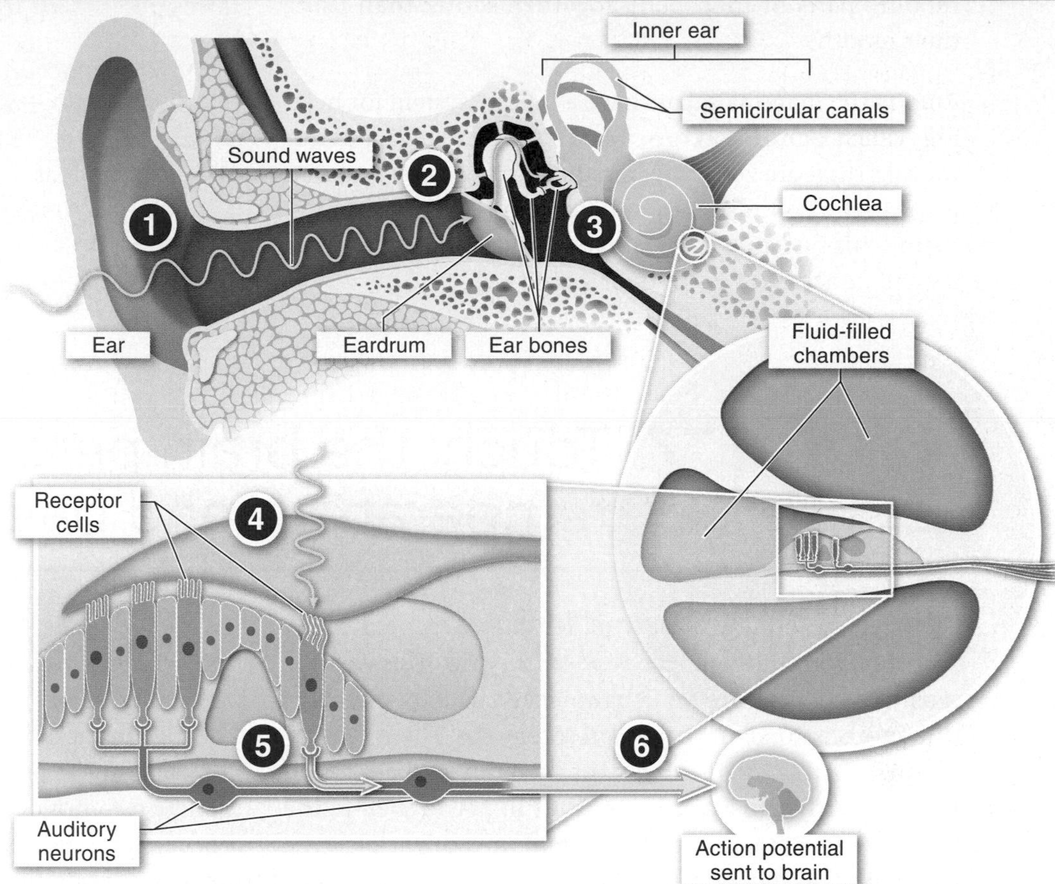

FIGURE 24-22 **The process of hearing.**

Hearing occurs when sound waves vibrate the eardrum, triggering action potentials that are interpreted by the brain as sound.

brain and lead to feelings of nausea. Some people experience similar effects when playing certain videogames.

Q How can loud music lead to hearing loss?

Long-term exposure to loud noises, including music, can be damaging to hearing, because such stimulation can wear out the cochlea of the inner ear. The hair cells in the inner ear are very fragile and are irreplaceable, so chronic over-stimulation due to loud noises can damage them. This reduces their ability to release neurotransmitters and stimulate auditory neurons, causing hearing loss (FIGURE 24-23).

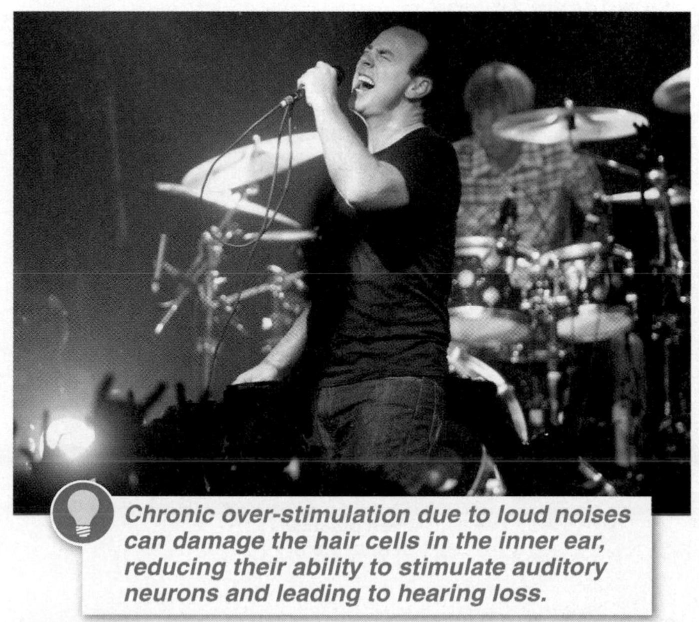

Chronic over-stimulation due to loud noises can damage the hair cells in the inner ear, reducing their ability to stimulate auditory neurons and leading to hearing loss.

FIGURE 24-23 **Loud noise can cause hearing loss.**

Are ears necessary for hearing? Not really. Most insects hear a wide range of sounds, with delicate hairs on their antennae or on other parts of the body. These insects often produce noises (that is, communicate by sound) by rubbing parts of their body together, rather than using their mouth.

Among the mammals, bats have a unique system for hearing, called **echolocation.** Most bats emit high-pitched squeaks (that are generally inaudible to humans) and wait to hear the sounds as they bounce off surfaces around them. This system is so sensitive that bats can detect even a small moth as it flies in their vicinity.

TAKE HOME MESSAGE 24.12

» Hearing occurs when sound waves cause the eardrum to vibrate, moving tiny bones that pass on the vibrations to the inner ear, where the vibrations bend hair cells and thus trigger a pattern of action potentials that varies with the wavelengths of the sound. This is interpreted by the brain as a particular sound.

24.13 Touch: the brain perceives pressure, temperature, and pain.

Touch is actually a class of sensations generated by numerous different types of sensory neurons that are sensitive to pressure (mechanoreceptors), temperature (thermoreceptors), or pain (pain receptors). These sensory receptors are located throughout the body and are found in particularly dense concentrations in places such as the fingertips in primates and the side of the body (called the lateral line) in most types of fish.

Whether they are mechanoreceptors or pain receptors or thermoreceptors—and any given part of the body usually has many of each of these sensory neuron types—they function similarly. External stimulation, such as the prick of a pin, the tickle of a feather, or the heat of a pan, can cause a change in the shape of the neuron's membrane, momentarily altering its permeability. This change in permeability then increases or decreases the rate at which ions enter the cell, converting the stimulus into action potentials and, ultimately, sensations perceived by the brain (FIGURE 24-24).

Eerily, most amputees report that they sometimes experience tingling, prickly sensations or even shooting pains

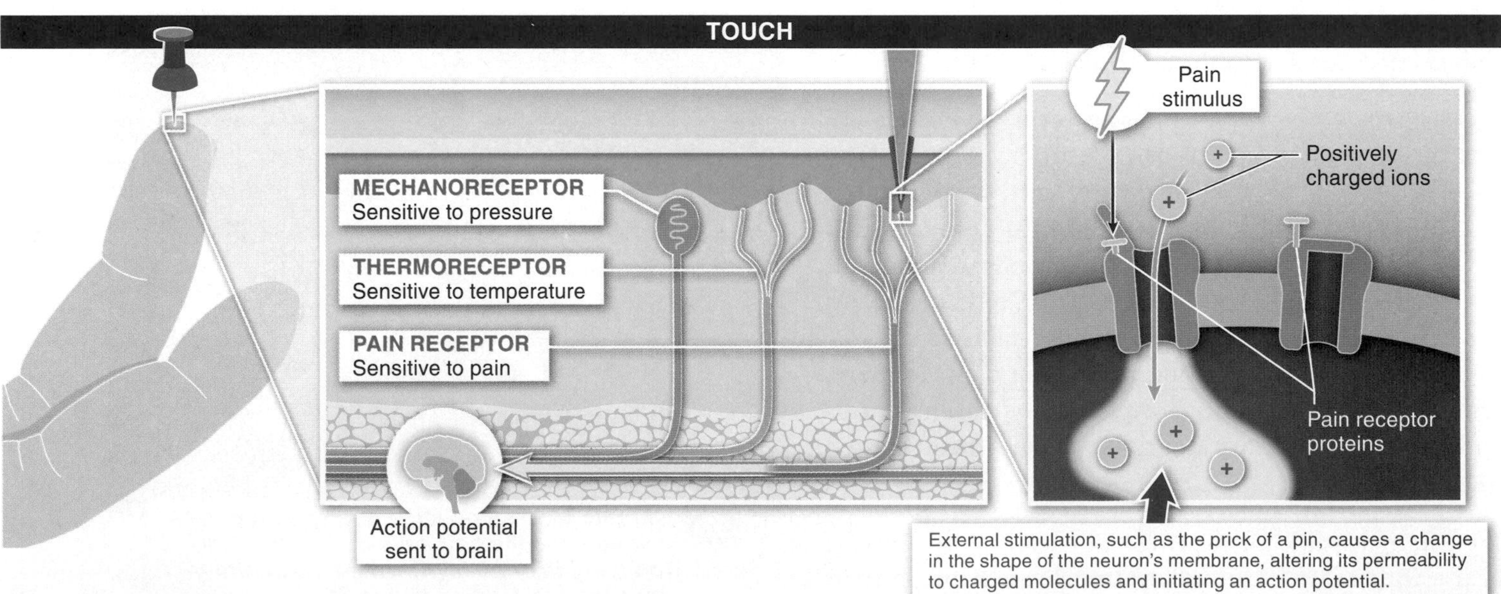

TOUCH
MECHANORECEPTOR
Sensitive to pressure
THERMORECEPTOR
Sensitive to temperature
PAIN RECEPTOR
Sensitive to pain
Action potential
sent to brain
Pain
stimulus
Positively
charged ions
Pain receptor
proteins
External stimulation, such as the prick of a pin, causes a change in the shape of the neuron's membrane, altering its permeability to charged molecules and initiating an action potential.

FIGURE 24-24 **How the sensation of touch is produced.**

PHANTOM PAIN

Many amputees experience "phantom pain" that seems to come from their missing limb, as neurons in the brain that had synapses with touch receptors in the lost limb are stimulated, usually by nearby neurons.

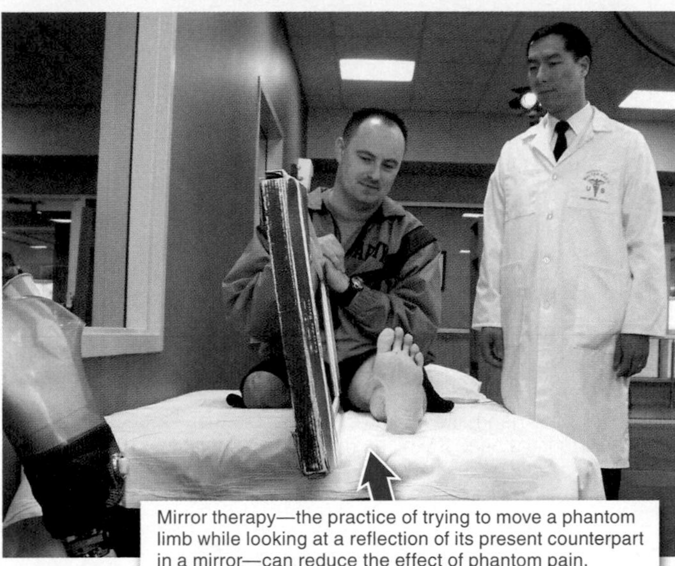

Mirror therapy—the practice of trying to move a phantom limb while looking at a reflection of its present counterpart in a mirror—can reduce the effect of phantom pain.

FIGURE 24-25 Sensory neurons may be gone, but real pain can remain.

Q Amputees sometimes feel real pain where their limb once was. How can this be?

that come from where their amputated limb once was. Although called "phantom pain," these sensations are very real and occur when neurons in the brain that used to receive information from touch receptors in the lost limb are firing, usually due to stimulation from other, nearby neurons (FIGURE 24-25).

TAKE HOME MESSAGE 24.13

» Touch is a class of sensations generated by mechanoreceptors, thermoreceptors, and pain receptors located throughout the body. Stimulation of these receptors causes a change in the shape of the sensory neuron's membrane, altering its permeability, generating action potentials, and causing the perception of touch by the brain.

22.14–22.15

The muscular and skeletal systems enable movement.

The Gentoo penguin is a masterful surfer.

24.14 Muscles generate force through contraction.

Running, flying, swimming, crawling, walking, digging, dancing. Animals, with very few exceptions, move. And in vertebrates, movement is generally initiated by cells of the nervous system that stimulate muscle contractions that pull on a rigid skeletal system.

As we saw in Section 21.4, muscle tissue is made up of elongated cells capable of generating force when they contract, and there are three types of muscle tissue.

1. *Skeletal muscle* is attached to bones by connective tissue and is controlled by individual neurons attached to each muscle fiber. The muscular system of humans contains about 700 skeletal muscles and makes up about 40% of human body weight.

2. *Cardiac muscle* causes the heart to pump blood through the body. Not initiated by the nervous system or under conscious control, cardiac muscle contractions and relaxations occur continuously throughout life, with the muscle cells containing more energy-releasing mitochondria than other types of muscle cells.

3. *Smooth muscle,* which also is not under conscious control, surrounds blood vessels and many internal organs. In some cases, smooth muscle contractions are stimulated by the autonomic nervous system, and in other cases, smooth muscle is able to contract without any stimulation from the nervous system. Compared with the other muscle types, smooth muscle generates slower contractions, which can gradually move blood, food, or other substances through vessels and organs.

Our focus here is on skeletal muscle. To facilitate movement, skeletal muscles are attached to bones by connective tissue. Often, a muscle tapers at each end into tendons, which attach the muscle at the two ends—the "insertion" and the "origin" of the muscle—to bones. The contraction of the muscle then pulls the attachment points closer together and causes a movement of the bone. The biceps muscle of the upper arm, for example, has its origin in the shoulder and its insertion in one of the bones of the forearm. Contraction can rotate the forearm or bring the lower part of the arm closer to the shoulder. Two muscles generally work in opposition, so that one contracts to move a bone in one direction, and the other—the triceps in the upper arm—has its insertion and origin located so that its contraction moves the bone in the opposite direction (FIGURE 24-26).

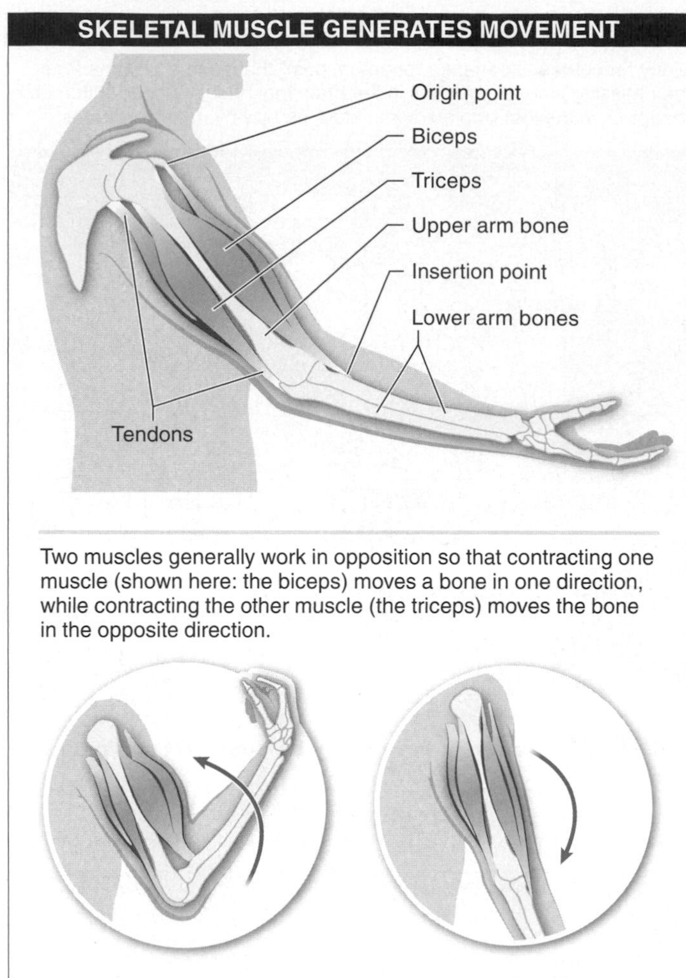

SKELETAL MUSCLE GENERATES MOVEMENT
Origin point
Biceps
Triceps
Upper arm bone
Insertion point
Lower arm bones
Tendons
Two muscles generally work in opposition so that contracting one muscle (shown here: the biceps) moves a bone in one direction, while contracting the other muscle (the triceps) moves the bone in the opposite direction.

FIGURE 24-26 Movement: the work done by skeletal muscles.

A skeletal muscle is made up of a bundle of fibers. Each fiber is a single cell, but has multiple nuclei and numerous **myofibrils,** cylindrical organelles that shorten when they contract. A myofibril contains repeating units called **sarcomeres,** which are where the contraction takes place.

The sarcomeres are composed of large numbers of long filaments, overlapping and parallel to each other. The filaments come in two types: thin filaments made mostly from the protein **actin,** and thick filaments made mostly from the protein **myosin.** There may be 100,000 sarcomeres in a biceps muscle cell.

The sarcomere shortens in a four-step process that resembles climbing a rope (FIGURE 24-27).

1. **Detach.** A link between a myosin filament and an actin filament parallel to it is broken as a molecule of ATP binds to the myosin.
2. **Reach.** As the ATP breaks down, energy released alters the shape of the myosin to form a higher-energy shape, much like bending a twig—the bent twig has a higher-energy shape because it can release energy as it snaps back to its original shape.
3. **Reattach.** In its altered shape, the myosin reaches farther down the actin filament, where it reattaches.
4. **Pull back.** The myosin then snaps back to its original shape, pulling the actin filament as it does so and thus shortening the fiber. This last step is considered the "power stroke."

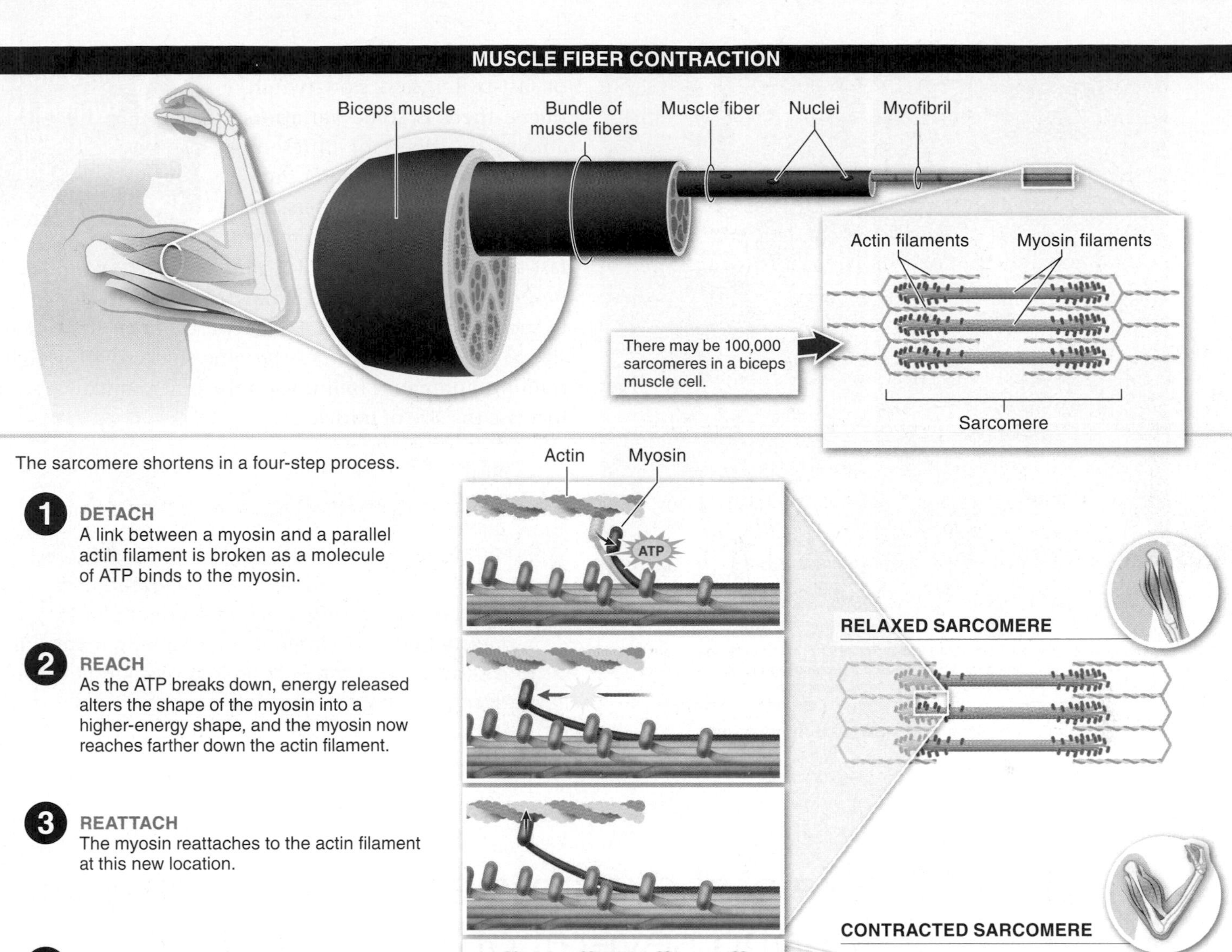

MUSCLE FIBER CONTRACTION
Biceps muscle
Bundle of muscle fibers
Muscle fiber
Nuclei
Myofibril
Actin filaments
Myosin filaments
There may be 100,000 sarcomeres in a biceps muscle cell.
Sarcomere
The sarcomere shortens in a four-step process.
1 DETACH
A link between a myosin and a parallel actin filament is broken as a molecule of ATP binds to the myosin.
2 REACH
As the ATP breaks down, energy released alters the shape of the myosin into a higher-energy shape, and the myosin now reaches farther down the actin filament.
3 REATTACH
The myosin reattaches to the actin filament at this new location.
4 PULL BACK
The myosin then snaps back to its original shape, pulling the actin filament as it does so and shortening the fiber.
Actin
Myosin
ATP
RELAXED SARCOMERE
CONTRACTED SARCOMERE

FIGURE 24-27 **Muscle fibers contract to generate force.**

As the contraction process occurs across numerous sarcomeres in a muscle cell, the entire muscle can shorten. From the perspective of energy, the potential energy of ATP is converted to the kinetic energy of sarcomere shortening, which can do work.

When a muscle contracts, the duration between a contraction and a relaxation is called a **twitch.** Muscle fibers within a muscle vary in how quickly they can twitch, with two general types. Fast-twitch fibers can contract and relax about 10 times faster than slow-twitch fibers, but have relatively weak endurance. The slow-twitch fibers, which are surrounded by rich oxygen-delivering capillary beds and large numbers of oxygen-storing myoglobin molecules, can sustain activity for much longer periods of time. Although most muscles have

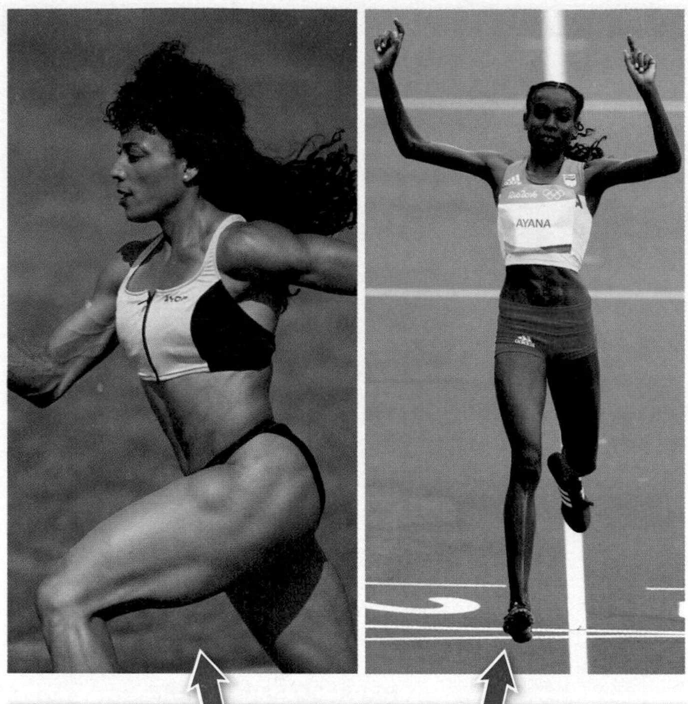
AYANA
World-class sprinters have more fast-twitch fibers (which enlarge in response to heavy use) in their leg muscles, while world-class long distance runners tend to have more slow-twitch fibers in their leg muscles.

FIGURE 24-28 Fast-twitch and slow-twitch muscles.

approximately equal proportions of fast-twitch and slow-twitch fibers, there is some variation among the fibers of different muscles, as well as among individuals. World-class sprinters, for example, have significantly more fast-twitch fibers in their leg muscles, while world-class long distance runners tend to have more slow-twitch fibers in their leg muscles (FIGURE 24-28). Muscle fibers increase in size as the fast-twitch fibers become thicker with weight training. Endurance training, on the other hand, does not increase the size of muscle cells.

Q Marathon runners tend to have smaller leg muscles than sprinters. Why?

TAKE HOME MESSAGE 24.14

» Muscle tissue—including skeletal, cardiac, and smooth muscle—is made up of elongated cells capable of generating force when they contract. A muscle fiber is a single cell containing myofibrils that shorten with the making and breaking of links between parallel actin and myosin filaments.

24.15 Skeletal systems enable movement, among several other important functions.

By serving as a structural frame, the skeletal system interacts with the muscular system to produce movement. But this is just one of many important functions of a skeleton. A skeletal system also provides support for the organism, affording shape and structure and a means of securing the organs and other structures within the animal's body. In addition, a skeletal system offers protection—the brain, for example, is protected from harm within the skull. In some organisms, including humans, the skeletal system is also the site for the production of important blood and immune cells, a process that occurs in bone marrow (see Section 27.3). Bones also serve as a reservoir of some minerals, such as calcium, absorbing excess quantities from the bloodstream or releasing minerals when their concentration in the blood is low (see Section 25.5).

There are three distinct types of skeletal systems (FIGURE 24-29). In a *hydrostatic skeleton*—such as is found in earthworms, jellyfish, and squid—pressure created within a fluid-filled cavity creates sufficient rigidity that the muscles surrounding it can generate movement. An *exoskeleton* is a rigid outer covering that supports and protects an animal's body. Examples include the shell of a turtle, the rigid, calcium-based covering of some mollusks, and the polysaccharide-based, chitinous exoskeleton of insects, crabs and lobsters, and spiders. The third type of skeleton—an *endoskeleton*—is a support structure of hard mineralized tissue, as is found in echinoderms and in all vertebrates, including humans.

Among the vertebrates, the jawless fishes and sharks have a flexible skeleton made from cartilage. All other

TYPES OF SKELETAL SYSTEMS

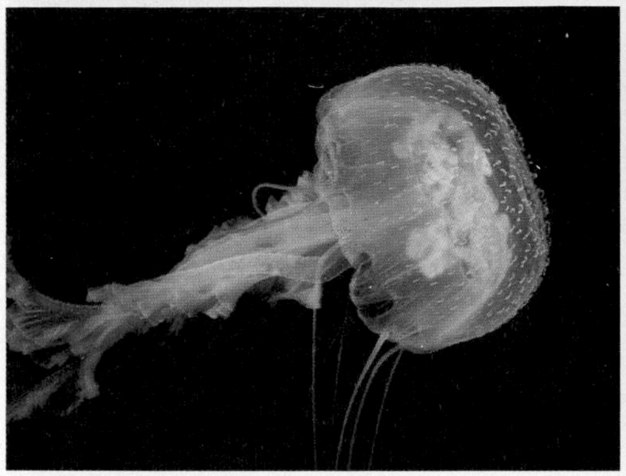

HYDROSTATIC SKELETON
A fluid-filled cavity with sufficient rigidity that muscles surrounding it can generate movement

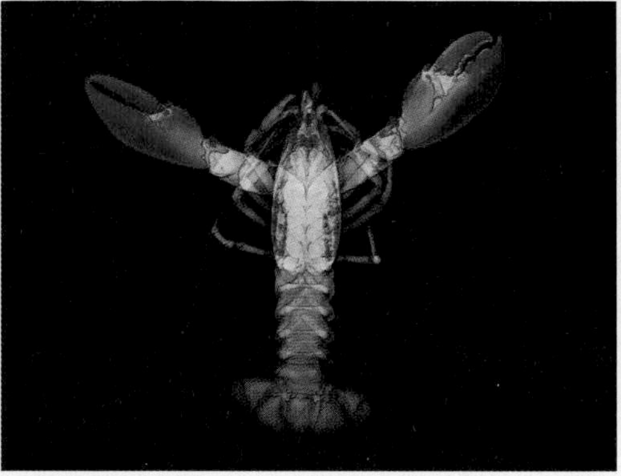

EXOSKELETON
A rigid outer covering that supports and protects an animal's body

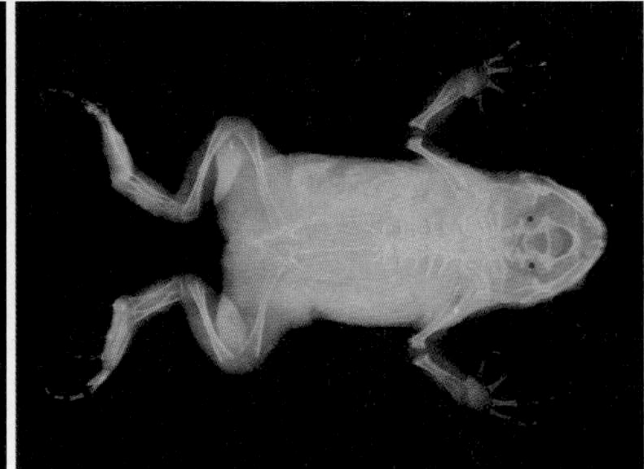

ENDOSKELETON
An internal support structure composed of hard, mineralized tissue

FIGURE 24-29 **There are three distinct types of skeletal systems in animals.** Shown (left to right): purple jellyfish, lobster, and toad.

vertebrates have a skeletal system made primarily from bone, a harder, but still resilient connective tissue material (described in Section 21.2).

At the locations in the vertebrate skeleton where bones meet, several types of joints occur. Some allow movement while maintaining support, such as in the shoulder, hip, knee, and elbow joints. Others enable little movement (such as the joints between vertebrae) or no movement at all (such as the "joints" that are the sutured connections between the bones of the skull). Arthritis is a condition in which joints have become damaged. There are several forms of arthritis. In the most common type, osteoarthritis, a breakdown of connective tissue, usually cartilage, reduces the cushioning of the bones and can lead to inflammation of tissue around the joint, with accompanying pain.

Another type of breakdown in the skeletal system is osteoporosis, a disease that occurs as the density of bone is reduced and the chemical composition of the bones changes (FIGURE 24-30). More common in women, particularly following menopause, osteoporosis increases the risk of bone fractures and falls. Dietary modifications, including increased intake of calcium and vitamin D that can strengthen bones, along with lifestyle changes such as increased exercise that can build bone mass and strengthen muscles, can reduce osteoporosis and the risks associated with it.

OSTEOPOROSIS

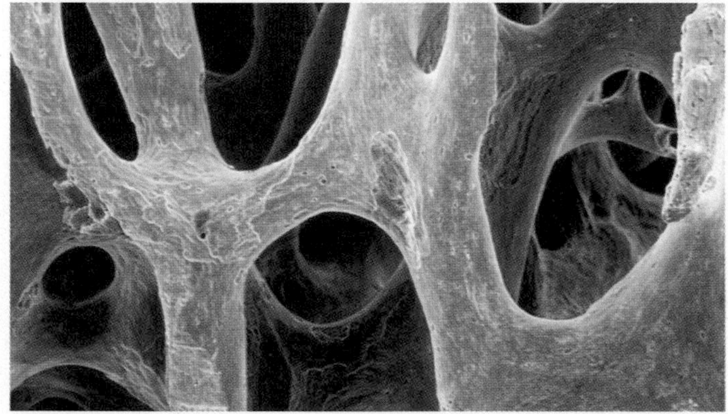

Healthy bone

Osteoporosis results in reduced bone density and a change in the chemical composition of bone, leading to an increased risk of bone fracture.

Osteoporotic bone

FIGURE 24-30 **Osteoporosis weakens bone.**

TAKE HOME MESSAGE 24.15

» There are three distinct types of skeletal systems: hydrostatic skeletons, exoskeletons, and endoskeletons. The vertebrate skeletal system, an endoskeleton, enables movement, provides support, offers protection, provides a site for production of important blood and immune cells, and serves as a reservoir of minerals, including calcium.

24.16–24.18

The brain is organized into distinct structures dedicated to specific functions.

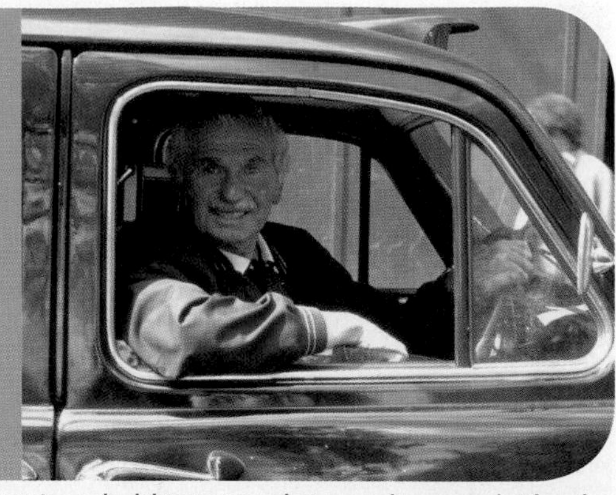

As they master complex street layouts, cab drivers experience an increase in the size of a brain structure—the posterior hippocampus—that assists in spatial memory.

24.16 The brain has several distinct regions.

From a cellular perspective, all animal brains function in the same general way. A neuron is a neuron is a neuron. But this isn't the whole picture. Each region of the brain is characterized by neurons with different axon and dendrite branching patterns and complexities, and using different neurotransmitters at their synapses. Each of these regions of the brain is the control center for particular activities in the body.

Evolutionary changes have resulted in differences among species in the relative amount of the brain dedicated to various body structures. These differences have, in turn, led to corresponding differences in abilities and behaviors. For example, humans have significantly more neurons in the cerebral cortex than do other species. The extra brain matter in this particular part of the brain (which we examine in more detail later in the chapter) is associated with many abilities that are particularly well developed in humans, including abstract thought, perception, and language. Similarly, a rat's brain has an unusually large number of smell-sensing neurons (relative to overall brain size). The abundance of these neurons gives the rat a particularly powerful sense of smell that contributes to its abilities to find food.

In this section and the next, we explore the three principal regions of the vertebrate brain—the hindbrain, the midbrain, and the forebrain—and the functions they control (FIGURE 24-31).

Hindbrain The central nervous system is made up of the brain and the spinal cord. The spinal cord extends from the brain down through the backbone (though not down the entire length of the backbone), protected by the vertebral column. Branching off from the spinal cord at regular intervals are bundles of axons, some carrying motor information to muscles and glands, others carrying in sensory information from the senses to the brain. The top of the spinal cord is the beginning of the brain. Here, the spinal cord expands into the **hindbrain.** Three important structures are located in the hindbrain: the **medulla,** the **pons,** and the **cerebellum.** Because these structures are not involved in higher thought, hindbrain damage can lead to paralysis without harming consciousness.

Medulla and pons. Running through these structures are motor neurons traveling from the brain to the spinal cord and sensory neurons traveling from all parts of the body to the brain. Additionally, within the medulla are collections of neurons that regulate basic physiological functions such as respiration, heart rate, and digestion. Consequently, injury to these structures—such as occurred when President John F. Kennedy was shot in 1963—nearly always leads to death.

Cerebellum. This structure coordinates numerous types of motor activity. When it works properly, graceful motion is possible. The coordination of many complex movements is overseen by the cerebellum. Knitting or typing, for example, involves a wide range of complex muscle movements that must be executed in precise sequences. Once learned, these movements run on "auto-pilot": we don't have to think consciously about these activities, because the cerebellum directs them (FIGURE 24-32).

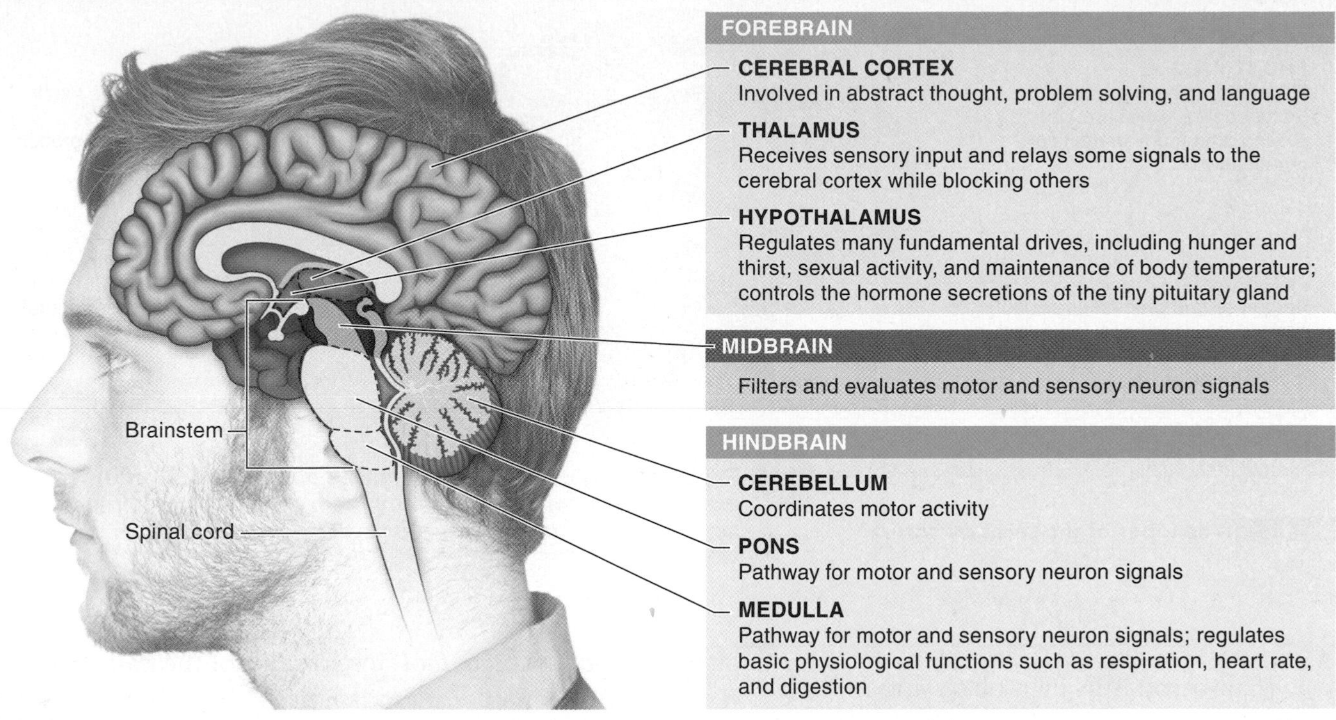

FIGURE 24-31 **The brain is divided into three main regions: forebrain, midbrain, and hindbrain.**

After complex muscle movements are learned, such as the precise sequences of knitting, they can then be done without much conscious direction, on a sort of "auto-pilot" as the cerebellum directs them.

FIGURE 24-32 **Coordinated by the cerebellum.** Once learned, knitting does not require conscious thought.

Midbrain The **midbrain,** along with the medulla and pons of the hindbrain, makes up the *brainstem.* Most of the sensory information and motor neuron connections passing through the brain travel through the midbrain, which helps filter and evaluate the importance of each signal.

Forebrain In humans, the **forebrain** is the largest region of the brain. It includes two control and relay structures, the thalamus and hypothalamus, as well as the cerebrum. The cerebral cortex, which makes up the bulk of the cerebrum, is responsible for most of what we consider "higher" thought, including perception, memory, language, intelligence, and personality.

Thalamus and hypothalamus. The **thalamus** is one of the primary switchboards in the brain. It receives most of the visual, auditory, and touch input and, like a secretary deciding which calls to let through, relays some signals to the cerebral cortex while blocking others. The **hypothalamus** (part of the limbic system, discussed later in this section) is one of the chief regulatory centers of the brain. It regulates many fundamental drives, including hunger and thirst, sexual activity, and maintenance of body temperature. It also controls the hormone secretions of the tiny pituitary gland, located right next to it, which play important roles in reproduction and development.

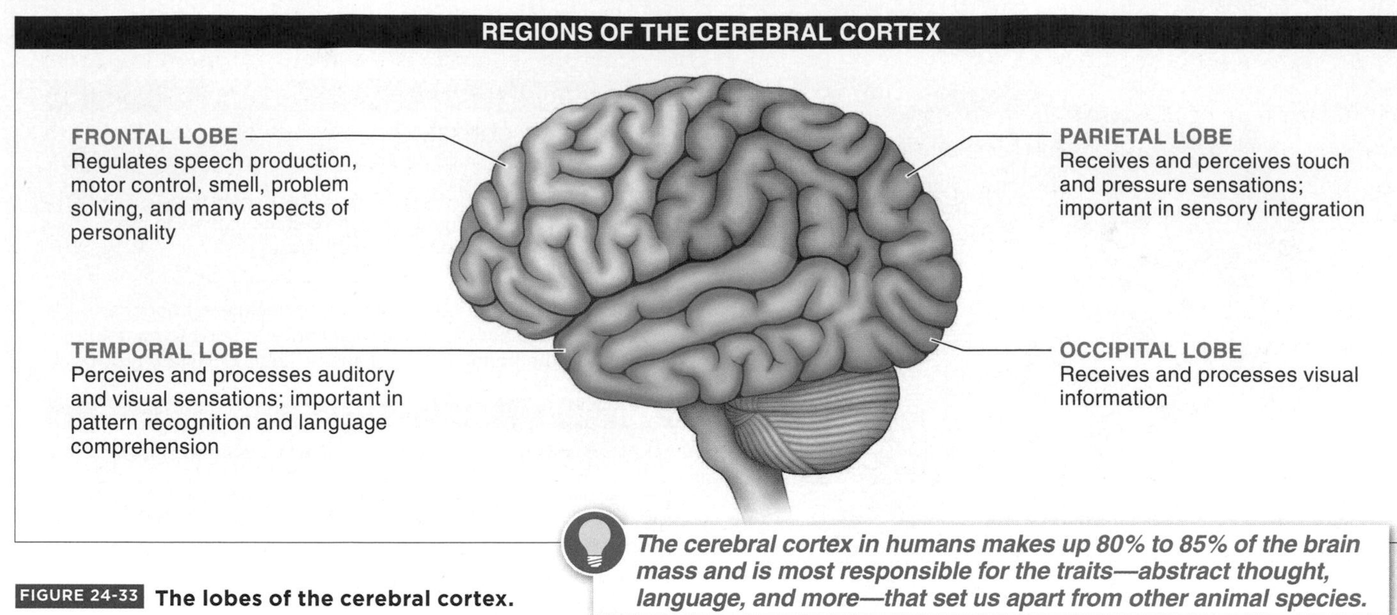
REGIONS OF THE CEREBRAL CORTEX
FRONTAL LOBE
Regulates speech production, motor control, smell, problem solving, and many aspects of personality
TEMPORAL LOBE
Perceives and processes auditory and visual sensations; important in pattern recognition and language comprehension
PARIETAL LOBE
Receives and perceives touch and pressure sensations; important in sensory integration
OCCIPITAL LOBE
Receives and processes visual information
The cerebral cortex in humans makes up 80% to 85% of the brain mass and is most responsible for the traits—abstract thought, language, and more—that set us apart from other animal species.

FIGURE 24-33 **The lobes of the cerebral cortex.**

> "The mind is its own place, and in itself
> Can make a heaven of hell, a hell of heaven."
>
> — JOHN MILTON
> *Paradise Lost, 1667*

In the 1950s, James Olds discovered that when he inserted a tiny electrode into the hypothalamus of a rat's brain, the animal seemed to experience great pleasure. Of course, Olds couldn't be certain that it was pleasure the rats were experiencing, but he hypothesized that this was so by noting how the rats would work very hard (for instance, by learning to run very complex mazes) to get the stimulation. When he set up the apparatus so that an animal could administer the stimulation to its own brain simply by pushing a lever, the rats became lever addicts, pushing the lever more than 700 times an hour for many hours. They would even choose to push the lever over eating food, with some rats starving to death while pushing the lever rather than taking a break to eat.

This discovery that the brain has "pleasure centers" or "do-it-again centers" was important because it revealed a mechanism by which animals could be motivated to engage in evolutionarily important behaviors. Drinking when thirsty or eating when hungry produces stimulation in this brain area, thereby reinforcing behaviors essential to survival and reproduction.

Cerebral cortex. For humans, the **cerebral cortex** is the most sophisticated part of the brain. Involved in abstract thought, problem solving, language, and more, this is the part of the brain most responsible for the traits that set us apart from other animal species.

In 1848, a railroad worker named Phineas Gage suffered and recovered from a tragic accident, providing insight into the functions of the cerebral cortex. An explosion blasted a three-and-a-half-foot iron bar right through his face, just below his left eye, through the front part of his head, and out through the top of his skull. Soon after his recovery, it became clear that Gage's personality had changed. Before the accident he was serious and industrious, but after the accident he became vulgar, unpleasant, uninhibited, and irresponsible. Gage died about 13 years after the accident, and his skull was preserved. Recent analyses of the skull indicate that Gage suffered from severe damage to part of his frontal cortex.

In other case studies of patients with damage to the same part of their frontal cortex, such as that caused by certain tumors, similar personality changes have been noted, with patients frequently exhibiting radically altered emotional responses.

The highly folded cerebral cortex is by far the largest brain structure, covering all of the brain except for the cerebellum (FIGURE 24-33). The cerebral cortex makes up 80% to 85% of the brain mass in humans, with about 10 billion neurons and hundreds of billions of synapses. Although the brain looks fairly symmetrical, functions are not divided in a symmetrical way. The brain is divided into a left and a right hemisphere, connected by a broad, thick band of neurons, called the **corpus callosum.** The **left hemisphere** in most, but not all, individuals is the site of areas specializing in language, logic, and mathematical skills, while the **right hemisphere** is more commonly home to larger areas dedicated to emotions, intuitive thinking, and artistic expression.

Each hemisphere of the cerebral cortex has four different lobes: frontal, temporal, parietal, and occipital. Each lobe is associated with specific functions (but each also has many other subareas that control a variety of additional functions; see Figure 24-33).

1. **Frontal lobe.** Important in speech production, motor control, smell detection, problem solving, and many aspects of personality.
2. **Temporal lobe.** Perceives and processes auditory and visual sensations; also important in pattern recognition and language comprehension.
3. **Parietal lobe.** Receives and perceives touch and pressure sensations; also important in **sensory integration,** the process of incorporating information from all of the senses to form a single perception of something.
4. **Occipital lobe.** Receives and processes visual information.

Limbic system. Within the forebrain, near the center of the brain, is a set of structures that together make up the **limbic system.** It includes parts of the cerebral cortex and two structures deeper within the center of the brain, the **hippocampus** and the **amygdala,** and is responsible for many of our physiological drives and instincts. Together, these structures also play an important role in emotions, learning, and memory, which we discuss below.

TAKE HOME MESSAGE 24.16

» The brain has three distinct regions. The hindbrain regulates basic physiological functions such as respiration, heart rate, and digestion, and coordinates numerous types of motor activity. The midbrain, part of the brainstem, helps filter and evaluate the importance of sensory and motor information. The forebrain, the largest region of the brain in humans, is responsible for most higher thought, including perception, memory, language, intelligence, and personality.

24.17 Specific brain areas are involved in the processes of learning, language, and memory.

Brains are capable of myriad amazing feats. Learning and memory, along with the use of complex language, are among the most sophisticated and enigmatic of these functions.

Language What is language, exactly? In its most broad interpretation, **language** is simply the means of communication between individuals of any species. In humans, language has evolved to its most complex level. Words and sentences allow us to convey not just simple information but also complex thoughts such as things and events distant in space or time, something rarely if ever seen in other species.

Several distinct brain structures, and two in particular, are necessary for the acquisition and use of language. One is required for the understanding of speech, including the linking of words with meaning (FIGURE 24-34). Located in the left temporal lobe, this region is called **Wernicke's area.** People with damage to Wernicke's area (from a stroke or head injury) are able to produce sounds properly, such as when repeating things they hear, but they cannot understand speech and can speak only gibberish. It's as if they've lost the ability to link words with their meanings. The second structure, **Broca's area,** controls the muscles involved in the actual process of speaking. It is located in the front part of the left frontal lobe.

Interestingly, women are much more likely than men to regain the ability to speak after a stroke. This may be related to the fact that, on average, women have a thicker corpus callosum, the band of fibers connecting the left and right hemispheres of the brain. The additional thickness may facilitate the ability of another part of the brain to take over language functions after the stroke-induced death of brain tissue.

Q Do men or women have a better chance of language recovery following a stroke? Why?

Another recent discovery suggests that children have a propensity to learn any language. In a study of children raised by parents who spoke using sign language only, observations of their children revealed that the infants "babbled" with their hands (FIGURE 24-35). Much as, from age six months to one year, babies with parents who speak begin to put together nonsense words made up from fragments of words and sounds, the babies of signing parents used their hands to produce fragments of hand signs. The babies of speaking parents did not exhibit such hand babbling.

Infants raised by parents who speak sign language "babble" only with their hands.

FIGURE 24-35 **"Babbling" baby.**

Learning and memory are understood even less well than language. Still, certain cellular changes seem to occur in neurons when we learn and remember things. Two parts of the limbic system are critical in storing memories: the amygdala and the hippocampus. The amygdala seems to associate emotional feelings and sensory input with memories (see Figure 24-34). This is why certain smells or tastes, for example, elicit memories. The hippocampus is also important in the storage and retrieval of memories, as well as in the transfer of short-term memories—things that are stored in our brain only briefly, such as a phone number or the name of a person we have just met—into long-term memories that can be recalled days, years, or even decades later. Memories and associations of sensations are stored throughout much of the cerebral cortex, too.

SPEECH AND MEMORY CENTERS IN THE BRAIN

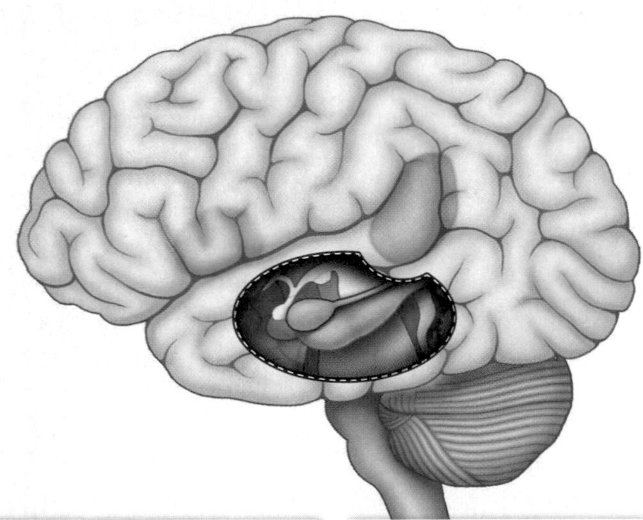

FIGURE 24-34 **Areas of the brain associated with memories.**

Humans don't have a monopoly on powerful memory. Each fall, for example, a red squirrel may hide more than 3,000 acorns and other nuts. Come winter, the squirrel is able to remember the locations of more than 80% of those nuts and recover them for food. Similarly, birds such as the black-capped chickadee store large numbers of nuts for the winter. Researchers looking for neurological changes in these birds found that the hippocampus is significantly larger in the fall, when the birds must remember all of their nut-stash locations, than in the spring, when they no longer need to rely on stored nuts for food.

What happens to the neurons at a cellular level during the processes of learning and forming memories? When repeated action potentials are generated by a neuron, changes occur at the synapse. There are two types of change that generally occur. One is called **long-term depression** and the other is called **long-term potentiation.** In each, as neurotransmitters are repeatedly released by the presynaptic membrane, the adjacent neuron responds by initiating an action potential, but also by becoming less likely (long-term depression) or more likely (long-term potentiation) to respond to future stimulation with an action potential. That's what constitutes the learning: future responses are modified by past experience. This makes sense when we consider that perhaps the simplest form of learning comes from repeated attempts at doing something, during which the individual becomes better and better at it.

TAKE HOME MESSAGE 24.17

» Learning, memory, and use of complex language are among the most sophisticated brain functions. We are learning about brain structures that are responsible for these functions—for language, structures in the left frontal and left temporal lobes, and for learning and memory, structures in the amygdala and hippocampus.

THIS IS HOW WE DO IT

Developing the ability to apply the process of science

24.18 Can intense cognitive training induce brain growth?

"Use it or lose it." Weightlifters often say this about muscles. Can the same be said about the memory-forming components of our brain? Scientists have long believed that no new neurons are added to the human brain once development is completed in early childhood. While this is largely true, more recent experimental approaches suggest that some parts of the hippocampus are an exception.

Brain scans called functional MRI, or fMRI, can visualize the size and activity of different brain structures. Using this scanning technology, researchers have investigated how individuals' brains change over time. They began, as many scientists do, with an interesting question: *Can environmental stimulation improve your brain's abilities?* This is an important question, but perhaps too general for tackling in a single, carefully controlled experiment.

How could you set about answering this question for a real-world situation?

The researchers refined the broad question into something more testable: *Does acquiring and mastering a large and complex body of knowledge measurably change your brain anatomy?* They identified a population well-suited to the study: London cab drivers. To become licensed as a cab driver in London, a trainee must master the layout of more than 25,000 streets within a six-mile radius of the central train station. This is known as acquiring "The Knowledge" and typically takes about four years, during which the trainee must pass a series of stringent exams. Among trainees, 50% to 60% fail to qualify for their license.

Armed with fMRI for measuring brain structures and an interesting pair of groups to compare—cabbies and non-cabbies—the researchers were able to make an exciting discovery. The posterior portion of the hippocampus was significantly larger among cabbies than among non-cabbies. (And its size was positively correlated with the number of years the cab driver had been on the job.)

Why would people who don't drive cabs be interested in this result?

These findings were consistent with the hypothesis that learning The Knowledge creates new neurons due to the greater demands placed on the posterior hippocampus as cabbies navigate and learn the streets of London. Furthermore, if enduring structural changes in the brain can be induced by behaviors, then there may be long-term benefits of educational interventions in a variety of areas, as well as for rehabilitation after illness or injury.

How much confidence can we have in this conclusion?

Many people were cautious about immediately drawing broad conclusions from the cab driver study. One issue in particular aroused skepticism. The study results are consistent with an alternative hypothesis. Maybe cab driving doesn't make your hippocampus bigger. Maybe, instead, people differ innately in posterior hippocampus size and those with a larger hippocampus are simply predisposed to be successful cab drivers.

How would a longitudinal study resolve uncertainty about why cabbies have a larger posterior hippocampus?

An experimental approach known as a longitudinal study could resolve this uncertainty. You could study individuals as they trained to become cab drivers, measuring their posterior hippocampus size before and after they acquired The Knowledge.

If posterior hippocampus size *is increased* as a result of acquiring The Knowledge, the hypothesis that brain changes can be induced by behaviors is supported. If posterior hippocampus size *is not increased* by acquiring The Knowledge, the hypothesis supported is that some people have a larger hippocampus and those people are predisposed to succeed as cab drivers.

Clearly, a longitudinal study is more useful than the initial approach. So why were the first "cross-sectional" studies even conducted?

We can answer this question with three words: faster, easier, cheaper. Based on the promising but ambiguous results from the initial experiment, however, the researchers decided to design and conduct a longitudinal study with 79 people training to become London cabbies (as well as 31 control participants, not undergoing the training). For each participant, the researchers recorded several measurements, including assessments of memory. They also determined the volume of each individual's hippocampal gray matter, using fMRI. They collected these data at two time points: T1, at the beginning of training, and T2, four years later, just after qualification for a license.

Which aspects of the longitudinal study made it more difficult to carry out than the earlier, cross-sectional study?

Of the original 79 trainees, 59 returned for the post-training T2 measurement: 39 who had qualified for their license and 20 who had not. Among the 20 original participants who did not return for the T2 measurement, 2 had qualified and the other 18 did not complete the training. All 31 of the control group returned for the T2 measurement.

Based on the trainees' license examination performance, the participants could be divided into three groups: group Q, trainees who *q*ualified for their license (having mastered The Knowledge); group F, trainees who *f*ailed to qualify for their license; and group C, *c*ontrols who did not undergo the training.

No hippocampus size differences were detected among the three groups before the start of training. In other words, the groups started out on equal terms. At T2, however, some definitive and dramatic results from the longitudinal study became clear.

1. Those who qualified—group Q—had a greater posterior hippocampus volume at T2 than at T1 (and a greater volume than participants in groups F and C at T2).

2. Those in groups F and C showed no change in posterior hippocampus volume at T2 relative to volume at T1.

Additionally, those who qualified (Q) were found to have spent an average of twice as many hours per week in training as did those who did not qualify (F). The qualifiers also scored significantly higher on tests of spatial memory at T2 than did groups F and C (which did not differ from each other).

The impact on hippocampus size of acquiring The Knowledge is quite clear when presented graphically (see the graph at the top of the next page).

What can you conclude from these results?

Based on the longitudinal study, the researchers were confident in concluding that acquisition of The Knowledge was associated with an increase in size of the posterior hippocampus and with better performance on memory assessments.

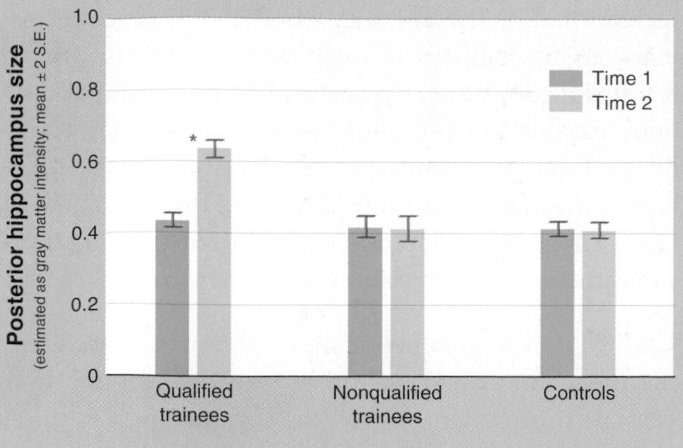

* Difference in size is statistically significant

GRAPHIC CONTENT
Thinking critically about visual displays of data
Turn to p. 871 for a closer inspection of this figure.

The researchers also thought a more general conclusion was warranted—that "specific, enduring, structural brain changes in adult humans can be induced by biologically relevant behaviors engaging higher cognitive functions such as spatial memory"—and speculated on the potential for memory-improvement training in educational and clinical settings.

The researchers also reported that *retired* cabbies' posterior hippocampus volume and performance on memory assessments were "normalized"—that is, indistinguishable from those of non-cabbies. Does this surprise you? Why or why not?

TAKE HOME MESSAGE 24.18

» Longitudinal and cross-sectional data reveal that London cab drivers' acquisition of The Knowledge—knowledge of the complex layout of ~25,000 streets in central London—is associated with both an increase in size of the posterior hippocampus and better performance on memory assessments.

Drugs can hijack pleasure pathways

The French poet Jacques Prévert tinkers with synaptic activity in his brain in Paris, 1955.

24.19 Our nervous system can be tricked by chemicals.

Stimulation of the reward centers in the brain spurs animals to behave in certain ways. We're built this way on purpose: when we behave in ways important to our evolutionary fitness, our brain releases neurotransmitters such as dopamine. Exposed to such neurotransmitters, our pleasure centers produce the sensation of bliss. And that bliss spurs us to seek out that feeling again once it fades. Usually this system works well.

REUPTAKE INHIBITORS

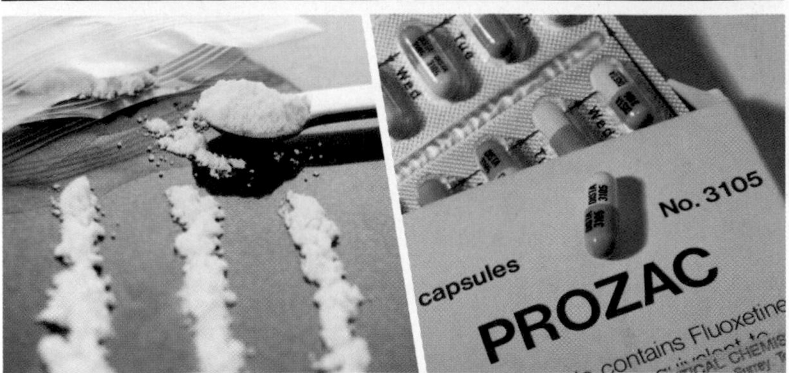

Reuptake inhibitors, such as cocaine and Prozac, cause prolonged action of a neurotransmitter in the synapse. Cocaine inhibits the reuptake of dopamine, while Prozac inhibits the reuptake of serotonin.

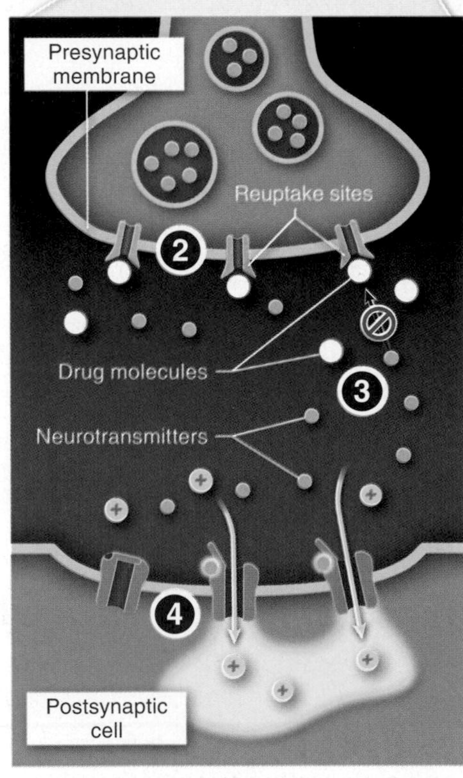

1. After the drug is taken, it crosses the blood-brain barrier and quickly reaches the brain's pleasure centers.
2. Drug molecules bind to sites on the presynaptic membrane where neuro-transmitters are normally reabsorbed by the cells that originally released them.
3. As long as reuptake sites are blocked by drug molecules, the neuro-transmitters previously released by that neuron remain in the synaptic cleft.
4. The neurotransmitters repeatedly stimulate the postsynaptic cell and cause non-stop activity in the pleasure centers of the brain.

FIGURE 24-36 Prolonging pleasure sensations.

Our brain's signaling system can be tricked. Drugs—whether recreational or therapeutic, whether found in nature or made in the laboratory—can work by mimicking neurotransmitters. When we take a pleasure-causing drug, our brain experiences the same sensations as if appropriately released neurotransmitters were flooding the system. The brain cannot distinguish this artificial high from a natural high. There are several different methods by which drugs can influence our nervous system.

Cocaine When someone snorts a bit of cocaine, it crosses the blood-brain barrier and quickly reaches the brain's pleasure centers. Once there, the cocaine binds to the sites on the presynaptic membrane where dopamine is normally reabsorbed from the synaptic cleft by the cells that originally released it. As long as these reuptake sites are blocked by the drug, dopamine released by that neuron remains in the synaptic cleft, repeatedly stimulating the postsynaptic cell and causing nonstop activity in the pleasure center (FIGURE 24-36).

Prozac and Zoloft Antidepressants work by a mechanism almost identical to that of cocaine. In addition to dopamine, another neurotransmitter important in our brain's pleasure centers is serotonin. The antidepressants Prozac and Zoloft block serotonin from being reabsorbed and recycled by the presynaptic cells that released it, prolonging its effect (see Figure 24-36). This is why these drugs are called **selective serotonin reuptake inhibitors (SSRIs).** The net result is that people taking these medications generally experience an elevated mood because of serotonin's prolonged stay in the synaptic cleft.

Morphine and Heroin Endorphins are our body's natural painkillers. Produced by our brain, endorphins block pain messages arriving from throughout the body. Under a variety of situations of extreme stress—such as if an individual has just been seriously injured in a fight or is in mile 12 of a half-marathon—the body responds by releasing endorphins.

The popular opiates morphine and heroin mimic endorphins and bind to their receptor sites. With a large enough dose, opiate users can give themselves an "endorphin rush" and feelings of euphoria far more intense than anything possible with their own natural supply of these pleasure compounds. The respiratory slowdown that comes with heroin use can occur quickly and, with large doses of heroin, can be fatal.

Nicotine One of the most commonly used drugs is nicotine. Shortly after entering the bloodstream, nicotine begins mimicking one of the body's most common and important neurotransmitters: acetylcholine. Fooled by the nicotine binding to acetylcholine receptors, cells release adrenaline and other stimulating chemicals, including pleasure-causing dopamine. Nicotine causes rapid surges, then rapid depletions, of these chemicals, leaving smokers happy for a short while but soon yearning for another cigarette.

When we consume drugs such as those described above, our bodies soon respond to the perception that there are larger amounts than usual of certain neurotransmitters. Usually the response is a reduction in sensitivity to the drug. Although tolerance is inevitable, its magnitude and consequences vary with the type of drug. In one study, volunteers were injected with a uniform daily dose of heroin and monitored for their level of euphoria. Initially they were ecstatic, but their bodies reacted by reducing the number of receptors that bind heroin. With fewer and fewer receptors, the euphoric heroin effects dropped almost to zero in just three weeks. Similar changes occur in response to using caffeine, nicotine, or alcohol.

Do human drug addictions and dependencies reflect differences in our genes? Recent data suggest that they might be influenced, at least, by the alleles a person carries. In one study of 283 individuals, a third of the people who smoked had a version (i.e., an allele) of an important gene that almost none of the non-smokers carried. This gene, labeled DRD4, codes for the receptor that enables the pleasure centers of our brain to respond to dopamine. The allele of the DRD4 gene implicated in this study of drug dependency is the same one we discussed in the Street Bio at the end of Chapter 9, a gene that appears to influence a variety of personality traits associated with risk-taking and novelty-seeking behavior.

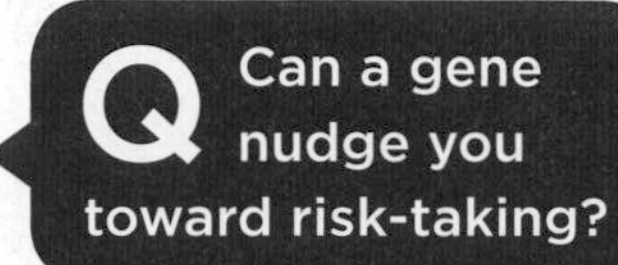

Q
Can a gene nudge you toward risk-taking?

TAKE HOME MESSAGE 24.19

» Our brain's signaling system can be tricked by drugs that mimic neurotransmitters. Such drugs—including cocaine, Prozac, heroin, and nicotine—can produce euphoric sensations, can reduce depression, and can block pain, but some carry significant health risks.

24.20 A brain slows down when it needs sleep. Caffeine wakes it up.

Caffeine has powerful effects on nearly every animal species. For instance, all rats can eventually be trained to race through mazes, but some learn quickly while others languish in remedial maze-running classes. What they all have in common, however, is that when they are given a caffeine pick-me-up before their maze lessons, they learn the solutions faster and remember them better.

> "If it weren't for coffee, I'd have no discernible personality at all."
>
> — DAVID LETTERMAN

How does caffeine work its magic? As long as we are awake, our brain is working hard and consuming about 20% of the energy we take in. Much as a running motor generates exhaust fumes, however, all of this neural activity leads to a buildup of cellular waste products. Neuron "exhaust" takes the form of a variety of molecules, including one called adenosine. As a cell releases adenosine, it fills adenosine receptors on nearby cells, particularly those in the brainstem. When a molecule of adenosine binds to an adenosine receptor, this reduces a neuron's likelihood of initiating an action potential. Adenosine has been described as putting a brake on neuron activity. Consequently, as the production of adenosine continues throughout our day, more and more receptors on more and more neurons are filled, and the neurons in our brain become less likely to fire, regardless of how strongly they are stimulated. We perceive this as tiredness. Then, as we sleep, the cellular waste products such as adenosine are reabsorbed and recycled. Upon awaking, we feel better because we are, literally, more clear-headed.

Caffeine has its effect by blocking the message to reduce activity and to sleep. Here's how it accomplishes this effect. The caffeine we ingest quickly circulates to our brain, and once there, it diffuses around the cells. Because the shape of the caffeine molecule is similar to that of the adenosine molecule, the caffeine slips into some of the receptors intended for adenosine (FIGURE 24-37). It doesn't have the same effect as adenosine, however. Adenosine receptors with bound caffeine molecules do not reduce a neuron's propensity to respond to other stimuli with an action potential. Thus, when many of the adenosine receptors are blocked by caffeine, the adenosine is unable to pass on the message that we need to sleep. Instead, we feel surprisingly alert, as caffeine interrupts one of the normal sleep-signaling systems.

Surprisingly, caffeine seems to be safe for most people. Despite considerable searching for ill effects, there is no clear evidence that moderate consumption of caffeine increases our risk of anything beyond the occasional jitters. For healthy people, there appears to be no increased risk of heart, lung, or kidney disease or even cancer.

TAKE HOME MESSAGE 24.20

» Normal neural activity leads to a buildup of cellular waste products. One of these, adenosine, fills adenosine receptors on nearby neurons, reducing the likelihood that a neuron will initiate an action potential and causing fatigue. Caffeine binds to adenosine receptors, but without reducing their likelihood of firing, thus blocking the fatigue-inducing message of adenosine.

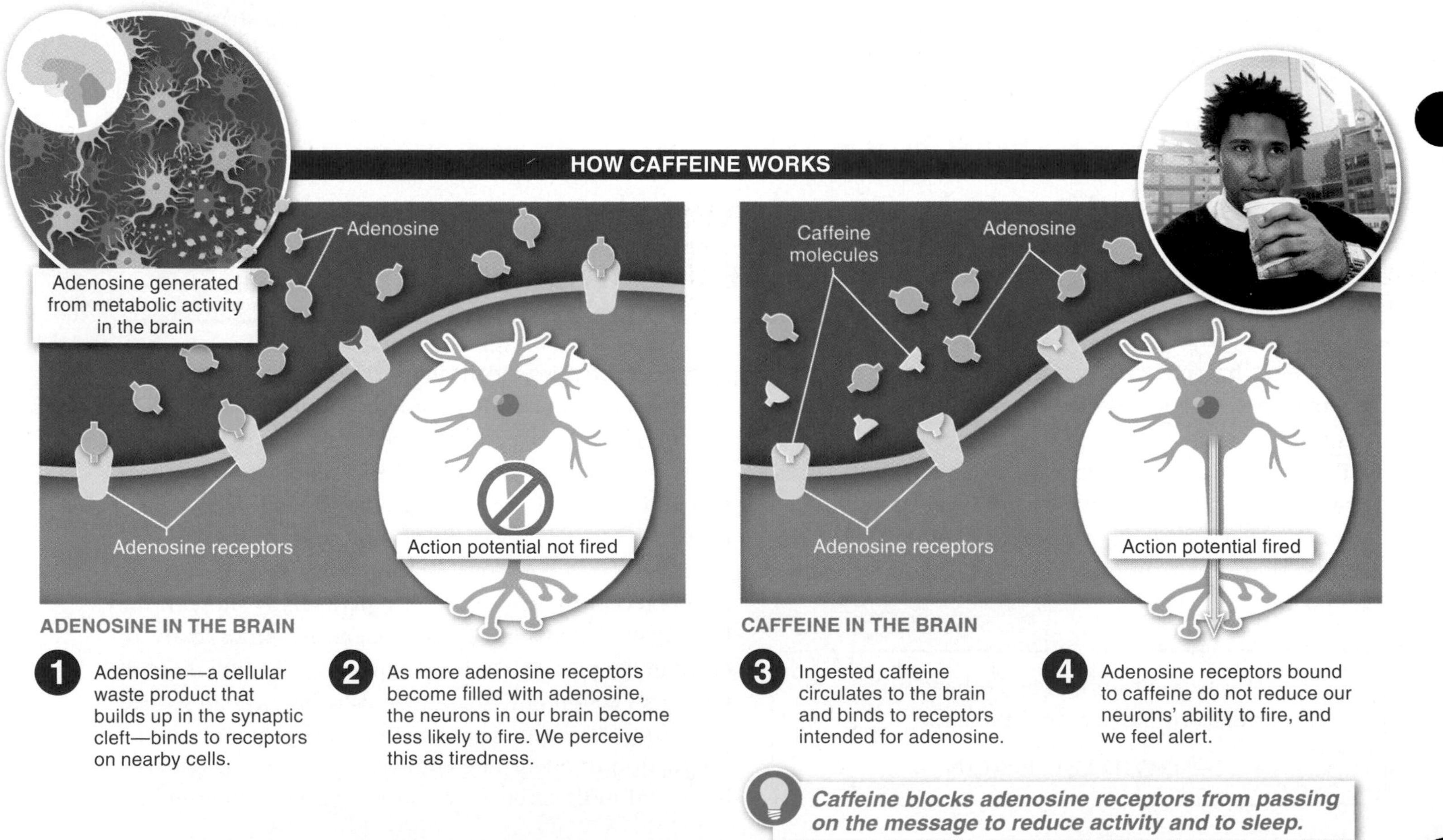

HOW CAFFEINE WORKS
Adenosine
Adenosine generated from metabolic activity in the brain
Adenosine receptors
Action potential not fired
ADENOSINE IN THE BRAIN
1 Adenosine—a cellular waste product that builds up in the synaptic cleft—binds to receptors on nearby cells.
2 As more adenosine receptors become filled with adenosine, the neurons in our brain become less likely to fire. We perceive this as tiredness.
Caffeine molecules
Adenosine
Adenosine receptors
Action potential fired
CAFFEINE IN THE BRAIN
3 Ingested caffeine circulates to the brain and binds to receptors intended for adenosine.
4 Adenosine receptors bound to caffeine do not reduce our neurons' ability to fire, and we feel alert.
Caffeine blocks adenosine receptors from passing on the message to reduce activity and to sleep.

FIGURE 24-37 **How caffeine reduces the feeling of fatigue.**

24.21 Alcohol interferes with many different neurotransmitters.

So far, we've seen that drugs frequently can alter our brain functioning by mimicking the neurotransmitters used by our neurons during normal functioning. Their specific effects are quite predictable, as long as we know the neurotransmitter that a particular drug mimics. Their work is like a surgical strike, altering our neurochemistry in one specific way.

But what happens if the drug is more of an everyman, looking enough like many different neurotransmitters to impersonate them all? This is the case with ethanol, the "alcohol" molecule in alcoholic beverages (FIGURE 24-38).

1. Alcohol slows us down, "relaxing" our neurons. By blocking receptors for glutamate, one of the brain's chief excitatory neurotransmitters, alcohol slows our reaction times and slurs our speech. It may also have more serious effects: many animals can't learn as well when their glutamate receptors are blocked.

2. Alcohol gives us a pleasant buzz. Acting like cocaine—but much weaker—alcohol blocks dopamine reuptake at synapses, increasing stimulation in the reward centers of our brain.

3. Alcohol blocks pain. By stimulating the release of endorphins—neurotransmitters that block pain messages—alcohol causes the same feeling that has been called "runner's high." Resembling morphine and heroin in this respect, but again at a greatly reduced magnitude, alcohol spurs our body to produce a little opiate-like high.

4. Alcohol makes us happier while it's in our system. Much like Prozac, alcohol modifies and increases the efficiency of our serotonin receptors. Interestingly, rats that have been bred for a high preference for alcohol turn out to have lower serotonin levels than normal rats. The alcohol-seeking animals may be trying to compensate for their lack of serotonin stimulation by looking for the alcohol effects.

For most people, moderate alcohol consumption is pleasant and does not have significant health risks. For some, however, alcohol abuse and alcoholism can lead to serious problems. Nearly 14 million Americans abuse alcohol, and the consequences can be serious. Alcohol has a variety of physiological effects in our bodies that can become harmful when alcohol is consumed in large amounts or for long periods of time. These include increasing the risk for some types of cancer, increasing the risk of liver disease (because the liver must metabolize and detoxify all the ethanol molecules consumed), increasing the likelihood of harm to the fetus during pregnancy, increasing the likelihood of taking part in risky sexual and other behaviors, and increasing the risk of automobile crashes.

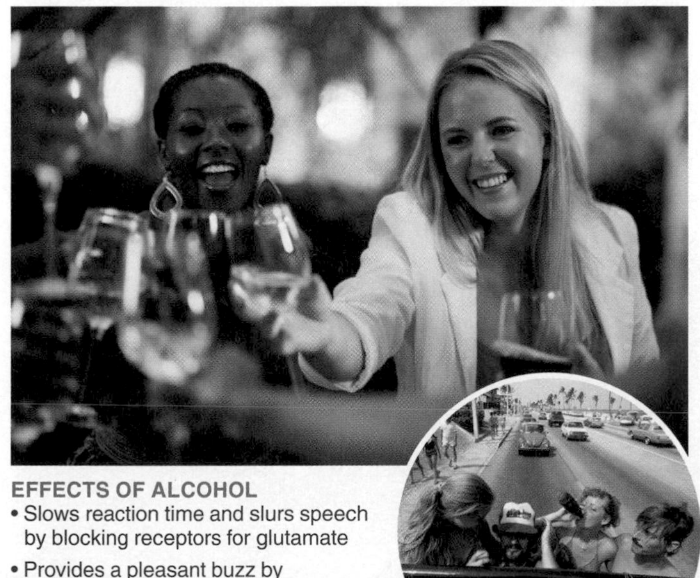

EFFECTS OF ALCOHOL
• Slows reaction time and slurs speech by blocking receptors for glutamate
• Provides a pleasant buzz by blocking dopamine reuptake
• Blocks pain by stimulating the release of endorphins
• Increases feelings of happiness by modifying and increasing the efficiency of our serotonin receptors
Alcohol intake lowers inhibitions and, in excess, can increase the likelihood of risky behavior.

FIGURE 24-38 Alcohol impersonates several neurotransmitters, with several consequences.

TAKE HOME MESSAGE 24.21

» Alcohol affects the functioning of multiple neurotransmitters—including glutamate, endorphins, dopamine, and serotonin—slowing reaction times, slurring speech, blocking pain, and increasing contentment.

STREET BIO

Using evidence to guide decision making in our own lives

"We use only 10% of our brain!" Fact or fiction?

Q: Well, do we? Besides being just plain silly, this pervasive myth is completely false. But it prompts some reasonable questions. Read on.

Q: How can we measure brain activity? Brain imaging technologies capture pictures of the brain as it functions; they reveal activity patterns, based on increased metabolism, in the different regions of the brain. One technique is called fMRI (functional magnetic resonance imaging). In an fMRI scan, a giant magnet detects the relative amounts of oxygenated blood and deoxygenated blood in various parts of the brain. When parts of the brain are active, more blood flows to them and the sensitive magnet can pick up the change.

Q: So what do these scans show? Do we use some parts of the brain more than others? Do mental tasks or physical sensations require the use of a specific part of the brain? Some recent findings include these:

Seeing your favorite foods activates brain reward centers. Subjects seeing images of their favorite foods show increased brain activity in the pleasure centers of the forebrain. And recovering drug addicts have a similar response to images of a syringe (or even an anti-drug advertisement), often prompting them to relapse. Both cases underscore the power of advertising.

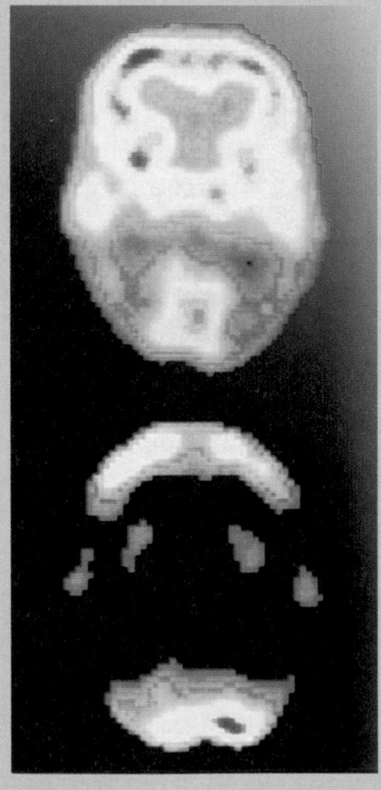

Altruistic behavior activates reward centers. When playing a game called "Prisoner's Dilemma," in which people gain rewards based on their decisions to cooperate or act selfishly, brain activity in the forebrain's reward center is increased when a player chooses to cooperate. This may indicate that our brains are built to produce a physiological "prodding" toward kindness in certain situations.

People with Alzheimer's disease exhibit significantly reduced brain activity. When compared with healthy individuals of the same age, patients with Alzheimer's show decreased activity throughout the parts of the brain associated with thinking and understanding, but normal activity in the areas active in movement and sensory perception.

Language tasks are localized in different parts of the brain. Language uses very different parts of the cerebral cortex, depending on whether someone is hearing, speaking, or reading. Among bilingual people, their second language is stored in a slightly different part of the brain if they acquired it past a certain age.

Q: So how much of our brain *are* we using? For any particular task, not all regions of the brain are called into action, but over the course of a day, all of your neurons are used.

GRAPHIC CONTENT

Thinking critically about visual displays of data

1 What is the "take home message" of this graph?

2 What does the asterisk above the two bars for "Qualified trainees" signify? Does the lack of an asterisk above the other two pairs of bars have any significance?

3 Prior to this study, an alternative explanation for cab drivers having a larger posterior hippocampus was proposed. What was that hypothesis? What evidence is supplied by this graph relevant to that hypothesis?

4 Each participant in the study was measured at two points in time. What differed between the two time periods?

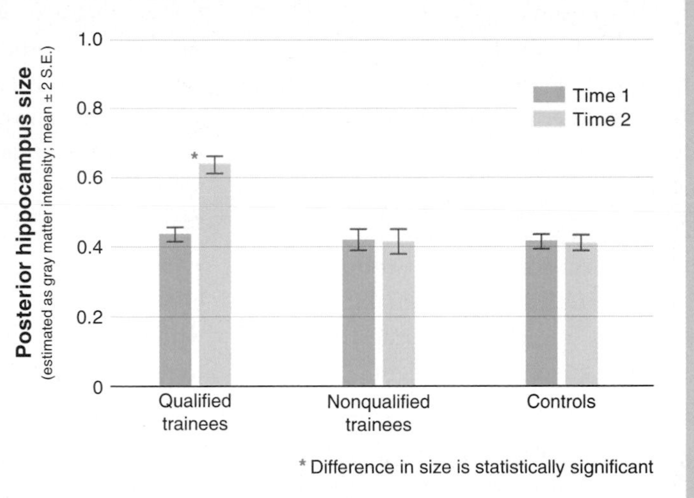

5 What can you learn from later measurements of the same cab drivers that you could not determine from a comparison of cab drivers versus non-cab drivers?

6 The *y*-axis indicates that posterior hippocampus size is measured as "gray matter intensity." Why do you think that measure is used rather than "number of brain cells"?

7 The blue-green bars (Time 1) for each of the three groups are all the same height. What does this tell us?

See answers at the back of the book.

KEY TERMS IN NERVOUS AND MOTOR SYSTEMS

actin, p. 854
action potential, p. 838
amygdala, p. 861
autonomic nervous system, p. 835
axon, p. 833
Broca's area, p. 861
cell body, p. 833
central nervous system (CNS), p. 835
cerebellum, p. 858
cerebral cortex, p. 860
chemoreceptor, p. 845
color-blind, p. 848
compound eye, p. 849
cone, p. 848
corpus callosum, p. 860
dendrite, p. 833
ear canal, p. 850
eardrum, p. 850
echolocation, p. 852
forebrain, p. 859
glial cell, p. 834
hindbrain, p. 858
hippocampus, p. 861
hypothalamus, p. 859
interneuron, p. 834
iris, p. 848
language, p. 861
left hemisphere, p. 860
limbic system, p. 861
long-term depression, p. 862
long-term potentiation, p. 862
mechanoreceptor, p. 850
medulla, p. 858
midbrain, p. 859
motor neuron, p. 834
myelin sheath, p. 838
myofibril, p. 854
myosin, p. 854
nerve, p. 833
nervous system, p. 832
neuron, p. 833
neurotransmitter, p. 841
optic nerve, p. 848
parasympathetic nervous system, p. 836
peripheral nervous system (PNS), p. 835
photoreceptor cell, p. 848
pons, p. 858
postsynaptic membrane, p. 841
presynaptic membrane, p. 841
reflex, p. 835
resting potential, p. 837
retina, p. 848
right hemisphere, p. 860
rod, p. 848
sarcomere, p. 854
selective serotonin reuptake inhibitors (SSRIs), p. 866
sensory integration, p. 861
sensory neuron, p. 834
single-lens eye, p. 848
somatic nervous system, p. 835
sympathetic nervous system, p. 836
synapse, p. 840
synaptic cleft, p. 841
terminal button, p. 838
thalamus, p. 859
twitch, p. 855
vesicle, p. 840
Wernicke's area, p. 861

BRIEF SUMMARY

What is the nervous system?

• Present in all multicellular animals other than sponges, the nervous system is a network of cells that collects information about the organism's internal and external environments, processes that information, and responds by sending signals to muscles and glands.

• The neuron is a cell specialized for carrying electrical signals. Groups of neurons bundled together enable us to sense light, sound, touch, tastes, and smells and to respond to them.

• In vertebrates, the nervous system is divided into the peripheral and central nervous systems. The central nervous system is made up of the spinal cord and brain. The peripheral nervous system carries signals to and from the sensory and motor pathways.

How do neurons work?

• Dendrites on motor neurons and interneurons generally connect with and receive signals from other neurons. Sensory neuron dendrites are modified to respond to a specific external stimulus.

• Each neuron has one axon extending from the cell body. At its end, the axon branches into numerous terminal buttons. In response to an action potential, the axon terminals release chemicals, potentially influencing adjacent cells.

• At the synapse, in response to an action potential, neurotransmitters are released into the synaptic cleft, diffuse, and may bind to receptors on an adjacent neuron, muscle cell, or gland, potentially stimulating an action potential, muscle contraction, or secretion. Neurotransmitters may then be recycled or broken down.

Our senses detect and transmit stimuli.

• The process by which each of our senses works is basically the same. A modified dendrite on a sensory neuron is stimulated, causing the sensory neuron either to fire an action potential or to alter its rate of neurotransmitter secretion.

• Taste buds on our tongues contain chemoreceptors, which are stimulated when particular chemicals in food bind to receptor proteins on the cell surface, triggering a signal to the brain.

• Neurons that can detect smells have dendrites modified with hair-like projections covered with chemoreceptors, densely packed within the nasal cavity.

• Vision results from the stimulation of light-sensitive sensory neurons, called photoreceptor cells, which have light-sensitive molecules embedded within their membranes. Signals are conveyed to the brain and interpreted as an image.

• Hearing occurs when sound waves cause the eardrum to vibrate, moving tiny bones that pass on the vibrations to the inner ear, where the vibrations bend hair cells and trigger a pattern of action potentials related to the wavelengths of the sound.

• Touch is a class of sensations generated by mechanoreceptors, thermoreceptors, and pain receptors. Stimulation of these receptors generates action potentials and causes the perception of touch.

The muscular and skeletal systems enable movement.

• Muscle tissue—including skeletal, cardiac, and smooth muscle—is made up of elongated cells capable of generating force when they contract.

• The vertebrate skeletal system enables movement, provides support, offers protection, provides a site for production of important blood and immune cells, and serves as a reservoir of minerals.

The brain is organized into distinct structures dedicated to specific functions.

• The brain has distinct regions. The hindbrain regulates basic physiological functions and coordinates motor activity. The midbrain helps filter and evaluate sensory and motor information. The forebrain, in humans, is responsible for most higher thought.

• Learning and memory, along with the use of complex language, are among the most sophisticated brain functions. The brain structures chiefly responsible for these functions include the left frontal and temporal lobes and the amygdala and hippocampus.

Drugs can hijack pleasure pathways.

• Our brain's signaling system can be tricked by drugs that mimic neurotransmitters.

• Neural activity leads to a buildup of cellular waste products. One of these, adenosine, fills receptors on nearby neurons, inhibiting them. Caffeine binds to adenosine receptors, but without inhibiting them, thus blocking the fatigue-inducing message of adenosine.

• Alcohol affects the functioning of multiple neurotransmitters.

CHECK YOUR KNOWLEDGE

Short Answer

1. Why is being born without the ability to feel pain a curse rather than a blessing?

2. Compare and contrast the structures and roles of neurons and glial cells in the nervous system.

3. Describe the two ways in which dendrites receive external information.

4. What is the effect of a substance that blocks acetylcholine from binding to its receptor cells?

5. From an evolutionary perspective, why have the brain and head become more pronounced during human evolution?

6. Why are the senses of taste and smell closely connected in humans?

7. Why can some insects see wavelengths of light in the ultraviolet spectrum that are invisible to us?

8. List the primary types of sensory receptor cells, their functions, and which of the five traditional senses they are associated with.

9. Describe the structure of a myofibril of skeletal muscle. What changes occur in the myofibril that make contraction possible?

10. Describe two functions of a skeletal system.

11. Which parts of the limbic system are crucial in storing memories? Describe evidence that supports your answer.

12. How does caffeine alter the functioning of adenosine?

Multiple Choice

1. Animal nervous systems have several principal features, which include all of the following except:

a) they coordinate the organism's long-term growth and development.

b) they receive input from the world around the organism.

c) they initiate responses to the information they receive from the world, when necessary.

d) they process the information from the world that the organism receives.

e) All of the above are principal features of animal nervous systems.

0 EASY — 46 — HARD 100

2. Which of the following statements about plants and nervous systems is correct?

a) Plants can feel pain in their trunks and stems, where they have neurons, but not in their leaves.

b) Plants can feel pain in their leaves, where they have neurons, but not in their trunks or stems.

c) Plants can feel pain because they have a nervous system that extends throughout their tissues.

d) Plants can feel pain in their trunks, where they have neurons, but not in their stems or leaves.

e) Plants cannot feel pain, because they do not have nervous systems.

0 EASY — 11 — HARD 100

3. In a neuron, the cell body:

a) contains the nucleus, mitochondria, and other cell organelles.

b) does not contain a nucleus.

c) has a membrane that is impermeable to water and intracellular solutes.

d) is highly modified and coated in a fatty myelin sheath.

e) is considered the primary structure of the "central" nervous system.

0 EASY — 20 — HARD 100

4. Dendrites:

a) conduct action potentials away from the cell body.

b) are present in mammalian nervous systems but not in the nervous systems of other vertebrates.

c) receive information from other neurons or from the external environment.

d) are coated in a fatty substance called the myelin sheath, which speeds up the rate at which signals are conducted.

e) are bundles of axons.

0 EASY — 12 — HARD 100

5. In individuals with multiple sclerosis (MS), myelin is gradually lost. What symptoms would you expect these individuals to have?

a) Their cell membranes have reduced numbers of sodium channels.

b) Their sensory neurons lose the ability to initiate action potentials.

c) Their brain becomes smaller.

d) They are able to continue athletic activity long after pain would cause most individuals to stop.

e) Their neurons gradually lose their ability to conduct electrical impulses.

0 EASY — 25 — HARD 100

6. Neurotransmitters in a synaptic cleft have all of the following possible fates except:

a) reuptake by the presynaptic neuron.

b) enzymatic breakdown in the synaptic cleft.

c) inactivation by acetylcholine in the synaptic cleft.

d) binding to a receptor in the postsynaptic cell membrane.

e) All of the above are possible fates of neurotransmitters in a synaptic cleft.

0 EASY — 49 — HARD 100

7. When you put a piece of chocolate on your tongue, your brain registers a sensation of sweetness. What aspect of a molecule is responsible for its having a particular taste?

a) the molecule's shape

b) the total number of protons in the molecule

c) the number of hydrogen bonds in the molecule (more hydrogen bonds = sweeter taste)

d) the ratio of covalent bonds to ionic bonds joining the atoms of the molecule

e) the speed of melting in your mouth

0 EASY — 22 — HARD 100

8. Because cone cells are less sensitive to light than are rod cells:

a) cones are better suited for daylight vision.

b) cones are better suited for night vision.

c) cones are better suited for peripheral vision.

d) Both a) and c) are correct.

e) Both b) and c) are correct.

0 EASY — 46 — HARD 100

9. The connective tissue that connects a muscle to a bone is known as:

a) collagen.

b) a ligament.

c) cartilage.

d) a tendon.

e) None of the above are correct.

0 EASY — 27 — HARD 100

10. The blood-brain barrier:

a) is formed from muscle tissue.

b) is a semi-permeable barricade that allows essential nutrients and gases to pass through.

c) prevents harmful molecules such as metabolic wastes from entering the brain.

d) restricts the passage of molecules to neurons but not to glial cells.

e) Both b) and c) are correct.

0 EASY — 17 — HARD 100

11. A boy is born with a brain defect in which the two hemispheres of his cerebrum are unable to "talk" to each other (they are not connected). This boy is born without a:

a) hippocampus.

b) brainstem.

c) medulla oblongata.

d) corpus callosum.

e) hypothalamus.

0 EASY — 38 — HARD 100

Ch25

Hormones are chemical messengers regulating cell functions.

Hormones are produced in glands throughout the body.

Hormones influence nearly every facet of an organism.

Environmental contaminants can disrupt normal hormone functioning.

Hormones such as testosterone activate processes that lead to increased muscle mass and influence physical performance. Shown here: Kachalka, a famous outdoor public gym in Ukraine.

Hormones

Mood, emotions, growth, and more: hormones as master regulators

25.1–25.3

Hormones are chemical messengers regulating cell functions.

A ring-tailed lemur with her baby.

25.1 The "cuddle" chemical: oxytocin increases trust and enhances pair bonding.

"Trust me." It's easy enough to say, but can be very difficult to do. Recently, however, some researchers discovered a way to actually make people more trusting. Here's how they did it.

Study participants, called investors, were given an amount of money. Each investor could keep the money or transfer some or all of it to another person, called a trustee. Any amount the investor chose to transfer to the trustee was tripled: the trustee could then return to the investor some portion of the significantly increased pot of cash, or the trustee could simply keep it all. For the investors, this created a dilemma—they could lose money, but trusting behavior could also reward them with a big payoff.

Using a double-blind experimental design, the researchers created two different groups of investors. Before making the decision whether to entrust some of their money to the trustee, one group of investors used a nasal spray to inhale a dose of **oxytocin**—a chemical normally found in humans and other vertebrates, but in smaller amounts than used in the study. Investors in the other group inhaled a placebo, rather than oxytocin.

The results were dramatic and unambiguous. The oxytocin group was significantly more trusting than the placebo group: more than twice as many of those receiving oxytocin entrusted the maximum amount of their money to the trustee, compared with investors who did not inhale oxytocin. And more than twice as many of those in the placebo group as in the oxytocin group transferred the smallest amounts of cash to the trustee, apparently exhibiting a greater fear of betrayal.

This result is consistent with some findings of research on other animals. In those studies, exposure to oxytocin reliably caused animals to let down their guard, initiate interaction, and form social attachments with other individuals at a much higher rate than they typically would.

In a clever follow-up investor-trustee study, the same methods were used but with one exception: the investors were told that the "trustee" in the game was a computer rather than another person. In this study, the investors receiving oxytocin did not transfer greater amounts of money to the trustee; they behaved exactly like those in the placebo group. This finding suggests that oxytocin specifically increased investors' trust of another human, rather than just increasing their willingness to take a financial risk.

Researchers believe that these findings give some insight into the biological basis for trust, which might even lead to effective treatments for people with extreme shyness. And the results provide a pretty clear demonstration that complex emotions and behaviors can be influenced by chemical signals produced in the body.

Oxytocin influences much, much more in humans than a propensity for trust (FIGURE 25-1). It plays significant roles in making milk available to nursing babies and in facilitating birth. In fact, the name "oxytocin" is from the Greek

FIGURE 25-1 **Oxytocin: the cuddle chemical.**

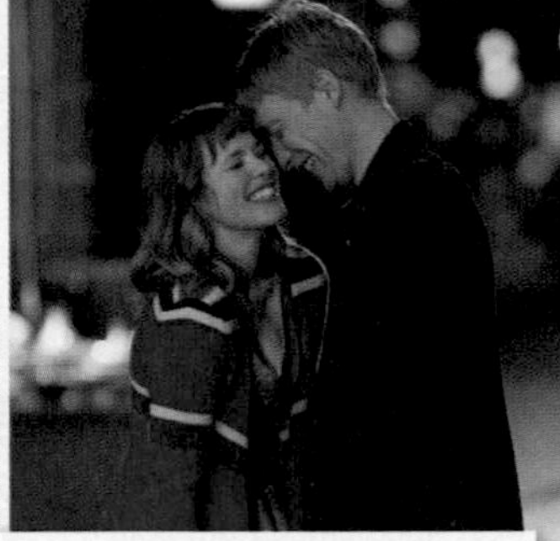

Oxytocin, a chemical signal produced in the body, can influence complex emotions and physiological processes, such as:

- *Increasing trust*
- *Making milk available to nursing babies*
- *Forming social attachments*
- *Stimulating uterine contractions*

word meaning "quick birth," based on the recognition of its role in stimulating uterine contractions. Since oxytocin has also been shown to increase the propensity to form social attachments, some researchers refer to it as the "cuddle drug."

In this chapter, we explore just what these chemical signals are, where they are produced, and how they have their effects. We also explore their pervasive influence in nearly every aspect of vertebrate life, from emotions to physical structures, behavior, and health and physiology.

TAKE HOME MESSAGE 25.1

» Complex emotions and behaviors can be influenced by chemical signals produced in the body. For example, exposure to oxytocin causes humans to be more trusting of others, as well as facilitating birth, making milk available to nursing babies, and increasing the propensity to form social attachments.

25.2 Hormones travel through the circulatory system to influence cells elsewhere in the body.

You wouldn't use the postal service to notify the fire department that your house is burning down. For all types of communication, depending on the message, some modes of delivery are more effective than others.

Animals need effective ways to respond to the outside world—ways of sending signals about specific conditions and giving instructions to tissues to respond to those conditions. They have two systems for carrying out this internal communication and regulation (FIGURE 25-2). As we saw in Chapter 24, the nervous system generally is responsible for controlling rapid movement and sensations in response to environmental changes. In multicellular animals, including humans, there is another system as well, with a special type of messenger. **Hormones** (from the Greek word meaning "that which sets in motion or urges forward") are a type of chemical messenger, secreted by cells into the circulatory system (the bloodstream or, in insects, the hemolymph), that influences the actions of cells elsewhere in the body, as part of an internal communication and regulation system. Oxytocin, described in Section 25.1, is an example of an animal hormone.

Cells that secrete hormones are called **endocrine cells,** and the cells that receive their signals are called **target cells.** Some hormone-secreting cells are individual cells, such

INTERNAL COMMUNICATION IN ANIMALS

Animals have two systems for carrying out internal communication in response to external conditions.

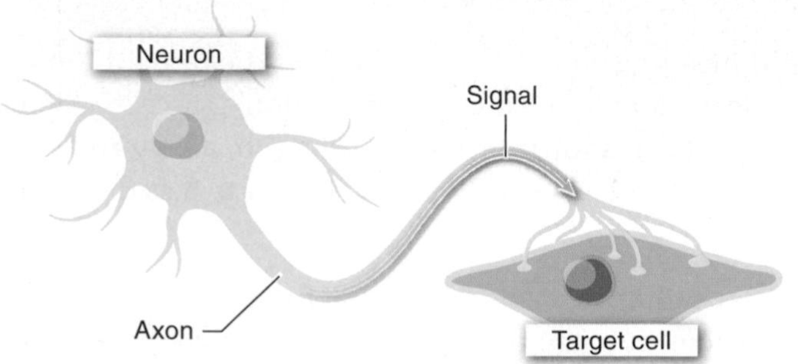

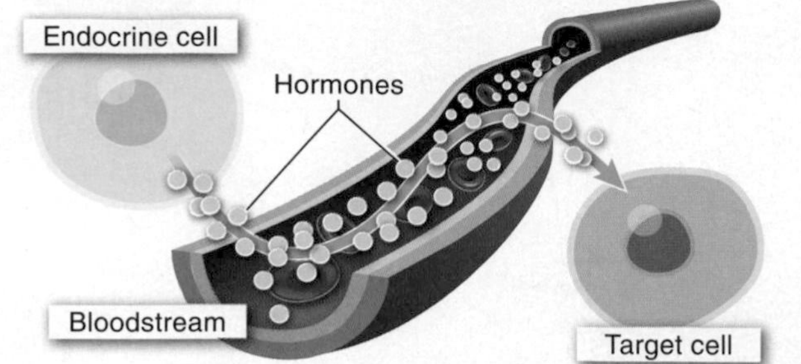

THE NERVOUS SYSTEM

- Messages sent via chemical and electrical signals down the axons of neurons and across synapses to target cells
- Generally responsible for controlling rapid movement and sensations

THE ENDOCRINE SYSTEM

- Messages sent via chemical signals secreted by endocrine cells, often into the circulatory system, to target cells throughout body
- Generally responsible for slower, longer-term regulation

FIGURE 25-2 The nervous and endocrine systems work together in animals.

as the endocrine cells in the lining of the stomach and small intestine that aid in digestion. Larger collections of hormone-secreting cells—including the pituitary gland, the pancreas, and the ovaries and testes—are called **endocrine glands.** Together, all of the hormone-secreting cells in an animal make up its **endocrine system** (FIGURE 25-3).

The endocrine system controls chemical signals that influence cells close to and far from the hormone-secreting cells. Whereas the nervous system controls rapid movements and sensations, the endocrine system is generally responsible for longer-term, slower regulation, such as growth and development, as well as a variety of secretions (see Figure 25-2). It's important to note, however, that there are some nervous system cells that secrete hormones, so the systems are not completely distinct but can overlap somewhat in function. Interestingly, although they don't have nervous systems, most plants do have hormones (see Chapter 20), suggesting that hormone systems predate nervous systems evolutionarily.

When a hormone gets to a target cell, it elicits a response. Commonly, the hormone's effect on a target cell is to alter the animal's physiology to help the organism maintain homeostasis. In doing so, the hormone may alter the target cell's metabolism or growth, spur cell division, or initiate a developmental pathway.

Another type of chemical messenger, called a **pheromone,** is considered an "ecto-hormone," because it is transported to the outside of the animal and can cause a behavioral or physiological change in another individual. Pheromones serve many different purposes. Territorial

OVERVIEW OF THE ENDOCRINE SYSTEM

The endocrine system consists of all the hormone-secreting cells—including larger collections of cells called endocrine glands—in an animal.

Hypothalamus
Pineal gland
Pituitary gland
Thyroid and parathyroid glands
Adrenal glands
Pancreas
Gonads
- Ovaries (in females)
- Testes (in males)

HORMONE SECRETION

Endocrine gland
Bloodstream
Hormones
Target cells

Hormones secreted into the bloodstream can influence cells far away in the body.

FIGURE 25-3 Organization of the endocrine system.

The dog marks its territory and also exaggerates its size by lifting its leg during urination.

FIGURE 25-4 **High hopes.**

pheromones, such as those found in dogs' urine, mark the boundaries of a territory (FIGURE 25-4); alarm pheromones, such as those that signal the presence of a predator, can trigger aggression or flight in a group of individuals; and trail-marking pheromones, such as those produced by ants, serve as a guide to the colony. The first pheromones discovered were those signaling sexual receptivity. In some butterfly species, for example, males can detect a female's sex pheromones from more than six miles away and fly in the direction of increasing pheromone concentration to locate the female.

TAKE HOME MESSAGE 25.2

» Hormones are chemical messengers that are secreted into the circulatory system by endocrine cells and endocrine glands. Hormones influence the actions of target cells elsewhere in the body, as part of an internal communication and regulation system.

25.3 Hormones can regulate target tissues in different ways.

If you turn on a radio, you can hear music. If you turn the radio off, the music stops. The signal from the radio station, however, is still there. But without the proper receiver, the signal goes undetected. Hormones function much like radio signals. Glands produce and release hormones, which then make their way through the organism, often distributed by the bloodstream, and bump into cells throughout the body. But, like a person without a radio, a cell doesn't respond to a hormone unless it has a specific receptor for that hormone. The hormone may even diffuse right through a cell that has no intracellular receptors, without any effect.

There are dozens of hormones and even more ways that they regulate target tissues, but the process by which hormones affect a particular cell is generally consistent.

1. **Signal is sent.** A hormone is released by a gland.
2. **Signal is received.** Although the hormone has no effect on most tissues it comes in contact with, cells with the proper receptor in their cytoplasm or on their plasma membrane receive the signal.
3. **Cell responds.** When the hormone binds to an appropriate receptor, it causes a response in the target cell. For some hormones, the response is a change in gene expression in the target cell's nucleus, which may cause the cell to start (or stop) producing a particular protein, or may alter the rate at which it produces the protein. A single hormone may cause different responses in different target cells.

Most hormones fall into one of three categories: (1) the amines (such as adrenaline), which are synthesized from single amino acids; (2) the polypeptide hormones (such as insulin), which are chains of amino acids; and (3) the steroid hormones (such as estrogen and testosterone), which are lipids, most of which are synthesized from cholesterol. The chemical structure of a hormone determines whether it can pass through a cell's plasma membrane (FIGURE 25-5). Most amines and polypeptide hormones are water-soluble, as opposed to lipid-soluble. They cannot pass through membranes, which are lipid-rich and hydrophobic. By contrast, steroid hormones are built from the lipid cholesterol and so are

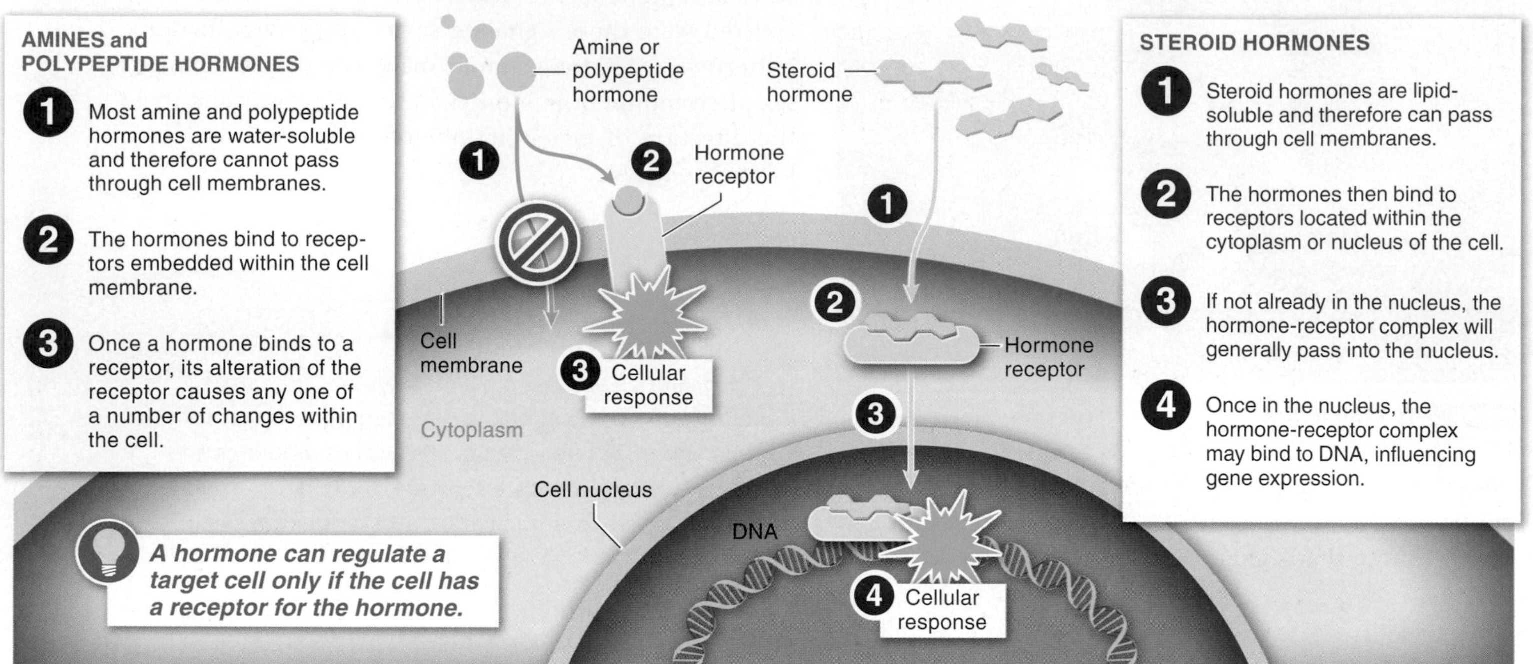

AMINES AND POLYPEPTIDE HORMONES vs. STEROID HORMONES
Hormones influence gene expression in different ways, depending on their chemical structure.
AMINES and POLYPEPTIDE HORMONES
1 Most amine and polypeptide hormones are water-soluble and therefore cannot pass through cell membranes.
2 The hormones bind to receptors embedded within the cell membrane.
3 Once a hormone binds to a receptor, its alteration of the receptor causes any one of a number of changes within the cell.
Amine or polypeptide hormone
1
2 Hormone receptor
Cell membrane
3 Cellular response
Cytoplasm
Steroid hormone
1
2
Hormone receptor
3
Cell nucleus
DNA
4 Cellular response
A hormone can regulate a target cell only if the cell has a receptor for the hormone.
STEROID HORMONES
1 Steroid hormones are lipid-soluble and therefore can pass through cell membranes.
2 The hormones then bind to receptors located within the cytoplasm or nucleus of the cell.
3 If not already in the nucleus, the hormone-receptor complex will generally pass into the nucleus.
4 Once in the nucleus, the hormone-receptor complex may bind to DNA, influencing gene expression.

FIGURE 25-5 **Most hormones belong to one of two main groups, differing in chemical properties.**

lipid-soluble. They diffuse right through cell membranes and into the cytoplasm.

Because hormones that are not lipid-soluble cannot pass through membranes, these hormones bind to a receptor on the outside of the cell and alter the receptor (perhaps changing its shape), thus causing any one of a number of changes within the cell. The hormone may activate an enzyme, initiating or speeding a reaction in the cell. Or it may alter the cell membrane's permeability, facilitating absorption or secretion of certain molecules. Or it may alter proteins within the cell, causing them to move into the nucleus and bind directly to the DNA, influencing the rate of transcription of one or more genes (for example, turning on or off the production of some proteins).

Steroid hormones, on the other hand, pass through the cell membrane and bind to receptors in the cytoplasm or nucleus of the target cell. If binding occurs in the cytoplasm, the hormone-receptor complex usually passes into the nucleus. In the nucleus, the hormone-receptor complex may bind to the DNA, influencing the rate of transcription of one or more genes.

By traveling through the bloodstream, hormones secreted by glands in one part of the body are able to regulate cell function in another part of the body. However, some glands produce regulators that act more locally. Called **paracrine regulators,** these molecules generally diffuse from the cells in which they are produced into nearby tissue, binding to receptors on or in neighboring cells and influencing their activity. *Prostaglandins* are one of the most important and common groups of paracrine regulators, produced by almost every cell and nearly every organ in an animal's body. Prostaglandins have numerous effects, including causing the dilation or constriction of blood vessels and affecting tissue inflammation.

One of the chief reasons that aspirin is such an effective pain reliever is that it inhibits an enzyme necessary for the production of prostaglandins. As a consequence, aspirin reduces inflammation—and the pain that often accompanies it—but it also inhibits another, similar enzyme that is involved in maintaining the lining of the stomach. The drug celecoxib, sold as Celebrex, has proved beneficial to people suffering from arthritis, who need regular, long-term anti-inflammatory

Q Why might long-term use of aspirin cause stomach bleeding?

medication. Celecoxib inhibits the prostaglandin-producing enzyme without inhibiting the stomach-lining-maintaining enzyme, making long-term use of this painkiller possible.

TAKE HOME MESSAGE 25.3

» Hormones can regulate the activities of a target tissue only if the cells have a receptor specific for that hormone. Water-soluble hormones bind to receptors on cell membranes, while lipid-soluble hormones bind to receptors within a cell. Both types of hormones, once bound to a receptor, cause changes in the target cell, such as influencing the rate of transcription of genes.

25.4–25.5 Hormones are produced in glands throughout the body.

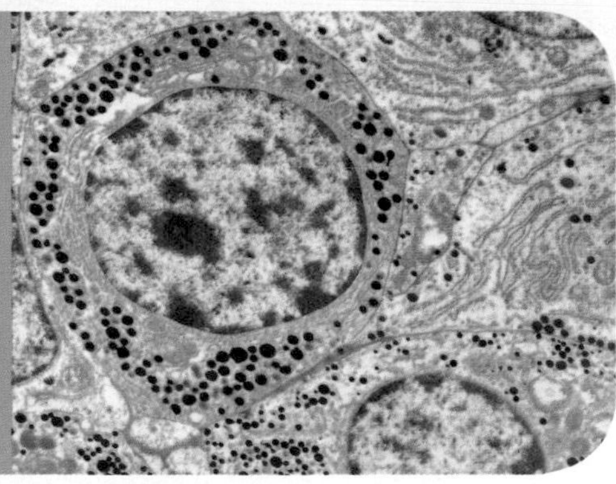

These hormone-secreting cells (stained green, with darker green nuclei) are from the anterior pituitary gland, at the base of the brain. The red granules in the cells contain hormones to be secreted.

25.4 The hypothalamus controls secretions of the pituitary.

As we have seen, animals have both nervous and endocrine systems, and there is considerable interaction between the two. The **hypothalamus,** part of the underside of the brain, functions as a liaison between the nervous and endocrine systems, and it receives input from neurons throughout the brain and the rest of the body (FIGURE 25-6). Using this information about the external environment and the physiological state of the body, the hypothalamus sends out the appropriate hormones (and nervous signals) to regulate nearly every aspect of the organism's physiology, including body temperature, hunger, thirst, and water balance.

The **pituitary gland,** which is about the size of a pea, is attached to the hypothalamus by a thin stalk. Signals from the hypothalamus directly influence the pituitary gland. The hypothalamus may release hormones that cause the pituitary to increase its production and release of hormones, or the hypothalamus may direct the pituitary to reduce or stop the release of hormones.

The pituitary gland is actually two separate glands fused together. The **posterior pituitary** has a fibrous appearance, because it contains a large amount of nervous tissue, particularly axons coming straight from the hypothalamus—from which the posterior pituitary develops. The posterior pituitary does not itself produce hormones, but it releases two important polypeptide hormones that are produced in neurons within the hypothalamus.

1. **Oxytocin** is the "cuddle" or "trust" hormone discussed in Section 25.1; it also directs the ejection

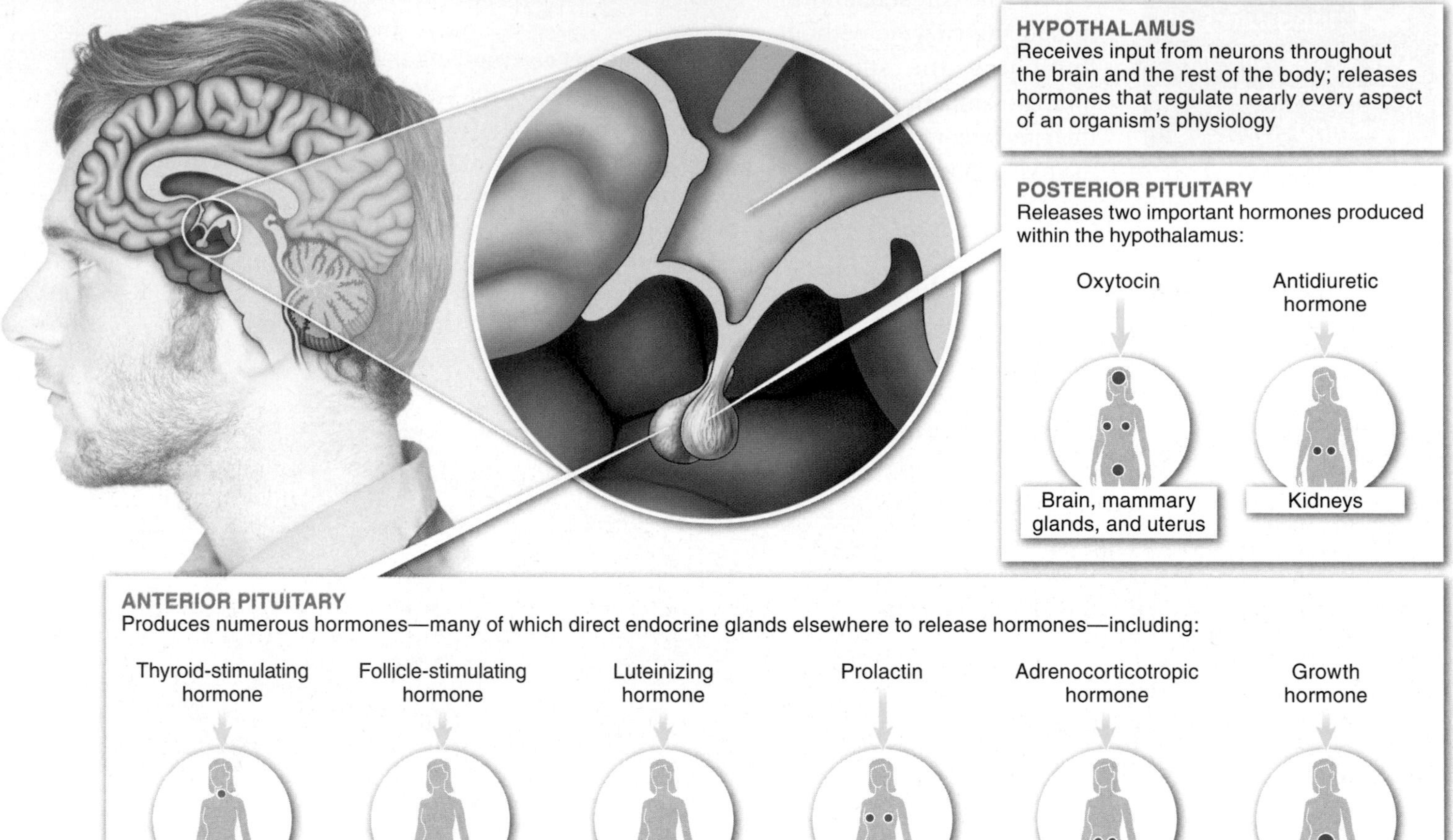

FIGURE 25-6 **Functions of the glands of the endocrine system: hypothalamus and pituitary.**

(or "let-down") of milk for nursing babies and stimulates muscle contractions in the uterus during childbirth.

2. **Antidiuretic hormone (ADH)** influences water retention by the kidneys. With increased levels of ADH, more water is saved, reducing the amount of urine produced while increasing its concentration.

The **anterior pituitary** produces numerous hormones in response to commands by the hypothalamus. Many of the anterior pituitary hormones direct endocrine glands elsewhere to release hormones. Some of the most important hormones produced by the anterior pituitary in mammals are the following:

1. **Thyroid-stimulating hormone (TSH)** causes the thyroid to produce thyroxine, important in cellular respiration (which we discuss further in Section 25.5).
2. **Follicle-stimulating hormone (FSH)** stimulates development of follicles in the ovaries, and **luteinizing hormone (LH)** triggers ovulation. In males, FSH stimulates sperm maturation and LH stimulates testosterone production.
3. **Prolactin** stimulates the mammary glands to produce milk.
4. **Adrenocorticotropic hormone (ACTH),** also known as **corticotropin,** stimulates the **adrenal glands** to produce cortisol and other stress-related hormones.
5. **Growth hormone** has several effects, including stimulating the liver to release chemicals that spur the growth of bones, cartilage, and many other tissues.

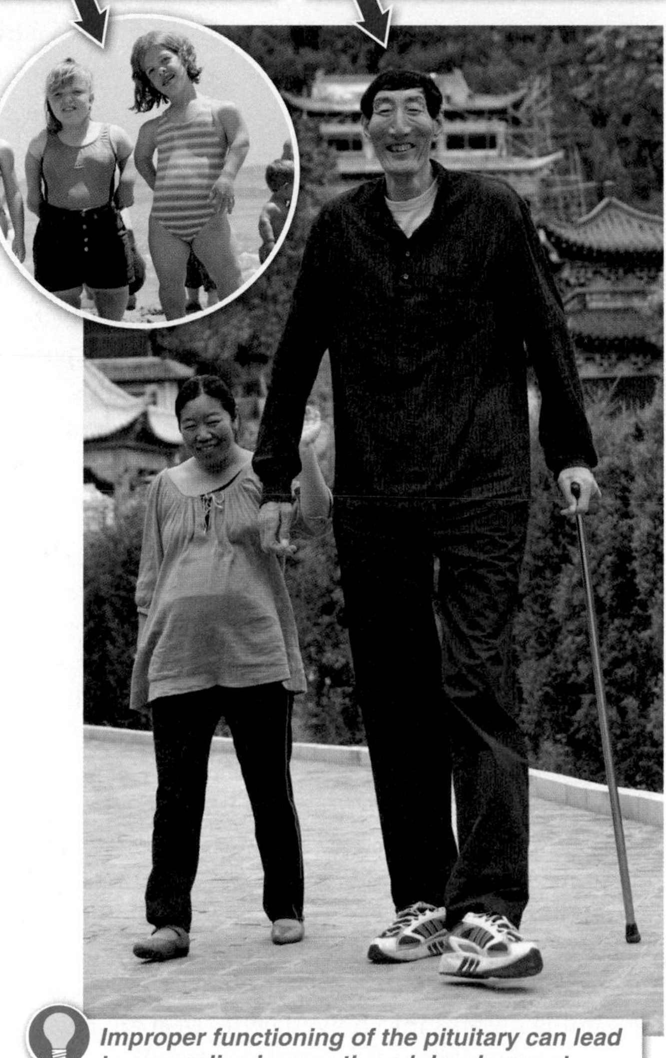
Reduced production or absence of growth hormone during childhood can lead to pituitary dwarfism.
Excessive production of growth hormone during childhood can cause extreme growth, with some individuals reaching 8 feet in height.
Improper functioning of the pituitary can lead to anomalies in growth and development.

FIGURE 25-7 **Pituitary malfunction can affect stature in humans.**

Improper functioning of the pituitary can lead to some anomalies in growth and development. Excessive production of growth hormone during childhood, for example, can cause extreme growth, called gigantism, with some individuals reaching 8 feet in height (FIGURE 25-7). If the increased exposure to growth hormone doesn't occur until adulthood, only the hands, face, and feet tend to respond with unusual growth. Individuals with reduced or no production of growth hormone during childhood develop a condition called pituitary dwarfism and may grow not more than 4 feet tall. Early diagnosis of this condition and treatment with human growth hormone can restore normal growth. (See Section 7.7 for a discussion of human growth hormone and recombinant DNA technology.)

Although the hypothalamus and pituitary gland control much of the hormone secretion in the body, there are several other endocrine glands with important regulatory roles, which we explore in the sections that follow. Keep in mind, too, that the hypothalamus and pituitary don't necessarily control all hormone secretions; they themselves are also regulated, in large part by the glands that they regulate, through numerous feedback loops. (See Section 21.9 for a discussion of feedback loops.)

TAKE HOME MESSAGE 25.4

» The hypothalamus receives input from neurons throughout the brain and the rest of the body, and using this information, it sends out the appropriate hormones (and nervous signals), often directing the pituitary gland to release hormones with important regulatory effects on body tissues.

25.5 Other endocrine glands also produce and secrete hormones.

Along with the hypothalamus and pituitary gland, many other glands produce the hormones that regulate physiology. From anxiety about public speaking, to changes in sleep patterns after flying to a new time zone, to the changes in metabolism that accompany aging, endocrine glands throughout the body are responsible for detecting and responding to signals reflecting an organism's internal and external environments. Here we explore

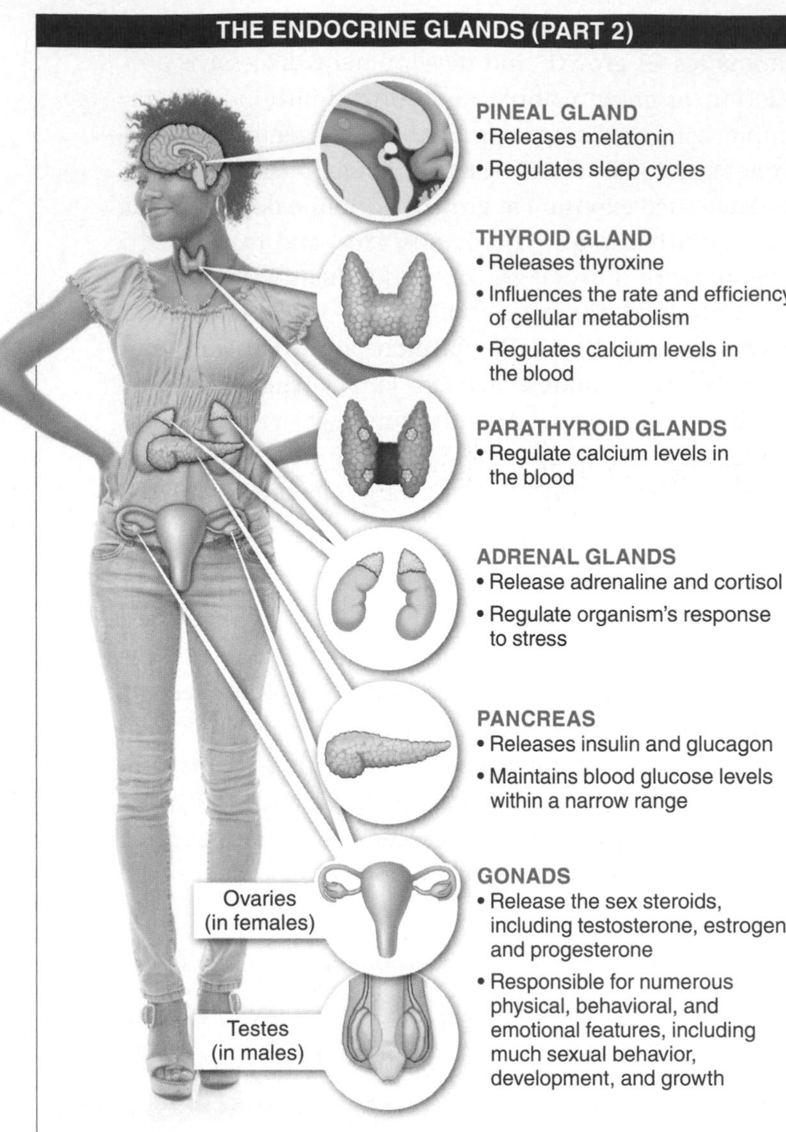

FIGURE 25-8 Functions of the glands of the endocrine system: pineal, thyroid, parathyroids, adrenals, pancreas, and gonads.

The sight of a predator is enough to initiate the "fight-or-flight" response in an animal, and within seconds, adrenal gland secretions of adrenaline and cortisol prepare the body for action.

FIGURE 25-9 Who will win in this conflict?

the signals of some of the most important of these other endocrine glands (FIGURE 25-8).

Adrenal Glands Regulating an organism's response to stress is largely a function of secretions of the two adrenal glands, which sit just above the kidneys and secrete the hormones cortisol and adrenaline (also called epinephrine), among others. Simply the sight of a predator is enough to initiate the "fight-or-flight" response—the secretion of adrenaline and cortisol that prepares the body for action (FIGURE 25-9). Within seconds, in a case of positive feedback, these secretions can cause goose bumps, an increased heart rate, an increased rate of glycogen breakdown in the liver and skeletal muscles, release of stored fatty acids, and dilation of bronchioles in the lungs, enabling greater absorption of oxygen for delivery to needy tissues.

> **Q** Why is adrenaline given to people experiencing an asthma attack?

The stress pathways are modulated by negative feedback loops. As an animal takes action in response to a stressful situation and the source of the stress is removed, the secretion of cortisol and adrenaline is reduced. This is how the stress response usually works in nature. But when there is no outlet through which an organism can deal with the stress, long-term consequences of a chronic stress response include ulcers, cardiovascular problems, decreased immune function, and other types of illness. With an increased understanding of the stress response and its function as a short-term physiological state that helps organisms quickly and effectively respond to stressful situations, researchers are gaining insights into how to better treat anxiety.

Pineal Gland The 17th century philosopher René Descartes believed that the **pineal gland** was the site where the soul connected with the body. He believed this largely because the pea-sized gland is located near the center of the brain, is singular (that is, there's just one, not one on the left side of the brain and another on the right), and is, so he thought, unique to humans. Scientists have abandoned such a view of the pineal gland (and we now know that the pineal gland is present in all vertebrates), but there's still considerable scientific interest in this gland. It has neuron connections with the retina of the eye, and it controls secretion of the hormone melatonin, which is derived from the amino acid tryptophan and affects diurnal-nocturnal wake and sleep patterns, called *circadian cycles*.

Although the exact mechanism by which melatonin influences circadian cycles is not understood, researchers have shown that melatonin has some benefits in synchronizing

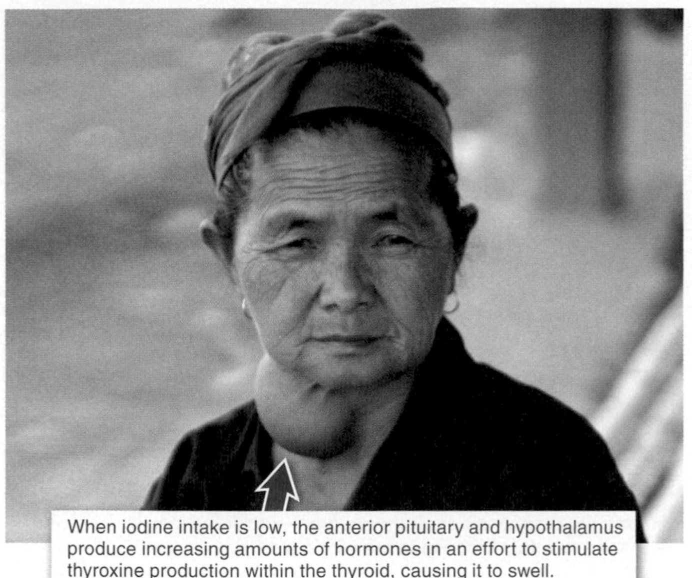

When iodine intake is low, the anterior pituitary and hypothalamus produce increasing amounts of hormones in an effort to stimulate thyroxine production within the thyroid, causing it to swell.

FIGURE 25-10 An enlargement of the thyroid gland in the neck is called a goiter.

individuals' sleep-wake cycles to the environment and in treating some types of insomnia. Melatonin is now sold over the counter throughout the United States.

Thyroid Gland One of the largest endocrine glands in humans is the **thyroid gland,** located in the neck, just below the Adam's apple. It secretes hormones—including **thyroxine**—that influence the speed and efficiency with which body cells break down macromolecules in our diet and use the energy released to produce proteins. In short, it controls most of what we think of as metabolism. As a consequence, poor thyroid function is believed to be at the root of many metabolic disorders, with underactive thyroid responsible for fatigue and weight gain, and overactive thyroid responsible for jitteriness, rapid heartbeat, weight loss, and irritability.

Goiter is a common health problem caused by enlargement of the thyroid gland (FIGURE 25-10). There are several causes of goiters, and they are particularly common in areas with low iodine consumption. When iodine intake is low, the thyroid is unable to produce thyroxine (which contains iodine). This causes thyroxine levels in the body to drop. In the absence of the normal negative feedback telling the body to slow its production of thyroxine, the hypothalamus and anterior pituitary produce increasing amounts, respectively, of thyroxine-releasing hormone and thyroxine-stimulating hormone. These cause the thyroid to swell into a visible lump—a goiter—as it tries unsuccessfully to make thyroxine. In the United States, the widespread use of iodine-fortified table salt prevents most iodine deficiencies, but they are common in Asia, Central Africa, and parts of South America.

The thyroid also regulates levels of calcium in the blood. Calcium is necessary for building and maintaining bones and teeth, and it influences the functioning of nerves and muscles. When there is too much calcium in the blood, the thyroid increases its release of **calcitonin**—a hormone particularly important in babies and children—which causes bones to take up the excess calcium. Additionally, embedded in the surface of the thyroid are the four small **parathyroid glands,** which produce parathyroid hormone, another hormone that plays a central role in regulating calcium levels in adults (FIGURE 25-11). Throughout

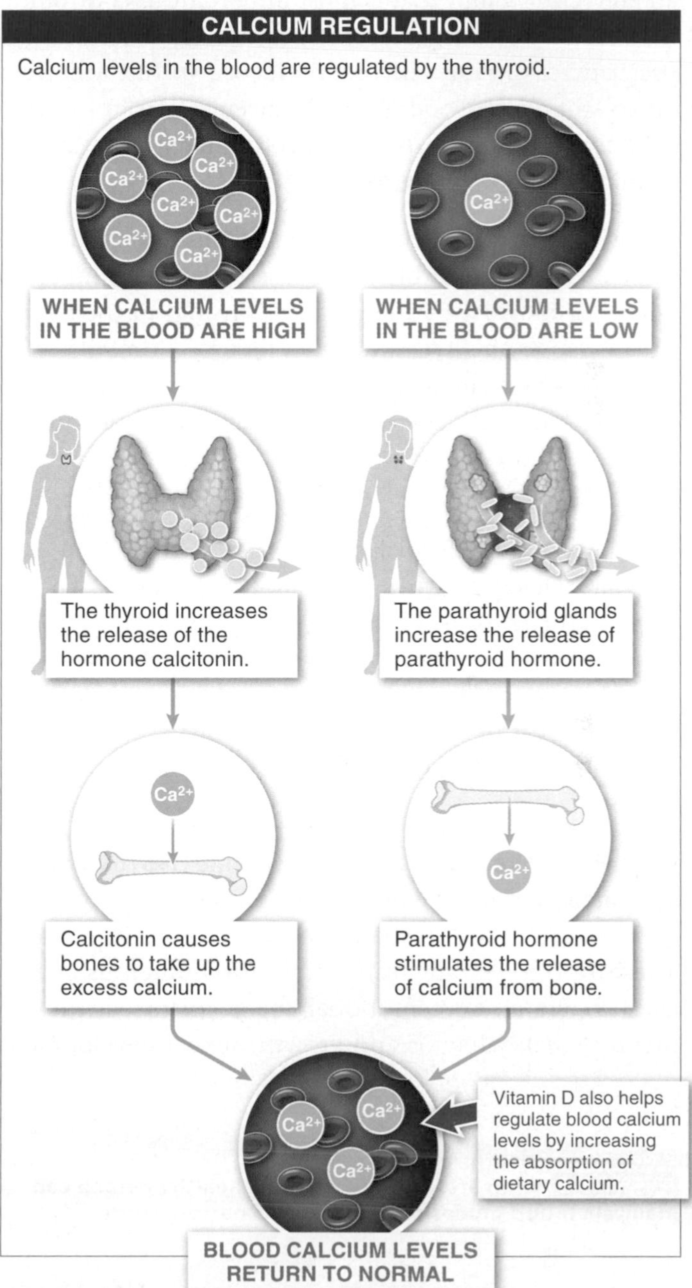

FIGURE 25-11 The body's use of calcium is regulated by the thyroid and parathyroid glands.

life, bone is continually broken down and remade as minerals are lost and added. Parathyroid hormone, working in concert with calcitonin, is important in stimulating much of this continued turnover of bone. Parathyroid hormone further reduces calcium loss in the urine, regulates the release of calcium from bone, and in conjunction with vitamin D (from the diet and produced in response to sun exposure), helps increase the body's ability to utilize the calcium supplied in the diet.

Pancreas Located next to the stomach and connected to the small intestine via a short duct, the **pancreas** is an endocrine gland that is most important in controlling the levels of blood glucose. As we saw in Section 23.18, the pancreas maintains blood glucose within a narrow range—a typical blood glucose concentration in humans is 90 mg/100 mL—through the coordinated secretions of insulin and glucagon (FIGURE 25-12).

Following a meal (particularly one rich in carbohydrates), the concentration of blood glucose rises. This stimulates release of **insulin** by the pancreas. Insulin in the bloodstream causes the liver and other tissues—primarily muscle—to take up glucose, which reduces the blood glucose level. As blood glucose levels fall, there is a reduction in insulin secretion.

Conversely, after a few hours of fasting, the blood glucose level gradually drops. The reduced blood glucose triggers release of **glucagon** by the pancreas. Glucagon has the reverse effect of insulin, causing the liver to convert stored glycogen into glucose, which is released into the bloodstream. Rising blood glucose concentration then causes the pancreas to reduce its glucagon secretion, maintaining homeostasis through negative feedback.

Gonads The sex steroids, including **testosterone, estrogen,** and **progesterone,** are produced largely by the gonads—the testes in males and the ovaries in females. These hormones are responsible for numerous physical, behavioral, and emotional characteristics, including much sexual behavior and growth, sexual development

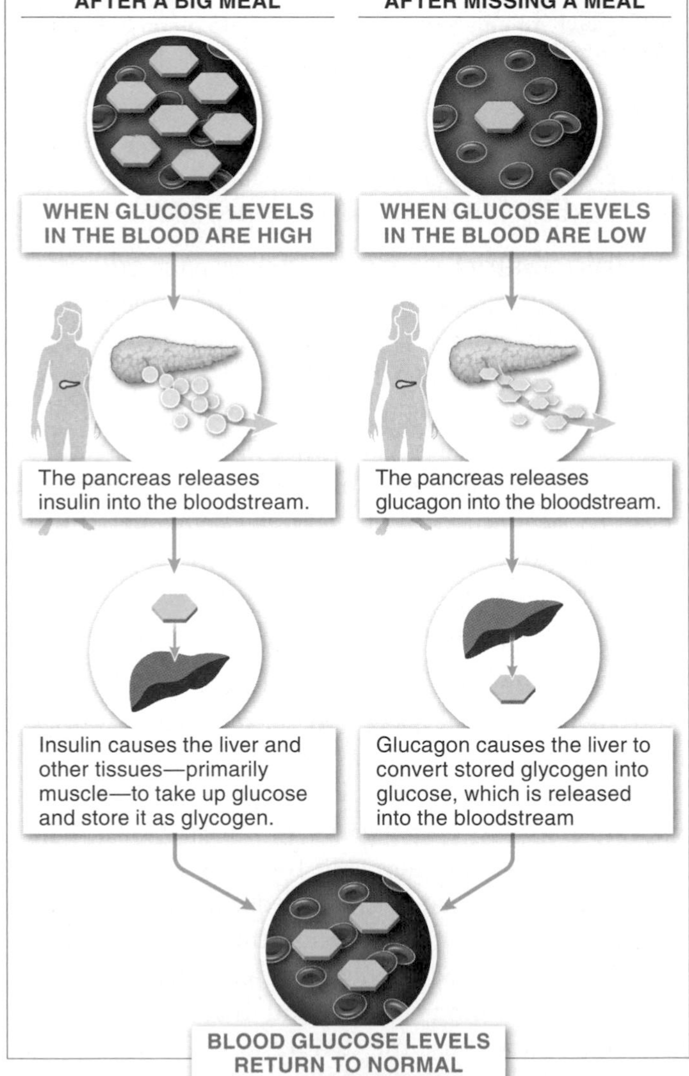

BLOOD GLUCOSE REGULATION
Hormone secretions from the pancreas—including insulin, following a meal—maintain blood glucose within a narrow range.
AFTER A BIG MEAL
AFTER MISSING A MEAL
WHEN GLUCOSE LEVELS IN THE BLOOD ARE HIGH
WHEN GLUCOSE LEVELS IN THE BLOOD ARE LOW
The pancreas releases insulin into the bloodstream.
The pancreas releases glucagon into the bloodstream.
Insulin causes the liver and other tissues—primarily muscle—to take up glucose and store it as glycogen.
Glucagon causes the liver to convert stored glycogen into glucose, which is released into the bloodstream
BLOOD GLUCOSE LEVELS RETURN TO NORMAL

FIGURE 25-12 Even after a heavy meal, a healthy person can maintain blood glucose levels within a narrow range.

(embryonically, in puberty, and continuing into adulthood), and maintenance of gamete production throughout an organism's reproductive life.

FIGURE 25-13 provides a summary of the major animal hormones and their actions.

OVERVIEW OF THE ENDOCRINE GLANDS AND THEIR HORMONES

ENDOCRINE GLAND AND HORMONE	TARGET AREA	PRIMARY FUNCTION
HYPOTHALAMUS		
Regulatory hormones	Anterior pituitary	Increase the production and release of hormones, or reduce or stop the release of hormones
Posterior pituitary hormones	Posterior pituitary	(See posterior pituitary hormones below)
ANTERIOR PITUITARY		
Thyroid-stimulating hormone	Thyroid	Stimulates the production of thyroxine, important in cellular respiration
Follicle-stimulating hormone	Ovaries (females), testes (males)	Stimulates follicles in the ovaries to begin development (females); stimulates sperm maturation (males)
Luteinizing hormone	Ovaries (females), testes (males)	Triggers ovulation (females); stimulates testosterone production (males)
Prolactin	Mammary glands	Stimulates the production of milk
Adrenocorticotropic hormone	Adrenal glands	Stimulates the production of cortisol and other stress-related hormones
Growth hormone	Liver and many other organs	Stimulates the release of chemicals that spur the growth of bones, cartilage, and many other tissues
POSTERIOR PITUITARY		
Oxytocin	Brain, mammary glands, uterus	Influences people's trust in others; increases propensity to form social attachments; directs the ejection of milk for nursing babies; stimulates uterine contractions during childbirth
Antidiuretic hormone	Kidneys	Influences water retention by the kidneys
PINEAL GLAND		
Melatonin	Brain	Regulates sleep cycles
THYROID GLAND		
Thyroxine	Cells throughout the body	Influences the rate and efficiency of cellular metabolism
Calcitonin	Bone	Regulates calcium levels in the blood
PARATHYROID GLANDS		
Parathyroid hormone	Bone, digestive system	Regulates the release of calcium from bone and, in conjunction with vitamin D, helps increase the body's ability to utilize calcium in the diet
ADRENAL GLANDS		
Adrenaline	Smooth muscle, cardiac muscle, skeletal muscle, blood vessels	Initiates the body's response to stress: increases heart rate, increases glycogen breakdown, initiates bronchiole dilation in the lungs, enabling greater absorption of oxygen
Cortisol	Cells throughout the body	Initiates and modulates the body's response to long-term stress; raises blood glucose level
PANCREAS		
Insulin	Liver, skeletal muscle, adipose tissue	Released in response to increasing blood glucose (typically following a meal), causes the liver, muscle, and other tissues to take up glucose
Glucagon	Liver, adipose tissue	Released in response to decreasing blood glucose, causes the liver to convert stored glycogen into glucose, which is released into the bloodstream
GONADS		
Estrogen	Cells throughout the body, female reproductive structures	Stimulates development of female secondary sex characteristics and growth of sex organs at puberty; stimulates monthly preparation of uterus for pregnancy
Progesterone	Uterus, breasts	Stimulates development of breasts; completes preparation of uterus for pregnancy
Testosterone	Cells throughout the body, male reproductive structures	Stimulates development of male secondary sex characteristics and growth spurt at puberty; stimulates development of sex organs; stimulates sperm production

FIGURE 25-13 **Summary of endocrine glands, hormones, and their primary functions.**

In the next few sections, we investigate some of the gonadal hormones and see how closely linked they are to athletic performance and the attributes necessary to excel physically. We also note the extreme health consequences that accompany the illegal use of some hormone supplements.

TAKE HOME MESSAGE 25.5

›› Endocrine glands throughout the body are responsible for detecting and responding to internal and external signals. The adrenal glands regulate responses to stress. The pineal gland regulates sleep cycles. The thyroid gland influences cellular metabolism. The pancreas maintains blood glucose levels. Hormones from the gonads are responsible for many physical, behavioral, and emotional characteristics.

25.6–25.10 Hormones influence nearly every facet of an organism.

Growth hormones influence stature.

25.6 Hormones can affect physique and physical performance.

The Tour de France is a grueling bike race that takes place over three weeks, as riders cover more than 2,200 miles (3,500 km), in 21 separate stages. In 2006, an American rider named Floyd Landis, in 11th place overall, turned in a performance in stage 17 that was so improbably fast that some commentators called it "the greatest performance ever." His win brought him within 30 seconds of the overall lead and propelled him, over the final four stages, to the Tour de France victory.

Less than a week after his victory, it was announced that he had tested positive for unusually high levels of testosterone, which a second test confirmed. The typical ratio of testosterone to the hormone epitestosterone in men is 1:1 or 2:1. While the Tour de France allows ratios of up to 4:1, Landis's level was 11:1. Moreover, the lab tests detected a synthetic form of testosterone in his bloodstream, in addition to the naturally produced hormone. Landis was stripped of his title and banned from the sport for two years.

While such revelations that competitors have resorted to illegal supplementation in an effort to boost physical performance are disappointing, the history of sports, including baseball, track and field, and football, is littered with drug scandals, frequently involving testosterone or its variants (FIGURE 25-14).

Testosterone and many structurally similar steroid hormones affect the composition of the body. Experimental studies of testosterone supplementation in men reported weight gains of 5–12.5 pounds within 10 weeks. The gains were due primarily to increases in lean muscle mass in the shoulders, upper chest, and upper arms, areas with the highest concentrations of testosterone receptors. In randomized, controlled studies, researchers also documented strength increases of 5% to 20% in both experienced and novice athletes.

Q How do steroids affect a person's physique and performance?

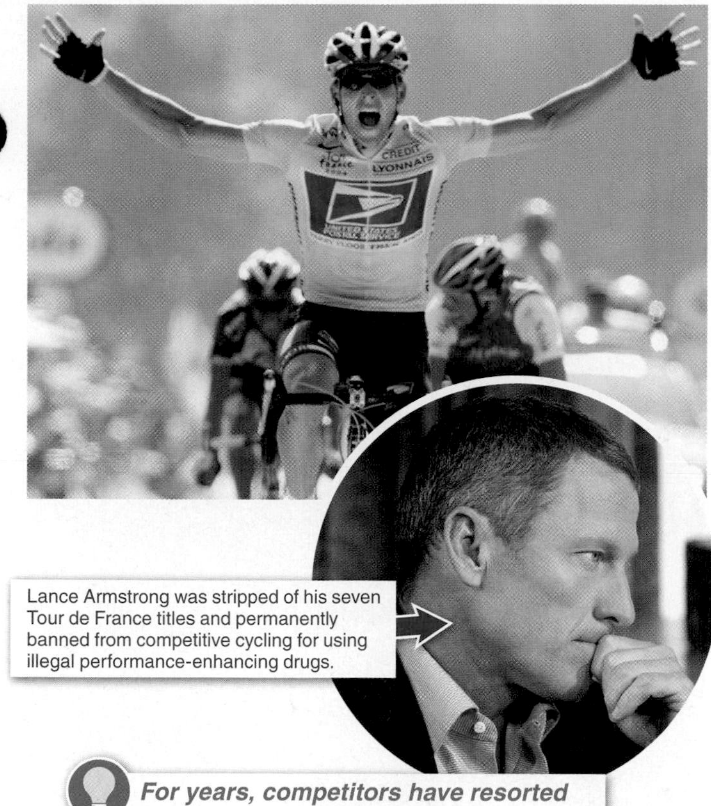

Lance Armstrong was stripped of his seven Tour de France titles and permanently banned from competitive cycling for using illegal performance-enhancing drugs.
For years, competitors have resorted to illegal steroid supplementation in an effort to boost physical performance.

FIGURE 25-14 Unsporting conduct. Many athletes have tested positive for illegal testosterone supplementation.

THE EFFECTS OF TESTOSTERONE

INCREASE IN LEAN MUSCLE MASS

IMPROVED SPEED AND STAMINA

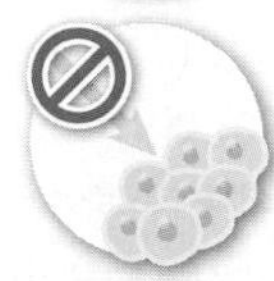
REDUCED PRODUCTION OF FAT-STORAGE CELLS

DEVELOPMENT OF SECONDARY SEX CHARACTERISTICS

THE ADVERSE EFFECTS OF SUPPLEMENTAL TESTOSTERONE

- Increased blood pressure and risk of heart disease
- Kidney and liver damage
- Development of breast tissue in males
- Atrophy (shrinkage) of the testes
- Increased risk of many types of cancer
- Potential undesirable behavioral changes

FIGURE 25-15 The normal effects of testosterone in the body, and the harmful effects of testosterone supplementation.

Testosterone has also been shown to influence locomotor performance—speed and endurance, in particular—in animals. In an experimental study on lizards (the northern fence lizard, *Sceloporus undulatus*), two to three weeks of supplemental testosterone caused a 24% increase in sprinting speed and a 17% increase in stamina.

Produced primarily by the testes in males (but also by the adrenal glands of both males and females), testosterone binds to receptors in target cells, then moves to the nucleus and influences gene expression (see Figure 25-5). The cells increase their protein synthesis and grow, resulting, in the case of muscle cells, in increased muscle mass (with a simultaneous reduction in the rate of production of fat-storage cells). Supplemental testosterone also increases cell division of satellite cells, small cells found in muscle tissue that aid in muscle repair. Additional responses to testosterone supplementation include the development of secondary sex characteristics, including chest and facial hair and a deepened voice—effects seen in both males and females.

There are numerous adverse effects caused by supplemental steroid hormones. They include increased blood pressure and risk of heart disease, damage to the kidneys and the liver (where the steroids are metabolized), development of breast tissue in males, and atrophy (shrinkage) of the testes as negative feedback leads to reduced production of testosterone in response to the increased levels of circulating testosterone (FIGURE 25-15). Also, because testosterone increases cell division—increasing muscle mass, for example—it increases the risk of unrestrained cell division in many tissues and, consequently, increases the risk of many types of cancer.

TAKE HOME MESSAGE 25.6

» Testosterone and similar hormones affect a person's physique and physical performance by influencing gene expression and thus protein synthesis in cells with the appropriate receptors. There are also numerous adverse effects caused by supplemental steroid hormones.

25.7 Hormones can affect mood.

In a randomized, controlled, double-blind study, researchers injected 47 men each week for six weeks with large amounts of testosterone and made numerous measurements of each participant's physical state. The testosterone caused more than just physical changes, though; after six weeks, one man said, "I feel great! I'm confident, happy, and productive. I want to be on this stuff forever." The researchers said that this was a common response among the study's participants.

Not only do hormones affect our physical traits and performance, they can also influence how we *feel*. Many hormones have pronounced effects on **moods,** defined as relatively long-lasting emotional states (shorter-lasting than a person's temperament, but longer-lasting and less specific than a single emotion). Here we describe just a few of the documented effects of hormones on mood in humans.

Estrogen In women, the levels of estrogen in the bloodstream change throughout life. And the incidence of depression follows a similar pattern. The large increases in estrogen levels at puberty, the sharp drop-off in estrogen levels after a woman gives birth, and the reduced estrogen levels at the onset of menopause are all associated with increased occurrence of depression (FIGURE 25-16). Estrogen as part of hormone replacement therapy is effective at reducing depression in women going through menopause.

The sharp drop in estrogen levels after a woman gives birth is associated with an increased occurrence of depression, sometimes referred to as the "baby blues."

FIGURE 25-16 **Drastic changes in estrogen levels affect mood.**

Testosterone Many studies have demonstrated a link between testosterone and mood in human males, including increased self-esteem and a sense of overflowing with new ideas. In the study discussed above, the men were described by their significant others as energetic, euphoric, confident, and charismatic. In a small number of cases, however, the men became uncharacteristically aggressive and exhibited verbal hostility. The researchers even had to withdraw one participant from the study when he became "alarmingly" aggressive—consistent with anecdotal claims of so-called 'roid rage resulting from supplemental testosterone.

Several studies of competitive male athletes have documented that testosterone levels—and the mood changes they influence—rise before a competition. After the contest, testosterone levels in the losers of the competition drop quickly and dramatically, while testosterone levels in the winners remain high and sometimes even increase (FIGURE 25-17).

Recent studies reveal that sports fans share in this biological "victory dance." Measurements of testosterone levels in male fans of Brazilian and Italian soccer teams during World Cup finals (and replicated in male fans of a college basketball team) revealed that the testosterone levels of the fans of the winning team rose, but not those of fans of the losing team.

Q How can the exhilaration of watching your favorite team win a game resemble the feeling you get from winning a sports contest yourself?

Melatonin In randomized, controlled, double-blind studies, researchers have demonstrated that under certain

FIGURE 25-17 **Biological victory dance.**

Prior to a competition, testosterone levels rise in the competitors. After the contest, levels in the losers drop quickly, while remaining high in the winners. Similar results have been reported for the fans of sports teams, too.

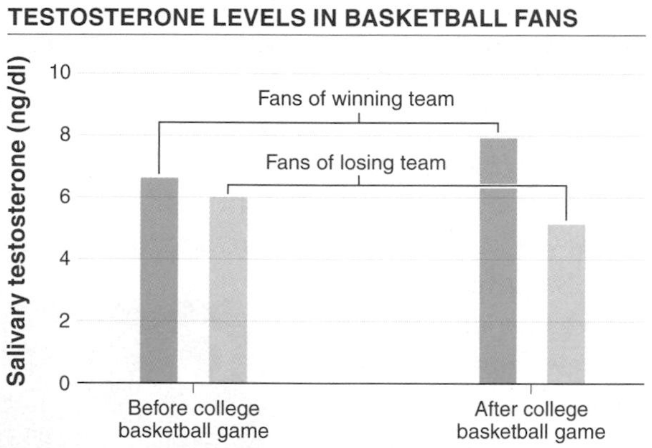

TESTOSTERONE LEVELS IN BASKETBALL FANS

Salivary testosterone (ng/dl)

Fans of winning team

Fans of losing team

Before college basketball game

After college basketball game

Testosterone levels surge among competitors as well as their fans. Victory keeps the levels high.

conditions, oral melatonin can induce hypnotic, sedative-like effects and improve sleep efficiency. Other studies showed that, in comparison with placebo treatment, melatonin supplementation significantly reduced participants' self-reported vigor, while increasing their fatigue and confusion.

Cortisol The majority of individuals suffering from Cushing's disease, a condition in which the adrenal glands release unusually large amounts of cortisol, experience depression and anxiety. Also, an atypically high number of individuals treated with cortisone (a chemical variant of cortisol) to reduce tissue inflammation suffer from anxiety and depression. As a consequence of these observations, researchers conducted a controlled study exposing rodents to cortisol in their drinking water. Two weeks or more of cortisol exposure caused the animals to take significantly longer to emerge from small, dark compartments into a brightly lit area—a measure of anxiety-like behavior in rodents—compared with animals not exposed to the cortisol.

TAKE HOME MESSAGE 25.7

» Many hormones, including estrogen, testosterone, melatonin, and cortisol, have pronounced effects on moods.

25.8 Hormones can affect behavior.

It can be hard to imagine how a little more or a little less of a chemical could influence an animal's behavior, but experimental research on hormones and behavior has documented literally *thousands* of such effects. The scholarly journal *Hormones and Behavior,* for example, in its 100th volume as of 2018, publishes 10 issues a year, each containing 25 or more articles describing laboratory and field studies on hormones and their influences on the development and expression of behaviors (FIGURE 25-18). In these studies, researchers take a variety of experimental approaches. Just two of these are described here.

Genetic manipulation of hormone levels. Researchers bred some lab mice that were unable to make the enzyme aromatase, which is essential to the production of estrogen. The mice with this deficiency had two striking behavioral differences from typical mice: they ran excessively on their exercise wheels and, when sprayed lightly with water, spent significantly longer grooming themselves. Both of these behaviors are indicators of obsessive-compulsive disorder in mice.

Physiological supplementation of hormone levels. Researchers implanted testosterone capsules in male juncos, a type of songbird. These males produced a song that was more attractive to females, and they produced more offspring than control-group males that received a capsule containing no testosterone. (In similar studies, male birds with supplemented testosterone tended to have increased muscle mass and were able to establish and maintain larger territories than non-supplemented males.) However, the male birds with testosterone implants spent less time with their offspring and gave the offspring less food. The testosterone-supplemented birds were also more susceptible to disease and had shorter life spans.

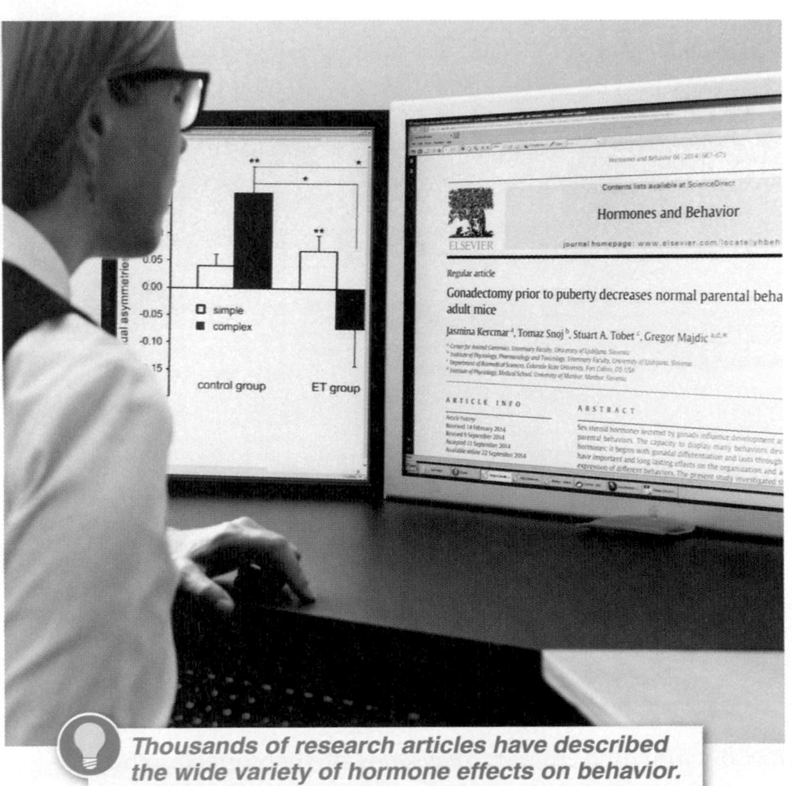

Thousands of research articles have described the wide variety of hormone effects on behavior.

FIGURE 25-18 The influence of hormones on behavior is an area of active scientific research.

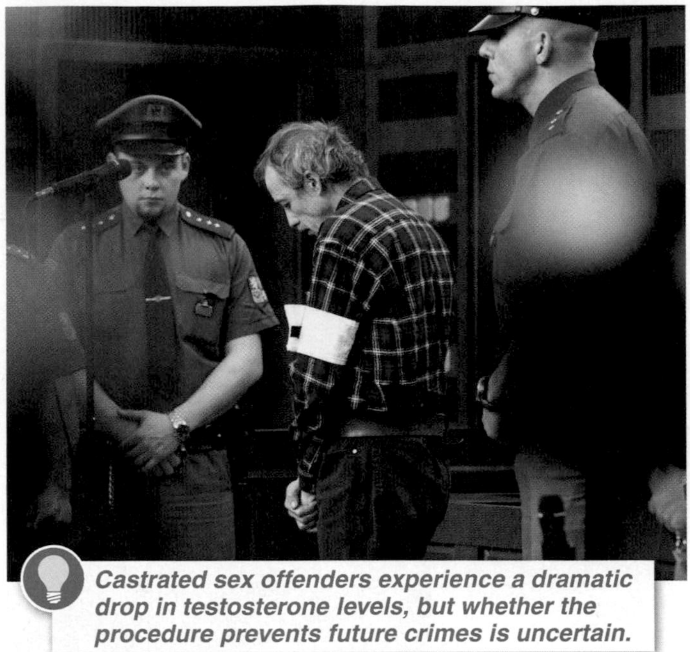

Castrated sex offenders experience a dramatic drop in testosterone levels, but whether the procedure prevents future crimes is uncertain.

FIGURE 25-19 Is castration of convicted sex offenders an ethical procedure?

The dramatic and close link between hormones and behavior has caused people to debate whether male sex offenders should have the choice to be castrated to modify their behavior (FIGURE 25-19). Whether or not castration—removal of the testes, with a resultant near-complete drop in testosterone levels—can reliably rehabilitate violent sex offenders is intensely debated.

Q Is castration an effective and humane treatment for sex offenders?

In the Czech Republic, one of the very few countries where sex offenders have the choice to be castrated as a treatment for their behavior and an alternative to life confinement in institutional care, 80 prisoners underwent castration between 2000 and 2010. Among these men, 4% were reported to commit further offenses. This is similar to data reported by the German government that 3 of 104 male sex offenders who were castrated committed further offenses, compared with 40% of non-castrated sex offenders. Similarly, in a Danish study of 900 castrated sex offenders, the rate of repeat offense dropped to approximately 2%, from close to 80% among non-castrated sex offenders.

Opponents of the castration of sex offenders argue that the evidence cannot be trusted, because it relies partly on self-reporting by the castrated men. Moreover, the opponents of castration argue that having such an option is coercive, because many convicted sex offenders will feel obligated to opt for the surgery, leading to a violation of their rights. Some argue for the reversible form of castration by injection of chemicals that block the effects of testosterone, but this is opposed by many on the grounds that it relies on the released offenders voluntarily undergoing treatment, which they may stop at any point.

TAKE HOME MESSAGE 25.8

» Many laboratory and field studies have demonstrated the influence of hormones on the development and expression of behaviors.

25.9 Hormones can affect cognitive performance.

Cognitive abilities—mental processes involving perception, memory, judgment, and reasoning—are influenced by hormones. Studies on the effects of hormones on cognitive performance have primarily focused on two types of measures, designed to reflect different aspects of cognition: (1) motor and verbal tasks, and (2) spatial tasks (FIGURE 25-20).

1. Motor and verbal tasks. These tasks include tests of articulation speed, such as a tongue twister, or complex wrist and hand movements and fine-muscle movements, such as those required in surgery and machine repair.

Most tests of hormone effects on motor and non-verbal tasks involve comparisons of the performances of men and women or of girls and boys on these tasks. There is consistent and significant evidence that females perform better than males and that the disparity between male and female performances is greatest at points of high estrogen levels in

THE EFFECTS OF HORMONES ON COGNITIVE TASKS

Repeat this sentence five times, as quickly as you can:

"A box of mixed biscuits in a biscuit mixer."

MOTOR AND VERBAL TASKS

- Females tend to perform better than males on tests of articulation speed and fine-muscle movement.
- The disparity between male and female performance is greatest at points of high estrogen levels in the female reproductive cycle and least at the points of lowest estrogen levels.

Identify which of the four shapes on the right represent(s) the target shape on the left in alternative positions.

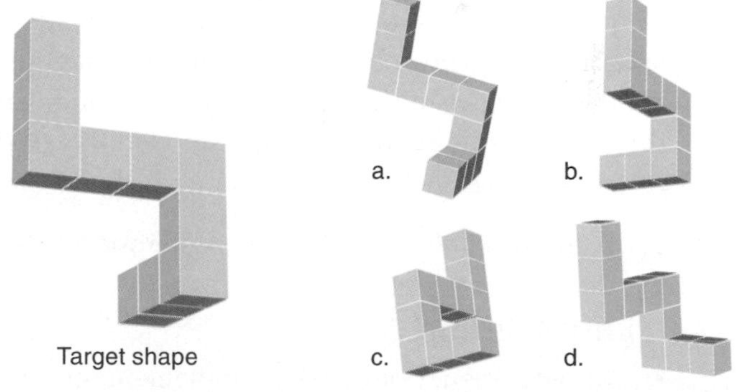

SPATIAL TASKS

- Males tend to perform better than females on tests of spatial ability, such as mental rotations of two- or three-dimensional objects.
- Evidence suggests that this performance difference reflects effects of testosterone—although the mechanisms are unclear.

Men and women have slightly different abilities in certain areas, based on the abundance or lack of the hormones testosterone and estrogen.

FIGURE 25-20 **Different talents.**

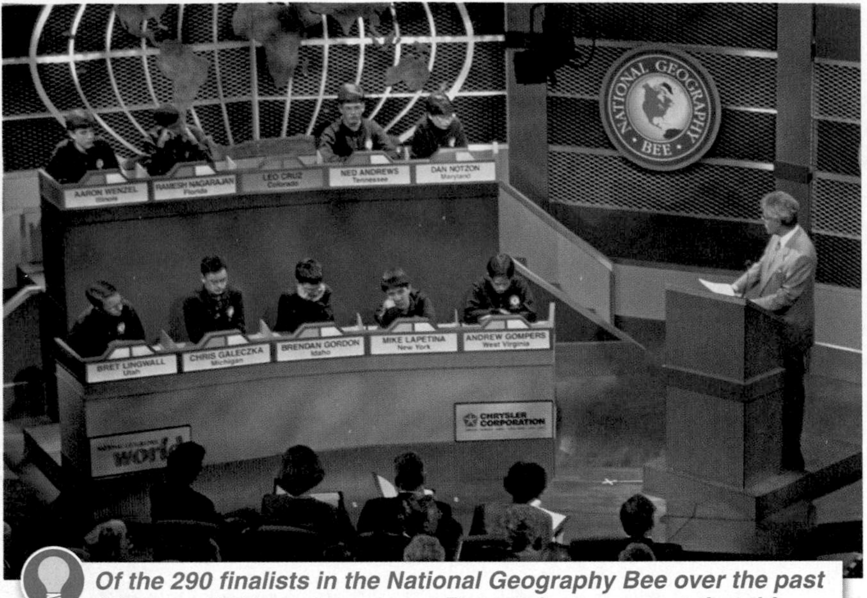

Of the 290 finalists in the National Geography Bee over the past 29 years, 273 have been boys. Researchers suggest that this disparity is linked to male-female differences in spatial abilities.

FIGURE 25-21 **Showing an interest in geography.**

the female reproductive cycle and smallest at the points of lowest estrogen levels.

2. Spatial tasks. These include tests such as predicting the shapes of boxes when they are unfolded and flat, and mental rotations, in which two drawings of two- or three-dimensional objects are compared to see whether they represent the same object, either rotated or as a mirror image.

In humans and other animal species, performance on tests of spatial ability is, on average, higher in males than in females. A large body of evidence suggests that this performance difference reflects effects of testosterone—although the mechanisms are poorly understood. Some of the suggestive findings include:

- As men get older and their testosterone levels decrease, so does their performance on tests of spatial ability.
- In older men, testosterone supplementation improves performance on spatial tasks, while administration of chemicals that reduce the production of testosterone (for the treatment of prostate cancer) causes a decrease in spatial abilities.
- Women with one form of a genetic condition (called congenital adrenal hyperplasia) that causes them to have higher than typical levels of circulating testosterone also have increased performance on tests of spatial ability.

As with tests of motor and verbal tasks, it is not clear exactly how—or whether—performance on tests of spatial ability affects individuals in contexts more relevant to everyday life. Each year since 1989, the National Geographic Society has held a "National Geography Bee" contest. In the 29 years of the contest (through 2017), 273 of the 290 finalists (and 27 of the 29 winners) have been boys (FIGURE 25-21). Interest in such competitions is unquestionably influenced by cultural and societal forces. But the researchers conducting this study suggested that the results stem from the effects of male-female differences in spatial abilities on reading and interpreting maps. An author of the study, Lynn Liben, commented that "it's not true that every woman is worse than every man or every girl is worse than every boy. But at the group level, it is true." Conversely, girls remain significantly more likely than boys to make the honor roll in school.

Beyond tests of motor and verbal tasks and tests of spatial ability, another component of cognitive performance is memory. Numerous studies report a strong role for estrogen in memory, suggesting that estrogen increases the growth and connections of neurons. A recent study in rodents, for example, showed that treatment with a high dose of estrogen improved performance on a maze-running task that tests memory.

The stress hormone cortisol has also been implicated in memory abilities. Most studies have reported a relationship that, when plotted as a graph, resembles an inverted "U": up to a point, increased cortisol improves memory, but an excess of cortisol and other stress hormones consistently reduces performance. (Removing the adrenal glands, which produce most of the body's cortisol, impairs performance on tests of memory, but abilities are restored when cortisol is supplemented to typical levels.) Optimum memory performance seems to require moderate levels of cortisol and other stress hormones.

TAKE HOME MESSAGE 25.9

» A great number of experimental studies demonstrate that cognitive abilities in humans are influenced by hormones, particularly the reproductive hormones estrogen and testosterone, as well as the stress hormone cortisol.

25.10 Hormones can affect health and longevity.

Sometimes the links between cause and effect in biological systems appear straightforward. Take, for example, the results from one of the most appalling "experiments" ever conducted. In the early 1900s, many men committed to sanitariums were castrated. In a case-controlled study, the castrated men were compared with non-castrated, institutionalized men, born in the same year and with the same estimated IQ. It turned out that the median longevity of the 297 castrated men, at 69.3 years, was almost 14 years longer than that of the 735 men in the control group. Similarly, data from analyses of more than 1,000 cats (male and female) showed that sterilized cats live significantly longer than those not sterilized. Among male cats, for example, those not castrated lived a mean of 5.3 years, while castrated males had a mean life span of 8.1 years (FIGURE 25-22).

Laboratory mice also have significantly lengthened life spans when sterilized. And in similar experiments, female fruit flies exposed to high temperature or X rays experienced a dramatic reduction in ovary size, a severe drop in egg-laying rate, and an accompanying 73% increase in life span. Other researchers found that the amounts of reproductive hormones in the bloodstream of mice were significantly reduced in calorie-restricted males and females compared with controls—animals given access to as much food as they wanted. Mice on the low-calorie diet lived more than 40% longer than those with unlimited access to food. All of these results suggest that reducing the levels of circulating reproductive hormones increases longevity.

Hormones probably have their significant impact on longevity because the rate of cancer occurrence—in rodents as well as in humans—is closely related to the concentrations of circulating reproductive hormones, such as estrogen.

- Increased lifelong exposure to estrogen is associated with increased rates of cell division and, consequently, increased cancer risk.
- High concentrations of estrogen in the bloodstream are associated with an increased incidence of endometrial cancer.
- Women taking oral contraceptives and women giving birth to higher numbers of children—both of which reduce a woman's lifetime exposure to estrogen and other hormones produced by developing follicles—have decreased incidences of ovarian cancer. An analysis of 20 studies reported that five years of taking oral contraceptives reduced women's lifetime risk of ovarian cancer by approximately 50%.

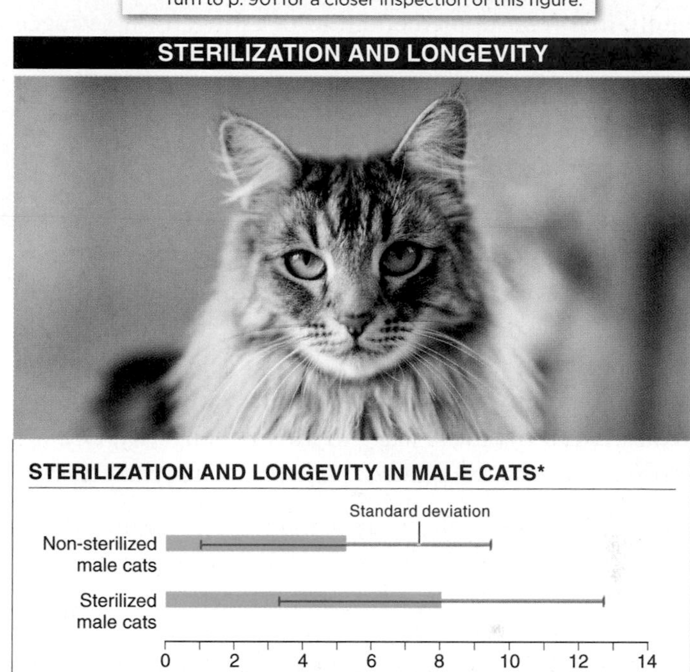

STERILIZATION AND LONGEVITY IN MALE CATS*

Standard deviation

Non-sterilized male cats

Sterilized male cats

0 2 4 6 8 10 12 14

Average longevity (years)

*Data included only for "protected" cats for which the cause of death was known and excluding deaths by trauma or poisoning, which are known to occur more commonly in non-castrated cats.

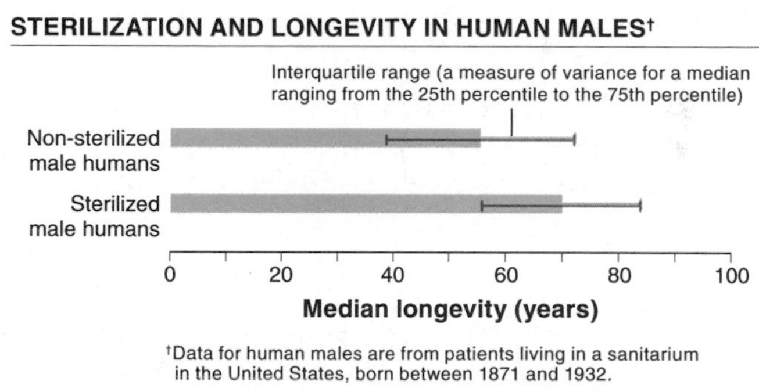

Research suggests that reducing the levels of circulating reproductive hormones increases longevity.

FIGURE 25-22 **Long-lived pets.** Sterilization increases longevity in cats and humans.

There is similar evidence of increased cancer risk linked to circulating levels of progesterone and testosterone.

These seemingly straightforward results, however, lead us to an example that illustrates the extreme complexity and interrelatedness of the body's physiological systems. It would seem, given the results above, that longevity should be decreased by hormone replacement therapy, which generally involves supplementation of estrogen (or estrogen and progesterone), following menopause. In fact, though, almost all studies of hormone replacement have found that the treatment reduces the annual risk of death by about 25% and leads to increased longevity.

How can increased longevity be reconciled with the observed adverse effects of estrogen? It turns out that estrogen reduces circulating levels of cholesterol, which can reduce the risk of death from heart disease, the top killer of elderly women. Heart disease kills significantly more women than do breast cancer and cancers of the reproductive system, so any treatment that reduces heart disease is likely to increase longevity, *even if it increases the risk of breast cancer and reproductive-system cancers.*

Nonetheless, as a consequence of the complex nature of these results, there is still confusion and concern when it comes to hormone replacement therapy, and the debate continues. In 2013, a review of studies on the mortality risks of hormone replacement therapy concluded that for younger postmenopausal women (under 60 years of age), there is a significant reduction in mortality risk associated with hormone therapy, but the benefit is reduced or absent among older women (over 70 years of age). And so, while there are several strong links between hormones and health and longevity, the complex and myriad ways in which hormones can affect health make it difficult to make simple predictions about the influence of any hormone on health and longevity.

TAKE HOME MESSAGE 25.10

» Hormones affect health and longevity in complex ways. Sterilization of animals, for example, reduces levels of circulating reproductive hormones and increases longevity, usually due to reduced cancer mortality. But in other cases, such as hormone replacement therapy in women under 60, the hormone treatment reduces the annual risk of death.

25.11–25.12

Environmental contaminants can disrupt normal hormone functioning.

Potentially hazardous chemicals are found not just in dry cleaning agents but also in most printed receipts.

25.11 Chemicals in the environment can mimic or block hormones, with disastrous results.

Beginning in 1939 and continuing for more than three decades, the chemical DDT was used as a pesticide, with widespread agricultural application. It is extremely effective in killing a variety of insects, including mosquitoes, which also made it popular in strategies against malaria—a disease transmitted by mosquitoes. Unfortunately, after DDT killed the intended insect pests, it remained in the environment. Animals that consumed insects exposed to DDT ingested the chemical, which was stored in their body tissues. And when those animals were consumed by others, the DDT was passed on. Because larger animals ate greater numbers of animals carrying DDT, they stored more and more of the toxic chemical.

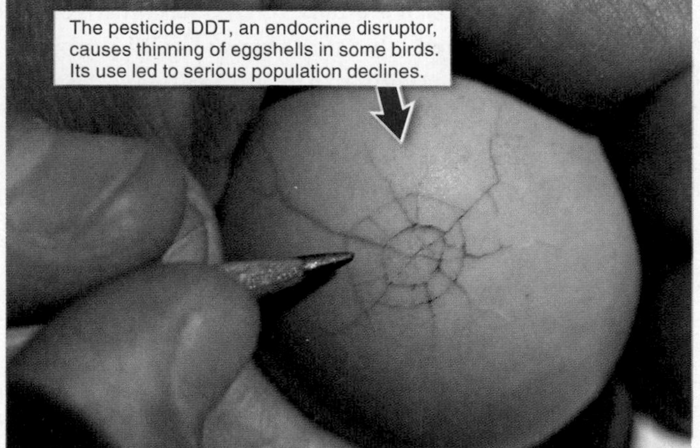

FIGURE 25-23 **Effects of DDT.**

In predatory birds such as bald eagles, peregrine falcons, and pelicans, DDT disrupted the development and functioning of the reproductive tract and impaired the birds' ability to produce normal eggshells—the shells would crack under the weight of the parent incubating the eggs (FIGURE 25-23). These problems led to serious population declines in many bird species and may have led to the extinction of some species.

The publication of *Silent Spring* by biologist Rachel Carson in 1962 called attention to the consequences of DDT use. Carson's book represented the beginning of the environmental movement in the United States. It incited such a public outcry that DDT was eventually banned in the United States—a ban that is cited as instrumental in the rebound of bald eagle populations.

DDT is a synthetic chemical called an **endocrine disruptor.** Endocrine disruptors are chemically similar to hormones, particularly estrogen. Many endocrine disruptors can directly bind to the same receptors as estrogen and can block or otherwise interfere with hormones, leading to harmful effects. In addition to DDT, endocrine disruptors include **PCBs (polychlorinated biphenyls),** which are used as industrial coolants and lubricants; **phthalates,** commonly found in soft toys and cosmetics; and **bisphenol A (BPA),** found in many plastics, including the lining of food cans, plastic water bottles, and baby bottles (FIGURE 25-24).

Endocrine disruptors are sometimes carried in runoff water from industrial manufacturing processes or as airborne pollutants. Exposure to endocrine disruptors has been implicated in a variety of adverse physiological effects, often related to the chemicals' feminizing effects. Many animal groups seem to be adversely affected by endocrine disruptors.

THE ADVERSE EFFECTS OF ENDOCRINE DISRUPTORS

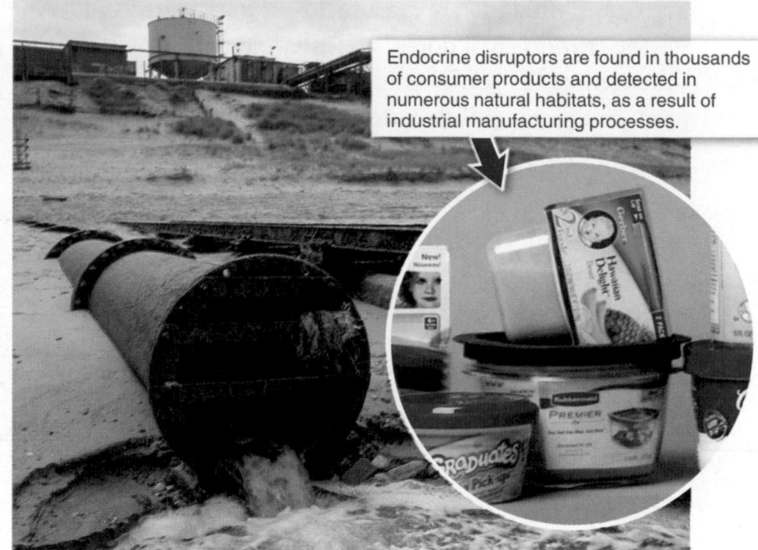

Endocrine disruptors are found in thousands of consumer products and detected in numerous natural habitats, as a result of industrial manufacturing processes.

A variety of animal groups appear to have been adversely affected by endocrine disruptors.

MAMMALS
In populations of Baltic ringed seals, endocrine disruptors have caused partial or complete sterility in 70% of the animals.

FISHES
In a variety of fish species, including carp, rainbow trout, and flounder, exposure to endocrine disruptors interferes with reproductive functioning.

INVERTEBRATES
In marine invertebrates, endocrine disruptors have led to the production of defective shells, as well as the masculinization of female genitals, reducing fertility.

FIGURE 25-24 **Some consumer products contain chemicals that act as endocrine disruptors in wildlife.**

1. Mammals. Populations of Baltic ringed seals have declined significantly over the past 100 years, due, in part, to the presence of large concentrations of pollutants, including PCBs and DDT, in their habitats. These pollutants have interfered with female reproductive functioning in many ways, leading to partial or complete sterility in 70% of the animals.

2. Fishes. In a variety of fish species, including carp, rainbow trout, and flounder, exposure to endocrine disruptors in sewage runoff is known to interfere with reproductive functioning. In some cases, the exposure causes males to produce egg proteins that are typically found only in females.

3. Invertebrates. A group of chemicals called tributyltin compounds, or TBTs, are used to reduce the growth of organisms on the hulls of ships. TBTs act as endocrine disruptors in several marine species. Oysters, for example, when exposed to TBTs, produce defective shells, and in numerous other marine invertebrates, females' genitals become masculinized and the animals have reduced fertility. These impacts have led to a worldwide decline in populations of gastropods such as sea snails and limpets.

Large numbers of synthetic and natural chemicals have endocrine-disrupting functions in both natural and laboratory populations of animals. It remains controversial, however, whether these chemicals cause health problems in humans. Many studies are under way—so far, with conflicting or inconsistent results—and the U.S. government has taken many steps to restrict the use of the endocrine disruptors described here.

TAKE HOME MESSAGE 25.11

» Endocrine disruptors are chemicals used by humans that can mimic, block, or otherwise interfere with hormones and can lead to a variety of adverse physiological effects. Although endocrine disruptors affect many animal species, it remains controversial whether these chemicals cause health problems in humans.

THIS IS HOW WE DO IT

Developing the ability to apply the process of science

25.12 Would you like your receipt? (Maybe not.)

Humans are masters of altering environments and creating goods to make life safer and easier. But as we create more chemicals, we must monitor and evaluate their effects. As we've seen, exposure to some of these chemicals, such as DDT, PCBs, phthalates, and TBTs, can disrupt normal hormone functioning and have harmful consequences for a wide variety of organisms—including ourselves.

Concerns over one such endocrine disruptor, bisphenol A (BPA), which can mimic human estrogen, have led to a ban on its use in reusable food and beverage containers in some U.S. states, and others are considering similar bans. In more than 50 studies in humans, exposure to concentrations of BPA that people might encounter in everyday life has been associated with reduced reproductive functioning in adults and impaired brain development in children.

Even so, BPA shows up in one item almost everyone in developed countries has contact with: *printed receipts*. More than half of all thermal paper—which is used for printing receipts from cash registers and ATMs—contains a layer of BPA that produces color when it is heated. Could BPA get into our bodies just from coming in contact with our skin?

How could we find out whether BPA is absorbed through our skin?

Here's one approach to answering that question. Researchers measured BPA levels (in micrograms per liter, μg/L) in 389 pregnant women, and those who routinely handled receipts had higher-than-average levels of BPA in their bodies. And while the average concentration of BPA in urine was 1.9 μg/L (SE = 1.0; SE is a measure of variation across study participants), the women cashiers had the highest concentrations of BPA, at 2.8 μg/L (SE = 1.1). Teachers and industrial workers, had levels of 1.2 μg/L and 1.8 μg/L, respectively.

Does this prove that touching receipts causes BPA to get into our bodies?

These results are consistent with the hypothesis that contact with receipts leads to higher levels of BPA in the body. But are there viable alternative hypotheses for these observations? Absolutely. Here's one: The population of people who become cashiers may differ from the population in other occupations in ways that—other than receipt handling—increase their exposure to BPA.

How could you test whether increased levels of BPA in the body are due to the handling of receipts?

Some researchers set about testing this idea, reporting their findings in a 2014 article published in the *Journal of the American Medical Association*. The researchers first measured BPA levels in the urine of 24 volunteers. Then the volunteers handled receipts continuously for 2 hours, while not wearing gloves, and their BPA levels were measured again. As a control, researchers conducted the procedure a week later, but with one change: participants wore gloves. Again, BPA levels were measured before and after 2 hours of handling receipts.

Here's what they found:

	BPA in urine, μg/L (95% confidence interval)	
	Before handling receipts	After handling receipts
No gloves	1.8 (1.3–2.4)	5.8 (4.0–8.4)
With gloves	1.3 (0.8–2.1)	1.8 (0.9–3.5)

In the group handling receipts without gloves, BPA levels were more than three times higher than for the group wearing gloves. (When researchers re-measured 12 of the non-glove participants after 8 hours, the BPA levels were five times higher!) Based on these results, the researchers concluded that BPA can and does pass through our skin and get into our bodies after handling receipts. Does the "control" procedure help rule out other means by which BPA on receipts might get into the body? How?

From these results, can we conclude that BPA from receipts is a health hazard?

In this case, it's reasonable to have high confidence that BPA can get into the body as a consequence of handling receipts on thermal paper. However, the researchers did not measure any health outcomes. (And, in fact, the increased BPA levels from receipt handling were relatively low. In another study, BPA levels after eating canned soup were considerably higher: 20.8 μg/L.)

Does this mean that this study has no value in addressing whether BPA from receipts is a health hazard?

First, we should answer this question: What experiment *would* generate the appropriate evidence for testing the hypothesis that "BPA from handling receipts carries a health risk"? You could recruit large numbers of individuals for a randomized, controlled, double-blind study in which the treatment group would be exposed to BPA for a long period of time so that researchers could detect any increase in negative health outcomes.

Of course, this experimental approach for testing a hypothesis may not be possible or ethical. By contrast, the smaller-scale, indirect investigative approach is doable and has value.

What should we conclude from our current state of knowledge about BPA, receipt handling, and health?

The researchers in the receipt-handling study, as well as other experts, conclude that people may be exposed to much higher levels of BPA than previously thought. They suspect, however, that most people's exposure to BPA in thermal paper may not be high enough to have significant adverse effects on health. Still, they suggest that people in certain occupations, such as cashiers and bank tellers, who may handle receipts for 40 hours a week, as well as pregnant women, might have more cause for concern.

TAKE HOME MESSAGE 25.12

» BPA is a chemical associated with adverse health outcomes. Exposure typically occurs through consumption of food or drink packaged in BPA-lined containers, but it can also occur through the handling of thermal receipts. Investigations of BPA exposure reveal an important role for experiments that, for practical and ethical reasons, cannot utilize randomized, controlled, double-blind approaches.

STREET BIO

Using evidence to guide decision making in our own lives

Are pheromones real-life "love potions"?

Q: What are sex pheromones? Sex pheromones, such as those produced by female gypsy moths, are chemical messengers released into the environment to announce to males, over a wide area, that a female is ready to mate.

Q: Do humans produce them? The jury is out on this question. In 1971, data were published suggesting that among women living in college dormitories, pheromones, probably released from underarms, could shorten or lengthen the reproductive cycle in other women (see Section 26.8). But, despite ample evidence of sex pheromones in other species, the evidence of their existence in humans is still hotly debated.

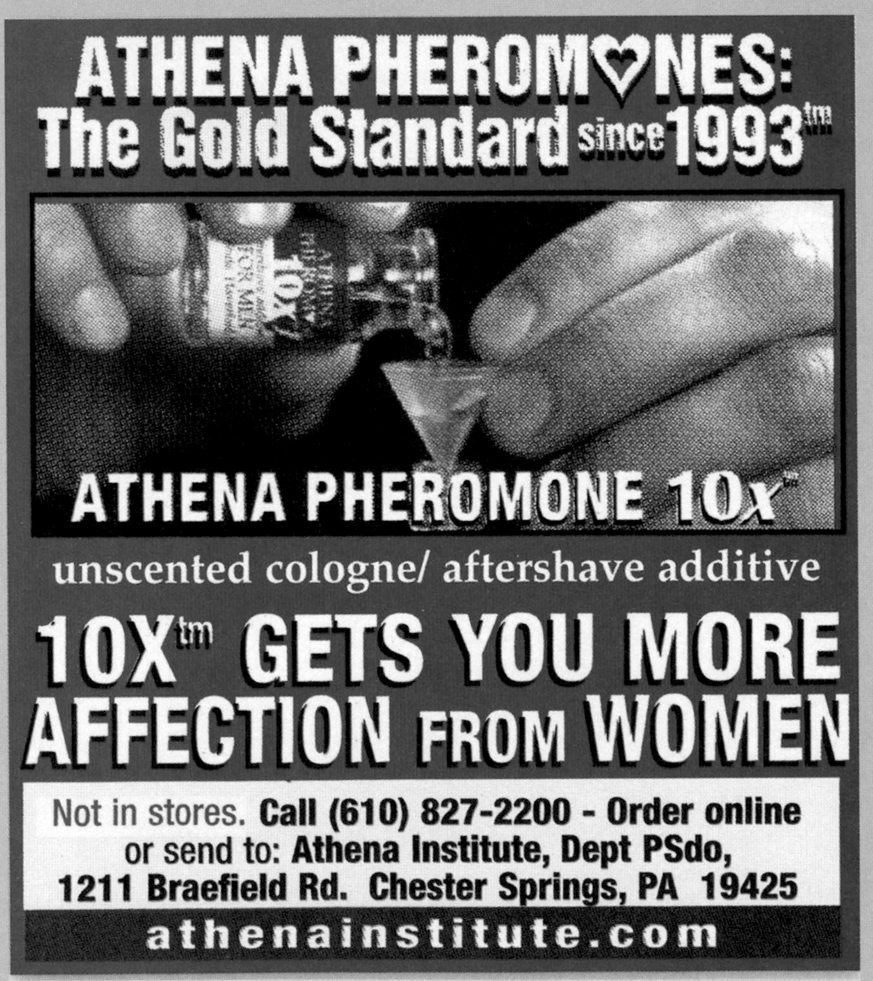

Q: What about the claims made in other studies? A study published in 2002 claimed that human pheromones could increase the sexual attractiveness of women to men. In the double-blind, placebo-controlled study, one group of women had a purported human sex pheromone added to their perfume, while women in a control group had a placebo. The women then recorded, over the next three months, their "socio-sexual" behaviors—including kissing, dating, sexual activity, and male approaches—which were then compared with recorded observations from the same women *prior* to receiving the pheromone or placebo. The results? More than three times as many pheromone users as placebo users (74% vs. 23%) experienced an increased frequency of socio-sexual behaviors. The researchers concluded that the pheromone acted as a sex attractant.

Q: Can we be certain of the claims of just one study? The double-blind and placebo-controlled aspects of this study represent important attempts at making a rigorous test of the authors' hypothesis. But the study has been criticized on several grounds. The authors tested only a small number of women ($n = 36$), within a limited age range (27.8 ± 6.7 years), over only a three-month period. Perhaps more important, they did not disclose in their published report the chemical preparation added to the perfumes. Additionally, one of the study's authors synthesized the proprietary chemical and currently markets it for profit. And re-analyses of the results challenge their significance. So while the results do suggest that there may be something in the air, we probably ought to wait for a bit more experimental evidence before drawing any strong conclusions.

GRAPHIC CONTENT

Thinking critically about visual displays of data

1 What is the "take home message" of these graphs? What additional data would be helpful? Why?

2 What are the scales used in the graphs? How would the graphs be perceived differently if the cat graph ranged from 0 to 100 years and the human graph ranged from 50 to 70 years? Would that be a misleading presentation? Why?

3 What is the percentage longevity increase resulting from sterilization in cats? What is it in humans?

4 In non-technical terms, what does "standard deviation" mean? The top error bar here extends farther than the "average longevity" indicated by the bar below it. What does that reveal?

5 For populations with the same average value for a trait, what is the difference between having a large and a small standard deviation?

6 In the lower graph, longevity is measured as median longevity. What does this mean? How does it differ from average longevity?

7 What does the 25th percentile mean? What does 75th percentile mean?

STERILIZATION AND LONGEVITY IN MALE CATS*

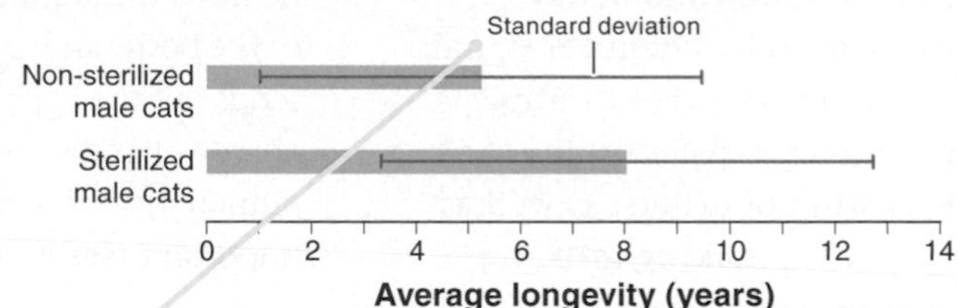

*Data included only for "protected" cats for which the cause of death was known and excluding deaths by trauma or poisoning, which are known to occur more commonly in non-castrated cats.

STERILIZATION AND LONGEVITY IN HUMAN MALES†

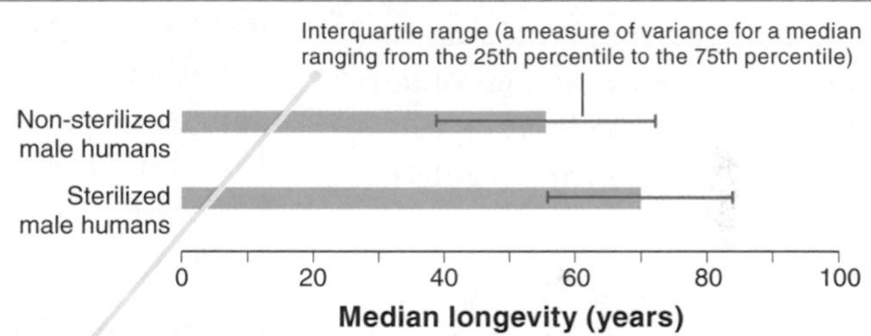

†Data for human males are from patients living in a sanitarium in the United States, born between 1871 and 1932.

8 Suggest why the data for the lower graph were collected from sanitarium patients.

See answers at the back of the book.

KEY TERMS IN HORMONES

adrenal glands, p. 882
adrenocorticotropic hormone (ACTH), p. 882
anterior pituitary, p. 882
antidiuretic hormone (ADH), p. 882
bisphenol A (BPA), p. 897
calcitonin, p. 885
corticotropin, p. 882
endocrine cell, p. 877
endocrine disruptor, p. 897
endocrine gland, p. 878
endocrine system, p. 878
estrogen, p. 886
follicle-stimulating hormone (FSH), p. 882
glucagon, p. 886
growth hormone, p. 882
hormone, p. 877
hypothalamus, p. 881
insulin, p. 886
luteinizing hormone (LH), p. 882
mood, p. 890
oxytocin, pp. 876, 881
pancreas, p. 886
paracrine regulator, p. 880
parathyroid glands, p. 885
PCBs (polychlorinated biphenyls), p. 897
pheromone, p. 878
phthalates, p. 897
pineal gland, p. 883
pituitary gland, p. 881
posterior pituitary, p. 881
progesterone, p. 886
prolactin, p. 882
target cell, p. 877
testosterone, p. 886
thyroid gland, p. 885
thyroid-stimulating hormone (TSH), p. 882
thyroxine, p. 885

BRIEF SUMMARY

Hormones are chemical messengers regulating cell functions.

• Complex emotions and behaviors can be influenced by chemical signals produced in the body. For example, exposure to oxytocin causes humans to be more trusting of others, as well as facilitating birth, making milk available to nursing babies, and increasing the propensity to form social attachments.

• Hormones are chemical messengers, secreted into the circulatory system by endocrine cells and endocrine glands. Hormones influence the actions of target cells elsewhere in the body, as part of an internal communication and regulation system.

• Hormones can regulate the activities of a target tissue only if the cells have a receptor specific to that hormone. Water-soluble hormones bind to receptors on cell membranes, while lipid-soluble hormones bind to receptors within a cell. Both types of hormones, once bound to a receptor, cause changes in the target cell, such as influencing the rate of transcription of genes.

Hormones are produced in glands throughout the body.

• The hypothalamus receives input from neurons throughout the brain and the rest of the body, and using this information, it sends out the appropriate hormones (and nervous signals), often directing the pituitary gland to release hormones with important regulatory effects on body tissues.

• Endocrine glands throughout the body are responsible for detecting and responding to internal and external signals. The adrenal glands regulate responses to stress. The pineal gland regulates sleep cycles. The thyroid gland influences cellular metabolism. The pancreas maintains blood glucose. Hormones from the gonads are responsible for many physical, behavioral, and emotional characteristics.

Hormones influence nearly every facet of an organism.

• Testosterone and other similar hormones affect a person's physique and physical performance by influencing gene expression and protein synthesis in cells with the appropriate receptors. There are numerous adverse effects associated with taking supplemental steroid hormones.

• Many hormones, including estrogen, testosterone, melatonin, and cortisol, have pronounced effects on moods.

• Many laboratory and field studies have demonstrated the influence of hormones on the development and expression of behaviors.

• A great number of experimental studies demonstrate that cognitive abilities in humans are influenced by hormones, particularly the reproductive hormones estrogen and testosterone, as well as the stress hormone cortisol.

• Hormones affect health and longevity in complex ways. Sterilization of animals, for example, reduces levels of circulating reproductive hormones and increases longevity, usually due to reduced cancer mortality.

Environmental contaminants can disrupt normal hormone functioning.

• Endocrine disruptors are chemicals used by humans that can mimic, block, or otherwise interfere with hormones and can lead to a variety of adverse physiological effects.

CHECK YOUR KNOWLEDGE

Short Answer

1. Describe the effects of the hormone oxytocin in humans.

2. Compare the endocrine system and the nervous system. How are they different?

3. Describe the general process by which hormones affect target cells.

4. Which part of the brain functions as a liaison between the nervous and endocrine systems?

5. The stress-response pathways are modulated by negative feedback loops. Describe how this mechanism works.

6. Compare and contrast polypeptide and amine hormones with steroid hormones.

7. How do the hypothalamus, posterior pituitary, and anterior pituitary interact, and what role do they play in the endocrine system?

8. Explain why castration of some male criminals may lead to decreases in criminal activity.

9. Performance on tests of spatial ability is, on average, higher in males than in

females. Describe how studies of females with congenital adrenal hyperplasia help strengthen these findings.

10. Why do women taking oral contraceptives and women giving birth to higher numbers of children have a decreased incidence of ovarian cancer?

11. What are the general effects of endocrine disruptors? Why can even small quantities in the environment be harmful to some organisms?

12. Describe the impact of the pesticide DDT on bird populations.

Multiple Choice

1. Target cells of the hormone ______ are located in a woman's uterus and breasts. This hormone stimulates its target cells to contract, which causes the uterine contractions that push out the newborn during labor and releases milk from the breast cells to feed the infant.

a) oxytocin

b) progesterone

c) estrogen

d) ACTH

e) prolactin

0 EASY 35 HARD 100

2. To operate effectively, the endocrine system is largely dependent on the:

a) muscular system.

b) circulatory system.

c) digestive system.

d) reproductive system.

e) immune system.

0 EASY 32 HARD 100

3. Which of the following statements, if true, best supports the suggestion that endocrine systems in living things might have evolved earlier than nervous systems?

a) Nervous systems are more complex than endocrine systems, so they obviously must have evolved later.

b) Endocrine systems can provide all of the same functions that nervous systems provide.

c) The oldest organism we know of was capable of producing hormones.

d) Plants have endocrine systems, but they don't have nervous systems.

e) Simple animals that lack brains do have endocrine systems.

0 EASY 48 HARD 100

4. Human sex hormones are which type of biological molecules?

a) enzymes

b) lipids

c) proteins

d) carbohydrates

e) nucleic acids

0 EASY 48 HARD 100

5. Injecting a female prairie vole with a molecule that blocks oxytocin will have which of the following effects?

a) The female will experience uterine contractions.

b) The female will not bond with the male that she mates with.

c) Male voles will reject the female and any of her offspring.

d) The female will be unable to produce offspring.

e) All of the above are correct.

0 EASY 63 HARD 100

6. All except one of the following pairs link a human gland with the hormone(s) it secretes. Which pairing is incorrect?

a) adrenal cortex: cortisol

b) pancreas: insulin

c) testis: testosterone

d) ovary: estrogen and progesterone

e) thyroid: oxytocin

0 EASY 27 HARD 100

7. What important issue has arisen about the evidence that castrating sex offenders is effective in stopping their criminal behavior?

a) Only males have been tested; we can't get a clear picture unless females are tested too (by having their ovaries removed).

b) The evidence on repeat offenders was self-reported by the castrated offenders themselves.

c) In some cases, castration actually increased the tendency to re-offend.

d) Only males from a particular culture and country were castrated, so it's hard to extend the evidence to other cultures.

e) In 100% of the cases, castration effectively stopped repeat offenses. The evidence is clear.

0 EASY 60 HARD 100

8. Which of the following is not associated with increased levels of estrogen resulting from hormone replacement therapy?

a) increased risk of heart disease

b) increased longevity

c) increased risk of endometrial cancer

d) increased risk of ovarian cancer

e) increased levels of cell division

0 EASY 86 HARD 100

9. Blood samples from an individual who has fasted for 24 hours would have:

a) high levels of insulin and low levels of glucagon.

b) high levels of both insulin and glucagon.

c) high levels of glucagon and standard levels of insulin.

d) low levels of both insulin and glucagon.

e) high levels of glucagon and low levels of insulin.

10. Many female bodybuilders take steroids (testosterone) to increase their muscle mass and achieve their desired body shape. The consequences can include all of the following except:

a) increased competitiveness and aggressiveness.

b) increased hair growth.

c) decreased performance on tests of object memory.

d) increased concentration.

e) decreased fertility.

0 EASY 37 HARD 100

11. If a high dose of time-released testosterone is given to a male songbird, which of the following is not a likely effect?

a) He would obtain a larger territory than males with lower levels of circulating testosterone.

b) He would get into many fights with larger males.

c) He would have a longer life span than most other males.

d) He would enjoy increased reproductive success.

e) He would develop a greater muscle mass than other males of similar body weight.

Ch26

How do animals reproduce?

26.1 Reproductive options (and ethical issues) are on the rise.

26.2 There are costs and benefits to having a partner: sexual versus asexual reproduction.

26.3 Fertilization can occur inside or outside a female's body.

Male and female reproductive systems have important similarities and differences.

26.4 Sperm are made in the testes.

26.5 There is unseen conflict among sperm cells.

26.6 *This is how we do it:* Can males increase sperm investment in response to the presence of another male?

26.7 Eggs are made in the ovaries (and the process can take decades).

26.8 Hormones direct the process of ovulation and the preparation for gestation.

Sex can lead to fertilization, but it can also spread sexually transmitted diseases.

26.9 In fertilization, two cells become one.

26.10 Numerous strategies can help prevent fertilization.

26.11 Sexually transmitted diseases reveal battles between microbes and humans.

Human development occurs in specific stages.

26.12 Early embryonic development occurs during cleavage, gastrulation, and neurulation.

26.13 There are three stages of pregnancy.

26.14 Pregnancy culminates in childbirth and the start of lactation.

Reproductive technology has benefits and dangers.

26.15 Assisted reproductive technologies are promising and perilous.

Red-and-green macaws (*Ara chloroptera*) generally mate for life.

Reproduction and Development

From two parents to one embryo to one baby

How do animals reproduce?

Parent and child King Penguin.

26.1 Reproductive options (and ethical issues) are on the rise.

Thirty-year-old Diane Blood wanted to have a child with her husband, Stephen. And then the High Court in England ruled that she could not. The problem: her husband was dead.

Stephen Blood had died two years earlier, but when he was in a coma, his wife asked the doctors to freeze a sperm sample from him, which they did, the day before he died. The Human Fertilisation and Embryology Authority, however, ruled that because he hadn't given written consent, his sperm could not be used.

Q Should frozen embryos be divided up like other marital assets at divorce?

This case illustrates a type of ethical and legal quandary that is becoming increasingly common (FIGURE 26-1). Technology is making pregnancy possible in many situations where it previously was not. But along with many happy outcomes, numerous complex legal and ethical questions stem from reproductive technologies. For example, issues related to surrogacy, ownership of frozen embryos, and rights related to sperm or egg donation have yet to be sorted out in the courts.

In this chapter, we describe how men produce sperm and women produce eggs, as well as the process by which fertilization occurs (or does not occur), and we explore the early stages of development, following fertilization. We also investigate the perils and promise of a variety of assisted reproductive technologies.

But first, a resolution to the case of Diane Blood. She was eventually allowed to take the sperm to another country to be inseminated (after re-mortgaging her house to pay for the expensive legal battles). The law in England still prevents storage of sperm from a man without his written consent, but an appeals court made an exception in her case, and Diane Blood was able to conceive a child (and, three and a half years later, another child) with her (deceased) husband's sperm.

FIGURE 26-1 In the headlines: ethical issues and reproductive technologies.

TAKE HOME MESSAGE 26.1

» **Technology is making pregnancy possible in many situations where it previously was not, but is simultaneously giving rise to complex legal battles and ethical dilemmas.**

26.2 There are costs and benefits to having a partner: sexual versus asexual reproduction.

The term "reproduction," the process by which new organisms are produced from existing organisms, usually conjures images of a male and a female, producing offspring together. And for most animals, this is how it's done. But recall from Section 8.13 that there are two fundamentally different ways in which organisms can reproduce. While the vast majority of plants and animals reproduce sexually, all prokaryotes and many plant and animal species reproduce asexually—and some can reproduce in both ways.

Sexual reproduction involves two individuals contributing genetic material to produce offspring (FIGURE 26-2). The genetic material is contained in gametes, the reproductive cells. The male gamete is called a **sperm** (or sperm cell) and the female gamete is called an egg or **ovum** (*pl.* **ova**). Recall from Section 8.10 that the cellular division process known as meiosis produces gametes, cells that contain only half as many sets of chromosomes as other body cells (somatic cells). When the male and female gametes fuse in **fertilization,** the full chromosome number is restored. Both the production of gametes and the combination of genetic material from two individuals tend to increase genetic diversity among offspring. We explore sexual reproduction in greater detail in the remainder of this chapter.

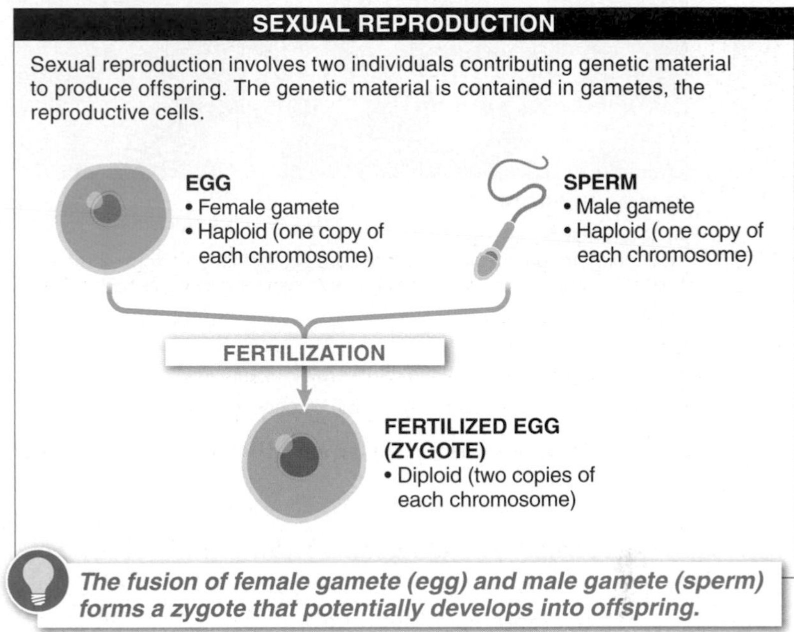

SEXUAL REPRODUCTION
Sexual reproduction involves two individuals contributing genetic material to produce offspring. The genetic material is contained in gametes, the reproductive cells.
EGG
• Female gamete
• Haploid (one copy of each chromosome)
SPERM
• Male gamete
• Haploid (one copy of each chromosome)
FERTILIZATION
FERTILIZED EGG (ZYGOTE)
• Diploid (two copies of each chromosome)
The fusion of female gamete (egg) and male gamete (sperm) forms a zygote that potentially develops into offspring.

FIGURE 26-2 **In sexual reproduction, two parents contribute genetic material to the offspring.**

An important feature of sexual reproduction is that it leads to offspring that differ genetically from each other and from either parent (see Figure 8-26). This genetic diversity can be an evolutionary adaptation, increasing fitness in changing environments. If an environment is gradually changing, individuals producing diverse offspring increase the likelihood that one of their offspring will be suited, genetically, to the new environment. There are, however, disadvantages to sexual reproduction. The two main drawbacks are that (1) finding a partner and mating can be time-consuming and challenging, and (2) each individual contributes only half of the alleles that its offspring will carry (FIGURE 26-3).

In contrast to sexual reproduction, **asexual reproduction** involves the production of offspring by a single individual without a contribution of genetic material from another individual. There are several types of asexual reproduction in animals, including parthenogenesis, budding, and fragmentation (FIGURE 26-4).

1. Parthenogenesis. In **parthenogenesis,** a female's egg develops into a new organism without having to be fertilized by a sperm cell. Some species, including the desert grassland whiptail lizard, are exclusively asexual, and all of the individuals in the species are female. Other species can reproduce either asexually or sexually, depending on environmental conditions; parthenogenesis allows them to utilize resources—particularly food—as quickly as possible. For example, in the spring, when food is plentiful, aphids produce diploid maternal cells that develop into normal adults without fertilization. When food is limited, they produce haploid eggs that must be fertilized before development. Hammerhead sharks and, occasionally, turkeys also have the ability to reproduce by parthenogenesis (although turkeys resulting from asexual reproduction tend to be less healthy).

Q Do female turkeys need males to reproduce?

SEXUAL vs. ASEXUAL REPRODUCTION

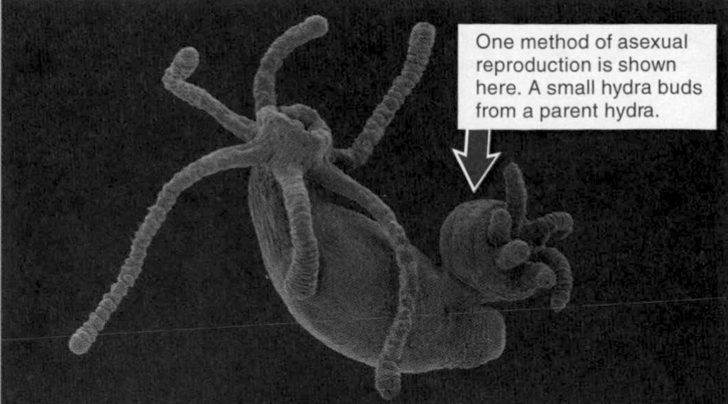

One method of asexual reproduction is shown here. A small hydra buds from a parent hydra.

SEXUAL REPRODUCTION

ADVANTAGES

- Offspring are genetically different from each other and from either parent—an evolutionary adaptation that can lead to increased fitness in changing environments.

DISADVANTAGES

- Finding a partner and mating can be difficult and time-consuming.
- Only half of an individual's alleles will be passed to its offspring.

ASEXUAL REPRODUCTION

ADVANTAGES

- Reproduction is fast and efficient.
- All of an individual's alleles are passed on to its offspring.

DISADVANTAGES

- With a changing environment, individuals producing genetically identical offspring are less likely to have offspring suited to the environment.

FIGURE 26-3 **Genetic variation versus efficiency: advantages and disadvantages of sexual and asexual reproduction.**

2. Budding. In budding, an offspring grows directly out of the body of its parent. Hydras, which are predatory cnidarians, reproduce by budding.

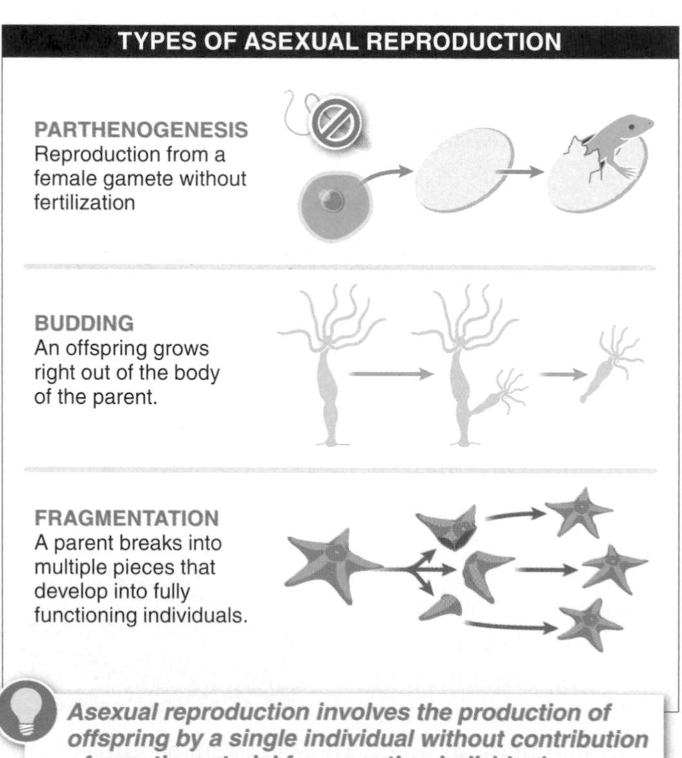

FIGURE 26-4 **Asexual reproduction: parthenogenesis, budding, and fragmentation.**

3. Fragmentation. In fragmentation, one individual breaks into multiple pieces, each of which develops into a fully functioning, independent individual. Fragmentation is seen among many species of flatworms, as well as some sea stars. In some sea stars, for example, if even a tiny part of one arm breaks off, it can develop into a complete individual.

Asexual reproduction can be fast and easy, because it involves only a single individual. And if an organism's environment is stable, it is beneficial for offspring to carry all of the genes that their parent carried. If an environment is changing, however, asexually reproducing organisms may be at a disadvantage.

TAKE HOME MESSAGE 26.2

» Organisms can reproduce sexually or asexually—or both. Sexual reproduction, which leads to offspring that are genetically different from each other and from both parents, occurs in the vast majority of plant and animal species. Asexual reproduction, which can be fast and efficient, leads to offspring genetically identical to the parent; it occurs in all prokaryotes and in many plant and animal species.

26.3 Fertilization can occur inside or outside a female's body.

Two general strategies for fertilization have evolved: **external fertilization,** in which the sperm and egg unite outside the male's and female's bodies, and **internal fertilization,** in which the sperm are deposited directly in the female's reproductive tract and unite with an egg inside her body.

The first vertebrates evolved in the oceans, where it was possible for females to release batches of eggs right into the water. Males could then release sperm into the water, and fertilization could take place. Many aquatic invertebrates—including sea urchins and clams—along with most fishes and amphibians use external fertilization to bring sperm and eggs together.

As we learned in Chapter 13, vertebrates' colonization of the land opened up a huge number of new niches but presented one big problem. External fertilization just doesn't work well on land. First, gametes cannot be moved around (and thus toward each other) without water. Second, and even more important, gametes quickly dry out. For these reasons, there was strong selective pressure on land-colonizing vertebrates for a new method to evolve: internal fertilization.

Used by nearly all terrestrial animals, internal fertilization solves the problem of gamete desiccation (drying out) by having the males deposit their sperm directly inside the females. This deposit is usually accomplished by the male placing his reproductive organ within the female's reproductive tract, in the act of **copulation.** Inside the female's body, there is sufficient moisture for the sperm to remain viable until fertilization can occur.

Once an egg is fertilized by a sperm, the fertilized egg is called a **zygote.** Following the first division into two cells (and continuing until approximately 8 weeks of development in humans), it is called an **embryo.** The offspring must, at some point in its development, leave the female's body. There are three different strategies for this (FIGURE 26-5).

1. Oviparity, a strategy in which the fertilized egg moves outside the body, where most embryonic development continues. The embryo is nourished by nutrients in the egg's yolk, and a live offspring emerges from the egg. This developmental strategy occurs in birds, most fishes, amphibians, reptiles, insects, and spiders.

2. Ovoviviparity, a less common strategy, in which most embryonic development takes place inside an egg, with the embryo nourished by the egg's yolk, but the egg itself remains in the female's body until hatched or near

EMBRYONIC DEVELOPMENT

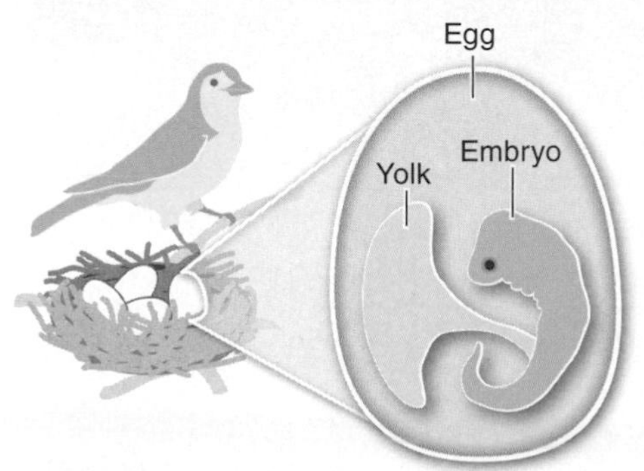

OVIPARITY
- Embryonic development: in egg outside the mother's body
- Nutrient source: egg yolk
- Examples: all birds; also some fishes, amphibians, reptiles, insects, and spiders

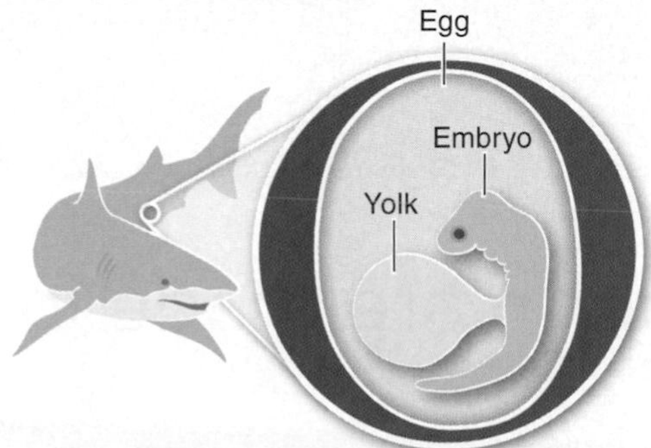

OVOVIVIPARITY
- Embryonic development: in egg inside the mother's body until hatching
- Nutrient source: egg yolk
- Examples: some sharks and other fishes, amphibians, reptiles, and invertebrates

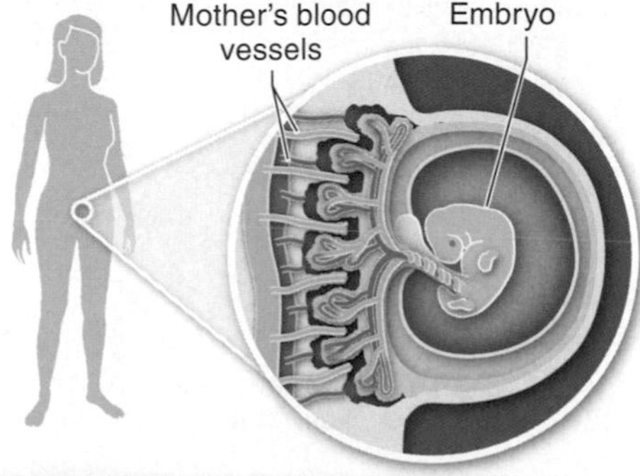

VIVIPARITY
- Embryonic development: inside mother until live birth
- Nutrient source: mother's bloodstream
- Examples: nearly all mammals; also some fishes, amphibians, and reptiles

FIGURE 26-5 **Three strategies for protecting and nourishing a developing embryo.**

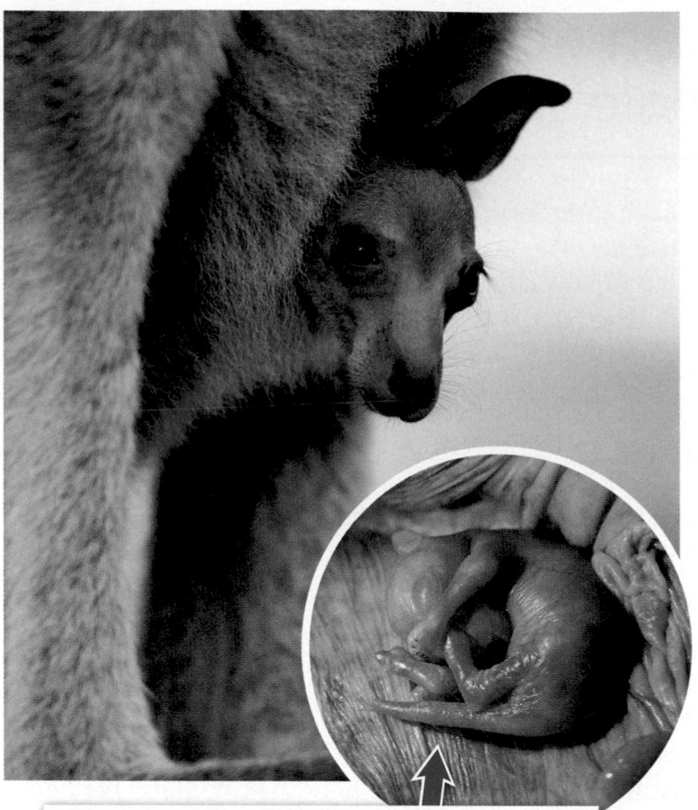
Kangaroo offspring emerge from the womb after about 33 days, blind, hairless, and just a few centimeters long. They continue developing (for many months) in the mother's protective pouch.

FIGURE 26-6 The kangaroo is a viviparous animal.

hatching. This strategy occurs in many aquatic organisms, including sharks and some other fishes, and in some species of amphibians, reptiles, and invertebrates.

3. Viviparity, a strategy in which the embryo develops inside the mother, nourished by nutrients in her blood, and a live offspring is born. This strategy occurs in all mammals (including marsupials; FIGURE 26-6) and among some reptiles, amphibians, and fishes.

TAKE HOME MESSAGE 26.3

» Sexual reproduction requires fertilization, which occurs externally or internally. The developing embryo can be nourished by yolk within an egg (that may or may not be retained within the female's body during development) or, remaining in the mother's body, by nutrients in her blood.

26.4–26.8

Male and female reproductive systems have important similarities and differences.

From Here to Eternity, 1953.

26.4 Sperm are made in the testes.

Beginning at puberty and continuing until death, men produce sperm, often more than 100 million per day. The process of sperm production, called **spermatogenesis,** is similar among most mammals and requires 9–10 weeks in humans. In this section, we examine the male reproductive system and the production of sperm and semen, a fluid expelled at ejaculation that usually contains sperm. We begin with the male reproductive structures and the role that each plays in the process of sperm formation and fertilization (FIGURE 26-7).

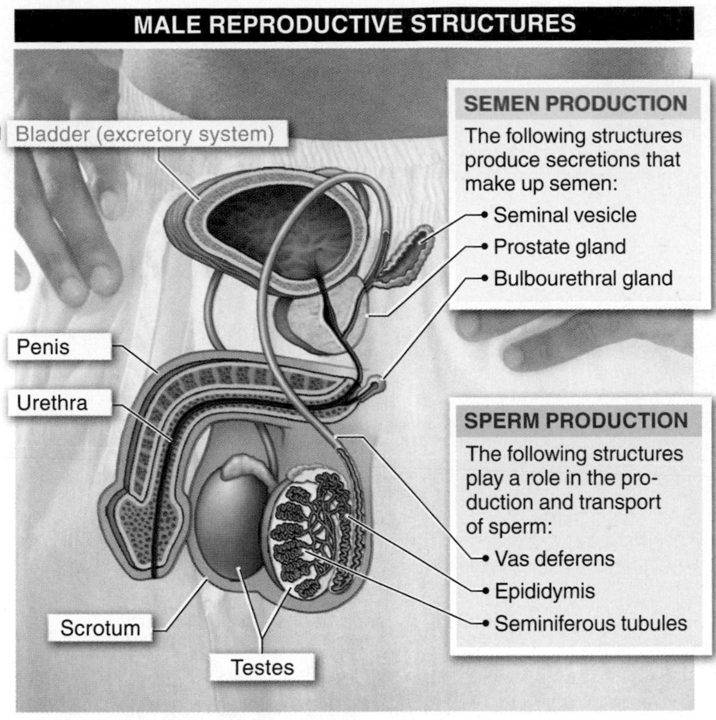

FIGURE 26-7 **The structures of the male reproductive system.**

Male Reproductive Structures

A male has just two reproductive structures, the penis and the scrotum. The **penis** has three columns of tissue—one along each side and a third underneath—that can become engorged with blood, causing erection and making copulation possible. In most animal species, including dogs, walruses, and most primates other than humans, a bone (known as a baculum) is present in the penis and contributes to its stiffness. The **scrotum,** generally on the outside of the body, is a sac containing the two **testes** (*sing.* **testis;** also called testicles), the site of sperm production. Each testis is made up of highly coiled **seminiferous tubules,** lined with cells, called spermatogonia, that are the site of sperm production. It is in cells between the seminiferous tubules that **testosterone,** the principal male sex hormone, and other androgens are produced. Testosterone stimulates sperm production.

Connected to the seminiferous tubules is the **epididymis,** a 15- to 20-foot-long coiled tube, in each testis, where sperm mature. The epididymis in each testis is linked to a **vas deferens,** a tube of smooth muscle tissue that passes from the testis into the body. The vas deferens from each testis connects to a single ejaculatory duct, which continues into the urethra, a duct passing through the penis and through which semen and urine are expelled. Sperm make up 1% to 5% of the semen volume in an ejaculate.

Besides the seminiferous tubules, three other structures produce secretions that make up semen. These include the **prostate gland,** located just below the urinary bladder, which secretes into the urethra a milky, basic (as opposed to acidic) fluid containing enzymes and sperm nutrients that makes up just under one-third of the volume of the ejaculate. A pair of **seminal vesicles** secrete into the semen nutrients for the sperm, as well as substances that increase sperm motility and make the female reproductive tract more hospitable to the sperm. These secretions make up about 65% of the volume of the ejaculate. And finally, a pair of **bulbourethral glands,** located just below the urethra near the base of the penis, contribute the remaining 1% or so of the ejaculate, as well as a mixture of mucus and sugar that lubricates the tip of the penis prior to copulation.

Gametogenesis in Males

Recall from Sections 8.9 through 8.13 that gametes are produced by cells that undergo meiosis. The spermatogonia within the seminiferous tubules are diploid cells (FIGURE 26-8). Each spermatogonium divides by mitosis to produce two cells. One of these cells is another spermatogonium, so the male never runs out of a store of sperm-producing cells; the other is a **primary spermatocyte,** which undergoes meiosis, in the first step of sperm production. Each primary spermatocyte produces two cells in the first meiotic division. These two cells are called **secondary spermatocytes,** and they then complete the second meiotic division, each producing two spermatids. The four spermatids produced by each spermatogonium, as products of meiosis, are haploid. As each spermatid matures into a sperm cell, it moves from the seminiferous tubules to the epididymis, where, over the course of approximately 18 hours, the sperm become motile.

Each sperm cell consists of three primary parts: (1) the head region, containing the nucleus with its DNA, plus a cap-like **acrosome** containing enzymes that can break down the protective layers surrounding an egg; (2) the body region, which contains many energy-generating mitochondria; and (3) the tail, a flagellum that propels the sperm through the fluid in the female reproductive tract (FIGURE 26-9).

Sperm production is sensitive to numerous factors, particularly hormones and temperature. The optimum temperature for sperm production is approximately two degrees lower than body temperature. This is why, in most mammals, the testes hang outside the body, where their temperature can be controlled by moving them closer to

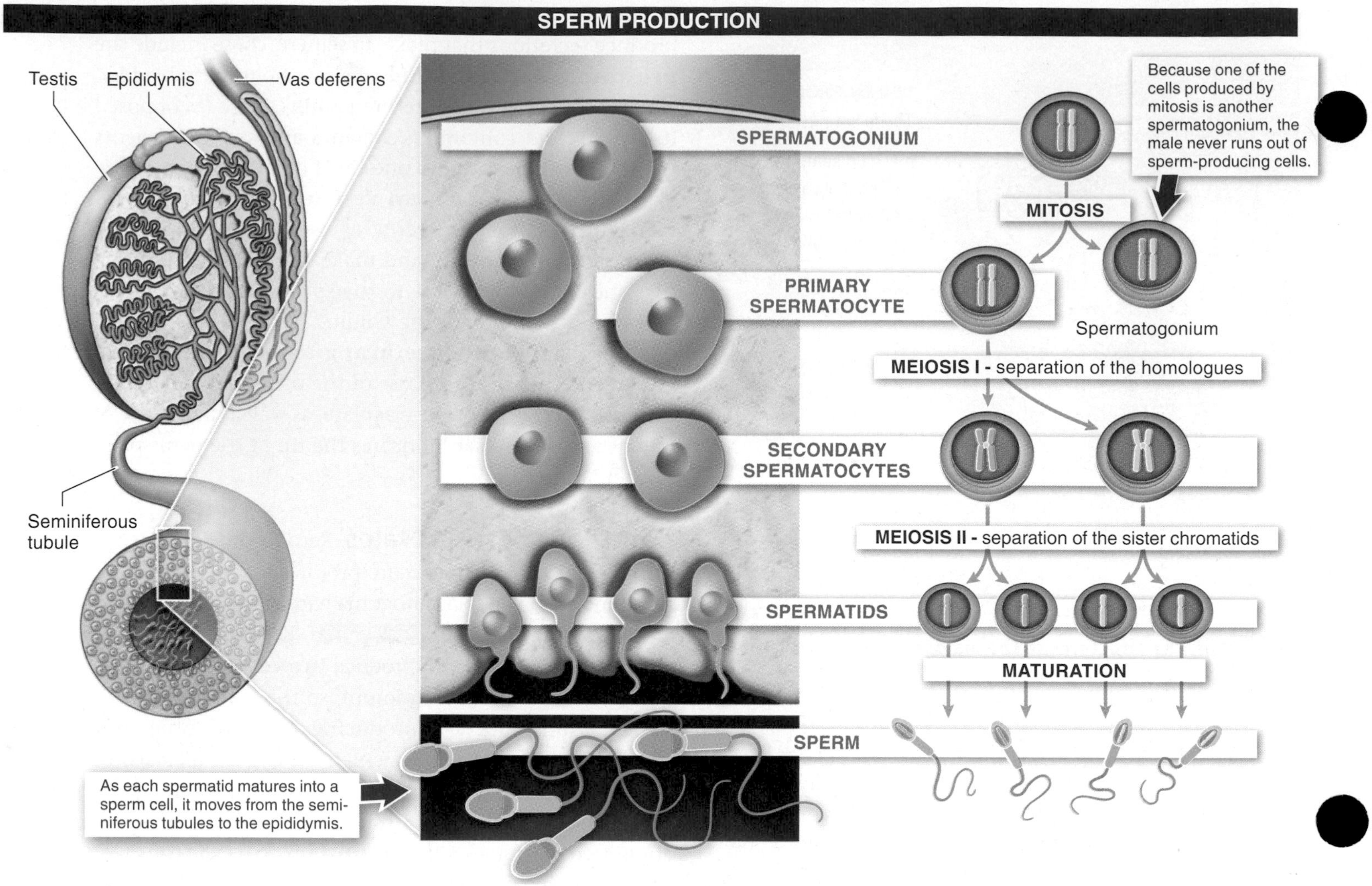

SPERM PRODUCTION
Testis
Epididymis
Vas deferens
Seminiferous tubule
SPERMATOGONIUM
Because one of the cells produced by mitosis is another spermatogonium, the male never runs out of sperm-producing cells.
MITOSIS
PRIMARY SPERMATOCYTE
Spermatogonium
MEIOSIS I - separation of the homologues
SECONDARY SPERMATOCYTES
MEIOSIS II - separation of the sister chromatids
SPERMATIDS
MATURATION
SPERM
As each spermatid matures into a sperm cell, it moves from the seminiferous tubules to the epididymis.

FIGURE 26-8 **Gametogenesis in males.** Sperm are continuously produced in the testis by meiosis.

STRUCTURE OF SPERM
HEAD
Contains the nucleus with DNA and is capped by an acrosome, which aids in breaking down the protective layers surrounding an egg
BODY
Contains many mitochondria, which generate energy
TAIL
A flagellum, which propels the sperm
Acrosome
Nucleus

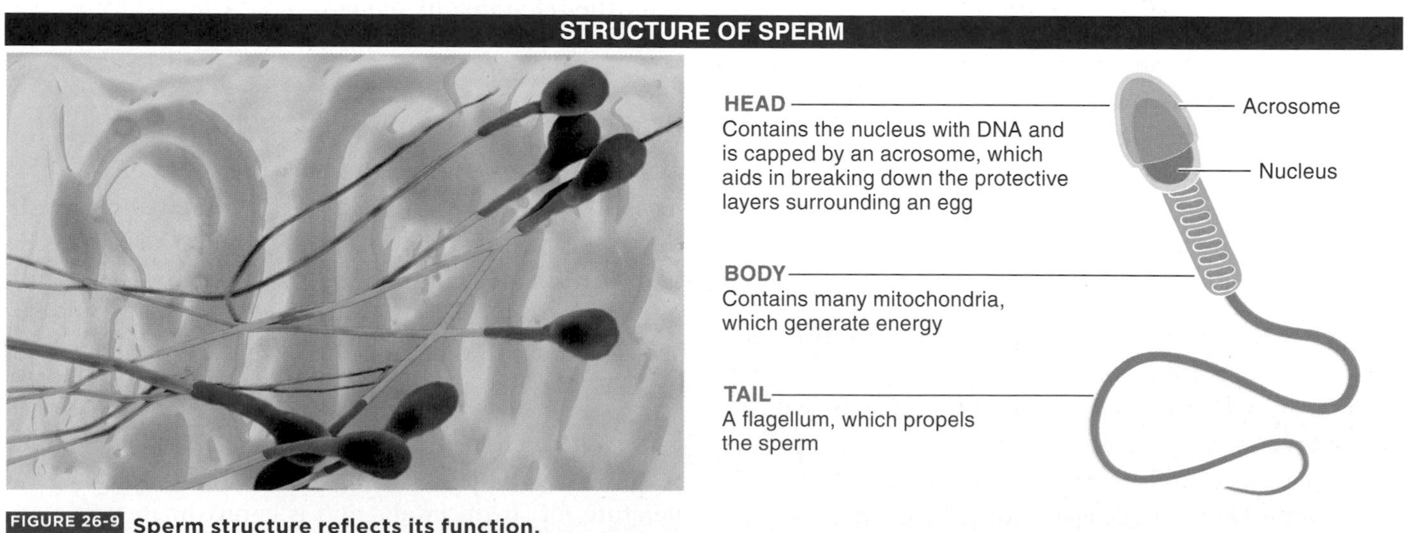

FIGURE 26-9 **Sperm structure reflects its function.**

FIGURE 26-10 **The pathway taken by sperm during ejaculation.**

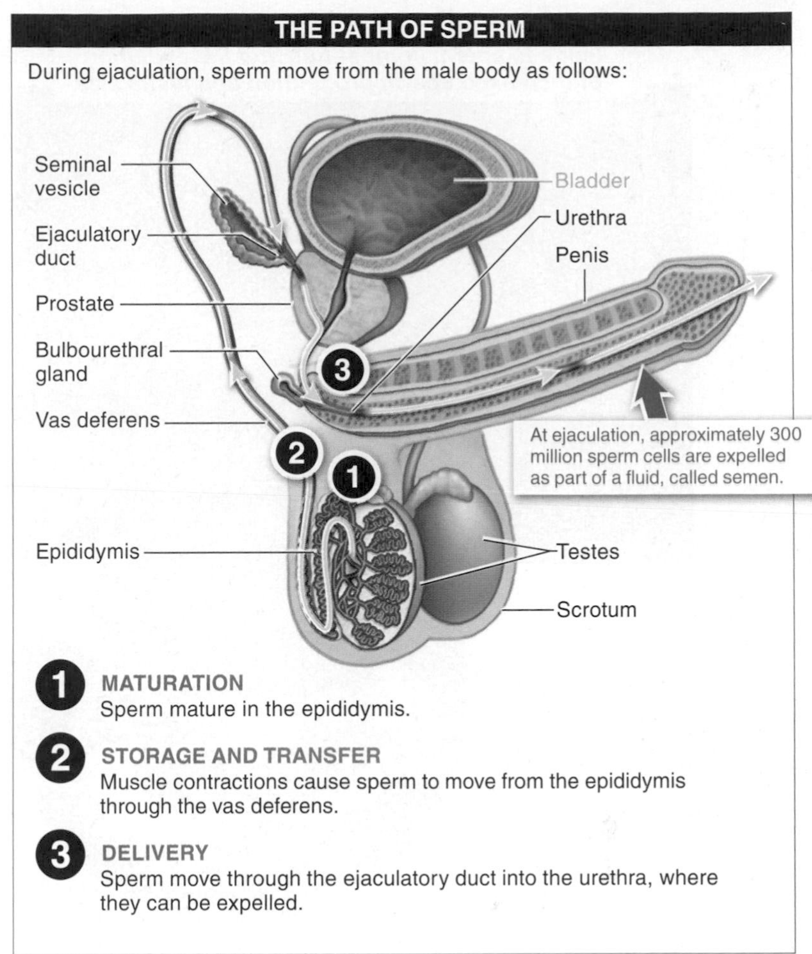

THE PATH OF SPERM
During ejaculation, sperm move from the male body as follows:
Seminal vesicle
Ejaculatory duct
Prostate
Bulbourethral gland
Vas deferens
Epididymis
Bladder
Urethra
Penis
At ejaculation, approximately 300 million sperm cells are expelled as part of a fluid, called semen.
Testes
Scrotum
1
2
3
1 MATURATION
Sperm mature in the epididymis.
2 STORAGE AND TRANSFER
Muscle contractions cause sperm to move from the epididymis through the vas deferens.
3 DELIVERY
Sperm move through the ejaculatory duct into the urethra, where they can be expelled.

Q How do hot tubs reduce a man's fertility?

or farther from the heat of the body. It's also why men exposed to hot tubs or hot baths for 30 minutes or more each week generally show signs of lower sperm count and motility. Fortunately, the condition can almost always be reversed by reducing exposure to the hot tubs or baths.

The path taken by sperm, from the male to the female, is as follows (FIGURE 26-10):

1. **Maturation.** Sperm mature in the epididymis, within each testis.
2. **Storage and transfer.** During ejaculation, sperm move from the epididymis (in each testis) through the vas deferens as a result of contractions of the muscular tissue of the vas deferens.
3. **Delivery.** Sperm moving from each vas deferens pass through the ejaculatory duct and into the urethra. During copulation, sperm are ejected from the urethra into the female reproductive tract. At ejaculation, approximately 300 million sperm cells are expelled as part of the semen.

TAKE HOME MESSAGE 26.4

» In adult men, sperm are continuously produced in the testes by meiosis. Semen—consisting of sperm cells and fluids that nurture and aid the sperm in fertilization—is ejaculated during copulation.

26.5 There is unseen conflict among sperm cells.

In *The Chimpanzees of Gombe: Patterns of Behavior,* Jane Goodall reported female chimpanzee copulatory rates of "an average of between five and six copulations per female per hour in the early morning, after which the rate dropped gradually to about two per hour in the midmorning, rose very slightly during the afternoon, and tapered off to one per hour in the evening." These observations were relevant to the realization reached by earlier researchers that if females (of any species, not just chimps) mate with more than one male, there may be competition among the males' sperm.

The idea of sperm competition, or "sperm wars" as this has been called, gives rise to several testable predictions. One simple prediction is this: when a female is likely to mate with more than one male, the males that produce more sperm are more likely to be successful at fertilizing

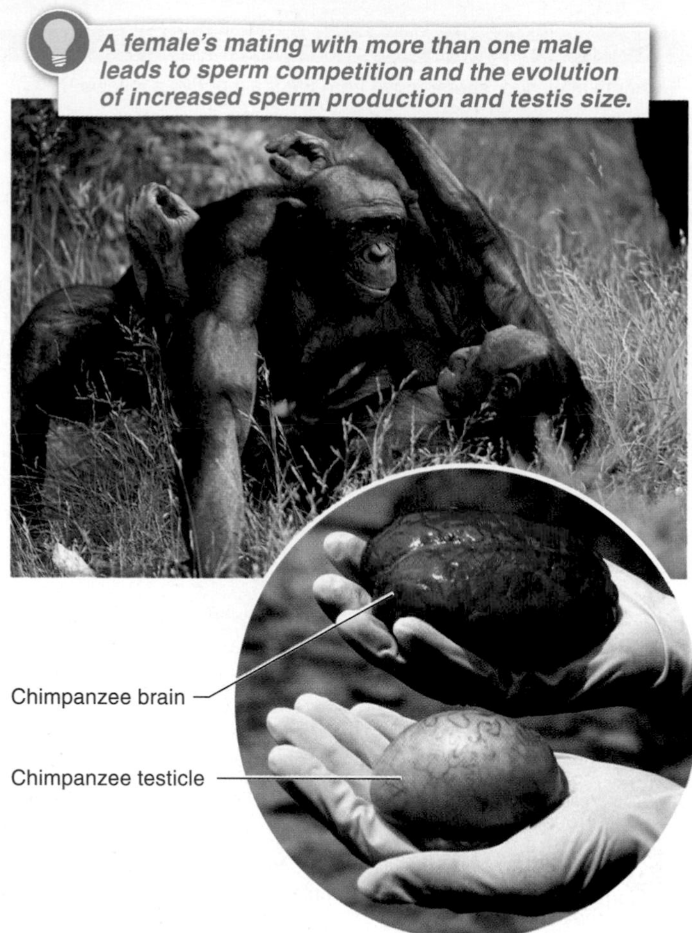

A female's mating with more than one male leads to sperm competition and the evolution of increased sperm production and testis size.
Chimpanzee brain
Chimpanzee testicle

FIGURE 26-11 **The evolutionary consequences of sperm competition among males.** The inset shows the size of a male chimp's brain (top) relative to one of its testicles.

Q A chimp's testicles are 15 times bigger, relative to body weight, than a gorilla's! Why?

the female's eggs. Some interesting observations support this prediction. For example, gorillas have golf-ball-sized testes, while chimpanzees' testes are closer in size to baseballs. Gorilla testicles account for only 0.02% of body weight, whereas chimp testicles account for 0.30%, 15 times as much. Why the huge difference? Gorilla groups are relatively small, and all females within a group mate with just one male, the dominant silverback. In contrast, as Jane Goodall noted, fertile chimpanzee females may have sex dozens of times a day with many different males. Consequently, the tiny testes of the gorilla are perfectly adequate, but for a chimp male to win these sperm competitions, he must produce significantly more sperm (FIGURE 26-11). Similarly, among fruit bats, males of species that live in large social groups have significantly larger testes than males of species that live in smaller groups.

Sperm competition has given rise to several other adaptations.

- Physical barriers to copulation, such as the mating plugs produced in many species (including some squirrels, mice, scorpions, and spiders) by the coagulation of semen, preventing other males' sperm from entering the female's reproductive tract.
- Toxic semen components, such as the protein in fruit fly semen that suppresses further mating by the female for several days (ensuring a male's paternity), as well as other fruit fly semen components that can incapacitate the sperm of other males (and even decrease the female's lifespan by about 10%).

Q Why do male fruit flies produce toxic semen?

- Genital morphology, such as the claspers and scrapers of dung flies, as well as the "shovel penis" of some dragonfly species that makes it possible for males to dislodge the sperm of other males that have already mated with a female.

TAKE HOME MESSAGE 26.5

» When females mate with more than one male, sperm competition occurs and can lead to a variety of adaptations, including increased sperm production and testis size, semen that can create a physical barrier to subsequent mating, toxic semen components, and penis morphology that aids in the displacement of rival males' sperm.

THIS IS HOW WE DO IT

Developing the ability to apply the process of science

26.6 Can males increase sperm investment in response to the presence of another male?

As we've seen, in a population of animals, when a female typically mates with more than one male, sperm competition occurs. This can result in the evolution of increased sperm production and testis size—as illustrated dramatically in chimps.

What impact does sperm competition have on the individual male? Researchers explored this question, beginning with the hypothesis "males respond to a risk of sperm competition with increased sperm investment."

Is this a testable hypothesis? How might we test it?

Like many hypotheses, this one may be a bit too broad. For example, it doesn't identify any particular species, or what qualifies as a "risk of sperm competition." In addition, for a hypothesis to be useful, it must be falsifiable. But it may be difficult to determine whether males are *unable to respond* to the risk of sperm competition with increased sperm investment, or whether males *are able to respond, but simply do not perceive a risk* of sperm competition.

These are issues to keep in mind when modifying the hypothesis and considering how to test it.

What is a good rule of thumb when crafting a hypothesis that will be testable?

Keep it simple. Try to control for as many potential sources of variation in the measure of interest—in this case, sperm investment—as possible. For example, males might differ from one another in their sperm investment even under similar conditions.

How could you control for the potential issue of male-to-male variability?

Controlling for differences among individuals (for a variety of traits) is a common goal when designing an experiment. An effective way to do this is to use a large number of individuals and to test each individual under both conditions. That way, regardless of whether a male has relatively high or low sperm investment (compared with other males), we can determine whether his sperm investment is *increased* when there is a risk of sperm competition. And that's the effect we're interested in.

Some researchers took this approach, but first they made a couple of refinements to the study. They decided to use one species—meadow voles (*Microtus pennsylvanicus*). They chose voles because, in the wild, litters are usually sired by more than one male, so males regularly experience sperm competition, and because the voles can be maintained under carefully controlled lab conditions. They also decided to simulate situations of sperm competition by exposing males to the odors of other male meadow voles. Can you suggest some other ways they might have simulated situations in which sperm competition occurred—and the pros and cons of those alternatives?

Their experimental approach was simple and straightforward.

The set-up The researchers paired a male vole with a sexually receptive female in two different contexts. In one, the voles were placed together in a cage that contained odors of another adult male vole (i.e., urine and feces from the other vole's cage). This was called the "risk of sperm competition" (RSC) context. In the other—the control context—the male and female were placed in a cage that did not contain odors from another male vole.

Ten males were used. Five were randomly assigned to the control context first, and then, 30 days later, to the RSC context. The other five were assigned to the RSC context first and, 30 days later, to the control context. This set-up minimized the possibility of any potential impact of the order in which the male voles experienced each context.

The measurements When the mating was completed, the researchers collected all of the semen from the female vole's reproductive tract, and determined the male's sperm investment—measured as the number of sperm present.

The results In all 10 voles, the sperm investment was greater in the RSC context. The average sperm investment in the control context was 98 (± 18) million sperm, and in the RSC context was 169 (± 18) million sperm—significantly higher. The researchers also noted that in the cage with odors from another male (RSC context), in every case the male vole first explored the cage, investigating the odor.

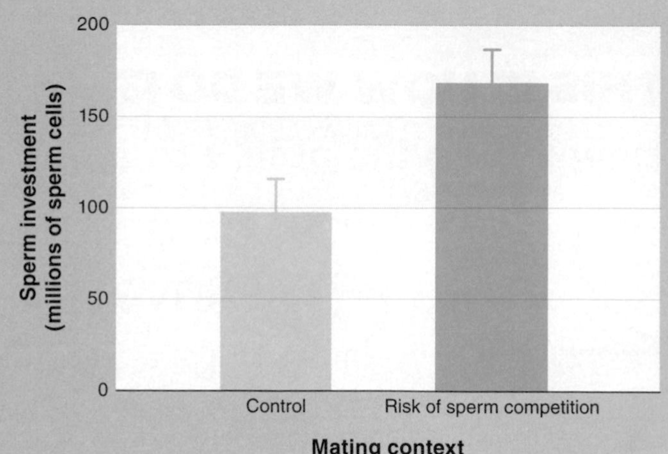

What other data might help evaluate the researchers' hypothesis?

In a subsequent experiment, the researchers found there was no increase in sperm investment when the male voles were exposed to the odors of males from another species. Why does using the odor of another species serve as a type of control in this study?

What can you conclude from these results?

The researchers concluded that male mammals may increase their sperm investment in response to an odor from another male of the same species. Could an individual's ability to alter his sperm investment in response to chemical cues in his environment have adaptive value? How might you explore this question further?

TAKE HOME MESSAGE 26.6

» Male voles respond to a risk of sperm competition with increased sperm investment. This physiological response can be triggered by odors present in the urine and feces of other males of the species.

26.7 Eggs are made in the ovaries (and the process can take decades).

From a genetic perspective, making eggs barely differs from making sperm. In the female gonads—the **ovaries**—diploid cells undergo meiosis to produce haploid eggs, and in the process, a great deal of variation is generated so that each haploid egg has a unique genetic makeup. When the haploid egg is fertilized by a sperm, the diploid condition is restored.

But the process of egg production differs from sperm production in several key ways. Eggs are produced in much smaller numbers; each egg is considerably larger than a sperm cell; and the production process can take decades rather than days, even though it begins at a much younger age. Let's explore the specifics, beginning with a description of the female reproductive system (FIGURE 26-12).

Female Reproductive Structures Externally, a female's clitoris and labia develop from the same embryonic tissue that, in males, produces the penis and scrotum. There is also a vaginal opening. Internally, the **vagina** is a tube-like chamber into which sperm are released during copulation. The vagina connects with the **uterus,** also called the womb, where an embryo develops throughout pregnancy. The lower, narrowest portion of the uterus is called the **cervix.** The lining of the uterus, rich with blood vessels, is the **endometrium,** where a fertilized egg implants and is nourished.

Connecting to the top of the uterus on both sides are the **Fallopian tubes,** or **oviducts.** The Fallopian tubes extend outward and are funnel-shaped near the end where the ovaries lie. Each ovary is about the size of a large olive.

Gametogenesis in Females The process of gametogenesis in females, called **oogenesis** (pronounced oh-oh-gen-eh-sis), starts in the ovaries while the female is still a fetus (FIGURE 26-13). Here, diploid cells called **oogonia** (*sing.* **oogonium**) multiply by mitosis. Each oogonium then begins meiosis, but stops at prophase I, at which point the cell is called a **primary oocyte** and is contained within a **follicle**—the small structure in which an egg will form. At birth, there are approximately one million follicles in a female's ovaries, each follicle containing a primary oocyte.

The primary oocytes remain in their hibernation-like state until puberty. At that point, periodic bursts of **follicle-stimulating hormone (FSH)** cause several primary oocytes to complete meiosis I. Unlike in the production of sperm, however, when the primary oocyte divides into two cells, although the pairs of homologous chromosomes split evenly, nearly all of the cytoplasm goes to just one of the daughter cells, now called a **secondary oocyte.** The other cell, known as a **polar body,** gets almost no cytoplasm and eventually disintegrates. At this point, meiosis again stops.

When females **ovulate,** generally a single follicle in one ovary ruptures, releasing the secondary oocyte—which still has not completed meiosis. The secondary oocyte, now called an egg, is swept by cilia into the Fallopian tube and carried down toward the uterus. If sperm have been deposited in the vagina during copulation, it is in the Fallopian tube that the sperm, swimming up from the vagina and through the uterus, are most likely to fertilize the egg. It is only *after* fertilization that a secondary oocyte is triggered

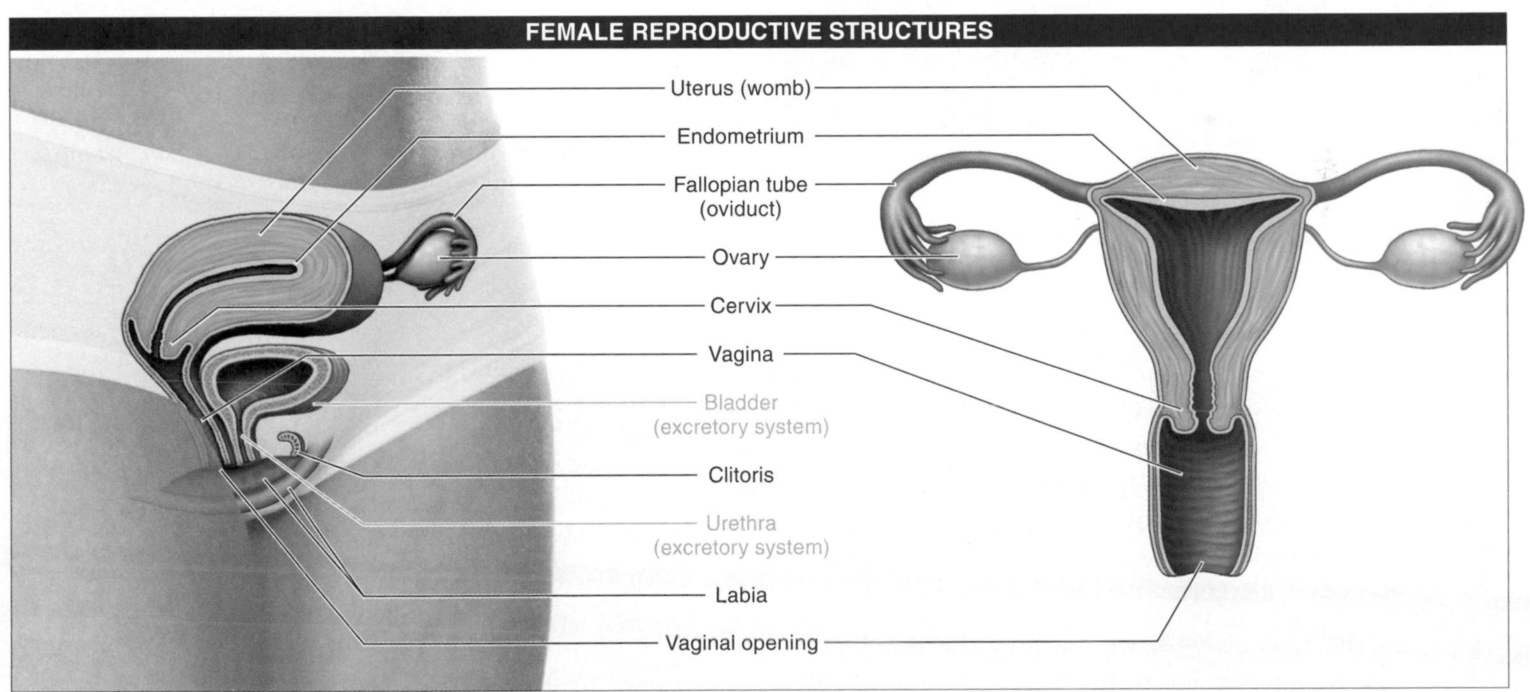

FIGURE 26-12 **The structures of the female reproductive system.**

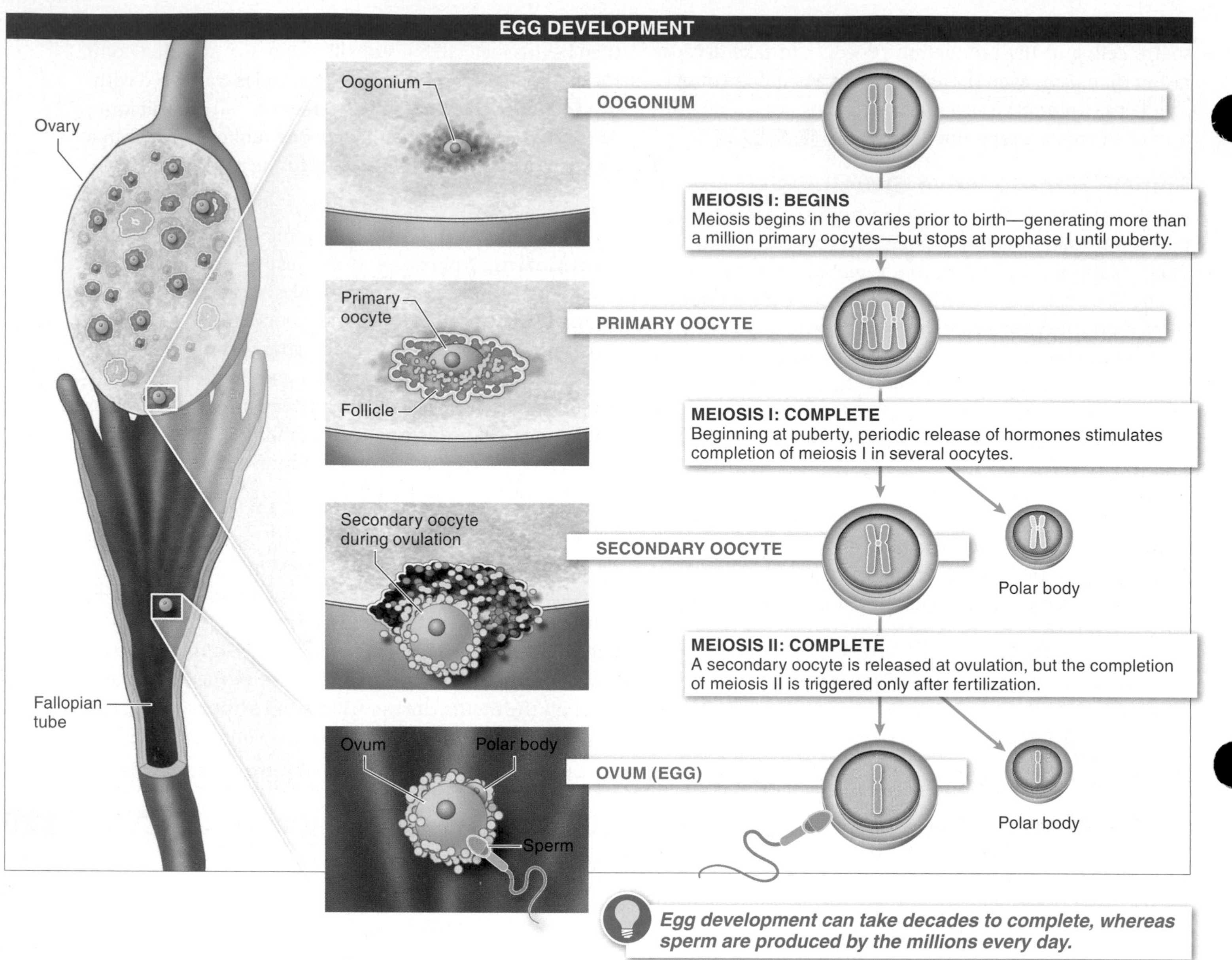
EGG DEVELOPMENT
Ovary
Fallopian tube
Oogonium
Primary oocyte
Follicle
Secondary oocyte during ovulation
Ovum
Polar body
Sperm
OOGONIUM
MEIOSIS I: BEGINS
Meiosis begins in the ovaries prior to birth—generating more than a million primary oocytes—but stops at prophase I until puberty.
PRIMARY OOCYTE
MEIOSIS I: COMPLETE
Beginning at puberty, periodic release of hormones stimulates completion of meiosis I in several oocytes.
SECONDARY OOCYTE
Polar body
MEIOSIS II: COMPLETE
A secondary oocyte is released at ovulation, but the completion of meiosis II is triggered only after fertilization.
OVUM (EGG)
Polar body
Egg development can take decades to complete, whereas sperm are produced by the millions every day.

FIGURE 26-13 Gametogenesis in females.

to finally complete meiosis—once again with an unequal division of cytoplasm. One of the resulting cells, the ovum, receives most of the cytoplasm, while the other receives almost none. This second polar body disintegrates just as the first polar body did. The haploid ovum now fuses with the haploid nucleus of the sperm, forming a diploid fertilized egg, called a zygote.

In the next section, we see how the development of follicles, preparation of the uterus for implantation, and ovulation are coordinated by hormone secretions.

TAKE HOME MESSAGE 26.7

» Genetically, the production of eggs barely differs from sperm production. In the ovaries, diploid cells begin to undergo meiosis, a process that continues in the Fallopian tubes following ovulation, producing genetically varied haploid gametes. A much smaller number of eggs than sperm are produced, however, and each egg is considerably larger than a sperm cell.

26.8 Hormones direct the process of ovulation and the preparation for gestation.

If you are female, here's something you may have noticed if you are living in a dormitory (or a prison): when women live in close proximity, their reproductive cycles become synchronized over time so that they menstruate and—perhaps less obviously—ovulate at approximately the same time.

This was first reported in 1971 by Martha McClintock, inspired by her own experiences and based on data from 135 female students living in a dormitory at Wellesley College in Massachusetts. Similar observations have been reported for other animal species, and subsequent studies implicate airborne chemicals, called **pheromones** (probably released from women's underarms), that can shorten or lengthen the reproductive cycle in other women (FIGURE 26-14).

It's not clear why such synchrony of reproductive cycles would occur, however. And other researchers have been critical of the original data analyses and have argued that the different lengths of women's cycles make it impossible for them to become truly synchronized. Results from recent studies are conflicting; several find evidence for menstrual synchrony, but others do not. The jury is still out on this hotly contested issue.

Hormones regulate the timing and development of egg production, called the **ovarian cycle,** which occurs approximately every 28 days. Hormones also regulate the **menstrual cycle,** during which the uterus prepares for the possible implantation and nurturing of a fertilized egg, and then sheds its lining when fertilization does not occur. We describe each of these cycles and how each influences the other (FIGURE 26-15).

In the normal monthly cycle, hormones direct the development of a follicle and the release of an egg (ovulation), while simultaneously preparing the uterus for

MENSTRUAL SYNCHRONY IN DORMS

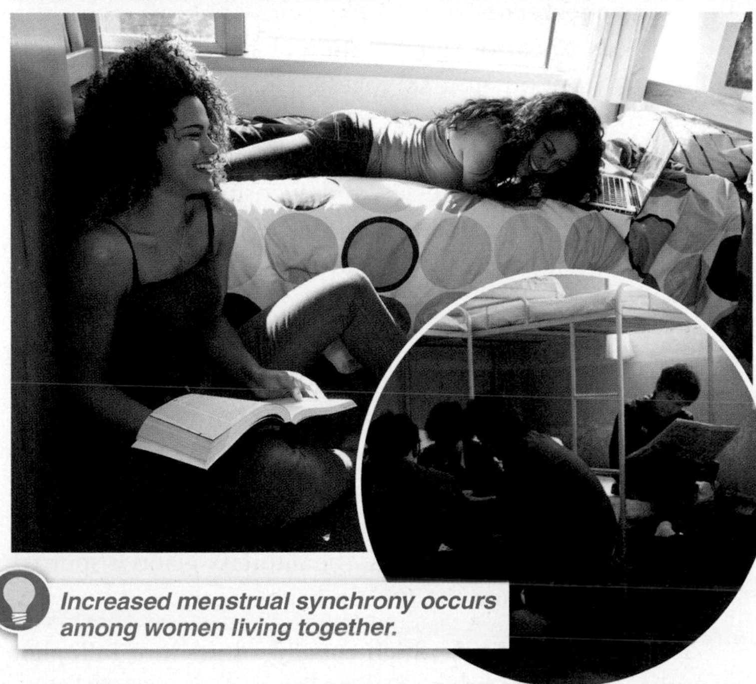

Increased menstrual synchrony occurs among women living together.

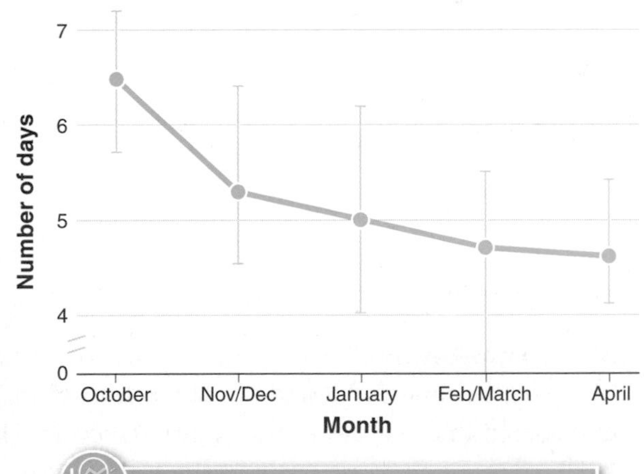

GRAPHIC CONTENT
Thinking critically about visual displays of data
Turn to p. 937 for a closer inspection of this figure.

FIGURE 26-14 **Synchronizing menstrual cycles.**

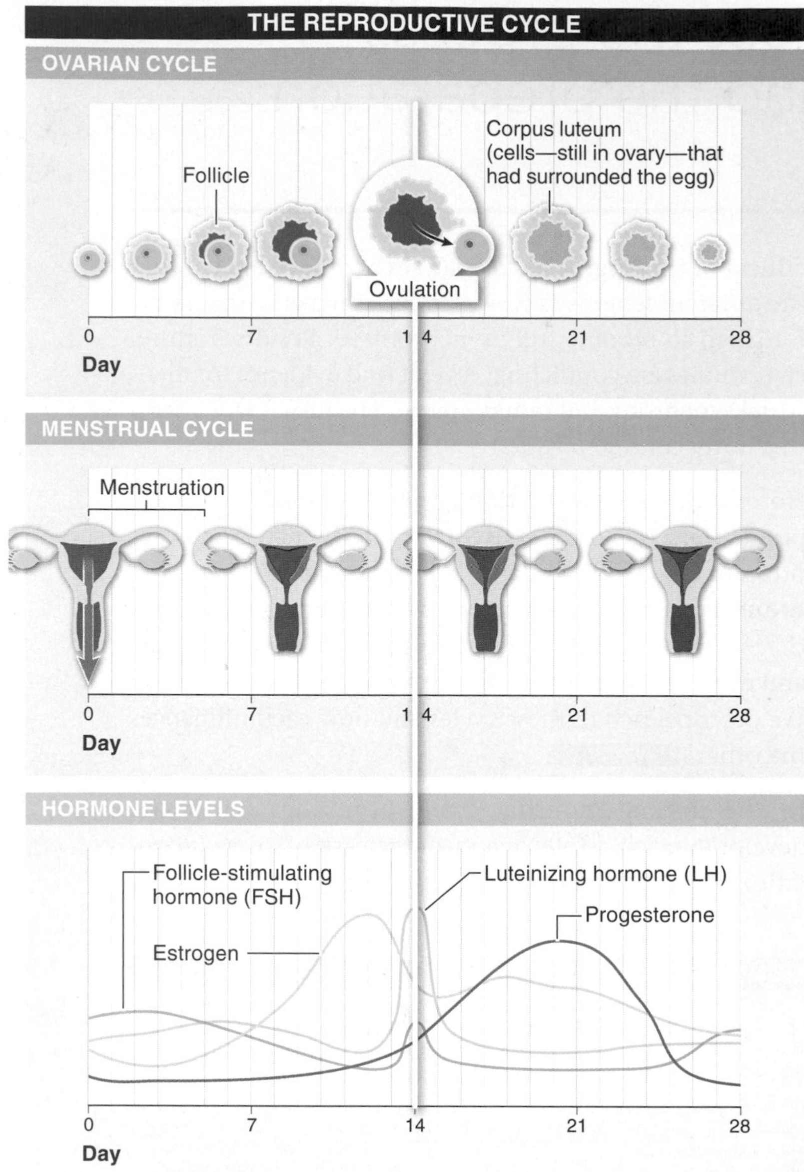

FIGURE 26-15 **The stages of the female reproductive cycle and associated hormonal changes.**

implantation. Let's follow the changes in hormone levels and their effects during the 28-day cycle (see Figure 26-15).

1. **Menstruation.** Traditionally, the first day of menstrual bleeding, a woman's "period," or **menstruation,** is considered the first day of the menstrual cycle. Three to five days of bleeding occur as the lining of the uterus (the endometrium) is sloughed off.
2. **FSH produced.** As the uterine lining sheds, the levels of **estrogen,** the chief female sex hormone, drop. Reduced levels of estrogen cause the pituitary gland to release follicle-stimulating hormone (FSH).
3. **Follicle develops.** FSH, just as its name indicates, causes a few follicles in the ovaries to grow and develop, although only one follicle reaches full maturity. Within this one follicle, the primary oocyte completes its first meiotic division and becomes a secondary oocyte.
4. **Estrogen produced.** As the follicles develop, they produce estrogen, gradually increasing the level of estrogen in the blood.
5. **LH released.** The high levels of estrogen trigger the release of a burst of luteinizing hormone (LH) and more FSH.
6. **Ovulation.** The burst of LH triggers ovulation, causing the secondary oocyte to erupt from the follicle and out of the ovary. This release occurs approximately halfway through the menstrual cycle, around day 14, and signals the end of the **follicular phase.**
7. **Progesterone produced.** As the second half, or **luteal phase,** of the reproductive cycle begins, the follicle cells that had surrounded the oocyte develop into a structure called the **corpus luteum** (Latin for "yellow body"). These cells begin secreting smaller amounts of estrogen, but increasing amounts of progesterone.
8. **Endometrium thickens.** Just as its name suggests, **progesterone** (*pro* = for; *gestare* = to bear) causes the body to prepare for gestation of an embryo, in case fertilization occurs. Progesterone's primary effect is to cause a thickening of the endometrium, or lining of the uterus. The endometrium becomes increasingly rich with blood vessels and deposits of glycogen that can nourish a developing embryo.

At this point, the process can go in one of two directions, depending on whether or not the egg is fertilized (**FIGURE 26-16**).

If the egg is not fertilized, the egg disintegrates and the corpus luteum degenerates, sloughs off, and is shed, along with the disintegrating egg. Abruptly removing this source of estrogen and progesterone causes the lining of the uterus to slough off, and menstruation begins. With continuing reduction in estrogen levels, the pituitary gland is spurred to release FSH, and the process begins again.

If the egg is fertilized, which usually occurs in a Fallopian tube, the zygote begins to develop and after several days begins to secrete **human chorionic gonadotropin (hCG).** This hormone prevents degradation of the

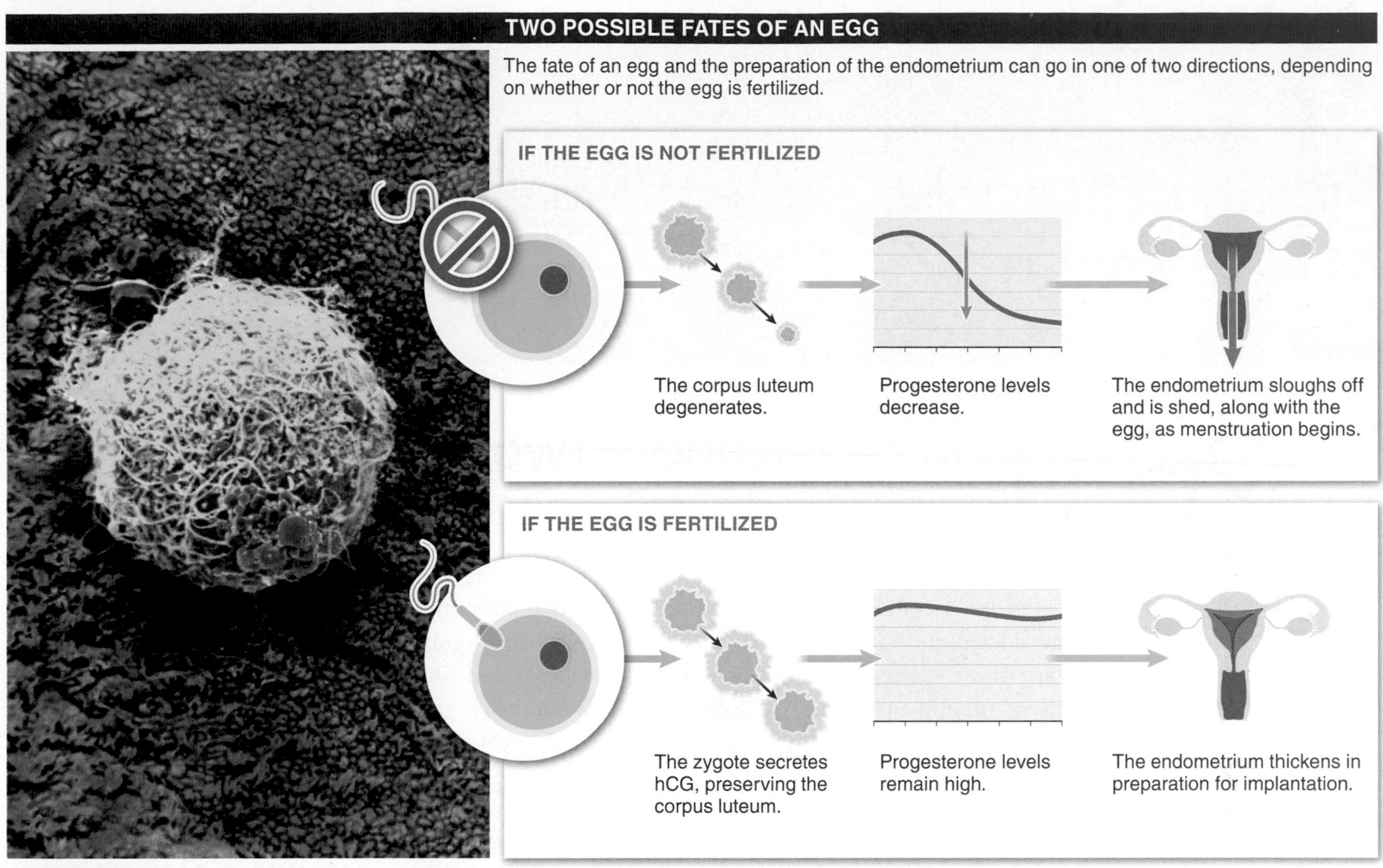

TWO POSSIBLE FATES OF AN EGG
The fate of an egg and the preparation of the endometrium can go in one of two directions, depending on whether or not the egg is fertilized.
IF THE EGG IS NOT FERTILIZED
The corpus luteum degenerates.
Progesterone levels decrease.
The endometrium sloughs off and is shed, along with the egg, as menstruation begins.
IF THE EGG IS FERTILIZED
The zygote secretes hCG, preserving the corpus luteum.
Progesterone levels remain high.
The endometrium thickens in preparation for implantation.

FIGURE 26-16 **An egg's fate can go in one of two ways.** The photo shows an egg moving down the oviduct.

corpus luteum, which continues to secrete progesterone, thereby maintaining the endometrium. In the next section, we investigate what happens next in development of the embryo.

Q What does the process of being an egg donor entail?

Females have about one million follicles, or potential eggs, when they are born, but most women ovulate fewer than 500 times over the course of their life. This leaves a lot of potential eggs lying around. In the normal course of events, these follicles just disintegrate, but in recent decades, modern medicine has made it possible for a woman to donate her eggs to another woman. The process involves several steps, beginning with suppression of the donor's cycle, followed by hormonal stimulation of one to two dozen follicles and extraction of the follicles from the ovaries.

Normal reproductive cycling continues from puberty until **menopause**—the cessation of ovulation and menstruation—usually between ages 45 and 55. It is not clear whether there is any evolutionarily adaptive value to menopause. In most species other than humans, females remain fertile throughout their entire adult life. Menopause may be a consequence of the relatively recent (on an evolutionary time scale) increase in human longevity. Regardless of the evolutionary explanation, oocyte depletion seems to influence the onset of menopause.

TAKE HOME MESSAGE 26.8

» In the ovaries, a cell within a follicle is stimulated to develop into a fertile egg by coordinated secretions of the hormones estrogen, FSH, and LH. Preparation of the uterus for implantation of a fertilized egg is coordinated by progesterone.

26.9–26.11

Sex can lead to fertilization, but it can also spread sexually transmitted diseases.

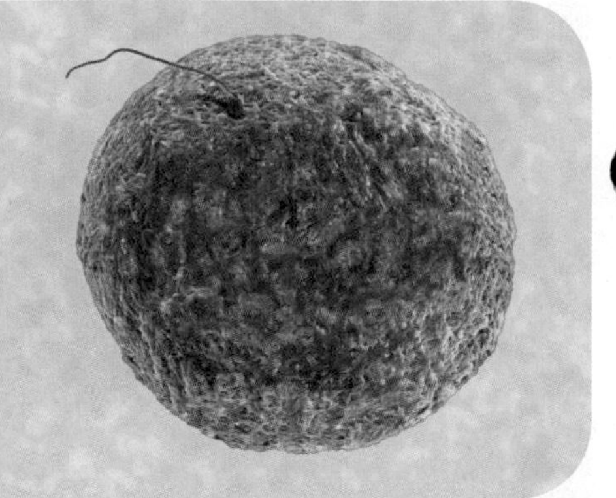

Sperm fertilizing an egg.

26.9 In fertilization, two cells become one.

For fertilization to occur, it's not enough for sperm cells to be in the same general location as an egg. In vertebrates, three separate steps take place: sperm activation, penetration, and fusion of the sperm and egg nuclei (FIGURE 26-17). And even before a sperm can take step one, it must find the egg. Recently, researchers discovered that sperm have receptors, like chemical sensors, that cause the sperm cells to swim toward a chemical attractant released by the egg. This attractant, in essence, says, "I'm over here!"

After ovulation, the egg is still surrounded by some smaller cells, called granulosa cells. These cells stand between the sperm and the egg, but they're not the only barrier. Between the granulosa cells and the egg's membrane, there is a glycoprotein layer called the **zona pellucida** (puh-loo-sih-duh). As sperm approach the egg, release of calcium by cells surrounding the egg seems to increase the sperms' swimming speed and tail movements, in a process known as sperm activation. A second phase of sperm activation

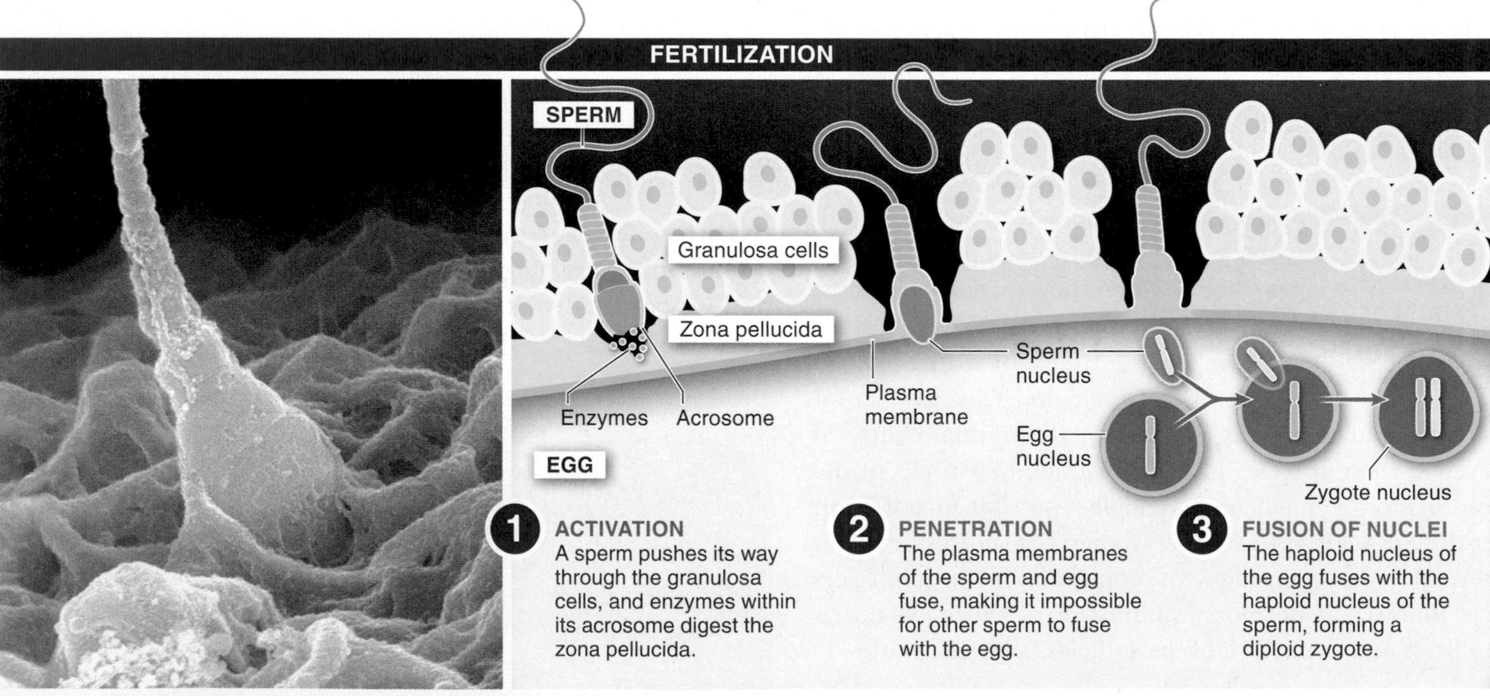

FERTILIZATION
SPERM
Granulosa cells
Zona pellucida
Enzymes
Acrosome
EGG
Plasma membrane
Sperm nucleus
Egg nucleus
Zygote nucleus
1 ACTIVATION
A sperm pushes its way through the granulosa cells, and enzymes within its acrosome digest the zona pellucida.
2 PENETRATION
The plasma membranes of the sperm and egg fuse, making it impossible for other sperm to fuse with the egg.
3 FUSION OF NUCLEI
The haploid nucleus of the egg fuses with the haploid nucleus of the sperm, forming a diploid zygote.

FIGURE 26-17 **Fertilization is the fusion of one sperm with an egg.**

occurs as a sperm cell pushes its way through the granulosa cells, and the acrosome (at the head of the sperm) dissolves and releases digestive enzymes that help the sperm make its way through the zona pellucida.

Penetration occurs as the sperm cell membrane fuses with the egg cell membrane, at which point the oocyte is activated. This activation and fusion of cell membranes has a couple of important consequences. First, it changes the egg's membrane in such a way that it is impossible for any other sperm to also fertilize the egg. After all, it would be a genetic disaster for the fertilized egg to have more than two sets of chromosomes. Second, activation of the oocyte triggers completion of its second meiotic division so that one haploid egg, the ovum, is formed, along with one smaller polar body. The latter disintegrates or is ejected from the egg.

Finally, the haploid nucleus of the egg fuses with the haploid nucleus of the sperm, creating a diploid cell—the zygote.

For many species, if the zona pellucida is stripped away (in the lab), it is possible for the egg to be fertilized by sperm from a different species. This suggests that something in the zona pellucida, perhaps recognition sites, ensures that the egg can be fertilized only by sperm from the same species.

TAKE HOME MESSAGE 26.9

» At fertilization, a sperm cell penetrates the protective zone around the egg, the egg blocks additional sperm entry, and the sperm and egg membranes fuse. The egg then completes its second meiotic division, and the haploid nuclei of the egg and sperm fuse, forming a diploid zygote.

26.10 Numerous strategies can help prevent fertilization.

Contraception, or birth control, is the attempt to prevent pregnancy. A wide variety of contraception strategies are used—with varying degrees of effectiveness—but the methods can be grouped into five general categories: (1) barrier methods, (2) hormonal methods, (3) intrauterine devices (IUDs), (4) "natural" methods, and (5) sterilization. FIGURE 26-18 summarizes the most commonly used methods of contraception and how they work.

Barrier Methods Barrier methods physically prevent conception by keeping sperm from coming in contact with an egg. The diaphragm or cervical cap is a dome-shaped piece of rubber placed in the vagina, blocking the cervix and thus preventing pregnancy by blocking sperm from reaching an egg. The use of spermicidal creams with a diaphragm or cervical cap (and with a condom as well) can increase the effectiveness of this contraceptive method. A condom, a thin rubber or natural membrane sheath placed on the penis or inside the vagina, covering the cervix, is another method of preventing contact between sperm and

METHODS OF CONTRACEPTION AND BIRTH CONTROL

METHOD	HOW IT WORKS
Barrier methods	• Physically prevent conception by keeping sperm from coming in contact with an egg • Include diaphragm, cervical cap, and both male and female condoms
Hormonal methods	• Alter the levels of hormones that influence the development and release of the egg; change cervical mucus, reducing sperm function or transport • Include birth control pills, hormone injections, patches, or implants, and emergency contraceptive pills ("morning-after" pills, such as Plan B and ella)
IUDs	• Small devices inserted into the uterus that change the conditions in the cervix and uterus to prevent pregnancy and/or inhibit the transit of sperm from the cervix to the Fallopian tubes • Include copper IUDs and hormonal IUDs
"Natural" methods	• Individuals refrain from sexual intercourse completely or during days when fertility is likely, based on analysis of the woman's menstrual cycle pattern, cervical secretions, and core body temperature
Sterilization	• Medical procedures that permanently alter the reproductive system to prevent the release of sperm or block the production or release of eggs or their movement down the Fallopian tubes

FIGURE 26-18 **Preventing pregnancy.**

egg. Condoms have the added benefit of being one of the only methods of contraception that also offer effective protection against sexually transmitted diseases, including HIV/AIDS. (See the next section for further discussion of this topic.)

Hormonal Methods Hormonal methods alter the levels of female hormones that influence the development and release of an egg. They may also change cervical mucus, thus reducing sperm function or transport. Available since 1960, birth control pills are among the most effective of all methods of contraception, with a failure rate of 1% to 8%; that is, among 100 women using birth control pills over the course of one year, there would be 1–8 pregnancies. (If taken every day at exactly the same time, the pills are more than 99% effective; the problem is that not all women are consistent in taking them.) Among 100 women of reproductive age using no contraception, there would be about 85 pregnancies during one year.

Q How do birth control pills work?

Two chief varieties of birth control pills, or oral contraceptives, are available, one that contains synthetic versions of estrogen and progesterone and one that contains only a synthetic progesterone. The estrogen-progesterone pill prevents ovulation by keeping estrogen levels just high enough that release of FSH by the pituitary gland is never triggered. As long as FSH is never released, eggs do not develop and ovulation does not occur. The progesterone component of the pill causes just enough development of the uterine lining that a plug of mucus forms at the connection between the vagina and uterus, blocking sperm from getting through.

Taking the pill at the same time every day is essential. If more than 24 hours go by between pills, the estrogen level in the body begins to drop; if it gets below a critical level, FSH release by the pituitary is triggered, which can lead to ovulation and the risk of pregnancy.

The long-term health consequences of using birth control pills include a slightly elevated risk of cardiovascular disease, particularly among women who smoke. However, because birth control users have a reduced risk of pregnancy—a condition that carries some significant health risks—and a reduced risk of ovarian and endometrial cancers, women's overall mortality risk is reduced while taking the pill (see Section 25.10).

Injections or capsule implants, just under the skin, of synthetic estrogen and progesterone, or progesterone only, are a slight variation on birth control pills. The prevention of pregnancy works in the same way, but with the added convenience of not having to take a pill every day. Some implants, in fact, need to be replaced only once every three years. These methods are slightly more effective than birth control pills, because much of the user error is eliminated.

Q What is the "morning-after" pill?

Not recommended as a long-term strategy for contraception, emergency contraceptive pills are available in several different types, containing different hormones or combinations of hormones, including progestin (a synthetic form of progesterone) and/or estrogen. Sometimes called "morning-after pills," and marketed under several different names (including Plan B and ella), emergency contraceptive pills can be effective when taken up to five days after sex. They have a failure rate of up to 25% and are recommended only for cases of rape or for emergencies in which other methods of contraception have failed.

Intrauterine Devices (IUDs) An IUD is a small, T-shaped plastic or metal device inserted by a doctor into the uterus (FIGURE 26-19). IUDs change the conditions in the

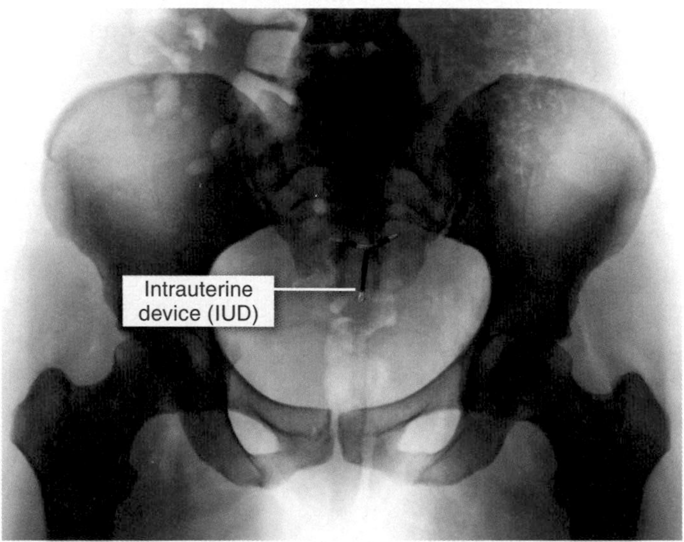

FIGURE 26-19 X ray of an intrauterine contraceptive device (IUD).

cervix and uterus to prevent pregnancy and/or inhibit the transit of sperm from the cervix to the Fallopian tubes. An IUD may contain copper or hormones (a type of progestin) and can be left in the uterus for three or four years. IUDs are very effective at preventing pregnancy and have a failure rate of just under 1%. Some women experience side effects from IUDs (including cramps and other pain) and may need to have the device removed.

"Natural" Methods "Natural" methods of contraception are those in which individuals refrain from sexual intercourse completely or on days when fertility is likely, based on analysis of the woman's menstrual cycle pattern, cervical secretions, and core body temperature.

Abstinence—not having sexual intercourse—is the most effective method of avoiding pregnancy, if practiced continuously. But, in reality, this is difficult to maintain, particularly in the context of a long-term relationship. Temporary abstinence during the times when conception is most likely (often called the "rhythm method"), while theoretically effective, has a failure rate as high as 25%, probably due to variation in the time of ovulation and the longevity of sperm in the female reproductive system (72 hours or more).

Sterilization Sterilization is the permanent alteration of the reproductive system to prevent the movement of eggs down the Fallopian tubes or the release of sperm. In a tubal ligation, the woman's oviducts are cut and tied so that eggs cannot reach the uterus. In a vasectomy, the man's vas deferens on each side is cut and tied so that sperm cannot reach the urethra, thereby causing the semen to carry no sperm. There are no side effects in either case. Each of these procedures, however, should be considered permanent (although, in rare cases, the procedures can be successfully reversed).

In contrast to the contraception methods described above, the drug RU486—a progesterone receptor antagonist—is used to induce abortion in the first seven weeks of pregnancy. Considered an anti-hormone drug, RU486 blocks progesterone receptors in the uterus, thereby causing the lining of the uterus to be sloughed off and ending a pregnancy. In addition to causing abdominal pain and cramping, RU486 can also have other adverse effects, including nausea, vomiting, and fever.

TAKE HOME MESSAGE 26.10

» Pregnancy can be prevented by many contraceptive methods, of five general types: barrier methods, hormonal methods, intrauterine devices (IUDs), "natural" methods, or sterilization.

26.11 Sexually transmitted diseases reveal battles between microbes and humans.

For many microbes, the human genitals and reproductive tract provide a desirable place to find shelter, nourishment, and opportunities for reproducing and dispersing. Unfortunately, these microbes can cause problems for humans in the form of **sexually transmitted diseases (STDs).** STDs produce symptoms of varying severity, from mild to extreme discomfort to sterility or even death. It is estimated that, worldwide, more than 300 million new cases of STDs occur each year.

Sexually transmitted diseases are caused by bacteria, viruses, fungi, protists, and even some arthropods. The organisms are passed from the mucous membranes (of the genitals, anus, or mouth) of one individual to those

SEXUALLY TRANSMITTED DISEASES			
CAUSE	EXAMPLES	SYMPTOMS	TREATMENT
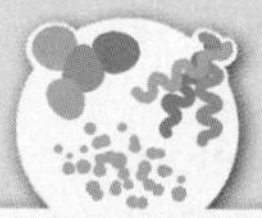 BACTERIUM	• Gonorrhea	Often none; sometimes painful urination, genital discharge, or irregular menstruation	Several antibiotics can successfully cure gonorrhea; however, drug-resistant strains are increasing.
	• Syphilis	Often no symptoms for years; eventual sores, skin rash, and if untreated, organ damage	Penicillin, an antibiotic, can cure a person in the early stages of syphilis.
	• Chlamydia	Often none; sometimes painful urination, genital discharge	Chlamydia can be easily treated and cured with antibiotics.
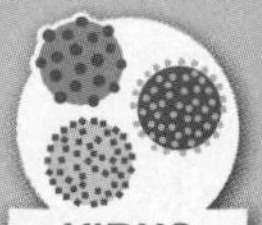 VIRUS	• HIV/AIDS	Initial symptoms range from none to flu-like; late stages involve severe infections and death	Currently no cure. Antiretroviral treatment can slow progression. Drug-resistant strains occur.
	• Genital herpes	Often none; outbreaks include sores on genitals, flu-like symptoms	Currently no cure. Antiviral medications can shorten and prevent outbreaks.
	• Human papilloma virus (HPV)	Often none; some types can lead to genital warts, others can cause cervical cancer	A vaccine prevents HPV, and is recommended for boys and girls 11–12. Warts and cancerous lesions can be removed.
	• Hepatitis	Initial symptoms range from none to fever, fatigue, and abdominal pain; can develop into a chronic disease that may cause liver cancer.	Antiviral drugs can help chronic disease; regular monitoring needed to identify liver damage or cancer.
 PROTIST	• Trichomoniasis	Painful urination and/or vaginal discharge in women; often no symptoms in men	Trichomoniasis can usually be cured with prescription drugs.
 FUNGUS	• Yeast infections	Genital itching or burning and/or vaginal discharge in women; genital itching in men	Yeast infections can usually be cured with antifungal suppositories or creams.
 ARTHROPOD	• Crab lice	Visible lice eggs or lice crawling or attached to pubic hair; itching in the pubic and groin area	Crab lice can be treated with over-the-counter lotions.

FIGURE 26-20 The most common STDs.

of another during sexual contact; sometimes they can also be transmitted by needles used for drug injections.

Some of the most common STDs, with their symptoms and treatments, are listed in FIGURE 26-20. Although most are curable with antibiotics, antifungal drugs, or anti-protozoan drugs, two characteristics of STDs make them nearly impossible to completely eradicate from a population: (1) their symptoms may be mild or completely absent at times, causing many people to unwittingly pass an infection to their partners, and (2) to prevent reinfection, both partners must be treated simultaneously. Furthermore, because most microbes have such high reproductive rates, populations of a microbe can evolve quickly and become resistant to existing drugs, reducing the long-term effectiveness of treatments. The treatment of STDs is one of the most pressing public health issues in the world today.

TAKE HOME MESSAGE 26.11

» Sexually transmitted diseases (STDs) are caused by a variety of organisms, including bacteria, viruses, protists, fungi, and arthropods. Worldwide, more than 300 million people are newly infected each year. The symptoms of STDs range from nonexistent to mild to extreme discomfort, sterility, or even death.

26.12–26.14

Human development occurs in specific stages.

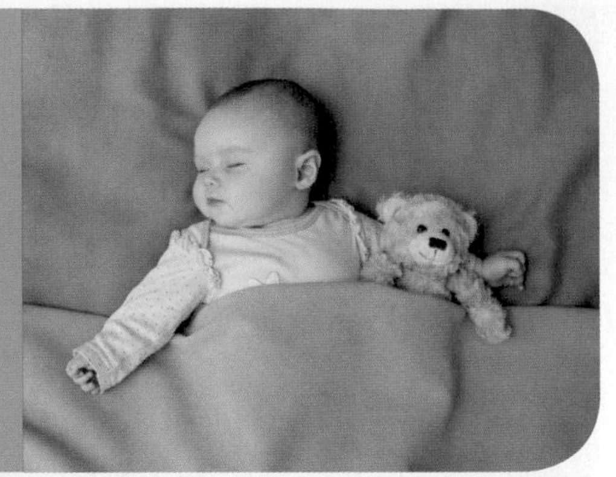

Multiple stages of development occur even before birth.

26.12 Early embryonic development occurs during cleavage, gastrulation, and neurulation.

How do you build a complete human, with specialized cells, tissues, and organs, when at fertilization there is just a zygote, a single cell that doesn't resemble a human at all? The field of developmental biology seeks to explain how a complex organism develops from a single cell. The genes in that initial cell and the environmental conditions in which the cell develops are critical. These factors interact as the development of humans and most other vertebrates proceeds in a carefully coordinated sequence of three stages: cleavage, gastrulation, and neurulation.

Cleavage The first phase, **cleavage,** is the early cell division, by mitosis, of the zygote, beginning shortly after fertilization (FIGURE 26-21). It starts about 30 hours after fertilization, as the zygote divides into two cells. After another 30 hours or so, it divides again, becoming four cells, and then again, through numerous divisions. Although the number of cells increases significantly during cleavage, there is little or no growth in overall size. Rather, the fertilized egg is partitioned into more and more cells that are smaller and smaller. These multiple cell divisions are fueled by nutrients that were in the egg. At the cleavage stage, it is difficult to distinguish between species as dissimilar as sea urchins and humans.

After about six days of continuous cell division, the cells form a hollow ball, called a **blastula**—the mammalian

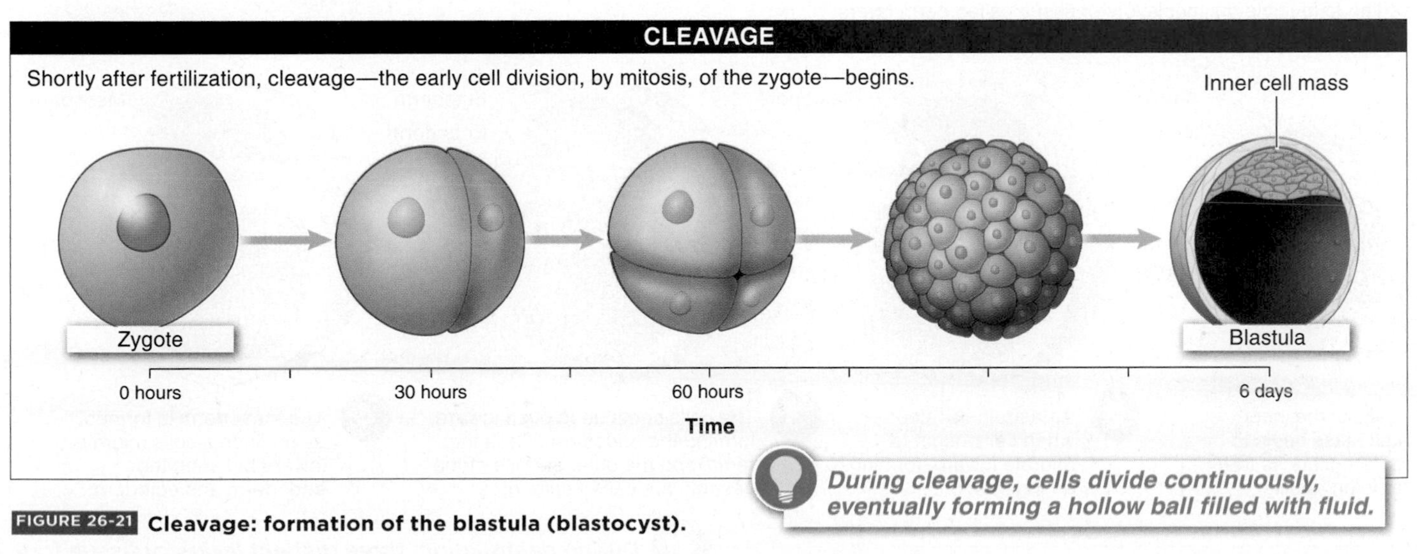

CLEAVAGE
Shortly after fertilization, cleavage—the early cell division, by mitosis, of the zygote—begins.
Inner cell mass
Zygote
Blastula
0 hours
30 hours
60 hours
6 days
Time
During cleavage, cells divide continuously, eventually forming a hollow ball filled with fluid.

FIGURE 26-21 **Cleavage: formation of the blastula (blastocyst).**

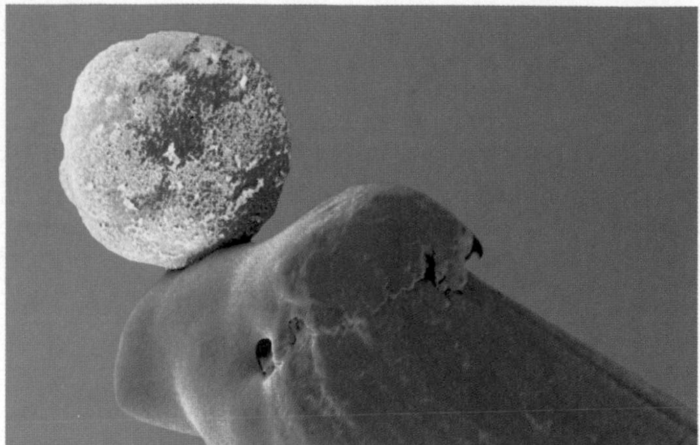

A human embryo in the blastocyst stage on the tip of a pin

FIGURE 26-22 **Tiny, but with tremendous potential.**

version is called a **blastocyst**—of approximately a thousand cells (FIGURE 26-22). The cells secrete fluid into the center of the ball, where there is an inner mass of cells that form the embryo. The outer cells also produce human chorionic gonadotropin (hCG), which keeps the corpus luteum from disintegrating. And the continued maintenance of the corpus luteum keeps progesterone levels high.

The blastocyst grows rapidly and forms membranes that surround and protect it. These include the **amnion,** which surrounds the embryo, and the **chorion,** which, along with the endometrium, forms the **placenta,** the structure that connects the developing embryo to the wall of the uterus (see Figure 26-26). The placenta is packed with blood vessels that bring nourishment to the embryo and remove wastes generated by the embryo.

Gastrulation In the second phase of development, **gastrulation,** three distinct **germ layers** of tissue form. Picture a beach ball—a hollow sphere. Now imagine making a fist and slowly pushing inward on the beach ball. This resembles the process of gastrulation. The entire mass of cells is called a gastrula (FIGURE 26-23), and the indentation is called the blastopore. The blastopore is located almost opposite the point at which the sperm entered the egg and, in vertebrates, becomes the anus.

Three distinct layers of tissue form during gastrulation. The cells in these layers have not yet differentiated (that is, they all look similar to one another), but they have become "determined," which means that the type of tissue they will become, their ultimate developmental fate, is irreversibly decided. Prior to determination, cells are referred to as "totipotent" and have the capacity to develop into any type of tissue in the body.

Here's a summary of what happens during gastrulation.

1. The blastocyst begins to fold inward, forming an indentation, the blastopore.
2. As the cells continue to push inward, they are called **endoderm,** and they form the lining of what will eventually become the digestive tract. Endoderm cells will also form the liver, the pancreas, and the

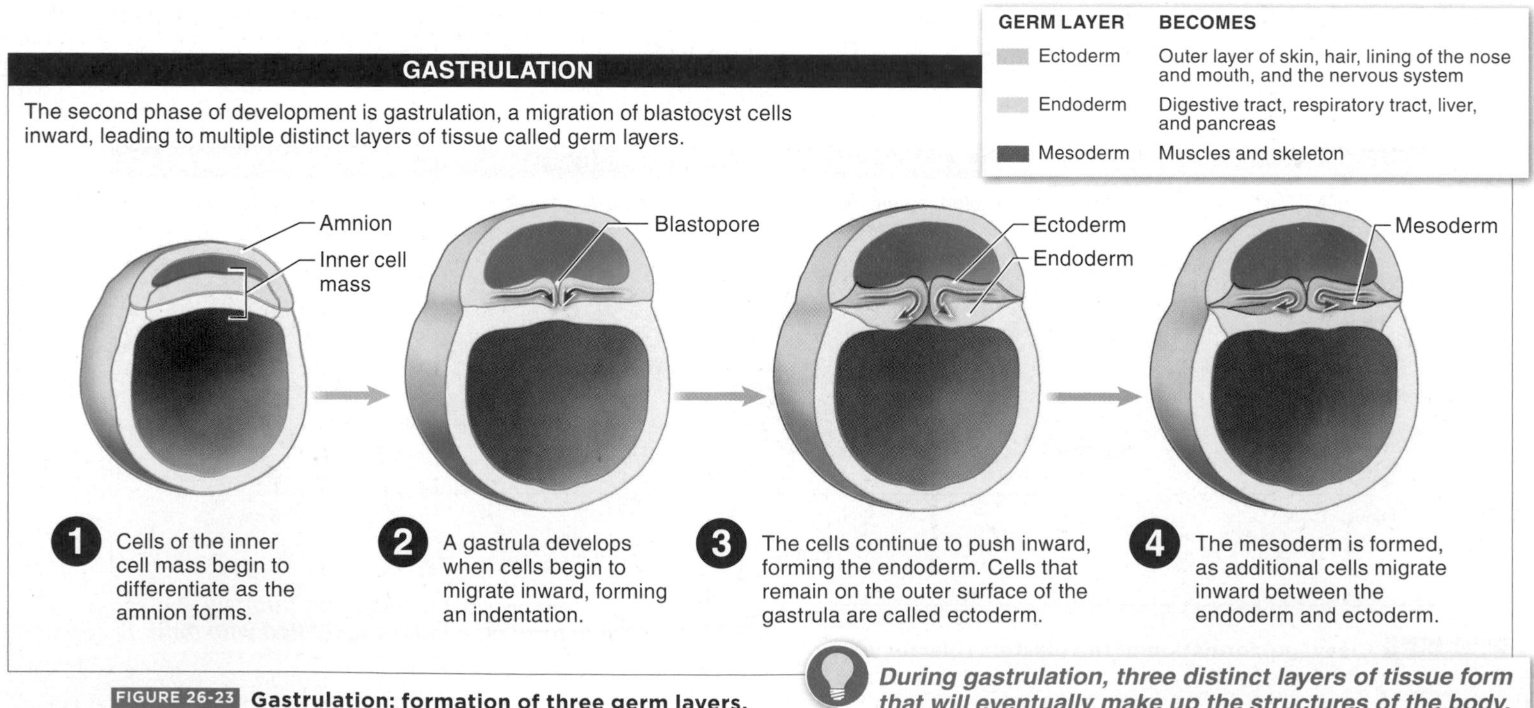

GASTRULATION
The second phase of development is gastrulation, a migration of blastocyst cells inward, leading to multiple distinct layers of tissue called germ layers.
GERM LAYER
BECOMES
Ectoderm
Outer layer of skin, hair, lining of the nose and mouth, and the nervous system
Endoderm
Digestive tract, respiratory tract, liver, and pancreas
Mesoderm
Muscles and skeleton
Amnion
Inner cell mass
Blastopore
Ectoderm
Endoderm
Mesoderm
1 Cells of the inner cell mass begin to differentiate as the amnion forms.
2 A gastrula develops when cells begin to migrate inward, forming an indentation.
3 The cells continue to push inward, forming the endoderm. Cells that remain on the outer surface of the gastrula are called ectoderm.
4 The mesoderm is formed, as additional cells migrate inward between the endoderm and ectoderm.
During gastrulation, three distinct layers of tissue form that will eventually make up the structures of the body.

FIGURE 26-23 **Gastrulation: formation of three germ layers.**

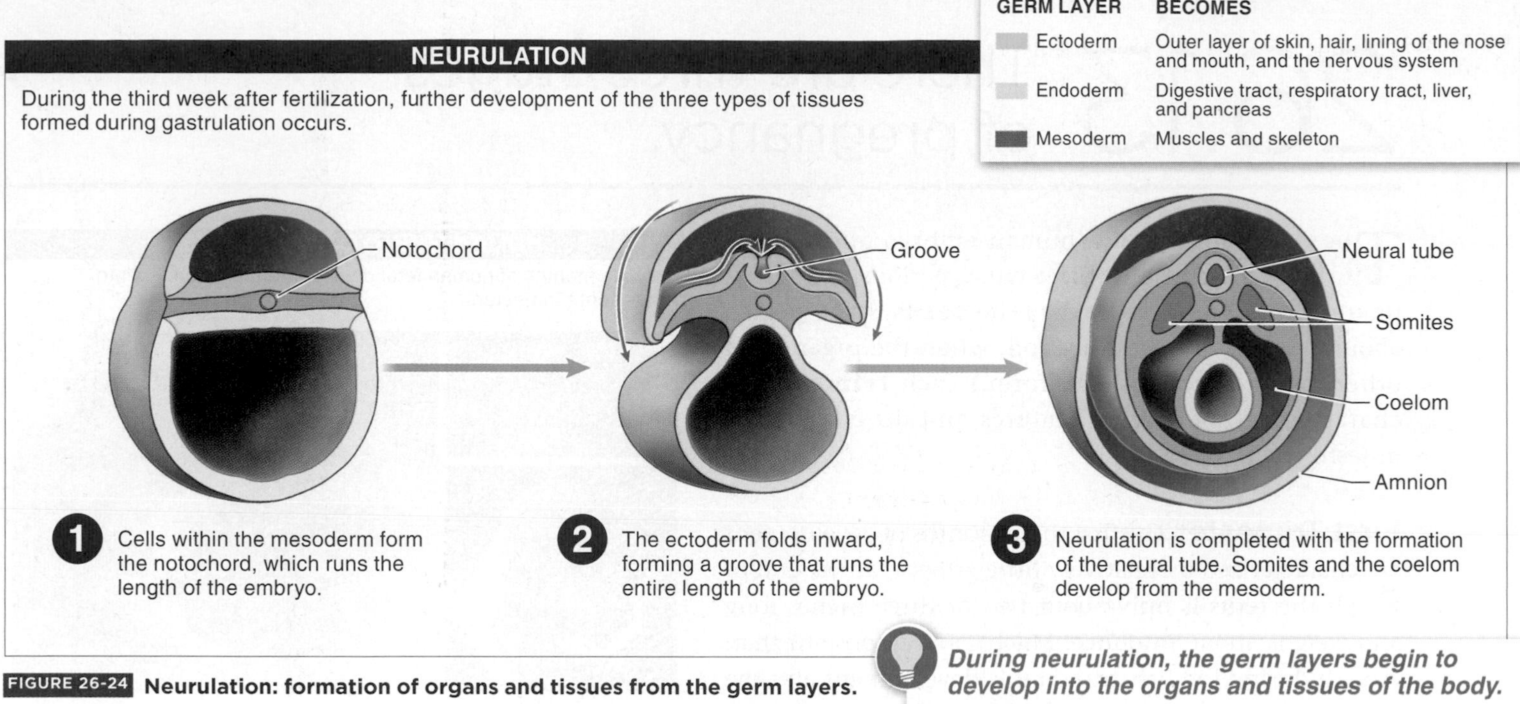

NEURULATION
During the third week after fertilization, further development of the three types of tissues formed during gastrulation occurs.
GERM LAYER
BECOMES
Ectoderm
Outer layer of skin, hair, lining of the nose and mouth, and the nervous system
Endoderm
Digestive tract, respiratory tract, liver, and pancreas
Mesoderm
Muscles and skeleton
Notochord
Groove
Neural tube
Somites
Coelom
Amnion
1 Cells within the mesoderm form the notochord, which runs the length of the embryo.
2 The ectoderm folds inward, forming a groove that runs the entire length of the embryo.
3 Neurulation is completed with the formation of the neural tube. Somites and the coelom develop from the mesoderm.
During neurulation, the germ layers begin to develop into the organs and tissues of the body.

FIGURE 26-24 **Neurulation: formation of organs and tissues from the germ layers.**

lining of the respiratory tract. At this point, the entire mass of cells is called a **gastrula.**

3. The cells that remain on the outer surface of the gastrula are called **ectoderm** and will ultimately form the outer layer of skin, the hair, the lining of the nose and mouth, and the nervous system.
4. Some additional cells migrate inward, between the endoderm and ectoderm, forming a third type of tissue, called **mesoderm,** which gives rise to the muscles and skeleton.

The specific details of gastrulation differ slightly from one species to the next. The general outcome, however—the formation of endoderm, mesoderm, and ectoderm—is the same throughout the vertebrates.

Neurulation In the third week of embryonic development, the three types of tissue formed during gastrulation begin to develop into the various organs and tissues. At this time, within the mesoderm and running the length of the embryo, just above the digestive tract, the notochord forms. The **notochord** is a flexible rod that develops in all chordates and gives support to surrounding tissue. In vertebrates, the notochord's function is ultimately taken over by the backbone.

Just above the notochord, ectoderm folds in for the entire length of the embryo, first forming a groove and then becoming a long hollow tube, called the **neural tube.** The process is referred to as **neurulation,** and the resulting neural tube ultimately develops into the entire nervous system, including the brain and the spinal cord (FIGURE 26-24).

As the neural tube is developing, blocks of mesoderm tissue, called somites, begin to form down the length of the embryo. These somites will form the vertebrae and the muscles. Next to the somites, spaces form that will become the body cavity, or **coelom** (see-lum), within which many of the organs will form. At this point, toward the end of the third week of development, the human embryo is only 2 millimeters (less than one-tenth of an inch) in length. Further development and specialization occur as some cells are programmed to die, in essence "sculpting" the body.

Other cells may be induced to develop into one particular cell type by neighboring cells that send signals, possibly proteins, to turn certain genes on or off.

TAKE HOME MESSAGE 26.12

» Soon after fertilization, cleavage takes place, and many rapid cell divisions occur without overall growth in size. Then, in gastrulation, a gut begins to form, along with three distinct germ layers with specific developmental fates. In neurulation, mesoderm forms a supporting rod called the notochord, and above that, an infolding of ectoderm forms a neural tube, which will become the brain and spinal cord.

26.13 There are three stages of pregnancy.

The development of a human embryo and fetus is usually divided into three equal periods of approximately three months each. (The **fetus** stage begins about 8 weeks after fertilization, when the organs and other major structures first form.) Each **trimester** is characterized by specific features and developmental milestones (FIGURE 26-25).

First Trimester The first three months of development are characterized by relatively little growth, at the end of which the fetus is only about two or three inches long and weighs about an ounce. Much more important than growth during the first trimester is development and the differentiation of cells into specialized types of tissues. Here are some of the chief milestones (times indicate how much time has passed since fertilization).

First month

30 hours: The first cleavage occurs.

60 hours: The second cleavage occurs. In both cleavages, there is no overall growth, just cell division.

6–7 days: The embryo reaches the uterus for implantation in the endometrium.

2nd week: Gastrulation occurs, and three distinct tissue layers develop. The placenta also forms, as two membranes grow around the blastocyst—the amnion around the embryo and the chorion around the amnion. The chorion ultimately branches out and surrounds the lining of the uterus and, with the endometrium, forms the placenta (FIGURE 26-26), a mass of tissue in which nutrients, respiratory gases, and waste products are transferred between the circulatory systems of the mother and the developing embryo. (Although there is a yolk sac in placental mammals, its size is reduced and the placenta takes over the yolk sac's role of nourishing the embryo.) The embryo is connected to the chorion by the body stalk, which will become the **umbilical cord.**

HUMAN EMBRYO DEVELOPMENT

The nine months of human fetal development are divided into three-month trimesters.

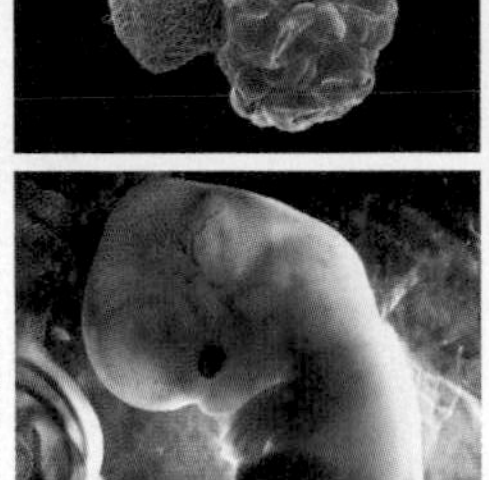

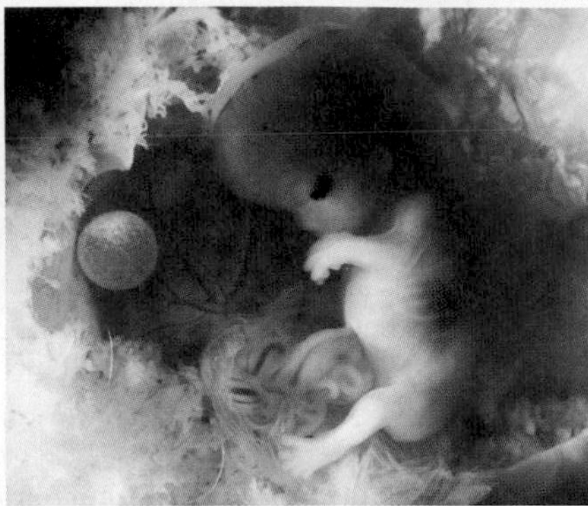

FIRST TRIMESTER: MONTHS 1–3
- Cells begin to differentiate into specialized types of tissues.
- Major organs and structures begin to form, including the eyes, heart, liver, pancreas, and gallbladder, as well as the limbs.

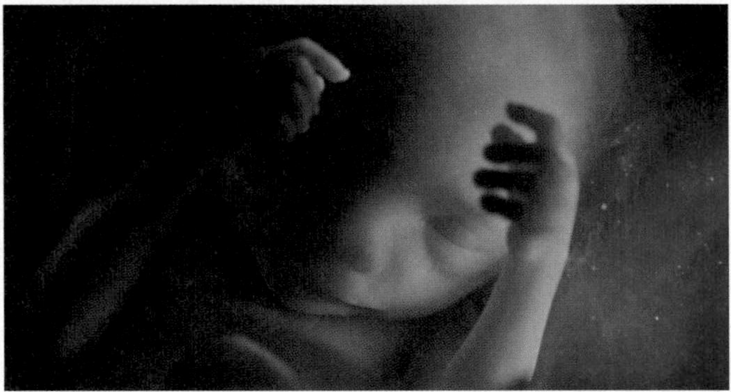

SECOND TRIMESTER: MONTHS 4–6
- Significant muscle and bone growth occurs, with less new development relative to the first trimester.

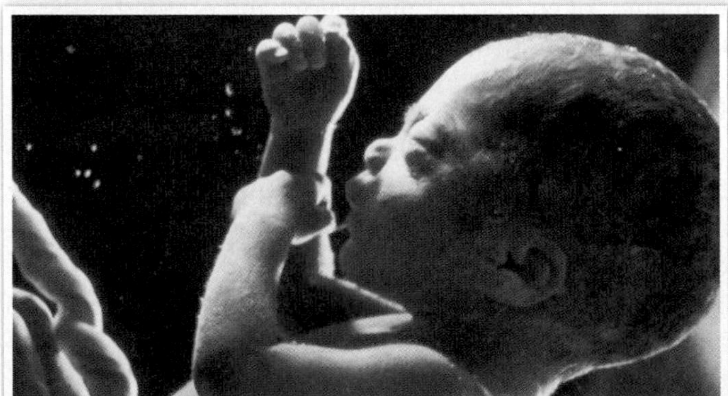

THIRD TRIMESTER: MONTHS 7–9
- Significant development of the nervous system takes place.

FIGURE 26-25 **Development of the human embryo and fetus.**

THE PLACENTA

The placenta is a mass of highly vascularized tissue in which nutrients, respiratory gases, and waste products are transferred between the mother and the developing embryo.

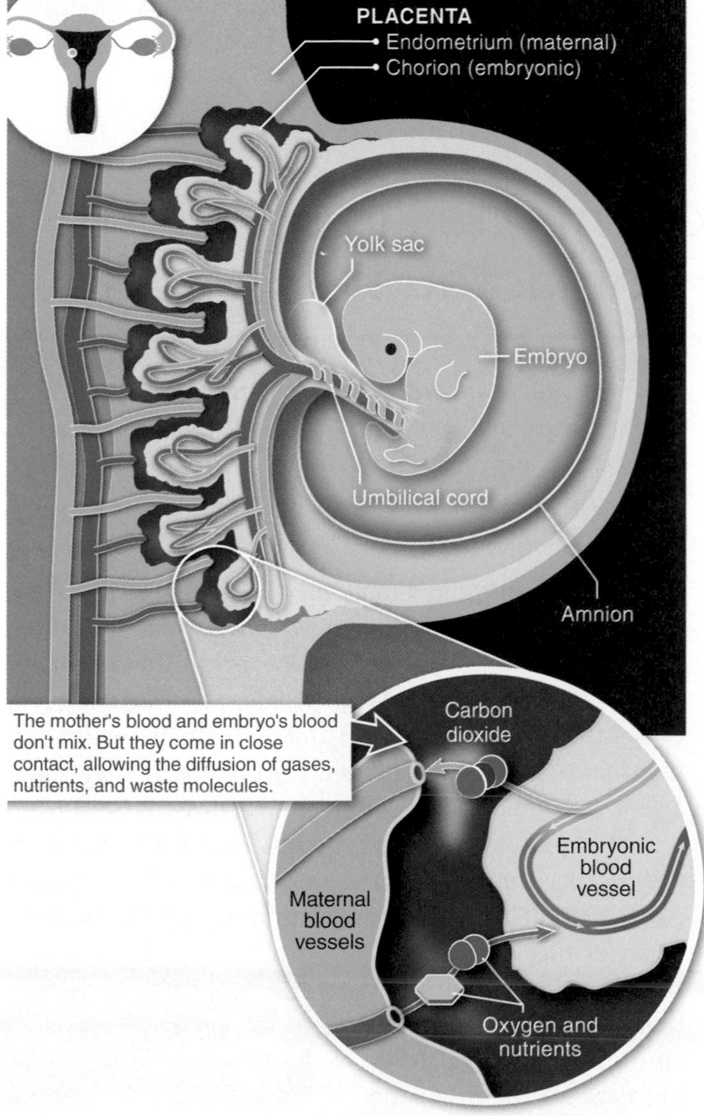
PLACENTA
Endometrium (maternal)
Chorion (embryonic)
Yolk sac
Embryo
Umbilical cord
Amnion
The mother's blood and embryo's blood don't mix. But they come in close contact, allowing the diffusion of gases, nutrients, and waste molecules.
Carbon dioxide
Embryonic blood vessel
Maternal blood vessels
Oxygen and nutrients

Although the mother's blood and the embryo's blood do not mix, they come in very close contact at the placenta, and oxygen diffuses from mother to embryo as carbon dioxide diffuses from embryo to mother. The placenta also nourishes the embryo and secretes hormones, including hCG, which maintains the corpus luteum, ensuring continued secretions of progesterone and, consequently, maintenance of the endometrium.

3rd week: Neurulation occurs.

4th week: Major structures and organs, including the eyes and heart, begin to form. At this point, it is also possible to see the arm and leg buds.

Second month

During this time, the limbs take shape. The major organs, including the liver, pancreas, and gallbladder, can also be seen. In spite of such dramatic development and differentiation, very little growth has occurred yet, and the weight of the embryo is about one gram—less than one-tenth of an ounce!

Third month

During the third month, the nervous system develops, as do some of the first reflexes, such as suckling. At this point, the corpus luteum finally degenerates. In the absence of the progesterone it has been producing, a new ovulatory cycle would typically (i.e., without pregnancy) occur, but because the placenta produces significant levels of estrogen and progesterone (which inhibit the production of FSH and LH), no ovulation occurs. Around the 10th week, the fetus's heartbeat usually becomes audible, and the development of mammary glands in the mother's breasts is also stimulated, in preparation for later milk production (lactation).

Second Trimester Months 4, 5, and 6 of a pregnancy are a time of significant growth and less new development relative to the first trimester. During this time, as the bones get bigger and the muscles develop, the fetus moves around and the mother can often feel it kick. The heartbeat becomes much louder, and its "whooshing" can be heard easily with a stethoscope. Although the fetus cannot survive outside the womb at the end of the second trimester,

FIGURE 26-26 **Structure and function of the placenta.**

its size has increased to about 600 grams (1.3 pounds) and 12 inches in length.

Third Trimester The third trimester is all about growth, as the fetus ultimately becomes large enough to survive outside the mother (the minimum size for survival is generally about 1.5 pounds, or 680 g). Lung maturation also occurs during the third trimester and is associated with the production of increasing amounts of surfactant, consisting of compounds that reduce the surface tension and increase the lungs' ability to inflate and deflate without collapsing. The lungs of premature infants commonly have insufficient surfactant and are structurally immature. As a consequence, the infant may have difficulty with gas exchange and suffer from infant respiratory distress syndrome, which is the leading cause of death in premature infants. In addition to growth and lung maturation, during the third trimester there is significant development of the nervous system. The addition of new neurons occurs at the explosive rate of 15 million per hour, a rate that continues even after birth.

TAKE HOME MESSAGE 26.13

» The nine-month development of a human embryo and fetus is divided into three equal periods, or trimesters. The first trimester is primarily a time of development and differentiation of cells into specialized types of tissues, as the embryo implants in the uterus and the placenta forms. The second and third trimesters are characterized mostly by significant growth and rapid development of the fetus's nervous system.

26.14 Pregnancy culminates in childbirth and the start of lactation.

Birth, also called parturition, is the culmination of pregnancy, and it occurs in three phases. The first is the initiation of contractions and the dilation, or opening, of the cervix. Toward the end of the third trimester, the fetus usually becomes positioned with its head down and its skull resting on the cervix. The weight and positioning of the fetus stretches the uterus. Although the exact process is not fully understood, fetal hormones cause the placenta to produce hormones, including estrogen and prostaglandin, that stimulate contractions. The stretching of the uterus causes the pituitary gland to release oxytocin, another hormone that causes contractions. These contractions cause a gradual dilation of the cervix and are referred to as **labor.** The rate of contractions increases from one or two per hour over many hours or even days before the birth, to about one every two to three minutes when birth is about to occur (FIGURE 26-27).

The second phase of birth is the delivery of the baby through the vagina, which is also called the birth canal. This generally occurs with the head passing first. After delivery of the baby, its umbilical cord is clamped and cut, and clotting quickly stops any bleeding. There is a brief relaxation of the contractions before the third and final phase of the birth process. When the contractions resume, the placenta is sheared from the wall of the uterus and expelled.

During pregnancy, mammary gland development is stimulated by estrogen, progesterone, and other hormones. After birth, the pituitary hormone prolactin stimulates milk production, or **lactation.** Suckling by the infant, too, causes the release of prolactin and of oxytocin, which further increases milk production. During the first few days, a yellowish fluid, called colostrum, is released. Colostrum is high in protein and contains numerous antibodies from the mother that protect the infant from some diseases (FIGURE 26-28). Gradually, the colostrum is replaced by milk, which is higher in fat and sugar, with less protein than colostrum. The median duration of breastfeeding varies a lot from one

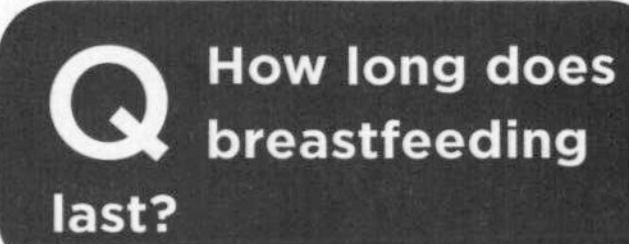
Q How long does breastfeeding last?

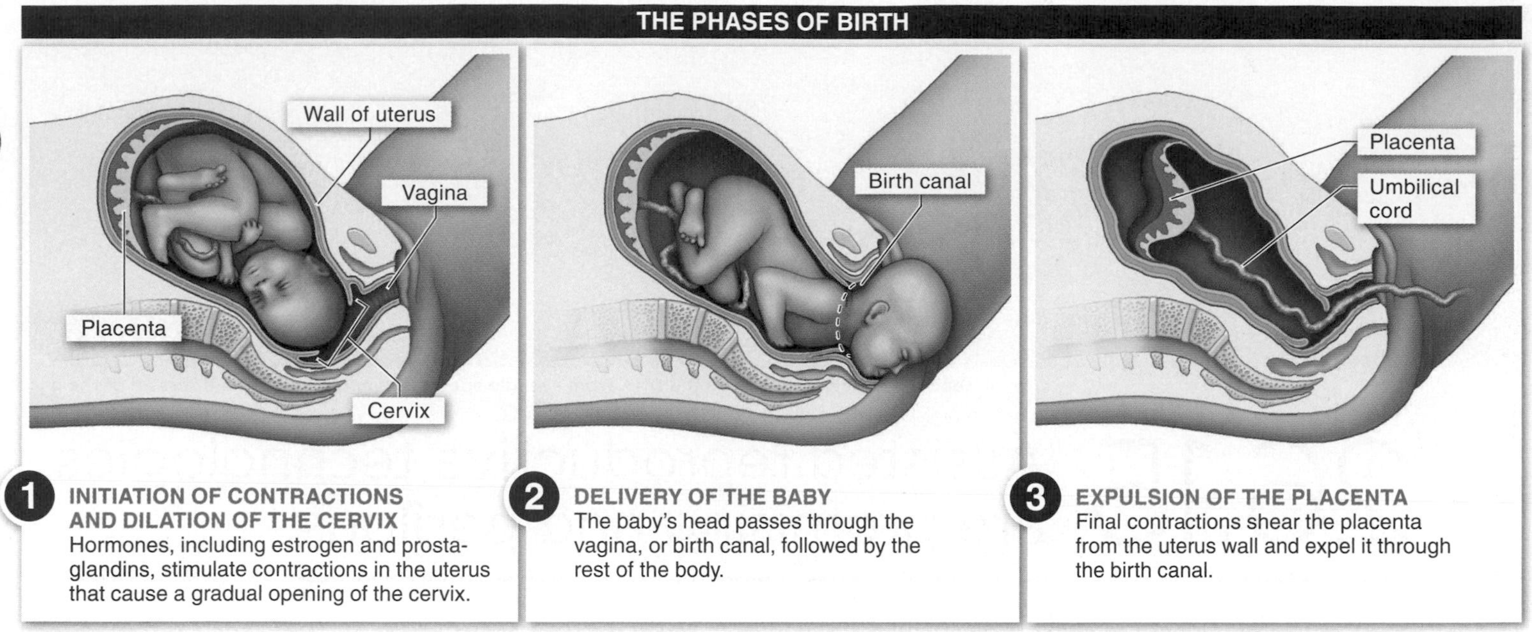

THE PHASES OF BIRTH

Wall of uterus
Vagina
Placenta
Cervix

1 INITIATION OF CONTRACTIONS AND DILATION OF THE CERVIX
Hormones, including estrogen and prostaglandins, stimulate contractions in the uterus that cause a gradual opening of the cervix.

Birth canal

2 DELIVERY OF THE BABY
The baby's head passes through the vagina, or birth canal, followed by the rest of the body.

Placenta
Umbilical cord

3 EXPULSION OF THE PLACENTA
Final contractions shear the placenta from the uterus wall and expel it through the birth canal.

FIGURE 26-27 **Labor and birth.**

country, and one culture, to another. Published reports by the journal *Pediatrics* and by the World Health Organization, for example, note that the median duration of "any breastfeeding" is approximately 3 months in the United States and 33 to 34 months in Bangladesh and Nepal. The American Academy of Pediatrics recommends exclusive breastfeeding for the first 6 months of a baby's life, and breastfeeding in combination with other foods until at least 12 months of age.

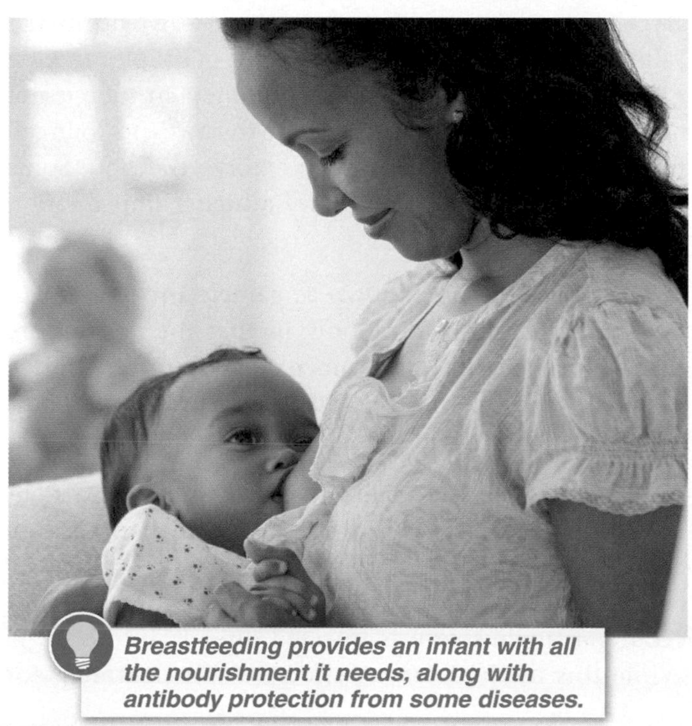

Breastfeeding provides an infant with all the nourishment it needs, along with antibody protection from some diseases.

FIGURE 26-28 **Complete nutrition.**

Q Can breastfeeding prevent pregnancy?

During lactation, the suckling of the infant prevents the pituitary from releasing a sufficient surge of LH to cause ovulation. For this reason, during lactation, a woman's fertility is significantly reduced. As a successful method of birth control, this "lactational amenorrhea" depends on several conditions, including that the mother is exclusively breastfeeding (i.e., the baby is not getting nourishment in any other way), the baby is less than six months old, and no more than six hours pass between any two feedings.

TAKE HOME MESSAGE 26.14

» Birth is the culmination of pregnancy and occurs in three phases. The first phase is initiation of contractions and dilation of the cervix. The second is delivery of the baby. The third is expulsion of the placenta. Lactation, stimulated by prolactin and other hormones, nourishes the infant with proteins, carbohydrates, and lipids and provides it with protective antibodies.

26.15

Reproductive technology has benefits and dangers.

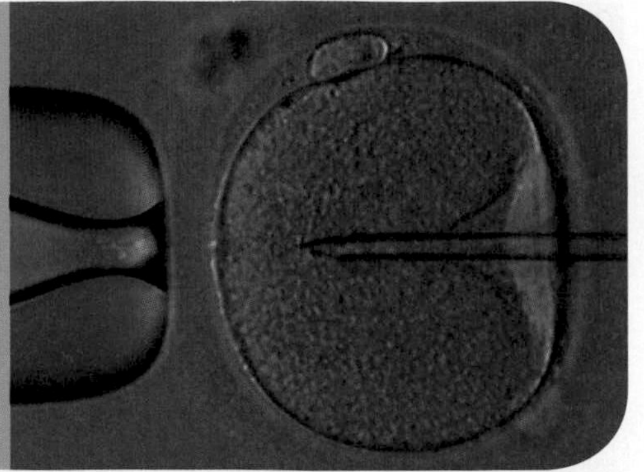

A needle penetrates an egg as the DNA from a single sperm is injected into the cytoplasm of the egg.

26.15 Assisted reproductive technologies are promising and perilous.

Many methods exist to help otherwise infertile couples to have children. In this section we examine some of these techniques.

Infertility is defined as the inability to get pregnant after one year of trying. It affects approximately 7% of married couples in the United States in any given year. There are many causes of infertility—evenly divided between males and females—including low sperm counts and insufficient sperm motility, Fallopian tube damage or blockage, and ovulation problems. In 10% of cases, no cause can be found.

The reproductive technologies available are varied and include such interventions as surgery to repair blocked Fallopian tubes and hormone treatments to increase sperm counts or egg production. Technologies also include a range of options called **assisted reproductive technology (ART),** the use of fertility treatments in which both sperm and egg are handled. These procedures typically involve removing eggs from a woman's ovaries, combining them with sperm to achieve fertilization in a Petri dish, and reinserting the fertilized eggs into the reproductive tract of that woman, or another woman. In many cases, the eggs, sperm, or fertilized eggs may be frozen for some period of time (FIGURE 26-29). ART includes several different methods.

1. IVF-ET. This stands for **i**n **v**itro **f**ertilization–**e**mbryo **t**ransfer. It is the most common of all ART methods. Several secondary oocytes are collected and combined with sperm in a Petri dish. After several days, the fertilized eggs, at the eight-cell stage, are inserted into the woman's uterus.

2. ZIFT. Zygote **i**ntra-**F**allopian **t**ransfer differs from IVF-ET in only two respects. After fertilization in a Petri dish, the fertilized eggs are transferred not to the uterus but to one of the Fallopian tubes instead. Also, they are transferred earlier, at the one-cell stage. Perhaps because surgery is needed to place a fertilized egg within the Fallopian tube, ZIFT is used by only 1% of couples seeking ART. In some cases, transfer to the Fallopian tube is not made until the fertilized egg is in the two- to four-cell stage (this is referred to as tubal embryo transfer). Some studies have shown ZIFT to have a higher implantation rate than IVF-ET.

3. GIFT. The method known as **g**amete **i**ntra-**F**allopian **t**ransfer differs from ZIFT only in that the oocytes are mixed with sperm and immediately transferred to a Fallopian tube, so that fertilization occurs in the body, rather than in a Petri dish.

In some cases, if the male's sperm counts are particularly low or if his sperm have low motility, a procedure called ICSI, or **i**ntra**c**ytoplasmic **s**perm **i**njection, may be used in conjunction with IVF-ET or ZIFT. In this procedure, a single sperm is injected directly into an egg. While this method can be effective as a treatment for

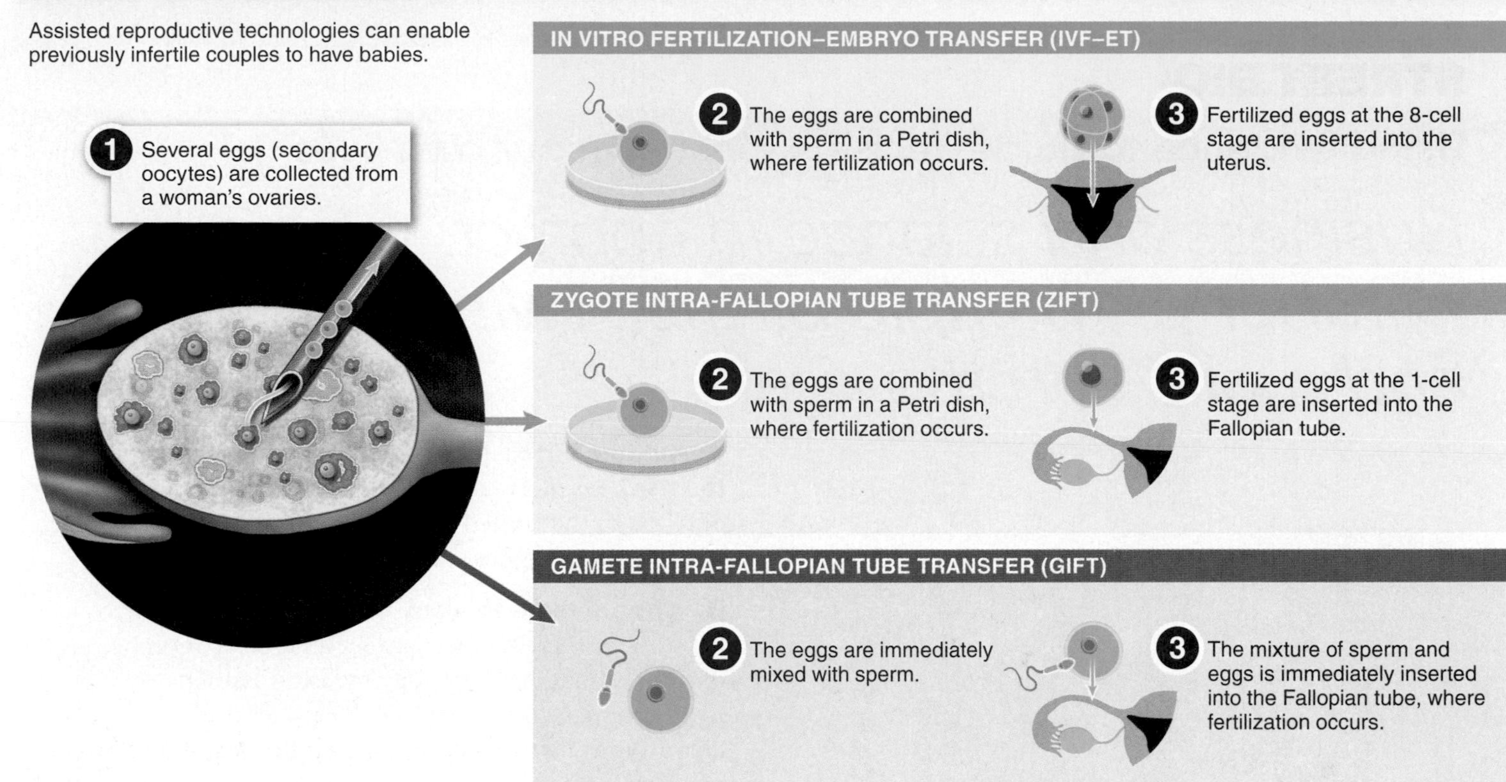

ASSISTED REPRODUCTIVE TECHNOLOGY (ART) METHODS
Assisted reproductive technologies can enable previously infertile couples to have babies.
1 Several eggs (secondary oocytes) are collected from a woman's ovaries.
IN VITRO FERTILIZATION–EMBRYO TRANSFER (IVF–ET)
2 The eggs are combined with sperm in a Petri dish, where fertilization occurs.
3 Fertilized eggs at the 8-cell stage are inserted into the uterus.
ZYGOTE INTRA-FALLOPIAN TUBE TRANSFER (ZIFT)
2 The eggs are combined with sperm in a Petri dish, where fertilization occurs.
3 Fertilized eggs at the 1-cell stage are inserted into the Fallopian tube.
GAMETE INTRA-FALLOPIAN TUBE TRANSFER (GIFT)
2 The eggs are immediately mixed with sperm.
3 The mixture of sperm and eggs is immediately inserted into the Fallopian tube, where fertilization occurs.

FIGURE 26-29 **ART procedures.**

male infertility, there are concerns because sperm selection is bypassed. That is, instead of one sperm "winning" the competition to fertilize the egg, a sperm is selected by the individual doing the procedure. This may be why there's an increased incidence of genetically carried birth defects with ICSI.

Outside the definition of ART is the increasingly effective process for selecting the sex of a baby, known as "sperm sorting." In this method, a fluorescent dye that binds to DNA is applied to a sperm sample. Because the X chromosome is larger than the Y chromosome (by about 3%), the X-chromosome-bearing sperm bind more dye. The dyed sperm are then given a small electric charge, positive or negative, depending on which chromosome they carry, and as they are passed through a tube in single file, the sperm cells are deflected to a collecting tube on one side or the other, depending on their charge. The selected sperm carrying an X or Y sex chromosome can then be used in any of the ART methods described above. The purities of the X versus Y samples of sperm range from 70% to 90%.

TAKE HOME MESSAGE 26.15

» Assisted reproductive technology (ART) procedures typically involve removing eggs from a woman's ovaries, combining them with sperm to achieve fertilization, and reinserting the fertilized eggs into the woman's uterus or Fallopian tube. These technologies can enable previously infertile couples to have babies.

STREET BIO

Using evidence to guide decision making in our own lives

Business and science in conflict: why is the number of multiple births—twins, triplets, and more—on the rise?

Q: Multiple births are in the news a lot. Are they on the rise? Yes! Between 1980 and 2008, the incidence of twins in the United States increased by 70%. The rate of triplets and higher multiples increased even more.

Q: Why are there so many more multiple births? For starters, there has been an increase in the number of older women having babies by using assisted reproductive technology (ART). Consider this: if you are running a fertility clinic, one of the most important pieces of information that will lead to the success or failure of your clinic is the percentage of treated couples who become pregnant. Unfortunately, in most ART procedures, this number is influenced by the number of embryos transferred into the woman: 5–10 transferred embryos are more likely to result in a pregnancy than 1–4 transferred embryos. When 5–10 embryos are transferred, however, the risk of "multiples"—twins, triplets, or more—also increases, as does the health risk to the mother and the babies. The decision regarding the number of embryos to transfer involves a trade-off between the perceived effectiveness of the clinic and the risk to the mother. This point was vividly illustrated in 2009 by the case of Nadya Suleman (called "Octomom" by the media), a California woman who gave birth to octuplets after eight embryos were implanted.

Q: Are fertility clinics bringing the rate back down to safer levels? In the past 10 years, the proportion of in vitro fertilization cycles in which four or more embryos were transferred dropped from 62% to 21%. This has significantly reduced the incidence of triplets, but the rate of twins has remained high. And although, for women under 35, the American Society of Reproductive Medicine now recommends the transfer of just a single embryo during ART procedures, for women above 35, it still recommends 3–5 embryos be transferred.

GRAPHIC CONTENT

Thinking critically about visual displays of data

1 In October, what was the average number of days between a woman's first day of menstruation and that of her 5–10 closest friends in the dorm? What was this number in April? By what percentage had it changed in six months?

2 If you assume that all women in the study had 28-day menstrual cycles, what is the highest value that would have been possible for the *y*-axis? Why?

3 Women on the pill were not excluded from this study. How might they influence the results? Why do you think they were included?

4 The data and the idea being tested were developed by a college undergraduate. Should that influence your interpretation of them? Why or why not?

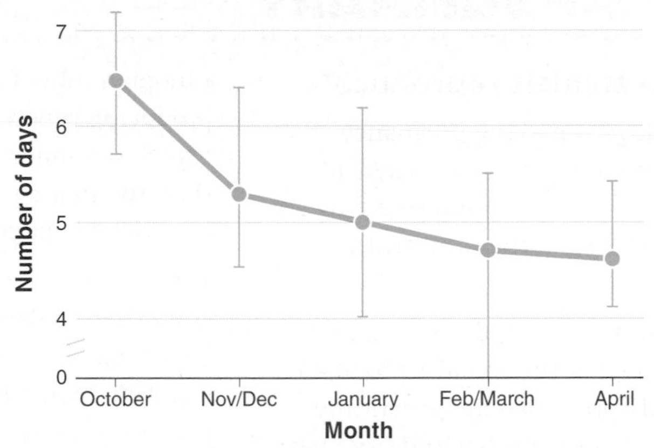

5 What does the break in the *y*-axis (near the bottom) indicate? How would the graph appear different if this break were not made?

6 What do the vertical lines through each data point indicate? How would your evaluation of this graph change if the vertical lines through the data points were twice as long?

7 What is the "take home message" of this graph? What additional data would be helpful? Why?

See answers at the back of the book.

KEY TERMS IN REPRODUCTION AND DEVELOPMENT

primary oocyte, p. 917
primary spermatocyte, p. 911
progesterone, p. 920
prostate gland, p. 911
scrotum, p. 911
secondary oocyte, p. 917
secondary spermatocyte, p. 911
seminal vesicle, p. 911
seminiferous tubule, p. 911
sexual reproduction, p. 907
sexually transmitted disease (STD), p. 925
sperm, p. 907
spermatogenesis, p. 910
testis, p. 911
testosterone, p. 911
trimester, p. 930
umbilical cord, p. 930
uterus, p. 917
vagina, p. 917
vas deferens, p. 911
viviparity, p. 910
zona pellucida, p. 922
zygote, p. 909

BRIEF SUMMARY

How do animals reproduce?

• Technology is making pregnancy possible in many situations where it previously was not, but is also giving rise to complex legal battles and ethical dilemmas.

• Organisms can reproduce sexually or asexually—or both. Sexual reproduction leads to offspring that are genetically different from each other and from both parents. Asexual reproduction, which can be fast and efficient, leads to offspring genetically identical to the parent.

• Sexual reproduction requires fertilization, which occurs externally or internally. Development of the embryo can be nourished by yolk within an egg or by nutrients in the mother's blood.

Male and female reproductive systems have important similarities and differences.

• In men, sperm are continuously produced in the testes by meiosis. Semen—consisting of sperm cells and fluids that nurture and aid the sperm in fertilization—is ejaculated during copulation.

• When females of some species mate with more than one male, sperm competition occurs and can lead to a variety of adaptations, including increased sperm production and testis size.

• In the ovaries, diploid cells undergo meiosis, a process that continues in the Fallopian tubes following ovulation, producing genetically varied haploid gametes. A much smaller number of eggs than sperm are produced and each egg is considerably larger than a sperm cell.

• In the ovaries, a cell within a follicle is stimulated to develop into a fertile egg by coordinated secretions of the hormones estrogen, FSH, and LH. Preparation of the uterus for implantation of a fertilized egg is coordinated by progesterone.

Sex can lead to fertilization, but it can also spread sexually transmitted diseases.

• At fertilization, a sperm cell penetrates the protective zone around the egg, the egg blocks additional sperm entry, and the sperm and egg membranes fuse. The egg then completes its second meiotic division, and the haploid nuclei of the egg and sperm fuse, forming a diploid zygote.

• Pregnancy can be prevented by barrier methods, hormonal methods, intrauterine devices (IUDs), "natural" methods, or sterilization.

• Sexually transmitted diseases (STDs) are caused by a variety of organisms, including bacteria, viruses, protists, fungi, and arthropods. Worldwide, more than 300 million people are newly infected each year.

Human development occurs in specific stages.

• Soon after fertilization, cleavage takes place, and many rapid cell divisions occur without overall growth. In gastrulation, a gut begins to form, along with three distinct germ layers with specific developmental fates. In neurulation, mesoderm forms a supporting rod and ectoderm forms a neural tube, which will become the brain and spinal cord.

• The nine-month development of a human embryo and fetus is divided into three equal periods, or trimesters. The first is primarily a time of development and differentiation of cells into specialized types. The second and third trimesters are characterized mostly by growth and development of the fetus's nervous system.

• Birth occurs in three phases. The first phase is initiation of contractions and dilation of the cervix. The second is delivery of the baby. The third is expulsion of the placenta. Lactation nourishes the infant with proteins, carbohydrates, and lipids and provides it with protective antibodies.

Reproductive technology has benefits and dangers.

• Assisted reproductive technology (ART) procedures typically involve removing eggs from a woman's ovaries, combining them with sperm to achieve fertilization, and reinserting the fertilized eggs into the woman's uterus or Fallopian tube.

CHECK YOUR KNOWLEDGE

Short Answer

1. Describe an unintended consequence of technology that makes pregnancy possible in situations where it once was not.

2. Asexual reproduction is not ideal when the environment is changing. Why?

3. In humans, how are male gametes formed? Where in the male reproductive system does this process occur?

4. Describe two adaptations that result from sperm competition.

5. Describe the key ways in which egg production differs from sperm production.

6. Why does degeneration of the corpus luteum result in the lining of the uterus sloughing off?

7. Describe the importance of activation of the oocyte and fusion of the plasma membranes of the sperm and egg.

8. In estrogen-progesterone birth control pills, what is the purpose of the estrogen component? What is the purpose of the progesterone component?

9. Describe two features of STDs that make them nearly impossible to eradicate completely.

10. What are germ layers? How many layers are there, what are they called, and what structures ultimately form from them?

11. Summarize the three stages of pregnancy and the major events that occur in each.

12. How is ovulation suppressed during pregnancy?

13. How is a woman's fertility greatly reduced during lactation and nursing?

Multiple Choice

1. Some animals are primarily asexual in their reproduction, but have the ability to switch to sexual reproduction under certain conditions. What might be the adaptive advantage to an animal that generally reproduces asexually to also be able to reproduce sexually?

a) to increase the genetic diversity of its offspring during periods of stress

b) to confuse its predators

c) to give the animal some variety in its life

d) to increase its own likelihood of survival

e) Both a) and d) are correct.

0 EASY — 55 — HARD 100

2. Which three sets of accessory glands add secretions to the semen in human males?

a) the seminal vesicles, the prostate gland, and the bulbourethral glands

b) the ejaculatory duct, the prostate gland, and the vas deferens

c) the seminal vesicles, the prostate gland, and the urethra

d) the seminal vesicles, the prostate gland, and the epididymis

e) the ejaculatory duct, the prostate gland, and the bulbourethral glands

0 EASY — 38 — HARD 100

3. When individual females mate with multiple males within short periods of time:

a) a male that produces more sperm than other males has an increased probability of reproductive success.

b) internal fertilization cannot occur.

c) there is selection for a reduction in size of the male testes.

d) the male gametes experience reduced desiccation.

e) All of the above are correct.

0 EASY — 32 — HARD 100

4. The tissue that in males develops into the penis, in females develops into the:

a) vagina.

b) labia.

c) clitoris.

d) oviduct.

e) ovary.

0 EASY — 31 — HARD 100

5. The completion of meiosis II in the oocyte of human females occurs:

a) before birth.

b) during menstruation.

c) after the oocyte is fertilized by a sperm.

d) a few hours before ovulation.

e) at ovulation.

0 EASY — 51 — HARD 100

6. The corpus luteum:

a) secretes chorionic gonadotropin.

b) is reabsorbed when fertilization occurs.

c) is the only source of the hormones that maintain pregnancy.

d) secretes progesterone but not estrogen.

e) is the initial source of progesterone during pregnancy.

0 EASY — 28 — HARD 100

7. The hormones present in birth control pills typically prevent pregnancy by:

a) preventing formation of the endometrial lining of the uterus.

b) preventing maturation of ovarian follicles.

c) preventing sperm from fertilizing the ovulated egg.

d) causing the endometrium to be shed once a fertilized egg has implanted.

e) preventing implantation of fertilized eggs in the endometrium.

0 EASY — 53 — HARD 100

8. Tying off the vas deferens leading from each testis will:

a) prevent formation of sperm.

b) decrease testosterone secretion.

c) reduce secretions from the seminal vesicles.

d) reduce secretions from the prostate gland.

e) None of the above.

0 EASY — 53 — HARD 100

9. When do cells begin to lose their totipotency?

a) anaphase of mitosis

b) gastrulation

c) late prophase I of meiosis

d) cleavage

e) neurulation

0 EASY — 49 — HARD 100

10. In humans, fetal growth is slowest during which trimester?

a) first

b) second

c) third

d) fourth

e) It occurs at the same rate in all trimesters.

0 EASY — 36 — HARD 100

11. In which of the following methods of assisted reproductive technology does fertilization occur within the woman's body?

a) intracytoplasmic sperm injection

b) in vitro fertilization–embryo transfer (IVF-ET)

c) zygote intra-Fallopian transfer (ZIFT)

d) gamete intra-Fallopian transfer (GIFT)

e) All of the above are correct.

0 EASY — 34 — HARD 100

Ch27

Your body has different ways to protect you against disease-causing invaders.

27.1 Three lines of defense prevent and fight pathogen attacks.

27.2 External barriers prevent pathogens from entering your body.

27.3 The non-specific division of the immune system recognizes and fights pathogens and signals for additional defenses.

27.4 The non-specific system responds to infection with the inflammatory response and with fever.

Specific immunity develops after exposure to pathogens.

27.5 The specific division of the immune system forms a memory of specific pathogens.

27.6 The structure of antibodies reflects their function.

27.7 Lymphocytes fight pathogens on two fronts.

27.8 Clonal selection helps in fighting infection now and later.

27.9 *This is how we do it*: Does contact with dogs make kids healthier?

27.10 Cytotoxic T cells and helper T cells serve different functions.

Malfunction of the immune system causes disease.

27.11 Autoimmune diseases occur when the body turns against its own tissues.

27.12 AIDS is an immune deficiency disease.

27.13 Allergies are an inappropriate immune response to a harmless substance.

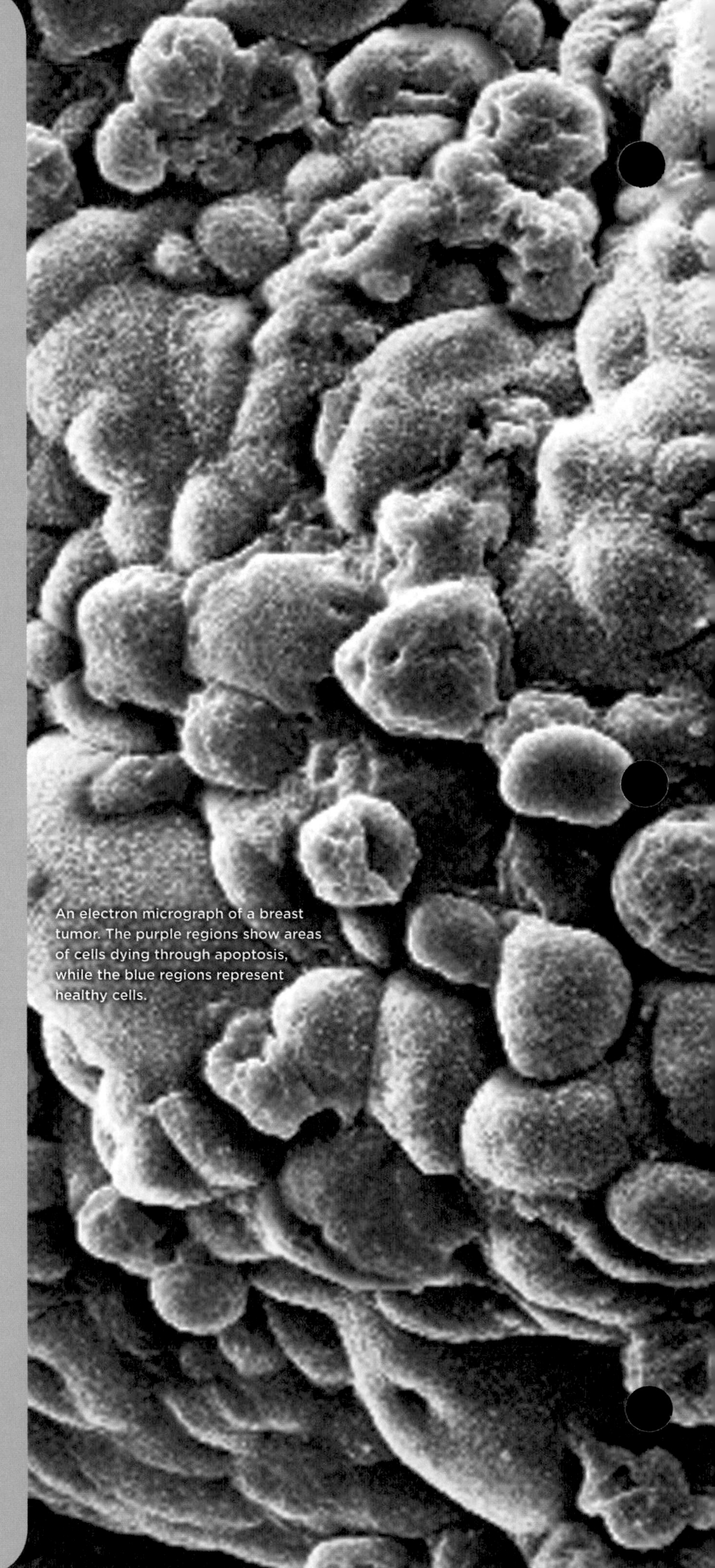

An electron micrograph of a breast tumor. The purple regions show areas of cells dying through apoptosis, while the blue regions represent healthy cells.

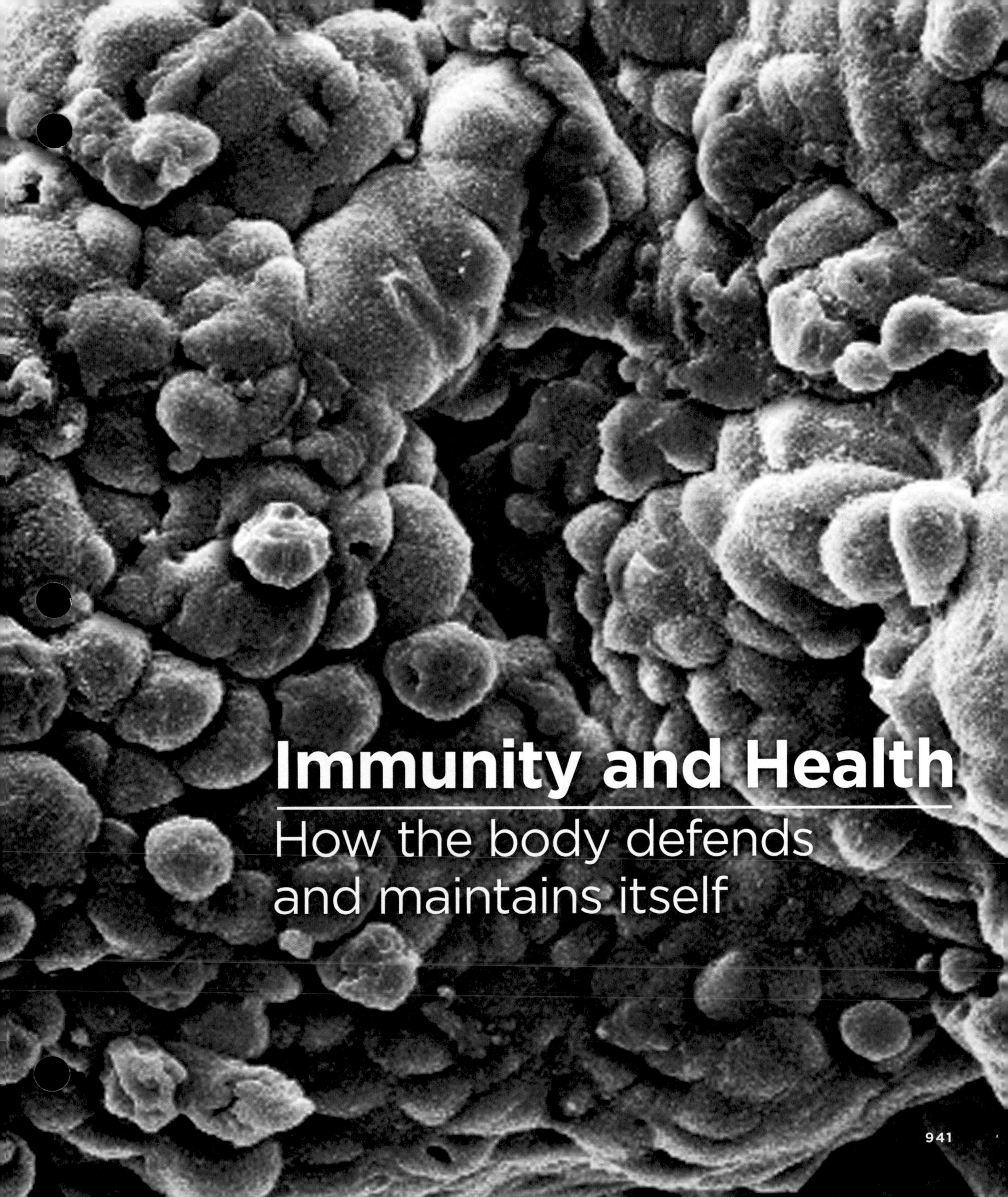

Immunity and Health

How the body defends and maintains itself

27.1–27.4

Your body has different ways to protect you against disease-causing invaders.

Sneezing is an effective way to spread disease-causing microorganisms.

27.1 Three lines of defense prevent and fight pathogen attacks.

Most of us don't think about the germs that constantly attempt to invade our bodies, until we hear that an illness is "going around." For example, in the spring of 2009, Mexico alerted the world to an outbreak of the H1N1 virus that caused numerous deaths—particularly among people under 25 years old and pregnant women. This outbreak was caused by a new strain of the influenza virus that the world's current population had never before encountered. The media showed footage of Mexico City residents wearing face masks and reported that schools and businesses were closing for extended periods of time. Within weeks, the virus had spread to the United States and around the world. Fortunately, as the flu strain began infecting people worldwide, researchers noted that although this influenza strain was strong, it was not as virulent as first suggested (FIGURE 27-1).

Unfortunately, as the risk of one illness recedes, the risk of another often increases. In 2014, for example, there was an outbreak of Ebola in West Africa that spread rapidly. Caused by the Ebola virus, which is transmitted through direct contact—through broken skin or mucous membranes—the disease initially manifests as flu-like symptoms, including fever, headache, and muscle and abdominal pain. With a rapid and strong immune response, a person may recover. But often, an infection becomes severe rapidly, and in 60% to 90% of cases it is fatal. As of April 2016, more than 15,000 cases had been confirmed, with 11,310 deaths.

For influenza, when you first hear about an outbreak of a particularly dangerous type, do you think about washing your hands more frequently, wearing a face mask, and staying away from people who are coughing? These are reasonable responses—influenza can be transmitted more easily than Ebola, including by inhaling airborne viruses or touching contaminated surfaces. The H1N1 flu example demonstrates that when we are aware of **pathogens,** disease-causing microorganisms, molecules, and viruses (sometimes referred to as "germs"), in our environment, we may take precautions to avoid them.

Think about the commonplace items that we all touch. Did you know that shopping cart handles rank as one of the "germiest" items around? A 2006 study by the Centers for Disease Control and Prevention (CDC) found that for infants, riding in a shopping cart of a market that sells meat was one of the biggest risk factors for *Salmonella* infection (just below exposure to reptiles and eating partially cooked eggs)! Following up on this report, a 2010 study reported that

Indian girls wearing protective masks outside a museum in the Indian city of Bangalore in August 2009, just after a 26-year-old woman died of swine flu (H1N1 virus) in a nearby hospital.

FIGURE 27-1 **Outbreak.**

Shopping cart handles have significantly more illness-causing bacteria than even public restroom toilet seats and flush handles.

FIGURE 27-2 **Everyday encounters with pathogens.** Shopping cart handles harbor germs.

72% of the shopping carts the researchers sampled (across four states) tested positive for fecal bacteria, and the average contamination levels exceeded those measured on toilet seats and flush handles in public restrooms (FIGURE 27-2). Not surprisingly, many stores now offer disinfectant wipes, and some states have even proposed legislation to make providing these wipes mandatory.

The vast majority of microscopic organisms found on everyday items will not cause disease, but some will. Our highly effective **immune system** sorts out benign from disease-causing microorganisms and provides protection against an enormous variety of pathogens, including many bacteria, fungi, viruses, and parasitic protists and worms (FIGURE 27-3). To protect the body from pathogens, the immune system must be able to rapidly recognize what belongs in the body and what doesn't, as well as distinguish among specific disease-causing microbes or viruses. Once a pathogen enters the human body, the immune system works within minutes to begin destroying it. The three major divisions of the immune system undertake this challenging task of protecting us against diverse threats (FIGURE 27-4).

1. Physical barriers provide the first defense against pathogens, keeping them out of the body. For example, skin acts as a nearly impenetrable barrier that keeps viruses, bacteria, and other pathogens from entering the body. In addition, skin surfaces and cells within mucous membranes secrete various chemicals that stop the growth of many pathogens.

2. Non-specific immunity is the part of the immune system that provides defense when pathogens make it past the physical barriers and into the body. The term "non-specific" indicates that these immune cells rapidly recognize and attack any and all pathogens that enter the body. The non-specific immune system is ready to function at birth and is sometimes called the *innate system* (meaning "inborn"). Non-specific immunity is found in virtually all multicellular organisms, including plants, nematode worms, fruit flies, and humans.

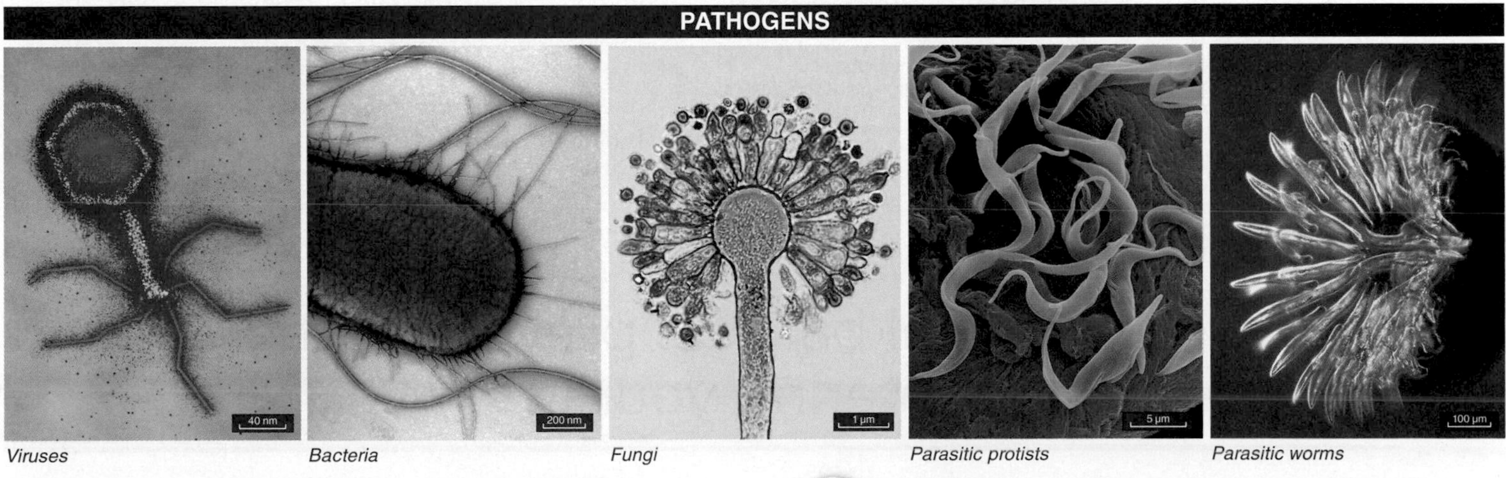

FIGURE 27-3 **Examples of the foreign, "non-self" microbes and viruses that can harm the body.**

DIVISIONS OF THE IMMUNE SYSTEM

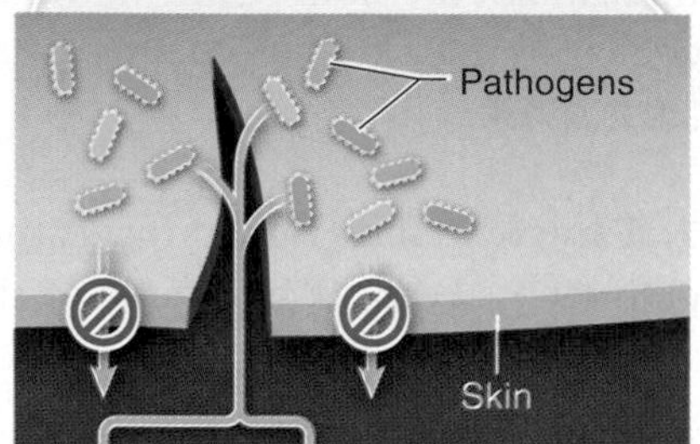

PHYSICAL BARRIERS (PREVENTION)
- Form a nearly impenetrable wall, keeping pathogens from entering body tissues
- Consist of skin, mucous membranes, and their associated anti-pathogen secretions

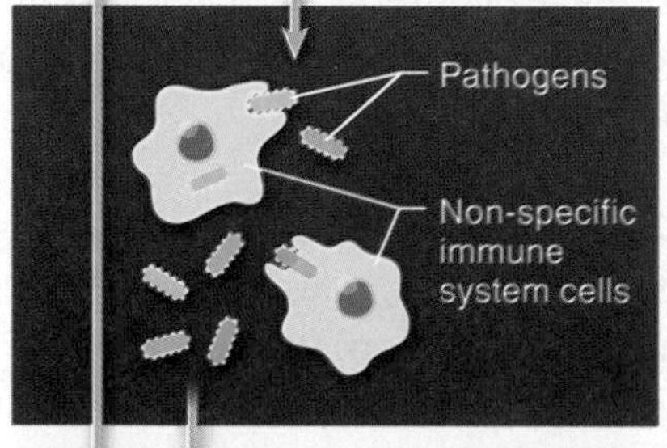

NON-SPECIFIC IMMUNITY (RAPID RESPONSE)
- Recognizes and destroys pathogens that breach external barriers
- Responds to all pathogens in the same way
- Responds to infection within minutes

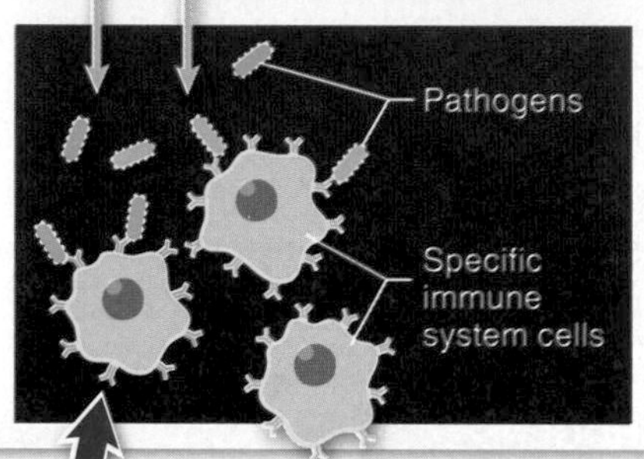

SPECIFIC IMMUNITY (LONG-TERM PROTECTION)
- Destroys pathogens that are not killed by non-specific defenses
- Recognizes specific pathogens and forms a memory of each
- Responds to infection in hours to days

After the specific immune system forms a "memory" of a pathogen, it fights off the same pathogen more quickly in the future, often resulting in lifelong protection from that pathogen.

FIGURE 27-4 Layered defenses of the immune system block entry and fight invaders.

3. Specific immunity, also called *adaptive immunity,* is the division of the immune system that deals with pathogens that have been missed by the non-specific system or cannot be overcome by that system. The specific division recognizes particular infectious agents—the bacterium *Staphylococcus aureus* that causes common skin infections, for example, or the virus varicella-zoster that causes chicken pox. Because the specific immunity part of your immune system must first recognize a specific threat and then manufacture specific weapons to combat it, there is a time lag of a few hours to a few days between your exposure to the pathogen and the response to it. Once activated, the cells of the specific division "remember" past infections and can then fight off the same infection more quickly in the future. As a result, an individual often has lifelong protection from a pathogen that he or she has previously encountered.

Working together, the three divisions of the immune system protect us against diseases and infections. Physical barriers prevent access to the interior of the body, but if entry does occur, a pathogen encounters an immediate response in the form of the non-specific system. Then, to defeat the pathogen for the long term, chemical signals trigger the cells of the specific immune system to retain information about the infection for an efficient defensive response in the future.

TAKE HOME MESSAGE 27.1

» The immune system, which protects us from a diverse group of pathogens, has three basic parts: physical barriers, non-specific immunity, and specific immunity. Physical barriers and non-specific immunity are the first lines of defense, serving to distinguish a substance as a pathogen. Cells in the specific immunity line of defense recognize individual pathogens and remember them, so they can fight them more effectively in the future.

27.2 External barriers prevent pathogens from entering your body.

Just as a high-security building may have walls and barbed wire to keep out potential intruders, your body has ways to prevent pathogens from entering your body, tissues, and cells. The **integumentary system** is made up of skin, mucous membranes (such as those lining the respiratory system), sweat glands, oil glands, hair, and nails, all of which protect you from stress or injuries caused by external forces such as friction, pressure, sunlight, dehydration, and, of course, pathogens (FIGURE 27-5).

Q How dangerous is it to sit directly on a toilet seat in a public restroom?

Have you ever sat down on a public toilet seat only to wonder if you were going to contract some horrible disease from it (FIGURE 27-6)? The good news is that most human pathogens cannot survive outside the body for very long, and a cold, smooth toilet seat is not an ideal place for them to survive. Furthermore, pathogens need some way to get into your body, and the skin on your buttocks is generally a strong barrier to entry for any pathogens on the toilet seat. Many public bathrooms offer toilet seat covers that can act like a second skin, further reducing the already very low risk.

Skin is not just a passively protective wall, however. It's more like a wall with armed guards. In addition to forming a nearly impenetrable barrier, skin cells actively secrete chemicals that inhibit bacterial growth (FIGURE 27-7). Sweat secreted onto the skin's surface, for example, contains **lysozyme,** an enzyme that kills bacteria by damaging their cell walls. The protection provided by skin becomes really clear in burn victims: lacking skin in certain areas, they are susceptible to infection by numerous pathogenic bacteria and fungi. The external and internal surfaces of the body, such as eyes, ears, and the linings of the digestive and respiratory tracts, also protect us from environmental stresses and forces. These body areas, too, are shielded by the integumentary system (and, as we'll see, by the cells of the non-specific defense system).

Consider, for example, how the digestive tract is protected. Imagine that minutes after you push a germ-laden shopping cart, you put your finger in your mouth to bite your nail. Are you guaranteed to get ill? No, because mucus (a major component of the saliva in your mouth) also contains the antibacterial chemical lysozyme. And even if lysozyme doesn't destroy potential invaders, the pathogens will quickly reach the stomach, a highly acidic environment flooded with digestive enzymes that can destroy a variety of pathogens.

In the respiratory tract, sticky mucus traps pathogens before they can reach the lungs. The epithelial cells that line the respiratory tract have hair-like extensions, called **cilia,** that continually move mucus-entrapped pathogens up and out of the lungs. As the mucus moves up into

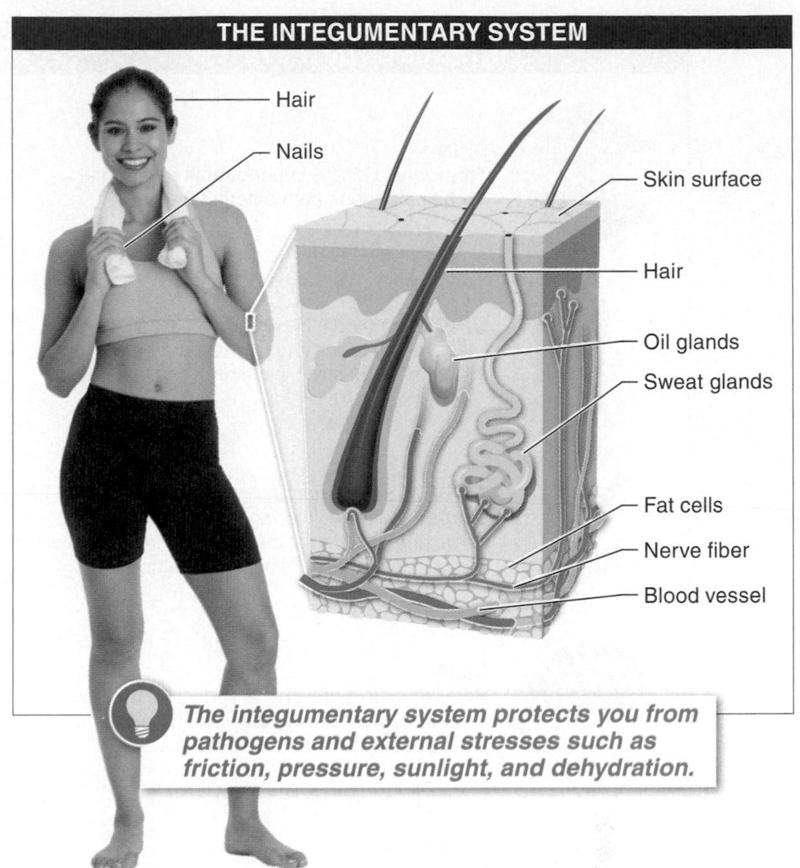

FIGURE 27-5 **Providing a nearly impenetrable barrier: the integumentary system.**

Pathogens do not pass through intact skin. So, while there may be germs on a toilet seat, your skin acts as a barrier that prevents entry.

FIGURE 27-6 **A safe place to sit.** Public restrooms are generally safe to use (but be sure to thoroughly wash your hands!).

DEFENSE MECHANISMS OF THE INTEGUMENTARY SYSTEM

SKIN
This forms a nearly impenetrable barrier that keeps pathogens from entering the body.

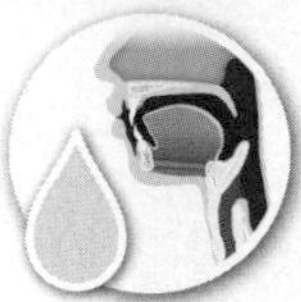

LYSOZYME AND OTHER ENZYMES
Lysozyme in saliva and tears, and digestive enzymes in the small intestine, kill many bacteria.

ACIDIC SECRETIONS
Stomach acids, acidic vaginal secretions, and acidic urine all protect the digestive, reproductive, and urinary tracts from bacterial pathogens.

CILIA
Hair-like extensions on the surface of the respiratory tract move mucus-entrapped pathogens up and out of the lungs.

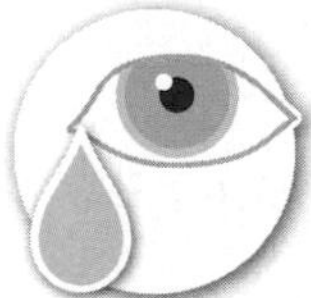

TEARS
Fluid containing antiviral and antibacterial chemicals washes away microorganisms from around the eyes.

EAR WAX
This sticky substance can trap microorganisms in the ear canal.

The integumentary system provides multiple defenses to protect you from pathogens.

FIGURE 27-7 **Security barrier.**

your throat, you continually swallow the germ-containing mucus and send it to your stomach, where the germs are destroyed. Mucus that's not swallowed continues up further into your throat until you feel the urge to "clear your throat" or cough.

The eyes produce tears that physically wash away microorganisms, and that also contain lysozyme and antiviral and antibacterial chemicals.

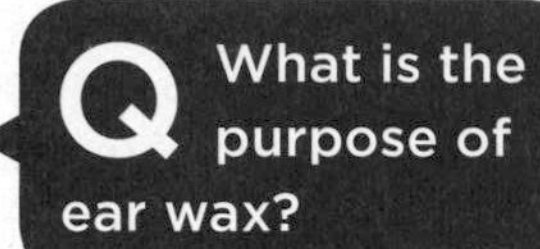
Q What is the purpose of ear wax?

Ear wax serves a protective function, too. Secretions produced by numerous glands inside the ear canal, combined with dead skin cells, make up this sticky substance that is effective at trapping microorganisms. The slight acidity of ear wax and the lysozyme the wax contains also serve to inhibit the growth of some microorganisms that find their way into the ear canal.

TAKE HOME MESSAGE 27.2

» Skin, part of the integumentary system, is a physical barrier that prevents pathogens from entering the body's cells. Cells that are not covered by skin but are exposed to the external environment are protected by defenses such as bacteria-destroying chemicals, acidic secretions, sticky mucus, and wax.

27.3 The non-specific division of the immune system recognizes and fights pathogens and signals for additional defenses.

Sometimes security measures fail. When intruders break through a defensive wall, the building's security guards need to (1) locate the security breach, (2) call for backup, and (3) attack and remove as many of the invaders as possible. Similarly, your body *recognizes* any security breaches, such as a wound to the skin or some spoiled food you've eaten, and then *responds* instantly to prevent an infection (FIGURE 27-8). Chemical signals are put out to call for backup cells, and cellular and molecular weapons are used to destroy both the pathogens and any cells they've infected.

RESPONDING TO PATHOGENS IN THE BODY

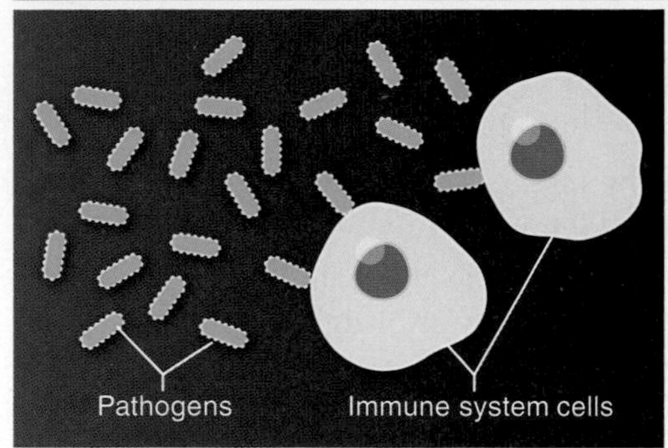

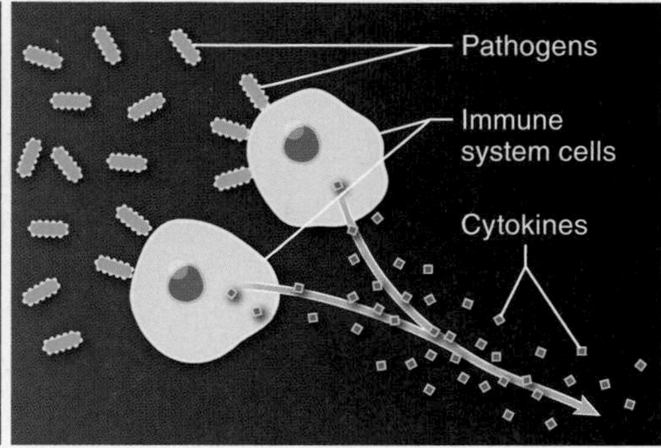

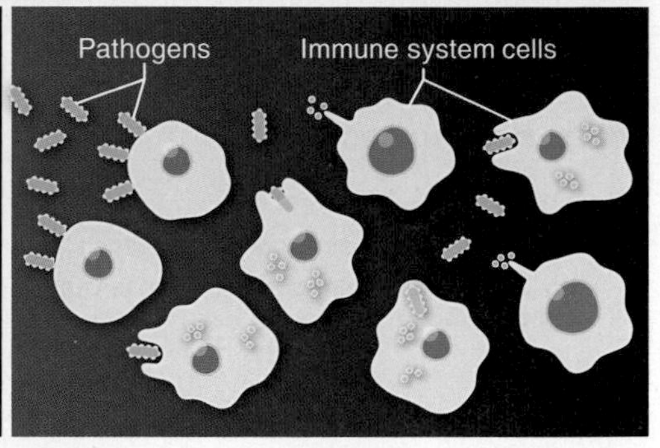

IDENTIFY THE INTRUDER
Non-specific immune system cells recognize molecules found only on the surface of pathogens and bind to them, marking the pathogens as invaders.

CALL FOR BACKUP
Immune system cells secrete signaling proteins called cytokines that recruit more immune cells to the site of the infection or warn other cells to protect themselves.

3 **ATTACK AND REMOVE**
Specialized immune system cells destroy, break down, and ingest both the pathogens and any cells they've infected.

FIGURE 27-8 **Sometimes the security measures fail.** The body recognizes and responds to invading pathogens.

In this section, we examine the cells that perform these duties and the tools that they use to combat infection. **White blood cells** (introduced in Section 22.8) are specialized cells that play roles in specific and non-specific immunity, but we focus here on non-specific immunity. These cells are made in the bone marrow and released into the bloodstream, where they patrol the body's tissues and search for invaders.

Patrolling white blood cells are able to distinguish the body's own cells ("self") from invaders ("non-self") because of unique molecules found on the surface of pathogens. The immune cell surface receptors recognize these distinctive molecules, usually proteins, and bind to them. Because your own cells do not contain these molecules, a cell that possesses any of them can be identified as an invader (non-self) by the cells of the non-specific part of the immune system.

This type of recognition is considered non-specific because immune cell receptors recognize and bind to classes of molecules shared by a wide array of organisms. For example, one receptor recognizes a protein contained in all bacterial flagella. Another receptor recognizes a molecule found in the cell walls of many types of fungi.

Once the immune system cells recognize these molecules and bind to them, the cells of the non-specific immune system begin to attack. There are four types of white blood cells that mount the non-specific defense (FIGURE 27-9).

1. Neutrophils circulate in the blood and exit blood vessels at sites of injury and infection in tissues. These cells are phagocytes, ingesting harmful particles or pathogens, primarily bacteria. Neutrophils normally make up between 50% and 70% of all white blood cells circulating in the bloodstream, and their numbers increase dramatically and rapidly during the first few days of some infections. Neutrophils destroy themselves as they ingest pathogens and live only 3–5 days, on average.

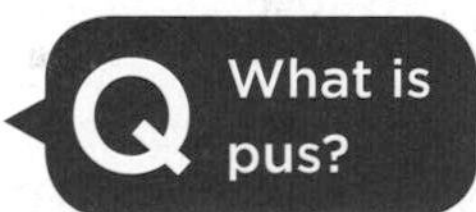

After self-destruction, the dead neutrophils become a major part of **pus,** the thick yellowish fluid we often notice at the site of infection. (Other components of pus include damaged cell debris from the invaded tissue and both live and dead pathogens.) Often, our bodies can clear infections on their own or with some help from an **antiseptic,** a solution used on a body surface to kill or discourage growth of microorganisms, such as rubbing alcohol or hydrogen peroxide. But when there is excessive pus, combined with signs of a more serious infection, antibiotics might be needed to kill bacteria inside the body.

2. Macrophages are large ("macro") phagocytic white blood cells that reside in and patrol the tissues (outside the blood vessels). They engulf and digest small pathogens whole. Macrophages also remove any pathogen and neutrophil debris that remains after the neutrophils have done their job. Unlike neutrophils, however, macrophages do not destroy themselves. Once macrophages have ingested a pathogen, they "present" digested pieces of the pathogen on their cell surface, "advertising" the infection to cells of the specific system. We'll learn more about the role of these "presenting cells" when we examine specific immunity (see Section 27.10).

WHITE BLOOD CELL DEFENDERS

The non-specific immune system consists of several types of white blood cells that are made in the bone marrow and released into the bloodstream.

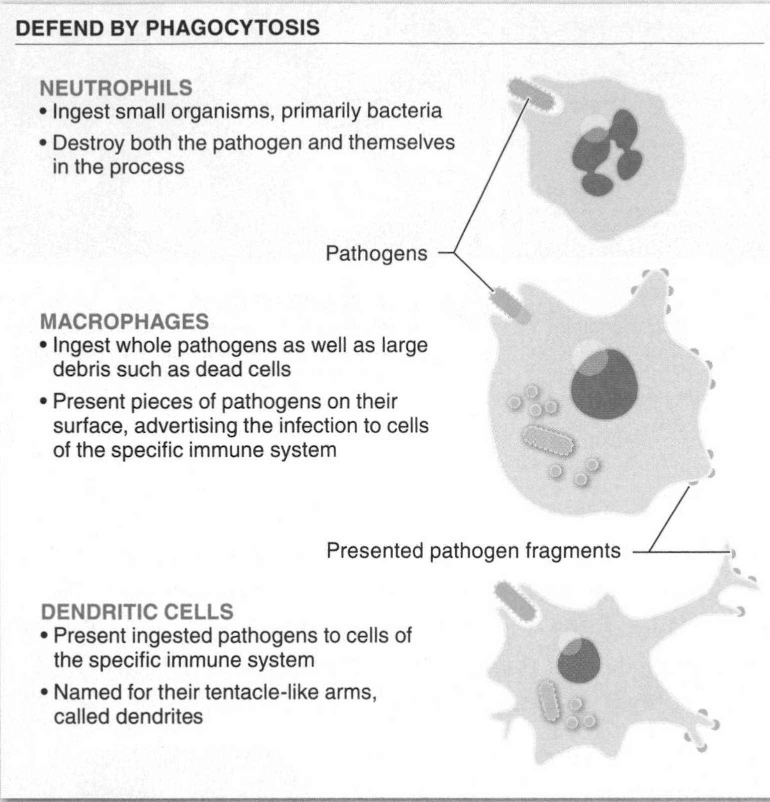

DEFEND BY PHAGOCYTOSIS
NEUTROPHILS
• Ingest small organisms, primarily bacteria
• Destroy both the pathogen and themselves in the process
Pathogens
MACROPHAGES
• Ingest whole pathogens as well as large debris such as dead cells
• Present pieces of pathogens on their surface, advertising the infection to cells of the specific immune system
Presented pathogen fragments
DENDRITIC CELLS
• Present ingested pathogens to cells of the specific immune system
• Named for their tentacle-like arms, called dendrites

DEFEND BY PUNCTURING CELLS

NATURAL KILLER (NK) CELLS

- Kill body cells infected by pathogens by poking holes in the cell membranes
- The first line of resistance to viruses
- Also play a role in recognizing and killing cancer cells

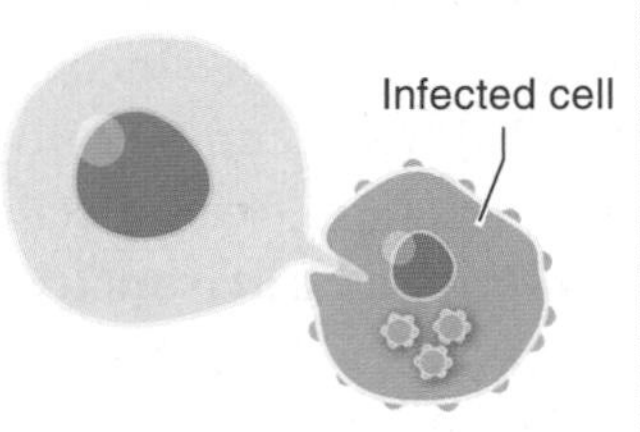

FIGURE 27-9 **Hardworking cells.** Several types of white blood cells defend the body in the non-specific division of the immune system.

Neutrophils and macrophages employ chemical warfare to destroy pathogens. For example, both types of cells can produce hydrogen peroxide and hypochlorous acid (also components of household bleach), two chemicals that are effective in killing bacteria and fungi.

3. Dendritic cells are phagocytes that link the non-specific and specific divisions of the immune system by "presenting" the cells of the specific system with foreign matter. Present in skin and mucous membranes, such as those lining the nose, lungs, stomach, and intestines, these phagocytic cells, once activated, migrate to the lymph nodes, where they interact with cells of the specific immune system.

4. Natural killer (NK) cells are the first line of resistance to viruses. They kill viruses not by ingesting the virus by phagocytosis but by killing a virus-infected cell by poking holes in it, thereby killing both the virus and the cell. The initial response of the NK cells may eliminate a viral infection, or at least slow it down until the specific system can respond to the virus in a more specific manner. NK cells also play a role in recognizing and killing cancer cells (FIGURE 27-10).

Although the cells of the non-specific system are the initial responders to a pathogen, there are never enough of them at the initial site of infection to ingest and kill all the invading pathogens. Thus, after recognizing the security breach and starting the response, the patrolling cells of the non-specific system sound the alarm and call for backup.

Additional support for the non-specific immune response is provided by several types of proteins. The chief way that cells of the immune system—both the specific and non-specific divisions—talk to one another is by secreting signaling proteins called **cytokines** in response to pathogens in the body. Cytokines secreted by one immune cell can bind to receptors on other immune cells and signal them to respond in various ways. For example, cytokine binding can cause immune cells to move closer to the site where the cytokines are being produced (*cyto* = cell; *kine* = to move).

Interferons are an important type of cytokine produced by cells infected by a virus. An interferon alerts nearby healthy cells to turn on protective measures that help them resist viruses. These alerted cells take on an "antiviral state"

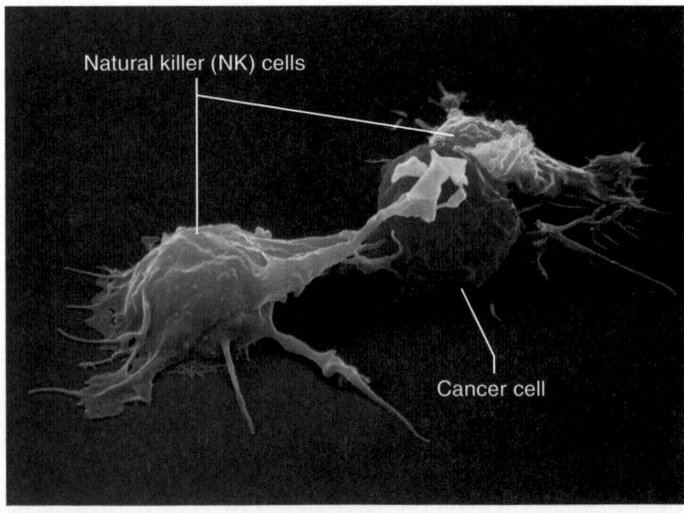

FIGURE 27-10 **Attacking cancer.** Natural killer cells are attaching to a cancer cell.

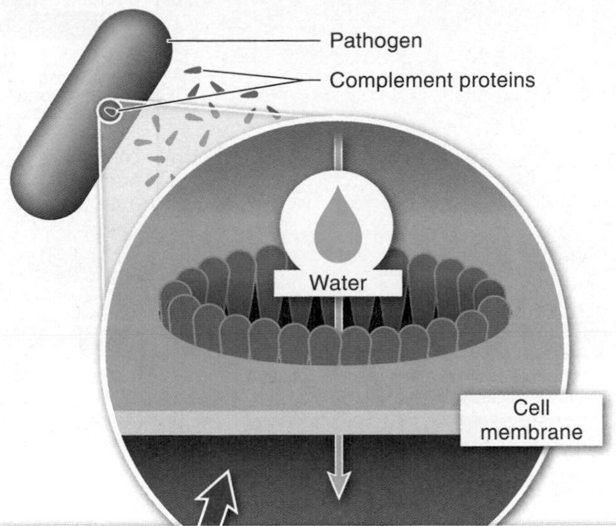

COMPLEMENT PROTEINS
A collective group of circulating proteins, called complement proteins, recognize pathogens and help fight them in various ways, including by sticking to them and increasing the ability of phagocytes to consume them, and by directly destroying them (shown here).
Pathogen
Complement proteins
Water
Cell membrane

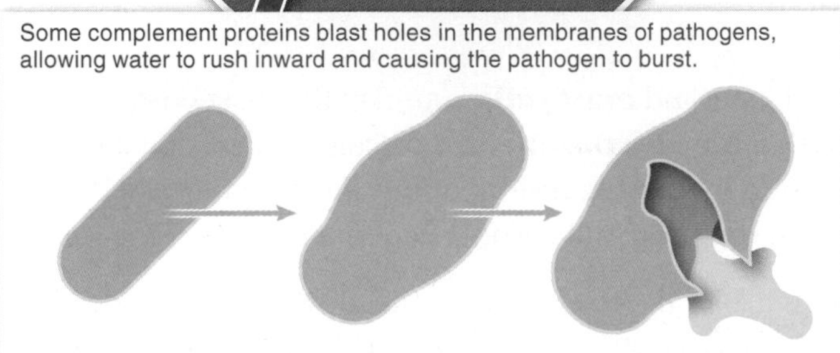

Some complement proteins blast holes in the membranes of pathogens, allowing water to rush inward and causing the pathogen to burst.

FIGURE 27-11 **Defensive proteins "complement" the other cells of the non-specific system to combat pathogens.**

in which they transcribe specific genes and translate them into proteins that can inhibit viral replication and can degrade viral RNA.

In addition to using the cells we've learned about here, the non-specific system can also make use of some circulating defensive proteins, collectively called **complement proteins,** to quickly recognize and destroy invaders. These proteins "complement" (work with) the activities of cells in the non-specific and specific immune systems and can link the two systems. Foreign molecules, such as the components of bacterial cell walls, activate the complement proteins (FIGURE 27-11). Once activated, the complement proteins fight pathogens in several ways. For example, a particular combination of complement proteins can blast holes in the cell membranes of pathogenic organisms, allowing water and ions, but not larger molecules such as proteins, to pass through. As a result, water rushes inward and the pathogen swells and bursts. Other complement proteins help destroy pathogens by sticking to them and forming a coating that enhances the ability of phagocytes to bind and engulf them.

TAKE HOME MESSAGE 27.3

» Non-specific immunity provides defenses against pathogens by recognizing molecules on their cell surfaces and by recruiting other cells to the site of infection or warning them to protect themselves. White blood cells of the non-specific system include phagocytes (neutrophils, macrophages, and dendritic cells) that ingest and kill pathogens, cells (macrophages and dendritic cells) that display pathogens to cells of the specific immune system, and cells (natural killer cells) that kill virus-infected cells and cancer cells. Complement proteins also non-specifically recognize invaders and help to destroy them.

27.4 The non-specific system responds to infection with the inflammatory response and with fever.

While we don't notice that our non-specific immune system is constantly on patrol for invaders, we definitely do notice when it responds at the site of an infection. A little response is quickly amplified into a noticeable one. Think about what happens when you get a splinter. Initially, you might feel only the pain of the splinter going into your finger. Within minutes, though, you feel pain and warmth, and see that the area around the splinter is red and swollen. The changes you are observing at the site of the splinter are the signs of an **inflammatory response.** The inflammatory response is a combination of events that leads to the recruitment of phagocytes and other immune cells to assist with pathogen destruction, followed by tissue healing. Whenever tissues are damaged by an invading pathogen such as a virus infecting throat cells, or by a physical injury such as a burn, bug bite, or other wound to the skin, the inflammatory response is triggered (FIGURE 27-12).

In the first century A.D., the Romans first described the four signs of inflammation: redness, heat, swelling, and pain—hence the term used today, "inflammation," which means, literally, "setting on fire." Whether it is a sore throat or wounded skin, these hallmarks of inflammation are always the same.

Let's imagine a scenario in which you cut your finger with a small knife while slicing a bagel. Three quick steps happen within a matter of minutes. First, invading pathogens from the knife and the skin surface are quickly engulfed by the macrophages that reside outside the blood vessels, in all tissues. Second, those macrophages release cytokines to recruit more phagocytes to the wound location. And, third, the cytokines also summon two other types of white blood cells, which initiate inflammation. **Basophils,** which circulate in the blood, and **mast cells,** found in the tissues (FIGURE 27-13), both release **histamine,** a molecule that produces inflammation by causing nearby non-injured blood vessels to (1) dilate (open up and become wider) and (2) become leaky.

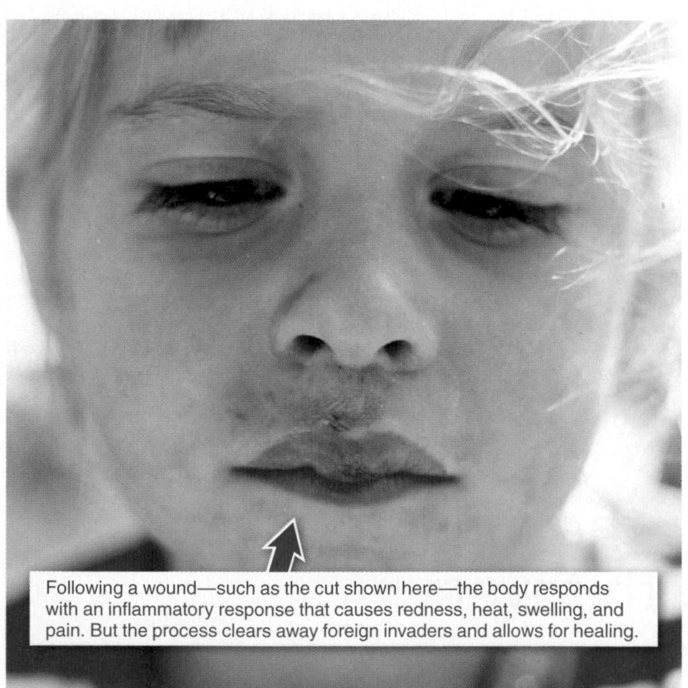

FIGURE 27-12 **An inflammatory response.**

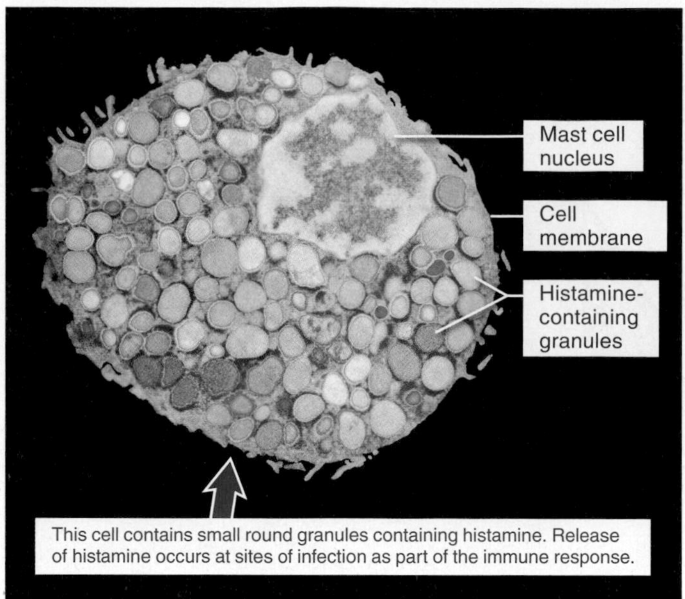

Mast cell nucleus
Cell membrane
Histamine-containing granules
This cell contains small round granules containing histamine. Release of histamine occurs at sites of infection as part of the immune response.

FIGURE 27-13 **A mast cell.**

When blood vessels dilate, the flow of blood increases and allows a faster, greater supply of defensive molecules and cells that can fight infection. The increased leakiness of the blood vessels makes it easier for neutrophils to exit the blood, enter tissue at the site of infection, and begin destroying any invading pathogens. The increased blood supply at the site of injury causes the *redness* and *heat* associated with inflammation. The leakiness is the reason for the *swelling,* because fluid leaking from the blood vessels accumulates in the tissue. And the increased pressure stimulates local *pain* receptors. Macrophages and neutrophils then begin an immediate response to destroy pathogens, and they also release cytokines that call more immune cells to the infection site (FIGURE 27-14).

Complement proteins also play a role in inflammation. Just as cytokines (secreted by macrophages) trigger the inflammatory response, activated complement proteins can also trigger initial inflammatory reactions. Activated complement proteins at the site of an infection cause mast cells to release histamine and further amplify the inflammatory response by attracting additional phagocytes. As mentioned in Section 27.3, complement proteins also directly affect

CALL FOR BACKUP

1. After you cut your finger with a knife, invading pathogens from the knife and the skin's surface enter your body.

2. Macrophages residing in tissues surrounding the cut begin engulfing the pathogens and releasing cytokines, recruiting more phagocytes and other white blood cells to the area.

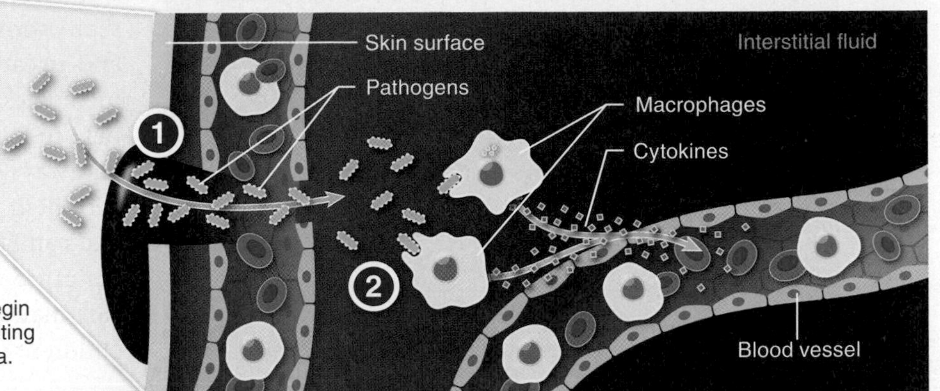

HISTAMINE RELEASE

3. Basophils, which circulate in the blood, and mast cells, found in the tissues, trigger the inflammatory reaction by releasing histamine.

4. Histamine causes nearby non-injured blood vessels to dilate, increasing the flow of blood and allowing a greater supply of defensive molecules and cells that can fight infection.

5. Histamine also causes the blood vessels to become more leaky, allowing neutrophils to more easily exit the blood and enter the site of infection.

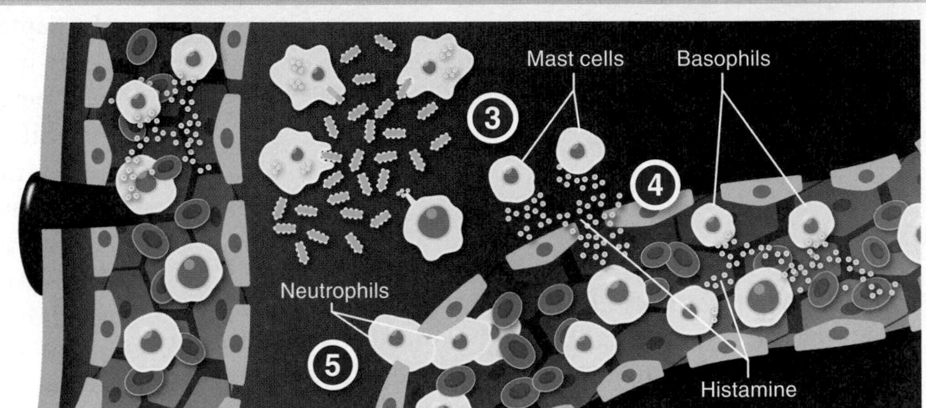

HEALING

6. The inflammatory response continues until the pathogens have been eliminated and the skin grows back, forming an impenetrable barrier once again.

An inflammatory response causes redness, heat, swelling, and pain, but ultimately leads to tissue healing.

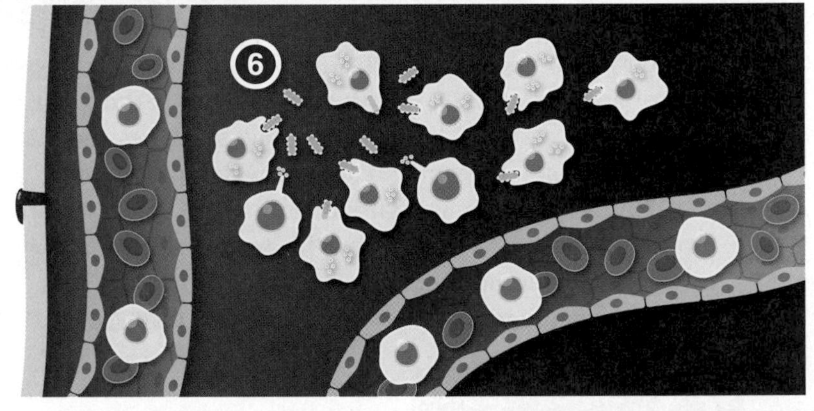

FIGURE 27-14 **Reacting to injury.**

pathogens by attacking their membranes and making it easier for phagocytes to engulf them.

So what about that cut on your finger? A scab—formed from dried, clotted blood and interstitial fluid—temporarily forms to cover the wound, much like a natural Band-Aid, and prevents more pathogens from entering. At the end of the inflammatory process, collagen fibers are secreted by nearby cells to close the wound in a more stable fashion, and the scab falls off. Skin begins to grow back, once again forming an impenetrable barrier. And you are more cautious about cutting bagels!

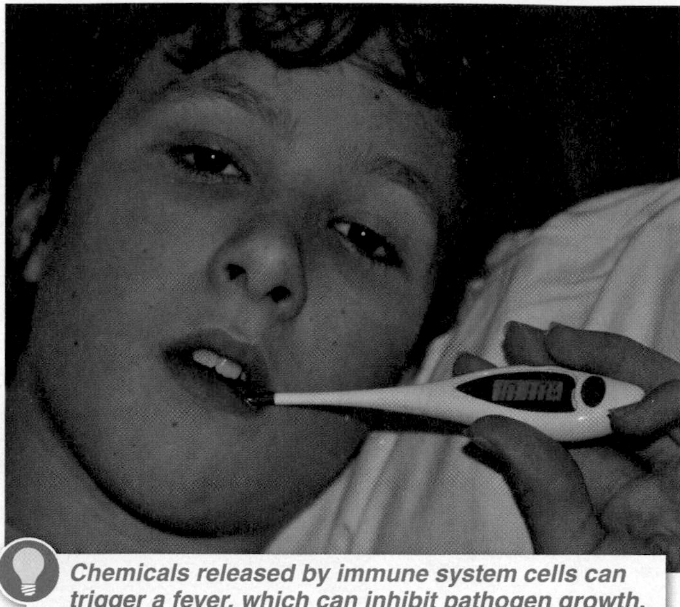

Chemicals released by immune system cells can trigger a fever, which can inhibit pathogen growth.

FIGURE 27-15 **Turning up the thermostat.** Fever slows the growth of pathogens.

Sometimes an infection is overcome quickly by the inflammatory response and is confined to the site of damage. However, if pathogens are not quickly destroyed, macrophages will be constantly stimulated and will continue to release cytokines. Some cytokines can cause a **fever,** an elevated body temperature, if their concentration is high enough—a situation that both stimulates the immune response and reduces the rate at which many pathogenic bacteria can divide. The fever-causing cytokines travel through the bloodstream and affect a region of the brain, called the hypothalamus, that functions as the body's thermostat (see Section 24.16). Just as you can change the settings on your home thermostat, your body's thermostat can also be set higher, thus leading to a fever (FIGURE 27-15).

Q Why do chills and sweating sometimes occur together when you have a fever?

When your body's thermostat is set higher with a fever, you feel cold and your body responds with rapid periods of muscle contractions and relaxations (shivers and chills). Chills are the body's way of trying to warm up to the higher temperature setting of the thermostat. When your fever "breaks," the thermostat is being reset to normal. You sweat because your body is now too hot, and sweating cools your body down. Anti-inflammatory medicines such as aspirin, acetaminophen, and ibuprofen help to lower a fever by blocking steps in the biochemical pathways promoted by the fever-producing cytokines. A very high fever (105° F or higher) is extremely dangerous and can be fatal. At such high temperatures, proteins can denature (unfold) and critical biochemical processes can malfunction, causing cellular stress and, in small children, seizures.

Q Is it always wise to treat a fever with aspirin or other medication?

In the Street Bio at the end of Chapter 21, we described "Darwinian medicine," a new perspective on when to treat and when not to treat symptoms. From this perspective, many protective responses that have evolved in organisms are recognized as having value in fighting illness. And consequently, treating symptoms such as fever, which is actually part of the immune system response to infection, can hinder our efforts at fighting infection. This idea was supported, for example, by the observation that chicken pox lasted longer, on average, in patients taking aspirin than in patients taking a non-fever-reducing placebo.

TAKE HOME MESSAGE 27.4

» Inflammation is a major way in which pathogens are eliminated by the non-specific immune system. The four recognizable signs of the inflammatory response (redness, heat, swelling, and pain) are related to the changes in blood vessels that enhance the recruitment of phagocytes and complement proteins to the site of inflammation. Fever-promoting cytokines help the body fight infection by stimulating immune responses and inhibiting the growth of some pathogens.

27.5–27.10

Specific immunity develops after exposure to pathogens.

Scanning electron micrograph of cell types found in blood, including white and red blood cells (leukocytes and erythrocytes).

27.5 The specific division of the immune system forms a memory of specific pathogens.

Often, a pathogen can be eliminated completely by the non-specific system's inflammatory response. Sometimes, however, an infection is persistent or the pathogen molecularly "disguises" itself in some way and is not recognized as a pathogen by the non-specific system. Luckily, the specific division of the immune system is also there to protect us.

Let's investigate how the immune system recognizes and responds to pathogens. An **antigen** is any molecule or fragment of a molecule that induces a specific immune response. A single pathogen often contains many different antigens (for example, the different glycoproteins, lipoproteins, and polysaccharides of a single bacterium can all be antigens). The body responds to antigens by making **antibodies,** circulating proteins that recognize specific antigens. (You can remember the difference between "antigen" and "antibody" by remembering that antigen is short for *anti*body *gen*erating.) Antibodies provide protection by enhancing the immune system's ability to recognize and destroy pathogens such as bacteria and viruses and the body cells they infect (FIGURE 27-16). But keep in mind that antibodies don't directly kill a pathogen. Instead, antibodies act as beacons that make pathogens conspicuous to patrolling phagocytes. Antibody levels peak within the first two weeks after encountering a pathogen or receiving a vaccine. We discuss the body's ability to recognize and destroy antigens in the next section.

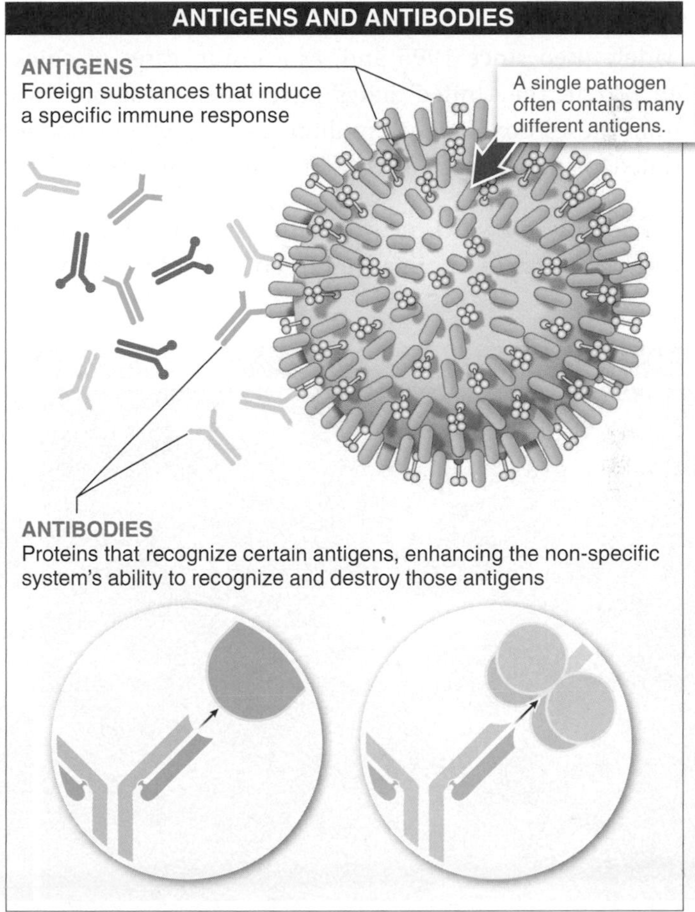

ANTIGENS AND ANTIBODIES
ANTIGENS
Foreign substances that induce a specific immune response
A single pathogen often contains many different antigens.
ANTIBODIES
Proteins that recognize certain antigens, enhancing the non-specific system's ability to recognize and destroy those antigens

FIGURE 27-16 Antibodies are proteins that recognize foreign molecules called antigens.

Even though antibody production eventually diminishes, the specific immune system retains a memory of the disease. If the same antigen is encountered again, the body takes only 1–2 days to begin producing large amounts of specific antibodies that aid in destruction of the pathogen; within just 7–9 days, there will be massive amounts of the antibodies. So, if a friend exposes you to chicken pox after you've already had the vaccine, your body won't need two weeks to respond: it will quickly kick into high gear and start producing antibodies immediately!

The memory of the specific division of the immune system results in **immunity,** a state of long-term protection against a specific pathogen. There are two ways in which we can acquire immunity to a particular pathogen. The first is to become sick with the disease (FIGURE 27-17). For example, people who contract chicken pox when they are young cannot get chicken pox again; they are immune to it. Fortunately, there is a second way to gain immunity, without having to suffer through a disease and its unpleasant symptoms. A vaccine for chicken pox has been widely used since 1995 and, as a result, cases of chicken pox in the United States have fallen by about 75% from the approximately 4 million cases in 1995 (which resulted in more than 10,000 hospitalizations and 100 deaths).

A **vaccine** is a weakened or harmless form of a specific pathogen that is administered to an individual to induce immunity, without subjecting the individual to the disease. Vaccines trick the body into thinking it has the full-blown disease, and the body mounts a specific immune response. An individual who receives the chicken pox vaccine and is later exposed to the virus will not get the disease, because the body already has a memory of this virus. It has manufactured a group of cells—called memory cells—that are already primed to recognize and rapidly respond to the virus. When the individual's immune system again encounters the virus, there will be no lag time in responding. Vaccine technology has been a major success in medicine.

Today we have vaccines against numerous viruses (including polio, measles, rubella, mumps, and rotavirus) and bacteria (including tetanus, cholera, and meningitis). These vaccines are less available in the developing world, and thus, worldwide, many people die each year from what are now preventable illnesses.

One of the most common infections fought off by the specific division of the immune system is influenza, an infectious disease caused by a virus. Each year there are many different versions, or strains, of the influenza virus, which are constantly changing. With each new flu season, the body encounters a slightly different form of an influenza virus and, therefore, a different set of antigens.

Because the influenza virus adapts to changing conditions and evolves, the best protection is to get a flu shot each year

EXPOSURE LEADS TO IMMUNITY

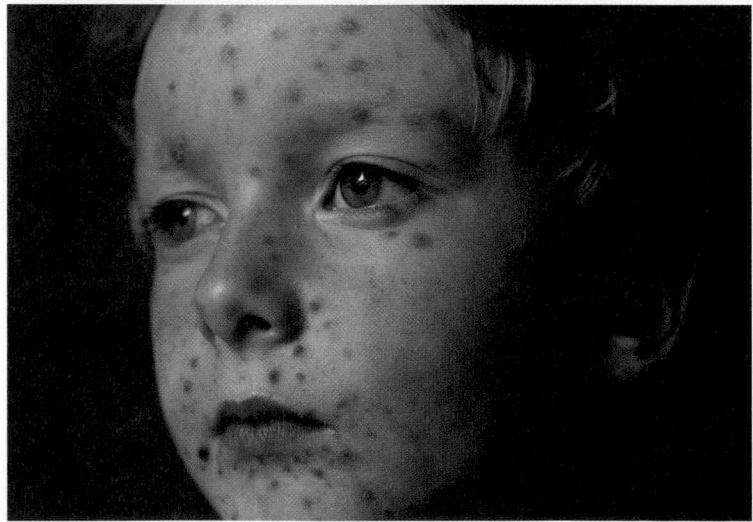

GETTING SICK
Exposure to a pathogen causes the body to form a memory of the pathogen, producing the cells necessary to rapidly respond to the pathogen if encountered again. Shown here: a toddler with chicken pox.

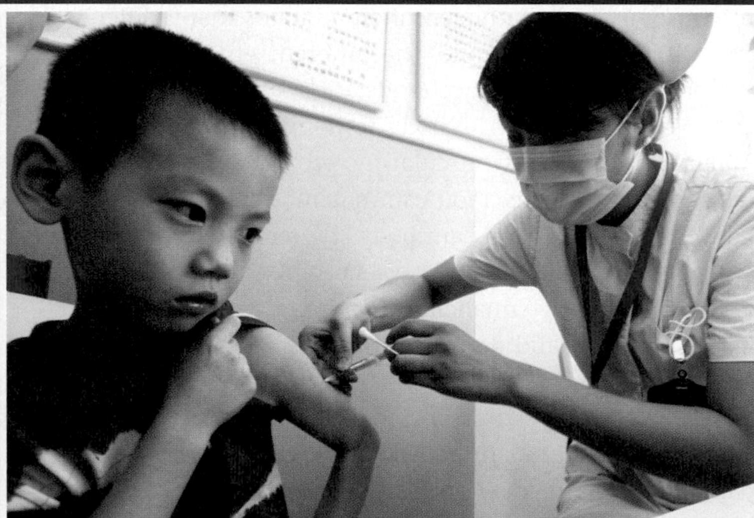

GETTING A VACCINATION
Exposure to a weakened or harmless form of a pathogen in a vaccine allows the body to form a memory of the pathogen without the risk of symptoms. The body then produces the cells necessary to rapidly respond to the pathogen if encountered again.

FIGURE 27-17 **Two paths to immunity: contracting and fighting an illness, or receiving a vaccine.**

THE INFLUENZA VIRUS IS CONSTANTLY CHANGING

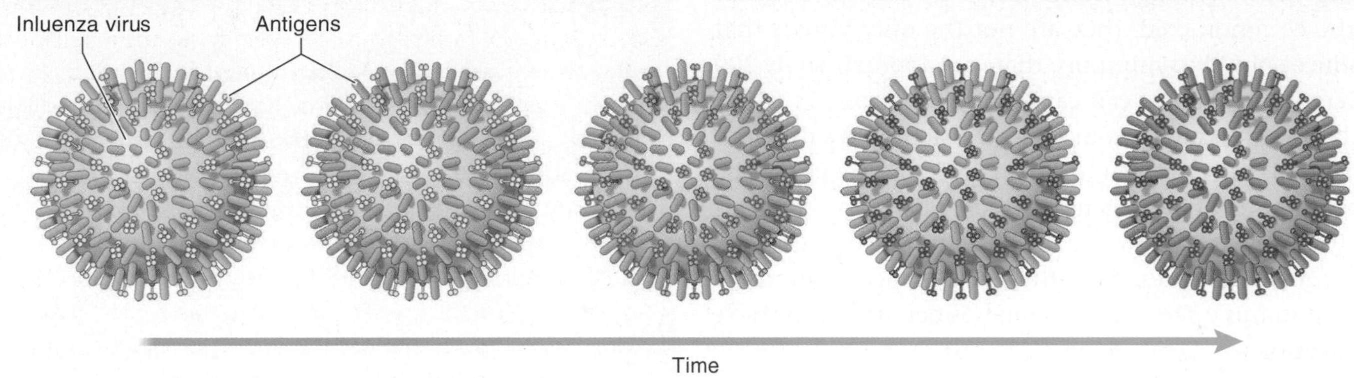

Strains of the influenza virus are constantly changing, so the body encounters a slightly different form with each new flu season, and therefore a different set of antigens.

FIGURE 27-18 Because the influenza virus changes so rapidly, flu vaccinations need to be given annually.

(FIGURE 27-18). And in the case of the flu vaccine, because any vaccine must contain an antigen, the vaccine is commonly made from an influenza virus (of the most common current strain or a mixture of strains) that has been inactivated by heat or from a live, weakened form of the virus that does not cause the flu. Although you can get the flu at any time of year, the flu season occurs in winter, so it makes sense to get vaccinated in autumn to give your body time to make antibodies that can protect you throughout the winter flu season. Every year a new flu vaccine is available in August or September.

Yet, even with these preventive measures, a person may still get the flu. Although it is commonly thought that the flu shot itself can give someone the flu, this is simply not true. So, why do individuals sometimes get the flu after having the vaccination? There are several possible reasons. First, scientists developing the vaccine must make predictions, well before the flu season, about how the influenza virus will change in the upcoming season. Sometimes the vaccine "matches" the current virus, and sometimes it may be less than perfect. Second, different strains of the flu virus circulate at the same time, and the vaccine may not protect an individual from all strains. Lastly, a person might already be infected with the flu at the time of vaccination.

Q Why don't people develop immunity to the common cold?

If the specific immune response can protect a person from illness during a future encounter with a specific pathogen, why is it that every winter you develop the runny nose, cough, and sore throat of the common cold? Why don't you develop immunity to this most annoying and recurrent disease? Just like the influenza virus, the rhinovirus ("nose virus"), one of the viruses that cause the common cold, continually changes (FIGURE 27-19). Your immune system mounts a response to each version

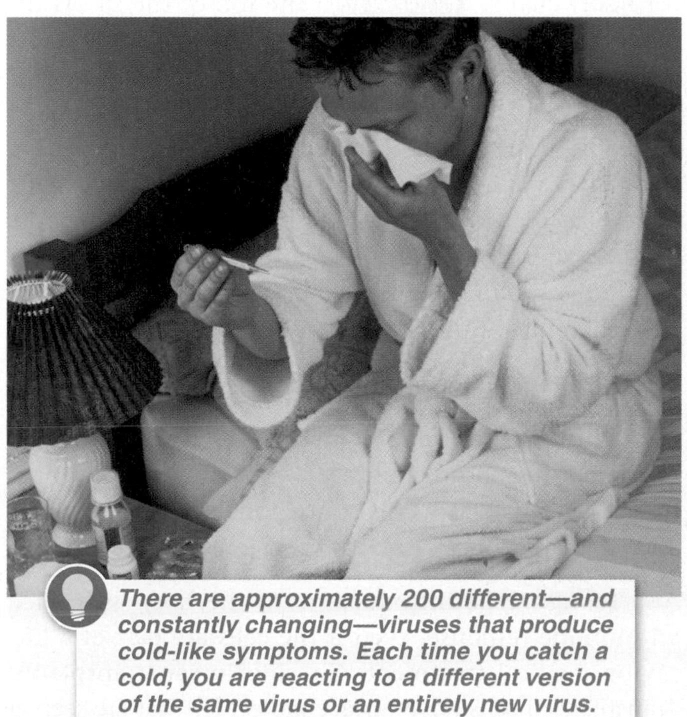

There are approximately 200 different—and constantly changing—viruses that produce cold-like symptoms. Each time you catch a cold, you are reacting to a different version of the same virus or an entirely new virus.

FIGURE 27-19 A constant challenge. Many different, rapidly changing viruses cause the common cold.

of the virus. Although rhinoviruses are a frequent cause of the common cold, they are not the only viruses that produce cold-like symptoms; there are approximately 200 different viruses that can cause the common cold. Thus, each time you catch a cold, you are reacting to a different version of the same virus or to an entirely new virus. The sheer number of pathogens and their ability to adapt and evolve over time are two challenges that the immune system constantly faces. Nonetheless, the specific immune system stands up to this challenge, as detailed in the next few sections.

TAKE HOME MESSAGE 27.5

» Antigens are molecules or fragments of molecules on the surfaces of pathogens that can be identified by disease-fighting proteins of the immune system, called antibodies. Antibodies are produced after exposure to a specific antigen. The specific immune system is continually responding to numerous pathogens, many of which change over time. Long-term protection from, or immunity to, a specific pathogen can form in two ways: exposure to the natural pathogen or exposure to an altered version of the pathogen in a vaccine.

27.6 The structure of antibodies reflects their function.

We now examine the structure and function of antibodies to learn exactly how they protect us against pathogens. All antibodies are Y-shaped, and their diversity and specificity come from variations in the top of the "Y."

Each antibody consists of four polypeptide chains joined together: two long (also called heavy) chains and two short (or light) chains (FIGURE 27-20). The top of the "Y" is the region that varies from antibody to antibody. But there are only about five variants for the lower part, the base of the "Y," and so this is referred to as the constant region of an antibody. At the top of the "Y," the variable region, each antibody has a unique three-dimensional shape that dictates which specific antigen it will "fit." An enormous diversity of antibodies is needed to recognize the correspondingly enormous group of antigens that the immune system may encounter. (And a given pathogen may have several different types of antigens on its surface.) Keep in mind that antibodies recognize *antigens,* not pathogens. An antigen is anything that an antibody recognizes as not being part of the body, as not being "self," including snake venom toxins, bacterial toxins that cause food poisoning, pollen, and various chemicals (such as those found in latex gloves).

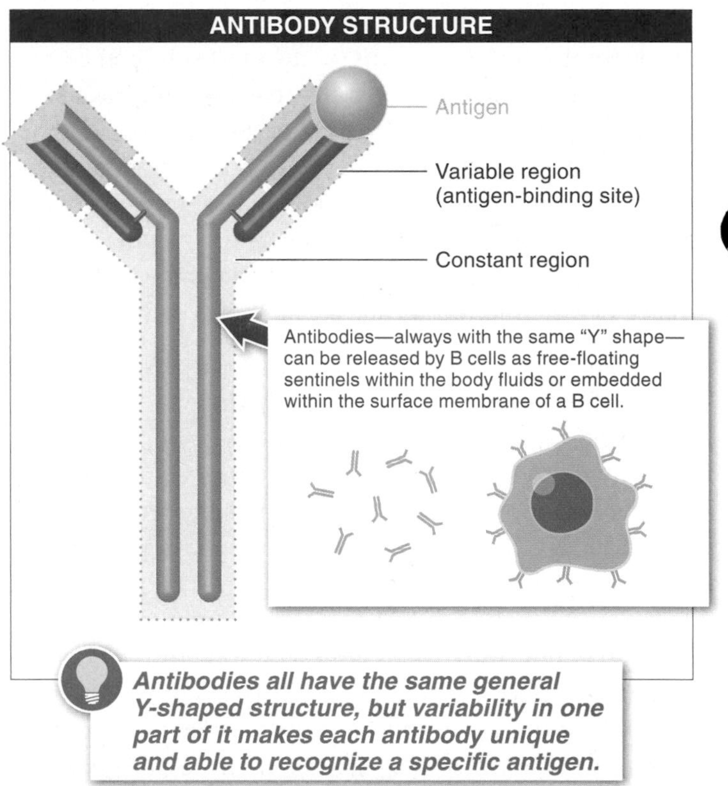

ANTIBODY STRUCTURE
Antigen
Variable region (antigen-binding site)
Constant region
Antibodies—always with the same "Y" shape—can be released by B cells as free-floating sentinels within the body fluids or embedded within the surface membrane of a B cell.
Antibodies all have the same general Y-shaped structure, but variability in one part of it makes each antibody unique and able to recognize a specific antigen.

FIGURE 27-20 **Variations on a molecular theme.** The shape at the top of the "Y" varies from one antibody to another and fits with only one antigen.

Any foreign cells introduced into the body will elicit an immune response, even if the cells are part of a lifesaving medical treatment. Consider organ transplants. Transplanted cells have surface molecules that are seen as antigens ("non-self") by the recipient's immune system. Although the transplanted cells are not pathogens, the recipient's immune system identifies them as invaders, and an immune response begins. Thus, medicines are needed to suppress the immune system's attack on and eventual destruction of a transplanted organ.

ANTIBODY FUNCTIONS

Antibodies function in several ways to help destroy pathogens and free-floating antigens.

PHAGOCYTE SIGNALING
Antibodies bind to antigens on the surface of pathogens, making it easier for phagocytes to find the pathogens and destroy them.

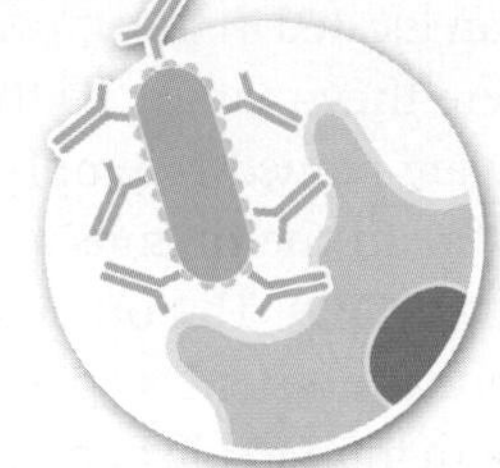

ANTIGEN CLUMPING
Antibodies make pathogens and soluble antigens clump together, making it easier for phagocytes to find them and destroy them.

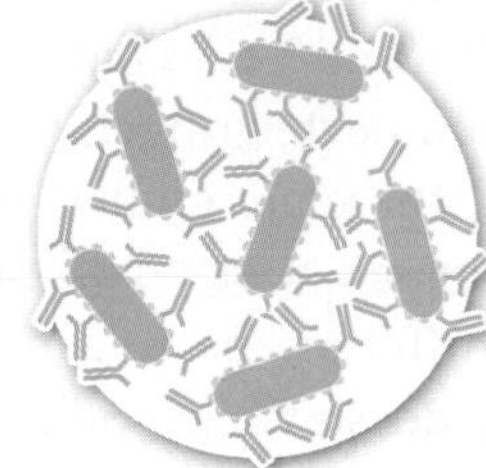

PREVENTION OF CELL ENTRY
Antibodies coating the surface of pathogens prevent the pathogens from entering body cells.

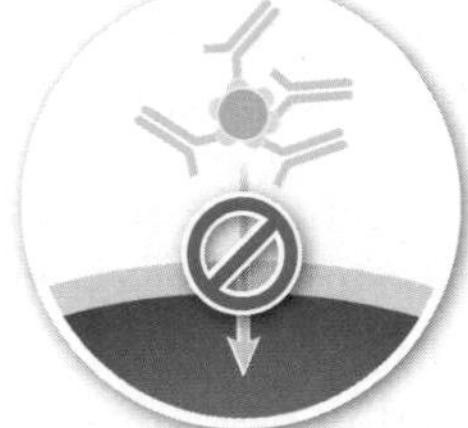

COMPLEMENT PROTEIN SIGNALING
Antibodies recruit complement proteins, which poke holes in pathogen membranes, causing the pathogen cells to burst.

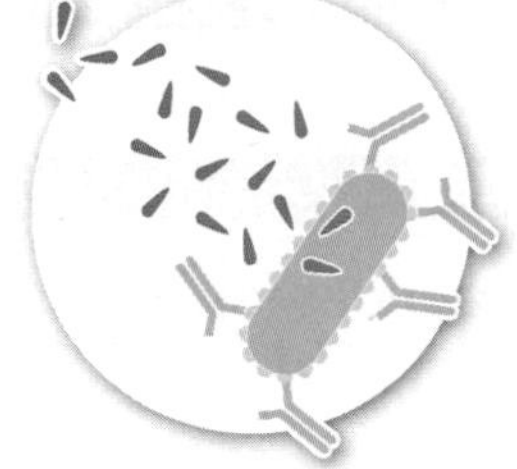

FIGURE 27-21 **Antibodies fight pathogens in several ways.**

The immune system is also the reason that people who need blood transfusions must have a proper match of blood type. As we saw in Section 9.10, if a transfusion is made between individuals with incompatible blood types, the recipient's antibodies will attach to antigens on the surface of the foreign (in this case, the donor's) red blood cells and mount an attack that can end in death.

Now let's consider the main ways in which antibodies function (FIGURE 27-21). The immune system pumps out tons of antibodies following exposure to a pathogen, and these antibodies attach to antigens. When the antigens on the surface of a pathogen have an antibody attached to them, the pathogen is easily ingested by phagocytes (such as macrophages and neutrophils) of the non-specific immune system.

Besides "marking" for destruction the antigens on a pathogen, or marking free-floating antigens (such as snake venom toxin), antibodies act in two other important ways. (1) They cause pathogens or antigens to clump together, making it easier for phagocytes to find and ingest them. And (2) they coat the surfaces of pathogens and prevent them from binding to and entering body cells, and thus they block the infection from spreading.

Lastly, when antibodies are bound to antigens on the surface of a pathogen, they recruit another player of the non-specific system: the complement proteins. Recall that some complement proteins can poke holes in membranes and cause the pathogenic cells to burst. Thus, by recruiting these complement proteins, antibodies indirectly destroy pathogens.

TAKE HOME MESSAGE 27.6

» Each antibody has a unique structure that recognizes a specific antigen. Antibodies are effective in helping to destroy pathogens and free-floating antigens in several ways: by enhancing ingestion by phagocytes, preventing the infection of more cells, and enhancing the destruction of pathogens by complement proteins.

27.7 Lymphocytes fight pathogens on two fronts.

Lymphocytes are the white blood cells responsible for the specific immune response. These cells can be found circulating in the blood and lymphatic systems, and they reside in lymphatic organs such as the lymph nodes and spleen. (See Section 22.11 to review the lymphatic system.) In contrast to blood, which contains a variety of white blood cells, 99% of the lymph is made up of just one type: the lymphocytes. Lymphocytes have **antigen receptors,** proteins on their plasma membranes that stick out from the cell surface and can bind to specific antigens—each

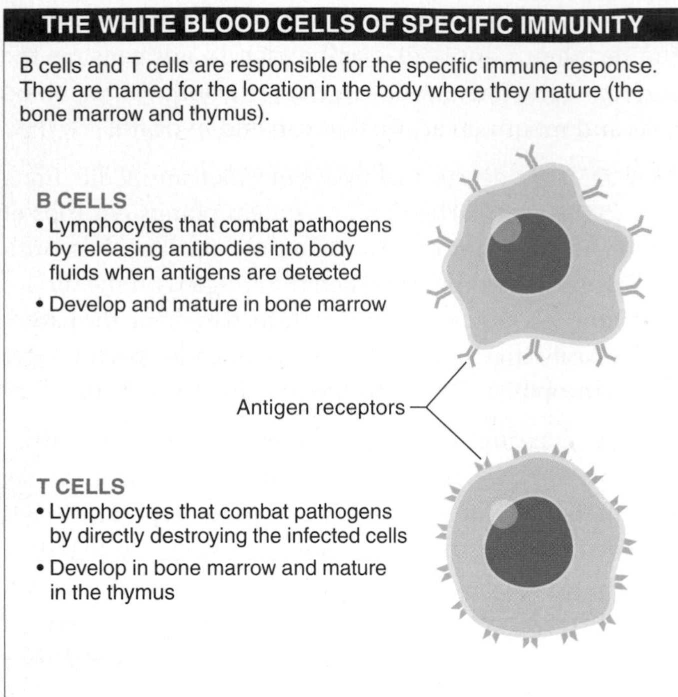

FIGURE 27-22 **B cells and T cells.**

lymphocyte can bind to just one type of antigen. The presence of antigen receptors allows lymphocytes, collectively, to recognize and react to a wide array of antigens.

There are two major types of lymphocytes. Both develop from stem cells in the bone marrow, but one type leaves the bone marrow and continues to mature in the thymus, a lymphatic organ located in the upper chest. Lymphocytes are named for where they mature, so those that mature in the bone marrow are named B lymphocytes, or **B cells,** and those that mature in the thymus are called T lymphocytes, or **T cells** (FIGURE 27-22). The other cells that normally circulate in the lymphatic system are dendritic cells, which use the lymphatic system for the sole purpose of meeting up with lymphocytes.

Both B and T cells have antigen receptors on their surfaces. Lymphocyte antigen receptors are structurally diverse. In fact, there are endless variations of these receptors. It is estimated that our lymphocytes can recognize billions of different pathogens. But each individual lymphocyte doesn't have billions of different antigen receptors. Instead, each lymphocyte has just one type of receptor that recognizes just one antigen, and it has many copies of this receptor on its surface.

Like an intense military operation on land and sea, the specific immune system operates on two fronts in order to win the war. Lymphocytes fight pathogens both in body fluids (humoral immunity) and within cells (cell-mediated immunity) (FIGURE 27-23).

SPECIFIC IMMUNE SYSTEM RESPONSES

HUMORAL IMMUNITY
Protection against pathogens and toxins found in body fluids, such as blood and lymph

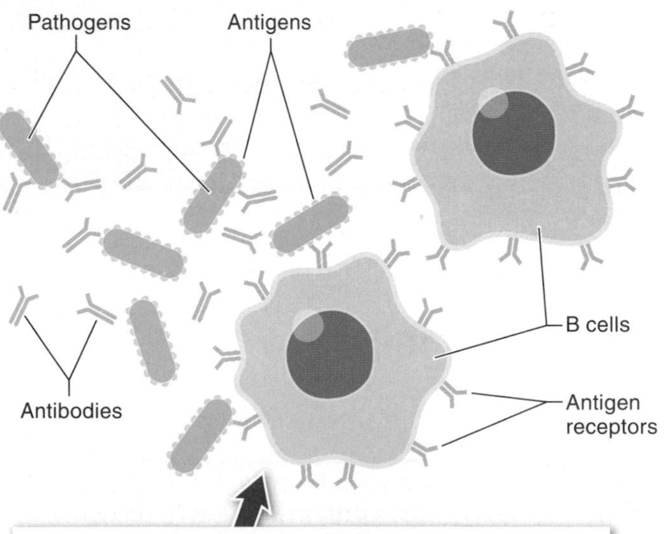

CELL-MEDIATED IMMUNITY
Protection against pathogens and toxins found within body cells

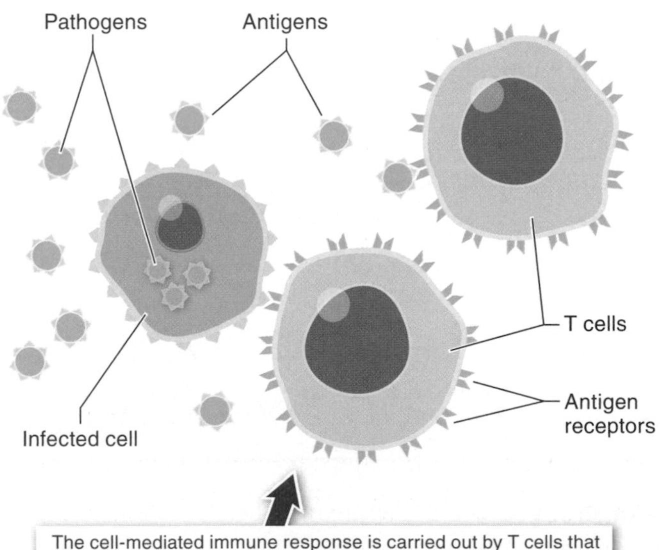

Pathogens
Antigens
T cells
Infected cell
• Develop and mature in bone marrow
Antigen receptors
The cell-mediated immune response is carried out by T cells that directly destroy invading pathogens as well as the infected cells.

FIGURE 27-23 **Fighting invaders in body fluids and within cells.** B cells are responsible for humoral immunity (antibody-mediated immunity), and T cells are responsible for cell-mediated immunity.

1. In *humoral immunity,* also called antibody-mediated immunity, the individual is protected against pathogens and toxins found in body fluids, such as blood and lymph. The word *humoral* comes from "humors"—the term that Greeks and Romans used for body fluids. Blood was one of the four humors (the other three were phlegm, yellow bile, and black bile). When an antigen is detected by a B cell with the matching antigen receptors on its surface, the B cell secretes antibodies into the blood, lymph, and other body fluids. The circulating antibodies defend against specific pathogens and toxins and make it easier for phagocytes to engulf them.

2. In *cell-mediated immunity,* the individual is protected against pathogens that have invaded and are located *inside* body cells. All your body cells have molecules on their cell membranes that identify them as "self" to the cells of your immune system—as cells that belong. Body cells that have been infected by a pathogen (and many cancer cells) present some different proteins (antigens) on their membranes. When a T cell recognizes one of these antigens (because the T cell has the matching antigen receptors on its surface), it binds to the antigen. This binding initiates an immune response that kills the infected cell and, usually, the pathogen within.

To defeat a specific pathogen, numerous identical lymphocytes (bearing the same antigen receptor) are needed. You may be starting to ask yourself, do we have trillions of immune cells? How does the body have room for all of these individual armies of lymphocytes ready to face any enemy it might—or might not ever—encounter? The answer is that the body doesn't store vast quantities of identical lymphocytes. Rather, for each of the billion or more different antigen-recognizing receptors, it has just a small number of lymphocytes with those receptors. It makes copies of these lymphocytes *only if* the antigen is encountered. In the next section, we see how the body does this.

TAKE HOME MESSAGE 27.7

» Lymphocytes are a type of white blood cell. Two types of lymphocytes are associated with the specific immune system: B cells and T cells. B cells are responsible for the humoral response, and T cells are responsible for the cell-mediated response. Because of the diversity and specificity of lymphocyte receptors, almost any pathogen can be recognized by the body's B and T cells.

27.8 Clonal selection helps in fighting infection now and later.

Just about every lymphocyte in an individual is unique, because of its specific antigen receptor. Although the immune system is *ready* to encounter any antigen, many lymphocytes will never meet the antigen they are capable of recognizing. But when a lymphocyte does come into contact with the antigen specific to its receptor, a sequence of events begins—and it doesn't end until the antigen is destroyed.

When a B or T cell binds to an antigen, the lymphocyte and its descendants rapidly divide numerous times to create a population of genetically identical cells (clones), all with the same antigen specificity. This process, known as **clonal selection,** ensures that there are enough B and T cells to recognize and respond to all of the specific pathogens that have invaded the body.

Recognizing a new pathogen is an important first step, but the specific immune system has two additional challenges: to *respond* to and *remember* the invader. Clonal selection generates many lymphocytes that can recognize the same pathogen, and these lymphocytes then function in one of two ways. Some attack the antigen (these are called the effector cells), and others are involved in creating a memory of the invasion (the memory cells). We discuss the immune system's initial interaction with a pathogen, called the **primary response,** first.

The cells of the primary immune response must first be generated through clonal selection. Once generated, these lymphocyte responders, or **effector cells,** recognize the antigen and immediately take some action that leads to its destruction. For example, **plasma cells,** derived from B cells, are the effector cells in the humoral response because they secrete antibodies. And T cells that directly kill infected cells are the effector cells in the cell-mediated response.

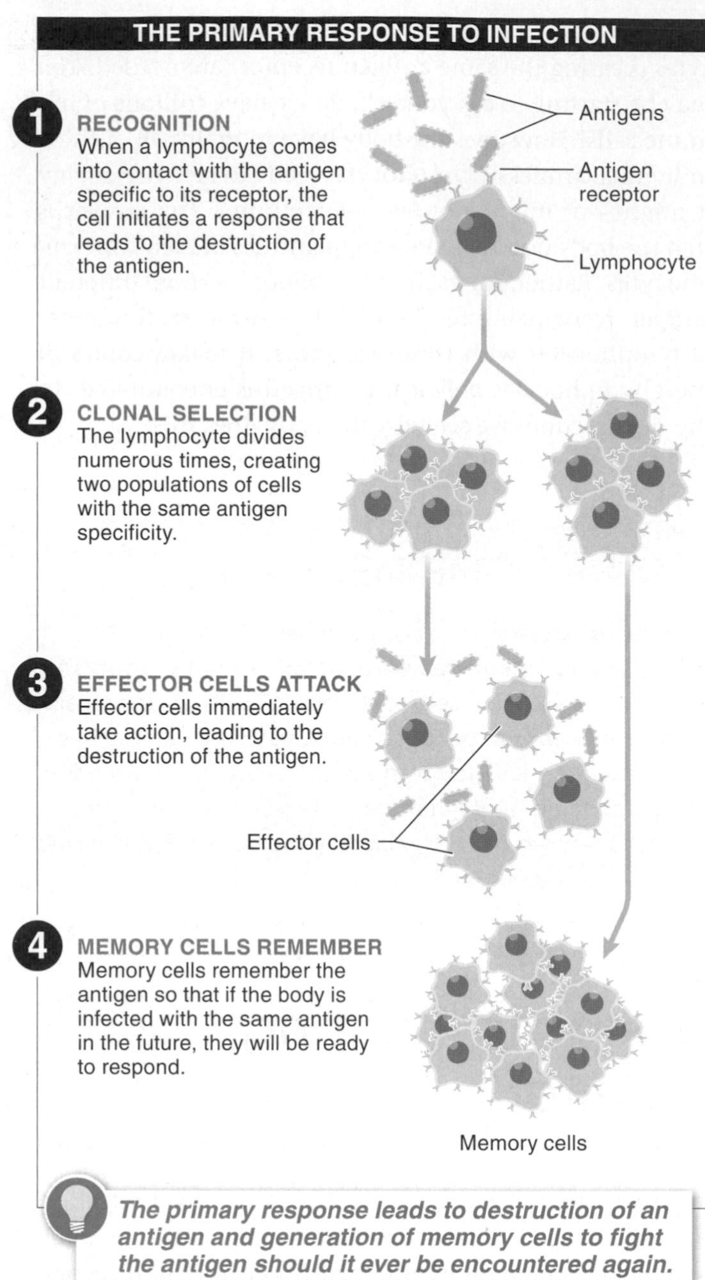

THE PRIMARY RESPONSE TO INFECTION
1 RECOGNITION
When a lymphocyte comes into contact with the antigen specific to its receptor, the cell initiates a response that leads to the destruction of the antigen.
Antigens
Antigen receptor
Lymphocyte
2 CLONAL SELECTION
The lymphocyte divides numerous times, creating two populations of cells with the same antigen specificity.
3 EFFECTOR CELLS ATTACK
Effector cells immediately take action, leading to the destruction of the antigen.
Effector cells
4 MEMORY CELLS REMEMBER
Memory cells remember the antigen so that if the body is infected with the same antigen in the future, they will be ready to respond.
Memory cells
The primary response leads to destruction of an antigen and generation of memory cells to fight the antigen should it ever be encountered again.

FIGURE 27-24 **Fighting now and later.**

On average, it takes two weeks to produce enough effector cells to combat and vanquish an infection (FIGURE 27-24).

Also produced during the primary response is the second type of lymphocyte, the **memory cells.** Like effector cells, memory cells are produced through clonal selection. The memory cells' job is to remember an antigen so that, if the body is infected with the same antigen in the future, it will be ready to attack the invader relatively quickly (in a matter of a couple of days rather than two weeks). Thus, memory cells remain in the lymph and blood and wait for the return of their specific antigen. There are both B and T memory cells, and they have the same antigen specificity as the effector cells produced in the primary response. While effector cells live less than a week, memory cells remain and circulate for a long time. The lifespan of memory cells is variable, but they may exist for years and possibly even for the individual's lifetime.

With this understanding of how effector cells and memory cells act in the primary immune response, let's look at what happens when a five-year-old kindergartener comes to school not knowing that he has a chicken pox infection. Imagine that the kindergarten teacher had chicken pox as a small girl. At that time, her specific immune system produced a population of chicken pox memory cells during her primary immune response. When the teacher interacts with the infected boy, these specific memory cells go through clonal selection and give rise to new B and T effector and memory cells armed to fight off the chicken pox virus. This **secondary response,** the creation of B and T effector cells, occurs rapidly (beginning after as little as 1–2 days) and produces a more intense response (peaking at 7–9 days) than the primary response that occurred in her childhood (FIGURE 27-25). Because the

THE SECONDARY RESPONSE TO INFECTION

Memory cells produced during a primary immune response (such as to chicken pox) enable you to mount a faster, stronger secondary response following exposure to the virus later in life, preventing illness.

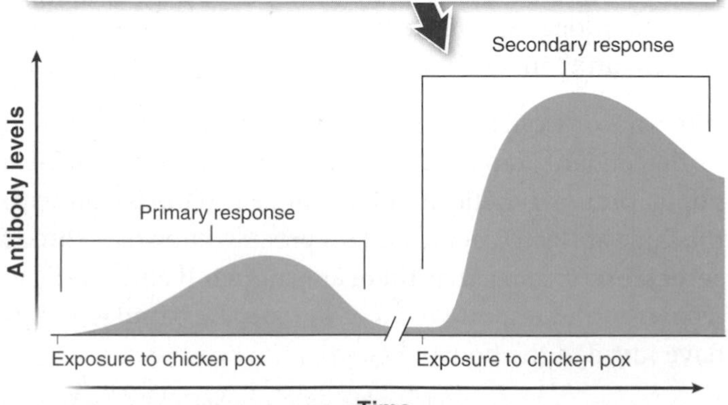

FIGURE 27-25 **Another encounter with the same antigen.**

B cells of humoral immunity and the T cells of the cell-mediated response work immediately to destroy the virus, the teacher won't even miss a day of work, because this secondary immune response attacks the chicken pox virus so rapidly and effectively.

TAKE HOME MESSAGE 27.8

» When the body encounters a specific antigen, the lymphocytes recognizing this antigen divide to produce many identical cells through clonal selection. During a primary response, effector cells respond to and attack the antigen. Memory cells produced during the primary response are ready to go through clonal selection if a secondary response is necessary.

THIS IS HOW WE DO IT

Developing the ability to apply the process of science

27.9 Does contact with dogs make kids healthier?

Infants get sick a lot. On average, a baby experiences three to six episodes of respiratory tract infection during the first year of life. There is a lot of variation, however, from one household to another. Numerous factors affect the frequency and severity of childhood illnesses, such as attending day care, having older siblings, and a parental history of asthma or smoking.

Recently, another factor was identified. Studies show that children exposed to animals—including pets—early in childhood seem to get sick *less* frequently and to experience *fewer* infections, allergies, and asthma attacks. These observations have even given rise to a name for this phenomenon: the "hygiene hypothesis," which states that reduced exposure to infectious agents—resulting from improved hygiene—suppresses development of the immune system in children. And this, in turn, leaves them more vulnerable to illness later in life.

Can it be true that our "reward" for cleaner living is that we are more likely to get sick?

Could a young child's exposure to infectious agents really be a good thing? As we saw in the previous section, exposure to pathogens (and the antigens that mark them) causes our immune system to be more effective at fighting later infections. Our immune system learns and remembers. Without early challenges, we may be depriving it of important "learning" opportunities.

This is an issue almost everyone is interested in, and one with significant public health implications, so it's worth exploring more. Of course, it would not be ethical to deliberately expose infants to pathogens in a randomized, controlled, double-blind type of study. So how could this problem be investigated?

In 2012, some researchers reported a study in which they examined the effect of contact with dogs and cats on the frequency of respiratory tract illnesses during the first year of life. Using a "prospective birth cohort study," they observed 397 children until the age of one year, beginning when their mother was pregnant. They collected information on the children through weekly diaries kept by the parents, including information about any contact with dogs or cats and any respiratory symptoms or infections.

Can you think of some difficulties in conducting a prospective cohort study? Some strengths?

Several features of prospective cohort studies make them a valuable tool for scientific thinking. Defining before the study period begins the groups—for example, infants having frequent or infrequent contact with pets—and the outcomes that will be measured can reduce potential sources of bias. And collecting data regularly throughout the study increases accuracy and reduces the likelihood of "recall error" (such as might happen if parents were simply asked about their child's illnesses over the past year).

How much healthier are kids living with a dog?

The results were dramatic. Children having dogs at home experienced 31% fewer respiratory infections and 44% fewer ear infections, and received 29% fewer antibiotic prescriptions. (The researchers found that cat ownership, too, provided a protective effect, but it was much weaker than for dog ownership.)

In their study, the researchers noted only the amount of animal exposure and the incidence of respiratory tract infections, so they could not offer a definitive explanation for their findings. They speculated, however, that animal contacts were responsible for strengthening the immune system. Specifically, they proposed that (1) animals bring dirt into the home, (2) the more dirt brought in, the greater the diversity of bacteria in the home, and (3) exposure to greater bacterial diversity has a positive effect on maturation of the child's immune system.

Does it matter whether the dog is indoors a lot or a little?

In some of the homes with dogs, the dog was only occasionally inside (less than 6 hours per day), while in others, the dog was inside "often" (6–16 hours per day) or "mostly" (more than 16 hours per day). Which condition do you think produced the greatest health benefit? As it turned out, the greatest benefit occurred when the dog was inside only occasionally. The researchers suggested that this was because dogs living mostly indoors probably brought in less dirt (and microbes) than the dogs spending most of their time outdoors.

Should we strive to create the most hygienic environment possible for a baby?

So, how clean should an infant's environment be? This is a hugely important question and one that the researchers weren't sure they could answer yet. Still, their finding adds to a growing body of evidence supporting the hygiene hypothesis and suggesting that there are benefits to an environment in which a developing child's immune system is challenged.

TAKE HOME MESSAGE 27.9

» Babies living in homes with dogs gain protection from respiratory illnesses during the first year of life, perhaps because challenges to the immune system make it more effective at fighting infections later.

27.10 Cytotoxic T cells and helper T cells serve different functions.

The specific immune response, as we saw in Section 27.7, has two parts: the humoral response (B cells) and the cell-mediated response (T cells). While antibodies operate in the realm of the blood and lymph, it is the role of the T cell-mediated response to fight pathogens that are already inside cells. We focus here on this T cell-mediated response, and begin by examining the two major types of T cells, both of which are lymphocytes.

1. **Cytotoxic T cells** are the effectors (the first responders) of the cell-mediated response; they directly kill cells infected with pathogens.

2. **Helper T cells** do not directly kill infected cells but, instead, stimulate other immune cells. Helper T cells stimulate B cells to produce antibodies and cytotoxic T cells to kill infected cells.

We'll follow a viral infection through the cell-mediated response to demonstrate one example of how T cells function (FIGURE 27-26).

When a pathogen enters the body, the phagocytes and NK cells of the non-specific system respond. At the same time, dendritic cells and macrophages call on the specific system by "advertising" the presence of antigens to circulating lymphocytes. Dendritic cells and macrophages (as well as B cells) that "present" digested particles of pathogens on their cell surfaces as a means of advertising are called **antigen-presenting cells.** Presenting cells are essential to all T cell functioning, because T cells can recognize infected cells only by the "non-self" antigens presented on their plasma membranes.

Circulating through a lymph node or the spleen, a macrophage may bump into helper T cells with specificity for the viral antigen being presented on the macrophage surface. The helper T cell receptor recognizes the antigen, and the T cell locks onto the antigen-presenting macrophage. Once this binding has taken place, the helper T cell has been activated. Activated helper T cells quickly ramp up an immune response by producing signaling molecules (cytokines) that activate the other type of T cell, the cytotoxic T cells. This is important because cytotoxic T cells are the only T cells that kill infected cells.

Both helper T cells and cytotoxic T cells then undergo clonal expansion, producing vast numbers of memory and effector cells with specificity for the viral antigen. Other cytokines produced by the activated helper T cells provide the final signals that make the cytotoxic T cells "mature" and ready to fight the pathogen at the site of infection. You may notice this process occurring when you've been sick: the rapid division of lymphocytes in the lymph nodes causes the nodes to swell (usually called "swollen glands"). The helper T cells and cytotoxic T cells then leave the lymph nodes or spleen and circulate throughout the body. (Helper T cells also are required to stimulate B cells to produce antibodies; B cell development occurs only in response to these signals from helper T cells.)

Mature cytotoxic effector cells are like newly trained soldiers, ready to fight the enemy, but first they need to detect where the enemy is hiding. Consider a cell in the throat that is infected with a respiratory virus. Is the cell revealing that it's under attack? Yes! The infected cell displays some of the pathogen's molecules on its surface receptors, advertising

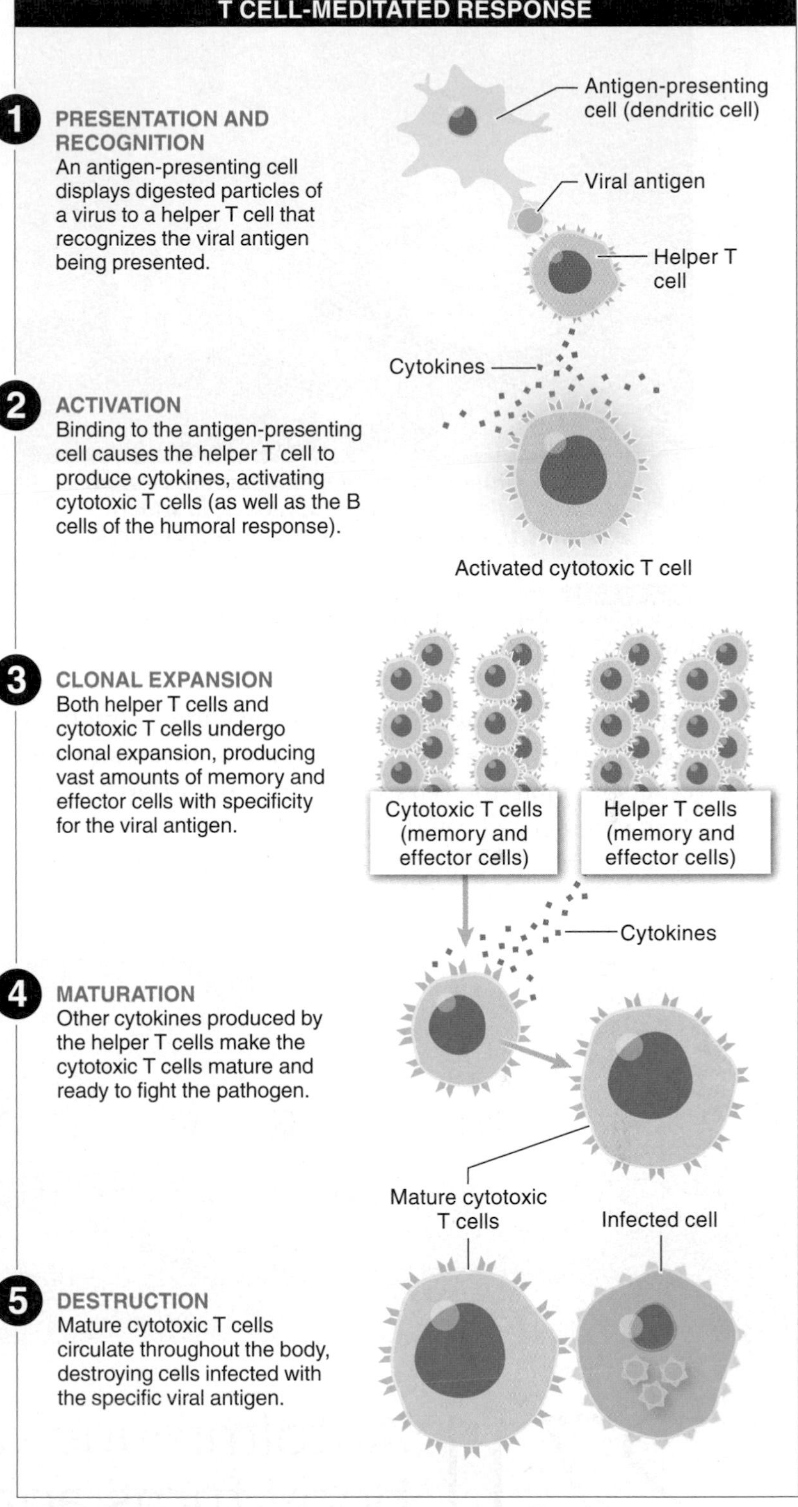

T CELL-MEDITATED RESPONSE
1 PRESENTATION AND RECOGNITION
An antigen-presenting cell displays digested particles of a virus to a helper T cell that recognizes the viral antigen being presented.
Antigen-presenting cell (dendritic cell)
Viral antigen
Helper T cell
Cytokines
2 ACTIVATION
Binding to the antigen-presenting cell causes the helper T cell to produce cytokines, activating cytotoxic T cells (as well as the B cells of the humoral response).
Activated cytotoxic T cell
3 CLONAL EXPANSION
Both helper T cells and cytotoxic T cells undergo clonal expansion, producing vast amounts of memory and effector cells with specificity for the viral antigen.
Cytotoxic T cells (memory and effector cells)
Helper T cells (memory and effector cells)
Cytokines
4 MATURATION
Other cytokines produced by the helper T cells make the cytotoxic T cells mature and ready to fight the pathogen.
Mature cytotoxic T cells
Infected cell
5 DESTRUCTION
Mature cytotoxic T cells circulate throughout the body, destroying cells infected with the specific viral antigen.

FIGURE 27-26 **Helper T cells and cytotoxic T cells work together to recognize and respond to pathogens.**

the infection to the now numerous cytotoxic T cells that can recognize that specific respiratory virus.

How do the cytotoxic effector cells fight the internal pathogen? They do something drastic—they kill the cell

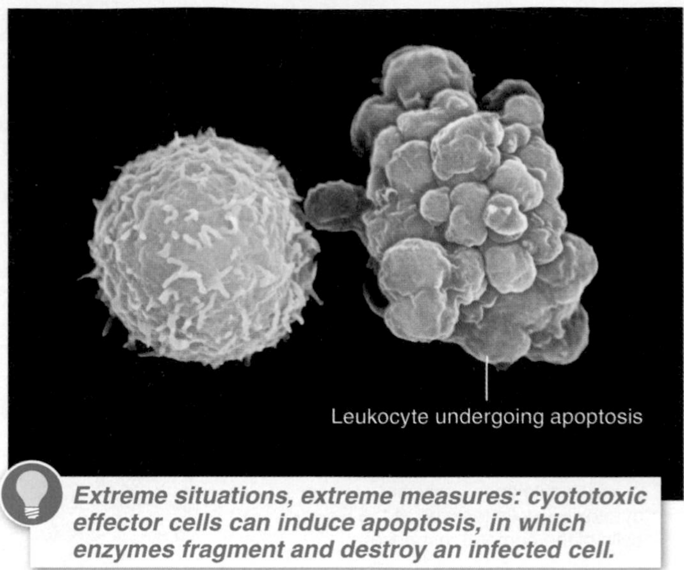

Extreme situations, extreme measures: cytotoxic effector cells can induce apoptosis, in which enzymes fragment and destroy an infected cell.

FIGURE 27-27 **Programmed cell death.** Shown are two leukocytes (white blood cells). The one on the right is undergoing apoptosis.

that is infected. The weapons that they use are proteins: some punch holes in the cell's plasma membrane, and others promote the cell's self-destruction, or **apoptosis.** This programmed cell death does not explode the cell—sending virus particles everywhere. Just the opposite: it is a neatly organized, well-orchestrated process carried out by enzymes that break down macromolecules inside the cell (including the pathogen), fragment the cell into smaller pieces, and package the pieces in vesicles (FIGURE 27-27). The vesicles are then engulfed and digested by phagocytes. So, while some body cells are lost in the battle, ultimately it is for the good of the entire organism.

Because memory helper T cells and memory cytotoxic T cells are also made during an infection, this same virus cannot cause illness again. Memory cells will leap into action to make more effector cells if the virus ever returns.

TAKE-HOME MESSAGE 27.10

» Antibodies, produced by B cells, cannot destroy pathogens that are inside cells. The specialization of cytotoxic T cells is required to kill infected cells. Antigen-presenting cells, such as macrophages, display antigens to circulating helper T cells and cytotoxic T cells, to alert the specific immune system that an infection is under way. In turn, helper T cells produce cytokines that instruct cytotoxic T cells to mature and respond to the infection.

27.11–27.13

Malfunction of the immune system causes disease.

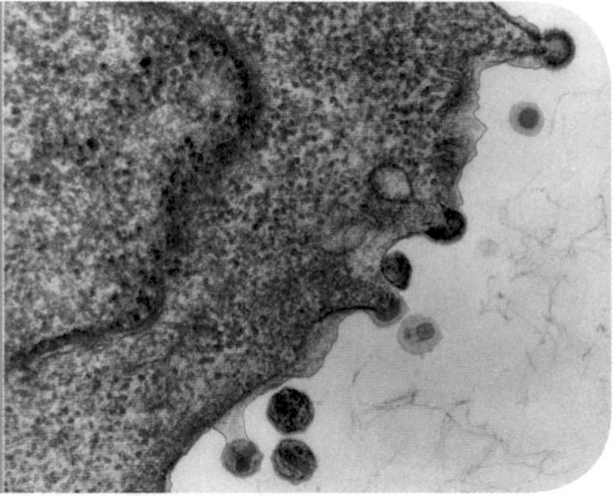

HIV viruses being released from lymph tissue.

27.11 Autoimmune diseases occur when the body turns against its own tissues.

What happens when the immune system malfunctions? Genetic defects, environmental influences, and even viruses can trigger immune dysfunction. And in many cases, such as autoimmune diseases, a combination of causes is suspected, but not completely understood.

Autoimmunity occurs when an individual's immune system responds inappropriately to the individual's own cells and tissues as if they were pathogens, mistaking "self" for "non-self." If immunity can be compared to warfare, then autoimmunity is equivalent to an army turning its weapons upon its own citizens. Lymphocyte receptors wrongly recognize an individual's own molecules or cellular structures as antigens, and the humoral and/or cell-mediated immune responses can be initiated by these antigens. Most autoimmune diseases result in significant tissue and organ damage, and some can even result in death.

Type 1 diabetes (sometimes called juvenile diabetes) is an example of an autoimmune disorder. It is diagnosed in more than 13,000 young people in the United States each year. Symptoms include high blood glucose levels, weight loss, unquenchable thirst, frequent urination, fatigue, and weakness. In this disease, cytotoxic T cells, complement proteins, and macrophages destroy the person's own pancreatic cells. Without a normal pancreas, a person cannot make insulin (FIGURE 27-28). Without insulin, cells in need of glucose are not able to take it up from the blood, even though glucose is plentiful. In turn, the body breaks down muscle and fat tissues to provide cells with needed energy, and blood glucose levels remain high. The exact cause of this autoimmune reaction is not understood. Treatment for individuals with type 1 diabetes is insulin injections or implantation of an insulin pump, which directly monitors blood glucose levels and releases insulin as needed.

In some disorders, including type 1 diabetes, the immune system destroys a single cell type or a single organ. In other disorders, the damage is spread throughout the body. When the insulation that surrounds nerve fibers of the brain and spinal cord is under immune attack, the resulting disorder is called **multiple sclerosis (MS)** (see Section 24.5). Individuals with MS usually begin to exhibit symptoms between 20 and 40 years of age, and these symptoms include blurred vision, weakness in the arms and legs, trouble with balance, and tingling sensations (FIGURE 27-29). This disorder tends to affect women more than men, and it affects more than 300,000 people in the United States. There is no cure for MS, but anti-inflammatory drugs can reduce the severity of some symptoms. Many new drugs are being investigated, including some that block the inflammatory response.

Multiple sclerosis (MS) results from an autoimmune attack on the insulation surrounding neurons. People with MS may experience arm and leg weakness and trouble with balance.

FIGURE 27-29 **An autoimmune disorder: multiple sclerosis.**

Rheumatoid arthritis is another autoimmune disease that affects women more frequently than men, and it usually strikes between ages 40 and 60. (This form of arthritis is a far less common cause of joint inflammation than osteoarthritis, which is not an autoimmune disease.) The linings of joints are the targets of this immune attack—leading to pain and severe joint swelling that can result in deformities. Various pharmaceuticals are available to help reduce symptoms, and joint replacement surgeries also provide relief for individuals suffering from rheumatoid arthritis.

AUTOIMMUNITY: TYPE 1 DIABETES

Type 1 diabetes is an autoimmune disorder in which cytotoxic T cells destroy the body's own pancreatic cells.

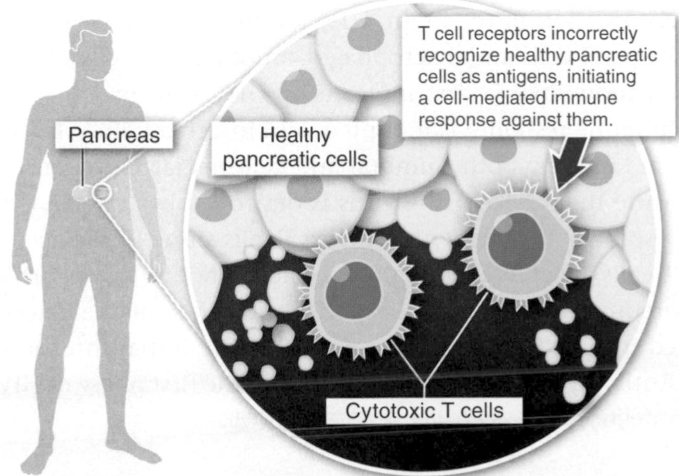

T cell receptors incorrectly recognize healthy pancreatic cells as antigens, initiating a cell-mediated immune response against them.
Pancreas
Healthy pancreatic cells
Cytotoxic T cells

FIGURE 27-28 **Type 1 diabetes is an all-too-common autoimmune disorder.**

TAKE HOME MESSAGE 27.11

» When lymphocytes bear receptors that inappropriately recognize structures of a person's own body as foreign invaders, autoimmunity develops. Autoimmune responses can do significant damage to specific organs or to tissues throughout the body, depending on where the antigens are located.

27.12 AIDS is an immune deficiency disease.

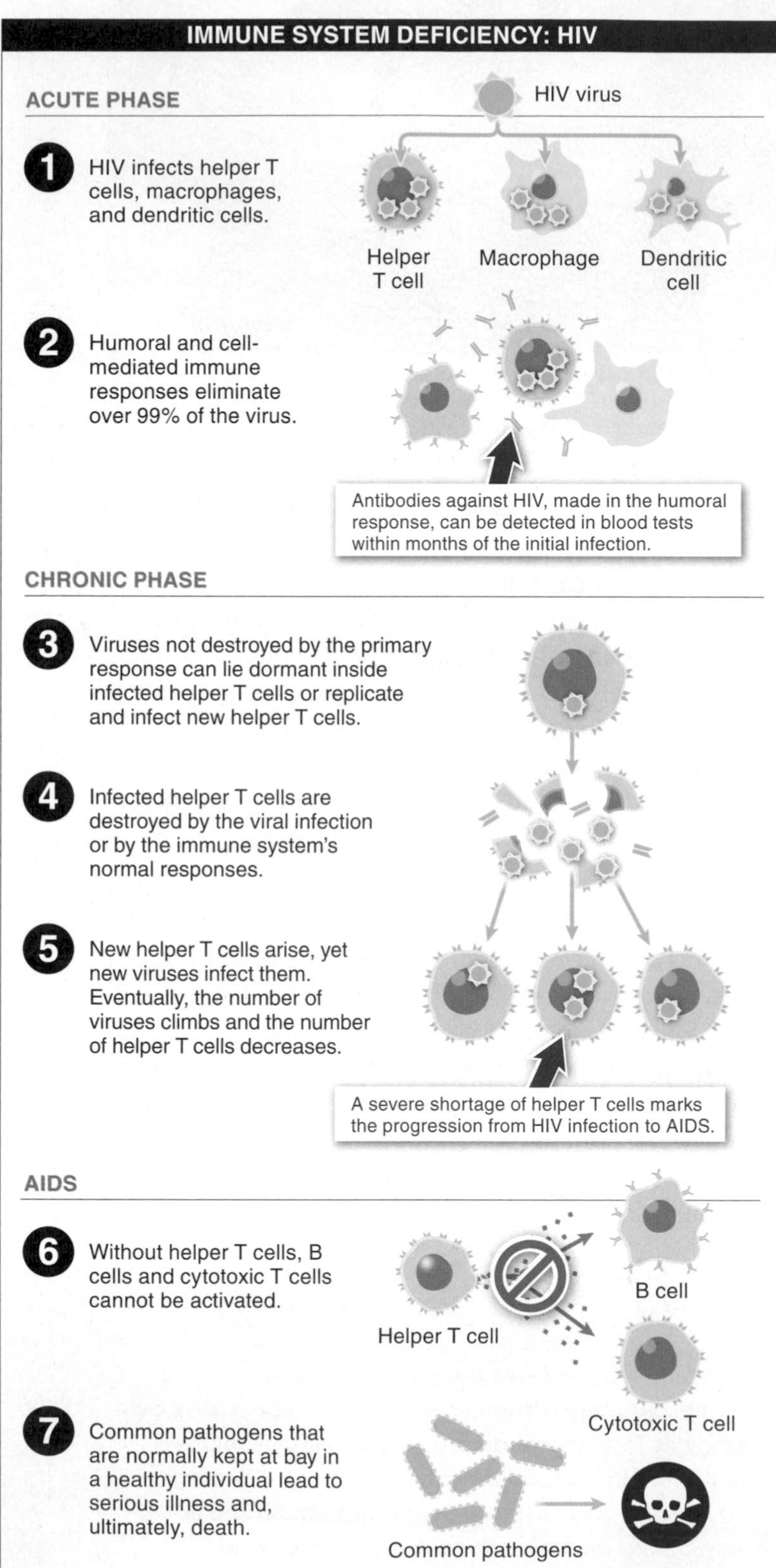

IMMUNE SYSTEM DEFICIENCY: HIV
ACUTE PHASE
HIV virus
1 HIV infects helper T cells, macrophages, and dendritic cells.
Helper T cell
Macrophage
Dendritic cell
2 Humoral and cell-mediated immune responses eliminate over 99% of the virus.
Antibodies against HIV, made in the humoral response, can be detected in blood tests within months of the initial infection.
CHRONIC PHASE
3 Viruses not destroyed by the primary response can lie dormant inside infected helper T cells or replicate and infect new helper T cells.
4 Infected helper T cells are destroyed by the viral infection or by the immune system's normal responses.
5 New helper T cells arise, yet new viruses infect them. Eventually, the number of viruses climbs and the number of helper T cells decreases.
A severe shortage of helper T cells marks the progression from HIV infection to AIDS.
AIDS
6 Without helper T cells, B cells and cytotoxic T cells cannot be activated.
Helper T cell
B cell
Cytotoxic T cell
7 Common pathogens that are normally kept at bay in a healthy individual lead to serious illness and, ultimately, death.
Common pathogens

Autoimmune diseases demonstrate how immune recognition can fail and have devastating effects on the body. Here we examine a case in which recognition of a true pathogen occurs, but the *response* is deficient. Normally, the immune system destroys body cells that are infected with a virus. But when a virus infects immune cells, the body's ability to respond to infection is impaired. The **human immunodeficiency virus (HIV),** which causes **AIDS (acquired immune deficiency syndrome),** infects immune system cells, including helper T cells—cells that are crucial to survival. With AIDS, a deficiency of helper T cells leads to complete failure of the specific immune response. As a result of this failure, individuals are prey to, and may die from, infections that a healthy immune system would have no trouble defeating.

HIV infection occurs through contact with blood, semen, vaginal fluid, or breast milk from an infected person. Thus, using clean (not shared) intravenous needles and using condoms are effective precautionary measures against HIV infection. People who become infected with HIV often do not have immediate symptoms and often do not know they are infected. Because individuals are contagious once they are infected, this initial lack of symptoms contributes to the spread of the disease. Let's examine the details of how immunity diminishes as the immune system loses its battle against HIV.

Different viruses infect different kinds of body cells. Parvovirus B19 infects heart muscle, for example, causing an inflammation of the heart, and West Nile virus attacks brain tissue and can cause encephalitis, a dangerous inflammation of the brain. HIV, on the other hand, infects helper T cells and macrophages, among other cells. When an individual is first infected by HIV, a normal specific immune response is mounted, and infected helper T cells are killed by both humoral and cell-mediated responses. More than 99% of the virus is destroyed in this primary response. Although many helper T cells are lost, new helper T cells arise. As you might expect, antibodies against HIV are made in a humoral response and can easily be detected in blood tests within months of the initial infection. Antibody detection marks the end of the first phase of HIV infection, the *acute phase* (FIGURE 27-30).

FIGURE 27-30 **The progression from HIV infection to AIDS.**

Unfortunately, the viruses that were not destroyed by the primary response can lie dormant inside infected helper T cells and macrophages. The infection then enters a period called the *chronic phase,* in which the person is infectious but doesn't have outward symptoms. This period is variable but rarely exceeds 12 years. Although the infection may seem inactive, a battle is going on within the immune system. Some of the HIV viruses lie dormant, but others replicate and infect new helper T cells. Helper T cells that are infected are killed by the viral infection or are destroyed by the immune system's normal responses. New helper T cells arise, and new viruses infect them. This cycle continues, but it is not a balanced situation. Sooner or later, the number of viruses climbs and the number of helper T cells decreases, and serious clinical symptoms appear. (This process is like an army valiantly fighting a battle but ultimately running out of soldiers.)

A severe shortage of helper T cells marks the progression from HIV infection to AIDS. Without helper T cells, B cells and cytotoxic T cells cannot be activated (recall from Section 27.10 that helper T cells stimulate the production of B cells and cytotoxic T cells). Common pathogens that are normally kept at bay in a healthy individual—such as a parasitic protist that causes diarrhea, a fungus that causes "thrush" on the tongue and throat, and viruses that can cause skin lesions and swelling—ultimately lead to debilitating illness and death.

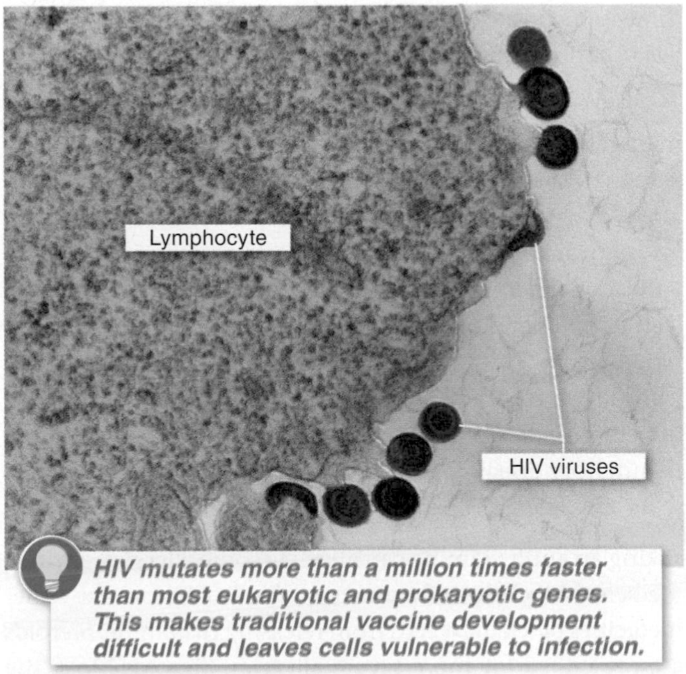

Lymphocyte

HIV viruses

HIV mutates more than a million times faster than most eukaryotic and prokaryotic genes. This makes traditional vaccine development difficult and leaves cells vulnerable to infection.

FIGURE 27-31 **An ever-changing foe.**

Q Why is there no vaccine against AIDS yet?

As we saw in Section 15.19, HIV is a rapidly mutating, RNA-containing virus, with surface proteins that can change with each new round of replication (FIGURE 27-31). This ever-changing nature of HIV makes development of vaccines against it challenging. Consider a vaccine that is made using a surface glycoprotein of HIV as the antigen. The vaccine becomes obsolete immediately, because the viral antigen is changing too quickly. More than 25 years after isolating the virus, the traditional method of making vaccines against viral antigens has failed because of the constantly mutating HIV. Interestingly, some people have a mutation in a T cell receptor that makes it more difficult for the virus to enter their helper T cells. Ongoing studies of individuals who are naturally resistant to HIV are helping scientists develop new potential vaccines and drug treatments.

Current therapies for HIV infection have been successful at increasing the length and quality of life for those infected. The regimen, called combination therapy, includes numerous drugs that affect HIV's ability to replicate or infect new cells. These drugs keep the number of viruses low and maintain a suitable number of helper T cells. HIV resistance to the drugs, however, is a fear shared by patients and clinicians. Nonetheless, the life span for individuals taking combination therapy is increasing and getting close to a normal length of life. Although there is not a cure yet, individuals who have access to treatment are able to live with HIV as a chronic condition. (Unfortunately, treatment is unavailable for millions of infected people in developing nations.) New therapies aim to find latent viruses that hide inside helper T cells, sometimes for decades, and eliminate them. In the future, HIV may be a disease that can be cured, but for now it continues to be a lifelong condition and a challenge to scientists.

Looking at a map depicting HIV/AIDS cases around the world, we can see just how widespread this virus is (FIGURE 27-32). Sub-Saharan Africa is the region most severely affected. In some countries in this region, such as Swaziland and Lesotho, close to 25% of the adult population is infected with HIV. In South Africa, more than 5 million people, or almost 20% of its adult population, are infected. In contrast, less than 1% of the U.S. population is infected with HIV. Not everyone who is HIV-infected has AIDS. For example, estimates indicate that of the more than one million people infected by HIV in the United States, in fewer than 40,000 cases has the infection progressed to AIDS.

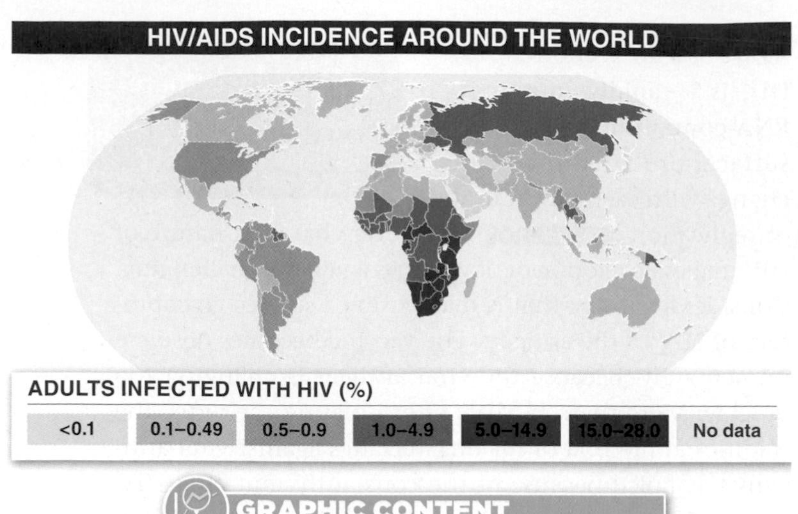

GRAPHIC CONTENT
Thinking critically about visual displays of data
Turn to p. 971 for a closer inspection of this figure.

FIGURE 27-32 **HIV/AIDS is most widespread in sub-Saharan Africa.**

TAKE HOME MESSAGE 27.12

» **AIDS is an immune system disease caused by the human immunodeficiency virus (HIV), which infects helper T cells—immune cells that are crucial to an individual's survival. As helper T cells are killed by the infection or by the body's own immune cells in response to the infection, the deficiency of helper T cells leads to complete failure of the specific immune response, resulting in illness and death from infections that a healthy immune system could defeat. There is currently no vaccine for HIV, because traditional vaccination methods do not work against this rapidly mutating virus, nor is there a cure for AIDS.**

27.13 Allergies are an inappropriate immune response to a harmless substance.

A three-year-old girl bites into a peanut butter sandwich. She immediately develops hives around her mouth and begins vomiting. She is having an allergic reaction. This is only the second time she's eaten peanut butter. Unfortunately, her story is not unusual. The CDC estimates that approximately 3 million children in the United States have food allergies, 90% caused by just eight foods: peanuts, tree nuts, milk, eggs, fish, shellfish, soy, and wheat. **Allergies** are the result of an inappropriate immune response to what should be a harmless substance.

In discussing antibodies earlier in the chapter, we did not focus on the groups of antibodies that differ in their constant (heavy) region at the base of the "Y." There are five classes of antibodies, differing in their chemical structure in this region. When antibodies of one of these classes bind to mast cells (the white blood cells found in tissues), they cause the mast cells to release histamine and cytokines, both of which cause local blood vessels to become dilated and leaky (as we've seen, standard parts of the inflammatory response), and recruit other immune cells to the site.

This mast cell-binding class of antibodies can cause much harm to the body if they are made in response to an **allergen,** an antigen that causes an allergic response. On first encounter, an allergen induces a normal humoral response: memory cells form, and plasma cells secrete antibodies specific to the allergen. The antibodies bind to the surface of mast cells and can remain bound for months and maybe even years. The mast cells are now "sensitized" to the allergen. In a second exposure, the allergen binds to the antibodies that are still attached to the mast cells—which are now "activated." This time, allergen binding causes the mast cells to release histamine, which causes the blood vessels to dilate and become leaky, and inflammation ensues (FIGURE 27-33).

The effect of different allergens reflects which particular mast cells have been exposed to the allergen and thus activated. Most allergic reactions occur in the digestive or respiratory tracts, because this is where the body first encounters an eaten or inhaled allergen. Allergic reactions associated with food often lead to vomiting or diarrhea, but can also lead to hives and other symptoms if the allergen enters the bloodstream. Respiratory allergens, such as pollen and dust, can cause a runny nose and teary eyes ("hay fever") in the upper respiratory tract, or can result in asthma, a chronic inflammatory disease of the lower respiratory tract that causes wheezing, coughing, and shortness of breath.

Taking an antihistamine can alleviate some allergies, as these medicines block the inflammatory effects of histamine. Other medicines block mast cells from releasing histamine. Steroids are prescribed for more severe allergies. Steroids block the production of the chemicals released from immune cells, thus

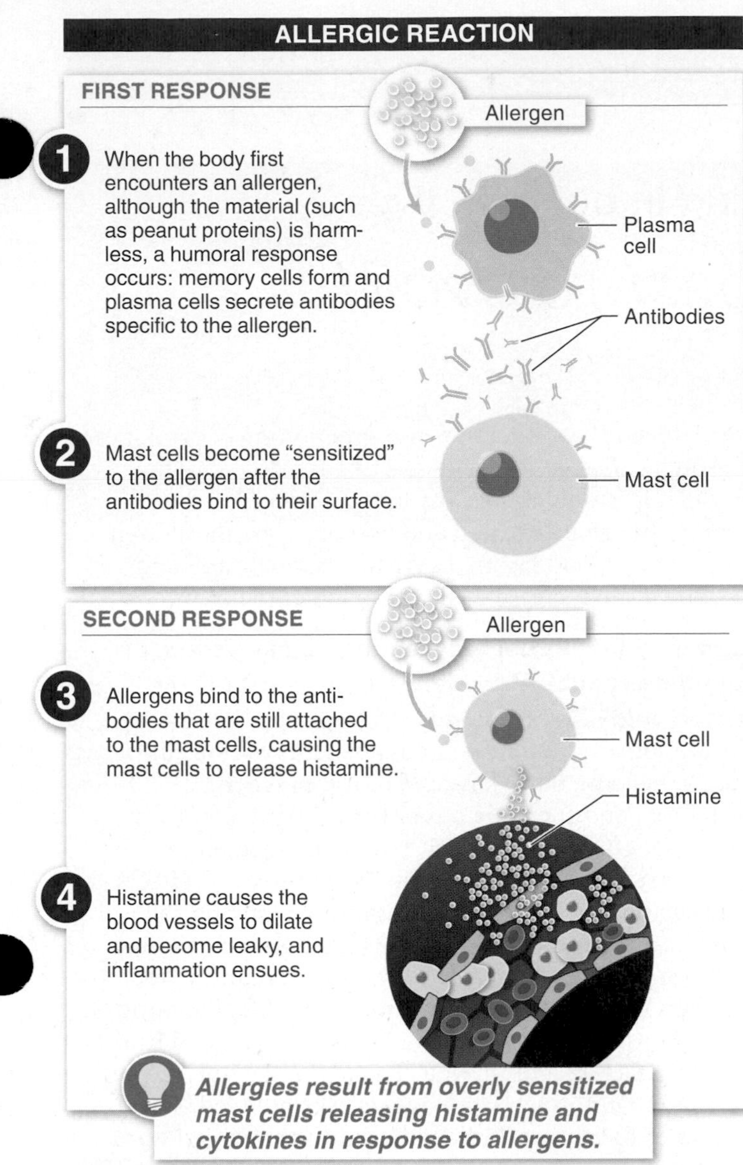

FIGURE 27-33 Allergens induce a humoral response in some individuals.

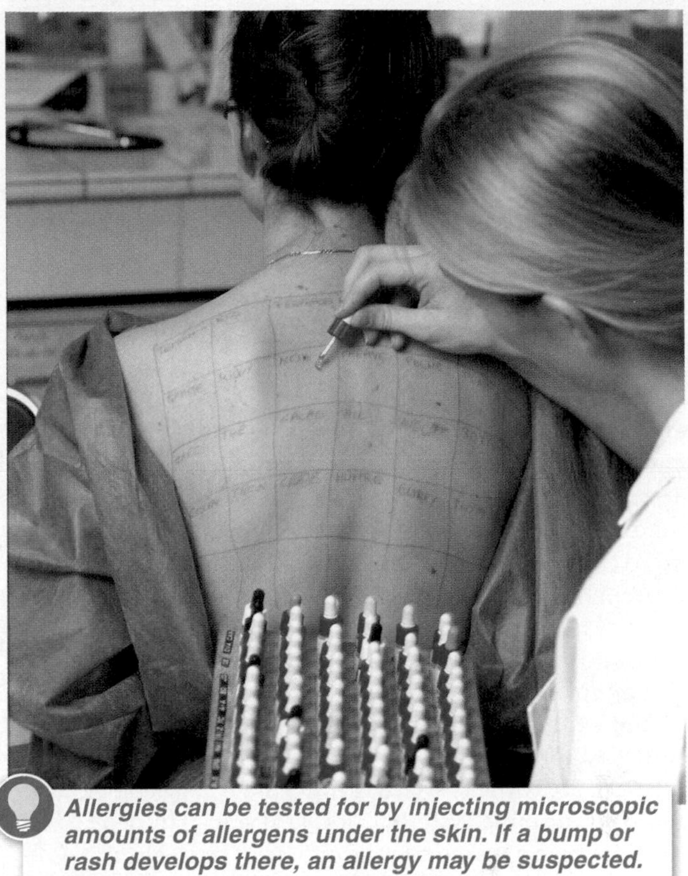

Allergies can be tested for by injecting microscopic amounts of allergens under the skin. If a bump or rash develops there, an allergy may be suspected.

FIGURE 27-34 Screening for allergies.

preventing the migration of more inflammatory cells, such as macrophages and neutrophils, to the site of inflammation. Additionally, steroids can block the killing ability of these phagocytes, resulting in less damage to the inflamed tissue.

An individual with a severe allergic response may experience **anaphylactic shock,** a life-threatening allergic reaction that is systemic, meaning that it is not localized to the site of exposure. Anaphylactic shock can lead to severe respiratory distress, as the throat swells and asthma develops. Swelling in various tissues means there is less fluid in the blood system, and this can lead to dangerously low blood pressure. Individuals with severe allergies often carry with them a dose of adrenaline, also called epinephrine, a chemical found naturally in our bodies (see Section 25.5) and also manufactured by several drug companies. Epinephrine, which can be injected if needed, reverses the effects of histamine on blood vessels, but it must be injected quickly, because death can occur rapidly from anaphylactic shock.

A common method of testing for allergies is called *skin allergy testing,* in which microscopic amounts of one or more potential allergens are injected under the skin—usually on the back or the forearm. The development of a rash can indicate a hypersensitivity to that allergen, enabling the person to take steps to reduce exposure to the allergen (FIGURE 27-34).

TAKE HOME MESSAGE 27.13

» In individuals who are sensitive, an allergen induces a humoral response in which one class of antibodies binds to and activates mast cells. With a second exposure to the allergen, activated mast cells release histamine and other chemicals, resulting in allergy-related symptoms. Swelling and inflammation can be localized or can be systemic and lead to anaphylactic shock.

STREET BIO

Using evidence to guide decision making in our own lives

Facts or fallacies? How best to avoid getting sick

Q: Can I catch a cold from going out with wet hair in the winter? No. Your grandmother was wrong about this one! To catch a cold, you must be infected with a pathogen (such as a rhinovirus) that causes the common cold. Colds do tend to be more common in colder weather, because the viruses are more stable in colder air with low humidity. And droplets of water and virus (such as those that escape when we sneeze and cough) remain airborne slightly longer in dry air than in humid summer air. Also, because people are indoors more often during winter, the increased population in confined quarters increases the incidence and ease of transmission. So, while the virus is affected by the colder weather, your immune system is not. Dry your hair before you leave the house if you don't want icicles in it, but not as a strategy for combating the common cold.

Q: Can I catch a cold because I am not sleeping enough? This gets an emphatic *yes!* In randomized, controlled studies in which individuals were given nasal drops containing rhinovirus, individuals getting less sleep were significantly more likely to develop colds than those sleeping 7 hours or more. And simply resting in bed or having an interrupted night's sleep doesn't count. Seven hours or more of good-quality sleep seems to be important to the immune system, particularly for the production and functioning of cytokines. Interestingly, there is also evidence that individuals who sleep 7–8 hours a night have the lowest rates of heart disease. Although eating well and exercising are most often associated with good health, sleep, too, is important, and neglecting it can increase your vulnerability to illness.

Q: Can drinking herbal teas, eating chicken soup, or taking the product Airborne help me ward off the common cold? *Herbal teas and chicken soup?* Not exactly. There are some symptom-reducing benefits of hot fluids, including keeping nasal passages moist, preventing dehydration, and soothing a sore throat. And one study showed that chicken soup with vegetables seemed to have an anti-inflammatory effect. But while these effects can make an infection less unpleasant by reducing the symptoms, they don't decrease the likelihood of catching a cold or shorten its duration. *Airborne*? For many years, the makers of Airborne made claims that a "double-blind, placebo-controlled study" showed their product had health benefits. In fact, as a result of a class-action lawsuit, it was discovered that no such study was ever conducted. There is no evidence to back up Airborne's claims that it can ward off colds or boost your immune system.

GRAPHIC CONTENT

Thinking critically about visual displays of data

1 Based on this figure, estimate the number of adults in the United States who are infected with HIV. Show your work.

2 Can you identify which countries have the largest number of adults infected with HIV? Explain your answer.

3 From this figure, can you conclude that there are twice as many adults infected with HIV in the United States as there are in Canada? Explain your answer.

4 What can you conclude about the HIV infection rates in South America relative to the infection rates in southern Africa?

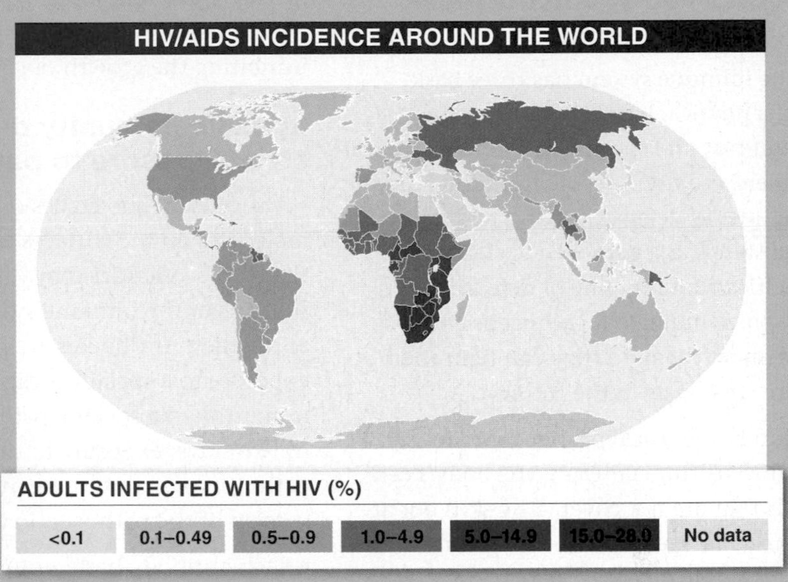

5 Each of the first three colored bars covers a range of less than 1 percentage point, while the darkest red bar covers almost 15 percentage points. Why? How might this difference mislead a reader? Why might this be a useful way to present the data?

See answers at the back of the book.

KEY TERMS IN IMMUNITY AND HEALTH

AIDS (acquired immune deficiency syndrome), p. 966
allergen, p. 968
allergy, p. 968
anaphylactic shock, p. 969
antibody, p. 953
antigen, p. 953
antigen-presenting cell, p. 963
antigen receptor, p. 957
antiseptic, p. 947
apoptosis, p. 964
autoimmunity, p. 965
B cell, p. 958
basophil, p. 950
cilia, p. 945
clonal selection, p. 959
complement protein, p. 949
cytokine, p. 948
cytotoxic T cell, p. 962
dendritic cell, p. 948
effector cell, p. 959
fever, p. 952
helper T cell, p. 963
histamine, p. 950
HIV (human immunodeficiency virus), p. 966
immune system, p. 943
immunity, p. 954
inflammatory response, p. 950
integumentary system, p. 944
interferon, p. 948
lymphocyte, p. 957
lysozyme, p. 945
macrophage, p. 947
mast cell, p. 950
memory cell, p. 960
multiple sclerosis (MS), p. 965
natural killer (NK) cell, p. 948
neutrophil, p. 947
non-specific immunity, p. 943
pathogen, p. 942
physical barriers (immune system), p. 943
plasma cell, p. 959
primary response, p. 959
pus, p. 947
secondary response, p. 960
specific immunity, p. 944
T cell, p. 958
vaccine, p. 954
white blood cell, p. 947

BRIEF SUMMARY

Your body has different ways to protect you against disease-causing invaders.

• The immune system has three basic parts: physical barriers, non-specific immunity, and specific immunity. Physical barriers and non-specific immunity are the first lines of defense, serving to distinguish a substance as a pathogen. Cells in the specific immunity line of defense recognize individual pathogens and remember them, so they can fight them more effectively in the future.

• Skin is a physical barrier that prevents pathogens from entering the body's cells. Cells that are not covered by skin but are exposed to the external environment are protected by defenses such as bacteria-destroying chemicals, acidic secretions, sticky mucus, and wax.

• Non-specific immunity provides defenses against pathogens by recognizing molecules on their cell surfaces and by recruiting other cells to the site of infection or warning them to protect themselves. White blood cells of the non-specific system include phagocytes, cells that display pathogens to cells of the specific immune system, and cells that kill virus-infected cells and cancer cells. Complement proteins also non-specifically recognize invaders and help to destroy them.

• Inflammation is a major way in which pathogens are eliminated by the non-specific immune system. Fever-promoting cytokines help fight infection by stimulating immune responses and inhibiting the growth of some pathogens.

Specific immunity develops after exposure to pathogens.

• Antigens are molecules or fragments of molecules on the surfaces of pathogens that can be identified by disease-fighting proteins of the immune system, called antibodies. Antibodies are produced after exposure to a specific antigen. Long-term immunity to a specific pathogen can form in two ways: exposure to the natural pathogen or exposure to an altered version of the pathogen in a vaccine.

• Each antibody has a unique structure that recognizes a specific antigen.

• Lymphocytes are a type of white blood cell. Two types of lymphocytes are associated with the specific immune system: B cells and T cells. B cells are responsible for the humoral response, and T cells are responsible for the cell-mediated response. Almost any pathogen can be recognized by the body's B and T cells.

• When the body encounters a specific antigen, the lymphocytes recognizing this antigen divide to produce many identical cells through clonal selection. Memory cells produced during the primary response are ready to go through clonal selection if a secondary response is necessary.

• Antibodies, produced by B cells, cannot destroy pathogens that are inside cells. The specialization of cytotoxic T cells is required to kill infected cells. Antigen-presenting cells, such as macrophages, display antigens to circulating helper T cells and cytotoxic T cells, to alert the specific immune system that an infection is under way.

Malfunction of the immune system causes disease.

• When lymphocytes bear receptors that inappropriately recognize structures of a person's own body as foreign invaders, autoimmunity develops.

• AIDS is an immune system disease caused by the rapidly mutating human immunodeficiency virus (HIV), which infects helper T cells. As helper T cells are killed by the infection or by the body's own immune cells in response to the infection, the deficiency of helper T cells leads to complete failure of the specific immune response, resulting in illness and death from infections that a healthy immune system could defeat.

• In individuals who are sensitive, an allergen induces a humoral response in which one class of antibodies binds to and activates mast cells. With a second exposure to the allergen, activated mast cells release histamine and other chemicals, resulting in allergy-related symptoms.

CHECK YOUR KNOWLEDGE

Short Answer

1. Describe the three general ways in which the immune system combats pathogens.

2. What is the function of lysozyme?

3. White blood cells can kill "invader cells" that have entered the body. Why do white blood cells not mount attacks on the body's own cells?

4. The signs of inflammation (following a cut, for example) are related to changes in which body structure(s)?

5. Why is it necessary to get only a single vaccination against the pathogen causing chicken pox, whereas a flu shot is necessary every year?

6. Describe two ways in which you can develop immunity to a disease.

7. Given that our bodies are good at recognizing and attacking "non-self" cells, how is it possible to receive a transplanted organ from another person?

8. Generally, the second time a person is exposed to a particular pathogen, his or her immune system is able to respond more quickly and effectively. Why?

9. Contrast the functions of effector cells and memory cells in fighting infections.

10. How can the self-destruction of an organism's own cells be effective when fighting pathogens?

11. In an autoimmune disease, which part of the immune system is malfunctioning?

12. What characterizes the progression from HIV infection to "full-blown" AIDS?

Multiple Choice

1. Which of the following statements about the integumentary system is incorrect?

a) It is one of the body's three lines of defense against infectious pathogens.

b) It confers specific immunity on an individual, because it confers protection against individual pathogens.

c) It includes sweat glands.

d) It is responsible for the production of tears and ear wax.

e) It helps protect the digestive system from pathogens.

2. Which of the following statements about complement proteins is incorrect?

a) Bound complement proteins are able to recruit other complement proteins to come to the site of a bacterial infection.

b) Complement proteins aggregate to form larger protein complexes that create holes in the bacterial cell surface.

c) Bacterial death results from membrane holes created by complexes formed between complement and surface proteins.

d) Complement proteins surround and bind to the entire surface of a bacterium, marking it to be engulfed by phagocytes and cytotoxic T cells.

e) Complement proteins are able to recognize and bind directly to the surface of bacteria's cell surface proteins.

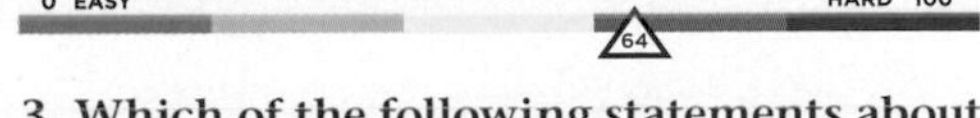

3. Which of the following statements about the inflammatory response is incorrect?

a) The release of histamine by mast cells initiates inflammation.

b) Cells lining blood vessels near inflamed tissue become "stickier" to neutrophils.

c) Pain can occur with inflammation, as increased pressure in an inflamed area stimulates local neurons.

d) The release of histamine causes blood vessels to constrict, reducing blood loss in an injured area.

e) Four signs of inflammation are swelling, redness, heat, and pain.

4. Vaccines:

a) activate a primary response in the vaccinated individual.

b) lead to the creation of memory B cells that provide a rapid secondary response against a specific infectious agent.

c) are sometimes made of just the unique antigens from a pathogen's surface.

d) All of the above are correct.

e) Only b) and c) are correct.

5. The major difference between T cells and B cells is that:

a) T cells develop in the thymus, whereas B cells develop in the bone marrow.

b) T cells do not interact with the non-specific immune system, but B cells do.

c) T cells bind directly to foreign antigens, whereas B cells produce proteins (antibodies) that are secreted and bind to the foreign antigens.

d) All of the above are major differences between T cells and B cells.

e) Only a) and c) are major differences between T cells and B cells.

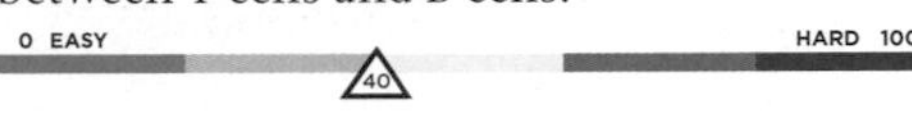

6. Why don't people develop immunity to the common cold?

a) Actually, many people do develop immunity to the common cold and get sick only from bacterial infections.

b) There are at least 200 different viruses that can cause the common cold, and they change over time.

c) Only non-specific immunity helps individuals fight the common cold; specific immunity is not an effective defense against viruses.

d) The rhinovirus that causes the common cold has no surface antigens and so can move through the body without being detected.

e) The pathogen that causes the common cold attacks and kills memory cells, continually "erasing" the immune system's memory of it.

0 EASY HARD 100 13

7. Antibodies:

a) are produced by helper B cells.

b) are produced by helper T cells.

c) are produced by B cells bound by a specific antigen.

d) are produced by plasma cells.

e) are made in the bone marrow.

0 EASY HARD 100 70

8. Which of the following is not a function of helper T cells?

a) They stimulate B cells to proliferate and produce antibodies.

b) They activate cytotoxic T cells.

c) They stimulate B cells to secrete cytokines.

d) They recognize antigens presented on macrophage surfaces.

e) All of the above are functions of helper T cells.

9. Autoimmunity is:

a) one of three lines of defense in the vertebrate immune system.

b) responsible for muscular dystrophy, which occurs when immune cells destroy the myelin sheath, reducing motor control and balance.

c) the culprit in rheumatoid arthritis, which occurs when immune attack on the linings of joints causes inflammation.

d) a consequence of a body's inability to produce antigens, leading it to attack its own cells as if they were pathogens.

e) All of the above statements about autoimmunity are correct.

10. As the human immunodeficiency virus (HIV) begins to kill the host's ________, the host's immune system begins to fail.

a) thymus cells

b) spleen cells

c) red blood cells

d) white blood cells

e) liver cells

11. Hay fever affects millions of individuals every year and is caused by an allergic reaction to pollen. Which of the following is not a step of the process in which a pollen grain causes an allergic reaction in an individual?

a) Mast cells burst when they encounter the pollen during the initial exposure, releasing histamine and other chemicals.

b) During the first exposure, B cells are activated by the pollen and differentiate into antibody-secreting plasma cells.

c) During the second exposure, the pollen grains are recognized by previously created antibodies on mast cells.

d) A type of antibody, specific to the pollen grain, is produced during the initial attack, and these antibodies adhere to mast cells, where they remain.

e) Histamine and other allergen chemicals cause itchy eyes and a runny nose.

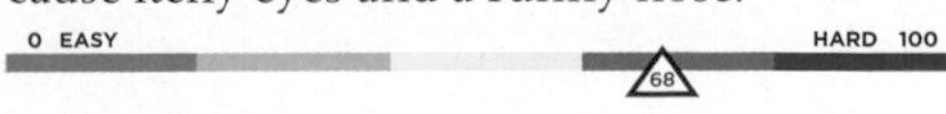

Periodic Table

THE PERIODIC TABLE OF ELEMENTS

ELEMENT CATEGORIES

- Alkali metals
- Alkaline earth metals
- Transition metals
- Lanthanides
- Actinides
- Post-transition metals
- Metalloids
- Other nonmetals
- Halogens
- Noble gases

Key (Be): 4 — Atomic number; Be — Element symbol; Beryllium — Element name; 9.0122 — Atomic weight

PERIOD \ GROUP	1	2	3	4	5	6	7	8	9	10	11	12	13	14	15	16	17	18
1	1 H Hydrogen 1.008																	2 He Helium 4.0026
2	3 Li Lithium 6.94	4 Be Beryllium 9.0122											5 B Boron 10.81	6 C Carbon 12.011	7 N Nitrogen 14.007	8 O Oxygen 15.999	9 F Fluorine 18.998	10 Ne Neon 20.180
3	11 Na Sodium 22.990	12 Mg Magnesium 24.305											13 Al Aluminum 26.982	14 Si Silicon 28.085	15 P Phosphorus 30.974	16 S Sulfur 32.06	17 Cl Chlorine 35.45	18 Ar Argon 39.948
4	19 K Potassium 39.098	20 Ca Calcium 40.078	21 Sc Scandium 44.956	22 Ti Titanium 47.867	23 V Vanadium 50.942	24 Cr Chromium 51.996	25 Mn Manganese 54.938	26 Fe Iron 55.845	27 Co Cobalt 58.933	28 Ni Nickel 58.693	29 Cu Copper 63.546	30 Zn Zinc 65.38	31 Ga Gallium 69.723	32 Ge Germanium 72.630	33 As Arsenic 74.922	34 Se Selenium 78.971	35 Br Bromine 79.904	36 Kr Krypton 83.798
5	37 Rb Rubidium 85.468	38 Sr Strontium 87.62	39 Y Yttrium 88.906	40 Zr Zirconium 91.224	41 Nb Niobium 92.906	42 Mo Molybdenum 95.95	43 Tc Technetium [98]	44 Ru Ruthenium 101.07	45 Rh Rhodium 102.91	46 Pd Palladium 106.42	47 Ag Silver 107.87	48 Cd Cadmium 112.41	49 In Indium 114.82	50 Sn Tin 118.71	51 Sb Antimony 121.76	52 Te Tellurium 127.60	53 I Iodine 126.90	54 Xe Xenon 131.29
6	55 Cs Caesium 132.91	56 Ba Barium 137.33	57–71	72 Hf Hafnium 178.49	73 Ta Tantalum 180.95	74 W Tungsten 183.84	75 Re Rhenium 186.21	76 Os Osmium 190.23	77 Ir Iridium 192.22	78 Pt Platinum 195.08	79 Au Gold 196.97	80 Hg Mercury 200.59	81 Tl Thallium 204.38	82 Pb Lead 207.2	83 Bi Bismuth 208.98	84 Po Polonium [209]	85 At Astatine [210]	86 Rn Radon [222]
7	87 Fr Francium [223]	88 Ra Radium [226]	89–103	104 Rf Rutherfordium [267]	105 Db Dubnium [268]	106 Sg Seaborgium [269]	107 Bh Bohrium [270]	108 Hs Hassium [269]	109 Mt Meitnerium [278]	110 Ds Darmstadtium [281]	111 Rg Roentgenium [280]	112 Cn Copernicium [285]	113 Nh Nihonium [286]	114 Fl Flerovium [289]	115 Mc Moscovium [289]	116 Lv Livermorium [293]	117 Ts Tennessine [294]	118 Og Oganesson [294]

57 La Lanthanum 138.91	58 Ce Cerium 140.12	59 Pr Praseodymium 140.91	60 Nd Neodymium 144.24	61 Pm Promethium [145]	62 Sm Samarium 150.36	63 Eu Europium 151.96	64 Gd Gadolinium 157.25	65 Tb Terbium 158.93	66 Dy Dysprosium 162.50	67 Ho Holmium 164.93	68 Er Erbium 167.26	69 Tm Thulium 168.93	70 Yb Ytterbium 173.05	71 Lu Lutetium 174.97
89 Ac Actinium [227]	90 Th Thorium 232.04	91 Pa Protactinium 231.04	92 U Uranium 238.03	93 Np Neptunium [237]	94 Pu Plutonium [244]	95 Am Americium [243]	96 Cm Curium [247]	97 Bk Berkelium [247]	98 Cf Californium [251]	99 Es Einsteinium [252]	100 Fm Fermium [257]	101 Md Mendelevium [258]	102 No Nobelium [259]	103 Lr Lawrencium [262]

Answers

CHAPTER 1

Graphic Content

1. The green portion represents the proportion of all papers published in the journal *Behavioral Ecology* that had a female first author. The gray portion represents the proportion of all papers published in *Behavioral Ecology* that had a male first author. Together, the green and gray portions, making up the whole pie, represent all papers published in *Behavioral Ecology* (that is, 100%).

2. The pie chart on the left shows the proportion of all papers published in *Behavioral Ecology* that had a female first author (and, in gray, the proportion with a male first author) during the years 1997–2001, when the journal had a "single-blind" review process in which reviewers knew authors' identities (and, presumably, their sex). The pie chart on the right presents the same information, but for the years 2002–2005, when the journal had a "double-blind" review policy in which reviewers did not know authors' identities or sex.

3. When the journal changed its review policy so that reviewers did not know the sex of authors, the proportion of accepted papers that had female first authors increased from 23.7% to 31.6%. This suggests that when reviewers did know the sex of authors, they were less likely to accept a paper with a female first author. This would seem to indicate that the reviewers for *Behavioral Ecology* at that time were biased against female authors.

4. It could be that, beginning in 2002, there was an increase in the proportion of submitted papers that had a female first author. This explanation seems unlikely, however, given the information in the text that a similar journal that did not change its review policy had no change in the proportion of published papers with a female first author. Another possible explanation is that the proportion of women in the field of behavioral ecology increased in the early 1990s, so that, during the second period measured, 2002–2005, there were more women at a more advanced stage in their careers, and the quality of the papers they submitted increased over time.

5. The increase in proportion of published papers with female first authors was 31.6% − 23.7%, or 7.9 percentage points. It could also be described as a 7.9%/23.7% × 100% = 33% increase (one-third of 23.7% added to 23.7% gives 31.6%).

6. The information in the figure tells us what percentage of the papers published had a female first author. We cannot tell from this information whether 50% of the papers submitted had female first authors. If that were true, it would be reasonable to be concerned that the proportion published wasn't also 50%. But, what if only 10% of behavioral ecologists are women? If that were the case, we might expect that only 10% of the papers published in the journal would have female first authors, and we might be concerned that with such a small proportion of female behavioral ecologists, more than one-quarter of the papers published in the journal had female first authors.

7. The graphs do not prove a general bias against female scientists. Because the data are drawn from just one journal in one field of science, we cannot be certain that they reflect a similar bias across all journals in biology, or in chemistry, physics, or other scientific fields. Nonetheless, the findings do suggest such a bias may exist. It would be valuable to collect additional data on this topic across all of the sciences.

Check Your Knowledge

Short Answer

1. Two claims were made, requiring two double-blind experiments, because only one variable can be changed or tested at a time. *Experiment 1:* Measure intestinal transit time (the time for consumed foods and other substances to travel through the digestive system) for two groups. *Group 1* (control group) does not receive Activia yogurt and instead receives an inert yogurt substitute. *Group 2* (experimental group) receives Activia yogurt. Intestinal transit times are measured for both groups. Neither the participants nor the administrators know which participants are consuming Activia and which are consuming the substitute. Data from group 2, the experimental group, are compared with data from group 1, the control group, to determine the effectiveness of Activia. If the data show that intestinal transit time is shorter in the experimental group, the statement by the Dannon Company is supported. *Experiment 2:* Determine whether DanActive dairy drink helps prevent colds and flu, again using two groups. *Group 1* (control group) does not receive DanActive dairy drink and instead receives an inert dairy drink substitute. *Group 2* (experimental group) receives DanActive dairy drink. The incidence of colds and flu is measured over a finite period of time, such as one year, for the two groups—the same period of time for both groups. Neither the participants nor the administrators know which participants are consuming DanActive dairy drink and which are consuming the substitute. Data collected from the experimental group are compared with data from the control group. If the incidence of colds and/or flu is lower in the experimental group, the statement made by the Dannon Company is supported.

2. Biological literacy can be defined in several ways: (a) The ability to use the process of scientific inquiry to think creatively about real-world issues that have a biological component. Examples could include using the scientific method to answer questions about which plant fertilizers are more effective or to evaluate whether companies' claims about herbal remedies are legitimate and supported by properly designed experiments. (b) The ability to communicate these thoughts to others. Examples could include discussing with a friend the importance of consuming adequate amounts of calcium or of making accurate tables and graphs for data collected in experiments. (c) The ability to integrate these ideas into decision making. Examples could include interpreting data from experiments and using this information to make decisions. For instance, you could read and evaluate an article about the benefits of eating a diet rich in fruits and vegetables for reducing certain cancers, then decide whether you and others would benefit from this.

3. "Empirical" refers to knowledge gathered through observations and experiments that are rational, testable, and repeatable. In contrast to this empirical approach, knowledge can also be gained through reason, reflection, and observation in the absence of experimentation.

4. Predictions are made based on a formulated hypothesis and are tested with experiments. If a hypothesis is not supported by the experimental results, it can be revised, new predictions made, and these predictions tested with new experiments. Experiments can be repeated many times by many different scientists, and the hypothesis can again be revised and tested if conflicting results arise. In this way, the scientific method, done properly, is self-correcting.

5. When a phenomenon can be measured, it can be tested through experimentation. To be testable by the scientific method, all examples would need to comply with this requirement. If a phenomenon cannot be measured, it cannot be tested using the scientific method; examples include determining the existence of God or the beauty of a Shakespeare sonnet.

6. The null hypothesis would state that exercise does not affect the amount of acne a person develops.

7. The prediction would be that a group of people who eat fresh fruits will have a lower rate of illness than another group of people who do not consume fresh fruits.

8. Key features of an experiment: *Treatment:* The condition that is changed for members of the experimental group. It is the variable (e.g., temperature, pressure, light) that is changed to test a hypothesis. Only the experimental group experiences this treatment; otherwise, comparison with the control group will not be possible and the hypothesis cannot be properly evaluated. *Experimental group:* The group that receives the treatment, which is the variable being manipulated to test the hypothesis. *Control group:* The group for which all variables, with the exception of the treatment, are the same as for the experimental group. This makes it possible to determine differences between the experimental group and control group that result from the variable being tested. *Variables:* Characteristics that can be changed, such as temperature, light, pressure, or quantity and type of food, and can potentially illicit a change. *Blind experimental design:* For studies involving human beings, a study design in which the placebo effect and bias can be minimized by preventing members of the experimental group from knowing whether they are part of the experimental (treatment) or control group. However, the experimenters know who is a member of the experimental group or the control group. *Double-blind experimental design:* For studies involving human beings, a study design in which the placebo effect and bias can be minimized by preventing members of the experimental and control groups *and* the experimenters from knowing who is part of the experimental (treatment) and control groups. *Randomized:* The random selection of participants to be in either the experimental or control group.

9. (a) Include a control group that does not drink green tea and instead drinks water or some other inert drink. (b) Conduct this as a double-blind experiment. (c) Include a large number of people who would be representative of the population advised to drink green tea as a weight-loss method. (d) Stipulate a certain amount of green tea of the same brand to be consumed by study participants. (e) Ensure that only one variable, green tea consumption, differs between the experimental and control groups. In neither group should participants alter their consumption of food and beverages or change their total calories burned or consumed. (f) Perform a statistical analysis of collected data to determine whether weight loss is significant.

10. If the experimental conclusion does not support the hypothesis, the hypothesis can be revised and retested. If the hypothesis is supported by the experimental conclusion, the experiment can be repeated by other scientists. This will ensure that the process is self-correcting.

11. Both theory and hypothesis are attempts to understand a natural phenomenon. A hypothesis is an idea that has been proposed but not yet widely tested, whereas a theory has been widely tested by many different scientists, with repeated results to support it. Both hypotheses and theories can be revised as new information arises from research and experimentation.

12. The study should be a double-blind, controlled experiment utilizing a sugar pill or other inert substance for the control group. All participants in the experiment should be informed of all the potential risks of the drug(s) being tested and the potential consequences of taking part in the study.

13. *Randomized:* Participants are randomly selected to be in either the experimental or control group. *Controlled:* There is a control group in which all variables, with the exception of the treatment, are the same as for the treatment group. This makes it possible to determine whether differences between outcomes for the experimental group and the control group result from the variable being tested. *Double-blind:* For studies involving human beings, the placebo effect and bias can be minimized by preventing members of the experimental and control groups and the experimenters from knowing who is part of the experimental (treatment) group and who is in the control group.

14. A randomized, controlled, double-blind experiment can be used to try to eliminate biases.

15. Statistics can help scientists determine whether differences between the control and experimental groups are a result of chance or are due to the experimental variable (treatment).

16. No conclusion can be drawn. The fact that the male students who signed up for French tutoring have blue eyes could be due to chance or to some biological differences. Without a randomized, controlled, double-blind experiment, there is no way to determine whether these observations result from chance or not.

17. This does not disprove that Scandinavians are usually fair-skinned and blonde. There is a greater likelihood of people from this region being fair-skinned and blonde, but there will always be some people with genetic variation that includes darker skin and/or hair. This could result from variation within the population or migration of individuals into the area.

18. The study of life is unified by the themes of hierarchical organization and the power of evolution.

Multiple Choice

1. b; 2. d; 3. c; 4. a; 5. a; 6. a; 7. d; 8. a; 9. b; 10. e; 11. d

CHAPTER 2

Graphic Content

1. A table would allow presentation of more data in less space—for example, it could show the percentage of every element known to be present in the human body. But conveying the maximum number of data within a given space isn't always the goal of using a figure. In this figure, for example, the goal is to convey, with maximum impact, that just four elements—oxygen, carbon, hydrogen, and nitrogen—make up most of the human body. This surprising fact might be less effectively conveyed in a table, in which data for every element present might be given equal space (one row, probably).

2. The *y*-axis indicates, for the elements making up a human body, the cumulative amount accounted for by each element listed in the figure. The gray portion of this visual display of quantitative information accounts for 18.5%, but you'll notice that the gray begins at about 17.5% and extends for 18.5%, which brings it to about 36% on the *y*-axis.

3. The data in this figure show the proportions of the human body that are made up of 10 elements. For the top four, the data include the name of each element and the percentage it accounts for. For the next six most prevalent elements, the names are given, along with the amount contributed by all six combined.

4. Bar graph:

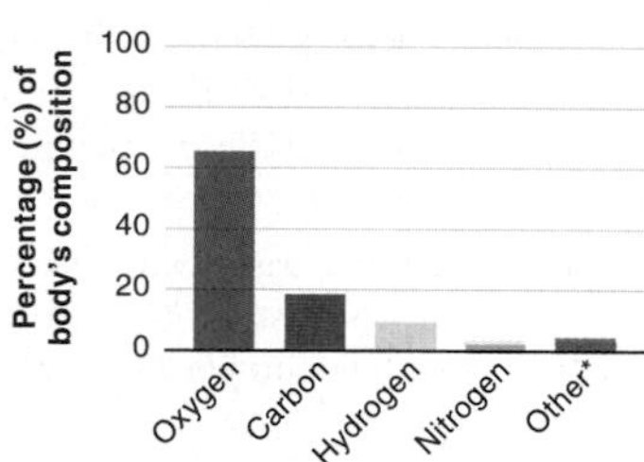

Strength: It's easy to make quick comparisons of the *relative* amounts of each element present. *Weakness:* It's less obvious that the top four elements together make up the vast majority of the human body.

Pie chart:

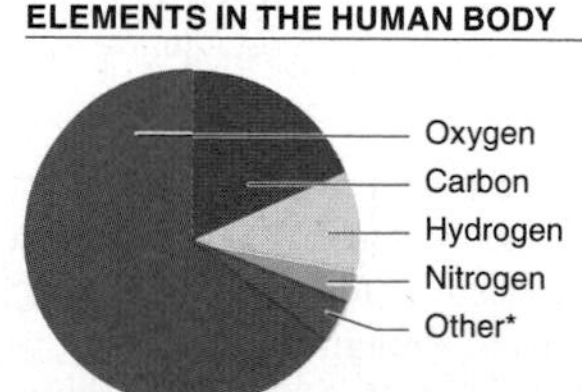

*Calcium, phosphorus, potassium, sulfur, sodium, chlorine, and trace amounts (less than 0.1%) of 15 other elements

Strength: It's immediately clear that the top four elements together make up the vast majority of the human body. It's also easy to make quick comparisons of the *relative* amounts of each element present. *Weakness:* On first viewing, a circle (the pie) does not convey that the information presented relates to amounts of various elements *making up the human body*. It's just a round pie and can be interpreted only by reading the title—unlike the version resembling a human body.

5. Yes. The positions of the colored regions for nitrogen, hydrogen, carbon, and oxygen could be reversed, as long as the amount of each color was still proportional to the percentage contributed by that element. In the given presentation, however, the elements are shown, from top to bottom, in order of most common to least common. Changing their positions might make their relative amounts just a bit less obvious.

6. There is no labeled *x*-axis, but the title makes it clear that the stacked color sections represent portions of the whole for a *human body.* And each color section is labeled with the element it represents. A graph is "appropriate" as long as the data presented are accurate and understandable and an appropriate interpretation can be made by examining the figure.

Check Your Knowledge

Short Answer

1. The mass of an electron is so small (less than one-twentieth of one percent of the mass of a proton) that its mass can be ignored when calculating the mass of an atom.

2. Atoms are most stable when their outermost shell, called the valence shell, is full. Bonding with another atom will not necessarily create a more stable atom, and it could potentially produce a more unstable atom with a partially full valence shell.

3. Atoms are most stable when their outermost shell, the valence shell, is full. Because the first shell holds a maximum of two electrons, a hydrogen bonded covalently to another hydrogen will fill its valence shell as each atom shares its single valence electron, forming more stable atoms within the H_2 molecule.

4. Covalent and ionic bonding involve the sharing of or transfer of electrons, respectively. Hydrogen bonding involves the attraction between hydrogen and another molecule as a result of weak opposite charges, which is a much weaker attraction.

5. Water is a polar molecule; the region of the hydrogen atoms is slightly positive and the region of the oxygen is slightly negative. Other polar molecules and ionic substances will readily dissolve in water because charged regions or ions are attracted to oppositely charged regions of water molecules.

6. Vomit tastes sour because the stomach produces hydrochloric acid, which causes the stomach contents to have a pH of 1–3.

Multiple Choice

1. a; 2. e; 3. d; 4. b; 5. a; 6. d; 7. d

CHAPTER 3

Graphic Content

1. Green dots indicate that an essential amino acid is present at optimal levels. White dots indicate that an essential amino acid is not present at optimal levels.

2. Some foods (primarily plant products) contain some of the essential amino acids, and others (primarily animal products) contain all the essential amino acids.

3. The shading indicates an example of how different foods (in this case, lentils and rice) that alone lack one or more essential amino acids can be combined to include all the essential amino acids.

4. Foods without optimal levels of one or more essential amino acids should not be avoided. Instead, these foods can be combined with other foods containing the missing amino acids to ensure that complete protein needs are met.

5. Although apples are missing several essential amino acids, when combined with white rice, which contains the missing amino acids, all the essential amino acids are provided. If lentils and almonds are combined for a meal, the essential amino acids methionine and cysteine—which are missing from both—will not be supplied in adequate amounts.

6. There is no information in this figure on optimal levels. We can assume that consuming these foods will provide amino acids as shown in the figure, but how much of each food is needed is not indicated.

7. It would be helpful to provide the serving sizes needed to meet the daily needs for each essential amino acid.

Check Your Knowledge

Short Answer

1. The chemical bonds of carbohydrates are readily broken, releasing energy that the cell can use to perform its activities. Excess carbohydrates can be stored for future use.

2. The three possible fates of glucose are: (a) breakdown for immediate energy needs, (b) storage as glycogen, or (c) conversion to fats.

3. An enzyme in saliva breaks the bonds of starch to produce sugar, which tastes sweet. Because it takes time for the enzyme to break these bonds, the degree of sweetness will increase over time as the sugar concentration increases.

4. Humans, as well as other mammals, do not produce the enzymes capable of breaking the chemical bonds of cellulose, preventing it from being digested and absorbed.

5. Lipids contain many more carbon-hydrogen bonds than carbohydrates. These bonds are rich in energy.

6. In animals, including humans, a strong taste preference for fats has evolved because fats contain so much stored energy.

7. Phospholipids form the lipid bilayer of the plasma membrane.

8. Any two of the following functions are acceptable answers. *Structural:* Proteins are used to build a variety of structures, such as hair and nails. *Protective:* Proteins are used to form antibodies in an immune response, as well as forming the components of blood clots. *Regulatory:* Proteins form some types of hormones (chemical messengers) and almost all enzymes (the biological catalysts that speed up chemical reactions). *Contractile:* Proteins form contractile fibers used in the contraction of muscle cells and the movement of cilia and flagella. *Transport:* Proteins assist with the transport of substances around the body, such as transport of oxygen by hemoglobin in the blood.

9. Essential amino acids cannot be synthesized by the body and must be consumed in the diet.

10. The amino acid sequence determines the three-dimensional shape of a protein, and the shape of a protein determines its function. As the shape of a protein changes, so does its function.

11. If the shape of an enzyme changes, the active site may no longer be able to bind its specific substrate, preventing a chemical reaction from occurring.

12. Nucleic acids store information in the sequence of nucleotide bases attached to the backbone of the DNA molecule. The specific sequence of nucleotides determines the particular protein produced.

13. The base sequence determines the genetic information contained in a segment of DNA; if the base sequence changes, so does the genetic information it contains. Researchers in the Human Genome Project present the sequence of only one DNA strand because the complementary base sequence of the other DNA strand can be inferred from this strand.

14. (a) The sugar in DNA is deoxyribose; the sugar in RNA is ribose. (b) DNA forms double strands; RNA generally forms single strands. (c) The base thymine is found only in DNA; the base uracil only in RNA.

Multiple Choice

1. a; 2. d; 3. c; 4. b; 5. a; 6. a; 7. a; 8. a; 9. e; 10. e

CHAPTER 4

Graphic Content

1. The variables are the different cell types observed and the number of mitochondria they contain.

2. It would be helpful to see data for more cell types, such as sperm cells, egg cells, or brain cells. Data on variation in the number of mitochondria in cells of a particular type in the same individual would also be helpful. For example, do all liver cells have the same number of mitochondria, or does the number vary a lot? And does the number of mitochondria in a specific cell type of an individual change over time—such as when the person gets older, or in response to exercise?

3. Cells with higher rates of metabolic activity seem to have more mitochondria.

4. First, perhaps "number of mitochondria per unit of cell volume" would be better, which would be a measure of mitochondrial density. This would make it easier to compare energy-generating capacity for cells that differ greatly in size. Second, perhaps mitochondria can shrink or swell. If that's the case (and some researchers believe it is), the energetic needs of a cell might be met by a few big mitochondria *or* by a large number of small mitochondria—in which case, the *number* of mitochondria wouldn't be the best measure. A measure of total mitochondrial *volume* might be better.

5. Some other cells with high metabolic activity might also have relatively large numbers of mitochondria. Sperm cells, for example, might have larger numbers. Similarly, for one cell type, such as muscle cells, more active cells (such as those in leg muscles) might have more mitochondria than less-used muscles (such as those controlling your smallest toe).

Check Your Knowledge

Short Answer

1. Cells are the smallest unit of life that can perform all the activities necessary for life, and all living organisms consist of one or more cells.

2. All prokaryotes share four structural features: (a) a plasma membrane composed of a phospholipid bilayer, which encloses the cellular contents, including the DNA, ribosomes, and cytoplasm; (b) cytoplasm, a jelly-like fluid contained within the plasma membrane; (c) DNA, organized into one or more circular loops, that contains the cell's genetic material; and (d) many ribosomes, the sites of protein synthesis (directed by the cell's genetic material).

3. *Endosymbiosis:* According to this theory, mitochondria and chloroplasts were originally prokaryotes capable of cellular respiration and photosynthesis, respectively, that were engulfed by a cell that developed into a eukaryotic cell. *Invagination:* In this theory, the cell membrane folded in on itself to form inner compartments that specialized to become the membranous organelles.

4. Phospholipids have a hydrophilic polar head containing a glycerol molecule attached to a phosphorus-containing molecule, resulting in regions of partial positive and partial negative charge. Phospholipids also have two nonpolar hydrophobic tails, each containing a chain of carbon and hydrogen. As a result of their chemical structure, phospholipids associate to form two layers (a bilayer), with the hydrophobic tails facing the center region and the hydrophilic heads facing the outer, water-containing regions. Phospholipids can float around within the same layer of a bilayer, but generally do not pass to the other side of the bilayer. The bilayer is semipermeable, allowing passage of some materials but not others.

5. The plasma membrane behaves like a *fluid* because many membrane molecules can move around within their side of the phospholipid bilayer, and it is described as a *mosaic* because it is composed of a variety of proteins, lipids, and carbohydrates.

6. Cystic fibrosis is a genetic disorder in which faulty transmembrane proteins improperly pump chloride ions across the cell membrane. This occurs primarily in cells lining the respiratory and digestive systems and results in accumulation of chloride ions in cells and the production of very thick mucus.

7. Common ways in which HIV can be transmitted include: (a) the transfer of blood, (b) the transfer of semen, (c) the transfer of vaginal fluid, and (d) the transfer of milk from infected mother to nursing child. In each case, to infect exposed cells, the virus must have access to cells that carry the CD4 marker.

8. None of these methods of transport require the expenditure of additional energy. All are limited to movement of substances down their concentration gradients, from regions of higher concentration to regions of lower concentration.

9. Energy is required to move substances against their concentration gradient.

10. Cells lining the small intestine have tight junctions between them, preventing the passage of bacteria and nutrients from the intestine into the body cavity.

11. Most of a cell's energy is produced in its mitochondria, which extract energy from food. This energy conversion requires oxygen.

12. Lysosomes contain acid and about 50 different enzymes that break down certain materials brought into the cell, as well as unwanted materials and worn-out organelles.

13. Smooth endoplasmic reticulum contains enzymes involved in detoxifying chemicals. This is one of the many important functions of the liver.

Multiple Choice

1. e; 2. b; 3. e; 4. d; 5. c; 6. d; 7. c; 8. b; 9. b; 10. e

CHAPTER 5

Graphic Content

1. The *y*-axis (vertical axis) indicates the amount of pigment molecules present in the leaves. The *x*-axis (horizontal axis) indicates the pigment types: chlorophyll *a,* chlorophyll *b,* and carotenoids.

2. (a) There are three variables: the type and relative amount of pigments—chlorophyll *a,* chlorophyll *b,* and carotenoids—and the season, either fall or spring. (b) The amounts of pigments in the leaves were measured. No numerical values are provided. The bars indicate only the relative amount of each pigment. (c) Each color represents a different pigment: chlorophyll *a* is represented by dark green, chlorophyll *b* by light green, and carotenoids by yellow.

3. The source of the data is not indicated; if it were, this might tell us something about the reliability of the information presented. Given that these data are presented in a textbook written and reviewed by experts in their fields, we can assume that the information is from a reliable source. If these same data were presented in a publication from a less reliable source, we might have some concerns about the reliability and accuracy of the data.

4. The two graphs indicate relative amounts of pigments during two different seasons: the graph on the left, the spring; the graph on the right, the fall.

5. The amounts of chlorophyll *a* and chlorophyll *b* are lower in the fall than in the spring, whereas the amount of carotenoids stays the same.

6. It would be helpful if the figure indicated: (a) units of measurement for the pigments; (b) types of plants in which the pigments were measured; (c) whether the amounts shown are averages or minimum/maximum amounts; (d) the source of the data; (e) the geographic region where samples were collected; and (f) the dates the samples were obtained.

7. This is an informational graph. No hypothesis has been stated, and the information is not presented or used to test a hypothesis. If the information were presented differently—especially if numerical values were given and a hypothesis stated—it could be viewed as data from an experiment.

Check Your Knowledge

Short Answer

1. Fossil fuels derive from plant and animal remains. They require millions of years to form, which essentially makes them a non-renewable resource: any fossil fuel used today will not be replaced for an immense period of time—possibly long after humans have become extinct.

2. The ball has the greatest amount of *potential energy* at the top of the ramp. As it rolls down the ramp, the amount of potential energy decreases. This energy is converted to the energy of motion, or *kinetic energy,* and to *heat,* from friction between the ramp and the ball and between the air and the ball. Once it reaches the bottom of the ramp and stops moving, the ball has the least amount of potential energy and no longer has kinetic energy.

3. Some energy is always converted to heat, which is the least useful form of energy. Energy conversions in which a high percentage of energy is converted to heat are considered less efficient than those in which a smaller percentage of energy is converted to heat.

4. Each of the three phosphate groups of ATP carries a negative charge, causing repulsion between the groups. Bonds between the phosphate groups can easily be broken to release energy that can be used by the cell to power energy-requiring processes. Cells can regenerate ATP by attaching a phosphate group to adenosine diphosphate (ADP), with the input of energy. These characteristics make ATP an ideal molecule to store, carry, and release energy.

5. Most chloroplasts are located in the leaves, particularly in cells close to the leaf surface. Leaves are designed to capture light. Many leaves have flat blades that face the sun, and most are located in the highest parts of the plant.

6. Chlorophyll, found in chloroplasts, is a green pigment, reflecting light in the green part of the electromagnetic spectrum. The observed color of any object is a result of the wavelengths of light it reflects. Pigments are chemicals with the ability to absorb light in specific portions of the electromagnetic spectrum.

7. Rubisco is the plant enzyme that plucks off (removes) the carbons from carbon dioxide for use in the building of carbohydrates and other organic molecules. The carbons are used in the Calvin cycle to synthesize glucose.

8. When stomata are closed to conserve water, the exchange of gases—carbon dioxide and oxygen—is disrupted. If CO_2 is prevented from entering the leaf, it will not be available for fixation in the Calvin cycle and photosynthesis will come to a halt.

9. Glycolysis takes place in the cytosol, the jelly-like fluid outside the organelles.

10. Mitochondria are the sites of aerobic cellular respiration, in which a majority of the cell's ATP is produced. Glycolysis occurs in the cell's cytoplasm. During glycolysis, only 2 ATP molecules are produced per glucose molecule broken down. Cellular respiration in the mitochondria, requiring oxygen, includes the Krebs cycle (citric acid cycle) and the electron transport chain. It produces an additional 30–32 ATP molecules per glucose molecule. Therefore, the vast majority of the cell's ATP is produced in the mitochondria, making them the "ATP factories."

11. In cellular respiration, electrons removed from glucose ultimately pass down the electron transport chain in the inner membrane of the mitochondria. The energy derived from electrons as they pass down the electron transport chain is used to concentrate hydrogen ions (protons) in the mitochondrion's intermembrane space. This proton gradient across the membrane is used to drive ATP synthesis, and the electrons that pass down the electron transport chain are accepted by oxygen, which combines with hydrogen to form water. Hydrogen ions can accumulate in the intermembrane space because mitochondrial membranes are impermeable to ions; the concentration gradient provides the potential energy necessary to make ATP.

12. Oxygen acts as an electron acceptor at the end of the electron transport chain. Imagine that the electron transport chain is a set of stairs. Every time an electron drops down one step, it loses some energy and this energy is used for ATP production. Once each electron has made its way to the end of the electron transport chain (the bottom of the stairs), it is accepted by oxygen, combining with hydrogen to form water. If no oxygen were available, electrons would have no place to go and would collect at the bottom of the stairs. This would prevent other electrons from moving down the electron transport chain, thus halting aerobic cellular respiration and the majority of the cell's ATP production. Glycolysis would continue, but would produce a net of only 2 ATP molecules per glucose molecule—not enough energy to support cellular activities.

Multiple Choice

1. e; 2. a; 3. b; 4. d; 5. b; 6. e; 7. b; 8. e

CHAPTER 6

Graphic Content

1. Among the species listed, humans have the largest percentage of "junk DNA," at 98%. *E. coli,* with 10% "junk DNA," has the smallest percentage.

2. The gray section of each bar represents the proportion of that species' DNA that does not code for proteins (sometimes referred to as "junk DNA"). The green section of each bar represents the proportion of DNA that does code for proteins.

3. No. You cannot rank the species in order of how many coding DNA molecules each species contains. This is because, for each species, the graph shows only the *percentage* of DNA that codes or does not code for proteins. The data do not give any information about the absolute amount of coding or noncoding DNA. Even though, for *E. coli,* a high percentage of DNA codes for proteins, the actual number of genes (~5,000) is much lower than for humans (~20,000).

4. For the species listed, there seems to be no relationship between the proportion of its DNA that codes for proteins and any indicators of that species' complexity. For example, humans do not seem less complex than fruit flies or roundworms. Nor do flies or roundworms seem more complex than *E. coli.*

5. The *x*-axis (horizontal axis) shows the percentage of DNA that codes for proteins. The *y*-axis (vertical axis) shows the species. Usually, the independent variable is on the *x*-axis and the dependent variable on the *y*-axis. In this bar graph, however, the species are arranged along the *y*-axis and the bars extend horizontally rather than vertically, as is more common. If the axes were transposed, this figure would look more like most typical bar graphs. But the dependent variable would still be the "percentage of DNA that codes for proteins," so our interpretation of the figure would not change. This presentation of the data incorporates a few features that make it more effective than a simple table. Most important, the figure of interest—the percentage of a species' DNA that codes for proteins—is shown in green, and the amount of green is proportional to that value. So our eye is immediately drawn to the differences in the amount of green (and the relative differences); we understand the data being presented before we read a single number. Having an image of each species, rather than just a name at the heading of a column in a table, also makes it easier to quickly identify the data being presented.

6. The presentation of the data controls for genome size. This allows us to evaluate the variation among different species in the percentage of DNA coding for proteins, without the comparison being confounded by the huge variation in overall genome size across these species. This means we can see, for example, that the greater the percentage of noncoding DNA, the more complex the species seems to be. (The issue of genome size and complexity is a different question—and an interesting one—but is not addressed in this figure.)

7. We can draw at least two important conclusions. First is the interesting and unexpected finding that for many species, a significant proportion of their genome (the majority of it, in fact) does not code for proteins. And second, for the species listed, there seems to be no relationship between the proportion of DNA that codes for proteins and any indicators of that species' complexity. In fact, if there is any relationship between these variables, it seems to be in the opposite direction (i.e., more complex species have a higher percentage of noncoding DNA).

Check Your Knowledge

Short Answer

1. When biological evidence is available, DNA can be used to identify criminals with great accuracy. DNA can also be used to clear suspects and exonerate persons mistakenly accused or convicted of crimes.

2. The DNA molecule would contain 30% thymine. The cytosine bases, at 20%, are complementary to guanine bases, which would make up an additional 20%, together totaling 40%. Adenine and thymine together make up the remainder, and because they are complementary bases, they occur in equal quantities. Therefore, the DNA has 30% adenine and 30% thymine.

3. The base sequence, which carries the genetic code, is held in place by the sugar-phosphate backbone.

4. The exact function of much of this "junk" DNA is not known. It consists of segments of DNA, called introns, found within genes, as well as noncoding DNA outside the genes. The sequences of noncoding DNA are made up of duplicate versions of genes, gene fragments, pseudogenes, and DNA segments that act as switches. "Junk" is not an appropriate term because scientists have yet to fully understand what functions, if any, this DNA has.

5. Transcription is the process in which DNA is used as a template for the production of a messenger RNA molecule within the nucleus. The nucleotide sequence of the mRNA is then used to direct the amino acid sequence of a protein, synthesized on ribosomes in the cytoplasm. Because DNA is restricted to the nucleus, mRNA is needed to transfer the genetic information to the cytoplasm for translation. Transcription provides the directions needed for translation, and the processes cannot occur in reverse order.

6. Messenger RNA must move out to the ribosomes, the organelles that are the site of protein synthesis. Ribosomes are located in the cytoplasm, either attached to the endoplasmic reticulum or free-floating in the cytosol.

7. Transfer RNA molecules carry specific amino acids, corresponding to each tRNA's anticodon sequence, to the growing amino acid chain being produced on an mRNA-ribosome complex. Each tRNA molecule, carrying a specific amino acid, binds to a complementary RNA sequence, called a codon, in the mRNA to ensure the correct sequence of amino acids in the new protein.

8. When *E. coli* inhabits an environment without glucose (its preferred fuel) but with lactose, lactose binds to the repressor protein produced by a regulatory gene located outside the lac operon. This causes a change in shape of the repressor that prevents it from binding to the lac operator. RNA polymerase can now bind to the lac promoter and begin transcription of the mRNA needed for producing the enzymes required for lactose metabolism.

9. Mutations allow evolution to occur. With new alleles added to the genome, new traits may provide reproductive advantages not present in the past.

10. This inherited disease is caused by a genetic mutation that affects an enzyme involved in lipid digestion in lysosomes. As a result, undigested lipid accumulates in lysosomes, which swell until they fill the cell, resulting in cell death. A person with Tay-Sachs disease will eventually die as a result of this process.

Multiple Choice

1. e; 2. e; 3. b; 4. c; 5. d; 6. d; 7. b; 8. c; 9. b; 10. b

CHAPTER 7

Graphic Content

1. According to the pie chart, 86% of the corn grown in the United States is genetically modified.

2. Because the pie charts indicate only the proportion of genetically modified crop, not the quantity, for each type of crop, it is impossible to determine from these charts whether more of one crop is grown than another. If the pie charts indicated the amount of each crop grown, we could determine whether more genetically modified cotton or corn is grown.

3. The majority of cotton, corn, and soybean crops grown in the United States are genetically modified. Opinions on the use of genetically modified crops should be based on your reading of available scientific data.

4. It is much easier to interpret the data by presenting the proportions of each crop than by showing their absolute amounts. A key purpose of these pie charts is to illustrate that a majority of each of these crops grown in the United States is genetically modified. Showing their magnitude would only detract from this point.

5. With this information, it might be interesting to investigate why a smaller proportion of these crops are genetically modified in other parts of the world.

6. Several different types of charts could be produced: (a) a series of pie charts showing the proportion of genetically modified crops for each year; (b) a bar graph with each bar representing the proportion for each year, with the *x*-axis representing the year and the *y*-axis representing the proportion of genetically modified crop grown; or (c) a line graph, with the *x*-axis representing the year and the *y*-axis representing the proportion of genetically modified crop grown, with a line connecting the data points for each year. The clearest conclusion from the data in the figure is the dramatic increase in the proportion of genetically modified corn grown in the United States over the years shown.

Check Your Knowledge

Short Answer

1. CRISPR is a gene-editing (DNA-editing) system that brings greater precision and efficiency to the editing of genes. CRISPR stands for "clustered regularly interspaced short palindromic repeats." The name describes the organization of a DNA that originally came from viruses but now exists within the genomes of bacteria and helps them thwart infection by potentially lethal viruses. Here's how CRISPR is used: (a) After identifying a particular DNA sequence of interest, researchers synthesize an RNA "guide" molecule with a sequence that matches the target gene to be sliced. (b) The sequences for the CRISPR RNA and the Cas9 enzyme are introduced into target cells by a plasmid. (c) Within the cells, the RNA leads the Cas9 enzyme to the desired location on the DNA, and Cas9 cuts the DNA there. (d) At the location where the DNA is cut, a sequence can be inserted that repairs or alters the host cell's DNA.

2. It is not possible to falsify the hypothesis "genetically modified foods are dangerous" because, although one study might show that a GMO does not, for example, cause cancer, that doesn't rule out the possibility that it might cause infertility. And if another study then shows that a GMO does not cause cancer or infertility, that doesn't rule out the possibility that it might cause autism. The possibilities are endless. But this is not a reasonable argument for permanently prohibiting the use of GMOs. Rather, with each study, we must ask a specific question about a potential risk and figure out a way to approach it. In this way, each such study adds to a body of evidence that, cumulatively, can give us confidence in GMO safety. This is the current approach taken in the United States.

3. Two examples: (a) Several types of crops have been genetically engineered by inserting insecticide-producing genes into their genomes, which has reduced the need for application of pesticides. This has lowered costs for farmers and reduced crop loss. (b) Some plants have been genetically engineered by insertion of a gene that provides resistance to herbicides. When farmers apply herbicides to fields of these crop plants, the crop is unharmed but the weeds are killed. The need for tilling is reduced and crop yield is increased.

4. As farmers use more genetically modified crops, there is much less genetic diversity in populations of these crop plants. The loss of genetic diversity makes crops more susceptible to environmental changes, such as drought, insects, or fungal infections. Plant populations no longer have the genetic potential—that is, the variety of alleles—to impart resistance to these stressors.

5. *Improvements in agriculture:* Plants and animals can be genetically altered to improve yield, to resist environmental pressures such as drought and insects, and to provide additional nutrients. The net result is the potential to provide more food and improve nutritional health for more people. *Treatment of diseases and production of medicines:* Many of today's medicines have resulted from biotechnology, improving or saving many lives. For example, the gene for human insulin has been inserted into the DNA of bacteria, which then produce insulin for use by people with diabetes. Gene therapy, though holding much promise, has not yet produced the benefits hoped for. *Cloning:* Agriculture has used cloning in animals to ensure that desired traits are maintained from one generation to the next. Cloning of tissues and organs provides the potential for treating many human ailments. When used responsibly, cloning helps improve human health and nutrition.

6. Because many genetic diseases occur only when an individual carries two recessive alleles for a particular gene, potential parents not showing symptoms of the disease might be carriers of the harmful allele. DNA testing can reveal whether they are carriers. If both potential parents are carriers, they can seek genetic counseling and medical alternatives.

7. Use of conventional breeding techniques to produce animals with the traits that farmers desire takes a long time and may not maintain the desired traits in successive generations. Cloning of animals ensures that the desired traits are maintained.

8. An STR region is a position in a gene with a repetitive sequence of 4–5 nucleotides, present in different numbers in different individuals or in two homologous chromosomes of the same individual. As with a structural gene, for an STR region, an individual carries two "alleles." Individuals inherit one allele from each of their parents. Similarly, they pass on one allele to each of their gametes, with half of their gametes having one allele and the other half having the other allele. Within a population, there is a huge amount of variation for an STR region. The region may have two or three dozen different alleles. For most structural genes, there is less variation: a population may have just one allele, with everyone having two copies of that allele, or sometimes two or more alleles for a gene are present in the population. Structural genes code for structural traits that are observable in an organism's phenotype. STR regions do not code for any protein products and so are not manifest in the phenotype.

Multiple Choice

1. e; 2. a; 3. e; 4. c; 5. d; 6. a; 7. b; 8. c; 9. a; 10. d; 11. a

CHAPTER 8

Graphic Content

1. According to the graph, the probability of a 35-year-old woman giving birth to a baby with Down syndrome is about 1 in 250 live births (assuming 4 per 1,000 live births). The graph could also be interpreted as indicating 3 babies with Down syndrome per 1,000 live births, with a probability of giving birth to a baby with Down syndrome of 1 in 333.3 live births.

2. For women aged 45, the incidence of live births of babies with Down syndrome is 20 per 1,000 live births, which means the woman is more likely to give birth to a baby without Down syndrome than a baby with Down syndrome. The probability of having a baby with Down syndrome is 1 in 50 live births.

3. No. The graph indicates only the incidence of babies having Down syndrome per 1,000 live births. The graph does not indicate the total number of babies born with Down syndrome for women of any age. (Note: Most babies with Down syndrome were born to women younger than 35, given that the total number of all babies born is greatest for this age group.)

4. No information is given for women over 45 years of age. The exclusion of women over 45 may be due to lack of data, because a much smaller number of women have children after the age of 45. (Note: Some research indicates that the risks are slightly reduced after age 47, but further studies of this are needed.)

5. The graph shows data for *live births*. Fetuses lost through early miscarriage, late miscarriage, or abortion that have an extra copy of chromosome 21 are not included in the data.

Check Your Knowledge

Short Answer

1. Rebuilding telomeres could result in cells becoming cancerous. Individuals undertaking therapy of this kind would have a greater chance of developing and dying from cancer.

2. Prokaryotes typically have circular chromosomes attached to the cell membrane, whereas eukaryotes have linear, free-floating chromosomes in the nucleus. Eukaryotic cells have more genetic material than prokaryotic cells. Eukaryotic cells typically have two (or sometimes more) copies of their chromosomes, contributed by maternal and paternal gametes. Because prokaryotic cells do not reproduce sexually, they do not typically have multiple copies of chromosomes.

3. The main phases of the eukaryotic cell cycle are *interphase,* during which the cell grows and prepares for division, and *mitosis* (or M phase), during which the nucleus and genetic material divide, followed by division of the rest of the cellular contents. Interphase consists of several sub-phases. During Gap 1 (G_1), the cell grows and performs normal functions such as protein synthesis and waste removal; cells spend most of their time in Gap 1. Some cells may enter a resting phase (sometimes for days or even years), called G_0, when no cell division occurs. Cells destined for mitosis enter S phase (DNA synthesis), in which chromosomes are replicated such that each has an exact copy. In Gap 2 (G_2), cells undergo a high rate of growth and protein synthesis in preparation for division. In M phase, or mitosis, the cell nucleus (including its chromosomes) divides. This generally is followed by cytokinesis, leading to division of the cytoplasm between the two daughter cells.

4. During the S phase (DNA synthesis) of interphase, the cell replicates its DNA, creating a duplicate copy of each chromosome (sister chromatids). Once this process is complete, the cell is ready to divide into two identical daughter cells.

5. Cancer cells are unlike normal cells in that: (a) they lack contact inhibition, (b) they can divide indefinitely, and (c) they have reduced "stickiness." These cancer cell properties are indirect causes of death. Often, death occurs when cancer cells crowd out normal, healthy cells, causing organs and organ systems to fail.

6. During meiosis, two cell divisions occur. The first division separates homologous chromosomes equally between two daughter cells. The second division separates sister chromatids equally into four daughter cells. The four daughter cells, called gametes, are haploid, meaning that they contain a single copy of each chromosome. When haploid gametes fuse at fertilization (the union of two gametes), the fertilized egg returns to the diploid state—in other words, it has two copies of each chromosome. If gametes were not haploid, fertilization would result in increasing numbers of chromosomes in each generation of offspring.

7. Sexual reproduction produces genetic variation without the creation of new alleles, for the following reasons: (a) Alleles from two different parents are contributed during fertilization, producing offspring with a genetic makeup different from that of either parent. (b) Crossing over during prophase I of meiosis results in a mixture of maternal and paternal genetic information on sister chromatids. (c) Homologous chromosomes are randomly distributed to daughter cells during meiosis I, resulting in a mixing of maternal and paternal chromosomes.

8. During crossing over, sister chromatids swap segments of equal size between the homologous pairs of maternal and paternal chromosomes, producing sister chromatids that are no longer genetically identical. This results in unique genetic combinations and offspring that are genetically different from their parents—creating genetic variation on which natural selection can act in the process of evolution.

9. Non-sex chromosomes contain genetic information for traits other than gender-specific characteristics. The sex chromosomes contain genetic information that instructs the body to develop as male or female.

10. Unlike the X chromosome, the Y chromosome lacks essential genetic information needed for normal development. Information contained on the much smaller Y chromosome is limited to directing the development of male gonads.

Multiple Choice

1. c; 2. b; 3. c; 4. a; 5. e; 6. d; 7. b; 8. d; 9. d; 10. d

CHAPTER 9

Graphic Content

1. The average novelty-seeking score for individuals with the *CC* genotype was 25. (For comparison, the average score for individuals with the *CT* or *TT* genotype was just above 20.)

2. *CC* refers to a genotype for the promoter polymorphism −521C/T, which alters the transcription rate of the dopamine receptor gene. This gene influences the activity of the brain's pleasure centers. There are two different alleles, *C* and *T*, and a person can have one of each or two copies of *C* or two copies of *T*. Having one or two copies of the *T* allele (*CT* or *TT*) seems to lead to reduced transcription of the dopamine receptor gene; the *CC* genotype is associated with greater receptor gene activity and thus greater novelty-seeking behavior.

3. The data reported for the initial study provided information on how much individuals in each of the two groups (first group: *CC;* second group: *CT* or *TT*) varied in their score on the novelty-seeking questionnaire (everyone in a group would not be expected to score exactly the same). This variation is conveyed by the light green bars. The smaller the variation observed in each group, the more likely it is that differences between the two groups reflect true differences in the measure between the groups.

4. When the vertical green bar is at 0, there is no difference in the novelty-seeking score of individuals with the *CC* genotype and those with the *CT* or *TT* genotype. If all the green bars were at 0, it would mean that in each of the 11 studies reported, the novelty-seeking score of individuals with the *CC* genotype was no greater than that of individuals with the *CT* or *TT* genotype. We would then have high confidence that this gene does not influence novelty-seeking.

5. The 11 different studies did not include exactly the same participants. In fact, they may have used groups of individuals that differed in significant ways, such as age, sex, profession, or ancestry. They may also have measured different numbers of individuals, or they may have used slightly different versions of novelty-seeking questionnaires. This variation is an important strength of meta-studies. By combining the results of many studies that address a related research hypothesis, we can get a more accurate picture of the phenomenon; a single study doesn't necessarily have enough power to reveal a true effect.

6. The results of any one study showing that individuals with the *CC* genotype score higher on a novelty-seeking questionnaire than individuals with the *CT* or *TT* genotype might be interesting and provocative, but not fulling convincing. Questions might remain about the population from which the study participants were chosen, the number of individuals used, the laboratory methods used in determining genotypes, the conditions under which the questionnaire was administered, and a whole range of unconscious, inadvertent, or otherwise hidden biases on the part of the researchers. But it is a powerful finding when the results of all 11 studies testing this hypothesis are reported and most show similar results. As a consequence, we can be much more confident in drawing our conclusion—that an individual's genotype for this gene influences novelty-seeking behavior—than we would be based only on the initial reported study.

Check Your Knowledge

Short Answer

1. Offspring resemble their biological parents because they receive genes from their parents. Genes code for traits that determine physical characteristics that parents and offspring may share.

2. Gregor Mendel wanted to understand how traits are passed from one generation to the next. For his research, he chose easily observable traits that he could categorize and count, in organisms (pea plants) that he could manipulate experimentally to understand their patterns of inheritance.

3. The cells of diploid organisms, with the exception of gametes, have two copies (alleles) of each gene, one allele on each of a pair of homologous chromosomes. For offspring to inherit the same (diploid) number of chromosomes and genes, they must receive only one copy in each parent's gamete. The two gametes combine to form a diploid zygote.

4. "Phenotype" describes the expression of alleles as physical, physiological, or behavioral characteristics. "Genotype" describes the alleles an organism has inherited, which determine the phenotypes expressed.

5. If an individual exhibits a dominant trait but has an unknown genotype, a test-cross can reveal the genotype (whether homozygous or heterozygous). The individual is crossed with an individual that is homozygous recessive for the trait. If all the offspring express the dominant trait, the unknown genotype is homozygous dominant. If half the offspring exhibit the recessive trait and half exhibit the dominant trait, the individual with the unknown genotype must be heterozygous for that trait.

6. A "carrier" is an individual who carries one copy of a recessive allele, which is not expressed in the phenotype. There is no outward evidence that the individual is carrying the recessive allele.

7. In incomplete dominance, an individual that is heterozygous for a trait expresses a phenotype intermediate between the phenotypes associated with the two alleles. An example is the snapdragon flower. If a true-breeding red-flowered plant is crossed with a true-breeding white- flowered plant, all their offspring have pink flowers—intermediate between red and white.

8. An individual who is heterozygous for the sickle-cell allele has resistance to malaria because the parasite that causes malaria does not survive well in sickled cells, the red blood cells with the defective hemoglobin.

9. Both the X and Y chromosomes are sex chromosomes. There are many more genes on the much larger X chromosome, with no corresponding alleles on the Y chromosome. If a recessive allele on the X chromosome is inherited by a male (XY), that allele will always be expressed.

10. An individual with PKU inherits a defective allele for an enzyme needed to convert phenylalanine to tyrosine; phenylalanine builds up in cells, eventually increasing to toxic levels. If the person modifies his or her diet—that is, modifies the environment—to largely exclude phenylalanine, this amino acid will not increase to toxic levels in the body.

11. Mendel's second law states that the inheritance of one trait does not influence the inheritance of another trait. However, this is not true for traits determined by alleles located on the same chromosome, especially if they are close together. When a chromosome is inherited, groups of alleles, especially those close together, are inherited together.

Multiple Choice

1. a; 2. e; 3. e; 4. a; 5. e; 6. c; 7. e; 8. d; 9. b; 10. a

CHAPTER 10

Graphic Content

1. The curvy line represents the average beak size (length) in millimeters.

2. Birds with longer/larger beaks are more likely to survive during drought conditions, whereas birds with shorter beaks are more likely to survive during very wet conditions, so average beak size in the bird population varies across the years.

3. This information shows the influence of drought and wet conditions on average beak size. The average beak size is larger during drought conditions than during wet conditions.

4. If data on average precipitation were presented, it would allow a qualitative prediction of average beak size. Beaks will be, on average, longer during drought conditions and shorter during wet conditions.

5. There is no optimal beak size, as illustrated by the graph. Beak size varies with level of precipitation. If there were a single, optimal beak size, there would be no variation in average beak size with changes in precipitation.

6. Average beak size varies according to precipitation on the Galápagos Islands for the finches surveyed in this study.

Short Answer

1. Fruit flies reach maturity in about two weeks, so a scientist can include many generations during an experiment on starvation resistance. With an organism having a late age of maturity, it would require many years to complete a study such as the starvation resistance experiments.

2. It was commonly thought that: (a) as in the Biblical account, the earth was only about 6,000 years old; (b) the earth had not changed very much over time, with the exception of the occasional earthquake, flood, or volcanic eruption; and (c) all species, including humans, were created at the same time, and species never change or die out.

3. Darwin found many fossils that were very similar to species still living in the same area. He also noticed that finches on the Galápagos Islands, which he assumed were all of the same species, had different physical characteristics on different islands. Biologists at the Zoological Society in London later found that there were 13 different finch species. All of these species closely resembled a single species of finch on the closest mainland, in Ecuador.

4. Evolution is a genetic change in a population over time. A change in allele frequency in the population occurs as a result of natural selection, mutations, genetic drift, and/or migration.

5. Fixation for an allele occurs as a result of genetic drift, when an allele frequency becomes 100% in a population. As a result, there is no genetic variation for that gene. This can be detrimental because the population may be less able to adapt to environmental changes.

6. The frequency of a recessive allele does not change in a population as long as the allele does not affect reproductive success. If a recessive allele does affect reproductive success, its frequency will change. Recessive alleles that adversely affect reproduction will decrease in frequency, and those that improve reproductive success will increase in frequency.

7. Fitness is a measure of the reproductive success of individuals. (a) An organism's fitness depends on the environment in which it lives. The organism may be fit in one environment, but less fit in another environment. (b) Fitness is relative to that of other organisms in the same environment. The fitness of one individual is measured against that of other individuals with specific genotypes and phenotypes. (c) The relative fitness of an individual is measured against that of other individuals of the same species in the same population. Those individuals that reproduce more frequently than others are more fit, irrespective of the age at which they die.

8. The use of an antibiotic creates environmental pressure, and bacteria with one or more alleles imparting resistance to that antibiotic are more likely to reproduce than those that do not possess such alleles. Over time, bacteria with antibiotic-resistance alleles become more frequent in the population. Should another antibiotic be introduced, other alleles present in the population that impart resistance to that antibiotic will become more frequent over time.

9. (a) Natural selection requires time, sometimes more time than is allowed by the rate at which changes are occurring in the environment. The world's environments are constantly changing and require continual evolutionary change if organisms are to maintain fitness. No single, perfect organism for all environments will ever evolve. (b) Because mutations occur regardless of the needs of the organism, new mutations (i.e., new alleles) may not be those required for the organism to adapt to changes in its environment. (c) More than one allele may be suitable for selection in an environment, and thus there is no single optimal adaptation.

10. Most babies weigh between 7 and 8 pounds at birth, as a result of stabilizing selection. Modern medicine has enabled very-low-weight, premature babies to survive, and very large babies to be delivered through Caesarean births. This has prevented selection against extremes in the size of babies born, and greater numbers of babies of extreme size (small or large) may become more prevalent in the future.

11. Scientists in the field of biogeography have observed that species migrate to nearby locations and adapt to their new habitats. Different species living less than 100 miles apart and occupying drastically different habitats resemble each other more than they resemble species many thousands of miles away that occupy very similar habitats. This suggests shared common ancestors among species that are close together geographically. An example is found on the continent of Australia. Many niches in Australia that are occupied by marsupials would elsewhere be occupied by placental mammals. For example, the Tasmanian wolf, now thought to be extinct, occupied the same type of habitat as gray wolves. However, the Tasmanian wolf is much more closely related to other marsupials living in Australia than to gray wolves found thousands of miles away.

12. *Analogous structures* are structures that have similar functions and developed as a result of convergent evolution, forming from different starting materials in unrelated species. For example, the wings of insects, birds, and bats are all used for flying, but developed independently in each group of animals. *Homologous structures* develop in related species in which the same structures are modified to perform different functions. An example is the modification of the bones making up the forelimbs of mammals. All mammals have a common ancestor, but the bones of the forelimbs have been modified for different functions, such as flying, swimming, and walking.

13. A molecular clock is a way for biologists to determine how long two different species have been evolving separately. The longer the two species have been evolving separately, the greater are the differences in their genetic sequences. For example, humans and chimpanzees share many genes, whereas fruit flies and humans share much fewer genes.

Multiple Choice

1. e; 2. e; 3. a; 4. e; 5. b; 6. c; 7. d; 8. b; 9. d; 10. c

CHAPTER 11

Graphic Content

1. A female with one mate produces 60 offspring. This is the same number as for a male mating with one female. (In both graphs, the number of offspring produced is recorded in the leftmost bar.) The two numbers are the same because they are measures of the same thing: one male mating with one female.

2. A female with two mates produces 50 offspring. It is not clear why females with two mates would produce fewer offspring than females with one mate. At least two explanations are possible. (a) Because the numbers are measured for different females—that is, a female's reproductive output can be measured when she has one mate or two mates, but not both—this may just be a consequence of natural variation among females. Maybe those in the two-mate group just happened to have fewer offspring, unrelated to the number of mates they had. (b) Perhaps the presence of two males created stress, which somehow reduced either the female's or the males' fertility. For additional data, it would be useful to have error bars. These would give us a sense of the extent of variation in reproductive output among females given access to two males and among females given access to one male. If there is much variation within each of the two groups, the lower number seen here may just be an artifact of the individual females measured. In that case, if greater numbers of flies were used in the study, the average reproductive output would be the same for females given access to one mate or two.

3. Females with four mates would most likely still have about 60 offspring, while males with four mates would have about 150 offspring. These numbers assume continuation of the trends shown in the graphs.

4. The females do not seem to be limited by access to mates. Rather, they seem to be limited by their own production of eggs. Males' reproductive output, however, does seem to be constrained by the number of mates they have. If a male has more mates, he has greater reproductive output.

5. To make comparisons between the data in the two graphs, not only must they be measuring the same thing, but the data must be presented using the same scale. Otherwise, a comparison between the two graphs could be misleading.

Check Your Knowledge

Short Answer

1. Egg retrieval is an example of an innate behavior that does not need to be learned. It is fixed, occurring in exactly the same pattern each time the behavior is started, and normally is not altered.

2. Behaviors that improved our ancestors' reproductive success and survival are more likely to behaviors that are easily learned.

3. The squirrel producing the alarm call is likely to be related to many of the other squirrels in the colony. When the squirrel sounds the alarm, it is helping to ensure the safety of others in the colony, many of whom share many of its genes. This is an example of inclusive fitness.

4. Any two of the following three conditions are correct: (a) Individuals must interact with others who could potentially be recipients or donors of altruistic-appearing behavior. (b) The benefit to the recipient of an altruistic-appearing behavior must be much greater than the cost to the donor. (c) Recipients of altruistic-appearing acts that do not reciprocate at a later time will be recognized and punished.

5. An example given in the text illustrates a maladaptive behavior. People often donate to a cause that benefits people who live in distant reaches of the planet, people the donor is unlikely to meet, so the behavior is unlikely to be reciprocated and, in that sense, is maladaptive. The origins of this behavior go back to when humans lived in small communities in which altruistic-like behaviors were directed to other members of the community and were likely to be reciprocated. Humans of today belong to much larger communities but behave as if they still lived in small ones.

6. No. Natural selection favors behaviors that benefit the individual, not necessarily the group. In group selection, the reproductive fitness of individuals is decreased, and this type of selection cannot be sustained.

7. The female produces and retains the fertilized egg, and the offspring develops within her body. The male initially contributes only sperm. After birth of the offspring, the male may play a large role—and in some cases, an equal role—in rearing young. Because offspring develop within the female, her energetic investment will, initially, always be larger than the male's.

8. The sex that makes the greater reproductive investment is choosier about selecting a mate. Females make the greatest reproductive investment initially, but their overall reproductive investment may or may not be greater than that of the male. In cases where the male has a greater overall reproductive investment, it is in his best interest to ensure that his mate is carrying offspring with his genes and that these offspring grow to maturity. If the female makes the greater overall reproductive investment, she will try to find a mate that best ensures the success of her offspring. The degree of investment determines the risk: the greater the investment, the greater the potential risk.

9. In these mating systems, females are choosy about which males they mate with, selecting males with preferred physical traits or superior resources. Males in these mating systems must compete with other males for available females and may mate with multiple females. Males that cannot compete for mates, due to undesirable physical characteristics or poor access to resources, will mate with few if any females.

10. In *monogamy,* two individuals mate and remain with each other. In polygamy, an individual has multiple mates: in *polygyny,* the male mates with multiple females; in *polyandry,* the female mates with multiple males.

11. The male moorhen probably provides the most care. The female makes a smaller overall reproductive investment; she must compete with other females for males and is physically larger and more aggressive than the males. Distinct differences between the sexes indicate sexual dimorphism, which is typically associated with polygamous mating systems.

Multiple Choice

1. b; 2. b; 3. c; 4. d; 5. a; 6. e; 7. a; 8. a; 9. b; 10. b

CHAPTER 12

Graphic Content

1. The background extinction rate ranged from a few percent to approximately 14 percent of all families of species during the time period represented.

2. Background extinction rates (yellow-green) are lower than rates of mass extinctions (blue-green). Mass extinctions occur over relatively short periods and are caused by sudden and unusual environmental changes. Background extinctions occur constantly but at much lower rates.

3. There are several possible reasons: (a) Fossils found thus far represent only a small percentage of all species that existed at any time in the distant past. However, fossils are representative of *families* of multiple species present at that time. (b) Scientists do not know how many species existed at any given time in the distant past. Even today, we don't know how many living species there are. (c) It may not be possible to determine whether multiple fossils are of the same species, but perhaps they can be assumed to belong to the same family.

4. Because the causes of extinction vary over time, background extinction rates change over time.

5. The fossil record may or may not represent the number of species that existed at any time in the distant past. Some species leave behind fossils, others do not. Only organisms that leave fossils can be identified as extinct organisms.

6. Scientists know what the background extinction rate should be, and any indication of current extinction rates higher than expected may be a sign that human activities are causing another mass extinction. Human activities could then be modified to halt the progression of extinctions.

Check Your Knowledge

Short Answer

1. The development of a membrane enabled the separation of a group of chemicals from the environment and thus the evolution of cells. Later, segregation of chemicals within cells by intracellular membranes allowed chemical reactions of different kinds to occur in different parts of the cell.

2. According to the biological species concept, a species is a population of organisms that interbreed or can possibly interbreed with each other under natural circumstances and cannot interbreed with organisms from other groups.

3. Reproductive isolation enables organisms to have independent evolutionary fates because there can no longer be genetic exchange between the two populations and each population can evolve differently, depending on the specific environmental conditions it encounters.

4. A phylogenetic tree also allows biologists to hypothesize about the evolutionary history of life on earth. As new knowledge is discovered, new branches are added to the tree and a more complete understanding of the evolutionary history of all life is constructed.

5. Analogous traits are produced through convergent evolution. They do not share a common ancestry but have developed as similar traits evolved to serve similar purposes, through natural selection. Problems arise when constructing evolutionary trees based on similarity of structures. Although similar structures are often a sign of shared heritage, for analogous structures this is not the case.

6. Microevolution occurs as a result of changes in allele frequencies in a population; this process modifies the genetic makeup of a population simply by changing allele frequency. Macroevolution results in the development of whole new groups of species and may involve many new alleles.

7. Fossils are an invaluable source of transitional species that lived in the past. Study of the physical features of fossilized organisms allows comparisons with other fossilized organisms and with existing organisms. More recently, genetic analysis allows researchers to examine the genetic relatedness of species, often presenting a clearer picture than anatomical comparisons alone.

8. Background extinctions occur at lower rates and typically result from natural selection. Mass extinctions do not result from natural selection; they are caused by extraordinary changes in the environment over relatively short periods of time. Typically, mass extinctions are global events, while background extinctions are more local.

9. Prior to the 1970s and 1980s, classification of organisms depended on comparisons of observable physical characteristics. When molecular sequencing methods became available, genetic information could be used for a more accurate classification.

10. Archaeal cell membranes contain polysaccharides not found in bacteria or eukaryotes, and archaeal ribosomes, enzymes, and membranes are more similar to those of eukaryotes than those of bacteria.

11. The domain eukarya contains single-celled and multicellular organisms with cells containing a membrane-enclosed nucleus.

Multiple Choice

1. a; 2. d; 3. d; 4. d; 5. e; 6. d; 7. d; 8. e; 9. b; 10. a; 11. c

CHAPTER 13

Graphic Content

1. We are of the species *Homo sapiens,* which has *Homo heidelbergensis* as an ancestor, which, in turn, may have had the following ancestors, in order of most recent to most distant: *Homo ergaster, Homo habilis, Australopithecus* (gracile), and *Ardipithecus*. We seem less likely to be direct descendants of *Homo neanderthalensis* and *Australopithecus* (robust), which are dead-end branches that split off from our common ancestors and became extinct.

2. *Homo* appeared about 2.5 million years ago. It seems to have had a significantly larger skull (and so brain size) than its *Australopithecus* ancestors.

3. The solid lines represent transitions from one species to another that are well supported by fossils showing gradual changes. The dotted lines represent potential relationships that are not as well supported by transitional fossils. The discovery of additional fossils may allow further clarification of the relationships.

4. About 1.6 million years ago, four different human species may have been living at the same time: *Australopithecus* (robust), *Homo erectus, Homo ergaster,* and *Homo habilis*.

5. It's not long ago that our species became the only living representative of the genus *Homo*. From this graph, it seems to be approximately 10,000 years ago.

Check Your Knowledge

Short Answer

1. The degree of complexity of an organism is not related to how well that organism is adapted to its environment. There is a tendency to judge animals that are more like humans as "higher organisms," but animals perfectly adapted to their environment, though they may not be at all like humans, are not "lower organisms."

2. The common ancestor of all animals is thought to have been an ancestral unicellular protist resembling a sperm in shape and size. This conclusion is based on DNA and RNA analyses.

3. Roundworms are probably the most abundant animals on earth. Some roundworms, also called nematodes, are parasites that cause damage to plants and animals.

4. Mollusks have a sandpaper-like tongue called a radula, which is used for feeding. The radula's rough surface assists with pulling food into the mouth.

5. It is difficult to define what intelligence is, and even more difficult to apply this concept to non-human animals. Animals have evolved in response to the selective forces in their environments, and our concept of intelligence cannot be applied objectively to non-human animals. In other words, all animals have adapted to manipulate their environment to meet their needs in ways that may not conform to our concept of intelligence.

6. *Benefits:* The exoskeleton provides protection, much like a suit of armor, and helps prevent water loss, keeping the organism from drying out. *Costs:* An exoskeleton prohibits growth; animals with exoskeletons must go through developmental stages or shed their exoskeleton and grow a new one as they increase in size.

7. (a) Insects' ability to fly enables them to escape from predators, find food and mates, and occupy a wide variety of habitats. (b) Their ability to pass through different life stages (larva, pupa, and adult) allows them to grow even though they have an exoskeleton.

8. (a) The notochord is a rod of tissue that extends from the head to the tail and stiffens when muscles contract during movement. It is present in all vertebrates during at least some stage of life. (b) The dorsal hollow nerve cord is located along the back, or dorsal side, of the animal and extends from the head to the tail. It forms the brain and spinal cord of the central nervous system. (c) Pharyngeal slits, located between the back of the mouth and the top of the throat, are present during at least some stage of vertebrate development. (d) A post-anal tail is present during at least some stage of vertebrate development; the tail is located behind the anus, the end of the digestive system.

9. (a) Long legs set vertically beneath the trunk of the body, enabling animals to run faster and for a longer time to capture prey. (b) Endothermy, having a warm-blooded body in which body temperature can be maintained, regardless of environmental temperature, through increased metabolic activity, which is due to the muscle contractions required for running.

10. Endotherms produce body heat as a result of cellular respiration. Ectotherms depend on outside sources of heat, such as from sunlight, to heat their body and do not depend on cellular respiration for this purpose. Endotherms use a great amount of energy to produce heat and maintain their body temperature, and must consume more food more frequently than ectotherms.

11. The placenta is rich in blood vessels and enables the transfer of nutrients, respiratory gases, and metabolic wastes between the mother's circulatory system and that of the developing fetus. Finger-like projections of the placenta enter the uterine wall, which is also rich in blood vessels, so blood vessels of both structures are in close proximity. This arrangement allows efficient exchange between the circulatory systems of the mother and developing fetus.

12. (a) The brain is larger in humans. (b) Humans walk on two legs, whereas chimpanzees mostly walk on four legs. (c) Body size is larger in humans.

Multiple Choice

1. e; 2. d; 3. c; 4. a; 5. b; 6. a; 7. c; 8. b; 9. e; 10. d

CHAPTER 14

Graphic Content

1. At some point between 20 and 47 days in this study, plants with mycorrhizal fungi had a significantly higher survival rate than those without mycorrhizal fungi.

2. To answer this question, two considerations must be kept in mind. First, the percentage survival rates are given in intervals of 20%, so between these intervals only estimations can be made. Second, error bars are given for survival data at 47 days. A best estimate is that the percentage survival rate is 57% or 58%, up to about 62%, for offspring of parents with mycorrhizal fungi, and about 21% to 25% or 26% for those having parents without mycorrhizal fungi.

3. At 10 and 20 days, no plants died, so there could be no errors or variations in measurement. At 47 and 96 days, some plant deaths had occurred and thus there was a chance of measurement errors or variations.

4. "Statistically significant difference" means a difference greater than a difference due only to error or variation in measurements.

5. A useful additional point of time would be about halfway between days 20 and 47. At some point between these dates, a difference in survival rates between the two groups arose. By adding an additional data point, it might be possible to determine exactly when this change occurred. However, many additional data points might be required to determine the actual time when a benefit was derived from this mutualistic plant-fungal relationship.

Check Your Knowledge

Short Answer

1. *Sexual reproduction:* Getting sperm to eggs for fertilization can be a challenge for many plants living on land, especially if the plants are large or live in dry locations. *Resisting predators:* As a result of their inability to move, plants cannot run from or otherwise escape from predators, and they have had to develop other means of defense, including thorns and poisonous chemicals.

2. Growth close to the ground minimizes the effects of gravity on the plant body—a major adaptation necessary for life on land. Buoyancy provided by water enables aquatic organisms to resist gravity; in organisms living on land, including the plants, traits to resist gravity had to evolve.

3. Mosses produce motile sperm with flagella, which require water so they can swim to the eggs and fertilize them. Mosses grow low to the ground, and the areas they inhabit become temporarily flooded as rain falls or snow melts, providing the needed environment for sperm to reach eggs.

4. Non-vascular plants do not have vascular tissue, which limits their height to only a few centimeters; their gametophyte is the dominant generation. Vascular plants have vascular tissue that transports water and nutrients over great distances within the plant body, enabling them to grow very tall; the sporophyte is the dominant generation. Vascular plants are more successful in areas where the surface of the ground dries out between rains. Non-vascular plants and mosses can grow in deserts.

5. The growing embryo and young seedling derive nutrients from the endosperm, a specialized nutritive tissue contained in the seed. The endosperm provides the nutrients and energy required for the embryo to grow before the seedling establishes roots and leaves and the ability to photosynthesize.

6. Seeds contain nutrients and an embryonic plant within the protective seed coat. Spores contain only a cell with DNA, RNA, and a small amount of nutrients. For seed plants, growing from an embryo provides a head start and can improve chances for success.

7. Plants and animals have evolved together, depending on each other for survival. If animals became extinct, there would be no bees or other pollinators, and many plant species would no longer be able to reproduce. And for many plants, there would be no way for seeds to be dispersed away from the parent plant. Altogether, this would result in many plant species becoming less common, if not extinct. Plants that do not depend on animals for pollination and/or seed distribution would dominate.

8. A pollen grain lands on the stigma, a pollen tube grows down through the style into the ovary, and two sperm travel down the pollen tube. One sperm fertilizes the egg to form a zygote, and the other fuses with the two central nuclei in the embryo sac to form endosperm. The endosperm grows into a food source for the developing embryo after germination. Thus, double fertilization forms not only a zygote but also a significant food source. Gymnosperms do not produce endosperm, so angiosperms have the advantage when it comes to nourishing the embryo.

9. The fleshy fruit provides a bribe to entice an animal into eating the fruit and transporting the fruit along with its seeds to another location, away from the parent plant. After the seeds have passed through the digestive system, they are deposited with some nutrient-rich manure to help nourish the seedling.

10. Unlike other eukarya, fungi have cell walls containing chitin and are decomposers, digesting food externally prior to absorption of nutrients.

11. Because fungi are not capable of photosynthesis, they depend on outside food sources for nutrients. An association with plant roots enables them to obtain food easily (without having to externally digest and absorb it), as the plant supplies them with sugar.

Multiple Choice

1. a; 2. b; 3. c; 4. e; 5. c; 6. c; 7. d; 8. b; 9. d; 10. d

CHAPTER 15

Graphic Content

1. Most of the cells in a human body are microbial cells, living in and on the human—about 90 trillion microbial cells, compared with 10 trillion human cells.

2. There are about 10 trillion human cells.

3. About 90 trillion microbial cells live in and on the human body.

4. Microbial cells are much smaller than the human cells that make up the body.

5. It is not clear how the data were obtained for this graph. It would be helpful to know the age, gender, and number of people in the represented sample. A breakdown of cell types might also be useful.

6. Any type of graph that compares the numbers of bacterial and human cells is acceptable. (a) A pie chart could be divided into two wedges to show the numbers of bacterial and human cells. A strength of this presentation is that it would convey nicely that most of the "pie" is bacterial rather than human. A weakness is that it would convey information on how each component represents some proportion of the whole, but no information on the absolute numbers of each cell type. (b) A bar graph could be used, with one bar representing the number of bacteria and the other bar representing the number of human cells. One strength of this approach is that it would make it easier to compare the two numbers. A weakness is that it would de-emphasize that the two components, together, make up the entire count of cells in a human body.

Check Your Knowledge

Short Answer

1. Amoebas are tiny cells. Their small size allows the exchange of gases, including oxygen and carbon dioxide, across their cell membrane at a high enough rate to meet their metabolic needs, so they do not require a respiratory system. Larger, multicellular organisms do require a respiratory system because, without it, gas exchange cannot occur at a rate fast enough to meet their metabolic needs.

2. Microbes are genetically very diverse. They live in just about every type of environment imaginable, including at extreme temperatures and pressures, and can live on just about any type of food. Thus, it is difficult to make any generalizations about microbes. There are huge numbers of microbes living on earth, with a huge number of species, many of which are still to be identified.

3. *Conjugation:* A copied piece of DNA is transferred between bacteria, even if they are not of the same species. *Transduction:* A virus inadvertently transfers bacterial DNA from one bacterium to another. This occurs as a virus picks up DNA from one host and, as it infects another host, transfers the DNA to that cell. *Transformation:* As bacteria die they release their DNA into the environment, and it can be scavenged by other bacteria. This new DNA may code for traits the host cell does not already have.

4. The characteristic taste and texture of yogurt result from the use of bacteria in yogurt production. The process uses bacteria that metabolize milk sugar, called lactose, for energy and produce lactic acid as a byproduct. Lactic acid has a low pH and causes milk proteins to denature, resulting in the characteristic taste and texture.

5. The source of this cholera outbreak was investigated in 1854 by Dr. John Snow, who identified the source as the Broad Street pump, which was delivering contaminated water. The handle of the pump was removed so no one else could use it as a source of water. It was Dr. Snow who persuaded the authorities to remove the handle, and as a result the number of new cholera cases fell dramatically.

6. Some strains of *Staphylococcus aureus* have become resistant to most of the antibiotics available today. Untreatable staph infections have spread from hospitals into the larger community and are more common than HIV infections, killing more people than HIV/AIDS.

7. (a) Many people are unaware that they are infected with an STD because symptoms may be mild or absent, and they may inadvertently pass the disease to another person. (b) It is nearly impossible to completely eliminate all STDs because both partners need to be treated simultaneously, and this is hard to arrange if people are unaware of their infection.

8. Any two of the following answers are possible. Bacteria and archaea have chemical and/or structural differences in: (a) their DNA, (b) their plasma membranes, (c) their cell walls, and (d) their flagella.

9. Many archaea live in extreme environments, and it can be challenging to replicate these conditions in the laboratory so as to successfully grow and study these organisms.

10. The ciliates, including *Paramecium,* are a group of protists considered to be animal-like because, like animals, they move and hunt other organisms (in this case, microorganisms).

11. *Plasmodium* goes through a series of developmental stages, each of which has different surface proteins. These proteins are what an immune system uses to identify foreign organisms. It takes time for an immune system to recognize a foreign organism by its surface proteins. The constant change in surface proteins with each stage of *Plasmodium* development means the immune system has to repeatedly identify and react to new proteins.

12. A virus is composed of a piece of DNA or RNA surrounded by a protein coat. Unlike living organisms, a virus is not made up of cells, cannot replicate on its own, carries out no metabolic activities within its protein coat, lacks the structures found in cells, and cannot detect and respond to stimuli in its environment.

Multiple Choice

1. b; 2. e; 3. a; 4. a; 5. d; 6. b; 7. a; 8. d; 9. e; 10. b

CHAPTER 16

Graphic Content

1. This graph illustrates the growth of a population over time, its maximum sustainable yield, and the environment's carrying capacity. With this information, it is possible to better manage a resource by understanding how much can be harvested/hunted/utilized sustainably.

2. The blue line becomes flat as the population reaches the environment's carrying capacity, the point at which all available resources are being utilized by the organisms in the population. No further increase in population size is possible.

3. A population can stay the same size (plateau) as long as the number of individuals arriving (through birth or immigration) is exactly the same as the number of individuals leaving (through death or emigration). This situation can persist indefinitely, even as organisms age and die.

4. The data contained in this graph include: (a) the size of the population, with no units given; (b) time, with no units given; (c) maximum sustainable yield, the population size at half the carrying capacity; and (d) the carrying capacity of the environment.

5. The following information could be added: (a) units of time and units of population size, to provide quantitative data; (b) the species that the population data represent; (c) the environment that the data represent; and (d) the season when the data were obtained.

6. The graph assumes that: (a) the population will grow exponentially until it reaches the environment's carrying capacity and will neither fall below nor exceed this rate over time, and (b) the environment's carrying capacity will not change over time due to degradation/reduction of resources or increased resource availability.

7. The graph will have an inverted "U" shape: growth rate starts out very fast; increases to a maximum rate, reaching maximum sustainable yield at half the carrying capacity ($K/2$); then slows down, approaching zero as the population size approaches the carrying capacity (K) of the environment.

Check Your Knowledge

Short Answer

1. A significant part of ecology deals with the interaction of populations of organisms with other organisms and with their environment. Study of a single individual does not meet the study objectives of ecology.

2. Density-dependent factors are pressures on a population's growth that result from population size itself. These factors include limitations on resources such as food, water, and space. As the population size increases, fewer resources are available to support the greater number of organisms. Density-independent factors do not result from population size; they usually result from geological or meteorological forces such as earthquakes or hurricanes, or from human activity that degrades the environment. These factors reduce the environment's potential to support populations of organisms.

3. Maximum yield is obtained by the harvesting or removal of all or almost all available organisms of a specific kind, to be used as food or for other purposes. Not enough organisms remain to allow normal population growth. Maximum *sustainable* yield is the maximum number of organisms that can be harvested or removed while allowing the normal growth of the population.

4. Organisms with a shorter breeding season produce more offspring per reproductive event than organisms with a longer breeding season. Generally, the closer an organism lives to the equator, the longer its breeding season.

5. Sickness and death early in life prevent reproduction and the passing of alleles to the next generation, and thus the prevalence of alleles that cause disease and death early in life will be reduced in a population over time.

6. Species that have evolved in environments that present a high risk of dying at a relatively early age live shorter lives, reproduce earlier, and have more offspring per reproductive event than species that have evolved in environments that present a relatively low risk of dying young. If a species that evolved in a high-risk environment were transferred to a low-risk environment, it could eventually evolve to live longer, breed later, and have fewer offspring.

7. Researchers began with a large number of fruit flies, and after 2 weeks they supplied them with a fresh dish of food and collected the eggs laid in the dish. They measured the longevity of some flies hatched from these eggs (average life span 33 days), and repeated the procedure with this new generation of flies—but this time waiting longer than 2 weeks to collect eggs. They repeated the procedure a number of times, gradually increasing the time until eggs were collected to 10 weeks, and in this way almost doubled the average life span of the flies to 63 days.

8. These demographic data can be used for planning and meeting the needs of people in different age groups, called cohorts. For instance, based on the current size of particular cohorts, the data would be useful for planning when and how many schools or other child services might be needed, or hospital maternity services, or assisted living facilities for the elderly.

9. "Demographic transition" refers to the change in the make-up of a region's population as it progresses from high birth rates and death rates, to high birth rates and low death rates, and finally to low birth rates and low death rates.

10. The ecological footprint of a population is a description of the amount of resources it uses. Average resource use by each person is so much greater in Japan than in India, which is a much poorer nation, that, overall, the smaller population of Japan has a larger ecological footprint than the larger population of India. Generally, as nations become richer, their ecological footprint increases.

Multiple Choice

1. a; 2. b; 3. d; 4. b; 5. c; 6. c; 7. e; 8. d; 9. a; 10. b

CHAPTER 17

Graphic Content

1. The drawings are roughly proportional in size to the relative biomass of each level in the energy pyramid—producer, primary consumer, secondary consumer, tertiary consumer—for this particular type of ecosystem. The tertiary consumers (top predators) are fewest in number and have the lowest biomass.

2. The size of the "pie" at each trophic level represents the biomass in an ecosystem of organisms fulfilling a particular role at that level. The pies are different sizes because the biomass of producers is typically 10 times the biomass of primary consumers, and the biomass of each successive level higher up in the pyramid is typically one-tenth of that at the previous level. This decreasing biomass is due to the inefficiencies of energy flow and conversion.

3. Each pie slice, typically 10% of the total pie, represents the conversion efficiency of biomass at one trophic level into biomass at the next level up the pyramid.

4. At the level of tertiary consumers, there is not sufficient biomass to support a population of quaternary consumers, given the 10% rule.

5. A level of secondary or tertiary consumers that also eat plants could have greater biomass than shown here, because these consumers would draw energy not just from the animals they eat, but also from plants. The expected biomass of such consumers would be 10% of the total biomass of the organisms they consume—both plants and animals. This would be much greater than for any of the consumers shown here.

6. There are likely to be many fewer predators than prey organisms (particularly herbivores), because the total biomass of predators is estimated to be one-tenth the total biomass of prey organisms.

Check Your Knowledge

Short Answer

1. The temperature generally decreases on moving from the equator toward the Poles. The type of biome located in a geographic region is determined by: (a) average temperature and whether it varies seasonally or is constant, and (b) average precipitation and whether it varies seasonally.

2. At lower latitudes, closer to the equator, the sun's rays hit the earth more directly and the solar energy spreads out over a smaller area (is more concentrated). Closer to the Poles, energy from the sun's rays is spread out over a greater area, The greater the concentration of sunlight reaching the earth, the higher the temperature.

3. By reducing the amount of solar radiation they absorb, cities can become cooler without the expenditure of energy. Steps that could be taken include: (a) making rooftops lighter in color to reflect sunlight, (b) planting trees and plants around buildings to provide shade and cool areas, and (c) creating rooftop green areas and gardens.

4. Many organisms are omnivores, which eat both plants and animals and thus can occupy more than one trophic level in a food chain. These multiple connections between organisms, not apparent in a food chain, are better represented by a food web.

5. Energy flows from producers to consumers through several levels called trophic levels. Producers, the photosynthesizers, are at the base of the food chain and outnumber all consumers. Primary consumers, the herbivores, feed on the producers. Energy then passes to secondary consumers as they feed on the primary consumers. Finally, at the very top of the food chain are the tertiary consumers: the top carnivores. Only about 10% of the energy at each trophic level is available for consumers at the next higher trophic level. Very little energy originally absorbed by producers is available to top carnivores, so a relatively small number of these animals can be supported.

6. Carbon dioxide levels reflect fluctuations in plant growth during different seasons. Photosynthetic rates decrease during the winter, resulting in higher atmospheric CO_2 levels. The CO_2 levels decrease during the summer as photosynthetic levels increase.

7. An organism's niche can be described as its place in a community. The features of a niche include the organism's space and food requirements; the time of year it reproduces; its temperature, moisture, and other living requirements; the plants or other organisms it feeds on; and how it influences competitors. If two species occupy the same niche, two outcomes are possible. Competitive exclusion occurs when the two species compete until the one that uses resources more efficiently drives the other species to extinction in that locality. Resource partitioning occurs when the two species alter their use of the niche through behavioral or structural changes, so that one or both species become restricted to just part of their niche.

8. *Competitive exclusion:* The two species compete until the one that uses resources more efficiently drives the other specie to extinction in that locality. *Resource partitioning:* The two species alter their use of the niche through behavioral or structural changes, so that one or both species become restricted to just part of their niche.

9. Unlike animals, plants are immobile and unable to move away from danger. They have developed chemical defenses that can injure or kill organisms that try to eat or otherwise harm them.

10. A crucial feature of a colonizer in primary succession is its ability to disperse far from its site of origin. Because of their ability to travel far, these species can reach remote environments that have been disturbed. Generally, they face few if any competitors at first and can flourish, but in later stages of succession they do not compete well with other species.

Multiple Choice

1. e; 2. a; 3. e; 4. d; 5. e; 6. a; 7. a; 8. a; 9. e; 10. b

CHAPTER 18

Graphic Content

1. There are two graphs, each presenting data on the number of species found at different latitudes, from the equator toward the Poles. The *x*-axis shows the number of species: in the top graph, the number of mammalian species; in the bottom graph, the number of marine copepod species. The *y*-axis shows the latitude. In the top graph, data are presented for the numbers of species at 40° and 20° south, the equator, and 20°, 40°, and 60° north. In the bottom graph, data are presented for the equator and 20°, 40°, and 60° north.

2. The independent variable is latitude. The dependent variable is species number. Typically, the *x*-axis is used for the independent variable and the *y*-axis for the dependent variable. In these graphs, the variables are placed on the opposite axes. This is done because the data are overlaid on a world map, which aids in identifying exactly where the selected latitudes fall. This makes it easy to see, for example, that 20° north includes Central America, while 60° north cuts across southern Alaska and northern Canada.

3. For mammalian species and for marine copepods, the largest numbers of species occur at 0°, which is the equator.

4. The value of having data both for mammalian species and for marine copepod species is that it enables us to see that the trend of decreasing species diversity toward the Poles holds true for *terrestrial* species and for *marine* species. This suggests that the phenomenon is robust and not just a quirk of one group of organisms. It would be helpful if data were presented for all species, but such data do not exist. (And, as we saw in Chapter 12, for prokaryotes and many other organisms it is difficult to even define what a species is.) Only a fraction of the world's species have been described. In this figure, it is more valuable to include animal groups for which most species have been described. Also important is that the figure uses large groups of organisms, for which any bias in the proportion of total species described at the equator versus the proportion described at higher latitudes is unlikely.

5. There are more mammalian species than marine copepod species at 20° north. Someone might answer this question incorrectly because the bar representing the number of copepod species is longer than the bar representing mammalian species—but this is because the scales for the axes differ. There are more than 300 mammalian species and about 70 copepod species at 20° north.

6. The bars here represent exact counts rather than data from repeated measures of a variable, such as height, that would have an associated variance.

Check Your Knowledge

Short Answer

1. The observed drop in oxygen level was correlated with the proliferation of bacteria using methane. This confirmed scientists' suspicion that bacteria were the cause of the disappearance of the methane. Scientists also discovered, through DNA testing, that the bacteria responsible for consuming oil were also found at other oil spills. These bacteria live in the Gulf at all times, because there are naturally occurring oil leaks in this area.

2. Biodiversity can be defined as the diversity of species in an area, the diversity of genes and alleles for a particular species, and the number of different ecosystems in an area.

3. Studying biodiversity can be difficult because more than just the number of species must be studied—including the diversity of genes and alleles and the diversity of ecosystems.

4. Three primary factors contribute to species richness. (a) *The solar energy available:* More available solar energy correlates with greater species diversity. The sun's solar energy is more concentrated in areas closer to the equator and less concentrated in areas farther away from the equator. The more energy available, the greater the numbers of organisms and species that can be supported. (b) *The evolutionary history of the area:* The greater the time that has elapsed in an undisturbed inhabited area, the greater the biodiversity. Climatic disasters in an area diminish the number of species and lower biodiversity. (c) *The rate of environmental disturbance:* Environments with an intermediate amount of disturbance have the greatest biodiversity. Species that would outcompete (have a competitive advantage over) many other species may be lost from the environment due to disturbance, enabling greater species diversity.

5. The fewer the individuals making up a species, the greater is the chances of the species becoming extinct. As the population of tigers has decreased, their genetic/allele diversity has decreased. Should tigers be faced with a severe environmental disturbance such as disease or other change, the population may lack the alleles that would permit some individuals to survive and reproduce.

6. The chief reason for species loss is habitat loss and habitat degradation due to human activities, particularly in tropical rain forests. It is estimated that half the world's tropical rain forests have been destroyed over the past 25 years. These areas are biologically very diverse, and their loss represents a significant loss in biodiversity. Overexploitation of species and the introduction of exotic species are additional reasons for the loss of so many species.

7. Once a species is lost, it can never exist again. Each species has formed through a unique, uninterrupted, never-to-be-repeated evolutionary history that gave rise to a genome unlike any other.

8. Fossil fuels, which include oil, coal, and natural gas, are composed primarily of carbon and hydrogen, with smaller amounts of nitrogen and sulfur. As fossil fuels are burned, these elements react with oxygen in the air to form carbon dioxide, water, nitrogen dioxide, and sulfur dioxide.

9. Carbon dioxide and methane in the air absorb infrared energy radiating from the earth, preventing its escape into space and thus heating up the atmosphere. With carbon dioxide and methane added to the atmosphere, more heat is retained and contributes to global rises in temperature. This effect is comparable to that of the glass panes of a greenhouse, which have the same effect as carbon dioxide and methane in retaining more infrared energy and warming the internal environment of the greenhouse.

10. (a) *Reducing biodiversity:* A disproportionate percentage of the world's species live in the rain forests. With the disappearance of these forests, many known and unknown species are lost. (b) *Reducing photosynthesis and increasing greenhouse gases:* Each year, an estimated 610 billion tons (550 trillion kg) of CO_2, a greenhouse gas, are removed from the atmosphere during photosynthesis by plants of the tropical rain forests—an amount that will be reduced as rain forests are destroyed. Additionally, when rain forests are burned, CO_2 is released, adding more greenhouse gas to the atmosphere.

11. A keystone species, such as kelp, has a disproportionate effect on the biodiversity of a community, as is revealed if that species should die out or otherwise be removed. When a keystone species is lost from a community, a relatively large number of other species, which depend on it, directly or indirectly, are also lost.

Multiple Choice

1. e; 2. c; 3. d; 4. d; 5. b; 6. d; 7. a; 8. e; 9. c; 10. a

CHAPTER 19

Graphic Content

1. Like all pie charts, this figure attempts to show the proportion of the whole that is contributed by each component. In this case, the whole pie represents all the material of which soil is composed. The pie "slices" represent the relative amounts of minerals, organic materials, and water and air.

2. The source of the data is not given. This can be problematic because the reader might assume that all soil has approximately the same composition as the soil sample used for this analysis.

3. No. There is no information in the pie chart to suggest that different types of soil differ significantly in composition. In other words, the reader might conclude that, whether sandy soil at the beach or moist soil around a lake, the soil will be 50% minerals, 1% to 5% organic materials, and 45% to 50% water and air. Given that different soils *do* differ in composition, the data for several different soil types could be presented as a group of pie charts, which would allow a quick comparison of the chief ways in which they differ. Alternatively, the data could be presented in a table, with each row representing a soil type and three columns to represent minerals, organic materials, and water and air.

4. There is no way to include an error bar reflecting variation in the size of a pie slice, so it can be difficult to present information about variability in a pie chart. For some phenomena, this could be important. It is possible (though not commonly done) to show fuzzy borders between pie slices, with the size of the fuzzy region proportional to the variation in that measure. Alternatively, the labels for the pie slices can indicate the proportion that each slice represents along with some measure of variation. In this pie chart, for example, the water and air slice represents 45% to 50%, although it is drawn as 45%; similarly, the organic materials slice represents 1% to 5%, although it is drawn as 5%.

5. The benefit of a pie chart rather than a bar graph is that it is immediately clear that the greatest proportion of soil is made up of minerals, with water and air making up slightly less, and organic materials making up much less. In general, a pie chart is effective when you wish to convey relative proportions of a whole. If the data were presented as a bar graph, the *y*-axis label would be "proportion of the whole," and each bar along the *x*-axis would represent "soil component."

6. A three-dimensional pie chart may be more aesthetically pleasing, but it can also be deceptive. The data do not have three dimensions, and the depth of the pie carries no meaning. A simple circle conveys exactly the same data. Moreover, in a three-dimensional pie, the slice at the front (in this chart, "water and air" and, to a lesser extent, "organic materials") is given more prominence because more of its color is visible than for other slices. This might cause a reader to misjudge the relative size of the slices, since the proportion of each color does not reflect the proportion contributed to the whole by that particular component. In this chart, the blue slice appears to be larger than the purple slice because more blue than purple is visible. This method of data presentation could be used to mislead readers into believing that one component is more or less significant than it actually is, and thus to persuade readers to believe something that is not true.

Check Your Knowledge

Short Answer

1. The three distinct parts of a vascular plant are leaves, stems, and roots. Leaves are the photosynthetic structures that absorb light and produce sugars; they are attached to the stem. Stems are support structures that carry the leaves and serve as conduits for water, minerals, sugars, and other nutrients. Roots absorb nutrients (especially minerals) and water from the soil, anchor the plant in the soil, and provide support for the plant body.

2. In monocots, the vascular tissue, which transports nutrients, is arranged in many, randomly scattered bundles in the interior of the stem, so the shallow cut caused by the wire would not damage all of the vascular tissue. In eudicots, the vascular tissue is arranged in a ring just under the stem surface, and the cut might damage all of this tissue.

3. The vascular tissue of plants functions much like the circulatory system in vertebrates; however, plants have no heart-like structure that acts as a pump to move water and nutrients around the plant body.

4. When parenchyma cells of the ground tissue of a root store starch—a form of energy that the plant can use later—the ground tissue increases in size and the root becomes enlarged and fleshy. Sweet potatoes, carrots, and radishes are examples of starch-filled roots.

5. Both types of meristems are made up of undifferentiated cells that eventually differentiate into epidermis, vascular tissue, or ground tissue. Growth at lateral meristems causes the plant part to become thicker. Growth at apical meristems causes the plant part to increase in height or length.

6. (a) The cuticle has a waxy coating that seals in water. (b) Leaf hairs on the surface reduce the temperature inside the leaf, decreasing evaporation. (c) Guard cells can close the stomata to prevent water loss.

7. Soil contains the minerals essential for plant growth and metabolism (primarily nitrogen, phosphorus, magnesium, potassium, sulfur, and calcium).

8. Nitrogen is often the limiting nutrient for plants, so adding nitrogen to the soil increases plant growth. Plants need nitrogen to build proteins, and nitrogen is required by nearly all plant cells and tissues.

9. Fungi (which have a large absorptive surface area) grow around and into plant roots and thus increase a plant's access to water and minerals; the fungi benefit by gaining access to energy and nutrients from the plant.

10. Cohesion between water molecules causes them to be pulled up through the plant, from the roots to the leaves, as water evaporates from the leaves. The height of trees is limited by the collective strength of the hydrogen bonds that cause the water molecules to stick together. Water can be pulled up to a height of approximately 400 feet, at the most.

11. Phloem delivers food to plant tissues. It transports sugars produced in the leaves throughout the plant body to where they are needed or can be stored.

Multiple Choice

1. b; 2. a; 3. d; 4. e; 5. d; 6. d; 7. b; 8. c; 9. a; 10. c

CHAPTER 20

Graphic Content

1. The fuzzy borders of each "slice" of the pie convey that the exact size of each slice is not certain—there is some variation. This variation could also be conveyed in several other ways. The pie slices could have sharp borders, with each slice labeled with an indication of its size range (3% to 7%, 20% to 25%, etc.). Or a bar graph could be used, with each component represented as a separate bar, with an error bar showing the uncertainty or variation around the estimate. Or the bar graph could have just a single bar (representing 100% of flowering plant species) divided into segments representing each type of pollinator, with fuzzy borders between segments to represent uncertainty. The labels for each region could indicate the size as a range (again, 3% to 7%, 20% to 25%, etc.).

2. The different shades of one color represent a broad category of plant pollinators. All of the green slices represent insect pollinators, with each shade of green representing a different group of insects. Similarly, each shade of blue represents a different group of non-insect animal pollinators. This allows quick comparisons of the relative proportions of non-insect animal pollinators and insect pollinators, while also allowing a comparison of the relative proportions of pollinators in each group of insects (bees versus beetles, for example).

3. Actual percentages for each slice would allow more precise comparisons between slice sizes. They may have been excluded here because such precise data do not exist, as suggested by the title above the pie chart ("approximate proportion").

4. Bees are the pollinators of the largest proportion of plant species: about 45% of flowering species. The second largest pie slice is for birds, indicating that they pollinate the second-largest proportion of flowering plant species.

5. It would be useful to know which habitats (and how many of each) were surveyed to produce the data, because the data may not apply to all types of habitats. For example, in some habitats, the majority of flowering plant species might be pollinated by birds, while in others, the majority might be pollinated by bees or beetles. Or perhaps the proportions *are* the same across all habitats. We cannot evaluate the variation across habitats (tropical versus arctic, for example) in the proportions of flowering plant species pollinated by each group of pollinators.

6. (a) Most flowering plant species are pollinated by animals rather than wind or water. (b) Among animals, insects are the most common pollinators. (c) More flowering plant species are pollinated by birds than by bats. (d) More flowering plant species are pollinated by bees than by any other group of animals. (e) Of the non-animal methods, wind pollination is much more common than water pollination.

7. An important detail in this figure is that the slices represent the proportion of flowering plant *species* rather than the proportion of *individual* flowering plants. All species do not have the same number of individuals. An extremely common plant species with millions of individuals that is pollinated by, say, flies counts as "one species, pollinated by flies," and a rare species with just a few hundred individuals that is pollinated by, say, bats counts as "one species, pollinated by bats." If you set out to survey just one habitat and count the number of individual plants pollinated by insects versus the number pollinated by non-insect animals, and if the species pollinated by non-insect animals had much smaller population sizes than the species pollinated by insects, you might rarely see any non-insect pollinators. If all flowering plant species had the same population size, however, you would see approximately 75% of the plants being pollinated by insects and approximately 20% by non-insect animals.

8. For data that are "loose approximations," we cannot have complete confidence in their accuracy. Any interpretations should be made with caution and with the understanding that they are subject to change. With the collection of additional observations on pollinators and, perhaps, in different types of habitats, it may turn out that the size of any pie slice is a bit larger or smaller than shown here. Loose approximations can still be valuable, however, especially for things that are difficult to count exactly. From this pie chart, it is probably safe to assume that bees really are the pollinators of the largest proportion of flowering plant species, that more flowering plant species are pollinated by animals than by non-animal methods, and that, among animals, insects are a more important group of pollinators than non-insects. Each of these conclusions can help us understand the general patterns of pollination. The presentation of loose approximations can also help guide future studies aimed at gaining a deeper and more accurate understanding of the subject under study.

Check Your Knowledge

Short Answer

1. By planting a potato adapted to its particular environment, a farmer can preserve the well-adapted set of alleles present in the parent plant. The farmer will choose for planting those potatoes that have the characteristics she wants to propagate in the next crop. The seeds produced from the parent plants would most likely have a different combination of alleles.

2. Sexual reproduction in plants allows the selective breeding of particular plants for sought-after features such as size, flavor, or speed of ripening.

3. Proteins projecting from the cell wall of each pollen grain increase the stickiness of the pollen, which helps in fertilization, but they can also act as allergens—proteins detected as foreign substances and producing an allergic reaction—when people breathe in the pollen grains.

4. During the coevolution of flowering plants and their animal pollinators, plants have become increasingly effective at attracting pollinators and deterring other species from visiting their flowers, while pollinators have become more effective at exploiting the resources offered by the plants.

5. Double fertilization involves two separate fusions of male nuclei with female nuclei—one fusion event producing the embryo and the other producing the endosperm. This process is efficient because when, and only when, an embryo is produced, so, too, is a ready-made food source.

6. Flowers commonly contain both male and female reproductive parts, so a pollinator can assist in sexual reproduction. This *hermaphroditism*—having both male and female reproductive parts—in flowering plants also allows self-fertilization. One of the pros of self-fertilization is that the plant can reproduce in the absence of pollen arriving from another plant. One of the cons is that the offspring will be less genetically diverse than those produced by cross-pollination and are more likely to express an undesirable gene (because of inbreeding).

7. Fruits do not ripen and become edible until the seeds are fully developed—only then are the seeds ready for dispersal. At that point, the fruit sweetens and turns a bright color to attract the attention of dispersers.

8. A plant will grow taller (undergo primary growth) if its lower branches are removed, or pruned (but not overly pruned), to reallocate resources to growth of the unpruned portions. This growth occurs through the activity of apical meristems.

9. Gibberellins are like "miracle grow" chemicals. They stimulate both cell division and cell elongation to cause rapid plant growth.

10. Ethylene has a role in the ripening of fruit. Fruit can be transported while still green, then ripened just before marketing. For example, ethylene can be applied to a large container of green bananas so that they ripen simultaneously and in conjunction with their arrival at grocery stores.

11. Temperature and rainfall amounts are not reliable markers of the seasons, as they can vary a lot from year to year—say, from one summer to the next. The number of hours of daylight or of darkness is a much better indicator of season.

12. Poinsettias are short-day plants and require extended, uninterrupted periods of darkness to form flowers. The conditions inside a home may not offer enough continuous darkness for blossoming.

Multiple Choice

1. a; 2. e; 3. d; 4. b; 5. b; 6. b; 7. a; 8. c; 9. b; 10. e

CHAPTER 21

Graphic Content

1. The curve representing the cat's body temperature (on the *y*-axis) as a function of ambient temperature (on the *x*-axis) shows a very slight increase, so you could say that a cat's body temperature is influenced by the ambient temperature. However, the effect is very small. Over an ambient temperature increase of about 17° C, the cat's body temperature rises by only one or two degrees. It would be more accurate to say that a cat's body temperature is almost completely unaffected by ambient temperature.

2. At 10° C, you would expect the lizard's body temperature to be 10° C, because at every other point on the graph showing the relationship between a lizard's body temperature and the ambient temperature, its body temperature is exactly the same as the ambient temperature.

3. Metabolic rate, as shown in the lower graph, is measured as the amount of oxygen consumed (in milliliters) per gram of body weight per hour. At any ambient temperature, the cat's metabolic rate is higher than the lizard's.

4. To make the comparison of metabolic rates meaningful, the rate is reported as a function of body weight (i.e., per gram of body weight). In other words, the difference in the animals' weight is factored out so that for both organisms, metabolism is measured as *rate per gram* per hour rather than as total consumption of oxygen per hour.

5. Between 27° and 36° C, the lizard's metabolic rate increases from about 0.15 ml O_2/g/hr to 0.35 ml O_2/g/hr. In other words, the lizard's metabolic rate more than doubles. Across the same temperature range, the cat's metabolic rate does not increase at all.

6. The lizard's metabolic rate is considerably lower than the cat's at every temperature reported. The lizard's metabolic rate, however, increases dramatically between 27° and 36° C, while the cat's remains constant. To make this difference more apparent, for lizards you could use a *y*-axis with a different scale, perhaps multiplying the rate by 10. This would make the slope much steeper, highlighting the greater influence of ambient temperature on metabolic rate in lizards than in cats. (Most textbooks do this.) A drawback to this method is that it might mislead readers into concluding that the *metabolic rate* is greater in lizards than in cats, when it is not. It is only the *rate of increase* in metabolic rate that is greater.

7. "Thermoneutral zone" refers to a range of ambient temperatures across which an animal's metabolic rate remains unchanged. Cats have a clear thermoneutral zone, but lizards do not: in lizards, any increase in ambient temperature is accompanied by an increase in metabolic rate.

8. In the ambient temperature range between 27° and 36° C, it is more efficient to be an endotherm. When the ambient temperature falls below 27° C or rises above 36° C, however, metabolic costs rise more dramatically in endotherms than in ectotherms, suggesting that, under such conditions, it is more efficient to be an ectotherm.

Check Your Knowledge

Short Answer

1. A chief benefit of multicellularity is that it allows a division of labor and specialization at the cellular level.

2. Connective tissue is like a "cellular glue" because it holds cells together in a matrix and gives shape, support, and structure to other tissues and organs throughout the body.

3. Epithelial tissue can form exocrine glands, which, depending on the type, produce and secrete saliva, sweat, or mucus.

4. (a) Skeletal muscles attach to bones and generate movement; most are under conscious control (some, such as those that control breathing, are not). (b) Cardiac muscle is located only in the heart and is not under conscious control. (c) Smooth muscle is found in the bladder and stomach and in many other organs; it generates slow, rhythmic contractions and is not under conscious control. All three types of muscle are similar in that they consist of cells that can contract.

5. Glial cells support, nourish, and protect nerve cells (neurons). Damage to glial cells could cause overall dysfunction of the nervous system, including impairment of signal propagation, of nervous system development, and of the repair of nervous system injury.

6. The circulatory system regulates the pumping of blood around the body.

7. Hyperthermia is a dangerously elevated body temperature. It is caused by a failure of the body to maintain homeostasis, a consistent internal environment, including consistent body temperature, which is necessary for the normal functioning of cells, tissues, and organ systems.

8. In large, multicellular organisms, the vast majority of cells are not in direct contact with the external environment and thus the internal environment takes on greater importance in influencing cell functioning. Internal variables such as pH, gas concentrations in the blood, and temperature are kept within a relatively constant range.

9. Negative feedback systems are more common because they oppose or reduce change in the internal environment. Positive feedback systems, in contrast, oppose homeostasis in response to a change, pushing the body away from normal conditions.

10. (a) Evaporation is the loss of heat that occurs as a liquid turns to gas (e.g., evaporation of sweat). (b) Radiation is the transfer of heat, without direct contact, from a warmer object to a colder object (e.g., the heat from a fireplace).

11. No. For an invertebrate osmoconformer living in salt water, the solute concentrations of its body fluids reflect those of the external environment.

12. The amount of urine excreted each day depends on the individual's fluid intake and water loss due to perspiration or exhalation.

Multiple Choice

1. c; 2. d; 3. d; 4. d; 5. c; 6. d; 7. a; 8. a; 9. e; 10. a

CHAPTER 22

Graphic Content

1. The graph illustrates the relationship between the partial pressure of oxygen and the degree to which hemoglobin is saturated with oxygen. As the partial pressure of oxygen increases, more oxygen binds to hemoglobin. As the partial pressure of oxygen decreases, oxygen bound to hemoglobin is released.

2. Red indicates a high hemoglobin saturation with oxygen, and blue indicates low hemoglobin saturation with oxygen. The color at any point along the S-shaped curve reflects the degree to which hemoglobin is saturated with oxygen at that partial pressure of oxygen.

3. The graph has no specific data points. The curve represents the general relationship between the partial pressure of oxygen and the oxygen saturation of hemoglobin. This relationship could be determined only by measuring the oxygen saturation of large amounts of hemoglobin across the range of oxygen partial pressure covered by the *x*-axis. We can assume that the graph is based on data points that were obtained in this way but are not individually shown.

4. The S-shaped curve shows the non-linear relationship between the partial pressure of oxygen and hemoglobin's saturation with oxygen. As the partial pressure decreases from 100 mm Hg, there is at first a gradual decrease in oxygen saturation. But as the partial pressure is further reduced, to approximately 50 mm Hg, the rate at which the oxygen saturation of hemoglobin drops increases rapidly. If the curve were a straight line, hemoglobin would release oxygen to tissues at a constant rate as the partial pressure of oxygen in a tissue decreased. With the S-shaped curve, hemoglobin stays relatively saturated with oxygen even as the partial pressure is halved; after that point, the rate of oxygen release increases rapidly. Thus, with physical exertion or other increased metabolic activity, the increased use of oxygen by cells is accompanied by a more rapid release of oxygen than would occur if the curve were linear.

5. At a partial pressure of 60 mm Hg, the oxygen saturation of hemoglobin is approximately 90%. This would not make sense if we were considering just a single molecule of hemoglobin, because the only possible states of saturation for a molecule are 0%, 25%, 50%, 75%, or 100%, corresponding to 0, 1, 2, 3, or 4 oxygen molecules bound to the hemoglobin (which has four binding sites). However, if numerous molecules in a sample of hemoglobin are analyzed for oxygen saturation, some may be at 75% saturation and others (most others) at 100% saturation. When the *average* of these is taken, it could be 90%.

6. The greatest change in oxygen saturation for any change in partial pressure of oxygen occurs where the curve is steepest. We can estimate this change as occurring between 25 and 45 mm Hg.

7. The partial pressure of oxygen in the air is lower at high elevations than at sea level. A climber's hemoglobin may not reach full saturation with each breath. As a consequence, the hemoglobin may not be able to deliver sufficient oxygen to the tissues, which may lead to shortness of breath and faster breathing. This oxygen deficiency can be exacerbated by the strenuous physical activity of climbing.

Check Your Knowledge

Short Answer

1. With increasing size of multicellular organisms, most of an organism's cells are not in direct contact with the external environment, so essential nutrients and wastes cannot diffuse into and out of cells adequately. A dedicated delivery and removal system (a circulatory system) is necessary.

2. In single-circuit cardiovascular systems, found in fishes, blood passes (once) through the heart to the gills, then through the body and back to the heart. Gill capillaries slow down the blood flow, so the body receives blood at a relatively low pressure, decreasing the rate of oxygen supply to the cells. In double circulation, found in amphibians, reptiles, birds, and mammals, blood flows from the heart to the lungs and back to the heart, then from the heart to the body and back to the heart. Blood leaves the heart at a relatively high pressure, especially in birds and mammals, which increases the rate of food and oxygen supply to cells and the rapid removal of wastes—a system better suited to high energy demand.

3. The pressure of blood in veins is only about one-tenth of the pressure in arteries. Friction in the capillaries slows the blood so that by the time it passes into the veins, its pressure is much diminished.

4. With high blood pressure, the heart must work harder, potentially weakening it, and arteries become less able to expand to accommodate increased blood flow during exertion. More cholesterol sticks to the more rigid artery walls, narrowing their diameter and further reducing the efficiency of the circulatory system and heart. These changes increase the risk of heart attack and stroke.

5. To reduce the risk of heart disease: (a) avoid all types of tobacco, (b) exercise, (c) achieve and maintain a healthy body weight, and (d) limit alcohol consumption.

6. Lymph travels through the body in lymphatic vessels. These vessels are squeezed as adjacent skeletal muscles contract, pushing the lymph toward the shoulder region, where the fluid enters the bloodstream.

7. In animals with gills, gas exchange occurs as water rushes across the gills, passing between the lamellae, which contain elaborately branched capillaries. Dissolved oxygen passes from the inhaled water to the blood by direct diffusion into the capillaries, and dissolved carbon dioxide passes by direct diffusion from the blood to the exhaled water. In animals with lungs, air moves down the trachea into the lungs, where oxygen diffuses into capillaries and thus into the bloodstream; carbon dioxide diffuses from the blood into air, which is exhaled.

8. Hemoglobin releases oxygen when it encounters a low partial pressure of oxygen in the tissues, such as in actively contracting muscle tissue.

9. By having vessels oriented so that the blood flows in the opposite direction to the flow of water, gills extract significantly more oxygen from the water than if blood and water flowed in the same direction.

10. Birds have highly efficient gas exchange systems, extracting more oxygen from the air (including at high altitudes) than do other terrestrial vertebrates. Air moves in one direction through the lungs, and during both inhalation and exhalation, oxygen-rich fresh air continues flowing through the lungs.

11. At higher altitudes, with lower air pressure, the amount of oxygen in the inhaled air is lower. It becomes harder to push air into the lungs and to drive oxygen into the blood.

12. Humans living under the low-oxygen conditions of higher elevations are likely to have hemoglobin that is more "sticky," with a higher affinity for oxygen. This enables their hemoglobin to become saturated with oxygen even though the air is relatively "oxygen-poor."

Multiple Choice

1. c; 2. b; 3. a; 4. c; 5. e; 6. b; 7. c; 8. a; 9. d; 10. a

CHAPTER 23

Graphic Content

1. Two aspects of the graph support the hypothesis that the judges' blood sugar level influenced their decision making. (a) There are three time periods during which the judges were not eating and, consequently, their blood sugar was dropping. During these periods, the percentage of times that parole was granted consistently declined. If blood sugar did not affect decision making, there would be no pattern here: the percentage of times that parole was granted would be independent of the time elapsed since the judges had eaten. (b) The second and third open circles represent the first decision judges made following a food break. In both cases, this data point is far higher than the point just before the break (and all subsequent points until the next meal). In other words, not only does the probability of granting parole decline as blood sugar declines, but it also *increases* with an increase in blood sugar, just after consumption of food. If blood sugar did not affect decision making, decisions made just after eating would be no different from any others with respect to percentage of times parole was granted.

2. If judges' blood sugar level *did not* affect their decision making, the graph would have a scattering of points with no clear pattern, with an average "percentage of times parole granted" of 35.8%. In some time periods it would be above this amount and in some it would be below, but there would be no relationship between the value of the number and its position on the *x*-axis.

3. The dotted green lines highlight the increase in "percentage of times parole granted" following the two food breaks during the judges' day.

4. There are no numbers on the "time of day" axis because each day—and each food break—did not necessarily begin at exactly the same minute by the clock. Instead, the graph shows the "percentage of times parole granted" at time periods *relative* to the beginning of the day and the beginning of decision-making sessions after a food break. This allows us to more clearly see the decisions in relation to the time elapsed since the judges' most recent food consumption.

5. The green circles represent the first decision of the day or the first decision after a food break. The number reflects the percentage of times that this decision was to grant parole, averaged across the 50 days for which the study collected and analyzed data.

6. There are no error bars above and below the data points because the points do not reflect a summary statistic with an associated variance. Each data point represents a single number: "for the 50 days of the study, what percentage of the cases decided at this exact point following a meal break granted parole?"

7. The steeper drop-off after the third green circle might be due to additional aspects of fatigue and mental exhaustion, given that the judges had been awake and working for a long period of time.

Check Your Knowledge

Short Answer

1. Herbivores spend much of their time eating because plant material, when compared with the meat consumed by carnivores, provides a lot more fiber and is more difficult to digest, and thus provides fewer kilocalories and nutrients (for the same weight of food).

2. Basal metabolic rate is the minimum amount of energy expended by an organism when it is not engaged in any activity, other than resting.

3. Drinking too much water can cause an imbalance of electrolytes in the blood, particularly a deficit of sodium.

4. A complete protein contains, among its amino acids, all nine essential amino acids—those that cannot be made by the body and must be supplied by food.

5. Eleven of the 20 amino acids essential for protein synthesis are made by the body; the other nine must come from food.

6. Fat-soluble vitamins are stored in fatty tissue and in the liver and are not excreted in the urine, as are water-soluble vitamins. Over-consumption of fat-soluble vitamins can lead to accumulation of toxic levels of these vitamins in the body.

7. (a) Ingestion: food is taken into the body. (b) Digestion: food is dismantled by physical breakdown into smaller pieces and by chemical breakdown into smaller molecules. (c) Absorption: molecules are taken up from the intestine into cells. (d) Elimination: unnecessary materials are swept from the body as waste.

8. Stomach acids kill bacteria.

9. The small intestine has a huge absorptive surface area. Its surface area is increased by its long length, its folded structure, and an interior lining with finger-like projections called villi, which are further lined with tiny projections called microvilli.

10. Some mammals have four-part stomachs that contain symbiotic cellulose-digesting bacteria. Other mammals have cellulose-digesting bacteria living in an outcropping of the digestive tract called the cecum.

11. Following the digestion of carbohydrates, the pancreas secretes insulin in response to the increased glucose concentration in the blood. Insulin stimulates the uptake of glucose by cells, such as those of muscle and fat tissue.

Multiple Choice

1. c; 2. d; 3. a; 4. b; 5. d; 6. a; 7. b; 8. b; 9. b; 10. e

CHAPTER 24

Graphic Content

1. Individuals who completed extensive training and acquired "The Knowledge" that qualifies them to be a London cab driver had a larger posterior hippocampus relative to (a) their pre-training posterior hippocampus size, (b) the posterior hippocampus size of individuals who attempted to become cab drivers but did not qualify, and (c) the posterior hippocampus size of people who were not drivers (controls).

2. The asterisk indicates that the difference in hippocampus size between time 1 and time 2 is statistically significant for qualified cabbies. In other words, the finding of a difference is reliable and unlikely to be due to chance. The lack of an asterisk for the other two pairs of bars indicates that there is no statistically significant difference in hippocampus size between the two time points for individuals in these two groups.

3. The alternative hypothesis was that cabbie training does not make the hippocampus bigger; rather, people differ innately in posterior hippocampus size, and those with a larger hippocampus are predisposed to be more successful in qualifying as cab drivers, due to their better spatial memory abilities. Those with a smaller posterior hippocampus, because they are less able to acquire "The Knowledge," do not qualify. The study results presented here, however, indicate that there was no difference in hippocampus size among the three groups at the start of the study period (time 1). Only *after* the training was completed was the posterior hippocampus larger among the qualified cabbies. This observation supports the hypothesis that it is the training itself that is responsible for the increase in posterior hippocampus size.

4. The two time points represent the time (time 1) before any individuals engaged in training (learning "The Knowledge") and the time, four years later (time 2), when training was complete for drivers qualifying to become cabbies.

5. When data *on the same individual* are recorded at two or more points in time, it is called a longitudinal study. This longitudinal study measured the effect of the variable—cabbie training—on posterior hippocampus size in the same individual at two time points, giving us greater confidence that any change in size is associated with the training (when compared with non-cabbie controls measured at the same time points). Comparisons between cabbies and non-cabbies at a single point in time would not allow reliable conclusions about the

effect of training on hippocampus size. A difference in size might mean just that the two populations are different (and that individuals with a larger posterior hippocampus are more likely to become cab drivers).

6. Gray matter is made up of dendrites and cell bodies, which do much of the processing in the brain. Because the number of dendrites or the extensiveness of their branching may increase (without the number of cell bodies—that is, neurons—increasing), it is more informative to measure the amount of gray matter rather than number of brain cells. (Additionally, because brain neurons are so small, with such complex interconnectedness and dense packing, it may not be possible to count them individually.)

7. The same height of the blue-green bars for all three groups at time 1 represents an important control, reflecting the fact that posterior hippocampus size was not significantly different across the three groups at the start of the study. Consequently, any difference in size between the groups at time 2 can be attributed to a change that occurred between the two time points.

Check Your Knowledge

Short Answer

1. Without the sensation of pain, an individual can be injured and not know the injury has occurred. The injury—from minor to serious or even life-threatening—might not produce the necessary reaction to take care of the injured body part and prevent future injury.

2. Neurons (nerve cells) are a type of cell specialized for generating and conducting electrical impulses. Each neuron is made up of dendrites, a cell body, and an axon. Glial cells support, protect, and insulate neurons. There are about nine times as many glial cells as neurons in the human brain. Glial cells regularly divide, whereas neurons do not.

3. (a) Dendrites of motor neurons and interneurons generally connect with and receive signals from other neurons. (b) Dendrites of sensory neurons are modified to respond to a specific external stimulus (through a sensory receptor), such as touch, sound, light, or a chemical.

4. When acetylcholine binding is blocked at the receptor site, this neurotransmitter cannot bind to muscle cells and the cells are unable to contract.

5. Evolutionary changes have resulted in differences among species in the relative amount of brain structure dedicated to various abilities and behaviors. Compared with other species, humans have a high proportion of neurons devoted to abstract thought, perception, and language. As decision making gets increasingly complex, the necessary brain circuitry becomes more complex, so the brain and head become more prominent structures.

6. In humans, the senses of smell and taste are closely connected because the same air circulates around the mouth, throat, and nasal passages.

7. Some insects have a broader range of light-sensing pigments in their photoreceptor cells than do humans, including pigments that allow them to perceive wavelengths of light in the ultraviolet region of the spectrum.

8. *Chemoreceptor* cells on the tongue are stimulated by chemicals in food and produce the sensation of taste. Chemoreceptors are also present in the nasal cavity, in smell receptors—dendrites with tiny hair-like projections covered with receptors—where they detect odors. *Photoreceptor* cells, in the eyes, have light-sensitive molecules embedded in their membranes and facilitate vision. *Mechanoreceptors,* specialized receptors that respond to mechanical pressure, are important for hearing; when sound waves stimulate mechanoreceptors in the ear, they initiate an action potential that passes along a series of neurons to the brain. The sense of touch, too, is generated through mechanoreceptors sensitive to pressure. The sense of touch is also associated with *thermoreceptors*, sensitive to temperature, and *pain receptors*, sensitive to pain.

9. Myofibrils are organelles that shorten when they contract. Within the myofibrils, contraction takes place in repeating units called sarcomeres, which are composed of long, overlapping, parallel filaments of two types: thin (actin) and thick (myosin). A sarcomere shortens with the making and breaking of links between actin and myosin: (a) a link between a myosin filament and an actin filament is broken, (b) energy released from ATP converts the myosin to a higher-energy shape, (c) the myosin reaches farther down the actin filament and reattaches, and (d) the myosin snaps back to its original shape, pulling on the actin filament and shortening the sarcomere, and thus the myofibril and the muscle fiber.

10. The skeletal system (a) provides shape and structure, (b) enables movement through the action of muscles connected to the bones, and (c) protects internal organs.

11. The amygdala and hippocampus are responsible for storing memory. The amygdala seems to associate emotional feelings and sensory input with memories, so that certain smells or tastes can elicit certain memories. Some evidence for the role of the hippocampus in memory storage comes from the study of birds. Black-capped chickadees store large numbers of nuts for the winter; researchers found that the birds' hippocampus is significantly larger in the fall, when the birds must remember their nut-stash locations, than in the spring, when they no longer need to rely on stored nuts for food.

12. Caffeine binds to receptors intended for the binding of fatigue-inducing adenosine. With adenosine binding blocked, we feel alert rather than tired.

Multiple Choice

1. a; 2. c; 3. a; 4. c; 5. e; 6. c; 7. a; 8 a; 9. d; 10. e; 11. d

CHAPTER 25

Graphic Content

1. Sterilization seems to increase longevity in male mammals. To have greater confidence in this conclusion, it would be helpful to have data on specific causes of mortality following sterilization. For example, if rates of death from cancer decreased similarly in cats and humans who were sterilized, that finding would support the conclusion on increased longevity, while also suggesting a possible mechanism for this effect.

2. The graphs use different scales because humans live much longer than cats, whether sterilized or not. Because both graphs have an *x*-axis that begins at time 0, it is possible to evaluate the *proportional* increase in life span for sterilized males relative to non-sterilized males. If the *x*-axis for the human graph covered only the range 50–70 years, the magnitude of the increase in life span would appear much more dramatic. That could be misleading, however, because it might suggest that the *proportional* increase in longevity for sterilized males relative to non-sterilized males is much greater than it actually is.

3. Sterilized male cats lived about 8 years and non-sterilized cats about 5 years—a 3/5, or 60%, increase in longevity. For humans, sterilized males lived about 70 years and non-sterilized males about 55 years—a 15/55, or 27%, increase in longevity.

4. Standard deviation is an indication of the extent of variation (or dispersion) in a particular measure (such as longevity) in a population from one individual to another relative to the population average for that measure. A high standard deviation indicates a broad range of possible values around the average, while a low standard deviation indicates that most data points are close to the average. Here, average longevity for non-sterilized male cats (top bar) is about 5 years, and average longevity for sterilized cats (second bar) is about 8 years. The standard deviation for the top bar extends from 0.5 years to about 9.5 years. We would express this as an average longevity of "5 years ± 4.5 years." It reflects the finding of a large amount of variation in longevity among the non-sterilized male cats in the study, with some even living longer than the average life span for the sterilized cats. When a data distribution is "normally distributed," approximately 68% of the data points occur within one standard deviation of the mean (average).

5. For populations with the same average value for a trait, the population with the larger standard deviation consists of individuals with values that deviate more from the average than do those in a population with a smaller standard deviation—in which individual values are clustered much more closely around the average. For example, imagine two populations, each having an average value of 5 for a particular trait. If one population has a standard deviation of 0, everyone has the same value for that trait: 5. If the other population has a high standard deviation, some individuals might have measures of, say, 8 or 9, but these would be balanced out by others with measures of 2 or 1, respectively, so any two individuals in that population would be more likely to differ (and to differ by a greater amount) for that measure.

6. A *median* value is the numerical value for which half of the measures in the study sample are greater than that value and half are less. A *mean* value is the arithmetical average, obtained by adding the values of all the measures and dividing by the number of measures. The median is less influenced by extreme values than is the mean.

7. Being in the "25th percentile" means that 25% of all individuals in the study sample fall below (and 75% of individuals are above) that individual for a particular measure under study. For example, if an individual is in the 25th percentile for longevity, 25% of individuals have a lower longevity and 75% have a higher longevity. Being in the "75th percentile" means that 75% of all individuals fall below (and 25% are above) that individual for that measure.

8. The data were collected from male sanitarium patients because, at one time, some of these patients were castrated. The data were available only because this practice once occurred in sanitariums. The practice is now considered unethical and is not performed. No such data could be obtained today.

Check Your Knowledge

Short Answer

1. Oxytocin causes humans to be more trusting of others, facilitates birth, makes milk available to nursing babies, and increases an individual's tendency to form social attachments.

2. Both the nervous and endocrine systems act as systems of communication and coordination within the body. The nervous system is capable of rapid responses in controlling movements and sensations. The endocrine system controls chemical signals called hormones that influence cells close to and far from the hormone-secreting cells. In contrast to the nervous system, the endocrine system is generally responsible for longer-term, slower regulation, such as that seen in growth and development, as well as a variety of secretions. Some nervous system cells also secrete hormones, so the systems overlap somewhat in function.

3. All hormones act by binding to receptors on or in target cells. The chemical structure of a hormone determines how it regulates activity in the target cell. Receptors for polypeptide and amine hormones are embedded within and extend through target cell membranes. Once the hormone binds to the receptor on the outside of the cell, it alters the receptor and causes some type of change within the cell. Steroid hormones pass through cell membranes and bind to receptors inside the cell, eventually affecting DNA transcription in the nucleus.

4. The hypothalamus, part of the underside of the brain, functions as a liaison between the nervous and endocrine systems. It receives input from neurons throughout the brain and the rest of the body.

5. In a stress-related negative feedback loop, an animal takes action in response to a stressful situation, and the source of stress is thus removed. The secretions of cortisol and adrenaline are first elevated in response to the stress event, then are reduced through negative feedback.

6. Polypeptide and amine hormones are water-soluble and cannot pass through cell membranes; they interact on the outside of a target cell with receptors embedded in the membrane, and this interaction causes specific types of change within the cell. Lipid-soluble steroid hormones can diffuse through cell membranes and interact with receptors in the cytoplasm, before passing into the nucleus, or with receptors in the nucleus, in either case causing changes in DNA transcription.

7. Signals from the hypothalamus directly influence the pituitary gland. The pituitary consists of two separate glands fused together. The *posterior pituitary* does not itself produce hormones; two important peptide hormones produced by neurons within the hypothalamus travel down axon tracts into the posterior pituitary, from which they are released. The *anterior pituitary* produces numerous hormones in response to commands by the hypothalamus. Many of the anterior pituitary hormones direct endocrine glands elsewhere in the body to release other hormones.

8. Castration almost completely eliminates production of the hormone testosterone and may reduce the criminal activity of violent sex offenders.

9. Women with congenital adrenal hyperplasia, a genetic condition that leads to higher than typical levels of circulating testosterone, show increased performance on tests of spatial ability when compared with women without this disorder.

10. Women who use oral contraceptives and women who give birth to higher numbers of children have a reduced lifetime exposure to estrogen and other hormones produced by developing follicles. This reduced exposure is associated with a decreased incidence of ovarian cancer.

11. Endocrine disruptors are chemicals with a close structural similarity to hormones. When ingested by organisms, these chemicals can mimic, block, or otherwise interfere with the body's hormones, leading (especially in the case of those mimicking estrogen) to harmful effects. For animals that consume other animals, even small amounts of endocrine disruptors taken up from the environment can become concentrated as they pass through the food chain.

12. DDT disrupts the development and functioning of a bird's reproductive tract and impairs its ability to produce normal eggshells; the shells crack under the weight of the parent incubating the eggs. During the years of DDT use, this caused serious reductions in bird populations.

Multiple Choice

1. a; 2. b; 3. d; 4. b; 5. b; 6. e; 7. b; 8. a; 9. c; 10. d; 11. c

CHAPTER 26

Graphic Content

1. In October, the average number of days between a woman's first day of menstruation and those of her 5–10 closest friends in the dorm was 6.5 days. In April, this average was about 4.7 days. This is a decrease of 1.8/6.5, or approximately 28%.

2. If all women had a 28-day menstrual cycle, the highest possible value on the *y*-axis would be 14. If a woman's first day of menstruation was more than 14 days *later* than her friend's, it would be considered some smaller number of days *earlier* than her friend's *next* first day of menstruation. For example, if the woman's first day of menstruation occurred 16 days later than her friend's first day, it would be reported as occurring 12 days *earlier* than her friend's next first day of menstruation.

3. Inclusion of women who were taking the contraceptive pill makes the results a bit more convincing. Because these women were likely to have a 28-day menstrual cycle that did not change, the observed menstrual synchrony must have occurred through changes in the cycles of the women *not* on the pill. It is hard to be certain why women on the pill were included in the study. One possibility is that, although their cycles would not change, they were still part of the environment of the other women and were influencing the cycle length of those women.

4. The fact that these ideas and data were developed by an undergraduate should not influence our interpretation of the findings. One of the strengths of scientific thinking and observation is that any study can be evaluated solely on its merits, regardless of the academic qualifications of the researcher. That said, more experienced researchers are more likely to anticipate difficulties in interpretation or to appreciate complex issues relating to the development of experimental controls—enabling them to design a better-controlled experiment or conduct a more powerful analysis of the results.

5. The break in the *y*-axis indicates that part of the scale (0–3) is omitted. This was probably done because none of the data points or their associated error bars fell within that range, and including that range in the graph would simply reduce the proportion of the graph devoted to actual data. Without this break, the magnitude of the change shown by the data would appear smaller. This is because, in its current form, the entire graph covers the range of 4–7 days. If it covered 0–7 days, the portion in which the data points appear would make up only about half of the graph: the magnitude of the change from October to April would appear less steep (although, of course, its actual value would remain unchanged).

6. The vertical lines represent the variation around the average (the data point plotted). In other words, although the average number of days between a woman's first day of menstruation and those of her 5–10 closest friends in the dorm was 6.5 days, in some cases it was larger and in other cases smaller. The size of the error bars, which probably represent the standard deviation around the mean, gives an indication of the extent of variation among individuals for this measure. If the error bars were twice as long, this would indicate much more variation in the number of days between a woman's first day of menstruation and those of her 5–10 closest friends—which would decrease our confidence that the observed menstrual synchrony represented a true physiological phenomenon rather than a change due simply to chance.

7. This graph shows that over the course of less than one academic year, women living together in dorms tend to become synchronized in their menstrual cycles, as evidenced by the decrease in number of days between the first days of their menstrual cycles. It would be helpful to see data obtained from women living together for longer periods, and to see data from these same women after they stopped living together at the end of the academic year. We would expect the average number of days between a woman's first day of menstruation and those of her 5–10 closest friends in the dorm (no longer living together) to increase again, returning to the same value as at the beginning of the study.

Check Your Knowledge

Short Answer

1. One consequence of assisted reproductive technologies is their legal and ethical complications. Just one example: children conceived by sperm or egg donation may wish to make contact with, and even seek a relationship with, the sperm or egg donor.

2. The genetically identical offspring resulting from asexual reproduction are less likely to produce a population with the genetic capability to adapt to, and to produce offspring adapted to, a changing environment.

3. In adult men, sperm are continuously produced in the testes by meiosis. In the seminiferous tubules, diploid cells called spermatogonia divide by mitosis, each spermatogonium producing two cells. One of these cells remains a spermatogonium, so the male never runs out of sperm-producing cells; the other is a primary spermatocyte, which undergoes meiosis. Each primary spermatocyte produces two cells in the first meiotic division—the secondary spermatocytes. These then complete the second meiotic division, each producing two spermatids, which mature into sperm.

4. (a) Physical barriers to copulation, such as mating plugs, prevent other males' sperm from entering the female's reproductive tract. (b) Toxic semen components can incapacitate the sperm of other males. (c) Genital morphology, such as claspers and scrapers, allow males to dislodge the sperm of other males that have already mated with the female.

5. Genetically, egg production is very similar to sperm production, but there are several key differences in the process. In females, meiosis begins in the ovaries before birth, producing primary oocytes, but stops at prophase I. Beginning at puberty, periodic release of hormones stimulates completion of meiosis I in several oocytes. A secondary oocyte is released at ovulation, but completion of meiosis II to form the ovum (egg) is triggered only after fertilization. A much smaller number of eggs than sperm are produced, and each egg is considerably larger than a sperm cell.

6. If fertilization does not occur, the egg disintegrates, and the corpus luteum in the ovary degenerates, abruptly removing this source of estrogen and progesterone. This drop in hormone levels causes the lining of the uterus to slough off, and it is shed, along with the egg, as menstruation begins.

7. When the plasma membranes of the sperm and egg fuse (activation), the membrane of the egg changes in such a way that no other sperm can fuse with and fertilize the egg. Fusion also induces the egg to complete its second meiotic division to form the ovum, along with a (much smaller) polar body, which disintegrates or is ejected from the egg. Finally, the haploid egg nucleus fuses with the haploid sperm nucleus, creating a diploid cell—the zygote.

8. The estrogen-progesterone pill prevents ovulation by keeping estrogen levels just high enough that release of follicle-stimulating hormone (FSH) by the pituitary gland is never triggered. As long as FSH is not released, eggs do not develop and ovulation does not occur. The progesterone component of the pill causes just enough development of the lining of the uterus to allow a mucous plug to form at the connection between the vagina and uterus, blocking sperm from getting through.

9. STDs are difficult to eradicate for several reasons. (a) Symptoms may be mild or completely absent, causing people to unwittingly pass on an infection to their partners. (b) To prevent reinfection, both partners must be treated simultaneously. (c) Populations of a microbe can evolve quickly and become resistant to existing drugs, reducing the long-term effectiveness of treatments.

10. Three distinct layers of tissue, called germ layers, form during gastrulation. The cells in these layers have not yet differentiated (that is, they look similar to one another), but they have become "determined," which means that the type of tissue they will become is already decided. The *ectoderm* becomes the outer layer of skin, hair, the lining of the nose and mouth, and the nervous system. The *endoderm* becomes the digestive tract, respiratory tract, liver, and pancreas. The *mesoderm* becomes the muscles and skeleton.

11. The *first trimester* of pregnancy is primarily a time of development and differentiation of cells into specialized tissues, as the embryo implants in the uterus and the placenta forms. The *second* and *third trimesters* are characterized mostly by significant growth and rapid development of the fetal nervous system.

12. Ovulation is suppressed during pregnancy through a hormone-mediated chain of events. Human chorionic gonadotropin (hCG) prevents degradation of the corpus luteum, which continues to secrete progesterone, thereby maintaining the lining of the uterus (the endometrium). With the endometrium intact, the pituitary gland does not release follicle-stimulating hormone (FSH), follicles do not grow and develop, and no luteinizing hormone (LH) is released.

13. During lactation, suckling by the infant prevents the pituitary gland from releasing a sufficient surge of LH to cause ovulation. For this reason, during lactation, a woman's fertility is significantly reduced.

Multiple Choice

1. a; 2. a; 3. a; 4. c; 5. c; 6. e; 7. b; 8. e; 9. b; 10. a; 11. d

CHAPTER 27

Graphic Content

1. According to the map, at the time the data were collected, 0.5% to 0.9% of adults in the United States were infected with HIV. To estimate the *number* of adults infected, we need to estimate the number of adults in the United States. Assuming a U.S. population of about 300 million at that time (it was actually 313 million), we can estimate that three-quarters, or 225 million, were adults (in fact, about 24% of the U.S. population of 313 million was under 18 years old, so 238 million were adults). If 0.5% to 0.9% of these 225 million adults were infected with HIV, that would be about 1.12 million for the low estimate and 2.02 million for the high estimate. If we had to give one number, it would be best to take the average of these two numbers: approximately 1.57 million adults infected with HIV.

2. Based only on the data in the map, we cannot determine which countries had the largest number of adults infected with HIV at the time the data were collected. Certain countries, shown in dark red, had a much higher *percentage* of adults infected with HIV than other countries—the data here convey information about *proportions* of a population infected—but to calculate which country had the most infected adults, we would need to multiply the rate of infection by the number of adults. Data on population size are not given.

3. No. Based on the same logic as in the answer to question 2, we cannot determine whether there were twice as many adults infected with HIV in the United States as in Canada at the time the data were collected. The rate of infection in Canada was between 0.1% and 0.49% (or, to take the midpoint, approximately 0.3%). And the rate of infection in the United States was between 0.5% and 0.9% (approximately 0.75%). This suggests that the *rate* of infection in the United States was about twice that in Canada. Given that the adult population of the United States is much greater than that of Canada, we can conclude that the number of HIV-infected adults in the United States was more than twice that in Canada.

4. From the data in the map, the rate of adult HIV infection in southern Africa was between 15% and 28%. (To take a midpoint, we could say 21.5%.) In South America, most of the countries had infection rates of 0.5% to 0.9%, although at least one country had a rate of 1.0% to 4.9%, and another had a rate of 0.1% to 0.49%. (To take a midpoint, we could say 0.75%.) We can conclude that the rate of adult HIV infection when these data were collected was significantly—almost 30 times—higher in southern Africa than in South America.

5. The different colors used to indicate ranges of adult HIV infection rates do not represent *consistent* increases from one color to the next. Rather, for the last three color bars, each range is considerably broader than the one before, and each is much larger than the ranges for the first three color bars. While this might confuse or mislead a reader who does not carefully note the infection rates designated by each color, it serves an important purpose. The map covers a very large range of infection rates worldwide: from <0.1% to 28.0%; it needs to show fine gradations at the low end, while showing meaningful, much larger gradations at the high end.

Check Your Knowledge

Short Answer

1. (a) Physical barriers (including skin and mucous membranes) and (b) non-specific immunity (non-specific immune system cells) are the first lines of defense, serving to distinguish a substance as a pathogen and provide some initial protection. (c) Cells of the specific immune system recognize and combat individual pathogens and remember them, so that the body can fight them more effectively in the future.

2. Lysozyme is an enzyme that kills bacteria by damaging their cell walls.

3. White blood cells can distinguish the body's own cells ("self") from invaders ("non-self") because of unique molecules on the surface of the foreign cells (pathogens). The immune cells' surface receptors recognize these distinctive molecules, usually proteins, and bind to them. The body's own cells do not have these molecules, but any cell that does possess any of these proteins is identified as an invader (non-self) by the cells of the nonspecific immune system.

4. The four recognizable signs of the inflammatory response (redness, heat, swelling, and pain) are related to the changes in blood vessels that enhance the recruitment of phagocytes and complement proteins to the site of inflammation.

5. An individual who receives the chicken pox vaccine and is later exposed to the virus will not get the disease, because the body already has a memory of this virus and puts up a rapid defense through the activity of the specific immune system. After the person was vaccinated, the specific immune system manufactured a group of cells—called memory cells—primed to recognize and rapidly respond to the virus on later exposure. In contrast, for the influenza virus, each year there are many different versions, or strains, which are constantly changing. With each new flu season, the body encounters a slightly different form of influenza virus and thus a different set of antigens. Because the influenza virus adapts to changing conditions and evolves, the best protection is to get a flu shot each year—tailored as much as possible to that season's flu strains.

6. Long-term protection from (immunity to) a specific pathogen can be developed in two ways: (a) exposure to the natural pathogen, or (b) exposure to an altered version of the pathogen in a vaccine.

7. After organ transplantation, an anti-rejection medication must be taken to suppress the recipient's immune system and protect the transplanted organ against rejection.

8. The second time a person is exposed to a particular pathogen, memory cells produced during the primary response are ready to produce a secondary response through clonal selection.

9. *Effector cells* are lymphocytes that recognize an antigen and immediately take some action that leads to its destruction. On average, it takes two weeks to produce enough effector cells (through clonal selection) to combat and vanquish an infection. A second group of lymphocytes, the *memory cells,* remember an antigen so that, if the body is infected with the same pathogen in the future, it will be ready to attack the invader relatively quickly (in a couple of days rather than two weeks). Memory cells remain in the lymph and blood, ready to respond if they encounter their specific antigen again.

10. When cytotoxic effector cells encounter body cells that are infected with an invading pathogen, they kill these cells through apoptosis—programmed cell death. Although some of the body's own cells are lost, the removal of these infected cells may be beneficial to the organism as a whole.

11. In autoimmune diseases, the body's ability to distinguish "self" from "non-self" is impaired. Lymphocyte receptors wrongly recognize an individual's own molecules or cellular structures as antigens, and the humoral and/or cell-mediated immune responses are initiated by these antigens.

12. A severe shortage of helper T cells marks the progression from HIV infection to AIDS. Without helper T cells, the body's B cells and cytotoxic T cells cannot be activated. Common pathogens that are normally kept at bay by a healthy individual can lead to debilitating illness and death in a person with full-blown AIDS.

Multiple Choice

1. b; 2. d; 3. d; 4. d; 5. e; 6. b; 7. d; 8. e; 9. c; 10. d; 11. a

Glossary

A note about notation: The word or phrase being defined is in boldface type; it is followed by the number (in parentheses) of the chapter or chapters where it is discussed. In some cases, derivations are given. Abbreviations: Gk., Greek; Lat., Latin; *sing.*, singular; *pl.*, plural; *dim.*, diminutive (a smaller version of the object named); *pron.*, pronounced.

A

abiotic (17) Relating to the physical and chemical components in an environment. These include the chemical resources of the soil, water, and air (such as carbon, nitrogen, and phosphorus) and physical conditions (such as temperature, salinity, moisture, humidity, and energy sources).

abscisic acid (20) A plant hormone that has the general effect of inhibiting growth and reproductive activities.

absorption (23) In the processing of food, the process in which energy-rich molecules are taken up from the digestive tract into the bloodstream and moved to cells throughout the body.

acid (2) Any fluid with a pH below 7.0, indicating the presence of more H^+ ions than OH^- ions in solution. [Lat., *acidus,* sour]

acquired immune deficiency syndrome (AIDS) (15, 27) Infectious human disease caused by a retrovirus, HIV (human immunodeficiency virus); it compromises the immune system by attacking T cells, leaving an individual susceptible to infections, as well as cancers.

acrosome (26) The cap-like structure covering the head region of a sperm; contains enzymes for penetrating the outer membrane of an egg, enabling fusion of the nuclei.

actin (24) A protein of muscle tissue; makes up the thin filaments.

action potential (24) An electrical signal that travels along an axon, from a neuron to another neuron, a muscle cell, or a gland cell.

activation energy (3) The minimum energy needed to initiate a chemical reaction (regardless of whether the reaction releases or consumes energy).

activator (3) A chemical within a cell that binds to an enzyme, altering the enzyme's shape or structure in a way that causes the enzyme to catalyze a reaction.

active site (3) The part of an enzyme to which reactants (or substrates) bind and undergo a chemical reaction.

active transport (4) Molecular movement that depends on the input of energy, which is necessary when the molecules (or ions) to be moved are large or are being moved against their concentration gradient.

adaptation (10) The process by which, as a result of natural selection, a population's organisms become better matched to their environment. Also, a specific feature, such as the quills of a porcupine, that makes an organism more fit.

adaptive radiation (12) The rapid diversification of a small number of species into a much larger number of species, able to live in a wide variety of habitats.

additive effects (9) Effects from alleles of multiple genes that all contribute to the ultimate phenotype for a given characteristic.

adenosine triphosphate (ATP) (5) A molecule that temporarily stores energy for cellular activity in all living organisms. ATP is composed of adenine, a sugar molecule, and a chain of three negatively charged phosphate groups.

adrenal glands (25) Two glands of the endocrine system that sit just above the kidneys and help regulate an organism's response to stress, by secreting the hormones cortisol, adrenaline (also called epinephrine), and other hormones.

adrenocorticotropic hormone (ACTH) (25) A hormone produced by the anterior pituitary in mammals that stimulates the adrenal glands to produce cortisol and other stress-related hormones; also known as corticotropin.

adult (13) In insect development, in complete metamorphosis, the third and final stage of development.

aging (16) An increased risk of mortality with increasing age; generally characterized by multiple physiological breakdowns.

alleles (6, 9) Alternative versions of a gene. [Gk., *allos,* another]

allergen (27) An antigen that causes an allergic response.

allergy (27) An inappropriate immune response to what should be a harmless substance.

allopatric speciation (12) Speciation that occurs as a result of a geographic barrier between groups of individuals that leads to reproductive isolation and then genetic divergence. [Gk., *allos,* another + *patris,* native land]

alternation of generations (20) In plants, the life cycle characterized by an extended period in which the plant has a multicellular haploid form and a period in which it has a multicellular diploid form.

altruistic behavior (11) A behavior that comes at a cost to the individual performing it and benefits another. [Lat., *alter,* the other]

amino acid (3) One of 20 molecules built of an amino group, a carboxyl group, and a unique side chain. Proteins are constructed of combinations of amino acids linked together.

amino group (3) A nitrogen atom attached by single bonds to hydrogen atoms.

amnion (26) During animal development, the membrane enclosing the fluid-filled sac surrounding an embryo.

amniotes (13) Terrestrial vertebrates—reptiles, birds, and mammals—that produce eggs (called amniotic eggs) that are protected by a water-tight membrane and a shell.

amphibians (13) Members of the class Amphibia; ectotherms (that is, they are cold-blooded), with a moist skin, lacking scales, through which they can fully or partially absorb oxygen. They were the first terrestrial vertebrates. The young of most species are aquatic, and the adults are true land animals. [Gk., *amphi,* on both sides + *bios,* life]

amygdala (24) Part of the brain's limbic system; associates emotional feelings and sensory input with memories.

analogous traits (12) Characteristics (such as bat wings and insect wings) that are similar because they were produced by convergent evolution, not because they descended from a common structure in a shared ancestor. [Gk., *analogos,* proportionate]

anaphase (8) A phase in mitosis and meiosis in which chromosomes separate. In mitosis, it is the third phase, in which the sister chromatids are pulled apart by the spindle fibers, with a full set of chromosomes going to opposite sides of the cell. In meiosis, the homologues separate in anaphase I and the sister chromatids separate in anaphase II. [Gk., *ana,* up + *phasis,* appearance]

anaphylactic shock (27) A life-threatening allergic reaction that causes severe respiratory distress, as the throat swells and asthma develops. With swelling in various tissues, there is less fluid in the bloodstream, and this can lead to dangerously low blood pressure.

anecdotal observation (1) Observation of one or only a few instances of a phenomenon.

angiosperms (14) Vascular, seed-producing flowering and fruit-bearing plants in which the seeds are enclosed in an ovule within the ovary. [Gk., *angeion,* vessel, jar + *sperma,* seed]

animals (13) Members of the kingdom Animalia; eukaryotic, multicellular, heterotrophic (that is, they cannot produce their own food) organisms. Many animals have body parts specialized for different activities and can move during some stage of life. [Lat., *animal,* a living being]

annelids (13) Phylum of worms having segmented bodies; protostomes with defined tissues, which grow by adding segments rather than by molting. There are about 13,000 identified species of segmented worms, including earthworms and leeches.

anterior pituitary (25) A gland of the endocrine system that produces numerous hormones, including thyroid-stimulating hormone (TSH), luteinizing hormone (LH), follicle-stimulating hormone (FSH), prolactin, adrenocorticotropic hormone (ACTH), and growth hormone. Many of the hormones of the anterior pituitary direct endocrine glands elsewhere to release other hormones.

anther (14, 20) The part of the stamen, the male reproductive structure of a flower, that produces pollen. [Gk., *anthos,* blossom]

antibody (27) A protein produced by the specific immunity system in response to an antigen; circulating antibodies recognize and bind to specific antigens.

antidiuretic hormone (ADH) (25) A peptide hormone, produced in neurons within the hypothalamus and released by the posterior pituitary, that influences water retention by the kidneys. With increased levels of ADH, more water is saved, reducing the amount of urine produced while increasing its concentration.

antigen (27) Any molecule that induces a specific immunity response.

antigen-presenting cells (27) Dendritic cells and macrophages that ingest bacteria and viruses, destroying them and "presenting" digested particles of the pathogens on their cell surface.

antigen receptor (27) A protein on the plasma membrane of a lymphocyte that sticks out from the cell surface and can bind one specific type of antigen.

antiseptic (27) A solution such as rubbing alcohol or hydrogen peroxide that discourages growth of microorganisms.

apical meristem (19) The growth region at the tip of a plant root or stem; division of cells in this meristem causes the root or stem to increase in length.

apoptosis (8, 27) Programmed cell death, which takes place particularly in parts of the body where the cells are likely to accumulate significant genetic damage over time and are therefore at high risk of becoming cancerous. [Gk., *apoptosis,* falling away]

archaea (*sing.* archaeon) (12) A group of prokaryotes that are evolutionarily distinct from bacteria and that thrive in some of the most extreme environments on earth. Archaea is one of the three domains of life. [Gk., *archaios,* ancient]

arteriosclerosis (22) A disease process, following the development of atherosclerosis, in which calcium deposits harden the arteries.

artery (22) A blood vessel that transports blood from the heart to the capillaries of the body; the blood in arteries is at a higher pressure than the blood in veins.

arthropods (13) Members of the invertebrate phylum Arthropoda; characterized by a segmented body, an exoskeleton, and jointed appendages. [Gk., *arthron,* joint + *pous,* foot]

artificial selection (10) A special case of natural selection; the three necessary and sufficient conditions for natural selection are satisfied, but the differential reproductive success is determined by humans rather than by nature and so is typically goal-oriented.

asexual reproduction (8, 26) A type of reproduction common in prokaryotes and plants, and also occurring in many other multicellular organisms, in which the offspring inherit their DNA from a single parent.

assisted reproductive technology (ART) (26) The use of fertility treatments in which both sperm and egg are handled.

atherosclerosis (22) A disease process in which plaques develop in the arteries, reducing the flow of blood and increasing the risk of blood clots.

atom (2) A particle of matter than cannot be further subdivided without losing its essential properties. [Gk., *atomos,* indivisible]

atomic mass (2) The mass of an atom; the combined mass of the protons, neutrons, and electrons in an atom (the mass of the electrons is so small as to be almost negligible).

atomic number (2) The number of protons in the nucleus of an atom of a given element.

atomic weight (2) An average of the atomic mass of an element's isotopes, taking into account their different abundances.

atrium (22) A chamber of the heart that collects blood returning from the lungs or from the rest of the body.

autoimmunity (27) An inappropriate response of an individual's immune system to the individual's own cells and tissues as if they were pathogens.

autonomic nervous system (24) One of the two major divisions of the motor pathways of the peripheral nervous system (the other is the somatic nervous system); relays signals—not under conscious control—to glands, smooth muscle tissue (such as the muscle that moves food through the digestive tract), and cardiac muscle.

auxins (20) In plants, a small group of naturally occurring hormones (and a larger group of synthetic variants) that promote cell division, stem elongation, and formation of roots. Auxins also influence plant orientation, making sure the correct ends are up and down.

axon (21, 24) A projection from a neuron that transmits impulses away from the cell body.

B

B cell (27) A type of lymphocyte, maturing in the bone marrow, that is involved in humoral immunity; fights invaders in body fluids.

background extinctions (12) Extinctions that occur at lower rates than at times of mass extinctions; mostly result from aspects of the biology and competitive success of the species, rather than catastrophe.

basal metabolic rate (BMR) (23) The amount of energy expended by a living organism at rest in a neutral-temperature environment.

base (chemistry) (2) Any fluid with a pH above 7.0—that is, with more OH^- ions than H^+ ions in solution.

base (nucleotide) (3, 6) One of the nitrogen-containing side-chain molecules attached to a sugar molecule in the sugar-phosphate backbone of DNA and RNA. The four bases in DNA are adenine (A), thymine (T), guanine (G), and cytosine (C); the four bases in RNA are adenine (A), uracil (U), guanine (G), and cytosine (C). The information in a molecule of DNA and RNA is determined by its sequence of bases.

base pair (6) Two nucleotides on complementary strands of DNA that form a pair, linked by hydrogen bonds. The pattern of pairing is adenine (A) with thymine (T), and cytosine (C) with guanine (G); the base-paired arrangement forms the "rungs" of the double-helix structure of DNA.

basophil (27) A type of white blood cell that initiates inflammation.

behavior (11) Any and all of the actions performed by an organism, often in response to its environment or to the actions of another organism.

behavioral defenses (17) Actions taken by animals to guard against predators; include both seemingly passive and active behaviors, such as hiding, escaping, alarm calling, or fighting back.

bilateral symmetry (13) A body structure with left and right sides that are mirror images. [Lat., *bi-,* two + *latus,* side]

bile (23) A juice that aids in the breakdown of fats. It is produced by the liver, sent to the gallbladder, and passed through a small duct into the small intestine, where it initiates the first step in fat digestion.

binary fission (8) A type of asexual reproduction in which the parent cell divides into two genetically identical daughter cells. Bacteria and other prokaryotes reproduce by binary fission. [Lat., *binarius,* consisting of two + *fissus,* divided]

biodiversity (12, 18) The variety and variability among all genes, species, and ecosystems. [Gk., *bios,* life + Lat., *diversus,* turned in different directions]

biodiversity hotspots (18) Regions of the world with significant reservoirs of biodiversity that are under threat of destruction.

biofuels (5) Fuels produced from plant and animal products.

biogeography (10) The study and interpretation of distribution patterns of living organisms around the world. [Gk. *bios,* life + *geo-,* earth + *graphein,* to write down]

biological literacy (1) The ability to use scientific inquiry to think creatively about problems with a biological component, to communicate these thoughts to others, and to integrate these ideas into one's decision making.

biological species concept (12) A definition of species described as populations of organisms that interbreed, or could possibly interbreed, with each other under natural conditions and that cannot interbreed with organisms outside their own group.

biology (1) The study of living things. [Gk., *bios,* life + *logos,* discourse]

biomass (17) The total mass of all the living organisms in a given area. [Gk., *bios,* life]

biomes (17) The major ecological communities of earth. Terrestrial biomes, such as rain forest or desert, are defined and usually described by the predominant types of plant life in the area, which are mostly determined by the weather. Aquatic biomes are usually defined by physical features such as salinity, water movement, and depth.

biotechnology (7) The modification of organisms, cells, and their molecules for practical benefits. [Gk., *bios,* life + *technologia,* systematic treatment]

biotic (17) Relating to living organisms; the biotic environment (or community) consists of all the living organisms in a given area. [Gk., *bios,* life]

bisphenol A (BPA) (25) A chemical found in some plastic water bottles and baby bottles and in the lining of food cans; a known endocrine disruptor.

bivalve mollusks (13) Mollusks with two hinged shells; examples are clams, scallops, and oysters. [Lat., *bi-*, two]

blastocyst (26) The mammalian blastula.

blastula (26) An early stage of embryonic development produced by cleavage of a fertilized egg (zygote). A spherical layer of cells encloses a large fluid-filled space where the embryo forms. [Gk., *blastos*, sprout]

blind experimental design (1) An experimental design in which the subjects do not know what treatment (if any) they are receiving.

blood (21, 22) A connective tissue consisting of a liquid extracellular matrix containing blood cells, contained in a closed circulatory system. Blood is important in the transport of respiratory gases, vitamins and minerals, nutrients, hormones, components of the immune system, and metabolic wastes.

blood pressure (22) The force with which blood flows through the arteries.

body mass index (23) Body weight in kilograms divided by height in meters squared (kg/m^2); often used as an index of a healthy weight for an individual based on the person's height.

bond energy (2) The strength of a bond between two atoms, defined as the energy required to break the bond.

bone (21) A rigid connective tissue that protects and provides support.

bottleneck effect (10) A phenomenon through which genetic drift can occur; a sudden reduction in population size (often due to famine, disease, or rapid environmental disturbance) that can lead to changes in the allele frequencies of a population.

Bowman's capsule (21) In the kidney, a ball-like structure surrounding the blood-filtering unit of a nephron. It is connected to a single, long, urine-collecting tube that excretes filtered fluid into a collecting duct.

Broca's area (24) The discrete part of the brain—located, in most people, in the front part of the left frontal lobe—that is used for controlling the muscles responsible for the process of speaking.

bryophytes (14) Three groups of plants (liverworts, hornworts, and mosses) that lack vascular tissue and move water and dissolved nutrients through the plant body by diffusion. [Gk., *bruon*, tree-moss, liverwort + *phytas*, plant]

buffer (2) A chemical that can quickly absorb excess H^+ ions in a solution (preventing it from becoming too acidic) or quickly release H^+ ions (to counteract increases in OH^- concentration).

bulbourethral gland (26) A male reproductive gland, one on each side of the body, that contributes mucus and sugar to the ejaculate and lubricant to the tip of the penis prior to copulation.

C

C4 photosynthesis (5) A method (along with C3 and CAM photosynthesis) by which plants fix carbon dioxide, using the carbon to build sugar. It serves as a more effective method than C3 photosynthesis for binding carbon dioxide under low CO_2 conditions, such as when plants in warmer climates close their stomata to reduce water loss.

calcitonin (25) A hormone released from the thyroid gland that causes bones to take up excess calcium from the bloodstream.

calorie (23) The energy required to raise the temperature of 1 gram of water by 1° C. The kilocalorie, which is 1,000 calories, is commonly used as a measure of the amount of energy in a particular food (and is often inaccurately referred to as a calorie).

Calvin cycle (5) In photosynthesis, a series of chemical reactions in the stroma of chloroplasts in which sugar molecules are assembled. [From the name of one of its discoverers, Melvin Calvin, 1911–1997]

CAM (crassulacean acid metabolism) (5) Energetically expensive photosynthesis in which the stomata are open only at night to admit CO_2, which is bound to a holding molecule and released to enter the Calvin cycle to make sugar during the day. In this type of photosynthesis, found in many fleshy, juicy plants of hot, dry areas, water loss is reduced because the stomata are closed during the day.

cancer (8) Unrestrained cell growth and division. [Lat., *cancer*, crab]

capillaries (22) Tiny blood vessels that bring blood close enough to cells to allow the diffusion of molecules into and out of the blood.

capsid (15) The protein container surrounding the genetic material (DNA or RNA) of a virus. [Lat., *capsa*, box, case]

capsule (15) A layer surrounding the cell wall of many bacteria; it may restrict the movement of water out of the cell and thus allow bacteria to live in dry places, such as the surface of the skin. The capsule contributes to the virulent characteristics of some bacteria, making them resistant to phagocytosis by the host's immune system. [Lat., *capsula*, small box or case]

carbohydrates (3) One of the four types of biological macromolecules, containing mostly carbon, hydrogen, and oxygen. Carbohydrates are the primary fuel for cellular activity and form much of the cell structure in all life forms. [Lat., *carbo*, charcoal + *hydro-*, pertaining to water]

carboxyl group (3) A functional group characterized by a carbon atom double-bonded to one oxygen atom and single-bonded to another oxygen atom. Amino acids are made up of an amino group, a carboxyl group, and a side chain.

cardiac muscle (21) A type of muscle tissue, located only in the heart, that causes the heart to pump blood through the body. Cardiac muscle cells, which are fused together and connected by gap junctions, contain many more mitochondria than other types of muscle cells.

cardiovascular disease (22) Progressive deterioration of the arteries and other degradations of the circulatory system; the leading cause of death and disability in the United States.

carnivores (17, 23) Predatory animals (and some plants) that consume only animals. [Lat., *carnis*, of flesh + *vorare*, to devour]

carotenoids (5) Pigments that absorb blue-violet and blue-green wavelengths of light and reflect yellow, orange, and red wavelengths of light. [Lat., *carota*, carrot]

carpel (14, 20) The female reproductive structure of a flower, including the stigma, style, and ovary. [Gk., *karpos*, fruit]

carrier (9) In genetics, an individual who carries one allele for a recessive trait and does not exhibit the trait; if two carriers mate, they may produce offspring that do exhibit the trait.

carrying capacity (K) (16) The ceiling on a population's growth imposed by the limitation of resources for a particular habitat over a period of time.

cartilage (21) A dense connective tissue with an extracellular matrix rich in collagen, elastin, and proteins bound to long carbohydrate chains. It has a hardness between that of bone and of tendons. Strong, but also flexible, cartilage is found in the ears and tip of the nose in humans, and it cushions the bones in joints throughout the body.

cartilaginous fishes (13) Fish species characterized by a skeleton made completely of cartilage, not bone. [Lat., *cartilago*, gristle]

cell (4) The smallest unit of life that can function independently; a three-dimensional structure, surrounded by a membrane and, in prokaryotes and most plants, by a cell wall, in which many of the essential chemical reactions of the life of an organism take place. [Lat., *cella*, room]

cell body (21, 24) The part of a cell, such as a neuron, that contains the nucleus and other organelles.

cell cycle (8) In a cell, the alternation of activities related to cell division and those related to growth and metabolism.

cell-cycle control system (8) A system—made up of molecules, mostly proteins—by which the events of the cell cycle are coordinated and cell division is regulated. At critical points in the cell cycle, called checkpoints, progress to the next phase may be blocked until specific signals trigger continuation of the process.

cell theory (4) A unifying and universally accepted theory in biology that holds that all living organisms are made up of one or more cells and that all cells arise from other, preexisting cells.

cell wall (4) A rigid structure, outside the cell membrane, that protects and gives shape to the cell; found in many prokaryotes and plants.

cellular respiration (5) The process by which all living organisms extract energy stored in the chemical bonds of molecules and use it for fuel for their life processes.

cellulose (3) A complex carbohydrate, indigestible by humans, that serves as the structural material for a huge variety of plant structures. It is the single most prevalent organic compound on earth. [Lat., *cellula*, *dim.* of *cella*, room]

central nervous system (CNS) (21, 24) The part of an organism's nervous system that includes the brain and spinal cord and coordinates all nervous activity.

central vacuole See **vacuole.**

centriole (8) A structure, located outside the nucleus in most animal cells, to which the spindle fibers are attached during cell division.

centromere (8) After replication, the region of contact between sister chromatids, located near the center of the two strands. [Gk., *centron,* the stationary point of a pair of compasses, thus the center of a circle + *meris,* part]

centrosome (8) An organelle in most animal cells, consisting of a pair of centrioles and a mass of proteins; produces and anchors spindle fibers, which spread out and attach to the chromosome centromeres.

cephalopods (13) Mollusks in which the head is prominent and the foot has been modified into tentacles; examples are octopuses and squids. Cephalopods have a reduced or absent shell and possess the most advanced nervous system of the invertebrates. [Gk., *kephale,* head + *pous,* foot]

cerebellum (24) The part of the vertebrate hindbrain that controls muscle coordination.

cerebral cortex (24) The part of the vertebrate brain involved in abstract thought, problem solving, language, and more; the brain structure most responsible for the traits that set humans apart from other animal species.

cervix (26) The low, narrow portion of the uterus leading to the vagina. [Lat., *cervix,* neck]

character displacement (17) An evolutionary divergence in one or more of the species that occupy the same niche that leads to a partitioning of the niche between the species. Changes in characteristics, such as behavior or body plan, of two or more very similar species that have overlapping geographic locations result in a reduction of competition between the species.

checkpoints (8) Several critical points in the cell cycle at which progress is blocked—and cells are prevented from dividing—until specific signals trigger continuation of the process.

chemical energy (5) A type of potential energy in which energy is stored in chemical bonds between atoms or molecules.

chemical reaction (2) A process, involving the forming and breaking of chemical bonds, in which molecules (called reactants) are transformed into different molecules (called products).

chemolithotrophs (15) Bacteria that can use inorganic molecules such as ammonia, hydrogen sulfide, hydrogen, and iron as sources of energy. [Gk., *lithos,* stone, rock + *troph* , food]

chemoorganotrophs (15) Bacteria that consume organic molecules, such as carbohydrates, as an energy source. [Gk., *troph* , food]

chemoreceptor (24) A type of sensory cell that detects chemical changes in the organism's internal or external environment.

chiasmata (*sing.* chiasma) (8) The sites at which chromatids exchange genetic material during recombination.

chitin (3) (*pron.* kite-in) A complex carbohydrate, indigestible by humans, that forms the rigid outer skeleton of most insects and crustaceans. [Gk., *chiton,* undershirt]

chlorophyll (5) A light-absorbing pigment molecule in chloroplasts. [Gk., *chloros,* pale green + *phyllon,* leaf]

chlorophyll a (5) The primary photosynthetic pigment. Chlorophyll *a* absorbs blue-violet and red light; because it cannot absorb green light and instead reflects those wavelengths, we perceive the reflected light as the color green.

chlorophyll b (5) A photosynthetic pigment similar in structure to chlorophyll *a*. Chlorophyll *b* absorbs blue and red-orange wavelengths and reflects yellow-green wavelengths.

chloroplast (4, 5) The organelle in plant cells in which photosynthesis occurs. [Gk., *chloros,* pale green + *plastos,* formed]

cholesterol (3, 4) One of the sterols, a group of lipids important in regulating growth and development; an important component of most cell membranes, helping the membrane maintain its flexibility. [Gk., *chole,* bile + *stereos,* solid + *-ol,* chemical suffix for an alcohol]

chorion (26) During animal development, the outer membrane surrounding an embryo, which, with the endometrium, forms the placenta.

chromatid (8) One of the two strands of a replicated chromosome. [Gk., *chroma,* color]

chromatin (4, 8) A mass of long, thin fibers consisting of DNA and proteins in the nucleus of the cell.

chromosomal aberration (6) A type of mutation characterized by a change in the overall organization of genes on a chromosome, such as the deletion of a section of DNA; or the moving of a gene from one part of a chromosome to elsewhere on the same chromosome or to a different chromosome; or the duplication of a gene, with the new copy inserted elsewhere on the chromosome or on a different chromosome. [Lat., *aberrare,* to wander]

chromosome (6, 8) A linear or circular strand of DNA with specific sequences of base pairs. The human genome consists of two copies of each of 23 unique chromosomes, one from the mother and one from the father. [Gk., *chroma,* color + *soma,* body]

chyme (23) A liquid mass of partially digested food that passes from the stomach through the small intestine.

cilia (*sing.* cilium) (4, 27) Short projections from the cell surface, often occurring in large numbers on a single cell, that beat against the extracellular fluid to move the fluid past the cell. [Lat., *cilium,* eyelid]

citric acid cycle (5) The second step of cellular respiration, in which energy is extracted from sugar molecules as additional molecules of ATP and NADH are formed; also called the Krebs cycle [from the name of the discoverer, Hans Adolf Krebs, 1900–1981].

class (12) In the hierarchical taxonomic system developed by Carolus Linnaeus (1707–1778), a classification of organisms consisting of related orders.

cleavage (26) In embryonic development, the early cell division of the zygote, which begins shortly after fertilization.

climax community (17) A stable and self-sustaining community that results from ecological succession.

clonal selection (27) A process in which a B cell or T cell binds to its antigen and divides numerous times to create a population of cells with the same antigen specificity. This process ensures that there are sufficient numbers of B and T cells to recognize and respond to a specific pathogen that has invaded the body.

clone (7) A genetically identical DNA fragment, cell, or organism produced from a DNA fragment or single cell or organism. [Gk., *klon,* twig]

clone library (7) A collection of cloned DNA fragments; also known as a gene library.

cloning (7) The production of genetically identical cells, organisms, or DNA molecules.

closed circulatory system (22) A circulatory system in which blood is contained in vessels and is separate from the interstitial fluid that bathes cells.

code (6) In genetics, the base sequence of a gene. Information encoded within the genetic information can be translated into proteins.

codominance (9) The case in which a heterozygote displays characteristics of both alleles.

codons (6) Three-base sequences in mRNA that link with complementary tRNA molecules, which are attached to amino acids that are specified by that codon. A codon with yet another sequence ends the process of assembling a protein from amino acids.

coelom (26) (*pron.* SEE-lum) The body cavity; forms from the mesoderm in embryonic development.

coevolution (17) The concurrent appearance and modification over time, through natural selection, of traits in interacting species that enable each species to become adapted to the other; an example is the 11-inch-long tongue of a moth that feeds from the 11-inch-long nectar tube of an orchid.

cohesion-tension mechanism (19) In vascular plants, the force driving fluid flow in the xylem; it results from evaporation of water from leaves, which pulls water up from the roots.

collagen (21) A protein in the matrix of connective tissue, synthesized and secreted by connective tissue cells (fibroblasts).

collenchyma cell (19) A long, stringy plant cell with unevenly thickened cell walls that contributes to a plant's flexibility. Collenchyma cells are a type of ground tissue.

colon See **large intestine.**

colonizers (17) Species introduced into an area that has been disturbed and is undergoing the process of primary or secondary succession. The identity of a colonizing species varies, depending on the stage and type of succession.

color-blind (24) In humans, describes the inability to distinguish between colors that most individuals can see—most commonly, the inability to distinguish between red and green; occurs mostly in males.

commensalism (17) A symbiotic relationship between species in which one benefits and the other neither benefits nor is harmed. [Lat., *com,* with + *mensa,* table]

communication (11) An action or signal on the part of one organism that alters the behavior of another organism.

community (17) In ecology, the biotic environment; a geographic area defined as a loose assemblage of species with overlapping ranges.

competitive exclusion (17) The case in which two species battle for resources in the same niche until the more efficient of the two wins and the other is driven to extinction in that location.

competitive inhibitor (3) A chemical that binds to the active site of an enzyme, blocking substrate molecules from the site and thereby reducing the enzyme's ability to catalyze a reaction.

complement protein (27) A type of circulating defensive protein that can create holes in the membranes of pathogens or destroy pathogens by sticking to them and forming a coating that enhances the ability of phagocytes to bind to and engulf them.

complementarity (8) A characteristic of double-stranded DNA in which the base on one strand always has the same pairing partner, or complementary base, on the other strand.

complementary base (8) A base on a strand of double-stranded DNA that is a pairing partner to a base on the other strand: adenine (A) is the complementary base to thymine (T), and guanine (G) is the complementary base to cytosine (C).

complete metamorphosis (13) A developmental process in which, after hatching, an organism's life is divided into three completely different stages: larva, pupa, and adult. It occurs in about 83% of insect species, including beetles and butterflies, and is an important factor in the broad diversification of insects.

complex carbohydrates (3) Carbohydrates that contain multiple simple carbohydrates linked together; examples are starch, which is the primary form of energy storage in plants, and glycogen, which is the primary form of short-term energy storage in animals.

composting (19) The deliberate mixing of decaying organic matter, from leaves, food scraps, and manure, used to provide nutrients to soil.

compound eye (24) A group of light-sensing cells found in many invertebrates, made up of dozens to thousands of separate light-sensing units, each with its own lens that focuses light onto about a dozen photoreceptors.

condensation (8) In the cell cycle, just before mitosis, the process in which sister chromatids coil tightly and become compact—in contrast to the uncondensed and tangled state of the chromosomes prior to replication, during most of interphase.

cone (24) In the retina, a highly light-sensitive cell type that enables color vision.

conformer (21) An organism that, for a given physiological variable, has no set point for the variable and allows it to fluctuate with external changes in the environment.

conjugation (15) In bacteria, the process by which a bacterium transfers a copy of some or all of its DNA to another bacterium of the same or another species. [Lat., *coniugatio,* connection]

connective tissue (21) A type of animal tissue that consists of cells embedded in a large amount of extracellular material, called matrix, which together contribute to body structure and support; includes tendons, ligaments, fat, blood, bone, and cartilage.

connective tissue proper (21) A type of connective tissue, including fat tissue, tendons, and ligaments, that has a semi-fluid, flexible matrix and functions like packing material.

conservation biology (18) An interdisciplinary field, drawing on ecology, economics, psychology, sociology, and political science, that studies and devises ways of preserving and protecting biodiversity and other natural resources.

contraception (26) The attempt to prevent pregnancy by preventing ovulation, fertilization, or implantation; also called birth control.

control group (1) In an experiment, the group of subjects not exposed to the treatment being studied but otherwise treated identically to the experimental group.

convergent evolution (10, 12) A process of natural selection in which features of organisms not closely related come to resemble each other as a consequence of similar selective forces. Many marsupial and placental species resemble each other as a result of convergent evolution. [Lat., *con-,* together with + *vergere,* turn + *evolvere,* to roll out]

coprophagy (23) A nutritional strategy, common in rabbits and rodents, in which animals consume some of their feces, thereby increasing their ability to extract energy from food by passing it through their digestive system twice.

copulation (26) The act of placing the male reproductive organ within the female reproductive tract.

coral reef (13) Underwater structures that are assemblies of giant calcium carbonate skeletons and corals, small cnidarians that secrete the calcium carbonate to create hard shells on which numerous individual polyps can live as a colony. Coral reefs provide an environment that is home to a greater diversity of species than any other marine habitat.

cork cambium (20) A cylinder of waxy lateral meristem cells close to the outer edge of the trunk in some plants. It plays an important role in maintaining the protective covering of bark, protecting against water loss, fire, and infection by microbes.

cork cell (19) A cell of the protective outer covering of some plants, containing a waxy substance impermeable to water and resistant to fire and decay.

coronary arteries (22) The blood vessels that deliver oxygen and nutrients to the cells of the heart.

corpus callosum (24) In the brain, the broad, thick band of neurons connecting the left hemisphere to the right hemisphere.

corpus luteum (26) A structure in the ovary that develops from a follicle after ovulation; secretes hormones to maintain pregnancy. [Lat., *corpus,* body + *luteum,* yellow]

cortex (19) In vascular plants, the ground tissue located between the epidermis and the xylem and phloem.

corticotropin (25) A hormone produced by the anterior pituitary in mammals that stimulates the adrenal glands to produce cortisol and other stress-related hormones; also known as adrenocorticotropic hormone (ACTH).

cotyledon (19, 20) Part of the plant embryo within the seed; this structure usually becomes the first embryonic leaf or leaves of the plant; also called a seed leaf.

covalent bond (2) A strong bond formed when two atoms share electrons; the simplest example is the H_2 molecule, in which each of the two atoms in the molecule shares its lone electron with the other atom. [Lat., *con-,* together + *valere,* to be strong]

CRISPR (7) A gene-editing system that allows the targeting and cutting of DNA at a specific sequence in almost any species. The name is an acronym for "clustered regularly interspaced short palindromic repeats" and refers to DNA that originally came from viruses but now exists within the genomes of bacteria and helps thwart viral infection.

critical experiment (1) An experiment that makes it possible to determine decisively between alternative hypotheses.

cross (9) The breeding of organisms that differ in one or more traits.

crossing over (9) The exchange of some genetic material between a paternal homologous chromosome and a maternal homologous chromosome, leading to a chromosome carrying genetic material from each; also called recombination.

cuticle (14, 19) In terrestrial plants, a waxy layer produced by epidermal cells and found on leaves and shoots, protecting them from drying out. [Lat., *cuticula, dim.* of *cutis,* skin]

cytokine (27) A type of signaling protein secreted by cells of the immune system; the chief way in which cells of the specific and non-specific immunity systems communicate with one another.

cytokinesis (8) In the cell cycle, the stage (following mitosis) in which cytoplasm and organelles duplicate and are divided into approximately equal parts, and the cell separates into two daughter cells. In meiosis, two diploid daughter cells are formed in cytokinesis following telophase I, and four haploid daughter cells are formed in cytokinesis following telophase II. [Gk., *kytos,* container + *kinesis,* motion]

cytokinins (20) In plants, a group of hormones that stimulate cell division throughout the body and throughout the lifetime of the plant.

cytoplasm (4) The contents of a cell contained within the plasma membrane, including a jelly-like fluid, called the cytosol, and, in eukaryotes, the organelles. [Gk., *kytos,* container + *plasma,* anything molded]

cytoskeleton (4) A network of protein structures in the cytoplasm of eukaryotes (and, to a lesser extent, prokaryotes) that serves as scaffolding, adding support and, in some cases, giving animal cells of different types their characteristic shapes. The cytoskeleton serves as a system of tracks guiding intracellular traffic flow and, because it is flexible and can generate force, gives cells some ability to control their movement.

cytosol (4) The jelly-like fluid that fills the inside of the cell.

cytotoxic T cell (27) A type of lymphocyte, acting in the cell-mediated response, that directly kills body cells infected with a pathogen.

D

daughter cells (8) Cells produced by the division of a parent cell.

decomposers (14, 17) Organisms, including bacteria, fungi, and detritivores, that break down and feed on once-living organisms.

decomposition (19) The process of breaking down organic material into the individual molecules of which it is made.

demographic transition (16) A pattern of population growth characterized by the progression from high birth and death rates (slow growth) to high birth rates and low death rates (fast growth) to low birth and death rates (slow growth).

denaturation (3) The disruption of protein folding in which secondary and tertiary structures are lost, caused by exposure to extreme conditions in the environment such as heat or extreme pH.

dendrite (21, 24) A branched projection from a neuron that receives signals from the external environment or from other neurons. [Gk., *dendron,* tree]

dendritic cell (27) A type of white blood cell that links the non-specific and specific divisions of the immune system by engulfing foreign matter and then "presenting" digested remnants of the pathogen on its cell surface.

density-dependent factors (16) Limitations on a population's growth that are a consequence of population density.

density-independent factors (16) Limitations on a population's growth unrelated to population size, such as floods, earthquakes, fires, and lightning.

deoxyribonucleic acid (DNA) (3, 6) A nucleic acid that carries information, in the sequences of its nucleotide bases, about the production of particular proteins.

dependent variable (1) A measurable entity that is created by the process being observed and the value of which cannot be controlled; generally represented on the *y*-axis in a graph and expected to change in response to a change in the independent variable.

dermal tissue (19) The outer covering that protects the surface of a plant.

desert (17) A type of terrestrial biome and a type of dry climate, with very little rainfall, in which water loss through evaporation exceeds water gain through precipitation; typically found at 30° north and south latitudes.

desmosomes (4) Irregularly spaced connections between adjacent animal cells that, much like Velcro, hold the cells together by multiple attachments but are not water-tight. They provide mechanical strength and are found in muscle tissue and in much of the tissue that lines the cavities of animal bodies. [Gk., *desmos,* bond + *soma,* body]

detritivores (17) Organisms that break down and feed on once-living organic matter; the group includes scavengers such as vultures, worms, and a variety of arthropods. [Lat., *detritus,* worn out + *vorare,* to devour]

deuterostomes (13) Bilaterally symmetrical animals with defined tissues in which the gut develops from back to front: the anus forms first, and the second opening formed becomes the mouth of the adult animal. [Gk., *deuteros,* second + *stoma,* mouth]

diabetes (23) A condition in which an individual does not adequately regulate his or her blood sugar levels. It usually results from insufficient insulin secretion by the pancreas in response to an increase in blood sugar or from an inadequate response of the body's cells to insulin in the bloodstream.

diaphragm (22) A large sheet of muscle that separates the chest cavity from the abdominal cavity; when contracted, it allows an increased volume of the chest cavity during inhalation. [Gk., *dia,* through + *phrasein,* to enclose]

diastolic pressure (22) The second blood pressure reading; a measure of the force that blood exerts on the artery walls while the heart is between beats.

differential reproductive success (10) The situation in which some individuals have greater reproductive success than other individuals in a population. Along with variation and heritability, it is one of the three conditions necessary for evolution by natural selection.

diffusion (4) Passive transport in which a particle (the solute) is dissolved in a gas or liquid (the solvent) and moves from an area of higher solute concentration to an area of lower solute concentration. [Lat., *diffundere,* to pour in different directions]

digestion (23) The physical and chemical breakdown of food into its fundamental macromolecular and smaller molecular components for absorption or elimination.

dihybrid (9) An individual that is heterozygous at two genetic loci.

dihybrid cross (9) A cross between two individuals that are heterozygous for the same two genetic loci.

diploid (8) Describes cells that have two copies of each chromosome. In many organisms, including humans, somatic cells are diploid. [Gk., *diplasiazein,* to double]

directional selection (10) Selection that, for a given trait, increases fitness at one extreme of the phenotype and reduces fitness at the other, leading to an increase or decrease in the mean value of the trait.

disaccharides (3) Carbohydrates formed by the union of two simple sugars; examples are sucrose (table sugar) and lactose (the sugar found in milk). [Gk., *di-,* two + *sakcharon,* sugar]

dispersers (17) Organisms able to move away from their original home.

disruptive selection (10) Selection that, for a given trait, increases fitness at both extremes of the phenotype distribution and reduces fitness at middle values.

DNA helicase (8) In DNA replication, the enzyme that unwinds the coiled DNA and separates the two complementary strands.

DNA polymerase (8) In DNA replication, the enzyme that adds new DNA nucleotides to the 3′ end of the primer as it builds a strand complementary to the template strand.

DNA synthesis phase (S phase) (8) In the cell cycle, the phase during which the cell prepares for cell division by creating an exact duplicate of each chromosome by replication.

domain (12) In modern classification, the highest level of the hierarchy. There are three domains, Bacteria, Archaea, and Eukarya.

dominant (9) In genetics, describes an allele that masks the phenotypic effect of the other, recessive allele for a trait. The phenotype shows the effect of the dominant allele in both homozygous and heterozygous genotypes. [Lat., *dominari,* to rule]

dorsal hollow nerve cord (13) The central nervous system of vertebrates, consisting of the spinal cord and brain. [Lat., *dorsum,* back]

double-blind experimental design (1) An experimental design in which neither the subjects nor the experimenters know what treatment (if any) the individual subjects are receiving.

double bond (2) The sharing of two electrons between two atoms; for example, the most common form of oxygen is the O_2 molecule, in which two electrons from each of the two atoms of oxygen are shared.

double fertilization (14, 20) In angiosperms, the fertilization process in which two sperm are released by a pollen grain, and one fuses with an egg to form a zygote, while the other fuses with two nuclei to form a triploid endosperm.

double helix (3) The spiraling ladder-like structure of DNA composed of two strands of nucleotides. The bases protruding from each strand like "half-rungs" meet in the center and bind to each other (via hydrogen bonds), holding the sides of the ladder together. [Gk., *heligmos,* wrapping]

E

ear canal (24) The channel from the outer ear to the middle ear that conducts sound waves.

eardrum (24) The thin membrane that divides the outer ear from the middle ear.

echolocation (24) The ability of some animals (including bats) to sense the location of objects by producing sounds and then detecting echoes bounced off the objects.

ecological footprint (16) A measure of the impact of an individual or population on the environment by calculation of the amount of resources—including land, food and water, and fuel—consumed.

ecology (16) The study of the interaction between organisms and their environments, at the level of individuals, populations, communities, and ecosystems. [Gk., *oikos,* home + *logos,* discourse]

ecosystem (17) A community of biological organisms and the non-living environmental components with which they interact.

ecosystem services (18) The utilitarian value of biodiversity to humans described in terms of four distinct categories of services: provisioning, regulating, habitat, and cultural services.

ectoderm (26) The outermost embryonic cell layer during gastrulation; it eventually forms skin, hair, the nervous system, and the lining of the nose and mouth. [Gk., *ektos,* outside + *derma,* skin]

ectotherms (13, 21) Organisms that rely on heat from an external source to raise their body temperature and seek the shade when the air is too warm. [Gk., *ektos,* outside + *therm* , heat]

effector (21) A structure that can be triggered in response to a stimulus, aiding in the maintenance of homeostasis by opposing or reducing changes in the internal environment in response to changes in the external environment.

effector cell (27) In the specific (adaptive) immunity system, a type of cell that recognizes an antigen and immediately takes some action that leads to its destruction. Plasma cells, which are antibody-secreting, modified B cells, are the effector cells of the cell-mediated response.

egg See **ovum.**

El Niño (17) A sustained surface temperature change in the central Pacific Ocean that occurs every two to seven years. This event can start a chain reaction of unusual weather across the globe that can result in flooding, droughts, famine, and a variety of extreme climate disruptions. [Spanish, *the child;* a reference to the Christ child, because of the appearance of the phenomenon at Christmastime]

elastin (21) A protein in the matrix of connective tissue, synthesized and secreted by connective tissue cells (fibroblasts), that gives the tissue flexibility.

electromagnetic spectrum (5) The range of wavelengths that produce electromagnetic radiation, extending (in order of decreasing energy) from high-energy, short-wave, gamma rays and X rays, through ultraviolet light, visible light, and infrared light, to very long, low-energy, radio waves. [Lat., *specere,* to look at]

electron (2) A negatively charged particle that moves around the atomic nucleus.

electron transport chain (5) The path of high-energy electrons moving from one molecule within a membrane to another, coupled to the pumping of protons across the membrane, creating a concentration gradient that is used to make ATP; occurs in mitochondria and chloroplasts.

element (2) A pure substance that cannot be broken down chemically into any other substances; all atoms of an element have the same atomic number. [Lat., *elementum,* element, or first principle]

elimination (23) The last phase in the breakdown of food; the absorption of water, salts, and some vitamins in the large intestine, followed by defecation of remaining waste material.

embryo (26) The multicellular, developing, fertilized egg of a eukaryote. In humans, at about eight weeks following fertilization the embryo is called a fetus.

embryo sac (20) In flowering plants, the part of the female reproductive structure of a flower that contains an egg and is the site of fertilization.

empirical (1) Describes knowledge that is based on experience and observations that are rational, testable, and repeatable. [Gk., *empeiria,* experience]

endangered species (18) As defined by the Endangered Species Act, species in danger of extinction throughout all or a significant portion of their range.

Endangered Species Act (ESA) (18) A U.S. law that defines "endangered species" and is designed to protect those species from extinction.

endemic species (18) Species peculiar to a particular region and not naturally found elsewhere. [Gk., *en,* in + *demos,* the people of a country]

endocrine cell (25) A hormone-secreting cell. Endocrine cells are part of the endocrine system.

endocrine disruptor (25) A chemical manufactured by humans that, when taken up by organisms, can mimic, block, or otherwise interfere with their hormones, leading to harmful effects.

endocrine gland (21, 25) A collection of epithelial cells that produces hormones and releases them into the bloodstream or other fluids of the body.

endocrine system (25) An organ system comprising glands and cells that secrete hormones, which are chemical messengers that act on target cells to regulate body functions and maintain homeostasis.

endocytosis (4) A cellular process in which large particles, solid or dissolved, outside the cell are surrounded by a fold of the plasma membrane, which pinches off to form a vesicle, and the enclosed particle moves into the cell. The three types of endocytosis are phagocytosis, pinocytosis, and receptor-mediated endocytosis. [Gk., *endon,* within + *kytos,* container]

endoderm (26) The innermost embryonic cell layer during gastrulation; it eventually forms the lining of the respiratory and digestive tracts, the liver, and the pancreas. [Gk., *endon,* within + *derma,* skin]

endomembrane system (4) A system of organelles (rough endoplasmic reticulum, smooth endoplasmic reticulum, and Golgi apparatus) that surrounds the nucleus. It produces and modifies necessary molecules, breaks down toxic chemicals and cellular by-products, and is thus responsible for many of the fundamental functions of the cell. [Gk., *endon,* within + Lat., *membrana,* a thin skin]

endometrium (26) The lining of the uterus, where a fertilized egg implants and is nourished.

endosperm (14, 20) Tissue of a mature seed that stores certain carbohydrates, proteins, and lipids that fuel the germination, growth, and development of the embryo and young seedling. [Gk., *endon,* within + *sperma,* seed]

endosymbiosis theory (4) Theory of the origin of eukaryotes. For photosynthetic eukaryotes, the theory holds that, in the past, two different types of prokaryotes engaged in a close partnership and eventually one, capable of performing photosynthesis, was subsumed into the other, larger prokaryote. The smaller prokaryote made some of its photosynthetic energy available to the host and, over time, the two became symbiotic and eventually became a single, more complex organism in which the smaller prokaryote had evolved into the chloroplast of the new organism. A similar scenario can be developed for the evolution of mitochondria. [Gk., *endon,* within + *symbios,* living together]

endotherms (13, 21) Organisms that use the heat produced by their cellular respiration to raise and maintain their body temperature above air temperature. [Gk., *endon,* within + *therm* , heat]

energy (5) The capacity to do work, which is the moving of matter against an opposing force. [Gk., *energeia,* activity]

energy pyramid (17) A diagram that illustrates the path of energy through the organisms of an ecosystem; each layer of the pyramid represents the biomass of a trophic level.

enzyme (3) A protein that initiates and accelerates a chemical reaction in a living organism. Enzymes are found throughout the cell; they also take part in chemical reactions on the inside and outside surfaces of the plasma membrane. [Gk., *en,* in + *zyme,* leaven]

epidermis (19) In plants, the layer of tightly packed, very thin cells that covers and protects roots, leaves, and stems.

epididymis (26) A tube in each testis where the sperm mature.

epithelial tissue (21) A thin tissue that covers and protects the surfaces of an animal's body; also called epithelium.

epithelium See **epithelial tissue.**

erythrocytes See **red blood cells.**

esophagus (23) The passageway from the throat to the stomach through which food travels.

essential amino acid (23) An amino acid that is not made by the body and so must be consumed in food.

estrogen (25, 26) One of the primary female sex hormones; important in female development and the female reproductive cycle.

estuary (17) A tidal water passage, linked to the sea, in which salt water and fresh water mix; characterized by exceptionally high productivity. [Lat., *aestus,* tide]

ethanol (5) The end product of fermentation of yeast; the alcohol in beer, wine, and spirits. [Contraction of the full chemical name, ethyl alcohol]

ethylene (20) In plants, a gas that functions as a hormone that speeds up fruit ripening.

eudicot (19) A large monophyletic subset of the dicots (plants in which two cotyledons form) and one of the two major groups of flowering plants (the other is the monocots).

eukaryote (4) An organism composed of eukaryotic cells. [Gk., *eu,* good + *karyon,* nut, kernel]

eukaryotic cell (4) A cell with a membrane-surrounded nucleus that contains DNA, membrane-surrounded organelles, and internal structures organized into compartments.

eutrophication (17) The process in which excess nutrients dissolved in a body of water lead to rapid growth of algae and bacteria, which consume much of the dissolved oxygen and, in time, can lead to large-scale die-offs of other species. [Gk., *eu,* good + *troph* , food]

evolution (10) A change in allele frequencies of a population. [Lat., *evolvere,* to roll out]

exocrine gland (21) A collection of epithelial cells that secretes products onto the surface of the epithelium.

exocytosis (4) A cellular process in which particles within the cell, solid or dissolved, are enclosed in a vesicle and transported to the plasma membrane, where the membrane of the vesicle merges with the plasma membrane and the material in the vesicle is expelled to the extracellular fluid for use throughout the body. [Gk., *ex,* out of + *kytos,* container]

exoskeleton (13) A rigid external covering such as is found in some invertebrates, including insects and crustaceans. [Gk., *ex,* out of]

exotic species (18) Species introduced by human activities to areas other than the species' native range; also called introduced species.

experimental group (1) In an experiment, the group of subjects exposed to a particular treatment; also known as the treatment group.

exponential growth (16) Growth of a population at a rate that is proportional to its current size.

external fertilization (26) The process in which sperm and egg unite outside the bodies of the male and female.

extinction (12) The complete loss of all individuals in a species. [Lat., *extinguere,* to extinguish]

extracellular fluid (21) A body fluid that is outside the cells; as distinct from intracellular fluid. See also **blood; interstitial fluid.**

extremophiles (15) Bacteria and archaea that can live in extreme physical and chemical conditions. [Lat., *extremus,* outermost + Gk., *philios,* loving]

F

facilitated diffusion (4) Diffusion of molecules through the phospholipid bilayer of a plasma membrane that takes place through a transport protein (a "carrier molecule") embedded in the membrane. Molecules that require the assistance of a carrier molecule are those that are too big to cross the membrane directly or are electrically charged and would be repelled by the middle layer of the membrane.

Fallopian tube (26) The tube that conveys eggs from an ovary to the uterus; also called the oviduct.

family (12) In the system developed by Carolus Linnaeus (1707–1778), a classification of organisms consisting of related genera.

fatty acid (3) A long hydrocarbon (a chain of carbon-hydrogen molecules); fatty acids form the tail region of triglyceride fat molecules.

female (11) In sexually reproducing organisms, a member of the sex that produces the larger gamete.

fermentation (5) The process by which glycolysis occurs in the absence of oxygen; the electron acceptor is pyruvate (in animals) or acetaldehyde (in yeast) rather than oxygen.

fertilization (8, 26) The fusion of two reproductive cells.

fetus (26) A developmental stage in humans, from the end of the embryonic period, approximately eight weeks after fertilization, until birth.

fever (27) An elevated body temperature in response to infection; enhances the immune response and slows the growth of pathogens.

fibrous roots (19) A root system in which numerous roots branch out directly from the plant stem; found primarily in monocots.

filament (14, 20) In angiosperms, the supporting stalk of the anther of a stamen in flowers. [Lat., *filum,* thread]

filtrate (21) The fluid that accumulates in Bowman's capsule in the vertebrate kidney; contains salts, sugars, amino acids, vitamins, and many other molecules, all at the same concentration as in the blood.

first law of thermodynamics (5) A physical law stating that energy cannot be created or destroyed; it can only change from one form to another.

fitness (10) A relative measure of the reproductive output of an individual with a given phenotype compared with the reproductive output of individuals with alternative phenotypes.

fixation (10) In genetics, the point at which the frequency of an allele in a population is 100%, and thus there is no more variation in the population for this gene.

fixed action pattern (11) An innate sequence of behaviors, triggered under certain conditions, that requires no learning, does not vary, and once begun runs to completion; an example is egg-retrieval in geese.

flagellum (*pl.* flagella) (4) A long, thin, whip-like projection from a cell that aids in cell movement. In prokaryotes, a projection of the plasma membrane that aids in the cell's movement through the medium in which it lives. In eukaryotes, a microtubule-based projection; in animals, the only cell with a flagellum is the sperm cell. [Lat., *flagellum,* whip]

flatworms (13) Worms with flat bodies that are members of the phylum Platyhelminthes; characterized by well-defined head and tail regions, with some having clusters of light-sensitive cells for eyespots. Most are hermaphroditic and are protostomes that do not molt. Examples are tapeworms and flukes.

fleshy fruit (14) A fruit that consists of the ovary and some additional parts of the flower; fleshy fruits are an attractive food, and when eaten by animals, the seeds may be widely dispersed.

flower (14) The part of an angiosperm that contains the reproductive structures; consists of a supporting stem with modified leaves (the petals and sepals) and usually contains both male and female reproductive structures.

fluid mosaic (4) A term that describes the structure of the plasma membrane, which is made up of several different types of molecules, many of which are not fixed in place but float, held in their proper orientation by hydrophilic and hydrophobic forces.

follicle (26) In an ovary, the small structure in which an egg forms.

follicle-stimulating hormone (FSH) (25, 26) A hormone that stimulates the production of eggs.

follicular phase (26) The first half of the reproductive cycle in women, culminating in ovulation.

food chain (17) The path of energy flow from producers to tertiary consumers.

food web (17) A more precisely described path of energy flow from producers to tertiary consumers than the food chain, reflecting the fact that many organisms are omnivores and occupy more than one position in the chain.

forebrain (24) The largest region of the vertebrate brain; includes two control and relay structures, the thalamus and hypothalamus, and the cerebrum.

fossil (10) The remains of an organism, usually its hard parts such as shell, bones, or teeth, that have been naturally preserved; also, traces of such an organism, such as footprints. [Lat., *fossilis,* that which is dug up]

fossil fuels (5) Fuels produced from the decayed remains of ancient plants and animals; include oil, natural gas, and coal.

founder effect (10) A phenomenon by which genetic drift can occur. The isolation of a small subgroup of a larger population can lead to changes in the allele frequencies of the isolated population, because all the descendants of the smaller group will reflect the allele frequencies of the subgroup, which may differ from those of the larger source population.

fruit (20) The mature ovary of a flower that houses seeds for dispersal. See also **fleshy fruit.**

fundamental niche (17) The full range of environmental conditions under which an organism can live.

G

G_0 (9) A quiescent or "resting" phase, outside the cell cycle, in which no cell division occurs.

gamete (8) A cell (often haploid) that will combine with another (haploid) cell at fertilization to produce offspring; also called a reproductive cell or sex cell. [Gk., *gamete,* wife]

gametophyte (14, 20) The structure in land plants and some algae that produces gametes (sperm and eggs); the haploid life stage of plants and some algae, which may be either male (producing sperm) or female (producing eggs). [Gk., *gamete,* wife + *phytas,* plant]

Gap 1 (G_1) (8) The phase of the cell cycle during which the cell may grow and develop, as well as performing its various cellular functions. Most cells spend most of their time in this phase.

Gap 2 (G_2) (8) The phase of the cell cycle that follows the DNA synthesis phase, characterized by significant growth and high rates of protein synthesis in preparation for cell division.

gap junction (4) A junction between adjacent animal cells in the form of a pore in each of the plasma membranes that is surrounded by a protein, linking the two cells and acting like a channel between them, thus allowing materials to pass between the cells.

gastropods (13) Mollusks that are members of the class Gastropoda. Most have a single shell, a muscular foot for locomotion, and a radula used for scraping food from surfaces; examples are snails and slugs. [Gk., *gaster,* belly + *pous,* foot]

gastrula (26) The mass of cells, made up of three layers, formed during the gastrulation phase of embryonic development.

gastrulation (26) The second phase of embryonic development, in which cells form distinct layers.

gene (6) The basic unit of heredity; a sequence of DNA nucleotides on a chromosome that carries the information necessary for making a functional product, usually a protein or an RNA molecule. [Gk., *genos,* race, descent]

gene expression (6) The process by which information in a gene's sequence is used to synthesize a gene product (commonly a protein, but also RNA).

gene flow (10) A change in the allele frequencies of a population due to movement of some individuals of a species from one population to another, changing the allele frequencies of the population they join; also known as migration.

gene library (7) A collection of cloned DNA fragments; also known as a clone library.

gene regulation (6) The processes by which cells "turn on" or "turn off" genes, influencing the amount of gene products formed.

gene therapy (7) The process of inserting, altering, or deleting one or more genes in an individual's cells to correct defective versions of the gene(s).

genetic drift (10) A random change in allele frequencies over successive generations; a cause of evolution.

genetic engineering (7) The manipulation of an organism's genetic material by adding or deleting genes or transplanting genes from one organism to another.

genetic recombination See **recombination.**

genetically modified organism (GMO) (9) An organism in which the genetic material has been altered using recombinant DNA technology.

genome (6) The full set of DNA present in an individual organism; also can refer to the full set of DNA present in a species.

genotype (6, 9) The genes that an organism carries for a particular trait; also, collectively, an organism's genetic composition. [Gk., *genos,* race, descent + *typos,* impression, engraving]

genus (*pl.* genera) (12) In the system developed by Carolus Linnaeus (1707–1778), a classification of organisms consisting of closely related species. [Lat., *genus,* race, family, origin]

germ layers (26) The three distinct layers of cells formed during gastrulation.

germinate (20) In plant spores or seeds, to begin growing after a period of dormancy.

gibberellins (20) In plants, a large group of hormones that regulate a plant's growth processes, primarily by stimulating cell division and cell elongation.

gills (22) Organs in fishes and other aquatic animals in which gases are exchanged between water and blood capillaries.

gland (21) A collection of epithelial cells that produces secretions for use elsewhere in the body.

glial cells (21, 24) Cells of nervous tissue that support and provide nutrients to neurons; also called neuroglia.

glomerulus (21) In the kidney, the blood-filtering unit of the nephron; a mass of capillaries, surrounded by Bowman's capsule and connected to a single, long, urine-collecting tube that excretes its filtered fluid into a collecting duct.

glucagon (25) A hormone produced by the pancreas and secreted in response to a low blood sugar level. It has the reverse effect of insulin, causing the liver to convert stored glycogen into glucose, which is released into the bloodstream.

glycemic index (23) A measure of the extent to which foods cause a surge in blood sugar and subsequent release of insulin.

glycerol (3, 4) A small molecule that forms the head region of a triglyceride fat molecule. [Gk., *glykys,* sweet + *-ol,* chemical suffix for an alcohol]

glycogen (3) A complex carbohydrate consisting of stored glucose molecules linked to form a large web, which breaks down to release glucose when it is needed for energy. [Gk., *glykys,* sweet + *genos,* race, descent]

glycolysis (5) In all organisms, the first step in cellular respiration, in which one molecule of glucose is broken into two molecules of pyruvate. For some organisms, glycolysis is the only means of extracting energy from food; for others, including most plants and animals, it is followed by the citric acid cycle and the electron transport chain. [Gk., *glykys,* sweet + *lysis,* releasing]

Golgi apparatus (4) (*pron.* gohl-jee) An organelle, part of the endomembrane system, structurally like a flattened stack of unconnected membranes, each known as a Golgi body. The Golgi apparatus processes molecules synthesized in the cell and packages those molecules that are destined for use elsewhere in the body. [From the name of the discoverer, Camillo Golgi, 1843–1926]

gonads (8) The ovaries and testes in sexually reproducing animals. [Gk., *gon ,* food offspring]

Gram stain (15) A dye test used by microbiologists in identifying an unknown bacterium. The dye stains the layer of peptidoglycan outside the cell wall purple, but bacteria in which the layer of peptidoglycan is covered by a membrane are not colored by the dye. Bacteria that take a Gram stain are known as Gram-positive bacteria; those that do not are known as Gram-negative bacteria. [From the name of the inventor, Hans Christian Gram, 1853–1938]

gravitropism (20) In plants, growth in response to the pull of gravity. [Lat., *gravidus,* heavy + Gk., *tropikos,* turn]

ground tissue (19) Those parts of a plant that are not dermal tissue or vascular tissue; makes up the bulk of the plant and is where most of the plant's metabolic activities take place.

group selection (11) The process, extremely uncommon in nature, that brings about an increase in the frequency of alleles for traits (e.g., behaviors) that are beneficial to the persistence of the species or population but are detrimental to the fitness of the individual possessing the trait (or engaging in the behavior).

growth factors (8) Chemical signals that trigger transitions to subsequent phases in the cell cycle, typically by providing feedback about the cell's environment.

growth hormone (25) A peptide hormone, secreted by cells in the anterior pituitary, that is a major participant in the control of several processes, including growth and metabolism.

growth rate (r) (16) In populations, the birth rate minus the death rate; the change in the number of individuals in a population per unit of time.

guard cell (19) One of the pair of plant cells that surround and control the opening of a stoma, through which carbon dioxide and water can pass.

gymnosperms (14) Vascular plants that do not produce their seeds in a protective structure; seeds are usually found on the surface of the scales of a cone-like structure. The gymnosperms include conifers, cycads, gnetophytes, and ginkgo. [Gk., *gymnos,* naked + *sperma,* seed]

H

habitat (17) The physical environment of organisms, consisting of the chemical resources of the soil, water, and air, and physical conditions such as temperature, salinity, humidity, and energy sources. [Lat., *habitare,* to dwell or inhabit]

hair (13) Dead cells filled with the protein keratin that collectively serve as insulation covering the body or a part of the body; present in all mammals.

haploid (8) Describes cells that have a single copy of each chromosome (in many species, including humans, gametes are haploid). [Gk., *haploeides,* single]

hazard factor (16) An external force acting on a population that increases the risk of death.

heart (22) A muscular pump that, with each contraction, propels blood at high pressure to lungs or gills and other body organs and tissues.

heart murmur (22) A medical condition—usually non-life-threatening—resulting from incomplete closure of the atrioventricular or semilunar valves of the heart, leading to some blood flowing back through the valves, making a swishing noise.

heat stroke (21) A potentially life-threatening medical condition caused by extreme hyperthermia, characterized by an inability of the organism to maintain homeostasis.

heliotropism (20) In plants, growth in response to a light source, in which leaves and flowers orient themselves in relationship to the sun's rays. [Gk., *helios,* sun + *tropikos,* turn]

helper T cell (27) A type of lymphocyte that stimulates B cells to produce antibodies and stimulates cytotoxic T cells to kill infected cells.

hematocrit (22) The proportion of blood that is made up of red blood cells; determined by spinning a blood sample in a centrifuge.

hemoglobin (22) An oxygen-carrying protein molecule in red blood cells. [Gk., *haima,* blood + Lat., *globus,* ball]

hemolymph (22) In an open circulatory system, the single fluid that surrounds all cells and transports nutrients, gases, and waste products.

herbivores (17, 23) Animals that eat plants; also known as primary consumers. [Lat., *herba,* grass + *vorare,* to devour]

heredity (9) The greater resemblance of offspring to their parents than to other individuals in the population, a consequence of the passing of characteristics from parents to offspring through their genes. [Lat., *heres,* heir]

heritability (10) The transmission of traits from parents to offspring via genetic information; also known as inheritance.

hermaphrodite (8) An organism that produces both male and female gametes. [From the names of the Greek god Hermes and goddess Aphrodite]

heterotherm (21) An animal that has a body temperature that fluctuates as the environmental temperature changes. [Gk., *heteros,* other + *therm*, heat]

heterozygous (9) Describes the genotype of a trait for which the two alleles an individual carries differ from each other. [Gk., *heteros,* other + *zeugos,* pair]

hibernate (21) To go into a state of reduced metabolic activity for days or weeks, during which the animal's body temperature can drop considerably.

hindbrain (24) The region of the vertebrate brain that includes the medulla, pons, and cerebellum.

hippocampus (24) Part of the brain's limbic system; functions in long-term memory formation.

histamine (27) A molecule released by mast cells that assists in the inflammatory response by causing non-injured blood vessels to dilate and become leaky.

histones (8) Proteins around which the long, linear strands of DNA are wrapped; serve to keep the DNA untangled and to enable orderly, tight, and efficient packing of the DNA within the cell.

homeostasis (21) The body's use of physical and chemical processes to maintain a consistent internal environment. [Gk., *homos,* same + *stasis,* standing]

homeotherm (21) An animal that maintains a relatively constant body temperature. [Gk., *homos,* same + *therm*, heat]

homologous feature (12) A feature inherited from a common ancestor. [Gk., *homologia,* agreement]

homologous pair (homologues) (8) The maternal and paternal copies of a chromosome.

homologous structure (10) Body structures in different organisms that, although they may have been modified over time to serve different functions in different species, are derived through inheritance from a common evolutionary ancestor.

homozygous (9) Describes the genotype of a trait for which the two alleles an individual carries are the same. [Gk., *homos,* same + *zeugos,* pair]

honest signal (11) A signal, which cannot be faked, that is given when both the individual making the signal and the individual responding to it have the same interests. It carries the most accurate information about an individual or situation.

horizontal gene transfer (12) The transfer of genetic material directly from one individual to another, not necessarily related, individual; common among bacteria.

hormone (20, 25) A chemical signal that responds to environmental variables, found in both plants and animals. In plants, hormones are produced in various locations and may have their effect in that location or may be transported to another part of the plant to regulate the plant's activities. In animals, hormones are usually secreted by endocrine glands and are transported by the bloodstream to target cells as part of an internal communication and regulation system. [Gk., *horma*, that which sets in motion or urges forward]

host (15, 17) An organism in or on which a parasite lives.

human chorionic gonadotropin (hCG) (26) A hormone secreted by the embryo that keeps the lining of the uterus thickened for implantation.

human immunodeficiency virus (HIV) (15, 27) The virus responsible for AIDS, a disease that destroys the human immune system. HIV is a retrovirus, an RNA-containing virus that is thought to have been introduced to humans from chimpanzees.

humus (19) (*pron.* HYOO-muss) The decomposing remains of plants and animals found in the uppermost region of soil; absorbs and releases water easily and releases many nutrients.

hybrids (12) Offspring of individuals of two different species. [Lat., *hybrida,* animal produced by two different species]

hybridization (12) The interbreeding of closely related species.

hydrogen bond (2) A type of weak chemical bond formed between the slightly positively charged hydrogen atoms of one molecule and the slightly negatively charged atoms (often oxygen or nitrogen) of another. Hydrogen bonds are important in building complex molecules, such as large proteins and DNA, and are responsible for many of the unique and important features of water.

hydrophilic (3) Attracted to water, as, for example, polar molecules that readily form hydrogen bonds with water. [Lat., *hydro-,* pertaining to water + Gk., *philios,* loving]

hydrophobic (3) Repelled by water, as, for example, nonpolar molecules that tend to minimize contact with water. [Lat., *hydro-*, pertaining to water + Gk., *phobos,* fearing]

hydroponically grown plant (19) A plant grown without soil, but in water enriched with essential plant nutrients.

hypertension (22) A medical condition characterized by consistently high blood pressure in the arteries; a potential health hazard due to loss of arterial elasticity and the associated increased risk of heart attacks and strokes.

hyperthermia (21) A condition in which high external environmental temperature and humidity overwhelm the body's ability to dissipate heat, and body temperature becomes abnormally high; can lead to death. [Gk., *hyper,* above + *therm*, heat]

hypertonic (4) Describes one of two solutions that has the higher concentration of solutes. [Gk., *hyper,* above + *tonos,* tension]

hyphae (*sing.* hypha) (14) (*pron.* HIGH-fee) Long strings of cells that make up the mycelium of a multicellular fungus. [Gk., *hypha,* web]

hypotension (22) The condition of consistently low blood pressure; symptoms may include dizziness, particularly just after a person stands up. It does not typically have long-term health risks.

hypothalamus (24, 25) A structure on the underside of the brain that functions as a liaison between the nervous and endocrine systems and serves as one of the chief regulatory centers of the brain. It regulates many fundamental drives, including hunger and thirst, sexual activity, and maintenance of body temperature. It also controls the hormone secretions of the pituitary gland, located next to it. [Gk., *hypo,* under + *thalamos,* inner room]

hypothermia (21) A condition in which body temperature becomes abnormally low; it can lead to death. [Gk., *hypo,* under + *therm*, heat]

hypothesis (*pl.* hypotheses) (1) A proposed explanation for an observed phenomenon. [Gk., *hypothesis,* a proposal]

hypotonic (4) Describes one of two solutions that has the lower concentration of solutes. [Gk., *hypo,* under + *tonos,* tension]

I

immune system (27) Body system consisting of disease-fighting white blood cells and external barriers such as skin, providing protection against an enormous variety of pathogens, including bacteria, fungi, some viruses, parasitic protists, and worms.

immunity (27) A state of long-term protection against a specific disease-causing microorganism or virus.

inclusive fitness (11) The sum of an individual's indirect and direct fitness.

incomplete dominance (9) The case in which the heterozygote has a phenotype intermediate between those of the two homozygotes; an example is pink snapdragons, with an appearance intermediate between that of a homozygote for white flowers and a homozygote for red flowers.

incomplete metamorphosis (13) In insects, a process of growth and development in which, rather than passing through completely different stages, an organism passes through several stages in which the juvenile form resembles a smaller version of the adult—with no pupal stage; occurs in about 17% of insects, including grasshoppers and crickets.

independent variable (1) A measurable entity that is available at the start of a process being observed and the value of which can be changed as required; generally represented on the *x*-axis in a graph.

infertility (26) The inability of a couple to get pregnant after one year of trying.

inflammatory response (27) The recruitment of phagocytes and other immune cells to assist with pathogen destruction, followed by tissue healing. Signs of the inflammatory response include redness, heat, swelling, and pain.

ingestion (23) Intake of food via the mouth, teeth, tongue, and esophagus.

inheritance (10) The transmission of traits from parents to offspring via genetic information; also known as heritability.

inhibitor (3) A chemical that binds to an enzyme or substrate molecule and in doing so reduces the enzyme's ability to catalyze a reaction.

innate behaviors (11) Behaviors that do not require environmental input for their development. These behaviors are present in all individuals in a population and do not vary much from one individual to another or over an individual's life span; also known as instincts. [Lat., *innatus,* inborn]

instincts (11) Behaviors that do not require environmental input for their development. Instincts are present in all individuals in a population and do not vary much from one individual to another or over an individual's life span; also known as innate behaviors. [Lat., *instinctus,* impelled]

insulin (25) A hormone produced in the pancreas and released in response to an increased concentration of blood glucose. Insulin in the bloodstream causes the liver, muscle, and other tissues to take up glucose.

integumentary system (27) Body system consisting of skin, sweat glands, oil glands, hair, and nails, all of which protect the body from mechanical stress caused by external forces such as friction or pressure, sunlight, dehydration, and pathogens. [Lat., *integere,* to cover]

interferon (27) An important cytokine produced by cells infected by a virus; alerts other cells to turn on protective measures that help them resist infection.

intermediate filaments (4) One of three types of protein fibers (the others are microtubules and microfilaments) that make up the eukaryotic cytoskeleton, providing it with structure and shape; a durable, rope-like system of numerous different overlapping proteins.

intermembrane space (4) In a mitochondrion, the region between the inner and outer membranes. [Lat., *inter,* between + *membrana,* a thin skin]

internal fertilization (26) The process in which sperm are deposited in the female reproductive tract and unite with one or more eggs.

interneuron (24) A type of neuron that acts as a middleman, integrating the signals coming in from sensory neurons and relaying them to a motor neuron; located in the brain and spinal cord.

interphase (8) In the cell cycle, the phase during which the cell grows and functions. During this phase, replication of DNA occurs in preparation for cell division. [Lat., *inter,* between + Gk., *phasis,* appearance]

interstitial fluid (21, 22) An extracellular fluid that surrounds and bathes cells; consists mainly of water and also contains nutrients, raw materials, and waste products.

introduced species See **exotic species.**

intron (6) A noncoding region of DNA.

invagination (4) The folding in of a membrane or layer of tissue so that an outer surface becomes an inner surface. [Lat., *in,* in + *vagina,* sheath]

invasive species (18) Species introduced by human activities, intentionally or accidentally, to areas other than the species' native range that cause economic or environmental harm, or harm to human health.

invertebrates (13) Animals that do not have a backbone; although a term commonly used in organizing the animals, invertebrates are not a monophyletic group.

ion (2) An atom that carries an electrical charge, positive or negative, because it has either lost or gained an electron or electrons from its normal, stable configuration. [Gk., *ion,* going]

ionic bond (2) A bond created by the transfer of one or more electrons from one atom to another; the resulting atoms, now called ions, are charged oppositely and so attract each other to form a compound.

ionic compound (2) A chemical compound in which ions of two or more elements are held together by ionic bonds.

iris (24) The colored portion of the vertebrate eye, with an opening, called the pupil, that controls the amount of light reaching the eye's interior.

isotonic (4) Refers to solutions with equal concentrations of solutes. [Gk., *isos,* equal to + *tonos,* tension]

isotopes (2) Variants of atoms that differ in the number of neutrons they possess. Isotopes do not differ in charge, because neutrons have no electrical charge, but the atom's mass changes with the loss or addition of a particle in the nucleus.

K

karyotype (8) A visual display of an individual's full set of chromosomes. [Gk., *karyon,* nut or kernel + *typos,* impression, engraving]

keystone species (17) A species that has an unusually large influence on the presence or absence of numerous other species in a community.

kilocalorie (kcal) (23) One thousand calories (cal).

kin selection (11) "Kindness" toward close relatives, which may evolve as apparently altruistic behavior toward them, but which in fact is beneficial to the fitness of the individual performing the behavior.

kinetic energy (5) The energy of moving objects, such as legs pushing the pedals of a bicycle or wings beating against the air. [Gk., *kinesis,* motion]

kinetochore (8) During metaphase of mitosis and metaphase II of meiosis, the disk-like group of proteins that develops at the centromeres of a pair of sister chromatids to which the spindle fibers attach and, in anaphase, will pull the chromatids to opposite poles of the cell.

kingdom (12) In the system developed by Carolus Linnaeus (1707–1778), one of the three categories—animal, plant, and mineral—into which all organisms and substances on earth were placed. In modern classification, there are six kingdoms: bacteria, archaea, protists, plants, animals, and fungi.

Krebs cycle See **citric acid cycle.**

L

La Niña (17) The counterpart of the El Niño phenomenon, characterized by the reverse effects, including a decrease in surface temperature and air pressure at the surface of the western Pacific Ocean; the resulting climate effects are approximately opposite to El Niño effects.

labor (26) During childbirth, a series of contractions of the uterus.

lactation (26) Milk production in the mammary glands.

lamella (*pl.* **lamellae**) (22) Disk-like structure in gills, with elaborately branched capillaries.

landscape conservation (18) The conservation of habitats and ecological processes, as well as species.

language (11, 24) A type of communication in which arbitrary symbols represent concepts and grammar; a system of rules dictates how the symbols can be manipulated to communicate and express ideas.

large intestine (23) The last part of the vertebrate digestive system, larger in diameter but shorter in length than the small intestine; also called the colon. It serves to absorb water, salts, and some vitamins and, in its final compartment (the rectum), to store the indigestible parts of consumed food and symbiotic bacteria (the feces), which can then be defecated.

larva (13) In complete metamorphosis, the first stage of insect development; the larva is hatched from the egg and eats to grow large enough to enter the pupal stage. The larva (for example, a caterpillar) looks completely different from the adult (a butterfly or moth). [Lat., *larva,* ghost]

lateral meristem (19) Growth region of a plant consisting of a layer of cells called cambium; division in this meristem causes a stem or root to become thicker, as opposed to longer.

leaf (19) The chief site of photosynthesis in most plants; leaves and stems form the shoot system.

learning (11) The alteration and modification of behavior over time in response to experience.

left hemisphere (24) The region of the brain that, in most (but not all) humans, specializes in language, logic, and mathematical skills.

leukocytes See **white blood cells.**

lichen (14) Symbiotic partnership between fungi and chlorophyll-containing algae or cyanobacteria, or both.

life (12) A physical state characterized by the ability to replicate and the presence of metabolic activity.

life cycle (20) The series of stages experienced by an individual of a species throughout its development, from formation of the embryo through maturity and reproduction.

life history (16) The vital statistics of a species, including age at first reproduction, probabilities of survival and reproduction at each age, litter size and frequency, and longevity.

life table (16) A table presenting data on the mortality rates within defined age ranges for a population; used to determine an individual's probability of dying during any particular year.

ligament (21) A connective tissue that binds bone to bone.

light energy (5) A type of kinetic energy made up of energy packets called photons, which are organized into waves.

lignin (19) A chemical in the cell walls of the sclerenchyma cells of woody plants; it is indigestible by nearly all organisms and gives sclerenchyma cells their strength.

limbic system (24) The brain region that includes parts of the cerebral cortex and two structures deeper in the center of the brain, the hippocampus and the amygdala; responsible for many physiological drives and instincts.

linked genes (9) Genes that are close to each other on a chromosome and so are more likely than others to be inherited together.

lipids (3) One of the four types of biological macromolecules, insoluble in water and greasy to the touch. Lipids are important in energy storage and insulation (fats), in membrane formation (phospholipids), and in regulating growth and development (sterols). [Gk., *lipos,* fat]

liver (23) An organ in vertebrates and some other animals that is important in the detoxification of toxic molecules, storage of glycogen, and synthesis of bile, hormones, and digestive enzymes.

lobe-finned fishes (13) Fish species characterized by two pairs of sturdy lobe-shaped fins on the underside of the body.

locus (*pl.* loci) (6) The location or position of a gene on a chromosome.

logistic growth (16) In populations. a pattern of growth in which initially exponential growth levels off as the environment's carrying capacity is approached.

long-term depression (24) A long-lasting decrease in the sensitivity of a neuron, resulting from a period of intense stimulation.

long-term potentiation (24) A long-lasting increase in the sensitivity of a neuron, resulting from a period of intense stimulation.

lungs (22) Internal organs in most land vertebrates, with highly branched, moist respiratory surfaces where gases are exchanged between air and blood.

luteal phase (26) The second half of the reproductive cycle in women, in which the follicle cell that had surrounded the ovum develops into the corpus luteum; this phase culminates in pregnancy or the sloughing off of the uterine lining.

luteinizing hormone (LH) (25) A hormone produced in the anterior pituitary; a surge in luteinizing hormone triggers ovulation in females and testosterone production in males.

lymph (22) A clear fluid formed from interstitial fluid as it is filtered into the lymphatic vessels in vertebrates.

lymph nodes (22) In the lymphatic system, patches of connective tissue filled with pathogen-fighting white blood cells, through which lymph passes.

lymphocyte (27) A type of white blood cell that carries out the response of the specific division of the immune system.

lysosome (4) A round, membrane-enclosed, enzyme- and acid-filled vesicle in the cell that digests and recycles cellular waste products and consumed material. [Gk., *lysis,* releasing + *soma,* body]

lysozyme (27) An enzyme produced by skin cells and in tears that kills bacteria by damaging their cell walls.

M

M phase See **mitotic phase.**

macroevolution (12) Large-scale evolutionary change involving the origins of new groups of organisms; the accumulated effect of microevolution over a long period of time. [Gk., *macros,* large + Lat., *evolvere,* to roll out]

macromolecule (3) A large molecule, made up of smaller building blocks or subunits. The four main types of biological macromolecules are carbohydrates, lipids, proteins, and nucleic acids. [Gk., *macros,* large + Lat., *dim.* of *moles,* mass]

macrophage (27) A type of large ("macro") white blood cell that resides in tissues and ingests small organisms and large debris such as dead cells, serving as an important link between the non-specific and specific divisions of the immune system.

male (11) In sexually reproducing organisms, a member of the sex that produces the smaller gamete.

mammary glands (13) Glands in all female mammals that produce milk for the nursing of young. [Lat., *mamma,* breast]

marsupials (13) Mammals in which, in most species, after a short period of embryonic life in the uterus, the young complete their development in a pouch in the female. [Gk., *marsipos,* pouch]

mass extinctions (12) Extinctions in which a large number of species become extinct in a short period of time, usually because of extraordinary and sudden environmental change. [Lat., *extinguere,* to extinguish]

mast cell (27) A type of white blood cell found in tissues that triggers the inflammatory response by releasing histamine.

mate guarding (11) Behavior by an individual that reduces the opportunity for its mate to interact with other potential mates.

mating system (11) The pattern of mating behavior in a species, ranging from polyandry to monogamy to polygyny. Mating systems are influenced by the relative amounts of parental investment by males and by females.

matrix (21) In connective tissue, a mass of non-living, extracellular material in which the connective tissue cells are embedded.

matter (2) Anything that has mass and takes up space.

maximum sustainable yield (16) The point at which the maximum number of individuals can be removed from a population without impairing its growth rate; it occurs at half the carrying capacity.

mechanoreceptor (24) A type of sensory cell that responds to mechanical pressure.

medulla (24) The part of the vertebrate hindbrain that connects to the spinal cord; it controls functions such as respiration, heart rate, and digestion.

meiosis (8) In sexually reproducing organisms, a process of nuclear division in the gonads that, along with cytokinesis, produces reproductive cells that have half as much genetic material as the parent cell and that differ from each other genetically. [Gk., *meioun,* to lessen]

membrane enzymes (4) Proteins embedded within a phospholipid bilayer membrane that catalyze chemical reactions.

memory cells (27) Long-lived B cells or T cells, produced in the primary immune response to a particular antigen, that enable rapid response to subsequent exposure to the same antigen.

Mendel's law of independent assortment (9) The principle that allele pairs for different genes separate independently in meiosis, so the inheritance of one trait generally does not influence the inheritance of another trait (the exception, unknown to Mendel, occurs with linked genes). [From the name of its discoverer, Gregor Mendel, 1822–1884]

Mendel's law of segregation (9) The principle that during the formation of gametes, the two alleles for a gene separate, so that half the gametes carry one allele, and half the gametes carry the other.

menopause (26) The permanent cessation of ovulation and menstruation, typically beginning in the late forties to early fifties in humans; characterized by a reduced production of estrogen and progesterone by the ovaries.

menstrual cycle (26) The cycle in which the uterus prepares for the implantation and nurturing of a fertilized egg, and sheds its lining if fertilization does not occur.

menstruation (26) The shedding of the uterine lining.

meristem (19, 20) A growth region of a plant, made up of undifferentiated cells that divide to form both additional undifferentiated cells and other cells that differentiate to form specific plant tissues.

mesoderm (26) The middle embryonic cell layer during gastrulation; it eventually forms the muscles and skeleton. [Gk., *mesos,* middle + *derma,* skin]

messenger RNA (mRNA) (6) The ribonucleic acid that "reads" the sequence for a gene in DNA and then carries the information from the nucleus to the cytoplasm, where the next stage of protein synthesis takes place.

metamorphosis (13) The rebuilding of molecules from the larva to the adult stage, resulting in a change of form. *Complete metamorphosis* is the division of an organism's life history into three completely different stages; *incomplete metamorphosis* is the pattern of growth and development in which an organism does not pass through separate, dramatically different life stages. [Gk., *metamorphoun,* to transform]

metaphase (8) The second phase of mitosis, in which the sister chromatids line up at the center of the cell; in meiosis, the homologues line up at the center of the cell in metaphase I, and the sister chromatids line up in metaphase II. [Gk., *meta,* in the midst of + *phasis,* appearance]

microbe (12, 15) A microscopic organism; not a monophyletic group, since it includes protists, archaea, and bacteria. [Gk., *micros,* small + *bios,* life]

microevolution (12) A slight change in allele frequencies in a population over one or a few generations. [Gk., *micros,* small + Lat., *evolvere,* to roll out]

microfilaments (4) One of three types of protein fibers (the others are intermediate filaments and microtubules) that make up the eukaryotic cytoskeleton, providing it with structure and shape, These are the thinnest elements in the cytoskeleton. The long, solid, rod-like fibers help generate forces, including those important in cell contraction and cell division.

microsphere (12) A membrane-enclosed, small, spherical unit containing a self-replicating molecule and carrying information, but no genetic material. Microspheres may have been an important stage in the development of life. [Gk., *micros,* small + *sphaira,* ball]

microtubules (4, 8) One of three types of protein fibers (the others are intermediate filaments and microfilaments) that make up the eukaryotic cytoskeleton, providing it with structure and shape. These are the thickest elements in the cytoskeleton. They resemble rigid, hollow tubes, functioning as tracks to which molecules and organelles within the cell may attach and be moved along; also help pull chromosomes apart during cell division.

midbrain (24) Part of the brainstem; a sensory integration and relay center through which most of the sensory information and motor neuron connections entering and leaving the brain pass.

migration (10) In populations, a change in allele frequencies due to the movement of some individuals from one population to another; an agent of evolutionary change caused by the movement of individuals into or out of a population. [Lat., *migrare,* to move from place to place]

mimicry (17) The evolution of an organism to resemble another organism or object in its environment to help conceal itself from predators. [Gk., *mimesis,* imitation]

mineral (23) A chemical element other than those commonly found in organic molecules (carbon, hydrogen, oxygen, and nitrogen); some minerals are required in the diet, in small amounts.

mismatch (11) The phenomenon of organisms finding themselves in a situation where the environment they are in differs from the environment to which they are evolutionarily adapted. If this occurs, we expect (and see) behaviors that appear to be (and are) not evolutionarily adaptive.

mitochondrial matrix (4, 5) The space enclosed by the inner mitochondrial membrane, where the carriers NADH and $FADH_2$ begin the electron transport chain by carrying high-energy electrons to molecules embedded in the inner membrane.

mitochondrion (*pl.* mitochondria) (4) The organelle in eukaryotic cells that converts the energy stored in food, in the chemical bonds of carbohydrate, fat, and protein molecules, into a form usable by the cell for its functions and activities. [Gk., *mitos,* thread + *chondros,* cartilage]

mitosis (8) The division of a nucleus into two genetically identical nuclei that, along with cytokinesis, leads to the formation of two identical daughter cells. [Gk., *mitos,* thread]

mitotic phase (M phase) (8) The phase of the cell cycle during which first the genetic material and nucleus and then the rest of the cellular contents divide.

molecule (2) A group of atoms held together by covalent bonds. [Lat., *dim.* of *moles,* mass]

monocot (19) One of the two major groups of flowering plants (the other is the eudicots); monocot seedlings have one cotyledon (seed leaf).

monogamy (11) A mating system in which most individuals mate with and remain with just one other individual. [Gk., *monos,* single + *gamete,* wife]

monophyletic (12) Describing a group containing a common ancestor and all of its descendants. [Gk., *monos,* single + *phylon,* race, tribe, class]

monosaccharides (3) The simplest carbohydrates and the building blocks of more complex carbohydrates, which cannot be broken down into other monosaccharides; examples are glucose, fructose, and galactose. Also known as simple sugars. [Gk., *monos,* single + *sakcharon,* sugar]

monotremes (13) Present-day mammals that retain the ancestral condition of laying eggs. Monotremes are so called because they have a single duct, the cloaca, into which the reproductive system, the urinary system, and the digestive system (for defecation) open. [Gk., *monos,* single + *trema,* hole]

mood (25) A relatively long-lasting emotional state (shorter-lasting than a person's temperament, but longer-lasting and less specific than a single emotion).

morphological species concept (12) A concept that defines species on the basis of physical features such as body size and shape. [Gk., *morph* , shape]

motor neuron (24) A type of neuron that stimulates action by conveying signals to muscles or glands and initiating a body's response to stimuli.

multiple allelism (9) The case in which a single gene has more than two possible alleles.

multiple sclerosis (MS) (27) An autoimmune disease in which the body attacks the myelin sheath surrounding nerve fibers of the brain and spinal cord, causing blurred vision, weakness in arms and legs, and trouble with balance.

muscle fiber (21) An individual muscle cell; typically contains multiple nuclei.

muscle tissue (21) A body tissue consisting of contractile cells (muscle fibers) that generate force and can facilitate movement.

mutation (6, 10) An alteration in the base-pair sequence of an individual's DNA; may arise spontaneously or following exposure to a mutagen. [Lat., *mutare,* to change]

mutualism (17) A symbiotic relationship in which both species benefit and neither is harmed. [Lat., *mutuus,* reciprocal]

mycelium (14) A mass of interconnecting hyphae that make up the structure of a multicellular fungus. [Gk., *mykes,* fungus]

mycorrhizae (14,19) (*pron.* MY-ko-RYE-zee) Root fungi; symbiotic associations between roots and fungi in which fungal structures are closely associated with fine rootlets and root hairs. [Gk., *mykes,* fungus + *rhiza,* root]

myelin sheath (24) A fatty coating that insulates axons.

myofibril (24) A cylindrical organelle within muscle cells that can contract; contains repeating units, called sarcomeres, in which the contraction takes place.

myosin (24) A protein of muscle tissue, making up the thick filaments.

N

NADPH (5) A molecule (nicotinamide adenine dinucleotide phosphate) that is a high-energy electron carrier involved in photosynthesis, which stores energy by accepting high-energy protons. It is formed when the electrons released from the splitting of water are passed to $NADP^+$.

natural killer (NK) cell (27) A type of white blood cell that kills body cells infected by a pathogen.

natural selection (10) A mechanism of evolution that occurs when, for a heritable variation of a trait, individuals with one version of the trait have greater reproductive success than individuals with a different version of that trait.

nectar (20) A solution produced by a flower, rich in sugars and amino acids, that can serve as an attractant to animal pollinators.

negative feedback (21) A control mechanism in which sensors detect changes in the internal environment and trigger effectors to counteract the change; one of the chief strategies by which organisms maintain homeostasis.

nephron (21) The chief functional unit in the vertebrate kidney, consisting of a tubule and a mass of blood vessels that work together to accomplish the tasks of filtration, reabsorption, and excretion.

nerve (24) A group of neurons bundled together with connective tissue.

nervous system (24) A network of cells (neurons), present in all multicellular animals other than sponges, that collects information about the organism's internal and external environments, processes that information, and sends signals to muscles and glands in response to the information.

nervous tissue (21) A body tissue that specializes in storing and transmitting information.

neural tube (26) A long hollow tube, formed during neurulation in embryonic development, that ultimately develops into the nervous system.

neuroglia See **glial cells.**

neuron (21, 24) The "excitable" cell within the nervous system that receives and transmits signals; made up of three distinctive elements: dendrites, a cell body, and an axon.

neurotransmitter (24) A chemical of the nervous system that transmits signals to adjacent cells.

neurulation (26) During embryonic development, the folding in of the ectoderm for the entire length of the embryo, first forming a groove and then becoming the neural tube.

neutron (2) An electrically neutral particle in the atomic nucleus. [Lat., *neutro,* in neither direction]

neutrophil (27) A type of white blood cell that circulates in the blood and ingests small organisms, primarily bacteria. Neutrophils make up 50% to 70% of all white blood cells.

niche (17) The way an organism utilizes the resources of its environment, including the space it requires, the food it consumes, and the timing of reproduction.

nitrogen fixation (19) The process by which the bonds between paired nitrogen atoms in molecules of nitrogen (N_2) are broken and the individual atoms are converted into a form, such as ammonia, that is usable by plants.

node (phylogeny) (12) The point on an evolutionary tree at which species diverge from a common ancestor. [Lat., *nodus,* knot]

node (plant structure) (19) A small lump of tissue on a stem from which a leaf, flower, cone, or additional stem (or branch) may grow.

nodule (19) In plant roots, a round lump containing plant cells and nitrogen-fixing bacteria.

non-amniotes (13) Those groups of terrestrial vertebrates (such as amphibians) that reproduce in water and do not have desiccation-proof amniotic eggs.

noncompetitive inhibitor (3) A chemical that binds to part of an enzyme away from the active site but alters the enzyme's shape so as to alter the active site, thereby reducing or blocking the enzyme's ability to bind with substrate.

nondisjunction (8) The unequal distribution of chromosomes during cell division; can lead to Down syndrome and other disorders caused by an individual's having too few or too many chromosomes.

non-essential amino acids (23) Amino acids that can be produced by the body and so are not needed in the diet.

nonpolar (4) Electrically uncharged.

non-specific immunity (27) A division of the immune system that, through various types of non-specific immune cells, rapidly recognizes as pathogens substances that breach external barriers and are not part of the body's own tissue, but does not distinguish between specific pathogens. Also called the innate (meaning "inborn") system, because it is functioning at birth and does not require any previous exposure to pathogens to act.

non-vascular plants (14) Plants that do not have vessels to transport water and dissolved nutrients, but instead rely on diffusion; bryophytes are non-vascular plants. [Lat., *vasculum, dim.* of *vas,* vessel]

notochord (13, 26) A rod of tissue extending from head to tail that stiffens the body when muscles contract during locomotion. Primitive chordates retain the notochord throughout life, but in advanced chordates it is present only in early embryos and is replaced by the vertebral column. [Gk., *notos,* back + Lat., *chorda,* cord]

nuclear membrane (4) A membrane enclosing the nucleus of a cell, separating it from the cytoplasm. It consists of two bilayers and is perforated by pores, enclosed in embedded proteins, that allow the passage of large molecules between the nucleus and the cytoplasm. Also called the nuclear envelope.

nucleic acids (3, 6) One of the four types of biological macromolecules, involved in information storage and transfer. The nucleic acids DNA and RNA store genetic information in unique sequences of nucleotides.

nucleolus (4) An area near the center of the nucleus where subunits of the ribosomes are assembled. [Lat., *nucleolus,* dim. of *nucleus,* kernel, small nut]

nucleotide (3, 6) A molecule containing a phosphate group, a sugar molecule, and a nitrogen-containing molecule called a base. Nucleotides are the individual units that together, in a unique sequence, constitute a nucleic acid.

nucleus (atom) (2) The central and most massive part of an atom, usually made up of two types of particles, protons and neutrons, which move about the nucleus.

nucleus (cell) (4) A membrane-enclosed structure in eukaryotic cells that contains the organism's genetic information as linear strands of DNA in the form of chromosomes. [Lat., *nucleus, dim.* of *nux,* nut]

null hypothesis (1) A hypothesis that proposes a lack of relationship between two factors. [Lat., *nullus,* none]

nuptial gift (11) A food item or other item presented to a potential mate as part of courtship. [Lat., *nuptialis,* of marriage]

nutrient (23) A substance taken in by an organisms and used for energy, raw materials, and maintenance of the body's systems.

O

omnivores (17, 23) Animals that eat both plants and other animals and thus can occupy more than one position in the food chain. [Lat., *omnis,* all + *vorare,* to devour]

oogenesis (26) The process of egg production in females.

oogonium (*pl.* **oogonia**) (26) A diploid cell in the ovary that multiplies by mitosis. Each oogonium begins meiosis but pauses at prophase I, at which point the cell is called a primary oocyte.

open circulatory system (22) A circulatory system in which a single fluid, hemolymph, circulates to transport nutrients, gases, and waste products and also surrounds all cells.

operator (6) The region of DNA to which a repressor protein can bind and, by doing so, block RNA polymerase from transcribing genes.

operon (6) A group of several genes, along with the elements that control their expression as a unit, all within one section of DNA.

optic nerve (24) The nerve that transports visual information from the retina of the eye to the thalamus and other parts of the brain.

order (12) In the system developed by Carolus Linnaeus (1707–1778), a classification of organisms consisting of related families.

organ (21) A structure that serves specialized functions and contains several types of tissue.

organ system (21) A group of organs that work together to accomplish physiological functions.

organelles (4) Specialized structures in the cytoplasm of eukaryotic cells, with specific functions; include, among others, the rough and smooth endoplasmic reticulum, Golgi apparatus, and mitochondria. [Gk., *organon,* tool]

organic molecules (3) Chemical compounds that contain carbon and usually hydrogen.

origin of replication (8) The site where the coiled, double-stranded DNA molecule unwinds and separates into two strands, like a zipper unzipping, and where duplication of the DNA molecule begins. In prokaryotes, there is a single origin of replication; eukaryotes may have multiple origin sites on each chromosome.

osmoconformer (21) An organism that regulates water loss and gain by maintaining the solute concentration of its body fluids at the solute concentration of its environment.

osmoregulation (21) The regulation of water content and of the concentrations of dissolved solutes that influence osmosis; an important component of homeostasis in animals.

osmoregulator (21) An organism that regulates water loss and gain by maintaining the solute concentration of its body fluids within a narrow range that differs from that of its environment.

osmosis (4) A type of passive transport in which water molecules move across a membrane, such as the plasma membrane of a cell. The direction of osmosis is determined by the relative concentrations of all solutes on either side of the membrane. [Gk., *osmos,* thrust]

ovarian cycle (26) The cycle in which a woman's hormones regulate the timing and development of egg production.

ovary (14, 20, 26) The female gonad. In plants, an enclosed chamber at the base of the carpel of a flower that contains the ovules. [Lat., *ovum,* egg]

oviduct See **Fallopian tube.**

oviparity (26) A reproductive strategy in which the fertilized egg moves outside the body for embryonic development.

ovoviviparity (26) A reproductive strategy in which most embryonic development takes place in the female's body until the egg hatches (or is released just before hatching).

ovulate (26) To release a secondary oocyte from an ovarian follicle into a Fallopian tube, from where it moves to the uterus.

ovule (14, 20) The structure within the ovary of flowering plants that gives rise to egg cells.

ovum (*pl.* **ova**) (26) A female gamete; also called an egg.

Oxygen Revolution (15) In earth's history, the accumulation in the atmosphere of oxygen released by cyanobacteria and other photosynthetic organisms.

oxytocin (25) A peptide hormone, produced in neurons within the hypothalamus and released by the posterior pituitary, that influences people's trust in others, increases the propensity to form social attachments, directs the ejection (or "let-down") of milk for nursing babies, and stimulates muscular contractions in the uterus during childbirth.

P

pair bond (11) A bond between an individual male and female in which they spend a high proportion of their time together, often over many years, sharing a nest or other refuge and contributing equally to the care of offspring.

pancreas (23, 25) An organ that secretes digestive juice into the small intestine and secretes hormones into the blood.

paracrine regulator (25) A chemical that is secreted from a cell and acts locally. Paracrine regulators generally diffuse through interstitial fluid from the tissue in which they are produced to nearby tissue, binding to receptors of neighboring cells and influencing their activity.

parasite (15, 17) An organism that lives in or on another organism, the host, and damages it. [Gk., *para,* beside + *sitos,* grain, food]

parasitism (17) A symbiotic relationship in which one organism (the parasite) benefits while the other (the host) is harmed.

parasympathetic nervous system (24) One of two components of the autonomic nervous system (the other is the sympathetic nervous system); responsible for unconsciously controlling activities related to digesting food and eliminating waste, tending to slow the heart and breathing rates.

parathyroid glands (25) Endocrine glands embedded in the surface of the thyroid gland that produce parathyroid hormone, which plays a central role in regulating calcium levels in adults. [Gk. *para,* beside + *thyra,* door]

parenchyma cell (19) A type of cell that carries out most of a plant's metabolic activities. Parenchyma cells are a type of ground tissue.

parent cells (8) Cells that divide to form daughter cells, which may be genetically identical to the parent cell or may differ in the amount and composition of genetic material.

parthenogenesis (26) The asexual reproductive process in which a female's egg develops into a new organism without fertilization by a sperm cell.

passive transport (4) Molecular movement that occurs spontaneously, without the input of energy. The two types of passive transport are diffusion and osmosis.

paternity uncertainty (11) In species with internal fertilization in the female, the inability of a male to be 100% certain that any offspring a female produces are his. [Lat., *pater,* father]

pathogen (22, 27) A disease-causing substance or organism, such as infectious bacteria, viruses, and fungi. Pathogens are sometimes referred to as "germs."

pathogenic (15) Disease-causing.

PCBs (polychlorinated biphenyls) (25) Chemicals used in industrial coolants and lubricants; known endocrine disruptors.

pedigree (9) In genetics, a type of family tree that maps the occurrence of a trait in a family, often over many generations.

penis (26) An external male reproductive structure.

pepsin (23) A protein-dismantling enzyme produced by cells in the stomach lining.

peptide bond (3) A bond in which the amino group of one amino acid is bonded to the carboxyl group of another. Two amino acids so joined form a dipeptide; several amino acids so joined form a polypeptide. [Gk., *peptikos,* able to digest]

peptidoglycan (15) A glycoprotein that forms a thick layer on the outside of the cell wall of a bacterium. In some bacteria, the layer of peptidoglycan is covered by a membrane and so is not colored by a Gram stain.

peripheral nervous system (PNS) (21, 24) The part of an organism's nervous system that transmits information to and from the central nervous system; includes sensory and motor neurons.

periodic table (2) A tabular display in which all the known chemical elements are arranged in the order of their atomic number and on the basis of other aspects of their atomic structure.

peristalsis (23) Waves of smooth muscle contractions that propel food along the digestive tract. [Gk., *peri,* around + *stellein,* to place]

petal (20) One of the usually brightly colored leaf-like structures of a flower, located inside the ring of sepals; the color, shape, and size vary greatly among the flowering plants. The petals help to attract animal pollinators, with different colors generally appealing to different pollinator species.

pH (2) A logarithmic scale that measures the concentration of hydrogen ions (H^+) in a solution, with decreasing values indicating increasing acidity. Water, in which the concentration of hydrogen ions (H^+) equals the concentration of hydroxyl ions (OH^-), has a pH of 7, the midpoint of the scale. [Abbreviation for "power of hydrogen"]

phagocytosis (4, 15) One of the three types of endocytosis, in which relatively large solid particles are engulfed by the plasma membrane, a vesicle is formed, and the particle is moved into the cell.

pharyngeal slits (13) Slits in the pharyngeal region, between the back of the mouth and the top of the throat, for the passage of water for breathing and feeding. [Gk., *pharynx,* throat]

phenotype (6, 9) The manifested structure, function, and behaviors of an individual; the expression of the genotype of an organism. [Gk., *phainein,* to cause to appear + *typos,* impression, engraving]

pheromones (11, 25, 26) Molecules released by an individual into the environment that trigger behavioral or physiological responses in other individuals. [Gk., *pherein,* to carry]

phloem (19) (*pron.* FLOW-uhm) In vascular plants, the tissue that conducts nutrients—sugar, in particular—throughout the plant.

phospholipids (3, 4) A group of lipids that are the major components of the plasma membrane. Phospholipids are structurally similar to fats, but contain a phosphorus atom and have two, not three, fatty acid chains.

phospholipid bilayer (4) The structure of the plasma membrane; two layers of phospholipids, arranged tail to tail (the tails are hydrophobic and so avoid contact with water), with the hydrophilic head regions facing the watery extracellular and intracellular fluids.

photoautotrophs (15) Chlorophyll-containing bacteria, or other organisms, that use the energy from sunlight to convert carbon dioxide to glucose by photosynthesis. [Gk., *phos,* light + *autos,* self + *troph* , food]

photon (5) The elementary particle that carries the energy of electromagnetic radiation of all wavelengths. [Gk., *phos,* light]

photoperiodism (20) A plant's response to the cycling of light and dark in a day; it regulates flowering time and numerous other responses to seasonal changes.

photoreceptor cell (24) A type of cell that detects the presence of light.

photosynthesis (5) The process by which some organisms, including plants and some protists and bacteria, are able to capture energy from the sun and store it in the chemical bonds of sugars and other molecules. [Gk., *phos,* light + *syn,* together with + *tithenai,* to place or put]

photosystems (5) Two arrangements of light-absorbing pigments, including chlorophyll, within the chloroplast that capture energy from the sun and transform it first into the energy of excited electrons and ultimately into ATP and high-energy electron carriers such as NADPH. [Gk., *phos,* light + *systema,* a whole compounded of parts]

phototropism (20) In plants, a growth response toward or away from a light source. [Gk., *phos,* light + *tropikos,* turn]

phthalates (25) Chemicals, commonly found in soft toys and cosmetics, that can act as an endocrine disruptor, blocking or interfering with hormones in humans and other animals, with potentially harmful effects.

phylogeny (12) The evolutionary history of organisms.

phylum (12) In the system developed by Carolus Linnaeus (1707–1778), a classification of organisms consisting of related classes. [Gk., *phylon,* race, tribe, class]

physical barriers (immune system) (27) The first defense against pathogens, to keep them out of the body. For example, skin acts as a nearly impenetrable barrier that keeps viruses and bacteria from entering the body, and various chemicals secreted by epithelial surfaces stop the growth of many pathogens. See also **integumentary system.**

physical defenses (17) Adaptations of prey in defense against predators, including mechanical and chemical structures, warning coloration, and camouflage.

physiology (21) The study of the complex functions carried out by organs and organ systems in living organisms.

pigment (5) In photosynthesis, molecules that are able to absorb the energy of light of specific wavelengths, raising electrons to an excited state in the process. [Lat., *pigmentum,* paint]

pilus (*pl.* **pili**) (4) A thin, hair-like projection that helps a prokaryote attach to surfaces. [Lat., *pilus,* a single hair]

pineal gland (25) A gland (pea-sized in humans) located near the center of the brain in vertebrates. It has neuron connections with the retina of the eye and controls secretion of the hormone melatonin, which affects circadian and seasonal cycles.

pinocytosis (4) One of the three types of endocytosis, in which dissolved particles and liquids are engulfed by the plasma membrane, a vesicle is formed, and the material is moved into the cell. The vesicles formed in pinocytosis are generally much smaller than those formed in phagocytosis. [Gk., *pinein,* to drink + *kytos,* container]

pith (19, 20) The ground tissue close to the center of a stem.

pituitary gland (25) An endocrine gland, consisting of anterior and posterior parts, attached to the hypothalamus by a thin stalk and regulated by secretions from the hypothalamus. See also **anterior pituitary; posterior pituitary.**

placebo (1) An inactive substance used in controlled experiments to test the effectiveness of another substance. The treatment group receives the substance being tested, and the control group receives the placebo. [Lat., *placebo,* I shall please]

placebo effect (1) A frequently observed and poorly understood phenomenon in which there is a positive response to treatment with an inactive substance.

placenta (8, 26) The organ formed during pregnancy (and expelled at birth) that connects the developing embryo to the wall of the uterus and allows the transfer of gases, nutrients, and waste products between mother and fetus; the placenta is so called from its shape. [Lat., *placenta,* a flat cake]

placentals (13) Mammals in which the developing fetus takes its nourishment from the transfer of nutrients from the mother through the placenta, which also supplies respiratory gases and removes metabolic waste products.

plants (14) Members of the kingdom Plantae; multicellular eukaryotes that have cell walls made primarily of cellulose, contain true tissues, and produce their own food by photosynthesis. Plants are sessile, and most inhabit terrestrial environments.

plasma (22) The liquid part of blood that contains dissolved metabolites and wastes, salts and ions, proteins, vitamins, and other chemicals and transports them to the tissues where they are required.

plasma cell (27) A type of white blood cell (an effector B cell) that secretes antibodies as part of the specific immune system's response to a pathogen. Plasma cells can secrete hundreds to thousands of antibodies per second.

plasma membrane (4) A complex, thin, two-layered membrane that encloses the cytoplasm of the cell, holding the contents in place and regulating what enters and leaves the cell; also called the cell membrane. [Gk., *plasma,* anything molded]

plasmid (7, 15) A circular DNA molecule found outside the main chromosome in bacteria.

plasmodesma (*pl.* **plasmodesmata**) (4) In plants, microscopic tube-like channels connecting the cells and enabling communication and transport between them. [Gk., *plassein,* to mold + *desmos,* bond]

platelets (22) Cellular fragments, components of the blood, formed by the pinching off of fragments from large cells (megakaryocytes) in the bone marrow. They lack organelles but are filled with enzymes and chemicals important for blood clotting.

pleiotropy (9) A phenomenon in which an individual gene influences multiple traits. [Gk., *pleion,* more + *tropos,* turn]

point mutation (6) A mutation in which one base pair in DNA is replaced with another, or a base pair is either inserted or deleted.

polar (4) Having an electrical charge.

polar body (8, 26) In gamete formation, one of the two cells formed when a primary oocyte divides; it gets almost no cytoplasm and eventually disintegrates.

pollen grain (14, 20) A structure that contains the male gametophyte of a seed plant. [Lat., *pollen,* fine dust]

pollen tube (20) A long, tube-like structure formed by the growth of a single cell of the pollen grain after germination; it forms within the style of the female reproductive structure and carries the male gametes to the ovule.

pollination (14, 20) The transfer of pollen from the anther of one flower to the stigma of another flower.

polyandry (11) A polygamous mating system in which individual females mate with multiple males. [Gk., *polys,* many + *aner,* husband]

polygamy (11) A mating system in which, for one sex, some individuals attract multiple mates while other individuals of that sex attract none; among the opposite sex, all or nearly all of the individuals are able to attract a mate. [Gk., *polys,* many + *gamete,* wife]

polygenic (9) Describes a trait that is influenced by many different genes. [Gk., *polys,* many + *genos,* race, descent]

polygyny (11) A mating system in which, among males, some individuals attract multiple mates while others attract none; among females, all or nearly all individuals are able to attract a mate. [Gk., *polys,* many + *gune,* woman]

polymerase chain reaction (PCR) (7) A laboratory technique in which a fragment of DNA can be duplicated repeatedly. [Gk., *polys,* many + *meris,* part]

polyploidy (12) The doubling of the number of sets of chromosomes in an individual. [Gk., *polys,* many + *ploion,* vessel]

polysaccharides (3) Complex carbohydrates formed by the union of many simple sugars. [Gk., *polys,* many + *sakcharon,* sugar]

pons (24) Part of the vertebrate hindbrain that acts with the medulla to control body functions related to respiration, heart rate, and digestion.

population (10) A group of organisms of the same species living in a particular geographic region.

population density (16) The number of individuals of a population in a given area.

population ecology (16) A subfield of ecology that studies the interactions between populations of organisms of a species and their environment.

positive correlation (1) A relationship between variables in which they increase (or decrease) together. [Lat., *com,* with + *relatio,* report]

positive feedback (21) A control mechanism in which a deviation from normal internal conditions causes an increase or acceleration of the change.

post-anal tail (13) A tail that extends beyond the end of the trunk, a point that is marked by the anus; a characteristic of chordates. [Lat., *post,* after]

posterior pituitary (25) A gland of the endocrine system that releases several important peptide hormones, including oxytocin and antidiuretic hormone, that are produced in neurons of the hypothalamus.

postsynaptic membrane (24) The cell membrane of a neuron on the receiving side of a synapse, across from the presynaptic membrane.

postzygotic barrier (12) A barrier to reproduction caused by the infertility of hybrid individuals or the inability of hybrid individuals to survive long after fertilization.

potential energy (5) Stored energy; the capacity to do work that results from an object's location, position, or composition, as in the case of water held behind a dam. [Lat., *potentia,* power]

predation (17) An interaction between two species in which one species eats the other. [Lat., *praedari,* to plunder]

prepared learning (11) Behaviors that are learned easily by all, or nearly all, individuals of a species.

pressure-flow mechanism (19) In vascular plants, the force driving the movement of sugar in the phloem; osmotic pressure moves material from sugar-production sites (sources) to sites where sugar is needed or stored (sinks).

presynaptic membrane (24) The cell membrane of an axon at the synapse.

prezygotic barrier (12) A barrier to reproduction caused by the physical inability of individuals to mate with each other, or the inability of the male's reproductive cell to fertilize the female's reproductive cell. [Lat., *prae,* before]

primary active transport (4) Active transport using energy released directly from ATP.

primary consumers (17) Herbivores, which consume the output of producers.

primary electron acceptor (5) In photosynthesis, a molecule that accepts excited, high-energy electrons from chlorophyll *a,* beginning the series of electron handoffs known as an electron transport chain.

primary growth (20) Plant growth that originates in the apical meristems of shoots and roots, increasing the height of the shoot, increasing the length of roots and branches, and producing new tissues. See also **secondary growth.**

primary oocyte (26) In female gametogenesis, a cell produced by meiosis, which later completes meiosis; it can become an egg (ovum).

primary productivity (17) The amount of organic matter produced by living organisms, called the producers, primarily through photosynthesis.

primary response (27) The immune system's first interaction with a pathogen.

primary spermatocyte (26) In sperm production, one of the two cells resulting from division of the spermatogonium; it undergoes meiosis in the first step of sperm production.

primary structure (3) The sequence of amino acids in a polypeptide chain.

probiotic therapy (15) A method of treating infections by introducing benign bacteria in numbers large enough to overwhelm harmful bacteria in the body. [Lat., *pro,* for, on behalf of + Gk., *bios,* life]

producers (17) The organisms responsible for primary productivity, such as grasses, trees, and agricultural crops, which convert light energy from the sun into chemical energy (that is, food) through photosynthesis.

progesterone (25, 26) A hormone, secreted by the corpus luteum of the ovary, that causes thickening of the endometrium to prepare for gestation. [Lat., *pro,* for + *gestare,* to bear]

prokaryote (4) An organism consisting of a prokaryotic cell (all prokaryotes are one-celled organisms). [Gk., *pro,* before + *karyon,* nut, kernel]

prokaryotic cell (4) A cell bound by a plasma membrane enclosing the cell contents (cytoplasm, DNA, and ribosomes); the cell has no nucleus or other organelles.

prolactin (25) A hormone, secreted by the anterior pituitary, that stimulates the production of milk in the mammary glands.

promoter site (6) Part of a DNA molecule that indicates where the sequence of base pairs that makes up a gene begins.

prophase (8) The first phase of mitosis, in which the nuclear membrane breaks down, sister chromatids condense, and the spindle forms. In meiosis, homologous pairs of sister chromatids come together and cross over in prophase I, and the chromosomes in daughter cells condense in prophase II. [Gk., *pro,* before + *phasis,* appearance]

prostate gland (26) A gland in males that secretes enzymes and nutrients into the semen.

proteins (3) One of the four types of biological macromolecules, constructed of unique combinations of 20 amino acids that result in unique structures and chemical behavior. Proteins are the chief building blocks of tissues in most organisms. [Gk., *proteion,* of the first quality]

protein synthesis (6) The construction of a protein from its constituent amino acids, by the processes of transcription and translation.

prothallus (14) The free-living haploid life stage of a fern; produces haploid gametes. [Gk., *pro,* before + *thallia,* twig]

protists (12) In modern classification, one of the four eukaryotic kingdoms, the Protista; includes all the single-celled eukaryotes. Phylogenetic analyses show that they are not a monophyletic group.

proton (2) A positively charged particle in the atomic nucleus; it is identical to the nucleus of the hydrogen atom, which lacks a neutron and has atomic number 1. [Gk., *protos,* first]

protostomes (13) Bilaterally symmetrical animals with defined tissues in which the gut develops from front to back; the first opening formed becomes the mouth of the adult animal. [Gk., *protos,* first + *stoma,* mouth]

pseudoscience (1) Hypotheses and theories not supported by trustworthy and methodical scientific studies. [Gk., *pseudes,* false + Lat., *scientia,* knowledge]

pulmonary circuit (22) The flow of blood from the heart to the lungs and back to the heart; the returning blood is oxygenated.

punctuated equilibrium (12) A theory in biology about the tempo of evolution, suggesting that in many taxa, long periods in which there is relatively little evolutionary change are punctuated by brief periods of rapid evolutionary change.

Punnett square (9) A diagram showing the possible outcomes of a cross between two individuals; the possible crosses are shown in the manner of a multiplication table. [From the name of its designer, Reginald C. Punnett, 1875–1967]

pupa (13) In complete metamorphosis, the second stage of insect development, in which the larva is enclosed in a case and its body structures are broken down into molecules that are reassembled into the adult form. [Lat., *pupa,* a little girl]

pus (27) The thick, yellowish fluid at the site of an infection, made up of neutrophils, damaged cell debris from the invaded tissue, and both live and dead pathogens.

pyruvate (5) The end product of glycolysis.

Q

quaternary structure (3) Two or more polypeptide chains bonded together in a single protein; an example is hemoglobin. [Lat., *quaterni,* four each]

R

radial symmetry (13) A body structure like that of a wheel, or pie, in which any cut through the center would divide the organism into identical halves. [Lat., *radius,* spoke of a wheel + Gk., *symmetria,* symmetry]

radioactive (2) The property of some elements or isotopes of having a nucleus that breaks down spontaneously, releasing tiny, high-speed particles that carry energy.

radiometric dating (10) A method of determining both the relative and the absolute ages of objects such as fossils by measuring both the radioactive isotopes they contain, which are known to decay at a constant rate, and their decay products.

rain shadow (17) An area in the lee of a mountain where there is no or reduced rainfall, because the air passing over the mountain falls, becomes warmer, and thus can hold more moisture.

random assortment (8) During anaphase of meiosis I, the process in which homologues are separated and pulled apart toward opposite poles of the cell, with either the maternal or the paternal homologue (consisting of two sister chromatids) going to one pole and the other homologue to the opposite pole. Because the process is random, the pairs of sister chromatids at each pole (which will separate in meiosis II) are a mix of maternal and paternal sister chromatids.

randomized (1) In a research study, describes a manner of choosing subjects and assigning them to groups on the basis of chance—that is, randomly.

ray-finned fishes (13) Fish species characterized by rigid bones and a mouth at the apex of the body; so called because their fins are lined with hardened rays.

realized niche (17) The environmental conditions in which an organism is living at a given time.

receptor-mediated endocytosis (4) One of the three types of endocytosis, in which receptors on the surface of a cell bind to specific molecules; the plasma membrane then engulfs both molecule and receptor and draws them into the cell.

receptor protein (4) A protein in the plasma membrane that binds to specific chemicals in the cell's external environment to regulate processes within the cell; for example, cells in the heart have receptor proteins that bind adrenaline.

recessive (9) Describes an allele with a phenotypic effect that is masked by a dominant allele for that trait. [Lat., *recessus,* retreating]

reciprocal altruism (11) Costly behavior directed toward another individual that benefits the recipient, with the expectation that, at some later time, the recipient will behave in a similar manner, "returning the favor." [Lat., *reciprocare,* to move backward and forward + *alter,* the other]

recognition protein (4) A protein in the plasma membrane that provides a "fingerprint" on the outside-facing surface of the cell, making it recognizable to other cells. Recognition proteins make it possible for the immune system to distinguish the body's own cells from invaders that may produce infection; they also help cells bind to other cells or molecules.

recombinant DNA technology (7) Technology that depends on the combination of two or more sources of DNA into a product; an example is the production of human insulin from fast-dividing transgenic *E. coli* bacteria that contain, through genetic manipulation, the human DNA sequence that codes for insulin.

recombination (8) The exchange of some genetic material between a paternal homologous chromosome and a maternal homologous chromosome, leading to a chromosome carrying genetic material from each; also called crossing over.

red blood cells (22) Hemoglobin-containing, oxygen-transporting blood cells, the most common type of blood cell; also called erythrocytes.

reflex (24) An automatic response that occurs when signals travel directly from sensory neurons to the spinal cord, where they connect with motor neurons to stimulate a response to a sensation that does not need to be processed through the brain.

regulator (21) An organism that, for a given physiological variable, maintains homeostasis for that variable, keeping it within a narrow range, even in the face of changes in the external environment.

regulatory gene (6) A gene that codes for a product—such as a repressor protein or activator protein—that influences the expression of one or more other genes.
replication (DNA) (8) The process in both eukaryotes and prokaryotes by which DNA duplicates itself in preparation for cell division.
replication (experiment) (1) The process of repeating a study (by the same researcher or others) so as to increase our confidence in the validity of the results.
replication fork (8) In DNA replication, the structure formed by the unwinding and separating of the two DNA strands, which will be used as templates.
reproductive cells (8) Haploid cells from two individuals that, as sperm and egg, will combine at fertilization to produce offspring; also called gametes or sex cells.
reproductive investment (11, 16) Energy and material expended by an individual in the growth, feeding, and care of offspring.
reproductive isolation (12) The inability of individuals from two populations to produce fertile offspring together.
reproductive output (16) The number of offspring an individual or population produces.
resource partitioning (17) A division of resources that occurs when species overlap some portion of a niche in which one or more species differ in behavior or body plan in a way that divides the resources of the niche between the species.
resting potential (24) In a neuron, the state when the neuron is not transmitting a signal and during which, due to the action of sodium-potassium pumps, a greater positive charge is maintained outside the cell than inside the cell.
restriction enzyme (7) An enzyme that recognizes and binds to a specific sequence of four to eight bases in DNA and cuts the DNA at that point. Restriction enzymes are important in biotechnology, because they permit the cutting of short lengths of DNA that can be inserted into other chromosomes or otherwise utilized.
retina (24) A nervous tissue containing light-sensitive cells, on the inner surface of the vertebrate eye, which transmits impulses to the vision centers of the brain.
retrovirus (15) A virus containing RNA and the enzyme reverse transcriptase, which uses the viral RNA as a template to synthesize a single strand of DNA. [Lat., *retro*, backward + *virus*, slime]
ribonucleic acid (RNA) (3) A nucleic acid that serves as a intermediary in the process of converting genetic information in DNA into protein, as well as other functions. Messenger RNA (mRNA) takes instructions for production of a given protein from DNA to another part of the cell; transfer RNA (tRNA) interprets the mRNA code and directs the construction of the protein from its constituent amino acids.
ribosomal subunits (6) The two structural parts of a ribosome, which function together to translate mRNA to build a chain of amino acids that will make up a protein.
ribosomes (4) Granular bodies in the cytoplasm, consisting of protein and RNA, that are involved in using information in messenger RNA (which was transcribed from DNA) to synthesize proteins.
right hemisphere (24) The region of the brain dedicated, in most humans, to emotions, intuitive thinking, and artistic expression.
ring species (12) Populations that can interbreed with neighboring populations but not with populations separated by larger geographic distances. Because the non-interbreeding populations are connected by gene flow through geographically intermediate populations, there is no clear point at which one species stops and another begins; for this reason, ring species are problematic for the biological species concept.
rod (24) In the retina, a type of cell that enables vision in dim light.
root (14, 19) The part of a vascular plant, usually below ground, that absorbs water and minerals from the soil and transports them through vascular tissue to the rest of the plant, and anchors the plant in place. The overall structure of a plant's roots is called the root system.
root meristem (20) A cluster of active, dividing embryonic cells at the tips of roots that generates new tissue and thus increases root length.
rough endoplasmic reticulum (rough ER) (4) An organelle, part of the endomembrane system, structurally like a series of interconnected, flattened sacs connected to the nuclear envelope; called "rough" because its surface is studded with ribosomes. [Gk., *endon*, within + *plasma*, anything molded; Lat., *reticulum, dim.* of *rete*, net]
roundworms (13) A worm phylum with members characterized by a long, narrow, unsegmented body and growth by molting. Roundworms, also called nematodes, are protostomes with defined tissues; there are some 90,000 identified species.
rubisco (5) An enzyme (ribulose 1,5-bisphosphate carboxylase/oxygenase) involved in photosynthesis. It fixes carbon atoms from CO_2 in the air, attaching them to an organic molecule in the stroma of the chloroplast. This fixation is the first step in the Calvin cycle, in which molecules of sugar are assembled. Rubisco is the most abundant protein on earth.
ruminant (23) A mammal characterized by having a four-chambered stomach. Ruminants chew a cud consisting of regurgitated, partially digested food. Cows, sheep, goats, and bison are examples.

S

S phase See **DNA synthesis phase.**
sap (19) In vascular plants, (1) the fluid in the xylem, consisting of water and dissolved minerals and sometimes sugars; or (2) the fluid in the phloem, consisting of water and dissolved sugars and other nutrients.
sarcomere (24) The fundamental unit of muscle contraction, made up of actin and myosin.
saturated fats (3) Fats in which each carbon in the hydrocarbon chain forming the tail region of the molecule is bound to two hydrogen atoms; solid at room temperature.
savanna (17) A type of terrestrial biome; a tropical or subtropical grassland with scattered woody plants, characterized by a hot climate and distinct wet and dry seasons (rainfall is less than in the tropical seasonal forest biome).
science (1) A body of knowledge based on observation, description, experimentation, and explanation of natural phenomena. [Lat., *scientia*, knowledge]
scientific literacy (1) A general, fact-based understanding of the basics of biology and other sciences, the scientific method, and the social, political, and legal implications of scientific information.
scientific method (1) A process of examination and discovery of natural phenomena that involves making observations, constructing hypotheses, testing predictions, experimenting, and drawing conclusions and revising them as necessary.
scientific theory (1) An explanatory hypothesis for a natural phenomenon that is exceptionally well supported by empirical data. [Gk., *theorein*, to consider]
scientific thinking (1) A highly flexible process for exploring natural phenomena that involves making observations, constructing hypotheses, testing predictions, experimenting, and drawing conclusions and revising them as necessary.
sclerenchyma cell (19) A type of plant cell that provides strength, enabling plants to grow tall; at maturity, the cell is not living. Sclerenchyma cells are a type of ground tissue.
scrotum (26) External sac that contains the testes.
second law of thermodynamics (5) A physical law stating that every conversion of energy is not perfectly efficient and invariably includes the transformation of some energy into heat.
secondary active transport (4) Active transport in which there is no direct involvement of ATP (adenosine triphosphate). The transport protein simultaneously moves one molecule against its concentration gradient while letting another flow down its concentration gradient.
secondary consumers (17) Animals that feed on herbivores; also known as carnivores.
secondary growth (20) In woody plants, plant growth that originates in lateral meristems, increasing the girth of stems and branches. It is the process by which eudicots and some other angiosperms produce wood, acquiring increased water-conduction capacity and becoming thicker and sturdier. See also **primary growth.**
secondary oocyte (26) One of the two cells formed by division of a primary oocyte; it completes meiosis after fertilization.

secondary phloem (20) In woody plants, tissue produced by the vascular cambium, toward the outer circumference of the trunk or branch, during secondary growth; the cells conduct sugars throughout the plant body.

secondary response (27) The creation of B and T effector cells in the immune system's response to a second exposure to a pathogen. The secondary response occurs rapidly and produces a more intense response than the primary response.

secondary spermatocyte (26) One of the two cells produced by the first meiotic division of a primary spermatocyte, which then complete the second meiotic division; each produces two spermatids.

secondary structure (3) The corkscrew-like twists or folds of a protein that are held in place by hydrogen bonds between amino acids in the polypeptide chain.

secondary xylem (20) In woody plants, tissue produced by the vascular cambium, toward the inside of the trunk or branch, during secondary growth; the cells conduct water and minerals, while also providing structural support; also called wood.

seed (14, 20) An embryonic plant with its own supply of water and nutrients, encased within a protective coating.

segmented worms (13) A worm phylum with members characterized by grooves around the body that mark divisions between segments. Segmented worms, also called annelids, are protostomes with defined tissues and do not molt; examples are earthworms and leeches.

selective serotonin reuptake inhibitors (SSRIs) (24) Antidepressant medications (including Prozac and Zoloft) that block serotonin from being reabsorbed and recycled by the cells (neurons) that released it, prolonging its effect. The net result for the individual is generally an elevated mood because of serotonin's prolonged stay in the synapse.

seminal vesicle (26) One of a pair of male reproductive glands that secrete sugar, enzymes, vitamin C, proteins, and immune suppressants into semen.

seminiferous tubules (26) Coiled tubes in the testes that are lined with the cells that divide to produce sperm.

sensory integration (24) The process of incorporating information from all of the senses to form a single perception.

sensory neuron (24) A type of neuron that collects information from an animal's environment; its dendrites are modified to respond directly to external stimulation—temperature, touch, taste, smell, light, or sound.

sepal (20) One of the leaf-like structures that grow in a ring around the outside of a flower, at the point where the flower is connected to the branch, stem, or stalk. Sepals protect the developing bud and usually are green, but sometimes are the same color as the petals.

sessile (13) Describes organisms that are fastened in place, such as adult mussels and barnacles. [Lat., *sedere,* to sit]

set point (21) A target value or range; in homeostasis, the normal range of values for a variable (such as temperature, pH, or solute concentration).

sex-linked trait (9) A trait controlled by a gene on a sex chromosome.

sexual dimorphism (11) The case in which the sexes of a species differ in size or appearance. [Gk., *di-,* two + *morph* , shape]

sexual reproduction (8, 26) A type of reproduction in which offspring are produced by the fusion of gametes from two distinct sexes.

sexual selection (10) The process by which natural selection favors traits, such as ornaments or fighting behavior, that give an advantage to individuals of one sex in attracting mating partners.

sexually transmitted disease (STD) (15, 26) A disease passed from one person to another through sexual activity.

shoot (14, 19) The above-ground part of a plant, consisting of stems and leaves, and sometimes flowers and fruits. The stem contains vascular tissue and supports the leaves, the main photosynthetic organ of the plant. Also called a shoot system.

shoot meristem (20) A cluster of active, dividing embryonic cells found at the tips of shoots that increases plant height and generates new tissues, including branches and buds.

sieve tube (19) A tube structure, formed from specialized cells in the phloem, that conducts sugar throughout the plant.

sign stimulus (11) An external signal that triggers the innate behavior called a fixed action pattern.

simple diffusion (4) Diffusion of molecules directly through the phospholipid bilayer of a plasma membrane, without the assistance of other molecules. Oxygen and carbon dioxide, because they are small and carry no charge that would cause them to be repelled by the middle layer of the membrane, can pass through the membrane in this way.

simple sugars (3) Monosaccharide carbohydrates, generally containing three to seven carbon atoms, which store energy in their chemical bonds and can be broken down by cells; they cannot be broken down into other simple sugars. Examples are glucose, fructose, and galactose.

single-gene trait (9) A trait that is determined by instructions on only one gene; examples are a cleft chin, a widow's peak, and unattached earlobes.

single-lens eye (24) The image-forming eye found in jellyfish and some mollusks.

sister chromatids (8) The two identical strands of a replicated chromosome.

skeletal muscle (21) A type of muscle tissue, usually attached to bones, that is responsible for generating most of the movement in animals; it accounts for about 40% of human body weight; also called voluntary muscle.

small intestine (23) A long, thin tube of the digestive tract between the stomach and large intestine; the part of the digestive system where most digestion and absorption take place.

smooth endoplasmic reticulum (smooth ER) (4) An organelle, part of the endomembrane system, structurally like a series of branched tubes; called "smooth" because its surface has no ribosomes. Smooth ER synthesizes lipids such as fatty acids, phospholipids, and steroids. [Gk., *endon,* within + *plasma,* anything molded; Lat., *reticulum, dim.* of *rete,* net]

smooth muscle (21) A type of muscle tissue found in the walls surrounding blood vessels, the stomach and intestines, bladder, and many other organs and inner "tubes" within the body. It generates slow, rhythmic contractions that can gradually move blood, food, or other substances through the organ; not under conscious control.

solute (4) A substance that is dissolved in a gas or liquid; for example, in a solution of water and sugar, sugar is the solute. [Lat., *solvere,* to loosen]

solvent (4) The gas or liquid in which a substance is dissolved; for example, in a solution of water and sugar, water is the solvent.

somatic cells (8) The (usually diploid) cells of the body of an organism (in contrast to the usually haploid reproductive cells). [Gk., *soma,* body]

somatic nervous system (24) One of the two major divisions of the motor pathways of the peripheral nervous system (the other is the autonomic nervous system); it relays signals to skeletal muscles, enabling contraction of those muscles and conscious movement of limbs.

special connective tissue (21) A type of connective tissue, including bone, cartilage, and blood, in which the matrix differs from that of connective tissue proper in that it is rigid or liquid.

speciation (12) The process by which one species splits into two distinct species. The first phase of speciation is reproductive isolation; the second is genetic divergence, in which two populations evolve over time as separate entities with physical and behavioral differences.

speciation event (12) A point in evolutionary history at which a given population splits into independent evolutionary lineages.

species (12) Natural populations of organisms that can interbreed and are reproductively isolated from other such groups. In the Linnaean system, the species is the narrowest classification for an organism. [Lat., *species,* kind, sort]

species richness (18) The number of different species in a given area.

specific epithet (12) In the system developed by Carolus Linnaeus (1707–1778), a noun or adjective added to the genus name to distinguish a species; For example, in the name *Homo sapiens, Homo* is the genus name and *sapiens* is the specific epithet. [Gk., *epithetos,* added]

specific immunity (27) A division of the immune system that recognizes and defends against very specific infectious agents—for example, not just all bacteria, but specific species, individually; also called adaptive immunity.

sperm (20, 26) A male gamete.

spermatogenesis (26) The process of sperm production.

sphincter (23) A ring of muscle that opens or closes a passage between two chambers in the body, such as between the esophagus and the stomach.

spindle (8) A part of the cytoskeleton of a cell, formed in prophase (in mitosis) or prophase I (in meiosis), from which extend the fibers that organize and separate the sister chromatids.

spindle fibers (8) Fibers that extend from one pole of a cell to the other, which pull the sister chromatids apart in anaphase of mitosis or anaphase II of meiosis.

sporangia (*sing.* **sporangium**) (14) In many ferns, the structures on the underside of the leaves in which the spores are produced. [Gk., *spora,* seed + *angeion,* vessel]

spore (14, 20) A reproductive structure of non-vascular and some vascular plants that have an alternation of generations. Spores are typically haploid and unicellular and develop into either a male (producing sperm) or female (producing eggs) gametophyte. The eggs and sperm produced by gametophytes unite to produce the diploid generation (sporophyte). [Gk., *spora,* seed]

sporophyte (14, 20) The multicellular diploid structure in non-vascular plants, some vascular plants, and some algae that produces asexual spores; the diploid life stage in organisms exhibiting alternation of generations.

stabilizing selection (10) Selection that, for a given trait, produces the greatest fitness at the intermediate point of the phenotypic range.

stamen (14, 20) The male reproductive structure of a flower, consisting of a head-like anther on a stalk-like filament. Most flowers have several stamens. [Lat., *stamen,* thread]

starch (3) A complex polysaccharide carbohydrate consisting of a large number of monosaccharides linked in line; in plants, the primary form of energy storage.

statistics (1) A set of analytical and mathematical tools designed to further the understanding of numerical data.

stem (19) A plant structure that supports and positions the leaves and flowers. The stem and leaves (and flowers) make up the shoot system.

stem cells (7, 22) Undifferentiated cells that have the ability to develop into any type of cell in the body, a property that makes stem cells useful in biotechnology.

sterols (3) Lipids important in regulating growth and development; include cholesterol and the steroid hormones testosterone and estrogen. All sterols are modifications of a basic structure of four interlinked rings of carbon atoms. [Gk., *stereos,* solid + *-ol,* chemical suffix for an alcohol]

stigma (14, 20) The region of the carpel, the female reproductive structure of a flower, that has a flat, sticky surface that functions as a landing pad for pollen.

stomach (23) A J-shaped digestive organ with thick, elastic walls; the part of the digestive system where food is mixed and partially digested.

stomata (*sing.* **stoma**) (5, 19) Small pores, usually on the undersides of leaves, that are the primary sites for gas exchange in plants. Carbon dioxide (for photosynthesis) enters and oxygen (a by-product of photosynthesis) exits through the stomata. [Gk., *stoma,* mouth]

stroma (4, 5) In the leaf of a green plant, the fluid in the inner compartment of a chloroplast, which contains DNA and protein-making machinery. [Gk., *stroma,* bed]

style (14, 20) The long, thin part of the carpel, the female reproductive structure of a flower, that holds up the stigma for pollen capture and through which a pollen tube will grow down into the ovary.

substrate (3) The molecule on which an enzyme acts. The active site on the enzyme binds to the substrate, initiating a chemical reaction; for example, the active site on the enzyme lactase binds to the substrate lactose, breaking it down into the two simple sugars glucose and galactose. [Lat., *sub-,* under + *stratus,* spread]

succession (17) In ecology, the change in the species composition of a community over time, following a disturbance.

superstition (1) The irrational belief that actions not related by logic to a course of events can influence an outcome.

surface protein (4) A protein that resides primarily on the inner or outer surface of the phospholipid bilayer that constitutes a cell's plasma membrane.

survivorship curves (16) Graphs showing the proportion of individuals of particular ages now alive in a population; indicate an individual's likelihood of surviving through a given age interval.

sympathetic nervous system (24) One of two components of the autonomic nervous system (the other is the parasympathetic nervous system), responsible for unconsciously controlling activities related to coordinating the body's fight-or-flight response to stress, including increasing the heart and breathing rates and providing muscles with additional blood flow.

sympatric speciation (12) Speciation that results not from geographic isolation but from polyploidy or hybridization and allopolyploidy; relatively uncommon in animals but common in plants. [Gk., *syn,* together with + *patris,* native land]

synapse (24) The site where a neuron communicates with another neuron, muscle cell, or gland. [Gk., *syn,* together with + *haptein,* to fasten]

synaptic cleft (24) The space at the synapse between the axon terminal and whatever is next to it (neuron, muscle cell, or gland cell).

systematics (12) The modern approach to classification, with the broader goal of reconstructing the evolutionary history, or phylogeny, of organisms. [Gk., *systema,* a whole compounded of parts]

systemic circuit (22) The flow of blood from the heart to the body (other than the lungs) and back to the heart; returning blood is low in oxygen and high in carbon dioxide.

systolic pressure (22) The first blood pressure reading; a measure of the pressure when the heart contracts, pumping blood into the arteries.

T

T cell (27) A type of lymphocyte, maturing in the thymus, that is involved in cell-mediated immunity; fights invaders in body cells.

taproot (19) A thick primary root that grows deep into the soil and has smaller roots branching out from it; found primarily in eudicots.

target cell (25) A cell that responds to a chemical regulatory signal, such as a hormone, typically when the regulatory chemical interacts with a receptor on or in the target cell.

technology (1) The application of research findings to various fields such as manufacturing and medicine to solve problems.

telomere (8) A noncoding, highly repetitive section of DNA at the tip of every eukaryotic chromosome that shortens with every cell division. If it becomes too short, additional cell division can cause the loss of functional, essential DNA and almost certain cell death. [Gk., *telos,* end + *meris,* part]

telophase (8) The fourth and last phase of mitosis, in which the chromosomes begin to uncoil and the nuclear membrane is reassembled around them. In meiosis, in telophase I, the sister chromatids arrive at the cell poles and the nuclear membrane reassembles around them; in telophase II, the sister chromatids have been pulled apart and the nuclear membrane reassembles around haploid numbers of chromosomes. [Gk., *telos,* end + *phasis,* appearance]

temperate grassland (17) A type of terrestrial biome, having grass and/or shrubs as the chief plants; characterized as a semi-arid to semi-humid area with a hot season and a cold season.

tendon (21) A connective tissue structure that joins muscle to bone.

tendril (20) A specialized thread-like leaf, stem, or branch that wraps around whatever it touches. Tendrils support the stem of the plant.

terminal buttons (24) Knob-like ends of an axon; also called axon terminals.

tertiary consumers (17) Animals that eat animals that eat herbivores; also known as top carnivores. [Lat., *tertius,* third]

tertiary structure (3) The unique and complex three-dimensional shape of a protein formed by multiple twists of its secondary structure as amino acids come together to form hydrogen bonds or covalent sulfur-sulfur bonds.

test-cross (9) A mating in which a homozygous recessive individual is bred with individuals of unknown genotype that have the dominant phenotype. This type of cross can reveal the unknown genotype by the observed characteristics, or phenotypes, of the offspring.

testis (*pl.* **testes**) (26) The male gonad.

testosterone (25, 26) The principal male sex hormone; it influences the development of an embryo as male and the production of male secondary sex characteristics.

tetrapod (13) An organism with four limbs; all terrestrial vertebrates are tetrapods. [Gk., *tetra-,* four + *pous,* foot]

thalamus (24) One of the primary switchboards in the brain. It receives most of the visual, auditory, and touch input, and relays some signals to the cerebral cortex while blocking others.

theory See **scientific theory.**

thermodynamics (5) The study of the transformation of energy from one type to another, such as from potential energy to kinetic energy. [Gk., *therm* , heat + *dynamis,* power]

thermoneutral zone (21) In endotherms, the temperature range (typically between 27° and 36° C) within which the organism is able to maintain a constant internal body temperature without any energy expenditures.

thermoregulation (21) The maintenance of body temperature within its normal range in homeostasis.

thigmotropism (20) In plants, growth that occurs in response to touch or physical contact with an object. [Gk., *thigma,* touch + *tropikos,* turn]

thylakoids (4, 5) Interconnected membranous structures in the stroma of a chloroplast, where light energy is collected and converted to chemical energy in photosynthesis. [Gk., *thylakis, dim.* of *thylakos,* bag]

thyroid gland (25) An endocrine gland, located in the neck in humans, that produces thyroxine, an iodine-containing hormone that influences growth and metabolism.

thyroid-stimulating hormone (25)A hormone produced by the anterior pituitary that causes the thyroid gland to produce thyroxine, which is important in regulating cellular respiration.

thyroxine (25) A hormone produced and secreted by the thyroid gland that influences metabolic speed and efficiency.

tight junction (4) A continuous, water-tight connection between adjacent animal cells. Tight junctions are particularly important in the small intestine, where digestion occurs, to ensure that nutrients do not leak between cells into the body cavity and so become lost as a source of energy.

tissue (21) A group of similar cells that act together to perform specific functions in the body.

tonicity (4) For a cell in solution, a measure of the concentration of solutes outside the cell relative to that inside the cell. [Gk., *tonos,* tension]

topography (17) The physical features of a region, including those created by humans. [Gk., *topos,* place + *graphein,* to write down]

total reproductive output (11) The lifetime number of offspring produced by an individual.

totipotent (20) Describes undifferentiated cells that have the developmental potential to become any type of cell and tissue in an organism.

trachea (23) A structure in the respiratory system that conducts air to the lungs; also called the windpipe.

trait (6, 10) Any characteristic or feature of an organism, such as red petal color in a flower.

trans fats (3) Unsaturated fats that have been partially hydrogenated (meaning that hydrogen atoms have been added to make the fat more saturated and to improve a food's taste, texture, and shelf-life). The added hydrogen atoms are in a trans orientation, which differs from the cis ("near") orientation of hydrogen atoms in the unsaturated fat. [Lat., *trans,* on the other side of]

transcription (6) The process by which a gene's base sequence is copied into mRNA.

transduction (15) A method of lateral transfer of DNA from one bacterial cell to another by means of a virus containing pieces of bacterial DNA that it picked up from a previous host. The virus infects the recipient bacterium and passes along new genes to it. [Lat., *trans,* on the other side of + *ducere,* to lead]

transfer RNA (tRNA) (6) The type of RNA molecules in the cytoplasm that link specific triplet base sequences on mRNA to specific amino acids.

transformation (15) A method of lateral transfer of DNA from one bacterial cell to another in which a bacterial cell scavenges DNA released from burst bacterial cells in the environment. [Lat., *trans,* on the other side of + *formare,* to shape]

transgenic organism (7) An organism that contains in its cells some DNA from another species.

translation (6) The process by which mRNA, which encodes a gene's base sequence, directs the production of a protein.

transmembrane protein (4) A membrane protein that penetrates the phospholipid bilayer of a cell's plasma membrane.

transport protein (4) A transmembrane protein that provides a channel or passageway through which large or strongly charged molecules can pass. Transport proteins are of various shapes and sizes, making possible the transport of a wide variety of molecules.

treatment (1) In a research study, any condition applied to subjects— those in the treatment group— that is not applied to subjects in a control group.

triglycerides (3) Fats having three fatty acids linked to the glycerol molecule. [Gk., *tri-* three + *glykys,* sweet]

trimester (26) In human development, one of the three three-month periods of pregnancy.

trophic level (17) A step in the flow of energy through an ecosystem. [Gk., *troph* , food]

tropical rain forest (17) A type of terrestrial biome, found between the Tropic of Cancer (23.5° north latitude) and Tropic of Capricorn (23.5° south latitude) and characterized by constant moisture and temperature that do not vary across the seasons; vegetation is dense.

tropical seasonal forest (17) A type of terrestrial biome characterized by a hot climate and distinct wet and dry seasons; trees shed their leaves in the dry season.

tropism (20) In plants, a growth response by which plants grow toward or away from environmental stimuli such as light, gravity, or physical impediments, by bending, curving, and twisting. [Gk., *tropikos,* turn] See also **gravitropism; heliotropism; phototropism; thigmotropism.**

true-breeding (9) Describes a population of organisms in which, for a given trait, the offspring of crosses of individuals within the population always show the same trait; for example, the offspring of pea plants that are true-breeding for round peas always have round peas.

turgor pressure (4) In plants, the pressure of the contents of the cell against the cell wall, which is maintained by osmosis as water flows into a cell that contains high concentrations of dissolved substances. Turgor pressure allows non-woody plants to stand upright, and its loss causes wilting. [Lat., *turgere,* to swell]

twitch (24) In a muscle contraction, the duration of the period between muscle contraction and relaxation.

U

umbilical cord (26) A mass of tissue that connects the embryo to the placenta and through which nutrients and wastes are exchanged.

unsaturated fats (3) Fats in which at least one carbon in the hydrocarbon chain forming the tail region of the molecule is bound to only one hydrogen atom; liquid at room temperature.

uterus (26) The reproductive organ in female mammals that is the site where an embryo develops; also called the womb.

V

vaccine (27) A weakened or harmless form of a specific pathogen that is administered to an individual to induce immunity without subjecting the individual to the disease.

vacuole (central) (4) A membrane-enclosed, fluid-filled, multipurpose organelle prominent in most plant cells (but also present in some protists, fungi, and animals). Its functions vary but can include storing nutrients, retaining and degrading waste products, accumulating poisonous materials, containing pigments, and providing physical support. [Lat., *vacuus,* empty]

vagina (26) A tube-like chamber connecting the female external genitals to the uterus; the place where sperm are released during copulation.

variables (1) The characteristics of an experimental system subject to change; for example, time (the duration of treatment) or specific elements of the treatment, such as the substance or procedure administered or the temperature at which it takes place. [Lat., *variare,* to vary]

vas deferens (26) A tube of smooth muscle tissue that passes from each testis into the body.

vascular cambium (20) In woody plants, the layer of lateral meristem cells that produces secondary xylem and secondary phloem.

vascular plants (14) Plants that transport water and dissolved nutrients by means of vascular tissue, a system of tubes that extends from the roots through the stem and into the leaves. [Lat., *vasculum, dim.* of *vas,* vessel]

vascular tissue (19) The part of a plant that conducts food, water, and nutrients throughout the plant body.

vein (22) A blood vessel that transports blood from capillaries in body tissues to the heart; the blood in veins is at a lower pressure than the blood in arteries,

ventricle (22) A chamber of the heart from which blood is pumped to the lungs or gills and the rest of the body.

vertebrates (13) Chordate animals that have a backbone (made of cartilage or hollow bones) and, at the front end of the organism, a head containing a skull, brain, and sensory organs.

vesicle (24) A small, membrane-enclosed sac within a cell. [Lat., *vesicula, dim.* of *vesica,* bladder]

vestigial structure (10) A structure that was once useful to the organism but has lost its function over evolutionary time; an example is molars in bats that now consume an exclusively liquid diet. [Lat., *vestigium,* footprint, trace]

viruses (12) Diverse and important biological entities that can replicate but can conduct metabolic activity only by taking over the metabolic processes of a host organism, and therefore fall outside the definition of life. [Lat., *virus,* slime]

vitamin (23) An organic compound that is an essential nutrient required by the body in small amounts.

viviparity (13, 26) The characteristic of bearing live young, giving birth to babies (rather than laying eggs). [Lat., *vivus,* alive + *parere,* to bear]

W

waggle dance (11) Behavior of scout honeybees that indicates, by the angle of the body relative to the sun and by physical maneuvers of various duration, the direction to a distant source of food.

waxes (3) Lipids similar in structure to fats but with only one long-chain fatty acid linked to the glycerol head of the molecule. Because the fatty acid chain is highly nonpolar, waxes are strongly hydrophobic.

Wernicke's area (24) The area of the brain responsible for understanding speech; located, in most people, in the left temporal lobe.

white blood cells (22, 27) Blood cells that defend against pathogens; the primary components of the immune response system; also called leukocytes.

X

X and Y chromosomes (8) The human sex chromosomes.

xylem (19) (*pron.* ZY-lum) In vascular plants, the tissue that conducts water and dissolved minerals throughout the plant.

Z

zona pellucida (26) In an egg, the glycoprotein layer between the granulosa cells and the egg's membrane.

zygote (20, 26) A diploid cell resulting when a sperm and egg fuse during fertilization.

Photo Credits

Legend: *L*, left; *C*, center; *R*, right; *T*, top; *M*, middle; *B*, bottom

FRONT MATTER p. iv: *(T)* Michael Nichols/National Geographic Creative; *(B)* Jay Phelan. **p. v:** *(T)* Miroslaw Swietek; *(B)* Keystone-France/Getty Images. **p. vi:** *(T)* Mary Gretchen Kaplan; *(B)* Betsie Van der Meer/Getty Images. **p. vii:** *(T)* Clark Sumida; *(B)* Marco Garcia/Getty Images. **p. viii:** *(T)* Eye of Science/Science Source; *(B)* Susan Chiang/Getty Images. **p. ix:** *(T)* © Uli Westphal 2008; *(B)* Marni Rolfes. **p. x:** *(T)* Courtesy HudsonAlpha Institute for Biotechnology; *(B)* © Photo copyright DNA Genotek Inc. All rights reserved. Used with permission. **p. xi:** *(T)* Steve Schapiro/CORBIS OUTLINE/Corbis via Getty Images; *(B)* Jade Brookbank/Getty Images. **p. xii:** *(T)* Imagno/Getty Images; *(B)* Julian Love/Getty Images. **p. xiii:** Media Drum World/Aurora Photos. **p. xiv:** *(T)* Juniors Bildarchiv GmbH/Alamy; *(B)* Zero Creatives/Getty Images. **p. xv:** *(T)* Kellar Autumn, Lewis & Clark College; *(B)* Reed Hutchinson. **p. xvi:** *(T)* Andrius Burba; *(B)* Ariel Skelley/Blend Images/Corbis. **p. xvii:** *(T)* Darlyne A. Murawski /National Geographic Creative; *(B)* AFP/Getty Images. **p. xviii:** *(T)* Dennis Kunkel Microscopy/Science Source; *(B)* David Woolley/Getty Images. **p. xix:** *(T)* Michael Wolf/Laif/Redux; *(B)* © Lane Oatey/Blue Jean Images/Corbis. **p. xx:** *(T)* LUMI Images/Getty Images; *(B)* NASA Image by Robert Simmon, based on Landsat data provided by the UMD Global Land Cover Facility. **p. xxi:** Gallo Images/Getty Images. **p. xxv:** *(T)* Cindy Kassab/Getty Images; *(B)* Annie Corrigan/WFIU. **p. xxvi** *(T)* Rob Kesseler; *(B)* BSIP/Photoshot. **p. xxvii** *(T)* Anandraj Thiyagarajaperumal; *(B)* © Image Source/Corbis. **p. xxviii** *(T)* Susumu Nishinaga/Science Source; *(B)* Wild Horizon/Getty Images. **p. xxix** *(T)* Copyright 2014 Francesco Tonelli; *(B)* © Ken Seet/Corbis. **p. xxx** *(T)* Joel Sartore/NGS Image Collection; *(B)* Science Source. **p. xxxi** *(T)* Kirill Golovchenko; *(B)* Editorial Image, LLC/Alamy. **p. xxxii** *(T)* Joel Sartore/National Geographic Creative; *(B)* Vincent Kessler/Reuters/Newscom. **p. xxxiii** *(T)* Khuloud T. Al-Jamal & Izzat Suffian, Wellcome Images; *(B)* M. Eric Honeycutt/Getty Images. **p. xxxv:** *(T, BR)* Photo by Thomas Jackson. *(BR)* The Daily Bruin. **p. xxxvi:** Kari MacRostie. **p. xxxvii:** *(L)* Dennis Kunkel Microscopy, Inc.; *(C)* Dennis Kunkel Microscopy/Science Source; *(R)* Ralph A. Clevenger/Getty Images.

CHAPTER 1 pp. 0–1: Michael Nichols/National Geographic Creative. **p. 2:** ©Tim Laman (www.TimLaman.com). **p. 3:** *Figure 1-1* Kristoffer Tripplaar/Alamy Stock Photo; *Figure 1-2 (L)* Sipa USA via AP, *Figure 1-2 (C)* Courtesy 23andme.com, *Figure 1-2 (R)* Universal Pictures/Photofest. **p. 4**: David Doubilet/National Geographic Creative. **p. 5:** *Figure 1-4 (L)* Tim Boyle/Getty Images, *Figure 1-4 (R)* Digital Vision Photography/Veer. **p. 6:** *Figure 1-5 (L)* Comstock Images/Getty Images, *Figure 1-5 (R)* AP Photo/Elaine Thompson. **p. 7:** *Figure 1-6* LWA/Dann Tardif/Blend Images/Corbis. **p. 12:** Mario Tama/Getty Images. **p. 14:** *Figure 1-11* Mary Evans Picture Library/Alamy Stock Photo. **p. 18:** B. Boissonnet/age fotostock. **p. 20:** *Figure 1-15* Jennifer Graylock/FilmMagic/Getty Images. **p. 21:** *Figure 1-16 (L)* Hero Images/Getty Images, *Figure 1-16 (R)* Jennifer Macmillan/Macmillan Learning. **p. 22:** *Figure 1-17* Jay Phelan. **p. 23:** *Figure 1-18* Every Child By Two. **p. 24:** *Figure 1-19* Philippe Psaila/Science Source. **p. 25:** Chip Clark/National Museum of Natural History, Smithsonian. **p. 26:** Jay Phelan.

CHAPTER 2 pp. 30–31: Miroslaw Swietek. **p. 32**: David Nicholls/Science Source; *Figure 2-1 (L)* Optimarc/Shutterstock, *Figure 2-1 (C)* Jeffrey Coolidge/Getty Images, *Figure 2-1 (R)* Chris Ratcliffe/Bloomberg via Getty Images. **p. 36:** *Figure 2-7* Michael Yarish/©CBS/Courtesy Everett Collection. **p. 39:** *Figure 2-10* Judith Collins/Alamy, *Figure 2-10 (inset)* Dennis Kunkel Microscopy/Science Source. **p. 41:** Jennifer Brower/National Geographic Creative; *Figure 2-13 (L)* Courtesy Steve Krichten, *Figure 2-13 (R)* R. B. Suter, Vassar College. **p. 43:** *Figure 2-16* Graham Eaton/Getty Images. **p. 49:** Keystone-France/ Getty Images.

CHAPTER 3 pp: 52–53: Mary Gretchen Kaplan. **p. 54:** *(T)* Зоряна Ивченко/500px; *(B)* deyangeorgiev/Getty Images. **p. 55:** *Figure 3-1* bitt24/Shutterstock. **p. 57:** *Figure 3-5 (L)* Tamelyn Feinstein/Getty Images, *Figure 3-5 (R)* TheCrimsonMonkey/Getty Images. **p. 58:** *Figure 3-6 (T)* Universal Images Group North America LLC (Lake County Discovery Museum)/Alamy, *Figure 3-6 (B)* D. Roberts/Science Source; *Figure 3-7* Chassenet/AgeFotostock. **p. 59:** Jeff Foott/Discovery Channel Images/Getty Images; *Figure 3-8* Fairfax Media via Getty Images. **p. 61:** *Figure 3-10* Gary Gladstone; *Figure 3-11 (L)* Ellen Isaacs/Alamy, *Figure 3-11 (R)* Masterfile. **p. 62:** *Figure 3-12* © Christine Schneider/Getty Images. **p. 64:** *Figure 3-14* Michael Ochs Archives/Getty Images. **p. 65:** Mirko_Rosenau/Getty Images. **p. 66:** *Figure 3-16 (from left to right)* Steve Bloom Images/Alamy, Steve Gschmeissner/Science Source, SPL/Science Source, Thomas Deerinck, NCMIR/Science Source, © Deco/Alamy. **p. 67:** *Figure 3-18* Ildi Papp/Shutterstock. **p. 69:** *Figure 3-21 (L)* Hero Images/Getty Images, *Figure 3-21 (R)* Peathegee Inc/Getty Images. **p. 73**: Vivek Mittal. **p. 76:** Betsie Van der Meer/Getty Images.

CHAPTER 4 pp. 80–81: Clark Sumida. **p. 82:** Biophoto Associates/Science Source; *Figure 4-1 (L)* Tim Gainey/Alamy Stock Photo, *Figure 4-1 (C)* Auscape/UIG via Getty Images, *Figure 4-1 (R)* David Scharf/Science Source. **p. 83**: *Figure 4-3 (TL)* John Springer Collection/CORBIS/Corbis via Getty Images, *Figure 4-3 (TR)* Rosenfeld/AgeFotostock, *Figure 4-3 (B)* Clouds Hill Imaging Ltd./Getty Images, *Figure 4-3 (inset)* Courtesy Reproductive Solutions. **p. 85:** *Figure 4-4* Alfred Pasieka/Science Source. **p. 86:** *Figure 4-5 (TL)* George F. Mobley/National Geographic Creative, *Figure 4-5 (TC)* Paul Kennedy/Getty Images, *Figure 4-5 (TR)* Matt Meadows/Peter Arnold Images/Getty Images; *Figure 4-6 (BL)* Dennis Kunkel Microscopy, Inc., *Figure 4-6 (BR)* James Cavallini/Science Source. **p 89:** © Toshiaki Ono/amanaimages/Corbis; *Figure 4-9 (L)* Kevin MacKenzie, University of Aberdeen. Wellcome Images, *Figure 4-9 (R)* Caro/Alamy. **p. 93:** *Figure 4-13 (TL)* Courtesy Cystic Fibrosis Foundation, *Figure 4-13 (TR)* Brendan Fitterer/ © Tampa Bay Times/Zuma Press; *Figure 4-14 (B)* Image Source/Veer. **p. 95:** *Figure 4-15 (T)* AP Photo/Jeffrey Boan; *(B)* Courtesy of Dr. G.M. Gaietta & T.J. Deerinck, CRBS/NCMIR, University of California, San Diego. **p. 100:** *Figure 4-20* Richard Keppel-Smith/Getty Images. **p. 101:** *Figure 4-21* Biophoto Associates/Science Source. **p. 102:** *Figure 4-23 (TL)* SPL/Science Source, *Figure 4-23 (TR)* SPL/Science Source; *Figure 4-24 (B)* SPL/Science Source. **p. 103:** Ed Burns/EyeEm/Getty Images; *Figure 4-25 (L)* Copyright 1997 Dennis Kunkel Microscopy, Inc., *Figure 4-25 (C)* Dennis Kunkel Microscopy, Inc., *Figure 4-25 (R)* Courtesy of Dr. G.D.Griffin, Oak Ridge National Laboratory. **p. 104:** Dennis Kunkel Microscopy, Inc. **p. 105:** *Figure 4-26 (L)* Mary Ellen Mark; *Figure 4-27 (R)* CNRI/Science Source. **p. 106:** *Figure 4-28 (T)* Biophoto Associates/Science Source, *Figure 4-28 (C)* Riccardo Cassiani-Ingoni/Science Source, *Figure 4-28 (B)* Dr. Jan Schmoranzer/Science Source. **p. 107:** *Figure 4-29 (L)* Dr. Yorgos Nikas/Phototake, *Figure 4-29 (R)* Juergen Berger/Science Source. **p. 108:** *Figure 4-30* Keith R. Porter/Science Source. **p. 111:** *Figure 4-32* Dr. Gopal Murti/Science Source. **p. 112:** *Figure 4-34* NIBSC/Science Source. **p. 113:** *Figure 4-35* Omikron/Science Source. **p. 114:** *Figure 4-36* Dennis Kunkel Microscopy, Inc. **p. 116:** *Figure 4-38* Omikron/Science Source. **p. 117:** *Figure 4-39* Biophoto Associates/Science Source. **p. 118:** *Figure 4-40* Biology Pics/Science Source. **p. 120:** Marco Garcia/Getty Images.

CHAPTER 5 pp. 124–125: Eye of Science/Science Source. **p. 126:** Sebastien Moreau/500px; *Figure 5-1* Lex Lieshout/AFP/Getty Images. **p. 128:** *Figure 5-3 (TL)* David Madison/Getty Images, *Figure 5-3 (TC)* © Brad Lenear, *Figure 5-3 (TR)* Bernhard Edmaier/Science Source, *Figure 5-3 (BL)* Copyright © Frans Lanting/www.lanting.com, *Figure 5-3 (BR)* Fredrik Von Erichsen /AFP/Getty Images; *Figure 5-4* © Halfdark/fstop/Corbis. **p. 129:** *Figure 5-5 (L)* Ed Gifford/Masterfile, *Figure 5-5 (R)* Courtesy Jay Phelan. **p. 130:** *Figure 5-6* Mark Rebilas/USA Today Sports Images. **p. 132**: Kari MacRostie. **p. 133:** *Figure 5-10 (L)* Dennis Kunkel Microscopy, Inc., *Figure 5-10 (C)* Dennis Kunkel Microscopy/Science Source, *Figure 5-10 (R)* Ralph A. Clevenger/Getty Images. **p. 134:** *Figure 5-12 (T)* Value Stock Images/Fotosearch, *Figure 5-12 (B)* Dennis Kunkel Microscopy, Inc. **p. 143:** *Figure 5-23* Al Seib/Getty Images; *Figure 5-24 (L)* Dr. Jeremy Burgess/Science Photo Library/Science Source, *Figure 5-24 (R)* Dr. Jeremy Burgess/Science Photo Library/Science Source. **p. 144:** *Figure 5-25* (L) Courtesy Jay Phelan, *Figure 5-25 (C)* Beth Perkins/Getty Images, *Figure 5-25 (R)* PBNJ Productions/Corbis. **p. 146**: Joe Petersburger/National Geographic Creative. *Figure 5-27 (L)* Anna Om/Shutterstock, *Figure 5-27 (C)* © Copyright

Frans Lanting/ www.lanting.com, *Figure 5-27 (R)* Volker Steger/Science Source. **p. 148:** *Figure 5-29 (L)* Courtesy Jay Phelan, *Figure 5-29 (C)* Yukihiro Fukuda/ Nature Picture Library, *Figure 5-29 (R)* A. Barry Dowsett/Science Source. **p. 152:** *Figure 5-33* Hybrid Medical Animation/Science Source. **p. 154:** *Figure 5-36 (L)* © Corbis, *Figure 5-36 (C)* © IMAGEMORE /Imagemore Co., Ltd./Corbis, *Figure 5-36 (R)* Cultura RM Exclusive/Brett Stevens. **p. 156:** Morsa Images/Getty Images; *Figure 5-37* Adam Pretty/Getty Images. **p. 157:** *Figure 5-39 (L)* © Cephas Picture Library/Alamy, *Figure 5-39 (R)* Yuriko Nakao/Bloomberg via Getty Images, *Figure 5-39 (inset)* SCIMAT/Science Source. **p. 158:** Susan Chiang/Getty Images.

CHAPTER 6 pp. 162–163: © Uli Westphal 2008. **p. 164:** Peter Blixt/Alamy Stock Photo; *Figure 6-1* Vasna Wilson/Virginian-Pilot. **p. 165:** *Figure 6-2* Jay Phelan. **p. 166:** *Figure 6-3 (L)* A. Barrington Brown/Science Source, *Figure 6-3 (R)* Science Source/Science Source. **p. 168:** *Figure 6-5 (L to R)* Courtesy Jay Phelan, Sian Irvine/Getty Images, Eye of Science/Science Source, Biophoto Associates/Science Source, Eric Carr/Alamy Stock Photo. **p. 170:** *Figure 6-8 (T to B)* Eye of Science/ Science Source, Courtesy Jay Phelan, Sian Irvine/Getty Images, Eric Carr/Alamy Stock Photo, Biophoto Associates/Science Source; *Figure 6-9 (T to B)* Courtesy Jay Phelan, Eye of Science/Science Source, Dennis Kunkel Microscopy/Science Source, Fancy/Veer/Corbis, © Dennis Kunkel Microscopy, Inc. **p. 172:** Deco Images II/ Alamy Stock Photo. **p. 178:** *Figure 6-16:* © Zave Smith/Getty Images. **p. 180:** Matthijs Kuijpers/Alamy Stock Photo. **p. 181:** *Figure 6-19 (L)* Andrew Syred/ Science Source, *Figure 6-19 (R)* Ted Kinsman/Science Source. **p. 183:** *Figure 6-21 (L)* Mika/Getty Images, *Figure 6-21 (C)* Andia/UIG via Getty Images, *Figure 6-21 (R)* © Ocean/Corbis. **p. 186:** Marni Rolfes. **p. 187:** *(T to B)* Courtesy Jay Phelan, Eye of Science/Science Source, Dennis Kunkel Microscopy/Science Source, Fancy/Veer/ Corbis, © Dennis Kunkel Microscopy, Inc.

CHAPTER 7 pp. 190–191: Courtesy HudsonAlpha Institute for Biotechnology. **p. 192**: Chung Sung-Jun/Getty Images. **p. 193:** *Figure 7-1 (L)* Douglas Peebles Photography/Alamy, *Figure 7-1 (C)* Peter DaSilva/The New York Times/Redux, *Figure 7-1 (R)* Jochen Tack/Getty Images. **p. 198:** *Figure 7-8 (T)* IanDagnall Computing/Alamy, *Figure 7-8 (C)* Steve Gschmeissner/Science Source, *Figure 7-8 (B)* Lara Jo Regan/Barcroft Media. **p. 199**: Thomas Barwick/Getty Images. **p. 200:** *Figure 7-9* Ed Young/Getty Images, *Figure 7-9 (inset)* Photo by John Doebley; *Figure 7-10 (TL)* Professor Peter Beyer/Humanitarian Board for Golden Rice/www. goldenrice.org, *Figure 7-10 (BL)* Professor Peter Beyer/Humanitarian Board for Golden Rice/www.goldenrice.org, *Figure 7-10 (inset, L)* Martin Ruegner/Getty Images, *Figure 7-10 (inset, R)* The James Hutton Institute. **p. 201:** *Figure 7-11 (L)* Philippe Huguen/AFP/Getty Images, *Figure 7-11 (C)* Lance Nelson/Getty Images, *Figure 7-11 (R)* Daniel Acker/Bloomberg via Getty Images. **p. 202:** *Figure 7-13* AFP/Getty Images. **p. 203:** *Figure 7-14* Photo by Dr. Garth Fletcher. *Figure 7-15* Havakuk/Reuters/Newscom. **p. 204:** *Figure 7-16* Gordon Chibroski/Portland Press Herald via Getty Images. **p. 207**: © Remi Benali and Stephen Ferry. **p. 208:** *Figure 7-17* Voisin/Phanie/Getty Images. **p. 209:** *Figure 7-18* MGM/courtesy Everett Collection. **p. 210:** *Figure 7-19* Rick Friedman/Getty Images; *Figure 7-20* Baylor Medical Center. **p. 212:** *Figure 7-22 (T)* AP Photo/John Chadwick, *Figure 7-22 (C)* AP Photo/Ben Margot, *Figure 7-22 (R)* © RIA Novosti/Alamy. **p. 213:** AP Photo/ Michael Probst. **p. 214:** *Figure 7-23 (L)* Neville Chadwick/Science Source, *Figure 7-23 (R)* David Parker/Science Photo Library/Science Source. **p. 216**: © Photo copyright DNA Genotek Inc. All rights reserved. Used with permission.

CHAPTER 8 pp. 220–221: Steve Schapiro/CORBIS OUTLINE/Corbis via Getty Images. **p. 222**: Steve Gschmeissner /© Science Photo Library/Alamy. **p. 223:** *Figure 8-1* Science Photo Library/Science Source; *Figure 8-2* Sven Dillen/Reporters/ Redux. **p. 225:** *Figure 8-4* Hazel Appleton, Health Protection Agency Centre for Infections/Science Source. **p. 231:** Indeed/Getty Images; *Figure 8-10 (L)* Dan Guravich/Science Source; *Figure 8-10 (R)* Alex Blackburn Clayton/Photolibrary/ Getty Images; *Figure 8-10 (inset, L)* Biology Pics/Science Source; *Figure 8-10 (inset, R)* SPL/Science Source. **p. 233:** *Figure 8-12* David Scharf/Science Source. **p. 234**: *Figure 8-14* Courtesy T. Wittmann. **p. 235**: *Figure 8-14* Courtesy T. Wittmann. **p. 237:** *Figure 8-16* GJLP/CNRI/Science Source. **p. 238**: Lennart Nilsson/TT; *Figure 8-17* Kevin Laubacher/Getty Images. **p. 240:** *Figure 8-19* AP Images. **p. 244:** *Figure 8-23* Don Fawcett/Science Photo Library/Science Source. **p. 247:** *Figure 8-26* Margot Granitsas/Science Source. **p. 248:** *Figure 8-27* James H. Robinson/Science Photo Library/Science Source; *Figure 8-28* Dr. Kari Lounatmaa/Science Source. **p. 249:** Dale Robinette/©Summit Entertainment/courtesy Everett Collection; *Figure 8-29* Biophoto Associates/Science Source. **p. 252:** Phanie/Alamy Stock Photo. **p. 253**: *Figure 8-32* L.Willatt, East Anglian Regional Genetics Service/ Science Source; *Figure 8-33* Saturn Stills/Science Source. **p. 254:** *Figure 8-34 (T)* Mika/Getty Images; *Figure 8-34 (B)* L. Willatt/East Anglian Regional Genetics Service/Science Source. **p. 258:** Jade Brookbank/Getty Images.

CHAPTER 9 pp. 262–263: Imagno/Getty Images. **p. 264**: Steve Bloom Images/ Alamy. **p. 265:** *Figure 9-3* Felix Wirth/Corbis. **p. 266:** *Figure 9-4 (L)* Jonathan Hordle/Rex Features; *Figure 9-4 (R)* Jason LaVeris/Getty Images; *Figure 9-5* Newspix/ Rex USA. **p. 267:** *Figure 9-6 (TL)* © Michael Weber/imagebroker/Corbis, *Figure 9-6 (TR)* Aspen Photo/ Shutterstock, *Figure 9-6 (BL)* Satapat/Shutterstock, *Figure 9-6 (BR)* Razvan Losif/Shutterstock. **p. 268:** *Figure 9-8* James King-Holmes/Science Source. **p. 270:** *Figure 9-9 (L)* R. L. Webber/Shutterstock, *Figure 9-9 (R)* Sally Reed/age fotostock. **p. 271:** *Figure 9-11* Thomas & Pat Leeson/Science Source. **p. 273:** Wally McNamee/Getty Images. **p. 275:** *Figure 9-15* Richard Ellis/Alamy. **p. 277:** *Figure 9-17* Courtesy Michael Comte. **p. 278:** The Photo Works/Alamy. **p. 279:** *Figure 9-19* Rene Maltete/RAPHO/Gamma-Rapho via Getty Images. **p. 283:** *Figure 9-23* Big Cheese Photo LLC/Alamy. **p. 284:** *Figure 9-24 (L)* Lennart Nilsson/TT, *Figure 9-24 (R)* Jackie Lewin, Royal Free Hospital/Science Source. **p. 287:** *Figure 9-26* Tommy Moorman. **p. 288:** *Figure 9-27 (T)* Lysandra Cook/Getty Images, *Figure 9-27 (B)* Juniors Bildarchiv/Alamy. **p. 289**: Joel Meyerowitz/Gallery Stock. **p. 291:** *Figure 9-29* Gregg DeGuire/Getty Images. **p. 292:** Julian Love/Getty Images.

CHAPTER 10 pp. 296–297: Media Drum World/Aurora Photos. **p. 298:** Lucas Bustamante; *Figure 10-1* Graphic Science/Alamy. **p. 300:** Scott Chandler. **p. 301:** *Figure 10-3 (L to R)* Pictorial Press Ltd/Alamy, Sheila Terry/Science Source, The Granger Collection, New York, The Granger Collection, New York, Mary Evans Picture Library/Alamy. **p. 302:** *Figure 10-5* imageBROKER/Alamy. **p. 304:** *Figure 10-7* Pictorial Press Ltd/Alamy. **p. 305:** Joe Petersburger/National Geographic Creative; *Figure 10-8* Photo by Louie Fasciolo/Bernard J. Shapero Rare Books. **p. 306:** *Figure 10-9* © Renee Lynn/Corbis/VCG/Getty Images. **p. 308:** *Figure 10-11 (L)* © Fabrice Lerouge/Onoky/ Corbis; *Figure 10-11 (R)* Arco Images GmbH/Alamy. **p. 310:** *Figure 10-13 (L)* EVARISTO SA/Getty Images, *Figure 10-13 (R)* EVARISTO SA/Getty Images; *Figure 10-14* Copyright © Frans Lanting/www.lanting.com. **p. 312:** *Figure 10-16* Dimitrios Kambouris/ Getty Images; *Figure 10-17* Max Nash/AFP/Getty Images. **p. 313:** *Figure 10-18* Trinity Mirror/Mirrorpix/Alamy. **p. 317**: Orsolya Haarberg/National Geographic Creative. **p. 318:** *Figure 10-22* Jason Edwards/National Geographic Creative. **p. 319:** *Figure 10-23* J. Sneesby/B. Wilkins/Getty Images. **p. 321:** *Figure 10-26* Courtesy Minnesota Historical Society, Photo by Ryan Howard. **p. 322:** *Figure 10-27* Owen Humphreys/ Press Association via AP Images; *Figure 10-28* Walter B. McKenzie/Getty Images. **p. 323:** *Figure 10-29* Tom & Pat Leeson/Science Source. **p. 326:** *Figure 10-30* Will & Deni McIntyre/Science Source; *Figure 10-31* Thegoodly/Getty Images. **p. 327:** Martin Harvey/Getty Images. **p. 328:** *Figure 10-32* Jason Edwards/National Geographic/ Getty Images. **p. 331:** *Figure 10-36 (T)* George Grall /National Geographic Creative, *Figure 10-36 (L)* Eric VanderWerf, *Figure 10-36 (C)* Eric VanderWerf, *Figure 10-36 (R)* Jack Jeffrey; *Figure 10-37 (TL)* ANT Photo Library/Science Source, *Figure 10-37 (CL)* Copyright © Frans Lanting/www.lanting.com, *Figure 10-37 (BL)* Dave Watts/Alamy, *Figure 10-37 (TR)* Nicholas Bergkessel, Jr./Science Source, *Figure 10-37 (CR)* Copyright © Frans Lanting/www.lanting.com, *Figure 10-37 (BR)* John W. Warden/age fotostock. **p. 332:** *Figure 10-38 (TL)* From: The Sharks of North America by José I. Castro, © 2011, *Figure 10-38 (TR)* Laboratory of Evolutionary Morphology, *Figure 10-38 (BL)* Omikron/Science Source, *Figure 10-38 (BR)* S. K. Ackerley, Department of Integrative Biology, University of Guelph, Guelph, ON, Canada. **p. 333:** *Figure 10-39 (L to R)* PBNJ Productions/PBNJ/AgeFotostock, Juniors Bildarchiv/AgeFotostock, © Malcolm Schuyl/Alamy, © Juniors Bildarchiv/Age Fotostock; *Figure 10-40* Bruce Dale/National Geographic Creative; *Figure 10-41 (T)* Robert C Nunnington/Getty Images; *Figure 10-41 (B)* CraigRJD/Getty Images. **p. 334**: *Figure 10-42 (L to R)* Jay Phelan, Arco Images GmbH/Alamy Stock Photo, Agency Animal Picture/Getty Images, © Danita Delimont/ Alamy, Ulrike Klenke and Zeev Pancer, Center of Marine Biotechnology, Baltimore, Md.

CHAPTER 11 pp. 342–343: Juniors Bildarchiv GmbH/Alamy. **p. 344:** C. S. Ling; *Figure 11-1* AP Photo/Independence Daily Reporter/Rob Morgan. **p. 345:** *Figure 11-2* Radius Images/Alamy; *Figure 11-3* Isabel M. Smallegange. **p. 346:** *Figure 11-4 (L)* Ralph Reinhold/Age Fotostock, *Figure 11-4 (TR)* © FLPA/Dave Pressland/Age Fotostock, *Figure 11-4 (BR)* A Hartl/Age Fotostock. **p. 348:** *Figure 11-6* Stacy Gold/ National Geographic Creative, *Figure 11-6 (Inset left)* Karn Bulsuk/EyeEm/Getty Images, *Figure 11-6 (Inset right)* Courtesy Everett Collection. **p. 349:** tbkmedia.

de/Alamy. **p. 350:** *Figure 11-8* Dr. Theodore Evans. **p. 351:** *Figure 11-9* Ron Wolf. **p. 353:** *Figure 11-11* Courtesy Gerald Carter. **p. 354:** *Figure 11-12* Courtesy Jay Phelan, Julia Phelan, Alon Ziv, Joshua Malina, Terry Burnham, Jennifer Yousem, Janvi Patel, Michael Cooperson and Jeffrey Tambor. **p. 355:** *Figure 11-13* Peter Biro/IRC. **p. 357**: Tim Laman/National Geographic Creative; *Figure 11-15* David M. Phillips/Science Source. **p. 358:** *Figure 11-17 (L)* © Frank Parker/Age Fotostock, *Figure 11-17 (C)* GFC Collection/Age Fotostock, *Figure 11-17 (R)* Copyright © Frans Lanting/www.lanting.com. **p. 360:** *Figure 11-18 (L)* ©Fox Searchlight/courtesy Everett Collection, *Figure 11-18 (R)* © Columbia/Courtesy Everett Collection; *Figure 11-19* George Grall/National Geographic/Getty Images. **p. 361:** *Figure 11-20 (T to B)* Anthony Mercieca/Science Source, Barrett Hedges/Getty Images, Graphic Science/Alamy, Copyright © Frans Lanting/www.lanting.com. **p. 363:** *Figure 11-21* Thomas Hinsche/Getty Images; *Figure 11-22* James H. Robinson/Science Source. **p. 366:** *Figure 11-23* Copyright © Frans Lanting / Frans Lanting Stock. **p. 367:** *Figure 11-24* M. Botzek/Age Fotostock. **p. 368:** *Figure 11-26 (L)* Design Pics Inc/National Geographic Creative; *Figure 11-26 (R)* John Cancalosi/National Geographic Creative. **p. 369:** Design Pics Inc/National Geographic Creative; *Figure 11-27 (L)* Darlyne A. Murawski/National Geographic Creative. *Figure 11-27 (C)* © DLILLC/Corbis, *Figure 11-27 (R)* Ocean/Corbis. **p. 370**: *Figure 11-28 (L)* Scott Camazine/Alamy, *Figure 11-28 (C)* AP Photo/J Pat Carter, *Figure 11-28 (R)* kali9/ Getty Images. **p. 371:** *Figure 11-29* Werner Bollmann/Age Fotostock. **p. 372:** Zero Creatives/Getty Images.

CHAPTER 12 pp. 376–377: Kellar Autumn, Lewis & Clark College. **p. 378:** Copyright © Frans Lanting/www.lanting.com; *Figure 12-1* Courtesy of Smithsonian Institution. Painting by P. Sawyer. **p. 379:** *Figure 12-2* Bettmann/Getty Images. **p. 380:** *Figure 12-3* Courtesy of Andrew Knoll. **p. 383:** © YellowPaul/Alamy; *Figure 12-5 (L to R)* Michael Regan/Getty Images, Jason Kempin/Getty Images, Plantography/Alamy, The Collection by Jan Smith Photography/Alamy, Copyright © Frans Lanting/www.lanting.com, Copyright © Frans Lanting/www.lanting.com. **p. 384:** *Figure 12-6* Maria Stenzel/National Geographic Creative. *Figure 12-7* Udo Richter/EPA/Newscom. **p. 385:** *Figure 12-8 (L)* © Juniors Bildarchiv/Alamy, *Figure 12-8 (R)* BlackDorianstock/Shutterstock. **p. 389:** *Figure 12-11 (L)* John Cancalosi/Age Fotostock; *Figure 12-11 (R)* R. Höelzl/Wildlife/Juniors/Age Fotostock; *Figure 12-12 (T)* Michael Woodruff; *Figure 12-12 (CL)* David Hosking/Alamy, *Figure 12-12 (C)* Copyright © Frans Lanting/www.lanting.com, *Figure 12-12 (CR)* Joel Sartore/National Geographic Creative, *Figure 12-12 (BL)* © D. Parer & E. Parer-Cook/ardea.com, *Figure 12-12 (BR)* Ralph Lee Hopkins/National Geographic Creative. **p. 391:** Melissa Farlow/National Geographic Creative. **p. 394:** *Figure 12-17 (L-R)* DLILLC/Corbis, Nigel Dennis/Age Fotostock, Altrendo/Getty Images, James Gritz/Getty Images. **p. 396:** *Figure 12-19* Aaron Corcoran. *Figure 12-20 (L)* David Hosking/Alamy, *Figure 12-20 (C)* Galen Rathbun, © California Academy of Sciences, www.afrotheria.net, *Figure 12-20 (R)* Birgit Koch-Wolf. **p. 397**: Courtesy Jonathan Sequeira Barboza; *Figure 12-21* Annie Griffiths/National Geographic Creative. **p. 401:** *Figure 12-26* NASA/REUTERS/Newscom. **p. 402:** Jill Schneider/ National Geographic Creative. **p. 405:** *Figure 12-29* Arthur Sasse/AFP/Getty Images; *Figure 12-30* Steve Gschmeissner/Science Source. **p. 406:** *Figure 12-31* Massimo Lupidi; *Figure 12-31 (inset)* Derek Lovley/Kazem Kashefi/Science Source. **p. 407:** *Figure 12-32 (T):* Joao Paulo Burini/Getty Images, *Figure 12-32 (CL)* 1Photodiva/Getty Images, *Figure 12-32 (CR)* Clive Padden/EyeEm/Getty Images, *Figure 12-32 (B)* Correia Patrice/Alamy. **p. 408:** Reed Hutchinson.

CHAPTER 13 pp. 412–413: Andrius Burba. **p. 414:** © Frans Lanting/www.lanting.com; *Figure 13-1 (L)* Picture Hooked/Malcolm Schuyl/Alamy, *Figure 13-1 (C)* Prasenjeet Yadav/National Geographic Creative, *Figure 13-1 (R)* Joel Sartore/ National Geographic Photo Ark/National Geographic Creative. **p. 415:** *Figure 13-2 (L)* Philippe Clement/Nature Picture Library, *Figure 13-2 (R)* Robin Moore/National Geographic Creative. **p. 418:** *Figure 13-4 (top row, L to R):* Alexis Rosenfeld/Science Photo Library/Science Source, Dirk Wiersma/Science Source, Marc Perkins, Wayne G. Lawler/Science Source; *Figure 13-4 (bottom row, L to R):* Rick Sass/ National Geographic Creative, CNRI/Science Source, © George Grall/National Geographic Creative, Mark Conlin/Alamy. **p. 419:** Poul Beckmann. **p. 420:** *Figure 13-5 (L)* Natural History Museum, London/Science Photo Library/Science Source, *Figure 13-5 (R)* Wolcott Henry/Getty Images; *Figure 13-6* Sam Hodge/Alamy. **p. 421:** *Figure 13-7 (T)* Melissa Fiene Photography, *Figure 13-7 (BL)* Geophoto/ Natalia Chervyakova/imagebroker/Alamy, *Figure 13-7 (BR)* © Chris Howes/ Wild Places Photography/Alamy. **p. 422:** *Figure 13-9 (T)* Paul Nicklen/National Geographic Creative, *Figure 13-9 (B)* Howard Hall Productions. **p. 424:** *Figure 13-10 (L)* Steve Gschmeissner/Getty Images, *Figure 13-10 (R)* Clouds Hill Imaging Ltd./ Science Source; *Figure 13-11(T)* London School of Hygiene & Tropical Medicine/ Science Source, *Figure 13-11(B)* Dennis Kunkel Microscopy, Inc. **p. 425:** *Figure 13-12 (T)* Leslie Newman & Andrew Flowers/Getty Images, *Figure 13-12 (C)* Stephan de Prouw/Getty Images, *Figure 13-12 (B)* ARCO/H. Reinhard/AgeFotostock. **p. 427:** *Figure 13-13 (TL)* David B. Fleetham/SeaPics.com, *Figure 13-13 (TR)* © Wayne Lynch/GettyImages, *Figure 13-13 (BL)* Foodcollection RF/Getty Images, *Figure 13-13 (BR)* © Hal Beral/Corbis. **p. 428:** *Figure 13-14 (L)* David B. Fleetham, *Figure 13-14 (C)* Jonathan Fromager/EyeEm/Getty Images, *Figure 13-14 (R)* © Image Source/Corbis. **p. 429:** *Figure 13-15* © Tomasz Machnik. **p. 430:** *Figure 13-16 (TL)* Tim Laman/National Geographic Creative, *Figure 13-16 (TR)* George Grall/Getty Images, *Figure 13-16 (BL)* Patrick Lynch/Alamy, *Figure 13-16 (BR)* © George Karbus Photography/Media Bakery. **p. 431:** *Figure 13-17 (L to R):* Alex Wild Photography, Simon D. Pollard/Science Photo Library/Science Source, Thomas Shahan/ Thomasshahan.com, Steve Allen/ Science Photo Library/Science Source. **p. 434:** *Figure 13-18 (L to r):* © Frans Lanting/www.lanting.com, George Grall/National Geographic Creative, Rick & Nora Bowers/Alamy, Ron Rowan Photography/ Shutterstock, Dirk Seifert. **p. 435:** *Figure 13-19 (TL)* Paul Sutherland/National Geographic Creative, *Figure 13-19 (TR)* George Grall/National Geographic Creative, *Figure 13-19 (BL)* Biosphoto/Superstock, *Figure 13-19 (BR)* © Frans Lanting/www.lanting.com. **p. 436**: © Frans Lanting/www.lanting.com. **p. 437:** *Figure 13-20 (TL)* © Justin Lo/Getty Images, *Figure 13-20 (TR)* Ross Armstrong/Getty Images. *Figure 13-20 (C)* Images&Stories/Alamy, *Figure 13-20 (BL)* Custom Life Science Images/Alamy, *Figure 13-20 (BR)* Mike Theiss/National Geographic Creative. **p. 439:** *Figure 13-23 (L)* ARCO/Reinhard, H/AGE Fotostock, *Figure 13-23 (inset)* Darlyne A. Murawski/Getty Images, *Figure 13-23 (R)* Masa Ushioda/SeaPics.com. **p. 440:** *Figure 13-24 (L)* Paul Souders/Getty Images, *Figure 13-24 (C)* Tim Laman/ National Geographic Creative, *Figure 13-24 (R)* © Rudie Kuiter/OceanwideImages.com. **p. 442**: © Frans Lanting/www.lanting.com. **p. 443:** *Figure 13-26 (L)* Paul Colangelo/National Geographic Creative, *Figure 13-26 (C)* George Grall/National Geographic Creative, *Figure 13-26 (R)* Danita Delimont/Getty Images. **p. 444:** *Figure 13-28 (L)* John Cancalosi/Age fotostock, *Figure 13-28 (R)* ©Tim Laman (www.TimLaman.com). **p. 445:** *Figure 13-29* Chuang Zhao & Lida Xing/AFP/ Newscom. **p. 446:** *Figure 13-31 (L)* Dave Watts/Alamy, *Figure 13-31 (C)* Martin Ruegner/Getty Images, *Figure 13-31 (R)* Denis-Huot/Nature Picture Library. **p. 449:** *Figure 13-34* © Paul Springett 07/Alamy. **p. 452:** Ariel Skelley/Blend Images/Corbis.

CHAPTER 14 pp. 456–457: Darlyne A. Murawski/National Geographic Creative. **p. 458:** James Forte/National Geographic Creative*; Figure 14-1* Wendy Foden. **p. 459:** *Figure 14-2 (L)* Victoria Snowber/Getty Images; *Figure 14-2 (C)* © Frans Lanting/www.lanting.com; *Figure 14-2 (Inset)* George Shepherd; *Figure 14-2 (R)* Jason Edwards/National Geographic Creative. **p. 460**: © Frans Lanting/www.lanting.com. **p. 461:** *Figure 14-4* Charles F. Delwiche; *Figure 14-5 (L)* Jouan & Rius/ Nature Picture Library, *Figure 14-5 (R)* Emil von Maltitz/Gallo Images/Getty Images. **p. 462:** *Figure 14-6 (T)* Pal Hermansen/Getty Images, *Figure 14-6 (BL)* blickwinkel/ Alamy, *Figure 14-6 (BR)* Nature Photographers Ltd/Alamy. **p. 463:** *Figure 14-7* Barry Turner/Alamy; *Figure 14-8* The Irish Image Collection/Design Pics/Corbis, *(Inset)* © Hemis/Alamy. **p. 464:** *Figure 14-9* David Wall/Alamy. **p. 465:** *Figure 14-10 (T)* © Frans Lanting/www.lanting.com, *Figure 14-10 (BL)* Pascal Goetgheluck/Science Source, *Figure 14-10 (BR)* Rudmer Zwerver/Shutterstock; *Figure 14-11 (L)* Snowflake Studio/Getty Images, *Figure 14-11 (R)* © Bob Gibbons/Alamy. **p. 466:** Rob Kesseler; *Figure 14-12* Kevin Phelan. **p. 468:** *Figure 14-13 (T)* Design Pics Inc/Alamy, *Figure 14-13 (CL)* Bob Gibbons/Alamy, *Figure 14-13 (CR)* Fletcher & Baylis/Science Source, *Figure 14-13 (B)* Bohemian Nomad Picturemakers/Getty Images; *Figure 14-14 (L)* Dr. Keith Wheeler/Science Source, *Figure 14-14 (R)* © Tetra Images/Corbis. **p. 470:** Jonathan Blair/National Geographic Creative; *Figure 14-16* Peter French/Getty Images, *Figure 14-16 (Inset)* Philippe Bourseiller/Getty Images. **p. 471:** *Figure 14-17 (TL)* Michael Wheatley/Getty Images, *Figure 14-17 (TR)* Roger Standen/ Science Photo Library/Science Source, *Figure 14-17 (BL)* Garry Gay/Getty Images, *Figure 14-17 (BR)* Frank Krahmer/Getty Images. **p. 472:** *Figure 14-19 (L)* Guenter Fischer/Getty Images, *Figure 14-19 (R)* Topic Photo Agency/AgeFotostock. **p. 473:** *Figure 14-20 (T, L to R)* Craig K. Lorenz/Science Source, Rick & Nora Bowers/Alamy, Amanda Jastremski/National Geographic Creative; DESIGN PICS INC/National Geographic Creative, *Figure 14-20 (BL)* © W. P. Armstrong, *Figure 14-20 (BR)* George

Grall/National Geographic Creative. **p. 475:** *Figure 14-21* Paul Souders/Getty Images. **p. 476:** Todd Gipstein /National Geographic Creative. **p. 477:** *Figure 14-23 (L)* Aubrey Huggins/Alamy, *Figure 14-23 (C)* Westend61/Getty Images, *Figure 14-23 (R)* © Lothar Lenz/Corbis. **p. 478:** *Figure 14-24 (TL)* John Burcham/National Geographic Creative, *Figure 14-24 (TR)* WILDLIFE GmbH/Alamy, *Figure 14-24 (C)* Howard Grey/Getty Images, *Figure 14-24 (BL)* Gertrud & Helmut Denzau/naturepl.com, *Figure 14-24 (BR)* Dr. Keith Wheeler/Getty Images. **p. 479:** *Figure 14-25* Ed Reschke/Getty Images. **p. 480**: Jason Edwards/National Geographic/Robert Harding Picture Library; *Figure 14-26* Ian Nichols/National Geographic Creative. **p. 481:** *Figure 14-28 (L)* Envision/Getty Images, *Figure 14-28 (C)* Dennis Kunkel Microscopy, Inc., *Figure 14-28 (C inset)* Gregory G. Dimijian/Science Source, *Figure 14-28 (R)* E. Gueho/CNRI/Science Source, *Figure 14-28 (R inset)* Julia Phelan. **p. 482:** *Figure 14-29* © Elliott Neep. **p. 483:** *Figure 14-30* ARCO/M Jörrn/AGE Fotostock; *Figure 14-31* D. Hurst/Alamy. **p. 484:** *Figure 14-32* Dr. Jeremy Burgess/Science Source. **p. 485:** *Figure 14-33* Paul Redfearn, the Ozarks Regional Herbarium, Missouri State University. *Figure 14-34* All Canada Photos/Alamy. **p. 488:** AFP/Getty Images.

CHAPTER 15 pp. 492–493: Dennis Kunkel Microscopy/Science Source. **p. 494:** Derren Ready; *Figure 15-1 (L)* Alfred Pasieka/Science Source, *Figure 15-1 (C)* Eye of Science/Science Source, *Figure 15-1 (R)* Marek Mis/Science Source, *Figure 15-1 (inset)* Eye of Science/Science Source. **p. 495:** *Figure 15-2 (T to B)* Astrid & Hanns-Frieder Michler/Science Source, *Figure 15-2* Dennis Kunkel Microscopy, Inc., *Figure 15-2* Copyright 2003 Dennis Kunkel Microscopy, Inc., *Figure 15-2* Copyright 2002 Dennis Kunkel Microscopy, Inc. **p. 496:** *Figure 15-3 (L)* CNRI/Science Source, *Figure 15-3 (C)* © Marc Moritsch/National Geographic Creative, *Figure 15-3 (R)* © Scott T. Smith/Getty Images, *Figure 15-3 (L inset)* Scott Camazine/Medical Images, *Figure 15-3 (C inset)* Copyright 2009 Dennis Kunkel Microscopy, Inc., *Figure 15-3 (R inset)* Dennis Kunkel Microscopy/Science Source; *Figure 15-4* Jay Phelan. **p. 497:** CNRI/Science Source. *Figure 15-5* Copyright 2003 Dennis Kunkel Microscopy, Inc. **p. 498:** *Figure 15-6 (L)* amediacolor/Alamy, *Figure 15-6 (R)* Biophoto Associates/Science Source. **p. 501:** *Figure 15-9 (L)* Dodie Ulery; *Figure 15-6 (C)* Ralph White/Getty Images, *Figure 15-9 (R)* Biophoto Associates/Science Source. **p. 502:** Dennis Kunkel Microscopy/Science Source. **p. 503:** *Figure 15-10* Whitebox Media/Alamy. **p. 505:** *Figure 15-11* Justin Cormack. **p. 506:** *Figure 15-12* ImageState Royalty Free/Alamy. **p. 508:** *Figure 15-15* Gaetan Bally/Keystone/AP. **p. 509:** © Frans Lanting/www.lanting.com; *Figure 15-16* Steve Gschmeissner/Science Source. **p. 510:** *Figure 15-17* © James Forte/National Geographic Society; *Figure 15-18* Chris Wilkins/Getty Images. **p. 511:** Steve Gschmeissner/Science Source; *Figure 15-19* Tim Mantoani/Masterfile. **p. 512:** *Figure 15-20* UCL Geology Collections. **p. 513:** *Figure 15-21 (L)* Copyright 2001 Dennis Kunkel Microscopy, Inc., *Figure 15-21 (C)* Ed Reschke/Getty Images, *Figure 15-21 (R)* Justin Lewis/Getty Images. **p. 514:** *Figure 15-22* Inaki Relanzon/Nature Picture Library/Getty Images, *Figure 15-22 (T inset)* David Littschwager/National Geographic Creative, *Figure 15-22 (B inset)* © Frans Lanting/www.lanting.com. **p. 515:** Lee D. Simon/Science Source; *Figure 15-23* Medical Images RM/Walter Reinhart. **p. 516:** *Figure 15-24 (L)* SPL/Science Source, *Figure 15-24 (R)* Hazel Appleton, Health Protection Agency Centre for Infections/Science Source. **p. 518:** *Figure 15-26 (L)* U.S. Army/Science Source; *Figure 15-26 (C)* AP Photo/Charles Tasnadi; *Figure 15-26 (R)* xPACIFICA/Getty Images. **p. 519:** *Figure 15-28* Bettmann/Getty Images. **p. 521:** *Figure 15-30* NIH/NIAID/Seth Pincus/Elizabeth Fischer/Austin Athman/Science Source. **p. 522:** David Woolley/Getty Images.

CHAPTER 16 pp. 526–527: Michael Wolf/Laif/Redux. **p. 528:** © Frans Lanting/www.lanting.com; *Figure 16-1* James L. Stanfield/National Geographic Creative. **p. 529:** *Figure 16-3* © Frans Lanting/www.lanting.com. **p. 531:** *Figure 16-4* Juan Carlos Muñoz/AgeFotostock. **p. 532:** *Figure 16-5* © Frans Lanting/www.lanting.com. **p. 533:** *Figure 16-7* John McColgan, Bureau of Land Management, Alaska Fire Service. **p. 534:** *Figure 16-8* George Steinmetz/National Geographic Creative; *Figure 16-9* Reuters/Pierre Holtz/Alamy. **p. 535:** *Figure 16-10* Jeffrey Lepore/Science Source. **p. 537:** *Figure 16-11* Dirk Anschutz/Getty Images; *Figure 16-12* © National Museum of Natural History, Smithsonian Institution. Photo by Chip Clark. **p. 538:** Ray Hennessy/Shutterstock. **p. 539:** *Figure 16-13 (L)* B.G. Thomson/Science Source, *Figure 16-13 (C)* Jim Doberman/Getty Images, *Figure 16-13 (R)* G Ronald Austing/Getty Images. **p. 540:** *Figure 16-14* Photo by Greg Katsoulis, courtesy Terry Burnham. **p. 541:** *Figure 16-15 (L)* Gunter Marx Photography/Getty Images, *Figure 16-15 (C)* Norbert Rosing/National Geographic Creative, *Figure 16-15 (R)* © Jack Goldfarb/Design Pics/Corbis, *Figure 16-15 (inset)* iStockphoto.com/jeridu. **p. 544:** *Figure 16-16* Michel Gunther/Science Source. *Figure 16-17 (L)* © Frans Lanting/www.lanting.com, *Figure 16-17 (C)* © Bernd Zoller/imagebroker/Corbis, *Figure 16-17 (R)* © Santiago Urquijo/Getty Images. **p. 545**: Arco Images/Robert Harding Picture Library. **p. 546:** *Figure 16-18 (T)* Them #6, photography, 78 x 200 cm, 1998-2000 (left-hand original, 197gmail.com3). © Hai Bo, Courtesy Pace Beijing, *Figure 16-18 (B)* Them #6, photography, 78 x 200 cm, 1998-2000 (left-hand original, 197gmail.com3). © Hai Bo, Courtesy Pace Beijing. **p. 549:** *Figure 16-20 (L)* Renaud Visage/Getty Images, *Figure 16-20 (R)* photoiconix/Shutterstock. **p. 551:** Michael Wolf/Laif/Redux. **p. 556:** *Figure 16-26 (T)* Yadid Levy/Getty Images, *Figure 16-26 (C)* Kletr/Shutterstock, *Figure 16-26 (B)* Michael Wolf/laif/Redux. **p. 557:** *Figure 16-27* Deng Peibo-Imaginechina via AP Images. **p. 558:** © Lane Oatey/Blue Jean Images/Corbis.

CHAPTER 17 pp. 562–563: LUMI Images/Getty Images. **p. 564:** © Frans Lanting/www.lanting.com; *Figure 17-1* ©Tim Laman (www.TimLaman.com). **p. 565:** *Figure 17-2* Steve Allen/Getty Images. **p. 566:** *Figure 17-3 (TL)* © Frans Lanting/www.lanting.com, *Figure 17-3 (TC)* © Andrey Nekrasov/imagebroker/Corbis, *Figure 17-3 (TR)* Fritz Poelking/AGE Fotostock, *Figure 17-3 (ML)* John Warburton-Lee Photography/Alamy, *Figure 17-3 (MC)* Michael Melford/Getty Images, *Figure 17-3 (MR)* Stephen P. Parker/Science Source, *Figure 17-3 (BL)* GeoStock/Getty Images, *Figure 17-3 (BC)* ajliikala/Getty Images, *Figure 17-3 (BR)* Paul Nicklen/National Geographic Creative. **p. 567:** *Figure 17-4 (TL)* James Randklev/Corbis, *Figure 17-4 (TR)* © Frans Lanting/www.lanting.com, *Figure 17-4 (BL)* Rich Reid/National Geographic Creative, *Figure 17-4 (BC)* Troy Plota/Getty Images, *Figure 17-4 (BR)* Mark Conlin/V&W/Image Quest Marine. **p. 568:** © Steve Besserman. **p. 569:** *Figure 17-6* Mike Boyatt/AGStockUSA; *Figure 17-7* © Frans Lanting/www.lanting.com; **p. 570:** *Figure 17-8* Jacques Descloitres, MODIS Land Rapid Response Team, NASA/GSFC. **p. 571:** *Figure 17-9 (L)* Image courtesy NASA/Goddard Space Flight Center Scientific Visualization Studio, *Figure 17-9 (R)* © Sandy Felsenthal /Corbis/VCG/Getty Images. **p. 574:** © Frans Lanting/www.lanting.com. **p. 575:** *Figure 17-13 (L)* Dr. Jeremy Burgess/Science Source, *Figure 17-13 (R)* James Hager/robertharding/Getty Images. **p. 581:** David Doubliet/National Geographic; *Figure 17-19* © Peter Essick/Aurora Photos, *Figure 17-19 (inset)* Dr. Jennifer L. Graham, U.S. Geological Survey. **p. 582:** *Figure 17-20* Klaus Nigge/National Geographic Creative. **p. 583:** *Figure 17-21* Hermann Eisenbeiss/Science Source. **p. 586:** *Figure 17-24 (L to R)* teekayu/Shutterstock, George Grall/National Geographic Creative, George Grall/National Geographic Creative, Medford Taylor/National Geographic Creative. **p. 587:** *Figure 17-25 (L)* David Doubilet/National Geographic Creative, *Figure 17-25 (R)* Annelies Leeuw/Alamy. **p. 588:** *Figure 17-26 (L)* blickwinkel/Alamy, *Figure 17-26 (R)* Eye of Science/Science Source; *Figure 17-27* Evan Kafka/Getty Images. **p. 590:** *Figure 17-28 (L)* Darlyne A Murawski/National Geographic Creative, *(R)* Arco Images GmbH/Alamy. **p. 593:** © Eric Raptosh Photography/Blend Images/Corbis. **p. 597:** NASA Image by Robert Simmon, based on Landsat data provided by the UMD Global Land Cover Facility.

CHAPTER 18 pp. 600–601: Gallo Images/Getty Images. **p. 602:** © Frans Lanting/www.lanting.com. **p. 602:** *Figure 18-1* Dennis Flaherty/Science Source, *Figure 18-1 (inset)* © Brad Nelson/Phototake. **p. 603:** *Figure 18-2 (L)* David Doubilet/National Geographic Creative, *Figure 18-2 (TC)* Ezra Bailey/Getty Images, *Figure 18-2 (TR)* © Ocean/Corbis, *Figure 18-2 (BC)* Javier Larrea/Getty Images, *Figure 18-2 (BR)* Jack Phelan. **p. 606:** *Figure 18-4* Dennis Macdonald/Age Fotostock. **p. 609:** *Figure 18-7 (L)* age fotostock/Robert Harding Picture Library, *Figure 18-7 (C)* © Frans Lanting/www.lanting.com, *Figure 18-7 (R)* WaterFrame/Alamy. **p. 610:** Karl Ammann/Getty Images. *Figure* 18-8 Courtesy of the Public Libraries of Saginaw, *Figure 18-8 (inset)* Tom Uhlman/Alamy. **p. 613:** *Figure 18-11* © Frans Lanting/www.lanting.com. *Figure 18-12* © Frans Lanting/www.lanting.com, *Figure 18-12 (inset)* Courtesy Bonobo Conservation Initiative. **p. 614:** Mike Clarke/AFP/Getty Images. **p. 615:** *Figure 18-13 (L)* USGS Photo by Harry Glicken, *Figure 18-13 (C)* USGS Photo by Harry Glicken, *Figure 18-13 (R)* USGS photo by Liz Westby; *Figure 18-14 (L)* E.J. Keller/National Zoo/Smithsonian Institution, *Figure 18-14 (C)* AP Photo/Vincent Yu, *Figure 18-14 (R)* Denis Scott/Getty Images. **p. 616:** *Figure 18-15* Jany Sauvanet/Science Source, *Figure 18-15 (inset)* Gale S. Hanratty/Alamy. **p. 617:** *Figure 18-17 (L)* William Mullins/Alamy, *Figure 18-17 (C)* Bruno Petriglia/Science Photo Library/Getty Images, *Figure 18-17 (R)* Peter Yates/Time & Life Pictures/Getty Images. **p. 618:** *Figure 18-18* Jack Picone/Alamy. **p. 619:** *Figure 18-19* Dennis MacDonald/Alamy. **p. 620:** *Figure 18-22* Will & Deni McIntyre/Getty Images. **p. 622:** *Figure 18-25 (T)* NASA/Goddard Space Flight Center Scientific Visualization Studio Thanks to Rob Gerston (GSFC) for providing the data, *Figure* 18-25 *(B)* NOAA. **p. 625:** *Figure 18-28b* Florian Kopp/© imagebroker/Alamy. **p. 626:**

Jason Lee/Reuters/Alamy. **p. 627:** *Figure 18-30 (L-R)* NASA Goddard Space Flight Center Scientific Visualization Studio, NASA Goddard Space Flight Center Scientific Visualization Studio, NASA Goddard Space Flight Center Scientific Visualization Studio, NASA; *Figure 18-31* AP Photo/Elise Amendola, *Figure 18-31 (inset)* Jim Spellman/Getty Images. **p. 628:** *Figure 18-32 (T)* REUTERS/Alamy, *Figure 18-32 (B)* Martin Oeggerli/Science Source. **p. 629:** *Figure 18-33 (L)* © Frans Lanting/www.lanting.com, *Figure 18-33 (R)* © Frans Lanting/www.lanting.com. **p. 630:** *Figure 18-34* Tommy Moorman.

CHAPTER 19 pp. 636–637: Cindy Kassab/Getty Images. **p. 638:** William Rugen. **p. 639:** *Figure 19-1 (L to R)* Danita Delimont/Getty Images, John McAnulty/Getty Images, DEA/C.DANI-I.JESKE/Getty Images, andDraw/Getty Images. **p. 641:** *Figure 19-4* David T. Webb (Ph.D), *Figure 19-4 (inset)* Dr. Keith Wheeler/Science Source. **p. 642:** *Figure 19-6 (T)* Eye of Science/Science Source, *Figure 19-6 (T inset)* Mark Bolton/Getty Images, *Figure 19-6 (C)* DR JEREMY BURGESS/Getty Images, *Figure 19-6 (B)* Dennis Kunkel Microscopy, Inc., *Figure 19-6 (B inset)* Doug Wilson/Getty Images. **p. 643:** *Figure 19-7* Steve Gschmeissner/Science Source. *Figure 19-8 (L)* Dennis Kunkel Microscopy/Science Source, *Figure 19-8 (L inset)* Image Source/Corbis, *Figure 19-8 (C)* Biophoto Associates/Science Source, *Figure 19-8 (C inset)* Joe Petersburger/National Geographic Stock, *Figure 19-8 (R)* Cheryl Power/Science Source, *Figure 19-8 (R inset)* Ed Buziak/Alamy. **p. 645:** Hand coloured Scanning Electron Micrograph. © Rob Kesseler. **p. 646:** *Figure 19-10 (L)* Jim Richardson/Getty Images, *Figure 19-10 (R)* Design Pics Inc/Getty Images; *Figure 19-11* Susumu Nishinaga/Science Source. **p. 647:** *Figure 19-12* apomares/Getty Images. **p. 648:** *Figure 19-15 (L to R)* Design Pics Inc/Getty Images, Matthew Egginton/Shutterstock, Gabor Geissler/Getty Images, Brandon Bourdages/Shutterstock. **p. 650:** *Figure 19-17 (TL)* blickwinkel/Alamy, *Figure 19-17 (TR)* Natalia van D/Shutterstock, *Figure 19-17 (CL)* Gill Orsman/Getty Images, *Figure 19-17 (CR)* © PBNJ Productions/Blend Images/Corbis, *Figure 19-17 (BL)* Custom Life Science Images/Alamy, *Figure 19-17 (BR)* flowerphotos/Alamy. **p. 651:** *Figure 19-18* Installation view of Claude Monet's Water Lilies (1914-26) Oil on canvas, three panels. The Museum of Modern Art. Mrs. Simon Guggenheim Fund. Photo © Jason Brownrigg. **p. 652:** *Figure 19-19 (L)* Elizabeth Given/Alamy, *Figure 19-19 (C)* GAP Photos/Jonathan Buckley, *Figure 19-19 (R)* Dennis Kunkel Microscopy, Inc. **p. 653:** © Frans Lanting/www.lanting.com. **p. 654:** *Figure 19-21* Julian Nieman/Getty Images. **p. 655:** *Figure 19-22* Andrey_Kuzmin/Shutterstock; *Figure 19-23* Darrell Gulin/Getty Images. **p. 656:** *Figure 19-24 (L)* Arco Images GmbH/Alamy, *Figure 19-24 (R)* Jane Legate/Getty Images; *Figure 19-25* Envision/Getty Images. **p. 658:** *Figure 19-28 (L)* Cathy Keifer/Shutterstock, *Figure 19-28 (R)* Karl-Heinz Jacobi/ullstein bild via Getty Images. **p. 660:** Kenneth Ginn/National Geographic Creative. **p. 661:** *Figure 19-29* Nigel Cattlin/Alamy. **p. 662:** *Figure 19-31* Audrius Menkis, *Figure 19-31 (inset)* Eye of Science/Science Source. **p. 663:** *Figure 19-32* 4FR/Getty Images. **p. 665:** *Figure 19-35 (L)* Jonathan Lesage/Getty Images, *Figure 19-35 (R)* Radius Images/Corbis. **p. 667:** *Figure 19-37* George D. Lepp/Getty Images. **p. 668:** Annie Corrigan/WFIU.

CHAPTER 20 pp. 672–673: Rob Kesseler. **p. 674:** Prasit Chansareekorn /Getty Images; *Figure 20-1* © Frans Lanting/www.lanting.com; *Figure 20-2* Greg Dale/National Geographic Creative, *Figure 20-2 (inset)* Copyright © Regents of the University of Minnesota. All rights reserved. **p. 676:** *Figure 20-3 (L)* Marc Vassallo © 2001 by The Taunton Press, Inc. *Figure 20-3 (C)* Ed Reschke/Getty Images, *Figure 20-3 (R)* chonrawit boonprakob/Shutterstock. **p. 677:** *Figure 20-4* All Canada Photos/Alamy. **p. 679:** Clive Nichols/Getty Images; *Figure 20-6 (L to R)* Joe Petersburger/National Geographic Creative, Kasia Biel/iStockphoto, Darlyne A. Murawski/Getty Images, Virginia Saunders. **p. 680:** *Figure 20-8:* Mauro Fermariello/Science Source, *Figure 20-8 (inset)* Cultura RM Exclusive/Ben Pipe Photography/Getty Images. **p. 681:** *Figure 20-10* Garry DeLong/Science Source. **p. 682:** *Figure 20-11* Dennis Kunkel Microscopy/Science Source, *Figure 20-11 (inset)* Innocenti/Getty Images. **p. 683:** schnuddel/Getty Images. **p. 684:** *Figure 20-13* Herbert Kehrer/Getty Images; *Figure 20-14 (T)* Hayashida Masayoshi, *Figure 20-14 (B)* Martin Harvey/Alamy, *Figure 20-14 (B inset)* Irene Terry, University of Utah. **p. 686:** *Figure 20-15* De Agostini Picture Library/Getty Images, *Figure 20-15 (inset)* Micro Discovery/Getty Images. **p. 691:** *Figure 20-20* Bill Beatty/AGPix, *Figure 20-20 (inset)* John Oeth/Alamy. **p. 692:** Grant Faint/Getty Images; *Figure 20-21* Steve Gschmeissner/Science Source. **p. 694:** *Figure 20-23* Dennis Flaherty/Science Source; *Figure 20-24(TL)* Joseph F. Gennaro Jr./Science Source, *Figure 20-24 (BL)* Patrick J. Lynch/Science Source, *Figure 20-24 (R)* James M. Bell/Science Source. **p. 695:** *Figure 20-25* Digital Vision/Getty Images. **p. 697:** *Figure 20-27 (T)* Ulrich Baumgarten/Getty Images, *Figure 20-27(C)* Bertrand Guay/Getty Images, *Figure 20-27(B)* Steve Babineau/Getty Images. **p. 698:** *Figure 20-29* © GYRO PHOTOGRAPHY/amanaimages/Corbis. **p. 699:** *Figure 20-30* Adrian Weston/Alamy. **p. 700:** Christopher Furlong/Getty Images. **p. 701:** *Figure 20-32* Yoshio Niikura/Getty Images. **p. 702:** *Figure 20-33* SWNS. **p. 703:** *Figure 20-34* Tim Laman/National Geographic Creative. **p. 705:** *Figure 20-36 (L)* Danny Lehman/Corbis/VCG/Getty Images, *Figure 20-36 (R)* Colin Underhill/Alamy; *Figure 20-37* Jimmy Villalta/Getty Images. **p. 706:** *Figure 20-38* Roberto Machado Noa/Getty Images; *Figure 20-39* Kenneth M. Highfill/Science Source. **p. 707:** Ed Reschke/Getty Images; *Figure 20-40* Graham Jordan/Science Source. **p. 708:** *Figure 20-41* Power and Syred/Science Source; *Figure 20-42* age fotostock/Superstock. **p. 709:** *Figure 20-43* Courtesy of Roger P. Hangarter, Chancellor's Professor of Biology, Indiana University. **p. 711:** *Figure 20-44* Foto Grebler/Alamy. **p. 713:** *Figure 20-46 (L)* Power and Syred/Science Source, *Figure 20-46 (C)* Pan Xunbin/Science Source, *Figure 20-46 (R)* Martin Shields/Science Source; *Figure 20-47* Voisin/Phanie/Science Source. **p. 714:** BSIP/Photoshot.

CHAPTER 21 pp. 718–719: Anandraj Thiyagarajaperumal. **p. 720:** Stephen Frink/Getty Images; *Figure 21-1* © Tim Davis/Corbis/VCG/Getty Images. **p. 721:** *Figure 21-3 (T)* Image Quest Marine/Alamy, *Figure 21-3 (L to R)* Dennis Kunkel Microscopy, Inc., Alfred Pasieka/Science Source, Copyright 1995 Dennis Kunkel Microscopy, Inc., Copyright 2009 Dennis Kunkel Microscopy, Inc. **p. 722:** *Figure 21-4* Chris Trotman/Getty Images. **p. 723:** *Figure 21-5* SPL/Science Source, *Figure 21-5 (inset)* Dr. Gopal Murti/Science Source; *Figure 21-6 (TL)* Quest/Science Source, *Figure 21-6 (BL)* Photo Quest Ltd/Science Photo Library/Corbis; *Figure 21-6 (C)* Robin Moore/National Geographic Creative, *Figure 21-6 (TR)* De Agostini Picture Library/Getty Images, *Figure 21-6 (CR)* Ed Reschke/Getty Images, *Figure 21-6 (BR)* David Spears FRPS FRMS/Getty Images. **p. 725:** *Figure 21-7 (L)* Steve Gschmeissner/Science Source, *Figure 21-7 (C)* Steve Gschmeissner/Science Source, *Figure 21-7 (R)* Steve Gschmeissner/Science Source. **p. 726:** *Figure 21-8 (T)* Joe McDonald/Getty Images, *Figure 21-8 (L)* Quest/Science Source, *Figure 21-8 (C)* Roger J. Bick & Brian J. Poindexter/UT-Houston Medical School/Science Source, *Figure 21-8 (R)* Roger J. Bick & Brian J. Poindexter/UT-Houston Medical School/Science Source. **p. 728:** *Figure 21-9* Robert S. McNeil/Baylor College of Medicine/Science Source; *Figure 21-10* Panoramic Images/Getty Images. **p. 729:** *Figure 21-11* Eric Grave/Science Source. **p. 730:** *Figure 21-12 (TL)* MIXA/Alamy, *Figure 21-12 (TC)* Stuart O'Sullivan/Getty Images, *Figure 21-12 (TR)* © Jose Luis Pelaez, Inc./Blend Images/Corbis, *Figure 21-12 (BL)* Rubberball/Mike Kemp/Getty Images, *Figure 21-12 (BC)* Hola Images/Getty Images, *Figure 21-12 (BR)* Radius Images/Alamy. **p. 731:** *Figure 21-12 (TL)* © Andersen Ross/Blend Images/Corbis, *Figure 21-12 (TC)* Rubberball/Mark Andersen/Getty Images, *Figure 21-12 (TR)* Radius Images/Alamy, *Figure 21-12 (BL)* Caroline Schiff/Getty Images, *Figure 21-12 (BC)* Andersen Ross/Getty Images. *Figure 21-12 (BR)* RubberBall/Alamy. **p. 732:** Anton Novoderezhkin/Newscom/ZUMA Press/Russian Federation; *Figure 21-13* AP Photo/Sue Ogrocki. **p. 733:** *Figure 21-14 (L)* AP Photo/Sergei Chuzavkov, *Figure 21-14 (R) Robert Harding Picture Library*/National Geographic Creative. **p. 734:** AP Photo/Andy Wong. **p. 737:** *Figure 21-18 (T)* Sarah Leen/National Geographic Creative, *Figure 21-18 (B)* Gary Bell/Getty Images. **p. 738:** *Figure 21-20 (L)* Reed Hoffmann/Getty Images, *Figure 21-20 (R)* Chris Johns/Getty Images. **p. 739:** *Figure 21-21 (L to R)* Paul Nicklen/National Geographic Creative, © Herbert Kratky/imagebroker/Corbis, Brian Gordon Green/National Geographic Creative, Chuck Savage/Getty Images; *Figure 21-22* TS Corrigan/Alamy. **p. 747:** *Figure 21-26* cglade/Getty Images, *Figure 21-26 (inset)* Tony Freeman/Photo Edit. **p. 748:** © Image Source/Corbis.

CHAPTER 22 pp. 752–753: Susumu Nishinaga/Science Source. **p. 754:** Bruce Dale/National Geographic Creative; *Figure 22-1 (L)* Susumu Nishinaga/Science Source, *Figure 22-1 (C)* Joel Sartore/National Geographic Creative, *Figure 22-1 (R)* Science Source. **p. 755:** *Figure 22-2* Chris Howes/Wild Places Photography/Alamy. **p. 756:** *Figure 22-3* A.S.Floro/Shutterstock. **p. 757:** *Figure 22-5* Kletr/Shutterstock; *Figure 22-6* © Frans Lanting/www.lanting.com. **p. 758:** *Figure 22-7* © Frans Lanting/www.lanting.com. **p. 759:** Photo Insolite Realite/Science Source; *Figure 22-8* © Blue Jean Images/Corbis. **p. 762:** *Figure 22-11* AP Photo/Sara D. Davis. **p. 763:** *Figure 22-12* Alfred Pasieka/Science Source, *Figure 22-12 (inset)* © Dennis Kunkel Microscopy, Inc. **p. 764:** *Figure 22-13 (L)* Eyecandy Images/Aurora Photos, *Figure 22-13 (R)* BSIP/Phototake. **p. 767:** *Figure 22-14* Heinrich van den Berg/Getty Images. **p. 768:** *Figure 22-15 (L)* © Dennis Kunkel Microscopy, Inc., *Figure 22-15 (C)* © Dennis Kunkel

Microscopy, Inc., *Figure 22-15 (R)* CNRI/Science Source; *Figure 22-16* Vanessa Price. **p. 769:** *Figure 22-17* AP Photo/Eckehard Schulz. **p. 770:** *Figure 22-18* Popperfoto/ Getty Images, *Figure 22-18 (inset)* Steve Percival/Science Source. **p. 771:** *Figure 22-19* © Frans Lanting/www.lanting.com. **p. 772:** *Figure 22-20 (L)* SPL/Science Source, *Figure 22-20 (R)* SPL/Science Source. **p. 773:** *Figure 22-21 (L)* © Simon Murrell/ Corbis, *Figure 22-21 (R)* PhotoCuisine RM/Alamy. **p. 774:** *Figure 22-22 (L)* © Blue Jean Images/Corbis, *Figure 22-22 (R)* Michael Abbey/Science Source. **p. 775:** © Frans Lanting/www.lanting.com. **p. 778:** *Figure 22-28 (T)* Jay Phelan, *Figure 22-28 (B)* Jay Phelan. **p. 779:** *Figure 22-29* Lennart Nilsson. **p. 781:** *Figure 22-31* Fancy/Veer/ Corbis. **p. 782:** *Figure 22-32* © Dennis Kunkel Microscopy, Inc. **p. 783:** *Figure 22-33* Tim Laman Photography. **p. 785:** *Figure 22-35 (L)* Pr. M. Brauner/Science Source, *Figure 22-35 (R)* Pr. M. Brauner/Science Source. **p. 786:** Art Twomey/Getty Images; *Figure 22-36* Jim Richardson/National Geographic Creative. **p. 787:** © Helen Jones-Florio. **p. 788:** Wild Horizon/Getty Images.

CHAPTER 23 pp. 792–793: Copyright 2014 Francesco Tonelli. **p. 794:** © Frans Lanting/www.lanting.com. **p. 795:** *Figure 23-2 (T)* Paul Nicklen/National Geographic Stock, *Figure 23-2 (C)* Don Johnston/Getty Images, *Figure 23-2 (BL)* Michael Sulis/Wikipedia Commons, *Figure 23-2 (BR)* Rolf Nussbaumer. **p. 796:** *Figure 23-3* Look at Sciences/Science Source, *Figure 23-3 (inset)* CDO Ranching and Development LLP. **p. 797:** © Frans Lanting/www.lanting.com. *Figure 23-4 (L)* Jay Phelan, *Figure 23-4 (R)* Jay Phelan. **p. 798:** *Figure 23-5* Michael Nichols/National Geographic Creative. **p. 799:** *Figure 23-6* John Carey/Getty Images. **p. 800:** *Figure 23-7 (L)* Bill Brennan/Getty Images, *Figure 23-7 (R)* Chris Ryan/Getty Images. **p. 801:** *Figure 23-8* Photo by David E. Scherman/The LIFE Picture Collection/Getty Images. **p. 802:** *Figure 23-9 (L)* Jutta Klee/Getty Images. *Figure 23-9 (C)* Brian Yarvin/ Getty Images, *Figure 23-9 (R)* Marie C Fields/Shutterstock. **p. 803:** *Figure 23-10* Goydenko Liudmila/Getty Images; *Figure 23-11(L)* elena moiseeva/Shutterstock, *Figure 23-11 (R)* Tim Hill/fabfoodpix.com. **p. 806:** *Figure 23-14* Callum Bennetts/ Alamy. **p. 807:** Beverly Joubert/National Geographic Creative. **p. 808:** *Figure 23-16* Image Source/Veer. **p. 810:** *Figure 23-18* Science Photo Library/Alamy. **p. 812:** *Figure 23-20 (L)* Steve Gschmeissner/Science Source, *Figure 23-20 (C)* SPL/Science Source, *Figure 23-20 (R)* SPL/Science Source. **p. 814:** *Figure 23-22* SPL/Science Source. **p. 815:** *Figure 23-23* Dr. Kari Lounatmaa/Science Source. **p. 816:** *Figure 23-24* David Coleman/Alamy. **p. 817:** © 2/Grove Pashley/Ocean/Corbis; *Figure 23-25* Olga Miltsova/Shutterstock, *Figure 23-25 (inset)* Mango Productions/Getty Images. **p. 818:** *Figure 23-26 (L)* Tommy Moorman, *Figure 23-26 (R)* Tommy Moorman. **p. 821:** *Figure 23-27* Robert Daly/Getty Images. **p. 822:** *Figure 23-29* H. Armstrong Roberts/Getty Images. **p. 823:** *Figure 23-30 (L)* AP Photo/Karl-Heinz Kreifelts, *Figure 23-30 (C)* Image Source/Corbis, *Figure 23-30 (R)* stefanolunardi/Shutterstock. **p. 825:** *Figure 23-32* Lea Paterson/Science Photo Library/Getty Images. **p. 826:** © Ken Seet/Corbis.

CHAPTER 24 pp. 830–831: Joel Sartore/NGS Image Collection. **p. 832:** Chadden Hunter; *Figure 24-1* Alex Livesey/Getty Images, *Figure 24-1 (inset)* AP Photo/Jim Mone. **p. 834:** *Figure 24-3 (L)* © Dennis Kunkel Microscopy, Inc., *Figure 24-3 (C)* Kent Wood/Science Source, *Figure 24-3 (R)* Professor Xiaoming Jin. **p. 835:** *Figure 24-4* Image Source/Alamy. **p. 836:** *Figure 24-6* Bianca Lavies/Getty Images. **p. 837:** Dr. Gopal Murti/Science Source. **p. 839:** *Figure 24-9* CNRI/Science Source. **p. 840:** *Figure 24-10* Tommy Moorman. **p. 842:** *Figure 24-12* Andrew D. Bernstein/Getty Images. **p. 843:** Paul Scott Thomas Photography; *Figure 24-13 (L)* Getty Images, *Figure 24-13 (R)* Lars Niki/Corbis via Getty Images. **p. 844:** *Figure 24-14 (TL)* up close with nature/Getty Images, *Figure 24-14 (TR)* peebert/ Getty Images, *Figure 24-14 (CL)* Franck Fife/AFP/Getty Images, *Figure 24-14 (CR)* Timm Schamberger/Getty Images, *Figure 24-14 (B)* © Frans Lanting/www.lanting.com. **p. 845:** *Figure 24-15* Tommy Moorman. **p. 847:** *Figure 24-18* humonia/ Getty Images. **p. 849:** *Figure 24-21 (L)* Bjorn Rorslett/Science Source, *Figure 24-21 (R)* Bjorn Rorslett/Science Source. **p. 851:** *Figure 24-23* Noel Vasquez/Getty Images. **p. 853:** Andy Rouse/Nature Picture Library; *Figure 24-25* Jeffrey Hopkins/ Uniformed Services University. **p. 856:** *Figure 24-28 (L)* Tony Duffy/Getty Images, *Figure 24-28 (R)* The Asahi Shimbun/Getty Images. **p. 857:** *Figure 24-29 (L)* Phillip Colla/SeaPics.com, *Figure 24-29 (C)* Scott Camazine/Science Source, *Figure 24-29 (R)* Dave Roberts/Science Source; *Figure 24-30 (T)* SPL/Science Source, *Figure 24-30 (B)* Dr. Tony Brain/Science Source. **p. 858:** VisitBritain/Britain on View/Getty Images. **p. 859:** *Figure 24-31* Image Source/Getty Images; *Figure 24-32* MGP/ Getty Images. **p. 862:** *Figure 24-34* © Pop! Studio Photography/www.alamy.com/ Corbis. **p. 865:** Robert Doisneau/Gamma-Rapho via Getty Images. **p. 866:** *Figure 24-36 (L)* Gustoimages/Science Source, *Figure 24-36 (R)* Dod Miller/Alamy. **p. 868:** *Figure 24-37* Steve Prezant/Getty Images. **p. 869:** *Figure 24-38* © Corbis, *Figure 24-38 (inset)* Kevin Fleming/Getty Images. **p. 870:** Science Source.

CHAPTER 25 pp. 874–875: Kirill Golovchenko. **p. 876:** © Frans Lanting/www.lanting.com. **p. 877:** *Figure 25-1 (T)* gradyreese/Getty Images, *Figure 25-1 (BL)* KidStock/Blend Images/Getty Images, *Figure 25-1 (BR)* Murray Close/©Universal Pictures/Courtesy Everett Collection. **p. 878:** *Figure 25-3* drbimages/Getty Images. **p. 879:** *Figure 25-4* Victor Lee. **p. 881:** Steve Gschmeissner/Science Source. **p. 882:** *Figure 25-6* Image Source/Getty Images. **p. 883:** *Figure 25-7* Tao Qi/VCG via Getty Images, *Figure 25-7 (inset)* Stewart Cohen/Gallery Stock. **p. 884:** *Figure 25-8* drbimages/Getty Images; *Figure 25-9* Gallo Images/Getty Images. **p. 885:** *Figure 25-10* Chris Hellier/Science Source. **p. 886:** *Figure 25-12* Alan Powdrill/Getty Images. **p. 888:** Sean Justice/Getty Images. **p. 889:** *Figure 25-14* Wolfgang Rattay/REUTERS/Newscom, *Figure 25-14 (inset)* George Burns/Handout/Corbis via Getty Images. **p. 890:** *Figure 25-16* Nick Vedros/Getty Images. **p. 891:** *Figure 25-17* AP Photo/The Canadian Press, Paul Chiasson, *Figure 25-17 (inset)* Justin Tafoya/Getty Images. **p. 892:** *Figure 25-18* Jay Phelan; *Figure 25-19* Alexandra Mlejnkova/CTK via AP Images. **p. 894:** *Figure 25-21* Todd Gipstein/National Geographic Creative. **p. 895:** *Figure 25-22* Purple Collar Pet Photography/Getty Images. **p. 896:** © moodboard/Corbis. **p. 897:** *Figure 25-23* James P. Blair/Getty Images; *Figure 25-24 (T)* Photofusion/ Getty Images, *Figure 25-24 (insets T to B)* Gary Porter/Milwaukee Journal Sentinel/MCT/Newscom, Arco Images GmbH/Alamy, Alvis Upitis/Getty Images, © Ryan Stuart/moodboard/Corbis. **p. 900:** Editorial Image, LLC/Alamy.

CHAPTER 26 p. 904–905: Joel Sartore/National Geographic Creative. **p. 906:** © Steve Bloom/stevebloom.com; *Figure 26-1* Courtesy Diane Blood/Rex Features/ Shutterstock. **p. 908:** *Figure 26-3 (L)* Juniors Bildarchiv GmbH/Alamy, *Figure 26-3 (R)* © Dennis Kunkel Microscopy, Inc. **p. 910:** John Springer Collection/Getty Images; *Figure 26-6* © Frans Lanting/www.lanting.com, *Figure 26-6 (inset)* ©Steve Downer/ardea/AGE Fotostock. **p. 911:** *Figure 26-7* Michael Stewart/Corbis. **p. 912:** *Figure 26-9* Dennis Kunkel Microscopy/Science Source. **p. 914:** *Figure 26-11* Martin Harvey/Biosphoto, *Figure 26-11 (inset)* Martin Muller. **p. 917:** *Figure 26-12* Michael Stewart/Corbis. **p. 919:** *Figure 26-14* Peathegee Inc/Getty Images, *Figure 26-14 (inset)* Peter Turnley/Getty Images. **p. 921:** *Figure 26-16* Dr. Yorgos Nikas/ Science Source. **p. 922:** Dennis Kunkel Microscopy/Science Source; *Figure 26-17* Clouds Hill Imaging Ltd./Science Source. **p. 924:** *Figure 26-19* George Mattei/ Science Source. **p. 927:** Emma Kim/Getty Images. **p. 928:** *Figure 26-22* Dr. Yorgos Nikas/Phototake/Diomedia. **p. 930:** *Figure 26-25 (First trimester TL)* Dr. Yorgos Nikas/Science Source, *(First trimester BL)* Petit Format/Science Source, *Figure 26-25 (TR)* Lennart Nilsson/TT *Figure 26-25 (C)* Lennart Nilsson/TT, *Figure 26-25 (B)* Petit Format/Science Source. **p. 931:** *Figure 26-26* CNRI/ Science Source. **p. 933:** *Figure 26-28* Blend Images/Alamy. **p. 934:** Phanie/Alamy. **p. 936:** Vincent Kessler/ Reuters /Newscom.

CHAPTER 27 pp. 940–941: Khuloud T. Al-Jamal & Izzat Suffian, Wellcome Images. **p. 942:** Custom Medical Stock Photo/Alamy; *Figure 27-1* JAGADEESH NV/EPA/Newscom. **p. 943:** *Figure 27-2* Mark Douet/Getty Images; *Figure 27-3 (L to R)* Lee D. Simon/Science Source, David Gregory & Debbie Marshall/Wellcome Images, Spike Walker, Copyright Dennis Kunkel Microscopy, Inc., Spike Walker. **p. 945:** *Figure 27-5* Tetra Images/Getty Images; *Figure 27-6* Jaminson J. Moses/AGE Fotostock. **p. 948:** *Figure 27-10* Eye of Science/Science Source. **p. 950:** *Figure 27-12* Stephanie Rausser/Getty Images; *Figure 27-13* University of Edinburgh, Wellcome Images. **p. 952:** *27-15* Julia Phelan. **p. 953:** Science Source - NIH/Getty Images. **p. 954:** *27-17 (L)* Dave Thompson/Press Association via AP Images, *27-17 (R)* Zheng Shuai /Imaginechina via AP Images. **p. 955:** *27-19* Cordelia Molloy/Science Source. **p. 960:** *27-25* Ian Boddy/Science Source. **p. 964:** Copyright Dennis Kunkel Microscopy, Inc.; *Figure 27-27* Dr. Gopal Murti/Science Source. **p. 965:** *Figure 27-29* Peter Macdiarmid/Getty Images. **p. 967:** *Figure 27-31* Copyright 1997 Dennis Kunkel Microscopy, Inc. **p. 969:** *Figure 27-34* Phanie/Science Source. **p. 970:** M. Eric Honeycutt/Getty Images.

Index

boldface indicates a definition, *italics* indicates a figure.

C

N

O

P

S